The Encyclopaedic Dictionary of Environmental Change

Edited by

John A. Matthews

Department of Geography, University of Wales Swansea, UK

E. Michael Bridges • Christopher J. Caseldine
Adrian J. Luckman • Geraint Owen • Allen H. Perry
Richard A. Shakesby • Rory P.D. Walsh
Robert J. Whittaker • Katherine J. Willis

8

Hardback edition first published in Great Britain in 2001.
This edition first published in Great Britain in 2003 by
Arnold, a member of the Hodder Headline Group,
338 Euston Road, London NW1 3BH

http://www.arnoldpublishers.com

Distributed in the United States of America by
Oxford University Press Inc.
198 Madison Avenue, New York, NY10016

British Library Cataloguing in Publication Data
A catalogue record for this book is available from the British Library

Library of Congress Cataloging-in-Publication Data
A catalog record for this book is available from the Library of Congress

ISBN 0 340 80976 0

1 2 3 4 5 6 7 8 9 10

Production Editor: Anke Ueberberg
Production Controller: Deborah Smith
Cover Design: Terry Griffiths

Typeset in 9 on 10 Plantin by Phoenix Photosetting, Chatham, Kent
Printed and bound in Great Britain by The Bath Press

What do you think about this book? Or any other Arnold title?
Please send your comments to feedback.arnold@hodder.co.uk

Contents

REVIEWS OF *THE ENCYCLOPAEDIC DICTIONARY OF ENVIRONMENTAL CHANGE*

'All in all, *The Encyclopaedic Dictionary of Environmental Change* helps to make sense of the babel that can be so characteristic of the literature and conferences in the environmental sciences and gives excellent support when reading across the disciplines. But beware: if you enjoy following up cross-references, this book may bind your attention longer than you had intended.'
                                                                                                                                      NATURE

'... a reference work which, because of its comprehensiveness and clarity of layout, prose and illustration, has something to offer everyone who, regardless of background, is concerned with environmental change and with the role played by people in it. This encyclopaedic dictionary can be thoroughly recommended as a gateway to deeper studies.'
                                                                                                                                      THE HOLOCENE

'The *Encyclopaedic Dictionary of Environmental Change* is a very valuable contribution to the literature on environmental change and the editors are to be congratulated for making an excellent job of such a large and difficult project.'
                                                                                                                          QUATERNARY SCIENCE REVIEWS

'A valuable feature is the intensity of reference citation throughout; even some brief items have reference lists citing mainly primary sources in the scientific literature. This will be of considerable value to those wishing to follow up particular topics in greater detail ... Most of the references cited have the advantage of being relatively accessible, increasing their usefulness to undergraduate students ... The quality and clarity of writing is extremely high, as are the line drawings and diagrams.'
                                                                                                                      BRITISH ECOLOGICAL SOCIETY BULLETIN

Preface

Environmental change is a major growth area of interdisciplinary science with a wide and developing terminology. Although there are an increasing number of disciplinary and even subdisciplinary dictionaries, which cover parts of the field of environmental change, no other dictionary covers the whole. *The Encyclopaedic Dictionary of Environmental Change* is an advanced work of reference for academics, teachers, students and others who may encounter environmental change in a wide range of disciplines and from a variety of perspectives. It also represents a work of scholarship and research synthesis. There is an academic niche for such a guide. In addition, there is a need to define, clarify and interpret the words employed by researchers in this field for a wider audience, increasingly aware of the existence of natural and anthropogenic environmental change. These are topical issues and have important social consequences. The drawing together herein of diverse material all relating to environmental change means that the reader interested in these topics need no longer to consult several, separate disciplinary dictionaries.

The field covered is environmental change in the broad sense, including environmental changes affecting the Earth over geological, intermediate and short timescales: pre-Quaternary, Quaternary (the last two million years or so) and Holocene (the last c. 11 500 years), historical, current and potential future environmental changes. This involves a very wide range of topics extending from the natural environmental sciences (such as physical geography, Earth observation, Quaternary geology, palaeoclimatology, environmental archaeology, ecology and soil science) into relevant areas of the physical, Earth, biological, archaeological and social sciences, and also into the applied sciences (such as engineering, agriculture and forestry). Indeed, the question of what to include and what to exclude presented a major problem to the editors in that almost everything can be included in the environment of some aspect of the Earth or solar system; all is interconnected; and all environments change.

Climatic change – past, present and future – is a major focus, but other topics relating to natural and human-induced changes to the lithosphere, hydrosphere, cryosphere, biosphere and pedosphere are also covered in detail. Concepts, theories and issues relating to environmental change are included, as are: the diverse sources of evidence of environmental change; the wide-ranging local, regional and global effects of environmental change from the polar regions to the tropics; and the approaches and techniques used for reconstructing, dating, monitoring, modelling and predicting change. The broader cosmosphere is included insofar as it impinges on the Earth (e.g. through meteorite impact). The basis and emphasis of *The Encyclopaedic Dictionary of Environmental Change* is the science of environmental change, but the social, economic and political dimensions of environmental issues, environmental problems, conservation, management and policy aspects are strongly represented. Thus, the content reflects the interdependence not only of the sciences but also of science and society.

The descriptor 'encyclopaedic' is used deliberately in the title to signify that this work is intermediate between a dictionary and an encyclopaedia. There are several levels of entries ranging from a definition alone (one or two sentences; there are 2200 entries of this type) to short reviews of about 1000 words or so (54 entries), and over 1000 entries at intermediate level. Hence the work is considerably more than a dictionary. It attempts to cover the terminology comprehensively, with a much larger number of entries than is usual in an encyclopaedia and it uses, where appropriate, a 'nested approach' to the definitions. The longest entries cover broad topics or areas of understanding that are particularly relevant to environmental change. Entries at intermediate levels cover terms that may appear in one or more of the longer entries, but the importance of which in the context of environmental change is deemed to warrant separate treatment. Shorter entries include terms likely to be encountered but not always in an environmental-change context. The nested approach means that explanations in the longer entries do not have to be interrupted to explain subsidiary terms.

Two other aims underlie this encyclopaedic dictionary. First, it develops the context of the terms, explaining and exemplifying the terminology of environmental change in relation to its concepts, techniques, methodology and philosophy. Even though the longest entries are of restricted length, because this is more than a dictionary, the reader must be able to find related entries easily. This is facilitated by extensive cross-referencing between entries and an index that includes terms that have been considered too specialised for inclusion as headwords. Second, this encyclopaedic dictionary can be used as a sourcebook for further information. A total of over 7000 terms are in fact defined in over 3450 entries. Sources listed after each entry include not only a wide range of specialist texts and review articles but also frequent examples of the cutting-edge research literature. We have striven to provide an up-to-date picture of the field incorporating appropriate recent references as well as some classics. In many respects, therefore, *The Encyclopaedic Dictionary of Environmental Change* provides an accessible gateway into a complex field.

This volume was inspired by the *Encyclopedic Dictionary of Physical Geography* (edited by Andrew Goudie with others and published by Blackwell), now in its third edition but regrettably having dispensed with the distinctive adjective. *The Encyclopaedic Dictionary of Environmental Change* was conceived originally as a millennial project amongst the physical geographers in the Department of Geography at the University of Wales Swansea, seven of whom are Editors. In order fully to cover such a diverse field, academics with important expertise outside that available at Swansea – from the School of Geography and Archaeology at the University of Exeter and the School of Geography and the Environment at the University of

Oxford – were invited to become involved. The areas covered by each Editor are listed below: each had responsibility for entries in particular areas, usually involving a combination of a particular environment, timescale and subject matter. We are all grateful to the large number of contributors (over a hundred), many of whom provided specialist inputs from such disciplines as archaeology, biology, chemistry, forestry, geology, geophysics, law, meteorology, oceanography and social science. Very few missed deadlines – at least the final one!

The end product represents a much enlarged and altered list of terms, headwords and levels to that we originally agreed upon. Each entry almost always begins with an explicit definition of the term; this is elaborated to a greater or lesser degree depending on the level. Though it did not at times appear to be so, there was an upper limit to the size of the work, the total amount of information included and hence the number and/or size of the entries at the various levels. This limit necessitated minimising any overlap between related entries, which means that some essential information may appear in cross-referenced entries. We feel that the overall balance between subject areas is about right but there will inevitably be room for improvements, which can be incorporated in a second edition, and we look forward to hearing about them along with any errors or omissions, which we recognise are most likely where terms have an alternative usage (a frequent occurrence in the field of environmental change). Occasionally, a particular usage has been emphasised or encouraged where we think it is necessary.

It was decided early on to include a full index to supplement the cross-referencing system, although the headwords themselves have not been repeated in the index. Several recent dictionaries have dispensed with an index but this assumes foolproof cross-referencing and/or inclusion of every possible term as a headword. In *The Encyclopaedic Dictionary of Environmental Change*, the index should be viewed as an indispensable aid to finding a topic or term if the headword is not the term with which the reader is most familiar. We have, however, downplayed if not dispensed completely with acronyms, the number of which, we feel especially in the field of environmental change, is unnecessary and fast becoming counter-productive.

English rather than American variants of terms have been used throughout (for example, palaeoenvironment is used in preference to paleoenvironment), as have SI units. Spellings that have fallen into disuse are generally avoided (e.g. Cenozoic is used rather than Cainozoic) but climatic is preferred to climate in climatic change and radiocarbon years have not always been calibrated. Where BP is employed without qualification, this refers to uncalibrated radiocarbon years.

Special thanks are due to those who have assisted, encouraged or otherwise put up with this very large publishing enterprise, especially: Gillian Young and Eileen Baker (word processing) and Nicola Jones and Anna Ratcliffe (diagrams) at Swansea; and Laura McKelvie, Luciana O'Flaherty, Elizabeth Gooster and Anke Ueberberg of Arnold in London as well as David Penfold, Judith Jenkins and Samantha Jones (in their various roles from commissioning to proofreading).

John A. Matthews
Swansea, 5 November 2000

Note

For this paperback edition of *The Encyclopaedic Dictionary of Environmental Change*, minor corrections and editing changes have been made to the text.

How to use *The Encyclopaedic Dictionary of Environmental Change*

1. First consult the dictionary entries for over 3450 terms and topics that are covered in alphabetical order.
2. If a term is not found as an entry in the body of the dictionary, consult the index, which directs the reader to the most appropriate entry.
3. In the text, words in SMALL CAPITALS indicate other entries in the dictionary where further related information can be found.
4. Words in *italics* in the text do not appear as separate entries, but they are nevertheless defined. Consult the index for the most appropriate entry.
5. The references listed after most entries point the reader to a wide range of sources and further reading. They include any sources cited in the text, figures and tables.

Editorial division of responsibilities

Co-ordinating editor

John A. Matthews　　Holocene environmental change; techniques, methodology and philosophy of environmental change; palaeoclimatology; conservation, management and policy aspects of environmental change; general co-ordination, final editing, cross-referencing, index.

Editors

E. Michael Bridges　　Soils and palaeosols; terrestrial pollution and land restoration.

Christopher J. Caseldine　　Quaternary environmental change; dating techniques; environmental archaeology; palaeoenvironmental reconstruction.

Adrian J. Luckman　　Earth observation; remote sensing; geographical information systems; environmental monitoring and environmental modelling.

Geraint Owen　　Geology and oceanography; plate tectonics; environmental change on geological timescales in terrestrial and marine environments.

Allen H. Perry　　Climate and climatic change; historical climatology, contemporary and future human impacts on climate.

Richard A. Shakesby　　Geomorphological change in mediterranean, temperate and cold environments; coastal environmental change; human impacts on geomorphology.

Rory P.D. Walsh　　Hydrology and palaeohydrology; fluvial geomorphology and human impacts; geomorphology and environmental change in tropical environments.

Robert J. Whittaker　　Ecology; biological aspects of contemporary environmental change; ecological change in tropical environments; human impacts on ecosystems.

Katherine J. Willis　　Biological aspects of environmental change on longer timescales; evolution and biogeography; palaeoecology; ecological change in temperate and cold environments; environmental archaeology and Holocene human impacts.

Biographical information on the Editors

Co-ordinating Editor

John A. Matthews is Professor of Physical Geography at the University of Wales, Swansea, where he is also Director of the Swansea Radiocarbon Dating Laboratory. After gaining a BSc and PhD in Geography from King's College, University of London, he worked at the Universities of Edinburgh and Cardiff before moving to Swansea in 1994. His research interests are in physical geography, ecology and environmental change, focusing on the geoecology and geomorphology of recently-deglaciated landscapes; and in the reconstruction and dating of glacier variations, climatic change and colluvial activity during the Holocene (the last c. 11 500 years). In pursuit of these studies he has planned and led 30 research expeditions to southern Norway, for which he received the Ness Award of the Royal Geographical Society in 1988. He has also carried out research in Finnish Lapland, on Mount Kenya, in the Austrian Alps and in the Brecon Beacons, Wales; he has been a Visiting Landsdowne Scholar at the University of Victoria, Canada, and was awarded the Bronze Medal of the University of Helsinki by the Science Faculty in 1996. Professor Matthews is the founding editor of *The Holocene*, an interdisciplinary journal focusing on recent environmental change, and co-ordinating editor of the book series *Key issues in environmental change*, published by Arnold. His publications include over 100 contributions in scientific journals and his books include: *Quantitative and statistical approaches to geography* (Pergamon, 1982); *The ecology of recently deglaciated terrain: a geoecological approach to glacier forelands and primary succession* (Cambridge University Press, 1992); *Solifluction and climatic variation during the Holocene* (Gustav Fischer, 1993); and *Rapid mass movement as a source of climatic evidence for the Holocene* (Gustav Fischer, 1997).

Editors

E. Michael Bridges is a Fellow of the Institute of Professional Soil Scientists and has specialised in the geography and genesis of soils. He gained a BSc and MSc from the University of Sheffield and a PhD from the University of Wales. He began his career as a soil surveyor, working in Derbyshire as a research scientist employed by Rothamsted Experimental Station. From 1961 he worked for the University of Wales at Swansea, where he became a Senior Lecturer in Geography and Sub Dean of Science. During his career, he was a Visiting Lecturer at the Universities of Auckland (New Zealand), New England, Armidale (NSW, Australia), Khartoum (Sudan) and the West Indies (Trinidad). In 1986 he was a Visiting Scientist with the Commonwealth Scientific and Industrial Research Organisation (CSIRO) at Brisbane, Australia. He joined the staff of the International Soil Reference and Information Centre (ISRIC) in The Netherlands in 1991 to work on the compilation of a world soil data set and retired from full-time employment in 1996. Dr Bridges has written over 120 contributions in scientific journals and edited or co-edited nine books including *Principles and applications of soil geography* (Longman, 1982) and *Response to land degradation* (Science Publishers Inc., 2000). He has authored or co-authored eight books including: *World soils*, (Cambridge University Press, 3rd edn, 1997); *The soils and land use of the district north of Derby* (Memoir of the Soil Survey of Great Britain, Harpenden, 1966); *Surveying derelict land* (Oxford University Press, 1988); *Soils of the Mary River alluvia* (CSIRO, 1990); *World geomorphology* (Cambridge University Press, 1990); and *Classic landforms of the Gower coast* (1987) and *The North Norfolk coast* (1991) for the Geographical Association.

Christopher J. Caseldine is Professor in Environmental Change at the University of Exeter. He gained an MA and PhD from the University of St Andrews and joined the Geography Department at the University of Exeter in 1976, where he was Head of Department from 1993 to 1999. He has been a Visiting Professor at the Universities of Munich and Innsbruck. His main interests relate to Late Quaternary environmental change in northwest Europe, and he has worked in a number of areas including Iceland, southern Norway, northwest Scotland, Ireland and southwest England. His research has involved studying both natural environmental change, principally climatic change, and the influence of past human communities on the landscape, often in collaboration with archaeologists. Dr Caseldine has published over 70 contributions in scientific journals and books, and edited *Environmental change in Iceland: past and present* (Kluwer, 1991).

Adrian J. Luckman is a Lecturer in Remote Sensing at the University of Wales, Swansea. He gained a BSc and DPhil in Electronic Engineering from York University and went on to work for the British National Space Centre at Monks Wood, Cambridgeshire, before joining the staff at Swansea in 1997. His interests and publications lie in the use of radar remote sensing, particularly interferometric techniques, to study various land applications including forestry and deforestation, glaciology (particularly surging glaciers in the Arctic) and urban change. His work includes the first complete interferometric observations of an entire glacier surge in Northern Spitsbergen and the development of the first multifrequency satellite radar image mosaic of Central Siberia. In 2000, Dr Luckman was awarded the Remote Sensing Society President's Prize for the best presentation at the 1999 Remote Sensing Society Annual Symposium.

Geraint Owen is a Geology Lecturer in the Department of Geography at the University of Wales Swansea. He gained his BSc in Geological Sciences at Leeds University and his PhD in Sedimentology at the University of Reading before joining the Department of Geology at the University of Wales Swansea as a Lecturer in 1984, and the Department of Geography in 1990. His research interests are in physical sedimentology and he has published on soft-sediment deformation processes and structures, fluvial sedimentology of the Old Red Sandstone and palaeo-environmental interpretation of Quaternary sands and gravels. Dr Owen is active in the Earth Science Teachers' Association and the South Wales Group of the Geologists' Association, and he is also European Book Review Editor of the interdisciplinary journal *The Holocene*.

Allen H. Perry is a Senior Lecturer in the Department of Geography, University of Wales Swansea, where he has worked since gaining a BA and PhD in Geography from the University of Southampton. His interests are in climatology and especially in synoptic and applied climate studies. He is a Fellow of the Royal Meteorological Society and a past Chairman of the Association of British Climatologists and has been a Visiting Lecturer at Rhodes University, Grahamstown, South Africa. Dr Perry is joint author of *Synoptic climatology* (Methuen, 1973), *The ocean–atmosphere system* (Longman, 1977), *Highway meteorology* (Spon, 1991) and *Applied climatology* (Routledge, 1998). He is also the author of *Environmental hazards in the British Isles* (Allen and Unwin, 1981) and of over 50 contributions in scientific journals. In addition to extensive consultancy work in applied climatology, he is a member of the Welsh Office Climate Change Impacts in Wales Steering Group, a contributing author to the Intergovernmental Panel on Climate Change Third Assessment and a member of the ACACIA project assessing the potential effects of climate change in Europe.

Richard A. Shakesby is a Senior Lecturer in the Department of Geography at the University of Wales Swansea. After gaining a BA in Geography from Portsmouth Polytechnic and a PhD in Glacial Geomorphology from Edinburgh University, he worked at Hereford College of Education before moving to Swansea in 1978. His interests range from glacial and periglacial geomorphology and glacial history to human impact on landforms and geomorphological processes with a particular emphasis on soil erosion, its causes and consequences. He has been a Visiting Lecturer at Khartoum and Zimbabwe Universities and, since 1988, a principal investigator on three European Union-funded projects concerned with land degradation and soil conservation issues in Portugal. In addition to Portugal, he has conducted research in the UK, Norway, Austria, Sudan and Zimbabwe. Dr Shakesby was Editor of the journal *Cambria* during the 1980s and is the author of over 80 contributions in scientific journals and books.

Rory P.D. Walsh is a Professor in Physical Geography at the University of Wales, Swansea. He is also the Research Co-ordinator of the Royal Society South-East Asia Rain Forest Research Programme, based at Danum Valley in Sabah (Malaysian Borneo), where he has been researching since 1990. After gaining an MA and PhD in Geography at St John's College, Cambridge, he worked for a year as Temporary Lecturer in Geography at Durham University before moving to a permanent lectureship at Swansea. His research interests lie in the fields of hydrology and tropical and Mediterranean physical geography and have included: drainage networks and hydrogeomorphological processes in the humid tropics; recent and historical changes in rainfall, tropical cyclone frequency and drought magnitude-frequency in tropical areas and their hydrological, geomorphological and ecological impacts; the influence of forest fires and land management on hydrology and soil erosion

in the Mediterranean; and acid waters and heavy-metal river pollution problems in the UK. In pursuit of these interests he has carried out field research in the Eastern Caribbean islands, the Sudan, Sarawak and Sabah, Thailand, Indonesia, Vietnam and Portugal. He was a member of the Royal Geographical Society Expedition to Mulu (Sarawak) in 1977–1978, a Visiting Lecturer at the University of Khartoum on several occasions over the period 1982–1990, and a Visiting Scholar at the University of Würzburg in his sabbatical year of 1990. Professor Walsh received the Back Award from the Royal Geographical Society in 1996 for contributions to tropical hydrology and geomorphology. His publications include over 60 contributions to scientific journals and books, including a major contribution to the book *The tropical rain forest* (Cambridge University Press, 1996).

Robert J. Whittaker is a Reader in Biogeography at the University of Oxford and a Fellow in Physical Geography at St Edmund Hall, Oxford. He gained a BSc in Botany and Geography from the University of Hull, an MSc in Ecology from the University College of North Wales and a PhD in Geography from University College Cardiff. After a brief spell working on a GIS project at Birkbeck College, he took up a post at Oxford in 1986, where he has since remained. His interests focus on island biogeography, dispersal ecology, vegetation succession, ecosystem development on the Krakatau Islands (Indonesia), vegetation–environment relationships and the relationships between climate and diversity. He has carried out ecological research in a range of tropical, temperate and arctic–alpine environments. Dr Whittaker is the Editor of the journal *Global Ecology and Biogeography - a Journal of Macroecology*, and is Deputy Editor of both the *Journal of Biogeography* and *Diversity and Distributions*. He is the author of some 50 publications in scientific journals and books, including the book: *Island biogeography: ecology, evolution and conservation* (Oxford University Press, 1998).

Katherine J. Willis is a University Lecturer in Physical Geography and Fellow of St Hugh's College, Oxford. She gained her first degree in Geography and Environmental Science at the University of Southampton, followed by a PhD in the Godwin Institute for Quaternary Research, Department of Plant Sciences, University of Cambridge. Her research focuses on the long-term relationship between vegetation dynamics and global environmental change. Projects have centred on the use of palaeoecological techniques to provide a detailed temporal perspective to ecological, archaeological and geological methods, ranging from questions on rates of postglacial soil development to processes responsible for the initiation of the Northern Hemisphere glaciation. Research sites have included lakes, bogs and volcanic maars in central and southeastern Europe. Before taking up her University Lectureship, Dr Willis held a Trevelyan Research Fellowship, an NERC Postdoctoral Fellowship and then a Royal Society University Research Fellowship in the University of Cambridge. Publications to date have included a number of primary research articles, review articles and book chapters. Her first book *Plant evolution: from the first cell to the flower in 2 billion years* is to be published shortly by Oxford University Press. She was the Book Review Editor and is now an Associate Editor of the journal *The Holocene* and a member of the Editorial Advisory Boards of *Global Ecology and Biogeography*, *The Norwegian Journal of Geography* and *Porocilo Slovenski*.

Contributors

The initials in parentheses after each contributor's address are those used in the entries to indicate authorship.

Professor Brigitta Ammann, Institute of Plant Sciences, University of Bern, Altenbergrain 21, CH-3013 Bern, Switzerland;
email: Brigitta.Ammann@ips.unibe.ch (*BA*)

Dr William E.N. Austin, School of Geography and Geosciences, University of St Andrews, St Andrews, Fife KY16 9AL, Scotland, UK;
email: wena@st-andrews.ac.uk (*WENA*)

Dr Heiko Balzter, Section for Earth Observation, Centre for Ecology and Hydrology, Monks Wood, Abbots Ripton, Huntingdon, Cambridgeshire PE28 2LS, UK; email: hbal@ceh.ac.uk (*HB*)

Dr Christopher J. Barrow, School of Social Sciences and International Development, University of Wales Swansea, Singleton Park, Swansea SA2 8PP, Wales, UK; email: C.J.Barrow@swansea.ac.uk (*CJB*)

Ir Niels H. Batjes, International Soil Reference and Information Centre, PO Box 353, 6700 A J Wageningen, The Netherlands; email: batjes@isric.nl (*NHB*)

Dr Douglas I. Benn, School of Geography and Geosciences, University of St Andrews, St Andrews, Fife KY16 9AL, Scotland, UK;
email: doug@st-andrews.ac.uk (*DIB*)

Professor Keith D. Bennett, Quaternary Geology, Department of Earth Sciences, Uppsala University, Villavägen 16, S-752 36 Uppsala, Sweden;
email: Keith.Bennett@geo.uu.se (*KDB*)

Professor Hilary H. Birks, Botanical Institute, University of Bergen, Allégaten 41, N-5007 Bergen, Norway; email: Hilary.Birks@bot.uib.no (*HHB*)

Professor H. John B. Birks, Botanical Institute, University of Bergen, Allégaten 41, N-5007 Bergen, Norway; email: John.Birks@bot.uib.no (*HJBB*)

Mrs Mary A. Boulton, Department of Geography, University of Tennessee, Knoxville, 304 Burchfield Geography Building, Knoxville, TN 37996-0925, USA; email: mboulton@utk.edu (*MAB*)

Dr Doreen S. Boyd, Centre for Earth and Environmental Science Research, School of Earth Sciences and Geography, Kingston University, Penrhyn Road, Kingston-upon-Thames KT1 2EE, Surrey, UK; email: d.boyd@kingston.ac.uk (*DSB*)

Dr E. Michael Bridges, 3 The Green, Hempton, Fakenham NR21 7LG, Norfolk, UK;
email: embridges@freenet.co.uk (*EMB*)

Professor Keith R. Briffa, Climatic Research Unit, School of Environmental Sciences, University of East Anglia, Norwich NR4 7TJ, Norfolk, UK;
email: K.Briffa@uea.ac.uk (*KRB*)

Mr Stephen J. Brooks, Department of Entomology, Natural History Museum, Cromwell Road, London SW7 5BD, UK; email: S.Brooks@nhm.ac.uk (*SJB*)

Dr Nick D. Brown, Department of Plant Sciences, University of Oxford, South Parks Road, Oxford OX1 3RB, UK;
email: nick.brown@plant-sciences.oxford.ac.uk (*NDB*)

Dr Paul Budd, Department of Archaeology, University of Durham, South Road, Durham DH1 3LE, UK;
email: p.d.budd@durham.ac.uk (*PB*)

Dr M. Jane Bunting, Department of Geography, University of Hull, Hull HU6 7RX, UK;
email: m.j.bunting@geo.hull.ac.uk (*MJB*)

Dr Mark B. Bush, Department of Biological Sciences, Florida Institute of Technology, 150 West University Boulevard, Melbourne, FL 32901, Florida, USA;
email: mbush@fit.edu (*MBB*)

Professor Chris J. Caseldine, School of Geography and Archaeology, University of Exeter, Amory Building, Exeter EX4 4RJ, UK; email: C.J.Caseldine@exeter.ac.uk (*CJC*)

Professor Frank M. Chambers, Centre for Environmental Change and Quaternary Research, GEMRU, University of Gloucestershire, Francis Close Hall, Cheltenham GL50 4AZ, Glos, UK; email: FChambers@Glos.ac.uk (*FMC*)

Dr Lesley Cherns, Department of Earth Sciences, Cardiff University, PO Box 914, Cardiff CF10 3YE, Wales, UK; email: cherns@cardiff.ac.uk (*LC*)

Dr Michèle L. Clarke, School of Geography, University of Nottingham, University Park, Nottingham NG7 2RD, UK;
email: Michele.clarke@nottingham.ac.uk (*MLC*)

Dr Michelle A. Clarke, National Soil Resources Institute, Cranfield University, Silsoe, Bedfordshire MK45 4DT, UK;
email: MichelleClarke@cranfield.ac.uk (*MAC*)

Dr Edward R. Cook, Tree-Ring Laboratory, Lamont-Doherty Earth Observatory, 61 Route 9W, Palisades, NY 10964, USA; email: drdendro@ldeo.columbia.edu (*ERC*)

Dr John C.W. Cope, Department of Earth Sciences, Cardiff University, PO Box 914, Cardiff CF10 3YE, Wales, UK; email: CopeJCW@Cardiff.ac.uk (*JCWC*)

Mr Martin G. Coulson OBE, Ministry of Defence, Defence Estates Agency, Durrington, Salisbury SP4 8AF, Wiltshire, UK; email: martin.coulson@de.mod.uk (*MGC*)

Dr Bryan T. Cronin, Department of Geology and Petroleum Geology, University of Aberdeen, King's College, Aberdeen AB24 3UE, Scotland, UK; email: cronin@abdn.ac.uk (*BTC*)

Dr S. Petra Dark, Department of Archaeology, University of Reading, Whiteknights, PO Box 218, Reading RG6 6AA, UK; email: S.P.Dark@reading.ac.uk (*SPD*)

Professor Alastair G. Dawson, Centre for Quaternary Science, School of Natural and Environmental Sciences, Coventry University, Coventry CV1 5FB, UK; email: A.Dawson@coventry.ac.uk (*AGD*)

Dr Mark H. Dinnin, School of Geography and Archaeology, University of Exeter, Amory Building, Exeter EX4 4RJ, UK; email: m.h.dinnin@exeter.ac.uk (*MHD*)

Dr Stefan H. Doerr, Land Surface Processes and Management Research Group, Department of Geography, University of Wales Swansea, Singleton Park, Swansea SA2 8PP, Wales UK; email: S.Doerr@swansea.ac.uk (*SHD*)

Dr P. Quentin Dresser, Swansea Radiocarbon Dating Laboratory, Department of Geography, University of Wales Swansea, Singleton Park, Swansea SA2 8PP, Wales, UK; email: P.Q.Dresser@swansea.ac.uk (*PQD*)

Professor Rose M. D'Sa, Economic and Social Committee of the European Communities, Rue Ravenstein 2, B-1000 Brussels, Belgium (*RMD*)

Dr Lisa Dumayne-Peaty, School of Geography and Environmental Sciences, University of Birmingham, Edgbaston, Birmingham B15 2TT, UK; email: Lpeaty@worcestershire.gov.uk (*LD-P*)

Dr Julia B. Edwards, Department of Humanities and Science, University of Wales College Newport, Caerleon Campus, PO Box 179, Newport NP18 3YG, Wales, UK; email: julia.edwards@newport.ac.uk (*JBE*)

Professor Derek M. Elsom, Department of Geography, Oxford Brookes University, Oxford OX3 0BP, UK; email: dmelsom@brookes.ac.uk (*DME*)

Ms Anette Engelmann, Centre for Environmental Change and Quaternary Research, GEMRU, Cheltenham and Gloucester College of HE, Francis Close Hall, Cheltenham GL50 4AZ, Glos, UK; email: AEngelmann@chelt.ac.uk (*AE*)

Dr John R. Evans, School of Geography and Geosciences, University of St Andrews, St Andrews, Fife KY16 9AL, Scotland, UK; email: jre1@st-andrews.ac.uk (*JRE*)

Mr Tim Fearnside, Department of Geography, University of Wales Swansea, Singleton Park, Swansea SA2 8PP, Wales, UK; email: T.Fearnside@swansea.ac.uk (*TF*)

Dr António J.D. Ferreira, Departamento de Ambiente e Ordenamento, Universidade de Aveiro, 3810-193 Aveiro, Portugal; email: vibrante@dao.ua.pt, aferreira@mail.esac.pt (*AJDF*)

Dr Katherine J. Ficken, Environmental Change Research Group, Department of Geography, University of Wales Swansea, Singleton Park, Swansea SA2 8PP, Wales, UK; email: K.J.Ficken@swansea.ac.uk (*KJF*)

Professor Giles M. Foody, Department of Geography, University of Southampton, Highfield, Southampton SO17 1BJ, UK; email: G.M.Foody@soton.ac.uk (*GMF*)

Professor Hugh M. French, Departments of Geography and Earth Sciences, University of Ottawa, PO Box 450 Station A, Ottawa, Ontario KIN 6N5, Canada; email: hfrench@science.uottawa.ca (*HMF*)

Dr Alejandro C. Frery, Universidade Federal de Pernambuco, Centro de Informática, CP 7851 Recife, PE, 50732-970 Brazil; email: frery@cin.ufpe.br (*ACF*)

Dr Janice L. Fuller, Department of Botany, National University of Ireland, Galway, Ireland; email: Janice.Fuller@nuigalway.ie (*JLF*)

Dr Adam R. Gardner, School of Geography, University of Nottingham, University Park, Nottingham NG7 2RD, UK; email: Adam.Gardner@nottingham.ac.uk (*ARG*)

Dr Brian D. Giles, Honorary Research Fellow, School of Geography and Environmental Sciences, University of Birmingham, 3/50 Onepoto Road, Takapuna, Auckland 1309, New Zealand; email: Gilesnz@ihug.co.nz (*BDG*)

Professor Dr Rüdiger Glaser, Geographisches Institut, Universität Würzburg, Am Hubland, 97074 Würzburg, Germany; email: ruediger.glaser@mail.uni-wuerzburg.de (*RG*)

Professor Michael J. Hambrey, Centre for Glaciology, Institute for Geography and Earth Sciences, University of Wales, Aberystwyth, Ceredigion SY23 3DB, Wales, UK; email: mjh@aber.ac.uk (*MJH*)

Ir Alfred E. Hartemink, International Soil Reference and Information Centre, PO Box 353, 6700 A J Wageningen, The Netherlands; email: Hartemink@ISRIC.nl (*AEH*)

Dr Dawn Hendon, Department of Geography, School of Geography and Archaeology, University of Exeter, Amory Building, Exeter EX4 4RJ, UK; email: D.Hendon@exeter.ac.uk (*DH*)

Professor David T. Herbert, Urban and Social Policy and Practice Research Group, Department of Geography, University of Wales Swansea, Singleton Park, Swansea SA2 8PP, Wales, UK; email: D.T.Herbert@swansea.ac.uk (*DTH*)

Dr Sheila P. Hicks, Institute of Geosciences, PL3000, 90401 University of Oulo, Finland; email: sheila.hicks@oulu.fi (*SPH*)

Dr Richard J. Huggett, School of Geography, University of Manchester, Oxford Road, Manchester M13 9PL, UK; email: richard.huggett@man.ac.uk (*RJH*)

Dr John B. Hunt, Centre for Environmental Change and Quaternary Research, GEMRU, Cheltenham and Gloucester College of HE, Francis Close Hall, Cheltenham GL50 4AZ, Glos, UK; email: jhunt@chelt.ac.uk (*JBH*)

Professor John L. Innes, FRBC Chair of Forest Management, University of British Columbia, Forest Resources Management, Second Floor, Forest Sciences Centre, 2045, 2424 Main Mall, Vancouver, BC V6T 1Z4, Canada; email: innes@interchg.ubc.ca (*JLI*)

Mr Paul R. Jepson, School of Geography and the Environment, Oxford University, Mansfield Road, Oxford OX1 3TB, UK;
email: Paul.jepson@geog.ox.ac.uk (*PRJ*)

Dr Stephen H. Jones, School of Geography and the Environment, University of Oxford, Mansfield Road, Oxford OX1 3TB, UK;
email: stephen.jones@blacksci.co.uk (*SHJ*)

Dr Jeremy T. Kerr, Department of Zoology, University of Oxford, South Parks Road, Oxford OX1 3PS, UK; email: jeremy.kerr@ccrs.nrcan.gc.ca (*JTK*)

Dr Damian M. Lawler, School of Geography and Environmental Sciences, University of Birmingham, Edgbaston, Birmingham B15 2TT, UK; email: D.M.Lawler@bham.ac.uk (*DML*)

Dr Mark V. Lomolino, Department of Environmental and Forest Biology, Illick Hall, SUNY College of Environmental Science and Forestry, Syracuse, NY 13210, USA; email: island@esf.edu (*MVL*)

Dr Adrian J. Luckman, Environmental Modelling and Earth Observation Group, Department of Geography, University of Wales Swansea, Singleton Park, Swansea SA2 8PP, Wales, UK; email: A.Luckman@swansea.ac.uk (*AJL*)

Dr Anson W. Mackay, Environmental Change Research Centre, Department of Geography, University College London, 26 Bedford Way, London WC1H 0AP, UK; email: amackay@geog.ucl.ac.uk (*AWM*)

Professor John A. Matthews, Environmental Change Research Group, Department of Geography, University of Wales Swansea, Singleton Park, Swansea SA2 8PP, Wales, UK; email: J.A.Matthews@swansea.ac.uk (*JAM*)

Dr Julian C. Mayes, Environment, Resources and Geographical Studies, School of Life Sciences, Whitelands College Campus, University of Surrey Roehampton, London SW15 3SN, UK;
email: J.Mayes@Roehampton.ac.uk (*JCM*)

Dr Danny McCarroll, Environmental Change Research Group, Department of Geography, University of Wales Swansea, Singleton Park, Swansea SA2 8PP, Wales, UK; email: D.McCarroll@swansea.ac.uk (*DMcC*)

Dr Jenny C. McElwain, Department of Geology, The Field Museum, 1400 S Lake Shore Drive, Chicago IL 60605, USA; email: mcelwain@fmnh.org (*JCMc*)

Dr T. Victor Mesev, School of Environmental Studies, University of Ulster at Coleraine, Coleraine BT52 1SA, Co Londonderry, Northern Ireland;
email: tv.mesev@ulst.ac.uk (*TVM*)

Professor Atle Nesje, Department of Geology, University of Bergen, Allégaten 41, N-5007 Bergen, Norway; email: atle.nesje@geol.uib.no (*AN*)

Professor Greg O'Hare, Geography Division, University of Derby, Kedleston Road, Derby DE22 1GB, UK; email: G.Ohare@derby.ac.uk (*GOH*)

Dr Colin P. Osborne, Department of Animal and Plant Sciences, University of Sheffield, Alfred Denny Building, Western Bank, Sheffield S10 2TN, UK; email: C.P.Osborne@sheffield.ac.uk (*CPO*)

Dr Geraint Owen, Earth Surface Processes and Management Research Group, Department of Geography, University of Wales Swansea, Singleton Park, Swansea SA2 8PP, Wales, UK; email: G.Owen@swansea.ac.uk (*GO*)

Dr Martin A. Pearce, Millenia Limited, Unit 3, Weyside Park, Newman Lane, Alton, Hampshire, GU34 2PJ, UK; email: MilleniaSC@msn.com (*MAP*)

Dr Allen H. Perry, Environmental Change Research Group, Department of Geography, University of Wales Swansea, Singleton Park, Swansea SA2 8PP, Wales, UK; email: A.H.Perry@swansea.ac.uk (*AHP*)

Dr Andrew J. Plater, Department of Geography, PO Box 147, University of Liverpool, Liverpool L69 7ZT, UK; email: Gg07@liv.ac.uk (*AJP*)

Mr Tristan Quaife, Environmental Modelling and Earth Observation Group, Department of Geography, University of Wales Swansea, Singleton Park, Swansea SA2 8PP, Wales, UK; email: T.Quaife@swansea.ac.uk (*TQ*)

Dr David A. Richards, School of Geographical Sciences, University of Bristol, Bristol BS8 1SS, UK; email: David.Richards@bristol.ac.uk (*DAR*)

Dr Iain Robertson, Quaternary Dating Research Unit, CSIR Environmentek, PO Box 395, Pretoria 0001, South Africa; email: irobertson@csir.co.za (*IR*)

Dr Deborah Z. Rosen, Environment Agency, Waterside House, Waterside North, Lincoln LN1 1LA, UK; email: debbierosen@hotmail.com (*DZR*)

Dr Paul J. Saich, Department of Geography, University College London, 26 Bedford Way, London WC1H 0AP, UK; email: psaich@geog.ucl.ac.uk (*PJS*)

Dr Danielle C. Schreve, Department of Geography, Royal Holloway, University of London, Egham TW20 0EX, Surrey, UK; email: danielle.schreve@rhul.ac.uk (*DCS*)

Dr Richard A. Shakesby, Land Surface Processes and Management Research Group, Department of Geography, University of Wales Swansea, Singleton Park, Swansea SA2 8PP, Wales, UK; email: R.A.Shakesby@swansea.ac.uk (*RAS*)

Professor John Shaw, Department of Earth and Atmospheric Sciences, University of Alberta, Edmonton, Alberta T6G 2E3, Canada;
email: John.Shaw@ualberta.ca (*JS*)

Mr Jamie G. Smith, Environmental Change Research Group, Department of Geography, University of Wales Swansea, Singleton Park, Swansea SA2 8PP, Wales, UK; email: ggsmithj@swansea.ac.uk (*JGS*)

Dr Geoffrey M. Smith, Section for Earth Observation, Centre for Ecology and Hydrology, Monks Wood, Abbots Ripton, Huntingdon PE28 2LS, Cambridgeshire, UK; email: gesm@ceh.ac.uk (*GMS*)

Professor Ian F. Spellerberg, Environmental Management and Design Division, PO Box 84, Lincoln University, Canterbury, New Zealand; email: spelleri@lincoln.ac.nz (*IFS*)

Mr Greg Spellman, School of Environmental Science, University College Northampton, Boughton Green Road, Northampton NN2 7AL, UK; email: greg.spellman@northampton.ac.uk (*GS*)

Dr Thomas Spencer, Cambridge Coastal Research Unit, Department of Geography, University of Cambridge, Downing Place, Cambridge CB2 3EN, UK; email: TS111@cam.ac.uk (*TSp*)

Dr Catherine E. Stickley, Environmental Change Research Centre, Department of Geography, University College London, 26 Bedford Way, London WC1H 0AP, UK; email: c.stickley@ucl.ac.uk (*CES*)

Dr Tazio Strozzi, Gamma Remote Sensing, Thunstrasse 130, CH-3074 Muri BE, Switzerland; email: strozzi@gamma-rs.ch (*TS*)

Dr John C. Sweeney, Department of Geography, National University of Ireland, Maynooth, Maynooth, Co Kildare, Ireland; email: John.Sweeney@May.ie (*JCS*)

Professor Michael R. Talbot, Department of Geology, University of Bergen, Allégaten 41, N-5007 Bergen, Norway; email: michael.talbot@geol.uib.no (*MRT*)

Mr Andrew R. Tallon, Urban and Social Policy and Practice Research Group, Department of Geography, University of Wales Swansea, Singleton Park, Swansea SA2 8PP, Wales, UK; email: ggtallon@swansea.ac.uk (*ART*)

Dr Kevin J. Tansey, Environmental Modelling and Earth Observation Group, Department of Geography, University of Wales Swansea, Singleton Park, Swansea SA2 8PP, Wales, UK; email: kevin.tansey@jrc.it (*KJT*)

Dr Andrew D. Thomas, School of Environment and Life Science, Telford Institute of Environmental Systems, University of Salford, Greater Manchester M5 4WT, UK; email: a.d.thomas@salford.ac.uk (*ADT*)

Dr Colin J. Thomas, Urban and Social Policy and Practice Research Group, Department of Geography, University of Wales Swansea, Singleton Park, Swansea SA2 8PP, Wales, UK; email: C.J.Thomas@swansea.ac.uk (*CJT*)

Dr D. Neil Thomas, Centre for Earth and Environmental Sciences Research, School of Geological Sciences, Kingston University, Penrhyn Road, Kingston-upon-Thames KT1 2EE, Surrey, UK; email: n.thomas@kingston.ac.uk (*NDT*)

Dr Russell D. Thompson, Honorary Fellow, University of Reading, 13 Badgers Brook Drive, Ystradowen, Cowbridge CF71 7TX, Vale of Glamorgan, Wales, UK; email: russellthompson@lineone.net (*RDT*)

Dr Colin E. Thorn, Department of Geography, University of Illinois at Champaign-Urbana, 220 Davenport Hall, 607 South Mathers Avenue, Urbana, IL 61801, USA; email: c-thorn@uiuc.edu (*CET*)

Dr John G. Tyrrell, Department of Geography, University College of Cork, Cork, Ireland; email: stgg8007@ucc.ie (*JGT*)

Professor H. Jesse Walker, Department of Geography and Anthropology, Louisiana State University, Baton Rouge, LA 70803, USA; email: hwalker@lsu.edu (*HJW*)

Professor Lawrence R. Walker, Department of Biological Sciences, University of Nevada, Las Vegas, Box 454004, NV 89154-4004, USA; email: walker@nevada.edu (*LRW*)

Dr Chris D. Walley, c/o Earth Resources Limited, University Innovation Centre, Singleton Park, Swansea SA2 8PP, Wales, UK; email: chris.walley@ntlworld.com (*CDW*)

Professor Rory P.D. Walsh, Land Surface Processes and Management Research Group, Department of Geography, University of Wales Swansea, Singleton Park, Swansea SA2 8PP, Wales, UK; email: R.P.D.Walsh@swansea.ac.uk (*RPDW*)

Dr Dennis A. Wheeler, Geography Department, University of Sunderland, Forster Building, Chester Road, Sunderland SR1 3SD, UK; email: dennis.wheeler@sunderland.ac.uk (*DAW*)

Miss M. Louise Whitehead, Ministry of Defence, Defence Estates Agency, Durrington, Salisbury SP4 8AF, Wiltshire, UK; email: whitehead400@hotmail.com (*MLW*)

Dr Robert J. Whittaker, School of Geography and the Environment, University of Oxford, Mansfield Road, Oxford OX1 3TB, UK; email: Robert.Whittaker@geog.ox.ac.uk (*RJW*)

Dr Katherine J. Willis, School of Geography and the Environment, University of Oxford, Mansfield Road, Oxford OX1 3TB, UK; email: Kathy.Willis@geog.ox.ac.uk (*KJW*)

Miss Lindsay J. Wilson, School of Geography and Geosciences, University of St Andrews, St Andrews, Fife KY16 9AL, Scotland, UK; email: ljw4@st-andrews.ac.uk (*LJW*)

Dr Peter Wilson, School of Biological and Environmental Studies, University of Ulster at Coleraine, Coleraine BT52 1SA, Co Londonderry, Northern Ireland; email: P.Wilson@ulst.ac.uk (*PW*)

Professor U. Barbara Wohlfarth, Department of Physical Geography and Quaternary Geology, Stockholm University, SE-106 91 Stockholm, Sweden; email: Wohlfarth@geo.su.se (*UBW*)

Dr Matthew J. Wooller, Geophysical Laboratory, Carnegie Institution of Washington, 5251 Broad Branch Road NW, Washington DC 20015-1305, USA; email: wooller@gl.ciw.edu (*MJW*)

Dr Tim Young, GeoArch, 54 Heol y Cadno, Thornhill, Cardiff CF14 9DY, Wales, UK; email: Tim.Young@GeoArch.demon.co.uk (*TY*)

A

a *Annum*: sometimes used as an abbreviation for years before the present. *GO*

[See also BEFORE PRESENT (BP), GA, KA, MA]

aa Blocky lava, from the Hawaiian for sore feet. A blocky, rubbly or clinkery LAVA FLOW, typical of slow-moving, viscous, BASALT lavas. *GO*

[See also PAHOEHOE]

abduction Reasoning that involves seeking the controlling states of affairs (causes) from the resulting states of affairs (effects) by application of a LAW (or law-like statement). Abduction is widely used in the explanation of past events. *JAM*

[See also DEDUCTION, INDUCTION, RETRODICTION, THEORY]

Von Englehardt, W. and Zimmermann, J. 1988: *Theory of Earth science*. Cambridge: Cambridge University Press.

abiotic Non-living physical and chemical components and factors of an ecosystem (mineral soil particles, rock, water, atmospheric gases, inorganic salts). Sometimes simple organic substances resulting from excretion or DECOMPOSITION may be included. *UBW*

[See also BIOTIC, ECOSYSTEM, ENVIRONMENTAL FACTORS]

ablation Ablation refers to the processes causing mass loss from a GLACIER, including *melting*, wind DEFLATION, avalanching from the front, CALVING of ICEBERGS, RUNOFF, EVAPORATION and SUBLIMATION. *AN*

ablation zone The lower part of a GLACIER where the annual ablation exceeds accumulation (see ACCUMULATION ZONE). *AN*

aboriginal Relating to the original inhabitants as opposed to the later colonisers of a country or region. The term is derived from the Latin *ab origine* (from the beginning). The term *aborigine* is reserved for the NATIVE inhabitants of Australia. *JAM*

Clark, J.S. and Royall, P.D. 1995: Transformation of a northern hardwood forest by aboriginal (Iroquois) fire: charcoal evidence from Crawford Lake, Ontario, Canada. *The Holocene* 5, 1–9. **Head, L.** 1999: *Second nature: the history and implications of Australia as aboriginal landscape*. New York: Syracuse University Press. **Horton, D.R.** 1982: The burning question: aborigines, fire and Australian ecosystems. *Mankind* 13, 237–251.

above-ground productivity The BIOMASS production rate in leaves, branches, stems, trunks and all parts of plants lying above the ground. *RJH*

[See also BELOW-GROUND PRODUCTIVITY, NET PRIMARY PRODUCTIVITY, PRODUCTIVITY]

abrasion The wearing down of rock surfaces by contact with particles undergoing gravitative transfer or transportation by water, wind, snow or ice. *Glacial abrasion*, for example, involves wearing down of rock surfaces by debris trapped between the moving ice and the rock surface. *Polishing* of rock surfaces between STRIATIONS is included by some experts as a separate abrasion mechanism involving removal of small protuberances by overriding rock particles and ice. On a large scale, COMMINUTION of rock particles enhances CHEMICAL WEATHERING, which draws down CARBON DIOXIDE, one of the prominent GREENHOUSE GASES. *JS*

[See also CORRASION, FLUVIAL PROCESSES, GLACIAL EROSION, WIND EROSION]

Benn, D.I. and Evans, D.J.A. 1998: *Glaciers and glaciation*. London: Arnold, 184–188. **Drewry, D.** 1986: *Glacial geologic processes*. London: Edward Arnold.

abrupt climatic change A sharp and sustained *discontinuity* in the climatic record. Abrupt climatic changes are associated with critical physical and biological THRESHOLDS between two different states and have played an important part in the long-term evolution of climate and EXTINCTION events. However, 'abruptness' is a concept relative to the timescale under consideration: abruptness on the geological timescale differs from abruptness from a human perspective. During the past 20 000 years there have been several major changes of short, 50–200 years, duration or less with temperature AMPLITUDES of < 1.0 to 10.0°C. Although there are examples during the historic period, the concept is usually associated with GLACIAL–INTERGLACIAL CYCLES as they experienced some of the most rapid and extreme CLIMATIC CHANGES on record.

The study of ICE CORES shows that at any time the Earth is in one of two basic states, glacial/stadial or interglacial/interstadial, and that the ocean–atmosphere system may switch between these modes in very short time periods. POLLEN ANALYSIS has shown that at the end of the last interglacial some TEMPERATE FORESTS were replaced by BOREAL FORESTS (pine–birch–spruce) possibly within 150 years, although the main climatic shift may have been even more rapid. The last DEGLACIATION occurred in two main steps. The warming around 13 000 radiocarbon years BP was abrupt, large and possibly concentrated within a span of 200 years. Possible explanations have included an abrupt shift in the ocean–atmosphere circulation, ALBEDO changes, increased CARBON DIOXIDE, VOLCANIC ASH and CLOUD. After a brief decline there was a second abrupt warming about 10 000 BP. ICE CORES and MARINE SEDIMENT CORES have shown that this was associated with a rapid change of AEROSOL concentration, possibly over 20 years.

On the shorter HOLOCENE TIMESCALE various mechanisms have been proposed for abrupt climatic changes. These can be divided into three types: natural causes, human causes and CHAOS THEORY. The natural causes include VOLCANIC ERUPTIONS, which add large quantities of VOLCANIC AEROSOLS to the atmosphere, which both reflects incoming SOLAR RADIATION and is warmed by it with the consequence that the Earth's surface is cooled and rainfall may be increased because of slow fallout of AEROSOLS. The eruption of Asama in Japan in AD 1783 and of Laki in Iceland in 1784 cooled the Northern Hemisphere by 1.3°C, while the Mount Pinatubo eruption in the Philippines in 1991 reduced global mean air temperatures by 0.5°C. A combination of the 11-year SUNSPOT CYCLE, the QUASI-BIENNIAL OSCILLATION and EL NIÑO–SOUTHERN OSCILLATION has also been shown to affect temperatures.

The main human activities that caused abrupt climatic changes during the Holocene concern the introduction of agriculture, which changed the reflectivity of the Earth's surface, and URBANISATION, which was associated with much POLLUTION, especially after the beginning of the INDUSTRIAL REVOLUTION of the mid-eighteenth century. The MEDIAEVAL WARM PERIOD may have been due to a change in the Northern Hemisphere atmospheric circulation patterns, whilst current thinking suggests that the LITTLE ICE AGE was a time of enhanced atmospheric dust load. *JGT/BDG*

[See also ATMOSPHERE–OCEAN INTERACTION, DANSGAARD–OESCHGER EVENTS, LATE GLACIAL ENVIRONMENTAL CHANGE, RAPID ENVIRONMENTAL CHANGE]

Adams, J., Maslin, M. and Thomas, E. 1999: Sudden climate transitions during the Quaternary. *Progress in Physical Geography* 23, 1–36. Alverson, K. and Oldfield, F. 2000: Abrupt climate change. *PAGES Newsletter* 8, 7–10. Anderson, D. 2000: Abrupt climatic change. *Geography Review* 13, 2–6. Bard, E., Arnold, M., Maurice, P. *et al.* 1987: Retreat velocity of the North Atlantic polar front during the last deglaciation determined by ^{14}C accelerator mass spectrometry. *Nature* 328, 791–794. Berger, W.H. and Labeyrie, L.D. (eds) 1987: *Abrupt climatic change: evidence and implications.* Dordrecht: Reidel. Broecker, W.S. 2000: Abrupt climate change: causal constraints provided by the paleoclimate record. *Earth Science Reviews* 51, 137–154. Broecker, W.S. and Denton, G.H. 1989: The role of ocean–atmosphere reorganisations in glacial cycles. *Geochimica Cosmochimica Acta* 53, 2465–2501. Dansgaard, W., White, J.W.C. and Johnsen, S.J. 1989: Abrupt termination of the Younger Dryas climate event. *Nature* 339, 532–534. Herron, M.M. and Langway, C.C. 1985: Chloride, nitrate, and sulfate in the Dye 3 and Camp Century, Greenland ice cores. In Langway, C.C., Oeschger, H. and Dansgaard, W. (eds), *Greenland ice core: geophysics, geochemistry, and the environment* [*Geophysical Monograph* 33]. Washington, DC: American Geophysical Union, 77–84. Hu, F.S., Slawinski, D., Wright Jr, H.E. *et al.* 1999: Abrupt changes in North American climate during early Holocene times. *Nature* 400, 437–440. Loon, A.J. van 1999: The meaning of 'abruptness' in the geological past. *Earth Science Reviews* 45, 209–214. McIntyre, A., Ruddiman, W.F. and Jantzen, R. 1976: Southward penetrations of the North Atlantic polar front: faunal and floral evidence of large scale surface water mass movements over the last 225,000 years. *Deep Sea Research* 19, 61–77. Oppo, D.W., McManus, J.F. and Cullen, J.L. 1998: Abrupt climatic events 500,000 to 340,000 years ago: evidence from subpolar North Atlantic sediments. *Science* 279, 1335–1338. Parker, D.E., Wilson, H., Jones, P.D. *et al.* 1996: The impact of Mount Pinatubo on world-wide temperatures. *International Journal of Climatology* 16, 487–497. Severinghaus, J.P., Sowers, T., Brook, E.J. *et al.* 1998: Timing of abrupt climate change at the end of the Younger Dryas interval from thermally fractionated gases in polar ice. *Nature* 391, 141–146. Woillard, G. 1979: Abrupt end to the last interglacial in north-east France. *Nature* 281, 558–562.

abscissa A scale value measured along the *x* axis of a Cartesian coordinate system from the origin of the system to the desired point location. *TF*

[See also CARTESIAN COORDINATES, ORDINATE]

absolute counting A method often used as part of POLLEN ANALYSIS and less frequently in other MICROFOSSIL analyses. The purpose is to obtain data in terms of numbers per unit volume of sediment or per unit area of sediment per unit of time (POLLEN INFLUX). This is achieved by adding a known amount of a marker (usually an exotic pollen type, from a plant not naturally occurring in the region under investigation; in Britain *Lycopodium* spores or *Eucalyptus* pollen are commonly used) to a known volume of sample before beginning processing for pollen analysis. This mode of data collection, also known as *absolute pollen analysis,* has an advantage over percentage data, because the value for a taxon is independent of the values of all other taxa, but calculations of pollen influx are dependent on knowing the sediment ACCUMULATION RATE, which can be uneven and is dependent in turn on the quality and reliability of RADIOCARBON DATING. *MJB*

[See also POLLEN SUM]

Berglund, B.E. and Ralska-Jasiewicsowa, M. 1986: Pollen analysis and pollen diagrams. In Berglund, B.E. (ed.), *Handbook of Holocene Palaeoecology and Palaeohydrology.* Chichester: Wiley, 455–483. Peck, R.M. 1974: A comparison of four absolute pollen preparation techniques. *New Phytologist* 73, 567–587. Stockmarr J. 1972: Tablets with spores used in absolute pollen analysis. *Pollen et Spores* 13, 614–621.

absolute sea level Global EUSTATIC sea level, excluding local influences on the relative heights of land and sea. The most important controls are changes in water volume (GLACIO-EUSTASY), changes in the volume of the world's oceans (*tectono-eustasy*) and changes in water distribution in the oceans (GEOIDAL EUSTASY). Less significant factors in controlling absolute sea level are ISOSTATIC DECANTATION and the addition of juvenile water from the Earth's interior. There is thought to be a detectable impact on absolute sea level caused by human interference with the hydrological cycle (e.g. GROUNDWATER extraction, river management schemes) as well as the possible impact of GLOBAL WARMING. *MJH*

[See also GEOID, RELATIVE SEA LEVEL, SEA-LEVEL CHANGE, SEA-LEVEL RISE]

Mörner, N.-A. 1987: Models of global sea level changes. In Tooley, M.J. and Shennan, I. (eds), *Sea level changes.* Oxford: Blackwell, 332–355.

absolute temperature Temperature measured on the *Kelvin* scale. Absolute zero is −273.16°C, the freezing point of water is 273 K and the boiling point 373 K (the Kelvin being the SI unit of temperature). *JAM*

absorption The process of incorporation or taking up of one substance by another. Gases may be absorbed by liquids or solids, liquids by solids. Absorbed substances do not necessarily exhibit ADSORPTION. *JAM*

absorption spectrum The array of ABSORPTION lines and absorption bands that results from the passage of radiant energy from a continuous source through a selectively absorbing medium. The absorption by atmospheric gases is highly selective. The ATMOSPHERE is virtually transparent to short-wave SOLAR RADIATION except for about 18% of this incoming energy, which is absorbed directly by OZONE and WATER VAPOUR. Ozone absorption is concentrated in three solar spectral bands (0.20–0.31, 0.31–0.35 and 0.45–0.85 μm) and water vapour absorbs in several bands between 0.9 and 2.1 μm. Most of the *long-wave radiation* emitted by the Earth is absorbed by water vapour and CARBON DIOXIDE (see GREENHOUSE EFFECT) with the remainder escaping to space through the *atmospheric window* between 8 and 13 μm.   *GS*

Ramanathan, V., Barkstrom, B.R. and Harrison, E.F. 1989: Climate and the Earth's radiation budget. *Physics Today* 42, 22–32.

abstraction The artificial removal of water from surface water bodies or GROUNDWATER sources, commonly for agricultural or industrial use or human consumption. Intensive abstraction of groundwater may cause, for example, WATER TABLES to fall, MINERALISATION of soil, impoverishment of vegetation and LAND SUBSIDENCE. Excessive surface water abstraction has been the major factor in the lowering of the level and changes in SALINITY of the Aral Sea since the 1960s. Groundwater abstraction from the chalk aquifer beneath London has caused a lowering of the *piezometric level* by over 60 m since the mid-nineteenth century.   *JAM*

Goudie, A.G. 2000: *The human impact on the natural environment*, 5th edn. Oxford: Blackwell.

abyssal Depths and environments between about 3000 and 6000 m in the OCEAN; although authors differ on precise depth limits, abyssal environments account for about 75% of the oceans. The ocean floor at these depths is typically represented by the *abyssal plains*, very flat areas (gradient less than 0.05°) at the base of the CONTINENTAL RISE. They are underlain by OCEANIC CRUST, the topography of which is subdued by a cover of sediment, largely deposited from TURBIDITY CURRENTS and DEBRIS FLOWS, but including also PELAGIC SEDIMENTS. About 75 true abyssal plains are recognised, ranging in size from smaller areas of the Mediterranean Sea (e.g. Alboran abyssal plain, 2600 km^2) to the Enderby abyssal plain in the Southern Ocean (3 703 000 km^2). They are least common in the Pacific Ocean, where OCEANIC TRENCHES trap land-derived sediment, and most common in the Southern Ocean, where sediment input is higher. The rapid accumulation of allochthonous sediment in deep water, distance from human populations and scarcity of indigenous life has encouraged their consideration as sites for RADIOACTIVE WASTE disposal.   *BTC*

[See also BATHYAL, CONTINENTAL MARGIN, HADAL, OCEAN BASIN]

accelerator mass spectrometry (AMS) dating
A RADIOCARBON DATING technique, in which a nuclear accelerator is used to enable the individual counting of ^{14}C ATOMS. Conventional *mass spectrometers* are able to separate the three ISOTOPES of carbon, ^{12}C, ^{13}C and ^{14}C, by virtue of their mass difference, but are not sensitive enough to detect natural concentrations of ^{14}C. The ^{14}C ion beam is also affected by the presence of IONS of the same mass, such as ^{14}N and ^{13}CH. A nuclear accelerator is required to enable the discrimination between the conta-

minating ions and to provide the ^{14}C ions with sufficient energy to enable their detection. The equipment required is expensive, but the sample throughput rate is greater than conventional techniques. The big advantage of AMS dating is the sample size requirement of c. 1 mg in contrast to the > 2 g required by most conventional laboratories.   *PQD*

Gove, H.E. 1992: The history of AMS, its advantages over decay counting: applications and prospects. In Taylor, R.E., Long, A. and Kra, R.S. (eds), *Radiocarbon after four decades: an interdisciplinary perspective*. New York: Springer, 214–229. **Hedges, R.E.M. and Gowlett, J.A.J.** 1986: Radiocarbon dating by accelerator mass spectrometry. *Scientific American* 254, 100–107. **Kirner, D., Southon, J.R., Hare, P.E and Taylor, R.E.** 1996: Accelerator mass spectrometry radiocarbon measurement of sub-milligram samples. In Orna, M.V. (ed.), *Archaeological chemistry: organic, inorganic and biochemical analysis*. Washington, DC: American Chemical Society, 434–442. **Lowe, J.J., Lowe. S., Fowler, A.J.** *et al.* 1988: Comparison of accelerator and radiometric radiocarbon measurements obtained from Late Devensian lateglacial lake sediments from Lake Gwernan, North Wales, UK. *Boreas* 17, 355–369. **Tuniz, C., Bird, J.R., Fink, D and Herzog, G.F.** 1998: *Accelerator mass spectrometry: ultrasensitive analysis for global science*. Boca Raton, FL: CRC Press.

accessory mineral A mineral present in a rock in small quantities but not a characteristic of the rock type.   *JAM*

acclimatisation The physiological adjustment of living organisms, especially animals and including humans, to changes in their contemporary environment. Rapid movement of humans to high altitudes can result in acute 'mountain sickness', accompanied by headaches, sickness and lethargy, as a result of lower oxygen concentrations at altitude. Drinking plenty of liquid and resting allows acclimatisation to occur over a period of days. Peoples living at high altitudes in the Andes or Himalayas are well acclimatised. A distinction should be made between acclimatisation, *acclimation* (whereby changes are induced rapidly under artificial, laboratory conditions) and genetic ADAPTATION.   *AHP*

accommodation space Space available for SEDIMENTARY DEPOSITS to accumulate; a concept central to SEQUENCE STRATIGRAPHY. If accommodation space is not available at one locality, sediment will be carried further along the transport path. In a subaqueous environment, accommodation space equates to water depth.   *GO*

[See also TRANSGRESSION]

Muto, T. and Steel, R.J. 1997: Principles of regression and transgression: the nature of the interplay between accommodation and sediment supply. *Journal of Sedimentary Research* 67, 994–1000.

accordant summits The tops of hills or mountains of approximately the same altitude, assumed to represent the remnants of a considerably dissected EROSION SURFACE, or the uniform reduction in the height of summits between evenly-spaced valleys.   *RAS*

[See also DENUDATION CHRONOLOGY]

accretion (1) In geomorphology, the accumulation of sediments. (2) In meteorology, the growth of a particle of ice by collision with a water drop.   *RAS*

[See also ACCRETIONARY COMPLEX, OROGENY]

accretionary complex An accumulation of OCEANIC SEDIMENTS scraped from the surface of a plate undergoing SUBDUCTION at a DESTRUCTIVE PLATE MARGIN, and accreted to the edge of the over-riding plate (see PLATE TECTONICS). It is also known as an *accretionary prism, accretionary wedge* or *subduction complex*. The geological structure is dominated by THRUST FAULTS, allowing accretionary complexes to be recognised in the GEOLOGICAL RECORD. Fragments of OCEANIC CRUST may be scraped off the subducting plate and preserved as *ophiolite complexes*. The accreted material may partly fill the OCEANIC TRENCH and may protrude above sea level to form a non-volcanic ISLAND ARC close to the trench (sometimes known as an *outer arc* or *outer-arc ridge*). Modern examples include Barbados in the West Indies, and the Andaman and Nicobar Islands on the Indian Ocean side of the Sunda Arc. *GO*

Corfield, R.I., Searle, M.P. and Green, O.R. 1999: Photang thrust sheet: an accretionary complex structurally below the Spontang ophiolite constraining timing and tectonic environment of ophiolite obduction, Ladakh Himalaya, NW India. *Journal of the Geological Society, London* 156, 1031–1044. **Karig, D.E., Lawrence, M.B., Moore, G.F. and Curray, J.R.** 1980: Structural framework of the forearc basin, NW Sumatra. *Journal of the Geological Society, London* 137, 77–91. **Leggett, J.K.** 1987: The Southern Uplands as an accretionary prism: the importance of analogues in reconstructing palaeogeography. *Journal of the Geological Society, London* 144, 737–752.

accumulated temperature A summation of mean daily temperature values above a chosen THRESHOLD value, expressed as DEGREE DAYS or heat units. The threshold value is often taken as 6°C, representing the minimum temperature necessary for the commencement of growth in temperate cereals and grasses. This is an index of the efficiency, as opposed to the length, of the thermal GROWING SEASON. Variants of the index may be used outside of agricultural applications to quantify heat stress and energy demand. Significant changes in accumulated temperature may accompany more modest changes in mean temperature. As many crops have known bio-climatic requirements, in terms of degree-days for germination/ripening, their responses to future CLIMATIC CHANGE SCENARIOS may be modelled. Similarly, the potential impacts of GLOBAL WARMING on heating/cooling energy demand may be examined using variants of the degree day concept. *JCS*

Soulé, P.T. and Suckling, P.W. 1995: Variations in heating and cooling degree days in the South Eastern U.S.A, 1960–1989. *International Journal of Climatology* 15, 335–368. **Tivy, J.** 1990: *Agricultural ecology*. London: Longman, 30–33.

accumulation area ratio (AAR) The ratio of the accumulation area to the total glacier area. The AAR of a GLACIER varies mainly as a function of its MASS BALANCE: ratios below 0.5 indicate negative mass balance; 0.5–0.8 corresponds to *steady-state* conditions; and values above 0.8 reflect positive mass balance regimes. An AAR of 0.6 ± 0.05 is generally considered to characterise steady-state conditions of valley/cirque glaciers. Ice caps and piedmont glaciers may, however, differ significantly from this ratio. The largest source of inaccuracy related to the AAR method of determining the EQUILIBRIUM-LINE ALTITUDE (ELA) on former glaciers is the reconstruction of the surface contours, especially if the glacier margins intersect valley-side topographic contours at small angles or coincide with them for some distance. However, this source of error is considered to be randomly distributed and is unlikely to introduce major deviations from representative conditions. *AN*

[See also EQUILIBRIUM-LINE ALTITUDE RECONSTRUCTION]

Andrews, J.T. 1975: *Glacial systems: an approach to glaciers and their environments*. North Scituate, MA: Duxbury Press. **Porter, S.C.** 1975: Equilibrium line altitudes of late Quaternary glaciers in the Southern Alps, New Zealand. *Quaternary Research* 5, 27–47.

accumulation rate The accumulation rate of SEDIMENT is usually expressed in cm y^{-1} and calculated on the basis of an AGE DEPTH MODEL, which may be based on RADIOCARBON DATING or other dating methods. Variations in the accumulation rate of, for example, PEAT can give useful information about the HYDROLOGY of the system, since the amount of organic material entering the permanently waterlogged lower layers depends on the decay rate in the upper layers; when the surface is relatively wet, decay is relatively slow and hence the sediment depth increases rapidly, compared with drier conditions when the decay rate is higher. Accumulation rate data are also an important parameter for ABSOLUTE COUNTING of POLLEN, SPORES and other MICROFOSSILS. Calculating the accumulation rate is dependent on establishing an independent CHRONOLOGY, and therefore affected by the problems common to age depth models. *MJB*

[See also GRENTZHORIZONT, RECURRENCE SURFACE]

Higgit, S.R., Oldfield, F. and Appleby, P.G. 1991: The record of land use change and soil erosion in the late Holocene sediments of the Petit Lac d'Annecy, eastern France. *The Holocene* 1, 14–28. **Snowball, I. and Thompson, R.** 1992: A mineral magnetic study of Holocene sediment yields and deposition patterns in the Llyn Geirionydd catchment, north Wales. *The Holocene* 2, 238–248. **Webb, R.S., and Webb III, T.** 1988: Rates of sediment accumulation in pollen cores from small lakes and mires of Eastern North America. *Quaternary Research* 30, 284–297.

accumulation zone The upper part of a GLACIER, where the annual accumulation exceeds ABLATION. *AN*

accuracy The notion of how close an approximation is to the real or true value, i.e. by analogy closeness of the shot to the bull's-eye on a target (see Figure). *CET*

[See also PRECISION]

Griffiths, J.C. 1967: *Scientific method in the analysis of sediments*. London: Longman. **Wagner, G.A.** 1998: *Age determination of young rocks and artifacts*. Berlin: Springer.

acid (1) A substance that dissociates in water producing H$^+$ ions and reacts with a BASE to form a salt. (2) Possessing a pH of less than 7.0 (see ACIDITY). *ADT*

[See also ACIDIFICATION, ACID RAIN, ALKALI, ALKALINITY]

acid deposition The transfer of acidic substances of anthropogenic origin (primarily derived from the combustion of FOSSIL FUELS) from the atmosphere to the Earth's surface, including DRY DEPOSITION, OCCULT DEPOSITION and WET DEPOSITION. *JAM*

[See also ACID PRECIPITATION, ACID RAIN]

Adams, D.D. and Page, W.P. (eds) 1985: *Acid deposition: environmental, economic and political issues*. New York: Plenum.

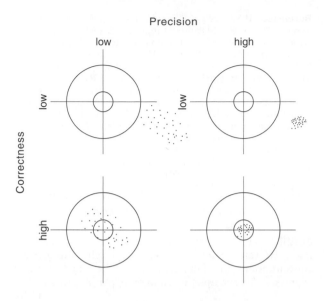

Precision

accuracy *A representation of precision and correctness: the target analogue. Accuracy implies the combination of correctness with precision (after Griffiths, 1967; Wagner, 1998)*

acid mine drainage   The seepage of water containing sulphuric acid from active or abandoned MINING operations, such as coal, nickel and copper mines. Resulting from the OXIDATION of sulphur-bearing minerals, such as iron sulphide (pyrites), it often contains toxic metals as well as being corrosive. Control measures include the sealing off of deep mines, LAND RESTORATION of surface strip mines and the chemical treatment of drainage water or backfill (such as the addition of an alkaline buffer).   *JAM*

Griggs, G.B. and Gilchrist, J.A. 1983: *Geologic hazards, resources and environmental planning.* Belmont, CA: Wadsworth.

acid precipitation   Rain, snow (*acid snow*) or other forms of PRECIPITATION, with a pH of < 5.6 – the equilibrium pH of atmospheric CARBON DIOXIDE (0.03%) in contact with water.   *JAM*

[See also ACID RAIN]

Norton, S.A., Lindberg, S.E. and Page, A.L. (eds) 1990: *Acidic precipitation.* Vol. 4. *Soils, aquatic processes and lake acidification.* New York: Springer.

acid rain   Precipitation with an artificially low pH due to industrial EMISSIONS. The pH of pure water in equilibrium with atmospheric CARBON DIOXIDE is 5.6 and thus the term 'acid rain' is reserved for precipitation of pH < 5.6. There has been a notable increase in the acidity of precipitation across North America and Europe since the INDUSTRIAL REVOLUTION as a result of the emission of sulphur and NITROGEN OXIDES from the burning of fossil fuels. New regions with significant acid rain problems are emerging with the spread of INDUSTRIALISATION (see Figure). The OXIDATION of sulphur and nitrogen oxides in the atmosphere to sulphuric acid (H_2SO_4) and nitric acid (HNO_3) creates relatively strong acids. It is estimated that 60–70% of the mean annual acidity of precipitation originates from sulphuric acid and the remainder from nitric

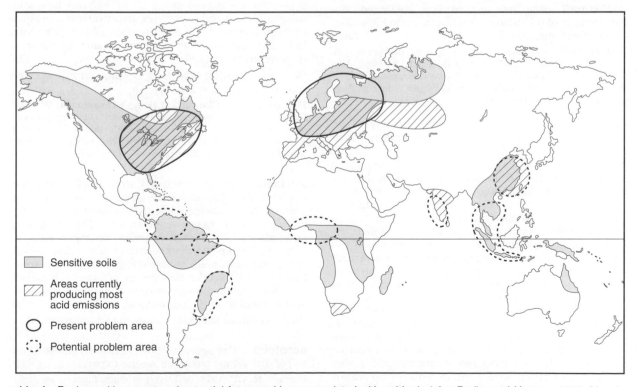

acid rain *Regions with current and potential future problems associated with acid rain (after Rodhe and Herrera, 1988; Mannion, 1999)*

acid. This reflects the much larger quantities of sulphur emitted and the fact that the sulphuric acid molecule in solution releases two H$^+$ ions and the nitric acid molecule only one. Precipitation in Scotland and North America with a pH as low as 2.1–2.4 has been recorded. Acid rain leads to accelerated leaching of base cations from soils, increased solubility of metals and the acidification of surface waters. *ADT*

[See also ACIDIFICATION, FOREST DECLINE]

Howells, G. 1990: *Acid rain and acid waters.* New York: Ellis Horwood. **Likens, G.E., Driscoll, C.T. and Buso, D.C.** 1996: Long-term effects of acid rain: response and recovery of a forest ecosystem. *Science* 272, 244–246. **Mannion, A.M.** 1999: Acid precipitation. In Pacione, M. (ed.), *Applied geography: principles and practice.* London: Routledge, 36–50. **McCormick, J.** 1997: *Acid Earth: the politics of acid pollution,* 3rd edn. London: Earthscan. **Park, C.C.** 1987: *Acid rain: rhetoric and reality.* London: Methuen. **Rodhe, H. and Herrera, R. (eds)** 1988: *Acidification in tropical countries.* Chichester: Wiley. **Schindler, D.W.** 1988: Effects of acid rain on freshwater ecosystems. *Science* 239, 149–157. **Veselý, J.** 1994: Effect of acidification on trace metal transport in fresh waters. In Steinberg, C.E.W. and Wright, R.F., *Acidification of freshwater ecosystems: implications for the future.* Wiley: Chichester, 141–151. **Wellburn, A.** 1994: *Air pollution and acid rain: the biological impact,* 2nd edn. Harlow: Longmans. **White, J.C. (ed.)** 1989: *Global climate change linkages: acid rain, air quality, and stratospheric ozone.* New York: Elsevier.

acid sulphate soils

Soils rich in sulphates occurring mainly in the tropics and associated with former mangrove swamps in the Far East, and coastal areas of West Africa and South America. Pyrite in peaty soils oxidises to sulphuric acid when soils are drained, making soils extremely acid (pH < 4.0), causing toxic levels of aluminium ions and restricting crop growth. These soils can be reclaimed using lime and fertiliser, combined with LEACHING out of the sulphate conditions, but in most cases the process is uneconomic. *EMB*

Dent, D. 1980: Acid sulphate soils: morphology and prediction. *Journal of Soil Science* 31, 87–99. **Van Breemen, N.** 1980: Acid sulphate soils. In *Land reclamation and water management: developments, problems and challenges.* Wageningen: International Institute for Land Reclamation, 53–57.

acidification

The process of increasing ACIDITY or, more usually, a decrease in pH. In the environmental context, acidification may be NATURAL or ANTHROPOGENIC. Anthropogenic acidification results from the inputs of ACID RAIN as well as from acidity generated within the catchment itself. Acidification of freshwaters is a widespread problem across North America, Scandinavia, Central Europe and the UK. The problems associated with acidification, however, do not necessarily reflect the distribution of acid rain. The degree of acidification will depend upon the underlying geology and soils. Calcareous lithologies are not sensitive to acidification, whereas areas of granite and gneiss and areas of thin soils are much more sensitive. This reflects the availability of base cations and hence the BUFFERING of soils and water from the effects of acidification by neutralising the effects of acidity. LANDUSE is also a crucial factor and increased acidification is often associated with coniferous AFFORESTATION. This is because conifers are efficient scavengers of atmospheric acidity and the pH of THROUGHFALL under coniferous forests is significantly lower than that in rainfall. In addition, the litter under such forests also tends to acidify the soils. *ADT*

Battarbee, R.W. 1994: Diatoms, lake acidification and the Surface Water Acidification Program (SWAP) – a review. *Hydrobiologia* 274, 1–70. **Battarbee, R.W., Flower, R.J., Stevenson, A.C. and Rippey, B.** 1985: Lake acidification in Galway: a paleoecological test of competing hypotheses. *Nature* 314, 350–352. **Battarbee, R.W., Flower, R.J., Stevenson, A.C. et al.** 1988: Diatoms and chemical evidence for reversibility of acidification of Scottish Lochs. *Nature* 332, 530–532. **Gee, A.S. and Stoner, J.H.** 1988: The effects of afforestation and acid deposition on the water quality of upland Wales. In Usher, M.B. and Thompson, D.B.A. *Ecological change in the uplands.* Blackwell: Oxford 273–287. **Henriksen, A.** 1982: Acidification of groundwater in Norway. *Nordic Hydrology* 13, 183–192. **Mannion, A.M.** 1999: Acidification and entrophication. In Mannion, A.M. and Bowlby, S.R. (eds), *Environmental issues in the 1990s.* Chichester: Wiley, 177–195. **Mason, C.F.** 1996: *Biology of freshwater pollution.* Harlow: Longman.

acidity

(1) The state or quality of being ACID, i.e. having a pH value below 7 (a relatively high hydrogen ion concentration in solution). (2) The acid content expressed, for example, in milligrams per litre (mg L^{-1}). *UBW*

[See also ACID RAIN, ACIDIFICATION, ALKALINITY, DIATOM ANALYSIS, ICE CORES, PALAEOLIMNOLOGY, PALAEOSALINITY, pH, VOLCANIC ASH]

Adams, F. (ed.) 1984: *Soil acidity and liming.* Madison, WI: American Society of Agronomy. **Lerman, A. (ed.)** 1978: *Lakes: chemistry, geology, physics.* New York: Springer.

acidity record of volcanic eruptions

Volcanic eruptions may inject large volumes of acidic AEROSOLS into the atmosphere, which are subsequently deposited in a range of environments. A detailed record of such acidic deposition is preserved in ICE CORES, and may be studied using ELECTRICAL CONDUCTIVITY MEASUREMENTS, pH measurements or chemical identification of particular IONS. Individual eruptions may be identified, allowing the age of ice layers to be checked. *DIB*

[See also GLACIOLOGICAL VOLCANIC INDEX, SULPHUR DIOXIDE, VOLCANIC AEROSOLS]

Hammer, C.U., Clausen, H.B. and Dansgaard, W. 1980: Greenland ice sheet evidence of post-glacial volcanism and its climatic impact. *Nature* 288, 230–235. **Zielinski, G.A.** 2000: Use of paleo-records in determining variability within the volcanism–climate system. *Quaternary Science Reviews* 19, 417–438.

Acrisols

Strongly leached humid tropical and subtropical soils with a subsurface ARGIC HORIZON rich in LOW ACTIVITY CLAYS, having a low CATION EXCHANGE CAPACITY and a BASE SATURATION of less than 50%. These soils occur in Southeast USA, South America, Southeast Asia and Africa on old EROSION SURRFACES or deposition surfaces and in PIEDMONT areas of humid regions (SOIL TAXONOMY: Kandic and Kanhaplic great groups of *Ultisols*). If Acrisols lose their topsoil through erosion the subsoil is strongly acid and fertility is greatly reduced. *EMB*

[See also WORLD REFERENCE BASE FOR SOIL RESOURCES]

acrotelm

The upper, shallow, seasonally AEROBIC horizon of diplotelmic MIRES, above the CATOTELM. *FMC*

Ingram, H.A.P. 1978: Soil layers in mires: function and terminology. *Journal of Soil Science* 29, 224–227. **Ivanov, K.E.** 1981: *Water movement in mirelands.* London: Academic Press.

activation function In the study of NEURAL NETWORKS, the activation function of a node is a mathematical TRANSFER FUNCTION that describes the relationship between the input values to the node and the output value.   *PJS*

active continental margin A CONTINENTAL MARGIN characterised by an abrupt transition from the coastline (usually backed by a high, coast-parallel, volcanically active mountain range) to a deep OCEANIC TRENCH and subject to frequent severe EARTHQUAKES (see Figure). Examples occur around the Pacific rim, most notably the 9000 km long western margin of South America, and they are also known as *Pacific-type* or *seismic* continental margins. An active continental margin corresponds to a convergent or DESTRUCTIVE PLATE MARGIN where SUBDUCTION causes the consumption of OCEANIC CRUST beneath CONTINENTAL CRUST, giving rise to andesitic VOLCANISM, UPLIFT and TECTONIC activity.   *CDW*

Burk, C.A. and Drake. C.L. (eds) 1974: *The geology of continental margins.* New York: Springer. Busby, C.J. and Ingersoll, R.V. (eds) 1995: *Tectonics of sedimentary basins.* Oxford: Blackwell. Pickering, K.T., Hiscott, R.N. and Hein, F.J. 1989: *Deep marine environments: clastic sedimentation and tectonics.* London: Unwin Hyman.

active layer The layer of ground that is subject to annual freezing and thawing in areas underlain by PERMAFROST. The thickness varies from as little as 15 cm in polar regions to over 1.0–1.5 m in SUBARCTIC regions. Thickness depends upon the interaction of factors such as ambient air temperature, degree and orientation of slope, vegetation, drainage, snow cover, soil and/or rock type and water content.   *HMF*

French, H.M. 1988: Active layer processes. In Clark, M.J. (ed.), *Advances in periglacial geomorphology.* Chichester: Wiley, 151–179.

active volcano A VOLCANO that has produced a VOLCANIC ERUPTION in very recent historical times, usually

within living memory, and is expected to produce further eruptions in the near future. An example is El Teide on Tenerife, in the Canary Islands, which last erupted in AD 1909 (the Chinyero eruption).   *GO*

[See also DORMANT VOLCANO, EXTINCT VOLCANO]

actualism The idea that the geological past should be interpreted by invoking processes that operate at the present day rather than by appealing to extra, unknown causes. The expression is essentially a synonym for the 'Uniformity of Process' element of Lyell's UNIFORMITARIANISM. Its greater precision and brevity makes it a useful term, but it has been less popular in the English-speaking world than in mainland Europe. This probably reflects the fact that in many European languages 'actual' (as in *actualisme, aktualismus*) means 'the present' rather than in English where, confusingly, it means 'real'. In the context of actualism and environmental change there must be a suspicion that the present Earth, in its immediate postglacial state, may be quite aberrant when compared with long periods of apparent stability that characterise much of the GEOLOGICAL RECORD.   *CDW*

[See also CATASTROPHISM, GRADUALISM, NEOCATASTROPHISM, PALIMPSEST]

Gould, S. J. 1987: *Time's arrow, time's cycle: myth and metaphor in the discovery of geological time.* Cambridge, MA: Harvard University Press. Goodman, N. 1967: Uniformity and simplicity. *Geological Society of America Special Paper* 89, 93–99.

adaptation The process of modifying to suit changed, or changing, conditions. The term is also used for the condition of being adapted. Darwinian NATURAL SELECTION proposes that organisms with a characteristic that improves the chances of leaving descendants relative to organisms without the characteristic will become better adapted to the ENVIRONMENT, and the characteristic may be termed an adaptation. Adaptation, as a process, is a continuing dynamic change.   *KDB*

[See also DARWINISM, ETHOLOGICAL (BEHAVIOURAL) ADAPTATIONS, EVOLUTION, GENETIC FITNESS, LAMARCKISM, VULNERABILITY]

adaptionist approach An approach to archaeological change and human evolution that assumes the changes are physiological or behavioural adjustments to contemporary environmental conditions rather than MIGRATION, DIFFUSION or internal cultural dynamics.   *JAM*

adaptive radiation The evolutionary diversification of a taxon in response to environmental change.   *JAM*

[See also ADAPTATION]

Schluter, D. 2000: *The ecology of adaptive radiation.* Oxford: Oxford University Press.

adhesion The trapping of wind-blown sand on a damp or sticky surface, producing BEDFORMS and SEDIMENTARY STRUCTURES that include *adhesion ripples, adhesion warts* and horizontal to inclined *adhesion laminations.* In the geological record these are indicators of aeolian sediment transport over a damp substrate as in, for example, an interdune setting.   *MRT*

Brookfield, M.E. 1992: Eolian systems. In Walker, R.G. and James, N.P. (eds), *Facies models: response to sea level change.* St

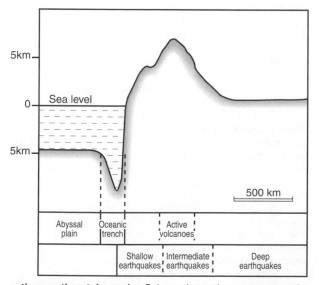

active continental margin *Schematic section across an active continental margin*

John's, Newfoundland: Geological Association of Canada, 143–156. **Kocurek, G. and Fielder, G.** 1982: Adhesion structures. *Journal of Sedimentary Petrology* 52, 1229–1241.

adjacency A POLYGON ANALYSIS operation used to determine whether two or more polygons share common boundary lines (see ARC). In graph theory, two vertices (see POINT) are adjacent if they are connected by an edge. *TVM*

adsorption The attachment of molecules to the outside or interior surfaces of a solid by physical or chemical bonding. The molecules of absorbed substances may be adsorbed onto interior surfaces. *JAM*

[See also ABSORPTION]

Harter, R.D. (ed.) 1986: *Adsorption phenomena*. New York: Van Nostrand Reinhold. **Ruthven, D.M.** 1984: *Principles of adsorption*. New York: Wiley.

advection The horizontal transport of energy in the atmosphere in contrast to the vertical transport by CONVECTION and the non-advective process of RADIATION. Advection is normally carried out by AIR MASSES: for example, warm air from the North Atlantic Ocean is regularly advected eastward into Europe in winter. The term is also used for the redistribution of moisture or POLLUTANTS in the air, and for similar transportation in water. *AHP*

aeolian Formed by the action of wind (from the Greek meaning 'God of the winds'). The American spelling is *eolian*. *SHD*

[See also AEOLIANITE, WIND EROSION].

Gill, T. 1996: Eolian sediments generated by anthropogenic disturbance of playas. *Geomorphology* 17, 207–228. **Goudie, A.S., Livingstone, I. and Stokes, S. (eds)** 1999: *Aeolian environments, sediments and landforms*. Chichester: Wiley.

aeolian sediments Sediments transported and deposited by the wind, principally sand as DUNES and silt as LOESS. Wind-blown clay contributes to PELAGIC SEDIMENT in the OCEAN. Aeolian deposits are significant in DESERTS and some coastal settings. *GO*

[See also COASTAL DUNES, DUST]

Nickling, W.G. 1994: Aeolian sediment transport and deposition. In Pye, K. (ed.), *Sediment transport and depositional processes*. Oxford: Blackwell Scientific, 293–350.

aeolian zone The *altitudinal zone* (see ALTITUDINAL ZONATION OF VEGETATION) above the ALPINE ZONE in mountain regions characterised by the absence of established plant life but the presence of organisms, such as arthropods and micro-organisms, transported there by wind. *JAM*

[See also NIVAL ZONE]

Sugg, P.M. and Edwards, J.S. 1998: Pioneer aeolian community development on pyroclastic flows after the eruption of Mount St. Helens, Washington, USA. *Arctic and Alpine Research* 30, 400–407. **Swan, L.W.** 1963: Aeolian zone. *Science* 140, 77–78.

aeolianite Lithified AEOLIAN sands (e.g. dunes) usually cemented by carbonates (the American spelling is *eolianite*). Aeolianites are most commonly formed near high-energy tropical (and Mediterranean) coasts, where carbonate sand is shifted shorewards by wave action, blown onshore and subsequently lithified by carbonate precipitation. Its development often, although not always, reflects glacio-eustatically lowered sea level, exposing sand unavailable for wind action during INTERGLACIALS. Because of rapid lithification, aeolianites are useful for PALAEOENVIRONMENTAL RECONSTRUCTION. PALAEOWIND directions may be determined, dating may be possible (e.g. LUMINESCENCE DATING) and the nature of environmental changes may be determined using techniques such as land snails (MOLLUSCA) and PALAEOSOLS. *SHD*

[See also SEA-LEVEL CHANGE: PAST EFFECTS ON COASTS]

Cook, L.M., Goodfriend, G.A. and Cameron, R.A.D. 1993: Changes in the land snail fauna of eastern Madeira during the Quaternary. *Philosophical Transactions of the Royal Society* B339, 83–103. **Gardner, R.A.M.** 1983: Aeolianite. In Goudie, A.S. and Pye, K. (eds) *Chemical sediments and geomorphology*. London: Academic Press, 265–300. **Yaalon, D.** 1967: Factors affecting the lithification of aeolianite and interpretation of its environmental significance in the coastal plain of Israel. *Journal of Sedimentary Petrology* 37, 1189–1199. **Zhou, L.P., Williams, M.A.J. and Peterson, J.A.** 1994: Late Quaternary aeolianites, paleosols and depositional environments on the Nepean Peninsula, Victoria, Australia. *Quaternary Science Reviews* 13, 225–239.

aerial photography The collection of remotely sensed information by recording it on photographic film using a camera mounted on an airborne platform. Aerial photography was the first form of REMOTE SENSING and was extensively developed during World War II as a source of intelligence information.

The majority of cameras used for aerial photography are conventional in design, but with special features to improve performance; large format film (frame size of typically 23 by 23 cm), low-distortion lenses, precision construction for both geometric and radiometric accuracy and mountings that take account of the orientation of the camera to the flight line and the motion of the aircraft while the shutter is open. More unusual cameras are also available, for instance *panoramic cameras* to see from horizon to horizon and *strip cameras* for continuous recording rather than collecting single frames.

Black-and-white film is widely used, with variants available for different applications: *mapping film* has a conventional *panchromatic response*; reconnaissance film has a reduced response to blue light to reduce the effects of atmospheric haze; *near-infrared film* has an extended response to include the large response of vegetation in that region. Colour films fall into two categories: *conventional colour film* records true colour photographs; on *false-colour infrared film* the near-infrared response appears red, the red response appears green and the green response appears blue. *False-colour infrared film* is particularly suited to studying vegetation, which appears a strong red colour.

Oblique aerial photographs, recorded when the surface is viewed at an angle, give a realistic view of objects, but scale changes across the photograph make measurements difficult. The majority of aerial photographs are therefore taken vertically (*vertical aerial photographs*), with overlap between adjacent photographs, to give a map-like view, a constant scale under the assumption of a flat surface and the potential for stereoscopic analysis.

Aerial photographs are usually interpreted manually using characteristics such as tone/colour, texture, shape, size, context and pattern. They are now regularly scanned into a digital format for computer-aided analysis, particularly for PHOTOGRAMMETRY. Even with the development of digital OPTICAL REMOTE SENSING INSTRUMENTS, large amounts of survey work are still undertaken with aerial photography. *GMS*

Falkner, E. 1994: *Aerial mapping: method and applications.* Boca Raton, FL: CRC Press. Ciciarelli, J.A. 1991: *Practical guide to aerial photography.* Dordrecht: Kluwer. Arnold, R.H. 1997: *Interpretation of airphotos and remotely sensed imagery.* New York: Prentice Hall.

aerobic   An ENVIRONMENT in which oxygen is available to organisms that require its presence for growth or a process that occurs only in the presence of oxygen. Aerobic organisms acquire oxygen during respiration. Aerobic decomposition involves OXIDATION. *UBW*

[See also ANAEROBIC, ORGANIC SEDIMENT, BIOGENIC SEDIMENT]

aerobiology   The microbiology of the atmosphere, including the study of micro-organisms and organic matter in the atmosphere. *JAM*

Jacobs, W.C. 1951: Aerobiology. In Malone, T.F. (ed.), *Compendium of meteorology.* Boston, MA: American Meteorological Society.

aerology   (1) A synonym for AERONOMY: the meteorology of the upper atmosphere. (2) Less commonly, a synonym for METEOROLOGY. *JAM*

Deland, R. 1987: Aerology. In Oliver, J.E. and Fairbridge, R.W. (eds), *The encyclopedia of climatology.* New York: Van Nostrand Reinhold.

aeronomy   The study of the physics and chemistry of the middle (above the TROPOSPHERE) and upper (above the mesosphere) ATMOSPHERE of Earth and the atmospheres of other planets. *JAM*

[See also AEROLOGY]

Brasseur, G. and Solomon, S. 1986: *Aeronomy of the middle atmosphere: chemistry and physics of the stratosphere and mesosphere.* Dordrecht: Reidel. Romick, G.J. 1996: Aeronomy. In Schneider, S.H. (ed.), *Encyclopedia of climate and weather.* New York: Oxford University Press.

aerosols   Small particles and liquid droplets suspended in a gaseous medium, such as the ATMOSPHERE, where they may be primary or secondary. Primary aerosols are directly derived from dispersal of solids from the Earth's surface; secondary aerosols form from chemical reactions in the atmosphere. *Primary aerosols* include soil and sand particles, sea salts, POLLEN, SPORES, microorganisms (bacteria, fungi), insect fragments and elemental carbon (soot) from combustion. *Secondary aerosols* include nitrate-containing aerosols formed by the OXIDATION of NITROGEN OXIDES and sulphate aerosols formed from OXIDATION of sulphur dioxide. Most aerosols of anthropogenic origin are found in the lower TROPOSPHERE (below 2 km) and typically have residence times of a few days, being removed largely by PRECIPITATION. Aerosols resulting from volcanic eruptions can reach the STRATOSPHERE, where winds transport them around the Earth over many months or even a few years. Aerosols scatter and absorb SOLAR RADIATION so influencing the RADIATION

budget of the Earth–atmosphere system and consequently the CLIMATE. High concentrations of aerosols in the ATMOSPHERE can reduce visibility, producing a haze. Aerosols are involved in the formation of CLOUDS and PRECIPITATION by acting as CONDENSATION NUCLEI or freezing nuclei for the formation of water droplets and ice crystals. *DME*

[See also PARTICULATES, VOLCANIC AEROSOLS]

Houghton, J.T., Filho, L.G.M., Callander, B.A. *et al.* (eds) 1996: *Climate change 1995: the science of climate change.* Cambridge: Cambridge University Press. Thompson, R.D. 1995: The impact of atmospheric aerosols on global climate: a review. *Progress in Physical Geography* 19, 336–350. Twomey, S. 1977: *Atmospheric aerosols.* Amsterdam: Elsevier.

aesthetic degradation   Deterioration of the visual quality of the environment including, for example, inappropriate siting of industry, buildings or power lines. It may be reduced by URBAN AND RURAL PLANNING, by LANDSCAPE ARCHITECTURE and by specific measures such as zoning. *JAM*

[See also LAND DEGRADATION, SOIL DEGRADATION]

Fisher, J.A. 2001: Aesthetics. In Jamieson, D. (ed.), *A companion to environmental philosophy.* Oxford: Blackwell, 264–276.

aestivation   A process by which some animals reduce their metabolic activity in order to survive hot and dry conditions; it may be manifest in summer inactivity. *JLI*

[See also HIBERNATION]

afforestation   An increase in the area of forest either by planting or by natural succession. A distinction is sometimes made between *forestation*, establishing forest for the first time (or after a long period without forest), and REFORESTATION, re-establishing forest that has been removed. In eastern North America, where 60–85% of forests were cleared for agriculture during European settlement in the seventeenth and eighteenth centuries, 60–85% of the land is currently forested following natural reforestation of agricultural land abandoned in the late nineteenth and early twentieth centuries. *JAM*

Motzkin, G., Wilson, P., Foster, D.R. and Allen, A. 1999: Vegetation patterns in heterogeneous landscapes: the importance of history and environment. *Journal of Vegetation Science* 10, 903–920.

afforestation: ecological impacts   Afforestation commonly involves the planting of trees on land not recently or previously covered by woodland, as in marginal upland zones and in disturbed sites including wasteland, derelict land and mining spoils. Many upland moorland sites in northern Europe have been afforested using tree plantations with foreign, ALIEN or EXOTIC coniferous tree species (e.g. Sitka Spruce and Douglas Fir from Canada). There have been many criticisms from environmentalists of such afforestation. Chief among these are: (1) loss of MOORLAND and other *semi-natural habitats* (see SEMI-NATURAL VEGETATION), such as WETLANDS; (2) ACIDIFICATION of soils and PEAT accumulation; (3) disruption to local HYDROLOGICAL BALANCES and HABITATS in and round plantations; (4) the use of PESTICIDES in plantations; (5) reduced BIODIVERSITY; and (6) AESTHETIC DEGRADATION. The low biodiversities of coniferous plantations are a

9

result of the INTRODUCTION of a few alien tree species without their associated biota and HABITAT LOSS consequent upon the creation of simplified tree MONOCULTURES with even-age stands. An estimated 51 local vascular plants, many from WETLAND HABITATS (e.g. *Drosera* spp.), and 22 native species of bird (e.g. the dipper), have been displaced or seriously reduced by plantations in the UK.

On the positive side, OLD-GROWTH FOREST can develop its own flora and fauna, e.g. the return of the Goshawk to spruce plantations in the UK. Forest plantations are also a SINK for CARBON DIOXIDE (both trees and soil) and may regulate water supplies better than moorland areas. Specific conservation measures to improve the ECOLOGICAL VALUE of coniferous plantations thus include long rotations (100 years) between commercial fellings to encourage build up of old wood and its associated high biodiversity, restocking by natural regeneration and the use of NATIVE species within woodland plantations.

The ECOLOGICAL RESTORATION of mining spoils and other toxic sites by afforestation has often been unsuccessful because of the difficulty of effectively establishing trees on such sites: soil compaction by machinery, water-logging, NUTRIENT deficiency and toxic minerals are some of the reasons. Attempts to establish tree plantations on coal spoil heaps in the uplands of the UK have often resulted in only low, stunted, ELFIN WOODLAND.   *GOH*

[See also REFORESTATION, REGENERATION, FOREST CLEARANCE]

Forestry Commission 1998: *The UK forestry standard: the government's approach to sustainable forestry.* Edinburgh: Forestry Commission. **Mannion, A.M.** 1997: *Global environmental change: a natural and cultural environmental history.* Harlow Addison-Wesley Longman. **Mather, A.S.** 1990: *Global forest resources.* London: Belhaven. **Peterken, G.** 1996: *Natural woodland: ecology and conservation in northern temperate regions.* Cambridge: Cambridge University Press. **Petty, S.J., Garson, P.J. and McIntosh, R.** 1995: Kielder, the ecology of a man-made spruce forest. *Forest Ecology and Management* 79(1–2) [Special Issue]. **Watkins, C. (ed.)** 1993: *The ecological effects of afforestation.* Wallingford, Oxfordshire: CAB International. **Usher, M.B. and Thompson, D.B.A.** 1988: *Ecological change in the uplands.* Oxford: Blackwell Scientific

afforestation: impacts on hydrology and geomorphology

Afforestation changes the *water balance* within a catchment as well as altering soil properties. Both these lead to impacts on hydrology and geomorphology. Forests have high INTERCEPTION and *transpiration* rates and thus reduce the fraction of rainfall reaching the surface. This reduces the amount of RUNOFF. *Eucalyptus* plantations have been used deliberately to lower high WATER TABLES, in order to control waterlogging and SALINITY problems in Western Australia. Pine afforestation in upland areas can also lead to ACIDIFICATION of soil and streamwaters as well as subsidence and increased depths to the water table. Afforestation also leads to changes in soil properties, notably increases in INFILTRATION rates and improved structure. Incidences of overland flow are thus reduced in forested catchments and storm peaks are lower. These conditions also reduce EROSION and sediment movement down slopes and in channels, although EROSION RATES can be high during preparations for planting when ground cover is minimal and artificial DISTURBANCE is high. Reduced suspended SEDIMENT concentrations (see SUSPENDED LOAD) in channels can lead to accelerated channel incision (see CHANNEL SHAPE), especially in ALLUVIUM.   *ADT*

Douglas, I., Greer, T., Sinun, W. *et al.* 1995: Geomorphology and rainforest logging practices. In McGregor, D. and Thompson, D. (eds), *Geomorphology and land management in a changing environment.* Chichester: Wiley, 309–320. **Shotbolt, L., Anderson, A.R. and Townend, J.** 1998: Changes to blanket peat bog adjoining forest plots at Bad a' Cheo, Rumster Forest, Caithness. *Forestry* 71, 311–324.

African Humid Period   The PLUVIAL in the latest PLEISTOCENE and early HOLOCENE (between about 14 500 and 5500 calendar years BP) when the Sahara was much more vegetated than today with abundant SAVANNA supporting a richer fauna, denser human populations and numerous perennial lakes. The African Humid Period is attributed to strengthening of the African MONSOON in response to a gradual increase in summer-season insolation; the onset and termination nevertheless appear to have been abrupt as a result of the crossing of a THRESHOLD in summer insolation receipt of about 4.2% greater than at present. It was punctuated by a relatively arid phase (corresponding to the YOUNGER DRYAS): the term *Holocene African Humid Period* should be reserved for the humid phase following this arid phase and prior to the change to late-Holocene hyper-aridity.   *JAM*

[See also ARIDLAND PAST ENVIRONMENTAL CHANGE, SAPROPEL]

deMenocal, P., Ortiz, J., Guilderson, T. *et al.* 2000: Abrupt onset and termination of the African Humid Period: rapid climatic responses to gradual insolation forcing. *Quaternary Science Reviews* 19, 347–361. **Ritchie, J.C., Eyles, C.H. and Haynes, C.V.** 1985: Sediment and pollen evidence for an early to mid-Holocene humid period in the eastern Sahara. *Nature* 330, 645–647. **Thinon, M., Ballouche, A. and Reille, M.** 1996: Holocene vegetation of the Central Saharan Mountains: the end of a myth. *The Holocene* 6, 457–462.

Afro-alpine zone   The tropical alpine zone on mountains in Africa. 'Tropical alpine' refers to the regions within the tropics between the upper limit of continuous, closed canopy forest (often around 3500–3900 m above sea level – m.a.s.l.) and the upper limit of plant life (often around 4600–4900 m.a.s.l.). The Afro-alpine environment is characterised by a cool climate and diurnal climatic variations in temperature greater than the seasonal variation, causing nocturnal soil frost heaving. Afro-alpine vegetation is characterised by tussock grasses (of the subfamily Pooideae), erect shrubs with leathery evergreen leaves and megaphytic plants of the genera *Senecio* and *Lobelia*, with physiognomy varying according to climatic and *edaphic factors*. The Afro-alpine zone may be subdivided into upper and lower zones on the basis of floristic variation (e.g. presence or absence of *Senecio* spp.). The movement of the Afro-alpine zone is used, like that of RAINFOREST zones and TREE-LINE VARIATIONS, in tropical palaeoenvironmental reconstruction.   *MJW*

[See also, TROPICAL VEGETATION, ALTITUDINAL ZONATION OF VEGETATION]

Coe, M. J. 1967: *The ecology of the alpine zone of Mount Kenya.* The Hague: Junk. **Hedberg, O.** 1955: Altitudinal zonation of the vegetation on East African Mountains. *Proceedings of the Linnean Society of London* 165, 1952–1953. **Rundel, P. W., Smith, A. P., and Meinzer, F. C.** 1994: *Tropical alpine environments: plant form and function.* Cambridge: Cambridge University Press.

Smith, A. P. 1987: Tropical alpine plant ecology. *Annual Review of Ecology and Systematics* 18, 137–158.

aftershock An EARTHQUAKE that occurs during the days, weeks or months after a major earthquake as a result of movements along the same FAULT surface. Despite usually having a lesser magnitude (see EARTHQUAKE MAGNITUDE) than the main event, aftershocks can cause significant further damage. After the Taiwan earthquake of 20 September 1999 (magnitude 7.6) there were five aftershocks with magnitudes of around 6 within 30 minutes of the main shock, and there had been 21 aftershocks of magnitude greater than 5 within a few days.   *DNT*

Kilb, D., Ellis, M., Gomberg, J. and Davis, S. 1997: On the origin of diverse aftershock mechanisms following the 1989 Loma Prieta earthquake. *Geophysical Journal International* 128, 557–570.

age calibration Conversion of ages or dates into forms approaching *calendrical years* (see CALENDAR/CALENDRICAL AGE/DATE) or the transformation of parameters considered to be proxies for age into a regular timescale using a number of points of known age. Most age calibration follows the latter form, as in LICHENOMETRIC DATING, where measurements of the diameters of lichen thalli on surfaces of unknown age can be calibrated using a curve based on observations of thalli diameters on surfaces of known age. RADIOCARBON DATING utilises data from TREE RINGS to allow calibration of radiocarbon determinations into years BEFORE PRESENT (BP) with a known error term.   *CJC*

[See also CALIBRATED-AGE DATING, CALIBRATION OF RADIO-CARBON DATES]

Colman, S.M., Pierce, K.L. and Birkeland, P.W. 1987: Suggested terminology for Quaternary dating methods. *Quaternary Research* 28, 314–319.

age depth models The age depth modelling of a SEDIMENT sequence may be of critical importance for all interpretations of an age or a rate of an individual horizon. Ages are typically determined at intervals along the sequence by the RADIOCARBON DATING method (or by other methods) and the problem is then to interpolate between such dates or, more rarely, to extrapolate from them to the ends of the sequence of interest. This can be done by any of several methods, including linear interpolation, spline interpolation and polynomial line-fitting. It is then possible to derive sediment ages and sediment ACCUMULATION RATES for any point along the sequence. The rates may also be used to help derive other quantities, such as POLLEN INFLUX. In general, best results are obtained if the ages are evenly spaced (in time) along the sequence, but this is difficult to achieve in practice. However, the choice of age depth model may make a considerable difference to the results and, although it is possible to assess the errors associated with any calculated age or rate, assessment of the error associated with the choice of model is not yet practicable.   *KDB*

[See also SEDIMENTATION, SEDIMENTOLOGICAL EVIDENCE OF ENVIRONMENTAL CHANGE]

Bennett, K.D. 1994: Confidence intervals for age estimates and deposition times in late-Quaternary sediment sequences. *The Holocene* 4, 337–348.

age An interval of geological time: a subdivision of an EPOCH in GEOCHRONOLOGY; equivalent to a STAGE in CHRONOSTRATIGRAPHY.   *GO*

aggradation The build-up of SEDIMENT such that the surface of deposition rises vertically, in contrast to the lateral migration of PROGRADATION. Aggradation can occur if subaqueous sedimentation keeps pace with a rising water level, or increase in ACCOMMODATION SPACE. The development and thickening of permanently frozen ground is termed *permafrost aggradation*.   *GO*

[See also DEGRADATION]

aggregate stability The structural stability of soil aggregates (aggregations of particles generally < 10 mm) regardless of changes in soil moisture. The stability of soil aggregates is related to the resistance of a soil surface to WATER EROSION. Aggregate stability can be reduced through cultivation (e.g. tillage and depletion of organic matter).   *SHD*

[See also SOIL EROSION]

Amezketa, E. 1999: Soil aggregate stability: a review. *Journal of Sustainable Agriculture* 14, 83–151.

aggressive water In the context of KARST processes, water capable of dissolving calcium carbonate. Water may be made more aggressive in respect to the CORROSION of limestone by being acidified through enhanced take-up of carbon dioxide or organic acids from a soil cover or by atmospheric pollution. *Aggressivity* can also be increased by the mixing of two water bodies, even though each may be saturated with respect to calcium carbonate.   *SHD*

[See also CORROSION]

agrarian civilisation A society or state, the economy of which is based on agricultural production and land cultivation. The term is often used to describe those early societies in which agriculture originated and developed, leading to early urban civilisations. Some also use the term *hydraulic civilisation* for those early civilisations, which utilised water in complex IRRIGATION systems.   *LD-P*

[See also AGRICULTURAL HISTORY, AGRICULTURAL ORIGINS, NEOLITHIC, URBANISATION]

Butzer, K.W. 1976: *Early hydraulic civilisation in Egypt.* Chicago, IL: Chicago University Press. **Maisels, C.K.** 1990: *The emergence of civilisation: from hunting and gathering to agriculture.* London: Routledge.

agricultural abandonment The complete withdrawal of energy subsidies and management practices from AGROECOSYSTEMS as a result of economic factors, drought, population migration, warfare, etc. Agricultural abandonment leads to BIOLOGICAL INVASIONS, SECONDARY SUCCESSION and LANDUSE CHANGE.   *GOH*

[See also AGRICULTURAL IMPACTS, AGRICULTURAL INTENSIFICATION]

agricultural history The origin, evolution, diffusion and development of agriculture on any geographical or temporal scale. It may include the study of different types of agriculture, agricultural methods, technology, production systems and crop and animal types. The changing role of the agriculture in the economy, society and culture of a community and its history are also important elements. *LD-P*

[See also AGRICULTURAL IMPACTS, AGRICULTURAL INTENSIFICATION, AGRICULTURAL ORIGINS, AGRICULTURAL REVOLUTION, DOMESTICATION, FOREST CLEARANCE, GENETIC ENGINEERING, GREEN REVOLUTION]

Piggott, S. 1981: *The agrarian history of England and Wales: prehistory.* Cambridge: Cambridge University Press.

agricultural impact on climate

Agriculture impacts on climate through aggregation of microscale alterations to ENERGY BALANCE/BUDGETS and HYDROLOGICAL BALANCE BUDGETS. A crop or vegetation canopy ameliorates CLIMATIC EXTREMES, modifies airflow and humidity and may assist with removal of dust and fog droplets. LANDUSE and LAND COVER changes may thus result in modifications to RADIATIVE TRANSFER or ATMOSPHERIC COMPOSITION.

DEFORESTATION and OVERGRAZING may double the previous ALBEDO value and may be associated with the prolongation of DROUGHTS in areas such as the SAHEL. Estimates have been made of a 15–30% reduction in regional rainfall consequent on the removal of TROPICAL RAIN FORESTS. Vegetation removal and/or burning may also increase the carbon pool in the atmosphere and contribute further to GLOBAL WARMING, as can increased METHANE emissions from paddy rice cultivation and enteric fermentation from livestock. Agriculture–climate interactions exhibit complex FEEDBACK MECHANISMS and some crops (see C-3, C-4 AND CAM PLANTS) may assist with the removal of extra carbon dioxide in the troposphere by growing more vigorously, thereby reducing global warming. *JCS*

Charney, J. 1975: Dynamics of deserts and drought in the Sahel. *Quarterly Journal of the Royal Meteorological Society* 101, 193–202. **Henderson-Sellers, A. and Gornitz, V.** 1984: Possible climatic impacts of land cover transformations, with particular emphasis on tropical deforestation. *Climatic Change* 6, 231–258. **Nicholson, S.** 1988: Land surface-atmosphere interaction: physical processes and surface changes and their effects. *Progress in Physical Geography* 12, 36–65. **Shukla, J., Nobic, C. and Sellers, P.** 1990: Amazon deforestation and climate change. *Science* 247, 1322–1325.

agricultural impact on geomorphology

Modifying the landscape to make it suitable for growing plants and raising animals, together with management of land for agricultural purposes, has led both to reshaping of LANDFORMS and altered geomorphological processes. The building of, for example, STRIP LYNCHETS and TERRACING has intentionally reshaped the landscape. Unintentional changes to landforms include: the formation of RILLS, GULLIES and the creation of BADLANDS; sedimentation of RESERVOIRS; subsidence due to GROUNDWATER extraction for IRRIGATION; AGGRADATION of river valleys and infilling of valleys with VALLEY FILL DEPOSITS; the creation of THERMOKARST in PERMAFROST areas; and the triggering of MASS MOVEMENTS.

FOREST CLEARANCE and PLOUGHING increase the ERODIBILITY of soils causing SOIL EROSION by water and wind to be up to many times higher than in areas characterised by natural vegetation and undisturbed soil. It has been estimated that some 2.7×10^9 tonnes of soil are removed annually from croplands in the United States. In Queensland, Australia, estimated soil losses of

70–150 t ha^{-1} y^{-1} (about 5–10 mm depth of soil removed) on agricultural land have been reported. On the Loess Plateau, China, the annual average EROSION RATE is estimated at 60 t ha^{-1} y^{-1}. Caution is needed in accepting such figures without question as they may be based on data that are not suitable for such extrapolation.

The onset of DOMESTICATION, cultivation and major technological improvements (e.g. introduction of the plough) are often marked in the HOLOCENE sedimentary record by large increases in SEDIMENTATION rates. Such information is derived from, for example, lake cores and sediments derived from other SEDIMENT TRAPS. Dates vary world-wide according to the timing of the introduction of agriculture and its technological innovations. *RAS*

[See also AGRICULTURAL HISTORY, AGRICULTURAL IMPACT ON SOILS, CAESIUM-137, DEFORESTATION: IN THE HOLOCENE, HUMAN IMPACT ON LANDFORMS AND GEOMORPHIC PROCESSES, TILLAGE]

Bell, M.G. and Boardman, J. (eds) 1992: *Past and present soil erosion* [Oxbow Monograph 22]. Oxford: Oxford Books. **Boardman, J.** 1998: An average soil erosion rate for Europe: myth or reality? *Journal of Soil and Water Conservation* 53, 46–50. **Boardman, J., Foster, I.D.L. and Dearing, J.A. (eds)** 1990: *Soil erosion on agricultural land.* Chichester: Wiley. **Dearing, J.** 1994: Reconstructing the history of soil erosion. In Roberts, N. (ed.), *The changing global environment.* Oxford: Blackwell, 242–261. **Mannion, A.M.** 1995: *Agriculture and environmental change: temporal and spatial dimensions.* Chichester: Wiley. **Pimentel, D. (ed.)** 1993: *World soil erosion and conservation.* Cambridge: Cambridge University Press.

agricultural impact on soils

Removal of natural vegetation for cropping interrupts natural MINERAL CYCLING between soils and plants, leaving the soil more vulnerable to WIND EROSION and WATER EROSION. Cultivation provides suitable conditions for seedling emergence and deeper rooting. It loosens the soil and causes an increased rate of OXIDATION of organic matter, and weakens SOIL STRUCTURE. *Liming* counters acidity of soils, and FERTILISERS increase yields by providing available plant nutrients. Some amendments cause acidity. Drainage schemes (see SOIL DRAINAGE) remove excess rainwater from soils, allow a longer period for cultivation and permit deeper rooting by crops. Cropping and GRAZING gradually reduce the fertility of soils (see Figure), and unless plant nutrients are replaced can lead to SOIL EXHAUSTION. In arid regions, IRRIGATION allows cropping where lack of rainfall prohibits it. Climatic change will influence the crops grown and their geographical distribution. Profound modification of soils by agricultural activities can lead to their classification as ANTHROSOLS. Such modifications have occurred since the NEOLITHIC TRANSITION. *EMB*

[See also AGRICULTURAL HISTORY, HUMAN IMPACT ON SOILS, LANDUSE IMPACTS ON SOILS, SOIL VULNERABILITY]

Bridges, E.M. and de Bakker, H. 1998: Soil as an artifact: human impact on the soil resource. *Land* 1, 197–215. **Greenland , D.J. and Szabolcs, I. (eds)** 1994: *Soil resilience and land use.* Wallingford: CAB International. **Macphail, R.I., Courty, M.A. and Gevhardt, A.** 1990: Soils and early agriculture. *World Archaeology* 22, 53–69. **Tilman, D.** 1999: The greening of the green revolution. *Nature* 396, 211.

agricultural intensification

(1) The addition of *energy subsidies* to AGROECOSYSTEMS to enhance the yield of

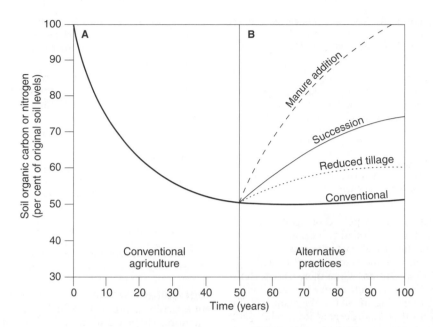

agricultural impact on soils *Typical effects of different agricultural practices on the total carbon or nitrogen content of soil: (A) the effects of conventional agriculture over about 50 years (B) the effects of alternative practices on the recovery of soil fertility (after Tilman, 1999)*

plant and animal products. Agricultural intensification includes *direct subsidies* (such as labour, fuel oil and electricity) and *indirect subsidies* (such as seeds, FERTILISERS, PESTICIDES, machinery and water).

(2) The progressive increase in the application of technology in general and the resulting enhanced production of agricultural systems since the onset of the NEOLITHIC TRANSITION. There have been periods when intensification was relatively rapid. Agricultural intensification has led to progressively greater AGRICULTURAL IMPACTS on environment and society. With a decreasing requirement for land and labour, indirect effects related to agricultural intensification include URBANISATION and the conversion of agricultural land to other uses.

Prior to the mid 1980s in Europe, '*productivist agriculture*' led to intensification, specialisation and concentration of agricultural production. This was supported in eastern Europe by the communist state and, in western Europe, by the Common Agricultural Policy of the European Community. In western Europe, since the 1980s, however, *de-intensification* has resulted from such factors as: the movement towards food quality rather than quantity; withdrawal of state subsidies; greater environmental regulation; reduced use of fertilisers and/or introduction of organic farming methods; the emphasis on SUSTAINABILITY, land management for public access and leisure; and increasing long-term SET-ASIDE SCHEMES for agricultural land.   *GOH/JAM*

[See also AGRICULTURAL REVOLUTION, GREEN REVOLUTION]

Ilbery, B. 1999: The de-intensification of European agriculture. In Pacione, M. (ed.), *Applied geography: principles and practice*. London: Routledge, 274–287. **Meyer, W.B. and Turner, B.L.** 1994: *Changes in land use and land cover: a global perspective*. Cambridge: Cambridge University Press. **Tivy, J.** 1990: *Agricultural ecology*. London: Longman. **Wolman, M.G. and**

Fournier, F.G.A. (eds), 1987: *Land transformation in agriculture*. Chichester: Wiley.

agricultural origins   How, when and where agriculture developed, the beginning of agricultural production or the transition from a HUNTING, FISHING AND GATHERING ECONOMY to permanent/semi-permanent agriculture (i.e. SEDENTISM). The term is often used to refer to the origin of domestic crops and animals (DOMESTICATION). Some of the earliest work on the initial development of agriculture was undertaken by Vavilov in 1926, who proposed that there were eight centres between Asia and Europe in which agriculture began. The areas which are currently recognised as being the most important for the origins of agriculture are the Near East, Meso-America, China, southeast Asia, and certain other parts of the Tropics (e.g. West Africa). There is significant debate as to whether agriculture was invented in one area and spread by processes of diffusion and migration of people and ideas to new areas or whether agriculture had an independent origin in different areas. Agricultural development was a gradual process of change that led to increased economic productivity, human population growth and new social organisations in pre-existing hunting–fishing–gathering communities. These initiated the emergence of AGRARIAN CIVILISATIONS.   *LD-P*

[See also AGRICULTURAL HISTORY, CEREAL CROPS, DOMESTICATION, NEOLITHIC]

Cohen, M.N. 1977: *The food crisis in prehistory: overpopulation and the origins of agriculture*. New Haven, CT: Yale University Press. **Harris, D. R.** 1990: Vavilov concept of centres of origin of cultivated plants, its genesis and its influence on the study of agricultural origins. *Biological Journal of the Linnean Society* 39, 7–16. **Harris, D.R. and Hillman, G.C. (eds)** 1989: *Foraging and farming: the evolution of plant exploitation*. London: Unwin

Hyman. **Moore, A., Hillman, G.C. and Legge, A.J.** 1998: *Abu Hureyra and the advent of agriculture.* New York: Oxford University Press. **Piperno, D.R. and Pearsall, D.M.** 1998: *The origins of agriculture in the lowland neotropics.* San Diego, CA: Academic Press. **Sage, R. F.** 1995: Was low atmospheric carbon dioxide during the Pleistocene a limiting factor for the origin of agriculture. *Global Change Biology* 1, 93–106. **Sherratt, A.** 1997: Climatic cycles and behavioural revolutions: the emergence of modern humans and beginning of farming. *Antiquity* 71, 271–287. **Smith, B.D.** 1995: *The emergence of agriculture.* New York: Scientific American Library. **Stemler, A.B.L.** 1980: The origins of plant domestication in the Sahara and Nile Valley. In Williams, M.A.J. and Faure, H. (eds), *The Sahara and the Nile.* Rotterdam: Balkema, 507–526. **Thorpe, I.J.** 1996: *The origins of agriculture in Europe.* London: Routledge.

agricultural revolution A period of significant agricultural change. The term is applied to a variety of periods in the prehistoric and historical past, the present and the future. It is commonly associated, however, with the period AD 1750–1850 in Britain, when there were technological advances leading to change from essentially subsistence to commercial farming involving the adoption of a mixed farming system that incorporated into arable CROP ROTATIONS, fodder crops such as turnips, and nitrogen-fixing plants such as clover. Changes in hand-tool technology and the introduction of machinery for harvesting and threshing grain facilitated change, as did parliamentary ENCLOSURE, which involved change in the land tenure system with private property rights replacing common property rights. The large open fields were replaced by smaller regular fields surrounded by hedges or walls. In northern and western parts of the country, previously open areas of rough pasture and wasteland became enclosed. An alternative view is that the 'revolution' occurred somewhat earlier, in the 150–250 years leading up to 1750, when in England rising output and productivity were achieved by open field farmers. Similar changes in agriculture also occurred in other parts of Europe and elsewhere in technologically advanced societies at around the same time.

At a global scale, at least three additional agricultural 'revolutions' have been identified: (1) the term has been used for the beginnings of agriculture that more or less coincided with the beginning of the HOLOCENE, when domestication of food plants and animals occurred (see NEOLITHIC TRANSITION). Traditionally this was thought to have involved simple diffusion of ideas from a few distinct agricultural hearths (e.g. in the FERTILE CRESCENT), but the process of innovation is now viewed as more complex. (2) It has been used to refer to a European Mediaeval revolution that took place between the sixth–ninth centuries and the late eighth–twelfth centuries, when significant changes were made to the field systems and plough technology. (3) It has also been used for the process of agricultural INDUSTRIALISATION (*agro-industry*), in which farming became increasingly dominated by the industrial model. Its beginnings are dated as the late 1920s by some researchers but a century earlier according to others. It has been associated with increased risk of animal diseases such as foot-and-mouth and bovine spongiform encephalitis (BSE), as experienced in the UK recently. GENETIC ENGINEERING could be seen as a major development in this industrialisation process.

Wherever or whenever major changes in agriculture have occurred, they have often effected considerable environmental change, most notably to patterns and types of fauna and flora and to soil characteristics (e.g. structure and organic matter), leading to many areas subject to SOIL DEGRADATION. *RAS*

[See also AGRICULTURAL INTENSIFICATION, GREEN REVOLUTION, HUMAN IMPACT ON ENVIRONMENT, HUMAN IMPACT ON TERRESTRIAL VEGETATION, HUMAN IMPACT ON SOIL, SOIL EROSION]

Allen, R.C. 1999: Tracking the agricultural revolution in England. *Economic History Review* 52, 209. **Chambers, J.D. and Mingay, G.E.** 1966: *The agricultural revolution, 1750–1880.* London: Batsford. **Harlan, J.R.** 1976: The plants and animals that nourish man. *Scientific American* 235, 88–97. **Lamberg-Karlovsky, C.C. and Sabloff, J.A.** 1979: *Ancient civilizations: the Near East and Meso-america.* Menlo Park, CA: Benjamin Cummings. **Kerridge, E.** 1967: *The agricultural revolution.* London: Allen and Unwin. **Roberts, N.** 1998: *The Holocene: an environmental history,* 2nd edn. Oxford: Blackwell. **Wallace, A.** 1984: The next agricultural revolution. *Communications in Social Science and Plant Analysis* 15, 191–197.

agricultural waste Unused materials and byproducts of agricultural production, including crop residues, animal manure, slurries and effluents and unused AGROCHEMICALS. Although not generally seen as an environmental problem, agriculture represents the single largest source of waste in many countries, especially those of the developing world. AGRICULTURAL INTENSIFICATION results in large quantities of *animal wastes*, such as faeces, urine and bedding material. In certain circumstances, such as highly permeable soils or a high GROUNDWATER table (see also WATER TABLE), large quantities disposed of upon the land as manure for crops causes *agricultural pollution*. In European countries, 20–50% of animal wastes are used as manure. In the Netherlands this has led to incineration of animal wastes. Leaks from silage pits and slurry ponds on farms with intensively reared livestock can have disastrous local effects upon rivers and groundwater. It is in the processing of agricultural products that many problems become evident. Large quantities of water are required in cleaning farm produce that may become contaminated with soil, organic materials and PESTICIDE residues. Residues from dairies and related industries, such as tanneries, fellmongers and animal processing factories, may also be rich in organic materials, detergents and disinfectants, all of which can have an undesirable environmental impact. *EMB*

[See also DOMESTIC WASTE, INDUSTRIAL WASTE, WASTE MANAGEMENT]

Wilcox, A. 1997: Wastes. In Brune, D., Chapman, D.V., Gwynne, M.D. and Pacna, J.M. (eds), *The global environment: science, technology and management.* Weinheim, Germany: VCH, 625–634.

agrochemicals Chemicals used to increase yields in agriculture, such as FERTILISERS and PESTICIDES. *JAM*

Cremlyn, R.J. *Agrochemicals.* Chichester: Wiley

agroclimatology/agrometeorology The study and application of CLIMATOLOGY/METEOROLOGY to agriculture, including animal husbandry and forestry. *JBE*

Chang, J.H. 1968: *Climate and agriculture.* Chicago, IL: Aldine. **Smith, L.P.** 1975: *Methods in agricultural meteorology.* Amsterdam: Elsevier. **Tivy, J.** 1990: *Agricultural ecology.* Harlow: Longman.

agro-ecological zone A land resource unit defined in terms of temperature regime, a rainfall:potential transpiration ratio and soil water-holding capacity, having a specific range of potentials and constraints for LANDUSE, expressed by the length of the growing period. *EMB*

United Nations Food and Agriculture Organization (FAO) 1996: *Agro-ecological zoning. Guidelines* [*Soils Bulletin* 73]. Rome: FAO.

agro-ecosystem An agricultural ecosystem for the extensive or intensive PRODUCTION of crops and/or livestock. While less structurally and functionally complex than SEMI-NATURAL VEGETATION, by using energy subsidies agro-ecosystems can achieve high levels of NET PRIMARY PRODUCTIVITY and yields. *GOH*

[See also AGRICULTURAL INTENSIFICATION]

Tivy, J. 1990: *Agricultural ecology.* London: Longman.

agroforestry The intimate association of trees and agriculture, whereby woody and non-woody components exhibit economic and ecological interactions. Although used elsewhere (e.g. in the Mediterranean), agroforestry is mainly practised in thè tropics. *AJDF*

[See also DEHESA, DRY FARMING, SOIL CONSERVATION, SUSTAINABILITY]

Sanchez, P.A. 1995: Science in agroforestry. *Agroforestry Systems* 30, 5–55. **Young, A.** 1986: *Agroforestry for soil conservation.* Wallingford: CAB International.

agronomy The study of agricultural land management and rural economy. *JAM*

[See also ENVIRONMENTAL MANAGEMENT, SUSTAINABILITY]

AIDS *Acquired immune deficiency syndrome* (AIDS) is a DISEASE caused by the *human immunodeficiency virus* (HIV), of which there are two main strains – HIV-1 and HIV-2 – both of which originated in Africa and are transmitted mainly by unprotected sexual activity with different partners. The disease or, more correctly, a set of conditions resulting from the collapse of the immune system, became a global *pandemic* in the last two decades of the twentieth century: over 20 million people have died of AIDS and over 30 million are currently living with HIV. The spread of HIV has been greater than predicted, as has been its demographic, social and economic impacts, especially in sub-Saharan Africa where > 10% of the adult population is infected in many countries. *JAM*

Gould, P. 1993: *The slow plague.* Oxford: Blackwell. **Piot, P., Bartos, M., Ghys, P.D.** *et al.* 2001: The global impact of HIV/AIDS. *Nature* 410, 968–973.

air mass An extensive body of air with near-uniform horizontal characteristics of temperature and humidity over a large area (generally hundreds to thousands of kilometres). Air masses tend to be characterised by their region of origin (a tropical maritime air mass, for example, originates in warm oceanic areas), but they undergo modification through time. FRONTS may mark their boundaries. *JBE*

[See also SYNOPTIC CLIMATOLOGY]

Bryson, R.A. 1966: Airmasses, streamlines and the Boreal forest. *Geographical Bulletin* 8, 228–269. **James, R.W.** 1970: Air mass climatology. *Meteorologische Rundschau* 23, 65–70. **Miller, A.A.** 1950: Air mass climatology. *Geography* 38, 55–67.

air pollution The addition of dust, gases, smoke or other PARTICULATE matter to the atmosphere in quantities sufficient to be injurious to the health or welfare of humans or other organisms, or to cause damage to property or ECOSYSTEMS. Although air pollution has affected humans since PREHISTORY, it was the large-scale use of FOSSIL FUELS (especially coal) during and since the INDUSTRIAL REVOLUTION that led to widespread and damaging effects. Damage to health and property was brought about particularly through the incomplete COMBUSTION of fossil fuels producing carbon monoxide, soot and partly pyrolised hydrocarbons, chains of free radical reactions leading to the formation of nitric oxide from atmospheric nitrogen and oxygen (the *Zeldovic cycle*), as well as compounds of sulphur and chlorine from impurities in the fuel.

Comprehensive clean-air legislation began in the mid-twentieth century with the UK Clean Air Act, 1956. This promoted air pollution abatement measures to control air pollution, including the change from coal to oil, gas and electricity, engineering solutions (such as tall chimneys, industrial scrubbers and filters and ABSORPTION and ADSORPTION techniques) and the zoning of landuses. The main source of air pollution in urban areas, where PHOTOCHEMICAL SMOG has replaced the SMOG of Victorian cities, is now the automobile. Air quality management is increasingly on the agenda in URBAN AND RURAL PLANNING, both in the outdoor and indoor environment. *JAM/AHP*

[See also ACID RAIN, ALTERNATIVE ENERGY, ARCTIC HAZE, LEAD, OZONE DEPLETION, PHOTOCHEMISTRY, POLLUTANT, POLLUTION, POLLUTION HISTORY]

Bridgman, H. 1990: *Global air pollution.* London: Belhaven. Press. **Brimblecombe, P.** 1987: *The big smoke.* London: Methuen. **Brimblecombe, P.** 1999: Air pollution. In Alexander, D.E. and Fairbridge, R.W. (eds), *Encyclopedia of environmental science.* Dordrecht: Kluwer, 13–15. **Elsom, D.M.** 1992: *Atmospheric pollution: a global problem,* 2nd edn. Oxford: Blackwell. **Innes, J.L. and Haron, A.H.** (eds) 2000: *Air pollution and the forests of developing and rapidly industrialising countries.* Wallingford: CABI Publishing. **Leslie, G.B. and Lanau, F.W.** 1992: *Indoor air pollution problems and priorities.* Cambridge: Cambridge University Press. **Lyons, T.J. and Scott, W.D.** 1990: *Principles of air pollution meteorology.* London: Belhaven. **Metcalfe, S.** Forthcoming: *Air pollution: an environmental change perspective.* London: Arnold. **Pacyna, J.M. and Ahmadzai, H.** 1997: Air pollution abatement. In Brune, D., Chapman, D.V., Gwynne, M.D. and Pacyna, J.M. (eds), *The global environment: science, technology and management.* Weinheim, Germany: VCH, 724–748. **Seinfeld, J.H.** 1986: *Atmospheric chemistry and physics of air pollution.* New York: Wiley. **Wayne, R.P.** 1991: *Chemistry of the atmosphere.* Oxford: Clarendon Press.

air subsidence Sinking motion in the atmosphere. Air subsidence occurs, for example, in the various types of ANTICYCLONES, in the tropical ocean subsidence zones associated with the WALKER CIRCULATION and as *downdrafts* in convective clouds and thunderstorms. It carries cold, dry air from aloft to the lower troposphere: during subsidence, the air is warmed by adiabatic processes (see LAPSE RATES). *JAM*

air temperature The temperature recorded in a Stevenson screen, normally at a height of 1 m above the ground surface. Often known as the *dry bulb temperature*, it is normally recorded in degrees Celsius. *AHP*

[See also MEAN ANNUAL AIR TEMPERATURE]

airborne remote sensing The collection of remotely sensed information from an instrument mounted on an airborne platform. This was the first form of remote sensing, which began when primitive cameras were carried aloft in balloons in the AD 1860s.

Airborne REMOTE SENSING is undertaken predominantly from fixed-wing aircraft, but microlights, balloons, helicopters and remotely piloted vehicles are not uncommon. The choice of platform depends on the specification of the data acquisition. Large-area coverage for *mapping* would use a fixed-wing aircraft, while small-area reconnaissance could use a microlight. Repeat measurements of the same location are undertaken most easily from a helicopter.

OPTICAL REMOTE SENSING INSTRUMENTS and RADAR REMOTE SENSING INSTRUMENTS are flown to collect remotely sensed information over a broad range of the ELECTROMAGNETIC SPECTRUM. GEOPHYSICAL SURVEY instruments are also flown to make measurements of the Earth's magnetic and gravity fields. The majority of remote sensing instruments provide information in an image format, either as a photograph or as a RASTER. Instruments with a single field of view, such as *spectroradiometers*, and active point-measurement instruments, such as LIDARS, are also used. Airborne platforms have often been the test beds for instruments that are later flown on orbital platforms.

Airborne remote sensing allows a large amount of flexibility in data collection. The suite of instruments used and their configuration can be tailored to the particular application. This may include the choice of film for a camera or the wavebands to be recorded by a digital scanner. It is now common to fly optical remote sensing instruments and LIDARS together to record surface reflectance and topography simultaneously. Adjustments can also be made during data acquisition in response to changing conditions and circumstances, so as to optimise system performance. The timing of data collection can be controlled to avoid cloud, optimise viewing conditions or collect data in response to important events. Although airborne remote sensing offers fine *spatial resolution*, it cannot provide the regional to global near-synoptic coverage of instruments on orbital platforms. In addition, changes in aircraft altitude as it passes through the turbulent ATMOSPHERE distort the recorded images, which then require GEOMETRIC CORRECTION. *GMS*

Falkner, E. 1994: *Aerial mapping: method and applications.* Boca Raton, FL: CRC Press. **Barrett, E.C. and Curtis, L.F.** 1992: *Introduction to environmental remote sensing.* Dordrecht: Kluwer. **Lillesand, T.M. and Kiefer, R.W.** 1994: *Remote sensing and image interpretation.* New York: Wiley.

airfall The gravitational settling of particulate material. The term is particularly used in relation to VOLCANIC ERUPTIONS to describe the process and products of PYROCLASTIC FALL. *JBH*

airflow types The range of atmospheric circulation types of specific areas is often described by airflow types. Unlike AIRMASSES, an airflow type usually has an explicit directional component expressing the flow pattern of a given level of the atmosphere. The WEATHER TYPE classification of H.H. Lamb allocated a single airflow type to the British Isles region for every day from AD 1861 to 1995 based largely (but not solely) on the characteristics of surface airflow. Classification of airflow types over larger regions typically refers to either the upper atmosphere flow or to the locations of pressure centres. Changes in type frequency influence the geographical distribution of shelter and exposure. Knowledge of the local effects of each type may become an important aspect of future CLIMATIC CHANGE SCENARIOS in assisting the '*downscaling*' from GENERAL CIRCULATION MODEL output to scenarios of local and REGIONAL CLIMATIC CHANGE. *JCM*

[See also SYNOPTIC CATALOGUES]

El-Kadi, A.K. and Smithson, P.A. 1996: An automated classification of pressure patterns over the British Isles, *Transactions of the Institute of British Geographers NS* 21, 141–56. **Lamb, H.H.** 1972: *British Isles weather types and a register of the daily sequence of circulation patterns, 1861–1971* [*Meteorological Office Geophysical Memoir* 116]. London: HMSO. **Wilby, R.** 1997: Non-stationarity in daily precipitation series: implications for GCM downscaling using atmospheric circulation indices, *International Journal of Climatology* 17, 439–454.

airshed The boundary of the 'catchment area' for POLLUTANTS around a source of EMISSIONS. It is a concept developed by analogy from WATERSHED, although the boundary of the airshed is not fixed by topography and may vary in space and time. *JAM*

alas A THERMOKARST depression, 3–40 m deep and 100 m to 15 km across. It was originally defined as a type of enclosed grassy depression in the BOREAL FOREST of central Siberia caused by the thawing of unconsolidated, ice-rich, silty sediments through disturbance of the ground surface from, for example, forest fire or CLIMATIC CHANGE. Distinctive *Alas thermokarst relief* and an *alas thermokarst cycle* have been identified in central Yakutia, Siberia. *HMF*

[See also PERMAFROST, PERIGLACIAL LANDFORMS]

Czudek, T. and Demek, J. 1970: Thermokarst in Siberia and its influence upon the development of lowland relief. *Quaternary Research* 1, 103–120. **Desyatkin, R.V.** 1991: Soil formation in alases. *Soviet Soil Science* 23, 9–19.

albedo The fraction of total incident SOLAR RADIATION that is reflected by a surface. Albedo values are expressed using the ranges 0.01 to 1.00 or 0 to 100%. Reflection does not change the *wavelength* of the incident RADIATION. Typical values of albedo include coniferous forest (5–15%), crops (15–25%), sand (20–30%), grass (25%), old snow (40–70%) and freshly fallen snow (80–90%). Water surfaces such as a calm sea normally have a very low albedo (around 5%), but this increases to around 70% when the Sun's elevation is low and the waves are high. The albedo of CLOUDS varies widely in relation to the type and thickness. The planetary ALBEDO, measured at the top of the ATMOSPHERE and representing the Earth–atmospheric system as a whole, is between 30 and 35%. CLOUDS are responsible for much of the planet's 'brightness' as seen from space. Venus and Mars have albedos of around 75% and 15%, respectively. *DME*

Oke, T.R. 1987: *Boundary layer climates*, 2nd edn. London: Methuen. **Sagan, C., Toon, O.B. and Pollock, J.B.** 1979: Anthropogenic albedo changes and the Earth's climate. *Science* 206, 1363–1368.

Albeluvisols Soils having a clay-enriched subsurface ARGIC HORIZON with a markedly irregular upper boundary resulting from tonguing of the overlying bleached ALBIC HORIZON into the clay-rich subsoil. A *perched watertable* develops upon the argic horizon after snow melt leading to periodic saturation and GLEYING in the albic horizon. These soils occur in cold-continental regions of North America and Eurasia and also under forest in cool moist temperate regions on transitional deposits between sands and loess (SOIL TAXONOMY: Glossic and Fragic great groups of *Alfisols*). If limed and grazed, or cultivated, biological activity is encouraged and restrictions upon water penetration are eased. *EMB*

[See also WORLD REFERENCE BASE FOR SOIL RESOURCES]

albic horizon

A pale-coloured upper mineral SOIL HORIZON with weak structural development from which clay and free iron oxides have been removed by, for example PODZOLISATION, GLEYING or LEACHING. Often sandy, the albic horizon is associated with PODZOLS, ALBELUVISOLS and SOLONETZ and is symptomatic of soil development in acid, moist conditions. *EMB*

[See also WORLD REFERENCE BASE FOR SOIL RESOURCES]

Food and Agriculture Organization of the United Nations 1998: *World reference base for soil resources* [*Soil Resource Report* 84]. Rome: FAO, ISRIC, ISSS.

algae

Unicellular or multicellular plants, which contain chlorophyll and carry out photosynthesis. Most are microscopic and found in AQUATIC ENVIRONMENTS. The main groups are *red algae*, *brown algae* (mainly marine), *green algae* (important as freshwater PLANKTON) and *diatoms*. *JAM*

[See also ALGAL ANALYSIS, DIATOM ANALYSIS]

Sze, P. 1997: *The biology of algae.* New York: McGraw Hill.

algal analysis

Algae comprise several groups of photosynthetic multi- or unicellular organisms, e.g. marine, terrestrial and freshwater blue–green algae, mainly marine red and brown algae, predominantly freshwater green algae and siliceous algae (DIATOMS, freshwater chrysophytes). Algae are excellent ENVIRONMENTAL INDICATORS and possess a wide range of application possibilities for environmental assessments, PALAEO-ENVIRONMENTAL RECONSTRUCTIONS and CLIMATIC RECONSTRUCTIONS (see also PALAEOCLIMATOLOGY). Algae are an important component of the global CARBON CYCLE and are known for ALGAL BLOOMS in freshwater and marine systems and for their antimicrobial activity. Blue–green, green and siliceous algae have, among others, long been recognised as a valuable source of information about changes in the TROPHIC state of lakes and WETLANDS. Their spores and coenobia can be found in SEDIMENTS and SOILS. Characean oospores may be among the main constituents of MARL. High-performance liquid chromatography has been used to quantify changes in FOSSIL pigments in LAKE SEDIMENTS and to reconstruct SEA-SURFACE TEMPERATURES (U^k_{37} index – see ALKENONES). *UBW*

[See also ANNUALLY LAMINATED SEDIMENTS, CHRYSOPHYTE CYST ANALYSIS, EUTROPHICATION, GYTTJA, LIPID STRATIGRAPHY, PALAEOLIMNOLOGY, POLLUTION, SEDIMENT TYPES]

Battarbee, R.W. 1986: Diatom analysis. In Berglund, B.E. (ed.), *Handbook of Holocene palaeoecology and palaeohydrology*. Chichester: Wiley, 527–570. **Cronberg, G.** 1986: Blue–green algae, green algae and crysophyceae in sediments. In Berglund, B.E. (ed.), *Handbook of Holocene palaeoecology and palaeohydrology*. Chichester: Wiley, 507–526. **Hamilton, W.D. and Lenton, T.M.** 1998: Spora and Gaia: how microbes fly with their clouds. *Ethology, Ecology and Evolution* 10, 1–16. **Livingstone, D.** 1984: The preservation of algal remains in recent lake sediments. In Haworth, E.Y. and Lund, J.W.G. (eds). *Lake sediments and environmental history.* Leicester: Leicester University Press, 191–202. **Meyers, P.A.** 1997: Organic geochemical proxies of paleoceanographic, paleolimnologic, and paleoclimatic processes. *Organic Geochemistry* 27, 213–250.

algal bloom

The rapid growth of algae producing a visible discolouration (such as so-called '*red tides*') in the surface waters of lakes and coastal areas. Algal blooms occur naturally in spring but are also induced on a grand scale by EUTROPHICATION. They may cause damage to wildlife and fisheries either by their decomposition, which raises the BIOCHEMICAL OXYGEN DEMAND (BOD), or by toxic excretions (*toxic bloom*). Algal blooms have been recognised in the stratigraphic record. *JAM*

Faure, K. and Cole, D. 1999: Geochemical evidence for lacustrine microbial blooms in the vast Permian Main Karoo, Parana, Falkland Islands and Huab basins of southeastern Gondwana. *Palaeogeography, Palaeoclimatology, Palaeoecology* 152, 189–213. **Pearce, F.** 1995: Dead in the water. *New Scientist* 145, 26–31. **Taylor, D.L. and Seliger, H.H.** 1979: *Toxic dinoflagellate blooms.* Amsterdam: Elsevier. **Thorsen, T.A., Dale, B. and Nordberg, K.** 1995: 'Blooms' of the toxic dinoflagellate *Gymnodinium catenatum* as evidence of climatic fluctuations in the late Holocene of southwestern Scandinavia. *The Holocene* 5, 435–446.

alien species

A 'foreign' species moved by humans to a region outside its natural, geographical RANGE or environment. It can exhibit BIOLOGICAL INVASION, causing extinctions in the native *biota*, especially in island HABITATS. *GOH*

[See also EXOTIC SPECIES, INTRODUCTION]

Cronk, Q.C.B. and Fuller, J.L. 1995: *Plant invaders: the threat to natural ecosystems.* London: Chapman and Hall. **Salisbury, Sir E.** 1964: *Weeds and aliens.* London: Collins. **Swanson, T.** 1997: *Global action for biodiversity.* Cambridge: Earthscan, IUCN.

Alisols

Soils of the humid tropical and subtropical regions with a subsurface, clay-enriched, ARGIC HORIZON of HIGH-ACTIVITY CLAYS and a high CATION EXCHANGE CAPACITY. Intense weathering of silicate clays releases large amounts of aluminium, making a very acid soil, rich (60%) in exchangeable aluminium. These soils occur in Southeast USA, South America, Southeast Asia and Africa. (SOIL TAXONOMY: Aquults, Humults and Udults within the *Ultisols*). Alisols are commonly strongly acid and susceptible to erosion. *EMB*

[See also WORLD REFERENCE BASE FOR SOIL RESOURCES]

alkali

A BASE that is soluble in water; usually a hydroxide of a metal, which in solution produces an excess of hydroxyl ions (OH^-). Alkalis can neutralise ACIDS as exemplified by the addition of calcium hydroxide to lakes affected by ACIDIFICATION. *JAM*

[See also ACID RAIN, ALKALINITY]

alkalinity

A measure of the total amount of weak acid anions (primarily bicarbonate, carbonate and hydroxyl ions) in a solution that are available to neutralise H⁺. High alkalinity (capacity to neutralise a strong acid) should be distinguished from high *basicity* (high pH). *UBW*

[See also ACIDITY, ALKALISATION, DIATOM ANALYSIS, PALAEOLIMNOLOGY, pH, SALINITY]

Lerman, A. (ed.) 1978: *Lakes: chemistry, geology, physics.* New York: Springer. **Wetzel, R.G. and Likens, G.E.** 1991: *Limnological analyses.* New York: Springer.

alkalisation A SOIL FORMING PROCESS in which sodium ions accumulate and dominate the CATION EXCHANGE CAPACITY of the soil forming a dense, subsurface NATRIC HORIZON. Sodium bicarbonate may be formed, raising soil pH in excess of pH 8.5 and effectively prohibiting the growth of crop plants. Unlike in SALINISATION, the soil loses its structure and becomes hard when dry but loses its bearing strength when wet. Alkalisation occurs both naturally, with change to a more arid climate (DESERTIFICATION), and also as a result of IRRIGATION. *EMB*

[See also SALINISATION]

alkenones Long-chain alkenones are organic compounds that occur ubiquitously in MARINE SEDIMENTS of CRETACEOUS to recent age. They commonly occur with 37, 38 or 39 carbon atoms with two, three or four double bonds in the biologically unusual E (*trans*) configuration. These alkenones are biosynthesised by several species of *phytoplankton*, especially *Emiliania huxleyi*, and are considered as BIOMARKERS for algae of the class Prymnesiophyceae. Long-chain alkenones have also been identified in some LACUSTRINE SEDIMENTS, although the source is as yet undetermined. The relationship between the relative abundances of alkenones and the temperature of the water in which they biosynthesised resulted in the U^k_{37} index, defined by the relative abundance of the $C_{37:4}$, $C_{37:3}$ and $C_{37:2}$ alkenones. This ratio enables palaeo SEA-SURFACE TEMPERATURES to be reconstructed. The U^k_{37} index covaries with OXYGEN ISOTOPE records and historical measurements of EL NIÑO—SOUTHERN OSCILLATION events indicating its effectiveness as an indirect measure of past climates. *KJF*

Brassell, S.C., Eglinton, G., Marlowe, I.T. *et al.* 1986: Molecular stratigraphy: a new tool for climatic assessment. *Nature* 320, 129–133. **Leeuw, J.W. de, van der Meer, F.W., Rijpstra, W.I.C. and Schenck, P.A.** 1979: On occurrence and structural identification of long chain unsaturated ketones and hydrocarbons in sediments. In Douglas, A.G. and Maxwell, J.R. (eds), *Advances in Organic Geochemistry 1979.* Oxford: Pergamon, 211–217. **Rosell-Mele, A., Carter, J.F., Parry, A.T. and Eglinton, G.** 1995: Determination of the U^k_{37} index in geological samples. *Analytical Chemistry* 67, 1283–1289. **Rechka, J.A. and Maxwell, J.R.** 1988: Characterisation of alkenone temperature indicators in sediments and organisms. *Organic Geochemistry* 13, 727–734.

allele One of two or more genes that occupy the same position on a chromosome, but which differ in form, presumably through genetic mutation. *KJW*

allelopathy In the broadest sense, biochemical interactions (primarily involving volatile terpenes and phenolic compounds) between plants, although the term is usually used in the context of adverse interactions (inhibition). *JLI*

[See also COMPETITION]

Rice, E.L. 1984: *Allelopathy*, 2nd edn. New York: Academic Press. **Souto X.C., Gonzalez L. and Reigosa M.J.** 1994: Comparative analysis of allelopathic effects produced by four forestry species during decomposition process in their soils in Galicia (NW Spain). *Journal of Chemical Ecology* 20, 3005–3015.

allergen A normally harmless substance that provokes an allergy or abnormal reaction of the human body, such as hay fever or asthma. Although allergenic reactions may be brought about by such outdoor ENVIRONMENTAL FACTORS as AIR POLLUTION, they appear to be more commonly caused by changes in the indoor environment, such as the house-dust mite allergen, or changes in diet. *JAM*

Allerød Interstadial The final interval of relatively warm conditions in Europe during the *Late Glacial Interstadial* (following the LAST GLACIAL MAXIMUM), which precedes the YOUNGER DRYAS STADIAL. Defined originally in Scandinavia, it has been dated to between c. 11 800 and 11 000 ¹⁴C years BP. Summer temperatures approached those of today. *CJC*

[See also LATE GLACIAL ENVIRONMENTAL CHANGE]

Lotter, A.F., Birks, H.J.B., Eicher, U. *et al.* 2000: Younger Dryas and Allerød summer temperatures at Gerzensee (Switzerland) inferred from fossil pollen and cladoceran assemblages. *Palaeogeography, Palaeoclimatology, Palaeoecology* 159, 349–361. **Lowe, J.J., Ammann, B., Birks, H.H.** *et al.* 1994: Climatic changes in areas adjacent to the North Atlantic during the last glacial–interglacial transition (14–9 ka BP). *Journal of Quaternary Science* 9, 185–198. **Mangerud, J., Andersen, S.T., Berglund, B.E. and Donner, J.J.** 1974: Quaternary stratigraphy of Norden: a proposal for terminology and classification. *Boreas* 10, 109–127.

allochthonous MINEROGENIC or ORGANIC material transported from its place of origin to form a SEDIMENT (or part of a sediment) at another locality. *UBW*

[See also AUTOCHTHONOUS, DETRITUS]

Ashmore, P., Brashay, B.A., Edwards, K.J. *et al.* 2000: Allochthonous and autochthonous mire deposits, slope instability and palaeoenvironmental investigations in the Borve Valley, Barra, Outer Hebrides, Scotland. *The Holocene* 10, 97–108.

allocyclic change A change driven by external factors; ALLOGENIC CHANGE is a synonym. For example AVULSION, or switching of a depositional lobe on a DELTA or ALLUVIAL FAN, may occur in response to CLIMATIC CHANGE, EUSTATIC change or TECTONIC effects (allocyclic controls) as opposed to occurring in response to the accumulation of sediment (an AUTOCYCLIC CHANGE).

Until the 1960s ENVIRONMENTAL CHANGE represented in the GEOLOGICAL RECORD of SEDIMENTARY DEPOSITS tended to be attributed to allocyclic mechanisms – a TRANSGRESSION, for example, might be related directly to a eustatic sea-level rise. With the development of FACIES ANALYSIS and FACIES MODELS in the 1960s and 1970s, it was recognised that such changes occur through autocyclic mechanisms in many environments: a transgression in a delta can result from PROGRADATION and abandonment of one lobe with avulsion to another. The Mississippi Delta has had at least five distinct lobes during the past 7000 years, so each change is unlikely to have had an allocyclic control. Through the 1980s and 1990s it became accepted that many abrupt facies changes are indeed driven by allocyclic mechanisms, and SEDIMENTOLOGY has been brought closer to STRATIGRAPHY in attempts to identify the relative roles of climate, tectonics, eustasy and autocyclic mechanisms such as sediment accumulation and progradation in controlling the GEOLOGICAL RECORD OF ENVIRONMENTAL CHANGE, leading to the development of disciplines such as EVENT STRATIGRAPHY, BASIN ANALYSIS and SEQUENCE STRATIGRAPHY. *GO*

[See also REGRESSION]

Lidz, B.H. and McNeill, D.F. 1998: New allocyclic dimensions in a prograding carbonate bank: evidence for eustatic, tectonic, and paleoceanographic control (late Neogene, Bahamas). *Journal of Sedimentary Research* 68, 269–282. **Plint, A.G., Eyles, N., Eyles, C.H. and Walker, R.G.** 1992: Control of sea level change. In Walker, R.G. and James, N.P. (eds), *Facies models: response to sea level change.* St John's, Newfoundland: Geological Association of Canada, 15–25. **Reading, H.G. and Levell, B.K.** 1996: Controls on the sedimentary rock record. In Reading, H.G. (ed.), *Sedimentary environments: processes, facies and stratigraphy,* 3rd edn. Oxford: Blackwell Science, 5–36. **Zaleha, M.J.** 1997: Intra- and extrabasinal controls on fluvial deposition in the Miocene Indo-Gangetic foreland basin, northern Pakistan. *Sedimentology* 44, 369–390.

allogenic change (1) Alteration of species composition during ECOLOGICAL SUCCESSION due to ABIOTIC forces that are *external* to or independent of the community. (2) The term is also used in the sedimentological context as a synonym for ALLOCYCLIC CHANGE. *LRW*

[See also AUTOGENIC CHANGE, PRIMARY SUCCESSION]

alluvial architecture The three-dimensional geometry of ALLUVIAL DEPOSITS, particularly where there is a thick succession of these deposits in the STRATIGRAPHICAL RECORD; it is also known as *fluvial architecture.* The principal components, or *architectural elements,* of alluvial deposits are channel deposits of relatively coarse GRAIN-SIZE, the deposits of various types of CHANNEL BARS, and finer grained FLOODPLAIN deposits. Alluvial architecture developed as a reaction against reconstructions of alluvial environments based on inadequate data from one-dimensional GRAPHIC LOGS of vertical profiles through alluvial deposits and stresses the importance of the geometry of BOUNDING SURFACES and architectural elements. Alluvial architecture can be applied to alluvial deposits of any age to reconstruct environmental characteristics such as channel dimensions, CHANNEL PATTERNS and aspects of PALAEOHYDROLOGY. Changes in these characteristics through time are commonly interpreted in terms of external controls (see ALLOCYCLIC CHANGE), although it may be difficult to distinguish changes driven by environmental or CLIMATIC CHANGE from those caused by TECTONIC effects. *GO*

[See also ALLUVIAL STRATIGRAPHY, ARCHITECTURAL ELEMENT ANALYSIS, FACIES ARCHITECTURE]

Alexander, J. and Leeder, M.R. 1987: Active tectonic control on alluvial architecture. In Ethridge, F.G., Flores, R.M. and Harvey, M.D. (eds), Recent developments in fluvial sedimentology, *Society of Economic Paleontologists and Mineralogists Special Publication* 39, 243–252. **Bridge, J.S.** 1985: Paleochannel patterns inferred from alluvial deposits: a critical evaluation. *Journal of Sedimentary Petrology* 55, 579–589. **Marriott, S.B.** 1999: The use of models in the interpretation of the effects of base-level change on alluvial architecture. In Smith, N.D. and Rogers, J. (eds), *Fluvial sedimentology VI. International Association of Sedimentologists Special Publication* 28. Oxford: Blackwell Science, 271–281.

alluvial deposits Also known as *alluvial sediments, fluvial deposits, fluvial sediments, fluviatile deposits* or *fluviatile sediments,* alluvial deposits are the SEDIMENTARY DEPOSITS of rivers, commonly known as ALLUVIUM. Their accumulation is controlled by FLUVIAL PROCESSES. Typical alluvial deposits include CLASTIC sediments of a range of GRAIN-SIZES, principally GRAVEL and SAND that accumulate in river channels, and SILT that accumulates on the FLOODPLAIN. Biological activity within alluvial deposits includes SOIL FORMING PROCESSES (see also PALAEOSOL) and organic remains may be preserved as FOSSILS. Many alluvial deposits in the GEOLOGICAL RECORD are red in colour as a result of OXIDATION during the early stages of burial and DIAGENESIS (see RED BEDS). Modern and ancient alluvial deposits host many RESOURCES such as PLACER DEPOSITS, COAL, uranium and PETROLEUM. Interpretation of the SEDIMENTOLOGY of alluvial deposits, including the analysis of SEDIMENTARY STRUCTURES, allows the reconstruction of former river conditions, that may have varied in response to environmental or climatic change. *GO*

[See also ALLUVIAL ARCHITECTURE, ALLUVIAL STRATIGRAPHY, FACIES ANALYSIS, GRAVEL-BED RIVER, SAND-BED RIVER, SUSPENDED-LOAD RIVER]

Collinson, J.D. 1996: Alluvial sediments. In Reading, H.G. (ed.), *Sedimentary environments: processes, facies and stratigraphy,* 3rd edn. Oxford: Blackwell Science, 37–82. **Grossman, S. and Gerson, R.** 1987: Fluviatile deposits and morphology of alluvial surfaces as indicators of Quaternary environmental changes in the southern Negev, Israel. In Frostick, L. and Reid, I. (eds), *Desert sediments: ancient and modern. Geological Society, London, Special Publication* 35, 17–29. **Miall, A.D.** 1996: *The geology of fluvial deposits.* New York: Springer.

alluvial fan A fan-shaped, low-angle depositional LANDFORM of predominantly fluvial origin, characterised by a DISTRIBUTARY pattern of channels, and composed of ALLUVIAL DEPOSITS (see also ALLUVIUM). Alluvial fans, also known as *alluvial cones,* tend to be located in the PIEDMONT zone at the foot of a mountain or upland where rivers and streams deposit BEDLOAD at a distinct break of slope, commonly FAULT-controlled. Unlike mountain-front river systems, alluvial fans are oriented transverse to the mountain front. They are well developed at the mouths of narrow valleys or canyons, where flow expansion occurs, leading to deposition. They typically show a concave longitudinal profile, a convex cross-profile, and down-fan fining of sediment GRAIN-SIZE. Fluvially dominated alluvial fans (*fluvial fans*) differ from DEBRIS FLOW-dominated COLLUVIAL FANS (*gravity-flow fans*) in that the former tend to be larger, lower-angled and associated with larger, less rugged DRAINAGE BASINS.

Many classic examples of alluvial fans occur in arid and semi-arid settings, such as Death Valley in California. They typically have a radius of kilometres to tens of kilometres from apex to toe, and their deposits may be several hundred metres thick. Adjacent fans along a mountain front may coalesce to form a BAJADA. Alluvial fans in humid settings (*humid fans*) include the Kosi Fan in northern India, with a radius of 150 km from apex to toe. Alluvial fans in *proglacial* settings are sometimes known as *outwash fans,* in contrast to *outwash plains* (see SANDUR). Some river systems terminate inland at a *terminal fan* due to infiltration or evaporation. Alluvial fans that advance into a body of standing water are sometimes described as *fan deltas.*

19

Alluvial fan deposits are well represented in the STRATI-GRAPHICAL RECORD, where they are an important component of the fill of many fault-controlled SEDIMENTARY BASINS. Their interpretation provides important SEDIMENTOLOGICAL EVIDENCE OF ENVIRONMENTAL CHANGE. The sedimentary FACIES of alluvial fan deposits can be confused with those of braided river deposits, and the term *fanglomerate* has been used to describe coarse-grained sedimentary deposits attributed to alluvial fan deposition.

GO/JAM

[See also SUBMARINE FAN]

Ballantyne, C.K. and Whittington, G. 1999: Late Holocene floodplain incision and alluvial fan formation in the central Grampian Highlands, Scotland: chronology, environment and implications. *Journal of Quaternary Science* 14, 651–671. **Blair, T.C. and McPherson, J.G.** 1984: Alluvial fans and their natural distinction from rivers based on morphology, hydraulic processes, sedimentary processes and facies assemblages. *Journal of Sedimentary Research* A64, 450–489. **Bull, W.B.** 1977: The alluvial fan environment. *Progress in Physical Geography* 1, 222–270. **Gupta, S.** 1997: Himalayan drainage patterns and the origin of fluvial megafans in the Ganges foreland basin. *Geology* 25, 11–14. **Harvey, A.M., Wigand, P.E. and Wells, S.G.** 1999: Response of alluvial fan systems to the late Pleistocene to Holocene climatic transition: contrasts between the margins of pluvial Lakes Lahontan and Mojave, Nevada and California, USA. *Catena* 36, 255-281. **Kochel, R.C. and Johnson, R.A.** 1984: Geomorphology, sedimentology and depositional processes of humid-temperate alluvial fans in central Virginia, U.S.A. *Canadian Society of Petroleum Geologists Memoir* 10, 108–122. **McCarthy, T.S. and Ellery, W.N.** 1995: Sedimentation on the distal reaches of the Okavango Fan, Botswana, and its bearing on calcrete and silcrete (ganister) formation. *Journal of Sedimentary Research* A65, 77–90. **Nemec, W. and Steel, R.J. (eds)** 1988: *Fan deltas: sedimentology and tectonic settings*. London: Blackie. **Rachocki, A.H. and Church, M. (eds)** 1990: *Alluvial fans: a field approach*. Chichester: Wiley.

alluvial fill A general term used to describe ALLUVIAL DEPOSITS, usually referring to those resulting from one phase of AGGRADATION by rivers, which may have been in response to an ALLOCYCLIC CHANGE such as BASE LEVEL rise, CLIMATIC CHANGE or TECTONIC effects in the source area. *GO*

[See also ALLUVIAL STRATIGRAPHY, VALLEY FILL]

alluvial stratigraphy The documentation and interpretation of the chronology and succession of ALLUVIAL DEPOSITS, with the principal aim of reconstructing the pattern of ENVIRONMENTAL CHANGE preserved in the record of past river environments. In contrast, ALLUVIAL ARCHITECTURE is concerned with the geometry of alluvial deposits, with the aim of reconstructing the alluvial environments themselves. Alluvial stratigraphy and alluvial architecture, together with approaches such as PALAEOHYDROLOGY, PALAEOSOL analysis and PALAEOECOLOGY, allow for the detailed reconstruction of past environmental change in continental environments. This record is sometimes referred to as a *fluvial archive* (see NATURAL ARCHIVES).

The STRATIGRAPHY of alluvial deposits comprises a succession of alternating channel and FLOODPLAIN deposits, recording alternating phases of incision (channel downcutting or DEGRADATION) and AGGRADATION. These environmental changes can be interpreted in terms of a variety of controls, and the challenge for alluvial stratigraphy is to interpret accurately the controls on past alluvial develop-

ment. During gradual sediment accumulation, AUTO-CYCLIC CHANGES such as channel migration and intermittent AVULSION can give rise to channel and floodplain deposits alternating in vertical succession. In many cases, however, ALLOCYCLIC CHANGES have been inferred to explain such successions, including CLIMATIC CHANGE, BASE LEVEL change and TECTONIC effects. It can be difficult to discriminate amongst these allocyclic controls on long-term fluvial development. A fruitful line of approach is the *correlation* of changes recorded in alluvial successions in adjacent basins. *GO*

[See also SEQUENCE STRATIGRAPHY]

Antoine, P., Lautridou, J.P. and Laurent, M. 2000: Long-term fluvial archives in NW France: response of the Seine and Somme rivers to tectonic movements, climatic variations and sea-level changes. *Geomorphology* 33, 183–207. **Gibbard, P.L.** 1988: The history of the great northwest European rivers during the past three million years. *Philosophical Transactions of the Royal Society of London, Series* B318, 559–602. **Mack, G.H. and Leeder, M.R.** 1998: Channel shifting of the Rio Grande, southern Rio Grande rift: implications for alluvial stratigraphic models. *Sedimentary Geology* 117, 207–219. **Mackey, S.D. and Bridge, J.S.** 1995: 3-dimensional model of alluvial stratigraphy – theory and application. *Journal of Sedimentary Research* B65, 7–31. **Taylor, M.P. and Lewin, J.** 1997: Non-synchronous response of adjacent floodplain systems to Holocene environmental change. *Geomorphology* 18, 251–264. **Vandenberghe, J. and Maddy, D.** 2000: Editorial: The significance of fluvial archives in geomorphology. *Geomorphology* 33, 127–130.

alluviation The build-up of ALLUVIAL DEPOSITS by a river system. Over long time spans rivers may alternate between periods of AGGRADATION (alluviation) and incision (channel downcutting or DEGRADATION). This behaviour may be controlled by ALLOCYCLIC CHANGES such as CLIMATIC CHANGE, BASE LEVEL change or other environmental factors. *GO*

[See also ALLUVIAL STRATIGRAPHY, ALLUVIUM, COLLUVIATION]

alluvium A general term for the SEDIMENTARY DEPOSITS of rivers, or ALLUVIAL DEPOSITS. *GO*

[See also COLLUVIUM, ELUVIUM]

along-track direction In the context of REMOTE SENSING, the direction parallel to that in which a moving sensor travels. It is also known as the *azimuth direction*. *PJS*

alp Pasture on a gently-sloping bench or shoulder typically above the steep trough form of a U-SHAPED VALLEY in *glaciated mountain terrain* (see GLACIER) of the European Alps. *RAS*

[See also ALPINE ZONE]

alpine zone The *altitudinal zone* (see ALTITUDINAL ZONATION OF VEGETATION) between the TREE LINE and the upper limit of plant life in mountain regions. Although the alpine zone is a geoecological zone with distinctive geomorphology as well as ecology, it is commonly divided into three belts or (sub-) zones, primarily on the basis of differences in the vegetation. In the Scandinavian mountains, the *low-alpine belt*, also known as the 'willow belt', has a more-or-less complete cover of mainly dwarf-shrub-dominated communities and alpine mires together with some tall-herb communities. In the *mid-alpine belt*, increasingly severe climatic conditions, more extensive SNOWBEDS and a shorter growing season lead to a mosaic of graminoid heath communities alternating with late-snowbed communities. In the *high-alpine belt*, low-growing perennial FORBS, which occur only in favourable habitats, predominate with lichens and bryophytes. Each of these belts is of the order of a few hundred metres wide. A more complex scheme is applied to the European Alps, as illustrated in the Figure, and this recognises a distinct NIVAL ZONE above the alpine zone. Especially in the low-alpine zone, alpine grassland ('alps' in the Alps; sætre grassland in Norway) may have developed in response to a long history of seasonal grazing by domesticated animals involved in TRANSHUMANCE systems.   *JAM*

[See also AFRO-ALPINE ZONE, MOUNTAIN REGIONS]

Dahl, E. 1986: Zonation in arctic and alpine tundra and fjellfield ecobiomes. In Polunin, N. (ed.), *Ecosystem theory and applications*. Chichester: Wiley, 35–62. **Ellenberg, H.** 1988: *Vegetation ecology of Central Europe*. Cambridge: Cambridge University Press. **Grabherr, G.** 1997: The high-mountain ecosystems of the Alps. In Wielgolasjki, F.E. (ed.), *Polar and Alpine Tundra. Ecosystems of the World*. Vol. 3. Amsterdam: Elsevier, 97–121. **Price, L.W.** 1981: *Mountains and man: a study of process and environment*. Berkeley, CA: University of California Press. **Troll, C. (ed.)** 1972: *Geoecology of the high-mountain regions of Eurasia*.

Proceedings of the Symposium of the International Geographical Union Commission on High-Altitude Geoecology, Mainz, Germany, 1969. *Erdwissenschaftliche Forschung* 4. Wiesbaden: Steier.

alternative energy   Energy generated from sources other than traditional HYDROCARBON materials (coal, oil or gas) and nuclear energy. Concerns over the finite nature of NON-RENEWABLE RESOURCES of HYDROCARBONS and a desire to reduce emissions of GREENHOUSE GASES, plus fears over the safety of nuclear power, have prompted many nations to develop alternative methods of generating heat and electricity. Examples of 'clean' and RENEWABLE ENERGY resources that can be harnessed include HYDROPOWER, wind, geothermal and solar energy, as well as tidal or wave power and BIOENERGY.   *MLW*

Drennen, T.E., Erickson, J.D. and Chapman, D. 1996: Solar power and climate change policy in developing countries: *Energy Policy* 24, 9–16. **Golob, R. and Brus, E.** 1993: *The almanac of renewable energy*. New York: H. Holt. **Gupta, H.K.** 1980: *Geothermal resources: an energy alternative* [*Developments in Economic Geology* 12]. Amsterdam: Elsevier. **Lee, S.** 1996: *Alternative fuels*: Washington, DC: Taylor and Francis. **Rosenberg, P.** 1992: *The alternative energy handbook*: Lilburn, GA: Fairmont Press. **Simeons, C.** 1980: *Hydro-power: the use of water as an alternative source of energy*: Oxford: Pergamon Press. **Street, P. and Miles, I.** 1996: Transition to alternative energy

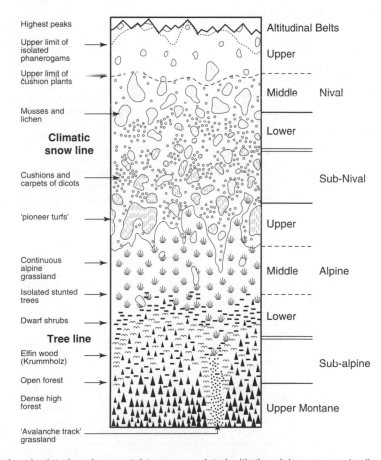

alpine zone *Vegetational and related environmental terms associated with the alpine zone and adjacent altitudinal zones in the European Alps (after Ellenberg, 1981)*

supply technologies – the case of windpower. *Energy Policy* 24, 413–425.

altimetry

altimetry Traditionally, the measurement of altitudes of 'benches' and summits in the landscape in order to identify the scattered remnants of former EROSION SURFACES. Identification is assisted by plotting the data as an *altimetric frequency distribution* (e.g. as an altimetric frequency curve), enabling any major clustering of frequencies to be identified. *RAS*

[See also DAVISIAN CYCLE OF EROSION, DENUDATION CHRONOLOGY]

altitudinal zonation of vegetation

altitudinal zonation of vegetation The recognition of more-or-less distinct zones or belts in phenomena (e.g. climate, landforms, geomorphic processes, vegetation or landscapes) with variation in altitude in mountain regions. The species composition of *plant communities* tends to show continuous variation with altitude, allowing VEGETATION to be classified into distinct zones along an altitudinal gradient (see Figure). Altitudinal variation in species composition and abundance is largely controlled by climate. In particular, many stages of plant life history are strongly limited by temperature, which declines with altitude and sets boundaries to the species RANGE (see EXTREME CLIMATIC EVENTS: IMPACT ON ECOSYSTEMS). However, a general decrease in the stature of vegetation with altitude offsets this limitation to some extent, by reducing aerodynamic conductance to sensible and LATENT HEAT loss. This reduction conserves heat and tends to decrease the LAPSE RATE of leaf temperature during the PHOTOPERIOD in comparison with that of air temperature. Conversely, the same mechanism causes leaf temperatures to be cooler than air temperatures during the night, necessitating low temperature TOLERANCE in alpine species.

Variation in vegetation structure with altitude is mediated by latitude, CONTINENTALITY, topography, *edaphic factors* and ASPECT, and depends on whether precipitation increases or decreases with elevation. Vegetation may also show changes in species composition, ADAPTATION or ACCLIMATISATION in response to wind exposure, SOLAR and *ultraviolet radiation* maxima, and diurnal variation in temperature, which all increase with altitude, as well as to partial pressures of atmospheric CARBON DIOXIDE and WATER VAPOUR, which decrease.

GLOBAL WARMING may cause a general shift in plant and associated *animal communities* to higher latitudes, with potential implications for CONSERVATION. For example, the range of the gelada baboon is likely to follow that of its montane grassland food source with future climatic warming, moving to higher altitudes and thereby becoming fragmented and contracted in area. Similarly, tropical montane CLOUD FOREST may be forced upslope by reduced cloud contact and increasing EVAPOTRANSPIRATION, which could accompany warming at lower altitudes. These changes will reduce the size of POPULATIONS and greatly increase the risk of future EXTINCTION, which will be particularly high in ENDEMIC and RARE SPECIES with limited DISPERSAL capabilities. *CPO*

[See also AFRO-ALPINE ZONE, ALPINE ZONE, CLINE, ECOCLINE, MOUNTAIN PERMAFROST, TREE LINE/LIMIT, ZONATION SPATIAL]

Archibold, O.W. 1995: *Ecology of world vegetation.* London: Chapman and Hall. **Beniston, M.** 2000: *Environmental change in mountains and uplands.* London: Arnold. **Beniston, M. and Fox, D.G.** 1996: Impacts of climate change in mountain regions. In Watson, R.T., Zinyowera, M.C., Moss, R.H. and Dokken, D.J. (eds), *Climate change 1995. Impacts, adaptations and mitigation of climate change: scientific-technical analysis.* Cambridge: Cambridge University Press, 191–213. **Dunbar, R.I.M.** 1998: Impact of global warming on the distribution and survival of the gelada baboon: a modelling approach. *Global Change Biology* 4, 293–304. **Fernandez-Palacios, F.M. and de Nicolas, J.P.** 1995: Altitudinal pattern of vegetation variation on Tenerife. *Journal of Vegetation Science* 6, 183–190. **Friend, A.D. and Woodward, F.I.** 1990: Evolutionary and ecophysiological responses of mountain plants to the growing season environment. *Advances in Ecological Research* 20, 59–124. **Harris, C.** 1982: The distribution and altitudinal zonation of periglacial landforms, Okstindan, Norway. *Zeitschrift für Geomorphologie* 26, 283–304. **Still, C.J., Foster, P.N. and Schneider, S.H.** 1999: Simulating the effects of climate change on tropical montane cloud forests. *Nature* 398, 608–610.

amber

amber A fossilised tree resin. Amber takes a fine polish and has been prized for jewellery since prehistoric times, especially TERTIARY amber from the Baltic region.

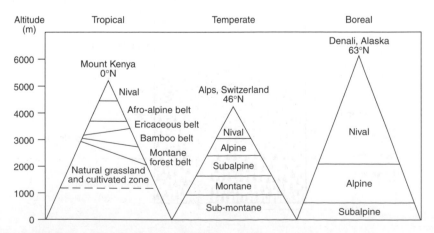

altitudinal zonation of vegetation *Schematic diagram of altitudinal zonation of vegetation in tropical, temperate and boreal mountains. Large aspect differences are indicated on Mount Kenya (after Beniston, 2000)*

Arthropods (e.g. insects and spiders) are commonly preserved as FOSSILS in amber: they represent a type of conservation FOSSIL LAGERSTÄTTEN. *GO*

Crichton, M. 1991: *Jurassic park*. London: Random Century Group. Poinar, G. 1995: *The quest for life in amber*. Oxford: Perseus Publishing. Poinar, G. Jr and Poinar, R. 1999: *The amber forest*. Princeton, NJ: Princeton University Press.

americium-241 dating

The americium-241 (^{241}Am) isotope, derived from nuclear weapons testing, provides additional RADIOMETRIC DATING marker horizons in near-surface sediments where degradation of the LEAD-210 and CAESIUM-137 *dating profiles* appears likely. *DZR*

Appleby, P.G., Richardson, N. and Nolan, P.J. 1991: ^{241}Am dating of lake sediments. *Hydrobiologia* 214, 35–42.

amino acid dating

A RELATIVE-AGE DATING technique based on the measurement of amino acid decay products in protein residues in FOSSIL bones and shells. The most widely used technique is based on the principle that only L-amino acid ISOMERS are found in living tissue, and on death these begin to transform to D-amino acid isomers by processes of *racemisation* or *epimerisation* until an equilibrium is reached (e.g. the DIAGENESIS of L-isoleucine to D-alloisoleucine). Since these processes are time-dependent, the D/L amino acid ratio of a sample reflects its RELATIVE AGE. The ranked relative ages of samples can be used to construct an *aminostratigraphy* for SEDIMENT sequences. Samples used in amino acid dating for the QUATERNARY have been principally Mollusca (see MOLLUSCA ANALYSIS), both terrestrial and marine; hence their particular value in correlating terrestrial and marine sequences and in the dating of past SHORELINES. Assumptions have to be made about the temperature history of the samples and different taxa epimerise at different rates. It is therefore necessary to produce a series of species-specific rates when determining ages and to estimate the temperature history of the sediment. Although of potential value for the whole Quaternary in cold environments with very low rates of epimerisation, amino acid dating has so far produced aminostratigraphies of most value for the last 0.5–1 million years. *MHD/CJC*

Bowen, D.Q. (ed.) 1999: *A revised correlation of Quaternary deposits in the British Isles* [Geological Society Special Report 23]. Bath: The Geological Society. Bowen, D.Q., Hughes, S., Sykes, G.A. and Miller, G.H. 1989: Land sea correlations in the Pleistocene based on isoleucine epimerisation in non-marine molluscs. *Nature* 350, 49–51. Brooks, A.S., Hare, P.E. Kokis, J.E. *et al.* 1990: Dating Pleistocene archeological sites by protein diagenesis in ostrich eggshells. *Science* 248, 60–64. Hare, P.E., Von Endt, D.W. and Kokis, J.E. 1997: Protein and amino acid diagenesis dating. In Taylor, R.E. and Aitken, M.J. (eds), *Chronometric dating in archaeology*. New York: Plenum, 261–296. Hearty, P.J. and Aharon, P. 1988: Amino acid chronostratigraphy of late Quaternary coral reefs: Huon Peninsula, New Guinea and the Great Barrier Reef, Australia. *Geology* 16, 579–583. Hearty, P.J. and Vacher, H.L. 1994: Quaternary stratigraphy of Bermuda: a high-resolution pre-Sangamonian rock record. *Quaternary Science Reviews* 13, 685–698. Johnson, B.J. and Miller, G.H. 1997: Archaeological applications of amino acid racemization. *Archaeometry* 39, 265–287. Miller, G.H., Magee, J.W. and Jull, A.J.T. 1997: Low latitude, low elevation Southern Hemisphere ice-age cooling deduced from amino acid racemization in emu eggshell. *Nature* 358, 241–244. Oches, E.A. and McCoy, W.D. 1995: Amino acid geochronology applied to the correlation and dating of central European loess deposits. *Quaternary Science Reviews* 14, 767–782. Sykes, G. 1991: Amino acid dating. In Smart, P.L. and Frances, P.D. (eds), *Quaternary dating methods – a users guide* [QRA Technical Guide 4]. Cambridge: Quaternary Research Association, 161–176.

amorphous

A MINERAL, or aggregate of genetically related minerals, lacking CRYSTALLINE structure. Examples include opal (amorphous silica) and some hydrated iron oxides. *MRT*

[See also PHYTOLITH]

amplitude

The magnitude of the variation of a given variable over time (for example, PRECIPITATION VARIATION) or space (for example, LONG WAVES) often applied to regular or cyclic variation. *JCM*

[See also PERIODICITIES, RESILIENCE]

anabranching

A river CHANNEL PATTERN consisting of multiple channels separated by stable islands, which are large relative to channel size and which divide the flow at discharges up to and including BANKFULL DISCHARGE. Anabranching differs from BRAIDING in which CHANNEL BARS are covered by water at high flows. Flow patterns of individual branch channels are thus largely independent of each other and the isolated islands (usually of a similar height to the floodplain) may persist for centuries. Anabranching rivers have been classified into six types, on the basis of flow, sediment and behavioural characteristics, of which those with cohesive-sediment *anastomosing channels* are the most common. Although anabranching rivers occur in a wide range of climatic environments, a highly seasonal or extremely episodic river regime is favourable to the AVULSION processes necessary for their formation. *RPDW*

Nanson, G.C. and Knighton, A.D. 1996: Anabranching rivers: their cause, character and classification. *Earth Surface Processes and Landforms* 21, 217–239.

anaerobic

An environment in which oxygen is absent; organisms which are able to grow only in the absence of oxygen; a process that can only occur in the absence of oxygen. Anaerobic processes include *denitrification, sulphate reduction*, METHANOGENESIS and FERMENTATION. *UBW*

[See also AEROBIC, ANOXIC, BIOGENIC SEDIMENT, GLEYSOLS, LAKE STRATIFICATION, NITROGEN CYCLE, ORGANIC SEDIMENT, REDUCTION, SULPHUR CYCLE]

Bates, T.S., Lamb, B.K., Guenther, A. *et al.* 1992: Sulfur emissions to the atmosphere from natural sources. *Journal of Atmospheric Chemistry* 14, 315–337. Holland, K.T., Knapp, J.S. and Shoesmith, J.G. 1987: *Anaerobic bacteria*. Glasgow: Blackie.

analogue method

In the context of environmental change, the use of the phenomena or events from one point in time to interpret the phenomena or events at another time. MODERN ANALOGUES use the diagnostic features from contemporary organisms or environments to interpret similar palaeoorganisms or PALAEOENVIRONMENTS. *Past analogues* are widely used in the interpretation of the present and the PREDICTION of the future. Thus, the mid-Holocene HYPSITHERMAL and the MEDIAEVAL WARM PERIOD have been suggested as past analogues for a warmer Earth in the twentyfirst century. Similarly, events at the Palaeocene/Eocene boundary (when GLOBAL WARM-ING appears to have occurred following a massive release of METHANE from GAS HYDRATES) suggest that enhanced planktonic CARBON SEQUESTRATION may return the Earth to cooler conditions following greenhouse warming. Use of analogues in these ways entails a uniformitarian approach (see UNIFORMITARIANISM), assuming that organisms and environments in the past were/are comparable to those existing today. The main problem with the analogue method is that most analogous situations are not perfect: they are *partial analogues* only. And even if the effects are the same, the causes may be different. *DH/JAM*

[See also MODERN ANALOGUE, ANALOGUE MODEL]

Bains, S., Norris, R.D., Corfield, R.M. and Faul, K.L. 2000: Termination of global warmth at the Palaeocene/Eocene boundary through productivity feedback. *Nature* 407, 171–174. Bryson, R.A. 1985: On climatic analogs in paleoclimatic reconstruction. *Quaternary Research* 23, 275–286. Delcourt, H.R. and Delcourt, P.A. 1991: *Quaternary ecology: a paleoecological perspective*. London: Chapman and Hall. Dickens, G.R. 1999: The blast from the past. *Nature* 401, 752–754. Schneider, S.H. 1989: The greenhouse effect: science and policy. *Science* 243, 771–781. Schmitz, B. 2000: Plankton cooled a greenhouse. *Nature* 407, 143–144.

analogue model (1) A model in which predictions and/or explanations are based on an analogous historical situation. These may range from the similar synoptic conditions used in *weather forecasting* to the use of analogues from the GEOLOGICAL RECORD. (2) The representation of a physical system by an analogous physical system that simulates the former. *AHP/JAM*

[See also ANALOGUE METHOD, MODERN ANALOGUE]

Norris, R.D. and Röhl, U. 1999: Carbon cycling and chronology of climatic warming during the Palaeocene/Eocene transition. *Nature* 401, 775–778.

analysis of covariance (ANCOVA) An extension of ANALYSIS OF VARIANCE that allows for the possible effects of continuous concomitant variables (*covariates*) on the response variable in addition to the effects of the factors. *HJBB*

[See also GENERALISED LINEAR MODELS]

analysis of variance (ANOVA) The separation of variance attributable to one cause from the variance attributable to others. By partitioning the total variance of a set of observations into parts due to particular factors and comparing variances (= mean squares) by F-tests, differences between means can be evaluated. A multivariate equivalent, *multivariate analysis of variance* (MANOVA) exists when there is more than one response variable. *HJBB*

[See also GENERALISED LINEAR MODELS, ANALYSIS OF COVARIANCE]

ancient woodland Woodland that has been continuously present on a site for hundreds if not thousands of years without major human disturbance. It is not synonymous with PRIMARY WOODLAND. In Europe, a distinction between ancient and recent woodland has been based on a threshold c. AD 1600 (Mediaeval vs. postmediaeval woodland). ECOLOGICAL INDICATORS of ancient woodlands have been developed, such that the presence of especially certain indicator plant species or communities may be used to determine the condition of the woodland and the changes that have taken place over time. *IFS*

[See also COMMUNITY CONCEPTS, PHYTOSOCIOLOGY, SECONDARY WOODLAND]

Peterken, G.F. 1974: A method for assessing woodland flora for conservation using indicator species. *Biological Conservation* 6, 239–245. Rackham, O. 1980: *Ancient woodland*. London: Edward Arnold.

Andosols Soils developed in volcanic PARENT MATERIAL occurring in all continents, characterised by low BULK DENSITY, high phosphate retention, ammonium acetate-extractable Fe and Al, and < 10% VOLCANIC GLASS. The surface horizon is often very dark coloured as a result of organic matter accumulation. Andosols are fertile soils if not strongly leached and characteristically the CATION EXCHANGE CAPACITY varies with the pH (SOIL TAXONOMY: Andisols). As volcanic eruptions are episodic, most Andosols are POLYGENETIC exhibiting successive phases of soil formation. *EMB*

[See also WORLD REFERENCE BASE FOR SOIL RESOURCES]

Mizota, C. and van Reeuwick, L.P. 1989: *Clay mineralogy and chemistry of soils formed in volcanic material in diverse climatic regions* [*Soil monograph* 2]. Wageningen: International Soil Reference and Information Centre. Quantin, P. 1985: Characteristics of the Vanuatu Andosols. *Catena supplement (Volcanic soils)* 7, 99–105. Shoji, S., Namzyo, M. and Dahlgren, R.A. 1993: *Volcanic ash soils: genesis, properties and utilization* [*Developments in Soil Science* 21]. Amsterdam: Elsevier.

angiosperms A subdivision of the Spermatophyta (*seed plants*), which may be distinguished by the protection of the ovule (seed) within a fruit. They are the *flowering plants* and are subdivided into *monocotyledons* and *dicotyledons*. After originating about 130 million years ago, they evolved rapidly and are today the dominant species in almost all vegetation FORMATIONS apart from the BOREAL FOREST. *JCMc*

[See also GYMNOSPERMS]

Friis, E.M., Chaloner, W.G. and Crane, P.R. 1987: *The origin of angiosperms and their biological consequences*. Cambridge: Cambridge University Press.

anhysteretic remanent magnetisation (ARM) Magnetisation induced by subjecting a sample to a strong alternating field that is smoothly decreased to zero in the presence of a small steady field. The susceptibility of anhysteretic remanent magnetisation (χ_{ARM}) is highly selective of stable single domain (SSD) ferrimagnetic grains in the size range of 0.02 to 0.4 μm. *AJP*

[See also MINERAL MAGNETISM]

animal remains The study of faunal remains, either FOSSIL or SUBFOSSIL, of which vertebrates (and particularly mammals) may be the most numerous. Analysis of fossil animal remains must first consider the effects of TAPHONOMY, taking into account potential biases that may have occurred during the transition from the *biocoenosis* (the faunal community in its natural living proportions) to the *thanatocoenosis* (or DEATH ASSEMBLAGE), which in turn may be affected by the fossilisation process and by sampling or recovery factors. Animal remains may be used in PALAEOENVIRONMENTAL RECONSTRUCTIONS and CLIMATIC RECONSTRUCTIONS (see also PALAEOCLIMATOLOGY); for example, on the basis of the presence or absence (or relative abundance) within an assemblage of animals with well defined ecological preferences. This approach, however, relies upon the assumption that the ecologies of extant taxa have not changed significantly over time. Past climatic or environmental information may further be deduced by changes in body size (*Bergmann's rule*, for example, states that individuals of warm-blooded animals are larger in relatively cold climates) or body shape.

CARBON ISOTOPE analysis of mammalian remains may be employed as a means of reconstructing *palaeodiet*, as may tooth microwear and morphology studies. Animal bones, preserved soft tissues, hair, claws or even dung may also be analysed for ancient genetic or biomolecular information. Many mammalian LINEAGES show marked patterns of significant morphological EVOLUTION and EXTINCTION through time, in addition to shifts in their geographical distribution brought about by past ENVIRONMENTAL CHANGE. Mammalian remains are therefore particularly suitable for dating and correlating deposits through the application of BIOSTRATIGRAPHY. With more recent remains, numerical age estimates may be obtained through direct RADIOCARBON DATING of animal remains, whereas more ancient remains may be subjected to other dating methods, such as LUMINESCENCE or URANIUM-SERIES DATING. The analysis of animal remains may also provide information on HOMINID subsistence strategies (for example, hunting or scavenging) or on the economies of early agricultural communities (see AGRARIAN CIVILISATIONS), through reconstruction of age and sex profiles of domestic animals. Indeed, analysis of the morphological changes visible in domesticated livestock is one of the major means of establishing when the transition to animal husbandry occurred (see DOMESTICATION). Examination of deer antlers, whether shed or unshed, and of tooth eruption patterns may reveal the season during which an archaeological site was occupied. *DCS*

[See also DNA ANCIENT, BEETLE ANALYSIS, COPROLITE, MOLLUSCA ANALYSIS, RODENT MIDDENS, SEASONALITY INDICATOR]

Behrensmeyer, A.K. and Kidwell, S.M. 1985: Taphonomy's contribution to paleobiology. *Paleobiology* 11, 105–119. **Coard, R. and Chamberlain, A.T.** 1999: The nature and timing of faunal change in the British Isles across the Pleistocene/Holocene transition. *The Holocene* 9, 372–376. **Davis, S.J.M.** 1987: *The archaeology of animals.* London: Batsford. **Hillson, S.** 1986: *Teeth.* Cambridge: Cambridge University Press. **Martin, R.A. and Barnosky, A.D.** (eds) 1993: *Morphological change in Quaternary mammals of North America.* Cambridge: Cambridge University Press. **Ringrose, T.J.** 1993: Bone counts and statistics: a critique. *Journal of Archaeological Science* 20, 121–157. **Stuart, A.J.** 1982: *Pleistocene vertebrates in the British Isles.* London: Longman. **Stuart, A.J.** 1991: Mammalian extinctions in the Late Pleistocene of northern Eurasia and North America. *Biological Reviews* 66, 453–562. **Sutcliffe, A.J.** 1985: *On the track of ice age mammals.* Cambridge, MA: Harvard University Press. **Vickers-Rich, P., Monaghan, J.M., Baird, R.F. and Rich, T.H.** (eds) 1991: *Vertebrate palaeontology of Australasia.* Melbourne: Pioneer Design Studio. **Yalden, D.** 1999: *The history of British mammals.* London: Poyser.

annotation Additional text information associated with graphic-image elements of a map or a digital image. It enhances the basic facts of a feature's existence, for example by describing status, condition or characteristics. *TF*

annual A plant that completes its entire life cycle within a single year. *JLI*

[See also EPHEMERAL, PERENNIAL]

annual mean air temperature The mean air temperature for a particular year; the average of twelve MONTHLY MEAN TEMPERATURES. It is used to show year-to-year variations, CLIMATIC FLUCTUATIONS and longer-term TRENDS in temperature, usually by way of ANOMALIES relative to the MEAN ANNUAL AIR TEMPERATURE (MAAT) of a CLIMATIC NORMAL period. It is not the same as the MAAT. *BDG/JAM*

Türkes, M., Sümer, U.M. and Kiliç, G. 1995: Variations and trends in annual mean air temperatures in Turkey with respect to climatic variability. *International Journal of Climatology* 15, 557–569.

annually resolved record An environmental archive where incremental growth rate and/or postdepositional preservation are such that physical, chemical or biological properties can be observed to vary in response to seasonal variation of environmental factors. Examples include the density variations in TREE RINGS (DENDROCLIMATOLOGY), intensity of luminescence signal in annually banded SPELEOTHEMS, the grey scale in VARVES and most signals in ICE CORES. *DAR*

[See also CORAL AND CORAL REEFS, DENDROCHRONOLOGY, HIGH-RESOLUTION RECONSTRUCTION, SCLEROCHRONOLOGY]

anomalies (1) Deviations of a given variable from mean values. These may be expressed as either absolute or percentage values or in terms of standard deviations from the mean over a stated period. (2) Values that depart so much from the mean value that they can be regarded as unlikely to belong to the same population as the rest of the dataset. (3) Events that depart greatly from the average or the usual. The term *climatic anomaly* is widely used in the sense of the first and third definitions, the latter being exemplified by *anomalous weather events* (see Figure). More generally, *statistical anomalies* conform to the second definition. *JCM/JAM*

[See also DIVERSITY ANOMALY, EXTREME CLIMATIC EVENT, EXTREME EVENT, RESIDUAL]

Frenzel, B., Pfister, C. and Gläser, B. 1994: Climatic trends and anomalies in Europe 1675–1715. *Paläoklimaforschung* 13, 1–479. **Glantz, M.H.** 2001: *Currents of change: impacts of El Niño and La Niña on climate and society,* 2nd edn. Cambridge: Cambridge University Press. **Hansen, D.V. and Bezdek, H.F.** 1996: On the nature of decadal anomalies in North Atlantic sea surface temperature. *Journal of Geophysical Research* 101, 9749–9758.

anoxia/anoxic An environment, such as BOTTOM WATER, devoid or with very low levels of free OXYGEN exhibits anoxia or is said to be anoxic. Anoxic sediments are produced during oxygen deficiency as a result of high organic PRODUCTIVITY and lack of oxygen replenishment to the water/sediment during stagnation or STRATIFICATION. Mechanisms thought to be responsible for *anoxic events* include either a lowering in the oxygen content of the atmosphere or a rise in atmospheric CO_2. Both result primarily from geological activity (e.g. UPLIFT, *burial*, FAULTING, VOLCANIC activity) altering the availability of rocks and minerals for weathering and the amount of

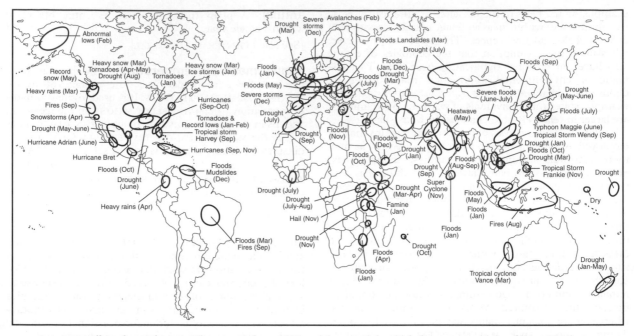

anomalies *Anomalous weather events in* AD *1999 and the month in which they occurred (after Glanz, 2001)*

ORGANIC CARBON in storage. Anoxia can occur in both TER-RESTRIAL and oceanic ENVIRONMENTS with devastating impacts. Ocean anoxia, for example, is thought to have been one of the major contributory factors leading to the end-Permian MASS EXTINCTION – a time when 90–96% of all durably skeletonised marine invertebrate species became extinct. *KJW/UBW*

[See also CARBON CYCLE, MASS EXTINCTIONS, OCEANIC ANOXIC EVENTS, TECTONICS]

Erwin, D.H. 1993: *The great Paleozoic crisis.* Columbia University Press: New York. **Fenchel, T. and Finlay, B.** 1995: *Ecology and evolution in anoxic worlds.* Oxford: Oxford University Press.

antagonism   Interaction between substances such that their combined toxic effect is less than the sum of the toxic effects that each exerts separately.   *JAM*

[See also ECOTOXICOLOGY, SYNERGISM]

Antarctic environmental change
Antarctica is a continent twice the size of Australia. With a 98% ice cover averaging 2 km in thickness, Antarctica influences both global circulation in the atmosphere and oceans and global sea level. Antarctica became thermally isolated from other GONDWANA continents 50–30 million years ago. Ice sheet-scale glaciation was underway by 36 million years ago, whilst a cold stable ice sheet, like that of today, was established sometime between 15 and 3 million years ago.

The modern ICE SHEET comprises 30 million km³ of ice, representing 90% of the world's ice and 70% of its fresh-water, equivalent to a 56 m rise in global sea level. It comprises the larger, mainly land-based, East Antarctic ice sheet and the West Antarctic ice sheet, which rests on the sea floor. The latter is thought by some glaciologists to be potentially unstable and may collapse under the influence of GLOBAL WARMING as the buttressing effect of ICE SHELVES is diminished. It is conceivable that the grounded ice could flow rapidly into the Southern Ocean, potentially causing a SEA-LEVEL RISE of 8 m. Already, in the Antarctic Peninsula, a number of ice shelves have collapsed in the last 30 years under the influence of a regional warming of climate and rising oceanic temperatures. Predictions are difficult, since the behaviour of ice masses is influenced by bed conditions and internal dynamics, as well as by climate.

The environmental record of ice sheets can be derived from ice cores, which extend to within a few tens of metres of the bed. In Antarctica, the Vostok ICE CORE, over 2 km long, has yielded a record back at least 500 000 years. Based on the OXYGEN ISOTOPIC RECORD and trapped gases, it is known that there is a positive correlation between the amount of CARBON DIOXIDE in the atmosphere and decreased global ice volume. For example, a doubling of CO_2 may lead to a 6°C temperature rise in Antarctica. However, because the Antarctic ice sheet is so cold, any initial warming is likely to cause increased snowfall and hence ice sheet growth. With such warming, the area covered by sea ice is likely to decline substantially and influence the distribution and population of marine animals. OZONE DEPLETION, which is linked to the release of CHLOROFLUOROCARBONS, was first recorded in Antarctica.

WHALING AND SEALING brought the first serious claims to sovereignty over parts of Antarctica, which until the twentieth century had no record of human habitation. With permanent occupation, largely in scientific research stations, came the first serious threats to the Antarctic terrestrial environment and the need for ENVIRONMENTAL PROTECTION. The Antarctic Treaty of 1961 has been followed by numerous generally successful CONVENTIONS to preserve this 'last WILDERNESS'.   *MJH*

Hansom, J.D. and Gordon, J.E. 1998: *Antarctic environments and resources*. Harlow: Longman. Harris, C. and Stonehouse, B. (eds) 1991: *Antarctica and global climatic change*. London: Belhaven. Herr, R.A., Hall, H.R. and Haward, M.E. (eds) 1990: *Antarctica's future: continuity or change*. Hobart, Tasmania: Antarctic Institute of International Affairs. Houghton, J.T., Jenkins, G.J. and Ephraums, J.J. (eds) 1990: *Climate change – the IPCC scientific assessment*. Cambridge: Cambridge University Press. Houghton, J.T., Filho, L.G.M., Callander B.A. *et al.* (eds) 1996: *Climate change 1995. The science of climate change*. Cambridge: Cambridge University Press. Intergovernmental Panel on Climate Change 1992: *Climate change*. World Meteorological Organisation and United Nations Environment Programme. International Union for the Conservation of Nature and Natural Resources (IUCN) 1991: *A strategy for Antarctic conservation*. Gland and Cambridge: IUCN. Laws, R. 1989: *Antarctica – the last frontier*. London: Boxtree. Walton, D.W.H. (ed.) 1987: *Antarctic science*. Cambridge: Cambridge University Press.

antecedent drainage A type of DISCORDANT DRAINAGE in which a river maintains its course despite localised UPLIFT and hence may cut across geological structures such as FOLDS and FAULTS and even mountain ranges. The development of antecedent drainage, which requires the rate of river downcutting to exceed the land uplift rate, is difficult to establish with certainty. Examples include parts of the Colorado River and its tributaries, and the major Indian rivers that cut through the Himalayas.
JAM

[See also SUPERIMPOSED DRAINAGE]

Powell, J.W. 1875: *Exploration of the Colorado River of the West and its tributaries*. Washington, DC: Government Printing Office.

antecedent soil moisture The soil moisture content prior to a rainfall event. The speed and magnitude of THROUGHFLOW, SATURATION OVERLAND FLOW and streamflow responses to rainfall (especially in humid vegetated areas) increase markedly with antecedent soil-moisture levels.
RPDW

[See also RIVER DISCHARGE VARIATIONS, RUNOFF PROCESSES, STORM HYDROGRAPH]

anthraquic horizon An ANTHROPEDOGENIC HORIZON comprising the *puddled layer* (mixed layer) and underlying compacted PLOUGH PAN produced in paddy fields for the growth of wetland rice.
EMB

anthropedogenic horizons Surface soil horizons that have been profoundly modified by human activity. In general, these horizons have been improved by human activity as their fertility has been increased. They are best developed in areas of intensive cultivation, such as urban gardens and terraced fields. They are the PLAGGIC, IRRAGRIC, TERRIC, HORTIC, ANTHRAQUIC and HYDRAGRIC HORIZONS (see Figure).
EMB

Food and Agriculture Organization of the United Nations 1998: *World reference base for soil resources* [*Soil Resource Report* 84]. Rome: FAO, ISRIC, ISSS. Macphail, R.I. 1994: The reworking of urban stratigraphy by human and natural processes. In Hall, A.R. and Kenward, H.K (eds), *Urban–rural connexions: perspectives from environmental archaeology* [*Symposium of the Association of Environmental Archaeology* 12]. Oxford: Oxford Books. Macphail, R.I., Courty, M.A. and Bebhardt, A. 1990: Soils and early agriculture. *World Archaeology* 22, 53–69.

anthropochore A non-native (plant) species that was introduced unintentionally by humans.
FMC

[See also APOPHYTE]

anthropogene A little-used equivalent to QUATERNARY: used, mostly in Russia, to emphasise the human presence in the latest part of the GEOLOGICAL TIMESCALE.
GO

anthropogenic Caused directly or indirectly by the actions of people. The distinction is often made between NATURAL and anthropogenic. In POLLEN ANALYSIS, ANTHROPOGENIC INDICATORS are those pollen taxa that reveal HUMAN IMPACT on the environment.
SPH

Behre, K.E. 1981: The interpretation of anthropogenic indicators in pollen diagrams. *Pollen et Spores* 23, 225–246. Goudie, A.S. 2000: *The human impact*, 5th edn. Oxford: Blackwell. Starkel, L. (ed.) 1987: *Anthropogenic sedimentological changes during the Holocene*. Striae 26. Uppsala: Societas Upsaliesis pro Geologica Quaternia.

anthropogenic indicators Plant species or pollen types indicative of human activity. Many pollen types originate from plants that have a strong positive association with HUMAN IMPACT ON VEGETATION (e.g. RUDERAL weeds and CEREAL CROPS). Increased representation of these types in a *pollen diagram* can be interpreted in terms of human activities (e.g. livestock grazing or arable farming) in the landscape around the sampled site, and therefore are referred to as 'anthropogenic indicators'. The definition and interpretation in terms of specific activities of these types is the subject of ongoing research and debate.
MJB

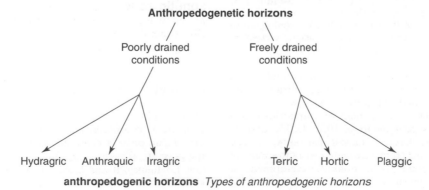

anthropedogenic horizons *Types of anthropedogenic horizons*

[See also ANTHROPOCHORE, APOPHYTE, HUMAN IMPACT ON ENVIRONMENT]

Behre, K.-E. (ed.) 1986: *Anthropogenic indicators in pollen diagrams*. Rotterdam: Balkema. Birks, H.J.B. 1990: Indicator values of pollen types from post-6000 B.P. pollen assemblages from southern England and southern Sweden. *Quaternary Studies in Poland* 10, 21–31. Chambers, F.M. (ed.) 1993: *Climate change and human impact on the landscape*. London: Chapman and Hall.

anthropology

The study of humanity, both in its biological sense and in the sense of its non-biological attributes (CULTURE). Three subdivisions are generally recognised: *physical* or *biological anthropology*, which investigates the nature and evolution of humans as biological phenomena; *social* or *cultural anthropology*, which analyses human culture based largely on experience of contemporary cultures; and ARCHAEOLOGY, which studies past societies primarily through reconstruction from the remains of their MATERIAL CULTURE. *JAM*

[See also ETHNO-ARCHAEOLOGY]

Renfrew, C. and Bahn, P. 1996: *Archaeology: theories, methods and practice*, 2nd edn. London: Thames and Hudson.

Anthrosols

Soils of all continents profoundly modified by HUMAN IMPACT to a depth of at least 50 cm; they are characterised by ANTHROPEDOGENIC HORIZONS. The original soil is no longer recognisable or has been buried. Usually, these soils have a thick human-made, organic-enriched, surface horizon and have widely differing physical and chemical properties. Already extensive and sometimes of considerable antiquity, Anthrosols are likely to increase substantially in area as human impact increases (SOIL TAXONOMY: *Anthric* subgroups). The buried soil below the Anthrosol can indicate environmental conditions before the land was modified. *EMB*

[See also AGRICULTURAL IMPACT ON SOILS, HUMAN IMPACT ON SOILS, TERRA PRETA DO INDIO, WORLD REFERENCE BASE FOR SOIL RESOURCES]

Bridges, E.M. 1997: The human factor in the World Reference Base: soil as an artifact. *Mitteilungen der Österreichische Bodenkundliche Geschellshaft* 55, 205–207. Deckers, J.A., Nachtergaele, F.O. and Spaargaren, O.C. (eds) 1998: *World reference base for soil resources: Introduction*. Leuven: Acco [ISSS Working Group RB]. Eidt, R.C. 1985: Theoretical and practical considerations in the analysis of anthrosols. In Rapp Jr, G. and Gifford, J.A. (eds), *Archaeological geology*. New Haven: Yale University Press, 135–190. Groenman-van Wateringe, W. and Robinson, M. (eds) 1988: *Man-made soils* [*BAR International Series* 410]. Oxford: BAR. Kosse, A. 1989: Anthrosols: proposals for a new soil order. *Transactions of the 13th International Congress of Soil Science, Kyoto* 3, 1175. Sandor, J.A. 1992: Long-term effects of prehistoric agriculture on soils: examples from New Mexico and Peru. In Holliday, V.T. (ed.), *Soils in archaeology: landscape evolution and human occupation*. Washington, DC: Smithsonian Institution, 217–245.

antibiotic

A complex chemical of microbial origin with antimicrobial properties. Although extracts from moulds have been used throughout history to combat infection, the modern development of antibiotics began with the discovery of *penicillin* in AD 1928. Unlike *antiseptics*, antibiotics are effective against BACTERIA, though widespread use has led to increasingly resistant strains of bacteria, which require the continual development of improved synthetic antibiotics. *JAM*

Coghan, A. 1996: Animal antibiotics threaten hospital epidemics. *New Scientist* 151, 4.

anticyclone

A region of high atmospheric pressure, also known as a *high pressure system* or '*high*', typically several thousand kilometres across, around which the air circulates in a clockwise direction in the Northern Hemisphere and counterclockwise in the Southern Hemisphere. Anticyclones are normally confined to the TROPOSPHERE and bring generally settled and usually dry WEATHER, which contributes to *phytochemical smog* and other problems of AIR POLLUTION. *Anticyclogenesis* refers to the formation of an anticyclone. *Cold anticyclones* are relatively shallow features resulting from winter radiative cooling over continental surfaces in temperate regions. *Warm anticyclones* are semipermanent features, such as the Azores High, maintained by subsiding and diverging air in the subtropical regions (see HADLEY CELL), and control the location of the major subtropical DESERTS. The smallest anticyclones (*mesohighs*, < 100 km across) are produced by cold thunderstorm downdrafts. *JAM*

[See also BLOCKING]

Alberta, T.L., Colucci, S.J. and Davenport 1991: Rapid mid-tropospheric cyclogenesis and anticyclogenesis. *Monthly Weather Review* 119, 1186–1204. Curry, J. 1987: The contribution of radiative cooling to the formation of cold-core anticyclones. *Journal of Atmospheric Sciences* 44, 2575–2592. Palmer, E. and Newton, C.W. 1969: *Atmospheric circulation patterns*. New York: Academic Press.

antitrades

The upper air winds that blow in some low latitude areas in the opposite direction to the TRADE WINDS, i.e. westerly and polewards. They do not have the same persistence or strength as the surface Trade Winds, tend to be more zonal in flow and are interrupted by other upper air winds, especially over the continents. In the classical three-cell model of the GENERAL CIRCULATION OF THE ATMOSPHERE they are the return flow of the HADLEY CELL. At about 30° latitude they merge into the subtropical JET STREAM. The antitrades or '*counter trades*' are replaced during the northern summer by the *tropical easterly jet* over the northern Indian Ocean and by the *African easterly jet* over West Africa. Over the Pacific Ocean the antitrades are subsumed by EL NIÑO–SOUTHERN OSCILLATION phenomena. *BDG*

McGregor, G.R. and Nieuwolt, S. 1998: *Tropical climatology*, 2nd edn. Chichester: Wiley.

aphelion

The point on the orbit of the Earth (or that of another planet) at which it is farthest from the Sun. *JAM*

aphotic zone

Depths in water bodies where there is not enough light for PHOTOSYNTHESIS to take place (usually deeper than 90 m). *BTC*

[See also EUPHOTIC ZONE, LAKE STRATIFICATION AND ZONATION]

apophyte

A native (plant) species favoured and spread directly or indirectly to newly established habitats by human cultural activity. Apophytes that penetrate new areas are ANTHROPOCHORES. Both terms are used in continental European POLLEN ANALYSIS. *FMC*

Behre, K.-E. 1988: The role of man in European vegetation history. In Huntley, B. and Webb III, T. (eds), *Vegetation history*. Dordrecht: Kluwer, 633–672.

apparent age An older *radiocarbon age* (see RADIOCARBON DATING) determined for contemporary samples from differing environments. Dating methods using *cosmogenic radionuclides* assume rapid (in terms of radionuclide HALF LIFE) exchange between the various parts of the geophysical reservoirs, leading to even distribution. This situation does not obtain, and for radiocarbon dating for instance, the *mean residence time* (MRT) (see RESIDENCE TIME) of carbon in the mixed layer of the oceans causes apparent ages of at least about 400 years for marine samples. Local factors, such as UPWELLING of deep (older) ocean water and the melting of (old) ice, increase this value, while it is reduced in shallow, coastal and Equatorial regions. For marine radiocarbon dates, there is a separate marine CALIBRATION dataset that displays none of the short-term variations of the terrestrial dataset as a result of the greater MRT. The phenomenon of apparent age is a RESERVOIR EFFECT. *PQD*

McCormac, F.G., Hogg, A.G., Higham, T.F.G. *et al.* 1998: Variations of radiocarbon in tree rings: Southern Hemisphere offset preliminary results. *Radiocarbon* 40, 1153–1159. Stuiver, M., Pearson, G.W. and Braziunas, T. 1986: Radiocarbon age calibration of marine samples back to 9000 cal year BP. *Radiocarbon* 28, 980–1021. Stuiver, M., Reimer, P.J. and Braziunas, T.F. 1998: High-precision radiocarbon age calibration for terrestrial and marine samples. *Radiocarbon* 40, 1127–1151.

apparent mean residence time (AMRT) A concept employed in SOIL DATING to account for the apparent age of soil organic material that is of mixed age, ranging from undecomposed plant litter (relatively young) to the inert or recalcitrant products of DECOMPOSITION (relatively old). The AMRT may be estimated with reference to the *radiocarbon age* (see RADIOCARBON DATING) of modern soils or by the radiocarbon dating of soils of known age. The radiocarbon age of a buried PALAEOSOL reflects the AMRT in addition to the time elapsed since burial. *JAM*

[See also SOIL AGE]

Campbell, C.A.E., Paul, E.A., Rennie, D.A. and McCallum, K.J. 1967: Factors affecting the accuracy of the radiocarbon-dating method in soil humus studies. *Soil Science* 104, 81–85. Matthews, J.A. 1993: Radiocarbon dating of arctic-alpine palaeosols and the reconstruction of Holocene palaeoenvironmental change. In Chambers, F.M. (ed.), *Climatic change and human impact on the landscape*. London: Chapman and Hall, 83–96.

apparent polar wander (APW) The apparent movement through time of the Earth's axis of rotation, as approximated by the palaeomagnetic north (or south) pole, with respect to any continent of observation (see PALAEOMAGNETISM). As the PLATE TECTONICS hypothesis became accepted, palaeomagnetists recognised that apparent polar wander was mainly due to the movement of lithospheric plates rather than the magnetic poles, bringing about changes in PALAEOLATITUDE: hence the description 'apparent' polar wander. There is, however, some 'true' polar wander of the magnetic pole (see GEOMAGNETISM, MAGNETIC EXCURSION). *Apparent polar wander paths* (APWPs) show successive positions of dated palaeomagnetic poles from individual continents, allowing their relative positions through GEOLOGICAL TIME to be determined: they are a major tool in the reconstruction of CONTINENTAL DRIFT. *DNT*

Fowler, C.M.R. 1990: *The solid Earth: an introduction to global geophysics*. Cambridge: Cambridge University Press. Sager, W.W. and Koppers, A.A.P. 2000: Late Cretaceous polar wander of the Pacific plate: evidence of a rapid true polar wander event. *Science* 287: 455–459. Zhang, H.M. 1998: Preliminary Proterozoic apparent polar wander paths for the South China Block and their tectonic implications. *Canadian Journal of Earth Sciences* 35, 302–320.

applied science The application of reliable knowledge and understanding (the fruits of *basic* or *pure science*) to the problems of individuals, organisations and societies in the real world; useful SCIENCE of immediate relevance to society. Pure and applied science are interdependent but successful applications require a sufficient knowledge base. Thus, knowledge and understanding of the pattern and timing of environmental changes, especially the development of reliable theories that provide explanations for change in terms of processes and mechanisms, are a necessary precondition to solving the applied problems associated with a wide range of ENVIRONMENTAL ISSUES. Applied science has considerable involvement with management, policy and ethical considerations. *JAM*

[See also ENVIRONMENTAL ETHICS, ENVIRONMENTAL MANAGEMENT, ENVIRONMENTAL POLICY, SCIENTIFIC METHOD]

Habermas, J. 1974: Theory and practice. London: Heinemann. Pacione, M. 1999: In pursuit of useful knowledge: the principles and practice of applied geography. In Pacione, M. (ed.), *Applied geography: principles and practice*. London: Routledge, 3–18. Trudgill, S. and Richards, K. 1997: Managing the Earth's surface: science and policy. *Transactions of the Institute of British Geographers NS* 22, 3–12.

aquaculture The culturing or 'farming' of fish and shellfish either in artificial systems, such as tanks, or in natural or semi-natural ecosystems, such as ponds, rivers or enclosed areas of the sea. There is a long history of freshwater aquaculture, especially in China; it is a relatively recent development in the sea (MARICULTURE). *JAM*

[See also FISHERIES CONSERVATION AND MANAGEMENT]

Beveridge, M.C.M., Ross, L.G. and Kelly, L.A. 1994: Aquaculture and biodiversity. *Ambio* 23, 497–502. Michael, R.G. (ed.) 1987: *Managed aquatic ecosystems* [*Ecosystems of the World*. Vol. 29]. Amsterdam: Elsevier. Seaman Jr, W. and Sprague, L.M. (eds) 1991: *Artificial habitats for marine and freshwater fishes*. New York: Academic Press. Stickney, R.R. 1996: *Aquaculture in the United States*. New York: Wiley.

aqualithic The 'non-Neolithic' communities and cultures of Africa from Senegal to Kenya, dating especially from the early HOLOCENE, whose livelihood and outlook were focused on lakes and rivers. *JAM*

Sutton, J.E.G. 1977: The African aqualithic. *Antiquity* 51: 25–34.

aquatic ape theory A controversial theory that interprets certain anatomical and behavioural features of HOMINIDS in terms of an early phase of HUMAN EVOLUTION that involved ADAPTATION to an aquatic environment. It disputes some aspects of the more widely accepted theory that human evolution involved adaptations to SAVANNA conditions. *GO*

Morgan, E. 1990: *The scars of evolution*. London: Souvenir Press.

aquatic environment ABIOTIC and BIOTIC conditions of lacustrine, riverine or marine ecosystems, where organisms live in water bodies that are stationary or flowing. An aquatic environment includes the physical (e.g. currents/circulation, light and temperature), chemical (e.g. nutrients, pH) and biological (e.g. primary producers such as algae and consumers such as zooplankton or fish) components and their interactions. BA

[See also TERRESTRIAL ENVIRONMENT]

Cushing, C.E., Cummins, K.W. and Minshall, G.W. (ed.) 1995: *River and stream ecosystems* [*Ecosystems of the World*. Vol. 22]. Amsterdam: Elsevier. **Ketchum, B.H. (ed.)** 1983: *Estuaries and enclosed seas* [*Ecosystems of the World*. Vol. 26]. Amsterdam: Elsevier. **Michael, R.G. (ed.)** 1987: *Managed aquatic ecosystems* [*Ecosystems of the World*. Vol. 29]. Amsterdam: Elsevier. **Taub, F.B. (ed.)** 1984: *Lakes and reservoirs* [*Ecosystems of the World*. Vol. 23]. Amsterdam: Elsevier.

aqueduct An artificial channel for transporting water, widely used throughout history, especially in facilitating IRRIGATION in SEMI-ARID environments and for conveying canals across valleys. JAM

[See also QANAT]

aquiclude An impermeable rock stratum overlying or underlying a water-bearing stratum or AQUIFER. JAM

[See also AQUITARD, GROUNDWATER]

aquifer A water-bearing stratum of rock or sediment. Aquifers contain and allow the passage of GROUNDWATER because they are PERMEABLE or PERVIOUS to water. A *confined aquifer*, located between two IMPERMEABLE or IMPERVIOUS strata, may be the source of ARTESIAN water; whereas the upper boundary of an *unconfined aquifer* (or *free-surface aquifer*) is defined by the WATER TABLE. JAM

[See also AQUICLUDE, AQUITARD, WELL]

Jones, J.A.A. 1997: *Global hydrology: processes, resources and environmental management*. Harlow: Longman.

aquitard A semi-permeable rock stratum through which GROUNDWATER flow is limited. JAM

[See also AQUICLUDE, AQUIFER]

arboreal Apertaining to trees; that is perennial plants with woody stems which grow to a height of more than 2 m. Arboreal animals live in trees. SPH

[See also ARBOREAL POLLEN]

arboreal pollen (AP) Pollen originating from ARBOREAL plants (trees). Pollen of SHRUBS can be included in either the AP or the NON-ARBOREAL POLLEN (NAP) for the purpose of the POLLEN SUM. SPH

[See also POLLEN ANALYSIS]

arborescent A GROWTH FORM that is tree-like. Some plants in the AFRO-ALPINE ZONE, for example, exceed 2 m in height (the conventional height of trees) but are not trees in other respects; notably they are not woody plants. JAM

arboriculture The cultivation of trees, including fruit and nut trees, and ornamental trees, but excluding trees in forests (SILVICULTURE). JAM

[See also HORTICULTURE]

arc A line, segment, chain, or string connecting a sequence of at least two POINTS, and represented in a GEOGRAPHICAL INFORMATION SYSTEM (GIS) by digital geometric COORDINATES. It is an Arc/Info term. TVM

ESRI 1997: *Understanding GIS: the Arc/Info method*. New York: Wiley.

Archaean An EON of geological time from the earliest known rocks (3960 million years old, from northwest Canada, although older dates have been obtained from individual MINERALS) to the beginning of the PROTEROZOIC. The first evidence of life on Earth is from the early Archaean, in the form of STROMATOLITES and filamentous microfossils in CHERT. GO

[See also BANDED IRON FORMATION, GEOLOGICAL TIMESCALE, GREENSTONE BELT, KOMATIITE, PRISCOAN]

Bowring, S.A., Williams, I.S. and Compston, W. 1989: 3.96 Ga gneisses from the Slave Province, NWT, Canada. *Geology* 17, 971–975. **McClendon, J.H.** 1999: The origin of life. *Earth-Science Reviews* 47, 71–93. **Nisbet, E.G.** 1987: *The young Earth: an introduction to Archaean geology*. London: Allen and Unwin. **Tajika, E. and Matsui, T.** 1993: Degassing history and carbon-cycle of the Earth – from an impact induced steam atmosphere to the present atmosphere. *Lithos* 30, 267–280.

archaeological chemistry An emerging subfield of chemistry that focuses on the application of the principles and techniques of chemistry to archaeological contexts, such as the chemical analysis of raw materials, ARTEFACTS and ECOFACTS. JAM

[See also ARCHAEOMETRY, ENVIRONMENTAL ARCHAEOLOGY, ENVIRONMENTAL CHEMISTRY, GEOARCHAEOLOGY, SCIENTIFIC ARCHAEOLOGY]

Goffer, Z. 1980: *Archaeological chemistry*. New York: Wiley. **Lambert, J.B. (ed.)** 1984: *Archaeological chemistry*. Washington, DC: American Chemical Society. **Pollard, A.M. and Heron, C. (eds)** 1996: *Archaeological chemistry*. Cambridge: Royal Society of Chemistry.

archaeological geology An emerging subdiscipline of geology where geological research is carried out at archaeological sites or on archaeological materials primarily for geological purposes but with archaeological implications. It is related to GEOARCHAEOLOGY, where the emphasis is on the application of geological techniques to the elucidation of problems in archaeology or ENVIRONMENTAL ARCHAEOLOGY. JAM

[See also ARCHAEOLOGICAL PROSPECTION, ENVIRONMENTAL GEOLOGY, SCIENTIFIC ARCHAEOLOGY]

Lasca, N.P. and Donahue, J. (eds) 1990: *Archaeological geology of North America* [*Centennial Special*. Vol. 4]. Boulder, CO: Geological Society of America. **McGuire, W.J., Griffiths, D.R., Hancock, P.L. and Stewart, I. (eds)** 2000: *The archaeology of geological catastrophes* [*Geological Society of London Special Publication* 171]. London: Geological Society of London. **Pollard, A.M.** 1999: Geoarchaeology: an introduction. In Pollard, A.M. (ed.), *Geoarchaeology: exploration, environments, resources*. London: The Geological Society, 7–14. **Rapp Jr, G. and Gifford, J.A.** 1982: Archaeological geology. *American Scientist* 70, 456–463. **Rapp Jr, G. and Hill, C.L.** 1998: *Geoarchaeology*. New Haven, CT: Yale University Press.

archaeological prospection Also known as *geoprospection*, archaeological prospection is essentially the discovery and exploration of archaeological sites before excavation, using geophysical, geochemical and/or REMOTE SENSING techniques. The most commonly employed techniques of *geophysical prospection (archaeogeophysical exploration)* include magnetic surveying, *electrical resistivity surveying, electromagnetic surveying* and GROUND-PENETRATING RADAR. SOIL PHOSPHATE ANALYSIS is the most widely used technique of *geochemical prospection*. Large quantities of the elements phosphorus, nitrogen and carbon are added to the soil in and around human habitations in various forms (originating from human and animal remains, faeces and other waste) but only the phosphorus accumulates as it largely remains fixed and insoluble. The presence of phosphorus concentrations can be detected long after visible evidence of habitation has disappeared. *JAM*

[See also AIRBORNE REMOTE SENSING, GEOPHYSICAL SURVEYING, SATELLITE REMOTE SENSING]

Garrison, E.G., Baker, J.G. and Thomas, D.H. 1985: Magnetic prospection and discovery of Mission Santa Catalina de Guale, Georgia. *Journal of Field Archaeology* 12, 299–313. **Spoerry, P. (ed.)** 1992: *Geoprospection of the archaeological landscape*. London: Oxbow Books. **Stickel, E.G. and Garrison, E.G.** 1988: New applications of remote sensing: geophysical prospection for underwater archaeological sites in Switzerland. In Purdy, B.A. (ed.), *Wet site archaeology*. Caldwell, NJ: Telford, 69–88.

archaeological timescale The division of time according to technological, cultural or political change used in archaeology to define distinct periods, which may be of local, regional or global application. These changes may be related to AGRICULTURAL ORIGINS, innovations in metallurgy or real (or perceived) socio-economic or sociopolitical developments, which are rarely synchronous over areas; hence most archaeological periods are TIME-TRANSGRESSIVE (see Table A). In the absence of written records or an accepted calendar, dating the boundaries that define the timescale relies on some form of RELATIVE-AGE DATING or *absolute-age dating*, the latter including RADIOCARBON DATING. Dates used in the archaeological timescale are usually expressed as either BC or AD, but these are often generalised rather than specifically calibrated ages.

The determination of detailed CHRONOLOGIES within broad archaeological periods was originally achieved through the determination of typological sequences for ARTEFACTS such as *pottery* or *flints*, based on assumed similarities or differences thought either to represent synchroneity (see SYNCHRONOUS) of production (thus of the same age) or to be due to gradual development through time (representative of different periods). This approach was originally developed in the nineteenth century by Oscar Montelius using tools and weapons of BRONZE AGE context in Europe and became the basis of much archaeological chronological work, particularly for changes within major periods. This has been further developed into the principle of SERIATION, in which assemblages of artefacts are placed within a succession indicative of their temporal sequence, under the assumption that certain styles come into common usage, peak and then become rare, forming what are known as *battleship curves* when plotted graphically in relation to time. This approach was originally developed using CERAMICS from the Mayan civilisation.

For periods with written records, such as the Egyptian,

Chinese and Mayan civilisations, it has been possible to develop fairly precise *archaeological chronologies*. Such records still require critical analysis of the original sources and also need tying to the current calendar. For Egypt this has been achieved by using the conquest by Alexander the Great in 332 BC as a known marker and then counting back or forward, also utilising astronomical observations of known date within the same records. There is, however, still uncertainty and debate about certain parts of the chronology, particularly where it appears to contradict independent dating evidence from other dating methods such as DENDROCHRONOLOGY. Chronologies developed from such sources originally provide a FLOATING CHRONOLOGY comparable to that found in dendrochronology. One of the most intricate examples is that of the extremely sophisticated *Mayan calendar* used between AD 300 and 900, which has only recently been interpreted and related to *calendar years* (see CALENDAR/CALENDRICAL AGE/DATE).

The earliest period recognised in the archaeological timescale is the PALAEOLITHIC, which is divided into the Lower, Middle and Upper (see Table B). The overall defining characteristic of this period that spans the evolution of HOMINIDS is a reliance on stone tools, at least since around 2.5 million years ago (Ma), hence the term '*Old Stone Age*'. This was followed by the MESOLITHIC ('*Middle Stone Age*'), which was still pre-agricultural, and eventually the NEOLITHIC ('*New Stone Age*'), when agriculture became established. The NEOLITHIC TRANSITION has been a phenomenon of considerable interest to both archaeologists and palaeoecologists and its dating is highly variable, with, for instance in Europe, DOMESTICATION first appearing in the Near East and nearby areas of Europe at least as early as 6500 BC, but only spreading across the continent to the west over the next 2000–3000 years. Following the Neolithic in Europe, the transition to the Bronze Age is defined at the first appearance of metal-working, to be followed by the IRON AGE, when iron tools and implements were developed. Although not as time-transgressive as earlier transitions, both periods vary in their dating from region to region. After the ROMAN PERIOD subsequent timescales are usually expressed purely in terms of CULTURAL CHANGE, reflecting the importance of dominant races or political systems. *CJC*

Baillie, M.G.L. 1995: *A slice through time: dendrochronology and precision dating*. London: Batsford. **Brainerd, G.W.** 1951: The place of chronological ordering in archaeological analysis. *American Antiquity* 16, 301–313. **Renfrew, C. and Bahn, P.** 1993: *Archaeology: theories, method and practice*. London: Thames and Hudson. **Robinson, W.S.** 1951: A method for chronologically ordering archaeological deposits. *American Antiquity* 16, 293–301.

archaeology The study of the human past through its material remains. *SPD*

[See also CONTEXTUAL ARCHAEOLOGY, ENVIRONMENTAL ARCHAEOLOGY, GEOARCHAEOLOGY, SCIENTIFIC ARCHAEOLOGY]

Ashmore, W. and Sharer, R.J. 1996: *Discovering our past: a brief introduction to archaeology*, 2nd edn. Mountain View, CA: Mayfield. **Barker, G. (ed.)** 1999: *Companion encyclopedia of archaeology*, two vols. London: Routledge. **Renfrew, C. and Bahn, P.** 1996: *Archaeology: theories, methods and practice*, 2nd edn. London: Thames and Hudson.

Years AD/BC	Near East	Egypt & Africa	Medi-terranean	North Europe	India	E. Asia & Pacific	Meso-America	South America	North America
1500		Great Zimbabwe	BYZANTINE EMPIRE	Mediaeval states		New Zealand settled	AZTEC TOLTEC	INCA CHIMU	Cahokia Chaco
1000	ISLAM	Towns (Africa) AXUM	ROMAN EMPIRE	ROMAN EMPIRE		States (Japan) Great Wall (China)	MAYA TEOTI-HUACAN	MOCHE	HOPEWELL PUEBLOS
500 AD BC	PERSIA BABYLON ASSYRIA	LATE PERIOD	CLASSICAL GREECE	IRON AGE	MAURYAN	Cast Iron (China) Lapita (Polynesia)	OLMEC	CHAVIN	Maize (Southwest)
500	HITTITES Iron	NEW KINGDOM MIDDLE KINGDOM	Iron MYCENAE	BRONZE Stonehenge AGE	Iron	SHANG (China)			
1000									
1500	SUMER	OLD (pyramids) KINGDOM	Minoan		INDUS	Walled villages (China)		Temple-mounds	
2000	Writing	EARLY DYNASTIC	LATE NEOLITHIC				Maize	Maize, Llama, cotton	
2500	Cities	Towns (Egypt)							
3000	Wheeled vehicles								
3500									
4000				Megaliths					
4500	Irrigation		Copper (Balkans)	Farming, pottery		Rice, millet (China)		Manioc	
5000			EARLY NEOLITHIC						
5500							Beans, squash, peppers?	Beans, squash, peppers?	
6000	Copper	Cattle (N. Africa)	Farming, pottery			Gardens (New Guinea)			
6500	Pottery				Farming			Pottery (Brazil)	
7000	Wheat, Rye etc.	Pottery (Sudan)							
7500									
8000			MESOLITHIC					Maize? (Argentina)	
8500	Sheep								
9000									
9500						Pottery (Japan) (14,000 BC)			
10,000									

archaeological timescale *(A) A summary of world-wide archaeological change during the Holocene, indicating some important cultural developments used in regional subdivisions of the archaeological timescale (after Renfrew and Bahn, 1996, with the addition of an approximate position of the Mesolithic, early Neolithic and late Neolithic in central Europe)*

archaeomagnetic dating The PALAEOMAGNETIC DATING of baked materials and archaeological sediments, using THERMOREMANENT MAGNETISATION and DETRITAL REMANENT MAGNETISATION as proxy records of the Earth's ambient geomagnetic field (see GEOMAGNETISM) at the time of the formation of the archaeological material.

MHD

Eighmy, J.L. and Sternberg, R.S. (eds) 1990: *Archaeomagnetic dating.* Tucson, AZ: University of Arizona Press.

Sternberg, R.S. 1997: Archaeomagnetic dating. In Taylor, R.E. and Aitken, M.J. (eds), *Chronometric dating in archaeology.* New York: Plenum, 323–356.

archaeomagnetism A subfield of PALAEOMAGNETISM that applies palaeomagnetic techniques to archaeological materials or artefacts, or to rocks in an archaeological context.

JAM

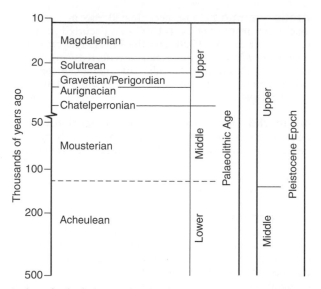

archaeological timescale *(B) Some archaeological subdivisions of the Palaeolithic in Europe (after Aitken, 1990)*

[See also ARCHAEOMAGNETIC DATING, ENVIRONMENTAL ARCHAEOLOGY, GEOARCHAEOLOGY]

Saribudak, M. and Tarling, D.H. 1993: Archaeomagnetic studies of the Urartian civilization, eastern Turkey. *Antiquity* 67, 620–668. **Sparks, R.S.J.** 1985: Archaeomagnetism, Santorini volcanic eruptions and fired destruction levels on Crete. *Nature* 313, 74–75. **Tarling, D.H.** 1985: Archaeomagnetism. In Rapp Jr, G. and Gifford, J.A. (eds), *Archaeological geology*. New Haven, CT: Yale University Press, 237–263.

archaeometry Strictly, the application of measurement to archaeological problems but, more broadly, the application of scientific techniques, especially DATING TECHNIQUES, site investigation using geophysical techniques (*archaeogeophysical exploration* or ARCHAEOLOGICAL PROSPECTION) and the analysis of the physical and chemical properties of archaeological materials, as exemplified by the journal, *Archaeometry*. *JAM*

[See also ARCHAEOLOGICAL CHEMISTRY, ARCHAEOLOGICAL GEOLOGY, ENVIRONMENTAL GEOLOGY, GEOARCHAEOLOGY, SCIENTIFIC ARCHAEOLOGY]

Leute, U. 1987: *Archaeometry*. Weinheim, Germany: VCH.

architectural element analysis An approach to FACIES ANALYSIS that groups FACIES into assemblages that represent elements of the depositional system, such as fluvial channels or *bars*. Together with the recognition of a hierarchy of BOUNDING SURFACES, architectural element analysis has proved a very successful methodology for interpreting river PALAEOENVIRONMENTS. *GO*

[See also FACIES ARCHITECTURE, FACIES SEQUENCE]

Allen, J.R.L. 1983: Studies in fluviatile sedimentation: bars, bar-complexes and sandstone sheets (low-sinuosity braided streams) in the Brownstones (L. Devonian), Welsh Borders. *Sedimentary Geology* 33, 237–293. **Miall, A.D.** 1985: Architectural element analysis: a new method of facies analysis applied to fluvial deposits. *Earth-Science Reviews* 22, 261–308.

Miall, A.D. 1996: *The geology of fluvial deposits: sedimentary facies, basin analysis, and petroleum geology*. Berlin: Springer.

archive A repository and source of information. The term was originally used for a place where DOCUMENTARY EVIDENCE is stored, but in the study of environmental change its use has broadened to include DATABASES and also NATURAL ARCHIVES. The explosion of data, the internationalisation of scientific activity and the necessity for sharing data in fields such as PALAEOCLIMATOLOGY and *earth-system science* have led to the growth of *data banks* or *data archives*, such as that provided by the World Data Center-A for Paleoclimatology based in Boulder, Colorado. *JAM*

[See also CLIMATIC ARCHIVE]

Anderson, D.M. and Webb, R.S. 2000: *The PAGES data guide: results from the Second Workshop on Global Paleoenvironmental Data*. Boulder, CO: World Data Center-A for Paleoclimatology. **Overpeck, J.T. and Pilcher, J.** 1995: *Global palaeoenvironmental data* [*Pages Workshop Report Series* 95–2]. Bern: PAGES Core Project Office.

Arctic All areas north of the Arctic Circle (latitude 66° 30′ N). This is the latitude at which the Sun does not rise in midwinter or set in midsummer. However, this simple definition is misleading since many areas distinctly 'arctic' in character are excluded and it obscures the range of environments present within the Arctic. A climatic definition of the Arctic, incorporating the concept of an *Arctic climate*, is the region where the warmest monthly mean air temperature does not exceed 10°C and the coldest is below 0°C. The term SUBARCTIC is then used to describe those areas where the mean monthly air temperatures do not exceed +10°C for more than four months and where the coldest is below 0°C. The boundary between the Arctic and Subarctic approximates to the northern limit of trees. *HMF*

[See also ARCTIC ENVIRONMENTAL CHANGE, LOW ARCTIC, HIGH ARCTIC, TUNDRA]

French, H.M. 1999: Arctic environments. In Alexander, D.E. and Fairbridge, R.W. (eds), *Encyclopedia of environmental science*. Dordrecht: Kluwer, 29–33.

Arctic environmental change The Arctic region differs from the Antarctic in having a 4000 m deep ocean located over the North Pole, surrounded by continental land masses, allowing only limited interchange of waters with other oceans.

The Arctic record of environmental changes through geological time is highly fragmentary, as the region today comprises different, formerly widely separated, tectonic TERRANES. Many parts of the Arctic originated south of the Equator and record marked climatic and depositional changes. Although some areas (North America, Scandinavia, Barents Shelf, parts of Siberia) were partly covered by successive ICE SHEETS during the last two million years, ice growth may have first begun eight million years ago. The extent of only the last ice sheets, however, is reasonably well known. Only the Greenland ice sheet survives today, covering 1.68 million km², with a volume of 2.95 million km³, equivalent to a 7 m SEA-LEVEL RISE. Apart from Greenland, ice cover is limited mainly to ice caps, highland icefields and valley glaciers. The potential contribution of these smaller ice masses to sea-level rise is about 0.35 m.

Most Arctic glaciers have generally receded since the early twentieth century. Some scientists argue that the Greenland ice sheet is growing; others that it is shrinking. A compromise view suggests increased melting in the south, but growth in the north from increased precipitation. Greenland ICE CORES provide an environmental record extending back through the last glacial–interglacial cycle. The positive link between CARBON DIOXIDE concentration and temperature is comparable to that from Antarctica, suggesting global synchroneity. Termination of the last glaciation, around 13 000 years ago, was abrupt and a temperature rise of several degrees in just one decade is recorded.

SEA ICE is increasingly being used as a monitor of global warming. Changes in the southern limits of PACK ICE, and of thickness of sea ice (15% decrease in the area north of Greenland since the mid 1970s) have been the main indicators used. Some GENERAL CIRCULATION MODELS predict an 8–15°C temperature rise for the Arctic with a doubling of atmospheric CO_2, enhancing the melting of ice masses and contributing to sea-level rise, reducing the thickness and extent of sea ice, melting the PERMAFROST and thereby having a profound effect on the indigenous fauna and flora. Other environmental problems in the Arctic are associated with water, air and soil POLLUTION, resource exploitation and development. *MJH*

[See also ARCTIC HAZE, ANTARCTIC ENVIRONMENTAL CHANGE]

Houghton, J.T., Jenkins, G.J. and Ephraums, J.J. (eds) 1990: *Climate change – the IPCC scientific assessment.* Cambridge: Cambridge University Press. **Intergovernmental Panel on Climate Change** 1992: *Climate change.* World Meteorological Organisation and United Nations Environment Programme. **Overpeck, J., Hughen, K., Hardy, D. et al.** 1997: Arctic environmental change of the last four centuries. *Science* 278, 1251–1256. **Sugden, D.E.** 1982: *Arctic and Antarctic.* Oxford: Blackwell. **Wadhams, P., Dowdeswell, J.A. and Schofield, A.N. (eds)** 1996: *The Arctic and environmental change.* Amsterdam: Gordon and Breach.

Arctic Front The semipermanent but discontinuous boundary in the atmosphere between the very cold, dense AIR MASS controlled by the cold Arctic ANTICYCLONE and the less cold polar air mass; it is a passive front, positional changes being the result of gravity and air mass weight. The stronger and more dynamic POLAR FRONT lies at somewhat lower latitudes between the cool polar air mass and the warmer tropical air mass. *JAM*

Binkley, M. 1987: Arctic Front. In Oliver, J.E. and Fairbridge, R.W. (eds), *The encyclopedia of climatology.* New York: Van Nostrand Reinhold, 91–92.

Arctic haze A reddish-brown atmospheric haze observed primarily in winter and spring (since the 1940s) when Arctic air is calm. Its existence is attributed to industrial AIR POLLUTION transported mostly from Europe and North America. By increasing surface and near-surface ABSORPTION, it could cause spring AIR TEMPERATURE in the Arctic to rise and this could result in changes in the hemispheric wind pattern. *GS*

[See also ARCTIC ENVIRONMENTAL CHANGE]

Barrie, L.A. 1986: Arctic air pollution: an overview of current knowledge. *Atmospheric Environment* 20, 643–663. **Kemf, E.** 1984: Air pollution in the Arctic. *Ambio* 13, 122–123.

Arctic warming The accentuated rise in the MEAN ANNUAL TEMPERATURE by 4–10°C that occurred in Arctic regions between the 1920s and the 1950s. The cause was mainly intrusions of warm ocean currents reducing SEA ICE formation. This moderated adjacent coastal climates with consequences for TUNDRA ecosystems and PERMAFROST DEGRADATION. *RDT*

ard marks Criss-crossing, linear scoring, mostly found beneath Neolithic barrows, caused by an early type of plough (the 'ard'), which broke the ground surface but did not turn the sod. *JAM*

Kristiansen, K. 1990: Ard marks under barrows: a response to Peter Rowley Conwy. *Antiquity* 64: 332–337.

area generalisation A simplification of detail, a resampling to larger spacing or a reduction in the number of POINTS in a line (see ARC) that make up an area. In GEOGRAPHICAL INFORMATION SYSTEM (GIS), this involves changes in CARTOGRAPHIC transformations and scale reductions. *TVM*

[See also AREA PATCH GENERALISATION, LINE REDUCTION, MAP GENERALISATION]

Muller, J.-C., Lagrange, J.-P. and Weibel, R. 1995: *GIS and generalisation.* London: Taylor & Francis.

area labels IDENTIFICATION tags describing some ATTRIBUTE, both name and value, of an area or any other ENTITY in a GEOGRAPHICAL INFORMATION SYSTEM (GIS). *TVM*

[See also LABELS]

area patch generalisation A type of generalisation that reduces the amount of detail to enable the effective representation of areal objects in two-dimensional space at varying scales. *TVM*

[See also AREA GENERALISATION]

Muller, J.C. and Zeshen, W. 1992: Area patch generalisation: a competitive approach. *Cartographic Journal* 29, 144–147.

arenaceous Adjective describing a SAND-rich SEDIMENT or SEDIMENTARY ROCK. *TY*

arenite A SANDSTONE with less than 15% MATRIX, typical of sediments deposited in a high ENERGY environment. *TY*

Pettijohn, F.J. 1975: *Sedimentary rocks*, 3rd edn. New York: Harper and Row. **Pettijohn, F.J., Potter, P.E. and Siever, R.** 1987: *Sand and sandstone.* New York: Springer.

Arenosols Sandy soils with weak or very weak soil development forming on sandstones or sand dunes, often marginal to deserts, where they may be indicative of CLIMATIC CHANGE. They do not have diagnostic horizons other than an OCHRIC or an ALBIC HORIZON. Arenosols are widespread in SEMI-ARID Africa, Australia and South America (SOIL TAXONOMY: Psammic and Psammaquic great groups of *Entisols*). With a low waterholding capacity these soils are used mainly for natural GRAZING, but if IRRIGATION water is available cropping can be successful. Arenosols are vulnerable to WIND EROSION and COMPACTION. *EMB*

[See also WORLD REFERENCE BASE FOR SOIL RESOURCES]

areography A little-used term referring to the study of the geographical RANGES of taxa. The factors determining the geographical range of a taxon are critical to the understanding of how a species will respond to environmental change. *JLI*

Rapoport, E.H. 1982: *Areography: geographical strategies of species.* Oxford: Pergamon Press.

argic horizon A diagnostic subsoil horizon with an increased clay content, ranging from > 3% in sandy soils to >8% in clayey soils, compared with overlying horizons (SOIL TAXONOMY: *Argillic horizon*). The argic horizon results from an *illuvial accumulation* of clay derived from the upper part of a SOIL PROFILE by the process of ELUVIATION. *EMB*

[See also ILLUVIATION, PERVECTION]

Bronger, A. 1978: Climatic sequences of steppe soils from eastern Europe and the USA with emphasis on the genesis of the 'argillic horizon'. *Catena* 5, 33–51. Bronger, A. 1991: Argillic horizons in modern loess soils in an ustic soil moisture regime: comparative studies in forest-steppe and steppe areas from eastern Europe and the USA. *Advances in Soil Science* 15, 41–90. Food and Agriculture Organization of the United Nations 1998: *World reference base for soil resources* [*Soil Resource Report* 84]. Rome: FAO, ISRIC, ISSS.

argillaceous Adjective describing a MUD-rich SEDIMENT or SEDIMENTARY ROCK. *TY*

argillite A lithified MUDROCK, although the term is not widely used. *TY*

[See also LITHIFICATION]

argon–argon (Ar-40/Ar-39) dating A RADIOMETRIC DATING technique based on the decay of argon-40 to the DAUGHTER ISOTOPE argon-39. The advantages of using Ar-40/Ar-39 over POTASSIUM–ARGON (K–Ar) DATING are: (1) abundance of Ar ISOTOPES is measured on the same sample, circumventing sample heterogeneity problems; and (2) step heating is used to release Ar from increasingly deeper, unaltered interiors of the crystal, enabling the assessment of alteration or contamination. *DAR*

Baksi, A.H., Hsu, V., McWilliams, M.O. and Farrar, E. 1992: ⁴⁰Ar/³⁹Ar dating of the Brunhes– Matuyama geomagnetic field reversal. *Science* 256, 356–357. Walter, R.C. 1994: Age of Lucy and the first family: single-crystal ⁴⁰Ar/³⁹Ar dating of Denen Dora and lower Kada Hadar members of the Hadar Formation, Ethiopia. *Geology* 22, 6–10.

aridity The characteristic of a CLIMATE having insufficient or inadequate precipitation to maintain a vegetation cover. This depends on soil factors as well as the interplay between temperature, evaporation, the amount of precipitation that falls annually and its seasonal distribution. Most aridity indices, which attempt to define aridity quantitatively, commonly take account of POTENTIAL EVAPOTRANSPIRATION (or EVAPORATION) and/or temperature as well as precipitation (i.e. *precipitation effectiveness*). *AHP/JAM*

[See also ARIDLAND, DESERT, DROUGHT, DRYLAND]

Durrenberger, R.W. 1987: Arid climates. In Oliver, J.E. and Fairbridge, R.W. (eds), *The encyclopedia of climatology*. New York: Van Nostrand Reinhold, 92–101. Stadler, S.J. 1987: Aridity indexes. In Oliver, J.E. and Fairbridge, R.W. (eds), *The encyclopedia of climatology*. New York: Van Nostrand Reinhold, 102–107.

aridland A DRYLAND environment defined by the United Nations Environment Program (UNEP) (1991) on the basis of the Thornthwaite index of *available humidity*, where the ratio of annual precipitation to POTENTIAL EVAPOTRANSPIRATION averages 0.05–0.20. This UNEP definition uses fixed time meteorological data for the years AD 1951–1980. From this, aridlands cover 12.1% of the world's land surface. Aridlands are usually too dry to support, unaided, dryland agriculture or grazing by domesticated livestock. *MLC*

[See also ARIDLAND VEGETATION, DESERT, DRYLAND]

Agnew, C. and Anderson, E. 1992: *Water resources in the arid realm*. London: Routledge. United Nations Environment Program 1991: *Status of desertification and implementation of the United Nations plan of action to combat desertification* [GCSS.III/3]. Nairobi: UNEP.

aridland: past environmental change SEDIMENTS that mantle the world's aridlands often preserve surficial and stratigraphical records of ENVIRONMENTAL CHANGE. Currently arid landscapes, including many DESERTS, contain significant lacustrine deposits, which may include whole shells of freshwater mollusc species. Death Valley in California, which is one of the hottest and driest places on earth (averaging 41 mm of rain annually) displays wavecut shorelines, the salt pan remnants of a large PLEISTOCENE lake and an impressive gravel beach bar, standing over 100 m above the valley floor. Death Valley is currently hyperarid, yet it has clearly contained at least one large and persistent lake during the Late PLEISTOCENE. TUFA-coated shorelines of past lakes exist across the aridlands of the southwest USA and further evidence for large PLUVIAL lakes has been described from the hyper arid Sahara and Namib deserts in Africa (see AFRICAN HUMID PERIOD). RADIOCARBON DATING of organic material from lake cores, tufa, shell and rock varnish has been used to define a CHRONOLOGY for past PLUVIAL periods, in which ENVIRONMENTS were wetter and colder than those that exist today.

Other evidence for past environmental change in aridlands, particularly of increased effective moisture, includes: areas of BADLAND erosion; ARROYO cutting and filling; ALLUVIAL FAN accretion and dissection; and the formation of carbonate-rich PALAEOSOL horizons. AEOLIAN deposits are mobilised by strong winds in areas of high sediment supply and low vegetation cover. Stratigraphical records within aeolian SAND DUNES and dust deposits (LOESS) provide evidence of episodic accretion, which has often been linked to past periods of intense aridity. Recent evidence has shown, however, that the relationship between sand dunes and climate change is not so simple, and dune accretion may relate to wetter periods providing an increased sand supply to DEFLATION areas.

Evidence for the former extension of deserts exists in the Permian and Triassic geological record within northern Europe, including the UK, where large-scale EVAPORITES, SABKHA and SAND SEA deposits are found and commercially exploited. *MLC*

[See also DESERT LANDSCAPE EVOLUTION, DESERT VEGETATION, DESERTIFICATION, LAKE-LEVEL VARIATIONS, PALAEOSALINITY, RODENT MIDDENS]

Clarke, M.L. and Rendell, H.M. 1998: Climate change impacts on sand supply and the formation of desert sand dunes in the south-west USA. *Journal of Arid Environments* 39, 517–531. Frostick, L.E. and Reid, I. 1987: *Desert sediments: ancient and* modern, *Geological Society Special Publication* 35. London: The Geological Society. **Goudie, A.S. and Eckardt, F.** 1999: The evolution of the morphological framework of the central Namib Desert, Namibia, since the early Cretaceous. *Geografiska Annaler* 81A, 443–458. **Hunt, C.B.** 1975: *Death Valley: geology, ecology, archaeology*. Berkeley, CA: University of California Press. **Stokes, S., Thomas, D.S. and Washington, R.** 1997. Multiple episodes of aridity in southern Africa since the last interglacial period. *Nature* 388, 154–158.

aridland vegetation The VEGETATION of aridland is characteristically sparse, with low BIOMASS, and dominated by drought-tolerant XEROPHYTES. Species vary according to environment factors such as temperature, rainfall, SALINITY and soil moisture. Adaptation to DROUGHT conditions varies between species and includes: extensive, near-surface rooting systems such that the majority of organic matter is below the surface; deep tap roots (which exceed 50 m in species of mesquite (*Prosopis* spp.); the ability to store water in tissues within roots, stems and leaves; stomata control; long dormant periods; and short growing and flowering phases that are closely linked to moisture availability. Annual PRIMARY PRODUCTIVITY is closely related to rainfall.

Aridland vegetation can be classified into *passive species*, which are inactive during the dry season, and *active species*, which are drought tolerant (xerophytic) and continue to photosynthesise throughout the dry season. Inactive species include many ANNUALS, which restrict their growth to rainy periods and survive drought by the production of a large number of desiccation-resistant seeds that lie dormant until sufficient moisture is available for germination. PERENNIALS in this class lose their leaves in the dry season and lie dormant until activated into growth by the rains. Active xerophytic species function throughout the dry period, using a variety of drought-resistant strategies. They are PERENNIAL, retain their biomass and are characterised by having a small number of stomata, which close during the day and open at night (the reverse of temperate species). They are often spiny, an adaptation that serves to minimise water loss. Changes in moisture availability in arid landscapes, as a result of short-term climatic change, have an immense impact on vegetation germination and growth.

Grasses and xerophytic shrubs are important for grazing animals and traditional PASTORALISM. Aridland environments are often too dry for rain-fed agriculture. The saline nature of many arid landscapes, particularly around salt pans, provides ideal habitat for HALOPHYTES such as salt grasses (*Distichlis* spp.) and dwarf shrubs such as *Sueda* spp. and saltbush (*Atriplex*). *MLC*

[See also BIOMES, DESERT VEGETATION]

Blume, H.P. and Berkowicz, S.M. 1995: *Arid ecosystems*. Cremingen: Catena. **Bowers, M.A.** 1987: Precipitation and the relative abundances of desert winter annuals: a 6 year study in the northern Mojave Desert. *Journal of Arid Environments* 12, 141–149. **Goodall, D.W. and Perry, R.W.** 1979: *Aridland ecosystems: structures, functioning and management*. Cambridge: Cambridge University Press.

arkose A SANDSTONE with a significant proportion (> 25%) of CLASTS of FELDSPAR, and with little or no (< 15%) MATRIX of MUD (i.e. an ARENITE). A *subarkose* has between 5 and 25% feldspar clasts. A *lithic arkose* has more than 25% feldspar clasts and a ratio of feldspar to LITHIC

clasts between 3:1 and 1:1. *Arkosic arenite* (or *feldspathic arenite*) covers both arkose and lithic arkose. Arkoses are derived from feldspar-rich source areas such as outcrops of *granite* and *gneiss* (see PROVENANCE). Because feldspars undergo rapid WEATHERING in humid conditions, arkoses commonly indicate source environments that favour PHYSICAL WEATHERING over CHEMICAL WEATHERING, such as ARIDLANDS or PERIGLACIAL settings, although rapid DEPOSITION may allow arkosic sediments to accumulate even under unfavourable weathering conditions. *TY*

Johnsson, M.J., Stallard, R.F. and Lundberg, N. 1991: Controls on the composition of fluvial sands from a tropical weathering environment – sands of the Orinoco River drainage basin, Venezuela and Colombia. *Geological Society of America Bulletin* 103, 1622–1647. **Kamp, P.C. van de, Helmold, K.P. and Leake, B.E.** 1994: Holocene and Paleogene arkoses of the Massif Central, France: mineralogy, chemistry, provenance, and hydrothermal alteration of the type arkose. *Journal of Sedimentary Research* A64, 17–33. **Pettijohn, F.J.** 1975: *Sedimentary rocks*, 3rd edn. New York: Harper and Row. **Stewart, A.D.** 1991: Geochemistry, provenance and palaeoclimate of the Sleat and Torridon groups in Skye. *Scottish Journal of Geology* 27, 81–95.

armouring A variety of processes by which potentially mobile sediment is protected from EROSION, usually by a surface layer of coarser or otherwise less readily transported material (*armour*). In the context of *environmental engineering* (or GEOENGINEERING) armouring is an important method of ENVIRONMENTAL PROTECTION in, for example, COASTAL (SHORE) PROTECTION or the stabilisation of river banks. Natural armouring of sediment surfaces occurs where a substrate is covered by a thin layer (commonly only one or two grains thick) of coarser material. In the context of river beds, particularly GRAVEL-BED RIVERS, a distinction is commonly made between an *armoured riverbed*, in which the coarse surface layer is mobile during FLOODS, allowing transportation of the substrate to occur, and a *paved riverbed*, in which the surface layer is considerably coarser than the substrate and is immobile, typically representing a LAG DEPOSIT in a channel undergoing DEGRADATION. *GO*

[See also BOULDER PAVEMENTS, DESERT PAVEMENT]

Dunkerley, D.L. 1990: The development of armour in the Tambo River, Victoria, Australia. *Earth Surface Processes and Landforms* 15, 405–412. **Evans, D.J.A.** 1991: A gravel diamicton lag on the south Albertan prairies, Canada – evidence of bed armoring in early deglacial sheet-flood spillway courses. *Geological Society of America Bulletin* 103, 975–982. **Gomez, B.** 1984: Typology of segregated (armoured/paved) surfaces: some comments. *Earth Surface Processes and Landforms* 9, 19–24. **Isla, F.I.** 1993: Overpassing and armoring phenomena on gravel beaches. *Marine Geology* 110, 369–376. **Sahayan, S.J.M. and Hall, K.R.** 1998: Optimum geometry for naturally armouring berm breakwaters. *Journal of Coastal Research* 14, 1293–1303.

arroyo A term of Spanish origin used in the American West for an ephemeral stream channel with a rectangular cross-section entrenched in ALLUVIUM. Arroyos are produced by CLIMATIC CHANGE and/or LANDUSE CHANGE. They differ from GULLIES, which tend to be V-shaped in cross-section and may be excavated also in soil and COLLUVIUM. *JAM*

[See also WADI]

Cooke, R.U. and Reeves, R.W. 1976: *Arroyos and environmental change in the American South-West*. Oxford: Clarendon Press.

Graf, W.L. 1983: The arroyo problem: paleohydrology and paleohydraulics in the short term. In Gregory, K.J. (ed.), *Background to palaeohydrology: a perspective*. Chichester: Wiley, 279–302.

artefact   An object that has been made or modified by humans.   *SPD*

[See also ECOFACT, GEOFACT, INTERFEROMETRY]

Slater, E. 1999: Studying artefacts. In Barker, G., *Companion encyclopaedia of archaeology*. London: Routledge, 344–388.

artesian   Relating to GROUNDWATER under HYDROSTATIC PRESSURE in a confined AQUIFER. *Artesian water* in *artesian basins* may flow at the Earth's surface from a natural, *artesian spring* or be tapped artificially by *artesian wells*. The Great Artesian Basin, where many of Australia's 400 000 groundwater-tapping boreholes are located, provides a classic example of the nature and use of artesian water.   *JAM*

Habermehl, M.A. 1985: Groundwater in Australia. *International Association of Hydrological Sciences Publication* 154, 31–52.

artificial   Made by humans; not NATURAL.   *JAM*

artificial shoreline   A shoreline constructed by human endeavours usually in an attempt to protect the existing shoreline from erosion by means of structures including SEAWALLS and RIPRAP, to change its configuration or to improve its recreational potential.   *HJW*

[See also COASTAL CHANGE, HUMAN IMPACT ON COASTS, LAND RECLAMATION]

Walker, H.J. (ed.) 1988: *Artificial structures and shorelines*. Dordrecht: Kluwer.

asbestos   A group of fibrous silicate minerals, the commonest of which, *chrysotile*, was widely used in electrical and thermal insulation. Health hazards, associated especially with the inhalation of asbestos fibres (e.g. *asbestosis* and *mesothelioma*) have been recognized since the AD 1930s. Its production, use and disposal have therefore been strictly regulated since the 1970s.   *JAM*

Selikoff, I.J. and Lee, D.H.K. 1978: *Asbestos and disease*. New York: Academic Press.

ash content   The residue of a sample after ignition (see LOSS-ON-IGNITION). When samples are burnt at 500°C, organic material is removed (see CARBON CONTENT) and what remains is the mineral component or ash. For PEAT or lake SEDIMENT samples, the ash content may represent: AUTOCHTHONOUS plant silica; ALLOCHTHONOUS mineral material brought in by streams or washed in by water due to erosion in the catchment area; or airborne mineral particles, including TEPHRA (volcanic ash). Temporal changes in ash content through a sediment profile often reflect different ENVIRONMENTAL IMPACTS within the sediment *catchment* area, while tephras provide valuable marker horizons, which can be dated and used for correlating the timing of events between profiles. When the sample is an animal, the ash content is a measure of that proportion of the body weight comprised of bone.   *SPH*

ash fall   Sometimes used as a synonym for PYROCLASTIC FALL, but the term should be restricted to deposits of GRAIN-SIZE less than 2 mm.   *GO*

ash flow   Sometimes used as a synonym for PYROCLASTIC FLOW, but the term should be restricted to deposits of GRAIN-SIZE less than 2 mm, or used as *block and ash flow*.   *JBH*

aspect   The direction in which a slope or LANDSCAPE faces, with environmental implications such as the amount of INSOLATION received or the degree of exposure. It is normally measured with a compass.   *JBE*

assemblage   (1) A set of ARTEFACTS recovered from the same archaeological site or from a particular archaeological context. Groups of assemblages may form INDUSTRIES, and sequences of industries may form TRADITIONS. (2) In an ecological context, a group of plants growing together at the same site. (3) In a palaeontological or palaeoecological context, a set of FOSSILS or SUBFOSSILS found together *in situ*.   *JAM*

assemblage zone   A unit in BIOSTRATIGRAPHY describing the dominant and/or characteristic species and their abundances (e.g. percentages) in a biologically more-or-less homogeneous sediment (including peat) section. Qualitative and quantitative description of compositional changes of FLORA or FAUNA at the lower and upper limit of the zone help to delimit the *local assemblage zone*, which is the basis for building up *regional assemblage zones*. It is independent of chronology.   *BA*

[See also PALAEOBOTANY, PALAEOECOLOGY, POLLEN ASSEMBLAGE ZONE]

Cushing, E.J. 1967: Late-Wisconsin pollen stratigraphy and the glacial sequence in Minnesota. *Proceedings of the 7th INQUA Congress* 7, 59–88. New Haven, CT: Yale University Press. **North American Commission on Stratigraphic Nomenclature** 1982: North American Stratigraphic Code. *American Association of Petroleum Geologists Bulletin* 67, 841–875.

association   A VEGETATION unit, with the term being used either in relation to the basic plant community derived from field samples (as in the Zürich–Montpellier classification system) or in relation to broad vegetation types, such as deciduous forest.   *JLI*

[See also COMMUNITY CONCEPTS, PHYTOSOCIOLOGY, VEGETATION FORMATION, VEGETATION FORMATION-TYPE]

Ellenberg, H. 1988: *Vegetation ecology of Central Europe*, 4th edn. Cambridge: Cambridge University Press.

asteroids   Solid, rocky bodies with low volatile content that range in size from about 5 m across (large METEOROIDS) to bodies such as *Ceres*, which has a diameter of about 1000 km. Most orbit the Sun between Mars and Jupiter in the *asteroid belt*. The term means 'star-like object'. *Near-Earth asteroids* (NEAs) together with short-period COMETS are the NEAR-EARTH OBJECTS that may threaten the Earth. Near-Earth asteroids are arbitrarily defined as objects that have a *perihelion distance* (closest distance to the Sun) of <1.0 AU (astronomical units based on the mean distance of the Earth from the Sun of about 150 000 km): around 700 had been identified by the year 2000 as a result of the NASA Near Earth Asteroid Tracking (NEAT) Program. Three groups of NEAs are recognised: (1) *Atens* – Earth-crossing NEAs named after 2062 Aten – which are generally closer to the Sun than the

Earth; (2) *Apollos* – Earth-crossing NEAs named after 1862 Apollo – which are generally farther from the Sun than the Earth with orbital periods greater than one year; and (3) *Amors* – Earth-approaching NEAs named after 1221 Amor – which have orbits beyond the Earth's but within that of Mars. *Potentially hazardous asteroids* (PHAs) are those > 150 m in diameter and with the potential to approach within 7 480 000 km of the Earth: around 250 PHAs had been identified by June 2000.   *JAM*

[See also METEORITE IMPACT, PLANETARY GEOLOGY]

Gehrels, T. (ed.) 1994: *Hazards due to comets and asteroids.* Tucson, AZ: University of Arizona Press. **Huggett, R.J.** 1997: *Environmental change: the evolving ecosphere.* London: Routledge. **McGuire, W.J., Kilburn, C.R.J. and Saunders, M.A.** 2001: *Natural hazards and environmental change.* London: Arnold.

asthenosphere   A weak, plastic layer in the outer MANTLE of the Earth (100–300 km depth below the surface) underlying the rigid LITHOSPHERE. The weakness of the asthenosphere, or *low velocity zone*, is attributed to a small amount (c. 5%) of partial melting, which is the source of MAGMA.   *GO*

[See also EARTH STRUCTURE, PLATE TECTONICS]

astrobleme   A large (diameter > 10 km) METEORITE IMPACT structure on the Earth's land surface. The term, literally meaning 'star wound', is sometimes restricted to old, eroded structures (*impact structures*).   *GO*

[See also NEAR-EARTH OBJECT]

Dietz, R.S. 1961: Astroblemes. *Scientific American* 205, 51–58.

astrogeology   The application of geological principles to the study of solid objects in the solar system (planets, satellites, asteroids, etc.) and beyond.   *GO*

[See also PLANETARY ENVIRONMENTAL CHANGE, PLANETARY GEOLOGY]

asymmetrical valley   A valley cross-profile in which one side is steeper than the other. The asymmetry may be the product of aspect-related differences in hillslope processes, differential erosion due to underlying geological structures (see UNICLINAL SHIFTING) and/or differential stream undercutting at the base of the slopes. Asymmetrical valleys may provide evidence of changed environmental conditions: in present-day TEMPERATE environments, for example, they may indicate former PERIGLACIAL conditions under which there were aspect-related differences in rates of FROST WEATHERING and SOLIFLUCTION.   *JAM*

Churchill, R.R. 1982: Aspect-related differences in hillslope processes. *Earth Surface Processes and Landforms* 7, 171–182. **Long, D. and Stoker, M.S.** 1986: Valley asymmetry: evidence for periglacial activity in the central North Sea. *Earth Surface Processes and Landforms* 11, 525–532. **Wende, R.** 1995: Drainage and valley asymmetry in the Tertiary Hills of Lower Bavaria, Germany. *Geomorphology* 14, 255–265.

atmosphere   The gaseous mass retained close to the Earth by gravity. There is also an atmosphere around other planets and stars. The reduction in atmospheric pressure and temperature variations with altitude define the vertical structure of the atmosphere summarised in the Figure. The lower layers, particularly the TROPOSPHERE and the STRATOSPHERE, are most important for environmental change. The Earth's ATMOSPHERIC COMPOSITION is predominantly nitrogen (average 78% by volume), oxygen (21%), argon (0.93%) and CARBON DIOXIDE (0.3%), but there are numerous other natural and anthropogenic gases present in very small amounts together with water vapour and suspended AEROSOLS, including some important POLLUTANTS.   *JAM*

Goody, R.M. and Walker, J.C.G. 1972: *Atmospheres.* Englewood Cliffs, NJ: Prentice-Hall. **Kasting, J.F.** 1996: Atmosphere: an overview. In Schneider, S.H. (ed.), *Encyclopedia of climate and weather.* New York: Oxford University Press, 55–58. **Walker, J.C.G.** 1977: *Evolution of the atmosphere.* New York: Macmillan.

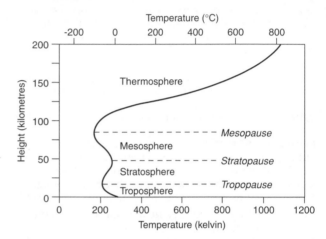

atmosphere   *Temperature variations with altitude and the vertical structure of the atmosphere up to 200 km (after Kasting, 1996)*

atmosphere–ocean interaction   As the oceans cover < 70% of the surface of the Earth, they are a major influence on the Earth's climates, especially through the exchange of moisture (see HYDROLOGICAL CYCLE), heat and momentum. The heat capacity of the oceans, SEA-SURFACE TEMPERATURES, OCEAN CURRENTS and UPWELLING all influence climate in important ways. Similarly, atmospheric conditions are a major control on the dynamics of the oceans. A major example of such interactions includes EL NIÑO–SOUTHERN OSCILLATION (ENSO) phenomena. Heat and momentum are transferred across the boundary between ocean and atmosphere but the dynamics of these transfers are only partially understood. Of particular interest is the sea surface temperature, which influences the rate and magnitude of latent and sensible heat fluxes and which is now regularly monitored by REMOTE SENSING.

Some particularly sensitive areas of the oceans, where a small change in the ocean circulation could have an amplifying effect on CLIMATIC CHANGE, include the following: (1) off the eastern tip of Brazil, where the branching of the Atlantic Equatorial Current affects the heat transfer to the Southern Hemisphere; (2) the southern limits of the polar seas in relation to the extent of SEA ICE; (3) the northern limit of the warm GULF STREAM, which creates temperature ANOMALIES in the north Atlantic; (4) the eastern Pacific area, where the extent of warm equatorial waters is responsible for ENSO. Ocean–atmosphere interactions play an important role in climatic change on many timescales and it has been suggested that stochastic changes in heat content of the ocean may initiate different atmospheric states. *AHP/JAM*

[See also COUPLED OCEAN–ATMOSPHERE MODELS, THERMOHALINE CIRCULATION, TROPICAL CYCLONES, WALKER CIRCULATION]

Bakun, A. 1990: Global climate change and intensification of coastal ocean upwelling. *Science* 247, 198–201. **Barrera, E. and Johnson, C.C.** 1999: Evolution of the Cretaceous ocean-climate system. *Geological Society of America Special Paper* 332, 1–446. **Cayan, D.R.** 1992: Latent and sensible heat flux anomalies over the northern oceans: the connection to monthly atmospheric circulation. *Journal of Climate* 5, 354–369. **Ganopolski, A., Kubarzki, C., Claussen, M.** *et al.* 1998: The influence of vegetation–atmosphere–ocean interaction on climate during the mid-Holocene. *Science* 280, 1916–1919. **Frankignoul, C.** 1985: Sea surface temperature anomalies, planetary waves and air-sea feedbacks in middle latitudes. *Reviews of Geophysics* 8, 233–246. **Kushnir, Y.** 1994: Interdecadal variations in north Atlantic sea surface temperature and associated atmospheric conditions. *Journal of Climate* 7, 141–157. **Perry, A.H. and Walker, J.M.** 1977: *The ocean–atmosphere system*. London: Longman. **Trenberth, K.E. and Hurrell, J.W.** 1994: Decadal atmosphere–ocean variations in the Pacific. *Climate Dynamics* 9, 303–319. **Wallace, J.M., Smith, C. and Jiang, Q.** 1990: Spatial patterns of atmosphere–ocean interaction in the northern winter. *Journal of Climate* 3, 990–998.

atmospheric composition

The chemical components of the Earth's ATMOSPHERE. The atmosphere acts as a medium of exchange between the GEOSPHERE, BIOSPHERE and HYDROSPHERE in the cycling of material on geological time scales. Hence, the composition of the Earth's atmosphere (see Table) has been determined from a combination of biological, physical and chemical processes. The effects of the biosphere on atmospheric composition are profound and the two main atmospheric components, nitrogen and oxygen, are almost entirely controlled by biological processes and the result of biological EVOLUTION. The early pre-life atmosphere consisted of a mixture of water vapour, carbon dioxide, carbon monoxide, nitrogen, methane and ammonia. The exact composition is still debated; however, the common consensus is that the early atmosphere was a reducing (see REDUCTION) one that favoured the evolution of life. The atmosphere today in an Earth devoid of life would contain only traces of oxygen and be characterised by nitrogen and CARBON DIOXIDE.

JCMc

[See also AEROSOLS, CARBON CYCLE, CARBON DIOXIDE VARIATIONS, GAIA HYPOTHESIS, NITROGEN CYCLE, NITROGEN OXIDES, NITROUS OXIDE VARIATIONS, OXYGEN, POLLUTANTS]

Francis, P. and Dise, N. 1997: *Atmosphere, Earth and life*. Milton Keynes: Open University Press. **Houghton, J.T., Meira Filho, L.G., Callander, B.A.** *et al.* 1996: *Climate change 1995*. Cambridge: Cambridge University Press.

atmospheric correction

A procedure required to remove the effects of the ATMOSPHERE from measurements of RADIANCE to convert them to surface reflectance. The atmosphere varies with time, location, wavelength, altitude and pressure and, even on a clear day, may be rich with *absorbing* and *scattering components*. AEROSOL scattering and absorption at wavelengths between 400 and 1000 nm are related to the size, shape, chemical composition and refractive index of the particles. *Rayleigh scattering* at wavelengths below 1000 nm, caused by molecules, is related to wavelength and pressure. Gas absorptions at wavelengths between 1000 and 2500 nm are dominated by WATER VAPOUR, CARBON DIOXIDE, METHANE, OZONE, OXYGEN, nitrous oxide and CARBON MONOXIDE. RADIATION is affected by the atmosphere both before and after it interacts with the surface. Atmospheric correction procedures range from complex models that attempt to simulate the interactions occurring in the atmosphere to simple empirical relationships. *GMS*

Asrar, G. 1989: *Theory and applications in optical remote sensing*. New York: Wiley.

atmospheric dust

Together with the finer AEROSOLS, atmospheric dust (about 5–20 μm) make up the PARTICULATE load of the atmosphere. Sources of dust can be natural, for instance from VOLCANIC ERUPTIONS and windblown sand and soil, or be the result of human activities, such as agriculture, forest fires and industry. Although most is removed by gravity and precipitation, dust can still affect the RADIATION BALANCE of the atmosphere. Incoming SOLAR RADIATION will be scattered or absorbed, thus increasing planetary ALBEDO and resulting in a cooling effect. On the other hand, the increased

atmospheric composition *Chemical components of the atmosphere*

Gas	Symbol	Atmospheric composition (percent by volume)	(parts per million)
Nitrogen	N_2	78.1	
Oxygen	O_2	20.9	
Argon	Ar	0.93	
Carbon dioxide	CO_2	0.036	360
Neon	Ne	1.8×10^{-3}	18
Helium	He	5.2×10^{-4}	5.2
Methane	CH_4	2×10^{-4}	2
Krypton	Kr	1×10^{-4}	1
Hydrogen	H_2	5×10^{-5}	0.5
Nitrous oxide	N_2O	3.2×10^{-5}	0.32
Xenon	Xe	9×10^{-6}	0.06
Ozone	O_3	1×10^{-6} to 1×10^{-5}	0.01–0.1
Water vapour	H_2O	Variable	
Pollutants, e.g. halocarbons		Variable	

absorption of outgoing *long-wave radiation* will enhance warming. Whether the net effect of atmospheric dust levels is a cooling or a warming on a global scale depends on the quantity and nature of the particles as well as the nature of the land or ocean surfaces below, and regional effects can be significant. *GS*

[See also DUST VEIL, SAHARAN DUST, VOLCANIC AEROSOLS, WET DEPOSITION]

Barry, R.G. and Chorley, R.J. 1998: *Atmosphere, weather and climate.* London: Routledge. **Goudie, A.S. and Middleton, N.J.** 1992: The changing frequency of dust storms through time. *Climatic Change* 20, 197–225. **Pye, K.** 1987: *Aeolian dust and dust deposits.* London: Academic Press. **Shaw, R.W.** 1987: Air pollution by particles. *Scientific American* 257, 96–103.

atmospheric loading The concentration of liquid or particulate impurities in the ATMOSPHERE, such as dust (e.g. VOLCANIC ASH) and smoke. *JCM*

[See also AEROSOLS, DUST VEIL]

atoll A 1–130 km diameter, often circular to elliptical, usually oceanic CORAL REEF, with a ring of low coral islands encircling a central lagoon of variable depth. *TSp*

atom The smallest constituent part of an ELEMENT, which retains its characteristic properties and can take part in a chemical reaction. It consists of a *nucleus*, made up of *protons* and *neutrons*, and *electrons*, which occupy surrounding orbitals. The nucleus contains most of the mass of the atom but occupies only a small fraction of its volume. The positive charges of the protons are balanced by the negative charges of the electrons. *JAM*

[See also ION]

atomic absorption spectrophotometry (AAS) A technique for elemental analysis used in GEOCHEMISTRY and related fields, involving measurement of the absorption of characteristic wavelengths of light (see ABSORPTION SPECTRUM) passing through a flame containing the vapourised sample. Instruments can typically analyse for around 50 elements sequentially (compare INDUCTIVELY COUPLED PLASMA ATOMIC EMISSION SPECTROMETRY), using a small sample size (200 mg of material in solution). The accuracy is about ±1% for major elements, and about ±15% for TRACE ELEMENTS. It is also known as *elemental absorption spectrophotometry (EAS)*. *TY*

[See also CHEMICAL ANALYSIS OF SOILS AND SEDIMENTS]

Fairchild, I.J., Hendry, G., Quest, M. and Tucker, M.E. 1988: Chemical analysis of sedimentary rocks. In Tucker, M.E. (ed.), *Techniques in sedimentology.* Oxford: Blackwell Scientific, 274–354. **Farmer, J.G., Eades, L.J. and Graham, M.C.** 1999: The lead content and isotopic composition of British coals and their implications for past and present releases of lead to the UK environment. *Environmental Geochemistry and Health* 21, 257–272.

attractor Attractors are points or areas within a phase space being used to depict the behaviour of a DYNAMICAL SYSTEM, which control the trajectory of the SYSTEM and to which all trajectories are drawn. *CET*

[See also FRACTAL ANALYSIS, STRANGE ATTRACTOR]

Phillips, J.D. 1999: *Earth surface systems, complexity, order and scale.* Oxford: Blackwell Publishers.

attribute A non-graphic description, IDENTIFICATION or value of an ENTITY (for example the name, length or velocity of a river) used in a GEOGRAPHICAL INFORMATION SYSTEM (GIS). It is also a FIELD entry in a RELATIONAL DATABASE. *TVM*

[See also AREA LABELS, LABELS, OVERLAY ANALYSIS]

auger A rod-like device for preliminary examination of sediments, soils or other subsurface materials. It is pushed or screwed into the substrate either by hand or with mechanical assistance. *JAM*

autecology The study of the ecology or life history of a single individual or species, as opposed to the study of groups of different species and their interactions (*synecology* or COMMUNITY ECOLOGY). *JLI*

authigenic Minerals and cements formed during or after the formation of the rock or SEDIMENT of which they are part as a result of chemical and biological reaction. *UBW*

[See also DIAGENESIS, GREIGITE]

autochthonous Rocks and deposits formed and deposited at the same place as they are found today; FOSSILS or SUBFOSSILS deposited at the place where the animal lived or the plant grew. Autochthonous features form *in situ*; they are not the products of transportation. *UBW*

[See also ALLOCHTHONOUS, COAL, PEAT]

autocompaction Pore-water expulsion and compression of PEAT (or ORGANIC-rich SEDIMENT) under its own mass. It can amount to 80% of the original thickness and be even greater when overlain by non-organic sediments. *UBW*

[See also COMPACTION, CONSOLIDATION]

Allen, J.R.L. 1999: Geological impacts on coastal wetland landscapes: some general effects of sediment autocompaction in the Holocene of northwest Europe. *The Holocene* 9. 1–12.

autocorrelation A measure that quantifies dependence between values in a dataset as a function of the spatial or temporal relationship between them. It can be used as a measure of IMAGE TEXTURE, for classification purposes. *ACF*

autocyclic change A change driven by internal factors, within a system; AUTOGENIC CHANGE is a synonym. For example AVULSION, or switching of a depositional lobe on a DELTA or ALLUVIAL FAN, may occur in response to the accumulation of sediment (an autocyclic control). In contrast, it could be driven by ALLOCYCLIC CHANGES such as CLIMATIC CHANGE, EUSTATIC change or TECTONIC effects. *GO*

[See also REGRESSION, TRANSGRESSION]

Mack, G.H., Love, D.W. and Seager, W.R. 1997: Spillover models for axial rivers in regions of continental extension: the Rio Mimbres and Rio Grande in the southern Rio Grande rift, USA. *Sedimentology* 44, 637–652. **Peper, T. and Cloetingh, S.** 1995: Autocyclic perturbations of orbitally forced signals in the sedimentary record. *Geology* 23, 937–940.

autogenic change　(1) Alteration of species composition during ECOLOGICAL SUCCESSION due to BIOTIC interactions and biotic modification of the environment. These causal factors are *internal* to the community. (2) The term is also used in the sedimentological context as a synonym for AUTOCYCLIC CHANGE.　　　　　　　　　　　*LRW*

[See also ALLOGENIC CHANGE, FACILITATION, INHIBITION MODEL]

Botkin, D.B. 1981: Causality and succession. In West, D.C., Shugart, H.H. and Botkin, D.B. (eds), *Forest succession: concepts and applications*. New York: Springer, 36–55.

automatic weather station　A device consisting of a range of environmental sensors providing automated observations of ambient meteorological conditions, usually via a DATA LOGGER.　　　　　　　　　　　*JCM*

autoregressive (AR) modelling　A linear method of estimating the *persistence structure* or memory in TIME-SERIES ANALYSIS. It is most easily understood in the context of linear REGRESSION ANALYSIS in which one or more *predictor variable* or independent VARIABLE is used to estimate a *predictand* or *dependent variable*. In the time-series context, this means that past values of the series are used to estimate its current value. In the simplest case of an AR model of order 1 (the *AR(1) model*), and if a time step of one year is assumed, the observed value for year t is estimated by its previous value (year $t-1$). This is equivalent to lagging the series on itself one year and regressing year t data on year $t-1$ data. AR modelling can easily be extended to include additional lags resulting in the *AR(p) model*, which regresses year t data on year $t-1, t-2, \ldots,$ $t-p$ data. AR models can be used for time-series FORECASTING and also for addressing the lack of serial independence that makes statistical tests of association between series difficult to apply. In the latter case, persistence reduces the number of independent observations and this reduces the effective sample size for hypothesis testing. AR modelling can be used to remove persistence in the time series being compared, resulting in a new set of serially random series (the residuals from the AR model) that do not have this problem.　　　　　　　　　　*ERC*

[See also AUTOREGRESSIVE MOVING AVERAGE MODELLING]

Box, G.E.P. and Jenkins, G.M. 1970: *Time series analysis: forecasting and control*. San Francisco, CA: Holden-Day. **Cook, E.R. and Kairiukstis, L.A.** 1990. *Methods of tree-ring analysis: applications in the environmental sciences*. Dordrecht: Kluwer.

autoregressive moving average (ARMA) modelling　A more general method of modelling persistence in TIME-SERIES ANALYSIS, as in for instance DENDROCHRONOLOGY, than that afforded by AUTOREGRESSIVE (AR) MODELLING alone. Whereas AR modelling only uses past observations from $t-1$ to $t-p$ years for estimating current observations, ARMA modelling, and its more restricted *moving average* (MA) *modelling* form, uses past model residuals from $t-1$ to $t-q$ to estimate current observations as well. This results in an ARMA(p, q) model fitted to the series. This generalisation allows for more complex persistence structure to be modelled and, thus, greatly increases the number of time series that can be adequately modelled. However, compared to *linear autoregressive* (AR) *models*, ARMA are much more difficult to estimate

because the coefficients of the MA part of the model are highly non-linear. Therefore, specialised non-linear estimation methods must be used. The determination of the correct order of the ARMA(p, q) model can also be more difficult than that of an AR(p) model. Given that AR modelling often provides an adequate representation of the persistence structure in time series used for CLIMATIC RECONSTRUCTION in PALAEOCLIMATOLOGY, ARMA modelling is not used as frequently.　　　　　　　　*ERC*

[See also FILTERING]

Box, G.E.P. and Jenkins, G.M. 1970: *Time series analysis: forecasting and control*. San Francisco, CA: Holden-Day. **Cook, E.R. and Kairiukstis, L.A.** 1990: *Methods of tree-ring analysis: application in the environmental sciences*. Dordrecht: Kluwer.

autotrophic organism　A 'self-nourishing' organism that makes its own food from simple inorganic carbon componds. *Photoautotrophs*, including most terrestrial plants and some bacteria utilise light energy through PHOTOSYNTHESIS; *chemoautotrophs* or *chemolithotrophs* comprise most bacteria, which obtain the necessary energy by OXIDATION of reduced inorganic compounds such as sulphides and ammonia. *Lithoautotrophic organisms* live within rocks.　　　　　　　　　　　*JAM*

[See also ENERGY FLOW, HETEROTROPHIC ORGANISM, TROPHIC LEVEL]

Stevens, T.O. and McKinley, J.P. 1995: Lithoautotrophic microbial ecosystems in deep basalt aquifers. *Science* 270, 450–454.

autovariation　Internally regulated sequences of behaviour exhibited by the atmospheric circulation whereby preferred recurring patterns are apparent. CLIMATIC CHANGE is caused by external FORCING FACTORS and autovariation generated within the CLIMATIC SYSTEM. *JCS*

[See also SELF-ORGANISATION]

Karl, T.R. (1988) Multi-year fluctuations of temperature and precipitation: the gray area of climate change. *Climatic Change* 12, 179–198.

avalanche　A rapid, gravity-driven MASS MOVEMENT of particulate material on a steep slope. Although frequently used specifically to refer to SNOW AVALANCHES, avalanches can also comprise rock debris or sediment (*rock falls*, *debris falls* and *grain flows*).　　　　　　　　　　　*GO*

[See also CROSS-STRATIFICATION, SEDIMENT GRAVITY FLOW, STURZSTROM]

Blikra, L.H. and Nemec, W. 1998: Postglacial colluvium in western Norway: depositional processes, facies and palaeoclimatic record. *Sedimentology* 45, 909–959.

avulsion　A relatively sudden abandonment of a major part of a depositional system. The behaviour of some rivers, for example, is characterised by the periodic avulsion of a long stretch of the channel to a new course on the FLOODPLAIN. This contrasts with more gradual shifting of the channel within a channel belt through processes such as lateral migration of a point bar of MEANDER CUTOFF. The avulsion or abandonment of a depositional lobe of a DELTA or ALLUVIAL FAN in favour of a new site is also known as *lobe switching*. Avulsion can occur in response to AUTOCYCLIC CHANGE or ALLOCYCLIC CHANGE, and can be recognised in SEDIMENTARY DEPOSITS through distinctive FACIES or FACIES ARCHITECTURE.   *GO*

Jones, L.S. and Schumm, S.A. 1999: Causes of avulsion: an overview. In Smith, N.D. and Rogers, J. (eds), *Fluvial sedimentology VI* [*International Association of Sedimentologists Special Publication* 28]. Oxford: Blackwell Science, 171–178. **Kraus, M.J. and Wells, T.M.** 1999: Recognizing avulsion deposits in the ancient stratigraphical record. In Smith, N.D. and Rogers, J. (eds), *Fluvial sedimentology VI* [*International Association of Sedimentologists Special Publication* 28]. Oxford: Blackwell Science, 251–268. **Perez-Arlucea, M. and Smith, N.D.** 1999: Depositional patterns following the 1870s avulsion of the Saskatchewan River (Cumberland Marshes, Saskatchewan, Canada). *Journal of Sedimentary Research* 69, 62–73. **Smith, N.D., Cross, T.A., Dufficy, J.P. and Chough, S.R.** 1989: Anatomy of an avulsion. *Sedimentology* 36, 1–23.

axiom A proposition that is assumed to be true and forms part of a THEORY that is the foundation for investigating a particular scientific research problem. Axioms are well established or self-evident truths, which are not questioned during the investigation.   *JAM*

azimuth The 360° plane at right angles to a fixed elevation angle. In REMOTE SENSING, it is used to denote position in the ALONG-TRACK DIRECTION.   *PJS*

azonal soil A weakly developed soil, dominated by the nature of the PARENT MATERIAL and lacking genetic SOIL HORIZONS, the distribution of which is unrelated to climate. A soil order in the system of classification used in the USA before 1965, but not used in SOIL TAXONOMY or the WORLD REFERENCE BASE FOR SOIL RESOURCES. Azonal soils are often vulnerable to SOIL DEGRADATION with environmental change.   *EMB*

Baldwin, M., Kellogg, C.E. and Thorp, J. 1938: Soil classification. In *Soils and men, yearbook of agriculture*. Washington, DC: US Department of Agriculture.

B

back-arc The area behind a volcanic ISLAND ARC on the over-riding plate at a DESTRUCTIVE PLATE MARGIN, characterised by a *fold-thrust belt* on CONTINENTAL CRUST (e.g. Canadian Rockies) or a back-arc MARGINAL BASIN in OCEANIC CRUST (e.g. Sea of Japan, Aegean Sea). *GO*

[See also FORE-ARC, PLATE TECTONICS, SEDIMENTARY BASIN]

background level (1) A measure of the quantity of a substance or any other environmental attribute in the absence of human effects. In relation to HEAVY METALS in the environment, for example, natural concentrations exist at background levels, which may be exceeded as a result of METAL POLLUTION. Background levels of RADIATION are also measures of this type. (2) A measure of the long-term state of an environmental phenomenon in relation to significant departures from the norm. Through most of geological time, for example, about four species per year became extinct (out of a total of some 10 million species): this background level of EXTINCTION was greatly exceeded during episodes of MASS EXTINCTION. *JAM*

[See also BACKGROUND RATE]

background rate A rate of change for a process, against which possibly significantly increased (or reduced) rates can be tested. For example, the significance of anthropogenically increased EXTINCTION rates cannot be assessed without information on the background rate: the rate of extinction in the absence of human activity. *KDB*

backscattering The scattering of RADIATION into the direction opposite to the incident direction. In the case of an active REMOTE SENSING instrument, for example, back to a sensor and in the case of a passive remote sensing instrument, back towards the Sun. *PJS*

[See also ENERGY BALANCE]

bacteria Unicellular micro-organisms that absorb soluble food and reproduce by binary fission. They perform important roles in natural and anthropogenic ecosystems, including in DECOMPOSITION and BIOGEOCHEMICAL CYCLES, POLLUTION control and WASTE MANAGEMENT, and the spread of DISEASES such as plague, tetanus and tuberculosis. *JAM*

[See also BIODEGRADATION, BIOREMEDIATION, OIL SPILLS]

Madigan, M.T., Martinko, J.M. and Parker, J. 1997: *Biology of micro-organisms*, 8th edn. Upper Saddle River, NJ: Prentice-Hall.

badlands A degraded landscape, especially in a SEMI-ARID location, comprising a maze of GULLIES cut into soft sediments including soil. The term was first applied to parts of South Dakota, North Dakota and Nebraska. The high *drainage densities* were originally attributed to intensive surface fluvial activity, but it was subsequently recognised that *piping* and MASS WASTING were also important. The term is now used to describe any intensely dissected landscape where vegetation cover is sparse or absent and the land is of no agricultural use. In many cases, SOIL EROSION caused by removal of the vegetation cover is blamed. SOIL DEVELOPMENT in association with geomorphic surfaces in badlands can be used to interpret environmental change. *RAS*

[See also LAND DEGRADATION]

Bryan, R. and Yair, A. 1982: *Badland geomorphology and piping*. Norwich: GeoBooks. Campbell, I.A. 1997: Badlands and badland gullies. In Thomas, D.S.G. (ed.), *Arid zone geomorphology*. Chichester: Wiley, 261–291. Howard, A.D. 1994: Badlands. In Abrahams, A.D. and Parsons, A.J. (eds), *Geomorphology of arid environments*. London: Chapman and Hall, 213–242. Kendrick, K.J. and McFadden, L.D. 1996: Comparison and contrast of processes of soil formation in the San Timoteo Badlands with chronosequences in California. *Quaternary Research* 46, 149–160.

bajada A gently sloping aggradational surface along the foot of a mountain front, typically in SEMI-ARID areas. In contrast to PEDIMENT slopes, bajadas (also known as *bahadas* or *alluvial bajadas*) are a product of deposition, formed by the coalescence of adjacent ALLUVIAL FANS. The term is of Spanish-American derivation from the southwestern USA and northern Mexico. *GO*

[See also AGGRADATION, PIEDMONT]

Milana, J.P. 2000: Characterization of alluvial bajada facies distribution using TM imagery. *Sedimentology* 47, 741–760.

balanced cross-section A cross-section through GEOLOGICAL STRUCTURES that allows the geometry of strata prior to DEFORMATION to be reconstructed. It is used particularly to determine the amount of crustal shortening caused by THRUST FAULTS. *GO*

[See also PALINSPASTIC MAP]

Davis, T.L. and Namson, J.S. 1994: A balanced cross-section of the 1994 Northridge earthquake, southern California. *Nature* 372, 167–169. Harris, C., Williams, G., Brabham, P. *et al.* 1997: Glaciotectonized Quaternary sediments at Dinas Dinlle, Gwynedd, North Wales, and their bearing on the style of deglaciation in the eastern Irish Sea. *Quaternary Science Reviews* 16, 109–127.

banded iron formation (BIF) Rock sequences made up of fine alternations of red iron oxides (haematite, magnetite) and grey silica (CHERT), effectively restricted to the PRECAMBRIAN geological record, between 3000 and 1800 million years ago. Banded iron formations are known from every continent, are up to 600 m thick, and extend for hundreds of kilometres. They are of major importance as iron ORES (e.g. Hamersley, West Australia). Their origin appears to have involved chemical or biochemical deposition of iron under oceanic conditions different from those of the present day. Overall oceanic oxygen levels may have been sufficiently low that iron passed into solution as the soluble ion Fe^{2+}, but locally enough oxygen was produced (perhaps by seasonal bacterial or ALGAL BLOOMS) to cause precipitation of the less soluble ion Fe^{3+}. BIFs represent a classic refutation of a strict ACTUALISM as they demonstrate that in the distant past Earth's environments were markedly different from those of today. *CDW*

[See also RED BEDS]

James, H.L. and Trendall, A.F. 1982: Banded iron formations: distribution in time and paleoenvironmental significance. In Holland, H.D. and Schidlowski, M. (eds), *Mineral deposits and the evolution of the biosphere.* Berlin: Springer, 199–218. **Holland, H.D.** 1984: *The chemical evolution of the atmosphere and oceans.* Princeton, NJ: Princeton University Press. **Kato, Y., Ohta, I., Tsunematsu, T.** *et al.* 1998: Rare earth element variations in mid-Archean banded iron formations: implications for the chemistry of ocean and continent and plate tectonics. *Geochimica et Cosmochimica Acta* 62, 3475–3497. **Trendall, A.F.** 2000: The significance of banded iron formation (BIF) in the Precambrian stratigraphic record. *Geoscientist* 10(6), 4–7. **Windley, B.F.** 1995: *The evolving continents.* Chichester: Wiley.

bank erosion

Detachment and entrainment of bank material as individual grains, aggregates or blocks by fluvial, geodynamic or subaerial processes. Fluid entrainment, mass failure and frost action mechanisms are especially important. Bank erosion rates vary from 0–1000 m y^{-1}, and increase with DRAINAGE BASIN area (see Figure). Rates may, however, peak in mid basin, where *stream power* and bank ERODIBILITY tend to achieve maximum values. Bank erosion rates may be increased by environmental changes such as increases in FLOOD MAGNITUDE-FREQUENCY resulting from a wetter climate or LANDUSE CHANGE or altered bank vegetation. Bank erosion processes are instrumental in the formation of MEANDERS and BRAIDING. *DML*

Darby, S.E. and Thorne, C.R. 1996: Development and testing of riverbank-stability analysis. *Proceedings of the American Society of Civil Engineers, Journal of Hydraulic Engineering* 122, 443–454. **Hooke, J.M.** 1980: Magnitude and distribution of rates of river bank erosion. *Earth Surface Processes* 5, 143–157. **Lawler, D.M.** 1993: The measurement of river bank erosion and lateral change. *Earth Surface Processes and Landforms, Technical and Software Bulletin* 18, 777–821. **Lawler, D.M., Thorne, C.R. and Hooke, J.M.** 1997: Bank erosion and instability. In Thorne, C.R., Hey, R.D. and Newson, M.D. (eds), *Applied fluvial geomorphology for river engineering and management.* Chichester: Wiley, 137–172. **Nanson, G.C. and Hickin, E.J.** 1986: A statistical analysis of bank erosion and channel migration in western Canada. *Geological Society of America Bulletin* 97, 497–504.

bankfull discharge

The rate of flow (normally m^3 s^{-1}) of a river flowing at CHANNEL CAPACITY, i.e. when exactly filling the river channel to bank-top level. It is sometimes considered the channel-forming discharge, or *dominant discharge.* Reported frequencies of such a flow (or greater) vary widely from more than once per year to once every 32 years. The *recurrence interval* or RETURN PERIOD of bankfull discharge averages 1.58 years for rivers in the USA and 2.33 years in the UK. *DML*

[See also DISCHARGE, FLOOD MAGNITUDE-FREQUENCY CHANGES]

Williams, G.P. 1978: Bankfull discharge of rivers. *Water Resources Research* 14, 1141–1158.

barrier beach

An elongated sand bar disconnected from and paralleling the main SHORELINE. It usually has a crest that is exposed at low water. It is also known as a *barrier bar.* A series of barrier beaches can constitute part of a BARRIER ISLAND. *HJW*

[See also BEACH DEPOSITS, STORM BEACH]

barrier island

An elongated, depositional LANDFORM parallel to the shore formed along gently sloping coasts with wide shelves and coastal plains (e.g. east coast of USA) and composed of sand or coarser SEDIMENT. Barrier islands are more or less permanently above high-TIDE level. On the seaward side, there is a BARRIER BEACH. Addition of beach ridges eventually forms a beach plain.

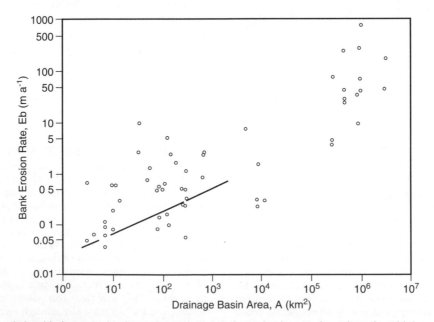

bank erosion *The relationship between bank erosion rate and drainage basin area for selected world rivers (after Hooke, 1980; Lawler* et al.*, 1997)*

Typically, there are also DUNE ridges backed by washover fans, tidal marshes and a LAGOON. Barrier islands are particularly common in low to middle latitudes and represent 10–15% of all the world's coastlines. Alternative terms are *offshore bar* or *barrier complex*. *RAS*

Oertel, G.F. and Leatherman, S.P. (eds) 1985: Barrier islands. *Marine Geology* 63(Special Issue), 1–396.

barrier reefs Coral structures separated from land by a lagoon. They form linear structures on continental margins (e.g. the Great Barrier Reef) but may be annular around mid-plate volcanoes. *TSp*

[See also CORAL REEF]

basalt A dark, fine-grained IGNEOUS ROCK composed of approximately equal proportions of pyroxene and plagioclase feldspar, sometimes with olivine, representing solidified LAVA of low-silica (basic) composition. *GO*

base A compound that dissociates in water to form hydroxyl ions (OH⁻) and reacts with ACIDS to form a salt. *ADT*

[See also ALKALI]

base level The theoretical, lower altitudinal limit to EROSION by subaerial geomorphological processes. Sea level provides a *general base level* but there are *local base levels*, such as lakes and valley floors, to which geomorphic processes, such as FLUVIAL PROCESSES and slope processes respectively, adjust. In the lower reaches of a river system, for example, a fall in base level tends to increase the rate of downcutting with headward erosion of a KNICKPOINT, whereas a rise in base level is likely to lead to AGGRADATION. However, simple EQUILIBRIUM CONCEPTS related to long periods of stable base level linked to the DAVISIAN CYCLE OF EROSION have been rejected in favour of modern ideas of a rapidly fluctuating general base level during QUATERNARY ENVIRONMENTAL CHANGE, to which the LANDSCAPE responds in a complex fashion. *JAM*

[See also REJUVENATION]

Davis, W.M. 1902: Base-level, grade and peneplain. *Journal of Geology* 1, 77–111. Schumm, S.A. 1993: River response to baselevel changes: implications for sequence stratigraphy. *Journal of Geology* 101, 279–294. Schumm, S.A. and Parker, R.S. 1973: Implications of complex response of drainage systems for Quaternary alluvial stratigraphy. *Nature* 243, 99–100.

base saturation The extent to which the CATION EXCHANGE CAPACITY of a soil is saturated with the cations Ca²⁺, Mg²⁺, Na⁺ and K⁺, expressed as a percentage of the total cation exchange capacity. To determine the base saturation, a soil sample is percolated with ammonium acetate solution at pH 7 and the cations determined in the percolate. *EMB*

Reeuwijk, L.P. van (ed.) 1992: *Procedures for soil analysis*, 3rd edn. [*Technical Paper* 9]. Wageningen: International Soil Reference and Information Centre.

baseflow The steady, slowly declining flow of streams and rivers between STORM events. Baseflow is provided by a range of delayed flow RUNOFF PROCESSES, notably GROUNDWATER, but also deeper THROUGHFLOW and PIPEFLOW in some environments. Baseflows are character-

istically low and invariant compared with QUICKFLOW processes, but fluctuate seasonally with changes in the WATER TABLE. *ADT/RPDW*

[See also STORM HYDROGRAPH]

basin analysis The interpretation of SEDIMENTARY ROCKS and FACIES in terms of the type of SEDIMENTARY BASIN in which they accumulated and its evolution through time. Basin analysis involves an integration of SEDIMENTOLOGY, STRATIGRAPHY, PALAEOGEOGRAPHY, GEOMORPHOLOGY and TECTONICS. The study of sedimentary basins sprang from the developments of PLATE TECTONICS in the 1960s (replacing GEOSYNCLINE concepts) and FACIES ANALYSIS in the 1970s and 1980s, driven in part by the search for PETROLEUM. Basin-wide studies of sedimentary rocks expanded in the 1990s to investigate ALLOCYCLIC controls on sediment accumulation, particularly the interrelationships of TECTONICS, EUSTASY and palaeoclimate (see PALAEOCLIMATOLOGY). *GO*

[See also SEQUENCE STRATIGRAPHY]

Allen, P.A. and Allen, J.R. 1990: *Basin analysis: principles and application.* Oxford: Blackwell. Leeder, M.R. 1999: *Sedimentology and sedimentary basins: from turbulence to tectonics.* Oxford: Blackwell Science. Leeder, M.R., Harris, T. and Kirkby, M.J. 1998: Sediment supply and climate change: implications for basin stratigraphy. *Basin Research* 10, 7–18. Miall, A.D. 2000: *Principles of sedimentary basin analysis*, 3rd edn. Berlin: Springer. Mitchell, A.H.G. and Reading, H.G. 1986: Sedimentation and tectonics. In Reading, H.G. (ed.), *Sedimentary environments and facies*, 2nd edn. Oxford: Blackwell, 471–519. Schlunegger, F., Slingerland, R. and Matter, A. 1998: Crustal thickening and crustal extension as controls on the evolution of the drainage network of the central Swiss Alps between 30 Ma and the present: constraints from the stratigraphy of the North Alpine Foreland Basin and the structural evolution of the Alps. *Basin Research* 10, 197–212.

bathyal Depths and environments between about 200 m and 3000 m in the OCEAN, corresponding to the CONTINENTAL SLOPE and CONTINENTAL RISE. *GO*

[See also ABYSSAL, CONTINENTAL MARGIN, HADAL, OCEAN BASIN]

bathymetry Measurement of the depth of water in large bodies of water and the information derived from such measurements. *HJW*

[See also ALTIMETRY]

bauxite An aluminium-rich RESIDUAL DEPOSIT formed in the HUMID TROPICS by LEACHING of WEATHERING products under strong SEASONALITY. Hydrated aluminium minerals predominate, including gibbsite [Al(OH)₃], diaspore [α-AlO(OH)] and boehmite [γ-AlO(OH)]. Aluminium-bearing rocks are weathered by wet season leaching and dry season drawing-up of leached material to the surface. CLAY MINERALS are generated initially, but these eventually become desilicified. Bauxite is closely related to the iron-rich residual deposit LATERITE. *TY*

[See also DURICRUST]

Schwarz, T. 1997: Lateritic bauxite in central Germany and implications for Miocene palaeoclimate. *Palaeogeography, Palaeoclimatology, Palaeoecology* 129, 37–50. Wilson, R.C.L. (ed.) 1983: *Residual deposits: surface related weathering processes and materials* [*Geological Society Special Publication* 11]. London: Geological Society.

Bayesian statistics A relatively new probabilistic approach to statistical analysis. Bayesian methods start with a *prior probability*, which is usually the initial probability of a hypothesis being true before data are considered. A *conditional probability*, or likelihood, is then formed from experimental or recorded data, which is effectively a TRAINING SET. The prior probability and conditional probability are combined using Bayes' theorem to produce a *posterior probability*; the probability of the initial hypothesis being true given the observed data. In itself, Bayes' theorem is uncontroversial; however, some specific applications have been viewed with suspicion arising from the derivation of prior probabilities. One advantage of the Bayesian approach is that it can be applied to continuous variables, using a smoothing technique, such as *kernel density estimation*. The Bayesian approach allows the successive revision of the posterior probability in the light of new data, as the posterior probability becomes the prior probability in the new hypothesis. Consequently, Bayesian methods lend themselves to many of the CALIBRATION and group-assignment problems encountered in the ENVIRONMENTAL SCIENCES. To date, they have been applied successfully to DENDROCLIMATOLOGY, SPATIAL ANALYSIS, RADIOCARBON DATING and the provenance of archaeological ARTEFACTS. *IR*

Bayliss, A., Groves, C., McCormac, G. *et al.* 1999: Precise dating of the Norfolk timber circle. *Nature* 402, 479. **Buck, C.E., Cavanagh, W.G. and Litton, C.D.** 1996: *Bayesian approach to interpreting archaeological data.* Chichester: Wiley. **Buck, C.E., Kenworthy, J.B., Litton, C.D. and Smith, A.F.M.** 1991: Combining archaeological and radiocarbon information: a Bayesian approach to calibration. *Antiquity* 65, 808–821. **Lucy, D., Aykroyd, R.G., Pollard, A.M., Solheim, T.** 1996: A Bayesian approach to adult human age estimation from dental observations by Johanson's age changes. *Journal of Forensic Sciences* 41, 189–194. **Robertson, I., Lucy, D., Baxter, L.** *et al.* 1999: A kernel-based Bayesian approach to climatic reconstruction. *The Holocene* 9, 495–500.

beach deposits When sediment transported along the shoreline comes to rest, it becomes a beach deposit. Such deposits may be ephemeral or stable, eventually becoming sedimentary rock (e.g. BEACHROCK). Most beach deposits range in texture from mud and fine sand to coarse gravels (*shingle*) and may be well sorted or mixed. The composition of beach deposits reflects that of source areas such as river basins, adjacent cliffs or offshore zones. Quartz sands predominate in beach deposits with a terrigenous origin, whereas calcareous sands are common along shores where carbonate-producing organisms, including corals, are abundant. *HJW*

[See also STORM BEACH]

Hardisty, J. 1990: *Beaches: form and process.* London: Unwin Hyman. **Komar, P.D.** 1976: *Beach processes and sedimentation.* Englewood Cliffs, NJ: Prentice Hall.

beach mining The process of removing materials, usually mineral, from a beach. Sand and gravel extraction for engineering projects is now generally illegal. Valuable minerals such as diamonds, gold and zircon are extracted from some beaches. *HJW*

beach nourishment A SOFT ENGINEERING technique developed in the USA during the 1930s and 1940s, whereby beach material (usually sand or gravel) replaces

sediment lost through BEACH EROSION. Beach nourishment may also lead to beach construction where none existed before. *HJW*

[See also BEACH SCRAPING, BRUUN RULE]

Møller, J.T. 1990: Artificial beach nourishment on the Danish North Sea Coast. *Journal of Coastal Research* 6(Special Issue), 1–9. **National Research Council** 1995: *Beach nourishment and protection.* Washington, DC: National Academy Press. **Trembanis, A.C., Pilkey, O.H. and Valverde, A.R.** 1999: Comparison of beach nourishment along the US Atlantic, Great Lakes, Gulf of Mexico, and New England shorelines. *Coastal Management* 27, 329–340.

beach ridges Usually subparallel, low, linear depositional ridges of sand, shells or CLASTS, ranging in size and spacing from centimetres to metres. Each ridge marks the position of a former shore and they are commonly found on shorelines exhibiting PROGRADATION. Beach ridges are produced through landward motion of water (*swash*) being greater than that returning as *backwash* flow and are caused by the percolation of water through the beach sediment. Beach-ridge formation is favoured by coarse material and low waves. In areas characterised by TECTONIC or GLACIO-ISOSTATIC UPLIFT, there may be *staircases* of beach ridges (*raised beaches*). For example, in Hudson Bay, Canada, flights of hundreds of emerged beach ridges up to 300 m above sea level testify to GLACIO-ISOSTATIC REBOUND during the HOLOCENE. In some cases, shell debris has been subjected to RADIOMETRIC DATING to reconstruct coastline change. *AGD*

[See also CHENIER, EUSTASY, GLACIO-ISOSTASY, ISOSTASY, RAISED SHORELINES, SEA-LEVEL CHANGE]

Dyke, A.S. and Prest, V.K. 1987: Late Wisconsinan and Holocene history of the Laurentide ice sheet. *Géographie Physique et Quaternaire* 41, 237–264. **Fletcher III, C.H., Fairbridge, R.W., Moller, J.J. and Long, A.J.** 1993: Emergence of the Varanger Peninsula, Arctic Norway, and climatic changes since deglaciation. *The Holocene* 3, 116–127. **Lichter, J.** 1997: AMS radiocarbon dating of Lake Michigan beach-ridge and dune development. *Quaternary Research* 48, 137–140. **Tanner, W.F.** 1993: An 8000-year record of sea level change from grain-size parameters: data from beach ridges in Denmark. *The Holocene* 3, 220–231. **Otavos, E.G.** 2000: Beach ridges: definitions and significance. *Geomorphology*, 32, 83–108.

beach scraping The process of smoothing a beach and adjusting its profile by mechanical means. It is usually regarded as a temporary, seasonal adjustment with no long-term effect. *HJW*

[See also BEACH NOURISHMENT]

Bruun, P. 1983: Beach scraping – is it damaging to beach stability? *Coastal Engineering* 7, 167–173.

beachrock Friable to indurated sands/gravels cemented by calcium carbonate, often forming seaward-dipping slabs, capable of rapid formation on intertidal tropical and subtropical beaches. *TSp*

Stoddart, D.R. and Cann, J.R. 1965: Nature and origin of beachrock. *Journal of Sedimentary Petrology* 35, 243–247.

beamwidth For a RADAR REMOTE SENSING INSTRUMENT, the finite angular width of the antenna pattern, which is the beam into which energy is transmitted (or received), characterised by the angular width at which the power density is half (3 dB lower than) the peak power on the beam axis. *PJS*

Beaufort Scale An empirical scale of wind speed estimates based on sea state. It has an integer range from 0 (calm) to 14 (*hurricane*) and was devised in AD 1805, originally to estimate storm damage potential for warships. Later, it was adapted for use on land. *JCS*

bed (1) A SEDIMENT surface (e.g. sea bed, gravel-bed river; see BEDFORM, BEDLOAD). (2) The smallest formal subdivision recognised in LITHOSTRATIGRAPHY. (3) A depositional unit of sediment thicker than 1 cm (see BEDDING, STRATIFICATION). *MRT*

bedding A type of STRATIFICATION characteristic of many sediment and sedimentary rock successions, comprising distinct units thicker than 1 cm (BEDS). Successive beds may or may not differ in LITHOLOGY from those above and below, and may be differentiated using features such as appearance (e.g. colour), TEXTURE, *mineralogy*, FOSSIL content or SEDIMENTARY STRUCTURES. Beds are separated from each other by *bedding planes* or, more generally, *bedding surfaces*. The orientation of bedding surfaces is described by their STRIKE and DIP. Some bedding surfaces represent changes in lithology within sequences that have accumulated gradually, without significant breaks in sedimentation. Others mark a HIATUS in accumulation or a major episode of EROSION (see DIASTEM). Irregular bedding surfaces in some LIMESTONE sequences are produced by KARSTic processes. Although bedding surfaces are essentially primary features, they may become accentuated by DIAGENESIS and in some homogeneous limestones are apparently entirely of secondary origin. *MRT*

[See also CROSS-BEDDING]

Bathurst, R.G.C. 1991: Pressure-dissolution and limestone bedding: the influence of stratified cementation. In Einsele, G. (ed.), *Cycles and events in stratigraphy*. Berlin: Springer, 450–463. **Collinson, J.D. and Thompson, D.B.** 1992: *Sedimentary structures*, 2nd edn. London: Chapman and Hall. **McKee, E.D. and Weir, G.W.** 1953: Terminology for stratification and cross-stratification in sedimentary rocks. *Geological Society of America Bulletin* 64, 381–390.

bedform A morphological feature produced by interaction between a flow and a surface (BED) of sediment or, occasionally, rock. *Depositional bedforms* are most commonly developed in SAND and include flat surfaces (PLANE BED) and a range of mainly flow-transverse sand ridges. These include: small-scale RIPPLES (*microforms*), including CURRENT RIPPLES, WAVE RIPPLES and WIND RIPPLES; medium-scale DUNES (*mesoforms*), both aqueous and aeolian; and large-scale *bars* (*macroforms*). Smaller bedforms may be superimposed on larger bedforms, forming *compound bedforms*. *Erosional bedforms* include CHANNELS and, on a smaller scale, SCOUR MARKS and potholes. Bedforms are important indicators of PALAEOCURRENTS and other palaeoenvironmental conditions. They can be preserved as SEDIMENTARY STRUCTURES in sediments and sedimentary rocks, where their geometry can be reconstructed using patterns of CROSS-STRATIFICATION and BOUNDING SURFACES. *GO*

[See also BEDFORM STABILITY DIAGRAM]

Allen, J.R.L. 1982: *Sedimentary structures: their character and physical basis*. Amsterdam: Elsevier. **Allen, J.R.L.** 1983: Studies in fluviatile sedimentation: bars, bar-complexes and sandstone sheets (low-sinuosity braided streams) in the Brownstones (L. Devonian), Welsh Borders. *Sedimentary Geology* 33, 237–293. **Allen, P.A.** 1997: *Earth surface processes*. Oxford: Blackwell Science. **Leeder, M.R.** 1999: *Sedimentology and sedimentary basins: from turbulence to tectonics*. Oxford: Blackwell Science. **Menzies, J. and Rose, J. (eds)** 1989: Subglacial bedforms: drumlins, rogen moraine and associated subglacial bedforms. *Sedimentary Geology* 62, 117–430.

bedform stability diagram A plot of flow strength against GRAIN-SIZE, which shows the conditions under which different depositional BEDFORMS can exist (see Figure). Such relationships are important for interpreting the significance of SEDIMENTARY STRUCTURES preserved in sediments and sedimentary rocks. It is also known as a *bedform phase diagram*. *GO*

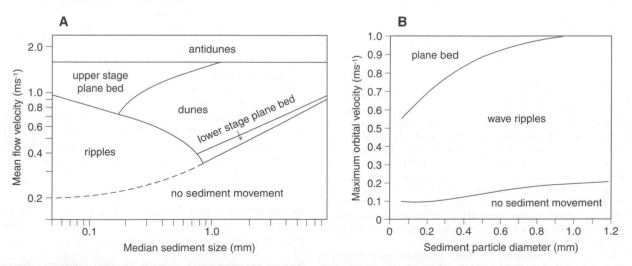

bedform stability diagram *Bedform stability diagrams for (A) unidirectional water flow (after Ashley, 1990) and (B) water waves (after Allen, 1982). Note that the horizontal axis (particle size) is logarithmic in (A) and linear in (B)*

Allen, J.R.L. 1982: *Sedimentary structures: their character and physical basis*. Amsterdam: Elsevier. **Ashley, G.** 1990: Classification of large scale subaqueous bedforms: a new look at an old problem. *Journal of Sedimentary Petrology* 60, 160–172. **Southard, J.B.** 1991: Experimental determination of bedform stability. *Annual Review of Earth and Planetary Sciences* 19, 423–455.

bedload Particulate sediment transported in contact with the basal surface (BED) of a flow, in contrast to SUSPENDED LOAD. Contact with the bed may be continuous (transport by *sliding* or *rolling*) or intermittent (transport by SALTATION), although some workers recognise *saltation load* as distinct from bedload. Under most conditions in rivers, bedload (also known as *traction load*) comprises particles of GRAVEL and SAND plus aggregates of finer material, such as fragments of MUD or SOIL. Bedload typically comprises less than 50% of the yield from a river's drainage basin. Bedload may be shaped into BEDFORMS, such as RIPPLES and DUNES. *GO*

Allen, P.A. 1997: *Earth surface processes*. Oxford: Blackwell Science.

bedrock channel A channel, typically a river channel, cut into rock, in contrast to an *alluvial channel* that is cut into ALLUVIUM. There are contrasts between the properties of bedrock channels and alluvial channels. Bedrock channels may have a less regular CHANNEL GRADIENT and CHANNEL SHAPE and they are characterised by erosional features such as WATERFALLS, *potholes*, GORGES and incised or entrenched MEANDERS. The processes of bedrock erosion by rivers are still poorly understood. *GO*

[See also CORRASION]

Grant, G.E., Swanson, F.J. and **Wolman, M.G.** 1990: Pattern and origin of stepped-bed morphology in high-gradient streams, western Cascades, Oregon. *Geological Society of America Bulletin* 102, 340–352. **Seidl, M.A.** and **Dietrich, W.E.** 1992: The problem of channel erosion into bedrock. *Catena Supplement* 23, 101–124.

beetle analysis The study of FOSSIL and SUBFOSSIL beetle remains lies within the wider field of PALAEOENTOMOLOGY (the study of fossil insects) and *archaeoentomology* (the analysis of insects from archaeological contexts). Although fossil beetles from TERTIARY and earlier deposits have been investigated, their main application as palaeoenvironmental indicators is in the QUATERNARY.

The pioneering work of G.R. Coope demonstrated that beetles (order *Coleoptera*) have remained evolutionarily stable over at least the Quaternary. Their value as palaeoenvironmental indicators arises also from the fact that: (1) beetles respond rapidly to ENVIRONMENTAL CHANGE; (2) they are well preserved and often abundant in waterlogged Quaternary sediments; and (3) fossil specimens may be matched with living species with restricted HABITAT or NICHE preferences (stenotopic), narrow climatic requirements (many are *stenothermic*) and known geographical ranges. Beetle analysis may therefore provide detailed information about a range of environmental variables including climate, vegetation composition and structure, WATER QUALITY and flow regimes, soil conditions and human activity.

Fossil beetle analysis provides quantitative *palaeoclimatic reconstructions* based on the overlap of modern climatic envelopes of carnivorous and scavenger beetles in fossil assemblages. This MUTUAL CLIMATE RANGE METHOD provides estimated mean temperatures for the warmest and coldest months of the year, as well as seasonality. Beetles respond more rapidly to climatic change than plant communities and have made an important contribution to the understanding of the rates, scale and timing of RAPID ENVIRONMENTAL CHANGE during the *Weichselian* stage in northern Europe. HOLOCENE beetle faunas provide detailed evidence for ANTHROPOGENIC impact on landscapes and ECOSYSTEMS, including the effects of DEFORESTATION and human settlement of previously undisturbed landscapes. Fossil beetles have also been used to address the BIOGEOGRAPHY and origins of BIOTAS, including those of isolated islands such as Iceland and Greenland.

Fossil beetles, together with the remains of other insects (e.g. flies, fleas and lice), are used in ENVIRONMENTAL ARCHAEOLOGY to interpret the origin of sediments, human living conditions and building function or usage. This has aided interpretation of human responses to environmental change during the *Mediaeval Period* in the North Atlantic islands.

Although much research has focused on Northern Europe, fossil beetle studies are increasingly widely used elsewhere in the world, most notably in North America. Attention has also widened to the examination of other insect orders, including true flies (Diptera) and non-biting midges (Chironomidae). *MHD*

[See also CHIRONOMID ANALYSIS]

Ashworth, A.C., Buckland, P.C., and **Sadler, J.P.** (eds) 1997: *Studies in Quaternary entomology: an inordinate fondness for insects* [*Quaternary Proceedings* No.5]. London: Quaternary Research Association. **Atkinson, T.C., Briffa, K.R.** and **Coope, G.R.** 1987: Seasonal temperatures in Britain during the last 22,000 years, reconstructed using beetle remains. *Nature* 325, 587–592. **Coope, G.R.** 1977: Fossil coleopteran assemblages as sensitive indicators of climatic changes during the Devensian (last) cold stage. *Philosophical Transactions of the Royal Society of London* B280, 313–340. **Coope, G.R.** 1994: Insect faunas in ice age environments: why so little extinction? In May, R. and Lawton, J. (eds), *Estimating extinction rates*. Oxford: Oxford University Press, 55–74. **Coope, G.R.** and **Lemdahl, G.** 1995: Regional differences in the Lateglacial climate of northern Europe based on coleopteran analysis. *Journal of Quaternary Science* 10, 391–395. **Dinnin, M.H.** and **Sadler, J.P.** 1999: 10,000 years of change: the Holocene entomofauna of the British Isles. *Quaternary Proceedings* 7, 545–562. **Elias, S. A.** 1994: *Quaternary insects and their environments*. Washington, DC: Smithsonian Institution Press.

before present (BP) In RADIOCARBON DATING, the 'present' is defined by convention as the year AD 1950. Thus, CONVENTIONAL RADIOCARBON AGES are all quoted by radiocarbon dating laboratories in 'radiocarbon years before present', 'years BP' or simply 'BP'. Archaeologists sometimes use 'bp' for conventional radiocarbon ages and 'BP' for true 'calendar years before present'. When radiocarbon ages are calibrated they may be expressed in 'calibrated years before present', or 'cal BP'. *JAM*

below-ground productivity The BIOMASS production rate of plant roots. *RJH*

[See also ABOVE-GROUND PRODUCTIVITY]

bench terrace A form of MECHANICAL SOIL CONSERVATION MEASURE comprising a series of flat or gently sloping shelf-like areas of ground linked by steeper slopes or 'risers' up to 30°. The riser is vulnerable to SOIL EROSION and is protected by a vegetation cover; alternatively, risers are sometimes faced with stone or concrete. Benches are made level and may have a raised lip at the outer edge, where retention of IRRIGATION water is required.   *RAS*

[See also LAND DEGRADATION, SOIL CONSERVATION, TERRACING]

Ternan, J.L., Williams, A.G., Elmes, A. and Fitzjohn, C. 1996: The effectiveness of bench terracing and afforestation for erosion control on Raña sediment in central Spain. *Land Degradation and Development* 7, 337–351. **Wadsworth, R. and Swetnam, R.** 1998: Modelling the impact of climate warming at the landscape scale: will bench terraces become economically and ecologically viable structures under changed climates? *Agriculture, Ecosystems and Environment* 68, 27–39.

benthic   Pertaining to the BED of a water body.   *GO*

[See also BENTHOS]

benthos   Organisms that live on or in the BED of a water body (e.g. the sea-floor). Benthic assemblages of FOSSILS or SUBFOSSILS are useful in palaeoenvironmental reconstruction (e.g. SILURIAN brachiopods and SEA LEVEL curves).   *LC*

[See also PELAGIC]

Johnson, M.E., Kaljo, D. and Rong, J.-Y. 1991: Silurian eustasy. In Bassett, M.G. and Edwards, D. (eds), *The Murchison Symposium. Special Papers in Palaeontology* 44, 145–163.

bias   (1) Systematic error in results due to flaws in EXPERIMENTAL DESIGN, measurement techniques or analytical method. (2) Lack of objectivity.   *JAM/RMD*

[See also OBJECTIVE KNOWLEDGE]

bi-directional reflectance distribution function (BRDF)   A description of the angular distribution of reflected electromagnetic radiation from a surface, with respect to the distribution and wavelength of the incident RADIATION. In REMOTE SENSING, knowledge of the BRDF is useful for several reasons. First, it provides a more accurate measure of ALBEDO than estimates derived by assuming that the reflectance of the surface is Lambertian. ALBEDO is an important input parameter to GENERAL CIRCULATION MODELS (GCMs). Second, uniform changes in brightness across remotely sensed images (see REMOTE SENSING) are often due to the directional reflectance properties of the surface and consequently, the more accurately the BRDF is known, the better these effects can be corrected. Finally, physical models of the BRDF of vegetation are typically parameterised by the biophysical properties of the canopy such as LEAF AREA INDEX, stand height and *leaf angle distribution* (LAD). Hence, with sufficient angular samples of the canopy reflectance it is possible mathematically to invert such models and retrieve estimates of these biophysical parameters. For remote sensing this offers a quantitative method of information retrieval not available from many of the more commonly used spectral or temporal methods.   *TQ*

Kimes, D.S. 1983: Dynamics of directional reflectance factor distributions for vegetation canopies. *Applied Optics* 22, 1364–1372. **Nicodemus, F.E., Richmond, J.C., Hsia, J.J. et al.** 1977: *Geometrical considerations and nomenclature for reflectance* [*NBS Monograph* 160]. Washington, DC: National Bureau of Standards. **Roujean, J-L., Leroy, M. and Deschamps, P-Y.** 1992: A bi-directional reflectance model of the Earth's surface for the correction of remote sensing data. *Journal of Geophysical Research* 97, 20455–20468.

big bang   A widely held theory in COSMOLOGY for the formation of the universe. The big bang theory holds that matter, energy, space and time came into being about 15 billion (15 000 000 000) years ago. The explosion of an 'initial singularity' resulted in expansion and cooling of the universe that has continued at an ever-decreasing rate to the present day. The acceptance of big bang theory, first developed in the late 1940s, was supported by the discovery in AD 1965 of background microwave COSMIC RADIATION. The solar system and the Earth formed about 4.6 billion years ago (see PRISCOAN).   *GO*

[See also ASTROGEOLOGY, PLANETARY GEOLOGY]

Barrow, J.D. 1994: *The origin of the universe.* London: Weidenfeld and Nicolson. **Hawking, S.W.** 1988: *A brief history of time: from the big bang to black holes.* London: Bantam.

binge–purge model   A model developed by D.R. MacAyeal to explain the origin of HEINRICH EVENTS. As a result of a combination of geothermal and frictional heat beneath the LAURENTIAN ICE SHEET, there is catastrophic failure of the ice sheet leading to rapid and massive iceberg discharge into the North Atlantic. Originally it was felt that this model failed to explain the parallel events leading to increased ICE-RAFTED DEBRIS (IRD) from other ice sheets, attributed by other workers to external climatic forcing. The apparent synchroneity of all the ice-sheet collapses has, however, recently been questioned.   *CJC*

Dowdeswell, J.A., Elverhoi, A., Andrews J.T. and Hebbeln, D. 1999: Asynchronous deposition of ice-rafted layers in the Nordic seas and North Atlantic Ocean. *Nature* 400, 348–351. **MacAyeal, D.R.** 1993: Binge/purge oscillations of the Laurentide ice sheet as a cause of the North Atlantic's Heinrich events. *Paleoceanography* 8, 775–784.

bio-accumulation   Selective accumulation by organisms of nutrients and harmful substances in cells and tissues. It can be further amplified by BIOLOGICAL MAGNIFICATION, as when humans eat oysters.   *GOH*

[See also MERCURY]

Connell, D.W. 1990: *Bioaccumulation of xenobiotic compounds.* Boca Raton, FL: CRC Press. **Cunningham, W.P. and Saigo, B.W.** 1997: *Environmental science: a global concern.* London: W.C. Brown. **McLachlan, M.S.** 1996: Bioaccumulation of hydrophobic chemicals in agricultural food chains. *Environmental Science and Technology* 30, 252–259.

bio-archaeology   The application of principles and techniques from the biological sciences to the study of the human past.   *SPD*

[See also ENVIRONMENTAL ARCHAEOLOGY, PALAEOETHNOBOTANY, ZOOARCHAEOLOGY]

biochemical oxygen demand (BOD)   The amount of oxygen (in $mg L^{-1}$) needed by AEROBIC micro-organisms over a period to oxidise organic material in water. BOD is a useful measure of the *organic loading* and hence *organic pollution* levels in watercourses. If the BOD value exceeds the available dissolved oxygen in the water, *oxygen depletion* occurs and aquatic organisms, including fish, are commonly killed. The international standard for tests is the amount of oxygen needed over a five-day period (the five-day BOD). High levels of BOD (i.e. in excess of 4 mg L^{-1}) indicate poor water quality associated with low levels of dissolved oxygen; natural clean streamwaters have a BOD of ≤ 2 mg L^{-1}. *ADT*

[See also WATER POLLUTION]

Adams, N. and Bealing, D. 1994: Organic pollution: biochemical oxygen demand and ammonia. In Calow, P. (ed.), *Handbook of ecotoxicology*. Vol. 2. Oxford: Blackwell Science, 264–285.

bioclimatology The study of the effects of CLIMATE on the life, health and distribution of people, plants and animals. It may be distinguished from *biometeorology*, which involves the study of the effects of weather. *JCM*

Kates, R.W., Ausubel, J.H. and Berberian, M. (eds) 1985: *Climate impact assessment* [*Scope Report* 27]. Chichester: Wiley. **Mintzer, I.M.** 1992: *Confronting climate change: risks, implications and responsibilities*. Cambridge: Cambridge University Press. **Tromp, S.W.** 1980: *Biometeorology*. London: Heyden.

biodegradation The DECOMPOSITION of organic materials by micro-organisms, largely under AEROBIC environmental conditions . Biodegradation is an essential part of BIOGEOCHEMICAL CYCLING and BIOREMEDIATION. Most naturally occurring and manufactured organic substances are biodegradable and therefore have low PERSISTENCE in the environment. However, non-biodegradable products, such as plastics made of synthetic polymers and PERSISTENT ORGANIC COMPOUNDS, remain in the environment as solid waste or may exhibit BIOACCUMULATION.

Naturally occurring inert or recalcitrant substances, such as some HUMUS substances, PHYTOLITHS and charcoal, are of value in PALAEOENVIRONMENTAL RECONSTRUCTION. Many other types of FOSSIL or SUBFOSSIL organic materials are preserved only because of ANAEROBIC or other conditions that are not conducive to biodegradation. *JAM*

Alexander, M. 1999: *Biodegradation and bioremediation*, 2nd edn. New York: Academic Press

biodiversity The variability of life from all sources, including within species (often termed *genetic diversity*), between species, and of ecosystems. Interactions between organisms, including humans, and their environment are part of biodiversity. Biodiversity is an abbreviation of *biological diversity*, now in common use.

At the genetic level, diversity can refer to the heritable molecular variation present in an individual or a population, or within higher taxa. Genetic diversity in a population can determine the ability of that population to respond to changing environmental conditions. Diversity is perhaps most commonly assessed at the level of species, because of the relative ease with which *species diversity* can be measured compared to genetic diversity. Global SPECIES RICHNESS (number of species) estimates range from five to 100 million species, with many suggesting 14 million, but only about 1.7 million species have been formally described. Most described species are insects, and global richness in this taxon is thought to exceed all others. The increasing rarity of systematicists renders complete biological inventories impractical. Shortcuts to assessing local or regional diversity are required, including the use of INDICATOR SPECIES or taxa.

Ecosystem or *ecological diversity* is the variety of ecosystems or HABITATS in a region. Measurement at this scale is not easily adapted to the common, fine scales of ecological studies. Ecosystem diversity is particularly relevant in larger-scale investigations, such as those relying on REMOTE SENSING, and to CONSERVATION. 'Latitudinal' gradients of diversity have been recognised from the earliest period of modern ecology. In general, diversity increases nearer the tropics, although this gradient is not a simple north–south trend, and many higher taxa are found nowhere else. The most diverse terrestrial habitats are generally believed to be TROPICAL RAIN FORESTS, while marine biodiversity is highest in CORAL REEFS. Latitude is a surrogate for environmental factors, ecological interactions and evolutionary/historical events that are variously thought to be responsible for geographical variation in biodiversity. The quality and quantity of evidence supporting these various explanations differs significantly. While no consensus explanation is yet accepted, strong evidence supports energy availability, either as POTENTIAL EVAPOTRANSPIRATION or PRIMARY PRODUCTIVITY, as a major determinant of biodiversity at a regional scale. Ecological, climatic/physical or historical factors may also influence biodiversity patterns in some regions or at some spatial scales. This is an area of active research.

Directly or indirectly, biodiversity provides indispensable ECOLOGICAL GOODS AND SERVICES to human societies, the estimated monetary value of which is far greater than the value of the entire human economy. It is fair to state that the atmosphere, soil fertility, medicines, agriculture, waste recycling and other products and services are indispensable benefits provided by biodiversity. Aesthetic, ethical, cultural and other non-consumptive benefits of biodiversity are intrinsically important, but also make tangible contributions to the economic well-being of many nations through different aspects of tourism (approximately US $2 trillion per year, globally). There is increasing evidence that species-level, and possibly genetic, diversity is important for ECOSYSTEM HEALTH and the maintenance of ecosystem services, though the design of experiments to test this hypothesis is very difficult. Aspects of ecosystem stability, such as resistance to DISTURBANCE, may also be related to diversity.

Biodiversity conservation activities are typically directed toward species, although ecosystem conservation and maintenance of genetic diversity are also advocated. One international conservation strategy is to focus on megadiverse countries in which biodiversity is concentrated. By prioritising at this scale, national and international organisations and governments are able to target funding to states with the most pressing conservation needs, such as Papua New Guinea and Venezuela. A complementary conservation strategy is the designation of biodiversity HOTSPOTS, based on local biodiversity and the degree to which species are ENDEMIC to the area. HABITAT LOSS, OVERHUNTING or other strong human influence in hotspot areas can cause many EXTINCTIONS in a short time. Several conservation organisations are concerned with conservation of hotspots, which frequently face particularly severe threat. Some particularly important hotspots include areas within the Philippines, tropical Andes and Madagascar. Centres of high biodiversity and endemism do not necessarily coincide, complicating conservation planning and management. GAP ANALYSIS can assist in resolving such difficulties when choosing which areas to protect.

Three primary international agreements are directly and primarily concerned with biodiversity conservation. While biodiversity conservation will often fail unless legitimate human welfare needs can also be met, these agreements play important roles in conservation. The *Ramsar Convention* (1971) identifies and protects WETLANDS of international importance. More than 40 million ha of wetland habitat have been protected under the auspices of this generally successful agreement. The *Convention on International Trade in Endangered Species of Wild Fauna and Flora* (CITES, 1973), restricts the export and import of species jeopardised by continued exploitation for trade. Trade may be completely banned for species that are currently endangered or strictly regulated for those that are at risk of becoming endangered. The *Convention on Biological Diversity* (1993) is the most comprehensive of these agreements. Parties to this convention are required to promote the conservation of biodiversity, and habitats, and to act to maintain viable populations of organisms in their natural surroundings. The establishment of PROTECTED AREAS, threatened species legislation and national biodiversity strategies are also mandated by this convention. The agreement recognises national sovereignty over genetic resources and implements procedures to facilitate more equitable sharing of financial and technological rewards between the developed and developing world.

Of increasing concern is the impact of CLIMATIC CHANGE on biodiversity patterns and maintenance. Historically, species have migrated in response to gradual climatic changes. Anthropogenic climate change is expected to be very rapid relative to geologically recent historical changes and species may be forced to migrate much more rapidly than they are able to without management intervention. This difficulty is enhanced because many species' populations have been reduced substantially, reducing their genetic diversity and potentially limiting their ability to adapt to new environmental conditions. Other forms of ENVIRONMENTAL CHANGE are very likely to exacerbate this problem. When species have migrated historically, they have typically done so across landscapes that had undergone little or no human modification. Presently, very few habitats at a global scale are reasonably free of human influence, and synergistic ENVIRONMENTAL DEGRADATION, such as HABITAT FRAGMENTATION, will reduce the natural capacity of species to migrate. Many species' likelihood of extinction will rise considerably in the face of combinations of negative human impacts. *JTK*

[See also BIO-INDICATORS, DIVERSITY ANOMALY, DIVERSITY CONCEPTS, DIVERSITY INDICES, SYNERGISMS]

Currie, D. J. 1991: Energy and large-scale patterns of animal and plant species richness. *American Naturalist* 137, 27–49. **Currie, D.J., Francis, A.R., and Kerr, J.T.** 1999: Some general propositions about the study of spatial patterns of species richness. *Ecoscience* 6, 392–399.. **Ehrlich, P.R. and Wilson, E.O.** 1991: Biodiversity studies: science and policy. *Science* 253, 758–762. **Gentry, A.** 1992: Tropical forest biodiversity: distributional patterns and conservation significance. *Oikos* 63, 19–28. **Heywood, V.H. (ed.)** 1995: *Global biodiversity assessment.* Cambridge: Cambridge University Press. **Huston, M.A.** 1994: *Biological diversity: the coexistence of species on changing landscapes.* Cambridge: Cambridge University Press. **Huston, M.A.** 1997: Hidden treatments in ecological experiments: re-evaluating the ecosystem function of biological diversity. *Oecologia* 110, 449–60.

Kerr, J.T. and L. Packer. 1997: Habitat heterogeneity as a determinant of mammal species richness in high energy regions. *Nature* 385, 252–254. **May, R.M.** 1988: How many species are there on Earth? *Science* 241, 1441–1449. **Mittermeier, R.A., Myers, N., Thomsen, J.B.** *et al.* 1998: Biodiversity hotspots and major tropical wilderness areas: approaches to setting conservation priorities. *Conservation Biology* 12, 516–520. **Pearce, D. and Moran, D.** 1994: *The economic value of biodiversity.* London: Earthscan. **Peters, R.L. and Lovejoy, T.E.** 1992: Global warming and biological diversity. New Haven, CT: Yale University Press. **Reid, W.V.** 1998: Biodiversity hotspots. *Trends in Ecology and Evolution* 13, 275–280. **Ricklefs, R.E. and Schluter, D. (eds)** 1993: *Species diversity in ecological communities: historical and geographical perspectives.* Chicago, IL: University of Chicago Press. **Rosenzweig, M.L.** 1995: *Species diversity in space and time.* Cambridge: Cambridge University Press. **Turner, J.R.G., Lennon, J.J. and Greenwood, J.J.D.** 1996: Does climate cause the global diodiversity gradient? In Hochberg, M., Claubert, J. and Barbault. R. (eds), *Aspects of the genesis and maintenance of biological diversity.* Oxford: Oxford University Press.

biodiversity loss The decline in the number of species, the loss in area and FRAGMENTATION of biotic communities, and the loss in *genetic diversity* are all examples of current BIODIVERSITY loss. Although EXTINCTION of species occurs naturally, the rate of loss of species currently far exceeds the natural BACKGROUND RATE. Most predictions of world extinction rates are dependent on uncertain estimates of species richness in, and deforestation rates of, TROPICAL RAIN FORESTS. Such predictions are of unknown accuracy and have led to some controversy. It seems better not to focus on precise rates of extinction but on measures to ensure the CONSERVATION of species rich areas (HOTSPOTS IN BIODIVERSITY). Monitoring the rate and extent of loss of diversity at different levels of biological organisation is undertaken on a global scale by the World Conservation Monitoring Centre, Cambridge, UK. *IFS*

[See also CONSERVATION BIOLOGY, EXTIRPATION, HABITAT LOSS]

Lawton, J.H. and May, R.M. (eds) 1995: *Extinction rates.* Oxford: Oxford University Press. **World Conservation Monitoring Centre** 1992: *Global biodiversity: status of the Earth's living resources.* London: Chapman and Hall.

bio-energy Bioenergy is produced by the burning or fermentation of organic material and most bioenergy sources are RENEWABLE and SUSTAINABLE. Wood, peat and dried dung have long been burned to generate heat in the home, and in many African countries over 70% of national energy consumption is derived from wood. Methane is one of the by-products of anaerobic fermentation of organic matter. Many modern sewage works burn methane emitted during digestion of sludge produced from primary and secondary treatment phases to generate electricity for use on site. It is also possible that methane produced in landfill sites could be tapped and used as a fuel. Ethanol, produced from the fermentation of sugar cane or molasses, is widely used in Latin American countries such as Brazil to fuel vehicles. Maize, wheat, barley and sugar beet could also be used to produce ethanol in temperate countries, and biofuel has the advantage of burning cleanly, without emitting lead or oxides of nitrogen and sulphur. *MLW*

[See also ALTERNATIVE ENERGY, FUELWOOD]

Bryant, R.R. 1986: US residential demand for wood. *Energy Journal* 7, 137–147. Fox, J. 1984: Firewood consumption in a Nepali village. *Environmental Management* 8, 243–50. International Energy Agency 1994: *Biofuels*. Paris: IEA. Solar Energy Research Institute 1982: *Ethanol fuels reference guide*. Washington, DC: SERI.

biofacies A FACIES of SEDIMENT or SEDIMENTARY ROCK characterised primarily by biological characteristics such as FOSSIL content (e.g. coral-stromatoporoid biofacies, graptolite biofacies). Biofacies are interpreted in terms of process, environment or PALAEOENVIRONMENT as, for example, reef biofacies or trough biofacies.   *LC*

[See also ICHNOFACIES, LITHOFACIES]

biogasification The ANAEROBIC digestion of organic waste and animal faeces for production of *biogas*, which contains about 50% METHANE. Small-scale biogas production is in widespread use in China and India and there is potential for wider use for cooking, heating and lighting in developing countries and rural areas where the residual sludge provides a nitrogen-rich fertiliser.   *JAM*

[See also ALTERNATIVE ENERGY, BIOENERGY]
International Energy Agency 1994: *Biofuels*. Paris: IEA.

biogenic   sediment Biochemical SEDIMENTS, ORGANIC SEDIMENTS or rocks formed through processes whereby living organisms contribute to and dominate the sediment's composition, e.g. plant and animal fragments, calcium carbonate and silica precipitated by PLANKTON.   *UBW*

[See also DIATOMITE, DY, GYTTJA, MARL, PEAT, SEDIMENT TYPES]

biogeochemical cycles The cyclical progress of elements essential to life (*bioelements*) through living things, air, rocks, soil and water. The cycles have BIOTIC (within organisms) and ABIOTIC (within the geochemical environment) phases. They involve SINKS (*pools* or *stores*) of various chemical species in the ATMOSPHERE, HYDROSPHERE, PEDOSPHERE and LITHOSPHERE and *fluxes* of chemical species between the stores. They are sometimes termed *nutrient cycles*.

The major cycles involve the storage and flux of hydrogen, carbon, nitrogen, oxygen, magnesium, phosphorus, potassium, sulphur and calcium. Cycles of carbon, hydrogen, oxygen and nitrogen are *gaseous cycles* – their component chemical species are gaseous for a leg of the cycle. Chemical species that do not readily volatilise and are exchanged between the BIOSPHERE and its environment in solution follow *sedimentary cycles*. The magnesium cycle is an example. There are also biogeochemical cycles of the MICRONUTRIENTS sodium and chlorine and of many other elements suspected of being micro-bioelements (e.g. boron, molybdenum, silica, vanadium and zinc).

Biogeochemicals cycle through ecosystems in three stages – uptake, turnover and DECOMPOSITION. Solutes and gases are taken up by green plants and are incorporated into phytomass (plant biomass), the rate of uptake broadly matching BIOMASS production rate (PRIMARY PRODUCTIVITY). Oxygen is released in photosynthesis. The remaining minerals are either passed on to consumers or returned to soil and water bodies when plants die or when leaves and other parts are shed. The minerals in the consumers eventually return to the soil, water bodies or atmosphere.

There is substantial evidence that the storage and, to a greater degree, the fluxes of many biogeochemicals – including carbon, sulphur, nitrogen and phosphorus – are being drastically changed by human activities. The geological effects of these human perturbations of the biogeochemical system are unclear. Some simulations suggest long-delayed, radical shifts in the state of the ECOSPHERE as increased SOIL EROSION transports carbon to the sea floor at a heightened rate. Of more immediate concern are changes now and in the next century. Humans are increasing the atmospheric carbon dioxide store and likely changes in the CARBON CYCLE are under investigation. Extra nitrogen in lakes causes accelerated EUTROPHICATION. Sulphur liberated by burning fossil fuels creates ACID RAIN.   *RJH*

[See also ECOSYSTEM CONCEPT, HYDROLOGICAL CYCLE, NITROGEN CYCLE, SULPHUR CYCLE]
Bolin, B. and Cook, R. B. 1983: *The major biogeochemical cycles and their interactions*. New York: Wiley. Butcher, S.S., Charleson, R. J., Orians, G. H. and Wolfe, G.V. (eds) 1992: *Global biogeochemical cycles*. London: Academic Press. Chameides, W. L. and Perdue, E. M. 1997: *Biogeochemical cycles: a computer-interactive study of Earth system science and global change*. New York: Oxford University Press.

biogeochemistry The study of chemical elements that are essential to life (bioelements) in both their BIOTIC (or biochemical) and ABIOTIC (geochemical) phases. *RJH*

[See also BIOGEOCHEMICAL CYCLES, ECOSYSTEM CONCEPT, LANDSCAPE GEOCHEMISTRY]
Degens, E.T., Kempe, S. and Richey, J.E. (eds) 1991: *Biogeochemistry of major world rivers*. Chichester: Wiley. Likens, G. E. and Bormann, F. H. 1995: *Biogeochemistry of a forested ecosystem*, 2nd edn. New York: Springer.

biogeocoenosis A community of organisms together with the physical environment supporting it in a LANDSCAPE context. Widely used in the former Soviet Union and Eastern Europe, the term is similar but not identical to ecosystem (see Figure).   *RJH/JAM*

[See also ECOSYSTEM CONCEPT, GEOECOLOGY, LANDSCAPE ECOLOGY]
Sukachev, V.N. and Dylis, N.V. 1964: *Fundamentals of forest biogeocoenology*, translated by J.M. Maclennan. Edinburgh: Oliver and Boyd.

biogeography The geography of life. The study of the variation in any biological attribute (e.g. species number, densities of populations or body size of individuals) across any spatial gradient (e.g. distance, latitude, longitude, depth or area).   *MVL*

[See also PHYTOGEOGRAPHY, ZOOGEOGRAPHY]
Brown, J.H. and Lomolino, M.V. 1998: *Biogeography*, 2nd edn. Sunderland, MA: Sinauer. Cox, C.B. and Moore, P.D. 1993: *Biogeography: an ecological and evolutionary approach*, 5th edn. Oxford: Blackwell Scientific.

biogeomorphology The study of the influence of plants, animals and micro-organisms on Earth surface processes and LANDFORM development. *Phytogeomorphology* and ZOOGEOMORPHOLOGY are recognised as subdivisions.   *JAM*

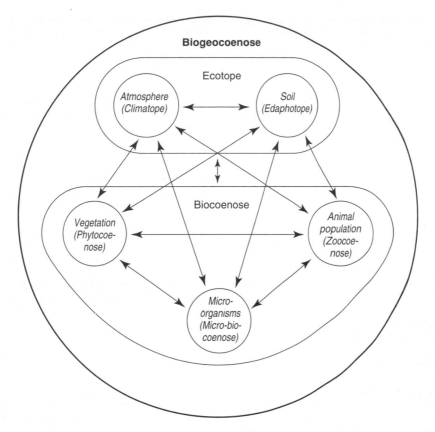

biogeocoenosis *The components of a biogeocoenosis and their interactions (after Sukachev and Dylis, 1964)*

[See also GEOMORPHOLOGY]

Howard, J.A. and Mitchell, C.W. 1985: *Phytogeomorphology.* New York: Wiley. Viles, H.A. (ed.) 1988: *Biogeomorphology.* Oxford: Basil Blackwell.

bioherm A lens-shaped deposit of BIOGENIC origin (see FOSSIL) in sedimentary rocks (usually LIMESTONE) that can be interpreted as having formed a mound on the sea-bed. *Reef mounds* are a particular type (see REEF). A similar deposit that is tabular in shape is termed a *biostrome.*

GO/JAM

James, N.P. and Bourque, P.-A. 1992: Reefs and mounds. In Walker, R.G. and James, N.P. (eds), *Facies models: response to sea level change.* St John's, Newfoundland: Geological Association of Canada, 323–347.

bio-indicators Species, the ecological requirements of which are indicative of either a general suite of environmental and biotic conditions or the specific presence of a single factor (synonymous with INDICATOR SPECIES). In the context of ENVIRONMENTAL CHANGE, *bio-indicator species* can be used: (1) to reconstruct PALAEOENVIRONMENTS and palaeoecosystems by using FOSSIL or SUBFOSSIL records (e.g. diatoms); and (2) to monitor contemporary change in ecosystem condition and environmental quality. Certain *lichens*, for example, only grow where atmospheric concentrations of SULPHUR DIOXIDE are low. Similarly, the presence of a top-carnivore may indicate the functional integrity of the supporting ECOSYSTEM, such as in the case of the northern spotted owl *Strix occidentalis* in OLD-

GROWTH FOREST in the Pacific northwest of America. The loss of bio-indicators can give early warning of declining ECOSYSTEM HEALTH and may be used for ecological evaluation and ENVIRONMENTAL MONITORING. Other uses are for geobotanical surveys to identify ore deposits (*geobotanical prospecting*), to infer the presence of another less observable species for census and CONSERVATION purposes, and in PHYTOSOCIOLOGY. *SHJ*

[See also BIO-ACCUMULATION, ECOLOGICAL INDICATOR, ENVIRONMENTAL INDICATORS, GEOINDICATORS, INDICATOR SPECIES, PHYTOINDICATION]

Cole, M.M. 1973: Geobotanical and biogeochemical investigations in the sclerophyllous woodland and shrub associations of the Eastern Goldfields area of western Australia, with particular reference to the role of *Hybanthus floribundus* (Lindl) F. Muell, as a nickel accumulator and indicator plant. *Journal of Applied Ecology* 10, 269–320.

biokarst Limestone landforms (see KARST) produced by biological activity. Biokarst is also known as *phytokarst.* Rock surfaces are usually covered by a layer of micro-organisms that can erode the surface or enhance its corrosion by producing organic acids. *SHD*

Ford, D.C. and Williams, P.W. 1989: *Karst geomorphology and hydrology.* London: Chapman and Hall. Viles, H.A. 1984: Biokarst: review and prospect. *Progress in Physical Geography* 8, 523–543.

biological control The INTRODUCTION of an organism to eliminate or reduce a PEST population. The organism is normally a natural enemy of the PEST species (i.e. a parasitoid or predator). The term can also be used for the natural regulation of organisms in the absence of manipulation by humans and for the use of PESTICIDES, but biological control is usually contrasted with the latter. Biological control in the narrow sense has been widely accepted historically as a 'green' approach to PEST MANAGEMENT. There are, however, possible negative indirect and non-target effects. *GOH*

Cory, J.S. and Myers, J.H. 2000: Direct and indirect ecological effects of biological control. *Trends in Ecology and Evolution* 15, 137–139. **Simberloff, D. and Stiling, P.** 1996: How risky is biological control? *Ecology* 77, 1965–1974. **Strobel, G.A.** 1992: Biological control of weeds. *Scientific American* 265, 50–60. **Van Emden, H.F.** 1989: *Pest control*. London: Arnold.

biological half life The time required for the elimination of one half of the mass of an ingested substance. The biological half life of MERCURY in humans, for example, is about 70 days, whereas it may be up to two years in large fish. *JAM*

[See also HALF LIFE]

biological invasion The spread, usually rapid and long term, of a species (plant, animal, insect, pathogen) in a new HABITAT or in an area in which it is not a NATIVE species. Biological invasion is commonly described as a form of ECOLOGICAL EXPLOSION. Some *invasive species* such as the bracken fern and the rabbit have greatly expanded their distributions because they are highly competitive and tolerant of the many new open HABITATS created by humans. Another reason for the success of many invasive species is that humans have introduced them to new areas lacking their natural PREDATORS, PESTS or PATHOGENS. This applies, for example, to rabbits, Oxford ragwort, Canadian pondweed and Japanese knotweed.

Invasions induced by humans, both deliberate and unintentional, have increased in area and intensity with time within the Old World (up to AD 1500), between Old and New Worlds (AD 1500–1850) and globally with intensive multicentred interactions (AD 1850–present). Many invasive species are regarded as pests. A number of them have been successfully reduced by BIOLOCAL CONTROL methods. Notable among the successes are the prickly pear cactus (*Opuntia*) and the aquatic fern *Salvinia molesta*. Populations of *Salvinia molesta* spread rapidly in Australia when introduced from southern Brazil in the 1950s, but were quickly brought under control when its natural herbivore, the black long-snouted weevil, was introduced to Australia in the early 1980s. Invasive species that have not been controlled successfully include the bracken fern, rabbit, starlings and the Oxford ragwort. The Oxford Ragwort (*Senecio jacobaea*), a native plant of the European mainland, spread very rapidly in the UK, especially along roads and railways lines after its 'escape' from the Oxford Botanic Gardens in the eighteenth century. This poisonous plant is a particular problem for farmers, causing the death of a large number of livestock in Europe each year. Invasions of ecosystems by *alien* or EXOTIC SPECIES now rank second to HABITAT LOSS as a major threat to BIODIVERSITY. Today 13% of vulnerable mainland vertebrates and 31% of those on islands are threatened in this way. *GOH*

[See also ISLAND BIOGEOGRAPHY, THREATENED SPECIES]

Di Castri, F. 1989: History of biological invasions with special emphasis on the Old World. In Clark, W.C. and Munn, R.E. (eds), *Sustainable development of the biosphere*. Cambridge: Cambridge University Press, 252–289. **Drake, J.A., Mooney, H.A., di Castri, F. *et al.* (eds)** 1989: *Biological invasions: a global perspective*. Chichester: Wiley. **Roberts, L. (ed.)** 1997: *World Resources 1998–99: a guide to the global environment*. Oxford: Oxford University Press. **Starfinger, U., Edwards, K., Kowarik, I. and Williamson, M. (eds)**, 1998: *Plant invasions: ecological mechanisms and human responses*. Leiden: Backhuys. **Williamson, M.** 1996: *Biological invasions*. London: Chapman and Hall.

biological magnification An increase in the concentration of non-biodegradable compounds (e.g. ORGANOCHLORIDES and HEAVY METALS) as they pass from one TROPHIC LEVEL to another upward through the FOOD CHAIN/WEB. Such compounds can increase to harmful concentrations in higher predators and humans. The organochloride insecticide, Dieldrin, may undergo an eightfold *biomagnification* from phytoplankton through the food chain to fish-eating birds. Biological magnification is more common in marine than in terrestrial environments: on land, it is most common in aquatic environments. *GOH*

[See also BIO-ACCUMULATION]

Nester, E.W., Evans, R.C. and Evans, M.T. 1995: *Microbiology: a human perspective*. Oxford: W.C. Brown. **Sijm, D., Seinen, W. and Opperhuizen, A.** 1992: Life cycle biomagnification study in fish. *Environmental Science and Technology* 26, 2162–2174. **Woodwell, G.M.** 1967: Toxic substances and ecological cycles. *Scientific American* 216, 24–31.

biological species 'Species are groups of actually or potentially interbreeding natural populations, which are reproductively isolated from other such groups' (Mayr, 1942: 120). *KDB*

[See also SPECIES CONCEPT]

Mayr, E. 1942: *Systematics and the origin of species from the viewpoint of a zoologist*. New York: Columbia University Press.

biological weathering Physical disintegration, dissolution and chemical alteration of rocks or minerals driven by biological processes. Important processes of biological or *biotic weathering* include enhanced CHEMICAL WEATHERING by biologically derived acids. *SHD*

[See also BIOKARST, CHELATION, WEATHERING]

biomagnetics Organisms may contain intracellular fine-grained MAGNETITE or GREIGITE, which they use to sense geomagnetic field lines. Magnetite crystals in such magnetotactic bacteria are used to characterise bacterial and other biogenic magnetite. *UBW*

[See also GEOMAGNETISM, PALAEOMAGNETISM]

Dunlopp, D.J. and Özdemir, Ö. 1997: *Rock magnetism*. Cambridge: Cambridge University Press.

biomantle The upper part of the soil produced or considerably modified by BIOTURBATION. *JAM*

Johnson, D.L. 1990: Biomantle evolution and the redistribution of earth materials and artifacts. *Soil Science* 149, 84–102.

biomarker An organic compound, the structure of which reveals an unambiguous link with a known biological natural product (precursor). Biomarkers, or chemical fossils, have been identified for vascular plants, *phytoplankton*, bacteria and archaebacteria. True biomarkers can be assigned to a specific organism or type of organism (e.g. botryococcenes are biomarkers for *Botryococcus braunii*). In SEDIMENTS, biomarkers can be used to identify the major sources of organic matter and hence, to deduce some characteristics of the *depositional environment*. These conditions include oxicity/anoxicity, hypersaline, freshwater LACUSTRINE and MARINE, which can be used in CLIMATIC RECONSTRUCTIONS. True environmental biomarkers are very rare as they depend upon a compound restricted to a single organism or group of related organisms and for this organism(s) to be restricted to a certain environment. Other applications of biological markers include thermal maturation and oil/oil or oil/source rock correlations, commonly used in PETROLEUM EXPLORATION.

KJF

Eglinton, G. and Calvin, M. 1967: Chemical fossils. *Scientific American* 216, 32–43. **Hedges, R.E.M. and Sykes, B.C.** 1992: Biomolecular archaeology: past, present and future. In Pollard, A.M. (ed.), *New developments in archaeological science* [*Proceedings of the British Academy* 77]. Oxford: Oxford University Press, 267–283. **Johns, R.B.** 1986: Biological markers in the sedimentary record. *Methods in Geochemistry and Geophysics* 24. Amsterdam: Elsevier. **Tissot, B.P. and Welte, D.H.** 1984: *Petroleum formation and occurrence*, 2nd edn. Heidelberg: Springer.

biomass

The total weight of living tissue of organisms accumulated over time. Biomass is usually measured as DRY WEIGHT of organic matter per unit area (e.g. in t ha^{-1}). Plants accumulate biomass by primary production using PHOTOSYNTHESIS. The rate at which energy is stored in plants is called PRIMARY PRODUCTIVITY. The rate at which animals, bacteria and fungi build up biomass by consuming plant tissue is called SECONDARY PRODUCTIVITY. Biomass in TEMPERATE FORESTS is subject to seasonal variation due to loss of leaves in autumn. The dynamics of global forest biomass is particularly important for BIOGEOCHEMICAL CYCLES, as well as CLIMATIC CHANGE. OPTICAL REMOTE SENSING INSTRUMENTS are used routinely to estimate *above-ground biomass* in forest inventories in Finland. BACKSCATTERING from RADAR is known to be directly related to forest biomass and *synthetic aperture radar* is frequently employed for *biomass mapping* over large regions, especially where regular cloud cover is limiting the use of optical sensors.

HB

[See also ABOVE-GROUND PRODUCTIVITY, BELOW-GROUND PRODUCTIVITY, RADAR REMOTE SENSING INSTRUMENTS]

Baker, J.R., Mitchell, P.L., Cordey, R.A. *et al.* 1994: Relationships between physical characteristics and polarimetric radar backscatter for Corsican pine stands in Thetford Forest, UK. *International Journal of Remote Sensing* 15, 2827–2849. **Beaudoin, A., Le Toan, T., Goze, S.** *et al.* 1994: Retrieval of forest biomass from SAR data. *International Journal of Remote Sensing* 15, 2777–2796. **Tomppo, E.** 1995: Finnish national forest inventory. *Paper and Timber* 77, 374–378.

biomass burning

The COMBUSTION of plant and other organic material either intentionally, in place of fossil fuels, or inadvertently. Biomass burning can be a form of RENEWABLE ENERGY because plants replace themselves. Biomass burning is a significant source of atmospheric METHANE and CARBON DIOXIDE, contributes to the store of GREENHOUSE GASES in the atmosphere and may be an important driver of CLIMATIC CHANGE. Reconstruction of the level of biomass burning during the HOLOCENE shows high levels of burning during relatively dry episodes. Biomass burning is thought to have increased by some 50% since about AD 1850 and TROPICAL FOREST FIRES and savannah burning has increased the burnt area to as much as 2.5% of the land area of the globe annually. In recent years widespread haze, especially in southeast Asia, has accompanied biomass burning and resulted in health problems.

AHP/JAM

Crutzen, P.J. and Andreae, M. 1990: Biomass burning in the tropics. *Science* 250, 1669–1678. **Levine, J.S.** 1991: *Global biomass burning: atmospheric, climatic and biosphere implications*. Cambridge, MA: MIT Press. **Penner, J.E., Dickinson, R.E. and O'Neill, C.A.** 1992: Effects of biomass burning on the global radiation budget. *Science* 256, 1432–1434. **Taylor, K.C., Mayewski, P.A., Twickler, M.S. and Whitlow, S.I.** 1996: Biomass burning recorded in the GISP2 ice core: a record from eastern Canada? *The Holocene* 6, 1–6.

biome

A community of animals and plants occupying a climatically relatively uniform area on a continental scale. Biomes are usually thought of as CLIMATIC CLIMAX communities whose FAUNA and FLORA are in balance with prevailing environmental, and especially the climatic, conditions at the continental scale. They are distinguished by their plant assemblages – forest, grassland, desert, tundra and so on – and not by their component animals. The climatic influence on biomes imparts a strong latitudinal and altitudinal element in their distribution. The modern terrestrial BIOSPHERE consists of nine zonal biomes (*zonobiomes*) (see Figure). In the marine realm, the chief biomes are the intertidal (ESTUARINE, LITTORAL marine, algal bed, CORAL REEF) biome, the open sea (PELAGIC) biome, the UPWELLING zone biome, the BENTHIC biome and the HYDROTHERMAL VENT biome. Biomes are resistant to natural disturbances but human activities are transforming large areas of them. The distribution and nature of biomes is likely to alter in response to GLOBAL WARMING.

RJH

[See also COMMUNITY CONCEPTS, ECOSYSTEM CONCEPT, NATURAL VEGETATION, POTENTIAL NATURAL VEGETATION, VEGETATION FORMATION]

Bailey, R. G. 1996: *Ecosystem geography*. New York: Springer. **Carpenter, J. R.** 1939: The biome. *American Midland Naturalist.* 21, 75–91. **Huggett, R. J.** 1998: *Fundamentals of biogeography*. London: Routledge. **Schultz, J.** 1995: *The ecozones of the world: the ecological divisions of the geosphere*. Hamburg: Springer. **Walter, H.** 1985: *Vegetation of the Earth and ecological systems of the geo-biosphere*, 3rd edn, translated from the 5th German edition by O. Muise. Berlin: Springer.

bioprospecting

The search for whole organisms, genes (*gene bioprospecting*), natural compounds (*chemical bioprospecting*) and designs (*bionic bioprospecting*) in wildlife with potential for product development without disrupting NATURE.

JAM

Mateo, N., Nader, W. and Tamayo, G. 2001: Bioprospecting. In Levin, S.A. (ed.), *Encyclopedia of biodiversity*. Vol. 1. San Diego, CA: Academic Press, 471–488.

bioregional management

Integrated management of distinct biophysical regions that aims to conserve critical ecological flows and linkages and the core biological, cultural and landscape character of the region through

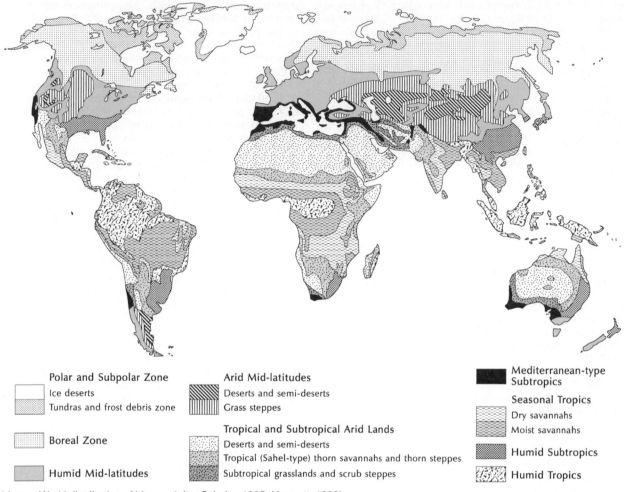

Polar and Subpolar Zone
Ice deserts
Tundras and frost debris zone

Boreal Zone

Humid Mid-latitudes

Arid Mid-latitudes
Deserts and semi-deserts
Grass steppes

Tropical and Subtropical Arid Lands
Deserts and semi-deserts
Tropical (Sahel-type) thorn savannahs and thorn steppes
Subtropical grasslands and scrub steppes

Mediterranean-type Subtropics

Seasonal Tropics
Dry savannahs
Moist savannahs

Humid Subtropics

Humid Tropics

biome *World distribution of biomes (after Schultz, 1995; Huggett, 1998)*

decentralised planning and strengthening people's 'sense of place'. *PJ*

[See also CONSERVATION BIOLOGY, ECOSYSTEM CONCEPT, ECOSYSTEM MANAGEMENT, LANDSCAPE ECOLOGY]

bioremediation Use of micro-organisms, usually bacteria but also fungi, in the DECOMPOSITION of organic chemical wastes and POLLUTANTS. The aim is to produce less hazardous chemicals and, where possible, achieve complete MINERALISATION. The process may involve contaminated soil or sediment, ground or surface water, and sewage or other WASTE materials. Bioremediation, which results in the destruction of contaminants *in situ*, is increasingly replacing other approaches to WASTE MANAGEMENT, such as the use of LANDFILL, which merely transfers the contaminants elsewhere. *JAM*

[See also BIODEGRADATION, BIOTECHNOLOGY]

Atlas, R.M. 1995: Bioremediation. *Chemical Engineering News* 73, 32–42. **Bragg, J.R., Prince, R.C., Harner, E.J. and Atlas, R.M.** 1994: Effectiveness of bioremediation for the Exxon Valdez oil spill. *Nature* 368, 413–418. **Chaparian, M.G.** 1999: Bioremediation. In Alexander, D.E. and Fairbridge, R.W. (eds), *Encyclopedia of environmental science*. Dordrecht: Kluwer, 53–58. **Devine, K.** 1992: *Bioremediation case studies: an analysis of vendor supplied data*. Washington, DC: US Environmental Protection Agency.

biosphere Coined by Eduard Suess, and sometimes used as a synonym for ECOSPHERE, the term biosphere is most appropriately used for the totality of life on Earth. It has also been used for the space occupied by living things. *RJH/JAM*

Huggett, R.J. 1999: Ecosphere, biosphere or Gaia? What to call the global ecosystem. *Global Ecology and Biogeography Letters* 8, 425–431. **Polunin, N. and Grinevald, J.** 1988: Vernadsky and biospheral ecology. *Environmental Conservation* 15, 117–122. **Suess, E.** 1875: *Die Entstehung der Alpen*. W. Baunmüller: Vienna. **Walker, B., Steffen, W., Canadell, J. and Ingram, J.** 1999: *The terrestrial biosphere and global change: implications for natural and managed ecosysems*. Synthesis Volume [International Geosphere-Biosphere Programme Series. Vol. 4] Cambridge: Cambridge University Press.

biosphere reserves Areas of terrestrial or coastal ecosystems and landscapes that are internationally recognised within the framework of Unesco's *Man and the Biosphere Programme* (MAB). *IFS*

[See also CONSERVATION, PROTECTED AREA]

Falk, D.A. 1998: Biosphere reserves. In Calow, P. (ed.), *The encyclopedia of ecology and environmental management*. Oxford: Blackwell Science, 93–94.

biostasy A relatively stable phase in soil evolution characterised by a well-developed vegetation cover, CHEMICAL WEATHERING and PEDOGENESIS with generally low-energy transformations is said to exhibit biostasy. Phases of biostasy contrast with phases of reduced vegetation cover, mechanical REDEPOSITION, SOIL EROSION and disequilibrium, which exhibit *rhexistasy*. Although biostasy and rhexistasy may alternate through time in response to environmental change, they may also co-exist in the same LANDSCAPE. *JAM*

[See also K-CYCLES, LOESS STRATIGRAPHY, PALAEOSOL]

Erhart, H. 1956: *La genèse des sols en tant que phénomène géologique*. Paris: Masson.

biostratigraphy The establishment of RELATIVE AGES for sediments and sedimentary rocks based on their FOSSIL or SUBFOSSIL content and patterns of change, including evolutionary changes within fossil groups. A *biozone* is a body of sediment or rock defined by its fossil content with reference to specific INDEX FOSSILS or fossil assemblages. If EVOLUTION is taken to occur rapidly on a geological timescale, then the boundaries of biozones can be taken to be time-equivalent horizons, making them valuable tools in CORRELATION. A good index fossil should be abundant and easily recognisable, showing rapid morphological evolution, wide geographical distribution, good preservation potential and tolerance of a range of sedimentary FACIES. The best index fossils show facies-independence as a result of free-swimming (NEKTONIC) or floating (PLANKTONIC) modes of life (e.g. graptolites, ammonites, foraminiferans) or as a result of dispersal by air (e.g. SPORES and pollen – see POLLEN ANALYSIS). BENTHIC (bottom-dwelling) fossil assemblages (e.g. brachiopods, corals, bivalves) are facies-dependent and zones based on such groups may show DIACHRONISM. However, such '*facies fossils*' can be useful locally in biostratigraphy, as in the coral zonation of the British Lower CARBONIFEROUS. The duration of individual biozones is variable, from about 10 million years for zones based on Lower CAMBRIAN trilobites to about 600 000 years for zones based on JURASSIC ammonites. In the QUATERNARY context, evolutionary changes are not common and '*assemblage biozones*' are defined on the basis of a characteristic mix of subfossils and their relative abundances. They typically have a duration of a few centuries or less. *LC*

Doyle, P. and Bennett, M.R. (eds) 1998: *Unlocking the stratigraphical record: advances in modern stratigraphy*. Chichester: Wiley. Nichols, G. 1999: *Sedimentology and stratigraphy*. Oxford: Blackwell Science. Whittaker, A., Cope, J.C.W., Cowie, J.W. et al. 1991: A guide to stratigraphical procedure. *Journal of the Geological Society, London* 148, 813–824.

biotechnology The artificial manipulation of living organisms, cells, or subcellular components in the production of useful goods (such as vaccines, new breeds or varieties, and foodstuffs) and services (such as the enzyme- or micro-organism-controlled chemical reactions used in POLLUTION abatement and WASTE disposal). Biotechnology has extended the range of useful organisms that benefit humans from plants and animals to BACTERIA, viruses and fungi. These are used, for example, in sewage treatment, in the BIOLOGICAL CONTROL of pests, in nitrogen fixation, in the production of protein-rich food substitutes, in the recovery of metals from ores (*bioleaching* or *biomining*), in the production of ethanol from sugar cane (a source of ALTERNATIVE ENERGY) and in the BIOREMEDIATION of OIL SPILLS at sea.

Aspects of biotechnology have a long history; for example in brewing and baking, and in the various purposeful and accidental attempts at plant and animal breeding. GENETIC ENGINEERING is one, very recent aspect of biotechnology that may well exceed all others in the scale of its environmental impacts. The implications and potential of biotechnology for *economic development* and SUSTAINABILITY in both the developed and developing world are immense. For example, the potential for increasing food production without the adverse direct and indirect environmental changes that accompanied many other innovations and 'revolutions' throughout AGRICULTURAL HISTORY is considerable. Developing countries have nevertheless much to lose in the conflicts of interest between those in control of biotechnological enterprise and those in greatest need of affordable applications. *JAM*

Agate, A.D. 1996: Recent advances in microbial mining. *World Journal of Microbiology and Biotechnology* 12, 487–495. Ginzburg, L.R. 1991: *Assessing ecological risks of biotechnology*. Boston: Butterworth-Heinemann. Gray, N.F. 1989: *Biology of wastewater*. Oxford: Oxford University Press. Lindsey, K. and Jones, M.K.G. 1989: *Plant biotechnology in agriculture*. Milton Keynes: Open University Press. Mannion, A.M. 1992: Sustainable development and biotechnology. *Environmental Conservation* 19, 297–306. Mannion, A.M. 1993: Biotechnology and global change. *Global Environmental Change* 3, 320–329. Nakas, P. and Hagerdorn, C. (eds) 1990: *Biotechnology of plant-microbe interactions*. New York: McGraw-Hill. Smith, J.E. 1988: *Biotechnology*, 2nd edn. London: Arnold.

biotic Living components and factors (*biota*) of the BIOSPHERE, or of an ECOSYSTEM, as distinct from the non-living ABIOTIC components. *UBW*

[See also BIOGENIC SEDIMENTS, ENVIRONMENTAL FACTORS, FOSSILS, ORGANIC SEDIMENT]

bioturbation The disturbance, disruption or mixing of soil or sediment by the activity of organisms. Examples include burrowing, root growth, tracks and trails associated with locomotion and feeding, and trampling. Bioturbation is particularly important near the soil surface and in the surficial sediments of lacustrine and marine deposits. It can be preserved in the geological record in the form of TRACE FOSSILS. *MRT*

[See also INFAUNA]

Bromley, R.G. 1996: *Trace fossils: biology, taphonomy and applications*. London: Chapman and Hall. Hole, F.D. 1981: Effects of animals on soil. *Geoderma* 25, 75–112.

birth rate The average number of births per 1000 people in the population. In 'westernised' and economically advanced countries birth rates tend to be considerably lower than in developing countries. *JAM*

[See also DEMOGRAPHIC CHANGE, DEMOGRAPHIC TRANSITION, POPULATION]

black-body radiation The electromagnetic energy emitted from an ideal body (known as a black body or *holeraum*), which would absorb all incident radiation and reflect none. Because of the KIRCHHOFF RADIATION LAW, a perfect absorber is also a perfect emitter and thus the radiant energy emitted by a black body is equal to the energy of the incident radiation. According to the *Stefan–Boltzmann law*, black-body radiation depends on its temperature alone and increases as the fourth power of the temperature. The black-body radiation spectrum is described by Planck's equation. This spectrum is used as a reference against which the radiation spectra of real materials (usually referred to as *grey bodies*) at the same KINETIC TEMPERATURE are compared. Grey bodies emit less than a black body does and do not necessarily absorb all the energy incident upon them. The emission properties of real materials characterise the media through their *spectral signature*. TS

[See also RADIOMETRIC DATING]

Schanda, E. 1986: *Physical fundamentals of remote sensing*. Berlin: Springer. Ulaby, F.T., Moore, R.K. and Fung, A.K. 1981: *Microwave remote sensing: active and passive*. Norwood, MA: Artech House.

black shale   A dark, fine-grained marine or lacustrine MUDROCK rich in organic carbon (total organic carbon, TOC, typically 1–15%). Some workers include black CHERTS, organic LIMESTONES and fine-grained TURBIDITES, although black shales should only include the finely laminated parts of such sediments, with quantities of reduced iron in the form of pyrite. Black shales are thought to form under ANAEROBIC conditions (less than 0.1 ml oxygen per litre of sea water) in a wide range of water depths and are particularly common in CRETACEOUS and PALAEOZOIC successions that accumulated during periods of sea-level HIGHSTAND. Current theories suggest that black shales formed during past episodes of GREENHOUSE CONDITIONS, when the absence of dense, cold polar OCEAN CURRENTS produced stratified oceans with poorly oxygenated waters at depth (see OCEANIC ANOXIC EVENTS). In equatorial regions, where surface waters are even warmer and less dense, oxygenation of bottom waters is less likely, enhancing black shale production. BTC

[See also METAL POLLUTION, OIL SHALE, PELAGIC SEDIMENT, SAPROPEL]

Arthur, M.A., Dean, W.E. and Stow, D.A.V. 1984: Models for the deposition of Mesozoic–Cenozoic fine-grained organic-carbon-rich sediment in the deep sea. In Stow, D.A.V. and Piper, D.J.W. (eds), *Fine-grained sediments: deep-water processes and facies*. Special Publication 15. London: Geological Society, 527–560. Arthur, M.A. and Sageman, B.B. 1994: Marine black shales: depositional mechanisms and environments of ancient deposits. *Annual Review of Earth and Planetary Sciences* 22, 499–551. Cramp, A. and O'Sullivan, G. 1999: Neogene sapropels in the Mediterranean: a review. *Marine Geology* 153, 11–28. Faure, K. and Cole, D. 1999: Geochemical evidence for lacustrine microbial blooms in the vast Permian Main Karoo, Parana, Falkland Islands and Huab basins of southwestern Gondwana. *Palaeogeography, Palaeoclimatology, Palaeoecology* 152, 189–213. Wignall, P.B. 1994: *Black shales* [Oxford Monographs in Geology and Geophysics 30]. Oxford: Oxford University Press.

black smoker   A submarine HYDROTHERMAL VENT that emits a jet of hot brine with finely dispersed black metal sulphide particles. These are precipitated as columns up to 10 m tall and provide HABITATS for life not nourished by sunlight. JBH

[See also CHEMOSYNTHESIS]

blanket mire   A form of extensive OMBROTROPHIC MIRE that blankets the landscape, especially on plateaux, but can spread on rising ground on slopes up to 15°. Blanket mires need high rainfall (*c.* 1250 mm per annum), which is well distributed through the year (> 160 wet days), and a cool-temperate climate. In the Northern Hemisphere, blanket MIRES are found in western Norway, the Faeroes, in upland areas of northern and western Britain and eastern Ireland and almost to sea level in the west of Ireland. Blanket mire types are also found in Newfoundland, in the Aleutian islands and in western Kamchatka, whilst 'alpine blanket bog complexes' have been described from the Central Alps, and 'peat cakes' on the fells of Finnish Lapland have been interpreted as continental fragments of blanket mires. In the Southern Hemisphere, blanket mires have been described from the Falkland Islands, Tierra del Fuego, Tasmania and New Zealand. Mires resembling blanket mires are claimed from upland equatorial Africa, in Uganda. Lindsay *et al.* (1988) suggest the world distribution of blanket mire is some 10 000 000 ha. The type region is the British Isles.

The vegetation of blanket mire in the British Isles was termed '*blanket bog*' – one of the recognised vegetational formations of ombrotrophic BOG – but 'blanket bog' is now frequently used synonymously to denote blanket mire. The 'blanket peat' produced from the incomplete decay of blanket bog species is typically highly humified and can be difficult to distinguish from heath peats. The term 'blanket peat' is therefore frequently used rather loosely in Britain to describe not just the deeper (> 1 m depth) peats that may derive from typical blanket bog species, but also areas of shallower hill peat that may derive largely from grass heath and from Callunetum.

It has recently been recognised that, despite the humified appearance of their peats, blanket mires are NATURAL ARCHIVES of environmental history and can be studied using a range of palaeoecological techniques, including POLLEN ANALYSIS, analysis of testate amoebae (RHIZOPOD ANALYSIS), STABLE-ISOTOPE ANALYSIS and determination of PEAT HUMIFICATION, to give various PROXY RECORDS of vegetational or climatic history, which assist in reconstructing environmental changes of the HOLOCENE.

Blanket mires have been most studied in Europe, especially in Britain and Ireland, where they developed in the Holocene and where their origins are claimed to result from a combination of prevailing climate, soil-forming processes and the influence of human activity in prehistory. Locally, tree remains are found at the peat base, or stratified within the peat. In Ireland, a distinction is made by Doyle (1997) between Atlantic (or western) blanket bogs below 200 m above sea level and mountain blanket bogs.

In Europe the largest extent of blanket mire is the Flow Country of Caithness and Sutherland, in northern Scotland. Extensive coniferous afforestation of the Flows commenced in 1978, and for the following decade posed a major threat to their conservation. Other European blanket mires have experienced different pressures. For example, in the English Pennines and in South Wales, their vegetation was formerly dominated by *Sphagnum* species, but their present vegetation is typically depauperate, with few *Sphagnum* species and with over-representation of *Eriophorum* or of *Molinia*, possibly as a result of human influence through a combination of burning, overgrazing and industrial atmospheric pollution.   *FMC*

Chambers, F.M., Barber, K.E., Maddy, D. and Brew, J. 1997: A 5500-year proxy-climate and vegetational record from blanket mire at Talla Moss, Borders, Scotland. *The Holocene* 7, 391–399. **Doyle, G.** 1997: Blanket bogs: an interpretation based on Irish blanket bogs. In Parkyn, L. Stoneman, R.E. and Ingram, H.A.P. (eds), *Conserving peatlands*. Wallingford: CAB International, 25–34. **Gore, A.J.P. (ed.)** 1983: *Mires: bog, fen, swamp and moor [Ecosystems of the World.* Vol. 4B]. Rotterdam: Elsevier. **Lindsey, R.A.** 1995: *Bogs: the ecology, classification and conservation of ombrotrophic mires*. Battleby: Scottish Natural Heritage. **Lindsey, R.A., Charman, D.J., Everingham, F. et al.** 1988: *The Flow Country: the peatlands of Caithness and Sutherland*. Peterborough: Nature Conservancy Council. **Luoto, M. and Seppälä, M.** 2000: Summit peats ('peat cakes') on the fells of Finnish Lapland: continental fragments of blanket mires? *The Holocene* 10, 229–241. **Moore, P.D.** 1993: Origin of blanket mires, revisited. In Chambers, F.M. (ed.), *Climate Change and Human Impact on the Landscape*. London: Chapman and Hall, 217–224. **Steiner, G.M.** 1997: Bogs of Europe. In Parkyn, L. Stoneman, R.E. and Ingram, H.A.P. (eds), *Conserving peatlands*. Wallingford: CAB International, 4–24.

blockfields AUTOCHTHONOUS or ALLOCHTHONOUS spreads of boulders on gently sloping terrain. Boulders tend to decrease in size with depth and fine-grained material may be present below the surface. Blockfields are also known as *felsenmeer*.   *PW*

[See also CLITTER, ROCK STREAM]

McCarroll, D. and Nesje, A, 1993: The vertical extent of ice sheets in Nordfjord, western Norway – measuring degree of rock surface weathering. *Boreas* 22, 255–265. **Rea, B.R., Whalley, W.B., Rainey, M.M. and Gordon, J.E.** 1996: Blockfields, old or new? Evidence and implications from some plateaus in northern Norway. *Geomorphology* 15, 109–121.

blocking An atmospheric WEATHER TYPE associated with an ANTICYCLONE or high-pressure area in mid-latitudes that blocks the movement of migratory DEPRESSIONS ('lows'), which are forced to move northwards or southwards around the block. Blocking is an extreme low index state of the *index cycle*, which is characterised by weakening of the westerlies throughout the troposphere and strong MERIDIONAL FLOW. Blocking, which tends to be more intense and long lived in the Northern Hemisphere, can cause anomalous weather such as DROUGHTS or severe cold while it persists. It is frequent over the eastern North Atlantic, especially in spring, and may last for a few days or several weeks. The frequency of blocking may change during CLIMATIC CHANGE; an increase in frequency was

likely in northwestern Europe during the HYPSITHERMAL or 'Climatic Optimum' of the early to mid Holocene.   *AHP/JAM*

Benzi, R., Saltzman, B. and Wiin-Nielsen, A.C. (eds) 1986: *Anomalous atmospheric flows and blocking [Advances in Geophysics* 29]. **Green, J.S.A.** 1977: The weather during July 1976: some dynamical considerations of the drought. *Weather* 32, 120–128. **Rex, D.F.** 1951: The effect of blocking action upon European climate. *Tellus* 3, 100–111.

blowout A bowl-shaped hollow caused by wind erosion in a SAND DUNE. It indicates loss of sand from the dune system.   *HJW*

blue-water refugium A REFUGIUM for marine organisms provided by an equatorial ocean that may have existed during episodes of global GLACIATION in the late PRECAMBRIAN according to a modification of the SNOWBALL EARTH hypothesis. The possible existence of blue-water refugia, or *pelagic refugia*, has implications for the CAMBRIAN EXPLOSION.   *JAM/GO*

Runnegar, B. 2000: Loophole for snowball Earth. *Nature* 405, 403–404.

Blytt–Sernander timescale The Norwegian, A.G. Blytt, and the Swede, R. Sernander, recognised a sequence of HOLOCENE ENVIRONMENTAL CHANGE in southern Scandinavia, based on PEAT STRATIGRAPHY. Changes in PEAT HUMIFICATION were used in CLIMATIC RECONSTRUCTION. Later, pollen analysis helped to correlate the Blytt–Sernander zones over geographical space. Before RADIOCARBON DATING, chronology was established using VARVE CHRONOLOGY and archaeological evidence. Mangerud et al. (1974) used the Blytt–Sernander terminology in their definition of CHRONOZONES for northwest Europe. However, because of possible confusion between CHRONOSTRATIGRAPHY, BIOSTRATIGRAPHY and CLIMATOSTRATIGRAPHY attached to these terms, their use is now not generally recommended (especially outside northwest Europe). The timescale, with equivalent radiocarbon dates, is given in the Table.   *BA*

[See also HOLOCENE TIMESCALE]

Mangerud, J., Andersen, S. T., Berglund, B. E. and Donner, J. J. 1974: Quaternary stratigraphy of Norden: a proposal for terminology and classification. *Boreas* 3, 109–128.

bog burst A spontaneous and catastrophic collapse of local MIRE structure in response to severe stress (and extreme rainfall events), leading to water release and PEAT slides, which can extend hundreds of metres.   *MJB*

Blytt–Sernander timescale *The timescale with the equivalent radiocarbon dates*

	Radiocarbon (ka BP)	Calibrated (y BP)	Inferred climate
Subatlantic	2.5–0	2600–0	Cool and wet
Subboreal	5–2.5	5700–2600	Warm and dry
Atlantic	8–5	7800–5700	Warm and wet
Boreal	9–8	10 500–7800	Warm and dry
Preboreal	10–9	11 500–10 500	Cool and dry

bog people The bodies of PREHISTORIC to MODERN people preserved in BOGS in northwestern Europe, especially Denmark, northern Germany, The Netherlands, Ireland and the UK. Most are skeletal but some, such as 'Tollund Man' and 'Grauballe Man' from Denmark, and 'Lindow Man' from Cheshire, UK, are extremely well preserved, allowing detailed analysis of cultural and environmental aspects of life especially in the BRONZE AGE, IRON AGE, ROMAN PERIOD and MEDIAEVAL WARM PERIOD. *JAM*

Glob, P.V. 1969: *The bog people: Iron-Age man preserved*, translated by R. Bruce-Mitford. London: Faber and Faber. **Turner, R.C. and Scaife, R.G. (eds)** 1995: *Bog bodies: new discoveries and new perspectives*. London: British Museum Press.

bolide An extraterrestrial object that impacts on a planetary or satellite surface or explodes in the lower ATMOSPHERE, including COMETS, ASTEROIDS and METEORITES. One of the most notable documented examples on Earth was the Tunguska explosion of AD 1908. The MASS EXTINCTION event at the K–T BOUNDARY and a decline in BRONZE AGE civilisation have been attributed to bolides.

JBH

[See also COSMOSPHERE, METEORITE IMPACT, PLANETARY ENVIRONMENTAL CHANGE]

Baillie, M.J.L. 1998: Tree-ring evidence for environmental disasters during the Bronze Age: causes and effects. In Peiser, B.J., Palmer, T. and Bailey, M.E. (eds), *Natural catastrophes during Bronze Age civilizations: archaeological, geological, astronomical and cultural perspectives* [*British Archaeological Reports* S72]. Oxford: Archaeopress. **Bronshten, V.A.** 1999: The nature of the Tunguska meteorite. *Meteoritics and Planetary Science* 34, 723–728. **Svetsov, W.** 1996: Total ablation of debris from the 1908 Tunguska explosion. *Nature* 383, 697–699.

Bølling Interstadial The earliest interval of relatively warm conditions in Europe during the *Late Glacial Interstadial* (following the LAST GLACIAL MAXIMUM), which precedes the OLDER DRYAS STADIAL. Defined originally in Scandinavia, the Bølling Interstadial has been dated to between c. 13 and 12 ka BP (^{14}C years). *MHD*

[See also LATE GLACIAL ENVIRONMENTAL CHANGE]

Lowe, J.J., Ammann, B., Birks, H.H. *et al.* 1994: Climatic changes in areas adjacent to the North Atlantic during the last glacial–interglacial transition (14–9 ka BP). *Journal of Quaternary Science* 9, 185–198. **Mangerud, J., Andersen, S.T., Berglund, B.E. and Donner, J.J.** 1974: Quaternary stratigraphy of Norden: a proposal for terminology and classification. *Boreas* 10, 109–127.

bolus A wad of chewed food or a food RESIDUE. For example, masticated plant tissues have been recovered as boluses from bison teeth and used for palaeoenvironmental reconstruction. *JAM*

Akersten, W.A., Foppe, T.M. and Jefferson, G.T. 1988: New source of dietary data for extinct herbivores. *Quaternary Research* 30, 92–97.

bomb effect The presence, initially in the atmosphere, of artificial or anthropogenic RADIONUCLIDES due to the atmospheric testing of nuclear weapons. In the context of RADIOCARBON DATING, this is also known as the *Libby effect*. Most notably, the testing of large weapons until the test ban treaty in AD 1962 released large quantities of ^{14}C into the atmosphere, causing a near doubling of the atmospheric ^{14}C concentration in 1963. Atmospheric CO_2 constitutes only a small part of the global carbon exchange reservoir and, since 1963, as shown in the Figure, the ^{14}C concentration has declined exponentially as the bomb carbon has moved into other parts of the exchange reservoir. *PQD*

[See also TRACER]

Levin, I. and Kromer, B. 1997: Twenty years of atmospheric $^{14}CO_2$ observations at Schauinsland station, Germany. *Radiocarbon* 39, 205–218. **Levin, I., Kromer, B., Schoch-Fischer, H.** *et al.* 1985: 25 years of tropospheric ^{14}C observations in central Europe. *Radiocarbon* 27, 1–19.

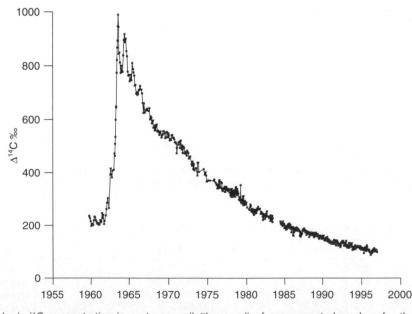

bomb effect *Atmospheric ^{14}C concentration in parts per mil (thousand) of age corrected modern for the sites at Vermunt and Schauinsland in central Europe since* AD *1960 (after Levin et al., 1985; Levin and Kromer, 1997)*

Bond cycle A cooling cycle or PERIODICITY of duration 6–15 ka, noted particularly during the part of the late Pleistocene between *c.* 80–10 ka BP (oxygen ISOTOPE STAGES 5a to 2). Each Bond cycle includes several DANSGAARD–OESCHGER EVENTS, culminates in the massive discharge of ICEBERGS across the North Atlantic (a HEINRICH EVENT) and ends with rapid warming to the next cycle (see Figure). *FMC/CJC*

Alley, R.B. 1998: Icing the North Atlantic. *Nature* 392, 335–337. Bond, G., Broecker, W., Johnsen, S. *et al.* 1993: Correlations between climate records from North Atlantic sediments and Greenland ice. *Nature* 365, 143–147.

Boolean logic A branch of deductive logic (which is the science dealing with the principles of valid reasoning and argument) founded by George Boole and Augustus de Morgan in the middle of the nineteenth century. It assumes that any well-formed assertion is either true or false and it can be used to express relations among objects in a symbolic and formal manner. The rules of Boolean logic allow the construction and validation of complex relationships, and many decisions in the study of ENVIRONMENTAL CHANGE rely on the use of this body of knowledge. *ACF*

boom and bust cycles A description of the responses of plants, animals and sometimes human populations in DRYLANDS to the characteristically highly variable environmental conditions. *RAS*

bootstrapping A general method of estimating standard errors and CONFIDENCE INTERVALS for virtually any statistic from an available data sample. It is based on the insight that the basic underlying PROBABILITY DISTRIBUTION of the statistic in question is contained in the data sample itself. Thus, bootstrapping is a powerful tool for generating confidence limits when there are no theoretical tests available and in cases where the assumptions of the theoretical tests may not be met by the data. There are a number of ways to generate bootstrap confidence intervals, but all are based on randomly sampling the data with replacement. This is done with a uniform random number generator that places equal probability on the selection of each datum for calculating the statistic of interest. The statistic for each bootstrap sample is calculated, as it would be from the original sample. This procedure is done many times to create an empirical probability distribution function that can be used to estimate the confidence intervals of the actual statistic. For example, if 1000 bootstrap statistics were estimated, then the empirical two-tailed 95% confidence intervals would be the 2.5% and 97.5% percentiles (equivalently the 25th and 975th ranked values) of the empirical bootstrap distribution. With the easy availability of fast computers today, this computationally intensive method can now be routinely applied in many cases. *ERC*

Ephron, B. and Tibshirani, R. 1986: Bootstrap methods for standard errors, confidence intervals, and other measures of statistical accuracy. *Statistical Science* 1, 54–77.

boreal forest Coniferous-dominated forest occupying an almost continuous c. 1000 km wide circumpolar zone (in Eurasia often referred to as *taiga*). The forest structure is simple with a dense EVERGREEN canopy, an understorey of low-growing ericaceous SHRUBS and a thick ground layer of *mosses* and *lichens*. SPECIES DIVERSITY is low and the dominant conifers are spruce (*Picea*) in North America, larch (*Larix*) in Asia and pine (*Pinus*) in Europe. WILDFIRES spread easily in these forests and are an essential element of forest REGENERATION. Birches (*Betula*), aspens and poplars (*Populus*) and alders (*Alnus*) play a role in the natural ECOLOGICAL SUCCESSION following fires and other disturbances, but they are also dominant in the extremely

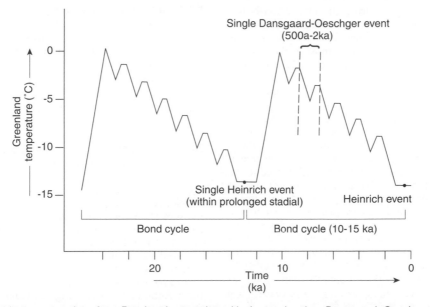

Bond cycle *Schematic representation of two Bond cycles together with shorter-duration, Dansgaard–Oescher events and infrequent but severe Heinrich events. The temperature scale is approximate mean annual temperature and the timescale, in thousands of years (ka), is arbitrary: zero years is not the present*

oceanic or extremely continental areas. Boreal forests are found where there is continuous snow cover in winter but not continuous permafrost. They are bordered to the south by TEMPERATE FORESTS and to the north by TUNDRA.

SPH

[See also BIOMES, FIRE IMPACTS: ECOLOGICAL, VEGETATION FORMATION-TYPE]

Engelmark, O. 1999: Boreal forest disturbances. In Walker, L.R. (ed.), *Ecosystems of disturbed ground* [*Ecosystems of the World*. Vol. 16]. Amsterdam: Elsevier, 161–186. **Hämet-Ahti, L.** 1981: The boreal zone and its biotic subdivision. *Fennia* 159, 69–75. **Heinselman, M.L. and Wright Jr, H.E. (eds)** 1973: The ecological role of fire in natural conifer forests of western and northern America. *Quaternary Research* 3(Special Issue), 317–482. **Kasischke, E.S. and Stocks, B.J. (eds)** 2000: *Fire, climate, and carbon cycling in the boreal forest*. Berlin: Springer. **Larsen, J.A.** 1980: *The boreal ecosystem*. New York: Academic Press. **Shugart, H.H., Leemanns, R. and Bonan, G.B. (eds)** 1992: *A systems analysis of the global boreal forest*. Cambridge: Cambridge University Press.

boreal forest: history

The history of the BOREAL FOREST has been traced primarily by POLLEN ANALYSIS. In Europe, the characteristic boreal trees, birch (*Betula*) and pine (*Pinus*), were the first ARBOREAL taxa to spread into the TUNDRA VEGETATION that colonised the land revealed by the retreating ice at the end of the LAST GLACIATION. These taxa dominated the forests of northern Europe during the Preboreal and Boreal CHRONOZONES and have remained dominant in northerly latitudes, where the short growing season precludes the establishment of THERMOPHILOUS forest. In North America, the same role was played by pine and spruce (*Picea*), which were later joined by fir (*Abies*). The role of spruce in European forests is still debated. There is some MACROFOSSIL evidence for individual trees being present at the treeline on the Scandes mountains early in the HOLOCENE, but spruce does not become an important forest component until much later, expanding westwards in Fennoscandia from 5000 radiocarbon years BP onwards. During the Holocene, boreal forests achieved their maximum northwards extent between 7000 and 5000 BP as evidenced by the presence of pine megafossils in the lakes of Fennoscandia far north of the present range of the species. After 5000 radiocarbon years BP, pine began to retreat southwards, reaching its present northern limit at around 3000 BP, at which time spruce reached its northern limit.

Over much of the boreal forest area human habitation is sparse and the economy is based on HUNTING, FISHING AND GATHERING or PASTORALISM, so that human impact has not changed the forest structure to the same degree as in the THERMOPHILOUS forests further south. In contrast, forest fires have been extensive and occurred at regular intervals. Where populations have practised agriculture, the *slash-and-burn* method has frequently been adopted and in regions where this continued for centuries and the burning cycle became very short, the forest has been drastically changed. In Europe, the damp richer soils occupied by spruce have been more subjected to slash-and-burn than the drier less fertile soils occupied by pine. *SPH*

[See also FIRE HISTORY, FIRE IMPACTS, HUMAN IMPACTS ON VEGETATION]

Eronen M. and Hyvärinen, H. 1982: Subfossil pine dates and pollen diagrams from northern Fennoscandia. *Geologiska Föreningens i Stockholm Förhandlingar* 103 (for 1981), 437–445. **Hicks, S.** 1993: Pollen evidence of localized impact on the vegetation of northernmost Finland by hunter–gatherers. *Vegetation History and Archaeobotany* 2, 137–144. **Mäkelä, E.** 1999: *The Holocene history of birch in northeastern Fennoscandia*. Helsinki: Helsinki University Press. **Solomon, A.M.** 1992: The nature and distribution of past, present and future boreal forests: lessons for a research and modeling agenda. In Shugart, H.H., Leemanns, R. and Bonan, G.B. (eds), *A systems analysis of the global boreal forest*. Cambridge: Cambridge University Press, 291–312.

borehole A hole drilled into soil, sediment or rock, usually with a mechanical drill, to provide subsurface information. Instruments can record data from the walls of the borehole, or a special drill-bit can be used to extract a CORE. *GO*

[See also CORER, DEEP-SEA DRILLING, ICE CORES, MARINE SEDIMENT CORE, NATURAL GAMMA RADIATION, SUBSURFACE TEMPERATURE]

bornhardt A large domed INSELBERG (isolated hill) produced by erosion. Bornhardts are found especially in granitic and gneissic regions and throughout a wide range of tropical climatic environments, from the HUMID TROPICS to SAVANNA areas. They may degenerate into residual *koppies* (see TOR) as a result of slope retreat or areal collapse. *MAC*

King, L.C. 1948: A theory of bornhardts. *Geographical Journal* 112, 83–87. **Twidale, C.R. and Bourne, J.A.** 1998: Multistage landform development, with particular reference to a cratonic bornhardt. *Geografiska Annaler* 80A, 79–94.

bottleneck A severe reduction in POPULATION numbers that is often associated with reduced genetic diversity and HETEROZYGOSITY and reduced adaptability of that population. *MVL*

[See also GENOTYPE]

bottom water Originating from the Southern Ocean, *Antarctic Bottom Water* (AABW) forms by a process of *brine exclusion* (see DOWNWELLING) during SEA ICE formation in the Weddell Sea and extends as far North as 40°N in the Atlantic. *WENA*

[See also DEEP WATER, WATER MASS]

boulder A SEDIMENT particle of GRAIN-SIZE greater than 256 mm. *TY*

boulder pavements In PERIGLACIAL areas, horizontal or gently sloping ground with closely packed stones (COBBLES and BOULDERS), the flat surfaces of which are uppermost; they are also termed *stone pavements*. Well developed examples are usually associated with SNOWBEDS/patches, although pavements are also found along shorelines. The stones may be of AUTOCHTHONOUS or ALLOCHTHONOUS derivation. Processes thought to be involved in pavement formation include: the upfreezing (FROST HEAVE) of stones; removal of fines by MELTWATER (SLOPEWASH) and/or DEFLATION; compression caused by the weight of the overlying SNOWBED or ICING together with ground saturation; and, along shorelines, ice-push and ice-pull processes associated with lake ice. Miniature examples of pavements have been termed *stone pockets*

(Jeffries, 1982: 21). Although widely distributed in the NIVAL ZONE, stone pavements are not necessarily of zonal significance and consequently fossil examples are of limited value for CLIMATIC RECONSTRUCTION. Boulder pavements may also form in the intertidal zone. *PW*

[See also BLOCKFIELDS, DESERT PAVEMENT, NIVATION]

Jeffries, M. 1982: Stone sorting phenomenon, Svartisen, Norway. *Geografiska Annaler* 64A, 21–24. **Hansom, J.D.** 1983: Ice-formed inter-tidal boulder pavements in the sub-Antarctic. *Journal of Sedimentary Petrology* 53, 135–145. **Hara, Y. and Thorn, C.E.** 1982: Preliminary quantitative study of alpine subnival boulder pavements, Colorado Front Range, U.S.A. *Arctic and Alpine Research* 14, 361–367. **Matthews, J.A., Dawson, A.G. and Shakesby, R.A.** 1986: Lake shoreline development, frost weathering and rock platform erosion in an alpine periglacial environment, Jotunheimen, southern Norway. *Boreas* 15, 33–50. **Washburn, A.L.** 1979: *Geocryology: a survey of periglacial processes and environments.* London: Edward Arnold. **White, S.E.** 1972: Alpine subnival boulder pavements in Colorado Front Range. *Geological Society of America Bulletin* 83, 195–200.

boundary conditions Initial conditions and external FORCING FACTORS that control or influence the operation of a SYSTEM. In the case of the global CLIMATIC SYSTEM, complex couplings exist, internal to the system, between its various components: atmosphere, hydrosphere, cryosphere, land surface and biosphere. These internal linkages are ultimately regulated by variations in external PARAMETERS, principally SOLAR RADIATION. Other parameters may act partly as external and partly as internal forcing factors: ocean basin shape, SALINITY, ATMOSPHERIC COMPOSITION, changes in land features, orography, vegetation, ALBEDO. Changes in boundary conditions force a system to adjust to a new EQUILIBRIUM state. Changes in solar radiation inputs to the climate system during the QUATERNARY, together with changes in global orography are now known to regulate the oscillating climates of GLACIAL–INTERGLACIAL CYCLES. Changes in boundary conditions may be mediated by a rearrangement of internal linkages (AUTO-VARIATION), such as alteration in the pattern of LONG WAVES.

The computer-based climatic simulations of GENERAL CIRCULATION MODELS (GCMs) require specification of boundary conditions such as ice sheet extent, land albedo, insolation, CO_2 and sea surface temperature to initiate and constrain model runs. These were often poorly parameterised by early models, especially with reference to ocean, land elevation and albedo. While these remain problematical, recent advances such as COUPLED OCEAN–ATMOSPHERE MODELS and nested models (which use the output from larger scale models to provide the boundary conditions for downscaled versions) have enabled a more realistic treatment of climate simulation both for past and possible future conditions. *JCS*

[See also CLIMATIC MODELS, PARAMETERISATION]

McGuffie, K. and Henderson-Sellers, A. 1997: *A climate modelling primer.* New York: Wiley. **Kondratyev, K. and Cracknell, A.** 1998: *Observing global climate change.* London: Taylor and Francis, 381–386.

boundary layer (1) In fluid mechanics, the zone at the junction between a fluid and a solid in which the velocity of the liquid is affected by motion of the solid. (2) In climatology, the lowest few hundred metres of the atmosphere, the properties of which (especially turbulence) are affected directly and rapidly by the properties of the Earth's surface. *JAM*

[See also MICROCLIMATE]

Brazel, A.J., Arnfield, A.J., Greenland, D.J. and Wilmott, C.J. 1991: Physical and boundary layer climatology. *Physical Geography* 12, 189–206. **Lenschow, D.H.** 1996: Planetary boundary layer. In Schneider, S.H. (ed.), *Encyclopedia of climate and weather.* New York: Oxford University Press, 590–593. **Oke, T.R.** 1987: *Boundary layer climates,* 2nd edn. London: Methuen.

bounding surface A prominent BEDDING surface in a succession of sediments or sedimentary rocks. In recent usage the term has been confined to surfaces that delineate genetically related packages. Some sediment bodies, particularly those characterised by CROSS-STRATIFICATION, contain a hierarchy of bounding surfaces that can be interpreted in terms of compound BEDFORMS, particularly in aeolian, fluvial and tidal deposits. Very large scale surfaces (major, first-order or SUPER BOUNDING SURFACES) are regional in extent and usually indicate a major environmental change. *MRT*

[See also FACIES ARCHITECTURE, SEDIMENTARY STRUCTURES, SEQUENCE STRATIGRAPHY]

Allen, J.R.L. 1983: Studies in fluviatile sedimentation: bars, bar-complexes and sandstone sheets (low-sinuosity braided streams) in the Brownstones (L. Devonian), Welsh Borders. *Sedimentary Geology* 33, 237–293. **Brookfield, M.E.** 1977: The origin of bounding surfaces in ancient aeolian sandstone. *Sedimentology* 24, 303–332. **Kocurek, G.A.** 1996: *Desert aeolian systems.* In Reading, H.G. (ed.), *Sedimentary environments: processes, facies and stratigraphy,* 3rd edn. Oxford: Blackwell, 125–153. **Miall, A.D.** 1987: Recent developments in the study of fluvial facies models. In Ethridge, F.G., Flores, R.M. and Harvey, M.D. (eds), *Recent developments in fluvial sedimentology* [*Special Publication* 39]. Tulsa, OK: Society of Economic Paleontologists and Mineralogists, 1–9. **Talbot, M.R.** 1985: Major bounding surfaces in aeolian sandstones – a climatic model. *Sedimentology* 32, 257–265.

brackish water Water with SALINITY intermediate between freshwater and seawater, commonly with a total dissolved solids concentration between 3000 and 20 000 mg L^{-1}. Brackish water is commonly found in ESTUARINE ENVIRONMENTS, in GROUNDWATER near coasts and in ARIDLANDS, where it may be the result of poor IRRIGATION practices. *MJB*

braiding A multichannel or multithread form of CHANNEL PATTERN, where the flow divides around active, low-relief CHANNEL BARS. Braiding is associated with high and variable discharges, high BEDLOAD transport rates, active BANK EROSION and steep channel slopes. It is found in high-energy river systems, especially in *proglacial* and SEMI-ARID environments. Occasional extreme floods are needed to maintain this inherently unstable channel pattern. *DML*

Ashworth, P.J. and Ferguson, R.I. 1986: Interrelationships of channel processes, changes and sediments in a proglacial braided river. *Geografiska Annaler* 68A, 361–371. **Best, J.L. and Bristow, C.S. (eds)** 1993: *Braided rivers* [*Special Publication* 75]. London: Geological Society. **Warburton, J. and Davies, T.** 1994: Variability of bedload transport and channel morphology in a braided river hydraulic model. *Earth Surface Processes and Landforms* 19, 403–421.

breakwater A barrier constructed from various materials (e.g. boulders, concrete), built in coastal locations to help prevent erosion and encourage sedimentation or to provide sheltered moorings. Breakwaters can be attached to the shore (e.g. Santa Barbara Harbor, California) or lie parallel in an offshore position. *RAS*

[See also COASTAL (SHORE) MANAGEMENT, COASTAL PROTECTION, WAVE ENERGY]

Bull, C.F., Davis, A.M. and Jones, R. 1998: The influence of fish-tail groynes (or breakwaters) on the characteristics of the adjacent beach at Llandudno, North Wales. *Journal of Coastal Research* 14, 93–105. **Komar, P.D.** 1983: Coastal erosion in response to the construction of jetties and breakwaters. In Komar, P.D. (ed.), *Handbook of coastal processes and erosion*. Boca Raton, FL: CRC Press, 191–204.

breccia A clastic SEDIMENTARY ROCK with a GRAIN-SIZE greater than 2 mm (a RUDITE or lithified GRAVEL) and angular CLASTS. *TY*

[See also CLASTIC CONGLOMERATE]

brickearth A deposit of the English chalklands, associated with COOMBE ROCK. It commonly comprises a matrix of silt-sized quartz grains and chalk pellets and is widely regarded as mainly LOESS reworked by SOLIFLUCTION. *RAS*

[See also AEOLIAN, PERIGLACIAL]

bristlecone pine The very long-lived five-needle pines (made up of the two species *Pinus aristata* and *P. longaeva*) that grow at high elevation in the western USA. These trees are renowned as the oldest living organisms, with living individuals proven, through *dendrochronological dating* (see DENDROCHRONOLOGY), to be greater than 4700 years old. By CROSS-DATING series of ring-width measurements from living trees with those from long dead trees, it has been possible to extend the continuous record of annual growth variations for bristlecone back more than 7000 years. It was the availability of absolutely dated bristlecone wood (along with oak samples from Europe) that enabled the CALIBRATION OF RADIOCARBON DATES, and ring-width and CARBON ISOTOPE data from bristlecones have been used in many studies of ENVIRONMENTAL CHANGE, particularly CLIMATIC RECONSTRUCTION over recent millennia. *KRB*

Ferguson, C.W. 1970: Dendrochronology of bristlecone pine, *Pinus aristata*: establishment of a 7484-year chronology in the White Mountains of eastern-central California, U.S.A. In Olsson, I.U. (ed.), *Radiocarbon variations and absolute chronology*. New York: Wiley, 237–259. **Graybill, D.A. and Idso, S.B.** 1993: Detecting the aerial fertilization effect of atmospheric CO_2 enrichment in tree-ring chronologies. *Global Biogeochemical Cycles* 7, 81–95. **LaMarche Jr, V.C.** 1974: Palaeoclimatic inferences from long tree-ring records. *Science* 183, 1043–1048. **Schulman, E.** 1958: Bristlecone pine, oldest known living thing. *National Geographic Magazine* 113, 345–372.

Bronze Age: landscape impacts The Bronze Age was the period between the Copper Age and the Iron Age, in which bronze, an alloy of copper and tin, became the primary material for the manufacture of tools and weapons. The Bronze Age started in Eurasia at around 5000 calendar years BP and the greater social order of the period led to the first civilisations and the earliest written records. The Bronze Age did not occur in the Americas.

In many areas of Europe, the remains of Bronze Age FIELD SYSTEMS are apparent in the modern landscape. These were constructed to regulate increased levels of agricultural activity. POLLEN ANALYSIS from Bronze Age mining sites in the British Isles has demonstrated that FOREST CLEARANCE was directly associated with bronze production and increased agricultural activity. The landscape instability resulting from forest clearance subsequently instigated the growth of BLANKET MIRES in upland areas. *ARG*

[See also COPPER AGE: LANDSCAPE IMPACTS, IRON AGE: LANDSCAPE IMPACTS, NEOLITHIC: LANDSCAPE IMPACTS]

Barrett, J.C. 1999: Rethinking the Bronze Age environment. *Quaternary Proceedings* 7, 493–500. **Buckland, P.C., Dugmore, A.J. and Edwards, K.J.** 1997: Bronze Age myths? Volcanic activity and human response in Mediterranean and North Atlantic regions. *Antiquity* 71, 581–593. **Cole, J.M. and Harding, A.F.** 1979: *The Bronze Age in Europe*. London: Methuen. **Fleming, A.** 1988: *The Dartmoor reaves: investigating prehistoric land divisions*. London: Batsford. **Mighall, T. and Chambers, F.M.** 1993: The environmental impact of prehistoric mining at Copa Hill, Cwmystwyth, Wales. *The Holocene* 3, 260–269. **Moore, P.D.** 1993: The origin of blanket mire, revisited. In Chambers, F.M. (ed.), *Climate change and human impact on the landscape*. London: Chapman and Hall, 217–224. **Willis, K.J., Sümegi, P. Braun, M. et al.** 1997: Prehistoric land degradation in Hungary: who, how and why? *Antiquity* 72, 101–113.

Brown Earth A soil type, occurring in nutrient-rich, mainly DECIDUOUS forest areas in TEMPERATE to semi-humid/SEMI-ARID climate regions of the Northern Hemisphere. The SOIL PROFILE is characterised by a thin litter horizon, a thick MULL horizon (a few cm to greater than a metre) and an underlying mineral soil, with a pH of c. 6–7. The mull horizon is weathered and leached (see WEATHERING and LEACHING) of, among others, iron and aluminium and its HUMUS content decreases with depth. A Brown Earth tends to be rich in earthworms and burrowing organisms, which effectively mix the ORGANIC and INORGANIC matter. The soil is therefore well mixed and oxygenated. Since Brown Earths occur over large areas with generally different climatic regimes, many different variants have been recognised, which has led to confusion regarding their classification. In the modern classification systems, e.g. the FAO (Food and Agriculture Organization) system, the different soil types have been renamed and are now classified within other soil types, for example, CAMBISOLS. *UBW*

[See also SOIL CLASSIFICATION, WORLD REFERENCE BASE FOR SOIL RESOURCES]

Andersen, S.T. 1979: Brown earth and podzol: soil genesis illuminated by microfossil analysis. *Boreas* 8, 59–73. **Andersen, S.T.** 1986: Palaeoecological studies of terrestrial soils. In Berglund, B.E. (ed.), *Handbook of Holocene palaeoecology and palaeohydrology*. Chichester: Wiley, 165–177. **Kubiena, W.L.** 1953. *The soils of Europe*. London: Thomas Murdy. **Stålfelt, M.G.** 1972: *Stålfelt's plant ecology: plants, the soil and man*, translated by M.S. Jarvis and P.G. Jarvis. London: Longman.

Brückner cycle One of the earliest climatic PERIODICITIES to be alleged. Recognised by E. Brückner in AD 1890, this 35-year cycle of alternating cool–damp and warm–dry periods was supposed to have occurred throughout north-western Europe during at least the eighteenth and nineteenth centuries. Although statistically inconclusive, recent advances in FILTERING techniques have hinted at potential connections between SOLAR CYCLES/lunar influences and climatic periodicities of this nature. *JCS*

Fairbridge, R.W. 1987: Brückner Cycle. In Oliver, J.E. and Fairbridge, R.W. (eds), *The encyclopedia of climatology*. New York: Van Nostrand Reinhold, 184.

Bruun Rule
A model proposed by P. Bruun in 1962, in which a beach profile remains in equilibrium during SEA-LEVEL RISE (see Figure). Sand eroded from the foreshore is deposited on the nearshore bottom, raising it by an amount equal to the rise in sea level. The rule is thought to apply where limitless supplies of sand are available. The significance for beaches under accelerated sea-level rise due to GLOBAL WARMING is that, according to the model, lateral retreat of beaches would be some one hundred times that of the vertical extent of sea-level rise. Thus, even a modest rise would mean substantial narrowing or even disappearance of beaches. Doubts have, however, been expressed about the applicability of this essentially two-dimensional model. *HJW*

Bird, E.C.F. 1993: *Submerging coasts*. Chichester: Wiley. **Bruun, P.** 1988: The Bruun Rule of erosion by sea-level rise – a discussion of large-scale two-dimensional and 3-dimensional usages. *Journal of Coastal Research* 4, 627–648. **Schwartz, M. L.** 1967: The Bruun theory of sea-level rise as a cause of shore erosion. *Journal of Geology* 75, 76–92.

bryophyte analysis
The identification of remains of bryophytes (macroscopic and SPORES) in SEDIMENTS. Bryophyte spores (in contrast to fern spores or pollen) are rather difficult to identify, even if well preserved, because they tend to be small and lack distinguishing features. Notable exceptions are the spores of *Sphagnum*, routinely recorded during POLLEN ANALYSIS, and spores of hornworts (*Anthoceros* and *Phaeoceros*). Macroscopic remains are readily extracted by careful sieving and may then be kept in liquid preservatives or mounted on microscope slides. Quaternary remains may be readily identified using standard bryophtye floras and reference material in a bryophyte herbarium. Bryophyte remains have been used extensively in understanding the development of *peatlands* (which may be composed largely of *Sphagnum* species) and in understanding details of past micro-environments.

Mosses are much more abundant than liverworts in the fossil record. *KDB*

Dickson, J. H. 1973: *Bryophytes of the Pleistocene: the British record and its chronological and ecological implications*. London: Cambridge University Press. **Dickson, J. H.** 1986: Bryophyte analysis. In Berglund, B.E. (ed.), *Handbook of Holocene palaeoecology and palaeohydrology*. Chichester: Wiley, 627–643.

buffer zone
A zone surrounding and separating a CONSERVATION area from a developed area, which helps to protect the biotic community from change. Buffer zones are used extensively in BIOSPHERE RESERVES. The ecology of the concept has not been researched extensively. *IFS*

[See also CONSERVATION BIOLOGY, PROTECTED AREAS]

buffer zone in GIS
An area of specified width around an ENTITY, commonly used in a GEOGRAPHICAL INFORMATION SYSTEM (GIS). This zone is essentially a polygon and is used in *proximity analysis* to determine, for example, the number of trees within 50 m of a logging road. *TVM*

[See also POLYGON ANALYSIS]

buffering
The resistance of a soil or water body to changes in pH. In soil, for example, BASES in solution and on ion-exchange sites (if available) act to neutralise ACID inputs and prevent ACIDIFICATION. Similarly, naturally occurring buffering agents, such as LIMESTONE, reduce the environmental impact of ACID RAIN on lake ecosystems. *ADT*

[See also ACID PRECIPITATION, WATER POLLUTION]

bulk density
The apparent density of a soil or sediment calculated from the oven-dried (105°C) mass divided by the volume occupied in the field. Loamy mineral soils normally have a bulk density of 1.3 Mg m^{-3}, sandy subsoils 1.6 Mg m^{-3}, indurated horizons and compacted soils 1.9 Mg m^{-3}, but volcanic soils (ANDOSOLS) may have a bulk density of only 0.5 Mg m^{-3}. Bulk density measurements are necessary in the estimation of rates of soil formation as well as SEDIMENT ACCUMULATION RATES and EROSION RATES. *EMB*

[See also PHYSICAL ANALYSIS OF SOILS AND SEDIMENTS]

McRae, S.G. 1988: *Practical pedology: studying soils in the field*. Chichester: Horwood/Wiley.

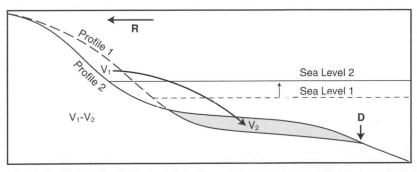

Bruun Rule *The Bruun rule: following sea-level rise, coastline retreat (R) results from sediment transfer from the shoreface sediment store (V1) to the nearshore zone (V2) to re-establish an equilibrium water depth as far as the closure depth (D), beyond which there is no profile adjustment (after Bird, 1993)*

buried soil A soil covered by later sedimentation from, for example, an alluvial, colluvial, aeolian or volcanic source, usually to a depth greater than 50 cm. *EMB*

[See also PALAEOSOL]

bush encroachment The spread of woody vegetation on sites cleared for agricultural use. It may be accentuated by AGRICULTURAL ABANDONMENT or by a reduction in the intensity of LANDUSE. *JLI*

[See also ECOLOGICAL SUCCESSION]

bush meat Wild animals used for food on a commercial, semi-commercial or subsistence basis, especially in the tropics. Small-scale hunting of wild animals as a food supplement has been a widespread and sustainable practice. The increased accessibility provided by LOGGING roads in TROPICAL AND SUBTROPICAL FORESTS, however, has led to a major increase in the hunting of wild animals, especially the larger vertebrates such as chimpanzee and gorilla, to the extent that their survival is threatened. Consequently, bush meat has recently become a major CONSERVATION issue, especially in tropical Africa. *JAM*

Moore, P.D. 2001: The rising cost of bushmeat. *Nature* 409, 775–776.

business-as-usual scenario One of several scenarios often used in considering likely future environmental change, especially CLIMATIC CHANGE. It implies predictions based on no change in ENVIRONMENTAL POLICY, specifically that no policies that limit emissions of GREENHOUSE GASES will be undertaken. The projection of future levels of these gases is fraught with enormous levels of uncertainty, since future emission levels depend on a wide range of economic, technical and political considerations, which are unpredictable. However, with increasing realisation of the likely detrimental environmental effects if business-as-usual scenarios are actually played out in the future, more environment-friendly scenarios (i.e. those based on various levels of reduced emissions) may be regarded as increasingly realistic. *AHP*

[See also CONVENTIONS, EMISSION CONTROL, ENVIRONMENTAL MANAGEMENT, INTERGOVERNMENTAL PANEL ON CLIMATE CHANGE]

Houghton, J.T., Jenkins, J.H. and Ephraums, J.J. (eds) 1990: *Climate change: the IPCC scientific assessment.* Cambridge: Cambridge University Press.

C

C-3, C-4 and CAM plants

C-3, C-4 and CAM plants   Plants that use the three carbon fixation pathways in plant PHOTOSYNTHESIS defined by the first products of their photosynthetic pathways. The advantages and disadvantages associated with these three different photosynthetic pathways are outlined in the Table. Also shown are their evolutionary and ecological history, their current distributions, and their speculated future survival.   *JCMc*

[See also CARBON CYCLE, CARBON DIOXIDE, GRASS CUTICLE ANALYSIS, RESPIRATION]

Cerling, T.E., Ehleringer, J.R. and Harris, J.M. 1998: Carbon dioxide starvation, the development of C4 ecosystems and mammalian evolution. *Philosophical Transactions of the Royal Society of London* 353, 159–171. Ehleringer, J.R., Sage, R.F., Flanagan, L.B. and Pearcy, R.W. 1991 Climate change and the evolution of C4 photosynthesis. *Trends in Ecology and*

C-3, C-4 and CAM plants   *Characteristics of C-3, C-4 and CAM plants*

Characteristics	C-3	C-4	CAM
Evolved (unequivocal evidence)	Mid-Silurian (400 Ma)	Mid-Miocene (11 Ma)	Quaternary
Evolved (equivocal evidence/ speculation)		Late Carboniferous (300 Ma) and Mid Cretaceous to Mid-Miocene (120–11 Ma)	Late Carboniferous (300 Ma) and Mid-Cretaceous to Holocene
Primary CO_2 acceptor	RuBP (ribulose-1,5-bisphosphate)	PEP (phosphenol pyruvate)	In light RuBP In dark PEP
First product of photosynthesis	3-Carbon acids: Phosphoglycerate (PGA)	4-Carbon acids: Oxaloacetate, Malate, Aspartate	In light PGA In dark Malate
Photosynthesis depression by O_2	Yes	No	No
CO_2-concentrating mechanism to avoid/reduce photorespiration	None. Rubisco, which catalyses the carboxylation of RUBP during photosynthesis, also catalyses the oxygenation of RUBP in the process of photorespiration, resulting in CO_2 release	10- to 20-fold elevation of CO_2 at Rubisco binding site from breakdown of C-4 acids (produced in the mesophyll cells from carboxylation of PEP) in spatially separated bundle sheath cells	Breakdown of Malate in the day to elevate CO_2 at Rubisco binding site. Malate is produced at night and stored in vacuoles
CO_2 release in light (photorespiration)	Yes	No	No
Net photosynthetic capacity	Slight to high	High to very high	In light: slight In dark: medium
Light saturation of photosynthesis	At intermediate irradiance	No saturation even at highest irradiance	At intermediate to high irradiance
Water use efficiency (WUE)	Low to moderate	High	Very high
Current distribution	Dominate most terrestrial ecosystems, especially cool climates	Dominate warm to hot open ecosystems. Tropical and temperate grasslands	Xeric ecosystems (deserts, epiphytic habit)
Composition of global flora	c. 85%	c. < 5%	c. 10%
Limits to distribution	High light and temperatures, low water availability	Low temperatures, very low water availability	Competition from C-3 and C-4 plants
Speculated future effects of elevated CO_2?	C-3 plants will have competitive advantage, leading to expansion of their ranges	C-4 plants will be less favoured as C-4 photosynthesis is saturated at current ambient CO_2 concentration	
Speculated future effects of elevated CO_2 plus global warming?	Contraction of C-3 plant range during drought due to competition from C-4 plants	Increased WUE conferring competitive advantage to C-4, leading to expansion of range during drought	

Evolution 6, 95–99. **Owensby, C.E., Ham, J.M., Knapp, A.** *et al.* 1996: Ecosystem-level responses of tallgrass prairie to elevated CO$_2$. In Koch, G.W. and Mooney, H.A. editors, *Carbon dioxide and terrestrial ecosystems*. San Diego, CA: Academic Press. **Raven, J.A. and Spicer, R.A.** 1996: The evolution of Crassulacean Acid Metabolism. In Winter Smith (eds), *Crassulacean acid metabolism* [*Ecological Studies* 114]. Berlin: Springer. **Sage, R.F. and Monson, R.K.** 1999: *C4 plant biology*. London: Academic Press.

cadmium (Cd)

A HEAVY METAL and a TRACE ELEMENT that is biologically non-essential. It is nevertheless available in soils, having been derived naturally from rocks and artificially via zinc mining, FERTILISERS, industrial EMISSIONS and sewage. An estimated 70% of environmental Cd has been produced anthropogenically over the last 30 years. Amongst other HUMAN HEALTH problems, it can cause kidney damage because it can displace zinc atoms in some enzymes. *JAM*

[See also METAL POLLUTION]

Campbell, I.T., Brand, U. and Morrison, J.O. 1999. Cadmium pollution and toxicity. In Alexander, D.E. and Fairbridge, R.W. (eds), *Encyclopaedia of Environmental Science*. Dordrecht: Kluwer, 69. **Nriagu, J.O. (ed.)** 1980: *Cadmium in the environment*. New York: Wiley.

cadmium : calcium ratio (Cd : Ca)

A measure of *palaeoproductivity* in the oceans derived from analyses of the ratio in the tests of BENTHIC FORAMINIFERA. Estimation of Cd:Ca ratios has allowed reconstruction of changing nutrient supply, especially as it relates to the formation of *North Atlantic Deep Water*. *CJC*

Boyle, E.A. 1988: Cadmium chemical tracer of deepwater. *Paleoceanography* 3, 471–489.

caesium-137

A radioactive ISOTOPE produced in the RADIOACTIVE FALLOUT from atmospheric testing of nuclear weapons since AD 1954 and used to assess the amount and patterns of SOIL EROSION since that time. In addition, the ratios of ^{137}Cs to ^{134}Cs can be used to assess EROSION RATES of soil in parts of Europe since the Chernobyl NUCLEAR ACCIDENT in April 1986. From the nuclear weapons testing era, ^{137}Cs has been distributed globally in the STRATOSPHERE and deposited on the Earth's surface in the rainfall (see WET DEPOSITION). Regionally, deposition is variable, but at the field scale it is reasonably uniform. The isotope has a HALF LIFE of 30 years and is well adsorbed to clay soil particles.

The principle of assessing soil erosion using this TRACER is relatively simple (see Figure). Comparatively undisturbed soils (e.g. in woodland or grassland) are used as reference sites to determine the variation in ^{137}Cs concentration throughout the soil profile and to derive a reference *loading* of ^{137}Cs for non-eroded sites. The isotope tends to be concentrated near the surface and total loading is moderately high in such locations. Some downward movement of ^{137}Cs occurs through biological disturbance. On eroded slopes, the isotope is typically more evenly spread throughout the soil profile, because of disturbance of the existing soil profile and some additions of soil material from upslope with variable concentrations of ^{137}Cs. At depositional sites, the total loading of ^{137}Cs is typically much higher than for either an undisturbed or eroded site (because of the net addition from upslope of eroded soil containing the tracer from upslope). In addition, the depth over which contaminated soil occurs is much greater than at either eroded or undisturbed sites. Where changes in isotope loading can be correlated with measured SEDIMENT YIELDS (e.g. using EROSION PLOTS), the method can be used to estimate medium-term WATER EROSION rates. Alternatively, a simple model can be applied in which it is assumed that net SOIL LOSS is directly proportional to the percentage loss of ^{137}Cs.

The technique has been used to gain an understanding of the spatial patterns of erosion on agricultural land, as well as estimates of total erosion at the field scale. It has been used also to gain a better understanding of the SEDIMENT BUDGET at the drainage basin scale, which is difficult using more traditional techniques. *RAS*

[See also AGRICULTURAL IMPACTS ON SOILS]

Branca, M. and Voltaggio, M. 1993: Erosion rate in badlands of central Italy – estimation by radiocesium isotope ratios from the Chernobyl nuclear accident. *Applied Geochemistry* 8, 437–445. **Chappell, A.** 1999: The limitations of using Cs-137 for estimating soil redistribution in semi-arid environments. *Geomorphology* 29, 135–152. **Quine, T.A.** 1999: Use of caesium-137 data for validation of spatially distributed erosion models: the implications of tillage erosion. *Catena* 37, 415–430. **Walling, D.E.** 1999: Linking land use, erosion and sediment yields in river basins. *Hydrobiologia* 410, 223–240. **Walling, D.E. and Quine, T.A.** 1990: Use of caesium-137 to investigate patterns and rates of soil erosion on arable fields. In Boardman, J., Foster, I.D.L. and Dearing, J.A. (eds), *Soil erosion on agricultural land*. Chichester: Wiley, 33–53.

calcic horizon

A diagnostic subsurface soil horizon with distinct calcium carbonate enrichment, at least 15 cm thick containing 15% (< 150 g kg^{-1}) or more CaCO$_3$ equivalent and at least 5% more (50 g kg^{-1}) than a horizon below. It is an important feature of soils in SEMI-ARID regions where calcareous DUST is washed into the soil in the rainy season and the calcium precipitated as carbonate during the dry season. Precipitation may also occur in the *capillary fringe* above the groundwater table. When continuously cemented, massive or indurated with secondary carbonate, it is a *petrocalcic horizon*. Erosion of the friable overlying soil to leave the calcic or petrocalcic horizon at the surface is a feature of DESERTIFICATION. *EMB*

[See also CALCISOL, CALCRETE]

Food and Agriculture Organization of the United Nations 1998: *World reference base for soil resources* [*Soil Resource Report* 84]. Rome: FAO, ISRIC, ISSS.

Calcisols

Soils with accumulation of secondary calcium carbonate in a CALCIC HORIZON or CALCRETE, characteristic of semi-arid areas of North America, North Africa, the Middle East and Australia (SOIL TAXONOMY: *Calciorthids*). *EMB*

[See also WORLD REFERENCE BASE FOR SOIL RESOURCES]

calcium-41

A *cosmogenic nuclide* with a HALF LIFE of 103×10^3 years formed in rocks by the thermal neutron capture of ^{40}Ca. Although it has potential as a SURFACE DATING technique, there are still a number of difficulties in its application, particularly the very low ^{41}Ca:^{40}Ca ratio in most rocks. *CJC*

[See also COSMOGENIC-ISOTOPE DATING]

Fink, D., Klein, J. and Middleton, R. 1990: ^{41}Ca: past, present and future. *Nuclear Instruments and Methods in Physics Research* B52, 1649–1662.

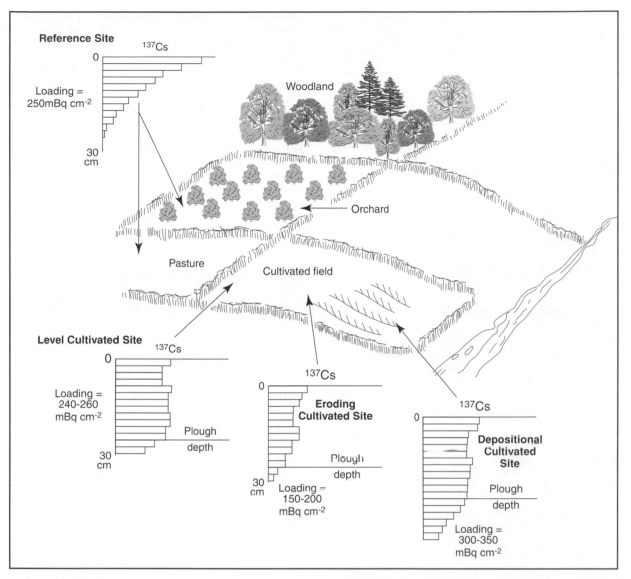

caesium-137 *Schematic representation of the effects of various agricultural practices on the loading and distribution within the soil profile of caesium-137 (after Walling and Quine 1990)*

calcrete   The accumulation of secondary calcium or magnesium carbonate as a hardened layer, normally within the soil or weathering zone and above the WATER TABLE, in SEMI-ARID regions. Calcretes are important as DURICRUSTS and PALAEOSOLS. Recently termed a *petrocalcic horizon* but also known as *caliche* and *kunkar,* calcrete may have a nodular or massive appearance and older deposits become capped by a laminar form, composed entirely of AUTHIGENIC carbonate. Calcrete does not slake in water and in the petrocalcic form cannot easily be penetrated by auger or spade.

Calcretes are widespread in semi-arid regions where mean annual precipitation is 200–600 mm – enough to enable upward movement of soil moisture by CAPILLARY ACTION without the loss of soluble minerals through LEACHING. Such *pedogenic calcretes* typically form close to the land surface where the carbonate may originate as dust

washed into the soil or be derived from calcium-rich groundwater. *Groundwater calcretes* may form at depths of several metres by direct CHEMICAL PRECIPITATION from GROUNDWATER. Calcretes associated with older geomorphic surfaces progressively become thicker with increasing age and may attain a thickness of several tens of metres. On stable sites in the New Mexico desert, calcretes dating back to early Pleistocene have accumulated up to 1840 kg m^{-2} at a rate of 2 to 3 kg m^{-2} per 1000 years.

EMB/MAC

[See also CALCIC HORIZON]

Gile, L.H., Hawley, J.W. and Grossman, R.B. 1981: *Soils and geomorphology in the Basin and Range area of southern New Mexico – Guidebook to the Desert Project* [*Memoir* 39]. Socorro: New Mexico Institute of Mining and Technology. **Goudie, A.S.** 1983: Calcrete. In Goudie, A.S. and Pye, K. (eds), *Chemical sed-*

iments and geomorphology: precipitates and residua in the near-surface environment. London: Academic Press, 93–131. **Nash, D.J. and Smith, R.F.** 1998: Multiple calcrete profiles in the Tabernas Basin, southeast Spain: their origins and geomorphological implications. *Earth Surface Processes and Landforms* 23,1009–1029. **Wright, V.P. and Tucker, M.E. (eds)** 1991: *Calcretes.* Oxford: Blackwell Scientific.

caldera A large CRATER (typically 5–20 km diameter) formed by the collapse of a VOLCANO following withdrawal of MAGMA from an underlying storage chamber. Caldera formation has accompanied many of the largest explosive VOLCANIC ERUPTIONS in the GEOLOGICAL RECORD, such as Krakatau (AD 1883) and Toba (74 000 years ago), both in Indonesia. The Toba eruption may have accelerated the LAST GLACIATION.   *JBH*

Francis, P. 1983: Giant volcanic calderas. *Scientific American* 248, 60–70. **Ninkovich, D., Sparks, R.S.J. and Ledbetter, M.J.** 1978: The exceptional magnitude and intensity of the Toba eruption, Sumatra: an example of the use of deep-sea tephra layers as a geological tool. *Bulletin of Volcanology* 41, 286–298.

calendar/calendrical age/date An age estimate or date based on the calendar system of years in relation to archaeological, historical or astronomical events; for example, calendar ages before or after Christ are quoted in years BC or AD respectively. CALIBRATION OF RADIOCARBON DATES converts ages to estimates of *calendrical, sidereal* or *solar years.*   *DAR*

[See also CONVENTIONAL RADIOCARBON AGE/DATE, SIDEREAL DATE]

calibrated-age dating Age estimation or *age calibration* where measurements of systematic changes resulting from exposure to environmental variables, such as climate and lithology, are calibrated by independent chronological control. The principal aim is to convert these changes into a reproducible form, with minimal error terms, to allow dating of surfaces or materials of unknown age. The term should not be confused with the term 'calibrated' as used, for example, in CALIBRATION OF RADIOCARBON DATES, although the ultimate aim of any dating technique should be to produce ages that as closely as possible estimate CALENDRICAL AGE.   *DAR/CJC*

Colman, S.M., Pierce, K.L. and Birkeland, P.W. 1987: Suggested terminology for Quaternary dating methods. *Quaternary Research*, 28, 314–319.

calibration The statistical operation of fitting a regression model to obtain a TRANSFER FUNCTION that may be used to provide estimates of one or more *predictand variables* (dependent VARIABLES) as a function of the variability over time in one or more *predictors* (independent variables). Various methods may be used, but the most common approach is to use linear LEAST-SQUARES REGRESSION.   *KRB*

[See also CROSS-CALIBRATION, VALIDATION]

Fritts, H.C., Guiot, J., Gordon, G.A. and Schweingruber, F.H. 1990: Methods of calibration, verification and reconstruction. In Cook, E.R. and Kairiukstis, L.A. (eds), *Methods of dendrochronology: applications in environmental sciences.* Dordrecht: Kluwer, 163–217.

calibration of radiocarbon dates The conversion of CONVENTIONAL RADIOCARBON DATES (AGES/DATES) into CALENDAR DATES (AGES/DATES) by the use of computer programs and a calibration dataset. The calibration curve displays short-term variations or 'wiggles', which in many instances have the undesired effect of a radiocarbon date having more than one equivalent calendar date, and also complicate the interpretation of the quoted PRECISION of a date. The Figure shows on the *y* axis the normal probability distribution of a radiocarbon date of 650 ± 40 BP between $\pm 2\sigma$ limits, and on the *x* axis the calibrated date PROBABILITY DISTRIBUTION. Arranged diagonally and linked with straight lines, individual dates from the INTCAL98 dataset are shown with $\pm 1\sigma$ error bars. Calibration in this case results in two nearly distinct ranges, which cannot be differentiated unless further analysis is carried out using techniques such as methods of BAYESIAN STATISTICS or curve fitting.   *PQD*

[See also WIGGLE MATCHING]

Bard, E., Hamelin, B., Fairbanks, R.G. and Zinder, A. 1990: Calibration of the ^{14}C timescale over the past 30,000 years using mass spectrometric U–Th ages from Barbados corals. *Nature* 345, 405–440. **Buck, C.E., Cavanagh, W.G. and Litton, C.D.** 1996: *Bayesian approach to interpreting archaeological data.* Chichester: Wiley. **Long, A. (ed.)** 1998: Calibration issue. *Radiocarbon* 40(3). **Pearson, G.W.** 1986: Precise calendrical dating of known growth-period samples using a 'curve fitting' technique. *Radiocarbon* 28, 292–299. **Stuiver, M., Reimer, P.J., Bard, E.** *et al.* 1998: INTCAL98: Radiocarbon age calibration, 24000–0 cal BP. *Radiocarbon* 40, 1041–1083.

calving The process of detachment of ice masses from a GLACIER into water. The resulting ICEBERGS may range from small so-called '*bergy bits*' a few metres across to large *tabular icebergs* covering thousands of square kilometres.   *MJH*

[See also HEINRICH EVENTS, ICE-RAFTED DEBRIS]

Kirkbride, M.P. 1993: The temporal significance of transitions from melting to calving termini at glaciers in the central Southern Alps of New Zealand. *The Holocene* 3, 232–240. **Kristensen, M.**

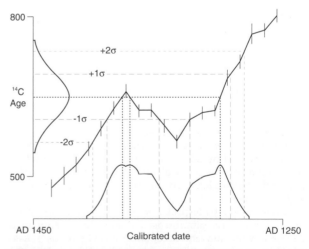

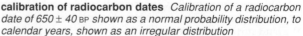

calibration of radiocarbon dates *Calibration of a radiocarbon date of 650 ± 40 BP shown as a normal probability distribution, to calendar years, shown as an irregular distribution*

1983: Iceberg calving and deterioration in Antarctica. *Progress in Physical Geography* 7, 313–328.

calving glaciers
Glaciers which terminate in water, with a significant amount of ABLATION by CALVING. Calving glaciers may terminate in a floating ICE SHELF or a non-floating grounded margin. Glaciers with non-floating, marine termini are known as *tidewater glaciers*. Calving glaciers commonly exhibit non-linear responses to *climatic forcing*. At times of positive MASS BALANCE, glacier growth may be limited or counteracted by calving losses and ice-front advances may depend on the build up of shoals of sediment at the margin, reducing water depth and calving rates. Rapid glacier retreat can occur if a glacier front withdraws into deep water, so that retreat rates may be independent of climatic triggers. Many glaciers in high latitudes terminate in water, including 77% of the Antarctic ice sheet. *DIB*

[See also ICEBERGS, ICE-RAFTED DEBRIS]

Brown C.S., Meier M.F. and Post A. 1982: Calving speed of Alaskan tidewater glaciers with applications to the Columbia Glacier, Alaska. *United States Geological Survey Professional Paper* 1258-C. **Warren, C.R.** 1992: Iceberg calving and the glacioclimatic record. *Progress in Physical Geography* 16, 253–282. **Warren, C.R., Glasser, N.F., Harrison, S.** *et al.* 1995: Characteristics of tide-water calving at Glaciar San Rafael, Chile. *Journal of Glaciology* 41, 273–289.

cambering
The movement of escarpment caprocks towards the valley axis. Cambering results from the DEFORMATION of underlying softer strata by the weight of the caprock. *PW*

[See also GULL, ROCK CREEP, VALLEY BULGING]

Hutchinson, J.N. 1992: Engineering in relict periglacial and extraglacial areas in Britain. *Quaternary Proceedings* 2, 49–65.

cambic horizon
A diagnostic mineral subsoil horizon of CAMBISOLS, recognised by differences in colour and structure from the horizon above and the PARENT MATERIAL below. It is formed by alteration relative to the underlying horizons so that rock structure is absent from at least half the volume, but it should contain more than 10% of weatherable minerals and have a CATION EXCHANGE CAPACITY of more than 16 $cmol_c kg^{-1}$ clay. LEACHING of constituents removes carbonates and CHEMICAL WEATHERING of iron minerals gives a stronger colour. The cambic horizon should have a thickness of at least 15 cm with its base at least 25 cm below the soil surface. *EMB*

Food and Agriculture Organization of the United Nations 1998: *World reference base for soil resources* [Soil Resource Report 84]. Rome: FAO, ISRIC, ISSS.

Cambisols
Weakly to moderately developed soils present in all continents characterised by a CAMBIC HORIZON. Under climatic change with increased leaching or introduction of acidophyllous vegetation, Cambisols may progress towards PODZOLS by development of ALBIC and SPODIC HORIZONS. Changes to wetter conditions could lead to the development of GLEYSOLS. Drier conditions on desert margins could result in carbonate or gypsum accumulation and a change towards CALCISOLS or GYPSISOLS. Cambisols possess a moderate to high natural fertility and make good agricultural lands (SOIL TAXONOMY: Eutrochrepts, Dystrochrepts). *EMB*

[See also BROWN EARTH, WORLD REFERENCE BASE FOR SOIL RESOURCES]

Food and Agriculture Organization of the United Nations, 1993: *World soil resources: an explanatory note on the FAO World Soil Resources Map at 1:25,000,000 scale* [*World Soil Resources Report* 66 Rev.1]. Rome: FAO.

cambium
The relatively thin layer of *meristem* (dividing) cells that lie just below the bark in the stems and roots of woody plants. These cells are capable of division to form either cork, phloem or xylem cells. In trees, the *phloem* cells, formed in the outside part of the vascular cambium, conduct food, whereas the *xylem*, formed on the inside, comprises the water-carrying cells. It is the formation of these xylem cells from year to year that makes up the wood of the tree and provides the basis for DENDROCHRONOLOGY in many subtropical regions of the world. *KRB*

Kozlowski, T.T. 1971: *Growth and development of trees*. II. *Cambial growth, root growth and reproductive growth*. New York: Academic Press. **Zimmermann, M.H. (ed.)** 1964: *The formation of wood in forest trees*. New York: Academic Press.

Cambrian
A SYSTEM of rocks and a PERIOD of geological time from 544 to 495 million years ago. Metazoan life first evolved exoskeletons in the early Cambrian (see CAMBRIAN EXPLOSION) and the first vertebrates appeared. *GO*

[See also GEOLOGICAL TIMESCALE]

Brasier, M.D. 1992: Background to the Cambrian explosion [Introduction to Special Issue, pp. 585–668]. *Journal of the Geological Society, London* 149, 585–587. **Tucker, M.E.** 1992: The Precambrian–Cambrian boundary: seawater chemistry, ocean circulation and nutrient supply in metazoan evolution, extinction and biomineralization. *Journal of the Geological Society, London* 149, 665–668.

Cambrian explosion
An informal description for the sudden diversification (*evolutionary radiation*) of multicellular life forms at the beginning of PHANEROZOIC time. It is also known as the *Cambrian radiation*. It is unclear whether environmental or genetic factors were responsible for this critical episode in the EVOLUTION of life on Earth. *GO*

Conway Morris, S. 1998: *The crucible of creation: the Burgess Shale and the rise of animals*. Oxford: Oxford University Press. **Derry, L.A., Brasier, M.D., Corfield, R.M.** *et al.* 1994: Sr-isotope and C-isotope in Lower Cambrian carbonates from the Siberian craton – a paleoenvironmental record during the Cambrian explosion. *Earth and Planetary Science Letters* 128, 671–681. **Gould, S.J.** 1989: *Wonderful life: the Burgess Shale and the nature of history*. London: Penguin. **Kimura, H., Matsumoto, R., Kakuwa, Y.** *et al.* 1997: The Vendian–Cambrian delta C-13 record, North Iran: evidence for overturning of the ocean before the Cambrian Explosion. *Earth and Planetary Science Letters* 147, E1–E7. **Knoll, A.H. and Carroll, S.B.** 1999: Early animal evolution: emerging views from comparative biology and geology. *Science* 284, 2129–2137.

canonical correlation analysis
A numerical technique for investigating the linear relationships between two sets of variables (e.g. biological response data and environmental predictor data) by finding linear combinations of one set of variables that maximise correlations with linear combinations of the variables in the other set. Canonical correlation analysis is limited to data-sets

where the number of variables (responses and predictors) in the two sets is less than the number of observations.

HJBB

[See also REDUNDANCY ANALYSIS]

Gittins, R. 1985: *Canonical analysis. A review with applications in ecology.* Berlin: Springer.

canonical correspondence analysis (CCA)

The canonical or constrained version of CORRESPONDENCE ANALYSIS where the ORDINATION axes are constrained to be linear combinations of predictor or explanatory variables. The response variables (e.g. species) are assumed to have a unimodal response to the explanatory variables (e.g. environmental variables). Canonical correspondence analysis provides a multivariate direct ORDINATION or GRADIENT ANALYSIS of objects, variables, and predictor variables.

HJBB

Braak, C.J.F. ter and **Verdonschot, P.F.M.** 1995: Canonical correspondence analysis and related multivariate methods in aquatic ecology. *Aquatic Sciences* 57, 153–187.

capillary action

Downward water movement through soil pores is mainly the result of gravity, but, in fine-textured materials, upward and lateral movement may also take place by capillary action through the forces of surface tension,. Water is drawn into soil pores mainly by SURFACE TENSION, but water is also held in the soil by ADSORPTION. It is not always possible to decide which of these mechanisms controls water retention in soils. Water held in a soil by capillary action above the water table is said to be in the *capillary fringe*.

EMB

Baver, L.D., Gardner, W.H. and **Gardner W.R.** 1991: *Soil physics.* New York:Wiley. **Wild, A. (ed.)** 1988: *Russell's soil conditions and plant growth,* 11th edn. Harlow: Longman.

captive breeding

Ex situ CONSERVATION of species, not in the natural habitat. *Zoos, botanical gardens* and *aquaria* undertake captive breeding programmes.

IFS

[See also CONSERVATION BIOLOGY, SPECIES TRANSLOCATION]

carbon assimilation

The incorporation of carbon from atmospheric CARBON DIOXIDE into organic compounds (sugars) during the process of PHOTOSYNTHESIS and the incorporation of carbon from plant and animal BIOMASS into animal biomass.

RJH

[See also CARBON CYCLE]

carbon balance/budget

An estimate of the gains and losses between carbon reservoirs (sources and sinks) in the CARBON CYCLE. Annual average rates of anthropogenic carbon emissions and uptake from various sources and sinks, respectively, are particularly important aspects. The main anthropogenic sources of CARBON DIOXIDE are the burning of FOSSIL FUELS and LANDUSE CHANGE. Over the period 1980 to 1989 the average emissions from fossil-fuel burning were 5.5 ± 0.5 gigatonnes of carbon per year (Gt C y^{-1}). Landuse changes cause both release and uptake of CARBON DIOXIDE. On average, carbon dioxide will be released to the atmosphere if the original ecosystem stored more carbon than the modified ecosystem that replaces it. Tropical DEFORESTATION as a carbon dioxide source accounted for an average emission to the atmos-

phere of 1.6 ± 1.0 Gt C y^{-1} in the 1980s. However, in Northern Hemisphere mid and high latitudes there are areas where forests are regrowing after clearance in the past and this provided a net CARBON SINK of 0.5 ± 0.5 Gt C y^{-1} in the 1980s. Consequently the net release due to global LANDUSE changes was 1.1 ± 1.2 Gt C y^{-1}. Carbon dioxide is readily soluble in water and the oceans act as a vast reservoir for the gas, estimated as an uptake of about 2.0 ± 0.8 Gt C y^{-1}. A terrestrial sink for anthropogenic CARBON DIOXIDE accounts for about 1.4 Gt C y^{-1} and is due to the fertilisation of plants by the application of nitrogen fertilisers and because of PHOTOSYNTHESIS being stimulated by increased levels of atmospheric CARBON DIOXIDE. Storage in the atmosphere accounted for 3.2 ± 0.2 Gt C y^{-1} in the 1980s, equivalent to an atmospheric CARBON DIOXIDE increase of about 1.5 ppmv y^{-1}. This represents about 50% of the anthropogenic CARBON DIOXIDE emissions over that same period.

DME

Intergovernmental Panel on Climate Change 1994: *Radiative forcing of climate change. The 1994 report of the Scientific Assessment Working Group of IPCC. Summary for Policymakers.* Geneva and Nairobi: World Meteorological Organisation and United Nations Environment Programme. **Houghton, J.T., Filho, L.G.M., Callander, B.A.** *et al.* **(eds)** 1996: *Climate change 1995: The science of climate change.* Cambridge: Cambridge University Press.

carbon content

Soils contain the largest terrestrial pool of *organic carbon* with global estimates ranging from 1115 to 2200 Pg [Pg (petagram) = 10^{15} g]. Reserves of *inorganic carbon* also occur in the form of coal, charcoal and carbonates. All soils contain some organic matter in the form of carbon compounds, but HISTOSOLS and CRYOSOLS and humic subgroups of the other soil units contain the greatest proportion. The mean soil organic carbon content in the upper 1 m of the various soil units ranges from 3.1 kg C m^{-2} for sandy ARENOSOLS to 77.6 kg C m^{-2} for HISTOSOLS. Release of this carbon as CARBON DIOXIDE or METHANE through GLOBAL WARMING or LANDUSE CHANGES could make a significant contribution to the GREENHOUSE EFFECT. Alternatively, enhanced sequestration of atmospheric CO_2 in the soil through adaptive management may help to mitigate it.

EMB

[See also CARBON SEQUESTRATION, ORGANIC CONTENT]

Batjes, N.H. 1996: Total carbon and nitrogen in the soils of the world. *European Journal of Soil Science* 47, 151–163. **Eswaran, H., van den Berg, E.** and **Reich, P.** 1993: Organic carbon in soils of the world. *Soil Science Society of America Journal,* 57, 192–194.

carbon cycle

The cyclical progress of carbon through living things, air, rocks, soil and water (see Figure). Atmospheric carbon, in the form of carbon dioxide, is converted to carbohydrates by photosynthesising plants. Some plant carbon returns to the atmosphere as CARBON DIOXIDE through producer RESPIRATION. Some is assimilated and metabolised by animals. A portion of animal carbon returns to the atmosphere as carbon dioxide released through consumer respiration. The rest enters the decomposer FOOD CHAIN and either returns to the atmosphere through DECOMPOSER respiration or accumulates as organic sediment (e.g. PEAT and COAL). Carbon dioxide is released into the atmosphere by combustion.

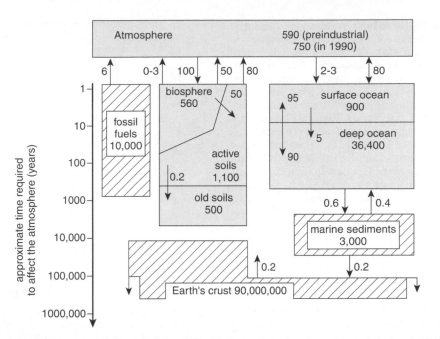

carbon cycle *Major reservoirs or stores in PgC (IPgC = 10¹⁵ gC), fluxes in PgC yr⁻¹, and timescale for exchange of carbon in the global carbon cycle (after Sunquist, 1993)*

Some is emitted by volcanoes. Agricultural, industrial and domestic practices alter the stores and fluxes in the carbon cycle. The burning of FUELWOOD and FOSSIL FUELS pumps some five billion tonnes of carbon dioxide into the atmosphere every year. The increasing carbon dioxide pool enhances the GREENHOUSE EFFECT and contributes to GLOBAL WARMING. Current knowledge of the carbon cycle within the oceans, on land and in the atmosphere is sufficient to conclude that, although natural processes can potentially slow the rate of increase in atmospheric CO_2, they are unlikely to assimilate all the anthropogenically produced CO_2 during the twenty-first century. Further understanding requires greater knowledge of the relationship between the carbon cycle, other BIOGEOCHEMICAL CYCLES and the CLIMATIC SYSTEM, together with an approach through EARTH-SYSTEM ANALYSIS. *RJH*

[See also METHANE, NITROGEN CYCLE, SULPHUR CYCLE]

Bolin, B. 1970: The carbon cycle. *Scientific American* 223, 124–132. **Falkowski, P., Scholes, R.J., Boyle, E.** *et al.* 2000: The global carbon cycle: a test of our knowledge of Earth as a system. *Science* 290, 291–296. **Holmén, K.** 1992: The global carbon cycle. In Butcher, S. S., Charlson, R.J., Orians, G.H. and Wolfe, G.V. (eds), *Global biogeochemical cycles*. London: Academic Press, 239–262. **Houghton, R.A.** 1995: Land-use change and the carbon cycle. *Global Change Biology* 1, 275–287. **Lal, R., Kimble, J.M. Follett, R. F. and Stewart, B. A. (eds)** 1998: *Soil processes and the carbon cycle*. Boca Raton, FL: CRC Press. **Wigley, T.M.L. and Schimel, D.S. (eds)** 2000: *The carbon cycle*. Cambridge: Cambridge University Press.

carbon dioxide A GREENHOUSE GAS released through the respiratory process of plant and animal life and by complete combustion of carbon-containing substances such as wood, COAL and oil. Carbon dioxide (CO_2) is removed from the ATMOSPHERE by plants (PHOTOSYNTHESIS) and by solution in water. Approximately half the carbon dioxide emissions remain in the atmosphere, with the remainder being taken up by the oceans and terrestrial sinks. Atmospheric concentrations of this trace gas (less than 1% in the atmosphere) have increased significantly in the past two centuries. In the pre-industrial era the annual average atmospheric carbon dioxide concentration was estimated to be 280 ppmv. In 1997 at Mauna Loa Observatory, Hawaii, it reached 364 ppmv, an increase of 30% over the pre-industrial level. Average annual atmospheric carbon dioxide concentrations increased by 1.2 ppmv during the 1990s, a slight reduction compared with around 1.4 ppmv during the 1980s. Monthly carbon dioxide concentrations display an annual cycle with a maximum around May and a minimum around September. This is related to the Northern Hemisphere vegetation GROWING SEASON and variations in the SEA-SURFACE TEMPERATURE, which affects the dissolved carbon dioxide content of the oceans. Analysis of air bubbles preserved in the deep ICE CORES from Antarctica indicate that atmosphere carbon dioxide concentrations decreased by as much as 100 ppmv during GLACIATIONS. The peak wavelength at which carbon dioxide absorbs TERRESTRIAL RADIATION is around 15 µm and it is considered the most important gas contributing to the human-enhanced GREENHOUSE EFFECT. When dissolved in water, carbon dioxide forms carbonic acid, which is important in CHEMICAL WEATHERING. *DME*

[See also CARBON DIOXIDE FERTILISATION, CARBON DIOXIDE VARIATIONS, CARBON CYCLE, CARBON MONOXIDE, CLIMATIC CHANGE]

Bach, W., Crane, A.J., Berger, A.L. and Longhetto, A. (eds) 1983: *Carbon dioxide*. Dordrecht: Reidel. **Houghton, J.T., Filho, L.G.M., Callander, B.A.** *et al.* **(eds)** 1996: *Climate change 1995: the science of climate change*. Cambridge: Cambridge University Press. **Trabalka, J.R. (ed.)** 1985: *Atmospheric carbon dioxide and the global carbon cycle*. Washington, DC: US Department of Energy.

carbon dioxide fertilisation The enhancement of plant growth as a result of elevated concentrations of atmospheric CARBON DIOXIDE. C-3 plants, including most trees and agricultural crops, such as rice, wheat, potatoes, soya bean and vegetables, show a more pronounced initial increase in productivity than C-4 plants, which are mainly of tropical origin, including maize, sugar cane, millet, sorghum and other tropical grasses. Reported (mostly laboratory) increases in plant net PRODUCTIVITY, BIOMASS and crop yield are countered by evidence of declines in the NUTRIENT and thus the food and forage quality of leaf tissue. Laboratory studies of the CO_2 fertilisation effect show enhanced plant growth rates and crop yields. But water and NUTRIENT shortages in many ecosystems may place a limit on increased plant PRODUCTION. Any reduction in the initial stimulation of photosynthesis by carbon dioxide fertilisation is termed *down regulation*. For the world's AGRO-ECOSYSTEMS, estimates suggest that while wheat yields in Scandinavia and rice yields in eastern Asia could rise, grain yields could fall in the USA and the former USSR.

GOH

[See also CARBON BALANCE/BUDGET, CLIMATIC CHANGE]

Bazzaz, F.A. 1990: The response of natural ecosystems to rising global carbon dioxide levels. *Annual Review of Ecology and Systematics* 21, 167–96. **Rosenzweig, C. and Hillet, D.** 1998: *Climate change and the global harvest: potential impacts of the greenhouse effect on agriculture.* Oxford: Oxford University Press. **Solomon, A.M. and Shugart, H.H. (eds)**, 1993: *Vegetation dynamics and global change.* London: Chapman and Hall.

carbon dioxide variations The concentration of carbon dioxide (CO_2) in the atmosphere has fluctuated markedly over geological time. Analysis of air trapped in Vostock polar ICE CORES has exposed a fluctuation in CO_2 during GLACIAL–INTERGLACIAL CYCLES, with levels of 180–200 ppm occurring during GLACIAL EPISODES and 280–300 ppm during INTERGLACIALS. The current ambient CO_2 concentration (360 ppm) is 25% higher than pre-industrial concentrations as a result of fossil-fuel burning and landuse change and is expected to reach levels of 560 ppm by the year 2100 according to projections by the INTERGOVERNMENTAL PANEL ON CLIMATE CHANGE (IPCC). Although the rate of anthropogenic-induced CO_2 rise is unprecedented, the magnitude of CO_2 change is not so in relation to CO_2 variations over geological time (see Figure), as indicated by geochemical models (GEOCARB) of the long-term CARBON CYCLE. These variations are caused by temporary imbalances between CARBON SINKS/SOURCES. The major CO_2 sources include decarbonation, VOLCANIC emissions, uplifting and WEATHERING of organic sediments, whilst the more minor sources include carbonate weathering and biosphere RESPIRATION. Conversely, the major sinks include weathering of silicate rocks (linked to burial of carbonate carbon in the sea) and photosynthesis (linked to burial of organic matter), whilst the shorter-term sinks include PHOTOSYNTHESIS and carbonate precipitation.

JCMc

[See also ATMOSPHERIC COMPOSITION]

Barnola, J.M., Raynaud, D., Korotkevich, Y.S., and Lorius, C. 1987: Vostock ice core provides 160,000-year record of atmospheric CO_2 *Nature* 329, 408–414. **Berner, R.A.** 1994: GEOCARB II. A revised model of atmospheric CO_2 over Phanerozoic time. *American Journal of Science* 294, 59–91. **Berner, R.A.** 1997: The rise of plants and their effect on weath-

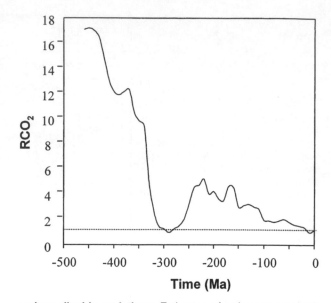

carbon dioxide variations *Estimates of palaeo-atmospheric carbon dioxide concentrations for the Phanerozoic from the GEOCARB II model: RCO_2 represents a ratio of atmospheric CO_2 concentration at some time in the past to that of the pre-industrial concentration of 300 ppm; the dashed horizontal line represents the current ambient CO_2 concentration of 360 ppm = 1.2 RCO_2 (after Berner, 1994)*

ering and atmospheric CO_2. *Science* 276, 544–546. **Houghton, J.T., Meira Filho, L.G., Callander, B.A. et al.** 1996: *Climate Change 1995.* Cambridge: Cambridge University Press. **Sigman, D.M. and Boyle, E.A.** 2000: Glacial/interglacial variations in atmospheric carbon dioxide. *Nature* 407, 859–869. **Street-Perrott, F.A., Yonsong Huang, Perrott, R.A. et al.** 1997: Impact of lower atmospheric carbon dioxide on tropical mountain ecosystems. *Science* 278, 1422–1426.

carbon isotopes Carbon has two STABLE ISOTOPES (natural abundances, $^{12}C = 98.90\%$ and $^{13}C = 1.10\%$), as well as one *radioisotope* (^{14}C or *radiocarbon*; half-life = 5730 y) caused by bombardment of nitrogen by high-energy COSMIC RADIATION in the upper atmosphere. The ISOTOPIC RATIO is expressed as $^{13}C{:}^{12}C$ relative to the Vienna Pee Dee Belemnite (VPDB) standard (see STANDARD SUBSTANCE). The $\delta^{13}C$ values vary as a result of physical, chemical and biological processes. The measurement of the $\delta^{13}C$ values of atmospheric CARBON DIOXIDE was started at Mauna Loa, Hawaii in AD 1956. Earlier values are estimated from the $\delta^{13}C$ analysis of TREE-RING cellulose, carbon dioxide trapped in ICE CORES, and C-4 PLANTS preserved in RODENT MIDDENS. Throughout the HOLOCENE, the $\delta^{13}C$ value of atmospheric carbon dioxide has been relatively constant until the increased burning of isotopically depleted FOSSIL FUELS following the INDUSTRIAL REVOLUTION. The $\delta^{13}C$ value of atmospheric carbon dioxide has been used to calculate the fate of ANTHROPOGENIC inputs into the CARBON CYCLE, with the 'missing' CARBON SINK allocated to increased CARBON SEQUESTRATION in the terrestrial biosphere.

ISOTOPIC FRACTIONATION occurs during the photosynthetic assimilation of atmospheric carbon dioxide as different isotopes have different rates of diffusion and carboxylation. Plant carbon is depleted in ^{13}C compared to atmospheric carbon dioxide with C-3 PLANTS being depleted more than C-4 plants; CAM plants may resemble either photosynthetic type. The utilisation of these different metabolic pathways may be detected from $\delta^{13}C$ values. Although $\delta^{13}C$ values from bulk LACUSTRINE SEDIMENTS have been used to infer changes in plant distribution, the analysis of species-specific $\delta^{13}C$ values is adopted to minimise the influence of DIAGENESIS. The predominance of C-4 plants during the LAST GLACIAL MAXIMUM was inferred from the COMPOUND-SPECIFIC CARBON ISOTOPIC ANALYSIS of individual BIOMARKERS in lacustrine sediments. The $\delta^{13}C$ of the carbon occluded in PHYTOLITHS may also be used to reconstruct vegetation assemblages. In non-arid environments, the $\delta^{13}C$ values from teeth or bone may be used to determine the contribution of C-3 or C-4 plants to the diet. TREE-RING $\delta^{13}C$ values have been used to infer that C-3 plants have adapted to increasing atmospheric CO_2 concentrations by increasing their intrinsic water-use efficiency. *IR*

[See also BIO-INDICATOR, C-3, C-4 AND CAM PLANTS, CORAL, MARINE SEDIMENT CORES, OSTRACODS, PEAT]

Coplen, T.B. 1996: Editorial: more uncertainty than necessary. *Paleoceanography* 11, 369–370. Farquhar, G.D., Ehleringer, J.R. and Hubick, K.T. 1989: Carbon isotope discrimination and photosynthesis. *Annual Review of Plant Physiology and Plant Molecular Biology* 40, 503–537. Friedli, H., Lötscher, H., Oescher, H. *et al.* 1986: Ice core record of the $^{13}C/^{12}C$ ratio of atmospheric CO_2 in the past two centuries. *Nature* 324, 237–238. Kelly, E.F., Amundson, R.G., Marino, B.D., and DeNiro, M.J. 1991: Stable isotope ratios of carbon in phytoliths as a quantitative method of monitoring vegetation and climate change. *Quaternary Research* 35, 222–233. Marino, B.D., McElroy, M.B., Salawitch, R.J. and Spaulding, W.G. 1992: Glacial-to-interglacial variations in the carbon isotopic composition of atmospheric CO_2. *Nature* 357, 461–465. Street-Perrott, F.A., Huang, Y., Perrott, R.A. *et al.* 1997: Impact of lower atmospheric carbon dioxide on tropical mountain ecosystems. *Science* 278, 1422–1426. Trolier, M., White, J.W.C., Tans, P.P. *et al.* 1996: Monitoring the isotopic composition of atmospheric CO_2: measurements from the NOAA Global Air Sampling Network. *Journal of Geophysical Research* 101(D20), 25897–25916.

carbon monoxide

Carbon monoxide (CO) is a colourless, odourless gas formed by the incomplete combustion of carbon-containing materials, which would otherwise, in the presence of sufficient oxygen, have formed CARBON DIOXIDE. It is a toxic gas which has an affinity for haemoglobin (Hb), the substance that carries oxygen in blood. It can combine with haemoglobin to form carboxyhaemoglobin (COHb) rather than oxyhaemoglobin, reducing the oxygen-carrying capability of the blood. Its accumulation in the blood leads to oxygen starvation, although the effects of brief exposure are reversible. Initial symptoms are impairment of vigilance and time-interval discrimination, drowsiness and headaches, but this can progress to loss of consciousness and death at very high COHb levels. The World Health Organisation recommends that the equilibrium blood concentration of COHb should not exceed 2.5% at a high level of physical activity. To meet this criterion in the non-smoking population and to include a safety margin of protection, atmospheric concentrations should not exceed 10 mg m^{-3} (8.5 ppmv) averaged over 8 h. Highest concentrations are found in enclosed spaces (such as garages and road tunnels) from petrol-engine vehicle emissions. Carbon monoxide levels may be high indoors as a result of emissions from faulty gas-fired heaters and gas cookers: smoking tobacco exposes the smoker to high concentrations too. Anthropogenic sources (mainly automobile exhaust) account for about half of total global carbon monoxide emissions. Catalytic converters fitted to vehicles reduce carbon monoxide emissions by converting them to CARBON DIOXIDE. *DME*

Expert Panel on Air Quality Standards 1994: *Carbon monoxide*. London: HMSO. Newell, R.E., Reichle, H.G. and Seiler, W. 1989: Carbon monoxide and the burning Earth. *Scientific American* 261, 58–64.

carbon sequestration

The processes involved in carbon storage, as opposed to release in the CARBON CYCLE. MARINE SEDIMENTS, MIRES and other WETLANDS, VEGETATION and SOILS are important sites of carbon sequestration. The amount of carbon stored in soils, for example, as fresh organic matter, stable humus or charcoal is two to three times higher than that contained in the above-ground BIOMASS of natural vegetation and standing crops. AFFORESTATION and REFORESTATION of cut-over areas can help to limit the increase of atmospheric CARBON DIOXIDE, but the potential for carbon sequestration in soils is even greater. Environmental changes that increase carbon sequestration would help reduce the GLOBAL WARMING caused by enhanced concentrations of atmospheric carbon dioxide. *EMB*

[See CARBON CONTENT]

Batjes, N.H. 1998: Options for mitigation of atmospheric CO_2 concentrations by increased carbon sequestration in the soil. *Biology and Fertility of Soils* 27, 230–235. Bouwman, A.F. (ed.) 1990: *Soils and the greenhouse effect*. Chichester: Wiley. Chisholm, S.W. 2000: Stirring times in the Southern Ocean. *Nature* 407, 685–687. Lal, R., Kimble, J.M. Follet, R.F., Stewart, B.A. 1998: *Management of carbon sequestration in soil*. Boca Raton, FL: CRC Press. Mooney, H.A., Vitousek, P.M. and Matson, P.A. 1987: Exchange of materials between terrestrial ecosystems and the atmosphere. *Science* 238, 926–932. Watson, A.J., Bakker, D.C.E., Ridgwell, A.J. *et al.* 2000: Effect of iron supply on Southern Ocean CO_2 uptake and implications for glacial atmospheric CO_2. *Nature* 407, 730–733.

carbon sink/source

A carbon sink or output is any *reservoir* to which carbon is added; conversely, a carbon source or input is any reservoir from which carbon is removed. Carbon sinks and sources and the *fluxes* between them represent the CARBON CYCLE. Alteration of the balance between carbon sources and sinks cause changes in the size of carbon reservoirs. For instance, the concentration of atmospheric CARBON DIOXIDE (CO_2) is controlled by the balance that exists between the major sources and sinks of carbon dioxide. Anthropogenic inputs of CO_2 from the burning of FOSSIL FUELS, LANDUSE CHANGE and the cement industry have altered the sink:source balance, causing the sinks to outweigh the sources of CO_2. Hence, atmospheric CO_2 has been increasing since the INDUSTRIAL REVOLUTION and is predicted to double the pre-industrial levels by AD 2100. There has been much debate as to whether natural ecosystems will be sources or sinks for carbon, which is of great importance to future atmospheric CO_2 concentrations. *JCMc*

[See also BIOMASS BURNING, CARBON ASSIMILATION, CARBON BALANCE, CARBON DIOXIDE VARIATIONS, DECOMPOSITION]

Houghton, J.T., Meira Filho, L.G., Callander, B.A. *et al.* 1996: *Climate Change 1995*. Cambridge: Cambridge University Press.

carbonate A group of MINERALS containing the anion [CO_3^{2-}], the most common of which are *calcite* [$CaCO_3$] and *dolomite* [$CaMg(CO_3)_2$]. LIMESTONE and associated rocks are commonly referred to as carbonate SEDIMENTARY ROCKS. *GO*

Marshall, J.D. 1992: Climatic and oceanographic isotopic signals from the carbonate rock record. *Geological Magazine* 129: 143–160. Sun, S.Q. 1994: A reappraisal of dolomite abundance and occurrence in the Phanerozoic. *Journal of Sedimentary Research* A64, 396–404.

carbonate compensation depth (CCD) The depth in the OCEANS at which the rate of supply of CARBONATE material (as PELAGIC SEDIMENT) is balanced by the rate of its dissolution and below which carbonate sediments do not accumulate because of the enhanced solubility of carbonate material with increasing pressure and decreasing temperature. The *calcite compensation depth* is typically about 1 km deeper than the *aragonite compensation depth* (ARD). The CCD ranges from 3000 m in polar regions to 5000 m in the warmest equatorial waters and is shallower closer to land. It typically lies a few hundred metres deeper than the LYSOCLINE. The CCD is analogous to the SNOW LINE, where rates of snow fall and snow melt are balanced. Studies of pelagic sediments (see DEEP-SEA DRILLING) have shown that the CCD has varied through time. For example, there was an abrupt deepening at the EOCENE–OLIGOCENE boundary, coincident with the initiation of major Antarctic ice cover. *BTC*

Andel, T.H. van 1975: Mesozoic/Cenozoic calcite compensation depth and the global distribution of calcareous sediments. *Earth and Planetary Science Letters* 26, 187–194. Berger, W.H. 1967: Foraminiferal ooze: solution at depth. *Science* 156, 383–385. Corliss, B.H. and Honjo, S. 1981: Dissolution of deep-sea benthonic foraminifera. *Micropaleontology* 27, 356–378.

Carboniferous A SYSTEM of rocks and a PERIOD of geological time from 354 to 295 million years ago. The first land vertebrates are known from the early Carboniferous and the first reptiles from the mid Carboniferous. A Southern Hemisphere ICE AGE developed in the late Carboniferous. The late Carboniferous is known for the extensive development of COAL-bearing sequences, preserved in northwest Europe and northeastern North America. In North American usage this interval of the GEOLOGICAL TIMESCALE is divided into the MISSISSIPPIAN and PENNSYLVANIAN periods. *GO*

Caputo, M.V. and Crowell, J.C. 1985: Migration of glacial centres across Gondwana during the Paleozoic Era. *Geological Society of America Bulletin* 96, 1020–1036. Eyles, N. and Young, G.M. 1994: Geodynamic controls on glaciation in Earth history. In Deynoux, M., Miller, J.M.G., Domack, E.W. *et al.* (eds), *Earth's glacial record*. Cambridge: Cambridge University Press, 1–28.

carcinogen A substance that causes cancer. DNA-reactive carcinogenic molecules may irreversibly damage cell DNA, which induces gene mutation, abnormal growth of cells and the production of cancerous tumours. Epigenetic carcinogens may lead to cancer indirectly, through altering the immune system. Examples of carcinogens include many synthetic PERSISTENT ORGANIC COMPOUNDS (POCs), asbestos, HEAVY METALS such as arsenic, CADMIUM, chromium and nickel, and RADIONUCLIDES from sources such as NUCLEAR ACCIDENTS, *nuclear waste* and RADON EMANATION. OZONE DEPLETION is responsible for increased amounts of *ultraviolet radiation* reaching Earth, raising the incidence of skin cancers. *MLW*

[See also ECOTOXICOLOGY, HUMAN HEALTH HAZARDS, MUTAGEN]

Longstreth, J. D., Degruijl, F.R., Kripke, M.L. *et al.* 1995: Effects of increased solar ultraviolet radiation on human health. *Ambio* 24, 153–165. Mohr, U., Schmal, D. and Tomatis, L. (eds) 1977: Air pollution and cancer in man. *Proceedings of the 2nd Hanover International Carcinogenesis Meeting, 22–24 October 1975*. Lyon: International Agency for Research on Cancer.

carnivore Any predatory or flesh-eating organism, such as a weasel, a tiger or an insectivorous plant. *Top carnivores*, such as the eagle, eat other carnivores and occupy the apex of a FOOD CHAIN/WEB. *RJH*

[See also HERBIVORE, TROPHIC LEVEL]

carrying capacity The maximum species population that the RESOURCES of an environment or place can sustain without jeopardising the species population, other species in the ECOSYSTEM and all necessary ecosystem processes. The idea of carrying capacity assumes a balance between populations and resources in an ecosystem. If populations should exceed the carrying capacity, there will be insufficient resources to support the excess numbers. The population will then fall until a new balance is attained (at the carrying capacity). If the population is lower than the carrying capacity, then it will normally grow until the carrying capacity is reached. In population models, carrying capacity is denoted by the letter K.

Carrying capacity applies to natural ecosystems, to AGROECOSYSTEMS and to recreational systems. In agriculture, carrying capacity determines the number of GRAZING animals that can be sustained. In recreational landuse, it determines the number of people and kinds of activity that can be accommodated without harming the environment.

Human activity often drives populations to levels above the carrying capacity and leads to ENVIRONMENTAL DEGRADATION. This may be the case when too many cattle are farmed in RANGE MANAGEMENT or when too many people are allowed to use a recreational area. On rangeland, the carrying capacity depends upon rainfall and soil fertility. In the USA, the carrying capacity of pasture and rangeland varies from about 200 cows per km^2 to 4 cows per km^2. In theory, the planet Earth has a carrying capacity. In practice, the planetary carrying capacity can only be calculated according to an agreed set of human values – what kind of life do humans want and what kind of environment? The highest estimate would be based on the amount of land and resources needed per person to provide basic necessities – food, water, clothing and waste disposal. A lower estimate would provide a better equality of life and allow greater space for such amenities as RECREATION and WILDERNESS. *RJH*

[See also ECOSYSTEM HEALTH, POPULATION, *k*-SELECTION, SUSTAINABILITY, SUSTAINABLE DEVELOPMENT, SUSTAINABLE YIELD, WILDLIFE CONSERVATION AND MANAGEMENT]

Campbell, D.E. 1998: Energy analysis of human carrying capacity and regional sustainability: an example using the State of Maine. *Environmental Monitoring and Assessment* 51, 531–569. **Cohen, J.E.** 1995: *How many people can the Earth support?* New York and London: W.W. Norton.

carse Extensive, relatively flat areas of estuarine silty clay sediments that occur above sea level as a result of GLACIO-ISOSTATIC REBOUND in Scotland. During the early decades of the eighteenth century, extensive areas of peat were stripped from the carse surfaces, which were subsequently converted to farmland. The highest carse clay sediments in Scotland reach up to about 14 m above sea level and attain maximum thicknesses of around 10 m. Thomas Jamieson, in the AD 1860s was the first to observe the occurrence of PEAT deposits *beneath* the carse clays and used this information to elucidate for the first time the theory of GLACIO-ISOSTASY. *AGD*

[See also ISOSTASY, SEA-LEVEL CHANGE]

Barras, B.F. and Paul, M.A. 1999: Sedimentology and depositional history of the Claret Formation ('carse clay') at Bothkennar, near Grangemouth. *Scottish Journal of Geology* 35, 131–143. **Sissons, J.B.** 1967: *The evolution of Scotland's scenery.* Edinburgh: Oliver and Boyd. **Smith, D.E., Cullingford, R.A., Harkness, D.D. and Dawson, A.G.** 1985: The stratigraphy of Flandrian relative sea level changes at a site in Tayside, Scotland. *Earth Surface Processes and Landforms* 10, 17–25.

Cartesian coordinates
Measurements taken along each of a set of orthogonal axes. The origin of the system occurs at the intersection of the axes. *TF*

[See also COORDINATE SYSTEM, ABSCISSA and ORDINATE]

cartography In general terms, cartography is associated with the making of maps. However, this simple description belies the range and coverage of a subject now enshrined in more technical definitions. The International Cartographic Association (ICA) submits that 'cartography is the science and technology of analysing, interpreting and communicating spatial relationships by means of maps' (Board, 1989), while the British Cartographic Society proposes that 'cartography is the organisation and communication of spatially related information in a graphic form. It can include all phases from data acquisition to presentation.' (Board, 1989). Using digital map data, all elements of map making from data collection through design and drafting to the production of print masters are now carried out on computers. Software is readily available for thematic mapping on personal computers. In this era, when there is more interest in mapping than at any previous time in history (Crampton, 1998), cartography is becoming entwined with other technologies such as GEOGRAPHICAL INFORMATION SYSTEMS (GIS), desktop mapping and visualisation. The boundaries of these topics do intersect, but cartographic principles of representation, recording, presentation and communication of information through spatially structured graphics still underlie these newer technologies.

The spatial datasets used in cartography are the basis of much of the environmental, economic and social information that underpins our politics and development. Maps are products for communication, recording, planning and decision making based on well established principles. They encompass the mathematical basis of cartography, principles of cartographic design, map type and construction and map proficiency (that is, the skills of graphic visualisation and understanding). Cartography is building on its knowledge base derived from conventional mapping and is expanding its concepts into GIS and visualisation to encompass their presentation potential in, for example, the display of dynamic and interactive map-type graphics. *TF*

[See also MAP PROJECTION, SURVEYING]

Board, C. 1989: ICA Working Group on Cartographic Definitions. *Cartographic Journal* 26, 175–176. **Crampton, J. 1998:** What's in a name? *Association of American Geographers, Cartography Specialty Group Newsletter* 18(2); http://www.csun.edu/~hfgeg003/csg/index.html. **Jones, C. 1997:** *Geographical information systems and computer cartography.* Harlow: Addison-Wesley Longman. **Robinson, A.H., Morrison, J.L., Muehrcke, P.C.** *et al.* **1995:** *Elements of cartography,* 6th edn. Chichester: Wiley.

cash cropping Growing crops for monetary return rather than subsistence. Short-term financial incentives may encourage inappropriate use and lead to SOIL DEGRADATION, especially of MARGINAL land. *JAM*

[See also SUBSISTENCE AGRICULTURE]

cast In the geological context, a cast or PSEUDOMORPH is a sedimentary structure reflecting the previous existence of an active form. When applied to ICE WEDGES, the thaw of PERMAFROST and subsequent infill of the wedge by adjacent sediment gives rise to an *ice-wedge cast, or wedge of secondary infilling.*

When applied to a *sand wedge,* however, the RELICT feature is better termed a *sand-wedge cast* rather than a 'wedge of secondary infilling'. This is because there has been no secondary infilling of the wedge consequent upon thaw. *HMF*

Harry, D.G. and Gozdzik, J.S. 1988: Ice wedges: growth, thaw transformation and palaeoenvironmental significance. *Journal of Quaternary Science* 3, 39–55

catalyst A substance that accelerates the rate of a chemical reaction without being altered chemically or being used up. It is recoverable and can be used repeatedly. Catalysts are widely used in industry, and in *catalytic converters* to convert harmful vehicle exhaust gases into less harmful ones. *Enzymes* are organic catalysts in living cells. *JAM*

Butt, J.B. and Peterson, E.E. 1988: *Activation, deactivation and poisoning of catalysts.* San Diego, CA: Academic Press.

catastrophe theory A topological mathematical language created by Réné Thom. In English the term is unfortunate because it involves neither THEORY nor catastrophes in the commonly used senses. Calculus is a mathematical language well suited to depicting smooth changes, while catastrophe theory was developed specifically for describing surfaces that exhibit abrupt changes. While catastrophe theory is not restricted in the number of dimensions that may be expressed, visualisation of more than three is very limited. Equilibrium surfaces of differing orders depicted using catastrophe theory reveal folds and cusps. Catastrophe theory is by no means easy to use in the environmental sciences; however, if such things as the correct control factors can be designated, it can demonstrate how a SYSTEM may exhibit smooth changes, abrupt changes, divergence, HYSTERESIS, bimodal behavior, instability or stability – all dependent upon antecedent conditions. *CET*

Thom, R. 1975: *Structural stability and morphogenesis: an outline of a general theory of models*. Reading, MA: W.A. Benjamin. Graf, W.L. 1979: Catastrophe theory as a model for change in fluvial systems. In Rhodes, D.D. and Williams, G.P. (eds), *Adjustments of the fluvial system*. Dubuque, IA: Kendell/Hunt, 13–32.

catastrophic event An event that is sudden and exceptionally large in scale, causing substantial damage to the system under consideration. A METEORITE IMPACT, for example, is a catastrophic event on a global scale. *KDB*

[See also EXTREME CLIMATIC EVENTS, MASS EXTINCTIONS]

catastrophism The principle that the GEOLOGICAL RECORD is primarily to be interpreted as the product of dramatic, short-lived processes, often of world-wide extent, that do not occur today. It was widely supported before the triumph of Lyell's UNIFORMITARIANISM from the 1830s onward. Catastrophism was, however, far more diverse and sophisticated than is commonly thought today and many proponents rejected direct supernatural intervention. The explanation of alleged flood deposits as of glacial origin (see GLACIAL THEORY) in the 1840s dealt traditional catastrophism a fatal blow. Since the 1960s there has been a renewed acceptance of the role of cataclysmic EVENTS within the EARTH SCIENCES (see NEOCATASTROPHISM). Since the 1950s there has also been a popular espousal (particularly within the USA) of a form of flood catastrophism by 'young Earth' creationists. Although claiming to be in continuity with pre-Lyell catastrophism, this is not the case: it has no support within mainstream science. *CDW*

[See also CREATIONISM, GRADUALISM]

Gould, S.J. 1987: *Time's arrow, time's cycle: myth and metaphor in the discovery of geological time*. Cambridge, MA: Harvard University Press. Huggett, R.J. 1989: *Cataclysms and Earth his-tory: the development of diluvialism*. Oxford: Clarendon Press. Huggett, R.J. 1997: *Catastrophism: asteroids, comets and other dynamic events in Earth history*. London: Verso. Numbers, R.L. 1992: *The creationists*. New York: Alfred A Knopf. Palmer, T. 1998: *Controversy, catastrophism and evolution: the ongoing debate*. New York: Plenum.

catena A topographic (hillside) sequence of soils of similar age but with different profiles developed at different positions on a slope from the same parent material (*simple catena*) or different parent materials (*composite catena*). The term was originally used by G. Milne to describe a related series of soils of different character from interfluve to valley bottom in East Africa (see Figure). Although proposed as a unit for mapping at a reconnaissance scale, it was rarely used. The catena provides however, a useful model for investigation of PEDOGENESIS when used in combination with the *clorpt equation* (see SOIL FORMING FACTORS) and stimulated development of the *nine-unit slope model* (see SLOPE ZONATION MODELS). *EMB*

[See also SOIL FORMING FACTORS]

Birkland, P.W. and Burke, R.M. 1988: Soil catena chronosequences on eastern Sierra Nevada moraines, California. *Arctic and Alpine Research* 20, 473–484. Dalrymple, J.B., Blong, R.J. and Conacher, A.J. 1968: An hypothetical nine unit landsurface model. *Zeitschrift für Geomorphologie* 12, 60–76. Milne, G. 1935: Some suggested units of classification and mapping, particularly for East African soils. *Soil Research* 4, 183–198. Ollier, C.D. 1976: Catenas in different climates. In Derbyshire, E. (ed.), *Geomorphology and climate*. London: Wiley, 137–170.

cathodoluminescence (CL) analysis *Cathodoluminescence* is the emission of photons in the visible range of the ELECTROMAGNETIC SPECTRUM after excitation by high-energy electrons in an electron microscope. A hot-cathode method is more widely used than low-temperature CL microscopy. Different colours originate from various impurities in the crystal lattice. The technique is important in the study of DIAGENESIS in sediments and sedimentary rocks and has been widely used in ARCHAEOLOGICAL GEOLOGY to determine the sources of white marbles, which have been divided into three major families

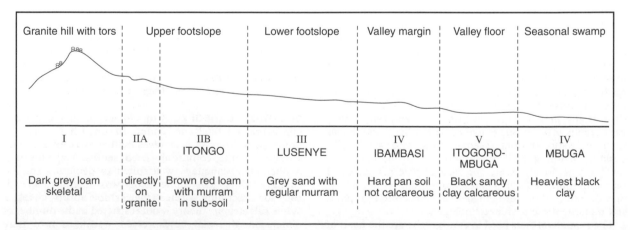

catena *A soil catena from East Africa: Itonga. Lusenye, Ibambasi, Itogoro-mbuga and Mbuga are colloquial soil names. Murram refers to ironstone (lateritic) gravel (after Milne, 1935)*

based on their luminescence colours: orange (mostly calcitic marbles), red (dolomitic marbles) and blue. *JAM*

[See also SCIENTIFIC ARCHAEOLOGY]

Barbin, V., Ramseyer, K., Decrouez, D. *et al.* 1992: Cathodoluminescence of white marbles: an overview. *Archaeometry* 34, 175–183. **Marshall, D.J.** 1988: *Cathodoluminescence of geological materials*. Boston: Unwin Hyman. **Miller, J.** 1988: Cathodoluminescence microscopy. In Tucker, M.E. (ed.), *Techniques in sedimentology*. Oxford: Blackwell, 174–190.

cation exchange capacity (CEC)

Clay and humus particles are able to attract, hold and exchange cations from the soil solution because they have an overall negative charge. The negative charge arises through unsatisfied charges at broken ends of crystals and isomorphic replacement of elements within the atomic structure of clays. This property of a soil is measured normally by percolation with an ammonium acetate solution at pH 7, the sample is then washed free of excess salt and percolated with potassium chloride. The content of ammonium is determined in the percolate to give a value for the exchange capacity. The sum of base cations, plus the soil acidity is the cation exchange capacity and the BASE SATURATION is the percentage of base cations held at an equilibrium condition of pH 7.0. The sum of the exchangeable bases and exchangeable aluminium is known as the *effective cation exchange capacity* (ECEC) and is a useful measure in soils where aluminium ions predominate. The base cations (Ca^{2+}, Mg^{2+}, Na^+, K^+) are displaced by hydrogen ions (H^+) and aluminium ions (Al^{3+}) in the process of LEACHING and ACIDIFICATION of the soil. Cation exchange, cation exchange capacity and effective cation exchange capacity are fundamental to understanding soil fertility and productivity. *EMB*

Bolt, G.H. and Bruggenwert, M.G.M. (eds), 1976: *Soil chemistry: A. basic elements* [*Developments in Soil Science* 5A]. Amsterdam: Elsevier. **van Reeuwijk, L.P. (ed.)** 1995: *Procedures for soil analysis*, 5th edn [*Technical Paper* 9]. Wageningen: International Soil Reference and Information Centre and FAO. **Wild, A. (ed.)** 1988: *Russell's soil conditions and plant growth*, 11th edn. Harlow: Longman

cation-ratio dating

The ratio between relatively mobile potassium (K^+) and calcium (Ca^{2+}) cations and the immobile titanium (Ti^{4+}) cation in ROCK VARNISH is considered to be a function of exposure to LEACHING, and hence time. Varnish development and the extent to which mobile cations are leached is related to accumulated environmental effects, which have considerable temporal variation. Attempts have been made to calibrate cation-ratio dating using RADIOCARBON DATING, but the results are controversial and the technique has been applied rarely in recent years. *DAR*

Lanteigne, M.P. 1991: Cation-ratio dating of rock engravings: a critical appraisal. *Antiquity* 65, 292–295. **Schnieder, J.S. and Bierman, P.R.** 1997: Surface dating using rock varnish. In Taylor, R.E. and Aitken, M.J. (eds), *Chronometric dating in archaeology*. New York: Plenum.

catotelm

In diplotelmic MIRES, the subsurface (i.e. beneath the ACROTELM) ANAEROBIC peat horizons, permanently waterlogged, in which decay of plant material continues but is much reduced. *FMC*

Ingram, H.A.P. 1978: Soil layers in mires: function and terminology. *Journal of Soil Science*, 29, 224–227. **Ivanov, K.E.** 1981: *Water movement in mirelands*. London: Academic Press.

causal relationship

A relationship between two phenomena or events where it is possible to distinguish between cause and effect. If there is a relationship between A and B such that B would not have occurred without A, then A is the cause of B. EXPERIMENT is the most appropriate approach used in SCIENCE to establish causal relationships. In the study of environmental change, many relationships are known that are suspected as being causal but where causality has not been established. Commonly, understanding of environmental change rests on *functional relationships* (numerical relationships based on techniques such as CORRELATION) that are not necessarily causal, imperfectly known causal chains and interactions between several causes with multiple effects. Thus, establishing causal relationships is an important aim but an all too rarely achieved outcome of research in this field. *JAM*

cave

From an anthropocentric perspective, commonly defined as a natural subterranean void sufficiently large to be enterable by a human. As a biological HABITAT, however, much smaller voids may be accepted as caves. KARST caves have been defined as solutional conduits large enough for turbulent water flow. Caves and CAVE PASSAGES in particular should be distinguished from excavated *tunnels*.

Caves may be exogenous, often known as ROCK-SHELTERS and little more than a short recess in a rock face, or endogenous, comprising chambers or passages, which may be interconnecting and run for many kilometers. Although characteristic of karst landscapes and normally found in LIMESTONE or CARBONATE rocks, caves also occur within SANDSTONES, EVAPORITES and ICE and are common in coastal environments and volcanic terrain. They are produced as a result of a number of interacting processes, often over long timescales, and their morphology reflects these interactions. The dominant formation process is dissolution of the limestone that depends on the availability of dissolved CARBON DIOXIDE in water, leading to the dissociation of *calcite* and *dolomite* in the rock. Rates of dissolution are higher in tropical climates where there is a higher concentration of carbon dioxide in the SOIL consequent upon bacterial activity and where there is higher rainfall. The form of the cave system is strongly influenced by LITHOLOGY in the form of JOINTS, FAULTS and FRACTURES. Cave formation is also affected by the hydrological characterisics of streams that flow into them and palaeo-discharges have been determined from morphometric analysis of systems and from the degree of scalloping seen on the surface of passages. Overall morphology may also be affected by TECTONIC activity and SEA-LEVEL CHANGE, which, either together or separately, lead to a raising or lowering of BASE LEVELS. In China, limestone towers have many caves presently situated well above the present base level as a result of long-term lowering (see TOWER KARST) and in the Bahamas the Blue Holes comprise cave systems now well below sea level (see DROWNED KARST).

Caves are valuable as sources of information about QUATERNARY ENVIRONMENTAL CHANGE because they can be of great age, their form and sedimentary inclusions representing changing conditions over many millennia or even millions of years. CAVE SEDIMENTS take a number of forms and are preserved with minimal alteration as a result of almost constant temperature and humidity, and hence very low rates of DIAGENESIS, with little erosion once base levels have been lowered. CLASTIC cave sediments can originate from both outside and within the cave system, reflecting such events as local GLACIATION or PERIGLACIATION and changing FLUVIAL regimes. SPELEOTHEMS within caves provide evidence of former surface conditions through a variety of ISOTOPES such as δD, $\delta^{18}O$ and $\delta^{13}C$, which derive from the overlying soil prior to being precipitated and locked within the calcite. Because of their longevity, caves have proved of considerable value in establishing long-term Quaternary environmental change in carbonate-rich environments. Caves in the Nullabar Plain in Australia can be shown to have formed originally in the TERTIARY, as can some of the highest caves in the Chinese tower karst around Guilin. Upper levels of the world's longest system, the Mammoth Cave system in Kentucky, formed prior to the *Brunhes–Matuyama palaeomagnetic reversal*, perhaps in the OLDUVAI EVENT.

The age of cave systems can be determined by the application of a number of DATING TECHNIQUES. URANIUM-SERIES DATING of speleothem has proved the most commonly used technique, sometimes supplemented by ELECTRON SPIN RESONANCE (ESR) and THERMOLUMINESCENCE DATING. RADIOCARBON DATING has also been used, but with limited success as a result of the availability of 'old' CARBON derived from the bedrock. Over the longer QUATERNARY TIMESCALE, PALAEOMAGNETISM provides the only real means of dating when applied to clastic deposits such as SANDS, SILTS and CLAYS.

Because they formed shelters, and occasionally traps, for both animals and humans, caves are rich sources of ORGANIC remains, particularly bones, and investigations of cave deposits have contributed significantly to the understanding of HUMAN EVOLUTION and more recent archaeological problems. Extensive excavation of fossiliferous cave deposits by antiquarians in the nineteenth and early twentieth centuries, notably in the British Isles, has led to the loss of many important records, as has the excavation of such sites for 'medicinal' use of bones in other areas.

CJC

[See also SALT CAVES]

Barba, P.L.A., Manzanilla, L., Chavez, R. *et al.* 1990: Caves and tunnels at Teotihuacan, Mexico: a geomorphological phenomenon of archaeological interest. In Lasca, N.P. and Donahue, J. (eds), *Archaeological geology of North America*. Boulder, CO: Geological Society of America. **Chapman, P.** 1993: *Caves and cave life*. London: HarperCollins. **Dreybrodt, W.** 1988: *Processes in karst systems: physics, chemistry and geology*. Berlin: Springer. **Farrant, A.R., Smart, P.L., Whitaker, F.F. and Tarling, D.H.** 1995: Long-term Quaternary uplift rates inferred from limestone caves in Sarawak, Malaysia. *Geology* 23, 357–360. **Ford, D.C. and Ewers, R.O.** 1978: The development of limestone cave systems in the dimensions of length and depth. *International Journal of Speleology* 10, 213–244. **Ford, D.C. and Williams, P.W.** 1989. *Karst geomorphology and hydrology*. London: Unwin Hyman. **Gillieson, D.** 1996: *Caves: processes, development and management*. Oxford: Blackwell. **Palmer, A.N.** 1991. Origin and morphology of limestone caves. *Geological Society of America Bulletin* 103, 1–21. **Sutcliffe, A.J.** 1985. *On the track of Ice Age mammals*. London: British Museum (Natural History). **Trudgill, S.T.** 1985: *Limestone geomorphology*. London: Longman. **Wood, C.** 1976: Caves in rocks of volcanic origin. In Ford, T.D. and Cullingford, C.H.D. (eds), *The science of speleology*. London: Academic Press, 127–150.

cave art Primitive drawings, paintings or engravings (PETROGLYPHS) in caves or rockshelters, especially on rock walls (PARIETAL ART). Upper PALAEOLITHIC cave art, as in the caves of Chauvet and Lascaux in southern France, usually depict single large mammals of interest to hunters; there are virtually no representations of vegetation, scenery or other context. Later examples, such as the art of shallow rock shelters in Levantine Spain, depict more complex scenes. Recent advances in dating techniques, such as CATION-RATIO DATING of petroglyphs and RADIOCARBON DATING based on ACCELERATOR MASS SPECTROMETRY (AMS) of charcoal used in the drawings, now permit the direct dating of cave art and hence a greater contribution to the study of environmental change is in prospect.

JAM

Beltrán, A. 1982: *Rock art of the Spanish Levant*. Cambridge: Cambridge University Press. **Jameson, R.J.** 1999: Cave art. In Shaw, I. and Jameson, R. (eds), *A dictionary of archaeology*. Oxford: Blackwell. **Layton, R.** 1994: *Australian rock art: a new synthesis*. Cambridge: Cambridge University Press. **Ucko, P. and Rosenfeld, A.** 1967: *Palaeolithic cave art*. London: World University Library.

cave passage A natural underground passage in rock large enough for people to enter. The longest system of connected cave passages explored to date exceeds 500 km. *SHD*

cave sediments Cave sediments can be classified into three main groups: CLASTIC or DETRITAL, ORGANIC and CHEMICAL. Clastic or detrital deposits comprise a wide range of material from angular BOULDERS to fine CLAYS and are predominantly allogenic or ALLOCHTHONOUS deriving from outside the cave system. Autogenic or AUTHIGENIC deposition occurs through the solutional breakdown of the country rock producing *cave earth* and by gravity fall from the roof or sides of passages to produce *thermoclastic scree*, often comprising extremely coarse blocks. The development of such material is enhanced near to cave entrances, where there is greater amplitude of temperature, leading in extreme cases to FREEZE–THAW CYCLES, but is also a normal result of solutional weakening of the country rock. Allochthonous material usually enters cave systems as water-lain sediments, but close to entrances wind-blown SILTS and SANDS may be found as well as TILL and PERIGLACIAL or COLLUVIAL SEDIMENTS. Clastic sediments may be dated by PALAEOMAGNETIC DATING and also by THERMOLUMINESCENCE DATING in appropriate material.

Organic material in sediments in caves can take many forms. Because caves have been used for shelter by animals, or have trapped animals that fell into chambers through surface openings, they can contain large quantities of animal bones, often of considerable age. Analysis of such ANIMAL REMAINS has had an important role to play in understanding QUATERNARY ENVIRONMENTAL CHANGE, for even if the sediments are not well dated, they can be stratigraphically related within sequences spanning many millennia. Further information can be obtained from COPROLITES left by animals and, because of their attraction for bats, caves can also contain sediments rich in guano.

Caves were important sites of early human occupation and hence both skeletal remains and occupation debris may be found interstratified in cave sediments. Analyses of MOLLUSCA from clastic sediments and POLLEN from both clastic sediments and SPELEOTHEMS has provided further information about Quaternary environmental change.

Chemical sediments predominantly take the form of CARBONATE precipitates, in the form of speleothems (*stalagmites* and *stalactites*) and TUFA (*travertine*). These sediments are made up of either calcite or aragonite and are formed by the loss of CARBON DIOXIDE (CO_2) or degassing and/or evaporation, the latter primarily close to cave entrances, where there is more airflow and exchange. Speleothems provide the base for ISOTOPIC analysis utilising δD, $\delta^{18}O$ and $\delta^{13}C$ as palaeoclimatic proxies for both temperature and precipitation. Recent studies have demonstrated the existence of annually banded speleothem, offering the opportunity for high-resolution CLIMATIC RECONSTRUCTION within chronologies constrained by URANIUM-SERIES DATING. FLUORESCENCE SPECTROSCOPY of the HUMIC ACIDS trapped within the calcite of speleothems may also be used to show changes in the soils and environment immediately overlying the cave system.

Overall, cave sediments have a number of advantages for Quaternary environmental reconstruction: they provide long records, especially in areas subject to glaciation and thus lacking in surface deposits; they contain a range of palaeoclimatic proxies; and they can be dated quite precisely over the Late Quaternary by the use of uranium-series dating. *CJC*

[See also SALT CAVES]

Baker, A., Smart, P.L., Edwards, R.L. and Richards, D.A. 1993: Annual growth banding in a cave stalagmite. *Nature* 364, 518–520. Burney, D.A., Brook, G.A.and Cowart, J.B. 1994: A Holocene pollen record for the Kalahari Desert of Botswana from a U-series dated speleothem. *The Holocene* 4, 225–232. Camacho, C.N., Carrión, J.S., Navarro, J. *et al.* 2000: An experimental approach to the palynology of cave deposits. *Journal of Quaternary Science* 15, 603–615. Carrión, J.S., Munuera, M. and Navarro, C. 1998: The paleoenvironment of Carihuela Cave (Granada, Spain): a reconstruction on the basis of palynological investigations of cave sediments. *Review of Palaeobotany and Palynology* 99, 317–340. Gascoyne, M. 1992: Palaeoclimate determination from cave calcite deposits. *Quaternary Science Reviews* 11, 609–632. Gillieson, D. 1996: *Caves: processes, development and management.* Oxford: Blackwell. Hill, C. and Forti, P. (eds) 1997: *Cave minerals of the world.* Huntsville, AL: National Speleological Society. Moriarty, K.C., McCulloch, M.T., Wells, R.T. and McDowell, M.C. 2000: Mid-Pleistocene cave fills, megafaunal remains and climate change at Naracoorle, S.Australia: towards a predictive model using U–Th dating of speleothems. *Palaeogeography, Palaeoclimatology and Palaeoecology* 159, 113–143.

cay An island, often elongate, of concentric to parallel sand ridges formed by wave refraction around an underlying REEF. Cays may be unvegetated and ephemeral or larger, vegetated and more stable forms. *TSp*

[See also CORAL REEF ISLANDS]

cement Mineral material precipitated in the pore spaces of a SEDIMENT during DIAGENESIS, bringing about LITHIFICATION to a SEDIMENTARY ROCK. Minerals forming cement are often termed AUTHIGENIC. The most common

are QUARTZ, calcite and CLAY MINERALS, but a very wide range of minerals and even HYDROCARBONS may form cement. *TY*

[See also CEMENTATION]

cementation The precipitation of minerals (CEMENT) in the pore spaces of a SEDIMENT during DIAGENESIS. The process may bind the GRAINS together as part of the process of LITHIFICATION, forming a SEDIMENTARY ROCK. *TY*

Cenozoic (*Cainozoic*) An ERA of geological time comprising two PERIODS: the TERTIARY and QUATERNARY. *GO*

[See also GEOLOGICAL TIMESCALE]

census enumeration district The basic data collection unit of the UK census and a basis for investigating DEMOGRAPHIC CHANGE. The districts are constructed differently in England and Wales, Scotland and Northern Ireland, respectively. Returns for each are the responsibility of one enumerator. *TF*

Openshaw, S. 1993: *Census users handbook.* Cambridge: Geoinformation International.

Central England Temperature Record (CET) The World's longest monthly mean HOMOGENEOUS SERIES starting in AD 1659, the year following the death of Oliver Cromwell. There is also a daily mean temperature series (the mean of the daily maximum and minimum temperatures) of the same name extending back to 1772. These series were originally prepared by Gordon Manley and are now maintained by the UK Meteorological Office. *DAW*

Jones, P. and Hulme, M. 1997: The changing temperature of 'Central England'. In Hulme, M. and Barrow, E. (eds), *Climates of the British Isles: present, past and future.* London: Routledge, 173–196. Manley, G. 1974: 'Central England Temperatures': monthly means 1659 to 1975. *Quarterly Journal of the Royal Meteorological Society* 100, 389–405.

centroid In a GEOGRAPHIC INFORMATION SYSTEM (GIS), the geometric (or weighted) centre of an area or polygon. The centroid is usually represented as a POINT, along with an ATTRIBUTE label or value, and located by geographical COORDINATES. The term is also used for the centre of a distribution in a non-geographical space, for example in ORDINATION. *TVM*

Martin, D.J. 1996: *Geographic information systems and their socioeconomic applications.* London: Routledge.

ceramics ARTEFACTS and archaeological materials, mostly *pottery* (earthenware, stoneware, porcelain, etc.) manufactured from clay with an addition of temper at high temperatures (*fired clay*). Fragments of ceramics (*potsherds*) are extremely common at archaeological sites and are relatively easily dated by stylistic differences and/or archaeometric dating techniques (see ARCHAEOMETRY). Ceramics may be investigated to yield a wide range of information on, for example, the origin of the materials and the nature and spread of cultures. *JAM*

Neff, H. (ed.) 1992: *Chemical characterization of ceramic pastes in archaeology* [*Monograph* 7]. Madison, Wisconsin: Prehistoric Press. Velde, B. and Druc, I.C. 1999: *Archaeological ceramic materials: origin and utilization.* Berlin: Springer.

cereal crops Corn or edible grain grown and consumed as food. Cereal crops were derived by DOMESTICATION of wild grasses and resulted in SEDENTISM. *Wheat, barley, oats* and *rye* were the earliest cereal crops, the domestication of which began in southwest Asia at least 13 000 calendar years ago (11 000 BP). These were followed by: *maize*, which was domesticated in Meso-America c. 8000 BP; sorghum and millet in west Africa c. 8000–6000 BP; and rice in southeast Asia and west Africa c. 7000 BP. Cultivation of such crops often required innovation in tool sets, IRRIGATION, hoeing, PLOUGHING, HARVESTING and crop processing techniques. Cereal crops are detectable in the palaeoecological record (PALAEOECOLOGY) by their remains as MACROFOSSILS and CEREAL POLLEN. Indirect evidence for growing of cereal crops can be obtained from archaeological ARTEFACTS such as quern stones and ploughs. *LD-P*

[See also AGRICULTURAL HISTORY, AGRICULTURAL ORIGINS, GREEN REVOLUTION]

Hillman, G., Hedges, R., Moore, A. *et al.* 2001: New evidence of lateglacial cereal cultivation at Abu Hureya on the Euphrates. *The Holocene* 11, 383–393. **Kroll, H.** 1998: Literature on the archaeological remains of cultivated plants 1996–1997. *Vegetation History and Archaeobotany* 7, 23–56.

cereal pollen The pollen of cultivated GRASSES can be distinguished from that of other Poaceae, largely on size grounds, and serves as a valuable ANTHROPOGENIC INDICATOR in POLLEN ANALYSIS. *MJB*

[See also AGRICULTURAL HISTORY, AGRICULTURAL ORIGINS]

Andersen, S.Th. 1979: Identification of wild grasses and cereal pollen. *Danmarks geologiske Undersøgele, Aarbog* 1978, 69–92.

chamber sampler A device for the collection of SEDIMENTS, which is pushed into the sediment to the sampling depth, then rotated in order to collect a core of sediment within the chamber. The most common type is the *Russian* or *Jowsey peat sampler.* *MJB*

[See also CORER, PISTON SAMPLER]

Jowsey, P.C. 1966: An improved peat sampler. *New Phytologist* 65, 245–248.

channel bar A BEDFORM within a channel, representing a site of DEPOSITION. Bars are particularly associated with river channels, but are also an important element of other systems, such as *tidal channels* or *submarine deepwater channels.* The scale of a bar is related to channel width (representing a *macroform*), in contrast to smaller bedforms that scale with channel depth (*mesoforms* such as DUNES) or flow parameters (*microforms* such as RIPPLES).

Bars can be classified according to their morphology, GRAIN-SIZE or position with respect to the channel. A *point bar* is a type of *bank-attached bar* on the inner bank of a MEANDER. An *in-channel bar* (channel bar in the strict sense) lies between the banks of the channel, and migrates downstream. Such bars are a distinctive feature of braided rivers (see BRAIDING) and are sometimes described as *braid bars.* There are many types: *longitudinal bars* are elongate parallel to the channel axis and are typically dominated by gravel; *transverse bars* are wider than they are long; and *linguoid bars* have an apex at the downstream end.

On modern or relict FLOODPLAINS, former bars can be reconstructed from their GEOMORPHOLOGY. Deposition on bars gives rise to distinctive SEDIMENTARY STRUCTURES that allow bars to be reconstructed from SEDIMENTARY DEPOSITS of any age. *GO*

[See also ARCHITECTURAL ELEMENT ANALYSIS, CHANNEL PATTERNS, FACIES ARCHITECTURE]

Allen, J.R.L. 1983: Studies in fluviatile sedimentation: bars, bar-complexes and sandstone sheets (low-sinuosity braided streams) in the Brownstones (L. Devonian), Welsh Borders. *Sedimentary Geology* 33, 237–293. **Knighton, D.** 1998: *Fluvial forms and processes: a new perspective.* London: Arnold. **Miall, A.D.** 1994: Reconstructing fluvial macroform architecture from two-dimensional outcrops: examples from the Castlegate Sandstone, Book Cliffs, Utah. *Journal of Sedimentary Research* B64, 146–158.

channel capacity The maximum area of cross-section of a river channel, i.e. channel cross-sectional area at BANKFULL DISCHARGE. Channel capacity, normally measured in m^2, can be difficult to identify precisely. Capacities can reduce or increase downstream of RESERVOIRS and areas of URBANISATION. *DML*

Knighton, A.D. 1998: *Fluvial forms and processes: a new perspective.* London: Arnold, Chapter 5.

channel change The change in river-channel form or position over time, as planform change, cross-sectional change or channel slope/longitudinal profile change. Despite substantial research on rates and patterns of river channel change, fewer studies on the dynamics, controls or mechanisms have emerged. This may reflect the difficulty of the problem and the absence, until recently, of techniques to monitor channel change continuously.

Rivers are amongst the most dynamic and sensitive elements in the landscape, and often the first to be affected by an environmental change that alters water and sediment supply from the basin. This has spawned numerous studies of the impact on channel stability and SUSPENDED LOADS of, for example, DEFORESTATION and URBANISATION. Furthermore, sedimentary sequences preserved in floodplains and other riverine depositional contexts as a legacy of channel change, especially if they contain dateable organic matter, may themselves be used to decipher past environmental changes. Understanding and quantifying 'natural', evolutionary, channel adjustment, and distinguishing it from channel response to imposed environmental change, is crucial to the development of a full understanding of fluvial systems. Channel instability also has key relevance to catchment sediment fluxes, the preservation in the geological (sedimentary) record of environmental events, aquatic organism population dynamics (e.g. freshwater invertebrates) and floodplain biological diversity. Applied implications of river-channel change, such as water resource development, bank stabilisation, security of channel and riparian structures and loss of agricultural land, have added further impetus to the field.

There are eight main methods of quantifying channel change, each with its own appropriate timescale of application and achievable temporal RESOLUTION. These can be grouped into techniques for deciphering change over: *long timescales* (c. 10–20 000 years), including sedimentological evidence, botanical evidence and historical sources; *intermediate timescales* (c. 1–30 years), including planimetric resurvey and repeated cross-profiling; and *short timescales* (c. 0.2–10 years), including the high-resolution methods that are often reserved for studies of the *dynamics* of channel change (such as terrestrial PHOTOGRAMMETRY, EROSION PINS and bedload TRACERS). Dateable sedimentological evidence is used to establish alluvial chronologies to reconstruct episodes of channel AVULSION or stability. Botanical, especially dendrochronological, evidence (see TREE RINGS) is used to quantify lateral channel change from rates of tree colonisation of newly created point bars. The use of HISTORICAL EVIDENCE may involve the quantitative comparison of sequences of early maps and aerial photographs to document lateral channel movements (see Figure). *Planimetric resurvey* is the repeated field surveying of channel planform to evaluate meander growth, bank erosion rates etc. Similarly, repeated *cross-profiling* in the field can record changes in river-channel cross-section size, shape and position. *Terrestrial photogrammetry* involves quantitative, now largely analytical, processing of overlapping, ground-based imagery taken with a phototheodolite to produce serial 3D models of the channel reach/bank: differencing of successive surfaces thus quantifies channel change. Erosion pins – networks of rods inserted into the channel boundary and remeasured at intervals to determine exposure differences and hence erosion or deposition – are a very common method for short timescale applications. More recently, the first *automatic* method of monitoring channel change has been developed in the form of the PHOTO-ELECTRONIC EROSION PIN (PEEP) monitoring system, based on a sensor containing an array of light-sensitive cells: changing exposure of the sensor by erosion or deposition is recorded by a datalogger to clarify, for example, BANK EROSION event magnitude, frequency and timing. Specific bed-scour detection methods (e.g. scour chain, bed profiler, MICRO-EROSION METER) have also been developed.

Channel change can be *autogenic* (part of a natural river development process, such as meander bend growth or downstream translation) or *allogenic* (adjustment to environmental change, such as CLIMATIC CHANGE, urbanisation or LANDUSE CHANGE): either can occur at timescales from a few days to millennia. Lateral channel change rates increase with stream power (although discharge is often used as a surrogate because data are more readily available) and decrease with bank resistance (indexed by, for example, percentage of silt-clay content in the bank material), though many bank erosion processes exist. River-bend migration rates can peak at *intermediate* values of bend curvature or 'tightness', when values of the r_m/w ratio lie between two and three (where r_m is the bend radius and w the channel width). Thus, in gently curving bends, flow asymmetry may not develop sufficiently to impose significant enough boundary shear stresses on the outer bank to cause bank erosion: in very tight bends, however, flow separation can take place, allowing a protective cell of low-velocity water to develop at the cutbank; this is a good example of a form–process feedback. Absolute migration rates can range from 0–1000 m y^{-1} (or 0–10% channel width per year) and tend to increase with drainage-basin area. Evidence has also emerged, however, that bank erosion rates peak in mid basin, in association with high *stream power* values. Significant channel change can occur in response to a wide range of flows, not just once THRESHOLDS for SEDIMENT TRANSPORT and channel boundary deformation (which themselves can change over various timescales) are crossed. If pattern change is extreme, it has been termed *channel metamorphosis*. Both negative (recoverable) and positive (irreversible) FEEDBACK may occur in response to environmental change. A tendency for channel migration rates to increase over the last 100 years has been observed, as discharge and/or sediment supply regimes have changed as a result of landuse change, urbanisation or climate change.

Studies of river cross-sectional change have revealed significant changes in CHANNEL CAPACITY, including increases downstream of urbanisation schemes and both increases and decreases downstream of RESERVOIRS. Fewer studies exist on channel-slope or longitudinal-profile change and these mainly address HUMAN IMPACTS, for example the effects of river regulation or CHANNELISATION. Bed scour rates are very low for rock-cut channels, but can be substantial for sand-bed rivers. This has generated the need for laboratory flume (hardware-model) experiments (e.g. of KNICKPOINT recession).

Prediction of channel change is at an early stage, as Richards and Lane (1997: 274–275) argue: 'successful modelling of both the transient and equilibrium channel responses to disturbance therefore await physically realistic models of the bank erosion and width adjustment processes'. Encouraging approaches embracing multiple channel adjustments, hydrodynamics, SEDIMENT TRANSPORT continuity and topographic feedback are currently being developed. QUATERNARY legacies (e.g. the nature of the alluvial materials, basin slope development and sediment supply inheritance effects), however, can impose constraints on modelling successes, as can scale, environment and uncertainty. *DML*

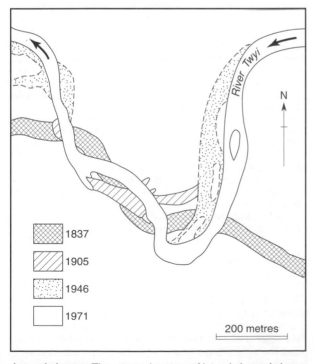

channel change *The rate and pattern of lateral channel change of the River Twyi at Llandeilo, South Wales, based on historical sources (after Lewin and Manton, 1975; Lawler et al., 1997)*

Gregory, K.J. (ed.) 1977: *River channel changes*. Chichester: Wiley. **Gurnell, A.M, Foster, I.J. and Petts, G.E. (eds)** 1995: *Changing river channels*. Chichester: Wiley. **Hickin, E.J.** 1983:

River channel changes: retrospect and prospect. In Collinson, J.D. and Lewin, J. (eds), *Modern and ancient fluvial systems* [*Special Publication of the International Association of Sedimentologists* 6]. Oxford: Blackwell, 61–83. **Hooke, J.M.** 1997: Styles of channel change. In Thorne, C.R., Hey, R.D. and Newson, M.D. (eds), *Applied fluvial geomorphology for river engineering and management*. Chichester: Wiley, 237–268. **Lawler, D.M.** 1993: The measurement of river bank erosion and lateral channel change: a review. *Earth Surface Processes and Landforms* 18, 777–821. **Lawler, D.M., Thorne, C.R. and Hooke, J.M.** 1997: Bank erosion and instability. In Thorne, C.R., Hey, R.D. and Newson, M.D. (eds), *Applied fluvial geomorphology for river engineering and management*. Chichester: Wiley, 137–172. **Leopold, L.B.** 1973: River channel change with time: an example. *Geological Society of America Bulletin* 84, 1845–1860. **Lewin, J.** 1987: Historical river channel changes. In Gregory, K.J., Lewin, J. and Thornes, J.B. (eds), *Palaeohydrology in practice*. Chichester: Wiley 161–175. **Lewin, J. and Manton, M.M.** 1975: Welsh floodplain studies: the nature of floodplain geometry. *Journal of Hydrology* 25, 37–50. **Macklin, M.G. and Lewin, J.** 1997: Channel, floodplain and drainage basin response to environmental change. In Thorne, C.R., Hey, R.D. and Newson, M.D. (eds), *Applied fluvial geomorphology for river engineering and management*. Chichester: Wiley, 15–45. **Petts, G.E., Möller, H. and Roux, R.L. (eds)** 1989: *Historical change of large alluvial rivers: Western Europe*. Chichester: Wiley. **Richards, K.S. and Lane, S.N.** 1997: Prediction of morphological changes in unstable channels. In Thorne, C.R., Hey, R.D. and Newson, M.D. (eds), *Applied fluvial geomorphology for river engineering and management*. Chichester: Wiley, 269–292.

channel gradient The concept of channel gradient, describing the slope of the floor of a channel, is usually applied to river channels. The variation in channel gradient along its length from its source is described as the *longitudinal profile*. Longitudinal profiles are typically concave-upwards and the downstream decrease in gradient is associated with an increase in DISCHARGE. Discontinuities in the gradual shape of a longitudinal profile, with a downstream increase in gradient, are referred to as KNICKPOINTS, which can be interpreted in terms of phases of channel downcutting downstream of the knickpoint related, for example, to changes in BASE LEVEL. Former channel gradients and longitudinal profiles can be reconstructed from the analysis of RIVER TERRACES and form an important tool in the reconstruction of PALAEOHYDROLOGY. *GO*

[See also GRADE CONCEPT]

Richards, K. 1982: *Rivers: form and process in alluvial channels*. London, New York: Methuen. **Seeber, L. and Gornitz, V.** 1983: River profiles along the Himalayan arc as indicators of active tectonics. *Tectonophysics* 92, 335–367.

channel pattern The planimetric form (planform) of a river reach, i.e. the two-dimensional nature of a river course as viewed from above. Channel patterns are different from *drainage patterns*, the larger-scale form of the river network at the basin scale. Most channel pattern classification schemes are based on morphological characteristics (usually planform geometry). Increasingly, however, classifications embrace river dynamism (e.g. whether MEANDERS are actively developing or are passive, i.e. inactive), channel instability, boundary sediments, channel slope, WIDTH–DEPTH RATIO and hydrological and geomorphological processes, including SEDIMENT TRANSPORT. Classification systems acknowledge a continuum of planform geometry. An early scheme (Leopold and Wolman,

1957) classified rivers as straight, meandering (highly sinuous rivers) or braided (high-energy, multithread), on the basis of planform sinuosity P (P is the ratio of channel length to straight-line valley length). They proposed a simple quasi-threshold based on slope s and BANKFULL DISCHARGE Q_b (two components of *stream power*), $s = 0.013Q_b^{-0.44}$, to separate meandering rivers from BRAIDING. Thus, at low stream powers, streams tend to be straight or inactively sinuous, though natural rivers are rarely straight for more than 10 channel widths. At higher powers, erosional energy is sufficient to deform the channel boundary and create MEANDERS; if stream power exceeds a critical value, then braiding is likely (see Figure).

Despite reservations about such fixed thresholds, most authors agree that the creation of a particular channel pattern is dependent on the total energy available. For example, Richards (1982) defines a direct positive relationship between sinuosity and stream power as $P = 2.64\Phi^{0.1}$, where Φ is a stream power index (discharge × channel slope). Recent classifications grade across a continuum of pattern, including transitional forms, between straight, sinuous, meandering, braided and anastomosed channels (see Figure). *Anastomosing channels*, occasionally defined as one type of ANABRANCHING channel, are low-energy multithread systems, which occur at lower slopes than braided forms, and are characterised by sinuous channels dividing around large, semi-permanent, vegetated islands at approximately floodplain height.

Quantitative prediction of channel pattern and CHANNEL CHANGES in response to environmental change usually address the degree of meandering–braided transition and are based on slope, discharge, bed material size or sediment supply variables. The implication is that those rivers plotting close to a pattern THRESHOLD should be more sensitive to environmental change and more susceptible to future channel-pattern change. CHANNELISATION of such rivers should be approached cautiously. *DML*

Ferguson, R. 1987: Hydraulic and sedimentary controls of channel pattern. In Richards, K.S. (ed.), *River channels: environment and process* [*Institute of British Geographers Special Publication* 18]. Oxford: Blackwell, 129–158. **Knighton, A.D.** 1998: *Fluvial forms and processes: a new perspective*. London: Arnold. **Leopold, L.B. and Wolman, M.G.** 1957: River channel patterns – braided, meandering and straight. *United States Geological Survey Professional Paper* 282B, 39–85. **Nanson, G.C. and Knighton, A.D.** 1996: Anabranching rivers: their cause, character and classification. *Earth Surface Processes and Landforms* 21, 217–239. **Richards, K.S.** 1982: *Rivers: form and process in alluvial channels*. London: Methuen, Chapter 7. **Rosgen, D.** 1994: A classification of natural rivers. *Catena* 22, 169–199. **Thorne, C.R.** 1997: Channel types and morphological classification. In Thorne, C.R., Hey, R.D. and Newson, M.D. (eds), *Applied fluvial geomorphology for river engineering and management*. Chichester: Wiley, 175–222.

channel shape The cross-sectional form of a river channel, defined by width (w) and depth dimensions (d). The WIDTH–DEPTH RATIO is a widely used shape index. Channel shape influences and reflects flow properties (see HYDRAULIC GEOMETRY), but also channel boundary sediment; and modern width explanations reflect this duality. Perimeter sediment properties (e.g. the proportion of fine particles in the bank material, which provide cohesion) influence resistance to BANK EROSION and hence width. Finer bank sediments lead to narrow and deep channels, while more easily entrained sandy bank materials lead to

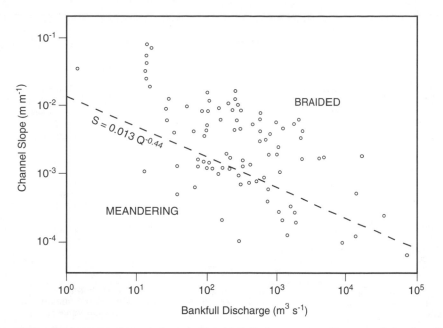

channel pattern *The relationship between channel slope and bankfull discharge, separating meandering from braided rivers (after Leopold and Wolman, 1957; Knighton, 1998)*

wide and shallow channels, e.g. $F = 255\ M^{-1.08}$, where F is the form or width–depth ratio and M is the weighted mean percentage of silt-clay in the boundary sediment. Bank resistance is also affected by its geotechnical characteristics (e.g. shear strength), composite structure and vegetation: tree-lined channels may be 30% narrower than expected from hydraulic geometry relations. Coarse bed materials inhibit bed scour, resulting in wider, shallow channels. Channel shape oscillates with CHANNEL PATTERN: asymmetry develops at MEANDER bends, where bed scour and bank retreat are maximised; more symmetrical cross-sections develop at inflection points. If river flows increase (e.g. with URBANISATION) and/or RIPARIAN trees are removed, channel widening would be expected: *in extremis*, BRAIDING, with abnormally wide and shallow channels, can develop. In KARST, the shape of cave channels can reflect creational environmental history. *DML*

Davies, T.R.H. 1987: Channel boundary shape – evolution and equilibrium. In Richards, K.S. (ed.), *River channels: environment and process* [*Institute of British Geographers Special Publication* 18]. Oxford: Blackwell, 228–248. **Knighton, A.D.** 1998: *Fluvial forms and processes: a new perspective.* London: Arnold, Chapter 5. **Miller, A.J. and Gupta, A. (eds)** 1999: *Varieties of fluvial form.* Chichester: Wiley.

Channeled Scablands

J. Harlen Bretz first used this term in 1923 to describe the area of eastern Washington State, USA where JÖKULHLAUPS occurred some 40 times as a result of the breaching of the ice-dammed Lake Missoula during the LAST GLACIATION and eroded deep canyons, left dry waterfalls and pockmarked the basalt surface with potholes. Deposits include fans with huge boulders and giant CURRENT RIPPLES up to 15 m high. Discharge during the jökulhlaup events has been estimated as reaching 10 times that of all the world's rivers combined. *JS*

[See also GLACIOFLUVIAL LANDFORMS, GLACIOFLUVIAL SEDIMENTS]

Baker, V.R. and Bunker, R.C. 1985: Cataclysmic Late Pleistocene flooding from Glacial Lake Missoula: a review. *Quaternary Science Reviews* 4, 1–41. **Waite Jr, R.B.** 1980: About forty last-glacial Lake Missoula jökulhlaups through southern Washington. *Journal of Geology* 88, 653–679.

channelisation

Deliberate physical modification of a river channel for the management of navigation, flood conveyance, land drainage, channel stability or erosion problems. Channelisation involves human alteration of: CHANNEL SHAPE; CHANNEL CAPACITY; CHANNEL PATTERN; long-profile regrading or alignment (e.g. removal of MEANDER bends); boundary materials; or riparian vegetation. It can be carried out for a short reach or for an entire river. Often, there has been a legislative requirement to carry out such '*channel improvement*' works (e.g. to reduce the cost of flooding). Many types of channelisation have been identified, depending on aims, extant legislation, agency protocols and available finance. They include *resectioning* (*channel widening* or *deepening*), *realigning*, *embanking*, *bank protection* and *clearing* (removal of aquatic weed, silt, etc). The principal effects are shown in the Figure.

Impacts depend on the nature and extent of scheme, sympathetic mediating measures taken, degree of basin-wide perspective adopted and efficacy of post-project appraisal. Key potential impacts (Brookes, 1988) can be grouped into four categories. First, channel instability may be introduced during the period of river readjustment to any newly imposed in-channel flows and sediment fluxes. Second, downstream (and occasionally upstream) effects can emerge, especially increased BANK EROSION and FLOODS from the higher velocities created by straightening further upstream. Third, channel cleaning, straightening and steepening, and removal of RIPARIAN vegetation and

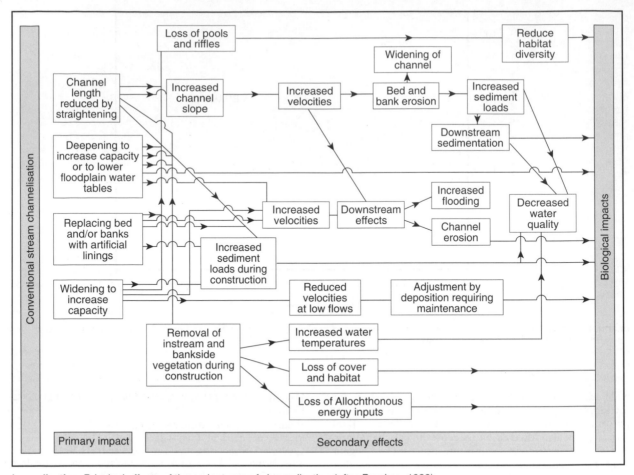

channelisation *Principal effects of the major types of channelisation (after Brookes, 1988)*

riffles, directly remove natural spatial variation from the stream system, especially in velocity distribution, water depth, geomorphology, bed sediments, light climate and water temperature (through lack of shaded sites). Such 'fluvial monotony' leads to a loss of habitat diversity, with well documented negative freshwater ecological effects, particularly on invertebrates and fish populations (which may be halved). Furthermore, consequent increases in bed shear stresses and SEDIMENT TRANSPORT, a lack of aquatic vegetation to provide organisms with shelter and food supply and reduced organic inputs from bankside vegetation may all serve to discourage recolonisation and species diversity. Finally, channelisation schemes can be visually intrusive and reduce aesthetic appeal. River behaviour subsequent to early channelisation works has helped to identify the nature of fluvial adjustment, the likely impact of future engineering works and potential mediating measures. However, because of many perceived disadvantages of channelisation, a recent tendency has emerged to 'reinstate' rivers through RESTORATION or REHABILITATION programmes, especially in North America and Europe (notably Denmark and the UK), and to devise river engineering schemes that make use of natural fluvial processes rather than imposing unnatural forms upon them. *DML*

Brookes, A. 1987: River channel adjustments downstream from channelisation works in England and Wales. *Earth Surface Processes and Landforms* 12, 337–351. **Brookes, A.** 1988: *Channelised rivers: perspectives for environmental management.* Chichester: Wiley. **Brookes, A.** 1997: River dynamics and channel maintenance. In Thorne, C.R., Hey, R.D. and Newson, M.D. (eds), *Applied fluvial geomorphology for river engineering and management.* Chichester: Wiley, 293–307. **Downs, P.W.** 1994: Characterisation of river channel adjustments in the Thames Basin, South-East England. *Regulated Rivers: Research and Management* 9, 151–175. **Leeks, G.J., Lewin, J. and Newson, M.D.** 1988: Channel change, fluvial geomorphology and river engineering: the case of the Afon Trannon, mid-Wales. *Earth Surface Processes and Landforms* 13, 207–223. **Simon, A.** 1989: A model of channel response in disturbed alluvial channels. *Earth Surface Processes and Landforms* 14, 11–26. **Simon, A. and Darby, S.E.** 1997: Disturbance, channel evolution and erosion rates: Hotphia Creek, Mississippi. In Wang, S.Y., Langendoen, E.J. and Shields Jr, F.D. (eds), *Management of landscapes disturbed by channel incision.* Oxford, MS: The University of Mississippi, 476–481.

chaos No single definition of chaos is universally accepted. One suggested definition is the complicated, aperiodic, attracting orbits of certain, usually low-dimensional DYNAMICAL SYSTEMS. Chaos addresses the path of a dynamical system through its state (or more commonly phase) space following an unambiguous rule specifying where to go next (the mathematicians' *vector field*). The long-term path of a classical dynamical system is drawn towards two possible ATTRACTORS – either to a fixed, *equilibrium point* or a *limit cycle* (a limited orbit). However, some *non-linear dynamical systems*, while DETERMINISTIC and frequently even simple in nature, exhibit complex, irregular and pseudo-random long-term behaviour, i.e. they are drawn to STRANGE ATTRACTORS. Such SYSTEMS are extremely sensitive to their initial conditions and those with even minutely different starting points diverge increasingly over time; they are also highly sensitive to PERTURBATION.

A test for the potentially chaotic behaviour of a non-linear dynamical system involves calculation of its *Lyapunov exponents*, of which there is one for each system dimension. If the average of the Lyapunov exponents is positive, the system will behave chaotically in general; if there is at least one positive exponent, the system will behave chaotically in at least some conditions. Chaos is an inherent system attribute, the origin of which should not be confused with STOCHASTIC complex behaviour, which is apparently random, system behaviour stemming from multiple external controls or the outcome of numerous individual phenomena (which may be individually deterministic). The nonlinear behaviour that is fundamental to chaos is equally fundamental to FRACTAL statistics. *Chaos theory* and chaotic behaviour are avowedly INTERDISCIPLINARY. *CET*

Casti, J.L. 1990: *Searching for certainty*. New York: Morrow and Company. Gleick, J. 1987: *Chaos making a new science*. New York: Penguin. Phillips, J.D. 1999: *Earth surface systems: complexity, order and scale*. Oxford: Blackwell. Turcotte, D.L. 1997: *Fractals and chaos in geology and geophysics*. Cambridge: Cambridge University Press.

charismatic species

Species with popular appeal and that tend to attract wide support for protection against environmental change. Typical examples include birds, mammals (such as the giant panda) and spectacular flowering plants (such as orchids). *IFS*

[See also CONSERVATION BIOLOGY]

charred-particle analysis

Charred-particle (*charcoal*) analysis forms part of investigations in PALAEOECOLOGY and ENVIRONMENTAL ARCHAEOLOGY into past fire regimes, where charcoal particles preserved in a stratigraphic setting are quantified to estimate the relative importance and/or use of fire in the past. Charcoal is also an appropriate material for RADIOCARBON DATING. An understanding of FIRE HISTORY is relevant to long-term studies of VEGETATION CHANGE, CLIMATIC CHANGE, ENVIRONMENTAL CHANGE and HUMAN IMPACT ON ENVIRONMENT. In ecosystems where fire is an important disturbance mechanism, an understanding of the fire regime is essential for land management and CONSERVATION purposes, as well as for understanding its role in NATURAL ecosystems. In many cases (especially forested ecosystems), the interval between fires can extend for decades and even centuries and therefore a retrospective approach is essential.

Fires in the natural environment are characterised by incomplete combustion and large quantities of charcoal are produced by most fires. A portion of this charcoal is carried by wind and water into sedimentary environments in which it can be preserved under ANAEROBIC conditions. The amount of charcoal produced is governed by the type of material being burnt and the nature of the fire (intensity, duration and temperature). The amount of charcoal quantified depends on the amount originally produced as modified by dispersal, deposition and preservation processes, as well as the amount detected by the sampling, preparation and counting method employed.

Both macroscopic and microscopic charred particles are analysed in palaeoecological and archaeological studies. Charcoal abundance can be quantified by a number of techniques: (1) counting particles on microscope slides prepared for pollen analysis or thin sections of embedded sediments; (2) sieving sediments using sieves of various sizes and categorising particles in several size classes; (3) elemental carbon analysis; (4) measuring the MINERAL MAGNETISM of sediments; (5) electron microscopy study; (6) spectroscopic study; and (7) *image analysis*.

Estimating the source area of charred particles preserved in sediments is highly problematic. Large particles (*macroscopic charcoal*) are thought primarily to reflect local fires, whereas small particles (*microscopic charcoal*) reflect fire in the greater region, sometimes extending a considerable distance from the sampling site. The size and intensity of a fire, its proximity to the sedimentary environment (from which charcoal samples are analysed) and the size of the sedimentary environment (lake or WETLAND) all affect the way the fire is represented in charcoal records. *JLF*

[See also BIOMASS BURNING, FIRE FREQUENCY, FIRE SCAR, SEDIMENT, SEDIMENT TRANSPORT, SEDIMENTOLOGICAL EVIDENCE OF ENVIRONMENTAL CHANGE]

Clark, J.S. and Hussey, T.C. 1996: Estimating the mass flux of charcoal from sedimentary records: effects of particle size, morphology, and orientation. *The Holocene* 6, 129–144. Clark, J.S. and Royall, P.D. 1995: Particle-size evidence for source areas of charcoal accumulation in late Holocene sediments of eastern North American lakes. *Quaternary Research* 43, 80–89. Clark, J.S. 1988: Particle motion and the theory of charcoal analysis: source area, transport deposition, and sampling. *Quaternary Research* 30, 67–80. MacDonald, G.M., Larsen, C.P.S, Szeicz, J.M., and Moser, K.A. 1991: The reconstruction of boreal forest fire history from lake sediments: a comparison of charcoal, pollen, sedimentological, and geochemical indices. *Quaternary Science Reviews* 10, 53–71. Ohlson, M. and Tryterud, E. 2000: Interpretation of the charcoal record in forest soils: forest fires and their production and deposition of macroscopic charcoal. *The Holocene* 10, 519–525. Patterson III, W.A. and Backman, A.E. 1988: Fire and disease history of forests. In Huntley, B. and Webb III, T. (eds), *Vegetation history*. Dordrecht: Kluwer, 603–632. Patterson III, W.A., Edwards, K.J. and Maguire, D.J. 1987: Microscopic charcoal as a fossil indicator of fire. *Quaternary Science Reviews* 6, 3–23. Tolonen, K. 1986: Charred particle analysis. In Berglund, B. (ed.), *Handbook of Holocene palaeoecology and palaeohydrology*. Chichester: Wiley, 485–496. Whitlock, C. and Millspaugh, S.H. 1996: Testing the assumptions of fire-history studies: an examination of modern charcoal accumulation in Yellowstone National Park, USA. *The Holocene* 6, 7–15.

chattermark

A crescent-shaped gouge found, for example, on beach cobbles, along geological faults and on sand grains. On glacially eroded bedrock surfaces, chattermarks form *en echelon* trails produced by the jerky stick–slip movement of CLASTS held in the sole of a GLACIER or ICE SHEET. They often occur within shallow grooves.

Chattermark trails on sand grains have mostly been associated with glacial environments, but can also be produced in marine intertidal environments. *JS*

[See also FRICTION CRACK, GLACIAL EROSION, QUARTZ GRAIN SURFACE TEXTURES]

Bull, P.A., Culver, S.J. and Gardner, R. 1980: Chattermark trails as palaeoenvironmental indicators. *Geology* 8, 318–322. **Chamberlin, T.C.** 1888: The rock scorings of the great ice invasions. *United States Geological Survey Seventh Annual Report*, 147–248. **Orr, E.D. and Folk, R.L.** 1983: New scents on the chattermark trail – weathering enhances obscure microfractures. *Journal of Sedimentary Petrology* 53, 121–129.

chelation The ability of organic breakdown products to form water-soluble chemical complexes with a metal, facilitating transport in solution through a soil. It helps to explain movement of iron and aluminium in the PODZOLISATION process. Chelates also enable certain MICRONUTRIENTS, such as iron, zinc and copper, to be made available to plants: these would be rendered unavailable if added, for example, to a calcareous soil. *EMB*

[See also BIOLOGICAL WEATHERING]

Bolt, G.H. and Bruggenwert, M.G.M. (eds) 1976: *Soil chemistry: A. basic elements* [*Developments in Soil Science* 5A]. Amsterdam: Elsevier. **Brady, N.C.** 1984: *The Nature and properties of soils*. New York: Macmillan.

chemical analysis of soils Analytical methods used to describe properties that can be measured in terms of their chemical composition. The most common chemical properties investigated in soils are: pH; organic carbon content; the available plant nutrients, nitrogen, phosphorus and potassium; extractable iron; calcium carbonate equivalent; gypsum; cation exchange properties, exchangeable bases, exchangeable acidity, base saturation; electrical conductivity; and trace elements. *EMB*

Allen, S.E., Grimshaw, H.M. Parkinson, J.A. and Quarmby, C. 1974: *Chemical analysis of ecological materials*. Oxford: Blackwell Scientific. **Bengtsson, L. and Enell, M.** 1986: Chemical analysis. In Berglund, B.E. (ed.), *Handbook of Holocene palaeoecology and palaeohydrology*. Chichester: Wiley, 423–451. **Hunt, D.T.E. and Wilson, A.L.** 1990: *The chemical analysis of water. General principles and techniques*, 2nd edn. Cambridge: Royal Society of Chemistry. **Van Reeuwijk, L.P. (ed.)** 1995: *Procedures for soil analysis*, 5th edn [*Technical Paper* 9]. Wageningen: FAO and ISRIC. **Van Reeuwijk, L.P.,** 1998: *Guidelines for quality management in soil and plant laboratories* [*Soils Bulletin* 74]. Rome: ISRIC and FAO.

chemical denudation The contribution to wearing down of the Earth's surface (DENUDATION) made by chemical processes. In order to determine rates and amounts of chemical denudation within landscapes, a catchment approach involving the monitoring and calculation of DISSOLVED LOAD is usually employed. The dissolved load (calculated in terms of unit area) is the product of the mean flow and mean solute concentration of a river divided by the catchment area. Estimations of load are subject to considerable error as both flows and concentrations are temporally variable and need to be monitored frequently in storms and at times of BASEFLOW. To estimate chemical denudation, the non-denudational component (largely precipitation and artificial inputs) needs to be subtracted from the total dissolved load. It has been suggested that,

for most lithological environments, after removing Na^+, Cl^- and NO_3^-, the remaining ions in streamwater will represent chemical denudation within a catchment. Rates of chemical denudation increase greatly with rock and soil solubility, and with annual rainfall and annual runoff (streamflow per unit area) and to a lesser extent with temperature.

Chemical denudation will be affected by environmental change on a number of levels. Over GEOLOGICAL TIME there may be a link through the CARBON CYCLE between rates of chemical denudation and climatic change. Rapid UPLIFT of the Tibetan Plateau in the late Cenozoic, for example, led to an increased intensity of the Asian MONSOON system and increased rates of chemical denudation of silicate rocks. This caused a decrease in the concentration of CARBON DIOXIDE in the atmosphere and contributed significantly to global cooling. Shorter-term fluctuations will occur with LANDUSE CHANGE within the catchment. This was shown in the Hubbard Brook Experimental Forest in New Hampshire, USA, where CLEAR CUTTING was shown to increase losses of nitrate and cations. *ADT/GO*

[See also CHEMICAL WEATHERING]

Filippelli, G.M. 1997: Intensification of the Asian monsoon and a chemical weathering event in the late Miocene early Pliocene: implications for late Neogene climate change. *Geology* 25, 27–30. **Janda, R.J.** 1971: An evaluation of procedures used in computing chemical denudation rates. *Geological Society of America Bulletin* 82, 67–80. **Likens, G.E., Bormann, F.H., Pierce, R.S. et al.** 1977: *Biogeochemistry of a forested ecosystem*. New York: Springer. **Raymo, M.E. and Ruddiman, W.F.** 1992: Tectonic forcing of late Cenozoic climate. *Nature* 359, 117–122. **Reesman, A.L. and Godfrey, A.E.** 1972: *Chemical erosion and denudation in the middle Tennessee. Water Resources Series*. Vol. 4. United States Department of Conservation. **Stallard, R.F. and Edmond, J.M.** 1987: Geochemistry of the Amazon: 3. Weathering chemistry and limits to dissolved inputs. *Journal of Geophysical Research* 92 (C8), 8293–8302.

chemical oxygen demand (COD) A measure of the non-microbial requirement of a waste for oxygen and hence the strength of an INDUSTRIAL WASTE that is toxic or resistant to BIODEGRADATION (biological oxidation). *JAM*

[See also BIOCHEMICAL OXYGEN DEMAND]

chemical precipitation The crystallisation of a dissolved component from a supersaturated solution, forming a *precipitate*. SPELEOTHEMS, such as stalactites and stalagmites for example, are common deposits in limestone caves and are formed by the precipitation of dissolved calcite under certain environmental conditions. *SHD*

[See also SUPERSATURATION]

chemical remediation Chemical methods for reducing the levels of POLLUTANTS in the environment including, for example, chemical conversion processes in WASTE MANAGEMENT, the chemical modification of pulp and paper mill bleaching processes and the reformulation of petrol and diesel fuels. Major reductions in TOXICANT emissions have been made in recent years, not only by chemical remediation but also by electrochemical, thermal and engineering remediation strategies. *JAM*

[See also BIOREMEDIATION]

Fishbein, L. 1997: Chemical remediation. In Brune, D., Chapman, D.V., Gwynne, M.D. and Pacyna, J.M. (eds), *The global environment: science, technology and management*. Weinheim, Germany: VCH, 908–926.

chemical time bomb

Up to certain critical concentrations, soils and sediments are able to accumulate toxic elements, HEAVY METALS or PESTICIDE residues without adverse environmental effects. Changes in LANDUSE, climate or soil conditions may trigger the sudden release of these toxic substances into the environment. This concept has been referred to as a 'chemical time bomb'. One of the best documented examples was the sudden release of MERCURY from Swedish soils into lakes as a result of ACID-IFICATION. This occurred long after industrial EMISSIONS had been brought under strict control and it produced toxic levels of mercury in fish, making them unfit for human consumption.   *EMB*

Stigliani, W.M. (ed.) 1991: *Chemical time bombs: definition, concepts and examples* [Executive Report 16]. Laxenburg: International Institute for Applied Systems Analysis. **Ter Meulen, G.B.R., Stigliani, W.M., Salomons, W.** *et al.* (eds) 1992: *Chemical time bombs. Proceedings of the European State of the Art Conference on Delayed Effects of Chemicals in Soils and Sediments*. Hoofddorp: Foundation for Ecodevelopment.

chemical weathering

The dissolution or chemical alteration of minerals *in situ*, usually in the presence of water. Important processes include HYDRATION, which is the absorption or adhesion of water molecules to a mineral, increasing the mass and volume of minerals and thus aiding the disintegration of rock. Solutional weathering (*solution*) is the dissolution of solid material. It is not necessarily a permanent chemical change, as substances often precipitate from solution back to their original chemical form. Although all rock types are soluble to some degree, solubility varies widely between rock types. OXIDATION is the chemical combination of minerals with oxygen and is most apparent with respect to iron compounds, especially in humid tropical climates. Tropical soils are often distinctively red, because they contain large quantities of iron oxides. *Carbonation* involves chemical reactions of substances with carbonic acid (H_2CO_3), which is the product of the dissolution of carbon dioxide in water and associated with HYDROLYSIS. Decaying organic matter in litter and soil is an important source of carbon dioxide. Thus carbonation is most pronounced in well vegetated environments. Carbonation results in the conversion of oxides into more soluble carbonates and therefore aids solutional weathering. It is most effective on limestone and very effective on feldspar minerals, which are major constituents of most igneous rocks.

Globally, chemical weathering is more important than PHYSICAL WEATHERING. Since chemical processes normally operate most effectively under warm and moist conditions, the HUMID TROPICS are usually associated with the most rapid and intense chemical weathering. Factors such as the chemical stability, texture and permeability greatly influence chemical weathering of rock. Chemical weathering under particular environmental conditions can produce distinctive weathering products. The CLAY MINERALS of soils, for example, usually differ between humid tropical and temperate climates and can thus be used as a sedimentary *palaeoclimatic indicator*.   *SHD*

[See also AGGRESSIVE WATER, BIOLOGICAL WEATHERING, KARST, WEATHERING, WEATHERING RATES]

Colman, S.M. and Dethier, D.P. (eds) 1986: *Rates of chemical weathering of rocks and minerals*. Orlando, FL: Academic Press. **Drever, J.E.** (ed.) 1985: *The chemistry of weathering*. Dordrecht: Reidel. **Gibbs, M.T., Bluth, G.J.S., Fawcett, P.J. and Kump, L.R.** 1999: Global chemical erosion over the last 250 my: variations due to changes in paleogeography, paleoclimate, and paleogeology. *American Journal of Science* 299, 611–651. **Tebbens, L.A., Veldkamp, A. and Kroonenberg, S.B.** 1998: The impact of climate change on the bulk and clay geochemistry of fluvial residual channel fillings: the Late Weichselian and Early Holocene River Meuse sediments (The Netherlands). *Journal of Quaternary Science* 13, 345–356. **Thomas, M.F.** 1994: *Geomorphology in the tropics*. Chichester: Wiley. **White, A.F. and Brantley, S.L.** 1995: Chemical weathering of silicate minerals: an overview. *Reviews in Mineralogy* 31, 1–22.

chemocline

The boundary or zone of rapid vertical change between WATER MASSES of contrasting chemical composition, most commonly the transition from oxic surface water to ANOXIC or SALINE deep water.   *MRT*

Beadle, L.C. 1981: *The inland waters of tropical Africa*. London: Longman. **Sinninghe Damsté, J.S., Wakeham, S.G., Kohnen, M.E.L.** *et al.* 1993: A 6,000 year sedimentary molecular record of chemocline excursions in the Black Sea. *Nature* 362, 827–829.

chemosphere

The altitudinal zone of the ATMOS-PHERE, mostly in the STRATOSPHERE and *mesosphere* between about 32 km and 92 km, characterised by photochemical reactions.   *JAM*

[See also OZONE, PHOTOCHEMISTRY]

chemostratigraphy

Chemostratigraphy or *chemical stratigraphy* is the definition of stratigraphic units according to the variations in their chemical properties. Although originally largely defined in terms of major elements, the increasing use of isotopes has led to the development of distinct isotopic stratigraphies such as that based on OXYGEN ISOTOPES.   *DH*

[See also MOLECULAR STRATIGRAPHY, STRATIGRAPHY]

chemosynthesis

The use by bacterial micro-organisms of inorganic and organic compounds as a source of energy in the absence of light. Chemosynthesis is important in the SULPHUR CYCLE and the NITROGEN CYCLE. The *chemo-autotrophic* or *chemolithotrophic bacteria* involved include those of HYDROTHERMAL VENT communities on the ocean floor and those used in the recovery of metals from WASTE.   *JAM*

[See also BIOTECHNOLOGY]

chenier

An isolated BEACH RIDGE of sand or shell fragments that overlies and is set in muds deposited by TIDES. *Chenier ridges* are commonly wooded (the term is from the French *chêne*, meaning oak) and form a *chenier plain* in which ridges are separated by SWAMPS (e.g. southwest Louisiana). Despite being usually less than 3 m high, cheniers have provided sites for coastal occupation.

GO/HJW

Davis, R.A. 1997: Regional coastal morphodynamics along the United States Gulf of Mexico. *Journal of Coastal Research* 13, 595–604. **Meldahl, K.H.** 1995: Pleistocene shoreline ridges from tide-dominated and wave-dominated coasts – northern Gulf of California and western Baja-California, Mexico. *Marine Geology* 123, 61–72. **Russell, R.J. and Howe, H.V.** 1935: Cheniers of southwestern Louisiana. *Geography Review* 25, 449–461. **Stanley, D.J. and Chen, Z.Y.** 1996: Neolithic settlement distributions as a function of sea level-controlled topography in the Yangtze delta, China. *Geology* 24, 1083–1086. **Vilas, F., Arche, A., Ferrero, M. and Isla, F.** 1999: Subantarctic macrotidal flats, cheniers and beaches in San Sebastian Bay, Tierra del Fuego, Argentina. *Marine Geology* 160, 301–326.

chernic horizon

A deep, dark-coloured, organic-rich and well-structured diagnostic surface soil horizon with high base status and supporting a rich biological population. It is typical of true CHERNOZEMS.   *EMB*

Food and Agriculture Organization of the United Nations 1998: *World reference base for soil resources* [*Soil Resource Report* 84]. Rome: FAO, ISRIC, ISSS.

Chernozem

Dark-coloured soils (*Black Earths*) with thick, organic-rich (4–16% organic carbon) CHERNIC or mollic surface horizons developed on LOESS parent materials occurring in central North America, South America and Eurasia. These soils are neutral in reaction with high BASE SATURATION and carbonates present in lower horizons. Environmental change and poor management are suggested as responsible for degraded chernozems in Russia and WIND EROSION is a hazard (SOIL TAXONOMY: Mollisols).   *EMB*

[See also WORLD REFERENCE BASE FOR SOIL RESOURCES]

chert

Fine-grained *silica* (SiO_2) made of microcrystalline QUARTZ and cryptocrystalline *chalcedony*. Chert occurs as nodules or BEDS, and varieties include *flint* and *jasper*. Processes that form chert in SEDIMENTARY ROCKS include the redistribution of biogenic silica during DIAGENESIS, the accumulation of siliceous micro-organisms (e.g. siliceous OOZE, DIATOMITE), chemical precipitation at hot springs associated with VOLCANISM and at SALINE LAKES, and the formation of SILCRETE. Many of the earliest traces of life on Earth are preserved as FOSSILS in chert.   *GO*

Golbubic, S. and Lee, S.J. 1999: Early cyanobacterial fossil record: preservation, palaeoenvironments and identification. *European Journal of Phycology* 34, 339–348. **Khadkikar, A.S., Sant, D.A., Gogte, V. and Karanth, R.V.** 1999: The influence of Deccan volcanism on climate: insights from lacustrine intertrappean deposits, Anjar, western India. *Palaeogeography, Palaeoclimatology, Palaeoecology* 147, 141–149. **Krainer, K. and Spötl, C.** 1998: Abiogenic silica layers within a fluvio-lacustrine succession, Bolzano Volcanic Complex, northern Italy: a Permian analogue for Magadi-type cherts? *Sedimentology* 45, 489–505. **Schopf, J.W. and Packer, B.M.** 1987: Early Archean (3.3-billion to 3.5-billion-year-old) microfossils from Warrawoona Group, Australia. *Science* 237, 70–73.

chironomid analysis

The identification and estimation of relative abundance of chironomid midge taxa recovered as larval head capsules from LAKE SEDIMENTS. Characteristic changes in the abundance and diversity of subfossil chironomid midge taxa sampled from lacustrine sediment cores are used to infer past ENVIRONMENTAL CHANGE. Chironomidae (Insecta: Diptera), or non-biting midges, are a family of two-winged flies with about 6000 species world-wide. The aquatic larvae are abundant, diverse, and ubiquitous in freshwater and many species are stenotopic (ecologically fastidious). The head capsules

of chironomid larvae are well preserved in lake sediments and they are readily identifiable. Typically, several hundred can be recovered from a few grams of sediment, thus enabling environmental reconstructions of high temporal resolution. The life cycle of most species is completed within a year and this, together with the abilities of the winged adults to disperse over a wide area, means that chironomids respond rapidly to environmental change.

Quantitative chironomid-inferred environmental reconstructions are achieved by developing CALIBRATION sets, based on the modern distribution of chironomid species along an environmental gradient, and using statistical methods such as *weighted averaging partial least squares regression* and calibration to develop quantitative TRANSFER FUNCTIONS, which interpret changes in the composition of subfossil chironomid assemblages as being driven by changes in a specific environmental variable. This technique has been used successfully to quantify LATE GLACIAL and HOLOCENE climate change expressed as mean July air temperature with a prediction error of about 1.5°C. In arid regions, climate change has been investigated by using chironomid analysis to reconstruct changes in lake salinity. The response of chironomids to EUTROPHICATION, following agricultural intensification, has also been studied by reconstructing chironomid-inferred values for total phosphorus, chlorophyll *a* and the duration of summer anoxia. Additionally, the impact of ACID RAIN can be inferred by the characteristic response of chironomid assemblages to changes in lake-water pH.   *SJB/HJBB*

[See also PALAEOLIMNOLOGY, CLIMATIC RECONSTRUCTION, NUMERICAL ANALYSIS]

Armitage, P., Cranston, P.S. and Pinder, L.C.V. (eds) 1995 *The Chironomidae. The biology and ecology of non-biting midges.* London: Chapman and Hall. **Brodin, Y.W. and Gransberg, M.** 1993: Responses of insects, especially Chironomidae (Diptera), and mites to 130 years of acidification in a Scottish lake. *Hydrobiologia* 250, 201–212. **Lotter, A.F., Birks, H.J.B., Hofmann, W. and Marchetto, A.** 1998: Modern diatom, Cladocera, chironomids and chrysophyte cyst assemblages as quantitative indicators for the reconstruction of past environmental conditions in the Alps. I. Climate. *Journal of Paleolimnology* 18, 395–420. **Quinlan, R., Smol, J.P. and Hall, R.I.** 1998: Quantitative inferences of past hypolimnetic anoxia in south-central Ontario lakes using fossil midges (Diptera: Chironomidae). *Canadian Journal of Fisheries and Aquatic Sciences* 55, 587–596. **Walker, I.R.** 1993. Paleolimnological biomonitoring using freshwater benthic macroinvertebrates. In: Resh, V.H. and Rosenberg, D.M. (eds) *Freshwater biomonitoring using benthic macroinvertebrates.* New York: Routledge, Chapman and Hall. **Walker, I.R., Wilson, S.E. and Smol, J.P.** 1995: Chironomidae (Diptera): quantitative palaeosalinity indicators for lakes of western Canada. *Canadian Journal of Fisheries and Aquatic Sciences* 52: 950–960.

chlorine-36

A *cosmogenic nuclide* with a HALF LIFE of 3.08×10^5 years, formed in rocks by the neutron activation of ^{35}Cl. ^{36}Cl has been used as a technique of SURFACE DATING as the level of activity increases over the time of exposure and has a potential dating range of 5 ka to 1 Ma.   *CJC*

[See also COSMOGENIC-ISOTOPE DATING]

Ballantyne, C.K., Stone, J.O. and Fifield, L.K. 1998: Cosmogenic Cl-36 dating of post glacial landsliding at The Storr, Isle of Skye, Scotland. *The Holocene* 8, 347–351. **Phillips, F.M., Leavey, B.D., Jannik, N.O. and Kubik, P.W.** 1986: The accumulation of cosmogenic chlorine-36 in rocks: a method for exposure dating. *Science* 231, 41–43. **Zreda, M.G. and Phillips, F.M.** 1994: Surface exposure dating by cosmogenic chlorine-36 accumulation. In Beck, C. (ed.), *Dating in exposed and surface contexts.* Albuquerque, NM: University of New Mexico Press, 161–183.

chlorofluorocarbons (CFCs) Compounds of carbon in which the hydrogen atoms have been replaced by chlorine and fluorine atoms. They are synthetic, human-produced chemicals that do not occur naturally and are used in various industrial processes as, for example, propellant gases, foaming agents, refrigerants and solvents. They have long lifetimes (up to about 400 years in the atmosphere) and act as GREENHOUSE GASES as well as depleting ozone. An international CONVENTION, the Montreal Protocol, deals with the phasing out of their use.
AHP

[See also HALOGENATED HYDROCARBONS, HYDROCHLORO-FLUOROCARBONS (HCFCs), OZONE DEPLETION, PERSISTENT ORGANIC COMPOUNDS]

chlorophyll The green and yellow pigments in plant cells, which facilitate PHOTOSYNTHESIS. There are two forms: chlorophyll-a and chlorophyll-b. *JAM*

[See also PLANT PIGMENT ANALYSIS]

chlorosis The yellowing or whitening of green plants reflecting a reduction in CHLOROPHYLL as a result of POLLUTION (such as ACID RAIN) or DISEASE. *JAM*

[See also DIEBACK]

cholera A DISEASE produced by ingestion of the water-borne bacterium *Vibrio cholerae*, which leads to rapid and potentially fatal dehydration through diarrhoea and vomiting. Cholera is believed to have originated in India and is now regarded as endemic in many tropical and subtropical regions. It is EPIDEMIC in Europe and it has reached *pandemic* proportions eight times in the last two centuries. Flooding, and the consequent contamination of water with sewage, has led to severe outbreaks around locations such as the Bay of Bengal. The bacterium thrives in warm waters and epidemics have frequently correlated with sea surface temperatures, elevating concerns that GLOBAL WARMING may enhance the future prevalence of the disease. *MLW*

[See also HUMAN HEALTH HAZARDS]

Banarjee, B. and Hazra, J. 1974: *Geoecology in West Bengal: a study in medical geography.* Calcutta: K.P. Bagchi. **Glass, R.I., Claeson, M., Blake, P.A. *et al.*** 1991: Cholera in Africa: lessons on transmission and control for Latin America. *The Lancet* 338, 791–795. **Reeves, P.R. and Lan, R.T.** 1998: Cholera in the 1990s. *British Medical Bulletin* 54, 611–623. **Wachsmuth, L., Blake, P. and Olsvik, Ø. (eds)** 1994: *Vibrio cholerae and cholera: molecular to global perspectives.* Washington, DC: American Society of Microbiology.

chott A large shallow fluctuating SALINE LAKE or the often tectonically formed lake basin or hollow in which the salt lake lies, located in North Africa. The word is Arabic in origin. *RAS*

Wadge, G., Archer, D.J. and Millington, A.C. 1994: Monitoring playa sedimentation using sequential radar images. *Terra Nova* 6, 391–396.

chronicles Historical written accounts, such as personal diaries, SHIP LOG BOOK RECORDS or agricultural records, which can be interpreted as PROXY EVIDENCE of past environmental conditions. *JCM*

Ingram, M.J., Underhill, D.J. and Farmer, G. 1981: The use of documentary sources for the study of past climate. In Wigley, T.M.L., Ingram, M.J. and Farmer, G. (eds), *Climate and History.* Cambridge: Cambridge University Press, 80–123.

chronofunction A quantitative relationship between a soil property and time based on soils of differing age (rather than from the development of a single soil). Chronofunctions may be derived by taking any soil characteristic and investigating it with respect to land surface age, keeping other factors as constant as possible. The principle may be applied to topics other than soils. *JAM*

[See also CHRONOSEQUENCE, ERGODIC HYPOTHESIS]

Jenny, H. 1946: Arrangement of soil series and types according to functions of soil forming factors. *Soil Science* 61, 375–391. **Yaalon, D.H.** 1975: Conceptual models in pedogenesis: can soil-forming functions be solved? *Geoderma* 14, 189–205.

chronology The assignment of dates to events or phenomena. In the ENVIRONMENTAL SCIENCES in general, chronology refers to the timescale associated with observations or measurements of various physical or biological processes (e.g. the accumulation of SEDIMENT in a lake or changes in the vegetation of a region). In PALAEOCLIMATOLOGY, many techniques are used to date the evidence of past changes. These have varying uncertainty associated with them. One specialised use of the word 'chronology' is found in DENDROCHRONOLOGY. Here the term refers to a continuous and annually resolved time series, where each value represents the mean of multiple measurements (or some transformation, e.g. into relative indices) of some tree-growth parameter (e.g. the width of the annual ring) taken from different trees. The values are averaged in their correct calendrical alignment so that the chronology represents the expression of some underlying forcing (see FORCING FACTOR, SOLAR FORCING) contained in the temporal pattern of variability of all of the trees. *KRB*

[See also CHRONOMETRY, TREE-RING INDEX, TREE RINGS]

Bradley, R.S. 1999: *Paleoclimatology: reconstructing climates of the Quaternary,* 2nd edn. London: Academic Press. **Fritts, H.C.** 1976: *Tree rings and climate.* London: Academic Press. **Plater, A.J.** 2000: Blind date: the importance of chronology in reconstructing the past. *Geology Today* 16, 63–70.

chronometric age An age estimate based on a dating system that refers to a specific point or range of time. Chronometric dates are not necessarily exact dates and they are often expressed as a range. *DAR*

chronometry The measurement of age. Linkage of STRATIGRAPHICAL RECORDS to chronometrically established ages (see CHRONOMETRIC AGE), establishes CHRONOLOGY. Thus: stratigraphy + chronometry = chronology. *JAM*

[See also TEPHROCHRONOLOGY, TEPHROCHRONOMETRY]

Bowen, D.Q. 1978: *Quaternary geology: a stratigraphic framework for multidisciplinary work.* Oxford: Pergamon. **Wagner, G.A.** 1998: *Age determination in young rocks and artifacts.* Berlin: Springer.

chronosequence A sequence of related soils that differ from each other in certain properties, primarily as a result of time as a SOIL FORMING FACTOR. Soils with different ages can occur in the same LANDSCAPE and they may overlap in age depending when soil formation began or ended. A fresh start to soil formation may be the result of human intervention or other change in environmental conditions. A flight of river terraces with different ages of alluvial, colluvial or aeolian deposits can produce soils that show a close relationship with the time taken in their formation. In most cases a chronosequence is built up by substituting space for time, taking examples from sites that illustrate a parallel increase in age and development. The concept is applicable to *vegetation chronosequences* and to whole landscapes as well as to *soil chronosequences*. Chronosequences have been classified by Vreeken (1975).
EMB

[See also CHRONOFUNCTION, K-CYCLE]

Dickson, B.A. and Crocker, R.L. 1953: A chronosequence of soils and vegetation near Mt Shasta, California. I. Definition of the ecosystem investigated and features of the plant succession. *Journal of Soil Science* 4,123–141. **Dickson, B.A. and Crocker, R.L.** 1953: A chronosequence of soils and vegetation near Mt Shasta, California. II. The development of the forest floors and the carbon and nitrogen profiles of the soils. *Journal of Soil Science* 4,142–154. **Dickson, B.A. and Crocker, R.L.** 1954: A chronosequence of soils and vegetation near Mt Shasta, California. III. Some properties of the mineral soils. *Journal of Soil Science* 5, 173–191. **Matthews, J.A.** 1992: *The ecology of recently deglaciated terrain: a geoecological approach to glacier forelands and primary succession.* Cambridge: Cambridge University Press. **Vreeken, W.J.,** 1975: Principal kinds of chronosequences and their significance in soil history. *Journal of Soil Science* 26:378–393.

chronostratigraphy The subdivision of the STRATIGRAPHICAL RECORD into units that have significance in terms of time, with the aim of establishing a standard relative time scale for global CORRELATION (the Global Standard Stratigraphy), represented on the *chronostratic scale* or STRATIGRAPHICAL COLUMN. Internationally agreed boundaries for units, primarily SYSTEMS, SERIES and STAGES, are co-ordinated through the International Commission on Stratigraphy of the International Union of Geological Sciences (IUGS). *Boundary stratotypes* are selected at *stratotype sections*, which should be accessible, have a continuous sediment or rock succession, be relatively undeformed, without marked FACIES changes, and be rich in FOSSILS or SUBFOSSILS, preferably with a varied fossil assemblage. A biostratigraphical datum marks the boundary level, the *Global Stratotype Section and Point* (GSSP) or '*golden spike*'. This represents a time point in the rock succession and a reference for correlation with other areas using all available methods of STRATIGRAPHY. Only the lower boundary of a unit is defined: the top is automatically defined by the base of the overlying unit. Independent geochronometric dating using radiometric isotopes (GEOCHRONOLOGY) provides absolute dates for chronostratigraphical units and boundaries (see GEOLOGICAL TIMESCALE). *LC*

Cowie, J.W. and Bassett, M.G. 1989: IUGS 1989 Global Stratigraphic Chart. *Episodes* 12, June 1989, Supplement. **Haq, B.U. and van Eysinga, W.B.** 1998: *Geological time table*, 5th edn. Amsterdam: Elsevier. **Harland, W.B., Armstrong, R.L., Cox, A.V.** et al. 1990: *A geologic time scale 1989.* Cambridge: Cambridge University Press. **Holland, C.H.** 1986: Does the

golden spike still glitter? *Journal of the Geological Society, London* 143, 3–21. **Whittaker, A., Cope, J.C.W., Cowie, J.W.** et al. 1991: A guide to stratigraphical procedure. *Journal of the Geological Society, London* 148, 813–824.

chronozone A non-hierarchial term used in CHRONOSTRATIGRAPHY applied to deposits formed during a designated interval of time based on the criteria of LITHOSTRATIGRAPHY, BIOSTRATIGRAPHY or MAGNETOSTRATIGRAPHY. Chronozones are often subject to uncertainty because the physical basis upon which they are defined may be TIME TRANSGRESSIVE. Definition of chronozones has played an important role in describing the sequence of events during LATE GLACIAL ENVIRONMENTAL CHANGE at the end of the LAST GLACIATION (as in the YOUNGER DRYAS or ALLERØD chronozones), although this approach has now been questioned, with researchers of the INTIMATE group favouring an approach based on EVENT STRATIGRAPHY.
DAR/CJC

Björck, S., Walker, M.J.C., Cwynar, L.C. et al. 1998: An event stratigraphy for the Last Termination in the North Atlantic region based on the Greenland ice-core record: a proposal by the INTIMATE group. *Journal of Quaternary Science* 13, 283–292. **Lowe, J.J. and Walker, M.J.C.** 1997: *Reconstructing Quaternary environments*, 2nd edn. Harlow: Longman. **Mangerud, J., Andersen, S.T., Berglung, B.E. and Donner, J.J.** 1974: Quaternary stratigraphy of Norden: a proposal for terminology and classification. *Boreas* 3, 109–127.

chrysophyte cyst analysis An ENVIRONMENTAL INDICATOR technique used to reconstruct pH, SALINITY and nutrients in lacustrine ecosystems. Chrysophyte cysts belong to two classes of algae, the Chrysophyceae and Synurophyceae, which occur mainly in freshwater ecosystems, although they can also be found in SALINE LAKES, MARINE SEDIMENTS and PEATlands. Chrysophytes produce hollow, siliceous, resting-stages called *stomatocysts*. Increasingly, these cysts are used in conjunction with other techniques to reconstruct *water quality*, e.g. DIATOM ANALYSIS, because they preserve well in the environment. Morphologically, stomatocysts are typically spherical or oval, 2–35 μm in diameter and have a single pore, frequently surrounded by a collar. Identifications are usually based on the structure of the stomatocyst (i.e. the cyst body and pore–collar complex). One major drawback of chrysophyte cyst analysis is that many cannot be identified to species level. Furthermore, many different species of chrysophyte produce identical cyst types, thereby limiting their potential in *environmental reconstruction*. *AWM*

Duff, K.E., Zeeb, B.A. and Smol, J.P. 1995: *Atlas of chrysophycean cysts.* Kluwer: Dordrecht. **Zeeb, B.A., Duff, K.E. and Smol, J.P.** 1996: Recent advances in the use of chrysophyte stomatocysts in palaeoecological studies. *Nova Hedwigia* 114, 247–252. **Facher, E. and Schmidt, R.** 1996: A siliceous chrysophycean cyst-based pH transfer function for Central European lakes. *Journal of Paleolimnology* 16, 275–321.

circadian rhythms The *diurnal rhythms* of behaviour exhibited by organisms in response to environmental stimuli, such as light or temperature, that oscillate with an approximately 24-hour period. Other *biological rhythms* operate over shorter (e.g. *tidal rhythms*, over 12.4 hours) or longer (e.g. *circannual rhythms* over one year) timescales. *JAM*

Saunders, D. 1977: *An introduction to biological rhythms.* London: Blackie.

cirque In uplands, a glacially eroded, typically semi-circular, steep-sided depression with an overdeepened, relatively flat floor. Cirques range from valley-side niches a few tens of metres across up to large features that are kilometres in width. Using an ERGODIC HYPOTHESIS, it is possible to view cirque development as the result of progressive growth from a SNOWBED occupying a NIVATION hollow through to a GLACIER in a true cirque, but development is probably more complex. The close relationships of cirque glaciers to climatic parameters, especially temperature, winter precipitation and PREVAILING WIND direction, and to orientation with respect to shading mean that the former existence of *cirque glaciers* (best indicated by MORAINES) can be of considerable value in CLIMATIC RECONSTRUCTION. Large cirques may have been re-occupied and excavated by glaciers throughout the PLEISTOCENE. Cirques are sometimes termed *corries* or *coires* in Scotland and *cwms* in Wales. *RAS*

[See also GLACIAL EROSION, U-SHAPED VALLEYS]

Jansson, P., Richardson, C. and Jonsson, S. 1999: Assessment of requirements for cirque formation in northern Sweden. *Annals of Glaciology* 28, 16–22. **Lewis, W.V. (ed.)** 1960: *Norwegian cirque glaciers* [*Research Series* 4]. London: Royal Geographical Society.

cirque altitude analysis Study of the spatial variation in the altitude of CIRQUE floors or thresholds across a region. Cirque altitude analysis is used to reconstruct former GLACIATION THRESHOLDS and to infer climatic parameters, such as precipitation gradients. However, cirques can rarely be assigned to any particular glacial event and their altitudes are unlikely to have a simple relationship with the EQUILIBRIUM-LINE ALTITUDES of the glaciers that formed them. Nevertheless, when used with caution, cirque altitude analysis can yield useful information on the average glacial conditions experienced during the PLEISTOCENE. *DIB*

Derbyshire, E. and Evans, I.S. 1976: The climatic factor in cirque variation. In Derbyshire, E. (ed.), *Geomorphology and climate.* New York: Wiley. **Richardson, C. and Holmlund, P.** 1996: Glacial cirque formation in northern Scandinavia. *Annals of Glaciology* 22, 102–106. **Robinson, G., Peterson, J.A. and Anderson, P.M.** 1971: Trend surface analysis of corrie altitudes in Scotland. *Scottish Geographical Magazine* 87, 142–146.

cladistics The analysis of ancestral LINEAGES through identification of groups of organisms with common ancestors, known as *clades*. Lineages are treated as a sequence of dichotomies representing the splitting of a parental species into two 'daughter' species. The ancestral species then cease to exist. This approach was established comprehensively by Willi Hennig, who called it *phylogenetic analysis*: it has since been renamed 'cladistics'. *KDB*

[See also CLADOGENESIS, SPECIATION]

Hennig, W. 1950: *Grundzüge einer Theorie der Phylogenetischen Systematik.* Berlin: Deutscher Zentralverlag.

Cladocera analysis A technique using the fossilised remains of a specific group of crustaceous *zooplankton* to reconstruct environmental changes in freshwater ecosystems, although some genera are marine. Remains of three families are usually used in Cladocera analysis: Chydoridae, Bosminidae and Daphniidae, each having different HABITAT preferences. Chydoridae live in mud or on macrophytes, whereas the latter two families are planktonic. Cladocera analysis provides detailed information on local freshwater habitat conditions, but provides less information on past climates and salinities. Cladoceran remains take the form of *exuviae*, or cast exoskeletons, which disassemble into their component parts, e.g. carapace and headshield. These remains, with the exception of those derived from Daphniidae, preserve quantitatively. Daphniidae remains preserve poorly overall, with only a few exoskeleton components, such as the mandibles, preserving in lake sediments. The geochemistry of exuviae is an important source of data in palaeoenvironmental reconstruction. *AWM*

Dodson, S.I. and D.G. Frey. 1991: Cladocera and other Branchipoda. In Thorpe, J.H. and A.P. Covich (eds), *Ecology and classification of North American freshwater invertebrates.* Toronto: Academic Press, 723–786. **Frey, D.G. 1986**: Cladocera analysis. In Berglund, B.E. (ed.), *Handbook of Holocene palaeoecology and palaeohydrology.* Chichester: Wiley. **Holmes, J.A.** 1992: Nonmarine ostracods as Quaternary palaeoenvironmental indicators. *Progress in Physical Geography* 16, 405–431. **Holmes, J.A., Hales, P.E. and Street-Perrott, F.A.** 1992: Trace element chemistry of nonmarine ostracods as a means of palaeolimnological reconstruction: an example from the Quaternary of Kashmir, northern India. *Chemical Geology* 95, 177–185.

cladogenesis The origin of new clades, or LINEAGES, by the splitting of an existing clade into two. *KDB*

[See also CLADISTICS, SPECIATION]

class intervals The range between the lower class limit and the upper class limit in classifying ordinal data into discrete classes (e.g. in IMAGE CLASSIFICATION). *HB*

[See also HISTOGRAM]

classification The process of putting entities, most commonly objects, into groups or clusters. Classifications can be subjective (e.g. as in some plant sociological procedures) or numerical and computer-based (e.g. CLUSTER ANALYSIS). Many classifications are hierarchical with groups nested within other groups. There are two main ways of deriving hierarchical classifications numerically. A divisive approach (e.g. TWO-WAY INDICATOR SPECIES ANALYSIS) starts with the entire set of objects and progressively divides into smaller and smaller groups to maximise some mathematical criterion of group uniformity. An agglomerative approach (e.g. CLUSTER ANALYSIS) starts with individual objects and progressively amalgamates them using a stated mathematical criterion of within-group variability into larger and larger groups until the entire data-set is grouped together. Other numerical approaches exist including non-hierarchical, overlapping, and fuzzy clustering. Classification and ORDINATION are the major approaches in the STRUCTURING of large, complex multivariate data-sets. The term classification is also used in the statistical literature to refer to DISCRIMINANT ANALYSIS and the assignment of unknown objects to *a priori* groups. Classification is mainly used nowadays to refer to the process of partitioning objects (or variables) into groups. *HJBB*

Gordon, A.D. 1999: *Classification,* 2nd edn. London: Chapman and Hall/CRC Press. **Legendre, P. and Legendre L.** 1998: *Numerical ecology.* Amsterdam: Elsevier.

clast A particle or GRAIN in a SEDIMENT or SEDIMENTARY ROCK. The term is commonly restricted to MINERAL and rock particles derived by WEATHERING and EROSION (i.e. DETRITAL particles): a particle of biological origin is a *bioclast*. *TY*

[See also GRAIN-SIZE, LIMESTONE, PYROCLASTIC]

clastic The fragmentary character of many SEDIMENTS and SEDIMENTARY ROCKS. Fragments (CLASTS) dominated by SILICATE MINERALS are commonly termed *siliciclastic*. TERRIGENOUS clastic material is derived from the WEATHERING of rocks on land and may include non-silicate material such as LIMESTONE or other CARBONATE debris as well as silicate minerals and silicate-rich rock fragments. *Bioclastic* material is of biological origin (e.g. shell fragments). Fragments of volcanic material (including PYROCLASTS and fragments of IGNEOUS ROCKS) are termed *volcaniclastic*. Clastic sediments are sometimes split into *epiclastic* (products of an extrinsic process of breakdown, such as weathering) and *autoclastic* (products of an intrinsic mechanism of breakage such as PYROCLASTIC material). *TY/GO*

[See also LITHIC, MINEROGENIC]

Nichols, G. 1998: *Sedimentology and stratigraphy.* Oxford: Blackwell.

clay (1) A material composed of CLAY MINERALS. (2) SEDIMENT particles of GRAIN-SIZE less than 0.004 mm. A SEDIMENTARY ROCK formed dominantly by particles of this size is a CLAYSTONE or clay SHALE. *TY*

[See also MUD, MUDROCK]

Chamley, H. 1989: *Clay sedimentology.* Berlin: Springer.

clay–humus complex A semi-stable, intimate mixture of silicate CLAY MINERALS and HUMUS having colloidal properties that largely determine the nature and fertility of soils. In temperate environments, the clay–humus complex results largely from the passage of soil and organic matter through the gut of EARTHWORMS. The resulting worm cast is slightly calcareous and has an intimate association of mineral and organic matter bound together in a stable *crumb structure*. In humid tropical environments the clay–humus complex is formed from a mixture of hydrous oxides of iron and aluminium as well as silicate clays and organic material. *EMB*

Bridges, E.M. 1997: *World soils.* Cambridge: Cambridge University Press.

clay lunettes Crescent-shaped dunes found on the downwind side of some ephemeral SALINE LAKES and pans and along tidal lagoons in coastal areas of arid and semi-arid regions. Generally, aggregates of CLAY constitute a much more important component than sand. They are formed by the DEFLATION of lagoonal and dried lake deposits. They are widespread in the SEMI-ARID region of southeast Australia, around pans in the north central Kalahari, in the coastal plain and High Plains of Texas and New Mexico, in the CHOTTS of north Africa and in the coastal plain of Argentina. RELICT examples can be useful in reconstructing CLIMATIC CHANGE with, for example, periods of dune stability indicated by PEDOGENESIS interrupting phases of dune SEDIMENTATION. *RAS*

[See also WIND EROSION]

Arbogast, A.F. 1996: Late Quaternary evolution of a lunette in the central Great Plains: Wilson Ridge, Kansas. *Physical Geography* 17, 354–370. **Bowler, J.M.** 1973: Clay dunes: their occurrence and environmental significance. *Earth-Science Reviews* 9, 315–338. **Holliday, V.T.** 1997: Origin and evolution of lunettes on the High Plains of Texas and New Mexico. *Quaternary Research* 47, 54–69.

clay minerals Phyllosilicate minerals with fine or very fine (< 2μm) platy, grains that can be identified only indirectly using a microscope, or by SCANNING ELECTRON MICROSCOPY. Clay minerals have a layered structure and chemically are hydrous silicates of aluminium. There are three main groups: the 1:1 KAOLINITE group, the 2:1 *smectite* group and the 2:1 *hydrous micas*, the numbers indicating the layers of silicon and aluminium atoms in a unit cell. Layers are held together by shared oxygen atoms and the unit cells by shared hydroxyl atoms. Silicate clays are formed by CHEMICAL WEATHERING of primary minerals, muscovite or mica, or by recrystallisation from solution or by substitution of elements in the mineral lattice. Clay minerals have the property of cation exchange (see CATION EXCHANGE CAPACITY), and possess many properties of COLLOIDS such as plasticity, cohesion, shrinkage and swelling, FLOCCULATION and dispersion. Some tropical soils are dominated by clay-sized particles of iron (*goethite*) and aluminium (GIBBSITE) oxide or hydroxide. Soils developing on volcanic materials contain non-crystalline colloidal matter such as allophane, imogolite and ferrihydrite. These different types of clay minerals may be used in PALAEOENVIRONMENTAL RECONSTRUCTION. *EMB*

[See also CATION EXCHANGE CAPACITY, ILLITE, MONTMORILLONITE, VERMICULITE]

Dixon, J.B., Weed, S.B., Kittrick, J.A., Milford, M.H. and White, L.J. 1977: *Minerals in soil environments.* Madison, WI: Soil Science Society of America.

claystone A SEDIMENTARY ROCK, with a dominant GRAIN-SIZE of CLAY grade (< 0.004 mm). Sometimes restricted to a rock of this grade that is not FISSILE. *TY*

[See also MUD, MUDROCK, MUDSTONE]

clear cutting Felling of all trees in a block of forest in a single operation. Clear cutting or *clear felling* is typically carried out in an even-aged tree plantation. *NDB*

[See also HARVESTING, SELECTIVE CUTTING]

Moffatt, A.S. 1993: Clearcutting's soil effects. *Science* 261, 1116.

climate The aggregate or totality of the WEATHER of a particular place or area over a protracted period, conventionally over an interval of at least 30 years. Climate is a complex phenomenon that is not static. The main *climatic elements* include temperature, precipitation, humidity, sunshine and wind, which are represented not only by average conditions, which may be summarised by climatic statistics, including daily, monthly, seasonal and annual averages (such as MEAN ANNUAL AIR TEMPERATURE), but also by the many aspects that reflect the varying nature of climate through time (such as CLIMATIC VARIABILITY, CLIMATIC EXTREMES, CLIMATIC FLUCTUATIONS, CLIMATIC VARIATIONS and EXTREME CLIMATIC EVENTS) and over space at local, regional and global scales. The term is derived from the Greek word *klima*, which refers to the angle of the Sun above the horizon. *JAM*

[See also CLIMATIC CLASSIFICATION, MACROCLIMATE, MESO-CLIMATE, MICROCLIMATE]

Barry, R.G. 1992: *Mountain weather and climate*, 2nd edn. London: Routledge. **Barry, R.G. and Chorley, R.J.** 1992: *Atmosphere, weather and climate*, 6th edn. London: Routledge. **Hare, F.K.** 1966: The concept of climate. *Geography* 5, 99–110. **Harvey, D.** 2000: *Climate and global environmental change.* Harlow: Longman. **Houghton, J.T. (ed.)** 1984: *The global climate.* Cambridge: Cambridge University Press. **Hulme, M. and Barrow, E.** 1997: *Climates of the British Isles: past, present and future.* London: Routledge. **Pearce, E.A. and Smith, C.G.** 1990: *The world weather guide*, 2nd edn. London: Hutchinson.

climatic archive

The main function of a climatic archive is to collate and standardise meteorological records over as long a period as possible to provide a consistent historical perspective on past climates. Many records prior to the twentieth century are irregular in both space and time. Extension and standardisation of these records is often by cross-referencing with historical DOCU-MENTARY EVIDENCE such as SHIP LOG BOOK RECORDS and personal WEATHER DIARIES (see CHRONICLES). Indicators of biophysical conditions such as ICE CORE analyses and DENDROCLIMATOLOGY can extend PROXY climatic records over hundreds or thousands of years.

Global climate archives providing estimates of land and sea surface temperatures extend back to the mid nineteenth century. The main data series for the British Isles is the CENTRAL ENGLAND TEMPERATURE RECORD of G. Manley, a database of mean monthly temperatures extending back to AD 1659, now updated on daily and monthly bases. The England and Wales Rainfall series extends back to 1727, though as with most such series, early accuracy is suspect. Climatic and palaeoclimatic datasets from all over the world may be archived at the National Geophysical Data Center, Boulder, Colorado, USA.   *JCM*

[See also ARCHIVE, DATABASE, NATURAL ARCHIVE]

Hulme, M. 1994: Historical records and recent climatic change, in Roberts, N. (ed.), *The changing global environment.* Oxford: Blackwell, 69–98. **Jones, P.D. and Bradley, R.S.** 1995: Climatic variations in the longest instrumental records, In Bradley, R.S. and Jones, P.D. (eds), *Climate since* AD *1500.* London: Routledge, 246–268.

climatic change

The world's climate is an inherently variable part of the wider environment. ICE AGES and GLACIAL–INTERGLACIAL CYCLES provide the most dramatic testimony to this natural CLIMATIC VARIABILITY (see Figure 1). In more recent times the MEDIAEVAL WARM PERIOD and the LITTLE ICE AGE also provide evidence of changes of a lower degree of magnitude and over shorter time intervals. The twentieth century began, however, with a general view that the world's climate had assumed a state of relative stability. This view, based on the limited volume of instrumental data then available, has long since been replaced by a view of a CLIMATIC SYSTEM being more dynamic, less predictable and subject to change by natural forces and by anthropogenic factors. Of these, the most widely publicised has been that of GREENHOUSE GASES and consequent problems with enhanced GLOBAL WARMING. This particular ANTHROPOGENIC IMPACT is closely followed in importance by that of OZONE DEPLETION, which, though related to the former, should be seen as a distinct aspect of ENVIRONMENTAL CHANGE.

Climatic change can be viewed as a largely RANDOM series, but one within which some TRENDS or PERIODICI-TIES can be detected. The sequence of climatic change over the past one million years reveals a degree of complex periodicity at the scale of tens of thousands of years. In particular the MILANKOVITCH THEORY indicates that the advances and retreats of the major ice caps during the PLEISTOCENE can be interpreted in the light of ORBITAL FORCING that results from regular variations in the Earth's orbit about the Sun and rotation about its own axis. It is important to note that, whilst such orbital variations produce only very small changes in the seasonal patterns of SOLAR INSOLATION receipts at the Earth's surface, their consequences are dramatic because of the possibility of FEEDBACK MECHANISMS within the atmosphere. At other time scales, the evidence of *cyclic phenomena* is more difficult to detect. In the longer term, such non-cyclic factors as the changing geography of ocean and land masses,

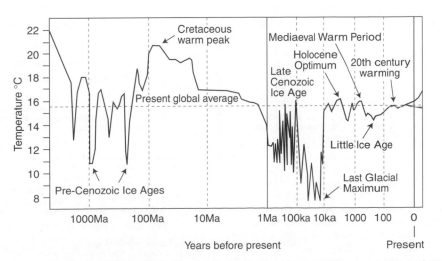

climatic change (1) *The average temperature of the Earth over geological time. Note that time is represented on a logarithmic scale with a break at 1.0 million years (after Bryant, 1997)*

which results from CONTINENTAL DRIFT, contribute significantly to climatic change through controlling oceanic and atmospheric circulations, and may obscure other patterns (see Figure 2).

At millennial and shorter time scales, much effort has been invested in trying to isolate evidence of the SOLAR CYCLE in the climatic record. An unambiguous demonstration of this influence remains, however, beyond the scope of current research. On the other hand, the QUASI-BIENNIAL OSCILLATION (QBO) offers greater possibilities for assessing cyclic patterns and many TIME-SERIES ANALYSIS studies of long-term temperature and rainfall series, such as the CENTRAL ENGLAND TEMPERATURE RECORD, suggest a recognisable signal at this critical frequency of just over two years.

Periodic or cyclical variations account, however, for only a small proportion of all climatic change and the majority of the natural signal, as it is understood at the close of the twentieth century, must be regarded as random and, in that sense, unpredictable. There are two exceptions to this, both of which reflect the monumental complexity of the atmospheric system. First, WEATHER FORECASTING is a process of prediction, but one that is limited to a period of some five to ten days ahead of the events themselves. Beyond that limit the natural chaos of the atmospheric system [see CHAOS THEORY] makes it impossible to foresee the day-to-day weather by conventional numerical modelling methods in which the atmosphere is replicated in computer programmes. Second, an impression of the overall climate, as opposed to daily 'weather', can be gained by employing quite different numerical representations of the atmosphere known as GENERAL CIRCULATION MODELS (GCMs). These models are used to predict the possible consequences of the feature of climatic change that has preoccupied scientists during the closing years of the twentieth century – that of anthropogenically induced GLOBAL WARMING. This is argued to be a result of the increasing concentrations of CARBON DIOXIDE and other greenhouse gases in the atmosphere. Whilst there is little doubt that global temperatures during the 1980s and 1990s are amongst the highest in the INSTRUMENTAL RECORD, the latter embraces little more than a century. Other sources such as PROXY CLIMATIC INDICATORS and documentary evidence point to earlier periods, such as the Mediaeval Warm Period and the HYPSITHERMAL (Holocene Climatic Optimum), when temperatures may have been higher than they are today. Both periods predate any possible anthropogenic interference and again emphasise the importance of natural variation. Arguably, the most important of the recent climatic phases was the Little Ice Age, which documentary evidence has shown caused major disruption to economic activity in the seventeenth and eighteenth centuries.

Such major phases as those noted above, whilst forming part of a non-cyclic, random series, may be a consequence of SOLAR FORCING. The MILANKOVITCH THEORY presupposes no change in the SOLAR CONSTANT in accounting for the Pleistocene Ice Ages, but there is evidence to connect more recent warm and cool phases of Earth history with periods of respectively higher and lower levels of solar

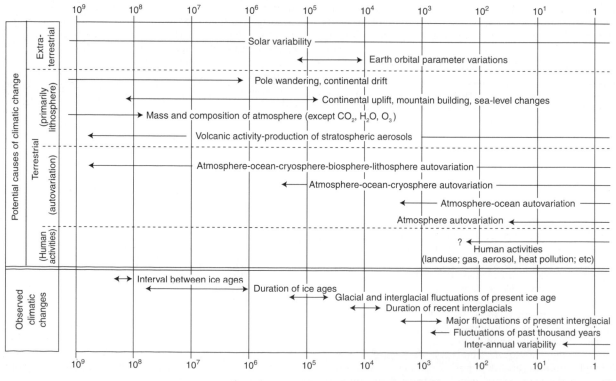

climatic change (2) *The characteristic timescales of (top) potential causes of climatic change and (bottom) observed climatic changes in nature (after Mitchell, 1965; Kutzbach, 1976)*

activity. In particular, the most intense period of the Little Ice Age coincides with a period of low SUNSPOT activity known as the MAUNDER MINIMUM. This conclusion is, however, obscured by the evidence that the Little Ice Age includes also some of the most significant volcanic eruptions of the millennium and the consequent VOLCANIC DUST VEILS would also have provided a global cooling explanation. Conversely, solar activity appears to have been much lower than today. On the basis of such evidence some scientists have gone so far as to challenge the whole concept of anthropogenically induced global warming and attribute much of the anomalous climatic behaviour of the late twentieth century to variations in the behaviour of the Sun.   *DAW*

[See also GLOBAL CHANGE, PALAEOCLIMATOLOGY]

Bradley, R.S. and Jones, P.D. 1993: 'Little Ice Age' summer temperature variations: their nature and relevance to recent global warming trends. *The Holocene* 3, 367–376. **Bryant, E.** 1997: *Climate process and change.* Cambridge: Cambridge University Press. **Calder, N.** 1997: *The manic sun: weather theories confounded.* London: Pilkington Press. **Eddy, J.A.** 1976: The Maunder Minimum. *Science* 192, 1189–1202. **Gribbin, J. (ed.)** 1978: *Climatic change: studying the climates of the past.* Cambridge: Cambridge University Press. **Hart, M.B. (ed.)** 2000: *Climate: past, present and future* [Geological Society Special Publication 181]. London: Geological Society. **Harvey, L.D.D.** 2000: *Climate and global environmental change.* Harlow: Pearson Education. **Houghton, J.** 1994: *Global warming: the complete briefing.* Cambridge: Cambridge University Press. **Intergovernmental Panel on Climate Change** 1996: *Climate change 1995: the science of climate change.* Cambridge: Cambridge University Press. **Jones, P.D., Bradley, R.S. and Jouzel, J. (eds)** 1996: *Climatic variations and forcing mechanisms of the last 2000 years.* Berlin: Springer. **Kutzbach, J.E.** 1976: The nature of climate and climatic variations. *Quaternary Research* 6, 471–480. **Lamb, H.H.** 1972: *Climate: present, past and future.* Vol. 1. *Fundamentals and climate now.* London: Methuen. **Lamb, H.H.** 1977: *Climate: present, past and future.* Vol. 2. *Climatic history and the future.* London: Methuen. **Lassen, K. and Friis-Christensen, E.** 1995: Variability of the solar cycle length during the past five centuries and the apparent association with terrestrial climate. *Journal of Atmospheric and Terrestrial Physics* 57, 835–845. **Mitchell Jr, J.M.** 1965: Theoretical paleoclimatology. In Wright, M. and Frey, D. (eds), *Quaternary of the United States.* Princeton, NJ: Princeton University Press, 881–901. **National Research Council** 1995: *Natural climatic variability on decade to century timescales.* Washington, DC: National Academy Press. **Thompson, R.D.** 1989: Short-term climatic change: evidence, causes, environmental consequences and strategies for action. *Progress in Physical Geography* 13, 315–347.

climatic change: past impact on animals

The possible effects of climatic change on animals range from TOLERANCE to ADAPTATION, MIGRATION and EXTINCTION. A recent example is provided by a poleward shift in the geographical RANGES of butterflies, apparently in response to GLOBAL WARMING. In a sample of 35 non-migratory European butterfly species, 63% have ranges that have shifted to the north by 35–240 km during the twentieth century. The greater mobility of animals compared with plants enables many animals to escape the vagaries of climatic change at local (HABITAT) to regional scales, although animals are dependent on plants and other aspects of the ECOSYSTEM, which restricts their ability to move. There are, moreover, great variations between animal types and species in their responses to similar climatic changes. Historical records indicate significant

effects of CLIMATIC VARIATIONS on animal distributions, sometimes with implications for humans. Catch records of the sockeye salmon in Alaska, for example, suggest an increased PRODUCTION during the twentieth century associated with increased SEA-SURFACE TEMPERATURES in the eastern North Pacific. Interpretation of such records is, however, complicated by commercial HARVESTING and HABITAT change.   *JAM*

[See also ANIMAL REMAINS, BEETLE ANALYSIS, CORAL BLEACHING, DISEASE, INSECT ANALYSIS, MASS EXTINCTION, MOLLUSCA ANALYSIS, RODENT MIDDEN]

Cushing, D.H. and Dickson, R.R. 1976: The biological response in the sea to climatic changes. *Advances in Marine Biology* 14, 1–122. **Dennis, R.L.H.** 1993: *Butterflies and climate change.* Manchester: Manchester University Press. **Grayson, D.K.** 1981: A critical review of the use of archaeological vertebrates in paleoenvironmental reconstruction. *Journal of Ethnobiology* 1, 28–38. **Finney, B.P., Gregory-Eaves, I., Sweetman, J. et al.** 2000: Impacts of climatic change and fishing on Pacific salmon abundance over the past 300 years. *Science* 290, 795–799. **Ford, M.J.** 1982: *The changing climate: responses of the natural fauna and flora.* London: George Allen and Unwin. **Parmesan, C., Ryrholm, N., Stefanescu, C. et al.** 1999: Poleward shifts in geographical ranges of butterfly species associated with regional warming. *Nature* 399, 579–583. **Vibe, C.** 1967: Arctic animals in relation to climatic fluctuations. *Meddelelser om Grønland* 170, 5.

climatic change: past impact on humans

The idea that climate influenced human society and culture was suggested in the writings of Mediaeval geographers, historians and philosophers. Influential enlightenment thinkers such as Abbé du Bos, Montesquieu and Hume compared ancient writings and their own WEATHER to link the rise and fall of creative historical eras to CLIMATIC CHANGE and promoted a brand of climatic DETERMINISM based on geographical location and the quality of the air. This culminated in Huntingdon's determinism that cast a long shadow into the twentieth century.

Since then, more rigorous studies, using historical and scientific methodologies, have identified both direct and indirect climatic impacts on human societies. These impacts have been mediated through water supplies (groundwater, soil moisture, rivers and glaciers), temperature (crop growth, human and animal comfort, fuel demand), sunshine, humidity and cloudiness (health and growth) and wind (structural damage, wind and wave power, crop and vegetation stress). With numerous FEEDBACK MECHANISMS and numerous possible non-climatic causes, these influences are very difficult to analyse and unravel.

There are, however, a limited number of ways that societies can respond to accumulated climatic stress, including movement (migration), social collaboration and innovation. In extreme cases, collapse is a possibility. Examples can be found throughout the history of civilisations, such as those of the ancient Egyptians, the Maya, the prehistoric Andean cultures and the NORSE GREENLAND SETTLEMENTS.   *JGT/JAM*

[See also AGRICULTURAL ORIGINS, DESERTIFICATION, DROUGHT, ENVIRONMENTAL ARCHAEOLOGY, FAMINE, HUMAN HEALTH HAZARDS]

Bell, B. 1971: The dark ages in ancient history I. The first dark age in Egypt. *American Journal of Archaeology* 75, 1–26. **Bell, B.** 1975: Climate and history in Egypt: the Middle Kingdom.

American Journal of Archaeology 79, 223–269. **Bos, A. du**1719: *Reflexions critiques sur la poesie et sur la peinture.* 2 vols. Paris. [Republished by Ecole nationale supérieure des Beaux-arts, Paris]. **Dunin-Wasowicz, T.** 1987: Climate as a factor affecting the human environment in the Middle Ages. *Journal of European Economic History* 4, 691–706. **Fagan, B.** 1999: *Floods, famines and emperors: El Niño and the fate of civilizations.* New York: Basic Books. **Harding, A. (ed.)** 1982: *Climatic change in later prehistory.* Edinburgh: Edinburgh University Press. **Hodell, D.A., Curtis, J.H. and Brenner, M.** 1995: Possible role of climate in the collapse of classic Maya civilization. *Nature* 375, 391–399. **Huntingdon, E.** 1914: *Civilisation and climate.* New Haven, CT: Yale University Press. **Issar, A.S. and Brown, N. (eds)** 1998: *Water, environment and society in times of climatic change.* Dordrecht: Kluwer. **Lamb, H.H.** 1982: *Climate, history and the modern world.* London: Methuen. **McAlpin, M.B.** 1983: *Subject to famine: food crisis and economic change in western India 1860–1920.* Princeton, NJ: Princeton University Press. **Nüzhet, H., Kukla, G. and Weiss, H. (eds)** 1994: *Third millennium B.C. climate change and Old World collapse.* Berlin: Springer. **Paulsen, A.C.** 1976: Environment and empire: climatic factors in pre-historic Andean culture change. *World Archaeology* 8, 121–132. **Shimata, I., Schaaf, C.B., Thompson, L.G. and Moseley-Thompson, E.** 1991: Cultural impacts of severe drought in the prehistoric Andes. *World Archaeology* 22, 247–270. **Wigley, T.M.L., Ingram, M.J. and Farmer, G. (eds)** 1981: *Climate and history: studies in past climates and their impact on man.* Cambridge: Cambridge University Press.

climatic change: past impact on landforms and geomorphological processes

Because of the substantial fluctuations in climate affecting the Earth during the comparatively recent geological past, the world's landscapes can be viewed essentially as geomorphological PALIMPSESTS, with LANDFORM and sediment traces of major climatic episodes varying in their impressiveness and clarity according to the recency, duration and landscape-modifying effectiveness of former geomorphological regimes. Repeated cooling of world climates during the PLEISTOCENE had its greatest effect geomorphologically in those parts of mid-latitude regions affected directly by GLACIER ICE. Here, landforms of GLACIAL and GLACIOFLUVIAL EROSION (e.g. CIRQUES, U-SHAPED VALLEYS, TUNNEL VALLEYS) and DEPOSITION (e.g. MORAINES, DRUMLINS, KAMES, *sandar*), together with large thicknesses of *glacigenic sediments*, bear witness to substantial geomorphological change, which altered comparatively little during the relatively stable geomorphological conditions pertaining during temperate climates of INTERGLACIAL times (including the HOLOCENE). Beyond the Pleistocene ice sheets, in mid-latitudes, PERIGLACIAL climates also caused substantial geomorphological change (e.g. PERMAFROST, slope instability, SOLIFLUCTION, freeze–thaw processes, *frost-shattering* of bedrock) though generally not as pronounced as in glaciated areas.

In lower latitudes, it was change in precipitation and/or evapotranspiration, rather than in temperature, during the Pleistocene that triggered geomorphological change. Much of the evidence is in the form of sediment accumulations reflecting phases of surface stability or instability. Paradoxically, it was often increased dryness rather than wetness that caused instability leading to increased sedimentation because of its effect in reducing the vegetation cover and thus making slopes vulnerable to erosion by water, particularly during the first rains after prolonged dry periods. Wetter phases in low latitudes are reflected in LAKE-LEVEL VARIATIONS and stabilisation of dunes by vegetation (see PALAEODUNES). Near-shore sea-floor sediments in equatorial regions have been used to help in the reconstruction of the environmental change on the adjacent land by the change in the nature of sediments (e.g. AEOLIAN or FLUVIAL sands denoting dry and wet conditions respectively and smectite and kaolinite clay minerals reflecting SAVANNA (dry) or RAIN FOREST (wet) conditions respectively). *RAS*

[See also CLIMATIC CHANGE: POTENTIAL FUTURE IMPACTS]

Büdel, J. 1982: *Climatic geomorphology.* Princeton, NJ: Princeton University Press. **Bull, W.B.** 1991: *Geomorphic responses to climatic change.* Oxford: Oxford University Press. **Derbyshire, E. (ed.)** 1976: *Geomorphology and climate.* London: Wiley. **Molnar, P. and England, P.** 1990: Late Cenozoic uplift of mountain ranges and global climate change: chicken or egg? *Nature* 346, 29–34. **Ollier, C.D.** 1992: Global change and long-term geomorphology. *Terra Nova* 4, 312–319. **Summerfield, M.A. and Kirkbride, M.P.** 1992: Climate and landscape response. *Nature* 355, 306. **Twidale, C.R.** 1976: On the survival of paleoforms. *American Journal of Science* 276, 77–95.

climatic change: past impact on soils

Soils provide a NATURAL ARCHIVE of past climatic environments. PEAT deposits, HISTOSOLS and acidic soils contain pollen grains that have fallen on the soil surface. POLLEN ANALYSIS, horizon by horizon, gives evidence of past vegetation and climate conditions that have influenced soil formation. Dating the various carbon compounds in soils, including STABLE ISOTOPES, provides information on the history of soil formation. Increased air temperatures, associated with relatively high CARBON DIOXIDE concentrations, may have induced soil processes similar to those under tropical conditions at present. Macro- and MICROMORPHOLOGICAL analysis of soils and the occurrence of carbonate or iron nodules in soils indicate specific climatic conditions that can be related to climatic changes in the past. Soils in arid or formerly arid regions contain carbon locked up as calcium carbonate or CALCRETE. Concretionary forms of iron are symptomatic of poor drainage conditions in the past. Pedological information, geographically referenced, can help researchers develop models of past climates and their extent. *EMB*

[See also PALAEOSOLS, SOIL DATING]

Dimbleby, G.W. 1961: Soil pollen analysis. *Journal of Soil Science* 12,1–11. **Matthews, J.A. and Caseldine, C.J.** 1987: Arctic-alpine Brown Soils as sources of palaeoenvironmental information: further ^{14}C dating and palynological evidence from Vestre Memurubreen, Jotunheimen, Norway. *Journal of Quaternary Science* 2, 59–71. **Rounsevell, M.D.A. and Loveland, P.J. (eds)** 1994: *Soil response to climate change.* Berlin: Springer.

climatic change: past impact on temperate vegetation

Climatic factors, principally temperature and PRECIPITATION, can be shown to be the major controls on the broad-scale distribution of BIOMES. Climate change would therefore be expected to have a major impact on vegetation and this is seen in the reconstructions from studies of the PALAEOECOLOGY of the TEMPERATE zone. During INTERGLACIALS, broad-leaved deciduous woodland (or *grassland* in drier continental interiors) was the dominant vegetation type, but during GLACIAL EPISODES much of the temperate zone was either covered by ICE SHEETS or ice-free but PERIGLACIAL and covered with TUNDRA-like vegetation. The trees characteristic of the interglacial vegetation seem to have persisted during glacial periods in

REFUGIA. Recolonisation of the region in successive inter-glacials led to distinctive patterns of arrival of different species and to the development of different competitive balances depending on the exact climatic character of the interglacial. These patterns are often used as a basis for biostratigraphic dating (see BIOSTRATIGRAPHY) of inter-glacial deposits. *MJB*

[See also CLIMATIC CHANGE: PAST IMPACTS ON SOILS, INTERGLACIAL CYCLE, LAST GLACIAL MAXIMUM, REFUGE THEORY IN TEMPERATE REGIONS]

Davis, M.B. 1981: Quaternary history and the stability of forest communities. In West, D.C., Shugart, H.H. and Botkin, D.B. (eds), *Forest succession: concepts and application*. New York: Springer, 132–153. **Huntley, B. and Birks, H.J.B.** 1983: *An atlas of past and present pollen maps for Europe 0–13 000 years ago*. Cambridge: Cambridge University Press. **Huntley, B. and Prentice, I.C.** 1993: Holocene vegetation and climates of Europe. In Wright Jr, H.E., Kutzbach, J.E., Webb III, T. *et al.* (eds), *Global climates since the last Glacial Maximum*. Minneapolis, MN: University of Minnesota Press, 136–168. **Prentice, I.C., Guiot, J. and Harrison, S.P.** 1992: Mediterranean vegetation, lake levels and palaeoclimate at the last glacial maximum. *Nature* 360, 658–660. **Webb III, T. and Bartlein, P.J.** 1992: Global changes during the last three million years: climatic controls and biotic responses. *Annual Review of Ecology and Systematics* 23, 141–173.

climatic change: past impact on tropical vegetation

Natural climatic changes, such as GLACIAL–INTERGLACIAL CYCLES profoundly affected tropical vegetation. The principal impact of GLACIATIONS was to lower tropical temperatures by 5–9°C in both montane and lowland areas. With this cooling, tropical MONTANE FOREST species were able to migrate 1000 to 1500 m vertically down-slope and invade the lowlands. The lowland RAIN FOREST species could not migrate down-slope since the 125 m drop in sea level was not sufficient to compensate for the temperature change. The cooling was accompanied by oscillations in precipitation sufficient to cause ECOTONE habitats to shift toward a wetter or drier vegetation type. However, the expansion of SAVANNA was much less than proposed in the REFUGE THEORY. The combination of cooler temperatures and lower, or more seasonal, precipitation resulted in novel species assemblages. As in the temperate region, species responded individualistically to these climatic changes and formed new communities. No MODERN ANALOGUE exists for many of these GLACIAL EPISODES or for early HOLOCENE communities. *MBB*

[See also SEA-LEVEL CHANGE, HOT SPOT IN BIODIVERSITY, TROPICAL RAIN FOREST, TROPICAL SEASONAL FOREST]

Colinvaux, P.A., De Oliveira, P.E., Moreno, J.E. *et al.* 1996: A long pollen record from lowland Amazonia: forest and cooling in glacial times. *Science* 274, 85–88. **Hooghiemstra, H.** 1989: Quaternary and upper-Pliocene glaciations and forest development in the tropical Andes: evidence from a long high-resolution pollen record from the sedimentary basin of Bogota, Colombia. *Palaeogeography, Palaeoclimatology, Palaeoecology* 72, 11–26. **Thompson, L.G., Davis, M.E., Mosley-Thompson, E.** *et al.* 1998: A 25,000-year tropical climate history from Bolivian ice cores. *Science* 282, 1858–1864.

climatic change: potential future ecological impacts

It is predicted that the greatest effects of GLOBAL WARMING will occur in high latitudes and the least in the tropics. The area of TUNDRA, cold and hot DESERTS will diminish. BOREAL and TEMPERATE FOREST TREE LINES will extend further polewards and to higher altitudes, respectively, and TROPICAL FORESTS and *grasslands* will also expand. The species composition of constituent ecosystems will also alter significantly. Changes in temperature and rainfall will modify the distribution of WETLANDS and the effects of rising sea levels (SALTWATER INTRUSION, more EROSION) will dramatically change coastal ecosystems including *salt marshes* and MANGROVES. Changes in sea level, OCEAN CURRENTS and nutrient availability will impact on marine ecosystems. The interplay of plants and animals will be altered. Migratory bird populations are already changing, as evidenced by BIOLOGICAL INVASIONS of the Arctic tundra. There will also be shifts in the prevalence of PESTS and DISEASES, with, for example, movement of tropical forms into existing temperate latitudes. *GOH*

[See also CARBON DIOXIDE FERTILISATION, CLIMATIC CHANGE, SEA-LEVEL RISE]

Lenihan, J.M. and Neilson, R.P. 1995: Canadian vegetation sensitivity to projected climate change at three organisation levels. *Climate Change* 30, 27–56. **Mannion, A.M.** 1997: *Global environmental change: a natural and cultural environmental history*. Harlow: Addison-Wesley Longman. **Moore, P.D., Chaloner, B. and Stott, P.A.** 1996: *Global environmental change*. Oxford: Blackwell Science. **Walker, B., Steffan, W., Canadell, J. and Ingram, J. (eds)** 1999: *The terrestrial biosphere and global change: implications for natural and managed ecosystems*. Cambridge: Cambridge University Press. **Watson, R.T., Zinyowera, M.C., Moss, R.H. and Dokken, D.J.** 1996: *Climate change 1995: impacts, adaptations, and mitigation of climate change, scientific and technical analyses*. Cambridge: Cambridge University Press.

climatic change: potential future economic impacts

As GREENHOUSE warming becomes more pronounced, its impacts will be felt widely so that the aggregate economic and ecological implications of climatic change will become increasingly important. Providing an accurate picture of likely future impacts of climate change and aggregating them into a tractable set of impact indicators is difficult and requires well defined underlying assumptions and professional judgement and skills. Regionally specific CLIMATIC CHANGE SCENARIOS and a broad base of underlying case studies and data form the core of CLIMATIC IMPACT ASSESSMENTS. Climatic change is complex and impacts are to a high degree region, sector and time specific. Despite these limitations, climatic impact assessment can inform ENVIRONMENTAL POLICY. In the UK, national as well as regional impact assessments have been made. *AHP*

Department of the Environment 1996: *Review of the potential effects of climate change in the UK*. London: HMSO. **Intergovernmental Panel on Climate Change (IPPC)** 1996: *Climate change 1995: economic and social dimensions of climate change: contribution of Working Group III to the Second Assessment Report*. Cambridge: Cambridge University Press. **Mintzer, I.M. (ed.)** 1992: *Confronting climate change: risks, implications and responses*. Cambridge: Cambridge University Press. **Parry, M.L. (ed.)** 1985: The sensitivity of natural ecosystems and agriculture to climatic change. *Climatic Change* 7(1), 1–152 [Special Issue]. **Parry, M.L. (ed.)** 1990: *The impact of climatic variations on agriculture*. 2 vols. Dordrecht: Kluwer. **Reddy, K.R. and Hodges, H.F. (eds)** 2000: *Climate change and global crop productivity*. Wallingford: CABI Publishing. **Whyte, I.D.** 1995: *Climatic change and human society*. London: Arnold.

climatic change: potential future geomorphological impacts

For many geomorphologists, anthropogenic modifications of the concentration of GREEN-HOUSE GASES in the atmosphere are expected to cause geomorphologically significant climatic changes in the future. Predicting geomorphological impacts has focused on landscapes expected to be particularly sensitive to relatively small amounts of climatic change. For example, the latitudinal limits of PERMAFROST have been predicted to be displaced polewards 100–250 km for every 1°C rise in mean annual temperature. Many CIRQUE and valley glaciers are expected to continue to shrink or even disappear. Depending on changes in precipitation and evaporation and the wind strength in desert margins, slopes could become more or less stable geomorphologically. If the change is to wetter conditions, currently dry stream courses could become reactivated and lakes could grow in size. The destructive geomorphological activity accompanying the passage of TROPICAL CYCLONES is expected to affect coastal areas that currently have sea surface temperatures below the threshold for cyclone activity (c. 27°C). Greatest attention, however, has been paid to the likely geomorphological impacts accompanying a greenhouse-induced SEA-LEVEL RISE. Even a relatively modest rise is expected to cause, for example, permanent coastal inundation, substantially increased erosion on unprotected coasts made of unconsolidated sediments, and narrowing or disappearance of sand beaches. With modest sea-level rises being predicted, impacts in some low-lying coastal environments (e.g. deltas, SABKHAS, marshes, MANGROVES) are less certain. Improved understanding of how geomorphological systems responded to HOLOCENE ENVIRONMENTAL CHANGE is increasingly viewed as critical for predicting their future responses to GLOBAL WARMING.　　*RAS*

[See also BRUUN RULE, CLIMATIC CHANGE: PAST IMPACTS, GLOBAL WARMING, SEA-LEVEL RISE: POTENTIAL FUTURE GEOMORPHOLOGICAL IMPACTS]

Beniston, M. 2000: *Environmental change in mountains and uplands.* London: Arnold. **Goudie, A.S.** 1990: The global geomorphological future. *Zeitschrift für Geomorphologie Supplementband* 79, 51–62. **Goudie, A.S.** 1996: Geomorphological 'hotspots' and global warming. *Interdisciplinary Science Reviews* 21, 253–259. **Haeberli, W. and Beniston, M.** 1998: Climatic change and its impact on glaciers and permafrost in the Alps. *Ambio* 27, 258–265. **Lavee, H., Imeson, A.C. and Sarah, P.** 1998: The impact of climate change on geomorphology and desertification along a Mediterranean–arid transect. *Land Degradation and Development* 9, 407–422.

climatic change: potential future impacts on soils

Changes in rainfall patterns and increased temperatures as a result of GLOBAL WARMING will have many implications for soils, natural vegetation and landuse. The SOIL FORMING PROCESSES will change with shifts of climate, and the effects will be felt in terms of soil physics, chemistry and biology. With warmer, drier conditions, the amount of water stored in soils is likely to decrease and SOIL MOISTURE DEFICITS increase. In wetter conditions, cultivation may become more difficult, demanding heavier machinery, and the period when cultivation is possible may be restricted, with implications for all stages of crop production. Loss of organic matter and decrease of soil biological activity would lead to reduced stability of SOIL STRUCTURE and greater SOIL VULNERABILITY to erosion.

Excessive drying of VERTISOLS will lead to additional structural damage to buildings as soils shrink and crack

widely. On these and other soils, deep cracks would allow rapid movement of water to the WATER TABLE, possibly carrying fertiliser, herbicides and pesticides rapidly to the GROUNDWATER and streams with consequent POLLUTION and EUTROPHICATION potential. Changes could be rapid or slow: rapid change could be associated with temperature and moisture regimes; slower change will affect organic matter content, salinity, alkalinity and loss of soil structure. The change of soil development from one genetic grouping to another is likely to occur over thousands of years.

HISTOSOLS will reduce in area as organic material is oxidised through cultivation, but poorly drained mineral soils could improve if climate becomes drier. In coastal lowlands, global warming, leading to SEA-LEVEL RISE, will lead to a rise in a brackish GROUNDWATER table (see also WATER TABLE) making the present drainage network less effective and possibly requiring costly investment if such land is to remain suitable for arable cropping. Saline intrusion is also a possibility that will affect the natural vegetation and crops grown, and even become a hazard for concrete foundations. Warmer temperatures in high latitudes could lead to large releases of the GREENHOUSE GASES from CRYOSOLS, as organic matter is oxidised releasing CARBON DIOXIDE, or reduced, releasing METHANE.　　*EMB*

Arnold, R.W., Szabolcs, I. and Targulian, V.O. (eds) 1990: *Global soil change.* Laxenburg: International Institute for Applied Systems Analysis. **Department of the Environment** 1991: *The potential effects of climate change in the United Kingdom,* Ch. 3. London: HMSO. **Food and Agriculture Organization of the United Nations,** 1993: *World soil resources: an explanatory note on the FAO world soil resources map at 1:25,000,000 scale.* Rome: FAO. **Scharpenseel, H.W., Schomaker, M. and Ayoub, A. (eds)** 1990: *Soils on a warmer earth: Effects of expected climate change on soil processes with emphasis on the tropics and sub-tropics* [*Developments in Soil Science* 20]. Amsterdam: Elsevier. **Watson, R.T. and Zinyowlka, R.H.** 1996: *Climate Change 1995. Impacts, adaptations and mitigation of climate change: scientific-technical analyses.* Cambridge: Cambridge University Press.

climatic change scenarios

Expectations of what the FUTURE CLIMATE may be like. A scenario is not a forecast, but an estimate of possible climatic developments normally based on the output of numerical CLIMATIC MODELS, statistical models and combinations of both. A range of future GREENHOUSE GAS emission scenarios and various assumed sensitivities of the CLIMATIC SYSTEM to these emissions are commonly considered. The scenarios present coherent, systematic and internally consistent descriptions of changing climates and are typically used as inputs into CLIMATIC CHANGE vulnerability, impact or adaptation assessments. In the UK four scenarios, referred to as UKCIP98 (United Kingdom Climatic Impact Programme 1998) are currently in use based on low, medium–low, medium–high and high future levels of greenhouse gases. The rate of warming assumed with these scenarios ranges from about 0.1°C to 3°C per decade over the next century.　　*AHP*

[See also BUSINESS-AS-USUAL SCENARIO]

Giorgi, F., Brodeur, C.S. and Baters, G.T. 1994: Regional climatic change scenario over the United States produced with a nested regional climate model. *Journal of Climate* 7, 375–399. **Hulme, M.** 1998: *Climate change scenarios for the United Kingdom* [*UK Climate Impacts Programme Technical Report* 1]. Norwich: Climatic Research Unit, University of East Anglia. **Palutikof,**

J.P. and Wigley, T.M.L. 1996: Developing climatic change scenarios for the Mediterranean region. In Leftić, L., Kečkeš, S. and Pernetta, J.C. (eds), *Climatic change and the Mediterranean.* Vol. 2. London: Arnold, 27–56. **Schaer, C., Frei, C., Lüthi, C. and Davies, H.C.** 1996: Surrogate climate change scenarios for regional climatic models. *Geophysical Research Letters* 23, 185–209.

climatic classification

The identification of macroscale climatic regions; the area generalisation of MACROCLIMATES. Climates can be classified into broad regimes using selected variables (in practice, this tends to be based primarily on temperature and precipitation). Climatic classifications may be termed *generic* where they are based on climatic characteristics alone, *empirical* where these classes are linked to the effects of climate and *genetic* where they are grounded in concepts of causation.

Relating NATURAL VEGETATION distribution to climate was the basis of the best known empirical classification by W. Köppen, who used monthly temperature to identify four major thermal regimes and precipitation to delimit a fifth regime of dry climates. Precipitation effectiveness and temperature efficiency formed the basis of C.W. Thornthwaite's classification, which popularised use of the concept of EVAPOTRANSPIRATION in climatic classification. Such early empirical classification schemes sought in vain to match perfectly climatic regions to soil and vegetation classifications. Single-purpose genetic classifications, often based on AIR MASS dominance concepts or derived from STATISTICAL ANALYSIS of large area climatic databases, have become more popular in recent years as the limitations of climatic classifications to provide multipurpose utility have been recognised. *JCS*

Davis, R.E. and Kalkstein, L.S. 1990: Development of an automated spatial synoptic climatological classification. *International Journal of Climatology* 10, 769–774. **Köppen, W.** 1923: *Die Klimate der Erde: Grundkriss der Klimakunde.* Berlin: de Gruyter. **Oliver, J.E.** 1970: A genetic approach to climate classification. *Annals of the Association of American Geographers* 60, 615–637. **Oliver, J.E.** 1991: History, status and future of climatic classification. *Physical Geography* 12, 231–251. **Thornthwaite, C.W.** 1948: An approach toward a rational classification of climate. *Geographical Review* 38, 55–94. **Wilcox, A.A.** 1968: Köppen after fifty years. *Annals of the Association of American Geographers* 58, 12–28.

climatic discontinuity

An abrupt and non-transitory change of a climatic VARIABLE from one average value to another. Climatic discontinuities include the effects of non-climatic influences such as changes or errors in observational practice. *JCM*

climatic extreme

An extreme value (high or low) in relation to a series of climatic data, conventionally of duration at least 30 years. Climatic extremes include such values as the highest or lowest temperature on record for a particular locality, the lowest MEAN ANNUAL AIR TEMPERATURE and the highest precipitation amount received during any winter month. *JAM*

[See also EXTREME CLIMATIC EVENT, EXTREME WEATHER EVENT]

Dukes, M. and Eden, P. 1997: 'Phew! What a scorcher': weather records and extremes. In Hulme, M. and Barrow, E. (eds), *Climates of the British Isles: present, past and future.* London: Routledge, 262–295. **Innes, J.L.** 1998: The impacts of climatic extremes on forests: an introduction. In Beniston, M. and Innes, J.L. (eds), *The impacts of climatic variability on forests.* Heidelberg: Springer, 1–18.

climatic fluctuation

A type of climatic inconstancy or short-term CLIMATIC VARIATION. Although some consider climatic variations and climatic fluctuations to be synonymous, the term fluctuation should be limited to the description of those systematic aspects of CLIMATIC VARIABILITY involving a climatic variable changing smoothly from one average value to another *and back again* over a relatively short period of time (conventionally, seasonal and interannual variations of less than about 30 years). Hence, the system more-or-less returns to its original condition after a short interval of time and a single fluctuation involves two maxima and one minimum, or vice versa, in the period of record displayed. Fluctuations do not include TRENDS, CLIMATIC DISCONTINUITIES or random variations, but they do include CLIMATIC OSCILLATIONS, CLIMATIC VACILLATIONS and regular, high-frequency PERIODICITIES, as well as irregular fluctuations. Climatic variations longer than about 30 years represent true CLIMATIC CHANGE, during which climatic fluctuations may occur; however, climatic fluctuations may also occur during phases of relatively stable climate. Some of these distinctions are summarised in the Figure. *JAM/AHP*

World Meteorological Organisation 1966: *Climatic change* [*Technical Note* 79]. Geneva: WMO.

climatic geomorphology

The scientific study of current LANDFORM development and geomorphological processes under different climates. Climatic geomorphology was developed at the turn of the twentieth century, with an emphasis on the definition of MORPHOCLIMATIC ZONES at a global scale. Such schemes assumed that climate is the main factor affecting landforms and chose climatic boundaries that were anticipated to be significant for landforming processes. The scheme proposed by Tricart and Cailleux uses this assumption and employs the major structural types of vegetation as indicators of morphoclimatic zones; its boundaries are chosen accordingly.

More recently, the literature has focused on problems with the climatic geomorphological approach. These include: (1) differentiating the influence of climatic factors, on the one hand, and non-climatic factors, such as tectonics, lithology and relief, on the other; (2) the fact that most areas have been subject to a history of CLIMATIC CHANGE and thus to a sequence of climates rather than a single climate operating over a long period; and (3) climatic transitions may be of greater geomorphological significance than periods of climatic stability. Above all, there remains a lack of basic knowledge about geomorphological processes under different climatic regimes. Traditional schemes lay emphasis on MEAN ANNUAL AIR TEMPERATURE and precipitation as the basis of links between climate, vegetation, soil, geomorphological process and landforms. Other climatic factors, however, such as the MAGNITUDE-FREQUENCY of large rainstorms, cyclones, frost or snow, which vary within morphoclimatic zones, may be of significance in influencing fundamental geomorphological variables such as drainage density. *MAC/RPDW*

[See also CLIMATOGENETIC GEOMORPHOLOGY, DAVISIAN CYCLE, LANDSCAPE EVOLUTION]

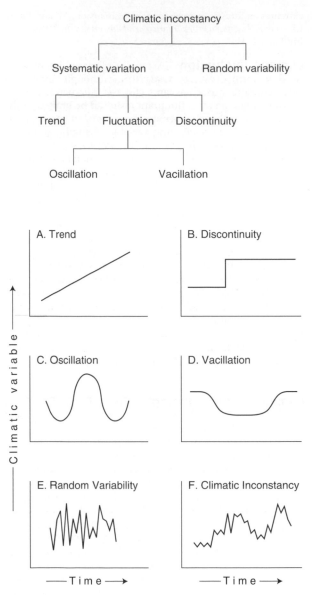

climatic fluctuation *Types of climatic inconstancy (partly after World Meteorological Organisation, 1966). Climatic change may be defined as all systematic climatic variations with a duration of >30 years and climatic fluctuations as a particular type of regular variation with a duration of < 30 years*

Birot, P. 1968: *The cycle of erosion in different climates*, translated by Jackson, C.I. and Clayton, K.M. London: Batsford. **Bremer, H.** 1985: Randschwellen: a link between plate tectonics and climatic geomorphology. *Zeitschrift für Geomorphologie NF Supplementband* 54, 11–21. **Bull, W.B.** 1991: *Geomorphic responses to climatic change.* Oxford: Oxford University Press. **Derbyshire, E. (ed.)** 1976: *Geomorphology and climate.* London: Wiley. **Milliman, J.D. and Syvitski, P.M.** 1992: Geomorphic/tectonic control of sediment discharge to the oceans: the importance of small mountain rivers. *Journal of Geology* 100, 525–544. **Stoddart, D.R.** 1969: Climatic geomorphology: review and re-assessment. *Progress in Geography* 1, 159–222. **Tricart, J. and Cailleux, A.** 1965: *Introduction à la géomorphologie climatique.* Paris: SEDES. **Walsh R.P.D.** 1993:

Problems of the climatic geomorphological approach in the humid tropics with reference to drainage density, chemical denudation and slopewash. *Würzburger Geographische Arbeiten* 87, 221–239.

climatic impact assessment   Assessing the impact of contemporary climate and the consequences of CLIMATIC CHANGE (and especially projected GLOBAL WARMING) on a wide range of socio-economic activities and on ecosystems and landscapes. This INTERDISCIPLINARY activity may focus on *first-order* or *primary impacts*, such as major droughts affecting biological productivity, or *second-order impacts*, such as yield decreases of crops, which can affect a wide spectrum of social and economic relationships. One of the major goals of climatic impact assessment, especially concerning aspects of FUTURE CLIMATES, is the prediction of future impacts. Other goals include: improving knowledge of climate–society interactions; identification of vulnerable areas, populations and activities; and providing a basis for mitigation efforts. *Impact assessment models*, also known as *integrated assessment models* (IAMs), which encompass both climatic change and economic impacts, are used to research policy options.   *AHP*

[See also ENVIRONMENTAL IMPACT ASSESSMENT]

Drake, F. 2000: *Global warming: the science of climate change.* London: Arnold. **Kates, R.W., Ausubel, J. and Berberian, M. (eds)** 1985: *Climate impact assessment* [SCOPE Report 27]. Chichester: Wiley. **Parry, M. and Carter, T.** 1998: *Climate impact and adaptation assessment.* London: Earthscan. **Riebsame, W.E.** 1988: *Assessing the social implications of climate fluctuations: a guide to climate impact studies.* Nairobi: United Nations Environment Programme. **Tegart, W.J.McG., Sheldon, G.W. and Griffiths, D.C. (eds)** 1990: *Climate change: the IPCC impacts assessment.* Canberra: Australian Government Publishing Service. **Weyant, J., Davidson O., Dowlatabadi, H. et al.** 1996: Integrated assessment of climatic change: an overview and comparison of approaches and results. In Bruce, J.P., Lee, H. and Haityes, E.F. (eds), *Climate change 1995: economic and social dimensions of climate change.* Cambridge: Cambridge University Press.

climatic models   Simplified mathematical representations of the CLIMATIC SYSTEM. Models provide climatologists with an improved understanding of the complex workings of the ATMOSPHERE and climatic processes as well as enabling FUTURE CLIMATE to be predicted. The accuracy of a model can be assessed (VALIDATION) by running it for many simulated decades and comparing the model's outputs with current and past observations of climate and/or reconstructed PROXY DATA. For the model to be considered a valid one, the annual, seasonal, geographical and vertical variations of appropriate parameters such as surface pressure, temperature and rainfall have to compare well with the independent evidence. Future climate can be predicted by altering one or more of the radiative forcing factors (GREENHOUSE GAS concentrations, SOLAR RADIATION output, VOLCANIC AEROSOLS) and comparing the model's output to the simulation with no change to the forcing factors (the control). This strategy is intended to simulate changes or perturbations to the CLIMATIC SYSTEM and partially overcome some imperfections in the model.

A key assessment in climatic model experiments is the sensitivity of global climate to the radiative forcing by the greenhouse-gas equivalent of a doubling of carbon dioxide concentrations. Early GENERAL CIRCULATION MODELS (GCMs) suggested an increase in global mean surface air temperature of 2.5–5.0°C, while more recent COUPLED OCEAN–ATMOSPHERE MODELS (CGCMs) suggest 1.0 to 2.0°C. Early GCMs were mainly concerned with equilibrium simulations and portrayed the final adjustment of the CLIMATE to a doubling of carbon dioxide concentration. More recent models attempt transient (time-dependent) simulations and employ CGCMs. Incorporation of an adequate representation of oceanic processes is vital since there is a time lag induced by the deep ocean circulation. The temporal evolution and the regional patterns of CLIMATIC CHANGE may depend significantly on the time dependence of the change in forcing, so the outputs from transient simulations are important in consideration of the likely impacts of climatic change over the coming decades before any equilibrium situation is reached. Coupled ocean–atmosphere models have progressed from treating the ocean as a simple 'swamp ocean', where sea-surface temperatures are calculated through an energy budget and no annual cycle is possible, through a slab (mixed-layer) ocean, where storage and release of energy can take place seasonally, to an ocean general circulation model that treats all key dynamic and thermodynamic processes and interactions.

The mathematical equations forming the models are solved numerically with large computers, using a three-dimensional grid over the globe. For climate, typical RESOLUTION is about 250 km in the horizontal and 1 km in the vertical in atmospheric GCMs, often with higher vertical resolution near the surface and lower resolution in the upper TROPOSPHERE and STRATOSPHERE. Many physical processes, such as those related to CLOUDS, take place on much smaller spatial scales and therefore cannot be properly resolved and modelled explicitly. Their average effects are included in the models in a simple way by taking advantage of physically based relationships with the larger scale variables and expressing these as empirical and statistical relationships (PARAMETERISATION). CGCMs may suffer 'drift' due to accumulating errors when used in long-term (century-scale) simulations. This tendency is sometimes constrained by arbitrary procedures. Increasingly, with improved computing facilities, modellers are not content to make climate predictions using a single run of a model, which may be subject to large random variability. Instead, they use ensembles of predictions in which each run of the model is started from slightly different initial conditions spanning a range of natural variability. Confidence in the predictions comes when the underlying longer-term climatic prediction from each run is the same and is not sensitive to the starting conditions.

Confidence in climatic change predictions is gained when different models from different international research groups provide agreement. However, current models contain many uncertainties concerning CLIMATE processes, such as FEEDBACK MECHANISMS associated with clouds, oceans, SEA ICE and VEGETATION CHANGE, which limit the accuracy of future climatic predictions. Positive feedbacks amplify an initial response, whereas negative feedbacks reduce it. Non-linear processes are especially difficult to model and unexpected changes in future climate cannot be ruled out. Nevertheless, climatic models

have continued to improve in respect of both their physical realism and their ability to simulate present climate on large scales, and increasingly at the regional scale. Climatic models are limited by our imperfect understanding of processes that control climate. Models have improved through fuller treatment of some previously neglected physical processes and improved spatial resolution. The accuracy of model predictions of climate is not only limited by the ability of the model to simulate climate processes accurately, but also because of uncertainties in future changes in the radiative forcing factors. *DME*

[See also BOUNDARY CONDITIONS]

Barry, R.G. and Chorley, R.J. 1998: *Atmosphere, weather and climate*, 7th edn. London and New York: Routledge. **Hecht, A.D. (ed.)** 1985: *Paleoclimate analysis and modeling*. New York: Wiley Interscience. **Houghton, J.T.** 1997: *Global warming: the complete briefing*, 2nd edn. Oxford: Lion. **Houghton, J.T., Filho, L.G.M., Callander, B.A.** *et al.* **(eds)** 1996: *Climate change 1995: the science of climate change*. Cambridge: Cambridge University Press. **Jones, P.D., Briffa, K.R., Barnett, T.P. and Tett, S.F.B.** 1998: High-resolution palaeoclimatic records for the last millennium: interpretation, integration and comparison with General Circulation Model control-run temperatures. *The Holocene* 8, 455–471. **Kohfeld, K.E. and Harrison, S.P.** 2000: How well can we simulate past climates? Evaluating the models using palaeoenvironmental datasets. *Quaternary Science Reviews* 19, 321–346. **North, G.R.** 1996: Models and modeling. In Schneider, S. (ed.), *Encyclopedia of climate and weather*. New York: Oxford University Press, 508–512. **Rind, D.** 1996: The potential for modeling the effects of different forcing factors on climate over the past 2000 years. In Jones, P.D., Bradley, R.S. and Jouzel, J. (eds), *Climatic variations and forcing mechanisms of the last 2000 years*. Berlin: Springer, 563–581. **Street-Perrott, F.A.** 1991: General circulation (GCM) modelling of palaeoclimates: a critique. *The Holocene* 1, 74–80.

climatic modification The intentional or inadvertent modification of atmospheric processes. Deliberate weather modification involves CLOUD SEEDING to promote precipitation, as well as hail and lightning suppression. Attempts have also been made to subdue TROPICAL CYCLONES (hurricanes) in the USA. The science of weather modification reached its peak in the 1960s and 1970s, since when environmental concerns and lack of state funding have curtailed activities. Inadvertent modifications include changes to URBAN CLIMATE, such as the *urban warming* that has accompanied the growth of cities and enhanced GLOBAL WARMING as a result of GREENHOUSE GAS emissions. *AHP*

Cottom, W.R. and Pielke, R. 1995: *Human impacts on weather and climate*. Cambridge: Cambridge University Press. **SMIC** 1971: *Inadvertent climate modification: report of the study of man's impact on climate*. Cambridge, MA: MIT Press.

climatic normals Average climatic data, computed over at least a 30-year period and corrected, where necessary. Such a timescale ensures that yearly WEATHER fluctuations are smoothed. The latest widely recognised *standard period* for the calculation of climatic normals was AD 1961–1990. *JBE*

Aune, B. 1993: *Temperatur normaler, normal periode 1961–90*. Oslo: Den Norske Meteorologiske Institutt.

climatic oscillation A type of CLIMATIC FLUCTUATION: when the average value of a climatic variable exhibits PERIODICITY or CYCLICITY with gradual transitions from peak to trough. A single oscillation is sometimes defined as the swing from one extreme to the other (a *half cycle* or *hemicycle*). Oscillations are said to be *damped* when they are constantly decreasing in amplitude, *unstable* when increasing in amplitude and *neutral* or *persistent* when maintaining a constant amplitude. Oscillations are different from, but may be superimposed on, a TREND. *JAM*

[See also EL NIÑO–SOUTHERN OSCILLATION, CLIMATIC VACILLATION, NORTH ATLANTIC OSCILLATION]

climatic prediction Forecasts of future climate are imperfect because we have an imperfect knowledge of the CLIMATIC SYSTEM and its workings and an imperfect understanding of future rates of GREENHOUSE GAS emissions, how these will change atmospheric concentrations and the response of climate to these changed concentrations. The sources and particularly the SINKS of the greenhouse gases are poorly understood. Climatic predictions also depend on CLIMATIC MODELS, which are themselves imperfect. Normally, a range of climatic predictions is given that reflects an estimate of the uncertainties due to model imperfections. Our imperfect understanding of climatic processes could make us vulnerable to surprises: for example, changes in ocean circulation in the North Atlantic Ocean could cool, rather than warm, European climates. There has been some success with shorter-term climatic prediction associated with the EL NIÑO–SOUTHERN OSCILLATION phenomenon. *AHP*

[See also FORECASTING]

Hastenrath, S., Greischar, L. and van Heerden, J. 1995: Prediction of the summer rainfall over South Africa. *Journal of Climate* 8, 1511–1518. **Hulme, M., Biot, Y., Borton, J. et al.** 1992: Seasonal rainfall forecasting for Africa. Part I: current status and future developments. *International Journal of Environmental Studies* 39, 245–256. **Washington, R. and Downing, T.E.** 1999: Seasonal forecasting of African rainfall: prediction, responses and household food security. *Geographical Journal* 165, 255–274.

climatic reconstruction Climate varies continuously in space and time and the reconstruction of past climatic change is a major aim of much of research in QUATERNARY ENVIRONMENTAL CHANGE. A wide range of geological, biological, physical, and historical evidence is available for climatic reconstruction at the many different temporal scales encompassed within the Quaternary ranging from geomorphological, lithological and biological evidence to historical and instrumental records. The interpretation of all such PROXY EVIDENCE in terms of past climates relies on the assumption of methodological UNIFORMITARIANISM (or ACTUALISM) that proposes that the nature of the geological, physical and biological processes observable today is the same as in the past, but that these processes may have occurred at different rates at different times. Catastrophes do occur but they all involve and follow basic laws of nature because the properties of matter and energy are invariant with time. These laws can be extended back in time and are thus applicable to the interpretation of past events. Methodological uniformitarianism is thus the basic logic and methodology by which all past climates can be reconstructed from proxy evidence. There is no way to prove methodological uniformitarianism, but there is no way to reject it.

Geomorphological evidence for climate change comes, for example, from GLACIAL LANDFORMS, indicating the extent and direction of ice-sheet movement and the altitude of CIRQUE GLACIERS, permitting the reconstruction of the EQUILIBRIUM-LINE ALTITUDE. PERIGLACIAL LANDFORMS can provide clear evidence of past climatic conditions, whereas coastal landforms may indicate past sea-level changes. RIVER TERRACES reflect changes in fluvial activity, which may be related to climatic change. PLUVIAL lake shorelines can provide striking evidence for past changes in precipitation and/or evaporation and SAND DUNES can indicate major changes in aeolian activity. By using modern landscapes as analogues, some aspects of past climate can be inferred from geomorphological evidence. Often such evidence cannot be precisely dated, it does not provide a continuous record of climatic change and it only gives a general picture of past climates. Exceptions are equilibrium-line altitude and pluvial lake shorelines, both of which can allow specific climatic parameters to be inferred.

Lithological evidence comes from stratigraphic sections and sediment cores. Such evidence can involve analysis of particle sizes, particle shapes, organic content, sediment chemistry, clay mineralogy, heavy minerals, magnetic properties and STABLE ISOTOPES (e.g. C, H, O, N). Some of these analyses can be applied to GLACIAL, PERIGLACIAL, CAVE, AEOLIAN, LACUSTRINE and MIRE, and MARINE SEDIMENTS, as well as to PALAEOSOLS and to ICE CORES. STABLE ISOTOPE analyses, particularly of $^{16}O/^{18}O$ of foraminifera preserved in marine CORES and of long ice cores have provided indirect palaeoclimatic records with high temporal resolution for the last 160 000 years or more. Stable-isotope analyses of speleothem deposits in caves and tree rings can provide annual or near annual indirect records of past climate for the last 3000–5000 years.

Biological evidence for climatic change largely takes the form of fossils preserved in sediments. Fossil groups that can provide a basis for climatic reconstruction from terrestrial sediments include POLLEN, PLANT MACROFOSSILS, BEETLES, CHIRONOMIDS, RHIZOPOD ANALYSIS, OSTRACODS and vertebrates. FORAMINIFERA, RADIOLARIA, COCCOLITHOPHORES and DIATOMS preserved in marine sediments can all be valuable sources of palaeoclimatic information. Other fossil groups, such as freshwater diatoms, CLADOCERA, CHRYSOPHYTES and MOLLUSCA, generally reflect local limnological conditions, which may be influenced by events within the lake's catchment (e.g. vegetation change), which, in turn, may result from climatic change. There can thus be complex organism–climate relationships.

Quantitative reconstructions of past climate from fossil assemblages usually involve the use of modern calibration or TRANSFER FUNCTIONS. These model numerically the relationship between biological assemblages preserved in surface sediments and modern climate. These quantitative relationships are established by inverse REGRESSION. The modern calibration functions are then used to transform fossil assemblages containing the same taxa into estimates of past climate, on the assumption that the modern calibration functions are applicable in the past.

Other biological evidence for climate change comes from changes in TREE RING thickness and density. Such evidence can provide annual records of temperature and/or precipitation for the last 3000–5000 years.

Historical records, such as weather notes in ships' logs,

trading and exploration company papers, personal diaries, annals, chronicles, sagas, and administrative records, can provide some information on past climate, particularly about extremes in the weather over the last 800 years (see HISTORICAL CLIMATOLOGY). Some historical records in China and Greece date back to 500 BC. Phenological information on crop harvests, flowering and fruiting of plants and times of animal migration can also provide indirect information about past climate.

INSTRUMENTAL RECORDS extend back to about AD 1660, although uncertainities about the accuracy of some early instruments, calibration with modern equipment and recording practices limit the reliability of some early instrumental records.

Reconstructing past climate from almost all proxy sources is not straightforward. Climate inferences are usually several stages removed from the primary geological, biological, physical or chemical evidence. An understanding of the modern processes influencing, for example, glacial equilibrium-line altitude, stable isotope ratios, chironomid assemblages or pollen assemblages is thus essential if past proxy evidence is to be interpreted reliably in terms of past climate. Different proxies differ in the timescales over which they respond to climate change. Some proxies (e.g. chironomids, beetles) appear to respond very rapidly to change, whereas other proxies (e.g. glaciers, terrestrial vegetation) may have a significant lag in their response. The duration of such lags may vary not only temporally and geographically, but also with the direction of change, for example if the climate is ameliorating or deteriorating. Slow sediment ACCUMULATION RATES may, despite fine-resolution sampling, result in short-lived climatic changes (< 50 years) being undetected. In the late Holocene, distinguishing between changes in fossil biological assemblages that result from climatic change or human impact is a major problem, particularly in Europe. *HJBB*

[See also MULTIPROXY APPROACH, RESOLUTION IN PALAEOENVIRONMENTAL RECONSTRUCTION]

Birks, H.J.B. 1981: The use of pollen analysis in the reconstruction of past climates: a review. In Wigley, T.M.L., Ingram, M.J., and Farmer, G. (eds), *Climate and history*. Cambridge: Cambridge University Press, 111–138. **Birks, H.J.B.** 1995: Quantitative palaeoenvironmental reconstructions. In Maddy, D. and Brew, J.S. (eds). *Statistical modelling of Quaternary science data*. Cambridge: Quaternary Research Association, 161–254. **Imbrie, J and Kipp, N.G.** 1971: A new micropaleontological method for quantitative paleoclimatology: application to a late Pleistocene Caribbean core. In Turekian, K.K. (ed.), *The Late Cenozoic glacial ages*. New Haven, CT: Yale University Press, 71–181. **Ingram, M.J., Underhill, D.J. and Farmer, G.** 1981: The use of documentary sources for the study of past climate. In Wigley, T.M.L., Ingram, M.J. and Farmer, G. (eds), *Climate and history*. Cambridge: Cambridge University Press, 180–213. **Lowe, J.J. and Walker, M.J.C.** 1997: *Reconstructing Quaternary environments*. London: Longman. **Roberts, N.** 1998: *The Holocene. An environmental history*. Oxford: Blackwell.

climatic system

The global climatic system consists of the atmospheric response to a variety of FORCING FACTORS, the fundamental one being solar energy. In general terms, the behaviour of the climatic system may be classified as exhibiting transitive, intransitive or almost intransitive responses to a given PERTURBATION (see Figure). A *transitive response* is one in which the system evolves

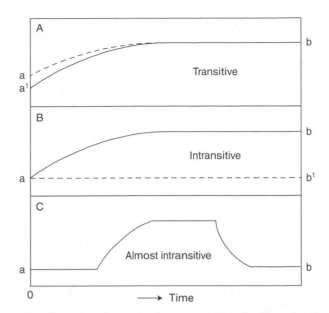

climatic system *Concepts of response of the climatic system to perturbations: (A) transitive response; (B) intransitive response; (C) almost intransitive response; (a, a') = starting positions; (b, b') = stable states (after Peixoto and Oort, 1992)*

towards the same stable state from different starting points (initial conditions); an *intransitive response* involves the possibility of evolution towards different stable states from the same initial conditions; whereas an *almost transitive response* is one that alternates between more than one stable state and suggests the existence of a non-linear, DYNAMICAL SYSTEM (as appears to be the case in GLACIAL–INTERGLACIAL CYCLES).

The climatic system derives energy from the translation of thermal contrasts (arising principally from latitude and oceanic influences) into atmospheric motion, both vertical and horizontal. The one significant energy source 'external' to the atmosphere is the Sun. Receipt of SOLAR RADIATION (in the form of visible, infrared and ultraviolet light) by the Earth's atmosphere is influenced by longer-term changes in the rotation and orbit of the Earth and by the output of the Sun. After losses by reflection and scattering, the remainder reaches the surface either as either direct or diffuse solar radiation. The proportion reflected by the surface is termed the ALBEDO, a key influence on surface temperature. Terrestrial or long-wave radiation influences the rate of heat loss, which is determined by the concentration of GREENHOUSE GASES in the atmosphere.

Imbalances between incoming and outgoing radiation at different latitudes act as a catalyst for the GENERAL CIRCULATION OF THE ATMOSPHERE. Latitudinal temperature gradients generate a westerly flow around each hemisphere due to the interaction of the poleward pressure gradient force and the CORIOLIS effect, which deflects moving air to the right (left) in the Northern (Southern) Hemisphere. The degree of sinuosity of the upper westerlies, denoted by LONG WAVES (Rossby waves), plays an important part in redistributing heat across different latitudes. Well developed Rossby waves are associated with periods when a weakened westerly flow is blocked by quasi-stationary high-pressure areas. Temperature ANOMALIES then result from anomalies in OCEANICITY and CONTINENTALITY due to changes in the prevalence of westerly winds.

Vertical motion is superimposed on this horizontal flow where warm air rises at the Equator (INTERTROPICAL CONVERGENCE ZONE) and at the convergence zone between contrasting air masses at the POLAR FRONT in the mid-latitudes. This leads to divergence of the air column in the upper atmosphere and the development of low-pressure systems. Rising air cools adiabatically, leading to condensation and the development of CLOUD and PRECIPITATION.

JCM

Coley, P.F. and Jonas, P.R. 1999: Back to basics: clouds and the Earth's radiation budget. *Weather* 54(3), 66–70. Drake, F. 2000: *Global warming: the science of climate change*. London: Arnold. Peixoto, J.P. and Oort, A.H. 1992: *Physics of climate*. New York: American Institute of Physics. Trenberth, K.E. and Soloman, A. 1994: The global heat balance: heat transports in the atmosphere and ocean. *Climate Dynamics* 10, 107–134. Trenberth, K.E., Houghton, J.T. and Meira Filho, L.G. 1996: The climate system: an overview. In Houghton, J.T., Meira Filho, L.G., Callander, B.A. *et al.* (eds), *Climate change 1995: the science of climate change*. Cambridge: Cambridge University Press, 51–64.

climatic vacillation
A type of CLIMATIC FLUCTUATION: when the average value of a climatic variable exhibits relatively long periods of near constancy separated by relatively short change-over periods. *JAM*

climatic variability
Climatic variability usually consists of natural or anthropogenic changes of a climatic element over inter-monthly to inter-decadal timescales. As timescale lengthens, climatic TRENDS may be observed if the change is unidirectional; such trends may develop into CLIMATIC CHANGE if sustained over several decades. At shorter timescales, the variability is itself the main feature, regardless of trend. The wider range of climatic states may then lead to an increased frequency of EXTREME WEATHER EVENTS.

The causes of climatic variations differ according to duration. Inter-monthly changes are typically due to changes in the *atmospheric circulation*. Occasionally, volcanic eruptions may increase the atmospheric aerosol loading sufficiently to effect a short-term climatic variation. Decadal climate variations are more likely to be related to Earth surface changes (ice-cover variations or DESERTIFICATION), which lead to changes in the surface radiation budget. Anomalies of SEA-SURFACE TEMPERATURE have a major influence upon the regional atmospheric circulation, especially where ocean–atmosphere processes are closely coupled (see, for example, EL NIÑO–SOUTHERN OSCILLATION and NORTH ATLANTIC OSCILLATION). Another cause of temperature is the 11-year solar cycle. Longer-term variations in climate may be caused by changes in the strength of the GREENHOUSE EFFECT.

M. Hulme has suggested that the warming from the late nineteenth to late twentieth centuries of roughly +0.5°C was at the upper limit of potential natural climate variability, citing work by T.M.L. Wigley indicating a typical upper limit of ±0.3°C per century. The observed rate of warming has accelerated markedly over the final 15 years of the twentieth century following cooling (mostly Northern Hemisphere) between the 1940s and about 1970. The current warming rate is roughly +0.2°C per

decade, close to the rate suggested by CLIMATIC CHANGE SCENARIOS of twentyfirst century warming.

These global averages mask large regional and SEASONALITY variations. High-latitude continental areas in the Northern Hemisphere warmed by over 1°C between the 1960s and 1990s and by several degrees in winter. Conversely, much of Greenland and adjacent parts of the North Atlantic have cooled over this period by as much as 0.5°C, changes related to the post-1980 positive phase of the North Atlantic Oscillation, which has led to REGIONAL CLIMATIC CHANGE over much of Europe.

Precipitation shows more variability over space and time than temperature. However, large-scale changes in precipitation have been observed in the twentieth century, notably increases over some high-latitude locations in the Northern Hemisphere and over the mid-latitudes in winter. Some low-latitude regions have been subject to drier conditions over the same period, notably the SAHEL in Africa. This has been linked to changes in the latitude of sea surface temperature anomalies off West Africa.

The identification of the 'signal' of climatic trends from the 'noise' of climatic variability involves many uncertainties related to the spatial and temporal sampling of the data. A further uncertainty concerns the consistency of the observations themselves as a result of changes in instrumentation or technology. Examples include the undersampling of snowfall in conventional rain gauges and the derivation of surface temperature from satellite observations. A non-climatic variable, which is now taken account of in CLIMATIC ARCHIVES is the warming effect associated with increasing URBANISATION. Despite these uncertainties, a GLOBAL WARMING signal is also being derived from glacier retreat, Northern Hemisphere snow cover and from other parts of the natural environment. ICE CORE records have also provided some evidence for abrupt climatic variations of as much as 7°C over a few decades during INTERSTADIALS within the last glaciation, possibly related to variations in the rate of DEEP WATER formation in the North Atlantic. Together with evidence for rapid temperature shifts during the last (*Eemian*) interglacial, the present (HOLOCENE) interglacial has been reinterpreted as a period of muted climatic variability.

There is little evidence of a consistent increase in global climatic variability, though this is not surprising as climatic variations are subject to such a wide range of influences. It is essential to take a global overview because increased variability in one region may be balanced by reduced variability elsewhere. D. Parker has found some evidence of a slight increase in seasonal variability in temperature.

A plausible cause of any increased climatic variability is greater variability of air pressure patterns, some evidence for which has been identified in the Atlantic sector. In other parts of the world, climate variability is related to regional or larger OCEAN CURRENT or temperature anomalies, which influence the atmospheric circulation by a series of teleconnections. The most profound effect of this is the El Niño–Southern Oscillation coupling in which changes in ocean temperature and circulation lead to temperature anomalies of more than 4°C over many months in the Eastern South Pacific, partly due to the capping of the 'normal' UPWELLING of cool deep water.

COUPLED OCEAN–ATMOSPHERE MODELS show some evidence for increases in seasonal variability in scenarios of future warming. One of the most widespread changes is an increase in winter precipitation in mid-latitude areas whilst that of summer is expected to decline, thus increasing seasonal variability of moisture availability (see figure).

One of the 'costs' to society of increased short-term climatic variability is an increase in the frequency of EXTREME CLIMATIC EVENTS. It is important that such scenarios also incorporate a measure of the change in variability in addition to that of the mean of a climatic variable. It is then possible to estimate the probability or return period of particular extreme events. For example, research at the Climatic Research Unit, University of East Anglia, UK has estimated that the return period of a year being as warm as 1990 in the British Isles will change from 65 years (1961–1990) to 1.6 years by the 2050s. *JCM*

[See also CLIMATIC FLUCTUATIONS, CLIMATIC VARIATIONS, PERIODICITIES, VOLCANIC IMPACTS ON CLIMATE]

Born, K. 1996: Tropospheric warming and changes in weather variability over the Northern Hemisphere, 1967–91. *Meteorology and Atmospheric Physics* 59, 201–215. **Bradley, R.J. and Jones, P.D.** 1993: 'Little Ice Age summer temperature variations: their nature and relevance to recent global warming trends. *The Holocene* 3, 367–376. **Bradley, R.J. and Jones, P.D. (eds)**, 1995: *Climate since AD 1500*. London: Routledge. **Conway, D.** 1998: Recent climate variability and future climate scenarios for Great Britain. *Progress in Physical Geography* 22(3), 350–374. **Heino, R.** 1994: *Climate variations in Europe, Academy of Finland publication* 3/94. Helsinki: Academy of Finland. **Hulme, M.** 1994: Historical records and recent climatic change. In Roberts, N. (ed.), *The changing global environment*, Oxford: Blackwell, 69–98. **Hulme, M. and Jenkins, G**. 1998: *Climate change scenarios for the U.K.* [*UKCIP Technical Report* 1]. Norwich: Climatic Research Unit, University of East Anglia Climate Impacts Programme. **Jones, P.D. and Briffa, K.R.** 1992: Global surface air temperature variations over the twentieth century: part 1. Spatial, temporal and seasonal details. *The Holocene* 2, 165–179. **Jones, P.D., Briffa, K.R., Barnett, T.P. and Tett, S.F.B.** 1998: High-resolution palaeoclimatic records for the last millennium: interpretation, integration and comparison with General Circulation Model control-run temperatures. *The Holocene* 8, 455–471. **Kattenberg, A., Giorgi, H., Grassl, H. *et al.*** 1996: Climatic models: projections of future climate. In Houghton, J.T., Meiro Filho, L.G., Callander, B.A. *et al.* (eds), *Climatic*

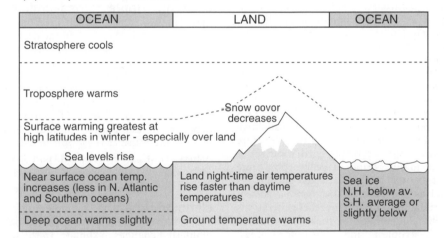

(A) Temperature indicators

| OCEAN | LAND | OCEAN |

Stratosphere cools

Troposphere warms

Snow cover decreases

Surface warming greatest at high latitudes in winter - especially over land

Sea levels rise

Near surface ocean temp. increases (less in N. Atlantic and Southern oceans)

Land night-time air temperatures rise faster than daytime temperatures

Sea ice N.H. below av. S.H. average or slightly below

Deep ocean warms slightly

Ground temperature warms

(B) Hydrological indicators

| OCEAN | LAND | OCEAN |

Tropical precipitation variability increases
High latitude precipitation increases year round in N.H.
Mid to high latitude precipitation increases in winter, may decrease in summer in N.H.

Water vapour increases

Evaporation in tropics increases

Mid-latitude soil moisture increases in winter, decreases in summer

climatic variability   *Environmental changes associated with scenarios of global warming in the twentyfirst century for some (A) temperature and (B) hydrological indicators (after Kattenberg et al. 1996)*

change 1995. Cambridge: Cambridge University Press, 285–357. **Lamb, H.H.** 1995: *Climate history and the modern world*, 2nd edn. London: Routledge. **Mitchell, J.M.** 1976: An overview of climatic variability and its causal mechanisms. *Quaternary Research* 6, 481–493. **National Research Council** 1995: *Natural climate variability on decadal to century timescales*. Washington, DC: National Academy Press. **Nicholls, N., Gruza, G.V., Jouzel, J.** *et al.* 1996: Observed climate variability and change. In Houghton, J.T., Meira Filho, G., Callander, B.A. *et al.* (eds), *Climate change 1995: the science of climate change*. Cambridge: Cambridge University Press, 133–192.

climatic variation

A synonym for CLIMATIC CHANGE: conventionally any systematic change in a climatic variable (e.g. mean value, other measure of central tendency or a measure of dispersion or CLIMATIC VARIABILITY) that persists for about 30 years or more. Some include shorter-term inconstancies in climate, such as CLIMATIC FLUCTUATIONS, i.e. all systematic aspects of climatic variability.
JAM

Urban, F.E., Cole, J.E. and Overpeck, J.T. 2000: Influence of mean climate change on climate variability from a 155-year tropical Pacific coral record. *Nature* 407, 989–993.

climatogenetic geomorphology

The scientific study of climatically controlled landform development over time using past as well as contemporary climatic influences. According to Büdel, the task of climatogenetic geomorphology is recognising, ordering and distinguishing relief generations within LANDSCAPES, in order to be able to analyse today's highly complex relief. If climatic phases are of sufficient duration to form characteristic LANDFORMS and deposits, the landscape may provide a record of this, with at least some parts of these features surviving through each subsequent period of distinctive morphogenesis. Most landscapes are therefore PALIMPSESTS of features engraved on the Earth's surface at different times. The major period of geomorphic time between the CRETACEOUS and the Mid-Pliocene, is described by Büdel as the time of the *tropicoid palaeo-Earth*, when some of the most important relief generations were formed by climates similar to those experienced today in seasonally tropical and subtropical areas. Following this, the relief generations of the late PLIOCENE, PLEISTOCENE, and HOLOCENE, were created.
MAC

[See also CLIMATIC GEOMORPHOLOGY, LANDSCAPE EVOLUTION]

Büdel J. 1982: *Climatic geomorphology*, translated by Fischer, L. and Busche, D. Princeton, NJ: Princeton University Press. **Derbyshire, E. (ed.)** 1976: *Geomorphology and climate*. London: Wiley. **Holzner, L. and Weaver, G.D.** 1965: Geographical evaluation of climatic and climato-genetic geomorphology. *Annals of the Association of American Geographers* 55, 592–602.

climatology

The study of the meteorology and weather for specified time periods and locations comprising not only averages but also the range, frequency and causes of atmospheric events.
JCM

[See also CLIMATE, CLIMATIC SYSTEM, CLIMATIC VARIABILITY]

Barry, R.G. and Chorley, R.J. 1992: *Atmosphere weather and climate*, 6th edn. London: Routledge.

climax vegetation

A mature plant community that is self-replacing and in equilibrium with environmental conditions. There are several different concepts of climax vegetation, including: *monoclimax* (a single type of vegetation towards which ecological succession converges); *polyclimax* (a number of types, each controlled by different ENVIRONMENTAL FACTORS); and the *climax pattern* concept (a continuum of vegetational variation corresponding with environmental gradients). Traditionally, such communities were considered to represent the terminal stage of ECOLOGICAL SUCCESSION but with the current emphasis on the importance of DISTURBANCE and ENVIRONMENTAL CHANGE in structuring plant communities, climax vegetation is now regarded as a useful concept, but not one with a great deal of applicability to natural situations.
LRW

[See also CONVERGENCE AND DIVERGENCE, IN SUCCESSION, INDIVIDUALISTIC CONCEPT]

Clements, F.E. 1916: Nature and structure of the climax. *Journal of Ecology* 24, 252–284. **Davis, M.B.** 1986: Climatic instability, time lags and community disequilibrium. In Diamond, J. and Case, T.J. (eds), *Community ecology*. New York: Harper and Row. **McIntosh, R.P.** 1985: *The background of ecology: concept and theory*. Cambridge: Cambridge University Press. **Whittaker, R.H.** 1953: A consideration of climax theory: the climax as a population and pattern. *Ecological Monographs* 23, 41–78.

climbing ripple cross-lamination

A style of CROSS-LAMINATION in which the boundaries between sets (TRANSLATENT STRATA) rise in the downstream direction as a result of sediment accretion during RIPPLE migration. Some climbing ripple cross-lamination can be used to calculate sediment ACCUMULATION RATES. Climbing ripple cross-lamination is a common SEDIMENTARY STRUCTURE in EVENT DEPOSITS, and beds of climbing ripple cross-laminated sand several tens of centimetres thick can be deposited in hours to tens of hours. Sometimes described in older literature as *ripple drift cross-lamination*.
GO

Ashley, G.M., Southard, J.B. and Boothroyd, J.C. 1982: Deposition of climbing-ripple beds: a flume simulation. *Sedimentology* 29, 67–79.

climostratigraphy

Climostratigraphy or *climatostratigraphy* is the division of the stratigraphic record on the basis of inferred changes in climate. In the study of QUATERNARY ENVIRONMENTAL CHANGE, these *geologic–climatic* or *climostratigraphic units* have formed the main elements of the chronostratigraphic subdivision of the QUATERNARY TIMESCALE into STAGES, which mostly correspond with GLACIAL EPISODES (*glacials*) and INTERGLACIALS.
DH

[See also GLACIAL–INTERGLACIAL CYCLE, INTERSTADIAL, STADIAL]

Lowe, J.J. and Walker, M.J.C. 1999: *Reconstructing Quaternary environments*, 2nd edn. Harlow: Addison-Wesley Longman.

cline

A variation in the genetic characteristics within or between species along a gradient, usually allied with a geographical transect such as altitude or latitude.
KJW

[See also ECOCLINE]

Clausen, J., Keck, D.D. and Hiesey, W.M. 1958: *Experimental studies on the nature of species*, III. *Carnegie Institution of Washington Publication* 581.

clitter

The accumulation of frost-weathered boulders surrounding a TOR.
PW

[See also BLOCKFIELDS, FROST WEATHERING]

closed system A SYSTEM that may exchange energy, but not matter, with its surroundings (environment). It is occasionally used as synonymous with an ISOLATED SYSTEM (i.e. no energy or matter exchange). *CET*

cloud forest A MONTANE FOREST formation found on TROPICAL and SUBTROPICAL mountains with persistent cloud cover and heavy rainfall caused either by lifting of moist air over mountains or by *stripping* of moisture from clouds by trees. *NDB*

[See also ALTITUDINAL ZONATION OF VEGETATION, ELFIN WOODLAND, PRECIPITATION, RAIN FOREST, TROPICAL FOREST, VEGETATION FORMATION-TYPE]

Hamilton, L.S., Juvik, J.O. and Scatena, F.N (eds) 1995: *Tropical montane cloud forests* [*Ecological Studies*.Vol. 110]. New York: Springer. Nadkarni, N.M. and Wheelright, N.T. 2000: *Monteverde: ecology and conservation of a tropical cloud forest.* Oxford: Oxford University Press. Sandved, K.B. and Emsley, M. 1979: *Rain forests and cloud forests.* New York: Harry N. Abrams.

cloud seeding The stimulation of condensation and precipitation by the introduction of particles into super-cooled clouds (see SUPERCOOLING) to act as CONDENSATION NUCLEI. These may include solid CARBON DIOXIDE and silver iodide crystals. A 10–15% increase in precipitation appears possible under optimal conditions, namely seeding of clouds at temperatures between -5.0 and $-15.0°C$. *JGT*

[See also CLIMATIC MODIFICATION]

Mason, B.J. 1975: *Clouds, rain and rainmaking.* Cambridge: Cambridge University Press.

clouds Assemblages of water droplets and/or ice particles in the atmosphere above the Earth's surface, which have formed around CONDENSATION NUCLEI. Generally limited to the TROPOSPHERE, a few special clouds do occur in the STRATOSPHERE. Clouds are formed when moist air cools and becomes saturated. Clouds are composed of either water droplets above freezing point, ice crystals at temperatures below about $-40°C$ or a mixture of supercooled water droplets and ice crystals at temperatures in between. A standard classification of clouds is based on their height (values are higher in the tropics) above the Earth's surface: high (ice crystal) clouds (above 5 km in mid-latitudes) include *cirrus*, *cirrocumulus*, and *cirrostratus*; medium (mixed) clouds (2 to 5 km) include *altocumulus*, *altostratus* and *nimbostratus*; whereas low (water droplet) clouds (0 to 2 km) include *cumulus*, *cumulonimbus*, *stratocumulus* and *stratus*. Three of these (nimbostratus, cumulus and cumulonimbus) extend beyond these height limits and are more accurately considered as *multilevel clouds*. *BDG*

Cotton, W.R. and Anthes, R.A. 1989: *Storm and cloud dynamics.* San Diego, CA: Academic Press. Fouquart, Y., Buriez, J.C. and Herman, M. 1990: The influence of clouds on radiation: a climate-modeling perspective. *Reviews of Geophysics* 28, 145–166. McIlveen, R. 1992: *Fundamentals of weather and climate.* London: Chapman and Hall. Rogers, R.R. and Yau, M.K. 1989: *A short course in cloud physics,* 3rd edn. Oxford: Pergamon. World Meteorological Organization 1956: *International cloud atlas.* Geneva: WMO.

cluster analysis A set of numerical methods for constructing a CLASSIFICATION of an initially unclassified set of data, using the variable values (attributes) observed for each object (individual). Many methods exist including TWO-WAY INDICATOR SPECIES ANALYSIS (TWINSPAN). *HJBB*

Everitt, B.S. 1993: *Cluster analysis.* London: Edward Arnold.

coal A *carbonaceous* SEDIMENTARY ROCK rich in organic carbon (< 33% inorganic material) derived from plant remains. *Humic coals* comprise *in situ* organic material and form from PEAT by COMPACTION and loss of volatiles. They are divided by *coal rank* (carbon content) into *lignite* and *brown coals* (approx. 60–70% carbon), *bituminous coals* (70–90% carbon) and *anthracite* (90–95% carbon). *Sapropelic coals* comprise transported organic material (SAPROPEL) and include *cannel coal* (derived mainly from higher plants) and *boghead coal* (derived from algae and fungae). Coal releases energy when burned and is one of the main FOSSIL FUELS, used as a domestic and industrial fuel, in electricity generation and in industrial processes such as iron smelting.

Coal occurs in BEDS (*coal seams*) that form a part (usually 5–20%) of coal-bearing rock successions that are usually dominated by other CLASTIC sedimentary rocks such as SANDSTONE and SHALE. Most coal seams in Britain are in Upper CARBONIFEROUS rocks and reach a maximum of a few metres thick, but elsewhere coals range from DEVONIAN to TERTIARY age and seams may reach over 100 m thick. Coal is extracted from *shallow mines* (excavations from tunnels driven from the ground surface), from *deep mines* (excavations from the base of a vertical shaft), and by *opencast* or *open pit* methods (also known as *strip mining*). In recent years advances have been made in the development of *coal-bed methane*, which involves the extraction of METHANE gas from *in situ* coal seams.

The environmental significance of coal falls into two fields. First, environmental problems accompanying the exploitation of coal include: ENVIRONMENTAL DEGRADATION associated with coal extraction; AIR POLLUTION associated with coal combustion and opencast extraction; slope stability problems in *spoil tips*; and WATER POLLUTION from mine drainage. Second, coal seams are important indicators of PALAEOCLIMATOLOGY and PALAEOENVIRONMENT in the GEOLOGICAL RECORD, commonly indicating sediment accumulation in waterlogged environments in the HUMID TROPICS. *GO*

Cook, A.C. and Sherwood, N.R. 1991: Classification of oil shales, coals and other organic-rich rocks. *Organic Geochemistry* 17, 211–222. Dimichele, W.A. and Phillips, T.L. 1994: Paleobotanical and paleoecological constraints on models of peat formation in the Late Carboniferous of Euramerica. *Palaeogeography, Palaeoclimatology, Palaeoecology* 106, 39–90. Gayer, R. and Petek, J. (eds) 1997: *European coal geology and technology* [*Special Publication* 125]. London: Geological Society. Haszeldine, R.S. 1989: Coal reviewed: depositional controls, modern analogues and ancient climates. In Whateley, M.K. and Pickering, K.T. (eds), *Deltas: sites and traps for fossil fuels* [*Special Publication* 41]. London: Geological Society. 289–308. McCabe, P.J. 1984: Depositional environments of coal and coal-bearing strata. In: Rahmani, R.A. and Flores, R.M. (eds), *Sedimentology of coal and coal-bearing sequences. Society of Economic Paleontologists and Mineralogists Special Publication* 7, 13–42. Montgomery, S.L. 1999: Powder River basin, Wyoming: an expanding coalbed methane (CBM) play. *American Association of Petroleum Geologists Bulletin* 83, 1207–1222.

coal balls Nodules of mineralised PEAT that occur in some COAL seams, posing problems in mining. They formed in the early stages of DIAGENESIS and preserve the three-dimensional structure of FOSSIL plants, providing important information about plant communities and PALAEOENVIRONMENTS. *GO*

DeMaris, P.J. 2000: Formation and distribution of coal balls in the Herrin Coal (Pennsylvanian), Franklin County, Illinois Basin, USA. *Journal of the Geological Society, London* 157, 221–228. **Lyons, P.C., Zodrow, E.L., Millay, M.A.** *et al.* 1997: Coal-ball floras of Maritime Canada and palynology of the Foord seam: geologic, paleobotanical and paleoecological implications. *Review of Palaeobotany and Palynology* 95, 31–50.

coastal bars and spits Deposits that parallel the shoreline, either submerged or exposed, usually as a sand bar (*offshore* or *longshore bar*) or attached at one end to the land as a *spit*. Spits are usually curved toward the shore at their distal end. A bar that extends from the mainland to a spit, offshore bar, island, stack or breakwater is a *tombolo*. Bars and spits form depositional sinks for sediment transported along a shoreline. *HJW*

Schwartz, M. L. (ed.) 1972: *Spits and bars*. Stroudsburg, PA: Dowden, Hutchinson, and Ross.

coastal change The coast is the junction between an unstable solid and a constantly moving liquid. It is the most conspicuous and yet most dynamic boundary on Earth. The three most critical natural characteristics of the coast and its SHORELINE are its constituent materials, the forms it assumes and the processes that affect it. All three of these conditions are subject to change, which may be: instantaneous or require millennia to be effective; cyclic (hourly, daily, seasonally) or random (earthquakes); and independent or interrelated. Shore materials may be: mineral or organic; solid, liquid or gaseous; such that they even undergo phase change. Shore forms also vary widely, e.g. from steep cliffs to extensive mudflats, barrier complexes to LAGOONS and coastal marsh, and from CORAL REEFS to MANGROVE swamps. The processes operating on these materials and forms include virtually all of the physical, chemical and biotic agents found in nature.

Because of PLATE TECTONICS and CONTINENTAL DRIFT, the present coastline is longer and more varied than at any other time in Earth's history. Coasts along the leading edge of a plate are usually more rugged than those formed along its trailing edge. The impact of tectonic activity on coastlines also influences the amount and type of sediment transported to the shoreline by rivers and therefore the formation of DELTAS, which are so important to human activities. The position of the coastline has fluctuated with changing sea level and is affected, not only by tectonic and sedimentary activity, but also by ISOSTASY and EUSTASY. For example, isostatic adjustment was (and continues to be) responsible for rebound along many coasts (e.g. Hudson Bay and Scandinavia) following the waning of ICE SHEETS, which also led to recovery of world eustatic sea level by mid-HOLOCENE times from its minimum of about –120 m at the maximum of the LAST GLACIATION around 18 000 years ago. A classification of coasts based on change is shown in the Figure.

Human activity directly affects present-day coastlines, especially those in densely populated areas, to a consider-able extent. Indirectly, SEA-LEVEL RISE may be caused (at least partially) on a global scale through GLOBAL WARMING. *HJW*

[See also COASTAL ENGINEERING STRUCTURES, HUMAN IMPACT ON COASTS, SEA-LEVEL CHANGE, SEA-LEVEL RISE: POTENTIAL FUTURE GEOMORPHOLOGICAL IMPACTS, SEA-LEVEL RISE: POTENTIAL FUTURE IMPACTS ON PEOPLE]

Davies, J. L. 1977: *Geographical variation in coastal development*. London: Longman. **Dubois, R.N.** 1988: Seasonal changes in beach topography and beach volume in Delaware. *Marine Geology* 81, 79–96. **Schwartz, M. L.** 1982: *The encyclopedia of beaches and coastal environments*. New York: Hutchinson Ross. **Shennan, I. and Andrews, J.E. (eds)** 1999: *Holocene land–ocean interaction and environmental change around the North Sea* [*Special Publication* 166]. Bath: Geological Society. **Spencer, T.** 1999: Coastal Erosion. In Pacione, M. (ed.), *Applied geography: principles and practice*. London: Routledge, 109–123. **Valentin, H.** 1952: Die Kusten der Erde. *Petermanns Geographische Mitteilungen* 46, 118. **Walker, H. J.** 1975: Coastal morphology. *Soil Science* 119, 1, 3–19. **Zenkovich, V. P.** 1967. *Processes of coastal development*. Edinburgh: Oliver and Boyd.

coastal dunes AEOLIAN sand DUNES that develop above high water in a coastal setting as sand is blown from BEACH DEPOSITS. Individual dunes and dune ridges are typically up to several tens of metres high and are commonly separated by wet, sometimes PEAT-filled hollows known as *dune slacks*, where the ground surface intersects the WATER TABLE. Coastal dunes are dominated by quartz sand, but most have a significant CARBONATE content from shell debris. Where the carbonate content is high, particularly in Mediterranean and tropical settings, dune sands readily lithify to become AEOLIANITES.

Coastal dune systems may extend inland for several kilometres. They are best developed in areas of strong

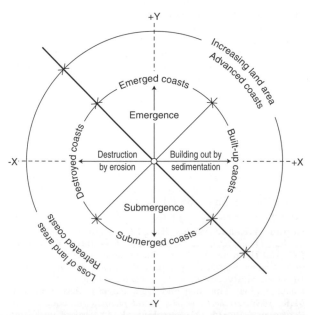

coastal change *Coastal classification in terms of (historical) emergence or submergence and (contemporary) erosion or sedimentation. The solid line separates all advancing coasts (top right) from all retreating coasts (bottom left) (after Valentin, 1952; Spencer, 1999)*

onshore winds, with plentiful sand supply due to factors such as extensive glacial or glaciofluvial offshore deposits, a large tidal range, and a suitable site for DEPOSITION. Coastal dunes differ from most inland active dunes and DESERT DUNES because moisture is normally plentiful, and the interplay of stabilising vegetation and wave action affects sediment supply and shapes the foredunes. There is typically a consistent variation from active, unvegetated dunes close to the shore to older, vegetated dunes further inland. Dune systems may be artificially vegetated and stabilised as a measure of COASTAL (SHORE) PROTECTION, and destabilisation through TRAMPLING is a major concern along coastlines popular with tourists.

The activity of coastal dune systems is strongly influenced by climatic change. There is DOCUMENTARY EVIDENCE, for example, of increased sand movement during periods of increased STORM frequency, leading to SAND DUNE ENCROACHMENT onto human settlements or agricultural land. In addition, entire dune systems may migrate seaward (PROGRADATION) or landward. A range of dating and environmental reconstruction techniques applied to relict coastal dunes (e.g. MOLLUSCA ANALYSIS, LUMINESCENCE DATING) have aided PALAEOENVIRONMENTAL RECONSTRUCTION, particularly of past SEA-LEVEL CHANGE.

RAS/GO

Borja, F., Zazo, C., Dabrio, C.J. et al. 1999: Holocene aeolian phases and human settlements along the Atlantic coast of southern Spain. The Holocene 9, 333–339. Bridges, E.M. 1997: Classic landforms of the Gower coast. Sheffield: The Geographical Association. Clemmensen, L.B., Pye, K., Murray, A. and Heinemeier, J. 2001: Sedimentology, stratigraphy and landscape evolution of a Holocene coastal dune system, Lodbjerg, NW Jutland, Denmark. Sedimentology 48, 3–27. Kamola, D.L. and Chan, M.A. 1988: Coastal dune facies, Permian Cutler Formation (White Rim Sandstone), Capitol Reef National Park area, southern Utah. Sedimentary Geology 56, 341–356. Nordstrom, K.F. 2000: Beaches and dunes of developed coasts. Cambridge: Cambridge University Press. Ruz, M.-H. and Allard, M. 1995: Sedimentary structures of cold-climate coastal dunes, Eastern Hudson Bay, Canada. Sedimentology 42, 725–734. Sherman, D.J. and Bauer, B.O. 1993: Dynamics of beach-dune systems. Progress in Physical Geography 17, 413–447. Wilson, P. and Braley, S.M. 1997: Development and age structure of Holocene coastal dunes at Horn Head, near Dunfanaghy, Co. Donegal, Ireland. The Holocene 7, 187–197.

coastal engineering structures Structures placed parallel or at angles to the shoreline primarily to control erosion. Shore parallel structures include REVETMENTS, SEAWALLS, bulkheads, dykes and BREAKWATERS. Breakwaters may also be placed offshore (detached) to impede wave action. Perpendicular or subperpendicular structures include GROYNES and JETTIES. Their main function is to intercept the lateral distribution of sediment by LONGSHORE DRIFT. These structures may be made of earth, RIPRAP, or fabricated concrete armour units in the form of, for example, TETRAPODS and HEXAPODS. *HJW*

[See also HARD ENGINEEERING, COASTAL (SHORE) PROTECTION]

Thorne, C.R. Abt, S.R., Barends, F.B.J. and Pilarczyk, K.W. 1995: River, coastal and shoreline protection. Chichester: Wiley. Walker, H.J. (ed.) 1988: Artificial structures and shorelines. Dordrecht: Kluwer.

coastal (shore) protection Measures to reduce the negative impact associated with erosion. The most com-

mon protection measures adopted are the stabilisation of the shoreline with HARD ENGINEERING structures (such as BREAKWATERS, SEAWALLS, *dykes*, LEVÉES, GROYNES, and JETTIES) or by SOFT ENGINEERING procedures (such as BEACH NOURISHMENT and DUNE stabilisation). *HJW*

Bird, E. C. F. 1996: Beach management. Chichester: Wiley. Kraus, N. C. (ed.) 1996: History and heritage of coastal engineering. New York: American Society of Civil Engineers. National Research Council 1990: Managing coastal erosion. Washington, DC: National Academy Press.

coastline The general line that forms the boundary between the land and sea. The *coastal zone* extends an ill-defined distance inland from the coastline. *HJW*

[See also SHORELINE, SHORE ZONE]

Bird, E.C.F. and Schwartz, M.L. (eds) 1985: The world's coastline. New York: Van Nostrand Reinhold.

cobble A SEDIMENT particle of GRAIN-SIZE between 64 and 256 mm. *TY*

coccolithophores Unicellular marine algae covered in light-intensifying calcium carbonate platelets (*coccoliths*). The commonest forms of calcareous *nannoplankton*, they are important in MICROPALAEONTOLOGY as INDEX FOSSILS. Construction and deposition of coccoliths also perturbs the ocean carbon system, thereby exerting an impact on atmospheric CARBON DIOXIDE levels. *DCS*

Winter, A. and Siesser, W.G. (eds) 1994: Coccolithophores. Cambridge: Cambridge University Press.

cockpit karst A KARST landscape with a high density of deep and large, often star-shaped, DOLINES ('cockpits'). Cockpit karst, known as *kegelkarst* in Germany, is usually associated with the HUMID TROPICS, although it has also been found in other climatic zones. The star-shape arises from small and usually dry valleys that incise the doline slopes. These have been attributed to high-intensity tropical rainfall leading to incision by overland flow, though solution by concentrated shallow subsurface flow is perhaps a more likely mechanism. The high density of dolines compared with temperate areas has also sometimes been attributed to the higher rainfall of the humid tropics.

SHD

[See also TROPICAL KARST]

Ford, D.C. and Williams, P.W. 1989: Karst geomorphology and hydrology. London: Chapman and Hall.

coefficient of variation A descriptive statistic that measures the *relative variability* in a dataset. It is the standard deviation expressed as a percentage of the mean and is often used in preference to the standard deviation because the latter, as a measure of *absolute variability*, is affected by the value of the mean. *JAM*

coevolution A process whereby EVOLUTION in two or more separate LINEAGES takes place in such a way that each affects the survival of the other. In extreme cases, coevolution may lead to close relationships between species of widely differing lineages, such as insects and flowers, where each species is completely dependent on the other. *KDB*

[See also ADAPTATION, SYMBIOSIS]

Thompson, J.N. 1994: *The coevolutionary process.* Chicago, IL: University of Chicago Press. Futuyma, D.J. and Slatkin, M. 1983: *Coevolution.* Sunderland, MA: Sinauer. Futuyma, D.J. 1998: *Evolutionary biology.* Sunderland, MA: Sinauer.

cohesion The tendency of particles of CLAY and fine SILT to stick together, giving *cohesive* sediment an inherent strength that is not possessed by *cohesionless* sediment, such as coarse silt, SAND and GRAVEL, in which particles are held together by friction alone. *GO*

[See also SEDIMENT GRAVITY FLOW]

cold-based glaciers Glaciers with basal ice entirely below the *pressure melting point* and therefore frozen to their beds. They occur in high-latitude and high-altitude regions with low mean annual air temperatures, especially where ice is thin or slowly moving. Cold-based glaciers do not cause significant erosion of their beds and may leave little evidence of their existence in the landscape. In some ARCTIC regions, TERTIARY land surfaces are thought to have been preserved beneath cold-based PLEISTOCENE *ice caps* and ICE SHEETS (see Figure). *DIB*

[See also GLACIER THERMAL REGIME]

Dyke, A.S. 1993: Landscapes of cold-centred Late Wisconsinan ice caps, Arctic Canada. *Progress in Physical Geography* 17, 223–247. Gellatly, A.F., Gordon, J.E., Whalley, W.B. and Hansom, J.D. 1988: Thermal regime and geomorphology of plateau ice caps in northern Norway: observations and implications. *Geology* 16, 983–986. Kleman, J. and Hättestrand, C. 1999: Frozen-based Fennoscandian and Laurentide ice sheets during the Last Glacial Maximum. *Nature* 402, 63–66.

collision zone The area affected by the convergence of two CONTINENTS in PLATE TECTONICS once all the inter-vening OCEANIC CRUST has been consumed at DESTRUCTIVE PLATE MARGINS. SEDIMENTS of the CONTINENTAL MARGINS experience compression, DEFORMATION and UPLIFT, producing a linear chain of high mountains, or OROGENIC BELT. The Himalayas are a classic example, arising from the collision between India and Asia. *GO*

[See also WILSON CYCLE]

Park, R.G. 1988: *Geological structures and moving plates.* Glasgow: Blackie.

colloid A substance that, by virtue of its particle size (usually finer than 0.01 µm), is capable of remaining indefinitely in suspension. Colloidal suspensions are commonly viewed as solids in liquids (e.g. colloidal CLAYS and HUMUS colloids), but the term also applies, for example, to liquids in gases (e.g. CLOUDS). *JAM*

[See also CLAY–HUMUS COMPLEX]

colluvial fan A fan-shaped deposit of COLLUVIUM at the foot of a mountain slope produced by slope processes, especially DEBRIS FLOWS and SNOW AVALANCHES. Colluvial fans tend to be smaller and steeper, and are associated with smaller and more rugged drainage basins, than ALLUVIAL FANS. *JAM*

colluvial history COLLUVIUM is clastic slope-waste material that is commonly coarse-grained and immature, deposited on the lower part of a mountain slope by sediment-gravity processes. Colluvial depositional systems have typically the form of relatively steep fans, commonly coalescing into *colluvial aprons.* Modern COLLUVIAL FANS are distinguishable from ALLUVIAL FANS due to the characteristic products of rapid MASS MOVEMENT PROCESSES such

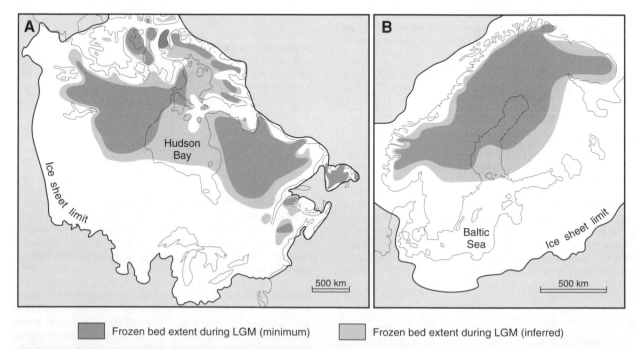

cold-based glaciers *The extent of frozen-bed conditions and hence cold-based glaciers in (A) North America and (B) Fennoscandia at the Last Glacial Maximum reconstructed from the distribution of relict landscapes and Rogen (ribbed) moraines. Dark grey areas indicate the minimum extent of frozen-bed conditions based directly on the geomorphological evidence; additional areas of inferred frozen-bed conditions are shown in light grey (after Kleman and Hättestrand, 1999)*

as DEBRIS FLOW and are well illustrated by postglacial colluvium in Norway. Such processes in mountainous terrain are generally controlled by climate and local slope conditions. Colluvial systems are characterised by highly episodic SEDIMENTATION as a response to major, episodic weather events. SNOW AVALANCHES are generally related to snowfall intensity and strong winds. Rock falls, which are directly related to bedrock weathering, are also likely to have been triggered by air temperature variations (FREEZE–THAW CYCLES). Debris flows are dependent on a wider range of factors, such as local slope conditions and high RAINFALL intensity.

Chronostratigraphic analyses of colluvial successions indicate that the colluvial deposits contain important PROXY EVIDENCE of regional palaeoclimatic variations. Studies in the Møre area, western Norway, suggest that stratigraphic changes in colluvial facies assemblages reflect significant climatic changes. The palaeoclimatic record obtained from the colluvial successions corresponds quite well to the record of alpine glacier variations in western Norway. Colluvial systems are also highly dependent upon local slope conditions. The palaeoclimatic record from colluvial facies data is most legible when data from a large number of sites are compiled. The regional HOLOCENE record of snow avalanches in the Møre area indicates a progressive climatic deterioration and increasingly colder winter conditions since at least c. 4700 calendar years BP with at least two subsequent pulses of slope activity. Some of the local colluvial systems with catchments at high altitudes show the first evidence of snow avalanches approximately 7000 calendar years BP. Evidence from LACUSTRINE SEDIMENTS in western Norway suggests increased snow avalanche activity even earlier, around 8700 calendar years BP. The Møre record shows an increase in snow avalanche activity between 3800 and 3000 calendar years BP, three significant pulses between 2500 and 1800 calendar years BP, a pulse around 1400 calendar years BP and historic evidence suggests increased activity during the LITTLE ICE AGE.   *AN*

Blikra, L.H. and Fjeldstad Selvik, S. 1998: Climatic signals recorded in snow avalanche-dominated colluvium in western Norway: depositional facies successions and pollen records. *The Holocene* 8, 631–658. **Blikra, L.H. and Nemec, W.** 1998: Postglacial colluvium in western Norway: depositional processes, facies and palaeoclimatic record. *Sedimentology* 45, 909–959. **Matthews, J.A., Dahl, S.O., Berrisford, M.S.** *et al.* 1997: A preliminary history of Holocene colluvial (debris-flow) activity, Leirdalen, Jotunheimen, Norway. *Journal of Quaternary Science* 12, 117–129.

colluvial processes Processes that operate in response to gravity acting on slopes. They are distinct from FLUVIAL PROCESSES, which involve water flow in channels, although there is some overlap with *alluvial processes*, which are associated more generally with the flow of freshwater. Colluvial processes involve the gravity-driven down-slope movement of rock and sediment, with or without water, and include DEBRIS FLOW, AVALANCHE and other LANDSLIDE processes. The products are known as COLLUVIUM and colluvial LANDFORMS include COLLUVIAL FANS and TALUS.   *GO*

[See also ALLUVIAL FAN, MASS MOVEMENT, REGOLITH]

Blikra, L.H. and Nemec, W. 1998: Postglacial colluvium in western Norway: depositional processes, facies and palaeoclimatic record. *Sedimentology* 45, 909–959.

colluviation The formation and, especially, the accumulation of COLLUVIUM. Colluviation is associated with hillside slopes everywhere but is particularly prominent in PERIGLACIAL ENVIRONMENTS, where thick colluvial deposits such as TALUS and COLLUVIAL FANS may accumulate at the foot of slopes to the extent that they dominate the landscape. Phases of colluviation and/or episodic REDEPOSITION of colluvium may alternate with relatively stable phases of PEDOGENESIS in response to environmental change.   *JAM*

[See also ALLUVIATION, BIOSTASY, LANDSCAPE EVOLUTION]

Curry, A.M. 2000: Holocene reworking of drift-mantled hillslopes in Glen Docherty, Northwest Highlands, Scotland. *The Holocene* 10, 509–518. **Innes, J.L.** 1983: Stratigraphic evidence of episodic talus accumulation on the Isle of Skye, Scotland. *Earth Surface Processes and Landforms* 8, 399–403. **Luckman, B.H. and Fiske, C.J.** 1997: Holocene development of coarse-debris landforms in the Canadian Rocky Mountains. In Matthews, J.A., Brunsden, D., Frenzel, B. *et al.* (eds), *Rapid mass movement as a source of climatic evidence for the Holocene.* Stuttgart: Gustav Fischer, 283–297.

colluvio-aeolian Sandy and/or silty SEDIMENT, the DEPOSITION of which is a result of both SLOPEWASH and wind action.   *PW*

[See also COLLUVIUM, COVERSAND, LOESS]

colluvium Sedimentary material deposited by slope processes, such as SNOW AVALANCHE, DEBRIS FLOW, rockfall, SLOPEWASH and SOLIFLUCTION. The characteristics of colluvium vary greatly, depending not only on the particular slope processes but also on the local bedrock and/or the deposits from which it is derived or reworked. It is extremely common but relatively poorly investigated.   *JAM*

[See also COLLUVIAL FAN, COLLUVIAL HISTORY, COLLUVIATION, HEAD]

Blikra, L.H. and Nemec, W. 1998: Postglacial colluvium in western Norway: depositional processes, facies and palaeoclimatic record. *Sedimentology* 45, 909–959. **Mason, J.A. and Knox, J.C.** 1997: Age of colluvium indicates accelerated Late Wisconsin hillslope erosion in the Upper Mississippi Valley. *Geology* 25, 267–270.

colonialism The exercise of controlling power and the exploitation of one state by another, often involving considerable environmental impact, direct and indirect. The *Romans* opened up large areas of the Mediterranean, Europe and North Africa to trade and agriculture, leading to widespread *forest degradation*. Industrial activity, such as LEAD mining in Europe, also damaged forests and left an air pollution 'signature' in the Greenland ice cap, one of the earliest identifiable effects of truly GLOBAL ENVIRONMENTAL CHANGE. Liberalisation of *trade* in the later Roman Empire led to overexploitation of land in many parts of the Mediterranean and consequent SOIL DEGRADATION, especially in northern Libya and Tunisia. Demand for exotic animals helped cause the EXTINCTION of large predators like the lion from Europe, the Middle East and North Africa and may have exterminated elephants north of the Sahara. Present-day cheetah populations south of the Sahara are genetically depauperate, endangering their survival, some suggest as a consequence of Roman over-hunting.

Portuguese and Spanish colonialism has had a huge impact on plant distributions with *introduced species* carried in both directions between Old and New Worlds. Their dispersal of human and livestock diseases has also had marked effect, not least helping in the conquest of Latin America, which in turn caused social breakdown and abandonment of large areas of farmland. Vast systems of hillside *terracing*, which gave their name to the Andes, were abandoned after the Spanish colonisation of western South America. The establishment of plantation agriculture in Brazil, the West Indies and Central America resulted in DEFORESTATION and the relocation, and often extinction, of Indian peoples.

Small islands have suffered world-wide through colonial expansion of plantation agriculture, and also as a consequence of the deliberate release and escape of goats, pigs, rats, cats and other animals. The impact on island birdlife has been considerable and continues. In Australia and New Zealand release of animals and plants from the colonists' home countries has resulted in immense damage, notably from rabbits, red deer (New Zealand) and cats and dogs. The latter have become FERAL and have driven many flightless bird and marsupial species to extinction in Australia, Tasmania and New Zealand. In Australia, New Zealand, Argentina and Brazil improved communications allowed cattle and sheep ranching with considerable impact on grasslands and forest that could be converted to pasturage. The expropriation of rubber (*Hevea brasiliensis*) seeds from *Amazonia* enabled the British and Dutch to establish thousands of square kilometres of MONOCULTURE *plantations* in Malaya and the Dutch East Indies, which not only decimated TROPICAL RAINFORESTS but also led to water POLLUTION through latex processing. One of the greatest impacts has been the establishment of communications infrastructures in colonies and the initiation of the process of INDUSTRIALISATION and exportation of primary products to the metropolitan countries. Hardwood forests in the West Indies, West Africa, India, Burma, Malaya and Brazil south of the Amazon have suffered, and still suffer, in supplying timber for Europe.

There have been some positive environmental impacts, albeit limited in comparison with the damage caused in colonial times. In South Africa, India and Malaysia, a number of Game Parks and RESERVES were established. Also, on several islands and steeper areas of mainland territories colonial authorities attempted REFORESTATION and maintained watershed protection forests. There was also progress with locust control in Africa and in veterinary medicine (with mixed impacts, some positive and some negative). *CJB*

[See also IMPERIALISM, NEOCOLONIALISM, POSTCOLONIALISM, ROMAN PERIOD LANDSCAPE IMPACTS]

Crosby, A.W. 1986: *Ecological imperialism: the biological expansion of Europe, 900–1900.* Cambridge: Cambridge University Press. Crosby, A.W. 1994: *Germs, seeds and animals.* Armonk, NY: Sharpe. Grove, R.H. 1995: *Green imperialism: colonial expansion, tropical island Edens and the origins of environmentalism 1600–1860.* Cambridge: Cambridge University Press. Guelke, L. and Kay, J. 1996: European settlement, 1450–1750. In Douglas, I., Huggett, R. and Robinson, M. (eds), *Companion encyclopedia of geography.* London: Routledge, 162–181. Williams, M. 1996: European expansion and land cover transformation. In Douglas, I., Huggett, R. and Robinson, M. (eds), *Companion encyclopedia of geography.* London: Routledge, 182–205.

colonisation The process of MIGRATION (dispersal) and ECESIS (establishment) of species either (1) following DISTURBANCE that creates open habitats (NUDATION) or (2) to a geographical region where the species did not previously occur. Establishment depends on the presence of viable propagules from dispersal to a site or on *in situ* seed banks, spore production or vegetative reproduction. The spatial and temporal characteristics of the environment interact with the physiological characteristics of the organisms to determine colonisation success. (3) In archaeological and historical context, it is the large-scale, long-distance migration of human populations, which is akin to (2) above. *LRW*

[See also ECOLOGICAL SUCCESSION, PRIMARY SUCCESSION]

Grubb, P.J. 1987: Some generalizing ideas about colonization and succession in green plants and fungi. In Gray, A.J., Crawley, M.J. and Edwards, P.J. (eds), *Colonization, succession and stability.* Oxford: Blackwell Scientific, 81–102. Lee, K. 2001: Colonization. In Jamieson, D. (ed.), *A companion to environmental philosophy.* Oxford: Blackwell, 486–497. Valk, A.G. van der 1992: Establishment, colonization and persistence. In Glenn-Lewin, D.C., Peet, R.K. and Veblen, T.T. (eds), *Plant succession: theory and prediction.* London: Chapman and Hall, 60–102.

colorimetry A method of chemical analysis by which the concentration of a compound in solution is measured by the intensity of its colour. *JAM*

colour measurement Colours of soil and rocks may be described accurately using the *Munsell Soil Colour Chart.* In the Munsell system, colour is analysed according to *hue* (the basic spectrum colour, e.g. red or yellow), *value* (lightness or darkness from white to black), and *chroma* (the strength of colour). The hues 10R, 2.5YR, 5YR, 7.5YR, 10YR, 2.5Y and 5Y are sufficient to describe the colours of the majority of soils. Each hue has a page with colour chips upon it that cover the range of chroma and value commonly met. The notation is written in the order: hue, value and chroma, and has a common name, for example 10YR 4/2 (dark greyish brown). Colour varies with moisture content, so both moist and dry colours are frequently given in soil descriptions. *EMB*

Bigham, J.M. and Ciolkosz, E.J. (eds), 1993: *Soil color.* Madison, Wisconsin: Soil Science Society of America. Munsell Color Company 1954: *Munsell soil color charts.* Baltimore: Munsell Color Company Inc.

combined sewer A sewer system that carries both wastewater and stormwater in the same pipes. Most modern systems involve *separate sewers* for these two functions. *JAM*

Tchobanoglous, G. and Burton, F.L. 1991: *Wastewater engineering: treatment, disposal and re-use,* 3rd edn. New York: McGraw-Hill.

combustion The rapid OXIDATION (burning) of a fuel with the production of heat. The *complete combustion* of a fuel containing only carbon and hydrogen oxidises to no other substances than CARBON DIOXIDE and water. *Incomplete combustion* may produce, in addition: (1) carbon particles, which are released as smoke or remain in the ash RESIDUE; (2) CARBON MONOXIDE gas; and/or (3) partially oxidised CONTAMINANTS. *JAM*

cometary impact Either the surface impact or atmospheric impact of debris from comets (or asteroids). Although largely ignored, and often derided, as an influence on Quaternary climatic change, there has recently been more interest shown in the possible effects of such an event. This has stemmed largely from an inability to link TREE RING records of poor climate with known forcing factors, such as VOLCANIC ERUPTIONS, and from the evidence of the effects of cometary (or asteroid) airburst above Tunguska, Siberia, in June 1908. *CJC*

[See also METEORITE IMPACT, NEAR-EARTH OBJECT]

Baillie, M.G.L. 1994: Dendrochronology raises questions about the nature of the AD 536 dust-veil event. *The Holocene* 4, 212–217. **Baillie, M.G.L.** 1999: *From Exodus to Arthur: catastrophic encounters with comets.* London: Batsford. **Velokovsky, I.** 1950: *Worlds in collision.* London: Victor Gollancz.

comets Diffuse, volatile-rich bodies that orbit the Sun and are generally less stable than ASTEROIDS. The term means 'long-haired star', after the 'tail' of partially vaporised volatiles induced by proximity to the Sun. The tail may be tens of millions of kilometres long but 'nuclei' of most comets are around 1 km across and contain up to 80% ice (H_2O) by mass. *Short-period comets* or *Halley-family comets* have orbital periods of < 20 years; *intermediate-period comets* or *Jupiter-family comets* have orbital periods of 20–200 years; and *long-period comets* or *parabolic comets* have orbital periods of > 200 years. Together with near-Earth ASTEROIDS (NEAs), the short-period comets constitute the NEAR-EARTH OBJECTS (NEOs) that pose a NATURAL HAZARD. The impact velocity of NEOs is typically in the range $16–22$ km s^{-1}, whereas long-period comets travel more rapidly and may impact at up to 55 km s^{-1}. Although they are relatively rare, the long-period comets comprise up to 25% of the impact hazard from space debris. *JAM*

[See also BOLIDE, METEORITE IMPACT, PLANETARY GEOLOGY]

Bailey, M.E., Clube, S.V.M. and Napier, W.M. 1990: *The origin of comets.* Oxford: Pergamon Press. **Gehrels, T. (ed.)** 1994: *Hazards due to comets and asteroids.* Tucson, AZ: University of Arizona Press. **McGuire, W.J., Kilburn, C.R.J. and Saunders, M.A.** 2001: *Natural hazards and environmental change.* London: Arnold.

commensalism A form of SYMBIOSIS whereby one organism benefits from living with another without affecting the 'host' either negatively (*parasitism*) or positively (MUTUALISM). The wild mammals and birds that live in and around human habitations may be regarded as *commensals* provided they do not become PESTS. *JAM*

[See also EPIPHYTE, PARASITE]

Somerville, E.M. 1999: Some aspects of the palaeoecology of commensals. *Quaternary Proceedings* 7, 605–613 [*Journal of Quaternary Science* 14, 605–613].

comminution The breakdown of rock fragments to smaller sizes by crushing, grinding and rubbing. *JS*

[See also ABRASION]

common land Land over which rights of use ('*rights of common*') apply to others than the owner. An estimated 566 580 hectares of common land remain in England and Wales as remnants of a land tenure system thought to have originated before the Norman Conquest. *Commons, com-*

mon fields and *waste* were ubiquitous features of the Mediaeval *open field system* (see FIELD SYSTEMS) and communal farming and the embodied principles were not restricted to those areas. Much land formerly 'held in common' disappeared during the long period of 'inclosures' that took land out of common holding into freehold (see ENCLOSURE). This was a legalistic process enabled by Acts of Parliament, but there were also many illegal encroachments upon common lands and waste. Common land that remains is often that of least agricultural value and may be owned privately or by a local authority. Others will have rights, such as access, sporting rights and GRAZING rights to common land. As the process of registration unfolds, common lands emerge with three main functions: first, the agricultural function is strongest in upland areas where common land is used for grazing; second, the public amenity function is strongest in lowland areas close to and within urban areas; and third, the CONSERVATION function is widely applied in an attempt to preserve historic components of the landscape. *DTH*

[See also TRAGEDY OF THE COMMONS]

Aitchison, J.W. and Hughes, E.J. 1988: The common lands of Wales. *Transactions of the Institute of British Geographers* 13, 96–108. **Hoskins, W.G. and Stamp, L.D.** 1963: *The common lands of England and Wales.* London: Collins. **Wilson, O.** 1993: Common lands in the Durham Dales: management and policy issues. *Area* 25, 237–245.

community assembly The coming together of species to form a COMMUNITY. In detail, community change involves new arrivals (through SPECIATION events and immigration), the persistence of existing species and species loss (through EXTINCTION and emigration). The process of community assembly is individualistic because each species has its own propensity for DISPERSAL, *invasion* and population expansion. Evidence that communities assemble (and disassemble) in this manner is seen in multidirectional ECOLOGICAL SUCCESSION, in which succession leads in several directions and does not end in a single climax state (CLIMAX VEGETATION). It is also seen in no-MODERN ANALOGUE communities, such as the PLEISTOCENE boreal grassland BIOME, and in 'disharmonious' communities such as existed in the southern Great Plains and Texas during the Pleistocene when present-day boreal mammals lived alongside present-day grassland and deciduous-forest mammals. The process of community assembly means that community response to GLOBAL WARMING is unlikely to involve a simple, zonal shift of biome boundaries. *RJH*

[See also BIOME, BOREAL VEGETATION, ISLAND BIOGEOGRAPHY]

Drake, J. A. 1990: Communities as assembled structures: do rules govern pattern? *Trends in Ecology and Evolution* 5, 159–163. **Morton, R.D., Law, R., Pimm, S.L. and Drake, J.A.** 1996: On models for assembling ecological communities. *Oikos* 75, 493–499.

community concepts Plant, animal and biotic communities consist of groups of plants and/or animals living at the same location and at the same time. The grouping may be viewed in different ways. Two extreme views are: (1) the community is the product of chance developments (the INDIVIDUALISTIC CONCEPT); and (2) the community is the result of a predictable development towards a CLIMAX VEGETATION community. Interactions occur within the community, with COMPETITION for resources being particularly important. However, other forms of interaction, such as SYMBIOSIS, MUTUALISM and parasitism (see PARASITE), may also occur. The extent to which communities are integrated, homogeneous, 'organismic' phenomena or arbitrary units within a continuum of variation has been long debated. *JLI*

[See also BIOMES, ECOSYSTEM CONCEPT]

Bowman W.D., Theodose T.A. and Fisk M.C. 1995: Physiological and production responses of plant growth forms to increases in limiting resources in alpine tundra: implications for differential community response to environmental change. *Oecologia* 101, 217–227. **Langford, A.N. and Buell, M.F.** 1969: Integration, identity and stability in the plant association. *Advances in Ecological Research* 6, 84–135. **McCook L.J.**, 1994: Understanding ecological community succession: causal models and theories, a review. *Vegetatio* 110, 115–147. **McIntosh, R.P.** 1967: The continuum concept of vegetation. *Botanical Review* 33, 99–187. **Richardson, J.L.** 1980: The organismic community: resilience of an embattled ecological concept. *Bioscience* 30, 465–471. **Whittaker, R.H.** 1962: Classification of natural communities. *Botanical Review* 28, 1–239.

community ecology
The study of the living organisms and their inter-specific relationships within an ECOSYSTEM. Also referred to as *synecology*. *JLI*

[See also AUTECOLOGY, COMMUNITY CONCEPTS, ECOSYSTEM CONCEPT, EMERGENT PROPERTIES]

Daubenmire, R. 1968: *Plant communities: a textbook of plant synecology.* New York: Harper and Row.

compaction
An increase in density of a soil or sediment, often caused by externally applied stresses (e.g. livestock or vehicles), but also by AUTOCOMPACTION processes such as drying or loss of organic matter. Compaction leads to a reduction in porosity, resulting in a reduced INFILTRATION CAPACITY and HYDRAULIC CONDUCTIVITY. *SHD*

[See also CONSOLIDATION, DESERTIFICATION, LAND DEGRADATION, SOIL DEGRADATION]

Dexter A.R.1988: Advances in the characteristics of soil structure. *Soil and Tillage Research* 11, 199–238.

competence
(1) The ability of a current to transport sediment, as measured by the largest GRAIN-SIZE that can be moved. Competence in this sense is a function of the strength of a flow and is used especially in the context of rivers and streams. It contrasts with the *capacity* of a river, which is the maximum load that can be carried. (2) A description of the relative strength of materials in response to DEFORMATION. A competent material, such as sandstone or limestone, is relatively strong and likely to undergo brittle deformation leading to FAULTS; an incompetent material such as shale is weaker and more likely to deform in a plastic manner, giving rise to FOLDS. *GO*

competition
Inter- or intra-specific interactions between individuals, whereby one or more of the interacting individuals is adversely affected. Species may avoid competition by the occupation of different NICHES. Competition for resources appears to be common in plant communities and may be an important process in ECOLOGICAL SUCCESSION and COEVOLUTION. *JLI*

composting
The process of rapid, controlled DECOMPOSITION of organic materials and WASTES, largely by AEROBIC micro-organisms. Domestic-scale and industrial-scale composting of BIODEGRADABLE waste is increasingly used in the production of a relatively stable compost material for agriculture and HORTICULTURE. *JAM*

Porteous, A. 2000: *Dictionary of environmental science and technology*, 3rd edn. Chichester: Wiley.

compound
A substance composed of two or more ELEMENTS joined by chemical bonds and requiring chemical reactions to separate them. Compounds have characteristic properties that are widely different from the properties of their constituent elements. *JAM*

compound-specific carbon isotope analysis
The stable carbon isotope ratios of individual organic BIOMARKERS present in complex mixtures are determined by *isotope ratio monitoring gas chromatography mass spectrometry* (irm-GC-MS). If the compounds are separated easily during gas chromatography the $^{13}C/^{12}C$ ratio of the individual compounds can be measured. The carbon isotopic composition of any naturally synthesised organic compound depends on the CARBON SOURCE utilised, isotopic effects associated with CARBON ASSIMILATION by the producing organism, isotope effects associated with metabolism and biosynthesis and cellular carbon budgets. The structure of the biomarker can identify the source organism, while the isotopic composition of the molecule can indicate the isotopic composition of the parent organism, which, in turn, can reveal the carbon source utilised by the producer. These factors are dependent on environmental conditions and so the distribution of ^{13}C among natural products is a sensitive palaeoenvironmental indicator and can provide a great deal of information about ancient biogeochemical processes.

Interpretation of carbon isotope data requires knowledge of ISOTOPE FRACTIONATION in organisms. The individual LIPID components are more depleted in ^{13}C than the total BIOMASS of an organism, while polysaccharides and proteins are somewhat enriched in ^{13}C. Terrestrial vegetation fixes atmospheric CO_2 *via* different enzymatic pathways (C-3, C-4, CAM), resulting in distinct $^{13}C/^{12}C$ ratios. Aquatic photosynthetic organisms also differ in their enzymatic fixation of CO_2 (HCO_3^- pumping, ribulose-1,5-bisphosphate carboxylase-oxygenase (Rubisco), reverse tricarboxylic acid cycle (TCA)) leading to major differences in $^{13}C/^{12}C$ ratios in their biomass. Methanogenic bacteria significantly discriminate against ^{13}C, leading to heavily depleted biomass and extremely low $^{13}C/^{12}C$ values of methane. Apart from these biosynthetically induced isotope variations, the concentration and ^{13}C content of atmosphere CO_2, water temperature, SALINITY, as well as many other physical and biochemical parameters ultimately determine the $^{13}C/^{12}C$ value of an organic compound.

Compound-specific carbon-isotope analysis is a useful tool in reconstructing PALAEOENVIRONMENTS as specific biomarkers, e.g. those representing specific algae, aquatic macrophytes, and terrestrial higher plants can be traced through a sediment core. As the ISOTOPIC signature of these source organisms depends on many factors, an overall view of the CARBON CYCLE in a particular environment can be established. *KJF*

[See also ISOTOPES AS INDICATORS OF ENVIRONMENTAL CHANGE, ISOTOPIC RATIO]

Ficken, K.J., Street-Perrott, F.A., Perrott, R.A. *et al.* 1998: Glacial/interglacial variations in carbon cycling revealed by molecular and isotope stratigraphy of Lake Nkunga, Mt. Kenya, East Africa. *Organic Geochemistry* 29, 1701–1719. Freeman, K.H., Hayes, J.M., Trendel, J-M. and Albrecht, P. 1990: Evidence from carbon isotope measurements for diverse origins of sedimentary hydrocarbons. *Nature* 343, 254–256. Hayes, J.M. 1993: Factors controlling ^{13}C contents of sedimentary organic compounds: Principles and evidence. *Marine Geology* 113, 111–125. Hayes, J.M., Freeman, K.H., Popp, B.N. and Hoham, C.H. 1990: Compound specific isotope analyses: A novel tool for reconstruction of ancient biogeochemical processes. *Organic Geochemistry* 16, 1115–1128. Rieley, G., Collister, J.W., Stern, B. and Eglinton, G. 1993: Gas chromatography/isotope ratio mass spectrometry of leaf wax *n*-alkanes from plants of differing carbon dioxide metabolisms. *Rapid Communications in Mass Spectrometry* 7, 488–491.

compression   In IMAGE PROCESSING, a set of techniques devised to reduce the physical amount of space required to store data in a computer system. Different compression techniques (based on information theory, statistical properties of the data, etc.) are employed in common image storage formats.   *ACF*

Murray, J.E. and van Ryper, W. 1994: *Encyclopaedia of graphic file formats*. Cambridge: O'Reilly.

conceptual model   A non-numerical, qualitative model based on knowledge of structures and processes of a system. Often the first stage in the development of a *mathematical model*, which uses a mathematical language, is based on quantification of the system and consists of one or more equations.   *HB*

concretion   A localised solid mass or aggregate of pedogenetic origin, formed in the soil by the concentration of a chemical compound, such as calcium carbonate or iron oxide, deposited in concentric layers within the concretion. The word *nodule* is used for similar but more irregular features without a concentric internal arrangement. Because the terms concretion and nodule had been used loosely, the term *glaebule* was introduced by Brewer for any aggregate of pedogenetic origin.   *EMB*

Brewer, R. 1964: *Fabric and mineral analysis of soils*. New York: Wiley. Hodgson, J.M. 1978: *Soil sampling and soil description*. Oxford: Oxford University Press.

condensation   The physical process by which a gas is transformed into a liquid accompanied by the release of LATENT HEAT. WATER VAPOUR, for example, is transformed into liquid water, either by air being cooled to its *dewpoint* or becoming saturated by EVAPORATION into it.   *JBE*

condensation level   The height at which rising air cools to its *dewpoint* temperature so that CONDENSATION occurs. It is marked by the base of CLOUDS formed by the uplift.   *JGT*

[See also ALTITUDINAL ZONATION]

condensation nuclei   The particles in the atmosphere upon which WATER VAPOUR condenses to form water droplets. They include sea salt, smoke, organic compounds, fine clays and dusts between 0.05 and 1.0 µm. *Aitken nuclei* are the smallest category (< 0.2 µm); *large nuclei* are > 0.2 µm and *giant nuclei* > 1 µm.   *BDG/JGT*

[See also AEROSOLS]

Mason, B.J. 1962: *Clouds, rain and rainmaking*. Cambridge: Cambridge University Press.

confidence interval   A range of values or random interval, calculated from the sample observations, that is believed, with a particular PROBABILITY or CONFIDENCE LEVEL to include the true value of the population parameter of interest. A 95% confidence interval implies that, if the estimation were repeated many times, 95% of the calculated intervals would be expected to contain the true value of the parameter. A confidence interval is a measure of the PRECISION of a sample statistic. If the interval is large, low reliance can be placed on the estimated value as an estimate of the true population parameter. The interval thus provides a measure of the degree of confidence in estimated parameters. The stated PROBABILITY level refers to the properties of the confidence interval and not to the parameter. The upper and lower extremes of the interval are the upper and lower *confidence limits*.   *HJBB*

Sokal, R.R. and Rohlf, F.J. 1995: *Biometry*. New York: W.H. Freeman.

confidence level   The PROBABILITY or level of confidence associated with a CONFIDENCE INTERVAL that quantifies the probability that the interval includes the true population parameter. Also known as the *confidence coefficient*.   *HJBB*

Sokal, R.R. and Rohlf, F.J. 1995: *Biometry*. New York: W.H. Freeman.

confirmation   The result of demonstrating that a HYPOTHESIS or MODEL is in agreement with independent evidence (observation). Confirmation of the hypothesis or model is only with respect to particular observations. It is not equivalent to *proof* or VERIFICATION. It means only that the hypothesis (1) remains in contention for truth and (2) is more probably true, or (3) that the model may be a good representation of some part of reality, not all aspects of reality. *Confirmation theory* is founded on the notion that SCIENCE is a hypothetico-deductive system in which attempts are made to refute hypotheses, which can never be verified. The more complex the hypothesis, the more obvious this conclusion becomes; it is equally applicable to models because they may be viewed as highly complex scientific hypotheses.   *JAM*

[See also CRITICAL RATIONALISM, DEDUCTION]

Oreskes, N., Shrader-Frechette, K. and Belitz, K. 1994: Verification, validation, and confirmation of numerical models in the Earth sciences. *Science* 263, 641–646.

confirmatory data analysis   Types of statistical data analysis in which the approach involves the formal *testing* of HYPOTHESES or *estimation* of parameters. It contrasts with EXPLORATORY DATA ANALYSIS.   *JAM*

[See also NUMERICAL ANALYSIS]

confounding   Failure to distinguish between alternative explanations for results. For example, a correlation between two variables (A and B) that suggests that A causes B may be the result of B causing A; another alternative is that they are not directly related but both are influenced by a third factor (C). A good EXPERIMENTAL DESIGN minimises the possibility of confounding. *JAM*

Benson, M.A. 1965: Spurious correlation in hydraulics and hydrology. *Journal of the American Society of Civil Engineers* 91, 35–42.

conglomerate A CLASTIC SEDIMENTARY ROCK with a GRAIN-SIZE greater than 2 mm (a RUDITE or lithified GRAVEL) and rounded CLASTS. *TY*

[See also BRECCIA]

connectivity The degree of complexity in the inter-linkage of lines in a network. High connectivity implies a node (see POINT) with many spatial, often linear, links with other nodes. *TVM*

[See also NETWORK ANALYSIS, SPATIAL ANALYSIS]

conservation The maintenance of species, ecosystems, landscapes and natural resources generally by both preservation and wise use. The term is also increasingly applied to human resources, such as archaeological ARTE-FACTS, historic buildings and HERITAGE. The emphasis, until recently, has been on preservation or protection of the *status quo* (as in traditional 'nature conservation' and the setting aside of PROTECTED AREAS), but this has been superseded by the concept of conservation as management for SUSTAINABILITY and, in some cases, for the ENHANCEMENT of environmental and resource quality.

Utilitarian and naturalistic ethics (see ENVIRONMENTAL ETHICS) may be seen as underlying the various concepts of conservation. Both have a long history. The former, which sees resources primarily for human use, was strongly developed in Judaism and in the Greek and Islamic cultures and is still the predominant western and world view. Its importance is emphasised by past failures to conserve resources, which are even implicated, directly or indirectly, in the collapse of many great civilisations. *Naturalism*, as reflected in Asia, Africa and the Americas and in such religions as Taoism, Buddhism and Hinduism, was less exploitative and humans were viewed as first amongst equals: not necessarily superior to other animals.

Conservation in both senses is clearly of major importance in the modern world. As exploitation of natural resources continues at an increasing rate and effects are increasingly felt globally, a unified, holistic approach to conservation is more necessary than ever. There are signs that such a need is being recognised locally, nationally and internationally; social attitudes are changing and ENVIRON-MENTAL LAW is becoming increasingly prominent, but institutional changes and international action are slow, possibly too slow to make much of an impact on the effects of seeking ever higher levels of economic productivity, development and wealth. *JAM*

[See also CONSERVATION BIOLOGY, DEEP ECOLOGY, ENVIRONMENTALISM, FISHERIES CONSERVATION AND MANAGEMENT, GEOLOGICAL CONSERVATION, RESERVES, WILDERNESS, WILDLIFE CONSERVATION AND MANAGEMENT]

Adams, W.M. 1996: *Future nature: a vision for conservation.* London: Earthscan. **Owen, O.S. and Chiras, D.D.** 1995: *Natural resource conservation*, 6th edn. Englewood Cliffs, NJ: Prentice-Hall. **Petulla, J.M.** 1977: *American environmental history: the exploitation and conservation of natural resources.* San Francisco, CA: Boyd and Fraser. **Ponting, C.** 1991: *A green his-tory of the world: the environment and the collapse of great civilisations.* New York: St Martin's Press.

conservation biology A multidisciplinary subject based on the integrated use of science and social disciplines to achieve CONSERVATION and sustainable use of nature. If conservation biology were to have a mission, it might be the conservation and sustainable use of the diversity of life (BIODIVERSITY). Conservation biology has emerged as a popular subject in the western world, where there have been expressions of concern about the degradation of nature caused by human impacts. The MULTI-DISCIPLINARY basis for conservation biology is particularly important and recognises that conservation cannot wholly be achieved by biology alone. The first annual meeting of the Society for Conservation Biology was held at Montana State University in 1987 and the journal, *Conservation Biology*, was launched in the same year. *IFS*

[See also EXTINCTION, SUSTAINABILITY]

Meffe, G. K. and Carroll, C.R. 1997: *Principles of conservation biology*, 2nd edn. Sunderland, MA: Sinauer Associates. **Primack, R.B.** 1995: *A primer of conservation biology.* Sunderland, MA: Sinauer Associates. **Spellerberg, I.F. (ed.)** 1996: *Conservation biology.* Harlow: Longman.

conservation laws, in science Regulative principles (i.e. unprovable assumptions) that state that the total of something is constant in a system. The conservation of energy and mass are the best known. *CET*

conservative plate margin A boundary between two plates (see PLATE TECTONICS) that are sliding past each other. LITHOSPHERE is neither created nor destroyed. There is thus normally no VOLCANIC activity, but friction between the plates generates EARTHQUAKES. The movement represents a type of strike-slip FAULT, but, because it links different types of PLATE MARGIN, it is known as a TRANSFORM FAULT. Examples include the San Andreas Fault in California, the Alpine Fault in New Zealand and the offsets of MID-OCEAN RIDGES. Bends in conservative plate margins (and other *strike-slip faults*) cause UPLIFT and SUBSIDENCE, giving rise to localised SEDIMENTARY BASINS known as PULL-APART BASINS. *GO*

Crowell, J.C. 1974: Origin of late Cenozoic basins in southern California. In Dickinson, W.R. (ed.), *Tectonics and sedimentation. Society of Economic Paleontologists and Mineralogists Special Publication* 22, 190–204. **Crowell, J.C.** 1979: The San Andreas fault system through time. *Journal of the Geological Society, London* 136, 293–302.

consolidation In the Earth sciences context: (1) a process by which loose, soft or liquid materials are solidified; or (2) the COMPACTION of soil or sediment in response to surface loading, often associated with the expulsion of water from void spaces in the soil or sediment. Normal consolidation occurs in the absence of an *overburden pressure* (loading on top); *overconsolidation* results from the previous existence of an overburden (e.g. LODGEMENT TILL becomes overconsolidated by the weight of an overlying glacier). *SHD*

constructive plate margin A divergent boundary (also known as a *divergent plate margin*) between two tectonic plates at which new LITHOSPHERE bearing OCEANIC

CRUST is formed by VOLCANISM and SEA-FLOOR SPREADING at MID-OCEAN RIDGES, causing an OCEAN to become wider. Examples include the Mid-Atlantic Ridge (MAR), where the North and South American plates are separating from the Eurasian and African plates at about 3 cm y^{-1}, and the East Pacific Rise (EPR), which is spreading at about 15 cm y^{-1}. The formation of new lithosphere is balanced by SUBDUCTION at DESTRUCTIVE PLATE MARGINS. *GO*

[See also PLATE TECTONICS]

containment A POLYGON ANALYSIS operation used in a GEOGRAPHICAL INFORMATION SYSTEM (GIS) for dealing with the association between two polygons where one is completely surrounded by the other. It also refers to a POINT within a polygon. *TVM*

[See also ADJACENCY]

contaminant In the natural environmental context, a substance introduced into a natural ecosystem by human agency. It may change the system in some way but, unlike a POLLUTANT, a contaminant does not necessarily impair or harm organisms. Exploitation of resources by organisms other than humans does not contaminate or pollute, nor do natural events such as volcanic ashfalls; natural agencies bring about *environmental anomalies* or cause DISTURBANCE rather than *contamination*. More generally, an impurity. *JAM*

contaminated land Land containing substances that when present in sufficient quantities or concentrations are likely to cause harm, directly or indirectly to humans, to the environment, or occasionally to other targets. Land occupied by industry, particularly, provides the majority of contaminated land sites: manufacture of coal gas, sewage works, scrapyards, chemical and metal works and LANDFILLS are all likely to cause gross contamination. Many derelict sites contain toxic or HAZARDOUS WASTE that can be injurious if taken up by plants and ingested by animals. The most common contaminants are the HEAVY METALS, COPPER, LEAD, zinc, CADMIUM and MERCURY; inorganic substances such as cyanides and phosphates; aromatic compounds such as benzene, toluene; polycyclic aromatic compounds like phenol and naphthalene, chlorinated organic compounds, PESTICIDES, asbestos, fuel and mineral oils. *Contaminative uses* cause contaminated land.

Land contamination can be dealt with in a number of ways depending upon the character, concentration and distribution on site of any contaminants. Treatment on site may involve leaching, separation by mechanical means or flotation, thermal decomposition, steam stripping, or chemical, microbial and stabilisation techniques. A smaller number of *in situ* treatments have been proposed, including cultivation, leaching, grouting, vitrification and electrokinetic techniques. Barriers and covering systems are the most commonly employed systems to safeguard health and the environment. The aim of all these techniques is to either isolate or stabilise the contamination so that it will not move out to contaminate surrounding areas. Excavation and removal elsewhere is not favoured as this simply spreads the contamination. Criteria to determine whether or not land is contaminated have been established and published by many governments. *EMB*

[See also CONTAMINANT, PERSISTENT ORGANIC COMPOUNDS, POLLUTION HISTORY, SOIL RECLAMATION, WASTE MANAGEMENT]

Fleming, G. 1996: *Recycling derelict land.* London: Thomas Telford. **Bridges, E.M.** 1987: *Surveying derelict land.* Oxford: Clarendon Press. **Rulkens, W.H., Assink, J.W. and van Gemmert, W.J.T.** 1985: On site processing of contaminated soil. In M.A. Smith (ed.), *Contaminated land: treatment and reclamation.* New York: Plenum. **Meeder, T., Versluijs, K., Bakker, R. and Sóczó, E.** 1998: Remedial action techniques for contaminated land. In Brune, D., Chapman, D.V., Gwynne, M.D. and Pacyna, J.M. (eds), *The global environment: science, technology and management.* Weinheim, Germany: VCH, 887–907. **Moen, J.E.T., Cornet, J.P. and Evers, C.W.A.** 1986: Soil protection and remedial actions: criteria for decision making and standardisation of requirements. In Assink, J.W. and van den Brink, W.J. (eds), *Contaminated soil.* Dordrecht: Martinus Nijhoff. **Inter-departmental Committee for the Redevelopment of Contaminated Land** 1983: *Guidance on the assessment and re-development of contaminated land.* London: Department of the Environment.

contamination of samples The inclusion of foreign or extraneous material in a sample at any time prior to analysis, which affects the result of the analysis. Excavation or recovery and storage of samples should be designed to preclude contamination and a cleaning or PRETREATMENT process should aim to remove any contamination. The problem of contamination is particularly severe where the concentration of the analysed species is close to BACKGROUND LEVELS, e.g. the RADIOCARBON DATING of old material where little ^{14}C remains. A 45 000 year old sample (c. eight HALF LIVES) will contain 1/256 of the modern level of ^{14}C. If the sample was contaminated with 0.4% of modern carbon, the ^{14}C concentration would be doubled and the 'age' reduced by one half life. *PQD*

Klinken, G.J. van and Hedges, R.E.M. 1998: Chemistry strategies for organic ^{14}C samples. *Radiocarbon* 40, 51–56.

contextual archaeology An approach to ARCHAEOLOGY, advocated especially by I. Hodder. It acknowledges the need for VALIDATION of THEORY by testing its coherence and correspondence against independent evidence, but emphasises that this must be supplemented by insights arising from personal experience and from an attempt to gain a deeper understanding of past events by imagining the motives of the individuals involved. *JAM*

Hodder, I. 1987: *The archaeology of contextual meaning.* Cambridge: Cambridge University Press. **Jameson, R.** 1999: Contextual archaeology. In Shaw, I. and Jameson, R. (eds), *A dictionary of archaeology.* Oxford: Blackwell.

continent A large land-mass on the Earth. The present-day continents are Africa, Antarctica, Australia, Asia, Europe, North America and South America. Continents are underlain by CONTINENTAL CRUST, which is older, thicker and less dense than OCEANIC CRUST. The boundary between the two types of crust underlies the outer edge of the CONTINENTAL SHELF, an area of shallow sea fringing the continents (see CONTINENTAL MARGIN). The land areas make up about 29% of the Earth's surface, the continental shelves 11% and the OCEANS cover the remaining 60%. Continental areas can be divided into: (1) CRATONS, which have been tectonically stable since PRECAMBRIAN times; (2) young OROGENIC BELTS at PLATE MARGINS or sites of recent continental collision; (3) older, denuded orogenic belts; (4) RIFT VALLEYS; (5) plateaus; and (6) present-day SEDIMENTARY BASINS. Because of CONTINENTAL DRIFT resulting from PLATE TECTONICS, the continents and their relative positions have changed through the history of the Earth, at times assembling as one or more SUPERCONTINENTS (e.g. PANGAEA). GO

O'Nions, K. 1992: The continents. In Brown, G.C., Hawkesworth, C.J. and Wilson, R.C.L. (eds), *Understanding the Earth: a new synthesis*. Cambridge: Cambridge University Press, 145–163. *The Times* 1992: *The Times atlas of the world*. London: Times Books. Windley, B.F. 1995: *The evolving continents*, 3rd edn. Chichester: Wiley.

continental crust

continental crust The outermost layer of the Earth that underlies the CONTINENTS and CONTINENTAL SHELF seas. Compared with OCEANIC CRUST, continental crust is relatively thick (average 35 km, maximum 70 km), low in density (typically 2700 kg m^{-3}), granitic in composition and old. The oldest continental crust is about 3800 million years old and the volume of continental crust has increased slowly through GEOLOGICAL TIME. GO

[See also EARTH STRUCTURE]

MacNiocaill, C. and Ryan, P.D. (eds), 1999: *Continental tectonics. Geological Society, London, Special Publication* 164. O'Nions, K. 1992: The continents. In Brown, G.C., Hawkesworth, C.J. and Wilson, R.C.L. (eds), *Understanding the Earth: a new synthesis*. Cambridge: Cambridge University Press, 145–163. Taylor, S.R. and McLennan, S.M. 1985: *The continental crust: its composition and evolution*. Oxford: Blackwell.

continental drift

continental drift The hypothesis that the relative positions of the CONTINENTS have changed through geological time (see Figure). Francis Bacon in AD 1620 noted the similarity in shape between the eastern coastline of the Americas and the western coastline of Europe and Africa. By the end of the nineteenth century geologists had recognised geological similarities across the oceans and began to construct persuasive arguments that the continents had once been joined. From 1915 to 1929, Alfred Wegener published descriptions of detailed geological and palaeontological similarities across the Atlantic Ocean and argued that all the present-day continents were once assembled in a single *supercontinent* that he named PANGAEA. Wegener's mechanism for continental drift, however, proved contentious. As there was clear evidence that the Earth responded to vertical forces by flowing vertically, he argued that it could flow horizontally in concert with forces produced by the rotation of the Earth and suggested that, as continents drifted apart, new ocean floor was created at the retreating margins. A major debate developed through the 1920s and 1930s as geologists gathered further evidence of continental drift, but physicists were adamant that there was no mechanism for it. Through the 1950s the study of PALAEOMAGNETISM and the construction of APPARENT POLAR WANDER paths provided further evidence for continental drift, now from GEOPHYSICS, but still with no indication of a mechanism.

Convincing evidence of a mechanism for continental drift came in the late 1950s and 1960s. In 1960 Hess established that there was abundant VOLCANISM at the MID-OCEAN RIDGES and proposed a theory of SEA-FLOOR SPREADING whereby new sea-floor formed at the ridges and spread outwards. Unlike Wegener, Hess proposed that the

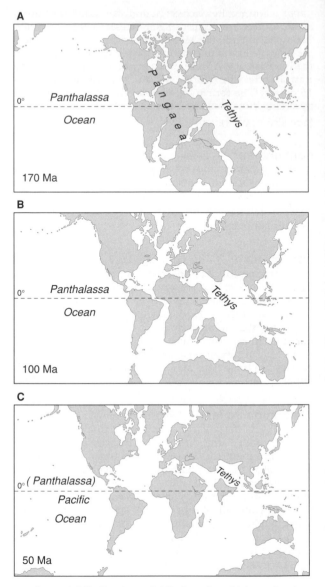

continental drift *Continental drift since the mid Jurassic, showing the break up of Pangaea. Note that former coastlines did not coincide with the present-day coastlines: (A) Jurassic, c. 170 million years ago; (B) Cretaceous, c. 100 million years ago; (C) Eocene, c. 50 million years ago (based on Open University Course Team, 1989)*

continents were part of a rigid system that included the ocean floors; sea-floor was not only generated at midoceanic spreading centres, but descended at OCEANIC TRENCHES. Geophysicists finally accepted that convection currents in the Earth's MANTLE could drive such a system of continental drift. The discovery of symmetrical MAGNETIC ANOMALIES on either side of the mid-ocean ridges led to the acceptance of sea-floor spreading and the development in the mid 1960s of PLATE TECTONICS. This maintains that lateral movement of the LITHOSPHERE takes place through the formation and destruction of lithosphere-bearing OCEANIC CRUST at the boundaries of rigid

plates. Since plates include both oceanic crust and CONTI-
NENTAL CRUST, continental drift is now seen as a necessary
consequence of plate tectonics.

An important implication of continental drift is that
areas of the Earth's surface have experienced different climatic and environmental conditions in the geological past,
for which evidence is preserved in the GEOLOGICAL
RECORD. For example, Upper Carboniferous COAL-bearing
rock sequences in Britain, northwest Europe and northeastern North America imply formation under tropical
conditions some 300 million years ago. The reconstruction of former continental positions for the past 200 million years can be achieved by removing ocean-floor
magnetic anomalies younger than the age of the reconstruction. Prior to the formation of Pangaea, PALAEOGEOG-
RAPHY is reconstructed using palaeomagnetism and
geological evidence. *JCWC/GO*

Andel, T.H. van 1994: *New views on an old planet: a history of
global change.* Cambridge: Cambridge University Press. Hallam,
A. 1990: *Great geological controversies,* 2nd edn. Oxford: Oxford
University Press. McKerrow, W.S. and Scotese, C.R. (eds)
1990: *Palaeozoic palaeogeography and biogeography,* Memoir 12.
London: Geological Society. Open University Course Team
1989: *The ocean basins: their structure and evolution.* Oxford:
Pergamon Press. Oresknes, N. 1999: *The rejection of continental
drift: theory and method in American Earth science.* New York:
Oxford University Press. Piper, J.D.A. 2000: The
Neoproterozoic Supercontinent: Rodinia or Palaeopangaea?
Earth and Planetary Science Letters 176, 131–146. Scotese, C.R.
1991: Jurassic and Cretaceous plate tectonic reconstructions.
Palaeogeography, Palaeoclimatology, Palaeoecology 87, 493–501.
Thompson, R. 1993: Palaeomagnetism and continental drift. In
Duff, P.McL.D. (ed.), *Holmes' principles of physical geology,* 4th
edn. London: Chapman and Hall, 616–640. Wegener, A. 1929:
The origin of continents and oceans, translated from the 4th revised
German edition by John Biram. New York: Dover. Wilson, J.T.
(ed.) 1976: *Continents adrift and continents aground.* San
Francisco, CA: Freeman.

continental glaciation
Glaciation mainly on terres-
trial land masses where the main ice body is or was located
on land. Examples include the Late Cenozoic Laurentide
and Scandinavian ICE SHEETS. *AN*

[See also LOCAL GLACIATION, GLACIAL EPISODES, ICE AGES]

continental island
An island formed as part of a con-
tinent and that has a nucleus of CONTINENTAL CRUST.
Although presently isolated by water, continental islands
were once connected to the mainland, for example during
glacial conditions of the Pleistocene when sea levels
dropped by over 100 m. An *oceanic island* is formed
unconnected to a continent. *MVL*

Whittaker, R.J. 1998: *Island biogeography: ecology, evolution and
conservation.* Oxford: Oxford University Press.

continental margin
The boundary zone between
CONTINENT and OCEAN, extending from the shoreline to
the deep ABYSSAL plain, spanning the CONTINENTAL SHELF,
CONTINENTAL SLOPE and CONTINENTAL RISE and representing an apparently sharp transition from CONTINENTAL
CRUST to OCEANIC CRUST. In terms of TECTONIC setting,
GEOMORPHOLOGY and economic interest there are profound differences between PASSIVE CONTINENTAL MARGINS,
which lie within plates, and ACTIVE CONTINENTAL MARGINS,
which correspond to DESTRUCTIVE PLATE MARGINS.

Continental margins are major sites for the accumulation of SEDIMENTARY DEPOSITS, and in many places up to
15 km thickness has built up over the last 200 million
years. Most of this is land-derived and transported onto
and across the continental shelves by currents associated
with waves, STORMS and TIDES. Continental margins may
be cut by SUBMARINE CANYONS, which act as major conduits of sediment transport. DEPOSITION on the continental margin in general, and on the shelf in particular, is very
sensitive to changes in SEA LEVEL, ocean chemistry, climate
and tectonic modification of the hinterland: hence, continental margin deposits are of great scientific interest
because of their potential to preserve such information.
Many depositional and erosional features of today's continental margins formed under low sea-level conditions
during PLEISTOCENE glacial maxima, when large parts of
the shelf would have been exposed. They are therefore
PALIMPSEST features. The importance of the continental
margins for PETROLEUM exploration is such that abundant
data are now available, particularly in the form of SEISMIC
STRATIGRAPHY and BOREHOLE records. *CDW*

[See also PLATE TECTONICS, OCEAN BASIN]

Andrews, J.T., Austin, W.E.N., Bergsten, H. and Jennings,
A.E. (eds) 1996: *Late Quaternary palaeoceanography of the North
Atlantic margins.* Special Publication 111. London: Geological
Society. Burk, C.A. and Drake, C.L. (eds) 1974: *The geology
of continental margins.* New York: Springer. Walker, R.G. and
Plint, A.G. 1992: Wave- and storm-dominated shallow marine
systems. In Walker, R.G. and James, N.P. (eds), *Facies models:
response to sea level change.* St John's, Newfoundland: Geological
Association of Canada, 219–238. Stoker, M.S., Evans, D. and
Cramp, A. (eds) 1998. *Geological processes on continental margins: sedimentation, mass-wasting and stability* [*Special Publication*
129]. London: Geological Society.

continental rise
The gently sloping part of the ocean
floor at BATHYAL depths between the foot of the CONTI-
NENTAL SLOPE and the ABYSSAL plain on PASSIVE CONTINEN-
TAL MARGINS. The continental rise is absent from ACTIVE
CONTINENTAL MARGINS. *CDW*

continental shelf
The area of shallow sea-floor (less
than 200 m depth) adjacent to the continents, underlain
by CONTINENTAL CRUST and effectively a submerged extension of the continents (see Figure). On PASSIVE CONTINEN-
TAL MARGINS shelves tend to be broad (up to 1500 km
wide); on ACTIVE CONTINENTAL MARGINS they are much
narrower. Modern continental shelves are important sites
for the accumulation of SEDIMENTARY DEPOSITS that hold
an important record of environmental change over the last
200 million years. Extensive areas of continental shelf
were exposed as land during periods of low SEA LEVEL.
CDW

Cook, P.J. and Carleton, C. 2000: *Continental shelf limits: the
scientific and legal interface.* Oxford: Oxford University Press.
Johnson, H.D. and Baldwin, C.T. 1996: Shallow clastic seas.
In Reading, H.G. (ed.), *Sedimentary environments: processes, facies
and stratigraphy.* Oxford: Blackwell, 232–280. Open University
Course Team 1989: *Waves, tides and shallow-water processes.*
Oxford: Pergamon Press. Skinner, B.J. and Porter, S.C.
1987: *Physical geology.* New York: Wiley. Tillman, R.W. and
Siemers, C.T. (eds) 1984: *Siliclastic shelf sediments* [*Special
Publication* 34]. Tulsa, OK: Society of Economic Paleontologists
and Mineralogists. Wright, V.P. and Burchette, T.P. 1996:

121

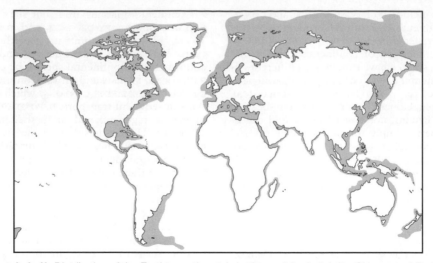

continental shelf *Distribution of the Earth's continental shelf seas (shaded) (after Skinner and Porter, 1987)*

Shallow-water carbonate environments. In Reading, H.G. (ed.), *Sedimentary environments: processes, facies and stratigraphy.* Oxford: Blackwell, 325–394.

continental slope The relatively steep (3–6°) part of the ocean floor at BATHYAL depths seaward of the CONTI-NENTAL SHELF. The continental slope passes into the CON-TINENTAL RISE on PASSIVE CONTINENTAL MARGINS and into a deep OCEANIC TRENCH on ACTIVE CONTINENTAL MARGINS. The junction between the continental shelf and the continental slope is typically an abrupt break of slope, termed the *shelf break.* *CDW*

Doyle, L.J. and Pilkey, O.H. 1979: *Geology of continental slopes* [*Special Publication* 27.] Tulsa, OK: Society of Economic Paleontologists and Mineralogists.

continentality The extent to which the CLIMATE of a location is influenced by land mass, as opposed to maritime influence (OCEANICITY). Continental climates tend to be dry, have extreme seasonal variations in temperature and a MEAN ANNUAL AIR TEMPERATURE below the latitudinal mean. Numerous indices have been developed to express continentality in numerical terms based on temperature range and latitude. The most continental climates are in east Asia. *GS*

Driscoll, D.M. and Yee Fong, J.M. 1992: Continentality: a basic climatic parameter re-examined. *International Journal of Climatology* 12, 185–192.

contingent drought A type of DROUGHT caused by irregular precipitation in areas that normally receive enough moisture for agricultural needs. The droughts of the UK in 1975–1976 and 1988–1992 were contingent droughts. *JAM*

[See also INVISIBLE DROUGHT, PERMANENT DROUGHT, SEASONAL DROUGHT]

Gregory, K.J. and Dornkamp, J.C. 1980: *Atlas of drought.* London: Institute of British Geographers.

contour A line joining sites of equal height value on a map or other graphical representation of the land surface.

Contour is an abbreviation of *contour line.* Interpretation of contour line patterns gives an indication of surface slope and steepness. *TF*

[See also ISOPLETH]

contour bund An earth bank typically 1.5–2.0 m wide, built across a cultivated slope of 1–7° in the tropics to act as an obstruction to *overland flow* and thus help to reduce SOIL EROSION and also store water. *SHD*

[See also CONTOUR RIDGING, MECHANICAL SOIL CONSERVA-TION MEASURES, TERRACING]

Morgan, R.P.C. 1995: *Soil erosion and conservation,* 2nd edn. Harlow: Longman, 138.

contour current A bottom current in the OCEAN that flows parallel to bathymetric contours as part of the THER-MOHALINE CIRCULATION (see OCEAN CURRENTS). Surface waters at high latitudes cool, forming dense WATER MASSES, which sink and flow towards lower latitudes. In northern latitudes, *Arctic Bottom Water,* or *North Atlantic Deep Water,* is sourced in the Norwegian and Greenland Seas. *Antarctic Bottom Water* forms in areas such as the Weddell Sea, circulates around the Antarctic and flows northwards into the South Atlantic, Indian and Pacific Oceans. These deep, cold, dense currents are influenced by the CORIOLIS FORCE, causing them to follow the western margins of the OCEAN BASINS as *Western Boundary Undercurrents.* Contour current velocities are highly variable (2–20 cm s^{-1}) as they are further affected by seasonal, topographic, surface storm and tidal factors. During periods of high velocity, sediment is transported in suspension (see SUSPENDED LOAD) and deposited as *contourite drifts.* Contour current deposits (*contourites*) may be difficult to distinguish from TURBIDITES. *BTC*

Neuman, G. 1968: *Ocean currents.* Amsterdam: Elsevier. **Stow, D.A.V.** 1994: Deep sea processes of sediment transport and deposition. In Pye, K. (ed.), *Sediment transport and depositional processes.* Oxford: Blackwell Scientific, 257–291. **Stow, D.A.V. and Faugeres, J.C. (eds)** 1998: Special issue: contourites, tur-bidites and process interaction. *Sedimentary Geology* 115, 1–386.

contour ploughing PLOUGHING of soil parallel to the contours. This practice can cause less SOIL EROSION on slopes than ploughing down slope. *SHD*

[See also MECHANICAL SOIL CONSERVATION MEASURES]

contour ridging The construction of soil ridges formed by PLOUGHING on slopes with silty or sandy soil and a shallow gradient (< 4.5°), leading to the storage of water rather than encouraging erosional *overland flow*, which causes SOIL EROSION. The resulting ridges are much less substantial than CONTOUR BUNDS. *SHD*

Morgan, R.P.C. 1995: *Soil erosion and conservation*, 2nd edn. Harlow: Longman.

control sites Sites that are used in field EXPERIMENTS as a standard for comparison with other sites at which specified changes have been induced or have occurred. Thus, they are comparable in all respects to the experimental sites except in that (those) parameter(s) which is (are) being manipulated. Because of the difficulties involved in explicit or implicit experimentation in many areas of ENVIRONMENTAL CHANGE, sites have to be accepted as controls that may not fulfil all the necessary criteria. *DH/CJC*

Karlén, W. and Matthews, J.A. 1992: Reconstructing Holocene glacier variations from glacial lake sediments: studies from Nordvestlandet and Jostedalsbreen-Jotunheimen, southern Norway. *Geografiska Annaler* 74A, 327–348.

controlled tipping The commonest method for the disposal of DOMESTIC WASTE. The refuse is tipped in layers, each of which is subjected to COMPACTION and sealed by a covering layer of inert material. Controlled tipping prevents, for example, unsightliness, offensive odours, the dispersal of refuse by wind or water and the breeding of PESTS. *JAM*

[See also LANDFILL]

convection Transfer of heat (and other properties) within a fluid by motion. In the atmosphere it is restricted to vertical motion, ADVECTION being used to describe horizontal heat transfer. Air, warmed by a heated ground surface, becomes lighter and rises, and is replaced by descending cooler air. Uplift leads to cooling and resultant cumulus cloud formation. *JBE*

[See also RADIATION]

Bejan, A. 1984: *Convection heat transfer*. New York: Wiley.

convention An international treaty or agreement concluded between states. Conventions comprise one dimension of international law and ENVIRONMENTAL LAW. There are probably over a thousand bilateral conventions and over a hundred multilateral ones. Some important conventions concerning environmental change and ENVIRONMENTAL PROTECTION are:

- Ramsar Convention on Wetlands of International Importance Especially as Wildfowl Habitats 1971.
- Oslo Convention for the Prevention of Marine Pollution by Dumping from Ships and Aircraft 1972.
- Washington Convention on International Trade in Endangered Species (CITES) 1973.
- Paris Convention for the Prevention of Marine Pollution from Land-Based Sources 1974.

- United Nations Convention on the Law of the Sea 1982 (a general convention with important environmental aspects).
- Vienna Convention for the Protection of the Ozone Layer 1985.
- Montreal Protocol on Substances that Deplete the Ozone Layer 1987.
- Basel Convention on the Control of Transboundary Movements of Hazardous Wastes and their Disposal 1989.
- Protocol on Environmental Protection to the Antarctic Treaty 1991.
- United Nations Framework Convention on Climate Change 1992.
- United Nations Convention on Biological Diversity 1992.

The Montreal Protocol was signed by many, but not all, industrial countries in 1987. Backed up by UNEP, it called for signatories to cut total CHLOROFLUOROCARBONS (CFCs) consumption by 1990 to their 1986 level and to freeze consumption of listed compounds. This convention was one of the first international agreements specifically intended to protect against GLOBAL ENVIRONMENTAL CHANGE (stratospheric OZONE DEPLETION). CFC manufacture has been phased out in developed countries, with developing countries following by 2010 (aided by funding from an Interim Multilateral Fund). *CJB/JAM*

Hass, P.M. 1991: Policy responses to stratospheric ozone depletion. *Global Environmental Change* 1, 224–234. **Von Molltke, K.** 1999: Conventions for environmental protection. In Alexander, D.E. and Fairbridge, R.W. (eds), *Encyclopedia of environmental science*. Amsterdam: Elsevier, 97–99.

conventional radiocarbon ages/dates The results of RADIOCARBON DATING are normally expressed in conventional radiocarbon years. These are based on (1) the LIBBY HALF-LIFE of 5 568 years; (2) the same modern reference standard; (3) the year AD 1950 as the present (the zero point from which to count ^{14}C time); and (4) normalisation of ^{14}C fractionation in the sample to a standard δ^{13}C value of −25‰ to correct for ISOTOPIC FRACTIONATION. CALIBRATION OF RADIOCARBON DATES corrects for changes that in fact have occurred in ^{14}C activity in the past. *JAM/PQD*

Stuiver, M. and Polach, H.A. 1977: Discussion: reporting of ^{14}C data. *Radiocarbon* 19, 355–363.

conventional radiocarbon dating RADIOCARBON DATING based on either *gas proportional counting* or *liquid scintillation* methods to measure ^{14}C activity (that is, the counting of beta particles from the decay of ^{14}C atoms). Sometimes termed *radiometric carbon dating*, it should not be confused with CONVENTIONAL RADIOCARBON AGES or the different and newer method of ACCELERATOR MASS SPECTROMETRY (AMS) DATING. *JAM*

convergence and divergence, in succession Convergence is the increasing similarity of initially different sites through successional time; divergence is the opposite of convergence. These terms are most frequently applied to community properties such as species composition, diversity or structure. Both convergence and divergence depend on the temporal and spatial scale under consideration and may occur simultaneously in the same

SERE for different properties (e.g. convergence of structure, divergence of composition) or sequentially for a single process (e.g. divergence of composition followed by convergence of composition). Biotic controls (e.g. low diversity, few immigrants or strong dominance by a few species) tend to favour convergence. Environmental controls (e.g. severe environmental conditions or frequent exogenous disturbances such as flooding or fire) favour divergence. *LRW*

Matthews, J.A. 1979: Refutation of convergence in a vegetation succession. *Naturwissenschaften* 66, 47–49. **Matthews, J.A.** 1992: *The ecology of recently-deglaciated terrain.* Cambridge: Cambridge University Press. **Peet, R.K.** 1992: Community structure and ecosystem function. In Glenn-Lewin, D.C., Peet, R.K. and Veblen, T.T. (eds), *Plant succession: theory and prediction.* London: Chapman and Hall. 103–151.

convergence, in evolution

Structures with similar functions and superficially similar appearance may have evolved independently from quite different ancestral structures in separate lineages of organisms. For example, fully functional eyes of different origins are found in octopuses, insects and vertebrates. This process is known as convergence. *KDB*

[See also CONVERGENCE AND DIVERGENCE, IN SUCCESSION, DIVERGENCE, EVOLUTION]

coombe rock

A silty chalk mud with flint and chalk CLASTS. It is an important PERIGLACIAL deposit of the English chalklands. *PW*

[See also DELL, HEAD, SOLIFLUCTION]

Ballantyne, C.K. and Harris, C. 1994: *The periglaciation of Great Britain.* Cambridge: Cambridge University Press.

coordinate

A value in a set measured from the origin of a location definition system. It may be an angle, distance or grid row/column value. *TF*

[See also COORDINATE SYSTEM]

coordinate system

A systematic method of defining points in space. The dimensionality of the space defines the number of values required to specify a unique location. For example, a two-dimensional plane surface requires two coordinate values. *TF*

Maling, D.H. 1992: *Coordinate systems and map projections.* Oxford: Pergamon Press.

copper (Cu)

A non-ferrous metal usually present in rocks as mineral veins and in small quantities in soils as a trace element (20–30 mg kg^{-1}). The metal is extracted from its ores relatively easily and its malleability and ease of combination with zinc as brass and with tin as bronze made it a useful metal. As a component of bronze, copper pollution is associated with early human culture. Copper is an essential MICRONUTRIENT for plants and grazing animals, but in higher concentrations is toxic. Elevated levels in soils are associated with point sources such as industrial metal-working activities, particularly since the INDUSTRIAL REVOLUTION, and power generation where pollution levels may be high but limited in area. Copper sulphate in *fungicide* sprays applied to vineyards, sewage sludge, municipal composts, as a contaminant in fertilisers and in pig slurry poses a wider risk to food and fibre pro-

duction where extensive applications have been made to crops or the land. *EMB*

[See also BRONZE AGE HUMAN IMPACTS, COPPER AGE, METAL POLLUTION]

Alloway, B.J. (ed.), 1990: *Heavy metals in soils.* Glasgow and London: Blackie.

Copper Age: landscape impacts

The Copper Age was a transitional phase between the NEOLITHIC and the BRONZE AGE, during which copper was used in the manufacture of tools, weapons and ornaments. The Copper Age occurred between 7000 and 5000 calendar years BP in Europe and is best known from Hungary, the Balkans and Greece, where the terms *Eneolithic* and *Chalcolithic* are also used. Agriculture flourished during this phase and could support a larger population, with the result that settlements became bigger. However, the most significant landscape impact was in the production of vast quantities of charcoal necessary for COPPER smelting. Clearance of woodland for use as fuel was widespread, although COPPICING of broad-leaved forest taxa such as *Corylus* (hazel) and *Carpinus betulus* (hornbeam) ensured a constant supply of wood for copper production. This is evident in some studies in PALAEOECOLOGY as peaks of microscopic charcoal, fluctuations in the pollen of broad-leaved forest taxa and an influx of copper minerals. *ARG*

[See also BRONZE AGE: LANDSCAPE IMPACTS, IRON AGE: LANDSCAPE IMPACTS, NEOLITHIC: LANDSCAPE IMPACTS, POLLUTION HISTORY]

Renfrew, C. 1969: The autonomy of the south-east European Copper Age. *Proceedings of the Prehistoric Society* 35, 12–47. **Simpson, D.D.A. (ed.)** 1971: *Economy and settlement in Neolithic and Bronze Age Britain and Europe.* Leicester: Leicester University Press. **Whittle, A.** 1996: *Europe in the Neolithic.* Cambridge: Cambridge University Press. **Willis, K.J., Sümegi, P., Braun, M. et al.** 1998: Prehistoric land degradation in Hungary: who, how and why? *Antiquity*, 72, 101–113.

coppicing

A WOODLAND MANAGEMENT PRACTICE whereby underwood trees are cut off close to ground level every 4–8 years (16–20-year cycles are also known) so that a number of new stems shoot up from the stool. The harvested stems provide strong, straight, supple poles for fencing or wattle and also fuel for burning. Within any one woodland different panels were coppiced at different time to ensure a constant supply of wood. The most commonly coppiced tree species is hazel (*Corylus avellana*), but hornbeam (*Caprinus betulus*), lime (*Tilia* spp.), ash (*Fraxinus excelsior*), oak (*Quercus* spp.), wych elm (*Ulmus glabra*) and elder (*Sambucus nigra*) coppices are also frequent. Coppicing is known from NEOLITHIC times: the Sweet Track that crossed the Somerset levels, England, is made of mixed coppice-wood. TRACKWAYS from the BRONZE and IRON AGE were usually constructed of wattle hurdles, which necessitated an elaborate coppicing system. Coppicing continued through Anglo-Saxon times and was important to the MEDIAEVAL economy. The practice has been revived in Britain in this century. *SPH*

Rackham, O. 1986: *The history of the countryside.* London: Dent.

coprolite

Fossilised faeces, which range in size from minuscule granular pellets produced by invertebrates to large, lump-like masses left by dinosaurs. *DCS*

[See also FAECES ANALYSIS, RODENT MIDDENS]
Bryant Jr, V.M. and Williams, D.G. 1975: The coprolites of man. *Scientific American* 232, 100–105. **Reinhard, K.J. and Bryant Jr, V.M.** 1992: Coprolite analysis: a biological perspective on archaeology. In Schiffer, M.B. (ed.), *Archaeological method and theory.* Vol. 4. Tucson, AZ: University of Arizona Press.

coquina A shell SAND or GRAVEL, or its lithified equivalent as a *bioclastic limestone.* GO

coral and coral reefs: environmental reconstruction

Coral reefs growing in shallow tropical waters create durable structures that both modify and respond to changing environmental forcing factors. Individual corals in REEFS contain several independent high-resolution NATURAL ARCHIVES of environmental variability in their aragonitic skeletons, with long, well constrained chronologies being matched to geochemical signatures of ocean water conditions.

On long timescales of > 1 million years (Ma) to 50 Ma, studies of ATOLL and BARRIER REEF stratigraphy from deep drilling yield information on long-term island SUBSIDENCE rates and, through studies of carbonate DIAGENESIS, the presence of PALAEOSOLS and modelling of UNCONFORMITY sequences, their interaction with large-scale SEA-LEVEL CHANGES. Mass spectrometric dating techniques allow precise dating (to within ± 1% of actual age) of coral samples from decades to several hundred thousand years; with the appropriate corrections, ^{14}C dating is also applicable on timescales up to 40 000 years. The combination of palaeoecological studies of fossil reef deposits with radiometric dating of individual coral colonies has been instrumental in establishing both the magnitude and form of Late Pleistocene GLACIAL–INTERGLACIAL sea-level fluctuations and the global variability of HOLOCENE sea-level change in the tropics and subtropics.

At short timescales, rapid (ca. 5–15 mm a^{-1} extension rates) but annually variable coral growth generates skeletal density banding, revealed by X-RADIOGRAPHY. This allows dating to subseasonal resolution (at times to weekly resolution) along the axis of maximum growth; study of such banding is akin to the study of tree rings and has been termed SCLEROCHRONOLOGY. This methodology is particularly valuable given that individual coral colonies can be remarkably long-lived (e.g. environmental records from AD 1587 in the Galapagos Archipelago and 1635 on the Great Barrier Reef).

The most useful of the skeletal signals are the STABLE ISOTOPE content ($\delta^{18}O, \delta^{13}C$) and the concentration of trace metals (particularly cadmium (Cd), barium (Ba) and manganese (Mn)). Coral $\delta^{18}O$ records reflect ambient seawater temperatures at subseasonal resolution; the $\delta^{18}O$ of calcium carbonate precipitated in equilibrium with seawater decreases by ca 0.22‰ for every 1°C rise in water temperature. Variations in the $\delta^{18}O$ of seawater obscure this simple relationship: these variations result from changes in EVAPORATION ($\delta^{18}O$ enrichment), PRECIPITATION ($\delta^{18}O$ depletion) and RUNOFF. If one or other of these thermal or hydrographic controls can be constrained, OXYGEN ISOTOPE variations provide a powerful means of environmental reconstruction (see Figure). The ratio of incorporation of strontium (Sr) to calcium (Ca) in coral skeletons is controlled by the *strontium:calcium activity ratio* in ocean waters (which can be regarded as invariant

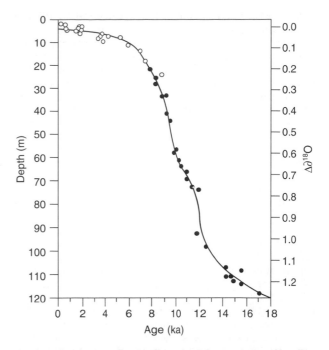

coral and coral reefs: environmental reconstruction *The rise in sea level over the last 17 000 years in Barbados reconstructed from the dated fossil coral (*Acropora palmata*), a species that lives in water depths of < 5 metres (after Fairbanks, 1989 – referenced in sea-level change entry)*

over 100 ka timescale) and the Sr.Ca distribution coefficient between aragonite and seawater, which depends largely upon water temperature and can be resolved, through THERMAL IONISATION MASS SPECTROMETRY (TIMS), to a measurement precision equivalent to < 0.5°C. *Magnesium:calcium thermometry* is being similarly explored. These techniques, singly or in combination, have been used to establish, for example: EL NIÑO—SOUTHERN OSCILLATION (ENSO) rainfall changes at Tarawa Atoll, equatorial West Pacific, over the last century; twentieth-century climate change in the western equatorial Pacific 'warm pool'; and ENSO-related coolings and long-term ocean warming on the Great Barrier Reef. On longer timescales, old records from the Galapagos Islands have identified low ocean temperatures during the early AD 1600s and 1800s and warmer conditions during the 1700s. Spectral and cross-spectral techniques applied to long time series of sea surface temperature change allow the establishment of dominant PERIODICITIES and identification of shifts in dominant cycles; they also form the basis for discussion of the changing dynamics of tropical ocean–atmosphere circulation systems. Extension of such methodologies to fossil corals has demonstrated abrupt ocean warming in the early Holocene and elevated mid-Holocene sea surface temperatures. Environmental signals from skeletal $\delta^{13}C$ records are more complex because variations are related not only to the isotopic composition of sea water, but also to insolation- and depth-dependent coral growth processes.

Oceanic distributions of certain metals in trace concentrations reflect specific environmental processes such as UPWELLING, ADVECTION, AEOLIAN dust transport and runoff. The variability of such processes can be derived from coral records because ambient seawater metal concentrations can be reconstructed from skeletally bound metal concentrations through known distribution coefficients. The modern distribution of both Cd and Ba follows that of marine nutrients, such as phosphate and silica; thus coral records may record ENSO-mediated variations in the upwelling of nutrient-rich deep water. Mn levels in coral skeletons may record not only variations in upwelling, but also fluctuations in AEROSOL deposition, advection of Mn-enriched shelf waters and local sediment fluxes.

Corals have been shown to fluoresce under ultraviolet light. In some cases there is clear banding, with 'bright' and 'dull' fluorescence associated with the 'light' and 'dark' component respectively of the annual skeletal density couplet. The fluorescence derives from the incorporation of HUMIC and FULVIC ACIDS from sediment-rich terrestrial runoff and is therefore best displayed where runoff shows strong seasonality and/or inter-annual variation. Where preserved in exceptionally old corals, fluorescent banding offers the possibility of reconstructing palaeo-discharges in the absence of historical records, but not all sites studied have shown a clear signal between fluorescence, rainfall events and inshore sediment plumes.

Understanding how environmental PERTURBATIONS are reflected in coral reef structure and individual coral skeletons may both help identify trends in the recent and present record of changing ocean conditions and aid prediction of near-future reef responses to global environmental change. *TSp*

Barnes, D.J. and Lough, J. 1996: Coral skeletons: storage and recovery of environmental information. *Global Change Biology* 2, 547–558. Beck, J.W., Edwards, R.L., Ito, E. *et al.* 1992: Sea-surface temperature from coral skeletal strontium/calcium ratios. *Science* 257, 644–647. Cole, J.E., Dunbar, R.B., McClanahan, T.R. and Muthiga, N.A. 2000: Tropical Pacific forcing of decadal SST variability in the western Indian Ocean over the past two centuries. *Science* 287, 617–619. Dunbar, R.B., Wellington, G.M., Colgan, M.W. and Glynn, P.W. 1994: Eastern Pacific sea surface temperatures since 1600 AD: the $\delta^{18}O$ record of climate variability in Galapagos corals. *Palaeoceanography* 9, 291–315. Edwards, R.L., Chen, J.H. and Wasserburg, G.J. 1987: $^{238}U-^{234}U-^{230}Th-^{232}Th$ systematics and the precise measurement of time over the last 500 000 years. *Earth and Planetary Science Letters* 81, 175–192. Isdale, P. 1984: Fluorescent bands in massive corals record centuries of coastal rainfall. *Nature* 310, 578–579. Knutson, D.W., Buddemeier, R.W. and Smith, S.V. 1972: Coral chronometers: seasonal growth bands in reef corals. *Science* 177, 270–272. Lincoln, J.M. and Schlanger, S.O. 1991: Atoll stratigraphy as a record of sea level change: problems and prospects. *Journal of Geophysical Research* 96: 6727–6752. Shen, G.T. and Boyle, E.A. 1988: Determination of lead, cadmium, and other trace metals in annually banded corals. *Chemical Geology* 67, 47–62. Tudhope, A.W., Shimmield, G.B., Chilcott, C.P. *et al.* 1995: Recent changes in climate in the far western equatorial Pacific and their relationship to the Southern Oscillation: oxygen isotope records from massive corals, Papua New Guinea. *Earth and Planetary Science Letters* 344, 575–590.

coral bleaching The discolouration of coral caused by death of the endosymbiotic dinoflagellates (*zooxanthellae*), which comprise up to 75% of coral tissue. Coral bleaching events may be extensive and sudden following a rise in sea-water temperature of 1–2°C. The highest SEA-

SURFACE TEMPERATURES ever recorded, related to both the 1997–1998 EL NIÑO–SOUTHERN OSCILLATION event and GLOBAL WARMING, caused severe bleaching of corals world-wide in 1998 (see Figure). *JAM*

[See also CONSERVATION, CORAL REEFS, HUMAN IMPACT]

Aronson, R.B., Precht, W.F., Macintyre, I.G. and Murdoch, T.J.T. 2000: Coral bleach-out in Belize. *Nature* 405, 36. Brown, B.E. 1990: Coral bleaching. *Coral Reefs* 8, 153–232. Goreau, T.J. and Hayes, R.L. 1994: Coral bleaching and ocean 'hot spots'. *Ambio* 23, 176–180.

coral reef A wave-resistant structure in tropical and subtropical waters (> 18°C monthly minimum temperature), composed of framework-building corals, coralline algae and associated organisms. *TSp*

coral reef islands Islands composed largely of CORAL REEF sediments. They are usually divided into 'high' and 'low' islands. 'High islands' with raised reefs up to 100 m above present sea level include raised ATOLLS (e.g. Henderson Island), raised BARRIER REEFS (e.g. New Georgia) and raised FRINGING REEFS (e.g. Vanuatu); they result from the secondary uplift of coral reefs initially formed by upward reef growth around subsiding volcanoes. Uplift results from plate reheating over asthenospheric 'bumps', lithospheric up-arching to compensate regional loading by young (ca. 1 Ma) volcanoes and plate buckling near subduction zones. These islands may exhibit Pleistocene limestones around an older Tertiary

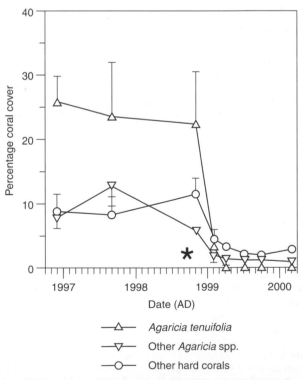

coral bleaching *Changes in coral cover at Channel Cay, Belize, associated with the high-temperature anomaly in August 1998. Almost all living colonies were completely bleached (after Aronson* et al.*, 2000)*

core, elevations representing a combination of tectonic UPLIFT and glacio-eustatic sea-level change (see EUSTASY). The most extensive of these Pleistocene raised reefs, up to about 6 m above present sea level in the absence of tectonism, are typically LAST INTERGLACIAL in age. In some locations (e.g. Bahamas, Bermuda) elevated islands are characterised by cemented oolitic shoals and AEOLIANITES. Slightly emergent islands, supporting fossil corals in growth position (including MICROATOLLS), cemented conglomeratic platforms, fossil algal ridges on reef flats and abandoned inter-island channels (or *hoa*) are typical of the South Pacific Ocean and record high sea levels between about 5 and 1.5 ka BP.

'Low islands' lie within a few metres of sea level and are composed of the disintegration products of reef organisms. They can be classified as: sand CAYS; sand cays with shingle ridges (or MOTU); shingle cays; MANGROVE cays; and low-wooded islands, typical of the Belize Barrier Reef and northern Great Barrier Reef. There are considerable differences in the abundance of low reef islands world wide (e.g. Tuamotu atolls: 100–300 islands each; Marshall Islands: ca. 20; Carolines: ca. 10). On many islands, long-term migration of reef islands across reef flats (and even into lagoons) is indicated by the presence of offshore BEACHROCK ridges recording former beach positions. Low islands have been shown to change in size and orientation with long-term shifts in wind patterns and may undergo major morphological change when impacted by TROPICAL CYCLONES. Tropical cyclones are viewed as essential, however, for island growth, particularly once vegetation has become established, but replacement of mangroves and natural thickets with coconuts can result in catastrophic island erosion and island disappearance.

TSp

Bayliss-Smith, T.P. 1988: The role of hurricanes in the development of reef islands, Ontong Java, Solomon Islands. *Geographical Journal* 154, 377–391. **Guilcher, A.** 1988: *Coral reef geomorphology*. Chichester: Wiley. **Spencer, T.** 1994: Geodynamics, ocean dynamics and island geomorphology on the Pacific plate. *Geo – Eco – Trop* 16, 49–99. **Stoddart, D.R. and Steers, J.A.** 1977: The nature and origin of coral reef islands. In Jones, O.A. and Endean, R. (eds), *Biology and geology of coral reefs*. Vol. IV. *Geology 2*. New York: Academic Press, 59–105.

coral reefs: human impact and conservation

Coral reef communities have a narrow range of tolerance of SALINITY, temperature, light and water quality and are extremely sensitive to ENVIRONMENTAL CHANGE. Changes in LANDUSE that lead to increased SEDIMENT discharge from rivers into coastal waters present the most serious threat. Hermatypic corals get much of their energy from PHOTOSYNTHESIS of their zooxanthellae (microscopic algae that form a symbiotic relationship with coral polyps) and thus require light. EUTROPHICATION enhances the growth of macro-algae at the expense of coral. OVERFISHING and damaging fishing methods, such as poisoning and dynamiting reefs, are also major causes of destruction. Coral is mined for road stone, lime and as souvenirs. CORAL BLEACHING, caused by the death of zooxanthellae, affected vast areas of coral reef around the globe from the end of 1997. The likely cause was increased water temperatures as a result of an extreme EL NIÑO event, but this may have been compounded by GLOBAL WARMING and other human impacts. Despite the importance given to marine and coastal BIODIVERSITY in the United Nations Convention on Biological Diversity, to date very few nations have implemented comprehensive programmes aimed at protecting coral reefs. As many of the organisms that inhabit reefs do so for only part of their life cycle, there is concern that protection of reefs alone will do little to conserve them.

NDB

[See also FISHERIES CONSERVATION AND MANAGEMENT]

Alcala, A.C. and Russ, G.R. 1990: A direct test of the effects of protective management on abundance and yield of tropical marine resources. *Journal de Conseil* 47, 40–47. **Pennisi, E.** 1997: Brighter prospect for the world coral reefs? *Science* 277, 491–493. **Smith, S.V. and Buddemeier, R.W.** 1992: Global change and coral reef ecosystems. *Annual Review of Ecology and Systematics* 23, 89–118.

coral reefs: impact of tropical cyclones

Reefs in the tropical storm belts suffer periodic damage from hurricane-generated waves and surge. Corals in water depths up to 20 m may be broken and removed; branching forms are more susceptible to damage than massive corals. Reef-crest and reef-front coral debris accumulates as talus on fore-reef slopes, in massive individual blocks and storm ridges on reef flats, as lobes in back-reef lagoons and as drapes of carbonate sands and muds on deeper fore-reefs and lagoons. Tropical cyclones with windspeeds of 120–150 km h^{-1} may trim reef topography, whereas windspeeds of 200 km h^{-1} may lead to massive devastation and a complete loss of reef morphology. The time interval between hurricanes influences, through renewed reef growth, the availability of material and the degree to which previous storm deposits have been stabilised. Many reef flat storm deposits cluster at 3–4 ka BP, probably related to a slight sea level fall rather than an increase in storminess. The production of coral debris via occasional hurricanes is seen as essential for the creation, growth and maintenance of CORAL REEF ISLANDS. Reductions in cyclone frequency could starve reef islands of sediment and lead to the decline or even disappearance of some MOTUS and CAYS.

TSp

[See also MANGROVE SWAMPS: IMPACT OF TROPICAL CYCLONES]

Done, T.P. 1993: On tropical cyclones, corals and coral reefs. *Coral Reefs* 12, 143–152. **Scoffin, T.P.** 1993: The geological effects of hurricanes on coral reefs and the interpretation of storm deposits. *Coral Reefs* 12, 203–221. **Stoddart, D.R.** 1985: Hurricane effects on coral reefs: Conclusion. *Proceedings of the 5th International Coral Reef Symposium, Tahiti* 3, 349–350.

coral terraces

Subaerial fossil reefs, often forming 'staircases' on uplifting coasts (e.g. Barbados, Huon Peninsula, Papua New Guinea) or submarine benches on coral fore-reef slopes.

TSp

Chappell, J. 1974: Geology of coral terraces, Huon Peninsula, New Guinea: a study of Quaternary tectonic movements and sea level change. *Geological Society of America Bulletin* 85, 553–570.

core

(1) A cylinder of soil, sediment, ice or rock extracted from a BOREHOLE by drilling or other means, allowing detailed examination of buried materials. Cores can be obtained from beneath the land surface, lake beds or the sea-floor.

(2) The innermost zone of the Earth, separated from the MANTLE by the *Gutenberg discontinuity*, 2900 km below the Earth's surface. The core is very dense, with an

iron–nickel composition. A molten *outer core* gives way to a solid *inner core* at 5150 km below the Earth's surface. Convection in the outer core is responsible for the Earth's magnetic field.

(3) In the archaeological context, a core is a stone ARTE-FACT from which flakes are removed; it can also be a tool or blank from which other tools are made. *GO/JAM*

[See also CORE CORRELATION, CORE SCANNING, CORER, DEEP-SEA DRILLING, EARTH STRUCTURE, GEOMAGNETISM, ICE CORE, LITHICS, LITHOSPHERE, MARINE SEDIMENT CORE, PISTON SAMPLER]

Bott, M.H.P. 1982: *The interior of the Earth: its structure, constitution, and evolution*, 2nd edn. London: Arnold.

core correlation
Sediment cores from the same or different deposits can be correlated using SEDIMENT TYPE, LOSS-ON-IGNITION, LAMINATIONS, TEPHRA layers, MAGNETIC SUSCEPTIBILITY, BIOSTRATIGRAPHY and RADIOCARBON DATING or, in young sediments, LEAD-210 DATING. Correlation is necessary to ensure that particular cores are representative and to identify features that are unique to individual cores and hence may have limited significance. *HHB*

core–satellite model
An ecological model which holds that assemblages or communities are composed of two classes of species, some that are abundant and found at many sites (*core species*) and others that are rare and restricted to a few sites (*satellite species*). *MVL*

[See also METAPOPULATION MODEL]

Hanski, I. 1982. Communities of bumblebees: testing the core–satellite model. *Annales Zoologici Fennici* 19, 65–73.

core scanning
Rapid, commonly NON-DESTRUCTIVE SAMPLING of a sediment core. Core scanning is often used in CORE CORRELATION of unextruded cores and/or as a preliminary to more detailed, destructive analysis of the material. MAGNETIC SUSCEPTIBILITY and X-RADIOGRAPHY are commonly used in this way. *JAM*

corer
A device used to retrieve samples of peat from MIRES or SEDIMENTS from lakes, oceans, peatlands or GLACIERS. Corers work on a variety of principles, dependent upon the type of material sampled. The primary goal is to get continuous, undisturbed and uncontaminated records (including fine structures such as varves) even if the material is either very loose (in which case FREEZE CORING is used) or very tough (in which case various mechanical or electrically powered systems are used). *BA*

[See also CHAMBER/PISTON SAMPLER]

Aaby, B. and Digerfeldt, G. 1986: Sampling techniques for lakes and bogs. In: Berglund, B.E., Birks, H.J.B., Ralska-Jasiewiczowa, M. and Wright, H.E. (eds), 1986: *Palaeoecological events during the last 15000 years. Regional syntheses of palaeoecological studies of lakes and mires in Europe* Chichester: Wiley. **Wright, H.E.** 1991: Coring tips. *Journal of Paleolimnology* 6, 37–49.

corestone
A blocky or spheroidal shaped, cobble-to-boulder size, relatively unweathered rock mass, currently or previously surrounded by weathered bedrock. Deep subsurface WEATHERING, followed by exhumation of corestones by various subaerial processes, has been used in a

two-stage theory of TOR formation in both tropical and temperate environments. *RAS*

[See also CLITTER, GROWAN, GRUSSIFICATION]

CORINE land cover map
The CORINE (coordination of information on the environment) land cover map is a European inventory of land cover in 44 classes. The production and management of the map is carried out by the European Topic Centre on Land Cover (ETC/LC in Kiruna, Sweden), established by the European Environment Agency (EEA) to provide accurate data on land cover in Europe, corresponding to needs across a wide range of applications. The production of the map started in 1985 and to date the ETC/LC is holding a DATABASE with LAND COVER data from 12 of the currently 18 EEA member states, covering 3.6 million km^2. Additionally, six of 13 Central and Eastern European Countries are included in the database. Switzerland has recently produced a CORINE land cover map. The update of the CORINE land cover map from AD 2000 will provide information on land-cover change. CORINE land cover has generally been produced by visual interpretation of hard-copy satellite images, followed by manual digitising to give computer maps in vector format at a scale of 1 : 100 000. The smallest mapping unit is 25 ha (500 m × 500 m). For some countries, such as Great Britain, Sweden and Finland, CORINE land cover was generated using semi-automated IMAGE CLASSIFICATION and generalisation approaches. The land-cover classes are ordered in a three-level hierarchy. *HB*

Cruickshank, M.M. and Tomlinson, R.W. 1996: Application of CORINE land cover methodology to the UK – Some issues raised from Northern Ireland. *Global Ecology and Biogeography Letters* 5, 235–248. **European Commission** 1994: *EUR12585 – CORINE Land Cover Project – Technical Guide*. Luxembourg: Office for Official Publications of the European Communities. **Rodwell, J.S., Pignatti, S., Mucina, L. and Schaminee, J.H.J.** 1995: European vegetation survey – update on progress. *Journal of Vegetation Science* 6, 759–762.

Coriolis force
An apparent force, named after Gaspard de Coriolis, which deflects surface airflow (responding to the initial pressure gradient force) as a result of the Earth's rotation. The magnitude of this deflection, which is to the right in the Northern Hemisphere and to the left in the Southern Hemisphere, is directly proportioned to the speed of the moving air mass and the sine of the latitude (being maximum at the poles). *RDT*

corrasion
Mechanical erosion of rock surfaces and cohesive beds by running water, especially rivers: it is normally restricted to the abrasive wearing away of the channel sides and channel bed by transported sedimentary particles. Corrasion differs from the chemical process of CORROSION, the purely *hydraulic action* of water (see QUARRYING), and also the process of *attrition* (the wear that the sedimentary particles undergo during transport). It is also used in the context of marine erosion by waves armed with rock fragments, and AEOLIAN erosion by sand particles. *JAM*

Stamp, L.D. (ed.) 1961: *A glossary of geographical terms*. London: Longman.

correlated-age dating
A category of techniques based on stratigraphic CORRELATION with deposits or

events that are independently dated. The techniques are essentially nominal and only demonstrate *age equivalence* between sites or sequences, as in the case of palaeomagnetic, lithostratigraphic or tephrochronological correlation. *DAR*

[See also LITHOSTRATIGRAPHY, MARKER HORIZON, PALAEOMAGNETIC DATING, TEPHROCHRONOLOGY]

Colman, S.M., Pierce, K.L. and Birkeland, P.W. 1987: Suggested terminology for Quaternary dating methods. *Quaternary Research* 28, 314–319.

correlation analysis A set of statistical techniques for measuring the extent to which two variables are related. For interval data, the strength of relationships can be expressed quantitatively by using a correlation coefficient such as r, which varies between -1.0 and $+1.0$. A value of the *correlation coefficient* of $+1.0$ indicates perfect positive correlation (in the sense that increases in one variable accompany increases in the other), whilst a value of -1.0 indicate that increases in one variable perfectly reflect decreases in the other (they are negatively related). A correlation coefficient of zero indicates no correlation. The degree of similarity does not allow one to infer that variations in one quantity actually cause variations in the other. *Multiple correlation* measures the extent to which more than one variable are related. *Partial correlation* measures the extent to which variables are related while holding constant the statistical effects of one or more other variables. *PJS*

[See also AUTOCORRELATION, NUMERICAL ANALYSIS, PARTIAL CORRELATION COEFFICIENT]

correlation (stratigraphical) The principle of establishing the time-equivalence of rock or sediment units in different places. Correlation is a central objective of STRATIGRAPHY and can be achieved using: FOSSILS or SUBFOSSILS (BIOSTRATIGRAPHY); widespread units of distinctive LITHOLOGY (LITHOSTRATIGRAPHY, e.g. TEPHRA layers, EVENT horizons); changes in magnetic polarity (MAGNETOSTRATIGRAPHY); PALAEOSOLS (soil stratigraphy or pedostratigraphy); landforms (MORPHOSTRATIGRAPHY); and, in some cases controversially, patterns of SEA-LEVEL CHANGE (SEQUENCE STRATIGRAPHY). *GO*

[See also CORE CORRELATION, EVENT STRATIGRAPHY, STRONTIUM ISOTOPE RATIO]

Dunay, R.E. and Hailwood, E.A. 1995: *Non-biostratigraphical methods of dating and correlation* [*Special Publication* 89]. London: Geological Society. **Miall, A.D.** 1997: *The geology of stratigraphic sequences*. Berlin: Springer. **Nichols, G.** 1999: *Sedimentology and stratigraphy*. Oxford: Blackwell Science. **Shackleton, N.J. and Turner, C.** 1967: Correlation between marine and terrestrial Pleistocene successions. *Nature* 216, 1079–1082.

correspondence analysis (CA) An indirect ORDINATION technique for simultaneously displaying similarities between objects and variables. It was originally developed for the analysis of categorical data but is now widely used with quantitative data. Each axis is selected to maximise the dispersion of the variable scores, subject to the constraint that the axes are uncorrelated to each other. It assumes a unimodal response of the variables to the underlying latent variables or gradients. Its main fault is the arch effect. This can usually be removed by detrending (*detrended correspondence analysis*). Also known as reciprocal averaging and optimal or dual scaling. *HJBB*

[See also PRINCIPAL COMPONENTS ANALYSIS, CANONICAL CORRESPONDENCE ANALYSIS, GRADIENT ANALYSIS]

Hill, M.O. 1974: Correspondence analysis: a neglected multivariate method. *Applied Statistics* 23, 340–354.

corrosion (1) The deterioration of metals by chemical action, such as the rusting of iron. (2) *Chemical erosion* of rock surfaces by running water: it includes the removal of rock by *solution* rather than the mechanical erosion brought about by rock particles transported in both fluvial and marine environments (CORRASION). Corrosion is a particularly important process in the context of LIMESTONE rocks and KARST landscapes, where the term *mixing corrosion* refers to the enhanced erosion made possible by the AGGRESSIVE WATER created when two streams saturated with respect to calcium carbonate are mixed. *JAM*

[See also CHEMICAL WEATHERING]

Bögli, A. 1971: Corrosion by mixing of karst water. *Transactions of the Cave Research Group of Great Britain* 13, 109–114. **Stamp, L.D. (ed.)** 1961: *A glossary of geographical terms*. London: Longman.

coseismic Geological changes that accompany an EARTHQUAKE. For example, land uplift that takes place during an earthquake is described as *coseismic uplift*. The term should be contrasted with *interseismic*, which describes the period between earthquakes. Interseismic and coseismic crustal movements are of particular interest in understanding the dynamics of FAULT activity. One of the largest postulated coseismic uplift events is thought to have taken place in western Crete during an earthquake in AD 365, when several areas were uplifted by up to 6 m. *AGD*

[See also PALAEOSEISMICITY, SEISMICITY]

Hunstad, I., Anzidei, M., Cocco, M. *et al.* 1999: Modelling coseismic displacements during the 1997 Umbri–Marche earthquake (central Italy). *Geophysical Journal International* 139, 283–295. **Pirazzoli, P.A., Stiros, S.C., Laborel, J.** *et al.* 1994: Late-Holocene shoreline changes related to palaeoseismic events in the Ionian Islands, Greece. *The Holocene* 4, 397–405. **Vita-Finzi, C.** 1985: *Neotectonics*. Chichester: Wiley.

cosmic radiation A combination of *solar cosmic rays*, and much higher energy *galactic cosmic rays*, which strike the Earth from all directions. The latter are the high-speed subatomic particles, mostly *protons*, that are responsible for the production of *cosmogenic nuclides* both in the atmosphere (e.g. ^{14}C) and at the Earth's surface (e.g. ^{10}Be and ^{26}Al). Cosmogenic isotopes are used in RADIOCARBON DATING and EXPOSURE-AGE DATING. *DMcC*

[See also COSMOGENIC-ISOTOPE DATING, SURFACE DATING]

cosmic ray flux The variable amount of COSMIC RADIATION reaching the Earth's surface. The Earth's MAGNETIC FIELD deflects the charged particles and filters out those with the lowest energy, with a much greater efficiency at the Equator than the poles. Variations in the SOLAR WIND and in the Earth's magnetic field mean that the cosmic-ray flux, and therefore the production rate of *cosmogenic nuclides*, has varied over time, which is important in RADIOCARBON DATING and COSMOGENIC-ISOTOPE DATING. *DMcC*

[See also GEOMAGNETISM]

129

cosmogenic-isotope dating A technique for estimating how long rock surfaces have been exposed to COSMIC RADIATION and thus a basis for EXPOSURE-AGE DATING. High-energy cosmic rays interact with atomic nuclei near the surface of rocks to produce *cosmogenic nuclides*, which may be stable or radioactive. If the production and decay rates can be estimated, then so can the exposure age of the surface. Different cosmogenic isotopes are appropriate for dating different materials over different timescales (see Table). The stable isotopes ^{3}He and ^{21}Ne are measured using conventional MASS SPECTROMETRY. The other isotopes require complex chemical separation and accelerator mass spectrometry (as used in ACCELERATOR MASS SPECTROMETRY DATING), so that dates are expensive. Production rates vary with altitude and latitude, as well as over time, and are reduced if the sample is not fully exposed on all sides (as on a mountain summit). Any surface loss, due to weathering and erosion, will influence the results.

The most widely used isotopes are ^{10}Be and ^{26}Al, often together, and ^{36}Cl. Surfaces that have been dated include METEORITE-IMPACT craters, glacial MORAINES and RIVER TERRACES. The same techniques can be used to estimate long-term DENUDATION RATES. *DMcC*

[See also CALCIUM-41, CHLORINE-36, SURFACE DATING]

Bierman, P.R., Marsella, K. A., Patterson, C. *et al.* 1999: Mid-Pleistocene cosmogenic minimum-age limits for pre-Wisconsinan glacial surfaces in southwestern Minnesota and southern Baffin Island: a multiple nuclide approach. *Geomorphology* 27, 25–39. **Cerling, T.E. and Craig, H.** 1994: Geomorphology and *in situ* cosmogenic isotopes. *Annual Review of Earth and Planetary Science* 22, 273–317. **Kurtz, M.D. and Brook, E.J.** 1994: Surface exposure dating with cosmogenic nuclides. In Beck, C. (ed.), *Dating in exposed and surface contexts.* Albuquerque, NM: University of New Mexico Press, 139–159. **Plummer, M.A., Phillips, F.M., Fabryka-Martin, J.** *et al.* 1987: Chlorine-36 in fossil rat urine: an archive of cosmogenic nuclide deposition during the past 40,000 years. *Science* 277, 538–540. **Stone, J. O., Ballantyne, C. K. and Fifield, L. K.** 1998: Exposure dating and validation of periglacial weathering limits, NW Scotland. *Geology* 26, 587–590. **Summerfield, M. A., Stuart, F. M., Cockburn, H. A. P.** *et al.* 1999: Long-term rates of denudation in the Dry Valleys, Transantarctic Mountains, southern Victoria Land, Antarctica based on in-situ-produced cosmogenic ^{21}Ne. *Geomorphology* 27, 113–129.

cosmology The study of the universe or *cosmos*, including its origin (see BIG BANG) and evolution. *GO*

[See also ASTROGEOLOGY, COSMOSPHERE, PLANETARY GEOLOGY]

Sagan, C. 1981: *Cosmos.* London: Macdonald.

cosmosphere The domain of all non-living things and forces, which, strictly speaking, includes the Earth, but is usually taken to include phenomena beyond the Earth's atmosphere. Environmental change on Earth is intimately associated with the cosmosphere through three main factors: (1) the force of gravity exerted by extraterrestrial bodies (Earth is a player in the so-called 'music of the spheres'); (2) energy received from stars, principally the Sun; and (3) *space debris*, which includes the ASTEROIDS, COMETS and METEOROIDS that may collide with the Earth. *JAM*

[See also COSMIC RADIATION, COSMOGENIC-ISOTOPE DATING, METEORITE IMPACT, MILANKOVITCH THEORY, NEAR-EARTH OBJECTS, PHOTOSYNTHESIS, PLANETARY ENVIRONMENTAL CHANGE]

Huggett, R.J. 1997: *Environmental change: the evolving ecosphere.* London: Routledge.

cost–benefit analysis An approach to assessing the value of a project based on the premise that the potential benefits accrued should exceed the monetary costs involved in its implementation. It is commonly used in *environmental engineering*, e.g. in the evaluation of alternative FLOOD CONTROL MEASURES, but also has broader applications throughout decision making, planning and resource development. The relative merits of two or more alternative strategies or solutions may be assessed by comparison of their *cost–benefit ratios*. In the context of environmental change the main limitations are: (1) the practical difficulty in assessing accurately all the costs and all the benefits; and (2) the conceptual difficulty in measuring such concepts as ECOLOGICAL VALUE, environmental value and human life in monetary terms. *JAM*

Layard, R. and Glaister, S. 1994: *Cost–benefit analysis*, 2nd edn. Cambridge: Cambridge University Press.

counterurbanisation A process of population deconcentration away from urban areas towards regions classified as rural. The dominant trend in urbanised countries until the end of the twentieth century was for the MIGRATION of rural dwellers to urban centres (see URBANISATION) with consequent widespread *rural depopulation*. In recent decades, however, this has tended to reverse due largely to urban–rural migration. Quality-of-life considerations have been critical here. In the UK, for example, counterurbanisation began in the 1960s and net outmigration from the main metropolitan areas averages about 90 000 people per year. Counterurbanisation has been an important factor in *rural regeneration*. *JAM*

[See also RURAL DECLINE]

Boyle, P. and Halfacree, K. (eds) 1998: *Migration into rural areas: theories and issues.* Chichester: Wiley. **Champion, A.** 1999: Urbanisation and counterurbanisation. In Pacione, M. (ed.), *Applied geography: principles and practice.* London, Routledge, 347–357.

cosmogenic-isotope dating *Some cosmogenic nuclides used for exposure-age dating (after Kurtz and Brook, 1994)*

Isotope	Half life	Age range	Comments
^{41}Ca	103×10^3	to 300 ka	Difficult to measure
^{3}He	Stable	1 ka to c. 3 Ma	Diffusive loss; requires large crystals
^{21}Ne	Stable	7 ka to 10 Ma?	Requires correction for inherited neon
^{36}Cl	3.08×10^5	5 ka to 1 Ma	Composition dependent; applicable to a wide range of minerals
^{10}Be	1.5×10^6	3 ka to 4 Ma	Subject to atmospheric contamination; quartz surfaces can be acid-cleaned
^{26}Al	7.16×10^5	5 ka to 2 Ma	Used on Al-poor minerals, e.g. quartz

coupled ocean–atmosphere models (CGCMs)
A type of CLIMATIC MODEL in which ocean dynamics forms an integral part of the model. Although CGCMs are in many ways highly sophisticated, they still exhibit fundamental weaknesses, especially in (1) their treatment of clouds and water vapour and (2) coupling the rapidly responding atmosphere with the relatively slow response of the ocean. Such coupled models have nevertheless superseded earlier generations of GENERAL CIRCULATION MODELS (GCMs) and produce more accurate climatic PREDICTION and RETRODICTION. *JAM*

[See also ATMOSPHERE–OCEAN INTERACTION]

Battisti, D.S. 1995: Decade-to-century time-scale variability in the coupled atmosphere–ocean system: modeling issues. In National Research Council, *Natural climate variability on decade to century time scales*. Washington, DC: National Academy Press, 419–431. **Manabe, S. and Stouffer, R.J.** 2000: Study of abrupt climate change by a coupled ocean–atmosphere model. *Quaternary Science Reviews* 19, 285–299. **Meehl, G.A. and Branstator, G.W.** 1992: Coupled climate model simulation of El Niño/Southern Oscillation: implications for paleoclimate. In Diaz, H.F. and Markgraf, V. (eds), *El Niño*. Cambridge: Cambridge University Press, 69–91.

covariance
A measure of the simultaneous variation of two quantities (characterised by the deviation from their respective mean values). If both variables are always above and below their means at the same time, the covariance is positive. If one variable is above its mean whenever the other is below its mean (or vice versa), the covariance is negative. If the variables are not associated, then covariance is zero. *PJS*

[See also ANALYSIS OF COVARIANCE]

coverage
A layer (or digital map) in a GEOGRAPHICAL INFORMATION SYSTEM (GIS) containing sets of ENTITIES representing geographic information, such as rivers, terrain and forested areas. Coverages are used in GIS operations, including OVERLAY ANALYSIS, and the calculation of BUFFER ZONES. *TVM*

coversand
Fine sand of AEOLIAN origin and PLEISTOCENE age; it differs from LOESS in that it is (a) coarser and (b) of more local origin. Coversand occurs widely over the western and central European lowlands, where it forms flat to gently undulating terrain. The source is assumed to have been the exposed southern North Sea Basin together with GLACIOFLUVIAL outwash channels and SANDUR plains. THERMOLUMINESCENCE dating has shown that the coversands in northwest Europe date from Late Glacial times. *HMF*

[See also PERIGLACIAL SEDIMENTS]

Bateman, D. 1998: The origin and age of coversand in north Lincolnshire, UK. *Permafrost and Periglacial Processes* 9, 313–325. **Kasse, C.** 1997: Cold-climate aeolian sand-sheet formation in North West Europe (*c*. 14–12.4 ka): a response to permafrost degradation and increased aridity. *Periglacial and Permafrost Processes* 8, 295–311. **Koster, E.A.** 1988: Ancient and modern cold-climate aeolian sand deposition: a review. *Journal of Quaternary Science* 3, 69–83.

crack propagation
In the geological context, the growth of cracks or fractures in bedrock. It is important in terms of rock failure. The existence of cracks causes pronounced stress concentration around the crack tip, which increases the tendency for crack growth and rock fractures. For any given stress, the likelihood of crack growth depends on the length of the crack, with longer cracks (*critical cracks*) tending to grow rather than shorter ones (*subcritical cracks*).

In terms of, for example, GLACIAL EROSION, over-riding ice tends to induce stress in rocks such that critical cracks can grow rapidly, which causes blocks of rock to become isolated and thus EROSION can occur. Subcritical cracks may grow slowly through CHEMICAL WEATHERING until they cross the crack-growth threshold to become critical cracks, at which point they too can grow rapidly. This may be a mechanism by which the erosion of intact rock beneath glaciers can occur. *RAS*

Atkinson, B.K. (ed.) 1987: *Fracture mechanics*. London: Academic Press. **Benn, D.I. and Evans, D.J.A.** 1998: *Glaciers and glaciation*. London: Arnold.

crag and tail
In glaciated terrain, a streamlined hill formed by GLACIAL EROSION. Crag and tail usually comprises resistant bedrock at the up-ice end and more easily eroded rock or sediment forming a down-ice tail. Classic examples occur in central Scotland (e.g. Edinburgh Castle and the Royal Mile), where the term originates. Originally viewed as an entirely erosional feature, depositional forms, where the tail comprises a cavity infill of debris in the lee of a resistant obstacle, are also recognised. The term has also been used for features of similar appearance eroded by other processes, such as PYROCLASTIC FLOWS and seafloor currents. *JS*

[See also ROCHE MOUTONNÉE]

Crag deposits
The Crag deposits of East Anglia, England, are mainly shallow-marine fossiliferous sediments that built up on the western edge of the southern North Sea basin during the PLIOCENE and Early PLEISTOCENE, representing the earliest QUATERNARY sediments in the British Isles. The Crags comprise three broad groups. Fossil assemblages in the Late Pliocene *Coralline Crag* indicate water temperatures of 18–24°C. The *Red Crag* and *Norwich Crag* rest unconformably on the Coralline Crag (*cf*. Reuveran of The Netherlands) and may span the Early Pleistocene (Pre-Ludhamian to Bramertonian stages), perhaps extending back into the Late Pliocene. Their faunas indicate fluctuations between cool and temperate conditions. The overlying *Cromer Forest Bed Formation* probably spans the Pre-Pastonian to Cromerian stages (i.e. pre-Oxygen Isotope Stage 12), and includes fossil evidence for fluctuating cold and temperate stage climates. *MHD*

Bowen, D.Q. (ed.) 1999: *A revised correlation of Quaternary deposits in the British Isles* [Geological Society Special Report 23]. Bath: The Geological Society. **Funnell, B.M. and West, R.G.** 1977: Preglacial Pleistocene deposits of East Anglia. In: Shotton, F.W. (ed.), *British Quaternary studies: recent advances*. Oxford: Clarendon, 247–265. **Zalasiewicz, J.A., Hughes, M.J., Gibbard, P.L.** *et al.* 1988: Stratigraphy and palaeoenvironments of the Red Crag and Norwich Crag formations between Aldeburgh and Sizewell, Suffolk, England. *Philosophical Transactions of the Royal Society of London* B322, 221–272.

crater
A roughly circular depression in the surface of the Earth or another planet, commonly with a raised rim, caused by VOLCANISM or METEORITE IMPACT. *JBH*

craton An area of CONTINENTAL CRUST that has been stable and experienced little or no TECTONIC disturbance since PRECAMBRIAN times. Examples include the Pilbara Craton (Australia) and the Yangtze Craton (China). Cratons form the ancient cores of the CONTINENTS. A craton may comprise *platform* areas that have a thin cover of essentially undeformed SEDIMENTARY ROCKS (e.g. Siberian Platform), and *shield* or *continental shield* areas where ARCHAEAN or PROTEROZOIC rocks are exposed at the surface (e.g. Canadian Shield and Fennoscandian Shield). *GO*

Kashirtsev, V.A., Kontorovich, A.E., Philp, R.P. *et al.* 1999: Biomarkers in crude oils of the eastern Siberian Platform as indicators of paleoenvironment of source-rock deposition. *Geologiya i Geofizika* 40, 1700–1710. **Mareschal, J.C., Jaupart, C., Cheng, L.Z.** *et al.* 1999: Heat flow in the Trans-Hudson Orogen of the Canadian Shield: implications for Proterozoic continental growth. *Journal of Geophysical Research – Solid Earth* 104, 29007–29024. **Rasmussen, B. and Buick, R.** 1999: Redox state of the Archean atmosphere: evidence from detrital heavy minerals in ca. 3250–2750 Ma sandstones from the Pilbara Craton, Australia. *Geology* 27, 115–118.

creationism The view that attributes the origin of matter and biological species to special acts of creation by one or more Gods. *KDB*

[See also DARWINISM, EVOLUTION, LAMARCKISM]

Eldridge, N. 2000: *The triumph of evolution and the failure of creationism.* New York: Macmillan. **Ruse, M.** 1982: Creation science: the ultimate fraud. In Cherfas, J. (ed.), *Darwin up to date.* London: New Science Publications, 7–11.

Cretaceous A SYSTEM of rocks and a PERIOD of geological time from 144 to 65 million years ago. The first placental mammals and the first flowering plants (angiosperms) appeared. The end of the Cretaceous is marked by a major MASS EXTINCTION event (see K–T BOUNDARY). *GO*

[See also GEOLOGICAL TIMESCALE]

critical level The threshold concentration of a gaseous POLLUTANT above which, according to current knowledge, significant harmful effects are likely on specific sensitive elements of the environment, such as vegetation. *JAM*

[See also CRITICAL LOAD]

critical load A quantitative estimate of an exposure to one or more POLLUTANTS below which significant long-term (50 years) harmful effects on specified sensitive elements of ecosystems do not occur. The concept is widely accepted as a basis for ENVIRONMENTAL POLLUTION control. It is ultimately dependent on scientific findings (both field and laboratory based) regarding dose–response relationships that define the TOLERANCE of ecosystems to pollutants. Operationally, it is difficult to define, being subject to technical, political, economic and social factors. Pollutants interact in complex, synergistic and anergistic ways (see SYNERGISM) in the environment, the effects of which are difficult to identify. Reducing critical loads to BACKGROUND LEVELS may be expensive, so THRESHOLDS are often pitched at levels where some damage will occur. For instance, critical loads of nitrogen compounds can be detrimental to water quality, yet they allow increased fish-ery yields. For many ecosystems there is no universally acceptable load level. *GOH*

Grennfelt, P. and Thornelof, E. (eds) 1992: *Critical loads for nitrogen: a workshop report.* Copenhagen: Nordic Council of Ministers, NORD. **Horning, M. and Skeffington, R. A.** 1993: *Critical loads: concepts and applications* [*ITE Symposium* 28]. London: HMSO. **Reynolds, B. and Skeffington, R. A.** 1999: Developments in critical loads. *Progress in Environmental Science* 1, 371–381.

critical rationalism Karl Popper's philosophy of science, developed originally to counter LOGICAL POSITIVISM. Most scientists emphasise Popper's principle of FALSIFICATION, in contrast to the principle of VERIFICATION promoted by the logical positivists. Popper emphasised the asymmetry between falsification and verification – a single instance of falsification is sufficient to falsify a hypothesis, while no number of verifications is truly definitive. He claimed that science progresses by scientists subjecting their HYPOTHESES to the most rigorous test available and engaging in an iterative procedure built upon error elimination (i.e. falsification of hypotheses followed by appropriate adjustment, restatement and fresh testing). While such an approach has much logical appeal, it does not reflect the way most scientists actually conduct research. Today, philosophers of science have generally moved away from the business of telling scientists how to conduct research (a *normative approach*) and examine the way scientists actually conduct research – so-called *naturalised approaches.* *CET*

[See also SCIENTIFIC METHOD, THEORY]

Popper, K. R. 1972: *The logic of scientific discovery*, 6th revised impression. London: Hutchinson. **Bird, J.H.** 1975: Methodological implications for geography from the philosophy of K.R. Popper. *Scottish Geographical Magazine* 91, 153–163. **Haines-Young, R.H. and Petch, J.R.** 1980: The challenge of critical rationalism for methodology in physical geography. *Progress in Physical Geography* 4, 43–78.

Cromerian Interglacial In Britain, an INTERGLACIAL of the Early Middle QUATERNARY preceding the *Anglian Glaciation*, now defined as an interglacial complex covering at least four warm stages, lying between oxygen ISOTOPIC STAGES 13 and 21 (500–800 ka BP). Most Cromerian deposits in Britain are found in Norfolk and Suffolk, resting unconformably on the underlying deposits reflecting an underlying HIATUS of up to one million years in the sequence. The later parts of the complex are equivalent to the *Elsterian* interglacial in continental Europe. *DH/CJC*

[See also QUATERNARY TIMESCALE]

Gibbard, P.L., West, R.G., Zagwijn, W.H. *et al.* 1999: Early and Middle Pleistocene correlations in the Southern North Sea Basin. *Quaternary Science Reviews* 10, 23–52.

crop rotation The cultivation of the same field with different crops in a regular sequence over a period of years. It allows the periodic use of certain crops, which, if cultivated continuously, would not sustain a high yield or might cause SOIL DEGRADATION through, for example, nutrient depletion or SOIL EROSION. Thus, legumes are commonly rotated with crops with a high nitrogen demand. *SHD*

[See also SUSTAINABILITY]

Foth, H.D. 1990: *Fundamentals of soil science.* Chichester: Wiley.

cropmarks Patterns in the landscape detectable by archaeologists and others because of the differential growth of crops. They are most commonly caused by spatial variation in soil moisture and have been instrumental in revealing that surviving archaeological features are often least common where past human activity and modification of the landscape have been most intense. *JAM*

[See also SOILMARKS]

Edis, J., MacLeod, D. and Bewley, R. 1989: An archaeologist's guide to classification of cropmarks and soilmarks. *Antiquity* 63: 112–126.

cross-bedding A SEDIMENTARY STRUCTURE formed by the downstream migration of DUNES in air or water and characterised by CROSS-STRATIFICATION with sets thicker than 10 cm. There is little overlap in size between dunes and current RIPPLES, and hence between cross-bedding and CROSS-LAMINATION. Cross-bedding is common in fluvial, aeolian and shallow marine SEDIMENTARY DEPOSITS. It is a major tool in the interpretation of PALAEOCURRENT directions, and the analysis of REACTIVATION SURFACES and BOUNDING SURFACES allows the reconstruction of compound BEDFORMS including *fluvial bars*, tidal SANDWAVES, aeolian dunes and desert DRAAS. It is sometimes described in older literature as *dune bedding, false bedding* or, in the case of trough cross-bedding, as *festoon bedding*. *GO*

Allen, J.R.L. 1980: Sand waves: a model of origin and internal structure. *Sedimentary Geology* 26, 281–328. **Chakraborty, T.** 1999: Reconstruction of fluvial bars from the Proterozoic Mancheral Quartzite, Pranhita-Godavari Valley, India. In Smith, N.D. and Rogers, J. (eds), *Fluvial sedimentology VI* [*International Association of Sedimentologists Special Publication* 28]. Oxford: Blackwell Science, 451–466. **Clemmensen, L.B.** 1987: Complex star dunes and associated aeolian bedforms, Hopeman Sandstone (Permo-Triassic), Moray Firth Basin, Scotland. In Frostick, L. and Reid, I. (eds), *Desert sediments: ancient and modern*, [*Special Publication* 35]. London: Geological Society, 213–231.

cross-calibration A technique used in REGRESSION ANALYSIS, where the aim is to establish the likely reliability of a predictive model that is applied outside the fitting (CALIBRATION) period and where there is a possibility of overfitting the calibration equation(s) by the use of too many predictors (e.g. in DENDROCLIMATOLOGY). Cross-calibration involves division of the available predictand (observation) series into two parts, fitting the regression equation over one of these and using it to estimate the predictand data over the other. The predictand estimates can then be compared with the withheld data (validated or 'verified') using a range of statistical tests. This process is then repeated with the periods reversed so that the model is now calibrated over the previously withheld data period and validated over the previous calibration period. This exercise generally provides a more reliable indication of the realism of a reconstruction than is suggested by the calibration-period statistics and also provides some indication of the stability or time dependence of the regression model (i.e. the extent to which the relative magnitude of the regression coefficients change through time). *KRB*

Briffa, K.R. and Schweingruber, F.H. 1992: Recent dendroclimatic evidence of northern and central European summer temperatures. In Bradley, R.S. and Jones, P.D. (eds), *Climate since A.D. 1500*. London: Routledge, 366–392.

cross-dating One of the fundamental principles that underlies the field of DENDROCHRONOLOGY. It is the process whereby the patterns of INTERANNUAL VARIABILITY in ring-width or some other growth parameter (e.g. ring density) in different trees are uniquely synchronised, sometimes in association with other distinguishing wood anatomical features, to identify the precise year of formation of each ring in each tree. After synchronisation, the individual tree records can then be amalgamated to form an average *master series* or MASTER CHRONOLOGY. This will be absolutely dated when it extends through to the present day or when it can also be cross-dated against a previously dated chronology. Where a chronology is not firmly anchored in time, it may nevertheless represent relative tree-growth variability for some unknown period. Such a chronology is said to be a FLOATING CHRONOLOGY. In practice, the ease with which cross-dating is achieved is dependent on the strength (and interannual variability) of the underlying growth influences and the length of the growth series being compared. For example, if the main factor limiting tree growth varies strongly and erratically from year to year, it will leave a much more distinct pattern of wide and narrow rings than will be seen in an area with little interannual variability. Also, if the growth records from different trees are long, say several centuries, the probability of achieving a strong correlation by chance will be very low. *KRB*

Douglass, A.E. 1941: Crossdating in dendrochronology. *Journal of Forestry* 39, 825–831. **Fritts, H.C.** 1976: *Tree rings and climate*. London: Academic Press. **Stokes, M.A. and Smiley, T.L.** 1968: *An introduction to tree-ring dating*. Chicago, IL: University of Chicago Press.

cross-lamination A SEDIMENTARY STRUCTURE formed by the migration of RIPPLES and characterised by CROSS-STRATIFICATION with sets thinner than 5 cm. Cross-lamination formed from the migration of CURRENT RIPPLES in unidirectional water flows has relatively simple geometry: it is a common and important indicator of PALAEOCURRENT strength and direction. There is little overlap in size between current ripples and DUNES and hence between cross-lamination and CROSS-BEDDING. Cross-lamination formed by WAVE RIPPLES has more complex geometry. If significant sedimentation accompanies ripple migration, the sediment surface will rise as cross-lamination extends downstream, forming CLIMBING RIPPLE CROSS-LAMINATION. *GO*

Allen, J.R.L. 1968: *Current ripples: their relation to patterns of water and sediment motion*. Amsterdam: North-Holland. **de Raaf, J.F.M., Boersma, J.R. and van Gelder, A.** 1977: Wave-generated structures and sequences from a shallow marine succession, Lower Carboniferous, County Cork, Ireland. *Sedimentology* 24, 451–483.

cross-stratification Layering (STRATIFICATION) in SEDIMENTS and SEDIMENTARY ROCKS which is at an angle (up to 35°) to the depositional horizontal, as represented by the BEDDING. Cross-stratification is a common SEDIMENTARY STRUCTURE that forms from the deposition of cohesionless granular sediment (coarse silt, sand or gravel) in water or air. Deposition generally occurs on the downstream surface (lee side or *slip face*) of a RIPPLE or DUNE, giving rise respectively to CROSS-LAMINATION and CROSS-BEDDING. Flow separation in the lee of the BEDFORM

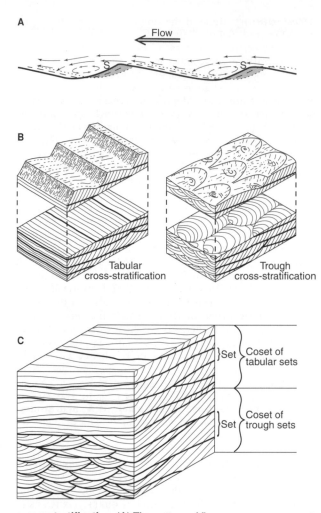

A

Flow

B

Tabular
cross-stratification

Trough
cross-stratification

C

}Set { Coset of
tabular sets

}Set { Coset of
trough sets

cross-stratification *(A) The pattern of flow across an asymmetrical bedform (ripple or dune) showing flow separation (S). Dashed lines represent two previous positions of the sediment surface, showing the formation of cross-stratification (shaded). (B) Tabular and trough cross-stratification resulting from the migration of straight-crested and curved-crested bedforms respectively. (C) Terminology for stacked cross-stratification (based on Allen, 1982)*

causes BEDLOAD particles to accumulate near its crest, steepening the lee slope. This fails intermittently as a cascade of grains moving down the lee slope as a *grain flow* or AVALANCHE, forming a distinct cross-stratum or *foreset* (see Figure). A unit of cross-stratification formed by the migration of one bedform is a *set*; stacked sets of similar character form a *coset*. Laterally extensive BOUNDING SURFACES separate cosets. Bedforms with straight crests in plan view produce *tabular* (or *tabular planar*) *cross-stratification*; those with curved crests (*linguoid* or *lunate* planforms) migrate to produce *trough cross-stratification*. Trough cross-bedding is sometimes described as *festoon bedding* in older literature. Cross-stratification is an important indicator of PALAEOCURRENT type (see also HERRINGBONE CROSS-STRATIFICATION), strength (see BEDFORM STABILITY DIAGRAM) and direction, and can be used to reconstruct bedform morphology.   *GO*

Allen, J.R.L. 1982: *Sedimentary structures: their character and physical basis.* Amsterdam: Elsevier. **Anastas, A.S., Dalrymple, R.W., James, N.P. and Nelson, C.S.** 1997: Cross-stratified calcarenites from New Zealand: subaqueous dunes in a cool-water, Oligo-Miocene seaway. *Sedimentology* 44, 869–891. **Collinson, J.D. and Thompson, D.B.** 1992: *Sedimentary structures*, 2nd edn. London: Chapman and Hall. **McKee, E.D. and Weir, G.W.** 1953: Terminology for stratification and cross-stratification in sedimentary rocks. *Geological Society of America Bulletin* 64, 381–390. **Potter, P.E. and Pettijohn, F.J.** 1977: *Paleocurrents and basin analysis*, 2nd edn. Berlin: Springer.

crude oil   The liquid form of PETROLEUM, exploited as a FOSSIL FUEL by drilling into buried rocks, usually SEDIMENTARY ROCKS, that hold the oil in pore spaces or other cavities, often beneath accumulations of NATURAL GAS. Crude oil must be refined to yield fuels and other materials of use to society. Oil is the world's main ENERGY RESOURCE. It accounted for about 40% of world energy consumption in 1999 and RESERVES are estimated to last until late in the twentyfirst century (see also SYNFUELS). The countries of the Middle East together hold about 65% of world reserves: Saudi Arabia has 25% and Iraq 11%, and they produce 12% and 4% respectively of the world's oil. In contrast, Europe and the USA each hold between 2% and 3% of reserves (UK 0.5%) and produce between 9% and 11% (UK 4%) of the world's oil. This imbalance in the distribution of reserves and consumption of the world's major fuel is the source of much international trade and geopolitical friction.   *GO*

[See also OIL SPILL]

BP Amoco 2000: *BP Amoco stastical review of world energy, June 2000.* London: BP Amoco.

crust   The outermost layer of the Earth, distinct in composition and density from the underlying MANTLE from which it is separated by the *Mohorovičić discontinuity* or *Moho*. CONTINENTAL CRUST is quite different from OCEANIC CRUST (see Figure). The terms *sial* and *sima*, referring respectively to the upper continental crust and the oceanic crust, are no longer used, and it is now clear that a layer with the characteristics of oceanic crust does not extend beneath the continents.   *GO*

[See also EARTH STRUCTURE]

Walton, E.K. 1993: A view of the Earth. In Duff, P.McL.D. (ed.), *Holmes' principles of physical geology*, 4th edn. London: Chapman and Hall, 9–21.

crustal subsidence   The depression of part of the Earth's CRUST, leading to its flooding by the sea, or to the accumulation of SEDIMENTARY DEPOSITS in a SEDIMENTARY BASIN. Crustal subsidence may be caused by stretching and thinning (*tectonic subsidence*), by heating and cooling (*thermal subsidence*), or by loading (ISOSTASY). Typical subsidence rates in sedimentary basins range from 0.1 to > 2.5 mm per year.   *GO*

[See also AIR SUBSIDENCE, LAND SUBSIDENCE]

Allen, P.A. and Allen, J.R. 1990: *Basin analysis: principles and applications.* Oxford: Blackwell. **Chen, Y.-G. and Liu, T.-K.** 2000: Holocene uplift and subsidence along an active tectonic margin southwestern Taiwan. *Quaternary Science Reviews* 19, 923–930. **McKenzie, D.P.** 1978: Some remarks on the development of sedimentary basins. *Earth and Planetary Science Letters* 40, 25–32.

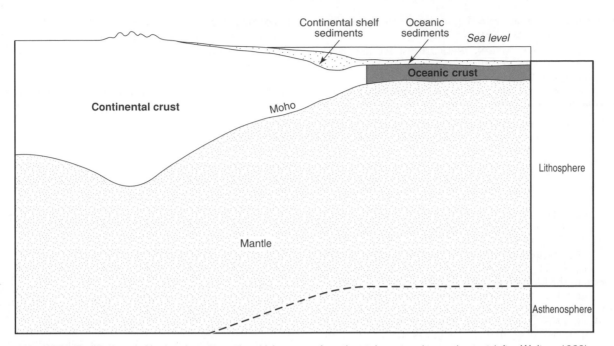

crust *The Earth's crust, showing the contrasting thicknesses of continental crust and oceanic crust (after Walton, 1993)*

crusting of soil The development of a thin crust on the soil surface, ranging in thickness from a few millimetres to centimetres; the crust is more compact, hard and brittle than the material immediately beneath it. Crusting can be caused by the mobilisation and deposition of dispersed fine particles (SILT size or smaller) by, for example, *rainsplash*. The development of a crust takes place when the soil dries out. The term *sealing* refers to the reorganisation of the surface soil layer prior to when the soil dries out.

Crusting can occur on most soils, except sands or soils with a high organic-matter content, and can lead to reduced INFILTRATION CAPACITY and increased *overland flow*. In ARIDLANDS and SEMI-ARID areas and elsewhere in the early stages of PRIMARY SUCCESSION, assemblages of algae, mosses and fungi can create a *microphytic* or *cryptogamic soil crust* that protects the soil from rainsplash, helps to promote water-stable aggregates and generally stabilises the ground surface. *SHD*

[See also AGGREGATE STABILITY, SOIL EROSION]

Eldridge, D.J. and Greene, R.S.B. 1994: Assessment of sediment yield and splash erosion on a semi-arid soil with varying cryptogam cover. *Journal of Arid Environments* 26, 221–232. **Poesen, J.W.A. and Nearing, M.A. (eds)** 1993: *Soil surface sealing and crusting* [Catena 24(Supplement)]. Cremlingen: Catena. **Worley, I.A.** 1973: The 'black crust' phenomenon in upper Glacier Bay, Alaska. *Northwest Science* 47, 20–29.

cryopedology The study of the effects of frost action upon soils and soil-forming processes. *HMF*

[See also CRYOSOLS, FREEZE–THAW CYCLES, PATTERNED GROUND]

Cailleux, A. and Taylor, G. 1954: *Cryopédologie, études des sols gelés: Expéditions Polaires Françaises, Missions Paul-Emile Victor* IV [*Actualités Scientifiques et Industrielles* 1203]. Paris: Hermann.

cryoplanation The flattening and lowering of a landscape by processes related to frost action and/or the presence of PERMAFROST. *HMF*

[See also PERIGLACIAL LANDSCAPE EVOLUTION, PERIGLACIAL LANDSCAPES]

Boch, S.G. and Krasnov, I.I. 1943: O nagornykh terraskh i drevnikh poverkhnostyakh vyravnivaniya na Urale i svyazannykh s nimi problemakh. *Vsesoyuznogo Geograficheskogo obshchestva, Izvestiya* 75, 14–25 (English translation by A. Gladunova, 1994: On altiplanation terraces and ancient surfaces of levelling in the Urals and associate problems. In Evans, D.J.A. (ed.), *Cold climate landforms*. Chichester: Wiley, 177–186). **Priesnitz, K.** 1988: Cryoplanation. In Clarke, M.J. (ed.), *Advances in periglacial geomorphology*. Chichester: Wiley, 49–67.

cryoplanation terraces Gently sloping bedrock surfaces, also referred to as *altiplanation terraces*. They range in size from several tens to hundreds of metres in width, and either bevel summits or form steps or benches cut into upper hillslopes. They occur within PERIGLACIAL landscapes but are problematic in origin. Some are thought to form through a combination of bedrock disintegration by FREEZE–THAW CYCLES and scarp recession. This process is generally referred to as CRYOPLANATION. However, many cryoplanation terraces occur in never-glaciated terrain and these may reflect nothing more than differential erosion of contrasting lithologies. *RAS/HMF*

[See also CRYOPLANATION]

Demek, J. 1969: Cryogenic processes and the development of cryoplanation terraces. *Biuletyn Peryglacjalny* 18, 115–125.

Cryosols Soils with PERMAFROST within 1 m of the soil surface, characterised by PATTERNED GROUND with ice wedges and convoluted (cryoturbated) subsoil materials. GLOBAL WARMING is potentially a major problem in Arctic and subarctic areas, releasing GREENHOUSE GASES from these soils (SOIL TAXONOMY: Gelisols). *EMB*

[See also CRYOTURBATION, INVOLUTIONS, WORLD REFERENCE BASE FOR SOIL RESOURCES]

Campbell, I.B. and Claridge, G.G.C. 1987: *Antarctica: soils, weathering processes and environment.* Amsterdam: Elsevier. **Permafrost and Periglacial Processes** 1999: Special Issue on Cryosols and Cryogenic Environments. *Permafrost and Periglacial Processes* 10(3), 209–307. **Rieger, S.** 1983: *The genesis and classification of cold soils.* New York: Academic Press. **Tarnocai, C., and Smith, C.A.S.** 1993: The formation and properties of soils in the permafrost regions of Canada. *Proceedings of the 1st International Conference on Cryopedology.* Puschino: Russian Academy of Sciences, 21–24.

cryosphere   That part of the Earth's surface covered by snow and ice limited to the polar areas and high mountains. It covers 14.9 million km^2 (2.9% of the Earth's surface). Most (98.4%) of the cryosphere is in the polar regions: 84.5% (12.6 million km^2) in the south polar region in the form of the Antarctic ICE SHEET (12.5 million km^2) and 13.9% (2.1 million km^2) in the Arctic, mostly comprising the Greenland ice sheet (1.7 million km^2). With an average thickness of 2.8 km in Antarctica and a central thickness of about 3 km in Greenland, these two ICE CAPS have volumes of 26 million km^3 and 4 million km^3 respectively. All the other GLACIERS in the world contain only 1.6% (0.2 million km^2). The SEA ICE, in the form of PACK ICE and ICE SHELF, covering the Arctic Ocean and surrounding the Antarctic continent is also part of the cryosphere, as is the winter land snow cover in both hemispheres. Normal usage (and the above values) excludes the additional areas of PERMAFROST and SEASONALLY FROZEN GROUND. *BDG*

Sugden, D.E. 1987: The polar and glacial world. In Clark, M.J., Gregory, K J. and Gurnell, A.M. (eds), *Horizons in physical geography.* Basingstoke: Macmillan, 214–231. **Ye, H. and Mather, J.R.** 1997: Polar snow cover changes and global warming. *International Journal of Climatology* 17, 155–162.

cryostatic pressure   The pressure thought to be generated during freeze-back in a confined, wet unfrozen pocket in the ACTIVE LAYER. The existence of cryostatic pressure in the field has not been convincingly demonstrated because the presence of voids in the soil, the occurrence of frost cracks in winter and the weakness of the confining soil layers lying above prevent high pressures from developing. On theoretical grounds, ice lensing at the top and bottom of the active layer will desiccate the unfrozen pocket so that pore water is under tension, not pressure. *HMF*

[See also PERMAFROST, FREEZE–THAW CYCLES]

Mackay, J.R. and MacKay, D.K. 1976: Cryostatic pressures in non-sorted circles (mud hummocks), Inuvik, Northwest Territories. *Canadian Journal of Earth Science* 13, 889–897.

cryoturbation   The process of cryogenic deformation of near-surface sediments in PERIGLACIAL areas as a result of the winter freezing and release of excess water during summer thaw consolidation; also the forms produced by such deformation. *RAS*

[See also INVOLUTION, PERMAFROST]

Vandenberghe, J. 1988: Cryoturbations. In Clark, M.J. (ed.), *Advances in periglacial geomorphology.* Chichester: Wiley, 179–198.

cryptotephra   TEPHRA layers that are invisible, in sediment core or section, to the naked eye. *JBH*

Hunt, J.B. and Lowe, D.L. 2001: Tephra nomenclatura. *Journal of Archaeological Sciences* (in press).

crystalline   A substance with a regular arrangement of atoms or molecules (crystal lattice) that may form crystals. *MRT*

[See also AMORPHOUS]

cuesta   An asymmetrical ridge with a steep scarp (*escarpment*) and more gently sloping *dip slope*. The latter is sometimes termed the *backslope*, which is more correct where the topographic surface does not exactly parallel the DIP of the strata. The word is of Spanish origin *AJDF*

Schmidt, K.H. 1989: Talus and pediment flatirons: erosional and depositional features on dryland cuesta scarps. *Catena Supplement* 14, 107–118.

cultivars   Varieties of cultivated plants produced by selective breeding (cf. *breed* in animals). Also known as *cultigens*, they are distinguished by heritable, uniform and different characteristics, as exhibited by broccoli, brussel sprouts and cauliflower within the cabbage family. The distinction has been made, however, between a 'cultigen' as a wild plant species that is tolerated and encouraged for use and a *domesticate*, which is dependent on human intervention for survival (Cowan and Watson, 1992:4). *GOH*

Cowan, C.W. and Watson, P.J. 1992: *The origins of agriculture: an international perspective.* Washington, DC: Smithsonian Institution. **Harlan, J.** 1995: *The living fields.* Cambridge: Cambridge University Press.

cultural change   Cultures can be 'fixed' in terms of their main components of language, religion and social organisation, but are neither residual nor static. The idea of cultural change acknowledges the fact that cultures have both a diversity and a dynamism that defies rigid definition. One view is that successive generations redefine CULTURE in ways that fit their ongoing lifestyles. Culture is part of people's everyday lives and its meanings are continuously being reinterpreted and socially constructed. From a position where cultures could be identified by one or two main indicators, usually language, the range has widened to embrace subcultures, distinguished by specific and perhaps temporal, sets of values, and notions such as popular culture, high culture and low culture. All of these diversities make the idea of cultural change, of shifting and transitory cultures, more real and recognisable.

An older concept of *culture history* has been used in archaeological research to examine culture change over long time periods in the past. Originally cultural change was conceived in terms of the evolution from primitive to advanced forms of culture and technology. It was later used in reference to the emergence of different cultural groupings. *Cultural geography* recognises cultural differences in place and space. LANDSCAPE as a PALIMPSEST usefully captures the idea of the overlaying of successive cultural imprints upon a segment of the Earth's surface. That concept, along with more traditional ideas on culture hearths, culture areas, culture regions and CULTURAL LANDSCAPE, sits easily with the broad idea of cultural change. These 'places' are anything but static. *DTH*

Crang, M. 1998: *Cultural geography*. London: Routledge. Flannery, K.V. 1972: The cultural evolution of civilizations. *Annual Review of Ecology and Systematics* 3, 399–426. Leighley, J. 1962: *Land and life: a selection of the writings of Carl Sauer*. Berkeley, CA: University of California Press. Rockwell, R.C. 1994: Culture and cultural change. In Meyer, W.B. and Turner II, B.L. (eds), *Changes in land use and land cover: a global perspective*. Cambridge: Cambridge University Press, 357–382. Storey, J. 1993: *Cultural theory and popular culture*. London: Harvester Wheatsheaf.

cultural landscape

In the broadest sense, those areas of LANDSCAPE that have been influenced by human activity. The term became common after World War II with increasing recognition of past and present HUMAN IMPACT on ecological and environmental systems. Contemporary cultural landscapes of Britain and northwest Europe have been studied intensively via vegetation survey, *field mapping*, AERIAL PHOTOGRAPHY and DOCUMENTARY EVIDENCE. Prehistoric and older historical cultural landscapes, which cannot be observed directly, have been investigated by the techniques of ENVIRONMENTAL ARCHAEOLOGY, such as CHARRED-PARTICLE ANALYSIS and POLLEN ANALYSIS. Significant cultural events identified in this way include evidence of initial MESOLITHIC activity, the onset of NEOLITHIC cultivation, the *hazel declines* due to COPPICING and the ELM DECLINE *circa* 5000 BP. Causal relationships between past LANDSCAPE change and cultural development nonetheless often remain difficult to establish because of the retrospective nature of the data and spatial and temporal diversity in both nature and human activity.

Human influence on the landscape was initially subtle and confined spatially as populations and technological capacities were limited. Landscape modification increased with the introduction of fire as a management tool, followed by technological and agricultural development, INDUSTRIALISATION and URBANISATION. The INDUSTRIAL REVOLUTION and associated increases in population and resource utilisation produced more substantial landscape change whilst political regimes and conflicts have also had considerable impacts. Since the eighteenth century, cultural rather than natural factors have probably been the major cause of LANDSCAPE change throughout Britain and northwest Europe. Around 99% of the almost continuous WILDWOOD that covered Britain during the early HOLOCENE has been destroyed and the British landscape appears to have been predominantly open by AD 1700, mainly as a result of FOREST CLEARANCE for timber and agricultural land, whilst lowland wetlands were also drained and cultivated. More recently, twentieth-century AGRICULTURAL INTENSIFICATION, urban–industrial growth and AFFORESTATION have produced particularly distinctive landscapes with many EXOTIC SPECIES (wild or cultivated), in which ECOLOGICAL TRANSITIONS are often sudden. Although small areas that are often perceived as 'NATURAL', such as the New Forest, may have escaped more recent impacts of cultural development, these too have in fact been altered by MANAGEMENT or non-intensive land-use. Many other distinctive SEMI-NATURAL VEGETATION types created by people over previous centuries have meanwhile been continually modified or destroyed. *DZR*

Berglund, B.E. 1986: The cultural landscape in a long-term perspective. Methods and theories behind the research on land-use and landscape dynamics. *Striae* 24, 79–87. Berglund, B.E.

(ed.) 1991: *The cultural landscape during 6000 years in southern Sweden: the Ystad Project* [*Ecological Bulletins* 41]. Copenhagen: Munksgaard. Birks, H.H., Birks, H.J.B., Kaland, P.E. and Moe, D., eds, 1988: *The cultural landscape- past, present and future*. Cambridge: Cambridge University Press. Chambers, F.M. (ed.) 1993: *Climate change and human impact on the landscape*. London: Chapman and Hall. Dimbleby, G.W. 1984: Anthropogenic changes from Neolithic through Medieval times. *New Phytologist* 98, 57–72. Fowler, P.J. 1983: *The farming of prehistoric Britain*. Cambridge: Cambridge University Press. Ratcliffe, D.A. 1984: Post-Medieval and recent changes in British vegetation: the culmination of human influence. *New Phytologist* 98, 73–100. Rowntree, L.B. 1996: The cultural landscape concept in American human geography. In Earle, C., Mathewson, K. and Kenzer, M.S., *Concepts in human geography*. Lanham, MD: Rowman and Littlefield, 127–160.

cultural resource management (CRM)

The CONSERVATION of archaeological HERITAGE by *site protection* and *salvage archaeology* (*rescue archaeology* immediately prior to construction projects). The legislative basis is particularly well developed in the USA. *JAM*

Cleere, H. (ed.) 1984: *Approaches to the archaeological heritage: a comparative study of world cultural resource management systems*. Cambridge: Cambridge University Press. Gummerman, G. and Schiffer, M.B. (eds) 1977: *Conservation archaeology*. New York: Academic Press.

culture

Most simply summarised as 'way of life', the word culture is difficult to define. The traditional components of culture include language, religion, family, social organisation and technology. More recently, the concept of social construction has entered the defining qualities of culture and Crang (1998) suggests that culture represents 'sets of beliefs or values that give meanings to ways of life and produce (or are reproduced through) material and symbolic forms'. *DTH*

[See also CULTURAL CHANGE, CULTURAL LANDSCAPE]

Crang, M. 1998: *Cultural geography*. London: Routledge.

current ripple

A small, asymmetrical, flow-transverse periodic BEDFORM in coarse silt or sand, with height less than 5 cm and spacing less than 50 cm, characterised by a gently sloping stoss side facing upstream and a more steeply sloping lee side facing downstream, sometimes described as an *asymmetrical ripple*. Downstream migration of the lee slope forms CROSS-LAMINATION. The crest lines of current ripples in plan change from straight to sinuous to linguoid (convex down-current) with increasing current strength. Current ripples are formed by near-bed turbulence in unidirectional water flows. They are important indicators of PALAEOCURRENT strength and direction. *GO*

[See also BEDFORM STABILITY DIAGRAM, DUNE, WAVE RIPPLE, WIND RIPPLE]

Allen, J.R.L. 1968: *Current ripples: their relation to patterns of water and sediment motion*. Amsterdam: North-Holland. Reid, I. and Frostick, L.E. 1994: Fluvial sediment transport and deposition. In Pye, K. (ed.), *Sediment transport and depositional processes*. Oxford: Blackwell Scientific, 89–155.

curvature

The shape, inclination and orientation of the Earth's surface. When generating maps, the Earth's curvature is an important consideration in MAP PROJECTIONS, scale and CARTESIAN COORDINATE systems. *TVM*

[See also CARTOGRAPHY]
Robinson, A.H., Morrison, J.L., Muehrcke, P.C. *et al.* 1995: *Elements of cartography*. New York: Wiley, 41–111.

cut-and-fill A close association between EROSION and DEPOSITION, especially in the context of river FLOOD-PLAINS. A distinction is sometimes made between cut-and-fill in the strict sense, in which erosion and deposition closely related in space allow a level surface to develop (for example, erosion on the outer bank of a river MEANDER, accompanied by deposition on the inner *point bar*) and *scour-and-fill* in which erosion and deposition alternate in time (for example, FLOOD erosion followed by AGGRADA-TION). *GO*

cuticle The thin waxy covering of the aerial parts of terrestrial plants that serves to protect the plant from water loss and the development of which was an essential prerequisite to life on land. *JCMc*

[See also EVAPOTRANSPIRATION, GRASS CUTICLE ANALYSIS]

cyanobacteria *Prokaryotic* organisms (i.e. with no distinct nucleus in the cell) formerly known as *blue-green algae*. Cyanobacteria are the earliest well-known form of life on Earth, with a FOSSIL record to at least 3500 million years ago (see ARCHAEAN) as filaments in CHERT and as STROMATOLITES. PHOTOSYNTHESIS by cyanobacteria is likely to have been the source of oxygen in the early ATMOSPHERE. *GO*

cyclic regeneration Cycles of vegetation change where two or more plant species (plant communities) repeatedly return to the original species composition. It has been recognised, for example, in TROPICAL and TEM-PERATE FORESTS, HEATHLAND and MIRES. Mechanisms that drive this process may include biological processes such as synchronised senescence of cohorts (similarly aged individuals of one species), auto-toxicity and fluctuations in soil microbial communities favouring a certain species or plant community. Alternatively, physical processes, such as shifting resource levels (light, nutrients or water) or differential responses of species to EROSION, WINDFALL or FIRE, may promote cyclic regeneration. Cyclic regeneration can be synonymous with direct replacement of adults by juveniles of the same species following AUTOGENIC or ALLOGENIC CHANGE, but can also involve several distinct plant communities. In-filling of gaps by canopy species or alternative species can also result in cyclic regeneration. Although cyclic regeneration may be driven by the same processes as ECOLOGICAL SUCCESSION, it represents a situation where there is no stable endpoint. *LRW*

[See also CLIMAX VEGETATION, CONVERGENCE AND DIVERGENCE IN SUCCESSION]

Burrows, C.J. 1990: *Processes of vegetation change*. London: Unwin Hyman. Gimingham, C.H. 1988: A reappraisal of cyclical processes in *Calluna* heath. *Vegetatio* 77, 61–64. Sprugel, D.G. 1976: Dynamic structure of wave-regenerated *Abies balsamea* forests in the north-eastern United States. *Journal of Ecology* 64, 889–911. Watt, A.S. 1947: Pattern and process in the plant community. *Journal of Ecology* 35, 1–22.

cyclicity A tendency towards regular PERIODICITY in any TIME SERIES. In climatic context cyclic phenomena include the long cycles of MILANKOVITCH THEORY as well as shorter cycles such as SUNSPOT CYCLES. As environmental phenomena are normally too irregular to be described as truly cyclic, the term periodicity is generally preferable. *AHP*

[See also CLIMATIC FLUCTUATION]

cyclogenesis The 'birth' and/or development of a DEPRESSION (English) or CYCLONE (American) or '*low*'. It refers to the anticlockwise (clockwise) circulation around an area of mid-latitude atmospheric low pressure in the Northern (Southern) Hemisphere. 'Development' means that the pressure at the centre of the depression/cyclone is falling or decreasing. These pressure changes and patterns produce travelling wind systems, which commonly originate as waves along the POLAR FRONT. Sometimes cyclogenesis takes place in a *vorticity centre*, which develops on the poleward side of a JET STREAM, *vorticity* being a measure of the rotation in a fluid. *Lee cyclogenesis* occurs downwind of mountain barriers, such as the Alps, Himalayas and Rocky Mountains. The term *cyclolysis* is used to denote the weakening or disappearance of a depression, i.e. the central pressure is increasing. Analogous terms are used for the formation/intensification of anticyclones or '*highs*' (*anticyclogenesis*) and their decline (*anticyclolysis*). *BDG*

McIlveen, R. 1992: *Fundamentals of weather and climate*. London: Chapman and Hall. Sturman, A.P. and Tapper, N.J. 1996: *The weather and climate of Australia and New Zealand*. Melbourne: Oxford University Press.

cyclone An organised circulation of air around an area of low pressure within which the winds blow counter-clockwise in the Northern Hemisphere and clockwise in the Southern Hemisphere, usually with a diameter of 2000–3000 km. TROPICAL CYCLONES (hurricanes) should be distinguished from *extratropical cyclones* (DEPRESSIONS). *JGT*

[See also CYCLOGENESIS]

Jones, D.A. and Simmonds, I. 1993: A climatology of Southern Hemisphere extratropical cyclones. *Climate Dynamics* 9, 131–145. Newton, C.W. and Holopainen, E.O. (eds) 1990: *Extratropical cyclones*. Boston, MA: American Meteorological Society. Palmén, E. and Newton, C.W. 1969: *Atmospheric circulation systems*. Orlando, FL: Academic Press.

cyclostratigraphy The use of the record of major global cycles in SEDIMENTS and SEDIMENTARY ROCKS to provide CORRELATION of stratigraphical units and successions. Cyclical processes that may be global in extent include those driven by astronomical influences (see MILANKOVITCH THEORY), and SEA-LEVEL CHANGES (cycles of TRANSGRESSION and REGRESSION) resulting from PLATE TECTONICS effects or GLACIO-EUSTASY. Cyclostratigraphy has the potential to resolve GEOLOGICAL TIME to tens or hundreds of thousands of years for parts of the GEOLOGI-CAL TIMESCALE. *GO*

[See also SEDIMENTOLOGICAL EVIDENCE OF ENVIRONMEN-TAL CHANGE, SEQUENCE STRATIGRAPHY, STRATIGRAPHY]

Fisher, A.G. and Bottjer, D.J. (eds) 1991: Orbital forcing and sedimentary sequences. *Journal of Sedimentary Petrology* 61, 1063–1252. Gale, A.S. 1998: Cyclostratigraphy. In Doyle, P. and Bennett, M.R. (eds), *Unlocking the stratigraphical record: advances in modern stratigraphy*. Chichester: Wiley, 195–220. House, M.R. and Gale, A.S. (eds) 1995: *Orbital forcing timescales and cyclostratigraphy* [*Special Publication* 85]. London:

Geological Society. **Lehmann, C., Osleger, D.A. and Montañez, I.P.** 1998: Controls on cyclostratigraphy of Lower Cretaceous carbonates and evaporites, Cupido and Coahuila platforms, northeastern Mexico. *Journal of Sedimentary Research* 68, 1109–1130.

cyclothem A recurring pattern of FACIES in a succession of sediments or sedimentary rocks. The term was originally applied to upward-coarsening FACIES SEQUENCES associated with COAL seams in Upper CARBONIFEROUS rocks, each attributed to PROGRADATION of a DELTA. *GO*

Carter, R.M., Abbott, S.T. and Naish, T.R. 1999: Plio–Pleistocene cyclothems from Wanganui Basin, New Zealand: type locality for an astrochronologic time-scale, or template for recognizing ancient glacio-eustasy? *Philosophical Transactions of the Royal Society of London Series A* 357, 1861–1872. **Hampson, G.J., Stollhofen, H. and Flint, S.** 1999: A sequence stratigraphic model for the Lower Coal Measures (Upper Carboniferous) of the Ruhr district, north-west Germany. *Sedimentology* 46, 1199–1231.

D

Dalton minimum The *modern* minimum of solar sunspot activity observed in the late eighteenth and early nineteenth centuries, which may have caused a relatively small fall in global temperature. *JAM*

[See also MAUNDER MINIMUM]

Daly level A terrace of marine ABRASION, up to 6 m above present sea level, thought to represent postglacial high eustatic sea level. *TSp*

Daly, R.A. 1925: Pleistocene changes of level. *American Journal of Science Ser.* 5 10, 281–313.

dambo A distinctive type of freshwater, seasonal WETLAND in a saucer-shaped depression representing a channelless extension of the drainage network. Dambos occur typically on gently sloping land surfaces, especially in the TROPICS (the term dambo originates from a Bantu word meaning 'meadow grazing'), and are underlain by deeply weathered and often lateritised terrain. In Africa, dambos are particularly well developed in a broad latitudinal belt from Angola to Tanzania. Alternative local terms for dambos or dambo-like features in Africa include *fadama* (Nigeria), *mbuga* (East Africa), *matoro* (Mashonaland), *vlei* (South Africa), *bolis* (Sierra Leone). Comparable names in Europe are *bas-fond* (France) and *Spültäl* (Germany). Dambos appear to be sensitive to CLIMATIC CHANGE and to human impacts on the vegetation cover and soils caused by cultivation and/or OVERGRAZING. Most dambos contain COLLUVIUM or ALLUVIAL FILL, which can be 10 m or more in depth. This can provide evidence of QUATERNARY ENVIRONMENTAL CHANGE in the form of phases of AGGRADATION and channel cutting by headward-eroding GULLIES. *RAS*

[See also LATERITE]

Boast, R. 1990: Dambos: a review. *Progress in Physical Geography* 14, 153–177. **Goudie, A.S.** 1996: The geomorphology of the seasonal tropics. In Adams, W.M., Goudie, A.S. and Orme, A.R. (eds), *The physical geography of Africa*. Oxford: Oxford University Press, 152–153. **McFarlane, M.J. and Whitlow, R.** 1990: Key factors affecting the initiation and progress of gullying in dambos in parts of Zimbabwe and Malawi. *Land Degradation and Rehabilitation* 2, 215–235. **Thomas, M.F. and Goudie, A.S.** (eds) 1985: Dambos: small channelless valleys in the tropics. *Zeitschrift für Geomorphologie Supplementband* 52, 1–222.

Dansgaard–Oeschger (D–O) events Sudden and short-lived warm INTERSTADIAL events within the last GLACIAL EPISODE were first recognised in ice cores by Hans Oeschger and Willy Dansgaard. Confirmation was gained from the OXYGEN ISOTOPE and ATMOSPHERIC DUST records of both the Greenland *GRIP* and *GISP2* ice cores. A series of alternating high-dust (cold) and low-dust (warm) conditions are recognised. Twentyfour of these Dansgaard–Oeschger warm events are identified between 115 000 and 14 000 years ago within the ICE CORES. Each of the warm interstadial events is linked to a colder interval (Dansgaard–Oeschger cycles), thus indicating a climatic

shift over relatively short periods of time. Warming takes place on a decadal scale and slower cooling conditions evolve over a number of centuries. The warm interstadials, lasting from 500 to 2000 years, exhibit a temperature variation of 7°C, but are typically 5°C cooler than present. Some of these interstadial events are recognised in long terrestrial records, as in the ARBOREAL/NON-ARBOREAL POLLEN record from La Grande Pile, France, which exhibits a number of warming events during the last cold stage.

Further evidence of these rapid warming events comes from HIGH-RESOLUTION RECONSTRUCTIONS based on North Atlantic MARINE SEDIMENT CORES. These records exhibit high-frequency variability in oxygen isotopes, ICE-RAFTED DEBRIS (IRD) and calcium carbonate content, which correlates with the Dansgaard–Oeschger cycles and shows them also to be linked to lower-frequency variations known as BOND CYCLES. Debate continues surrounding a causal mechanism or driving force for these cycles. The rapidity of the warming event suggests that retreat and advance of ICE SHEETS cannot be the major triggering feature for this CLIMATIC CHANGE, for the RESPONSE TIME would be too long. However, some correlate with HEINRICH EVENTS. It is likely that the changes in the rates of heat transport within the ocean's THERMOHALINE CIRCULATION account for some of the variation, although variations in solar output have also been implicated. *LJW/WENA*

Bell, R.J. and Kennett, J.P. 1996: Brief interstadial events in the Santa Barbara basin, NE Pacific, during the past 60 kyr. *Nature* 379, 243–246. **Bond, G.C. and Lotti, R.** 1995: Iceberg discharges into the North Atlantic on millennial time scales during the last glaciation. *Science* 267, 1005–1010. **Johnsen, J., Clausen, H.B., Dansgaard, W.** *et al.* 1992: Irregular glacial interstadials recorded in a new Greenland ice core. *Nature* 359, 311–313. **Porter, S.C. and Zisheng, A.** 1995: Correlation between climate events in the North Atlantic and China during the last glaciation. *Nature* 375, 305–308.

Darwinism The biological theory of Charles Robert Darwin (1809–1882) concerning the EVOLUTION of species by means of NATURAL SELECTION. Darwinism has been used in this sense since 1871, but it was used earlier to refer to the doctrines of Erasmus Darwin (1731–1802). *KDB*

[See also ADAPTATION, CREATIONISM, LAMARCKISM, SPECIES CONCEPT]

Darwin, C.R. 1859: *On the origin of species by means of natural selection, or the preservation of favoured races in the struggle for life.* London: John Murray.

data acquisition The phase of data handling that begins with the sensing of variables and ends with a record of raw data. *TS*

data logger A micro-electronic device for automatically storing digital data generated in ENVIRONMENTAL

MONITORING and in laboratory and field EXPERIMENTS. Data loggers are widely used in, for example, geology, geomorphology, pedology, hydrology and climatology. Periodically, the stored data can be transferred (downloaded) to a computer for display and analysis. Some data loggers operate for over a year unattended. *RAS*

data quality An important aspect of any dataset, which depends on many factors. For instance, in the context of a GEOGRAPHICAL INFORMATION SYSTEM, data quality depends on the age of the data, areal coverage, map scale, density of observations, relevance, format, accessibility and positional and thematic ACCURACY. *HB*

database A collection of inter-related information, usually stored in some form of mass-storage system, such as magnetic tape or the hard disk of a computer. For soils or other environmental data, the system must be capable of handling datasets of both quantitative and qualitative information. A database is managed by a *database management system* (DBMS) comprising a set of computer programs for organising the information. It typically contains routines for data input, verification, storage, retrieval and combination. Examples of databases include the INTERNATIONAL TREE-RING DATABASE (ITRDB), the *European Pollen Database* (EPD), the *North American Pollen Database* (NAPD), the FAO–Unesco Soil Map of the World digital database that is available on CD-ROM, the World Soils

and Terrain database (SOTER; see Figure), and the soil databases of the European Union and the National Resource Conservation Service of the United States.

SOTER is an ambitious project to prepare an up-to-date global soil resource database utilising available information from topographic, soil and ecological surveys. The geographic data is handled by GEOGRAPHICAL INFORMATION SYSTEM software and the soil and terrain data are held in attribute files handled by a RELATIONAL DATABASE management system. There is space available for 13 terrain attributes, 19 attributes of terrain components and 78 attributes of the soil component at each site recorded. Data for South America, to update the Soil Map of the World, have already been compiled and some countries such as Hungary have used SOTER to hold their soil and terrain data. *EMB*

[See also ARCHIVE]

Batjes, N.H. 1997: A world dataset of derived soil properties by FAO–UNESCO soil unit for global modelling. *Soil Use and Management* 13, 9–16. **Baumgardner, M.F. (ed.)** 1999: Soil databases. In: *Handbook of soil science*. Boca Raton, F: CRC Press. **Burrough, P.A.** 1986: *Principles of geographical information systems for land resources assessment* [*Monographs on Soil and Land Resources Survey* 12]. Oxford: Clarendon Press. **FAO, ISRIC, UNEP and CIP** 1998: *Soil and terrain database for Latin America and the Caribbean* [*Land and Water Media Series* 5]. Rome: FAO. **Grissino-Mayer, H.D. and Fritts, H.C.** 1997: The International Tree-Ring Data Bank: an enhanced global database

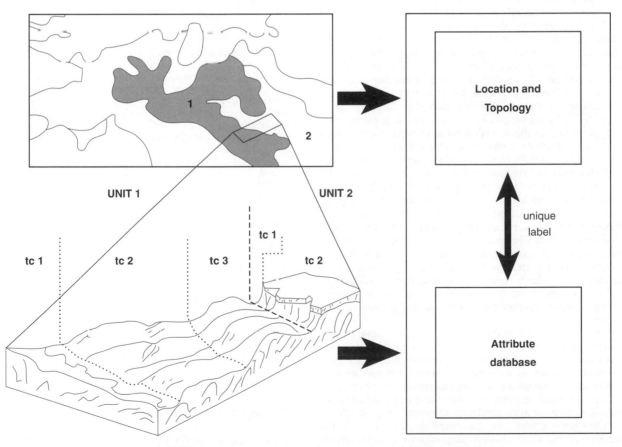

database *SOTER database units, their terrain components (tc), attributes and location*

serving the global scientific community. *The Holocene* 7, 235–238. **Le Bas, C., King, D., Jamagne, M. and Daroussin, J. (eds)** 1998: The European soil information system. In Heineke, H.J., Echelmann, W., Thomasson, A.J. *et al. Land information systems – developments for planning the sustainable use of land resources.* Ispra: European Soil Bureau, European Commission, 43–49. **Nachtergaele, F.O.** 1996: *From the soil map of the world to the digital global soil and terrain database: 1960–2000* [*AGLS Working Paper*]. Rome: FAO. **Van Engelen, V.W.P. and Wen, T.T.** 1993: *Global and national soils and terrain digital databases (SOTER).* Wageningen: UNEP, ISSS, ISRIC, FAO.

dating techniques Dating provides the chronological framework for ENVIRONMENTAL CHANGE: it permits the measurement and quantification of time. Many of the major advances made in the study of environmental change have been contingent on the development of appropriate dating techniques. A large battery of dating techniques is now available for dating a wide range of materials from all sorts of contexts on long and short timescales (see, for example, GEOLOGICAL, QUATERNARY and HOLOCENE ENVIRONMENTAL CHANGE). Although they may differ considerably in terms of their individual characteristics, dating techniques in general contribute to knowledge and understanding of the following specific aspects of environmental change: (1) the age of particular objects and events; (2) the timing of sequences of events; (3) rates of change or the speed at which changes occur; (4) the synchroneity of events in time (including leads and lags); and (5) the correlation of events in different places.

Four overlapping types of dating techniques may be recognised, based on the level of measurement achieved and the degree of confidence that can be placed in the age estimate. First, NUMERICAL-AGE DATING techniques produce results on a ratio scale; that is, they produce quantitative estimates of age and uncertainty, the ratios of which can be compared. *Absolute-age dating* or *absolute dating* are terms that have been used for this type of technique. Particular techniques that fall into this category range from those based on precise ARCHAEOLOGICAL or HISTORICAL EVIDENCE to the so-called INCREMENTAL DATING techniques of DENDROCHRONOLOGY, SCLEROCHRONOLOGY and VARVE CHRONOLOGY, each of which is capable of providing age estimates in SIDEREAL or CALENDAR years and enables HIGH-RESOLUTION RECONSTRUCTIONS of environmental change. Also included are many RADIOMETRIC DATING techniques (also known as *isotopic dating* techniques), based on the measurement of ISOTOPES affected by or produced by radioactive decay, such as COSMOGENIC-ISOTOPE DATING, POTASSIUM-ARGON DATING, RADIOCARBON DATING and URANIUM-SERIES DATING, and the *radiogenic dating* techniques, which measure cumulative non-isotopic effects of radioactive decay, such as ELECTRON SPIN RESONANCE DATING, FISSION-TRACK DATING, and LUMINESCENCE DATING.

The second type of dating technique, CALIBRATED-AGE DATING, measures systematic changes in a parameter with time but must be calibrated by independent chronological control in order to account for various ENVIRONMENTAL FACTORS that affect the rates of change in the parameter. Many biological (e.g. LICHENOMETRIC DATING), biochemical (e.g. AMINO ACID DATING), chemical (e.g. OBSIDIAN HYDRATION DATING), or geomorphological (e.g. dating based on rates of erosion, deposition or weathering) processes can provide the basis of calibrated-age dating

techniques, provided the rates of the underlying processes can be calibrated. If independent chronological control is not available, however, then such techniques may still provide a basis for RELATIVE-AGE DATING or *relative dating*. This third type of dating technique provides ordinal-scale information; that is, the rank-order or age sequence may be established but not the precise amount of time between objects or events.

The fourth type of dating technique, CORRELATED-AGE DATING, which establishes *age equivalence* to independently dated objects or events, is essentially providing ages on a nominal scale. Examples include PALAEOMAGNETIC DATING, TEPHROCHRONOLOGY and ORBITAL TUNING based on ISOTOPIC RATIOS. These techniques are particularly useful when linked to the precise chronological control provided by the first two categories above.

Choice of a dating technique for a particular problem involves three major considerations: applicability (is dateable material present?); time range (is the technique appropriate to the probable age of the material?); and RESOLUTION (is the PRECISION and ACCURACY of the technique commensurate with the temporal resolution required to solve the problem?). There is also the strong possibility that the practical and theoretical complexities encountered in the reconstruction of environmental change in a particular context will require the application of more than one dating technique. *JAM*

[See also CHRONOMETRIC AGE, MARKER HORIZON, SURFACE DATING]

Aitken, M.J. 1990: *Science-based dating in archaeology.* London: Longman. **Beck, C.** 1994: Introduction. In Beck, C. (ed.), *Dating in exposed and surface contexts.* Albuquerque, NM: University of New Mexico Press, 1–13. **Colman, S.M., Pierce, K.L. and Birkeland, P.W.** 1987: Suggested terminology for Quaternary dating methods. *Quaternary Research* 28, 314–319. **Geyh, M.A. and Schleicher, H.** 1990: *Absolute age determination: physical and chemical dating methods and their application.* Berlin: Springer. **Rutter, N.W. (ed.)** 1985: *Dating methods of Pleistocene deposits and their problems.* Toronto: Geoscience Canada. **Smart, P.L.** 1991: General principles. In Smart, P.L. and Frances, P.D. (eds), *Quaternary dating methods: a user's guide* [*QRA Technical Guide* 4]. Cambridge: Quaternary Research Association, 1–15. **Spence, K.** 2000: Ancient Egyptian chronology and the astronomical orientation of pyramids. *Nature* 408, 320–324. **Taylor, R.E. and Aitken, M.J. (eds)** 1997: *Chronometric dating in archaeology.* New York: Plenum. **Stratton-Noller, J., Sowers, J.M. and Lettis, W.R. (eds)** 2000: *Quaternary geochronology: methods and applications.* Washington, DC: American Geophysical Union. **Wagner, G.A.** 1998: *Age determination of young rocks and artifacts: physical and chemical clocks in Quaternary geology and archaeology.* Berlin: Springer. **Zeuner, F.E.** 1952: *Dating the past,* 3rd edn. London: Methuen.

daughter isotope Daughter isotopes are derived from *parent isotopes*: unstable parent *radioisotopes* undergo spontaneous *radioactive decay* to form daughter isotopes. As the decay rate is time-dependent, this forms a basis for forms of RADIOMETRIC DATING such as URANIUM-SERIES DATING. *IR*

[See also HALF LIFE]

Davisian cycle of erosion W.M. Davis proposed the idea that LANDFORMS may be viewed as representing various stages in a developmental sequence from 'youth' to 'maturity' and 'old age', and in so doing provided the leading PARADIGM for GEOMORPHOLOGY in the first half of the twentieth century. Although he also stated that landforms were the product of structure, process and stage, it was the emphasis on 'stage' that was paramount as relatively little was known about geomorphological processes at that time. His theoretical model – the so-called *cycle of erosion* or *geographical cycle* – was said to be initiated by UPLIFT (a change in BASE LEVEL) and proceeded, largely by SLOPE DECLINE, towards a PENEPLAIN. Landforms such as those produced by GLACIATION, which could not be fitted into the so-called '*normal cycle*' that was developed and elaborated largely in the temperate regions of the USA and Europe, tended to be viewed as '*climatic accidents*'. Different cycles of erosion were later proposed for different climatic regions. There are two main limitations of this concept of a cycle of erosion: first, its simplicity in relation to the complexities of landscape evolution in space and time. In particular, there is the implicit assumption of climatic stability: climatic changes have been too rapid and too frequent, especially during the QUATERNARY, for even a small part of the cycle to be completed without interruption. The second problem is that uplift and tectonic changes are not generally confined to 'short periods' of landscape initiation, but are operative while landscapes are being reduced. *JAM*

[See also CLIMATIC GEOMORPHOLOGY, CLIMATOGENIC GEOMORPHOLOGY, DENUDATION CHRONOLOGY, EQUILIBRIUM CONCEPTS IN GEOMORPHOLOGICAL/LANDSCAPE CONTEXTS, ROCK CYCLE, QUATERNARY ENVIRONMENTAL CHANGE]

Chorley, R.J. 1965: A re-evaluation of the geomorphic system of W.M. Davis. In Chorley, R.J. and Haggett, P. (eds), *Frontiers in geographical teaching*. London: Methuen. Chorley, R.J., Beckinsale, R.P. and Dunn, A.J. 1973: *The history of the study of landforms*. Vol. 2. *The life and work of W.M. Davis*. London: Methuen. Cotton, C.A. 1942: *Climatic accidents in landscape-making*. Christchurch, New Zealand: Whitcombe and Tombs. Davis, W.M. 1899: The geographical cycle. *Geographical Journal* 14, 481–504.

De Geer moraines

Subaqueous, transverse MORAINES deposited at or close to the grounding lines of GLACIERS terminating in water. Different terms have been suggested for these subaqueous moraines, including *morainal bank*, *composite grounding line fans*, cross-valley moraines and *washboard moraines*. The term 'De Geer moraines' is normally used for relatively small, separate, narrow and sharp-crested ridges of varying length and shape occurring in fields of closely spaced subparallel moraines formed subglacially at an oblique angle to the general ice-movement direction. The morphology and sedimentology of subaqueous moraines are highly variable and a large number of depositional MODELS have been proposed. Lenses of layered, water-lain sediments may occur in the ridges. There has been some discussion whether De Geer moraines are ANNUAL MORAINES or not. In plan form, individual subaqueous moraines may be linear or have sinuous crestlines. In some cases they are slightly concave upglacier, apparently reflecting calving bays along the ice front. *AN*

Boulton, G.S. 1986: Push moraines and glacier contact fans in marine and terrestrial environments. *Sedimentology* 33, 677–698. Boulton, G.S. 1990: Sedimentary and sea level changes during glacial cycles and their control on glacimarine facies architecture. In Dowdeswell, J.A. and Scourse, J.D. (eds), *Glaciomarine environments: processes and sediments* [*Special Publication* 53]. London: Geological Society, 15–52. Hoppe, G. 1959: Glacial morphol-

ogy and inland ice recession in northern Sweden. *Geografiska Annaler* 41A, 193–212. Larsen, E., Longva, O. and Follestad, B.A. 1991: Formation of De Geer moraines and implications for deglaciation dynamics. *Journal of Quaternary Science* 6, 263–277.

de Vries effect

The non-concordance of RADIOCARBON ages with CALENDAR AGES discovered by Hessel de Vries when dating timber of known age. Radiocarbon ages can be converted to calendar ages by CALIBRATION. Such terms as '*wiggles*', '*kinks*' and '*warps*' have been used in the past for de Vries effects. *PQD*

[See also WIGGLE MATCHING]

Vries, H. de 1958: Variations in concentration of radiocarbon with time and location on Earth. *Proceedings, Nederlandsche Akademie van Wetenschappen* B61, 94–102.

dead ice

Glacier ice that no longer flows actively but is downwasting; also known as stagnant ice. *Dead-ice topography* typically comprises a hummocky surface formed of irregular mounds of glacigenic debris left following surface melting of the ice. *JS*

[See also GLACIAL SEDIMENTS, HUMMOCKY MORAINE]

death assemblage

An assemblage of FOSSILS or SUBFOSSILS that has become modified by processes such as reworking and time-averaging, in contrast to a LIFE ASSEMBLAGE (see TAPHONOMY). The *fidelity* of a death assemblage refers to comparison with the living COMMUNITY from which it derived. Its *allochthoneity*, or degree of displacement from its original HABITAT, is assessed through population features such as the amount of shell breakage and abrasion, disarticulation, valve ratios and size-frequency distributions. Faunal similarity with adjacent BEDS and evidence from SEDIMENTOLOGY are also useful in analysing assemblages. Death assemblages that are little displaced may be termed neighbourhood or indigenous; more extensively transported assemblages are termed ALLOCHTHONOUS or *exotic*; those reworked from older rocks are *remanié*. *Neighbourhood death assemblages* commonly record an accumulation of shells from species living in an area through successive generations over a period of time. They are typical of the shelly community for that habitat and hence are useful for palaeocommunity reconstructions (while accepting the limitation of the general absence of soft-bodied organisms from the fossil record). *LC*

Kenward, S.M. and Bosence, D.W.J. 1991: Insect communities and death assemblages, past and present. In Hall, A.R. and Kenward, H.K. (eds), *Environmental archaeology in the urban context* [*Research Report* 43]. London: Council for British Archaeology. Kidwell, S.M. and Bosence, D.W.J. 1991: Taphonomy and time-averaging of marine shelly faunas. In Allison, P.A. and Briggs, D.E.G. (eds), *Taphonomy: releasing the data locked in the stratigraphic record*. New York: Plenum, 115–209.

death rate

The average number of deaths per 1000 people in the POPULATION. *JAM*

[See also BIRTH RATE, DEMOGRAPHIC TRANSITION]

debris cone

A cone-shaped accumulation of rock debris at the foot of a gully, talus slope or tributary valley; commonly a product of repeated DEBRIS FLOW activity. *PW*

[See also ALLUVIAL FAN, COLLUVIAL FAN]

debris flow A laminar, SEDIMENT GRAVITY FLOW with a high debris concentration. This type of rapid MASS MOVEMENT occurs in a wide range of environments from semi-arid to arctic–alpine when unconsolidated sediments are mobilised. They may be triggered, for example, by intense rainfall or rapid melting of snow, or by the thawing of PERMAFROST or SEASONALLY FROZEN GROUND. Debris-flow events produce characteristic landform–sediment assemblages including *lateral levées* and *terminal lobes* composed of DIAMICTON; but *distal fans* composed of fine, predominantly sandy sediments may also be present. The debris-flow fans are produced by water flow and intermediate-type flow (wet mudflow or *hyperconcentrated flow*) in the later stages of debris-flow events and individual debris-flow pulses. Deposits from numerous debris-flow events may accumulate to form COLLUVIAL FANS and they also contribute to some ALLUVIAL FANS. Debris-flow landforms and sediments have potential in reconstructing EXTREME CLIMATIC EVENTS and in identifying debris-flow hazards. *JAM*

[See also COLLUVIAL HISTORY, NATURAL HAZARDS]

Costa, J.E. 1984: Physical geomorphology of debris flows. In Costa, J.E. and Fleischer, P.J. (eds), *Developments and applications of geomorphology*. Berlin: Springer, 268–317. **Jackson, L.E., Kostaschuk, R.A. and MacDonald, G.M.** 1987: Identification of debris flow hazard on alluvial fans in the Canadian Rocky Mountains. *Geological Society of America Reviews in Engineering Geology* 7, 115–124. **Matthews, J.A., Dahl, S.-O., Berrisford, M.S. et al.** 1997: A preliminary history of Holocene colluvial (debris-flow) activity, Leirdalen, Jotunheimen, Norway. *Journal of Quaternary Science* 12, 117–129. **Matthews, J.A., Shakesby, R.A., McEwen, L.J. et al.** 1999: Alpine debris-flows in Leirdalen, Jotunheimen, Norway, with particular reference to distal fans, intermediate-type deposits and flow types. *Arctic, Antarctic and Alpine Research* 31, 421–435. **Rebetez, M., Lugon, R. and Baeriswyl, P.A.** 1997: Climatic change and debris flows in high mountain regions. *Climatic Change* 36, 371–389. **Zimmermann, M. and Haeberli, W.** 1992: Climatic change and debris flow activity in high-mountain areas: a case study in the Swiss Alps. *Catena Supplement* 22, 59–72.

debris-mantled glaciers Glaciers with substantial coverings of debris in their ABLATION ZONE, caused by either high rates of debris delivery from valley sides or NUNATAKS, or the melt-out of ENGLACIAL debris. Debris-mantled glaciers are common in high mountain environments such as the Andes and Himalayas and may exhibit a delayed response to climate due to the insulating properties of thick debris cover. An abrupt transition to rapid ABLATION can occur in association with the growth of SUPRAGLACIAL lakes dammed behind large terminal MORAINES. Such lakes are susceptible to catastrophic outburst floods or JÖKULHLAUPS and present major HAZARDS in some mountain regions. *DIB*

Kirkbride, M.P. 1993: The temporal significance of transitions from melting to calving termini at glaciers in the central Southern Alps of New Zealand. *The Holocene* 3, 232–240. **Llibouträy, L., Arnao, B.M., Pautre, A. and Schneider, B.** 1977: Glaciological problems set by the control of dangerous lakes in Cordillera Blanca, Peru. I. Historical failures of morainic dams, their causes and prevention. *Journal of Glaciology* 18, 239–254. **Nakawo, M., Raymond C.F. and Fountain, A. (eds)** 2000: *Debris-Covered Glaciers* [Publication 264]. Budapest: International Association for Scientific Hydrology.

deciduous Perennial plant species that seasonally shed their leaves in order to avoid environmental stress. Within TEMPERATE latitudes this is during the winter months and in lower latitudes, especially in the seasonal tropics, it is during the dry season. *BA*

[See also EVERGREEN]

decision making The process of finding a solution to a problem given a set of objectives and constraints. This is the optimal or least-cost path in NETWORK ANALYSIS or in making and supporting decisions in OVERLAY ANALYSIS within a GEOGRAPHICAL INFORMATION SYSTEM (GIS). *TVM*

[See also SPATIAL ANALYSIS]

declination (*D*) The angle between *magnetic north*, as indicated by a compass needle, and *true* or *geomagnetic north* of the Earth's MAGNETIC FIELD. It is a parameter of the geomagnetic field used in PALAEOMAGNETIC DATING. *MHD*

[See also INCLINATION]

Thompson, R. 1991: Palaeomagnetic dating. In Smart, P.L. and Frances, P.D. (eds), *Quaternary dating methods – a users guide* [QRA Technical Guide 4]. Cambridge: Quaternary Research Association, 177–198.

décollement A surface along which shearing or sliding has taken place, which may separate highly deformed from relatively undeformed material. Some types of LITHOLOGY readily act as décollement surfaces during DEFORMATION, such as EVAPORITES or CLAY. Décollement surfaces are important in THRUST FAULTS and MASS MOVEMENTS such as SLUMPS. *GO*

decolonisation (1) The ending of colonial rule. (2) The departure of the colonisers from a colonised territory. *JAM*

[See also COLONIALISM, COLONISATION]

Cooper, F. 1997: *Decolonization and African society*. Cambridge: Cambridge University Press. **Grimal, H.** 1978: *Decolonization*. London: Routledge.

decommissioning 'The final closing down and putting into a state of safety of a nuclear reactor or other industrial plant or device when it has come to the end of its useful life' (Porteous 2000: 143). *JAM*

Porteous, A. 2000: *Dictionary of environmental science and technology*, 3rd edn. Chichester: Wiley.

decomposer An organism that takes part in the DECOMPOSITION subsystem of ECOSYSTEMS. *JLI*

[See also BIOGEOCHEMICAL CYCLES, FOOD CHAIN]

decomposition The breakdown of organic matter by animals, bacteria, fungi and abiotic processes, without which BIOGEOCHEMICAL CYCLES would be impossible. The term does not include the breakdown of inorganic materials (WEATHERING). Decomposition is temperature-sensitive, making it an important process in many ecological models developed for environmental-change impact studies. Where PRODUCTIVITY exceeds decomposition rate, organic matter accumulates, as in *peatlands*.   *JLI*

Ågren, G.I. and Bosatta, E. 1996: *Theoretical ecosystem ecology. Understanding element cycles*. Cambridge: Cambridge University Press. **Pastor, J. and Post, W.M.** 1988: Response of northern forests to CO_2 induced climate change. *Nature* 334, 55–58.

deduction   In a properly formed deductive argument the truth of the premises ensures the truth of the conclusion: more generally, to reason from general truth to a particular instance of that truth.   *CET*

West, A. 1992: *A rulebook for arguments*, 2nd edn. Indianopolis, IN: Hackett.

deep ecology   An ethical and moral philosophical position within ECOCENTRISM. One form emphasises the intrinsic right of all species to exist in the NATURAL ENVIRONMENT and the immorality of human actions that interfere in any way with this right. Another emphasises the interdependence of species in the ECOSYSTEM and considers immoral any human action that disturbs the HOMEOSTASIS of natural ecosystem interactions. Deep ecologists maintain that the alternative 'shallow ecology', which espouses such concepts as ECOLOGICAL MANAGEMENT, ENVIRONMENTAL MANAGEMENT and SUSTAINABLE DEVELOPMENT, does not take sufficient account of ecological principles and inevitably leads to ENVIRONMENTAL DEGRADATION.   *JAM*

[See also ENVIRONMENTAL ETHICS, ENVIRONMENTALISM, GREEN POLITICS]

Devall, B. and Sessions, G. 1985: *Deep ecology* Salt Lake City, UT: Peregrine Smith Books. **Mathews, F.** 2001: Deep ecology. In Jamieson, D. (ed.), *A companion to environmental philosophy*. Oxford: Blackwell, 218–232. **Naess, A.** 1989: *Ecology, community and lifestyle: outline of an ecosophy*. Cambridge: Cambridge University Press.

deep water   Dense water masses, such as *North Atlantic Deep Water* (NADW), that flow within the deep ocean basins and form largely in polar regions, either by winter cooling or by *brine exclusion* during SEA ICE formation.   *WENA*

[See also BOTTOM WATER, DOWNWELLING, WATER MASS]

Dickson, R.R. and Brown, J. 1994: The production of North Atlantic deep water: sources, rates and pathways. *Journal of Geophysical Research* 99, 12319–12341.

deep-sea drilling   The recovery of long MARINE SEDIMENT CORES from the OCEAN floor. Drilling in water depths of more than 3000 m is carried out from ships and involves state-of-the-art technology. The first core from the ocean floor was drilled in the South Atlantic by the *Glomar Challenger* in AD 1968. It demonstrated that the oldest sediments overlying igneous OCEANIC CRUST were only 180 million years old and led to further data collection to test theories of SEA-FLOOR SPREADING and PLATE TECTONICS. Between 1968 and 1983, under the direction of the *Deep Sea Drilling Project* (DSDP) based at Texas A&M University, the *Glomar Challenger* travelled over 600 000 km and drilled over 1000 holes at 624 sites in water depths greater than 6000 m, to recover and curate over 96 km of deep-sea cores, each up to 1.5 km long.

The initial aims of the Deep Sea Drilling Project included improved understanding of the processes of plate tectonics, the Earth's crustal structure and composition, ocean conditions and climate changes through time. In 1985 the work continued through the *Ocean Drilling Program* (ODP) using the ship *JOIDES Resolution*, which undertakes six scientific expeditions per year, is capable of drilling in water depths up to 8.2 km and has drilled over 160 km of core since 1985. The objectives were expanded to include the evolution of oceanic crust, marine sedimentary sequences and CONTINENTAL MARGINS, with a significantly expanded environmental agenda, which included long-term changes in the Earth's ATMOSPHERE, oceans, polar ice caps and marine life. International participation in DSDP began with the *International Phase of Ocean Drilling* (IPOD) from 1974 to 1976. Scientific planning for ODP is now provided by the *Joint Oceanographic Institutions for Deep Earth Sampling* (JOIDES), an organisation of advisory committees and panels representing scientists and research institutions in over 20 countries.
                   BTC/GO

[See also OCEANIC SEDIMENTS]

Carter, R., Carter, L., McCave, N. and the Leg 181 Shipboard Scientific Party 1999: The DWBC sediment drift record from Leg 181: drilling in the Pacific gateway for the global thermohaline circulation. *JOIDES Journal* 25, 8–13. **Hsü, K.J.** 1983: *The Mediterranean was a desert: a voyage of the Glomar Challenger*. Princeton, NJ: Princeton University Press. **Kemp, A.E.S. and Baldauf, J.G.** 1993: Vast Neogene laminated diatom mat deposits from the eastern equatorial Pacific Ocean. *Nature* 362, 141–144. **Warme, J.E., Douglas, R.G. and Winterer. E.L. (eds)** 1981: *The Deep Sea Drilling Project: a decade of progress* [*Special Publication* 32]. Tulsa, OK: Society of Economic Paleontologists and Mineralogists.

deflation   The entrainment and removal of loose particles by the wind. Deflation constitutes one of the two erosive effects of wind in deserts (the other being ABRASION). Localised deflation involves sand-sized particles (which may be used to form DUNES in nearby areas), but silt and, more rarely, clay-sized material can be transported over much greater distances and lead to the development of LOESS deposits. The deflation of large areas leaves behind a gravel or stone-strewn LAG DEPOSIT (known in the Sahara as *hamada*) and to the development of a DESERT or DESERT PAVEMENT.   *MAC*

[See also DUST STORM]

defoliant   A chemical that causes plants to lose leaves, such as so-called 'agent orange', a mixture of the herbicides 2,4,5-trichlorophenoxyacetic acid (2,4,5-T) and 2,4,-dichlorophenoxyacetic acid (2,4-D) used in the Vietnam War by the US military.   *JAM*

[See also PESTICIDES]

deforestation   The loss of forest canopy. Deforestation is mainly caused by human activities like logging and burning (see BIOMASS BURNING), associated with human POPULATION growth. Tropical deforestation is considered as a major cause of global CLIMATIC CHANGE, as it releases large quantities of carbon into the atmosphere. In Brazil alone, 1.5 million ha had been deforested in AD 1978, increasing to 4.26 million ha in 1991. At this rate of vegetation destruction, forest is not able to regenerate itself and SOIL EROSION is an unavoidable consequence.

To monitor the extent of deforestation on large scales, REMOTE SENSING techniques are employed. *LANDSAT Thematic Mapper* (TM), *AVHRR* and *JERS*-1 *synthetic aperture radar* have been used for the mapping of tropical deforested areas and FOREST REGENERATION stages. The remotely sensed data can be used for evaluating socio-economic and ecological MODELS designed to assess different economic scenarios and likely consequences of different policy options for the ENVIRONMENT. *HB*

[See also CARBON CYCLE, CARBON SINK/SOURCE, FOREST CLEARANCE, FOREST DECLINE, FOREST MANAGEMENT, HUMAN IMPACT ON CLIMATE, HUMAN IMPACT ON VEGETATION, IMAGE CLASSIFICATION, TROPICAL FOREST FIRES]

Frohn, R.C., McGwire, K.C., Dale, V.H. and Estes, J.E. 1996: Using satellite remote sensing analysis to evaluate a socio-economic and ecological model of deforestation in Rondonia, Brazil. *International Journal of Remote Sensing* 17, 3233–3255. **Gilruth, P.T., Hutchinson, C.F. and Barry, B.** 1990: Assessing deforestation in the Guinea Highlands of West Africa using remote-sensing. *Photogrammetric Engineering and Remote Sensing* 56, 1375–1382. **Jusoff, K. and Manaf, M.R.A.** 1995: Satellite remote sensing of deforestation in the Sungai Buloh Forest Reserve, Peninsular Malaysia. *International Journal of Remote Sensing* 16, 1981–1997.

deforestation: climatic impacts

There are several potential impacts on climate through BIOGEOCHEMICAL CYCLES and the resulting effects, particularly via the CARBON CYCLE on CARBON DIOXIDE concentrations in the atmosphere. It is estimated that each year about 2 Gt C is released to the atmosphere as a result of tropical DEFORESTATION. This is the largest change in LAND COVER currently occurring on Earth. In recent years the burning of rain forests (see BIOMASS BURNING) is estimated to have contributed 15–30% of the carbon dioxide added to the atmosphere by humans. One estimate indicates that SHIFTING CULTIVATION (slash-and-burn) and scavenging for FUELWOOD world-wide accounts for 40–50% of deforestation. Deforestation also releases significant amounts of METHANE and NITROUS OXIDE. If all tropical forests were to be removed, it is estimated that the input to the atmosphere would be 150–240 Gt C, which might increase atmospheric carbon dioxide by about 35–60 ppm. Forests are able to store carbon via photosynthesis (see CARBON SEQUESTRATION); they also affect climate by increasing ALBEDO and via the HYDROLOGICAL CYCLE by decreasing EVAPOTRANSPIRATION. Model results suggest that regional rainfall reductions of about 15% might occur if the forest cover to the north of 30°S in South America was removed and replaced by grassland. *AHP*

Gornitz, V. 1987: Climatic consequences of anthropogenic vegetation changes from 1880–1980. In Rampino, M.R., Sanders, J.E., Newman, W.S. and Königsson, L.K. (eds), *Climate: periodicity and predictability*. New York: Van Nostrand Reinhold, 47–69. **Myers, N.** 1988: Tropical deforestation and climatic change. *Proceedings of the Symposium on Resources, Development and Planning*. Madras: Madras University, 1–19. **Shukla, J., Nobic, C. and Sellers, P.** 1990: Amazon deforestation and climate change. *Science* 247, 1322–1325.

deforestation: ecological impacts

Deforestation is the CLEAR FELLING of forest and the conversion of the land to some other use. Forests that are not cleared may be subject to degradation through the SELECTIVE CUTTING of trees for timber or fuel. The impacts of deforestation depend on the extent of clearance and the type of LANDUSE that replaces forest. The most common cause of deforestation is FOREST CLEARANCE for agriculture. Conversion of PRIMARY RAIN FORESTS for agriculture represents one of the most profound changes to the global environment of the present era. Commercial logging of forests for timber is often an indirect cause of deforestation as it provides access routes along which colonisation and clearance take place. In Amazonia between AD 1991 and 1996, 73% of deforestation occurred less than 50 km from three major road networks.

Deforestation is typically a process of forest FRAGMENTATION that leaves small blocks of forest in a matrix of pasture and crops. An important area for ecological research is the investigation of the MINIMUM CRITICAL SIZE OF ECOSYSTEMS. In fragments of < 100 ha, rates of tree mortality and damage increase dramatically as a consequence of EDGE EFFECTS, particularly desiccation and wind disturbance. As a consequence, such fragments are particularly vulnerable to FIRE IMPACTS. There are also changes in species composition, with PIONEER PLANTS becoming more abundant at the expense of species characteristic of the forest interior. These factors, in combination, often result in a decline in forest BIOMASS. Many species in tropical rain forests have very restricted or patchy distributions and they may be poorly represented in residual forest fragments. In contrast, as many TEMPERATE forest species are well dispersed and abundant, even widespread deforestation has resulted in very low rates of species EXTINCTION.

In intact forest, small local populations are often maintained by high rates of immigration within a larger METAPOPULATION. Forest fragmentation not only reduces the total habitat area for many animals; it also isolates small populations, some of which may be below the MINIMUM VIABLE POPULATION SIZE. Habitat and ecological specialists are sensitive to the changes in forest MICROCLIMATE and species composition. *NDB*

[See also BIODIVERSITY, DISPERSAL, HABITAT LOSS]

Brown, K. and Pearce, D.W. 1994: *The causes of tropical deforestation*. Vancouver: University of British Columbia Press. **Grainger, A.** 1993: *Controlling tropical deforestation*. London: Earthscan. **Grainger, A.** 1993: Rates of deforestation in the humid tropics: estimates and measurements. *Geographical Journal* 159, 33–44. **Food and Agriculture Organization (FAO)** 1997: *State of the World's forests 1997*. Rome: FAO. **Laurance, W.F. and Bierregaard, R.O. (eds)** 1997: *Tropical forest remnants: ecology, management and conservation of fragmented communities*. Chicago, IL: University of Chicago Press. **Sayer, J.A. and Whitmore, T.C.** 1991: Tropical moist forests: destruction and species extinction. *Biological Conservation* 55, 199–213. **Whitmore, T.C. and Sayer, J.A. (eds)** 1992: *Tropical deforestation and species extinction*. London: Chapman and Hall.

deforestation: hydrological and geomorphological impacts

Removal of a forest cover by CLEAR FELLING or fire causes the underlying soil to be much less well protected from the direct erosive effects of rainfall. RUNOFF also tends to be generated far more readily and to be greater in volume. Water chemistry of streams may also be affected. With a forest cover intact, SOIL EROSION is usually low because of the reduction in the kinetic energy of raindrops under most tree canopies and the presence of soil humus, which is highly permeable so that INFILTRATION CAPACITY is high. The humus also tends to reduce *rainsplash erosion* and macropores formed by roots and the abundant soil fauna help transmission of water beneath the soil surface. AGGREGATE STABILITY is high as a result of the high organic content and EARTHWORM activity. Tree roots also help to stabilise steep slopes. Once the forest cover is removed, and particularly if the ground is left bare, rates of soil loss will tend to increase and MASS MOVEMENT will be greater and/or more frequent. Evidence of long-term rates of soil erosion derived from, for example, SEDIMENT TRAPS show that DEFORESTATION of mature forest in historical and prehistoric times led to accelerated rates of soil erosion substantially higher than those experienced prior to deforestation.

Experiments in forested and deforested catchments have shown that runoff amounts tend to increase many times if the forest cover is removed. The reasons include a lower INTERCEPTION rate and reduced OVERLAND FLOW in mature forests, as well as a better infiltration capacity and structure of forest soils, both of which tend to reduce overland flow. Where present, soil WATER REPELLENCY may exacerbate this effect. If vegetation REGENERATION is allowed to occur after deforestation, soil erosion and runoff tend to revert back towards levels typical of mature forest, but recovery may take several decades, depending on environmental conditions. *RAS*

[See also FOREST CLEARANCE, HUMAN IMPACT ON HYDROLOGY, HUMAN IMPACT ON LANDFORMS AND GEOMORPHOLOGIC PROCESSES]

Baumler, R. and Zech, W. 1999: Effects of forest thinning on the streamwater chemistry of two forested watersheds in the Bavarian Alps. *Forest Ecology and Management* 116, 119–128. Goudie, A.S. 2000: *The human impact on the natural environment*, 5th edn. Oxford: Blackwell. Lal, R. 1997: Deforestation effects on soil degradation and rehabilitation in western Nigeria. 4. Hydrology and water quality. *Land Degradation and Development* 8, 95–126.

deforestation: in the Holocene

The clearing of forests, inferred especially from *pollen diagrams*: from increasing NON-ARBOREAL POLLEN (NAP) (especially ANTHROPOGENIC INDICATORS); from decreasing ARBOREAL POLLEN (AP); and – if the temporal resolution permits – from the increased abundance of early- versus late-successional woody taxa. Together with increasing charcoal and erosional input, deforestation can indicate early farming (*agricultural deforestation*). To estimate the degree of the deforestation that has occurred, the characteristics of the pollen collecting source and the distance from the clearance/archaeological site are important considerations that must be taken into account. *BA*

[See also AGRICULTURAL HISTORY, FOREST CLEARANCE, HUMAN IMPACT ON VEGETATION HISTORY, LANDUSE CHANGE, LANDNAM, NEOLITHIC: LANDSCAPE IMPACTS]

Gaillard, M.J. and Berglund B.E. (eds) 1998: *Quantification of land surfaces cleared of forests during the Holocene: modern pollen/vegetation/landscape relationships as an aid to the interpretation of fossil pollen data* [Paläoklimaforschung 27]. Stuttgart: Gustav Fischer. Sugita, S., Gaillard, M.-J. and Broström, A. 1999: Landscape openness and pollen records: a simulation approach. *The Holocene* 9, 409–421.

deformation

Change in shape, size or orientation resulting from a system of applied stresses. It is commonly applied to permanent changes in sediments and rocks resulting from TECTONIC stresses (see PLATE TECTONICS), preserved as GEOLOGICAL STRUCTURES such as FAULTS and FOLDS. *GO*

[See also SOFT-SEDIMENT DEFORMATION]

Ramsay, J.G. and Huber, M.I. 1983: *The techniques of modern structural geology*. Vol.1. *Strain analysis*. London: Academic Press.

degassing

The release of dissolved gases, commonly triggered by a decrease in pressure or some physical disturbance. Degassing is an important process in many VOLCANIC ERUPTIONS, in which the release of gases such as CARBON DIOXIDE from the MANTLE to the ATMOSPHERE can contribute to CLIMATIC CHANGE. Controlled degassing of certain crater lakes is a possible mechanism for controlling the hazard represented by LIMNIC ERUPTIONS. Volcanic degassing in the ARCHAEAN is the origin of the Earth's early atmosphere and its surface waters. Degassing of METHANE locked up in GAS HYDRATES is a potential cause of climatic change. Degassing of GROUNDWATER may cause problems for the supply of WATER RESOURCES. *GO*

Jarsjo, J. and Destouni, G. 2000: Degassing of deep groundwater in fractured rock around boreholes and drifts. *Water Resources Research* 36, 2477–2492. Johnson, J.B. and Lees, J.M. 2000: Plugs and chugs – seismic and acoustic observations of degassing explosions at Karymsky, Russia and Sangay, Ecuador. *Journal of Volcanology and Geothermal Research* 101, 67–82. Kusakabe, M., Tanyileke, G.Z., McCord, S.A. and Schladow, S.G. 2000: Recent pH and CO_2 profiles at Lakes Nyos and Monoun, Cameroon: implications for the degassing strategy and its numerical simulation. *Journal of Volcanology and Geothermal Research* 97, 241–260.

deglaciation

The process of ICE SHEET decay and recession that occurs in response to climatic amelioration at the end of a GLACIAL EPISODE. Deglaciation may be driven either by a decrease in accumulation or an increase in ablation, or a combination of both. In these circumstances, the ice sheet has a negative MASS BALANCE. Deglaciation may be caused by a decrease in precipitation as snow, increasing temperatures and (where a glacier terminates in the sea) SEA-LEVEL RISE. Deglaciation may be extremely rapid, for example through rapid *draw-down* of interior ice into the sea via fast-flowing ice streams. Alternatively, on land, deglaciation is accomplished by *downwasting* and *areal stagnation* or by *ice-margin recession*. A wide range of sediments and depositional landforms are the products of deglaciation, including DIAMICTONS, sands and gravels, which make up a range of ice-contact features, including ESKERS, KAMES and MORAINES. *MJH*

[See also DEGLACIERISATION, GLACIATION]

Bauch, H.A. and Weinelt, M.S. 1997: Surface water changes in the Norwegian Sea during last deglacial and Holocene times. *Quaternary Science Reviews* 16, 1115–1124. Bennett, M.R. and Glasser, N.F. 1996: *Glacial geology – ice sheets and landforms*. Chichester: Wiley.

deglacierisation

The process whereby a LANDSCAPE progressively loses its cover of GLACIER ice. *PW*

[See also DEGLACIATION, GLACIERISATION]

degradation (1) The lowering of a land surface or a river bed by erosional processes. In rivers, it is the opposite of AGGRADATION. (2) A deterioration in ENVIRONMENTAL QUALITY in the context of ENVIRONMENTAL DEGRADATION, LAND DEGRADATION and SOIL DEGRADATION.
JAM/RPDW

[See also DENUDATION, EROSION, PERMAFROST DEGRADATION]

degree days A measure of the duration and intensity of temperatures above or below stated reference temperatures or thresholds, as used, for example, in agriculture and space heating/cooling requirements. It is essentially a measure of ACCUMULATED TEMPERATURE (surplus or deficit in relation to the defined threshold temperature). In relation to plant growth, degree days above 6°C are commonly used; for calculating space heating needs, the threshold temperature is commonly 15.5°C. *JCM*

[See also FREEZING INDEX].

dehesa In Spain and Portugal (where it is referred to as *montado*), a land-use practice traditionally comprising pasture and cultivation in combination with widely spaced trees (10–80 per hectare). Dehesa is a type of DRY FARMING. The tree species are oaks (*Quercus suber, Q. ilex, Q. pyrenaica* and *Q. rotundifolia*) grown for cork, timber, fuel, charcoal, tannin and acorns (for fodder). In addition to tree products, there is shifting cultivation of cereals and grazing of pigs, sheep, goats and cattle on FALLOW land. Dehesa/montado systems still cover 5 million ha in southwestern Spain and about 1 million ha in Portugal, but since the AD 1970s there has been change in management practices, and in some circumstances abandonment of systems caused by changing socio-economic conditions, leading to reversion to shrubs and increased risk of WILDFIRE. *RAS*

[See also CROP ROTATION, HUMAN IMPACT ON TERRESTRIAL VEGETATION, LANDUSE CHANGE, LANDUSE IMPACTS ON SOILS, MEDITERRANEAN ENVIRONMENTAL CHANGE AND HUMAN IMPACT, MEDITERRANEAN-TYPE VEGETATION, SAVANNA]

Joffre, R. 1992: The dehesa: does this complex ecological system have a future? In Teller, A., Mathry, P. and Jeffers, J. (eds), *Responses of forest ecosystems to environmental changes*. London: Elsevier Applied Science, 381–388. **Pinto-Correira, T. and Mascarenhas, J.** 1999: Contribution to the extensification/intensification debate: new trends in the Portuguese *montado*. *Landscape and Urban Planning* 46, 125–131.

de-industrialisation A decline in industry, especially a local or regional decline in the numbers employed in MANUFACTURING INDUSTRY as a result of plant closures, usually in response to changing industrial conditions globally. *JAM*

[See also POSTINDUSTRIALISATION]

Martin, R. and Rowthorn, B. (eds) 1986: *The geography of deindustrialization*. Basingstoke: Macmillan.

Delaunay triangulation An interpolation technique employing a TRIANGULAR IRREGULAR NETWORK, which requires that: (1) triangles be as near equilateral as possible; and (2) the circumcircle of a triangle may not contain any other triangulation point. *TF*

Worboys, M.F. 1995: *GIS: A computing perspective*. London: Taylor and Francis.

dell A small, shallow and usually DRY VALLEY floored with SOLIFLUCTION debris. *PW*

[See also COOMBE ROCK, SAPPING]

Tuckfield, C.G. 1986: A study of dells in the New Forest, Hampshire, England. *Earth Surface Processes and Landforms* 11, 23–40.

delta A partly subaerial accumulation of sediment at the mouth of a river where it enters a body of standing water (a lake or the sea). Deltas are highly variable in morphology, depending largely on the relative importance of river and basinal processes. Basinal processes attempt to redistribute sediment deposited at the river mouth and include TIDES, waves and nearshore currents. DELTA DEPOSITS contain an important record of ENVIRONMENTAL CHANGE. *GO*

[See also GILBERT-TYPE DELTA]

Elliott, T. 1986: Deltas. In Reading, H.G. (ed.), *Sedimentary environments and facies*, 2nd edn. Oxford: Blackwell, 113–154. **Wright, L.D.** 1978: River deltas. In Davis Jr, R.A. (ed.), *Coastal sedimentary environments*. New York: Springer, 5–68.

delta deposits SEDIMENTARY DEPOSITS that accumulate in a DELTA. They are an important source of information on environmental change because the river mouth is an area of rapid SEDIMENT accumulation and the coastal location is sensitive to changes in SEA LEVEL. The GILBERT-TYPE DELTA represents a model for delta deposits in their simplest form, but larger deltas give rise to more complex deposits characterised by repeated upward-coarsening FACIES SEQUENCES (CYCLOTHEMS) that record PROGRADATION of a depositional lobe at a river mouth. Periodic AVULSION, or lobe-switching, gives rise to a shift in the locus of sediment accumulation. The old lobe is starved of sediment but continues to undergo SUBSIDENCE, resulting in deposition of a characteristic *abandonment facies*, followed by a marine TRANSGRESSION. This essentially AUTOCYCLIC model for delta deposits has undergone major revisions in the light of SEQUENCE STRATIGRAPHY interpretations that emphasise the role of ALLOCYCLIC controls such as eustatic (see EUSTASY) or TECTONIC effects. Delta deposits are important repositories of ENERGY RESOURCES: several present-day deltas contain subsurface accumulations of PETROLEUM, and many COAL deposits occur in ancient delta deposits. *GO*

Bhattacharya, J.P. and Walker, R.G. 1992: Deltas. In Walker, R.G. and James, N.P. (eds), *Facies models: response to sea level change*. St John's, Newfoundland: Geological Association of Canada, 157–177. **Leeder, M.R.** 1999: *Sedimentology and sedimentary basins: from turbulence to tectonics*. Oxford: Blackwell Science. **Maynard, J.R. and Leeder, M.R.** 1992: On the periodicity and magnitude of Late Carboniferous glacio-eustatic sea-level changes. *Journal of the Geological Society, London* 149, 303–311. **Reading, H.G. and Collinson, J.D.** 1996: Clastic coasts. In Reading, H.G. (ed.), *Sedimentary environments: processes, facies and stratigraphy*, 3rd edn. Oxford: Blackwell Science, 154–231. **Whateley, M.K.G. and Pickering, K.T. (eds)** 1989: *Deltas: sites and traps for fossil fuels* [Geological Society, London, Special Publication 41]. Oxford: Blackwell.

dematerialisation A decrease in the materials used per unit of output. The concept can be applied at any level from the manufacture of a particular product to the world economy. Examples include the use of less metal in beverage cans (the mass of an aluminium can has been reduced by 25% since 1973), *miniaturisation* in the electronics industry and the stabilised material use of some bulk materials such as steel and copper relative to industry value added or gross domestic product (GDP). Dematerialisation does not mean an absolute decline in material use. *JAM*

[See also MANUFACTURING INDUSTRY]

Wernick, I.K., Herman, P., Govind, S. and Ausubel, J.H. 1996: Materialization and dematerialization. *Dædalus* 125, 171–198.

demic diffusion The spread of innovations or cultural traits by cumulative small-scale movements of people rather than by cultural DIFFUSION or the large-scale MIGRATION of populations (COLONISATION). *JAM*

[See also DIFFUSIONISM]

demographic change World population has exceeded 6000 million people and is estimated to reach between 7.3 and 10.7 billion people by the year AD 2050 (Smil, 1999). Throughout prehistory and history the human population can be viewed as exhibiting phased EXPONENTIAL GROWTH, with phases of growth and long periods of relative stability associated with successive advances in technology and social organisation (see Figure). Shorter-term fluctuations have been common and are linked to specific events.

From Malthus onwards there has been speculation about ultimate limits to this growth in population and the ability of the resource base to support it. More recently, *rates of growth* have begun to decline and it is doubtful whether the world's population will again double. During the late 1960s the *relative growth rate* peaked at just over 2% per year and fell below 1.5% per year for the first time in the 1990s. As a consequence the *absolute annual growth* in numbers peaked at 85 million people per year in the late 1980s and in 1995 was down to 80 million. Future projections are uncertain, mainly because of unknown future fertility rates and changing life expectancies. The transition to low fertility has occurred in all developed countries, is well advanced in most of Asia and Latin America and it is underway in sub-Saharan Africa.

Factors such as population migration, war, genocide, epidemics, acquired immune deficiency syndrome (AIDS) and social attitudes to family size affect not only population numbers but also demographic age structures. The growth of the human population and the increase in population density within the ECUMENE are major aspects of human environmental change that underlie increasing HUMAN IMPACTS on the natural environment. *JAM*

[See also DEMOGRAPHIC TRANSITION, OVERPOPULATION]

Clarke, J.I. 1996: Changes in global demography. In Douglas, I., Huggett, R. and Robinson, M. (eds), *Companion encyclopedia of geography*. London: Routledge, 249–273. **Ness, G.D.** 1997: World population growth. In Brune, D., Chapman, D.V., Gwynne, M.D. and Pacyna, J.M. (eds), *The global environment: science, technology and management*. Vol. 2. Weinheim, Germany: VCH, 637–656. **Smil, V.** 1999: How many billions to go? [Millennial Essay] *Nature* 401, 429. **Whitmore, T.M., Turner II, B.L., Johnson, D.L. et al.** 1990: Long-term population change. In Turner II, B.L., Clark, W.C., Kates, R.W. *et al.* (eds), *The Earth as transformed by human action*. Cambridge, Cambridge University Press, 25–39.

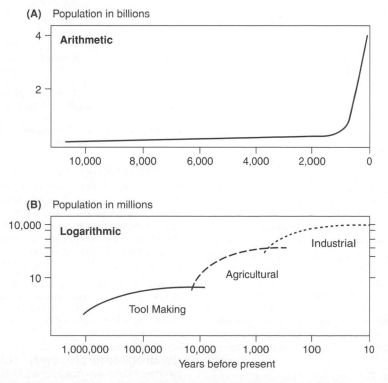

(A) Population in billions

(B) Population in millions

Years before present

demographic change *Global human population growth (A) over the last 10 000 years and (B) over the last one million years. Note the varying units and scales in the two graphs and the effects of major technological innovations in (B) (after Whitmore et al., 1990)*

demographic transition A concept of human population change from high to low birth and death rates, which accompanies *economic development* from a rural–agrarian to an urban–industrial society. There are notable differences in pattern and cause between the past transitions of the developed nations and those currently occurring in the developing world. *JAM*

[See also DEMOGRAPHIC CHANGE]

Szreter, S. 1993: The idea of the demographic transition and the study of fertility change: a critical intellectual history. *Population and Development Review* 19, 659–702.

demography The study of the human population, especially its vital statistics – numbers, fertility (birth rates), mortality (death rates), age structure – and so on. *Ecological demography* involves a similar approach to populations of other organisms. *JAM*

denature To render something unpalatable by addition of a noxious substance. Thus grain may be denatured to ensure that it is used as seed rather than human food. *JAM*

dendrochronology The study of annual TREE RINGS. The term derives from the Greek *dendron* (tree) and the Latin *chronos* (time). In its widest sense it encompasses many subfields concerned with different branches of *dendro-indication*, such as DENDROCLIMATOLOGY and DENDRO-GEOMORPHOLOGY. All have a common foundation: an expression of various phenomena in the growth of trees, recorded on a rigid, continuous and absolute calendrical timescale that is provided by the CROSS-DATING of multiple tree-ring time series.

In a narrow sense dendrochronology can be considered the technique by which a long, continuous (usually ring-width) CHRONOLOGY is constructed by averaging indices of interannual ring widths, aligned in their correct calendar years, to form long '*master series*' for a particular tree species in some site or region. Such chronologies can be composed of data from living trees or a combination of material from living and dead wood. Dendrochronology also involves the process of dating otherwise undated wood samples by uniquely matching the pattern of year-to-year variability in the sample wood to that of an established MASTER CHRONOLOGY. When a well replicated chronology is available, many wooden ARTEFACTS (such as historical building timbers, archaeological wood samples) may all be dated with PRECISION and ACCURACY, provided they contain sufficient rings reflecting the same overall growth forcing represented in the master chronology.

It should be stressed that dendrochronology is distinct from simple ring counting in as much as it has the fundamental requirement to compare and synchronise the ring patterns in multiple trees. This cross-dating process ensures the veracity of the absolute timescale by resolving problems associated with the following: unclear ring boundaries; locally missing (partial) rings that can occur in parts of the CAMBIUM during years when tree growth is very suppressed; and seemingly double rings when seasonal growth is interrupted, for example by DROUGHT. *KRB*

Bauch, J. and Eckstein, D. 1981: Wood biological investigations on panels of Rembrandt paintings. *Wood Science and Technology* 15, 251–263. **Baillie, M.G.L.** 1995: *A slice through time: dendrochronology and precision dating.* London: Batsford. **Cook, E.R., Briffa, K.R., Meko, D.M. et al.** 1995: The segment length curse in long tree-ring chronology development for palaeoclimatic studies. *The Holocene* 5, 229–237. **Cook, E.R. and Kairiukstis, L.A.** 1990: *Methods of dendrochronology: applications in environmental sciences.* Dordrecht: Kluwer. **Cook, E.R., Meko, D. M., Stahle, D.W. and Cleveland, M.K.** 1999: Drought reconstructions for the continental United States. *Journal of Climate* 12, 1145–1162. **Dean, J.S.** 1997: Dendrochronology. In Taylor, R.E. and Aitken, M.J. (eds), *Chronometric dating in archaeology.* New York: Plenum, 31–64. **Fritts, H.C.** 1976: *Tree rings and climate.* London: Academic Press. **Hillam, J.** 1998: *Dendrochronology: guidelines on producing and interpreting dendrochronological dates.* London: Ancient Monuments Laboratory, Conservation and Technology, English Heritage. **Jacoby, G.C.** 1997: Application of tree-ring analysis to paleoseismology. *Reviews of Geophysics* 35, 109–124. **Schweingruber, H.C.** 1998: *Tree rings and environment: dendroecology.* Berne: Haupt.

dendroclimatology A subfield of DENDROCHRONOL-OGY, within which the relationships between annual tree growth and climate are studied. This encompasses exploration of the detailed effects of climate variability on the temporal and spatial variability of tree growth, as measured and recorded in terms of numerous ring-width, tree-ring densitometric and isotopic parameters. Dendroclimatology also encompasses the use of tree-ring data for estimating past climate variability, principally by the application of tree-ring/climate TRANSFER FUNCTIONS, calibrated against modern climate data, and applied to past tree-growth measurements. *KRB*

[See also RESPONSE FUNCTION, DENSITOMETRY, ISOTOPE DENDROCHRONOLOGY, MICRODENDROCLIMATOLOGY]

Briffa, K.R. 2000: Annual climate variability in the Holocene: interpreting the message of ancient trees. *Quaternary Science Reviews* 19, 84–105. **Briffa, K.R., Schweingruber, F.H., Jones, P.D. et al.** 1998: Reduced sensitivity of recent tree growth to temperatures at high latitudes. *Nature* 391, 678–682. **Cook, E.R. and Kairiukstis, L.A.** 1990: *Methods of dendrochronology. Applications in environmental sciences.* Dordrecht: Kluwer/IIASA. **Fritts, H.C.** 1976: *Tree rings and climate.* London, Academic Press. **Hughes, M.K., Kelly, P.M., Pilcher, J.R. and LaMarche Jr, V.C.** (eds) 1982: *Climate from tree rings.* Cambridge: Cambridge University Press. **Jacoby, G.J., D'Arrigo, R.D. and Davaajamts, Ts.** 1996: Mongolian tree rings and 20th century warming. *Science* 273, 771–773.

dendroecology The study of various ecological processes using the evidence of their effect on the growth of trees, as recorded in their annual tree-rings. Hence the term includes studies of diverse subjects such as AIR POLLUTION, SOIL ACIDIFICATION, HEAVY-METAL POLLUTION and tree-population dynamics. It includes the other general subdiscipline of DENDROCLIMATOLOGY. *KRB*

[See also DENDROCHRONOLOGY]

Fritts, H.C. and Swetnam, T.W. 1989: Dendroecology. A tool for evaluating variations in past and present forest environments. *Advances in Ecological Research* 19, 111–188. **Schweingruber, F.H.** 1996: *Tree rings and environment: dendroecology.* Bern: Haupt.

dendrogeomorphology The branch of DENDROCHRONOLOGY dealing with Earth surface processes (including *dendroglaciology* – the study of past glacier behaviour). Many geomorphological processes may leave an impression on the growth of trees. Events such as EARTHQUAKES, LANDSLIDES, AVALANCHES or even glacier movements may be deduced from the pattern of ring-width growth in trees that were affected but not killed. The SYNCHRONOUS death of many trees may also indicate events, while the germination dates of others (e.g. on a glacial MORAINE) may provide a minimum age for the formation of landscape features. *KRB*

[See also CROSS-DATING]

Alestalo, J. 1971: Dendrochronological interpretation of geomorphic processes. *Fennia* 105, 1–140. **Jacoby, G.C., Sheppard, P.R. and Sieh, K.E.** 1988: Irregular recurrence of large earthquakes along the San Andreas Fault: Evidence from tree-rings. *Science* 241, 196–199. **LaMarche, V.C. Jr.** 1961: Rate of slope erosion in the White Mountains, California. *Geological Society of America Bulletin* 72, 1579–1580. **Luckman, B.H.** 1995: Calendar-dated, early 'Little Ice Age' glacier advance at Robson Glacier, British Columbia, Canada. *The Holocene* 5, 149–159. **Shroder Jr, J.F.** 1980: Dendrogeomorphology: review and new techniques of tree-ring dating. *Progress in Physical Geography* 4, 161–188.

densitometry A subdiscipline within DENDROCHRONOLOGY and DENDROCLIMATOLOGY based on the measurement of tree-ring density. The study of TREE RINGS was established originally on the basis of measurements of the widths of the annual rings seen in cross-section through the stems of extratropical trees. The patterns of relative-width variability were then associated with changing environmental factors, particularly the variability of climate. Tree-ring densitometry measures the intra-annual variability in ring density (e.g. as measured by X-ray penetration of sample wood) to enable the quantification of seasonal tree growth in terms of various densitometric parameters (e.g. maximum latewood density or mean latewood density). The strong contrast in the density of spring and late-summer wood also allows the more traditional ring-width measurements to be extracted by this technique. *KRB*

[See also X-RADIOGRAPHY]

Polge, H. 1970: The use of X-ray densitometric methods in dendrochronology. *Tree-Ring Bulletin* 30, 1–10. **Schweingruber, F.H.** 1988: *Tree rings: basics and applications of dendrochronology.* Dordrecht: Reidel. **Schweingruber, F.H.** 1990: Radiodensitometry. In Cook, E.R. and Kairiukstis, L.A. (eds), *Methods of dendrochronology: applications in the environmental sciences.* Dordrecht: Kluwer/IIASA, 55–63.

density The dry mass of mineral particles of a soil or sediment divided by the volume they occupy, sometimes referred to as *particle density*. Particle density of soil minerals varies between 2.6 and 2.7 Mg m^{-3}. *EMB*

[See also BULK DENSITY]

density dependence The concept that a (demographic) population parameter (e.g. numbers, death rate, migration rate) is determined by the density of individuals per unit area. All populations are density-dependent to some degree. Density dependence implies regulation by FEEDBACK MECHANISMS. *JLI*

[See also CARRYING CAPACITY, COMPETITION, POPULATION DYNAMICS]

Begon, M., Harper, J.L. and Townsend, C.R. 1996: *Ecology: individuals, populations and communities*, 3rd edn. Oxford: Blackwell Scientific.

density flow A current, also known as a *density current*, driven by a density contrast between the flow and the ambient fluid. This can be caused by differences in temperature (e.g. at a FRONT in the ATMOSPHERE), sediment concentration (TURBIDITY CURRENT, PYROCLASTIC FLOW) or salinity (CONTOUR CURRENT). Density flows occur in the atmosphere, in association with VOLCANIC ERUPTIONS, and in sea- and freshwater where they are important in the distribution and reworking of deep-water sediments. An aqueous density flow typically moves as a bottom current (*underflow*), but in a stratified water body it can spread along the interface (THERMOCLINE or CHEMOCLINE) between two WATER MASSES to form an *interflow*. Less dense flows in water bodies can spread as a surface *plume*. *MRT*

[See also SEDIMENT GRAVITY FLOW]

Allen, P.A. 1997: *Earth surface processes.* Oxford: Blackwell Science. **Hall, I.R., McCave, I.N., Chapman, M.R. and Shackleton, N.J.** 1998: Coherent deep flow variation in the Iceland and American basins during the last interglacial. *Earth and Planetary Science Letters* 164, 1521. **Talbot, M.R. and Allen, P.** 1996: Lakes. In Reading, H.G. (ed.), *Sedimentary environments: processes, facies and stratigraphy*, 3rd edn. Oxford: Blackwell, 83–124.

density slicing A point-wise non-linear IMAGE PROCESSING operation that maps ranges of values into fewer shades of grey. This transformation, which reduces detail, is used to enhance (see IMAGE ENHANCEMENT) particular features, to reduce noise and to outline distinct classes in the image. *ACF*

denudation The loss of material from both surface and subsurface parts of a landscape by both PHYSICAL and CHEMICAL WEATHERING and by EROSION. Denudation is a broader concept than erosion and can occur without erosion. It is usually considered in terms of whole drainage basins or regions and includes also subsurface material losses caused by leaching or groundwater movement. *Subaerial denudation* occurs at the interface of the LITHOSPHERE and ATMOSPHERE (i.e. land surface), thus excluding the marine denudational process operating beyond the coastal zone and normally excluding GLACIAL EROSION. *SHD*

[See also DENUDATION CHRONOLOGY, DENUDATION RATES]

Leeder, M. 1999: *Sedimentology and sedimentary basins: from turbulence to tectonics.* Oxford: Blackwell. **Sparks, B.W.** 1971: *Rocks and relief.* London: Longman.

denudation chronology The reconstruction of the erosional history of an area of the Earth's surface based primarily on the recognition of EROSION SURFACES and, where available, their RESIDUAL DEPOSITS and secondly on the morphological evidence of drainage networks. Denudation chronology is epitomised by the influential 'British school' of geomorphology developed in particular by S.W. Wooldridge around the mid-twentieth century. *JAM*

[See also LANDSCAPE EVOLUTION]

Brown, E.H. 1960: *The relief and drainage of Wales: a study in geomorphological development.* Cardiff: University of Wales Press. **Wooldridge, S.W. and Linton, D.** 1955: *Structure, surface and drainage in south-east England*, 2nd edn. London: Philip.

denudation rate The rate at which material is removed from the landscape. Denudation rates consist of the combined mechanical (sediments) and chemical (solutes) losses from a drainage basin or region (see EROSION RATES). For most climatic conditions and rock types, the mechanical component is greater than the chemical component in total DENUDATION (globally, the ratio is about 6:1). Calculation may involve a geological approach (using, for example, lowering experienced by a dated surface or using RADIOMETRIC DATING and FISSION-TRACK DATING to determine the specific depth below the surface of a layer of rock at a particular time in the past) or estimates based on the present-day operation of processes.

Globally, much of the sediment entering the oceans comes from the monsoonal, tectonically active river basins of south and southeast Asia and Oceania. In contrast, in the comparatively tectonically stable continents of Africa and Australia, denudation rates have apparently been low over prolonged geological time. Controls on calculated denudation rates include: relief (mountainous areas experience higher sediment loads than plains); scale of measurement (smaller drainage basins yield higher rates per unit area than larger ones); the impact of CLIMATIC CHANGE on vegetation cover and RUNOFF; and, for the HOLOCENE, the type and intensity of human impact (e.g. DEFORESTATION and river channel modification). In terms of climate and its relation to SEDIMENT YIELD, there are three climatic zones where rates may be especially high: mediterranean seasonal climates; monsoonal areas with large amounts of tropical rain; and semi-arid areas.

At a global scale, there is a broad correspondence between estimated present-day and long-term geological rates of denudation, despite the influence on the former of Quaternary oscillations of eustatic sea level (see EUSTASY) and of CLIMATIC CHANGE. The extent to which this correspondence has any significance or merely reflects fortuitous inaccuracies in the estimates is not clear.

The complexity of the denudation process has led some workers to develop models to provide forecasts of erosion potential. An example is the *cumulative seasonal erosion potential* (CSEP) *model*, which suggests, for example, that relative denudation rates were about 25% higher at the LAST GLACIAL MAXIMUM in the Mediterranean area than they are today. *RAS*

[See also MEDITERRANEAN LANDSCAPE EVOLUTION]

Kirkby, M.J. and Cox, N.J. 1995: A climatic index for soil erosion potential (CSEP) including seasonal and vegetation factors. *Catena* 25, 333–352. **Leeder, M.R., Harris, T. and Kirkby, M.J.** 1998: Sediment supply and climate change: implications for basin stratigraphy. *Basin Research* 10, 7–18. **Meade, R.H.** 1969: Errors in using modern stream-load data to estimate natural rates of denudation. *Bulletin of the Geological Society of America* 80, 1265–1274. **Summerfield, M.A.** 1991: *Global geomorphology*. Harlow: Longman, 371–402. **Walling, D.** 1987: Rainfall, runoff and erosion of the land: a global view. In Gregory, K.J. (ed.), *Energetics of physical environment*. Chichester: Wiley, 89–117.

deoxyribonucleic acid (DNA) The self-replicating complex molecule that carries genetic information, which is located in the chromosomes of the cell nucleus in organisms. It is a polymer composed of chains of *nucleotides* arranged in a double helix, each nucleotide consisting of a sugar, a phosphate group and a nitrogenous base. Inheritable characteristics are chemically coded in segments of DNA termed *genes*, of which there are around 100 000 in humans located on 23 chromosomes. The *genome* is the complete complement of genes in an organism, which characterises and defines the organism. *JAM*

[See also DNA, ANCIENT, GENETIC ENGINEERING]

Hecht, S.M. (ed.), 1996: *Bio-organic chemistry: nucleic acids*. New York: Oxford University Press.

deposition The letting-down of material transported by water, wind, snow, ice or tides and currents in the sea. Deposition includes CHEMICAL PRECIPITATION as well as *mechanical deposition*. *AJDF*

[See also DRY DEPOSITION, EROSION, WET DEPOSITION]

depositional environment An environment characterised by DEPOSITION of sediment rather than EROSION. Sedimentary rocks accumulated in former depositional environments, the characteristics of which can be identified using FACIES ANALYSIS and PALÆOENVIRONMENTAL RECONSTRUCTION. *GO*

Reading, H.G. (ed.) 1996: *Sedimentary environments: processes, facies and stratigraphy*, 3rd edn. Oxford: Blackwell.

depression An area of low atmospheric pressure, usually referring to mid-latitude features (*extratropical cyclone*) evident on surface synoptic maps associated with convergence and uplift of AIR MASSES at FRONTS. *JCM*

[See also CYCLONE]

Carlson, T.N. 1991: *Mid-latitude weather systems*. New York: Wiley.

derived fossil A FOSSIL reworked from its original context of preservation into a younger SEDIMENT or SEDIMENTARY ROCK. *TY*

[See also TAPHONOMY]

desalination The removal of salts from saline water (e.g. sea water or GROUNDWATER) to render it suitable for domestic, agricultural or industrial use. The first desalination plant, based on distillation, was built by the British government in Aden, Yemen, in 1869. New methods have been developed since the 1960s, based on reverse OSMOSIS, ion exchange, electrodialysis, liquid extraction or freeze separation, but distillation remains the most common method in use. Desalination plants are in use in > 50 countries and are particularly important in the Middle East, California and Florida, where they have brought about a transformation of the local environment, but they still provide no more than 0.1% of the Earth's freshwater. *JAM*

Porteous, A. 1983: *Desalination technology*. London: Applied Science. **Shahin, M.** 1989: Review and assessment of water resources in the Arab Region. *Journal of Water International* 14, 206–219.

desalinisation The removal of salts from salty soils. This is effectively the reversal of the SALINISATION process that occurs naturally in semi-arid environments and artificially as a result of ill-founded IRRIGATION schemes. Desalinisation may also be natural or artificial. Improved drainage, irrigation and use of salt-tolerant plants (HALOPHYTES) are techniques used in artificial desalinisation. Desalinisation occurs naturally on EMERGENT COASTS; it is also necessary during LAND RECLAMATION from the sea. In the empoldered areas of The Netherlands, however, there are few salinity problems because the soils become desalinised by DIFFUSION of salts into fresh water. *JAM*

descriptive statistic A quantitative summary of the measurements made on a set of objects. Measures of *central tendency* (e.g. mean and median), *dispersion* (e.g. standard deviation and interquartile range), sums and percentages are all examples of descriptive statistics. *JAM*

[See also INFERENTIAL STATISTIC]

desert An ambiguous term used by local communities in DRYLAND ENVIRONMENTS to describe areas of perceived ARIDITY. Deserts are complex and diverse landscapes, commonly found in environments ranging from hyperarid to SEMI-ARID, which may be characterised by sparseness of VEGETATION, dominance of bare rock surfaces, angular rock debris and, often, the presence of AEOLIAN deposits, including SAND DUNES. A key feature of many deserts is a large diurnal temperature range, which inspired early explorers and travellers to write of the efficiency of INSOLATION WEATHERING in deserts. Deserts are associated with certain landforms and vegetation communities, which, while not unique to deserts, are nevertheless found in most hyper-arid and arid environments. Features include: ROCK VARNISH; *playas* and SALINE LAKES; DESERT PAVEMENTS (also known as GIBBER, *hamada* or *reg*); and WADIS. Fog and dew are an important source of moisture for landscape evolution, soil development and vegetation communities. The popular view of deserts involves marching sand dunes, implying a dominance of aeolian processes in both contemporary deserts and ancient palaeodeserts that are found within the geological record. However, episodic FLUVIAL and COLLUVIAL PROCESSES respond rapidly to a change in rainfall regime and many desert features may reflect short-term changes in moisture availability, rather than long-term aridity. Given the ephemeral nature of many desert processes, an understanding of the importance of ENVIRONMENTAL CHANGE in driving desert landscape evolution is a study currently in its infancy.

Evidence of past environmental change exists in the form of sedimentary remnants of former lakes visible in the hyperarid Sahara and Namib deserts of Africa and the arid deserts of the American southwest. Many of these lakes are believed to have formed in PLUVIAL periods sometimes associated with the LAST GLACIAL MAXIMUM. *Aridisols* are characteristic soils of deserts; these may indicate palaeoenvironments with wetter climates, as revealed by the presence of carbonate precipitation at depth within the soil profile. The presence of PALAEOSOLS in ALLUVIAL FANS, sand dunes and windblown DUST deposits is also believed to reflect persistent periods of landscape stability, which may result from environmental change. *MLC*

[See also ARIDLAND: PAST ENVIRONMENTAL CHANGE, DESERT VEGETATION, ERG, GEOLOGICAL RECORD OF ENVIRONMENTAL CHANGE]

Abrahams, A.D. and Parsons, A.J. 1994: *Geomorphology of desert environments*. London: Chapman and Hall. Cooke, R.U., Warren, A. and Goudie, A.S. 1993: *Desert geomorphology*. London: Longman. Goudie, A.S., Allchin, B. and Hegde, K.T.M. 1973: The former extension of the Great Indian sand desert. *Geographical Journal* 139, 243–257. Goudie, A.S. and

Wilkinson, J.C. 1977: *The warm desert environment*. Cambridge: Cambridge University Press. Mabbutt, J.A. 1977: *Desert landforms*. Canberra: ANU Press. Sarnthein, M. 1978: Sand deserts during the last glacial maximum. *Nature* 272, 43–46. Street, F.A. and Grove, A.T. 1976: Environmental change and climatic implications of late Quaternary lake-level fluctuations in Africa, *Nature* 261, 385–390. Thomas, D.S.G., Stokes, S. and Shaw, P.A. 1997: Holocene aeolian activity in the southwestern Kalahari Desert, southern Africa: significance and relationships to late-Pleistocene dune-building events. *The Holocene* 7, 273–281. Vogel, J.C. 1989: Evidence of past climate change in the Namib Desert. *Palaeogeography, Palaeoclimatology, Palaeoecology* 70, 355–366.

desert pavement A LAG DEPOSIT of coarse particles set on or in finer material comprising varying mixtures of sand, silt or clay. Most common in unvegetated areas, pavements are typically caused by the processes of DEFLATION and *sheetwash*, which remove the smaller particles from the ground surface. Desert pavements, or *stone pavements*, provide a barrier of protection for desert soils, help control surface stability and influence surface infiltration and runoff. *MAC*

[See also GIBBER, SAND SEA,]

Cooke, R.U. 1970: Stone pavements in deserts. *Annals of the Association of American Geographers* 60, 560–577. Wainwright, J., Parsons, A.J. and Abrahams, A.D. 1995: A simulation study of the role of raindrop erosion in the formation of desert pavements. *Earth Surface Processes and Landforms* 20, 277–291.

desert vegetation Typically, the vegetation consists of XEROPHTYE species, which are characteristic of hyperarid and ARIDLANDS. Whilst the term DESERT is used to cover a range of DRYLANDS, perceptions of desert vegetation communities are dominated by succulents, such as cacti and agaves, which have adapted to conditions of low rainfall and high EVAPOTRANSPIRATION by storing water in specially adapted tissues in their roots, stems and leaves. There is no clear distinction between desert vegetation and ARIDLAND VEGETATION, in that both communities populate land where EVAPORATION exceeds PRECIPITATION. In addition to shrubs and grasses, DESERTS contain a wide variety of lichen and algae, which colonise areas of shade and, with bacteria, can contribute to surface bio- and mycophytic crusts.

In some deserts, such as the Mojave Desert in California, specific plant communities exist in zones clearly related to altitude, lithology and moisture availability. The Piñon pine-juniper (*Pinus-Juniperius*) zone consists of woodland occurring on mountain slopes above 1200 m, creosote bush (*Larrea tridentata*) scrub occurs on well drained alluvial fans and gentle slopes, upper sagebrush-grass (*Artemisia tridentata*) scrub occurs on valley sides and upland valleys and the zone occupied by shadscale (*Atriplex convertifolia*) is in valley bottoms. Past changes in environmental conditions in the PLEISTOCENE and HOLOCENE have changed both the species found in these locations and the altitude of the vegetation zones.

Unique evidence of past ENVIRONMENTAL CHANGE in deserts derives from the habits of desert packrats (*Neotoma* spp.), which collect a wide variety of plant material, including seeds, POLLEN, twigs, leaves and cactus spines, from the local desert environment to create their middens. These *packrat middens* build up over time and are cemented by urine, creating an archive of desert vegetation that can be dated using RADIOCARBON techniques. The vegetation record in such RODENT MIDDENS can be used to determine changing environmental conditions, identifying past episodes when currently arid deserts have been wetter and colder than today.   MLC

[See also ARIDLAND VEGETATION, BIOMES]

Betancourt, J.L., Van Devender, T.R. and Martin, P.S. 1990: *Packrat middens: the last 40,000 years of biotic change.* Tucson, AZ: University of Arizona Press. **Evenari, M., Noy-Meir, I. and Goodall, D.W. (eds)** 1985 and 1986: *Hot deserts and arid shrublands* [*Ecosystems of the World*. Vols 12A and 12B]. Amsterdam: Elsevier. **Grayson, D.K.** 1993: *The desert's past: a natural prehistory of the Great Basin.* Washington, DC: Smithsonian Institution Press. **Louw, G.N.and Seely, M.K.** 1982: *Ecology of desert organisms.* London: Longman. **Ritchie, J.C.** 1994: Holocene pollen spectra form Oyo, northwestern Sudan: problems of interpretation in a hyperarid environment. *The Holocene* 4, 9–15. **Spaulding, W.G.** 1991: A middle Holocene vegetation record from the Mojave Desert of North America and its paleoclimatic significance. *Quaternary Research* 35, 427–437.

desertification   The process of LAND DEGRADATION in the context of DRYLANDS. The principal processes of desertification are: degradation of the vegetation cover; accelerated water and wind erosion; and SALINISATION and waterlogging. For most authorities, the term *desertisation* is regarded as synonymous with desertification. There are more than 100 definitions, which partly explains why desertification is an extremely controversial environmental issue.

The term was first used, though not formally defined, by Aubréville, a French forester, in AD 1949 in referring to the way in which he considered that the Sahara Desert was expanding to incorporate desert-marginal SAVANNA grasslands. The term reached a much wider audience in the 1970s when international attention became focused on the drought-ridden SAHEL zone of Africa. A result of this attention was the United Nations Conference on Desertification (UNCOD) held in Nairobi in 1977. The world map of desertification arising out of the conference (see Figure) suggested that desertification was occurring in drylands throughout the world, and not just in Africa, although the map is based on opinion rather than fact. The definition that emerged from the conference proposed that desertification represented the 'diminution or destruction of the biological potential of land that can lead ultimately to desert-like conditions'. The definition was criticised for being too broad. In 1984, a joint UNEP and FAO Desertification Hazards Map was produced and the revised definition drew attention to the role of socio-economic factors in causing desertification. This was later emphasised in the 1992 UN official definition, which stated that desertification was 'land degradation in arid, semiarid and dry subhumid areas resulting mainly from adverse human impact'. In the 1992 United Nations Conference on Environment and Development (UNCED) in Rio de Janeiro, an intergovernmental committee negotiated the Convention to Combat Desertification, which was enforced on 26 December 1996. In this definition, acknowledgement of the role of climatic factors in causing desertification was reinstated ('land degradation in arid, semiarid and dry subhumid areas resulting from various factors, including climatic

variations and human activities'). Hence the term can be used in the context of Mediterranean regions. Hyperarid zones, the true deserts, are not usually considered to be prone to desertification because of their naturally very low biological activity; this is why they are not included in the last two climatically specific definitions. The term desertification has, however, been applied to non-dryland areas (e.g. Iceland) when severe land degradation is described.

Various attempts have been made to assess the global extent of the world's land areas at risk from desertification. These are based on estimates of desertification HAZARD, rather than the results of ENVIRONMENTAL MODELLING of its extent. As a consequence, the figures vary. According to how drylands are defined, some 70–79% (c. 3.5–4.0 billion ha) are estimated to be affected by desertification. Furthermore, apart from recent applications of REMOTE SENSING, there has been no accurate assessment of the world-wide extent of desertification or of its rate of increase and/or decrease. Desertification is thought to be a direct threat to some 200 million people world-wide and an indirect threat to an additional 700 million people, although figures vary from source to source.

The causes of desertification are very controversial. In terms of climatic factors, the possible roles of temporary intense DROUGHT periods, long-term CLIMATIC CHANGE towards aridity (either in terms of the now discredited PROGRESSIVE DESICCATION concept or as part of a 200-year cycle) and human-induced climatic change have been considered. Certainly, severe droughts do take place and have taken place in susceptible drylands and their effects are aggravated by the increased pressure on the land induced by increases in human and domestic animal populations. The devastating Sahel Drought, which lasted (with six 'wet' years) from AD 1968 to 1984, caused considerable ecological stress, but it was not unprecedented. Broadly comparable droughts occurred in the periods AD 1910–1915 and 1944–1948, but it was an increase in population numbers that made the impact of the most recent period so serious. There is some suggestion that climate may have varied since the LITTLE ICE AGE in the Sahel, but there is no clear evidence of a general global trend towards a drier climate this century. Whether current human impacts on the vegetation cover can have a positive feedback on rainfall remains speculative and contentious. A typical, though not universally accepted, present-day view is that the low annual totals and high annual variability of rainfall with periodic successions of dry years provide a suitable *milieu* in which injudicious human action can cause land degradation, rather than climate being a primary cause of desertification.

There is a range of recognised 'classic' anthropogenic impacts that are thought to cause desertification, but they are often based on subjective assessment rather than long-term environmental monitoring. OVERGRAZING by livestock is one such cause. Vegetative destruction is caused by TRAMPLING, disturbance of root systems and COMPACTION causing reduced INFILTRATION, as well as by eating. Livestock-based livelihoods are an important component of both subsistence and commercial economic activities in drylands. In recent decades, ranching has spread onto many dryland areas formerly only utilised on a nomadic basis (see NOMADISM). Increasing livestock numbers have been linked in some areas to the provision

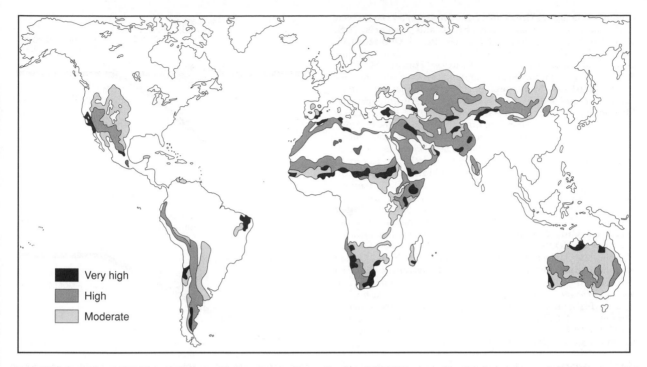

desertification *The 1977 United Nations Conference on Desertification (UNCOD) desertification hazard map (after Thomas and Middleton, 1994)*

of boreholes and wells to tap GROUNDWATER supplies. It has been argued that continual pressure around each watering point leads to the formation of PIOSPHERES, which become highly degraded. Coalescence of growing piospheres supposedly leads to widespread destruction of the vegetation cover, but this simple view has been questioned recently. Similarly, a simplistic interpretation of the concept of environmental degradation being caused by overgrazing exceeding the potential CARRYING CAPACITY of the land may be flawed. For example, livestock numbers may actually decline through mortality, low birth rates and, in some cases, sales and thus reduce the potential impacts of large livestock numbers at the point when the vegetation cover is under most stress.

Some researchers believe that OVERCULTIVATION is the main cause of desertification. It is thought to occur as a result of the shortening of periods when the land is left FALLOW, leading to nutrient depletion, or from the use of mechanical techniques, which damage the soil structure and reduce AGGREGATE STABILITY, leading to widespread loss of soils by wind or water. The use of mechanical techniques was certainly the reason for the DUST BOWL of the USA in the 1930s when ignorance led to the use of inappropriate deep ploughing, which rendered the soils of the semi-arid grassland prone to WIND EROSION when drought occurred. However, the reasons for the similar large-scale wind erosion in the 1970s were more complex. Although the need to avoid ploughing was then recognised, economic incentives overcame the disincentive to plough.

Dryland irrigated cropland increased in area by 34% in the period 1961–1978 and the mismanagement and misuse of irrigation schemes can lead to SALINISATION, ALKALISATION and the accumulation of sodium, all of which cause reduced crop productivity.

DEFORESTATION and the removal of woody material represent the third main commonly recognised anthropogenic cause of desertification. Acceleration of woodland and FOREST CLEARANCE in drylands has occurred in recent decades. In the late 1980s, it was estimated that only 4–6% of Ethiopia was forested, whereas there had formerly been a 40% cover. In developing countries, many people depend on wood for domestic uses (construction, cooking, heating, brick manufacture, etc.). Demand for wood has come increasingly from urban populations. For example, in Burkino Faso, the capital Ougadougou accounts for as much as 95% of national forest consumption.

There are several other contributory causes of desertification including: soil COMPACTION and CRUSTING caused by trampling, heavy machinery or raindrop impact; disruption of fragile surfaces and vegetation cover by vehicular access; quarrying for construction materials; and disturbance due to desert warfare, such as the Iraqi invasion of Kuwait in 1990 (see MILITARY IMPACTS ON ENVIRONMENT).

As regards GLOBAL WARMING and its possible impact on desertification in the world's drylands, whilst there are differences in the predictions of the different CLIMATIC MODELS, there was an overall warming trend in the twentieth century and warmer years may be associated with drier conditions. The implication is that global warming is likely to reduce further the already limited availability of moisture, increasing the risk of desertification. *RAS*

[See also AGRICULTURAL IMPACT ON SOILS, SOIL DEGRADATION]

Aubréville, A. 1949: *Climats, forêts et désertification de l'Afrique tropicale.* Paris: Société d'Editions Géographiques, Maritimes et

Coloniales. **Dregne, H.E.** 1983: *Desertification of arid lands*. New York: Harwood Academic. **Fantechi, R. and Margaris, N.S.** 1986: *Desertification in Europe*. Dordrecht: Reidel. **Glantz, M.H. (ed.)** 1977: *Desertification: environmental degradation in and around arid lands*. Boulder, CO: Westview. **Hulme, M.** 1996: Recent changes in the world's drylands. *Geophysical Research Letters* 23, 61–64. **Mainguet, M.** 1991: *Desertification: natural background and human mismanagement*. Heidelberg: Springer. **Middleton, N. and Thomas, D. (eds)** 1997: *World atlas of desertification*, 2nd edn. London: Arnold. **Perez-Trejo, F.** 1994: *Desertification and land degradation in the European Mediterranean*. European Commission, Directorate General Science, Research and Development, Report EUR 14850. Luxembourg: European Commission. **Thomas, D.S.G. and Middleton, N.J.** 1994: *Desertification: exploding the myth*. Chichester: Wiley. **Tucker, C.J. and Nicholson, S.E.** 1999: Variations in the size of the Sahara Desert from 1980 to 1997. *Ambio* 28, 587–591. **United Nations** 1977a: *Desertification: its causes and consequences*. Oxford: Pergamon. **United Nations** 1977b: *World map of desertification* [*Document* A/CONF.74/2]. UN Conference on Desertification, Nairobi, 29 August–9 September 1977. New York: United Nations. **United Nations Environment Programme (UNEP)** 1994: *United Nations Convention to Combat Desertification in countries experiencing serious drought and/or desertification, particularly in Africa*. Elaborated by the Intergovernmental Negotiating Committee (INCD), signed June 1994. UN General Assembly A/AC.241/15/Rev.5. **Verstraete, M.M.** 1986: Defining desertification: a review. *Climatic Change* 9, 5–18.

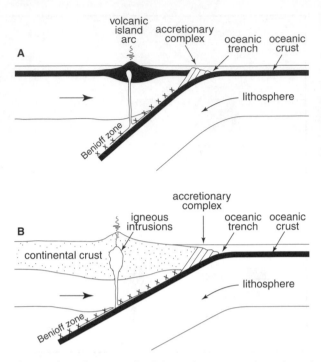

destructive plate margin *Schematic cross-sections through destructive plate margins where (A) plates on both sides comprise lithosphere bearing oceanic crust, and (B) one plate carries continental crust. Crosses mark sites of earthquake foci along the inclined Benioff zone*

desiccation (1) The removal of water from a substance and/or its drying out. (2) In DRYLANDS, the process of longer-term reductions in moisture availability resulting from a dry period at the decadal scale. A period of long-term rainfall deficit can have a major impact on rainfed crop production and LAND DEGRADATION. In Sudan, for example, rainfed cereal production is estimated to have declined by 14% in the period AD 1970–1980, when rainfall totals declined. In southern Africa, an 18-year cycle of rainfall fluctuations extending back to the 1840s in the summer rainfall zone has been identified. The apparently inherent rainfall variability of drylands means that defining and delimiting their extent using simple climatic measures is extremely difficult. *RAS*

[See also DESERTIFICATION, DROUGHT, MUD CRACKS, PROGRESSIVE DESICCATION]

Hare, F.K. 1987: Drought and desiccation: twin hazards in a variable climate. In Wilhite, D. and Easterby, W (eds), *Planning for drought*. London: Westview Press, 3–10. **Thomas, D.S.G. and Middleton, N.J.** 1994: *Desertification: exploding the myth*. Chichester: Wiley, 106–108.

destructive plate margin A convergent boundary between two tectonic plates (see Figure) at which LITHOSPHERE is destroyed (also known as a *convergent plate margin*). One plate sinks beneath the other and its lithosphere is absorbed into the MANTLE in the process of SUBDUCTION. The junction between the plates is marked at the surface by a deep OCEANIC TRENCH and friction along the upper surface of the subducting plate generates EARTHQUAKES along an inclined *Benioff zone*. Granitic INTRUSIONS and *volcanic activity* of andesitic composition develop above the subducting plate. Because only lithosphere bearing OCEANIC CRUST can be subducted, on account of its higher density, the WILSON CYCLE envisages changing types of destructive plate margin as an OCEAN is consumed. An *ocean–ocean destructive plate margin* is associated with a volcanic ISLAND ARC on the over-riding plate (e.g. Aleutian

Islands). At an *ocean–continent destructive plate margin* the plate bearing oceanic crust is subducted beneath an ACTIVE CONTINENTAL MARGIN and may form a *continental margin orogen* (e.g. Andes). If the subducting plate eventually brings CONTINENTAL CRUST, subduction ceases and a continent–continent COLLISION ZONE gives rise to an OROGENIC BELT (e.g. Himalayas). *GO*

[See also PLATE TECTONICS]

detachment-limited erosion Where agents (e.g. wind or water) have the capacity to transport more soil material than is available by *detachment* (i.e. loosening of soil particles from the soil mass). *SHD*

[See also EROSION, TRANSPORT-LIMITED EROSION, WATER EROSION]

Morgan, R.P.C. 1995: *Soil erosion and conservation*, 2nd edn. Harlow: Longman.

deteriorated pollen FOSSIL and SUBFOSSIL pollen grains can show various signs of damage of the exine (outer layer of pollen grain wall) caused by physical, chemical or biological processes. Various types of damage may be distinguished: *corroded* (ektexine – layers making up the pollen grain wall – only affected); *degraded* (structural change of the whole exine, elements 'blurred' or 'diffuse'); *broken* (split); and *crumpled* (folded). Physical (mechanical) corrosion and crumpling typically prevail in SILTS, chemical corrosion in PEATS. *BA*

[See also MICROFOSSIL, POLLEN ANALYSIS]

Cushing, E.J. 1967: Evidence of differential pollen preservation in Late Quaternary sediments in Minnesota. *Review of Palaeobotany and Palynology* 4, 87–101. **Havinga, A.J.** 1984: A 20-year investigation into differential corrosion susceptibility of pollen and spores in various soil types. *Pollen et Spores* 26, 541–558.

determinism A simplified and subjective attempt to explain human activities as a direct response to the prevailing environmental factors. It emphasised the control of climate (including CLIMATIC CHANGE) on the progress of history and the rise and fall of civilisations. This approach to human–climate relationships was popular at the end of the nineteenth century and continued until the 1940s, although the extreme and oversimplified theories became unacceptable to most geographers and others by the late 1920s.

Ellsworth Huntington vigorously propounded the case for *climatic determinism* with the publication of *Civilisation and Climate*, which remained a 'best-seller' for some 30 years, with a third edition published in 1924. Ellen Churchill Semple was an equally famous advocate of the broader *environmental determinism*. Huntington received support from other climatologists into the 1940s, when the decline of descriptive, static determinism coincided with the growth of a more pertinent dynamic climatology – a 'synthesis of weather' – with the dominance shifting to advective controls of atmospheric dynamism.

Current renewed interest in relationships between people and environment (including climate) are altogether more complex and cautious. Detailed consideration of human environments in ENVIRONMENTAL ARCHAEOLOGY, for example, is justified as the background to the conditions of life in which human choices and decisions are made. It does not entail *deterministic* interpretations. Environmental effects upon human communities are mediated through technology and cognition, the distinctively human means of ADAPTATION. Contextual richness and specificity have been reinstated in understanding human-environmental interaction. *RDT/JAM*

[See also CLIMATIC CHANGE: IMPACTS ON HUMANS, CONTEXTUAL ARCHAEOLOGY, ENVIRONMENTALISM, POSSIBILISM]

Binford, M.W., Kolata, A.L., Brenner, M. *et al.* 1997: Climate and the rise and fall of an Andean civilization. *Quaternary Research* 47, 235–248. **Cullen, H.M., deMenocal, P.B., Hemming, S.** *et al.* 2000: Climate change and the collapse of the Akkadian empire: evidence from the deep sea. *Geology* 28, 379–382. **Dincauze, D.F.** 2000: *Environmental archaeology: principles and practice.* Cambridge: Cambridge University Press. **Huntington, E.** 1924: *Civilisation and climate,* 3rd edn. Boston, MA: Yale University Press. **Manley, G.** 1958: The revival of climatic determinism. *Geographical Review* 48, 98–105. **Markham, S.F.** 1944: *Climate and the energy of nations.* New York: Oxford University Press. **Mills, C.A.** 1942: *Climate makes the man.* New York: Harper. **Semple, E.C.** 1935: *Influences of the geographical environment.* London: Constable. **Siddiqui, A.H. and Oliver, J.E.** 1987: Determinism, climatic. In Oliver, J.E. and Fairbridge, R.W. (eds), *The encyclopedia of climatology.* New York: Van Nostrand Reinhold.

deterministic model A model in which no component is inherently uncertain. Thus no PARAMETER in the model is characterised by a PROBABILITY DISTRIBUTION and, for fixed starting values, a deterministic model always produces the same result. Because of the complexity and uncertainty of environmental processes and their inter-

actions, deterministic models are generally less realistic than *stochastic models* in the context of environmental change. *JAM*

[See also STOCHASTICITY]

detrital A CLAST, SEDIMENT or SEDIMENTARY ROCK derived by WEATHERING, often used synonymously with TERRIGENOUS clastic and *siliciclastic* (see CLASTIC). *GO*

detrital remnant magnetisation The magnetisation of ferromagnetic sediment particles settling out of air, water or saturated sediments, usually preserved in LAKE SEDIMENTS or MARINE SEDIMENTS, providing a proxy record of the ambient geomagnetic field (see GEOMAGNETISM) at the time of formation. *CJC*

[See also PALAEOMAGNETIC DATING]

Thompson, R. and Oldfield, F. 1986: *Environmental magnetism.* London: Allen and Unwin.

detritus (1) Weathered and eroded mineral or rock fragments transported by natural processes to their place of deposition. (2) ALLOCHTHONOUS and AUTOCHTHONOUS fragments of plants and animals deposited in water. *UBW*

[See also GYTTJA, SEDIMENT TYPES]

Devensian The last GLACIAL EPISODE identified in the British Isles, originally defined as beginning at the end of the last full INTERGLACIAL, the *Ipswichian*, and terminating at the onset of the present interglacial, the HOLOCENE, at 10 ka BP (^{14}C years). More recently the opening has been defined by the boundary between oxygen ISOTOPIC STAGES 5a and 4 (c. 74 ka BP). It is equivalent to the *Wisconsinan* in North America and the *Weichselian* in Northern Europe. *DH*

[See also QUATERNARY TIMESCALE]

Bowen, D.Q. (ed.) 1999: *A revised correlation of Quaternary deposits in the British Isles* [*Special Report* 23]. Bath: The Geological Society.

devolution The concept of devolving political powers, normally to local or regional levels of government, as exemplified by the recently created Scottish Parliament and Welsh Assembly in the UK. The term can also be used for the devolution of powers from a supranational organisation to nation states. In the case of the European Union, the principle of *subsidiarity*, though complex, may be viewed as complementary to the concept of devolution. *JAM/RMD*

Deacon, R. 1996: New Labour and the Welsh Assembly: 'Preparing for a new Wales', or updating the Wales Act 1978? *Regional Studies* 30, 689–693. **Weatherill, S. and Beaumont, P.** 1999: *EU Law: the essential guide to the legal workings of the European Union.* Harmondsworth: Penguin.

Devonian A SYSTEM of rocks and a PERIOD of geological time from 416 to 354 million years ago. *GO*

[See also GEOLOGICAL TIMESCALE]

diachronism An EVENT that occurred at different times in different places exhibits diachronism. The concept is most commonly applied in the GEOLOGICAL RECORD to changes in LITHOLOGY or FACIES in SEDIMENTS and SEDIMENTARY ROCKS. An example is the change from marine

157

to non-marine deposits caused by the PROGRADATION of a DELTA into a marine basin: the base of the non-marine deposits would be progressively younger, the further the delta advanced. It may only be possible to demonstrate that changes are diachronous where CHRONOSTRATIGRAPHY is precise, for example as established by EVENT STRATIGRAPHY or high-resolution BIOSTRATIGRAPHY. A classic example from the UK, first demonstrated over a century ago, used ammonite faunas to demonstrate that the Lower JURASSIC Upper Lias Sands became progressively younger when traced from the Cotswold Hills southwards to the Dorset coast. *JCWC*

Brown, A.G. 1990: Holocene floodplain diachronism and inherited downstream variations in fluvial processes – a study of the River Perry, Shropshire, England. *Journal of Quaternary Science* 5, 39–51. **Buckman, S.S.** 1889: On the Cotteswold, Midford and Yeovil Sands, and the division between Lias and Oolite. *Quarterly Journal of the Geological Society of London* 45, 440–474. **Callomon, J.H. and Cope, J.C.W.** 1995: The Jurassic geology of Dorset. In Taylor, P.D. (ed.), *Field geology of the British Jurassic*. London: Geological Society, 51–103. **Tremblay, A., Ruffet, G. and Castonguay, S.** 2000: Acadian metamorphism in the Dunnage zone of southern Quebec, northern Appalachians: Ar-40/Ar-39 evidence for collision diachronism. *Geological Society of America Bulletin* 112, 136–146.

diachronous A TIME TRANSGRESSIVE unit, EVENT or boundary that varies spatially in age. QUATERNARY glacigenic sequences, for example, are related to diachronous ice margin fluctuations and their boundaries are not SYNCHRONOUS. *DAR*

[See also CHRONOSTRATIGRAPHY]

diagenesis Physical and chemical changes, such as AUTOCOMPACTION, COMPACTION, CONSOLIDATION and CEMENTATION, that affect a SEDIMENT during burial – i.e. between the time of DEPOSITION and the onset of metamorphism, under relatively low temperature and pressure. One important consequence is LITHIFICATION to form a SEDIMENTARY ROCK. Chemical changes are associated with the interaction of sediment and its porewater. Minerals formed during diagenesis are termed AUTHIGENIC, and may replace existing GRAINS, or grow into pore space as a CEMENT. The early stages of diagenesis are influenced by surface environments, and diagenetic alteration of geological and archaeological materials must be taken into account when chemical analysis is undertaken. *TY/GO*

Burley, S.D., Kantorowicz, J.D. and Waugh, B. 1985: Clastic diagenesis. In Brenchley, P.J. and Williams, B.P.J. (eds), *Sedimentology: recent developments and applied aspects* [*Special Publication* 18]. London: Geological Society, 189–226. **Dickson, J.A.D.** 1985: Diagenesis of shallow-marine carbonates. In Brenchley, P.J. and Williams, B.P.J. (eds), *Sedimentology: recent developments and applied aspects* [*Special Publication* 18]. London: Geological Society, 173–188. **Harwood, G.M.** 1988: Microscopic techniques: II. Principles of sedimentary petrography. In Tucker, M.E. (ed.), *Techniques in sedimentology*. Oxford: Blackwell Scientific, 108–173. **Quattropani, L., Charlet, L., de Lumley, H. and Menu, M.** 1999: Early Palaeolithic bone diagenesis in the Arago cave at Tautavel, France. *Mineralogical Magazine* 63, 801–812. **Trewin, N.H.** 1988: Use of the scanning electron microscope in sedimentology. In Tucker, M.E. (ed.), *Techniques in sedimentology*. Oxford: Blackwell Scientific, 229–273.

diagnostic soil horizons SOIL HORIZONS with stipulated criteria including thickness that are used to allocate SOIL PROFILES to the appropriate category in modern systems of SOIL CLASSIFICATION. The MOLLIC, UMBRIC and HISTIC HORIZONS are examples of *diagnostic surface horizons*; CAMBIC, ARGIC, FERRALIC, CALCIC, SPODIC and *petrogypsic* HORIZONS are examples of *subsurface diagnostic horizons*. Diagnostic or *reference soil horizons* are used in most current systems of soil classification, including the SOIL TAXONOMY of the United States Department of Agriculture and the WORLD REFERENCE BASE FOR SOIL RESOURCES. *EMB*

Food and Agriculture Organization of the United Nations 1998: *World reference base for soil resources* [*Soil Resource Report* 84]. Rome: FAO, ISRIC, ISSS. **Soil Survey Staff** 1999: *Soil taxonomy: a basic system of soil classification for making and interpreting soil surveys* [*Agriculture Handbook* 436]. Washington, DC: Natural Resources Conservation Service.

diamagnetism An extremely weak magnetic property that arises from the interaction of an applied MAGNETIC FIELD with the orbital motion of electrons such that the electron orbits become aligned so as to oppose the external field. It is a very weak negative MAGNETISATION and becomes lost once the magnetic field is removed. *AJP*

diamict, diamictite, diamicton Diamict is the general term for non-sorted terrigenous sediment or rock containing a wide range of particle sizes, commonly ranging from clay to boulder. The term embraces *diamicite* for the rock (lithified) equivalent and *diamicton* for the unconsolidated sediment. *MJH*

[See also FABRIC ANALYSIS, GLACIAL SEDIMENTS, PERIGLACIAL SEDIMENTS, SEDIMENT GRAVITY-FLOWS]

Flint, R.F., Sanders, J.E. and Rodgers, J. 1960: Diamictite: a substitute term for symmictite. *Geological Society of America Bulletin* 71, 1809–1810. **Hambrey, M.J.** 1994: *Glacial environments*. London: UCL Press. **Harland, W.B., Herod, K.N. and Krinsley, D.H.** 1966: The definition and identification of tills and tillites. *Earth-Science Reviews* 2, 225–256.

diapir A gravity-driven dome-like INTRUSION of low-density material into a denser cover. Examples include some intrusions of MAGMA, SALT DOMES, MUD diapirs, LOAD STRUCTURES and movements of thawed material within PERMAFROST. *GO*

[See also INVOLUTIONS, SOFT-SEDIMENT DEFORMATION]

Ramberg, H. 1981: *Gravity, deformation and the Earth's crust*, 2nd edn. London: Academic Press. **Swanson, D.K., Ping, C.-L. and Michaelson, G.J.** 1999: Diapirism in soils due to thaw of ice-rich material near the permafrost table. *Permafrost and Periglacial Processes* 10, 349–367.

diastem A gap in the GEOLOGICAL RECORD of lesser duration than an UNCONFORMITY; for example, the gap represented by an erosion surface at the base of an EVENT bed. The concept is important in stressing that only a very small proportion of geological time is actually represented in preserved SEDIMENTARY DEPOSITS. *GO*

Ager, D.V. 1993: *The nature of the stratigraphical record*, 3rd edn. Chichester: Wiley. **Barrell, J.** 1917: Rhythms and the measurement of geologic time. *Geological Society of America Bulletin* 28, 745–904.

diastrophism A little-used term for DEFORMATION of the CRUST on a large scale, including OROGENY, EPEIROGENY and PLATE TECTONICS. *GO*

diatom analysis Diatoms are aquatic, unicellular algae belonging to the class Bacillariophycea and are unique in that they possess siliceous cell walls composed of two ornate valves. These valves are used taxonomically to differentiate between many species. There are several direct applications for diatom analysis, such as ENVIRON-MENTAL ARCHAEOLOGY, forensic medicine and oil and gas exploration. Diatoms are however most commonly used as ENVIRONMENTAL INDICATOR species.

Diatoms are sensitive to changes in the physical, chemical and biological variables in AQUATIC ENVIRONMENTS, and are thus widely used to monitor or reconstruct environmental conditions, such as ACIDITY, NUTRIENT STATUS, SALINITY and CLIMATE. Traditionally, diatom analysis has been based on the concept of CLASSIFICATION according to species preferences to specific environmental conditions, especially pH. Over the last decade, the TRANSFER FUNC-TION concept has gained popularity, and uses numerical techniques, including calibration and various methods of REGRESSION ANALYSIS.

Diatom analysis played a crucial role in the ACID RAIN debate during the 1980s, demonstrating that since the INDUSTRIAL REVOLUTION, POLLUTION has had an acidifying impact on freshwater ecosystems (see Figure). More recently, however, the technique has been used to monitor and manage acidified ecosystems, as well as helping to determine natural ACIDIFICATION processes throughout the HOLOCENE. Diatom analysis has also been widely used to: (1) reconstruct EUTROPHICATION histories in freshwater and coastal habitats, especially phosphorus levels; and (2) reconstruct past hydrology and salinity of lakes from LACUSTRINE SEDIMENTS, especially in arid and semi-arid regions. Increasingly, the technique is also being used to reconstruct CLIMATIC CHANGE in continental and marine environments, especially in conjunction with other techniques such as STABLE ISOTOPE ANALYSIS. Although in marine ecosystems diatom analysis is not as widely used as in lacustrine environments, the technique has been used to: (1) track the strength and volume of deep ocean currents through GLACIAL–INTERGLACIAL CYCLES; (2) infer the position of oceanic fronts and their migration through time; (3) determine palaeoproductivity histories; (4) determine SEA-SURFACE TEMPERATURES; (5) infer the extent of SEA-ICE at polar regions during glacial–interglacial cycles; and (6) infer SEA-LEVEL CHANGES at coastal sites.

AWM

[See also ALGAL ANALYSIS, INDICATOR SPECIES, BIO-INDICA-TOR]

Hustedt, F. 1937–1939: Systematische und ökologische Untersuchungen über den Diatomeen-Flora vo Java, Bali Sumatra. *Archiv für Hydrobiologie* (Suppl.) Nos. 15 and 16. **Battarbee, R.W.** 1984: Diatom analysis and the acidification of lakes. *Philosophical Transactions of the Royal Society of London* B305, 451–477. **Battarbee, R.W.** 1986: Diatom analysis. In Berglund, B.E. (ed.), *Handbook of Holocene palaeoecology and palaeohydrology.* Chichester: Wiley, 527–570. **Dixit, S.S., Smol, J.P., Kingston, J.C. and Charles, D.F.** 1992: Diatoms: powerful indicators of environmental change. *Environmental Science and Technology* 26, 22–33. **Round, P.E., Crawford, R.M. and Main, D.G.** 1990: *The diatoms.* Cambridge: Cambridge University Press. **Stoermer, E.F. and Smol, J.P.** 1999: *The diatoms: applications for the environmental and Earth sciences.* Cambridge: Cambridge University Press.

diatomite A light-coloured soft, MARINE or LACUSTRINE SEDIMENT or SEDIMENTARY ROCK, consisting mainly of opa-

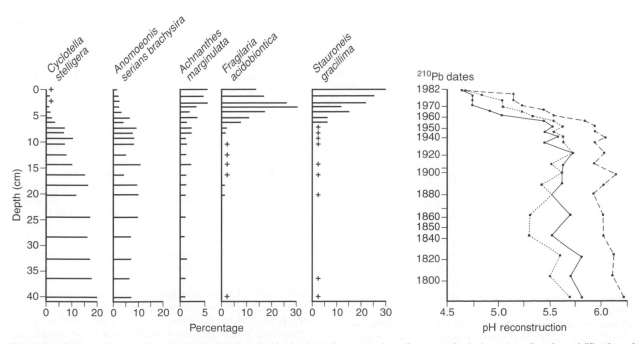

diatom analysis *Changing diatom assemblages and pH reconstructions based on diatom analysis demonstrating the acidification of Big Moose Lake, New York State, since the* AD *1950s (after Battarbee, 1984)*

line frustules of diatoms, a unicellular aquatic plant related to the algae. *UBW*

[See also ALGAL ANALYSIS, DIATOM ANALYSIS, LACUSTRINE SEDIMENTS, OOZE, SEDIMENT TYPES, SILICA ANALYSIS]

dieback Visible injury to leaves and other growing tissues by *chlorosis* (yellowed areas), *necrosis* (blackened tissues) and *defoliation*. It may be caused, for example, by ACID RAIN, OZONE, DROUGHT, *disease* attack by PESTS or *nutrient deficiency*. *GOH*

[See also FOREST DECLINE]

Barrow, C.J. 1994: *Land degradation: development and breakdown of terrestrial environments*. Cambridge: Cambridge University Press. **Goodman, P. J. and Williams, W. J.** 1961: Investigations into 'die-back' in *Spartina townsendii* agg. III. Physiological correlates of 'die-back'. *Journal of Ecology* 49, 391–398.

dielectric constant The dielectric constant of a material is the ratio of the capacity of a condenser (or capacitor) with that material as dielectric to the capacity when the dielectric is a vacuum. The dielectric constant of an object helps to determine the strength with which it scatters electromagnetic radiation and is a parameter used in MODELS of RADAR REMOTE SENSING. *PJS*

diffusion (1) The random movement of particles from an area of high concentration to one of low concentration. In gases and liquids diffusion continues until the concentration becomes uniform. (2) The spread of organisms, people, ideas and technology. *JAM*

[See also DEMIC DIFFUSION, DIFFUSIONISM, OSMOSIS]

diffusionism An approach to CULTURAL CHANGE that assumes technologies and other traits are adopted by neighbours or trading partners rather than spread by migrating peoples. *JAM*

[See also DEMIC DIFFUSION]

digital chart of the world (DCW) A map of the world stored in digital VECTOR FORMAT, generated at a scale of $1:10^6$. It contains 17 layers of thematic information derived from government mapping agencies around the world. *TF*

digital elevation model (DEM) A set of data values based on a network of points, for example, raster or *TIN*, representing elevation with respect to a defined datum. The data may be displayed graphically as a surface using appropriate computer software. By manipulation in the computer, this surface may be viewed from any azimuth or height. The model generated may also be used as the setting for a second image, for example land cover or geology, which is 'draped' over it.

DEM is increasingly applied as the standard term for such a dataset or model. However, other descriptions, such as *digital ground model, digital height model, digital surface model* and DIGITAL TERRAIN MODEL, while representing a similar form, may imply more specific surface information. *TF*

Petrie, G. and Kennie, T.J.M. 1990: *Terrain modelling in surveying and civil engineering*. London: Whittles in association with Thomas Telford. **Johnston, C.A.** 1998: *GIS in ecology*. Oxford: Blackwell Science, chapter 5.

digital number (DN) The discrete representation of a number, using a binary representation, suited for storage and processing in digital computers. *ACF*

digital terrain model (DTM) A set of data values, based on a network of points, for example, raster or *TIN*, representing a surface. In some applications, the term TERRAIN may indicate the inclusion of ground surface features, whilst in others terrain is considered to be the uncovered ground. Enhanced information about terrain, such as slope, aspect and drainage nets, can nowadays be derived from digital elevation data. Although DTM is still applied synonymously with DIGITAL ELEVATION MODEL (DEM), the latter is increasingly applied as the standard term. *TF*

dikaka A sand accumulation on which scrub or grasses occur in DRYLANDS (see NEBKHA). The term is also used to refer to root cavities in dunes. *RAS*

dilatancy The expansion of sediment (especially TILL in the SUBGLACIAL environment), caused by a change from a close-packed to a more open-packed structure, involving an increase in pore volume. The term DILATION is also used to describe this process. The concept has been used to explain the development of DRUMLINS. *MJH*

[See also GLACIAL LANDFORMS, GLACIAL SEDIMENTS]

Smalley, I.J. and Unwin, D.J. 1968: The formation and shape of drumlins and their distribution and orientation in drumlin fields. *Journal of Glaciology* 7, 377–390.

dilation/dilatation (1) A form of EXFOLIATION that involves the expansion of freshly exposed rock due to the removal of confining pressures. Dilat(at)ion bedrock joints or sheeting, especially on massive rocks, can be conspicuous on exposed bedrock surfaces in U-SHAPED VALLEYS and CIRQUES. (2) Deformation of sediment with volume change, although the term DILATANCY is more frequently used. (3) In REMOTE SENSING and IMAGE PROCESSING, dilation is one of the basic operations on which mathematical morphology is based. When dealing with binary images, where '1' is seen as white and '0' as black, it consists of enlarging (dilating) white spots. *JS/ACF*

Diluvial Theory A historical, scientific PARADIGM underpinned by religious belief in the biblical flood and accounting for many landforms and deposits (DILUVIUM) now recognised as glacial DRIFT. It was associated with CATASTROPHISM and belief in the recent creation of the Earth (about 6000 years ago according to Archbishop Ussher). It was strongly advocated by the Reverend William Buckland and other leading geologists of the early nineteenth century, many of whom were later 'converted' to the GLACIAL THEORY. *JAM*

Davies, G.L. 1969: *The Earth in decay: a history of British geomorphology 1578–1878*. London: Macdonald.

diluvium Deposits believed, according to the DILUVIAL THEORY, to have been deposited by Noah's flood. Prior to acceptance of the GLACIAL THEORY, most unconsolidated sedimentary deposits overlying bedrock in temperate latitudes were classified as *diluvial, antediluvial* (dating from before the deluge) or *postdiluvial* (dating from after the flood and later termed *alluvial*, hence ALLUVIUM). With the acceptance of the Glacial Theory, much diluvium was reinterpreted as GLACIAL, GLACIOFLUVIAL and PERIGLACIAL in origin. It appears that the biblical account of Noah's flood was inspired by the flooding of the Black Sea basin by sea water from the Mediterranean Sea about 7600 years ago, a large scale but regional natural event that has been the subject of detailed recent research. *JAM*

[See also DRIFT, ICEBERG DRIFT THEORY]

Ryan, W. and Pitman, W. 1998: *Noah's flood: the new scientific discoveries about the event that changed history.* New York: Simon and Schuster.

dimethyl sulphide (DMS) A compound of sulphur emitted during planktonic ALGAL BLOOMS, which is oxidised in the atmosphere to SULPHUR DIOXIDE, *methane sulphonic acid* (MSA) and sulphate AEROSOLS. Through FEEDBACK MECHANISMS that reduce incoming SOLAR RADIATION, it has potential as a natural countermeasure to GLOBAL WARMING. *JAM*

Charlson, R.J., Lovelock, J.E., Andreae, M.O. and Warren, S.G. 1987: Oceanic plankton, atmospheric sulphur, cloud albedo and climate. *Nature* 326, 655–661.

Dimlington Stadial A Late Devensian STADIAL, occurring between c. 22 and 18 ka BP (^{14}C years) and named after the stratotype site at Dimlington (East Yorkshire, UK). It is equivalent to the LATE GLACIAL MAXIMUM and marks the maximum expansion of ice during the DEVENSIAN (*Weichselian*) in the British Isles. *MHD*

[See also QUATERNARY TIMESCALE]

Eyles, N., McCabe, A.M. and Bowen, D.Q. 1994: The stratigraphic and sedimentological significance of Late Devensian ice sheet surging in Holderness, Yorkshire, U.K. *Quaternary Science Reviews* 13, 727–759. **Rose, J.** 1985: The Dimlington Stadial Chronozone: a proposal for naming the main glacial episode of the Late Devensian in Britain. *Boreas* 14, 225–237.

dinoflagellate cyst analysis The study of a constituent of *phytoplankton* used in BIOSTRATIGRAPHY (stratigraphic range: Late Triassic to Recent) and palaeoenvironmental reconstruction. Dinoflagellates are primarily single-celled Protists, most are heterotrophs or autotrophs (see HETEROTROPHIC and AUTOTROPHIC ORGANISMS); a few are parasites or symbiotes. They occupy most AQUATIC ENVIRONMENTS and are a major component of the marine PLANKTON with highest abundances over the outer CONTINENTAL SHELF. During their lifecycle, some dinoflagellates produce preservable resting cysts that are believed to be an integral stage of their development and which function as a mechanism for surviving adverse environmental conditions. Encystment is understood to take place from early summer to late autumn following exponential population growth (see ALGAL BLOOMS). SALINITY variation, proximity to shoreline, UPWELLING intensity, ORBITAL FORCING cycles and relative water temperature and depth can be deduced from dinoflagellate cyst species. However, for studies concerning environments of increasing geological age, more emphasis is placed on assemblages and morphogroups of cysts as modern representatives are rare. *MAP*

Harland, R. and Howe, J.A. 1995: Dinoflagellate cysts and Holocene oceanography of the northeastern Atlantic Ocean. *The Holocene* 5, 220–228. **Stover, L.E., Brinkhuis, H., Damassa,**

S.P. *et al.* 1996: Mesozoic–Tertiary Dinoflagellates, Acritarchs and Prasinophytes. In Jansonius, J. and McGregor, D.C. (eds), *Palynology: principles and applications*. Vol. 2. Dallas, TX: American Association of Stratigraphic Palynologists Foundation, 641–750. **Taylor, F.J.R.** 1987: Ecology of dinoflagellates. In Taylor, F.J.R. (ed.), *The Biology of Dinoflagellates* [*Botanical Monographs*. Vol. 21]. Oxford: Blackwell Scientific, 398–402.

dip The maximum angle between the horizontal and a sloping surface, used (along with STRIKE) to describe the orientation of sediment or rock strata and other geological features, including GEOLOGICAL STRUCTURES. *GO*

disaster A major event causing loss of human life, damage to property or loss of livelihood. Various minimum thresholds have been suggested to give precision to the concept, such as at least 100 people dead or at least one million dollars damage. *JAM*

[See also ENVIRONMENTAL DISASTER, NATURAL DISASTER, NATURAL HAZARD]

Quarantelli, E.L. (ed.) 1998: *What is a disaster? Perspectives on the question*. London: Routledge.

discharge The volume of flow per unit time. In the context of rivers, it is sometimes termed RUNOFF or *stream flow*. The term is also used in the context of GROUNDWATER. *River discharge* can be measured using a range of methods, such as the velocity/area method using a current meter, dilution gauging, control structures (weirs and flumes), ultrasonics, electromagnetic gauging and rated cross-sections. River discharge is a fundamental variable in hydrology and geomorphology. It varies spatially with catchment size, vegetation and landuse, topography, rainfall, soils and geology and temporally with weather and season. Accurate discharge measurements of RUNOFF are important for a variety of environmental purposes including the calculation of DISSOLVED LOAD and SEDIMENT YIELD, water supply and FLOOD FREQUENCY analysis. In geomorphology, the concept of a *dominant discharge* is used to describe flows (usually at or close to BANKFULL DISCHARGE) at which most sediment transport occurs. Similar dominant discharges have been calculated for SOLUTE transport and for channel formation. *ADT*

[See also RIVER DISCHARGE VARIATIONS, RIVER REGIMES]

Carling, P.A. 1988: The concept of dominant discharge applied to two gravel-bed streams in relation to channel stability thresholds. *Earth Surface Processes and Landforms* 13, 355–367. **Shaw, E.M.** 1988: *Hydrology in practice*. London: Chapman and Hall.

discordant drainage Rivers or streams that 'cut across' and hence appear unadjusted to the underlying geological structure. Examples include the rivers Avon, Test and Itchen in southern England, which transect the axes of anticlines and synclines (see FOLDS). Such discordance may be explicable in terms of ANTECEDENT DRAINAGE, SUPERIMPOSED DRAINAGE, RIVER CAPTURE or *glacial diversion* of drainage. *JAM*

Small, J. and Witherick, M. 1995: *A modern dictionary of geography*, 3rd edn. London: Arnold.

discriminant analysis A numerical technique to find the linear combination of variables, the discriminant function, that maximises the separation between two or more *a priori* defined groups of objects. If there are three or more groups, it is *multiple discriminant analysis* or *canonical variates analysis*. This is a special case of CANONICAL CORRELATION ANALYSIS where one variable set consists of nominal variables defining group membership. It can be used as an ORDINATION technique to display patterns of variation within and between groups in relation to the discriminating variables. Unknown objects can be assigned to the groups using the discriminant functions and associated decision rules. This is sometimes called CLASSIFICATION. *HJBB*

James, M. 1985: *Classification algorithms*. London: Collins. Matthews, J.A. 1979: A study of the variability of some successional and climax plant assemblage types using multiple discriminant analysis. *Journal of Ecology* 67, 255–271.

disease An unhealthy condition in an organism, including plants, animals and humans, such that functioning is impaired and death may ultimately result. *Acute diseases* have a sudden onset and a short duration; *chronic diseases* have a long duration and may recur frequently. *Infectious diseases* are spread by micro-organisms, *contagious diseases* by contact. Infectious and contagious diseases in particular have been linked to conditions in the environments in which disease sufferers live. Whereas many diseases in the developing world are still linked with conditions of the natural environment, more developed societies now experience more diseases that originate in built environments and social behaviour. For infectious diseases, VECTORS or carriers remain critical factors and, amongst these, the mosquito reigns supreme.

Chronic diseases, such as *leprosy* and *yaws*, may affect small populations and take a long time to kill victims, who stay alive for a long time acting as a reservoir for new infection. These seem to be the most ancient diseases of humanity, whereas the infectious diseases that give rise to EPIDEMICS and spread rapidly probably evolved later following the growth of dense human populations (hence the term *crowd diseases*). The first attested dates for many of the well known infectious diseases are relatively recent: around 1600 BC for *smallpox* (based on evidence from Egyptian MUMMIES), 400 BC for *mumps*, 200 BC for leprosy, AD 1840 for epidemic *polio*, and AD 1959 for AIDS.

Early agricultural communities not only sustained dense populations; they also practised a sedentary lifestyle (see SEDENTISM) and often lived in unsanitary conditions, which provided breeding grounds for disease-transmitting rodents, with rats in particular acting as 'reservoirs' for germs that cause contagious diseases. Such communities also lived in close proximity to domesticated animals with PATHOGENS that would later evolve into human diseases, spread faeces on fields as fertiliser and practised IRRIGATION, which provided habitats for WATER-RELATED VECTORS and encouraged the spread of WATER-BORNE and WATER-BASED DISEASES.

The even more unsanitary conditions that accompanied URBANISATION were an even greater spur to the spread of disease. Prior to the twentieth century, urban populations were not self-sustaining: they were dependent on constant immigration from rural areas to replace the victims of disease. The development of world trade routes likewise encouraged epidemics and pandemics. For example, smallpox reached Rome between AD 165 and 180 (the Plague of Antoninus) and *bubonic plague* first appeared in Europe in AD 542–543 (the Plague of Justinian). The later

impact of COLONIALISM through the direct or indirect introduction of disease was immense: far more non-European peoples were killed by Eurasian germs than by guns or steel weapons. Eurasian germs played a key role in decimating native peoples in the Americas, the Pacific Islands, Australia and southern Africa. However, the colonisers were to some extent affected in their colonies, especially by the diseases of southeast Asia (e.g. CHOLERA and MALARIA) and tropical Africa (e.g. malaria and *yellow fever*). The application of *spatial diffusion models* to the spread of disease has enabled both a descriptive and a predictive tool to be added to the study of disease ecology. The technique has been applied to a number of diseases including the spread of HIV and AIDS.

Two mechanisms can be distinguished whereby environmental change affects the likelihood of disease. First, it operates through changes in the ecology of the vectors in vector-borne diseases, the occurrence of which is largely determined by vector abundance, the abundance of intermediate and reservoir hosts, the abundance of disease-causing pathogens and parasites adapted to the vectors, the human or animal host, and local environmental conditions, especially temperature and humidity. Second, environmental change may operate through the direct modification of human risk factors, such as the availability and quality of water for domestic and agricultural purposes. Present and future climatic change, for example, is likely to affect the distribution and frequency of many diseases by both mechanisms. These mechanisms have operated in the past and appear to be coming into play in response to GLOBAL WARMING with the spread of tropical vector-borne diseases such as malaria, *dengue fever* and yellow fever to higher latitudes and altitudes. As disease is also closely linked to the social environment, which affects the conditions under which people live, the late twentieth century witnessed the return of diseases such as *tuberculosis* in Western inner cities as the disparities between living environments of rich and poor became greater.

JAM/DTH

[See also EPIDEMIOLOGY, HUMAN HEALTH HAZARDS]

Anderson, R. and May, R. 1992: *Infectious diseases of humans*. Oxford: Oxford University Press. Baker, J.R. and Brothwell, D.R. 1980: *Animal diseases in archaeology*. London: Academic Press. Beniston, M. 2000: *Environmental change in mountains and uplands*. Arnold: London. Burnet, M. 1953: *Natural history of infectious disease*. Cambridge: Cambridge University Press. Bushnell, O.A. 1993: *The gifts of civilization: germs and genocide in Hawaii*. Honolulu, HI: University of Hawaii Press. Cockburn, A. 1967: *Infectious diseases: their evolution and eradication*. Springfield, IL: Thomas. Cockburn, A. and Cockburn, E. (eds) 1983: *Mummies, diseases and ancient cultures*. Cambridge: Cambridge University Press. Crosby, A. 1972: *The Columbian exchange: biological consequences of 1492*. Westport, CT: Greenwood. Diamond, J. 1998: *Guns, germs and steel: a short history of everybody for the last 13,000 years*. London: Vintage. Haggett, P. and Cliff, A.D. 1988: *An atlas of disease distributions: analytical approaches to epidemiological data*. Oxford: Blackwell. Ingram, D. and Robertson, N. 1999: *Plant disease*. London: HarperCollins. Roberts, C. and Manchester, K. (eds) 1995: *The archaeology of disease*, 2nd edn. Ithaca, NY: Cornell University Press. Rothschild, B.M. and Martin, L. 1992: *Paleopathology: disease in the fossil record*. Boca Raton, FL: CRC Press. Zinsser, H. 1935: *Rats, lice and history*. Boston: Little Brown.

disjunct distribution A distribution pattern exhibited by taxa that have POPULATIONS that are geographically separated, as shown, for example, by the southern beeches (*Nothofagus*), which occur in South America and Australasia. The explanation of such distributions is an important element of the subdiscipline of BIOGEOGRAPHY. Many such distributions are thought to have been caused by past environmental change.   *JLI*

[See also RELICT]

Brown, J.H. and Lomolino, M.V. 1998: *Biogeography*, 2nd edn. Sunderland, MA: Sinauer.

dispersal   The movement of an organism away from its natal site and a basic life-history trait. While dispersal occurs in ecological time, it is also a major focus of historical BIOGEOGRAPHY in that contemporary species distributions may result from past dispersal patterns. *Passively dispersing organisms* rely on the natural environment or other species for their movement. Plants, animals, their parasites and microbes frequently employ *passive dispersal*. *Actively dispersing organisms* do so under their own power (i.e. they exhibit *active dispersal*). Dispersal may be inhibited by human environmental modification, such as through habitats suffering from FRAGMENTATION, although *habitat corridors* may partially alleviate this problem. Successful dispersal requires the organism to overcome ecological and physiological barriers. For example, many birds will not cross small water bodies, while a variety of plants are inhibited from dispersing across salt water because of poor tolerance of saline environments.   *JTK*

[See also NOMADISM, MIGRATION, RANGE ADJUSTMENT, VICARIANCE]

Brown, J.H. and Lomolino, M.V. 1998: *Biogeography*, 2nd edn. Sunderland, MA: Sinauer. **Whittaker, R.J. and Jones, S.H.** 1994: The role of frugivorous bats and birds in the rebuilding of a tropical forest ecosystem, Krakatau, Indonesia. *Journal of Biogeography* 21, 689–702.

dissolved load   The total amount of SOLUTES transported in RUNOFF by a river or stream per unit time; also often called *solute load*. It can be expressed in absolute units (for example, t y^{-1}) or as a *specific load* (per unit catchment area) so that catchments of different sizes can be compared (t km^{-2} y^{-1}). The dissolved component is sometimes equal to or greater than the amount of material transported as PARTICULATES in some humid vegetated regions of soluble lithology. The dissolved load encompasses both denudational and non-denudational components. The *denudational component* comprises the solutes derived from the weathering of bedrock and soil within the catchment. Artificial solute sources, such as sewage effluent and industrial pollutants (see WATER POLLUTION), as well as naturally derived atmospheric inputs and solutes derived from the MINERALIZATION of organic matter, all represent *non-denudational components*.

Gibbs (1970) classified the natural solute loads of rivers into: (1) rivers dominated by atmospheric inputs of Na$^+$ and Cl$^-$ with low solute concentrations (20–30 mg L^{-1}); (2) rivers with intermediate solute concentrations dominated by solutes derived from rock and soil WEATHERING; and (3) rivers with high solute concentrations (> 1000 mg L^{-1}) dominated by EVAPORATION and subsequent precipitation of salts. The contribution of weathering to the dissolved load depends upon the solubility of the bedrock (with rock solubility decreasing from evapor-

ites, through carbonates, volcanics and clastic sediments to crystalline rocks).

There is usually an inverse relationship between solute concentrations and DISCHARGE, such that, at low flows, solute concentrations are highest (because solute-rich groundwater dominates and the evaporation–concentration effect is at a maximum) and, at high flows, concentrations are lowest (because of the increased contribution from dilute *overland flow* and shallow subsurface RUNOFF PROCESSES). High discharge, however, more than outweighs reduced solute concentrations in solute budgets and high dissolved loads are associated with high flows. Thus, the global pattern of dissolved load reflects the global distribution of runoff and is greatest in the HUMID TROPICS and lowest in ARIDLANDS and SEMI-ARID regions. It also follows that dissolved loads of rivers over the Earth's surface have fluctuated greatly with changes in PRECIPITATION and RIVER DISCHARGE VARIATIONS during the Quaternary.   *ADT*

Gibbs, R.J. 1970: Mechanisms controlling world water chemistry. *Science* 170, 1088–1090. **Meybeck, M.** 1987: Global chemical weathering of surficial rocks estimated from river solute loads. *American Journal of Science* 287, 401–428. **Walling, D.E. and Webb, B.W.** 1983: *The dissolved load of rivers: a global overview* [*Publication* 141]. Budapest: International Association for Scientific Hydrology, 3–20.

distributary   A channel that branches from others in the downstream direction. Distributary systems develop where there is a downstream reduction in transporting capacity and characterise fan systems such as ALLUVIAL FANS, DELTAS and SUBMARINE FANS.   *GO*

[See also TRIBUTARY]

disturbance   'A disturbance is any relatively discrete event in time that disrupts ecosystem, community, or population structure and changes resources, substrate availability, or the physical environment' (Pickett and White, 1985: 1). This general definition encompasses both ABIOTIC events, such as the destruction of trees by *windfall* or *lightning strike*, and BIOTIC events, for instance PREDATION, GRAZING or DISEASE, irrespective of whether they are EXOGENETIC or ENDOGENETIC processes. CLIMATIC CHANGE may impact on disturbance directly, via the frequency and intensity of EXTREME CLIMATIC EVENTS, and indirectly, through its influence on biotic interactions. Disturbance is the rule rather than the exception where the world's ecosystems are considered (see Table).

Since natural disturbances may be viewed as STOCHASTIC, their frequency can be described in terms of probability and VARIANCE. SELECTION PRESSURE on populations for ADAPTATION to disturbance increases with its frequency, which can therefore have important consequences for EVOLUTION. For example, FIRE FREQUENCY in MEDITERRANEAN-TYPE VEGETATION can be high in relation to plant life spans and has led to numerous specialised mechanisms for survival and regeneration after fire (see FIRE IMPACTS). In addition, disturbance frequency may have important implications for SPECIES RICHNESS (see INTERMEDIATE DISTURBANCE HYPOTHESIS).

The magnitude of disturbance may be measured in terms of its intensity, the physical force exerted or its ecological impacts, and the association between these measures depends on the ability of individuals to resist

disturbance *Percentage of the Earth's surface regularly affected by some major disturbances (after Walker and Willig, 1999)*

Element	Primary disturbance	Percentage*
Earth (tectonic)	Earthquake	1
	Erosion	> 50
	Volcano	1
Air	Hurricane	15
	Tornado	< 1
	Tree-fall	nd
Water	Drought	30
	Flood	15
	Glacier	10
Fire	Fire	> 50
Biota – non-human	Herbivory	nd
	Invasion	nd
	Other animal activity (includes building, excavating, waste products, movement, death, diseases, parasites)	1
Biota – human	Agriculture	45
	Forestry	10
	Mineral extraction	1
	Military activity	1–40 (USA 1%; Vietnam 40%)
	Transportation (includes motorised and non-motorised transportation)	5
	Urban	3

*nd = no data available

disturbance. Destruction of plants by intense disturbance creates gaps in the VEGETATION CANOPY, allowing the REGENERATION of new individuals, and is a key element of community and POPULATION DYNAMICS.

Disturbance regimes are an important determinant of heterogeneity in ecosystems, because of the discrete nature of disturbance events (see PATCH DYNAMICS). The significance of each event depends on the areal extent of disturbed patches in relation to the spatial scale of interest (see SCALE CONCEPTS IN ENVIRONMENTAL CHANGE). For example, heavy rainfall may destroy a significant fraction of a grass seedling population occupying a small patch of otherwise bare ground, but cause minimal disturbance to the *grassland* ecosystem. CATASTROPHIC EVENTS cause significant ecological disturbance on a very large spatial scale. For example, a METEORITE IMPACT at the K–T BOUNDARY has been linked with MASS EXTINCTION at this time.   *CPO*

[See also DISTURBED ECOSYSTEMS, TROPICAL CYCLONES]

Binford, M.W., Brenner, M., Whitmore, T.J. *et al.* 1987: Ecosystems, paleoecology and human disturbance in subtropical and tropical America. *Quaternary Science Reviews* 6, 115–128. **Pausas, J.G.** 1999: Mediterranean vegetation dynamics: modelling problems and functional types. *Plant Ecology* 140, 27–39. **Pickett, S.T.A. and White P.S. (eds)** 1985: *The ecology of natural disturbance and patch dynamics*. Orlando, FL: Academic Press. **Sousa, W.P.** 1984: The role of disturbance in natural communities. *Annual Review of Ecology and Systematics* 14, 353–391. **Walker, L.R. and Willig, M.R.** 1999: An introduction to terrestrial disturbances. In Walker, L.R. (ed.), *Ecosystems of disturbed ground*. Amsterdam: Elsevier, 1–16.

disturbed ecosystems   Those altered from their 'mean' equilibrium (stable) or non-equilibrium (naturally unstable) condition. Disturbances can be NATURAL or ANTHROPOGENIC and external (fire, storm) or internal (pre-

dation, competition) to the ecosystem. Disturbances vary in scale, intensity and frequency: compare a single CATASTROPHIC EVENT (such as a TROPICAL CYCLONE) with regular or continuous disturbance by ATMOSPHERIC POLLUTION. Durable ecosystems resist change better than fragile systems and resilient ecosystems recover quickly from disturbance. Three types of disturbed ecosystem may be recognised: (1) disturbance-controlled ecosystems, where regular disturbance by physical factors, such as TIDES or FROST HEAVE, maintains the system at an early stage of ECOLOGICAL SUCCESSION; (2) disturbance-dependent systems, where periodic external disturbance is necessary for the system's existence, such as the creation of natural openings in the TROPICAL RAIN FOREST, which maintain BIODIVERSITY; (3) disturbance-maintained systems, where the intensity of disturbance is so great or continuous that the original system has been replaced by another, e.g. AGROECOSYSTEMS, TROPICAL GRASSLAND and mediterranean scrubland (see mediterranean-type VEGETATION).   *GOH*

[See also DISTURBANCE, EQUILIBRIUM CONCEPTS, RESILIENCE]

Begon, M., Harper, J.L. and Townsend, C.R. 1996: *Ecology: individuals, populations and communities*, 3rd edn. Oxford: Blackwell Science. **Griggs, R.F.** 1934: The problem of Arctic vegetation. *Journal of the Washington Academy of Sciences* 25, 153–175. **Walker, L. R. (ed.)** 1999: *Ecosystems of disturbed ground* [*Ecosystems of the world*. Vol. 16]. Amsterdam: Elsevier. **White, P.S. and Jentsch, A.** 2001: The search for generality in studies of disturbance and ecosystem dynamics. *Progress in Botany* 62, 399–450.

diurnal changes   The changes that are observed in environmental variables over the 24-hour SOLAR CYCLE. AIR TEMPERATURE and relative humidity display a systematic diurnal cycle albeit of an opposite phase. Mountain winds and sea breezes commonly exhibit a diurnal cycle. A diurnal cycle of wind is also a chief feature of low-latitude climates. *GS*

divergence As lineages evolve over time, organisms' structures may become altered and may have quite different appearances and functions in descendant lineages. For example, the mammalian fore limb has evolved into the fore legs of running quadrupeds, arms of bipeds, fins of whales and wings of bats. This process is known as divergence. *KDB*

[See also CONVERGENCE IN EVOLUTION, CONVERGENCE AND DIVERGENCE IN SUCCESSION, EVOLUTION]

divergent weathering A principle (*divergierende Verwitterung*) developed by H. Bremer, maintaining that a differential in CHEMICAL WEATHERING rate between hillslopes stripped of their soil cover (low rate) and valley plains covered by a deep SAPROLITE (high rate) will promote relief enhancement of a landscape by leading to preferential lowering of plains (ETCHPLAINS). Such a mechanism may play a role in the development of TORS and INSELBERGS. Bremer considered this to be particularly effective in the *humid tropics*, where rainfall is sufficient to promote high rates of chemical weathering in valley bottoms enhanced by the supply of overland flow from the hillslopes; however, her ideas are probably more applicable in the seasonally wet tropics. In the ever-wet tropics, moreover, the principle may not apply, as slopes are actually covered by deep and permeable soils; in such a climate, high rates of chemical weathering are characteristic of both slopes and valley bottoms and landsliding may be the dominant process in landscape development. Bremer saw this divergence as a reversal of the situation in arid or cold climates, where PHYSICAL WEATHERING (such as frost shattering) is enhanced on the exposed hillslopes. *MAC*

[See also DOUBLE PLANATION SURFACES, LANDSCAPE EVOLUTION]

Bremer, H. 1972: Flussarbeit, Flächen – und Stufenbildung in den feuchten Tropen. *Zeitschrift für Geomorphologie Supplementband* 14, 21–38. **Bremer, H.** 1990: *Allgemeine Geomorphologie. Methodik – Grundvorstellungen – Ausblick auf den Landschaftshaushalt.* Berlin: Bornträger.

diversity anomaly A difference in BIODIVERSITY within similar HABITATS between different regions. For example, TEMPERATE zone genera of plants that have DISJUNCT DISTRIBUTIONS between similar environments in eastern Asia and eastern North America have twice as many species in Asia as in North America. Diversity anomalies may arise through different rates of SPECIATION or EXTINCTION and are affected by topographic heterogeneity. *JAM*

Qian, H. and Ricklefs, R.E. 2000: Large-scale processes and the Asian bias in species diversity of temperate plants. *Nature* 407, 180–182.

diversity concepts BIODIVERSITY may be viewed from different perspectives, and can be assessed at three general levels: genetic, species (community) and ecological (ecosystem). It includes the interactions between organisms, including humans, and their environment. Many DIVERSITY INDICES are available to measure diversity, especially at the species level. These indices may

include assessments of species abundance, *equitability* and phylogenetic interrelationships as well as SPECIES RICHNESS. Diversity concepts are complicated by the availability and use of various SPECIES CONCEPTS, which use different criteria to define a species. *Genetic diversity* data describing aspects of the heritable molecular variability of the organisms may be required to assess species diversity. *Species diversity* is measurable at the scale of individual habitats or locales (*alpha diversity*), at the scale of regions (*gamma diversity*) or as the turnover between different habitats (*beta diversity*). *Ecosystem* or *ecological diversity* may be measured from the scale of HABITATS and LANDSCAPES to BIOMES at a global scale. *JTK*

[See also CLADISTICS, ENDEMIC]

Heywood, V.H. (ed.) 1995: *Global biodiversity assessment.* Cambridge: Cambridge University Press. **Huston, M.A.** 1994: *Biological diversity: the coexistence of species on changing landscapes.* Cambridge: Cambridge University Press.

diversity indices Species diversity can be measured most simply by counting the number of species present in a defined area. This is SPECIES RICHNESS or *species density*, perhaps the most commonly used estimator of BIODIVERSITY. More sophisticated techniques exist that incorporate species abundance as well as richness. The *Shannon–Wiener* (H^i) and *Simpson* (S) indices incorporate information on the abundances of species as well as richness. These indices suggest that diversity is maximised for a given number of species when those species are all equally abundant. While species abundances provide valuable information about a community, interpretation of these indices may be difficult. A general problem with incorporation of abundance data is that these are often impractical to collect, limiting their use to relatively well studied locales. Other diversity indices measure the *evenness* or *equitability* of species abundances, but exclude species richness information, and scale more intuitively than H^i and S. *JTK*

Heywood, V.H. (ed.) 1995: *Global biodiversity assessment.* Cambridge: Cambridge University Press. **Magurran, A.** 1988: *Ecological diversity and its measurement.* Princeton, NJ: Princeton University Press.

diversity–stability hypothesis This hypothesis proposes that ECOSYSTEMS become increasingly able to withstand DISTURBANCE as local species diversity increases. Ecosystem functions are more likely to be maintained when BIODIVERSITY is high because of *functional redundancy* and the larger number of interactions among species at different TROPHIC LEVELS. If several different species recycle nitrogen in an ecosystem, for example, EXTIRPATION of a few of these species is less likely to reduce nitrogen recycling in the ecosystem. This is particularly important in the long term, as ecosystems are more likely to undergo extreme environmental disturbances that extirpate species from the system. Functional redundancy buffers ecosystem function against these changes and allows more rapid ecosystem recovery. Biogeochemical processes in an ecosystem become increasingly efficient as species diversity increases, but evidence from tropical communities shows that efficiency does not improve in ecosystems containing 100 or more species. This result may vary geographically and does not take into account long-term ecosystem stability. *JTK*

[See also DIVERSITY CONCEPTS, INTERMEDIATE DISTURBANCE HYPOTHESIS, STABILITY CONCEPTS]

Ewel, J.J., Mazzarino, M.J. and Berish, C.W. 1991: Tropical soil fertility changes under monocultures and successional communities of different structure. *Ecological Applications* 1, 289–302. **McCann, K.S.** 2000: The diversity–stability debate. *Nature* 405, 228–233. **McCann, K., Hastings, A. and Huxel, G.R.** 1998: Weak trophic interactions and the balance of nature. *Nature* 395, 794–798. **Naeem, S., Thompson, L.J., Lawler, S.P.** *et al.* 1994: Declining biodiversity can alter the performance of ecosystems. *Nature* 368, 734–737. **Vitousek, P.M.** 1982: Nutrient cycling and nutrient use efficiency. *American Naturalist* 119, 553–572.

DNA, ancient

In the study of ENVIRONMENTAL CHANGE, DNA preserved in FOSSIL material is used to reconstruct the genetic code of ancient organisms. Remains containing ancient DNA have been obtained from herbarium specimens, charred seeds and cobs, mummified seeds and embryos, fossil leaves, bones, mummified skin, hair and amber. Some estimates suggest that material as old as 17–20 million years contains ancient DNA. However, there has been much debate as to whether DNA can be preserved for this length of time without significant damage to its structure through degradation, particularly through exposure to water (hydrolitic) and air (oxidative). Studies have indicated that the most successful preservation of ancient DNA occurs in environments where there is rapid and directed DEPOSITION, thus reducing physical damage and minimising externally and internally induced wound reactions. In addition, rapid burial is needed to reduce the exposure of the tissue to biotic DEGRADATION. One particularly good medium for the long-term preservation of ancient DNA sequences is AMBER. Amber is a fossil resin formed by members of the coniferous family Araucariaceae and the angiosperm family. When fauna or flora become trapped in the sticky resin, in the process of fossilisation, the sugars withdraw the moisture from the original tissue and initiate the process of *inert dehydration*. Thus the DNA is largely dehydrated, protected from atmospheric oxygen, and not exposed to microbial contamination. Successful studies utilising ancient DNA tend to be from material less than 100 000 years old and range from specific taxonomic and phylogenetic studies, to archaeological work on AGRICULTURAL ORIGINS and genetic analyses of PREHISTORIC populations. *KJW*

[See also PHYLOGEOGRAPHY]

Austin, J.J., Smith, A.B. and Thomas, R.H. 1997: Palaeontology in a molecular world: the search for authentic ancient DNA. *Trends in Ecology and Evolution* 12, 303–306. **Brown, T.A.** 1999: How ancient DNA may help in understanding the origin and spread of agriculture. *Philosophical transactions of the Royal Society of London* B354, 89–97. **DeSalle, R.** 1994: Implications of ancient DNA for phylogenetic studies. *Experientia* 50, 543–550. **Golenberg, E.M., Giannasi, D.E., Clegg, M.T.** *et al.* 1991: Chloroplast DNA sequence from a Miocene *Magnolia* species. *Nature* 344, 656–658. **Jones, M.K. and Brown, T.** 2000: Agricultural origins: the evidence of modern and ancient DNA. *The Holocene* 10, 769–777. **Lindahl, T.** 1993: Instability and decay of the primary structure of DNA. *Nature* 362, 709–715.

documentary evidence

Written records of many aspects of everyday life, of government, administration and commerce, especially agriculture, have survived from the past 1000 years. In some exceptional instances older records have also come down to the present day. Some are of particular importance in HISTORICAL CLIMATOLOGY and, whilst traditionally being regarded as source material for the historian, anthropologist or sociologist, it is now recognised that a significant number of these documents have a wider application. They lack the apparent objectivity of the INSTRUMENTAL RECORD or of PROXY CLIMATIC INDICATORS, but, when used cautiously, have been shown to reveal important evidence of CLIMATIC CHANGE and other aspects of past ENVIRONMENTAL CHANGE. They can also be used in the CONFIRMATION and CALIBRATION of evidence from other sources. This latter role is particularly important for pre-instrumental episodes such as the MEDIAEVAL WARM PERIOD.

Some of the oldest documentary evidence dates from 3050 BC, when records of the height of the annual NILE FLOOD were first made. In those distant times before the invention of paper and parchment, 'documents' exist in other forms. Clay and stone tablets are obvious examples while oracle bones record aspects of Chinese climate from before 1000 BC. Elsewhere, the famous CAVE ART of cattle herding and hippopotamus in Aounrhet from 3500 BC in what is today the Sahara Desert have been used as evidence to support theories that this area was once wetter than today and sustained an environment similar to the savannah lands further south beyond the present day arid zone.

Past societies, particularly those with organised governmental systems, often had an interest in the weather, not for any scientific reason but because contemporary agricultural economies depended so heavily upon its vagaries. EXTREME EVENTS often appear in such records, but the more mundane annual records of crop prices, harvest dates and tax yields also cast an indirect light on the prevailing climate, although it has been recognised that factors other than the climate or physical environment might act on these indicators. The accounts of individual farms and estates or of specially commissioned works serve a similar purpose. The Domesday Book, for example, reveals clearly that much of the deforestation of Lowland England had already occurred by the late eleventh century. Later documents, both civic and ecclesiastical, have allowed scientists to plot detailed aspects of CLIMATIC CHANGE as varied as the frequency and intensity of North Sea STORMS from the sixteenth century and of FLOODS and DROUGHTS in Mediterranean Spain from as early as the fourteenth century – both long before formal weather recording began.

Many European countries possess important collections of documents and many more documents can be found in China and Japan. Other areas of the World are, however, less fortunate in this respect and Africa and America have little comparable material. On the other hand, oceanic areas are well represented from the mid-seventeenth onwards by the SHIP LOG BOOK RECORDS that are preserved in the archives of the once powerful imperial nations of France, Spain, England and Holland.

Broader themes have also been investigated using documentary sources. A good example is provided by London's air POLLUTION problems, the nature of which have been studied using Mediaeval and later documents, including diaries, letters, reports and accounts of the coal trade. More generally, documentary evidence can help place a wide range of so-called environmental problems or ENVIRONMENTAL ISSUES in perspective.

Given the value to be derived from such sources, it is remarkable to note the resistance previously offered by the scientific community to their use. The two most prominent figures responsible for overcoming these objections are E. Le Roy Ladurie and H.H. Lamb. The former is a historian while the latter is a climatologist. Le Roy Ladurie used French documents of the dates of VINE HARVESTS to construct a proxy temperature series for some French regions from the sixteenth to the nineteenth century. Lamb's work was even more wide ranging in this respect. He successfully used sources as diverse as the Bible and Norse folklore and more recent sources such as the famous *Landnámabók*, which records many aspects of the environment of Iceland from the earliest days of its occupation shortly before AD 1000. He also assisted in reconstructing the unsettled weather of the summer of 1588 using the letters and documents of the storm-tossed survivors of the Spanish Armada. Lamb has also used documents to construct decadal indices of 'winter severity' and 'summer wetness' for European climate starting as long ago as AD 1000.

Although success has been achieved using documentary sources, they present a number of problems: the records were not prepared in any systematic fashion; the handwriting and the language may also present difficulties; and most contain much additional information that may be of little or no relevance to the environmental scientist, but which needs nonetheless to be checked. They are also scattered across a large number of archives and museums in different countries. The biggest problem with documentary sources, although not unique to this *genre*, is that of VERIFICATION. In this respect, methods of qualitative investigation, such as *content analysis* of written terms, have been of great assistance. This approach was pioneered by a number of scientists including A.J.W. Catchpole and D.W. Moodie, who studied the records of the Hudson's Bay Company to extract an annual series of freeze and thaw dates from the early eighteenth to the late nineteenth century.   *DAW/RG*

[See also EARLY INSTRUMENTAL RECORDS, ROGATION INDEX, WEATHER DIARIES, WEATHER TYPES]

Alcoforada, M.J., de Fátima Nunes, M., Garcia, J.C. and Taborda, J.P. 2000: Temperature and precipitation reconstruction in southern Portugal during the Maunder Minimum (AD 1675–1715). *The Holocene* 10, 333–340. **Baker, A.R.H., Hamshere, J.D. and Langton, J.** (eds) 1970: *Geographical interpretations of historical sources.* Newton Abbott: David and Charles. **Brimblecombe, P.** 1987: *The big smoke: a history of air pollution in London since medieval times.* London: Routledge. **Catchpole, A.J.W. and Moodie, D.W.** 1978: Archives and the environmental scientist. *Archivaria* 6, 113–136. **Cuesta, M.J.D., Sánchez, M.J. and García, A.R.** 1999: Press archives as temporal records of landslides in the north of Spain: relationships between rainfall and instability slope events. *Geomorphology* 30, 125–132. **Douglas, K.S., Lamb, H.H. and Loader, C.** 1978: *A meteorological study of July to October 1588: the Spanish Armada storms* [*University of East Anglia, Climatic Research Unit, Report* 6]. Norwich: University of East Anglia. **Frenzel, B., Pfister, C. and Gläser, B.** (eds) 1992: European climate reconstructed from documentary data: methods and results. *Paläoklimaforschung* 24, 1–182. **Ingram, M.J., Underhill, D.J. and Farmer, G.** 1981: The use of documentary sources for the study of past climates. In Wigley, T.M.L., Ingram, M.J. and Farmer, G. (eds), *Climate and history: studies in past climates and their impact on man.* Cambridge: Cambridge University Press, 180–213. **Lamb, H.H.** 1977: *Climate: present past and future.* Vol. 2. *Climatic history and the future.* London: Methuen. **Lamb, H.H.** 1991: *Historic Storms of the North Sea, British Isles and north-west Europe.* Cambridge: Cambridge University Press. **Le Roy Ladurie, E.** 1971: *Times of feast, times of famine: a history of climate since the year 1000.* London: George Allen and Unwin. **Moodie, D.W. and Catchpole, A.J.W.** 1976: Valid climatological data from historical sources by content analysis. *Science* 1993, 51–53. **Pfister, C.** 1984: *Das Klima der Schweiz von 1525 bis 1860 und seine Bedeutung in der Geschichte von Bevölkerung und Landwirtschaft.* Bern: Haupt. **Rotberg, R.I. and Rabb, T.K.** (eds) 1981: *Climate and history: studies in interdisciplinary history.* Princeton, NJ: Princeton University Press. **Vallve, M.B. and Martin-Vide, J.** 1998: Secular climatic oscillations as indicated by catastrophic floods in the Spanish Mediterranean coastal area (14th–19th centuries). *Climatic Change* 38, 473–491. **Vogel, C.H.** 1989: A documentary-derived climatological chronology for South Africa, 1820–1900. *Climatic Change* 14, 291–307.

doline A typically circular- or oval-shaped depression, a few metres up to several hundreds of metres in width, found in KARST landscapes. Dolines are usually formed by the solution and/or collapse of underlying limestone bedrock. Dolines are often the locations of water *sink holes* or *swallow holes* and may coalesce to form larger depressions (*uvalas*).   *AJDF*

[See also POLJE]

Gibbard, P.L., Bryant, I.D. and Hall, A.R. 1986: A Hoxnian Interglacial doline infilling at Slade Oak Lane, Denham, Buckinghamshire, England. *Geological Magazine* 123, 27–43. **Jennings, J.N.** 1985: *Karst geomorphology.* Oxford: Blackwell.

domestic waste Household wastes are collected in many countries by municipal authorities and are either incinerated or more usually consigned to LANDFILL. In the developed countries, nineteenth-century domestic waste was largely composed of cinder and ash from domestic fires and kitchen wastes arising from household activities. Early in the twentieth century, in Europe, this amounted to around 17 kg per household per week. Towards the end of the twentieth century domestic waste has become more varied and consists of greater amounts of plastic, paper and putrescent material, amounting to between 10 and 11 kg per household per week. Larger items and garden waste are also collected, but local authorities may charge for the service. Increasingly, local authorities are promoting schemes for WASTE RECYCLING. Bottles, glass items, paper, cardboard and cans are now generally collected separately, and in The Netherlands, for example, putrescent kitchen waste is collected for COMPOSTING.   *EMB*

[See also INDUSTRIAL WASTE, LANDFILL, WASTE MANAGEMENT]

Coggins, P.C., Cooper, A.D. and Brown, R.W. 1992: Civic amenity waste disposal sites: the Cinderella of the waste disposal system. In Clarke, M., Smith, D. and Blowers, A. (eds), *Waste location: spatial aspects of waste management, hazards and disposal.* London: Routledge. 79–104. **Jones, B.F. and Tinzmann, M.** 1990: *Too much trash?* Columbus, OH: Zaner-Bloser. **National Research Council** 1984: *Disposal of industrial and domestic wastes: land and sea alternatives.* Washington, DC: National Academy Press.

domestication The naturalisation of wild plants and animals under human control, whereby the genetic make-up of plants and animals changes, making them more suited to the needs of people and/or to the environment within a deliberately managed agricultural system. This can mean that the plant or animal is no longer able to survive outside of the managed system. Domestication of plants involves the planting of harvested and selected seed and the process can lead to increased size and yield of the plant, greater robustness, loss of natural defence mechanisms (e.g. thorns, bitterness) and the loss of the ability to disseminate itself. Domestication of animals can lead to a change in body size that is often smaller in pigs and cattle and larger in horses. Such changes may eventually lead to significant differences between wild and domesticated types.

Domestication is thought to have occurred independently at different times and places (see Figure). The most important animals to have been domesticated (c. 10 000–8000 radiocarbon years BP) include dogs, sheep, cattle, goats and pigs. The earliest plants to have been domesticated include *wheat*, *barley*, lentils, peas and vetch (c. 10 000–8500 BP). Several weeds of cultivation (*oats*, *rye*) became domesticated between c. 4000 and 3000 BP. Sorghum, sesame and rice were later domesticates (c. 3000 BP). The New World was also an area in which domestication occurred from 10 000 BP (*maize*, potatoes, tomatoes, avocados, sunflowers and capsicums), but the process was slower to develop.

Domesticates spread from their centres of origin to the rest of the world by diffusion and migration of ideas and people. ENVIRONMENTAL CHANGE during the HOLOCENE may also have been a stimulus to domestication and the spread of agriculture. As domestication occurred and spread to new areas, there were major changes in society, settlement patterns and ARTEFACT production. The most important change was from nomadic HUNTING, FISHING AND GATHERING societies to SEDENTISM. *LD-P*

[See also AGRARIAN CIVILISATION, AGRICULTURAL HISTORY, AGRICULTURAL ORIGINS, CEREAL CROPS, NOMADISM, OASIS HYPOTHESIS, PASTORALISM]

Andersen, K. 1998: Animal domestication in geographic perspective. *Society and Animals* 6, 119–135. **Barigozzi, C. (ed.)** 1986: *The origin and domestication of cultivated plants*. Amsterdam: Elsevier. **Bell, M. and Walker, M.J.C.** 1992: *Late Quaternary environmental change*. Harlow: Longman. **Clutton-Brock, J.** 1987: *A natural history of domesticated mammals*. Cambridge: Cambridge University Press. **Hillman, G.C. and Davies, M.S.** 1990: Domestication rates in wild-type wheats and barley under primitive cultivation. *Biological Journal of the Linnean Society* 79, 39–78. **Hodder, I.** 1990: *The domestication of Europe*. Oxford: Blackwell. **Kajale, M. D.** 1996: Archaeology and domestication of crops in the Indian subcontinent. *World Archaeology* 28, 5–19. **Loftus, R.T., MacHugh, D.E., Cunningham, P and Bradley, D.G.** 1999: Animal domestication. *Science* 283, 329–330. **Moore, P.D.** 1998: Plant domestication – getting to the roots of tubers. *Nature* 395, 330–331. **O'Connor, T.** 1997: Working at relationships: another look at animal domestication. *Antiquity* 71, 149–156. **Olsen, S.J.** 1979: Archaeologically, what constitutes an early domestic animal? *Advances in Archaeological Method and Theory* 2, 175–197. **Thorpe, I.J.** 1996: *The origins of agriculture in Europe*. London: Routledge.

dominant species A species that exerts functional control (e.g. through COMPETITION) over other species in communities and ECOSYSTEMS. It need not necessarily be the most abundant or conspicuous species. *JAM*

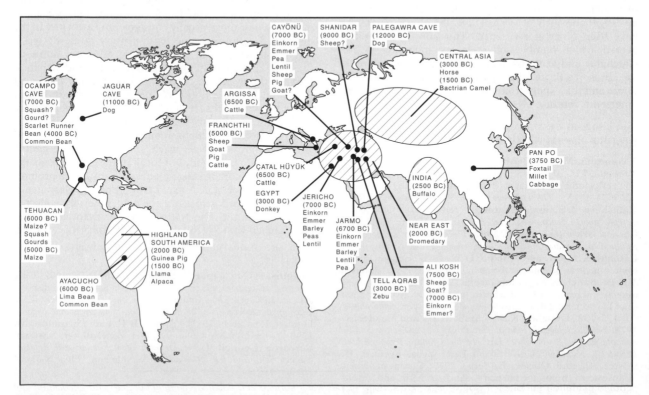

domestication *The locations and dates at which some of the most important domestications are likely to have taken place (after Bell and Walker, 1992)*

dominant wind The wind direction that has the strongest winds or the greatest effects, in contrast to the highest frequency of occurrence (PREVAILING WIND). *JAM*

donga A southern African term for a small, narrow, steep-sided gully, subject to turbulent water flow, or a BADLAND area subject to severe gully erosion. A donga is usually dry except in the rainy season. Dongas are common in areas of COLLUVIUM and erodible deeply weathered bedrock, where the mean annual rainfall is 600–800 mm. Dongas are sometimes discontinuous features within otherwise channelless washes (DAMBOS) and in Zimbabwe have been observed to develop from road collapse of pre-existing soil pipes (see PIPEFLOW). In Australia, the term has been used to describe circular depressions, of varying size, on the Nullarbor Plain, caused by the roof collapse of subterranean chambers. *MAC*

[See also EPHEMERAL STREAMS]

Rienks, S.M., Botha, G.A. and Hughes, J.C. 2000: Some physical and chemical properties of sediments exposed in a gully (donga) in northern KwaZulu-Natal, South Africa and their relationship to the erodibility of the colluvial layers. *Catena* 39, 11–31. **Shakesby, R.A. and Whitlow, R.** 1991: Perspectives on prehistoric and recent gullying in central Zimbabwe. *Geojournal* 23, 49–58.

Doppler effect The effect of the relative motion of a source and receiver on a signal passing between them. Relative motion along the line of sight causes a frequency shift in the propagating wave. If a transmitter and receiver are moving towards (away from) one another, the frequency of the signal passing between them is shifted to shorter (longer) wavelengths. *PJS*

dormant volcano A VOLCANO that is presently inactive, but is thought likely to produce a VOLCANIC ERUPTION at some time in the future. *GO*

[See also ACTIVE VOLCANO, EXTINCT VOLCANO]

Capra, L., Macias, J.L., Espindola, J.M. and Seibe, C. 1998: Holocene Plinian eruption of La Virgen volcano, Baja California, Mexico. *Journal of Volcanology and Geothermal Research* 80, 239–266.

double planation surfaces The two surfaces of levelling (the *wash surface* and the subsurface *basal surface of weathering*) considered by Julius Büdel (*Doppelten Einebnungsflächen*) to be operating concurrently in landscapes in seasonally wet tropical regions. Along the basal surface of weathering (WEATHERING FRONT) CHEMICAL WEATHERING works downwards, while in the rainy season finely worked material is correspondingly removed from above by highly effective *sheetwash* (see Figure). The separation of the upper wash surface from the weathering front by a thick layer of decomposed rock (SAPROLITE) creates a major contrast with TEMPERATE areas, where WEATHERING and EROSION are sometimes considered to proceed together. Tropical plains exhibiting such surfaces form gently concave INSELBERG-studded ETCHPLAINS over long geologic periods. Büdel considered that rivers played little role in such landscapes, but later workers regard fluvial action as an integral part in the lowering of the wash surface and in the erosion of such landscapes to form INSELBERG-studded ETCHPLAINS. *MAC*

[See also DIVERGENT WEATHERING, LANDSCAPE EVOLUTION]

Büdel J. 1982: *Climatic geomorphology*, translated by Fischer, L. and Busche, D. Princeton, NJ: Princeton University Press. **Büdel J.** 1957: Die 'Doppelten Einebnungsflächen' in den feuchten Tropen. *Zeitschrift für Geomorphologie NF* 1, 201–231.

downscaling The methods by which large-scale results are related to smaller-scale situations. Downscaling is a major problem in relating the global results of GENERAL CIRCULATION MODELS (GCMs) to regional patterns of CLIMATIC CHANGE and is approached in two ways. First, *empirical downscaling* involves the use of statistical techniques to relate locally observed climatic VARIABLES to the 'large-scale variables' available from GCM output. The second approach uses a high-resolution *regional climatic model* (RCM) nested within the GCM. *Upscaling* involves the reverse process: the application of results derived at the local scale to regional or global problems. Upscaling is a major task in relating GROUND MEASUREMENTS to the results of REMOTE SENSING at the global scale. *JAM*

[See also REGIONAL CLIMATIC CHANGE, SCALE CONCEPTS IN ENVIRONMENTAL CHANGE]

Wilbey, R.L. and Wigley, T.M.L. 1997: Downscaling general circulation model output: a review of methods and limitations. *Progress in Physical Geography* 21, 530–548.

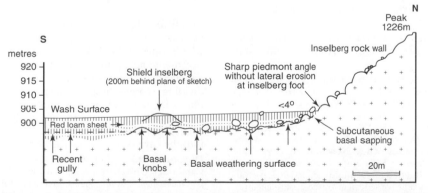

double planation surfaces *A double planation surface (upper wash surface and basal weathering surface) and associated landforms with the concentration of weathering at the piedmont angle and beneath the plain that surrounds the inselberg (after Büdel, 1982)*

downwelling The vertical movement of water from the sea surface towards the ocean depths. It may occur, for example, where surface OCEAN CURRENTS move towards the shore with the result that nearshore surface water is transported downwards and in the offshore direction, or as a result of density-driven DEEP WATER formation. Two major processes are responsible for the latter: (1) *open-ocean convection* as warm salty water reaches high latitudes, is cooled and sinks following reduced buoyancy; and (2) *brine exclusion* or *brine rejection*, which occurs in regions of large-scale SEA ICE formation, where surface water increases in density and sinks as salt is 'rejected' during freezing. *JAM*

[See also UPWELLING, THERMOHALINE CIRCULATION]

Dickson, R.R. and Brown, J. 1994: The production of North Atlantic deep water: sources, rates and pathways. *Journal of Geophysical Research* 99, 12319–12341.

draa A large compound DUNE (up to 450 m high) formed by the aggregation of smaller BEDFORMS in an AEOLIAN setting. *MRT*

Clemmensen, L.B. and Hegner, J. 1991: Eolian sequence and erg dynamics: the Permian Corrie Sandstone, Scotland. *Journal of Sedimentary Petrology* 61, 768–774. **Lancaster, N.** 1988: The development of large aeolian bedforms. *Sedimentary Geology* 55, 69–89.

drainage basin An area of the Earth's surface sloping and hence draining towards a particular point or having a common outlet. Most LANDSCAPES consist of a mosaic of nested drainage basins, also termed *catchments* (and, in America, *watersheds*), at various scales, each delimited by a *water divide* (in Europe, WATERSHED). River drainage basins may be considered fundamental units of HYDROLOGY and fluvial GEOMORPHOLOGY, while lake drainage basins have been proposed as fundamental units in both LANDSCAPE GEOCHEMISTRY and PALAEOECOLOGY, at least in temperate environments. River drainage basins have been widely used (as so-called *representative* or *experimental drainage basins*) in monitoring processes and assessing short-term EROSION RATES at the catchment scale. Lake drainage basins have been used to provide spatially integrated records of long-term erosion and DENUDATION RATES, and stratigraphic records as a basis for basin-wide palaeoenvironmental reconstruction. The drainage basin is also a meaningful unit for management and planning, including so-called *integrated drainage basin management*. *JAM*

[See also ENVIRONMENTAL MONITORING, LACUSTRINE SEDIMENTS, PALAEOLIMNOLOGY, SEDIMENT TRANSPORT, SEDIMENT YIELD]

Chorley, R.J. 1969: The drainage basin as the fundamental geomorphic unit. In Chorley, R.J. (ed.), *Water, Earth and man*. London: Methuen, 77–99. **Downs, P.W., Gregory, K.J. and Brookes, A.** 1991: How integrated is river basin management? *Environmental Management* 15, 299–309. **Gregory, K.J. and Walling, D.E.** 1973: *Drainage basin form and process: a geomorphological approach*. Chichester: Wiley. **Newson, M.** 1992: *Land, water and development: river basin systems and their sustainable development*. London: Routledge. **Oldfield, F.** 1978: Lakes and their drainage basins as units of sediment-based ecological study. *Progress in Physical Geography* 1, 460–504. **Toebes, C. and Ouryvaev, V. (eds)** 1970: *Representative and experimental basins: an international guide for research and practice*. Paris: Unesco.

draw down (1) The spatial and vertical extent to which the WATER TABLE is lowered due to water ABSTRACTION from WELLS. (2) The extent to which the water level in a reservoir is lowered by abstraction. *JAM*

[See also DEGLACIATION, GROUNDWATER]

dredging The removal of bottom material from the bed of a water body. At the coast, it is often carried out to improve navigation and to provide sediment for BEACH NOURISHMENT. *HJW*

drift A general term used to describe all unconsolidated sediments, especially on geological *drift maps*. The term has its origins in the nineteenth century, when many surface sediments were believed to have been deposited by drifting ICEBERGS. *DIB*

[See also GENETIC DRIFT, ICEBERG DRIFT THEORY, OLDER DRIFT]

driftwood Wood transported by rivers and oceans. In the HIGH ARCTIC, driftwood has occasionally been used, in the absence of other datable organic materials, to radiocarbon date the age of *raised beach* deposits and, hence, to establish rates of ISOSTATIC UPLIFT and regional patterns of DEGLACIATION. In the Canadian Arctic islands, driftwood consists usually of spruce (*Picea* sp.) fragments derived from the northern BOREAL FORESTS to the south, but other drifted materials such as charred fat, driftwood charcoal (*Betula nana*; *Juniperus* sp.; *Salix* sp.), whale bone and PEAT can also be radiocarbon dated.

Along the northern coastlines of the western North American Arctic and central Siberia, where large rivers discharge into the Arctic Ocean from the south, drift-transported tree trunks commonly occur at the high-tide level. The wood is usually either spruce (*Picea* sp., North America) or larch (*Larix* sp., Siberia) that has entered the river system following fluvio-thermal river-bank erosion. *HMF*

[See also GLACIO-ISOSTASY, RADIOCABON DATING]

Blake Jr, W. 1970: Studies of glacial history in Arctic Canada. 1. Pumice, radiocarbon dates and differential postglacial uplift in the eastern Queen Elizabeth Islands. *Canadian Journal of Earth Science* 7, 634–664. **Czudek, T. and Demek, J.** 1970: Thermokarst in Siberia and its influence on the development of lowland relief. *Quaternary Research* 1, 103–120. **Johansen, S.** 1998: The origin and age of driftwood on Jan Mayen. *Polar Research* 17, 125–146.

dripline The narrow zone at the mouth of a CAVE or ROCKSHELTER directly below the overhanging cliff edge where water and debris from the cliff face hit the ground. Sediment accumulations at the dripline, sometimes enhanced by built structures, tend to form a watershed, dividing contrasting sedimentary environments and microclimates and directing falling sediments and water either towards the outside (where they may contribute to TALUS) or inside. Through time, the dripline may migrate into the cave or shelter as the cliff face retreats. *JAM*

Dincauze, D.F. 2000: *Environmental archaeology*. Cambridge: Cambridge University Press. **Waters, M.R.** 1992: *Principles of geoarchaeology*. Tucson, AZ: University of Arizona Press.

dropstone Isolated CLASTS deposited onto the ocean floor or a lake bottom, rafted actively by ICEBERGS, SEA ICE or lake ice and passively by marine algae. Clasts range from CLAY to BOULDERS, depending on the source material and rafting agent. *LJW/WENA*

[See also ICE-RAFTED DEBRIS]

Bennett, M.R., Doyle, P. and Mather, A.E. 1996: Dropstones: their formation and significance. *Palaeogeography, Palaeoclimatology, Palaeoecology* 121, 331–339. **Gilbert, R.** 1990: Rafting in glacimarine environments. In Dowdeswell, J. and Scourse, J. (eds), *Glacimarine environments: processes and sediments* [*Special Publication* 53]. London: The Geological Society, 105–120. **Luckman, B.H.** 1975: Drop stones resulting from snow-avalanche deposition on lake ice. *Journal of Glaciology* 14, 186–188.

drought

Commonly defined as a period of RAINFALL deficiency or exceptionally dry weather; it must be distinguished from ARIDITY. It is a relative concept that needs to be explained in particular contexts. Four kinds of drought may be recognised: *meteorological drought* is when there is little rain compared with normal, note being taken of the rainfall *variability* at the place in question. Until 1961, an absolute drought in the UK was defined as 15 consecutive days, none of which received more than 0.2 mm of rainfall. In Australia, a drought occurs if the annual rainfall is within the range of the driest 10% of years at a particular place. In the United States, it is defined in terms of the *Palmer Drought Severity Index*. This depends on comparisons of estimated values of EVAPOTRANSPIRATION and measured rainfall with normal values.

Agricultural drought occurs when a crop failure is due to insufficient rainfall. In this case the drought is specific to each crop because it depends on the SOIL MOISTURE conditions at different stages in the crop's growth. Thus, it depends on previous RUNOFF and EVAPORATION as well as rainfall. In France, a disaster is declared if a drought affects at least 25% of the projected yield and 12% of a farm's total production. *Hydrological drought* is the reduction in streamflow below a specified level for a particular period of time – essentially the drying up of streams. It is affected by all the RUNOFF PROCESSES and is usually revealed by *water balance/budget* techniques. Finally, *socioeconomic drought* is when there is a water shortage due to management decisions. This can be in a variety of contexts. Rural–urban migration may produce an urban water shortage if there has been insufficient planning of the water supply infrastructure. OVERGRAZING and/or overstocking will also result in this kind of drought. It is clear from this discussion that there is considerable overlap and interaction between the various kinds of drought, but they all depend on a relative deficiency of rainfall in relation to specified environmental factors, whether natural or anthropogenic. Furthermore, drought is a recurrent feature in virtually all climatic regimes. *BDG*

[See also CONTINGENT DROUGHT, DESERTIFICATION, INVISIBLE DROUGHT, PERMANENT DROUGHT, SEASONAL DROUGHT]

Alley, W.M. 1984: The Palmer Drought Severity Index: limitations and assumptions. *Journal of Climate and Applied Meteorology* 23, 1100–1109. **Briffa, K.R., Jones, P.D. and Hulme, M.** 1994: Summer moisture variability across Europe, 1892–1991: an analysis based on the Palmer Drought Severity Index. *International Journal of Climatology* 14, 475–506. **Doornkamp, J.C. and Gregory, K.J.** (eds) 1980: *Atlas of drought in Britain 1975–76*. London: Institute of British Geographers. **Gantz, M.H.** (ed.) 1994: *Drought follows the plough: cultivating marginal areas*. Cambridge: Cambridge University Press. **Karl, T.R. and Koscielny, A.J.** 1982: Drought in the United States, 1895–1981. *Journal of Climatology* 2, 313–329. **Laird, K.R., Fritz, S.C., Maasch, K.A. and Cumming, B.F.** 1996: Greater drought intensity and frequency before AD 1200 in the northern Great Plains, U.S.A. *Nature* 384, 552–554. **Palmer, W.C.** 1965: *Meteorological drought* [*Research Paper* 45]. Washington, DC: US Weather Bureau. **Riebsome, W.E., Changnon Jr, S.A. and Karl, T.R.** 1991: *Drought and natural resource management in the United States: impacts and implications of the 1987–89 drought*. Boulder, CO: Western Press. **Rosenberg, N.J.** (ed.) 1978: *North American droughts*. Boulder, CO: Western Press. **Scian, B. and Donnari, M.** 1997: Retrospective analysis of the Palmer Drought Severity Index in the semi-arid pampas region, Argentina. *International Journal of Climatology* 17, 313–322. **Steila, D.** 1983: Quantitative versus qualitative drought identification. *Professional Geographer* 35, 192–194. **Van Royen, W.** 1937: Prehistoric droughts in the central Great Plains. *Geographical Review* 27, 637–650. **Wilhite, D.A.** (ed.) 1999: *Drought: a global assessment*. 2 vols. London: Routledge. **Wilhite, D.A. and Glantz, M.H.** 1987: Understanding the drought phenomenon: the role of definitions. In Wilhite, D.A. and Easterling, W.E. (eds), *Planning for drought*. Boulder, CO: Westview Press, 11–27.

drought: impact on ecosystems

DROUGHT events have significant consequences for ecosystem function, limiting PRIMARY PRODUCTIVITY and CARBON and NITROGEN CYCLES, and, in terms of ecosystem structure, controlling the LEAF AREA INDEX and height of VEGETATION CANOPIES and acting as a strong filter for species presence. Drought is an important factor controlling species survival in ARIDLAND ecosystems, exerting a strong SELECTION PRESSURE for ADAPTATION to low water availabilities. XEROPHYTES show physiological ADAPTATION and ACCLIMATISATION for conservation of water or TOLERANCE of dry conditions, or they have life histories in which development and reproduction are completed within rainy seasons and drought is spent in a dormant state. Since aridland ecosystems often experience high SOLAR RADIATION, morphological and physiological features that promote efficient heat dissipation may also be important in avoiding thermal damage.

Future responses of vegetation structure and function to drought are likely to be mediated by rising atmospheric CARBON DIOXIDE. STOMATAL CONDUCTANCE to water loss exerts an important control over *transpiration* from individual leaves, usually declining in response to CO_2 and plant, atmospheric and SOIL MOISTURE DEFICITS and thereby conserving water. Stomatal closure in high CO_2 concentrations may therefore reduce rates of transpiration in vegetation canopies, hence improving plant water economy and slowing the development of drought conditions. However, NEGATIVE FEEDBACK mechanisms may offset this benefit and the operation of each will govern the extent to which stomatal closure in high-CO_2 conditions improves water relations of vegetation during future drought.

While CARBON DIOXIDE FERTILISATION is likely to improve water economy in some temperate tree species, it may exacerbate the impacts of drought in others. For example, stomata of the beech *Fagus sylvatica* are relatively insensitive to CO_2 enrichment and the leaf area tends to increase, leading to greater transpiration in this drought-sensitive species. Furthermore, stomatal sensitivity to CO_2 declines during drought and the response is therefore lost at precisely the time when it would be most effective. Differing species responses to atmospheric CO_2

could result in differences in their survival of drought events, ultimately leading to altered species ranges and alteration in the plant species composition of ecosystems subjected to periodic drought.   *CPO*

Beerling, D.J., Heath, J., Woodward, F.I., Mansfield, T.A. 1996: Drought–CO₂ interactions in trees: observations and mechanisms. *New Phytologist* 134, 235–242. **Cloudsley-Thompson, J.L.** 1991: *Ecophysiology of desert arthropods and reptiles*. Berlin: Springer. **Field, C.B., Jackson, R.B., Mooney, H.A.** 1995: Stomatal responses to increased CO₂: implications from the plant to the global scale. *Plant, Cell and Environment* 18, 1214–1225. **Heath, J.** 1998: Stomata of trees growing in CO₂-enriched air show reduced sensitivity to vapour pressure deficit and drought. *Plant, Cell and Environment* 21, 1077–1088. **Smith, S.D., Monson, R.K., Anderson, J.E.** 1997: *Physiological ecology of North American desert plants*. Berlin: Springer.

drowned karst   KARST landforms and landscapes developed above sea level and subsequently submerged during a rise in sea level or during land SUBSIDENCE. Dated CAVE SEDIMENTS from drowned karst systems can provide indirect information on RELATIVE SEA LEVEL fluctuations.   *SHD*

[See also SEA-LEVEL CHANGE]

drumlin   A SUBGLACIAL landform in the form of a streamlined hill, typically like an inverted spoon and often grouped into extensive fields (*basket-of-eggs topography*) and with long axes parallel to the ice-flow direction. Drumlins usually comprise unconsolidated glacigenic sediments, but can also consist almost entirely of bedrock. There are many theories of origin for the components of the subglacial LANDSYSTEM. Drumlins are a type of longitudinal (i.e. parallel to ice-flow direction) subglacial bedform, which also includes FLUTED MORAINE (or flutings) and *megaflutes* (or megaflutings), the latter being classified as a particularly large form of fluted moraine (> 100 m in length).   *JS*

[See also GLACIAL LANDFORMS, ROCHE MOUTONNÉE]

Menzies, J. and Rose, J. (eds) 1987: *Drumlin symposium*. Rotterdam: Balkema. **Patterson, C.J. and Hooke, R.L.** 1995: Physical-environment of drumlin formation. *Journal of Glaciology* 41, 30–38. **Rose, J.** 1987: Drumlins as part of a glacier bedform continuum. In Menzies, J. and Rose, J. (eds), *Drumlin symposium*. Rotterdam: Balkema, 45–67. **Shaw, J., Kvill, D. and Rains, B.** 1989: Drumlins and catastrophic subglacial floods. *Sedimentary Geology* 62, 177–202.

dry deposition   Used in the context of deposition from the atmosphere of POLLUTANTS, especially ACID PRECIPITATION, dry deposition includes two classes of materials: PARTICULATES deposited by gravitational settling; and gases adsorbed (see ADSORPTION) by vegetation, soil and surface water.   *JAM*

[See also WET DEPOSITION, OCCULT DEPOSITION]

Davidson, C.I. and Wu, Y.L. 1990: Dry deposition of particles and vapors. In Lindberg, S.E., Page, A.L. and Norton, S.A. (eds), *Acid precipitation*. Vol. 3. *Sources, deposition and canopy interactions*. New York: Springer, 103–216. **Rustad, L., Kahl, J.S., Norton, S.A. and Fernandez, I.J.** 1994: Multi-year estimates of dry deposition at the Bear Brook Watershed in eastern Maine. *Journal of Hydrology* 162, 319–336.

dry farming   Agriculture in SEMI-ARID or other DROUGHT-prone environments without the use of IRRIGA-

TION. Water is conserved by dry-farming techniques, such as the breaking down of soil aggregates and the maintenance of a fine-textured *tilth* in the surface layer by TILLAGE (which reduces evaporation from subsurface layers while allowing rainwater to percolate downwards from the surface and acts as a seedbed), MULCHING (addition of organic residues, often plant litter and/or dung, to the soil surface), use of species adapted to seasonal rains and/or periodic drought (such as deciduous trees in some forms of ARBORICULTURE and cereals), and not necessarily cropping in every year.   *JAM*

[See also DEHESA]

Rees, J. 1987: Agriculture and horticulture. In Wacher, J. (ed.), *The Roman world*. Vol. 2. London: Routledge, 481–503.

dry ravel   A form of movement of loose, dry surface soil material on steep slopes, which may collect as DEBRIS CONES at the bases of slopes. The process may be encouraged by soil WATER REPELLENCY as a result of the lower bulk density of the water-repellent soil.   *RAS*

[See also SEDIMENT GRAVITY FLOW]

Krammes, J.S. and Osborn, J. 1968: Water-repellent soils and wetting agents as factors influencing erosion. *Proceedings of the Symposium on Water Repellent Soils. Riverside, University of California, May 1968*, 177–187.

dry valley   A valley without a permanent stream. Dry valleys are a feature particularly associated with limestone, chalk and sandstone rocks of southern England, although they also occur in a wide range of other locations and on other rock types. Origins for dry valleys range from uniformitarian ideas (not involving CLIMATIC CHANGE or BASE LEVEL change) to marine hypotheses (associated with base level change) and palaeoclimatic hypotheses (linked primarily to PLEISTOCENE climatic change and especially to the flow of water over impermeable PERMAFROST).   *RAS*

Doran, P.T., Berger, G.W., Lyons, W.B. *et al.* 1999: Dating Quaternary lacustrine sediments in the McMurdo Dry Valleys, Antarctica. *Palaeogeography, Palaeoclimatology, Palaeoecology* 147, 223–239. **Espejo, J.M.R., Catt, J.A. and Mackney, D.** 1992: The origin of very flinty dry-valley deposits in the Marlow area, Buckinghamshire, England. *Journal of Quaternary Science* 7, 227–234. **Goudie, A.** 1989: *The nature of the environment*, 2nd edn. Oxford: Blackwell, 101–104.

dry weight   Soils and sediments contain a variable amount of water, so results of chemical and physical analysis are usually reported on the basis of a sample dried at 105°C.   *EMB*

drylands   The arid, SEMI-ARID and dry subhumid regions of the world that are susceptible to DROUGHT and DESERTIFICATION. These regions are characterised by low but variable rainfall and very high rates of potential EVAPOTRANSPIRATION, which restrict plant growth to a minimum without IRRIGATION and generally prevent other than a nomadic or extremely low-density occupance. Drylands represent about 5.1 billion ha or one-third of the world's land area and support about one-fifth of the world's population. For some authorities, *hyperarid* areas (e.g. Atacama and Sahara Deserts) would also be included in this term, although many use the term without definition, so that the climatic and spatial boundaries may be difficult to judge.   *RAS*

[See also ARIDLAND]

Beaumont, P. 1993: *Drylands: environmental management and development.* Routledge: London. **Hulme, M.** 1996: Recent changes in the world's drylands. *Geophysical Research Letters* 23, 61–64. **United Nations Environment Programme (UNEP)** 1992: *Status of desertification and implementation of the United Nations Plan of Action to combat desertification* [*Report of the Executive Director*]. Nairobi: UNEP. **Williams, M.A.J. and Balling Jr, R.C.** 1996: *Interactions of desertification and climate.* London: Arnold.

dune A medium scale, flow-transverse periodic BED-FORM in sand or gravel with height greater than 10 cm (up to tens of metres) and spacing greater than 1 m. Dunes develop in response to processes in the turbulent BOUND-ARY LAYER of moderately strong currents of water (aqueous dunes) or air (aeolian dunes). Dune migration forms CROSS-BEDDING, which is a valuable indicator of PALAEOCURRENT type, strength and direction. *GO*

[See also BEDFORM STABILITY DIAGRAM, COASTAL DUNES, PALAEODUNES, SAND DUNES]

Allen, J.R.L. 1982: *Sedimentary structures: their character and physical basis.* Amsterdam: Elsevier. **Collinson, J.D. and Thompson, D.B.** 1992: *Sedimentary structures*, 2nd edn. London: Chapman and Hall. **Lancaster, N.** 1995: *Geomorphology of desert dunes.* London: Routledge.

duricrust A hard crust formed on the surface of, or within, the upper horizons of a soil during the processes of WEATHERING and soil formation. In French literature a duricrust is sometimes termed a *cuirasse*. Typically, it is formed by the accumulation of soluble material precipitated by mineral-bearing waters that move upward by CAPILLARY ACTION and evaporate during the dry season. Climatic factors controlling duricrust formation are difficult to isolate because most deposits have experienced repeated environmental changes during their development. They cover wide areas of the tropics and subtropics and occur less extensively elsewhere. The indurated nature of duricrusts means they are often relatively resistant to erosion and tend to encourage the formation of positive relief features; indeed, they provide the main relief features of low-relief deserts in central Australia, southern Africa and the central Sahara. They therefore have a profound effect on slope hydrology, sediment transport processes and slope evolution.

Duricrusts often cap residual hills (see INSELBERG) and plateaux and encourage the development of hillslopes by PARALLEL SLOPE RETREAT. Undercutting and collapse of summital duricrust cappings may result in the formation of boulder fields and PSEUDOKARST topography. If duricrust formation has taken place preferentially in low-lying areas over long periods of time, relief and drainage inversion may take place. INDURATION of the near-surface layers of a sediment body also helps to preserve otherwise ephemeral forms such as aeolian dunes and river terrace deposits.

Duricrusts can provide evidence of palaeoenvironmental conditions. Formation and accumulation of thick sequences of chemical precipitates usually require substantial periods of landscape stability. Fossil deposits may therefore indicate relatively stable tectonic conditions and ineffective erosion processes. In addition, although also affected by the factors of lithology, topography and time,

to some extent their chemical composition indicates the degree of wetness of the climate that prevailed at their time of formation with CALCRETE, GYPCRETE, FERRICRETE and *alcrete* (indurated BAUXITE) representing progressively wetter environments. SILCRETE, however, can form in both arid and humid tropical environments. *MAC*

[See also LATERITE]

Beauvais, A. and Colin, F. 1993: Formation and transformation processes of iron duricrust systems in tropical humid environment. *Chemical Geology* 106, 77–101. **Goudie, A.S.** 1973: *Duricrusts in tropical and sub-tropical landscapes.* Oxford: Oxford University Press. **Goudie, A.S. and Pye, K. (eds)** 1983: *Chemical sediments and geomorphology: precipitates and residua in the near-surface environment.* London: Academic Press. **McNally, G.H., Clarke, G and Weber, B.W.** 2000: Porcellanite and the urban geology of Darwin, Northern Territory. *Australian Journal of Earth Sciences* 47, 35–44. **Nash, D.J., Shaw P.A. and Thomas D.S.G.** 1994: Duricrust development and valley evolution – process-landform links in the Kalahari. *Earth Surface Processes and Landforms* 19, 299–317. **Nash D.J., Thomas D.S.G. and Shaw P.A.** 1994: Siliceous duricrusts as paleoclimatic indicators – evidence from the Kalahari desert of Botswana. *Palaeogeography, Palaeoclimatology, Palaeoecology* 112, 279–295. **Tardy, Y.** 1997: *Petrology of laterites and tropical soils.* Rotterdam: Balkema. **Twidale, C.R. and Bourne, J.A.** 1998: The use of duricrusts and topographic relationships in geomorphological correlation: conclusions based in Australian experience. *Catena* 33, 105–122. **Watson, A. and Nash, D.J.** 1997: Desert crusts and varnishes. In Thomas, D.S.G. (ed.), *Arid zone geomorphology: process, form and change in drylands.* Chichester: Wiley, 69–107.

Durisols Soils with accumulation of secondary silica in a *duric* or *petroduric horizon*, occurring in USA, South Africa and Australia. Accumulation of opaline or microcrystalline silica occurs in the subsurface of soils in ARIDLANDS and SEMI-ARID regions where sufficient weathering occurs to break down silicate CLAY MINERALS, releasing silica. The accumulation may be in the form of nodules (*durinodes*) or a continuously cemented *duripan*, fragments of which do not disintegrate in water or concentrated hydrochloric acid (SOIL TAXONOMY: Durorthic or Durargic great groups of *Aridisols* and other Orders). If the HARDPAN is broken, followed by IRRIGATION or heavy rains, SALINISATION may occur. *EMB*

[See also DURICRUST, WORLD REFERENCE BASE FOR SOIL RESOURCES].

Thompson, C.H., Bridges, E.M. and Jenkins, D.A. 1993: An exploratory examination of some relict hardpans in the coastal lowlands of southern Queensland. In Ringrose-Voase A.J. and Humphries, G.S. (eds), *Soil micromorphology: studies in management and genesis* [*Developments in Soil Science.* Vol. 22]. Amsterdam: Elsevier. **Thompson, C.H., Bridges, E.M. and Jenkins, D.A.** 1996: Pans in humus podzols (Humods and Aquods) in coastal southern Queensland. *Australian Journal of Soil Research* 34, 161–182.

dust Solid, irregular microscopic particles less than about 0.08 mm in size in suspension in a gas, or a deposit of such particles. Grains above about 0.02 mm settle back to the surface quite quickly when the turbulence associated with strong winds decreases, but smaller particles can remain in suspension for days or even weeks unless washed out by rain. Dust can be derived from a range of natural and artificial sources (e.g. volcanic eruption, saltspray from the oceans, POLLEN and bacteria, smoke and ash from forest fires and industrial combustion processes, soil particles). It travels as suspension load in wind transport and can be lifted to considerable heights in the atmosphere. DUST STORMS are associated strongly with DROUGHTS in DRYLANDS. *RAS*

[See also ATMOSPHERIC DUST, LOESS, SAHARAN DUST, VOLCANIC AEROSOLS]

Middleton, N. 1997: Desert dust. In Thomas, D.S.G. (ed.), *Arid zone geomorphology*, 2nd edn. London: Wiley, 413–436. **Péwé, T.L. (ed.)** 1981: *Desert dust: origin, characteristics and effect on man* [*Special Paper* 186]. Boulder, CO: Geological Society of America. **Pye, K.** 1987: *Aeolian dust and dust deposits*. London: Academic Press.

Dust Bowl A spectacular example of WIND EROSION that occurred during the AD 1930s in America's Great Plains region (see Figure). It has been estimated that the amount of DUST blown from the area in 1934 was equivalent to between a third and half of the annual sediment load of the Mississippi River. By 1937, the US Soil Conservation Service estimated that 43% of a 6.5 million ha area at the core of the region affected had been seriously damaged by DEFLATION. The Dust Bowl was caused by a combination of climatic, edaphic and anthropogenic factors. Rainfall amounts were relatively high in the period 1900–1930, when farmers settled the area and were able to harvest good crops. PLOUGHING of the grassland cover meant that SOIL ERODIBILITY was high when annual rainfall totals declined to low levels in successive years during the 1930s. The resulting LAND DEGRADATION led to one of the largest migrations in American history, with an estimated 3.5 million people moving to nearby towns or towards the west coast in search of alternative work. Written in the lower case (i.e. dust bowl), the term is also applied to any region where deflation of cultivable land occurs. *RAS*

[See also DESERTIFICATION, DUST STORM]

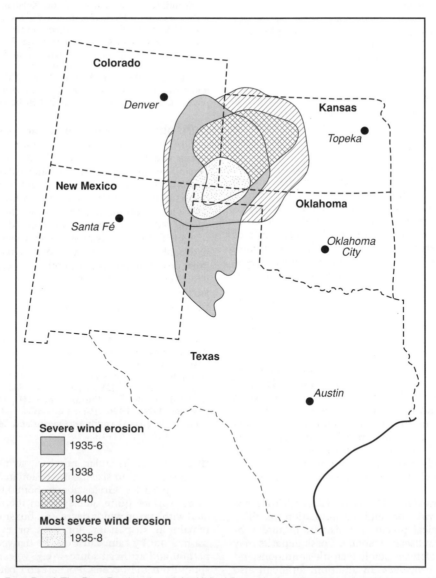

Dust Bowl *The Dust Bowl region of the United States in the AD 1930s (after Worster, 1979)*

Goudie, A.S. 1983: Dust storms in space and time. *Progress in Physical Geography* 7, 502–530. **Rosenberg, N.J. (ed.)** 1978: *North American droughts.* Boulder, CO: Westview Press. **Worster, D.** 1979: *Dust Bowl: the southern plains in the 1930s.* New York: Oxford University Press.

dust storm A condition of the atmosphere characterised by DUST concentrations sufficient to reduce visibility below 1000 m. The *haboob* (derived from the Arabic word *habb*, to blow) is a severe dust storm in North Africa characterised by a wall of dust, which may rise to over 1000 m, moving at speeds up to 60 km h^{-1}. Dust storms are one of the main mechanisms on the surface of the Earth for the transport of fine sediments. Agriculture and transport by road and air can be affected, together with human health. The frequency of dust storms in some of the DESERT areas of Algeria and Libya is over 15 days per year, but they were probably more frequent during cold, dry INTERPLUVIALS of the QUATERNARY. As a result of reduced rainfall and increased LANDUSE pressures, the frequency of dust storms in the SAHEL and Sudan zones has increased by four- to six times since the mid 1960s.
JAM/AHP

[See also ATMOSPHERIC DUST, DUST BOWL, SAHARAN DUST]

Goudie, A.S. 1983: Dust storms in space and time. *Progress in Physical Geography* 7, 502–530. **Idso, S.B.** 1976: Duststorms. *Scientific American* 235, 108–114. **Middleton, N.J.** 1985: Effect of drought on dust production in the Sahel. *Nature* 316, 431–434.

dust veil A layer of fine particles (AEROSOLS < 0.6 mm diameter, especially sulphates), normally of volcanic origin, ejected into the upper TROPOSPHERE or lower STRATOSPHERE, changing the Earth's ENERGY BALANCE/BUDGET for several weeks or months. It alters atmosphere TURBIDITY and is associated with global cooling.
JGT

[See also DUST VEIL INDEX, VOLCANIC AEROSOLS]

dust veil index (DVI) A semi-quantitative assessment of the ATMOSPHERIC LOADING of *volcanic dust* in the atmosphere first developed by H.H. Lamb. The AD 1883 eruption of Krakatoa was given a DVI of 1000 and hence used as a standard against which all other eruptions could be measured. In an analysis extending back to 1680, the highest DVI was attained by the eruption of Tambora in 1815, which led to the following 'year without a summer' in Europe. Lamb's Index has been superseded by others.
AHP

[See also GLACIOLOGICAL VOLCANIC INDEX, VOLCANIC EXPLOSIVITY INDEX]

Bradley, R.S. and Jones, P.D. 1992: Records of explosive volcanic eruptions over the last 500 years. In Bradley, R.S. and Jones, P.D. (eds), *Climate since A.D. 1500.* London: Routledge, 606–622. **Lamb, H.H.** 1970: Volcanic dust in the atmosphere with a chronology and assessment of its meteorological significance. *Philosophical Transactions of the Royal Society of London* A266, 425–533. **Robock, A.** 1981: A latitudinally dependent volcanic dust veil index, and its effect on climate simulations. *Journal of Volcanological and Geothermal Research* 11, 67–80.

dy Acidic, brown/black SEDIMENT composed mainly of precipitated, colloidal, humic substances (gel-mud), transported by surface or groundwater into a lake. It is deposited in the profundal to lower sublittoral zone.
UBW

[See also DYSTROPHIC, GYTTJA, LACUSTRINE SEDIMENTS, SEDIMENT TYPES]

dynamic climatology The study of climate in terms of thermodynamics and the energy processes of the atmosphere.
JCM

Hare, F.K. 1957: The dynamic aspects of climatology. *Geografiska Annaler* 39, 87–104. **Raynor, J.N. and Hobgood, J.S.** 1991: Dynamic climatology: its history and future. *Physical Geography* 12, 207–219.

dynamical systems Dynamic is used to mean that the SYSTEM changes and evolves (i.e. it is not static), while dynamical specifically means that the system(s) may change state. Simple examples of dynamical behaviour in the Earth sciences would be variation between deposition and erosion in a river, advance and retreat in glacier. Many, if not most, Earth systems are dynamical and their behaviour may be modelled using mathematical models that invoke dissipative structures, multiple equilibria, bifurcations and catastrophes – in short the concepts associated with CHAOS theory. The essential elements for construction of a dynamical systems model are description of the system using a set of state variables and the use of one or more differential equations to describe change over time. Model evaluation often focuses upon *sensitivity analysis.*
CET

Huggett, R.J. 1985: *Earth surface systems.* Berlin: Springer. **Phillips, J.D.** 1999: *Earth surface systems: complexity, order and scale.* Oxford: Blackwell.

dystrophic Lakes in acid PEAT areas, poor in plant nutrients (OLIGOTROPHIC), with low productivity, low water transparency and deposition mainly of ALLOCHTHONOUS humic substances.
UBW

[See also DY, LAKES AS INDICATORS OF ENVIRONMENTAL CHANGE, LAKE STRATIFICATION, SEDIMENT TYPES]

E

early instrumental meteorological records

Meteorological readings from instruments made before the introduction of official meteorological station networks and the establishment of standardised procedures and instruments. The transition between descriptive weather records and instrumental measurements is often blurred and WEATHER DIARIES often contain a mixture of the two. Early instrumental record series often contain both systematic and other errors that result from technical problems and setting-up or reading-off difficulties. They can often be identified using homogeneity tests because of sudden changes in data-series values. Additional problems occur because of the variety of scales that were used in the past and, in the case of barometer readings, a lack of correction for altitude and gravity. If these early and valuable data are to be used for scientific purposes, then these series must be homogenised and calibrated. *Homogenising* is the attempt at removing breaks and inconsistencies in the single data series. *Calibration* is the STANDARDISATION and transfer of early scale values into Celsius or Kelvin scales as used today.

The first thermometer is traced back to Galileo Galilei in AD 1597. The first liquid thermometer was described in the year 1632 by J. Rey. The thermometer was improved on and further developed by Fahrenheit, Réaumur and Celsius (hence the three early scales of those names). Long historical European record series have been extensively analysed and interpreted. Especially worth mentioning are the oldest measurements between 1654 and 1670 in Florence and Pisa, Manley's CENTRAL ENGLAND TEMPERATURE RECORD and the measurement series still being maintained at De Bilt, The Netherlands. In connection with the instrument series the possibilities for measuring networks was discussed early on. Essential impulses came from the *Royal Society of London*, which started its service in 1667. As a first result of this synoptic research, the wind map by Halley – the oldest meteorological map of all – appeared in the year 1688, published as a supplement to an article on the trade winds and monsoons in *Philosophical Transactions*. With the same aims, the *Société Royale de Médecin* was founded in France shortly after the Royal Society. In the same tradition was the foundation in 1780 of the *Societas Meteorologica Palatina* (SMP) in Mannheim by Elector Karl Theodor with considerable cooperation from his court chaplain Hemmer. In retrospect, this society has proved to be probably the best conceived scientifically, as well as the most fruitful of such institutions. The SMP not only confined itself to the collection of data, but also developed from the data new concepts on synopsis; thus it may be considered to be the foundation of world-wide climatic observation. *RG*

[See also INSTRUMENT CALIBRATION, INSTRUMENTAL DATA, INSTRUMENTAL RECORDS]

Brazdil, R. (ed.) 1990: *Climatic change in the historical and the instrumental periods.* [*Proceedings of the International Conference of the Commission on Climatology of the International Geographical Union, Brno, June 1989*]. Available from the Editor, Brezinova 6, 61600 Brno, Czechoslovakia. **Ball, T.F. and Kingsley, R.A.** 1984: Instrumental temperature records at two sites in Canada 1768 to 1910. *Climatic Change* 6, 39–56. **Demarre, G.R., Van Engelen, A.F.V. and Geurts, H.A.M.** 1994: Les observations météorologiques de Théodore-Augustin Mann effectuées à Nieuport en 1775, 1776 et 1777. *Ciel et Terre* 110, 41–48. **Jones, P.D. and Bradley, R.S.** 1992: Climatic variations in the longest instrumental records. In Bradley, R.S. and Jones, P.D. (eds), *Climate since A.D. 1500*. London: Routledge, 246–268. **Kington, J.A.** 1988: *The weather of the 1780s over Europe*. Cambridge: Cambridge University Press. **Legrand, J.P. and Legoff, M.** 1994: *Les observations météorologiques de Louis Morin* [*Monographie 6*]. Paris: Direction de la Météorologie Nationale. **Rudloff, H. von** 1967: *Die Schwankungen und Pendlungen des Klimas in Europa seit Beginn der regelmäßigen Instrumentenbeobachtungen (1670).* Braunschweig: Vieweg. **Van den Dool, H.M., Krijnen, H.J. and Schuurmans, C.J.E.** 1978: Average winter temperatures at De Bild Netherlands: 1634–1977. *Climatic Change* 1, 319–330. **Wang, S.** 1991: Reconstruction of palaeo-temperature series in China from the 1380s to the 1980s. *Würzburger Geographische Arbeiten* 80, 1–20.

Earth observation (EO)

The REMOTE SENSING of the land surface and oceans. It has often and recently been used synonymously with the term 'remote sensing', especially in environmental science. *AJL*

earth pillar

A pinnacle, typically many metres high, of unconsolidated material protected from *rainsplash erosion* by the presence of a stone or boulder (caprock) at the top. Alternative terms for earth pillars include, in North America, *hoodoos* or *tepee buttes* and, in the French Alps, *demoiselles* or *cheminées de fées* (fairy chimneys). *SHD*

Francek, M. 1988: Earth pillar formation on the Mountain Pine Ridge, Belize. *Earth Surface Processes and Landforms* 13, 183–186.

Earth science

Although commonly used as a synonym for GEOLOGY, the Earth sciences are all those scientific disciplines concerned with the Earth that are not normally considered as physical, chemical or biological sciences. The field includes CLIMATOLOGY, GEOCHEMISTRY, GEOMORPHOLOGY, GEOPHYSICS, HYDROGEOLOGY, METEOROLOGY and OCEANOGRAPHY as well as geology. *CDW*

[See also ASTROGEOLOGY, GEOSCIENCE, PLANETARY GEOLOGY]

Earth structure

The shape of the Earth (see GEOID) is a flattened sphere with an equatorial radius of 6378 km and a polar radius of 6357 km. The relief on the Earth's surface is about 20 km, from 8848 m above sea level (Everest, Himalayas) to 11 524 m below sea level (Mindanao Trench, Pacific Ocean). Internally the Earth can be divided into concentric zones that vary in composition and density (see Figure). The outermost layer is the CRUST, which occurs as OCEANIC CRUST and CONTINENTAL CRUST. The *Mohorovičić discontinuity* (*Moho*) separates the

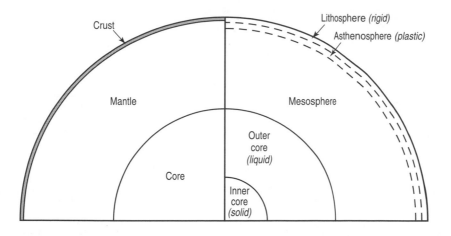

Earth structure *The gross structure of the Earth's interior (after Skinner and Porter, 1987). The left side shows the compositional zones of crust, mantle and core; the right side shows the rheological zones of lithosphere, asthenosphere, mesosphere, outer core and inner core*

crust from the denser, rocky MANTLE, which is separated from the very dense, iron-rich CORE at the *Gutenberg discontinuity*, 2900 km below the Earth's surface. In terms of RHEOLOGY, the crust and outermost mantle to a depth of 100 km below the surface together comprise the rigid LITHOSPHERE, underlain by the weak, plastic ASTHENOSPHERE (100–300 km). Direct sampling has only been undertaken on the crust. Samples of the mantle can be studied in nodules expelled in some VOLCANIC ERUPTIONS and in *ophiolite complexes* – fragments of oceanic lithosphere caught up in OROGENIC BELTS. The structure of the Earth's interior is mainly known from the geophysical analysis of SEISMIC WAVES. *GO*

Anderson, D.L. 1992: The Earth's interior. In Brown, G.C., Hawkesworth, C.J. and Wilson, R.C.L. (eds), *Understanding the Earth: a new synthesis*. Cambridge: Cambridge University Press, 44–66. **Bott, M.H.P.** 1982: *The interior of the Earth: its structure, constitution, and evolution*, 2nd edn. London: Arnold. **Poirier, J.-P.** 2000: *Introduction to the physics of the Earth's interior*, 2nd edn. Cambridge: Cambridge University Press. **Skinner, B.J. and Porter, S.C.** 1987: *Physical geology*. New York: Wiley. **Vasco, D.W., Johnson, L.R. and Marques, O.** 1999: Global Earth structure: inference and assessment. *Geophysical Journal International* 137, 381–407.

Earth-surface system

A SYSTEM at and near the surface of the Earth, such as the global HYDROLOGICAL CYCLE or an ECOSYSTEM. An Earth-surface systems approach involves investigating the interactions between the component parts and the behaviour of the system as a whole rather than describing, classifying and analysing the individual elements (REDUCTIONISM). Earth-surface systems are inherently complex and orderly but may be unstable and chaotic, exhibit SELF-ORGANISATION and behave, at least in part, as nonlinear DYNAMICAL SYSTEMS. *JAM*

[See also CHAOS]

Phillips, J.D. 1999: *Earth surface systems: complexity, order and scale*. Oxford: Blackwell.

Earth-system analysis (ESA)

The scientific investigation of the ECOSPHERE and HOMOSPHERE (or *anthroposphere*) as a whole with particular reference to (1) understanding the complex response of the total Earth system to major environmental perturbations and (2) sustainable global management of GLOBAL ENVIRONMENTAL CHANGE. Schellnhuber (1999) suggests there are three distinct ways to perceive and achieve understanding of the Earth system in a holistic way: the *bird's eye approach* (viewing Earth from space); the *digital-mimicry approach* (development of SIMULATION MODELS); and the *Lilliput approach* (building experimental microcosms). If a BUSINESS-AS-USUAL scenario is assumed to be unacceptable in relation to HUMAN IMPACTS ON ENVIRONMENT, there are several principles from which humanity may choose to guide future actions in pursuit of SUSTAINABILITY.

Three principles for such '*geodesign*' include the optimisation, stabilisation and pessimisation principles, which may be elaborated and operationalised using 'Earth-system models of intermediate complexity' (EMICs). The *optimisation principle* aims for the best solution in the sense of maximising an 'aggregated ecosphere–homosphere welfare function': for example, temperate regions could specialise in producing global food supplies, the subtropics producing renewable energies and the tropics preserving biodiversity. The *stabilisation principle* aims to achieve and maintain a 'desirable state' by using technology, such as BIOTECHNOLOGY and GEOENGINEERING, to modify human actions. The *pessimisation principle* avoids the worst outcomes of human actions in terms of adverse effects on the ecosphere by not allowing irreversible transgression of critical thresholds. Present intergovernmental actions in relation to CLIMATIC CHANGE and OZONE DEPLETION fall in this category. Earth-system analysis is clearly an INTERDISCIPLINARY 'grand challenge' to scientific endeavour in the twentyfirst century. *JAM*

[See also GAIA HYPOTHESIS, GLOBAL ENVIRONMENTAL CHANGE, INTERMEDIATE–COMPLEXITY MODELLING]

Graedel, T.E. and Crutzen, P.J. 1993: *Atmospheric change: an Earth system perspective*. New York: Freeman. **Jacobsen, M., Charleson, R.J. and Rodhe, H.** 2000: *Earth system science: from biogeochemical cycles to global change*. Sidcup, Kent: Academic

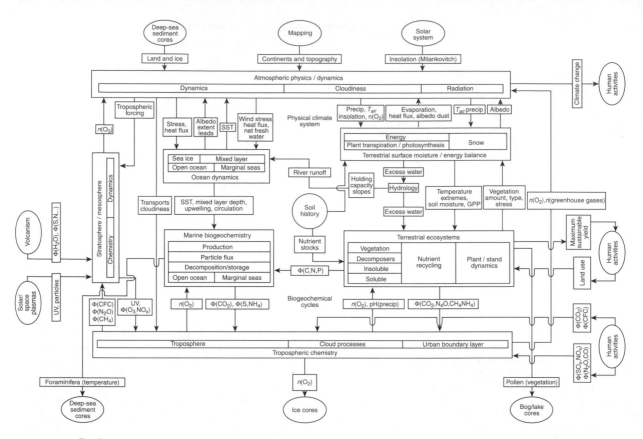

Earth-system analysis *A simple conceptual model of the whole Earth system (after Schellnhuber, 1999)*

Press. **Schellnhuber, H.-J.** 1999: 'Earth system' analysis and the second Copernican revolution. *Nature* 402 (Supplement), C19–23. **Schellnhuber, H.-J. and Wenzel, V. (eds)** 1998: *Earth system analysis: integrating science for sustainability.* Berlin: Springer.

earth tremor   An informal term used to describe an EARTHQUAKE of low *intensity*. This may be either because it was of low magnitude or because it was felt at a large distance from the EPICENTRE. Many AFTERSHOCKS are described as tremors. *GO*

[See also EARTHQUAKE MAGNITUDE, MICROSEISM]

earthquake   Shaking of the Earth caused by the sudden release of strain energy in the form of shock waves (SEISMIC WAVES) in the Earth's CRUST or outer MANTLE. Earthquakes can be generated by METEORITE IMPACT, by the movement of MAGMA beneath a VOLCANO (*volcanic earthquake*) or by the instantaneous release of strain energy that has accumulated gradually across a FAULT (*tectonic earthquake*). Shocks resulting from meteorite impact are the cause of *moonquakes* detected on the Moon. The monitoring of volcanic earthquakes can show the movement of magma beneath a volcano and provide warning of a VOLCANIC ERUPTION. The vast majority of earthquakes are TECTONIC earthquakes, and most major destructive earthquakes are generated at faults associated with PLATE MARGINS. The geographical distribution of large earth-

quakes (see SEISMICITY) describes well defined, narrow linear zones such as the *Circum-Pacific belt*, which experiences about 90% of the world's large earthquakes, and the *Alpide belt* from the Mediterranean through Turkey and Iran to northern India. These SEISMIC BELTS define plate boundaries (see PLATE TECTONICS) and earthquakes along them tend to recur at time intervals of decades to centuries (see SEISMIC GAPS). In a severe earthquake, shaking may last for tens of seconds. *Surface rupture* or displacement may or may not be evident. Earthquakes are also known as *shocks*, *quakes*, *seisms* or *macroseisms*. Minor earthquakes that are barely felt are sometimes described as EARTH TREMORS; those that are detectable only by instruments are MICROSEISMS. The study of earthquakes is SEISMOLOGY.

The point at which seismic waves are generated is the *focus* or *hypocentre* and the point on the Earth's surface directly above the focus is the EPICENTRE. *Shallow focus earthquakes* (focal depth < 100 km) occur at CONSTRUCTIVE, DESTRUCTIVE and CONSERVATIVE PLATE MARGINS; *intermediate focus earthquakes* (focal depth 100–300 km) and *deep focus earthquakes* (focal depth < 300 km) occur only at destructive plate margins, where they are generated along an inclined surface known as a *Benioff zone* or *Wadati–Benioff zone*, which represents a SUBDUCTION zone. Seismic waves are transmitted in all directions from the focus, as *body waves* through the Earth's interior and as *surface waves* along the surface. Surface waves cause the damage associated with large earthquakes. Body waves can spread around the entire Earth from a large earthquake. Their detection provides information about the earthquake (e.g. location of the epicentre, magnitude, type of fault displacement, depth of focus) and about EARTH STRUCTURE. Earthquake vibrations are recorded as a *seismogram* by a *seismograph*, in which a SEISMOMETER detects the arrival of seismic waves. The severity or strength of an earthquake is measured in terms of the energy released (EARTHQUAKE MAGNITUDE, measured on the *Richter scale*) or the effects caused (EARTHQUAKE INTENSITY).

There are essentially three types of earthquake effect: primary, secondary and tertiary. The main primary effect is ground shaking, while secondary effects include LIQUE-FACTION, surface displacement and SUBSIDENCE, slope instability and MASS MOVEMENTS, FLOODS, fires and TSUNAMIS. The intensity and frequency of secondary effects can depend on, and directly affect, the environmental stability of the region affected by an earthquake. Tertiary effects develop some time after an earthquake and can include loss of livelihood, FAMINE and disease. Many lives may be lost in the aftermath of a severe earthquake: recent devastating earthquakes include those at Tangshan, China (1976, estimated 250 000 deaths), Izmit, Turkey (August 1999, more than 17 000 deaths) and Gujarat, India (January 2001, more than 16 000 deaths). EARTHQUAKE PREDICTION and EARTHQUAKE ENGINEERING attempt to limit the impact of severe earthquakes. *DNT/GO*

[See also AFTERSHOCK, FORESHOCK, GEOPHYSICAL SURVEY, GEOPHYSICS, NATURAL HAZARDS, PALAEOSEISMOLOGY, SEISMIC SURVEY]

Bishop, I., Styles, P. and Allen, M. 1993: Mining-induced seismicity in the Nottinghamshire Coalfield. *Quarterly Journal of Engineering Geology* 26, 253–279. **Bolt, B.A.** 1999: *Earthquakes*, 4th edn. New York: Freeman. **Gutenberg, B. and Richter, C.F.** 1954: *Seismicity of the Earth and associated phenomena*. Princeton, NJ: Princeton University Press. **Hubert-Ferrari, A., Barka, A., Jacques, E.** *et al.* 2000: Seismic hazard in the Marmara Sea region following the 17 August 1999 Izmit earthquake. *Nature* 404, 269–273. **Ishihara, K.** 1993: Liquefaction and flow failure during earthquakes. *Géotechnique* 43, 351–415. **Koukouvelas, I., Mpresiakas, A., Sokos, E. and Doutsos, T.** 1996: The tectonic setting and earthquake ground hazards of the 1993 Pyrgos earthquake, Peloponnese, Greece. *Journal of the Geological Society, London* 153, 39–49. **Rymer, M.J. and Ellsworth, W.L. (eds)** 1990: The Coalinga, California, earthquake of May 2, 1983. *United States Geological Survey Professional Paper* 1487. **Tibaldi, A., Ferrari, L. and Pasquare, G.** 1995: Landslides triggered by earthquakes and their relations with faults and mountain slope geometry: an example from Ecuador. *Geomorphology* 11, 215–226. **Walker, B.** 1982: *Planet Earth: earthquake.* Amsterdam: Time-Life Books. **Yeats, R.S., Sieh, K. and Allen, C.R.** 1997: *The geology of earthquakes.* Oxford: Oxford University Press.

earthquake engineering

Technology employed to mitigate against the damaging effects of large ground accelerations on human activities and structures during an EARTHQUAKE. Steps that can be taken to minimise damage include the design of structures to prevent resonance with ground shaking or to sway with ground motion, reinforcement of masonry to prevent collapse, improved planning and the preparation of reliable *seismic hazard maps*. Even in areas where strict engineering codes of conduct exist, however, structures may fail unexpectedly as a result of lapses in the implementation of regulations, or of unexpected behaviour of the ground. *DNT*

[See also EARTHQUAKE PREDICTION]

Ambraseys, N.N. 1988: Engineering seismology. *Earthquake Engineering and Structural Dynamics* 17, 1–105. **Severn, R.T.** 1999: European experimental research in earthquake engineering for Eurocode 8. *Proceedings of the Institution of Civil Engineers – Structures and Buildings* 134, 205–217. **Shedlock, K.M.** 1999: Seismic hazard map of North and Central America and the Caribbean. *Annali di Geofisica* 42, 977–997. **Theodulidis, N., Lekidis, V., Margaris, B.** *et al.* 1998: Seismic hazard assessment and design spectra for the Kozani-Grevena region (Greece) after the earthquake of May 13, 1995. *Journal of Geodynamics* 26, 375–391.

earthquake intensity

A measure of the strength of an EARTHQUAKE in terms of the effects of the shaking and the degree and extent of damage caused. Earthquake intensity, or *seismic intensity*, is determined on the basis of eyewitness accounts and surveys after an earthquake. *Isoseismal lines* are lines enclosing points of equal intensity on an *isoseismal map*: they normally form elliptical patterns centred about the earthquake EPICENTRE. Seismic intensity depends on local ground conditions (e.g. GEOLOGY, SOIL type), depth of the *focus* and distance from the epicentre as well as on the absolute size, or MAGNITUDE, of the earthquake.

Different qualitative scales are used in North America, Europe, Japan and elsewhere to express seismic intensity. The *Modified Mercalli Intensity Scale*, or *Mercalli Scale*, is widely used in North America. This ranks damage on a descriptive scale from I to XII, where I is 'not felt except by a very few', IV is 'during the day felt indoors by many, outdoors by a few' and XII is 'damage total'. The *European Macroseismic Scale* (*EMS*) was adopted in 1998 to replace the MCS Scale (Mercalli–Cancani–Sieberg Scale) and the MSK Scale (Medvedev, Sponheuer and Karnik Scale) that had previously been used in Europe. The EMS scale is also a 12-point scale, ranging from 1 (not felt) through 3 (weak) and 7 (damaging) to 9 (destructive), 11 (devastating) and 12 (completely devastating – practically all structures above and below ground are heavily damaged or destroyed). The EMS is considered to be the first intensity scale with clear instructions to ensure its consistent application. A seven-point scale is used in Japan (the Japanese Meteorological Agency or JMA Scale). *GO*

Bakun, W.H. and Wentworth, C.M. 1997: Estimating earthquake location and magnitude from seismic intensity data. *Bulletin of the Seismological Society of America* 87, 1502–1521. **Bolt, B.A.** 1999: *Earthquakes*, 4th edn. New York: Freeman. **Grünthal, G.** 1998: *European Macroseismic Scale 1998, EMS-98*. Luxembourg: Council of Europe. **Kozak, J. and Ebel, J.E.** 1996: Macroseismic information from historic pictorial sources. *Pure and Applied Geophysics* 146, 103–111.

earthquake magnitude

The strength of an EARTHQUAKE, expressed in terms of the amount of energy released and calculated from the maximum displacement on a *seismogram* (see SEISMOMETER) with a correction for the distance from the EPICENTRE. Magnitude (also known as *seismic magnitude* or *Richter magnitude*) is measured on the logarithmic *Richter Scale*, developed by Charles Richter in the 1930s (see Figure).

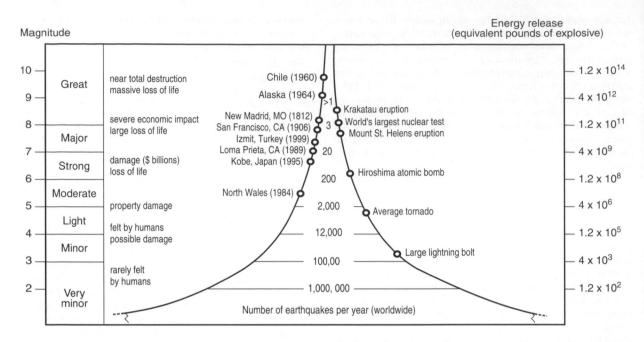

earthquake magnitude *The approximate number of earthquakes per year of a given magnitude, together with descriptive terms for earthquake magnitude, the effects of earthquakes of different magnitudes, sizes of some well-known earthquakes and other events, and the equivalent explosive energy release (after Bolt, 1999)*

The largest recorded earthquake is generally agreed to have been the AD 1960 Chile earthquake, with magnitude 9.6. It is likely that an upper limit of magnitude for earthquakes generated by slip along FAULTS is determined by the strength of the rocks of the Earth's CRUST. On average there is one earthquake per year of magnitude greater than 8, 18 between magnitudes 7.0 and 7.9, and 120 between magnitudes 6.0 and 6.9. Many of these are not felt, occurring in sparsely populated areas or beneath the OCEANS. There are an estimated 6200 earthquakes each year with magnitudes between 4.0 and 4.9, 49 000 between magnitudes 3.0 and 3.9, and about 8000 per day between magnitudes 1 and 2; none of these are felt (see MICROSEISM). The UK mainland experiences an earthquake of magnitude 5.0 or greater on average once every eight years, and an earthquake of magnitude 4.0 to 4.9 once every two years. The largest earthquake in Britain was in the Dogger Bank area of the North Sea in AD 1931, with a magnitude of 6.1. The AD 1984 North Wales earthquake of magnitude 5.4 was the largest UK onshore earthquake of the last century.

The effects of an earthquake are expressed by EARTHQUAKE INTENSITY, which depends not only on the magnitude, but also on local ground conditions, building quality, the depth of the *focus*, and the distance from the epicentre. Earthquake intensity is measured on scales such as the *Mercalli scale* and the *European Macroseismic Scale*. A large, but deep earthquake (focal depth 637 km) of magnitude 8.3 in Brazil in 1994 was felt from Argentina to Canada, but caused just five deaths and little damage, whereas an earthquake in Newcastle, New South Wales, in AD 1989 of magnitude 5.4 and focal depth 4 km caused 12 deaths and an earthquake in Cairo in AD 1992 of magnitude 5.9 and focal depth 22 km caused 552 deaths.

GO/DNT

[See also EARTHQUAKE ENGINEERING, EARTHQUAKE PREDICTION]

Bolt, B.A. 1999: *Earthquakes*, 4th edn. New York: Freeman. **Gutenberg, B. and Richter, C.F.** 1954: *Seismicity of the Earth and associated phenomena*. Princeton, NJ: Princeton University Press.

earthquake prediction Attempts to forecast the time, place, magnitude and effects of an EARTHQUAKE. Only a few major earthquakes have been successfully predicted, but many have been 'postdicted' by the retrospective analysis of precursory phenomena (see RETRODICTION). There are three basic approaches to earthquake prediction: statistical, geological and geophysical. Statistical methods involve calculating the probability of an event occurring either on a global scale or in any given region. Relevant data include knowledge of SEISMIC GAPS and earthquake recurrence intervals, obtained from records of past earthquakes (see PALAEOSEISMICITY). It is estimated, for example, that there is a 67% probability of an earthquake of magnitude greater than 6.7 striking the San Francisco Bay area in the next 30 years. Geological methods include monitoring changes in ground tilt prior to an event and recognising *asperities* (rough spots) along FAULT surfaces: where asperities are large and separated by great distances, earthquakes tend to be infrequent and large. Geophysical methods rely on identifying physical changes in rocks in a seismically active zone prior to an event.

Geophysical and some geological approaches provide the potential for short-term earthquake prediction. A major earthquake in Haicheng, China in February 1975 was apparently successfully predicted by a combination of geophysical analysis and local reports of anomalous observations such as changes in water levels in wells and unusual behaviour of animals. Large-scale evacuations preceded the magnitude 7.3 earthquake and there seem to have been few casualties, although there was widespread destruction of buildings. However, a magnitude 7.6 earthquake struck Tangshan, China in July 1976 with no prior warning, causing an estimated 250 000 deaths, while in other cases evacuations have been ordered but no earthquake has occurred. Earthquake prediction is notoriously difficult and efforts are better concentrated on hazard mitigation (see EARTHQUAKE ENGINEERING).   *DNT/GO*

Adams, R.D. 1976: The Haicheng, China, earthquake of 4 February 1975; the first successfully predicted major earthquake. *Earthquake Engineering and Structural Dynamics* 4, 423–437. **Jackson, D.D., Aki, K., Cornell, C.A.** *et al.* 1995: Seismic hazards in southern California – probable earthquakes, 1994 to 2024. *Bulletin of the Seismological Society of America* 85, 379–439. **Kagan, Y.Y. and Jackson, D.D.** 1994: Long-term probabilistic forecasting of earthquakes. *Journal of Geophysical Research – Solid Earth* 99, 13685–13700. **Sykes, L.R., Shaw, B.E. and Scholz, C.H.** 1999: Rethinking earthquake prediction. *Pure and Applied Geophysics* 155, 207–232. **Wesnousky, S.G., Prentice, C.S. and Sieh, K.E.** 1991: An offset Holocene stream channel and the rate of slip along the northern reach of the San-Jacinto fault zone, San-Bernardino-Valley, California. *Geological Society of America Bulletin* 103, 700–709.

earthworms   Soil-inhabiting invertebrates of the order Oligochaeta. Earthworms are important for BIOTURBATION, DECOMPOSITION and MINERAL CYCLING in tropical and temperate soils, and also help to promote a high INFILTRATION CAPACITY and reduce overland flow and EROSION. There are about 200 native species in Europe, of which 26 occur in Britain.   *JAM*

Darwin, C. 1881: *The formation of vegetable mould through the action of worms.* London: John Murray [Reissued in 1945 as *Darwin on humus and the earthworms.* London: Faber and Faber.] **Enckell, P.H. and Rundgren, S.** 1988: Anthropochorous earthworms (Lumbricidae) as indicators of abandoned settlements in the Faroe Islands. *Journal of Archaeological Science* 15, 439–452. **Lavelle, P.** 1988: Earthworm activities and the soil system. *Biology and Fertility of Soils* 6, 237–251.

éboulis ordonnés   A type of coarse-grained stratified slope deposit formed in PERIGLACIAL environments. Such deposits are distinguished from GRÈZES LITÉES in that they are coarser and form by the accumulation of rockfall material rather than accumulating by SLOPEWASH and SOLIFLUCTION.   *PW*

[See also MASS WASTING]

Pappalardo, M. 1999: Observations on stratified slope deposits, Gesso Valley, Italian maritime Alps. *Permafrost and Periglacial Processes* 10, 107–111.

eccentricity   One of the three main orbital parameters in the MILANKOVITCH THEORY of GLACIAL–INTERGLACIAL CYCLES, the 'eccentricity of the orbit' refers to the shape of the Earth's orbit around the Sun (specifically its variation from near-circular to various degrees of ellipticity). *JAM*

[See also OBLIQUITY, PRECESSION]

ecesis   The successful establishment, growth and persistence to reproduction of organisms on a disturbed site; the establishment phase of ECOLOGICAL SUCCESSION. *LRW*

ecliptic   The apparent path of the Sun through the sky. Its plane tilts 60.5° to the Earth's axis. The OBLIQUITY of

the ecliptic of 23.5° from the equator varies between 22.0 and 24.5° with a PERIODICITY of 40 000 years.   *JGT*

ecocentrism   An environmental philosophy that is environment- or Earth-centred. Ecocentrism or *biocentrism* stresses that the interests of humanity should not be placed above those of other species, and that the continued existence of the human species depends on treating the environment with respect if not humility. It contrasts with and provides a counterweight to the excesses of *anthropocentrism*.   *JAM*

[See also DEEP ECOLOGY, ENVIRONMENTAL ETHICS, ENVIRONMENTAL MANAGEMENT, ENVIRONMENTALISM, TECHNOCENTRISM]

O'Riordan, T. 1981: *Environmentalism*, 2nd edn. London: Pion. **Pepper, D.** 1984: *The roots of modern environmentalism.* London: Croom-Helm.

ecocline   (1) A gradient in both the biotic communities (a *coenocline*) and physical environmental factors (a complex gradient), such as occurs with increasing altitude, forming a sequence of ecosystem types. (2) A CLINE. *JLI*

[See also ECOTONE]

Whittaker, R.H. 1975: *Communities and ecosystems.* London: Macmillan.

ecofact   In the archaeological context, an inorganic or organic object that has cultural relevance but is not an ARTEFACT. Ecofacts are *nonartefactual* (natural) remains, such as bones and seeds, which may shed light on, for example, past environmental conditions, diet or resource use.   *JAM*

[See also GEOFACT, MATERIAL CULTURE]

Ashmore, W. and Sharer, R.J. 1996: *Discovering our past: a brief introduction to archaeology.* Mountain View, CA: Mayfield.

ecofeminism   'A series of theoretical and practical positions bringing feminist insight to environmental philosophy' (Davion 2001: 233). Ecofeminists emphasise the historical, conceptual, empirical, epistemological, ethical, political and theoretical links between the domination of women and the domination of NATURE.   *JAM*

Davion, V. 2001: Ecofeminism. In Jamieson, D. (ed.), *A companion to environmental philosophy.* Oxford: Blackwell, 233–247.

ecological energetics   The study of energy flow within or through ecosystems. Solar energy flowing into the system is converted to chemical energy through PHOTOSYNTHESIS and is eventually lost as heat energy.   *RJH*

[See also BIOGEOCHEMICAL CYCLES, ECOSYSTEM CONCEPT, TROPHIC LEVELS]

Phillipson, J. 1966: *Ecological energetics.* London: Edward Arnold.

ecological engineering   'The design of human society with its natural environment for the benefit of both' (Mitsch and Jørgensen, 1989). There are two essential components: (1) a basis in ecological theory; and (2) a reliance on the self-designing nature of ecosystems. It differs from *environmental engineering* in the absence of major engineering structures and the deployment of only small inputs of human energy and resources. Ecological engineering, together with its synonym *ecotechnology*, is used to describe situations where the main driving force is not the human inputs. Ecological engineering also differs from BIOTECHNOLOGY and GENETIC ENGINEERING, which involve the manipulation of organisms, their cells or their subcellular components. *JAM*

Ma Shijun 1985: Ecological engineering: application of ecosystem principles. *Environmental Conservation* 12, 331–335. **Mitsch, W.J. and Jørgensen, S.E. (eds)** 1989: *Ecological engineering: an introduction to ecotechnology.* New York: Wiley.

ecological evaluation

The process of assessing the CONSERVATION importance of an area especially in connection with a PROTECTED AREA APPROACH, as in the establishment of nature reserves. *IFS*

[See also ECOLOGICAL VALUATION, ECOLOGICAL VALUE]

ecological explosion

The massive and rapid increase in the POPULATION of an organism when the various natural checks on its growth and reproduction, such as limited food supply or predation, are absent. Ecological or *population explosions* can be localised and short-lived, taking place within the HABITAT of a NATIVE SPECIES (e.g. ALGAL BLOOMS in the ocean and lemmings in the Arctic TUNDRA) or spread temporarily beyond the original habitat (e.g. locusts in semi-arid Africa). Delayed DENSITY DEPENDENCE relationships allow such population explosions to overshoot the CARRYING CAPACITY of their habitat. They are eventually controlled by NEGATIVE FEEDBACK mechanisms or HOMEOSTASIS. Algal blooms, for example, subside because of NUTRIENT shortages. Such explosions are also reversed by heavy predation (e.g. lemming numbers may be reduced by weevil attack as well as by poor quality forage). Ecological explosions can also be widespread and long-lasting, as in BIOLOGICAL INVASIONS. *GOH*

Begon, M., Harper, J.L. and Townsend, C.R. 1996: *Ecology: individuals, populations and communities*, 3rd edn. Oxford: Blackwell Science. **Elton, C. S.** 1958: *The ecology of invasions by animals and plants.* London: Methuen. **Krebs, C. J.** 1978: *Ecology: the experimental analysis of distribution and abundance.* New York: Harper and Row. **Mackenzie, A., Ball, A.S. and Virdee, S.R.** 1998: *Instant notes in ecology.* New York: Bios.

ecological goods and services

Ecological goods (resources) are attributes and processes in nature that support human production and consumption. *Biological goods* provide materials, including for food, building, shelter, pharmaceuticals and energy. Ecological or *ecosystem services* are those attributes, processes and qualities of ecosystems (e.g. water supply, CO_2–O_2 balance, BIOGEOCHEMICAL CYCLES and SOIL formation) that support life and provide services for humans and human activity. WETLANDS, for example, provide a variety of goods (e.g. food, fuel and building materials) and services (e.g. buffering coastal regions from floods, maintaining water quality, waste absorption, recreational area). In another sense, ecosystems can be said to provide *ecological sources* (of goods) and *ecological sinks* (for waste materials). *IFS*

[See also ECOSYSTEM COLLAPSE, ECOSYSTEM HEALTH, SUSTAINABILITY]

Barbier, E.B., Burgess, J.C. and Folke, C. 1994: *Paradise lost? The ecological economics of biodiversity.* London: Earthscan. **Ewel, K.C., Twilley, R.R. and Ong, J.E.** 1998: Different kinds of mangrove forests provide different goods and services. *Global Ecology and Biogeography Letters* 7, 83–94. **McNeely, J.A.**

1988: *Economics and biological diversity: developing and using economic incentives to conserve biological resources.* Gland: International Union for the Conservation of Nature.

ecological group

A group of species with similar ecological requirements. Ecological groups may be used as a basis for the interpretation of past environmental conditions (under an assumption of UNIFORMITARIANISM). Examples include acidophilous DIATOMS or nitrophilous plants. Such groups may contain both EURY- and STENOtypic species. *BA*

[See also MODERN ANALOGUE, PALAEOECOLOGY, PALAEOENVIRONMENT, TRANSFER FUNCTION]

ecological indicator

(1) An organism, the presence of which signifies a particular set of environmental conditions. (2) A biological criterion for measuring ECOSYSTEM HEALTH. (3) A measurable biotic or abiotic phenomenon that can be used to determine the presence or extent of environmental change. A range of different indicators have been used for the last purpose, ranging from the basic (e.g. measurements of AIR TEMPERATURE) to the highly sophisticated (e.g. measurements of changes in species interactions). To be useful, an ecological indicator must be clearly related to some aspect of the environment, must be relatively easy to assess and must be sensitive to environmental change. *JLI*

[See also BIO-INDICATOR, ENVIRONMENTAL INDICATOR, PROXY DATA/PROXY EVIDENCE/PROXY RECORDS, INDICATOR SPECIES]

ecological response to environmental change

Environmental changes have numerous, complex and interacting impacts on the structure and function of ECOSYSTEMS. However, groups of species may share similar responses to a particular change, allowing generalisations to be made about COMMUNITY responses (see FUNCTIONAL TYPES). Rates of change in the species composition of a community will depend on the life span of individuals, DISTURBANCE regimes and rates of MIGRATION or DISPERSAL.

Ecosystem components may be directly affected by ENVIRONMENTAL CHANGE, as in the response of DECOMPOSITION to temperature, or indirectly influenced, e.g. changes in HERBIVORE activity resulting from altered leaf tissue quality with rising atmospheric CARBON DIOXIDE. Immediate responses of organisms to changes in the environment may be offset by ACCLIMATISATION during the lifetime of an individual and by ADAPTATION on evolutionary timescales.

Ecosystem structure and function may be significantly modified by the following: CARBON DIOXIDE FERTILISATION; changes in average and extreme temperatures (see GLOBAL WARMING and EXTREME CLIMATIC EVENTS: IMPACT ON ECOSYSTEMS); changes in DROUGHT frequency, duration and intensity resulting from altered precipitation patterns or changes in POTENTIAL EVAPOTRANSPIRATION caused by temperature variation (see DROUGHT: IMPACT ON ECOSYSTEMS); tropospheric OZONE excursions, elevated atmospheric SULPHUR DIOXIDE and NITROGEN OXIDES, nitrogen and sulphur deposition (see AIR POLLUTION, ACID RAIN and ACIDIFICATION); METAL POLLUTION; increases in *ultraviolet radiation* resulting from OZONE DEPLETION in the STRATOSPHERE; changes in STORM frequency or intensity; POLLU-

TION of watercourses by agricultural FERTILISERS, industrial EFFLUENT, THERMAL POLLUTION or *sewage disposal* (see EUTROPHICATION).

In many cases, ecological systems do not just respond to environmental changes, but may also mediate the changes themselves via FEEDBACK MECHANISMS. For example, the future rise in atmospheric carbon dioxide could be moderated by greater CARBON SEQUESTRATION in the terrestrial biosphere. *CPO*

[See also SCALE CONCEPTS IN ENVIRONMENTAL CHANGE].

Bigg, G.R. 1996: *The oceans and climate*. Cambridge: Cambridge University Press. **Mason, C.F.** 1996: *Biology of freshwater pollution*. Harlow: Longman. **Walker, B. and Steffen, W. (eds)** 1996: *Global change and terrestrial ecosystems*. Cambridge: Cambridge University Press. **Walker, B., Steffen, W., Canadell, J. and Ingram, J. (eds)** 1999: *The terrestrial biosphere and global change: implications for natural and managed ecosystems*. Cambridge: Cambridge University Press. **Watson, R.T., Zinyowera, M.C., Moss, R.H. and Dokken, D.J. (eds)** 1996: *Climate change 1995. Impacts, adaptations and mitigation of climate change: scientific–technical analysis*. Cambridge: Cambridge University Press. **Wellburn, A.** 1994: *Air pollution and climate change. The biological impact*. Harlow: Longman.

ecological restoration RESTORATION of biotic communities and habitats on the basis of ecological principles, but also taking into consideration human values. *IFS*

[See also BIOREGIONAL MANAGEMENT, CONSERVATION BIOLOGY, LAND RESTORATION, SOIL REGENERATION]

Higgs, E.S. 1997: What is good ecological restoration? *Conservation Biology* 11, 338–348.

ecological succession A sequential change in species composition or other ecosystem characteristics. Vegetation succession typically occurs over time spans ranging from about 1 year to 1000 years. Insect or microbial succession may occur more rapidly (days to months) because the organisms involved have relatively short life cycles. Much longer-term changes (> 1000 y) are often due to widespread climatic shifts and are called *secular succession* or *geohistorical changes*. Some argue that the term ecological succession is too loaded with connotations of directionality (to CLIMAX VEGETATION) or CONVERGENCE of multiple pathways and should be discarded. If not defined too broadly or too narrowly, ecological succession is still a functional term that is widely recognised and perhaps more helpful than the alternatives (e.g. vegetation dynamics, vegetation change). However, such broad terminology results from recognition that there are many non-linear pathways of species change including CYCLIC REGENERATION, direct replacement of dominant species following a disturbance, and other, less predictable pathways.

PRIMARY SUCCESSION and SECONDARY SUCCESSION form endpoints on a continuum between clearly primary SERES (e.g. a newly exposed rock surface), through seres with some nutrients, organic matter and propagules (e.g. a floodplain with seeds and sediments from upstream communities but no organised soil layers), to clearly secondary seres (e.g. a forest clearing), where soils remain intact but above-ground vegetation is removed by wind, harvest or disease. PRIMARY SUCCESSION is the biological response to an extreme ALLOGENIC CHANGE. In contrast, SECONDARY SUCCESSION may occur following allogenic disturbance

(e.g. fire) or AUTOGENIC CHANGE (e.g. herbivory). Because primary succession embodies the complete assembly of biotic communities, many patterns and processes differ between primary and secondary succession. For example, primary seres may be less predictable, with more potential pathways, in part because of the important yet largely unpredictable role of DISPERSAL of propagules to the site. Similarly, vascular species with nitrogen-fixing symbionts may be more important in soil development and FACILITATION of the growth of other species during primary succession. Both primary and secondary succession can occur on land or in the water (*hydrarch succession*). Soil structural development and the availability of nutrients and light generally limit plant and microbial COLONISATION on land; a surface to attach to and sufficient light often limit colonisation in water.

The complexity of changes that occur during ecological succession makes both definitions and predictions difficult. Site history determines the initial template upon which change will occur. The type of disturbance that altered the site enough to initiate change can strongly influence the rate and direction of succession. SOIL DEVELOPMENT is dependent on the parent material, topography, climate and biological activity. The vegetation strongly influences soil formation (e.g. through leaf litter fall, root growth and decay and nitrogen fixation), and the soil influences which plants can colonise. Animal and soil microbial populations both depend on and influence plant colonisation.

Several basic processes have been identified that drive successional change in plant communities. NUDATION is the creation of space for colonisation by disturbance. Colonisation is the process of filling in that space. Dispersal to a recently opened site is usually from the plants and animals that are nearby. Long-distance dispersal is not common, but there are RUDERALS that specialise in the colonisation of open habitats and have adaptations for long-distance dispersal. ECESIS is the establishment of individuals and involves the processes of dispersal, colonisation, competition and FACILITATION. Winners in this process are often the species that arrive first and monopolise the existing resources. Dense mats or thickets of vegetation often result and these may delay species change, as described by the INHIBITION MODEL of succession. Ultimately, the relative longevities of each species may determine the overall pattern of succession, while the relative balance of competitive and facilitative interactions determines the rate of species change.

Despite the agreement on what basic processes drive ecological succession, no one model successfully predicts successional change across a variety of sites. Conclusions about succession are still mostly site-specific. However, a variety of models have been successful in refining those variables that are most likely to determine successional trajectories and have provided guidance for field measurements. A traditional emphasis on deterministic pathways to a self-replacing climax vegetation has been replaced by an emphasis on the influence, interaction and fate of individuals (the biotic component) in a rapidly changing ABIOTIC environment. Future research on ecological succession will undoubtedly continue to look for general principles, but also focus on collection of data from long-term observations of plots, continue to examine the mutual influences of plants and soils, explore the importance of soil microbial communities, study the influences of plant/animal interactions and standardise the collection of data to facilitate comparisons across sites.

Understanding ecological succession is not just an intellectual exercise. Soaring human populations have led to massive destruction of natural habitats and are contributing to rapidly shifting environmental conditions. To the extent that we understand successional processes, we can manipulate these conditions to our advantage for the restoration of disturbed habitats. *LRW*

[See also ECOLOGICAL RESTORATION, LAND RESTORATION, SOIL RECLAMATION]

Burrows, C.J. 1990: *Processes of vegetation change.* Unwin Hyman. **Clements, F.E.** 1928: *Plant succession and indicators.* New York: Wilson. **Connell, J.H. and Slatyer, R.O.** 1977: Mechanisms of succession in natural communities and their role in community stability and organization. *The American Naturalist* 111, 1119–1144. **Gleason, H.A.** 1937: The individualistic concept of the plant association. *American Midland Naturalist* 21, 92–110. **Glenn-Lewin, D.C., Peet, R.K. and Veblen, T.T. (eds)** 1992: *Plant succession: theory and prediction.* London: Chapman and Hall. **Luken, J.O.** 1990: *Directing ecological succession.* London: Chapman and Hall. **Pickett, S.T.A. and White, P.S.** 1985: *The ecology of natural disturbance and patch dynamics.* London: Academic Press. **Tilman, D.** 1988: *Plant strategies and the dynamics and structure of plant communities.* Princeton, NJ: Princeton University Press. **Walker, L.R. (ed.)** 1999: *Ecosystems of disturbed ground* [*Ecosystems of the World.* Vol. 16]. Amsterdam: Elsevier. **Watt, A.S.** 1947: Pattern and process in the plant community. *Journal of Ecology* 35, 1–22.

ecological transition The reduced dependency of farmers on the land, which accompanies the change from SUBSISTENCE AGRICULTURE to CASH CROPPING and may lead to OVERCULTIVATION. *JAM*

Bennett, J.W. 1976: *The ecological transition: cultural anthropology and human adaptation.* Oxford: Pergamon Press.

ecological valuation Placing a monetary value on an ecological entity, such as a species, a habitat or a landscape. Ecological valuations are necessary for COST–BENEFIT ANALYSIS, but do not measure non-monetary ECOLOGICAL VALUE. *JAM*

ecological value (1) The attributes or features of biological communities and ecosystems that are valued by humans for any reason. Such reasons include, for example, food, shelter, fuel, medicine, enjoyment, science and morals. (2) The value in the sense of a measurable ecological attribute of a biological community or ecosystem (e.g. BIODIVERSITY). *IFS*

[See also ECOLOGICAL GOODS AND SERVICES, SUSTAINABILITY]

ecology The study of the interactions between organisms and their environment, including other organisms and any factors or processes influencing the distributions, abundances and behaviours of organisms. *Applied ecology* includes the CONSERVATION and management of ecological systems. *MVL*

[See also AUTECOLOGY, BIOGEOGRAPHY, COMMUNITY ECOLOGY, ECOSYSTEM CONCEPT]

McIntosh, R.P. 1985: *The background to ecology: concept and theory.* Cambridge: Cambridge University Press. **Worster, D.** 1994: *Nature's economy: a history of ecological ideas,* 2nd edn. Cambridge: Cambridge University Press.

ecosphere The global ecosystem; the BIOSPHERE plus its life-support systems. It includes all living beings, together with those parts of the ATMOSPHERE, CRYOSPHERE, HYDROSPHERE, LITHOSPHERE and PEDOSPHERE with which the BIOSPHERE interacts. It has also been used in relation to other planetary environments (extraterrestrial ecospheres) that could support life as we know it. The age of the earliest fossils (STROMATOLITES) of *ca* 3.5×10^9 years indicates that an ecosphere has existed throughout most of geological time. *JAM*

[See also ECOSYSTEM, GAIA HYPOTHESIS, NOÖSPHERE]

Cole, L.C. 1958: The ecosphere. *Scientific American* 198, 83–96. **Edwards, D. and Selden, P.A.** 1992: The development of early terrestrial ecosystems. *Botanical Journal of Scotland* 46, 337–366. **Huggett, R.J.** 1997: *Environmental change: the evolving ecosphere.* London: Routledge. **Huggett, R.J.** 1999: Ecosphere, biosphere or Gaia? What to call the global ecosystem. *Global Ecology and Biogeography Letters* 8, 425–431.

ecosystem collapse The degradation of ecosystems (caused by human activities) to the stage that characteristics and functions are grossly changed and no longer maintained. Typically this includes EXTINCTION of species, loss of biomass and nutrient content, and major disruption of ecological processes. In reality an ecosystem cannot collapse in the sense of falling in or tumbling down, but, after a certain level of degradation has been reached, further degradation may follow even after impacts have been removed. Some research suggests that ecosystems exhibit complex SELF-ORGANISATION, but ecosystems vary in terms of their stability (ability to recover from disturbance) and RESILIENCE (rate at which former states and functioning are achieved). *IFS*

[See also DIVERSITY–STABILITY HYPOTHESIS, ECOLOGICAL GOODS AND SERVICES, SOIL DEGRADATION, STABILITY CONCEPTS]

Schulze, E.-D. and Mooney, H.A. (eds) 1994: *Biodiversity and ecosystem function.* New York: Springer. **Rapport, D.J. and Whitford, W.G.** 1999: How ecosystems respond to stress. *BioScience* 49, 193–203.

ecosystem concept An ecosystem (*ecological system*) was defined by Tansley as a self-sustaining community of organisms together with the physical environment that supports it. The wide acceptance of the ecosystem concept led to the rise of modern ECOLOGY and a framework for major research programmes, such as the International Biological Programme (IBP) and its successors. Ecosystems range in size from the smallest units that can sustain life (consisting of several species and a fluid medium) to the global ecosystem or ECOSPHERE. They are characterised by exchanges of energy and materials between living organisms and their supporting environment. The living organisms form an interacting set of microbes, animals and plants organised into FOOD CHAINS/WEBS and TROPHIC LEVELS. The ABIOTIC part consists of inorganic materials, organic by-products of biotic activity and physical environmental factors (winds, tides, light, heat and so on). Ecosystems are real units – the ecological community (the set of interacting species living in the same place) and its non-living environment function as a unified, if complex and dynamic, whole. *RJH*

[See also BIOGEOCHEMICAL CYCLES, BIOGEOCOENOSIS, COMMUNITY CONCEPTS, GAIA HYPOTHESIS]

Dickinson, G. and Murphy, K. 1998: *Ecosystems: a functional approach.* London: Routledge. **Golley, F. B.** 1993: *A history of the ecosystem concept in ecology.* New Haven, CT: Yale University Press. **Odum, E. P.** 1969: The strategy of ecosystem development. *Science* 164, 262–270. **Tansley, A.G.** 1935: The use and abuse of vegetational concepts and terms. *Ecology* 16, 284–307. **Willis, A. J**. 1997: The ecosystem: an evolving concept viewed historically. *Functional Ecology* 11, 268–271.

ecosystem health The state or condition of 'health' of ecosystems and their constituent entities of biota and processes. The concept (often the subject of disagreement) is used in connection with 'State of the Environment' reporting and ENVIRONMENTAL MANAGEMENT. *IFS*

[See also ECOSYSTEM CONCEPT, SUSTAINABILITY]

Constanza, R. (ed.) 1993: *Ecosystem health: new goals for environmental management.* London: Earthscan.

ecosystem indicator An ENVIRONMENTAL INDICATOR that tracks the state and trends of ecosystem conditions. It provides a basis for assessment of ECOSYSTEM HEALTH and the effects of human-imposed stress. *JAM*

[See also INTERACTION INDICATOR, SYNTHESIS INDICATOR]

ecosystem management Management at ecosystem levels. In practice, it means the integrated management of multiple attributes and processes (over time) of wetlands, forests, agricultural ecosystems, etc. *IFS*

[See also BIOGEOCHEMICAL CYCLES, ECOSYSTEM CONCEPT, ENVIRONMENTAL MANAGEMENT, LANDSCAPE MANAGEMENT]

Grumbine, R.E. 1994: What is ecosystem management. *Conservation Biology* 8, 27–38.

ecotone The border or spatial transition zone between two vegetation communities (such as the TREE LINE ecotone), characterised by having elements of both communities. Ecotones are thought to be particularly useful for the study of environmental change impacts as they represent tension zones between two community types. *JLI*

[See also COMMUNITY CONCEPTS, ECOCLINE, EDGE EFFECTS, INDIVIDUALISTIC CONCEPT]

Kent, M., Gill, W.J., Weaver, R.E. and Armitage, R.P. 1997: Landscape and plant community boundaries in biogeography. *Progress in Physical Geography* 21, 315–353. **Risser P.G.** 1995. The status of the science examining ecotones. *BioScience* 45, 318–325.

ecotope The ABIOTIC part of a BIOGEOCOENOSIS and/or the physical space occupied by it. The other part is the *biocoenosis*, which consists of the plants, animals and microorganisms present in the system. Although widely used in Central Europe, the term can be replaced by HABITAT. The term is also used for the combined factors of the habitat and NICHE, which influences the survival of species in the context of evolution. *JLI*

ecotourism In contrast to traditional *mass tourism*, ecotourism, *environmental tourism, nature tourism* or *wilderness tourism* refers to 'environmentally friendly' tourism. It involves experiencing wildlife and landscapes at first hand, but is non-destructive and non-intrusive and requires little or no infrastructure; it is ecocentric and contributes positively to CONSERVATION and SUSTAINABILITY, and has

greater educational value than other forms of tourism. It excludes most adventure tourism, hunting, fishing, boating and climbing. It includes such activities as bird watching, hiking, wildlife safaris and the appreciation of giant trees, some of which are becoming large-scale businesses and may threaten its *raison d'être*. *JAM*

[See also RECREATION]

Ceballos-Lascurain, H. 1987: *Estudio prefactibilidad socioeconomica del turismo ecologico y anteproyecto arquitectonico y urbanistico del Centro de Tourismo Ecologico de Sian Ka'an Roo.* Mexico City: SEDUE. **Dowling, R.** 2000: *Ecotourism.* Harlow: Longman. **Fennell, D.A.** 1999: *Ecotourism: an introduction.* London: Routledge. **Honey, M.** 1999: *Ecotourism and sustainable development.* Washington, DC: Island Press.

ecotoxicology Ecotoxicology or *environmental toxicology* is the study and prevention of the effects of POLLUTANTS on living organisms. It grew out of toxicology with the recognition that pollution affects other organisms than humans within ECOSYSTEMS and that such effects have implications for humans. An important aspect is the definition of toxicity threshold levels above which pollutants produce, for example, detectable adverse (*chronic*) effects or lethal (*acute*) effects. ENVIRONMENTAL CHEMISTRY, which investigates the environmental distribution and fate of contaminants, including pollutants, is a related field. *JAM*

[See also CONTAMINANT, POLLUTION]

Connell, D.W., Lam, P. and Wu, R. 1999: *Ecotoxicology: an introduction.* Oxford: Blackwell Scientific. **Renzoni, A. and Bacci, E.** 1999: Environmental toxicology. In Alexander, D.E. and Fairbridge, R.W. (eds), *Encyclopedia of environmental science.* Dordrecht: Kluwer, 230–235. **Truhaut, R.** 1975: Ecotoxicology – a new branch of toxicology: a general survey of its aims, methods and prospects. In McIntyre, A.D. and Mills, C.F. (eds), *Ecological toxicology research.* New York: Plenum, 3–23.

ecotype A morphologically distinct type within a species, reflecting genetic variability. Ecotypes enable a species to adapt to environmental gradients. Taxa with ecotypes are EURY-typic and do not make good INDICATOR SPECIES. *BA*

[See also ECOCLINE, GENOTYPE, NATURAL SELECTION, SPECIATION]

Turesson, G. 1922: The genotypic response of the plant species to the habitat. *Hereditas* 6, 147–236.

ecozone A large-scale ecological subdivision of the Earth's LANDSCAPES characterised by a similar VEGETATION FORMATION-TYPE and *zonobiome* (see BIOME), which operate as a SYSTEM under broadly similar environmental constraints. *JAM*

Bailey, R.G. 1998: *Ecoregions: the ecosystem geography of the oceans and continents.* London: Springer. **Schultze, J.** 1995: *The ecozones of the world: the ecological divisions of the geosphere.* Berlin: Springer.

ecumene That part of the Earth's surface permanently inhabited by humans. Now, arguably the whole world, the ecumene (or *œcumene*) was much more restricted for most of prehistory and history. The major expansion of the ecumene occurred at the expense of WILDERNESS following the NEOLITHIC REVOLUTION. *JAM*

[See also HOMOSPHERE, NOÖSPHERE]

Roberts, N. 1996: The human transformation of the Earth's surface. *International Social Science Journal* 48, 493–510.

edge detection IMAGE PROCESSING techniques that aim at finding boundaries between regions. They are typically based on FILTERS and they are usually applied to grey-level (or gray-scale) images. *ACF*

edge effects Ecologists often refer to two types of edge effects, both of which are associated with the interface of two relatively distinct HABITATS and biotic communities. On the one hand, the edge effect may refer to the relatively high BIODIVERSITY often encountered along the edge between two habitats (ECOTONE) because of the co-occurrence of species from both communities, along with some species that may specialise for edges. On the other hand, edge effects may refer to the deleterious effects of ABIOTIC forces (e.g. winds and fire) or the movement of undesirable species inward from the edges of nature reserves or habitat fragments deep into their interior. *MVL*

Whittaker, R.J. 1998: *Island biogeography: ecology, evolution and conservation.* Oxford: Oxford University Press.

effective population size The estimated number of individuals that would result in the same degree of GENETIC DRIFT as detected in the *actual population* (higher levels of drift, lower effective population size). Effective population size is almost invariably lower than the total number of individuals in a population, especially for populations with substantially high variability in numbers over time, skewed sex ratios or age distributions. *MVL*

[See also MINIMUM VIABLE POPULATION, POPULATION VIABILITY ANALYSIS]

Lacy, R.C. 1992: The effects of inbreeding on isolated populations: are minimum viable population sizes predictable? In Fielder, P.L.and Jain, S.K. (eds), *Conservation biology: the theory and practice of nature conservation and management.* New York: Chapman and Hall, 276–320.

effective precipitation According to Horton's theory of RUNOFF, all the water that remains on the surface after INFILTRATION and any losses in surface detention, INTERCEPTION and EVAPORATION. It is the proportion of the *total precipitation* that remains on the surface. *GS*

Horton, R.E. 1933: The role of infiltration in the hydrological cycle. *Transactions of the American Geophysical Union* 14, 446–460.

effluent Gaseous or liquid WASTE discharged from point sources into the environment as a result of human activities. Agricultural, industrial and residential areas are all sources of effluent, which may be treated or untreated. *JAM*

[See also EMISSIONS, WASTE MANAGEMENT]

eigenvalue Eigenvalues are the roots, $\lambda_1, \lambda_2, ..., \lambda_m$ of the mth-order polynomial defined by $|\mathbf{A} - \lambda \mathbf{I}|$ where $\mathbf{A}$ is a $m \times m$ square matrix and $\mathbf{I}$ is an identity matrix of order m. Associated with each eigenvalue is a non-zero eigenvector. Both eigenvalues and eigenvectors are important in ORDINATION techniques such as PRINCIPAL COMPONENTS ANALYSIS and CORRESPONDENCE ANALYSIS. The eigenvalue gives the strength of the ordination axes, usually as the variance of a linear combination of the variables. The elements of the *eigenvector* define a linear function of the variables with particular mathematical and geometrical properties. An eigenvalue is also known as a *characteristic root*. *HJBB*

Healy, M.J.R. 1986: *Matrices for Statistics.* Oxford: Clarendon Press.

ejecta Solid material thrown from a VOLCANIC ERUPTION (PYROCLASTIC and VOLCANICLASTIC material as opposed to liquid LAVA or gaseous products) or a METEORITE IMPACT. *GO*

Ekman motion The motion of water in the surface layers of the OCEAN and large lakes in response to wind stress. Because of the effect of the CORIOLIS FORCE, the topmost layer flows in a direction at 45° to the wind direction. This angular difference increases with depth, although the velocity decreases. The resultant pattern of motion is described as the *Ekman spiral* and the layer in which it occurs, which is typically 50–100 m thick, as the *Ekman layer*. The net transport of water (*Ekman transport*) is at right angles to the wind direction, to the right in the Northern Hemisphere and to the left in the Southern Hemisphere. *GO*

[See also GEOSTROPHIC CURRENT, GYRE, OCEAN CURRENTS, UPWELLING]

Allen, P.A. 1997: *Earth surface processes.* Oxford: Blackwell Science. **Leeder, M.R.** 1999: *Sedimentology and sedimentary basins: from turbulence to tectonics.* Oxford: Blackwell Science. **Open University Course Team** 1989: *Ocean circulation.* Oxford: Pergamon Press.

El Niño Originally a warm OCEAN CURRENT that periodically flows southwards along the coast of Ecuador and northern Peru during the Southern Hemisphere summer when the Southeast TRADE WINDS are weakest. The term is Spanish for 'Christ Child', which is associated with its occurrence shortly after Christmas. Current usage associates the term with the climatic ANOMALIES over a wide area of the tropical Pacific Ocean in those years when the current penetrates exceptionally far south (as in an 'El Niño year'). Such years were traditionally known as *años de abundancia*, which brought rainfall and abundant vegetative growth to the arid or semi-arid coast. As a result of the associated breakdown in the UPWELLING of cold, nutrient-rich waters, El Niño is a major determinant of periodic failures in the Peruvian and northern Chilean archoveta fishery. *JAM/AHP*

[See also EL NIÑO–SOUTHERN OSCILLATION]

Enfield, D.B. 1989: El Niño, past and present. *Reviews of Geophysics* 27, 159–187. **Glantz, M.H.** 2000: *Currents of change: impacts of El Niño and La Niña on climate and society,* 2nd edn. Cambridge: Cambridge University Press. **Grove, R.** 1998: Global impact of the 1789–1793 El Niño. *Nature* 393, 318–319. **Idyll, C.P.** 1973: The anchovy crisis. *Scientific American* 228, 22–29. **Laws, E.A.** 1997: *El Niño and the Peruvian anchovy fishery.* Sausalito, CA: University Science Books. **Navarra, A.** 1999: *Beyond El Niño.* New York: Springer. **Pezet, F. A.** 1895: The countercurrent 'El Niño' on the coast of northern Peru. *Boletines del Sociedad Geográfico Lima* 11, 603–606. **Quinn, W.H. and Neal, V.T.** 1983: Long-term variations in the Southern Oscillation, El Niño, and Chilean sub-tropical rainfall. *Fishery Bulletin* 81, 363–374. **United Nations Environment Program (UNEP)** 1992: *The El Niño phenomenon.* Nairobi: UNEP.

El Niño–Southern Oscillation (ENSO)

The SOUTHERN OSCILLATION (SO) was described in the 1920s by Sir Gilbert Walker. He noted that when pressure is high in the Pacific it tends to be low in the Indian Ocean from Africa to Australia. These conditions are associated with low temperatures in both these areas and rainfall varies in the opposite direction to pressure. The difference in pressure between Tahiti and Darwin is frequently used as an index of the state of the Southern Oscillation (SOI). EL NIÑO is the invasion from time to time of warm surface waters from the western part of the equatorial Pacific Basin to the eastern part. This disrupts UPWELLING along the western coast of South America, causing a rise in water temperature, heavy rainfall and widespread mortality of plankton, fish and birds. In the 1960s it became clear that annual variations in SEA-SURFACE TEMPERATURE and consequent El Niño events were closely linked to the Southern Oscillation: hence the coupling between El Niño and the Southern Oscillation to derive the term El Niño–Southern Oscillation (ENSO) now in common use.

ENSO is a prime example of ATMOSPHERE OCEAN INTERACTION operating over an interannual timescale of several years with world-wide impacts. Indeed, the ENSO phenomenon is the largest single source of interannual CLIMATIC VARIABILITY on a global scale. During El Niño episodes extremely heavy precipitation occurs in coastal Ecuador and Peru, while droughts are noted in Indonesia, eastern Australia, southern India and southern Africa. Severe STORMS and FLOODS also occur in California and the Gulf States of the United States, but the impact in Europe is minor. Injection of cold water into the eastern Pacific leads to a LA NIÑA episode. Strong south Pacific high pressure and Southeast TRADE WINDS off the South American coast accompany these events. La Niña has been suggested as a causative factor in causing hot dry summers in North America.

El Niño events occur on average about every five years, although the interval between events is irregular: the largest of recent times occurred in AD 1982–1983 and 1997–1998. Improved understanding of the ENSO phenomenon has led, since the 1980s, to successful CLIMATIC PREDICTIONS around the Pacific Basin. Further research is required, however, on its effects outside the tropical Pacific and on the Asian MONSOON, the mechanisms behind the observed TELECONNECTIONS and the interactions of ENSO with the other controlling factors on the general circulation of the atmosphere. The Southern Oscillation Index (SOI) explains, for example, about 45% of the variance of Indian summer rainfall.

Recently El Niño events appear to have become more frequent, suggesting the existence of *persistent El Niño sequences* and prompting concern that there may be a link with the GREENHOUSE EFFECT. Similar, patterns on decadal to century timescales have been detected in NATURAL ARCHIVES, which provide sources of information about ENSO prior to the instrumental period. MARINE SEDIMENTS, fisheries data (including fish scales in coastal marine sediments) and CORAL all act as ENVIRONMENTAL INDICATORS of ENSO-related changes in the ocean environments. Tropical ICE CORE records from southern Peru and TREE RINGS in the southwestern United States also exhibit ENSO signals. *AHP/JAM*

Allan, R.J. and D'Arrigo, R.D. 1999: 'Persistent' ENSO sequences: how unusual was the 1990–1995 El Niño? *The Holocene* 9, 101–118. **Bigg, G.R.** 1990: El Niño and the Southern Oscillation. *Weather* 45, 2–8. **Cane, M.A., Eshel, G. and Buckland, R.W.** 1994: Forecasting Zimbawean maize yields using eastern equatorial Pacific sea surface temperatures. *Nature* 370, 204–205. **Changnon, S. (ed.)** 2000: *El Niño, 1997–1998: the climate event of the century.* New York: Oxford University Press. **Diaz, H.F. and Markgraf, V. (eds)** 1992: *El Niño: historical and paleoclimatic aspects of the Southern Oscillation.* Cambridge: Cambridge University Press. **Diaz, H.F. and Markgraf, V. (eds)** 2000: *El Niño and the Southern Oscillation: multiscale variability and global and regional impacts.* Cambridge: Cambridge University Press. **Enfield, D.B.** 1988: Is El Niño becoming more common? *Oceanography* 1, 23–27. **Glantz, M., Katz, R. and Krenz, M. (eds)** 1987: *The societal impacts associated with the 1982–1983 worldwide climatic anomalies.* Boulder, CO: National Center for Atmospheric Research. **Glynn, P.W. (ed.)** 1990: *Global consequences of the 1982–1983 El Niño–Southern Oscillation.* Amsterdam: Elsevier. **Mooley, D.A. and Parasarathy, B.** 1984: Indian summer monsoon and El Niño. *Pure and Applied Geophysics* 121, 339–352. **National Research Council** 1994: *Learning to predict El Niño: accomplishments and legacies of the TOGA Program.* Washington, DC: National Academy Press. **Philander, S.G.H.** 1990: *El Niño, La Niña and the Southern Oscillation.* Orlando, FL: Academic Press. **Prieto, M. del R., Herrera, R. and Dussel, P.** 1999: Historical evidences of streamflow fluctuations in the Mendoza River, Argentina, and their relationship with ENSO. *The Holocene* 9, 473–481. **Quinn, T.M. and Neal, V.T.** 1992: The historical record of El Niño events. In Bradley, R.S. and Jones, P.D. (eds), *Climate since A.D. 1500.* London: Routledge, 623–648. **Rasmusson, E.M and Wallace, J.M.** 1983: Meteorological aspects of the El Niño/Southern Oscillation. *Science* 222, 1195–1202. **Rodbell, D.T., Seltzer, G.O., Anderson, D.M. et al.** 1999: An ~15,000-year record of El Niño-driven alluviation in southeastern Ecuador. *Science* 283, 516–520. **Takayabu, Y.N., Iguchi, T., Kachi, M. et al.** 1999: Abrupt termination of the 1997–98 El Niño in response to a Madden–Julian oscillation. *Nature* 402, 279–282. **Zhang, Y., Wallace, J.M. and Battisti, D.S.** 1997: ENSO-like decade-to-century scale variability: 1900–93. *Journal of Climatology* 10, 1004–1020.

El Niño–Southern Oscillation: impacts on ecosystems

Significant changes in rainfall, temperature, humidity and storm patterns over much of the tropics and subtropics disrupt many ecosystems and provide conditions that opportunistic organisms may exploit. The best-known biological impact is the reduction in PRIMARY PRODUCTIVITY that occurs in the eastern Pacific. This is caused by the decline in coastal UPWELLING and an increase in rainfall bringing large amounts of sediment to coastal waters. Fish populations collapse, with consequences for marine birds, mammals and predatory fish at higher TROPHIC LEVELS. Periodic population declines in the Antarctic Weddell seal population and of sea birds in Peru and the Galápagos have been correlated with ENSO events. The AD 1997–1998 ENSO event caused massive CORAL BLEACHING and mortality among coral-reef communities in the eastern Pacific. Substantial increases in rainfall in the eastern Pacific induce prolific plant growth in otherwise arid ecosystems. An increase in invasive species (see BIOLOGICAL INVASIONS) has been noted in the Galápagos during these unusually wet periods.

In contrast, a decrease in rainfall in the western Pacific may inflict severe DROUGHT on ecosystems. ENSO-related droughts extend to India, west Africa, Brazil and parts of central and north America. Rates of tropical tree mortality have been observed to increase as a result of drought stress, especially among *canopy* emergents. A dense forest canopy buffers the forest MICROCLIMATE against temporary dry periods and thus protects understorey species. Populations of animals, especially *frugivores* (fruit eaters), decline as a result of reduced food availability.

Most ecosystems are robust in the face of periodic natural DISTURBANCE such as ENSO. However, the impact of ENSO events may be compounded by other, human-inflicted stress. In recent decades, ENSO droughts have been strongly associated with extensive TROPICAL FOREST FIRES. In 1997–1998 large areas of TROPICAL FORESTS in southeast Asia and Brazil were burned. Many fires were started deliberately for DEFORESTATION and the fire hazard was unequivocally exacerbated by forest degradation caused by logging and SHIFTING CULTIVATION. There are impacts too on human populations. ENSO events are associated with increases in diseases such as MALARIA, typhoid and CHOLERA. Failure of crops across southeast Asia has caused FAMINE. *NDB*

[See also EXTREME CLIMATIC EVENTS, CLIMATIC VARIABILITY, CLIMATIC VARIATIONS, CORAL REEFS, HUMAN IMPACTS, CONSERVATION]

Glynn, P.W. and De Weerdt, W.H. 1991: Elimination of two reef-building hydrocorals following the 1982–83 El Niño warming event. *Science* 253, 69–71. **Glynn, P.W. (ed.)** 1990: *Global ecological consequences of the 1982–83 El Niño* [*Elsevier Oceanographic Series* 52]. Amsterdam: Elsevier. **Polis, G.A., Hurd, S.D., Jackson, C.T. and Pinero, F.S.** 1997: El Niño effects on the dynamics and control of an island ecosystem in the Gulf of California. *Ecology* 78, 1884–1897. **World Health Organization** 1997: El Niño and its health impacts. *Journal of Communicable Diseases* 29, 375–377.

electrical conductivity measurements (ECM)

A rapid means of estimating the amount of dissolved impurities (solutes) in water, ice or snow. *Specific conductance* is a measure of SOLUTE concentrations in rivers, for which continuous sampling is now possible. In ICE CORES electrical conductivity measurements have provided valuable palaeoclimatic information (see PALAEOCLIMATOLOGY), as enhanced deposition of impurities onto ICE SHEETS occurs as a result of strengthened *atmospheric circulation* patterns during GLACIAL EPISODES. *DIB/CJC*

Collins, D.N. 1979: Hydrochemistry of meltwaters draining from an Alpine glacier. *Arctic and Alpine Research* 11, 307–324. **Foster, I.D.L., Grieve, I.C. and Christmas, A.D.** 1982: The use of specific conductance in studies of natural waters and solutions. *Hydrological Sciences Bulletin* 26, 257–269. **Taylor, K.C., Hammer, C.U., Alley, R.B. et al.** 1993: Electrical conductivity measurements from the GISP2 and GRIP Greenland ice cores. *Nature* 366, 549–552.

electromagnetic distance measurement (EDM)

A technique for determination of distance between two points. Distance is computed from measurement of a time or phase difference between transmission and return of a signal pulse. Normally the target point must be marked by a reflector to return the transmitter signal, but in some cases, namely short-range laser measurers, no special reflector is required. *TF*

Banniser, A., Raymond, S. and Baker, R. 1998 *Surveying*, 7th edn. Harlow: Longman, chapter 5.

electromagnetic spectrum

The total array of wavelengths or frequencies of electromagnetic RADIATION, extending from the longest radio waves to the shortest known cosmic rays. Electromagnetic waves propagate outward from any electric charge that is accelerated and consist of oscillating electric and magnetic fields; they move at the speed of light and at right angles to each other and to the direction of motion. The relationship between the frequency v, the wavelength λ and the velocity of the wave c (in the air 3×10^8 m s^{-1}) is $v = c/\lambda$. The energy E of a PHOTON (or light quantum) is given by $E = hv$, where h is the Planck constant with a value of about 6.6×10^{-34} W s^2. Electromagnetic waves with different wavelengths have different names and properties, interact in different ways with materials and are used for different applications. The spectrum is usually divided into seven sections: *radio, microwave, infrared, visible, ultraviolet, X-ray* and *gamma-ray* (see Figure). Exact frequency limits for the division of the spectrum are not defined.

The frequency range of radio waves extends approximately from 10 kHz to 300 MHz. Above this frequency, the short waves used, for example, by RADARS and *radiometers* but also for radiation ovens are called microwaves. Still higher we do not need an instrument to detect the electromagnetic wave; it can be seen with the human eye. The lowest frequency that the human eye can see appears as red light. The highest visible frequency has a value almost double that of red light and appears as violet. OPTICAL REMOTE SENSING uses electromagnetic waves of this frequency range. The electromagnetic waves with frequencies lower than that of red light constitute the infrared. These waves are used in THERMAL REMOTE SENSING. The

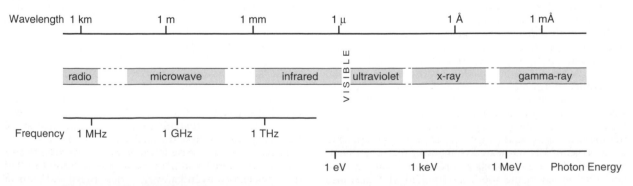

electromagnetic spectrum *Wavelength and frequency of the components of the electromagnetic spectrum*

electromagnetic waves with frequencies higher than that of violet are known as ultraviolet. Still higher frequencies and energies are carried by X- and gamma-rays. *TS*

Feynman, R.P., Leighton, R.B. and Matthew, S. 1963: *The Feynman lectures on physics.* Reading, MA: Addison-Wesley. **Hewitt, P.G.** 1987: *Conceptual physics.* Reading, MA: Addison-Wesley. **Schanda, E.** 1986: *Physical fundamentals of remote sensing.*: Heidelberg: Springer. **Ulaby, F.T., Moore, R.K. and Fung, A.K.** 1981: *Microwave remote sensing: active and passive.* Norwood, MA: Artech House.

electron microprobe analysis (EMPA)

A rapid, non-destructive technique for the quantitative elemental analysis of solid materials, in which samples are excited by a beam of electrons from an electron gun. Alternatively known as *electron probe microanalysis (EPMA)*, the technique is of particular value for determining the provenance of TEPHRAS, which tend to have characteristic suites of major elements. It is also widely used in the investigation of archaeological materials, especially metals and CERAMICS. *DH*

Hunt, J.B. and Hill, P.G. 1993: Tephra geochemistry: some persistent analytical problems. *The Holocene* 3, 271–278. **Kane, W.T.** 1973: Applications of the electron microprobe in ceramics and glass technology. In Anderson, C.A. (ed.), *Microprobe analysis.* New York: Wiley, 241–270. **Reed, S.J.B.** 1993: *Electron microprobe analysis,* 2nd edn. Cambridge: Cambridge University Press.

electron paramagnetic resonance

The resonant absorption of electromagnetic radiation by a paramagnetic substance containing unpaired spinning electrons, simultaneously subjected to a constant, strong magnetic field (also called *electron spin resonance* or *electron magnetic resonance*). This is detected by microwave spectroscopy and forms the basis of ELECTRON SPIN RESONANCE DATING. *MHD*

Thompson, R. 1991: Palaeomagnetic dating. In: Smart, P.L. and Frances, P.D. (eds), *Quaternary dating methods – a users guide* [QRA Technical Guide 4]. Cambridge: Quaternary Research Association, 177–198.

electron spin resonance (ESR) dating

A form of LUMINESCENCE DATING, electron spin resonance (ESR) dating measures in a non-destructive form the electrons which, after excitation, become trapped in the crystal lattice of certain minerals, particularly in *calcite*. The intensity measured is a function of the average radiation flux at the trap site, the time elapsed since the ELECTRON TRAPS were last emptied, or zeroed, and the sensitivity of the mineral to radiation, which derives principally from radioactive decay of uranium and thorium in the surrounding rock, sediment or soil. Whilst the technique has the advantage of requiring only small samples, < 2 g, it has low PRECISION, at best ±2% but in some cases up to ±100%. ESR dating can be applied to calcite in SPELEOTHEMS and CORALS, as well as to aragonite in molluscs, and hydroxyapatite in tooth enamel. It is of less value in porous, open-system conditions characteristic of TUFA and bones, but has been used to date QUARTZ in VOLCANIC rocks and *flints* that were zeroed following firing. It is most commonly used to date material up to 800 000 years old, being particularly valuable in testing and extending timescales derived from URANIUM-SERIES DATING, but can be used to two million years. As well as dating QUATERNARY sequences in CAVES and from corals, ESR dating has contributed to the understanding of later HOMINID evolution. *CJC*

Aitken, M.J. 1985: *Thermoluminescence dating.* London: Academic Press. **Grün, R.** 1990: Electron spin resonance (ESR) dating. *Quaternary International* 1, 65–109. **Grün, R.** 1997: Electron spin resonance dating. In Taylor, R.E. and Aitken, M.J. (ed.), *Chronometric dating in archaeology.* New York: Plenum, 217–260. **Ikeya, M.** 1975: Dating a stalactite by electron spin paramagnetic resonance. *Nature* 255, 48–50. **Ikeya, M.** 1993: *New applications of electron spin resonance dating, dosimetry and microscopy.* Singapore: World Scientific. **Jonas, M.** 1997: Concepts and methods of ESR dating. *Radiation Measurements* 27, 943–973. **Kailath, A.J., Rao, T.K.G., Dhir, R.P.** *et al.* 2000: Electron spin resonance characterization of calcretes from the Thar desert for dating applications. *Radiation Measurements* 32, 371–383. **Porat, N., Schwarcz, H.P., Valladas, H.** *et al.* 1994: Electron spin resonance dating of burned flint from Kebara Cave, Israel. *Geoarchaeology* 9, 393–407. **Rink, W.J.** 1997: Electron spin resonance (ESR) dating and ESR applications in Quaternary science and archaeometry. *Radiation Measurements* 27, 975–1025. **Skinner, A.R.** 2000: ESR dating: is it still an 'experimental' technique? *Applied Radiation and Isotopes* 52, 1311–1316. **Smart, P.L.** 1991: Electron spin resonance (ESR) dating. In Smart, P.L. and Frances, P.D. (eds), *Quaternary dating methods – a users guide* [QRA Technical Guide 4]. Cambridge: Quaternary Research Association, 128–160.

electron trap

A defect in the crystal lattice of minerals (particularly QUARTZ or FELDSPARS), as in a negative-ion valency, to which *electrons* are attracted. Energy absorbed from ionising radiation frees electrons to move through the crystal lattice. Electron traps form paramagnetic centres detected in ELECTRON SPIN RESONANCE (ESR) DATING. Light emitted when electrons are evicted from traps by heat or light is measured in LUMINESCENCE DATING. *DAR*

element

A substance that cannot be separated into simpler materials by physical or chemical means. The properties of an element are determined by its *atomic number*, which refers to the number of *protons* in the nucleus of ATOMS of the element (also equal to the number of *electrons*). Although each element is unique, patterns in their structures lead to similarities in their properties, which can be understood with reference to their position in the *periodic table.* *JAM*

elfin woodland

A MONTANE FOREST formation of stunted stature (probably due to wind stress in exposed locations) that occurs at high altitude on TROPICAL and SUBTROPICAL mountains. It is sometimes used as a synonym for CLOUD FOREST. *NDB*

[See also ALTITUDINAL ZONATION OF VEGETATION, KRUMMHOLZ, TROPICAL FOREST]

Howard, R.A. 1968: The ecology of an elfin forest in Puerto Rico. *Journal of the Arnold Arboretum* 49, 381–418.

elm bark beetle

The elm bark beetle (*Scolytus scolytus* and *S. multistratus*) forms breeding galleries in the bark of elm trees affected by the fungal PATHOGEN *Ceratocystis ulmi.* The galleries are suitable places for the production of fruiting bodies by the fungi, which are spread to other trees by the beetle. *JLF*

[See also ELM DECLINE]

Gibbs, J.N. 1974: *The biology of Dutch elm disease, Ceratocystis ulmi.* [*Forest Record* 94]. Harpenden: Forestry Commission.
Gibbs, J.N. 1978: Development of the Dutch elm disease epidemic in southern England, 1971–76. *Annals of Applied Biology* 88, 219–228.

elm decline A mid-HOLOCENE decline in the abundance of trees of the genus *Ulmus* (elm) throughout north-western Europe. Pollen data from numerous locations record a decline in the abundance of elm about 5100 radiocarbon years BP. Several hypotheses have been proposed to explain the decline in this DECIDUOUS tree taxon, including CLIMATIC CHANGE, human activity and/or a disease outbreak. Although recent opinion favours the *pathogenic hypothesis*, possibly exacerbated by human activity, the cause of the decline is still highly debatable. Elm appears to have been the only tree taxon to decline significantly at this time in most cases, and at many sites indications of human activity associated with Neolithic agricultural activity did not precede the decline.   *JLF*

[See also ELM BARK BEETLE, HEMLOCK DECLINE, HUMAN IMPACT ON VEGETATION HISTORY, NEOLITHIC: LANDSCAPE IMPACTS]

Girling, M.A. and Greig, J. 1985: A first fossil record for *Scolytus scolytus* (F.) (elm bark beetle): its occurrence in elm decline deposits from London and the implications for Neolithic elm disease. *Journal of Archaeological Science* 12, 347–352.
Molloy, K.M. and O'Connell, M. 1987: The nature of the vegetation changes at about 5000 B.P. with particular reference to the elm decline: fresh evidence from Connemara, western Ireland. *New Phytologist* 107, 203–220. **Peglar, S.M.** 1993: The mid-Holocene *Ulmus* decline at Diss Mere, Norfolk, UK: a year-by-year pollen stratigraphy from annual laminations. *The Holocene* 3, 1–13. **Peglar, S.M. and Birks, H.J.B.** 1993: The mid-Holocene *Ulmus* fall at Diss Mere, South-East England – disease and human impact. *Vegetation History and Archaeobotany* 2, 61–68. **Perry, L. and Moore, P.D.** 1987: Dutch elm disease as an analogue of Neolithic elm decline. *Nature* 326, 72–73. **Whittington, G., Edwards, K.J. and Cundill, P.R.** 1991: Palaeoecological investigations of multiple elm declines at a site in north Fife, Scotland. *Journal of Biogeography* 18, 71–87.

eluviation The removal of fine-grained soil material, especially clay and organic matter, in aqueous suspension from the upper horizons of a soil. Eluviation is best expressed in humid climates with a dry season. The results of eluviation may be observed as clay or silt coatings on soil structure faces. Currently, the term LEACHING refers to removal of soluble material only, but in older literature it is included in eluviation.   *EMB*

[See also ILLUVIATION, LEACHING, LUVISOLS, PERVECTION]

Buol, S.W., Hole, F.D. and McCracken, R.J. 1988: *Soil genesis and classification.* Ames, IA: Iowa State University Press.

eluvium Sedimentary material produced by rock weathering *in situ*, in contrast to that deposited by rivers (ALLUVIUM) or slope processes (COLLUVIUM).   *JAM*

[See also REGOLITH]

emergent coast A coast that has been raised relative to sea level, usually because of TECTONIC movement, *isostatic rebound* and/or falling SEA LEVEL.   *HJW*

[See also SUBMERGENT COAST and EUSTASY]

emergent properties The appearance of synergistic or organisational qualities, structures or properties at higher levels in a model or system that cannot be predicted from relationships between isolated components.   *CET*

[See also HOLISM]

emission The release of waste, especially POLLUTANTS, to the ATMOSPHERE from point sources. The *emissions* released may be gaseous, such as CARBON DIOXIDE and other GREENHOUSE GASES, liquid droplets or PARTICULATES.   *JAM*

Nakicenovic, N. and Swart, R. (eds) 2000: *Emissions scenarios: special report of the Intergovernmental Panel on Climate Change.* Cambridge: Cambridge University Press.

emission control A general term that refers to the range of options open to a regulatory body to reduce ambient levels of a wide range of POLLUTANTS in order to conform to air quality standards. These can take the form of general policy strategies designed to influence the behaviour of polluters, for instance, urban planning, investment in public transport and various disincentives for private car use, or they can be enforced by legislation. The latter strategy includes regulations related to vehicle design (e.g. the fitting of catalytic converters and diesel particulate traps), the regular inspection and certification of vehicles by government, the use of reformulated fuels or the prohibition of high polluting fuel types in certain areas.

Reductions of industrial emissions may be achieved by improving the efficiency of furnaces, boilers and operating processes to use less fuel, using less polluting fuel types and fitting pollution-control equipment (e.g. electrostatic precipitators or flue-gas desulphurisation systems). In addition, emission controls include financial incentives (subsidies) to industry to reduce pollution, the introduction of TRADEABLE EMISSION PERMITS and the installation of fees and charges related to the amount of pollutants emitted (the *polluter-pays principle*). Emissions from residential areas can be reduced using more efficient cooking and heating equipment, switching from kerosene, coal and lignite to cleaner fuels such as natural gas and electricity, and improving household insulation.   *GS/DME*

Elsom, D.M. 1992: *Atmospheric pollution.* Oxford: Blackwell.
Elsom, D.M. 1996: *Smog alert: managing urban air quality.* London: Earthscan. **Holmes, G., Theodore, L. and Singh, B.** 1993: *Handbook of environmental management and technology.* New York: Wiley.

empiricism Empiricism in philosophy comes in many forms and degrees. However, empiricist approaches always place heavy emphasis upon experience, observation and sensory perception as the source(s) of knowledge.   *CET*

enclosure The process by which a system of subdivisions with clearly marked boundaries is imposed on an agricultural LANDSCAPE. Many boundaries are historic and tend to follow natural features; others are much affected by forms of social organisation and patterns of inheritance. As a legalistic process, enclosure has been modifying rural landscapes for a long period of time with the broad aim of clarifying land ownership or land tenancy. Enclosure has also been used to achieve change, such as the breaking up of large tracts of land and the reverse process of consolidating smaller holdings.

The great period of enclosures or '*inclosures*' (the latter has precise legal meaning) in the UK occurred between AD 1760 and 1830. Between 1760 and 1810, for example, there were 2765 separate Acts of Enclosure. Many of these were targeted at the Mediaeval *open field system* and signified the major social change from the communal holding of land to individual ownership. There were large-scale encroachments of enclosures upon COMMON LAND and *common fields* that played a part in the outmigration of land-less labourers from rural areas. One definition of 'inclosures' is the process by which common rights over a piece of land are extinguished and the land turned into ordinary freehold. Enclosures were justified in terms of economic efficiency but did represent a major societal change. *DTH*

Royal Commission on Common Land, 1955–58, Command 462, 1958: London: HMSO.

endangered species A species in danger of extinction and a category in the International Union for the Conservation of Nature RED DATA BOOK, including species with at least 20% probability of EXTINCTION within 20 years or 10 generations. *IFS*

[See also RARE SPECIES, THREATENED SPECIES, VULNERABLE SPECIES]

World Conservation Monitoring Centre 1992: *Global biodiversity: status of the Earth's living resources.* London: Chapman and Hall.

endemic A taxon that is restricted to a given area. This area may range considerably in size, from a single water hole in the Nevada desert (for the Devil's hole pupfish, *Cyprinidon diabolis*) to an entire continent. *Narrow* or *local endemics* are confined to relatively small areas (up to the area of a mountain range for example); *broad endemics* have more extensive distributions. Biogeographers recognise two major categories: *neo-endemics* and *palaeo-endemics*. The former are relatively recently evolved arrays of closely related species, while the latter are relictual higher taxa, often containing few species. The relationship between richness and endemism is variable: *oceanic islands*, for example, are often species-poor but rich in endemics. There are many areas of extreme endemism, such as Lake Victoria, home to several hundred endemic cichlid fish. The degree to which endemic species contribute to BIODIVERSITY is an important consideration in designating HOTSPOTS. The most common use of 'endemic' in the CONSERVATION literature refers to neo-endemic taxa. Palaeo-endemics may constitute a small proportion of overall biodiversity in a region, but they often receive special conservation attention in recognition of their distinctiveness. *JTK*

[See also ADAPTIVE RADIATION, DIVERSITY CONCEPTS, ISLAND BIOGEOGRAPHY]

Gentry, A.H. 1991: Biological extinction in western Ecuador. *Annals of the Missouri Botanical Garden,* 78, 273–295. **May, R.M.** 1990. Taxonomy as destiny. *Nature* 347, 129–130. **Mittermeier, R.A., Myers, N., Thomsen, J.B.** *et al.* 1998: Biodiversity hotspots and major tropical wilderness areas: approaches to setting conservation priorities. *Conservation Biology* 12, 516–520. **Paulay, G.** 1994: Biodiversity on oceanic islands: its origin and extinction. *American Zoologist* 34, 134–144. **Richardson, I.B.K.** 1978: Endemic taxa and the taxonomist. In Street, H.A. (ed.), *Essays in plant taxonomy.* London: Academic Press, 245–262.

endemic bird area (EBA) An area with notable levels of ENDEMIC birds, defined as areas containing two or more bird species of restricted range (< 50 000 km²). A global database of EBAs has been developed by BirdLife International. *IFS*

Long, A.K., Crosby, M.J., Stattersfield, A.J. and Wege, D.C. 1996: Towards a global map of biodiversity: patterns in the distribution of restricted-range birds. *Global Ecology and Biogeography Letters* 5, 281–304.

endogenetic Pertaining to the internal agencies of formation, particularly the processes originating from within the Earth's interior (e.g. those of EPEIROGENY, OROGENY, PLATE TECTONICS and VOLCANISM) that affect the formation of LANDFORMS at the Earth's surface. The term is also used for the landforms, such as RIFT VALLEYS and VOLCANOES produced by endogenetic processes. *JAM*

[See also EXOGENETIC, POLYGENETIC]

endorheic lake A closed lake basin, fed by inflow from surface streams and direct precipitation. As there is no outlet, fluctuations in water level within endorheic lakes can often be used as an index of climatic variation. They are characteristic of ARIDLANDS and DESERTS. *MLC*

[See also CLIMATIC CHANGE, LAKE-LEVEL VARIATIONS, SABKHA, SALINE LAKES].

Street-Perrott, F.A. and Roberts, N. 1983: Fluctuations in closed lakes as an indicator of past atmospheric circulation patterns. In Street-Perrott, F.A., Beran, M.A. and Ratcliffe, R.A.S. (eds), *Variations in the global water budget.* Dordrecht: Reidel, 331–345.

energy (of environment) A concept loosely used in the interpretation of SEDIMENTS and SEDIMENTARY ROCKS to indicate the strength of currents characteristic of a DEPOSITIONAL ENVIRONMENT. *High-energy* environments frequently experience strong currents (e.g. GRAVEL-BED RIVER); *low-energy* environments are characterised by weak currents (e.g. lake). Proxy indicators of the energy of an environment include GRAIN-SIZE and SEDIMENTARY STRUCTURES. *GO*

[See also WAVE ENERGY]

energy balance/budget The exchange of energy between a surface, object or SYSTEM and its environment over a specified time period. The concept has wide application, for example to ECOSYSTEMS, LANDSCAPES, the ATMOSPHERE and the whole Earth. The energy balance of the Earth–atmosphere system is summarised in the Figure. *AHP/JAM*

[See also ECOLOGICAL ENERGETICS, ENERGY FLOW, HEAT BALANCE, RADIATION BALANCE]

Gregory, K.J. (ed.) 1987: *Energetics of physical environment.* Chichester: Wiley. **Trenberth, K.E., Houghton, J.T. and Meira Filho, L.G.** 1996: The climate system: an overview. In Houghton, J.T., Filho, L.G.F., Callander, B.A. *et al.* (eds), *Climate change 1995: the science of climate change.* Cambridge: Cambridge University Press, 51–64.

energy-dispersive spectrometry (EDS) A technique for the simultaneous and rapid determination of multiple elements from their X-ray emission spectra. It is used in some X-RAY FLUORESCENCE (XRF) and ELECTRON

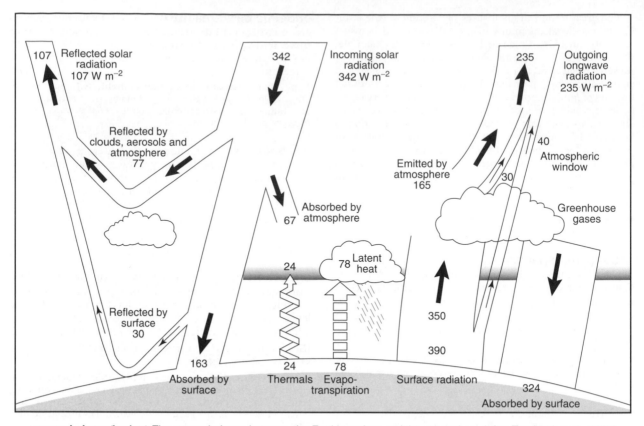

energy balance/budget *The energy balance between the Earth's surface and the atmosphere (after Trenberth et al., 1996)*

MICROPROBE ANALYSES. Processing of spectra to determine elemental composition follows the recognition of particular elemental lines in the collected spectrum. It is, however, slower with higher detection limits than *wavelength dispersive spectrometry* (WDS), which is also used for X-ray fluorescence analysis (WD-XRF) and in electron microprobes.   *TPY*

Winsome, T. and McColl, J.G. 1998: Changes in chemistry and aggregation of a Californian forest soil worked by the earthworm *Argilophilus papillifer* Eisen (Megascolecidae). *Soil Biology and Biochemistry* 30, 1677–1687.

energy flow   An ECOSYSTEM CONCEPT and an integral part of the TROPHIC–DYNAMIC APPROACH to ecology and of ECOLOGICAL ENERGETICS. Energy fixed by PHOTOSYNTHESIS in green plants (*autotrophs*) flows through a small number of TROPHIC LEVELS occupied by HERBIVORES and CARNIVORES (*heterotrophs*). Only a small fraction of the energy reaching each level is available for the next as most is dissipated during RESPIRATION, either in the so-called *grazing food chain* or in the *decomposer food chain* (see FOOD CHAIN). The latter often predominates, especially in forest ecosystems, whereas the grazing food chain is relatively more important in grasslands. There are many other environmental contexts in which concepts of energy flow have been found useful.   *JAM*

[See also ENVIRONMENTAL ENERGETICS, MINERAL CYCLING]

energy resources   Materials that can be used by society to obtain energy (see Figure), including: FUEL-WOOD; dung and biofuels; non-renewable FOSSIL FUELS (COAL; PETROLEUM); NON-RENEWABLE RESOURCES with effectively unlimited RESERVES (GEOTHERMAL energy, NUCLEAR ENERGY); and RENEWABLE ENERGY sources (hydroelectric, solar, tidal and wind power). With the exception of nuclear, geothermal and tidal power, the ultimate source of energy in all these forms is the Sun, either at the present day or, in the case of fossil fuels, in the geological past. The sixfold increase in global *per capita* energy consumption over the last 200 years is a major factor in such global environmental issues as ACID RAIN, AIR POLLUTION and the enhanced GREENHOUSE EFFECT.   *GO*

[See also ALTERNATIVE ENERGY, INDUSTRIAL REVOLUTION, POLLUTION HISTORY, RENEWABLE RESOURCES]

BP Amoco 1999: BP *Amoco statistical review of world energy, June 1999*. London: BP Amoco. **Byrne, J. and Rich, D.** 1992: *Energy and environment: the policy challenge*. New Brunswick, New Jersey: Transaction Publishers. **Edwards, D.J.** 1997: Crude oil and alternate energy production forecasts for the twenty-first century: the end of the hydrocarbon era. *American Association of Petroleum Geologists Bulletin* 81, 1292–1305. **LeBel, G.P.** 1982: *Energy economics and technology*. Baltimore, MD: Johns Hopkins University Press. **Prest, J.** 1984: *Energy: principles, problems, alternatives*. Reading, MA: Addison-Wesley. **Ricketts, J.** 1995: *Competitive energy management and environmental technologies*. Lilburn, GA: Fairmont Press.

englacial   Relating to the interior of a glacier. *Englacial debris* is derived either from a SUPRAGLACIAL position through debris being buried or falling into

A

Million tonnes oil equivalent

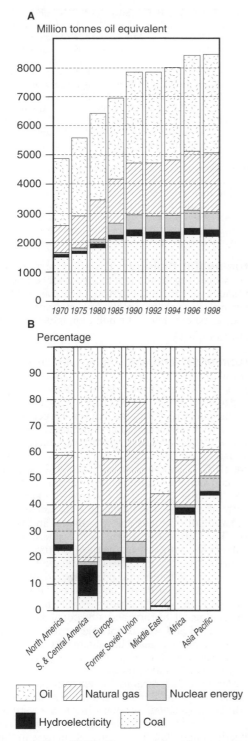

B

Percentage

Oil | Natural gas | Nuclear energy
Hydroelectricity | Coal

energy resources *(A) World consumption of energy resources from 1970 to 1998, according to energy type. (B) Regional consumption of different energy types for 1998 (after BP Amoco, 1999)*

crevasses, or from a SUBGLACIAL position by folding and thrusting processes. An *englacial* stream is one that has penetrated below the glacier surface and is *en route* to the bed.   *MJH*

enhancement   The improvement of an environmental SYSTEM measured in terms of diversity, stability, ENVIRONMENTAL QUALITY or SUSTAINABILITY.   *JAM*

[See also CONSERVATION]

National Research Council 1992: *Restoration of aquatic ecosystems: science, technology and public policy.* Washington, DC: National Academy Press.

entities   Fundamental units of information in a GEOGRAPHICAL INFORMATION SYSTEM (GIS). They can be POINTS, lines (see ARCS) or areas, each identified by LABELS and ATTRIBUTES and spatially positioned in terms of CARTESIAN COORDINATES.   *TVM*

[See also AREA LABELS, FIELDS, OVERLAY ANALYSIS]

entrainment   The process by which stationary particles are picked up and set in motion in a flowing medium, such as air, water or ice. Entrainment, which occurs when the *critical erosion velocity* is exceeded by the flow, involves the forces of lift and drag overcoming the weight of particles and friction with the substrate.   *JAM*

[See also GLACIAL EROSION, TURBULENT FLOW]

environment   The complex interaction of elements and processes that surround and support life. All organisms exist in an environment (usually on the Earth's surface) that may be considered as consisting of ABIOTIC and BIOTIC FACTORS. There are reciprocal interactions between an organism and its environment, which normally includes other organisms. 'Environment' is an organism-centred concept: it has little meaning if divorced from organisms. If the concept is extended from one of a 'NATURAL' to a 'human' environment, the concept becomes both more subjective and complex. A human environment is socially constructed and may be perceived and experienced in different ways. In the general context of any SYSTEM, the environment may be defined as what lies outside the system.   *JAM/DTH*

[See also ECOSPHERE, ENVIRONMENTALISM, NATURE]

Simmons, I.G. 1993: *Interpreting nature: cultural constructions of the environment.* London: Routledge.

environmental accident   A relatively small-scale event or incident, caused by human agency, that results in environmental damage. Examples include some forest fires, industrial explosions, OIL SPILLS and NUCLEAR ACCIDENTS, although major events of these types in terms of either intensity of damage or areal extent may be classified as ENVIRONMENTAL DISASTERS.   *JAM*

environmental archaeology   Broadly, the use of biological, palaeoecological, sedimentological, geophysical and other methods to study and interpret the environment in which humans lived. Deriving in part from the pioneering work of G.W. Dimbleby and of K.W. Butzer, the term was introduced in Europe (UK) in the 1970s, to emphasise the additional information that could be obtained from linked scientific investigation of the palaeoenvironment of archaeological sites, as opposed to merely the excavation of archaeological site structures or the classification of inorganic artefacts (such as flint typologies, *pottery*, coins, etc.). It encompasses BIOARCHAEOLOGY (which to a degree it has as a term superseded) and GEOARCHAEOLOGY. Advocates of a broader definition might include also the archaeological use of 'ancient BIOMARKERS' (biochemical detection of biological lineage or marker molecules), chemical and mineral analysis of artefacts and the wide range of dating applications in archaeology. However, environmental archaeology is perhaps more appropriately regarded as part of 'science-based archaeology' (SCIENTIFIC ARCHAEOLOGY), in which a range of dating, forensic and other skills are used to help reconstruct, provenance and date the past environments, economic base and lineage of human groups. In recent years, environmental archaeology has been regarded as being equivalent to *human palaeoecology*.

Environmental archaeology can be practised either 'on-site', in which the various contexts or layers of an archaeological site excavation might be subjected to a range of field or laboratory analyses, or 'off-site', in which the sediments of a nearby lake or the peat of a MIRE might be cored to produce, for example, a vegetational history for the locality or region. Although, for decades, palaeoecologists had practised off-site methods such as POLLEN ANALYSIS, they had tended to work separately from archaeologists; where collaboration did exist, environmental evidence had previously been relegated to appendices in site excavation reports. One of the best examples of its early practice (as BIOARCHAEOLOGY) was in England at Star Carr, where a range of specialists combined to produce a more complete picture of a Mesolithic site (Clark, 1954). This degree of co-operation and integration was then unusual. Improved techniques, including FLOTATION – designed to recover PLANT MACROFOSSILS (especially cereal remains) from archaeological sites – were developed in the 1970s and have since been applied at a range of sites, particularly in Europe and the Near East. The formation of the Association for Environmental Archaeology in the UK in the late 1970s was intended to give a higher profile to environmental evidence in archaeology and to educate classical, finds and field archaeologists in the value of integrating palaeoenvironmental research.

Current best practice would include a combination of 'on-site' and 'off-site' records, to arrive at a more complete interpretation of the archaeological site environment and its (pre-)historical context, but in cases where disturbance of the archaeological site is either not possible or not desirable, then 'off-site' methods might be used alone or in combination with non-destructive (e.g. geophysical) methods, on site. The methods that are used by environmental archaeologists can also be used extensively to contribute to landscape archaeology through reconstruction of the CULTURAL LANDSCAPE over time.

The subdiscipline has spawned a number of journals, including *Journal of Archaeological Science*, *Vegetational History and Archaeobotany* and *Environmental Archaeology* (formerly *Circaea*) and is populated by geophysicists, sedimentological geoarchaeologists and a range of specialists, particularly from the biological sciences (for example 'bone' people, pollen analysts, (palaeoethno)archaeobotanists, palaeoethnoparasitologists, etc.), so making the subject MULTIDISCIPLINARY. Environmental archaeology as practised in the future (either as human palaeoecology, or as CULTURAL LANDSCAPE history) may exhibit more of the characteristics of INTERDISCIPLINARY RESEARCH. *FMC*

Bell, M. and Walker, M.J.C. 1992: *Late Quaternary environmental change: physical and human perspectives*. London: Longman. **Berglund, B.E. (ed.)** 1991: *The cultural landscape during 6000 years in southern Sweden – the Ystad Project* [*Ecological Bulletins* 41]. Copenhagen: Munksgaard. **Butzer, K. W.** 1971: *Environment and archaeology*. London: Methuen. **Clark, J.D.** 1954: *Excavations at Star Carr, an Early Mesolithic site at Seamer, near Scarborough, Yorkshire*. Cambridge: Cambridge University Press. **Dimbleby, G.W.** 1978: *Plants and archaeology*, 2nd edn. London: Baker. **Dimbleby, G.W.** 1985: *The palynology of archaeological sites*. London: Academic Press. **Dincauze, D.F.** 2000: *Environmental archaeology: principles and practice*. Cambridge: Cambridge University Press. **Evans, J.G.** 1993: The influence of human communities on the English chalklands from the Mesolithic to the Iron Age: the molluscan evidence. In Chambers, F.M. (ed.), *Climate change and human impact on the landscape*. London: Chapman and Hall, 147–156. **Evans, J.G. and O'Connor, T.** 1999: *Environmental archaeology: principles and methods*. Stroud: Alan Sutton. **Harris, D.R. and Thomas, K.D. (eds)** 1991: *Modelling ecological change*. London: Institute of Archaeology, University of London. **Shackley, M.** 1981: *Environmental archaeology*. London: George Allen and Unwin.

environmental biology The study of the environment in which organisms live and reproduce and of the ENVIRONMENTAL FACTORS, physical, chemical and biological, that influence them. In practice environmental biology tends to be dominated by pollution issues. *GOH*

[See also ECOLOGY]

environmental change At present, there is great concern about future GLOBAL ENVIRONMENTAL CHANGES. Problems with emission of GREENHOUSE GASES, depletion of the OZONE layer, DEFORESTATION, LAND DEGRADATION and marine and terrestrial POLLUTION are receiving special attention from both the scientific community and the public. Human-induced GLOBAL WARMING, which according to the latest INTERGOVERNMENTAL PANEL ON CLIMATE CHANGE (IPCC) report may be under way, will have a range of consequences for the global CLIMATIC SYSTEM, for example for the frequency of STORMS, FLOODS and DROUGHTS, the MIGRATION of animal and plant species, the MASS BALANCE of GLACIERS and ICE SHEETS, and SEA-LEVEL CHANGE. The Earth–atmosphere–ocean system has always been subject to NATURAL ENVIRONMENTAL CHANGE, including CLIMATIC CHANGE, and CLIMATIC VARIABILITY. Reliable predictions of environmental change based on scientific assessments are therefore difficult to make. For example, the concentration of GREENHOUSE GASES has fluctuated through geological history and ICE CORE data obtained from Greenland and Antarctica suggest that the concentration of greenhouse gases has fluctuated in phase with global temperature changes during the last glacial/interglacial cycle (< 150 000 years).

In order to make meaningful PREDICTIONS of future environmental changes, it is of great importance to investigate different NATURAL ARCHIVES in order to study past changes in the Earth–atmosphere–ocean system. Investigations of global environmental changes in the past suggest that restricted areas may be critical in triggering large-scale changes. The build up and decay of the large QUATERNARY ice sheets, for example, were restricted to a few sensitive areas like Fennoscandia, North America, Greenland and Antarctica. The snow and ice ALBEDO *positive feedback* is of special interest in this case. Future predictions are, however, complicated by the HUMAN IMPACTS on several Earth systems, the rate of change of which may affect the capacity of these systems to recover. Long-range

INDUSTRIAL EMISSIONS have in some cases affected the most remote ECOSYSTEMS, for example in the Arctic and Antarctic. As a result, in some ENVIRONMENTAL SYSTEMS it is, therefore, difficult to know what the pre-industrial, 'NATURAL' conditions were like. Manipulation of RIVER DISCHARGE VARIATIONS, LAND DEGRADATION and FOREST CLEARANCE may individually have serious impacts on local and regional environments, and cause serious disturbance to the complex global environmental system as a result of cumulative smaller-scale changes.

The distinction between GLOBAL CHANGE and *global environmental problems* is important. Global changes are world-wide in character and must be studied at a global scale: climatic change is a good example. Global environmental problems, on the other hand, include regional phenomena, such as ACID RAIN from industrial POLLUTION, affecting large regions in Eurasia and North America. It is therefore necessary to study environmental changes at different scales – global, regional and local. The effects of environmental change on the Earth are also uneven spatially. The ability to make specific predictions of future environmental change is therefore a major limitation. ECOTONE boundaries that separate different vegetation zones at, for example, alpine or polar TREE LINES or the woodland/grassland transition in SAVANNA ECOSYSTEMS are examples of areas/zones sensitive to global warming. TROPICAL FORESTS, high-latitude TUNDRA soils and peat and the vast ocean carbon reservoir play key roles in this issue. In addition, the COASTLINE is vulnerable to possible future SEA-LEVEL RISE, for example MANGROVE SWAMPS, RIVER DELTAS and CORAL REEFS. The majority of the Earth's largest river deltas are cultivated and have dense populations. In addition to the projected global sea-level rise, these large river deltas are subject to natural and accelerated SUBSIDENCE related to GROUNDWATER abstraction and a decrease in FLUVIAL SEDIMENT influx due to upstream *dam* construction.

Global warming is, if not compensated by increased winter precipitation, likely to cause melting and retreat of minor VALLEY GLACIERS, mostly located in the Himalayas, Rocky Mountains, the European Alps, western Scandinavia, and New Zealand. Shrinking glaciers may put WATER RESOURCES and IRRIGATION SYSTEMS, intimately linked to glacier meltwater, at risk.

To understand the interaction between different components in the global environmental system, *modelling* provides the essential theoretical basis for understanding the complex processes involved in changes of the Earth's environmental system. Models in use are both conceptual and numerical and vary in spatial scale from local to global. Numerical computer models function over a wide range of scales, from local to global. Possible interactions between different components in the Earth–atmosphere–ocean system make the global models extremely complex. GENERAL CIRCULATION MODELS (GCMs) form the basis, for example, for the predictions of the IPCC relating to future greenhouse-gas-induced warming. All the GCMs predict future warming, but the magnitude of temperature rise varies between 1.9° and 5.2°C for a doubling of the CO_2 content. The different models do not show a consistent regional pattern, but they suggest that the land areas will warm more rapidly than the oceans and that the northern latitudes will warm more than the global mean, especially during winter. The inconsistent results

from the different GCMs are due to the fact that the models are oversimplifications of the real world and that the spatial resolution is too low to take account of the effects of topography. Non-climatic factors, such as the ocean heat flux, certainly affect the climatic system and should be included in model experiments. The effects of clouds are also difficult to model. Up to now, most effort has been spent on getting realistic models of modern climate and weather situations. In the future, the challenge will be to use COUPLED ATMOSPHERE–OCEAN MODELS with high spatial resolution to model climatic change or climatic conditions different from the prevailing modern situation. The combination of numerical modelling and empirical testing based on field evidence is an important way to achieving progress in the understanding of the complex relation between the different components of the Earth's environmental system.

Several international scientific research programmes have been launched to study the Earth's environmental systems. The International Geosphere-Biosphere Programme (IGBP) of the International Council of Scientific Unions (ICSU) was established to study the nature and dynamics of the global environmental system and the effect of human impact. This project has addressed several key questions about environmental change. *AN*

[See also FUTURE CLIMATE, HOLOCENE ENVIRONMENTAL CHANGE, QUATERNARY ENVIRONMENTAL CHANGE]

Alverson, K.D., Oldfield, F. and Bradley, R.S. (eds) 2000: Past global changes and their significance for the future. *Quaternary Science Reviews* 19(Special Issue), 1–5. **Bell, M.** and **Walker, M.J.C.** 1992: *Late Quaternary environmental change*. London: Longman. **Beniston, M.** 2000: *Environmental change in mountains and uplands*. London: Arnold. **Bradley, R.S.** 1999. *Paleoclimatology: reconstructing climates of the Quaternary*, 2nd edn. San Diego, CA: Academic Press. **Goudie, A.** 1992: *Environmental change*. Oxford: Clarendon Press. **Kemp, D.D.** 1994: *Global environmental issues: a climatological approach*. London: Routledge. **Lowe, J.J.** and **Walker, M.J.C.** 1997: *Reconstructing Quaternary environments*, 2nd edn. London: Longman. **Mannion, A.M.** 1991: *Global environmental change*. London: Longman. **Nesje, A. and Dahl, S.O.** 2000: *Glaciers and environmental change*. London: Arnold. **Roberts, N.** (ed.) 1994: *The changing global environment*. Oxford: Blackwell.

environmental chemistry The study of the nature and implications of chemical reactions in nature, including those involved in natural processes and the distribution and fate of CONTAMINANTS in the environment. *JAM*

[See also AIR POLLUTION, ARCHAEOLOGICAL CHEMISTRY, ECOTOXICOLOGY, PHOTOCHEMISTRY]

Baird, C. 1999: *Environmental chemistry*, 2nd edn. New York: Freeman. **Brimblecombe, P.** 1995: *Air composition and chemistry*, 2nd edn. Cambridge: Cambridge University Press. **Huztinger, O.** (ed.) 1990: *The handbook of environmental chemistry*. Berlin: Springer. **McBride, M.B.** 1994: *Environmental chemistry of soils*. Oxford: Oxford University Press.

environmental crisis A severe decline in biological PRODUCTIVITY and associated SOIL DEGRADATION resulting from extensive and persistent adverse human impacts on ecosystems and landscapes. Environmental crises threaten SUSTAINABILITY and may be evident in a deterioration in the health of human populations. Such crises represent a stage in the development of *creeping environmental problems* (CEPs) defined by Glantz (1999: iv) as 'long-term, low-grade, incremental but cumulative environmental problems'. Beyond *irreversibility thresholds*, environmental crises become *environmental catastrophes*, which are characterised by major disruptions to ecosystems and dependent economic and social systems.   *JAM*

[See also ENVIRONMENTAL ACCIDENT, ENVIRONMENTAL DISASTER]

Glantz, M.H. (ed.) 1999: *Creeping environmental phenomena in the Aral Sea Basin.* Cambridge: Cambridge University Press. **Saiko, T.** 2001: *Environmental crises: geographical case studies in post-socialist Eurasia.* Harlow: Pearson Education.

environmental degradation

A temporary or permanent decrease in the capability of the ENVIRONMENT to support the resource demands of the organisms that inhabit it. Such a broad definition is necessary to encompass the various elements of the environment that may be affected by degradation and to stress that the impacts of degradation are normally expressed in terms of the environment as a resource base for the biological population, including humans.

In terms of the Earth's surface, SOIL DEGRADATION and LAND DEGRADATION are the predominant forms. The former is expressed by a decrease in the productive capacity of the soil due to changes in the nutrient budget, erosion processes and chemical or biological attributes; the latter is a composite term to signify decreasing productivity of the land as a whole, thus implying a socio-economic dimension in parallel with environmental changes. SOIL EROSION, SALINISATION and DESERTIFICATION are common causes of land degradation.

In terms of the HYDROSPHERE, degradation is primarily in terms of the declining productivity potential of water bodies, owing to POLLUTION. ACIDIFICATION of lakes due to ACID RAIN and METAL POLLUTION caused by contaminated runoff from waste heaps and derelict smelting works are two examples; both are known to cause long-term productivity decreases in both floral and faunal assemblages. In addition, the pollution of water bodies used to provide drinking water is a form of degradation, because, without specialist treatment, polluted water is incapable of sustaining its current usage.

Similarly, AIR POLLUTION from a range of sources (including CO_2 and dust from the combustion of fossil fuels and carbon monoxide from vehicle exhaust emissions) has resulted in generally decreased *air quality*, which poses a threat to human health and causes productivity decline in the ecosystems of certain areas. The growing threat of GLOBAL WARMING caused by an enhanced GREENHOUSE EFFECT also has the potential to degrade the environment on a global scale. The shifting of vegetation zones, possible changes to world SEA LEVEL and modification of the HYDROLOGICAL and BIOGEOCHEMICAL CYCLES may locally change the suitability of the environment to particular purposes. Stronger zonal winds in response to steepened temperature gradients, for example, may cause accelerated soil erosion in susceptible areas such as the Chinese loess plateau, which would contribute markedly to soil and land degradation and cause a decrease in agricultural productivity.   *JGS*

[See also HUMAN IMPACT ON ENVIRONMENT]

Lowe, M.S. and Thompson, R.D. 1992: Pollution and development. In Bowlby, S.R and Mannion, A.M. (eds),

Environmental issues in the 1990s. Chichester: Wiley, 197–210. **Roberts, N.** 1996: The human transformation of the Earth's surface. *International Social Sciences Journal* 48, 493–510. **Runnels, C.** 1995: Environmental degradation in ancient Greece. *Scientific American* 272, 72–75. **Stocking, M.** 2000: Soil erosion and land degradation. In O'Riordan, T. (ed.), *Environmental science for environmental management*, 2nd edn. Harlow: Prentice Hall, 287–322.

environmental disaster

A relatively large-scale event, caused by human agency, that results in high levels of environmental damage over a large area. There are consequences for ECOSYSTEM HEALTH and possibly for human health. Examples include major OIL SPILLS, OIL FIRES and NUCLEAR ACCIDENTS.   *JAM*

[See also DISASTER, ENVIRONMENTAL ACCIDENT, NATURAL DISASTER]

environmental disobedience

A radical response, practised by some activists, to the ENVIRONMENTAL POLICIES that they consider unacceptable. Environmental disobedience is essentially illegal activity motivated by environmental concern. It may include protests, boycotts and civil disobedience, such as blockades and violence against people and property.   *JAM*

Hettinger, N. 2001: Environmental disobedience. In Jamieson, D. (ed.), *A companion to environmental philosophy.* Oxford: Blackwell, 498–509.

environmental economics

The study of environmental aspects of economics, especially the direct and indirect dependence of economies on ECOLOGICAL GOODS AND SERVICES supplied by the BIOSPHERE. Traditionally focused on 'economic' approaches, such as COST–BENEFIT ANALYSIS, some recognise the emergence of *ecological economics*, which focuses on more 'ecological' concepts such as ECOSYSTEM HEALTH and SUSTAINABILITY.   *JAM*

Turner, R.K., Pearce, D.W. and Bateman, I.J. 1994: *Environmental economics: an elementary introduction.* Hemel Hempstead: Harvester Wheatsheaf.

environmental energetics

The study of energy in the context of environmental systems. Concepts of ENERGY FLOW and ENERGY BALANCE, for example, are fundamental concepts in the understanding of patterns, processes and change throughout the Earth–atmosphere–ocean system, including the BIOSPHERE.   *JAM*

[See also ECOLOGICAL ENERGETICS]

Broda, E. 1975: *The evolution of the bioenergetic process.* Oxford: Pergamon. **Gregory, K.J. (ed.)** 1987: *Energetics of physical environment: energetic approaches to physical geography.* Chichester: Wiley. **Odum, H.T. and Odum, E.C.** 1976: *Energy basis for man and nature.* New York: Wiley. **Smil, V.** 1991: *General energetics: energy in the biosphere and civilisation.* Chichester: Wiley. **Warburton, J.** 1993: Energetics of alpine proglacial geomorphic processes. *Transactions of the Institute of British Geographers NS* 18, 197–206.

environmental ethics

The moral philosophical guiding principles of decision making in the fields of CONSERVATION and ENVIRONMENTAL MANAGEMENT. *Ethics* are a system of cultural values motivating people's behaviour, which draw upon human reasoning, morals, knowledge and goals to shape our worldview. Ethics operate at the level of individuals, institutions and societies, as well as internationally. Some blame western Judeo–Christian ethics for overexploitation of nature since Classical times (White, 1967). Interest in developing appropriate environmental ethics aiming for sustainable development has grown since the 1980s, especially as reflected in DEEP ECOLOGY and GREEN POLITICS. However, the ethical component of many environmental problems is often still avoided by environmental scientists, managers and public policy makers. *CJB*

Attfield, R. 1991: *The ethics of environmental concern*, 2nd edn. Athens, GA: University of Georgia Press. **Attfield, R.** 1999: *The ethics of the global environment*. Edinburgh: Edinburgh University Press. **Carley, M. and Christie, I.** 1992: *Managing sustainable development*. London: Earthscan. **Cheny, J.** 1989: Post-modern environmental ethics: ethics as bioregional narrative. *Environmental Ethics* 11, 117–134. **Dower, N. (ed.)** 1989: *Ethics and environmental responsibility*. Aldershot: Gower. **White Jr, L.** 1967: The historical roots of our ecologic crisis. *Science* 155, 1203–1207 [reproduced in Barr, J. (ed.), *The environmental handbook*. London: Ballantine/Friends of the Earth, 3–16].

environmental factors Conditions in the environment that influence life and/or soils. Examples include radiation and light, temperature, moisture, wind, fire, gravity, salinity, currents, plants, animals, microorganisms, topography, substrate and human activity. Environmental factors are commonly classified as ABIOTIC or BIOTIC, but this division is arbitrary and they are interactive. *RJH*

[See also ALLOGENIC, SOIL FORMING FACTORS]

environmental geology The scientific study of the interaction between humans and the geological environment. Four components may be identified: (1) managing the use of geological RESOURCES such as FOSSIL FUELS, including the organisation of their exploration and exploitation to minimise environmental impact; (2) adjusting planning, engineering and construction procedures (GEOENGINEERING) to the constraints of particular geological environments; (3) planning appropriate use of the geological environment for WASTE MANAGEMENT; and (4) identifying and quantifying the potential effects of NATURAL HAZARDS and, where possible, mitigating their effects (see ENVIRONMENTAL IMPACT ASSESSMENT). These are bold aims and their attainment would be difficult in a steady-state environment. In a changing environment most, if not all, become very difficult. Most flood prevention schemes, for example, rely on hazard maps based on FLOOD FREQUENCY analysis: if rainfall or runoff patterns change these become invalid. Understanding of environmental change is fundamental to environmental geology. *CDW*

Bennett, M.R. and Doyle, P. 1997: *Environmental geology: geology and the human environment*. New York: Wiley. **Pickering, K.T. and Owen, L.A.** 1997: *An introduction to global environmental issues*, 2nd edn. London: Routledge. **Smith, K.** 1993: Riverine flood hazard. *Geography* 78, 182–185.

environmental impact assessment (EIA) A process that seeks to improve a development through precautionary studies designed to identify potential unwanted and beneficial impacts. A wide variety of assessment methods may be used for an EIA, such as checklists, impact matrices and systems approaches. If applied carefully and early enough in the development process, an EIA can reduce mistakes and highlight the best alternatives for achieving a goal. It is therefore an important tool for SUSTAINABLE DEVELOPMENT in that it offers a way of avoiding damaging and possibly irreversible impacts. EIA should be a systematic, objective and independent approach, which improves policy making and planning. Some argue it is more than a planning and decision making aid: it can integrate environmental management with development; it can offer a way of involving the public in decision making; it can make planners and decision-makers more accountable for their actions, and therefore more careful. It involves value judgements and the integration of planning and politics, so it is unavoidably subjective and is not a precise science.

The origins of EIA lie with the 1969 United States National Environmental Policy Act, passed in 1970, which required its application to any federal development likely to affect the environment significantly. The Act also specified that the EIA should cover physical and socio-economic impacts and thus helped establish social impact assessment. It overlaps, uses or runs parallel with RISK ASSESSMENT, hazard assessment and *eco-audit*. From roughly the mid-1970s EIA has been required before the provision of US foreign aid funds. Project-focused EIA is increasingly applied at programme and policy levels.

The EIA process typically consists of: initial screening to determine whether it is required; scoping to decide the terms of reference (i.e. time available, funding, approach, etc.); identification of possible impacts; assessment of impacts to determine their significance; presentation of findings – which should flag problems, especially irreversible impacts and an assessment of available alternative ways of meeting development goals. Following presentation, the public may provide opinions that can be used in final decision making based on the assessment. The document (and possibly a public meeting) that presents the findings of an EIA is known as an *environmental impact statement* (EIS).

The EIA process should not give planners, decision-makers or the public a sense of false security, because some possible impacts are likely to be missed. In particular, there are problems with cumulative impacts that result from a number of causes, which may originate in different ways and interact after delays, and indirect impacts, which are felt at the end of a chain of causation that is difficult to trace. There is also a risk that the process may be mishandled, with the findings distorted by special-interest-group pressures, poor quality study or lack of adequate data. Without effective institutional support, planners and decision-makers may misuse or ignore an EIA. To combat these problems there is a trend toward assessors becoming independent accredited professionals, 'policed' by a body like the International Association for Impact Assessment, which has the power to strike off those who fail to provide an adequate standard of assessment.

As development proceeds, the situation changes; there is also likely to be local and global environmental and socio-economic change. A single 'snapshot-view' EIA is therefore unsatisfactory. There should be repetition to identify problems at construction, completion, ongoing activity, closure and decommissioning stages. EIA should therefore integrate with and help shape ongoing monitoring. In the European Union impact assessment was established by the 1985 'Environmental Impact Assessment

Directive'; however, it was not until 1992 that all European countries adopted its rules. EIA is now established for project planning and some countries are moving to adopt strategic environmental assessment procedures to try to integrate policy (global), programme (regional) and project (local) assessment.   *CJB*

[See also CLIMATIC IMPACT ASSESSMENT, ENVIRONMENTAL MANAGEMENT, ENVIRONMENTAL PROTECTION]

Barrow, C.J. 1997: *Environmental and social impact assessment: an introduction.* London: Arnold. **Gilpin, A.** 1995: *Environmental impact assessment (EIA): cutting edge for the twenty-first century.* Cambridge: Cambridge University Press. **Glasson, J., Therivel, R. and Chadwick, A.** 1994: *Introduction to environmental impact assessment.* London: UCL Press. **Wathern, P. (ed.)** 1988: *Environmental impact assessment: theory and practice.* London: University Hyman. **Wood, C.** 1995: *Environmental impact assessment: a comparative review.* Harlow: Longman.

environmental indicator   (1) Ideally, any measure of the 'integrity, stability and SUSTAINABILITY of the biological and physical environment, especially those aspects that can, like miners' canaries, warn of impending rapid change' (Berger, 1996: 3). Such 'state-of-the-environment indicators' take many forms (e.g. BIO-INDICATORS, ECOSYSTEM INDICATORS and GEOINDICATORS) and relate to the many important issues concerning both NATURAL ENVIRONMENTAL CHANGE and changes in the environment due to HUMAN IMPACTS on local, regional and global scales.

(2) More generally, the term is used for any biological or physical indicator of environmental conditions, either at present or in the past. INDICATOR SPECIES are often used in this way. In the PALAEOSCIENCES, environmental indicators provide the basis for inferring and reconstructing past environmental conditions.   *JAM*

[See also BIOMARKER, ECOLOGICAL INDICATOR, GLOBAL ENVIRONMENTAL CHANGE, INTERACTION INDICATOR, PALAEOENVIRONMENTAL INDICATOR, SEASONALITY INDICATOR, SYNTHESIS INDICATOR]

Berger, A.R. 1996: The geoindicator concept and its application: an introduction. In Berger, A.R. and Iams, W.J. (eds), *Geoindicators: assessing rapid environmental changes in Earth systems.* Rotterdam: Balkema, 1–14. **Environment Canada** 1991: *A report on Canada's progress towards a national set of environmental indicators* [*State of the Environment (SOE) Report* 91-1]. Ottawa: Environment Canada. **National Research Council** 1999: *Measures of environmental performance and ecosystem condition.* Washington, DC: National Academy Press.

environmental issue   A topic of concern to society because of the actual or potential HUMAN IMPACT ON ENVIRONMENT and/or the impact of the NATURAL ENVIRONMENT on human society. Examples include ACID RAIN, ALTERNATIVE ENERGY, CLIMATIC CHANGE, DESERTIFICATION, METEORITE IMPACT and NATURAL HAZARDS.   *JAM*

[See also ENVIRONMENTAL CHANGE]

Kemp, D.D. 1994: *Global environmental issues: a climatological approach.* London: Routledge. **Middleton, N.J.** 1999: *The global casino: an introduction to environmental issues,* 2nd edn. London: Arnold.

environmental law   The topics encompassed by this subject are very wide: it has been described as ranging from the street corner to the stratosphere. Law is also only one element of a major cross-disciplinary topic such as the

environment. A basic definition might cover laws concerning the air, space, water, land, plants and wildlife. Alternatively, the subject may be divided into two groups of laws: (1) those dealing with natural resource CONSERVATION; and (2) those dealing with HUMAN HEALTH issues involving human contact with POLLUTANTS in the air, water or land. The wide scope of the subject is also evidenced by the fact that even the Antarctic – which has been described as a land beyond normal law – is in fact governed by, for example, *international environmental law.*

There are diverse sources of environmental law at national, regional and international level. Most western countries have introduced legislation to protect the environment and some of these laws date back to the nineteenth century (e.g. in England and the USA). In the United States, environmental laws can be passed at the federal, state and local levels and are currently numbered in hundreds.

The GLOBALISATION of environmental problems was reflected by, for example, the United Nations Conference on Environment and Development held in Rio de Janeiro in June 1992 and other bodies such as the World Commission on Environment and Development (the Brundtland Commission), which contributed to the development of the concept of SUSTAINABLE DEVELOPMENT.

In Europe, the significance of European Community (EC) Law relating to the environment must be emphasised. The ENVIRONMENTAL POLICY of the European Communities (as set out in the EC Treaty, Article 174) pursues various objectives: preserving and protecting the quality of the environment; protecting public health; prudent and rational utilisation of NATURAL RESOURCES; and promoting measures at international level to deal with regional or world-wide environmental problems. This policy is based on three key principles, which have been applied by the Court of Justice of the European Communities, namely: precaution and prevention; rectifying environmental damage at source; and the *polluter-pays principle* – but this may be subject to exceptions, e.g. under EC rules on state aid and on environmental projects. Principles such as the 'polluter pays' may arguably be among the legal concepts that have been developed specifically in the context of environmental law.

In general, the scope of measures covered by environmental law may include laws relating to human health, such as measures concerning the quality of water intended for human consumption, urban wastewater treatment, the quality of bathing water, as well as laws concerned with other animal life, such as fish and shellfish breeding water, GROUNDWATER POLLUTION and MARINE POLLUTION. Relevant legislation also covers AIR POLLUTION generally, such as emissions from industrial plants, or air quality limits regarding EMISSION CONTROL of gaseous pollutants from diesel and other motor vehicle engines. Other laws concern INDUSTRY, including BIOTECHNOLOGY and chemical substances (such as the classification, packaging and labelling of HAZARDOUS WASTES or the prevention and reduction of environmental pollution from ASBESTOS), industrial risks and NOISE POLLUTION. The last includes the control of sound levels of motor vehicles, motorcyles, recreational craft (e.g. motorboats), subsonic aircraft and even the noise of lawnmowers and the airborne noise of household appliances and specific items of industrial and road-making machinery.

Within the EC, one of the most widely cited legal instruments relating to matters of the environment is Directive 85/337/EEC on the assessment of certain public and private projects on the environment, which includes a requirement for an ENVIRONMENTAL IMPACT ASSESSMENT. Laws also concern WASTE MANAGEMENT, notably measures for the disposal of toxic, hazardous (e.g. the Basel Convention on the control of transboundary movements of hazardous wastes and their disposal) and general waste, which fall generally within the ambit of public health measures.

Environmental law also covers ENVIRONMENTAL PROTECTION and the conservation of wild species, such as the Convention on International Trade in Endangered Species of Wild Fauna and Flora (concluded in Washington in 1973) and BIODIVERSITY (including issues concerning the deliberate release into the environment of GENETICALLY MODIFIED ORGANISMS). Laws have also been enacted concerning the protection of animal uses for experimental and other scientific purposes. Environmental law may also include issues of RADIOACTIVE WASTE and nuclear safety and civil protection, such as basic safety standards for health protection against ionising radiation.

Environmental law has grown considerably in recent times. The European Union's informal Network for the Implementation and Enforcement of Environmental Law (IMPEL) has been established to improve legal policy and implementation on the one hand and to enhance inspection, practical application and enforcement issues on the other. Another aspect of environmental law of topical interest is that of ENVIRONMENTAL LIABILITY (e.g. civil liability for damage caused by waste) and also issues of freedom of access to information on the environment. However, it has been pointed out that environmental law is a political discipline and is just one tool alongside others such as fiscal policy, education, research and voluntary solutions. *RMD*

Bell, S. 1998: *Environmental law*, 4th edn. London: Blackstone Press. **D'Sa, R.M.** 1998: *European Community law on state aid*. London: Sweet and Maxwell. **Epstein, S., Brown, L. and Pope, C.** 1982: *Hazardous waste in America*. San Francisco, CA: Sierra Club Books. **Findley, R.W. and Farber, D.A.** 1992. *Environmental law*, 3rd edn. St Paul, Minnesota: West. **Joyner, C.C.** 1999: *Governing the frozen commons: the Antarctic regime and environmental protection*. Charleston, SC: University of South Carolina Press. **Kramer, L.** 2000: *European Community environmental law*. London: Sweet and Maxwell.

environmental liability

Environmental liability aims at making the polluter pay for remedying the damage caused. Most states have laws that deal with liability for damage to goods or persons and many have introduced laws to deal with the liability for the clean-up of CONTAMINATED LAND. However, most national legal frameworks have not addressed the issue of liability to the whole environment. The introduction of liability for damage to BIODIVERSITY and the strengthening of laws relating to the contamination of land are likely to bring about improvements in POLLUTION prevention and environmental care. New laws will probably require greater disclosure of environmental liabilities and demand that companies begin declaring environmental liabilities that may be associated with their activities prior to the time that a plant is closed. As a result, many companies now have increased motivation to sell environmentally impaired sites. *MGC*

[See also ENVIRONMENTAL LAW]

Bartsch, E. 1999: *Liability for environmental damages: incentives for precaution and risk allocation*. Tubingen: JCB. Mohr. **Hollins, M. and Percy, S.** 1998: Environmental liability for contaminated land – towards a European consensus. *Land Use Policy* 15, 119–133.

environmental management

A field of study concerned with guiding human–environment interactions. More specifically, environmental management can be pursued as a multilayered process with different 'environmental managers' interacting with the environment and with each other; or as an approach co-ordinated by a single environmental manager, or more likely a well co-ordinated group. However it is practised, environmental management may be described as an approach to *environmental stewardship* that integrates ecology, policy making, planning and socio-economic development. It is a broad and ambitious discipline, which goes beyond natural resource management to encompass political and socio-economic issues and develop environmentally sound development strategies. *Natural resources management* focuses on specific components of the environment, mainly those with utility, and tends more to seek short-term gains for specific interests, often governments or companies. Environmental management is more proactive and precautionary, involves a wider range of disciplines, is increasingly holistic in approach and is likely to act on behalf of a much wider range of interests. Environmental management is more concerned with stewardship than exploitation and, because it faces environmental change and social issues, which are unpredictable, must be flexible and adaptive.

Environmental management overlaps and supports *environmental planning*, but the focus is on understanding human–environment interactions and on coping with problems, rather than theoretical planning. There are two extreme stances, anthropocentric and ecocentric. The first, places human welfare before that of the environment and biota; the second, if need be, places the environment above human welfare. Mainstream environmental management is mainly anthropocentric. The growth of environmental management was prompted by public concern in western countries since the 1960s over loss of BIODIVERSITY, perceived deterioration of environmental quality and fears of ecologically damaging accidents. The process of environmental management is: to identify problems; to determine goals; to decide on appropriate action; to implement action; and to put in place ongoing management procedures and monitoring. Many *environmental problems* appear suddenly or are insidious. Environmental managers should seek to warn of problem 'thresholds' in time for them to be avoided or for remedial action to be taken, and so prevent breakdowns that endanger people or biota, quality of life and SUSTAINABLE DEVELOPMENT. The discipline is still developing. Having developed in western democracies where there is relative freedom of information, reasonable access to data and fewer pressing problems of poverty manifest elsewhere in the world, environmental management has to be adapted to suit other political, socio-economic and environmental conditions.

Critics of environmental management argue that it is a poorly defined discipline, lacking established theory, that it is too prescriptive, is insufficiently analytical and is subjective and unscientific. Nevertheless, with increasing global change and growing numbers of people, some form of environmental management is vital; since 1972 most countries have established ministries or departments charged with the task (e.g. the United States Environmental Protection Agency). Increasingly, interest in environmental management is coming from within commerce and from non-governmental bodies. The United Nations has established the *UN Environment Programme* (UNEP) to oversee environmental care and the European Union has established an *Environmental Management and Audit System* (EMAS). *CJB*

[See also ENVIRONMENTAL IMPACT ASSESSMENT, ENVIRON-MENTAL PROTECTION]

Barrow, C.J. 1999: *Environmental management: principles and practice.* London: Routledge. **O'Riordan, T. (ed.)** 1997: *Environmental science for environmental management.* Harlow: Addison-Wesley Longman. **Owen, L. and Unwin, T.** 1997: *Environmental management: readings and case studies.* Oxford: Blackwell. **Theodore, L., Dupont, R.R. and Baxter, T.E.** 1998: *Environmental management: problems and solutions.* Boulder, CO and New York: CRC Press and Springer. **Wilson, G.A. and Bryant, R.L.** 1997: *Environmental management: new directions for the twenty-first century.* London: UCL Press.

environmental modelling The development and investigation of SIMULATION MODELS of environmental processes to allow the prediction of environmental SYSTEM behaviour under varying *scenarios*. In general, environmental systems and processes are highly complex in nature, often involving many interacting subsystems at different spatial or temporal scales. To predict the way that different parts of the environment may react to differ-ent scenarios of input conditions, aspects of HUMAN IMPACT and FORCING FACTORS, it is necessary to develop simplified MODELS that have none of the time, space or complexity constraints of the real system. These models can then be used as an analogue of the system to under-stand its current dynamic behaviour, test its future behav-iour and/or serve as an early warning for any potential detrimental outcomes.

Model types include those that are descriptive or con-ceptual, empirical or black-box, and theoretical. Environmental models generally fall into the latter two categories with the emphasis on *theoretical models* based on equations representing the relationships between system elements. These models are generally DETERMINISTIC MODELS, i.e. the model relationships are fixed and non-random, as opposed to stochastic or statistical models (see STOCHASTICITY) in which a random element is introduced to the system. *Empirical* or *semi-empirical models* are also used for environmental modelling where the relationship between system elements is too complex to derive deter-ministically, so data derived experimentally is used instead. For example, the height of a tree (which is diffi-cult to measure) is dependent on many complex factors related to its environment, but the diameter of the trunk (which is easy to measure) may be used in a regressive model to predict tree height.

Elements of an environmental model include *reservoirs* or *stocks* (e.g. carbon within tropical forests, heat energy in the world's oceans, number of individuals in an ecosystem population), *processes* or *activities* (e.g. exchange of carbon between vegetation and atmosphere), *rates* or *converters*

(e.g. the rate of deforestation in the tropics) and *interrela-tionships* (e.g. global temperature influences snow cover, which influences albedo, which influences global temper-ature). In all cases, the principle of conservation is used (e.g. carbon is neither gained by nor lost from the bios-phere, it is merely exchanged between the land, atmos-phere and oceans).

Environmental models are generally used in one of two modes. First, a *predictive model* may be employed to esti-mate the future value of a system variable (such as mean annual global temperature) based on current system para-meters and likely environmental change. Such predictions tend, at best, to be only as good as the data that goes into the model and hence the quality of measured or estimated inputs must be carefully considered. Second, many envi-ronmental models are also used to understand the dynam-ics of the system being modelled. Such '*system dynamic models*' are used to understand how one parameter is sen-sitive to changes in another (e.g. the population of foxes compared to the population of rabbits) and are generally only critically dependent on the description of the rela-tionships between system elements.

Environmental models vary in scale and scope from simple models of, for example, population dynamics in small ECOSYSTEMS (see BOOM AND BUST CYCLES) and GLAC-IER MODELLING, to global models of the flux of energy and matter within the whole ECOSPHERE. These latter models include GENERAL CIRCULATION MODELS and models of BIO-GEOCHEMICAL CYCLES, the CARBON CYCLE, the NITROGEN CYCLE and the HYDROLOGICAL CYCLE. Other large-scale processes that environmental modelling has provided an insight into include DESERTIFICATION, the THERMOHALINE CIRCULATION and the EL NIÑO–SOUTHERN OSCILLATION. Environmental modelling has been particularly influential in studies of GLOBAL WARMING, CLIMATIC CHANGE and ENVIRONMENTAL CHANGE since these processes have potentially large impacts on humans and modelling is the most appropriate method of predicting future behaviour. In these cases the systems in question are large and com-plex but on a global scale can generally be reduced to rel-atively simple models with measurable input and outputs. One example of how environmental modelling has influ-enced general thinking is that of NUCLEAR WINTER. Although there is no direct experience of such an event, models of atmospheric dust transport and energy exchange following a nuclear war have led to the wide-spread understanding of the ways in which all nations would suffer from such a conflict. *AJL*

Ford, A. 1999: *Modeling the environment.* Washington, DC: Island Press. **Harte, J.** 1988: *Consider a spherical cow.* Sausalito, CA: University Science Books. **Huggett, R.J.** 1993: *Modelling the human impact on nature.* Oxford: Oxford University Press. **Hulme, M., Barrow, E.M., Arnell, N.W.** *et al.* 1999: Relative impacts of human-induced climate change and natural climate variability. *Nature* 397, 688–691. **Hyde, W.T., Crowley, T.J., Baum, S.K. and Peltier, W.R.** 2000: Neoproterozoic 'snow-ball Earth' simulations with a coupled climate/ice-sheet model. *Nature* 405, 425–429. **Kirkby, M.J., Naden, P.S., Burt, T.P. and Butcher, D.P.** 1993: *Computer simulation in physical geogra-phy.* New York: Wiley. **Saether, B.E., Tufto, J., Engen, S.** *et al.* 2000: Population dynamical consequences of climate change for a small temperate songbird. *Science* 287, 854–856.

environmental monitoring The repeated data col-lection of environmental variables to understand ENVIRON-

MENTAL CHANGE and aid DECISION MAKING. Environmental monitoring is carried out by the United Nations, intergovernmental, governmental, non-governmental, academic and commercial organisations. It is gaining increasing world-wide importance through international environmental conferences like the Earth Summit in Rio de Janeiro 1992 and in Kyoto 1997. An example is the monitoring of global CARBON BALANCES/BUDGETS, especially absorption and release of CARBON DIOXIDE, which has been given a high priority by the United Nations Framework CONVENTION on Climate Change. GLOBAL CHANGE and its effects on human life are perhaps the greatest challenge in the new millennium. GREENHOUSE GASES and CLIMATIC CHANGE, stratospheric OZONE DEPLETION, POLLUTION of air, soil and water resources, tropical DEFORESTATION, DESERTIFICATION and BIODIVERSITY LOSS are the main causes of concern. Each of these phenomena is affected by processes on local and global scales. To describe, understand and predict these potentially threatening processes, repeated measurements on large scales are needed.

The central importance of environmental monitoring arises from its function as a key step in the iterative process of DECISION MAKING. Given the aim of a more SUSTAINABLE DEVELOPMENT, it is crucial to review continuously the effects of economic and political decisions on the environment in order to assess whether they have the desired impacts.

Techniques for continuous monitoring and prediction of environmental key variables (ENVIRONMENTAL INDICATORS) have been developed and are still being improved to tackle these problems. On national, continental and global scales, REMOTE SENSING plays a major role in delivering environmental data. In the past, remote sensing data were collected on coarse SPATIAL RESOLUTION, but advances in remote sensing techniques have made it possible to acquire data on continental scales with improved spatial and spectral resolution in an automated way. Particularly satellite remote sensing can be employed to derive key variables to monitor environmental change. Optical and microwave sensors are delivering complementary information on global processes. Among the optical sensors (OPTICAL REMOTE SENSING INSTRUMENTS), *Landsat Thematic Mapper* (TM), *Système pour l'Observation de la Terre* (SPOT) and the *Advanced Very High Resolution Radiometer* (AVHRR) of the National Oceanographic and Atmospheric Administration (NOAA) are the most important. The last of these provides daily coverage on a 1.1 km grid, acquiring five channels in the 580–12 500 nm ELECTROMAGNETIC SPECTRUM, i.e. thermal, infrared and visible light. Especially the red and infrared bands are widely used for VEGETATION monitoring, as they directly measure the amount of the PHOTOSYNTHETICALLY ACTIVE RADIATION absorbed by the VEGETATION CANOPY. The daily acquisition ensures availability of cloud-free images. Ehrlich *et al.* (1994) gave a review of AVHRR applications to environmental monitoring. Main applications include LAND COVER mapping, vegetation dynamics, monitoring of TROPICAL RAIN FOREST, vegetation PRODUCTIVITY estimation, fire risk assessment and biophysical parameter estimation. A widely used data product is the *normalised difference vegetation index* (NDVI), which is a linear combination of two sensor channels in the red and near-infrared spectrum. Correlations between NDVI and SPECIES RICH-

NESS of trees and perennial herbs, as well as with LEAF AREA INDEX, moisture content of the vegetation and BIOMASS have been found. NDVI composites are also used for continuously monitoring locust HABITATS in the Sahel, providing a means of PEST control and an early warning system. AVHRR data can furthermore be used for the estimation of greenhouse gas release from forest fires. However, environmental monitoring by repeated mapping has been found difficult, as IMAGE CLASSIFICATION techniques are generally not robust over time because of changing atmospheric and moisture conditions.

Studies specifically designed for change detection are generally more reliable than the direct repeated comparison of images. In the microwave spectrum (RADAR REMOTE SENSING INSTRUMENTS), the *European Research Satellites* ERS-1 and ERS-2, the Canadian *RADARSAT*, and the *Japanese Earth Resources Satellite JERS*-1 (data acquisition until October 1998: to be replaced by the *Advanced Land Observation Satellite*, *ALOS*, in 2003) are used for environmental monitoring. Because of their independence of cloud cover and daylight, active microwave sensors can provide images at any time and guarantee regular repetitions. Microwaves are able to penetrate the vegetation canopy to a WAVELENGTH-dependent extent and are used to derive structural and biophysical variables such as BIOMASS, soil moisture, forest type, tree age, Leaf Area Index and expected crop yield. Applications have been tropical deforestation monitoring, CARBON CYCLE quantification, global CLIMATIC CHANGE, LAND COVER mapping and the monitoring of SEA ICE, GLACIER VARIATIONS and FLOOD events. The new technique of SAR interferometry has significantly improved the applicability of microwave remote sensing to environmental problems. However, because of the dependence of BACKSCATTERING on moisture conditions, serious calibration problems have made direct comparison of repeated imagery difficult.

For monitoring pollutants in the environment on small scales, repeated assessments of BIO-INDICATORS can be used as an alternative to analytical chemistry. To assess the potential impacts of human activities on the environment, an ENVIRONMENTAL IMPACT ASSESSMENT can be carried out.

SIMULATION MODELS play an important role in predicting likely changes of environmental key variables. Different *scenarios* of complex environmental systems can be simulated by using different model parameters as input. Models may be used to temporally and spatially interpolate and extrapolate environmental measurements.

HB

[See also ENVIRONMENTAL PROTECTION, ECOTOXICOLOGY, ENVIRONMENTAL POLICY, ENVIRONMENTAL MANAGEMENT, TEMPORAL CHANGE DETECTION]

Belward, A.S. 1992: Spatial attributes of AVHRR imagery for environmental monitoring. *International Journal of Remote Sensing* 13, 193–208. **Burt, T.P.** 1994: Long term study of the natural environment: perceptive science or mindless monitoring. *Progress in Physical Geography* 18, 475–496. **Cowen, D.J., Jensen, J.R., Bresnahan, P.J. et al.** 1995: The design and implementation of an integrated geographic information-system for environmental applications. *Photogrammetric Engineering and Remote Sensing* 61, 1393–1404. **Eastwood, J.A., Plummer, S.E., Wyatt, B.K. and Stocks, B.J.** 1998: The potential of SPOT-vegetation data for fire scar detection in boreal forests. *International Journal of Remote Sensing* 19, 3681–3687. **Ehrlich, D., Estes, J.E. and**

Singh, A. 1994: Applications of NOAA-AVHRR 1 km data for environmental monitoring. *International Journal of Remote Sensing* 15, 145–161. **Hart, G.F. and Trinder, J.C.** 1991: ISPRS Commission VII Symposium – global and environmental monitoring – techniques and impact – 17–21 September 1990, Victoria, Canada. *ISPRS Journal of Photogrammetry and Remote Sensing* 46, 176–177. **Leshkevich, G.A., Schwab, D.J. and Muhr, G.C.** 1993: Satellite environmental monitoring of the Great-Lakes – a review of NOAA's Great-Lakes Coastwatch Program. *Photogrammetric Engineering and Remote Sensing* 59, 371–379. **Maselli, F., Conese, C., Petkov, L. and Gilabert, M.A.** 1992: Use of NOAA-AVHRR NDVI data for environmental monitoring and crop forecasting in the Sahel – preliminary results. *International Journal of Remote Sensing* 13, 2743–2749. **Phinn, S.R.** 1998: A framework for selecting appropriate remotely sensed data dimensions for environmental monitoring and management. *International Journal of Remote Sensing* 19, 3457–3463. **Rauste, Y., Herland, E., Frelander, H.** *et al.* 1997: Satellite-based forest fire detection for fire control in boreal forests. *International Journal of Remote Sensing* 18, 2641–2656. **Stow, D., Hope, A., Nguyen, A.T.** *et al.* 1996: Monitoring detailed land surface changes using an airborne multi-spectral digital camera system. *IEEE Transactions on Geoscience and Remote Sensing* 34, 1191–1203. **Tripathi, N.K., Venkobachar, C., Singh, R.K. and Singh, S.P.** 1998: Monitoring the pollution of river Ganga by tanneries using the multi-band ground truth radiometer. *ISPRS Journal of Photogrammetry and Remote Sensing* 53, 204–216.

environmental movement

Any notable upsurge in environmental awareness incorporated into political, social, economic, educational or recreational activity. The environmental movement of the 1960s culminated in the *Blueprint for Survival* and the Club of Rome's LIMITS TO GROWTH, with ensuing popular concern leading to the formation of several prominent PRESSURE GROUPS and European 'Green' political parties. An unprecedented upsurge in environmental awareness occurred in the late 1980s and early 1990s, with the promotion of sustainable, less damaging industrial practices of particular concern worldwide, mainly as a result of the growing prospect of GLOBAL WARMING and its associated effects.   *JGS*

[See also ENVIRONMENTAL CONSERVATION, ENVIRONMENTAL MANAGEMENT, GREEN POLITICS, SUSTAINABILITY]

Bowlby, S.R. and Lowe, M.S. 1992: Environmental and green movements. In Bowlby, S.R. and Mannion, A.M. (eds), *Environmental issues in the 1990s*. Chichester: Wiley, 161–174. **Connors, L. and Hutton, D.** 1999: *History of the Australian environmental movement*. Cambridge: Cambridge University Press. **Shabecoff, P.** 1993: *A fierce green fire: the American environmental movement*. New York: Farrar, Straus and Giroux.

environmental policy

A national environmental policy or strategy is made up of the stated opinions and the apparent attitude of a government and its officials towards ENVIRONMENTAL ISSUES. It is necessarily based on government statements, laws, regulations, guidelines and case law. It is also reflected in government spending priorities and in other policies that may make little direct mention of the environment, for example in transport policy. However, as public attitudes to GENETICALLY MODIFIED ORGANISMS and international opinions on nuclear power have shown, environmental policy development can be unclear and subject to strong influences from outside the national government.

The formal development of international environmental policy can be said to date from the *World Conservation Strategy* prepared by the International Union for the Conservation of Nature. Following the 1987 World Commission on Environment and Development (Bruntland Commission) the ideal of SUSTAINABLE DEVELOPMENT became the political orthodoxy. The definition from Bruntland was: 'development that meets the needs of the present without compromising the ability of future generations to meet their own needs'. Although the United States had developed ENVIRONMENTAL LAW with the National Environmental Policy Act (NEPA) in 1969, national governments in North America and Europe eventually prepared comprehensive environment strategies in the 1990s. In the UK, the policy titled *This Common Inheritance* referred to a 'moral duty to look after our planet and hand it on in good order to future generations'. The United Nations Conference on Environment and Development (Rio Conference) in 1992 was an opportunity for leaders and representatives from over 150 states collectively to adopt a declaration committing themselves to the principles of SUSTAINABILITY. Their *Programme of Action on Environment and Development (Agenda 21)* is a non-treaty agreement designed to serve as the basis for nationally oriented sustainable management policies. Other similar international policy coalitions in the environmental field have influenced environment policies, for example the 1987 Convention on the Protection of the Ozone Layer (Montreal Protocol).

United States and Canadian environmental policy has moved from the purely regulatory approach to an encouragement of market-based incentive systems and the uptake of comprehensive ENVIRONMENTAL MANAGEMENT systems. The European Community's *Fifth Environmental Action Plan* is the Community's programme for sustainable development up to the year 2000 and beyond. Environmental laws and regulations impact on national economic activity and business costs and must be implemented equally across borders to ensure fair competition. Hence, there is a need for general consensus and discussion over policy development. For example, the options for laws on ENVIRONMENTAL LIABILITY are contained in a European Commission White Paper for debate (February 2000).

Environmental policies are implemented at the local level and the UK strategy document *Sustainable Development* views the role of local authorities in Britain as crucial in putting the UK on a path towards sustainable development. It outlines the role and importance of local authorities in implementing the principles of Agenda 21. Here it is recognised that: 'as the level of governance closest to the people, they [local authorities] play a vital role in educating, mobilising and responding to the public to promote sustainable development'. Regional energy and environmental action policy initiatives for businesses can be delivered through a variety of advice schemes partially funded by central government, to complement the local authority work.

Most countries have a thriving voluntary sector of environment and CONSERVATION groups. Central and local government policies often support local environmental initiatives and many cities now have charitably funded Environment Centres to provide information, advice, education and training with the aim of encouraging mass participation in ENVIRONMENTAL PROTECTION, conservation and green purchasing initiatives.   *MGC*

[See also CONVENTIONS, ENVIRONMENTAL MOVEMENTS, ENVIRONMENTALISM, GREEN POLITICS, PROTECTED AREAS]

Caldwell, L.K. 1990: *International environmental policy: emergence and dimensions*, 2nd edn. Durham, NC: Duke University Press. **Commission of the European Communities** 1996: *Draft Council Resolution on Waste Policy. Commission of the European Communities paper COM(96) 399 final of 30.07.96.* Brussels: European Commission. **Department of the Environment** 1995: *Making waste work: a strategy for sustainable waste management in England and Wales, Cm 3040.* London: HMSO. **Her Majesty's Government** 1990: *This common inheritance: Britain's environmental strategy, Cm 1200.* London: HMSO. **Her Majesty's Government** 1994: *Sustainable development: the UK strategy, Cm 2426.* London: HMSO. **Her Majesty's Government** 1997: *This common inheritance: UK annual report 1997, Cm 3556.* London: HMSO. **International Union for the Conservation of Nature (IUCN)** 1980: *World conservation strategy: living resources for sustainable development.* Gland, Switzerland: IUCN. **Kirby, J., O'Keefe, P. and Timberlake, L.** 1996: *Sustainable development.* London: Earthscan. **Lester, J.M. (ed.)** 1994: *Environmental politics and policy*, 2nd edn. Durham, NC: Duke University Press. **O'Riordan, T.** 2000: *Environmental science for environmental management.* Harlow: Longman Scientific. **UNCED** 1992: *Earth summit '92.* London: Regency Press. **Vig, N.T. and Kraft, M.E. (eds)** 1994: *Environmental policy in the 1990s: towards a new agenda.* Washington, DC: CQ Press. **WCED** 1987: *Our common future.* Oxford: Oxford University Press.

environmental protection

The protection of human populations from POLLUTANTS and, more generally, the CONSERVATION or stewardship of natural resources. This function is carried out by environmental protection agencies (regulatory authorities) at various levels of government – local, national and international – and with varying degrees of integration, guided by environmental protection policies and regulations. Some regulations reduce *environmental nuisances* (such as NOISE POLLUTION) rather than HUMAN HEALTH HAZARDS; others may enhance aesthetic or historical value. *JAM*

[See also CONVENTIONS, PROTECTED AREAS]

environmental quality

'A measure of the condition of an environment relative to the requirements of one or more species and/or to any human need or purpose.' (Johnson *et al.* 1997: 586). Thus, mandatory or non-mandatory environmental quality standards may be established, in terms of *air quality*, *soil quality* or *water quality* for example, to avoid or limit harm to humans. These tend to be specific minimum concentrations of substances that are considered harmful. *JAM*

Curran, P.J. 1998: Environmental quality standard (EQS). In Calow, P. (ed.), *The encyclopedia of ecology and environmental management.* Oxford: Blackwell Science, 244. **Godish, T.** 1991: *Air quality.* Chelsea, MI: Lewis. **Gray, N.F.** 1994: *Drinking water quality.* London:Wiley. **Johnson, D.L., Ambrose, S.H., Bassett, T.J. et al.** 1997: Meanings of environmental terms. *Journal of Environmental Quality* 26, 581–589. **Maybeck, M., Chapman, D.V. and Helmer, R. (eds)** 1990: *Global freshwater quality: a first assessment.* Cambridge, MA: WHO/UNEP/Blackwell.

environmental science

The recently emerging, INTERDISCIPLINARY field of scientific study examining the complex interactions of human beings with the NATURAL ENVIRONMENT in which they live. Environmental scientists usually work in MULTIDISCIPLINARY teams to address specific problems caused by humans and are thus a response to the growth of ENVIRONMENTALISM in the late twentieth century. Because modern environmental problems cannot be satisfactorily remedied by the application of any one discipline, environmental science is based on a number of scientific disciplines (including chemistry, biology, physics, geography, geology, hydrology, ecology, meteorology and oceanography) and social-science disciplines such as economics and social policy. In this sense environmental science is a truly APPLIED SCIENCE driven by topics of public interest and concern (such as the release of unwanted processed materials into the environment) and fuelled by theoretical advances and insights from a wide array of traditional sciences. *JGS*

[See also ENVIRONMENTAL MOVEMENTS, ENVIRONMENTAL ETHICS, GREEN POLITICS]

Groot, W. de 1992: *Environmental science theory: concepts and methods in a one world, problem oriented paradigm.* Dordrecht: Kluwer. **Jorgensen, S.E.** 1989: *Principles of environmental science.* New York. Elsevier. **Mannion, A.M. and Bowlby, S.R. (eds)** 1992: *Environmental issues in the 1990s.* Chichester: Wiley. **O'Riordan, T. (ed.)** 2000: *Environmental science for environmental management*, 2nd edn. Harlow: Prentice Hall. **Trudgill, S.T. and Richards, K.S.** 1997: Environmental science and policy: generalizations and context sensitivity. *Transactions of the Institute of British Geographers NS* 22, 5–12. **Wakeford, T. and Walters, M. (eds)** 1995: *Science for the Earth: Can science make the world a better place?* New York: Wiley.

environmental security

Broad definitions of environmental security reflect two main conceptual views: first, that environmental security is about relative safety from environmental dangers, whether these are caused by natural or human processes or are due to ignorance, mismanagement or design. Second is the view that environmental security is the state of human-environment dynamics that includes ENVIRONMENTAL DEGRADATION, military damage or biological threats that could lead to social disorder and conflict. *Environmental threats* often involve cross-border or global impacts and require international co-operation by national governments. Environmental security and SUSTAINABLE DEVELOPMENT can be viewed as mutually reinforcing concepts. However, in contrast to the benign emphasis of sustainable development, environmental security focuses on the potential for conflict from environmental factors.

The expansion of national security to include non-military factors has been recognised for some time. Environmental degradation and resource shortages have affected relationships between states in the past. Poverty, food insecurity, poor health conditions, displacement and migration and the disruption of social and political institutions are regarded as the most important consequences of ENVIRONMENTAL STRESS. However, the great majority of these issues do not become security problems because they are resolved through compromise within a favourable political, economic and social context. It can be seen, for example, that a water problem between Turkey and Iraq is in a very different security context to a similar dispute between the USA and Mexico. By using contextual elements, it has been argued that deterioration of environmental conditions in the USSR was a significant factor in its break-up. Although democracies rarely go to war, the example of the 'cod war' between the UK and Iceland shows that resource disputes have the capacity to mobilise nationalistic feelings and mask political best interests.

The tracking and monitoring of both environmental and contextual indicators is essential in order to warn of the potential for conflict or analyse the potential for conflict escalation. Environmental stress poses a threat to security at all geographic levels, so taking preventive action is the most appropriate approach to reducing the potential for environmental conflicts. International agreements and regional environmental agreements could play an important role in preventing conflict. However, there is a need to consider how ENVIRONMENTAL POLICY at the national level can lead to security concerns. At present there is little evidence of the integration of environmental considerations and security concerns. Security institutions also need to develop a capacity to link their potential ability to respond to threat to the greater damage that conflict itself induces.

Environmental concerns are valuable topics for establishing dialogue and co-operation between security institutions. Collaboration on commonly shared environmental issues, such as CONTAMINATED LAND through defence activities, can open lines of communication across borders that might be valuable in times of stress. In the international context there is a need to develop methods of mediation and dispute resolution that can address environmental conflicts and produce regional agreements to reflect the political, economic, social and environmental context as a whole. Furthermore, once conflict has been prevented the post-crisis phase needs mechanisms to address and monitor changes in ENVIRONMENTAL FACTORS that could lead once again to stress and an escalation of the political temperature. *MGC*

[See also ENVIRONMENTAL LIABILITY]

Baechler, G. and Spillman, K.R. 1996: *Environmental degradation as a cause of war* [*Final report of the environment and conflict project ENCOP*]. Chur, Zurich: Ruegger. **Brundtland, G.H.** 1993: The environment, security and development. In *SIPRI yearbook 1993: world armaments and disarmament*. Oxford: Oxford University Press. **Duedney, D.** 1990: The case against linking environmental degradation and national security. *Millennium* 19, 461–476. **Gleik, P.H.** 1993: Water and conflict: fresh water resources and international security. *International Security* 18, 79–112. **Hillel, D.** 1994: *Rivers of Eden: the struggle for water and the quest for peace in the Middle East*. New York: Oxford University Press. **Homer-Dixon, T.F.** 1994: Environmental scarcities and violent conflict. *International Security* 19, 5–40. **Imber, M.F.** 1994: *Environment, security and UN reform*. New York: St Martin's Press. **Lietzmann, K.M. and Vest, G.D.** 1999: *Environment and security in an international context* [*Pilot study summary report*]. Brussels: NATO CCMS. **Myers, N.** 1993: *Ultimate security: the environmental basis of political stability*. New York: Norton.

environmental stress 'An action, agent or condition that impairs the structure or function of a biological system' (Cairns, 2001: 315). *JAM*

[See also DISTURBANCE, DISTURBED ECOSYSTEM, FOREST DECLINE, PERTURBATION]

Cairns Jr, J. 2001: Stress, environmental. In Levin, S.A. (ed.), *Encyclopedia of biodiversity*. Vol. 5. San Diego, CA: Academic Press, 515–522.

environmentalism A set of mediating values representing human conduct in relation to the environment and a core concept that creates critical connections between PHYSICAL and HUMAN GEOGRAPHY. In essence, environmentalism is concerned with the reflexive relationships between people and environment. A conceptual focus was the emergence of the idea of environmental or climatic DETERMINISM as part of the impact of DARWINISM in the latter part of the nineteenth century. The determinist position, in common with its possibilist and probabilist derivatives, fell into the trap of attempting to generalise from a false dichotomy when the nature of people–environment interactions was infinitely more various. Environmentalism has moved on to more tangible concerns with the need to conserve the frail balance of ECOSYSTEMS, to advance the concept of SUSTAINABILITY, to monitor and control the impact of human activities on LANDSCAPE and to find more efficient ways of predicting catastrophic change. Environmentalism has become part of the political agenda and the practice of geography has much to offer its understanding. Today, those calling themselves 'environmentalists' are a very diverse group, although there are theorists who suggest that a shift has taken place to *postenvironmentalism*. *DTH/CJB*

Goudie, A. and Viles, H. 1997: *The Earth transformed*. Oxford: Blackwell. **Nash, R.F.** 1990: *American environmentalism: readings in conservation history*. New York: McGraw Hill. **Pepper, D.** 1986: *The roots of modern environmentalism*. London: Routledge. **O'Riordan, T.** 1981: *Environmentalism*, 2nd edn. London: Pion. **O'Riordan, T.** 1996: Environmentalism on the move. In Douglas, I., Huggett, R. and Robinson, M., *Companion encyclopedia of geography: the environment and human kind*. London: Routledge, 449–476. **Simmons, I.G.** 1997: *Humanity and environment: a cultural ecology*. Harlow: Addison-Wesley Longman.

Eocene An EPOCH of the TERTIARY period, from 56.5 to 35.4 million years ago. *GO*

[See also GEOLOGICAL TIMESCALE]

eoliths Rock fragments produced by natural agency, which were mistaken for crude pre-Palaeolithic tools in the late nineteeth century. *JAM*

[See also GEOFACT]

Barnes, A.S. 1939: The differences between natural and human flaking on prehistoric flint implements. *American Anthropologist* 41: 99–112. **Warren, S.H.** 1905: On the origin of eolithic flints by natural causes, especially by the foundering of drifts. *Journal of the Royal Anthropological Institute* 35, 337–364.

eon A unit of GEOLOGICAL TIME representing a first-order subdivision of the GEOLOGICAL TIMESCALE. The history of the Earth can be divided into four eons: PRISCOAN, ARCHAEAN, PROTEROZOIC and PHANEROZOIC. The equivalent unit in CHRONOSTRATIGRAPHY is an *eonothem* (although by definition no rocks remain on Earth from the Priscoan eon). *GO*

[See also GEOCHRONOLOGY]

epeirogeny The warping of large areas (thousands of kilometres across) of the Earth's crust without significant deformation. *SHD*

[See also OROGENY]

Ollier, C.D. 1981: *Tectonics and landforms*. London: Longman.

ephemeral Short-lived, or every now and again; commonly referring to less than a year. *Ephemeral plants* can complete their life cycle in less than a year: some produce more than one generation in a year; in extreme cases, plants in DESERTS require only a few weeks to complete their life cycle after rain. *Ephemeral habitats* exist for a short period after DISTURBANCE. *Ephemeral streams* flow only temporarily during and after a rainstorm, typically for a matter of days. *JAM/RPDW*

[See also ARROYO, WADI]

Patton, P.C. and Schumm, S.A. 1981: Ephemeral stream processes: implications for studies of Quaternary valley fills. *Quaternary Research* 15, 24–43.

epicentre The point on the Earth's surface directly above the origin (*focus* or *hypocentre*) of an EARTHQUAKE. *DNT*

epidemic A large-scale OUTBREAK of a communicable DISEASE, such as typhoid or CHOLERA, that is characterised by a large number of cases affecting a wide geographical area in a short period of time. To qualify as an epidemic, an outbreak might be expected to reach regional scale, affecting hundreds to thousands of people over a period of weeks. This temporary prevalence differs from *endemic* diseases, which are continually present. When an epidemic reaches global, or at least continental proportions, the disease is termed *pandemic*. The Bubonic Plague (or Black Death) of the AD 1330s killed around 25 million people in Europe (one third of the population) in five years. More recently, the 1918 influenza pandemic left up to 50 million people dead world wide. Such DISASTERS are less likely in the modern world because of improved medical care, increased water sanitation and food hygiene and the existence of preventative measures such as vaccines. *MLW*

[See also AIDS, HUMAN HEALTH HAZARDS]

Post, J.D. 1985: *Food shortage, climatic variability and epidemic disease in pre-industrial Europe.* New York: Cornell University Press. **Scott, M.E. and Smith, G. (eds)** 1992: *Parasitic and infectious diseases.* New York: Academic Press.

epidemiology A branch of medical science that studies the distribution of DISEASE in a human population and the factors determining that distribution. Common study techniques include: using demographic surveys to identify which sections of the community are particularly affected by a disease; monitoring the progression of a disease over time in order to identify temporal variations in incidence or mortality; and monitoring the geographical patterns of OUTBREAKS. The results of epidemiological analyses can be used to identify susceptible sections of populations, and times of year or spatial locations that display increased disease prevalence. *MLW*

[See also EPIDEMIC, HUMAN HEALTH HAZARDS]

Learmouth, A. 1987: *Disease ecology: an introduction to ecological medical geography.* Oxford: Basil Blackwell. **Timmreck, T.C.** 1994: *An introduction to epidemiology.* Boston, MA: Jones and Bartlett.

epifauna Bottom-dwelling aquatic organisms (BENTHOS) that live on the surface of the SEDIMENT. They may have fixed (*sessile*) or free-living (*vagile*) life modes and various life habits, such as suspension-feeding, deposit-feeding, scavenging, predatory, herbivorous or, in the case

of plants, AUTOTROPHIC. Through the PHANEROZOIC, epifaunal life strategies developed a TIERING, or vertical spatial separation above the sediment–water interface, which reached its maximum of 1 m (stalked crinoids) from mid-PALAEOZOIC to MESOZOIC times. Ecological resource partitioning within suspension-feeding communities exploits the abundance of organic material in the boundary layer close to the sediment, where drag reduces current velocity. Sepkoski's Paleozoic Fauna was dominated by suspension-feeding epifauna, particularly brachiopods, corals, bryozoans and crinoids, but among the 'Modern Fauna' (since CRETACEOUS times) infaunal benthos, including suspension- and deposit-feeders, and predators, have been dominant. *LC*

[See also INFAUNA, TRACE FOSSIL]

Bottjer, D.J. and Ausich, W.I. 1986: Phanerozoic development of tiering in soft substrata suspension-feeding communities. *Paleobiology* 12, 400–420. **Sepkoski Jr, J.J.** 1984: A kinetic model of Phanerozoic taxonomic diversity. III Post-Paleozoic families and mass extinctions. *Paleobiology* 10, 246–267.

epigenetic An ICE WEDGE that grows in pre-existing PERMAFROST and is usually much younger than the host materials is said to be epigenetic. The wedge grows progressively wider rather than higher or deeper. *PW*

[See also SUPERIMPOSED DRAINAGE, SYNGENETIC, PERIGLACIAL]

epiphyte A plant that uses another as a physical support but, in contrast to a PARASITE, does not draw any nutrition from it. Many ferns, orchids and bromeliads are epiphytes; mistletoe species are parasites. Epiphytes reach their maximum diversity and abundance in TROPICAL RAIN FORESTS and MONTANE (cloud) FORESTS, where moisture conditions are particularly favourable. *JLI*

episodic events Related EVENTS that occur at irregular time intervals as distinct from regular, *periodic events*. Extreme FLOODS in DRAINAGE BASINS and the passage of a severe TROPICAL CYCLONE directly over a coastal or island locality are examples of episodic events. The comparative roles and rates of such episodic events and more frequent, so-called 'normal' events in influencing the geomorphological and ecological characteristics of landscapes and ecosystems form an important but often unresolved focus of research. Although regular PERIODICITIES have been widely sought in the study of ENVIRONMENTAL CHANGE, many such changes are too occasional, irregular, dissimilar and/or exhibit too much STOCHASTICITY to be regarded as truly periodic. *JAM/RPDW*

[See also MAGNITUDE-FREQUENCY CONCEPTS, EXTREME EVENTS]

epistemology Central issues in epistemology are the nature and derivation of knowledge, the scope of knowledge and the reliability of claims to knowledge. The term is usually contrasted with ONTOLOGY. Epistemology seeks to connect questions of content with structures of belief and issues of authority. There are many epistemologies within the environmental sciences. *Rationalists* have argued that ideas of reason intrinsic to the mind are the only source of knowledge. In contrast, *empiricists* argue that experience and observation are the primary source of our ideas and knowledge.   *ART*

[See also EMPIRICISM, RATIONALISM].

Dancy, J. 1985: *Introduction to contemporary epistemology*. Oxford: Blackwell. **Gregory, D.** 1978: *Ideology, science and human geography*. London: Hutchinson.

epoch   An interval of geological time: a subdivision of a PERIOD in GEOCHRONOLOGY; equivalent to a SERIES in CHRONOSTRATIGRAPHY.   *GO*

epsilon cross-bedding   A style of large-scale CROSS-BEDDING that occurs in solitary sets, usually 1–5 m thick, with differences in LITHOLOGY between foresets, and a PALAEOCURRENT direction (as indicated, for example, by CROSS-LAMINATION) approximately perpendicular to the dip direction of the epsilon foresets. It is attributed to LATERAL ACCRETION on the inner bank of a bend in a channel (e.g. migration of the *point bar* of a river meander) and is a key element in a FACIES MODEL for meandering river deposits in the geological record.   *GO*

Allen, J.R.L. 1965: The sedimentation and palaeogeography of the Old Red Sandstone of Anglesey, North Wales. *Proceedings of the Yorkshire Geological Society* 35, 139–185. **Collinson, J.D. and Thompson, D.B.** 1992: *Sedimentary structures*, 2nd edn. London: Chapman and Hall.

equifinality   The concept that identical outward form may be achieved by a variety of differing pathways or processes, and/or from differing initial conditions. Usually called equifinality in North America and the UK – often called *convergence* elsewhere. Equifinality has been an important concept in geomorphology where the explanation of landform(s) is central, but determination of a one-to-one relationship between form and process has been equivocal. Recent examinations of equifinality suggest that it stems largely from inadequacy or imprecision in the scientific approach, or may be seen as a reflection of disciplinary immaturity that will be accommodated by embracing non-linear *dynamical systems*. At present, model equifinality is an important issue; in this context the ability of different models with a wide variety of parameter sets to produce comparable goodness of fit to the available data hinders inference and prediction.   *CET*

Beven, K. 1996. Equifinality and uncertainty in geomorphological modelling. In Rhoads, B.L. and Thorn, C.E. (eds), *The scientific nature of geomorphology*. Chichester: Wiley, 289–313. **Culling, W.E.H.** 1987: Equifinality: modern approaches to dynamical systems and their potential for geographical thought. *Transactions of the Institute of British Geographers NS* 12, 57–72. **Haines-Young, R.H. and Petch, J.R.** 1983: Multiple working hypotheses: Equifinality and the study of landforms. *Transactions of the Institute of British Geographers NS* 8, 458–466.

equilibrium concepts   Equilibrium concepts in all disciplines derive from physics, excepting those having a purely mathematical origin. In dynamics, three 'conditions of equilibrium' underpin individual concepts, e.g. static, dynamic, rotational, neutral, stable, unstable and metastable equilibrium. However, dynamics is limited in requiring that each particle or molecule be treated discretely. Thermodynamics overcomes this limitation by treating groups of particles or molecules using ideas developed in statistical mechanics.

Analysis in both dynamics and thermodynamics requires prior identification of the SYSTEM. Virtually anything can be identified as a system: however, the system and its boundary must be predetermined so that it can be distinguished from its environment (that is everything outside the system by definition). The concept of a system is much older than the notion of GENERAL SYSTEM(S) THEORY.

The concepts of thermodynamic equilibrium and entropy were initially developed for ISOLATED SYSTEMS and depend on the principle of the conservation of energy, but they have been extended to CLOSED and OPEN SYSTEMS where the definition of equilibrium differs in each type. Thermodynamic studies focus on three broad zones: equilibrium, linear (near-to-equilibrium), and non-linear (far-from-equilibrium). Furthermore, thermodynamic equilibrium is defined in a number of ways, but the most widely applicable form is Boltzmann's statistical version founded on PROBABILITY theory. In an isolated system thermal equilibrium, the most probable statistical state and the ATTRACTOR state are synonymous. In a closed system true thermal equilibrium is unattainable and the attractor state is defined in terms of minimum free energy, or as the statistical tendency to minimise deviations from the most probable macrostate. Open systems are inherently in non-equilibrium but evolve towards a non-equilibrium stationary state in the linear zone and may exhibit many stationary states in the non-linear zone. Non-linear behaviour of an open system implies order rather than disorder and it is open-system behaviour far-from-equilibrium to which some forms of CHAOS theory pertain.

Various orders of difference equations may produce exotic behaviour and have generated an equilibrium vocabulary including stable points, stable cycles, and STRANGE ATTRACTORS. This behaviour is purely mathematical in origin.   *CET*

Gleick, J. 1987: *Chaos: making a new science*. New York: Penguin Books. **May, R.M.** 1976: Simple mathematical models with very complicated dynamics. *Nature* 261, 459–467. **Prigogine, I. and Stengers, I.** 1984: *Order out of chaos*. Toronto: Bantam Books. **Thorn, C.E. and Welford, M.R.** 1994: The equilibrium concept in geomorphology. *Annals of the Association of American Geographers* 84, 666–696.

equilibrium concepts in ecological and evolutionary contexts   The concept of equilibrium is closely connected with that of CLIMAX VEGETATION. It involves the idea that, through ECOLOGICAL SUCCESSION, a stable community will develop. This view has been increasingly questioned, initially through the INDIVIDUALISTIC CONCEPT and more recently through the growing recognition of both the role played by DISTURBANCES in ecosystems and evidence from PALAEOECOLOGY for the short-lived nature of many modern species combinations. The idea of an equilibrium remains popular in other contexts, for example in relation to the GAIA HYPOTHESIS. Such an approach is essentially HOLISTIC. Different types of equilibrium are possible (e.g. dynamic, static), with some forms allowing for progressive changes in the SYSTEM through time. On both ecological and evolutionary time-scales, equilibrium has been advocated in relation to ISLAND BIOGEOGRAPHY. In this case, the equilibrium is achieved between colonisation (and speciation) and extinction rates on islands. Such a view appears today as rather simplistic, with the processes occurring on islands being extremely complex.   *JLI*

[See also EQUILIBRIUM CONCEPTS, GENERAL SYSTEMS THE-
ORY, STABILITY CONCEPTS]
Davis, M.B. 1986: Climatic instability, time lags and commu-
nity disequilibrium. In Diamond, J. and Case, T.J. (eds),
Community ecology. New York: Harper and Row. **Foster, D.R.,
Schoonmaker, P.K. and Pickett, S.T.A.** 1990: Insights from
palaeoecology to community ecology. *Trends in Ecology and
Evolution* 5, 119–122. **Whittaker, R.J.** 1998: *Island biogeography:
ecology, evolution, and conservation*. Oxford: Oxford University
Press.

equilibrium concepts in geomorphological and landscape contexts

A system or feature (such as a
river, landform, slope profile or landscape) is said to be in
a state of equilibrium when it exhibits relatively stable
characteristics, to which it tends to return after DISTUR-
BANCE of less than a critical (THRESHOLD) magnitude or
frequency. Equilibrium is not necessarily a static state and
processes (such as slope erosion and sediment transport)
may be very active, but it is considered that there is an
overall balance between form over a period and the factors
and processes controlling it.

The concept has a long history, particularly in slope and
river studies. In some schemes of LANDSCAPE EVOLUTION
certain features of slope form, such as the scarp and pediment slope segments, are maintained at the same equilibrium angle during most of landscape reduction. In river
studies, the concept developed from earlier ideas of GRADE
and the achievement of a smooth concave longitudinal
profile. Equilibrium concepts became more widely used as
the SYSTEMS approach was increasingly adopted across
PHYSICAL GEOGRAPHY from the late 1960s. Chorley and
Kennedy (1971) distinguished and defined a variety of
equilibrium states. An alternative scheme, identifying a
range of equilibrium, disequilibrium and non-equilibrium
types, was produced by Renwick (1992). A modified version combining elements of both schemes is given in the
Figure.

Static equilibrium (Type 1) describes a situation where
the system or feature of interest does not change appreciably over a reasonable length of time; slope profile form
within some landscapes may constitute an example. *Stable
equilibrium* (left-hand part of Type 3) is somewhat similar,
but in this situation the system has a capacity for recovery
to the same equilibrium state following any disturbances
to the system. An example might be beach and nearshore
profiles within a bay; these may be radically altered by the
waves of a severe storm with abnormal wind direction, but
return quickly to the normal pattern in the weeks and
months following the storm. In many cases, particularly
with river channels, a more realistic situation is a *steady-
state equilibrium* (Type 2), in which an equilibrium can
only be recognised as an average state but where features
alter with individual process events around that mean
condition. Thus, provided the controlling variables governing the magnitude-frequency of discharge, sediment
supply and sediment transport remain constant in the
mean, natural rivers may develop characteristic or equilibrium forms that are recognisable as statistical averages and
are related to the control variables; examples are meander
wavelength, riffle-pool spacing and channel dimensions
and form.

Metastable equilibrium (Type 3 right) describes a systems
change in which a disturbance or EXTREME EVENT exceeds
a threshold magnitude or character, leading to an irreversible change in the value of the parameter or system in
question. An example would be the increase in drainage
density that a gullying episode might produce following
deforestation; once formed, a gullying network tends to
remain or maintain itself long after the landscape might
have stabilised with revegetation by pasture or reversion to
forest.

The concepts of *dynamic equilibrium* and *dynamic
metastable equilibrium* (Types 4 and 5) incorporate the
ideas of steady-state equilibrium and long-term progressive change (in the case of dynamic equilibrium) and in
addition a threshold change (in the case of metastable
dynamic equilibrium). Classifying these types as '*equilibrium*' has proven somewhat controversial and they could
be viewed, using the terminology of Renwick (1992), as
forms of '*disequilibrium*', because a progressive change in
mean value or state is involved.

Renwick (1992) has also suggested various types of
'*non-equilibrium*' state, ranging from a system or landform
lurching from one state to another in response to *episodic
threshold events* (Type 8) to a completely *chaotic state* (Type
9). What constitutes non-equilibrium and equilibrium,
however, is often unclear and will vary with the timescale
envisaged by, and the perception of, the worker. What
constitutes acceptable variations around a steady state is
rarely specified. It has been suggested that mathematical
criteria should be applied, but very often information on
variability is restricted to human timescales and is insufficient to define objective criteria.

There are also important conceptual issues. Type 6
shows schematically a temporal pattern in which *occasional episodic events* play an important part in governing a
system or feature, but where between events the system or
feature tends towards contrasting characteristics. There
are many examples in both geomorphology and ecology of
this type of situation. In fluvial geomorphology, Werritty
and Ferguson (1981) demonstrated how BRAIDING in the
Feshie tributary of the River Spey in Scotland is dependent for its survival on occasional extreme floods. They
demonstrated that an extreme flood in AD 1960 led to a
great increase in braiding index, but that lesser, but still
competent, flood events in the preceding (1946–1960)
and succeeding (1960–1976) periods led to a progressive
simplification of the drainage system, a reduction in the
braiding index and increased meandering of component
braids. The braided river pattern is characterised therefore
by occasional increases in braiding separated by long periods of decline. Bayliss-Smith (1988) views the dynamics
of CORAL REEF ISLANDS in a similar manner, with MOTUS
and CAYS in complex and delicate relationships with TROP-
ICAL CYCLONE frequency. Tropical cyclones are seen as
essential for the growth or maintenance of motus via
storm ridge development, with lesser wave activity
between cyclones leading to beach erosion. If tropical-
cyclone frequency declines, erosion phases dominate and
islands may disappear. Cays, although reduced in area by
erosion during cyclones, are dependent upon cyclones for
the production of coral debris for their construction
between cyclones; a decline in cyclone frequency could
therefore starve them of sediment. Finally, tropical rain
forests affected by occasional tropical cyclones or drought
(with or without fire) may also be characterised by sudden
impacts separated by periods of recovery, with some fea-

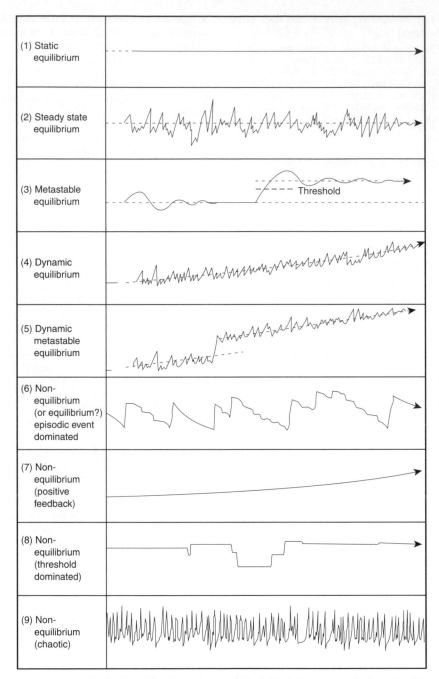

equilibrium concepts in geomorphological and landscape contexts *A graphical representation of various equilibrium concepts (based on Chorley and Kennedy, 1971; Renwick, 1992)*

tures (such as size distribution, structure and species composition) dependent on such occasional extreme events for their survival. Whether such situations are viewed as equilibrium or non-equilibrium is in the eyes of the beholder.

Kennedy (1994) views the future of the term – like that of its predecessors, grade and PENEPLAIN – as limited. This is in part because systems are so often disturbed by changes in climate, geological movements and base level that attaining equilibrium is an unrealistic idea. In addition, there are the problems of the unavailability of objec-

tive time-series data and a lack of consensus on appropriate timescales and mathematical criteria for distinguishing different types of equilibrium, disequilibrium and non-equilibrium states.   *RPDW*

[See also MAGNITUDE-FREQUENCY CONCEPTS]

Bayliss-Smith, T.P. 1988: The role of hurricanes in the development of reef islands. *Geographical Journal* 154, 377–391. **Brokaw, N.V.L. and Walker, L.R.** 1991: Summary of the effects of Caribbean hurricanes on vegetation. *Biotropica* 23, 442–447. **Chorley, R.J. and Kennedy, B.A.** 1971: *Physical geography: a systems approach*. London: Prentice-Hall. **Howard, A.D.** 1988:

Equilibrium models in geomorphology. In Anderson, M.G. (ed.), *Modelling geomorphological systems*. Chichester: Wiley, 49–72. **Kennedy, B.A.** 1992: Hutton and Horton: views of sequence, progression and equilibrium in geomorphology. *Geomorphology* 5, 231–250. **Kennedy, B.A.** 1994: Requiem for a dead concept. *Annals of the Association of American Geographers* 84, 702–705. **Renwick, W.H.** 1992: Equilibrium, disequilibrium, and non-equilibrium landforms in the landscape. *Geomorphology* 5, 265–276. **Schumm, S.A. and Lichty, R.W.** 1965: Time, space and causality in geomorphology. *American Journal of Science* 263, 110–119. **Thorn, C.E. and Welford, M.R.** 1994: The equilibrium concept in geomorphology. *Annals of the Association of American Geographers* 84, 666–696. **Walsh, R.P.D.** 1996: Drought frequency changes in Sabah and adjacent parts of northern Borneo since the late nineteenth century and possible implications for tropical rain forest dynamics. *Journal of Tropical Ecology* 12, 385–407. **Walsh, R.P.D.** 1996: Climate. In P.W. Richards (ed.), *The Tropical Rain Forest*, 2nd edn. Cambridge: Cambridge University Press, 159–206. **Walsh, R.P.D. and Newbery, D.M.** 1999: The ecoclimatology of Danum, Sabah, in the context of the world's rainforest regions, with particular reference to dry periods and their impact. *Philosophical Transactions of the Royal Society of London* B354, 1869–1883. **Werrity, A. and Ferguson, R.I.** 1981: Pattern changes in a Scottish braided river over 1, 30 and 200 years. In Cullingford, R.A., Davidson, D.A. and Lewin, J. (eds), *Timescales in geomorphology*, Chichester: Wiley, 53–68.

equilibrium concepts in soils

In an open system, a state may exist where inputs of energy and matter are balanced by outputs (a *steady-state equilibrium*). Receipt of organic matter at the soil surface can reach equilibrium with organic breakdown such that a surface humus horizon is present. It has been shown that organic matter in cultivated soils is gradually reduced until it reaches equilibrium at around 1%. After a change of environment during PEDOGENESIS, a soil will reach equilibrium in its upper horizons first, retaining evidence of previous conditions in the subsoil. Many northern European and North American soils, for example, reveal CRYOTURBATION in their subsoils, a relic from the last cold phase of the Pleistocene. Moreover, some processes, such as LEACHING of carbonates, phosphorus depletion or iron and aluminium accumulation, may tend towards a *terminal state* rather than a steady state. Thus, for some soil properties, and for soils in extreme cold desert environments in particular (where soil properties may require > 250 000 years to equilibrate), the concept of equilibrium has little value.

EMB/JAM

Bockheim, J.G. 1990: Soil development rates in the Transantarctic Mountains. *Geoderma* 47, 59–77. **Johnson, D.L. and Watson-Stegner, D.** 1987: Evolution model of pedogenesis. *Soil Science* 143, 349–366. **Messer, A.C.** 1988: Regional variations in rates of pedogenesis and the influence of climatic factors on moraine chronosequences, southern Norway. *Arctic and Alpine Research* 20, 31–39. **Smeck, N.E., Runge, E.C.A. and Mackintosh, E.E.** 1983: Dynamics and genetic modelling of soil systems. In Wilding, L.P., Smeck, N.E. and Hall, G.F. (eds), *Pedogenesis and soil taxonomy I. Concepts and interactions*. Amsterdam: Elsevier, 51–81.

equilibrium line

The line on a GLACIER separating the ACCUMULATION ZONE from the ABLATION ZONE and marking where ANNUAL accumulation and ablation are equal. The *equilibrium-line altitude* is a sensitive indicator of CLIMATIC CHANGE (see EQUILIBRIUM-LINE ALTITUDE (ELA) RECONSTRUCTION). The *steady-state equilibrium-line altitude* delimits the elevation where the *glacier net balance* is zero. The *climatic ELA* is the annual mean ELA over a standard 30-year period.

AN

Benn, D.I. and **Evans, D.J.A.** 1998: *Glaciers and glaciation*. London: Arnold.

equilibrium-line altitude (ELA) reconstruction

Processes influencing the equilibrium-line altitude (ELA) on GLACIERS commonly involve ABLATION (mainly determined by summer temperature) and ACCUMULATION (winter snow precipitation). In addition, however, wind transport of dry snow is an important factor for the glacier MASS BALANCE. On plateau glaciers, snow deflation and drifting dominate on the windward side, while snow accumulates on the leeward side. By calculating the mean ELA in all glacier quadrants, the influence of wind on plateau glaciers can be neglected. The resulting ELA is therefore defined as the TP-ELA (*temperature–precipitation ELA*). The TP-ELA reflects the combined influence of the regional ablation-season temperature and accumulation-season precipitation.

In deeply incised cirques and valleys surrounded by wide, wind-exposed mountain plateaux the snow may deflate from the plateaux and accumulate in the cirques and valleys, either by direct accumulation on the cirque/valley glaciers or by avalanching from the mountain slopes. This may thereby increase significantly the accumulation on the cirque/valley glaciers, where the ELA is commonly influenced by wind-transported snow, giving the TPW-ELA (*temperature–precipitation–wind ELA*). Thus the TPW-ELA is commonly lower than the TP-ELA.

The most common approaches in reconstructing palaeo-ELAs are to use: (a) the *maximum elevation of lateral moraines* (MELM); (b) the *median elevation of glaciers* (MEG); (c) the *toe-to-headwall altitude ratio* (THAR); and (d) the ACCUMULATION AREA RATIO (AAR). Because of the nature of glacier flow towards the centre and the margin of the glacier above and below the ELA, respectively, lateral moraines are theoretically only deposited in the ablation zone below the ELA. As a result, the maximum elevation of lateral moraines reflects the corresponding ELA. The AAR is normally based on the assumption that the steady-state AAR of former glaciers is approximately 0.6.

ELA reconstruction can be the key to reconstructing records of both summer temperatures and winter precipitation, provided one of the parameters can be estimated from independent evidence.

AN

[See also HOLOCENE ENVIRONMENTAL CHANGE]

Dahl, S.O. and **Nesje, A.** 1992: Palaeoclimatic implications based on equilibrium-line altitude depressions of reconstructed Younger Dryas and Holocene cirque glaciers in inner Nordfjord, western Norway. *Palaeogeography, Palaeoclimatology, Palaeoecology* 94, 87–97. **Lowe, J.J.** and **Walker, M.J.C.** 1997: *Reconstructing Quaternary environments*. London: Longman. **Nesje, A.** 1992: Topographical effects on the equilibrium-line altitude on glaciers. *GeoJournal* 27.4, 383–391. **Paterson, W.S.B.** 1994: *The physics of glaciers*. Oxford: Pergamon. **Sutherland, D.G.** 1984: Modern glacier characteristics as a basis for inferring former climates, with particular reference to the Loch Lomond Stadial. *Quaternary Science Reviews* 3, 291–309.

equinox

One of two periods each year when the noon-day sun is directly overhead at the equator and astronomical day and night are equal in length at all latitudes. It occurs on or about 21 March (spring or vernal equinox) and 22 September (autumnal equinox). *JBE*

[See also PRECESSION]

era A unit of GEOLOGICAL TIME comprising several PERIODS (e.g. MESOZOIC era). The equivalent unit in CHRONO-STRATIGRAPHY is an *erathem*. *GO*

[See also GEOLOGICAL TIMESCALE]

erg An Arabic word referring to an extensive tract of sandy desert deeply covered with shifting sand and commonly occupied by a variety of complex SAND DUNES, especially in the Sahara. Major examples include the Iguidi Erg (south of Morocco), the Western Erg (south of Algeria) and the Eastern Erg (adjoining Tunisia). Ergs are more generally known as SAND SEAS. Many modern ergs appear to represent different generations of dune deposits that have accumulated episodically in response to periods of increased ARIDITY and/or SEA-LEVEL CHANGE during the Quaternary. *MAC*

[See also PALAEODUNES]

Lancaster, N. 1995: *Geomorphology of desert dunes.* London: Routledge. **Stamp, L.D. (ed.)** 1961: *A glossary of geographical terms.* London: Longman.

ergodic hypothesis A methodological assumption, widely used in the natural environmental sciences, that sampling of patterns in space can be a basis for the inference of changes through time. Fundamental to the so-called *geographical* or *comparative method*, such *space-for-time substitution* is one approach to investigating those environmental changes (such as ECOLOGICAL SUCCESSION, LANDSCAPE EVOLUTION and SOIL DEVELOPMENT) that operate on timescales too long for the application of direct observation and experiment. Problems include variations in initial conditions and differences in environmental histories between sites, establishing site age and the risk of circular argument. *JAM*

[See also CHRONOSEQUENCE]

Paine, A.D.M. 1985: 'Ergodic' reasoning in geomorphology: time for a review of the term? *Progress in Physical Geography* 9, 1–15. **Pickett, S.T.A.** 1988: Space-for-time substitution as an alternative to long-term studies. In Likens, G.E. (ed.), *Long-term studies in ecology: approaches and alternatives.* New York: Springer, 110–135. **Savigear, R.A.G.** 1952: Some observations on slope development in South Wales. *Transactions of the Institute of British Geographers* 18, 31–51.

erodibility The degree to which a material is susceptible to EROSION. The erodibility of, for example, a soil will depend on its resistance to both detachment and transport. Soils or sediments of fine sand or SILT texture, or soils with low organic-matter content, are generally the most erodible. The disturbance or removal of vegetation generally leads to an increased erodibility of the soil. *SHD*

[See also AGGREGATE STABILITY, EROSIVITY, SOIL EROSION, WATER EROSION, WIND EROSION]

Bryan, R.B. 2000: Soil erodibility and processes of water erosion on slopes. *Geomorphology* 32, 385–415. **Rowell, D.L.** 1994: *Soil science: methods and application.* Harlow: Longman.

erosion (1) The group of processes by which material is removed from any part of the Earth's surface by water, wind, ice, snow or other agents. It differs from WEATHERING in that material is transported away from the site. Erosion at a particular site is defined with reference to coordinates fixed beneath the surface so that it can be directly and locally measured.

(2) In REMOTE SENSING and IMAGE PROCESSING, erosion is one of the basic operations on which mathematical morphology is based. When binary images are being processed, where '1' is seen as white and '0' as black, erosion consists of reducing (eroding) whitespots. *SHD/ACF*

[See also DENUDATION, ERODIBILITY, EROSIVITY, GLACIAL EROSION, HEADCUT EROSION, INTERRILL EROSION, SOIL EROSION, WATER EROSION, WIND EROSION]

erosion pin A rod of rigid plastic, steel or wood inserted and anchored in soil or SEDIMENT, so that it can be used as a stationary reference point from which the EROSION or DEPOSITION of material around the pin can be measured. *SHD*

[See also EROSION PLOT, GERLACH TROUGH, MICRO-EROSION METER, SEDIMENT TRAP, SOIL MICRO-PROFILING DEVICE]

Haigh, M.J. 1977: The use of erosion pins in the study of slope evolution. *British Geomorphological Research Group, Technical Bulletin* 18, 31–49.

erosion plot A physically isolated piece of land of known size, slope and soil characteristics from which SOIL LOSS is monitored. A standard size is 22 m long by 1.8 m wide, although other plot sizes are used. Erosion, *runoff* or *bounded plots* are very useful in determining soil losses under different LANDUSES and/or land management scenarios. They do not, however, always provide reliable estimates of soil losses at slope or DRAINAGE BASIN scales, since the plot boundary interferes with processes operating over the entire slope (e.g. *overland flow* and THROUGHFLOW). *SHD*

[See also EROSION PIN, GERLACH TROUGH, SEDIMENT TRAP, SOIL MICRO-PROFILING DEVICE]

Shakesby, R.A., Walsh, R.P.D. and Coelho, C.O.A. 1991: New developments in techniques for measuring soil erosion in burned and unburned forested catchments. *Zeitschrift für Geomorphologie Supplementband* 83, 161–174. **United States Department of Agriculture** 1979: *Field manual for research in agricultural hydrology,* USDA Agricultural Handbook. US Department of Agriculture, 224.

erosion rate The rate at which material is removed from the Earth's surface by water, wind, snow, ice or other agency. Contemporary erosion rates can be monitored at points or for areas. On soil-mantled slopes, for example, EROSION PINS, SOIL MICRO-PROFILING DEVICES, or the projection of tree roots above the surrounding ground provide techniques for monitoring erosion at points. On rock surfaces, a MICRO-EROSION METER can be used to provide comparable data. Areal rates on soil-mantled slopes may be monitored by, for example, EROSION PLOTS, GERLACH TROUGHS or CAESIUM-137); and, for drainage basins, areal rates may be determined using the SUSPENDED SEDIMENT and BEDLOAD of streams. Rates of erosion have also been determined for agents other than WATER EROSION on hillslopes. For example, rates of GLACIAL ABRASION have been assessed by recovering overridden rock and metal plates attached at an earlier date to abraded rock surfaces.

Erosion rates on vertical surfaces (e.g. river banks, cliffs) have been determined using, for example, erosion pins, automatic electronic distance measurement devices, maps or plans of different dates, (see DOCUMENTARY EVIDENCE) and recent terrestrial or AERIAL PHOTOGRAPHY.

For longer-term assessments of rates of lowering of the ground surface, the depth of sediment at a point or series of points above a chronostratigraphic marker (see CHRONOSTRATIGRAPHY) or datable material have been used, as has the volume of sediment in a SEDIMENT TRAP, assuming its contributing area and starting date as a sediment sink are known (see Figure). It may be possible to derive approximate erosion rates if there are reasonable time constraints for (1) the removal by erosion for a calculable volume of rock or deposit or (2) for the construction of a depositional LANDFORM for which the source area and age of its constituent sediments can be determined. An example of the former is provided by estimates of the amount of PLEISTOCENE glacial erosion involved in excavating U-SHAPED VALLEYS along preglacial river courses. The derivation of erosion rates for a CIRQUE glacier based on GLACIAL MORAINE volume provides an example of the latter. *RAS*

[See also DENUDATION RATES]

Biersman, P.R. 1994: Using in-situ produced cosmogenic isotopes to estimate rates of landscape evolution – a review from the geomorphic perspective. *Journal of Geophysical Research – Solid Earth* 99, 13885–13896. **Boulton, G.S.** 1974: Processes and patterns of glacial erosion. In Coates, D.R. (ed.), *Glacial geomorphology*. London: Allen and Unwin, 41–87. **Clayton, K.M.** 1996: Quantification of the impact of glacial erosion on the British Isles. *Transactions of the Institute of British Geographers NS* 21, 124–140. **Golonbek, M.P. and Bridges, N.T.** 2000: Erosion rates on Mars and implications for climatic change: constraints from the Pathfinder landing site. *Journal of Geophysical Research – Planets* 105, 1841–1853. **Goudie, A.** 1995: *The changing Earth: rates of geomorphological processes*. Oxford: Blackwell. **Hughes, R.J., Sullivan, M.E. and Yok, D.** 1991: Human induced erosion in a Highlands catchment in Papua New Guinea: the prehistoric and contemporary records. *Zeitschrift für Geomorphologie Supplementband* 83, 227–239. **Judson, S.** 1968: Erosion rates near Rome, Italy. *Science* 160, 1444–1445. **Nesje, A., Dahl, S.O., Valen, V. and Øvsterdal, J.** 1992: Quaternary erosion in the Sognefjord drainage basin, western Norway. *Geomorphology* 5, 511–520.

erosion surface   A flattish plain or so-called *planation surface* (PEDIPLAIN, PENEPLAIN or ETCHPLAIN) produced by EROSION. Identification of their remnants in the modern landscape forms a major part of the study of DENUDATION CHRONOLOGY. *RAS*

[See also ALTIMETRY, DAVISIAN CYCLE OF EROSION]

Adams, G.F. (ed.) 1975: *Planation surfaces: peneplains, pediplains and etchplains*. Stroudsburg, PA: Dowden, Hutchinson and Ross.

erosivity   The ability of an eroding agent (e.g. wind or water) to erode. For example, rainstorms of high intensity and/or long duration often have a higher erosivity than low-intensity and/or short-duration rainfall events. SOIL EROSION is considered to be a function of both erosivity and ERODIBILITY. *SHD*

Leek, R. and Olsen, P. 2000: Modelling climatic erosivity as a factor for soil erosion in Denmark: changes and temporal trends. *Soil Use and Management* 16, 61–65. **Morgan, R.P.C.** 1995: *Soil erosion and conservation*, 2nd edn. Harlow: Longman.

erratic   A fragment of rock transported from its source by a GLACIER or ICE SHEET and deposited in an area of differing geology (from Latin, *errare* meaning 'to go astray'). A so-called *erratics train* forms a fan-shaped distribution of erratics aligned in the direction of ice flow down-ice of individual source rock outcrops. Erratics are particularly useful when reconstructing the extent and flow patterns of former glaciers or ice sheets. The Foothills erratics train in the USA extends over 1000 km in the Rocky Mountain valleys and demonstrates Cordilleran ice sheet flow between the LAURENTIDE ICE SHEET and the Rocky Mountain Foothills. It represents a medial MORAINE and helps in reconstructing ice-sheet coalescence, critical in understanding the timing of MIGRATION of plants and

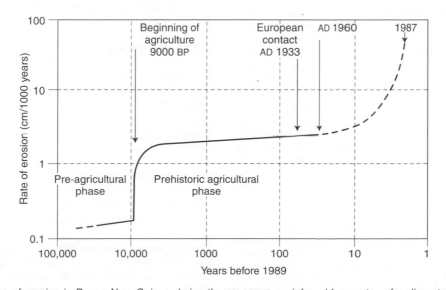

erosion rate *Rates of erosion in Papua New Guinea during the* HOLOCENE *as inferred from rates of sedimentation in Kuk Swamp (after Hughes* et al., *1991)*

animals during DEGLACIATION. Since erratics trains can provide tangible, regional representations of ice flow, they may be useful in validating ice-sheet reconstructions. They are also used in mineral prospecting in formerly glaciated areas. Recognition of the glacial origin of erratics was seminal in the rise of the GLACIAL THEORY. *JS*

[See also GLACIAL SEDIMENTS]

Agassiz, L. 1842: On glaciers and the evidence of their having once existed in Ireland and England. *Proceedings of the Geological Society of London* 3, 327–332. **Prest, V.K.** 1990: Laurentide ice flow patterns: a historical review, and implications of the dispersal of Belcher Island erratics. *Géographie Physique et Quaternaire* 44, 113–136. **Salonen, V.P.** 1987: Observations on boulder transport in Finland. *Geological Survey of Finland*, Special Paper 3, 103–110. **Shakesby, R.A.** 1979: Glacial dispersal of rock fragments and mineral grains from two point sources. *Zeitschrift für Gletscherkunde und Glazialgeologie* 15, 31–45. **Stalker, A.MacS.** 1956: The erratics train, foothills of Alberta. *Geological Survey of Canada, Bulletin* 37, 5–28.

esker An elongate winding ridge of glaciofluvial sand and gravel formed as infill to an ice-walled river channel and deposited by subglacial, englacial or supraglacial meltwater streams. The word is an anglicised version of the Irish *eiscir*, meaning ridge. *JS*

Bannerjee, I. and McDonald, B.C. 1975: Nature of esker sedimentation. In Joplin, A.V. and McDonald, B.C. (eds), *Glaciofluvial and glaciolacustrine sedimentation* [*SEPM Special Publication* 23]. Tulsa, OK: Society of Economic Paleontologists and Mineralogists, 132–154.

establishment, agencies of Those agencies (for example, fire or human disturbance) that *affect* or alter local substrate and light conditions, so making conditions suitable for a plant taxon's germination and growth. *FMC*

[See also ECESIS, SPREAD, VECTORS OF]

Chambers, F.M. 1993: Late Quaternary climatic change and human impact: commentary and conclusions. In Chambers, F.M. (ed.), *Climate change and human impact on the landscape.* London: Chapman and Hall, 247–259.

estuarine environments The transition zones between rivers and the ocean within which the river broadens and discharges into the sea and the TIDES operate in diurnal cycles, mixing saline and fresh water. Estuarine environments are very dynamic, being continually modified by EROSION and DEPOSITION in the short term and in the long term by, for example, sea-level fluctuations, tectonic activity and glaciation. Estuaries are usually characterised by a distinct ocean inlet, which defines the point of entry of incoming tides. The Severn Estuary, UK, and the Bay of Fundy, Nova Scotia, Canada, have the highest tidal ranges in the world caused primarily by topographic factors, especially their funnel-like shape. Estuarine environments exhibit steep sedimentary, salinity, hydrochemical, hydrophysical and ecological gradients over a relatively short distance. Water circulation and fluctuating salinity maintain highly productive ecosystems at a variety of scales. Estuarine plant assemblages (ranging from microscopic algae to seagrasses) provide abundant food and shelter to an equally productive and diverse fauna. Many estuarine ecosystems are strictly protected and monitored to ensure that conflicting usage by fishing and shipping industries, for example, are not detrimental to the ecosystem. Estuaries are nevertheless often the sites of high-level MARINE POLLUTION. *JGS*

[See also AQUATIC ENVIRONMENT, COASTAL ZONATION, TIDES]

Dyer, K.R. 1986: *Coastal and estuarine sediment dynamics.* Chichester: Wiley. **Grant, A. and Jickells, T.** 2000: Marine and estuarine pollution. In O'Riordan, T. (ed.), *Environmental science for environmental management*, 2nd edn. Harlow: Prentice Hall, 263–282. **Kennish, M.J.** 1992: *Ecology of estuaries: anthropogenic effects.* Boca Raton, FL: CRC Press. **McLusky, D.S.** 1990: *Estuarine ecology.* Glasgow: Chapman and Hall.

etchplain A type of EROSION SURFACE, often INSELBERG-studded, found particularly in shield areas of the seasonal tropics. Etchplains are believed to develop in the long term by both deep CHEMICAL WEATHERING and stripping, which operate as a DOUBLE PLANATION SURFACE. The feature was originally described as an 'etched plain' by Wayland, but the concept has been developed by Büdel. Erosion only operates on the upper soil surface of most of the plain, so that the basal and wash surfaces have different functions: SAPROLITE development at the basal surface of weathering (or WEATHERING FRONT) and soil erosion at the *wash surface*. Weathering is active all year at the basal surface, which remains moist throughout the dry season, and is intensified by the high levels of soil CO_2 and the occurrence of organic acids. GRUNDHOECKER typify the relief of the basal surface and larger rock masses form both inselbergs and more extensive rock pavements. Isolated CORESTONES can be detached by progressive lowering of the weathering surface as weathering penetrates from wash depressions (DAMBOS). According to Büdel, it is this characteristic etchplain–inselberg relief that forms the prime evidence for the hypothesised gradual landsurface reduction or *etchplanation*. *MAC*

Büdel, J. 1982: *Climatic geomorphology*, translated by Fischer, L. and Busche, D. Princeton, NJ: Princeton University Press. **Thomas, M. F., Thorp, M. and McAlister, J.** 1999: Equatorial weathering, landform development and the formation of white sands in north western Kalimantan, Indonesia. *Catena* 36, 205–232. **Thomas, M.F.** 1994: *Geomorphology in the tropics: a study of weathering and denudation in low latitudes.* Chichester: Wiley.

ethno-archaeology Studying contemporary societies in order to understand the production and use of their MATERIAL CULTURE so that the necessary knowledge is available for the interpretation of past cultures in archaeological contexts. *JAM*

[See also ANTHROPOLOGY, ETHNOGRAPHY]

Binford, L.R. 1978: *Nunamiut ethnoarchaeology.* New York: Academic Press.

ethnography The anthropological study of contemporary societies and cultures, and a basis for the ethno-archaeological interpretation of prehistory. *JAM*

[See also ANTHROPOLOGY, ETHNO-ARCHAEOLOGY]

ethological (behavioural) adaptations Behavioural (rather than structural) characteristics that have improved the chances that an organism will leave descendants. *KDB*

[See also ADAPTATION]

Euclidean distance measure A dissimilarity index that measures the distance between two points in Euclidean space calculated using Pythagoras' theorem. In the two-dimensional case, it is given by the square root of the sum of the squares of the separations in the two dimensions. *PJS*

euphotic zone Depths in water bodies at which light penetration is sufficient for PHOTOSYNTHESIS to take place. The euphotic zone is usually 50 to 150 m thick, and shallowest in CONTINENTAL SHELF areas, where sediment concentrations in the water column are greatest. *BTC*

[See also APHOTIC ZONE, LAKE STRATIFICATION AND ZONATION]

eury- A prefix used to describe organisms that have a broad tolerance for one or more environmental factors. *JLI*

[See also STENO-]

eustasy The shape and level of the surface of the world's oceans in equilibrium with the gravitational field of the Earth. Rise and fall in this surface may be due to water volume changes (mainly as a result of GLACIO-EUSTASY), to changes in the shape of the equipotential surface (GEOIDAL EUSTASY) and to changes in the shape of ocean basins (*tectono-eustasy*). The best-known *eustatic changes* are *glacio-eustatic changes* caused by the addition of water to and removal of water from the total ocean volume as a result of climate induced variations in the dimensions of the world's GLACIERS and ICE SHEETS. *Geoidal-eustatic changes* are less well known and arise because the surface of the world's ocean is uneven as a result of regional variations in the Earth's gravitational field. The surface of oceans represents an equipotential surface of the Earth's gravitational field, known as the GEOID.

Tectono-eustatic changes are long-term alterations in the shape of ocean basins as a result of plate motion (see PLATE TECTONICS). For example, it has been argued that as a result of SEA-FLOOR SPREADING, ocean basins have been widening at an average rate of 16 cm y⁻¹ during the late QUATERNARY. If correct, this would mean that the volume of the world's ocean basins has increased by 6% since the LAST INTERGLACIAL. The situation is complex, however, since it can be argued that during the Quaternary, the total amount of global sea-floor spreading has been offset by reductions in ocean volume caused by SUBDUCTION processes. *AGD*

[See also ISOSTASY, MEAN, ABSOLUTE AND RELATIVE SEA LEVEL, SEA-LEVEL CHANGE]

Bard, E., Hamelin, B., Arnold, M. *et al.* 1996: Deglacial sea-level record from Tahiti corals and the timing of global meltwater discharge. *Nature* 382, 241–244. **Dawson, A.G.** 1992: *Ice Age Earth: Late Quaternary geology and climate.* London: Routledge. **Fairbridge, R.W.** 1983: Isostasy and eustasy. In Smith, D.E. and Dawson, A.G. (eds), *Shorelines and isostasy.* London: Academic Press, 3–28. **Harmon, R.S., Mitterer, R.M., Kriansakal, N.** *et al.* 1983: U-series and amino-acid racemisation geochronology of Bermuda: implications for eustatic sea-level fluctuations over the past 250,000 years. *Palaeogeography, Palaeoclimatology, Palaeoecology* 44, 41–70. **Mörner, N.-A.** 1976: Eustasy and geoid changes. *Journal of Geology* 84, 123–152.

eutrophic Nutrient-rich; used with reference to river, lake and estuarine waters when they are enriched in nitrogen and/or phosphorus. Eutrophic waters are often associated with high levels of PRIMARY PRODUCTIVITY and depleted dissolved oxygen. *ADT*

[See also EUTROPHICATION, OLIGOTROPHIC]

eutrophication The enrichment of waters by inorganic plant nutrients, usually nitrogen and phosphorus, resulting in an increase in PRIMARY PRODUCTIVITY. Although the process occurs naturally as lakes age, accelerated nutrient loading to freshwater bodies may result from inputs of DOMESTIC WASTE or AGRICULTURAL WASTE or from events such as forest fires. Inputs can be from either *point sources* (such as sewage outfalls or storm drains) or *diffuse sources* (originating from agricultural fields and arriving in streamwater via *overland flow* or THROUGH-FLOW). The problem is exemplified by the Great Lakes of North America since the AD 1970s and extends to enclosed seas, such as the Mediterranean, and ESTUARINE ENVIRONMENTS.

PHOSPHORUS (P) is usually the LIMITING FACTOR to algal growth in freshwater because the relative proportion of available P in water is less than that in ALGAE. Phosphorus from detergents often makes up over half the P in sewage EFFLUENT and because it is in an immediately available form it can lead to rapid increases in algal productivity. Agricultural sources of P tend to have a less immediate impact on productivity because P is largely insoluble and adsorbed onto eroded sediments. Once in the water, however, it may subsequently become available for utilisation by algae. In contrast, NITRATE is soluble and easily leached from agricultural fields (see LEACHING, NITRATE POLLUTION). For example, agriculture is the source of 71% of the nitrogen (N) load in the River Great Ouse, England, but only 6% of the P (the remainder is from sewage).

Water affected by eutrophication is characterised by an initial increase in plant and animal BIOMASS, but a decrease in SPECIES DIVERSITY. This is associated with an increase in TURBIDITY, SEDIMENTATION rate and ultimately the development of ANOXIC conditions because of the depletion of dissolved oxygen. These changes cause a number of problems associated with water supply, recreation and the management of watercourses. *ADT*

[See also ALGAL BLOOM, WATER POLLUTION]

Ashworth, W. 1986: *The late, Great Lakes: an environmental history.* New York: Knopf. **Chapman, D.V.** 1997: Eutrophication. In Brune, D., Chapman, D.V., Gwynne, M.D. and Pacyna, J.M. (eds), *The global environment: science, technology and management.* Weinheim: VCH, 532–549. **Damiani, A.** 1992: Aspects of eutrophication in the Venice Lagoon. *Journal of the Institution of Water and Environmental Management* 6, 62–72. **Hanson, S. and Ruslam, L.G.** 1990: Eutrophication and Baltic fish communities. *Ambio* 19, 123–125. **Harper, D.** 1992: *Eutrophication of freshwaters.* London: Chapman and Hall. **Mason, C.F.** 1996: *Biology of freshwater pollution.* Harlow: Longman, Chapter 5 (Acidification). **Nixon, S.W.** 1990: Marine eutrophication: a growing international problem. *Ambio* 19, 101. **Smith, S.J., Sharpley, A.N. and Ahuja, L.R.** 1993: Agricultural chemical discharge in surface runoff. *Journal of Environmental Quality* 22, 474–449. **Sutcliffe, D.W. and Jones, J.G. (eds)** 1992: *Eutrophication: research and applications to water supply.* Ambleside: Freshwater Biological Association. **Vollenweider, R.A.** 1968: *Scientific fundamentals of the eutrophication of lakes and flowing water with special reference to nitrogen and phosphorus as fac-*

tors in eutrophication. Paris: Organisation for Economic Co-operation and Development.

evaporation The process by which a liquid changes into a gas. It depends upon the vapour pressure of the air, its temperature, wind conditions and the nature of the ground surface. Evaporation involves absorption of energy from the environment (LATENT HEAT), which is released when the process is reversed (CONDENSATION). *JGT*

[See also EVAPOTRANSPIRATION]

evaporite A MINERAL, SEDIMENT or SEDIMENTARY ROCK that forms by chemical precipitation upon the evaporation of salty water (a *salt*). As sea-water evaporates, a sequence of evaporite minerals forms with increasing salinity. First to precipitate is *calcite* ($CaCO_3$), which makes up 0.3% of the total precipitates. This is followed by calcium sulphate, as *anhydrite* ($CaSO_4$) and GYPSUM ($CaSO_4.2H_2O$), which accounts for 3.5% of the precipitates; then *halite* or *rock salt* (NaCl), which makes up 78% of the precipitates; and finally a variety of potassium salts, which make up the remaining 18%. It has been estimated that complete evaporation of the OCEANS would yield an evaporite deposit with an average thickness of 60 m, but many evaporite successions in the GEOLOGICAL RECORD exceed 1000 m in thickness and are dominated by gypsum or anhydrite: some mechanism must have operated to keep the brines concentrated to form such deposits. Models include: evaporation from a shallow embayment that is continually replenished with sea-water of normal salinity; repeated evaporation of a marine basin, as may have happened during the MESSINIAN salinity crisis in the Mediterranean; SABKHA processes on arid coastlines; or evaporation from SALINE LAKES (e.g. Dead Sea) or *playas*. Evaporite minerals also form within SOILS (see DURICRUSTS). Evaporite deposits in the geological record are good indicators of a palaeoclimate that was hot and arid (see PALAEOCLIMATOLOGY). Many evaporite minerals are exploited as RESOURCES with important industrial applications and evaporite deposits commonly trap accumulations of PETROLEUM. Evaporites in the subsurface are liable to solution by GROUNDWATER and areas where evaporites form part of the local rock succession experience problems of ground collapse or SUBSIDENCE. *GO*

[See also SEDIMENTOLOGICAL EVIDENCE OF ENVIRONMENTAL CHANGE]

Hardie, L.A. 1996: Secular variations in seawater chemistry: an explanation for the coupled secular variation in the mineralogies of marine limestones and potash evaporites over the past 600 my. *Geology* 24, 279–283. **Hsü, K.J.** 1972: Origin of saline giants: a critical review after the discovery of the Mediterranean evaporite. *Earth-Science Reviews* 8, 371–396. **Kendall, A.C.** 1992: Evaporites. In Walker, R.G. and James, N.P. (eds), *Facies models: response to sea level change.* St John's, Newfoundland: Geological Association of Canada, 375–409. **Testa, G. and Lugli, S.** 2000: Gypsum-anhydrite transformations in Messinian evaporites of central Tuscany (Italy). *Sedimentary Geology* 130, 249–268.

evapotranspiration The EVAPORATION of water from the Earth's surface plus *transpiration* from vegetation, represented either by actual observed amounts (*actual evapotranspiration*) or the POTENTIAL EVAPOTRANSPIRATION. *JCM*

event An intermittent, short-lived process, commonly of high magnitude or energy, with characteristics that vary markedly from normal background conditions. On a geological timescale events are instantaneous, synchronous happenings, although their actual duration may be hours or days. *Physical events* are 'short (hours to days), usually rare intervals of rapid deposition within a system of relatively slow background sediment accumulation' (Einsele *et al.*, 1996). Examples include river FLOODS, STORMS, *hurricanes*, TSUNAMIS, some SEA-LEVEL CHANGES, EARTHQUAKES, VOLCANIC ERUPTIONS and METEORITE IMPACTS. Their products are EVENT DEPOSITS. *Chemical events* include oxygen events in the ʼATMOSPHERE and salinity events in the OCEANS: these may be more subtly represented in the stratigraphical record. *GO*

[See also EVENT STRATIGRAPHY, NEOCATASTROPHISM]

Ager, D.V. 1993: *The nature of the stratigraphical record*, 3rd edn. Chichester: Wiley. **Dressler, B.O., Grieve, R.A.F. and Sharpton, V.L. (eds)** 1994: Large meteorite impacts and planetary evolution. *Geological Society of America Special Paper* 293. **Einsele, G., Ricken, W. and Seilacher, A. (eds)** 1991: *Cycles and events in stratigraphy.* Berlin: Springer.

event deposit A BED of sediment or sedimentary rock deposited by a short-lived EVENT of high energy. *Event beds* (*event horizons*) are 'discrete and frequently widely extended sediment layers which normally deviate in texture, structure, and fossil content from their host sediments' (Einsele *et al.*, 1996). Common characteristics include a sharp base, often with SCOUR MARKS, GRADED BEDDING and other SEDIMENTARY STRUCTURES that indicate deposition from a waning flow. Each event bed represents a short episode of sedimentation separated by a long period of low-energy conditions characterised by a much slower SEDIMENT ACCUMULATION RATE. Examples of event beds include deposits from river FLOODS, STORMS (*tempestites*), TSUNAMIS (*tsunamiites*), DEBRIS FLOWS (*debrites*) and TURBIDITY CURRENTS (TURBIDITES). In some cases it is possible to identify deposits modified by the effects of EARTHQUAKES (*seismites*). Other event deposits include the deposits of JÖKULHLAUPS, PYROCLASTIC products of VOLCANIC ERUPTIONS (see TEPHRA) and debris such as shocked quartz grains or *microtektites* resulting from METEORITE IMPACTS. The recognition and interpretation of event deposits is important in understanding the relative roles of cataclysmic versus gradual processes in Earth history (see ACTUALISM, GRADUALISM, NEOCATASTROPHISM, UNIFORMITARIANISM). They provide a long-term record of short-lived events that were regional or global in extent and may have contributed to ENVIRONMENTAL CHANGE. Because they tend to be synchronous horizons covering a large area, their CORRELATION forms the basis of EVENT STRATIGRAPHY. *GO*

Dott Jr, R.H. 1983: Episodic sedimentation – how normal is average? How rare is rare? Does it matter? *Journal of Sedimentary Petrology* 53, 5–23. **Einsele, G., Chough, S.K. and Shiki, T.** 1996: Depositional events and their records – an introduction. *Sedimentary Geology* 104, 1–9 (and thematic set of 15 papers, pp. 11–255).

event stratigraphy The use of the products of discrete, relatively rare EVENTS of high magnitude or energy that are widespread and synchronous in their effects, to provide CORRELATION of successions of SEDIMENTS and SEDIMENTARY ROCKS. *GO/BTC*

[See also CYCLOSTRATIGRAPHY, SEQUENCE STRATIGRAPHY, STRATIGRAPHY]
Bosellini, A., Morsilli, M. and Neri, C. 1999: Long-term event stratigraphy of the Apulia platform margin (Upper Jurassic to Eocene, Gargano, southern Italy). *Journal of Sedimentary Research* 69, 1241–1252. **Dott Jr., R.H.** 1996: Episodic event deposits versus stratigraphic sequences – shall the twain never meet? *Sedimentary Geology* 104, 243–247. **Einsele, G.** 1998: Event stratigraphy: recognition and interpretation of sedimentary event horizons. In Doyle, P. and Bennett, M.R. (eds), *Unlocking the stratigraphical record: advances in modern stratigraphy.* Chichester: Wiley, 145–193. **Walker, M.J.C., Björck, S., Lowe, J.J. et al.** 1999: Isotopic 'events' in the GRIP ice core: a stratotype for the Late Pleistocene. *Quaternary Science Reviews* 18, 1143–1150.

evergreen A plant species that has at least some green leaves throughout the year, e.g. conifers in BOREAL FORESTS, *laurisilva* in humid SUBTROPICAL zones or TROPICAL RAIN FOREST species. *BA*

[See also DECIDUOUS]

evolution The process by which species and lineages give rise to new, descendant, species and lineages. The term was originally used in the sense of the development of organisms from embryo to mature organism, but was first used with reference to species by Charles Lyell in 1832. During the nineteenth century there was vigorous debate about the fact of evolution with supporters of CREATIONISM and about the mechanism, between supporters of LAMARCKISM and DARWINISM. Lamarck brought the topic to the fore, but mid-nineteenth century debate over Darwinism was mostly about the fact of evolution. In 1859, Darwin's theory of evolution by means of NATURAL SELECTION was the only significant contender as a mechanism. The study of CREATIONISM is no longer considered part of science. Darwin's theory has dominated thinking on the subject since the mid-nineteenth century. Natural selection is essentially an ecological argument, as Darwin had no notion of the underlying GENETICS. During the early twentieth century, genetics was incorporated into Darwinian theory and in the mid-century three influential books, by the geneticist Theodosius Dobzhansky, the systematist Ernst Mayr and the palaeontologist George Gaylord Simpson, produced a view of evolution (known as the '*Modern Synthesis*') that appeared consistent across the whole of geology and biology. This view, that evolution proceeded more-or-less continually because of the mechanisms of genetic inheritance, mediated by the interaction between individuals and their environment, was not seriously challenged until the publication of the hypothesis of PUNCTUATED EQUILIBRIA, defined as a contrast to PHYLETIC GRADUALISM, as the Modern Synthesis view has been termed.

The debate over 'punctuated equilibria' versus 'phyletic gradualism', heated at times, has brought about a radical rethink of the process of evolution and some new ideas have come to the fore, some originating within the debate and some being drawn in. In particular, there has been a focus on the interaction between macroevolutionary patterns and ENVIRONMENTAL CHANGE. Stephen Jay Gould proposed that MASS EXTINCTIONS, on time-scales of 26–30 million years, reset the evolutionary clock. For shorter timescales, perhaps millions of years, Elisabeth Vrba has suggested that evolutionary changes may take place simultaneously across several lineages, forced by CLIMATIC CHANGE: the '*turnover-pulse hypothesis*'. Species tend to persist for periods of time of 1–30 million years, depending on lineage; this is clearly a few orders of magnitude longer than the time scales of orbitally forced climatic change. This persistence of species in the face of significant environmental change must be a permanent feature of Earth history. The main response of species to these changes is through distribution and abundance, both of which may vary individualistically by several orders of magnitude. This view of species constancy through environmental change is hard to reconcile with classic Darwinian theory, as Gould pointed out for MASS EXTINCTIONS. It may be that climatic changes on the timescales of MILANKOVITCH THEORY are also partly responsible for undoing accumulated change from ecological timescales and resetting the evolutionary clock.

The debate has come full-circle, in the sense that Darwin argued that environmental changes brought about evolution through differential survival of members of populations. But increased knowledge of timescales shows that this is probably not the case and it is now being argued that environmental change actually blocks the accumulation over time of evolution by means of natural selection. On the other hand, a dynamic Earth on timescales of thousands of years and longer continually brings about conditions under which populations may become separated and then develop into new species in ISOLATION. *KDB*

[See also GENETIC FITNESS]

Bennett, K.D. 1997: *Evolution and ecology: the pace of life.* Cambridge: Cambridge University Press. **Darwin, C.** 1859: *On the origin of species by means of natural selection, or the preservation of favoured races in the struggle for life.* London: John Murray. **Dobzhansky, T.** 1937: *Genetics and the origin of species.* New York: Columbia University Press. **Eldredge, N. and Gould, S.J.** 1972: Punctuated equilibria: an alternative to phyletic gradualism. In Schopf, T.J.M. (ed.), *Models in paleobiology.* San Francisco, CA: Freeman, Cooper, 82–115. **Gould, S. J.** 1985: The paradox of the first tier: an agenda for paleobiology. *Paleobiology* 11, 2–12. **Lyell, C.** 1832: *Principles of geology, being an attempt to explain the former changes of the Earth's surface, by reference to causes now in operation.* London: John Murray. **Mayr, E.** 1942: *Systematics and the origin of species from the viewpoint of a zoologist.* New York: Columbia University Press. **Simpson, G.G.** 1944: *Tempo and mode in evolution [Columbia Biological Series XV].* New York: Columbia University Press. **Stebbins, G.L.** 1977: *Processes of organic evolution,* 3rd edn. Englewood Cliffs, NJ: Prentice-Hall. **Vrba, E.S.** 1993: Turnover-pulses, the Red Queen, and related topics. *American Journal of Science* 293A, 418–452.

excavation An approach to FIELD RESEARCH involving the physical uncovering of subsurface phenomena. It ranges from relatively small pits (common in soil survey) to trenches and the uncovering of whole settlements and landscapes (common in archaeological excavation). It tends to be more informative, more costly and more damaging to the site and the environment than alternatives such as REMOTE SENSING, ARCHAEOLOGICAL PROSPECTION and the use of CORERS. Its particular advantages include the stratigraphic detail revealed in exposures where horizontal as well as vertical variation can be observed, described and sampled *in situ*. *JAM*

Barker, P. 1993: *Techniques of archaeological excavation.* London: Batsford.

exfoliation The peeling off of a surface layer or layers of rock; a type of PHYSICAL WEATHERING process. Exfoliation can be subdivided into: (subsurface) SPHEROIDAL WEATHERING (in which the volume is held constant); UNLOADING (in which expansion due to removal by DENUDATION of overlying rock occurs); and (subaerial) *flaking* (which takes place due to external forces). It is also known as *spalling, scaling* and *onion-skin weathering*. On a large scale, DILATION joints may be produced by UNLOADING and the resulting larger-scale form of exfoliation is sometimes termed *sheeting*. It can lead to slabs of rock up to tens of metres in size being removed. Exfoliation tends to affect medium- to coarse-grained rocks. *SHD*

exhumed soil A soil that was once buried by the deposition of sediments (see BURIED SOIL), and possibly lithified (see FOSSIL SOIL, GEOSOL), but has since been exposed in the LANDSCAPE by erosion. *JAM*

exogenetic Pertaining to the external agencies of formation, especially the geological or geomorphological processes that originate at or near the Earth's surface. *Exogenetic landforms*, such as SAND DUNES and V-SHAPED VALLEYS, are those produced by *exogenetic processes*. Thus, most processes of WEATHERING and EROSION are exogenetic, in contrast to the ENDOGENETIC processes that originate within the Earth's interior. *JAM*

[See also POLYGENETIC]

exotic species Individuals and populations of a species that occur outside their natural geographic ranges. Exotic species are also known as ALIEN SPECIES. *MVL*

[See also NATIVE SPECIES, NATURALISED SPECIES]

experiment In LABORATORY SCIENCE, especially physics, chemistry and molecular biology, experiment involves simplification and manipulation of phenomena in order to understand the underlying mechanisms and causes. In particular, *controlled experiments* enable causal factors to be identified and investigated by holding constant the many other factors that obscure explanation, typically involving a closely specified EXPERIMENTAL DESIGN, REPLICATION and the direct and quantitative testing of theoretical PREDICTIONS. *Treatments* (manipulations of one factor or a small number of factors) and *controls* (not affected by treatments) may be analysed at the same time on different samples (*synchronic contrast*) or at different times on the same sample in before (control) and after (treatment) situations (*diachronic contrast*). In the NATURAL ENVIRONMENTAL SCIENCES, archaeological sciences, historical sciences and SOCIAL SCIENCES, other forms of experiment are often used which, although less well controlled, remain an important part of the SCIENTIFIC METHOD. Examples include the *comparative method* and *natural experiments*, in which spatial and temporal patterns in nature are used as surrogates for true laboratory conditions. The use of nature in this way – as a *natural laboratory* – involves immense skill. *JAM*

[See also CHRONOSEQUENCE, ERGODIC HYPOTHESIS, FIELD RESEARCH]

Deevey, E.S. 1969: Coaxing history to conduct experiments. *Bioscience* 19, 40–43. **Diamond, J.** 1986: Overview: laboratory experiments, field experiments and natural experiments. In Diamond, J. and Case, T. (eds), *Community ecology*. New York: Harper and Row, 3–22. **Robinson, D. (ed.)** 1990: *Experimentation and reconstruction in environmental archaeology*. Oxford: Oxford Books. **Slaymaker, H.O. (ed.)** 1991: *Field experiments and measurement problems in geomorphology*. Rotterdam: Balkema.

experimental archaeology An approach to archaeology that uses explicit experiments, especially *field experiments*. It is an aspect of SCIENTIFIC ARCHAEOLOGY. Five categories have been recognised: constructions, process investigations, simulation trials, probability trials and technological innovation. Large-scale projects, such as the experimental earthworks at Overton Down and Bascomb Down and crop trials at Butser Ancient Farm, UK, may involve several of these categories simultaneously. *JAM*

Bell. M., Fowler, P.J. and Hilson, S.W. 1996: *The experimental earthwork project: 1960–1992* [*Report* 100]. York: Council for British Archaeology. **Coles, J.** 1973: *Archaeology by experiment*. London: Hutchinson. **Reynolds, P.J.** 1994: *Experimental archaeology: a perspective for the future*. Leiden: Brill.

experimental design The term is most commonly found in a statistical context (e.g. a Latin Square) where an experimental design is a matter of precise conceptualisation and economy so that the statistical results are precisely focused upon the research question(s) being evaluated, provide answers at the specified confidence level and do both with optimal economy. It should also be noted that the word EXPERIMENT commonly implies human manipulation of at least some part of the subject matter; however, in the environmental sciences '*field experiments*' are often 'natural', with little if any human manipulation. In both statistical and *natural experiments*, the primary requirement of the experimental design is to isolate effectively individual factors wherever possible, and thus produce a critical or sharp experiment or test. There is an entire literature on statistical experimental design; the literature on the nature of field experiments is more limited. *CET*

[See also SAMPLING]

Church, M. 1984: On experimental method in geomorphology. In Burt, T.P. and Walling, D.E. (eds), *Catchment experiments in fluvial geomorphology*. Norwich: Geo Books, 563–580.

exploration In many ways, the very seed of GEOGRAPHY is found in the concept of exploration with its related qualities of discovery, expeditions, FIELD RESEARCH and map making. The first 'geographers' were the explorers who ventured into the unknown, kept their records and journals, constructed their maps and generated an understanding of the wider world for all its inhabitants. In Britain, the meeting rooms of the Royal Geographical Society witnessed historic expositions as the great names returned to report their success in exploring new lands. By its nature, the practice of map making, SURVEYING and CARTOGRAPHY needed to be a precise science and one that saw the early appearance of measurement and systematic method in the discipline. The appeal of exploration remains strong in its modern manifestations of REMOTE SENSING, imagery from SATELLITES and the geographies of distant worlds. The continuance of expeditions and fieldwork ensures that the spirit of exploration remains integral to modern geography, its vitality and attractiveness.

Expeditions and fieldwork are no longer the preserve of remote mountains or other distant places; they extend to the environments of the great cities and diverse cultural milieux. *DTH*

Allen, J.L. 1997: *North American exploration* (3 vols). Vol. 2. *A continent defined.* Lincoln, NB: University of Nebraska Press. **Baker, J.N.L.** 1937: *A history of geographical discovery and exploration.* London: Harrap. **Livingstone, D.** 1992: *The geographical tradition.* Oxford: Blackwell. **Nansen, F.** 1911: *In northern mists: Arctic exploration in early times* (2 vols). London: Heinemann.

exploratory data analysis

An approach to NUMERICAL ANALYSIS, especially MULTIVARIATE ANALYSIS, in which patterns are sought in datasets without the formal statistical constraints of hypothesis testing or parameter estimation. This does not mean that such analysis is carried out in an intellectual vacuum or without a basis in theory, but it does mean that some of the requirements of INFERENTIAL STATISTICS are relaxed. Exploratory analysis may lead to the refining of hypotheses that can later be tested more formally by CONFIRMATORY DATA ANALYSIS. *JAM*

Tukey, J.W. 1977: *Exploratory data analysis.* Reading, MA: Addison-Wesley.

exponential growth

Common in animal populations where there is unlimited availability of resources, exponential growth is characterised by a constant *rate of increase* in numbers. On a graph in which population numbers are plotted against time, with time on a logarithmic scale, exponential growth is represented as a straight line. *JAM*

[See also DEMOGRAPHIC CHANGE]

exposure-age dating

An alternative term for SURFACE DATING. *CJC*

[See also CATION-RATIO DATING, COSMOGENIC-ISOTOPE DATING]

Beck, C. (ed.) 1994: *Dating in exposed and surface contexts.* Albuquerque, NM: University of New Mexico Press.

extinct volcano

A VOLCANO that is considered unlikely to produce any future VOLCANIC ERUPTIONS. Activity may continue in the form of FUMAROLE or HYDROTHERMAL processes. *GO*

[See also ACTIVE VOLCANO, DORMANT VOLCANO]

extinction

The loss of all populations of a species or higher taxon. Extinction may also refer to the loss of a local population of a species within a certain area, an event more formally known as EXTIRPATION. Except during periods of MASS EXTINCTION, extinctions in the geological record were relatively infrequent. Human activities have accelerated extinction rates by perhaps a factor of 10^4. For CONSERVATION purposes, the World Conservation Union (IUCN) defines an *extinct species* as one not definitely located in the wild during the past 50 years.

Several methods for estimating extinction rates exist. The most commonly used technique measures the extent of habitat loss and extrapolates the resulting decline in BIODIVERSITY from the SPECIES–AREA RELATIONSHIP. An alternative method examines how quickly species pass between different categories of threat, but no method claims to be accurate. By AD 1995, there were little more than 1000 documented extinctions, two thirds of which were of plants. However, as most species remain undescribed, this is considered a major underestimation of recent numbers of extinctions. For example, there are about 115 documented bird species extinctions, but humans are thought to have caused perhaps 2000 bird species extinctions on tropical Pacific Islands alone. From 5% to 20% of the species in many groups are currently threatened with extinction (see Figure).

Assessing the causes of extinction at a global scale is complicated by the interdependence of socioeconomic factors affecting human welfare and proximate causes of extinction, such as HABITAT LOSS, FRAGMENTATION and OVERHUNTING. CLIMATIC CHANGE, a major cause of extinction in the geological record, will likely cause many future extinctions. The populations and geographical ranges of many species have been reduced drastically because of human activities. These species may be ecologically 'extinct', in that they are no longer able to modify the environment significantly or function to provide ecosystem services. Many of these species are the 'living dead', committed to extinction by small population demographics, susceptibility to environmental catastrophes, genetic problems or a combination of these factors. There remains no accepted technique for calculating the rate at which such species will become extinct, but if populations and geographical ranges are permitted to recover, many such species may avoid extinction. *JTK*

[See also ENDANGERED SPECIES, EXTINCTION CRISIS]

Chapin III, S.F., Zavaleta, E.S., Eviner, V.T. *et al.* 2000: Consequences of changing biodiversity. *Nature* 405, 234–242. **Heywood, V.H. (ed.)** 1995: *Global biodiversity assessment.* Cambridge: Cambridge University Press. **Hughes, J.B., Daily, G.C. and Ehrlich, P.R.** 1997: Population diversity: its extent and extinction. *Science,* 278, 689–692. **Kerr, J.T. and Currie, D.J.** 1995: Effects of human activity on global extinction risk. *Conservation Biology,* 9, 1528–1538. **Lawton, J.H. and May, R.M. (eds)** 1995: *Extinction rates.* Oxford: Oxford University Press. **Pimm, S.L., Russell, G.J., Gittleman, J.L. and Brooks, T.M.** 1995: The future of biodiversity. *Science* 269, 347–350. **Soulé, M.E. (ed.)** 1986: *Conservation biology: the science of scarcity and diversity.* Sunderland, MA: Sinauer. **Whittaker, R.J.** 1998: *Island biogeography: ecology, evolution, and conservation.* Oxford: Oxford University Press.

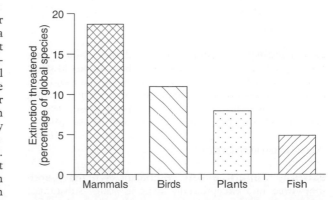

extinction *The proportion of the global number of species of mammals, birds, plants and fish that are currently threatened with extinction (after Pimm* et al., *1995; Chapin* et al., *2000)*

extinction crisis The elimination of large numbers of species resulting in important implications for nature in relation to EVOLUTION and for humans in relation to ECOLOGICAL GOODS AND SERVICES. EXTINCTION is a natural process and periodically there have been MASS EXTINCTIONS. Human activity has steadily contributed to increasing losses of species. The current rate of loss of species is considered by many to be far greater than has ever occurred before. Some estimates suggest species losses occurring 100 to 1000 times prehuman, BACKGROUND RATES. *IFS*

[See also CONSERVATION BIOLOGY]

Pearce, D.W. and Moran, D. 1997: The economics of biodiversity. In Tietenberg, T. and Folmer, H. (eds), *International Yearbook of Environmental and Resource Economics: A Survey of Current Issues.* Cheltenham: Edward Elgar, 82–113. **Wilson, E.O.** 1992: *The diversity of life.* London: Allen Lane/Penguin.

extirpation The loss of a local population, not implying global loss of the species. Thus extirpation differs from true EXTINCTION. *MVL*

extractive industry The *primary sector* of industry, especially the exploitation of NON-RENEWABLE RESOURCES, such as MINING, quarrying and oil drilling; some would include the exploitation of RENEWABLE RESOURCES, such as fisheries and forestry. Extractive industries provide raw materials, including materials for construction (building stone, sand, gravel and limestone), ferrous and non-ferrous metals, a wide range of other non-metallic resources and fuels for energy, transport and MANUFACTURING INDUSTRY. Environmental issues facing the extractive industries include: POLLUTION from present extraction and inadequate past WASTE MANAGEMENT; pollution caused during transport of bulky materials (e.g. OIL SPILLS); future (possibly imminent in some cases) exhaustion of RESERVES; CONSERVATION problems of prospecting activities; and LAND RESTORATION after extraction. *JAM*

[See also EXTRACTIVE RESERVE, FISHERIES CONSERVATION AND MANAGEMENT, FOREST MANAGEMENT, FOSSIL FUELS, LIMITS TO GROWTH, MINING]

Brookins, D.G. 1993: *Mineral and energy resources: occurrence, exploitation and environmental impact.* New York: Macmillan.

extractive reserve A NATURE RESERVE where local people are permitted to extract and sell non-timber forest products, such as latex and nuts, but not trees. *NDB*

[See also HARVESTING]

extrapolation Estimating from a data-set values lying beyond the range of the available data. In REGRESSION ANALYSIS, a value of a response variable may be estimated from the fitted equation for a new observation having values of the predictor variables beyond those used in deriving the equation. It is often a dangerous procedure. *HJBB*

[See also INTERPOLATION]

extreme climatic event A climatic event that is both short-lived and exceptionally large in scale. Substantial damage may be caused to the system under consideration. Such an event may be an extreme of natural CLIMATIC VARIABILITY (such as 'the one in a thousand years storm') or may originate in some external perturbation of the climatic system (or CATASTROPHIC EVENT). At the outer limits of the wide range of variations of contemporary climate, MAGNITUDE-FREQUENCY CONCEPTS are used to describe such events. Relatively small changes in climatic variability or mean CLIMATE may be accompanied by large changes in the frequency of short-term weather events (e.g. flash floods and TROPICAL CYCLONES) or somewhat longer-term climatic events (e.g. heat waves and DROUGHTS).

International reinsurance has shown an upward trend in economic losses due to extreme climatic events in recent years. Around 95% of the deaths that occur annually due to extreme climatic events occur in *developing countries*, especially in tropical Asia and Central America. The 1990s 'International Decade for Natural Hazard Reduction' attempted to reverse the rising tide of such climatic DISASTERS. The long-term importance of extreme climatic events in such fields as LANDSCAPE EVOLUTION and species EVOLUTION is poorly understood. *AHP/KDB*

[See also CLIMATIC EXTREMES, EVOLUTION, EXTREME WEATHER EVENTS, MASS EXTINCTION, NATURAL HAZARDS]

Morgan, M.D. and Moran, J.M. 1997: *Weather and people.* Englewood Cliffs, NJ: Prentice-Hall. **Vrba, E. S.** 1993: Turnover-pulses, the Red Queen, and related topics. *American Journal of Science* 293A, 418–452. **Wigley, T.M.** 1985: Impact of extreme events. *Nature* 316, 106–107.

extreme climatic events: impact on ecosystems Responses to extreme climatic events are key determinants of the geographic limits (RANGE) of plant species, and govern the species present in many BIOMES. The ability to survive extreme low temperatures differs markedly between plant species and is an important criterion in the classification of global VEGETATION into FUNCTIONAL TYPES. In addition, extreme DROUGHT, low temperatures or high temperatures can reduce competitive ability, for example by defoliating the leaf canopy, and may lead to competitive exclusion by species that are able to tolerate or avoid these extremes (see COMPETITION and TOLERANCE). Changes in patterns of extreme climate events accompanying CLIMATIC CHANGE therefore have important biogeographical and ecological implications.

Extreme climatic events are also important causes of DISTURBANCE in ECOSYSTEMS. In particular, STORMS destroy vegetation by WINDTHROW and lightning, and the latter is also important in igniting WILDFIRES. *CPO*

[See also EL NIÑO–SOUTHERN OSCILLATION (ENSO): IMPACTS ON ECOSYSTEMS, FIRE IMPACTS, TROPICAL CYCLONES: IMPACT ON ECOSYSTEMS]

Whittaker, R.J. 1998: *Island biogeography: ecology, evolution and conservation.* Oxford: Oxford University Press. **Woodward, F.I.** 1987: *Climate and plant distribution.* Cambridge: Cambridge University Press.

extreme event An event that is exceptional in terms of low frequency and high magnitude: it tends to be of short duration and differs substantially from the mean in the context of a specified timescale. However, the frequency and/or magnitude of extreme events may themselves change, for example in response to an increase in the mean value (see Figure, part A) the number of extreme high-magnitude events would be expected to increase and parallel a decrease in the number of low-magnitude events; whereas the numbers of both extreme

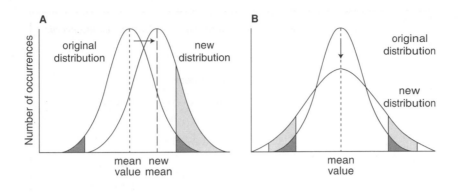

extreme event *Expected changes in the number of extreme events following changes in the frequency distribution: (A) the response to an increase in the mean value; (B) the response to an increase in variability (after Mitchell et al., 1990)*

high-magnitude and extreme low-magnitude events increase in response to an increase in variability (see Figure, part B).

Because geophysical systems and any associated ecosystems and socio-economic systems may be adjusted to higher-frequency, lower-magnitude events, extreme events cause disruptions, pose NATURAL HAZARDS and may result in DISASTERS unless such systems have the capacity sufficiently to absorb or buffer the environmental impact.

There is a growing body of evidence, however, that extreme events are an integral part of some natural EARTH SURFACE SYSTEMS and may play a disproportionately large role in accounting for their attributes. Thus, LANDSLIDES are now viewed as more important in the erosion of high-relief humid tropical environments than all other physical processes of EROSION combined, and the occasional extreme DROUGHT accounts for important aspects of the structure, dynamics and species composition of some rain forests. *JAM/RPDW*

[See also DISTURBANCE, DISTURBED ECOSYSTEMS, EXTREME WEATHER EVENT]

Alexander, D. 1993: *Natural disasters.* London: UCL Press, 5. **Ely, L.L.** 1997: Response of extreme floods in the southwestern United States to climate variations in the late Holocene. *Geomorphology* 19, 175–201. **Katz, R.W. and Brown, B.G.** 1992: Extreme events in a changing climate: variability is more important than averages. *Climatic Change* 21, 289–302. **Larsen, M.C., Torres-Sánchez, A.J. and Conceptión, I.M.** 1999: Slopewash, surface runoff and fine-litter transport in forest and landslide scars in humid-tropical steeplands, Luquillo Experimental Forest, Puerto Rico. *Earth Surface Processes and Landforms* 24, 481–502. **Mitchell, J.F.B., Manabe, S., Meleshko, V. and Tokioka, T.** 1990: Equilibrium climate change – and its implications for the future. In Houghton, J.T., Jenkins, G.J. and Ephraums, J.J. (eds), *Climate change: the IPCC scientific assessment.* Cambridge: Cambridge University Press. **Walsh, R.P.D.** 1996: Drought frequency changes in Sabah and adjacent parts of northern Borneo since the late nineteeth century and possible implications for tropical rain forest dynamics. *Journal of Tropical Ecology* 12, 385–407. **Walsh, R.P.D. and Newbery, D.M.** 1999: The ecoclimatology of Danum, Sabah, in the context of the world's rainforest regions, with particular reference to dry periods and their impact. *Philosophical Transactions of the Royal Society of London* B354, 1869–1883.

extreme weather event A weather event (duration hours to months) that is exceptional in terms of high magnitude and low frequency. Because the concept of an EXTREME EVENT depends on the timescale, there is a continuum from short-term extreme weather events to longer-term EXTREME CLIMATIC EVENTS. However, extreme weather events may also be classified as either absolute or relative (see Figure): *absolute extreme events* are regarded as extreme because of high magnitude, even in places or regions where they are quite frequent occurrences (e.g. TROPICAL CYCLONES, STORMS and hailstorms); *relative extreme events* are regarded as extreme on the basis of their rarity, or probability of occurrence at the location concerned (e.g. extremes of heat, cold or drought).

Extreme weather events, and the weather-related events that they commonly trigger (such as FLOODS, LANDSLIDES, SNOW AVALANCHES and STORM SURGES), present several distinctive problems in their analysis: first, their low frequency limits the sample size for investigation; second, they tend to be localised spatially and of variable magnitude; third, these attributes present difficulties with prediction and forecasting; fourth, impacts on landscapes and societies vary not only with event characteristics but also with numerous human variables such as LANDUSE, the way HAZARDS are perceived and the nature of mitigation strategies. There are several forms of statistical *extreme-value analysis* (e.g the use of Gumbel extreme-value distributions) that specialise in extreme events, but the nature of the data (low-frequency events) means that results are often tentative and unreliable. *RPDW/JAM*

[See also CLIMATIC EXTREMES, NATURAL HAZARDS]

Gumbel, E.V. 1958: *Statistics of extremes.* New York: Columbia University Press. **Smith, K.** 1997: Climatic extremes as a hazard to humans. In Thompson, R.D. and Perry, A.H. (eds), *Applied climatology: principles and practice.* London: Routledge, 304–316. **Walsh, R.P.D.** 1999: Extreme weather events. In Pacione, M. (ed.), *Applied geography: principles and practice.* London: Routledge, 51–65.

extremophile An organism that thrives in an environment that is extreme in terms of the range of environments found on Earth. If the environment is extreme in relation

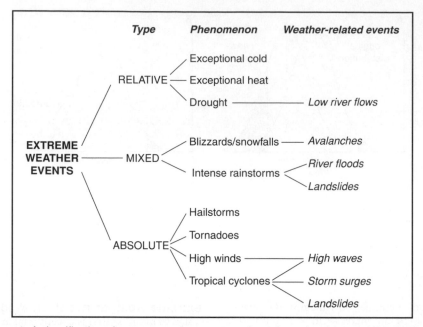

Type	Phenomenon	Weather-related events
RELATIVE	Exceptional cold	
	Exceptional heat	
	Drought	Low river flows
MIXED	Blizzards/snowfalls	Avalanches
	Intense rainstorms	River floods
		Landslides
ABSOLUTE	Hailstorms	
	Tornadoes	
	High winds	High waves
	Tropical cyclones	Storm surges
		Landslides

extreme weather event *A classification of extreme weather events and weather-related events (after Smith, 1997; Walsh, 1999)*

to more than one ENVIRONMENTAL FACTOR or VARIABLE, the organism is termed a *polyextremophile*. The term normally implies physical (e.g. temperature or pressure) or geochemical (e.g. salinity or pH) extremes. Many but not all extremophiles are BACTERIA or relatively simple plants and animals. Extremophiles are widely used in industry and BIOTECHNOLOGY, and are also of interest for possible insights into life in the COSMOSPHERE, beyond Earth. *JAM*

Hirikoshi, K. and Grant, W. 1998: *Extremophiles: microbial life in extreme environments*. New York: Wiley-Liss. **Rothschild, L.J. and Mancinelli, R.L.** 2001: Life in extreme environments. *Nature* 409, 1092–1101.

F

fabric The spatial arrangement of the constituents ('fabric elements') of a SEDIMENT or rock (an aspect of TEXTURE). Important attributes in SEDIMENTARY ROCKS are the alignment, packing and contact relationships of GRAINS or CLASTS, including FOSSIL fragments (*bioclasts*).

(1) *Alignment.* Where particles are spherical or have no preferred orientation, the fabric is *isotropic*. *Anisotropic* fabrics are caused by the alignment of particles in a force field, which may occur during DEPOSITION (e.g. a PALAEOCURRENT), DIAGENESIS or DEFORMATION. Depositional alignments are sometimes referred to as *orientation* in plan view and IMBRICATION in cross-section. They are commonly expressed by the orientation of the long (a), intermediate (b) and short (c) axes of clasts, or their a–b, b–c and a–c planes. In the study of SAND and SANDSTONE (e.g. in THIN SECTION) it may be possible only to measure the apparent long-axis alignment.

(2) *Packing.* The packing of clasts is expressed by the *fractional volume concentration* (C = ratio of volume occupied by solid particles to total volume) or *porosity* (p = ratio of pore volume to total volume), where $C + p = 1$. The *void ratio*, commonly used in an engineering context, is the ratio of pore volume to the volume occupied by solids. Porosity is affected by GRAIN-SIZE, GRAIN SHAPE, SORTING and the process of deposition. The porosity of most freshly deposited sand varies between 34% and 48%, while that of freshly deposited MUD can be as high as 80%. Lithified sedimentary rocks have much lower values owing to compaction and cementation. Porosity affects the BULK DENSITY of granular materials and is one influence on *permeability*, the ability of a fluid to flow through a porous medium.

(3) *Contact relationships.* The contacts between GRAVEL particles (CLASTS) and their relationship to any MATRIX gives an indication of depositional processes. Where clasts are in contact, with or without a matrix between them, the fabric is described as *clast-supported* (or *grain-supported* or *framework-supported – orthoconglomerate*); where the clasts are not in contact, the fabric is *matrix-supported*. This latter term is usually restricted to gravel clasts dispersed in a mud matrix (as opposed to a poorly sorted pebbly sand): such sediments are described as DIAMICT or *paraconglomerate* and are normally indicative of non-aqueous deposition by, for example, DEBRIS FLOW or as glacial TILL.

Fabric analysis is also important in engineering and the study of SOIL and GEOLOGICAL STRUCTURES, in which contexts fabric elements include features such as crystal alignment and joint systems, and their analysis provides information on SLOPE STABILITY analysis or palaeo-stress fields. *GO/TY*

[See also GLACIOTECTONICS, MICROMORPHOLOGICAL ANALYSIS OF SEDIMENTS, MICROMORPHOLOGICAL ANALYSIS OF SOILS, TILL FABRIC ANALYSIS]

Allen, J.R.L. 1982: *Sedimentary structures: their character and physical basis.* Amsterdam: Elsevier. **Bertran, P., Hétu, B., Texier, J.-P. and Van Steijn, H.** 1997. Fabric characteristics of subaerial slope deposits. *Sedimentology* 44, 1–16. **Collinson, J.D. and Thompson, D.B.** 1992: *Sedimentary structures*, 2nd edn. London: Chapman and Hall. **Graham, J.** 1988: Collection and analysis of field data. In Tucker, M.E. (ed.), *Techniques in sedimentology.* Oxford: Blackwell, 5–62. **Harris, C.** 1998: The micromorphology of paraglacial and periglacial slope deposits: a case study from Morfa Bychan, west Wales, UK. *Journal of Quaternary Science* 13, 73–84. **Pettijohn, F.J.** 1975: *Sedimentary rocks*, 3rd edn. New York: Harper & Row. **Potter, P.E. and Pettijohn, F.J.** 1977: *Paleocurrents and basin analysis*, 2nd edn. Berlin: Springer.

fabric analysis Strictly speaking, analysis of the structural aspects of framework, matrix and planar elements of a sediment. Common usage limits the term to the three-dimensional disposition of the particles, sand-size up to CLAST-size. The technique is widely applied to sediments in order to reconstruct palaeoflow direction (especially DIAMICTONS of GLACIAL and PERIGLACIAL origin). Usually, the azimuth and plunge (or angle of dip) of the longest axes of a sample of elongate clasts are measured. It is assumed that the depositional flow direction can be reconstructed from the preferred orientation of clast long axes (e.g. ice-flow direction for LODGEMENT TILL (*till fabric analysis*) and down-slope flowage direction for SOLIFLUCTION deposits). More sophisticated three-dimensional analysis (using *Eigenvectors*) together with other sedimentary characteristics, can be used to infer depositional processes and to reconstruct DEPOSITIONAL ENVIRONMENTS. *JS*

[See also TILL]

Dowdeswell, J.A. and Sharp, M.J. 1986: Characterisation of pebble fabrics in modern terrestrial glacigenic sediments. *Sedimentology* 33, 699–710. **Holmes, C.D.** 1941: Till fabric. *Geological Society of America Bulletin* 52, 1299–1354. **Mark, D.M.** 1973: Analysis of axial orientation data including till fabrics. *Geological Society of America Bulletin* 84, 1369–1374. **Lawson, D.E.** 1981: Distinguishing characteristics of diamictons at the margin of the Matanuska Glacier, Alaska. *Annals of Glaciology* 2, 78–74.

facet (1) A flat surface of a CLAST formed by ABRASION. (2) An element of the surface of a crystal. (3) In the classification of LANDSYSTEMS, groupings of similar land elements. (4) A component of a hillslope profile, usually relatively short and of uniform nature. *RAS*

facies A body of SOIL, SEDIMENT or rock with characteristics that are distinct from those of adjacent bodies. The term is widely used with regard to sediments and SEDIMENTARY ROCKS, in which a facies can be characterised by its LITHOLOGY, geometry, SEDIMENTARY STRUCTURES, PALAEOCURRENTS and FOSSILS (see also BIOFACIES, ICHNOFACIES, LITHOFACIES). FACIES ANALYSIS and FACIES MODELS are used to identify DEPOSITIONAL ENVIRONMENTS in the GEOLOGICAL RECORD. The term is also used with the same general sense in other contexts, such as metamorphic facies or TECTONIC facies. *GO*

[See also FACIES ARCHITECTURE, FACIES ASSOCIATION, FACIES SEQUENCE]

Nichols, G.J. 1999: *Sedimentology and stratigraphy*. Oxford: Blackwell Science. Reading, H.G. (ed.) 1986: *Sedimentary environments and facies*, 2nd edn. Oxford: Blackwell. Robertson, A.H.F. 1994: Role of the tectonic facies concept in orogenic analysis and its application to Tethys in the Eastern Mediterranean region. *Earth-Science Reviews* 37, 139–213.

facies analysis

The interpretation of FACIES in sediments and sedimentary rocks, with the principal aim of determining the DEPOSITIONAL ENVIRONMENT or PALAEOENVIRONMENT. Vertical and lateral relationships of LITHOLOGY, BEDDING, SEDIMENTARY STRUCTURES, PALAEOCURRENTS and FOSSILS are examined in surface outcrops or CORES and recorded on GRAPHIC LOGS. In *subsurface facies analysis*, much used in the petroleum industry, lithological changes, marker beds, UNCONFORMITIES, key BOUNDING SURFACES and stratal geometries are inferred from BOREHOLE log data or SEISMIC profiles (see SEISMIC STRATIGRAPHY). Facies analysis typically proceeds from the subdivision of deposits into facies and the subsequent interpretation of these in terms of sedimentary or biological processes. Typical assemblages of facies (FACIES ASSOCIATIONS and FACIES SEQUENCES) are diagnostic of particular depositional environments, and their characteristics can be summarised in FACIES MODELS. The analysis of lateral and vertical facies successions (FACIES ARCHITECTURE), and their relationships to key bounding surfaces and regional SEQUENCE STRATIGRAPHY, can reveal changes in sedimentary environments through time. *MRT/GO*

Reading, H.G. (ed.) 1996: *Sedimentary environments: processes, facies and stratigraphy*, 3rd edn. Oxford: Blackwell. Selley, R.C. 1996: *Ancient sedimentary environments and their sub-surface diagnosis*. London: Chapman and Hall. Walker, R.G. and James, N.P. (eds) 1992: *Facies models: response to sea level change*. St John's: Geological Association of Canada.

facies architecture

The three-dimensional geometry of FACIES and BOUNDING SURFACES in SEDIMENTARY DEPOSITS, used particularly with respect to fluvial and aeolian deposits. Facies architecture is important in the analysis of DEPOSITIONAL ENVIRONMENTS. *Alluvial architecture*, for example, is concerned with the reconstruction of *bar* morphology and channel geometry and behaviour. *GO*

[See also ARCHITECTURAL ELEMENT ANALYSIS, FACIES SEQUENCE]

Clemmensen, L.B. and Dam, G. 1993: Aeolian sand-sheet deposits in the Lower Cambrian Neksø Sandstone Formation, Bornholm, Denmark: sedimentary architecture and genesis. *Sedimentary Geology* 83, 71–85. Glover, B.W. and Powell, J.H. 1996: Interaction of climate and tectonics upon alluvial architecture: Late Carboniferous Early Permian sequences at the southern margin of the Pennine Basin, UK. *Palaeogeography, Palaeoclimatology, Palaeoecology* 121, 13–34. Miall, A.D. 1993: The architecture of fluvial–deltaic sequences in the Upper Mesaverde Group (Upper Cretaceous), Book Cliffs, Utah. In Best, J.L. and Bristow, C.S. (eds), *Braided rivers, Geological Society, London, Special Publication* 75, 305–332. Vincent, S.J. 1999: The role of sediment supply in controlling alluvial architecture: an example from the Spanish Pyrenees. *Journal of the Geological Society, London* 156, 749–759.

facies association

A grouping of FACIES in a succession of SEDIMENTS or SEDIMENTARY ROCKS. If each facies is interpreted as the product of a particular process or subenvironment, the association enables the overall DEPOSITIONAL ENVIRONMENT to be identified. The concept is similar to that of architectural elements (see ARCHITECTURAL ELEMENT ANALYSIS), although the latter places more emphasis on the geometry of facies associations. *GO*

facies model

An idealised summary of the characteristics of a DEPOSITIONAL ENVIRONMENT in terms of sediment FACIES in the deposits. Facies models may be represented as block diagrams or GRAPHIC LOGS. *GO*

Walker, R.G. 1992: Facies, facies models and modern stratigraphic concepts. In Walker, R.G. and James, N.P. (eds), *Facies models: response to sea level change*. St. John's: Geological Association of Canada, 1–14.

facies sequence

A grouping of FACIES that tend to occur in a particular order in a vertical succession of SEDIMENTS or SEDIMENTARY ROCKS, with gradational boundaries between them. A distinction can be made between *cyclic sequences* (e.g. 1234321) and *rhythmic sequences* (e.g. 12341234). Successive facies in a rhythmic sequence can commonly be interpreted as the products of laterally adjacent subenvironments that migrated laterally with PROGRADATION of the overall environment, a concept commonly summarised as WALTHER'S LAW. Some workers prefer the term *facies succession* to avoid potential confusion with the usage of *sequence* in SEQUENCE STRATIGRAPHY. *GO*

[See also FACIES ARCHITECTURE]

Reading, H.G. and Levell, B.K. 1996: Controls on the sedimentary rock record. In Reading, H.G. (ed.), *Sedimentary environments: processes, facies and stratigraphy*, 3rd edn. Oxford: Blackwell Science, 5–36.

facilitation

In the context of ECOLOGICAL SUCCESSION, the positive influence of one species on another. It is a central theme in the *facilitation model*, whereby early colonists improve the environment for later colonists, resulting in the replacement of the early colonists. Facilitation can apply to all life history stages of a species, including DISPERSAL (perch trees promote bird dispersal of defaecated seeds), germination (build-up of organic matter can promote necessary soil nutrients), growth (species that add nitrogen to the soil can increase growth of associated or subsequent species) or reproduction (shared pollinators). Facilitation can involve direct species interactions (species A positively affects species B), indirect effects (species A inhibits species C and the status of species B improves) or long-term habitat amelioration (increased shade or nutrients promote species change). Facilitation is presumably more important in harsh physical environments such as those occurring in PRIMARY SUCCESSION, but little evidence exists that facilitation is obligatory for species change to occur. Instead, when it does occur, the rate of ecological succession may be accelerated by facultative (non-obligatory) facilitation. In contrast, the rate of successional change may be slowed by the inhibitory effects of COMPETITION. *LRW*

[See also AUTOGENIC CHANGE, COLONISATION, INHIBITION MODEL, NURSE PLANT, TOLERANCE MODEL]

Callaway, R.M. and Walker, L.R. 1997: Competition and facilitation: a synthetic approach to interactions in plant communities. *Ecology* 78, 1958–1965. Connell, J. H. and Slatyer, R.O. 1977: Mechanisms of succession in natural communities and their role in community stability and organization. *American Naturalist* 111, 1119–1144. **Matthews, J.A.** 1992: *The ecology of recently-deglaciated terrain.* Cambridge: Cambridge University Press.

factor analysis A group of statistical techniques related to PRINCIPAL COMPONENTS ANALYSIS that attempts to resolve complex relationships by identifying causal factors (latent variables) and ranking them according to the extent to which each factor accounts for the variance in the data. It differs from PRINCIPAL COMPONENTS ANALYSIS in that correlations between variables are only of interest if they reflect putative underlying causal factors. Sometimes but erroneously equated with principal components analysis or ORDINATION in general.   *HJBB*

Everitt, B.S. 1984: *An introduction to latent variable methods.* London: Chapman and Hall.

facultative Pertaining to an ENVIRONMENTAL FACTOR that is tolerated but not required by an organism for survival.   *JAM*

[See also OBLIGATE]

faeces analysis A valuable means of interpreting PALAEODIET through the identification of preserved vegetable matter or fragmented bone and also of reconstructing the palaeoenvironment from identifiable remains. The study of modern faeces also serves as a useful proxy for animal population size and habits in terrestrial environments, as well as providing a source of DNA.   *DCS*

[See also COPROLITE, MICROVERTEBRATE ACCUMULATIONS, RODENT MIDDENS]

Akeret, Ö., Haas, J.N., Leuzinger, U. and Jacomet, S. 1999: Plant macrofossils and pollen in goat/sheep faeces from the Neolithic lake-shore settlement Arbon Bleiche 3, Switzerland. *The Holocene* 9, 175–182. **Greig, J.R.A.** 1982: Garderobes, sewers, cesspits and latrines. *Current Archaeology* 85, 49–52. **Hellwig, M.** 1997: Plant remains from two cesspits (15th and 16th century) and a pond (13th century) from Göttingen, southern Lower Saxony, Germany. *Vegetation History and Archaeobotany* 6, 105–116.

failure The loss of strength (DEFORMATION) of rock or SOIL as a result of compressive, tensile or shearing forces.   *PW*

[See also LANDSLIDE, LIQUID LIMIT, MASS MOVEMENT PROCESSES, PLASTIC LIMIT]

fallout The deposition of PARTICULATES from the atmosphere onto the Earth's surface. It may be natural, such as volcanic fallout, or anthropogenic, such as RADIOACTIVE FALLOUT.   *JAM*

[See also DRY DEPOSITION, WET DEPOSITION]

fallow Agricultural fields left uncultivated. Leaving fields fallow allows nutrient reserves in the soil to be replenished. It is an integral part of SHIFTING CULTIVATION and, in more advanced agricultural systems, is often combined with CROP ROTATION.   *SHD*

Schmidt-Vogt, D. 1999: *Swidden farming and fallow vegetation in northern Thailand.* Stuttgart: Franz Steiner.

false colour composite A representation of a REMOTE SENSING image using colours of the ELECTROMAGNETIC SPECTRUM other than those used to acquire the image. For instance the infrared band of an image may be assigned to the red parts of a representation of that image (e.g. on a computer screen) to highlight vegetation within the scene.   *AJL*

false rings These are encountered in DENDROCHRONOLOGY and refer to a disruption in the normal gradation of larger, thinner-walled cells to smaller, thicker-walled cells that characterise the usual annual ring formed during the growing season of trees in many extra-tropical regions of the world. This change in the cell structure mimics the appearance of normal latewood growth and suggests that two rings were formed, whereas in fact there is only one true annual ring. These false rings can often be identified by close inspection of the cell structure and by rigorous CROSS-DATING of the ring patterns in contemporaneous ring measurements on other trees.   *KRB*

[See also TREE RINGS]

Telewski, F.W. and Lynch, A.M. 1991: Measuring growth and development of stems. In Lassoie, J.P. and Hinckley, T.M. (eds), *Techniques and approaches in forest tree ecophysiology.* Boca Raton, FL: CRC Press, 503–555.

falsification To prove a HYPOTHESIS untrue, as opposed to attempting to corroborate or verify it. The possibility of falsification is the critical test of the scientific nature of something in Popper's CRITICAL RATIONALISM.   *CET*

famine Widespread starvation following from food shortages usually associated with NATURAL DISASTERS, especially DROUGHT and FLOOD, but also triggered or accentuated by CLIMATIC CHANGE, political and social unrest and OVERPOPULATION. Famine should be differentiated from *hunger*, the consumption of a diet that is inadequate to sustain good health, normal activity and growth, which is less acute but even more ubiquitous.

Today, the populations of tropical developing countries, where BACKGROUND LEVELS of hunger and malnutrition are also high, tend to be the most vulnerable to famine, whereas those in the industrialised countries have balanced diets and food surpluses. In the developing countries famines are alleviated to a variable extent by highly organised, international, governmental and non-governmental food aid. Nevertheless, major famines continue to occur: it is thought that more than one million people died during the AD 1984–1985 famine in the SAHEL. In historical times, Europe suffered frequent famines, especially due to grain harvest failures: in France, for example, between AD 1371 and 1791 there were 111 famines, 16 of them in the eighteenth century. There was a major reduction in famine frequency in Europe with the introduction of the potato and other developments, although overdependence on this vegetable led to its vulnerability to disease (potato blight), the Irish famine of 1845, the death of 800 000 people and the MIGRATION of 1.6 million people in the following six years. Arguably the greatest famine of modern times was the Chinese famine of AD 1958 to 1962, which is believed to have accounted for at least 25 million deaths.   *JAM*

Fagan, B. 1999: *Floods, famines and emperors.* New York: Basic Books. **Glantz, M.H. (ed.)** 1987: *Drought and hunger in Africa: denying famine a future.* Cambridge: Cambridge University Press. **Jowett, J.** 1989: China: the population of the People's Republic.

SAGT Journal 18, 38–49. **Le Roy Ladurie, E.** 1972: *Times of feast, times of famine: a history of climate since the year 1000.* London: George Allen and Unwin. **Messer, E. and Uvin, P.** 1999: Hunger and food supply. In Alexander, D.E. and Fairbridge, R.W. (eds), *Encyclopedia of environmental science.* Dordrecht: Kluwer, 328–330. **Seaman, J. and Holt, J.** 1980: Markets and famines in the third world. *Disasters* 4, 289–299. **Wisner, B., Weiner, P. and O'Keefe, P.** 1982: Hunger: a polemical review. *Antipode* 14, 1–16.

fault A GEOLOGICAL STRUCTURE formed by brittle DEFORMATION causing a fracture (*fault surface* or *fault plane*) across which there has been relative displacement. Different types of faults (see Figure) form in response to tensional stresses (*normal faults*), compressional stresses (*reverse faults* and THRUST FAULTS) and shearing stresses (*strike-slip faults*). Total displacements vary from millimetres to hundreds of kilometres, with very large horizontal displacements associated with CONSERVATIVE PLATE MARGINS. Incremental movements on active faults generate EARTHQUAKES. Active and inactive faults have an important influence on GEOMORPHOLOGY, LANDSCAPE EVOLUTION and SEDIMENTATION. *GO*

[See also FAULT SCARP, FOLD, GRABEN, HORST, JOINT, RIFT VALLEY, TECTONICS]

Alexander, J. and Leeder, M.R. 1987: Active tectonic control on alluvial architecture. In Ethridge, F.G., Flores, R.M. and Harvey, M.D. (eds), *Recent developments in fluvial sedimentology. Society of Economic Paleontologists and Mineralogists Special Publication* 39, 243–252. **Ballance, P.F. and Reading, H.G. (eds)** 1980: *Sedimentation in oblique slip mobile zones. International Association of Sedimentologists Special Publication* 4. **Crowell, J.C.** 1974: Sedimentation along the San Andreas Fault, California. In Dott Jr, R.H. and Shaver, R.H. (eds), *Modern and ancient geosynclinal sedimentation. Society of Economic Paleontologists and Mineralogists Special Publication* 19, 292–303. **Paton, S.** 1992: Active normal faulting, drainage patterns and sedimentation in southwestern Turkey. *Journal of the Geological Society, London* 149, 1031–1044. **Ramsay, J.G. and Huber, M.I.** 1987: *The techniques of modern structural geology.* Vol. 2. *Folds and fractures.* London: Academic Press.

fault propagation The increase in size with time of a fracture in the Earth's crust due to relative movement across it. *DNT*

[See also CRACK PROPAGATION, FAULT, JOINT]

fault scarp A step on the Earth's surface formed by *surface rupture* and displacement across a FAULT. The formation of a fault scarp up to several metres high accompanies some, but not all, EARTHQUAKES. They are normally worn down rapidly by the agents of DENUDATION, but successive increments of movement may accumulate large steps where fault displacement is rapid, denudation is slow or there is a marked contrast in the resistance to denudation of the material on either side: fault scarps may then represent major topographical features, such as the margins of the East African RIFT VALLEY. *GO*

fauna The total animal life of a particular place or geographical region, or at a particular time or geological time period. The *palaeofauna* is not the same as the fossil animal assemblage or DEATH ASSEMBLAGE, since the latter is rarely representative of the living community. *DCS*

[See also ANIMAL REMAINS]

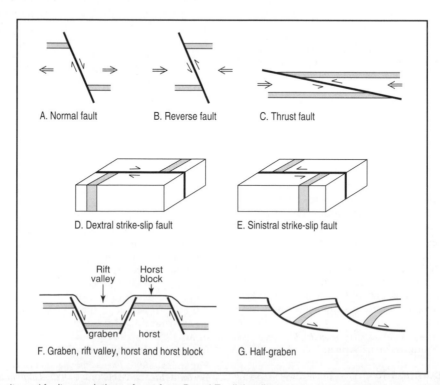

A. Normal fault   B. Reverse fault   C. Thrust fault

D. Dextral strike-slip fault   E. Sinistral strike-slip fault

F. Graben, rift valley, horst and horst block   G. Half-graben

fault *Varieties of faults and fault associations. Apart from D and E, all the diagrams represent vertical cross-sections. Note that, in the case of normal, reverse and thrust faults, the ground surface may or may not experience displacement*

faunal provinces Large geographical areas or regions characterised by the presence of certain animal species and higher taxonomic groups that are more-or-less unique to them. Differences between faunal provinces depend on various limiting factors, including climate, topography and the existence of barriers to migration. A minimum of six faunal provinces or regions are recognised: *Palaearctic* (Eurasia north of the tropics); *Nearctic* (most of North America); *Neotropical* (Central and South America), *Ethiopian* (Africa south of the Sahara); *Oriental* (tropical Asia); and *Australian* (including parts of southeast Asia and Oceania). *DCS*

[See also FLORAL PROVINCES, LAND BRIDGES, ZOOGEOGRAPHY]

Schmidt, K.P. 1954: Faunal realms, regions and provinces. *Quarterly Review of Biology* 29, 322–331.

faunation 'The assemblance of animal individuals of all species occurring at a locality' (Udvardy 1969: viii). The animal equivalent of VEGETATION. *JAM*

Udvardy, M.D.F. 1969: *Dynamic zoogeography with special reference to land animals.* New York: Van Nostrand Reinhold.

feedback mechanisms One of the fundamental aspects of the behaviour of SYSTEMS: the processes involved when a change in one variable is transmitted through the system so that observed effects are eventually nullified or accentuated. The feedback is said to be negative or positive in nature. With NEGATIVE FEEDBACK, the responses to change tend to remove or negate the trends and are generally recognised as self-regulating loops. For example, GLOBAL WARMING, could initiate (in theory, at least) more effective EVAPOTRANSPIRATION (especially in mid-latitudes), which could increase the atmosphere's moisture capacity. Associated LAPSE RATE steepening and free convection could lead to more CONDENSATION (given sufficient hygroscopic nuclei). The resultant, increased

cumiliform cloud cover could reflect more SOLAR RADIATION back to space, so depleting INSOLATION and lowering surface temperatures. This could initiate a global cooling response and a return to the original thermal state. Conversely, *positive feedback* tends to reinforce or accentuate the changes set in motion. For example, within the HYDROLOGICAL CYCLE, DEFORESTATION will expose the soil surface, reduce INFILTRATION and accelerate RUNOFF. There is a consequent increase in SOIL EROSION and, if this denudation exposes impermeable soil, percolation rates are reduced even further, which accentuates runoff, flooding and even greater soil erosion. Self-regulation is then out of the question and the situation deteriorates, perhaps until remedial action is taken through AFFORESTATION and/or the adoption of a variety of RUNOFF control measures and soil conservation techniques.

Negative and positive *feedback loops* commonly interact as illustrated in the Figure, which shows the likely effects of temperature, atmospheric carbon dioxide and soil moisture on terrestrial carbon storage. If the product of all the signs (plus and minus) in a given loop is positive, then the feedback loop is positive. *Fast feedbacks* in the climatic system operate over timescales of days to months. *RDT*

[See also AUTOVARIATION, HOMEOSTASIS, RESPONSE TIME, SELF-ORGANISATION]

Bennett, R.J. 1978: *Environmental systems.* New York and London: Methuen. Cess, R.D., Potter, G.L., Blanchet, J.P. *et al.* 1990: Intercomparison and interpretation of climatic feedback mechanisms in 19 atmospheric general circulation models. *Journal of Geophysical Research* 95(D10), 16601–16615. Chorley, R.J. and Kennedy, B.A. 1971: *Physical geography: a systems approach.* London and Englewood Cliffs, NJ: Prentice Hall. Drury, G.H. 1981: *An introduction to environmental systems.* London and Exeter, NH: Heinemann. Harvey, L.D.D. 1996: Development of a risk-hedging CO₂-emissions policy, part 1: risks of unrestrained emissions. *Climatic Change* 34, 1–40. Rampino, M.R. and Self, S. 1993: Climate–volcanism feedback and the Toba eruption of ~74,000 years ago. *Quaternary Research* 40, 269–280.

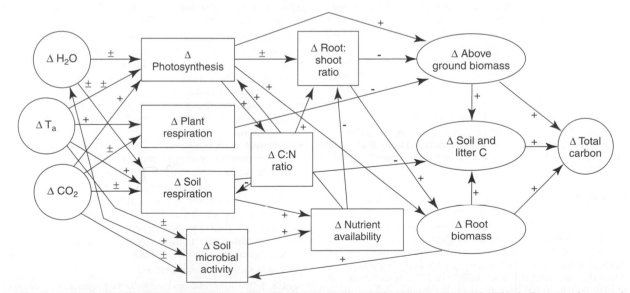

feedback mechanisms *Feedback mechanisms and feedback loops in terrestrial carbon storage: a plus sign indicates an increase in the upstream quantity, which produces an increase in the downstream quantity; a negative sign indicates an increase in the upstream quantity, which produces a decrease in the downstream quantity (after Harvey, 1996)*

feldspar A group of SILICATE MINERALS in which some silica is replaced by aluminium. Feldspars are the most common minerals in the Earth's CRUST. *GO*

Fennoscandian ice sheet The ice sheet that formed over large areas of Finland and Scandinavia on many occasions during the QUATERNARY. The last Fennoscandian ice sheet formed during the WEICHSELIAN glaciation. *DIB*

Donner, J. 1995: *The Quaternary history of Scandinavia.* Cambridge: Cambridge University Press.

feral Pertaining to domesticated plants and, especially, animals that have escaped and become wild, such as feral cats and goats. *JAM*

fermentation The breakdown of carbohydrates by micro-organisms under ANAEROBIC conditions. The production of ethanol in this way from sugar cane for example provides a source of ALTERNATIVE ENERGY. The *fermentation layer* (or *F layer*) occurs in the organic surface horizon of soils (see HUMUS). *JAM*

Ward, O.P. 1989: *Fermentation biotechnology: principles, processes and products.* Englewood Cliffs, NJ: Prentice-Hall.

ferralic horizon A soil horizon having medium or fine texture, a low CATION EXCHANGE CAPACITY (CEC) and lacking weatherable minerals as a result of prolonged and intense CHEMICAL WEATHERING (see FERRALITISATION). These horizons are associated with old stable land surfaces, particularly in the HUMID TROPICS. *EMB*

Food and Agriculture Organization of the United Nations 1998: *World reference base for soil resources* [*Soil Resource Report* 84]. Rome: FAO, ISRIC, ISSS.

ferralitisation A process of prolonged and intense CHEMICAL WEATHERING in a warm humid environment, resulting in the residual development of an acid subsurface horizon, rich in the oxides and hydroxides of aluminium, iron and manganese, from which the more soluble weathering products have been removed. The process leads to the development of a FERRALIC HORIZON. Ferralitisation is described also as a process of *desilicification*. Kaolinitic clays are formed at depth in the soil, and *haematite*, *goethite* and GIBBSITE accumulate relative to the other minerals present in the soil. *EMB*

[See also CLAY MINERALS, FERROLYSIS, KAOLINITE]

Bridges, E.M. 1997: *World soils*, 3rd edn. Cambridge: Cambridge University Press. **Buol, S.W., Hole, F.D. and McCracken, R.J.** 1988: *Soil genesis and classification.* Ames, IA: Iowa State University Press.

Ferralsols Soils having a FERRALIC HORIZON; deep, strongly weathered soils with LOW-ACTIVITY CLAYS, which are physically stable, occurring mainly in the HUMID TROPICS of South America and Africa. Ferralsols have yellowish or reddish colours, diffuse horizons and contain micro-aggregations of clay referred to as *pseudosand* or *pseudosilt*. They do not have the irreversible hardening character of PLINTHOSOLS, but ironstone nodules inherited from these older soils are common (USDA: Oxisols). *EMB*

[See also WORLD REFERENCE BASE FOR SOIL RESOURCES]

ferricrete An iron-rich DURICRUST, associated with deep weathering profiles, especially under a seasonal tropical or subtropical climate with a mean annual precipitation of 1500 mm. Weathering and soil-forming processes such as FERRALITISATION lead to an accumulation of iron oxides and hydroxides in the upper zone of the WEATHERING PROFILE. Ferricrete can also result from reprecipitation of iron-rich solutes at the base of slopes in valley bottom areas as a SECONDARY LATERITE. It is often called LATERITE or described as an *iron pan*. *MAC*

[See also CALCRETE, GYPCRETE, HARDPAN, SILCRETE]

Goudie, A.S. 1985: Duricrusts and landforms. In Richards, K.S., Arnett, R.R. and Ellis, S. (eds), *Geomorphology and soils.* London: Allen and Unwin, 37–57. **Ollier, C.D.** 1991: Laterite profiles, ferricrete, and landscape evolution. *Zeitschrift für Geomorphologie NF* 35, 165–173. **Phillips, J.D.** 2000: Rapid development of ferricretes on a subtropical valley side slope. *Geografiska Annaler* 82A, 69–78.

ferrolysis A weathering process, occurring in tropical climates with strongly alternating wet and dry seasons. It is driven by energy derived from bacterial decomposition of organic matter and involves cycles of OXIDATION and REDUCTION that cause the breakdown of CLAY MINERALS. Cations, liberated by exchange with hydrogen ions in the reduced phase, are leached away. In the oxidised phase, acid weathering of clay minerals occurs and as iron minerals re-oxidise they liberate hydrogen ions to begin the cycle again. The process is associated with PLANOSOLS and ALBELUVISOLS, and with GLEYING in other soils. *EMB*

[See also FERRALITISATION]

Brinkman, R. 1969: Ferrolysis, a hydromorphic soil-forming process. *Geoderma* 3,199–206.

ferromagnetism Spontaneous MAGNETISATION occurs where the spins of unpaired adjacent atoms become aligned in the absence of an applied MAGNETIC FIELD. In crystals where atoms are fixed, the spontaneous magnetisation forms in one particular direction. Different forms of ferromagnetic behaviour result from the arrangement of atoms in the crystal lattice. These are:

- *Ferromagnetic*: parallel coupling of all unpaired electrons, resulting in a strong magnetisation.
- *Antiferromagnetic*: alternate layers of the crystal lattice become paired in opposite directions, giving no overall ferromagnetic behaviour.
- *Ferrimagnetic*: the atomic magnetic moments of two sublattices are unequal, giving a net spontaneous magnetisation.
- *Imperfect antiferromagnetic*: sublattices in an otherwise antiferromagnetic crystal may not be perfectly anti-parallel, giving a small residual spontaneous magnetisation.

These types of magnetic behaviour have been used successfully in the identification of different magnetic minerals, such as MAGNETITE and HAEMATITE, and estimation of their relative contributions to sedimentary environments, thus providing information on sediment supply, provenance, processing and diagenesis. *AJP*

[See also MINERAL MAGNETISM]

Smith, J.P. 1999: An introduction to the magnetic properties of natural materials. In Walden, J., Oldfield, F. and Smith, J.P. (eds), *Environmental magnetism: a practical guide* [*QRA Technical Guide* 6]. London: Quaternary Research Association, 5–25.

Fertile Crescent An arc-shaped area extending from the alluvial lowlands of the Tigris and Euphrates rivers in Mesopotamia through Syria and Lebanon to the southeast Mediterranean coast. It is thought to be the core of one of the main areas of the world (others include Meso-America and southeast Asia) where agriculture made its earliest developments. By about 9000 years ago, crops such as einkorn, emmer, barley, pea and lentil had been domesticated in the area. *RAS*

[See also AGRICULTURAL HISTORY, AGRARIAN CIVILISATIONS, AGRICULTURAL ORIGINS, CEREALS, DOMESTICATION]

Harris, D.R. 1990: Vavilov's concept of centres of origin of cultivated plants: its genesis and its influence on the study of agricultural origins. *Biological Journal of the Linnean Society* 39, 7–16. **Simmons, I.G.** 1996: *Changing the face of the Earth: culture, environment and history*, 2nd edn. Oxford: Basil Blackwell, 87–134.

fertilisers Substances added to soil to provide essential nutrients for plants with the aim of increasing the yield of crops. Relatively low yields, if not characteristic deficiency symptoms, encourage the use of fertilisers. Chemical fertilisers are commonly used to provide required MACRONUTRIENTS, especially nitrogen, phosphorus and potassium, and sometimes MICRONUTRIENTS. Although lime (calcium carbonate) is often added to the soil to increase yields, it is not strictly a fertiliser as it normally affects nutrition *indirectly* through its effects on soil pH. Farm manure and other traditional forms of organic fertiliser perform a dual function of providing nutrients and also improving SOIL STRUCTURE.

Excessive use of fertilisers, combined with failing to coincide applications with the GROWING SEASON of crops, has accentuated losses from the soil of soluble nutrients, particularly nitrates, by LEACHING. Large-scale use of fertilisers, especially those containing NITRATES and PHOSPHATES, has led to the contamination (*agricultural pollution*) of GROUNDWATER supplies and the EUTROPHICATION of rivers, lakes and estuaries. *JAM*

[See also AGRICULTURAL IMPACTS, AGRICULTURAL REVOLUTION, BIOGEOCHEMICAL CYCLES, GREEN REVOLUTION, WATER POLLUTION]

Bacon, P.E. 1995: *Nitrogen fertilization in the environment*. New York: Dekker. **Engelstad, O.P. (ed.)** 1985: *Fertilizer technology and use*. Madison, WI: Soil Science Society of America.

fetch The distance over which wind blows resulting in the generation of waves. *HJW*

feudalism A hierarchical political and economic system, prevalent in Mediaeval England for example, in which use of the land was exchanged for military or labour services. The system was based around the Lord and his manor with characteristic FIELD SYSTEMS (open fields divided into strips). *JAM*

[See also MIDDLE AGES: LANDSCAPE IMPACTS]

Christie, N. 1999: Europe in the Middle Ages. In Barker, G. (ed.), *Companion encyclopedia of archaeology*. London: Routledge, 1040–1076.

field capacity A measure of the maximum water content that a saturated soil can hold under conditions of unrestricted drainage, which is determined on *in situ* field soils. *SHD*

Baize, D. 1993: *Soil science analysis: a guide to current use*. Chichester: Wiley.

field drainage The artificial acceleration of the drainage of agricultural land by enhancement of subsurface flow from soils. Alternatively known as *underdrainage*, field drainage is distinct from the enhancement of surface flow and subsurface flow using ditches. The main approach is *tile drainage*, which involves the burial of PERMEABLE pipes in the soil. A survey in 1969 suggested that about one quarter of the 11 million hectares of agricultural land in England and Wales had already been drained and a similar area could benefit from tile drainage. Government grants resulted in 250 000 acres per year being drained between AD 1970 and 1980, but the emphasis has changed with the scaling down of agricultural subsidies in the UK and the European Union. *JAM*

[See also LAND DRAINAGE, LAND RECLAMATION]

Davis, B., Walker, N., Ball, D. and Fitter, A. 1992: *The soil*. London: Harper Collins. **Green, F.H.W.** 1978: Field drainage in Europe. *Geographical Journal* 114, 171–174.

field research Environmental field research or *fieldwork* consists essentially of recording field observations or collecting materials according to an established scientific framework. In the study of ENVIRONMENTAL CHANGE, field research complements laboratory-based or computer-based approaches. The identification of suitable sites or areas for investigation is followed by the sampling and use of such sites as sources of information. Attempts to reconstruct or monitor environmental change may be based directly on the field evidence or upon subsequent data analysis using numerical techniques and ENVIRONMENTAL MODELLING. The vast number of techniques encompassed by the term field research, and the great variation in type, sophistication and accuracy, reflect their evolution over many years, the diversity of environments under investigation and the nature of the investigation. Many of the methods used to reconstruct QUATERNARY or HOLOCENE ENVIRONMENTAL CHANGE rely upon the collection of suitable and intact sedimentary sequences for later analysis and increasingly use automated techniques. Common techniques used in the study of more recent environmental change also include ENVIRONMENTAL MONITORING, REMOTE SENSING and *field mapping*. Such field-based data are of fundamental importance in the testing of GLOBAL CIRCULATION MODELS and hence in achieving reliable predictions about future environmental change. *DZR*

[See also CORER, FREEZE CORING, CHAMBER/PISTON SAMPLER, ENVIRONMENTAL ARCHAEOLOGY]

Berglund, B.E. (ed.) 1986: *Handbook of Holocene palaeoecology and palaeohydrology*. Chichester: Wiley. **Drewett, P.L.** 1999: *Field archaeology: an introduction*. London: UCL Press. **Gardener, V. and Dackcombe, R.** 1983: *Geomorphological field manual*. London: George Allen and Unwin. **Hester, T.N., Shafer, H.J. and Heizer, R.F.** 1987: *Field methods in archaeology*, 7th edn. Palo Alto, CA: Mayfield. **Kuklick, H. and Kohler R.** 1996: *Science in the field*. Ithaca, NY: Cornell University Press. **Lounsbury, J.F. and Aldrich, F.T.** 1986: *Introduction to geographic field methods and techniques*. Columbus, OH: Charles E. Merrill.

field systems The ways in which agricultural land is organised and subdivided for arable or pastoral use. There is variation in practice on regional, national and international bases that reflects differences in conditions of the physical environment, cultural practices and traditions. The open grid of the North American prairies, for example, is the product of a specific land allocation system and the nature of the land. It offers a sharp contrast to the small, intensively cultivated TERRACING that has been created in Mediterranean countries or the sharply demarcated farms of upland Europe.

The existence of field systems can be traced back to PREHISTORIC times and the field systems are discernible in LANDSCAPE through both AERIAL PHOTOGRAPHY and ARCHAEOLOGICAL PROSPECTION. Much interest has focused on the boundary markers of old field systems, best preserved in MOORLANDS, which have been little used for agriculture since prehistoric times. Evidence has been found for *walls* (such as the parallel *reaves* of Dartmoor), *dykes* and LYNCHETS and also for the *banks* and *ditches* designed to control livestock. *Celtic fields* (the often compact fields that were in use before the Saxon strip field) offer an example of a discernible field system associated with a specific prehistoric period and society. Similarly, at Engaruka at the foot of the escarpment of the Great Rift Valley in northern Tanzania is found one of the few examples of sub-Saharan IRON AGE field systems (dating from 300–600 years ago) that can be discerned in the landscape.

Historically, the main distinction in Western Europe in particular has been between the open and enclosed field systems. The *open field systems* had close ties with Mediaeval FEUDALISM and were cultivated communally. Found throughout lowland Europe with their distinctive linear strips and nucleated villages, they remain a clear part of the landscape. COMMON LAND, *common fields* and *waste* were key original features of this form of land organisation. The arable associated with each village was worked in two or more large open, common fields (one of which normally lay fallow for a year), whereas the waste was the little-used common land. The *enclosed field systems* were originally the characteristic form throughout upland areas, reflecting both topography and a social organisation that was based on dispersed, independently held farms. One variant in Europe was the *infield–outfield system*: the *infield* lying close to the farm or settlement was kept in fertile condition and used for arable, whereas the *outfield* was largely maintained for pasture (although, in Scotland, a type of SHIFTING CULTIVATION was widespread in the outfields).

Field systems change over time and may also indicate patterns of land inheritance that often led to subdivisions of a holding among surviving children. This led to widescale fragmentation into smaller plots and subsequent attempts to consolidate these into larger and more economic units. The twentieth-century French process of *remembrement* is specifically of this character. Modern trends have been to organise farming land on market principles and both ENCLOSURES and the application of ranch farming reflect these. *DTH*

Baker, A.R.H. and Butlin, R.A. (eds) 1973: *Studies of field systems in the British Isles*. Cambridge: Cambridge University Press. Beckett, J.V. 1989: *A history of Laxton: England's last open-field village*. Oxford: Blackwell. Dodgson, R.A. 1980: *The origin of British field systems: an interpretation*. London: Academic Press. Johnston, R. 1999: Prehistoric field systems in Northwest Europe. *The Prehistoric Society, Past* 31, 12–14. Mercer, R. 1981: *Farming practice in British Prehistory*. Edinburgh: Edinburgh University Press. Sutton, J.E.G. (ed.) 1989: History of African agricultural technology and field systems. *Azania* 24 [Special Issue].

fields   Subsets of a record in a DATABASE. A field contains one or more ATTRIBUTES, which describe an inter-related set of information. *TVM*

[See also RELATIONAL DATABASE]

filtering   Methods for *smoothing* or enhancing the signal in a data series, for example in TIME-SERIES ANALYSIS or REMOTE SENSING. The simplest type of filter commonly employed is a *moving-window average*, which calculates the mean of the data values in a locale and replaces the central value with that mean. In effect, this is a method for smoothing the data and reduces the impact of any noise, accentuating the low frequencies in the series. In the context of a time series, this is also known as a *running mean*. More complex filters of this sort assume a model of the data and remove data that deviate too greatly from this model (e.g. GAUSSIAN FILTERS). Moving-window filters may also be used to boost high frequencies and perform more complex operations such as EDGE DETECTION. *High-pass filters* are used to eliminate relatively low frequencies; *low-pass filters* enhance the low frequencies by eliminating relatively high frequencies. FOURIER ANALYSIS may be used to filter out specific frequency components in a signal: by decomposing a signal into its frequency components, it is possible to exclude the undesired frequencies when the inverse transform is applied. *TQ*

[See also KALMAN FILTER, SPATIAL ANALYSIS, SPATIAL FILTERING]

fire frequency   The average annual probability of a site being burned. More loosely, fire frequency is the number of fires that occur within a given time period or the interval between fires. *JLF*

Larsen, C.P.S. 1997: Spatial and temporal variation in boreal forest fire frequency in northern Alberta. *Journal of Biogeography* 24, 663–673.

fire history   Fire history is a record of the fire regime within an area in the past. The main components of a fire regime include FIRE FREQUENCY, size, duration and intensity. The impacts of fire can range from killing the local FLORA and FAUNA, to instances where fire can pass through an area without having much of an effect, depending on environmental conditions, fuel loading, fire tolerance of the flora and mobility of the fauna. In areas where fire is an important DISTURBANCE mechanism (e.g. in the BOREAL FOREST), an understanding of the fire regime is essential to understand ecosystem composition, structure and function, and for ENVIRONMENTAL MANAGEMENT and CONSERVATION purposes. Where natural fires occur, the interval between fires can span from years to decades to centuries and therefore long-term records are essential. As fire is controlled largely by climatic conditions, by fuel type and loading and by ignition frequencies, information on past fire regimes can provide indirect evidence for CLIMATIC CHANGE in the past (see Figure). An understanding of the relationship between fire and climate is essential to predict the impacts of projected climatic change on disturbance factors such as fire.

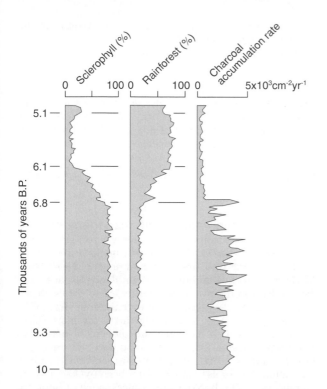

fire history *An illustration of the effect of changing fire frequency on vegetation history at Lake Barrine, northeastern Australia. This summary pollen diagram indicates the replacement of fire-adapted sclerophyllous vegetation by tropical rain forest between 6800 and 6100 BP following an abrupt decline in fire frequency caused by an abrupt climatic change with increased precipitation (Walker and Chen, 1987; Delcourt and Delcourt, 1991)*

Fire history within an area can be determined from charred-particle abundances in sedimentary environments, FIRE SCAR studies, DENDROECOLOGY, historic accounts and modern records. Stratigraphic CHARRED-PARTICLE ANALYSIS can provide information on past fire frequency and intensity (fine-resolution studies), although in most cases it provides data on the relative importance of fire in the past. FIRE SCAR studies and DENDROCHRONOLOGY can provide an accurate chronology of past fires within the lifetime of the tree sampled. Historical accounts can be useful for determining fire frequency, but often fire size and location are not recorded accurately and the records do not extend far back in time. Data from studies of modern fire regimes may be extrapolated back into the past; however, these projections may not be reliable under conditions of changing climate, vegetation and land-use practices. In most cases, therefore, it is only possible to determine the relative importance of fire in the past. *JLF*

Delcourt, H.R and Delcourt, P.A. 1991: *Quaternary ecology: a palaeoecological perspective.* London: Chapman and Hall. **Heinselman, M.L.** 1973: Fire in the virgin forests of the Boundary Waters Canoe Area, Minnesota. *Quaternary Research* 3, 329–382. **Johnson, E.A. and Gutsell, S.L.** 1994: Fire frequency models, methods, and interpretations. *Advances in Ecological Research* 25, 239–287. **Johnson, E.A.** 1992: *Fire and vegetation dynamics: studies from the North American boreal forest.* Cambridge: Cambridge University Press. **Mohr, J.A.,**

Whitlock, C. and Skinner, C.N. 2000: Postglacial vegetation and fire history, eastern Klamath Mountains, California, USA. *The Holocene* 10, 587–601. **Patterson III, W.A. and Backman, A.E.** 1988: Fire and disease history of forests. In Huntley, B. and Webb III, T. (eds), *Vegetation History.* Dordrecht: Kluwer, 603–632. **Patterson III, W.A., Edwards, K.J. and Maguire, D.J.** 1987: Microscopic charcoal as a fossil indicator of fire. *Quaternary Science Reviews* 6, 3–23. **Pitkänen, A. and Huttunen, P.** 1999: A 1300-year forest-fire history at a site in eastern Finland based on charcoal and pollen records in laminated lake sediment. *The Holocene* 9, 311–320. **Tinner, W., Conedera, M., Ammann, B.** *et al.* 1998: Pollen and charcoal in lake sediments compared with historically documented fires in southern Switzerland since AD 1920. *The Holocene* 8, 31–42. **Walker, D. and Chen, Y.** 1987: Palynological light on rainforest dynamics. *Quaternary Science Reviews* 6, 77–92. **Whelan, R.J.** 1995: *The ecology of fire.* Cambridge: Cambridge University Press.

fire impacts: ecological Fire is a powerful SELECTION PRESSURE and one of the most important processes of DISTURBANCE that determine the structure and composition of ecosystems. The flammability and fuel load of an ecosystem control FIRE FREQUENCY and intensity and the relationship of these to SPECIES DIVERSITY is described by the INTERMEDIATE DISTURBANCE HYPOTHESIS. Fires of low intensity but intermediate frequency are thought to be crucial to the maintenance of high BIODIVERSITY in MEDITERRANEAN-TYPE VEGETATION. Many short-lived graminoids (see GRASSES) and FORBS are able to co-exist with large numbers of shrub species as fire prevents competitively dominant species taking over. Patchy fires are also important in creating spatial heterogeneity. In contrast, in many TEMPERATE coniferous forests large-scale intense wildfire may maintain the dominance of a small number of tree species. In ecosystems that are less flammable, fires are less frequent. As a consequence there is often time for a greater quantity of BIOMASS to accumulate and for the intensity of a WILDFIRE, when it occurs, to be greater. In such ecosystems, few species have ADAPTATIONS to fire and most perish. In contrast to *pyrophytic ecosystems* (where frequent fire may promote structural and compositional stability), high-intensity, low-frequency fires may destabilise an ecosystem, leading to SECONDARY SUCCESSION.

It is believed that many invertebrates but few vertebrates are killed by wildfire. However, the indirect effects of fire, through the modification of HABITAT, may have substantial impacts on vertebrate populations. Fire typically has a lethal effect on soil micro-organisms, but their populations recover quickly once plant REGENERATION has commenced. The sanitising effect of fire on soil is a justification for an ancient agricultural practice.

Sudden changes in fire frequency can lead to dramatic changes in ecosystems. Large areas of the HUMID TROPICS that have been subject to DEFORESTATION have been transformed into fire-climax *grasslands*. Fire plays an important ecological role in a number of human-created, PLAGIOCLIMAX ecosystems such as HEATHLAND and tallgrass *prairie*, which are now perceived to have high CONSERVATION value. Their management includes the application of prescribed burning at regular intervals in order to favour characteristic species. *NDB*

[See also BIOMASS BURNING, TROPICAL FOREST FIRES]

Daubenmire, R. 1968: Ecology of fire in grassland. *Advances in Ecological Research* 5, 209–266. **Kimmins, J.P.** 1996: *Forest ecol-*

ogy: a foundation for sustainable management, 2nd edn. Englewood Cliffs, NJ: Prentice Hall. **Kozlowski, T.T. and Ahlgren, C.E. (eds)** 1974: *Fire and ecosystems.* New York: Macmillan. **Kutiel, P. and Inbar, M.** 1993: Fire impact on soil nutrients and soil erosion in a Mediterranean pine forest. *Catena* 20, 129–139. **Pyne, S.J., Andrews, P.L. and Laven, R.D.** 1996: *An introduction to wildland fire,* 2nd edn. New York: Wiley. **Tyler, C.M.** 1995: Factors contributing to postfire seedling establishment in chaparral: direct and indirect effects of fire. *Journal of Ecology* 83, 1009–1020.

fire impacts: geomorphological

Fire impact will depend largely on the intensity and the type of burn, which governs the extent of modifications to vegetation cover and soil properties. To a greater or lesser extent, however, vegetation cover is reduced or killed during fire and, as a result, EVAPOTRANPIRATION losses are temporarily reduced or cease altogether. Consequently, RUNOFF regimes in fire-affected catchments are altered, usually exhibiting a temporary increase in RUNOFF and storm peak flows. Soil properties are also affected by fire, commonly with reductions in organic matter and INFILTRATION. The reduction in vegetation cover results in more effective rainsplash detachment of soil and, in combination with higher occurrences of overland flow due to the reduced infiltration, erosion rates are significantly higher after fire (Imeson *et al.*, 1992). Enhanced EROSION RATES and runoff can occur for several weeks to years following fire, largely depending on vegetation recovery but also on post-fire land management practices.   *ADT*

[See also TROPICAL FOREST FIRES]

Imeson, A.C., Verstraten, J.M., Van Mulligen, E.J. and Sevink, J. 1992: The effects of fire and water repellency on infiltration and runoff under Mediterranean type forest. *Catena* 19, 345–361. **Kutiel, P., Lavee, H., Segev, M. and Beyamini, Y.** 1995: The effect of fire-induced surface heterogeneity on rainfall–runoff–erosion relationships in an eastern Mediterranean ecosystem, Israel. *Catena* 25, 77–87. **Walsh, R.P.D., Boakes, D., Coelho, C. de O.A.** *et al.* 1995: Post-fire landuse and management and runoff responses to rainstorms in Northern Portugal. In McGregor, D. and Thompson, D. (eds), *Geomorphology and Land Management in a Changing Environment.* Chichester: Wiley, 283–308.

fire scar

Low-intensity fires will sometimes kill or damage part of the CAMBIUM of a tree, or group of trees, near ground level, but the trees will survive and continue growing. A characteristic inverted V-shaped fire scar at the base of a tree is often evidence of past fire. If cross-sections are taken through the base of such trees, the damage caused by many fires may be apparent at the margins of the main scar, separated by healthy overgrown wood tissue. By CROSS-DATING the ring-width measurements from a number of trees at the site, the exact dates of the fires can be established.   *KRB*

[See also DENDROCHRONOLOGY, FIRE FREQUENCY]

Lehtonen, H. and Huttunen, P. 1997: History of forest fires in eastern Finland from the fifteenth century A.D. – the possible effects of slash-and-burn cultivation. *The Holocene* 7, 223–228. **Swetnam, T.W.** 1993: Fire history and climate change in giant sequoia groves. *Science* 262, 885–889.

firn

Snow that has survived one melt season and has begun transformation; it has a density of 0.4–0.83. The transition from firn to ice occurs when interconnected air passages become sealed off.   *AN*

Paterson, W.S.B. 1994: *The physics of glaciers.* Oxford: Pergamon.

firn line

The boundary between FIRN and ice on the GLACIER surface at the end of the ABLATION season. On TEMPERATE GLACIERS, the firn line and the EQUILIBRIUM-LINE ALTITUDE will be more-or-less coincident.   *AN*

[See also SNOW LINE]

Benn, D.I. and **Evans, D.J.A.** 1998: *Glaciers and glaciation.* London: Arnold.

fisheries conservation and management

The necessity for conservation of fisheries was recognised during the 1880s when depletion of fish stocks, especially by steam trawlers, was first detected in the North Sea, Mediterranean Sea and off Newfoundland. Expansion of fishing fleets in the twentieth century led to the establishment of numerous international organisations devoted to fisheries conservation and management, the main scientific basis for which has been the concept of maximum SUSTAINABLE YIELD. Despite this, and partly as envisaged in the TRAGEDY OF THE COMMONS, the practical result of the exploitation of fish resources has been OVERFISHING on a grand scale. Technological advances, such as sophisticated acoustical equipment for locating fish, which was developed for the detection of submarines during the Second World War, and the monofilament nylon drift net, which permitted even the fish of the open ocean to be caught in large numbers, meant that almost all the major fish stocks of the world could be exploited. Thus, collapse of many of the most important traditional fisheries has occurred since the 1960s. By 1990, the total world fish catch was fluctuating around 100 million tonnes, which, according to one estimate, is the maximum sustainable yield. Establishment of Exclusive Economic or Fishing Zones from the 1970s was a significant event because there are now signs of better conservation of fish stocks under national control than has been evident from the many years of broken international agreements.

Overfishing, along with HABITAT degradation (from HYDROLOGICAL change, LANDUSE IMPACTS, POLLUTION and EUTROPHICATION) has lead to shifts in the relative importance of the wild species caught and an overall reduction in fish diversity. Although most fish species are marine, the greatest losses are amongst the freshwater fishes. It has been estimated that some 20% of the world's 9500 freshwater fish species are extinct or in danger of extinction; indeed, in Mediterranean regions the figure may be as high as 65%. These figures reflect the comparative neglect of fish diversity in particular and of aquatic BIODIVERSITY in general.

Developments in AQUACULTURE and MARICULTURE are partly responsible for maintaining the global fish catch. Global production of farmed fish increased to about 30 million tonnes by 1997 (around 25% of fish harvested). However, the farming of carnivorous species requires large inputs of wild fish as feed. In many cases the ratio of wild-fish biomass used to farmed fish produced is between 2.5:1 and 5:1, which is having a detrimental effect on some wild species. Other detrimental effects have resulted from introduced species (see INTRODUCTION) and HABITAT LOSS or transformation (especially of MANGROVE swamps). *JAM*

[See also CLIMATIC CHANGE PAST IMPACT ON ANIMALS, WHALING AND SEALING]

Anderson, L.G. 1986: *The economics of fisheries management.* Baltimore, MD: Johns Hopkins University Press. **Bruton, M.N.** 1995: Have fishes had their chips? The dilemma of threatened fishes. *Environmental Biology of Fishes* 43, 1–27. **Densen, W.L.T. and Morris, M.J. (eds)** 2000: *Fish and fisheries of lakes and reservoirs in Southeast Asia and Africa.* Otley: Westbury. **Erickson, C.L.** 2000: An artificial landscape-scale fishery in the Bolivian Amazon. *Nature* 408, 190–193. **Hall, S.** 1999: *The effects of fishing on marine ecosystems and communities.* Oxford: Blackwell Science. **Moyle, P.B. and Moyle, P.R.** 1995: Endangered fishes and economics: intergenerational obligations. *Environmental Biology of Fishes* 43, 29–37. **Naylor, R.L., Goldburg, R.J., Primavera, J.H. et al.** 2000: Effect of aquaculture on world fish supplies. *Nature* 405, 1017–1024. **Pauly, D. and Christensen, V.** 1995: Primary production required to sustain global fisheries. *Nature* 374, 255–257. **Ryther, J.H.** 1969: Photosynthesis and fish production in the sea. *Science* 166: 72–76. **Smith, T.M.** 1994: *Scaling fisheries: the science of measuring the effects of fishing 1855–1955.* Cambridge: Cambridge University Press. **Voigtlander, C.W. (ed.)** 1994: *The state of the World's fisheries resources.* New Delhi: IBH Publishing.

fissile, fissility The tendency of some fine-grained rocks (e.g. SHALE) to break into sheets. *GO*

fission-track analysis The analysis of damage trails (*fission tracks*) within single crystals caused by the spontaneous decay of the radioactive isotope ^{238}U. The method is usually applied to crystals of apatite or zircon. The density of fission tracks is related to the age of a sample in FISSION-TRACK DATING. *Fission track thermochronology* relates the lengths of fission tracks to the thermal history of a sample. By assuming appropriate values of the GEOTHERMAL gradient, this can be converted to a trajectory of burial depth through time. Regional variations can be interpreted in terms of histories of UPLIFT and SUBSIDENCE, yielding information about PALAEOGEOGRAPHY, DENUDATION CHRONOLOGY and long-term LANDSCAPE EVOLUTION, particularly when used in conjunction with PALAEOMAGNETISM or other techniques of *palaeothermometry* such as VITRINITE REFLECTANCE. The method can also be applied to determining the PROVENANCE of sedimentary deposits or the MATURITY of a sedimentary basin that may potentially yield petroleum. *GO*

Carter, A. 1999: Present status and future avenues of source region discrimination and characterization using fission track analysis. *Sedimentary Geology* 124, 31–45. **O'Sullivan, P.B., Gibson, D.L., Kohn, B.P. et al.** 2000: Long-term landscape evolution of the Northparkes region of the Lachlan Fold Belt, Australia: Constraints from fission track and paleomagnetic data. *Journal of Geology* 108, 1–16. **Steinmann, M., Hungerbuhler, D., Seward, D. and Winkler, W.** 1999: Neogene tectonic evolution and exhumation of the southern Ecuadorian Andes: a combined stratigraphy and fission-track approach. *Tectonophysics* 307, 255–276. **Thomson, K., Underhill, J.R., Green, P.F. et al.** 1999: Evidence from apatite fission track analysis for the post-Devonian burial and exhumation history of the northern Highlands, Scotland. *Marine and Petroleum Geology* 16, 27–39.

fission-track dating A RELATIVE-AGE DATING technique based on the spontaneous radioactive decay of ^{238}U by nuclear fission. In suitable minerals, particularly zir-

con, but also mica, sphene and apatite, and in glasses such as obsidian, this fission takes the form of the splitting of the nucleus into two smaller nuclei that recoil at 180° to each other. In this process the fission tracks are left within the atomic lattice. Although very short, < 0.01 mm, these tracks can be identified microscopically following chemical etching and counted. The number of tracks is dependent upon the amount of uranium originally present in the sample and the rate of spontaneous fission. Tracks are lost following heating, which causes zeroing of the signal. The number of tracks found in a sample are compared with the number of extra tracks induced by placing it within a nuclear reactor, and are expressed as number of tracks per cm^2. Results are converted to age either by CALIBRATION against a sample of known age or by reference to known rates of fission.

The age range covered by fission-track dating depends on the uranium content but, although dates considerably younger than 20 ka have been produced, this is a realistic lower limit due to the lack of tracks and the difficulty of identification in younger material. Because of the paucity of tracks in some archaeological material as a result of low rates of fission, tracks left by more frequently emitting alpha particles have been used, although they are more difficult to identify. Some of the oldest rocks on the EARTH have been dated by this method as it is an important technique over the longer GEOLOGICAL TIMESCALE. It is particularly useful for dating VOLCANIC GLASS in TEPHROCHRONOLOGY, having proved valuable in early HOMINID studies by dating volcanic layers in the fossil-bearing STRATIGRAPHIES, and more directly in archaeology for dating ARTEFACTS or waste flakes made of OBSIDIAN. *CJC*

Bellot-Gurlet, L., Bigazzi, G., Dorighel, O. et al. 1999: The fission-track analysis: an alternative technique for provenance studies of prehistoric obsidian artefacts. *Radiation Measurements* 31, 639–644. **Espizua, L.E. and Bigazzi, G.** 1998: Fission-track dating of the Punta de Vacas glaciation in the Rio Mendoza valley, Argentina. *Quaternary Science Reviews* 17, 755–760. **Fleischer, R.L.** 1965: Fission track dating of Bed I, Olduvai Gorge. *Science* 148, 72–74. **Fleischer, R.L., Price, P.B. and Walker, R.M.** 1975: *Nuclear tracks in solids: principles and applications.* Berkeley, CA: University of California Press. **Gogen, K. and Wagner, G.A.** 2000: Alpha-recoil track dating of Quaternary volcanics. *Chemical Geology* 166, 127–137. **Guo, S.L., Huang, W., Hao, X.H. and Chen, B.L.** 1997: Fission track dating of ancient man site in Baise, China, and its significance in space research, paleomagnetism and stratigraphy. *Radiation Measurements* 28, 565–570. **Huang, W.H. and Walker, R.M.** 1967: Fossil alpha-particle recoil tracks: a new method of age determination. *Science* 155, 1103–1106. **Hurford, A.J.** 1991: Fission track dating. In Smart, P.L. and Frances, P.D. (eds), *Quaternary dating methods – a users guide* [QRA Technical Guide 4]. Cambridge: Quaternary Research Association, 84–107. **Naeser, C.W. and Naeser, N.D.** 1988: Fission-track dating of Quaternary events. *Geological Society of America Special Paper* 227, 1–11. **Wagner, G.A. and Van den Haute, P.** 1992: *Fission-track dating.* Stuttgart: Ferdinand Enke. **Walter, R.C.** 1989: Application and limitation of fission-track geochronology to Quaternary tephras. *Quaternary International* 1, 35–46. **Westgate, J., Sandhu, A. and Shane, P.** 1997: Fission track dating. In Taylor, R.E. and Aitken, M.J. (eds), *Chronometric dating in archaeology.* New York: Plenum, 127–158.

fissure eruption A VOLCANIC ERUPTION along a linear fracture, characterised by LAVA FLOWS of BASALT and emission of VOLATILES. On a large scale, FLOOD BASALTS may form. Most notable during the HOLOCENE was the Laki eruption of AD 1783, with a volume of 12.3 km³, which caused 35% human and 75% sheep mortality in Iceland and AIR POLLUTION over much of Europe. Eruptions at MID-OCEAN RIDGES are of this type. *JBH*

Grattan, J. and Brayshay, M. 1995: An amazing and portentous summer: environmental and social responses in Britain to the 1783 eruption of an Icelandic volcano. *Geographical Journal* 161, 125–134.

fjard A Swedish term for a submerged valley in low-lying glaciated terrain, usually shallow. *HJW*

[See also FJORD]

fjord A Norwegian–Danish term for a glacial U-SHAPED VALLEY partially submerged beneath SEA LEVEL. *HJW*

[See also FJARD, RIA]

Flandrian A local (British) term for the present INTER-GLACIAL, more commonly known as the HOLOCENE. *JAM*

[See also HOLOCENE TIMESCALE]

flaser bedding A SEDIMENTARY STRUCTURE comprising sand with CROSS-LAMINATION interrupted by thin, discontinuous lenses of mud, indicating deposition under conditions of repeatedly alternating sediment transport by water currents, and still water. Flaser bedding is commonly taken as an indicator of sedimentation by TIDES. Different ratios of sand to mud characterise *wavy bedding* (interbedded mud and cross-laminated sand) and *lenticular bedding* (lenses of cross-laminated sand encased in mud). *GO*

[See also MUD DRAPES, TIDAL BUNDLES, TIDAL RHYTHMITES]

Allen, J.R.L. 1982: *Sedimentary structures: their character and physical basis.* Amsterdam: Elsevier. **Reineck, H.-E. and Singh, I.B.** 1986: *Depositional sedimentary environments.* Berlin: Springer.

flash flood A short-lived river or stream FLOOD characterised especially by rapid onset. Flash floods are common in ARIDLANDS in response to high-intensity rainfall, but *flashy regimes* may occur elsewhere, particularly in areas of high relief and low INFILTRATION CAPACITY. LAND-USE CHANGES (such as DEFORESTATION, OVERGRAZING and URBANISATION) and attendant SOIL EROSION and GULLY erosion may exacerbate flash flooding if the river channel system is unable to accommodate the extra RUNOFF generated. *RPDW/JAM*

[See also DRAINAGE BASIN, RAPID-ONSET HAZARD]

Schick, A.P. 1995: Fluvial processes on an urbanizing alluvial fan: Eilat, Israel. In Costa, J.E., Miller, A.J., Potter, K.W. and Wilcock, P.R. (eds), *Natural and anthropogenic influences in fluvial geomorphology* [*Geophysical Monographs* 89]. Washington, DC: American Geophysical Union.

floating bog/mire A WETLAND, also termed *Schwingmoor*, consisting of a quaking mat of plants (particularly SPHAGNUM moss) growing over a water body. *JAM*

Warner, B.G. 1993: Palaeoecology of floating bogs and landscape change in the Great Lakes drainage basin of North America. In Chambers, F.M. (ed.), *Climate change and human impact on the landscape: studies in palaeoecology and environmental archaeology.* London: Chapman and Hall, 237–245.

floating chronology In DENDROCHRONOLOGY, a ring-width CHRONOLOGY that is not fixed to a calendar timescale. It is possible to cross-date the time series of year-to-year ring width values, as measured in a number of different trees that were not living at the time of sampling (see CROSS-DATING). The goodness of fit between the series can be sufficient to establish that the trees grew contemporaneously and the measurement series may therefore be averaged with the correct relative alignment. However, unless the resulting chronology can be reliably cross-dated against an existing, firmly dated chronology, it is said to be floating. Often, the absolute date of such chronologies can be estimated fairly accurately (to within a decade or so) by WIGGLE MATCHING a series of high-precision radiocarbon dates of known relative position (from wood samples whose ring widths were measured to build the chronology) against the published radiocarbon anomaly curve. Until it is firmly tied to the dendrochronological (absolute) timescale, however, such a chronology is still deemed to be floating. *KRB*

Baillie, M.G.L. 1995: *A slice through time: dendrochronology and precision dating.* London: Routledge.

flocculation A process causing CLAY particles to form aggregates. It occurs, for example, when a river carrying electrically charged colloidal clays mixes with saline water, which carries electrically charged particles in solution. *UBW*

[See also SEDIMENT TYPES, SOIL DEVELOPMENT]

flood A *coastal flood* is the incoming (rising) wind-induced and/or astronomical tide along a shoreline. *River floods* are the higher stages of a river regime, especially the OVERBANK FLOWS. The four main causes of river floods are: (1) intense and/or prolonged rainfall; (2) snow melt; (3) failure of a natural structure; and (4) failure of an artificial dam. The four main methods of reducing the impacts of flood hazard are: (1) flood modification, by reservoir construction; (2) flood containment, by engineering works such as CHANNELISATION; (3) flood loss reduction by, for example, structural measures or LANDUSE zoning; and (4) flood loss bearing (through insurance). HISTORICAL EVIDENCE can be used to investigate past FLOOD FRE-QUENCY–MAGNITUDE CHANGES and the RETURN PERIODS or *recurrence intervals* of floods of given magnitudes. Geomorphological and stratigraphical evidence can be used to investigate PALAEOFLOODS on longer timescales. *RAS/JAM*

[See also FLOOD HISTORY, PALAEOHYDROLOGY, RESERVOIRS: ENVIRONMENTAL EFFECTS]

Baker, V.R., Kochel, R.C. and Patton, P.C. (eds) 1988: *Flood geomorphology.* New York: Wiley. **Beven, K. and Carling, P.A.** 1989: *Floods: hydrological, sedimentological and geomorphological perspectives.* Chichester: Wiley. **Kale, V.S., Ely, L.L., Enzel., Y. and Baker, V.R.** 1996: Palaeo and historical flood hydrology, Indian Peninsula. In Branston, J., Brown, A.G., and Gregory, K.J. (eds), *Global continental changes: the context of palaeohydrology* [*Special Publication* 115]. Bath: Geological Society, 155–163. **Ward, R.C.** 1978: *Floods: a geographical perspective.* London: Macmillan. **Wohl, E.E. (ed.)** 2000: *Inland flood hazards: human, riparian and aquatic communities.* Cambridge: Cambridge University Press.

flood basalt An extensive sheet of basalt LAVA from a FISSURE ERUPTION. Flood basalts represent some of the largest single eruptive events known, some individual sheets exceeding 1500 km³ in volume (mid-MIOCENE Columbia River Plateau, USA). Repeated eruptions may build up a LAVA PLATEAU. Flood basalt eruptions may be associated with the rise of a MANTLE PLUME to the Earth's surface (see HOTSPOT), and may have occurred at regular intervals during Earth history, with major VOLCANIC IMPACTS ON CLIMATE.   *JBH*

[See also K–T BOUNDARY]

Johnston, S.T. and Thorkelson, D.J. 2000: Continental flood basalts: episodic magmatism above long-lived hotspots. *Earth and Planetary Science Letters* 175, 247–256. **Kent, R.W., Thomson, B.A., Skelhorn, R.R.** *et al.* 1998: Emplacement of Hebridean Tertiary flood basalts: evidence from an inflated pahoehoe lava flow on Mull, Scotland. *Journal of the Geological Society, London* 155, 599–607. **Rampino, M.R. and Stothers, R.B.** 1988: Flood basalt volcanism during the past 250 million years. *Science* 241, 663–668.

flood history Historical DOCUMENTARY EVIDENCE and *flood-level markers* can be used to establish chronologies of river flood events for a location, reach or basin of a river. Flooding is today still considered a great natural catastrophe and/or NATURAL HAZARD. The question of repetition, height and trends of such events is regularly posed, particularly with predictions by GENERAL CIRCULATION MODELS of increased FLOOD MAGNITUDE-FREQUENCY with GLOBAL WARMING. Flood-level markers, a common feature on buildings, bridges and other riverine structures in the older towns and cities in Europe have yielded *flood chronologies* back to the sixteenth century. Documentary accounts of individual floods can extend records back even further. In the case of the River Nile, heights of NILE FLOODS at the Roda gauge in Lower Egypt have been recorded back to the seventh century. Such data series not only give direct chronologies of flood heights, but, if allowance is made for changes in the river cross-section (through progressive sedimentation in the case of the Nile) and any buildings or structures in the vicinity of the site, they can sometimes be converted into flood DISCHARGE series. Often, indicators as to climatic or underlying anthropogenic causes (such as changes in LANDUSE) of floods or changes in flood frequency can be derived from descriptive historical data. In some cases, flood-level series are available from a hierarchy of sites within larger drainage basins, allowing: (1) spatial patterns of flooding and rainfall during individual flood events to be established; and (2) temporal changes in flood magnitude-frequency for basins of different sizes or locations to be differentiated.   *RG*

[See also PALAEOFLOOD, PALAEOHYDROLOGY]

Changnon, S.A. 1983: Trends in floods and related climate conditions in Illinois. *Climatic Change* 5, 341–358. **Evans, T.** 1990: History of Nile flows. In Howell, P.P. and Allen, J.A. (eds), *The Nile.* London: School of Oriental and African Studies. **Jones, R.D., Ogilvie, A.E.J. and Wigley, T.M.L.** 1984: *Riverflow data for the United Kingdom: reconstructed data back to 1844 and historical data back to 1556.* Norwich: Climatic Research Unit, University of East Anglia. **Prieto, M de R., Herrera, R. and Dussel, P.** 1999: Historical evidences of streamflow fluctuations in the Mendoza River, Argentina, and their relationship with ENSO. *The Holocene* 9, 473–481. **Walsh, R.P.D., Davies, H.R.J. and Musa, S.B.** 1994: Flood frequency and impacts at

Khartoum since the early nineteenth century. *Geographical Journal* 160, 266–279.

flood magnitude-frequency changes Fluctuations over time in the depth, discharge or recurrence of FLOODS adjacent to rivers, i.e. changes to the temporal distribution of OVERBANK FLOWS, which spill out and inundate land adjacent to a river. Much work has been based on analysis of instrumental records at river gauging stations, with discharge data sometimes extending back 100 years. Records can sometimes be extended back to earlier centuries using dated historic flood height marks or other DOCUMENTARY EVIDENCE such as newspapers or journals (e.g. Rivers Severn and Nile). Prior to the historical period, or in ungauged catchments, flood chronologies have to be established from dated OVERBANK DEPOSITS or from complete FLOODPLAIN sedimentary and terrace sequences (e.g. the upper Mississippi basin in Wisconsin, USA, dated with RADIOCARBON methods).

Many studies in the Severn basin, South Wales and Scottish rivers, based on RETURN PERIODS, flood frequency analysis or other methods of TIME-SERIES ANALYSIS, have demonstrated significant increases in flooding since the AD 1920s or 1940s. This has been ascribed primarily to an increase in heavy daily rainfall events (e.g. >50 or 63 mm), despite largely invariant annual precipitation totals. Such changes appear to be driven by *atmospheric circulation* changes and, especially for upland areas, increases in the incidence of weather of a westerly or cyclonic type. A secondary control is thought to be LAND-USE CHANGE, especially the expansion of LAND DRAINAGE systems that route rainfall to streams more efficiently.

Work on flow time series in the upper Mississippi basin has shown declines in flood frequencies since about 1950 (see Figure), which, in the presence of unchanging rainfall event distributions, have been explained by improvements in LAND MANAGEMENT practices (e.g. shifts to contour ploughing). At longer timescales, ^{14}C-dated VALLEY FILL chronologies established for the past 25 000 years, combined with palaeohydraulic modelling, have revealed several episodes of increased or decreased flood magnitude and frequency of the order of ±15–20%. Increases are associated with periodic, yet modest, switches to cooler and wetter conditions. Substantial flood increases followed replacement of prairie and forest by cultivated crops in the early nineteenth century. Globally, G.H. Dury postulated huge increases in flood magnitude as part of his MISFIT MEANDER hypothesis.

Defining the cause(s) of any changes in flood frequency and/or magnitude is often difficult, especially disentangling the competing hypotheses of CLIMATE CHANGE and LANDUSE CHANGE. Even when these are resolved, an additional intriguing question is which specific climate change or landuse shift, of the several candidates that often emerge (e.g. concurrent URBANISATION and AGRICULTURAL change in the basin), is responsible. Just as a FLOOD event can have many causes (e.g. catastrophic lake drainage, JÖKULHLAUPS and dam-bursts), so too may FLOOD magnitude-frequency changes.   *DML*

[See also NILE FLOODS]

Enzel, Y. 1992: Flood frequency of the Mojave River and the formation of late Holocene playa lakes, southern California. *The Holocene* 2, 11–18. **Higgs, G.** 1987: Environmental change and hydrological response: flooding in the Upper Severn catchment.

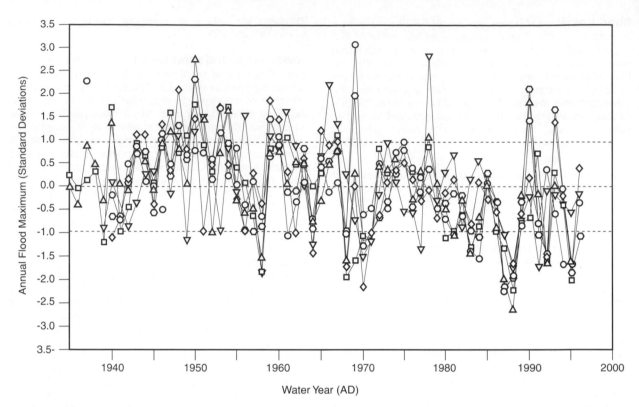

flood magnitude-frequency changes *Time series relating to six rivers in southwestern Wisconsin, USA, showing a tendency for decline in the frequency of high-magnitude floods since* AD *1950 (after Knox, 1999)*

In Gregory, K.J., Lewin, J. and Thornes, J.B. (eds), *Palaeohydrology in practice: a river basin analysis.* Chichester: Wiley, 131–159. **Jones, J.A.A., Changming Liu, Ming-Ko Woo and Hsiang-Te Kung (eds)** 1996: *Regional hydrological response to climate change.* Dordrecht: Kluwer. **Knox, J.C.** 1993: Large increases in flood magnitude in response to modest changes in climate. *Nature* 361, 430–432. **Knox, J.C.** 1999: Long-term episodic changes in magnitudes and frequencies of floods in the Upper Mississippi River Valley. In Brown, A.G. and Quine, T.A. (eds), *Fluvial processes and environmental change.* Chichester: Wiley, 255–282. **Merrett, S.P. and Macklin, M.G.** 1999: Historic river response to extreme flooding in the Yorkshire Dales, Northern England. In Brown, A.G. and Quine, T.A. (eds), *Fluvial processes and environmental change.* Chichester: Wiley, 345–360. **Walsh, R.P.D., Hudson, R.N. and Howells, K.** 1982: Changes in the magnitude-frequency of flooding and heavy rainfalls in the Swansea valley since 1875. *Cambria* 9, 36–90.

floodplain The relatively flat LANDFORM adjacent to a river channel, composed of ALLUVIUM and subject to periodic flooding. Relief features may include natural *levées* (low banks alongside the river channel produced by OVERBANK DEPOSITION), MEANDER CUTOFFS and RIVER TERRACES (often the result of a change in RIVER REGIME). WETLANDS commonly occupy the poorly drained areas. Regular OVERBANK FLOW and associated deposition maintains the fertility of floodplain soils, which, combined with their flatness and the availability of a water supply and water transport, has led to the extensive use of floodplains for agriculture and settlement. Early AGRARIAN CIVILISATIONS in Egypt and Mesopotamia depended on the annual cycle of SILTATION to maintain productivity. Human occupation

of floodplains combined with inadequate *floodplain management* has, however, especially in modern times, led to enhancement of the *flood hazard* and increasing losses. ALLUVIAL DEPOSITS and ALLUVIAL STRATIGRAPHY associated with floodplains are important in ENVIRONMENTAL ARCHAEOLOGY and for palaeoenvironmental reconstruction generally. *JAM*

[See also FERTILE CRESCENT, FLOOD, FLOOD MAGNITUDE-FREQUENCY CHANGES, FLOODWAY, NATURAL HAZARD, PALAEOFLOOD, PALAEOHYDROLOGY]

Bailey, R.G., Jose, P.V. and Sherwood, B.R. (eds) 2000: *UK floodplains.* Otley: Westbury. **Nanson, G.C. and Croke, J.C.** 1992: A genetic classification of floodplains. *Geomorphology* 4, 459–486.

floodway That part of the FLOODPLAIN where there is rapidly flowing water in times of FLOOD. The remaining part of the floodplain is occupied by *slackwater*. The floodway is not only a particularly hazardous location but, if crossed or occupied by buildings, roads and railway embankments, obstructions to flow may cause and exacerbate flooding in adjacent areas and upstream, but reduce flooding downstream. It follows that improvements to the floodway will reduce flooding in that vicinity but exacerbate it downstream. *JAM/RPDW*

Ward, R.C. 1978: *Floods: a geographical perspective.* London: Macmillan, 122–128.

flora The sum (or list) of the plant species in an area. *BA*

[See also ANGIOSPERMS, BIOGEOGRAPHY, BIODIVERSITY, C-3, C-4 AND CAM PLANTS, FAUNA, GYMNOSPERMS, VEGETATION, VEGETATION INDICES]

floral provinces/realms/regions
A division of the Earth into geographical regions delimited on the basis of the composition of their flora.   *MJB*

[See also FAUNAL PROVINCES, PHYTOGEOGRAPHY]

flotation
A simple technique for the physical separation of materials based on whether they float. Charcoal particles and other MACROFOSSILS, for example, may be separated from MINEROGENIC sediments by flotation in water. Flotation is commonly used in combination with wet sieving.   *JAM*

[See also CHARRED-PARTICLE ANALYSIS, PLANT MACROFOSSIL ANALYSIS]

Toll, M.S. 1988: Flotation sampling: problems and some solutions, with examples from the American Southwest. In Hastorf, C.A. and Popper, V.S. (eds), *Current paleoethnobotany: analytical methods and cultural interpretations of archaeological plant remains.* Chicago, IL: Chicago University Press, 36–52.

flow slide
A tongue-shaped form of DEBRIS FLOW emanating from unstable, poorly compacted or otherwise generally loose debris, usually on a steep slope. Characteristically, flow slides have long RUNOUT DISTANCES and move at high velocity. They can occur in certain natural sediments but their association with artificial spoil heaps such as colliery waste, fly ash and tailings is probably better known. Such spoil heaps often lie in close proximity to people, leading to loss of life and damage to buildings and other structures, in addition to possible burial of crops and destruction of trees. They differ from STURZSTROMS, which develop through the fall or slide of a rock body or through the mobilisation of debris caused either by a fall of an overhanging rock mass or because of seismic shock. Typically, the debris is in a metastable state and initiation of the flow slide is triggered by its change from a drained to an undrained state such that the internal stress becomes transferred to the fluid in the structure, which is usually water or air.   *RAS*

[See also LAHAR, LANDSLIDE, MUDSLIDE]

Bishop, A.W. 1973: The stability of tips and spoil heaps. *Quarterly Journal of Engineering Geology* 6, 335–376. **Ibsen, M.-L., Brunsden, D., Bromhead, E. and Collison, A.** 1996: Flow slide. In Dikau, R., Brunsden, D., Schrott, L. and Ibsen, M.-L. (eds), *Landslide recognition: identification, movement and causes.* London: Wiley, 202–211. **Shakesby, R.A. and Whitlow, J.R.** 1991: The failure of a gold mine waste dump, Arcturus, Zimbabwe: causes and geomorphological consequences. *Environmental Geology and Water Resources* 18, 143–153.

flow till
Material originally transported by glacier ice, which is subsequently translocated by gravitational processes (e.g. mud flow, DEBRIS FLOW). Most researchers now consider that the term TILL is inappropriate and that, in view of the redepositional mechanisms in their formation, tills that have flowed are better classified as a MASS WASTING sediment.   *JS*

[See also GLACIAL SEDIMENTS, LODGEMENT TILL]

Benn, D.I. and Evans, D.J.A. 1998: *Glaciers and glaciation.* London: Arnold.

fluidization
The loss of strength of a cohesionless granular material (e.g. SAND or GRAVEL) caused by the upward movement of a fluid through the pore spaces at a rate such that the upward-directed fluid drag equals the downward-acting particle weight. Fluidized sediment is unable to support a load and may be described as *quicksand*. Fluidization can develop in areas of upwelling GROUNDWATER and is an important process in the mobility of PYROCLASTIC FLOWS, which may be fluidized by escaping VOLATILES. Evidence of fluidization in the past may be preserved in the form of SOFT-SEDIMENT DEFORMATION structures such as *water escape structures* or *sand volcanoes*. Fluidization differs from LIQUEFACTION in that an external source of fluid is needed, and the reduction in strength can continue for as long as fluid is supplied.   *GO*

[See also LIQUIDIZATION]

Guhman, A.I. and Pederson, D.T. 1992: Boiling sand springs, Dismal River, Nebraska: agents for formation of vertical cylindrical structures and geomorphic change. *Geology* 20, 8–10. **Lowe, D.R.** 1976: Subaqueous liquefied and fluidized sediment flows and their deposits. *Sedimentology* 23, 285–308. **Owen, G.** 1996: Anatomy of a water-escape cusp in Upper Proterozoic Torridon Group sandstones, Scotland. *Sedimentary Geology* 103, 117–128. **Wilson, C.J.N.** 1984: The role of fluidization in the emplacement of pyroclastic flows. 2. Experimental results and interpretation. *Journal of Volcanology and Geothermal Research* 20, 55–84.

fluorescence spectroscopy
A technique, also known as *luminescence spectroscopy*, for measuring the luminescence properties of a range of sources. It has recently been applied to SPELEOTHEMS and to PEATS. When excited by a high-energy source, molecules release energy as luminescent light. The light takes the form of wavelength pairs of excitation : emission wavelengths depending on the nature of the source. In speleothems and in peats such emissions can be defined according to varying organic acid sources and used in CLIMATIC RECONSTRUCTION.   *CJC*

Caseldine, C.J., Baker, A., Charman, D.J. and Hendon, D. 2000: A comparative study of optical properties of NaOH extracts: implications for humification studies. *The Holocene* 10, 649–658. **McGarry, S.F. and Baker, A.** 2000: Organic acid fluorescence: applications to speleothem palaeoenvironmental reconstruction. *Quaternary Science Reviews* 19, 1087–1101.

fluorine dating
A technique used to estimate the RELATIVE AGE of skeletal material up to several million years old, based on the rate of fluorine uptake from natural waters during fossilisation. *Age calibration* is required because of considerable temporal variation of fluorine concentration in GROUNDWATER and because antlers, dentine and enamel yield different calibration curves. URANIUM–THORIUM DATING provides a means of AGE CALIBRATION, but assumptions also have to be made about the open-system behaviour of uranium.

Fluorine-profile dating is a variant of the technique based on the gradient in the fluorine content with increasing distance from the surface of the bone. *Fluorine–uranium–nitrogen (FUN) dating* involves the combined use of the contents of fluorine, uranium and nitrogen, as used in the case of the 'Piltdown man' forgery. *Fluorine-diffusion dating*, based on the rate of diffusion of fluorine into the chipped surfaces of stone objects, has also been suggested.   *DAR/JAM*

[See also NITROGEN DATING]
Aitken, M.J. 1990: *Science-based dating in archaeology.* London: Longman. **Haddy, A. and Hanson, A.** 1982: Nitrogen and fluorine dating of Moundville skeletal samples. *Archaeometry* 24, 37–44. **Schurr, M.R.** 1989: Fluoride dating of prehistoric bones by ion selective electrode. *Journal of Archaeological Science* 16, 265–270. **Taylor, R.E.** 1975: Fluorine diffusion: a new dating method for chipped lithic material. *World Archaeology* 7, 125–135. **Wagner, G.A.** 1998: *Age determination of young rocks and artifacts: physical and chemical clocks in Quaternary geology and archaeology.* Berlin: Springer. **Weiner, J.S., Oakley, J.P. and Le Gros Clark, W.E.** 1963: The solution of the Piltdown problem. *Bulletin British Museum Natural History (Geology)* 2, 139–146.

fluted moraine   Elongate glacially streamlined ridges, also known as *flutings* or *flutes*, generally up to a few metres high and wide and up to about 100 m in length. They comprise DIAMICTON and usually form a series of ridges parallel to ice-flow direction. Various origins have been suggested. They are SUBGLACIAL landforms that are formed close to the margins of retreating GLACIERS and are most readily associated with recently deglacierised terrain.   *RAS*

[See also DRUMLIN]
Benn, D.I. and Evans, D.J.A. 1996: The interpretation and classification of subglacially-deformed materials. *Quaternary Science Reviews* 15, 23–52. **Hoppe, G. and Schytt, V.** 1953: Some observations on fluted moraine surfaces. *Geografiska Annaler* 35, 105–115.

fluvial processes   Fluvial processes are those relating to rivers and streams that flow in channels. In practice, the terms fluvial and *alluvial* are commonly used interchangeably to refer to rivers and associated processes and environments, although *alluvial processes* relate more widely to the flow of freshwater on land surfaces.

Erosional processes include fluvial incision and lateral erosion. Incision gives rise to channels and channel downcutting. It may occur during FLOODS or in response to ALLOCYCLIC CHANGE, such as CLIMATIC CHANGE or a lowering of BASE LEVEL, and may be associated with AVULSION. Lateral erosion occurs, for example, on the outer, cut banks of MEANDERS, giving rise to channel migration.

Processes of sediment transportation include the movement of sediment particles as BEDLOAD and SUSPENDED LOAD, and the transport of salts in solution as DISSOLVED LOAD. DEPOSITION gives rise to ALLUVIAL DEPOSITS if the current strength decreases, the sediment load becomes too great, or through CHEMICAL PRECIPITATION (see also COMPETENCE).

Within the wider remit of alluvial, rather than strictly fluvial, processes are those that occur on a river's FLOODPLAIN (such as non-channelised *overland flow*, postdepositional alteration to form SOIL, and some AEOLIAN processes) and COLLUVIAL PROCESSES involving water flow (such as SLOPEWASH).   *GO*

[See also ALLUVIAL FAN, CORRASION, CORROSION, GULLY, HYDROLOGY]
Brown, A.G. and Quine, T.A. (eds), 1999: *Fluvial processes and environmental change.* Chichester: Wiley. **Reid, I. and Frostick, L.E.** 1994: Fluvial sediment transport and deposition. In Pye, K. (ed.), *Sediment transport and depositional processes.* Oxford: Blackwell, 89–155.

fluviokarst   A type of terrain developed by a combination of fluvial erosion and KARST processes. Typical features are DRY VALLEYS and GORGES.   *SHD*

Fluvisols   Young soils, developed in Holocene ALLUVIUM of all continents. They have weakly developed horizons, but retain stratification from the mode of deposition of their parent material in the subsoils. The content of organic carbon in Fluvisols has an erratic distribution down the profile (SOIL TAXONOMY: Fluvents).   *EMB*

[See also WORLD REFERENCE BASE FOR SOIL RESOURCES]
Bridges, E.M. and Creutzberg, D. 1994: Leptosols and Fluvisols. *Transactions of the 15th International Congress of Soil Science, Acapulco* 6a, 868–872. **Klimek, K.** 1999: A 1000 year alluvial sequence as an indicator of catchment/floodplain interaction: the Ruda Valley, Sub-Carpathians, Poland. In Brown, A.G. and Quine, T.A., *Fluvial processes and environmental change.* Chichester: Wiley, 329–343.

fly ash   The particulate fraction emitted by high-temperature FOSSIL FUEL combustion. Fly ash consists of SPHEROIDAL CARBONACEOUS PARTICLES and inorganic ash spheres formed by the fusing of mineral inclusions. *DZR*

Renberg, I. and Wik, M. 1985: Carbonaceous particles in lake sediments – pollutants from fossil fuel combustion. *Ambio* 14, 161–163.

flysch   A term originally applied in the early nineteenth century to sparsely fossiliferous, thick sequences of interbedded SANDSTONE and SHALE of marine origin that accumulated from the late Cretaceous to the early Tertiary in the European Alps. Flysch represents material that accumulated rapidly in SEDIMENTARY BASINS adjacent to a rising mountain belt and was itself deformed in later stages of the OROGENY (see also MOLASSE). As such, the concept can be considered as a 'tectono-stratigraphic facies' – a sedimentary FACIES with a particular relationship to TECTONIC events. The term has since been applied to other Alpine-type OROGENIC BELTS. Some workers have applied the name to a sedimentary facies, regardless of tectonic setting, and in some cases it has been used almost as a synonym for TURBIDITE. Because of this confusion in terminology, the term is now seldom used outside the European Alps.   *BTC/GO*

Bouma, A.H. 1962: *Sedimentology of some flysch deposits: a graphic approach to facies interpretation.* Amsterdam: Elsevier. **Dzulynski, S. and Walton, E.K.** 1965: *Sedimentary features of flysch and greywackes* [*Developments in Sedimentology* 7]. Amsterdam: Elsevier. **Hsü, K.J.** 1995: *The geology of Switzerland.* Princeton, NJ: Princeton University Press.

fold   A GEOLOGICAL STRUCTURE formed by the ductile bending of strata under compression and shortening. Folds form on a range of scales from millimetres to kilometres, with varied geometry and orientations (see Figure). The principal varieties are *antiforms* (arch-shaped – strata dip away from the centre), *synforms* (inverted arch – strata dip towards the centre), *anticlines* (oldest strata in the centre) and *synclines* (youngest strata in the centre). Most commonly, antiforms are anticlines and synforms are synclines, but in areas of complex deformation (e.g. associated with THRUST FAULTS) antiforms may be synclines and synforms anticlines: this situation is described as downwards-*facing*. Fold *vergence* describes the sense of

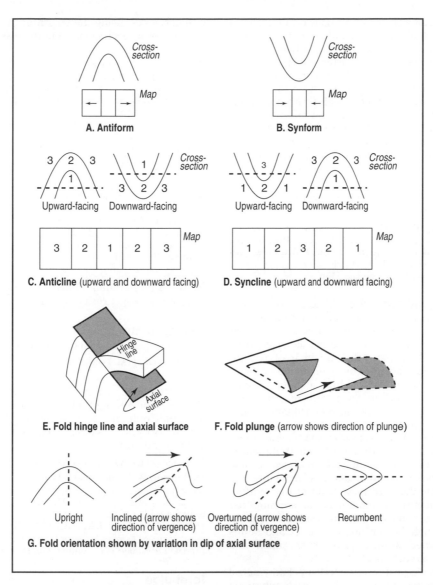

fold *Varieties of folds and their orientation*

asymmetry in minor folds of unequal limb length, which can be used to map out very large-scale structures.

Fold orientation is described by means of the *hinge line* (line of maximum curvature on each folded surface) and *axial surface* or *axial plane* (surface containing successive hinge lines). The axial surface runs through the hinge region of the fold and separates two *limbs*. Hinge lines may be horizontal or tilted (*plunging fold*). The axial surface may be upright, inclined or horizontal (*recumbent fold*). An inclined fold with one limb overturned (i.e. rotated through more than 90° from the horizontal) is commonly described as an *overfold* or *overturned fold*. Hinge lines and axial surfaces may themselves be curved, and folds may become refolded in a subsequent episode of DEFORMA-TION. In three-dimensions, folds may be continuous (*cylindrical fold* – shape can be formed by a straight line, the *fold axis*, moved parallel to itself) or discontinuous

(hinge lines die out in each direction, forming a *pericline* or *periclinal fold*). *GO*

[See also FAULT]

Bell, A.M. 1981: Vergence: an evaluation. *Journal of Structural Geology* 3, 197–202. **Carey, P.J.** 1998: Field interpretation of complex tectonic areas. In Doyle, P. and Bennett, M.R. (eds), *Unlocking the stratigraphical record: advances in modern stratigraphy*. Chichester: Wiley, 81–121. **Lisle, R.J.** 1995: *Geological structures and maps*. Oxford: Butterworth-Heinemann. **Ramsay, J.G.** 1967: *Folding and fracturing of rocks*. New York: McGraw-Hill. **Ramsay, J.G. and Huber, M.I.** 1987: *The techniques of modern structural geology*. Vol. 2. *Folds and fractures*. London: Academic Press.

food chain/web Concepts relating to the structure of ecological communities in terms of the feeding relationships amongst organisms. Food and energy passes along a food chain or through a food web along two fundamental pathways, the so-called *grazing food chain* and *decomposer food chain*: the former tends to be particularly well developed in grassland ecosystems, the latter in forests. A food chain envisages a linear pattern with each organism at one TROPHIC LEVEL feeding on an organism from the trophic level below. A food web is closer to reality with, for example, some predators sharing the same prey and DECOMPOSITION affecting organic remains from all levels. Although food webs may be complex, there are normally only a small number of trophic levels. The complexity of food webs appears to be related to community and ecosystem stability. *JAM*

[See also ENERGY FLOW, STABILITY CONCEPTS, TROPHIC–DYNAMIC APPROACH]

Hall, S.J. and Raffaelli, D.G. 1993: Food webs: theory and reality. *Advances in Ecological Research* 24, 187–239. **Pimm, S.L.** 1982: *Food webs*. London: Chapman and Hall. **Polis, G.A. and Winemiller, K.O. (eds)** 1995: *Food webs: integration of pattern and dynamics*. New York: Chapman and Hall.

Foraminiferal analysis

Foraminifera are single-celled PROTISTA that inhabit nearly all marine and marginal environments. They are either BENTHIC (bottom dwelling) or planktonic (surface dwelling; see PLANKTON) in nature and have shells (*tests*) that are made of calcium carbonate (*calcite* and *aragonite*) or agglutinated particles or are naked. Their small size and high abundance, plus high preservation potential in the FOSSIL record, make Foraminifera useful tools in BIOSTRATIGRAPHY and PALAEOENVIRONMENTAL RECONSTRUCTION.

To analyse Foraminifera it is necessary to separate them from the SEDIMENT in which they are found. Sediment is normally washed over a 63 μm sieve and the coarse residue obtained is examined under a stereo binocular microscope. Foraminifera are picked out of the residue, normally with a fine artist's paintbrush, and are mounted on cardboard slides, which are often divided into cells. Specimens are arranged according to species and abundance values recorded. Counts of > 300 specimens are typically made per sample.

The modern ecological characteristics, bathymetric ranges, assemblage diversities and species distributions are amongst some of the factors used to infer past environmental and oceanographic conditions. An example of this is the type and abundance of Foraminifera found in MARINE SEDIMENT CORES and the variation of OXYGEN ISOTOPES and CARBON ISOTOPES recorded within their shells.

Oxygen ISOTOPE RATIOS of ^{16}O and ^{18}O in the tests of calcareous Foraminifera can be used to determine the *palaeotemperature* of sea water, as in the CLIMAP project, which made estimates of SEA-SURFACE TEMPERATURES for the LAST GLACIAL MAXIMUM (18 000 BP) using TRANSFER FUNCTIONS. These are based upon the modern relationships that exist between an assemblage and its surrounding environment; in this case the planktonic Foraminifera and SEA-SURFACE TEMPERATURE. This relationship is first determined for MODERN ANALOGUE conditions and then MULTIVARIATE ANALYSES are used to apply the transfer function to a fossil dataset.

Known first and last occurrence datums of benthic and planktonic Foraminifera are used in biostratigraphy to determine age and level within a stratigraphic section. These sections can be subdivided into biostratigraphic units or zones, using the length of time that a particular Foraminifera species existed in the fossil record. *JRE/WENA*

Austin, W.E.N. and Evans, J.R. 2000: NE Atlantic benthic foraminifera: modern distribution patterns and palaeoecological significance. *Journal of the Geological Society* 157, 679–691. **Ganssen, G.M. and Kroon, D.** 2000: The isotopic signature of planktonic foraminifera from NE Atlantic surface sediments: implications for the reconstruction of past oceanic conditions. *Journal of the Geological Society* 157, 693–699. **McIntyre, A., Kipp, N.G., Bé, A. et al.** 1976: Glacial North Atlantic 18,000 years ago: a CLIMAP reconstruction. *Geological Society of America Memoir* 145, 43–47. **Murray, J.W.** 1991: *Ecology and palaeoecology of benthic Foraminifera*. Harlow: Longman. **Shackleton, N. J. and Kennett, J.P.** 1975: Late Cenozoic Oxygen and Carbon isotopic changes at DSDP Site 284: Implications for glacial history of the Northern Hemisphere and Antarctica. In Kennet, J.P., Houtz, R.E. *et al.* (eds), *Initial reports of the deep sea drilling project*. Washington, DC: US Government Printing Office, 801–807.

forb A non-graminoid herb: a non-woody plant that is not a grass, sedge, rush or other grass-like plant. *JAM*

forcing factor Any *external* ENVIRONMENTAL FACTOR that causes a directional change in a SYSTEM. *JAM*

[See also ORBITAL FORCING, SOLAR FORCING]

fore-arc The area between a volcanic ISLAND ARC and the plate undergoing SUBDUCTION at a DESTRUCTIVE PLATE MARGIN. Components of the fore-arc zone include a fore-arc SEDIMENTARY BASIN, ACCRETIONARY COMPLEX and OCEANIC TRENCH. *GO*

[See also BACK-ARC, PLATE TECTONICS]

Dickinson, W.R. and Seely, D.R. 1979: Structure and stratigraphy of forearc regions. *Bulletin of the American Association of Petroleum Geologists* 63, 2–31. **Hamilton, W.B.** 1988: Plate tectonics and island arcs. *Geological Society of America Bulletin* 100, 1503–1527. **Leggett, J.K. (ed.)** 1982: *Trench-forearc geology: sedimentation and tectonics on modern and ancient active plate margins* [*Special Publication* 10]. London: Geological Society.

forebulge A compensatory UPLIFT of the Earth's CRUST beyond the margin of an area subjected to an additional load. Forebulges may involve uplift of several hundred metres. Examples include uplift in areas adjacent to major ice sheets, where *glacial forebulges* are an aspect of GLACIO-ISOSTASY, and uplift adjacent to areas of *tectonic loading* associated with FORELAND BASIN thrust sheets. *JAM/GO*

Barnhardt, W.A., Gehrels, W.R., Belknap, D.F. and Kelley, J.T. 1995: Late Quaternary relative sea-level change in the western Gulf of Maine: evidence for a migrating glacial forebulge. *Geology* 23, 317–320. **Fjeldskaar, W.** 1994: The amplitude and decay of the glacial forebulge in Fennoscandia. *Norsk Geologisk Tidsskrift* 74, 2–8. **Lorenzo, J.M., O'Brien, G.W., Stewart, J. and Tandon, K.** 1998: Inelastic yielding and forebulge shape across a modern foreland basin: North West Shelf of Australia, Timor Sea. *Geophysical Research Letters* 25, 1455–1458. **Straeten, C.A.V. and Brett, C.E.** 2000: Bulge migration and pinnacle reef development, Devonian Appalachian foreland basin. *Journal of Geology* 108, 339–352.

forecasting The anticipation and projection of future short- to medium-range developments and needs in the context of society. It is an integral part of the planning process and involves the application of statistical, modelling and decision making techniques. Examples range from predictions of POPULATION growth on annual to decadal timescales to *weather forecasting* on a daily to monthly basis. Increasingly accurate *seasonal forecasting* of rainfall and crop yields may revolutionise the management of agricultural resources in Africa. *JAM*

[See also CLIMATIC PREDICTION]

Anderson, M.G. and Burt, T.P. (eds) 1985: *Hydrological forecasting.* Chichester: Wiley. **Cane, M.A., Eshel, G. and Buckland, R.W.** 1994: Forecasting Zimbabwean maize yield using eastern equatorial Pacific sea surface temperature. *Nature* 370, 204–205. **Gilchrist, A.** 1986: Long-range forecasting. *Quarterly Journal of the Royal Meteorological Society* 112, 567–592. **Mielke Jr, P.W., Berry, K.J., Landsea, C.W. and Gray, W.M.** 1996: Artificial skill and validation in weather forecasting. *Weather Forecasting* 11, 153–169. **Pierce, C., Dikes, M. and Parker, G.** 1997: Forecasting the British Isles weather. In Hulme, M. and Barrow, E. (eds), *Climates of the British Isles: present, past and future.* London: Routledge, 299–325. **Shukla, J.** 1998: Predictability in the midst of chaos: a scientific basis for climate forecasting. *Science* 282, 728–731. **Washington, R. and Downing, T.E.** 1999: Seasonal forecasting of African rainfall: prediction, responses and household food security. *Geographical Journal* 165, 255–274.

foreland basin

A SEDIMENTARY BASIN that develops adjacent to an OROGENIC BELT. SUBSIDENCE is caused by loading of the CRUST through the stacking up of sheets of rock bounded by THRUST FAULTS in a *fold-thrust belt*, the EROSION of which supplies SEDIMENT to the basin. The migrating locus of DEFORMATION as shortening continues may result in the deformation of earlier foreland basins. Foreland basins contain an important record of the growth of major mountain chains. Present-day examples include the MOLASSE basin of the Alps and the Ganges basin south of the Himalayas. *GO*

Allen, P.A., England, P.C., Grotzinger, J. and Sinclair, H. (eds), 1992: Thematic set on foreland basins. *Basin Research* 4, 169–352. **Burgess, P.M. and Gayer, R.A.** 2000: Late Carboniferous tectonic subsidence in South Wales: implications for Variscan basin evolution and tectonic history in SW Britain. *Journal of the Geological Society, London* 157, 93–104. **Evans, M.J. and Elliott, T.** 1999: Evolution of a thrust-sheet-top basin: the Tertiary Barreme basin, Alpes-de-Haute-Provence, France. *Geological Society of America Bulletin* 111, 1617–1643. **Najman, Y. and Garzanti, E.** 2000: Reconstructing early Himalayan tectonic evolution and paleogeography from Tertiary foreland basin sedimentary rocks, northern India. *Geological Society of America Bulletin* 112, 435–449. **Ori, G.-G. and Friend, P.F.** 1984: Sedimentary basins formed and carried piggyback on active thrust sheets. *Geology* 12, 475–478.

foreshock

A minor EARTHQUAKE that occurs in the weeks or days before a major earthquake, caused by preliminary movements along the same FAULT surface. Although foreshocks can be useful in EARTHQUAKE PREDICTION, most major earthquakes are not heralded by foreshocks. *GO*

Reasenberg, P.A. 1999: Foreshock occurrence rates before large earthquakes worldwide. *Pure and Applied Geophysics* 155, 355–379.

foreshortening

A geometrical distortion of TERRAIN when viewed obliquely. The term is applied in particular to the distortion inherent in REMOTE SENSING images formed by a side-looking RADAR, which arises where the relative distance from the radar to elevated parts of the terrain (mountain tops) is shorter than when projected along the ground. Foreshortened areas appear bright since the pixels cover a larger area on the ground and therefore scatter a larger amount of energy. The extreme case is *layover*, when the top of the mountain appears in the image to be closer than the bottom. *PJS*

forest clearance

A decrease in the extent of tree-dominated vegetation communities, as a consequence (either intentional or accidental) of human activity. Although hunter–gatherer populations seem to have manipulated the forest ENVIRONMENT and have had some effect on the openness of the canopy, especially in marginal forest environments such as upland Britain, significant, permanent forest clearances only really began with the introduction of agriculture.

NEOLITHIC culture spread across Europe from the near East between 9000 and 4500 radiocarbon years BP, although early agricultural practices, at least in the Balkans, were not necessarily associated with forest clearance as detected by POLLEN ANALYSIS. In northwestern Europe, forest clearance and the arrival of Neolithic agriculture are usually detected simultaneously, often coincident with the ELM DECLINE. The mechanisms of clearance include suppression of regeneration as a result of GRAZING pressure or understorey burning and direct assault, usually in order to clear ground for agricultural crops. Iversen's classic experiments at the Draved forest demonstrated the effectiveness of Neolithic axes, although the exact nature of Neolithic forest agriculture is still debated. The overall pattern from the arrival of the first farmers to the present day has been one of progressive forest clearance, although the timing and rate has varied from place to place: lighter, sandy or chalky soils were cleared first and the heavy, waterlogged clay soils of major northwest European river FLOODPLAINS were only substantially cleared in the IRON AGE.

The situation in northwest Europe is in marked contrast to the record from North America, where, despite the strong evidence that native populations were using fire to manipulate the forest understorey for hunting and agriculture, forest canopies seem to have remained largely intact, probably largely because of a lack of domesticated livestock. The first European settlers came into a landscape in which forest cover remained at its natural, climatically controlled extent. Massive forest clearance for European-style agriculture followed, particularly during the mid–late nineteenth century. Many areas are currently reafforesting rapidly, since agriculture in the forested east became uneconomic once the western prairie *grasslands* began to be exploited efficiently and infrastructure links were established, enabling affordable transhipment of agricultural produce. *MJB*

[See also AGRICULTURAL HISTORY, DEFORESTATION, FIRE IMPACTS, FOREST REGENERATION, GRAZING HISTORY, HUMAN IMPACT, LANDUSE CHANGE, WILDWOOD]

Brown, A.G. 1997: Clearances and clearings: deforestation in Mesolithic–Neolithic Britain. *Oxford Journal of Archaeology* 16, 133–146. **Edwards, K.J.** 1989: Meso–Neolithic vegetational impact in Scotland and beyond: palynological considerations. In Bonsall, C. (ed.), *The Mesolithic in Europe.* Edinburgh: Donald,

143–155. **Edwards, K.J.** 1993: Models of mid-Holocene forest farming for north-west Europe. In Chambers, F.M. (ed.), *Climate change and human impact on the landscape.* London: Chapman & Hall, 133–145. **Iversen, J.** 1973: The development of Denmark's nature since the last glacial. *Danmarks Geologiske Undersøgelse,* V Raekke, 7-C, 1–126. **McAndrews, J.H.** 1988: Human disturbance of North American forests and grasslands: the fossil pollen record. In Huntley, B. and Webb, T. (eds), *Vegetation history.* Dordrecht: Kluwer, 674–697. **Whitney, G.G.** 1994: *From wilderness to fruited plain: a history of environmental change in temperate North America from 1500 to the present.* Cambridge: Cambridge University Press. **Willis, K.J. and Bennett, K.D.** 1994: The Neolithic Transition – fact or fiction? Palaeoecological evidence from the Balkans. *The Holocene* 4, 326–330.

forest decline Specifically, the deterioration in the quality of forest (health, genetic diversity and age profile), although reductions in quantity (see also FOREST CLEARANCE) may also be included. Forest decline or *Waldsterben* may be considered as a result of environmental stress, forest clearance, and natural processes.

(1) *Environmental stress:* Air pollution by sulphur dioxide and nitrous oxides (ACID DEPOSITION) and low-level ozone produce both visible (e.g. defoliation, chlorosis) and invisible (loss of yield) injury to trees. Such injuries are often similar to damage from other stresses such as DROUGHT, PEST attack, DISEASE and NUTRIENT deficiency or excess. Moreover, trees weakened by air pollution are more susceptible to injury by drought, disease and lack of nutrients. Acid deposition can also cause nutrient deficiency in trees from accelerated LEACHING of Ca and Mg from soils. Air-pollution damage is especially evident in Europe, North America and Asia, and in cities throughout the world. More than one quarter of Europe's trees show moderate to severe defoliation (25% leaf loss) from air pollution and other stresses, according to regular surveys by the UN Economic Commission for Europe.

(2) *Forest clearance:* A 1997 World Resources Institute Report estimates that 57% of the world's remaining original forests consist of highly fragmented (see FRAGMENTATION) and degraded human-modified woodlands. Though TROPICAL FORESTS are suffering the greatest current losses in forest area and quality, TEMPERATE FORESTS have been affected more than other forest types by FRAGMENTATION and DISTURBANCE. For instance, 95–98% of US forests have been logged at least once and in Europe less than 1% of OLD-GROWTH FOREST remains. Compared with original forests, human-disturbed forested ecosystems in temperate and tropical areas have simplified structures, lower ENERGY FLOW, drastically depleted BIODIVERSITY and greater AESTHETIC DEGRADATION. Attempts at restoring forest cover through AFFORESTATION programmes have often created simple-structured, even-aged tree plantations of low biodiversity.

(3) *Natural processes:* Forest quality may decrease during the later stages of ECOLOGICAL SUCCESSION when progressive soil leaching, and increasing forest shade, can remove 'climax' trees and replace them with earlier successional types (e.g. oak by pine in the UK). *GOH*

[See also ACID RAIN, DEFORESTATION, DIEBACK, RETROGRESSIVE SUCCESSION]

Brown L., Flavin, C. and French, H. 1998: *State of the world* [*Worldwatch Institute Report*]. London: Earthscan. **Roberts, L. (ed.),** 1997: *World resources 1998–99: a guide to the global environ-*

ment. Oxford: Oxford University Press. **Innes, J.L.** 1992: Forest decline. *Progress in Physical Geography* 16, 1–64. **Innes, J.L.** 1993: *Forest health: its assessment and status.* Wallingford: CAB International. **Wellburn, A.** 1994: *Air pollution and climate change: the biological impact.* London: Longman.

forest fallow Forest which, in connection with SHIFTING CULTIVATION and/or *wood pasture,* has been partially cleared and/or used and then left and allowed to return to forest. It is therefore land under regenerating SECONDARY WOODLAND. *SPH*

[See also FOREST CLEARANCE, FOREST MANAGEMENT, WOODLAND MANAGEMENT PRACTICES].

forest management The management of forest resources for particular purposes, such as ENVIRONMENTAL PROTECTION, provision of a facility for RECREATION, maximisation of timber yield, maintenance of water quality and WILDLIFE CONSERVATION. *JAM*

[See also NEW FORESTRY, SILVICULTURE]

Davis, L.S. and Johnson, K.N. 1987: *Forest management,* 3rd edn. New York: McGraw-Hill.

formation (stratigraphical) The basic unit of LITHOSTRATIGRAPHY, defined as the smallest mappable unit. It is defined during geological mapping and described using gross characteristics of LITHOLOGY. The thickness of a formation may vary from less than 1 m to thousands of metres. Its boundaries may be defined arbitrarily in a gradational sequence, or at a sharp change in lithology. A formation should be, or have been, laterally continuous, and traceable because of its lithological characteristics in surface outcrop, or in the subsurface using GEOPHYSICAL SURVEYING or BOREHOLE data. Generally, a formation is restricted to one SEDIMENTARY BASIN, and it is debatable whether the same name should be applied to similar formations of equivalent age in other basins (e.g. Upper JURASSIC Kimmeridge Clay Formation of the Wessex Basin and equivalent OIL SHALES in North Sea basins). Distinctive local subunits of formations are termed MEMBERS. *LC*

[See also VEGETATION FORMATION]

Brown, S. 1986: Jurassic. In Glennie, K.W. (ed.), *Introduction to the petroleum geology of the North Sea,* 2nd edn. Oxford: Blackwell Scientific, 133–159. **Cox, B.M. and Sumbler, M.G.** 1998: Lithostratigraphy: principles and practice. In Doyle, P. and Bennett, M.R. (eds), *Unlocking the stratigraphical record: advances in modern stratigraphy.* Chichester: Wiley, 11–27. **Whittaker, A., Cope, J.C.W., Cowie, J.W. et al.** 1991: A guide to stratigraphical procedure. *Journal of the Geological Society, London* 148, 813–824.

fossil The preserved remains of life in the GEOLOGICAL RECORD. Most commonly only hard parts – mineralised skeletons, shells, or resistant organic material – are preserved as *body fossils* in PHANEROZOIC sedimentary rocks, having been variously altered through taphonomic processes (see TAPHONOMY). The likelihood of preservation (*preservation potential*) depends on the abundance of the organism, composition and structure of its hard parts, the enclosing sediment and the postdepositional environment. Moulds (internal or external) in the sediment, and CASTS formed by replacement of a shell following dissolution, represent indirect fossil preservation. The preservation of soft-bodied organisms is exceptional (see FOSSIL

LAGERSTÄTTEN), although evidence of their activity may be preserved as TRACE FOSSILS. Chemical fossils (BIOMARKERS) are organic molecules that record biological activity. BIOSTRATIGRAPHY relies on INDEX FOSSILS, but fossils are also important for studies of EVOLUTION (systematics), PALAEOECOLOGY and PALAEOENVIRONMENTAL RECONSTRUCTION. The earliest fossils are PRECAMBRIAN (ARCHAEAN) microbes and STROMATOLITES in the Warrawoona Group from Australia and the Onverwacht Group from South Africa, dated at about 3500 million years old. They provide evidence for the build-up of oxygen in the Earth's early ATMOSPHERE.   *LC*

[See also BIOFACIES, DEATH ASSEMBLAGE, DERIVED FOSSIL, FOSSIL RECORD, LIFE ASSEMBLAGE, MICROFOSSIL, SUBFOSSIL]

Benton, M.J. and Harper, D.A.T. 1997: *Basic palaeontology.* Harlow: Longman. **Goldring, R.** 1999: *Field palaeontology.* Harlow: Longman. **Schopf, J.W. and Packer, B.M.** 1987: Early Archaean (3.3 billion to 3.5 billion year old) microfossils from Warrawoona Group, Australia. *Science* 237, 70–73.

fossil fuels: as non-renewable resources

Fuels made from fossil organic material trapped in SEDIMENTARY ROCKS. Fossil fuels include COAL, NATURAL GAS and PETROLEUM, which are the principal sources of industrial and domestic energy in the industrialised world. When plants or animals die and decompose, atmospheric oxygen combines with carbon and hydrogen stored in organic compounds in the decomposing BIOMASS and releases a small amount of energy, which effectively reverses the photosynthetic reaction. This process is not ubiquitous and spatial variations in the relative efficiency of organic production and DECOMPOSITION means that in some areas a small proportion of organic matter is trapped and stored before it is completely decayed. In this way, secondary solar energy becomes stored in rocks. Accumulating organic matter is converted into fossil fuel by a combination of further decomposition, overlying pressure and heat. The kind of organic matter trapped in the sediment and the location of the sediments (oceanic, riverine or terrestrial) determines whether coal, oil or gas is produced.

Fossil-fuel combustion has become a contentious issue because of the effect of emissions of GREENHOUSE GASES and ACID RAIN production, leading to ACIDIFICATION of water bodies down-wind.   *JGS*

[See also NON-RENEWABLE RESOURCES]

Laudon, R.C. 1995: *Principles of petroleum geology.* Englewood Cliffs, NJ: Prentice Hall. **Perry, H.** 1983: Coal in the United States: a status report. *Science* 222, 377–384. **Skinner, B.J. and Porter, S.C.** 1987: *Physical geology.* New York: Wiley. **Thomas, L.** 1992: *Handbook of practical coal geology.* New York: Wiley.

fossil fuels: in climatic change

The burning of fossil fuels (coal, petroleum, natural gas) is the primary source of increasing trace gas concentrations that contribute to the anthropogenically enhanced GREENHOUSE EFFECT. Fossil fuels used in heating, transportation and industrial activities generate 5×10^{20} J y^{-1} of energy. Some 95% of burning takes place in the Northern Hemisphere, where, in many industrialised countries, per capita annual releases reach about 5 t C in contrast to most developing countries, where releases are in the range 0.2–0.6 t C.

From the beginning of the INDUSTRIAL REVOLUTION, the concentration of CARBON DIOXIDE in the atmosphere has risen by 25% from 280 ppmv (around AD 1750) to over 350 ppmv. Half of this increase has taken place since the mid-1960s. The primary net source of CO_2 is fossil-fuel combustion accounting for about 5×10^{12} kg C y^{-1}. However, the consumption of fossil fuels should actually have produced an increase almost twice as great as that observed. The difference is attributed to the uptake and dissolution in the oceans and terrestrial biosphere. Carbon dioxide has a significant impact on global temperature by its ABSORPTION and re-emission of radiation from the Earth and atmosphere.   *GS*

[See also CARBON CYCLE, CARBON SEQUESTRATION, GREENHOUSE GASES]

Callendar, G.S. 1938: The artificial production of carbon dioxide and its influence on temperature. *Quarterly Journal of the Royal Meteorological Society* 64, 223–240. **Intergovernmental Panel on Climate Change** 1996: *Climate change 1995: the science of climate change.* Cambridge: Cambridge University Press.

fossil Lagerstätten

Deposits where the quantity and/or quality of preservation of FOSSILS is exceptional. Exceptional preservation can commonly be attributed to ENVIRONMENTAL FACTORS and the quality of preservation may allow important interpretations to be made about PALAEOENVIRONMENTS. Unusual abundance of fossils in *concentration Lagerstätten* can result from condensed sedimentation (e.g. hiatus COQUINAS), LAG DEPOSITS (*bone beds*) or SEDIMENT TRAPS (caves, fissures). Unusual preservation of fossils in *conservation Lagerstätten* may include soft-part preservation if microbial decay and scavenging were limited or halted. This may result from water-body stagnation (Jurassic Solnhofen Limestone; Devonian Hunsrück Slate; Tertiary Messel OIL SHALES), OBRUTION (smothering by rapid burial as in the Cambrian Burgess Shale), sealing (Tertiary insects in AMBER, Pleistocene mammoths in PERMAFROST) or CONCRETIONS (Upper Carboniferous Mazon Creek fauna). Fossil Lagerstätten are known from the late Precambrian (Ediacara Fauna) and through the Phanerozoic. The preservation of soft tissues has vastly enhanced knowledge of the early diversification and PALAEOBIOLOGY of fossil biotas (e.g. conodont animal).   *LC*

[See also TAPHONOMY]

Allison, P.A. and Briggs, D.E.G. 1991: Taphonomy of non-mineralised tissue. In Allison, P.A. and Briggs, D.E.G. (eds), *Taphonomy: releasing the data locked in the stratigraphic record.* Plenum: New York, 25–70. **Brenchley, P.J. and Harper, D.A.T.** 1998: *Palaeoecology: ecosystems, environments and evolution.* London: Chapman and Hall. **Briggs, D.E.G., Clarkson, E.N.K. and Aldridge, R.J.** 1983: The conodont animal. *Lethaia* 16, 1–14. **Gould, S.J.** 1990: *Wonderful life: the Burgess Shale and the nature of history.* London: Hutchinson Radius. **Seilacher, A.** 1970: Begriff und Bedeutung der Fossil-Lagerstätten. *Neues Jahrbuch für Geologie und Paläontologie* 1, 34–39. **Whittington, H.B. and Conway Morris, S. (eds)** 1985: Extraordinary fossil biotas: their ecological and evolutionary significance. *Philosophical Transactions of the Royal Society of London* B311, 1–192.

fossil record

The record through time of life on Earth, as recorded in FOSSILS, from the origins of life on Earth through its EVOLUTION to the present day. Recent analysis has shown that, although the fossil record is incomplete, its quality is equally good throughout PHANEROZOIC time, and is adequate to demonstrate the history of life.   *GO*

[See also GEOLOGICAL RECORD, STRATIGRAPHICAL RECORD]
Benton, M.J., Wills, M.A. and Hitchin, R. 2000: Quality of the fossil record through time. *Nature* 403, 534–537. **Conway Morris, S.** 1998: *The crucible of creation: the Burgess Shale and the rise of animals.* Oxford: Oxford University Press. **Donovan, S.K. and Paul, C.R.C. (eds)** 1998: *The adequacy of the fossil record.* Chichester: Wiley. **Gould, S.J.** 1989: *Wonderful life: the Burgess Shale and the nature of history.* London: Penguin. **McClendon, J.H.** 1999: The origin of life. *Earth-Science Reviews* 47, 71–93. **Seilacher, A.** 1999: Earth history seen as a long-term experiment: the great revolutions in the development of life. *Eclogae Geologicae Helvetiae* 92, 73–79.

fossil soil   A soil not formed in the present phase of formation and, strictly speaking, preserved by LITHIFICATION in a geological deposit. Such soils are usually buried by later sedimentation (BURIED SOIL), but may appear at the surface where erosion has removed the overlying material (EXHUMED SOIL).   *EMB*

[See also PALAEOSOL, GEOSOL]

founder effect   When populations of new species become established, there may be very few 'founders', carrying with them only a reduced, randomly selected, proportion of the total GENE POOL of the population from which they originate. In the absence of exchange of genetic material with other populations of the species, such a population may develop into a new species.   *KDB*

[See also GENETIC FITNESS, ISLAND BIOGEOGRAPHY, ISOLATION]

Fourier analysis   In TIME-SERIES ANALYSIS or IMAGE PROCESSING, the Fourier transform is applied to a dataset in order to visualise *frequency* rather than *spatial* information. In the Fourier transform of a digital image, noise and edges are seen as high-frequency components and periodic features can be easily detected and removed. Spatial frequency FILTERING is often achieved through Fourier analysis.   *ACF*

Richards, J.A. 1986: *Remote sensing digital image analysis: an introduction.* Berlin: Springer.

fractal analysis   The study of curves and shapes that have complex, self-similar geometry and, in particular, the determination of the FRACTAL DIMENSION of these shapes. Many objects in nature have complex forms that are not well represented by simple shapes and conventional geometry, examples being clouds, mountains and plant structures. Fractals are rough or fragmented shapes that have complicated structures on all length scales and are characterised by *self-similarity* in that patterns of the shape are repeated at all scales. Therefore, when divided into smaller parts, fractals are (at least approximately) reduced copies of the original shape. The classic application of fractal analysis was to the determination of the 'true' length of the British coastline, where it was shown that the result depends upon the length of the measuring instrument (RESOLUTION scale). At higher resolutions, finer details can be discerned and the total length increases accordingly.

Fractals are usually defined as structures for which the fractal dimension *(Hausdorff–Besicovich dimension)* is greater than the topological dimension. The fractal dimension measures the degree of roughness of a shape and, roughly speaking, can be defined as the ratio:

log (change in object size) : log (change in measurement scale)

If one considers a smooth line (which has a topological dimension of unity), then the length of the line is independent of the length of the instrument that is used to measure it. In this case, the fractal dimension is also unity (and the object is therefore not a fractal). However, in the case of a rough curve (such as a coastline), finer details may be apparent at higher resolutions, so that, as one uses a smaller measurement scale, the total length becomes larger. In this case, the fractal dimension is larger than unity and the object is therefore a fractal. As well as applications to spatial patterns (objects in nature, the shapes of voting districts), fractal analysis has also been applied to temporal patterns ranging from the time series of brain signals to financial markets.   *PJS*

Barnsley, M. 1993: *Fractals everywhere*, 2nd edn. New York: Academic Press. **De Cola, L., and Lam, N.S.** 1993: Introduction to fractals in geography. In Lam, N.S. and De Cola, L. (eds), *Fractals in geography.* Englewood Cliffs, NJ: Prentice Hall. **Gao, J. and Xia, Z.** 1996: Fractals in physical geography. *Progress in Physical Geography* 20, 178–191. **Mandelbrot, B.** 1983: *The fractal geometry of nature.* San Francisco, CA: W.H. Freeman. **Xiu, T.B., Moore, I.D. and Gallant, J.C.** 1993: Fractals, fractal dimensions and landscapes: a review. *Geomorphology* 8, 245–262.

fractal dimension   A measure of the dimensionality of an object that falls between the four topological dimensions of traditional Euclidean Geometry: 0-D (points); 1-D (lines); 2-D (planes) and 3-D (volumes). For example, an irregular line may have a fractal dimension of 1.8 (falling between straight lines and planes) or an image, such as that produced by REMOTE SENSING, may have a fractal dimension of 2.3. The fractal, or *Hausdorff–Besicovich dimension* may be used as a measure of IMAGE TEXTURE in remote sensing.   *AJL*

[See also FRACTAL ANALYSIS]

fracture   A physical discontinuity or break in a material. Fractures in rocks (and some sediments) are divided into FAULTS, where there has been relative displacement across the fracture, and JOINTS, where there has not. *DNT*

fracture zone   A step or scarp in the OCEAN floor that cuts the flanks of a MID-OCEAN RIDGE as a continuation of a TRANSFORM FAULT, beyond the zone where the ocean floor is moving in opposite relative directions. Because the two sides of the fracture zone lie within the same plate, there is no lateral sliding between them, but because they differ in age and therefore temperature, they are at different elevations.   *GO*

[See also PLATE TECTONICS]

fragipan   A lower soil layer that is physically compacted so that when dry it has a brittle consistence and a high bulk density, but fragments placed in water will disintegrate. Freeze–thaw activity in the past with added bonding agents such as silica, iron and aluminium is thought to have been responsible for the dry strength of the fragipan. It has been redefined for the WORLD REFERENCE BASE FOR SOIL RESOURCES as a *fragic horizon* and is commonly found in CAMBISOLS and GLEYSOLS in areas formerly marginal to continental ice sheets.   *EMB*

[See also HARDPAN, INDURATION]

Bridges, E.M. and Bull, P.A. 1981: The role of silica in the formation of compact and indurated horizons in the soils of South Wales. In Bullock, P. and Murphy, C.P. *Soil micromorphology*. Vol. 2. *Soil genesis*. Berkhamsted: AB Academic Publishers. Brooks, S.M., Anderson, M.G. and Crabtree, K. 1995: The significance of fragipans to early Holocene slope failure: application of physically based modelling. *The Holocene* 5, 293–303. Grossmann, R.B. and Carlisle, F.J. 1969: Fragipan soils of the eastern United States. *Advances in Agronomy* 21, 237–279. Habecker, M.A., McSweeney, K. and Madison, F.W. 1990: Identification and genesis of fragipans in Ochrepts in north central Wisconsin. *Soil Science Society of America Journal* 54, 139–146.

fragmentation

When human societies colonise new regions, they typically transform native landscapes into agricultural systems and other environments that are more capable of supporting human populations, at least for the short term. As a result, the total surface area of native ecosystems decreases (HABITAT LOSS) and they are converted to archipelagos of increasingly isolated, or fragmented, habitats (the process of fragmentation). Almost invariably, these two phenomena occur simultaneously and should be viewed as outcomes of the same, more general phenomenon – *anthropogenic transformation* of native landscapes. *MVL*

[See also DEFORESTATION, RELAXATION]

Harris, L.D. 1984: *The fragmented forest: island biogeography theory and the preservation of biotic diversity*. Chicago, IL: University of Chicago Press. Laurance, W.F. and Bierregard, R.O. (eds) 1997: *Tropical forest remnants: ecology, management and conservation of fragmented communities*. Chicago, IL: University of Chicago Press. Meffe, G.K. and Carroll, C.R. 1997: *Principles of conservation biology*, 2nd edn. Sunderland, MA: Sinauer. Skole, D. and Tucker, C. 1993: Tropical deforestation and habitat fragmentation in the Amazon: satellite data from 1978 to 1988. *Science* 260, 1905–1910.

freeze coring

Obtaining a lake-sediment core by means of freezing the sediment to the *outside* of a tube, box or wedge filled with dry ice (a so-called '*frozen finger*'). The device is dropped or lowered into the sediment and pulled out once the sediment has frozen on. It is particularly useful for obtaining an undisturbed sample of delicate sedimentary structures, such as VARVES, in the uppermost soft sediments, which tend to be disturbed by more conventional CHAMBER or PISTON SAMPLERS during either sampling or transportation or core extrusion. *JAM*

Lotter, A.F., Renberg, I., Hansen, H. *et al.* 1997: A remote controlled freeze corer for sampling unconsolidated surface sediments. *Aquatic Science* 59, 295–303. Renberg, I. 1981: Improved methods for sampling, photographing and varve-counting of varved lake sediments. *Boreas* 10, 255–258. Saarnisto, M. 1986: Annually laminated lake sediments. In Berglund, B.E. (ed.), *Handbook of Holocene palaeoecology and palaeohydrology*. Chichester: Wiley, 343–370.

freeze–thaw cycles

The frequency of freezing and thawing, based on either air or ground temperatures, is a commonly used quantitative index with respect to cold-climate (cryogenic) WEATHERING. Freeze–thaw cycles can range from long-term (i.e. annual) to short-term (i.e. diurnal) in nature. The annual cycle is relatively easily characterised in terms of its time of occurrence, duration and intensity. Short-term freeze–thaw cycles are difficult to characterise. According to the literature, the annual frequency of short-term cycles at the ground surface, however defined, is relatively low, between 50 and 100. These cycles are twice as numerous as air cycles and, with increasing depth, there is a rapid drop in frequency such that, below 10–20 cm, only the annual cycles usually occur.

Oceanic PERIGLACIAL environments and mid-latitude alpine environments are commonly regarded as being the most suited for freeze–thaw processes, but the relatively low frequency of freeze–thaw cycles must cast doubt upon the assumption that freeze–thaw weathering is the dominant process of rock disintegration in these environments. Thermal stress is also important, and the complexities of cryogenic weathering are still not fully understood. *HMF*

[See also CRYOSOLS]

Cook, F.A. and Raiche, V. 1962: Freeze–thaw cycles at Resolute, N. W. T. *Geographical Bulletin* 18, 64–78. Matsuoka, N. 1991: A model of the rate of frost shattering: application to field data from Japan, Svalbard and Antarctica. *Permafrost and Periglacial Processes* 2, 271–280.

freezing front

The boundary between frozen and unfrozen soil in the soil profile. It should be distinguished from the *cryofront*, which is the 0°C boundary, and the *thawing front*, which is the boundary between seasonally frozen and seasonally thawed soil. *HMF*

[See also PERMAFROST, ACTIVE LAYER]

Associate Committee on Geotechnical Research 1988: *Glossary of permafrost and related ground-ice terms* [*Technical Memorandum* 142]. Ottawa: Permafrost Subcommittee, National Research Council of Canada.

freezing index

The cumulative total of temperature degrees (i.e. DEGREE DAYS) below zero for any one year; a measure of the severity and duration of freezing. The opposite, the *thawing index*, is the cumulative total of degree-days above zero for any one year. Mean daily air temperatures in °C are normally used. *HMF*

[See also ACCUMULATED TEMPERATURE, FREEZE–THAW CYCLES]

frequency

(1) A statistical term for the number of occasions a variable takes a certain value or lies in a certain range of values. (2) In a time series, the number of cycles of a periodic phenomenon per unit time. *GS*

[See also HISTOGRAM, PERIODICITIES, TIME-SERIES ANALYSIS]

friction cracks

A fracture mark or crack in bedrock caused by the removal of flakes of rock by subglacial plucking. Friction cracks include *crescentic gouges*, *conchoidal fractures*, *lunate fractures* and CHATTERMARKS. They differ from STRIATIONS in being formed by intermittent, as opposed to continuous, contact between bedrock and clasts in subglacial traction. They can be used to provide an indication of former ice-flow direction. *RAS*

[See also GLACIAL EROSION]

fringing reefs

Coral structures lying on relatively shallow foundations, close to shore, with a shorter history than BARRIER REEFS and ATOLLS. *TSp*

front A transition zone between AIR MASSES that differ in terms of temperature and/or humidity, which develops during *frontogenesis*. A *cold front* is where cold air is displacing (undercutting) warmer air: its position on a weather map indicates where the front reaches the Earth's surface, where triangles point in the direction of movement of the front. A *warm front* is where cold air is being displaced (overridden) by warm air and is represented by halfmoon symbols. The *warm sector* lies between the cold and warm fronts in an extratropical CYCLONE or DEPRESSION. An *occluded front* or occlusion, represented by alternating cold- and warm-front symbols, may occur during the later stages of CYCLOGENESIS, occlusion being defined by the warm sector having been uplifted by converging cold and warm fronts to the extent that it is no longer in contact with the ground. A *frontal system* is a complex of fronts associated with CYCLONES or DEPRESSIONS. Fronts also exist between water masses (see OCEANIC POLAR FRONT). *JAM*

[See also ARCTIC FRONT, FREEZING FRONT, INTERTROPICAL CONVERGENCE ZONE, POLAR FRONT, WEATHERING FRONT, WETTING FRONT]

Carlson, T.N. 1991: *Mid-latitude weather systems*. London: HarperCollins.

frost boil A type of non-sorted circle or mud hummock. The term *boulder-cored frost boil* has been used for a small hummock produced by the vertical frost heaving of single boulders. *HMF*

[See also PATTERNED GROUND]

Harris, C. and Matthews, J.A. 1984: Some observations on boulder-cored frost boils. *Geographical Journal* 150, 63–73.

frost crack A *thermal contraction crack*, i.e. a fissure formed as a result of stresses generated in frozen soil or rock. In PERMAFROST regions, cracking of the ground usually results from thermal contraction. Frost cracks in rocks and boulders are also thought to result because of the 9% volume expansion of water upon freezing when water penetrates the cracks, joints and bedding planes. Differential thermal stresses generated at subzero temperatures may also cause cracking. *HMF*

[See PERIGLACIAL LANDFORMS, PERIGLACIAL LANDSCAPE EVOLUTION, PERIGLACIAL SEDIMENTS, ICE WEDGE]

French, H.M. 1996: *The periglacial environment*, 2nd edn. London: Addison-Wesley Longman, 39–40.

frost heave The predominantly upward movement of mineral soil during freezing caused by the migration of water to the freezing plane and its subsequent expansion upon freezing. Frost heaving is usually associated with the ACTIVE LAYER above PERMAFROST or with seasonally frozen ground. Frost heaving results in upfreezing of objects, the heave of bedrock blocks, tilting of stones, formation of NEEDLE ICE and the sorting and migration of soil particles. Frost heaving presents problems in the construction of roads, buildings, pipelines and other structures in cold regions. *HMF*

[See also PATTERNED GROUND]

frost mound Any mound-shaped LANDFORM produced by ground freezing combined with groundwater movement or the migration of soil moisture. Frost mounds can be distinguished on the basis of (a) their structure and duration and (b) the character of the ice contained within them. *Frost blisters* are seasonal frost mounds produced by the doming of seasonally frozen ground by a subsurface accumulation of water under high hydraulic potential during progressive freezing of the active layer. An *icing blister* is a seasonal frost mound consisting of ice only and formed at least in part through lifting of one or more layers of an ICING by injected waters. An *icing mound* is a seasonal frost mound consisting exclusively of thinly layered ice formed by the freezing of successive flows of water issuing from the ground or from below river ice. PALSAS are peaty PERMAFROST mounds (up to several metres in height) possessing a core of alternating layers of segregated ice and PEAT or mineral soil material. Those existing without a peaty overburden are sometimes called *mineral permafrost mounds*, LITHALSAS or *mineral palsas*. PINGOS (also known as *bulgannyakh* and *hydrolaccoliths*) are perennial frost mounds consisting of a core of massive ice, produced primarily by injection of water and covered with soil and vegetation. They are usually circular or oval in plan form, up to 600 m in diameter and 50 m high. It is common to distinguish between *closed system pingos*, formed by the freezing of injected water supplied by expulsion of pore water during aggradation of permafrost in the closed TALIK beneath a former water body, and *open system pingos*, formed by the freezing of injected water supplied by groundwater moving down-slope through taliks to the site of the pingo, where it moves towards the surface.

Remnants of perennial frost mounds (i.e. pingos and palsas), if found in non-permafrost environments today, provide evidence of former permafrost conditions. So-called *pingo scars* (many of which may be remnants of other types of ground-ice depressions or even not of PERIGLACIAL origin at all) typically comprise more or less circular or elongated depressions surrounded or partly surrounded by low ramparts. Many such depressions contain peat and, through the application of POLLEN ANALYSIS and RADIOCARBON DATING, likely minimum dates for their formation can be determined. Where such techniques have been used, they have generally indicated a LATE GLACIAL STADIAL age. *HMF*

[See also PERMAFROST, ACTIVE LAYER, ICING]

Degroot, T., Cleveringa, P. and Klijnstra, B. 1987: Frost mound scars and the evolution of a Late Dryas environment (northern Netherlands). *Geologie en Mijnbouw* 66, 239–250. **Mackay, J.R.** 1986: Frost mounds. In French, H.M. (ed.), Focus: permafrost geomorphology. *The Canadian Geographer* 30, 363–364. **Marsh, B.** 1987: Pleistocene pingo scars in Pennsylvania. *Geology* 15, 945–947. **Matthews, J.A., Dahl, S.O., Berrisford, M.S. and Nesje, A.** 1997: Cyclic development and thermokarstic degradation of palsas in the mid-alpine zone at Leirpullan, Dovrefjell, southern Norway. *Permafrost and Periglacial Processes* 8, 107–122. **Seppälä, M.** 1988: Palsas and related forms. In Clark, M.J. (ed.), *Advances in periglacial geomorphology*. Chichester: Wiley, 247–278. **Sollid, J.L. and Sorbel, L.** 1998: Palsa bogs as a climatic indicator – examples from Dovrefjell, southern Norway. *Ambio* 27, 287–291. **Worsley, P., Gurney, S.D. and Collins, P.E.F.** 1995: Late Holocene mineral palsas and associated vegetation patterns – a case-study from Lac-Hendry, northern Québec, Canada and significance for European Pleistocene thermokarst. *Quaternary Science Reviews* 14, 179–192.

frost weathering A general term to describe the predominantly mechanical WEATHERING of rocks by FREEZE–THAW CYCLES in non-glacial cold climates. Expansion of ice upon freezing within rocks and in cracks and fissures leads to the process of *frost shattering* or *frost splitting*. The magnitude of frost weathering is influenced by both daily and annual cycles of freezing, but only the latter appears capable of releasing large blocks following relatively deep freezing and thawing. Frost weathering within continental areas tends to be influenced mainly by annual freeze cycles, whilst daily freeze-thaw cycles are more effective in more maritime locations. The process is particularly effective along cold-climate shores, where water is more easily able to penetrate rock joints. *AGD*

[See also ROCK PLATFORM, PHYSICAL WEATHERING, POLAR SHORE EROSION, STRANDFLAT]

Ballantyne, C.K. and Harris, C. 1994: *The periglaciation of Britain*. Cambridge: Cambridge University Press. **Dredge, L.A.** 1992: Breakup of limestone bedrock by frost shattering and chemical weathering, eastern Canadian Arctic. *Arctic and Alpine Research* 24, 314–323. **Matthews, J.A., Dawson, A.G. and Shakesby, R.A.** 1986: Lake shoreline development, frost weathering and rock platform erosion in an alpine periglacial environment. *Boreas* 15, 33–50.

fruits and seeds A true fruit is the dry or fleshy case formed from the wall of a plant ovary after fertilisation, which contains the *seeds* (the fertilised ovules), but the term fruit or *fruiting body* can be applied to any plant organ containing propagules. Fruits include *berries* (succulent fruits, normally with more than one seed and seeds lacking a stony coat, e.g. raspberries), *drupes* (succulent or spongy fruits, usually with one, stony, coated seed, e.g. peaches), and *nuts* (dry, indehiscent one-seeded fruits with a hard woody wall, e.g. acorns). Fruits serve to protect seeds and often aid seed dispersal (e.g. by attracting birds, which eat the fruit and pass the seed through the gut, depositing it later in a location that may be far removed from the parent tree). They also provide a useful food source for humans. Parts of dry fruits are often encountered in MACROFOSSIL assemblages (e.g. hazelnuts), but fleshy fruits rarely survive intact and are usually represented only by their seeds. *MJB*

[See also PLANT MACROFOSSIL ANALYSIS]

Minnis, P.E. 1981: Seeds in archaeological sites: sources and some interpretive problems. *American Antiquity* 46, 143–152.

fuelwood Wood derived from woodlands and forests for the purposes of heating, cooking, metal smelting and charcoal production. Its removal in DRYLANDS is regarded as one of the classic human (as opposed to natural environmental) causes of DESERTIFICATION. *RAS*

[See also LAND DEGRADATION]

Fox, J. 1984: Firewood consumption in a Nepali village. *Environmental Management* 8, 243–250. **Sefe, F., Ringrose, S. and Matheson, W.** 1996: Desertification in north central Botswana: causes, processes, and impacts. *Journal of Soil and Water Conservation* 51, 241–248. **Soussan, J., O'Keefe, P. and Munslow, B.** 1990: Urban fuelwood: challenges and dilemmas. *Energy Policy* 18, 572–582.

fulgurite Fused sand in irregular, branching, tube-shaped structures a few centimetres in diameter and up to several metres long caused by lightning striking the ground. Distribution patterns of fulgurites are indicative of thunderstorm distribution and have significance for PALAEOENVIRONMENTAL RECONSTRUCTION, especially in ARIDLANDS and DESERTS. *JAM*

Sponholz, B., Baumhauer and Felix-Henningsen, P. 1993: Fulgurites in the southern Central Sahara, Republic of Niger and their palaeoenvironmental significance. *The Holocene* 3, 97–104.

fulvic acids Complex organic substances present in HUMUS that remain in solution after an alkali soil extract has been acidified. Fulvic acids are high molecular weight hydroxyl-carboxylic acids that have a light yellow colour and lower solubility in water compared with HUMIC ACIDS. In the RADIOCARBON DATING of soil or PEAT, the *fulvic acid fraction* tends to be avoided as it is relatively mobile and hence susceptible to inclusion of modern contaminants. *EMB*

[See also SOIL DATING]

Schnitzer, M. and Khan, U. 1972: *Humic substances in the environment*. New York: Marcel Dekker.

fumarole A small high-temperature volcanic vent, or opening in the Earth's surface, that emits gases (e.g. CO_2, SO_2) to the ATMOSPHERE, often with the SUBLIMATION of less VOLATILE components such as sulphur, but does not emit LAVA or EJECTA. Fumaroles may be associated with DORMANT or EXTINCT VOLCANOES. *JBH*

Kaneko, T. and Wooster, M.J. 1999: Landsat infrared analysis of fumarole activity at Unzen Volcano: time-series comparison with gas and magma fluxes. *Journal of Volcanology and Geothermal Research* 89, 57–64.

functional factorial approach An approach to soil study in particular, in which soil properties are examined as functions of a small number of SOIL FORMING FACTORS. It emphasises the possibility of isolating the effects of individual soil forming factors, such as relief through *toposequences* (CATENAS) and time through CHRONOSEQUENCES or CHRONOFUNCTIONS. It has also been applied to vegetation studies and illustrates an approach widely used in environmental research. *JAM*

Jenny, H. 1941: *Factors of soil formation*. New York: McGraw-Hill. [Republished as *Factors of soil formation: a system of quantitative pedology* 1994: New York: Dover.] **Jenny, H.** 1980: *The soil resource: origin and behaviour*. New York: Springer.

functional type A group of organisms that respond to environmental PERTURBATION in a similar manner, and through the same mechanism. Functional types are used particularly in MODELS of ECOLOGICAL RESPONSE TO ENVIRONMENTAL CHANGE. *CPO*

Lavorel, S. and Cramer, W. (eds) 1999: *Plant functional types and disturbance dynamics. Journal of Vegetation Science* 10, 603–730 [Special Issue]. **Smith, T.M., Shugart, H.H. and Woodward, F.I. (eds)** 1997: *Plant functional types. Their relevance to ecosystem properties and global change*. Cambridge: Cambridge University Press.

fungal hyphae analysis The nature of fungal hyphae in soils, peat and wood may be of value in PALAEOENVIRONMENTAL RECONSTRUCTION. In soils, living fungal hyphae occur mainly near the surface in the litter layer, where they decompose rapidly. If decomposition is retarded, dark-coloured (pigmented) hyphae develop, but they become fragmented by soil fauna and characteristically decrease in frequency with depth (macroarthropods cut the hyphae into fragments of length 20–50 µm; microarthropods produce fragments around 10–30 µm). Hyphae fragments in BROWN EARTHS are produced by the extant arthropods, whereas in PODZOLS the fragmentation is due to former animal communities. Thus, the former existence of a Brown Earth stage has been recognised by an increase in the frequency of hyphae fragments with depth in Podzols. *JAM*

Andersen, S.T. 1984: Stages in soil development reconstructed by evidence from hypha fragments, pollen, and humus contents of soil profiles. In Haworth, E.Y. and Lund, J.W.G. (eds), *Lake sediments and environmental history*. Leicester: Leicester University Press, 295–316.

future climate

Climates of the future will depend on the natural and anthropogenic causes of CLIMATIC CHANGE in both the long and short terms. Predictions about future climate are best made by CLIMATIC MODELS that have been tested against data from PALAEOCLIMATOLOGY. Models based on ORBITAL FORCING predict the end of the present interglacial in a few thousand years, cold episodes around 5, 23 and 60 ka AP (thousands of years After Present) with the last as severe as the Last Glacial Maximum (LGM). This pattern is unlikely to be changed very much as a result of enhanced GLOBAL WARMING from carbon dioxide forcing (see Figure). Various more extreme CLIMATIC SCENARIOS have, however, been proposed ranging from the *superinterglacial scenario* (the present interglacial is made warmer) to the *delayed-glaciation scenario* (warming is more effective, resulting in delayed onset of the next glaciation) and the extreme, *irreversible-greenhouse scenario* (prevention of the next glaciation).

In the shorter-term future, there are fewer uncertainties but predictions are far from certain because of (1) the use of simple models, combined with (2) imperfect understanding of the likely natural background CLIMATIC VARIATIONS and (3) uncertainty over the extent of government action in limiting emissions of GREENHOUSE GASES. It would appear that the GLOBAL WARMING predicted by GENERAL CIRCULATION MODELS is now discernible in global temperature measurements, but it is likely that some of the recent variations in mean global air temperature are partly controlled by natural decadal-scale variations in climate. Following the recommendations of the INTERGOVERNMENTAL PANEL ON CLIMATE CHANGE (IPCC), EMISSION CONTROLS are being implemented, but what will replace the BUSINESS-AS-USUAL SCENARIO is at present unclear.

JAM

[See also CLIMATIC MODIFICATION, CLIMATIC PREDICTION, DEFORESTATION: CLIMATIC IMPACTS, MILANKOVITCH THEORY, OZONE DEPLETION]

Allen, M.R., Stott, P.A., Mitchell, J.F.B. *et al.* 2000: Quantifying the uncertainty in forecasts of anthropogenic climate change. *Nature* 407, 617–620. **Berger, A. and Loutre, M.F.** 1996: Modelling the climate response to astronomical and CO_2 forcings. *Comptes Rendus de l'Académie des Sciences Serie 2 Fascicule A*, 1–16. **Broecker, W.S.** 1998: The end of the present interglacial: how and when? *Quaternary Science Reviews* 17, 689–694. **Conway, D.** 1998: Recent climatic variability and future climate change scenarios for Great Britain. *Progress in Physical Geography* 22, 350–374. **Goodess, C.M., Palutikof, J.P. and Davies, T.D.** 1992: *The nature and causes of climate change: assessing the long term future*. London: Belhaven. **Houghton, J., Meira Filho, L.G., Callander, B.A.** *et al.* (eds) 1996: *Climate change 1995: the science of climate change*. Cambridge: Cambridge University Press. **Hulme, M., Barrow, E.M., Arnell, N.W.** *et al.* 1999: Relative impacts of human-induced climatic change and natural climate variability. *Nature* 397, 688–691. **Nesje, A. and Dahl, S.O.** 2000: *Glaciers and environmental change*. Arnold: London. **Santer, B.D., Taylor, K.E., Wigley, T.M.L.** *et al.* 1995: Towards the detection and attribution of an anthropogenic effect on climate. *Climate Dynamics* 12, 79–100. **Santer, B.D., Taylor K.E., Wigley, T.M.L.** *et al.* 1996: A search for human influences on the thermal structure of the atmosphere. *Nature* 382, 39–46. **Tett, S.F., Mitchell, J.F., Parker, D.E. and Allen, M.R.** 1996: Human influences on the atmospheric vertical temperature structure: detection and observations. *Science* 274, 1170-1173.

fuzzy logic

A mathematical concept based on the theory of *fuzzy sets*. Instead of an element belonging either to set A (membership 1) or not (membership 0), an element may belong to several sets at a time with graded membership degrees $\mu(A)$, $\mu(B)$ etc. The sum of all membership

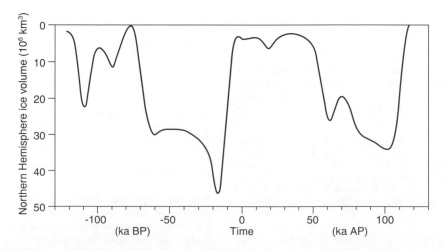

future climate *The possible course of future climate for the next 120 000 years: the curve shows predicted ice volume in the Northern Hemisphere resulting from astronomical and carbon dioxide forcing (after Berger and Loutre, 1996; Nesje and Dahl, 2000)*

degrees of an element is one. Fuzzy logic has been applied to REMOTE SENSING, mainly for fuzzy IMAGE CLASSIFICATION. Given n classes, a classified PIXEL in the image has n membership degrees. Fuzzy logic is sometimes confused with PROBABILITY theory. A *maximum likelihood classification* estimates *a posteriori* probabilities of each pixel belonging to the classes, but it is assumed that there is a single 'true' class for each pixel. Fuzzy logic allows more than one class membership for each pixel. The membership degrees may be interpreted, for instance, as within-pixel variation in LAND COVER type. The fuzzy classification image can be transformed to a traditional classified image by selecting the class with the highest membership degree for each pixel. *HB*

Foody, G.M. 1996: Fuzzy modelling of vegetation from remotely sensed imagery. *Ecological Modelling* 85, 3–12. **Klawonn, F. and Kruse, R.** 1995: Clustering methods in fuzzy control. In Gaul, W. and Pfeifer, D. (eds), *From data to knowledge: theoretical and practical aspects of classification, data analysis and knowledge organization.* Berlin: Springer, 195–202. **Wang, F.** 1990: Fuzzy supervised classification of remote sensing images. *IEEE Transactions on Geoscience and Remote Sensing* 28, 194–201.

G

Ga   A *giga-annum*: an abbreviation for thousands of millions (billions) of years before the present. For example, 4.6 Ga is 4600 million years ago.   *GO*

gabion   A mesh basket (usually of wire) filled with rocks to prevent erosion or to serve as a foundation. It is usually an inexpensive, temporary structure, much used in COASTAL (SHORE) PROTECTION.   *HJW*

Gaia hypothesis   A concept suggesting that the Earth is a complex entity in a state of equilibrium maintained by FEEDBACK MECHANISMS. The physical and chemical 'health' of the planet, especially ATMOSPHERIC COMPOSITION, is seen as the result of self-regulation throughout geological time and, contrary to more conventional ideas that see life as adapting to the ABIOTIC environment, the Gaia concept stresses control by the BIOSPHERE. The concept was developed by James E. Lovelock, who noted that the lower atmosphere of the Earth is an integral, regulated and necessary part of life itself. The Gaia hypothesis may be viewed as a recent manifestation of the HOLISTIC APPROACH to knowledge: *weak Gaia* asserting a substantial influence of life over the abiotic environment, *strong Gaia* emphasising the *organismic analogue* (see COMMUNITY CONCEPTS).   *JAM/AHP*

[See also EARTH-SYSTEM ANALYSIS, GLOBAL ENVIRONMENTAL CHANGE, HOMEOSTASIS]

Lovelock, J.E. 1995: *Gaia: a new look at life on Earth*, 2nd edn. Oxford: Oxford University Press. **Lovelock, J.E. and Margulis, L.** 1974: Atmospheric homeostasis by and for the biosphere. *Tellus* 26, 1–10. **Schneider, S.H. and Boston, P.J. (eds)** 1991: *Scientists on Gaia*. Cambridge, MA: MIT Press.

gallery forest   A narrow strip of EVERGREEN or semi-deciduous closed-canopy forest along a water course draining an area of DECIDUOUS or SAVANNA vegetation. In the tropics, gallery forests are often similar to TROPICAL RAIN FOREST in structure and physiognomy and some have been regarded as the relics of more extensive evergreen forests in the past. However, continuously high soil moisture permits an evergreen habit in regions that are seasonally dry. Gallery forests are often very species-rich ecosystems and fire is believed to promote this diversity. The boundaries of gallery forests are fire-prone, but they contain core zones into which fire very rarely intrudes. Fire-tolerant trees in the outer zone may protect a forest interior of low flammability. Although of high CONSERVATION value, many gallery forests are threatened by irrigated agriculture and flooding for hydroelectric power generation.   *NDB*

[See also DISTURBANCE, FIRE FREQUENCY, FIRE IMPACTS, RIPARIAN]

Kellman, M. and Meave, J. 1997: Fire in the tropical gallery forests of Belize. *Journal of Biogeography* 24, 23–34. **Kellman, M., Tackaberry, R. and Rigg, L. 1998:** Structure and function in two tropical gallery forest communities: implications for forest conservation in fragmented systems. *Journal of Applied Ecology* 35, 195–206.

game management   The management of essentially wild animals primarily for economic (HARVESTING) purposes. It contrasts with the raising of domesticated animals on farms, although the distinction is blurred in *game ranching*.   *JAM*

[See also WILDLIFE CONSERVATION AND MANAGEMENT]

Leopold, A. 1933: *Game management*. Madison, WI: University of Wisconsin Press.

gamma correction   In IMAGE ENHANCEMENT, a non-linear point-wise transformation used to control the overall brightness in images. It is based on the relationship between the actual values in the pixels, the voltage into which these values will be translated in the display device and the brightness that will be shown to the user for every voltage.   *ACF*

Jain, A.J. 1989: *Fundamentals of digital image processing*. Englewood Cliffs, NJ: Prentice-Hall.

gamma ray attenuation porosity evaluator (GRAPE)   A rapid, non-destructive method for assessing the water content of SEDIMENTS by measuring the attenuation of a beam of gamma rays projected through the sediment. It is widely used for CORE logging, as GRAPE records from some MARINE SEDIMENT CORES display cyclicities characteristic of MILANKOVITCH THEORY.   *MRT*

Mayer, L.A., Jansen, E., Backman, J. and Takayama, T. 1993: Climatic cyclicity at Site 806: the GRAPE record. *Proceedings of the Ocean Drilling Program, Scientific Results* 130, 623–639.

gap analysis   A technique used to identify biota and biotic communities that are not adequately represented in an existing series or network of PROTECTED AREAS. The technique can help to locate areas for CONSERVATION and can also be used to prioritise planning with respect to HABITAT protection and management. Elements of the analysis typically include vegetation classification and the mapping of the geographical distribution of vegetation types, plant and animal species, of areas set aside for protection of nature and of the habitats of greatest importance to THREATENED SPECIES. GEOGRAPHICAL INFORMATION SYSTEMS are often employed in gap analysis.   *IFS*

[See also CONSERVATION BIOLOGY, PROTECTED AREA APPROACH, SINGLE LARGE OR SEVERAL SMALL RESERVES]

Scott, J.M., Davis, F., Csuti, B. *et al.* 1993: Gap analysis: a geographic approach to protection of biological diversity. *Wildlife Monographs* 123, 1–41. **Spellerberg, I.F. and Sawyer, J.W.D.** 1999: *An introduction to applied biogeography*. Cambridge: Cambridge University Press.

gas hydrates   CRYSTALLINE mixtures of METHANE and water stable at high ambient pressure (> 50 bar) and low temperature ($< 7°C$), which are widespread in CONTINENTAL SLOPE and CONTINENTAL RISE sediments and in

PERMAFROST areas. They are also known as *methane hydrates* or *clathrates*. The gas is thought to have been released from organic-rich sediments at shallow burial depths and is trapped in sediment pore spaces within about 1 km of the sea-floor.

Degassing is the sudden release of large volumes of methane gas into the oceans and atmosphere, where it is a potent GREENHOUSE GAS, as a result of the disturbance of sea-floor sediments or a lowering of hydrostatic pressure, triggered by processes such as EARTHQUAKES, SEA-LEVEL CHANGE or mass movements of sediment, for example TURBIDITY CURRENTS. Sudden degassing produces large pockmarks on the sea-floor and is thought by some to be responsible for the sinking of ocean-going vessels in the 'Bermuda Triangle', an area above the Blake Ridge and Carolina Rise where gas hydrate production in marine sediments is known to be very high. The release of gas hydrates may have contributed to the global firestorm at the K–T BOUNDARY. Abrupt BOTTOM WATER warming at the end of the PALAEOCENE may have caused widespread methane release, resulting in a world-wide negative carbon isotopic excursion (see CARBON ISOTOPES). The change from solid to gas phase also raises sediment pore pressure, which may destabilise slope deposits, causing large-scale mass movement. This may be particularly prevalent at times of falling sea-level and there is some evidence of increased frequency of slumping in hydrate-prone areas during sea-level lowstands. Degassing may cause a fall in sea-level, through a reduction in the volume of sea-bed sediments.

Gas hydrates may be underlain by sediment containing large volumes of gaseous methane, which could be exploited as an ENERGY RESOURCE. It has been estimated that gas hydrates are associated with at least twice as much combustible carbon as in all other FOSSIL FUELS.

BTC/MRT

Bratton, J.F. 1999: Clathrate eustasy: methane hydrate melting as a mechanism for geologically rapid sea-level fall. *Geology* 27, 915–918. Dickens, G.R., O'Neil, J.R., Rea, D.K. and Owen, R.M. 1995: Dissociation of oceanic methane hydrate as a cause of the carbon isotope excursion at the end of the Paleocene. *Paleoceanography* 10, 965–971. Henriet, J.-P. and Mienert, J. (eds) 1998: *Gas hydrates: relevance to world margin stability and climatic change* [*Special Publication* 137]. London: Geological Society. Max, M.D., Dillon, W.P., Nishimura, C. and Hurdle, B.G. 1999: Sea-floor methane blow-out and global firestorm at the K–T boundary. *Geo-Marine Letters* 18, 285–291. Max, M.D. and Lowrie, A. 1996: Oceanic methane hydrates: a 'frontier' gas resource. *Journal of Petroleum Geology* 19, 41–56. Paull, C.K., Buelow, W., Ussler, W. and Borowski, W.S. 1996: Increased continental-margin slumping frequency during sea-level lowstands above gas hydrate-bearing sediments. *Geology* 24, 143–146. Rothwell, R.G., Thomson, J. and Kahler, G. 1998: Low-sea-level emplacement of a very large Late Pleistocene 'megaturbidite' in the western Mediterranean Sea. *Nature* 392, 377–380.

Gauss–Gilbert geomagnetic boundary

The geomagnetic polarity boundary between the Gilbert (reverse polarity) and Gauss (NORMAL POLARITY) POLARITY CHRONS, dated by POTASSIUM–ARGON (K–Ar) DATING to c. 3.58 Ma BP. *DH*

[See also GEOMAGNETIC POLARITY REVERSAL, GEOMAGNETIC POLARITY TIMESCALE]

Løvlie, R. 1989: Palaeomagnetic stratigraphy: a correlation method. *Quaternary International* 1, 129–149.

Gauss–Matuyama geomagnetic boundary

An important geomagnetic boundary between the Gauss (NORMAL POLARITY) and Matuyama *reverse polarity* POLARITY CHRONS or Epochs, K–Ar dated to 2.48 Ma BP and astronomically tuned to 2.6 Ma BP (see ORBITAL TUNING). This boundary has been suggested as a MARKER HORIZON for determining the PLIOCENE–PLEISTOCENE TRANSITION. *MHD*

[See also GEOMAGNETIC POLARITY REVERSAL, GEOMAGNETIC POLARITY TIMESCALE]

Thompson, R. 1991: Palaeomagnetic dating. In Smart, P.L. and Frances, P.D. (eds), *Quaternary dating methods – a users guide* [*QRA Technical Guide* 4]. Cambridge: Quaternary Research Association, 177–198. Valet, J.-P. and Meynadier, L. 1993: Geomagnetic field intensity and reversals during the past four million years. *Nature* 366, 234–238.

Gaussian filter

In IMAGE PROCESSING, an implementation in the FILTERING process where the values in the neighbouring pixels are multiplied by constants proportional to the exponential of the square of the distance to the central pixel (i.e. the *Gaussian function*). Use of the Gaussian filter is a noise-reduction technique, which induces a blur in the original image. *ACF*

Gaussian model

The most widely used probability distribution function for real-world data including image data from OPTICAL REMOTE SENSING INSTRUMENTS. Conformity to this model, which is manifested as a bell-shaped curve, is a prerequisite for many parametric statistical techniques such as the *maximum likelihood* classification. *ACF*

gelifluction

In PERIGLACIAL ENVIRONMENTS, the slow down-slope flow of soil or surface sediments through the release of MELTWATER from thawing ice lenses. It can occur on seasonally frozen ground as well as on ground underlain by PERMAFROST. Together with frost creep, it constitutes SOLIFLUCTION. *RAS*

Washburn, A.L. 1967: Instrumental observations of mass wasting in the Mesters Vig district, Northeast Greenland. *Meddelelser øm Grønland* 176, 1–303.

gene pool

A collective term for all the genes in the organisms of a population. *KDB*

genecology

The study of genetic differences between individuals in populations, in relation to their environment. *KDB*

[See also GENETIC FITNESS]

Turesson, G. 1922: The genotypic response of the plant species to the habitat. *Hereditas* 3, 211–350.

general circulation models (GCMs)

Mathematical models of the GENERAL CIRCULATION OF THE ATMOSPHERE based on fundamental principles and developed using the largest and fastest digital computers. They were first developed in the 1960s. The central element of a GCM is a three-dimensional, time-evolving model of the atmosphere, usually represented as an array of 'grid boxes'. Beyond this there is a set of BOUNDARY CONDITIONS at the top and bottom of the model atmosphere, which include a specification of the physical character of the Earth's surface and the ocean. The mathematical formulation of the model involves a set of time-dependent governing equations, which describe the dynamics of the atmosphere. The better the spatial resolution of the model, the greater the requirement for computer power. Despite their complexity, different GCMs continue to yield widely different predictions from the same questions, such as the character of past climates or the likely scale of enhanced GREENHOUSE warming in the near future, indicating the need for further improvements.   *AHP*

[See also CLIMATIC MODELS, FUTURE CLIMATE, GLOBAL WARMING]

Randall, D.J. 2000: *General circulation model development: past, present and future.* Sidcup: Academic Press–Harcourt Publishers.

general circulation of the atmosphere

The average long-term state of the atmosphere, especially the planetary wind systems, which are driven by the distribution of heat received from the Sun and the Earth's rotation. The general circulation is three-dimensional and includes such planetary-scale phenomena as the INTERTROPICAL CONVERGENCE ZONE, HADLEY CELLS, LONG WAVES, MONSOONS, POLAR FRONTS, TRADE WINDS and the WESTERLIES.

AHP/JAM

Lorenz, E.N. 1967: *The nature and theory of the general circulation of the atmosphere.* Geneva: World Meteorological Organization. Palmén, E. and Newton, C.W. 1969: *Atmospheric circulation systems.* New York: Academic Press. Smagorinsky, J. 1972: The general circulation of the atmosphere. In McIntyre, D.P. (ed.), *Meteorological challenges: a history.* Ottawa: Information Canada, 3–42.

general system(s) theory (GST)

A concept developed by Ludwig von Bertalanffy and introduced to the English-speaking world in 1951, although his publications in German predate 1951. Von Bertalanffy had two primary objectives: (1) to resolve the problems biologists were having accommodating the OPEN SYSTEMS of physicists (necessitating their invocation of 'supposedly vitalistic characteristics'); and (2) to minimise the inefficiencies associated with repetitive discoveries stemming from the isolated nature of individual scientific disciplines. General System(s) Theory (GST) was promoted as a discipline akin to logic and mathematics, and some general principles (sometimes called LAWS by von Bertalanffy) such as the exponential and logistic curves were emphasised. Von Bertalanffy dismissed analogy as an appropriate transfer mechanism from one discipline to another, while emphasising homologues and isomorphs, which he conflated. SYSTEMS concepts predate GST and are pervasive in science; consequently, the benefits of invoking GST (as opposed to systems in general) has been questioned. *CET*

Bertalanffy, L. von 1951: An outline of general system theory. *British Journal for the Philosophy of Science* 1: 134–165. Bertalanffy, L. von 1951. General system theory: 5. Conclusion. *Human Biology* 23, 336–345. Smalley, I.J. and Vita-Finzi, C. 1969. The concept of 'system' in the Earth sciences, particularly geomorphology. *Geological Society of America Bulletin* 80, 1591–1594.

generalised additive models (GAMs)

A class of statistical models that are NON-PARAMETRIC extensions of GENERALISED LINEAR MODELS (GLMs). They use smoothing techniques such as locally weighted regression or SPLINE functions to identify and represent possible non-

linear relationships between the response (*y*) and the explanatory variables as an alternative to the PARAMETRIC models implicit in GLMs. In GAMs the link function of the expected value of the response function is modelled as the sum of a number of smooth functions of the explanatory variables rather than in terms of the explanatory variables themselves as in GLMs. The basic GAM equation is thus:

$$\textit{link function } (y) = \text{(explanatory variables)} \times \text{(smooth functions)} + \text{error function}$$

GAMs allow the data to determine the shape of the response function. As a result bimodality and pronounced SKEWNESS can be easily detected. GAMs often provide a better tool for initial data exploration than GLMs. *HJBB*

Hastie, T.J. and Tibshirani, R.J. 1990: *Generalized additive models.* London: Chapman and Hall. Yee, T.W. and Mitchell, N.D. 1991: Generalized additive models in plant ecology. *Journal of Vegetation Science* 2, 587–602.

generalised linear models (GLMs)

GLMs provide a framework for all REGRESSION ANALYSIS and associated statistical *modelling* that involves linear models, namely equations that contain mathematical variables, parameters and random variables that are *linear* in the parameters and the random variables. A GLM consists of an error function, a linear predictor, and a *link function*.

Least-squares REGRESSION ANALYSIS assumes a NORMAL DISTRIBUTION error function. Many kinds of environmental data have non-normal errors. In GLMs the error function can be expressed as one of the members of the exponential function of PROBABILITY DISTRIBUTIONS (e.g. gamma, exponential, normal for continuous probability distributions; binomial, multinomial, Poisson for discrete probability distributions). The structure of the GLM relates each observed response value (*y*) to a predicted value. The predicted value is obtained by transforming the value derived from the linear predictor, which is a linear combination of one or more explanatory variables. There are as many terms in the linear predictor as there are parameters to be estimated from the data. To determine the fit of a given model, the linear predictor is evaluated for each *y* value and the predicted value is then compared with a transformed value of the response variable. The transformation applied is specified by the link function. This relates the mean value of the response variable to its linear predictor. The use of a non-linear link function allows the model to use response and explanatory variables measured in different scales as it maps the linear predictor onto the scale of the response variable. The basic or core equation of a GLM is thus:

$$\textit{link function } (y) = \text{linear predictor} + \text{error function}$$

With an appropriate choice of link function (e.g. log, logit, identity, reciprocal, exponential, probit) and error function, a wide range of statistical models can be developed that are appropriate for different data types (e.g. counts, percentages or proportions, binary). Techniques such as least-squares REGRESSION ANALYSIS, ANALYSIS OF VARIANCE, ANALYSIS OF COVARIANCE, logit regression, probit regression, *multiple linear regression* and contingency table analysis are all types of GLM.

Parameter estimation is by maximum likelihood estimation. Goodness-of-fit is assessed by *deviance*, a measure of the extent to which a particular model differs from the full or saturated model for a data-set where there is one parameter for every data point. Deviance is analogous to residual sum-of-squares in least-squares regression and has a chi-squared distribution. Care is required to find the minimal adequate model, namely the model that produces the minimal residual deviance, subject to the constraint that it has the smallest number of statistically significant parameters.

Advantages of GLMs are: (1) they have great versatility; (2) they can cope with different data types and associated error distributions; (3) they provide a common and powerful framework for REGRESSION ANALYSIS, ANALYSIS OF VARIANCE and, if covariates are included, ANALYSIS OF COVARIANCE; and (4) it is not necessary to transform the data as the regression is transformed through the link function. GLMs can be fitted by software packages such as GLIM, GENSTAT, SAS and S-PLUS.   *HJBB*

[See also GENERALISED ADDITIVE MODELS]

Brew, J.S. and Maddy, D. 1995: Generalised linear modelling. In Maddy, D. and Brew, J.S. (eds), *Statistical modelling of Quaternary science data.* Cambridge: Quaternary Research Association, 125–160. **Crawley, M.J.** 1993: *GLIM for ecologists.* Oxford: Blackwell. **Dobson, A.J.** 1990: *An introduction to generalized linear models.* London: Chapman and Hall. **McCullagh, P. and Nelder, J.A.** 1989: *Generalized linear models.* London: Chapman and Hall. **O'Brian, L.** 1992: *Introducing quantitative geography. Measurement, methods and generalised linear models.* London: Routledge.

genetic algorithms

Algorithms that carry out simulated EVOLUTION on a population of numbers, which are called chromosomes. They were first postulated by Holland in 1975. Simulated evolution takes place by modifying the chromosomes, which are made up of sequences of genes. During evolution, NATURAL SELECTION favours 'successful' chromosomes, which reproduce more frequently than others. Evolution is mainly based on the reproduction process, whereas MUTATIONS occur relatively rarely.

A genetic algorithm has at least five components: chromosome/gene representation; initialisation of the population; evaluation function determining the fitness for survival; genetic operators altering chromosomes (reproduction, crossover and mutation) and parameters for population size; and PROBABILITIES of genetic operators. Genetic algorithms provide robust estimates in complex spaces. They have been applied in REMOTE SENSING (e.g. for automatic identification of spectral signatures).   *HB*

Clark, C. and Canas, A. 1995: Spectral identification by artificial neural-network and genetic algorithm. *International Journal of Remote Sensing* 16, 2255–2275. **Davis, L.** 1987: *Genetic algorithms and simulated annealing.* London: Pitman. **Zhou, J. and Civco, D.L.** 1996: Using genetic learning neural networks for spatial decision making in GIS. *Photogrammetric Engineering and Remote Sensing* 62, 1287–1295.

genetic drift

Changes in the GENE POOL of a population by random processes. These include the spread of an allele because it is neutral (and hence neither eliminated nor favoured) or changes in gene frequency as populations diminish. Such processes are only likely to be significant in small populations, being swamped by non-random processes in larger populations.   *KDB*

Futuyma, D.J. 1998: *Evolutionary biology.* Sunderland, MA: Sinauer.

genetic engineering

The artificial manipulation of the genetic make-up or genome of an organism; the molecular aspect of BIOTECHNOLOGY. Genetic engineering, also known as *recombinant DNA technology, gene cloning* and *'in vivo* (in cell) genetic manipulation', has already been widely used to introduce desirable traits in crops and domesticated animals and to produce antibiotics and hormones. Some of the most important applications of genetic engineering relate to *genetically modified (GM) crops*, which improve productivity either directly or indirectly through engineered resistance to, for example, PESTS or frost. Genetic engineering modifies organisms to suit the environment rather than vice versa: the latter being the traditional approach to AGRICULTURAL INTENSIFICATION. The potential benefits are clear but the environmental risks are largely unknown. There are lessons to be learnt from the release of introduced species (see INTRODUCTION) in the past, but conclusions from these may not be readily transferred, as species behave individualistically and many genetically modified organisms are *transgenic* (derived from more than one species), which adds to the uncertainty in their behaviour when released into the environment.   *JAM*

Fincham, J.R.S. and Ravetz, J.R. 1991: *Genetically-engineered organisms: benefits and risks.* New York: Wiley. **Gasser, C.S. and Fraley, R.T.** 1991: Transgenic crops. *Scientific American* 266, 34–39. **Lindow, S.E.** 1990: Use of genetically altered bacteria to achieve plant frost control. In Nikas, J. and Hagerdorn, C. (eds), *Biotechnology of plant–microbe interaction.* New York: McGraw-Hill, 85–110. **Lycett, G. and Grierson, D. (eds)** 1990: *Genetic engineering of crop plants.* London: Butterworth. **Mannion, A.M.** 1992: Biotechnology and genetic engineering: new environmental issues. In Mannion, A.M. and Bowlby, S.R. (eds), *Environmental issues in the 1990s.* Chichester: Wiley, 147–160. **Mannion, A.M.** 1999: Biotechnology, environmental impact. In Alexander, D.E. and Fairbridge, R.W. (eds), *Encyclopedia of environmental science.* Amsterdam: Elsevier, 59–64. **Primrose, S.B.** 1991: *Molecular biotechnology.* Oxford: Blackwell.

genetic fitness

Typically, the *genetic diversity* of a population, which, in turn, is assumed to be a measure of its ability to adapt to a variable environment. Conservation biologists often worry about the loss of genetic fitness in captive or otherwise small populations of ENDANGERED SPECIES.   *MVL*

[See also INBREEDING DEPRESSION, POPULATION VIABILITY ANALYSIS]

genetic pollution

The process by which genes from GENETICALLY MODIFIED ORGANISMS become incorporated into wild species. Genes conferring resistance to PESTICIDES or DISEASE in genetically modified crops may, for example, be transferred in pollen from the crop plants to related WEED species.   *JAM*

Porteous, A. 2000: *Dictionary of environmental science and technology,* 3rd edn. Chichester: Wiley.

genetically modified organism (GMO)

'Any biological entity capable of replication or of transferring genetic material, in which genetic material has been altered in a way that does not occur naturally by mating and/or natural recombination' (European Economic Community Deliberate Release Directive, 1990, Article 2). Genetically modified organisms are increasingly produced by *recombinant DNA technology* for use in *genetically modified foods* or for other reasons. Most scientists agree that fears about the safety of such foods are unfounded.

JAM

[See also BIOTECHNOLOGY, DEOXYRIBONUCLEIC ACID, GENETIC ENGINEERING, GENETIC POLLUTION]

European Economic Community Deliberate Release Directive 90/220, 1990: EEC Council Directive on the deliberate release to the environment of genetically modified organisms. *Official Journal of the European Communities* L117, 15–27. **McHughen, A.** 2000: *Pandora's picnic basket: the potential and hazards of genetically modified foods.* Oxford: Oxford University Press.

genetics   The part of biological science that deals with the study of heredity and variation.   *KDB*

genotype   The total genetic constitution of an individual.   *KDB*

[See also PHENOTYPE]

gentrification   The replacement of traditional inner-city, working-class, residential areas with improved housing, more affluent inhabitants and up-market services.

JAM

Smith, N. 1992: Gentrification and uneven development. *Economic Geography* 58, 139–155.

geoarchaeology   The application of principles and techniques from the Earth sciences to the study of the human past.   *SPD*

[See also ARCHAEOLOGICAL GEOLOGY, ENVIRONMENTAL ARCHAEOLOGY, LANDSCAPE ARCHAEOLOGY]

Brown, A.G. 1997: *Alluvial geoarchaeology: floodplain archaeology and environmental change.* Cambridge: Cambridge University Press. **Butzer, K.W.** 1982: *Archaeology as human ecology.* Cambridge: Cambridge University Press. **Davidson, D.A. and Shackley, M.L. (eds)** 1976: *Geoarchaeology: Earth science and the past.* London: Duckworth. **Rapp Jr, G. and Hill, C.L.** 1998: *Geoarchaeology.* New Haven, CT: Yale University Press. **Renfrew, C.** 1983: Geography, archaeology and environment. *Geographical Journal* 149, 316–322. **Waters, M.R.** 1992: *Principles of geoarchaeology: a North American perspective.* Tuscon, AZ: University of Arizona Press.

geochemistry   The study of the chemistry of Earth materials and processes. Within SEDIMENTOLOGY much geochemical study is focused on DIAGENESIS. Approaches include the study of MINERAL and aqueous phase chemistry, STABLE ISOTOPE ANALYSIS (e.g. for palaeotemperature determination) and *radioisotope* investigation (e.g for RADIOMETRIC DATING). Analytical techniques include ATOMIC ABSORPTION SPECTROPHOTOMETRY, ENERGY-DISPERSIVE SPECTROMETRY, INDUCTIVELY COUPLED PLASMA ATOMIC EMISSION SPECTROMETRY, INDUCTIVELY COUPLED PLASMA MASS SPECTROMETRY, OPTICAL EMISSION SPECTROSCOPY and X-RAY FLUORESCENCE ANALYSIS. A key technique for mineralogical determination is X-RAY DIFFRACTION ANALYSIS, and microanalytical techniques include use of the ELECTRON MICROPROBE and SCANNING ELECTRON MICROSCOPY.   *TY*

[See also CHEMICAL ANALYSIS OF SOILS]

Brownlow, A.H. 1979: *Geochemistry.* Englewood Cliffs, NJ: Prentice-Hall. **Chester, R.** 1999: *Marine geochemistry*, 2nd edn. Oxford: Blackwell. **Gill, R. (ed.)** 1977: *Modern analytical geochemistry.* Harlow: Longman. **Holland, H.D.** 1984: *The chemical evolution of the atmosphere and oceans.* Princeton, NJ: Princeton University Press. **Killops, S.D. and Killops, V.J.** 1993: *An introduction to organic geochemistry.* Harlow: Longman. **Tucker, M.E. (ed.)** 1988: *Techniques in sedimentology.* Oxford: Blackwell Scientific.

geochronology   The study of GEOLOGICAL TIME. Successions of SEDIMENTS and SEDIMENTARY ROCKS clearly record a history of events comprising the GEOLOGICAL RECORD, and during the nineteenth century a timescale of RELATIVE AGES was built up without any idea of the actual time involved (see STRATIGRAPHICAL COLUMN). CHRONOSTRATIGRAPHY is concerned with the application of time to sediment and rock successions, usually through the use of FOSSILS that provide divisions between chronostratigraphical units. Work continues on the establishment of a series of international standard reference sections that will ultimately provide a boundary marker point at a *Global Stratotype Section and Point (GSSP)* for world-wide geochronological CORRELATION for each chronostratigraphical unit.

Geochronometry, the direct measurement of geological time, is an important application of geochronology. Early attempts, for example comparing the thickness of a sedimentary succession with an 'average' rate of sedimentation, failed because of the incomplete nature of rock successions and widely varying sediment ACCUMULATION RATES in different environments. Calculations by the physicist William Thomson (Lord Kelvin) in the 1890s, assuming that the Earth's internal heat was a relic from a once-molten state, gave a maximum age for the Earth of 60 million years. Kelvin's stature assured respect for his calculations and provided comfort for anti-evolutionists because Darwin had stated that EVOLUTION of life would require at least 120 million years. The discovery of RADIOACTIVITY in the early 20th century showed that the Earth had its own internal heat source, thus negating Kelvin's calculations. The discovery that the rate of radioactive decay was constant provided the potential for accurate geochronometry, and the results of the first *absolute-age dating* using radioactive minerals were published in 1907. Ages obtained in this way are referred to as *isotopic dates* or *radiometric dates* (see RADIOMETRIC DATING) and are quoted in millions of years (Ma) or thousands of millions of years (Ga) before the present.

ISOTOPES suitable for dating PRECAMBRIAN and most PHANEROZOIC rocks include rubidium (^{87}Rb), which decays to strontium (^{87}Sr) with a HALF-LIFE of 48 800 million years; potassium (^{40}K), which decays to argon (^{40}Ar) with a half-life of 11 930 million years; and uranium (^{238}U), which decays to lead (^{206}Pb) with a half-life of 4 469 million years. Different dating methods may yield different ages and the experimental error associated with radiometric dates are commonly millions or tens of millions of years. The minerals that contain these elements are mostly found in IGNEOUS ROCKS, in which case the age obtained represents the cooling of the rock through some critical temperature. Where rocks have subsequently experienced metamorphism the radiometric 'clock' may be reset and care must be taken over the meaning of any radiometric age. Rock successions in which igneous rocks such as LAVA FLOWS are interbedded with fossiliferous sedimentary rocks have been used to calibrate the stratigraphical column, producing the GEOLOGICAL TIMESCALE. The age of the Earth, currently estimated at between 4500 and 4600 Ma., cannot be obtained directly from terrestrial rocks (see PRISCOAN), but is deduced from the range of ages given by chondritic meteorites landing on Earth.

More recently developed methods of geochronology, some of which are directly applicable to sediments and sedimentary rocks, include FISSION-TRACK DATING, COSMO-GENIC-ISOTOPE DATING and ELECTRON SPIN RESONANCE DATING. Dates of late PLEISTOCENE and HOLOCENE organic samples may be determined using RADIOCARBON DATING. Their age is usually quoted as years before the present (years BP), using 1950 as the baseline. *JCWC/GO*

[See also DATING TECHNIQUES]

Berggren, W.A., Kent, D.V., Aubry, M.-P. and Hardenbol, J. (eds) 1995: *Geochronology, time scales and global stratigraphic correlation* [Special Publication 54]. Tulsa, OK: Society of Economic Paleontologists and Mineralogists. **Dunay, R.E. and Hailwood, E.A. (eds)** 1995: *Non-biostratigraphical methods of dating and correlation* [Special Publication 89]. London: Geological Society. **Geyh, M.A. and Schleicher, H.** 1990: *Absolute age determination: physical and chemical dating methods and their application.* Berlin: Springer. **Harland, W.B., Armstrong, R.L., Cox, A.V. et al.** 1990: *A geologic time scale 1989.* Cambridge: Cambridge University Press. **Wagner, G.A.** 1998: *Age determination in young rocks and artifacts: physical and chemical clocks in Quaternary geology and archaeology.* Berlin: Springer. **Wells, J.W.** 1963: Coral growth and geochronometry. *Nature* 197, 948–950.

geocryology The study of perennially frozen ground, or PERMAFROST science. *HMF*

Ershov, E. 1998: *Geocryology.* Cambridge: Cambridge University Press. **Williams, P.J. and Smith, M.J.** 1989: *The frozen Earth. Fundamentals of geocryology.* Cambridge: Cambridge University Press.

geodesy The study of the Earth's gravitational field and related topics including the shape of the Earth (see GEOID), rotation of the Earth, TIDES and precise measurement and mapping of points on the Earth's surface (*geodetic surveying*). *DNT*

Argus, D.F., Peltier, W.R. and Watkins, M.M. 1999: Glacial isostatic adjustment observed using very long baseline interferometry and satellite laser ranging geodesy. *Journal of Geophysical Research – Solid Earth* 104, 29077–29093. **Bomford, G.** 1980: *Geodesy.* Oxford: Clarendon Press. **Ivins, E.R. and James, T.S.** 1999: Simple models for late Holocene and present-day Patagonian glacier fluctuations and predictions of a geodetically detectable isostatic response. *Geophysical Journal International* 138, 601–624. **Vanyo, J.P. and Awramik, S.M.** 1985: Stromatolites and Earth–Sun–Moon dynamics. *Precambrian Research* 29, 121–142.

geodynamics The study of dynamic processes affecting the Earth, particularly those relating to TECTONICS, including PLATE TECTONICS. *GO*

geoecology The interdisciplinary area of GEOGRAPHY and ECOLOGY that focuses on the interrelations and interactions of BIOTIC and ABIOTIC components of NATURAL and CULTURAL LANDSCAPES. *JAM*

[See also GEOECOSYSTEM, LANDSCAPE ECOLOGY, LANDSCAPE SCIENCE]

Huggett, R.J. 1995: *Geoecology: an evolutionary approach.* London: Routledge. **Matthews, J.A.** 1992: *The ecology of recently-deglaciated terrain: a geoecological approach to glacier fore-*

lands and primary succession. Cambridge: Cambridge University Press. **Troll, C.** 1971: Landscape ecology (geoecology) and bio-geocenology – a terminological study. *Geoforum* 8, 43–46.

geoecosphere The global geoecosystem: the totality of those landscapes of the Earth in which organisms are present. As a concept of LANDSCAPE ECOLOGY, it is the landscape sphere, not only the terrestrial BIOSPHERE but also the interacting parts of the upper LITHOSPHERE and lower ATMOSPHERE. *JAM*

[See also GEOECOLOGY, GEOECOSYSTEM, GEOSPHERE]

geoecosystem A landscape system including the ecosystem in its landscape context. Geoecosystems are dynamic spatial entities, the components of which interact and are continually responding to environmental change, but the behaviour of which is holistic and therefore not predictable from the behaviour of their component parts. *JAM*

[See also BIOGEOCOENOSIS, ECOSYSTEM, GEOECOLOGY, GEOECOSPHERE, HOLISTIC APPROACH, LANDSCAPE ECOLOGY, LANDSCAPE GEOCHEMISTRY]

geoengineering (1) The application of geological knowledge and techniques to dealing with environmental problems (e.g. the safe disposal of RADIOACTIVE WASTE). (2) The deliberate manipulation at the global scale of the Earth's environment; also referred to as *environmental engineering*. The term has become associated particularly with GLOBAL WARMING: how the FORCING FACTORS might be abated; how people and economic systems might be adapted to reduce the impact; and how the climate system might be manipulated to counteract anthropogenic impact (e.g. spreading DUST in the STRATOSPHERE to mitigate expected global temperature rise). *RAS*

Aoki, K. and Shiogama, Y. 1993: Geoengineering techniques used in the construction of underground openings in jointed rocks. *Engineering Geology* 35, 167–173. **Schneider, S.H.** 1996: Geoengineering: could we or should we do it? *Climatic Change* 33, 291–302. **Watts, R.G.** 1998: *Engineering response to global climate change.* London: CRC Press.

geofact A pseudo-artefact. Geofacts are objects created by natural processes (e.g. frost-shattered pebbles), which may be mistaken for archaeological ARTEFACTS (e.g. primitive axe heads). *JAM*

[See also ECOFACT, EOLITH]

Haynes, C.V. 1973: The Calico site: artefacts or geofacts? *Science* 181: 305–310.

geoglyph Ground figures carved from rock, such as those on Easter Island (although they need not be of such an immense size). Such archaeological features, also known as *intaglios* or *earthen art*, should be distinguished from PETROGLYPHS, which are carvings into rock surfaces. *JAM*

Bahn, P. and Flenley, J. 1992: *Easter Island, Earth island.* London: Thames and Hudson.

geographical information system (GIS) A technique developed in the mid-1960s, during the so-called *quantitative revolution* in GEOGRAPHY, by governmental agencies in response to a new awareness in dealing with complex environmental issues. At the time a GIS was designed to automate the process of handling large amounts of spatial data, but with the growth in computer technology a GIS can now not only capture and store data, but also manipulate and display any digital data that are spatially referenced to the Earth. These four stages, or subsystems, not only define a GIS, but also describe what a GIS does. The Figure illustrates the basic components of these subsystems and how they interact with the main computer (shown in bold) and each other (DBMS refers to a database management system).

Like all information systems a GIS is centred around the use of computers for processing data, but what makes a GIS different is that the processed data are explicitly geographical. Geographical information (for example, a spot height, river or forest) is held in digital form in one of three ENTITIES: POINTS, lines (see ARCS) or areas (sometimes known as polygons). A spot height would be represented by a digital point, a river by a digital line and a forest by a digital polygon. Entities are identified by LABELS, which can be either names (e.g. spot height of Ben Nevis, the River Severn and the New Forest), or values (e.g. spot height of 1335 m, river velocity of 5 m s^{-1} and forest area of 350 km^2). Names and values of entities are often termed ATTRIBUTES.

Collections of related entities and their associated attributes are stored in a GIS as layers or COVERAGES. For example: elevation samples would be combined to produce a relief coverage (known as a DIGITAL ELEVATION MODEL); roads would contribute to a transport network coverage; and lakes a hydrology coverage. A coverage can best be viewed as a digital map and, like all maps, it is bound by conventional cartographic rules that dictate that geographical information is symbolised, generalised to scale and referenced to commonly accepted cartographic COORDINATE and MAP PROJECTION systems. Coverages can then be stored in one of three main formats, vector, RASTER and QUAD-TREE. Coverages based on the VECTOR DATA MODEL use precise geometric coordinates to locate all points: lines are represented by interconnected points; and polygons are lines that connect points which start and end at the same point. Raster-based coverages, on the other hand, are composed of a matrix of GRID CELLS, where points are represented by a single occupied cell; lines are a series of adjacent occupied cells; and polygons are a mass of neighbouring cells. Lastly, quad-trees are less popular and use a hierarchical structure based on quadrants to pinpoint occupied cells recursively.

Associated coverages representing the same geographical area or thematic features (for example, agricultural LANDUSE CHANGE), are combined and stored in a digital database (see RELATIONAL DATABASE) and controlled by a *database management system* (DBMS). The most important strength of a GIS is its ability to interrogate simultaneously coverages held in a database, using SPATIAL

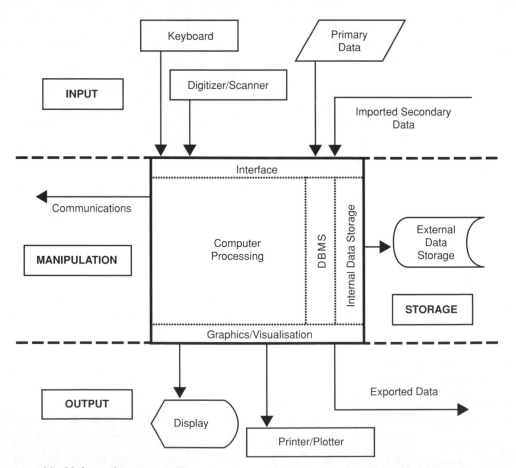

geographical information system *The main stages and components of a geographical information system*

ANALYSIS routines, including OVERLAY ANALYSIS (for a diagram), POLYGON ANALYSIS and the generation of BUFFER ZONES. The efficient handling and manipulation of digital geographical information can answer many questions dealing with ENVIRONMENTAL CHANGE. For instance, HABITAT LOSS assessments may involve questions such as: what impact will the location of a new power station have on an environmentally sensitive area? Questions such as the impact of effluents on water supplies, soil and CONTAMINATED LAND can be answered by a GIS within a DECISION MAKING process. Coverages representing information on water bodies, soil types, woodland land cover, agricultural landuse, relief and prevailing winds would be captured, spatially registered to common geometric coordinates, and held within a database. A QUERY LANGUAGE would then be used to interrogate these coverages and reveal answers to the likely effects of a power station on the environment. The process would start with the generation of buffer zones to determine the proportion of land within some distance of the power plant deemed to be at risk. Polygon analysis would then calculate affected areal units of land and overlay analysis would finally compare all at-risk land at common locations, along with some indications of the degree of risk to each parcel of land and water. Over time, further data will reveal changes in water and soil acidity, tree damage and any fall in crop yield.

For most people a GIS is proprietary computer software. There are many GIS packages available on the market, including Arc/Info (from the Environmental Systems Research Institute, USA), and Idrisi (from Clark Laboratories, USA). A broader definition of a GIS would include data and people. Data and DATA ACQUISITION are becoming increasingly more important to a GIS as many environmental-change projects are now global in nature and data can be transferred around the world via the Internet. As computer software and hardware costs have decreased, DATA QUALITY is now seen as the most critical aspect in GIS environmental measurement and monitoring. Data can be obtained in the following ways: by digitising paper maps; as transferable secondary digital data, surveys, GROUND MEASUREMENTS and pin-point locations from a GLOBAL POSITIONING SYSTEM; and from AIRBORNE and SATELLITE REMOTE SENSING. People too are fast becoming valuable components of a successful GIS. Skilled analysts, efficient managers and focused users are essential for a structured and coordinated GIS. A GIS is also known as a *land information system* (LIS), particularly when applied solely to the monitoring of ENVIRONMENTAL INDICATORS.

The Association of Geographic Information (AGI) is a non-profit, non-governmental organisation geared towards the promotion and awareness of GIS in the UK and includes vendors, government, the private sector and academia. AGI's annual Source Book is a good starting point on the use of data for environmental applications.

Idrisi is a type of desktop geographical information system (GIS), capable of offering educational and professional functions. Usually there is a suite of software programs with standard GIS operations, such as the ability to evaluate coverages for many environmental risk assessments and modelling exercises. *TVM*

[See also BOOLEAN LOGIC, CARTOGRAPHY]

AGI 1998: *The AGI source book for GIS 1998/99.* London: Taylor and Francis. **Burrough, P.A.** and **McDonnell, R.A.** 1998: *Principles of geographical information systems.* Oxford: Oxford University Press. **ESRI**, 1997: *Understanding GIS: the Arc/Info method.* New York: Wiley. **Larsen, L.** 1999: GIS in environmental monitoring and assessment. In Longley, P.A., Goodchild, M.F., Maguire, D.J. and Rhind, D.W. (eds), *Geographical information systems: principles, techniques, applications, and management.* Chichester: Wiley, 999–1007. **Maidment, D.R.** 1996: Environmental modeling with GIS. In Goodchild, M.F., Steyaert, L.T., Parks, B.O. *et al.* (eds), *GIS and environmental modeling: progress and research issues.* Fort Collins, CO: GIS World Books, 315–324. **Malanson, G.P.** and **Armstrong, M.P.** 1990: Issues in spatial representation: effects of number of cells and between-cell step size on models of environmental processes. *Geographical and Environmental Modelling* 1, 47–64. **Mitasova, H., Hofierka, J., Zlocha, M.** and **Iverson, L.R.** 1996: Modelling topographic potential for erosion and deposition using GIS. *International Journal of Geographical Information Systems* 10, 629–642. **Welch, R., Marguerite, M.** and **Doren, R.F.** 1999: Mapping the Everglades. *Photogrammetric Engineering and Remote Sensing* 65, 163–170.

geography The study of the surface of the Earth. It involves the phenomena and processes of the Earth's natural and human environments and landscapes at local to global scales. Its basic division is between PHYSICAL GEOGRAPHY, which is unambiguously a science and analyses the physical make-up of the Earth's surface (including the BIOSPHERE and lower ATMOSPHERE), and HUMAN GEOGRAPHY, where the focus is upon the human occupance of this area. There are key unifying themes across the physical and human aspects of geography. One is the link between the NATURAL ENVIRONMENT, with its LANDFORMS, VEGETATION, SOILS and CLIMATES, and the patterns and processes of human settlement. ENVIRONMENTALISM focuses on the nature of this link and its study takes many forms, including investigating the consequences of the human use and transformation of the Earth. Another connecting theme is the cartographic tradition and the geographer as mapmaker; these draw together the many elements of the Earth's surface into a visible and unified form and exemplify the central importance of location, spatial variation and spatial relationships to the discipline. With the advent of satellite imagery, EARTH OBSERVATION has added new dimensions to the cartographic tradition and, allied with the new technologies of GEOGRAPHICAL INFORMATION SYSTEMS (GIS), lays claim to be regarded as a subdiscipline of geography in its own right.

Geography carries forward its historic identifiers – exploration, discovery, FIELD RESEARCH – together with its central interests in concepts such as landscape, region and place. Its INTERDISCIPLINARY character is seen in its interactions with allied disciplines in the physical, biological and EARTH SCIENCES, HISTORY, the HUMANITIES and SOCIAL SCIENCES. There is a strong applied dimension that ranges from ENVIRONMENTAL IMPACT ASSESSMENT and ENVIRONMENTAL MANAGEMENT to URBAN AND RURAL PLANNING and the definition of political space. Geography addresses many of the major issues of our time that are associated with ENVIRONMENTAL CHANGE, such as POLLUTION, GLOBAL WARMING, CONSERVATION, SUSTAINABILITY and the manifestations of GLOBALISATION. The surface of the Earth is complex and ever-changing; geography provides the methodologies needed to understand, portray and predict these complexities. *DTH/JAM*

[See also CARTOGRAPHY, REMOTE SENSING]

Douglas, I., Huggett, R. and Robinson, M. (eds) 1996: *Companion encyclopedia of geography: the environment and humankind.* London: Routledge. **Haggett, P.** 1995: *The geographer's art.* Oxford: Blackwell. **Makinson, D. (ed.)** 1996: Geography: state of the art I – the environmental dimension. *International Social Science Journal* 48, 439–536 [Special Issue 150]. **Makinson, D. (ed.)** 1997: Geography: state of the art II – societal processes and geographic space. *International Social Science Journal* 49, 1–104 [Special Issue 151]. **Martin, G.J. and James, P.E.** 1993: *All possible worlds: a history of geographical ideas*, 3rd edn. New York: Wiley. **National Research Council** 1997: *Rediscovering geography: new relevance for science and society.* Washington, DC: National Academy Press. **Pacione, M. (ed.)** 1999: *Applied geography: principles and practice. An introduction to useful research in physical, environmental and human geography.* London: Routledge.

geohazard A category of NATURAL HAZARDS relating to the Earth's internal or surface processes. Geohazards, or *geological hazards*, include EARTHQUAKES, VOLCANIC IMPACTS ON PEOPLE, TSUNAMIS, magnetic storms, LANDSLIDES, ground SUBSIDENCE, river FLOODS and SEA-LEVEL CHANGE. *Meteorological hazards* (e.g. TROPICAL CYCLONES, TORNADOES, STORMS, DROUGHT), *extraterrestrial hazards* (METEORITE IMPACT) and *biological hazards* (e.g. PESTS, EPIDEMICS) are excluded. *GO*

[See also RAPID-ONSET HAZARD]

Maund, J.G. and Eddleston, M. (eds) 1998: *Geohazards in engineering geology* [Engineering Geology Special Publication 15]. London: Geological Society. **Pickering, K.T. and Owen, L.A.** 1997: *An introduction to global environmental issues*, 2nd edn. London: Routledge.

geoid The shape of the Earth's sea surface without tides, currents, water-density variations and atmospheric effects. It is an *equipotential surface* that is uneven as a result of regional variations in the Earth's gravity field. REMOTE SENSING has shown that there are considerable differences in geoidal sea surface altitudes, with a 180 m difference in sea level between the low level off the Maldive Islands (Indian Ocean) and the high level off New Guinea. *RAS/SHD*

[See also GEOIDAL EUSTASY]

Mörner, N.-A. 1976: Eustasy and geoid changes. *Journal of Geology* 84, 123–152.

geoidal eustasy Variations in SEA LEVEL caused by the unevenness of the sea surface caused by regional variations in the Earth's gravitational field. Ocean surfaces represent an *equipotential surface* of the Earth's gravitational field, known as the GEOID. There are large regional differences in geoidal sea-surface altitudes. For example, there is a 180 m difference in sea level between the low level off the Maldive Islands in the Indian Ocean and the high level near New Guinea. In the Gulf of Corinth, Greece, the geoidal sea surface varies by as much as 12 m over a distance of only c. 150 km. During ICE AGES, the distribution of ICE SHEETS caused considerable changes in the Earth's gravity field and therefore also large changes in the topography of the geoidal sea surface. *AGD*

[See also GLACIO-ISOSTASY, ISOSTASY]

Devoy, R.J.N. (ed.) 1987: *Sea surface studies.* London: Croom Helm. **Mörner, N.-A.** 1980: Eustasy and geoid changes as a function of core/mantle changes. In Mörner, N.-A. (ed.), *Earth rheology, isostasy and eustasy.* Chichester: Wiley, 535–553.

geoindicator A high-resolution measure of short-term change in the geological environment, which is important for ENVIRONMENTAL MONITORING and ENVIRONMENTAL IMPACT ASSESSMENT. Geoindicators (or *geoenvironmental indicators*) are the magnitudes, rates and trends of the near-surface geological processes and phenomena that vary appreciably over timescales of < 100 years. Most involve local (0.1–10 km scale) and mesoscale (10–100 km) landscapes although some, such as RELATIVE SEA LEVEL and volcanic activity, have regional to global dimensions. Among the characteristics of a good geoindicator are: scientific validity, geographic scope, responsiveness to change, relevance and utility to users and capability of forward projection. These attributes mean that they should be sensitive to HUMAN IMPACT on the environment and also relevant to PALAEOENVIRONMENTAL RECONSTRUCTION. *JAM*

[See also BIO-INDICATOR, ENVIRONMENTAL INDICATOR, INTERACTION INDICATOR]

Berger, A.R. and Iams, W.J. (eds) 1996: *Geoindicators: assessing rapid environmental changes in Earth systems.* Rotterdam: Balkema.

geological conservation The CONSERVATION of sites and features of significance to GEOLOGY and GEOMORPHOLOGY. MINERALS, FOSSILS, rock types, GEOLOGICAL STRUCTURES, SOILS, LANDFORMS, LANDSCAPES and sites of educational value or historical significance need to be conserved against damage, over-collecting, EROSION or development. In the UK, geological conservation comes under the remit of the national conservation bodies and is incorporated into the system of SITES OF SPECIAL SCIENTIFIC INTEREST (SSSIs). A further, lower status of conservation is awarded to REGIONALLY IMPORTANT GEOLOGICAL AND GEOMORPHICAL SITES (RIGs). *GO*

Ellis, N., Bowen, D.Q., Campbell, S., Knill, J. *et al.* 1996: *An introduction to the Geological Conservation Review.* Peterborough: Joint Nature Conservation Committee. **Nature Conservancy Council** 1990: *Earth science conservation in Great Britain: a strategy.* Peterborough: Nature Conservancy Council.

geological controls on environmental change Geological processes exert a major influence on the environment at all scales. As controls on environmental change, they may be divided into two categories, external controls that bring about ALLOCYCLIC CHANGE and internal controls that bring about AUTOCYCLIC CHANGE. The latter are typically local in scope and include events such as DELTA lobe switching and river-channel AVULSION.

Many large-scale, allocyclic controls are related to PLATE TECTONICS. At the global scale, the migration of continental areas across climatic zones (CONTINENTAL DRIFT) leads to a gradual change in climate, as happened in the case of Spitsbergen as it moved from equatorial latitudes in the DEVONIAN period to its present Arctic position. The formation and breakup of supercontinents such as PANGAEA also has major environmental consequences. The interiors of these enormous land areas develop extreme continental climates (see CONTINENTALITY), while breakup involves the formation of RIFT VALLEYS and ultimately seaways, opening up new routes for moisture supply. The collision of continental fragments leads to mountain building (OROGENESIS). Physical UPLIFT results in environmental change, but in addition the creation of relief can influence the GENERAL CIRCULATION OF THE ATMOSPHERE. Uplift of the Tibetan Plateau, for example, may have affected the course of the subtropical JET STREAM, changing the extent and intensity of the Asian MONSOON. Global tectonics also influences SEA LEVEL by changing the volume of the ocean basins (*tectono-eustasy*). Formation and decay of the MID-OCEAN RIDGE systems and continental collision are especially significant.

Regional TECTONICS affect the environment in a variety of ways. Tectonic closure of the Strait of Gibraltar in the Late Miocene isolated the Mediterranean, leading to the MESSINIAN salinity crisis and ultimately complete desiccation. Formation of the Isthmus of Panama separated the Atlantic and Pacific Oceans, changing the equatorial OCEAN CURRENT system, with dramatic consequences for the North Atlantic. Rift-shoulder uplift resulting from continental extension has dammed and diverted major drainage systems in East Africa.

Catastrophic EVENTS can bring about profound local change. LANDSLIDES, DEBRIS FLOWS, LAVA or PYROCLASTIC deposits may dam valleys or fill them in. Subaquatic SLUMPS can cause loss of coastline. VOLCANIC ERUPTIONS cause physical and chemical changes to the atmosphere, which may lead to CLIMATIC CHANGE. Longer-term variations in the composition of the atmosphere are also subject to geological control, particularly the abundance of two of the most important GREENHOUSE GASES, CARBON DIOXIDE and METHANE. The former is added to the atmosphere by volcanic activity and probably reaches its highest concentrations when SEA-FLOOR SPREADING is at a maximum. CHEMICAL WEATHERING and LIMESTONE formation remove carbon dioxide from the atmosphere. Large volumes of methane are trapped in deep-sea sediments as GAS HYDRATE. Episodic gas release from this repository could increase the greenhouse effect and may be a cause of ABRUPT CLIMATIC CHANGE.   *MRT*

Berner, R.A. 1994: GEOCARB II: A revised model of atmospheric CO_2 over Phanerozoic time. *American Journal of Science* 294, 56–91. **Fort, M.** 1996: Late Cenozoic environmental changes and uplift on the northern side of the central Himalayas: a reappraisal from field data. *Palaeogeography, Palaeoclimatology, Palaeoecology* 120, 123–145. **Haq, B.U.** 1998: Natural gas hydrates: searching the long-term climatic and slope-stability records. In Henriet, J.P. and Mienert, J. (eds), *Gas hydrates: relevance to world margin stability and climate change* [*Special Publication* 137]. London: Geological Society, 303–318. **Keefer, D.K.** 1999: Earthquake-induced landslides and their effects on alluvial fans. *Journal of Sedimentary Research* 69, 84–104. **Kutzbach, J.E.** 1994: Idealized Pangean climates: sensitivity to orbital change. In Klein, G.D. (ed.), *Pangea: paleoclimate, tectonics and sedimentation during accretion, zenith and breakup of a supercontinent* [*Special Paper* 288], 41–55. Boulder, CO: Geological Society of America. **Plint, A.G., Eyles, N., Eyles, C.H. and Walker, R.G.** 1992: Controls of sea level change. In Walker, R.G. and James, N.P. (eds), *Facies models: response to sea level change*. St. John's: Geological Association of Canada, 15–25. **Reading, H.G. and Levell, B.K.** 1996: Controls on the sedimentary rock record. In Reading, H.G. (ed.), *Sedimentary environments: processes, facies and stratigraphy*, 3rd edn. Oxford: Blackwell, 5–36.

geological evidence of environmental change

The GEOLOGICAL RECORD is the primary source of information on prehistoric environmental change. Information about PALAEOENVIRONMENTS is recorded in many different ways. The most widespread is SEDIMENTOLOGICAL EVIDENCE, which reflects the response of DEPOSITIONAL ENVIRONMENTS to changing conditions of sediment formation and accumulation. Changes in LITHOLOGY, ACCUMULATION RATE, FACIES or PALAEOCURRENT direction may be a result of local or global environmental change. FOSSILS can be excellent indicators of environmental change, as many organisms are sensitive to variations in their HABITAT. Fossil evidence may be preserved as solid remnants (shell, bone, wood, pollen, etc.) of the original animal or plant, as moulds or CASTS of these, or as TRACE FOSSILS.

The mineralogical and chemical composition of sediments, fossils and mineral deposits also provides important palaeoenvironmental information. DETRITAL sediment mineralogy may reflect the relative effectiveness of physical and chemical WEATHERING processes in the source area, which are functions of climate and relief. Syndepositional minerals typically reflect the composition of the water from which they precipitate. Variations in the mineralogy of, for example, CARBONATE or EVAPORITE minerals in a sequence of lagoonal or lacustrine deposits are commonly the result of changes in water chemistry and salinity, which may in turn be related to changes in HYDROLOGICAL BALANCE. Fluid inclusions trapped within minerals can be used to determine the temperature of precipitation as well as water composition. GEOCHEMISTRY has been central to many modern studies of environmental change. Cyclic variations in the OXYGEN ISOTOPIC composition of marine fossils, for example, have been used to confirm the MILANKOVITCH THEORY of climatic change. Other important types of geochemical evidence include TRACE ELEMENTS, which may reveal variations in environmentally significant parameters such as the temperature or oxygen content of waterbodies, and organic compounds, particularly BIOMARKERS.

Geomorphological features such as former glaciated valleys, overflow channels and *wave-cut platforms* are significant indicators of environmental change and have considerable historical importance as some of the earliest recognised evidence of PLEISTOCENE glaciation and sea-level change (see GLACIAL THEORY). Ancient examples of GLACIAL LANDFORMS provide vital evidence in support of postulated ICE AGES of PRECAMBRIAN and PALAEOZOIC age.
                    MRT

Broecker, W.S. 1995: *The glacial world according to Wally.* Palisades: Eldigio Press. **Bromley, R.G.** 1996: *Trace fossils: biology, taphonomy and applications.* London: Chapman and Hall. **Hinnov, L.A. and Park, J.** 1998: Detection of astronomical cycles in the stratigraphic record by frequency modulation (FM) analysis. *Journal of Sedimentary Research* 68, 524–539. **Martini, I.P. (ed.)** 1997: *Late Glacial and postglacial environmental changes: Quaternary, Carboniferous–Permian, and Proterozoic.* Oxford: Oxford University Press. **Maynard, J.R. and Leeder, M.R.** 1992: On the periodicity and magnitude of Late Carboniferous glacio-eustatic sea-level changes. *Journal of the Geological Society, London* 149, 303–311. **Pettijohn, F.J., Potter, P.E. and Siever, R.** 1987: *Sand and sandstone.* Berlin: Springer. **Reading, H.G. (ed.)** 1996: *Sedimentary environments: processes, facies and stratigraphy*, 3rd edn. Oxford: Blackwell. **Tyson, R.V.** 1995: *Sedimentary organic matter: organic facies and palynofacies.* London: Chapman and Hall.

geological record

The total of all materials that preserve information about the Earth's past, including particularly SEDIMENTS and SEDIMENTARY ROCKS of all ages, but also IGNEOUS ROCKS, METAMORPHIC ROCKS, FOSSILS and GEOLOGICAL STRUCTURES. The geological record, particularly the STRATIGRAPHICAL RECORD of materials that accumulated on the Earth's surface, is the main source of information about past environmental change, and virtually the only source of information about pre-QUATERNARY environmental change.   *GO*

[See also FOSSIL RECORD, GEOLOGICAL EVIDENCE OF ENVIRONMENTAL CHANGE, GEOLOGICAL RECORD OF ENVIRONMENTAL CHANGE, NATURAL ARCHIVES, SEDIMENTOLOGICAL EVIDENCE OF ENVIRONMENTAL CHANGE]

Ager, D.V. 1993: *The nature of the stratigraphical record*, 3rd edn. Chichester: Wiley. **Taylor, S.R. and McLennan, S.M.** 1985: *The continental crust: its composition and evolution. An examination of the geological record preserved in sedimentary rocks.* Oxford: Blackwell Science.

geological record of environmental change

Geological evidence is vital in understanding environmental change because of the time limitations of other data sources. HISTORICAL EVIDENCE and DOCUMENTARY EVIDENCE provide data for only the very recent historical past; ENVIRONMENTAL ARCHAEOLOGY extends the record to not much further than 20 000 years. In contrast, GEOLOGICAL EVIDENCE OF ENVIRONMENTAL CHANGE extends as far back as the oldest rocks on Earth, about 3800 million years ago (Ma); only the first 700–800 million years of Earth history are without evidence (see PRISCOAN). Additionally, the rapid climatic fluctuations of the late PLEISTOCENE and HOLOCENE suggest that the time period covered by nongeological evidence may not be typical of Earth history generally (see ACTUALISM).

SEDIMENTOLOGICAL EVIDENCE OF ENVIRONMENTAL CHANGE is considerable, as many SEDIMENTARY DEPOSITS record evidence of their DEPOSITIONAL ENVIRONMENT and, in the case of TERRIGENOUS clastic sediments and sedimentary rocks, that of the hinterland from where the sediments were eroded.

FOSSILS represent some of the most unequivocal and potent evidence of environmental change. The FOSSIL RECORD, however, poses many problems for PALAEOENVIRONMENTAL RECONSTRUCTION. Some of these are palaeobiological, such as the loss of soft-tissued organs and organisms (see TAPHONOMY). Other problems are the mixing together of fossils from environments that were separated in space (see DEATH ASSEMBLAGE) or time (see DERIVED FOSSIL). Some groups of fossils are good palaeoenvironmental indicators. For example, colonial scleractinian corals are confined to warm, shallow marine waters. TRACE FOSSILS provide direct evidence of the activity of organisms. Plants are excellent environmental indicators for non-marine settings and POLLEN ANALYSIS in particular is a sensitive tool for defining palaeoclimates. In general the confidence and precision with which a fossil type can be used for palaeoenvironmental analysis decreases with age; the environmental significance of a Cenozoic mollusc is easier to interpret than a Cambrian one. Equally the PALAEOECOLOGY of fossils with close relations to living taxa can be more confidently interpreted than those without.

A correct palaeoenvironmental interpretation is critical if rocks and fossils are to be used as data for environmental change. This is sometimes difficult. Sand bodies formed in the deep OCEAN and in fluvial systems may be superficially similar. Furthermore, DIAGENESIS can remove, modify or disguise environmental information in both sediments and fossils. Ideally sedimentological, palaeontological and FACIES information obtained at the THIN SECTION, hand specimen and outcrop scales should be integrated in any palaeoenvironmental analysis. A necessary component is the unravelling of lateral, vertical and temporal relationships: this is the province of STRATIGRAPHY. Increasingly, such data is tied in, at a still higher level, with the general depositional framework of the SEDIMENTARY BASIN in the context of BASIN ANALYSIS.

On a regional scale, any geological interpretation of environmental change must take into account the PLATE TECTONICS context. Replacement of warm-water sediments over time by a glacial TILLITE, for example, could indicate either global cooling or a regional poleward shift in PALAEOLATITUDE as a result of CONTINENTAL DRIFT.

A major problem with interpreting environmental change in the geological record is that the RESOLUTION of DATING TECHNIQUES is such that the timescale of change is often imprecisely known. In rocks older than about 30 Ma the greatest precision using BIOSTRATIGRAPHY and RADIOMETRIC DATING is around 0.5 million years and frequently it is no better than 2–3 million years. One implication is that it is hard to determine the exact duration of most geological events unless they are many hundred times longer than all recorded human history. Another is that it is difficult to be certain that particular short-lived EVENTS were truly synchronous across wide areas.

The GEOLOGICAL RECORD sets limits on the range of environmental fluctuations that have occurred on the Earth. Thus, at the most extreme level, the continuous fossil record of life from about 3800 Ma onwards indicates that at no time did all surface water become either ice or steam as appears to have happened on Mars and Venus respectively (see PLANETARY ENVIRONMENTAL CHANGE). Given the variability of SOLAR RADIATION as the Sun evolved, this is remarkable and is one line of evidence for some sort of planetary HOMEOSTASIS as proposed by the GAIA HYPOTHESIS. Within this gross stability the geological record does, however, indicate considerable variation. For instance, during the Jurassic and Cretaceous periods there is very limited evidence for polar *ice caps*, suggesting that global temperatures were such that GREENHOUSE CONDITIONS prevailed. In contrast, during the late Proterozoic, ice sheets appear to have been widespread, extending to low latitudes, and the Earth experienced ICEHOUSE CONDITIONS (see SNOWBALL EARTH).

The geological record also shows evidence of environmental conditions and processes very different from those prevailing at present. One example is oceanic circulation patterns that allowed the development of widespread OCEANIC ANOXIC EVENTS. Geology also suggests that short-term biotic and environmental crises have occurred periodically in MASS EXTINCTIONS (see also K–T BOUNDARY).

Finally, geological evidence is vital in both the creation and VALIDATION of GENERAL CIRCULATION MODELS for the ancient atmosphere and reconstructions of PALAEOCEANOGRAPHY. These models are developed from regional geological studies and based on PLATE TECTONICS, and their outputs can be tested by examining sedimentological and palaeontological data in contemporaneous rocks. Geology similarly provides data for testing models of the past and future evolution of the atmosphere. Thus the idea that the Earth's atmosphere became oxygen-rich during the PRECAMBRIAN is supported by the change from

BANDED IRON FORMATIONS to RED BEDS from 2000 to 1500 Ma.   *CDW*

Allen, J.R.L., Hoskins, B.J., Sellwood, B.W, *et al.* **(eds)** 1994: *Paleoclimates and their modelling.* London: Chapman and Hall. **Brasier, M.D.** 1989: Global ocean–atmosphere change across the Precambrian–Cambrian transition. *Geological Magazine* 129, 161–168. **Einsele, G., Ricken W. and Seilacher, A. (eds)** 1991: *Cycles and events in stratigraphy.* Berlin: Springer. **Eyles, N. and Young, G.M.** 1994: Geodynamic controls of glaciation in Earth history. In Deynoux, M., Miller, J.M.G., Domack, E.W. *et al.* (eds), *Earth's glacial record.* Cambridge: Cambridge University Press, 1–28. **Fischer, A.G.** 1981: Climatic oscillations in the biosphere. In Nitecki, M. (ed.), *Biotic crises in ecological and evolutionary time.* New York: Academic Press, 103–131. **Francis, J.E.** 1998: Interpreting palaeoclimates. In Doyle, P. and Bennett, M.R. (eds), *Unlocking the stratigraphical record: advances in modern stratigraphy.* Chichester: Wiley, 471–490. **Hambrey, M.J. and Harland, W.B.** 1981: *Earth's pre-Pleistocene glacial record.* Cambridge: Cambridge University Press. **Pirrie, D.** 1998: Interpreting the record: facies analysis. In Doyle, P. and Bennett, M.R. (eds), *Unlocking the stratigraphical record: advances in modern stratigraphy.* Chichester: Wiley, 395–420. **Price, G.D., Valdes, P.J. and Sellwood, B.W.** 1997: Quantitative palaeoclimate GCM validation: Late Jurassic and mid-Cretaceous case studies. *Journal of Geological Society, London* 154, 769–772. **Reading, H.G. and Levell, B.K.** 1996: Controls on the sedimentary rock record. In Reading, H.G. (ed.), *Sedimentary environments: processes, facies and stratigraphy,* 3rd edn. Oxford: Blackwell, 5–36.

geological structures Features produced by the permanent DEFORMATION of sediments or rocks, mainly in response to TECTONIC stresses while rocks are buried in the Earth's CRUST. FAULTS and JOINTS are *brittle* structures produced by the elastic deformation and fracturing of rocks in tension or compression. FOLDS are *ductile* structures formed by the bending of rock strata, usually in compression. Geological structures range in scale from microscopic features to regional structures tens or hundreds of kilometres across. They are responsible for the irregularities of outcrop patterns on a geological map. The orientation of geological structures is described by the STRIKE and DIP of planar features (e.g. tilted BEDS, fault surfaces, fold axial surfaces) and the *plunge* of linear features (e.g. fold hinge lines).

The analysis of geological structures can provide information about the magnitude and orientation of stresses in the geological past, allowing the deformation history of an area to be reconstructed. This is important in reconstructing past PLATE TECTONICS, unravelling the STRATIGRAPHY of an area and understanding the RHEOLOGY of the CRUST and LITHOSPHERE. *Structural geology* is the branch of GEOLOGY concerned with the interpretation of geological structures. It overlaps with tectonics, GEODYNAMICS, rheology and materials science.   *GO*

[See also SEDIMENTARY STRUCTURES, SOFT-SEDIMENT DEFORMATION, UNCONFORMITY]

Carey, P.J. 1998: Field interpretation of complex tectonic areas. In Doyle, P. and Bennett, M.R. (eds), *Unlocking the stratigraphical record: advances in modern stratigraphy.* Chichester: Wiley, 81–121. **Hobbs, B.E., Means, W.D. and Williams, P.F.** 1976: *An outline of structural geology.* New York: Wiley. **Lisle, R.J.** 1995: *Geological structures and maps.* Oxford: Butterworth Heinemann. **Park, R.G.** 1988: *Geological structures and moving plates.* Glasgow: Blackie. **Park, R.G.** 1997: *Foundations of structural geology,* 3rd edn. Glasgow: Blackie.

geological time Geological time is the same as time in the conventional sense, but there are difficulties in measuring the immense spans of time represented in the GEOLOGICAL RECORD, which extends over 4500 million years. The ages of geological materials and events are expressed in two distinct ways: as RELATIVE AGES, in which they are placed in sequence; and as *absolute ages* measured in years. *Absolute-age dating* only became possible on geological timescales with the discovery of radioactivity in the early twentieth century. Relative dating, primarily using FOSSILS (see STRATIGRAPHICAL COLUMN), is the only method directly applicable in fieldwork and is capable of identifying more precise time 'slices' than absolute dating methods, although those slices may be of uncertain absolute age.   *GO*

[See also GEOCHRONOLOGY, GEOLOGICAL TIMESCALE, STRATIGRAPHY]

Gould, S.J. 1987: *Time's arrow, time's cycle: myth and metaphor in the discovery of geological time.* London: Penguin.

geological timescale An absolute time scale, or *geochronologic scale,* for the GEOLOGICAL RECORD, expressed in years (*chronometric units*). The scale has been developed by calibrating the RELATIVE AGES of the STRATIGRAPHICAL COLUMN (*chronostratic scale*) with *absolute-age dating* from appropriate sediments and rocks, using RADIOMETRIC DATING or other methods of GEOCHRONOLOGY. The calibration of these two distinct methods of measuring GEOLOGICAL TIME is subject to revision and refinement, so dates applied to specific boundaries vary in different versions of the geological time scale. For example, the beginning of the CAMBRIAN period was placed at 600 million years ago (Ma) in the 1960s, 570 Ma in the 1980s, and 544 Ma in 1998. The position of the boundary in terms of BIOSTRATIGRAPHY, or its physical position in a rock succession, have not, however, changed.

A relative timescale is more appropriate than an absolute scale with regard to field observations, where a reference to rocks as 'early Jurassic' is more appropriate than a statement that they are 'approximately 200 million years old' since absolute-age dating can only be carried out through complex laboratory analyses of certain types of rock. This is analogous to the usage of terms such as 'Bronze Age' on the ARCHAEOLOGICAL TIMESCALE. Moreover, radiometric dates are generally not as precise as some CORRELATION using BIOSTRATIGRAPHY. For example, the correlation of some JURASSIC ammonite faunas allows the discrimination of rock units representing intervals of the order of 50 000 years, which is much more precise than radiometric dates can provide for the same period.

A parallel geological timescale is provided by the GEOMAGNETIC POLARITY TIMESCALE (*magnetostratigraphic timescale*), which uses the pattern of GEOMAGNETIC POLARITY REVERSALS recorded in rock successions. Such changes are synchronous and can be correlated world-wide, although some overall biostratigraphical or radiometric control is also usually necessary.

The principal divisions of geological time and dates for the boundaries are shown in the Table. No rocks are preserved from PRISCOAN time. The ARCHAEAN and PROTEROZOIC *eonothems* were originally classed together as PRECAMBRIAN and are still often referred to as such. This major division separates rocks in which FOSSIL content is

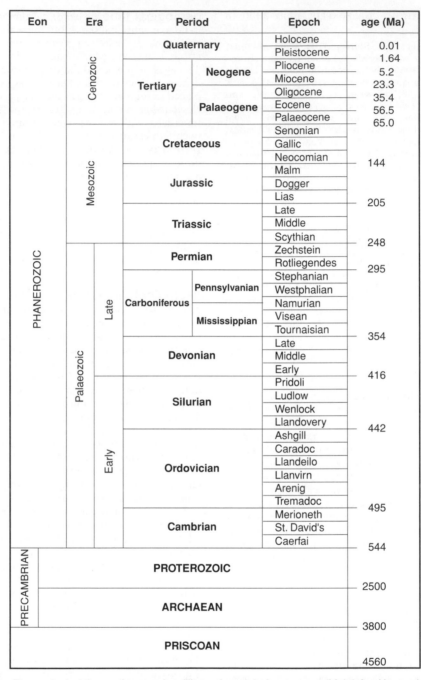

Eon	Era	Period		Epoch	age (Ma)
PHANEROZOIC	Cenozoic	Quaternary		Holocene	0.01
				Pleistocene	1.64
		Tertiary	Neogene	Pliocene	5.2
				Miocene	23.3
			Palaeogene	Oligocene	35.4
				Eocene	56.5
				Palaeocene	65.0
	Mesozoic	Cretaceous		Senonian	
				Gallic	
				Neocomian	144
		Jurassic		Malm	
				Dogger	
				Lias	205
		Triassic		Late	
				Middle	
				Scythian	248
	Palaeozoic	Permian		Zechstein	
				Rotliegendes	295
	Late	Carboniferous	Pennsylvanian	Stephanian	
				Westphalian	
				Namurian	
			Mississippian	Visean	
				Tournaisian	354
		Devonian		Late	
				Middle	
				Early	416
	Early	Silurian		Pridoli	
				Ludlow	
				Wenlock	
				Llandovery	442
		Ordovician		Ashgill	
				Caradoc	
				Llandeilo	
				Llanvirn	
				Arenig	
				Tremadoc	495
		Cambrian		Merioneth	
				St. David's	
				Caerfai	544
PRECAMBRIAN		PROTEROZOIC			2500
		ARCHAEAN			3800
		PRISCOAN			4560

geological timescale *The geological timescale: ages in millions of years before present (Ma) (after Haq and van Eysinga, 1998, for the Precambrian, Palaeozoic and Mesozoic; and after Harland et al., 1990, for the Cenozoic)*

obvious (PHANEROZOIC, Greek for 'visible life') from those in which it is much less obvious and often absent. The boundary between the Archaean and the Proterozoic is taken arbitrarily at 2500 Ma. The PALAEOZOIC (Greek for 'ancient life') begins rapidly with the CAMBRIAN EXPLOSION – a sudden appearance of invertebrate taxa, many with hard parts. Palaeozoic faunas are dominated by such fossil groups as trilobites, brachiopods and graptolites, together with crinoids and rugose and tabulate corals. Following the end-Permian MASS EXTINCTION, in which at least 75%

of species became extinct, the MESOZOIC (Greek for 'middle life') began. Mesozoic marine faunas are dominated by molluscs, especially ammonoids and belemnoids, as well as large marine reptiles such as ichthyosaurs, plesiosaurs and pliosaurs. On land, dinosaurs were the dominant vertebrates, but the first mammals and birds appeared in the TRIASSIC and Jurassic respectively. The first flowering plants (angiosperms) appeared in the CRETACEOUS. At the end-Cretaceous mass extinction (K–T BOUNDARY) the ammonites, belemnites, large marine reptiles and dinosaurs disappeared. CENOZOIC (Greek for 'recent life') marine invertebrate faunas are dominated by molluscs, especially bivalves and gastropods. Mammals evolved rapidly on land, the evolution of the GRAZING habit following on from the appearance of grasses in the PALAEOGENE. HOMINIDS appeared in the PLIOCENE. *JCWC/GO*

Hailwood, E.A. 1989: *Magnetostratigraphy* [*Special Report* 19]. London: Geological Society. **Hailwood, E.A. and Kidd, R.B.** 1993: *High resolution stratigraphy* [*Special Report* 70]. London: Geological Society. **Haq, B.U. and van Eysinga, W.B.** 1998: *Geological time table*, 5th edn. Amsterdam: Elsevier. **Harland, W.B., Armstrong, R.L., Cox, A.V.** *et al.* 1990: *A geologic time scale 1989*. Cambridge: Cambridge University Press. **International Union of Geological Sciences.** 1989: 1989 Global Stratigraphic Chart. *Episodes* 12(2), Supplement. **Palmer, A.R.** 1983: The decade of North American geology 1983 geologic time scale. *Geology* 11, 503–504.

geology The investigation of the solid Earth, its composition, structure, processes and history, through the study of materials such as rocks, MINERALS and FOSSILS – the GEOLOGICAL RECORD. The discipline can be usefully split into *physical geology* (the study of present-day processes and products) and *historical geology* (study of the Earth's past). The principal subdivisions of geology are crystallography, mineralogy, PETROLOGY (including IGNEOUS, SEDIMENTARY and METAMORPHIC ROCKS), PALAEONTOLOGY, STRATIGRAPHY, SEDIMENTOLOGY, structural geology, TECTONICS, GEOPHYSICS, GEOCHEMISTRY, HYDROGEOLOGY, MARINE GEOLOGY, applied geology and ENVIRONMENTAL GEOLOGY. Unlike EARTH SCIENCE, geology does not normally cover OCEANOGRAPHY or CLIMATOLOGY. In recent years the methods of geology have been applied to other bodies in the solar system (see PLANETARY GEOLOGY). *CDW*

[See also GEOLOGICAL EVIDENCE OF ENVIRONMENTAL CHANGE, GEOLOGICAL RECORD OF ENVIRONMENTAL CHANGE]

Duff, P.McL.D. (ed.) 1993: *Holmes' principles of physical geology*, 4th edn. London: Chapman and Hall. **Lyell, C.** 1830–1833: *Principles of geology, being an attempt to explain the former changes of the Earth's surface by reference to causes now in operation.* London: John Murray. **Rothery, D.A.** 1997: *Teach yourself geology*. London: Hodder and Stoughton.

geomagnetic polarity reversal A change in the orientation of the Earth's magnetic field (see GEOMAGNETISM) such that north and south magnetic poles swap positions. Geomagnetic polarity reversals (also known as *field reversals, geomagnetic reversals, magnetic reversals, magnetic polarity reversals* or *polarity reversals*) have occurred roughly every few hundred thousand years over the past 100 million years. Polarities like that existing today are known as *normal polarity*, in contrast to episodes of *reversed polarity* (see POLARITY CHRON). During a reversal the intensity (strength) of the Earth's magnetic field gradually declines, the polarity suddenly flips and the intensity gradually increases, the entire event apparently lasting a few thousand years. Reversals are attributed to fluctuations in flow in the Earth's outer CORE. Reversals are globally synchronous, providing a means for CORRELATION in MAGNETOSTRATIGRAPHY, and their dating provides the basis for the GEOMAGNETIC POLARITY TIMESCALE. *GO*

[See also MAGNETIC ANOMALY, PALAEOMAGNETISM]

Cox, A. (ed.) 1973: *Plate tectonics and geomagnetic reversals.* San Francisco, CA: Freeman. **Kageyama, A., Ochi, M.M. and Sato, T.** 1999: Flip-flop transitions of the magnetic intensity and polarity reversals in the magnetohydrodynamic dynamo. *Physical Review Letters* 82, 5409–5412. **Tarling, D.H.** 1983: *Palaeomagnetism*. London: Chapman and Hall. **Willis, D.M., Holder, A.C. and Davis, C.J.** 2000: Possible configurations of the magnetic field in the outer magnetosphere during geomagnetic polarity reversals. *Annales Geophysicae – Atmospheres, Hydrospheres and Space Sciences* 18, 11–27.

geomagnetic polarity timescale (GPTS) A GEOLOGICAL TIMESCALE based on GEOMAGNETIC POLARITY REVERSALS, calibrated by *absolute-age dating* of reversals (see Figure). The geomagnetic polarity timescale, or *magnetostratigraphic timescale*, is principally used where a distinctive set of reversals in a rock sequence can be reliably matched to a section of the timescale. It provides a detailed record of geomagnetic polarity reversals from c.180 million years ago to the present through the dating of OCEAN-floor MAGNETIC ANOMALIES. The numbers given to ocean-floor anomalies and the names given to the most recent POLARITY CHRONS are also applied to the timescale. Calibration of the reversal pattern with BIOSTRATIGRAPHY and RADIOMETRIC DATING is only available for the past 80 million years (i.e. latest Cretaceous and Tertiary) and there are considerable uncertainties in its pre-Mesozoic definition. *DNT*

Harland, W.B., Armstrong, R.L., Cox, A.V. *et al.* 1990: *A geologic time scale 1989*. Cambridge: Cambridge University Press. **Nichols, G.** 1999: *Sedimentology and stratigraphy*. Oxford: Blackwell Science.

geomagnetism The study of the Earth's magnetic field (*geomagnetic field*). This is best approximated by a bar magnet at the centre of the Earth, its axis aligned with the Earth's axis of rotation. This *geocentric axial dipole* (GAD) model represents the time-averaged geomagnetic field throughout geological time. The geomagnetic field originates in the iron-rich fluid outer CORE of the Earth (see EARTH STRUCTURE) and is maintained by convective motions in this region. Changes in the *intensity* (strength), *declination* (azimuth: orientation relative to true north) and *inclination* (DIP: orientation in a vertical plane relative to the Earth's surface) define the *secular variation* (variations in time) of the geomagnetic field and are attributed to physical and chemical changes in the core and the core–MANTLE boundary region. Over longer time scales the field experiences GEOMAGNETIC POLARITY REVERSALS when north and south magnetic poles reverse. *DNT*

[See also PALAEOMAGNETISM]

Juarez, M.T. and Tauxe, L. 2000: The intensity of the time-averaged geomagnetic field: the last 5 Myr. *Earth and Planetary Science Letters* 175, 169–180. **Merrill, R.T., McElhinny, M.W. and McFadden, P.H.** 1998: *The magnetic field of the Earth: paleomagnetism, the core, and the deep mantle*. London: Academic Press. **Robinson, E.S. and Çoruh, C.** 1988: *Basic exploration geophysics*. New York: Wiley.

geometric correction A mathematical transformation applied in the rectification of errors and restoration of images to geographical correctness or specific map projections. Each type of collection instrumentation, whether a film camera, sensor or scanner, has its own characteristics of production and presentation of raw data. Errors and

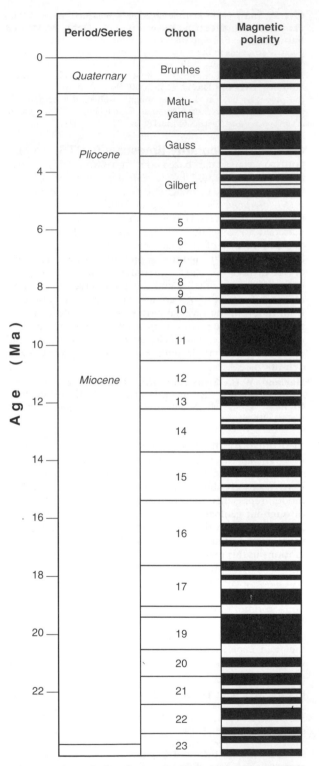

Period/Series	Chron	Magnetic polarity
Quaternary	Brunhes	
	Matu-yama	
Pliocene	Gauss	
	Gilbert	
	5	
	6	
	7	
	8	
	9	
	10	
	11	
Miocene	12	
	13	
	14	
	15	
	16	
	17	
	19	
	20	
	21	
	22	
	23	

Age (Ma)

geomagnetic polarity timescale *The geomagnetic polarity timescale for the past 24 million years (Ma). Black bands are episodes of normal polarity; white bands are episodes of reversed polarity (after Nichols, 1999)*

distortion (e.g. FORESHORTENING), which occur during the collection of such data, may be constant, systematic or random. These are generally checked and adjusted prior

to the imagery being released for wider use and analysis. The corrected image is then supplied in a recognised COORDINATE SYSTEM format. Further computation and adjustment may be required to match other images generated on different coordinate systems or lacking well defined geometry. *TF*

Jones, C. 1997: *Geographical information systems and computer cartography.* Harlow: Addison-Wesley Longman.

geomorphology The study of LANDFORMS from micro-scale to global scale, including their formative ENDOGENETIC and, especially, EXOGENETIC processes at the Earth's surface, the evolution of landforms in the face of environmental change, and relationships between landforms, geomorphological processes and human society (*applied geomorphology*). *JAM*

Bloom, A.L. 1998: *Geomorphology.* Upper Saddle River, NJ: Prentice-Hall. Chorley, R.J., Schumm, S.A. and Sugden, D.A. 1984: *Geomorphology.* London: Methuen. Hails, J.R. (ed.) 1977: *Applied geomorphology.* Amsterdam: Elsevier. Summerfield, M.A. 1991: *Global geomorphology.* Harlow: Longman. Tinkler, K.J. 1985: *A short history of geomorphology.* London: Croom Helm.

geomorphology and environmental change Since many geomorphic processes are conditioned by changes in CLIMATE and in the nature of the ground surface (both natural and human-induced), GEOMORPHOLOGY is intimately linked with ENVIRONMENTAL CHANGE. The Earth's land surface comprises POLYGENETIC LANDSCAPES reflecting the influence of processes that either no longer operate or operate at spatial or temporal scales different from earlier ones. Some geomorphic systems and processes are more sensitive to change than others, making it difficult to generalise about LANDFORM development in response, for example, to a particular magnitude of change in temperature or precipitation.

The importance of environmental change in geomorphological research is now recognised more widely than formerly, but much remains to be understood about the nature and speed of geomorphic response to natural and/or human-induced change. Public awareness of the concept of GLOBAL WARMING and concern about the consequences of *human population growth* have generated unprecedented interest in the ENVIRONMENT amongst the general public as well as amongst politicians and scientists. This reflects as much a desire to know the possible socio-economic impacts of environmental (including geomorphic) response as it does a wish to acquire greater scientific understanding.

Attempts to assess the impact of environmental change on geomorphic systems and processes have frequently adopted one or more of the following three distinct approaches. First, PROXY EVIDENCE and historical records of past changes, ranging from long to short timescales, have been investigated. The resulting palaeogeomorphological data may enable RESPONSE TIME and RECURRENCE interval to be assessed, but the *temporal resolution* is often comparatively coarse (i.e. century-to-millennial scale). The decadal and subdecadal resolution needed for future predictions is seldom achieved. Second, *monitoring* of current geomorphic processes has been carried out (e.g. using EROSION PLOTS, EROSION PINS or a MICRO-EROSION METER), providing valuable information at a higher temporal resolution about the physical basis of processes and site-specific quantifiable data. Such data inevitably suffer, however, from the brevity of the monitoring periods and limited scope for spatial and temporal EXTRAPOLATION (except by using ERGODIC principles) and are likely to omit extreme events with a long RECURRENCE interval. Third, *modelling* and simulation of geomorphic systems and processes in order to assess impact–process–response relationships have been applied. Data input can be drawn from the first two approaches, although substitute parameters and data may have to be used if field data are unavailable. Criticisms of models include oversimplification because of the limited number of parameters and the strong dependence of the output on the quality of the input. Ideally, more than one approach may be appropriate in evaluating geomorphic response to environmental change, but understanding it also requires recognition and comprehension of the interrelationships existing between various processes and individual system components. This is more likely to be achieved through the adoption of a HOLISTIC APPROACH than of a strategy based on REDUCTIONISM.

SENSITIVITY of geomorphic systems and processes to change is complex. It can be defined in different ways: as the recognisable response of a system to a change in system controls; or as the susceptibility of a system to DISTURBANCE. Whether geomorphic systems and processes change following an environmental PERTURBATION depends on the spatial and temporal balance of the resisting and disturbing forces. Landscape stability is maintained if the resisting forces prevent environmental perturbations from having any persistent effect.

Landscape response to environmental change varies through time and space and operates at different scales. Geomorphic systems and processes may respond slowly, adjusting to change over several millennia. Alternatively, response may be sudden, even catastrophic, with recognisable change occurring in minutes only. In reality, both levels of response condition all systems. For much of the time, geomorphic systems evolve slowly through processes of moderate intensity and FREQUENCY. Major modifications are often associated with low-frequency/high-intensity events to the extent that it has often been assumed that most change is effected during these events. This is not necessarily so because geomorphic sensitivity is not simply a function of the magnitude of the change, but depends on whether, be it large or small, it crosses important geomorphic THRESHOLDS.

An important focus for geomorphology in the last decade or so, as with other environmental sciences, has been the prediction of sensitivity of systems and processes to CLIMATIC CHANGE induced by GLOBAL WARMING. Projected magnitudes and likely impacts of temperature and precipitation changes from GENERAL CIRCULATION MODELS are critical, as are any predictions of increased frequency of high-magnitude extreme climatic events. Many palaeo-studies documenting geomorphic response to climate shifts since the Late PLEISTOCENE provide a useful basis for developing and exploring future scenarios, although they usually relate to times and/or situations in which human agency in environmental change was either lacking or inconsequential. For most areas of the world, the history of HUMAN IMPACT ON LANDFORMS AND GEOMORPHIC PROCESSES is well documented, but the response

of geomorphic systems and processes to future climatic change in landscapes currently strongly affected by human action is by no means clearly understood.   *PW*

[See also EQUILIBRIUM CONCEPTS, FEEDBACK MECHANISMS, LAG TIME, LANDFORM EVOLUTION, lead time, MAGNITUDE-FREQUENCY CONCEPTS, PHYSIOGRAPHY, RATES OF ENVIRONMENTAL CHANGE, RELAXATION TIME, SCALE CONCEPTS]

Baker, V.R. and C.R. Twidale. 1991: The reenchantment of geomorphology. *Geomorphology* 4, 73–100. **Boer, D.H. de** 1992: Hierarchies and spatial scale in process geomorphology: a review. *Geomorphology* 4, 303–318. **Brunsden, D.** 1990: Tablets of stone: towards the Ten Commandments of geomorphology. *Zeitschrift für Geomorphologie Supplementband* 79, 1–37. **Brunsden, D. and Thornes, J.B.** 1979: Landscape sensitivity and change. *Transactions of the Institute of British Geographers* 4, 463–484. **Bull, W.B.** 1991: *Geomorphic responses to climatic change.* New York: Oxford University Press. **Eybergen, F.A. and Imeson, A.C.** 1989: Geomorphological processes and climatic change. *Catena* 16, 307–319. **Scheidegger, A.E.** 1987: The fundamental principles of landscape evolution. *Catena* 10 (Supplement). Cremlingen: Catena, 199–210. **Schumm, S.A.** 1979: Geomorphic thresholds: the concept and its applications. *Transactions of the Institute of British Geographers* 4, 485–515. **Thomas, D.S.G. and Allison, R.J.** (eds), 1993: *Landscape sensitivity.* Chichester: Wiley.

geophysical surveying   The practical application of GEOPHYSICS to investigate the Earth's interior, particularly shallow levels in the CRUST. Geophysical surveying (*geophysical exploration* or *exploration geophysics*) is analogous to REMOTE SENSING in that measurements made at the Earth's surface are used to investigate the subsurface. Techniques of geophysical surveying include *electrical surveying, electromagnetic surveying, gravity surveying* (see GRAVITY ANOMALY), GROUND-PENETRATING RADAR, *magnetic surveying* (see MAGNETIC ANOMALY) and SEISMIC SURVEYING (including SEISMIC REFLECTION SURVEYING and SEISMIC REFRACTION SURVEYING). Geophysical surveys can detect objects or GEOLOGICAL STRUCTURES in the subsurface that have contrasting physical properties to their surroundings and, together with data from surface exposures and BOREHOLES, are important in the study of EARTH STRUCTURE, in ARCHAEOLOGICAL PROSPECTION, in the exploration for geological NATURAL RESOURCES such as PETROLEUM, and in surveys to assess ground conditions for engineering.   *GO*

Armadillo, E., Massa, F., Caneva, G. *et al.* 1998: Modelling of karst structures by geophysical methods. An example: the doline of S. Pietro dei Monti (Western Liguria). *Annali di Geofisica* 41, 389–397. **Gowda, B.M.R. Ghosh, N., Wadhwa, R.S.** *et al.* 1998: Seismic refraction and electrical resistivity methods in landslide investigations in the Himalayan foothills. *Environmental and Engineering Geoscience* 4, 130–135. **Kearey, P. and Brooks, M.** 1991: *An introduction to geophysical surveying,* 2nd edn. Oxford: Blackwell Science. **Milsom, J.** 1996: *Field geophysics,* 2nd edn. Chichester: Wiley. **Reynolds, J.M.** 1997: *An introduction to applied and environmental geophysics.* Chichester: Wiley.

geophysics   The application of principles of physics to the study of the Earth. Within EARTH SCIENCE, geophysics involves studies of the Earth's gravity, magnetism, (GEOMAGNETISM), behaviour with respect to elastic waves (SEISMOLOGY), heat flow and radioactivity, to investigate the structure, physical and environmental condition and geological evolution of the Earth. *Pure geophysics* (*global geophysics*) is the study of processes occurring within the whole planet (see EARTH STRUCTURE). Some methods of pure geophysics can also be applied to the study of PLANETARY GEOLOGY. *Applied geophysics* (*exploration geophysics* or GEOPHYSICAL SURVEYING) is the study of the Earth's CRUST, particularly at shallow depths, for practical purposes such as mineral exploration, engineering and archaeology. The relatively new discipline of *environmental geophysics* is the study of near-surface physical and chemical interactions and the evaluation of their implications for ENVIRONMENTAL MANAGEMENT.   *DNT*

Fowler, C.M.R. 1990: *The solid Earth: an introduction to global geophysics.* Cambridge: Cambridge University Press. **Jones, E.J.W.** 1999: *Marine geophysics.* Chichester: Wiley. **Lowrie, W.** 1997: *Fundamentals of geophysics.* Cambridge: Cambridge University Press. **Reynolds, J.M.** 1997: *An introduction to applied and environmental geophysics.* Chichester: Wiley.

geophyte   A terrestrial plant that can survive environmental stress (e.g. summer drought or winter frosts) through underground food-storage organs, such as bulbs and tubers.   *JLI*

[See also LIFE FORM]

geoscience   A synonym for EARTH SCIENCE sometimes criticised on etymological, if not other grounds.   *CDW*

Edwards, D. and King, C. 1999: *Geoscience: understanding geological processes.* London: Hodder and Stoughton.

geosols   In part synonymous with the term PALAEOSOL, geosols include all soil stratigraphical units (see SOIL STRATIGRAPHY) formed in past environments. Recent geological research has shown that soils are preserved in the FOSSIL RECORD with examples from most geological formations, and particularly in the Quaternary. *Seat earths* have long been recognised in the Carboniferous succession in association with coals, but other soil characteristics also survive LITHIFICATION and enable an explanation and interpretation of otherwise inexplicable sedimentary details. *Petrocalcic horizons*, mottle patterns and vertic structures are recognised in beds from the Carboniferous to the Tertiary.   *EMB*

[See also BURIED SOILS, FOSSIL SOILS, PALAEOSOLS].

Morrison, R.B. 1978: Quaternary soil stratigraphy – concepts, methods and problems. In Mahaney, W.C. (ed.), *Quaternary soils.* Norwich: Geo Abstracts, 77–108. **Wright, V.P. (ed.)** 1986: *Palaeosols: their recognition and interpretation.* Oxford: Blackwell Scientific.

geosphere   (1) The totality of geophysical spheres from Earth's core to the outer layers of the atmosphere. (2) Any one of these spheres, such as the ATMOSPHERE, BIOSPHERE, CRYOSPHERE, HYDROSPHERE, LITHOSPHERE or PEDOSPHERE.   *JAM*

[See also ECOSPHERE]

Young, G.I. and Bartuska, T.J. 1974: Sphere: term and concept as an integrative device towards understanding environmental unity. *General Systems Yearbook* 19, 219–230.

geostrophic current   An OCEAN CURRENT driven by pressure gradients related to variations in water surface level, balanced by the CORIOLIS FORCE so that the direction

of flow is at right angles to the horizontal pressure gradient.   *GO*

[See also EKMAN MOTION, GEOSTROPHIC WIND, THERMOHALINE CIRCULATION]

Allen, P.A. 1997: *Earth surface processes.* Oxford: Blackwell Science. **Open University Course Team** 1989: *Ocean circulation.* Oxford: Pergamon.

geostrophic wind   A horizontal wind blowing parallel to the isobars, which is determined by the balance between the CORIOLIS FORCE and the pressure gradient force. The speed of the geostrophic wind is inversely related to the spacing of the isobars. It is an approximation to the actual wind, except at the Equator (where the Coriolis force is zero) and near the ground (where frictional effects dominate producing a *gradient wind* that crosses the isobars). The geostrophic wind may be considered the top (non-frictional) level of the *Ekman spiral* (see EKMAN MOTION). A similar concept applies to ocean currents (see GEOSTROPHIC CURRENT).   *JAM*

Hess, S.L. 1957: *Introduction to theoretical meteorology.* New York: Holt.

geosyncline   A long-lived, regional- to continental-scale depression in the Earth's CRUST in which thick piles of SEDIMENTARY ROCKS accumulated before DEFORMATION, UPLIFT and the formation of an OROGENIC BELT. Geosyncline theory has been abandoned since the 1960s as a result of developments in the understanding of PLATE TECTONICS, CONTINENTAL MARGINS and SEDIMENTARY BASINS.   *GO*

Dott Jr, R.H. and Shaver, R.H. (eds) 1974: *Modern and ancient geosynclinal sedimentation* [*Special Publication* 19]. Tulsa, OK: Society of Economic Paleontologists and Mineralogists. **Mitchell, A.H.G. and Reading, H.G.** 1986: Sedimentation and tectonics. In Reading, H.G. (ed.), *Sedimentary environments and facies*, 2nd edn. Oxford: Blackwell, 471–519.

geothermal   Associated with heat from the Earth's interior. The *geothermal gradient* is the increase in temperature with depth in the Earth's CRUST. Typical values are 20–30°C km^{-1}, with a range from less than 10°C km^{-1} (e.g. CRATONS) to over 300°C km^{-1} (e.g. MID-OCEAN RIDGES). *Geothermal energy* is obtained by transferring underground heat to the surface using heated GROUNDWATER (see HYDROTHERMAL activity) or by pumping water down from the surface. It is considered a relatively clean source of energy, but there are technical problems in its extraction and its use is restricted to areas of high heat flow, particularly areas of active or recently active VOLCANISM, such as Iceland and North Island, New Zealand. *GO*

[See also ALTERNATIVE ENERGY, EARTH STRUCTURE]

Murphy, H. and Niitsuma, H. 1999: Strategies for compensating for higher costs of geothermal electricity with environmental benefits. *Geothermics* 28, 693–711.

Gerlach trough   A type of trough or container dug into the surface on a hillslope to catch *overland flow* and trap most of the sediment being transported. There have been various versions of this type of SEDIMENT TRAP since the original device was first described by T. Gerlach. Most versions comprise some form of metal or plastic box or trough (typically up to 50 cm wide) with, importantly, a lip (separate or integral) on the upslope side, flush with the ground surface to ensure that overland flow and transported sediment are guided into the container. A lid prevents splashed material from entering the box if only overland flow-transported sediment is required. As the contributing area of overland flow is not usually known (see EROSION PLOTS), Gerlach troughs tend to be used to indicate relative amounts of EROSION during a given period. They have the advantage over erosion plots of not interfering with upslope overland flow processes. *RAS*

[See also EROSION PIN, SOIL EROSION, SOIL MICROPROFILING DEVICE, WATER EROSION]

Gerlach, T. 1967: Hillslope troughs for measuring sediment movement. *Revue de Géomorphologie Dynamique* 17 (64), 132–140. **Shakesby, R.A., Walsh, R.P.D. and Coelho, C.de O.A.** 1991: New developments in techniques for measuring soil erosion in burned and unburned forested catchments, Portugal. *Zeitschrift für Geomorphologie Supplementband* 83, 161–174.

Ghijben–Herzberg principle
In a coastal AQUIFER, where the difference in density between freshwater and saltwater ensures that a freshwater GROUNDWATER lens overlies saltwater at depth, the Ghijben–Herzberg principle defines the quantitative relationship between the height of the fresh WATER TABLE above sea level to the depth of the interface between freshwater and saltwater. Because the difference in density between freshwater and saltwater is so small, ABSTRACTION of fresh groundwater leading to a DRAW DOWN of the water table will be compensated by a relatively large upward SALTWATER INTRUSION. Thus, for every 1 m draw-down in a well, for example, there is likely to be a corresponding rise of around 40 m in the saltwater table beneath the well.
JAM/RPDW

Thomas, D.S.G. and Goudie, A. (eds) 2000: *The dictionary of physical geography*, 3rd edn. Oxford: Blackwell. [They use the spelling Ghyben–Herzberg.] **Ward, R.C. and Robinson, M.** 1990: *Principles of hydrology*, 3rd edn. Maidenhead: McGraw-Hill.

gibber
An ABORIGINAL Australian term for a desert plain covered with a layer of pebbles or boulders. It is a type of DESERT PAVEMENT or *stone pavement* thought to be formed from the break-up of a siliceous surface crust (SILCRETE). *MAC*

[See also BOULDER PAVEMENTS, DURICRUST]

gibbsite
A crystalline aluminium hydroxide found in soils, particularly in highly weathered soil (FERRALSOLS, ACRISOLS and PLINTHOSOLS) material of tropical regions. It is mined as an ore of aluminium. *EMB*

[See also CLAY MINERALS, FERRALITISATION]

gigantism
The tendency for some isolated populations to undergo significant increases in body size in comparison to their conspecifics on the mainland. Gigantism is especially common in insular populations of small mammals. *MVL*

[See also EVOLUTION, ISLAND BIOGEOGRAPHY, NANISM]

Brown, J.H. and Lomolino, M.V. 1998: *Biogeography*, 2nd edn. Sunderland, MA: Sinauer. **Foster, J.B.** 1964: Evolution of mammals on islands. *Nature* 202, 234–235. **Lomolino, M.V.** 1985: Body size of mammals on islands: the island rule re-examined. *American Naturalist* 125, 310–316.

Gilbert-type delta
A rivermouth DELTA characterised by coarse-grained SEDIMENT and a steep (up to 35°) delta front slope (*foreset*) overlain and underlain by more gently sloping *topset* and *bottomset* beds. Gilbert-type deltas develop at steep basin margins and are typical of relatively small, freshwater deltas. *GO*

Colella, A. and Prior, D.B. (eds) 1990: *Coarse-grained deltas. International Association of Sedimentologists Special Publication* 10. Oxford: Blackwell Scientific. **Dorsey, R.J., Umhoefer, P.J. and Falk, P.D.** 1997: Earthquake clustering inferred from Pliocene Gilbert-type fan deltas in the Loreto basin, Baja California Sur, Mexico. *Geology* 25, 679–682. **Gilbert, G.K.** 1885: The topographic features of lake shores. *Annual Reports of the United States Geological Survey* 5, 75–123.

gilgai
An Australian aboriginal word used to describe the undulating micro-relief found on clay-rich VERTISOLS. The undulations are produced by shrink–swell activity of *smectite* clays during *wetting and drying cycles*. In the dry season the soil cracks as the clays lose moisture and crumbs of soil fall down the cracks. When the soil is re-wetted the clays expand but the additional material at depth causes heaving and the development of trapezoidal structures, often separated by curved, slickensided, thrust planes. *EMB*

Hallsworth, E.G., Robertson, G.K. and Gibbons, F.R. 1955: Studies in pedogenesis in New South Wales. VII. The 'gilgai' soils. *Journal of Soil Science* 6, 1–34 **Verger, F.** 1964: Mottureux et gilgais. *Annales de géographie* 73, 413–430. **Wilding, L.P. and Puentes, R. (eds)** 1988: *Vertisols: their distribution, properties, classification and management* [*Technical Monograph* 18]. Austin, TX: Soil Management Support Services, Texas A&M University.

glacial deposition
The release of sediments by GLACIER ice or in close proximity to a glacier. A wide range of processes are involved in glacial deposition, reflecting not only the interaction of glaciers with rivers, lakes and the sea, but also processes of reworking (e.g. by gravity flowage) following deposition. As a result, the deposits resulting from glacial activity are among the most complex to be found in the natural environment. The processes and products of glacial deposition in different environments are summarised in the Figure. *MJH*

Benn, D.I. and Evans, J.A. 1998: *Glaciers and glaciation.* London: Arnold. **Bennett, M.R. and Glasser, N.F.** 1996: *Glacial geology – ice sheets and landforms.* Chichester: Wiley. **Ehlers, J., Gibbard, P.L. and Rose, J. (eds)** 1991: *Glacial deposits in Great Britain and Ireland.* Rotterdam: Balkema. **Hambrey, M. J.** 1994: *Glacial environments.* London: UCL Press. **Hambrey, M.J.** 1999: The record of the Earth's glacial climates over the last 3000 Ma. In Barrett, P.J. and Orombelli, J. (eds), *Geological records of global change* [*Special Publication*]. Siena: Terra Antarctica. **Menzies, J.** 1995: Glaciers and ice sheets. In Menzies, J. (ed.), *Modern glacial environments – processes, dynamics and sediments.* Oxford: Butterworth Heinemann.

glacial episode
A protracted cold phase or interval of CONTINENTAL GLACIATION marked by the expansion of ICE SHEETS and glaciers. Glacial episodes, *glacial periods*, GLACIATIONS or *glacials* are subdivided into STADIALS and INTERSTADIALS. The latest glacial or LAST GLACIATION terminated about 11 500 years ago and was known in the British Isles as the DEVENSIAN, in northern Europe as the *Weichselian* and in America as the *Wisconsinan*. *DH/JAM*

(a) Terrestrial temperate/polythermal glacier

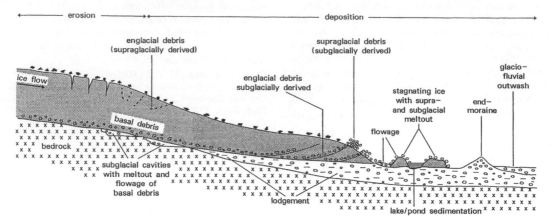

(b) Temperate tidewater glacier in fjord

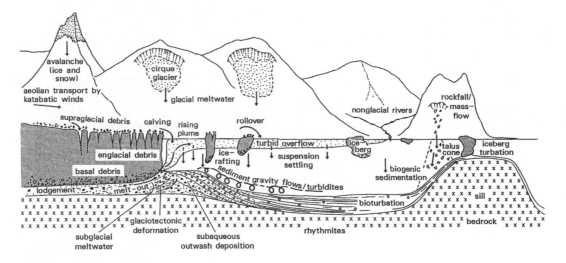

(c) Ice shelf and continental shelf (Antarctica)

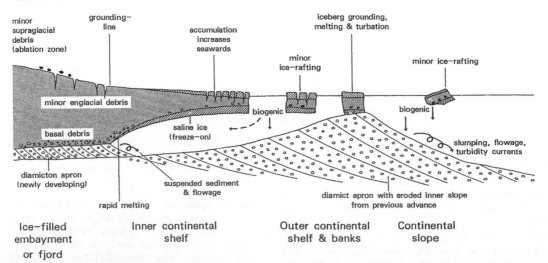

glacial deposition *Examples of processes and facies in three contrasting glacial environments: (A) a temperate glacier terminating on land; (B) a temperate glacier terminating in an Alaskan fjord; and (C) a cold-based glacier/ice shelf terminating on the continental shelf of Antarctica (after Hambrey, 1999)*

[See also ICE AGE, ISOTOPIC STAGE, QUATERNARY ENVIRONMENTAL CHANGE, TERMINATION]

glacial erosion Glaciers erode their beds by three principal processes: ABRASION; QUARRYING; and freezing on at the glacier bed by slab ENTRAINMENT. Abrasion produces STRIATIONS, FRICTION CRACKS and polished surfaces. Quarrying or *plucking* involves the removal of blocks of rock, commonly defined by joints, by hydraulic forces and polythermal conditions at the ice bed. Entrainment by a freeze-on process is inferred from the presence in tills of slabs of bedrock and surficial deposits (up to kilometres wide and tens of metres thick) from modern glacier beds. The mechanism for entrainment is not clear, although high pore-water pressures beneath a permanently frozen ice substrate are suspected of reducing rock strength and allowing failure by glacier shearing. Accretion of ice below the slabs may also contribute to their entrainment. Thus, large-scale slabs are likely indicators of ice advance over PERMAFROST. *JS*

[See also OVERDEEPENING, U-SHAPED VALLEYS]

Benn, D.I. and Evans, D.J.A. 1998: *Glaciers and glaciation.* London: Arnold. **Bennett, M.R. and Glasser, N.F.** 1996: *Glacial geology: ice sheets and landforms.* Chichester: Wiley. **Boulton, G.S.** 1979: Processes of glacial erosion on different substrates. *Journal of Glaciology* 23, 773–799. **Brodzikowski, K. and van Loon, A.J.** 1991: *Glacigenic sediments.* Amsterdam: Elsevier. **Drewry, D.J.** 1986: *Glacial geologic processes.* London: Arnold. **Hallet, B., Hunter, L. and Bogen, J.** 1996: Rates of erosion and sediment evacuation by glaciers: a review of field data and their implications. *Global and Planetary Change* 12, 213–235. **Menzies, J. (ed.)** 1995: *Modern glacial environments.* Oxford: Butterworth Heinemann. **Röthlisberger, H. and Iken, A.** 1981: Plucking as an effect of water pressure variations at the glacier bed. *Annals of Glaciology* 2, 57–62.

glacial–interglacial cycle The repeated alternation of cold (GLACIAL) and warm (INTERGLACIAL) stages within an ICE AGE. These cycles appear to have a PERIODICITY of c. 100 000 years between the major interglacials and also exhibit periodicities at c. 43 ka and 23–19 ka. The MILANKOVITCH THEORY of variations in global insolation receipt due to the changing astronomical position of the Earth relative to the Sun is generally acknowledged as the dominant driving force behind the cycles. *DH*

[See also INTERGLACIAL CYCLE, ORBITAL FORCING, ORBITAL TUNING]

glacial landforms Morphological features on a micro, meso or macro scale produced by erosion or deposition by the direct action of glacier ice. Micro-scale erosional forms include STRIATIONS and FRICTION CRACKS (including CHATTERMARKS); meso-scale features include ROCHES MOUTONNÉES, whalebacks, rock grooves, rock basins and KNOCK AND LOCHAN TOPOGRAPHY; and macro-scale features include CIRQUES, glacial troughs (U-SHAPED VALLEYS) and *flyggbergs* (large roches moutonnées). Depositional LANDFORMS include those that lie transverse to ice-flow direction (end and recessional MORAINES, ROGEN MORAINES and subaqueous moraines), those that are parallel to the ice-flow direction (FLUTED MORAINE, DRUMLINS) and those with no particular orientation with respect to ice flow (e.g. HUMMOCKY MORAINE). All the meso- and macro-scale erosional and depositional features are hills, valleys, ridges, depressions and plains, which can be used to reconstruct both the lateral and vertical extent of former GLACIERS and ICE SHEETS. They contribute to reconstructions of GLACIER THERMAL REGIME and glacier form, dynamics and HYDROLOGY. GLACIAL LANDFORMS of former continental ICE SHEETS are commonly arranged in regional-scale fields (c. 10^4 km²) and landforms within such fields show gradations in form and scale. There is debate on the origin, glacial or glaciofluvial, of many subglacial landforms of the PLEISTOCENE, mid-latitude ice sheets. *JS*

[See also GLACIAL SEDIMENTS, GLACIOFLUVIAL LANDFORMS]

Menzies, J. and Rose, J. (eds) 1989: Subglacial bedforms – drumlins, rogen moraine and associated subglacial bedforms. *Sedimentary Geology* 62, 117–430. **Patterson, C.J.** 1998: Laurentide glacial landscapes: the role of ice streams. *Geology* 26, 643–646. **Sugden, D.E. and John, B.S.** 1976: *Glaciers and landscape.* London: Arnold.

glacial maximum The maximum extent of ICE SHEET coverage during a GLACIAL EPISODE, usually defined by MORAINE limits. The LAST GLACIAL MAXIMUM, which occurred some 20 000 years ago, was so extensive in many regions that earlier limits were overridden; hence, defining the maximum extent of ice reached during some earlier glaciations is proving difficult. However, the timings of glacial maxima are now well established from the continuous sedimentary record preserved in MARINE SEDIMENT CORES and ICE CORES. *DH*

glacial protection concept An early view that GLACIERS protect the landscape from EROSION, which is no longer seen as generally applicable. A modern view is that COLD-BASED GLACIERS and ICE SHEETS (or cold-based parts of ice sheets with warm-based sections) may be protective. This is suggested by the preservation of such delicate landforms as TORS (e.g. Cairngorms, UK) and sorted PATTERNED GROUND in formerly glaciated areas. The thermal conditions at the ice bed have important implications for ice-sheet reconstruction and dynamics, which, in turn, are essential elements of CLIMATIC MODELS for the period of GLACIATION. *JS*

Dyke, A.S. 1993: Landscapes of cold-centred late Wisconsinan ice caps, Arctic Canada. *Progress in Physical Geography* 17, 223–247. **Kleman, J. and Stroeven, A. P.** 1997: Preglacial surface remnants and Quaternary glacial regimes in northwest Sweden. *Geomorphology* 19, 35–54. **Lagerbäck, R.** 1988: The Veiki Moraines in northern Sweden, widespread evidence of an early Weichselian deglaciation. *Boreas* 17, 487–499.

glacial sediments Materials that are being or have been transported by GLACIER ice, usually excluding debris that is subsequently reworked (see REDEPOSITION). *Glacigenic sediment* is usually used in a broader sense to include sediments with a greater or lesser component derived from glacier ice. A comprehensive classification of terrestrially deposited glacigenic sediments has been developed by the Till Work Group of the International Union for Quaternary Research, based on the present understanding of the processes of transport and deposition. Sediment beneath the glacier, where it is commonly modified during ice-flow, is referred to as SUBGLACIAL. Basal glacial debris is the term used if sediment is incorporated into the basal ice zone of the glacier. SUPRAGLACIAL debris is transported on the glacier surface, and ENGLACIAL debris is ingested either from the surface *via* crevasses or meltwater conduits, or from the bed by folding and thrusting.

The principal genetic types of deposited glacigenic sediment are as follows. Sediment deposited by uniquely glacial processes without subsequent disaggregation and reworking is termed TILL. The term 'till' has also been applied to glacial sediments reworked by MASS MOVEMENTS and to sediments that accumulate in the glaciomarine environment. Although some workers prefer to include these types of sediment as tills, the emerging consensus is towards the more restrictive definition. In the former category, LODGEMENT TILL results from active plastering onto the bed of sediment by the glacier, while MELTOUT TILL is released by melting from DEAD ICE or relatively inactive debris-rich ice. DEFORMATION TILL comprises weak rock and sediment that has been detached by the glacier from its source and its original structural integrity destroyed. Together, all these are grouped under the name *subglacial* or *basal till. Secondary tills* are the product of reworking of saturated glacial sediment, especially by debris flowage, which is a very common process close to ice margins. The name FLOW TILL, or *glacigenic sediment flow*, is often applied to this type of sediment. Sediment released directly by glaciers into water has commonly been referred to as *waterlain till*, although the terms *ice-proximal* GLACIOMARINE (or GLACIOLACUSTRINE) SEDIMENT are preferred today. Farther from the ice in the water body can be found *ice-distal glaciomarine (or glaciolacustrine) sediment*.

All these sediments are generally poorly sorted, with material potentially ranging in size from clay to boulders. Except for the last, all these types of glacigenic sediment usually lack bedding and can only be distinguished on the basis of detailed textural analyses, e.g. GRAIN-SIZE analysis, *particle shape*, clast-surface features (e.g. STRIATIONS) and FABRIC ANALYSIS. Ice-distal deposits include material derived directly from icebergs, mixed with sediment formerly in suspension, and other types of MARINE or LACUSTRINE SEDIMENT.

In practice, until the mode of deposition has been clearly established, use of a non-genetic term (e.g. DIAMICTON) for poorly sorted sediment, irrespective of origin, is preferred. *MJH*

[See also GLACIATION, GLACIAL DEPOSITION, GLACIOFLUVIAL SEDIMENTS]

Benn, D.I. and Evans, J.A. 1998: *Glaciers and glaciation.* London: Arnold. **Bennett, M.R. and Glasser, N.F.** 1996: *Glacial geology – ice sheets and landforms.* Chichester: Wiley. **Brodzikowski, K. and van Loon, A.J.** 1991: *Glacigenic sediments.* Amsterdam: Elsevier. **Dowdeswell, J.A. and Scourse, J.D.** (eds) 1990: *Glacimarine environments: processes and sediments* [*Special Publication* 53]. London: Geological Society. **Hambrey, M.J.** 1994: *Glacial environments.* London: UCL Press. **Menzies, J.** (ed.) 1995: *Modern glacial environments – processes, dynamics and sediments.* Oxford: Butterworth Heinemann.

Glacial Theory

A theory developed in the early- to mid-nineteenth century largely by members of the Helvetic Society, including Perraudin, Venetz, de Charpentier and Louis Agassiz. Their theory proposed the existence of a former ICE AGE when greatly expanded glaciers and ice sheets produced many of the erosional landforms and especially the sedimentary deposits in and around the Alps. Critical evidence in establishing the Glacial Theory by the application of the principles of UNIFORMITARIANISM included the recognition of (1) ERRATICS, (2) STRIATIONS, grooves and glacially polished rock surfaces, (3) MORAINES, (4) unsorted drift (TILL), and (5) oversteepened 'U-shaped' valleys and, in Norway, FJORDS. Soon afterwards, multiple GLACIATIONS and their global implications were recognised (see Table). The Glacial Theory provided a new PARADIGM, which replaced the DILUVIAL THEORY, and can now be seen as the beginning of modern ideas on ENVIRONMENTAL CHANGE and QUATERNARY science. *JAM*

Bowen, D.Q. 1978: *Quaternary geology: a stratigraphic framework for multidisciplinary work.* Oxford: Pergamon. **Flint, R.F.** 1957: *Glacial and Pleistocene geology, first edition.* New York: Wiley. **Imbrie, J. and Imbrie, K.P.** 1979: *Ice ages: solving the mystery.* London: Macmillan. **North, F.J.** 1943: Centenary of the Glacial Theory. *Proceedings of the Geologists Association* 54, 1–28.

glaciation

The term glaciation is used in several senses. At its most general, it refers to occupancy of part or all of a landscape by GLACIERS and/or ICE SHEETS. Glaciation is also used to refer to a glacier advance–retreat cycle, sometimes synonymously with glacial period or GLACIAL EPISODE, although this is not correct formal usage. The term, however, may also refer to periods of ice occupancy on both longer and shorter timescales, in this case synonymous with *stade* or STADIAL. Periods of glaciation normally occurred during even-numbered ISOTOPIC STAGES in the OXYGEN ISOTOPE timescale, although they should more appropriately be seen simply as cold stages. In METEOROLOGY, the term glaciation refers to the formation of ice crystals in clouds.

Glaciations (ICE AGES) have occurred on Earth during many geological periods, notably the Early and Late PROTEROZOIC, ORDOVICIAN, CARBONIFEROUS–PERMIAN and CENOZOIC, persisting for many millions of years on each occasion. The reasons why major Ice Ages occurred at these times are still unclear, although the distribution of land masses and mountain belts, and concentrations of greenhouse gases are likely to have been major factors. The most recent Ice Age is the CENOZOIC, which is still ongoing. Significant ice cover began to form in the late MIOCENE about 10 million years ago and has been a particular characteristic of the QUATERNARY.

Numerous glaciations (GLACIAL EPISODES or *glacials*) during the Quaternary have been interspersed with INTERGLACIALS in the form of GLACIAL–INTERGLACIAL CYCLES. Each glacial was probably composed of a number of relatively short-lived cold stages or stadials (not always involving extensive glaciation), with intervening periods of comparatively mild climate as *interstades* or INTERSTADIALS. During the most recent glacial period, large ice sheets formed over North America (the LAURENTIDE or LAURENTIAN ice sheet), northern Europe (the FENNOSCANDIAN and British ice sheets) and the islands and shelves surrounding the ARCTIC (e.g. the Barents Sea ice sheet). Considerable ice cover also developed on all of the world's great mountain ranges, including the North American Cordillera, the Andes, the European and New Zealand Alps and the Himalayas.

ICE-SHEET GROWTH and decay during glaciation is asso-

Glacial Theory *Some personalities and their landmark contributions in the development of the Glacial Theory, before and after* AD *1840 (after Flint, 1957; Bowen, 1978; and others)*

Date (AD)	Personality	Contribution
1779	De Saussure	Coined the term ERRATIC; attributed them to floods
1787	Kuhn	Interpreted erratics as evidence of ancient GLACIATION
1795	Hutton	Recognized erratics are glacially transported
1802	Playfair	Supported Hutton
1815	Perraudin	Recognized an Alpine glaciation
1821	Venetz	First scientific account of an Alpine glaciation
1823	Buckland	Publication of '*Reliquae Diluvianae*' (see DILUVIAL THEORY)
1824	Esmark	Recognized Norwegian MOUNTAIN GLACIATION
1829	Venetz	Recognized extensive European glaciation beyond the Alps
1832	Bernhardi	Recognized separate CONTINENTAL GLACIATION in Germany
1834	de Charpentier	Recognized glacial STRIATIONS
1837	Agassiz	Discourse of Neuchâtel; strongly advocates the Glacial Theory
1838	Buckland	'Converted' to the Glacial Theory
1839	Conrad	First acceptance of the theory in America
1840	Agassiz	Publication of *Études sur les glaciers*; elaborates the theory
1840	Agassiz	Visits Britain; lectures with Buckland and Lyell
1840	Schimper	Coined the term 'Eizeit' (ICE AGE)
1841	de Charpentier	Publication of *Essai sur les glaciers*
1841	Maclaren	Recognized GLACIO-EUSTASY (sea-level fall)
1847	Collomb	Recognized two glaciations in the Vosges Mountains
1851	Godwin-Austen	Recognized the PERIGLACIAL origin of much COLLUVIUM
1852	Ramsay	Recognized two glaciations in Wales
1853	Chambers	Recognized two glaciations in Scotland
1856	Morlot	Recognized two glaciations in Switzerland
1858	Heer	Recognized INTERGLACIAL deposits
1863	Jamieson	Recognized high lake levels in ARIDLANDS during glacial episodes
1865	Jamieson	Recognized GLACIO-ISOSTASY (depression of the Earth's crust)
1868	Tylor	Coined the term PLUVIAL
1868	Whittlesey	Calculated approximate value of glacio eustatic sea level fall
1872	von Richthofen	Recognized the AEOLIAN origin of LOESS
1882	Penck	Recognized three glaciations in the Alps
1905	Howorth	Last major scientific opposition to the Glacial Theory in Britain
1909	Penck & Bruckner	Recognized four glaciations in the Alps

ciated with major environmental changes on global and regional scales, including major reorganisations of the atmospheric and oceanic circulation. *DIB*

[See also GLACIERISATION, GLACIERS AND ENVIRONMENTAL CHANGE, ICEHOUSE CONDITION, QUATERNARY ENVIRONMENTAL CHANGE, TERMINATION]

Benn, D.I. and Evans, D.J.A. 1998: *Glaciers and glaciation.* London: Arnold. **Lowe, J.J. and Walker, M.J.C.** 1997: *Reconstructing Quaternary environments*, 2nd edn. Harlow: Addison-Wesley Longman. **Nesje, A. and Dahl, S.O.** 2000: *Glaciers and environmental change.* London: Arnold.

glaciation threshold The glaciation threshold, *glaciation limit* or *glaciation level* is the critical elevation above which GLACIERS can exist. The concept is operationalised as the mean elevation between the highest topographically suited summit without a glacier and the lowest summit hosting a glacier in the same geographical region. *AN*

Østrem, G. 1966: The height of the glaciation limit in southern British Columbia and Alberta. *Geografiska Annaler* 48A, 126–138.

glaciel A term of French-Canadian origin covering PERIGLACIAL phenomena formed by floating ice in fluvial, lacustrine and marine environments. For example, glaciel STRIATIONS (*stries glacielles*) formed by drift ice contrast with glacial striations formed by GLACIER ice. Other glaciel phenomena include BOULDER PAVEMENTS, *ice-shove ridges*, ICEBERG PLOUGH MARKS and ICE-RAFTED DEBRIS. *JAM*

[See also ALPINE PERIGLACIOFLUVIAL SYSTEM, FROST WEATHERING, POLAR SHORE EROSION]

Dionne, J.-C. 1973: Distinction entre stries glacielles et stries glaciaires. *Revue de Géographie de Montréal* 27, 185–190. **Hamelin, L.-E.** 1976: La famille du mot 'glaciel'. *Revue de Géographie de Montréal* 30, 233–236.

glacier A mass of ice, irrespective of size, derived largely from snow and continuously moving from higher to lower ground or spreading over the sea. Glaciers may be classified according to their size and morphology. The largest are ICE SHEETS, defined as more than 50 000 km², and so today are limited to Antarctica and Greenland. Morphologically similar, but covering < 50 000 km² are *ice caps*. Together ice sheets and ice caps are referred to as ICE DOMES, in view of their shape. Within ice sheets are zones of fast-flowing ice called *ice streams*, which sometimes continue into the sea as so-called *ice tongues*. Ice sheets also feed large floating slabs of ice called ICE SHELVES, although these are mainly restricted to Antarctica today. Extensive areas of undulating ice that mirror the underlying topography, and through which mountains project as NUNATAKS, are referred to as *highland ice fields* or *plateau glaciers*. Glaciers flowing between rock walls, often within a U-SHAPED VALLEY (or more strictly a valley that has a parabolic cross-sectional form) are known as *valley glaciers*, whether or not they have a well defined local accumulation area or emanate as OUTLET GLACIERS from an ice cap or ice sheet. Where glacier ice spreads out as a wide lobe onto flat land beyond the confines of a valley, *piedmont glaciers* are formed. Small glaciers occupying hollows carved out high in the mountains or uplands are referred to as CIRQUE glaciers. Steep areas of ice clinging to precipitous mountainsides are known as *ice aprons* or *niche glaciers*. Glaciers that form at the foot of a slope from ice and snow avalanche debris are known as *rejuvenated* or *regenerated glaciers*. *Hanging glaciers* occupy HANGING VALLEYS. *Transection glaciers* flow in several directions in highly dissected landscapes. *MJH*

Benn, D.I. and Evans, J.A. 1998: *Glaciers and glaciation*. London: Arnold. **Hambrey, M.J. and Alean, J.C.** 1992: *Glaciers*. Cambridge: Cambridge University Press. **Hambrey, M.J.** 1994: *Glacial environments*. London: UCL Press. **Jansson, P., Richardson, C. and Jonsson, S.** 1999: Assessment of requirements for cirque formation in northern Sweden. *Annals of Glaciology* 28, 16–22. **Knight, P.G.** 1999: *Glaciers*. Cheltenham: Stanley Thornes. **Nesje, A. and Dahl, S.O.** 2000: *Glaciers and environmental change*. London: Arnold. **Robin, G. de Q.** 1975: Ice shelves and ice flow. *Nature* 253, 168–172.

glacier foreland The recently deglacierised zone in front of a retreating glacier. Derived from the German 'Gletschervorfeld', and introduced into the English language literature by R.E. Beschel, the term glacier foreland is usually restricted to the area deglacierised in historical time since the maximum glacier extent of the LITTLE ICE AGE. Thus, it is a visually distinct zone of relatively bare terrain, a generally immature landscape and the site of PRIMARY SUCCESSION of vegetation. Glacier forelands are increasingly used as field laboratories for investigating ecological and other aspects of LANDSCAPE EVOLUTION over timescales of decades to centuries and possibly millennia. The landscapes of increasing age with distance from the glacier can be used to infer temporal change using the CHRONOSEQUENCE approach. *JAM*

Beschel, R.E. 1961: Dating rock surfaces by lichen growth and its application to glaciology and physiography (lichenometry). In Raasch, G.O. (ed.), *Geology of the Arctic*, Vol. 2. Toronto: University of Toronto Press, 1044–1062. **Engstrom, D.R., Fritz, S.C., Almendinger, J.E. and Juggins, S**. 2000: Chemical and biological trends during lake evolution in recently deglaciated terrain. *Nature* 408, 161–166. **Matthews, J.A.** 1992: *The ecology of recently-deglaciated terrain: a geoecological approach to glacier forelands and primary succession*. Cambridge: Cambridge University Press.

glacier milk The sediment-laden MELTWATER emanating from GLACIERS. The cloudy appearance of the water is due to the SUSPENDED LOAD, especially silt (ROCK FLOUR). *MJH*

[See also GLACIOFLUVIAL SEDIMENTS]

Keller, W.D. and Reesman, A.L. 1963: Glacier milks and their laboratory-simulated counterparts. *Geological Society of America Bulletin* 74, 61–76.

glacier modelling The numerical simulation of GLACIERS or ICE SHEETS, or some aspect of their behaviour. Because glaciers and ice sheets evolve over long timescales, cycles of growth and decay cannot be observed directly. Understanding of such cycles, therefore, largely depends upon numerical modelling, in which glaciers are simulated (see SIMULATION MODEL) by sets of equations. Furthermore, models are widely used in developing theories of SUBGLACIAL processes, which generally operate in inaccessible environments. Several types of glacier model have been developed, with varying degrees of detail and sophistication.

A widely used modelling procedure attempts to reconstruct the climatic inputs required to 'grow' a glacier or ice sheet of a given size. Inputs for such models include: a representation of topography, annual temperature and PRECIPITATION values (and their variation with altitude), and equations describing ice flow; and yield outputs such as glacier thickness, extent and velocity. Successive model runs with varying climatic inputs can be compared to derive the most likely climatic conditions associated with observed glacier limits. Alternatively, such models can be used to explore the future behaviour of ice masses under given climatic inputs to determine; for example, glacier response to GLOBAL WARMING. At a more detailed level, glacier ABLATION rates can be modelled by simulating the ENERGY BALANCE at the ice surface, using inputs such as SOLAR RADIATION, air temperature, HUMIDITY and windspeed. Other types of glacier model explore the evolution of ice temperatures and GLACIER THERMAL REGIMES, HYDROLOGY, surging behaviour (see GLACIER SURGE), SEDIMENT TRANSPORT and EROSION RATES. *DIB*

Boulton, G.S. 1996: Theory of glacial erosion, transport and deposition as a consequence of subglacial sediment deformation. *Journal of Glaciology* 42, 43–62. **Hubbard, A., Blatter, H., Nienow, P. et al.** 1998: Comparison of a three-dimensional model for glacier flow with field data from Haut Glacier d'Arolla, Switzerland. *Journal of Glaciology* 44, 368–378. **Hulton N., Sugden, D.E., Payne, A.J. and Clapperton, C.M.** 1994: Glacier modelling and the climate of Patagonia during the last glacial maximum. *Quaternary Research* 42, 1–19. **Kamb, B.** 1987: Glacier surge mechanism based on linked cavity configuration of the basal water conduit system. *Journal of Geophysical Research* 92, 9083–9100. **Paterson, W.S.B.** 1994: *The physics of glaciers*, 3rd edn. Oxford: Elsevier. **Schmeits, M.J. and Oerlemans, J.** 1997: Simulation of the historical variations in length of Unterer Grindelwaldgletscher, Switzerland. *Journal of Glaciology* 43, 152–164.

glacier surge A short-lived phase of accelerated GLACIER flow (up to 10–100 times faster than previously), during which the surface becomes broken into a maze of crevasses and the terminus advances rapidly. Only about 4% of glaciers globally surge and they tend to be concentrated geographically (e.g. Svalbard, Alaska). One trigger mechanism is thought to be a change in the SUBGLACIAL drainage system, leading to increased basal water pressure. The pattern of surging glaciers world-wide may well be altered by GLOBAL WARMING in the future. *MJH*

Meier, M.F. and Post, A. 1969: What are glacier surges? *Canadian Journal of Earth Sciences* 6, 807–817. **Porter, S.C.** 1989: Late Holocene fluctuations of the fjord glacier system in Icy Bay, Alaska, U.S.A. *Arctic and Alpine Research* 21, 364–379. **Sharp, M.** 1988: Surging glaciers: behaviour and mechanisms. *Progress in Physical Geography* 12, 349–370.

glacier thermal regime   The state of a GLACIER as determined by its temperature distribution. Temperature is one of the most important parameters controlling glacier behaviour. Temperature affects glacier morphology, ice flow, water flow, debris ENTRAINMENT, SEDIMENTATION and LANDFORMS. Based on thermal regime, there are two main types of glacier: (1) *warm* or *temperate glaciers*, in which the ice is mainly at the *pressure melting point* (PMT) throughout, except for a cold surface layer (about 10–15 m thick) that develops in winter; and (2) *cold* or *polar glaciers*, in which the bulk of the ice is below the PMT. A transitional form, in which the ice in the upper and marginal parts of the glacier is below the PMT, whereas ice at depth is at the PMT, is referred to as a *polythermal glacier* (also commonly known ambiguously as a *subpolar glacier*). The fundamental difference between these glacier types is that temperate glaciers slide on their beds and therefore erode them, whereas cold glaciers are frozen to their beds and only erode them in exceptional circumstances. Polythermal glaciers are *wet-based glaciers* in parts and dry in others and thus have a more complex interaction with the substrate. Since the thermal regime is influenced by climate, it is important to understand the sediments and landforms produced by the different types of glacier, in order to make sound judgements about past glacial climates.   *MJH*

[See also COLD-BASED GLACIERS, GLACIAL EROSION, GLACIAL PROTECTION CONCEPT, ICE-SHEET GROWTH]

Bennett, M.R. and Glasser, N.F. 1996: *Glacial geology – ice sheets and landforms*. Chichester: Wiley. **Knight, P.G.** 1999: *Glaciers*. Cheltenham: Stanley Thornes. **Menzies, J.** 1995: Glaciers and ice sheets. In Menzies, J. (ed.), *Modern glacial environments – processes, dynamics and sediments*. Oxford: Butterworth Heinemann, 101–138. **Paterson, W.S.B.** 1994: *The physics of glaciers*, 3rd edn. Oxford: Pergamon.

glacier variations   Changes in the size of glaciers, commonly their frontal variations; see the Figure for an example. Glacier variations provide information about CLIMATIC CHANGE and rates of change with respect to short- and long-term energy fluxes at the GLACIER surface. Historical and longer-term glacier variations reconstructed from direct measurements, paintings, written sources and MORAINES indicate that the glaciers in many mountain regions have fluctuated considerably in extent. During the HOLOCENE, for example, the range of variabil-

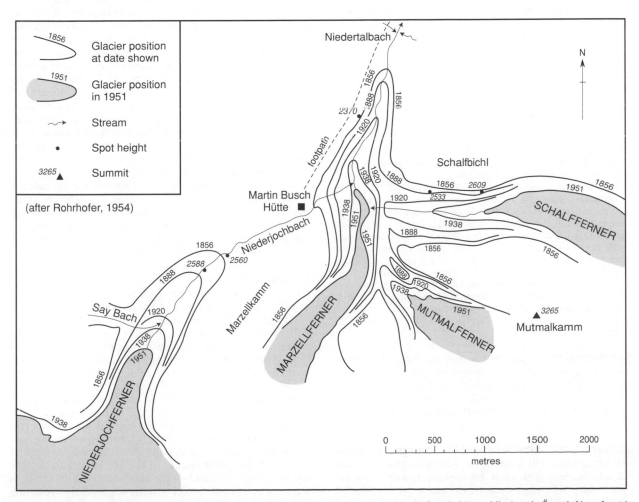

glacier variations *The pattern of glacier retreat AD 1856–1951 in the vicinity of Martin Busch Hütte, Niedertal, Ötztal Alps, Austria; further retreat has occurred in the last 50 years (after Rohrhofer, 1954)*

ity is defined by the early-Holocene HYPSITHERMAL ('Climatic Optimum') and the maximum extent of glaciers during the LITTLE ICE AGE.

Glacier margins *advance* or *retreat*, with variable time lags, in response to variations in GLACIER MASS BALANCE. ABLATION removes ice from the glacier and the horizontal velocity component carries ice forward. A glacier margin remains in the same position when the horizontal velocity component is equal to the horizontal component of ablation. Although the frontal position is stationary, the ice is in motion, but is removed from the glacier at a rate equal to the velocity. Frontal retreat takes place when the horizontal velocity component is less than the horizontal ablation component, whereas glacier advance occurs when the horizontal velocity component is larger than the horizontal ablation component. During the winter season, glacier sliding velocities of temperate glaciers tend to be low as a result of there being little meltwater at the glacier base. Commonly, winter advances start late in the ablation season when melting at the margin does not exceed the forward flow of glacier ice. Commonly, the horizontal ablation component is low in late winter, causing the small winter flow velocities to produce small glacier advances. Despite higher summer flow velocities than in winter, high summer ablation rates cause net retreat of the glacier.

Advance and retreat of the glacier front normally lags behind the *climatic forcing* because the signal must be transferred from the ACCUMULATION AREA to the snout. This is referred to as the LAG TIME or RESPONSE TIME, which is longest for long, low-gradient and slowly moving glaciers and shortest on short, steep and fast-flowing glaciers. Kinematic wave theory has been applied to calculating response times. However, physically based flow models may help determine the response times more precisely. *AN*

[See also GLACIERS AND ENVIRONMENTAL CHANGE]

Frenzel, B., Boulton, G.S., Gläser, B. and Huckreide, U. (eds) 1997: *Glacier fluctuations during the Holocene.* Stuttgart: Gustav Fischer. **Johannesson, T., Raymond, C. and Waddington, E.** 1989: Timescale for adjustment of glaciers to changes in mass balance. *Journal of Glaciology* 35, 355–369. **Kaiser, G.** 1999: A review of the modern fluctuations of tropical glaciers. *Global and Planetary Change* 22, 93–103. **Nesje, A. and Dahl, S.O.** 2000: *Glaciers and environmental change.* London: Arnold. **Nye, J.F.** 1960: The response of glaciers and ice sheets to seasonal and climatic changes. *Proceedings of the Royal Society of London* A256, 559–584. **Paterson, W.S.B.** 1994: *The physics of glaciers,* 3rd edn. Oxford: Pergamon. **Rohrhofer, F.** 1954: Untersuchungen am Ötztaler Gletschern über den Ruckgang 1850–1950. *Geographischer Jahresbericht aus Österreich* 25 (1953/1954), 57–84.

glacierisation The progressive covering of a landscape by glacier ice. Glacierisation and its antonym, DEGLACIERISATION, should not be confused with GLACIATION and DEGLACIATION, which refer to related processes associated with GLACIAL EPISODES during the QUATERNARY. *JAM*

glaciers and environmental change GLACIERS and ICE SHEETS will grow if the ACCUMULATION of snow and ice exceeds ABLATION by melting and/or CALVING. Conversely, shrinkage and retreat occur when ablation exceeds accumulation. Thus, glaciers and ice sheets will respond, sometimes dramatically, to any climatic or other environmental changes that alter the local MASS BALANCE. This, in turn, may precipitate further environmental changes, including CLIMATIC CHANGE due to the action of FEEDBACK MECHANISMS. Consequently, it is important to understand glacier–environmental relationships to be able to predict the role of glaciers and ice sheets in current and future environmental change. Furthermore, the past fluctuations of glaciers and ice sheets can be used to reconstruct past climates, providing an important source of evidence for long-term environmental change.

The rate at which a glacier or ice sheet advances or retreats following climatic change depends on its RESPONSE TIME, which generally increases with the dimensions of the ice mass. Although in many cases glacier response will be predictable, some glaciers may behave non-linearly, with delayed or disproportionate responses to climatic inputs. Non-linear behaviour is particularly characteristic of *surging glaciers*, CALVING GLACIERS and DEBRIS-MANTLED GLACIERS. Complex responses are also thought to be characteristic of ICE SHEETS, as a result of feedbacks involving global SEA-LEVEL CHANGES and changes in surface ALBEDO.

The *glacier advance* of valley glaciers can result in the destruction of farmland and settlements, as happened in Norway, Iceland and the European Alps during the LITTLE ICE AGE. Although *glacier retreat* is widespread at present, glacier advance is occurring in some mountain regions, mainly as a result of increased snowfall. Glacier retreat may trigger wide-ranging environmental changes. First, the frequency of slope failures tends to increase as a result of the destabilisation of rock slopes by the removal of ice support and the exposure of unconsolidated sediments on steep hillsides. An interval of accelerated sediment yield, known as a PARAGLACIAL stage, thus tends to be associated with glacier retreat. Second, river DISCHARGE will increase as snow and ice are removed from storage. With the disappearance of glacier ice from a catchment, river discharges may dramatically decrease, especially in summer, with potentially damaging consequences for agriculture. Third, unstable moraine-dammed or ICE-DAMMED LAKES may form, which are prone to catastrophic outburst floods or JÖKULHLAUPS. Dangerous lakes associated with retreating glaciers pose a particularly severe environmental hazard on DEBRIS-MANTLED GLACIERS in high mountain regions. Fourth, the return of MELTWATER to the oceans will contribute to a rise in SEA LEVEL. It is estimated that the retreat of valley glaciers and ice caps may cause a rise in global sea level of up to 18 cm during the next century.

Changes in the magnitude of ice sheets have a potentially large environmental impact. ICE-SHEET GROWTH has, on several occasions during the QUATERNARY, engulfed large areas of the mid-latitudes, removing huge volumes of water from the oceans and lowering global sea level by up to about 120 m (*eustatic* SEA-LEVEL CHANGE). The formation of successive LAURENTIAN and FENNOSCANDIAN ICE SHEETS was particularly important in this respect. Large ice sheets also affect regional sea levels through ISOSTATIC effects, causing the underlying LITHOSPHERE to sag into the MANTLE because of the additional imposed weight of the ice. Subsequent retreat of the ice results in *isostatic rebound*, which, in much of the area within the ice limits, is large enough to counteract eustatic sea-level rise, resulting in local sea-level fall. Some areas of northern Canada and Scandinavia are still experiencing such a fall in RELATIVE

SEA LEVEL as a result of residual isostatic uplift. The retreat and disappearance of the mid-latitude ice sheets was associated with a complex series of environmental changes with both regional and global consequences.

In recent years, much attention has focused on the possible response of the Antarctic and Greenland ice sheets to recent GLOBAL WARMING. In particular, it has been suggested that rapid ice-sheet collapse may occur in the near future, precipitating significant SEA-LEVEL RISE. While there is evidence that parts of the high-latitude ice sheets are undergoing thinning and retreat, other parts appear to be thickening, probably in response to increased snowfall from warmer air masses that can carry greater amounts of moisture. On balance, therefore, the Greenland and Antarctic ice sheets appear to be rather stable features of the CRYOSPHERE, a view that is supported by most interpretations of the QUATERNARY geological record.

Evidence for former glacier limits, such as MORAINES and GLACIOMARINE and GLACIOLACUSTRINE sediments can provide information on long-term climatic fluctuations, provided that the glacier limits can be dated and glacier–climate relationships are known. The climatic significance of former glaciers is commonly estimated using reconstructed EQUILIBRIUM-LINE ALTITUDES or the elevation at which glacier ABLATION and ACCUMULATION are in balance. Such reconstructions are most accurate in mountain regions, where the three-dimensional form of glaciers can be determined with greatest ACCURACY and where some form of independent palaeoclimatic data (see PALAEOCLIMATOLOGY) are available, e.g. from faunal or pollen evidence. *DIB*

[See also GLACIER VARIATIONS, GLACIOSEISMOTECTONICS]

Ballantyne, C.K. and Benn, D.I. 1996: Paraglacial slope adjustment during recent deglaciation: implications for slope evolution in formerly glaciated terrain. In Brooks, S. and Anderson, M.G. (eds), *Advances in hillslope processes*. Chichester: Wiley. **Benn, D.I. and Evans, D.J.A.** 1998: *Glaciers and glaciation*. London: Arnold. **Bentley, C.R.** 1998: Rapid sea-level rise from a West Antarctic ice-sheet collapse: a short-term perspective. *Journal of Glaciology* 44, 157–163. **Grove, J.M.** 1988: *The Little Ice Age*. London: Routledge. **Hughes, T.J.** 1992: Abrupt climatic change related to unstable ice-sheet dynamics: toward a new paradigm. *Palaeogeography, Palaeoclimatology, Palaeoecology* 97, 203–234. **Krabill, W., Frederick, E., Manizade, S.** *et al.* 1998: Rapid thinning of parts of the southern Greenland Ice Sheet. *Science* 283, 1522–1524. **Nesje, A. and Dahl, S.O.** 2000: *Glaciers and environmental change*. London: Arnold. **Oerlemans, J.** (ed.) 1989: *Glacier fluctuations and climatic change*. Dordrecht: Kluwer. **Oescher, H. and Langway Jr, C.C. (eds)** 1989: *The environmental record in glaciers and ice sheets*. New York: Wiley. **Sugden, D.E., Marchant, D.R. and Denton, G.H.** 1993: The case for a stable East Antarctic Ice Sheet: the background. *Geografiska Annaler* 75A, 151–154. **Warrick, R.A., Le Provost, C., Meier, M.F.** *et al.* 1996: Changes in sea level. In Houghton, J.T., Meira Filho, L.G., Callander, B.A. *et al.* (eds), *Climate change 1995*. Cambridge: Cambridge University Press. **Yamada, T. and Sharma, C.K.** 1993: Glacier lakes and outburst floods in the Nepal Himalaya. *International Association for Scientific Hydrology (IAHS) Publication* 218, 319–330.

glacio-eustasy

The addition of water to and removal of water from the total ocean volume as a result of climate-induced changes in the dimensions of GLACIERS and ICE SHEETS. A reasonably accurate indirect record of glacio-eustatic ocean volume changes has been derived from OXYGEN ISOTOPE records using FORAMINIFERAL ANALYSIS of ocean-floor sediment cores since the measured oxygen isotope changes mostly record long-term changes in the isotopic composition of seawater. These changes, in turn, are dependent on temporal changes in the volume of ice locked up in the world's ice sheets and glaciers. The difference in *sea level* between GLACIAL EPISODES and INTERGLACIALS caused by glacio-eustasy was up to c. 120 m.

AGD

[See also EUSTASY, GLACIO–ISOSTASY, SEA-LEVEL CHANGE]

Bard, E., Hamelin, B. and Fairbanks, R.G. 1990: U–Th ages obtained by mass spectrometry in corals from Barbados: sea level during the past 130,000 years. *Nature* 346, 456–458. **Mörner, N.-A.** 1995: Sea level and climate – the decadal to century signals. In Finkl Jr, C.W. (ed.), *Holocene cycles: climate, sea levels and sedimentation. Journal of Coastal Research* Special Issue, 261–268.

glaciofluvial landforms

MELTWATER, created mainly by glacial ABLATION, is organised into veins, channels, films, cavities linked by narrow necks (or *throttles*) and broad sheets, which produce LANDFORMS in SUBGLACIAL, SUPRAGLACIAL, ice-marginal and *proglacial* positions. Glaciofluvial erosional forms include meltwater channels (ice-marginal, submarginal, subglacial, lake overflow or overspill and *proglacial*) and P-FORMS (various sculpted bedrock features). Depositional forms include valley trains, outwash plains, KAMES, kame terraces and ESKERS.

Glaciofluvial subglacial deposits are commonly found in regional-scale systems of TUNNEL VALLEYS, eskers and arcuate, ice-marginal MORAINES. Such drainage systems correlate with major changes in climate. For example, large esker networks in Canada and Scandinavia, terminating in arcuate 'glaciofluvial moraines', mark stable positions of YOUNGER DRYAS ice margins. These positions may reflect re-equilibration of ICE SHEETS following the quiescent stages of major surges, which redistributed large ice volumes. JÖKULHLAUPS involving rapid evacuation of subglacial meltwater storage may have contributed to events forming 'equilibration moraines'.

Large-scale catastrophic drainage ('megafloods') is advocated by J. Shaw and coworkers to explain landscapes of bedrock erosion, HUMMOCKY MORAINE and ROGEN MORAINE, FLUTED MORAINE and DRUMLINS in fields of about 10^5 km^2 in area, which characterise much of the land occupied by former mid-latitude ICE SHEETS. Meltwater EROSION on this scale would require reservoirs of water, in the ice-sheet hydrological system, equivalent to several metres of SEA-LEVEL RISE. The rapid drainage of these reservoirs would have major implications for CLIMATIC CHANGE, CLIMATIC MODELS and DANSGAARD–OESCHGER and HEINRICH EVENTS.

Tunnel valleys are formed by pressurised subglacial meltwater in areas of low relief. Such channels may show convex-upwards long profiles, are commonly eroded into uplands and contain mostly streamlined bedforms (drumlins, fluted moraine, hummocky moraine and Rogen moraine). Eskers within tunnel valleys commonly have up-and-down long profiles indicating subglacial formation under hydrostatic pressure in conditionally stagnant ice. Tunnel valleys may or may not be infilled with sediment, depending on events following their erosion.

Both erosional and depositional glaciofluvial landforms occur in proglacial positions. Changing conditions may cause incision (degradation), producing *outwash terraces* or aggradation (vertical accumulation) of outwash plains.

KETTLE HOLES may form through subsequent melting of buried ice blocks. *JS*

[See also GLACIOFLUVIAL SEDIMENTS, GLACIAL LANDFORMS, GLACIAL SEDIMENTS]

Brennand, T.A. 1994: Macroforms, large bedforms and rhythmic sedimentary sequences in subglacial eskers south-central Ontario: implications for esker genesis and meltwater regime. *Sedimentary Geology* 91, 9–55. **Fyfe, G.** 1990: The effect of water depth on ice-proximal glaciolacustrine sedimentation. *Boreas* 19, 147–164. **Kor, P.S.G., Shaw, J. and Sharpe, D.R.** 1991: Erosion of bedrock by subglacial meltwater. *Canadian Journal of Earth Sciences* 28, 623–642. **Maizels, J.K.** 1989: Sedimentology, paleoflow dynamics and flood history of jökulhlaup deposits: paleohydrology of Holocene sediment sequences in southern Iceland sandur deposits. *Journal of Sedimentary Petrology* 59, 204–223. **Miall, A.D.** 1983: Glaciofluvial transport and deposition. In Eyles, N. (ed.), *Glacial geology*. Oxford: Pergamon Press, 168–183. **Price, R.J.** 1973: *Glacial and fluvioglacial landforms*. Edinburgh: Oliver and Boyd.

glaciofluvial sediments

MELTWATER discharging from a GLACIER carries a large volume of sediment, including sand, silt and clay in suspension (known as ROCK FLOUR), as well as rock fragments up to boulder size, which roll or bounce (see SALTATION) on the bed. This sediment is deposited in the *proglacial* area, where it forms extensive *braidplains*. Glacial meltwater streams generally cause reworking and SORTING of glacigenic sediment as it is transported downstream, with the finer fraction being carried farthest. The seasonal and diurnal variations in discharge from glaciers control the nature of sedimentation on, and the morphology of, braidplains. Modifications of the fluvial system occur as a result of the transport and burial of ice blocks. Although glaciofluvial sediments are best preserved in the *proglacial* area, they are also deposited in contact with the glacier, including supraglacially, englacially, subglacially and ice-marginally.

Braidplains resulting from glaciofluvial sedimentation are known as OUTWASH plains, sandar (singular: SANDUR) or, where constrained by steep mountain-sides, *valley trains*. The channel system is constantly changing in braidplains. Characteristic forms are bars, which are ridges of sediment that form between channels (see CHANNEL BAR). Bars include longitudinal, point and linguoid types. Cross-bedding may be developed in bars and channel-fill sequences, while smaller-scale SEDIMENTARY STRUCTURES, such as ripples, cross-lamination, desiccation cracks, mud films and scour structures, may also be found. Organic matter may be trapped in the sediment.

The principal sedimentary types are well sorted sands and gravels, often arranged in fining-upward cycles. In backwater areas, finer sediments may also be deposited. In many outwash plains, ice blocks carried by the river from collapsed ice tunnels may become buried. Later, as the blocks slowly melt, depressions called KETTLE HOLES form in the braidplain, giving rise to *pitted outwash plains*.

Beyond some GLACIERS, major catastrophic flood events can heavily modify the braidplain. Such floods are generated by ICE-DAMMED LAKE outbursts or subglacial volcanic eruptions (see JÖKULHLAUPS). *MJH*

[See also GLACIAL SEDIMENTS]

Gurnell, A.M. and Clark, M.J. (eds) 1987: *Glacio-fluvial sediment transfer: an alpine perspective*. Chichester: Wiley. **Hambrey, M.J.** 1994: *Glacial environments*. London: UCL Press. **Joplin, A.V. and McDonald, B.C. (eds)** 1975: *Glaciofluvial and glacio-lacustrine sedimentation*. [*Special Publication* 23]. Tulsa, OK: Society of Economic Palaeontologists and Mineralogists. **Königsson, L.K. (ed.)** 1985: Glaciofluvium Special Issue. *Striae* 22. **Maizels, J.** 1995: Sediments and landforms in modern proglacial terrestrial environments. In Menzies, J. (ed.), *Modern glacial environments – processes, dynamics and sediments*. Oxford: Butterworth Heinemann.

glacio-isostasy

Deformation of the LITHOSPHERE due to the loading and unloading of ICE SHEETS, which was a prominent process during the QUATERNARY. *AGD*

[See also FOREBULGE, GLACIO-ISOSTATIC REBOUND, ISOSTASY, SEA-LEVEL CHANGE]

Andrews, J.T. (ed.) 1974: *Glacio isostasy*. Stroudsburg, PA: Dowden, Hutchinson and Ross. **Chappell, J.** 1974: Late Quaternary glacio- and hydro-isostasy on a layered Earth. *Quaternary Research* 4, 405–428. **Lambeck, K.** 1995: Late Devensian and Holocene shorelines of the British Isles and North Sea from models of glacio–hydro-isostatic rebound. *Journal of the Geological Society of London* 152, 437–448. **Sabadini, R., Lambeck, K. and Boschi, E. (eds)** 1991: *Glacial isostasy, sea level, and mantle rheology*. Dordrecht: Kluwer.

glacio-isostatic rebound

Upward deformation of the LITHOSPHERE due to the unloading of ICE SHEETS as a result of DEGLACIATION. Investigations of patterns of change in RELATIVE SEA LEVEL in areas affected by GLACIO-ISOSTATIC REBOUND have been used to reconstruct former patterns of isostatic deformation and regional patterns of ice thinning and retreat (see Figure). In areas previously covered by late Quaternary ice sheets, glacio-isostatic rebound is still taking place. For example, in Angermanland in northern Scandinavia, the present rate of glacio-isostatic rebound is about 9 mm y^{-1}. *AGD*

[See also ISOSTATIC UPLIFT, ISOSTASY, SEA-LEVEL CHANGE]

Goudie, A.S. 1992: *Environmental change*, 3rd edn. Oxford: Blackwell. **Gray, J.M.** 1995: Influence of Southern Uplands ice on glacio-isostatic rebound in Scotland – the Main Rock Platform in the Firth of Clyde. *Boreas* 24, 30–36. **Shennan, I.** 1989: Holocene crustal movements and sea level changes in Great Britain. *Journal of Quaternary Science* 4, 77–89.

glaciolacustrine deposits

Sediments deposited in lake water, with a significant input of glacially derived material. Glaciolacustrine deposits can be: (1) *glacier-fed*, in which glacial debris is transported to the lake via subaerial meltstreams; or (2) *glacier-contact*, in which sediment is delivered directly to the lake from glacier ice. Deposits laid down in ice-proximal environments (i.e. closest to the glacier) typically have complex geometries and may take the form of DELTAS, MORAINES or subaqueous outwash fans. Distal glaciolacustrine deposits tend to have more uniform geometry, forming extensive blankets of fine-grained sediments with thickness depending on basin topography and proximity to sediment influx points. They may contain a record of GLACIER VARIATIONS. Annual cycles of sedimentation in distal environments may be recorded in VARVES, rhythmically bedded sands, silts and clays that can be used to establish VARVE CHRONOLOGIES, allowing correlation of successions within and between large basins. Studies of MICROFOSSILS in distal glaciolacustrine deposits provide important data on climatic and other environmental changes. *DIB*

Ashley, G.M. 1995: Glaciolacustrine environments. In Menzies J. (ed.), *Glacial environments*. Vol. 1. *Modern glacial environments:*

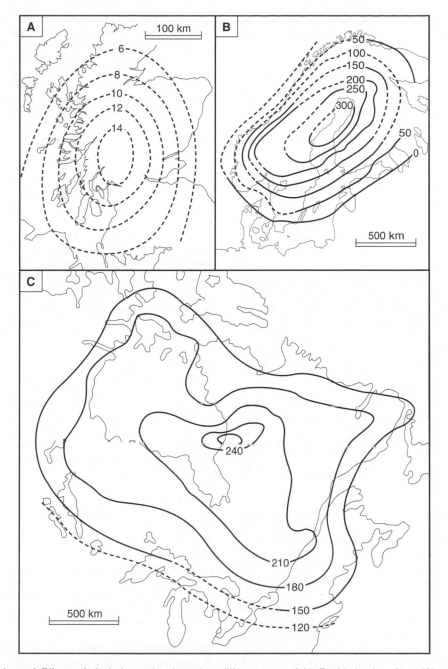

glacio-isostatic rebound *Effects of glacio-isostatic rebound on different parts of the Earth's land surface: (A) generalised isobases (m) for the main postglacial raised shoreline in Scotland; (B) the amount of isostatic recovery (m) of Scandinavia during the Holocene; and (C) the maximum postglacial rebound (m) of northeastern North America (based on Goudie, 1992)*

processes, dynamics and sediments. Oxford: Butterworth Heinemann. **Benn, D.I. and Evans, D.J.A.** 1998: *Glaciers and glaciation.* London: Arnold. **Donnelly, R. and Harris, C.** 1989: Origin and sedimentology of glacio-lacustrine deposits from a small ice-dammed lake, Leirbreen, Norway. *Sedimentology* 36, 581–600. **Matthews, J.A., Dahl, S.O., Nesje, A.** *et al.* 2000: Holocene glacier variations in central Jotunheimen, southern Norway based on distal glaciolacustrine sediment cores. *Quaternary Science Reviews* 19, 1625–1647. **Nesje, A., Dahl, S.O., Andersson, C. and Matthews, J.A.** 2000: The lacustrine sedimentary sequence in Sygneskardvatnet, western Norway: a continuous, high-resolution record of the Jostedalsbreen ice cap during the Holocene. *Quaternary Science Reviews* 19, 1047–1065.

glaciological volcanic index (GVI) A measure of the importance of volcanic eruptions for PALAEOCLIMATOLOGY derived from ELECTRICAL CONDUCTIVITY (ECM) measurements on ICE CORES. The index has the potential to provide a continuous record of volcanic eruptions and, unlike the DUST VEIL INDEX or the VOLCANIC EXPLOSIVITY

INDEX, it is directly related to atmospheric acidity and hence to the addition of volcanic acids to the atmosphere.

JAM

[See also VOLCANIC AEROSOLS]

Bradley, R.S. and Jones, P.D. 1992: Record of explosive volcanic eruptions over the last 500 years. In Bradley, R.S. and Jones, P.D. (eds), *Climate since A.D. 1500*. London: Routledge, 606–622. **Legrand, M. and Delmas, R.J.** 1987: A 220-year continuous record of volcanic H_2SO_4 in the Antarctic Ice Sheet. *Nature* 327, 671–676. **Robock, A. and Free, M.P.** 1995: Ice cores as an index of global volcanism from 1850 to the present. *Journal of Geophysical Research* 100, 11549–11567.

glaciomarine deposits

Sediments deposited in the sea, containing a significant amount of glacially derived material. Glaciomarine sediments laid down in close proximity to glacier ice may be similar in overall geometry to GLACIOLACUSTRINE DEPOSITS, although sediment tends to be more widely distributed because of the buoyancy of glacial meltwater in sea water and the action of tides. Distal glaciomarine deposits commonly take the form of extensive MUD DRAPES, which may contain a rich archive of MICROFOSSILS. ICEBERGS can carry glacial debris considerable distances in high-latitude oceans. ICE-RAFTED DEBRIS may be preserved in distinctive layers in MARINE SEDIMENT CORES, allowing periods of heavy iceberg passage to be identified. Several extensive layers of ice-rafted debris below the North Atlantic record a succession of intense ice-rafting events (HEINRICH EVENTS) that occurred during the Wisconsinan–Weichselian GLACIAL EPISODE. *DIB*

Dowdeswell, J.A. and Scourse, J.D. (eds) 1990: *Glaciomarine environments: processes and products* [*Special Publication* 53]. London: Geological Society. **Powell, R.D. and Domack, E.** 1995: Modern glaciomarine environments. In, Menzies, J. (ed.), *Glacial environments*. Vol. 1. *Modern glacial environments: processes, dynamics and sediments*. Oxford: Butterworth Heinemann, 445–486.

glaciomarine hypothesis

The disputed hypothesis that many sediments previously interpreted as glacial TILL are glaciomarine MUDS. The best known examples are in the Irish Sea Basin, where DIAMICTONS containing marine shells and FORAMINIFERA occur on both coasts. Though first interpreted as evidence of the Biblical Flood, these have long been regarded as tills, with the mud and shells being reworked from deposits that accumulated in the Irish Sea before the LAST GLACIATION. They were reinterpreted as glaciomarine sediments deposited both proximal and distal to the ice margin as it retreated northwards. During the Last Glaciation, world sea levels were eustatically depressed, so the glaciomarine hypothesis requires very substantial isostatic depression to raise local relative sea levels to far above the present height (see GLACIO-ISOSTASY). Similar sediments in East Anglia and elsewhere have been reinterpreted in the same way.

The main evidence used in support of the glaciomarine hypothesis is sedimentological. The marine shells and foraminifera show mixed assemblages, but the pre-Quaternary forms and those with warm affinities are regarded as the product of REDEPOSITION or *reworking* whilst the cold-water forms are considered to be *in situ*. Sand and gravel deposits around the Irish Sea basin have been interpreted as deltas marking RELATIVE SEA LEVEL during deglaciation. The varying altitudes, which do not increase from south to north, are explained using the concept of '*piano-key tectonics*', whereby different fault-bounded parts of the crust were depressed by varying amounts and rebounded at varying speeds.

All of the evidence on which the glaciomarine hypothesis is based has been disputed. At many sites, alternative sedimentological models have been used to interpret the DEPOSITIONAL ENVIRONMENTS of the sediments. It has been argued that all of the shells and foraminifera are derived and reworked from marine sediments that predate the Last Glaciation. The delta deposits may represent local ponding of water at the ice margin or where glaciers from different source areas uncoupled. There is no evidence for substantial movements on reactivated faults during the Quaternary. No consensus has been reached. *DMcC*

[See also DILUVIAL THEORY, GLACIOMARINE DEPOSITS, QUATERNARY TIMESCALE]

Eyles, N. and McCabe, A.M. 1989: The Late Devensian (< 22,000 BP) Irish Sea Basin: the sedimentary record of a collapsed ice sheet margin. *Quaternary Science Reviews* 8, 307–351. **McCabe, A.M.** 1997: Geological constraints on geophysical models of relative sea-level change during deglaciation of the western Irish Sea Basin. *Journal of the Geological Society of London* 154, 601–604. **McCarroll, D.** 2001. Deglaciation of the Irish Sea basin: a critique of the glacimarine hypothesis. *Journal of Quaternary Science* 16: 393–404.

glacioseismotectonics

The interplay between glacial dynamics, structural deformation and SEISMICITY. Glacioseismotectonics, also termed *deglaciation seismotectonics*, operates at the scale of the Earth's CRUST as distinct from the largely surficial direct effects of glacier movement and loading at the base or margin of a glacier. Important aspects include (1) the role of GLACIO-ISOSTATIC REBOUND in modulating crustal deformation, FAULTS and EARTHQUAKE generation, (2) the possible triggering of HEINRICH EVENTS, (3) the broader interactions with climate, TECTONICS and topography, and (4) the implications for the disposal of RADIOACTIVE WASTE, in relatively 'stable' continental regions. *JAM*

[See also FOREBULGE, GLACIO-ISOSTASY, GLACIOTECTONICS, NEOTECTONICS]

Hunt, A.G. and Malin, P.E. 1998: Possible triggering of Heinrich events by ice-load-induced earthquakes. *Nature* 393, 155–158. **Stewart, I.S., Sauber, J. and Rose, J.** 2000: Glacioseismotectonics: ice sheets, crustal deformation and seismicity. *Quaternary Science Reviews* 19, 1367–1389.

glaciotectonics

The dislocation of sediment or rock masses under glacially applied stresses. This may occur beneath a glacier (*subglacial tectonics*) or at the ice margin (*proglacial tectonics*) and can involve extensional or compressional DEFORMATION. The term is normally restricted to the surficial deformation as a direct result of glacial movement or loading in the SUBGLACIAL and proximal *proglacial* domains, rather than the deeper and more far-reaching crustal effects (see GLACIOSEISMOTECTONICS). Subglacial tectonics may result in the detachment and transport of intact *megablocks* or rafts, the excavation of ice-scooped basins, the deformation of bedrock or SOFT-SEDIMENT DEFORMATION. Proglacial tectonic deformation typically occurs in response to high subsurface porewater pressures, steep stress gradients and rapid loading associated with glacier advances or *surging glaciers*. Submarginal excavation combined with proglacial upthrusting produces *thrust moraines*, which may be several tens of metres in height. *DIB*

Aber, J.S. 1985: The character of glaciotectonism. *Geologie en Mijnbouw* 64, 389–395. **Aber, J.S., Croot, D.G. and Fenton, M.M.** 1989: *Glaciotectonic landforms and structures.* Dordrecht: Kluwer. **Harris, C., Williams, G., Brabham, P.** *et al.* 1997: Glaciotectonized Quaternary sediments at Dinas Dinlle, Gwynedd, north Wales, and their bearing on the style of glaciation in the North Sea. *Quaternary Science Reviews* 16, 109–127. **Hart, J.K. and Boulton, G.S.** 1991: The inter-relation of glaciotectonic and glaciodepositional processes within the glacial environment. *Quaternary Science Reviews* 10, 335–350. **Thorson, R.M.** 2000: Glacial tectonics: a deeper perspective. *Quaternary Science Reviews* 19, 1391–1398. **Wateren, D.F.M. van der** 1995: Processes of glaciotectonism. In Menzies, J. (ed.), *Glacial environments.* Vol. 1. *Modern glacial environments: processes, dynamics and sediments.* Oxford: Butterworth Heinemann, 309–335.

glacis A French term for a low-angle slope segment found at the foot of a hillslope, used especially to describe PIEDMONT slopes in ARIDLANDS and SEMI-ARID regions. It may be composed of bedrock, COLLUVIUM and/or ALLUVIUM. *MAC*

[See also PEDIMENT]

gleying A process of REDUCTION of iron and its segregation into mottles and CONCRETIONS (or complete leaching) that occurs in poorly drained soils as a result of ANAEROBIC conditions. The result is that the soil has a grey appearance when saturation is continuous and a mottled character when saturation alternates with AEROBIC conditions. These features of gleying persist after drainage has removed the cause of their development. Features of reduction are described as *stagnic* if they occur within 50 cm of the surface and *gleyic* if they relate to saturation associated with a high level of groundwater. *EMB*

[See also GLEYSOLS]

Buol, S.W., Hole, F.D. and McCracken, R.J. 1988: Soil genesis and classification. Ames, IA: Iowa State University Press. **Crompton, E.** 1952: Some morphological features associated with poor soil drainage. *Journal of Soil Science* 3, 277–289. **Food and Agriculture Organization of the United Nations** 1998: *World reference base for soil resources* [*Soil Resource Report* 84]. Rome: FAO, ISRIC, ISSS.

Gleysols Soils present in all continents with temporary or permanent wetness within the soil profile. As a result of the GLEYING process iron compounds may be segregated into mottles or concretions of ferrihydrite, lepidocrocite or goethite, most likely where there is temporary wetness and hence OXIDATION for some of the time (*oxymorphic* conditions), or in extreme cases leached from the profile leaving a (reduced) grey soil matrix (SOIL TAXONOMY: Aquic great groups of *Entisols, Afisols, Inceptisols* and *Mollisols*). Climate change with increased wetness could lead to extension of gleying, as could higher sea levels in coastal areas. *EMB*

[See also WORLD REFERENCE BASE FOR SOIL RESOURCES].

Department of the Environment 1991: *The potential effects of climate change in the United Kingdom.* London: HMSO.

global capitalism The notion that the economic system of the world is predominantly a unitary, global one in which national controls play a decreasing role. Only half of the 100 largest economic entities in the world are nation states: the rest are *international corporations*, which are increasingly influential in the political, social, cultural, scientific and environmental fields. Notable adverse environmental impacts include those of emissions from the FOSSIL FUEL industry on AIR POLLUTION and CLIMATIC CHANGE, and the inequitable use of tropical genetic resources in agriculture. International corporations are able to play off nation states against one another in seeking lowest costs, thus putting pressure on countries to reduce the attention they give to ENVIRONMENTAL PROTECTION. There are, however, growing constraints on the inherent lack of social and environmental responsibility of global capitalism, which are exemplified by the parallel growth of *non-governmental organisations* (NGOs): in 1975 there were only 1400 NGOS; in 2000 there were over 30 000. *JAM*

[See also GLOBALISATION, NATURAL CAPITALISM]

Hutton, W. and Giddens, A. (eds) 2000: *On the edge: living with global capitalism.* London: Jonathan Cape.

global change 'Transformation processes that operate at a truly planetary scale plus processes that operate at smaller spatial scales (local, regional and continental) but that are so ubiquitous and pervasive as to assume global proportions' (Grübler 1998: 3). Global change research began as a field of the NATURAL ENVIRONMENTAL SCIENCES but it is now recognised that the 'global environment' includes not only (1) NATURAL processes of the environment, their impacts on humans and the ANTHROPOGENIC causes of change in these processes, but also (2) a wide range of human processes, such as those directly and indirectly affecting TECHNOLOGICAL CHANGE, and the many economic, social and political factors controlling production and consumption.

Global change is synonymous with GLOBAL ENVIRONMENTAL CHANGE, although the latter is sometimes confined to global aspects of NATURAL ENVIRONMENTAL CHANGE (a position that is no longer tenable). *JAM*

[See also GLOBALISATION, INTERNATIONALISATION]

Grübler, A. 1998: *Technology and global change.* Cambridge: Cambridge University Press. **Mather, J.R. and Sdasyuk, G.V. (eds)** 1991: *Global change: geographical approaches.* Tucson, AZ: University of Arizona Press. **National Research Council** 1999: *Global environmental change: research pathways for the next decade* [*Report of the Committee on Global Change Research*]. Washington, DC: National Academy Press. **Organisation for Economic Cooperation and Development (OECD)** 1994: *Global change of planet Earth* [*Report from the OECD Megascience Forum*]. Paris: OECD. **Proctor, J.D.** 1998: The meaning of global environmental change: retheorising culture in human dimensions research. *Global Environmental Change* 8, 227–248. **Taylor, P.J., Watts, M.J. and Johnston, R.J.** 1995: Global change at the end of the twentieth century. In Johnston, R.J., Taylor, P.J. and Watts, M.J. (eds), *Geographies of global change.* Oxford: Blackwell, 1–10.

global environmental change Directional ENVIRONMENTAL CHANGES that are experienced in most regions of the world at approximately the same time and so might be considered 'global' in extent or scale, such as GLACIAL–INTERGLACIAL CYCLES in climate. Possible examples since the LAST GLACIATION include the YOUNGER DRYAS STADIAL and the LITTLE ICE AGE, for which there is evidence of global TELECONNECTIONS. The term is *not* usually applied to regional, continental or hemispherical phenomena, such as EL NIÑO–SOUTHERN OSCILLATION (ENSO) events or to the NORTH ATLANTIC OSCILLATION (NAO), nor to short-lived climatic fluctuations associated with single explosive volcanic eruptions. Causes, triggers or FORCING FACTORS of global NATURAL ENVIRONMENTAL CHANGE include variations in receipt of SOLAR RADIATION (see SOLAR FORCING, MILANKOVITCH THEORY), changes in the THERMOHALINE CIRCULATION of the ocean-heat conveyor, reorganisation of atmospheric circulation (AUTOVARIATION), and solar variability on SUB-MILANKOVITCH timescales.

It is well established that natural environmental change has affected the Earth since its formation and that the effects of environmental change are ubiquitous. Many disciplines, especially the NATURAL ENVIRONMENTAL SCIENCES, investigate such changes at a range of temporal and spatial scales. The concept of global environmental change is newer and has become increasingly important for two main reasons. First, there is the realisation that the complex, interactive, physical, chemical and biological processes of the total Earth system are poorly understood. This means there are fundamental weaknesses in our ability to explain not only the long-term environmental changes that are prominent on GEOLOGICAL and QUATERNARY TIMESCALES but also HOLOCENE ENVIRONMENTAL CHANGE and the related short-term changes that are likely to affect human society in the immediate future. Second, there is the realisation that HUMAN IMPACTS on the environment, which have increased during the Holocene, have the potential to affect the Earth system and, indeed, are already visible at the global scale.

Some recognise a distinction between *systemic* global change and *cumulative* global change, at least in the context of human impacts: systemic referring to operation at the global scale (such as enhanced GLOBAL WARMING); cumulative implying accumulation of local effects until the extent is global (such as DEFORESTATION and URBANISATION). *FMC/JAM*

[See also BIOGEOCHEMICAL CYCLES, CLIMATIC CHANGE, EARTH-SYSTEM ANALYSIS, GAIA HYPOTHESIS, GLOBAL CHANGE, HUMAN IMPACT ON ENVIRONMENT]

Bradley, R.S. 2000: Past global changes and their significance for the future. *Quaternary Science Reviews* 19, 391–402. **Eddy, J.A. and Oeschger, H. (eds)** 1993: *Global change in the perspective of the past.* Chichester: Wiley. **Graham, R.W. and Grimm, E.C.** 1990: Effects of global climatic change on the patterns of terrestrial biological communities. *Trends in Ecology and Evolution* 5, 289–292. **Human Dimensions of Global Environmental Change Programme (HDP)** 1996: *A framework for research on the human dimensions of global environmental change* [HDP Report 1]. Barcelona: HDP Secretariat. **Mannion, A.M.** 1991: *Global environmental change: a natural and cultural environmental history.* Harlow: Longman Scientific and Technical. **Moore, P.D., Chaloner, W. and Stott, P.** 1996: *Global environmental change.* London: Blackwell Science. **National Research Council** 1999: *Global ocean science.* Washington, DC: National Academy Press. **Roberts, N. (ed.)** 1994: *The changing global environment.* Oxford: Blackwell. **Slaymaker, O. and Spencer, T.** 1998: *Physical geography and global environmental change.* Harlow: Addison-Wesley Longman. **Stern, P.C., Druckman, D. and Young, O.R. (eds)** 1992: *Global environmental change: understanding the human dimension.* Washington, DC: National Academy Press. **Turner, B.L., Clark, W.C. et al. (eds)** 1990: *The Earth as transformed by human action.* Cambridge: Cambridge University Press. **Turner, B.L., Kasperson, R.E., Meyer, W.B. et al.** 1990: Two types of global environmental change.

Definitional and spatial-scale issues in their human dimensions. *Global Environmental Change* 1, 14–22.

global navigation satellite system (GNS)
A generic term applied to the civil requirement and potential for applications based on world-wide position, velocity and time determination using satellite systems. The European Community Galileo system will be in this category and operated by a public–private (funding) partnership, in contrast to the military motivated GPS. *TF*

[See also GLOBAL POSITIONING SYSTEM (GPS)]

global positioning system (GPS)
A location determination system using a radio receiver to translate signals transmitted from at least four orbiting satellites and hence compute position, speed and time at the receiver. Developed by the US Department of Defense, there are currently 27 *Navstar* satellites in orbit at about 20 000 km. Initially intended for military and other selected users, the accuracy of GPS was limited for civil use by data degradation and encryption, but full accuracy was granted for civilian users on 1 May 2000. The GLONASS system, preferred by some scientific users and operated by the Ministry of Defence of the Russian Federation, also offers precise data from satellites for location and navigation purposes. A third system, the *Galileo* system, is under definition by the European Community. This intended constellation of 24 orbiting satellites, with three geostationary space vehicles covering Europe, is expected to be operational by AD 2008.

The result of the signal measurement is that the geographic coordinates of any location on the Earth can be rapidly computed. Results may be provided in any of a selection of coordinate systems, for example, latitude and longitude, UK Ordnance Survey National Grid or Universal Transverse Mercator (UTM) co-ordinates. Removal of *selective availability*, the degraded signal supplied to non-special users, has provided a considerable increase in positional accuracy, to better than 20 m, for many civilian applications and users. Location fixes can be improved to practical survey levels by drawing on differential GPS methods. This technique uses a fixed control station in conjunction with a mobile unit. Readings and timings are compared and adjusted with respect to the fixed receiver, providing an accuracy, following processing, to within 1 cm.

Applications of this technology now include in-vehicle navigation systems linked to digital map display, geodetic control, PLATE TECTONICS investigation and site survey measurement. Height information is available, but accuracy is considerably less than for planimetric detail. *TF*

Kennedy, M. 1996: *The GPS and GIS; an introduction.* Ann Arbor, MI: Press Incorporated. **Husti, G. and de Jong, K.** 2000: GPS, GLONASS and Galileo status and future plans. *GeoInformation* 3, 26–27.

global warming
The recent rise in global air temperature caused, at least in part, by an anthropogenically enhanced GREENHOUSE EFFECT is frequently referred to as global warming. Over the past century and a half the Earth's mean surface air temperature has increased by about 0.5–0.6°C. Most of this rise occurred in two steps: first, between about AD 1910 and 1940, and second after 1970, with the 1980s and 1990s being the warmest

decades (see Figure). Global CLIMATIC MODELS suggest that, with a BUSINESS-AS-USUAL SCENARIO, global temperatures will continue to rise by between 1.5 and 3.5°C over the next century. This rise of temperature exceeds the fastest global temperature trend since the INSTRUMENTAL RECORD began and is probably greater than any that has occurred over the last 10 000 years. Global warming has become one of the most important environmental issues of our time. The United Nations INTERGOVERNMENTAL PANEL ON CLIMATE CHANGE (IPCC) concluded in 1995 that the balance of evidence suggests a discernible human influence on climate exists. A consensus now exists amongst scientists that global warming is taking place, although a small but vocal minority continue to question its reality.

Predictions suggest that the world will not warm uniformly. The climatic models suggest that warming will be greatest in the polar regions and over land masses. The predicted warming is likely to be sufficient overall to cause a SEA-LEVEL RISE of about 50 cm. In the next century it could trigger large and unpredictable changes to large scale climatic phenomena such as the EL NIÑO–SOUTHERN OSCILLATION and the GULF STREAM. Confidence in the climatic predictions is less at the regional level than at the global scale, and less in relation to associated changes in PRECIPITATION. The impacts of global warming are complex and far from uniform over the globe or seasonally (see REGIONAL COOLING).   *AHP/JAM*

[See also ANALOGUE METHOD, CLIMATIC CHANGE, FUTURE CLIMATE, GREENHOUSE CONDITION, HUMAN IMPACT ON CLIMATE]

Cox, P.M., Betts, R.A., Jones, C.D. et al. 2000: Acceleration of global warming due to carbon-cycle feedbacks in a coupled climate model. *Nature* 408, 184–187. **Drake, F.** 2000: *Global warming: the science of climate change.* London: Arnold. **Harvey, D.** 1999: *Global warming: the hard science.* London: Prentice-Hall. **Houghton, J.** 1997: *Global warming: the complete briefing.* Cambridge: Cambridge University Press. **Houghton, J., Meira Filho, L.G., Callander, B.A. et al. (eds)** 1996: Climate change 1995: the science of climate change. Cambridge: Cambridge University Press. **Huber, B.T., Macleod, K.G. and Wing, S.L.** 1999: *Warm climates in Earth history.* Cambridge: Cambridge University Press. **Hughes, L.** 2000: Biological consequences of global warming: is the signal already apparent? *Trends in Ecology and Evolution* 15, 56–61. **Tett, S.F.B., Stott, P.A., Allen, M.R. et al.** 1999: Causes of twentieth-century temperature change near the Earth's surface. *Nature* 399, 569–572. **Warrick, R.A.** 1993: Slowing global warming and sea level rise: the rough road from Rio. *Transactions of the Institute of British Geographers NS* 18, 140–148. **Whyte, I.D.** 1995: *Climatic change and human society.* London: Arnold.

globalisation The term has moved from a simple descriptor of a process that becomes world-wide or 'makes global' to a more complex concept. This transformation arises in large part from the rapid development of systems of communications and the impact of trends, policies and role models on a global scale. Robertson (1992) used the phrase, 'the scope and depth of consciousness of the world as single place' and this usefully captures the potential universality of many global processes. There are key actors in globalisation. Pragmatically, the multinational companies have major influences on economies at all scales, marketing organisations export brand images to many parts of the world, the media invade even the remoter parts of the Earth's surface and global forums (such as the United Nations) regard the world as their parish. Conceptually, the globalisation theme raises key questions. Giddens (1990) theorised ways in which space and time were compressed by what he termed 'distanciation' and 'disembedding'. The significance of locality is, however, strongly contested. Cultural diversity is a strong and resilient quality, and community has its roots in tradition and history as well as in the practice of everyday life. All the conditions for greater globalisation are present; the evidence to measure its impacts remains to be realised.   *DTH*

[See also GLOBAL CHANGE]

Eade, J. (ed.) 1997: *Living the global city: globalisation as a local process.* London: Routledge. **Giddens, A.** 1990: *The consequences of modernity.* Cambridge: Polity. **Kiely, R. and Marfleet, P. (eds)** 1999: *Globalisation and the third world.* London: Routledge. **Robertson, R.** 1992: *Globalisation.* London: Sage.

glow curve A curve showing the light emission or thermoluminescence given off by SEDIMENTS or *pottery* as

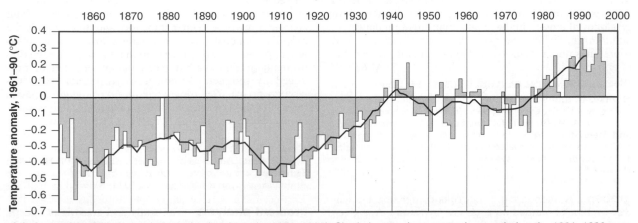

global warming *Global warming over land and sea (AD 1851–1996). Shaded areas show years above or below the 1961–1990 average; the solid curve is a ten-year running mean (after Houghton, 1997)*

they are heated rapidly to 500°C and measured by a photomultiplier. Glow curves are used in THERMOLUMINESCENCE DATING (TL), because on first heating the curve represents both the red-hot glow due to the heating and the inherent thermoluminescence derived from radiation from surrounding radioactive impurities in minerals over the period since deposition or firing, whereas on second heating only the glow due to heating is recorded. The amount of thermoluminescence represented in the glow curve is a function of the radiation flux, the *dose rate* and the susceptibility of the minerals to acquire thermoluminescence.   *CJC*

Aitken, M.J. 1985: *Thermoluminescence dating.* London: Academic Press.

Gondwana A former SUPERCONTINENT comprising South America, Africa, Antarctica, Australia and peninsular India. Its migration across the South Pole in CARBONIFEROUS and PERMIAN times gave rise to a major ICE AGE. During the Permian period Gondwana formed the southern part of PANGAEA. It is named after a region of India.   *JCWC*

[See also CONTINENTAL DRIFT, PLATE TECTONICS]

Audley-Charles, M.G. and Hallam, A. (eds) 1988: *Gondwana and Tethys* [*Special Publication* 37]. London: Geological Society. **Caputo, M.V. and Crowell, J.C.** 1985: Migration of glacial centers across Gondwana during Paleozoic Era. *Geological Society of America Bulletin* 96, 1020–1036. **Crowley, T.J., Mengel, J.G. and Short, D.A.** 1987: Gondwanaland's seasonal cycle. *Nature* 329, 803–807. **Smith, A.G.** 1999: Gondwana: its shape, size and position from Cambrian to Triassic times. *Journal of African Earth Sciences* 28, 71–97.

gorge An incised valley or small *canyon* characterised by a relatively high depth:width ratio, usually with precipitous rocky walls and steep channel gradient. Gorges are generally considered to originate where fluvial downcutting greatly exceeds valley widening. In some environments, however, gorge excavation by purely FLUVIAL PROCESSES may be augmented by other processes, such as FROST WEATHERING in alpine PERIGLACIOFLUVIAL SYSTEMS. The deepest gorges have been excavated by rivers in regions of rapid tectonic UPLIFT. In some areas, gorge sections of rivers have been attributed to the headward erosion of KNICKPOINTS marking REJUVENATION of rivers following uplift. Some gorges have been attributed to glaciofluvial erosion in the SUBGLACIAL environment where MELTWATER, highly charged with coarse sediment and sometimes under HYDROSTATIC PRESSURE, may be capable of enhanced incision.   *JAM/RPDW*

[See also SUBMARINE CANYON, V-SHAPED VALLEY]

Hantke, R. and Scheidegger, A. 1993: On the genesis of the Aare gorge, Berner Oberland, Switzerland. *Geographica Helvetica* 48, 120–124. **McEwen, L.J., Matthews, J.A., Shakesby, R.A. and Berrisford, M.S.** 2002: Holocene gorge excavation and boulder-fan formation related to frost weathering in a Norwegian alpine periglacio-fluvial system. *Arctic, Antarctic and Alpine Research* 34.

graben A GEOLOGICAL STRUCTURE comprising a downthrown block bounded on each side by FAULTS. At a given level the rock between the faults is younger than that on each side. A RIFT VALLEY is the corresponding landform:

the topographic expression of a graben. Grabens usually result from extensional TECTONICS. A *half-graben* is a tilted downthrown block adjacent to a single extensional fault.   *GO*

[See also HORST]

Nelson, W.J., Denny, F.B., Follmer, L.R. and Masters, J.M. 1999: Quaternary grabens in southernmost Illinois: deformation near an active intraplate seismic zone. *Tectonophysics* 305, 381–397. **Peakall, J.** 1998: Axial river evolution in response to half-graben faulting: Carson River, Nevada, USA. *Journal of Sedimentary Research* 68, 788–799.

grade concept An old term in fluvial geomorphology used to indicate that a river channel is in equilibrium or balance with its environment. A graded stream or stream reach was defined by Mackin (1948: 471) as one 'in which, over a period of years, slope is delicately adjusted to provide, with available discharge and prevailing channel characteristics, just the velocity required for the transportation of the load supplied from a DRAINAGE BASIN'. The concept was closely associated with the idea that a smooth concave longitudinal profile indicated a landscape in which its drainage system was adjusted to the stage of LANDSCAPE EVOLUTION and that conversely irregularities in a longitudinal profile indicated interruptions to an erosion cycle (such as uplift or sea-level change) and a river that was unadjusted to the landscape. Such irregularities were widely used in DENUDATION CHRONOLOGY in reconstructions of the long-term history of landscapes.   *RPDW*

[See also EQUILIBRIUM CONCEPTS IN GEOMORPHOLOGICAL AND LANDSCAPE CONTEXTS]

Chorley, R.J. 2000: Classics in physical geography revisited: 'Mackin, J.H. 1948: Concept of the graded river. *Geological Society of America Bulletin* 59, 463–512.' *Progress in Physical Geography* 24, 563–578. **Knox, J.C.** 1975: Concept of the graded stream. In Mulhorn, W.N. and Flemel, R.C. (eds), *Theories of landform development.* London: George Allen and Unwin, 169–198. **Mackin, J.H.** 1948: Concept of the graded river. *Geological Society of America Bulletin* 59, 463–512.

grade (of particles) A general term for GRAIN-SIZE in a SEDIMENT or SEDIMENTARY ROCK, such as 'a sediment of medium sand grade'.   *GO*

graded bedding A SEDIMENTARY STRUCTURE characterised by a consistent variation in GRAIN-SIZE from bottom to top of a single BED. *Normal grading* passes from coarser at the base to finer at the top, and is commonly the product of DEPOSITION from a waning flow, such as a TURBIDITY CURRENT or other depositional EVENT, although thin normally graded laminae (see LAMINATION) can be produced by sediment settling from suspension (see SUSPENDED LOAD). *Reverse grading* or *inverse grading* passes from fine up to coarse within a single bed and is commonly characteristic of the deposits of DEBRIS FLOWS.   *BTC*

Allen, J.R.L. 1982: *Sedimentary structures: their character and physical basis.* Amsterdam: Elsevier. **Kuenen, Ph.H.** 1953: Significant features of graded bedding. *Bulletin of the American Association of Petroleum Geologists* 37, 1044–1066.

gradient analysis The study and analysis of species occurrences and abundances along gradients. The gradients may be unknown *a priori* and are derived from the data by ORDINATION using CORRESPONDENCE ANALYSIS or PRINCIPAL COMPONENTS ANALYSIS (*indirect gradient analysis*) or they are known and measured (*direct gradient analysis*). The latter involves REGRESSION ANALYSIS (one species) or canonical ordination techniques such as CANONICAL CORRESPONDENCE ANALYSIS or REDUNDANCY ANALYSIS (many species). *HJBB*

Braak, C.J.F. ter and Prentice, I.C. 1988: A theory of gradient analysis. *Advances in Ecological Research* 18, 271–317.

gradualism The concept that the GEOLOGICAL RECORD is dominated by the products of slow, steady and gradual processes, as opposed to the cataclysmic EVENTS preferred by CATASTROPHISM. A component of Lyell's UNIFORMITARIANISM, gradualism alone is now considered an inadequate explanation for the geological record. The modern consensus on the nature of Earth history could be described by analogy with biology's PUNCTUATED EQUILIBRIUM as 'punctuated gradualism', with a background of slow, long-term processes periodically affected by dramatic, short-lived, major interruptions. *CDW*

[See also NEOCATASTROPHISM, PHYLETIC GRADUALISM]

Ager, D.V. 1993: *The nature of the stratigraphical record*, 3rd edn. Chichester: Wiley. **Dott Jr, R.H.** 1983: Episodic sedimentation – how normal is average? How rare is rare? Does it matter? *Journal of Sedimentary Petrology* 53, 5–23. **Gould, S.J.** 1984: Towards the vindication of punctuational change. In Berggren, W.A. and Van Couvering, J.A. (eds), *Catastrophes and Earth history: the new uniformitarianism*. Princeton, NJ: Princeton University Press.

grain (1) A general term for a particle forming part of a rock or soil, particularly used for a SEDIMENT particle or CLAST and sometimes restricted to particles smaller than a few millimetres. (2) More specifically, a grain is a particle in a sediment or SEDIMENTARY ROCK that is of the nominal or larger GRAIN-SIZE (i.e. GRAVEL, SAND or MUD) in contrast to MATRIX, which is of finer grain-size. (3) 'Grain' is also sometimes used as a synonym for 'texture' in the broader context of landforms and LANDSCAPES, referring to the scale of the components of the landscape (for example, landscapes with much detail may be termed 'fine-grained'). *TY*

grain-size The size (diameter) of particles (particularly CLASTS) in a SOIL, SEDIMENT or SEDIMENTARY ROCK; one of the most important attributes of sediment TEXTURE. Grain-size tends to diminish with distance of sediment transport, and at any place variations in grain-size reflect the COMPETENCE of agents of sediment transport: it can therefore be used as a proxy indicator of the ENERGY of a DEPOSITIONAL ENVIRONMENT. Patterns of grain-size frequency distributions – mean sizes and other moment measures of the distribution such as SORTING, SKEWNESS and KURTOSIS – have been used to characterise different processes of SEDIMENT TRANSPORT and DEPOSITION and attempts have been made to relate grain-size characteristics to depositional environments in SEDIMENTARY DEPOSITS.

The most commonly used scale of grain-size among Earth scientists is the *Udden–Wentworth scale* (see Figure 1), which recognises GRAVEL, SAND, MUD and subdivisions. Other scales are in use, notably for PYROCLASTIC material and in disciplines such as engineering. A modification of the Udden–Wentworth scale has recently been proposed to encompass particles with diameters up to 1075 km (= very coarse megalith). Figure 2 shows the names given to mixtures of gravel, sand and mud.

The size of *gravel* particles can be directly measured. It is common to measure the long, intermediate and short axes of a number of clasts, giving both a size frequency distribution and a measure of particle SHAPE. Grain-size of unconsolidated *sand* is commonly determined by *sieve analysis* – passing the sediment through a stack of sieves with mesh diameter decreasing downwards. The mass retained on each sieve after shaking gives a mass frequency distribution. Grain-size in lithified SANDSTONES can be estimated visually against a reference scale to give a representative GRADE on the Udden–Wentworth scale, or else clasts are measured in THIN-SECTION ANALYSIS to give a number frequency distribution. *Mud* particles are difficult to resolve by eye. In unconsolidated or poorly consolidated samples, the ratio of SILT to CLAY can be estimated by grinding a small sample between the fingers or teeth: silt is gritty whereas clay is smooth. In the laboratory a grain-size distribution can be obtained by SEDIMENTATION

particle diameter		SEDIMENT	SEDIMENTARY ROCK
mm	ø units		
		BOULDER	CONGLOMERATE (rounded CLASTS) BRECCIA (angular CLASTS)
— 256	— -8		
		COBBLE	
— 64	— -6	GRAVEL	RUDACEOUS
		PEBBLE	
— 4	— -2		
— 2	— -1	GRANULE	
— 1.0	— 0	very coarse	
— 0.5	— 1	coarse	SANDSTONE
— 0.25	— 2	medium	SAND
— 0.125	— 3	fine	ARENACEOUS
— 0.063	— 4	very fine	
		SILT	MUDROCK (MUDSTONE, SHALE)
— 0.004	— 8		MUD
		CLAY	ARGILLACEOUS

grain-size (1) *The Udden–Wentworth scale for grain-size*

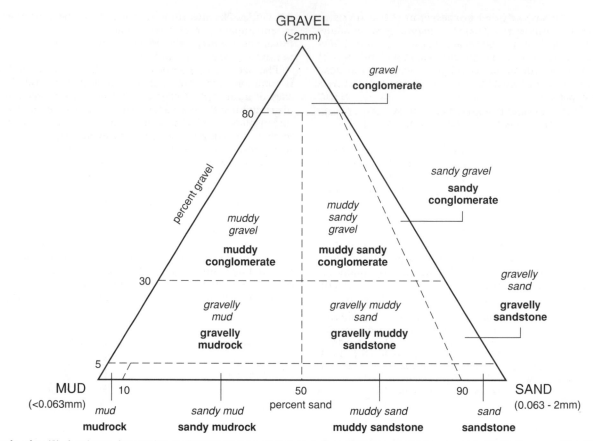

grain-size (2) *A scheme for naming sediments (names in italics) and sedimentary rocks (names in bold) that are combinations of gravel, sand and mud (after Nichols, 1999)*

methods that use the relationship between settling velocity and particle mass. Grain-size of unconsolidated sand and mud can also be determined using a *Coulter counter*, which uses the electrical properties of particles carried in a fluid, or a *Sedigraph*, which uses X-rays, or by *laser diffraction* techniques. There are difficulties in comparing frequency distributions obtained by different methods, particularly where particles are non-spherical.

Grain-size can be expressed as a typical value on the Udden–Wentworth scale, as a histogram or curve of the frequency distribution, or as a cumulative percentage curve. When plotted on probability graph paper, cumulative percentage curves of grain-size distributions commonly approximate to a straight line or to a line made up of straight segments. Attempts have been made to relate these straight-line segments to depositional processes.

GO/TY

[See also SOIL TEXTURE]

Bianchi, G.G., Hall, I.R., McCave, I.N. and Joseph, L. 1999: Measurement of the sortable silt current speed proxy using the Sedigraph 5100 and Coulter Multisizer IIe: precision and accuracy. *Sedimentology* 46, 1001–1014. **Blair, T.C. and McPherson, J.G.** 1999: Grain-size and textural classification of coarse sedimentary particles. *Journal of Sedimentary Research* 69, 6–19. **Ehrlich, R.** 1983: Editorial: size analysis wears no clothes, or have moments come and gone? *Journal of Sedimentary Research* 53, 1. **Folk, R.L.** 1980: *Petrology of the sedimentary rocks.* Austin, TX: Hemphill. **Friedman, G.M. and Sanders, J.E.** 1978:

Principles of sedimentology. New York: Wiley. **Harrell, J. and Eriksson, K.A.** 1979: Empirical conversion equations for thin-section and sieve derived size distribution parameters. *Journal of Sedimentary Petrology* 49, 273–280. **Konert, M. and Vandenberghe, J.** 1997: Comparison of laser grain size analysis with pipette and sieve analysis: a solution for the underestimation of the clay fraction. *Sedimentology* 44, 523–535. **McManus, J.** 1988: Grain size determination and interpretation. In Tucker, M.E. (ed.), *Techniques in sedimentology.* Oxford: Blackwell, 63–85. **Syvitski, J.P.M. (ed.)** 1991: *Principles, methods and applications of particle size analysis.* Cambridge: Cambridge University Press. **Wentworth, C.K.** 1922: A scale of grade and class terms for clastic sediments. *Journal of Geology* 30, 377–392.

granule A SEDIMENT particle of GRAIN-SIZE 2–4 mm.

TY

[See also GRAVEL]

graphic log A diagrammatic representation of a vertical succession of SEDIMENTS, SEDIMENTARY ROCKS or PYROCLASTIC ROCKS in which the vertical scale represents thickness, the horizontal scale represents GRAIN-SIZE and the resulting column of variable width is ornamented to illustrate features such as SEDIMENTARY STRUCTURES, contacts between BEDS, FOSSILS and sampling points. The value of a graphic log is that grain-size reflects the ENERGY of the DEPOSITIONAL ENVIRONMENT and the log gives a clear visual impression of ENVIRONMENTAL CHANGE through time. There has been some criticism of over-reliance on graphic logs as a tool in FACIES ANALYSIS because their one-dimensional form detracts from a careful three-dimensional reconstruction of PALAEOENVIRONMENT. *GO*

Bouma, A.H. 1962: *Sedimentology of some flysch deposits: a graphic approach to facies interpretation.* Amsterdam: Elsevier. **Bridge, J.S.** 1985: Paleochannel patterns inferred from alluvial deposits: a critical evaluation. *Journal of Sedimentary Petrology* 55, 579–589. **Nichols, G.** 1998: *Sedimentology and stratigraphy.* Oxford: Blackwell. **Graham, J.** 1988: Collection and analysis of field data. In Tucker, M.E. (ed.), *Techniques in sedimentology.* Oxford: Blackwell, 5–62.

graphical user interface (GUI)

A set of programs that act as mediators between the user and other computer programs (usually referred to as 'commands'). GUIs usually rely on windows (different information can be displayed simultaneously), iconic representation of entities (such as files, processes, etc.), pull-down or pop-up menus (commands or other options are selected from lists rather than from their actual names) and pointing devices (the mouse can be used for selection). These programs are intended to help the user in the pursuit of tasks, and they should take into account cognitive factors, such as the size of short-term memory, colour perception, etc. The interface should be based on user-oriented terms and concepts rather than on computer concepts, and typical users should be involved in the design of GUIs. *ACF*

grass cuticle analysis

The use of SUBFOSSIL grass cuticles, extracted from SEDIMENTS, for PALAEOENVIRONMENTAL RECONSTRUCTION. SUBFOSSIL grass cuticles are the resistant surface remains of grass leaves, retaining the micromorphological features of the cells they once covered, and consist principally of complex polymers (cutin, cutan and suberin), cuticular leaf waxes and PHYTOLITHS embedded in the cuticle complex. Grass cuticles are relatively inert and well preserved in SEDIMENTS, because of the chemical stability of their constituent hydroxy mono-carboxylic acids – phloinic acid ($C_{18}H_{34}O_{16}$), cutinic acid ($C_{13}H_{22}O_3$), and cutic acid ($C_{26}H_{50}O_6$) – and retain micromorphological features that can aid identification to taxonomic levels below family, thus providing valuable BIO-INDICATORS. Grass cuticle analysis therefore helps to overcome the limited taxonomic resolution associated with grass pollen (often only identifiable to family level). Grass cuticle analysis is particularly suited to tropical environments where the BIODIVERSITY of grasses is high. Grass cuticles are also used for monitoring shifts in woodland–grassland ECOTONES and for determining the photosynthetic pathways used by past grasslands. *MJW*

[See also CHARRED-PARTICLE ANALYSIS, TROPICAL, PALAEO-ENVIRONMENT, PALAEOECOLOGY]

Dugas, D.P. and Retallack, G.J. 1993: Middle Miocene fossil grasses from Fort Ternan, Kenya. *Journal of Paleontology* 67, 113–128. **Mworia-Maitima, J.** 1997: Prehistoric fires and land-cover change in Western Kenya: evidence from pollen, charcoal, grass phytoliths, and grass cuticle analyses. *The Holocene* 7, 409–417. **Palmer, P.G.** 1976: Grass cuticles: a new paleoecological tool for East African lake sediments. *Canadian Journal of Botany* 54, 1725–1734.

grasses

Fossil evidence suggests that grasses (Poaceae) are one of the most recent ANGIOSPERM families to evolve. The first unequivocal MACROFOSSIL evidence for grasses has been found in FOSSIL deposits from North America dated to approximately 60 Ma BP. *Grassland* BIOMES are more recent with palaeoecological evidence suggesting that open TEMPERATE GRASSLANDS and SAVANNA did not come into existence until approximately 15 Ma. There are a number of suggestions to account for their relatively late EVOLUTION and radiation. These include increased global ARIDITY, increased FIRE FREQUENCY and an increase in FAUNA (e.g. hoofed mammals) physiologically adapted to a diet rich in cellulose and silica. Despite their late appearance, the radiation of grasslands has been rapid. Presently there are more than 10 000 species of grasses on Earth, with estimates suggesting that modern grasslands cover more than 30% of the land surface, providing up to 52% of the protein in human diets world-wide. *KJW*

Axelrod, D.I. 1985: Rise of the grassland biome, central North America. *Botanical Review* 51, 163–201. **Crepet, W.L. and Feldman, G.D.** 1991: The earliest remains of grasses in the fossil record. *Americal Journal of Botany* 78, 1010–1014. **Janis, C.M.** 1993: Tertiary mammal evolution in the context of changing climates, vegetation and tectonic events. *Annual Review of Ecology and Systematics* 24, 467–500. **Stebbins, G.L.** 1981: Coevolution of grasses and herbivores. *Annals of the Missouri Botanical Gardens* 68, 75–86.

graticule

The pattern generated by intersecting lines of meridians and parallels on a globe or map. The pattern will vary according to the MAP PROJECTION employed and the map area covered. *TF*

Maling, D.H. 1992: *Coordinate systems and map projections.* Oxford: Pergamon.

gravel

A SEDIMENT with a dominant GRAIN-SIZE greater than 2 mm. The term is occasionally used as a synonym for GRANULE. *TY*

gravel-bed river

A river with a sediment load dominated by GRAVEL moving as BEDLOAD, although such rivers also transport significant quantities of sediment of finer GRAIN-SIZE. Gravel-bed rivers tend to dominate under conditions of high CHANNEL GRADIENT and high, often variable, DISCHARGE. They have unstable, mobile channels due to a lack of cohesive bank material. BEDFORMS are dominated by longitudinal CHANNEL BARS and most gravel-bed rivers have a low-sinuosity, braided CHANNEL PATTERN, although some meandering rivers have gravel beds. Many streams in glacial outwash settings (see SANDUR) and in SEMI-ARID settings are gravel-bed rivers. *GO*

[See also BEDROCK CHANNEL, BRAIDING, SAND-BED RIVER, SUSPENDED-LOAD RIVER]

Alonso, A. and Garzón, G. 1994: Quaternary evolution of a meandering gravel bed river in central Spain. *Terra Nova* 6, 465–475. **Ashmore, P.** 1991: How do gravel-bed rivers braid? *Canadian Journal of Earth Sciences* 28, 326–341. **Billi, P., Hey, R.D., Thorne, C.R. and Tacconi, P. (eds)** 1992: *Dynamics of gravel-bed rivers.* Chichester: Wiley. **Brasington, J., Rumsby, B.T. and McVey, R.A.** 2000: Monitoring and modelling morphological change in a braided gravel-bed river using high resolution GPS-based survey. *Earth Surface Processes and Landforms* 25, 973–990. **Carson, M.A.** 1986: Characteristics of high-energy 'meandering' rivers: the Canterbury Plains, New Zealand. *Geological Society of America Bulletin* 97, 886–895.

gravity anomaly

A local variation in the strength of the Earth's gravity field that can be attributed to variations in the density of buried materials. In GEOPHYSICAL SURVEYING a *Bouguer anomaly* is a gravity reading that has been corrected for the predictable effects of altitude and latitude: residual variations in Bouguer anomalies can be interpreted in terms of buried geological or archaeological features. *GO*

grazing: contemporary impacts on ecosystems

Grazing is the consumption of green plants by vertebrates, invertebrates and some micro-organisms. Throughout EVOLUTION, grazing has led to interactions between grazing animals and plants. Interactions between herbivorous insects and plants include toxic chemicals being produced by the plants and selective grazing by the insects. Plants grazed by vertebrates range from the palatable to those that are avoided. *Selective grazing* affects the floristic composition of pastures, with some species declining and some becoming abundant. For example, CONSERVATION of floristic diversity of chalk grasslands in the UK is dependent on the selective grazing of domestic herbivores. Grazing by wild herbivores (ALIEN SPECIES) has halted ECOLOGICAL SUCCESSION in some communities and elsewhere has had major implications for conservation of NATIVE or indigenous flora. *IFS*

[See also OVERGRAZING]

Crawley, M.J. 1983: *Herbivory: the dynamics of animal–plant interactions.* Oxford: Blackwell Scientific. **Hodgson, J. and Illius, A.W. (eds)** 1996: *The ecology and management of grazing systems.* Wallingford: CAB International.

grazing history

Although grazing animals are often believed to be one of the primary mechanisms of FOREST CLEARANCE, detecting their presence from the palaeoecological record can be problematic. Some ANTHROPOGENIC INDICATOR species are closely associated with grazed HABITATS, and EROSION often increases as a result of increased TRAMPLING by animals. During the MESOLITHIC, palaeoecological evidence for woodland DISTURBANCE and that from FIRE HISTORY are often interpreted in terms of human activity designed to open up the canopy and encourage plentiful new growth of ground-level vegetation, in order to make hunting easier either by concentrating populations of native grazing animals (e.g. red deer in Britain) in one location or by increasing the CARRYING CAPACITY of a given tract of woodland. The Mesolithic–NEOLITHIC TRANSITION includes two major changes with regard to grazing animals, the domestication of native species (e.g. cattle) and the introduction of new species (e.g. sheep, goats). Once the location and nature of grazing herds was controlled to some extent by human activity, landscape impacts often become significant.

The nature and severity of grazing-related ENVIRONMENTAL CHANGE varies in both space and time and depends on the herbivore, since different grazing animals have different tastes and requirements, and thus different effects on the landscape. Intensification of grazing activity is often believed to have been an important factor in the initiation of *blanket peat* growth and the development of MOORLAND and HEATHLAND, and therefore in profound changes in the use and value of LANDSCAPES for prehistoric communities. In upland Britain, heather MOORLAND is an important resource for grazing sheep, red deer and grouse and this use seems to have continued sustainably for several thousand years. However, intensification of burning (designed to improve the grazing) and higher stocking levels of sheep in the last 200 years are leading to a reduction in heather moorland and show that grazing pressure is an important factor in environmental change. *MJB*

[See also OVERGRAZING, PASTORALISM, TRANSHUMANCE]

Mellars, P.A. 1976: Fire ecology, animal populations and man: a study of some ecological relationships in prehistory. *Proceedings of the Prehistoric Society* 42, 15–45. **Moore, P.D.** 1975: Origin of blanket mires. *Nature* 256, 267–269. **Moore, P.D.** 1993: The origin of blanket mire, revisited. In Chambers, F.M. (ed.), *Climate change and human impact on the landscape*. London: Chapman & Hall, 133–145. **Stephenson, A.C. and Thompson, D.B.A.** 1993: Long term changes in the extent of heather moorland in upland Britain and Ireland: palaeoecological evidence for the importance of grazing. *The Holocene* 3, 70–76.

green lists

Those species that are known to be secure and not in need of CONSERVATION. If a species is not in a RED DATA BOOK, it does not necessarily mean that it is not in need of conservation. The use of green lists has been suggested as an alternative to red lists. *IFS*

green manure

Plant material incorporated with the soil while green, or soon after maturity, to improve the soil. *RAS*

[See also SOIL CONSERVATION]

Nyberg, G., Ekblad, A., Buresh, R.J. and Hogberg, P. 2000: Respiration from C-3 plant green manure added to a C-4 plant carbon dominated soil. *Plant and Soil* 218, 83–89.

green politics

The injection of increased environmental awareness into political thinking and policy making since the late 1980s. Although environmental concerns have a long history, the realisation that humanity must learn to conserve and protect the NATURAL ENVIRONMENT, rather than exploit it in a non-sustainable way, only became widespread through the publicity of the ENVIRONMENTAL MOVEMENTS and radical PRESSURE GROUPS. Recognition of the very real threat to humanity posed by increased GREENHOUSE GAS emissions, for example, led to shifts in the political agenda with environmental issues taking a high profile for the first time.

In the UK, the government environmental White Paper of 1990 prompted a restructuring of central government to include a Minister for the Environment and two 'green' cabinet committees responsible for issuing guidance documents for civil servants. It also encouraged corporate efforts to integrate environmental concerns and sustainable practices into day-to-day organisation of business, industrial and domestic life. As well as government policy changes and restructuring world wide, green politics incorporates political parties like the *Green Party*, the principal objective of which is to 'green society' and promote environmental awareness and sustainable business practices. *JGS*

[See also ENVIRONMENTAL POLICY, ENVIRONMENTAL PROTECTION, GREENING OF SOCIETY]

Jordan, A. 1998: The construction of a multi-level environmental governance system. *Environment and Planning* 17, 227–235. **Lester, J.M. (ed.)** 1994: *Environmental politics and policy*, 2nd edn. Durham, NC: Duke University Press. **Weale, A.,** 1998: Environmental policy. In Budge, I., Crewe, I., McKay, D. and Newton, K. (eds), *The new British politics*. Harlow: Longman, 171–129.

Green Revolution The rapid increase in crop yields of the AD 1950s brought about by the introduction of high-yielding varieties of grain, particularly rice and wheat, combined with heavy inputs of artificial FERTILIS-ERS. Especially in the developing world, where the 'revolution' has supported rapid *population growth*, the SUSTAINABILITY of such systems is open to question for both ecological and economic reasons. For example, a large number of resistant varieties of grain have been replaced by a small number of varieties that are vulnerable to PESTS and DISEASE; and the price of fertilisers and PESTI-CIDES is both high and variable.   *JAM*

[See also AGRICULTURAL REVOLUTION]

Brown, L.R. 1970: *Seeds of change: the Green Revolution and development in the 1970s.* New York: Praeger/Overseas Development Council. **Gordon, R.C. and Barbier, E.R.** 1990: *After the Green Revolution: sustainable agriculture for development.* East Haven, CT: Earthscan.

greenhouse condition A term used to describe a long interval of global warmth in the Earth's past, associated with high SEA LEVEL, an absence of ICE SHEETS or *ice caps* and a prominence of sedimentary FACIES that indicate a warm palaeoclimate (see PALAEOCLIMATOLOGY). At least five oscillations between greenhouse and cooler ICEHOUSE CONDITIONS can be recognised in the GEOLOGICAL RECORD of the late PROTEROZOIC and PHANEROZOIC. Greenhouse conditions characterised late Cambrian to late Devonian and mid-Triassic to early Cretaceous times.   *GO*

[See also GEOLOGICAL RECORD OF ENVIRONMENTAL CHANGE, GREENHOUSE EFFECT, SEQUENCE STRATIGRAPHY]

Brenchley, P.J., Marshall, J.D., Carden, G.A.F. *et al.* 1994: Bathymetric and isotopic evidence for a short-lived Late Ordovician glaciation in a greenhouse period. *Geology* 22, 295–298. **Fischer, A.G.** 1981: Climatic oscillations in the biosphere. In Nitecki, M. (ed.), *Biotic crises in ecological and evolutionary time.* New York: Academic Press, 103–131. **Pope, M. and Read, J.F.** 1998: Ordovician metre-scale cycles: implications for climate and eustatic fluctuations in the central Appalachians during a global greenhouse, non-glacial to glacial transition. *Palaeogeography, Palaeoclimatology, Palaeoecology* 138, 27–42. **Sun, S.Q.** 1994: A reappraisal of dolomite abundance and occurrence in the Phanerozoic. *Journal of Sedimentary Research* A64, 396–404. **Veevers, J.J.** 1990: Tectonic climatic supercycle in the billion-year plate-tectonic eon – Permian Pangean icehouse alternates with Cretaceous dispersed-continents greenhouse. *Sedimentary Geology* 68, 1–16.

greenhouse effect The imperfect analogy between the Earth's atmosphere and a greenhouse. Incoming SOLAR RADIATION penetrates the atmosphere whereas outgoing, long-wave TERRESTRIAL RADIATION heats the atmosphere as it is absorbed by trace gases (GREENHOUSE GASES). This NATURAL greenhouse effect results in the air temperature close to the Earth's surface being some 21°C higher than it would otherwise be and should be differentiated from any *enhanced* greenhouse effect being produced by anthropogenic emissions of greenhouse gases.   *JAM*

[See also GLOBAL WARMING]

Jones, M.D.H. and Henderson-Sellers, A. 1990: History of the greenhouse effect. *Progress in Physical Geography* 14, 1–18.

greenhouse gases Atmospheric trace gases which allow short-wave SOLAR RADIATION to pass through unaf-fected, but which absorb long-wave TERRESTRIAL RADIA-TION. The net effect is to trap part of the *infrared radiation* that would otherwise escape to space and to warm the Earth's surface and lower ATMOSPHERE. Greenhouse gases naturally present in the ATMOSPHERE raise the average global surface temperature by about 21°C more than it would be if the natural greenhouse gases were not present. This effect is known as the GREENHOUSE EFFECT. Human activities are causing concentrations of some greenhouse gases to increase and this is causing an additional anthropogenic or enhanced greenhouse effect estimated by GENERAL CIRCULATION MODELS to be currently around 0.3°C per decade.

Greenhouse or radiatively active gases include principally WATER VAPOUR (H_2O) but also CARBON DIOXIDE (CO_2), METHANE (CH_4), NITROUS OXIDE (N_2O), OZONE (O_3), HALOGENATED HYDROCARBONS and halocarbon substitutes. Halogenated hydrocarbons include CHLOROFLUO-ROCARBONS (CFCs), bromofluorocarbons (BFCs), perfluorocarbons (PFCs), sulphur hexafluoride (SF_6), methyl chloroform (CH_3CCl_3) and carbon tetrachloride (tetrachloromethane – CCl_4). Halocarbon substitutes include HYDROCHLOROFLUOROCARBONS (HCFCs) and *hydrofluorocarbons* (HFCs). Carbon dioxide is produced by the combustion of FOSSIL FUELS and LANDUSE change (especially tropical DEFORESTATION). Methane is released by agriculture (rice paddies, animal husbandry), waste disposal (LANDFILL), BIOMASS BURNING and FOSSIL FUEL production and use. Nitrous oxide sources are mainly agriculture (development of pasture in tropical regions), biomass burning and some industrial processes (nitric acid production). Tropospheric ozone is a secondary pollutant formed by the action of sunlight on NITROGEN OXIDES and volatile organic compounds. The increase in carbon dioxide concentrations is held responsible for about two-thirds of the enhanced radiative forcing by greenhouse gases.   *DME*

[See also CARBON BALANCE]

Boag, S., White, D.H. and Howden, S.M. 1994: Monitoring and reducing greenhouse gas emissions from agricultural, forestry and other human activities. *Climatic Change* 27, 5–11. **Bolle, H.J., Seiler, W. and Bolin, B.** 1986: Other greenhouse gases and aerosols. In Bolin, B., Doos, B.R., Jager, J. and Warrick, R.A. (eds), *The greenhouse effect, climatic change and ecosystems* [*SCOPE Report* 29]. New York: Wiley. **Houghton, J.T.** 1997: *Global warming: the complete briefing*, 2nd edn. Oxford: Lion. **Houghton, J.T., Filho, L.G.M., Callander, B.A.** *et al.* (eds) 1996: *Climate change 1995: the science of climate change.* Cambridge: Cambridge University Press. **Raynaud, D., Barnola, J.-M., Chappellaz, J.** *et al.* 2000: The ice record of greenhouse gases: a view in the context of future changes. *Quaternary Science Reviews* 19, 9–17.

greening of society The realignment of societal values since the 1970s in response to the growing threat of ENVIRONMENTAL DEGRADATION on a global scale. The greening of society has been exemplified at many levels, but mainly through consumption habits, which, in response partly to government legislation and more to the influence of ENVIRONMENTAL MOVEMENTS and PRESSURE GROUPS, has trended towards low consumption and sustainable practices such as paper and glass RECYCLING.

The unprecedented success of 'environmentally friendly' produce and the encouragement of environmental awareness expressed through government policy, has added to the growing sense of responsibility felt by society for limiting further damage to the environment. In addition, environmental education has grown in status at all levels, with the next generation of adults being particular targets for extensive media coverage of issues such as DEFORESTATION and OZONE DEPLETION. Also, the expansion of environmental awareness into the political arena has seen the development of green legislation and ENVIRONMENTAL LAW to monitor compliance with regulations for potentially damaging practices.   *JGS*

[See also ENVIRONMENTALISM, GREEN POLITICS, SUSTAINABILITY]

Carley, M. and Spapens, M. 1998: *Sharing the world: sustainable living and global equality in the 21st century*. London: Earthscan. Sachs, W., Laske, R. and Linz, M. 1998: *Greening the north: a post-industrial blueprint for ecology and equity*. London: Zed Books.

greenstone belt   An area of deformed and metamorphosed sedimentary and volcanic rocks (including KOMATIITE lavas) of ARCHAEAN age. Many greenstone belts have been intruded by granite, forming *granite–greenstone belts*, which form the cores of many CRATONS. It is now generally agreed that greenstone belts formed through SEA-FLOOR SPREADING and provide evidence for the initiation and nature of PLATE TECTONICS on the early Earth, which may have operated in rather different ways than in the PROTEROZOIC and PHANEROZOIC (see UNIFORMITARIANISM). Greenstone belts demonstrate the existence in the Archaean of many of the aqueous processes and environments that operate today, and preserve the earliest traces of life on Earth in the form of STROMATOLITES and filamentous microfossils in CHERT, both with a record to 3500 million years ago.   *GO*

Coward, M.P. and Ries, A.C. (eds) 1995: *Early Precambrian processes* [*Special Publication* 95]. London: Geological Society. de Wit, M.J. 1998: On Archean granites, greenstones, cratons and tectonics: does the evidence demand a verdict? *Precambrian Research* 91, 181–226. Hollings, P., Wyman, D. and Kerrich, R. 1999: Komatiite–basalt–rhyolite volcanic associations in Northern Superior Province greenstone belts: significance of plume–arc interaction in the generation of the proto continental Superior Province. *Lithos* 46, 137–161. Windley, B.F. 1995: *The evolving continents*, 3rd edn. Chichester: Wiley.

greigite   Ferrimagnetic iron sulphide (Fe_3S_4) occurring in various forms in freshwater and BRACKISH or MARINE SEDIMENTS, as well as in GLEYSOLS and PEAT.   *UBW*

[See also LAKE SEDIMENTS, MAGNETIC INTENSITY, MAGNETIC SUSCEPTIBILITY, MAGNETITE, GEOMAGNETISM, PALAEOMAGNETISM]

Ariztegui, D. and Dobson, J. 1996: Magnetic investigations of framboidal greigite formation: a record of anthropogenic environmental changes in eutrophic Lake St.Mortiz, Switzerland. *The Holocene* 6, 235–241. Hilton, J. 1990: Greigite and the magnetic properties of sediments. *Limnology and Oceanography* 35, 509–520. Walden, J., Oldfield, F. and Smith, J. (eds) 1999: *Environmental magnetism, a practical guide* [*QRA Technical Guide* 6]. London: Quaternary Research Association.

Grentzhorizont   An abrupt reduction in *Sphagnum* DECOMPOSITION around 2500 radiocarbon years BP reported from northwestern and central European MIRES. Grentzhorizont was originally used to link peatlands from different geographical areas to the same fluctuations in cli-

mate, but PEAT stratigraphic evidence and RADIOCARBON DATING shows that the surfaces are neither as regular or as synchronous as was originally believed, and the term can be misleading.   *MJB*

[See also RECURRENCE SURFACE]

Weber, C.A. 1900: Über die Moore, mit besondere Berücksichtigung der zwischen Underweser und Underelbe liegenden. *Jahresbericht der Manner von Morgenstern* 3, 3–23.

grey-scale analysis   A measure of variations in the relative reflectivity of SEDIMENTS, normally obtained by scanning the freshly cut surface of a CORE. The digital records can be treated statistically and are a powerful tool in the analysis and CORRELATION of VARVES and other laminated sediments.   *MRT*

Hughen, K.A., Overpeck, J.T., Peterson, L.C. and Tumbore, S. 1996: Rapid changes in the tropical Atlantic region during the last deglaciation. *Nature* 380, 51–54.

greywacke   A largely outdated term for a WACKE (i.e. a MUD-rich SANDSTONE, with > 15% MATRIX). The term, sometimes spelled 'graywacke', became closely associated with sandstones interpreted as having been deposited by TURBIDITY CURRENTS and hence was too tainted with process implications for continued employment as a formal descriptive term.   *TY*

Dott Jr, R.H. 1964: Wacke, graywacke and matrix – what approach to immature sandstone classification? *Journal of Sedimentary Petrology* 34, 625–632. Dzulynski, S. and Walton, E.K. 1965: *Sedimentary features of flysch and greywackes* [*Developments in Sedimentology* 7]. Amsterdam: Elsevier.

grèzes litées   A regional name given to the type of rhythmically stratified slope-waste deposit, of PLEISTOCENE age, that occurs in the limestone region of Charente, France. The more general term for this type of sediment is *stratified slope deposit*. Stratified slope deposits occur mostly in mid-latitudes, but are also known to be forming today in cold humid regions. They are interpreted as the result of intense frost acting upon frost-susceptible bedrock assisted by snow-melt runoff processes. There has been some confusion with the term ÉBOULIS ORDONNÉS, which some view as synonymous with grèzes litées, but which originally referred to a coarse TALUS deposit formed of rockfall debris.   *HMF*

[See also PERIGLACIAL SEDIMENTS]

Dewolf, V. 1988: Stratified slope deposits. In Clark, M.J. (ed.), *Advances in periglacial geomorphology*. Chichester: Wiley, 91–110. Guillien, Y. 1951: Les grèzes litées de Charente. *Revue de Géographie des Pyrénées et du Sud-Ouest* 22, 154–162. Ozouf, J.C., Coutard, J.P. and Lautridou, J.P. 1995: Grèzes, grèzes litées: historique des définitions. *Permafrost and Periglacial Processes* 6, 85–87. Vansteijn, H., Bertran, P., Hetu, B. and Texier, J.P. 1995: Models for the genetic and environmental interpretation of stratified slope deposits – review. *Permafrost and Periglacial Processes* 6, 125–146.

grid   A square network formed by two sets of intersecting equidistant parallel straight lines. Grids are used as the basis of reference location system such as the UK Ordnance Survey National Grid or the US State Plane system.   *TF*

[See also GRID CELL]

grid cell   A single enclosed element in a GRID structure. Grid cells are associated with RASTER-type images, where a grid cell may equate to a PIXEL (picture element).   *TF*

[See also GRID]

grit   Formerly used in an ill-defined sense to describe a 'gritty' SANDSTONE; in most cases a tough, coarse-grained sandstone with angular CLASTS. The term is obsolete in a descriptive sense because of a lack of clarity and consistency in its use. However, it is retained in many formal names in STRATIGRAPHY (e.g. Ystrad Meurig Grits FORMATION).   *GO*

gross primary productivity (GPP)   The total fixation rate of energy by PHOTOSYNTHESIS. GPP is the production rate before respiratory heat losses are accounted; it is the *photosynthetic production rate* for plants and *metabolisable production rate* for animals.   *RJH*

[See also NET PRIMARY PRODUCTIVITY, PRODUCTIVITY]

ground ice   A general term used to refer to all types of ice formed, or preserved, in freezing and frozen ground. There are many types of ground ice. Ground ice is one of the most important components of PERMAFROST and is the main cause of THERMOKARST.   *HMF*

Associate Committee on Geotechnical Research 1988: *Glossary of permafrost and related ground-ice terms* [*Technical Memorandum* 142]. Ottawa: Permafrost Subcommittee, National Research Council of Canada.

ground-based remote sensing   REMOTE SENSING of environmentally relevant variables carried out from the ground. Ground-based remote sensing of the atmosphere with LIDARS, RADARS, *radiometers* and other instruments is used to retrieve the distribution and profile of different atmospheric quantities such as temperature, precipitation, water vapour, oxygen, ozone and AEROSOLS. Ground-based remote sensing is also relevant for the establishment of useful relationships between specific terrestrial parameters and the energy received by OPTICAL, THERMAL or MICROWAVE REMOTE SENSING instruments and for the proper choice of the instrument's parameters (e.g. wavelength, incidence angle and polarisation) to be used from space.   *TS*

Schanda, E. 1986: *Physical fundamentals of remote sensing.* Heidelberg: Springer. **Ulaby, F.T., Moore, R.K. and Fung, A.K.** 1981: *Microwave remote sensing: active and passive.* Norwood, MA: Artech House.

ground measurement   Measurement of variables on the Earth's surface in order to calibrate data derived by REMOTE SENSING. Such measurements should be taken at the same time as the data acquisition from the sensor, at accurate sampling locations.   *HB*

ground-penetrating radar (GPR)   A method of GEOPHYSICAL SURVEYING that uses electromagnetic radiation to provide detailed images of the subsurface structure of SOILS and SEDIMENTS to depths of several metres.   *GO*

[See also RADAR REMOTE SENSING INSTRUMENTS]

Bridge, J.S., Alexander, J., Collier, R.E.Ll. *et al.* 1995: Ground-penetrating radar and coring used to study the large-scale structure of point-bar deposits in three dimensions.

Sedimentology 42, 839–852. **Bristow, C.S., Skelly, R.L. and Ethridge, F.G.** 1999: Crevasse splays from the rapidly aggrading, sand-bed, braided Niobrara River, Nebraska: effect of base-level rise. *Sedimentology* 46, 1029–1047. **Davis, J.L. and Annan, A.P.** 1989: Ground penetrating radar for high resolution mapping of soil and rock stratigraphy. *Geophysical Prospecting* 37, 531–551.

ground temperature   The temperature (1) at the ground surface or (2) near the ground surface at specified shallow depths. At climatological stations minimum temperatures are measured at the ground surface: *grass minimum temperature* is recorded a few millimetres above a grass surface. At such a height, temperatures below freezing are known as *ground frosts*. In addition, *soil temperatures* are recorded at depths of 50 mm, 100 mm and 200 mm.

Ground temperature may differ considerably from the AIR TEMPERATURE above, which may be a poor predictor of it, even though smoothing of data from daily to monthly scales may improve the correlation between the two. Ground temperatures tend to be more extreme and variable than the corresponding air temperature. *Mean annual ground temperature (MAGT)* estimated from measurements based on a thermistor cable to 10 m depth in an area of MOUNTAIN PERMAFROST in southern Norway was about 2.5°C lower than the MEAN ANNUAL AIR TEMPERATURE (MAAT), but this may vary greatly between sites as a result of such factors as vegetation cover, topography and snow depth.   *JGT/JAM*

[See also MICROCLIMATE, SUBSURFACE TEMPERATURE]

Ødegård, R.S., Sollid, J.L. and Liestøl, O. 1992: Ground temperature measurements in mountain permafrost, Jotunheimen, southern Norway. *Permafrost and Periglacial Processes* 3, 231–234. **Thorn, C.E., Schlyter, J.P.L., Darmody, R.G. and Dixon, J.C.** 1999: Statistical relationships between daily and monthly air and shallow-ground temperatures in Kärkevagge, Swedish Lapland. *Permafrost and Periglacial Processes* 10, 317–330.

grounding line   The boundary between the part of an ICE SHEET that is floating and the part that is in contact with bedrock: it lies below sea level.   *JAM*

[See also GLACIOMARINE DEPOSITS]

groundwater   The part of the HYDROSPHERE beneath the ground surface. Groundwater, under half of which is *fresh groundwater*, comprises about 1.6% of all the Earth's water. Fresh groundwater comprises about 30% of the Earth's freshwater (excluding frozen groundwater in PERMAFROST environments). Most is contained within PERMEABLE rocks (AQUIFERS) in the saturated or PHREATIC ZONE, where it forms contiguous bodies of water that move. Groundwater is a geological and geomorphological agent (particularly apparent in KARST landscapes), is an important supplier of BASEFLOW to streams and rivers, is an important *reservoir* or *store* within the global HYDROLOGICAL CYCLE and may have a considerable age in deep aquifers (commonly hundreds of years and in some cases thousands of years or more) but often a much shorter residence time in shallow aquifers (especially in relatively wet environments).

Groundwater is a major source for urban and rural water supplies and hence is a major natural RESOURCE. In the developed world, URBANISATION is a major cause of

287

GROUNDWATER DEPLETION through excessive ABSTRACTION, which often leads to a lowering of the WATER TABLE. The impact and use of water in cities in ARIDLANDS, in contrast and somewhat perversely, may lead to a localised rise in the water table as surplus water descends through the VADOSE ZONE to the phreatic zone by PERCOLATION. Elsewhere in aridlands, however, abstraction of groundwater for IRRIGATION generally leads to a lowering of the water table. Natural recharge of the depleted aquifers (*groundwater recharge*) may take a very long time to the extent that, in some areas at least, it should be regarded as a NON-RENEWABLE RESOURCE. In coastal areas, abstraction may also lead to SALTWATER INTRUSION and *groundwater salinisation*. In areas affected by AGRICULTURAL INTENSIFICATION, groundwater POLLUTION from FERTILISERS is an additional concern. *JAM/RPDW*

[See also ARTESIAN, DRY VALLEYS, GROUNDWATER MINING, HYDROGEOLOGY, METEORIC WATER]

Brown, A.G. (ed.) 1995: *Geomorphology and groundwater*. Chichester: Wiley. **Domenico, P.A. and Schwartz, P.W.** 1990: *Physical and chemical hydrogeology*. New York: Wiley. **Price, M.** 1996: *Introducing groundwater*, 2nd edn. London: Chapman and Hall.

groundwater depletion   Reduction in the volume of water stored in an AQUIFER and consequent lowering of the WATER TABLE through water ABSTRACTION. Groundwater depletion may be the direct effect of *groundwater abstraction* or a more complex, indirect effect of the abstraction of river water, or the result of LANDUSE CHANGE or a drier climate, leading to reduced *groundwater recharge*. It results, therefore, when abstraction and natural losses to rivers, the sea or adjacent basins exceed groundwater recharge (from PERCOLATION of soil water or infiltration of river water into beds and banks). *JAM/RPDW*

[See also GROUNDWATER MINING]

groundwater mining   Excessive groundwater ABSTRACTION, used in particular in situations where (1) GROUNDWATER recharge fails to keep pace with current abstraction and hence GROUNDWATER DEPLETION occurs and/or (2) the groundwater is ancient and hence it can be considered as a NON-RENEWABLE RESOURCE. The Ogallala Aquifer 1800 m beneath the Great Plains, USA, provides a good example of groundwater mining because the abstraction rate appears to be 100 times the current recharge rate and much of the water in the aquifer dates from the LAST GLACIATION. The term is increasingly used in ARIDLANDS and SEMI-ARID regions, such as the Middle East, where groundwater abstraction is increasing and recharge is slow. *JAM*

Sloggett, G. and Dickason, C. 1986: *Groundwater mining in the United States*. Washington, DC: Government Printing Office.

group   A unit in LITHOSTRATIGRAPHY comprising two or more FORMATIONS that have similar characteristics of LITHOLOGY or origin (e.g. Torridon Group, Lias Group). In SEQUENCE STRATIGRAPHY, a group may link FORMATIONS between major bounding UNCONFORMITIES. *LC*

Woodcock, N.H. 1990: Sequence stratigraphy of the Palaeozoic Welsh Basin. *Journal of the Geological Society, London* 147, 537–547.

growan   *In situ* decomposed granite as found on Dartmoor, southwest England. It is also referred to as *gruss*. *PW*

[See also GRUSSIFICATION, SAPROLITE]

Ballantyne, C.K. and Harris, C. 1994: *The periglaciation of Great Britain*. Cambridge: Cambridge University Press.

growing season   The period of a year when plants will grow, commonly operationalised as the period when mean daily temperatures exceed the temperature at which plants will grow. It determines the mix of natural vegetation and types of crop that will grow. In middle and high latitudes, the growing season is commonly limited by spring and autumn frosts. In DRYLANDS, moisture may be the LIMITING FACTOR. *RJH*

[See also PHENOLOGY, PHOTOSYNTHESIS]

growth form   A group of organisms (usually plants) classified according to their morphological characteristics. Examples include broad-leaved evergreen trees, palms, succulents, dwarf shrubs and cushions. Vegetation classification based on growth form or LIFE FORM is complementary to taxonomic classification. *JLI*

[See also LIFE FORM]

groyne   A structure built perpendicular or subperpendicular to a SHORELINE to intercept sediment moving laterally along the beach by LONGSHORE DRIFT. Groynes are built to reduce EROSION and/or to widen a beach. There is often enhanced erosion beyond the downdrift end of a groyne (American spelling *groin*) or series of groynes. *HJW*

[See also JETTY, COASTAL ENGINEERING STRUCTURES]

Grundhoecker   A German term coined by Julius Büdel to describe a joint-controlled convexity or *basal knob* of the basal surface of weathering, forming a shield INSELBERG with knobbly small-scale relief when cropping out at the surface of the weathering mantle. *MAC*

Büdel, J. 1982: *Climatic geomorphology*, translated by Fischer, L. and Busche, D. Princeton, NJ: Princeton University Press.

grussification   The process of *in situ* granular disintegration of coarse-grained crystalline rock (e.g. granite) with little or no decomposition. *PW*

[See also GROWAN, SAPROLITE]

Gulf Stream   A warm OCEAN CURRENT that originates in the eastern Gulf of Mexico, travels north along the east coast of the USA to about 40°N, then crosses the Atlantic Ocean to reach the British Isles at 50°N as the *North Atlantic Drift* (NAD), which flows on to the northern reaches of the Atlantic. The Gulf Stream and its continuation as the NAD has a considerable influence on the climate of Western Europe. *GS*

Duplessy, J.-C. 1999: Climate and the Gulf Stream. *Nature* 403, 594–595. **Lynch-Stieglitz, J., Curry, W.B. and Slowey, N.** 1999: Weaker Gulf Stream in the Florida Straits during the Last Glacial Maximum. *Nature* 402, 644–648.

gull   An expanded bedrock joint, parallel to an escarpment, that results from CAMBERING of valley-side caprocks. *PW*

Ballantyne, C.K. and Harris, C. 1994: *The periglaciation of Great Britain.* Cambridge: Cambridge University Press.

gully A steep-sided ephemeral channel, often formed in poorly consolidated sediments or soils by concentrated *overland flow* or ephemeral streams. On agricultural land the term is used for features that are too large to be removed by PLOUGHING. Gullies are often initiated by the reduction or complete removal of protective vegetation and may enlarge rapidly and extend to form gully networks. *SHD*

[See also ARROYO, BADLANDS, DONGA, EROSION, HEADCUT EROSION, RILL, WADI]

Bocco, G. 1991: Gully erosion – processes and models. *Progress in Physical Geography* 15, 392–406. **Heede, B.H.** 1976: *Gully development and control: the status of our knowledge* [*USDA Forest Service Research Paper* RM-196]. Fort Collins, CO: US Department of Agriculture.

gully control dam A small structure, usually up to about 2 m in height, constructed from wood, rocks, GABIONS or other materials and aligned across a GULLY to trap sediment and control EROSION. *SHD*

Heede, B.H. 1976: *Gully development and control: the status of our knowledge* [*USDA Forest Service Research Paper* RM-196]. Fort Collins, TX: US Department of Agriculture.

guyot A flat-topped SEAMOUNT. The flat top is considered to have been eroded by waves when the guyot was an island, close to the MID-OCEAN RIDGE, that subsided and moved away from the ridge due to SEA-FLOOR SPREADING. *CDW*

[See also ATOLL]

gymnosperms A subdivision of the Spermatophyta (seed plants), which literally means 'naked' (gymno) 'seed' (sperm), in which the seeds are borne in an exposed position on the cone scale or equivalent structure. Originating at least 350 million years ago, they dominated the vegetation on Earth before the origination of ANGIOSPERMS (flowering plants). The most successful today are the coniferous trees of the BOREAL FOREST and certain other environments. *JCMc*

gypcrete A GYPSUM-cemented DURICRUST, largely confined to very arid regions where the mean annual precipitation is below 250 mm. *MAC*

[See also CALCRETE, FERRICRETE, SILCRETE]

Watson, A. 1983: Gypsum crusts. In Goudie, A.S. and Pye, K. (eds), *Chemical sediments and geomorphology: precipitates and residua in the near-surface environment.* London: Academic Press.

gypsic horizon A SOIL HORIZON with an accumulation of secondary gypsum as thin, filamentous *pseudomycelia*, separate crystals or *crystalaria* that is at least 15 cm thick and contains 5% more gypsum than the underlying parent material. When continuously cemented it is a *petrogypsic horizon* or GYPCRETE. *EMB*

Food and Agriculture Organization of the United Nations 1998: *World reference base for soil resources* [*Soil Resource Report* 84]. Rome: FAO, ISRIC, ISSS.

Gypsisols Soils with a secondary accumulation of gypsum in a GYPSIC HORIZON or *petrogypsic horizon*, char-

acteristic of arid areas in North Africa and the Middle East (SOIL TAXONOMY: *Gypsiorthids*). There are two potential sources for gypsum in soils, the most common of which is for it to be precipitated from gypsum-rich GROUNDWATER; the alternative source is as DUST blown out of sedimentary basins, as for example, the Great Chott Lake in Tunisia. IRRIGATION of gypsisols can lead to solution of gypsum and collapse of irrigation structures. *EMB*

[See also WORLD REFERENCE BASE FOR SOIL RESOURCES]

Alphen, J.G. and de los Rios, F. 1991: *Gypsiferous soils.* Wageningen: International Institute for Land Reclamation and Improvement (ILRI). **Herrero, J. and Poch, R.M. (eds)** 1998: Soils with gypsum. *Geoderma* 87, 1–135 [Special Issue]. **Porta, J. and Herrero, J.** 1990: Micromorphology and genesis of soils enriched with gypsum. In Douglas, L.A. (ed.), *Soil micromorphology: a basic and applied science* [*Developments in Soil Science* 19]. Amsterdam: Elsevier. **Watson, A.** 1983: Gypsum crusts. In Goudie, A.S. and Pye, K. (eds), *Chemical sediments in geomorphology.* London: Academic Press, 133–161.

gypsum A widely distributed soft mineral consisting of hydrous calcium sulphate ($CaSO_4.2H_2O$). It is the most common sulphate mineral and is frequently associated with halite and anhydrite in EVAPORITES, forming thick, extensive beds interstratified with limestone, shale and clay. Where evaporites of gypsum harden, they are termed GYPCRETE. *MAC*

gyre A large-scale ocean circulatory system generated by the *atmospheric circulation*. Gyres circulate clockwise in the Northern Hemisphere and counterclockwise in the Southern Hemisphere. The most important are the five, mid-ocean *subtropical gyres*, each bounded by an *equatorial current*, a *western boundary current* and an *eastern boundary current*, and located between latitudes 20° and 40°N or S. The position of the oceanographic boundary between the North Atlantic subtropical gyre and subpolar North Atlantic water (the OCEANIC POLAR FRONT) is particularly sensitive to GLACIAL–INTERGLACIAL CYCLES and shorter-term CLIMATIC CHANGES. *Subpolar gyres* exist only in the North Atlantic and North Pacific Oceans. *JAM*

[See also OCEAN CURRENTS]

Pediosky, J. 1990: The dynamics of the oceanic subtropical gyres. *Science* 248, 316–322.

gyttja A term introduced by H. von Post to describe a LACUSTRINE SEDIMENT formed mainly under ANAEROBIC and EUTROPHIC conditions and with an ORGANIC CONTENT of > 30%. It consists of a mixture of ALLOCHTHONOUS and decomposed plant and animal fragments, minerogenic particles (see MINEROGENIC SEDIMENT) and AUTOCHTHONOUS material. *Coarse detrital gyttja* is rich in MACRO- and MICROFOSSILS and occurs in shallow water, whereas homogenous *fine detrital gyttja* is deposited in deeper water. *Drift gyttja* with larger, rounded plant fragments and sand is found close to the shore. *Algal gyttja* is mainly composed of AUTOCHTHONOUS algae detritus and to a minor extent of minerogenic particles and forms in shallow, high-productivity lakes. *Calcareous gyttja*, which is rich in shell fragments, has a $CaCO_3$ content of 20–80% and deposits in shallow water. *Shell gyttja* is mainly composed of shell fragments, but may have a matrix of algae or fine detritus.

Gyttja clay has an organic CARBON CONTENT of 3–6%, which rises to 6–30% in *clay gyttja*. *UBW*

[See also DETRITUS, DIATOMITE, DY, MARL, SEDIMENT TYPES]

Lundqvist, G. 1925: Utvecklingshistoriska insjöstudier i Sydsverige. *Sveriges Geologiska Undersöknings Årsbok* 18, 4–129.
Post, H. von 1862: Studier öfver Nutidens koprogena Jordbildningar, Gyttja, Dy, Torf och Mylla. *Kungliga Svenska Vetenskapsakademiens Handlingar* 4(B), 1. Stockholm.

H

habitat The physical space, place, or ECOTOPE, occupied by organisms during one or more parts of their life cycle. Habitats are characterised by distinct suites of ENVIRONMENTAL FACTORS. *Habitat structure* refers to the physical arrangement or 'architecture' of the environment of organisms. *JLI*

[See also HABITAT DEGRADATION, HABITAT LOSS, NICHE]

Bell, S.S., McCoy, C.D. and Mushinsky, H.R. 1991: *Habitat structure: the physical arrangement of objects in space.* London: Chapman and Hall. **Morrison, M.L., Marcot, B.G. and Mannan, R.W.** 1992: *Wildlife–habitat relationships: concepts and applications.* Madison, WI: University of Wisconsin Press.

habitat degradation Decline in the quality and distribution (FRAGMENTATION) of BIOTIC environments. It is measured as a quality reduction in ecosystem structural (e.g. BIODIVERSITY, VEGETATION, SOILS) and functional (e.g. ENERGY FLOW, NUTRIENT cycling) components. *GOH*

[See also HABITAT LOSS]

Jeffries, M.J. 1997: *Biodiversity and conservation.* London: Routledge.

habitat island A patch or fragment of a formerly more extensive and relatively continuous ecosystem. Habitat islands may be formed by natural processes, such as those associated with ENVIRONMENTAL CHANGE during the Pleistocene, or by anthropogenic activity in more recent times. They include, for example, isolated woodlands in a 'sea' of grassland, and mountains surrounded by lowland. *MVL*

[See also CONSERVATION BIOLOGY, FRAGMENTATION, ISLAND BIOGEOGRAPHY, SINGLE LARGE OR SEVERAL SMALL RESERVES]

habitat loss Reduction in the area of NATURAL and SEMI-NATURAL ecosystems as a result of human activities, especially through FRAGMENTATION. This reduces the number and diversity of habitats available for NATIVE species, is a major factor in the loss of BIODIVERSITY and may also precipitate BIOLOGICAL INVASIONS by ALIEN SPECIES. An example is shown in the Figure, which highlights the rapid loss of hedges, woodland patches and individual trees from an area of The Netherlands over a 20-year interval. *JAM*

[See also DEFORESTATION, HEDGEROW REMOVAL, LANDUSE CHANGE, WETLAND CONSERVATION]

Arnold, R. and Villain, C. 1990: *New directions for European agricultural policy* [CEPS Paper 49]. Brussels: Centre for European Policy Study (CEPS). **Brothers, T.S. and Spingarn, A.** 1992: Forest fragmentation and alien plant invasion of central Indiana old-growth forests. *Conservation Biology* 6, 191–200.

hadal Depths and environments in the OCEAN deeper than ABYSSAL (i.e. > 6000 m), corresponding principally to OCEANIC TRENCHES. The term is derived from the French, *Hadés* (hell). *BTC*

[See also OCEAN BASIN]

Hadley cell A direct, thermally driven, meridional circulation, which consists of rising air in low latitudes, at the INTERTROPICAL CONVERGENCE ZONE, and sinking air in the subtropics. Part of the GENERAL CIRCULATION OF THE ATMOSPHERE and of MACROCLIMATE, it was first proposed by George Hadley in AD 1735. Poleward transport aloft carries latent and sensible heat from the tropics to mid-latitudes. There is one Hadley cell in each hemisphere and their intensity and position can vary seasonally, with the one in the winter hemisphere being the stronger. Based on zonally averaged observations, a thermally indirect *Ferrel cell* (named after William Ferrel) is also recognised polewards of each Hadley cell with ascending air in the mid-latitude region of the POLAR FRONT. *AHP*

[See also ANTITRADES, TRADE WINDS]

Hadley, G. 1735: Concerning the cause of the general trade wind. *Philosophical Transactions of the Royal Society of London* 39, 58–73. **Hastenrath, S.L.** 1968: On mean meridional circulations in the tropics. *Journal of Atmospheric Science* 25, 979–983.

haematite An iron oxide(α-Fe_2O_3) found in oxidised igneous rocks and sediments formed under oxidising conditions, often giving a blood-red colour. Adjacent iron layers in haematite are coupled antiferromagnetically, resulting in imperfect antiferromagnetic behaviour and low values of MAGNETIC SUSCEPTIBILITY between c. 0.3 and 2.0×10^{-6} m^3 kg^{-1}. *AJP*

half life A measure of the rate of disintegration of a RADIONUCLIDE or radioactive ISOTOPE. It is the time taken for a quantity of a radioactive isotope to reduce by a half by *radioactive decay*. Radioactive decay is exponential; after two half lives a quarter of the original remains, and after 10 half lives 0.1% remains. Half lives of individual radionuclides vary from small fractions of a second to 10^{10} years. *PQD*

[See also BIOLOGICAL HALF LIFE, LIBBY HALF LIFE]

halocline A steep gradient in salinity in the ocean. There is a halocline in the upper part of the Arctic Ocean where relatively fresh water overlies saltier water. *JAM*

halogenated hydrocarbons (halocarbons) Hydrocarbons into which one or more of the *halogen* elements – chlorine (Cl), bromine (Br) or fluorine (Fl) – have been introduced. In *chlorinated hydrocarbons*, for example, the halogens are *added* to hydrocarbon molecules. In CHLOROFLUOROCARBONS (CFCs – or *freons*), *bromofluorocarbons* (*halons*), *methyl chloride* (CH_3Cl) and *methyl bromide* (CH_3Br), the halogens *replace* hydrogen atoms.

Both categories are important POLLUTANTS: chlorinated hydrocarbons have been extensively used as synthetic PESTICIDES; CFCs, widely used in refrigerants and air-conditioning systems, are important GREENHOUSE GASES and probably the major cause of OZONE DEPLETION; halons, used mainly in fire extinguishers are actually more effec-

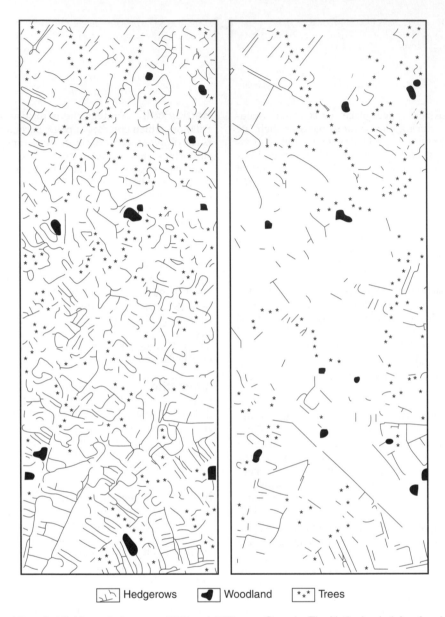

Hedgerows ⬛ Woodland `*.*` Trees

habitat loss *Loss of three habitat types between* AD *1950 and 1970 near Groenlo, The Netherlands (after Arnold and Villain, 1990)*

tive than freons in the breakdown of ozone; methyl bromide, used as a fumigant in the fruit and vegetable industry since the 1960s, has a comparable *ozone depletion potential* (ODP) to many CFCs.

Measurements of trace gases trapped in FIRN (consolidated snow) from sites in the Antarctic and Greenland have demonstrated that natural sources of chlorofluorocarbons, bromofluorocarbons and most other halogenated hydrocarbons are minimal or non-existent. Reconstructions from these firn samples of atmospheric concentrations back to the late nineteenth century are consistent with anthropogenic emission rates and their known PERSISTENCE in the atmosphere. The firn samples also show that methyl chloride and methyl bromide were the only species of halogenated hydrocarbons to exhibit unequivocal natural BACKGROUND LEVELS prior to the twentieth cen-

tury. It appears that most methyl chloride in the atmosphere is of natural origin, whereas only half of the methyl bromide can be attributed to natural sources. *JAM*

[See also CONVENTION, HYDROCHLOROFLUOROCARBONS, PERSISTENT ORGANIC COMPOUNDS]

Butler, J.H., Battle, M., Bender, M.L. *et al.* 1999: A record of atmospheric halocarbons during the twentieth century from polar firn air. *Nature* 399, 749–755. **Lovelock, J.E., Maggs, R.J. and Wade, R.J.** 1973: Halogenated hydrocarbons in and over the Atlantic. *Nature* 241, 194–196.

halophyte A salt-tolerant plant species defined as having an ability to grow and complete its life cycle at salt concentrations in excess of 100–200 mM NaCl. Halophytes are commonly found in SALTMARSH, DRYLANDS and SALINE LAKE environments. *MLC*

Flowers, T.J., Hajibaghri, M.A. and Clipson, N.J.W. 1986: Halophytes. *Quarterly Review of Biology* 61, 313–317. **Reimold, R.J. and Queen,W.H. (eds)** 1974: *Ecology of halophytes.* New York: Academic Press.

haloturbation   The disturbance of soils or sediments by salt crystal growth.   *JAM*

hanging valley   In glaciated terrain, a tributary valley, the mouth of which ends abruptly part way up the side of a trunk valley, formed as a result of the greater amount of GLACIAL EROSION of the latter.   *MJH*

haptic interface   A device that allows a user to interact with a computer-generated virtual world by receiving tactile feedback; it orients users to the location and nature of objects in a virtual space.   *ACF*

Haralick texture measures   A means of quantifying the spatial relationships among grey levels (or IMAGE TEXTURE) in a digital image, such as those formed by REMOTE SENSING. These measures, which are one of many methods used to quantify image texture, are derived from the grey-level co-occurrence matrix of the input image and can be used successfully as new features for IMAGE CLASSIFICATION.   *ACF*
Haralick, R.M., Shanmugam, K. and Dinstein, I. 1973: Textural features for image classification. *IEEE Transactions on Systems, Man and Cybernetics* 3(6), 610–621.

hard engineering   The emplacement of structures (e.g. SEAWALLS, GROYNES) along a shoreline with the intention of stabilising the coastline.   *HJW*
[See also COASTAL ENGINEERING STRUCTURES, COASTAL (SHORE) PROTECTION, SOFT ENGINEERING]
Walker, H. J. (ed.) 1988: *Artificial structures and shorelines.* Dordrecht: Kluwer.

hardpan   A soil horizon or layer that has become hardened by CEMENTATION of the soil particles by organic matter, silica, sesquioxides or calcium carbonate. The hardness does not change with varying moisture content and fragments do not disintegrate when placed in water. Development of hardpans may follow FOREST CLEARANCE on sandy soils that subsequently are podzolised with an *iron–humus pan*. In Alaska, development of a hardpan has led to vegetation change as woodland turns to MUSKEG.   *EMB*
[See also FRAGIPAN, INDURATION, PODZOLISATION]
Chartres, C.J. 1985: A preliminary investigation of hardpan horizons in north-west New South Wales. *Australian Journal of Soil Research* 23, 325–337. **Thompson, C.H., Bridges, E.M. and Jenkins, D.A.** 1993: An exploratory examination of relict hardpans in the coastal lowlands of southern Queensland. In Ringrose-Voase, A.J. and Humphreys, G.S. (eds), *Soil micromorphology: studies in management and genesis.* Amsterdam: Elsevier, 233–245. **Ugolini, F.C. and Mann, D.H.** 1979: Biopedological origin of peatland in south east Alaska. *Nature* 281, 366–368. **Van Dijk, D.C. and Beckmann, G.G.** 1978: The Yeluba hardpan, and its relationship to soil geomorphic history, in the Yeluba-Tara region, south-east Queensland. In Langford-Smith T. (ed.), *Silcrete in Australia.* Armidale, New South Wales: Department of Geography, University of New England, 73–91.

hard-water effect/error   This effect or error in a radiocarbon date results from dating material derived from aquatic plants that have assimilated 'dead' (inert) carbon that has dissolved in the lake water from ancient carbonate-bearing rocks (e.g. limestone). The contemporary ^{14}C is diluted and the resulting age is too great. Lake mud (GYTTJA) contains decayed plant material from macrophytes and phytoplankton that have utilised 'dead' carbon to a greater or lesser degree. The 'dead' carbon effect can be estimated by measuring ^{14}C activity in lake water and aquatic plants. Submerged plants are more likely to have a reservoir age than emergent aquatics that photosynthesise atmospheric CO_2. A correction can be made for dates from a lake whose modern reservoir age has been estimated. The effect may be avoided by ACCELERATOR MASS SPECTROMETRY DATING terrestrial MACROFOSSILS that have been incorporated in the gyttja, ensuring that no aquatic material is included.   *HHB*
[See also RADIOCARBON DATING, MARINE RESERVOIR EFFECT]
Andrée, M., Oeschger, H., Siegenthaler, U. et al. 1986: ^{14}C dating of plant macrofossils in lake sediment. *Radiocarbon* 28, 411–416. **Deevey, E.S., Gross, M.S., Hutchinson, G.E. and Henry, L.** 1954: The natural C^{14} content of materials from hardwater lakes. *Proceedings of the National Academy of Sciences* 40, 285–288. **Olsson, I.U.** 1986: Radiometric dating. In Berglund, B.E. (ed.), *Handbook of Holocene Palaeoecology and Palaeohydrology.* Chichester: Wiley, 273–312.

harvest records   Quantitative or qualitative records of harvest yields for crops, vineyards or fruit that are of potential use as PROXY DATA for reconstructions of past climate. Harvest yields fluctuate from year to year and these fluctuations can be interpreted as an expression of the yield forming factors that mould them. Yield can be seen as a function of work, soil, technology, capital and climate. Although the weighting of these factors can vary greatly both in space and through time, climate plays an important role. It becomes noticeable in the INTERANNUAL VARIATIONS, while economic innovations tend to become effective in the longer term and changes in field boundaries must be taken into consideration. In addition, if the yields in MARGINAL AREAS decrease, then fields may be abandoned, as happened for example at northerly latitudes and high altitudes in the Alps and other mountain areas in the LITTLE ICE AGE. The decoding of fluctuations is a difficult operation, which must be solved by a variety of statistical methods. Results both between and within studies do not always agree. Inconsistencies in the flora–climate relationships can, however, often be explained by regional or even local models.   *RG*
Neumann, J. and Sigrist, R.M. 1978: Harvest dates in ancient Mesopotamia as possible indicators of climatic variations. *Climatic Change* 1, 239–252. **Oram, P.A.** 1985: Sensitivity of agricultural production to climatic change. *Climatic Change* 7, 129–152. **Parry, M.L.** 1981: Climatic change and the agricultural frontier: a research strategy. In Wigley, T., Ingram, M. and Farmer, G. (eds), *Climate and history: studies in past climates and their impact on man.* Cambridge: Cambridge University Press, 319–376. **Rowntree, L.B.** 1985: A crop-based rainfall chronology for preinstrumental record in southern California. *Climatic Change* 7, 327–341.

harvesting   Exploitation of biological RENEWABLE RESOURCES as practised, for example, in agriculture, forestry (SILVICULTURE), fish farming (AQUACULTURE), WHALING AND SEALING, and GAME MANAGEMENT. A central idea is that of maximum SUSTAINABLE YIELD, which, if exceeded, will lead to population decline and possible harvest failure. *JAM*

hazard A phenomenon that presents a RISK of harm or damage to humans, other organisms or their environment. *JAM*

[See also DISASTER, NATURAL HAZARD, RAPID-ONSET HAZARD]

hazardous substance A substance with properties capable of causing harm to humans or other organisms, or damage to ecosystems. Such substances may be harmful because they are toxic, flammable, explosive, corrosive or have high chemical reactivity. They are often WASTE products. In the USA, the most important sources of HAZARDOUS WASTE, in decreasing order by volume, are: industrial organics; chemical manufacturing; petroleum refining; explosives manufacture; plastics and resins; refuse, agricultural chemicals; inorganic pigments; and alkaline substances. Hazardous WASTE MANAGEMENT may involve recovery and re-use, incineration, detoxification, BIOREMEDIATION, CHEMICAL REMEDIATION or long-term storage. *JAM*

[See also RADIOACTIVE WASTE]

Nemerow, N.L. and Dasgupta, A. 1991: *Industrial and hazardous waste.* New York: Van Nostrand.

hazardous waste As new chemical compounds are being continuously produced, it is impossible to produce a definitive list of hazardous wastes. Hazardous waste may be defined broadly as any waste substance with the potential to have an adverse effect upon human beings, plant and animal life or the environment. Throughout the nineteenth century, before the rapid growth of the chemical industry, hazardous waste was viewed as a localised issue. It became a 'second-generation' environmental concern in all sectors of industrialised societies by the late 1970s. This can be attributed partly to the number and diversity of potentially hazardous substances in common use: approximately half of the c. 70 000 synthetic chemicals in common use have some associated characteristic (e.g. toxicity, flamability or corrosivity) that renders them hazardous. Other factors include their association with some of the most feared HUMAN HEALTH concerns (such as cancer and birth defects) and the CHEMICAL TIMEBOMB effect stemming from the lack of WASTE MANAGEMENT in the past.

Categories of hazardous waste include: inorganic acids; organic acids; alkalis; toxic metal compounds; elemental metals; metal oxides; inorganic compounds; other inorganic materials, e.g. asbestos; organic compounds; polymeric materials; fuel, oil and greases; pharmaceutical chemicals and biocides; miscellaneous chemical wastes; filter materials; treatment sludge and contaminated rubbish; tars, paints, dyes and pigments from interceptor pits; wastes from tanneries and fellmongers, cellulose and paper production; timber preservatives; soap and detergents; and animal processing wastes. *EMB/JAM*

[See also INDUSTRIAL WASTE, PERSISTENT ORGANIC COMPOUNDS, RADIOACTIVE WASTE]

Budd, W.W. 1999: Hazardous waste. In Alexander, D.E. and Fairbridge, R.W. (eds), *Encyclopedia of environmental science.*

Dordrecht: Kluwer, 311–312. **Cope, C.B., Fuller, W.H. and Willetts, S.L.** 1983: *The scientific management of hazardous wastes.* Cambridge: Cambridge University Press. **Department of the Environment** 1976: *Waste management paper* 4. London: HMSO. **Epstein, S., Brown, L. and Pope, C.** 1982: *Hazardous waste in America.* San Francisco: Sierra Club Books. **Nemerow, N.L. and Dasgupta, A.** 1991: *Industrial and hazardous waste.* New York: Van Nostrand Reinhold.

head A term originally used to describe relict PERIGLACIAL MASS MOVEMENT deposits (typically DIAMICTONS) capping many coastal cliff sections in parts of southwest England. Head is usually regarded as a term restricted to SOLIFLUCTION, but it has been broadened to include a wider range of slope mass movement processes (COLLUVIUM). In North America, head deposits are included within the more general term '*surficial deposits*', used to describe all unconsolidated Quaternary-age deposits. *HMF*

[See also PERIGLACIAL SEDIMENTS]

Dines, H.G., Hollingworth, S.E., Edwards, W. *et al.* 1940: The mapping of head deposits. *Geological Magazine* 77, 198–226. **Harris, C.** 1987: Solifluction and related periglacial deposits in England and Wales. In Boardman, J. (ed.), *Periglacial processes and landforms in Britain and Ireland.* Cambridge: Cambridge University Press, 209–224. **Harris, C.** 1998: The micromorphology of paraglacial and periglacial slope deposits: a case study from Morfa Bychan, west Wales, UK. *Journal of Quaternary Science* 13, 78–84.

headcut erosion Water EROSION at the upslope limit of a GULLY system, resulting in a steep wall, which gradually migrates upslope. Headcut erosion may be aided by *piping.* *SHD*

[See also KNICKPOINT]

Archibold, O.W., DeBoer, D.H. and Delanoy, L. 1996: A device for measuring gully headward morphology. *Earth Surface Processes and Landforms* 21, 1001–1005. **Crouch, R.J.** 1990: Erosion processes and rates for gullies in granitic soils, Bathurst, New South Wales, Australia. *Earth Surface Processes and Landforms* 15, 169–173.

headward erosion The extension of the channel network or an individual first-order channel by erosion of the channel head. In studies of long-term LANDSCAPE EVOLUTION and DENUDATION CHRONOLOGY, headward erosion is seen as an important process in accomplishing RIVER CAPTURE. The processes involved in headward erosion may include erosion by *overland flow*, GULLY erosion, SAPPING and *piping.* *RPDW*

[See also HEADCUT EROSION]

Bishop, P. 1995: Drainage rearrangement by river capture, beheading and diversion. *Progress in Physical Geography* 19, 449–473. **Dohrenwend, J.C., Abrahams, A.D. and Turrin, B.D.** 1987: Drainage development in basaltic lava flows, Cima Volcanic Field, southeast California, and Lunar Crater Volcanic Field, south-central Nevada. *Bulletin of the Geological Society of America* 99, 405–413.

heat balance/budget The balance of gains and losses of heat for a system over a specified time period. For example, the heat balance forms part of the ENERGY BALANCE of the ATMOSPHERE. *AHP*

Sellers, W.D. 1965: *Physical climatology.* Chicago, IL: University of Chicago Press.

heat island The area of localised higher temperatures that occurs, especially under clear, calm conditions and at night, over urban areas. The effect is caused by ANTHROPOGENIC heat and the heat absorbed by structures and urban surfaces. There is a relationship between city size and heat-island intensity. The difference in MEAN ANNUAL AIR TEMPERATURE from the surrounding rural areas is typically 1.0°C for small cities up to about 4.0°C for large cities. *AHP*

[See also URBAN CLIMATE]

Goward, S.N. 1981: Thermal behaviour of urban landscapes and the urban heat island. *Physical Geography* 2, 19–33. **Oke, T.R.** 1982: The energetic basis of the urban heat island. *Quarterly Journal of the Royal Meteorological Society* 108, 1–24.

heathland A vegetation community and/or landscape developed on relatively dry, acidic, sandy and/or podzolic soils, dominated by low shrubs such as heathers (*Calluna vulgaris* and *Erica* spp.) and gorse (*Ilex europaeus*). It usually occurs < 200 m above sea level in Britain, where, in previous INTERGLACIAL CYCLES, heathland development occurred towards the end of the INTERGLACIAL (in the telocratic phase). SOIL EXHAUSTION and climatic cooling contribute to the development of acid podzols, which favour colonisation by coniferous trees (e.g. pine, *Pinus*) and by heathland vegetation. In the HOLOCENE, human activities, especially GRAZING of livestock, have had a marked effect on the timing and extent of heathland and MOORLAND development in northwest Europe. *MJB*

[See also GRAZING HISTORY, HUMAN IMPACT ON VEGETATION HISTORY, PODZOLS]

Odgaard, B.V. 1988: Heathland history in western Jutland, Denmark. In Birks, H.H., Birks, H.J.B., Kaland, P.E. and Moe, D. (eds), *The cultural landscape: past, present and future*. Cambridge: Cambridge University Press, 311–319. **Rodwell, J.S.** (ed.) 1991: *British plant communities*. Vol. 2. *Mires and heaths*. Cambridge: Cambridge University Press. **Specht, R.L.** (ed.) 1979: *Heathlands and related shrublands: descriptive studies* [*Ecosystems of the World*. Vol. 9A]. Amsterdam: Elsevier. **Specht, R.L.** (ed.) 1981: *Heathlands and related shrublands: analytical studies* [*Ecosystems of the World*. Vol. 9B]. Amsterdam: Elsevier. **Thompson, D.B.A.** (ed.) 1995: *Heaths and moorland: cultural landscapes*. Edinburgh: HMSO.

heavy metals A term widely used for metallic TRACE ELEMENTS in soils and elsewhere in the environment. Some elements such as arsenic, CADMIUM, chromium, LEAD, MERCURY, nickel and uranium may be described as *toxic metals*, but others, such as cobalt, COPPER, manganese, selenium and zinc are biologically essential at low concentrations (MICRONUTRIENTS), but in all cases high concentrations are toxic and reduce the activity of soil organisms. *EMB*

[See also METAL POLLUTION, SOIL POLLUTION, TOXIN]

Alloway, B.J. (ed.) 1990: *Heavy metals in soils*. Glasgow and London: Blackie. **Nieboer, E. and Richardson, D.H.S.** 1980: The replacement of the nondescript term 'heavy metals' by a biologically and chemically significant classification of metal ions. *Environmental Pollution* 1, 3–26.

heavy mineral analysis The investigation of those MINERALS in a sample of SEDIMENT or SEDIMENTARY ROCK that are denser than, and can be separated using, bromoform (relative density = 2.9). Heavy minerals usually comprise less than 1% of the sample and include apatite, epidote, garnet, rutile, staurolite, tourmaline and zircon. Their study is particularly useful in determining the PROVENANCE of SANDSTONES. Several minerals of economic importance are heavy minerals (e.g. diamond, gold and the tin ore cassiterite) and can be exploited from PLACER DEPOSITS. *TY*

Blatt, H., Middleton, G. and Murray, R.C. 1980: *Origin of sedimentary rocks*, 2nd edn. Englewood Cliffs, NJ: Prentice-Hall. **Dill, H.G.** 1998: A review of heavy minerals in clastic sediments with case studies from the alluvial-fan through the nearshore-marine environments. *Earth-Science Reviews* 45, 103–132. **Els, B.G.** 1998: The auriferous late Archaean sedimentation systems of South Africa: unique palaeo-environmental conditions? *Sedimentary Geology* 120, 205–224. **Lubke, R.A. and Avis, A.M.** 1998: A review of the concepts and application of rehabilitation following heavy mineral dune mining. *Marine Pollution Bulletin* 37, 546–557. **Rasmussen, B. and Buick, R.** 1999: Redox state of the Archean atmosphere: evidence from detrital heavy minerals in ca. 3250–2750 Ma sandstones from the Pilbara Craton, Australia. *Geology* 27, 115–118.

heavy oil A form of PETROLEUM that is too viscous to flow, causing difficulties in extraction. Many heavy oil deposits have formed from a reaction between CRUDE OIL and GROUNDWATER. The Orinoco Oil Belt in Venezuela is a major source, producing an oil–water emulsion known as *orimulsion*. Toxic waste products from the burning of orimulsion have led to environmental objections to its use. *GO*

Martinez, A.R. 1987: The Orinoco Oil Belt, Venezuela. *Journal of Petroleum Geology* 10, 125–134. **Niu, J.Y. and Hu, J.Y.** 1999: Formation and distribution of heavy oil and tar sands in China. *Marine and Petroleum Geology* 16, 85–95.

hedgerow removal Hedges are valuable for conserving BIODIVERSITY, providing shelter and reducing SOIL DEGRADATION. In many countries they are being removed to facilitate mechanised farming. Government incentives in the UK and Europe have encouraged the loss in recent decades. Eastern England has been especially badly affected and in the UK as a whole roughly one quarter of hedgerows were lost between AD 1945 and 1985 (the loss has continued since). *CJB*

Dowdeswell, W.H. 1987: *Hedgerows and verges*. London: Allen and Unwin. **Hooper, M.D.** 1979: Hedges and small woodlands. In Davidson, J. and Lloyd R. (eds), *Conservation and agriculture*. Chichester: Wiley. **Pollard, E., Hooper, M.D. and Moor, N.W.** 1974: *Hedges*. London:Collins.

Heinrich events Short-lived cold events characterised by periods of ice-rafting related to DISCHARGE predominantly from the LAURENTIDE ICE SHEET. Sharp basal contacts, low foraminiferal abundance and ICE-RAFTED DEBRIS (IRD) characterise these events in marine records. Six events have been defined between 70 and 14 ka BP, labelled H6 to H1, and the IRD peak in the YOUNGER DRYAS is sometimes defined as H0. *LJW/WENA*

[See also BOND CYCLE, DANSGAARD–OESCHGER EVENT]

Andrews, J.T. 1998: Abrupt changes (Heinrich events) in late Quaternary North Atlantic marine environments: a history and review of data and concepts. *Journal of Quaternary Science* 13, 3–16. **Broecker, W.S., Bond, G., Klas, M. et al.** 1992: Origin of the northern Atlantic's Heinrich events. *Climate Dynamics* 6,

265–273. **Dowdeswell, J.A., Maslin, M.A., Andrews, J.T. and McCave, I.N.** 1995: Iceberg production, debris, rafting, the extent and thickness of Heinrich layers (H-1, H-2) in North Atlantic sediments. *Geology* 23, 301–304. **Heinrich, H.** 1988: Origin and consequences of cyclic ice-rafting in the Northeast Atlantic Ocean during the past 130,000 years. *Quaternary Research* 29, 142–152.

hemlock decline A marked decline in the abundance of the tree *Tsuga canadensis* L. (hemlock) throughout eastern North America in the mid-HOLOCENE. FOSSIL pollen data from numerous sites indicate that this tree species declined sharply in abundance about 5000 calendar years BP (c. 4700 radiocarbon years BP). As the hemlock decline appears to have been rapid, more-or-less synchronous, species specific and over a large area, it is thought to have been the result of a pathogenic outbreak. Recent evidence from Québec suggests that insect activity may have been responsible for the decline. Several other forest taxa appear to have increased in abundance following the hemlock decline, suggesting that the demise of this long-lived, highly shade-tolerant conifer may have had a major impact on regional forest dynamics. This dramatic decline of a highly competitive tree species indicates the importance of pathogenic outbreaks as a rare, but potentially catastrophic, disturbance mechanism in forested ecosystems. *JLF*

[See also ECOSYSTEM COLLAPSE, ECOSYSTEM CONCEPT, ELM DECLINE, PATHOGENS, POLLEN ANALYSIS]

Bhiry, N. and Filion, L. 1996: Mid-Holocene hemlock decline in eastern North America linked with phytophagous insect activity. *Quaternary Research* 45, 312–320. **Davis, M.B.** 1981: Outbreaks of forest pathogens in Quaternary history. In *Proceedings of the IV International Palynological Conference.* Vol. 3. Lucknow: Birbal Sahni Institute of Paleobotany, 216–227. **Fuller, J.L.** 1998: Ecological impact of the mid-Holocene hemlock decline in southern Ontario, Canada. *Ecology* 79, 2337–2351. **Webb III, T.** 1982: Temporal resolution in Holocene pollen data. In *Proceedings of the Third North American Paleontological Convention.* Vol. 2. Toronto: Business and Economic Service, 569–572.

herb Any non-woody terrestrial plant, either annual, biennial or perennial. In POLLEN ANALYSIS, pollen from herbs is often summarised as NON-ARBOREAL POLLEN (NAP), in contrast to ARBOREAL POLLEN (AP) from trees and shrubs. *BA*

[See also FORB]

herbivore Any animal that eats plants. Specialised herbivore NICHES include grazers, browsers, grain and seed eaters (*granivores*), leaf eaters (*folivores*), fruit and berry eaters (*frugivores*), nut eaters (*nucivores*), nectar eaters (*nectarivores*), and root eaters. *RJH*

[See also CARNIVORE, FOOD CHAIN/WEB, TROPHIC LEVEL]

heritage Something that is inherited from the past or which a past generation has preserved and handed on to the present and which a significant part of the population wishes to hand on to the future. Heritage takes many forms, but the broad division is between natural and cultural. *Natural heritage* refers to valued environments that can be found in LANDSCAPE, scenery and areas or SITES OF SPECIAL SCIENTIFIC INTEREST. There are 'heritage coastlines', NATIONAL PARKS and other types of

PROTECTED AREAS, all designated to preserve environments that a society classifies as valued. *Cultural heritage* centres on the imprint of people on landscape and includes physical remains, such as archaeological sites or historic monuments, and significant places, such as battlefields or the residences of major figures in history. These forms of heritage are found in '*heritage places*' and have a geographical expression but other forms of heritage, such as art, cultural events and rituals, are more aspatial.

All forms of heritage raise issues of preservation and CONSERVATION and many societies have laws enshrined in their constitutions to support these functions. Concepts such as STEWARDSHIP and custodianship are often linked with the responsibilities of government and special agencies are created to preserve, conserve and manage heritage. *DTH*

Herbert, D.T. (ed.) 1995: *Heritage, tourism and society.* London: Cassells. **Lowenthal, D.** 1996: *Possessed by the past: the heritage crusade and the spoils of history.* New York: The Free Press. **Prentice, R.** 1993: *Tourism and heritage attractions.* London: Routledge.

herpetology The study of amphibians and reptiles; from the Greek for 'crawling thing'. *JAM*

Beebee, T.J.C. and Griffiths, R.A. 2000: *Amphibians and reptiles: a natural history of the British herpetofauna.* London: HarperCollins.

herringbone cross-bedding A style of CROSS-BEDDING in which closely-associated *sets* indicate PALAEOCURRENT directions approximately 180° apart (bipolar), commonly taken as a good indicator of sediment movement by TIDES. The presence of MUD DRAPES and TIDAL BUNDLES can provide important information about past tidal conditions. *GO*

[See also FLASER BEDDING, TIDAL RHYTHMITES]

Johnson, H.D. and Levell, B.K. 1995: Sedimentology of a transgressive, estuarine sand complex: the Lower Cretaceous Woburn Sands (Lower Greensand), southern England. In Plint, A.G. (ed.), *Sedimentary facies analysis: a tribute to the research and teaching of Harold G. Reading* [*International Association of Sedimentologists Special Publication* 22]. Oxford: Blackwell Science, 17–46.

heterotrophic organism An organism that cannot make its own organic food. Heterotrophic organisms or *heterotrophs* therefore obtain their requirements from other living (e.g. HERBIVORES, CARNIVORES, PARASITES) or dead (SAPROPHYTES) organisms. *JAM*

[See also AUTOTROPHIC ORGANISM]

heterozygosity The proportion of genetic loci, or genes, that are heterozygous (possessing different alleles on paired chromosomes of the same individual) in the average individual. *MVL*

[See also GENOTYPE, POLYMORPHISM]

heuristic algorithm A computational procedure that uses trial and error or probabilistic methods to approximate a solution for statistically difficult problems. Heuristic algorithms are sometimes used in a GEOGRAPHICAL INFORMATION SYSTEM (GIS). *TVM*

hexapod A standard concrete form with six 'legs', designed to be used in numbers to protect a shoreline from erosion. *HJW*

[See also COASTAL ENGINEERING STRUCTURES, RIPRAP, TETRAPOD]

hiatus A cessation of deposition, resulting in a gap in the STRATIGRAPHICAL RECORD that is not represented by any SEDIMENTS. In the QUATERNARY of the British Isles there is a major hiatus in deposition between c. 1.6 million years ago and c. 0.65 Ma BP. *DH*

[See also UNCONFORMITY]

hibernation The reduction of metabolic rates that enables some mammals to survive cold conditions (generally over an entire winter season) without (true hibernation), or with, very limited food supplies. *JLI*

[See also AESTIVATION]

hidden surface removal A technique in the projection of a three-dimensional object onto a two-dimensional computer screen that entails displaying only those parts of surfaces that would be naturally visible to the user. Hidden surface removal is part of the study of computer graphics for the modelling and visualisation of objects such as DIGITAL ELEVATION MODELS. *ACF*

Foley, J.D., van Dam, A., Feiner, S.K. *et al.* 1994: *Introduction to computer graphics.* Reading, MA: Addison-Wesley.

hierarchy theory A theoretical framework proposed for (eco)system analysis, based on the idea that organisation results from differences in process rates. SYSTEMS can be decomposed into different units based on differences in process rates. Each stratum that is defined can be divided into a number of units, termed *holons*, which are separated by gradients in process rates. The method is highly applicable to studies of environmental change because of the emphasis that it places on SCALE CONCEPTS. It views ecosystems as hierarchical structures, with each stratum responding differently to changes in the environment. The overall response of the system can then be determined by examining the cumulative effects of changes in individual strata. *JLI*

[See also ECOSYSTEM CONCEPT, FRACTAL ANALYSIS]

Allen, T.F.H. and Starr, T.B. 1982: *Hierarchy: perspectives for ecological complexity.* Chicago, IL: University of Chicago Press. May, R.M. 1989: Levels of organization in ecology. In: Cherrett J.M. (ed.), *Ecological concepts.* Oxford: Blackwell Scientific, 339–63. O'Neill, R.V. 1988: Hierarchy theory and global change. In Rosswall, T., Woodmansee, R.G. and Risser, P.G. (eds), *Scales and global change.* London, Wiley, 29–45. O'Neill, R.V., DeAngelis, D.L., Waide, J.B. and Allen, T.F.H. 1986: *A hierarchical concept of ecosystems.* Princeton, NJ: Princeton University Press.

hieroglyph A character or 'letter', often pictorial, in early Egyptian (*ca* 3200 BC–AD 400) and Mesoamerican writing. Also called *glyphs* in the context of Mesoamerican archaeology, hieroglyphs are mostly known from the decoration carved or painted on architecture and ARTEFACTS. *JAM*

Rice, P. 1999: Hieroglyphs. In Shaw, I. and Jameson, R. (eds), *A dictionary of archaeology.* Oxford: Blackwell, 275–277.

high-activity clays Clays with a high CATION EXCHANGE CAPACITY (> 24 cmol$_c$ kg^{-1}clay), usually developed over base-rich rocks. Such clays are of considerable importance in soils of the HUMID TROPICS as they have greater potential for sustainable agricultural use than other soils with low-activity clays. *EMB*

[See also LOW-ACTIVITY CLAYS]

High Arctic Ecologists use this term to differentiate it from the LOW ARCTIC. It refers to the various islands within the ARCTIC Basin, such as the Canadian Arctic islands, Svalbard, Franz Josef Land, northern Novaya Zemblya and northern Greenland. It is characterised by a desert-like environment with a sparse vegetation cover, POLAR DESERT or semi-desert. *HMF*

[See also SUBARCTIC]

French, H.M. 1999: Arctic environments. In Alexander, D.A. and Fairbridge, R.W. (eds), *Encyclopedia of environmental science.* Dordrecht: Kluwer, 29–33.

high-precision dating A small number of RADIOCARBON DATING laboratories world-wide employ techniques producing results with high precisions (small ± error terms). Conventional RADIOMETRIC measurement techniques measure residual radioactivity and the precision is dependent on the number of decay events recorded. To obtain higher precisions more events must be recorded, either by using larger samples or by measuring the sample for a longer time. For routine dating laboratories, sample size is often limited and greater counting times would reduce sample throughput rates, increasing expense. Most specialist high-precision dating laboratories use techniques where larger sample sizes are used and these laboratories have been involved in the establishment of the calibration dataset (see CALIBRATION OF RADIOCARBON DATES). *PQD*

[See also HIGH-RESOLUTION RECONSTRUCTIONS, WIGGLE MATCHING]

Pearson, G.W. 1980: High precision radiocarbon dating by liquid scintillation counting applied to radiocarbon timescale calibration. *Radiocarbon* 22, 337–345. Stuiver, M., Robinson, S.W. and Yang, I.C. 1979: ^{14}C dating to 60,000 years BP with proportional counters. In Berger, R. and Suess, H.E. (eds), *International ^{14}C Conference, 9th Proceedings.* Berkeley, CA: University of California Press, 202–215. Tans, P.P. and Mook, W.G. 1978: Design, construction and calibration of a high accuracy carbon-14 counting set up. *Radiocarbon* 21, 22–40.

high-resolution reconstructions The phrase 'high-resolution reconstruction' normally means the reconstruction of different environmental parameters with high temporal RESOLUTION (annual, decadal or centennial). As the general concern with the issues surrounding GLOBAL ENVIRONMENTAL CHANGE increases, the pressure on the scientific community to produce MODELS and PREDICTIONS of CLIMATIC CHANGE increases. While recent meteorological and oceanographic observations have paid attention to the processes and mechanisms of atmospheric and oceanographic circulation, this has produced only short-term perspectives (in general less than 200–300 years) of global change, limited by the range of INSTRUMENTAL and HISTORICAL records. Studies in PALAEOCLIMATOLOGY and PALAEOCEANOGRAPHY have, on the other hand, been mainly on coarser (in general greater than century to millennial) timescales. The palaeorecords that have yielded the highest temporal (interannual to decadal) resolution are HISTORICAL records, TREE RINGS, ICE CORES, CORALS and laminated MARINE and LACUSTRINE SEDIMENTS.

High-resolution palaeoclimatic records from high latitudinal regions are important for understanding the behaviour of the GENERAL CIRCULATION OF THE ATMOSPHERE. The patterns and causes of climatic change at these timescales are poorly understood, even though the magnitude and frequency of change recorded are significant. ANNUAL to decadal records are necessary because FUTURE CLIMATIC CHANGE will take place on these timescales. We must therefore know the range of natural CLIMATIC VARIABILITY against which to measure future change. Several initiatives, including the IGBP-PAGES (International Geosphere-Biosphere Programme – Past Global Changes) and NSF-PALE (Paleoclimate of Arctic Lakes and Estuaries), focus on obtaining high-resolution records of climate change for the last 1000–2000 years. The instrumental climate record of the past 50–150 years underestimates the range of natural climate variability across all temporal scales. It is becoming apparent that significant regional, spatial and temporal variability characterised climate change of the last millennium. Networks of annually dated time series from trees, sediments, corals, ice cores, and historical documents reveal that globally synchronous cold periods longer than a decade or two have been difficult to detect within the last 500 years. Emerging data also suggest that one of the largest and most extensive temperature shifts of the past 500 to 1000 years occurred between AD 1850 and today. Greenland ICE-CORE data and instrumental records have revealed large decadal climatic variations over the North Atlantic, which can be related to a major source of high-frequency variability, the NORTH ATLANTIC OSCILLATION (NAO). *AN*

[See also ANNUALLY RESOLVED RECORD, NATURAL ARCHIVES, SUB-MILANKOVITCH]

Bradley, R.S. (ed.) 1991: *Global changes of the past*. Boulder, Colorado: UCAR/Office for Interdisciplinary Earth Studies. Jones, P.D. and Bradley, R.S. 1992: Climatic variations over the last 500 years. In Bradley, R.S. and Jones, P.D. 1992: *Climate since AD 1500*, 649–665. London: Routledge. Lowe, J.J. and Walker, M.J.C. 1997: *Reconstructing Quaternary environments*. London: Longman. National Research Council 1995: *Natural climate variability on decade-to-century time scales*. Washington, DC: National Academy Press. Oldfield, F. (ed.), 1998: *Past global changes (PAGES): study report and implementation plan* [*International Geosphere-Biosphere Programme (IGBP) Report* 45]. Stockholm: The Royal Swedish Academy of Sciences.

highstand The maximum of an environmental cycle, fluctuation or oscillation. It may leave visible evidence in the LANDSCAPE, such as a RAISED SHORELINE at a SEA-LEVEL highstand, or in SEDIMENTARY DEPOSITS (see TRANSGRESSION). There may be more than one highstand close in time to the maximum, such as the three LITTLE ICE AGE highstands in glacier extent recognised in the Alps. The complementary term, LOWSTAND, is used for the minimum point in the cycle: for example, the sea-level lowstand at the LAST GLACIAL MAXIMUM was some 120 m below present sea level. *JAM*

[See also CLIMATIC FLUCTUATION, LAKE TERRACE, SEQUENCE STRATIGRAPHY]

Himalayan uplift The Himalayan mountain range is the result of the ongoing collision between the Indo-Australian and Asian tectonic plates. The resulting rapid uplift of 1000–2500 m in a few million years had created a mountain barrier by the late Miocene (11–7.5 Ma). The elevation of the Himalayas by 5 km resulted in complex atmosphere–ocean–lithosphere interlinkages that have been a FORCING FACTOR in global cooling and the onset of the Late Cenozoic ICE AGE. GENERAL CIRCULATION MODELS have been used to model the complex FEEDBACKS initiated by Himalayan uplift.

It is suggested that the creation of the Himalayas and Tibetan plateau caused perturbations of the JET STREAM, which resulted in hemispheric changes in the GENERAL CIRCULATION OF THE ATMOSPHERE and climate. Himalayan uplift may have induced strong convective atmospheric circulation on the Tibetan Plateau and increased temperature differences between summer and winter, leading to a regionally intense eastern Asia MONSOON circulation. LAND UPLIFT and a wet humid climate on the southern and eastern flanks of the Himalayas would have induced accelerated WEATHERING and EROSION RATES, leading to a reduction in atmospheric CARBON DIOXIDE concentrations, forcing global cooling. *MHD*

Molnar, P. and England, P. 1990: Late Cenozoic uplift of mountain ranges and global climatic change: chicken or egg? *Nature* 346, 29–34. Raymo, M.E. 1994: The initiation of Northern Hemisphere glaciation. *Annual Review of Earth and Planetary Science* 22, 353–383. Raymo, M.E. and Ruddiman, W. F. 1992: Tectonic forcing of late Cenozoic climate. *Nature* 359, 117–122. Ruddiman, W.F. and Kutzbach, J.E. 1991: Plateau uplift and climatic change. *Scientific American* 264, 42–50.

histic horizon An organic horizon at the soil surface that contains > 20% organic matter and is water saturated for at least one month per year. *EMB*

Food and Agriculture Organization of the United Nations (FAO) 1998: *World reference base for soil resources* [*Soil Resource Report* 84]. Rome: FAO, ISRIC, ISSS.

histogram A graph designed to quantify the frequency of occurrence of data values within specific class intervals. A histogram is used to visualise the probability density function of large datasets (e.g. to determine if they follow the GAUSSIAN MODEL). Histograms are often used in IMAGE ENHANCEMENT of images acquired through REMOTE SENSING. *ACF*

historical archaeology The archaeological study of people in historical times, i.e. since the appearance of the earliest written records and extending to the present day. It is distinguished by focusing on the post-PREHISTORIC past, utilising a greater diversity of sources than 'prehistoric' archaeology and impinging on the modern, globalised world. The subdiscipline is well developed in America, where it is sometimes known as *historic sites archaeology* and the combined archaeological/anthropological/historical study of the effects of the spread of European culture is a strong theme. *JAM*

Falk, L. (ed.) 1991: *Historical archaeology in global perspective*. Washington, DC: Smithsonian Institution. Little, J. (ed.) 1992: *Text-aided archaeology*. Boca Raton, FL: CRC Press. Orser Jr, C.E. and Fagan, B. 1995: *Historical archaeology*. New York: Harper Collins. Schuyler, R.L. 1970: Historical archaeology and historic sites archaeology as anthropology: basic definitions and relationships. *Historical Archaeology* 4, 83–89.

historical climatology

The time period of relevance to 'historical' climatology has not been precisely defined. A conservative interpretation regards it as covering the period of written history of the last 2000 years or so. Scientists such as H.H. Lamb have enjoyed success in using the sparse historical records of both the written and oral traditions to deduce broad features of climates in 'pre-instrumental' times. Many descriptions and accounts of weather provide DOCUMENTARY EVIDENCE of environmental change.

Weather-dependent *natural phenomena*, both physical (PARAMETEOROLOGY) and biological (PHENOLOGY), are also used in historical climatology. VINE HARVEST dates, for example, provide good phenological indicators of summer heat and rainfall. Weather information for oceanic areas can be found in SHIP LOG BOOKS, which date back to the earliest years of European voyages of discovery. Looking back yet further, Mediaeval estate accounts, diaries and grain price records all shed further light on climate through its influence on agriculture. Such sources need, however, to be used with care as other, non-climatic, elements can influence agriculture. By these means information has been gathered to provide a more detailed picture of the two most distinctive periods of historical times, the MEDIAEVAL WARM PERIOD and the LITTLE ICE AGE.

PROXY CLIMATIC INDICATORS provide helpful indications of climatic change for those millennia for which written evidence is sparse or non-existent. DENDROCHRONOLOGY, VARVE and ICE and MARINE SEDIMENT CORE data provide information for much of the HOLOCENE and help to corroborate documentary evidence from more recent centuries.

Only since the middle of the nineteenth century have reliable instrumental data been available from organised networks. Some INSTRUMENTAL DATA exist for earlier times but are of more limited scientific value because of the lack of standardisation of instruments and of their exposure. The British Isles have some of the oldest INSTRUMENTAL RECORDS and these have been used to reconstruct the CENTRAL ENGLAND TEMPERATURE RECORD. Rainfall records present more problems because of variations of catch over short distances, but both Holland and England have continuous series that date from the start of the eighteenth century. Wind data, which require no complex instruments, exist in various forms in England from Mediaeval times. *DAW*

Bradley, R.S. and Jones, P.D. (eds) 1992: *Climate since AD 1500*. London: Routledge. **Lamb, H.H.** 1995: *Climate, history and the modern world*. London: Routledge. **Landsberg, H.E.** 1985: Historic weather data and early meteorological observations. In Hecht, A.D. (ed.), *Paleoclimate analysis and modelling*. New York: Wiley, 27–70. **Peterson, T.C. and Vose, R.S.** 1997: An overview of the Global Historical Climatology Network temperature database. *Bulletin of the American Meteorological Society* 78, 2837–2849. **Rabb, T.K.** 1983: Climate and society in history: a research agenda. In Chen, R.S., Boulding, E. and Schneider, S. H. (eds), *Social science research and climate change: an interdisciplinary appraisal*. Dordrecht: Reidel, 62–76. **Rotberg, R.I. and Rabb, T.K.** (eds) 1981: *Climate and history: studies in interdisciplinary history*. Princeton, NJ: Princeton University Press. **Wigley, T.M.L., Ingram, M.J. and Farmer, G.** (eds) 1981: *Climate and history: studies in past climates and their impact on man*. Cambridge: Cambridge University Press.

historical ecology

The study of the changes to ecosystems that have occurred over the historical period. It is perhaps best exemplified by studies of the status and history of British woodlands based on DOCUMENTARY EVIDENCE and FIELD RESEARCH, including investigation of the structure and composition of remaining fragments of ANCIENT WOODLAND. Such woodlands may be located on early maps and remain in areas unsuitable for agriculture, possess sinuous boundaries and trees that are irregularly-spaced with straight trunks; they may also be recognised by the mixed-age stands, a rich epiphytic lichen flora and an absence of light-demanding species, though none of these characteristics are conclusive. *JAM*

[See also CULTURAL LANDSCAPE]

Rackham, O. 1986: *The history of the countryside*. London: Dent.

historical evidence

The term historical evidence is used in at least three ways in the context of investigating past environmental phenomena: (1) as a synonym for DOCUMENTARY EVIDENCE, written, graphical and numerical; (2) for evidence from the historical period, irrespective of its type, and especially including ARCHAEOLOGICAL evidence; and (3) for any evidence from the past, irrespective of age or context. Here, we prefer the second use of the term, of which documentary evidence is an important component, and propose that historical evidence in the strict sense occupies the niche between the direct measurements of the INSTRUMENTAL RECORD and the proxy data from NATURAL ARCHIVES, the latter being the only source relating to PREHISTORIC times. Historians, archaeologists, scientists and others have developed rigorous methodologies to check for the veracity and reliability of historical evidence from a variety of sources, both human and natural.

The reconstruction, analysis and interpretation of climate in historical times prior to standardised instrumental meteorological records (HISTORICAL CLIMATOLOGY) is the field in which the use of historical evidence is most highly developed. Historical written records, or documentary evidence, are the main data source. Widely used sources include: EARLY INSTRUMENTAL METEOROLOGICAL RECORDS; WEATHER DIARIES and PROXY DATA yielding climatic information such as HARVEST RECORDS, TREE RINGS, cherry blossom dates and other phenological indicators; data on FLOOD HISTORY; SEA ICE information; and dates of FAMINE.

An important aim of historical climatology has been to extend the modern INSTRUMENTAL DATA time series backwards through the construction of quantitative, continuous and HOMOGENEOUS SERIES of climatic elements that ideally overlap with and have been correlated and calibrated with modern data. Because of the nature, comprehensiveness, timespan and degree of continuity of the historical information available, this is often not possible. Considerable attention has been given to detailed reconstruction of the LITTLE ICE AGE and the preceding MEDIAEVAL WARM PERIOD in Europe and North America, but much longer reconstructions have been undertaken using Chinese, Japanese and Arabian historical sources. Although most attention has focused on the development of time series of monthly or seasonal indices of temperature and precipitation, some studies have focused on EXTREME CLIMATIC EVENTS, such as FLOODS and TROPICAL CYCLONES. The methodology of historical climatology draws upon the disciplines of history and language in the translation and evaluation of historical source material and on HISTORICAL GEOGRAPHY in considerations of the impact upon and interrelationships with society, as well as statistics and climatology in their climatic evaluation and analysis. *RG/JAM*

[See also CALIBRATION, FIELD RESEARCH, PALAEOCLIMATOL-OGY, PHENOLOGY]

Bradley, R.S. and Jones, P.D. (eds) 1992: *Climate since AD 1500*. London: Routledge. **Glaser, R.** 1997: Data and methods of climatological evaluation in historical climatology. *Historical Social Research* 22, 59–87. **Glaser, R. and Walsh, R.P.D. (eds)** 1991: *Historical climatology in different climatic zones – Historische Klimatologie in verschiedenen Klimazonen.* [*Würzburger Geographische Arbeiten* 80]. Würzburg: Geographisches Institut, Universität Würzburg. **Hook, J.M. and Kain, R.J.P.** 1982: *Historical change in the physical environment*. London: Butterworth. **Lamb, H.H.** 1977: *Climate: present, past and future*. Vol. 2. London: Methuen. **Pfister, C.** 1992: Monthly temperature and precipitation in Central Europe 1525–1979: quantifying documentary evidence on weather and its effects. In Bradley, R.S. and Jones, P.D. (eds), *Climate since AD 1500*. London: Routledge, 118–192. **Wigley, T.M.L., Ingram, M.J. and G. Farmer, G. (eds)** 1981: *Climate and history: studies in past climates and their impact on man*. Cambridge: Cambridge University Press.

historical geography

Traditionally defined as the study of the geographies of past periods and, as such, historical geography has played a large part in the development of HUMAN GEOGRAPHY. Geographers have always recognised the central importance of the interplay between time and place. Key concepts such as LANDSCAPE and ENVIRONMENT can only be understood as products of ongoing processes of change and interaction. For some aspects of PHYSICAL GEOGRAPHY, the GEOLOGICAL TIME-SCALE is relevant; other aspects focus on much shorter timescales.

Historical geographers have traced the evolution of forms of settlement and LANDUSE from the earliest archaeological evidence to the changing form of the later industrial city. Beyond the descriptive, there are central questions on the forces that drive change and the contrasts from one kind of society to another. As with many other aspects of human geography, facts are not given and the past is seen differently by its variety of interpreters. Historical geography remains distinctive, though perhaps mainly through its method, rather than by its status as a subdiscipline. *Time–space studies* are different and tend to focus on the interaction of these variables in the modern world. *DTH*

Butlin, R.A. 1993: *Historical geography: through the gates of space and time*. London: Arnold. **Pacione, M. (ed.)** 1987: *Historical geography: progress and prospect*. London: Croom Helm.

history

In essence the academic discipline of understanding and interpreting past events. Maintaining the historical record is a key role for historians. The events normally relate to people, nations and global change but in one sense everything on the Earth's surface has a history that is essential to its understanding. *DTH*

Histosols

Histosols form in organic materials such as upland or lowland peat MIRES characteristically found in wet climates or in wet declivities in the landscape. Upland peats and RAISED MIRES/BOGS are normally extremely acid and are rarely cultivated, but lowland fen peats are neutral or even calcareous in reaction. Histosols, soils with more than 40 cm depth of organic matter, are most extensive in SUBARCTIC areas, but there is a scatter of these soils throughout the world in association with other soil groups. The diagnostic feature of Histosols is the presence of a HISTIC HORIZON. Drainage and aeration of these soils results in loss of peat through OXIDATION and shrinkage. When cultivated, Histosols are also subject to WIND ERO-SION after drying out. As a result the area of valuable lowland Histosols is decreasing through human activities. (SOIL TAXONOMY: Histosols). *EMB*

[See also PEAT, MIRES, WORLD REFERENCE BASE FOR SOIL RESOURCES]

Andriesse, J.P. 1988: *Nature and management of tropical peat soils* [*FAO Soils Bulletin* 59]. Rome: FAO. **Avery, B.W.** 1990: *Soils of the British Isles*. Wallingford: CAB International. **Driessen, P.M.** 1980. Peat soils. In *Land reclamation and water management: developments, problems and challenges*. Wageningen: International Institute for Land Reclamation, 49–53.

holistic approach

An approach to science that treats phenomena (such as ECOSYSTEMS) as entire entities, without trying to identify the individual processes operating within the SYSTEM. A holistic approach may be justified in terms of EMERGENT PROPERTIES that cannot be predicted from the behaviour of the component parts of the system. *JLI*

[See also GENERAL SYSTEMS THEORY, REDUCTIONISM]

Holocene

Also known as the *Postglacial, Recent, Flandrian,* or *Present Interglacial*, the Holocene is the latest epoch of the GEOLOGICAL TIMESCALE, which has so far lasted about 11 500 years (10 000 radiocarbon years). Taken literally, the term means 'wholly recent', which originally referred to Holocene fossil assemblages containing only modern species. Together with the preceding PLEISTOCENE EPOCH, it completes the QUATERNARY. *JAM*

[See also HOLOCENE ENVIRONMENTAL CHANGE, HOLOCENE TIMESCALE]

Gulliksen, S., Birks, H.H., Possnert, G. and Mangerud, J. 1998: A calendar age estimate of the Younger Dryas–Holocene boundary at Kråkenes, western Norway. *The Holocene* 8, 249–258.

Holocene environmental change

Major global environmental changes occurred at the start of the Present Interglacial. ABRUPT CLIMATIC CHANGE saw withdrawal of ICE SHEETS from the mid-latitude continental land masses and triggered important changes in the oceans and on land, such as an increase in ocean volume, SEA-LEVEL CHANGE, marine TRANSGRESSION, PARAGLACIAL geomorphological activity, plant and animal MIGRATION, ECOLOG-ICAL SUCCESSION and SOIL DEVELOPMENT. Beyond the limits of the ice sheets, tropical and subtropical regions were also affected. The Sahara DESERT, for example, which had been more extensive than at present in the late PLEIS-TOCENE, was effectively replaced by SAVANNA in the early Holocene and TROPICAL RAIN FORESTS expanded from their late-Pleistocene REFUGIA as conditions became warmer and wetter.

For the remainder of the Holocene, the Earth's present landscapes evolved under climatic conditions comparable to those of today with relatively small-scale natural climatic changes, such as temperature variations with an amplitude of a few degrees Celsius and frequencies of decades to millennia. In addition to this NATURAL background, however, the Holocene has witnessed unique and marked ANTHROPOGENIC environmental changes with increasing magnitude, frequency and complexity towards the present day. Indeed, it is the presence of the human agency that provides an important justification for differentiating the Holocene from the other INTERGLACIALS that characterised the PLEISTOCENE. Identifying and separating the natural from the anthropogenic effects are important research foci in the study of Holocene environmental change.

After the attainment of an early- to mid-Holocene 'Climatic Optimum' or HYPSITHERMAL when, for at least some of the time, temperate latitudes were 2–3°C warmer than today, the major trend in Holocene climate has been a LATE-HOLOCENE CLIMATIC DETERIORATION when temperatures have tended to decline and precipitation to increase in temperate latitudes. Irregular and/or quasi-cyclic periodicities superimposed on these relatively long-term trends include the prominent, cold event (termed the *Finse Event* in southern Norway) at about 8200 years ago, which has been recognised in ICE CORES, MARINE SEDIMENT CORES and other records, the MEDIAEVAL WARM PERIOD and the LITTLE ICE AGE. Both long- and short-term patterns can be seen in the CLIMATIC RECONSTRUCTION of summer temperatures and winter precipitation for southern Norway, based on evidence from GLACIER VARIATIONS and TREE-LINE VARIATIONS (see Figure). Orbital forcing, as explained by the MILANKOVITCH THEORY, accounts for the long-term trend but the shorter-term SUB-MILANKOVITCH variations are poorly understood: SOLAR FORCING may be an important cause, but VOLCANIC IMPACTS and AUTOVARIATION, including fluctuations in the THERMOHALINE CIRCULATION of the oceans, may also be implicated.

Prior to the Holocene, *Homo sapiens* had little more effect on the landscape than any other animal: environment affected humans far more than humans affected their environment. With the onset of the Holocene, and particularly after the *Neolithic Revolution*, this balance began to change. Major anthropogenic effects on vegetation and soils were evident first in the Old World over 7000 years ago. Technological innovation has since continually improved the ability of an expanding human population to occupy, use and transform more of the Earth's surface, reversing the balance. However, some areas remained relatively immune from human impacts, as illustrated by the dramatic discovery that dwarf mammoths survived on Wrangel Island in the Siberian Arctic until about 3000 years ago. HUMAN IMPACTS reached their apotheosis after the advent of European IMPERIALISM and especially since the INDUSTRIAL REVOLUTION. Only recently has the extent of HUMAN IMPACTS on the environment been acknowledged in the wide recognition of a need for CONSERVATION to be taken seriously on a global scale, together with the exploitation of resources associated with *economic development*.

An increasing recognition of the importance of the study of Holocene environmental change can be attributed to what may be termed the *social imperative* and the *scientific imperative*. First, there are reciprocal relations between human societies and the natural environment, which affect wealth creation and quality of life considerations. This social imperative requires an understanding of both natural BACKGROUND LEVELS and likely future human impacts on environment, which can be illuminated by a knowledge of what has happened in the recent past. Second, there are fundamental aspects of the science of the natural environment that can only be fully understood by investigating environmental change over the Holocene timescale.

This scientific imperative includes the unique opportunities afforded for developing theory about events that are rare spatially, exhibit a low-frequency of occurrence through time or have a duration of decades to millennia. The variety and quality of NATURAL ARCHIVES available for the Holocene is higher than for earlier parts of the Quaternary or longer geological timescales. Many sources of PROXY DATA for the Holocene, such as ICE CORES, TREE RINGS, LAKE SEDIMENTS, PEAT STRATIGRAPHY and CORALS are capable of yielding continuous records with high (often annual or seasonal) temporal RESOLUTION. By adopting a MULTIPROXY APPROACH to environmental reconstruction in the Holocene, the combined information abstracted from the evidence covers an impressive range of environmental variables in unprecedented detail. Spatial coverage on local, regional and global scales for the Holocene may approach that available for the present, and Holocene records overlap with observations, the INSTRUMENTAL RECORD and DOCUMENTARY EVIDENCE, which are invaluable for CALIBRATION and CONFIRMATION ('validation') of environmental reconstructions. These attributes also confer on the Holocene considerable potential as a source of MODERN ANALOGUES for interpreting the pre-Holocene. Lastly, they also render Holocene data invaluable for testing the GENERAL CIRCULATION MODELS (GCMs) that are currently the most effective basis for predicting FUTURE CLIMATIC CHANGE. Thus, the scientific importance of the environmental changes of the Holocene is immense as, in several respects, Holocene environmental change holds the key to a fundamental understanding of the past, present and future environments of the Earth.

JAM

[See also ENVIRONMENTAL ARCHAEOLOGY, NEOGLACIATION, VEGETATION HISTORY]

Alley, R.B., Mayewski, P.A., Sowers, T. et al. 1997: Holocene climatic instability: a prominent, widespread event 8200 yr ago. *Geology* 25, 483–486. **Berglund, B.** (ed.) 1986: *Handbook of Holocene palaeoecology and palaeohydrology*. Chichester: Wiley. **Bond, G., Showers, W., Cheseby, M.** et al. 1997: A pervasive millennial-scale cycle in north Atlantic Holocene and Glacial climates. *Science* 278, 1257–1266. **Broecker, W.S.** 1998: The end of the present interglacial: how and when? *Quaternary Science Reviews* 17, 689–694. **Chambers, F.M., Ogle, M.I. and Blackford, J.J.** 1999: Palaeoenvironmental evidence for solar forcing of Holocene climate: linkages to solar science. *Progress in Physical Geography* 23, 181–204. **Dahl, S.-O. and Nesje, A.** 1996: A new approach to calculating Holocene winter precipitation by combining glacier equilibrium-line altitudes with pine-tree limits: a case study from Hardangerjøkulen, central southern Norway. *The Holocene* 6, 381–398. **Edwards, K.J. and Sadler, J.P.** 1999: *Holocene environments of prehistoric Britain*. [*Quaternary Proceedings* 7 – *Journal of Quaternary Science* 14: 477–635]. Chichester: Wiley. **Karlén, W., Bodin, A., Kuylenstierna, J. and Näslund, J.-O.** 1995:

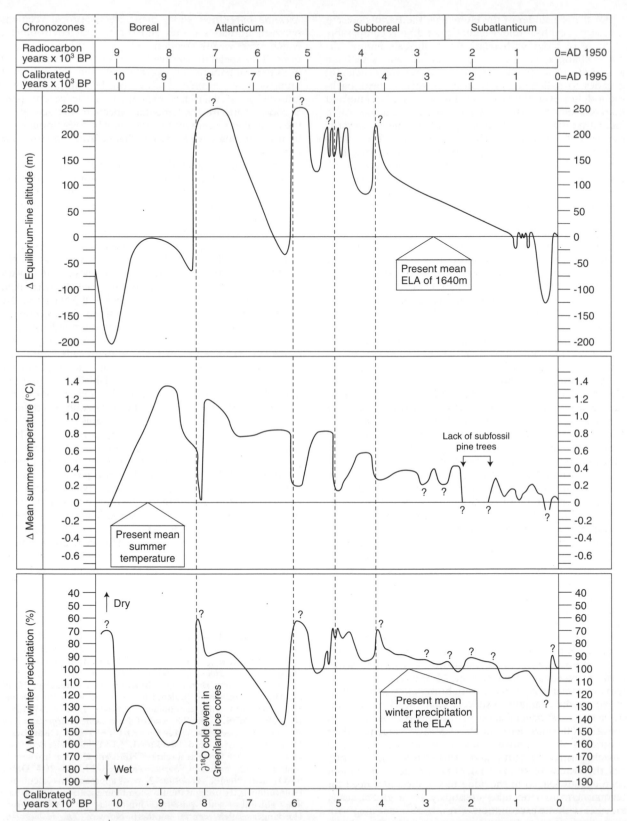

Holocene environmental change *Holocene glacier and climatic changes reconstructed at the Hardangerjøkulen ice cap, southern Norway: variations in equilibrium-line altitude (ΔELA) on Hardangerjøkulen (upper graph) were combined with variations in mean summer temperature derived from tree-line variations (middle) to derive the variations in mean winter precipitation (lower graph) (after Dahl and Nesje, 1996)*

Climate of northern Sweden during the Holocene. *Journal of Coastal Research, Special Issue* 17, 49–54. **Matthews, J.A.** 1998: The scientific and geographical importance of Holocene environmental change. *Swansea Geographer* 33, 1–6. **O'Brien, S.R., Mayewski, P.A., Meeker, L.D.** *et al.* 1995: Complexity of Holocene climate as reconstructed from a Greenland ice core. *Science* 270, 1962–1964. **Roberts, N.** 1998: *The Holocene: an environmental history*, 2nd edn. Blackwell: Oxford. **Vartanyan, S.L., Garutt, V.E. and Sher, A.V.** 1993: Holocene dwarf mammoths from Wrangel Island in the Siberian Arctic. *Nature* 362, 337–340. **Wright Jr, H.E. (ed.)** 1984: *The Holocene [Late-Quaternary Environments of the United States*, Vol. 2]. London: Longman. **Wright Jr, H.E., Kutzbach, J.E., Webb III, T.** *et al.* **(eds),** 1993: *Global climates since the last glacial maximum.* Minneapolis, MN: University of Minnesota Press.

Holocene timescale

The Holocene has been formally defined as a separate geological EPOCH following the PLEISTOCENE, although now it is commonly viewed as merely the most recent INTERGLACIAL of the QUATERNARY. The opening of the Holocene is conventionally drawn at 10 000 years BP (in radiocarbon years) approximating to 11 500–11 700 calendar years BP, as defined in ICE CORES, VARVES and other NATURAL ARCHIVES.

The traditional subdivision of the Holocene was based on the BLYTT–SERNANDER TIMESCALE, originating in the latter part of the nineteenth century. Using evidence from MACROFOSSIL remains in *peat bogs*, Blytt and Sernander defined an alternating series of zones reflecting climatic changes through the period: Pre-Boreal (increasing warmth but no trees); Boreal (warm and dry); Atlantic (warm and wet); Sub-Boreal (warm and dry); and Sub-Atlantic (cool and wet). This pattern was later set against the divisions of Iversen's INTERGLACIAL CYCLE, a pattern assumed to have been characteristic of all recent interglacials. Furthermore, early studies of vegetation history in the British Isles by Godwin and Iversen also utilised these periods as a basis for a scheme of pollen zonation applicable across the whole area, as shown in the Table.

In the absence of RADIOCARBON DATING, the ages of the boundaries between zones were approximate, based in the later Holocene on archaeological correlations. As radiocarbon dating developed, ages were derived from key sites and adopted over wide areas; hence, the timescale was based on a very limited series of determinations. As the number of horizons dated increased, it was realised that the vegetational changes on which the boundaries were founded were highly DIACHRONOUS. Because the environmental characteristics and changes described by the zones were, however, still considered largely valid, the terms were retained and are still used to describe the broad subdivisions of the Holocene as CHRONOZONES. A widely used scheme in northwestern Europe is that of Mangerud *et al.* (1974).

Defining ages within the Holocene is still very largely based on radiocarbon, although DENDROCHRONOLOGY and ICE-CORE DATING now provide much more precise dating techniques and THERMOLUMINESCENCE (TL), OPTICALLY STIMULATED LUMINESCENCE DATING (OSL dating), URANIUM-SERIES DATING, TEPHROCHRONOLOGY and even some forms of COSMOGENIC NUCLIDE DATING have been used, albeit usually with far less PRECISION. In the later Holocene, dating associated with archaeological evidence

has also been used and eventually human observation and documentary records can be used to improve or provide a timescale. In the absence of such evidence and because radiocarbon dating is influenced by the SUESS EFFECT and BOMB CARBON, dating events over the last 200 years has often proved relatively imprecise, relying on such techniques as LICHENOMETRIC and LEAD-210 DATING. Over the latter half of the last century nuclear weapons testing and nuclear accidents such as Chernobyl have allowed use of CAESIUM-137 for dating.

Definition of regional terms for periods is not such a problem in the Holocene as it is over the QUATERNARY TIMESCALE, although a number of ill-defined terms are widely used. The period of maximum warmth experienced in the Holocene is defined as the *Climatic Optimum* or HYPSITHERMAL, although the timing and length of such a period varies considerably. The *Neoglacial* (see NEOGLACIATION), a term originally coined in North America, is sometimes used for the period after the thermal optimum when glaciers redeveloped in mountain areas, culminating in the LITTLE ICE AGE, when glaciers in many areas reached their maximum Holocene limits. The Little Ice Age is difficult to define and is very variable in timing and length globally, although it tends to centre on the middle and later part of the last millennium. On the continent of Europe the term *Neuzeitlich* is also used in a chronological sense to identify renewed glaciation.

Because the Holocene has been a time of relative climatic stability, at least within a Quaternary perspective, changes have not been marked, making boundaries difficult to relate to specific dates. 'Events' have been identified and suggested to represent relatively RAPID ENVIRONMENTAL CHANGE, notably at 8200 calendar years BP and 2700 calendar years BP, but the extent to which these provide widely identifiable, precise time markers is uncertain. The recognition and CORRELATION of such SUB-MILANKOVITCH (decadal to millenial) events represents one of the main tasks of those researching environmental changes on the Holocene timescale.

With expanding human populations, especially in the later Holocene, and the increasing anthropogenic influences on the environment, it has become difficult to separate such effects from natural, largely *climatic forcing*. The development of an ARCHAEOLOGICAL TIMESCALE has meant that division of time according to cultural parameters is of importance for most populated areas. Thus, the timescale originally developed on CLIMATOSTRATIGRAPHIC grounds utilising biological evidence exists in parallel with a timescale based entirely on TECHNOLOGICAL CHANGE or CULTURAL CHANGE criteria for recent millennia. *CJC*

Bond, G., Showers, W, and Cheseby, M. *et al.*1997: A pervasive millennial-scale cycle in North Atlantic Holocene and glacial climates. *Science* 278, 1257–1266. **Burga, C.A.** 1988: Swiss vegetation history during the last 18,000 years. *New Phytologist* 110, 581–602. **Godwin, Sir H.** 1975: *History of the British flora: a factual basis for phytogeography*, 2nd edn. Cambridge: Cambridge University Press. **Hedberg, H.D. (ed.)** 1976: *International stratigraphic guide*. London: Wiley. **Mangerud, J., Andersen, S.T., Berglund, B.E. and Donner, J.J.** 1974: Quaternary stratigraphy of Norden: a proposal for terminology and classification. *Boreas* 3, 109–127. **Mangerud, J., Birks, H.J.B. and Jäger, K.-D.** 1982: Chronostratigraphical subdivisions of the Holocene: a review [Introduction to Special Issue]. *Striae* 16, 1–6. **Patzelt, B.** 1974: Holocene variations of glaciers in the Alps. *Colloques*

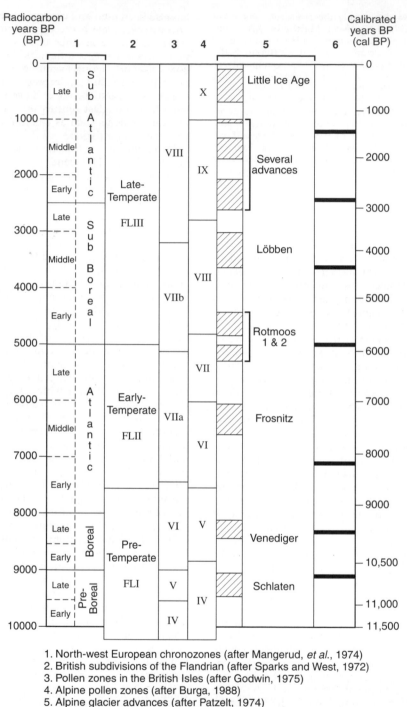

1. North-west European chronozones (after Mangerud, *et al.*, 1974)
2. British subdivisions of the Flandrian (after Sparks and West, 1972)
3. Pollen zones in the British Isles (after Godwin, 1975)
4. Alpine pollen zones (after Burga, 1988)
5. Alpine glacier advances (after Patzelt, 1974)
6. N. Atlantic ice rafting events (after Bond *et al.*, 1997)

Holocene timescale *The Holocene timescale in radiocarbon years (left,* BP *) and in calendrical or calendar years (right, cal. bp): (1) Northwest European chronozones (after Mangerud* et al., *1974); (2) British subdivisions of the Flandrian (after Sparks and West, 1972); (3) Pollen zones in the British Isles (after Godwin, 1975); (4) Alpine pollen zones (after Burga, 1988); (5) Alpine glacier advances (after Patzelt, 1974); (6) North Atlantic ice-rafting episodes (after Bond et al., 1997)*

Internationaux du Centre National de la Recherche Scientifique 219, 51–59. **Sparks, B.W. and West, R.G.** 1972: *The Ice Age in Britain.* London: Methuen. **Watson, R.A. and Wright Jr, H.E.** 1980: The end of the Pleistocene: a general critique of chronostratigraphic classification. *Boreas* 9, 153–163.

holokarst An area of pure limestone KARST that contains the full range of karst landforms, as exemplified by the Dinaric area of the former Yugoslavia. It contrasts with the relatively poorly developed MÉROKARST. *RAS*

Cvijić, J. 1925: Types morphologiques des terrains calcaires. Le holokarst. *Comptes Rendus de l'Académie des Sciences* 180, 592–594.

holomictic lake A lake that has unrestricted circulation through the whole water column after autumn overturn (cf. MEROMICTIC LAKE). *HHB*

[See also OVERTURNING].

homeostasis A *steady state* in which negative FEEDBACK MECHANISMS ensure that inputs and outputs counterbalance. In organisms, homeostasis is a physiological equilibrium produced by a balance of functions and of chemical compositions. *RJH*

[See also GAIA HYPOTHESIS, NEGATIVE FEEDBACK, SYSTEM]

hominid Of the Family Hominidae. The term Hominidae was previously used to refer only to humans and their immediate FOSSIL ancestors and, although scientifically questionable, the term 'hominid' is still frequently applied in this narrower sense with the term '*hominoid*' reserved for the broader grouping. The Hominidae in fact comprise two subfamilies, the Ponginae and the Homininae. The Ponginae migrated into Asia and are now represented solely by the orangutan, whereas the Homininae are represented by humans, chimpanzees and gorillas. This latter group remained in Africa, dividing again into African apes and the human lineage. Humans and their fossil ancestors are distinguished from other apes by larger brain sizes permitting complex communication such as speech, by an upright bipedal gait and by a slow rate of postnatal growth and development, favouring the development of a more elaborate social organisation with concomitant complex technological and cultural behaviour. *DCS*

[See also HOMINISATION, HUMAN EVOLUTION]

hominisation The progression from apes to MODERN humans whereby more 'human-like' characteristics, both biological and cultural, obtain. *JAM*

[See also HOMINID, HUMAN EVOLUTION]

Klein, R.G. 1999: *The human career: human biological and cultural origins.* Chicago, IL: Chicago University Press. **Tattersal, I.** 1998: *Becoming human: evolution and human uniqueness.* New York: Harcourt Brace. **Tobias, P.V.** 1979: Men, minds and hands: cultural awakenings over two million years of humanity. *South African Archaeological Bulletin* 34, 92–95. **Williams, M.A.J.** 1985: On becoming human: geographical background to cultural evolution [11th Griffith Taylor Memorial Lecture]. *Australian Geographer* 16, 175–184.

homogeneous series A climatic data series drawn from a single population. It may consist of a sequence of values recorded at one meteorological station or a sequence constructed from observations made under closely similar conditions that have been corrected statistically (homogenised) for any differences that may be attributable to changes of station location or of instrument. *AHP/JAM*

[See also CENTRAL ENGLAND TEMPERATURE RECORD]

Tabony, R.C. 1980: *A set of homogeneous European rainfall series.* Bracknell: UK Meteorological Office.

homoiothermic organism A 'warm-blooded' organism (mammal or bird) that, unlike *poikilothermic organisms* (such as invertebrates, fishes and reptiles) has the ability to vary and to some extent maintain its body temperature independent of the temperature of its environment. *JAM*

homosphere The homosphere or *anthroposphere* is the ECOSPHERE as modified by human presence and impacts. *JAM*

[See also ECUMENE, NOÖSPHERE]

Svoboda, J. 1999: Homosphere. In Alexander, D.E. and Fairbridge, R.W. (eds), *Encyclopedia of environmental science.* Dordrecht: Kluwer, 324–325.

horst A GEOLOGICAL STRUCTURE comprising an uplifted block bounded on either side by FAULTS. At a given level the rock between the faults is older than that on either side. Examples of topographically expressed horst blocks include the upland areas of the Vosges and the Black Forest, separated by the GRABEN of the Rhine Valley. *GO*

[See also RIFT VALLEY]

hortic horizon A surface ANTHROPEDOGENIC HORIZON that is thoroughly mixed by TILLAGE and earthworm activity, the casts of which comprise > 25% of the volume. Human ARTEFACTS are common and a BURIED SOIL may be preserved beneath the horizon resulting from human activity. *EMB*

horticulture The cultivation of garden plants, including fruit, vegetables, flowers and ornamental plants. *JAM*

[See also ARBORICULTURE, SILVICULTURE]

Piperno, D.R. Ranere, A.J., Holst, I. and Hansell, P. 2000: Starch grains reveal early root crop horticulture in the Panamanian tropical forest. *Nature* 407, 894–897.

Hortonian overland flow Water that flows over the ground surface, but not in defined channels. It occurs when rainfall intensity exceeds the INFILTRATION capacity of the surface soil, sediment or rock. First identified by R.E. Horton, it is sometimes termed *infiltration-excess overland flow*. Hortonian overland flow tends to be important in ARIDLANDS and SEMI-ARID environments, urban areas and overgrazed or eroded areas. *ADT*

[See also RUNOFF PROCESSES, SATURATION OVERLAND FLOW]

Horton, R.E. 1933: The role of infiltration in the hydrologic cycle. *Transactions of the American Geophysical Union* 14, 446–460.

hotspot, in biodiversity Areas with exceptional concentrations of ENDEMIC species and where HABITAT LOSS is a threat to their survival. Thus, it may be argued that hotspots are where there is the greatest need for BIOLOGICAL CONSERVATION and where conservation efforts might be appropriately concentrated in proportion to the share of the world's species at risk. An estimated 44% of all vascular plant species and 35% of four vertebrate animal groups (mammals, birds, reptiles and amphibians) are confined to 25 hotspots occupying only 1.4% of the Earth's land surface (see Figure). These hotspots have been defined on the basis of two criteria: first, they contain at least 0.5% of the world's plant species as endemic (i.e. at least 1500 endemic species); second, they have lost at least 70% or more of their primary vegetation (indeed, 11 have lost at least 90%).

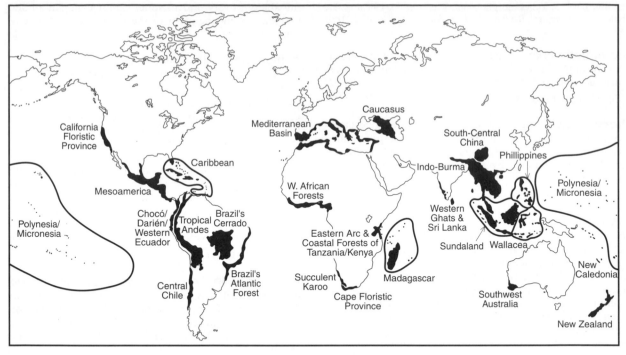

hotspot, in biodiversity *The 25 leading diversity hotspots (after Myers et al., 2000)*

TROPICAL FORESTS and MEDITERRANEAN regions account for 15 and five hotspots, respectively: nine are islands or mainly islands. The five leading hotspots – the tropical Andes, Sundaland (centred on the Sunda Straits of Indonesia), Madagascar, the Atlantic forests of Brazil and the Caribbean – each contain endemic plants and vertebrates amounting to at least 2.0% and together account for about 20% and 16%, respectively, of all plant and vertebrate species. Some 38% by area of the 25 hotspots is already protected in NATIONAL PARKS or other types of RESERVES, but this leaves 62% unprotected.

In 1995, more than 1.1 billion people (around 20% of the world's population) were living within biodiversity hotspots and three extensive tropical forest regions (covering a total of 12% of the land surface). The population growth rate in the hotspots (1995–2000) is 1.8% per year, higher than the global average rate of 1.3% and even that of developing countries as a group (1.6%). These data suggest that HUMAN IMPACTS are likely to continue in the hotspots and that DEMOGRAPHIC CHANGE is an important factor in the conservation of global biodiversity. Indeed, with possibly >60% of terrestrial biodiversity concentrated in the hotspots, they are undoubtedly the principal *Holocene refugia* for the Earth's biota. *JAM*

[See also BIODIVERSITY, REFUGIUM]

Cincotta, R.P., Wisnewski, J. and Engelman, R. 2000: Human population in the biodiversity hotspots. *Nature* 404, 990–992. **Mittermeier, R.A., Myers, N. and Mittermeier, C.G.** 2000: *Hotspots: Earth's biologically richest and most endangered terrestrial ecoregions.* Chicago, IL: University of Chicago Press. **Myers, N.** 1988: Threatened biotas: 'hot spots' in tropical forests. *The Environmentalist* 8, 187–208. **Myers, N., Mittermeier, R.A., Mittermeier, C.G.** *et al.* 2000: Biodiversity hotspots for conservation priorities. *Nature* 403, 853–858. **Prendergast, J.R., Quinn, R.M., Lawton, J.H.** *et al.* 1993: Rare species, the coincidence of diversity hotspots and conservation strategies. *Nature* 365, 335–337.

hotspot, in geology A localised, long-lived area of VOLCANISM, commonly characterised by FLOOD BASALTS. Hotspots overlie MANTLE PLUMES and are thus not part of the PLATE TECTONICS system. Therefore most hotspots do not coincide with PLATE MARGINS, producing *intraplate volcanism* (e.g. Hawaii, Yellowstone). Where a hotspot does coincide with a plate margin, it gives rise to greater melting of the MANTLE and more voluminous volcanic products than would otherwise occur (e.g. Iceland hotspot at a CONSTRUCTIVE PLATE MARGIN). As plates move, VOLCANOES formed at a hotspot move away from the site of high heat flow, so that an active hotspot such as Hawaii lies at one end of a chain of extinct VOLCANOES, SEAMOUNTS and GUYOTS. Dating these *hotspot tracks* provides a means of determining rates and directions of plate movements, and their distribution provides a fixed frame of reference for determining the absolute motion of the Earth's plates. Hotspots on thicker CONTINENTAL CRUST cause regional UPLIFT, influence patterns of EROSION and may lead to the break-up of continents and the development of SEA-FLOOR SPREADING. *GO*

[See also RIFT VALLEY, WILSON CYCLE]

Cope, J.C.W. 1994: A late Cretaceous hotspot and the southeasterly tilt of Britain. *Journal of the Geological Society, London* 151, 905–908. **DiVenere, V. and Kent, D.V.** 1999: Are the Pacific and Indo-Atlantic hotspots fixed? Testing the plate circuit through Antarctica. *Earth and Planetary Science Letters* 170, 105–117. **Duncan, R.A. and Richards, M.A.** 1991: Hotspots, mantle plumes, flood basalts, and true polar wander. *Reviews of Geophysics* 29, 31–50. **White, R. and McKenzie, D.** 1989: Magmatism at rift zones: the generation of volcanic continental margins and flood basalts. *Journal of Geophysical Research* 94, 7685–7729.

hotspot, in remote sensing The direction in which light is scattered most strongly from a surface relative to the source of illumination. For a VEGETATION CANOPY, this is typically back in the direction of illumination as it is in this direction that all shadow will be obscured from view.

TQ

[See also BI-DIRECTIONAL REFLECTANCE DISTRIBUTION FUNCTION]

Hoxnian Interglacial Named after the Hoxne type site in Suffolk, England, the Hoxnian Interglacial followed the Anglian and preceded the Wolstonian Glaciations in Britain. Occurring somewhere between c. 300 and 430 ka BP it broadly corresponds to the *Holsteinian Interglacial* of Northern Europe and to oxygen ISOTOPIC STAGE (OIS) 11, although in the British Isles, because of the difficulties in dating interglacials, biostratigraphical records believed to be Hoxnian may date from more than one interglacial or OIS.

MHD

[See also QUATERNARY TIMESCALE]

Bowen, D.Q. 1999: *A revised correlation of Quaternary deposits in the British Isles* [*Geological Survey Special Report* 23]. Bath: Geological Society.

hum A positive residual limestone (KARST) landform (hill) produced by the solutional removal of surrounding rock material.

SHD

human ecology The investigation of how humans relate to, interact with and impact upon their surroundings. Human ecology has also been defined in terms of the complex interactions between ecological systems and human social systems. *Cultural ecology* is a synonym. The study of dynamic human environmental relationships may involve the use of ecological methods and approaches. Carl Sauer (1889–1975) first proposed the term during his studies of how each intercommunicating and interbreeding human population ordered itself, and how humans had progressively displaced and disturbed more of the world around them to become the *ecological dominant.*

MLW

[See also ENVIRONMENTALISM, HUMAN GEOGRAPHY, HUMAN IMPACT ON ENVIRONMENT]

Butzer, K.W. 1989: Cultural ecology. In Gaile, G.L. and Wilmott, C.J. (eds), *Geography in America*. Columbus, OH: Merrill Publishing Company, 192–208. **Butzer, K.W.** 1990: The realm of cultural–human ecology: adaptation and change in historical perspective. In Turner, B.C., Clark, W.C., Kates, R.W. et al. (eds), *The Earth as transformed by human action*. New York: Cambridge University Press, 685–701. **Hawley, A.H.** 1986: *Human ecology: a theoretical essay*. Chicago: University of Chicago Press. **Ingold, T.** 1986: *The appropriation of nature: essays on human ecology and social relations*. Manchester: Manchester University Press. **Kates, R.W.** 1971: Natural hazard in human ecological perspective: hypotheses and models. *Economic Geographer* 47, 438–451. **Young, G.L.** 1974: Human ecology as an interdisciplinary concept: a critical enquiry. *Advances in Ecological Research* 8, 1–105.

human evolution A long sequence of events leading to the emergence of anatomically MODERN people (see Figure). Divergence of the Hominidae from ancestral gibbons is placed at 17 Ma ago, with the earliest fossil evidence suggesting an African origin. The genetic proximity of humans to other great apes suggests a recent separation (5–6 Ma ago), with *Ardpithecus ramidus* thought to lie close to the point of divergence. The appearance of *Australopithecus afarensis* 4 Ma ago heralded the development of bipedal locomotion, followed by diversification of the HOMINID lineage and the appearance of new genera *Paranthropus* and *Homo*, the latter the first to evolve a large brain and use tools. Migration from Africa into Asia and Europe was undertaken by *Homo erectus* (1.7 Ma) with descendant populations in the Old World ('archaic *sapiens*') evolving into Neanderthals in Europe and western Asia before world-wide replacement of ancestral populations by anatomically modern *Homo sapiens*.

DCS

[See also HOMINISATION]

Chamberlain, A.T. 1999: Human evolution. In Barker, G. (ed.), *Companion encyclopaedia of archaeology*. London: Routledge, 757–796. **Jones, S., Martin, R. and Pilbeam, D. (eds)** 1995: *The Cambridge encyclopedia of human evolution*. Cambridge: Cambridge University Press. **Stringer, C. and McKie, R.** 1996: *African exodus : the origins of modern humanity*. London: J. Cape.

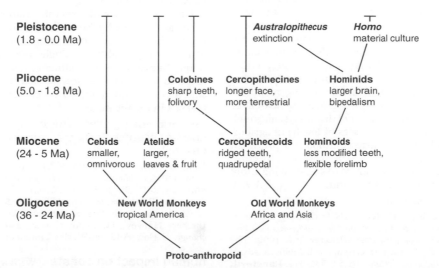

human evolution *Human evolution in the context of the divergence of anthropoid primates since the Oligocene and an indication of the major morphological adaptations in each lineage (after Chamberlain, 1999)*

human geography That part of GEOGRAPHY that studies the ways in which people occupy and interact with the surface of the Earth. Drawing upon the arts, HUMANITIES and SOCIAL SCIENCES, human geography is concerned with the patterns and processes that typify people on Earth. Its older PARADIGMS have been EXPLORATION, ENVIRONMENTALISM and REGIONALISM, all permeated by the cartographic tradition. Key modern concepts include the understanding of relationships between people, environments and landscapes, the meanings of place and the geometry of space and spatial interactions. The many subdivisions include urban geography, economic geography, cultural geography and political geography, although these 'adjectival' subdivisions have become overwritten by broad unifying themes such as GLOBALISATION, locality effects and the complexities of human relationships. Human geography has always had applied qualities that range from its early formative links with URBAN AND RURAL PLANNING, to its methodologies of SPATIAL ANALYSIS and GEOGRAPHICAL INFORMATION SYSTEMS (GIS), and its wider inputs to policy formulation and the understanding of modern patterns of consumption. *DTH*

[See also PHYSICAL GEOGRAPHY]

Agnew, J., Livingstone, D.N. and Rogers, A. 1996: *Human geography: an essential anthology.* Oxford: Blackwell. **Cloke, P., Crang, P. and Goodwin, M.** 1999: *Introducing human geographies.* London: Arnold. **Cloke, P., Philo, C. and Sadler, D.** 1991: *Approaching human geography: an introduction to contemporary theoretical debates.* London: Chapman.

human health hazards Predicted environmental changes such as GLOBAL WARMING, OZONE DEPLETION and a greater frequency of EXTREME CLIMATIC EVENTS have the potential to pose both direct and indirect hazards to human health. Direct impacts can be imposed by extreme events, whereby abnormally hot or cold spells can increase the risk of heat stroke and hypothermia. An increase in the frequency of storms and typhoons (with consequent FLOODS, landslides, population displacement or disease OUTBREAKS) can also lead to injury, illness, stress or fatality. An increase in ambient temperature and moisture level often facilitates the reproduction and spread of PATHOGENS and the vectors involved in DISEASE transmission. CHOLERA and MALARIA distributions are strongly linked to temperature and wetness and are both becoming more widespread in tropical and subtropical zones. Pathogens that cause diseases in crops and livestock may also proliferate, reducing the volume and quality of food production. Further reduction in food output (with accompanying malnutrition) is also a consequence of DESERTIFICATION. SEA-LEVEL RISE could lead to a greater number of drownings, although population displacements and losses of low-lying agricultural land will probably be more common and prevalent causes of stress and discomfort. Skin cancers are becoming more common as OZONE DEPLETION is allowing more ultraviolet radiation to reach the Earth. *MLW*

[See also CARCINOGEN, EPIDEMIC]

Abel, M.E. and Levin, S.K. (eds) 1981: *Climate's impact on food supplies: strategies and technology for climate-defensive food production.* Boulder, CO: Westview Press. **Baxter, P.J.** 1990: The medical consequences of volcanic eruptions: I. Main causes of death and injury. *Bulletin of Volcanology* 52, 532–544. **Langford, I.H. and Bentham, G.** 1995: The potential effects of climate change on winter mortality in England and Wales. *International Journal of Biometeorology* 38, 141–147. **Lines, J.** 1995: The effects of climate change and landuse changes on insect vectors of human disease. In Harrington, R. and Stork, N.E. (eds), *Insects in a changing environment.* London: Academic Press. **Loevinsohn, M.** 1994: Climate warming and increased malaria incidence in Rwanda. *The Lancet* 343, 714–718. **Martens, W.J.M., Niessen, L.W., Rotmans, T.H. and McMichael, A.J.** 1995: Potential impact of global climate change on malaria risk. *Environmental Health Perspectives* 103, 458–464. **Martin, P. and Lefebvre, M.** 1995: Malaria and climate: sensitivity of potential transmission to climate. *Ambio* 24, 200–257. **McMichael, A.J., Haines, A., Sloof, R. and Kovats, S. (eds)** 1996: *Climate change and human health.* Geneva: World Health Organisation. **Patz, J.** 1996: Global climate change and emerging infectious diseases. *Journal of the American Medical Association* 275, 217–223.

human impact on climate The impacts of humans on climate can be intentional or inadvertent. The former are usually small-scale and include the construction of walled gardens in temperate regions to create an artificial MICROCLIMATE to grow plants, such as fruit trees, that would not normally flourish. Unintentional impacts include the effects of urbanisation on URBAN CLIMATES and the effects of large reservoirs on land and water breezes. Such CLIMATIC MODIFICATION has occurred since the beginning of human occupancy of the Earth, particularly since the earliest settled communities and the NEOLITHIC TRANSITION. A very substantial body of evidence now exists from CLIMATIC MODELS, which suggests that relatively small PERTURBATIONS at the surface can impact on regional-scale climatology.

Changes in LAND COVER and LANDUSE, from natural Savanna to cropland for example, can alter soil moisture and runoff balances and affect precipitation amounts. Changes in surface characteristics may also have important climatic consequences in ARIDLANDS: removal of vegetation decreases soil-water storage and increases ALBEDO, affecting surface temperatures. DEFORESTATION CLIMATIC IMPACTS, for example in the Amazon Basin, may influence the HYDROLOGICAL BALANCE and hence climate on a regional scale. Furthermore, model results suggest that lower rates of EVAPOTRANSPIRATION feed less moisture into the WALKER CIRCULATION, which can weaken and produce extra-regional effects.

On a global scale, the modification of the composition of the atmosphere by the addition of GREENHOUSE GASES to create the enhanced GREENHOUSE EFFECT and GLOBAL WARMING is probably the single most important contemporary human impact on climate. Even with global agreement to limit the EMISSION of such gases, there is every sign that the human impact on climate will become more marked in the future. *AHP*

[See also AGRICULTURAL IMPACT ON CLIMATE, CLIMATIC CHANGE, FUTURE CLIMATE, GLOBAL ENVIRONMENTAL CHANGE]

Bonan, G.B. 1997: Effects of land use on the climate of the United States. *Climatic Change* 37, 449–486. **Frenzel, B. (series ed.)** 1992–1999: *European Science Foundation Programme: European Palaeoclimate and Man Since the Last Glaciation* [Paläoklimaforschung Special Issues 1–20]. Stuttgart: Gustav Fischer. **Harvey, D.** 2000: *Climate and global environmental change.* Harlow: Addison-Wesley Longman.

human impact on coasts With approximately 60% of the world's population living in the coastal zone, exploitation of coastal resources has been extensive. Many competing uses have drastically changed coastal morphology. These include: mining; reclamation; agriculture; industrial, residential and transportation structures; tourism; waste disposal; and COASTAL (SHORE) PROTECTION. In many countries, such as Belgium and Japan, erosion protection structures have converted natural shorelines into artificial ones. *HJW*

[See also HARD ENGINEERING, SEA-LEVEL RISE, SOFT ENGINEERING]

Kelletat, D. 1989: Biosphere and man as agents in coastal geomorphology and ecology. *Geoökodynamik* 10, 215–252. **Walker, H. J. (ed.)** 1988: *Artificial structures and shorelines*. Dordrecht: Kluwer.

human impact on environment

The actions of humans have dramatically modified the NATURAL ENVIRONMENT and have affected all parts of the Earth on a variety of temporal and spatial scales. ENVIRONMENTAL CHANGE due to human activity has diversified and intensified in parallel with DEMOGRAPHIC CHANGE and became a major focus of public concern in the late twentieth century. CULTURAL CHANGE has been accompanied by technological advances that have increased human potential to modify the ENVIRONMENT and have demanded more energy to sustain. Human induced environmental change may be conceptualised on a spatially hierarchical basis (see Figure 1), whereby local impacts act collectively to determine regional impacts, which, in turn, contribute to the planetary FEEDBACK MECHANISMS that drive GLOBAL ENVIRONMENTAL CHANGE. The three major subdivisions of human activity (agricultural, industrial and recreational) are convenient units of study for considering a potentially vast subject area.

The development of agriculture since its inception over 10 000 years ago (see AGRICULTURAL HISTORY) has resulted in a huge increase in the land area occupied by AGROECOSYSTEMS, which are typically low-diversity, artificial systems sustained to keep up with food demands from the growing world population. In the developed world, agricultural impacts have stemmed mainly from the gradual mechanisation and labour intensification of farming over the past two centuries. The most obvious expression of environmental change in the temperate mid-latitudes, for example, is the FRAGMENTATION of natural vegetation cover with the result that British vegetation in particular can be viewed as completely cultural in origin. Injudicious practices in Europe, parts of Australia and the USA have caused an alarming decline in the organic component of soils, resulting in SOIL DEGRADATION and major SOIL EROSION problems. The addition of synthetic FERTILISERS and PESTICIDES has led to significant WATER POLLUTION, with nitrate runoff from agricultural land into rivers and GROUNDWATER aquifers particularly problematic.

Agricultural impacts have been even more pronounced in the *developing countries*, where concern for the environment remains secondary to the necessity to produce food for an expanding population. Money for ENVIRONMENTAL PROTECTION and CONSERVATION measures is severely limited and land management strategies are non-sustainable because of a lack of education. DEFORESTATION in the tropics and humid subtropics is perhaps the largest cause for concern, not least because of the highly publicised effects on the CARBON CYCLE and possible contributions to GLOBAL WARMING. Locally, increased frequency of downstream flooding, accelerated soil erosion through unsuitable felling practices and declining productivity through removal of the BIOMASS nutrient store are the dominant impacts. It should be noted that agriculture is not the only source of deforestation; the timber industry is a major secondary source. AFFORESTATION and REFORESTATION,

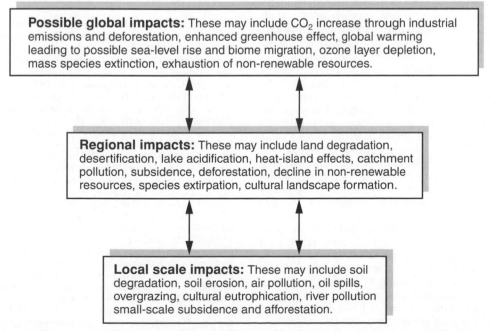

Possible global impacts: These may include CO_2 increase through industrial emissions and deforestation, enhanced greenhouse effect, global warming leading to possible sea-level rise and biome migration, ozone layer depletion, mass species extinction, exhaustion of non-renewable resources.

Regional impacts: These may include land degradation, desertification, lake acidification, heat-island effects, catchment pollution, subsidence, deforestation, decline in non-renewable resources, species extirpation, cultural landscape formation.

Local scale impacts: These may include soil degradation, soil erosion, air pollution, oil spills, overgrazing, cultural eutrophication, river pollution small-scale subsidence and afforestation.

human impact on environment (1) *The spatially hierarchical basis of human impacts on environment* although conservational and sustainable in nature, not only cause changes in flora and fauna but also modify water, sediment and nutrient transfer mechanisms. Ecologically speaking, the creation of managed AGRO-FORESTRY drastically modifies the functioning of the existing ecosystem.

In many arid and semi-arid areas ENVIRONMENTAL DEGRADATION, notably DESERTIFICATION, has emerged as a major problem in land management, as has SALINISATION and water-logging of irrigated agricultural land. The advent of BIOTECHNOLOGY and GENETIC ENGINEERING is an additional recent agent of environmental change, which permanently affects species composition and diversity. The overall impact of agriculture has been immense and has affected all aspects of the environment.

The exponential growth of industry since the INDUSTRIAL REVOLUTION has been accompanied by diverse and damaging environmental impacts on a global scale. The provision of accommodation and services for the growing number of industrial employees during the INDUSTRIAL REVOLUTION was, in itself, an agent of environmental change. The emplacement of artificial land surfaces and buildings altered albedo, wind patterns, vegetation composition and runoff. The principal effects of industry, however, have been environmental POLLUTION and ACIDIFICATION. The emission of particulate matter from FOSSIL FUEL consumption has resulted in high DUST concentrations in the troposphere, which contribute to the formation of PHOTOCHEMICAL SMOG and HEAT ISLAND effects. Similarly, CARBON DIOXIDE released into the atmosphere is a significant factor in global warming. The formation of dilute nitric and sulphuric acids in the atmosphere has been cited as the dominant cause of world-wide lake and stream acidification. ACID RAIN and runoff have caused a significant decrease in productivity in AQUATIC ENVIRONMENTS.

METAL POLLUTION is another significant human impact in the industrialised world. The dumping of waste metal products has caused widespread industrial LAND DEGRADATION and serious water pollution through HEAVY METAL contamination. Similarly, contaminated runoff from agricultural land, which transports artificial fertilisers, pesticides and plant nutrients into lake basins, has been the dominant factor in cultural EUTROPHICATION. Initially, eutrophication may stimulate productivity, but it eventually leads to serious oxygen depletion and declining productivity. In many areas the discharge of sewage and industrial waste into the sea has been responsible for MARINE POLLUTION and OIL SPILLS (e.g. the *Exxon Valdez*, Alaska, 1989, and the *Sea Empress*, Milford Haven, 1996), which have caused long-term damage to marine, shoreline and ESTUARINE ENVIRONMENTS. The nuclear energy industry has also been highly publicised as a source of environmental pollution, particularly in relation to NUCLEAR ACCIDENTS.

Although pollution is normally judged to be the main industrial impact, others do exist. Ground SUBSIDENCE caused by dewatering operations in AQUIFERS is common in mining areas and often results in the formation of large surficial chasms. The provision of transport routes is another example: the growing demand for roads and railways since the industrial revolution has meant that more and more land has been given over to this type of landuse. Roads in particular often require the destruction of nat-

ural habitat and modifications to the landscape in the form of hillside cuttings and tunnels. COASTAL (SHORE) PROTECTION works are a further example with the emplacement of groynes, jetties, breakwaters, sea-walls and riprap often causing disruption to sediment transport patterns and erosion through wave reflection. Similarly, modification of river channels by artificial straightening, concrete channelling, dams, barrages, weirs, diversions and dredging may significantly alter flow regime and discharge characteristics and may upset the balance between erosion and deposition.

As the standard of living in the developed world has increased, it has been paralleled by higher disposable income and more leisure time. As a result, the diversity of leisure activities has increased and recreation is viewed as an additional agent of environmental change. Vegetation TRAMPLING is the most obvious direct impact of recreation. Skiing, walking, boating, camping, fishing, riding and picnicking have all been cited as causes of trampling, which may result in local extinction of species and patch generation, as well as accelerated soil erosion. Motorcross, rallying, mountain-biking and military training practices (although the last is not recreational) may also result in denuded vegetation cover, soil compaction, enhanced nutrient cycling, intensified runoff, waterlogging or deflation.

The diversification and intensification of human impact on environment leads to speculation over the sustainability of human activities in all walks of life. In the future, further population growth and the increasing sophistication and pervasiveness of technological advances will surely further intensify and magnify human impacts to the point where they outweigh natural environmental changes, as illustrated in Figure 2.   JGS

[See also AGRICULTURAL IMPACT ON CLIMATE, AGRICULTURAL IMPACT ON GEOMORPHOLOGY, AGRICULTURAL IMPACT ON SOILS, AIR POLLUTION, CLIMATIC CHANGE, ENVIRONMENTAL CHANGE, GREENHOUSE EFFECT, WATER POLLUTION]

Andrew, R.W. and Jackson, J.M. 1996: *Environmental science: the natural environment and human impact.* Harlow: Longman. **Bell, M. and Walker, J.C.** 1992: *Late Quaternary environmental change.* Harlow: Longman. **Bowlby, S.R. and Mannion, A.M.** 1992: Perspective and prospect. In Bowlby, S.R. and Mannion, A.M. (eds), *Environmental issues in the 1990s.* Chichester: Wiley, 327–336. **Goudie, A.G.** 2000: *The human impact on the natural environment,* 5th edn. Oxford: Blackwell. **Goudie, A. and Alexander, D.E. (eds)** 1997: *The human impact reader: readings and case studies.* Oxford: Blackwell. **Goudie, A. and Viles, H. (eds)** 1996: *The earth transformed: an introduction to the human impact on the environment.* Oxford: Blackwell. **Messerli, B., Grosjean, M., Hofer, T. et al.** 2000: From nature-dominated to human-dominated environmental changes. *Quaternary Science Reviews* 19, 459–579. **Meyer, W.B.** 1999: *Human impact on the Earth.* Cambridge, Cambridge University Press. **Roberts, N.,** 1996: The human transformation of the earth's surface. *International Social Science Journal* 48, 493–510. **Roberts, N.** 1998: *The Holocene: an environmental history,* 2nd edn. Oxford: Blackwell. **Simmons, I.G.** 1996: The modification of the Earth by humans in pre-industrial times. In Douglas, I., Huggett, R. and Robinson, M. (eds), *Companion encyclopedia of geography: the environment and humankind.* London: Routledge, 137–156. **Wackernagel, M. and Rees, E. (eds)** 1996: *Our ecological footprint: reducing human impacts on the Earth.* Philiadelphia, PA: New Society. **Wright, L.,** 1993: *Environmental systems and human impact.* Cambridge: Cambridge University Press.

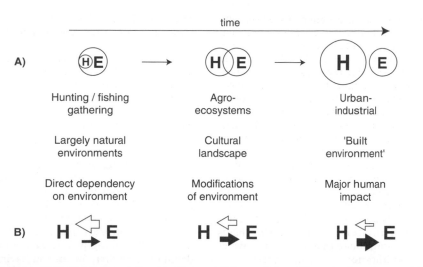

human impact on environment (2) *Changing relationships between humans (H) and the natural environment (E) over the course of the Holocene, including (A) the nature of interactions and (B) relative impacts (after Roberts, 1998)*

human impact on hydrology The direct and indirect effects of human activities on hydrology are diverse. All components of the HYDROLOGICAL CYCLE are potentially subject to human modification, with the speeds, magnitude and quality of water following different hydrological pathways all affected.

Landuse changes have had radical effects on: EVAPOTRANSPIRATION and its main component processes, INTERCEPTION and *transpiration*; INFILTRATION and RUNOFF PROCESSES; and RIVER DISCHARGE and the STORM HYDROGRAPHY. DEFORESTATION and URBANISATION tend to lead to higher streamflow, reduced interception and transpiration, increased streamflow and larger and quicker streamflow responses to rainstorms and increased flood frequency. URBANISATION, industrial activity and mining tends to lead to WATER POLLUTION. Forest fires (often the result of human action) can also lead to enhanced *overland flow* and storm peaks, as well as to a flush of nutrient losses. Agriculture is a major way in which humans have affected hydrology. ABSTRACTIONS from GROUNDWATER and rivers reduce flows, especially in summer. *Overland flow* and THROUGHFLOW from agricultural land can also contain high levels of nutrients and pesticides, leading to potential water pollution and EUTROPHICATION problems.

Structural changes within catchments can also induce important hydrological changes. Perhaps the most visible example of this is *dams*. Dam construction increased markedly across the world during the second half of the twentieth century in order to provide water for industry, HYDROPOWER generation, domestic consumption and agriculture. Dams also allow regulation of river flows, particularly reduced peak flows and maintenance of stable low flows. However, the modification of flow regime may lead to changes in the downstream channel, in part due to the reduction in sediment load (see RESERVOIRS: ENVIRONMENTAL EFFECTS). Many channels have been subject to *flood protection measures* with the construction of levées or through river CHANNELISATION, though frequently such measures merely transfer flooding problems further downstream.   *ADT*

[See also AFFORESTATION: IMPACTS ON HYDROLOGY AND GEOMORPHOLOGY, DEFORESTATION: HYDROLOGICAL AND GEOMORPHOLOGICAL IMPACTS]

Canter, L. 1985: *Environmental impact of water resources projects.* Boca Raton, FL: Lewis. **Cerdà, A., Imeson, A.C. and Calvo, A.** 1995: Fire and aspect induced differences on the erodibility and hydrology of soils at La Costera, Valencia, southeast Spain, *Catena,* 24, 289–304. **Goudie, A.** 1993: Human impact on the natural environment. Oxford: Blackwell, 143–158 **Goulding, K.W.T., Matchett, L.S., Heckrath, G.** *et al.* 1996: Nitrogen and phosphorus flows from agricultural hillslopes. In Anderson, M.G. and Brooks, S.M. (eds), *Advances in hillslope processes.* Vol. I. Chichester: Wiley, 231–247. **Lillehammer, A. and Saltveit, S.V. (eds)** 1984: *Regulated rivers.* Oslo: Universitetsforlaget. **National Research Council** 1993: *Hydrology, hydraulics and water quality.* Washington, DC: National Academy Press. **Nilsson, C., Ekblad, A., Gardfjell, M. and Carlberg, B.** 1991: Long-term effects of river regulation on river margin vegetation. *Journal of Applied Ecology* 28, 963–987. **Peiry, J.-L., Givel, J. and Panton, G.** 1999: Hydroelectric developments, environmental impact. In Alexander, D.E. and Fairbridge, R.W. (eds), *Encyclopaedia of environmental science.* Dordrecht: Kluwer, 332–336. **Shaw, E.M.** 1988: *Hydrology in practice.* London: Chapman and Hall. **Williams, G.E. and Wolman, M.G.** 1984: Downstream effects of dams on alluvial rivers. *United States Geological Survey Professional Paper* 1286, 1–83.

human impact on landforms and geomorphic processes Geomorphological impact by our ancestors for most of the five million years or so of human development was negligible. From about one million years ago and particularly from the beginning of the HOLOCENE, the scope and magnitude of impacts have multiplied as a result of population growth and technological developments. The earliest significant impacts by hunter–gatherer societies resulted from deliberate burning of vegetation, chiefly to encourage regrowth of new shoots and hence attract grazing animals for easier killing. This burning led to accelerated SOIL EROSION. The development of agriculture in different subtropical centres (e.g. FERTILE CRESCENT) in the early Holocene coincided broadly with a substantial increase in world population. In NEOLITHIC times, the hafted polished stone axe allowed efficient clearance of forest and woodland for cultivation. Development of the plough in the Old World meant that soils hitherto too heavy for agriculture could now be cultivated. Infills of valleys with sediment dating from BRONZE and IRON AGES in Europe indicate the accelerated erosion brought about by AGRICULTURAL INTENSIFICATION. Requirements of large amounts of wood for metal smelting doubtless contributed significantly to the loss of woodland and increased soil erosion. Cultivation requires a sedentary lifestyle, which leads to settlements and URBANISATION. The emergence of urban civilisations in Egypt, Mesopotamia, India and China in mid to late Holocene times was conditioned by specific physical conditions – major alluvial river valleys with low and/or unreliable rainfall. Here, successful agriculture on these fertile lands could only be achieved with major manipulation of water resources, involving modification of stream courses.

From the fifteenth century onwards, European COLONIALISM caused the rapid transformation of large tracts of the world (e.g. North America, Australia) from often a relatively unmodified state to one highly altered through intensive agriculture by technologically advanced societies. The most notable geomorphological consequence was a rapid and marked increase in soil erosion. The mechanical technology available over the last two centuries has enabled the Earth's land surface to be transformed on a scale and at a rate not possible under pre-industrial societies. Such changes (e.g. COASTAL modification, river management), impressive though they may be, are often localised, or at most regional, in their extent. Anthropogenic atmospheric GREENHOUSE GAS modifications, on the other hand, are anticipated to make future indirect human impact on geomophology potentially global in extent, as the precipitation and temperature changes resulting from CLIMATIC CHANGE cross geomorphological thresholds in certain environmental conditions.

RAS

[See also AGRICULTURAL IMPACT ON GEOMORPHOLOGY, CLIMATIC CHANGE: POTENTIAL FUTURE GEOMORPHOLOGICAL IMPACTS, CLIMATIC CHANGE: PAST IMPACT ON LANDFORMS AND PROCESSES, PERMAFROST DEGRADATION]

Brown, E.H. 1970: Man shapes the Earth. *Geographical Journal* 136, 74–85. **Goudie, A.** 1993: Human influence in geomorphology. *Geomorphology* 7, 37–59. **Jennings, J.N.** 1966: Man as a geological agent. *Australian Journal of Science* 28, 150–156. **Marsh, G.P.** 1864: *Man and nature*. New York: Scribner. **Nir, D.** 1983: *Man, a geomorphological agent: an introduction to anthropic geomorphology*. Jerusalem: Keter Publishing. **Roberts, N.** 1996: The human transformation of the Earth's surface. *International Social Science Journal* 150, 493–510. **Slaymaker, O.** (ed.) 2000: *Geomorphology, human activity and global environmental change*. Chichester: Wiley.

human impact on soil

Soils are the basis of all terrestrial ecosystems and are the main medium for human food production. They also provide fodder for animal husbandry, as well as timber for shelter and fuel. As a consequence, human impact on soils ranges from minimal changes to profound alteration. In most areas of the inhabited world there has been a history of human soil modification over thousands of years. In terms of site characteristics, the land surface may be altered by levelling or TERRACING. The land surface may have been raised or lowered through sedimentation or erosion or embanked to prevent flooding. In some places, soils have been created where none existed previously by importation of soil material; in others pre-existing soils have been buried, producing PALAEOSOLS.

The SOIL HORIZONS of the SOIL PROFILE may have been mixed by cultivation. The surface soil may be enriched or depleted of organic matter; lime and FERTILISERS may have been added to enhance fertility. *Contamination* by benign substances (see CONTAMINANT) or POLLUTION with HAZARDOUS WASTE may occur, especially in urban areas. Heavy machinery may be used to shatter compacted or indurated soil horizons, but increasing use of heavy implements is causing soil COMPACTION. For wetland rice cultivation, subsoils have been compacted to retain water in the paddy field and the surface soil *puddled* to provide suitable conditions for the young plants. Soils profoundly altered to a depth of more than 50 cm and having ANTHROPEDOGENIC HORIZONS composed of additions of earthy or turfy material, irrigation sediments, organic wastes and possessing the puddled and mottled layers of rice paddy soils are classified as ANTHROSOLS.

EMB

[See also AGRICULTURAL IMPACT ON SOILS, CONTAMINATED LAND, METAPEDOGENESIS]

Bell, M. and Boardman, J. (eds) 1992: *Past and present soil erosion: archaeological and geographical perspectives*. Oxford: Oxford Books. **Bidwell, O.W. and Hole, F.D.** 1965: Man as a factor in soil formation. *Soil Science* 99, 65–72. **Bridges, E.M.** 1978: Interaction of soil and mankind. *Journal of Soil Science* 29,125–139. **Bridges, E.M. and de Bakker, H.** 1998: Soil as an artifact: human impact on the soil resource. *Land* 1,197–215. **Lahmar, R.** 1998: *Des sols et des hommes: récits authentiques de gestion de la ressource sol*. Paris: Éditions Charles Léopold Mayer. **Yaalon, D.H. and Yaron, B.** 1966: Framework for man-made soil changes – an outline for metapedogenesis. *Soil Science* 102, 272–277.

human impact on terrestrial vegetation

Contemporary impacts vary in scale and intensity. The most intense impacts are found in highly managed AGROECOSYSTEMS where the original vegetation has been replaced by CULTIVARS. Next in intensity are areas with highly modified vegetation, such as the TROPICAL GRASSLANDS that have replaced TROPICAL FOREST. Finally, there are environments where the vegetation has not been substantially altered, as in certain remote BOREAL FOREST regions in northern latitudes. Impact can be direct (cutting) or indirect (fire, grazing and pollution). One of the main characteristics of human disturbed vegetation is structural and functional simplification. When PRIMARY WOODLAND is removed (see FOREST CLEARANCE), it is replaced, in different HABITATS, by SECONDARY WOODLAND, wet and dry grasslands and HEATHLANDS, and semi-arid SCRUB. Vegetation replacement in this way typically involves a reduction in plant DIVERSITY, plant height and layering, as well as simplified age profiles. The same applies to the replacement of vegetation by plant CULTIVARS and the human degradation of grass and SCRUB vegetation in semi-arid areas to DESERT landscapes in the process of DESERTIFICATION.

Many other aspects of the environment are transformed by the interference with the vegetation cover, including associated FAUNA, SOIL, CLIMATE, the HYDROLOGICAL

CYCLE and land surface features. Degraded vegetation with reduced BIODIVERSITY and more open, simplified HABITATS has encouraged a greater number/intensity of BIOLOGICAL INVASIONS. SOIL ORGANIC CONTENT is lost if grassland is degraded to desert SCRUB in semi-arid environments (e.g. the SAHEL). When the moisture recycling capacity of forest is lost during FOREST CLEARANCE (as in parts of the TROPICAL RAIN FOREST in Brazil), rainfall amounts can be reduced by 50%. The removal of the protective vegetation cover can accelerate SOIL LOSS by wind and WATER EROSION within and outside the deforested area. In the humid tropics, vegetation removal can lead to other sterile landscapes (e.g. DURICRUSTS).

In terms of ECOLOGICAL SUCCESSION, humans have altered CLIMAX VEGETATION communities and maintained many of them by FIRE, GRAZING and direct cutting at earlier PLAGIOCLIMAX stages. The maintenance by FIRE and GRAZING of well adapted vegetation types, including TEMPERATE and TROPICAL GRASSLANDS, Mediterranean SCRUB (*maquis, garrigue*) and cool temperate MOORLAND and heathland, are examples of this mechanism.   *GOH*

[See also DEFORESTATION: HUMAN IMPACT ON VEGETATION HISTORY, LANDUSE IMPACTS]

Barrow, C. J. 1991: *Land degradation: development and breakdown of terrestrial environments.* Cambridge: Cambridge University Press. **Goudie, A.** 2000: *The human impact on the natural environment,* 5th edn. Oxford: Blackwell Science. **Holzner, W., Werger, M.J.A. and Ikusima, I. (eds)** 1983: *Man's impact on vegetation.* The Hague: Junk. **Tivy, J. and O'Hare, G.** 1981: *Human impact on the ecosystem.* Edinburgh: Oliver and Boyd. **Tomaselli, R.** 1977: The degradation of the Mediterranean maquis. *Ambio* 6, 356–362.

human impact on vegetation history Human activity has been a driving force in vegetation change in many parts of the world in the latter half of the HOLOCENE. Human impacts have included *burning*, FOREST CLEARANCE, FOREST MANAGEMENT, agriculture (see AGRICULTURAL IMPACTS), DRAINAGE, IRRIGATION and the INTRODUCTION of plants to areas beyond their original geo-graphical range (see Table). The main sources of evidence for human impacts on long-term vegetation change are analysis of POLLEN, PLANT MACROFOSSILS and *charcoal*.

HUNTING, FISHING AND GATHERING peoples have had relatively minor effects, involving, for example, periodic burning of vegetation or creation of small temporary forest clearings. Agricultural peoples produced more pronounced changes, connected with the need to create fields for crop cultivation. In northwest Europe the original postglacial woodland began to be cleared in the NEOLITHIC period, although much of the landscape remained densely wooded until later prehistory. The spread of agriculture was accompanied not only by the expansion in the geographical range of domesticated plants and animals, but also by weeds, creating a variety of new plant communities of periodically disturbed soils. Shade-intolerant plants such as grasses, docks (*Rumex*) and ribwort plantain (*Plantago lanceolata*) were favoured by the creation of openings in the woodland, and those with distinctive pollen have been used in POLLEN ANALYSIS as ANTHROPOGENIC INDICATORS to assist in identifying human impacts in the pollen record.

Woodland clearance allowed an expansion of communities of open ground, such as *grassland*, HEATHLAND and MOORLAND, often maintained by grazing and/or burning. Sometimes it is difficult to disentangle the roles of human activity and natural factors, such as CLIMATIC CHANGE and soil maturation, in the creation of such communities. For example, BLANKET MIRES in northwest Europe began to form at various dates, from the MESOLITHIC period onward. While in the highest-rainfall areas PEAT formation probably began naturally, at some sites woodland clearance may have caused the ground-water level to rise, triggering the onset of peat accumulation.

It is clear that many of today's vegetation communities result from an interaction of human activity and natural ENVIRONMENTAL CHANGE operating on a variety of time scales. Understanding the long-term history of these communities is of particular importance for conserving them and in appreciating the potential effects of future changes in climate and human activity.   *SPD*

human impact on vegetation history *Relative human impact on vegetation and other aspects of the environment in the uplands of England and Wales since early Mesolithic times (after Simmons, 1996)*

Period	Impact	Category
Early Mesolithic	Apparently transient, but hints of locally severe impacts on woodland and soils	1
Later Mesolithic	Widespread management of woodlands and their edges; consequent paludification	2
Agricultural prehistory	Loss of woodland, introduction and expansion of field systems temporary and permanent, domestic cattle, etc; expanded population's need for wood. PALUDIFICATION	5
Roman	Demands for corn, road-building	3
Mediaeval	Permanent parcelling of landscape: common lands and grazing management systems emplaced; monasteries add to grazing, ironworking; also remove some settlements. Some evidence for continued paludification	6
Early modern	Steady state with cumulative effects of grazing and metal extraction; sheep progressively replace cattle	4
Nineteenth century	Industrialisation, especially of extractive processes; moor management for sport in east. Much vegetation acidified from rain-out	10
Twentieth century	Collapse of some industries; forestry and recreation gain in importance; sheep grazing progressively on economic knife-edge; expansion of bracken very rapid	7

[See also AGRICULTURAL HISTORY, BOREAL FOREST: HISTORY, BRONZE AGE: LANDSCAPE IMPACTS, COPPER AGE: LANDSCAPE IMPACTS, DEFORESTATION, FIRE IMPACTS, FOREST CLEARANCE, GRAZING HISTORY, IRON AGE: LANDSCAPE IMPACTS, LANDNAM, LANDUSE, MESOLITHIC LANDSCAPE IMPACTS, MIDDLE AGES: LANDSCAPE IMPACTS, NEOLITHIC LANDSCAPE IMPACTS, PALAEOLITHIC HUMAN—ENVIRONMENTAL RELATIONS, ROMAN PERIOD: LANDSCAPE IMPACTS, TUNDRA VEGETATION: HUMAN IMPACTS, VEGETATION HISTORY]

Birks, H.H., Birks, H.J.B., Kaland, P.E. and Moe, D. (eds) 1988: *The cultural landscape – past, present and future.* Cambridge: Cambridge University Press. **Chambers, F.M. (ed.)** 1993: *Climate change and human impact on the landscape.* London: Chapman & Hall. **Goudie, A.** 1993: *The human impact on the natural environment,* 4th edn. Oxford: Blackwell Science. **Huntley, B. and Webb III, T. (eds)** 1988: *Vegetation history.* Dordrecht: Kluwer. **Roberts, N.** 1998: *The Holocene: an environmental history,* 2nd edn. Oxford: Blackwell Science. **Simmons, I.G.** 1996: *Changing the face of the Earth: culture, environment, history,* 2nd edn. Oxford: Blackwell Science. **Simmons, I.G.** 1996: *The environmental impact of later Mesolithic cultures: the creation of moorland landscape in England and Wales.* Edinburgh: Edinburgh University Press.

humanism   A philosophical viewpoint that emphasises the distinctively human value, quality and subjectivity in people's lives. This is an approach made distinctive by the central and active role given to human awareness, human agency and human creativity. The rise of humanism in the 1970s resulted especially from a dissatisfaction with mechanistic models of social science and HUMAN GEOGRAPHY. Humanism offers another way of understanding people and their environments in contrast to more structural and positivist approaches. Principal humanist philosophies include IDEALISM, PHENOMENOLOGY and REALISM.

ART

Buttimer, A. 1999: Humanism and relevance in geography. *Scottish Geographical Journal* 115, 103–116. **Daniels, S.J.** 1985: Arguments for a humanistic geography. In Johnston, R. J. (ed.), *The future of geography.* London: Methuen, 143–158.

humanities   That set of academic disciplines belonging to the 'arts' and typically including HISTORY, languages and philosophy. The term is most commonly used in relation to the organisation of subjects or disciplines within university faculties.

DTH

Cartermill 1996: *Current research in Britain: humanities.* London: Cartermill.

humic acids   A complex mixture of dark-coloured organic substances precipitated by acidification of a dilute alkali extract from soil. Humic acids have a higher molecular weight than FULVIC ACIDS.

EMB

[See also SOIL DATING]

Flaig, W., Beutelspacher, H., and Rietz, E. 1975: Chemical composition and physical properties of humic substances. In Giesking, J.E. (ed.), *Soil components.* Vol. 1. Berlin: Springer, 1–211.

humid tropics: ecosystem responses to environmental change   Many coastal ecosystems in the TROPICAL environments are vulnerable to SEA-LEVEL RISE and an increase in WAVE ENERGY that may result from GLOBAL WARMING. Rapid rates of rise may reduce the extent of MANGROVE swamps. Many mangrove species are extremely sensitive to variation in their hydrological or tidal regimes. CORAL REEFS have been shown to be extremely sensitive to changes in water temperature. Increased water temperatures during the AD 1997–1998 EL NIÑO–SOUTHERN OSCILLATION (ENSO) event resulted in widespread CORAL BLEACHING.

Evidence of significant changes in the distribution and composition of tropical forests during periods of CLIMATIC CHANGE during the QUATERNARY suggests that present-day forests are unlikely to be resistant to future climatic change. Humid tropical ecosystems are likely to be more sensitive to changes in soil water availability than to temperature. Large areas of southern and eastern Amazonia have a climate that is marginal for TROPICAL RAIN FOREST and survive by having very deep root systems that can tap GROUNDWATER sources. Forest FRAGMENTATION and degradation exacerbate the vulnerability to DROUGHT and fire. The high resistance of humid tropical ecosystems to BIOLOGICAL INVASION is now believed to be attributable primarily to rapid rates of recovery after disturbance. Significant ENVIRONMENTAL CHANGE observed in forest fragments has made them much more vulnerable to invasions.

Tropical rain forests may respond to rising CARBON DIOXIDE levels by exhibiting faster turnover. Undisturbed tropical forest may be an important CARBON SINK for carbon, rather than simply a store, and there is some evidence that sink strength may be increasing as a result of the fertilisation effect of elevated levels of atmospheric carbon dioxide (see CARBON DIOXIDE FERTILISATION). As humid tropical plantations are among the most productive and tropical rain forests contain the largest total stock of biomass carbon, there is growing interest in the potential for managing humid tropical forests for CARBON SEQUESTRATION.

Changes inflicted on humid tropical ecosystems may also have a profound effect on the global environment. Annual global emissions of carbon from tropical DEFORESTATION are estimated to be about 1.9 Gt C, about one third of anthropogenic emissions. In most humid tropical countries, low incomes, high national debt and a dependency on agriculture make large human populations vulnerable to climatic variation.

NDB

[See also CARBON CYCLE, REFUGIA]

Field, C.D. 1995: Impact of expected climate change on mangroves. *Hydrobiologia* 295, 75–81. **Grace, J. and Malhi, Y.** 1999: The role of rain forests in the global carbon cycle. *Progress in Environmental Science* 1, 177–193. **Phillips, O.L. and Gentry, A.H.** 1994: Increasing turnover through time in tropical forests. *Science* 263, 954–958.

humidity   Relating to the WATER VAPOUR content of air. The *absolute humidity* is the mass of the water vapour in a given volume of air. The *specific humidity* is the ratio of the mass of the water vapour in the air to the combined mass of the water vapour and the air; it is approximated by the *mixing ratio* (the ratio of the mass of water vapour to the mass of dry air). The *relative humidity* (usually expressed as a percentage) is the ratio of the water-vapour content of the air to the maximum amount the air could hold at that temperature and volume. This is the common usage of the term 'humidity': 0% is 'dry' air; 100% is totally saturated with water vapour. Relative humidity may also be expressed as the ratio of the mixing ratio of the air to the saturation mixing ratio at the same temperature and pressure. Humidity is measured using a *hygrometer*. *JAM*

Fairbridge, R.W. and Oliver, J.E. 1987: Humidity. In Oliver, J.E. and Fairbridge, R.W. (eds), *The encyclopedia of climatology*. New York: Van Nostrand Reinhold, 479–483.

humification The processes of breakdown and synthesis of organic residues through biological activity, microbial synthesis and chemical reactions to HUMUS. In soils, as plant material is utilised by successive groups of animals for their nutrition, it is broken into smaller fragments and finally invaded by fungi and bacteria that complete the breakdown. In the process, the organic material is changed from green to brown and eventually to black *amorphous humus*. Once formed, the humus breaks down relatively slowly, offering the possibility of CARBON SEQUESTRATION in soils, which helps mitigate GLOBAL WARMING by abstracting carbon dioxide from the atmosphere and improves overall SOIL QUALITY. *EMB*

[See also PEAT HUMIFICATION]

Bolt, G.H. and Bruggenwert, M.G.M. (eds) 1976: *Soil chemistry*: A. *Basic elements* [*Developments in Soil Science* 5A]. Amsterdam: Elsevier. **Wild, A. (ed.)** 1988: *Russell's soil conditions and plant growth*, 11th edn. Harlow: Longman.

hummocky cross-stratification (HCS) A SEDIMENTARY STRUCTURE comprising gently undulating LAMINATION in fine- to medium-grained sand, characterised by convex-upward laminae with no preferred direction of DIP, that define hummocks and troughs (swales) with an amplitude of a few centimetres and spacing of a metre or so. It is generally agreed that HCS forms in response to combined flows. Most is reported from deposits that accumulated below fairweather WAVE BASE in response to STORMS, in which a storm-induced unidirectional current is combined with a storm-wave-induced oscillatory current, although there has been no unequivocal recognition of appropriate BEDFORMS in modern sediments. A related structure in which the swales are more prominent than the hummocks is sometimes recognised as *swaley cross-stratification (SCS)*. *GO*

Cheel, R.J. and Leckie, D.A. 1993: Hummocky cross-stratification. In Wright, V.P. (ed.), *Sedimentology review* 1. Oxford: Blackwell, 103–122. **Harms, J.C., Southard, J.B., Spearing, D.R. and Walker, R.G.** 1975: *Depositional environments as interpreted from primary sedimentary structures and stratification sequences* [*Short Course* 2]. Tulsa, OK: Society of Economic Paleontologists and Mineralogists. **Ito, M., Ishigaki, A., Nishikawa, T. and Saito, T.** 2001: Temporal variation in the wavelength of hummocky cross-stratification: implications for storm intensity through Mesozoic and Cenozoic. *Geology* 29, 87–89.

hummocky moraine An apparently chaotic assemblage of MORAINE mounds and ridges. The term has been used in a wide range of senses to refer to glacial landforms of different origins, although most commonly it is used to refer to moraines deposited by the ABLATION of DEBRIS-MANTLED GLACIERS. *DIB*

Benn, D.I. and Evans, D.J.A. 1998: *Glaciers and Glaciation*. London: Arnold.

humus The relatively resistant, usually dark brown to black fraction of soil organic matter that results from the biological breakdown and synthesis of organic residues

(HUMIFICATION). It is composed mainly of carbon and nitrogen in complex organic molecules, based on a chemical structure similar to the benzene ring; other elements are present in low amounts. Humus is capable of forming an intimate association with clay minerals to form the CLAY–HUMUS COMPLEX, the most reactive part of the soil. In acid conditions, beneath needle-leaved, coniferous BOREAL FOREST or HEATHLAND, collembola and mites are the most important animals, together with fungi, involved and the process takes place above the mineral soil. The resulting layers of *litter, fermentation* and *humus* are called MOR. In base-rich conditions, beneath broad-leaved deciduous TEMPERATE FOREST and particularly grassland, the organic matter is broken down and the humus is incorporated into the surface soil as MULL. Bacteria are more important than fungi and earthworms actively draw organic matter down into the soil so there is no surface accumulation. MODER is an intermediate form, often occurring on acid and poorly drained sites, having litter and fermentation layers of approximately equal thickness. *EMB*

[See also FULVIC ACIDS, HUMIC ACIDS]

Kononova, M.M. 1975: Humus of virgin and cultivated soils. In Gieseking, J.E. (ed.), *Soil Components*. Vol. 1. *Organic components*. Berlin: Springer. **Krosshavn, M., Bjorgum, J.O., Krane, J. and Steinnes, E.** 1990: Chemical structure of terrestrial humus materials formed from different vegetation characterised by solid-state ^{13}C NMR with CP-MAS techniques. *Journal of Soil Science* 41, 371–377. **Wild, A. (ed.)** 1988: *Russell's soil conditions and plant growth*, 11th edn. Harlow: Longman.

hunting, fishing and gathering A subsistence economy and lifestyle based on hunting wild animals (e.g. deer) for meat, hide and bone, fishing in coastal and fresh waters and collecting wild plants (e.g. nuts, seeds and fungi) for food. The balance between each type of activity in the economy can vary (e.g. seasonally or geographically). Unwooded areas were more likely to have been used for hunting (e.g. Late Pleistocene STEPPE of Europe) whilst gathering probably dominated in wooded landscapes (e.g. MESOLITHIC lowland Britain) where mobility would have been more restricted. The success of such an economy depends on an intimate knowledge of the environment and the resources available. Such communities could have had a significant impact on the environment and Late Pleistocene extinctions of MEGAFAUNA have been attributed to OVERHUNTING. Extensive burning of woodland was often undertaken to drive game, to ease movement by increasing the extent of open land, to increase the quality of browse or to encourage certain food plants (e.g. hazel nuts). *LD-P*

[See also NOMADISM, PASTORALISM, SEDENTISM]

Bettinger, R.L. 1991: *Hunter-gatherers: archaeology and evolutionary theory*. New York: Plenum. **Bunn, H.T. and Ezzo, J.A.** 1993: Hunting and scavenging by Plio-Pleistocene hominids: nutritional constraints, archaeological patterns and behavioural implications. *Journal of Archaeological Science* 20, 365–398. **Burch, E.S. and Ellanna, L.J. (eds)** 1994: *Key issues in hunter-gatherer research*. Oxford and Providence, RI: Berg. **Frison, G.C.** 1978: *Prehistoric hunters of the High Plains*. New York: Academic Press. **Head, L.** 1989: Prehistoric Aboriginal impacts on Australian vegetation: an assessment of the evidence. *Australian Geographer* 20, 37–46. **Ingold, T.** 1980: *Hunters, pastoralists and ranchers*. Cambridge: Cambridge University Press. **West, D.** 1997: *Hunting strategies in Central Europe during the last glacial* *maximum* [*BAR International Series* 672]. Oxford: British Archaeological Reports.

hybridisation The process of crossing between individuals of genetically distinct populations (such as species or subspecies), resulting in individuals (*hybrids*) that may be fertile or sterile. The offspring are more likely to be sterile with greater genetic distance between the parents.
KDB

hydragric horizon A subsurface ANTHROPEDOGENIC HORIZON lying beneath the ANTHRAQUIC horizon associated with wet cultivation of rice having iron–manganese segregations or evidence of reducing conditions in pores with a yellowish colour. *EMB*

hydration The ADSORPTION of water by certain minerals causing expansion and setting up stresses within a rock. It is a type of CHEMICAL WEATHERING process. *RAS*

hydraulic conductivity A measure of the ability of rock or soil to transmit water at a specified state of wetness. It is affected by porosity, the number and size of structural fractures, and fluid viscosity. The term is normally applied to flow in saturated soil (*saturated hydraulic conductivity*); if used for flows in unsaturated soil, then values of hydraulic conductivity will vary with the soil water content. *ADT*

[See also INFILTRATION, PERMEABLE, THROUGHFLOW]

hydraulic geometry The hydraulic geometry concept of L.B. Leopold and T. Maddock summarises the relationships between river-channel form adjustment and changing stream discharge and sediment load. *At-a-station hydraulic geometry* defines how width (w), mean depth (d) and mean velocity (v) at a specific cross-section change over time as discharge (Q) rises or falls. *Downstream hydraulic geometry* describes how, along a given river and for a given flow frequency (e.g. mean annual discharge), these three properties adjust in a downstream direction to accommodate an ever-increasing discharge. The relations are power functions:

$$w = aQ^b; \; d = cQ^f; \; v = kQ^m$$

where the coefficients a, c, k and the exponents (slopes) b, f, m are today derived from least-squares fits to the empirical data. Given that $Q = w . d . v$, then:

$$b + f + m = 1 \text{ and } a . c . k = 1$$

Catchment area is often used as a surrogate for discharge in downstream relations.

A certain consistency in the values implies that channels attain an equilibrium size and shape which minimises total work. For the downstream case, $b \approx 0.55$, $f \approx 0.35$ and $m \approx 0.1$. The WIDTH–DEPTH ratio therefore increases downstream and width adjusts most effectively to increasing flow (and is the most responsive to environmental change). A novel finding ($m \approx +0.1$) of the approach was that mean velocity increases downstream, a declining channel slope being more than offset by decreases in channel roughness and increases in channel depth and efficiency. The approach has now been extended to numerous other fluvial variables, such as channel roughness, stream power, shear stress, boundary sediment, BANK EROSION and, as the *river continuum concept*, to fresh-

water ecology. The downstream hydraulic-geometry approach has also been widely and very successfully used to detect, assess and predict downstream impacts of reservoirs, urbanisation and other catchment changes on river channels. *DML*

[See also RESERVOIRS: ENVIRONMENTAL EFFECTS, URBANISATION IMPACTS ON HYDROLOGY]

Ferguson, R.I. 1986: Hydraulics and hydraulic geometry. *Progress in Physical Geography* 10, 1–31. **Leopold, L.B. and Maddock, T.** 1953: The hydraulic geometry of stream channels and some physiographic implications. *United States Geological Survey Professional Paper* 252, 1–57. **Richards, K.S.** 1977: Channel and flow geometry. *Progress in Physical Geography* 1, 65–102.

hydrocarbons Complex organic compounds of carbon and hydrogen that make up PETROLEUM. The burning of hydrocarbons as FOSSIL FUELS releases into the atmosphere GREENHOUSE GASES that were formerly locked up in SEDIMENTARY ROCKS, potentially fuelling GLOBAL WARMING. Incomplete COMBUSTION and the presence of impurities, such as sulphur, contribute to other environmental problems such as PHOTOCHEMICAL SMOG and ACID RAIN. Hydrocarbons are also the basis for petrochemical industries producing, for example, FERTILISERS, PESTICIDES, pharmaceuticals and plastics, each with associated POLLUTION and WASTE MANAGEMENT problems. *GO/JAM*

[See also CHLOROFLUOROCARBONS, ORGANOCHLORIDES, PERSISTENT ORGANIC COMPOUNDS]

Schobert, H.H. 1991: *The chemistry of hydrocarbon fuels.* London: Newnes. **Selley, R.C.** 1997: *Elements of petroleum geology*, 2nd edn. New York: Freeman.

hydrochlorofluorocarbons (HCFCs) Compounds of carbon in which some of the hydrogen atoms have been replaced by chlorine and fluorine. They are less stable than CHLOROFLUOROCARBONS (CFCs) with limited atmospheric lifetimes of 1–20 years and tend to break down in the TROPOSPHERE before diffusing into the STRATOSPHERE. They are therefore around 95% less damaging to the OZONE layer and are widely used as substitutes for CFCs, but they are nevertheless to be phased out following on from the *Montreal Protocol. Hydrofluorocarbons (HFCs)* are viewed as more appropriate substitutes because they contain no chlorine and hence do not attack ozone. *JAM*

[See also HALOGENATED HYDROCARBONS]

Kanakidou, M., Dentener, F.J. and Crutzen, P.J. 1995: A global three-dimensional study of the fate of HCFCs and HFC-134a in the troposphere. *Journal of Geophysical Research* 100, 18781–18801.

hydroclimatology The study of the interaction between climatic processes and the atmospheric components of the HYDROLOGICAL CYCLE. *JBE*

Wendland, W.M. 1987: Hydroclimatology. In Oliver, J.E. and Fairbridge, R.W. (eds), *The encyclopedia of climatology.* New York: Van Nostrand Reinhold, 497–502.

hydrogen isotopes Hydrogen has two STABLE ISOTOPES (natural abundances, $^1H = 99.985\%$ and ^2H or *deuterium*, $D = 0.015\%$) and one *radioisotope* (^3H or *tritium*, T; half-life = 12.5 y). The stable ISOTOPE RATIO is expressed as $^2H{:}^1H$ relative to the Vienna Standard Mean Ocean Water (VSMOW) standard (see STANDARD SUBSTANCE). As hydrogen stable isotopes have the largest relative mass difference, hydrogen exhibits the largest variation in its stable isotope ratio. The hydrogen isotope record from the Vostok ICE CORE (Antarctica) has been used to demonstrate the influence of ORBITAL FORCING on global temperature variations over the last four GLACIAL–INTERGLACIAL CYCLES. The hydrogen and OXYGEN ISOTOPE values of METEORIC WATER are strongly associated. PHOTOSYNTHESIS causes a large ISOTOPIC DEPLETION in ^{2}H of non-exchangeable hydrogen relative to metabolic water. Post-photosynthetic processes cause a large ISOTOPIC ENRICHMENT. Small changes in metabolic activity can therefore strongly influence plant δD values. TREE-RING δD values of non-exchangeable hydrogen have been used to reconstruct both spatial and temporal variations in climate.   *IR*

[See also ISOTOPES AS INDICATORS OF ENVIRONMENTAL CHANGE]

Becker, B., Kromer, B. and Trimborn, P. 1991: A stable-isotope tree-ring timescale of the Late Glacial/Holocene boundary. *Nature* 353, 647–649. **Petit, J.R., Jouzel, J., Raynaud, D. et al.** 1999: Climate and atmospheric history of the past 420,000 years from the Vostok ice core, Antarctica. *Nature* 399, 429–436. **White, J.W.C.** 1989: Stable hydrogen isotope ratios in plants: a review of current theory and some potential applications. In Rundel, P.W., Ehleringer, J.R. and Nagy, K.A. (eds), *Stable isotopes in ecological research* [*Ecological Studies*. Vol. 68]. Berlin: Springer, 142–162. **Yakir, D.** 1992: Variations in the natural abundance of oxygen-18 and deuterium in plant carbohydrates. *Plant, Cell and Environment* 15, 1005–1020.

hydrogeology
The branch of GEOLOGY dealing with GROUNDWATER. Groundwater supplies are very sensitive to environmental change.   *GDW*

[See also CONTAMINANT, SALINISATION, SALTWATER INTRUSION]

Domenico P.A. and Schwartz, F.W. 1998: *Physical and chemical hydrogeology*, 2nd edn. New York: Wiley.

hydrograph
A graphical representation of channel DISCHARGE from a DRAINAGE BASIN over time. Analysis of the low-flow and storm-flow characteristics of hydrographs is useful in assessing impacts of LANDUSE CHANGE, CLIMATIC CHANGE, river CHANNELISATION or RESERVOIRS on the catchment WATER balance and responses to rainfall.   *ADT/RPDW*

[See also AFFORESTATION: IMPACTS ON HYDROLOGY AND GEMORPHOLOGY, STORM HYDROGRAPH]

hydro-isostasy
Isostatic loading and unloading by water upon ocean floors and CONTINENTAL SHELVES due to changes in RELATIVE SEA LEVEL. A similar process may occur in response to LAKE-LEVEL VARIATIONS, producing warped shorelines.   *AGD*

[See also ISOSTASY, SEA-LEVEL CHANGE]

Smith, D.E. and Dawson, A.G. 1983: *Shorelines and isostasy.* London: Academic Press.

hydrological balance/budget
An account-book approach to gains and losses (or inputs, storages and outputs) in any hydrological SYSTEM. It may be applied at any spatial scale, from water exchange at a small area of the Earth's surface, to the inputs and outputs of a LAKE or DRAINAGE BASIN, and the global water balance (see Figure). Seasonal fluctuations in rainfall, temperature and

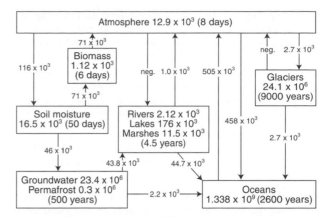

hydrological balance/budget *The global water balance: fluxes are shown as arrows (km^3 y^{-1}); storages (km^3) and turnover times (years) are indicated in the boxes (after Shiklomanov and Sokolov, 1983; Henshaw et al., 2000)*

EVAPOTRANSPIRATION result in parallel changes in storage of soil water and GROUNDWATER and lags between rainfall inputs and streamflow outputs, and hence at the catchment scale meaningful hydrological budgets, or *water budgets*, tend to be calculated on annual timescales.   *JAM/RPDW*

[See also HYDROLOGICAL CYCLE, LAKE-LEVEL VARIATIONS]

Baumgartner, A. and Reichel, E. 1975: *The world water balance: mean annual global, continental and maritime precipitation and runoff.* Amsterdam: Elsevier. **Henshaw, P.C., Charlson, R.J. and Burges, S.J.** 2000: Water and the hydrosphere. In Jacobson, M.C., Charlson, R.J., Rodhe, H. and Orians, G.H. (eds), *Earth system science: from biogeochemical cycles to global change.* San Diego, CA: Academic Press. **Shiklomanov, I.A. and Sokolov, A.A.** 1983: Methodological basis of world water balance investigation and computation. *International Association for Hydrological Sciences Publication* 148 [*Proceedings of the Hamburg Symposium*]. **Street-Perrott, A.F., Beran, M.A. and Ratcliffe, R.A.S. (eds)** 1983: *Variations in the global water budget.* Dordrecht: Reidel.

hydrological cycle
The complex of processes by which water circulates through the Earth–atmosphere–ocean SYSTEM. Also known as the *water cycle*, the hydrological cycle is not only a central concept of HYDROLOGY, but also constitutes a BIOGEOCHEMICAL CYCLE. In general, PRECIPITATION from the atmosphere may generate streamflow, may be stored for varying times in soils, lakes and GROUNDWATER or may be returned to the ATMOSPHERE by EVAPOTRANSPIRATION, but there are many complexities to any simple model. Because most of the Earth's surface is water, a high proportion of the global hydrological cycle occurs over the OCEANS involving only evaporation and precipitation.   *JAM/RPDW*

[See also HYDROLOGICAL BALANCE]

Berner, E.K. and Berner, R.A. 1987: *The global water cycle: geochemistry and environment.* Englewood Cliffs, NJ: Prentice-Hall. **Chahine, M.T.** 1992: The hydrologic cycle and its influence on climate. *Nature* 359, 373–380. **Elahire, E.A.B. and Bras, R.L.** 1996: Precipitation recycling. *Reviews of Geophysics* 34, 367–378.

hydrology
The study of water in all its phases in the Earth–atmosphere–ocean system, especially the nature of

hydrological processes (*physical hydrology*) and their application (*applied hydrology*). It includes not only RUNOFF on the Earth's surface but also GROUNDWATER and water in the ATMOSPHERE and CRYOSPHERE. *JAM*

[See also HYDROGEOLOGY, HYDROLOGICAL BALANCE, HYDROLOGICAL CYCLE, HYDROSPHERE, PALAEOHYDROLOGY]

Jones, J.A.A. 1997: *Global hydrology: processes, resources and environmental management.* Harlow: Longman. Shaw, E.M. 1988: *Hydrology in practice*, 2nd edn. New York: Van Nostrand Reinhold. Ward, A.C. and Elliot, W.J. (eds) 1995: *Environmental hydrology.* Boca Raton, FL: CRC. Ward, R.C. and Robinson, M. 2000: *Principles of hydrology*, 4th edn. London: McGraw-Hill.

hydrolysis (1) A type of CHEMICAL WEATHERING involving the formation of both an acid and a base from a salt when it dissociates with water. Hydrolysis particularly affects feldspar and mica minerals and is thus an especially important process affecting igneous rocks. In humid tropical environments, hydrolysis-induced weathering of igneous rocks can proceed to depths of several tens of metres. (2) The disintegration of organic compounds through their reaction with water. *SHD*

White, A.F. and Brantley, S.L. (eds) 1995: *Chemical weathering rates of silicate minerals.* Washington, DC: Mineralogical Society of America.

hydrophyte A waterplant, either submerged (e.g. *Chara, Najas* and *Potamogeton* spp.) or with floating leaves (e.g. *Potamogeton natans, Nymphaea* spp., *Nuphar* spp.). According to water depth, different species may form belts, the MACROFOSSILS of which can be used for the reconstruction of LAKE-LEVEL VARIATIONS. *BA*

[See also AQUATIC ENVIRONMENT, HYGROPHYTE, PALAEOLIMNOLOGY]

hydropower *Hydroelectric power* generated by flowing water; a form of ALTERNATIVE ENERGY or RENEWABLE ENERGY. Although it provides only a small minority of human needs at the global scale, it meets almost all the requirements of certain countries, such as Norway and Switzerland, where most of the available capacity has been developed. *JAM*

hydrosphere The sum total of all the Earth's water, including the OCEANS, liquid water on the continents (lakes, rivers and GROUNDWATER), SEA ICE and land ice (ICE SHEETS, GLACIERS and snowbeds), and atmospheric moisture. The oceans comprise most of the hydrosphere (some 97% by volume) and around 2% of the remainder is frozen as GLACIER ice. Some would exclude water in the ATMOSPHERE and CRYOSPHERE in a narrower definition of the term focusing on the liquid water. *JAM*

[See also HYDROLOGICAL CYCLE]

Henshaw, P.C., Charlson, R.J. and Burges, S.J. 2000: Water and the hydrosphere. In Jacobson, M.C., Charlson, R.J., Rodhe, H. and Orians, G.H. (eds), *Earth system science: from biogeochemical cycles to global change.* San Diego, CA: Academic Press.

hydrostatic pressure The pressure exerted in a fluid that depends on the depth or vertical head of the fluid and its density. It is important in understanding the flow of water from ARTESIAN springs and wells, and other aspects of GROUNDWATER movement. *JAM*

hydrothermal Activity and processes associated with the movement of GROUNDWATER heated by GEOTHERMAL activity, usually due to proximity to MAGMA in areas of active or dormant VOLCANISM. Hydrothermal activity causes the chemical alteration of rocks, precipitation of minerals (including ORE minerals), and surface features such as hot springs, *geysers* and submarine HYDROTHERMAL VENTS. *JBH*

hydrothermal vent An opening in the Earth's CRUST that emits a jet of hot water containing dissolved compounds. The term is most commonly used in relation to hydrothermal vents on the OCEAN floor which were discovered in the late 1970s along the axial regions of MID-OCEAN RIDGES. Hydrothermal vents include BLACK SMOKERS, in which the jet, at a temperature greater than 350°C, is blackened by the precipitation of metal sulphides that can build chimneys or columns up to several metres high, and *white smokers* at temperatures between 100 and 350°C, which precipitate minerals such as silica (SiO_2) and baryte $(BaSO_4)$. The areas of deep sea floor around many hydrothermal vents are inhabited by communities of chemosynthetic organisms (see CHEMOSYNTHESIS) that obtain energy ultimately from a GEOTHERMAL source. Hydrothermal vent processes have important implications for EVOLUTION, the origin of life on Earth and ORE genesis. *GO/JBH*

Huber, H., Jannasch, H., Rachel, R. *et al.* 1997: *Archaeoglobus veneficus sp. nov.*, a novel facultative chemolithoautotrophic hyperthermophilic sulfite reducer, isolated from abyssal black smokers. *Systematic and Applied Microbiology* 20, 374–380. Little, C.T.S. and Cann, J.R. 1999: Late Cretaceous hydrothermal vent communities from the Troodos ophiolite, Cyprus. *Geology* 27, 1027–1030. Parson, L.M., Walker, C.L. and Dixon, D.R. (eds) 1995: Hydrothermal vents and processes. *Geological Society, London, Special Publication* 87. Prieur, D., Erauso, G. and Jeanthon, C. 1995: Hyperthermophilic life at deep-sea hydrothermal vents. *Planetary and Space Science* 43, 115–122. Russell, M.J. 1996: The generation at hot springs of sedimentary ore deposits, microbialites and life. *Ore Geology Reviews* 10, 199–214. Tunnicliffe, V., McArthur, A.G. and McHugh, D. 1998: A biogeographical perspective of the deep-sea hydrothermal vent fauna. *Advances in Marine Biology* 34, 353–442.

hydrovolcanic eruption A category of explosive VOLCANIC ERUPTION driven by the interaction between hot MAGMA and water, including GROUNDWATER and ice, which is converted explosively to steam. *Phreatic eruptions* produce steam and may eject fragmented country rock (VOLCANICLASTIC debris). *Phreatomagmatic eruptions* also eject magmatic products (PYROCLASTIC debris). *GO*

[See also MAAR]

Chester, D. 1993: *Volcanoes and society.* London: Edward Arnold. Muffler, L.J.P., White, D.E. and Truesdell, A.H. 1971: Hydrothermal explosion craters in Yellowstone National Park. *Geological Society of America Bulletin* 82, 723–740. Orton, G.J. 1996: Volcanic environments. In Reading, H.G. (ed.), *Sedimentary environments: processes, facies and stratigraphy*, 3rd edn. Oxford: Blackwell Science, 485–567. Wohletz, K.H. and Sheridan, M.F. 1983: Hydrovolcanic explosions 2. Evolution of basaltic tuff rings and tuff cones. *American Journal of Science* 283, 385–413.

hygrophyte A WETLAND plant or a plant of humid habitats; hygrophytes often grow at transitions between water (inhabited by HYDROPHYTES) and land plants on drier ground. Wetland plants can survive ANOXIA of the roots. Some form PEAT in fens or bogs. *BA*

[See also AQUATIC ENVIRONMENT, ECOTONE, MANGROVE, MIRE, TERRESTRIAL ENVIRONMENT

hypothesis A conjecture that possesses generality and is testable or has the potential to be tested. In SCIENCE, progress is made by testing hypotheses or attempting to refute hypotheses (FALSIFICATION). The more general the hypothesis and the more severe the test, the greater is the likelihood of scientific advance. Hypotheses are an essential part of SCIENTIFIC METHOD, which may be conceived as a continuous cycle of comparing hypotheses (ideas) against observations (data), often involving the progressive elimination of *multiple working hypotheses*. *JAM*

[See also CONFIRMATION, VERIFICATION]

Baker, V.R. 1996: The pragmatic roots of American Quaternary geology and geomorphology. *Geomorphology* 16, 197–215. Battarbee, R.W., Flower, R.J., Stevenson, J. and Rippy, B. 1985: Lake acidification in Galloway: a palaeoecological test of competing hypotheses. *Nature* 314, 350–352. Carpinter, S.R., Cole, J.T., Essington, T.E. *et al.* 1998: Evaluating alternative explanations in ecosystem experiments. *Ecosystems* 1, 335–344. Chamberlin, T.C. 1965: The method of multiple working hypotheses. *Science* 148, 754–759. Turner, R.E. 1997: Wetland loss in the northern Gulf of Mexico: multiple working hypotheses. *Estuaries* 20, 1–13.

Hypsithermal Various terms have been used to describe the interval of relatively mild climate that characterised the first part of the HOLOCENE following the LAST GLACIATION, including *Climatic Optimum*, *Thermal Maximum*, *Megathermal*, *Xerothermic*, *Altithermal* and Hypsithermal. Some include the whole of the early- and mid-Holocene; others restrict these terms to a shorter time interval when thermal conditions were 'optimal', especially for vegetation development. Average summer temperatures may have reached 2–3°C warmer than today during parts of the Hypsithermal, depending on location. *JAM*

[See also HOLOCENE ENVIRONMENTAL CHANGE, LATE-HOLOCENE CLIMATIC DETERIORATION]

Deevey, E.S. and Flint, R.F. 1957: Postglacial Hypsithermal interval. *Science* 125, 182–184.

hysteresis Partial or incomplete reversibility. A situation where the condition is partly dependent on the previous state(s); e.g. at a given moisture content it matters whether a soil is in a drying or wetting phase. *CET*

[See also RESILIENCE]

I

ice ages Phases in Earth's history when ICE SHEETS expanded to cover large areas of the globe in the form of large, continental glaciers. The term was first used by K. Schimper in AD 1837 and is commonly used to describe time intervals on two different scales. First, it describes long, generally cool intervals of Earth history (tens to hundreds of millions of years), during which GLACIERS repeatedly expanded and contracted. Second, the term is also used to describe shorter time intervals (tens of thousands of years), during which glaciers were at times at, or near, their maximum extent. These shorter intervals are more appropriately known as GLACIATIONS, GLACIAL EPISODES or, more correctly, STADIALS. Ice ages in the first sense have affected the Earth on numerous occasions during its history. These glaciations have resulted in significant lowering of temperatures and are not randomly distributed in time. There are records of Early and Late PROTEROZOIC (650–700 million years ago), ORDOVICIAN (c. 450 Ma BP), CARBONIFEROUS–PERMIAN (250–300 Ma BP) and CENOZOIC (last 15 Ma) ice ages.

The QUATERNARY has been considered to be synonymous with the 'Ice Age'. Sir Edward Forbes wrote in 1846 that the PLEISTOCENE equated to the 'Glacial Epoch'. If ice age is used to refer to long, generally cool, intervals during which the CRYOSPHERE waxes and wanes, we are still in one today. Our modern climate represents a short, warm interval between glacial advances. An ice age comprises several glacials interspersed with INTERGLACIAL phases. The GLACIAL–INTERGLACIAL CYCLE appears to have a periodicity of approximately 100 000 years. It is now accepted that variations in SOLAR RADIATION receipt according to the MILANKOVITCH THEORY are responsible for global cooling as the Earth enters a GLACIAL EPISODE and for the fluctuations in ice extent during ice ages, but not for the existence of ice ages. CONTINENTAL DRIFT, PLATE TECTONICS, LAND UPLIFT and the reduction of CARBON DIOXIDE in the atmosphere may be cited as causal factors in the onset of ice ages. The evidence for ancient ice ages is clear in many localities and can be seen from striations and grooves in rock and from the presence of TILLITES. Basal temperatures and water conditions can be inferred from the types of TILL identified in Quaternary stratigraphic sequences, whilst SEA-LEVEL CHANGES can be inferred from rhythmic MARINE SEDIMENTS.

Penck and Bruckner, in their major work published in 1909 in which they defined four glaciations, used the term *Eiszeitalter* to describe the glacial episodes, and the term ice age remains in common usage to define a single period within which cold and/or cool conditions prevailed. Here the recommendation is to restrict the term ice age to the first sense defined above (i.e. an ice age includes many glacial–interglacial cycles). *DH/CJC/JAM*

[See also ICEHOUSE CONDITION, SNOWBALL EARTH]

Crowell, J.C. 1999: *Pre-Mesozoic Ice Ages: their bearing on understanding the climate system, United States Geological Survey Memoir* 192. Frakes, L.A., Francis, J.E. and Syktus, J.I. 1992: *Climate modes of the Phanerozoic*. Cambridge: Cambridge University Press. Imbrie, J. and Imbrie, K.P. 1979: *Ice ages: solving the mystery*. London: Macmillan. John, B.S. (ed.) 1979: *The winters of the world: Earth under the ice ages*. Newton Abbot: David and Charles. Murdoch, T.Q., Weaver, A.J. and Fanning, A.F. 1997: Palaeoclimatic response of the closing of the Isthmus of Panama in a coupled ocean–atmosphere model. *Geophysical Research Letters* 24, 253–256. Penck, A. and Bruckner, E. 1909: *Die Alpen in Eiszeitalter*. Leipzig: Tachnitz.

ice-contact slope Retreat of a glacier or the melting of stagnant ice commonly involves collapse of sediment shored up by the ice. This process can leave an abrupt ice-contact slope with faulting in associated sediments. Where ice-marginal deposition is dominated by glaciofluvial processes, ice-contact slopes may be the main evidence for former glacial limits. *JS*

[See also GLACIER VARIATIONS, MORAINES]

ice cores Cylindrical cores of ice drilled from ICE SHEETS and GLACIERS, which preserve records of annual snow and ice ACCUMULATION, providing a wealth of palaeoenvironmental data with up to annual, and even subannual, resolution. Dating of ice layers is straightforward in the upper parts of cores, where annual layers are clearly distinguishable from dirt or ICE-MELT LAYERS. In the deeper parts of cores, ice layers may be hard to detect and dating must be achieved by indirect means, such as models of ice-column thinning by compaction and flow. Long cores have been obtained from Antarctica (e.g. Vostok, Dome C and Byrd), Greenland (e.g. the *Greenland Ice-core Project* [GRIP] and *Greenland Ice Sheet Project* [GISP] cores), which span the last GLACIAL–INTERGLACIAL CYCLE, including the LAST (*Eemian*) INTERGLACIAL. Shorter cores have been drilled from Arctic *ice caps*, such as the Agassiz Ice Cap, Ellesmere Island, and low-latitude high-altitude ice masses, such as the Dunde Ice Cap, Tibet and the Quelccaya Ice Cap, Peru.

The recent completion of drilling at Vostok station in East Antarctica has allowed the extension of the ice-core record of atmospheric composition and climate over the last four glacial–interglacial cycles (see Figure). Ice was recovered from a depth of 3623 m, about 120 m above a subglacial lake to prevent contamination of the lake water. Data from the ice cores has revealed detailed patterns through each of the last four glacial TERMINATIONS indicating comparable sequences of CLIMATIC CHANGE, whereby the effects of ORBITAL FORCING are amplified by GREENHOUSE GASES and ice–albedo feedbacks leading to full INTERGLACIAL conditions.

Many types of information can be obtained from ice cores. First, the ice layers are composed of a range of ISOTOPES of oxygen and hydrogen, the relative abundances of which reflect global reservoirs and local air temperatures at the time of their deposition as snow. Variations in the isotopic composition of ice within a core thus provide detailed records of climatic change, at a much higher res-

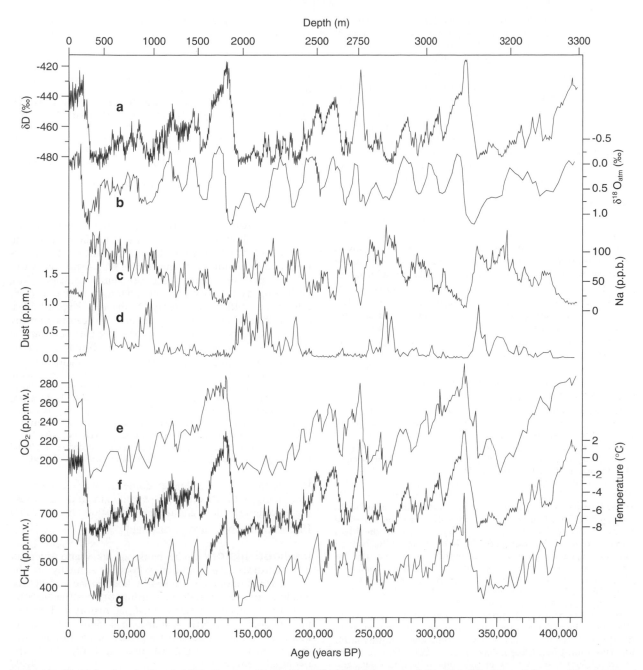

ice cores *Records of environmental change reconstructed from the Vostok ice core, Antarctica. The ice core is over 3 km long and covers the last 400 000 years. The last four glacial–interglacial cycles can be clearly see in these data relating to: (a) deuterium; (b) oxygen-18; (c) sodium; (d) dust; (e) carbon dioxide; (f) isotopic temperature; and (g) methane (after Petit et al., 1999)*

olution than almost all other types of evidence. The ice-core record of OXYGEN ISOTOPES has revealed patterns of climatic change during the last glacial cycle, including the quasi-periodic DANSGAARD–OESCHGER EVENTS. The high resolution of the record has revealed extremely rapid rates of climatic warming at the onset of the HOLOCENE, of the order of 5°C within a few decades.

Second, glacier ice contains air bubbles, sealed off from the atmosphere during compaction of snow layers, providing samples of the atmosphere at the time of sealing

(within a few decades for high-latitude cores, but possibly over 4000 years during glacial maxima in Antarctica). Bubbles in Antarctic and Greenland ice cores have provided long records of trace greenhouse gases such as CARBON DIOXIDE and METHANE, giving a unique record of changing atmospheric composition over the last glacial–interglacial cycle. This evidence has shown that greenhouse-gas concentrations and the oxygen isotope record are closely correlated, with high carbon dioxide and methane levels occurring at times of high atmospheric temperatures (see Figure). This correlation has suggested that greenhouse gases may participate in natural FEEDBACK MECHANISMS serving to amplify climatic cycles and has prompted the search for causal mechanisms.

A third type of evidence contained in ice cores is the concentrations of windblown PARTICULATES, such as DUST, SODIUM CHLORIDE and TEPHRA, contained within the ice. Such impurities can reveal changing patterns of atmospheric circulation and provide evidence of PALAEOWIND erosion, STORMS and VOLCANIC ERUPTIONS. In the upper parts of cores, industrial POLLUTANTS and RADIONUCLIDES can be used to reconstruct POLLUTION HISTORY and dispersal and fallout patterns. Additionally, ACIDITY levels can be detected using ELECTRICAL CONDUCTIVITY MEASUREMENTS (ECM). High concentrations of acids such as HNO_3 and H_2SO_4 in ice cores are generally indicative of volcanic AEROSOLS and therefore provide a record of past eruptions.

The presence of many forms of evidence within a single, high-resolution record means that ice cores are among the most valuable of all palaeoclimatic records, allowing many aspects of environmental changes to be interrelated. Furthermore, the relative ease of dating, at least for the upper parts of cores, places these environmental changes within an absolute timescale, independent of RADIOMETRIC DATING methods. As a result of these advantages, many now regard ice cores as standard yardsticks of climatic change, with which other forms of evidence – such as OCEAN CORES and POLLEN records – can be correlated, compared and calibrated. *DIB*

Alley, R.B. 2000: *The two-mile time machine: ice cores, abrupt climate change and our future*. Princeton, N.J.: Princeton University Press. Bond, G., Broecker, W., Johnsen, S. *et al.* 1993: Correlations between climate records from North Atlantic sediments and Greenland ice. *Nature* 365, 143–147. Dansgaard, W., Johnsen, S.J., Clausen, H.B. *et al.* 1993: Evidence for general instability of past climate from a 250-kyr ice-core record. *Nature* 364, 218–220. Dansgaard, W. and Oeschger, H. 1989: Past environmental long-term records from the Arctic. In Oeschger, H. and Langway, C.C. (eds), *The environmental record in glaciers and ice sheets*. Chichester: Wiley, 287–317. Etheridge, D.M., Steele, L.P., Langelfelds, R.L. *et al.* 1996: Natural and anthropogenic changes in atmospheric CO_2 over the last 1000 years from air in Antarctic ice and firn. *Journal of Geophysical Research* 101(4), 115–128. Hammer, C.U., Clausen, H.B. and Dansgaard, W. 1980: Greenland ice-sheet evidence of post-glacial volcanism and its climatic impact. *Nature* 288, 230–235. Petit, J.R, Jozel, J., Raynaud, D. *et al.* 1999: Climate and atmospheric history of the past 420,000 years from the Vostok ice core, Antarctica. *Nature* 399, 429–436. Raynaud, D., Barnola, J.M., Chappellaz, J. *et al.* 2000:The ice record of greenhouse gases: a view in the context of future climates. *Quaternary Science Reviews* 19, 9–17. Thompson, L.G. 1995: Ice core evidence from Peru and China. In Bradley, R.S. and Jones, P.D. (eds), *Climate since AD 1500*. London: Routledge, 517–548. Thompson, L.G. 2000: Ice core evidence for climate change in the Tropics: implications for our future. *Quaternary Science Reviews* 19, 19–35.

ice-dammed lake

A lake in glaciated terrain created where a GLACIER or ICE SHEET has moved against the regional slope or across a valley, forming a barrier to drainage. Ice-dammed lakes are noted for their sudden drainage, producing JÖKULHLAUPS. Evidence for former ice-dammed lakes includes shorelines, lake-floor RHYTHMITES or VARVES, DELTAS and overflow channels and associated deposits. Ice-dammed lakes may contain an important stratigraphic record of daily, intraseasonal, seasonal and interseasonal fluctuations in sediment input. *JS*

[See also CHANNELED SCABLANDS, GLACIOLACUSTRINE DEPOSITS]

Johnson, P.G. 1997: Spatial and temporal variability of ice-dammed lake sediments in alpine environments. *Quaternary Science Reviews* 16, 635–647. Röthlisberger, H. and Land, H. 1987: Glacial hydrology. In Gurnell, A.M. and Clark, M.J. (eds), *Glacio-fluvial sediment transfer*. Chichester: Wiley, 207–284. Shakesby, R.A. 1985: Geomorphological effects of jökulhlaups and ice-dammed lakes, southern Norway. *Norsk Geografisk Tidsskrift* 39, 1–16. Sissons, J.B. 1977: Former ice-dammed lakes in Glen Moriston, Inverness-shire, and their significance in upland Britain. *Transactions of the Institute of British Geographers* NS 2, 224–242.

ice-directional indicators

Erosional or depositional forms that show former directions of ice flow, including STRIATIONS, FLUTED MORAINES and DRUMLINS. Such features occur at widely differing spatial scales and have varying regional significance. Ice-directional indicators are commonly TIME TRANSGRESSIVE, relating to different periods of ice build up and decay. At the very largest scales, several generations of ice-directional indicators may be distinguished by SATELLITE REMOTE SENSING and can be used to establish aspects of ICE SHEET evolution. *DIB*

Boulton, G.S. and Clark, C.D. 1990: The Laurentide Ice sheet through the last glacial cycle: drift lineations as a key to the dynamic behaviour of former ice sheets. *Transactions of the Royal Society of Edinburgh, Earth Sciences* 81, 327–347. Clark, C.D. 1993: Mega-scale lineations and cross-cutting ice-flow landforms. *Earth Surface Processes and Landforms* 18, 1–29. Kleman, A., Hättestrand, H., Borgström, I. and Stroeven, A. 1997: Fennoscandian palaeoglaciology reconstructed using a glacial geological inversion model. *Journal of Glaciology* 43, 283–299.

ice dome

A symmetrical dome-shaped mass of ice, embracing both ICE SHEETS ($> 50\,000$ km^2 and up to several kilometres thick) and *ice caps* ($< 50\,000$ km^2). *MJH*

[See also GLACIER, OUTLET GLACIER]

ice-flood history

The banks of lakes and rivers at high latitudes are commonly flooded annually in the spring when damage is caused by floating ice and debris. Analysis of ice-scarred trees using DENDOCHRONOLOGY enables the reconstruction of ice-flood history and, in certain regions of Arctic Canada, has demonstrated an increase in magnitude and frequency of ice floods since the end of the LITTLE ICE AGE and, especially during the twentieth century. *JAM*

Tardif, J. and Bergeron, Y. 1997: Ice-flood history reconstructed with tree-rings from the southern boreal forest limit, western Québec. *The Holocene* 7, 291–300.

ice foot

An ephemeral feature that forms during winter as wave spray and swash freezes along a shoreline. Its thickness and width depend on tidal range and storm wave activity. It may be a locus for POLAR SHORE EROSION. *HJW*

ice-margin indicators

Geomorphological evidence that indicates the former position of the margin of a GLACIER or ICE SHEET including, for example, many types of MORAINES, KAME terraces, sandar (see SANDUR) and TRIMLINES. Ice-margin indicators are used in the reconstruction of the maximum extent of former glaciers and also subsequent recessional stages. *JAM*

Lowe, J.J. and Walker, M.J.C. 1997: *Reconstructing Quaternary environments*, 2nd edn. Harlow: Addison-Wesley Longman.

ice-melt layer

A layer of refrozen ice marking a former summer ABLATION surface in glacier ice, lying between winter ACCUMULATION layers. On Himalayan glaciers, two ice-melt layers are formed each year, during the pre- and post-monsoon ablation seasons. Ice-melt layers or *melt layers* typically contain few bubbles and high concentrations of wind-blown dust, allowing annual layers to be easily distinguished in ICE CORES. *DIB*

Alley, R. and Anandrakrishnan, S. 1995: Variations in melt-layer frequency in the GISP2 ice core: implications for Holocene summer temperatures in central Greenland. *Annals of Glaciology* 21, 64–70.

ice-rafted debris (IRD)

Particles or CLASTS deposited in a lacustrine or glacimarine environment, either from ICEBERGS or, passively or actively, by lake or SEA ICE. The size of the ice-rafted debris will depend on the agent by which it has been deposited. In the lacustrine context, ice-rafted debris is normally relatively coarse-grained, angular COLLUVIUM transported onto lake ice by DEBRIS FLOWS or SNOW AVALANCHES. In the glacimarine context, icebergs enable the transportation of particles from CLAY size particles (< 0.002 mm) to BOULDER size; particles transported by active sea ice do not normally include boulders, whereas passive sea ice rarely includes clay or SILT. SEDIMENT may be deposited as a single particle (DROPSTONE), agglomerations of more than one particle (*dump*), frozen aggregates or sediment-laden ice.

Deep-ocean sediments are rich in TERRIGENOUS sediment with estimates of 40% for the amount of sediment deposited in the oceans during the QUATERNARY consisting of IRD (see Figure). High-resolution MARINE SEDIMENT CORES show numerous episodes of North Atlantic ice-rafted debris deposition within the last GLACIAL EPISODE. These are seen as layers of coarse-grained lithic particles and clasts within the marine sediment, which were transported from Quaternary ice sheets to the North Atlantic by CALVING mechanisms. These clasts therefore allow PALAEOCEANOGRAPHIC studies to be conducted and source region identification on the basis of mineral composition. IRD in palaeoceanographic studies is often defined as lithic material > 125 μm.

Marine records of the last glaciation exhibit two notable PERIODICITIES of increased IRD content, at intervals of 2000–3000 years and 7000–10 000 years. These frequencies correspond respectively to DANSGAARD–OESCHGER EVENTS and HEINRICH EVENTS. IRD-defining Heinrich events in the mid-latitude North Atlantic contain lithic material found to originate predominantly from the LAURENTIAN ICE SHEET and to a lesser extent from the FENNOSCANDIAN, Icelandic and British ice sheets.

This multiple-source origin of the Heinrich events suggests that the discharge from the ice sheets may have been triggered by a common mechanism, although asynchroneity in their timing in different parts of the ocean is a complication. Debate continues over the triggering mechanism for widespread IRD on these millennial

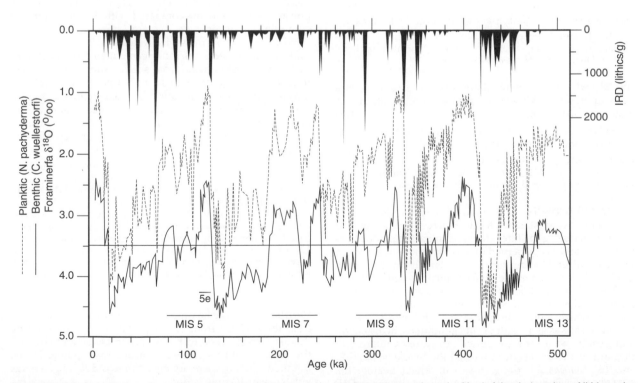

ice-rafted debris *Variations in an index of ice-rafted debris over the last 500 000 years from the North Atlantic (number of lithic grains larger than 150 mm per bulk sample weight, solid black) in relation to oxygen isotope variations in planktic (dashed) and benthic (lower solid line) foraminifera. Interglacial marine isotope stages (MIS) are numbered (after McManus et al., 1999)*

timescales; these include internal ice sheet dynamics (the BINGE–PURGE MODEL), eustatic SEA-LEVEL CHANGE and a linked MELTWATER–THERMOHALINE CIRCULATION system. External ORBITAL FORCING may provide a driving mechanism, but the *power spectra* are weak for the relevant PERIODICITIES. *LJW/WENA*

[See also EUSTASY, GLACIOMARINE HYPOTHESIS, HEINRICH EVENTS]

Bond, G.C. and Lotti, R. 1995: Iceberg discharges into the North Atlantic on millennial time scales during the last glaciation. *Science* 267, 1005–1010. **Bond, G., Heinrich, H., Broecker, W.** *et al.* 1992: Evidence for massive discharges of icebergs into the North Atlantic during the last glacial period. *Nature* 360, 245–249. **Bond, G., Showers, W., Cheseby, M.** *et al.*1997: A pervasive millenial-scale cycle in North Atlantic Holocene and glacial climates. *Science* 278, 1257–1266. **Dowdeswell J.A., Elverhoi, A., Andrews J.T. and Hebbeln, D.** 1999: Asynchronous deposition of ice-rafted layers in the Nordic seas and North Atlantic Ocean. *Nature* 400, 348–351. **Gilbert, R.** 1990: Rafting in glacimarine environments. In Dowdeswell, J. and Scourse, J. (eds), *Glacimarine environments: processes and sediments, Geological Society Special Publication* 53. London: The Geological Society, 105–120. **Lowe, J.J. and Walker, M.J.C.** 1997: *Reconstructing Quaternary environments*, 2nd edn. Harlow: Addison-Wesley Longman. **McManus, J.F., Oppo, D.W. and Cullen, J.L.** 1999: A 0.5 million year record of millennial-scale climatic variability in the North Atlantic. *Science* 283, 971–975.

ice scars Evidence of physical damage seen on the trunks of trees growing on the margins of lakes or rivers, caused by ice killing part of the CAMBIUM of the tree. These can often be dated to the precise year by dendrochronological study of the year or years of cambial injury. *KRB*

[See also TREE RINGS, DENDROCHRONOLOGY, FIRE SCAR]

Tardif, J. and Bergeron, Y. 1997: Ice-flood history reconstructed with tree rings from the southern boreal forest limit, western Quebec. *The Holocene* 7, 291–300.

ice sheet An extensive dome-shaped mass of ice and snow up to several kilometres thick and occupying an area of at least 50 000 km². *MJH*

[See also GLACIATION, GLACIER, GLACIER THERMAL REGIME, ICE DOME, ICE-SHEET GROWTH, ICE SHELF]

Denton, G.H. and Hughes, T. (eds) 1981: *The last great ice sheets*. New York: Wiley. **Oerlemans, J. and Van der Veen, C.J.** 1984: *Ice sheets and climate*. Dordrecht: Reidel.

ice-sheet growth The build up of an ice sheet due to persistent positive MASS BALANCE. As a result of erosion and deposition during the later stages of an ice sheet, little geological evidence is preserved of periods of ice-sheet growth and current understanding is based largely on theoretical models. Three main models have been proposed: (1) '*highland origin*', in which ice sheets are nucleated in mountain areas and flow out to lower regions; (2) '*instant glacierisation*', in which large continental areas become ice ACCUMULATION ZONES as a result of cold conditions, high ALBEDO and positive FEEDBACK processes (an extreme interpretation of this theory has been termed the '*snowblitz theory*'); and (3) '*marine ice transgression*', in which floating ICE SHELVES thicken and become grounded ice sheets. *DIB*

Denton, G.H. and Hughes, T.J. 1981: The Arctic Ice Sheet: an outrageous hypothesis. In Denton G.H. and Hughes T.J.,

(eds), *The last great ice sheets*. New York: Wiley, 437–467. **Hughes, T.J.** 1986: The marine ice transgression hypothesis. *Geografiska Annaler* 69A, 237–250. **Ives, J.D., Andrews, J.T. and Barry, R.G.** 1975: Growth and decay of the Laurentide ice sheet and comparisons with Fenno-Scandinavia. *Naturwissenschaften* 62, 118–125.

ice shelf A large slab of ice floating on the sea, up to several hundred metres thick, but remaining attached to, and partly fed by, land-based ice. *MJH*

[See also ANTARCTIC ENVIRONMENTAL CHANGE]

Thomas, R.H. 1974: Ice shelves: a review. *Journal of Glaciology* 24, 273–286.

ice wedge A wedge-shaped body of ice present in near-surface PERMAFROST caused by thermal contraction cracking of the ground in winter and the subsequent infilling of these FROST CRACKS in early summer by water, which then freezes. If this process is repeated for a number of years, a wedge-shaped body of foliated ice develops. The size of ice wedges depends largely upon the availability of water and the age of the ice wedge. In parts of central Siberia and the western North American Arctic, wedges in excess of 1–3 m in width and 5–10 m in depth can be observed. Most wedges are EPIGENETIC, but some may be SYNGENETIC. Where AEOLIAN transport is locally dominant, or in extremely arid regions such as Antarctica, the thermal-contraction cracks are filled largely with mineral soil particles rather than with ice and the features are termed *sand wedges*. In plan, ice wedges join predominantly at right angles and form polygonal, chiefly tetragonal, nets of PATTERNED GROUND that cover large areas of the Arctic and subarctic. The average dimensions of the polygons are 15–40 m, compared with 10–30 m for less common sand wedges.

When permafrost thaws, ice wedges melt and the void is infilled with mineral soil and organic material that slumps down from the sides and from above. Such structures are termed ice-wedge CASTS or PSEUDOMORPHS. *Thermal-contraction-cracking* has climatic and palaeoclimatic significance. In general, it requires a mean annual air temperature of −6°C or colder. Provided they are correctly identified, pseudomorphs provide incontrovertible evidence of the former existence of permafrost. *HMF*

French, H.M. and Gozdzik, J. 1988: Pleistocene epigenetic and syngenetic frost fissures, Belchatow, Poland. *Canadian Journal of Earth Sciences* 25, 2017–2027. **Gemmell, A.M.D. and Ralston, I.B.M.** 1984: Some recent discoveries of ice-wedge cast networks in northeast Scotland. *Scottish Journal of Geology* 20, 115–118. **Mackay, J.R.** 1974: Ice-wedge cracks, Garry Island, Northwest Territories. *Canadian Journal of Earth Sciences* 11, 1336–1383. **Mackay, J.R.** 1995: Air temperature, snow cover, creep of frozen ground and the time of ice wedge cracking, western Arctic coast. *Canadian Journal of Earth Sciences* 30, 1720–1729.

iceberg A large floating mass of ice that has broken off from the seaward front of a GLACIER or ICE SHELF during the CALVING process. Icebergs vary in shape and size and the larger examples can be a hazard to shipping. In the Southern Ocean, there are an estimated 300 000 icebergs greater than 10 m in width. *AHP*

[See also HEINRICH EVENT, ICEBERG DRIFT THEORY, ICE-RAFTED DEBRIS, SEA ICE]

Broecker, W. 1994: Massive iceberg discharges as triggers for global climate change. *Nature* 372, 421–424.

Iceberg Drift Theory   A 'hybrid' between the DILU-VIAL THEORY and the GLACIAL THEORY, which merged aspects of the catastrophic origin of DILUVIUM with the existence of a relatively cold climate. The theory, championed by Charles Lyell and others for a time in the mid to late nineteenth century, attributed DRIFT to the deposition of debris transported by icebergs during marine submergence. It delayed acceptance of the Glacial Theory, partly because of greater compatibility with religious beliefs. It also appeared to be supported by geological evidence of SEA-LEVEL CHANGE and reports of icebergs in high-latitude oceans from whalers, explorers and Charles Darwin's voyage in the Beagle.   *JAM*

iceberg plough marks   Elongated furrows formed by the keels of grounded icebergs dragged across the soft sediments of a lake or sea floor by tidal and wind action. Such marks are up to 20 m deep and 250 m wide and may extend for several kilometres. They are also known as *iceberg scour marks* or TOOL MARKS. Their form depends on the nature of the sediment, the shape of the iceberg keel(s) and the nature of the motion of the iceberg. In soft sediments, plough marks may be regular and continuous, curved or straight, flat-bottomed troughs or furrows. The microtopography of the plough marks tends to vary with the characteristics of movement of the iceberg. Linear *scour berms* (i.e. ridges) of displaced blocky material at the sides of the plough marks can attain heights of 6 m. The term *iceberg grounding structures* is used to refer to the deformed sediments associated with the plough marks as well as the marks themselves. Positive identification of fossil iceberg plough marks may provide evidence of PALAEOCURRENT or PALAEOWIND directions and, where applicable, the GLACIOMARINE HYPOTHESIS.   *RAS*

Bennett, M.R. and Bullard, J.E. 1991: Iceberg tool marks: an example from Heinabersjökull, South East Iceland. *Journal of Glaciology* 37, 181–183. **Delage, M. and Gangloff, P.** 1993: Relict iceberg marks near Montreal, Quebec. *Géographie Physique et Quaternaire* 47, 69–80. **Dowdeswell, J.A., Villinger, H., Whittington, R.J. and Marienfeld, P.** 1993: Iceberg scouring in Scoresby Sund, east Greenland. *Sedimentology* 41, 21–35. **Thomas, G.S.P. and Connell, R.J.** 1985: Iceberg drop, dump and grounding structures from Pleistocene glaciolacustrine sediments, Scotland. *Journal of Sedimentary Petrology* 55, 243–249.

icehouse condition   A term used to describe long intervals of global cooling in the Earth's past, characterised by relatively low SEA LEVEL, a scarcity of sedimentary FACIES characteristic of hot palaeoclimates (see PALAEOCLIMATOLOGY) and evidence of ICE AGE conditions for some of the time. At least five oscillations between icehouse and warmer GREENHOUSE CONDITIONS can be recognised in the GEOLOGICAL RECORD of the PHANEROZOIC and late PROTEROZOIC. Icehouse conditions characterised the late Proterozoic to late Cambrian, late Devonian to mid Triassic and early Cenozoic to the present day.   *GO*

[See also GEOLOGICAL RECORD OF ENVIRONMENTAL CHANGE]

Fischer, A.G. 1981: Climatic oscillations in the biosphere. In Nitecki, M. (ed.), *Biotic crises in ecological and evolutionary time.* New York: Academic Press, 103–131. **Price, G.D., Valdes, P.J. and Sellwood, B.W.** 1998: A comparison of GCM simulated Cretaceous 'greenhouse' and 'icehouse' climates: implications for the sedimentary record. *Palaeogeography, Palaeoclimatology, Palaeoecology* 142, 123–138. **Séranne, M.** 1999: Early Oligocene stratigraphic turnover on the west Africa continental margin: a signature of the Tertiary greenhouse-to-icehouse transition? *Terra Nova* 11, 135–140.

Iceman   The Alpine Iceman is the well preserved body of a prehistoric man that was discovered in AD 1991 melting out of glacier ice on the Tisa Pass (3280 m) close to the border between Austria and Italy. RADIOCARBON DATING revealed a NEOLITHIC age (4500 radiocarbon years BP; about 5200 years ago), which was older than expected given the accompanying clothing and equipment, including a copper axe, bow, arrows and backpack. Nicknamed 'Ötzi', a conflation of 'Ötztal' and 'yeti', the find is of considerable environmental as well as archaeological importance. The degree of preservation means that the body must have been buried very rapidly and remained buried in an environment conducive to its preservation until discovery: this is consistent with a rapid CLIMATIC CHANGE leading to NEOGLACIATION and with the glacier remaining no smaller than today throughout the period of burial. Soils at the periphery of the site have yielded dates as old as 5600 radiocarbon years and are indicative of the more favourable climate prior to burial of the Iceman. The lack of distortion of the corpse is explained by an absence of glacier flow or creep, which in turn indicates a thin ice mass frozen to the ground (*cold-based ice*) in a PERMAFROST environment.   *JAM*

[See also HOLOCENE ENVIRONMENTAL CHANGE, GLACIER VARIATIONS]

Baroni, C. and Orombelli, G. 1996: The Alpine 'Iceman' and Holocene climatic change. *Quaternary Research* 46, 78–83. **Bonani, G., Ivy, S.D., Niklaus, T.R. and Suter, M.** 1994: AMS ¹⁴C determinations of tissue, bone and grass samples from the Ötztal Ice Man. *Radiocarbon* 36, 247–250. **Bortenschlager, S. and Oeggl, K. (eds)** 2000: *The iceman in his natural environment: palaeobotanical results* [*The man in the ice.* Vol. 4]. Vienna and New York: Springer. **Rom, W., Golser, R., Kutschera, W. et al.** 1999: AMS ¹⁴C dating of equipment from the iceman and of spruce logs from the prehistoric salt mines of Hallstatt. *Radiocarbon* 41, 183–197. **Spindler, K.** 1994: *The man in the ice.* New York: Harmony Books.

ichnofacies   A body of sediment or rock characterised by its TRACE FOSSIL assemblage. Ichnofacies analysis can be combined with more conventional sedimentary FACIES ANALYSIS to enhance PALAEOENVIRONMENTAL RECONSTRUCTION.   *GO*

[See also BIOFACIES, FACIES, LITHOFACIES]

Smith, R.M.H., Mason, T.R. and Ward, J.D. 1993: Flash-flood sediments and ichnofacies of the Late Pleistocene Homeb Silts, Kuiseb River, Namibia. *Sedimentary Geology* 85, 579–599.

icings   Sheet-like, tabular masses of ice, also known as *naleds* and *Aufeis*, that form at the surface in winter wherever water issues from the ground. They occur in the ARCTIC and SUBARCTIC. Most are small, but some in central Alaska and Siberia associated with perennial springs and sub-PERMAFROST waters assume considerable dimensions, exceeding 10 km in length and 10 m in thickness. River icings form at localities where the river freezes to its bottom, thereby forcing water out of the river bed. Evidence for the possible previous existence of icings in former periglacial environments is difficult to identify with any certainty. *HMF*

[See also FROST MOUND]

Bennett, M.R., Huddart, D., Hambrey, M.J. and Ghienne, J.F. 1998: Modification of braided outwash surfaces by Aufeis: an example of Pedersenbreen, Svalbard. *Zeitschrift für Geomorphologie NF* 42, 1–20. **Hu, X.G. and Pollard, W.H.** 1997: The hydrologic analysis and modelling of river icing growth, North Fork Pass, Yukon Territory, Canada. *Periglacial and Permafrost Processes* 8, 279–294. **Van Everdingen, R.O.** 1990: Ground-water hydrology. In Prowse, T.D. and Ommanney, C.S.L. (eds), *Northern hydrology, Canadian perspectives* [*Report* 1]. Saskatoon: National Hydrology Research Institute, 77–101.

idealism A group of philosophical theories in which the central idea is that the 'real world' as defined by common standards is created by the human mind. Reality is seen as dependent on a mind and its processes for its existence; there is no such thing as a detached observer. The term should be associated primarily with 'ideas', rather than with 'ideals'. The primary contrast in philosophy is with REALISM, but contrasts with other 'isms' exist. *CET*

Harrison, S. and Dunham, P. 1998: Decoherence, quantum theory and their implications for the philosophy of geomorphology. *Transactions of the Institute of British Geographers NS* 23, 501–514.

identification (ID) A tag, label or ATTRIBUTE identifying an ENTITY or COVERAGE in a GEOGRAPHICAL INFORMATION SYSTEM (GIS). Also the primary FIELD for a tuple, or record, in a RELATIONAL DATABASE. *TVM*

[See also OVERLAY ANALYSIS]

idiographic science Science that is concerned with developing comprehensive explanations of individual cases, as opposed to general LAWS (NOMOTHETIC SCIENCE). A somewhat dated concept. *CET*

igneous rocks Rocks formed by the cooling and solidification of molten MAGMA, either below the Earth's surface as INTRUSIONS or at the surface through VOLCANIC ERUPTIONS of LAVA and PYROCLASTIC material. Their characteristic feature is a crystalline TEXTURE. Igneous rocks are subdivided into four categories according to their chemistry: *ultrabasic* (silica content, SiO_2 < 45%), *basic* (SiO_2 45–53%), *intermediate* (SiO_2 53–66%) and *acidic* (or silicic – SiO_2 > 66%). Mineral content, or percentage of dark minerals, is a proxy for chemistry. Crystal size ranges from *fine* (< 1 mm – *aphanitic*; fast cooling as lava or high-level intrusion) through *medium* to *coarse* (> 3 mm – *phaneritic*; slow-cooling as large, deep plutonic intrusion). VOLCANIC GLASS is a product of very rapid cooling. Common igneous rocks are defined in the Table. Pyroclastic rocks are named according to different criteria.

Minerals in igneous rocks are important for RADIOMETRIC DATING. Studies of igneous rocks contribute to understanding the behaviour and evolution of VOLCANOES, which constitute a major NATURAL HAZARD. Sites of igneous activity are closely related to PLATE TECTONIC activity (see also HOTSPOTS) and igneous rocks in the GEOLOGICAL RECORD can help to reconstruct past plate positions and movements. Pyroclastic BEDS form important, synchronous marker horizons of wide extent (see TEPHRA) and igneous rocks provide evidence of past episodes of VOLCANISM, which may have contributed to environmental change on a variety of time-scales. *JBH*

[See also METAMORPHIC ROCKS, SEDIMENTARY ROCKS, VOLCANIC IMPACT ON CLIMATE]

Cas, R.A.F. and Wright, J.V. 1987: *Volcanic successions: modern and ancient.* London: Allen and Unwin. **Cox, K.G., Bell, J.D. and Pankhurst, R.J.** 1979: *The interpretation of igneous rocks.* London: Allen and Unwin. **Le Maitre, R.W. (ed.)** 1989: *A classification of igneous rocks and glossary of terms: recommendations of the International Union of Geological Sciences Subcommission on the Systematics of Igneous Rocks.* Oxford: Blackwell. **McBirney, A.R.** 1984: *Igneous petrology.* San Francisco, CA: Freeman. **Middlemost, E.A.K.** 1997: *Magmas, rocks and planetary development: a survey of magma/igneous rock systems.* Harlow: Longman.

ignimbrite A deposit of a PYROCLASTIC FLOW. Because of confusion in past use, the term is best restricted to PUMICE-rich deposits. Some of the largest single eruptive units known (> 1000 km³) are ignimbrites, associated with major explosive VOLCANIC ERUPTIONS such as Santorini (1470 BC), Krakatau (AD 1883) and Mount St Helens (AD 1980). Many ignimbrite deposits are preserved as extensive sheets of WELDED TUFF. *JBH*

Branney, M.J. and Kokelaar, P. 1992: A reappraisal of ignimbrite emplacement – progressive aggradation and changes from particulate to non-particulate flow during emplacement of high-grade ignimbrite. *Bulletin of Volcanology* 54, 504–520. **Legros, F. and Druitt, T.H.** 2000: On the emplacement of ignimbrite in

igneous rocks *Characteristics of the principal igneous rock types*

Crystal size	Igneous setting	Igneous form	Rock type Silica content < 45% Ultrabasic	45–53% Basic	53–60% Intermediate	60–66%	> 66% Acidic
Fine (glassy)	Volcanic	Lava	—	Basalt	Andesite	Dacite	Rhyolite (obsidian)
Medium	Hypabyssal	Dykes, sills	—	Dolerite	—	—	Microgranite
Coarse	Plutonic	Deep intrusions, batholiths	Peridotite	Gabbro	Diorite	Granodiorite	Granite
Dominant mineral phases	—		Olivine, pyroxene	Pyroxene, plagioclase	Plagioclase, hornblende	Plagioclase, K-feldspar, quartz	K-feldspar, quartz

shallow-marine environments. *Journal of Volcanology and Geothermal Research* 95, 9–22.

illite A 2 : 1 CLAY MINERAL of the mica family having only slight expansion characteristics, as a result of potassium ions occupying sites in the interlayer space. It is sometimes called fine-grained mica. ISOMORPHIC SUBSTITUTION in the clay lattice gives a CATION EXCHANGE CAPACITY between that of KAOLINITE and *smectite*. *EMB*

Brady, N.C. 1984: *The nature and properties of soils*. New York: Macmillan.

illuviation Following ELUVIATION from the upper A and E horizons of a soil, the process by which colloidal material (clay and humus) is redeposited in a lower, B horizon. Illuviation is involved in the formation of the ARGIC HORIZON present in LUVISOLS, ALBELUVISOLS, ACRISOLS, ALISOLS, LIXISOLS, NITISOLS and the SPODIC HORIZON of PODZOLS. *EMB*

Buol, S.W., Hole, F.D. and McCracken, R.J. 1988: *Soil genesis and classification*. Ames, IA: Iowa State University Press.

image classification The procedure carried out to generate a thematic map from remotely sensed data. An image usually consists of a large number of PIXELS, each of which is characterised by a vector of spectral reflectance values (OPTICAL REMOTE SENSING INSTRUMENTS) and/or microwave BACKSCATTERING values (RADAR REMOTE SENSING INSTRUMENTS). *Per-pixel classification* assigns a class to individual pixels, while *per-parcel classification* assigns a class to individual polygons comprised of pixels regarded as belonging to the same class. Linear features in the landscape can be used to create parcels from a pixel-based image. A *supervised classification* uses user-defined TRAINING AREAS to estimate spectral signatures of individual classes. The classification assigns a class to each pixel/polygon by comparing the spectral information of that pixel/polygon with the estimated signature. An *unsupervised classification* uses the spectral content of the whole image and a user-defined number of classes to estimate spectral signatures automatically.

A widely used classification algorithm is the *maximum likelihood classification* (MLC). A pixel is assigned the class with the highest likelihood of causing the observed *reflectance* or backscatter given this class. BAYESIAN STATISTICS has provided the *maximum a posteriori classifier* (MAP). According to Bayes' theorem, the *a posteriori* PROBABILITY is calculated from the likelihood and the *a priori* probability. The likelihood contains information about spectral similarity, while the *a priori* probability includes previous knowledge about the region. A pixel is assigned the class with the highest *a posteriori* probability of belonging to this class, given the observed reflectance or backscatter.

Different methods of contextual classification have been developed in order to correct pixels that have been misclassified as a result of random variability in the data, by using information about spatially neighbouring pixels. FUZZY LOGIC has also been applied to image classification, as it enables quantification of the pixel-wise uncertainty of the thematic map. Because of the variability in the data, a classification always has errors associated with it. An ACCURACY assessment of the thematic map, in which the map is compared with GROUND MEASUREMENTS, is the final step of image classification. *HB*

[See also ELECTROMAGNETIC SPECTRUM, IMAGE ENHANCEMENT, IMAGE PROCESSING, IMAGE TEXTURE]

Augusteijn, M.F. and Warrender, C.E. 1998: Wetland classification using optical and radar data and neural network classification. *International Journal of Remote Sensing* 19, 1545–1560. Binaghi, E., Madella, P., Montesano, M.G. and Rampini, A. 1997: Fuzzy contextual classification of multisource remote sensing images. *IEEE Transactions on Geoscience and Remote Sensing* 35, 326–340. Durden, S.L., Haddad, Z.S., Morrissey, L.A. and Livingston, G.P. 1996: Classification of radar imagery over boreal regions for methane exchange studies. *International Journal of Remote Sensing* 17, 1267–1273. Li, W., Benie, G.B., He, D.C. *et al.* 1998: Classification of SAR images using morphological texture features. *International Journal of Remote Sensing* 19, 3399–3410. Miranda, F.P., Fonseca, L.E.N., Carr, J.R. and Taranik, J.V. 1996: Analysis of JERS-1 (Fuyo-1) SAR data for vegetation discrimination in north-western Brazil using the semivariogram textural classifier (STC). *International Journal of Remote Sensing* 17, 3523–3529. San Miguel Ayanz, J. and Biging, G.S. 1996: An iterative classification approach for mapping natural-resources from satellite imagery. *International Journal of Remote Sensing* 17, 957–981.

image enhancement Operations that are intended to augment the visibility of desired features within an image (e.g. in REMOTE SENSING). These features may be the contrast, some or all of the colours, the borders, large areas, etc., and the particular techniques to be used depend upon the input data and on the desired effect. Image enhancement is usually applied during IMAGE INTERPRETATION and some of the techniques it relies on are contrast stretching, HISTOGRAM equalisation and FILTERING. *ACF*

Jain, A.J. 1989: *Fundamentals of digital image processing*. Englewood Cliffs, NJ: Prentice-Hall.

image interpretation The transformation of data, presented as images, into information. This transformation can be performed by means of two (complementary rather than disjoint) approaches: quantitative analysis and photo-interpretation. In the former the digital essence of the data is exploited, through the use of computer-based techniques, while the latter presents the data to the human interpreter in the form of images, so that a visual inspection can be carried out. Quantitative analysis is optimum at PIXEL level; it is accurate for area estimation, can perform true multi-spectral analysis and can make use of all available brightness levels. Photo-interpretation is best suited for shape and contextual (spatial) determination and assessment. *ACF*

Richards, J.A. 1986 *Remote sensing digital image analysis: an introduction*. Berlin: Springer.

image processing The manipulation of pictures by means of digital computers. In order to be able to use digital machines, pictures have to be in a proper format: a finite array of real, complex or multidimensional data, represented by a finite number of bits. In this manner, a digital image is defined in a discrete domain and as having discrete values in each position. Three important applications of digital image processing are FILTERING, ENHANCEMENT and restoration. Image processing techniques can be classified into: linear or non-linear; point-wise (single-valued functions); local (functions of a few variables); or global (functions that use the whole image as input). Image processing is one of the most successful melting pots of disciplines: it uses concepts, techniques and ideas coming from such diverse areas as statistics, theoretical computing, artificial intelligence, psychometry and information theory, to name a few. Its success owes much to the advent of fast computers with massive storage capacity. A current trend in image processing is the development of systems that ease the user's burden, putting at his or her fingertips sharp tools that require little or no specialised training to obtain the desired results. *ACF*

Mather, P. M. 1999: *Computer processing of remotely-sensed images*, 2nd edn. New York: Wiley.

image texture   Syncretism between vision and touch; that is, the tactile response analogy to a visual stimulus. Areas with similar mean colour properties can be distinguished by their roughness properties (e.g. forest and non-forest in a REMOTE SENSING image) since texture is related to the spatial organisation of colours. Many techniques have been devised aiming at the quantification of this perceptual feature, HARALICK TEXTURE MEASURES being among the most well known and useful ones. Two main categories of textures are usually considered: structural (where repetition of exact patterns is observed) and statistical (where the similarity is in the distributional sense). *ACF*

Luckman, A.J., Frery, A.C., Yanasse, C.C.F. and Groom, G.B. 1997: Texture in airborne SAR imagery of tropical forest and its relationship to forest regeneration stage. *International Journal of Remote Sensing* 18(6), 1333–1349.

imbrication   (1) A FABRIC in SEDIMENTS and SEDIMENTARY ROCKS in which particles overlap like tiles; their long axes have a consistent DIP. As a result of shear stresses exerted by a current during DEPOSITION, the maximum projection planes (a–b plane), or apparent long axes in cross-section, of non-spherical particles are inclined gently (usually 10°–30°) in the up-current direction relative to the BEDDING surface. It is a useful indicator of PALAEOCURRENT direction, particularly for GRAVEL, which commonly lacks other SEDIMENTARY STRUCTURES. (2) In tectonically deformed rocks, *imbricate structure* describes a series of high-angle reverse FAULTS between two THRUST FAULTS. *TY/GO*

Allen, J.R.L. 1982: *Sedimentary structures: their character and physical basis.* Amsterdam: Elsevier. **Harms, J.C., Southard, J.B. and Walker, R.G.** 1982: *Structures and sequences in clastic rocks* [*Short Course Notes* 9]. Tulsa, OK: Society of Economic Paleontologists and Mineralogists. **Potter, P.E. and Pettijohn, F.J.** 1977: *Paleocurrents and basin analysis*, 2nd, corrected and updated, edn. Berlin: Springer.

impactite   A rock that formed as a result of a METEORITE IMPACT, with properties indicative of *shock metamorphism* or *impact metamorphism* (see METAMORPHIC ROCKS), such as shocked quartz grains, TEKTITES or included diamonds. An example of an impactite is *suevite*, a BRECCIA formerly interpreted as a TUFF, originally described from the Ries impact crater at Nördlingen, southern Germany. Suevite is similar in characteristics to lunar REGOLITH. *GO*

Poag, C.W. and Aubry, M.-P. 1995: Upper Eocene impactites of the U.S. east coast: depositional origins, biostratigraphic framework, and correlation. *Palaios* 10, 16–43. **Reimold, W.U.,**

von Brunn, V. and Koeberl, C. 1997: Are diamictites impact ejecta? No supporting evidence from South Africa Dwyka Group diamictite. *Journal of Geology* 105, 517–530. **Vonengelhardt, W., Arndt, J., Fecker, B. and Pankau, H.G.** 1995: Suevite breccia from the Ries Crater, Germany – origin, cooling history and devitrification of impact glasses. *Meteoritics* 30, 279–293.

imperialism   The pursuit of empire involving the control and exploitation of one state by another. The term *new imperialism* is used to distinguish the phase of territorial conquest beyond Europe in the latter half of the nineteenth century from the earlier phase of European expansion beginning in the fifteenth century. In a remarkably short time, Britain, France and Russia (and to a lesser extent Belgium, Germany and Italy) took control of much of the globe. *JAM*

[See also COLONIALISM]

impermeable   Having a structure and/or texture that does not allow the transmission of liquids or gases. The term can be applied to a rock strata, sediments or layers within a soil profile that do not allow (or slow considerably) the passage of water through pores or fissures. There is, however, a more restricted usage of the term to materials, including rocks and soils, that do not allow the DIFFUSION of fluids through the pores (see PERMEABLE). *JAM/ADT*

[See also HYDRAULIC CONDUCTIVITY, IMPERVIOUS, INFILTRATION]

impervious   (1) A synonym for IMPERMEABLE. (2) The term is sometimes used in a more restricted sense of materials, such as rocks, sediments or soils, that do not permit the flow of fluids through fissures, joints or cracks. (3) Urban areas within *catchments* where there is little or no INFILTRATION of rainfall are also said to be impervious. *ADT/JAM*

[See also PERMEABLE, PERVIOUS]

impulse response   The changes in a SYSTEM following from a sudden input to the system. An example is the variation through time in the amount of CARBON DIOXIDE that remains in the ATMOSPHERE following a sudden injection of CO_2 into the atmosphere. *JAM*

inbreeding depression   The loss in *genetic diversity* and GENETIC FITNESS that may result from breeding among close relatives. Inbreeding depression is especially problematic for populations with relatively few individuals, which unavoidably must be comprised of close relatives. *MVL*

[See also BOTTLENECK, POPULATION VIABILITY ANALYSIS]

Caughley, G. and Gunn, A. 1996: *Conservation biology in theory and practice.* Oxford: Blackwell Science.

inclination (*I*)   The angle of dip between the lines of force of the Earth's MAGNETIC FIELD and the horizontal. Inclination is a parameter of the geomagnetic field used in PALAEOMAGNETIC DATING, which varies between 90° at the magnetic equator and 0° at the magnetic poles. *MHD*

[See also DECLINATION]

Thompson, R. 1991: Palaeomagnetic dating. In: Smart, P.L. and Frances, P.D. (eds), *Quaternary dating methods – a users guide*

[*QRA Technical Guide* 4]. Cambridge: Quaternary Research Association, 177–198.

incremental dating methods

Techniques based on incremental growth of inorganic and organic deposits. Dating may use ANNUALLY RESOLVED RECORDS (see, for example, CORAL analysis, DENDROCHRONOLOGY, ICE CORES, SPELEOTHEMS, VARVE CHRONOLOGY) or growth rates calibrated by independent dating methods (for example, LICHENOMETRIC DATING, WEATHERING RINDS). Annually resolved records can provide age estimates of greater PRECISION than non-incremental techniques, such as ±9 years for dendrochronology and ±1% or less for ice-core years over the HOLOCENE, whereas incremental techniques requiring some form of AGE CALIBRATION, as in the case of lichenometry, are usually of relatively low precision, especially with increasing age. *DAR/CJC*

Baillie, M.G.L. 1995: *A slice through time: dendrochronology and precision dating.* London: Batsford. **Meese, D.A., Gow, A.J., Alley, R.B.** *et al.* 1997: The Greenland Ice Sheet Project 2 depth-age scale: methods and results. *Journal of Geophysical Research* 102, 26411–26423.

indeterminacy

In a large SYSTEM it becomes impossible to specify all of the exact causal chains relating variables to one another (indeterminacy of state); also measurement becomes inexact or impossible (indeterminacy of measurement). *CET*

[See also UNCERTAINTY PRINCIPLE]

index fossil

A FOSSIL or SUBFOSSIL species that characterises a biozone in BIOSTRATIGRAPHY. *GO*

Whittaker, A., Cope, J.C.W., Cowie, J.W. *et al.* 1991. A guide to stratigraphical procedure. *Journal of the Geological Society, London* 148, 813–824.

indicator species

Species that are indicative of a particular set of environmental conditions. Certain plants are widely used as indicator species for determining land suitability for forestry and other uses. Certain microorganisms, such as coliform bacteria, are useful indicators of the potential for WATER-BORNE DISEASE outbreaks. Subfossil diatoms, pollen and other organic remains are used as indicators of past environmental conditions. *JLI*

[See also BIO-INDICATORS, ENVIRONMENTAL INDICATORS, NICHE]

Brenniman, G.R. 1999: Water-borne diseases. In Alexander, D.E. and Fairbridge, R.W. (eds), *Encyclopedia of environmental science.* Dordrecht: Kluwer, 682–685. **Klinka, K., Krajina, V.J., Ceska, A. and Scagel, A.M.** 1989: *Indicator plants of coastal British Columbia.* Vancouver: UBC Press.

individualistic concept

The idea that individual species within a community behave differently, such that any community is essentially a chance combination of species and individuals. The idea was put forward by H.A. Gleason, and contrasts with the CLIMAX VEGETATION concept of F.E. Clements, which saw communities as eventually developing a stable and repeatable species composition determined principally by climate. The difference in approach is very important when the possible responses of plants and plant communities to environmental change are modelled. *JLI*

[See also COMMUNITY CONCEPTS]

Clements, F.E. 1916: *Plant succession. An analysis of the development of vegetation* [*Publication* 242]. Washington, DC: Carnegie Institution. **Gleason, H.A.** 1917: The structure and development of the plant association. *Bulletin of the Torrey Botanical Club* 44, 463–481. **Gleason, H.A.** 1926: The individualistic concept of the plant association. *Bulletin of the Torrey Botanical Club* 53, 7–26. **Gleason, H.A.** 1927: Further views on the succession concept. *Ecology* 8, 299–326. **Matthews, J.A.** 1996: Classics in physical geography revisited: Gleason, H.A. 1939: The individualistic concept of the plant association. *Progress in Physical Geography* 20, 193–203. **McIntosh, R.P.** 1995: H.A. Gleason's 'individualistic concept' and theory of animal communities: a continuing controversy. *Biological Reviews* 70, 317–357.

induction

To reason from particular instances to a general truth, or from individuals to universals. Induction develops a conclusion that is not a logical necessity from the premises. *CET*

inductively coupled plasma atomic emission spectrometry (ICP-AES)

A technique for elemental analysis used in GEOCHEMISTRY and related fields involving OPTICAL EMISSION SPECTROSCOPY of the sample as a plasma (a gas produced from the sample, held at around 10 000 K by a radio-frequency generator). Instruments can typically analyse for over 50 spectral lines simultaneously (compare ATOMIC ABSORPTION SPECTROPHOTOMETRY) with a very small sample size (1–10 mg of material in solution) and the accuracy is about ±5%. *TY*

[See also INDUCTIVELY COUPLED PLASMA MASS SPECTROMETRY (ICP-MS)]

Butler, O.T. and Howe, A.M. 1999: Development of an international standard for the determination of metals and metalloids in workplace air using ICP-AES: evaluation of sample dissolution procedures through an interlaboratory trial. *Journal of Environmental Monitoring* 1, 23–32. **Fairchild, I.J., Hendry, G., Quest, M. and Tucker, M.E.** 1988: Chemical analysis of sedimentary rocks. In Tucker, M.E. (ed.), *Techniques in sedimentology.* Oxford: Blackwell Scientific, 274–354. **Linderholm, J. and Lundberg, E.** 1994: Chemical characterization of various archaeological soil samples using main and trace-elements determined by inductively-coupled plasma-atomic emission-spectrometry. *Journal of Archaeological Science* 21, 303–314. **Szaloki, I., Somogyi, A., Braun, M. and Toth, A.** 1999: Investigation of geochemical composition of lake sediments using ED-XRF and ICP-AES techniques. *X-Ray Spectrometry* 28, 399–405.

inductively coupled plasma mass spectrometry (ICP-MS)

A technique for elemental analysis used in GEOCHEMISTRY and related fields involving mass spectrometry of a plasma (a gas produced from the sample, and held at around 10 000 K by a radio-frequency generator). Instruments can typically analyse for over 50 elements simultaneously, with accuracy to parts per billion level for many, with a small sample size (routinely 100–200 mg of material in solution is used to allow repeat measurements). *TY*

[See also INDUCTIVELY COUPLED PLASMA ATOMIC EMISSION SPECTROMETRY (ICP-AES); MASS SPECTROMETER, TEPHRA ANALYSIS]

Dai, X.X., Chai, Z.F., Mao, X.Y. *et al.* 2000: An alpha-amino pyridine resin preconcentration method for iridium in environmental and geological samples. *Analytica Chimica Acta* 403, 243–247. **Fairchild, I.J., Hendry, G., Quest, M. and Tucker, M.E.** 1988: Chemical analysis of sedimentary rocks. In Tucker, M.E. (ed.), *Techniques in sedimentology.* Oxford:

Blackwell Scientific, 274–354. **Golub, M.S., Keen, C.L., Commisso, J.F.** *et al.* 1999: Arsenic tissue concentration of immature mice one hour after oral exposure to gold mine tailings. *Environmental Geochemistry and Health* 21, 199–209. **Jarvis, K.E., Gray, A.L. and Houk, R.S.** 1991: *Handbook of inductively coupled plasma mass spectrometry.* London: Blackie.

induration A soil horizon, or part of a soil horizon, that is cemented with calcium carbonate or the oxides of silicon, iron, aluminium or humus, so that it is difficult to dig and is difficult for plant roots to penetrate. It also refers to the results of the processes of CEMENTATION and COMPACTION. Induration may occur in a number of soils, including PODZOLS and GLEYSOLS, but particularly in DURISOLS. *EMB*

[See also HARDPAN, DURICRUST]

Thompson, C.H., Bridges, E.M. and Jenkins, D.A. 1993: An exploratory examination of relict hardpans in the coastal lowlands of southern Queensland. In Ringrose-Voase, A.J. and Humphreys, G.S. (eds), *Soil micromorphology: studies in management and genesis* [*Developments in Soil Science* 22]. Amsterdam: Elsevier.

industrial archaeology The study of the legacy of the INDUSTRIAL REVOLUTION from an archaeological perspective. It includes investigation of abandoned industrial workings, their landscape settings, environmental impacts and broader social significance. The subdiscipline developed in postwar Britain following the recognition of a need to preserve some relics of the industrial HERITAGE. *JAM*

Industrial Archeology Review 1988: Special Issue on Textile Mills. *Industrial Archaeology Review* 10. **Palmer, M.** 1999: The archaeology of industrialization. In Barker, G. (ed.), *Companion encyclopedia of archaeology.* London: Routledge, 1160–1197. **Trinder, B. (ed.)** 1992: *The Blackwell encyclopedia of industrial archaeology.* Oxford: Blackwell.

industrial ecology An approach to industrial systems that emphasises ECOSYSTEM analogies. The emphasis is on the efficient use of RESOURCES, extending the life of products, POLLUTION prevention, WASTE RECYCLING and re-use and the establishment of *eco-industrial parks* (such as Kalundborg, Denmark), in which the outputs of several industrial processes serve as inputs for others. *JAM*

Porteous, A. 2000: *Dictionary of environmental science and technology* , 3rd edn. Chichester: Wiley.

Industrial Revolution The period of rapid transition from an agricultural to an industrial society, which is generally considered to have begun in Great Britain around the middle of the eighteenth century. It continued late into the nineteenth century and was characterised especially by the use of coal as an energy source powering heavy engineering industries dominated by steam, railways and steel. *JAM*

[See also INDUSTRIALISATION]

Ashton, T.S. 1967: *The Industrial Revolution – 1760–1830.* Oxford: Oxford University Press. **Hudson, P.** 1992: *Industrial Revolution.* London: Arnold.

industrial waste Industrial activities produce a wide variety of *natural wastes* from MINING and *by-products* from the processes of INDUSTRY. Some of these wastes are reusable; others are classified as HAZARDOUS WASTE and require special treatment to isolate them from the environment or render them innocuous. Reject material from COAL, china clay, slate and gravel extraction amount to many millions of tonnes and tipping is mainly on land adjacent to the site worked. Material from these industries can be used for brick making, light aggregate and concrete blocks, as well as for road-building material or as a filter medium for sewage works. Wastes from power stations includes *pulverised fuel ash* and *clinker*; some of the former is used in cement as a filler and, with the latter, can be made into concrete blocks and aggregate. Similarly, *slag* from blast furnaces and steel making can be used for aggregate and basic slag is used as a fertiliser for pastures. Mining spoil and slag from non-ferrous industries usually contain *toxic metal* elements that limit their usefulness, but tin and copper slags have been used for sand-blasting. Red mud from aluminium manufacture has been used as pigment in paints and plastics. Sludges from the petroleum industry are usually consigned to special LANDFILLS or incinerated. Carbonisation plants produce gas, coke, coal tar and ammoniacal liquor, all of which are useful materials, but in the process of gas purification, ferrous iron oxides were used and these became highly acidic and contaminated with cyanide. NATURAL GAS has made most of these plants redundant. Chemical and pharmaceutical production also results in sludges that require special treatment to make them innocuous. Nineteenth-century dumps from alkali and chromate works still remain as visual intrusions in the British landscape. Production of munitions and explosives for two World Wars and radioactive leakage from atomic sites have caused environmental problems and HUMAN HEALTH HAZARDS. LAND CONTAMINATION by scrapyards, sewage works and landfills are a legacy of poor WASTE MANAGEMENT during the past 100 years. *EMB*

[See also AGRICULTURAL WASTE, DOMESTIC WASTE, WASTE MINING, WASTE RECYCLING]

Bridges, E.M. 1987: *Surveying derelict land.* Oxford: Clarendon Press. **Brunner, C.R.** 1991: *Handbook of incineration systems.* New York: McGraw-Hill. **Eckenfelder Jr, W.W.** 1989: *Industrial water pollution control*, 2nd edn. New York: McGraw-Hill. **Gutt, W., Nixon, P.J., Smith, M.A.** *et al.* 1974: *A survey of the locations, disposal and prospective uses of the major industrial by-products and waste materials.* London: Building Research Establishment. **Nemerow, N.C.** 1984: *Industrial solid wastes.* Cambridge, MA: Ballinger.

industrialisation The process of change from a predominantly agricultural society to one dominated by industrial activity, particularly EXTRACTIVE and MANUFACTURING INDUSTRY. It began with the INDUSTRIAL REVOLUTION in Great Britain, which is dated, conventionally, from around the mid-eighteenth century. This was the culmination of many innovations initiated in the late seventeenth and early eighteenth centuries. Rapid advances in engineering had occurred in the water-based textile industry, and the charcoal-based iron smelting industry had resulted in experiments in the use of coal for smelting, while coal had also been used in the early development of the non-ferrous metal industry. Together, such advances combined to set the scene for the period of rapid change, which followed from the middle of the eighteenth century onwards. The second phase of industrialisation occurred around the mid-nineteenth century and had a much greater environmental impact, saw the application of the steam engine powered by coal to pumps, machines, ships and then railways, which spread rapidly throughout Europe. Industrialisation continued in the late nineteenth century with the rise of steel and chemical industries and the application of electrical power: Germany and the United States surpassed Great Britain in industrial output at this time, while Russia and Japan began to industrialise. These countries enhanced their position at the expense of Europe in the fourth phase, which saw a further revolution in transport, the invention of the internal combustion engine, its application to shipping and railways and the ascendancy of road and air transport. GLOBALISATION continues in the present phase, which began after World War II, with major centres of industrial activity developing around the Pacific rim, in the Middle East and in Asia based less on 'heavy engineering' and more on electronics and SERVICE INDUSTRIES.

Differences remain between the older centres of industrialisation and those that are 'catching up' but overall industrial expansion has been enormous. This has been possible only through the successive replacement of technologies during TECHNOLOGICAL CHANGE. These changes have yielded major productivity gains in energy use, materials and labour, which have sustained the increasing levels of output and incomes, eased demands on natural resources, reduced traditional environmental impacts such as AIR POLLUTION and created more leisure time and a RECREATION industry. However, new environmental concerns have arisen, such as enhanced GLOBAL WARMING from burning FOSSIL FUELS, EUTROPHICATION and PESTICIDES from the industrialisation of agriculture, synthetic chemicals leading to OZONE holes and the uncertain environmental impact of BIOTECHNOLOGY and GENETIC ENGINEERING.

Since the eighteenth century, labour productivity has risen by a factor of 200 in industry and at least a factor of 20 in agriculture, while productivity in the use of natural resources and in energy use per unit of economic output has risen by a factor of ten. Historically, such economic or *technological productivity* gains have been outpaced by output and consumption leading to heavier *environmental burdens*. However, in the last two decades, demand for bulk materials and energy use per unit of economic output has stabilised in the most advanced industrialised countries. *Environmental productivity* gains can therefore be seen to be an unplanned side effect of recent technological productivity gains and it has been suggested that there is a large potential for further environmental productivity gains in the future. *JAM*

Bairoch, P. 1982: International industrialization levels from 1750 to 1980. *Journal of European Economic History* 11, 269–333. **Cameron, R.** 1989: *A concise economic history of the world: from Paleolithic times to the present.* Oxford: Oxford University Press. **Grübler, A.** 1998: *Technology and global change.* Cambridge: Cambridge University Press. **Headrick, D.R.** 1990: Technological change. In Turner II, B.L., Clark, W.C., Kates, R.W. *et al.* (eds), *The Earth transformed by human action.* Cambridge: Cambridge University Press, 55–86. **Ilbery, B.W. and Bowler, I.R.** 1996: Industrialization and world agriculture. In Douglas, I., Huggett, R. and Robinson, M. (eds), *Companion encyclopedia of geography.* London: Routledge, 228–248.

industry (1) The production of goods and services including, for example, MANUFACTURING INDUSTRY and SERVICE INDUSTRIES; characteristric of developed societies. (2) In the context of archaeology, a set of ARTEFACTS drawn from different ASSEMBLAGES but related in terms of technology, style or other context: LITHIC industries and CERAMIC industries are examples. *JAM*

[See also CULTURE, INDUSTRIALISATION, TRADITION]

infant mortality rate The average number of deaths of children under one year of age per 1000 live births. Currently and historically, high infant mortality rates tend to reflect limited medical services. *JAM*

infauna Bottom-dwelling aquatic organisms (BENTHOS) that live in rather than on the SEDIMENT. Life habits include suspension-feeders that exploit food resources immediately above the sediment–water interface, deposit-feeders that mine the sediment for organic particles, predators and scavengers. Infauna inhabit soft (burrowers) and hard (borers) substrates. In soft substrates, controls on burrowing depths include the *redox* boundary, sediment firmness, biological constraints on movement and the development of specialised feeding structures (e.g. the siphons of bivalves). Organic material is concentrated in near-surface sediment (< 5 cm) and decreases rapidly downwards. Vertical partitioning of resources (TIERING) shows a Phanerozoic history of infauna to a depth of about 1 m from late Palaeozoic to Recent times. Sepkoski's Vendian, Cambrian and Paleozoic Faunas had only shallow infaunas, before tiering increased through deep-burrowing bivalves and soft-bodied organisms represented only by TRACE FOSSILS. The 'Modern Fauna' maintained well developed tiering, through evolutionary radiations of burrowing bivalves, echinoids, crustaceans and soft-bodied organisms. Important controls on the diversification of infaunal life habits may have been PREDATION and COMPETITION for space. *LC*

[See also EPIFAUNA]

Bottjer, D.J. and Ausich, W.I. 1986: Phanerozoic development of tiering in soft substrata suspension-feeding communities. *Paleobiology* 12, 400–420. **Sepkoski Jr, J.J.** 1984: A kinetic model of Phanerozoic taxonomic diversity. III Post-Paleozoic families and mass extinctions. *Paleobiology* 10, 246–267.

inferential statistic A statistic that involves the concept of PROBABILITY. Inferential statistics or *probabilistic statistics* provide a means of measuring the uncertainty associated with SAMPLING from statistical populations and a precise measure of the statistical confidence that can be placed in results based on representative samples. *JAM*

[See also DESCRIPTIVE STATISTICS, NUMERICAL ANALYSIS, STATISTICAL ANALYSIS]

infiltration The process by which rainfall enters the soil surface. *ADT*

[See also HORTONIAN OVERLAND FLOW, INFILTRATION CAPACITY, RUNOFF PROCESSES, SOIL]

Knapp, B.J. 1978: Infiltration and storage of soil water. In Kirkby, M.J. (ed.), *Hillslope hydrology.* Chichester: Wiley, 43–72.

infiltration capacity The maximum rate at which rainfall can enter the surface soil under given conditions.

If it is exceeded, excess water ponds on the surface and HORTONIAN OVERLAND FLOW is generated. Infiltration capacity under natural conditions only tends to be lower than characteristic rainfall intensities in some arid and semi-arid areas, where the lack of vegetation allows soil surface CRUSTING to develop, and in areas with soils rich in silt and clay. In most vegetated areas, infiltration capacities are very high and rarely, if ever exceeded. Infiltration capacities may be radically reduced by many forms of human interference, notably DEFORESTATION, overgrazing, SOIL EROSION and URBANISATION. *RPDW*

[See also RUNOFF PROCESSES]

infochemical A chemical that, in the natural context, conveys information between organisms and evokes a behavioural or physiological response. A *pheromone* evokes the response in an individual of the same species, whereas an *allelochemical* affects a different species (see ALLELOPATHY). Interactions between organisms are also mediated by other classes of chemicals, such as NUTRIENTS and TOXINS. *JAM*

Dicke, M. 1998: Infochemicals. In Calow, P. (ed.), *The encyclopedia of ecology and environmental management*. Oxford: Blackwell Science, 363–364.

infrared-stimulated luminescence (IRSL) dating
A form of LUMINESCENCE DATING based on stimulation by infrared wavelengths in FELDSPARS. Its advantages over OPTICALLY STIMULATED LUMINESCENCE (OSL) DATING are that (1) a wider wavelength region is available for detection and (2) IRSL in minerals is more effectively bleached during deposition. *DAR*

[See also THERMOLUMINESCENCE DATING]

Hutt, G., Jaek, I. and Tchonka, J. 1988: Optical dating: K-feldspars optical response stimulation spectra. *Quaternary Science Reviews* 7, 381–385. **Lang, A. and Wagner, G.A.** 1996: Infrared stimulated luminescence dating of archaeosediments. *Archaeometry* 38, 129–141. **Wiggenhorn, H., Lang, A. and Wagner, G.A.** 1994: Infrared stimulated luminescence – dating tool for archaeosediments. *Naturwissenschaften* 81, 556–558.

inhibition model A model suggesting that the net effect of early colonising species on later arrivals is negative, delaying establishment and growth of the next stage in ECOLOGICAL SUCCESSION. *LRW*

[See also COLONISATION, FACILITATION, TOLERANCE MODEL]

inlier An area of older rock completely surrounded by younger, contrasting with the normal situation where a rock outcrop is bounded by older rocks on one side and younger on the other. Inliers can form as a result of EROSION or DEFORMATION. *GO*

[See also OUTLIER]

inorganic matter Substances that are devoid of plant or animal material or ORGANIC compounds. *UBW*

[See also MINEROGENIC SEDIMENT, SEDIMENT TYPES]

insect analysis Insect remains are relatively rare in the fossil record. The majority of FOSSIL insect remains (except those in AMBER) consist of compressed and sometimes carbonised fragments or, more rarely, whole insects.

Material may be found in a wide range of situations where preservation conditions are favourable, such as LACUSTRINE SEDIMENTS, CAVES, ARIDLANDS and PERMAFROST environments, as well as in archaeological sites, such as wells and refuse middens. Aspects of insect analysis may be related to archaeology, particularly in the recognition of ANTHROPOGENIC modification of the natural environment. Assemblages associated with early occupation sites may provide information concerning insect infestations in ancient granaries or on cultivated crops, or even suggest directions of early trading systems or human migrations. A major use of fossil insects, particularly beetles, is also in PALAEOENVIRONMENTAL RECONSTRUCTION and CLIMATIC RECONSTRUCTION and in ZOOGEOGRAPHY, on account of their rapid colonising abilities, apparent evolutionary stability and specific habitat and temperature preferences. *DCS*

[See also BEETLE ANALYSIS, CHIRONOMID ANALYSIS, MUTUAL CLIMATIC RANGE METHOD]

Coope, G.R., 1970: Interpretations of Quaternary insect fossils. *Annual Review of Entomology*, 15, 97–120. **Coope, G.R.** 1994: The response of insect faunas to glacial–interglacial climatic fluctuations. *Philosophical Transactions of the Royal Society of London* B344, 19–26. **Elias, S.A.** 1994: *Quaternary insects and their environments*. Washington, DC: Smithsonian Institution. **Kenward, H.K.** 1975: Pitfalls in the environmental interpretation of insect death assemblages. *Journal of Archaeological Science* 2, 85–94.

inselberg A prominent, isolated, residual hill, usually smoothed and rounded. Inselbergs rise abruptly from, and are surrounded by, extensive lowland EROSION SURFACES and are characteristic of ARIDLANDS or SEMI-ARID landscapes in a late stage of evolution. The term was proposed by Bornhardt in AD 1900 to describe the abrupt, rocky hills that commonly interrupt tropical plains, but use of the term was soon extended to include a variety of isolated hill forms and its exact meaning became confused.

Theories concerning the origin of inselbergs can be divided into four groups: (1) non-climatic theories that lay greater emphasis on lithology or structural factors or on general relief development via scarp or PARALLEL SLOPE RETREAT under a range of possible climates; (2) single-climate theories that link inselberg development to a particular climate (see CLIMATIC GEOMORPHOLOGY); (3) POLYGENETIC theories that invoke CLIMATIC CHANGE or a sequence of climates, in which an usually long period of deep weathering and ineffective erosion perhaps under a hot wet climate, is followed by a stripping phase under a climate (perhaps less humid) that promotes erosion (see Figure); and (4) polygenetic theories that invoke a combination of climates and geological factors of uplift or tilting.

Single-climate theories have suggested that contrasting climates followed inselberg formation. Büdel's DOUBLE PLANATION SURFACE theory favoured the wet–dry tropics, whereas Bremer's DIVERGENT WEATHERING theory considered the wetter tropics as more favourable for inselberg formation. Similarly, authors proposing climatic change theories have also disagreed about the sequence of climates that may be involved, some suggesting dissection and stripping during more humid phases, whereas others suggest that semi-arid phases favour stripping. With really large inselbergs (BORNHARDTS) a polygenetic origin over geological time would seem the only logical explanation. *MAC*

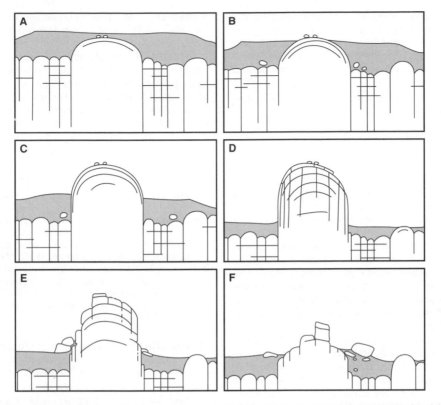

inselberg *The evolution of an inselberg: each stripping phase, during which the products of deep weathering (shaded) are removed from around the domed inselberg or bornhardt, is accompanied by lowering of the basal weathering surface, culminating in the formation of a residual koppie or tor (after Thomas, 1994)*

[See also ETCHPLAINS, LANDSCAPE EVOLUTION, PEDIMENT, TOR]

Bremer H. and Jennings J. (eds) 1978: Inselbergs. *Zeitschrift für Geomorphologie Supplementband* 31 [Special Issue]. **Ollier, C.D.** 1960: The inselbergs of Uganda. *Zeitschrift für Geomorphologie NF* 4, 43–52. **Thomas M.F.** 1994: *Geomorphology in the tropics: a study of weathering and denudation in low latitudes.* Chichester: Wiley. **Twidale, C.R.** 1990: The origin and implications of some erosional landforms. *Journal of Geology* 98, 343–364.

insolation The flow of direct SOLAR RADIATION intercepted by a unit area of a horizontal surface either at or above the Earth's surface. It varies with season, latitude, slope, ASPECT and atmospheric TURBIDITY and also on longer time scales due to SOLAR CYCLES, ORBITAL FORCING and other factors. *JCS*

[See also MILANKOVITCH THEORY]

instantaneous field of view (IFOV) The cone-shaped region of space from which radiation is collected for each discrete measurement, or PIXEL. The IFOV is defined by the solid angle subtended by a combination of the detector and the instrument focusing system. The sensitivity of an instrument is not usually uniform over the maximum extent of the IFOV, resulting in a point spread function. The nominal IFOV is usually reported for some fraction of the maximum of the point spread function. The field of view of an instrument is the sum of the IFOVs recorded for a single line of the RASTER image. The size of

the IFOV footprint on the surface is proportional to the distance between the point on the surface where the radiation is reflected and the instrument. The size of the IFOV footprint will therefore increase as it moves away from the NADIR POINT. *GMS*

Schowengerdt, R.A. 1997: *Remote sensing: models and methods for image processing.* New York: Academic Press.

instrument calibration The act of associating an instrument's measurement to a sensor-independent geophysical quantity. For the comparison of measurements of the same instrument gathered at different times or of values from different parts of the same environment or image, *relative calibration* may be sufficient. *TS*

[See also CALIBRATION]

instrumental data Derived from appliances with a standardised construction, exposure and observation times that respond to environmental conditions in a constant manner in order to acquire data that is precise and comparable from place to place and over periods of time. They normally have a scale of measurement that is calibrated against a known standard and often internationally agreed, as in the Système Internationale d'Unités (SI UNITS) agreed in AD 1948. Where instruments have a source of power (mechanical or electrical), they may record the environment in remote areas such as mountains, or continuously over long periods of time (e.g. AUTOMATIC WEATHER STATIONS). They contrast with visual observations and PROXY records, which depend on the responses of humans or organisms to environmental conditions that are more difficult to standardise. The data from early instruments lack the PRECISION and comparability of modern data because designs were unique and experimental. *JGT*

[See also EARLY INSTRUMENTAL METEOROLOGICAL RECORDS, INSTRUMENTAL RECORD]

instrumental record A record of an environmental variable measured on a quantitative scale using an instrument. In the context of ENVIRONMENTAL CHANGE, instrumental records are important for understanding *variability*, for detecting relatively short-term, interannual and decadal PERIODICITIES and TRENDS, and for the CALIBRATION of PROXY DATA in PALAEOENVIRONMENTAL RECONSTRUCTION.

The instrumental record of climate is particularly important and illustrative. The network of regular meteorological observations dwindles as one goes back in time. The British Isles have the longest systematic record of temperature in the world and precipitation was first measured at Kew in AD 1697. Atmospheric pressure was first recorded at Trondheim, Norway, in 1762 and in London in 1787. By 1750 several meteorological records were available from northwestern Europe and the first early records have survived from 1850 in the eastern USA as well as from eastern Europe and parts of Asia. From most other parts of the world records stretch back for about a century and a half (e.g. Australasia, Auckland 1853; South America, Rio de Janeiro 1832). EARLY INSTRUMENTAL METEOROLOGICAL RECORDS were often produced in unrepresentative locations: for example, temperature readings in Britain were recorded for a time in the eighteenth century in north-facing rooms. Factors such as changes in instrumentation, exposure, location and methodological practices can all cause inhomogeneities in records. *AHP/JAM*

[See also CENTRAL ENGLAND TEMPERATURE RECORD, CLIMATIC VARIABILITY, CLIMATIC VARIATIONS, HISTORICAL CLIMATOLOGY, HOMOGENEOUS SERIES]

Jones P.D and Bradley R.S. 1992: Climatic variations in the longest instrumental records. In Bradley, R.S. and Jones, P.D. (eds), *Climate since A.D.1500*. London: Routledge, 246–268. **Lamb, H.H.** 1966: *The changing climate*. London: Methuen. **Middleton, W.E.K.** 1969: *The history of the thermometer and its use in meteorology*. Baltimore, MD: Johns Hopkins Press.

interaction indicator ENVIRONMENTAL INDICATORS that signal the nature of the interaction between people and ecosystems, and hence enable an assessment of how human activities should be modified to avoid undesirable outcomes. *JAM*

[See also ECOSYSTEM INDICATOR, GEOINDICATOR, SYNTHESIS INDICATOR]

Hodge, R.A. 1996: Indicators and their role in assessing progress towards sustainability. In Berger, A.R. and Iams, W.J. (eds), *Geoindicators: assessing rapid environmental changes in earth systems*. Rotterdam: A.A. Balkema, 19–24.

Interactive Graphics Retrieval System (INGRES) A proprietary *database management system* for the efficient handling of RELATIONAL DATABASES. It is used in education and business, and capable of supporting structured QUERY LANGUAGES (such as SQL). *TVM*

Stonebraker, M. 1986: The design and implementation of INGRES. *ACM Transactions on Database Systems* 1, 189–222.

interannual variability Between-year variation in a time series. Interannual variability is an important characteristic of CLIMATE and of HIGH-RESOLUTION RECONSTRUCTIONS of climatic and environmental change, where the mean value of climatic variables (as expressed, for example, in the MEAN ANNUAL AIR TEMPERATURE) may mask great variations from year to year (as demonstrated, for example, by successive ANNUAL MEAN AIR TEMPERATURES). *JAM*

[See also ANOMALIES, CLIMATIC NORMAL]

Baumgartner, T.R., Michaelsen, J., Thompson, L.G. *et al.* 1989: The recording of interannual climatic change by high-resolution natural systems: tree-rings, coral bands, glacial ice layers and marine varves. In Peterson, D. (ed.), *Climatic change in the eastern Pacific and western Americas*. Washington, DC: American Geophysical Union, 1–14. **Cayan, D.R., Miller, A.J., Barnett, T.P.** *et al.* 1995: Seasonal-to-interannual fluctuations in surface temperature over the Pacific: effects of monthly winds and heat fluxes. In National Research Council (eds), *Natural climate variability on decade-to-century timescales*. Washington, DC: National Academy Press, 133–150.

interception The temporary storage of a proportion of PRECIPITATION, usually by vegetation, prior to it either evaporating back to the atmosphere or falling to the ground as *throughfall* or STEMFLOW. *Interception loss* is the amount (in mm) or proportion (as a percentage) of precipitation that fails to reach the ground and is evaporated back to the atmosphere. Interception is greatly affected by LANDUSE CHANGES, with significant consequences for RUNOFF, RIVER DISCHARGE, water resources and water supply. *ADT/RPDW*

Durocher, M.G. 1990: Monitoring spatial variability of forest interception. *Hydrological Processes* 4, 215–219.

interdecadal variability Between-decade variation in a time series. Interdecadal climatic variability includes those CLIMATIC VARIATIONS likely to affect society in the medium to long-term (defined with reference to a few human generations). Variations are often difficult to detect, and their causes and controlling mechanisms, such as radiative transfer, the GENERAL CIRCULATION OF THE ATMOSPHERE and oceans, the processes of PHOTOCHEMISTRY and the BIOGEOCHEMICAL CYCLES of TRACE GASES and NUTRIENTS, are still imperfectly understood. Current understanding of environmental change on interdecadal timescales is based on insufficient observations and imperfect models. *JAM*

[See also CLIMATIC VARIABILITY, PERIODICITIES]

National Research Council 1995: *Natural climate variability on decade-to-century time scales*. Washington, DC: National Academy Press. **Navarra, A. (ed.)** 1999: *Beyond El Niño: decadal and inter-decadal climate variability*. Berlin: Springer.

interdisciplinary research Research that exists between, or that transcends the boundaries of, conventional disciplines but that involves the merging of knowledge, the development of common concepts and the devising of unified methodologies to investigate and solve problems. True interdisciplinary research may be distinguished from the usual practice for issues that transcend conventional discipline boundaries, which would involve a range of specialists bringing their expertise and their own methodologies to bear: this is MULTIDISCIPLINARY research. O'Riordan (1995: 4) argued that 'true interdisciplinarity has probably never existed, because the phenomenon involves the unification of concepts that are designed to be conceived as separate entities'. Nevertheless, he identified four concepts that 'embrace both the social and natural sciences' and so 'look promising': CHAOS theory; social learning; dynamic equilibria; and CARRYING CAPACITY, which he saw as akin to SUSTAINABILITY. This latter term, or concept, has come increasingly to the fore over recent years and might be the foundation for an interdisciplinary futurology.   *FMC*

O'Riordan, T. (ed.) 1995: *Environmental science for environmental management*. London: Longman.

interferometry   The process of combining two coherent measurements or images to detect differences in phase. Measurements or images that are formed coherently contain both amplitude (brightness) and phase information. The phase information derives from interactions between ELECTROMAGNETIC RADIATION and scattering elements in the scene. It is often noise-like because of the large number of scattering elements and does not usually carry any information in itself. Under certain conditions, however, the relative phase of two measurements obtained at slightly different positions does carry useful information. REMOTE SENSING applications of interferometry are addressed using complex images acquired with space-borne *synthetic aperture radar* (SAR) instruments (in a technique known as SAR interferometry or INSAR). In this technique, two images of the same scene are acquired from slightly different positions (by a platform carrying two antennae in either across- or along-track configuration) or by a single antenna SAR imaging at two different times *(repeat-pass interferometry)*. Provided the two images are registered to sub-PIXEL accuracy, they can be combined to form an *interferogram*, which is a map of the phase differences of the two images.

The phase differences can be related directly to the differences in the distances to scatterers in the scene and it is therefore possible, in principle, to map the topography of the surface (by *phase unwrapping*). Applications of this have included topographic mapping over land (for DIGITAL ELEVATION MODEL generation) and monitoring glacier flows. The relative changes in topography can be mapped in a technique known as *differential interferometry*, which has been used to monitor land subsidence due to mining, water extraction and seismic activity.

Successful interferometry often has to compete with *decoherence*, which degrades the quality of the phase information. Decoherence is caused by *spatial decorrelation* (where the terrain is viewed from different directions and therefore has slightly different scattering properties) and also by *temporal decorrelation* (where changes in the scattering surface result from different weather conditions or movement of the surface). In some cases, propagation effects in the atmosphere may also randomise phases and lead to patches of low coherence (*artefacts*) in the image.   *PJS*

[See also RADAR REMOTE SENSING INSTRUMENTS, MICROWAVE REMOTE SENSING]

Massonnet, D., Rossi, M., Carmona, C. et al. 1993. The displacement field of the Landers earthquake mapped by radar interferometry. *Nature* 364, 138–142. **Wegmuller, U., Werner, C.L., Nuesch, D. and Borgeaud, M.** 1995: Land surface analysis using ERS-1 SAR interferometry. *European Space Agency Bulletin* 81, 30–37. **Zebker, H.A., Werner, C.L., Rosen, P.A. and Hensley, S.** 1994. Accuracy of topographic maps derived from ERS-1 interferometric radar. *IEEE Transactions on Geoscience and Remote Sensing* 32, 823–836.

interglacial   A period of thermal improvement separating cold GLACIAL EPISODES, when climatic conditions were similar to, or warmer than, those experienced today. The current interglacial, known as the HOLOCENE, has lasted approximately 11 500 years (10 000 radiocarbon years). Although a large number of interglacials are recognised for the QUATERNARY period in MARINE SEDIMENT CORES and given odd numbers in terms of oxygen ISOTOPIC STAGES, fewer interglacial periods are precisely identified and dated in the terrestrial record and some longer periods encompassing several interglacials are defined as *interglacial complexes*, for example the CROMERIAN INTERGLACIAL in the British Isles. Interglacials are usually considered to have been periods of relatively stable climate showing a consistent pattern of climatic warming and subsequent cooling, as defined in the INTERGLACIAL CYCLE, although recent evidence from ICE CORES from Greenland has suggested that they were punctuated by brief cooling episodes. This is however a matter of current debate.   *CJC*

[See also QUATERNARY TIMESCALE]

Bowen , D.Q. 1999: *A revised correlation of Quaternary deposits in the British Isles* [*The Geological Survey Special Report* 23]. Bath: Geological Society. **Dansgaard, W., Johnsen, S.J., Clausen, H.B. et al.** 1993: Evidence for general instability of past climate from a 250-kyr ice-core record. *Nature* 364, 218–220. **Lowe, J.J. and Walker, M.J.C.** 1997: *Reconstructing Quaternary environments*, 2nd edn. Harlow: Longman.

interglacial cycle   The cycle of responses of FLORA, VEGETATION and SOILS to changes in seasonal insolation due to MILANKOVITCH parameters. Originally proposed by Iversen (1958), and modified by Andersen (1966), the idea has been reviewed by Birks (1986). Iversen proposed four stages to the cycle: *cryocratic, protocratic, mesocratic* and *telocratic*. There are three key concepts: a rise and subsequent fall in mean annual temperature; progressive vegetation development (MIGRATION and ECOLOGICAL SUCCESSION) in the early stages; and RETROGRESSIVE SUCCESSION in the later stages (see Figure).   *BA*

Andersen, S. Th. 1966: Interglacial succession and lake development in Denmark. *Palaeobotanist* 15, 117–127. **Birks, H. J. B.** 1986: Late-Quaternary biotic changes in terrestrial and aquatic environments, with particular reference to north-west Europe. In Berglund, B.E. (ed.), *Handbook of Holocene palaeoecology and palaeohydrology*. Chichester: Wiley, 3–65. **Iversen, J.** 1958: The bearing of glacial and interglacial epochs on the formation and extinction of plant taxa. *Uppsala Universiteit Årsskrift* 6, 210–215.

Intergovernmental Panel on Climate Change (IPCC)   The IPCC was established jointly by the World Meteorological Organisation (WMO) and the United Nations Environment Programme (UNEP) in 1988. It was chaired by B. Bolin (Sweden); other members of the Panel were A. Al Gain (Saudi Arabia), who was vice chairman, J. A. Adejokun (Nigeria), who was rapporteur, and N. Sundararaman (WMO), who was secretary. The Panel

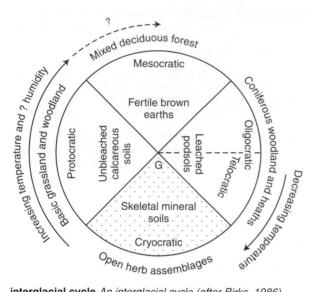

interglacial cycle *An interglacial cycle (after Birks, 1986)*

had two objectives: (1) to assess the scientific information related to the issue of CLIMATIC CHANGE and to evaluate its environmental and socio-economic consequences; and (2) to formulate realistic response strategies for the management of climate change.

To carry out these objectives, the Panel set up three Working Groups to assess, in relation to climate change, the available scientific information (Working Group I , chaired by J.T. Houghton, UK) and its impacts (Working Group II, Y. Israel, USSR) and to formulate policy responses (Working Group III, F. Bernthal, USA). In addition, a special committee (chaired by M.J. Ripert, France) was established to promote the participation of developing countries in these activities.

In 1990, the initial reports of Working Groups I and II were published in time for the Second World Climate Conference held in Geneva. Working Group I concluded that there is a natural GREENHOUSE EFFECT, which keeps the Earth warmer than it would otherwise be and that human activities result in emissions that have enhanced, and will increasingly enhance, this effect. It confidently calculated that immediate reductions by 60% in the EMISSIONS of long-lived gases (CARBON DIOXIDE, nitrous oxide and CHLOROFLUOROCARBONS) were necessary to keep concentrations at current levels and predicted that under the BUSINESS-AS-USUAL SCENARIO (i.e. no steps will be taken to limit emissions) global mean temperatures would rise by 0.3°C per decade over the next 100 years. Working Group II concluded that changes in climate would have important effects on agriculture, livestock, natural ecosystems, water resources, sea levels and a variety of other human activities. Working Group III found that climate change is a global issue and will require a global effort, but the onus will be on the developed countries. Updates are ongoing (e.g. Houghton *et al.*, 1996) and the third assessment was published in mid-2001 (see http://uk.cambridge.org for details). *BDG*

Houghton, J. T., Jenkins, G.J. and Ephraums, J.J. (eds) 1990: *Climate change. The IPCC scientific assessment.* Cambridge: Cambridge University Press. **Houghton, J.T., Meira Filho,**

L.G., Callender, B.A. *et al.* 1996: *Climate change 1995. The science of climate change.* Cambridge, Cambridge University Press. **Tegart, W.J.McG., Sheldon, G.W. and Griffiths, D.C. (eds)** 1990: *Climate change. The IPCC impacts assessment.* Canberra: Australian Government Publishing Service.

intermediate-complexity modelling

An approach to *modelling* any complex SYSTEM that aims to avoid the problems of both oversimplification and oversophistication. In EARTH-SYSTEM ANALYSIS, it provides a pragmatic solution to modelling the complexities of planet Earth as a whole. Such models have been termed *Earth-system models of intermediate complexity* (EMICs). *JAM*

Ganopolski, A., Rahmstorf, S., Petoukhov, V. and Claussen, M. 1998: Simulation of modern and glacial climates with a coupled global model of intermediate complexity. *Nature* 391, 351–356. **Schellnhuber, H.J.** 1999: 'Earth system' analysis and the second Copernican revolution. *Nature* 402 (Supplement), C19–23.

intermediate disturbance hypothesis

Originally proposed by J.H. Connell and M. Huston, this hypothesis states that species richness in a habitat is highest when disturbances occur with intermediate frequency or intensity . As DISTURBANCE frequency or intensity increases beyond this level, richness declines because only species that colonise and grow very rapidly are able to establish or maintain themselves successfully. At low levels of disturbance, competitive exclusion reduces local species richness. This hypothesis may apply at a local scale, but recent evidence suggests that it is infrequently important. There is no evidence to suggest that it can explain large-scale patterns such as the 'latitudinal' BIODIVERSITY gradient. Managing disturbance frequency is important in CONSERVATION. When humans alter natural disturbance regimes, ecosystems may undergo major structural changes, such as when fire-suppression causes prairie habitats to change into forests through *succession*. Such HABITAT LOSS is a cause of EXTINCTION. CLIMATIC CHANGE may change disturbance regimes at a very large scale. *JTK*

[See also DIVERSITY CONCEPTS, DIVERSITY–STABILITY HYPOTHESIS]

Connell, J.H. 1978: Diversity in tropical rain forests and coral reefs. *Science* 199, 1302–1310. **Currie, D.J.** 1991: Energy and large-scale patterns of animal and plant species richness. *American Naturalist* 137, 27–49. **Huston, M.** 1979: A general hypothesis of species diversity. *American Naturalist*, 113, 81–101. **Mackey, R.L. and Currie, D.J.** 2000: A re-examination of the expected effects of disturbance on diversity. *Oikos* 88, 483–492. **McNaughton, S.J.** 1989: Ecosystems and conservation in the twenty-first century. In Western, D. and Pearl, M.C. (eds), *Conservation for the twenty-first century.* Oxford: Oxford University Press, 109–120.

International Tree-Ring Database (ITRDB)

Part of the World Data Center-A for Paleoclimatology at the National Geophysical Data Center (NGDC) housed at Boulder, Colorado, USA, which comprises many different types of palaeoclimate data besides those derived from TREE RINGS (e.g. ICE CORE and POLLEN records). The ITRB grew out of a databank originally begun at the Laboratory of Tree-Ring Research in Tuscon, Arizona, but today it provides a central storage and distribution centre for tree-ring data from thousands of locations from around the world. These data are freely searchable and downloadable to anyone, simply by connecting to http://www.ngdc.noaa.gov/paleo/ftp-treering.html. A useful guide can also be found at http://web.utk.edu/~grissino/itrdb.html. *KRB*

Grissino-Mayer, H.D. and Fritts, H.C. 1997: The International Tree-Ring Data Bank: an enhanced global database serving the global scientific community. *The Holocene* 7, 235–238.

internationalisation of research

As the nature of ENVIRONMENTAL CHANGE research has itself changed from a largely local or regional pursuit to encompass GLOBAL CHANGE, it has become increasingly internationalised. Internationalisation of environmental change research is affecting disciplines, institutions, resources and methodologies. For example, the International Council of Scientific Unions (ICSU) recognises four GLOBAL ENVIRONMENTAL CHANGE research programmes: the *World Climate Research Programme* (WCRP), which investigates the physical CLIMATIC SYSTEM; the *International Geosphere–Biosphere Programme* (IGBP), which is concerned with interactions between living and non-living systems; the *International Human Dimensions Programme on Global Environmental Change* (IHDP), which considers interactions between human society and the environment at a global scale; and the *Diversitas Programme*, focused on the structure and function of BIOLOGICAL DIVERSITY. Such *supra-national science programmes* are a relatively new dimension of environmental change research, and raise important questions about how research problems are identified, pursued and utilised. While co-ordination is increasingly at the international level, most funding and operations are still carried out at the national level. *JAM*

Ehlers, E. 1999: Environment and geography: international programmes on global environmental change. *International Geographical Union Bulletin* 49, 5–18. **Jasanoff, S. and Wynne, B.** 1998: Science and decision making. In Rayner, S. and Malone, E.L. (eds), *Human choice and climate change.* Vol. 1. *The societal framework.* Columbus, OH: Battelle Press, 1–88.

interpluvial

A relatively dry time interval between two PLUVIAL phases. *JAM*

interpolation

Methods that estimate intermediate values of a dependent variable as a function of independent variables. These techniques are used, for instance, to ZOOM images and, more generally, to derive missing data values. *ACF*

interrill erosion

Removal of soil in areas between RILLS, i.e. most of the hillslope. It is generally caused by the combined action of *overland flow* and *rainsplash.* *SHD*

[See also SOIL EROSION]

Luk, S.H., Abrahams, A.D. and Parsons, A.J. 1993: Sediment sources and sediment transport by rill flow and interrill flow on a semiarid piedmont slope, southern Arizona. *Catena* 20, 93–111. **Morgan, R.P.C.** 1995: *Soil erosion and conservation,* 2nd edn. Harlow: Longman.

interstadial

A relatively short-lived period of thermal improvement during a GLACIAL EPISODE. Although considered to reflect amelioration to conditions not as warm as at present, some interstadials probably experienced rapid changes to temperatures as high as during the current INTERGLACIAL, albeit for very short periods. Whilst terrestrial records had suggested about eight interstadials within the last glacial phase (oxygen ISOTOPIC STAGES 5d–2), 24 interstadials are now recognised from the Greenland ICE CORE record for the same period. *CJC/DH*

[See also ALLERØD INTERSTADIAL, DANSGAARD–OESCHGER EVENT, LATE GLACIAL ENVIRONMENTAL CHANGE, QUATERNARY TIMESCALE, WINDERMERE INTERSTADIAL]

Fronval, T., Janssen, E., Bloemendal, J. and Johnsen, S. 1995: Oceanic evidence for coherent fluctuations in Fennoscandian and Laurentide ice sheets on millennium timescales. *Nature* 374, 443–446. **Johnsen, S.J., Clausen, H.B., Dansgaard, W.** *et al.* 1992: Irregular glacial interstadials recorded in a new Greenland ice core. *Nature* 359, 311–314.

intertropical convergence zone (ITCZ)

The belt of low pressure found near the Equator and formerly known as the *equatorial low* and, by mariners, as the *doldrums.* The ITCZ is a region of ascending air, visible on cloud images as a band of discontinuous convective cloud between the convergent TRADE WIND systems. Rainfall typically occurs from cloud clusters about 100 km across. Over the Western Hemisphere the ITCZ moves only a few degrees of latitude during the year, but in the Eastern Hemisphere variable seasonal shifts up to some 40° of latitude contribute to the problem of DROUGHT, especially in the African SAHEL and associated with the Asian MONSOON.

Moisture and ATMOSPHERE–OCEAN INTERACTIONS are important in the formation of the ITCZ: hence there is no exact correspondence between its location and the warmest land and SEA-SURFACE TEMPERATURES of the tropics. In the eastern Pacific, UPWELLING causes the formation of a northern and southern branch of the ITCZ and there are complex relationships during EL NIÑO–SOUTHERN OSCILLATION events. *Easterly waves* (also known as *tropical waves* or *African waves*) propagate along the ITCZ, travel towards the west and occasionally intensify into TROPICAL CYCLONES (hurricanes or typhoons). Longer-term variations in the average position of the ITCZ are very important in understanding ENVIRONMENTAL CHANGE in the tropics. *JAM/AHP*

[See also HADLEY CELL]

Hamilton, R.A. and Archbold, J.W. 1945: Meteorology over Nigeria and adjacent territory. *Quarterly Journal of the Royal Meteorological Society* 71, 231–262. **Nicholson, S.E. and Flohn, H.** 1980: African environmental and climatic changes and the general atmospheric circulation during the Late Pleistocene and Holocene. *Climatic Change* 2, 313–348. **Riehl, H.** 1979: *Climate and weather in the tropics.* New York: Academic Press. **Street-Perrott, F.A. and Perrott, R.A.** 1993: Holocene vegetation, lake levels, and climate of Africa. In Wright Jr, H.E., Kutzbach, J.E., Webb III, T., Ruddiman, W.F., Street-Perrott, F.A. and Bartlein, P.J. (eds), *Global climates since the Last Glacial Maximum.* Minneapolis, MN: University of Minnesota Press.

intrazonal soil

Soils with more-or-less well developed characteristics that reflect the dominating influence of a local factor of relief, PARENT MATERIAL or age over the 'normal' zonal effect of climate and vegetation. It was also a *soil order* in the system of classification used in the USA before 1965, but is not used in SOIL TAXONOMY or the WORLD REFERENCE BASE FOR SOIL RESOURCES. *EMB*

[See also ZONAL SOIL]

introduction The human-assisted movement of individual species into a site that lies outside their natural, historical range. It may be purposeful or accidental. Examples include the introduction of rabbits and *Opuntia* into Australia. *Reintroduction* involves the similar movement of species into sites inside their natural range, from which, however, they have been previously excluded by direct or indirect human impacts. *MVL*

[See also BIOLOGICAL CONTROL, BIOLOGICAL INVASIONS, SPECIES TRANSLOCATION]

intrusion A structural feature produced by the movement of MAGMA, ice or SEDIMENT beneath the Earth's surface, driven by contrasts in density or temperature between the intruding mass and its host (often termed 'country rock'). *Igneous intrusions* are classified according to their geometry. Near-surface, small intrusions (*hypabyssal*) solidify as medium-grained IGNEOUS ROCKS. *Dykes* cut across BEDDING and are commonly vertical or nearly so. *Sills* are intruded parallel to bedding. *Plugs* are the central remnants of eroded VOLCANOES and *ring dykes* are concentric intrusions around volcanic centres, which may lead to faulting and CALDERA formation. Deep-seated, large intrusions (*plutonic, plutons*) are preserved as coarse-grained igneous rocks. They may take thousands of years to cool completely. *Magma chambers* form reservoirs for VOLCANIC ERUPTIONS. *Batholiths* are very large granitic bodies, often several hundred kilometres long, intruded along destructive PLATE MARGINS. *Non-igneous intrusions* include salt or mud DIAPIRS or domes and the injection of ice to form PERMAFROST mounds. *JBH*

Johnson, A.M. 1970: *Physical processes in geology*. San Francisco, CA: Freeman and Cooper. **Ramberg, H.** 1981: *Gravity, deformation and the Earth's crust*, 2nd edn. London: Academic Press.

inversion A layer of air in which temperature rises with height; an inversion of the normal positive atmospheric LAPSE RATE. Inversions may form at the surface by nocturnal radiative cooling, above the surface as a result of air subsidence and adiabatic heating in ANTICYCLONES, or associated with the passage of a FRONT. Persistent inversions are usually associated with stable anticyclonic conditions, restriction of atmospheric dispersion mechanisms and thus AIR POLLUTION episodes. *JCS*

Voloshin, V. 1973: Duration of inversions and light wind periods in the boundary layer of the atmosphere. In *Atmosfernaya diffuziyai zagrryaznenie vozdukha, vypusk 293*. St Petersburg: Gidrometeoizdat [Translated in Berlyand, M. (ed.) 1974: *Air pollution and atmospheric diffusion*. New York: Halsted Press].

invisible drought A type of DROUGHT characterised by suboptimal crop yields in areas where moisture is normally sufficient for crop growth; it can be eliminated easily by IRRIGATION. *JAM*

[See also CONTINGENT DROUGHT, PERMANENT DROUGHT, SEASONAL DROUGHT]

involutions Disturbed, distorted and deformed structures occurring in unconsolidated, usually PLEISTOCENE-age, sediments. There are two types. One type is associated with PERIGLACIAL climatic environments, where involutions are thought to form through frost action within a seasonally frozen layer (or the ACTIVE LAYER if

PERMAFROST is present). The other type is of THERMOKARST origin, where the involution is produced by loading and density differences that develop during the thaw of ice-rich permafrost. *HMF*

[See also CRYOTURBATION]

French, H.M. 1996: *The periglacial environment*, 2nd edn. London: Addison-Wesley Longman, 240–243. **Murton, J.B. and French, H.M.** 1993: Thermokarst involutions, Summer Island, Pleistocene Mackenzie Delta, western Canadian Arctic. *Permafrost and Periglacial Processes* 4, 217–229.

ion An electrically charged particle, which may be atomic or polyatomic (formed from an atom or a group of atoms). Ions are formed by *ionisation* processes whereby electrons are added or removed from particles. In solution, many compounds become dissociated into their component ions, which may be positively (*anions*) or negatively (*cations*) charged, depending on whether an electron is removed or added, respectively. The process of *ion exchange*, used for example in water purification, water softeners, some DESALINATION plant and sewage-treatment works, involves the removal or replacement of anions or cations as a solution passes through a medium or filter. Gases may become ionised when an electrical charge passes through them or by *ionising radiation*, such as ultraviolet RADIATION and X-rays from the Sun; this commonly occurs in the *ionosphere* (above about 80 km in the upper ATMOSPHERE). Ions formed from metals are generally cations. *JAM/AHP*

Boekker, E. and Van Grondelle, R. 1995: *Environmental physics*. Chichester: Wiley. **Ratcliffe, J.A.** 1972: *An introduction to the ionosphere and magnetosphere*. Cambridge: Cambridge University Press. **Slater, M.J.** 1991: *Principles of ion exchange technology*. Oxford: Butterworth.

Iron Age: landscape impacts The Iron Age was the period following the BRONZE AGE, in which tools and weapons were primarily produced from iron. In Eurasia the Iron Age began c. 3000 calendar years BP and persisted until the ROMAN PERIOD, although iron continued to be the primary tool-making material up to recent times. Iron appeared in the Americas only after European settlement.

FOREST CLEARANCE was widespread in the Iron Age because of the requirement of wood for fuel and increasing agricultural activity in the larger, more permanent settlements. In some areas of central Europe, these activities favoured faster-growing forest taxa such as *Quercus* (oak) and *Carpinus betulus* (hornbeam). In addition, the construction of *hillforts* and *earthworks* during the Iron Age required large quantities of timber and left scars on the landscape that remain visible in the modern age. *ARG*

[See also BRONZE AGE: LANDSCAPE IMPACTS, COPPER AGE: LANDSCAPE IMPACTS, NEOLITHIC: LANDSCAPE IMPACTS]

Armit, I. and Ralston, I.B.M. 1997: The Iron Age. In Edwards, K.J. and Ralston, I.B.M. (eds), *Scotland: environment and archaeology 8000 BC–AD 1000*. Chichester: Wiley, 169–193. **Champion, T., Gamble, C., Shennan, S. and Whittle, A.** 1984: *Prehistoric Europe*. London: Academic Press. **Cunliffe, B.W.** 1991: *Iron Age communities in Britain*, 3rd edn. London: Routledge. **Küster, H.** 1997: The role of farming in the postglacial expansion of beech and hornbeam in the oak woodlands of central Europe. *The Holocene* 7, 239–242. **Ralston, I.B.M.** 1999: The Iron Age: aspects of human communities and their environments. *Quaternary Proceedings* 7, 501–512. **Turner, J.**

1981. The Iron Age. In Simmons, I. and Tooley, M. (eds), *The environment in British prehistory*. London: Duckworth, 250–281. **Willis, K.J., Sümegi, P., Braun, M.** *et al.* 1998: Prehistoric land degradation in Hungary: who, how and why? *Antiquity* 72, 101–113.

irradiance The amount of radiant flux per unit of surface that flows across or into a surface. The usual unit is watt per square metre (W m⁻2). *TS*

[See also RADIANCE]

irragric horizon An ANTHROPEDOGENIC soil horizon, raising the surface of the soil associated with IRRIGATION sediment, frequently containing fragments of brick and tile and showing evidence of considerable biological activity. *EMB*

irrigation The addition of water to enhance crop production where water shortage is a LIMITING FACTOR. Thus irrigation is a most common cultivation technique in DRY-LANDS. By AD 1989, there were c. 233×10^6 ha of irrigated land world-wide, of which 73% was located in the developing world. Global expansion of irrigation in the period 1950–1980 was implemented mainly by the construction of large dams. The environmental costs of irrigation may be high and depend on the efficiency and management of the irrigation system. Loss of nutrients, SALINISATION, ALKALISATION, waterlogging and GROUNDWATER POLLUTION are some of the main detrimental environmental impacts. In the worst cases, LAND DEGRADATION may be so far developed that land is abandoned. Dam construction often involves the displacement of people. For example, the impoundment of Lake Nasser behind the Aswan Dam, completed in 1964, required the displacement of about 100 000 people. In addition, if control measures are not adhered to in tropical countries, canals and ditches may help to spread various DISEASES (e.g. MALARIA, filariasis, yellow fever, dengue, river blindness, sleeping sickness, schistosomiasis or bilharzia, liver fluke and guinea worm). *RAS*

[See also AQUIFER, DESERTIFICATION, DRY FARMING]

Agnew, C. and Anderson, E. 1992: *Water resources in the arid realm*. London: Routledge. **Barrow, C.J.** 1987: *Water resources and agricultural development in the tropics*. London: Longman. **Hansen, V.E., Israelsen, O.W. and Stringham, G.E.** 1980: *Irrigation principles and practices*, 4th edn. New York: Wiley. **Hillel, D.** 1987: *The efficient use of water in irrigation: principles and practice for improving irrigation in arid and semiarid regions*. Washington, DC: World Bank. **Postel, S.** 1993: Water and agriculture. In Gleick, P.H. (ed.), *Water in crisis: a guide to the world's fresh water resources*. Oxford: Oxford University Press, 56–59. **Samad, M., Merrey, D., Vermillion, D.** *et al.* 1992: Irrigation management strategies for improving the performance of irrigated agriculture. *Outlook on Agriculture* 21, 279–286. **Zimmerman, J.D.** 1966: *Irrigation*. New York: Wiley.

island arc An arcuate chain of islands in the OCEAN, usually of volcanic origin and associated with an OCEANIC TRENCH, formed above a SUBDUCTION zone at a DESTRUCTIVE PLATE MARGIN (e.g. Aleutian Islands). *CDW*

Hamilton, W.B. 1988: Plate tectonics and island arcs. *Geological Society of America Bulletin* 100, 1503–1527. **Sigimura, A. and Uyeda, S.** 1973: *Island arcs: Japan and its environs*. Amsterdam: Elsevier.

island biogeography The study of the ecological and evolutionary characteristics of isolated biotas. Most science historians agree that island studies played a central role in the development of Darwin and Wallace's theory of NATURAL SELECTION. Their observations, along with those of others studying insular ecosystems, also became the foundations of the fields of BIOGEOGRAPHY and ECOLOGY. The *equilibrium theory of island biogeography* is an important theory put forward in 1963 by R.H. MacArthur and E.O. Wilson, which posits that island species number represents a dynamic equilibrium between immigration to an island and extinction from it. Island biogeographers study an impressive diversity of patterns, but most of them deal with the relationships between physical characteristics of islands (e.g. size and degree of isolation) and the characteristics of biotic communities (e.g. number and types of species, their distributions among islands and variation in their morphological, behavioural or genetic characteristics). Island biogeography continues to play an important role in providing insights for evolutionary biologists, ecologists and biogeographers in general. Perhaps just as important, because NATURE RESERVES and fragments of native ecosystems share many island-like characteristics, island biogeography theory has become an especially important tool for conserving BIODIVERSITY. *MVL*

[See also CONTINENTAL ISLAND, EQUILIBRIUM CONCEPTS, METAPOPULATION MODEL, RELAXATION, SINGLE LARGE OR SEVERAL SMALL RESERVES (SLOSS DEBATE), TAXON CYCLE]

Brown, J.H. and Lomolino, M.V. 1998: *Biogeography*, 2nd edn. Sunderland, MA: Sinauer. **Gilbert, F.S.** 1980: The equilibrium theory of island biogeography: fact or fiction? *Journal of Biogeography* 7, 209–235. **MacArthur, R.H. and Wilson E.O.** 1967: *The theory of island biogeography* [*Monographs in Population Biology* 1]. Princeton, NJ: Princeton University Press. **Whittaker, R.J.** 1998: *Island biogeography: ecology, evolution and conservation*. Oxford: Oxford University Press.

isochron technique A method used in RADIOMETRIC DATING to correct for the effect of initial DAUGHTER ISOTOPES (e.g. inherited argon in POTASSIUM–ARGON and ARGON–ARGON DATING or detrital thorium in URANIUM-SERIES DATING). Coeval samples from a deposit are analysed to produce a straight-line relationship between ISOTOPE RATIOS such that the gradient can be used to calculate the age. *DAR*

Faure, G. 1986: *Principles of isotope geology*. New York: Wiley.

isochrone A line joining points of equal age on a map. It is a particular kind of ISOPLETH. Isochrones provide a means of representing an *areal chronology* and are used, for example, to depict areal patterns of DEGLACIATION, coastal emergence and TREE MIGRATION. *JAM*

isolated system A SYSTEM in which the boundaries are closed to the import and export of both mass and energy. It is a concept most applicable to LABORATORY SCIENCE. *JAM*

[See also CLOSED SYSTEM, OPEN SYSTEM]

Chorley, R.J. and Kennedy, B.A. 1971: *Physical geography: a systems approach*. London: Prentice-Hall.

isolation The process of separation of populations of species to the extent that interbreeding is no longer possible. Isolation may arise in a number of ways and may have a geographic or ecological basis. It is probably the process that leads to most SPECIATION events. 'Rapidly evolving peripherally isolated populations may be the place of origin of many evolutionary novelties. Their isolation and comparatively small size may explain phenomena of rapid evolution and lack of documentation in the FOSSIL record, hitherto puzzling to the paleontologist' (Mayr, 1954: 179). *KDB*

[See also FOUNDER EFFECT, ISLAND BIOGEOGRAPHY]

Mayr, E. 1954: Change of genetic environment and evolution. In Huxley, J. Hardy, A.C. and Ford, E.B. (eds), *Evolution as a process* . London: George Allen and Unwin, 157–180.

isolation basin

A terrestrial basin that has previously been submerged below sea-level. MARINE SEDIMENTS accumulate in isolation basins while they are submerged below sea level. Subsequently, a fall in RELATIVE SEA LEVEL results in the isolation of the basin from marine influences and a shift to accumulation of freshwater LACUSTRINE SEDIMENTS. Isolation basins may later be resubmerged and re-emerge. Dating the changes from freshwater to marine or BRACKISH sedimentation in isolation basins may provide SEA-LEVEL *index points* for reconstructing regional patterns of SEA-LEVEL CHANGE. *MHD*

Long, A.J., Roberts, D.H. and Wright, M.R. 1999: Isolation basin stratigraphy and Holocene relative sea-level change on Arveprinsen Ejland, Disko Bugt, West Greenland. *Journal of Quaternary Science* 14, 323–345. **Shennan, I., Green, F., Innes, J.** *et al.* 1996: Evaluation of rapid relative sea-level changes in north-west Scotland during the last glacial–interglacial transition: evidence from Ardtoe and other isolation basins. *Journal of Coastal Research* 12, 862–874. **Svendsen, J.E. and Mangerud, J.** 1987: Late Weischselian and Holocene sea-level history for a cross section of western Norway. *Journal of Quaternary Science* 2, 113–132.

isomer

Two or more identical molecular counterparts of a COMPOUND, each with different structural or constitutional arrangements of the atoms. Different isomers of the same compound are represented by the same chemical formula but have different structural formulae. AMINO ACID DATING is based on the time-dependent transformation of amino acid molecules from one isomeric form to another. *MHD*

isomorphic substitution

The substitution of one atom for another of similar size but lower valence in a crystal without disrupting or changing the structure of the mineral. In soils, Al^{3+} is frequently substituted for Si^{4+} and Mg^{2+} for Al^{3+}, which leaves a deficit of positive charge on the crystal and allows the development of the CATION EXCHANGE CAPACITY. *EMB*

isopach

A line on a map joining points of equal thickness, usually of a distinctive BED or other stratigraphical unit (see LITHOSTRATIGRAPHY). The analysis of isopachs, which are also known as *isopachytes*, is important in the reconstruction of PALAEOGEOGRAPHY. *GO*

isopleth

(1) A line on a map connecting points of equal value in relation to a particular VARIABLE; see Table for examples. (2) A graph illustrating the frequency or intensity of a particular phenomenon as a function of two variables. *JCM/JAM*

Stamp, L.D. (ed.) 1961: *A glossary of geographical terms.* London: Longman.

isopleth *Examples of isopleths*

isobar	line of equal pressure
isobase	line of equal land uplift
isobath	line of equal water depth
isochrone	line of equal time lapse
isohaline	line of equal salinity
isohyet	line of equal precipitation
isohypse	line of equal altitude (a contour line)
isoneph	line of equal cloudiness
isonif	line of equal snowfall
isonomaly	line of equal anomalies
isopach	line of equal (geological) bed thickness
isophene	line of equal seasonal (botanical) phenomena
isophyte	line of equal vegetation height
isopoll	line of equal percentage of a pollen taxon
isoterp	line of equal (human) comfort
isotherm	line of equal temperature

isostasy

The condition of equilibrium caused by the Earth's LITHOSPHERE, which essentially 'floats' on the less dense underlying ASTHENOSPHERE. Deformation of the lithosphere due to the loading and unloading of ice sheets during the QUATERNARY is termed GLACIO-ISOSTASY. Owing to spatial variations in former ICE SHEET thickness, patterns of crustal rebound rate varied for different areas, the greatest rates occurring in areas where the ice had been thickest. During ice-sheet thinning and melting, however, most *isostatic rebound* takes place while rapidly thinning ice still covers the landscape.

Isostatic loading and *unloading* by water on ocean floors and continental shelves due to changes in RELATIVE SEA LEVEL is termed HYDROISOSTASY. For example, a long-term rise in sea level during DEGLACIATION of the last ICE SHEET was associated with a compensatory depression of the crust beneath major water bodies. This process is also greatly affected by the tectonic setting of individual regions and by the flexural rigidity of the lithosphere. That the magnitude of changes can be considerable is demonstrated by the fact that during the last 7000 years there has been an overall depression of 8 m in the ocean floors and an average uplift of adjacent continents by 16 m. It has also been argued, however, that, whereas hydro-isostatic loading during eustatic SEA-LEVEL RISE may induce ocean floor SUBSIDENCE, it does not follow that hydro-isostatic unloading due to regional sea-level fall would cause uplift of the ocean floor. *AGD*

[See also EUSTASY]

Chappell, J. 1974: Late Quaternary glacio- and hydro-isostasy on a layered Earth. *Quaternary Research* 4, 405–428. **Dawson, A.G.** 1992: *Ice Age Earth: Late Quaternary geology and climate.* London: Routledge. **Lyustikh, E.N.** 1960: *Isostasy and isostatic hypotheses.* New York: American Geophysical Union. **Pirazzoli, P.A.** 1996: *Sea level changes: the last 20,000 years.* Chichester: Wiley. **Smith, D.E. and Dawson, A.G. (eds)** 1983: *Shorelines and isostasy.* London: Academic Press. **Walcott, R.J.** 1972: Past sea levels, eustasy, and deformation of the Earth. *Quaternary Research* 2, 1–14.

isostatic decantation

Displacement (or decanting) of sea water into the oceans from shallow seas or continental shelves as a result of the unloading (isostatic) effect that follows melting of an ice mass. *MJH*

[See also GLACIO-EUSTASY, HYDROISOSTASY, ISOSTASY, SEA-LEVEL CHANGE]

Goudie, A.S. 1992: *Environmental change*, 3rd edn. Oxford: Blackwell, 218.

isothermal remanent magnetism (IRM)

One of the magnetic components comprising the NATURAL REMANENT MAGNETISM (NRM) of ROCKS and SEDIMENTS, it is magnetisation that remains in the absence of an applied magnetic field, i.e. remanent magnetisation after the application and subsequent removal of a magnetic field. IRM varies according to the strength of the applied field but the maximum remanence that can be produced is called the SATURATION ISOTHERMAL REMANENT MAGNETISATION (SIRM), which is normally considered to be that remaining after exposure to a magnetic field of 1 T (tesla). *AJP/CJC*

[See also MINERAL MAGNETISM]

Thompson, R. 1991: Palaeomagnetic dating. In: Smart, P.L. and Frances, P.D. (eds), *Quaternary dating methods – a users guide* [*QRA Technical Guide* 4]. Cambridge: Quaternary Research Association, 177–198.

isotope

The isotopes of an element are ATOMS having the same number of *protons* in the nucleus, but a different number of *neutrons* and hence a different atomic mass. Different isotopes of an element have the same chemical properties because they contain the same number of *electrons*. The variation in atomic mass causes isotopes to have slightly different physical properties. Isotopes can be divided into two forms, STABLE ISOTOPES and unstable *radioisotopes* (see RADIONUCLIDE). *IR*

[See also COSMIC RADIATION, ISOTOPES AS INDICATORS OF ENVIRONMENTAL CHANGE]

Choppin, G.R. and Rydberg, J. 1980: *Nuclear chemistry: theory and application*. New York: Pergamon.

isotope dendrochronology

A branch of TREE-RING studies concerned with the variations either of STABLE or radioactive ISOTOPES contained in the rings of trees. The absolute time scale that underpins dendrochronological studies means that measurements of the isotopic composition of wood may be assigned accurate dates. Perhaps the best known example of the value of such studies is the *radiocarbon anomaly curve* – a record of short (decadal) and longer (millennial) variations in the ^{14}C content of the atmosphere, assembled from measurements of the ^{14}C:^{12}C ratios in thousands of wood samples spanning the last 10 000 years. The variations must be accounted for when interpreting ^{14}C dates on organic matter. STABLE ISOTOPE ANALYSIS (e.g. studies of the ISOTOPE RATIOS ^{18}O:^{16}O, ^{1}H:^{2}H and ^{13}C:^{12}C) also provide evidence of changing physiological activity in trees through time and are often interpreted as evidence of CLIMATIC CHANGE. *KRB*

[See also DENDROCHRONOLOGY, MICRODENDROCLIMATOLOGY, RADIOCARBON DATING]

Bradley, R.S. 1999: *Paleoclimatology: reconstructing climates of the Quaternary*, 2nd edn. London: Academic Press. **Leavitt, S.W.** 1993: Environmental information from ^{13}C/^{12}C ratios of wood. *Geophysical Monographs* 78, 325–331. **Stuiver, M. and Braziunas, T.F.** 1987: Tree cellulose ^{13}C/^{12}C isotope ratios and

climatic change. *Nature* 328, 58–60. **Wilson, A.T. and Grinsted, M.J.** 1977: ^{13}C/^{12}C in cellulose in lignin as palaeothermometers. *Nature* 265, 133–135.

isotope ratio

Usually the relationship between the two main STABLE ISOTOPES of an element expressed as the proportion of the 'heavier' to the 'lighter' isotope relative to a REFERENCE STANDARD. The differences between samples and reference standards are usually small and expressed as parts per thousand or *per mille* (‰). *IR*

[See also ISOTOPES, ISOTOPIC FRACTIONATION, STABLE ISOTOPE ANALYSIS]

isotopes as indicators of environmental change

Changes in the natural abundance of CARBON ISOTOPES, HYDROGEN ISOTOPES, NITROGEN ISOTOPES, OXYGEN ISOTOPES and SULPHUR ISOTOPES are those most frequently used in studies of environmental change. ISOTOPIC FRACTIONATION is caused by different kinetic rate constants during physical processes and chemical reactions. Rather than measure absolute values, the difference between a sample and a REFERENCE STANDARD enables the determination of ISOTOPE RATIOS to a high degree of PRECISION.

The theoretical basis for the use of STABLE ISOTOPES as indicators of environmental change was established in the late AD 1940s. Carbon isotope values can be used to elucidate plant photosynthetic pathways as the δ^{13}C values of C-3 PLANTS exhibit greater ISOTOPIC DEPLETION than those of C-4 PLANTS. Reconstructions of PALAEOECOLOGY may be inferred from δ^{13}C values from LACUSTRINE SEDIMENTS, PALAEOSOLS, RODENT MIDDENS and MIRES. Concerns over anthropogenic CARBON DIOXIDE emissions have led to the establishment of a world-wide network to monitor the concentration and δ^{13}C values of atmospheric CO_2. Weekly air samples are analysed from remote sites to represent the large, well mixed air masses of the TROPOSPHERE. These values have revealed that there is a 'missing' CARBON SINK, which has been allocated to increased CARBON SEQUESTRATION in the terrestrial BIOSPHERE. The response of plants to increasing atmospheric carbon dioxide concentration has also been estimated using growth chamber experiments and theoretical models. The problems of upscaling growth-chamber results to represent an ECOSYSTEM may be overcome using intrinsic water-use efficiency values derived from δ^{13}C values in TREE RINGS to provide a time-integrated measure. These results confirm that the water-use efficiency of C-3 plants has increased together with the atmospheric carbon dioxide concentration.

The oxygen isotope value of FORAMINIFERA in MARINE SEDIMENT CORES represents one of the most frequently used indirect measures of long-term climatic change. PERIODICITIES in δ^{18}O values can be related to the changes in ORBITAL FORCING with frequencies at about 23 ka, 41 ka and 100 ka. ORBITAL TUNING may be used to date cores indirectly by matching properties with the predictable changes in the Earth's orbit and axis. Global changes in δ^{18}O values are well replicated and a composite 780 ka record forms the basis of the widely adopted SPECTRAL MAPPING PROJECT (SPECMAP) timescale. Terrestrial records of CLIMATIC CHANGE derived from POLLEN records, LOESS STRATIGRAPHY, LAKE-LEVEL VARIATIONS and ICE CORES are compared to this standard.

Stable isotope records have been obtained from ice

cores taken from Antarctica, the Greenland ICE SHEET and high-altitude ice caps. The first continuous ice-core record spanning a full GLACIAL–INTERGLACIAL CYCLE was obtained from Vostok in Antarctica. Global temperature variations have been inferred from the hydrogen isotope record of ice from the Vostok core, the oxygen isotope values of air occluded in this core and the oxygen isotope record from the Greenland ice cores (GRIP and GISP2). These cores retain a signal of climatic change covering the last 420 ka. The presence of high-frequency CHRONOSTRATIGRAPHIC markers in the ice-core oxygen isotope records and HEINRICH EVENTS in the marine sediment cores indicates that there are also climatic changes operating at a SUB-MILANKOVITCH scale, which are probably influenced by SOLAR FORCING and changes in the THERMOHALINE CIRCULATION.

The hydrogen isotope and oxygen isotope values of METEORIC WATER are strongly associated and represented by the *meteoric water line*. EVAPORATION of source water causes an ISOTOPIC DEPLETION in the 'heavier' isotopes of cloud water (^2H and ^{18}O) with CONDENSATION causing an ISOTOPIC ENRICHMENT of the 'heavier' isotopes of the resulting precipitation. Initial PRECIPITATION will have an isotopic composition similar to the source, but this will become progressively depleted in the 'heavier' isotopes and therefore precipitation is usually isotopically depleted compared to ocean water. As there is no isotopic fractionation during water uptake through the roots of terrestrial plants, the interpretation of non-exchangeable δD and $\delta^{18}O$ values from organic matter requires an understanding of isotopic fractionation during PHOTOSYNTHESIS and post-photosynthetic exchanges.

Unlike sulphur isotopes, nitrogen isotope values are isotopically enriched at each TROPHIC LEVEL in the food chain. In non-arid environments, dietary inputs from C-3 or C-4 plants can be established from $\delta^{13}C$ values. Marine inputs to the diet can be estimated from a combination of nitrogen, sulphur and STRONTIUM ISOTOPE values. The isotope ratio of atmospheric SULPHUR DIOXIDE (SO_2) originating from ANTHROPOGENIC activities is related to the source from which it was originally derived and can be used to monitor AIR POLLUTION. Similarly, LEAD ISOTOPE values have been used to evaluate changing sources of human exposure. Although the $\delta^{15}N$ values of nitrate have been used to demonstrate pollution originating from the application of nitrogen-based fertilisers, the use of $\delta^{15}N$ of nitrate in pollution studies is limited as the largest isotopic fractionation of nitrogen isotopes is from metabolic reactions. *IR*

[See also COMPOUND-SPECIFIC CARBON ISOTOPE ANALYSIS, ISOTOPE, RADIONUCLIDE]

Anderson, W.T., Bernasconi, S.M., McKenzie, J.A. and Saurer, M. 1998: Oxygen and carbon isotopic record of climatic variability in tree ring cellulose (*Picea abies*): an example from central Switzerland (1913–1995). *Journal of Geophysical Research* 103(D24), 31625–31636. **Barrie, A., Brookes, S.T., Prosser, S.J. and Debney, S.** 1995: High productivity analysis of ^{15}N and ^{13}C in soil/plant research. *Fertilizer Research* 42, 43–59. **Dawson, T.E. and Ehleringer, J.R.** 1991: Streamside trees that do not use stream water. *Nature* 350, 335–337. **Ehleringer, J.R. and Rundel, P.W.** 1989: Stable isotopes: history, units, and instrumentation. In Rundel, P.W., Ehleringer, J.R. and Nagy, K.A. (eds), *Stable isotopes in ecological research* [*Ecological Studies*. Vol. 68]. Berlin: Springer, 1–15. **Heaton, T.H.E.** 1986: Isotopic studies of nitrogen pollution in the hydrosphere and atmosphere:

a review. *Chemical Geology* 59, 87–102. **Imbrie, J., Hays, J.D., Martinson, D.G.** *et al.* 1984: The orbital theory of Pleistocene climate: support from a revised chronology of the marine $\delta^{18}O$ record. In Berger, A.., Imbrie, J., Hays, J. *et al.* (eds), *Milankovitch and climate: understanding the response to astronomical forcing*. Part I. Dordrecht: Reidel, 269–305. **Jouzel, J., Waelbroeck, C., Malaize, B.** *et al.* 1996: Climatic interpretation of the recently extended Vostok ice records. *Climate Dynamics* 12, 513–521. **Petit, J.R., Jouzel, J., Raynaud, D.** *et al.* 1999: Climate and atmospheric history of the past 420,000 years from the Vostok ice core, Antarctica. *Nature* 399, 429–436. **Shackleton, N.J. and Opdyke, N.D.** 1973: Oxygen isotope and palaeomagnetic stratigraphy of equatorial Pacific core V28-238: oxygen isotope temperatures and ice volumes on a 10^5 year and 10^6 year scale. *Quaternary Research* 3, 39–55. **Trolier, M., White, J.W.C., Tans, P.P.** *et al.* 1996: Monitoring the isotopic composition of atmospheric CO_2: measurements from the NOAA Global Air Sampling Network. *Journal of Geophysical Research* 101(D20), 25897–25916. **Tzedakis, P.C., Andrieu, V., de Beaulieu, J.-L.** *et al.* 1997: Comparison of terrestrial and marine records of changing climate of the last 500,000 years. *Earth and Planetary Science Letters* 150, 171–176. **Urey, H.C.** 1947: The thermodynamic properties of isotopic substances. *Journal of the Chemical Society of London* 85, 562–581.

isotopic depletion The reduction in the abundance of the 'heavier' ISOTOPE during both thermodynamic and kinetic reactions as a result of ISOTOPIC FRACTIONATION. *IR*

[See also ISOTOPIC ENRICHMENT]

isotopic enrichment The enhancement in the abundance of the 'heavier' ISOTOPE during both thermodynamic and kinetic reactions as a result of ISOTOPIC FRACTIONATION. *IR*

[See also ISOTOPIC DEPLETION]

isotopic fractionation The separation of ISOTOPES of an ELEMENT during physico-chemical processes due primarily to differences in their relative masses. In chemical systems, isotopic fractionation may be due to isotopic exchange reactions operating under equilibrium conditions or to unidirectional processes. In isotopic exchange reactions there is no net reaction, but a redistribution of isotopes amongst different chemicals, between different phases or between individual MOLECULES. Unidirectional processes are caused by unequal kinetic rate constants for the different isotopes of the reactants and usually result in ISOTOPIC DEPLETION of the 'heavier' isotope in the product. In a CLOSED SYSTEM, the reactant concentration falls progressively, causing the isotopic concentration of reactant and product to change with time. If the kinetic rate constant remains constant and the product is removed immediately under equilibrium conditions, the isotopic separation process may be modelled by the Rayleigh equation. The main physical processes that influence isotopic fractionation are DIFFUSION, EVAPORATION and CONDENSATION. *IR*

[See also ISOTOPES AS INDICATORS OF ENVIRONMENTAL CHANGE]

Chambers, L. A. and Trudinger, P. A. 1979: Microbiological fractionation of stable sulfur isotopes: a review and critique. *Geomicrobiology Journal* 1, 249–293. **Gat, J.R.** 1981: Isotopic fractionation. In Gat, J.R. and Gonfiantini, R. (eds), *Stable isotope hydrology: deuterium and oxygen-18 in the water cycle*. Vienna: IAEA, 21–33. **Urey, H.C.** 1947: The thermodynamic properties of isotopic substances. *Journal of the Chemical Society of London* 85, 562–581.

isotopic stage An interval defined by inflections in the global OXYGEN ISOTOPE stratigraphic record derived from MARINE SEDIMENTS. The oxygen ISOTOPE RATIO of

FORAMINIFERA in deep-ocean sediments is influenced primarily by global continental ice volume. Isotopic stages are assigned to GLACIAL EPISODES (even numbers) and INTERGLACIAL (odd numbers) periods (see Figure). These stages may be further subdivided as higher-resolution records become available, thus the last Interglacial, stage

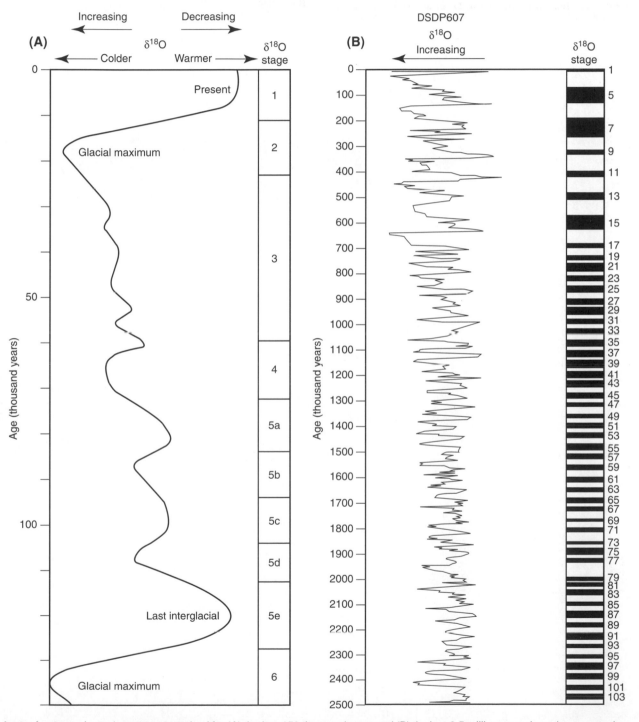

isotopic stage *Isotopic stages recognised for (A) the last 150 thousand years and (B) the last 2.5 million years based on oxygen isotope variations in marine sediment cores. Odd numbers indicate warm stages (interglacials shown as black bars in B); even numbers are cold stages (after Wilson et al., 2000)*

5, is divided into 5.1–5.5 or 5a-e, 'odd' letters (a, c, e etc.) relating to 'warm' *substages* and 'even' letters (b, d etc.) to cold substages. During the last 2.73 million years, 116 stages have been recognised.   *IR/CJC*

[See also ISOTOPES AS INDICATORS OF ENVIRONMENTAL CHANGE, MARINE SEDIMENT CORES, QUATERNARY TIME-SCALE]

Petit, J.R., Jouzel, J., Raynaud, D. *et al.* 1999: Climate and atmospheric history of the past 420,000 years from the Vostok ice core, Antarctica. *Nature* 399, 429–436. **Shackleton, N.J.** 1987: Oxygen isotopes, ice volume and sea level. *Quaternary Science Reviews* 6, 183–190. **Shackleton, N.J. and Opdyke, N.D.** 1973: Oxygen isotope and palaeomagnetic stratigraphy of equatorial Pacific core V28-238: oxygen isotope temperatures and ice volumes on a 10^5 year and 10^6 year scale. *Quaternary Research* 3, 39–55. **Wilson, R.C.L., Drury, S.A. and Chapman, J.L.** 2000: *The great ice age: climate change and life*. London: Routledge.

J

Jaramillo polarity event A NORMAL POLARITY event during the Matuyama Chron of the palaeomagnetic timescale, dated by POTASSIUM–ARGON DATING to between 0.90 and 0.97 Ma BP, and orbitally tuned (see ORBITAL TUNING) to between 0.99 and 1.07 Ma BP. *MHD*

[See also GEOMAGNETIC REVERSAL, MARKER HORIZON]

jet stream A strong, narrow air current in the TROPOSPHERE near the *tropopause*. It extends longitudinally for thousands of kilometres (length), latitudinally for hundreds of kilometres (width) and vertically for several kilometres (depth). Vertical wind shear ranges from 5–10 m s^{-1} km^{-1}, horizontal wind shear is 5 m s^{-1} (100 km)$^{-1}$. Arbitrary minimum speeds of 30 m s^{-1} are used to define the three-dimensional extent of the jet stream, whilst core speeds ('jet maxima') are often in excess of 70 m s^{-1}. Jet streams are analogous to a series of concentric hosepipes that follow LONG WAVE patterns. They are areas of clear air turbulence and usually avoided by aircraft. The *polar front jet* flows from west to east in the mid-latitudes and its position varies widely from day to day. It is found near the 300 hPa level (c. 10 km). The *subtropical jet* flows from west to east at about 30° latitude and its position is more constant in a given season. It is found near 200 hPa (c. 12 km). There are also jet streams near the stratopause and mesopause and sometimes low-level jets occur at inversions above the planetary boundary layer (see ATMOSPHERE). *BDG*

Giles, B.D. 1972: A three-dimensional model of a front. *Weather* 27, 352–363. **Reiter, E.R.** 1963: *Jet-stream meteorology*. Chicago, IL: University of Chicago Press.

jetty An artificial structure built out from the shore adjacent to the mouth of a river or tidal inlet to prevent shoaling by LONGSHORE DRIFT and to enhance hydraulic flushing. Jetties are mainly constructed at navigational entrances. *HJW*

[See also BREAKWATER, COASTAL ENGINEERING STRUCTURES, GROYNE, HARD ENGINEERING]

join operation A procedure for connecting two or more relations (or tables) in a DATABASE on the basis of common FIELDS or ATTRIBUTES. *TVM*

[See also RELATIONAL DATABASE]

joint A fracture in rocks with no relative lateral displacement of the two sides. Joints are ubiquitous in rocks exposed at the Earth's surface. They can develop because of stresses during DEFORMATION, contractional stresses associated with the cooling of a LAVA (*columnar jointing*) or the release of stress due to EROSION of overburden or melting of valley-filling GLACIERS (DILATION or pressure-release jointing). Joints influence the WEATHERING of rock masses, and their distribution and orientation can affect the stability of slopes. MINERAL material may grow in the cavity where the walls of a joint have separated, forming a *vein*. *GO*

jökulhlaup A catastrophic flood resulting from the sudden release of stored water associated with a GLACIER or ICE SHEET. Jökulhlaups can be caused by: the sudden drainage of an ICE-DAMMED LAKE below or through the ice dam; water overflow accompanied by rapid fluvial downcutting of ice, bedrock or sediment barriers; the build-up and release of SUBGLACIAL reservoirs. Jökulhlaup discharges can exceed ABLATION-related flows by several orders of magnitude and, in so doing, effect considerable geomorphological change, damage to buildings and other structures, as well as deaths. The retreat of glaciers in the European Alps following their maximum extents during the LITTLE ICE AGE caused a number of them to reach critical positions in tributary valleys such that they could provide effective barriers to streams in the main valleys. The timing of the catastrophic drainage and refilling of the resulting ice-dammed lakes is well documented for a number of glaciers in the Alps because of the concern generated by jökulhlaup destruction. Jökulhlaup size tended to attenuate through time until the retreating glacier could no longer form an effective barrier, reflecting the decreasing effectiveness of the ice dam. The term jökulhlaup is Icelandic and means 'glacier flood'. Alternative terms are *aluvión* (a moraine-dammed lake outburst) in South America and *débâcle* in Europe. *RAS*

[See also CHANNELED SCABLANDS, ICE-DAMMED LAKE]

Benn, D.I. and Evans, D.J.A. 1998: *Glaciers and glaciation*. London: Arnold, 117–123. **Fowler, A.C.** 1999: Breaking the seal at Grimsvötn. *Journal of Glaciology* 45, 506–516. **Grove, J.M.** 1988: *The Little Ice Age*. London: Routledge. **Reynolds, J.M.** 1992: The identification and mitigation of glacier-related hazards: examples from the Cordillera-Blanca, Peru. In McCall, G.J.H., Laming, D.J.C. and Scott, S.C. (eds), *Geohazards: natural and man-made*. London: Chapman and Hall, 143–157. **Shakesby, R.A.** 1985: Geomorphological effects of jökulhlaups and ice-dammed lakes, southern Norway. *Norsk Geografisk Tidsskrift* 39, 1–16. **Thorarinson, S.** 1939: The ice-dammed lakes of Iceland, with particular reference to their values as indicators of glacier oscillations. *Geografiska Annaler* 21, 216–242. **Tufnell, L.** 1984: *Glacier hazards*. London: Longman. **Tweed, F.S. and Russell, A.J.** 1999: Controls on the formation and sudden drainage of glacier-impounded lakes: implications for jökulhlaup characteristics. *Progress in Physical Geography* 23, 79–110.

jump dispersal A form of long distance dispersal in which a species colonises a patch of suitable habitat that is geographically separated from the area from which it dispersed. *JTK*

[See also DISPERSAL, RANGE ADJUSTMENT, VICARIANCE]

Kullman, L. 1998: Palaeoecological, biogeographical and palaeoclimatological implications of early Holocene immigration of *Larix sibirica* Ledeb. into the Scandes Mountains, Sweden. *Global Ecology and Biogeography Letters* 7, 181–188.

Junge layer A layer in the STRATOSPHERE at around 20–25 km above the Earth's surface characterised by a permanent high concentration of sulphates, mainly sulphuric acid droplets. The sulphates are VOLCANIC AEROSOLS, the concentration of which increases markedly after large, explosive, volcanic eruptions influence the temperature of the stratosphere and affect the Earth's RADIATION BALANCE.   *JAM*

Jurassic A SYSTEM of rocks, and a PERIOD of geological time from 205 to 144 million years ago. SEA-FLOOR SPREADING began in the Atlantic Ocean.   *GO*

[See also GEOLOGICAL TIMESCALE]

juvenile water Water originating from the Earth's interior that has not previously participated in the HYDROLOGICAL CYCLE. The term was proposed by O.E. Meinzer (1923), who contrasted juvenile water with both surface-derived METEORIC WATER and *connate water*, which was trapped in the interstices of SEDIMENTARY ROCKS during their formation.   *JAM*

Meinzer, O.E. 1923: Outline of ground-water hydrology. *United States Geological Survey Water-Supply Paper* 494.

K

K-cycles A concept relating to soils and LANDSCAPE, in which there is an alternation of stable and unstable phases of SOIL DEVELOPMENT as observed on slopes in Australia. Soils are formed in stable phases and eroded or covered by sedimentation in unstable phases; each couplet of instability (Ku) followed by stability (Ks) constitutes a single K-cycle and they are numbered consecutively. Thus, soils of different age may co-exist on the same landscape as a result of the redistribution of slope deposits. *EMB*

Butler, B.E. 1959: *Periodic phenomena in landscapes as a basis for soil studies* [*CSIRO Soil Publication* 14]. Melbourne: CSIRO. Vreeken, W.J. 1996: A chronogram for postglacial soil-landscape change from Palliser Triangle, Canada. *The Holocene* 6, 433–438.

K-selection A conceptual, evolutionary 'strategy' that emphasises competitive abilities to maintain populations near CARRYING CAPACITY (*K*) in stable habitats. *LRW*

[See also *R*-SELECTION]

Pianka, E.R. 1970: On r- and K-selection. *American Naturalist* 104, 592–597.

K–T boundary The boundary between the CRETACEOUS and TERTIARY periods (65 million years ago) is marked by a major MASS EXTINCTION event. Groups of organisms that became extinct include, most famously, the dinosaurs, but also other reptile groups, including the flying pterosaurs and the marine mosasaurs and plesiosaurs, together with marine molluscs, including the ammonites, belemnites and the rudist bivalves.

The cause of the extinctions has long been a matter of controversy. Explanations that have been put forward include SEA-LEVEL CHANGE and the effects of VOLCANIC ERUPTIONS, notably those of the Deccan traps in northwest India (see LAVA PLATEAU). Explanations specific to the extinction of the dinosaurs include disease, poisoning, suicide, constipation and sterility. Detailed studies of rock sections with continuous sedimentation from Cretaceous to Tertiary times, notably at Gubbio in central Italy, have revealed the presence of a distinct *K–T boundary layer*. Alvarez *et al.* (1980) demonstrated that CLAY in this layer was enriched in the rare-earth element iridium and attributed this *iridium anomaly* to an extraterrestrial source, specifically a major METEORITE IMPACT event that would also have caused the extinctions. This *impact hypothesis* was greeted with much controversy by the geological community. Further evidence to support the hypothesis has since been collected, including TSUNAMI deposits in the boundary layer, shocked QUARTZ grains and the identification of an impact site at Chicxulub on the Yucatán peninsula in southern Mexico. The theory of a major impact event at the K–T boundary is now widely accepted.

The problem remains, however, of whether or not the impact caused the extinctions. A major review of biotas across the boundary by MacCleod *et al.* (1997) revealed a surprising lack of detailed information for many groups. Some were in decline through the latest Cretaceous, before their final extinction (which in some cases occurred before, rather than at, the boundary), while others crossed the boundary with little apparent change. The nature of events at the K–T boundary remains a topic of lively debate. *JCWC/GO*

Alvarez, L.W., Alvarez, W., Asaro, F. and Michel, H.V. 1980: Extraterrestrial cause for the Cretaceous–Tertiary extinction. *Science* 208, 1095–1108. Alvarez, W. 1997: *T. rex and the crater of doom*. London: Penguin. Chaloner, W.G. and Hallam, A. (eds) 1989: *Evolution and extinction*. London: The Royal Society. Fortey, R. 1997: *Life: an unauthorised biography*. London: HarperCollins. Frankel, C. 1999: *The end of the dinosaurs: Chicxulub crater and mass extinctions*. Cambridge: Cambridge University Press. Hildebrand, A.R., Penfield, G.T., Kring, D.A. *et al.* 1991: Chicxulub crater: a possible Cretaceous/Tertiary boundary impact crater on the Yucatán Peninsula, Mexico. *Geology* 19, 867–871. Hudson, J.D. and MacLeod, N. 1998: Discussion on the Cretaceous–Tertiary biotic transition. *Journal of the Geological Society, London* 155, 413–419. MacLeod, N., Rawson, P.F., Forey, P.L. *et al.* 1997: The Cretaceous–Tertiary biotic transition. *Journal of the Geological Society, London* 154, 265–292. Norris, R.D., Huber, B.T. and Self-Trail, J. 1999: Synchroneity of the K–T oceanic mass extinction and meteorite impact: Blake Nose, western North Atlantic. *Geology* 27, 419–422. Ryder, G., Fastovsky, D. and Gartner, S. (eds) 1996: *The Cretaceous–Tertiary event and other catastrophes in Earth history*, Special Paper 307. Boulder, CO: Geological Society of America. Ward, P.D. 1995: After the fall: lessons and directions from the K/T debate. *Palaios* 10, 530–538.

ka A *kilo-annum*: an abbreviation for thousands of years before the present. *GO*

Kalman filter A powerful statistical method for FILTERING time series, used widely in detecting and characterising time-dependent patterns of behaviour in TREE RINGS. In the form developed for DENDROCLIMATOLOGY, Kalman filtering is essentially a multiple regression modelling procedure in which the regression coefficients of the predictor variables are allowed to vary with time. This is done in a recursive fashion by casting the regression problem into a *state-space model*, which explicitly allows for timewise changes in the coefficients to be modelled. In essence, a series of one-step-ahead predictions are made using the available predictors and their coefficients are allowed to change to improve the overall predictive ability of the model results. This is accomplished in an objective fashion using *maximum likelihood estimation*. The principal application of the Kalman filter in DENDROCHRONOLOGY has been in the detection of FOREST DECLINE through modelled changes in the response of tree growth to climate. The major difficulty in using the Kalman filter method lies in the interpretation of any time dependence detected by the technique. Unless there are strong *a priori* reasons to expect a certain pattern or direction of time dependence between tree growth and some predictor variable(s), the mere presence of time dependence may be difficult to interpret in a causal sense. However, when applied to a well constructed EXPERIMEN-

TAL DESIGN, the Kalman filter can be extremely useful for detecting changes in tree growth that may be induced by climatic and environmental change. *ERC*

[See also REGRESSION ANALYSIS, TIME-SERIES ANALYSIS]

Cook, E.R. and Johnson, A.H. 1989. Climate change and forest decline: a review of the red spruce case. *Water, Air, and Soil Pollution* 48, 127–140. **Harvey, A.C.** 1984: A unified view of statistical forecasting procedures. *Journal of Forecasting* 3, 245–275. **Van Deusen, P.C.** 1989. A model-based approach to tree ring analysis. *Biometrics* 45, 763–779. **Van Deusen, P.C.** 1990: Evaluating time-dependent tree ring and climate relationships. *Journal of Environmental Quality* 19, 481–488. **Visser, H. and Molenaar, J.** 1988: Kalman filter analysis in dendroclimatology. *Biometrics* 44, 929–940. **Visser, H. and Molenaar, J.** 1992: Air pollution stress in the Bavarian Forest? *Forest Science* 38, 870–872.

kame A steep-sided mound or short irregular ridge comprising mainly sand and gravel and formed by SUPRAGLACIAL or ice-contact glaciofluvial deposition. The deposits are typically bedded with faulting and folding structures found in fossil examples, particularly in the margins, reflecting removal of supporting ice. *RAS*

[See also ESKER, GLACIOFLUVIAL SEDIMENTS]

Benn, D.I. and Evans, D.J.A. 1998: *Glaciers and glaciation.* London: Arnold, 487–489.

Kampfzone Literally, the 'battlezone'. The term is widely used in the European Alps for the altitudinal zone (see ALTITUDINAL ZONATION OF VEGETATION) between the TREE LINE/LIMIT and the *tree-species line*. In the Kampfzone, tree species 'struggle' to attain tree growth-form, and are usually stunted (KRUMMHOLZ). *HHB*

kaolinite A 1:1 CLAY MINERAL having a basic structure of one sheet of silicon atoms and one sheet of aluminium atoms held together by shared oxygen atoms in a fixed lattice structure that does not expand when wetted. There is little ISOMORPHIC SUBSTITUTION and the CATION EXCHANGE CAPACITY is low. Kaolinite is a widely distributed clay mineral, but is dominant in moderately to strongly weathered soil environments where LUVISOLS, ACRISOLS and FERRALSOLS have developed. *EMB*

Brady, N.C. 1984: *The nature and properties of soils.* New York: Macmillan.

karoo The BIOME of the SEMI-ARID region of southern Africa bordering the Kalahari Desert and containing wide environmental and biotic diversity. The *succulent karoo*, characterised by relatively short-lived shrubs, occupies a broad coastal belt with a less variable annual rainfall but sparse winter rains and summer drought ameliorated by coastal fog. The *Nama-karoo*, characterised by more grasses and by longer-lived shrubs, lies further inland where the inter-annual rainfall variability is more extreme and low summer rainfall is received from the east and north. *JAM*

Dean, W.R.J. and Milton, S.J. 1999: *The Karoo: ecological patterns and processes.* Cambridge: Cambridge University Press. **Thomas, D.S.G. and Shaw, P.A.** 1991: *The Kalahari environment.* Cambridge: Cambridge University Press.

karren A term of German origin for small-scale (centimetre to metre) solutional sculptures on rock outcrops.

They are typical of KARST areas and include surface and subsurface pits, steps, grooves and channel forms. They are most commonly developed on carbonate and sulphate rocks, but have also been found on sandstone, quartzite and granite. *SHD*

Crowther, J. 1997: Surface roughness and the evolution of karren forms at Lluc, Serra de Tramuntana, Mallorca. *Zeitschrift für Geomorphologie NF* 41, 393–407. **Ford, D.C. and Williams, P.W.** 1989: *Karst geomorphology and hydrology.* London: Chapman & Hall. **Mottershead, D.** 1996: A study of solution flutes (Rillenkarren) at Lluc, Mallorca. *Zeitschrift für Geomorphologie Supplementband* 103, 215–241.

karst The processes, LANDFORMS and landscapes characterised by solutional WEATHERING along surface and subsurface pathways, typically leading to the progressive replacement of surface with underground drainage (*karstification*). Karst is usually associated with rocks of high solubility and well developed secondary porosity (e.g. limestone, gypsum). An estimated 7–10% of the Earth's ice-free land surface comprises karst terrain. Key elements of karst landscapes are solutional features such as CAVE PASSAGES, DOLINES, POLJES, KARREN, *sinkholes* and depositional forms such as SPELEOTHEMS. Datable CAVE SEDIMENTS can be used to reconstruct climatic, groundwater and SEA-LEVEL CHANGES. The term PSEUDOKARST is sometimes used for similar landforms produced in non-carbonate rocks. *SHD*

[See also AGGRESSIVE WATER, BIOKARST, COCKPIT KARST, DROWNED KARST, THERMOKARST, TOWER KARST, TROPICAL KARST, CHEMICAL WEATHERING, PRECIPITATION, RESURGENCE]

Ford, D. C. and Williams, P.W. 1989: *Karst geomorphology and hydrology.* London: Chapman & Hall. **Jennings, J.N.** 1985: *Karst geomorphology.* Oxford: Basil Blackwell. **Sweeting, M.M.** 1972: *Karst landforms.* London: Macmillan.

Kastanozems Soils with a thick dark-brown topsoil, a MOLLIC HORIZON, rich in organic matter and having a calcareous or gypsiferous subsoil. Kastanozems are common in central areas of North and South America, and Eurasia (SOIL TAXONOMY: Ustic and Boric great groups of Mollisols). Typically, Kastanozems develop in a warmer climate on the drier side of CHERNOZEMS under short-grass *prairie* or STEPPE. In earlier literature these soils are referred to as *Chestnut Soils*. *EMB*

[See also WORLD REFERENCE BASE FOR SOIL RESOURCES].

kettle hole A kettle or kettle hole is a self-contained bowl-shaped depression, formed as a result of melting of a buried ice block, especially common on glacier OUTWASH plains or in MORAINE complexes. The ice blocks may be derived as remnants of a GLACIER snout or as icebergs transported onto an outwash plain by floods. Kettle holes are commonly occupied by lakes (*kettle lakes*) and the sediments they contain are often used in PALAEOENVIRONMENTAL RECONSTRUCTION. The sediments in a kettle hole are younger than the surrounding deposits and conditions for preservation of organic material tend to be favourable in the kettle hole. In some cases, such stratigraphical archives have survived subsequent GLACIATION. *MJH*

Lagerbäck, R. and Robertsson, A.-M. 1988: Kettle holes – stratigraphical archives for Weichselian geology and palaeoenvironment in northernmost Sweden. *Boreas* 17, 439–468.

keystone species Many ecologists believe that a few species have a disproportionately strong influence on others in their community, such that, if they were removed, the structure of that community would change dramatically (ECOSYSTEM COLLAPSE), much like the collapse of the stones forming a Roman arch once the central stone ('keystone') has been removed. While there is much debate over this concept, examples of likely keystone species include nitrogen-fixing bacteria, decomposers, some top carnivores, beavers, prairie dogs and, at least in recent times, humans. *MVL*

[See also COMMUNITY CONCEPTS]

Pimm, S.L. 1991: *The balance of nature? Ecological issues in the conservation of species and communities.* Chicago, IL: The University of Chicago Press.

kinetic temperature The temperature that indicates the average kinetic energy of the molecules or atoms of a substance (i.e. that measurable by an instrument in direct contact with an object). It is also known as the *physical temperature*. *TS*

[See also RADIANT TEMPERATURE]

Kirchhoff radiation law The law that under conditions of local thermodynamic equilibrium, thermal emission has to be equal to absorption. A system is in thermodynamic equilibrium if mechanical, chemical and thermal equilibriums are held. *TS*

Ulaby, F.T., Moore, R.K. and Fung, A.K. 1981: *Microwave remote sensing: active and passive.* Norwood, MA: Artech House.

knickpoint A sharp break of slope in a river long profile characterised by a steepening of channel gradient. It may be marked by rapids or a WATERFALL and a GORGE. Also known as a *nickpoint* (American) or a *rejuvenation head*, a knickpoint tends to form and move upstream by headward erosion following a negative change in BASE LEVEL. Some, however, represent the occurrence of resistant geological structures. The term has usually been used in the context of long-term LANDSCAPE EVOLUTION and DRAINAGE EVOLUTION studies, but more recently more subtle knickpoints have been recognised in response to human interference in river channels. RIVER CHANNELISATION involving straightening of MEANDERS introduces an increased channel gradient, leading to DEGRADATION, which progresses upstream as a knickpoint. *RPDW/JAM*

[See also DENUDATION CHRONOLOGY, REJUVENATION]

Brookes, A. 1985: River channelization: traditional engineering methods, physical consequences and alternative practices. *Progress in Physical Geography* 9, 44–73. **Knighton, D.A.** 1998: *Fluvial forms and processes: new perspectives.* London: Arnold.

knock and lochan topography A Scottish term for an area of glacially scoured bedrock terrain comprising eroded rock knobs and rock basins, many of which are occupied by lakes. *RAS*

[See also GLACIAL EROSION]

Rea, B.R. and Evans, D.J.A. 1996: Landscapes of areal scouring in NW Scotland. *Scottish Geographical Magazine* 112, 47–50.

knowledge The sum of all that is known, including theoretical and practical knowledge. Knowledge may increase through time, but what is regarded as true, reliable or OBJECTIVE KNOWLEDGE may change in the light of new evidence or new perspectives. *JAM*

Ziman, J. 1978: *Reliable knowledge: an exploration of the grounds for belief in science.* Cambridge: Cambridge University Press.

knowledge-based systems Computer programs capable of advising and problem solving in a particular field of expertise on a level comparable to a human expert. During the last 20 years two paradigms can be distinguished: the knowledge *transfer paradigm* states that human knowledge is transferred rapidly into implemented computer systems and made available in the DECISION MAKING process. These knowledge-based systems could be developed quickly but failed to be reliable and maintainable. The *modelling paradigm* regards knowledge acquisition, not as the elicitation and collection of existing knowledge, but as the process of creating a knowledge model that did not exist beforehand. The design of such a knowledge-based system requires several steps: elicitation, interpretation, formalisation and operationalisation, design and implementation. *HB*

Angele, J. Fensel, D., Landes, D. and Studer, R. 1998: Developing knowledge-based systems with MIKE. *Journal of Automated Software Engineering* 5, 389–418.

komatiite An IGNEOUS ROCK with unusually high magnesia (MgO) content, representing very high-temperature and low-viscosity LAVA FLOWS, named after the Komati River in Swaziland. Komatiites are restricted to the GEOLOGICAL RECORD of the ARCHAEAN, forming an important constituent of many GREENSTONE BELTS. They pose problems for the strict application of UNIFORMITARIANISM and date from a period in the Earth's history when more internal heat was generated, with important implications for the nature of early PLATE TECTONICS. *GO*

Parman, S.W., Dann, J.C., Grove, T.L. and de Wit, M.J. 1997: Emplacement conditions of komatiite magmas from the 3.49 Ga Komati Formation, Barberton Greenstone Belt, South Africa. *Earth and Planetary Science Letters* 150, 303–323. **Williams, D.A., Wilson, A.H. and Greeley, R.** 2000: A komatiite analog to potential ultramafic materials on Io. *Journal of Geophysical Research – Planets* 105, 1671–1684.

koniology The scientific study of ATMOSPHERIC DUST and other suspended PARTICULATES in the atmosphere, such as soot, pollen and spores. The *konisphere* or *staubosphere* is the part of the atmosphere where such particles occur. *AHP*

[See also AEROBIOLOGY]

kriging A family of statistical techniques for optimal spatial INTERPOLATION between discrete measurement points, minimising the error VARIANCE. Kriging was developed by the mining engineer D.G. Krige. The interpolation of a random variable $Z(X)$ is based on the empirical VARIOGRAM $2\gamma(h)$. A theoretical variogram model is then fitted to the empirical variogram. From this model, local weights λ_i are estimated. These are used to predict $Z(x_0)$ at the unsampled location x_0 by calculating a local weighted mean. Kriging techniques include ordinary kriging, simple kriging, block kriging, universal kriging, disjunctive kriging, indicator kriging and cokriging. *HB*

Atkinson, P. M., Webster, R. and Curran, P.J. 1994: Cokriging with airborne MSS imagery. *Remote Sensing of*

Environment 50, 335–345. **Cressie, N.A.C.** 1993: *Statistics for spatial data*. New York: Wiley. **Rosenbaum, M.S. and Söderström, M.** 1996: Cokriging of heavy metals as an aid to biogeochemical mapping. *Acta Agriculturae Scandinavica* B46, 1–8. **Rossi, R.E., Dungan, J.L. and Beck, L.R.** 1994: Kriging in the shadows: geostatistical interpolation for remote sensing. *Remote Sensing of Environment* 49, 32–40.

Krummholz   The stunted growth form of tree species beyond the TREE LINE/LIMIT. Buds projecting beyond the climatically warmer boundary-layer or the snowpack are killed. Trees can develop if conditions ameliorate (e.g. flag-form trees of *Picea* spp.).   *HHB*

[See also KAMPFZONE]

kurtosis   The extent to which the peak of a unimodal frequency or PROBABILITY DISTRIBUTION departs from the shape of a NORMAL DISTRIBUTION by either being more pointed (*leptokurtic*) or flatter (*platykurtic*). The coefficient of kurtosis is the fourth moment about the mean divided by the variance squared (often with 3 subtracted so that a NORMAL DISTRIBUTION has zero kurtosis).   *HJBB*

[See also SKEWNESS]

Sokal, R.R. and Rohlf, F.J. 1995: *Biometry*. New York: W.H. Freeman.

L

La Niña A period of strong TRADE WINDS and unusually low SEA-SURFACE TEMPERATURE in the central and eastern tropical Pacific. The negative, opposite or 'cold' phase of EL NIÑO–SOUTHERN OSCILLATION (ENSO) to EL NIÑO.
AHP/JAM

Philander, S.G.H. 1985: El Niño and La Niña. *Journal of Atmospheric Science* 42, 2652–2662.

label A textual description of a geographic feature or object on a paper map. Also the name and description (ATTRIBUTE) of an ENTITY in a GEOGRAPHICAL INFORMATION SYSTEM (GIS).
TVM

[See also IDENTIFICATION, OVERLAY ANALYSIS]

laboratory science The use of laboratories for analysis and EXPERIMENT. Predominantly but not entirely the domain of the physical and biological sciences, the distinctive quality of the use of laboratories is the ability to restrict 'outside' influences, to reduce contamination and to conduct tightly controlled experiments. Thus, for example, in conducting laboratory experiments, CAUSAL RELATIONSHIPS may be firmly established by varying one causal factor at a time, an approach that is difficult, if not impossible, to achieve by FIELD RESEARCH.
JAM

lacustrine sediments Lake sediments; important temporal archives (see NATURAL ARCHIVES) of terrestrial environmental change. They can be sampled by a variety of CORERS used from open water, winter ice or the surface of an overgrown or drained lake. SEDIMENTARY STRUCTURES include laminations resulting from rhythmic deposition cycles. Annual laminations are VARVES. Minerogenic varves produced by glacial meltwater are widespread around the Baltic Sea and have been used to construct the Swedish Varve Timescale. Organic and chemical (e.g. carbonate) laminations are produced in certain lake types and are best preserved in MEROMICTIC LAKES, where BIOTURBATION is minimal. They can also be used to build a chronology, which can be used to calibrate the radiocarbon chronology.

ALLOCHTHONOUS sediments originate from the *catchment*. They may be organic (terrestrial macrofossils, soil, peat) or minerogenic and are derived by catchment erosion. High minerogenic content indicates catchment disturbance that can result from GLACIAL EROSION, PERIGLACIAL activity, ARIDITY (LOESS), HUMAN IMPACT and CATASTROPHIC EVENTS (AVALANCHES, LANDSLIDES, FLOODS, fires). In special circumstances a *trash layer* of terrestrial material is deposited during lake formation at the melting of a buried ice block that had supported vegetation and soil.

AUTOCHTHONOUS sediments are the net deposition of lacustrine biogenic material less decay and outflow loss. LOSS-ON-IGNITION (LOI) at 550°C reflects the balance between minerogenic and organic material. It can be a sensitive synthetic environmental indicator. LOI at 950°C estimates carbonate in calcareous sediments, including biogenic MARL, shells and clastic carbonates.

Sediment chemistry can be used to infer environmental change. Ca and Mg are indicators of mineral erosion in the catchment. Pb, Zn, Cu, Hg, etc. are indicators of HEAVY METAL POLLUTION. N, P, C and organic compounds (pigments derived from chlorophyll, lipids etc.) are indicators of algal types and EUTROPHICATION.

STABLE ISOTOPES (δ^{13}C, δ^{18}O, δ^{2}H) can indicate temperature and precipitation/humidity. Radioactive isotopes are used for dating: ^{14}C covers about 50 000 years; ^{210}Pb the last 200 years. Peaks of ^{137}Cs are time markers for nuclear testing culminating in AD 1963 and the 1986 Chernobyl accident. TEPHRA horizons are also chronological markers. All these can be used for CORE CORRELATION.

Lacustrine sediments contain animal and plant remains, the analysis of which results in a variety of environmental reconstructions through time. Ideally a study of environmental change should use evidence from both sediments and biota in a MULTIPROXY APPROACH.
HHB

Appleby, P.G. 1993: Forward to the ^{210}Pb dating anniversary series. *Journal of Paleolimnology* 9, 155–160. **Kitagawa, H. and van der Plicht, J.** 1998: Atmospheric radiocarbon calibration to 45,000 yr B.P.; late glacial fluctuations and cosmogenic isotope production. *Science* 279, 1187–1190. **Wohlfarth, B., Björck, S. and Possnert, G.** 1995: The Swedish Time Scale: a potential calibration tool for the radiocarbon time scale during the late Weichselian. *Radiocarbon* 37, 347–359.

lag deposit A sedimentary deposit of relatively coarse clasts produced *in situ* by the removal of finer interstitial particles. Examples include DESERT PAVEMENT formed from the winnowing away of fines by wind and periglacial BOULDER PAVEMENTS produced in pronival locations by snowmelt. Once formed, such pavements may be relatively stable and resistant to erosion.
JAM

lag time The time lapse between the occurrence of an event (such as a CLIMATIC CHANGE, LANDUSE CHANGE, a human intervention or an EXTREME CLIMATIC EVENT) and the resulting effect. The concept is important, for example, in understanding the CAUSAL RELATIONSHIPS involved in environmental change and in the prediction and mitigation of NATURAL HAZARDS. In the context of GLACIER VARIATIONS in response to a climatic change, it is the time interval between the mass balance change and the maximum (or minimum) of the resulting glacier advance (or retreat).
MAB/JAM

[See also LEAD AND LAG, REACTION TIME, RELAXATION TIME, RESPONSE TIME]

Allen, J.R.L. 1974: Reaction, relaxation and lag in natural sedimentary systems: general principles, examples and lessons. *Earth-Science Reviews* 10, 263–342.

lagoon A shallow body of water, connected permanently or intermittently to a larger body of water. Many lagoons lie parallel to the coast and are separated from it by a BARRIER ISLAND or barrier spit (connected to the mainland). Others are associated with CORAL REEF ISLANDS.
HJW

lahar A DEBRIS FLOW composed of volcanic debris. Such flows are common around VOLCANOES because of the combination of steep slopes and abundant, unconsolidated PYROCLASTIC debris. Lahars may occur during or immediately after a VOLCANIC ERUPTION or by secondary mobilisation weeks to years later. Lahars are a potentially devastating NATURAL HAZARD: over 25 000 people died in a lahar following the AD 1985 eruption of Nevado del Ruiz (Colombia). Some authors advocate abandoning the term because the flow processes and characteristics of the deposits are no different from debris flows in non-volcanic settings.   *JBH*

[See also LANDSLIDE HAZARD]

Lowe, D.R., Williams, S.N., Leigh, H. *et al.* 1986: Lahars initiated by the 13 November 1985 eruption of Nevado del Ruiz, Colombia. *Nature* 324, 51–53. **Pierson, T.C. and Scott, K.M.** 1985: Downstream dilution of a lahar: transition from debris flow to hyperconcentrated streamflow. *Water Resources Research* 21, 1511–1524. **Smith, G.A.** 1986: Coarse-grained nonmarine volcaniclastic sediment: terminology and depositional process. *Geological Society of America Bulletin* 97, 1–10. **Voight, B.** 1990: The 1985 Nevado del Ruiz Volcano catastrophe: anatomy and retrospection. *Journal of Volcanology and Geothermal Research* 42, 151–188.

lake A body of (normally fresh) water occupying a depression in the Earth's continental surface.   *JAM*

lake-level variations Past lake levels are an important source of PROXY DATA for reconstruction of palaeoclimates (see PALAEOCLIMATOLOGY). Lake-level variations reflect the HYDROLOGICAL BALANCE of the lake and its catchment and may be a particularly valuable indicator of changes in *effective moisture*. Regional- or continental-scale synchroneity in lake-level variations points to a climatic cause. Water-level changes can be reliably reconstructed from deep, *closed-basin lakes*, where water level may be strongly correlated with water SALINITY and the lake does not dry out. Reconstructions from *open-basin lakes*, where water is lost through surface outflows or groundwater seepage, may be more problematic. Nevertheless, water-level changes can be reconstructed from open basins by analysing: (1) PLANT MACROFOSSILS to identify changes in the distribution of lake-shore vegetation; (2) coarse minerogenic sedimentary composition to identify changes in the distribution of near-shore sediments affected by waves; and (3) changes in the position of the organic *sediment limit* (the highest level for permanent deposition of predominantly organic sediments or GYTTJA).   *JAM*

[See also LACUSTRINE SEDIMENTS, PALAEOLIMNOLOGY]

Almendinger, J.E. 1993: A groundwater model to explain past lake levels at Parkers Prairie, Minnesota, U.S.A. *The Holocene* 3, 105–115. **Digerfeldt, G.** 1986: Studies on past lake-level fluctuations. In Berglund, B.E. (ed.), *Handbook of Holocene palaeoecology and palaeohydrology*. Chichester: Wiley, 127–143. **Guiot, J., Harrison, S.P. and Prentice, I.C.** 1993: Reconstruction of Holocene precipitation patterns in Europe using pollen and lake-level data. *Quaternary Research* 40, 139–149. **Harrison, S.P. and Digerfeldt, G.** 1993: European lakes as palaeohydrological and palaeoclimatic indicators. *Quaternary Science Reviews* 12, 233–248. **Harrison, S.P., Frenzel, B., Huckriede, U. and Weiss, M.M. (eds)** 1998: *Palaeohydrology as reflected in lake-level changes as climatic evidence for Holocene times* [Paläoklimaforschung. Vol. 25]. Stuttgart: Gustav Fischer. **Street-Perrott, F.A. and Harrison, S.P.** 1985: Lake levels and climatic reconstruction. In Hecht, A.D. (ed.), *Paleoclimatic analysis and modeling*. New York: Wiley, 291–340.

lake stratification and zonation Water-column stratification occurs typically in temperate regions where water temperatures range either side of 4°C. Surface water warms in spring and, when wind action becomes insufficient to mix the water column, stratification develops. The warm, oxygenated *epilimnion* overlies the denser, cool, deoxygenated *hypolimnion*. Autumn cooling induces OVERTURNING. Winter stratification, with ice and water at 0°C overlying water at 4°C, overturns in spring (see Figure).

Macrophytes and phytoPLANKTON inhabit the *photic zone*. Macrophytes occupy the *littoral zone*, depth zonation being determined mainly by light penetration and distribution by wave exposure, sediment type and deposition rate. *Emergent plants* grow nearest the shore. Turf-like isoetids prefer wave-exposed shores and can extend (e.g. *Isoetes lacustris*) to a depth of about 6 m. Floating-leaved

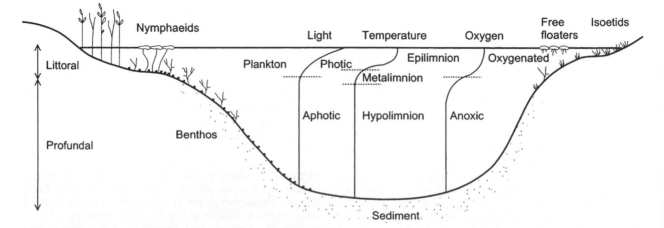

lake stratification and zonation *Zones of stratification of light, temperature, and oxygen in a lake, and vegetation zonation*

nyphaeids are restricted to a depth of about 2 m, but submerged elodeids can reach 11 m or deeper in clear water. Free-floating plants, (e.g. *Lemna, Azolla*) can cover a lake. Animal diversity is greatest in the epilimnion, but several invertebrate types thrive in the anoxic hypolimnion. *HHB*

[See also LIMNOLOGY, OVERTURNING, PALAEOLIMNOLOGY, PLANKTON]

Hutchinson, G.E. 1957: *Treatise on limnology.* Vol. 1. *Geography, physics, and chemistry.* New York: Wiley. **Lewis, W.M.** 1983: A revised classification of lakes based on mixing. *Canadian Journal of Fisheries and Aquatic Science* 40, 1779–1787. **Spence, D.H.N.** 1964: The macrophytic vegetation of lochs, swamps and associated fens. In Burnett, J.H. (ed.), *The vegetation of Scotland.* Edinburgh: Oliver and Boyd, 306–425.

lake terrace   A *terrace* at a lake margin formed during a regressive phase when there is a reduced volume of water in the lake. Lake terraces mark former SHORELINES and, if mapped and dated, can provide information on changing precipitation levels, particularly in SEMI-ARID regions and ARIDLANDS. In PERIGLACIAL environments they can be formed or enhanced by FROST WEATHERING of underlying bedrock, as in the case of the *parallel roads* of Glen Roy. *DH*

[See also LAKE-LEVEL VARIATIONS, POLAR SHORE EROSION]

Benson, L. and Thompson, R.S. 1987: The physical record of lakes in the Great Basin. In Ruddiman, W.F. and Wright Jr, H.E. (eds), *North America and adjacent oceans during the last deglaciation.* [*The Geology of North America.* Vol. K-3]. New York: Geological Society of America, 241–260. **Sissons, J.B.** 1978: The parallel roads of Glen Roy and adjacent glens. *Boreas* 7, 229–244.

lakes as indicators of environmental change
Lakes indicate environmental changes over time, and these can be observed directly or reconstructed using a MULTIPROXY APPROACH. *Closed-basin lakes* indicate precipitation:evaporation balance by changes in water level, detectable by former shorelines, by sediment transects from shallow to deep water associated with changes in aquatic macrophyte distribution and by SALINITY changes inferred, for example, from diatoms and sediment chemistry.

Lakes respond to changes in water chemistry. Diatoms reflect changes in pH and dissolved organic carbon resulting from ACID RAIN. EUTROPHICATION is indicated by changes in sediment composition, chemistry (e.g. C, N, P, pigments and lipids) and organism fossils (e.g. chironomids, cladocera, algae, cyanobacteria and aquatic macrophytes). HEAVY METAL concentrations in sediment indicate METAL POLLUTION levels. Lakes also respond to temperature changes. Sensitive organisms and stable isotopes in lake sediments provide a record of water and air temperature, ice cover, light penetration, productivity etc. (e.g. chironomids, cladocera, coleoptera and diatoms).

Lakes indicate *catchment* changes. MINEROGENIC SEDIMENT input reflects EROSION. This may be climatic (e.g. YOUNGER DRYAS and GLACIER VARIATIONS), catastrophic (e.g. AVALANCHES, DEBRIS FLOWS, FLOODS and FIRES) or human-induced SOIL EROSION (DEFORESTATION, AGRICULTURAL and landscape modification) exacerbated by ARIDITY or IRRIGATION. Lakes indicate terrestrial ecosystem changes by sedimentary records of terrestrial biota. *HHB*

[See also LIMNOLOGY, PALAEOLIMNOLOGY, LACUSTRINE SEDIMENTS, LAKE-LEVEL VARIATIONS, MULTIPROXY

APPROACH, SEDIMENTOLOGICAL EVIDENCE OF ENVIRONMENTAL CHANGE, CHIRONOMID ANALYSIS, CLADOCERA ANALYSIS, BEETLE ANALYSIS, DIATOM ANALYSIS, POLLEN ANALYSIS, MACROFOSSIL, STABLE ISOTOPES]

Battarbee, R.W. 2000: Palaeolimnological approaches to climate change, with special regard to the biological record. *Quaternary Science Reviews* 19, 107–124. **Fritz, S.C.** 1989: Lake development and limnological response to prehistoric and historic land-use in Diss, Norfolk, UK. *Journal of Ecology* 77, 182–202. **Fritz, S.C.** 1996: Paleolimnological records of climatic change in North America. *Limnology and Oceanography* 41, 882–889. **Gaillard, M.-J. and Digerfeldt, G.** 1991: Palaeohydrological studies and their contribution to palaeoecological and palaeoclimatic reconstructions. In Berglund, B.E. (ed.), *The cultural landscape during 6000 years in southern Sweden – the Ystad project* [*Ecological Bulletins* 41]. Copenhagen: Munksgaard, 275–282. **Laird, K.R., Fritz, S.C., Grimm, E.C. and Mueller, P.G.** 1996: Century-scale paleoclimate reconstruction from Moon Lake, a close-basin lake in the northern Great Plains. *Limnology and Oceanography* 41, 890–902. **Psenner, R. and Schmidt, R.** 1992: Climate driven pH control of remote alpine lakes and effects of acid deposition. *Nature* 356, 781–783.

Lamarckism   The view of Jean Baptiste Pierre Antione de Monet, Chevalier de Lamarck, (1744–1829) ascribing EVOLUTION to inheritable modification in the individual by habit, behaviour and the ENVIRONMENT. *KDB*

[See also CREATIONISM, DARWINISM]

laminar flow   Flow dominated by fluid viscosity and characterised by individual fluid elements (e.g. molecules) moving parallel to the flow direction. At a value of *Reynolds number* (ratio of inertial to viscous forces) between 500 and 2000, laminar flow (also known as *viscous flow*) transforms into TURBULENT FLOW. Examples of laminar flows include GLACIERS, DEBRIS FLOWS and LAVA FLOWS. Water flows are laminar only at very low velocities or very shallow depths. *GO*

Leeder, M.R. 1999: *Sedimentology and sedimentary basins: from turbulence to tectonics.* Oxford: Blackwell Science.

lamination   The smallest scale of STRATIFICATION commonly visible in sediments and sedimentary rocks. A *lamina* (plural: *laminae*) defines a sedimentation unit thinner than 1 cm. Horizontal lamination is formed by the vertical accumulation of sediment (see PARALLEL LAMINATION); inclined lamination is produced by deposition on laterally accreting surfaces such as the downstream faces of migrating RIPPLES (see CROSS-LAMINATION). The term *microlamination* is sometimes used for submillimetre-scale lamination present in some very fine-grained sediments. *MRT*

[See also HIGH-RESOLUTION RECONSTRUCTIONS, LACUSTRINE SEDIMENTS]

Collinson, J.D. and Thompson, D.B. 1992: *Sedimentary structures*, 2nd edn. London: Chapman and Hall. **Dickman, M.** 1987: Lake sediment microlaminae and annual mortalities of photosynthetic bacteria in an oligomictic lake. *Freshwater Biology* 18, 151–164. **Segall, M.P. and Kuehl, S.A.** 1994: Sedimentary structures on the Bengal shelf – a multiscale approach to sedimentary fabric interpretation. *Sedimentary Geology* 93, 165–180.

land   The FAO (1995:6) defines a land as 'a delineable area of the Earth's terrestrial surface, encompassing all attributes of the biosphere immediately above or below this surface, including those of the near-surface climate, the soil and terrain forms, the surface hydrology (including shallow lakes, rivers and swamps), the near-surface sedimentary layers and associated groundwater reserve, the plant and animal populations, the human settlement pattern and physical results of past and present human activity (terracing, water storage or drainage structures, roads, buildings etc). This definition refers to a natural unit (not an administrative area) and conforms to LAND-SCAPE UNITS, *land system units* or *landscape-ecological units* as building blocks of a WATERSHED (catchment area) or a phytogeographic unit (BIOME) and is useful for LANDUSE planning purposes. *EMB*

Food and Agriculture Organization of the United Nations (FAO) 1995: *Planning for sustainable use of land resources: towards a new approach* [*Land and Water Bulletin* 2]. Rome: FAO. **United Nations** 1994: *Convention on desertification* [*Information Programme on Sustainable Development*]. New York: United Nations.

land bridge A land connection between continents, between parts of continents or between islands. The fluctuating presence of land bridges has played a crucial biogeographical role in the history of Earth's FAUNA (particularly mammals) and FLORA. The availability of such 'highways' provides corridors for MIGRATION and DISPERSAL and directly affects terrestrial SPECIES DIVERSITY in the various FAUNAL PROVINCES through faunal interchange and COMPETITION. Conversely, the absence of land bridges has a profound influence on the EVOLUTION of geographically isolated populations (see ISOLATION). Prior to the Mesozoic, biotic homogeneity was largely maintained as a result of the presence of land connections throughout PANGAEA. The subsequent disruption of these terrestrial links led to a veritable explosion of faunal diversity in the Tertiary period through SPECIATION of isolated groups on continental landmasses and the evolution of orders such as marsupials and edentates. During the Pliocene, the formation of the Central American (Panamanian) land connection permitted faunal interchange between North and South America, but concomitantly led to the EXTINCTION of many South American groups through the dominance of placental mammals from the north. Thus, the presence of land bridges may facilitate dispersal and accordingly increase overall species diversity in any given area, but may also ultimately precipitate the extinction of other species in the face of newly arrived competitors.

The effects of ice build up and associated lowered sea level during the various glaciations of the PLEISTOCENE led to the periodic reconnection of North America and Eurasia across the land bridge of *Beringia*. This permitted the migration of many plants and animals, including humans, into the Americas from the Old World. The disappearance of Pleistocene land bridges as a result of SEA-LEVEL RISE following deglaciation also affected populations on small islands, leading to either dwarfism (NANISM) of some large mammal species or GIGANTISM of micromammals. In the Mediterranean, 16 islands (or former islands) have been found to contain fossils of dwarf ENDEMIC mammals, including elephants, hippopotami and antelope-like bovids. Further afield, fossil dwarf elephants are known from many islands in southeast Asia, while miniature MAMMOTHS have been found on Wrangel Island off

Siberia and on the Californian channel islands. Cases of gigantism, such as that of dormice on Malta, have also been noted. *DCS*

[See also BIOGEOGRAPHY, HABITAT, ISLAND BIOGEOGRAPHY, WILDLIFE CORRIDOR]

Darlington, P.J. 1957: *Zoogeography : the geographical distribution of animals.* New York: Wiley. **Hopkins, D.M., Matthews Jr, J.V., Schweger, C.E. and Young, S.B. (eds)** 1982: *Paleoecology of Beringia.* New York: Academic Press. **Lister, A.M.** 1995: Sea-levels and the evolution of island endemics: the dwarf red deer of Jersey. In Preece, R.C. (ed.), *Island Britain: a Quaternary perspective* [*Special Publication* 96]. London: Geological Society, 151–172. **Szalay, F.S., Novacek, M.J. and McKenna, M.C. (eds)** 1993: *Mammal phylogeny.* Vol. 1. *Mesozoic differentiation, multituberculates, monotremes, early therians, and marsupials.* Berlin: Springer. **Szalay, F.S., Novacek, M.J. and McKenna, M.C.** (eds), 1993: *Mammal phylogeny.* Vol. 2. *Placentals.* Berlin: Springer.

land cover A classification of features of the land surface. Rhind and Hudson (1980) defined vegetation and artificial construction as constituting land cover. However, Meyer and Turner (1994) limit land cover to 'the physical state of the land' of particular interest to natural scientists. Land-cover change is an important element in ENVIRONMENTAL MONITORING and is subject to *conversion* where a change of classification category is involved. Modification occurs only where an alteration of state within a classification category is found. Land-cover classification is often a primary goal of REMOTE SENSING. *TF*

[See also LANDUSE]

Meyer, W.B. and Turner, B.L. (eds) 1994: *Changes in land use and land cover: a global perspective.* Cambridge: Cambridge University Press. **Rhind, D. and Hudson, R.** 1980: *Land use.* London: Methuen.

land degradation A temporary or permanent decline in the productive capacity or resource potential of the land. Some types of degradation are irreversible, such as severe GULLY erosion or extreme SALINISATION; other types are reversible and can be changed by improved farming practices. Land degradation includes SOIL DEGRADATION, through SOIL EROSION or loss of fertility, the removal of natural vegetation, accompanied by HABITAT LOSS, and reduced BIODIVERSITY combined with a general deterioration of the landscape. It can be caused by both natural and anthropogenic factors, although the latter are normally viewed as the main cause. Land degradation may result from the effects of OVERGRAZING, excessive TILLAGE, FOREST CLEARANCE, disposal of INDUSTRIAL WASTES and other wastes causing soil contamination, and the effects of plants and animals that are capable of exploiting the degraded land situation. *EMB*

[See also CONTAMINATED SOILS, DESERTIFICATION]

Abel, J.O.J. and Blakie, P.M. 1989: Land degradation, stocking rates and conservation policies in the communal rangelands of Botswana and Zimbabwe. *Land Degradation and Rehabilitation* 1, 101–123. **Johnson, D.L. and Lewis, L.A.** 1995: *Land degradation: creation and destruction.* Oxford: Blackwell. **Lal, R.** 1995: Erosion–crop productivity relationships for soil in Africa. *Soil Science Society of America Journal* 59, 661–667. **Scherr, S.J. and Yadev, S.** 1996: *Land degradation in the developing world: implications for food, agriculture, and the environment to 2020* [*Discussion*

Paper 14]. Washington, DC: International Food Policy Research Institute (IFPRA).

land drainage

The removal of water from the land by artificial drainage systems for a variety of reasons, including: (1) conversion of WETLANDS to agricultural use; (2) improvement of existing agricultural land (FIELD DRAINAGE); (3) preservation of irrigated land from water-logging and SALINISATION; (4) *dewatering* and DESALINISATION of empoldered land reclaimed from the sea; (5) preparation of land for AFFORESTATION; (6) disposal of sewage and surface water from urban areas; (7) *flood control*; and (8) protection from GROUNDWATER CONTAMINATION.

Historically, land drainage has been seen as part of a process of progressive land improvement, but detrimental environmental effects are increasingly recognised. Particularly in relation to loss of wetlands, conflicts often arise between drainage interests and CONSERVATION interests. It has been estimated that the USA has lost 54% of its original wetlands since European settlement, whereas in Italy about 94% has been lost since Roman times. Much of the remaining wetland has suffered alteration, DEGRADATION and loss of functional integrity. The AGRICULTURAL INTENSIFICATION that often follows wetland drainage can produce *secondary ecological impacts* on the remaining wetland area, such as reduced WATER quality and changes in species composition. This has occurred, for example, in the Everglades National Park, Florida, following elevated phosphorus levels draining from the Everglades Agricultural Area. *JAM*

[See also SOIL DRAINAGE]

Armstrong, A.C. and Garwood, E.A. 1991: Hydrological consequences of artificial drainage of grassland. *Hydrological Processes* 5, 157–174. **Baldock, D.** 1984: *Wetland drainage in Europe.* London: International Institute for Environment and Development. **Framji, K.K. and Mahajan, I.K.** 1969: *Irrigation and drainage in the world: a global view* [two vols]. New Delhi: International Commission on Irrigation and Drainage. **Hill, A.R.** 1976: The environmental impact of agricultural land drainage. *Journal of Environmental Management* 4, 251–274. **Smedema, L.K. and Rycroft, D.W.** 1983: *Land drainage: planning and design of agricultural drainage systems.* Ithaca, NY: Cornell University Press.

land evaluation

As decisions about LANDUSE are a policy-driven activity, it is essential that all the factors involved are discussed with the stakeholders before landuse is changed. Land evaluation is a semi-quantitative process in which site and soil characteristics are assessed for specific purposes. In a land evaluation, LAND is considered to be more than soil; it includes all the reasonably stable attributes of the LANDSCAPE above and below an area, including geology, hydrology, plant and animal populations, and the results of past and present human activity. Soil maps are commonly interpreted in the light of these other environmental features to indicate the relative suitability (Highly, Moderately, Marginally) or unsuitability (Currently not suitable, Permanently not suitable) for a particular use or crop under well defined conditions of management. *EMB*

[See also LANDUSE CAPABILITY]

Food and Agriculture Organization of the United Nations (FAO) 1976: *A framework for land evaluation* [*Soils Bulletin* 32].

FAO: Rome. **Food and Agriculture Organization of the United Nations (FAO)** 1993: *FELSM, an international framework for evaluating sustainable land management* [*World Soil Resources Report* 73]. Rome:FAO. **McRae, S.G. and Burnham, C.P.** 1981: *Land evaluation.* Oxford: Clarendon Press.

land reclamation

In historical times, the process of bringing land under cultivation from a natural or semi-natural vegetation cover; the draining of marshes, clearing of heathland and woodland was spoken of as 'reclamation' of land for agriculture. Currently, the term is used for the REHABILITATION of derelict, contaminated or otherwise despoiled land, mainly resulting from industrial activity. However, the term is also used for the 'reclamation' of land from the sea, e.g. *empoldering* in The Netherlands, whereby new land with new uses results.

Remedial measures should always attempt to solve the problem once and for all as temporary solutions inevitably result in further work and greater expense at a later date. The process of RESTORATION will in many cases be determined by the future use of the land. Reuse of land for industrial purposes, where a solid concrete floor and tarmac parking places around the factory seal the ground surface, does not require as expensive reclamation as for housing or for a return to productive agriculture. A frequently employed alternative use for former derelict lands is to provide amenity open spaces in urban areas. During the past 50 years, considerable experience has been gained reclaiming land despoiled by open-cast coal and ironstone MINING or gravel quarrying. Such schemes require recreating a soil from the available geological materials to enable plants to make satisfactory growth.

Reclamation of irrigated land that has become salinised in areas of DESERTIFICATION is a growing problem in many countries with a SEMI-ARID environment. Calcium salts can usually be leached from permeable soils, but the presence of sodium salts causes greater problems and both must be dealt with in association with improved drainage. A rise of sea level through global warming may lead to increased problems of sea flooding in low-lying areas, in which case reclamation of the salt-affected land must take place before crops can be grown. *EMB*

[See also CONTAMINATED LAND, LAND DEGRADATION, LAND DRAINAGE, LAND RESTORATION, SALTWATER INTRUSION, SOIL DEGRADATION, SOIL RECLAMATION]

Bradshaw, A.D. 1987: The reclamation of derelict land. In Jordan, W.R., Gilpin, M.E. and Aber, J.E. (eds), *Restoration ecology.* Cambridge: Cambridge University Press, 53–74. **Bradshaw, A.D. and Chadwick, M.J.** 1980: *The restoration of land.* Oxford: Blackwell Scientific. **Bridges, E.M.** 1987: *Surveying derelict land: the ecology and reclamation of derelict and degraded land.* Oxford: Clarendon Press. **Hebbink, A.J.** 1999: Reclamation, polders. In Alexander, D.E. and Fairbridge, R.W. (eds), *Encyclopedia of environmental science.* Dordrecht: Kluwer, 367–369. **Schaller, F.W. and Sutton, P.** 1978: *Reclamation of drastically disturbed lands.* Madison, WI: American Society of Agronomy. **Wild, A. (ed.)** 1988: *Russell's soil conditions and plant growth,* 11th edn. Harlow: Longman Scientific and Technical.

land restoration

Land restoration in the strict sense may be described as the process of bringing back disused or CONTAMINATED LAND to a pre-existing LANDUSE. Whereas LAND RECLAMATION is the general term for bringing back to use, 'land restoration' implies full reconditioning and 'land rehabilitation' implies only partial success.

RESTORATION may be for either 'hard' or 'soft' uses. *Hard use* includes uses where people are in close daily contact with a site, such as children playing, or where there is a high dependence on garden produce grown on the site; *soft use* is where land is restored to amenity areas, public open spaces and playing fields, where contact is less intensive. Plans for site restoration would place considerable emphasis upon the after use of the site. Thus a restoration for agricultural use would imply a complete cleansing of the site to provide healthy conditions for plants and animals. The restored land would maximise gently sloping surfaces and a topsoil would be carefully replaced that was freely draining and composed of loam, sandy loam, sandy clay loam or silt loam. Special care should be taken to avoid loss of soil by erosion and graded waterways should be provided for the safe disposal of excess water. Similar qualifications are imposed where land is restored to forestry.

Restoration of derelict land and contaminated land for housing raises many problems as human beings are in close contact with the ground around their houses. It is essential that all toxic materials are removed and that EMISSIONS of potentially dangerous gases are eliminated. Where houses have been built upon contaminated land (Lekerkerke, The Netherlands) or even near industrial toxic waste dumps (Love Canal, USA), human health suffers, especially that of children, and expensive remedial measures become necessary. If an industrial after use is planned, it may be that hazardous materials can be surrounded by barriers below car parks or roadways and sealed by tarmacadam. Where land is restored to amenity use, less rigorous standards can be applied as human contact is not so intense; however, the well-being of the NATURAL ENVIRONMENT must still be considered.   *EMB*

Bradshaw, A.D. 1998: Land reclamation. In Calow, R. (ed.), *The encyclopedia of ecology and environmental management.* Oxford: Blackwell Science, 394–396. **Bridges, E.M.** 1987: *Surveying derelict land.* Oxford: Clarendon Press. **Paignen, B., Goldman, L.R., Highland, J.H.** *et al.* 1985: Prevalence of health problems in children living near Love Canal. *Hazardous Waste and Hazardous Materials* 2, 23–43. **Schuuring, C.** 1981: Dutch dumps. *Nature* 289, 340. **Smith, M.A. (ed.)** 1985: *Contaminated land: treatment and reclamation.* New York: Plenum.

land subsidence

The sinking or foundering of an area of the Earth's surface. Tectonic effects, such as downwarping, are normally excluded (see TECTONIC SUBSIDENCE). Land subsidence, or *ground subsidence*, can be the result of natural processes (such as the thawing of PERMAFROST, solution in KARST landscapes and the desiccation of *peatlands*), but the term is more widely applied to the results of human activities (such as the extraction of groundwater, oil or natural gas and the mining of coal, ores or salt). Hence subsidence can be either an ANTHROPOGENIC or NATURAL HAZARD.   *JAM*

Johnson, A.L. (ed.) 1991: Land subsidence. *International Association of Scientific Hydrology (IAHS) Publication* 200, 1–690.

land transformation

Changes in LAND COVER and LANDUSE CHANGE: the former involves the physical state of the land, changes of which may be caused by NATURAL and/or human agency; the latter involves changes in human use of the land.   *JAM*

Richards, J.F. 1991: Land transformation. In Turner II, B.L., Clark, W.C., Kates, R.W. *et al.* (eds), *The Earth as transformed by human action.* Cambridge: Cambridge University Press, 163–173. **Turner II, B.L. and Meyer, W.B.** 1994: Global land-use and land-cover change: an overview. In Meyer, W.B. and Turner II, B.L. (eds), *Changes in land use and land cover.* Cambridge: Cambridge University Press, 3–10.

land uplift

Land elevation occurs on a variety of spatial and temporal scales. Uplift may arise from tectonic OROGENESIS (e.g. collision and rifting of tectonic plates) and *tectono-isostasy* (e.g. unloading of crust by erosion or melting ice). ENVIRONMENTAL CHANGES that may be triggered are interlinked and largely determined by the nature of uplift. RAISED SHORELINES may result from lowering RELATIVE SEA LEVEL. River basin adjustment to falling BASE LEVEL may lead to enhanced EROSION, incision and RIVER TERRACE formation.

It is argued that land uplift may lead to local, regional and global CLIMATIC CHANGE, due to: OROGRAPHIC effects (such as RAIN SHADOW or WIND SHADOW); major perturbations of the ocean and GENERAL CIRCULATION OF THE ATMOSPHERE; and reduced atmospheric CARBON DIOXIDE levels resulting from enhanced WEATHERING and EROSION RATES. Tectonic uplift around the North Atlantic and in the Himalayas during the CENOZOIC may be an important driving force in global cooling at the onset of the late Cenozoic ICE AGE and the development of MONSOON climates in the Northern Hemisphere.   *MHD*

[See also HIMALAYAN UPLIFT, ISOSTASY, TECTONICS]

An, S.Z., Wang, S.M., Wu, X.H. *et al.* 1999: Eolian evidence from the Chinese Loess Plateau: the onset of the Late Cenozoic Great Glaciation in the Northern Hemisphere and Qinghai–Xizang Plateau uplift forcing. *Science in China Series* D42, 258–271. **Raymo, M.E.** 1994: The initiation of Northern Hemisphere glaciation. *Annual Review of Earth and Planetary Sciences* 22, 353–383. **Raymo, M.E. and Ruddiman, W.F.** 1992: Tectonic forcing of late Cenozoic climate *Nature* 359,117–122.

landfill

A method of waste disposal, known as *sanitary landfill* in North America, that involves dumping above or below ground. Disposal of domestic and industrial waste on land usually takes place at designated sites, the purpose of which is to contain the wastes without contaminating the surrounding environment. Older landfill sites were uncontrolled and so may contain hazardous substances, but environmental legislation during the past 20 years in most European and other industrialised countries has segregated HAZARDOUS WASTES from relatively harmless materials for disposal at designated facilities, where they can be effectively made harmless.

Accepted procedure is that ordinary DOMESTIC WASTE should be dumped in compartments, preferably on a puddled clay floor, in layers that are level and no more than 2.5 m deep. Each layer should be compacted and covered as soon as possible with inert material or subsoil. The capping of the landfill should comprise a metre of soil material. Co-disposal of industrial wastes and domestic waste is not considered to be a satisfactory means of dispersion of hazardous materials. INDUSTRIAL WASTE more often than not includes toxic materials (see TOXICANT, TOXIN) and these have to be treated or consigned to lined landfill sites, the aim of which is to contain the toxicity within the site. The lining of these landfill sites is thick polypropylene material and a suitable capping should be in place to limit ingress of rainwater. LEACHATE from both domestic and special landfill sites should be collected and treated before release into the environment. Emission of METHANE and other gases also takes place to the atmosphere unless these are collected and burnt off, or used for local heating schemes. Where landfill sites are badly sited, such as on very permeable rocks or below the GROUNDWATER table, there will be a rapid transfer of POLLUTION into the rivers or the groundwater. Where a slowly draining, unsaturated zone lies beneath a landfill, there is attenuation of the plume of pollution with distance from the site. *EMB*

[See also WASTE MANAGEMENT]

Al-Omer, M.A., Faiq, S.Y., Kitto, F.A. and Bader, N. 1987: Impact of sanitary landfill on air quality in Baghdad. *Water, Air and Soil Pollution* 32, 55–61. **Bagchi, A.** 1994: *Design, construction and monitoring of a sanitary landfill*, 2nd edn. New York: Wiley. **Cope, C.B., Fuller, W.H. and Willetts, S.L.** 1983: *The scientific management of hazardous wastes*. Cambridge: Cambridge University Press. **Department of the Environment** 1976: *Guidelines for the preparation of a waste disposal plan* [*Waste Management Paper* 3]. London: HMSO. **Department of the Environment** 1976: *The licensing of waste disposal sites* [*Waste Management Paper* 4]. London: HMSO. **Lema, J.M., Mendez, R. and Blazquez, R.** 1988: Characteristics of landfill leachates and alternatives for their treatment: a review. *Water, Air and Soil Pollution* 40, 223–250. **Lisk, D.J.** 1991: Environmental effects of landfills. *Science of the Total Environment* 100, 415–468. **Sumner Report** 1978: *Co-operation programme of research on the behaviour of hazardous wastes in landfill sites*. London: HMSO. **Wong, M.H.** 1999: Landfill, leachates, landfill gases. In Alexander, D.E. and Fairbridge, R.W. (eds), *Encyclopedia of environmental science*. Dordrecht: Kluwer, 356–361. **Wong, M.H. and Leung, C.K.** 1989: Landfill leachates as irrigation water for tree and vegetable crops. *Waste Management Research* 7, 311–324.

landform The three-dimensional form and nature of a particular feature ranging from micro-scale (STRIATIONS) to macro-scale (mountain ranges) of the Earth's land surface. *RAS*

[See also GEOMORPHOLOGY]

landnam Danish for taking possession of the land, landnam refers to the first FOREST CLEARANCE in PREHISTORY. The first DEFORESTATIONS coincided with the first findings of CEREAL POLLEN, implying a change from MESOLITHIC hunting and gathering to NEOLITHIC farming. The term has been widely applied beyond its area of first use in northwest Europe. *BA*

Binford, M.W., Brenner, M., Whithmore T.J. *et al.* 1987: Ecosystems, paleoecology and human disturbance in subtropical and tropical America. *Quaternary Science Reviews* 6, 115–128. **Buckland, P.C., Dugmore, A.J., Perry, D.W.** *et al.* 1991: Holt in Eyjafjallsveit, Iceland: a palaeoecological study of the impact of landnam. *Acta Archaeologica* 61, 252–271. **Dodson, J. (ed.)** 1992: *The native lands: prehistory and environmental change in Australia and the Southwest Pacific*. Harlow: Longman. **Iversen, J.** 1941: Landnam i Danmarks Stenalder (Landnam in Denmark's Stone Age). *Danmarks Geologiske Undersøgelse* II Raekke 66, 1–68.

landscape In the scientific and environmental sense, the landscape is a spatial concept relating to the interacting complex of systems on and close to the Earth's surface, including parts of the ATMOSPHERE, BIOSPHERE, HYDROSPHERE, LITHOSPHERE and PEDOSPHERE. The NATURAL LANDSCAPE may be differentiated from the CULTURAL LANDSCAPE, the latter encompassing the modifications and creations of human activities. *Farmscape, townscape* and *wildscape* may also be recognised as components of the landscape, depending on whether the LANDUSE is predominantly rural, urban or 'unproductive', respectively. *JAM*

[See also LANDSCAPE SCIENCE, LANDSCAPE UNITS]

Aston, M. 1985: *Interpreting the landscape: landscape archaeology and local history*. London: Routledge. **Atkins, P., Simmons, I. and Roberts, B.** 1998: *People, land and time: an historical introduction to the relations between landscape, culture and environment*. London: Arnold. **Chambers, F.M. (ed.)** 1993: *Climatic change and human impact on the landscape*. London: Chapman and Hall. **Head, L.** 2000: *Cultural landscapes and environmental change*. London: Arnold. **Isachenko, A.G.** 1977: L.S. Berg's landscape: geographical ideas, their origins and their present significance. *Soviet Geography* 18, 13–18. **Larkham, P.** 1999: Townscape conservation. In Pacione, M. (ed.), *Applied geography: principles and practice*. London: Routledge, 333–343. **Muir, R.** 1999: *Approaches to landscape*. London: Macmillan Press. **Rossignol, J.A. and Wandsnider, L. (eds)** 1992: *Space, time and archaeological landscapes*. New York: Plenum. **Wagstaff, J.M. (ed.)** 1987: *Landscape and culture: geographical and archaeological perspectives*. Oxford: Blackwell.

landscape archaeology An approach to archaeology that emphasises the topographic setting and environmental characteristics of the archaeological site, including the ways in which such sites were perceived by people in the past. *JAM*

[See also GEOARCHAEOLOGY, SITE-CATCHMENT ANALYSIS]

Aston, M. and Rowley, T. 1974: *Landscape archaeology: an introduction to fieldwork techniques on post-Roman landscapes*. Newton Abbott: David and Charles. **Eveson, P. and Williamson, T. (eds)** 1998: *The archaeology of landscape*. Manchester: Manchester University Press.

landscape architecture The modification of landscapes to make them more aesthetically pleasing, enjoyable or useful. Early examples are seen in the layout of extensive gardens and parklands around stately homes in England. Modern landscape architecture is an integral part of designing the built environment: the broader environmental context in the design of buildings, highways and golf courses. *JAM*

landscape ecology The ECOLOGY and management of distinct areas up to regional scale. There are different schools of landscape ecology, including: (1) visual and spatial arrangements of landscape elements and the ecological and cultural mechanisms that result in ecological change at a landscape scale; (2) the study of the form, structure, function and evolution of the visual aspects of landscapes, the attributes and spatial arrangements of attributes in landscapes, and the landscape as an ecosystem. Landscape ecology has both a strong theoretical basis and an applied aspect used in such areas as planning and natural resource management. European studies of regional geography and vegetation science led to the use of the term by Troll in AD 1939. *IFS*

[See also AUTECOLOGY, COMMUNITY ECOLOGY, ECOSYSTEM CONCEPT, GEOECOLOGY, LANDSCAPE GEOCHEMISTRY, LANDSCAPE SCIENCE, NATURAL AREAS CONCEPT]

Forman, R.T.T. and Godron, M. 1986: *Landscape ecology*. Chichester: Wiley. **Naveh, Z. and Lieberman, A.S.** 1990: *Landscape ecology: theory and application*. Berlin: Springer. **Turner, M.G. and Gardner, R.H. (eds)** 1991: *Quantitative* methods in landscape ecology: the analysis and interpretation of landscape heterogeneity. New York: Springer.

landscape evaluation

landscape evaluation Quantitative or semi-quantitative evaluation of the 'qualities' of landscape for planned development or CONSERVATION purposes. Landscape evaluation is primarily concerned with the visual, aesthetic, cultural and HERITAGE values of landscape rather than the ecological aspects, such as biodiversity, rarity and complexity. *JAM*

[See also LAND EVALUATION]

Brabyn, L. 1996: Landscape classification using GIS and national digital databases. *Landscape Research* 21, 277–287.
Burton, R. 1999: Landscape evaluation. In Pacione, M. (ed.), *Applied geography: principles and practice*. London: Routledge, 236–245.

landscape evolution

landscape evolution The nature and speed by which a LANDSCAPE changes through time. The knowledge and understanding of how landscapes evolve over a GEOLOGICAL TIMESCALE under different climatic and geological boundary conditions formed the main objective of GEOMORPHOLOGY until the 1960s and it remains an important aim, particularly in continental Europe. In the English-speaking world, timescales of study shortened considerably from the 1950s (in the USA) and from the late 1960s (in the UK) with the increased emphasis on process measurement and attempts to relate form to process; and in long-term studies an increased emphasis on late QUATERNARY and HOLOCENE landscape development.

Studies up to the 1960s sought to reconstruct the history of landscape development (or DENUDATION CHRONOLOGY) of parts of the Earth's surface using morphological (and increasingly, later, sedimentological) evidence in the current landscape. The timescale encompassed by such evidence (and ultimately the age of the landscape) largely determined the timescale covered by such studies. Thus, in Europe the timescales involved generally went back to the TERTIARY, but studies in parts of Africa and Australia encompassed much longer timescales. Morphological evidence used in denudation chronological studies included details of drainage patterns (such as orientation of river valleys, elbows of RIVER CAPTURE, longitudinal profiles, KNICKPOINTS, GORGES and *misfit streams*), the heights and extents of EROSION SURFACES and dry cols in scarps. In Britain, one of the reasons for the abandonment of such studies was the absence of means of dating and hence proving or disproving the chronological schemes that were proposed. In continental Europe, in contrast, the availability of TEPHRA (distinctive marker horizons associated with particular major volcanic eruptions) provided a means of RELATIVE-AGE DATING during the Tertiary and this partly explains the continued dominance of long-term geomorphology in Germany into the 1980s.

The first 60 years of the twentieth century were dominated by models of landscape evolution seeking to explain how initially high-relief terrain was progressively reduced during periods of crustal stability to produce plains (e.g. ETCHPLAINS, PENEPLAINS or PEDIPLAINS), before renewed OROGENY produced new high-relief terrain. The most influential scheme was that of W.M. Davis, who described a 'normal' cycle (DAVISIAN CYCLE OF EROSION) operating in climatic environments dominated by fluvial activity, but

with *glacial* and *arid cycles* applying in terrain dominated by ice and wind action respectively. A *periglacial cycle* was added by Peltier (1950), who also envisaged that the relative importance of different processes of WEATHERING and EROSION would vary with annual temperature and precipitation within the 'normal' cycle area. The alternative schemes of Walther Penck and Lester King differed from that of Davis in terms of the SLOPE EVOLUTION MODELS involved. The main problem with the cyclical concept, however, was the assumption that short periods of orogeny or crustal instability (i.e. landscape construction) alternated with long periods of crustal stability, in which landscapes could evolve under the influence of climate and lithology and inherited structure.

Schumm (1963) demonstrated this assumption to be false and that a more realistic scenario is that significant mountain building (often exceeding erosion) is not confined to short periods, but is characteristic of landscapes most of the time. Likewise, assumptions of a single climate or a narrow range of climates operating unchanged on a land surface throughout a cycle (or even for long periods during a cycle) have had to be rejected, even for the inner tropics, with the emergence of evidence in the latter part of the twentieth century of relatively frequent and large-scale climatic change over most of the Earth's surface.

Although landscape evolution is influenced by the interplay of a range of factors, notably climate and climatic history, lithology, structure and the history of earth movements (uplift, subsidence, tilt, folding, etc.), geomorphologists have tended to approach the issue by placing one of the factors in a primary position and treating the other factors as subsidiary, thus leading to CLIMATIC GEOMORPHOLOGY, *lithological geomorphology* and *structural geomorphology* respectively. Büdel (1963) made an early conceptual attempt to broaden such a view in proposing CLIMATOGENETIC GEOMORPHOLOGY, in which climate was seen to be operating within a long-term wider framework incorporating the other factors to produce generations of relief development.

In climatic geomorphology, many attempts have been made to define morphogenetic regions or MORPHOCLIMATIC ZONES on the basis of climatic and ecological controls over geomorphic processes and landforms and some of these have associated particular types of landscape evolution with particular climates or groups of climates (e.g. Peltier's periglacial cycle). These early attempts, however, were largely based upon a mixture of deduction about process on the basis of climatic parameters and simple (and often unwarranted) linkages between landforms and landscape and current process. This is demonstrated most strikingly by the various attempts to link INSELBERG-and-plain landscapes to climate, with different workers linking its development at different times to semi-arid, seasonal tropical and humid tropical conditions. More recent studies of inselbergs have acknowledged that the timescales required for development of such gross features of the landscape are much longer than the period of operation of any single climate or any period of crustal stability. Theories incorporating alternating deep weathering and STRIPPING PHASES linked to either a combination of sequences of different climates or tectonic changes have thus become more prevalent. More recent climatic geomorphology has thus tended to focus on the influence of climatic factors and climate-linked processes on current and past landforms, landscape development and rates of erosion in the context of a frequently changing climate. It has also focused more on the influence of particular elements of climate, the magnitude-frequency of climatic events, and the relative roles of EXTREME EVENTS and more frequent events, rather than on crude climatic mean values.

Equilibrium concepts within long-term landscape evolution, which became important in the late 1960s and 1970s, have also had to be revised with the realisation that landscapes are subject to more frequent climatic change and geological disruption than previously thought. Landscapes are probably in the process of adjusting to changed conditions (a state of disequilibrium) far more of the time than they are in a state of adjustment to a particular climate. Also, it is arguable that far more of a landscape and many of its individual landforms are the product of changes in climate rather than of the individual climates themselves. The shortening of timescales of interest – and in particular the growth in interest and focus on the impacts of human activities – has reinforced the importance of concepts of landscapes in transition (or adjustment) and therefore often in disequilibrium or non-equilibrium rather than equilibrium. Also important is the concept that there is a hierarchy of adjustment times of different components of a landscape. In rivers, bed configuration and channel width and depth may respond to changed discharge or sediment transport conditions almost immediately and meander wavelengths within a century, but longitudinal profile gradient and concavity may take 1000–10 000 years to respond. Similarly, evidence from stream networks developed on volcanic centres of contrasting age in the eastern Caribbean demonstrates how the speed of adjustment (RELAXATION TIME) of basin shape, bifurcation ratios and stream length ratios from the initial radial volcanic values to more typical values of dendritic networks falls with basin order, with higher-order basins and ratios retaining atypical values for much longer than first- or second-order basins.

Thus attention is increasingly focused on how, by how much and how quickly landscapes and processes respond to changes in climate, geological movements and human disturbance. Brunsden (1980) considered that geomorphic time can be divided into (a) REACTION TIMES (times for a landscape or individual landscape components to react to a change in conditions), (b) relaxation times (times for a landscape or component landforms to attain a new characteristic equilibrium state) and (c) *characteristic form times* (times over which those new states may be expected to persist). *RPDW*

[See also EQUILIBRIUM CONCEPTS IN GEOMORPHOLOGICAL AND LANDSCAPE CONTEXTS, MAGNITUDE-FREQUENCY CONCEPTS]

Brunsden, D. 1980: Applicable models of long term landform evolution. *Zeitschrift für Geomorphologie Supplementband* 36, 16–26. **Büdel, J.** 1948: Das System der klimatischen Geomorphologie. *Verhandlungen Deutscher Geographie* 27, 65–100. **Büdel, J.** 1963: Klimagenetische geomorphologie. *Geographische Rundschau* 15, 269–285. **Chorley, R.J., Schumm, S.A. and Sugden, D.E.** 1984: *Geomorphology*. London: Methuen. **Davis, W.M.** 1899: The geographical cycle. *Geographical Journal* 14, 481–504. **Hack, J.T.** 1960:

Interpretation of erosional topography in humid temperate regions. *American Journal of Science* 258A, 80–97. **King, L.C.** 1953: Canons of landscape evolution. *Bulletin of the Geological Society of America* 64, 721–752. **Knighton, D.A.** 1998: *Fluvial forms and process*. London: Arnold. **Peltier, L.C.** 1950: The geographic cycle in periglacial regions as it is related to climatic geomorphology. *Annals of the Association of American Geographers* 40, 214–236. **Penck, W.** 1924: *Die morphologische Analyse*. Stuttgart: Geographische Abhandlungen. **Schumm, S.A.** 1963: The disparity between present rates of denudation and orogeny. *United States Geological Survey Professional Paper* 454-H, 13pp. **Tricart, J. and Cailleux, J.** 1965: *Introduction à la géomorphologie climatique*. Paris: SEDES. **Walsh, R.P.D.** 1996: Drainage density and network evolution in the humid tropics: evidence from the Seychelles and the Windward Islands. *Zeitschrift für Geomorphologie Supplementband* 103. 1–23.

landscape geochemistry An approach to LANDSCAPE description and evolution developed from within the GEOSCIENCES in the former Soviet Union, notably by B.B. Polynov and A.I. Perel'man. Central ideas include the importance of geochemical and biological WEATHERING during LANDSCAPE EVOLUTION controlled by the position of the WATER TABLE. *JAM*

[See also BIOGEOCHEMISTRY, LANDSCAPE SCIENCE]

Fortescue, J.A.C. 1980: *Environmental geochemistry: a holistic approach*. Berlin: Springer. **Perel'man, A.I.** 1966: *Landscape geochemistry*. Moscow: Vysshaya Shkola; Geological Survey of Canada, Translation 676, 1972. **Snytko, V.A., Semenov, Yu.M. and Davydova, N.D.** 1981: A landscape-geochemical evaluation of geosystems for purposes of rational nature management. *Soviet Geography* 22, 569–578.

landscape management Applied LANDSCAPE ECOLOGY, or the integrated management of whole LANDSCAPE UNITS. *JAM*

Vink, A.P.A. 1983: *Landscape ecology and land use*. London: Longman.

landscape mosaic The mosaic of patches (woods, fields, ponds, rock outcrops, houses), corridors (roads, hedgerows, rivers) and matrices (background ecosystems or landuse types) that form landscapes. *RJH*

[See also LANDSCAPE ECOLOGY, PATCH DYNAMICS]

Remmert, H. (ed.) 1991: *The mosaic-cycle concept of ecosystems*. Berlin: Springer.

landscape science The study of LANDSCAPE in the scientific sense. Whether considered as a part of ECOLOGY or PHYSICAL GEOGRAPHY (LANDSCAPE ECOLOGY or GEOECOLOGY) or of the GEOSCIENCES (LANDSCAPE GEOCHEMISTRY), it is characterised by a HOLISTIC APPROACH to the interacting environmental processes at the surface of the Earth. *JAM*

Fortescue, J. 1996: Guidelines for a 'systematic landscape geoscience'. In Berger, A.R. and Iams, W.J. *Geoindicators: assessing rapid environmental changes in earth systems*. Rotterdam: A.A. Balkema, 351–364. **Isachenko, A.G.** 1973: *Principles of landscape science and physical-geographical regionalisation* [translated from Russian]. Melbourne: Melbourne University Press. **Kupfer, J.A.** 1995: Landscape ecology and biogeography. *Progress in Physical Geography* 19, 18–34. **Zonneveld, I.S.** 1979: *Land evaluation and landscape science*. Enschede: Enschede International Training Centre.

landscape sensitivity The magnitude of the response of a landscape to change in an external ENVIRON-MENTAL FACTOR or DISTURBANCE. High sensitivity implies a large response to relatively small natural disturbances and/or human impacts. *JAM*

[See also RESILIENCE, RESPONSE TIME]

Brundsen, D. and Thornes, J. 1979: Landscape sensitivity and change. *Transactions of the Institute of British Geographers NS 4*, 463–484.

landscape units Various systems of units have been proposed, mostly with an underlying hierarchical structure, for use in the description and investigation of the LANDSCAPE MOSAIC. Several based on morphology or PHYSIOGRAPHY are summarised in the Table. *JAM*

[See also LAND, LANDSCAPE, LANDSCAPE GEOCHEMISTRY, LANDSCAPE SCIENCE]

Fenneman, N.M. 1916: Physiographic divisions of the United States. *Annals of the Association of American Geographers 6*, 19–98. **Huggett, R.J.** 1995: *Geoecology: an evolutionary approach*. London: Routledge. **Linton, D.L.** 1949: The delimitation of morphological regions. *Transactions of the Institute of British Geographers 14*, 86–87. **Whittlesey, D.** 1954: The regional concept and the regional method. In James, P.E. and Jones, C.F. (eds), *American geography, inventory and prospect*. Syracuse, NY: Syracuse University Press, 19–68.

landslide Although the term landslide is in popular usage, a variety of definitions and classifications exist. Because few of these are clear and unambiguous, there are conflicting applications of this and other terms. Strictly speaking a landslide is a type of MASS MOVEMENT in which FAILURE occurs on a distinct zone of sliding (a *shear plane*) and the displaced material moves with uniform velocity throughout its mass. This excludes *fall*, *topple*, *flow* and *creep*. However, with the exception of creep, these latter mechanisms are frequently included in schemes of landslide classification, resulting in more broadly based definitions in which process is not inferred. Thus, a landslide is often defined as a perceptible down-slope displacement of rock or REGOLITH under the influence of gravity. As such, the term encompasses most forms of mass movement. This type of definition can be justified on the grounds that processes initiating movement are often complex and difficult to identify and that debris often undergoes transformation during movement, for example from slide to flow or AVALANCHE, depending on water content and degree of debris break-up.

Falls are free-fall movements of material from steep slopes or cliffs; *topples* involve a pivoting action at the base

of the FAILURE; and *flows* occur when the displaced units of material move as viscous substances in which air or water are significant components. *Slides* are subdivided into *rotational* and *translational* depending on the form of the shear plane; the former involves a curved (concave-upwards) shear surface, whilst the latter has a planar slip face roughly parallel to the ground surface. *Complex landslides* are a combination of two or more of these movements acting either simultaneously in different parts of the feature or sequentially down-slope. Many landslides begin as slides but become flows in their terminal zones.

Several factors influence landslide activity. A triggering factor may be recognised, but seldom can a landslide be attributed to a single cause. Important factors may include vegetation, SEISMICITY, water content, WEATHERING, climate and human impact. The probability of land sliding changes in response to CLIMATIC CHANGE and increasing levels of human activity, and it is important to recognise potential events. *PW*

[See also DEBRIS FLOW, LANDSLIDE FREQUENCY, LANDSLIDE HAZARD, RUNOUT DISTANCE, STURZSTROM]

Alexandrowicz, S.W. and Alexandrowicz, Z. 1999: Recurrent Holocene landslides: a case study of the Krynica landslide in the Polish Carpathians. *The Holocene 9*, 91–99. **Brunsden, D. and Prior, D.B.** (eds), 1984: *Slope instability*. Chichester: Wiley. **Crozier, M.J.** 1986: *Landslides: causes, consequences and environment*. London: Croom Helm. **Dikau, R., Brunsden, D., Schrott, L. and Ibsen, M-L.** (eds), 1996: *Landslide recognition: identification, movement and causes*. Chichester: Wiley. **Matthews, J.A., Brunsden, D., Frenzel, B., Gläser, B. and Weiss, M.M.** (eds), 1997: *Rapid mass movement as a source of climatic evidence for the Holocene* [Paläoklimaforschung.Vol. 19]. Stuttgart: Gustav Fischer. **Selby, M.J.** 1993: *Hillslope materials and processes*. Oxford: Oxford University Press.

landslide frequency The total geomorphic effect of an erosional process or event is determined partly by how often it occurs. Landslide frequency is therefore an important consideration in estimating the degree of risk involved and is fundamental to CLASSIFICATION of stability conditions. FREQUENCY is usually stated in terms of the PROBABILITY of recurrence – expressed as a probability that an event will occur in a stated number of years. Unfortunately, the detailed history of landslide activity in a given area is rarely well enough known for assessments of *recurrence interval* to be made by direct means. The factors that influence landslide frequency in the short term are those that vary substantially, such as MICROCLIMATE, soil moisture conditions and LANDUSE changes. Over longer

landscape units *Scales and terminology of landscape units (after Huggett 1995)*

Scale	Approximate area (km²)	Fenneman (1916)	Linton (1949)	Whittlesey (1954)
Micro (small)	$< 10^0$	–	Site	–
Meso (medium)	$10^0–10^1$	–	–	–
	$10^1–10^2$	–	Stow	Locality
	$10^2–10^3$	District	Tract	District
	$10^3–10^4$	Section	Section	–
Macro (large)	$10^4–10^5$	Province	Province	Province
	$10^5–10^6$	Major division	Major division	Realm
Mega (very large)	$> 10^6$	–	Continent	–

time periods WEATHERING and CLIMATIC CHANGE become significant determinants of frequency. *PW*

[See also LANDSLIDE, LANDSLIDE HAZARD]

Crozier, M.J. 1984: Field assessment of slope instability. In: Brunsden, D. and Prior, D.B. (eds), *Slope instability*. Chichester: Wiley, 103–142. **Selby, M.J.** 1993: *Hillslope materials and processes*. Oxford: Oxford University Press. **Van Asch, T.W.J.** 1998: The temporal activity of landslides and its climatological signals. In Matthews, J.A., Brunsden, D., Frenzel, B. *et al.* (eds), *Rapid mass movement as a source of climatic evidence for the Holocene*. Stuttgart: Gustav Fischer, 7–16.

landslide hazard A LANDSLIDE is only considered a HAZARD if people, property and/or utilities are at risk of injury, damage or loss. Recognition of a landslide hazard is not confined to the more rapid and violent events; relatively slow movements of SOIL and rock may cause significant disruption to the societies affected. The landslide scale, the terrain in which it occurs and the antecedent and prevailing climatic conditions are important factors influencing the level of impact. Increasingly, the landslide hazard is seen as being as much a human-induced as a natural one. In some areas modification of the landsurface has decreased slope stability and allowed natural events (e.g. rainfall) to trigger FAILURE. It is now a requirement in many areas for urban or economic development to be preceded by *landslide hazard mapping*. This entails investigation and evaluation of slope stability, a statement concerning the severity of the hazard and recommendations to reduce the hazard. A LAHAR (a volcanic mudflow) is a particularly hazardous landslide: a *cold lahar* results from heavy rains during an eruption, whereas a *hot lahar* results from the emptying of a crater lake. *PW*

[See also LANDSLIDE, LANDSLIDE FREQUENCY]

Hansen, A. 1984: Landslide hazard analysis. In Brunsden, D. and Prior, D.B. (eds), *Slope instability*. Chichester: Wiley, 523–602.

landsystem A large-scale grouping of FACIES associations and SEDIMENT–LANDFORM associations, together with broadly contemporaneously produced erosional forms. At the lowest level are *land elements* or individual landforms; at the intermediate level are land FACETS or groups of land elements; landsystems are composites of linked land facets. The term can be applied to a range of geomorphological systems, including glacial environments. For example, a landscape comprising areas of ice-scoured bedrock, subglacial TILLS, DRUMLINS, FLUTED MORAINE and ESKERS belongs to the *subglacial landsystem*. *JS*

[See also GLACIAL LANDFORMS, GLACIAL SEDIMENTS]

Boulton, G.S. and Eyles, N. 1979: Sedimentation by valley glaciers: a model and genetic classification. In Schlüchter, C. (ed.), *Moraines and varves*. Rotterdam: Balkema, 11–23. **Boulton, G.S. and Paul, M.A.** 1976: The influence of genetic processes on some geotechnical properties of tills. *Journal of Engineering Geology* 9, 159–194. **Eyles, N.** 1983: Glacial geology: a landsystem approach. In Eyles, N. (ed.), *Glacial geology*. Oxford: Pergamon, 1–18.

landuse A term applied to the 'human employment of the land' (Meyer and Turner, 1994). Landuse has at times been an overriding term that included LAND COVER, but in a strict sense is now applied only to human applications of the land. While it is entirely possible for the two to be associated on a one-to-one category basis, it is also feasible for landuse to extend over a variety of land covers. For example, a new development site may incorporate former built-up land and extend into green-field areas. *TF*

Meyer, W.B. and Turner, B.L. (eds) 1994: *Changes in land use and land cover: a global perspective*. Cambridge: Cambridge University Press. **Rhind, D. and Hudson, R.** 1980: *Land use*. London: Methuen.

landuse capability An approach to LAND EVALUATION for general use is called *land capability* or landuse capability. In this system, land is placed in classes depending upon the severity of the limitations on its use. The limitations used in the system include: erosion hazard, excess water, soil limitations in the rooting zone and climatic limitations. Thus, in the UK, *class I land* has no limitations and can be employed for any form of landuse, whereas *class VIII land* has several limitations and may have only the capability for supporting wildlife and watershed protection. *EMB*

Bibby, J.S. and Mackney, D. 1969: *Land use capability classification* [*Technical Monograph* 1]. Harpenden: Soil Survey of Great Britain. **Klingebiel, A.A. and Montgomery, P.H.** 1958: *Land-capability classification* [*USDA Agriculture Handbook* 210]. Washington, DC: United States Department of Agriculture.

landuse change Changes in human use of the land are caused largely by a variety of social, economic and political factors. For most of the time, the long history of landuse change involved impacts on vegetation and soil as, first, humans occupied most of the Earth's surface and, second, use of biotic resources was intensified. The degree and rate of change of this transformation has varied greatly in different regions. Agricultural DEFORESTATION, for example, began first in the Near East and East Asia, affected North Africa and Central Europe at 7000–8000 radiocarbon years ago, but affected the Americas much later (see Figure).

The net loss of forested area since pre-agricultural times is about 8 million km², of which more than 75% has been cleared since the end of the seventeenth century. Over the last three centuries 19% of the forests remaining in AD 1700 were removed: 8% of the world's grasslands and pastures were removed over the same period while there was a > 400% increase in the area of cropland. This agricultural expansion and related processes, such as LAND DRAINAGE and IRRIGATION of grassland, has for the most part accelerated under European political and economic control. In Europe itself, a common pattern of intensifying landuse developed around the major cities: intensive market gardening (HORTICULTURE) pushed outwards extensive cereal growing and livestock rearing. Shorter-term landuse changes have been many and varied. In England and Wales since the World War II, for example, there has been loss of ANCIENT WOODLAND and SEMI-NATURAL VEGETATION (heathland, wetlands, dunes and bracken), while the total area of woodland has increased as a result of AFFORESTATION (mainly coniferous plantations). Most recently, the area of agricultural land has decreased as a result of SET-ASIDE SCHEMES and changing economic conditions.

Although replacement of forests by agricultural systems and subsequent landuse changes throughout history have been viewed as 'land improvement', they have led in many

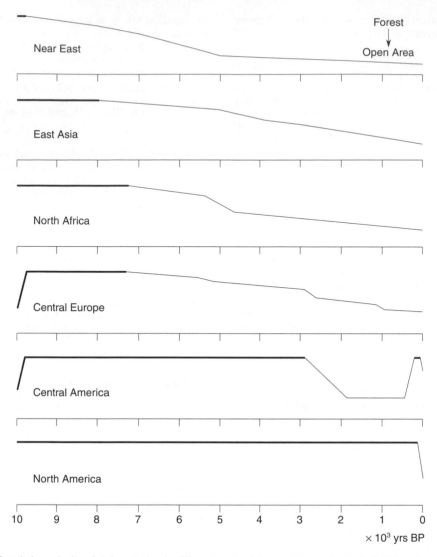

landuse change *Trends in agricultural deforestation in different parts of the world over the last 10 000 radiocarbon years. The thin line indicates human-induced deforestation (after Berglund, 1994)*

cases to SOIL DEGRADATION. Where land has remained productive for a long period with increasing yields this has only been possible with large inputs of human energy and materials. Landuse change accelerated and the direct and indirect environmental impacts increased in scale and complexity with AGRICULTURAL INTENSIFICATION and INDUSTRIALISATION. Almost all the world's lands are now used to some extent and landuse change is a major aspect of GLOBAL ENVIRONMENTAL CHANGE. Cumulative effects, such as reduced BIODIVERSITY and modification of the CARBON SEQUESTRATION capacity of the BIOSPHERE, have become global in extent. The historical conversion of forest to agricultural and urban landuse has resulted in release of carbon dioxide comparable in amount to that released by FOSSIL FUELS. Only since the mid-twentieth century has the rate of carbon release from fossil fuels exceeded that from land transformation.   *JAM*

[See also AGRICULTURAL HISTORY, AGRICULTURAL IMPACTS, AGRICULTURAL REVOLUTION, DEFORESTATION, FOREST CLEARANCE, GREEN REVOLUTION, HUMAN IMPACT ON ENVIRONMENT]

Bell, M. 1983: Valley sediments as evidence of prehistoric landuse on the South Downs. *Proceedings of the Prehistoric Society* 49, 119–150. **Berglund, B.E.** 1994: Methods for quantifying prehistoric deforestation. In Frenzel, B., Andersen, S.T., Berglund, B.E. and Gläser, B. (eds), *Evaluation of land surfaces cleared from forests in the Roman Iron Age and the time of migrating German tribes based on regional pollen diagrams* [*Paläoklimaforschung*. Vol. 12]. Stuttgart: Gustav Fischer, 5–11. **Bolin, B.** 1977: Changes in land biota and their importance for the carbon cycle. *Science* 196, 613–615. **Mannion, A.M.** 1995: *Agriculture and environmental change: temporal and spatial dimensions.* Chichester: Wiley. **Meyer, W.B. and Turner II, B.L. (eds)** 1994: Changes in land use and land cover: a global perspective. Cambridge: Cambridge University Press. **Renberg, I., Korsman, T. and Birks, H.J.B.** 1993: Prehistoric increases in the pH of acid-sensitive Swedish lakes caused by land-use changes. *Nature* 362, 824–826. **Richards, J.F.** 1990: Land transformation. In Turner II, B.L., Clark, W.C., Kates, R.W. *et al.* (eds), *The Earth as trans-*

formed by human action. Cambridge: Cambridge University Press, 163–178. **Thomas Jr, W.L. (ed.)** 1956: *Man's role in changing the face of the Earth.* Chicago, IL: University of Chicago Press. **Walker, B., Steffen, W., Canadell, J. and Ingram, J. (eds)** 1999: *The terrestrial biosphere and global change: implications for natural and managed ecosystems.* Cambridge: Cambridge University Press. **Walker, D. and Singh, G.** 1994: Earliest palynological records of human impact on the world's vegetation. In Chambers, F.M. (ed.), *Climate change and human impact on the landscape.* London: Chapman and Hall, 101–108. **Watson, R.T., Noble, I.R., Bolin, B.** *et al.* **(eds)** 2000: *Land use, land-use change, and forestry: special report of the Intergovernmental panel on climate change.* Cambridge: Cambridge University Press. **Wolman, M.G. and Fournier, F.G.A. (eds)** 1987: *Land transformation in agriculture.* Chichester: Wiley.

landuse impacts on hydrology

Changes in water, solute and sediment movement through catchments, produced deliberately or inadvertently by shifts in LANDUSE. Impacts have been evaluated at many timescales, using before-and-after, paired catchment, palaeoenvironmental reconstruction or modelling approaches. Disentangling the effects of landuse shift from CLIMATIC CHANGE impacts is challenging, however, especially given cyclical landuse changes and complex hydrological RESPONSE TIMES. Typical foci include: URBANISATION IMPACTS ON HYDROLOGY; the effects of artificial LAND DRAINAGE; AFFORESTATION; DEFORESTATION and shifts to intensive arable cultivation; and palaeohydrological effects of long-term agricultural change. Hydrological effects of landuse change partly depend on the location with respect to drainage networks and hydrologically sensitive areas. Impacts of field drainage include faster runoff response, higher flood peaks and frequencies, increased erosion rates and sediment yields and reduced solute fluxes. Afforestation can reduce basin water and sediment yields and modulate flows, though pre-afforestation ditching may enhance stream sediment loads and flood responses initially. Deforestation leads to enhanced erosion rates and flooding downstream. Shifts to intensive arable cultivation methods, if associated with bare soils stripped of protective surface vegetation, with regular ploughing and with application of fertilisers and pesticides, may result in higher flood risk, soil erosion rates and pollutant loads (e.g. nitrates). *DML*

Brown, A.G. and Quine, T.A. 1999: *Fluvial processes and environmental change.* Chichester: Wiley. **Newson, M.D.** 1997: *Land, water and development,* 2nd edn. London: Routledge, 70–80. **Pizzuto, J.E.** 1997: Combining geological and statistical methods with numerical modelling to evaluate the impacts of land use changes on rivers. In Wang, S.Y., Langendoen, E.J. and Shields Jr, F.D. (eds), *Management of landscapes disturbed by channel incision.* Oxford, MS: University of Mississippi Press, 647–652.

landuse impacts on soils

Interruption of natural cycling of nutrients; cultivation reduces organic matter content and fertility; soil is laid bare for erosion; irrigation may raise the WATER TABLE and cause SALINISATION in arid regions; crop removal reduces the content of plant nutrients; roads and buildings mean biological use of soil is ended. Positive impacts include increasing fertility and organic matter content; reduction of acidity by liming; drainage lowers the water table and helps with the removal of salts from the land in arid areas. However, increasing population in many parts of the developing world forces farmers onto MARGINAL AREAS for subsistence. With greater production risks in these areas, SOIL DEGRADATION is encouraged. *EMB*

[See also AGRICULTURAL IMPACT ON SOILS, HUMAN IMPACT ON SOIL, SOIL EROSION]

Bridges, E.M. and de Bakker, H. 1998: Soil as an artifact: human impact on the soil resource. *Land* 1,197–215.

lapilli

PYROCLASTIC fragments between 2 and 64 mm in GRAIN-SIZE. The name is from the Italian 'lapillus', meaning little stone. *GO*

[See also ASH, TEPHRA]

lapse rate

The variation of an atmospheric variable with height in the ATMOSPHERE; most commonly the temperature lapse rate. In the TROPOSPHERE, temperature decreases with height (average rate of 6.5°C km^{-1}): this is a positive lapse rate with cooler air above. A negative lapse rate (air warmer above) is an INVERSION, whilst an isothermal layer occurs when there is no change of temperature with height. A RADIOSONDE measures the *environmental* (or actual) *lapse rate.* This is compared with two theoretical lapse rates to assess the stability of an atmospheric layer: the *dry adiabatic lapse rate* (9.8°C km^{-1}; the rate at which a non-saturated pocket of air will cool if it is forced to rise) and the *saturated adiabatic lapse rate.* The latter depends on humidity but varies from 5°C km^{-1} in the lower troposphere to about the dry adiabatic lapse rate in the dry upper troposphere. An '*adiabatic*' *process* is one in which heat does not enter or leave a system. The STRATOSPHERE is either isothermal or has a negative lapse rate. *BDG*

Linacre, E. and Geerts, B. 1997: *Climates and weather explained.* London: Routledge. **McIlveen, R.** 1992: *Fundamentals of weather and climate.* London: Chapman and Hall.

large igneous province (LIP)

A thick pile of LAVA, commonly dominated by FLOOD BASALTS, erupted over a relatively short interval of geological time (probably 1 to 2 million years), forming a LAVA PLATEAU in the ocean or a *Continental Flood Basalt Province* (*CFBP*) on land. Examples include the flood basalts of the Deccan Traps of India and the Kerguelen Plateau in the southern Indian Ocean, both formed during the Cretaceous period. A large igneous province may form when the head of a MANTLE PLUME first reaches the base of the LITHOSPHERE. The plume tail is then responsible for HOTSPOT volcanoes, which remain active over a long period. Volcanic activity related to the formation of large igneous provinces may have implications for climatic change (see VOLCANIC IMPACTS ON CLIMATE). *GO*

Arndt, N. 2000: Hot heads and cold tails. *Nature* 407, 458–461. **Barley, M.E., Pickard, A.L. and Sylvester, P.J.** 1997: Emplacement of a large igneous province as a possible cause of banded iron formation 2.45 billion years ago. *Nature* 385, 55–58. **Todal, A. and Edholm, O.** 1998: Continental margin off Western India and Deccan Large Igneous Province. *Marine Geophysical Researches* 20, 273–291.

laser

A device (*light amplification by stimulated emission of radiation*) that emits a high-intensity beam by exciting electronic, ionic or molecular transitions to higher energy levels and then allows these to fall to lower energy levels. Generally, the beam is coherent (in phase), has narrow spectral width (or is monochromatic) and is highly directional. *PJS*

laser-induced fluorescence (LIF)

Fluorescence is the emission of radiation (usually at optical wavelengths of the ELECTROMAGNETIC SPECTRUM) from objects. Molecules within the object are excited into higher energy levels using a LASER and, when the molecules relax back to the lower energy level, they emit radiation (fluoresce). This phenomenon may be used in REMOTE SENSING to detect the presence of certain materials using LIDAR.   *PJS*

Last Glacial Maximum (LGM)

The oxygen isotope signal in DEEP-SEA SEDIMENT CORES and ICE CORES indicates that the maximum ICE SHEET volume occurred during oxygen ISOTOPIC STAGE (OIS) 2, towards the end of the LAST GLACIATION, and coincided with an estimated global SEA LEVEL of c. −120 m OD (*ordnance datum* – reference level of the Ordnance Survey defined as the mean level of the sea at Newlyn, Cornwall, England) and low atmospheric CO_2 levels below 190 ppmv. The LGM is approximately bracketed by HEINRICH EVENTS 2 and 3 in the North Atlantic, dated to 21–19 ka BP (^{14}C years) and 28–27 ka BP, respectively. According to recent sea-level data, the Last Glacial Maximum terminated at $19\,000 \pm 250$ years ago with a rapid decrease in ice volume by about 10% within a few hundred years. These events apparently coincide with low SEA-SURFACE TEMPERATURES and a reorganisation in the THERMOHALINE CIRCULATION.

At this time ice sheets spread across large areas in the high latitudes and were accompanied by significant modification of ecological, geomorphological and circulation systems. For example, the LGM is associated with increased LOESS mobilisation and deposition on the Loess Plateau of China and PERIGLACIATION and AEOLIAN activity in northwest Europe (see Figure). Low levels of pluvial lakes in tropical Africa and Australasia suggest regional ARIDITY in low latitudes during the LGM, although the question of whether Equatorial regions such as the Amazon Basin experienced aridity is still disputed, with recent studies suggesting continued high precipitation, at least in some areas.

In North America during the LGM the confluent LAURENTIAN and *Cordilleran ice sheet* covered c. 16 million km^2, resulting in the developments of ICE-DAMMED LAKES (e.g. Lake Agassiz) and superfloods beyond the ice margins (see CHANNELLED SCABLANDS)). The Laurentian ice sheet split the JET STREAM, enhancing the cooling of the continent and extensive periglaciation. LAKE-LEVEL VARIATIONS indicate that displacement of storm tracks (see STORMS) by the jet stream caused wetter and stormier conditions in the southwestern United States.

The CLIMAP and COHMAP projects originally placed the LGM at 18 ka BP, implying a global synchroneity in maximum ice-sheet extent. It is now clear that there were marked geographical variations in the timing of maximum ice-sheet development and phases of readvance were non-synchronous. For example, in North America the maximum extent of the Cordilleran ice sheet appears to have been c. 15 ka BP. In contrast, the maximum southern extent of the Laurentide ice sheet occurred between 21 and 18 ka BP, while the maximum northern extent was during OIS stage 5 in the early part of the last glaciation.   *MHD*

Bond, G. Broecker, W., Johnsen, S. *et al.* 1993: Correlations between climate records from North Atlantic sediments and Greenland ice. *Nature* 365, 143–147. **CLIMAP Project Members** 1981: Seasonal reconstructions of the Earth's surface at the last glacial maximum. *Geological Society of America Map and Chart Series*, MC-36. **COHMAP Members** 1988: Climatic changes of the last 18,000 years: observation and model simulations. *Science* 421, 1043–1052. **Colinvaux, P.A., De Oliveira, P.E. and Bush, M.B.** 2000: Amazonian and neotropical plant communities on glacial time-scales: the failure of the aridity and refuge hypotheses. *Quaternary Science Reviews* 19, 141–169. **McIntyre, A., Kipp, N.G., Bé, A.** *et al.* 1976: Glacial North Atlantic 18,000 years ago: a CLIMAP reconstruction. *Geological Society of America Memoir* 145, 43–47. **Prentice, I.C., Sykes, M.T., Lautenschlager, M.** *et al.* 1993: Modelling global vegetation patterns and terrestrial carbon storage at the Last Glacial Maximum. *Global Ecology and Biogeography Letters* 3, 67–76. **Wilson, R.C.L., Drury, S.A. and Chapman, J.L.** 2000: *The great Ice Age: climate change and life*. London: Routledge. **Wright Jr, H.E., Kutzbach, J.E., Webb III, T.** *et al.* **(eds)** 1993: *Global climates since the Last Glacial Maximum*. Minnesota, MN: University of Minneapolis Press. **Yokoyama, Y., Lambeck, K., De Deckker, P.** *et al.* 2000: Timing of the Last Glacial Maximum from observed sea-level minima. *Nature* 406, 713–716.

Last Glaciation

The last glaciation was the DEVENSIAN in Britain, otherwise known as the *Weichselian* in Northern Europe, the *Würm* in the Alps, the *Valdai* in Russia and the *Wisconsinan* in North America. It comprised oxygen ISOTOPIC STAGES 4–2 (*ca.* 74–11.5 ka calendar years BP).   *DH*

[See also QUATERNARY TIMESCALE, LAST GLACIAL MAXIMUM, PLENIGLACIAL]

Last Interglacial

The penultimate interglacial, i.e. before the present interglacial (the HOLOCENE), also known as the *Ipswichian* (Britain), *Eemian* (northern Europe) and *Sangamon* (North America) and correlated with Oxygen Isotope Substage 5e. The transition from the SAALIAN glaciation to the last interglacial is dated to c. 128 ka BP and the end to c.115 ka BP on the *SPECMAP* timescale (see SPECTRAL MAPPING PROJECT and QUATERNARY TIMESCALE). This chronology is, however, debated. Both terrestrial and marine proxy data indicate that during the Last Interglacial, mid and high latitudes were generally at least 2°C warmer than at present, enabling species such as hippopotamus and lion to colonise Britain. A range of recent evidence suggests that the interglacial climate was more unstable than previously thought. The interglacial climate may have experienced one or more abrupt SUB-MILANKOVITCH climatic oscillations, with maximum temperature decreases of 10–14°C for periods lasting 70 to 750 years. The existence of these events is however still uncertain and relies heavily on ICE CORE data and some BIOSTRATIGRAPHICAL evidence.   *MHD*

[See also DANSGAARD–OESCHGER EVENTS , RAPID ENVIRONMENTAL CHANGE]

Aalbersberg, G. and Litt, T. 1998: Multiproxy climate reconstructions for the Eemian and Early Weichselian. *Journal of Quaternary Science* 13, 367–390. **Dansgaard, W., Johnsen, S.J., Clausen, H.B.** *et al.* 1993: Evidence for general instability of past climate from a 250-kyr ice-core record. *Nature* 364, 218–220. **GRIP Members** 1993: Climate instability during the last interglacial period recorded in the GRIP ice core. *Nature* 364, 203–207. **McManus, J.F., Bond, G.C., Broecker, W.S.** *et al.* 1994: High-resolution climatic records from the North Atlantic during the last interglacial. *Nature* 371, 326–329. **Petit, J.R., Jouzel, J., Raynaud, D.** *et al.* 1999: Climate and atmospheric

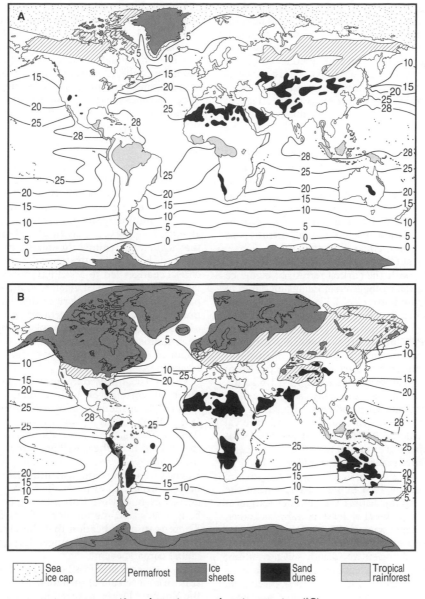

Last Glacial Maximum *The Earth (A) today and (B) at the Last Glacial Maximum, c. 18 ka* BP: *on land the distribution of ice sheets, perennially frozen ground (permafrost), active desert sand dunes and tropical rainforest are shown; sea surface temperatures are shown in the oceans. Note that the reduced area of tropical rainforest in (B) is probably overestimated (after Wilson et al., 2000)*

history of the past 420,000 years from the Vostok ice core, Antarctica. *Nature* 399, 429–436.

Late Glacial environmental change

The last GLACIAL–INTERGLACIAL transition, the *Late Weichselian* or *Late Devensian Lateglacial*, the boundary between marine oxygen ISOTOPIC STAGES 1 and 2 (TERMINATION I), often just referred to as the *Late Glacial*, marks the transition between the PLEISTOCENE epoch and the HOLOCENE. Although traditionally dated to about 13–10 ka BP (^{14}C years), Late Glacial environmental changes span a wider time period, both in *calendrical years* (15–11.5 ka BP) and in ^{14}C years. The period following Termination I was characterised by a series of rapid environmental changes between STADIAL and INTERSTADIAL climatic regimes, accompanied by a major reorganisation of atmospheric and ocean circulation, proxy evidence for which is best found in the HIGH-RESOLUTION RECONSTRUCTIONS based on, for example, ICE CORES, MARINE SEDIMENT CORES and LACUSTRINE SEDIMENTS. A relatively robust temporal framework is provided by, for example, RADIOCARBON DATING, VARVE CHRONOLOGIES, DENDROCHRONOLOGY and annually accumulating ice layers.

In Northern Europe the Late Glacial comprises four main biozones (see BIOSTRATIGRAPHY) that have subsequently been defined as CHRONOZONES and climatostratigraphic subdivisions (^{14}C years): BØLLING INTERSTADIAL (13–12 ka BP); OLDER DRYAS STADIAL (12–11.8 ka BP); ALLERØD INTERSTADIAL (11.8–11 ka BP); and YOUNGER DRYAS STADIAL (11–10 ka BP). The Younger Dryas ends with rapid warming into the Holocene interglacial (10 ka BP to present). The TIME TRANSGRESSIVE nature of environmental changes means that this system is rather unsatisfactory and a more informal division is often adopted: initial post LAST GLACIAL MAXIMUM warming (prior to 13 ka BP); LATE GLACIAL INTERSTADIAL (LGI: 13–11 ka BP); and Younger Dryas (11–10 ka BP).

Terrestrial evidence for Late Glacial environmental change includes: biological remains (see, for example, BEETLE ANALYSIS, POLLEN ANALYSIS and PLANT MACROFOSSIL ANALYSIS); AEOLIAN SEDIMENTS (e.g. COVERSAND and LOESS); LACUSTRINE SEDIMENTS (e.g. DIATOMITE, VARVES and RHYTHMITES); PEAT and various types of landforms (e.g. ice directional indicators, ICE-MARGIN INDICATORS and PERIGLACIAL LANDFORMS); and PALAEOSOLS.

The timing of DEGLACIATION was strongly dependent on latitude and altitude, with the earliest terrestrial evidence for climatic amelioration in the Iberian Peninsula dated to c. 15 ka BP, compared to c. 14.5 ka BP in Britain. Deglaciation was generally more widespread and marked by 13 ka BP. Although ICE SHEETS are thought to have been largely absent from Britain, Denmark and the southern Baltic Basin by 13 ka BP, there were periods of diachronous ice advances and retreats during the whole period of the Late Glacial. For the period preceding 13–13.5 ka BP, PROXY CLIMATIC INDICATORS generally suggest POLAR DESERT and steppe TUNDRA environments, with a cold and continental climate regime for northern and central Europe, including widespread PERIGLACIAL activity facilitated by winter temperatures of –20 to –25°C.

Fossil beetle evidence indicates very rapid warming at c. 13 ka BP, estimated at 2.6–7.2°C per century, with the thermal maximum of the LGI being reached within the first 500 years. During the early LGI, summer temperatures in much of Europe were at least as warm as today, but with more continental winters. This led to the replacement of open-ground communities by scrub vegetation and ultimately probably open woodland, soil formation, restricted SOLIFLUCTION and a shift towards organic sediment accumulation in lakes. Increased river discharge, resulting from enhanced MELTWATER, and slope stabilisation led to a transition from BRAIDING to MEANDER stream patterns, *channel downcutting* and RIVER TERRACE formation.

From c. 12.5 ka BP there was a step-wise climatic downturn (*revertence episodes*) and an abrupt fall at c. 12 ka BP (equivalent to the Older Dryas) to maximum summer temperature in northwest Europe of less than 11–15°C, leading to increased aeolian sand activity, minerogenic inwash into lakes, vegetation disturbance and glacial re-advances.

The onset of the Younger Dryas saw marked climatic cooling in Europe from c. 11 ka BP, with a 4–8°C fall in temperature from the LGI maximum, mean July temperatures below 10°C and extremely continental winters (i.e. at least c. –20°C). This was accompanied by landscape instability and evidence of SEA ICE off the west coast of Ireland. The Younger Dryas was characterised by a pronounced periglacial regime throughout northwest Europe, with extensive discontinuous PERMAFROST, ice-sheet

expansion, LOCAL GLACIATION (development of mountain glaciers and local icecaps) and ICE-DAMMED LAKES (e.g. the Baltic Ice Lake). The brevity of the Younger Dryas, together with a less severe climate than the LGM, restricted the build up of ice. Wood and heathland communities of the LGI were replaced by TUNDRA or alpine scrub, while THERMOPHILOUS beetles were replaced by Arctic *stenothermic* species. Renewed aeolian activity in the lowlands (e.g. coversand deposition) and LAKE-LEVEL VARIATIONS suggest the later part of the Younger Dryas (10.5–10 ka BP) was perhaps markedly colder and more arid.

The timing of the Late Glacial/Younger Dryas–Holocene transition is geographically variable in Europe between c. 10.2–9.5 ka BP. Biological evidence indicates a 6–8°C increase in mean summer temperatures during the first few hundred years of the Holocene (estimated rate of change: 1.7–2.8°C per century). All proxy records suggest that a thermal regime equitable with that of today was reached by 9.5 ka BP throughout northwest Europe. These changes led to a reduction in mineral inwash into lake basins, melting of glaciers and permafrost, restricted periglaciation, increased river discharge and soil formation. Cold-adapted, open ground and heath communities of the Younger Dryas were replaced by various woodland communities, themselves also changing as species migrated from warmer REFUGIA during the early Holocene.

Late Glacial environmental changes in Europe have potential correlatives with climatic oscillations reconstructed throughout the North Atlantic region and beyond (e.g. Older Dryas ≈ Killarney ≈ Gerzensee ≈ Amphi-Atlantic Oscillation). An EVENT STRATIGRAPHY in the isotope record from the GRIP ice core has been suggested as the *stratotype section* for the Last Termination. Interstadials and stadials identified from Greenland have been correlated with fossil beetle-based temperature curves in northwest Europe, oxygen isotope signals in Swiss lakes and marine cores from the North Atlantic. There is some indication that the climate cooling during the Late Glacial (i.e. Younger Dryas) may have been a global climatic event.

Both ice core and fossil beetle evidence suggest that the shift from interstadial to stadial conditions at the onset of the Younger Dryas occurred in under a century, while climate warming at the Younger Dryas/Holocene boundary occurred within a few decades. The rapid and almost synchronous changes in ocean circulation, sea surface temperatures, snow accumulation rates, atmospheric temperatures over ice sheets and terrestrial climatic regimes during the Late Glacial are best explained by the 'open and shut door' modes of the OCEANIC POLAR FRONT. Marine sediment evidence indicates that, during the LGM, the southerly limit of the Oceanic Polar Front and winter pack ice was on the same latitude as Lisbon, Portugal (20–13 ka BP). The front migrated rapidly (estimated rate c. 2 km y^{-1}) northwards to a position off Iceland during the Late Glacial Interstadial (13–11 ka BP), resulting in a 7–9°C warming in sea surface temperatures. An equally rapid southerly shift in position to the same latitude as southwest Ireland for the duration of the Younger Dryas (11–10 ka BP), resulted in a 5–8°C decline in SEA-SURFACE TEMPERATURES during the Younger Dryas. By 10 ka BP, the Oceanic Polar Front had retreated northwards (estimated rate c. 5 km y⁻¹) to its current position between Greenland and Iceland.

Variation in the strength of the THERMOHALINE CIRCULATION has been invoked to explain the shift in the Oceanic Polar Front and resulting changes in Late Glacial climate. The onset of the Younger Dryas apparently coincides with HEINRICH EVENT 0 and there is evidence for a series of cooling episodes during the LGI corresponding to ice melting pulses. It is suggested that enormous meltwater and iceberg discharges from collapsing ice sheets may have stalled the THERMOHALINE CIRCULATION and triggered the climatic down turn of the Younger Dryas. Whether or not Heinrich events are ultimately climatically forced or result from internal ice sheet dynamics is uncertain.   *MHD*

Andersen, E. 1997: Younger Dryas research and its implications for understanding of abrupt climatic change. *Progress in Physical Geography* 21, 230–249. **Bjørk, S., Kromer, B., Johnsen, S.** *et al.* 1996: Synchronised terrestrial–atmospheric deglacial records around the North Atlantic. *Science* 274, 1155–1160. **Bjørk, S., Walker, M.J.C., Cwynar, L.C.** *et al.* 1998: An event stratigraphy for the Last Termination in the North Atlantic region based on the Greenland ice-core record: a proposal by the INTIMATE group. *Journal of Quaternary Science* 13, 283–292. **Coope, G.R. and Lemdahl, G.** 1995: Regional differences in the Lateglacial climate in northern Europe based on coleopteran analysis. *Journal of Quaternary Science* 10, 391–395. **Dansgaard, W., Johnsen, S.J., Clausen, H.B.** *et al.* 1993: Evidence for general instability of past climate from a 250-kyr ice-core record. *Nature* 364, 218–220. **Gulliksen, S., Birks, H.H., Possnert, G. and Mangerud, J.** 1998: The calendar age of the Younger-Dryas–Holocene transition at Kråkenes, western Norway. *The Holocene* 8, 249–260. **Lowe, J.J., Ammann, B., Birks, H.H.** *et al.* 1994: Climatic changes in areas adjacent to the North Atlantic during the last glacial–interglacial transition (14–9 ka BP). *Journal of Quaternary Science* 9, 185–198. **Lowe, J.J., Coope, G.R., Sheldrick, C.** *et al.* 1995: Direct comparison of UK temperatures and Greenland snow accumulation rates, 15,000–12,000 yr ago. *Journal of Quaternary Science* 10, 175–180. **Markgraf, V., Baumgartner, T.R., Bradley, J.P.** *et al.* 2000: Paleoclimate reconstruction along the Pole–Equator–Pole transect of the Americas (PEP 1). *Quaternary Science Reviews* 19, 125–140. **Walker, M.J.C.** 1995: Climatic changes in Europe during the last glacial–interglacial transition. *Quaternary International* 28, 63–76.

late-Holocene climatic deterioration

The concept of a climatic cooling trend and/or increasing wetness following the Climatic Optimum or HYPSITHERMAL of the early HOLOCENE. Terms for the time interval encompassed include *Hypothermal, Katathermal, Medithermal* and *Neoglacial*. It may be accounted for by reduced insolation receipt on Earth as predicted by MILANKOVITCH THEORY.
                                                                      JAM

[See also HOLOCENE ENVIRONMENTAL CHANGE, NEOGLACIATION]

latent heat

The energy absorbed or released when a substance changes its phase: *latent heat of vaporisation* is involved during changes between gaseous and liquid state; *latent heat of fusion* is involved during liquid–solid transformations. Latent heat is released during the change from gas to liquid and from liquid to solid states: latent heat is absorbed in overcoming intermolecular bonds during the change from solid to liquid or liquid to gas. During SUBLIMATION, the change from a solid to a gas, the latent heat released is the sum of the latent heat of fusion and the latent heat of vaporisation. In the environmental context, latent heat is particularly important in understanding EVAPORATION, CONDENSATION, freezing and thawing processes, LAPSE RATES, ENERGY BALANCES/BUDGETS and heat transfer in the atmosphere and at the Earth's surface.
                                                                      AHP/JAM

Lock, G.S.H. 1994: *Latent heat transfer: an introduction to fundamentals.* New York: Oxford University Press.

lateral accretion

The accumulation of sediment on an inclined surface, resulting in the lateral migration of that surface, in contrast to the vertical AGGRADATION of a near-horizontal surface. The term is commonly applied to the inner bank (*point bar*) of a meander in a river channel, which migrates laterally across the FLOODPLAIN. In the geological record, EPSILON CROSS-STRATIFICATION has been interpreted as evidence of lateral accretion.   *GO*

Turner, B.R. and Eriksson, K.A. 1999: Meander bend reconstruction from an Upper Mississippian muddy point bar at Possum Hollow, West Virginia, USA. In Smith, N.D. and Rogers, J. (eds), *Fluvial Sedimentology VI, International Association of Sedimentologists Special Publication* 28. Oxford: Blackwell Science, 363–379.

laterite

A formerly widely used term for a highly weathered red and often mottled subsoil, rich in secondary oxides of iron, aluminium or both, nearly devoid of bases and primary silicates and commonly containing quartz and KAOLINITE. The term has largely been replaced by the term *plinthite* or FERRICRETE, though the latter is only the indurated form of laterite. Laterite develops in seasonal tropical or SUBTROPICAL climates and is a RESIDUAL product of large-scale CHEMICAL WEATHERING, where silica and bases are leached from the parent material, creating a concentration of iron and aluminium sesquioxides. It is associated mainly with mature surfaces of low relief and occurs in three forms: (1) a soft lateritic clay at depth which hardens on exposure to air; (2) a tough indurated layer forming a HARDPAN or DURICRUST at the surface, where the superficial material has been stripped off; and (3) a horizon of nodules or lenses. *Laterisation* is a general term for the process that converts a rock to laterite. SECONDARY LATERITES tend to form in valley-bottom areas through the reprecipitation of iron that has been transported down-slope by THROUGHFLOW. Laterite crusts may play a major role in long-term LANDSCAPE development.
                                                                      MAC

[See also PLINTHIC HORIZON, PLINTHOSOL, SAPROLITE, WEATHERING PROFILE]

Bourman, R.P. 1993: Perennial problems in the study of laterite: a review. *Australian Journal of Earth Science* 40, 387–401. **McFarlane, M.J.** 1976: *Laterite and landscape.* London: Academic Press. **McFarlane, M.J.** 1991: Some sedimentary aspects of lateritic weathering profile development in the major bioclimatic zones of tropical Africa. *Journal of African Earth Sciences* 12, 267–282. **Tardy, Y.** 1992: Diversity and terminology of laterite profiles. In Martini, I.P. and Chesworth, W. (eds), *Weathering, soils and paleosols.* Amsterdam: Elsevier, 379–405. **Tardy, Y.** 1997: *Petrology of laterites and tropical soils.* Rotterdam: Balkema. **Thomas, M.F.** 1994: *Geomorphology in the tropics.* Chichester: Wiley. **Young, A.** 1976: *Tropical soils and soil survey.* Cambridge: Cambridge University Press.

Laurentian/Laurentide ice sheet The ice sheet that formed over the Canadian *shield* (see CRATON) on many occasions during the QUATERNARY. It is named after the St Lawrence River. The Laurentide ice sheet had two major ACCUMULATION AREAS, located over Keewatin and Labrador, from which it advanced. Lobes of the last (Wisconsinan) Laurentide Ice Sheet extended over the Great Lakes and into the Northern American Plains. *DIB*

[See also GLACIATION]

Clark, P.U., Licciardi, J.M., MacAyeal, D.R. and Jenson, J.W. 1996: Numerical reconstruction of a soft-bedded Laurentide ice sheet during the last glacial maximum. *Geology* 24, 679–682. **Fulton, R.J. and Prest, V.K.** 1987: The Laurentide ice sheet and its significance. *Géographie Physique et Quaternaire* 41, 181–186. **Ives, J.D., Andrews, J.T. and Barry, R.G.** 1975: The growth and decay of the Laurentide ice sheet and comparisons with Fenno-Scandia. *Naturwissenschaften* 62, 118–125.

lava The liquid component of MAGMA emitted at the surface of the Earth (or other planet) in a VOLCANIC ERUPTION. Lava properties and behaviour depend on the magma composition: *low-silica magma* has a low viscosity, forming fast, mobile LAVA FLOWS of BASALT. Hazardous activity is associated with *high-silica magma*, which is highly viscous and inhibits the release of VOLATILES. Flows are very slow and *lava domes* or plugs develop. These can collapse because of volatile pressure or SEISMICITY, leading to *Plinian* eruptions or PYROCLASTIC FLOWS. *JBH*

[See also IGNEOUS ROCKS]

Chester, D. 1993: *Volcanoes and society*. London: Edward Arnold. **Sparks, R.S.J., Young, S.R., Barclay, J.** *et al.* 1998: Magma production and growth of the lava dome of the Soufriere Hills volcano, Montserrat: November 1995 to December 1997. *Geophysical Research Letters* 25, 3421–3424.

lava dome A build-up of very viscous, high-silica LAVA around the vent of a VOLCANO during a VOLCANIC ERUPTION. The collapse of a lava dome can cause a PYROCLASTIC FLOW. *GO*

Sparks, R.S.J., Young, S.R., Barclay, J. *et al.* 1998. Magma production and growth of the lava dome of the Soufriere Hills volcano, Montserrat: November 1995 to December 1997. *Geophysical Research Letters* 25, 3421–3424.

lava flow A flow of liquid LAVA from a VOLCANIC ERUPTION at the surface of the Earth or another planet. Lava flows vary from metre-scale effusions to volumes measured in tens of cubic kilometres (see FLOOD BASALT). Flows of basalt are rarely hazardous to life, although a basaltic flow from Nyiragongo (Zaire) in AD 1977 travelling at 11 ms^{-1} overran 400 houses, killing 72 people. Flow diversion is often attempted to prevent structural damage. Calculations suggest that a 3 m wall thickness is needed to deflect each 1 m depth of lava, as attempted successfully in the AD 1983 eruption of Etna (Sicily). Blocking of the harbour on Heimay (Iceland) in AD 1973 was prevented by spraying 1200 L s^{-1} of sea water onto the 1055°C basaltic flow. Bombing of lava flows has also been attempted. Surfaces of basaltic lava flows are commonly described by the Hawaiian terms AA (blocky or rubbly) or PAHOEHOE (smooth or rope-like). *JBH*

[See also IGNEOUS ROCKS]

Chester, D. 1993: *Volcanoes and society*. London: Edward Arnold. **Lockwood, J.P. and Torgerson, F.A.** 1980: Diversion of lava flows by aerial bombardment: lessons from Mauna Loa, Hawaii. *Bulletin Volcanologique* 43, 727–741. **Williams, R.S. and Moore, J.G.** 1973: Iceland chills a lava flow. *Geotimes* 18, 14–17.

lava plateau An elevated tableland mainly constructed by FLOOD BASALTS, with intercalated TUFFS and PALAEOSOLS. The CRETACEOUS Deccan Traps in northwest India cover an area greater than 500 000 km^2 and have a volume over 1 million km^3. Lava plateaux are constructed over a geologically short time interval and may affect ATMOSPHERIC COMPOSITION through the release of VOLATILES. This has been suggested as a mechanism for MASS EXTINCTIONS such as at the K–T BOUNDARY. *JBH*

Courtillot, V., Jaupart, C., Manighetti, I. *et al.* 1999: On causal links between flood basalts and continental breakup. *Earth and Planetary Science Letters* 166, 177–195. **Khadkikar, A.S., Sant, D.A., Gogte, V. and Karanth, R.V.** 1999: The influence of Deccan volcanism on climate: insights from lacustrine intertrappean deposits, Anjar, western India. *Palaeogeography, Palaeoclimatology, Palaeoecology* 147, 141–149. **Ray, J.S. and Pande, K.** 1999: Carbonatite alkaline magmatism associated with continental flood basalts at stratigraphic boundaries: cause for mass extinctions. *Geophysical Research Letters* 26, 1917–1920.

lava tube An elongated CAVE in a solidified LAVA FLOW that formed when the surface of the flow solidified but the LAVA beneath was still molten and drained away. *GO*

Martin, J.L. and Oromi, P. 1986: An ecological study of Cueva-de-los-Roques lava tube (Tenerife, Canary Islands). *Journal of Natural History* 20, 375–388. **Sarkar, P.K., Friedman, G.M. and Karmalkar, N.** 1998: Speleothem deposits developed in caves and tunnels of Deccan-Trap basalts, Maharashtra, India. *Carbonates and Evaporites* 13, 132–135.

law Scientific or *natural laws* (there are other kinds) are universally accepted generalisations believed to be true, i.e. they represent the expectations of scientists over broad realms. They are frequently falsifiable, but may refer to non-existent things (e.g. a frictionless body). A philosophically, universally acceptable, definition does not exist. At their indefinite lower limit, laws degenerate into HYPOTHESES and at their upper indefinite limit grow into THEORY (a web of laws) for many, but not all, scientists. *CET*

leachate A solution containing elements and compounds in solution (or suspension) that are taken up on its passage through a substance, such as a soil or a soil sample. Rainwater leaches soluble substances from soils and, in the laboratory, various solutions are used to extract elements from soil or sediment samples. *EMB*

leaching The process in which soluble materials are dissolved and carried in solution out of the soil profile. After soluble salts, gypsum and carbonates have been removed, leaching continues as base cations are removed from the CLAY–HUMUS COMPLEX and replaced by hydrogen ions. *EMB*

[See also ACIDIFICATION, SOIL FORMING PROCESSES]

Buol, S.W., Hole, F.D. and McCracken, R.J. 1988: *Soil genesis and classification*. Ames, IA: Iowa State University Press.

lead (Pb) A TRACE ELEMENT, HEAVY METAL and POLLUTANT, lead has been locally important in the environment and as a factor in human health since the onset of lead mining in the Roman period. There has been major release of lead into the environment and extensive pollution even of remote regions since the INDUSTRIAL REVOLUTION. Soils have been polluted by lead from industrial activity and sewage sludge, and there is long-term release from lead mine spoil into GROUNDWATER. A further escalation of ATMOSPHERIC POLLUTION by lead followed the widespread use of lead additives in petrol, but atmospheric concentrations fell slightly by the AD 1990s with the introduction of unleaded petrol. Human health has also been directly and indirectly affected by release of lead into the environment through industrial activity.

The normal range of lead in soil is between 30 and 100 mg kg^{-1} Pb. Uncontaminated soils in remote areas may contain less than 30 mg kg^{-1} Pb. In areas rich in metalliferous deposits, soils may contain up to 300 mg kg^{-1}. Lead links firmly with organic substances and mineral silicates in soils and so has a long residence time compared with other metal pollutants. It accumulates mainly in the organic-rich surface horizons and there is little evidence that it is leached out naturally. Alongside major roads soils are contaminated with lead emitted from vehicle engines, and soils in the gardens of old houses usually contain lead from paints. Despite the value of plant nutrients in sewage sludge, the presence of lead limits its disposal to farmland and forests. *EMB*

[See also METAL POLLUTION, LEAD-210 DATING, LEAD ISOTOPES, POLLUTION HISTORY]

Boutron, C.F., Gorlach, U., Candelone, J.P. *et al.* 1991: Decrease in anthropogenic lead, cadmium and zinc in Greenland snows since the late 1960s. *Nature* 353, 153–156. **Kramers, J.D., Reese, S. and Van Der Knaap, W.O.** 1998: History of atmospheric lead deposition since 12,370 ^{14}C yr BP from a peat bog, Jura Mountains, Switzerland. *Science* 281, 1635–1640. **Needleman, H.L. and Bellinger, D.** 1991: The health effects of low level exposure to lead. *Annual Review of Public Health* 12, 111–140. **Nriagu, J.O. (ed.)** 1978: *The biogeochemistry of lead in the environment.* Amsterdam: Elsevier.

lead and lag

Leads and lags are central to understanding the processes and causal mechanisms of environmental change. In a chain of events, causes occur before effects. If, therefore, it can be demonstrated that a particular phenomenon consistently leads another, then it is closer in time to the cause and may itself be a potential cause or connected in some way to the cause. Similarly, if a phenomenon lags behind another, it can be ruled out as a potential causal factor. For example, based on cross-dated oxygen isotope records from the last glacial episode in Greenland and Antarctic ice cores, it has been suggested that rapid warming events in the Southern Hemisphere lead those in the Northern Hemisphere by a little more than a millennium. This, however, has been disputed, which emphasises the need for accurate DATING TECHNIQUES and HIGH-RESOLUTION RECONSTRUCTIONS. *JAM*

Blunier, T., Chapellaz, J., Schwander, J. *et al.* 1998: Asynchrony of Antarctica and Greenland climate change during the last glacial period. *Nature* 394, 739–743. **White, J.W.C. and Steig, E.J.** 1998: Timing is everything in a game of two hemispheres. *Nature* 394, 717.

lead-210 dating

A RADIOMETRIC DATING technique used to establish chronologies for sediments aged less than about 150 years and therefore too young for RADIOCARBON DATING. Lead-210 (^{210}Pb) is present in the ATMOSPHERE as a DAUGHTER ISOTOPE of sedimentary radon. This ^{210}Pb is deposited to accumulating sediments in excess of any *in situ* ^{210}Pb and decays according to radiometric law. Reliable estimates of initial sedimentary ^{210}Pb activity plus measurements of current isotope activity are used to date the sediments.

There are two approaches to ^{210}Pb dating. The *constant initial concentration* (CIC) model is suited to homogeneous sediments with little variation in sediment accumulation rate and provides linear ^{210}Pb profiles that are fairly simple to interpret. Linear ^{210}Pb profiles are, however, uncommon as a result of changes in sediment accumulation rate, bulk density or other factors. The *constant rate of supply* (CRS) model is therefore normally preferred, in which initial ^{210}Pb activity does not necessarily decline monotonically with depth.

The interpretation of ^{210}Pb profiles may be complicated by their distortion or remobilisation through HYDROLOGICAL change within both peat and LACUSTRINE SEDIMENTS. The ^{210}Pb record in PEAT is particularly susceptible to degradation and the accuracy of *peatland* ^{210}Pb inventories is often significantly lower than those of lakes. The CRS ^{210}Pb dating model can often account for such discrepancies, although the ^{210}Pb profile in lake sediments may be further affected by SEDIMENT FOCUSING or LAKE-LEVEL VARIATIONS. Ambiguities in ^{210}Pb dating can nonetheless often be resolved by independent radiometric dating techniques (e.g. CAESIUM-137 or AMERICIUM-241 DATING). *DZR*

Appleby, P.G. and Oldfield, F. 1992: Application of lead-210 to sedimentation studies. In Ivanovich, M. and Harmon, R.S., eds, *Uranium series disequilibrium.* Oxford: Oxford University Press, 731–778. **Oldfield, F., Richardson, N. and Appleby, P.G.** 1995: Radiometric dating (^{210}Pb, ^{137}Cs, ^{241}Am) of recent ombrotrophic peat accumulation and evidence for changes in mass balance. *The Holocene* 5, 141–148. **Varvas, M. and Punning, J.M.** 1993: Use of the ^{210}Pb method in studies of the development and human impact history of some Estonian lakes. *The Holocene* 3, 34–44.

lead isotopes

Three *radiogenic isotopes* of lead, ^{206}Pb, ^{207}Pb and ^{208}Pb, occur naturally in the environment as DAUGHTER ISOTOPES of ^{238}U, ^{235}U and ^{232}Th, respectively. The partitioning of U/Th and Pb between different minerals and variations in the age of their formation give rise to systematic differences in the ratio of radiogenic isotopes and non-radiogenic ^{204}Pb between ore bodies and rock units throughout the GEOSPHERE. Lead ISOTOPE RATIO measurements have been used to record the sources of lead fluxes into the ATMOSPHERE, the OCEANS, SOILS and organisms. Particular attention has been given to the identification of ANTHROPOGENIC (ore body) lead in PROXY EVIDENCE for past atmospheric composition from the analysis of polar ICE CORES, LACUSTRINE SEDIMENTS and MIRES. Lead isotope measurements of contemporary and archaeologically preserved tissue have been used to assess changing sources of human exposure. *PB*

[See also ISOTOPES AS INDICATORS OF ENVIRONMENTAL CHANGE, LEAD-210 DATING]

Faure, G. 1986: *Principles of isotope geochemistry,* 2nd edn. New York: Wiley. **Rosman, K.J.R., Chrisholm, W., Boutron, C.F.** *et al.* 1993: Isotopic evidence for the source of lead in Greenland snows since the 1960s. *Nature* 362, 333–335. **Shotyk,**

W., Weiss, D., Appleby, P.G. *et al.* 1998: History of atmospheric lead deposition since 12,370 ^{14}C yr BP from a peat bog, Jura Mountains, Switzerland. *Science* 281, 1635–1640.

leaf area index (LAI)

The total one-sided (or one half of the total all-sided) green leaf area per unit ground surface area. This is an important biological parameter of vegetation because it defines the area that interacts with SOLAR RADIATION and provides the REMOTE SENSING signal that can be measured easily. It is also a measure of the surface that is responsible for carbon absorption and exchange with the atmosphere.   *KJT*

leaf physiognomy

Leaf or *foliar physiognomy* describes the character of leaves from their general shape and form. Various classification schemes have been proposed, most notably by Raunkiaer (1934), later modified by Webb (1959). Leaf physiognomy is strongly controlled by environment and plays an important role in whole plant ADAPTATION and survival, as the size, thickness, toughness and shape of leaves can affect the rate at which plants take up carbon, exchange heat and lose water. Broad patterns of leaf physiognomy have been observed in vegetation, demonstrating that generally large leaves occur in wet, hot environments and small leaves in hot or cold dry environments. Hence, analysis of FOSSIL leaf physiognomy has emerged as an important technique in PALAEOBOTANY in the reconstruction of palaeoclimates (see PALAEOCLIMATOLOGY) and PALAEOENVIRONMENTS, based on strong correlations that have been defined between extant ANGIOSPERM leaf characteristics and present-day climatic variables.

As early as 1915, Bailey and Sinott demonstrated that, as one moves from warmer to colder climate regimes, there is a marked decrease in the proportion of tree and shrub species with entire-margined leaves. This characteristic, along with leaf size, cuticle thickness and the presence or absence of a drip tip, were the first leaf physiognomy characteristics used to reconstruct Cretaceous and Tertiary palaeoclimates from angiosperm fossil floras. Since the 1940s the method has been much refined and MULTIVARIATE ANALYSIS has been applied to leaf physiognomy and modern climatic datasets to define regression models with which to reconstruct climatic variables, such as mean annual temperatures, mean range of temperatures and mean annual precipitation of Tertiary and Cretaceous times.   *JCMc*

[See also EVAPOTRANSPIRATION, PHOTOSYNTHESIS, RESPIRATION]

Bailey, I.W. and Sinnott, E.W. 1915: The climate distribution of certain types of angiosperm leaves. *American Journal of Botany* 3, 24–39. Chaloner, W.G. and Creber, G.T. 1990: Do fossil plants give a climate signal? *Journal of the Geological Society London* 147, 343–350. Givinish, T.J. 1979: On the adaptive significance of leaf form. In Solbrig, O.T., Jain, S. and Raven, P.H. (eds), *Topics in plant population biology*. New York: Columbia University Press, 375–407. Jacobs, B.F. 1999: Estimation of rainfall variables from leaf characters in tropical Africa. *Palaeogeography, Palaeoclimatology, Palaeoecology* 145, 231–251. Raunkiaer, C. 1934: *The life-forms of plants and statistical plant geography*, translated by H. Gilbert Carter. Oxford: Oxford University Press. Webb, L.J. 1959: A physiognomic classification of Australian rain forests. *Journal of Ecology* 47, 551–570. Wolfe, J.A. 1993: A method of obtaining climatic parameters from leaf assemblages. *United States Geological Survey Bulletin* 2040, 71.

least-squares regression

A method of fitting a statistical MODEL, in the simplest case a line, to data. Parameters of the model equation are estimated so that the sum of squared differences between the model and the observations of the dependent variable is minimised. Least-squares regression may be used as the basis for the prediction of the dependent variable, or merely to describe the relationship between dependent and independent variables.   *HB*

[See also NUMERICAL ANALYSIS, REGRESSION ANALYSIS]

Leptosols

Shallow soils of all regions over hard rock or in unconsolidated, very gravelly material, mainly occurring in mountainous areas and deserts (USDA: Lithic subgroups of *Entisols, Rendolls*). It is the most extensive of the major soils of the world and subject to erosion, particularly in mountainous areas.   *EMB*

[See also RANKER, RENDZINA, WORLD REFERENCE BASE FOR SOIL RESOURCES]

Bridges, E.M. and Creutzberg, D. 1994: Leptosols and Fluvisols. *Transactions of the 15th International Congress of Soil Science* 6a, 868–872.

levée

A ridge of fine-to-coarse SEDIMENT deposited on a FLOODPLAIN or alongside a DEBRIS FLOW track as a result of channel overtopping. Levées may be created artificially for FLOOD protection.   *PW*

[See also ALLUVIAL DEPOSITS]

Libby half life

The HALF LIFE of ^{14}C, determined by W.F. Libby to be 5568 years and used in the calculation of conventional RADIOCARBON dates even though a revised figure of 5730 years has since been determined.   *PQD*

[See also RADIOCARBON DATING]

Stuiver, M. and Polach, H.A. 1977: Discussion: reporting of ^{14}C data. *Radiocarbon* 19, 355–363.

lichenometric dating

The use of lichen size or related indices of lichen growth for dating rock surfaces. The technique, also known as *lichenometry*, was first developed and applied by Roland Beschel in the 1950s in the context of dating MORAINES in the Austrian Alps. It is still most widely used in alpine and polar environments where crustose lichens, especially those of the relatively slow-growing, yellow-green *Rhizocarpon* subgenus (widely known as the *Rhizocarpon geographicum* group), commonly dominate on rock outcrops and boulders. In principle, any abundant species that colonises rock surfaces rapidly after exposure, and grows steadily in an approximately circular fashion, is potentially useful for lichenometric dating. Applications have included dating a wide range of glacial, periglacial, lacustrine and coastal landforms but the technique has also been used on archaeological structures.

The maximum diameter of the largest thallus growing on a surface has traditionally been used on the grounds that it is the oldest specimen growing under optimal environmental conditions. In order to avoid the pitfall of anomalous single thalli, the average size of several 'largest lichens' (commonly five) is often employed instead of the single largest. Other, more time-consuming approaches have, however, been developed. As older surfaces tend to be characterised by larger lichens, lichen size can be used as a method of RELATIVE-AGE DATING and, given an adequate number of surfaces of known age to establish a *lichenometric-dating curve* or numerical relationship between lichen size and surface age, CALIBRATED-AGE DATING. This constitutes the so-called *indirect approach* to lichenometric dating, which has been most effective in dating surfaces up to about 500 years old to an accuracy of up to about 10% where accurate control points are available and lichen growth rates approach 1.0 mm y^{-1}. The *direct approach*, which has been less frequently used and less successful in practice, involves the construction of a *lichen growth curve* based on direct measurement of lichen growth rates. *JAM*

Beschel, R.E. 1961: Dating rock surfaces by lichen growth and its application to glaciology and physiography (lichenometry). In Raasch, G.O. (ed.), *Geology of the Arctic*. Vol. 1. Toronto: University of Toronto Press, 1044–1062. **Bull, W.B. and Brandon, M.T.** 1998: Lichen dating of earthquake-generated regional rockfall events, Southern Alps, New Zealand. *Geological Society of America Bulletin* 110, 60–84. **Innes, J.I.** 1985: Lichenometry. *Progress in Physical Geography* 9, 187–254. **Matthews, J.A.** 1994: Lichenometric dating: a review with particular reference to 'Little Ice Age' moraines in southern Norway. In Beck, C. (ed.), *Dating in exposed and surface contexts*. Albuquerque, NM: University of New Mexico Press, 185–212. **McCarroll, D.** 1994: A new approach to lichenometry: dating single-age and diachronous surfaces. *The Holocene* 4, 383–396. **McCarthy, D.P.** 1999: A biological basis for lichenometry. *Journal of Biogeography* 26, 379–386. **Worsley, P.** 1990: Lichenometry. In Goudie, A. (ed.), *Geomorphological techniques*, 2nd edn. London: Unwin Hyman, 422–428.

LIDAR An active REMOTE SENSING system (*Light Detection and Ranging*) that utilises pulses of light, transmitted usually from an aircraft, to illuminate the ground and build a picture of the three-dimensional structure of the terrain from the strength and delay time of the backscattered light. The LIDAR can operate in both profiling and scanning modes. The principal applications of LIDAR remote sensing systems are for the profiling of water depths and the heights of tree stands. This is achieved by transmitting pulsed LASER light such that the first strong measured return is from the water surface, followed by a weaker return from the bottom of the water body. The depth is determined from the two-way travel time that the pulse is in the water. Determining the heights of tree stands is possible because the light is scattered at the top of the trees and in clearings the light is scattered from the ground surface. Estimates of forest biomass are possible given tree height and species data. A further application of LIDAR is its use in measuring the LASER-INDUCED FLUORESCENCE (LIF) properties of the Earth's surface, which can be used to discriminate different materials and has been used operationally to detect chlorophyll concentration and oil slicks. *KJT*

Maclean, G.A. and Krabill, W. 1986: Gross merchantable timber volume estimate using an airborne LIDAR system. *Canadian Journal of Remote Sensing* 12, 7–18.

life assemblage Preservation of AUTOCHTHONOUS fossils or subfossils in life orientations with, for example, hinged or plated skeletons intact (see also DEATH ASSEMBLAGE). The FOSSIL assemblage is modified from the original COMMUNITY through the effects of TAPHONOMY and the loss of soft-bodied organisms, but its close similarity is

important for palaeocommunity reconstruction. Life assemblages are commonly preserved by rapid burial in EVENT DEPOSITS generated by STORMS or SLUMPS, which prevents later disturbance of shells by burrowing (BIOTURBATION) and disarticulation. Cemented organisms such as oysters or modern reef corals may also be preserved *in situ*. Life assemblages are important for PALAEOECOLOGY and PALAEOBIOLOGY. *LC*

[See also FOSSIL LAGERSTÄTTEN]

Kidwell, S.M. 1991: Stratigraphy of shell concentrations. In Allison, P.A. and Briggs, D.E.G. (eds), *Taphonomy: releasing the data locked in the stratigraphic record*. New York: Plenum, 211–261.

life form The characteristic overall form, or morphology or physiognomy, of an organism, most widely used in relation to mature plants. A particular well known scheme, Raunkiaer's life-form classification, which is based on the position of perennating buds in relation to the ground surface, includes such categories as *phanerophytes* (trees), *chamaephytes* (dwarf schrubs) and GEOPHYTES. *JLI*

[See also GROWTH FORM]

Raunkiaer, O. 1934 *The life forms of plants and statistical plant geography*. Oxford: Clarendon Press.

life zone A subdivision of the Earth's surface based on the major climatic controls on life. The best known is the *Holdridge system*, in which life zones are defined in terms of MEAN ANNUAL AIR TEMPERATURE, PRECIPITATION and POTENTIAL EVAPOTRANSPIRATION. Life zones correspond broadly with BIOMES and VEGETATION FORMATION-TYPES, but include the whole ecosystem and are more precisely defined in climatic terms. *JAM*

Holdridge, L.R. 1967: *Life zone ecology*. San José, Costa Rica: Tropical Science Center. **Kendeigh, S.C.** 1954. History and evaluation of various concepts of plant and animal communities in North America. *Ecology* 35, 152–171.

light rings A dendrochronological feature representing years of very little summer wood production, often in temperature-sensitive trees growing at high latitudes. These are equivalent to annual TREE RINGS of low maximum-latewood density and may indicate years when summer temperature was relatively low and/or light levels were significantly reduced. They have been shown to correspond in some cases to the dates of large explosive VOLCANIC ERUPTIONS. They may also represent years of heavy insect defoliation. *KRB*

[See also DENDROCHRONOLOGY, DENSITOMETRY]

Lavoie, C. and Payette, S. 1997: Late-Holocene light-ring chronologies from sub-fossil black spruce stems in mires of sub-arctic Quebec. *The Holocene* 7, 129–137. **Szeicz, J.M.** 1996: White spruce light rings in northwestern Canada. *Arctic and Alpine Research* 28, 184–189.

lignin phenols Lignin is a major biopolymer in vascular plants, formed by random polymerisation of three main monomers: *p*-coumaryl alcohol, coniferyl alcohol and sinapyl alcohol. The phenolic compounds of lignin origin produced by oxidation of SEDIMENTS are often used to evaluate terrestrial inputs into AQUATIC ENVIRONMENTS as the ratios of syringyl, vanillyl and cinnamyl phenols can be used to differentiate between the main vascular plant types: non-woody angiosperms, non-woody gymnosperms, woody angiosperms and woody gymnosperms. VEGETATION changes resulting from CLIMATIC CHANGES since the LAST GLACIAL MAXIMUM can therefore be deduced from specific lignin phenols. The $\delta^{13}C$ variations of lignin phenols reflect the differences in their origins and enable an estimate of C-4 plant abundances (see C-3, C-4 AND CAM PLANTS) to be calculated.   *KJF*

Goñi, M.A., Ruttengerg, K.C. and Eglinton, T.I. 1997: Sources and contribution of terrigenous organic carbon to surface sediments in the Gulf of Mexico. *Nature* 389, 275–278. **Hedges, J.I. and Mann, D.C.** 1979: The characterisation of plant tissues by their cupric oxide oxidation products. *Geochimica et Cosmochimica Acta* 43, 1803–1818. **Huang, Y., Freeman, K.H., Eglinton, T.I. and Street-Perrott, F.A.** 1999: $\delta^{13}C$ analyses of individual lignin phenols in Quaternary lake sediments: A novel proxy for deciphering past terrestrial vegetation changes. *Geology* 27, 471–474.

liman   A shallow bay with a muddy bottom.   *HJW*

limestone   A SEDIMENTARY ROCK composed entirely or dominantly of CARBONATE minerals, principally calcite and dolomite. Shells of macroscopic and microscopic marine invertebrate FOSSILS form an important constituent of many limestones, and biological and chemical processes are important in understanding their origins. This contrasts with the importance of physical processes to the formation of CLASTIC sediments and other sedimentary rocks. Limestones are important indicators of shallow-water PALAEOENVIRONMENTS.   *TY*

Bathurst, R.G.C. 1975: *Carbonate sediments and their diagenesis* [*Developments in Sedimentology* 12]. Amsterdam: Elsevier. **Kindler, P. and Hearty, P.J.** 1996: Carbonate petrography as an indicator of climate and sea-level changes: new data from Bahamian Quaternary units. *Sedimentology* 43, 381–399. **Tucker, M.E. and Wright, V.P.** 1990: *Carbonate sedimentology.* Oxford: Blackwell Science. **Wilson, J.L.** 1975: *Carbonate facies in geologic history.* Berlin: Springer. **Wright, V.P. and Burchette, T.P.** 1996: Shallow-water carbonate environments. In Reading, H.G. (ed.), *Sedimentary environments: processes, facies and stratigraphy*, 3rd edn. Oxford: Blackwell Science.

limiting factors   ENVIRONMENTAL FACTORS that slow down or stop the productivity, growth or reproduction of a population. Limiting factors may be physical (such as temperature and moisture levels) or chemical (such as NUTRIENT levels). They may operate at lower and upper extremes. For example, a population will normally be subject to a lower temperature limit and to a higher temperature limit. Some limiting factors partly determine species distributions; for example, the small-leafed lime tree (*Tilia cordata*) has a northern limit in England and Scandinavia corresponding to the 19°C mean July isotherm. Limiting factors also help to shape ecosystems. Broad-leaved deciduous forests grow in humid and mesic regions of Asia where the coldest-month mean is less than 1°C and the warmest-month mean is less than 20°C. Limiting factors affect agricultural systems. Successful agriculture commonly depends on recognising limiting factors and employing techniques to accommodate or surmount them. Once a limiting factor is modified, this triggers adjustments within the ecosystem, which may lead a different environmental factor to become limiting.   *RJH*

[See also ABIOTIC, ECOSYSTEM CONCEPT]

Kennedy, A.D. 1993: Water as a limiting factor in the Antarctic terrestrial environment: a biogeographical synthesis. *Arctic and Alpine Research* 25, 308–315.

limits to growth   In 1972 the Club of Rome, an informal international group concerned for the future predicament of humankind, published a non-technical report to promote public concern (entitled *The Limits to Growth*). This report was based on *system dynamics* modelling developed at the Massachusetts Institute of Technology (MIT). The MIT 'world model' tried to predict likely future behaviour of the global environment/economy between AD 1900 and 2100, concluding that if trends in population, INDUSTRIALISATION, POLLUTION, resource depletion and food production apparent in 1972 continued unchanged, the Earth's limits to growth would be reached within 100 years with dire effects on human welfare. The report was widely attacked for being based on poor data and computation and on simplistic 'neo-Malthusian' reasoning. Nevertheless, valuable debates were prompted. In 1992 the same researchers repeated the exercise with improved data, better computers and updated software. Their conclusion was that the original warnings were valid, but that appropriate action could avert catastrophe.   *CJB*

[See also DEMOGRAPHIC CHANGE, SUSTAINABILITY]

Kahn, H., Brown, W. and Martel, L. (ed.) 1978: *The next 200 years.* London: Abacus. **Meadows, D.H., Meadows, D.L., Randers, J. and Behrens, W.W.** 1972: *The limits to growth (a report for the Club of Rome on the predicament of mankind).* New York: Universal Books. **Meadows, D.H., Meadows, D.L. and Randers, J.** 1992: *Beyond the limits: global collapse or a sustained future?* London: Earthscan.

limnic eruption   In August 1986 at least 1700 people died of asphyxiation due to a massive release of CARBON DIOXIDE gas from Lake Nyos, a CRATER lake in Cameroon, West Africa (the *Lake Nyos gas disaster*). Subsequent investigations showed that the source of the gas was the sudden decompression of gas-rich BOTTOM WATER in the stratified lake (see LAKE STRATIFICATION), leading to the catastrophic *exsolution* of CO_2 gas as a self-sustaining fountain. Although the CO_2 was of magmatic origin (see MAGMA), its release was not associated with a VOLCANIC ERUPTION, and the terms *limnic eruption* or *eruptive outgassing* have been applied to the presumed process, which could have been triggered by a LANDSLIDE, EARTHQUAKE or STORM. The eruption, driven by the exsolution of CO_2, has similarities with water-driven HYDROVOLCANIC ERUPTIONS. Measurements have shown that CO_2 concentrations are again building up in the bottom water of Lake Nyos and *controlled degassing* has been suggested as a means of reducing the NATURAL HAZARD represented by Lake Nyos and similar lakes.   *GO*

Kling, G.W., Clark, M.A., Compton, H.R. *et al.* 1987: The 1986 Lake Nyos gas disaster in Cameroon, West Africa. *Science* 236, 169–175. **Rice, A.** 2000: Rollover in volcanic crater lakes: a possible cause for Lake Nyos type disasters. *Journal of Volcanology and Geothermal Research* 97, 233–239.

limnogeology   The study of modern and ancient LACUSTRINE SEDIMENTS and lake basins from an EARTH SCIENCE perspective.   *MRT*

[See also LAKES AS INDICATORS OF ENVIRONMENTAL CHANGE]

limnology The study of lakes, ponds and other standing waters, including physical and chemical characteristics of water and sediments, the biota and the relationship of the aquatic ecosystem to the catchment. *HHB*

[See also AQUATIC ENVIRONMENT, LAKES AS INDICATORS OF ENVIRONMENTAL CHANGE, LAKE STRATIFICATION AND ZONATION, LACUSTRINE SEDIMENTS, LAKE-LEVEL VARIATIONS, PALAEOLIMNOLOGY]

Horne, A.J. and Goldman, C.R. 1994: *Limnology* , 2nd edn. New York: McGraw- Hill. **Hutchinson, G.E.** 1957: *A treatise on limnology*. Vol. 1. *Geography, physics and chemistry*. New York: Wiley. **Hutchinson, G.E.** 1967: *A treatise on limnology*. Vol. 2. *Introduction to lake biology and limnoplankton*. New York: Wiley. **Hutchinson, G.E.** 1975: *A treatise on limnology*. Vol. 3. *Limnological botany*. New York: Wiley. **Hutchinson, G.E.** 1993: *A treatise on limnology*. Vol. 4. *The zoobenthos*. New York: Wiley.

line reduction A clipping technique used in computer graphics, which aims to determine the visible portion of lines that may be partially or totally occluded by other graphical objects. These algorithms are useful for HIDDEN SURFACE REMOVAL, for example in the visualisation of DIGITAL ELEVATION MODELS. *ACF*

Foley, J.D., van Dam, A., Feiner, S.K. *et al*. 1994: *Introduction to computer graphics*. Reading, MA: Addison-Wesley.

lineage A line of evolutionary descent, either of a single evolving species or of several species descended from a common ancestor. *KJW*

linear interpolation A technique for determining the value of a function at a point, given the value of the function at two neighbouring points, by assuming that the function varies linearly in this neighbourhood. *PJS*

linear mixture modelling A technique used for spectral unmixing of optical REMOTE SENSING data. The technique assumes the spectral signatures of PIXELS containing different LAND COVER types to be made up of a weighted linear sum of the component land-cover signatures. The weights are determined directly from the relative proportions of land cover types in the pixel. Given k land cover classes covering proportions f_i in a pixel with signature vectors μ_i, the resulting expected linearly mixed signature of that pixel in the absence of sensor noise is:

$$f_1\mu_1 \ldots + f_i\mu_i + \ldots + f_k\mu_k$$

The noise caused by the sensor and the natural *variability* on the ground can be added as a vector of errors. The linear mixture model assumes that each PHOTON hitting the ground is interacting with only one land-cover type before being reflected. It is thus only an approximation to reality. A prerequisite for the model is that there are more significant PRINCIPAL COMPONENTS in the satellite data than land-cover classes. *HB*

[See also OPTICAL REMOTE SENSING INSTRUMENTS]

Bastin, L. 1997: Comparison of fuzzy c-means classification, linear mixture modelling and MLC probabilities as tools for unmixing coarse pixels. *International Journal of Remote Sensing* 18, 3629–3648.

linguistic dating RELATIVE-AGE DATING of languages using their degree of linguistic similarity and based on the principle that the longer groups of people are separated,

the more different their languages become. *Glottochronology* is a form of *absolute-age dating* of languages, which assumes a quantifiable rate of divergence between languages. *JAM*

Renfrew, C. and Bahn, P. 1996: *Archaeology: theories, methods and practice*, 2nd edn. London: Thames and Hudson.

lipids Fats (animal, vegetable oils and waxes) that are insoluble in water, but soluble in organic solvents. Lipids are an organic chemical constituent in the BIOMASS. All organisms are composed of lipids, proteins and carbohydrates. The resistant parts of organisms (e.g. waxes, membranes) are comprised of lipids and hence lipids tend to survive decomposition and are the most important constituent in the formation of PETROLEUM. Up to a third of the chemical composition of organisms may be composed of lipids and lipid-like components. *KJF*

Cranwell, P.A. 1982: Lipids of aquatic organisms as potential contributors to lacustrine sediments II. *Organic Geochemistry* 11, 513–527. **Goossens, H., Düren, R.R., de Leeuw, J.W. and Schenck, P.A.** 1989: Lipids and their mode of occurrence in bacteria and sediments II. Lipids in the sediment of a stratified, freshwater lake. *Organic Geochemistry* 14, 27–41. **Tissot, B.P. and Welte, D.H.** 1984: *Petroleum formation and occurrence*, 2nd edn. Heidelberg: Springer.

liquefaction A temporary loss of strength that may affect cohesionless granular materials (e.g. SAND) related to an increase in pore fluid pressure to a level that equals the overburden pressure so that GRAINS are no longer supported at grain contacts, but 'float' in the pore fluid. The sediment then behaves as a fluid, or *quicksand*. Localised FLUIDIZATION may develop in the upper parts of BEDS that have liquefied, as a result of the focusing of escaping pore water. Unlike fluidization, no external source of fluid is required for liquefaction and grain contacts immediately begin to be re-established, restoring strength over time spans of tens of seconds to tens of minutes. Liquefaction is most commonly reported in loosely packed sediments and SOILS of SILT to fine sand grade as a consequence of large EARTHQUAKES, giving rise to failure of foundations and triggering slope failures. Liquefaction can also be triggered by waves during STORMS, by breaking waves or by the onset of FLOOD surges. Features such as *sand volcanoes* (*sand blows*) and other SEDIMENTARY STRUCTURES developed as a consequence of liquefaction may be preserved as SOFT-SEDIMENT DEFORMATION structures and can be used as evidence of PALAEOSEISMICITY. It has been suggested that the term *sand boils* be restricted to similar features with a demonstrably aseismic origin. Such studies have suggested a potential threat from large earthquakes along the eastern seaboard of North America, although there is no historical record of large earthquakes there. *GO*

[See also LIQUIDIZATION]

Amick, D. and Gelinas, R. 1991: The search for evidence of large prehistoric earthquakes along the Atlantic seaboard. *Science* 251, 655–658. **Hibsch, C., Alvarado, A., Yepes, H.** *et al*. 1997: Holocene liquefaction and soft-sediment deformation in Quito (Ecuador): a paleoseismic history recorded in lacustrine sediments. *Journal of Geodynamics* 24, 259–280. **Ishihara, K.** 1993: Liquefaction and flow failure during earthquakes. *Géotechnique* 43, 351–415. **Li, Y., Craven, J., Schweig, E.S. and Obermeier, S.F.** 1996: Sand boils induced by the 1993 Mississippi River flood: could they one day be misinterpreted as earthquake-induced liquefaction? *Geology* 24, 171–174. **Mörner,**

N.-A. and Tröften, P.-E. 1993: Palaeoseismotectonics in glaciated cratonal Sweden. *Zeitschrift für Geomorphologie NF Suppl.* 94, 107–117. **Obermeier, S.F.** 1996: Use of liquefaction-induced features for paleoseismic analysis – an overview of how seismic liquefaction features can be distinguished from other features and how their regional distribution and properties of source sediment can be used to infer the location and strength of Holocene paleo-earthquakes. *Engineering Geology* 44, 1–76.

liquid limit The minimum moisture content at which a soil passes from the plastic to liquid state and can then flow under its own weight. A sediment with a low liquid limit is likely to be susceptible to down-slope movement by GELIFLUCTION. *PW*

[See also FAILURE, PLASTIC LIMIT, SOLIFLUCTION]

Ballantyne, C.K. and Harris, C. 1994: *The periglaciation of Great Britain*. Cambridge: Cambridge University Press, 134–135.

liquidization Any mechanism that causes a sudden reduction in strength of a SEDIMENT, including LIQUEFACTION and FLUIDIZATION in cohesionless materials (e.g. SAND) and THIXOTROPY in cohesive MUD (see also SENSITIVE CLAYS). While liquidized, a sediment can be deformed by weak stresses and may develop SOFT-SEDIMENT DEFORMATION structures or induce slope failure (as considered in *slope stability analysis*). *GO*

Allen, J.R.L. 1982: *Sedimentary structures: their character and physical basis*. Amsterdam: Elsevier.

lithalsa A mesoscale FROST MOUND composed entirely of minerogenic material and SEGREGATED ICE, found in areas of PERMAFROST. Unlike PALSAS, to which they are genetically related, peat plays no part in their origin or development. *JAM*

Harris, S.A. 1993: Palsa-like mounds developed in a mineral substrate, Fox Lake, Yukon Territory. *Proceedings of the 6th International Conference on Permafrost, Beijing, China*. Vol. 1. Wushan Guangzhou: South China University Press, 238–243. **Pissart, A.** 2000: Remnants of lithalsas of the Hautes Fagnes, Belgium: a summary of present-day knowledge. *Permafrost and Periglacial Processes* 11, 327–355.

lithic Made of rock. In GEOLOGY, the term is used in the description of CLAST composition in SEDIMENTS and SEDIMENTARY ROCKS to distinguish rock fragments (*lithic clasts*) from MINERAL fragments. In ARCHAEOLOGY, lithics are stone materials worked by prehistoric humans, including both the useful ARTEFACTS (especially *core tools* but also *flake tools*) and workshop debris or *debitage* (made up of unused flakes). *GO/JAM*

[See also PROVENANCE, SANDSTONE]

Ashmore, W. and Sharer, R.J. 1996: *Discovering our past: a brief introduction to archaeology*. Mountain View, CA: Mayfield.

lithification The change from unconsolidated SEDIMENT to coherent SEDIMENTARY ROCK during DIAGENESIS, brought about principally by CEMENTATION and COMPACTION. INDURATION, meaning hardening, is commonly used in an equivalent sense. *TY*

lithofacies A FACIES in SEDIMENTS or SEDIMENTARY ROCKS defined primarily using characteristics of LITHOLOGY, which may include integral faunal characteristics (e.g. graded sandstone lithofacies, cross-bedded crinoidal grainstone lithofacies). Interpretations of descriptive lithofacies may indicate process, environment or PALAEOENVIRONMENT as, for example, TURBIDITE lithofacies or REEF flank lithofacies, although it is often advisable to avoid such genetic usage. *LC*

[See also BIOFACIES, ICHNOFACIES]

lithology The general characteristics of a rock, particularly those observable in field observations and hand-specimen, including rock type, composition and TEXTURE. *TY*

[See also LITHOFACIES]

lithosphere The rigid outer layer of the Earth, comprising the CRUST and the MANTLE to a depth of about 100 km. The lithosphere consists of discrete slabs (see PLATE TECTONICS), at the edges of which lithosphere bearing OCEANIC CRUST is created and destroyed (see PLATE MARGINS). *GO*

[See also EARTH STRUCTURE]

Walcott, R.I. 1970: Flexural rigidity, thickness and viscosity of the lithosphere. *Journal of Geophysical Research* 75, 3941–3954.

lithostratigraphy The subdivision of the STRATIGRAPHICAL RECORD using characteristics of LITHOLOGY or LITHOFACIES and the determination of the spatial relationships of the defined units through geological mapping. This is fundamental to all other types of STRATIGRAPHY. The primary unit for geological mapping is the FORMATION and the hierarchy of lithostratigraphical units is SUPERGROUP, GROUP, FORMATION, MEMBER and BED. *Lithostratigraphical units* are three-dimensional bodies of sediment or sedimentary, igneous or metamorphic rocks. A unit is defined and described from a *stratotype section*, or reference sections. Boundaries are drawn at horizons of lithological change, which may be abrupt, or at an arbitrary point in a gradational sequence. A lithostratigraphical unit may be characteristic of a particular DEPOSITIONAL ENVIRONMENT and may therefore migrate laterally and vertically within a rock succession (i.e. in space and time), indicating changes to the PALAEOENVIRONMENT. Hence lithostratigraphical units may show DIACHRONISM as demonstrated by BIOSTRATIGRAPHY or EVENT horizons. *LC*

Cox, B.M. and Sumbler, M.G. 1998: Lithostratigraphy: principles and practice. In Doyle, P. and Bennett, M.R. (eds), *Unlocking the stratigraphical record: advances in modern stratigraphy*. Chichester: Wiley, 11–27. **Nichols, G.** 1999: *Sedimentology and stratigraphy*. Oxford: Blackwell Science. **Whittaker, A., Cope, J.C.W., Cowie, J.W. et al.** 1991: A guide to stratigraphical procedure. *Journal of the Geological Society, London* 148, 813–824.

Little Ice Age Although originally used by Matthes to describe the renewed glaciation associated with NEOGLACIATION earlier in the HOLOCENE, the term Little Ice Age is now used for the cold interval during the last few centuries when GLACIERS were considerably larger than today and many attained their Holocene maxima. The Little Ice Age appears to have been a global event, but probably began at different times in different regions. In the European Alps, where recent GLACIER VARIATIONS and CLIMATIC CHANGE are known in greatest detail from DOCUMENTARY EVIDENCE, PALAEOENVIRONMENTAL RECONSTRUCTION and DATING, the Little Ice Age began around

AD 1300 as the MEDIAEVAL WARM PERIOD terminated. Subsequently, Alpine glaciers attained maximum advance positions (HIGHSTANDS) around AD 1350, 1650 and 1850 (see Figure). These three phases of the Little Ice Age may not have occurred elsewhere: in Scandinavia, for example, the evidence suggests one major glacier expansion episode, which began in the seventeenth century and peaked around AD 1750.

In the Little Ice Age, summer temperatures were likely to have been 1.0 to 2.0°C colder than today, at least in Europe, but this average value conceals CLIMATIC VARIATIONS at different temporal and spatial scales. Winter precipitation levels also varied, so that the climate of the Little Ice Age should not be viewed simply as a cold period. During the Little Ice Age, polar waters extended from Iceland, across the Norwegian Sea and south to the vicinity of the Shetland Islands, displacing warmer surface waters of the *North Atlantic Drift* farther south. This indicates that the average positions of the OCEANIC POLAR FRONT and associated atmospheric storm tracks lay south

of their present average positions. Outside Europe there is evidence of EXTREME CLIMATIC EVENTS, such as DROUGHTS in the SAHEL, being correlated to phases of the Little Ice Age, while recent research from equatorial east Africa has demonstrated a relatively wet climate at this time. Thus, the cause of the Little Ice Age can be linked to changes in the GENERAL CIRCULATION OF THE ATMOSPHERE. Possible ultimate cause(s), which are poorly understood, include FORCING FACTORS, such as variations in solar radiation and volcanic eruptions, and AUTOVARIATION within the Earth–atmosphere–ocean system.

The Little Ice Age is generally considered to have ended during the late nineteenth or early twentieth century at the latest. However, with a strong world-wide trend towards glacier recession dominant during the 1930s and 1940s but slowing subsequently, natural causes of climatic change may be affecting current rates of GLOBAL WARMING. Thus, in understanding the future course of global change we must consider the natural causes of climatic change associated with the Little Ice Age and similar

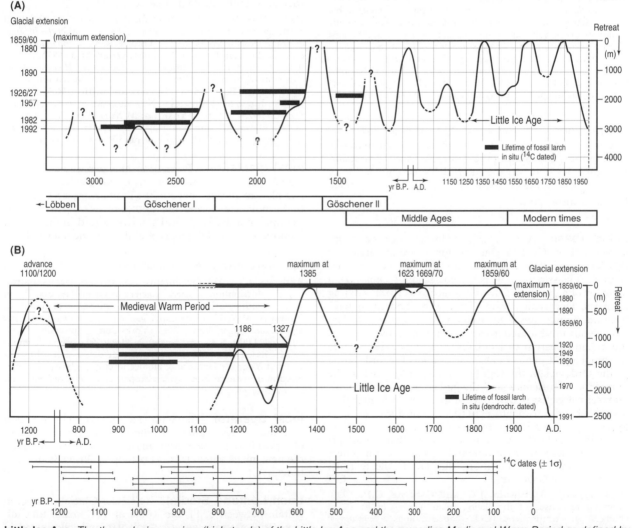

Little Ice Age *The three glacier maxima (highstands) of the Little Ice Age and the preceding Mediaeval Warm Period as defined by glacier variations of (A) the Grosser Aletsch Glacier and (B) the Gorner Glacier in the Swiss Alps. A multi-proxy approach has been used in the reconstruction, including documentary evidence and radiocarbon-dated fossil larch stumps (after Holzhauser, 1997)*

events earlier in the Holocene, as well as HUMAN IMPACT ON CLIMATE. *JAM*

[See also NORSE GREENLAND SETTLEMENTS]

Bradley, R.S. and Jones, P.D. 1993: 'Little Ice Age' summer temperature variations: their nature and relevance to recent global warming trends. *The Holocene* 3, 367–376. **Grove, J.M.** 1988: *The Little Ice Age.* London: Methuen. **Holzhauser, H.** 1997: Fluctuations of the Grosser Aletsch Glacier and the Gorner Glacier during the last 3200 years: new results. In Frenzel, B., Boulton, G.S., Gläser, B. and Huckriede, U. (eds), *Glacier fluctuations during the Holocene* [*Paläoklimaforschung.* Vol. 24; Special Issue 16 of the ESF/EPC Project 'European Palaeoclimate and Man']. Stuttgart: Gustav Fischer, 35–58. **Keigwin, L.D.** 1996: The Little Ice Age and Medieval Warm Period in the Sargasso Sea. *Science* 274, 1504–1508. **Kreutz, K.J., Mayewski, P.A., Meeker, L.D. et al.** 1997: Bipolar changes in atmospheric circulation during the Little Ice Age. *Science* 277, 1294–1296. **Luckman, B.H.** 2000: The Little Ice Age in the Canadian Rockies. *Geomorphology* 32, 357–384. **Matthes, F.** 1939: Report of the Committee on Glaciers. *Transactions of the American Geophysical Union* 20, 518–523. **Matthews, J.A.** 1991: The late Neoglacial ('Little Ice Age') glacier maximum in southern Norway: new ^{14}C-dating evidence and climatic implications. *The Holocene* 1, 219–233. **Mikami, T. (ed.)** 1992: *Proceedings of the International Symposium on the Little Ice Age Climate.* Tokyo: Department of Geography, Tokyo Metropolitan University. **Nicholson, S.E.** 1978: Climatic variations in the Sahel and other African regions during the past five centuries. *Journal of Arid Environments* 28, 13–30. **Swan, S.L.** 1981: México in the Little Ice Age. *Journal of Interdisciplinary History* 11, 633–648. **Verschuren, D., Laird, K.R. and Cumming, B.F.** 2000: Rainfall and drought in equatorial east Africa during the past 1,100 years. *Nature* 403, 410–414.

littoral zone   The intertidal zone along a SHORELINE. In Europe, the term is sometimes synonymous with *coastal zone*, but it is also used in the context of lakes. The *sublittoral zone* lies between the intertidal zone and the outer edge of the CONTINENTAL SHELF. *HJW*

[See also SHORE ZONE]

living fossil   A popular but loosely defined term, introduced by Charles Darwin, for an extant organism that has a long FOSSIL RECORD showing little evolutionary change. A classic example is the coelacanth (*Latimeria*), a member of a group of fish with a record as FOSSILS dating from the Devonian period. They were thought to have become extinct about 80 million years ago in the late Cretaceous until a modern specimen was recovered from the Indian Ocean in AD 1938. Other 'living fossils' include the brachiopod genus *Lingula*, with a fossil record dating to the Ordovician, the reptile tuatara, or *Sphenodon*, the horseshoe crab *Limulus* and the ginkgo tree. Living fossils represent organisms that have undergone little morphological change through EVOLUTION but have escaped EXTINCTION. A variety of factors seem to be responsible, but living fossils include some organisms that are adaptable to a range of environmental conditions and others that are adapted to conditions that are hostile to most organisms (see ECOLOGY). *GO*

Beerling, D.J., McElwain, J.C. and Osborne, C.P. 1998: Stomatal responses of the 'living fossil' Ginkgo biloba L. to changes in atmospheric CO$_2$ concentrations. *Journal of Experimental Biology* 49, 1603–1607. **Ward, P.D.** 1992: *On Methuselah's trail: living fossils and the great extinctions.* New York: Freeman.

Lixisols   Soils of the seasonally dry tropical, subtropical and warm temperate regions having an accumulation of LOW-ACTIVITY CLAYS in an ARGIC horizon and high BASE SATURATION. These soils occur in Brazil, Africa and India. (USDA: oxic subgroups of *Alfisols*). Erosion greatly reduces their already low fertility, so the thin surface horizon should be carefully conserved. When dry, the eluvial horizon of Lixisols may become very hard, a condition referred to as *hard setting.* *EMB*

[See also WORLD REFERENCE BASE FOR SOIL RESOURCES]

local climate   A small-scale CLIMATE, where climatic conditions are clearly different from those of nearby surrounding areas. Local climate may be the result of topography (TOPOCLIMATE), vegetation cover, soil type, water availability or the presence of anthropogenic features. It includes the climate in and around mountains, hills, lakes, coasts, forests and cities. Some view it as intermediate between MICROCLIMATE and MESOCLIMATE (see Figure); to others it is the equivalent of mesoclimate. A precise scale is difficult to define. *JBE/JAM*

[See also MACROCLIMATE, URBAN CLIMATE]

Barry, R.G. 1970: A framework for climatological research with particular reference to scale concepts. *Transactions of the Institute of British Geographers* 49, 61–70. **Oke, T.R.** 1987: *Boundary layer climates*, 2nd edn. London: Methuen. **Orlanski, I.** 1975: A subdivision of scales for atmospheric processes. *Bulletin of the American Meteorological Society* 56, 527–530. **Sturman, A.P., McGowan, H.A. and Spronken-Smith, R.A.** 1999: Mesoscale and local climates in New Zealand. *Progress in Physical Geography* 23, 611–635. **Yoshino, M.M.** 1987: Local climatology. In Oliver, J.E. and Fairbridge, R.W. (eds), *The encyclopedia of climatology.* New York: Van Nostrand Reinhold, 551–558.

local glaciation   Formation and persistence of individual, isolated GLACIERS, mostly cirque glaciers, ice fields and plateau glaciers, beyond the limit of larger ice masses. In Scandinavia during the YOUNGER DRYAS STADIAL, for example, hundreds of local glaciers existed in the coastal mountains beyond the margin of the continental ice sheet. *AN*

[See also CONTINENTAL GLACIATION]

Andrews, J.T. 1971: Quantitative analysis of the factors controlling the distribution of corrie glaciers in Okoa Bay, east Baffin Island (with particular reference to global radiation). In Morisawa, M. (ed.), *Quantitative geomorphology: some aspects and applications.* New York: New York State University Press, 223–241. **Dahl, S.O. and Nesje, A.** 1992: Palaeoclimatic implications based on equilibrium-line altitude depressions of reconstructed Younger Dryas and Holocene cirque glaciers in inner Nordfjord, western Norway. *Palaeogeography, Palaeoclimatology, Palaeoecology* 94, 87–97.

local operator   In IMAGE PROCESSING, any function that uses a few values around a pixel (usually, those observed in the NEAREST NEIGHBOURS) in order to compute the new image value. FILTERS belong to this class of functions. *ACF*

Loch Lomond Stadial   The Late Glacial STADIAL in Britain between c. 11 and 10 ka BP (^{14}C years) that followed the WINDERMERE INTERSTADIAL and culminated in the termination of the DEVENSIAN. It is equivalent to the YOUNGER DRYAS STADIAL of Scandinavia. *MHD*

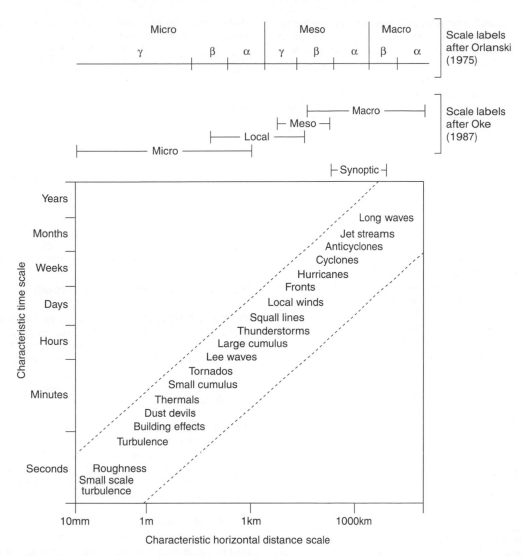

local climate *The temporal and spatial scales of atmospheric phenomena and the definition of microclimate, local climate, mesoclimate, macroclimate and synoptic climatology (from Sturman et al., 1999)*

[See also LATE GLACIAL ENVIRONMENTAL CHANGE]

Sissons, J.B. 1979: The Loch Lomond Stadial in the British Isles. *Nature* 280, 199–202.

lodgement till   TILL deposited at the base of a sliding GLACIER by plastering onto a rigid or semi-rigid bed.   *JS*

[See also GLACIAL SEDIMENTS, MELTOUT TILL]

Dreimanis, A. 1989: Tills, their genetic terminology and classification. In Goldthwait, R.P. and Matsch, C.L. (eds). *Genetic classification of glacigenic deposits*. Rotterdam: Balkema, 17–84.

loess   Well sorted, sedimentary deposits of predominantly SILT-sized particles, first correctly identified as being of AEOLIAN origin by Ferdinand van Richthofen. The material is derived from ARIDLANDS and/or outwash surfaces and has often travelled long distances from freshly deglaciated and PERIGLACIAL areas during periods of glaciation, being deposited beyond the ice margin.

Loess has therefore been entrained, transported, modified and sorted by aeolian transport processes. Loess deposits cover about 10% of the Earth's continental surfaces and are most extensive in the Loess Plateau region of central China, where the loess has a high carbonate content (sometimes exceeding 40% by weight). The Chinese loess deposits can be > 300 m thick and represent deposition which began as early as 2.5 million years ago.

Successions of loess deposits and interbedded PALAEOSOLS are interpreted as representing full GLACIAL conditions and warmer INTERGLACIAL or INTERSTADIAL episodes, respectively. In the deepest Chinese deposits almost 40 distinct palaeosol units have been identified. Loess deposits are also widespread in Central Europe, central Asia and the Great Plains in North America, and in the Southern Hemisphere they form the Pampas of Argentina and Uruguay. CLIMATIC RECONSTRUCTIONS and PALAEOENVIRONMENTAL RECONSTRUCTIONS can be inferred from lithological analysis of loess sequences and from a range of FOSSIL and SUBFOSSIL evidence, including POLLEN, MOLLUSCA and *vertebrates*. Loess is commonly unstratified, but can show well developed jointing and is amenable to LUMINESCENCE DATING. *DH*

[See also LOESS STRATIGRAPHY]

Derbyshire, E. (ed.) 1995: Wind blown sediments in the Quaternary record. *Quaternary Proceedings* 4. Chichester: John Wiley. **Derbyshire, E, Meng, X.M. and Kemp, R.A.** 1998: Provenance, transport and characteristics of modern aeolian dust in western Gansu Province, China, and interpretation of the Quaternary loess record. *Journal of Arid Environments* 39, 497–516. **Liu, T.S. (ed.)** 1991: *Loess, environment and global change*. Beijing: Science Press. **Wintle, A.G.** 1990. A review of current research on TL dating of loess. *Quaternary Science Reviews* 9, 385–397.

loess stratigraphy

Loess sequences can be characterised and correlated using a range of sedimentary properties, including heavy minerals (see HEAVY MINERAL ANALYSIS), CLAY MINERALS, micromorphology (see MICRO-MORPHOLOGICAL ANALYSIS) and MINERAL MAGNETISM. Loess sections usually display little visual evidence of stratification, although faint bedding may be seen with careful examination. SCANNING ELECTRON MICROSCOPY and optical MICROSCOPY of particle fabric can distinguish differences in TEXTURE, which are important for PALAEOEN-VIRONMENTAL RECONSTRUCTION. On the basis of such analyses two types of loess can be defined: one from areas that were arid during glacial periods with humid INTER-GLACIALS, as in China; the second from areas with persistently high humidities, as in Western Europe.

MAGNETIC SUSCEPTIBILITY analysis of loess deposits has shown that certain sedimentary units are more strongly characterised than others and can be correlated between individual loess sections across regions. Often the most distinctive features in loess sections are the darker interbedded PALAEOSOL units. Variations within palaeosol units appear to reflect CLIMATIC CHANGE, with relatively long periods of PEDOGENESIS during interglacial stages reflected by well developed, complex palaeosols. Shorter episodes of soil formation during INTERSTADIAL periods produce less prominent soil units. The successions of loess and interbedded palaeosols that have been found in Eastern Europe, China and former Soviet Central Asia provide evidence of alternating cold (loess) and warm (palaeosols) stages extending back over at least 17 GLACIAL–INTERGLACIAL CYCLES. *DH*

Derbyshire, E. 1995: Wind blown sediments in the Quaternary record. *Quaternary Proceedings* 4. Chichester: John Wiley. **Derbyshire, E. Kemp, R.A.** and **Meng, X.M.** 1997: Climate change, loess and palaeosols: proxy measures and resolution in North China. *Journal of the Geological Society* 154, 793–805. **Ding, Z., Yu, Z., Rutter, N.W. and Liu, T.** 1994: Towards an orbital time scale for Chinese loess deposits. *Quaternary Science Reviews* 13, 39–70. **Forman, S.L., Oglesby, R., Markgraf, V. and Stafford, T.** 1995: Paleoclimatic significance of Late Quaternary eolian deposition on the Piedmont and High Plains, central United States. *Global and Planetary Change* 11, 35–55. **Kemp, R.A. and Derbyshire, E.** 1998: The loess soils of China as records of climatic change. *European Journal of Soil Science* 49, 525–539. **Kukla, G. and An, Z.** 1989: Loess stratigraphy in central China. *Palaeogeography, Plaeoclimatology and Palaeoecology* 72, 203–225. **Liu, X., Rolph, T., Bloemendal, J.** *et al.* 1994: Quantitative estimates of palaeoprecipitation at Xifeng, in the Loess Plateau of China. *Palaeogeography, Palaeoclimatology and Palaeoecology* 113, 243–248. **Oches, E.A. and McCoy, W.D.**

1995: Amino acid geochronology applied to the correlation and dating of central European loess deposits. *Quaternary Science Reviews* 14, 767–782. **Verosub, K.L., Fine, P., Singer, M.J. and TenPas, J.** 1993: Pedogenesis and paleoclimate: interpretation of the magnetic susceptibility of Chinese loess–paleosol sequences. *Geology* 21, 1011–1014.

log

In the study of SEDIMENTS and rocks, particularly SEDIMENTARY ROCKS, a descriptive record of the characteristics of a vertical succession. A log can be compiled visually from a natural or artificial exposure or from a CORE or, using instruments, from a BOREHOLE. Information is recorded in the form of a columnar diagram or a GRAPHIC LOG. *GO*

logging

In the context of reconstructing environmental change, the recording of a continuous record (LOG) of stratigraphic units as a function of depth, according to visible variations in the rocks, SEDIMENTS or SOILS encountered. *DH*

[See also CLEAR CUTTING, DEFORESTATION]

logical positivism

A strongly empiricist, normative approach to the philosophy of science most closely identified with the Vienna Circle of philosophers. The VERIFICATION or verifiability principle is the most important attribute. *CET*

Suppe, F. 1977: *The structure of scientific theories*, 2nd edn. Urbana, IL: University of Illinois Press.

long waves

A hemispheric-scale wave pattern in the upper-air WESTERLIES, also known as *Rossby waves*. Preferred positions for troughs and ridges (wavelengths typically 3000–6000 km) strongly influence the formation and direction of motion of disturbances. Departures from quasi-stationary long-wave conditions are described by the *index cycle* (see ZONAL INDEX) and are frequently associated with *climatic anomalies*. Related long waves also develop in association with ocean currents. *JCS*

Dickinson, R.E. 1978: Rossby waves: long-period oscillations of oceans and atmospheres. *Annual Review of Fluid Mechanics* 10, 159–195. **Harman, J.** 1991: *Synoptic climatologies of the Westerlies: processes and patterns*. Washington: Association of American Geographers. **Platzman, G.W.** 1968: The Rossby wave. *Quarterly Journal of the Royal Meteorological Society* 94, 225–248.

longitudinal dune

A long, narrow SAND DUNE, parallel to the DOMINANT WIND (or perhaps two obliquely converging PREVAILING WINDS) responsible for its formation in arid areas with a moderate sand supply. Such dunes are also referred to as *linear dunes* and they can reach lengths of tens of kilometres, such as the QOZ dune systems of the Sudan. A *seif dune* is a longitudinal dune with a sinuous crest along which the slip face alternates from one side of the dune to the other (in response to seasonal differences in wind direction). In arid phases of the Quaternary longitudinal dunes were active over much larger areas than currently. *MAC*

[See also PALAEODUNE]

Brislow, C.S., Bailey, S.D. and Lancaster, N. 2000: The sedimentary structure of linear sand dunes. *Nature* 406, 56–59. **Lancaster, N.** 1995: *Geomorphology of desert dunes*. London: Routledge.

longshore drift   The movement of sand and other material along the shore by CURRENTS or WAVES. *Beach drifting* is the movement of beach material induced by the *swash* and *backwash* of waves. Longshore drift includes suspended sediment and BEDLOAD, as well as the water itself.   *HJW*

[See also BEACH, GROYNE]

Komar, P.D. 1971: The mechanism of sand transport on beaches. *Journal of Geophysical Research* 76, 713–721.

look-up table   In IMAGE PROCESSING, a memory-resident table that associates colours and numbers; used to convert DIGITAL NUMBERS in an image into colours on a display device. It is an economic way of colouring objects in computational systems and, usually, the user has access to this object in order to specify the behaviour of the display and other graphic devices.   *ACF*

loss-on-ignition (LOI)   A simple, crude but useful measure of the ORGANIC CONTENT of soils and sediments, obtained by weight difference before and after a soil sample is subjected to high-temperature ignition. Unless carried out at sufficiently low temperatures, ca 550°C, the method includes weight losses through breakdown of carbonates. At higher temperatures (ca 800°C) it forms the basis of a method for determination of carbonate content of rocks and sediments.   *EMB*

Baize, D. 1988: *Soil science analyses: a guide to current use.* Paris: Institute Nationale de Reserche Agronomique [English translation: 1993. Chichester: Wiley]. **Ball, D.F.** 1967: Loss-on-ignition as an estimate of organic matter and organic carbon in non-calcareous soils. *Journal of Soil Science* 15, 84–92. **Dean, W.E.** 1974: Determination of carbonate and organic matter in calcareous sediments and sedimentary rocks by loss on ignition: comparison with other methods. *Journal of Sedimentary Petrology* 44, 242–248.

low-activity clays   Clays with a low CATION EXCHANGE CAPACITY (< 24 cmol$_c$ kg^{-1}), associated with ACRISOLS, FERRALSOLS and LIXISOLS.   *EMB*

[See also CLAYS, HIGH-ACTIVITY CLAYS]

Low Arctic   The TUNDRA environment in the Northern Hemisphere north of the TREE LINE, where there is a predominantly complete vegetation cover, dominated by flowering plants including shrubby growth and dwarf woodland up to 2.0 m high in places. The Low Arctic has a richer plant and animal assemblage than does the HIGH ARCTIC.   *HMF*

[See also ARCTIC, SUBARCTIC]

French, H.M. 1999: Arctic environments. In Alexander, D.E. and Fairbridge, R.W. (eds), *Encyclopedia of environmental science.* Dordrecht: Kluwer, 29–33.

lowstand   The minimum of an environmental cycle, fluctuation or oscillation as reflected in, for example, GLACIER VARIATIONS, LAKE-LEVEL VARIATIONS and SEA-LEVEL CHANGE.   *JAM*

[See also HIGHSTAND]

Cross, S.L., Baker, P.A., Seltzer, G.O. *et al.* 2000: A new estimate of the Holocene lowstand level of Lake Titicaca, central Andes, and implications for tropical palaeohydrology. *The Holocene* 10, 97–108.

luminescence dating   A method used to determine the time elapsed since deposition of mineral grains (mainly QUARTZ and FELDSPAR) in sediments, including AEOLIAN (e.g. LOESS), FLUVIAL and GLACIOFLUVIAL SEDIMENTS. Luminescence dating is also used in ARCHAEOLOGY to date pottery, baked clay and flints from fire hearths.

In any geological environment, natural RADIATION induces free *electrons* in minerals that can be trapped in lattice defects (ELECTRON TRAPS). *Luminescence* is the light emitted when trapped electrons are subjected to heat (THERMOLUMINESCENCE or TL), a beam of light of visible wavelengths (OPTICALLY STIMULATED LUMINESCENCE or OSL) or infrared wavelengths (INFRARED-STIMULATED LUMINESCENCE or IRSL). The number of trapped electrons, and hence the luminescence signal, is proportional to the strength of the radioactive field (*dose rate*) and the length of time since the signal was last set to zero by exposure to light or heat (a process known as '*bleaching*'). The effective dating limit is the time elapsed between exposure and saturation of defect sites and depends on the radiation dose and capacity to accumulate electrons, but it is generally accepted to be about 0.25 Ma. The intensity of the luminescence emitted from a natural sample is compared with that from the same mineral separately irradiated in the laboratory by a dose of *beta* and/or *gamma radiation* to calculate the PALAEODOSE or *equivalent dose*. The age of deposition is determined using the equation *age = palaeodose/dose-rate*, where dose-rate is derived from the concentration of radioactive components in the sediment and surrounding matrix (uranium, thorium and potassium, and to a lesser degree, rubidium) and the COSMIC RAY FLUX.

One of the major assumptions of luminescence dating is that the signal is set to zero immediately prior to burial. However, zeroing will be incomplete in some environments, such as fluvial or glaciofluvial sediments where exposure to sunlight may be limited. The advantage of OSL and IRSL dating is that the electron traps utilised in these techniques are more sensitive to light than traps stimulated by heating in TL dating, and only brief exposure to sunlight is likely to reduce luminescence to zero. OSL and IRSL dating are therefore applicable to a wider variety of deposits, although IRSL only occurs in feldspars. Deposits as young as a few decades in age can be dated using OSL and IRSL.   *DAR*

Aitken, M.J. 1990: *Science based dating in archaeology.* Oxford: Oxford University Press. **Aitken, M.J.** 1998: *An introduction to optical dating: The dating of Quaternary sediments by the use of photon-stimulated luminescence.* Oxford: Oxford University Press. **Duller, G.A.T.** 1996: Recent developments in luminescence dating of Quaternary sediments. *Progress in Physical Geography* 20, 127–145. **Feathers, J. K.** 1997: Application of luminescence dating in American archaeology. *Journal of Archaeological Method and Theory* 4, 1–66. **Roberts, R., Walsh, G., Murray, A.** *et al.* 1997: Luminescence dating of rock art and past environments using mud-wasp nests in northern Australia. *Nature* 387, 696–699. **Stokes, S.** 1999: Luminescence dating applications in geomorphological research. *Geomorphology* 29, 153–171. **Wintle, A.G.** 1993: Luminescence dating of aeolian sands: an overview. In Pye, K. (ed.), *The dynamics and environmental context of aeolian sedimentary systems* [*Geological Society Special Paper* 72]. London: Geological Society, 49–58.

lunar cycles   Cycles due to the gravitational pull of the Moon on the Earth and its atmosphere. They range from the semi-diurnal TIDES to the 18.6 year lunar cycle. The atmospheric lunar tide is too small (< 0.2 hPa) to be of practical importance but, when combined with the effect of the Sun, the semi-diurnal pressure variation ranges from 4 hPa in the tropics to 1 hPa in mid-latitudes. As a result of its orbit characteristics in relation to the Earth, the Moon's declination relative to the ECLIPTIC varies with a PERIODICITY of 18.6 years and its tidal force is greatest when its perigee position coincides with maximum declination. Using maximum entropy spectrum analysis, the 18.6 year cycle has been found in a variety of long-term climatic records: AIR TEMPERATURE, atmospheric pressure, Chinese DROUGHTS, European VINE HARVEST dates and fish catches, TREE RINGS, NILE FLOODS, North Atlantic CYCLONES and thunderstorm occurrences. However, the physical CAUSAL RELATIONSHIP is difficult to establish. *BDG*

Currie, R. G. 1996: M_n and S_c signals in North Atlantic tropical cyclone occurrence. *International Journal of Climatology* 16, 427–439.**Lamb, H.H.** 1972: *Climate: present, past and future.* Vol. 1. *Fundamentals and climate now.* London: Methuen.

Lusitanean floral element Plant species with their main range in Portugal (latitude: Lusitania), some of which exhibit disjunct ranges extending to western Ireland and/or southwest England. *BA*

Luvisols Soils with subsurface accumulation of HIGH-ACTIVITY CLAYS in an ARGIC HORIZON, occurring in North America, Europe and Australia where there is seasonally a dry period (SOIL TAXONOMY: *Alfisols*). The presence of Luvisols is an indication of a stable land surface and some Luvisols occur in environments that are no longer conducive to clay eluviation. With change to a wetter climate these soils could become GLEYED and suffer ACIDIFICATION. They are widely used for crop production, having high activity clays and a good rooting environment. *EMB*

[See also WORLD REFERENCE BASE FOR SOIL RESOURCES]

lynchet A so-called *'cultivation terrace'* along the edge of a field boundary, which is formed by SOIL EROSION and deposition, especially by SLOPEWASH, of cultivated fields. Erosion tends to be most obvious on the down-slope side of walls or banks (*negative lynchet*) whereas deposition occurs against the obstacle on the upslope side (*positive lynchet*). Lynchets are important indicators of rates of erosion and of prehistoric FIELD SYSTEMS. *JAM*

[See also STRIP LYNCHET]

lysocline The depth in the OCEANS at which the solution rate of calcium carbonate increases markedly. It is thought to correspond to the upper boundaries of cold, corrosive BOTTOM WATERS. The lysocline is found a few hundred metres above the CARBONATE COMPENSATION DEPTH. *BTC*

M

Ma A *mega-annum:* an abbreviation for millions of years before the present. *GO*

maar A broad, low-rimmed CRATER caused by a VOLCANIC ERUPTION in which MAGMA encountered GROUNDWATER, converting it to steam with explosive expansion (see HYDROVOLCANIC ERUPTION). Maars are typically around 1 km in diameter and are commonly occupied by a *maar lake,* usually 10–500 m deep. These fill with SEDIMENT of organic, local slope wash and aeolian origin, providing excellent opportunities for CLIMATIC RECONSTRUCTION and PALAEOENVIRONMENTAL RECONSTRUCTION. The term is originally derived from the Eifel region of Germany. *Tuff rings* are similar features, and *tuff cones* are taller. *JBH*

[See also LACUSTRINE SEDIMENTS]

Allen, S.R., Bryner, V.F., Smith, I.E.M. and Ballance, P.F. 1996: Facies analysis of pyroclastic deposits within basaltic tuff-rings of the Auckland volcanic field, New Zealand. *New Zealand Journal of Geology and Geophysics* 39, 309–327. **Brauer, A.** 1999: High resolution sediment and vegetation response to Younger Dryas climate change in varved lake sediments from Meerfelder Maar, Germany. *Quaternary Science Reviews* 18, 321–329. **Mingram, J.** 1998: Laminated Eocene maar-lake sediments from Eckfeld (Eifel region, Germany) and their short-term periodicities. *Palaeogeography, Palaeoclimatology, Palaeoecology* 140, 289–305. **Negendank, J.F.W. and Zolitschka, B. (eds)** 1993: *Palaeolimnology of European maar lakes.* Berlin: Springer. **Schettler, G., Rein, B. and Negendank, J.F.W.** 1999: Geochemical evidence for Holocene palaeodischarge variations in lacustrine records from the Westeifel Volcanic Field, Germany: Schalkenmehrener Maar and Meerfelder Maar. *The Holocene* 9, 381–400. **Watts, W.A., Allen, J.R.M. and Huntley, B.** 1996: Vegetation history and palaeoclimate of the last glacial period at Lago Grande di Monticchio, Southern Italy. *Quaternary Science Reviews* 15, 133–153.

machair Calcareous COASTAL DUNE pastures of the highlands and islands of northwestern Scotland. Though currently treeless, these landscapes may have had a significant woodland cover in the early Holocene. *JAM*

Ritchie, W. 1976: The meaning and definition of machair. *Transactions of the Botanical Society of Edinburgh* 42, 431–440. **Whittington, G. and Edwards, K.J.** 1997: Evolution of a machair landscape: pollen and related studies from Benbecula, Outer Hebrides, Scotland. *Transactions of the Royal Society of Edinburgh: Earth Sciences* 87, 515–531.

macroclimate The distinguishing climatic features of large areas. Horizontal scales are typically above 100 km^2, at least regional and frequently subcontinental to global. Macroclimate incorporates and is controlled by the large-scale features of the GENERAL CIRCULATION OF THE ATMOSPHERE. *JCS*

[See also LOCAL CLIMATE, MICROCLIMATE, MESOCLIMATE]

macroecology An approach to ECOLOGY focusing on the emergent statistical phenomena exhibited by assem-

blages of species, e.g. body size, geographical RANGE and abundance, and on mechanisms concerning how species use and divide energy, space and other resources. *RJW*

[See also: AUTECOLOGY, BIOGEOGRAPHY, COMMUNITY ECOLOGY, ECOLOGY]

Brown, J.H. 1995: *Macroecology.* Chicago, IL: University of Chicago Press. **Brown, J.H.** 1999: Macroecology: progress and prospect. *Oikos* 87, 3–14.

macrofossil A FOSSIL or SUBFOSSIL of a size visible to the naked eye. For identification purposes a microscope or strong binocular loupe is usually needed. Examples include leaves, FRUITS AND SEEDS. Macrofossils often have the advantage over MICROFOSSILS, particularly pollen, because identification can more readily be made to the species instead of genus or family level. Macrofossils also experience less long-distance transport than microfossils. *BA*

[See also PALAEOBOTANY, PALAEOECOLOGY, PLANT MACRO-FOSSIL ANALYSIS]

macronutrients Elements other than carbon, hydrogen and oxygen, which are required by organisms in substantial amounts, including calcium, chlorine, magnesium, nitrogen, phosphorus, potassium and sulphur. *JAM*

[See also MICRONUTRIENTS, TRACE ELEMENTS]

Madden–Julian oscillation (MJO) An intraseasonal fluctuation of 30–60 days within the zonal, WALKER CIRCULATION characterised by an eastward shift in convectional activity and cloud clusters from the Indian Ocean to the central Pacific Ocean. The Madden–Julian oscillation occurs throughout the year but its amplitude and frequency are related to the MONSOON circulation and EL NIÑO–SOUTHERN OSCILLATION. In turn it may influence the evolution of extratropical weather on timescales of months to seasons. *JAM*

Lau, K.M. and Chan, P.H. 1986: The 40–50 day oscillation and the El Niño/southern oscillation: a new perspective. *Bulletin of the America Meteorological Society* 67, 533–534. **Madden, R. and Julian, P.R.** 1971: Detection of a 40–50 day oscillation in the zonal wind in the tropical Pacific. *Journal of Atmospheric Science* 28, 702–708.

magma Molten material generated in the Earth's CRUST and upper mantle, which solidifies to form an IGNEOUS ROCK. Magma is a mixture of silicate melt, crystals and gas. Gas is exsolved as magma decompresses on approaching the Earth's surface, releasing VOLATILES to the ATMOSPHERE. Magma temperatures in VOLCANIC ERUPTIONS range from 700–1200°C as the composition ranges from silica-rich (*acidic*) to silica-poor (*basic*). *JBH*

Middlemost, E.A.K. 1997: *Magmas, rocks and planetary development: a survey of magma/igneous rock systems.* Harlow: Longman.

magnetic anomaly A localised variation in the Earth's magnetic field, superimposed on the ambient regional magnetic field, which results from the deep geological character (see GEOMAGNETISM). Magnetic anomalies can be attributed to the magnetic properties of near-surface materials, and GEOPHYSICAL SURVEYING can be used to detect features such as GEOLOGICAL STRUCTURES, archaeological structures and ORE deposits. The interpretation of ocean-floor magnetic anomalies, which occur as bands parallel to the MID-OCEAN RIDGE and are symmetrical either side of the ridge, as a record of GEOMAGNETIC POLARITY REVERSALS preserved by SEA-FLOOR SPREADING, was critical to the development of PLATE TECTONICS theory in the early 1960s. Ocean-floor magnetic anomalies form the basis of the GEOMAGNETIC POLARITY TIMESCALE. They are numbered from 1, the most recent, such that those with dominantly normal polarity have odd numbers and those with dominantly reversed polarity have even numbers. *DNT*

[See also ARCHAEOLOGICAL PROSPECTION]

Cox, A. (ed.) 1973: *Plate tectonics and geomagnetic reversals.* San Francisco, CA: Freeman. **Robinson, E.S. and Çoruh, C.** 1988: *Basic exploration geophysics.* New York: Wiley. **Vine, F.J.** 1966: Spreading of the ocean floor: new evidence. *Science* 154, 1405–1415.

magnetic excursion A magnetic EVENT in which the geomagnetic pole (see GEOMAGNETISM) moves from one geographical pole to a position close to the other, but continues moving back towards the original pole rather than stabilising in the opposite position as in a true GEOMAGNETIC POLARITY REVERSAL. Excursions are evident in the record of PALAEOMAGNETISM, particularly in sequences of SEDIMENTARY ROCKS, and are a valuable tool in CORRELATION (see MAGNETOSTRATIGRAPHY). *DNT*

magnetic field The change in energy generated in a given volume of space such that the energy gradient created produces a force that can be detected, e.g. the torque induced on a compass needle. A magnetic field (H) is produced by a permanent magnet or whenever there is electrical charge in motion, and is measured in ampere/metre (A m^{-1}). *AJP*

magnetic grain size Ferrimagnetic materials can be divided into different regions or cells of MAGNETISATION, which are known as *domains*. Above ~110 µm, grains are referred to as *multi-domain* (MD) because energetically it is favourable to have more than one magnetic domain. In small grains < 0.2 µm only one domain will form and these are then *single domain* (SD) or *stable single domain* (SSD). Grains between 0.2 and 110 µm are large enough to have more than one domain but exhibit the magnetic properties of SD grains; such grains are termed *pseudo-single domain* (PSD). Very small (0.001–0.01 µm) ferro- or ferrimagnetic grains have thermal vibrations at room temperature that are of the same order of magnitude as their magnetic energies. Consequently, these *superparamagnetic* (SP) materials do not exhibit stable remanent magnetisation. Under certain circumstances, magnetic grain size can be used as a rapid, non-destructive proxy for physical grain-size. *AJP*

[See also MINERAL MAGNETISM]

Dearing, J. 1999: Magnetic susceptibility. In Walden, J., Oldfield, F., and Smith, J.P. (eds), *Environmental magnetism: a*

practical guide [*Technical Guide* 6]. London: Quaternary Research Association, 35–62.

magnetic induction When a MAGNETIC FIELD H has been generated in a medium, the response of the medium is its magnetic induction (B, also known as *magnetic flux density*). Magnetic induction is measured in tesla (T). *AJP*

magnetic intensity The strength (F) of the Earth's MAGNETIC FIELD. Natural materials can acquire a remanence in the Earth's magnetic field and, hence, the fossil record can be used to study ancient direction and intensity. *AJP*

[See also PALAEOMAGNETIC DATING]

magnetic moment The torque produced in the presence of a MAGNETIC FIELD; given by the equation $m = \tau_{max}/B$, where B is the MAGNETIC INDUCTION. The units of magnetic moment are ampere metre^2 (A m^2). *AJP*

magnetic remanence Upon removal of an applied MAGNETIC FIELD, the MAGNETISATION of a given material may not return to zero, i.e. the material is no longer unmagnetised but has a remanent magnetisation. *AJP*

magnetic secular variation The Earth's MAGNETIC FIELD can be described according to its strength and direction. In addition to MAGNETIC INTENSITY, the direction of the field is expressed as an angle of dip below the horizontal plane, i.e. the INCLINATION (I), and the angle between the horizontal component of the field and the true geographical north, i.e. the DECLINATION (D). Secular variation refers to changes in the direction of the geomagnetic field over time, reflecting shifts in the location of the Earth's geomagnetic poles. The first measurement of declination at London was in c. AD 1570, when it was found that the compass needle pointed 11° east. In AD 1660 the compass pointed due north and by AD 1820 it had swung around to 24° west. Since then, declination has decreased and is now approximately 3° west of true north, decreasing by about 0.5° every four years. *AJP*

[See also PALAEOMAGNETIC DATING]

Thompson, R. and Oldfield, F. 1986: *Environmental magnetism.* London: Allen and Unwin.

magnetic susceptibility The extent to which a sediment or other material can be magnetised. It is dependent on the magnetic mineralogy of the sediment. Various measures of magnetic susceptibility (see MINERAL MAGNETISM) are widely used in CORE CORRELATION, in the characterisation of sediments and in PALAEOENVIRONMENTAL RECONSTRUCTION. *JAM*

Dearing, J.A. 1999: *Environmental magnetic susceptibility.* Kenilworth: Chi Publishing. **Maher, B.A. and Thompson, R.** 1992: Palaeoclimatic significance of the mineral magnetic record of the Chinese loess and palaeosols. *Quaternary Research* 37, 155–170.

magnetisation The MAGNETIC MOMENT per unit volume of a solid: $M = m/V$ and is measured in ampere/metre (A m^{-1}). *AJP*

magnetite One of the most abundant iron oxide minerals, Fe_3O_4, found in the majority of igneous rocks, many metamorphic and sedimentary rocks and nearly all soils. It is a common magnetic mineral, which exhibits ferrimagnetic behaviour. Low frequency MAGNETIC SUSCEPTIBILITY values for magnetite lie in the range of c. 400–1000 × 10^{-6} m^3 kg^{-1}. *AJP*

[See also FERROMAGNETISM]

magnetometer A device used for measuring the strength and direction of a MAGNETIC FIELD. Magnetometer measurements of variations in the Earth's geomagnetic field (see GEOMAGNETISM) and magnetisation of materials can be used in PALAEOMAGNETIC DATING. *MHD*

Thompson, R. 1991: Palaeomagnetic dating. In: Smart, P.L. and Frances, P.D. (eds), *Quaternary dating methods – a users guide* [*QRA Technical Guide* 4]. Cambridge: Quaternary Research Association, 177–198.

magnetostratigraphy The use of variations in the geomagnetic field (see GEOMAGNETISM) through GEOLOGICAL TIME, preserved as GEOMAGNETIC POLARITY REVERSALS, MAGNETIC EXCURSIONS and variations in MAGNETIC SUSCEPTIBILITY, to subdivide sequences of sediments and rocks in CORRELATION and STRATIGRAPHY. The ABSOLUTE-AGE DATING of specific geomagnetic polarity reversals has given rise to the GEOMAGNETIC POLARITY TIMESCALE, or *magnetostratigraphic timescale*. *DNT*

Hailwood, E.A. 1989: *Magnetostratigraphy, Special Publication* 19. London: Geological Society. Roperch, P., Herail, G. and Fornari, M. 1999: Magnetostratigraphy of the Miocene Corque basin, Bolivia: implications for the geodynamic evolution of the Altiplano during the late Tertiary. *Journal of Geophysical Research – Solid Earth* 104, 20415–20429. Taberner, C., Dinares-Turell, J., Gimenez, J. and Docherty, C. 1999: Basin infill architecture and evolution from magnetostratigraphic cross-basin correlations in the southeastern Pyrenean foreland basin. *Geological Society of America Bulletin* 111, 1155–1174. Whittaker, A., Cope, J.C.W., Cowie, J.W. et al. 1991: A guide to stratigraphical procedure. *Journal of the Geological Society, London* 148, 813–824.

magnetozone A unit of rock characterised by a specific magnetic polarity (i.e. normal or reversed) or other palaeomagnetic property (see PALAEOMAGNETISM). The corresponding interval of time is a POLARITY CHRON. *GO*

magnitude-frequency concepts Used in relation to geomorphological, hydrological and biological processes operating in landscapes, the term refers to the frequency of process events of different size and the relative work or impact that they achieve. The overall concept also proposes that geomorphological (or vegetational) features may be influenced disproportionately by processes or events of a particular size and frequency (or range of sizes and frequencies), rather than by the whole range of events. Some features may be the result of relatively moderate, frequent events, whereas other features may be influenced only by high-magnitude events of rare frequency (see EXTREME EVENTS).

The concept is well exemplified with respect to fluvial processes, in which diagrams can be constructed of the relative proportions of the total geomorphic work done by different process magnitudes (where work is the product of the magnitude and the frequency of events). Thus analysis of the SUSPENDED LOADS of rivers has often demonstrated that most of the load is carried by a few storm events or flows of high magnitude but low frequency, though the RECURRENCE INTERVAL of the flow varies between rivers with climatic and other factors. Other studies have focused on trying to identify which flow (or range of flows) is responsible for producing, maintaining or changing the cross-sections (HYDRAULIC GEOMETRY) and CHANNEL PATTERNS of rivers.

As the magnitude-frequency characteristics of climatic variables, such as large STORMS, frost cycles and TROPICAL CYCLONES, have important bearings in turn on the magnitude-frequency of hydrological and geomorphological processes such as LANDSLIDES, FLOOD FREQUENCY and FREEZE–THAW CYCLES, it is important to include them in more realistic schemes relating climate to geomorphology and Ahnert (1987), for example, has argued that a magnitude-frequency index of EXTREME EVENTS should be developed and included in future, more refined schemes of morphogenetic regions (see MORPHOCLIMATIC ZONES). The lack of long-term records of climatic extremes and landscape processes and, where they do exist, the difficulties often of separating anthropogenic from natural components of magnitude and frequency prevent this objective being realised.

There is growing evidence of changes through time in the magnitude-frequency of climatic events with climatic change even over the timescale of the last two centuries. Rainstorms capable of generating WADI flows and shallow GROUNDWATER recharge in the White Nile Province of the Sudan have been significantly smaller and less frequent since AD 1965 than in the mid-twentieth century, resulting in reduced rural water supplies and population migration to the cities. An increase in flood frequency in the rivers of South Wales since 1925 has been linked to an increase in the magnitude-frequency of large daily rainfalls. Dry periods in the otherwise perhumid equatorial environment of Borneo, for example, were longer and more frequent in the late nineteenth and early twentieth centuries, and again in recent decades, than in the period 1916–1967 with implications for the dynamics of the local TROPICAL RAIN FORESTS.

Human interference in the landscape can have radical impacts on the magnitude-frequency of geomorphological and hydrological processes and events. LANDUSE CHANGES such as URBANISATION, DEFORESTATION, LAND DRAINAGE schemes and AFFORESTATION (if accompanied by ditching) lead to an increase in the flood magnitude-frequency, whereas RESERVOIRS usually lead to a reduction in flood flows. Such changes can lead to changes in channel size, shape and pattern. *RPDW*

Ahnert, F. 1987: An approach to the identification of morphoclimates. In Gardiner, V. (ed.), *International geomorphology 1986*. Part II [*Proceedings of the First International Conference on Geomorphology*]. Chichester: Wiley, 159–188. Douglas, I., Bidin, K., Balamurugan, G. et al. 1999: The role of extreme events in the impacts of selective tropical forestry on erosion during harvesting and recovery phases at Danum Valley, Sabah. *Philosophical Transactions of the Royal Society of London* B354, 1749–1761. Walsh, R.P.D. 1996: Drought frequency changes in Sabah and adjacent parts of northern Borneo since the late nineteenth century and possible implications for tropical rain forest dynamics. *Journal of Tropical Ecology* 12, 385–407. Walsh R.P.D., Hulme M. and Campbell, M. 1988: Recent rainfall changes and their impact on hydrology and water supply in the semi-arid zone of the Sudan. *Geographical Journal* 154, 181–198. Webb, B.W. and Walling, D.E. 1982: The magnitude and frequency characteristics of fluvial transport in a Devon drainage basin and some geomorphological implications. *Catena* 9, 9–24. **Wolman, M.G. and Miller, J.P.** 1960: Magnitude and frequency of forces in geomorphological processes. *Journal of Geology* 68, 54–74.

makatea island A South Pacific mid-plate island, with uplifted Tertiary to Late Pleistocene reefal limestones (sometimes encircling an old volcanic core), suggesting regional flexure of the lithosphere. *TSp*

Stoddart, D.R., Woodroffe, C.D. and Spencer, T. 1990: Mauke, Mitiaro and Atiu: geomorphology of makatea islands in the Southern Cooks. *Atoll Research Bulletin* 341, 1–66.

malacophyllous Soft-leaved, chiefly deciduous, low shrubs characteristic of MEDITERRANEAN-TYPE VEGETATION communities, which are subject to seasonal DROUGHT. Unlike the hard-leaves of SCLEROPHYLLOUS plants, the leaves of malacophyllous plants (e.g. *Cistus* spp.) wilt during drought: the plants tolerate reduced water content in the tissues rather than reduce water loss. *MLC/JAM*

[See also XEROPHYTE]

malaria A DISEASE endemic to most of the tropical world spread by the female *Anopheles* mosquito, which injects a PARASITE into the host during feeding. According to Stamp (1964: 38) 'it has been the greatest killer of all and when it does not kill it reduces mankind almost to incompetence'. LAND DRAINAGE, PESTICIDE use and drug treatment have reduced its extent and impact. It still kills more people each year than any other infectious disease except AIDS and tuberculosis. Billions of people in tropical countries are at risk of infection: > 400 million become infected each year, of whom between one and three million die. Starting in Asia in the AD 1960s, the parasite developed immunity to chloroquine, an old and reliable drug, and now resistance to the newer drugs is beginning to be developed (see Figure). *JAM*

[See also HUMAN HEALTH HAZARDS, WATER-RELATED VECTOR]

Marshall, E. 2000: A renewed assault on an old and deadly foe. *Science* 290, 428–430. **National Research Council** 1991: *Malaria: obstacles and opportunities*. Washington, DC: National Academy Press. **Stamp, L.D.** 1964: *The geography of life and death*. London: Collins.

mammoth An extinct elephant of the genus *Mammuthus*, closely related to modern elephants, although differentiated by twisted tusks and other features. The ancestral mammoth, *Mammuthus meridionalis*, evolved from a branch of the Elephantidae that migrated out of Africa approximately 2.5 Ma ago. Two separate lineages subsequently evolved, *Mammuthus trogontherii* (the steppe mammoth) in Eurasia and *Mammuthus colombi* (the Colombian mammoth) in North America. The word 'mammoth' is often used particularly to describe the descendant of *M. trogontherii*, the woolly mammoth *Mammuthus primigenius*, which arose in Eurasia around 500 000 years ago and became widespread in the Northern Hemisphere. Along with other elements of the Pleistocene MEGAFAUNA, the last mammoths became extinct from the continental landmasses by 10 000 years ago, although RADIOCARBON DATING of mammoth remains from Wrangel Island in the Arctic Ocean has demon-

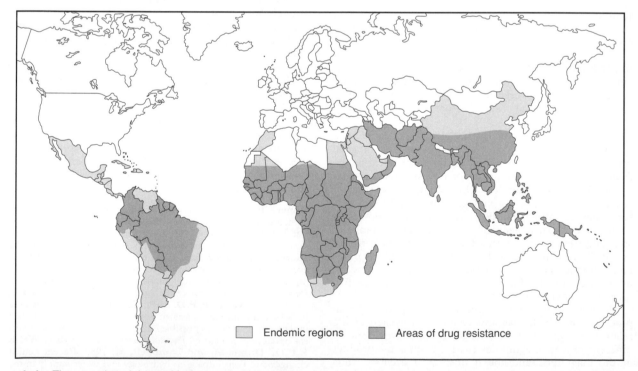

malaria *The countries of the world where malaria is endemic and those areas where the parasite has developed drug resistance (after Marshall, 2000)*

strated the survival of a relict dwarf race until about 3700 years ago. *DCS*

Haynes, G. 1991: *Mammoths, mastodonts, and elephants: biology, behaviour, and the fossil record.* Cambridge: Cambridge University Press. **Lister A. and Bahn, P.,** 1994: *Mammoths.* Macmillan: New York. **Vartanyan, S.L., Garutt, V.E. and Sher, A.V.** 1993: Holocene dwarf mammoths from Wrangel Island in the Siberian Arctic. *Nature* 362, 337–340.

managed retreat   A suggested strategy for dealing with SEA-LEVEL RISE due to GLOBAL WARMING along low-lying coasts, in which the coastline will be allowed to recede to a new position to restore natural coastal systems, such as mudflats, saltmarshes and dunes.   *HJW*

Brooke, J.S. 1992: Coastal defence: the retreat option. *Journal of the Institute of Water and Environmental Management* 6, 151–157. **Maddrell, R.J.** 1996: Managed coastal retreat, reducing flood risks and protection costs, Dungeness nuclear power station, UK. *Coastal Engineering* 28, 1–15.

management   The human control of a system with a view to directing it for economic gain or for some other purpose such as CONSERVATION.   *JAM*

[See also ENVIRONMENTAL MANAGEMENT]

mangrove   Highly productive tropical to subtropical ecosystems of taxonomically unrelated, shallow-rooted trees and shrubs on muddy coasts, associated with estuaries, CORAL REEFS and SALTMARSHES. Mangrove swamps occupy about two-thirds of the Earth's tropical coastline.   *TSp*

Tomlinson, P.B. 1986: *The biology of mangroves.* Cambridge: Cambridge University Press.

mangrove succession   Classical ECOLOGICAL SUCCESSION theory, in which the gradient concept and 'physiographical ecology' are used to explain persistence and change in MANGROVE communities. Successional models emphasise the active role of plants in inducing both organic and mineral accumulation. Increased surface elevation drives community transitions until a 'climax community' is reached (see CLIMAX VEGETATION). Temporal succession at a point may be mirrored in the spatial ZONATION of mangrove communities. The somewhat complementary gradient concept relates surface inundation and gradients in environmental factors, notably salinity, to plant physiology to establish spatial limits to species groups. *Physiographical ecology* argues that the evolution of mangrove vegetation patterns is closely related to the dynamics of shoreline change and that mangroves opportunistically colonise surfaces that develop at appropriate elevations. This approach predicts mangrove community mosaics in both terrigenous and carbonate environments and argues that the successional model is one case of a more general model, where sediment inputs drive shoreline PROGRADATION.   *TSp*

Snedaker, S.C. 1982: Mangrove species zonation: why? In Sen, D.N. and Rajpurohit, K.S. (eds), *Tasks for Vegetation Science,* The Hague: Junk, 111–125. **Thom, B.G.** 1984: Coastal landforms and geomorphic processes. In Snedaker, S.C. and Snedaker, J.G. (eds), *The mangrove ecosystem: research methods.* Paris: Unesco, 3–17. **Watson, J.G.** 1928: Mangrove forests of the Malay Peninsula. *Malaya Forest Records* 6, 1–275.

mangrove swamps: human impacts   Traditionally, MANGROVE ecosystems have been a renewable resource, sustaining local populations with building materials, firewood, charcoal and medicine and, from surrounding waters, fish and shellfish. Recently, however, large areas of mangrove have been lost to LOGGING, AGRICULTURAL IMPACTS, land conversion for industrial, urban and tourist resort developments, and replacement by brackish water AQUACULTURE. In Indonesia, 6% of the former mangrove area has been converted to fish and shrimp ponds. Furthermore, the removal of protective barriers, the interruption to natural sediment supply through *reclamation* and control of waterways, acceleration of SUBSIDENCE through oil and gas extraction and the direct removal of mangrove peats may lead to wetland loss through erosion and ecosystem degradation. Valuation of mangrove, and policy formulation for mangrove use, is complex and problematic.   *TSp*

Christiansen, B. 1983: Mangroves – what are they worth? *Unasylva* 35, 2–15. **Kamaludin, B.H.** 1993: The changing mangrove shorelines in Kuala Kuran, Peninsula Malaysia. *Sedimentary Geology* 83: 187–193. **Spalding, M., Blasco, F. and Field, C.** 1997: *World mangrove atlas.* Okinawa: International Society for Mangrove Ecosystems.

mangrove swamps: impact of tropical cyclones
MANGROVE forests in the tropical storm belts (7°–25° N and S) are vulnerable to hurricane damage because storm frequencies are often well within the lifespan of an individual tree. Impacts, some of which may be delayed, include: defoliation by winds and/or waves; shearing of branches and trunks; uprooting, often preferentially of large trees; and deposition of released sediments and organic debris, the latter leading to nutrient 'flushes' in mangrove-rimmed bays. REGENERATION is through sprouting, refoliation and the dispersal of propagules. Regrowth depends upon species characteristics, topography, sedimentation and hydrology and proximity to storm track. Periodic destruction of Caribbean mangroves by hurricanes may explain their low structural complexity as well as a lack of 'climax' components in the vegetation community. Mangrove forests are important buffers of wave energy and they help to promote net sedimentation on coral islands during cyclone events and aid their growth. Conversely, their removal may lead to more severe hurricane impacts on cleared coasts in terms of coastline erosion and retreat and island decline and disappearance.   *TSp*

[See also CORAL REEF ISLANDS, CORAL REEFS: IMPACT OF TROPICAL CYCLONES]

Roth, L.C. 1992: Hurricanes and mangrove regeneration: effects of Hurricane Joan, October 1988, on the vegetation of Isla del Venado, Bluefields, Nicaragua. *Biotropica* 24, 375–384. **Stone, G. and Finkl, C.** 1995: Impacts of Hurricane Andrew on the coastal zones of Florida and Louisiana: 22–26 August 1992. *Journal of Coastal Research, Special Issue* 21, 1–364. **Conner, W., Day, J., Baumann, R. and Randall, J.** 1989: Influence of hurricanes on coastal ecosystems along the northern Gulf of Mexico. *Wetlands Ecology and Management* 1: 45–56.

mantle   The part of the Earth's interior that underlies the CRUST, from which it is separated by the *Mohorovičić discontinuity (Moho),* and overlies the CORE, from which it is separated by the *Gutenberg discontinuity.* The mantle comprises over 80% of the Earth's volume. It is essentially solid and rocky, but its RHEOLOGY is such that it undergoes slow convective movements. A partially molten layer of the outer mantle – the ASTHENOSPHERE – allows PLATE TECTONICS to operate and is the source of most MAGMA.

GO

[See also EARTH STRUCTURE]

Jackson, I. (ed.) 1998: *The Earth's mantle: composition, structure and evolution.* Cambridge: Cambridge University Press.

mantle plume

A long-lived column of high heat flow from a deep level in the MANTLE. It causes enhanced partial melting of the ASTHENOSPHERE, giving rise to VOLCANISM at a HOTSPOT on the Earth's surface. *GO*

Sheth, H.C. 1999: Flood basalts and large igneous provinces from deep mantle plumes: fact, fiction, and fallacy. *Tectonophysics* 311, 1–29.

manufacturing industry

The *secondary sector* of industry, which converts the raw materials of EXTRACTIVE INDUSTRY into fabricated products. It includes the making and assembly of parts and components and it plays the major role in INDUSTRIALISATION. The two most important developments in the history of manufacturing from the viewpoint of environmental change were probably: first, the invention of the steam engine powered by coal which, combined with the introduction of the factory system, led to the INDUSTRIAL REVOLUTION; and, second, diversification of energy sources and energy end-use technologies. Particularly important in this respect were the far-reaching effects of electricity, the internal combustion engine and oil on the transport industry, employment and international trade. Although local, regional and global environmental impacts of manufacturing industry have been and remain considerable, several generic strategies for impact reduction can be recognised. These include DEMATERIALISATION (decrease in materials used per unit of output), MATERIALS SUBSTITUTION (use of different or new materials and alternative energy sources), WASTE RECYCLING in the strict sense (processing of discarded artefacts), and WASTE MINING (processing of manufacturing waste). Current world recycling rates for lead and steel are 45–50%. *JAM*

[See also WASTE MANAGEMENT]

Ausubel, J.H. 1991: Does climate still matter? *Nature* 350, 649–652. **Ayres, R.U. and Ayres, L.W.** 1996: *Industrial ecology: towards closing the materials cycle.* Cheltenham: Edward Elgar.

map generalisation

Also known as *cartographic generalisation*, map generalisation encompasses the essential process of reducing and symbolising map information from a real-world scale to a level of detail appropriate to the map. The smaller the scale of the map, the greater the level of generalisation applied. *TF*

[See also AREA GENERALISATION]

Robinson, A.H., Morrison, J.L., Muehrcke, P.C. et al. 1995: *Elements of Cartography*, 6th edn. Chichester: Wiley.

map projection

A graphical or mathematical transformation used in map construction to translate the curved surface of the Earth to the plane surface of a sheet map. There exist an infinite number of map projections, but many have little practical use. As it is impossible to retain all the characteristics of the curved global surface when translating to a plane surface, the objective of any map projection is to minimise distortion in, or retain one or more of, a set of desirable characteristics. Hence projections display specific properties (see Table).

Projections can vary in their aspect to the globe (effectively, the way in which the projection plane impinges on the global surface): hence projections are classed as *normal*, *transverse* or *oblique* (see Figure). *TF*

Kraak, M.J. and Ormeling, F.J. 1996: *Cartography: visualization of spatial data*, Harlow: Addison-Wesley Longman. **Maling, D.H.** 1992: *Coordinate systems and map projections*. Oxford: Pergamon.

map projection *Types of map projection and their characteristics*

Projection type	Property	Features
Equivalent	Equal area	Relative size of features is retained (shape is lost)
Conformal or orthomorphic	Shape retention	Shape of small elements is retained (size is not)
Azimuthal or zenithal	Directions from the central point are correct	All angles and directions from the central point are retained

Normal   Transverse   Oblique

map projection *Cylindrical class map projection showing aspect*

marching desert concept A simplistic, outdated and emotive view that LAND DEGRADATION in the form of DESERTIFICATION occurs through progressive advance of desert-like conditions across agriculturally useful land. *RAS*

Thomas, D.S.G. and Middleton, N.J. 1994: *Desertification: exploding the myth.* Chichester: Wiley, 100.

marginal areas Those areas where types of agriculture are or were practised at or near their climatic limits. This is marginality in a spatial/climatic sense, but the concept of marginality can also be considered in economic (when returns barely exceed costs) and social (population pressure) senses. Climatic marginal areas may be considered in terms of both horizontal and vertical (altitudinal) distributions. The marginality is generally due to insufficient water (either from precipitation or irrigation) being available or to either too much or too little heat for the type of agriculture being practised. The main global marginal areas are found in developing countries, where insufficient rainfall is combined with population pressure due either to war or to increasing birth rates (e.g. the SAHEL, southwest and southeast Asia, the Horn of Africa and the Andes). Marginal areas may increase in size as a result of GLOBAL WARMING. *BDG*

Carter, T.R., Parry, M. L., Harasawa, H. and Nishioka, S. 1994: *IPCC technical guidelines for assessing climate change impacts and adaptations.* London: UCL Press. Kates, R.W., Ausubel, J.H. and Berberian, M. (eds) 1985: *Climate impact assessment: studies of the interaction of climate and society* [*SCOPE report* 27]. Chichester: Wiley. Parry, M.L. 1977: *Climate change, agriculture and settlement.* Folkestone: Dawson. Parry, M.L. 1990: *Climate change and world agriculture.* London: Earthscan.

marginal basin One of several types of small SEDIMENTARY BASINS associated with DESTRUCTIVE PLATE MARGINS and ISLAND ARCS, including BACK-ARC basins, FORE-ARC basins and OCEANIC TRENCHES. *GO*

[See also PLATE TECTONICS]
Kokelaar, B.P. and Howell, M.F. (eds) 1984: *Marginal basin geology* [*Special Publication* 16]. London: Geological Society. Tamaki, K. and Honza, E. 1991: Global tectonics and formation of marginal basins: role of the Western Pacific. *Episodes* 14, 224–230.

mariculture The culturing of organisms, especially fish and shellfish, in marine environments. Expensive, high-quality seafoods such as shrimp and molluscs are much cultured in the tropics, and salmon in temperate seas and fjords. Mariculture may involve one or more of the following: feeding at various stages of the life history; construction of enclosures; provision of artificial habitats, including reefs and floating structures; breeding and release programmes, including *ocean ranching* in the open sea. *JAM*

[See also AQUACULTURE]

marine band A thin BED rich in marine FOSSILS within a succession of sediments or sedimentary rocks dominated by deposits lacking marine fossils. The concept is exemplified by goniatite-bearing SHALE horizons within COAL-bearing CYCLOTHEMS in Upper CARBONIFEROUS rocks. These successions represent DELTA DEPOSITS and the marine bands record periods of marine TRANSGRESSION, or *flooding surfaces* (see SEQUENCE STRATIGRAPHY). *GO*

Martinsen, O.J., Collinson, J.D. and Holdsworth, B.K. 1995: Millstone Grit cyclicity revisited, II: sequence stratigraphy and sedimentary responses to changes of relative sea-level. In Plint, A.G. (ed.), *Sedimentary facies analysis: a tribute to the research and teaching of Harold G. Reading* [*International Association of Sedimentologists Special Publication* 22]. Oxford: Blackwell Science, 305–327. Spears, D.A. and Sezgin, H.I. 1985: Mineralogy and geochemistry of the Subcrenatum Marine Band and associated coal-bearing sediments, Langsett, South Yorkshire. *Journal of Sedimentary Petrology* 55, 570–578.

marine geology The branch of GEOLOGY that deals with the OCEAN floor and CONTINENTAL MARGIN. Marine geology encompasses studies of: submarine relief; PLATE MARGINS; the influence of physical processes such as TIDES, waves, STORMS and CONTOUR CURRENTS on the sea-floor; the movement of SEDIMENTS on the CONTINENTAL SHELF, and to and in the deep ocean; the SEDIMENTOLOGY of SUBMARINE FANS and other deep-water CLASTIC systems; the GEOCHEMISTRY of rocks at and beneath the sea-floor; and fluids moving through the CRUST. Marine geology has expanded rapidly since the advent of DEEP-SEA DRILLING in the 1960s and as a result of recent technological advances such as SIDE-SCAN SONAR and SEISMIC STRATIGRAPHY. Thirty years ago the ocean floor was thought to be an inert landscape with little geological or biological activity: marine geology has shown it to be an important, dynamic, responsive and varied environment. *BTC*

[See also OCEANOGRAPHY]
Kennett, J.P. 1982: *Marine geology.* Engelwood Cliffs, NJ: Prentice-Hall. Pickering, K.T., Hiscott, N., Smith, R. and Kenyon, N.H. (eds) 1995: *Atlas of deep water environments – architectural style in turbidite systems.* London: Chapman and Hall.

marine pollution 'The introduction by man, directly or indirectly, of substances or energy into the marine environment (including estuaries) resulting in such deleterious effects as harm to living resources, hazards to human health, hindrance to marine activities including fishing, impairment of quality for use of seawater and reduction of amenities' (GESAMP, 1989). This definition from the IMO/FAO/Unesco/WMO/WHO/IAEA/UN/UNEP Joint Group of Experts on the Scientific Aspects of Marine Pollution summarises an environmental problem that has become worse since the 1980s, especially in coastal and ESTUARINE ENVIRONMENTS. Some 60% of the world's population live within 100 km of the shoreline and more than half the population of developing countries obtain ≥30% of their animal protein from marine fish. Once polluted discharge from the world's rivers reach the sea, it is diluted along with the POLLUTANTS from other sources: however, it is then not only relatively difficult to remove but also becomes an international problem.

Types of marine pollution include oil pollution (OIL SPILLS), synthetic persistent organic chemicals (such as HALOGENATED HYDROCARBONS), METAL POLLUTION (especially HEAVY METALS), microbial pollution (mostly from sewage WASTE), THERMAL POLLUTION and radioactive waste pollution. Artificial RADIOACTIVE ISOTOPES are new to the marine environment. *Dumping* of liquid or solid radioactive substances at sea is now banned by international agreement, but discharge from the land-based nuclear industry continues as it is regulated nationally. Particular concern has been expressed over dumping of RADIOACTIVE WASTE by the former Soviet Union in the Barents and Kara Seas of the Arctic Ocean. Findings suggest that contamination was restricted to the immediate vicinity of the radioactive material. *JAM*

[See also FISHERIES CONSERVATION AND MANAGEMENT, POLLUTION]

Bishop, P.L. 1983: *Marine pollution and its control.* New York: McGraw-Hill. **Clark, R.B.** 1992: *Marine pollution.* Oxford: Clarendon Press. **Føyn, L.** 1997: Marine pollution. In Brune, D., Chapman, D.V., Gwynne, M.D. and Pacyna, J.M. (eds), *The global environment: science, technology and management.* Vol. 1. Weinheim, Germany: VCH, 515–531. **GESAMP** 1989: *Report of Working Group 26 on the State of the Marine Environment [Joint Group of Experts on the Scientific Aspects of Marine Pollution, GESAMP XIX/6.1].* Nairobi, Kenya: United Nations Environment Programme. **Gorman, M.** 1993: *Environmental hazards: marine pollution.* Santa Barbara, CA: ABC-CLIO. **Jickells, T.D., Carpenter, R. and Liss, P.S.** 1991: Marine environment. In Turner, B.L., Clark, W.C., Kates. R.W. *et al.* (eds), *The Earth transformed by human action.* Cambridge: Cambridge University Press, 313–334. **Saliba, L.J.** 1992: Protection of the Mediterranean marine environment. *Journal of the Institution of Water and Environmental Management.* 6, 79–88. **Tippie, V.K. and Kester, D.R. (eds)** 1982: *Impact of marine pollution on society.* New York: Praeger.

marine sediment cores

Vertical sections of SEDIMENT collected from the sea floor. They represent the net accumulation of sediment over time at one location and cores can vary in length from tens of centimetres to tens of metres. Marine sediment cores are used to improve our understanding of the changes in physical, biological and chemical processes that occur within the oceans and at the sea floor. A wide range of scientific disciplines benefits from the collection of marine sediment cores, including PALAEOCEANOGRAPHY, GEOCHEMISTRY, MICROPALAEONTOLOGY, PALAEOMAGNETISM and SEDIMENTOLOGY.

Marine sediment cores are collected using different specially designed coring equipment. Murdoch and Macknight (1991) indicate that: 'CORERS are fundamental tools for obtaining sediment samples for geological and geotechnical surveys and, recently, for the investigation of historical inputs of contaminants to aquatic systems.' Typically, the collection and retrieval of marine sediment cores from the sea floor requires the use of a vessel equipped with appropriate winches and cranes. This can range from small boats to large research vessels, for example the drilling ship of the Ocean Drilling Program (ODP), the *JOIDES Resolution.* The main types of marine sediment cores that can be obtained are box, gravity, piston, vibro- and drilled cores.

Box corers, for example a *Kastenlot corer*, collect large rectangular sediment cores (about $30 \times 60 \times 22$ cm), into which plastic sediment-core tubes can be inserted to retrieve sediment once recovered. A *gravity corer* (see Figure) has a simple, large weighted head with a tubular or box section barrel. Once deployed from a ship, the corer free-falls through the water column and the barrel, often with an internal plastic liner, is driven into the sediment with the aid of a cutting head. The cutting head is placed on the end of the core barrel, in order to achieve better penetration into the sediment. Cutting heads are normally made out of stainless steel, brass or plastic, commonly with a screw or bayonet fitting. A catcher mechanism, often a series of spring-loaded metal fingers, closes as the corer is pulled out of the sediment and the sediment core is retained. A tubular shaped core of up to 3 m of marine sediment can be collected with a gravity corer; however, problems of sediment compression are encountered within the core barrel. PISTON CORERS are often used for studies of deep-ocean bottom sediment and range from 3 to 50 m in length. The corer consists of a weighted stabilised head, core barrel with plastic core liner, a piston, core retainer, cutting head and trigger mechanism. The piston within the corer creates a partial vacuum in the core barrel and sucks the sediment into the barrel. In recent years large piston corers have been developed, e.g. the French giant piston or *Calypso corer*, which has been utilised during the International Marine Global Change Studies (IMAGES).

Recovering loose and unconsolidated SAND and GRAVEL is far more problematic. The *vibrocorer* was developed to recover the non-cohesive sediments, which often dominate energetic, shallow marine environments. Vibrocorers over-

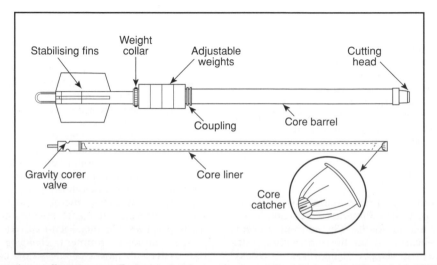

marine sediment cores *Typical parts of a gravity corer (after Murdoch and Macknight, 1991)*

come the resistance of sediment by a vibration action of the core barrel to retrieve marine sediment cores similar in nature to piston cores. Marine sediment cores can also be obtained from the sea floor by mechanical drilling. Improved drilling technology over the last 30 years, since the first site drilled by the *Glomar Challenger* in 1968, has enabled pioneering research to be carried out on the *JOIDES Resolution* for the ODP. This drilling ship is able to drill through soft sediment or MUD using an *advanced hydraulic piston corer* (APC) and, where more resistant sediment or rock is encountered, an *extended core barrel* (XCB) or *rotary core barrel* (RCB) method is applied. The drilling system utilised on the *JOIDES Resolution* is able to handle 9150 m of drill pipe in water depths up to 8235 m. The great revolution within ODP came with the advent of triple APC, allowing the construction of a continuous composite section. This allows '*metres composite depth*' (MCD) *scale*, accommodating core expansion and drilling gaps through interhole correlation using continuous measurements of core physical properties (e.g. MAGNETIC SUSCEPTIBILITY).

In the case of biological and geochemical analysis, a *multicore system* is often employed for the collection of marine sediment. This consists of a series of plastic core tubes mounted on a metal frame that is lowered slowly to the seabed. On contact with the sea floor the core tubes are pushed into the sediment, generally producing an excellent contact at the sediment–water interface.

Marine sediment cores have provided the fundamental data for understanding change in the global oceanographic system over the QUATERNARY period and earlier, as examination of important palaeoenvironmental proxies, such as FORAMINIFERAL ANALYSIS, ICE-RAFTED DEBRIS (IRD) and OXYGEN ISOTOPE analysis, have provided detailed continuous evidence of changes in major parameters, such as SEA-SURFACE TEMPERATURES, ocean volume, ocean PRODUCTIVITY and the nature of the THERMOHALINE CIRCULATION. These data can be placed within a uniform global chronology utilising the OXYGEN ISOTOPE STRATIGRAPHY, which can then be correlated to terrestrial evidence through PALAEOMAGNETIC DATING and the definition of occasional MARKER HORIZONS, as in the case of TEPHRA horizons. Correlation of the relative volumes of water held in the world's oceans and ice caps has proved possible through the reciprocal relationship seen in the oxygen isotope record of marine sediment cores and ICE CORES.

JRE/WENA

[See also MARINE SEDIMENTS, OCEANIC SEDIMENTS]

Curry, W.B., Shackleton, N.J., Richter, C. *et al.* 1995: *Proceedings of the ocean drilling programme (ODP), Initial Reports* 154. College Station, TX: Ocean Drilling Program. Davis, E.E., Mottl, M.J., Fisher, A.T. *et al.* 1992: *Proceedings of the ocean drilling programme (ODP), Initial Reports* 139. College Station, TX: Ocean Drilling Program. Murdoch, A. and Macknight, S.D., 1991. *CRC handbook of techniques for aquatic sediments sampling*. Boca Raton, FL: CRC Press, 29–95. Summerhayes, C.P. and Thorpe, S.A., 1996: *Oceanography – an illustrated guide*. London: Manson. Vernal, A. de and Hillaire-Marcel, C. 2000: Sea-ice cover, sea-surface salinity and halo-/thermocline structure of the Northwest North Atlantic: modern versus full glacial conditions. *Quaternary Science Reviews* 19, 65–85.

marine sediments SEDIMENTARY DEPOSITS in the SEAS and OCEANS, including OCEANIC SEDIMENTS and those that accumulate on the CONTINENTAL SHELF. *GO*

[See also CORAL REEFS, DEEP-SEA DRILLING, ICE-RAFTED DEBRIS, MARINE SEDIMENT CORES, OOZE, PALAEOCEANOGRAPHY, PELAGIC SEDIMENT, SAPROPEL, TURBIDITE]

marker horizon A position in a STRATIGRAPHIC sequence defined by a particular horizon found over a large area and representing a SYNCHRONOUS event. Marker horizons are useful for stratigraphical CORRELATION and RELATIVE-AGE DATING. The commonest marker horizons over the QUATERNARY timescales are: (1) TEPHRAS originating from single VOLCANIC ERUPTIONS; (2) PALAEOSOLS; and (3) changes in the polarity of the Earth's geomagnetic field. In the last case this provides a horizon found on the global scale. *CJC*

[See also GAUSS–MATUYAMA GEOMAGNETIC BOUNDARY]

Markov process A STOCHASTIC process in which the time parameter is discontinuous and the probability of an event is dependent only upon the state of the system and the probabilities of a change from one state to another (*transition probabilities*). In practice, this means that the state at any future time is dependent on the present state and the transition probabilities but is unaffected by any additional knowledge of the past history of the system. The usefulness of the concept and corresponding *Markov chain analysis* depends especially on the reliability of the established transition probabilities used and whether they actually remain constant through time. *AHP/JAM*

Caskey, J.E. 1963: A Markov chain model for the probability of precipitation occurrence in intervals of various length. *Monthly Weather Review* 91, 298–301. Caskey, J.E. 1964: Markov chain model of cold spells at London. *Meteorological Magazine* 93, 136–138. Waggoner, P.E. and Stephens, G.R. 1970: Transition probabilities for a forest. *Nature* 225, 1160–1161

marl A calcareous, slightly clayey LACUSTRINE SEDIMENT with >80% $CaCO_3$ deposited in the sublittoral and eulittoral zones of a lake and formed through the activity of calcareous algae. *UBW*

[See also ALGAL ANALYSIS, SEDIMENT TYPES, STABLE ISOTOPE ANALYSIS]

Kelts, K. and Hsü, K.J. 1978: Freshwater carbonate sedimentation. In Lerman, A. (ed.), *Lakes, chemistry, geology, physics*. New York: Springer, 295–323.

mass attraction The gravitational effect of large continental and ice masses on SEA LEVEL causing sea level to rise in their vicinity. *JS*

[See also EUSTASY, ISOSTASY, SEA-LEVEL CHANGE]

Mörner, N.-A. 1987: Models of global sea-level changes. In Tooley, M.J. and Shennan, I. (eds), *Sea level changes*. Oxford: Blackwell, 332–355.

mass balance The change in mass at any point on the surface of a GLACIER at any time. Commonly it means the change in mass of the entire glacier in a standard unit of time (the balance year or measurement year). The mass balance is the result of variations in ABLATION and ACCUMULATION, which cause volume changes. The *net balance* is the sum of the *winter balance* (positive) and the *summer balance* (negative). If ablation exceeds accumulation, the net balance is negative: the opposite situation produces a positive net balance. *AN*

Nesje, A. and Dahl, S.O. 2000: *Glaciers and environmental change*. London: Arnold.

mass extinctions

There have been a number of episodes in the history of the Earth when extinctions have occurred at rates far greater than the BACKGROUND RATE. These episodes are sufficiently prominent that they form the basis of much of the classic subdivision of the geological column (e.g. the extinction of the dinosaurs at the Cretaceous–Tertiary boundary). Some episodes are more pronounced than others and the severity, in terms of proportion of species that become extinct, varies widely between different groups. FAUNA, especially marine fauna, tend to be more severely affected than plants. Marine BIODIVERSITY tends to 'rebound' after a mass extinction, extinction rates being correlated with *origination rates* with a time lag of around 10 million years (see Figure). There has been, and still is, debate about the nature of these episodes, whether they are caused by a CATASTROPHIC EVENT and, in particular, whether they are periodic. They tend to occur every 26–30 million years, but it is not clear whether this is a genuinely periodic frequency.

It is also possible, but not established, that mass extinction may have an extraterrestrial explanation, through a METEORITE IMPACT, for example, and this might explain a PERIODICITY. The Earth is certainly subjected to all manner of extraterrestrial phenomena. So, for example, the number and size of known ASTEROIDS in orbits that approach the Earth means that on average one collision every 1.4 million years with objects greater than 2 km in diameter can be expected, one every 330 million years with objects greater than 8 km in diameter, and so on.

Whatever the explanation, a high proportion of the Earth's fauna, and some FLORA, become extinct during these episodes, with far-reaching consequences for EVOLUTION, because the continuance of life on Earth depends on the survivors, which may be a small, randomly selected, proportion of the original fauna.

It may be that mass extinctions solve what has been termed the '*paradox of the first tier*': 'our failure to find any clear vector of fitfully accumulating progress, despite expectations that processes regulating the first tier should yield such advance, represents our greatest dilemma for the study of pattern in life's history' (Gould, 1985: 4). Mass extinctions resolve the paradox because evolutionary processes at the first tier (i.e. ecological processes) cannot extend to the longest time scales. Mass extinctions, occurring randomly with respect to these shorter-term processes, prevent the extension of evolutionary trends. They may thus be of considerable importance in determining which LINEAGES comprise modern faunas and (possibly to a lesser extent) floras. *KDB*

[See also EXTREME CLIMATIC EVENTS, K–T BOUNDARY, NEAR-EARTH OBJECT]

Courtillot, V. 1999: *Evolutionary catastrophes: the science of mass extinction*. Cambridge: Cambridge University Press. **Erwin, D.H.** 1994: The Permo-Triassic extinction. *Nature* 367, 231–236. **Frankel, C.** 1999: *The end of the dinosaurs: Chicxulub crater and mass extinction*. Cambridge: Cambridge University Press. **Glen, W.** 1994: What the impact/volcanism/mass-extinction debates are about. In W. Glen (ed.), *The mass-extinction debates: how science works in a crisis*. Stanford, CA: Stanford University Press, 7–38. **Gould, S. J.** 1985: The paradox of the first tier: an agenda for paleobiology. *Paleobiology* 11, 2–12. **Kirchner, J.W. and Weil,**

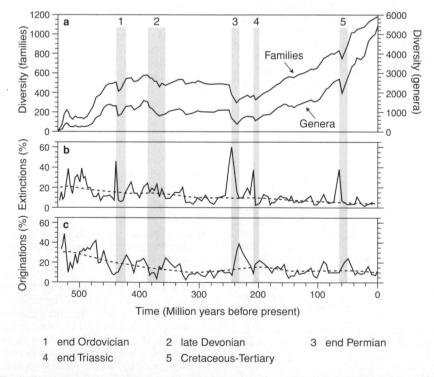

1 end Ordovician   2 late Devonian   3 end Permian
4 end Triassic   5 Cretaceous-Tertiary

mass extinctions *The five largest mass extinctions of the geological record (shaded and numbered columns) as exemplified by the fossil record of marine animal biodiversity: (a) number of families and genera; and corresponding percentages of (b) extinction and (c) origination. Dotted lines are background rates estimated by curve-fitting (after Kirchner and Weil, 2000)*

A. 2000: Delayed biological recovery from extinctions throughout the fossil record. *Nature* 404, 177–180. **Raup, D.M. and Sepkoski Jr, J.J.** 1984: Periodicity of extinctions in the geologic past. *Proceedings of the National Academy of Sciences of the USA* 81, 801–805. **Sepkoski Jr, J.J.** 1989: Periodicity in extinctions and the problem of catastrophism in the history of life. *Journal of the Geological Society, London* 146, 7–19. **Shoemaker, E.M., Wolfe, R.F. and Shoemaker, C.S.** 1990: Asteroid and comet flux in the neighbourhood of the Earth. In Sharpton, V.L. and Ward, P.D. (eds), *Global catastrophes in Earth history: an interdisciplinary conference on impacts, volcanism, and mass mortality* [Special Paper 247]. Boulder, CO: Geological Society of America, 155–170. **Stigler, S.M. and Wagner, M.J.** 1987: A substantial bias in non-parametric tests for periodicity in geophysical data. *Science* 215, 1501–1503. **Traverse, A.** 1988: Plant evolution dances to a different beat: plant and animal evolutionary mechanisms compared. *Historical Biology* 1, 277–301. **Willis, K.J. and Bennett, K.D.** 1995: Mass extinction, punctuated equilibrium and the fossil plant record. *Trends in Ecology and Evolution* 10, 308–309.

mass movement processes

Mass movement processes, involving the down-slope displacement of rock and REGOLITH under the influence of gravity, are of global significance in the development of hillslopes. They are also of economic and social importance as they may disrupt public utilities and cause loss of life. Although not confined to a particular environment, they probably reach their greatest intensity and efficacy in steep and mountainous PERIGLACIAL regions. A range of materials and processes is involved in mass movement and gives rise to a great variety of movement types. Criteria for recognition of these types include: velocity and mechanism of movement; physical properties of the material; mode of DEFORMATION; geometry of both the source area and the displaced mass; and water content. The most common groupings of movement are rebounds, saggings, slumps, falls, topples, slides, flows and creeps. The type of movement occurring in an area depends on the local climatic factors, lithology and structure and local topography. *PW*

[See also AVALANCHE, CAMBERING, DEBRIS FLOW, GELIFLUCTION, LANDSLIDE, PARAGLACIAL, PLOUGHING BLOCK, ROCK CREEP, ROCK STREAM, SOIL CREEP, SOLIFLUCTION, STURZSTROM, TALUS, UNLOADING, VALLEY BULGING]

Brunsden, B. and Prior, D.B. (eds) 1984: *Slope instability.* Chichester: Wiley. **French, H.M.** 1996: *The periglacial environment,* 2nd edn. Harlow: Addison-Wesley Longman. **Selby, M.J.** 1993: *Hillslope materials and processes.* Oxford: Oxford University Press.

mass spectrometer

An instrument designed to measure ISOTOPE RATIOS by separating positively charged IONS on the basis of their *mass-to-charge ratio.* *IR*

[See also ISOTOPES AS INDICATORS OF ENVIRONMENTAL CHANGE, STABLE ISOTOPE ANALYSIS]

Preston, T. 1992: The measurement of stable isotope natural abundance variations. *Plant, Cell and Environment* 15, 1091–1097.

mass wasting

A collective term for a range of processes causing down-slope movement of REGOLITH under the influence of gravity. Mass wasting occurs in all climatic regions, but is probably of greatest intensity in PERIGLACIAL environments. *PW*

[See also MASS MOVEMENT PROCESSES]

Harris, C. 1981: *Periglacial mass wasting: a review of research* [BGRG Research Monograph No. 4]. Norwich: Geobooks.

Massenerhebung effect

The effect of large mountain masses, particularly plateaux, in raising temperatures above the values found at similar heights in the free atmosphere and on isolated peaks. The effect arises because large upland surfaces present extensive areas for heating by SOLAR RADIATION receipts, which increase with altitude, whereas isolated mountains merely protrude into the free atmosphere and temperatures fall according to the local atmospheric LAPSE RATE. The term was first used in the European Alps, where the SNOW LINE, TREE LINES and the ALTITUDINAL ZONATION of vegetation are all at higher altitude on the large central massifs, such as the Pennine and Engadine Alps, than in the more coastal Alpine ranges. The telescoping of altitudinal limits in coastal mountains compared with further inland in tropical and temperate areas was formerly ascribed erroneously to *Massenerhebung* inland and its absence on coastal hills, whereas the main reason for the contrast is usually linked to higher humidity and hence cloud formation at lower altitudes in coastal areas. *RPDW*

Brockmann-Jerosch, H. 1913: Der Einfluss des Klimacharakters auf die Verbreitung der Pflanzen und Pflanzengesellschaften. *Botanisches Jahrbuch (Beiblätter)* 49, 19–43. **Ellenberg, H.** 1988: *Vegetation ecology of central Europe.* Cambridge: Cambridge University Press. **Walsh R.P.D.** 1996: Climate. In Richards, P.W. with Walsh R.P.D., Baillie I. and Greig-Smith P. *The tropical rain forest,* 2nd edn. Cambridge: Cambridge University Press, 159–205.

massive

BEDS or bodies of SEDIMENT lacking obvious STRATIFICATION or variations in TEXTURE. *MRT*

master chronology

An average time series of TREE RING data from many trees in a region. All of the annual values are aligned in the precise calendar year of their growth and averaged, so that the resulting series expresses the pattern of common year-to-year variability contained in the trees from that region. The master chronology thus represents a 'templet' of INTERANNUAL VARIABILITY in growth, against which series of ring widths in other wood samples from that area may be compared, matched and so dated with absolute PRECISION. The term is also used in the context of VARVE CHRONOLOGY. *KRB*

[See also CHRONOLOGY, CROSS-DATING, DENDROCHRONOLOGY, TREE-RING INDEX]

Baillie, M.G.L. 1995: *A slice through time: dendrochronology and precision dating.* London: Routledge.

material culture

In the archaeological context, all the physical expressions of a culture, including not only the portable objects (ARTEFACTS) but also nonportable human-made remains ('features') such as buildings, FIELD SYSTEMS, MIDDENS, burials and roads. *JAM*

Schlereth, T.J. (ed.) 1982: *Material culture studies in America.* Nashville, TN: American Association for state and local history. **Thomas, N.** 1991: *Entangled objects: exchange, material culture, and colonialism in the Pacific.* Cambridge, MA: Harvard University Press.

materials substitution

The substitution of one material for another in MANUFACTURING INDUSTRY and a core phenomenon of INDUSTRIALISATION. Classic examples include: the successive use of charcoal, coal, oil and natural gas in energy production, both during and since the

INDUSTRIAL REVOLUTION; and the displacement of natural materials by synthetic fibres, rubber, plastics and fertilisers. Materials substitution can overcome resource constraints, it can result in products that are more economically efficient and/or function better, it can promote new applications, it can replace harmful substances by environmentally friendly ones and it can contribute to reduced materials use. *JAM*

Grübler, A. 1998: *Technology and global change.* Cambridge: Cambridge University Press. **Wernick, I.K.** 1996: Consuming materials: the American way. *Technological Forecasting and Social Change* 53, 111–122.

matrix Particles in a SOIL, SEDIMENT or SEDIMENTARY ROCK that are finer grained than the nominal GRAIN-SIZE – e.g. MUD matrix in a SANDSTONE, SAND or mud matrix in a CONGLOMERATE or DIAMICTON. Some dispute surrounds the definition of matrix in sandstones: it is commonly defined as particles finer than 0.03 mm (i.e. medium to coarse SILT). Matrix can be formed by DEPOSITION of poorly sorted sediment, by INFILTRATION of fines after deposition or by COMPACTION of mechanically weak LITHIC fragments during DIAGENESIS. Larger particles or objects are embedded in the matrix: the matrix may support larger CLASTS or fill interstices between them. A mud matrix in a RUDITE may indicate non-aqueous deposition (e.g. by DEBRIS FLOW or as glacial TILL), particularly where it forms a *matrix-supported* FABRIC. In an archaeological context, the matrix is the physical medium (usually a soil or sediment) that surrounds and/or supports an archaeological find. *GO/TY*

[See also FLYSCH, GREYWACKE, MICROMORPHOLOGICAL ANALYSIS, PROVENIENCE, WACKE]

Pettijohn, F.J. 1975: *Sedimentary rocks*, 3rd edn. New York: Harper & Row. **Walker, T.R., Waugh, B. and Crone, A.J.** 1978: Diagenesis in first-cycle desert alluvium of Cenozoic age, southwestern United States and northwestern Mexico. *Geological Society of America Bulletin* 89, 19–32.

maturity (1) In the context of LANDSCAPES, maturity describes a stage in evolution. It is used especially as rivers develop towards the low gradient, wide, typically highly sinuous morphology found near BASE LEVEL (see also RIVER REGIME).

(2) In the context of SEDIMENTS and SEDIMENTARY ROCKS, maturity is a concept that combines several parameters of TEXTURE or composition of a SANDSTONE. The concept can be used descriptively or can be interpreted in terms of a more mature sediment having experienced more extensive WEATHERING, a longer path of SEDIMENT TRANSPORT or a DEPOSITIONAL ENVIRONMENT of higher ENERGY. However, the concept needs to be applied with caution. Source inheritance is a particular problem: for example, polycyclic or POLYGENETIC sediment derived from a quartz-rich source will necessarily be compositionally mature (see PROVENANCE). Such problems can be minimised by comparing different types of maturity for a given sediment.

Textural maturity describes the tendency towards being well sorted (see SORTING), having well rounded GRAINS (see ROUNDNESS) and lacking a MATRIX. *Compositional* (or *mineralogical*) *maturity* describes the tendency towards the occurrence of only the most stable MINERAL grains (typically QUARTZ) with the loss of the more degradeable

(labile) mineral and LITHIC grains, which are chemically altered to CLAY minerals. A *maturity index* expressed as the ratio (quartz + chert):(feldspars + rock fragments) is sometimes used. *Chemical maturity* describes the evolution towards a more inert chemical composition during CHEMICAL WEATHERING (e.g. during loss of soluble components by LEACHING). The chemical index of alteration (CIA) is calculated (on a CARBONATE-free basis) as the ratio of aluminium oxide to total oxide content.

(3) In the context of organic sediments, maturity refers to the degree to which organic material has been converted during burial to kerogen and to HYDROCARBONS (*maturation*). *TY/GO*

Cox, R. and Lowe, D.R. 1995: A conceptual review of regional-scale controls on the composition of clastic sediment and the co-evolution of continental blocks and their sedimentary cover. *Journal of Sedimentary Research* A65, 1–12. **Leeder, M.R.** 1982: *Sedimentology: process and product.* London: Unwin Hyman. **Pettijohn, F.J.** 1975: *Sedimentary rocks*, 3rd edn. New York: Harper and Row. **Selley, R.C.** 1985: *Elements of petroleum geology.* New York: Freeman. **Summerfield, M.A.** 1991: *Global geomorphology.* Harlow: Longman.

Maunder minimum The period of very low SUNSPOT activity from AD 1645 to 1715. According to Schove (1983), AD 1693 was the lowest trough in a SUNSPOT CYCLE ever recorded. The coincidence of the Maunder minimum in solar activity with the coldest period of the LITTLE ICE AGE provides evidence for SOLAR FORCING of SECULAR VARIATIONS in climate. The Maunder minimum, together with other historical sunspot minima (DALTON MINIMUM, OORT MINIMUM, SPÖRER MINIMUM, and WOLF MINIMUM) and the *contemporary solar activity maximum* (CSM) and *Mediaeval solar activity maximum* (MSM), have been detected in various PROXY DATA sources, such as the COSMOGENIC ISOTOPE record in ICE CORES (see Figure). *JAM/AHP*

Damon, P.E., Eastoe, C.J., Hughes, M.K. *et al.* 1998: Secular variation of Δ^{14}C during the Medieval Solar Maximum: a progress report. *Radiocarbon* 40, 343–350. **Eddy, J.A.** 1976: The Maunder minimum. *Science* 192, 1189–1202. **Frenzel, B., Pfister, C. and Gläser, B. (eds)** 1994: *Climatic trends and anomalies in Europe 1675–1715* [*Paläoklimaforschung.* Vol. 13]. Stuttgart: Gustav Fischer Verlag. **Schove, D.J.** 1983: *Sunspot cycles.* Stroudsburg, PA: Hutchinson Ross.

meadow A *grassland* maintained by mowing as distinct from PASTURE maintained by grazing. Meadows are ANTHROPOGENIC ecosystems and part (sometimes an ancient part) of the CULTURAL LANDSCAPE. Distinct *flood meadows* (flooded naturally), *water meadows* (flooded artificially), *hay meadows*, '*cornfields*' and *alpine meadows* are amongst the types that may be recognised. All are characterised by a rich flora of tall HERBS. Because of the decline in traditional agricultural practices, meadows are in urgent need of CONSERVATION. In Britain, for example, 97% of meadows have disappeared since the 1930s. *JAM*

[See also HABITAT LOSS]

Feltwell, J. 1992: *Meadows: a history and natural history.* Stroud, Gloucestershire: Alan Sutton. **Cook, H. and Williamson, T.** 1999: *Water management in the English landscape: field, marsh and meadow.* Edinburgh: Edinburgh University Press.

mean annual air temperature (MAAT) Usually calculated as the average ANNUAL MEAN AIR TEMPERATURE

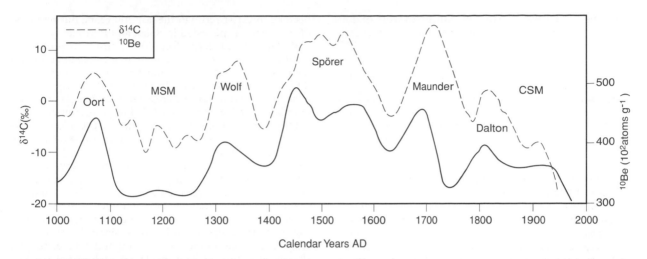

Maunder minimum *Trends in δ¹⁴C and ¹⁰Be over the last millennium from ice cores compared with named minima in solar activity, such as the Maunder minimum, the contemporary solar activity maximum (CSM) and the Mediaeval solar activity maximum (MSM) (after Damon et al., 1998)*

over a standard time period of 30 years (i.e. the average of 30 annual mean temperature values). The years 1961–1990 are commonly recognised as the latest *standard period*). *BDG/AHP*

[See also CLIMATIC NORMAL]

Hatch, D. 1984: *Weather around the world*. Amsterdam: Donald Hatch. **Pearce, E.A. and Smith, C.G.** 1990: *The world weather guide*, 2nd edn. London: Hutchinson. **Wernstedt, F.L.** 1972: *World climatic data*. Lemont, PA: Climatic Data Press.

mean lethal dose (L_{50}) The dose of a TOXIN required to kill one half of a target population. *JAM*

mean monthly temperature Usually calculated as the average MONTHLY MEAN TEMPERATURE over a standard time period of 30 years. *AHP*

mean sea level The average elevation of the sea surface along a SHORELINE over a given period of time. *HJW*

[See also RELATIVE SEA LEVEL, SEA-LEVEL CHANGE]

Pugh, D.T. 1987: *Tides, surges and mean sea level*. Chichester: Wiley. **Tooley, M.J.** 1993: Long term changes in eustatic sea level. In Warrick, R.A., Barrow, E.M. and Wigley, T.M.L. (eds), *Climate and sea level change: observations, projections and implications*. Cambridge: Cambridge University Press, 81–107.

meander A sinuous winding bend in a river. The term originates from the Maiandros River in northwestern Turkey. Meanders occur in moderately active systems and bend creation can be related to sufficient energy for systematic bed deformation and BANK EROSION. *Incised meanders* have cut down into bedrock. OCEAN CURRENTS and JET STREAMS also meander. *DML*

[See also CHANNEL PATTERN, MEANDER CUTOFF, MISFIT MEANDER]

Ikeda, S. and Parker, G. (eds) 1989: *River meandering* [*Water Resources Monograph* 12]. Washington, DC: American Geophysical Union.

meander cutoff A MEANDER bend abandoned by the main river channel after AVULSION or progressive erosion

of the meander neck. A meander cutoff appears on the floodplain as a horseshoe-shaped depression, sometimes water-filled; hence the term *oxbow lake*. A sudden increase in FLOOD MAGNITUDE-FREQUENCY with climatic or LANDUSE CHANGE may produce a large number of cutoffs over a short period. Sediment infill, vegetation succession and soil development follow their formation. *DML*

[See also CHANNEL CHANGE]

Lewis, G.W. and Lewin, J. 1983: Alluvial cutoffs in Wales and the Borderland. In Collinson, J.D. and Lewin, J. (eds), *Modern and ancient fluvial systems* [*Special Publication of the International Association of Sedimentologists* 6] . Oxford: Blackwell, 145–154.

mechanical soil conservation measures Structures or land-shaping techniques intended to prevent or reduce SOIL EROSION on slopes. On cultivated land, the measures include land formation techniques such as CONTOUR BUNDS and TERRACING. There are also stabilisation structures, including GABIONS and GULLY CONTROL DAMS. The other forms of SOIL CONSERVATION measures on cultivated land are agronomic measures (MULCHING and crop management) and soil management, which involves various types of conservation TILLAGE. *RAS*

[See also SOIL CONSERVATION]

Hudson, N.W. 1971: *Soil conservation*. Batsford: London.

Mediaeval Warm Period The *Little Climatic Optimum, Mediaeval Warm Epoch* or Mediaeval Warm Period extended from approximately the late ninth to the early fourteenth centuries and was characterised by a climate at least as warm as today. The Viking settlements flourished in Greenland and Iceland at this time, and agriculture expanded in MARGINAL AREAS of the uplands of northwest Europe only to decline during the subsequent LITTLE ICE AGE. The Mediaeval Warm Period was not uniformly warm: evidence from such sources as GLACIER VARIATIONS and DENDROCLIMATOLOGY indicates interruptions, to such an extent that some authors doubt its existence. Recent results from ice-sheet temperature profiles indicate temperatures of about 1.0°C warmer than today. It repre-

sents one of the short-term, SUB-MILANKOVITCH or cen-
tury- to millennial-scale climatic variations that characterised the HOLOCENE.   *JAM*

[See also SUBSURFACE TEMPERATURE]

Lamb, H.H. 1965: The early Medieval Warm Epoch and its
sequel. *Palaeogeography, Palaeoclimatology and Palaeoecology* 1,
13–37. **Hughes, M.K. and Diaz, H.F. (eds)** 1994: *The
Medieval Warm Period*. Dordrecht: Kluwer; reprinted from
Climatic Change 26, 109–342. **Stine, S.** 1994: Extreme and persistent drought in California and Patagonia during medieval time.
Nature 369, 546–549.

Mediterranean environmental change and human impact

During the PLEISTOCENE, CLIMATIC
CHANGE caused phases of slope instability and stability in
Mediterranean lands. In general, INTERGLACIAL episodes
were characterised by low sedimentation amounts, a
dense vegetation cover, relatively high INFILTRATION and
moderate river discharges. In contrast, GLACIAL phases
tended to be cold and dry with increased SEASONALITY of
rainfall, a sparse vegetation cover, extensive areas of bare
ground and loose SEDIMENTS, leading to extensive slope
EROSION and 'flashy' river regimes. Coastline development
and fluvial activity were affected by SEA-LEVEL RISE and fall
(see MEDITERRANEAN LANDSCAPE EVOLUTION). After fluctuations around 14 000 years ago, the climate became rather
wetter than hitherto and forest characterised the higher
rainfall and SAVANNA the low rainfall areas.

The Mediterranean landscape has a long history of
human occupance and modification of the landscape,
people, plants and animals. The mainland has been continuously occupied since the PALAEOLITHIC with islands
(e.g. Cyprus, Crete) not inhabited until about 8500 years
ago. The construction of settlements and agriculture
began at around this time in Anatolia, followed by
Palestine and southern Greece.

Around 3500 years ago, in the late BRONZE AGE, human
impact on the landscape was strongly in evidence with
TERRACING and IRRIGATION works being constructed and a
range of crops grown (e.g. winter wheat, barley, vines and
olives) and animals domesticated (e.g. cattle, sheep, goats
and pigs). Much of what survived in terms of WETLANDS
and forest was managed. Large population levels were
reached, with numbers in Greece peaking in the
Hellenistic period (about 2200 years ago), but later in
Italy during the ROMAN PERIOD (about 1800 years ago).
The Romans developed IRRIGATION methods, with storage
dams and aqueducts. After the Roman Period, there was
the Muslim expansion into the Mediterranean to include
certain Mediterranean islands (Crete, Sicily, Sardinia)
and most of the Iberian Peninsula. Persistence of Arabic
influence lasted longest in southeast Spain, where new
crops were introduced (e.g. mulberry, rice, sugar cane,
cotton, citrus fruits and aubergine), irrigation works were
elaborated and extended to include terraced hillsides, and
SEDIMENT TRAPS for soil eroded from BADLANDS were
devised.

By Mediaeval times, a typical landscape in most
Mediterranean countries comprised scattered trees in
grassland and heath (or SAVANNA). The agricultural use of
such terrain comprised (and still comprises in many areas)
grazing of animals (sheep, goats and pigs) combined with
woodcutting, shifting cultivation and the harvesting of
cork. Most savannas in Italy and the Balearic Islands have
become forest or scrub in the last 200 years, but retention
of savannas has persisted in Spain and Portugal.

Population increases from the eighteenth and nineteenth centuries led to an increase in the area of cultivated
land, with DEFORESTATION and terracing on increasingly
difficult terrain. From AD 1830 to 1950, however, most
European Mediterranean areas suffered periods of overpopulation and poverty, which led to programmes of tree
planting in some countries in order to provide employment, to recreate (supposedly) a former forest cover or in
an attempt to prevent soil erosion and flooding. Modern
agricultural practices, making use of machinery, artificial
fertilisers, greenhouses, piped irrigation water and pesticides, have encouraged cultivation on plains and on
drained WETLANDS that had been left uncultivated because
of malaria. These developments, together with the expansion of tourism in coastal locations since the 1960s, has
led, for example, to the construction of dams, leading to:
storage of sediments that would otherwise have reached
the coast to supply beach sediment; BEACH MINING for
construction materials causing increased coastal erosion;
excessive extraction of groundwater in coastal areas causing SALINISATION; and widespread abandonment of TER-
RACING, leading to reversion to flammable semi-natural
vegetation. In some places, the vegetation has been supplemented by highly flammable pines and eucalyptus.
With minimal management, these areas have been subject
to an increased frequency of WILDFIRES, which has led to
increased SOIL EROSION.

The long human occupance of Mediterranean landscapes and their predisposition to erosion as a result of
their subtropical location with summer drought, intense
rainfall, often steep terrain and often highly erodible soils
and/or unconsolidated SEDIMENTARY ROCKS has led to
speculation about the extent to which indications of
HOLOCENE erosion in the landscape reflect anthropogenic
or natural agencies. Such evidence includes hillslopes
denuded of soil, BADLANDS, FLOODPLAIN deposits, ALLU-
VIAL FANS and coastal SEDIMENTATION, with ancient ports
stranded far inland (e.g. Troy and Ephesus).

Spectacular GULLIES and badlands in southeast Spain
would seem to indicate a recent anthropogenic origin, but
archaeological evidence has shown that many of the gullies were already established some 4000 years ago. Later
vegetation clearance seems to have altered little the geomorphological patterns already in existence in the area.

An important milestone in the human versus natural
debate concerning Mediterranean landscape change was
the publication in 1969 of *The Mediterranean valleys* by C.
Vita-Finzi. It was argued that in many Mediterranean valleys there were terraces reflecting two types of alluvial fill
(an Older Fill and Younger Fill) (see Figure in the entry
on VALLEY FILL DEPOSITS). The Older Fill was thought to
date from the LAST GLACIAL MAXIMUM, when increased
FROST WEATHERING and more intense seasonal rainfall led
to removal of sediment from slopes and AGGRADATION in
valleys. Frequently 'nested' within the Older Fill is the
Younger Fill, which was interpreted as broadly synchronous throughout the region and ranging in age (based
largely on archaeological evidence) from about AD 400 to
1500 or even 1800. The apparent synchroneity, emerging
evidence of CLIMATIC CHANGE in northern Europe and
North America and the supposed inadequacy of an
anthropogenic origin led Vita-Finzi to favour a climatic-

change explanation. Subsequently, alternative views have been expressed for the origin of the Older Fill (possibly still under formation as late as the early Holocene), but there has been debate especially in respect of the Younger Fill, with suggestions of a wholly anthropogenic or a combined natural (extreme climatic events providing the necessary trigger) and anthropogenic (extensive removal of the vegetation cover) origin, as well as more complex climatic origins.

Irrespective of the cause(s), the phases of removal of soil from the steep slopes have had beneficial as well as negative effects. The extensive alluvial soils on the valley floors represent an important landuse resource. Nevertheless, soil erosion remains a major environmental problem in the Mediterranean and is aggravated by the gradual abandonment of terraces and a high incidence of forest fires. Estimated annual EROSION RATES of 10–165 t ha^{-1} commonly occur with instances of rates as high as 300 t ha^{-1}. Successive annual losses of 150 t ha^{-1} for Mediterranean soils can be expected to cause serious SOIL DEGRADATION in as few as five years. It is argued that many of the drier regions of the Mediterranean are suffering, or are prone to, DESERTIFICATION.   *RAS*

[See also MEDITERRANEAN LANDSCAPE EVOLUTION, VALLEY FILL DEPOSITS]

Bintliff, J.-L. 1992: Erosion in the Mediterranean lands: a reconsideration of pattern, process and methodology. In Bell, M. and Boardman, J. (eds), *Past and present soil erosion* [*Oxbow Monograph* 22]. Oxford: Oxbow. **Brückner, H.** 1986: Man's impact on the evolution of the physical environment in the Mediterranean region in historical times. *GeoJournal* 16, 7–17. **Butzer, K.W.** 1980: Holocene alluvial sequences: problems of dating and correlation. In Cullingford, R.A., Davidson, D.A. and Lewin, J. (eds), *Timescales in geomorphology*. Chichester: Wiley, 131–142. **Grove, A.T. and Rackham, O.** 1998: History of Mediterranean land use. In Mairota, P., Thornes, J.B. and Geeson, N. (eds), 1998: *Atlas of Mediterranean environments in Europe: the desertification context*. Chichester: Wiley, 76–78. **Grove, A.T. and Rackham, O.** 2000: *The nature of Mediterranean Europe: an ecological history*. London: Yale University Press. **Lewin, J., Macklin, M.G. and Woodward, J.C. (eds)** 1995: *Mediterranean Quaternary river environments*. Rotterdam: Balkema. **Roberts, N., Kuzucuoglu, C. and Karabiyikoglu, M. (eds)** 1999: The Late Quaternary in the eastern Mediterranean. *Quaternary Science Reviews* 18, 497–516. **Prentice, I., Guiot, J. and Harrison, S.P.** 1992: Mediterranean vegetation, lake levels and palaeoclimate at the Last Glacial Maximum. *Nature* 360, 658–660. **Rose, J., Meng, X. and Watson, C.** 1999: Palaeoclimatic and palaeoenvironmental responses in the western Mediterranean over the last 140 ka: evidence from Mallorca, Spain. *Journal of the Geological Society of London* 156, 435–448. **Vita-Finzi, C.** 1969: *The Mediterranean valleys*. Cambridge: Cambridge University Press. **Vita-Finzi, C.** 1976: Diachronism in Old World alluvial sequences. *Nature* 250, 568–570. **Wagstaff, J.M.** 1981: Buried assumptions: some problems in the interpretation of the 'Younger Fill' raised by recent data from Greece. *Journal of Archaeological Science* 8, 247–264. **Wise, S.M., Thornes, J.B. and Gilman, A.** 1982: How old are the badlands? A case study from south-east Spain. In Bryan, R and Yair, A. (eds), *Badland geomorphology and piping*. Norwich: GeoBooks, 259–277.

Mediterranean landscape evolution

The geological and geomorphological evolution of the modern Mediterranean began early in the Cenozoic era (65 Ma to the present) as the northerly-drifting African Plate collided with the main Eurasian Plate to the north and a mosaic of small plates throughout the Mediterranean (Arabian, Adriatic and Iberian plates). Initially, the TETHYS ocean was narrow and characterised by intense deformation and volcanicity in the Apennines, Greece and Turkey. Continental impact through PLATE TECTONICS resulted in mountain-building episodes that formed the Pyrenees, Carpathians and Alps. Continuing movements at plate boundaries account for the pattern of SEISMICITY throughout the Mediterranean Basin. The major EARTHQUAKE zone associated with plate collision lies along the Hellenic Arc, stretching from the west coast of Greece to southern Turkey.

There are strong links between GEOLOGICAL STRUCTURE and topography in landscapes around the Mediterranean Sea. In essence, the Mediterranean can be viewed as a ring of mountains surrounding the sea. At a large scale, the configuration of mountain belts, island chains and mountain ranges reflects collisional tectonics. These areas are separated by resistant microplates such as Iberia, or by areas such as the western Mediterranean and Adriatic and Dead Seas. The present plan form of the Mediterranean coastline is influenced strongly by fluvial processes with extensive alluvial fan systems (e.g. Rhône, Ebro, Po and Nile Deltas) and marine processes (e.g. deep-water channels in the Straits of Gibraltar and the Bosphorus).

In addition to the impacts of Pleistocene SEA-LEVEL CHANGE, CLIMATIC CHANGE and of changes to the vegetation cover (caused naturally or through human impact), drainage systems in the Mediterranean show strong links to geological structure and NEOTECTONICS. There are examples of disrupted drainage systems in the form of river reversal, capture, diversion or ponding (e.g. Greek islands, Turkey and Italy). Mature rivers tend to drain parallel to major mountain belts (e.g. the Po) and minor extensional basins (e.g. Rhône, Nile).

Whereas geological effects in the Mediterranean can be seen best at the large scale, the effects of Pleistocene climatic changes on Mediterranean landscapes are best displayed at the medium or small scale. Fluctuations in sea levels impacted on coasts as well as rivers, with raised *shore platforms* common on the predominantly rocky coasts (70–75% of the Mediterranean coastline is rocky) often reflecting high *eustatic* sea levels attained during INTERGLACIALS with their altitudes affected by subsequent neotectonic action. In many places, glacio-eustatically low sea levels are recorded along present-day coastlines by AEOLIANITE formation.

Inland, away from high mountains, Pleistocene climatic fluctuations in general led to phases of slope stability during interglacials and instability during glaciations, producing sequences of COLLUVIUM and alluvial SEDIMENTS, sometimes interdigitating with PALAEOSOLS. Human action during the HOLOCENE also caused slope instability leading to SOIL EROSION. Where eroded sediments reached valley floors, they often formed VALLEY FILL DEPOSITS.

RAS

[See also MEDITERRANEAN ENVIRONMENTAL CHANGE AND HUMAN IMPACT]

Butzer, K.W. 1975: Pleistocene littoral–sedimentary cycles of the Mediterranean basin: a Mallorquin view. In Butzer, K.W. and Isaac, G.L. (eds), *After the Australopithecines: stratigraphy, ecology and culture*. The Hague and Paris: Mouton, 25–71. **Dixon, J.E. and Robertson, A.H.F.** 1985: *The geological evolution of the Eastern Mediterranean* [*Special Publication* 17]. London:

Geological Society. **Macklin, M.G., Lewin, J. and Woodward, J.C.** 1992: Quaternary fluvial systems in the Mediterranean Basin. In Lewin, J, Macklin, M.G. and Woodward, J.C. (eds), *Mediterranean Quaternary river environments*. Rotterdam: Balkema, 1–24. **Rose, J. and Meng, X.** 1999: River activity in small catchments over the last 140 ka, northeast Mallorca, Spain. In Brown, A.G. and Quine, T.A. (eds), *Fluvial processes and environmental change*. Chichester: Wiley, 91–102. **Ruffell, A.** 1997: Geological evolution of the Mediterranean Basin. In King, R., Proudfoot, L. and Smith, B. (eds), *The Mediterranean: environment and society*. London: Arnold, 12–29.

mediterranean-type vegetation

True mediterranean-type vegetation is characterised by SCLEROPHYLLOUS woodland and SHRUBS, which are well adapted to periodic fire and the climate of the mediterranean regime, which has hot, dry summers and mild, humid winters. Mediterranean-type vegetation is characteristic of a VEGETATION FORMATION-TYPE found in five regions between latitudes 30°N and 40°S that have a mediterranean climate: the Mediterranean basin, southern Africa, southwestern and south-central Australia, central Chile and southern California.

All five mediterranean regions have been occupied by humans for millennia, but the greatest human impacts are apparent in the Mediterranean basin. The mediterranean climatic regime is young, having developed during the PLEISTOCENE, and a certain degree of COEVOLUTION is claimed for mediterranean-type vegetation and PALAEOLITHIC humans in the Mediterranean basin. The effects of NEOLITHIC and BRONZE AGE agriculture and PASTORALISM, in particular fire-clearance and goat-herding, initiated the transformation of the mediterranean vegetation from dense forest to low-lying SCRUB, in which highly valuable crops such as *Olea* (olive) and *Vitis* (vine) are now cultivated. Human activity has similarly altered the other mediterranean regions, although this has occurred only within the last 500 years since European settlement and has involved invasion by weedy European plant species better adapted to grazing. Although pockets of sclerophyllous woodland remain evident, the modern vegetation consists of SHRUB and HEATHLAND rarely exceeding 2 m in height. The thorny *garrigue* (*phrygana* in Greece, *chapparal* in California, *mallee* in Australia, and *mattoral* in Spain and Chile) consists of scattered chew-resistant sclerophyllous shrubs and is best represented in the Mediterranean basin by *Quercus coccifera* (kermes oak) with a ground layer of herbaceous PERENNIALS. The less common maquis (*fynbos* in southern Africa) heathlands are dominated by Ericaceous scrub of *Erica arborea* (tree heath) and *Arbutus unedo* (strawberry tree) in the Mediterranean basin and *Protea-Erica* heathland in southern Africa. *ARG*

Archibold, O.W. 1995: *Ecology of world vegetation* London: Chapman and Hall. **Arroyo, M.T.Z., Zedler, P.H. and Fox, M.D.** 1994: *Ecology and biogeography of Mediterranean ecosystems in Chile, California and Australia* [*Ecological Studies* 108]. New York: Springer. **Cowling, R.** 1992: *The ecology of fynbos: nutrients, fire and diversity* . Oxford: Oxford University Press. **Davis, G.W. and Richardson, D.M. (eds)** 1995: *Mediterranean-type ecosystems: the function of biodiversity*. Berlin: Springer. **Greater, W.** 1994: Extinctions in Mediterranean areas. *Philosophical Transactions of the Royal Society of London* B344, 41–46. **Groves, R.H. and Di Castri, F. (eds)**, 1991: *Biogeography of Mediterranean invasions*. Cambridge: Cambridge University Press. **Kruger, F.J., Mitchell, D.T. and Jarvis, J.U.M. (eds)**, 1983: *Mediterranean-type ecosystems*. Berlin: Springer.

megafauna

The larger animal species within a FAUNA, encompassing those species attaining a body mass exceeding about 40 kg (90 lb). *DCS*

[See also MASS EXTINCTIONS]

megafossil

A very large FOSSIL or SUBFOSSIL, such as a tree trunk preserved in a mire beyond the TREE-LINE. Unlike the MACROFOSSILS in PLANT MACROFOSSIL ANALYSIS, they are too large to be manipulated by hand. Because of their size, they are normally found *in situ* and hence are likely to be particularly reliable indicators of changing site conditions. *JAM*

[See also MICROFOSSIL, MICROFOSSIL ANALYSIS]

Eronen, M. and Huttunen, P. 1993: Pine megafossils as indicators of Holocene climatic changes in Fennoscandia. *Paläoklimaforschung* 9, 29–40.

megaripple

Sometimes used as a synonym for DUNE, but best avoided because it clouds the hydrodynamic distinction between dunes and RIPPLES. *GO*

Ashley, G. 1990: Classification of large scale subaqueous bedforms: a new look at an old problem. *Journal of Sedimentary Petrology* 60, 160–172.

megaturbidite

An extremely thick TURBIDITE of regional extent. Megaturbidites, or *megabeds*, may be many tens of metres thick, implying deposition from a major EVENT, such as a large EARTHQUAKE, sediment instability caused by a fall in SEA LEVEL, a catastrophic FLOOD or a VOLCANIC ERUPTION. Megaturbidites can be used in CORRELATION within a SEDIMENTARY BASIN (see EVENT STRATIGRAPHY). *GO*

Brunner, C.A., Normark, W.R., Zuffa, G.G. and Serra, F. 1999: Deep-sea sedimentary record of the late Wisconsin cataclysmic floods from the Columbia River. *Geology* 27, 463–466. **Rothwell, R.G., Thomson, J. and Kahler, G.** 1998: Low-sea-level emplacement of a very large Late Pleistocene 'megaturbidite' in the western Mediterranean Sea. *Nature* 392, 377–380. **Séguret, M., Labaume, P. and Madariago, R.** 1984: Eocene seismicity in the Pyrenees from megaturbidites in the South Pyrenean basin (North Spain). *Marine Geology* 55, 117–131.

meltout till

TILL deposited by debris melting out from stagnant glacier ice. *JS*

Boulton, G.S. 1970: The deposition of subglacial and melt-out tills at the margins of certain Svalbard glaciers. *Journal of Glaciology* 9, 231–245.

meltwater

Water derived from melting snow or ice. *Glacial meltwater* is produced in large quantities (except in Antarctica) during the summer season. In addition to the large annual variations in meltwater discharge from GLACIERS, strong diurnal cycles may be evident, resulting from variations in solar radiation. *MJH*

[See also GLACIOFLUVIAL LANDFORMS, GLACIOFLUVIAL SEDIMENTS, SNOWMELT]

member

A unit in LITHOSTRATIGRAPHY which is a locally distinct subdivision of a FORMATION. Members commonly wedge out through lateral FACIES changes. *LC*

mercury (Hg)

A silver-grey HEAVY METAL (quicksilver), liquid at normal temperatures and highly toxic in some compounds, especially as *methyl mercury*. It was relatively rare in the environment prior to extraction from its ore, cinnabar (HgS). Environmental levels first rose significantly above background levels in the Roman Period or possibly even earlier (see Figure). BIO-ACCUMULATION of mercury in organisms, especially fish, is a potential environmental hazard. *JAM*

[See also CHEMICAL TIME BOMB]

Martínez-Cortizas, A., Pontevedra-Pombal, X., García-Rodeja, E. *et al.* 1999: Mercury in a Spanish peat bog: archive of climatic change and atmospheric metal deposition. *Science* 284, 939–942. **Bargagli, R.** 1999: Mercury in the environment. In Alexander, D.E. and Fairbridge, R.W. (eds), *Encyclopedia of environmental science*. Dordrecht: Kluwer, 402–405.

meridional flow   A prevalence of north–south AIR-FLOW TYPES often associated with weakened upper westerly winds, LONG WAVES of large AMPLITUDE and atmospheric BLOCKING. This term is also used for the north–south (latitudinal) component of the atmospheric circulation, as distinct from the zonal (easterly or westerly) component. Meridional flow, parallel to the meridians (lines of longitude) is important for the transport of energy and moisture towards the poles. *JCM/JAM*

[See also GENERAL CIRCULATION OF THE ATMOSPHERE, HADLEY CELL, ZONAL INDEX]

mérokarst   A KARST landscape developed on thin sequences of limestones interbedded with other rocks as well as on pure carbonate formations. Also known as '*half karst*', mérokarst lacks the development of classical karst landforms. *RAS*

[See also HOLOKARST]

Cvijić, J. 1925: Le mérokarst. *Comptes Rendus de l'Académie de Sciences* 180, 757–758.

meromictic lake   A lake with a permanently stratified water column maintained by a density gradient caused by chemical stratification (including salinity). Meromictic lakes often contain annually laminated sediments (cf. HOLOMICTIC LAKE). *HHB*

[See also LAKE STRATIFICATION, LAMINATION, SEDIMENTS, VARVE]

mesoclimate   The climatic characteristics of a small or local area between about 10 and 100 km^2, such as a valley or a city. It differs from the wider MACROCLIMATE in terms of spatial scale. *JGT*

[See also LOCAL CLIMATE, MICROCLIMATE, URBAN CLIMATE]

Atkinson, B.W. 1981: *Meso-scale atmospheric circulations.* London: Academic Press. **Orlanski, I.** 1975: A rational subdivision of scales for atmospheric processes. *Bulletin of the American Meteorological Society* 56, 527–530.

mesocratic phase   The early-temperate phase of an INTERGLACIAL, characterised in northern Europe by the expansion of deciduous mixed oak woodland (typically *Quercus, Ulmus, Fraxinus* and *Corylus*) on rich forest soils. *MJB*

[See also INTERGLACIAL CYCLE]

Iversen, J. 1958: The bearing of glacial and interglacial epochs on the formation and extinction of plant taxa. *Uppsala Universitet Årsskrift* 6, 210.

Mesolithic: landscape impacts   MESOLITHIC peoples lived by HUNTING, FISHING AND GATHERING; their main impacts on the landscape were the creation of small temporary clearings and the use of fire. These have been detected by fine-resolution POLLEN ANALYSIS combined with CHARRED-PARTICLE ANALYSIS. Some types of vegetation seem to have been burned systematically, possibly as a deliberate management strategy, such as the burning of

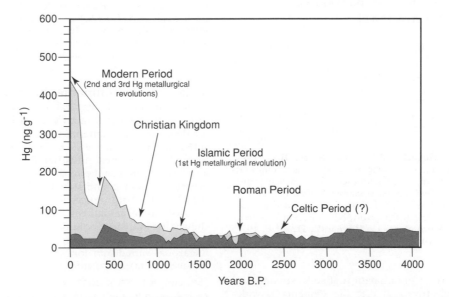

mercury *Total mercury and its natural (darker area) and anthropogenic (lighter area) components during the last 4000 years, as recorded in a Spanish peat bog. Major prehistoric and historic phases of mercury exploitation in Spain are indicated (after Martinez-Cortizas* et al.*, 1999)*

reedswamp at the early Mesolithic site of Star Carr, in northeast England.

In upland areas, such as the North York Moors in England, burning of MOORLAND and open woodland in the later Mesolithic may have lowered the height of the TREE LINE and even triggered BLANKET MIRE formation. It has been suggested that burning by Mesolithic peoples encouraged the spread of hazel (*Corylus avellana*) and alder (*Alnus glutinosa*) in the early HOLOCENE. There is little evidence to support this for hazel, but the population expansion of alder is sometimes associated with evidence of burning, suggesting that alder may have colonised areas of Mesolithic woodland disturbance. **SPD**

[See also VEGETATION HISTORY]

Bonsall, C. (ed.) 1989: *The Mesolithic in Europe.* Edinburgh: Edinburgh University Press. **Brown, A.G.** 1997: Clearances and clearings: deforestation in Mesolithic – Neolithic Britain. *Oxford Journal of Archaeology* 16, 133–146. **Mellars, P. and Dark, P.** 1998: *Star Carr in context.* Cambridge: McDonald Institute. **Mithen, S.J.** 1994: *The Mesolithic Age.* In Cunliffe, B. (ed.), *Prehistoric Europe.* Oxford: Oxford University Press, 79–135. **Mithen, S.J.** 1999: Mesolithic archaeology, environmental archaeology and human palaeoecology. *Quaternary Proceedings* 7, 477–483. **Simmons, I.G.** 1996: *The environmental impact of later Mesolithic cultures.* Edinburgh: Edinburgh University Press. **Simmons, I.G., Dimbleby, G.W. and Grigson, C.** 1981: The Mesolithic. In Simmons, I.G. and Tooley, M. (eds), *The environment in British prehistory.* London: Duckworth, 82–124. **Smith, A.G., Whittle, A., Cloutman, E.W. and Morgan, L.** 1989: Mesolithic and Neolithic activity and environmental impact on the south-east fen edge in Cambridgeshire. *Proceedings of the Prehistoric Society* 55, 207–249. **Smith, C.** 1992: *Late Stone Age hunters of the British Isles.* London: Routledge. **Vermeersch, P.M. and Van Peer, P. (eds)** 1990: *Contributions to the Mesolithic in Europe.* Leuven: Leuven University Press. **Zvelebil, M. (ed.)** 1986: *Hunters in transition: Mesolithic societies of temperate Eurasia and their transition to farming.* Cambridge: Cambridge University Press.

Mesozoic

An ERA of geological time comprising three PERIODS: the TRIASSIC, JURASSIC and CRETACEOUS. **GO**

[See also GEOLOGICAL TIMESCALE]

Messinian

The uppermost STAGE of the MIOCENE series (6.7 to 5.2 million years ago). The *Messinian salinity crisis* refers to the initially controversial theory, derived from DEEP-SEA DRILLING observations, that the Straits of Gibraltar were closed by TECTONIC processes, allowing evaporation of water from the enclosed Mediterranean basin and the precipitation of thick sequences of EVAPORITES. **JCWC**

Butler, R.W.H., McClelland, E. and Jones, R.E. 1999: Calibrating the duration and timing of the Messinian salinity crisis in the Mediterranean: linked tectonoclimatic signals in thrust-top basins of Sicily. *Journal of the Geological Society, London* 156, 827–835. **Hsü, K.J.** 1983: *The Mediterranean was a desert: a voyage of the Glomar Challenger.* Princeton, NJ: Princeton University Press. **Hsü, K.J., Montadert, L., Bernoulli, D.** *et al.* 1977: History of the Mediterranean salinity crisis. *Nature* 267, 399–403.

metal pollution

Contamination of soils, streams and groundwater with traces of metallic elements is widespread, but, as these so-called HEAVY METALS occur naturally in the rocks that are weathered to provide the parent materials for soils, all soils contain natural BACKGROUND LEVELS. As a rule of thumb, any soil with metallic elements in excess of the BACKGROUND LEVELS shown in the Table may be considered as polluted.

metal pollution *Natural background levels for heavy metals in soils*

Element	Median soil content (mg kg^{-1}) and range
As	6 (0.1–40)
B	20 (2–270)
Cd	0.35 (0.01–2)
Co	8 (0.05–65)
Cu	30 (2–250)
Hg	0.06 (0.01–0.5)
Mn	1000 (20–10 000)
Pb	35 (2–300)
Se	0.4 (0.1–2)
Zn	90 (1–900)

Small quantities of manganese, boron and molybdenum are essential for normal plant growth and ruminant animals require cobalt. However, the other metals are toxic, particularly if the background concentrations are enhanced by ANTHROPOGENIC sources. CADMIUM concentrations in Silurian and Lower Lias BLACK SHALE are abnormally high, leading to *teart pastures* and the use of phosphatic fertilisers containing cadmium as a contaminant has led to enhanced concentrations in agricultural soils. High LEAD concentrations from mining and smelting, sewage sludge and vehicle exhausts are widespread, the latter particularly affecting roadside verges. All urban areas and particularly sites where metal-working has taken place have enhanced metallic concentrations. The effect of toxic metals in soils is long-lasting as the metal ions are held by the CLAY–HUMUS COMPLEX and inhibit the beneficial activities of soil microfauna. **EMB**

Blum, W.E.H. 1990: *Soil pollution by heavy metals.* Strasbourg: Council of Europe. **Bridges, E.M.** 1988: *Surveying derelict land* [*Monographs on Soil and Resources Survey* 13]. Oxford: Clarendon Press. **Hutchinson, T.C. (ed.)** 1987: *Lead, mercury, cadmium and arsenic in the environment.* New York: Wiley. **Nriagu, J.O.** 1990: Global metal pollution: poisoning the biosphere. *Environment* 32, 7–32.

metals in environmental history

Metals released to the atmosphere by natural processes (such as EROSION) or by ANTHROPOGENIC activities (such as ore processing) are deposited in sedimentary sequences and can be measured using *geochemical analysis*. Much research has focused on the postindustrial flux of metals into the atmosphere amid concerns over POLLUTION, and high concentrations of LEAD and MERCURY have been discovered in the Greenland ICE CORES for this period. The ANTHROPOGENIC flux of metals released since the COPPER AGE has shown that industrial pollution has occurred for millennia, yet new research has demonstrated that the natural flux of metals has varied over longer timescales and can be used as a PROXY CLIMATIC INDICATOR. **ARG**

[See also POLLUTION HISTORY]

Craddock, P.T. (ed.) 1980: *Scientific studies in early mining and extractive metallurgy.* London: British Museum. **Lantzy, R.J. and Mackenzie, F.T.** 1979: Atmospheric trace metals: global cycles and assessment of man's impact. *Geochimica et*

Cosmochimica Acta 43, 511–525. **Maddin, R.** 1988: *The beginning of the use of metals and alloys.* Cambridge, MA: MIT Press. **Shepherd, R.** 1980: *Prehistoric mining and related industries.* London: Academic Press. **Shotyk, W., Weiss, D., Appleby, P.G.** *et al.* 1998: History of atmospheric lead deposition since 12,370 ^{14}C yr BP from a peat bog, Jura Mountains, Switzerland. *Science* 281, 1635–1640. **Tylecote, R.F.** 1987: *The early history of metallurgy in Europe.* London: Longman.

metamorphic rocks

Rocks that have been altered in the solid state (recrystallised) as a result of changes in temperature, pressure and/or chemical environment (e.g. HYDROTHERMAL effects). Recrystallisation involves changes to MINERALS and to their TEXTURE (shape or arrangement). Most metamorphic rocks have a CRYSTALLINE texture and a distinctly anisotropic FABRIC. *Contact metamorphism* converts rocks in a *metamorphic aureole* adjacent to an igneous INTRUSION into *hornfels. Regional metamorphism* results from heat and stress during OROGENESIS. MUDSTONE is converted into *phyllite, slate, schist* and *gneiss* with increasing degree of metamorphism. Basic IGNEOUS ROCKS are altered to *amphibolite,* SANDSTONE to *psammite* or, if QUARTZ-rich, to *metaquartzite,* and LIMESTONE to *marble. Dynamic metamorphism* results from rock DEFORMATION and forms rocks such as *mylonite* in FAULT zones. Other categories of metamorphism are summarised in the Table.

Much of the GEOLOGICAL RECORD, particularly of the PRECAMBRIAN, is preserved in metamorphic rocks and their interpretation yields important information about the evolution of environments, the ATMOSPHERE, climate and life on Earth. *Orogenic metamorphism* in the past may have been associated with the release of CARBON DIOXIDE into the atmosphere: although the significance of this process is controversial, it suggests that metamorphism may have a long-term effect on CLIMATIC CHANGE. Although usually considered in relation to rocks, metamorphic processes underlie the crystalline transitions between snow, FIRN, *nevé* and GLACIER ICE. *JBH*

[See also GREENSTONE BELT, SEDIMENTARY ROCKS]

Blatt, H. and **Tracy, R.** 1996: *Petrology: igneous, sedimentary, and metamorphic,* 2nd edn. New York: Freeman. **Graham, C.** 1992: Metamorphism and metamorphic rocks. In Duff, P.McL.D. (ed.), *Holmes' principles of physical geology,* 4th edn. London: Chapman and Hall, 147–175. **Kerrick, D.M. and**

Caldeira, K. 1998: Metamorphic CO_2 degassing from orogenic belts. *Chemical Geology* 145, 213–232.

metapedogenesis

Changes in soil characteristics resulting from human activity. The outcome may be beneficial or detrimental and results in changes to the natural SOIL PROFILE. If these changes are profound, and the original soil profile is completely altered or buried, soils are classified as ANTHROSOLS. *EMB*

[See also HUMAN IMPACT ON SOILS]

Yaalon D.H. and Yaron, B. 1966: Framework for man-made soil changes – an outine for metapedogenesis. *Soil Science* 102, 272–277.

metaphysics

The part of philosophy that raises questions about the nature of reality, and about being and knowing, and claims to deal with questions that are beyond the ability of SCIENCE to solve. It can be seen as an attempt to characterise existence or reality as a whole, instead of, as in the NATURAL ENVIRONMENTAL SCIENCES, particular parts or aspects of reality. Materialism and IDEALISM are examples of metaphysics in this sense. *ART*

metapopulation model

Derived from ISLAND BIOGEOGRAPHY theory of the AD 1960s, the fundamental assumption of the metapopulation model is that the characteristics of isolated communities and populations result from a dynamic balance between the opposing forces of immigration (which adds individuals and species) and extinction (which removes them). Metapopulation theory typically focuses on individual species. While this theory has developed into a diversity of models over the past two decades, all of these models assume that each species exists as a collection of interdependent populations, which comprise the metapopulation. Each population is assumed to be isolated enough to allow differentiation among populations, but dependent to some degree on immigration among populations to stave off EXTIRPATION and eventual EXTINCTION of the entire metapopulation. *MVL*

[See also FRAGMENTATION, POPULATION DYNAMICS]

Gilpin, M.E. and Hanski, I. 1991: *Metapopulation dynamics: empirical and theoretical investigations.* San Diego, CA: Academic

metamorphic rocks *Types of metamorphism associated with the formation of metamorphic rocks*

Location	Process	Description
Local metamorphism	Contact metamorphism	Metamorphic rocks adjacent to, and clearly related to IGNEOUS ROCKS
	Dynamic metamorphism	Metamorphic rocks associated with severe deformation along FAULT or shear zones
	Impact metamorphism	Metamorphic rocks associated with high-pressure–temperature regimes caused by METEORITE IMPACT
	Micro-contact metamorphism	Small-scale changes due to high-temperature lightning strikes (creating FULGURITES)
Regional metamorphism	Orogenic metamorphism	Metamorphic rocks formed in association with SUBDUCTION and collision-related zones of OROGENESIS
	Burial metamorphism	Metamorphic rocks buried in SEDIMENTARY BASINS, where higher pressures and temperatures have formed new minerals
	Oceanic metamorphism	Metamorphic rocks altered by circulating heated seawater driven by HYDROTHERMAL activity at MID-OCEAN RIDGES

Press. **Hanski, I. and Gilpin, M.E. (eds)** 1997: *Metapopulation biology: ecology, genetics, and evolution.* San Diego, CA: Academic Press. **Lennon, J.J., Turner, J.R.G. and Connell, D.** 1997: A metapopulation model of species boundaries. *Oikos* 78, 486–502.

meteoric water   Water at the Earth's surface or in the near-surface environment, which originated in the ATMOSPHERE and reached the Earth's surface by PRECIPITATION. It contrasts with JUVENILE WATER, which originated in the Earth's interior.   *JAM*

meteorite   A small ASTEROID or METEOROID that survives passage through the ATMOSPHERE and strikes the Earth's surface.   *JAM*

meteorite impact   The collision of debris from space (including ASTEROIDS and COMETS) with the surface of a planet, satellite or other object in the solar system, producing an *impact crater*. Impact CRATERS are well known from other bodies in the solar system, but most were thought to have formed early in its history and evidence of any that had formed on Earth was thought to have been removed by WEATHERING and EROSION, a notable exception being Meteor Crater (Barringer Crater) in Arizona, which is 1.2 km in diameter and formed about 49 000 years ago. The frequency of impacts is inversely proportional to the size of the impacting object (see Figure) and has declined through GEOLOGICAL TIME.

A renewed acceptance of the role of rare events in the history of the Earth (see EVENT, NEOCATASTROPHISM), driven partly by the *impact hypothesis* for the K–T BOUNDARY, has led to the successful search for more meteorite impact

craters on Earth. These can be distinguished from volcanic craters by details of their morphology, preservation of fragments of the impacting BOLIDE and evidence of shock-induced metamorphism of the surrounding rocks. They can be divided into *meteorite craters*, which are relatively fresh and uneroded, and *impact structures*, where the surface features have been eroded and deeper levels are now exposed. Large craters (> 10 km diameter) are sometimes called ASTROBLEMES, or *hydroblemes* if they formed in water: such large meteorite impact events are capable of inducing devastating ENVIRONMENTAL CHANGE through the shock waves at impact, huge TSUNAMI waves that would be generated if the impact were in the OCEAN and the ejection of dust and gases into the ATMOSPHERE (see also NUCLEAR WINTER).

Over 150 meteorite impact events have now been identified on Earth, ranging from the Sudbury structure in Canada, some 1850 million years old and 180 km across, which is a major nickel ore deposit, through the Chicxulub crater in Mexico identified as the 'smoking gun' for the end-Cretaceous MASS EXTINCTION, to the Sikhote Alin craters in Russia that formed in AD 1947. Additional incentive was given to the study of meteorite impacts by the observed impact of Comet Shoemaker–Levy 9 with Jupiter in 1994 and research is now being undertaken into the possibility of collisions between the Earth and NEAR-EARTH OBJECTS.   *GO*

[See also EARTHQUAKE MAGNITUDE, K–T BOUNDARY]

Chapman, C.R. and Morrison, D. 1994: Impacts on the Earth by asteroids and comets: assessing the hazard. *Nature* 367, 33–40. **Grady, M.M., Hutchison, R., McCall, G.J.H. and**

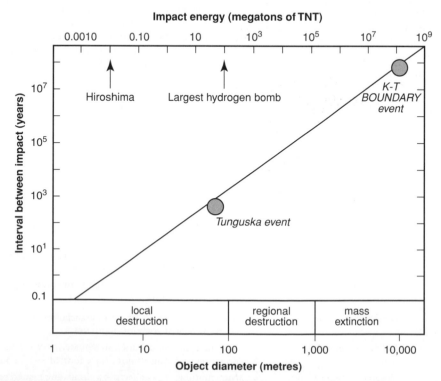

meteorite impact *Estimated recurrence intervals for impacts with Near Earth Objects of different sizes, with some key events for comparison (after Jewitt, 2000)*

Rothery, D. (eds) 1998: *Meteorites – flux with time and impact effects* [*Special Publication* 140]. London: Geological Society. **Grieve, R.A.F. and Pesonen, L.J.** 1992: The terrestrial impact cratering record. *Tectonophysics* 216, 1–30. **Hodge, P.** 1994: *Meteorite craters and impact structures of the Earth.* Cambridge: Cambridge University Press. **Jewitt, D.** 2000: Eyes wide shut. *Nature* 403, 145–148. **Levy, D.H.** 1995: *Impact Jupiter: the crash of comet Shoemaker–Levy 9.* New York: Plenum. **Melosh, H.J.** 1988: *Impact cratering: a geologic process.* Tucson, AZ: University of Arizona Press. **Silver, L.T. and Schultz, P.H. (eds)** 1982: *Geological implications of impacts of large asteroids and comets on the Earth* [*Special Publication I190*]. Boulder, CO: Geological Society of America.

meteoroids Natural solid objects moving in interplanetary space that are smaller than about 100 m in diameter but larger than a molecule. They are thought to be fragments of ASTEROIDS. Meteoroids that impact on the Earth's surface are termed METEORITES. Around 160 meteoroid *impact craters* are known world-wide (see Figure), most of which have been detected by SATELLITE REMOTE SENSING. It has been estimated, however, that only about 6% of PHANEROZOIC terrestrial impact structures with diameters > 10 km and about 16% of those with diameters > 20 km have been discovered.   *JAM*

[See also METEORITE IMPACT]

Grieve, R.A.F. and Shoemaker, E.M. 1994: The record of past impacts on Earth. In Gehrels, T. (ed.), *Hazards due to comets and asteroids.* Tucson, AZ: University of Arizona Press, 417–462. **Huggett, R.J.** 1997: *Environmental change: the evolving ecosphere.* London: Routledge. **Trefil, J.S. and Raup, D.M.**1990: Crater taphonomy and bombardment rates in the Phanerozoic. *Journal of Geology* 98, 385–398.

meteorological satellites Orbital platforms, also known as WEATHER SATELLITES, for making observations of the Earth's surface and ATMOSPHERE for WEATHER FORECASTING and CLIMATOLOGY. Observations must be made regularly over large areas. Weather satellites in *geostationary* orbit continuously view a hemisphere of the Earth. They allow regular observations of large-scale features and have a SPATIAL RESOLUTION of between 2.5 and 5 km at the Equator. A network of satellites, such as *METEOSAT* and *GOES* (*Geostationary Operational Environmental Satellites*), give global coverage. Weather satellites in Sun-synchro-nous polar orbits pass over the same place on the Earth's surface a number of times each day and can record images with a spatial resolution down to 1 km. Groups of satellites, such as the NOAA (North American Atmospheric Administration) series, collect images of the same location every two or three hours. Weather satellites carry OPTICAL REMOTE SENSING INSTRUMENTS to identify cloud formations, THERMAL REMOTE SENSING instruments to measure SEA-SURFACE TEMPERATURES and PASSIVE MICROWAVE REMOTE SENSING instruments to measure atmospheric properties.   *GMS*

Kidder, S.Q. and Vondar Haar, T.H. 1995: *Satellite meteorology.* New York: Academic Press.

meteorology The science of atmospheric phenomena and processes observed in the ATMOSPHERE, principally involving heat, motion and moisture. It includes but is not exclusively the study of WEATHER. Its origins go back to the Greeks and it now includes the *atmospheric science* of not only the Earth but also the other planets.   *JCM/AHP*

[See also CLIMATOLOGY, MICROCLIMATE]

Ahrens, C.D. 1994: *Meteorology today*, 5th edn. St Paul, MA: West Publishing. **Aristotle** 1952: *Meteorologica*, translated by H.D.P. Lee. Cambridge, MA: Harvard University Press. **Crutzen, P.J. and Ramanathan, V.** 2000: The ascent of atmospheric sciences. *Science* 290, 299–304. **National Research Council** 1999: *Atmospheric sciences entering the twenty-first century.* Washington, DC: National Academy Press. **Whiteman, C.D.** 2000: *Mountain meteorology: fundamentals and applications.* New York: Oxford University Press.

methane variations Atmospheric methane (CH_4) is one of the most important atmospheric trace gases because of its GREENHOUSE role in trapping RADIATION re-emitted by the Earth. Although methane only accounts for 0.2% of the total budget of the ATMOSPHERE, it has a very short RESIDENCE TIME (nine years compared, for example, with the 50–200 years residence time for atmospheric CARBON DIOXIDE) and therefore small variations can have a rapid effect on the global CLIMATIC SYSTEM. Major sources of atmospheric methane include emissions from FOSSIL FUELS, WETLANDS, TROPICAL PEATLANDS, domestic animals, rice fields, BIOMASS BURNING and TERMITES (see Figure). Evidence of past values of atmospheric methane has been

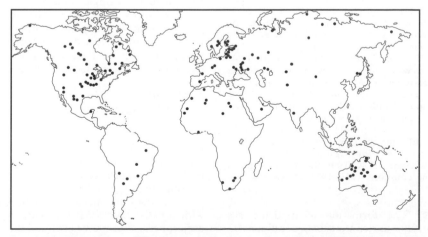

meteoroids *Sites of known terrestrial impact craters (after Grieve and Shoemaker, 1994)*

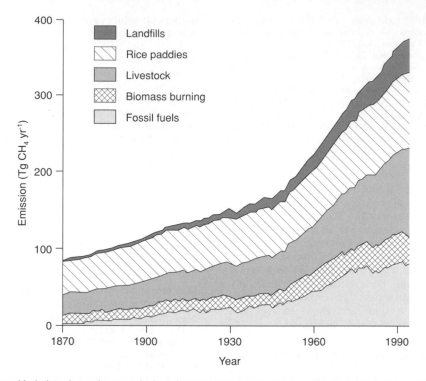

methane variations *Variations in methane emissions from various sources since* AD *1870 (after Stern and Kaufmann, 1996)*

recorded in the air trapped in the Arctic and Antarctic ICE CORES. Results from these studies indicate that levels of atmospheric methane have varied considerably through GLACIAL–INTERGLACIAL CYCLES with higher concentrations occurring during interglacials. During the last 150–200 years, concentrations of atmospheric methane have, however, reached unprecedented levels as a result of anthropogenic activity. Present estimates suggest that atmospheric methane is now 220% of its eighteenth century level and increasing on average 1% per year. This is raising concerns about its effect on GLOBAL WARMING.

KJW

Blake, D.R. and Rowland, F.S. 1988: Continuing worldwide increase in tropospheric methane. *Science* 239, 1129–1131. **Blunier, T., Chappellaz, J., Schwander, J.** *et al.* 1995: Variations in atmospheric methane concentrations during the Holocene Epoch. *Nature* 374, 46–49. **Bradley, R.S.** 1999: *Paleoclimatology – reconstructing climates of the Quaternary*, 2nd edn. San Diego, CA: Harcourt Academic Press. **Cronin, T.M.** 1999: *Principles of paleoclimatology*. New York: Columbia University Press. **Schlesinger, W.H.** 1997: *Biogeochemistry – an analysis of global change*. London: Academic Press. **Severinghaus, J.P. and Brooke, E.J.** 1999: Abrupt climate change at the end of the last glacial period inferred from trapped air in polar ice. *Science* 286, 930–934. **Smith, A.T.** 1995: Environmental factors affecting the global atmospheric methane concentration. *Progress in Physical Geography* 19, 322–335. **Stern, D.I. and Kaufmann, R.K.** 1996: Estimates of global anthropogenic methane emissions 1860–1993. *Chemosphere* 33, 159–176.

methanogenesis The formation of methane gas (CH$_4$) by DECOMPOSITION under ANAEROBIC conditions.

JAM

[See also METHANE VARIATIONS, RUMINANT]

Smith, A.T. 1995: Environmental factors affecting global atmospheric methane concentrations. *Progress in Physical Geography* 19, 322–335.

methodology The study of method, both the techniques of investigation and the approach to problem solving, including the setting of standards in scientific investigation.

JAM

[See also EXPERIMENTAL DESIGN, SCIENTIFIC METHOD]

Ford, E.D. 2000: *Scientific method for ecological research*. Cambridge: Cambridge University Press.

microatoll A single coral colony, usually massive and circular, with a dead, flat surface and living lateral margins. Growth banding and morphology records annual fluctuations in water depth over intertidal reef flats. *TSp*

Woodroffe, C.D. and McLean, R.F. 1990: Microatolls and recent sea level change on coral atolls. *Nature* 344, 531–534.

microbiology The scientific study of micro-organisms including, for example, ALGAE, BACTERIA, protozoa, viruses and yeasts.

JAM

microclimate The CLIMATE of a small space with dimensions 10^{-2} m to 10^3 m horizontally and within 10 m of the Earth's surface. The timescale of measurements for investigation of the microclimate is usually less than 24 h. These dimensions mean that there is an overlap in scale with LOCAL CLIMATE. Microclimatology is the study of the atmospheric BOUNDARY LAYER and the interaction between the ATMOSPHERE and the Earth's surface in terms of the

ENERGY BALANCE. A slightly looser definition includes the study of the climate of the BIOSPHERE: from the crowns of trees to the bottom of the root zone. This depends on the concept of the *active surface*, which is the level where the majority of the radiant energy is absorbed, reflected and emitted; where transformations of energy (radiant to thermal, sensible to latent) and mass (change of state of water) take place; where PRECIPITATION is intercepted; and where most frictional drag on airflow occurs. *BDG*

Geiger, R., Aron, R.H. and Todhunter, P. 1995: *The climate near the ground.* Wiesbaden: Vieweg. Oke, T. R. 1990: *Boundary layer climates,* 2nd edn. London: Routledge. Rosenberg, N.J., Blad, B.L. and Verma, S.B. (eds) 1983: *Microclimate: the biological environment,* 2nd edn. New York: Wiley.

microdendroclimatology
The intra-ring study of TREE RINGS, particularly their isotopic content, for HIGH-RESOLUTION RECONSTRUCTION of palaeoclimates. The major advantage of a microdendroclimatic approach is that SEASONALITY effects can be analysed or controlled. *JAM*

[See also DENDROCLIMATOLOGY, PALAEOCLIMATOLOGY]

Loader, N.J., Switsur, V.R. and Field, E.M. 1995: High-resolution stable isotope analysis of tree rings: implications of 'microdendroclimatology' for palaeoenvironmental research. *The Holocene* 5, 457–460. McCarroll, D. and Pawellek, F. 1998: Stable carbon isotope ratios of latewood cellulose in *Pinus sylvestris* from northern Finland: variability and signal strength. *The Holocene* 8, 675–684.

micro-erosion meter
An instrument for measuring accurately the change in the surface level of rock over time. Its three legs rest on metal studs fixed in the rock providing a stable platform from which measurements to the rock surface are made with a spring-loaded probe connected to an engineer's dial gauge. *RAS*

[See also EROSION, WEATHERING RATES]

Drysdale, R. and Gillieson, D. 1997: Micro-erosion meter measurements of travertine deposition rates: a case study from Louie Creek, northwest Queensland, Australia. *Earth Surface Processes and Landforms* 22, 1037–1051. Wayne, J.S. and Kirk, R.M. 1998: Rates and patterns of erosion on inter-tidal shore platforms, Kaikoura Peninsula, South Island, New Zealand. *Earth Surface Processes and Landforms* 23, 1071–1085.

microfossil
A FOSSIL or SUBFOSSIL of such a small size that a microscope is needed for inspection (e.g. pollen 10–100 μm). *BA*

[See also MACROFOSSIL, MICROFOSSIL ANALYSIS, PALAEOBOTANY, PALAEOECOLOGY, POLLEN ANALYSIS, SPORES]

Brasier, M.D. 1980: *Microfossils.* London: Allen and Unwin. Van Geel, B. 1986: Application of fungal and algal remains and other microfossils in palynological analyses. In Berglund, B.E. (ed.), *Handbook of Holocene palaeoecology and palaeohydrology.* Chichester: Wiley, 497–505.

microfossil analysis
The investigation of a wide range of microscopic FOSSILS or SUBFOSSILS, including POLLEN, SPORES, DIATOMS, OSTRACODS, TESTATE AMOEBAE, CHIRONOMIDS, CLADOCERA and FORAMINIFERA. Microfossil analysis is an important tool for studies in BIOSTRATIGRAPHY (where the occurrence of taxa or assemblages is used to establish the relative age of units), PALAEOECOLOGY (where microfossils are used to reconstruct past environ-

ments, plant and animal communities and interactions between them) and *evolutionary biology* (establishing the evolutionary descent of different LINEAGES of organisms). *MJB*

[See also MICROFOSSIL, MICROPALAEONTOLOGY]

Brasier, M.D. 1980. *Microfossils.* London: Chapman and Hall.

micromorphological analysis of sediments
Micromorphology is the term used to describe the distinctive arrangement of MATRIX particles and voids making up a sediment FABRIC. This can be established by THIN-SECTION ANALYSIS under a microscope. Micromorphological analysis of sediments can reveal evidence of: WEATHERING; alteration of MINERALS; orientation and packing of CLAY particles; arrangement and concentration of voids; presence and type of *calcite* crystal growth, animal excrement, clay coatings, rootlet PSEUDOMORPHS and other features. It can be used as both a descriptive and a diagnostic tool in palaeopedology and is regarded as one of the most reliable methods for detecting evidence of PEDOGENESIS, from which it is possible to distinguish sequential phases of soil formation and infer changes in environmental conditions. *DH*

[See also MICROMORPHOLOGICAL ANALYSIS OF SOILS, PALAEOSOLS]

Courty, M.A., Goldberg, P. and Macphail, R. 1989: *Soils and micromorphology in archaeology.* Cambridge: Cambridge University Press. Kemp. R.A. 1985: *Soil micromorphology and the Quaternary* [*QRA Technical Guide* 2]. London: Quaternary Research Association. Kemp, R.A. 1998: Role of micromorphology in palaeopedological research. *Quaternary International* 51, 133–141. Kemp, R.A. 1999: Micromorphology of loess–paleosol sequences: a record of paleoenvironmental change. *Catena* 35, 179–196.

micromorphological analysis of soils
Examination of soils using the microscope was introduced by Kubiena in the AD 1930s. Techniques developed for optical mineralogy have been adapted to study the fabric of soil in THIN SECTION under plain and polarised light. Soil samples with known orientation are taken from the field, dried and set in polyester resin and subsequently cut into thin slices 0.30 μm thick, mounted on microscope slides and examined under the polarising microscope. The *soil fabric* can be seen to comprise skeleton (mineral) grains, plasma and voids. The disposition of mineral grains, voids and clay domains and the location and appearance of various segregations and concentrations of clay, iron, manganese and organic matter enable the pedologist to assess the results of SOIL FORMING PROCESSES at the microscopic level. The three-dimensional arrangement of these components provides an insight into the detailed processes that operate in soils and can supply evidence of environmental change. SCANNING ELECTRON MICROSCOPY has carried the examination of soil fabrics into even greater detail. Micromorphology is used to confirm the presence of clayskins (*cutans*) in, for example, LUVISOLS and the presence of organic-iron-coated mineral grains in PODZOLS. *EMB*

[See also MICROMORPHOLOGICAL ANALYSIS OF SEDIMENTS].

Brewer, R. 1976: *Fabric and mineral analysis of soils,* 2nd edn. Huntington, NY: Krieger. Bullock, P., Fedoroff, N., Jongerius, A. *et al.* 1985: *Handbook for the description of thin sections of soils.* Wolverhampton: Waine Research Publishing. FitzPatrick, E.A. 1984: *Micromorphology of soils.* London:

Chapman and Hall. **Goldberg, P.** 1992: Micromorphology, soils and archaeological sites. In Holliday, V. (ed.), *Soils in archaeology: landscape evolution and human occupation.* Washington, DC: Smithsonian Institution, 145–167. **Kubiena, W.L.** 1938: *Micropedology.* Ames, IA: Collegiate Press.

micronutrients
Elements, mostly TRACE ELEMENTS, that are required by organisms in small amounts, such as boron, copper, iron, manganese, molybdenum and zinc by plants and cobalt, copper, fluorine, iodine, iron, manganese, selenium and zinc by animals. Deficiencies usually induce reduced growth; surpluses are often toxic. *JAM*

[See also MACRONUTRIENTS]

Marschner, H. 1995: *Mineral nutrition of higher plants.* London: Academic Press.

micropalaeontology
A branch of GEOLOGY encompassing the TAXONOMY and EVOLUTION of MICROFOSSILS for high-resolution BIOSTRATIGRAPHY and PALAEOENVIRON-MENTAL RECONSTRUCTION. *CES*

[See also DIATOM ANALYSIS, DINOFLAGELLATE CYST ANALYSIS, FORAMINIFERAL ANALYSIS, MICROFOSSIL ANALYSIS and OSTRACOD ANALYSIS, MICROSCOPY, RADIOLARIA].

Lipps, J.H. (ed.) 1993: *Fossil prokaryotes and protists.* Oxford: Blackwell Scientific.

microresidual fraction
The very fine PARTICULATE fraction remaining after PRETREATMENT of a sample for dating. For example, in the RADIOCARBON DATING of buried soils beneath MORAINES in New Zealand, the Himalaya, South America and Alaska, the microresidual fraction tended to be the oldest soil organic fraction and consisted of the resistant remains of pioneer plants. *JAM*

[See also SOIL DATING]

Geyh, M.A., Röthlisberger, F. and Gellatly, A. 1985: Reliability tests and interpretation of ${}^{14}C$ dates from palaeosols in glacier environments. *Zeitschrift für Gletscherkunde und Glazialgeologie* 21, 275–281.

microscopy
The use of the light microscope and/or electron microscopy for the resolution of microscopic objects and parts of larger objects. *CES*

[See also SCANNING ELECTRON MICROSCOPY].

microseism
A very minor EARTHQUAKE that is detectable only by instrumentation and is not necessarily generated by TECTONIC processes such as slip along a FAULT. Microseismic vibrations generate weak, almost continuous background SEISMIC WAVES or 'Earth noise' detected by SEISMOMETERS. They can be generated by ocean waves, surf, wind and human activities such as mining, quarrying, traffic and reservoir construction. The monitoring of changes in microseismic activity (*microseismic analysis*) can be used to predict sudden events, such as MINING hazards and LANDSLIDES. *GO*

Alcott, J.M., Kaiser, P.K. and Simser, B.P. 1998: Use of microseismic source parameters for rockburst hazard assessment. *Pure and Applied Geophysics* 153, 41–65. **Friedrich, A., Kruger, F. and Klinge, K.** 1998: Ocean-generated microseismic noise located with the Grafenberg array. *Journal of Seismology* 2, 47–64. **Rouse, C., Styles, P. and Wilson, S.A.** 1991: Microseismic emissions from flowslide-type movements in South Wales. *Engineering Geology* 31, 91–110. **Rushforth, I.M., Styles, P., Manley, D.M.J.P. and Toon, S.M.** 1999: Microseismic inves-

tigations of low frequency vibrations and their possible effects on populations. *Journal of Low Frequency Noise Vibration and Active Control* 18, 111–121.

microvertebrate accumulations
Accumulations of microvertebrate remains, composed predominantly of the remains of small mammals such as voles, mice and shrews, although small reptiles, amphibians and bird bones may also be present. The accumulations are the result of indigestible prey remains regurgitated as pellets by diurnal or nocturnal birds of prey (*raptors*) and are preserved most often in sediments within CAVES where the birds have eaten or roosted. The microvertebrate remains show traces of digestion that are frequently predator-specific, according to the degree of fragmentation and corrosion observed, thereby permitting the agent of accumulation to be identified. The microvertebrates themselves provide an important means of interpreting past environmental and climatic conditions and of dating the deposits, particularly in the case of small mammals. *DCS*

[See also RODENT MIDDENS]

Andrews, P. (1990) *Owls, caves and fossils.* London: The Natural History Museum.

microwave radiometer
A highly sensitive receiver for detecting microwave radiation by material media. In the microwave region of the ELECTROMAGNETIC SPECTRUM, the spectral RADIANCE of a BLACK BODY is directly related to its KINETIC (or physical) TEMPERATURE (*Rayleigh–Jeans approximation* of *Planck's law*). This has led to the interchangeable use of the two terms and the radiance of the scene observed by radiometers is usually characterised by the RADIANT (or brightness) TEMPERATURE. The brightness temperature may vary from zero Kelvin (0 K – for a non-emitting medium) to a maximum equal to the physical temperature of the scene (for a perfect emitter or blackbody). Equivalently, the emissivity, defined as the ratio of the brightness to the physical temperature, varies between zero and unity. The emissivity characterises the medium as a signature.

Through proper choice of the radiometer parameters (wavelength, polarisation and incidence angle) it is possible to establish useful relationships between the energy received by the radiometer and specific terrestrial or atmospheric parameters of interest. Microwave radiometry is therefore used for ENVIRONMENTAL MONITORING and has found extensive use in HYDROLOGY, OCEANOGRAPHY and METEOROLOGY. Monitoring of snow and soil moisture from satellite-based radiometers is useful in hydrology. Applications of microwave radiometry in oceanography include the monitoring of SEA-SURFACE TEMPERATURE, salinity and wind speed. In meteorology, satellite, airborne or ground-based radiometers are used, for instance, to measure water vapour, oxygen, OZONE and temperature profiles. The most limiting factor in SATELLITE REMOTE SENSING with radiometers is the coarse resolution of several kilometres per PIXEL. *TS*

Schanda, E. 1986: *Physical fundamentals of remote sensing.* Heidelberg: Springer. **Ulaby, F.T., Moore, R.K. and Fung, A.K.** 1981: *Microwave remote sensing: active and passive.* Norwood, MA: Artech House.

microwave remote sensing (MRS)
Any of a set of techniques for remotely gathering information using

instruments that operate in the microwave region of the ELECTROMAGNETIC SPECTRUM (wavelengths approximately from 1 mm to 1 m). Microwave sensors include *passive instruments* (which only receive natural radiation from the scene that is being imaged) and *active instruments* (which transmit energy to the Earth's surface and measure the amount scattered back to the sensor). Space-borne microwave instruments may have to compete against propagation effects associated with the atmosphere (which include absorption and scattering losses and possible rotation of the POLARISATION plane of the electromagnetic field by the Earth's magnetic field). At lower frequencies in the microwave range, the propagation is essentially unaffected by the atmosphere, meaning that the measurements are insensitive to weather and to the presence of clouds. Additionally, active instruments (because they transmit energy and then measure the amount that is reflected back) operate during both day and night.   *PJS*

[See also RADAR REMOTE SENSING INSTRUMENTS, RADAR, PASSIVE MICROWAVE REMOTE SENSING]

Ulaby, F.T., Moore, R.K. and Fung, A.K. 1981, 1982, 1986: *Microwave remote sensing: Active and passive* (three vols). Norwood, MA: Artech House.

midden   An archaeological deposit, essentially a refuse tip, often from food preparation (*kitchen midden*). Examples include the MESOLITHIC *shell middens* at coastal sites from Scandinavia to the Mediterranean, which provide important evidence of HUNTING, FISHING AND GATHERING economies.   *JAM*

[See also OCCUPATION LAYER, RODENT MIDDEN]

Bonsall, C. (ed.), 1985: *The Mesolithic in Europe*. Edinburgh: Edinburgh University Press. Roosevelt, A.C., Housley, R.A., Imazio da Silveira, M. *et al.* 1991: Eighth millennium pottery from a prehistoric shell midden in the Brazilian Amazon. *Science* 254, 1621–1624. Sullivan, M. and O'Connor, S. 1993: Middens and cheniers: implications of Australian research. *Antiquity* 67, 776–788. Waselkov, G.A. 1987: Shellfish gathering and shell midden archaeology. *Advances in Archaeological Method and Theory* 10, 93–210.

Middle Ages: landscape impacts

In the MIDDLE AGES much of the European landscape consisted of agricultural land or semi-natural communities, which were carefully managed to maximise their provision of useful resources. Arable activity was widespread, evidence for which survives as the remains of RIDGE AND FURROW extending into upland areas now suitable only for rough grazing. The cultivation of MARGINAL uplands may have been favoured by the MEDIAEVAL WARM PERIOD, while LAND DRAINAGE schemes allowed agriculture to extend onto former WETLANDS. The expansion of arable land seems to have increased SOIL EROSION, resulting in ALLUVIATION in river valleys across much of northern Europe. *Grassland* was widespread and much of this was grazed by sheep to supply the wool trade, while other areas were managed as hay MEADOWS. WOODLAND MANAGEMENT was also commonplace, often involving COPPICING or WOOD PASTURE. HEATHLAND expansion was encouraged by heavy GRAZING, *burning* and the practice of *plaggen*, a man-made surface probably produced by heavy manuring over a long period. Hunting was a popular pastime amongst the aristocracy, who created forests and parks for deer and warrens in which to keep rabbits.   *SPD*

[See also AGRICULTURAL HISTORY, VEGETATION HISTORY]

Astill, G. and Grant, A. (eds) 1988: *The countryside of Medieval England*. Oxford: Blackwell Science. Astill, G. and Langdon, J. (eds) 1997: *Medieval farming and technology*. Leiden: Brill. Christie, N. 1999: Europe in the Middle Ages. In Barker, G. (ed.), *Companion encyclopaedia of archaeology*. London: Routledge, 1040–1076. Rackham, O. 1986: *The history of the countryside*. London: Dent. Yeoman, P. 1994: *Medieval Scotland*. London: Batsford.

mid-ocean ridge (MOR)   A chain of submarine mountains extending some 84 000 km through the Atlantic, Arctic, Indian, and South Pacific OCEANS (see Figure). The mid-ocean ridges rise 1–3 km above the ABYSSAL plains, are about 1500 km wide and are characterised by very rugged topography. In places (e.g. Iceland) they rise above sea-level. The centre, or axis, of the ridge is offset laterally by TRANSFORM FAULTS and experiences shallow SEISMICITY. It is a site of active basaltic VOLCANISM, and related HYDROTHERMAL activity gives rise to submarine HYDROTHERMAL VENTS at which communities of chemosynthetic organisms were discovered in the 1970s. According to PLATE TECTONICS theory, the mid-ocean ridge is a constructive PLATE MARGIN and SEA-FLOOR SPREADING occurs at its centre. The ridge is elevated because its young OCEANIC CRUST is hot and buoyant. By the time it is about 110 million years old, ocean crust has spread from the ridge axis, lost excess heat and subsided to the level of the abyssal plains. Subsidence is accommodated along active FAULTS, which are responsible for the rugged topography. Where spreading is rapid, the ridge is narrow and typically has a central RIFT VALLEY (e.g. Mid-Atlantic Ridge, MAR); faster spreading leads to a wider ridge (e.g. East Pacific Rise, EPR).   *BTC*

[See also HOTSPOT (GEOLOGY)]

Cann, J.R. Elderfield, H. and Laughton, A.S. (eds) 1999: *Mid-ocean ridges: dynamics of processes associated with the creation of new oceanic crust*. Cambridge: Cambridge University Press. Decker, R. and Decker, B. 1997: *Volcanoes*, 3rd edn. Basingstoke: Freeman. Gebruk, A.V., Galkin, S.V., Vereschchaka, A.L. *et al.* 1997: Ecology and biogeography of the hydrothermal vent fauna of the Mid-Atlantic Ridge. *Advances in Marine Biology* 32, 93–144. Kelley, D.S. and Fruh-Green, G.L. 1999: Abiogenic methane in deep-seated mid-ocean ridge environments: insights from stable isotope analysis. *Journal of Geophysical Research – Solid Earth* 104, 10439–10460.

migration   The movement of an organism from one location to another. Migration may take many forms. Many species migrate on an annual basis, including a wide array of insects and birds, in response to seasonal climatic changes or changes in food supply. While most species migrate to and from a destination, annual migration patterns may involve more than one generation, such as occurs with the Monarch butterfly (*Danaus plexippus*). Migrations may also follow long-term CLIMATIC CHANGE. Following QUATERNARY ENVIRONMENTAL CHANGE, species migrated in response to shifting climatic zones or suffered EXTINCTION. Such long-term species migrations occur individualistically because of inter-specific variation in migration speed and climatic tolerance. Species associations and ECOSYSTEMS did not necessarily remain intact. Anthropogenic climate change is likely to force species to migrate to remain within climatically suitable areas, but species capacity to migrate successfully may be seriously

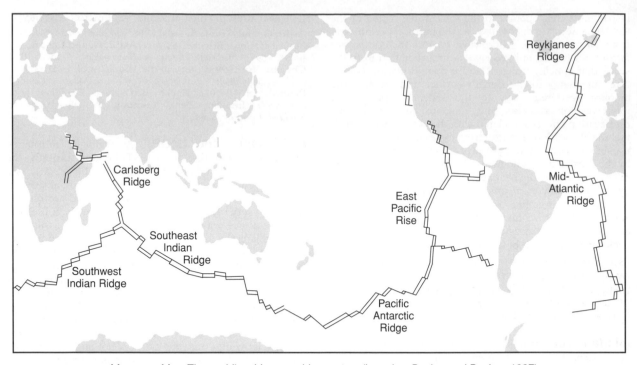

mid-ocean ridge *The world's mid-ocean ridge system (based on Decker and Decker, 1997)*

reduced by widespread habitat FRAGMENTATION or other ENVIRONMENTAL DEGRADATION.   *JTK*

[See also DISPERSAL, NOMADISM, RANGE ADJUSTMENT]

Baker, R.R. 1978: *The evolutionary ecology of animal migration.* London: Hodder and Stoughton. **Baker, R.R.** 1982: *Migration: paths through time and space.* London: Hodder and Stoughton. **Huntley, B. and Webb III, T.** 1989: Migration: species' response to climatic variations caused by changes to the Earth's orbit. *Journal of Biogeography* 16, 5–19. **Parmesan, C.** 1996: Climate and species range. *Nature 382,* 765–766. **Peters, R.L. and Lovejoy, T.E. (eds)** 1992: *Global warming and biological diversity.* New Haven, CT: Yale University Press.

Milankovitch theory   A mathematical theory that accounts for the PERIODICITY and timing of the CLIMATIC CHANGES responsible for GLACIAL–INTERGLACIAL CYCLES in terms of ORBITAL FORCING. Milutin Milankovitch (1879–1955) was a Serbian astronomer whose theory was disputed during his lifetime. It was not until the AD 1960s that the theory was re-examined and later fully tested and largely accepted. The theory incorporated and expanded on the earlier ideas of Adhémar (1842) and James Croll (1875).

The Milankovitch theory, also known as the *astronomical theory* of climatic change, suggests that glacial–interglacial cycles are linked to regular variations in the Earth's orbit around the Sun, which produce regular changes in the distance of the Earth from the Sun and hence the quantity and/or distribution of SOLAR RADIATION received by the Earth–atmosphere system. It has been estimated that about 60% of the VARIANCE in the record of global average temperature over the past million years occurs close to frequencies identified in the Milankovitch theory. This suggests that the CLIMATIC SYSTEM is not strongly chaotic but is responding in a predictable way to orbital forcing and hence that the Milankovitch theory provides a valid explanation of long-term climate changes.

The theory is based on the fact that the position and configuration of the Earth in relation to the Sun changes in a predictable way. Thus, the receipt of INSOLATION at the Earth's surface is equally predictable. There are three main orbital parameters that describe cyclical variations in the Earth's orbit with respect to the Sun: first, the *precession of the equinoxes* (PRECESSION) with a periodicity of about 21 000 years; second, the *obliquity of the ecliptic* (OBLIQUITY) with a periodicity of about 41 000 years; and, third, the *eccentricity of the orbit* (ECCENTRICITY) with a periodicity of about 96 000 years. These three parameters may be considered as representing, respectively, the 'wobble', 'tilt' and 'stretch' of the Earth's orbit. The 'composite curve' produced by combining the variations in these three parameters is the basis of the Milankovitch theory (see Figure 1).

The distance of the Earth from the Sun varies from about 147 to 152 million miles. Precession controls the time of year when the Earth is closest to the Sun. Summers receive more insolation and are relatively warm when the Earth is close to the Sun and the effects are greatest at low latitudes. We now have relatively cool summers and conditions are conducive to ICE SHEET growth; about 10 500 years ago conditions were ripe for ice sheet decay. Obliquity reflects changes in the Earth's axial tilt from about 21.5° to 24.5° and controls SEASONALITY (the contrast between winter and summer conditions) with relatively low tilt leading to lower seasonality. Obliquity effects are greatest at high latitudes. Eccentricity may be viewed as departures in ellipticity of up to 6% from a circular orbit (0% ellipticity); it accentuates the effects of

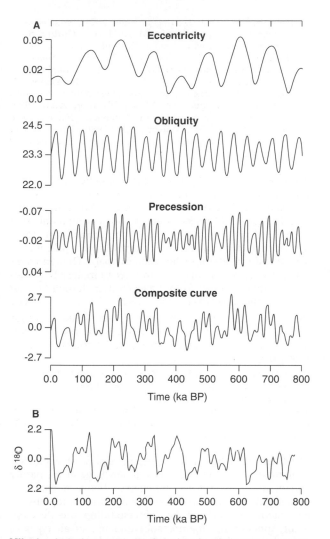

Milankovitch theory (1) *(A) Calculated variations in eccentricity, obliquity and precession over the last 800 000 years and the 'composite curve' produced by combining the three indices together after statistical normalisation; the basis of Milankovitch theory. (B) An observed (normalised and smoothed) palaeoenvironmental record of oxygen isotope variations from five marine sediment cores (after Imbrie et al., 1984)*

precession and it is the only one of the three parameters to affect the total heat received by the Earth.

These variations in the Earth's orbit interact and do not affect different latitudes in precisely the same way: for example, the 21 000-year precession cycle is more important at low latitudes and the 41 000-year obliquity cycle at high latitudes. GLACIAL EPISODES and INTERGLACIAL conditions are nevertheless expected when appropriate combinations of the three elements coincide and lead to greatest cooling or warming of the Earth, respectively.

Ever since Milankovitch proposed the theory, the predictions of the theory have been compared with PROXY DATA from the field of PALAEOCLIMATOLOGY. The first decisive test by Hays, Imbrie and Shackleton (1976) utilised data from MARINE SEDIMENT CORES, specifically the OXYGEN ISOTOPE ratios from the shells of microscopic plankton (FORAMINIFERA). The isotope ratio reflects the volume of water in the oceans and hence glacial–interglacial cycles. Use of SPECTRAL ANALYSIS identified dominant frequencies in the palaeoenvironmental data that were in very close agreement with the predictions of the theory. More recently, there have been further tests on data from CORAL REEF sequences, long vegetation records reconstructed by POLLEN ANALYSIS, LOESS sequences and ICE CORES. For example, spectral analyses of a variety of data from the Vostok ice core (Antarctica), all demonstrate significant effects of the 'Milankovitch periodicities' in the palaeoenvironmental record (see Figure 2).

Milankovitch theory is clearly of major importance in the explanation of climatic change on timescales of 10^4–10^5 years. The timing of cold and warm stages predicted by the theory and identified in palaeoenvironmental records provides a chronological and theoretical framework for studies of QUATERNARY ENVIRONMENTAL CHANGE and in many ways can be viewed as a PARADIGM for the field. It must be emphasised, however, that it does not explain the onset of ICE AGES or the quantitative scale of cooling during the glacial episodes. Changes in the Earth's geography and AUTOVARIATION, respectively, are likely to have played a role here. Neither does the theory account for the change from a dominant periodicity of about 41 000 years to one of about 100 000 years at around 750 000 years ago, when ice-sheet volumes also became larger. It is, nevertheless, aptly described as the 'pacemaker' or 'pulsebeat' of Ice Age climate. *JAM/AHP*

[See also SPECTRAL MAPPING PROJECT TIMESCALE]

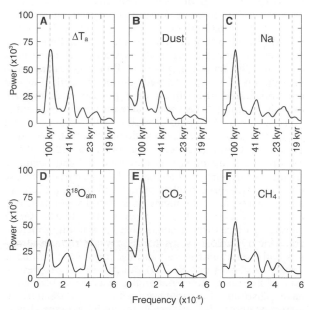

Milankovitch theory (2) *Results of spectral analysis (Blackman–Tukey technique) of time series from the Vostok ice core, Antarctica: (A) isotopic temperature of the atmosphere; (B) dust content; (C) sodium concentration; (D) $\delta^{18}O$; (E) carbon dioxide concentration; (F) methane concentration. Note correspondence of the spectral peaks with periodicities of 100 ka, 41 ka, 23 ka and 19 ka predicted by Milankovitch theory (after Petit et al., 1999)*

Hays, J.D., Imbrie, J. and Shackleton, N.J. 1976: Variations in the Earth's orbit: pacemaker of the Ice Ages. *Science* 194, 1121–1132. Hooghiemstra, H., Melice, J.L., Berger, A. and Shackleton, N.J. 1993: Frequency spectra and palaeoclimatic variability of the high-precision 30–1450 ka Funza I pollen record (Eastern Cordillera, Columbia). *Quaternary Science Reviews* 12, 141–156. Imbrie, J., Berger, A. and Shackleton, N.J. 1993: Role of orbital forcing: a two-million-year perspective. In Eddy, J.A. and Oescher, H. (eds), *Global changes in the perspective of the past.* New York: Wiley, 263–277. Imbrie, J., Hays, J.D., Martinson, D.G. *et al.* 1984: The orbital theory of Pleistocene climate: support from a revised chronology of the marine $\delta^{18}O$ record. In Berger, A., Imbrie, J., Hays, J., Kukla, G. and Salzman, B. (eds), *Milankovitch and climate.* Dordrecht: Reidel, 269–306. Imbrie, J., Berger, A., Boyle, E. *et al.*1993: On the structure and origin of major glaciation cycles. 2. The 100,000-year cycle. *Paleoceanography* 8, 699–735. Imbrie, J., Boyle, E.A., Clemens, S.C. *et al.* 1992: On the structure and origin of major glaciation cycles. 1. Linear responses to Milankovitch forcing. *Paleoceanography* 7, 701–738. Imbrie, J. and Imbrie, K.P. 1979: *Ice Ages: solving the mystery.* London: Macmillan. Kutzbach, J.E. and Street-Perrott, F.A. 1985: Milankovitch forcing of fluctuations in the level of tropical lakes from 18–0 kyr BP. *Nature* 317, 1301–1304. Milankovitch, M. 1941: *Canon of insolation and the Ice-Age problem.* Belgrade: Royal Serbian Academy [reprinted Jerusalem: Israel Program for Scientific Translations, 1968]. Palutikov, J.P., Goodess, C.M., Watkins, S.J. and Burgess, P.E. 1999: Developments in long-term climate change. *Progress in Environmental Science* 1, 89–96. Petit, J.R., Jouzel, J., Raynaud, D. *et al.* 1999: Climate and atmospheric history of the past 420,000 years from the Vostok ice core, Antarctica. *Nature* 399, 429–436. Rutherford, S. and D'Hondt, S. 2000: Early onset and tropical forcing of 100,000-year Pleistocene glacial cycles. *Nature* 408, 72–75.

military impacts on environment

Military forces have the ability to exert substantial impacts on air, land and water, during both peacetime and conflict. It is anticipated that a modern full-scale conflict would be an ENVIRONMENTAL DISASTER, as ENVIRONMENTAL PROTECTION is not a priority for warring troops. Pollution arising from the ignition of Kuwaiti oil wells during the Gulf War and the anticipated after-effects of a nuclear exchange are notable examples. Peacetime activities such as training exercises and the maintenance of infrastructure and equipment also contribute to ENVIRONMENTAL DEGRADATION, POLLUTION and ecological DISTURBANCE. Military training areas are often very extensive and can occupy a significant proportion of a nation's land resource. The use of heavy, armoured tracked vehicles in training exercises can be particularly damaging to soil and vegetation and impacts are akin to those of severe TRAMPLING. Other damaging or disturbing consequences of training include cratering by explosives, discharge of waste from naval vessels and the noise of aircraft and live firing. North Atlantic Treaty Organisation (NATO) policy and the progressive loss of immunity from national conservation legislation have ensured that ENVIRONMENTAL MANAGEMENT is assuming greater importance in the military sector. A positive consequence of the military ownership of large training areas is that agriculture, urban development and public access have been restricted on extensive areas of land. Thus, some such sites have become nationally or even internationally renowned as havens for rare species, habitats and archaeological features. *MLW*

[See also NUCLEAR WINTER, OIL FIRES]

Al-Hassan, J.M. 1992: *The Iraqi invasion of Kuwait: an environmental catastrophe.* Kuwait: Jassim M. Al-Hassan. Cuddy, S.M. 1990: Modelling the environmental effects of training on a major Australian army base. *Mathematics and Computers in Simulation* 32, 83–88. Prendergast, N.H.D. 1989: Military training areas and the countryside. *British Army Review* 93, 38–46. Vertegaal, P.J.M. 1989: Environmental impact of Dutch military activities. *Environmental Conservation* 16, 54–63. Westing, A.H. 1992: Protected natural areas and the military. *Environmental Conservation* 19, 343–347. Wilson, S.D. 1986: The effects of tank traffic on prairie: a management model. *Environmental Management* 12, 397–403.

mineral

A naturally occurring inorganic solid with a chemical composition that is fixed or varies within a limited range and a CRYSTALLINE structure. Minerals are the building blocks of rocks. Some geological materials that do not strictly conform to this definition are commonly described as minerals or, more correctly, *mineraloids* (e.g. AMBER and AMORPHOUS silica such as opal). The term *mineral deposit* is commonly used to refer to materials of economic value obtained from the ground, including OIL, *gas* and even GROUNDWATER. *GO*

[See also CARBONATE, EVAPORITE, FELDSPAR, IGNEOUS ROCKS, METAMORPHIC ROCKS, MINERALISATION, MINEROGENIC SEDIMENT, QUARTZ, SEDIMENTARY ROCKS, SILICATE MINERALS]

Deer, W.A., Howie, R.A. and Zussman, J. 1992: *An introduction to rock forming minerals,* 2nd edn. Harlow: Longman. Gaines, R.V., Skinner, H.C.W., Foord, E. et al. 1997: *Dana's new mineralogy,* 8th edn. Chichester: Wiley. Rutley, F. 1988: *Elements of mineralogy,* 27th edn, revised by C.D. Gribble. London: Unwin Hyman.

mineral cycling

An important aspect of ecological systems at local to global scales. Knowledge of the nature and quantity of minerals transferred between compartments and the rate of mineral cycling within and between ecosystems is essential for understanding the development, functioning, maintenance and use of all types of natural and anthropogenic ecosystems. *JAM*

[See also BIOGEOCHEMICAL CYCLES, ENERGY FLOW]

mineral magnetism

In measuring the 'magnetisability' of a material, MAGNETIC SUSCEPTIBILITY provides evidence primarily about the iron-bearing minerals found in rocks, sediments, soils and dusts. The data may, therefore, enable the following: identification of the different types of mineral present; estimation of their concentrations and relative inputs; classification of different types of materials; identification of their formation and/or processing; and the determination of sediment PROVENANCE. As a result of its non-destructive nature and its broad applicability in a range of environments, mineral or environmental magnetism is ideal as a reconnaissance tool where a large sample set is needed to characterise an environment or to reveal changes in input, source and/or environmental conditions. Indeed, it is rare to find examples where mineral magnetism is not used in parallel with other palaeoenvironmental or analytical techniques.

Both field- and laboratory-based measurements of magnetic susceptibility and MAGNETIC REMANENCE can be used to investigate the magnetic properties of environmental samples, i.e. DIAMAGNETISM, PARAMAGNETISM and FERROMAGNETISM. In addition, measurements of ISOTHERMAL

REMANENT MAGNETISM (IRM) and ANHYSTERETIC REMA-NENT MAGNETISATION (ARM) may be used to determine the *magnetic mineralogy*, e.g. the relative contributions from magnetic minerals, such as MAGNETITE or HAEMATITE, or the MAGNETIC GRAIN SIZE of a particular sample (see Table).

Mineral magnetic measurements have been used successfully in the study of lake sedimentation, with particular emphasis on temporal changes in sediment supply and provenance. Indeed, these studies have investigated the influence of both climate and humans on lake sediment budgets, catchment sediment yields and water quality. Long-term palaeoclimatic data have also been obtained from the mineral magnetic properties of MARINE SEDIMENT CORES and LOESS STRATIGRAPHY, which can be interpreted in the context of orbital forcing and/or major temporal and spatial changes in world climate patterns. The more recent impacts of humans in the form of POLLUTION have also been investigated in peat bog, lake, coastal and urban sites. *AJP*

[See also NATURAL REMANENT MAGNETISM]

Dearing, J.A., Alstrom, K., Bergman, A. *et al.* 1990: Recent and long-term records of soil erosion from southern Sweden. In Boardman, J., Foster, I.D.L. and Dearing, J.A. (eds), *Soil erosion on agricultural land.* Chichester: Wiley, 173–191. **Jiles, D.** 1991: *Introduction to magnetism and magnetic minerals.* London: Chapman and Hall. **Maher, B.A. and Thompson, R.** 1991: Mineral magnetic record of the Chinese loess and palaeosols. *Geology* 19, 3–6. **Robinson, S.** 1986: The late Pleistocene palaeoclimatic record of North Atlantic deep-sea sediments revealed by mineral magnetic measurements. *Physics of the Earth and Planetary Interiors* 42, 22–47. **Stoner, J.S., Channell, J.E.T. and Hillaire-Marcel, C.** 1996: The magnetic signature of rapidly deposited detrital layers from the deep Labrador Sea: relationship to North Atlantic Heinrich layers. *Palaeoceanography* 11(3), 309–325. **Thompson, R. and Oldfield, F.** 1986: *Environmental magnetism.* London: Allen and Unwin. **Walden, J., Oldfield, F. and Smith, J.P. (eds)** 1999: *Environmental magnetism: a practical guide* [*QRA Technical Guide* 6]. London: Quaternary Research Association.

mineral resources A loosely defined term covering MINERALS and rocks that are of value to society, either as obtained from the ground through MINING and quarrying or after processing. ENERGY RESOURCES are usually excluded. Mineral resources include metal ORE deposits, gemstones, construction materials (e.g. sand, gravel, clay, building stone) and *industrial minerals* such as gypsum, graphite and PHOSPHORITE. *GO*

mineral magnetism *Mineral magnetic parameters and their interpretation (after Stoner et al., 1996)*

Parameter	Characteristics
Volumetric magnetic susceptibility (κ) and **specific magnetic susceptibility** (κ) (measured at high and low frequency).	A first-order measure of the amount of ferrimagnetic material (e.g. magnetite). Susceptibility is enhanced by superparamagnetic (SP) magnetite (< 0.03 µm) and by large magnetite grains. When the concentration of ferrimagnetic material is low, susceptibility responds to antiferromagnetic (e.g. haematite), paramagnetic and diamagnetic materials.
Frequency-dependent magnetic susceptibility, χ_{fd} (or κ_{fd}), is the ratio of low-frequency to high-frequency susceptibility expressed as a percentage of the low-frequency susceptibility.	χ_{fd} indicates the presence of ultrafine ferrimagnetic (superparamagnetic – SP) material, produced by bacteria or by chemical processes mainly in soil. SP material in high concentrations complicates the grain-size interpretations using the interparametric ratios.
Isothermal remanent magnetisation (IRM), commonly expressed as a saturation IRM (or SIRM) when a field of 1 T is used. A backfield IRM is that magnetisation acquired in a reversed field after SIRM acquisition.	SIRM depends primarily on the concentration of magnetic, mainly ferrimagnetic, material. It is grain-size dependent, being particularly sensitive to magnetite grains smaller than a few tens of µm.
Anhysteretic remanent magnetisation (ARM), commonly expressed as susceptibility of ARM (χ_{ARM}) when normalised using the biasing DC field.	χ_{ARM} is a measure of the concentration of ferrimagnetic material, but is also strongly grain-size dependent. χ_{ARM} is highly selective of stable single-domain (SSD) ferrimagnetic grains in the range of 0.02 to 0.4 µm.
The **'hard' IRM** (HIRM) is derived by imparting a backfield IRM, typically 100–300 mT, on a sample previously given an SIRM.	HIRM is a measure of the concentration of magnetic material with a higher coercivity than the backfield, which commonly gives the concentration of antiferromagnetic (e.g. haematite and goethite) or very fine-grained ferrimagnetic grains.
The **'S ratio'** is derived by imparting a backfield of –100 mT on a sample previously given an SIRM from the ratio IRM/SIRM.	The S ratio can be used to estimate magnetic mineralogy. Values close to -1 indicate lower coercivity and a ferrimagnetic mineralogy (e.g. magnetite), whilst values closer to zero indicate a higher coercivity, possibly an antiferromagnetic mineralogy (e.g. haematite).
χ_{ARM}/χ (or κ_{ARM}/κ)	Providing the magnetic mineralogy is dominated by magnetite, χ_{ARM}/χ varies inversely with magnetic grain size, particularly 1–10 µm.
SIRM/χ (or SIRM/κ)	Providing the magnetic mineralogy is dominated by magnetite, SIRM/χ varies inversely with magnetic grain size. SIRM/χ is more sensitive to changes in the proportion of large (> 10 µm) grains.
SIRM/χ_{ARM} (or SIRM/κ_{ARM})	SIRM/χ_{ARM} increases with increasing magnetic grain size.

[See also NON-RENEWABLE RESOURCES]

Evans, A.M. 1997: *An introduction to economic geology and its environmental impact.* Oxford: Blackwell Science. **Woodcock, N.H.** 1994: *Geology and environment in Britain and Ireland.* London: UCL Press.

mineralisation

(1) The transformation of organic substances into inorganic ones. In soil it is the process by which NUTRIENT elements (such as nitrogen and phosphorus) are transformed from immobile organic forms (unusable by plants) to mobile inorganic ions that are available for root uptake. In BIOREMEDIATION, mineralisation occurs during the conversion of hazardous organic POLLUTANTS and WASTE into harmless inorganic compounds. It may also occur in this sense during the formation of FOSSILS.

(2) Mineralisation is used more generally in geology for the precipitation of minerals, especially in relation to useful minerals. *JAM/GO*

[See also BIOGEOCHEMICAL CYCLES, NITROGEN CYCLE, TAPHONOMY]

minerogenic sediment

Used especially in contradistinction to ORGANIC SEDIMENT, minerogenic sediment is either (1) a sediment composed wholly or dominantly of mineral particles or (2) the mineral component of a sediment. *JAM*

[See also CLASTIC]

minimum critical size of ecosystems

One of the most fundamental challenges for developing effective strategies to conserve BIODIVERSITY is to determine the minimum size of NATURE RESERVE necessary to maintain populations of particular ENDANGERED SPECIES. One of the earliest and most insightful attempts to estimate empirically the 'minimum critical size of ecosystems' was initiated by T.E. Lovejoy and colleagues during the early 1980s. In the TROPICAL RAIN FOREST of Brazil, just north of Manaus, Lovejoy's team carved out fragments of rain forest that varied in size from 1 to 1000 ha. By following the fate of populations of NATIVE SPECIES over the past two decades, these scientists have provided many important insights into the effects of habitat reduction and FRAGMENTATION on biological diversity. *MVL*

[See also SINGLE LARGE OR SEVERAL SMALL RESERVES]

Laurance, W.F. and Bierregaard Jr, R.O. 1997: *Tropical forest remnants: ecology, management and conservation of fragmented communities.* Chicago, IL: University of Chicago Press. **Lovejoy, T.E. and Bierregaard Jr, R.O.** 1990: Central Amazonian forests and the Minimum Critical Size of Ecosystems Project. In Gentry, A.H. (ed.), *Four Neotropical rainforests.* New Haven, CT: Yale University Press, 60–71.

minimum viable population

The minimum number of individuals that will ensure the survival of that population, not just in the short term, but in the long term, often cast in terms such as '95% probability of persistence for 100 or for 1000 years'. *RJW*

[See also EFFECTIVE POPULATION SIZE, POPULATION VIABILITY ANALYSIS]

Fielder, P.L. and Jain, S.K. (eds) 1992: *Conservation biology: the theory and practice of nature conservation, preservation and management.* London: Chapman and Hall.

mining

Extraction of mineral resources from underground (*deep mining*) or near the surface (*opencast* or *strip mining*). Surface mines and *quarries* (a term sometimes reserved for the mining of building stone) are known from prehistory but organised underground mining for salt, amber, metals and other materials was an activity developed by early civilisations. Underground mining is more hazardous than surface mining, but the latter generally presents greater LAND RECLAMATION and RESTORATION problems. These include the volume, calibre, chemical composition and instability of dumped mine waste, the problems of ACID MINE DRAINAGE and the toxicity of water seeping from mine waste tips long after mines have been abandoned. *JAM*

[See also EXTRACTIVE INDUSTRY, QUARRYING]

Carlsom, C.L. and Swisher, J.H. (eds) 1987: *Innovative approaches to mined land reclamation.* Carbondale, IL: Southern Illinois University Press. **Graf, W.L.** 1979: Mining and channel response. *Annals of the Association of American Geographers* 69, 262–275. **Gregory, C.E.** 1980: *A concise history of mining.* New York: Pergamon Press. **Hester, R.E. and Harrison, R.M.** 1994: *Mining and its environmental impact.* London: Royal Society of Chemistry. **Kelly, M.** 1988: *Mining and the freshwater environment.* Amsterdam: Elsevier. **Shepherd, R.** 1993: *Ancient mining.* Amsterdam: Elsevier. **Temple, J.** 1972: *Mining: an international history.* New York: Praeger.

Miocene

An EPOCH of the TERTIARY period, from 23.3 to 5.2 million years ago. *GO*

[See also GEOLOGICAL TIMESCALE]

mires

Used to denote peat-forming environments, mires include BOG, *fen*, SWAMP and *carr*, and may also include *moor*. Mire is all-embracing: it includes the peatland, the PEAT and the peat-forming vegetation. In this regard it is a particularly useful (though not precise) term. *Diplotelmic* (two-layered) mires have an anaerobic CATOTELM below the ACROTELM.

Classification of mires can be based on one or more of a range of features. Moore (1984, p. 2) considered these under seven headings: (i) floristics; (ii) vegetation structure and physiognomy; (iii) mire morphology; (iv) hydrology; (v) stratigraphy; (vi) chemistry; and (vii) peat characteristics. However, Moore noted that even such a classification of taxonomic criteria is far from perfect, for many of these features are themselves closely interrelated. This wide range of criteria has resulted in a diverse range of classifications, from detailed classifications based principally on PHYTOSOCIOLOGY, such as are used in central Europe and in Ireland, to the more generalised classification based on nutrient source, such as is sometimes used in parts of northwest Europe, particularly by palaeoecologists (e.g. OMBROTROPHIC MIRE and minerotrophic or RHEOTROPHIC MIRE). The difficulty with phytosociological classifications of mires is that they are based on the current vegetation. This can be a rather static view of what is inherently a dynamic system. Such classifications may not always acknowledge a mire's ontogeny, nor what might be growing there naturally but for the considerable human influence over recent centuries or decades (as, for example, the *under*-representation of carr habitats, the *over*-representation of depauperate BLANKET MIRE and the relative emphasis given to lowland wet heath in Britain's National Vegetation Classification of mires and heaths), nor what might be able to grow there were the climate to shift perceptibly. It is clear from analysis of PEAT STRATIGRAPHY that the surface vegetation of some mires has responded to climatic changes of the past, and might be expected to do so in the future. A simpler classification, which was formerly used in Britain, is one based on trophic status: OLIGOTROPHIC bogs and EUTROPHIC fens. However, 'poor fen' vegetation is at best mesotrophic, whereas some valley and basin mires are mesotrophic but support 'bog' rather than 'fen' communities.

Mires are NATURAL ARCHIVES of environmental history and can be examined using a range of palaeoecological techniques, including POLLEN ANALYSIS, PLANT MACROFOSSIL ANALYSIS, analysis of testate amoebae (RHIZOPOD ANALYSIS), STABLE-ISOTOPE ANALYSIS and determination of PEAT HUMIFICATION, to give various PROXY RECORDS of vegetational or climatic history, which assist in reconstructing environmental changes of the Late Quaternary.

The major PEAT formers in circum-boreal mires of the Northern Hemisphere are the bog mosses (*Sphagnum* spp.), Ericaceous shrubs and some members of the Cyperaceae (sedge) and Poaceae (Gramineae – grass) families. Further south, and particularly in Southern Hemisphere mires, a wider range of taxa may be major peat formers, including members of the Restionaceae. The peat of tropical bog forests may largely be composed of tree remains.

In some parts of northwest Europe, the deliberate drainage of mires, the cutting of peat for fuel and for horticultural use and the afforestation of bogs has led to a rapid and catastrophic loss of mire habitats. The loss of RAISED MIRE habitats through drainage was particularly great in The Netherlands – losses that accumulated over recent centuries and were paralleled in Britain to a lesser degree – but in the last 50 years, mechanised peat cutting has led to increased loss of raised mire habitats in both Britain and Ireland such that mire conservation and restoration has become a major conservation issue. There are even growing concerns for mire habitats in countries with abundant and extensive mires, such as Canada, Estonia and Finland. The fragility of mire vegetation is well recognised and there is increasing concern over the loss of WETLAND habitats world-wide (see WETLAND CONSERVATION).

Although geologically, most north-temperate bogs and fens are relatively young, having developed within the HOLOCENE, some TROPICAL PEATLANDS were initiated earlier. The present state of intact mires is the culmination of thousands of years of development, in which they may have passed through several stages of ECOLOGICAL SUCCESSION. The ontogeny of mires therefore needs to be considered carefully in plans for mire CONSERVATION and management, and particularly in those cases where restoration of damaged and cutover mires is attempted.

FMC

Brooks, S. and Stoneman, R. 1997: *Conserving bogs: the management handbook*. Edinburgh: The Stationery Office. **Gore, A.J.P. (ed.)** 1983: *Mires: bog, fen, swamp and moor* [*Ecosystems of the World*. Vol. 4B]. Rotterdam: Elsevier. **Heathwaite, A.L. and Gottlich, K.L. (eds)** 1993: *Mires: process, exploitation and conservation*. Chichester: Wiley. **Moore, P.D. (ed.)** 1984: *European mires*. London: Academic Press. **Parkyn, L. Stoneman, R.E. and Ingran, H.A.P. (eds)** 1997: *Conservating peatlands*. Wallingford: CAB International. **Rodwell, J.S. (ed.)** 1991: *British plant communities*. Vol. 2. *Mires and heaths*. Cambridge: Cambridge University Press.

misfit meander A meander apparently too small for its valley or drainage basin. Together with *misfit streams*, misfit meanders were studied extensively by G.H. Dury, who argued that they were globally widespread and reflected substantial postglacial shifts in hydrological regime, driven by CLIMATIC CHANGE. During cool, wetter, 'pluvial' periods, Dury estimated by RETRODICTION from HYDRAULIC GEOMETRY relationships for valley dimensions that discharges up to one hundred times present-day values were generated, which carved large 'valley meanders'. When rainfalls and temperatures assumed present-day levels, ALLUVIATION of valley meanders and 'shrinkage' of the river channels occurred as an adjustment to the much reduced discharge. These ideas proved highly controversial, especially in the 1970s, with some arguing that valley meanders were cut in one of the following ways: (1) by active estuarine processes at times of higher sea level to create *tidal palaeomorphs*; (2) during times of higher runoff caused by frozen ground in a PERIGLACIAL ENVIRONMENT; (3) under glacial river conditions characterised by a much greater seasonal concentration of flow (MELTWATER) in the summer months; (4) at times of higher discharge before RIVER CAPTURE had 'beheaded' part of the contributing catchment; (5) by catastrophic glacial meltwater releases (perhaps as JÖKULHLAUPS).

DML

[See also UNDERFIT STREAM]

Dury, G.H. 1983: Osage-type underfitness on the river Severn near Shrewsbury, Shropshire, England. In Gregory, K.J. (ed.), *Background to palaeohydrology*. Chichester: Wiley, 399–412. **Dury, G.H.** 1985: Attainable standards of accuracy in the retrodiction of palaeodischarge from former channel dimensions. *Earth Surface Processes and Landforms* 10, 205–213. **Williams, G.P.** 1988: Paleofluvial estimates from dimensions of former channels and meanders. In Baker, V.R., Kochel, R.C. and Patton, P.C. (eds), *Flood geomorphology*. Chichester: Wiley, 321–334.

mission to planet Earth (MTOP) An international research programme to understand the Earth's environment as a system, largely through the use of SATELLITE REMOTE SENSING. A major challenge is to observe, understand, model, assess and eventually predict GLOBAL CHANGE.

TS

[See also EARTH-SYSTEM ANALYSIS]

Mississippian In North American usage, a SYSTEM of rocks and a PERIOD of geological time from 354 to 324 million years ago. It is approximately equivalent to the early CARBONIFEROUS in Europe.

GO

[See also GEOLOGICAL TIMESCALE, PENNSYLVANIAN]

model A simplified representation of a (natural) phenomenon, developed to predict a new phenomenon or to provide insights into existing phenomena. Models may be seen as complex hypotheses and there are many types including semantic, conceptual, graphic, hardware, mathematical and statistical.

HB

[See also CLIMATIC MODELS, DETERMINISTIC, ENVIRONMENTAL MODELLING, SIMULATION MODEL, STOCHASTIC, VALIDATION]

moder An acid HUMUS form, in which there is incomplete breakdown of the organic material. It is loose and contains many animal droppings, mineral particles and brown-stained plant remains. F and H layers are of approximately equal thickness and moder is found in moist conditions under both deciduous and coniferous forests. See Figure under MULL. *EMB*

[See also MOR]

Kubiena, W.L. 1953: *The soils of Europe* London: Thomas Murby.

modern In the context of environmental change, 'modern' refers to similarity to and continuity with the present day in terms of conditions, characteristics or age. A radiocarbon date, for example, is said to be modern if the difference in age from the present (AD 1950 by convention) is not statistically significant. A fossil flora may be considered modern if the species are all extant. *JAM*

Kaser, G. 1999: A review of the modern fluctuations of tropical glaciers. *Global and Planetary Change* 22, 93–103.

modern analogue A modern analogue is the contemporary organism, assemblage or condition used as a benchmark for comparison with similar FOSSIL or SUBFOSSIL forms or conditions. Modern analogues are the basis for the reconstruction and interpretation of past conditions. A fundamental difficulty with geological analogues is whether strictly comparable analogues occur at present. In BIOSTRATIGRAPHY for instance, although the same species may have survived for many thousands, or even millions, of years, they may have changed in their ecological or climatic requirements. So-called *non-analogue conditions* or *non-analogue assemblages* are frequently encountered when investigating the past. Nevertheless, careful examination of modern analogues provides an invaluable tool in PALAEOENVIRONMENTAL RECONSTRUCTION. When TRANSFER FUNCTIONS were developed in PALAEOCEANOGRAPHY, where no analogues could be found, the *modern analogue technique* (MAT) was developed to define the degree of dissimilarity between modern and fossil faunal assemblages within defined temperature ranges. *DH*

[See also ANALOGUE METHOD]

Hutson, W.H. 1977: Transfer functions under no-analogue conditions: experiments with Indian Ocean planktonic Foraminifera. *Quaternary Research* 8, 355–367. **Kullman, L.** 1998: Non-analogous tree flora in the Scandes Mountains, Sweden, during the early Holocene – macrofossil evidence of rapid geographic spread and response to palaeoclimate. *Boreas* 27, 153–161. **Mock, C.J. and Brunelle-Daines, A.R.** 1999: A modern analogue of western United States summer palaeoclimate at 6000 years before present. *The Holocene* 9, 541–546. **Overpeck, J.T., Webb III, T. and Prentice, I.C.** 1985: Quantitative interpretation of fossil pollen spectra: dissimilarity coefficients and the method of modern analogues. *Quaternary Research* 23, 87–108. **Pflaumann, U., Duprat, J., Pujol, C. and Labeyrie, L.** 1996: SIMMAX: a modern analog technique to deduce Atlantic sea surface temperatures from planktonic foraminifera in deep-sea sediments. *Paleoceanography* 11, 15–35.

modernism The eighteenth- to twentieth-century artistic, architectural and intellectual movements that challenged the conventions of REALISM and romanticism by exploring ideas of 'newness' and expressing them as 'new' aesthetics. Ideas of modern are most commonly defined through their opposition to the old and the traditional. Hence 'modern' is synonymous with 'newness' and

'modernity' refers to the 'post-traditional' historical epoch within which 'newness' is produced and valued, as well as to the economic, social, political and cultural formations characteristic of that period. Modernity is generally associated with ideas of progress, order, science and rationality, which had arisen since the Age of Enlightenment in the late Middle Ages. From approximately the 1970s onwards there has been the rise of POST-MODERNISM as a critique of modernism, representing the end of 'progress'. *ART*

Blaikie, P.M. 1996: Post-modernism and global environmental change. *Global Environmental Change: Human and Policy Dimensions* 6, 81–85. **Harvey, D.** 1989: *The condition of postmodernity: an enquiry into the origins of cultural change.* Oxford: Blackwell.

Mohs' scale A relative scale of surface hardness for MINERALS, developed by F. Mohs (1773–1839). Common or distinctive minerals represent each of 10 points on the scale: 1 (softest) = talc, 2 = gypsum, 3 = calcite, 4 = fluorite, 5 = apatite, 6 = orthoclase FELDSPAR, 7 = QUARTZ, 8 = topaz, 9 = corundum, 10 (hardest) = diamond. The hardness of an unknown material can be determined relative to a known mineral or to common materials like finger nail ($H \sim 2.5$), copper ($H \sim 3.5$) and steel nail or knife blade ($H \sim 5.5$). *GO*

molasse A term originally applied in the early nineteenth century to thick sequences dominated by SANDSTONE and CONGLOMERATE of OLIGOCENE to MIOCENE age in Switzerland. Molasse represents material eroded from the then recently formed Alpine mountains and deposited in adjacent FORELAND BASINS in a range of shallow marine, freshwater and continental environments. Deposits of the Ganges–Brahmaputra SEDIMENTARY BASIN adjacent to the Himalayas represent a modern analogue. The term can be considered to represent a postorogenic tectono-stratigraphic FACIES (see also FLYSCH). It has since been applied more generally to postorogenic deposits of other OROGENIC BELTS. *BTC/GO*

Allen, J.R.L. 1962: Lower Old Red Sandstone of the southern British Isles: a facies resembling the Alpine Molasse. *Nature* 193, 1148–1150. **Allen, P.A.** 1984: Reconstruction of ancient sea conditions with an example from the Swiss Molasse. *Marine Geology* 60, 455–473. **Hsü, K.J.** 1995: *The geology of Switzerland.* Princeton, NJ: Princeton University Press.

molecular stratigraphy Molecular information such as BIOMARKERS, CARBON CONTENT, nitrogen content, bulk $\delta^{13}C$ values and COMPOUND-SPECIFIC CARBON ISOTOPE values can be analysed in MARINE, LACUSTRINE, PEAT BOG and SOIL cores. Molecular stratigraphic analyses, including LIPID distributions and compound-specific $\delta^{13}C$ measurements, represent changes in specific compound classes or individual compounds with depth or age. These molecular parameters form a stratigraphy that can be interpreted in terms of CLIMATIC CHANGE, organic matter source and atmospheric CO_2 concentration.

For the biomarker approach to be applicable to the PALAEOECOLOGY of the ENVIRONMENT *via* its molecular stratigraphy, biomarker proxies need to be assigned for the expected major inputs of organic matter, such as terrestrial higher plants, aquatic macrophytes and algae. Compound-specific isotope analysis by gas chromatography combined with isotope ratio mass spectrometry provides a tool with which to study the CARBON CYCLE at the molecular level. This technique allows the determination of molecule specific $\delta^{13}C$ values of the organic compounds in sediments, which will, in turn, clarify understanding of the carbon fluxes within the ecosystem. By measuring the relative abundances and the $\delta^{13}C$ values for individual compounds specific to higher plant leaf waxes, algae or aquatic macrophytes, changing inputs to the environments can be followed. Depth profiles of $\delta^{13}C$ values and biomarker distributions can be interpreted in terms of changes in atmospheric CO_2 concentration and local climate conditions. This approach is more discriminatory than conventional combustion isotope ratio mass spectrometry of total organic carbon (TOC) or conventional biomarker analyses alone. In effect, carbon cycling can now be studied using combined biomarker, stable isotope and biological data. *KJF*

Brassell, S.C., Eglinton, G., Marlowe, I.T. *et al.* 1986: Molecular stratigraphy: a new tool for climatic assessment. *Nature* 320, 129–133. Evershed, R.P. 1993: Biomolecular archaeology and lipids. *World Archaeology* 25, 74–93. Farrimond, P. and Flanagan, R.L. 1996: Lipid stratigraphy of a Flandrian peat bog (Northumberland, UK): comparison with the pollen record. *The Holocene* 6, 69–74. Farrimond, P., Poynter, J.G. and Eglinton, G. 1990: A molecular stratigraphic study of Peru Margin sediments, Hole 686B, Leg 112. In Suess, E., con Huene, R., Emeis, K.-C. *et al.* (eds), *Proceedings of the Ocean Drilling Program, Scientific Results* 112, 547–553. Ficken, K.J., Barber, K.E. and Eglinton, G. 1998: Lipid biomarker, $\delta^{13}C$ and plant macrofossil stratigraphy of a Scottish montane peat bog over the last two millennia. *Organic Geochemistry* 28, 217–237. Ficken, K.J., Street-Perrott, F.A., Perrott, R.A. *et al.* 1998: Glacial/interglacial variations in carbon cycling revealed by molecular and isotope stratigraphy of Lake Nkunga, Mt. Kenya, East Africa. *Organic Geochemistry* 29, 1701–1719. Huang, Y., Street-Perrott, F.A., Perrott, R.A. *et al.* 1999: Glacial–interglacial environmental changes inferred from molecular and compound-specific $\delta^{13}C$ analyses of sediments from Sacred Lake, Mt. Kenya. *Geochimica et Cosmochimica Acta* 63, 1383–1404.

molecule

The smallest constituent part of an ELEMENT or COMPOUND that retains its characteristic chemical properties. Molecules range from single ATOMS to very large numbers of atoms bonded together in complex chemical structures (e.g. HUMIC ACIDS or DEOXYRIBONUCLEIC ACID). *JAM*

mollic horizon

A dark coloured, humus-rich horizon with a *base saturation* of more than 50%. This horizon is found in KASTANOZEMS and PHAEOZEMS and in some CHERNOZEMS, ANDOSOLS and CAMBISOLS. It is characteristic of humus accumulation in a mid-continental neutral or calcareous soil environment. *EMB*

Food and Agriculture Organization of the United Nations (FAO) 1998: *World reference base for soil resources* [*Soil Resource Report* 84]. Rome: FAO, ISRIC, ISSS.

Mollusca analysis

Mollusca are common FOSSILS and SUBFOSSILS in calcareous sediments of PLEISTOCENE age. Terrestrial and freshwater Mollusca are valuable in both PALAEOENVIRONMENTAL RECONSTRUCTION and CLIMATIC RECONSTRUCTION based upon analogy with their present-day ecological and climatic preferences. The grouping of species into ecological categories provides a useful means of observing changes in the local environment, for example a transition from swamp to open-water conditions or woodland to grassland. Certain species of BRACKISH WATER may also be used as indicators of former marine TRANSGRESSIONS. Marine Mollusca are not particularly sensitive to changes in water temperature, although the appearance of genuine Arctic and southern species in Pleistocene deposits may be viewed as significant. Their application stems principally from the reconstruction of past water depths, SALINITY and energy conditions. Mollusca are less suitable as chronostratigraphic than palaeoenvironmental indicators, since they appear to have undergone little evolutionary change. Nevertheless, some diagnostic features, particularly extinctions, may be significant. *DCS*

[See also SCLEROCHRONOLOGY]

Evans, J.G., 1972: *Land snails in archaeology*. London: Seminar Press. Goodfriend, G.A. 1992: The use of land snail shells in palaeoenvironmental reconstruction. *Quaternary Science Reviews* 11, 665–685. Keen, D.H., 1990: Significance of the record provided by Pleistocene fluvial deposits and their included molluscan faunas for palaeoenvironmental reconstruction and stratigraphy: cases from the English Midlands. *Palaeogeography, Palaeoclimatology, Palaeoecology* 80, 25–34. Peacock, J.D. 1989: Marine molluscs and Late Quaternary environmental studies with particular reference to the Late-glacial period in north-west Europe: a review. *Quaternary Science Reviews* 8, 179–192. Rousseau, D.D., 1992: Terrestrial molluscs as indicators of global aeolian dust fluxes during glacial stages. *Boreas* 21, 105–110. Rousseau, D.D., Limondin, N., Magnin, F. and Puissegur, J.J. 1994: Temperature oscillations over the last 10,000 years in western Europe estimated from terrestrial mollusc assemblages. *Boreas* 23, 66–73. Sparks, B.W., 1961: The ecological interpretation of Quaternary non-marine mollusca. *Proceedings of the Linnean Society of London* 172, 71–80. Thomas, K.D. 1985: Land snail analysis in archaeology in theory and practice. In Fieller, N.J.R., Gilbertson, D.D. and Ralph, N.G.A. (eds), *Palaeobiological investigations: research design, methods and data analysis* [*BAR International Series* 266]. Oxford: British Archaeological Reports.

monochromatic

Any system or process using a very small range of colours or wavelengths of the ELECTROMAGNETIC SPECTRUM. The term is often used in reference to black-and-white photographic film (even though this system is really *panchromatic*) and REMOTE SENSING systems that utilise only a narrow range of wavelengths. *TS*

[See also ELECTROMAGNETIC SPECTRUM]

monoculture

The practice of cultivating single species rather than many species together (*polyculture*). Monoculture may be justified in terms of short-term economic return but tends to be unjustified in the long term or ecologically as there is loss of BIODIVERSITY and ecosystem complexity with concomitant vulnerability to PESTS and DISEASE and to downturns in market prices. *JAM*

monogenetic

An entity, such as a soil, sediment body, landform or landscape that was formed in a single time interval and environment (representing a single 'generation') is said to be monogenetic. Because of environmental change, monogenetic entities are the exception rather than the rule in nature. *JAM*

[See also POLYGENETIC]

monolith sampling The sampling of a vertical column of material (often PEAT, but can be SOIL, PALAEOSOL, LOESS or even lacustrine sediments, if accessible on land) by means of a metal (or wooden) three-sided box (usually open-ended). The intact column of sediment enclosed by the monolith box then needs to be severed from the vertical section by cutting behind and beneath. Special monolith cutters have been devised for peat monolith sampling.

FMC

Lageard, J.G.A., Chambers, F.M. and Grant, M.E. 1994: Modified versions of a traditional peat cutting tool to improve field sampling of peat monoliths. *Quaternary Newsletter*, 74, 10–15.

monsoon Derived from the Arabian word 'mausim', meaning season, the monsoon refers to large-scale seasonal winds in the Afro-Asian region from around 40°N to 20°S and 30°W to 180°E (see Figure, parts A and B).

Four criteria delimit these monsoon areas: PREVAILING WIND direction shifts by at least 120° between January and July; average frequency of prevailing directions in January and July exceeds 40%; mean resultant wind in at least one of the months exceeds 3 m s^{-1}; less than one CYCLONE–ANTICYCLONE alternation every two years in any one month in a 5° latitude-longitude rectangle.

Three factors account for the existence of monsoons: (1) differential seasonal heating of oceans and continents, which results in atmospheric pressure changes; (2) moisture processes in the tropical atmosphere; and (3) the Earth's rotation and the CORIOLIS effect. Four regional monsoon systems are recognised: Indian, east Asian, Australian and African. Over 55% of humankind live in these areas and the economies of tropical countries are closely linked to the annual cycle, especially the seasonal occurrence of monsoon rains. An alternative, wider and less traditional definition is given by Leroux (1998).

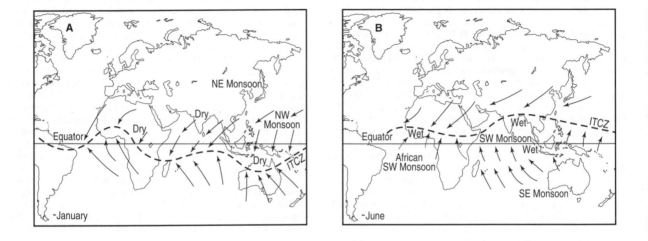

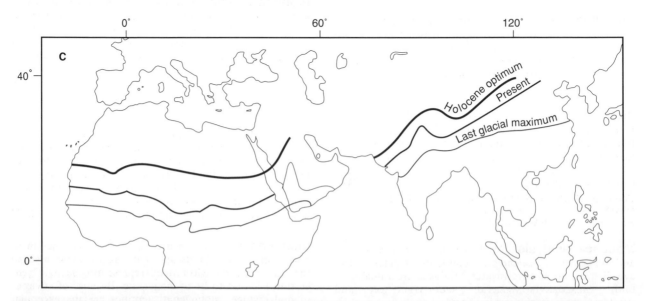

monsoon *(A, B) Seasonal patterns in monsoon winds at present (ITCZ = intertropical convergence zone) and (C) changes in the average northern limits of monsoon rains at the last glacial maximum and since the Holocene 'climatic optimum' (after Brown, 1994; Van Andel, 1994; Wilson et al., 2000)*

Variations in the strength of monsoonal atmospheric circulation have been recognised and the history of the circulation has been investigated on various timescales. Evidence from Chinese LOESS, for example, suggests a three-step evolution of the east Asian monsoon: (1) initiation about 2.6 million years ago (Ma); progressively greater variation at (2) 1.2 Ma and (3) 0.6 Ma between the relatively dry–cold periods dominated by the northerly monsoon and the humid–warm conditions characteristic of the southerly monsoon. During the last GLACIAL–INTERGLACIAL CYCLE there appear to have been at least six episodes of strengthened northerly monsoon, which may correspond with HEINRICH EVENTS in the north Atlantic Ocean; and millennial-scale variations have been identified in the HOLOCENE. Part C of the Figure suggests that these changes in the strength of the monsoon have been accompanied by appreciable changes in the areas receiving monsoon rains, with greater penetration of monsoon rains into the continental interiors during INTERGLACIALS.

BDG/JAM

Brown, G.M. 1994: Understanding the Earth's environment. *Nature and Resources (UNESCO)* 30, 21–33. **Chang, C.P.** 1987: *Monsoon meteorology.* Oxford: Oxford University Press. **Chun Chang Huang, Jiangli Pang and Jingpo Zhao** 2000: Chinese loess and the evolution of the east Asian monsoon. *Progress in Physical Geography* 24, 75–96. **Fein, J.S. and Stephens, P.L.** (eds) 1987: *Monsoons.* New York: Wiley. **Kutzbach, J.E. and Liu, Z.** 1997: Response of the African monsoon to orbital forcing and ocean feedbacks in the middle Holocene. *Science* 278, 440–443. **Leroux, M.** 1998: *Dynamic analysis of weather and climate.* Chichester: Wiley. **McGregor, G.R. and Nieuwolt, S.** 1998: *Tropical climatology,* 2nd edn. Chichester: Wiley. **Ramage, C.S.** 1971: *Monsoon meteorology.* New York: Academic Press. **Sirocko, F., Sarnthein, M., Erlenkeuser, H.** *et al.* 1993: Century scale events in monsoon climate over the past 24,000 years. *Nature* 364, 322–324. **Sontakke, N.A. and Singh, N.** 1996: Longest instrumental regional and all-India summer monsoon rainfall series using optimum observations: reconstruction and update. *The Holocene* 6, 315–331. **Van Andel, T.H.** 1994: *New views on an old planet.* Cambridge: Cambridge University Press. **Wilson, R.C.L., Drury, S.A. and Chapman, J.L.** 2000: *The great Ice Age: climate change and life.* London: Routledge.

montane forest A forest formation found on mountains in TROPICAL and SUBTROPICAL regions, characterised by short, gnarled, mesophyllous trees and abundant bryophytes. The formation is often split into lower and upper montane forest, depending on tree physiognomy. The altitude at which the formation begins is determined by the size of the mountain mass and its proximity to the sea (sometimes termed the MASSENERHEBUNG EFFECT). The term *montane* is sometimes used for any high-elevation forest , growing close to the TREE LINE. *NDB*

[See also ALTITUDINAL ZONATION OF VEGETATION, CLOUD FOREST, ELFIN WOODLAND, LEAF PHYSIOGNOMY, TROPICAL RAIN FOREST, VEGETATION FORMATION-TYPE]

Hamilton, L.S., Juvik, J.O and Scatena, F.N. (eds) 1995: *Tropical montane cloud forests.* Berlin: Springer. **Marchant, R. and Taylor, D.** 1998: Dynamics of montane forest in central Africa during the late Holocene: a pollen-based record from western Uganda. *The Holocene* 8, 375–381.

Monte Carlo methods Named after the famous casino in the principality of Monaco, Monte Carlo methods use random numbers to study either real data-sets or the behaviour of statistical methods through computer simulations. They are computer-intensive methods for finding solutions to mathematical and statistical problems by simulation or permutation. They are most commonly used when the analytical solution is intractable or very time-consuming. Within environmental statistics, Monte Carlo methods include *randomisation tests, permutation tests,* and, in some instances, BOOTSTRAPPING. They can also be used to estimate CONFIDENCE INTERVALS and limits for population parameters by using computer-generated data to estimate the expected amount of variation in the sample statistics of interest.

In many applied statistical applications, Monte Carlo methods are used to assess the statistical significance of an observed test statistic by comparing it with a large number of test statistics obtained by generating random data-sets *under some assumed model.* If the assumed model implies that all data orderings are equally likely, the test is a randomisation test with random sampling of the randomisation distribution. The reliability of any Monte Carlo test totally depends on the generation of data-sets that are *equally likely* under the relevant null hypothesis. Completely random permutations (as in a randomisation test) will yield invalid results if the observations are structured as a result of the way the data were collected, e.g. line transects, spatial grids, stratigraphical time series, split-plot designs, etc. Restricted permutation tests using model-based permutations are thus required in such instances to obtain reliable results.

The basic steps in all such randomisation and permutation tests are:

1. Calculate the test statistic (T_0) for the observed data-set.
2. Generate K new data-sets that *are equally likely* under the null hypothesis being tested.
3. Calculate the test statistic for each new data-set, giving estimates of $T_1, T_2, \ldots, T_K$.
4. Derive the exact Monte Carlo SIGNIFICANCE LEVEL by determining the proportion of values greater than or equal to T_0. The Monte Carlo level is thus the rank of T_0 among all values of T divided by $K + 1$ (one is added because T_0 is included in the null distribution).

Monte Carlo permutation tests provide distribution-free methods for statistical testing and are widely used in evaluating results of CANONICAL CORRESPONDENCE ANALYSIS and REDUNDANCY ANALYSIS. *HJBB*

[See also NON-PARAMETRIC STATISTICS]

ter Braak, C.J.F. and Šmilauer, P. 1998: *CANOCO Reference Manual and User's Guide to Canoco for Windows: Software for Canonical Community Ordination (version 4).* Ithaca, NY: Microcomputer Power. **Manly, B.F.J.** 1997: *Randomization, bootstrap and Monte Carlo methods in biology.* London: Chapman and Hall.

monthly mean temperature The mean of the daily maximum and minimum temperatures of a particular month. The most frequent expression of general temperature and the basis, for example, of the CENTRAL ENGLAND TEMPERATURE RECORD. *JCM*

[See also CLIMATIC ARCHIVE, MEAN MONTHLY TEMPERATURE]

montmorillonite A 2:1 clay mineral of the *smectite* group having a basic structure of two sheets of silicon atoms and one sheet of aluminium atoms held together by shared oxygen atoms. There is little attraction between the oxygen atoms in the top of one sheet and those in the bottom sheet of the next unit; consequently the lattice can expand when wetted. ISOMORPHIC SUBSTITUTION of Mg^{2+} for Al^{3+} within the lattice increases the negative charge so that it is 10 to 15 times that of KAOLINITE. Soils in which montmorillonite is the dominant CLAY MINERAL occur upon base-rich parent materials, particularly in warm temperate continental areas where VERTISOLS, CHERNOZEMS, *krasnozems* and some LUVISOLS are common.

EMB

Kittrick, J.A. 1971: Montmorillonite equilibria and the weathering environment. *Proceedings of the Soil Science Society of America* 35, 815–823.

moorland   A vegetation community and/or landscape formed on waterlogged, PEAT soils, which can accumulate to considerable depths, generally > 200 m above sea level in Britain, dominated by dwarf shrubs (e.g. heather, *Calluna vulgaris*, and bilberry, *Vaccinium myrtillus*), grasses and sedges (e.g. cotton-grass, *Eriophorum* spp.) adapted to wet, acidic environments.

MJB

[See also HEATHLAND, MIRE, WETLAND]

Pearsall, W.H. 1971: *Mountains and moorlands*, 2nd edn. London: Collins. **Simmons, I.G.** 1990: The mid-Holocene ecological history of the moorlands of England and Wales and its relevance for conservation. *Environmental Conservation* 17, 61–69.

mor   A strongly acid HUMUS form that has three clearly recognisable layers. At the surface is an L (*litter*) layer of plant debris accumulated from several years as decomposition is slow. Below is an F (*fermentation*) layer of increasingly fragmented plant material, matted together with fungal hyphae but containing few animal droppings. A thin H (*humus*) layer of completely humified material lies on the surface of the mineral soil. Incorporation of organic matter and conversion to mull can be achieved by liming and cultivation. See Figure under MULL.

EMB

[See also MODER]

Kubiena, W.L. 1953: *The soils of Europe* London: Thomas Murby.

moraines   Formerly used to describe GLACIAL SEDIMENTS as well as LANDFORMS, this term should only be used for depositional ridges formed by a range of processes (e.g. pushing, dumping, squeezing, thrusting) at the margins and termini of GLACIERS and for SUPRAGLACIAL debris in transport, originating largely by mass wasting of debris from slopes adjacent to a glacier. LATERAL MORAINES form along the flanks of a CIRQUE or valley glacier and an *end* or *terminal moraine* at its maximum extent downvalley. *Recessional moraines* form during marked stillstands or stages of renewed advance of a glacier. Supraglacial debris in transport forms a large component of lateral moraines along glacier margins and *medial moraines* usually form along lines of glacial confluence in the ABLATION zone. Moraines in alpine environments are usually composed of a combination of deposits originating in debris slope processes and in ENGLACIAL and basal transport. Dating of end and recessional moraines by a variety of techniques (e.g. LICHENOMETRY, RADIOCARBON DATING of incorporated or buried biogenic material, DENDROCHRONOLOGY,

SCHMIDT HAMMER *R*-VALUE) enables reconstruction of periods of glacial advance and retreat. The largest moraines of continental glaciers are made up of thick GLACIOFLUVIAL SEDIMENTS and GLACIOLACUSTRINE DEPOSITS with some TILL. They record STILLSTAND or re-equilibration positions of the ICE SHEETS and appear together with large ESKER and TUNNEL VALLEYS in complex landscapes.

JS

[See also DE GEER MORAINES, FLUTED MORAINE, GLACIAL LANDFORMS, GLACIOFLUVIAL LANDFORMS, GLACIER VARIATIONS, HUMMOCKY MORAINE, ROGEN MORAINES]

Boulton, G.S., van der Meer, J.J.M., Beets, D.J. *et al.* 1999: The sedimentary and structural evolution of a recent push moraine complex: Holmstrombreen, Spitsbergen. *Quaternary Science Reviews* 18, 339–371. **Embleton, C. and King, C.A.M.** 1975: *Glacial geomorphology.* London: Edward Arnold. **Goldthwait, R.P.** 1988: Classification of glacial morphological features. In Goldthwait, R.P. and Matsch, C.L. (eds), *Genetic classification of glacigenic deposits.* Rotterdam: Balkema, 267–277. **Röthlisberger, F., Haas, P., Holzhauser, H.** *et al.* 1980: Holocene climate fluctuations: radiocarbon dating of fossil soils and woods from moraines and glaciers in the Alps. *Geographica Helvetica*, 35, 21–52.

morphoclimatic zones   Regions of the Earth's surface governed by similar climatic relief-forming mechanisms. Each zone experiences dominant and subsidiary processes: the dominant processes largely determine the landform characteristics; and the subsidiary processes are more limited in their action, only affecting certain rock types or only acting at certain times. According to J. Tricart and A. Cailleux, delimitation of the world's morphoclimatic zones should be based primarily on present-day phenomena (see CLIMATIC GEOMORPHOLOGY). Their original classification divides the Earth's surface into 12 such zones, including glacial, periglacial, mid-latitude, desert and tropical forest zones. An alternative earlier approach was Peltier's system of *morphogenetic regions*, which were defined in terms of dominant geomorphic processes based on the climatic parameters of mean annual temperature and mean annual precipitation. The main drawback of morphoclimatic zone systems is that so many landforms and landscapes are the product of a past climate, a sequence of different climates and/or the transition between climates, rather than particular climates *per se.*

MAC

[See also CLIMATOGENETIC GEOMORPHOLOGY, LANDSCAPE EVOLUTION]

Derbyshire, E. (ed.) 1976: *Geomorphology and climate.* London: Wiley. **Peltier, L.C.** 1950: The geographic cycle in periglacial regions as it is related to climatic geomorphology. *Annals of the Association of American Geographers* 40, 214–236. **Tricart, J. and Cailleux, A.** 1965: *Introduction à la géomorphologie climatique.* Paris: SEDES.

morphometry   The quantitative description of form. Morphometric investigations exist in many disciplines and may be used as a basis for classification and for the inference of processes and change. In *geomorphometry* (measurement of the Earth's form), for example, *general geomorphometry* (analysis of the form of the entire landsurface) and *specific geomorphometry* (analysis of particular LANDFORMS) have been recognised.

JAM

Evans, I.S. 1990: General geomorphometry. In Goudie, A. (ed.), *Geomorphological techniques*, 2nd edn. London: Unwin Hyman, 31–37.

morphostratigraphy Originally developed by geologists and geomorphologists in the mapping of glacial STRATIGRAPHY, morphostratigraphy can be defined as part of the LITHOSTRATIGRAPHIC record seen in the form of geomorphological features (i.e. LANDFORMS). Although complexes of such features can represent temporal relationships, as in the case of a retreating ice margin, it is necessary to use other forms of evidence to provide a full record of chronological relationships between *morphostratigraphic units*. *CJC*

Lowe, J.J. and Walker, M.J.C. 1997: *Reconstructing Quaternary environments*, 2nd edn. Harlow: Addison-Wesley Longman.

mosaicing The process of combining and matching adjoining images, such as those acquired by AERIAL PHOTOGRAPHY or REMOTE SENSING. It is important that elements such as image geometry, edge feature matching and colour similarity are addressed. *TF*

motu An island typical of the inner reef flats of Indian and Pacific Ocean atolls, possessing a seaward shingle ridge of coral and molluscan fragments, an interior depression and a lagoonward sand ridge. *TSp*

[See also CORAL REEF ISLANDS]

mountain permafrost Perennially frozen ground that occurs in mountainous terrain, usually but not always at elevations above the TREE LINE. It is distinguished from *latitudinal permafrost* occurring at high latitudes (e.g. in ARCTIC and SUBARCTIC regions), and may also be distinguished from *plateau permafrost* occurring at high elevations but with relatively low slope gradients, e.g. on Quinghai-Xizang (Tibet) Plateau. Mountain permafrost is also referred to as *alpine permafrost*. *HMF*

[See also PERMAFROST]

Cheng, G. and Dramis, F. 1993: Distribution of mountain permafrost and climate. *Permafrost and Periglacial Processes* 3, 83–93. **Péwé, T.L.** 1983: Alpine permafrost in the contiguous United States: a review. *Arctic and Alpine Research* 15, 145–156.

mountain regions: environmental change and human impact Mountains are usually defined as natural elevations of the Earth's surface exceeding 600 m above sea level and rising abruptly above the surrounding terrain. They invariably possess several or all of the following characteristics: low temperatures; moderate to high precipitation; high wind speeds; steep slopes; and distinct summits or extensive plateaux. They may occur as single, isolated peaks or form substantial ranges, and many display evidence for past or present GLACIATION and/or PERIGLACIATION.

Mountain regions cover about 20% of the Earth's land-surface and about 50% of the human population are dependent, to some degree, on mountain environments for water supply, agricultural produce, mineral resources or power generation. The last 100 years has seen accelerated use of mountain areas for tourism, communications, MINING and quarrying, agriculture and timber supply and *hydroelectric power*. The pace of change has been such that it poses a serious threat to the SUSTAINABILITY of mountain ecosystems. This is particularly evident in parts of the Himalaya, where population pressures have resulted in forest destruction for both FUELWOOD and arable *cultiva-*

tion. As a consequence, SOIL EROSION has increased on deforested steep slopes and increased RUNOFF has caused a greater incidence of FLOOD events and SEDIMENTATION in the lower reaches of river systems. The natural component of soil erosion and hence the soil component in suspended sediment yields is, however, also important in mountain regions throughout the world (see Figure).

ENVIRONMENTAL DEGRADATION in mountain regions is also a product of developments associated with tourism and RECREATION. The growth in popularity of skiing and other mountain-based activities in, for example, western Europe has resulted in the construction of chairlifts, cablecars and restaurants, and bulldozing of hillsides to create pistes, often in ecologically and geomorphologically sensitive areas. Footpaths have multiplied and some are severely eroded. Associated impacts include visual intrusion, noise, litter, waste disposal, a decline in BIODIVERSITY and EROSION. However, these must be set against the socio-economic benefits that such developments can bring to small mountain communities.

In many mountain areas climatic warming during the last 100–150 years has caused significant GLACIER retreat and PERMAFROST DEGRADATION and has increased the incidence of slope FAILURE and outburst floods (JÖKULHLAUPS) from glacier- and moraine-dammed lakes. Predictions of continued warming imply that these types of geomorphic events will increase in FREQUENCY and become a more serious threat to human activities. *PW*

[See also CLIMATIC CHANGE, DEBRIS FLOW, LANDSLIDE, MASS MOVEMENT, STURZSTROM, UNLOADING]

Beniston, M. 2000: *Environmental change in mountains and uplands*. London: Arnold. **Dedkov, A.P. and Mozzheim, V.I.** 1992: *Erosion and sediment yields in mountain regions of the world* [*Publication* 209]. Budapest: International Association for Scientific Hydrology, 29–36. **Evans, S.G. and Clague, J.J.** 1994: Recent climatic change and catastrophic geomorphic processes in mountain environments. *Geomorphology* 10, 107–128. **Fox, J.** 1999: Mountaintop removal in West Virginia: an environmental sacrifice zone. *Organisation and Environment* 12, 163–183. **Gerrard, A.J.** 1990: *Mountain environments*. London: Belhaven. **Haeberli, W.** 1992: Construction, environmental problems and natural hazards in periglacial mountain belts. *Permafrost and Periglacial Processes* 3, 111–124. **Kalvoda, J. and Rosenfeld, C.L.** (eds), 1998: *Geomorphological hazards in high mountain areas*. Dordrecht: Kluwer. **Messerli, B. and Ives, J.D.** (eds), 1997: *Mountains of the world: a global priority*. New York: Parthenon. **Price, L.W.** 1981: *Mountains and man: a study of process and environment*. Berkeley, CA: University of California Press. **Rai, S.C. and Sundriyal, R.C.** 1997: Tourism and biodiversity conservation: the Sikkim Himalaya. *Ambio* 26, 235–242.

mud Fine-grained SEDIMENT (CLASTIC or CARBONATE) with a GRAIN-SIZE less than 0.063 mm, including both SILT and CLAY. Particles of fine silt and clay tend to stick together through COHESION. *TY*

[See MUDROCK, MUDSTONE]

mud cracks A SEDIMENTARY STRUCTURE comprising a polygonal network of downward-tapering cracks on a surface of cohesive sediment (MUD), that may be preserved in SEDIMENTARY ROCKS by an infilling of SAND. Most are DESICCATION cracks (*sun cracks*), formed by shrinkage as the mud dried; they may be associated with EVAPORITE minerals or PSEUDOMORPHS and are a valuable indicator of exposure in a PALAEOENVIRONMENT. Cracks can, however, form

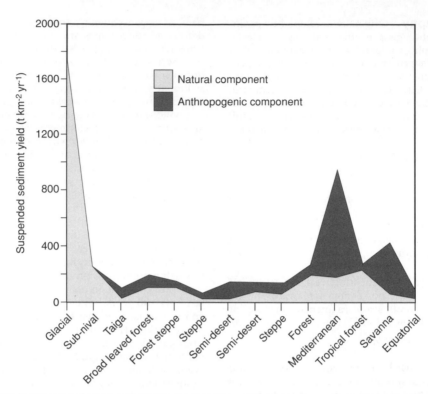

mountain regions: environmental change and human impact *The relative importance of natural and anthropogenic contributions to suspended sediment yields of mountain river basins in various climatic regions (after Dedkov and Mozzheim, 1992)*

in association with the expulsion of porewater in subaqueous settings; these *synaeresis cracks* are most commonly found in lacustrine deposits and can be confused with desiccation cracks.   *GO*

Allen, J.R.L. 1982: *Sedimentary structures: their character and physical basis.* Amsterdam: Elsevier. **Astin, T.R. and Rogers, D.A.** 1991: 'Subaqueous shrinkage cracks' in the Devonian of Scotland reinterpreted. *Journal of Sedimentary Petrology* 61, 850–859. **Pratt, B.R.** 1998: Syneresis cracks: subaqueous shrinkage in argillaceous sediments caused by earthquake-induced dewatering. *Sedimentary Geology* 117, 1–10.

mud drape   A thin layer of fine sediment (MUD) covering a BEDFORM, usually preserved within CROSS-BEDDING and indicative of an interruption to flow by, for example, a slack-water period between ebb and flood TIDES.   *GO*

[See FLASER BEDDING, TIDAL BUNDLE]

Allen, J.R.L. 1981: Lower Cretaceous tides revealed by cross-bedding with mud drapes. *Nature* 289, 579–581.

mudbelt   A zone of fine-grained sediments, for example on the CONTINENTAL SHELF where mudbelts are mostly of TERRIGENOUS origin and thicken in the vicinity of the mouths of major river systems.   *JAM*

Birch, G.F., Day, R.W. and du Plessis, A. 1991: Nearshore Quaternary sediments on the west coast of southern Africa. *Bulletin of the Geological Survey of South Africa* 101, 1–14. **Rogers, J. and Bremner, J.M.** 1991: The Benguela ecosystem. Part VII: marine geological aspects. *Oceanography and Marine Biology Annual Review* 29, 1–85. **Meadows, M.E., Dingle, R.V., Rogers, J. and Mills, E.** 1997: Radiocarbon chronology of

Namaqualand mudbelt sediments: problems and prospects. *South African Journal of Science* 93, 321–327.

mudlump   A small diapiric structure that forms at the mouth of a rapidly aggrading river where clays are extruded into overlying sands following loading. Mudlumps are especially common near the mouth of the Mississippi River, where they may interfere with navigation.   *HJW*

mudrock   A SEDIMENTARY ROCK (or its low-grade metamorphic derivative) entirely or dominantly comprising material with a GRAIN-SIZE less than 0.063 mm. The term embraces MUDSTONE, SHALE, SILTSTONE and CLAYSTONE.
                                                                    TY

mudslide   A type of MASS MOVEMENT in which fine-grained sediment moves down-slope under gravity mainly by sliding on discrete shear surfaces. The movement is usually slow (1–25 m y^{-1}), although there may be episodes of more rapid movement, commonly triggered by extreme rainfall. Many of the well documented coastal landslips or LANDSLIDES of west Dorset, UK, are of this type. There is some concern that CLIMATIC CHANGE might modify the activity of known mudslides.

The term is also used informally, for example by the media, for any damaging slide or flow of muddy water, including DEBRIS FLOWS, LAHARS and sediment-laden river FLOODS. Flows described as mudslides were responsible for over 30 000 deaths in Venezuela in December 1999 following heavy rainfall. DEFORESTATION and unsuitable URBAN development have been identified as factors contributing to the increasing frequency and severity of such NATURAL HAZARDS in recent years. *GO*

[See also FLOW SLIDE]

Allison, R.J. (ed.) 1992: *The coastal landforms of west Dorset* [*Geologists' Association Guide* 47]. London: The Geologists' Association. **Brunsden, D. and Ibsen, M.-L.** 1996: Mudslide. In Dikau, R., Brunsden, D., Schrott, L. and Ibsen, M.-L. (eds), *Landslide recognition* [*International Association of Geomorphologists Publication* 5]. Chichester: Wiley, 103–119. **Dehn, M., Burger, G., Buma, J. and Gasparetto, P.** 2000: Impact of climate change on slope stability using expanded downscaling. *Engineering Geology* 55, 193–204.

mudstone A SEDIMENTARY ROCK, entirely or dominantly comprising material with a GRAIN-SIZE less than 0.063 mm, including both CLAYSTONE and SILTSTONE. Sometimes restricted to a rock of this grade that is not FISSILE. Although the restriction to NON-FISSILE MUDROCKS is useful in distinguishing mudstone from SHALE, the term mudstone is commonly applied to any unmetamorphosed mudrock. *TY*

Potter, P.E., Maynard, J.B. and Pryor, W.A. 1980: *The sedimentology of shale: study guide and reference sources*. New York: Springer.

mulching The spreading of a material such as straw, sawdust, leaves or plastic film onto the soil surface to prevent soil and plant roots from freezing or to protect the soil surface against raindrop impact or excessive evaporation. However, some effects may be detrimental with reduced soil temperatures in cold climates or enhanced soil moisture conditions leading to GLEYING or ANAEROBIC conditions. *SHD*

[See also DRY FARMING]

Anderson, D.F., Garisto, M.A., Bourrut, J.C. *et al.* 1995: Evaluation of a paper mulch made from recycled materials as an alternative to plastic film mulch for vegetables. *Journal of Sustainable Agriculture* 7, 39–61. **Foth, H.D.** 1990: *Fundamentals of soil science*. Chichester: Wiley. **Zuzel, J.F. and Pikul, J.L.** 1993: Effects of straw mulch on runoff and erosion from small agricultural plots in northeastern Oregon. *Soil Science* 156, 111–117.

mull A form of HUMUS that is fully incorporated into the mineral soil (see Figure). Plant material is completely decomposed by the action of the soil fauna. EARTHWORMS actively bring together clay and humus in their alimentary canal to make a neutral, crumb-structured, humus-rich topsoil. Human activity may transform mull to MOR, as occurred in many areas of northwest Europe during prehistory when forests were replaced by HEATHLAND. *EMB*

[See also MOR]

Andersen, S.T. 1979: Brown earth and podzol: soil genesis illuminated by microfossil analyses. *Boreas* 8: 59–73.

multiconvex landscape Low relief formed by the presence of thick zones of partially altered SAPROLITE within weathered rock, found especially in TROPICAL areas of high precipitation and rapid weathering. *MAC*

[See also CHEMICAL WEATHERING, ETCHPLAIN, LANDSCAPE EVOLUTION]

multidimensional scaling A general term for methods that attempt to construct as accurately as possible a low-dimensional representation of a distance, dissimilarity, or proximity matrix between objects and/or variables. *HJBB*

[See also ORDINATION, PRINCIPAL COMPONENTS ANALYSIS, PRINCIPAL CO-ORDINATES ANALYSIS, CORRESPONDENCE ANALYSIS, NON-METRIC MULTIDIMENSIONAL SCALING]

Everitt, B.S. and Rabe-Hesketh, S. 1997: *The analysis of proximity data*. London: Edward Arnold.

multidisciplinary Involving a combination of several academic disciplines, with each discipline retaining its own methodologies. *FMC*

[See also ENVIRONMENTAL SCIENCE, ENVIRONMENTAL ARCHAEOLOGY, INTERDISCIPLINARY RESEARCH].

multiplicative model A set of distributions used to model data obtained with coherent illumination (for example in REMOTE SENSING), as is the case of *synthetic aperture radar*, sonar and laser images. Most of these distributions are quite different from the GAUSSIAN MODEL. *ACF*

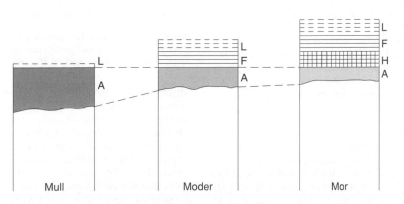

mull *Mull, moder and mor: the thin surface forms of soil organic matter. The litter (L), fermentation (F) and humus (H) layers make up the O horizon. Fully incorporated organic matter is a component of the A horizon*

Frery, A.C., Müller, H.-J., Yanasse, C.C.F. and Sant'Anna, S.J.S. 1997: A model for extremely heterogeneous clutter. *IEEE Transactions on Geoscience and Remote Sensing* 35, 648–659.

multipolarisation Characterising a set of measurements made using electromagnetic radiation that is transmitted and received in different POLARISATIONS. *PJS*

multiproxy approach The reconstruction of environmental change using many types of PROXY EVIDENCE from the same site or sediment core. Thus, the PROXY DATA derived from one source can be evaluated and extended by independent reconstructions from different proxies. PROXY CLIMATIC INDICATORS can be physical or biological parameters. Multiproxy evidence is also used to reconstruct environmental variables other than climate. The multiproxy approach is advocated as the best approach to climatic and environmental reconstruction, as well as to assessing the impact of environmental changes on the ecosystem.

Diverse multiproxy investigations have used LACUSTRINE SEDIMENTS. Important late-glacial environmental physical indicators include sediment composition and chemistry, GRAIN-SIZE, TEPHRA layers, VARVES, organic content (LOSS-ON-IGNITION), and PALAEOMAGNETISM. Strong LATE GLACIAL climate oscillations (Allerød, Younger Dryas, Holocene) are reflected by aquatic and terrestrial biota. Comparative quantitative climate reconstructions were made from a biostratigraphical sequence using pollen, chironomids, cladocera, and coleoptera, and vegetation analogues from plant macrofossil data (Battarbee, 2000). Examples of late-glacial multiproxy studies are Birks et al. (2000), and van Geel et al. (1989).

Multiproxy evidence for Holocene GLACIER VARIATIONS in lake sediments has been used to reconstruct summer temperature and winter precipitation (Dahl and Nesje, 1996). Vegetation history and TREE-LINE VARIATIONS have been reconstructed using pollen, plant macrofossils, and soils (e.g. Tinner *et al.*, 1996). EUTROPHICATION studies use multiproxy evidence from diatoms, cladocera, chironomids, aquatic macrophytes, cyanobacteria, sediment chemistry and fossil pigments (e.g. Birks *et al.*, 1976). SALINITY changes reflecting precipitation:evaporation balance, have been reconstructed using diatoms, mollusca, $\delta^{18}O$, sediment chemistry and organic content (e.g. Fritz *et al.*, 1994)

The reconstruction of environmental change from ICE CORES is typically multiproxy. Frequently used parameters are $^{18}O:^{16}O$ ratio, dust content, electrical conductivity, temperature of the borehole, annual layers, ^{14}C dating, and measurement of CO_2, CH_4 and isotopes in included air bubbles (e.g. Thompson *et al.*, 1998). Multiproxy investigations of marine sediment cores have used $\delta^{18}O$, ^{14}C, sedimentology, diatoms, foraminifera, tephra, ice-rafted detritus (IRD), etc.

ENVIRONMENTAL ARCHAEOLOGY uses multiproxy evidence. Plant material gives evidence of vegetation, building material, crops and other food, rope, etc. Animal remains demonstrate animal hunting/husbandry, diet, textiles and clothing, and insects include pests and parasites. The lifestyle of a community is well reflected by remains in a latrine (Greig, 1981). Archaeology is very diverse and past environments can also be reconstructed from human artefacts, tools, buildings, pottery, carving, writing, painting and grave contents, as well as from human bodies and skeletons. *HHB*

[See also CLIMATIC RECONSTRUCTION, NATURAL ARCHIVES]

Battarbee, R.W. 2000: Palaeolimnological approaches to climate change, with special regard to the biological record. *Quaternary Science Reviews* 19, 107–124. **Birks, H.H., Battarbee, R.W. and Birks, H.J.B.** 2000: The development of the aquatic ecosystem at Kråkenes Lake, western Norway, during the late glacial and early Holocene – a synthesis. *Journal of Paleolimnology* 23: 91–114. **Birks, H.H., Whiteside, M.C., Stark, D. and Bright, R.C.** 1976: Recent paleolimnology of three lakes in northwestern Minnesota. *Quaternary Research* 6, 249–272. **Dahl, S.O. and Nesje, A.** 1996: A new approach to calculating Holocene winter precipitation by combining glacier equilibrium-line altitudes and pine-tree limits: a case study from Hardangerjøkulen, central southern Norway. *The Holocene* 6, 381–398. **Fritz, S.C., Engstrom, D.R. and Haskell, B.J.** 1994: 'Little Ice Age' aridity in the North American Great Plains: a high resolution reconstruction of salinity fluctuations from Devil's Lake, North Dakota, USA. *The Holocene* 4, 69–73. **van Geel, B., Coope, G.R., and van der Hammen, T.** 1989: Palaeoecology and stratigraphy of the lateglacial type section at Usselo (The Netherlands). *Review of Palaeobotany and Palynology* 60, 25–129. **Greig, J.** 1981: The investigation of a medieval barrel-latrine from Worcester. *Journal of Archaeological Science* 8, 265–282. **Thompson, L.G., Davis, M.E., Mosley-Thompson, E. et al.** 1998: A 25,000-year tropical climate history from Bolivian Ice cores. *Science* 282, 1858–1864. **Tinner, W., Ammann, B. and Germann, P.** 1996: Treeline fluctuations recorded for 12,500 years by soil profiles, pollen, and plant macrofossils in the Central Swiss Alps. *Arctic and Alpine Research* 28, 131–147.

multitemporal analysis A technique for analysing REMOTE SENSING data, which makes use of several images acquired at different dates. It is used to detect TEMPORAL CHANGE of environmental variables on large spatial scales. Multitemporal analysis is also able to yield additional thematic information about LAND COVER and vegetation type. By combining images from winter and summer, coniferous and deciduous forests can easily be distinguished from the loss of leaves. Multitemporal images can be used for repeat-pass synthetic aperture radar (SAR) INTERFEROMETRY, which estimates the coherence and phase difference between two SAR images separated by a temporal baseline (usually 1–44 days). DIGITAL ELEVATION MODELS can be generated from the phase difference. *HB*

Schmidt, H. and Glaesser, C. 1998: Multitemporal analysis of satellite data and their use in the monitoring of the environmental impacts of open cast lignite mining areas in Eastern Germany. *International Journal of Remote Sensing* 19, 2245–2260. **Wegmüller, U. and Werner, C.** 1997: Retrieval of vegetation parameters with SAR interferometry. *IEEE Transactions on Geoscience and Remote Sensing* 35, 18–24.

multivariate analysis A general term for the many numerical techniques now available for the analysis of large, complex data-sets consisting of many observations and many variables. Each observation usually consists of values for more than one random response variable. In some cases there may also be one or more predictor variables. Examples include CLUSTER ANALYSIS, PRINCIPAL COMPONENTS ANALYSIS, *multiple discriminant analysis* and CANONICAL CORRESPONDENCE ANALYSIS. *HJBB*

Krzanowski, W.J. and Marriot, F.H.C. 1994–95: *Multivariate Analysis*. Parts 1, 2. London: Edward Arnold.

mummy A corpse deliberately preserved by human agency (*mummification*), as opposed to being preserved under natural waterlogged, arid or frozen conditions (e.g. the Alpine ICEMAN). Mummification has been practised by many different cultures at various times but especially in extreme environments such as the Sahara, the Andes and Siberia. *JAM*

Cockburn, A. and Cockburn, E. (eds), 1980: *Mummies, disease and ancient cultures.* Cambridge: Cambridge University Press.

muskeg An Indian term for the Canadian Arctic and especially subarctic *peatlands*. It is also the name of a RIVER REGIME that is characteristic of catchments where extensive WETLANDS act to increase water storage, reduce discharge rates and extend flood peaks. *DCS*

Radford, N.W. and Branner, C.O. (eds) 1977: *Muskeg and the northern environment in Canada.*

mutagen Any physical (e.g. ultraviolet light) or chemical (e.g. cigarette smoke) agent that alters *DNA* and that markedly increases the frequency of mutational events above an 'average' spontaneous MUTATION rate. *GOH*

[See also CARCINOGEN]

Russell, P.J. 1998: *Genetics.* Menlo Park, CA: Benjamin/Cummings.

mutation An abrupt alteration of chromosomal DNA, which occurs at random and is inheritable. Mutations are often deleterious, but such changes are the basis of genetic variability and hence the raw material for NATURAL SELECTION. *KDB*

mutual climatic range (MCR) method A method permitting the quantitative reconstruction of past climates, using biological proxies with precise thermal requirements, such as beetles or herpetofauna. For each species in an assemblage, a *climatic tolerance range* is deduced from present-day climatic distributions, commonly based on the temperature of the warmest month (T_{max}) and the temperature range between the warmest and coldest month (T_{range}). When the various climatic ranges of an assemblage of species are considered, an area of overlap will be expected within the climate space where all the species can co-exist (see Figure). This part of the climate space is known as the mutual climatic range of that assemblage. Where assemblages may be directly dated (for example by RADIOCARBON DATING), a palaeotemperature curve may then be constructed. *DCS*

[See also BEETLE ANALYSIS, INSECT ANALYSIS]

Atkinson, T.C., Briffa, K.R. and Coope, G.R. 1987: Seasonal temperatures in Britain during the past 22 000 years, reconstructed using beetle remains. *Nature* 325, 587–592. **Atkinson, T.C., Briffa, K.R., Coope, G.R. et al.** 1986: Climatic calibration of coleopteran data. In Berglund, B.E. (ed.), *Handbook of Holocene palaeoecology and palaeohydrology.* Chichester: Wiley, 851–858. **Lowe, J.J. and Walker, M.J.C.**

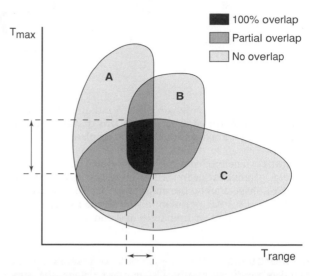

mutual climatic range method *Schematic representation of the mutual climatic range as the area of overlap between climatic tolerance ranges of three species (A, B and C) based on ecological knowledge: T_{max} is the mean monthly temperature of the warmest month; T_{range} is the difference between the mean monthly temperatures of the warmest and coldest months. Arrows indicate the reconstructed climate for any site where this species assemblage is found (after Lowe and Walker, 1997)*

1997: *Reconstructing Quaternary environments*, 2nd edn. Harlow: Addison-Wesley Longman.

mutualism A form of SYMBIOSIS in which all interacting organisms benefit from the association, such as the interaction between higher-order plants and MYCORRHIZA. An increasing number of cases of mutualism are being recognised, particularly between animals and plants, as in the case of those ant species that keep a tree stem free of PARASITES and EPIPHYTES. Environmental change may affect one species in an ASSOCIATION more than another, creating the potential for unpredictable effects within the ECOSYSTEM as a whole. *JLI*

Bronstein, J.L. 1994: Our current understanding of mutualism. *Quaterly Review of Biology* 69, 31–51. **Lavelle P., Lattaud C., Trigo D. and Barois I.** 1995: Mutualism and biodiversity in soils. *Plant and Soil* 170, 23–33.

mycorrhiza A symbiotic association between a fungus and the roots of a higher plant that enhances plant uptake of certain NUTRIENTS (sometimes water) from the soil. Two main types are *ectomycorrhizae* (sheathing the roots) and *endomycorrhizae* (inside the roots). *NDB*

[See also BELOW-GROUND PRODUCTIVITY, NUTRIENT, NUTRIENT POOL, SYMBIOSIS]

Brundrett, M. 1991: Mycorrhizas in natural ecosystems. *Advances in Ecological Research* 21, 171–313.

N

nadir point The point on the ground directly beneath (measured as plumb line) a camera lens or digital sensor. It forms a right angle with the recording plane. *TF*

nanism The tendency for some isolated populations to undergo significant decreases in body size in comparison to their conspecifics on the mainland. Nanism is especially common in insular populations of elephants and mammoths, deer and other ungulates and canids. *MVL*

[See also GIGANTISM]

nannofossils Microscopic FOSSIL or SUBFOSSIL calcite platelets produced by unicellular marine algae, the commonest of which are *coccoliths*. They are 2–15 μm in size and are PLANKTONIC (*nannoplankton*) or PELAGIC in origin, being restricted to the *photic zone* in the upper 200 m of the oceans. They are of particular value in indicating thermal conditions, although their distribution is also affected by light and salinity, and they provide one of the major constituents of MARINE SEDIMENT CORES covering the last 150 million years. *DH/CJC*

[See also COCCOLITHOPHORES]

McIntyre, A. and Ruddiman, W.F. 1972: North-east Atlantic post-Eemian palaeoceanography: a predictive analog for the future. *Quaternary Research* 2, 350–354.

National Parks Extensive areas of countryside that have been protected by national legislation from urban and corporate development, in order to protect and conserve the natural and cultural landscape. Yellowstone National Park (United States) became the first example in AD 1872. The first British examples were designated under the National Parks and Access to the Countryside Act (1949), 'to preserve and enhance the natural beauty of the areas and promote their quiet enjoyment by the public'. The degree of protection afforded by National Park status varies across different countries and states. *MLW*

[See also CONSERVATION, ENVIRONMENTAL PROTECTION, PROTECTED AREAS, RECREATION]

Lowry, W.R. 1994: *The capacity for wonder: preserving National Parks*. Washington, DC: Smithsonian Institution. **MacEwan, A. and MacEwan, M.** 1982: *National Parks: conservation or cosmetics?* London: Allen and Unwin. **Wright, R.G. (ed.)** 1996: *National Parks and protected areas: their role in environmental protection*. Cambridge, MA: Blackwell Science.

native Individuals and populations of a species that occur within the species' natural, historic range. Native species are also known as *indigenous species*. *MVL*

[See also ALIEN SPECIES, EXOTIC SPECIES, NATURALISED SPECIES]

natural Not made or caused by human agency. A broader definition would include 'not modified' by human agency. *JAM*

[See also NATURE]

Graf, W.L. 1996: Geomorphology and policy for restoration of impounded American rivers: what is 'natural'? In Rhoades, B.L. and Thorn, C.E. (eds), *The scientific nature of geomorphology*. Chichester: Wiley, 443–473.

natural archives Sediments or naturally occurring biological remains that have accumulated continuously and chronologically as a natural repository, which, when cored and examined with appropriate techniques, can be 'read' as an archive of, for example, vegetational history, catchment history and climate. Further examples are shown in the Table. *FMC*

Bradley, R.S. and Eddy, J.A. 1991: Introduction. In Bradley, R.S. (ed.), *Global changes of the past*. Boulder, CO: UCAR/Office for Interdisciplinary Earth Studies, 5–9. **Godwin, H.** 1981: *The archives of the peat bogs*. Cambridge: Cambridge University Press.

natural areas concept The concept refers to the intactness or integrity of HABITATS and ecosystems that

natural archives *Characteristics of natural archives (based on Bradley and Eddy 1991*

Archive	Best temporal resolution*	Temporal range (y)	Information derived†
Historical records	Hour/day	10^3	T, H, B, V, M, L, S
Tree rings	Season/year	10^4	T, H, C_a, B, V, M, S
Peat (especially ombrotrophic and estuarine)	(Sub-)decadal	10^3–10^4	T, H, C_a, C_w B, V, L, S
Lake sediments	Season/year, if varved	10^4–10^6	T, H, C_w, B, V, M
Ice cores	Year	10^5	T, H, C_a, B, V, M, S
Loess	100 years	10^6	H, B, M
Ocean cores	100s – 1000 years	10^7	T, C_w, B, M
Corals	Year	10^4	C_w, L
Palaeosols	100 years	10^5	T, H, C_s, V
River sediments	100 years	10^6	H, L, C_s, C_w
Geomorphic features	100 years	10^7	T, H, V, L
Sedimentary rocks	Year	10^7	H, C_s, V, M, L

* Minimum sampling interval in most cases
† T = temperature; H = humidity or precipitation; C = chemical composition of air (C_a), water (C_w) or soil (C_s); B = biomass and vegetation patterns; V = volcanic eruptions; M = geomagnetic field variations; L = sea level; S = solar activity

have not been changed or affected by human activity. There are few such areas remaining on Earth, but there are many areas that have been called *semi-natural*, that is, relatively unmodified by human activity. A framework for evaluating naturalness of ecosystems has been described by J.A. Anderson (1991), involving three indices: (1) the degree to which the system would change if humans were removed; (2) the amount of cultural energy required to maintain the functioning of the ecosystem as it currently exists; (3) the complement of NATIVE or indigenous species in the area compared to that which previously existed. In the UK, it refers to an approach developed by *English Nature* based on geographical integration of floras, geological maps, landscape accounts and so on, in order to derive distinct areas for CONSERVATION planning.   *IFS*

[See also CONSERVATION BIOLOGY, LANDSCAPE UNITS, NATURAL VEGETATION, SEMI-NATURAL VEGETATION, WILDERNESS CONCEPT]

Anderson, J.A. 1991: A conceptual framework for evaluating and quantifying naturalness. *Conservation Biology* 5, 347–352. **Spellerberg, I.F.** 1992: *Evaluation and assessment for conservation*. London: Chapman and Hall.

natural capitalism A strategy for sustainable economies, which safeguards the natural resource base that ultimately underpins all economic activity. Conventional capitalism is wasteful: in the United States, materials used by the 'metabolism' of industry involve over 20 times the total weight of the American population on a daily basis. Natural capitalism advocates reduction of materials and energy used *per unit output* and the eventual elimination of waste by emulating the recycling that characterises natural ECOSYSTEMS. It has been suggested that to achieve sustainable economies world wide, a 50% reduction in the intensity of materials and energy use is required and that, given the inability of developing countries to achieve this target, developed nations should aim for a 90% reduction.   *JAM*

[See also GLOBAL CAPITALISM, SUSTAINABILITY]

Hawken, P., Lovins, A.B. and L. Hunter Lovins 1999: *Natural capitalism: the next industrial revolution*. London: Earthscan.

natural change Environmental change that is not influenced by any human activity. In the context of HUMAN IMPACT ON CLIMATE, it is often referred to as the natural CLIMATIC VARIABILITY or the natural CLIMATIC VARIATION. A major concern is to identify when climatic change exceeds this natural BACKGROUND LEVEL and hence when a human imprint can be discerned in the climatic record.   *AHP*

[See also NATURAL, NATURAL ENVIRONMENTAL CHANGE]

natural disaster The outcome of a NATURAL HAZARD where major damage is inflicted on ecosystems, landscapes or the human environment by a natural agency such as storm, flood or volcanic eruption. In the human context, the recovery process may be characterised as a sequence of four overlapping stages: *emergency*, RESTORATION, *replacement reconstruction* and *developmental reconstruction* (see Figure).   *JAM*

[See also DISASTER, ENVIRONMENTAL DISASTER]

Alexander, D. 1993: *Natural disasters*. London: UCL Press. **Kates, R.W. and Pijawka, D.** 1977: From rubble to monu-

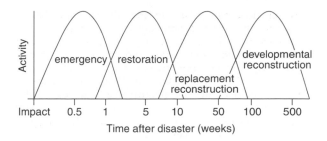

natural disaster *A model of the stages of recovery following a natural disaster (after Kates and Pjawka, 1977)*

ment: the pace of reconstruction. In Haas, J., Kates, M. and Bowden, M. (eds), *Disaster and reconstruction*. Cambridge, MA: MIT Press.

natural environment Those aspects of the environment that are not made by humans (but which humans may influence).   *JAM*

[See also NATURAL ENVIRONMENTAL CHANGE, NATURAL ENVIRONMENTAL SCIENCES, NATURE]

natural environmental change (1) Those aspects of ENVIRONMENTAL CHANGE caused by natural forces rather than human agency. Natural environmental change has occurred throughout Earth's history. Before the human species evolved, all environmental changes were NATURAL: in more recent times, natural environmental changes may be considered as a background and baseline to human-induced environmental changes. (2) Any change to the NATURAL ENVIRONMENT, irrespective of cause: changes to the natural environment are nevertheless increasingly caused by human activity.   *JAM*

Hulme, M., Barrow, E.M., Arnell, N.W. *et al.* 1999: Relative impacts of human-induced climate change and natural climate variability. *Nature* 397, 688–691. **Mannion, A.M.** 1999: *Natural environmental change*. London: Routledge.

natural environmental sciences Sciences such as ECOLOGY, ENVIRONMENTAL SCIENCE, GEOLOGY and PHYSICAL GEOGRAPHY that investigate the NATURAL ENVIRONMENT including HUMAN IMPACTS.   *JAM*

natural gamma radiation The emission of *gamma rays* by naturally occuring radioactive MINERALS, notably those containing potassium, uranium and thorium. It is commonly measured as a continuous record from CORES and BOREHOLES, indicating the presence of CLAY MINERALS and organic-rich SHALES.   *MRT*

Hoppie, B.W., Blum, P., Mountain, G. *et al.* 1994: Natural gamma-ray measurements on ODP cores: Introduction to procedures with examples from Leg 150. *Proceedings of the Ocean Drilling Program, Initial Reports* 150, 51–59.

natural gas The gaseous form of PETROLEUM, exploited as a FOSSIL FUEL by drilling into buried rocks, usually SEDIMENTARY ROCKS, that hold the gas in their pore spaces or other cavities, commonly overlying accumulations of CRUDE OIL. Chemically, natural gas is dominated by METHANE (CH_4). In the past, low energy prices and distance from markets led to natural gas being flared off in many oilfields, particularly in the Middle East. Rising prices since the AD 1970s and new developments in its liquid storage have led to increasing use. Natural gas is a cleaner ENERGY RESOURCE than crude oil and its share of world energy consumption increased to about 25% in the late 1990s (37% in the UK in 1999). The Russian Federation held about 33% of natural gas reserves in 1999 and Iran 16%. *GO*

BP Amoco 2000: *BP Amoco statistical review of world energy, June 2000.* London: BP Amoco. **Hay, N.E. (ed.)** 1992: *Guide to natural gas cogeneration.* Englewood Cliffs, NJ: Prentice-Hall.

natural hazards Elements of the physical environment potentially harmful to humankind and caused by natural forces. Generally speaking they are extreme cases of phenomena and have a short timescale. This definition excludes POLLUTION, which is normally considered to be caused only by humankind, and biological events such as infestation by PESTS (plants and animals). Consequently, natural hazards are usually considered to be either meteorological or geological/geomorphological.

The most important *meteorological natural hazards* are FLOODS, TROPICAL CYCLONES, DROUGHTS and TORNADOES, but other extreme weather events are also natural hazards: *blizzards* (a combination of near gale force winds with heavy snow); *fog* (defined internationally as visibility reduced to less than 1 km by suspended water droplets – in the UK it generally refers to visibility of less than 180 m); *frost* (when the air temperature is below 0°C); *hailstorms* (solid PRECIPITATION in the form of pieces of ice falling from cumulonimbus CLOUDS); *heat waves* (extremely high temperatures accompanied by high humidity, which cause physiological stress); and *lightning strikes* (lightning is a giant spark between unlike electrical charges in clouds and on the ground). These meteorological hazards can be predicted over short time periods (hours or a day or two) in a particular location.

The main *geomorphological natural hazards* are EARTHQUAKES, VOLCANIC ERUPTIONS, AVALANCHES, LANDSLIDES and TSUNAMIS. These natural hazards cannot be predicted in time with accuracy or location, although some places are more prone to them than others. In AD 1998, worldwide NATURAL DISASTERS caused damage of £40 billion and killed nearly 35 000 people. *BDG*

[See also RAPID ONSET HAZARD, WEATHER FORECASTING]

Blong, R.J. 1984: *Volcanic hazards: a sourcebook on the effects of eruptions.* New York: Academic Press. **Blong, R.** 1997: A geography of natural perils. *Australian Geographer* 28, 7–28. **Bryant, E.A.** 1991: *Natural hazards.* Cambridge: Cambridge University Press. **Burton, I, Kates, R. W., and White, G.F.** 1993: *The environment as hazard,* 2nd edn. New York: Guilford Press. **Kates, R.W.** 1978: *Risk assessment of environmental hazard* [*SCOPE Report* 8]. Chichester: Wiley. **Perry, A.H.** 1981: *Environmental hazard in the British Isles.* London: Allen and Unwin. **Smith, K.** 1996: *Environmental hazards: assessing risk and reducing disaster,* 2nd edn. London: Routledge. **Tilling, R.I.** 1989: Volcanic hazards and their mitigation: progress and problems. *Reviews of Geophysics* 27, 237–269. **United States Geological Survey** 1981: *Facing geological and hydrologic hazards: Earth Science considerations* [*Professional Paper* 1240-B]. Reston, VA: US Geological Survey. **White, G.F. (ed.)** 1974: *Natural hazards: local, national, global.* New York: Oxford University Press. **Whyte, A.V. and Burton, L. (eds)** 1980: *Environmental risk assessment. International working seminar on environmental risk assessment, 1977.* Chichester: Wiley.

natural remanent magnetism (NRM) Magnetism acquired by rocks and unconsolidated SEDIMENTS during formation as a result of alignment of contained ferromagnetic particles with the Earth's ambient geomagnetic field. Measurement of NRM is used in PALAEOMAGNETIC DATING. *MHD*

Thompson, R. 1991: Palaeomagnetic dating. In: Smart, P.L. and Frances, P.D. (eds), *Quaternary dating methods – a users guide* [*QRA Technical Guide* 4]. Cambridge: Quaternary Research Association, 177–198.

natural selection An evolutionary process identified and described, independently, by Alfred Russel Wallace (1823–1913) and Charles Robert Darwin (1809–1882). Both noted that individuals differ and that differential survival means that subsequent generations will have characteristics more suited to the ENVIRONMENT. 'This preservation of favourable variations and the rejection of injurious variations, I call Natural Selection' (Darwin, 1859: 81). *KDB*

[See also DARWINISM, EVOLUTION, SELECTION PRESSURE]

Darwin, C. 1859: *On the origin of species by means of natural selection, or the preservation of favoured races in the struggle for life.* London: John Murray. **Wallace, A.R.** 1895: *Natural selection and tropical nature.* London: Macmillan.

natural vegetation Vegetation that has become established without human intervention and that continues to remain undisturbed by human actions. *IFS*

[See also CLIMAX VEGETATION, POTENTIAL NATURAL VEGETATION, PRIMARY WOODLAND, SEMI-NATURAL VEGETATION]

naturalisation (1) The process by which an introduced species (see INTRODUCTION) adapts to local conditions and becomes integrated with the NATURAL or semi-natural ecosystem into which it has been introduced. (2) Human intervention in an environmental SYSTEM with the effect of driving the system towards greater diversity, stability and/or SUSTAINABILITY by natural processes. *JAM*

Rhodes, B.L., Wilson, D., Urban, M. and Hendricks, E. 1999: Interaction between scientists and non-scientists in community-based watershed management: emergence of the concept of stream naturalization. *Environmental Management* 24, 297–308.

naturalised species Individuals and populations of a species that are descendants of those introduced to a site outside their natural historic range. If these populations have persisted for a sufficient period to allow local adaptation, they may sometimes be viewed as integral components of the NATURAL ecosystem. *MVL*

[See also EXOTIC, INTRODUCTION, NATIVE SPECIES, NATURALISATION]

nature 'Perhaps the most complex word in the [English] language' (Williams, 1976: 184), but currently most commonly used to signify the natural world and the physical environment as opposed to CULTURE. However, 'nature' has both concrete and abstract meanings and its interpretations have changed throughout history. In the western world five important categories of meanings can be recognised: (1) nature as a physical place, notably places 'unspoiled' by human modification; (2) nature as the Earth or the universe as a collective phenomenon, including or excluding humans; (3) nature as an essence or quality that pervades the functioning of the Earth or universe; (4) nature as an inspiration and guide to human affairs; and (5) nature as the conceptual opposite to culture. *JAM*

[See also CONSERVATION, ENVIRONMENTAL ETHICS, ENVIRONMENTALISM, HUMAN IMPACTS]

Coates, P. 1998: *Nature: western attitudes since ancient times.* Cambridge: Polity Press. **Marsh, G.P.** 1864 (1965 edition, Lowenthal, D. (ed.)): *Man and nature: physical geography as modified by human action.* Cambridge, MA: Harvard University Press. **Macnaghten, P. and Urry, J.** 1998: *Contested natures.* London: Sage Publications. **Williams, P.** 1976: *Keywords: a vocabulary of culture and society.* Oxford: Oxford University Press.

nature reserve A haven for wildlife where nature CONSERVATION is prioritised. Land is usually owned by (or subject to the regulations of) international, national or non-governmental conservation organisations. Management plans and conservation practices are tailored towards both individual species and extensive rare habitats. *MLW*

[See also ENVIRONMENTAL PROTECTION, PROTECTED AREAS]

NBS oxalic acid Standard reference materials produced by the United States National Bureau of Standards, Washington for RADIOCARBON DATING laboratories. The first batch of material has run out and a second was prepared in 1974. The nomenclature of the first standard is variously given as NBS Oxalic acid I, NBS OxI, HoxI and SRM4990B; 0.95 times the activity is equivalent to AD 1950 with the δ^{13}C fractionation normalised to −19.0‰. The new standard is described as NBS OxII, HoxII or SRM4990C; 0.7459 times the activity is equivalent to AD 1950 with the δ^{13}C fractionation normalised to −25.0‰. *PQD*

[See also REFERENCE STANDARD, STANDARD SUBSTANCE]

Stuiver, M. 1983: International agreements and the use of the new oxalic standard. *Radiocarbon* 25, 793–795.

neap tide The TIDE of comparatively small range occurring twice each lunar month. *RAS*

[See also SPRING TIDE]

near-Earth object (NEO) Any asteroid or cometary debris with a solar orbit approaching within 1.3 Astronomical Units of the Earth. NEOs are classed as *Earth Crossing* or *Earth Approaching*, or as *Potentially Hazardous Asteroids* (PHAs), if their minimum potential Earth distance is less than 7 480 000 km. NEOs and PHAs are potential BOLIDES. *JBH*

[See also METEORITE IMPACT]

Rabinowitz, D., Helin, E., Lawrence, K. and Pravdo, S. 2000: A reduced estimate of the number of kilometre-sized near-Earth asteroids. *Nature* 403, 165–166.

near infrared reflectance spectroscopy (NIRS) A rapid, NON-DESTRUCTIVE SAMPLING technique for the acquisition of data on the chemical constituents of organic materials (such as carbon, nitrogen, phosphorus, lignin and ash content) and for obtaining PROXY DATA on a potentially wide range of biological and environmental variables including, for example: the digestibility of for-

age; the bulk density, HUMIFICATION, moisture content and MACROFOSSIL content of PEAT; and lake-water pH, carbon, nitrogen and phosphorus concentration, DIATOM content and inferred mean July air temperature from LACUSTRINE SEDIMENTS. Samples are scanned using a SPECTROPHOTOMETER producing reflectance spectra, the CALIBRATION of which may involve use of a TRAINING SET of samples that have been analysed by conventional laboratory techniques and/or correlation with environmental variables. *JAM*

Bokobza, L. 1998: Near infrared spectroscopy. *Journal of Near Infrared Spectroscopy* 6, 3–17. **Malley, D.F., Rönicke, H., Findlay, D.L. and Zippel, B.** 1999: Feasibility of using near-infrared reflectance spectroscopy for the analysis of C, N, P and diatoms in lake sediment. *Journal of Palaeolimnology* 21, 295–306. **McTiernan, K.B., Garnett, M.H., Mauquoy, D. et al.** 1998: Use of near-infrared reflectance spectroscopy (NIRS) in palaeoecological studies of peat. *The Holocene* 8, 729–740. **Rosén, P., Dåbakk, E., Renberg, I. et al.** 2000: Near-infrared spectrometry (NIRS): a new tool for inferring past climatic changes from lake sediments. *The Holocene* 10, 161–166.

nearest neighbour In IMAGE PROCESSING, local techniques such as FILTERING require the specification of the positions that will have influence in the computation of new values. The pixels in those positions are called the nearest neighbours and they are usually specified using distance criteria. Nearest-neighbour techniques are also widely used in ecology for pattern analysis. *ACF*

nebkha A small sand accumulation or DUNE formed around vegetation (also known as a *phytogenetic dune* or *nabkha*) rather than vegetated following their formation (see DIKAKA). *SHD*

Markley, W.G. and Wolfe, N.A. 1994: The morphology and origin of nabkhas, region of Mopti, Mali, West Africa. *Journal of Arid Environments* 28, 13–30.

needle ice Long, thin, needle-like ice crystals that form perpendicular to the ground surface. Needle ice, also known as *pipkrake*, forms at night when there is excessive radiation cooling, causing ice segregation in the surface soil layers. *HMF*

Branson, J., Lawler, D.M. and Glen, J.W. 1996: Sediment inclusion events during needle ice growth: a laboratory investigation of the role of soil moisture and temperature fluctuations. *Water Resources Research* 32, 459–466. **Lawler, D.M.** 1988: Environmental limits of needle ice: a global survey. *Arctic and Alpine Research* 20, 137–159. **Mackay, J.R. and Mathews, W.H.** 1974: Needle ice striped ground. *Arctic and Alpine Research* 6, 79–84.

negative feedback An interaction within a SYSTEM that causes a reduction or dampening of the response of the system to a force. Thus, a negative feedback has a stabilising effect. Conversely, *positive feedback* tends to reinforce or accentuate the changes set in motion. *AHP*

[See also FEEDBACK MECHANISMS]

nekton/nektonic/nektic Free-swimming organisms. *GO*

[See also PELAGIC, PLANKTON]

nemoral forest The TEMPERATE FORESTS dominated by broadleaved deciduous trees of genera such as

Fraxinus, Acer, Ulmus, Tilia, Fagus and *Quercus*. This is an alternative term for the deciduous summer forest VEGETATION FORMATION-TYPE with three formations (in North America, Europe and the Far East). The *boreo-nemoral forest*, which occupies the transition zone between the nemoral forest and the BOREAL FOREST, is characterised by a mixture of deciduous and coniferous tree dominants.

JAM

Sjörs, H. 1963: Amphi-Atlantic zonation, nemoral to Arctic. In Löve, Á. and Löve, D. (eds), *North Atlantic biota and their history*. Oxford: Pergamon, 109–125.

neocatastrophism

The principle that the GEOLOGICAL RECORD includes the products of rapid, short-lived natural processes (EVENTS), sometimes operating on scales that have not been observed in historical times. For example, most geologists accept that METEORITE IMPACTS have influenced Earth history and the observed impact of Comet Shoemaker–Levy 9 on Jupiter in 1994 elevated cometary impact (albeit on another planet) to the status of an actualistic process (see ACTUALISM). On a regional scale, catastrophic JÖKULHLAUPS are accepted as having been a factor in the creation of many *proglacial* landforms such as the CHANNELED SCABLANDS. Neocatastrophism can be distinguished from a mere revival of CATASTROPHISM by its moderation (it allows for a background GRADUALISM and is cautious about world-wide events) and its rejection of direct supernatural intervention. In some ways neocatastrophism simply involves the recognition that over geological time the rare event will happen. *CDW*

Ager, D.V. 1993: *The nature of the stratigraphical record*, 3rd edn. Chichester: Wiley. Ager, D.V. 1993: *The new catastrophism: the importance of the rare event in geological history*. Cambridge: Cambridge University Press. Baker, V.R. and Bunker, R.C. 1985: Cataclysmic Late Pleistocene flooding from glacial Lake Missoula: a review. *Quaternary Science Reviews* 4, 1–41. Dott Jr, R.H. 1996: Episodic event deposits versus stratigraphic sequences – shall the twain never meet? *Sedimentary Geology* 104, 243–247. Dressler, B.O., Grieve, R.A.F. and Sharpton, V.L. (eds) 1994: *Large meteorite impacts and planetary evolution* [*Special Paper* 293]. Boulder, CO: Geological Society of America.

neocolonialism

The concept of continuing outside economic and political control over independent developing countries. Control may be direct or indirect and exerted, for example, through the provision of various types of aid from governments, intergovernmental organisations, or multinational companies. *JAM*

[See COLONIALISM, POSTCOLONIALISM]

neo-Darwinism

The theory of EVOLUTION combining DARWINISM with modern ideas from GENETICS. *JAM*

Berry, R.J. 1982: *Neo-Darwinism*. London: Edward Arnold.

neo-ecology

The study of the interactions of extant organisms and their present-day environment (cf. PALAEOECOLOGY). *SHJ*

[See also COMMUNITY ECOLOGY, ECOLOGY, MACROECOLOGY]

neo-evolutionism

A PARADIGM relating to human behaviour and cultural change. It holds that CULTURAL CHANGE is characterised by distinct linear or multilinear sequences controlled largely by deterministic evolutionary processes involving uncontrollable factors of the human environment, such as demography, economics and technology. Concepts of the *neolithic* and URBAN REVOLUTIONS are neo-evolutionist. *JAM*

Trigger, B.G. 1989: *A history of archaeological thought*. Cambridge: Cambridge University Press.

Neogene

A subperiod of the TERTIARY, conventionally taken from 23.3 to 1.64 million years ago, comprising the MIOCENE and PLIOCENE epochs. *GO*

[See also GEOLOGICAL TIMESCALE]

neoglaciation

The concept of glacier recrudescence or regrowth following glacier disappearance (DEGLACIERISATION) during the HYPSITHERMAL or *Climatic Optimum* of the early HOLOCENE. Neoglaciation occurred at different times in different regions in response to LATE-HOLOCENE CLIMATIC DETERIORATION. Different models of neoglaciation have been proposed, one of which recognises an early Holocene when alpine glaciers, including the largest icecap on mainland Europe (Jostedalsbreen, southern Norway), were largely absent. Another model, which appears appropriate for at least the Austrian Alps and northern Sweden, recognises glacier expansion episodes at intervals throughout the Holocene but a larger number of more extensive advances in the late Holocene. *JAM*

[See also GLACIER VARIATIONS, LITTLE ICE AGE]

Nesje, A. and Kvamme, M. 1991: Holocene glacier and climatic variations in western Norway: evidence for early Holocene glacier demise and multiple Neoglacial events. *Geology* 19, 610–612. Matthews, J.A. and Karlén, W. 1992: Asynchronous neoglaciation and Holocene climatic change reconstructed from Norwegian glacio-lacustrine sequences. *Geology* 20, 991–994. Porter, S.C. and Denton, G.H. 1967: Chronology of Neoglaciation in the North American Cordillera. *American Journal of Science* 265, 177–210.

Neolithic: landscape impacts

The Neolithic is the period in which agriculture based on DOMESTICATION of animals and plants replaced HUNTING, FISHING AND GATHERING as the primary source of food. An agricultural economy required a sedentary lifestyle that created the stability for further social and technological developments, such as village societies and pottery firing. Originating in the Near East at c. 10 000 calendar years BP, the Neolithic 'package' spread throughout Eurasia and northern Africa, reaching the British Isles c. 5000 calendar years BP.

The first landscape-scale impacts resulting from ANTHROPOGENIC activity occurred during the Neolithic as woodland was cleared to create fields for agriculture. In some *pollen diagrams* this is indicated by a decline in ARBOREAL POLLEN and the appearance of herbaceous ANTHROPOGENIC INDICATOR species and the first appearance of CEREAL POLLEN grains, often with high concentrations of microscopic charcoal, which may indicate that fire was used in FOREST CLEARANCE. Studies that employ sedimentary and *geochemical analyses* have also shown a mineral SEDIMENT INFLUX, which is interpreted as SOIL EROSION from clearance activity.

Early Danish studies revealed an apparent cyclicity in the pollen records of Neolithic LANDNAM clearance phases that was related to SHIFTING CULTIVATION. In northwestern Europe, the appearance of anthropogenic indicator and cereal pollen often coincided with the ELM DECLINE.

Elsewhere, particularly in southeast Europe, such clearly defined pollen responses are not evident and Neolithic impacts on the landscape are assumed to be minor, perhaps due to the use of naturally available open land. *ARG*

[See also AGRARIAN CIVILISATIONS, AGRICULTURAL HISTORY, AGRICULTURAL ORIGINS, BRONZE AGE: LANDSCAPE IMPACTS, COPPER AGE: LANDSCAPE IMPACTS, FIRE IMPACTS, IRON AGE: LANDSCAPE IMPACTS, NEOLITHIC TRANSITION]

Andel, T.H. van and Runnels, C.N. 1995: The earliest farmers in Europe. *Antiquity,* 69, 481–500. **Edmonds, M.** 1999: Inhabiting Neolithic landscapes. *Quaternary Proceedings* 7, 485–492. **Simmons, I.G. and Innes, J.B.** 1996: The ecology of an episode of prehistoric cereal cultivation on the North York Moors, England. *Journal of Archaeological Science,* 23, 613–618. **Smith, A.G.** 1981: The Neolithic. In Simmons, I.G. and Tooley, M.M.J. (eds), *The environment in British Prehistory.* London: Duckworth, 125–209. **Thomas, J.** 1991: *Rethinking the Neolithic.* Cambridge: Cambridge University Press. **Willis, K.J. and Bennett, K.D.** 1994: The Neolithic Transition – fact or fiction? Palaeoecological evidence from the Balkans. *The Holocene,* 4, 326–330. **Whittle, A.** 1996: *Europe in the Neolithic.* Cambridge: Cambridge University Press.

Neolithic Transition

The period in which a mobile MESOLITHIC lifestyle based on HUNTING, FISHING AND GATHERING was succeeded by a sedentary NEOLITHIC lifestyle based on the production of food from the DOMESTICATION of plants and animals. From origins in the Near-East around 10 000 calendar years BP, Neolithic technology radiated outwards to north Africa, southwest Asia and Europe, reaching Greece c. 9000 calendar years BP and the British Isles by c. 5000 calendar years BP. Recent discoveries in China have indicated that the NEOLITHIC may have evolved independently at the same time in more than one location. *ARG*

[See also OASIS HYPOTHESIS]

Ammerman, A.J. and Cavalli-Sforza, L.L. 1984: *The Neolithic Transition and the genetics of population in Europe.* Princeton, NJ: Princeton University Press. **Armit, I. and Finlayson, W.** 1992: Hunter–gatherers transformed: the transition to agriculture in northern and western Europe. *Antiquity* 66, 664–676. **Normile, D.** 1997: Yangtze seen as earliest rice site. *Science* 275, 309. **Thomas, J.** 1988: Neolithic explanations revisited: the Mesolithic–Neolithic transition in Britain and south Scandinavia. *Proceedings of the Prehistoric Society* 54, 59–66. **Whittle, A.** 1996: *Europe in the Neolithic.* Cambridge: Cambridge University Press. **Zohary, D. and Hopf, M.** 1993: *Domestication of plants in the Old World,* 2nd edn. Oxford: Oxford University Press.

neotectonics

TECTONIC activity occurring during the relatively recent geological past. Recent vertical crustal movements are characteristic of areas of plate collision and island arcs. Detailed studies in the Mediterranean have enabled the reconstruction of numerous neotectonic movements taking place during recent centuries. *AGD*

[See also PLATE MARGIN, PLATE TECTONICS]

Fenart, P., Cat, N.N., Drogue, C. *et al.* 1999: Influence of tectonics and neotectonics on the morphogenesis of the peak karst of Halong Bay, Vietnam. *Geodinamica Acta* 12, 193–200. **Pirazzoli, P.A., Stiros, S.C., Arnold, M.** *et al.* 1994: Episodic uplift deduced from Holocene shorelines in the Perachora Peninsula (Corinth area, Greece) *Tectonophysics* 229, 201–209.

nepheloid layer

A turbid, near-bottom layer in parts of the OCEANS and some large lakes, with a relatively high concentration of very fine-grained sediment in suspension (average grain-size 0.012 mm). The nepheloid layer is best developed in deep water with strong bottom currents, particularly along the western margins of OCEAN BASINS, where it may be several hundred metres thick. *GO*

[See also CONTOUR CURRENT, PELAGIC SEDIMENT, WATER MASSES]

Bout-Roumazeilles, V., Cortijo, E., Labeyrie, L. and Debrabant, P. 1999: Clay mineral evidence of nepheloid layer contributions to the Heinrich layers in the northwest Atlantic. *Palaeogeography, Palaeoclimatology, Palaeoecology* 146, 211–228. **Open University Course Team** 1989: *Ocean chemistry and deep-sea sediments.* Oxford: Pergamon. **Ransom, B., Shea, K.F., Burkett, P.J.** *et al.* 1998: Comparison of pelagic and nepheloid layer marine snow: implications for carbon cycling. *Marine Geology* 150, 39–50. **Sly, P.G.** 1994: Sedimentary processes in lakes. In Pye, K. (ed.), *Sediment transport and depositional processes.* Oxford: Blackwell Scientific, 157–191.

net primary productivity (NPP)

GROSS PRIMARY PRODUCTIVITY (GPP) less the chemical energy used in plant RESPIRATION. It is usually 80% to 90% of GPP. Global mean NPP is $440 \, \mathrm{g \, m^{-2} \, y^{-1}}$. This is potentially available to primary consumers. *RJH*

Vitousek, P.M., Ehrlich, P.R., Ehrlich, A.H. and Matson, P.A. 1986: Human appropriation of the products of photosynthesis. *BioScience* 36, 368–373.

net radiation

The difference between the radiation travelling downwards to the Earth's surface (direct and diffuse shortwave SOLAR RADIATION together with atmospheric counter-radiation) and upwards from the Earth's surface (reflected shortwave radiation together with TERRESTRIAL RADIATION). Net radiation, the RADIATION BALANCE at a point on the Earth's surface, is measured using a *net radiometer* and is positive when the downwards radiation exceeds the upwards radiation. *JAM*

network analysis

An examination of a high-order system (or arrangement) of nodes interconnected by linear direction and magnitude. Network analyses are used to represent and MODEL the transport and flow of materials, for instance water (in rivers and pipes), natural gas, sewage or vehicles from one place to another. A river network tends to form a tree (dendritic) structure, controlled by changes in terrain and water availability. Over time, direction, discharge and velocity of a river system may be calculated to determine the balance and efficiency of flow within the network. GEOGRAPHICAL INFORMATION SYSTEMS (GIS) allows these measurements to be monitored, modelled and displayed more conveniently; they also facilitate many other functions, such as *minimum cost paths*. By calculating a three-dimensional surface (see DIGITAL ELEVATION MODEL), a GIS can determine the least-resistance route for say, a pipeline, with respect to gravity and land-use constraints. Network analysis also describes the operations of a type of DATABASE structure. It employs software pointers that explicitly link data items from one FIELD to another. More recently, network analysis has referred to distributed systems of interlinked computers around the world. *TVM*

[See also NEURAL NETWORKS, RELATIONAL DATABASE]

Band, L.E. 1993: Extraction of channel networks and topographic parameters from digital elevation data. In Kirkby, M.J.

and Bevan, K. (eds), *Channel network hydrology*. Chichester: Wiley, 13–42. **Lupien, A.E., Moreland, W.H. and Dangermond, J.** 1987: Network analysis in geographic information systems. *Photogrammetric Engineering and Remote Sensing* 53, 1417–1421.

neural network

A form of artificial intelligence that mimics aspects of biological neural systems. It consists of a relatively large number of processing units that are highly interconnected. The most widely used network is the *multilayer perceptron*, which comprises a set of units arranged in a layered architecture with each unit connected to every unit in adjacent layers. This network learns by example to convert a set of input data (e.g. on past climate) into an output (e.g. future climate). To achieve this, a sample of cases with known inputs and outputs is used with a learning algorithm to iteratively adjust the network's internal properties until it successfully predicts the output given the inputs. While each unit performs simple tasks, the entire network can solve problems that are complex, non-linear and poorly understood without making any assumptions about the data. Each type of network has its own application domain, but, as a whole, neural networks are general-purpose computing tools capable of application to almost any problem. They are particularly popular in REMOTE SENSING applications for CLASSIFICATION or as a non-parametric alternative to REGRESSION ANALYSIS. *GMF*

Bishop, C. M. 1995: *Neural networks for pattern recognition*. Oxford: Oxford University Press.

neutron activation analysis (NAA)

A non-destructive method of analysis in which a sample bombarded with neutrons gives off gamma rays with an energy characteristic of the original isotope. NAA is an analytical method in GEOCHEMISTRY that can be used to identify major elements and trace elements. The technique has also been used to identify trace elements in ores, pottery and metal ARTEFACTS, and hence to identify their sources and affinities. A *neutron activation log* or *neutron log*, obtained using a tool in a BOREHOLE, can be used to characterise LITHOLOGY and porosity and to distinguish water from oil. *GO*

[See also ARCHAEOLOGICAL GEOLOGY]

Hughes, M.J., Cowell, M.R. and Hook, D.R. (eds) 1991: *Neutron activation and plasma emission spectrometric analysis in archaeology* [*British Museum Occasional Paper* 82]. London: British Museum. **Porat, N., Yellin, J. and Heller-Kallai, L.** 1991: Correlation between petrography, NAA and ICP analyses: application to early bronze Egyptian pottery from Canaan. *Geoarchaeology* 6, 133–149. **Selley, R.C.** 1997: *Elements of petroleum geology*, 2nd edn. New York: Freeman. **Wilson, D.** 2000: Provenance of the Hillsboro Formation: implications for the structural evolution and fluvial events in the Tualatin Basin, northwest Oregon. *Journal of Sedimentary Research* 70, 117–126.

new forestry

Forest practices that retain greater BIO-DIVERSITY and ecosystem complexity than traditional, intensive, monocultural practices of SILVICULTURE. *JAM*

Swanson, F.J. and Franklin, J.F. 1992: New forestry principles from ecosystem analysis of Pacific northwest forests. *Ecological Applications* 2, 262–274.

niche

Originally used to describe the role of an organism within a community as opposed to the HABITAT in which it lives, the niche of an organism may also be defined by an array of environmental factors. The *fundamental niche* is that part of the RESOURCE availability field in which an organism can survive and which it may utilise. The *realised niche* may be limited by inter-specific interactions, such as COMPETITION, and is that part of the fundamental niche that is actually occupied by a species. *JLI*

Giller, P.S. 1984: *Community structure and the niche*. London: Chapman and Hall. **Grinnell, J.** 1917: The niche relations of the California thrasher. *Auk* 34, 364–82. **Schoener, T.W.** 1989: The ecological niche. In Cherrett, J.M. (ed.), *Ecological concepts*. Oxford: Blackwell Scientific.

Nile floods

The oldest year-by-year record of FLOOD levels is for the River Nile in Egypt. The earliest known records in Cairo date from the Early Dynastic Period, recording the height of floods back to about 3090 BC. The most reliable measurements date from AD 622 onwards. These are continuous up to 1470 and then, bar a few gaps, run up to the present day. This record is the longest continuous annual climatic series, monitoring the rainfall in a large DRAINAGE BASIN. It is particularly sensitive to precipitation in the Ethiopian Highlands, which are drained by the Blue Nile. Because of the river's links to other climatic zones, it is believed to hold key evidence of the global nature of CLIMATIC VARIABILITY and the existence of TELE-CONNECTIONS. Major, long-term variations can be identified with periods of low discharge between AD 630 and 1071 and between AD 1180 and 1350. High-discharge episodes occurred from AD 1070 to 1180 and from AD 1350 to 1470. The annual flood is related to the summer monsoonal rains in Ethiopia, which are associated with a northward shift in the INTERTROPICAL CONVERGENCE ZONE (ITCZ). *GS*

Fraedrich, K., Ziang, J., Gerstengarbe, F.-W. and Werner, P.C. 1997: Multiscale detection of abrupt climatic changes: applications to River Nile flood levels. *International Journal of Climatology* 17, 1301–1315. **Hassan, F. and Stucki, B.R.** 1987: Nile floods and climatic change. In Rampino, M.R., Sanders, J.E., Newman, W.S. and Königsson, L.K. (eds), *Climate: periodicity and predictability*. New York: Van Nostrand Reinhold, 37–46. **Popper, W.** 1951: *The Cairo Nilometer*. Berkeley, CA: University of California Press. **Said, R.** 1993: *The River Nile: geology, hydrology and utilization*. Oxford: Pergamon.

Nitisols

Soils with a deeply extended ARGIC HORIZON having a nut-shaped SOIL STRUCTURE with shiny ped faces, occurring in Africa, South America and India on base-rich PARENT MATERIAL (SOIL TAXONOMY: kandic groups of *Alfisols* and *Ultisols*). Nitisols resemble FERRALSOLS, but they are developed from parent materials rich in bases and, as a result, these soils are far more productive and stable under agriculture than other tropical soils. *EMB*

[See also WORLD REFERENCE BASE FOR SOIL RESOURCES]

Sombroek, W.G. and Siderius, W. 1982: Nitisols, a quest for significant diagnostic criteria. *Annual Report, International Soil Reference and Information Centre (Wageningen), 1982*. Wageningen: International Soil Reference and Information Centre

nitrate (NO_3^-)

The main source of nitrogen for plants, an important FERTILISER, the highest oxidation state for nitrogen in wastewater and a POLLUTANT in surface waters and GROUNDWATER. *Nitrifying bacteria* oxidise ammonia to nitrite (NO_2^-) and nitrate: *nitroso-bacteria* convert ammonia to nitrite, *nitro-bacteria* convert nitrite to nitrate. A high concentration of nitrate in wastewater is generally considered to represent a relatively stable effluent. However, excess nitrate contributes to EUTROPHICATION and causes potentially lethal *methaemoglobinaemia* (reduced ability of haemoglobin to carry oxygen) in infants. *JAM*

[See also NITROGEN CYCLE]

Burt, T.P., Heathwaite, A.L. and Trudgill, S.T. (eds), 1993: *Nitrate: processes, patterns and management.* Chichester: Wiley.

nitrogen cycle The cyclical progress of nitrogen through living things, air, rocks, soil and water. Relatively inactive atmospheric nitrogen either forms inorganic compounds in rainwater or is fixed by nitrogen-fixing bacteria (*nitrifiers*). It is then assimilated and metabolised by animals and plants. It returns to the soil in nitrogenous animal wastes and in dead organisms. Nitrogen in the soil is subject to *nitrification* (conversion to nitrates and nitrites by nitrifying micro-organisms), to MINERALISATION or *ammonification* (the release of ammonia and ammonium from dead organic matter by decomposers) and to *denitrification* (the reduction of nitrate to gaseous nitrogen forms that return to the atmosphere). Agricultural and industrial practices modify the stores and fluxes in the nitrogen cycle. Nitrogen FERTILISERS add to the soil nitrogen pool. Growing human and livestock populations increase nitrogenous waste volumes. Industrial activities release NITROGEN OXIDES into the atmosphere. These growing nitrogen pools create environmental problems, including EUTROPHICATION and OZONE DEPLETION. *RJH*

[See also ACID RAIN, BIOGEOCHEMICAL CYCLES, CARBON CYCLE, NITRATE, SULPHUR CYCLE]

Delwiche, C. C. 1970: The nitrogen cycle. *Scientific American* 223, 148–158. **Jaffe, D. A.** 2000: The nitrogen cycle. In Jacobsen, M.C., Charlson, R.J., Rodhe, H. and Orians, G.H. (eds), *Earth system science: from biogeochemical cycles to global change.* London: Academic Press, 322–342. **Sprent, J.I.** 1987: *The ecology of the nitrogen cycle.* Cambridge: Cambridge University Press.

nitrogen dating A RELATIVE-AGE DATING technique based on the *post mortem* decrease in the nitrogen content of bone and teeth as the protein, collagen, decomposes. *In vivo* bone and dentine contain about 5% nitrogen in the collagen and the decomposition rate is influenced by soil or sediment moisture content, pH and temperature. Nitrogen dating is mainly used in parallel with the accumulating uranium and fluorine content of the bone (see FLUORINE DATING). *Nitrogen-profile dating* is used in archaeological contexts on the carved surface of non-porous stone, such as jade and flint, to estimate the time elapsed since carving. The gradient in the nitrogen content with increasing distance from the surface of the stone is related to the rate of diffusion of nitrogen into the stone. *JAM*

Bowman, S. 1999: Nitrogen profiling. In Shaw, I. and Jameson, R. (eds), *A dictionary of archaeology.* Oxford: Blackwell, 428–429. **Ettinger, K.V. and Frey, E.L.** 1980: Nitrogen profiling: a proposed dating technique for difficult artefacts. *Proceedings of the 16th International Symposium on Archaeometry and Archaeological Prospection, Edinburgh, 1976,* 293–311. **Oakley, K.B.** 1980: Relative dating of the fossil hominids of Europe. *Bulletin British Museum Natural History (Geology)* 34, 1–63. **Ortner, D.J., von Endt, D.W. and Robinson, M.S.** 1972: The effect of temperature on protein decay in bone: its significance in nitrogen dating of archaeological samples. *American Antiquity* 37, 514–520. **Wagner, G.A.** 1998: *Age determination of young rocks and artefacts: physical and chemical clocks in Quaternary geology and archaeology.* Berlin: Springer.

nitrogen isotopes Nitrogen has two stable isotopes (natural abundances, $^{14}N = 99.64\%$ and $^{15}N = 0.36\%$). The ISOTOPIC RATIO is expressed as $^{15}N{:}^{14}N$ relative to atmospheric N_2. Most nitrogen is located in the ATMOSPHERE or dissolved in the ocean. The cycling of nitrogen through an ECOSYSTEM may be followed through the NITROGEN CYCLE. Metabolic processes fractionate nitrogen isotopes. Soil nitrogen usually exhibits ISOTOPIC ENRICHMENT in terms of ^{15}N relative to atmospheric N_2. Non-N_2-fixing plants, whose primary source of nitrogen is from the SOIL, have higher $\delta^{15}N$ values than plants, such as legumes, that have the ability to fix atmospheric N_2. The $\delta^{15}N$ values of nitrate have been used to identify POLLUTION from nitrogen-based FERTILISERS. However, pollution studies are limited since the largest ISOTOPIC FRACTIONATION of nitrogen isotopes is from metabolic reactions. Nitrogen in animal tissues is isotopically enriched in ^{15}N relative to dietary inputs, owing to the preferential excretion of ^{14}N in urea. This enrichment in ^{15}N is propagated through the FOOD CHAIN because $\delta^{15}N$ values increase at each TROPHIC LEVEL. In non-arid environments, marine and terrestrial diets may be distinguished using $\delta^{15}N$ and $\delta^{13}C$ values.

The ratio of ^{15}N to ^{14}N found in ICE CORES has recently been used in association with other isotopes (notably $^{40}Ar{:}^{39}Ar$) as confirming estimates of rapid temperature change originally based on OXYGEN ISOTOPE results. Thus, such RAPID ENVIRONMENTAL CHANGES have been shown not to be artefacts of the incorporation of such isotopes into the ice or of offsets in age between different proxy signals resulting from gas diffusion within the ice as it accumulates. *IR/CJC*

Heaton, T.H.E. 1986: Isotopic studies of nitrogen pollution in the hydrosphere and atmosphere: a review. *Chemical Geology* 59, 87–102. **Owens, N.J.P.** 1987: Natural variations in ^{15}N in the marine environment. *Advances in Marine Biology* 24, 389–451. **Sealy, J., Armstrong, R. and Schrire, C.** 1995: Beyond lifetime averages: tracing life histories through isotopic analysis of different calcified tissues from archaeological human skeletons. *Antiquity* 69, 290–300. **Severinghaus, J.P. and Brook, E.J.** 1999: Abrupt climate change at the end of the last glacial period inferred from trapped air in polar ice. *Science* 286, 930–934.

nitrogen oxides The reactive species of nitrogen and oxygen. The most important are nitric oxide (NO) and nitrogen dioxide (NO_2), which are major contributors to urban AIR POLLUTION, and nitrous oxide (N_2O), which is a major GREENHOUSE GAS. Nitrogen oxides (known informally as NO_x gases) are generated when combustion occurs at high temperatures ($> 1000°C$) so that naturally occurring nitrogen and oxygen combine to form nitric oxide. NO is relatively innocuous but is rapidly oxidised by OZONE in the air to form nitrogen dioxide: it accounts for $> 50\%$ of the natural destruction of stratospheric ozone. NO_x gases are primary POLLUTANTS in low air quality episodes in winter, and in summer they promote the formation of secondary pollutants in PHOTOCHEMICAL

SMOG. In urban areas, acute exposure to NO_2 can lead to coughing and sore throats, and will aggravate emphysema and other respiratory ailments. Peak hourly levels of NO_2 of 314 ppb or higher have been recorded in Amsterdam, Athens, Brussels, London, Los Angeles and Munich. *GS*

[See also NITROGEN CYCLE, NITROUS OXIDE VARIATIONS, OZONE DEPLETION]

Bouwman, A.F., Van der Hoek, K.W. and Oliver, J.G. 1995: Uncertainties in the global source distribution of nitrous oxide. *Journal of Geophysical Research* 100, 2785–2790. **Elsom, D.E.** 1996: *Smog alert*. London: Earthscan. **Lee, S.D.** 1980: *Nitrogen oxides and their effects on health*. Ann Arbor, MI: Ann Arbor Science.

nitrogen/nitrate analysis

A method in SOIL science to determine the nitrate/ammonium/nitrite content; in AQUATIC and marine ENVIRONMENTS to measure productivity changes and in SEDIMENTS to characterise the amount of ORGANIC material (see ORGANIC CONTENT). Dominant forms of nitrogen in freshwater include: dissolved molecular N_2, ammonia nitrogen (NH_4^+), nitrite (NO_2^-), nitrate (NO_3^-), and organic compounds. The organic *carbon:nitrogen* (C:N) *ratio* indicates an approximate state of resistance of complex mixtures of organic compounds to DECOMPOSITION. Organic compounds from ALLOCHTHONOUS and WETLAND sources commonly have C:N ratios from 45:1 to 50:1 and contain mainly humic compounds of low nitrogen content. AUTOCHTHONOUS organic matter produced by the decomposition of PLANKTON tends to have higher protein content and C:N ratios of about 12:1. Changing *carbon:phosphorus* (C:P) *ratios* reflect shifts in algal species. Increased loading of inorganic nitrogen to aquatic/marine ecosystems frequently results from agricultural activities, sewage and ATMOSPHERIC POLLUTION. *UBW*

[See also ALLOCHTHONOUS, AUTOCHTHONOUS, NITRATE, NITROGEN CYCLE, NITROGEN ISOTOPES, NITROGEN OXIDES, TROPHIC LEVEL]

Bengtsson, L. and Enell, M. 1986: *Chemical analysis*. In Berglund, B.E. (ed.), *Handbook of Holocene palaeoecology and palaeohydrology*. Chichester: Wiley, 423–451.

nitrous oxide variations

Nitrous oxide (N_2O) is an atmospheric TRACE GAS that is produced naturally by SOILS and OCEANS. Current estimates suggest that N_2O is increasing at the rate of 0.3% per year and that each molecule has the potential to contribute 300 times to the GREENHOUSE EFFECT relative to each molecule of CARBON DIOXIDE. Its impact on GLOBAL WARMING is therefore of serious concern. Evidence from air trapped in the ICE CORES indicate that variations in atmospheric nitrous oxide have mirrored the GLACIAL–INTERGLACIAL CYCLES. Lower levels during GLACIALS are thought to reflect either reduced soil activity or less output from oceanic sources. Variation between GLACIAL and INTERGLACIALS is considerable; during the last GLACIAL–INTERGLACIAL CYCLE, for example, nitrous oxide levels increased by 30%. Over the past 200 years, levels of atmospheric nitrous oxide have been strongly influenced by ANTHROPOGENIC ACTIVITY and in particular EMISSIONS resulting from the use of nitrogen-rich FERTILISERS. Current estimates suggest that this has increased levels of atmospheric nitrous oxide to 8% higher than pre-industrial levels. *KJW*

Battle, M., Bender, M. and Sowers, T. 1996: Atmospheric gas concentrations over the past century measured in air from the firn at the South Pole. *Nature* 383: 231–235. **Bradley, R.S.** 1999: *Paleoclimatology – reconstructing climates of the Quaternary*, 2nd edn. San Diego, CA: Harcourt Academic Press. **Cronin, T.M.** 1999: *Principles of paleoclimatology*. New York: Columbia University Press. **Machida, T., Nakazawa, T., Fujii, Y.** *et al.* 1995: Increase in atmospheric nitrous oxide concentrations during the last 250 years. *Geophysical Research Letters* 22, 2921–2924. **Naqvi, S.W.A., Jayakumar, D.A., Narrekar, P.V.** *et al.* 2000: Increased marine production of N_2O due to intensifying anoxia on the Indian Ocean shelf. *Nature* 408, 346–349. **Schlesinger, W.H.** 1997: *Biogeochemistry – an analysis of global change*. London: Academic Press.

nival zone

An ALTITUDINAL zone in high mountains above the SNOW LINE or above the altitude of the lowest permanent SNOWBEDS averaged over a number of years. *JAM*

[See also ALPINE ZONE, AEOLIAN ZONE]

nivation

A collective term for the erosive and depositional processes associated with semi-permanent and late-lying SNOWBEDS that give rise to such landforms as hillside nivation hollows, benches and CRYOPLANATION TERRACES. *PW*

[See also CHEMICAL WEATHERING, FROST WEATHERING, PRONIVAL RAMPART, SLOPEWASH, SNOWMELT, SOLIFLUCTION]

Berrisford, M.S. 1991: Evidence for enhanced mechanical weathering associated with seasonally late-lying and perennial snow patches, Jotunheimen, Norway. *Permafrost and Periglacial Processes* 2, 331–340. **Christiansen, H.H.** 1998: 'Little Ice Age' nivation activity in northeast Greenland. *The Holocene* 8, 719–728. **Thorn, C.E.** 1988: Nivation: a geomorphic chimera. In Clark, M.J. (ed.), *Advances in periglacial geomorphology*. Chichester: Wiley, 3–31.

niveo–aeolian deposits

Wind-blown silty sand deposited together with snow and subsequently reworked by snowmelt water. As seen in snow banks in winter and early summer, niveo–aeolian deposits consist of stratified snow and sediment layers. In summer, as the snow banks melt, the sediment is further stratified and locally reworked by wash processes. *HMF*

[See also PERIGLACIAL SEDIMENTS]

Koster, E.A. and Dijkmans, J.W.A. 1988: Niveo-aeolian deposits and denivation forms, with special reference to the Great Kobuk Sand Dunes, Northwestern Alaska. *Earth Surface Processes and Landforms* 13, 153–170.

noise pollution

Although the intensity of sound can be measured in decibels, and other aspects of sound (such as its frequency and duration) can also be measured precisely, noise (unwanted sound) is a subjective property. It is questionable whether noise should be described as a POLLUTANT or as causing POLLUTION. In certain environments, such as in some factories, in and around airports and close to urban motorways, noise may be more than an irritant or nuisance and becomes a HUMAN HEALTH HAZARD. Noise abatement can be achieved at source, during transmission or by receiver control. *JAM*

Foreman, J.E.K. 1990: *Sound analysis and noise control.* New York: Van Nostrand Reinhold. **Gjestland, T.** 1999: Sound and noise. In Brune, D., Chapman, D.V., Gwynne, M.D. and Pacyna, M. (eds), *The global environment: science, technology and management.* Vol. 1. Weinheim, Germany: VCH, 610–624.

nomadism A mobile community lifestyle involving more-or-less continued shifting of residence, which is often adapted to environmental constraints, particularly in marginal areas (e.g. high altitudes, ARIDLANDS). It is often associated with PASTORALISM, where the population relocates to new pasture. *LD-P*

[See also DOMESTICATION, MIGRATION, SEDENTISM]

Bonte, P., Guillaum, H. and **Zecchin, F.** 1996: Nomads: changing societies and environments. *Nature and Resources* 32, 2–10.

nomothetic science Science that is concerned with developing general ideas or LAWS, as opposed to a detailed explanation of individual cases (IDIOGRAPHIC SCIENCE). A somewhat dated concept. *CET*

non-arboreal pollen (NAP) Pollen originating from dwarf-shrubs and HERBS but not trees. The NAP percentage has been used as a rough indicator of non-forested land. *SPH*

[See also ARBOREAL POLLEN, POLLEN ANALYSIS]

non-destructive sampling Scientific investigation often influences the phenomenon under investigation and sometimes results in its destruction. It is good practice, particularly in environmentally conscious disciplines, to minimise such effects and non-destructive sampling is a case in point. Examples include satellite REMOTE SENSING and various types of CORE SCANNING used in palaeoenvironmental investigation. *JAM*

Caseldine, C., Baker, A. and **Barnes, W.L.** 1999: A rapid, non-destructive scanning method for detecting distal tephra layers in peats. *The Holocene* 9, 635–638.

non-metric multidimensional scaling A form of MULTIDIMENSIONAL SCALING in which the ranks of the distances or dissimilarities between objects are used to produce the required low-dimensional representation of the distance matrix. *HJBB*

Kruskal, J.B. 1964: Multidimensional scaling by optimising goodness of fit to a nonmetric hypothesis. *Psychometrika* 29, 1–27. **Matthews, J.A.** 1978: An application of non-metric multidimensional scaling to the construction of an improved species plexus. *Journal of Ecology* 66, 157–173.

non-parametric statistics Statistical procedures for testing hypotheses or estimating parameters that make no assumptions about the underlying PROBABILITY DISTRIBUTION of the variables (such as normality or linearity). They commonly involve the ranks of the observations rather than the observations themselves. They are often only marginally less powerful than their parametric counterparts, even when the underlying assumptions of the latter are true. They are also known as *distribution-free methods*. *HJBB*

[See also PARAMETRIC STATISTICS]

non-renewable resource A RESOURCE that, once used, cannot be replaced, at least within a timescale to be useful. Typically, non-renewable, *depletable, exhaustible* or *stock resources*, such as FOSSIL FUELS, MINERAL RESOURCES and biological species that, once extinct, cannot be replaced, are formed over geological timescales. At any time, the stock or RESERVES of a non-renewable resource are finite and, with use, will eventually be exhausted within a period of time that can be estimated. However, as a non-renewable resource is exploited and reserves are depleted, demand may change due to such factors as *product substitution* (use of substitutes) and WASTE RECYCLING, thus delaying exhaustion. *JAM*

[See also RENEWABLE RESOURCES, SUSTAINABILITY]

Fisher, C.A. 1981: *Resources and environmental economics.* Cambridge: Cambridge University Press. **Nordhaus, W.D.** 1974: Resources as a constraint to growth. *American Economics Review* 64, 22–26. **Rees, J.** 1990: *Natural resources: allocation, economics and policy,* 2nd edn. London: Routledge.

noösphere The realm or sphere of influence of the human mind in the context of the Earth and especially the ECOSPHERE. Unlike other GEOSPHERES, the noösphere is an abstract phenomenon without physical existence. It is, however, an increasingly pervasive influence on the ecosphere and may be viewed as the summation of the mental activity behind inadvertent human impacts as well as the conscious human use of the Earth, management and conservation. *JAM*

[See also HOMOSPHERE]

Samson, P.R. and **Pitt, D. (eds)** 1999: *The biosphere and noosphere reader: global environment, society and change.* London: Routledge. **Teilhard de Chardin, P.** 1956: The antiquity and world expansion of human culture. In Thomas Jr, W.L. (ed.), *Man's role in changing the face of the Earth.* Chicago, IL: University of Chicago Press, 103–114. **Teilhard de Chardin, P.** 1964: *The future of man.* London: Collins. **Vernadsky, V.I.** 1945: The biosphere and the noösphere. *American Scientist* 33, 1–12.

normal distribution A continuous PROBABILITY DISTRIBUTION, used to describe continuous random variables, that is assumed by many PARAMETRIC STATISTICAL methods. It is symmetrical, unimodal and bell-shaped. Its shape and distribution are defined by the mean and standard deviation. In a normal distribution, 95% of all observations lie within the mean ± 1.96 standard deviations and 99% within the mean ± 2.576 standard deviations. It is also known as a *Gaussian distribution.* *HJBB*

Sokal, R.R. and **Rohlf, F.J.** 1995: *Biometry.* New York: W.H. Freeman.

normal polarity The present day orientation of the dipole component of the Earth's geomagnetic field. Normal polarity and the opposite, *reverse polarity,* when the orientation changes through 180°, are fundamental components of the GEOMAGNETIC POLARITY TIMESCALE. *MHD*

[See also GEOMAGNETIC POLARITY REVERSAL]

Thompson, R. 1991: Palaeomagnetic dating. In: Smart, P.L. and Frances, P.D. (eds), *Quaternary dating methods – a users guide* [*QRA Technical Guide* 4]. Cambridge: Quaternary Research Association, 177–198.

normalisation A statistical procedure for making a non-normal distribution, such as a distribution exhibiting SKEWNESS, into a NORMAL DISTRIBUTION. Normalisation is commonly used in order to apply PARAMETRIC STATISTICS to data that do not exhibit a normal distribution. An example is the use of a *logarithmic transformation* in relation to a distribution that exhibits positive skew. Data are frequently converted into *z-scores* to approximate a standard normal distribution (STANDARDISATION to zero mean and unit variance). This involves subtracting the arithmetic mean from each value of the series and dividing the resulting values by the series standard deviation.

KRB/JAM

Read, C.B. 1985: Normal distribution. In Kotz, S. and Johnson, N.L. (eds), *Encyclopedia of statistical sciences*. Vol. 6. New York: Wiley, 347–359.

Norse Greenland Settlements

The Norse colony in Greenland was founded by settlers from Iceland around AD 985 and lasted for about 500 years. The Western Settlement lasted until the mid-fourteenth century; the more southerly Eastern Settlement disappeared towards the end of the late fifteenth century. CLIMATIC CHANGE associated with the LITTLE ICE AGE was certainly involved in their decline in these MARGINAL AREAS but the precise cause(s) have not been established despite intense interdisciplinary investigation. Other factors that have been implicated include Inuit competition, increasing crop failure and soil erosion, declining trade with Europe, and congenital infertility. *JAM*

Barlow, L.K., Sadler, J.P., Ogilvie, A.E.J. *et al.* 1997: Interdisciplinary investigations of the end of the Norse Western Settlement in Greenland. *The Holocene* 7, 489–499. **Gad, F.** 1973: *The history of Greenland*. Vol. 1. *Earliest times to 1700*. London: C. Hurst.

North Atlantic Oscillation (NAO)

A predominantly wintertime, decadal-scale, regional feature of the atmospheric circulation, the influence of which extends beyond the North Atlantic area and is of importance to the weather and climate of the British Isles and Europe. The NAO is defined as the sea-level pressure difference between Ponta Delgada (37°7–N, 25°7–W) in the Azores and Stykkisholmur (65°0–N, 22°8–W) in Iceland. The resultant standardised monthly index values provide a useful measure of the strength and frequency of air circulation across the eastern North Atlantic upwind of western Europe. Positive values indicate strong WESTERLIES with tracks of STORMS pushed farther south into Europe. Large negative values suggest BLOCKING and an easterly component to circulation. The NAO series begins in AD 1865, the start of the Azores record. Significant statistical associations exist between the NAO and certain climatic parameters (for instance winter precipitation and the CENTRAL ENGLAND TEMPERATURE RECORD), the THERMOHALINE CIRCULATION and cod fisheries of the North Atlantic, and the mass balance of European glaciers. *GS*

[See also PACIFIC–NORTH AMERICAN TELECONNECTION, QUASI-BIENNIAL OSCILLATION, TELECONNECTIONS]

Cook, E.R., D'Arrigo, R.D. and Briffa, K.R. 1998: A reconstruction of the North Atlantic Oscillation using tree-ring chronologies from North America and Europe. *The Holocene* 8, 9–17. **Dickson, R.R., Meincke, J., Malmberg, S.-A. and Lee, A.J.** 1988: The 'great salinity anomaly' in the northern North Atlantic, 1968–1982. *Progress in Oceanography* 20, 103–151. **Hurrell, J.W.** 1995: Decadal trends in the North Atlantic Oscillation: regional temperature and precipitation. *Science* 269, 676–679. **Hurrell, J.W. and van Loon, H.** 1997: Decadal variations in climate associated with the North Atlantic

Oscillation: climatic change at high-elevation sites. *Climatic Change* 36, 301–326. **Nesje, A., Lie, Ø. and Dahl, S.O.** 2000: Is the North Atlantic Oscillation reflected in Scandinavian glacier mass balance records? *Journal of Quaternary Science* 15, 587–601. **Perry, A.** 2000: The North Atlantic Oscillation: an enigmatic see-saw. *Progress in Physical Geography* 24, 289–294. **Pohjola, V.A. and Rogers, J.C.** 1997: Atmospheric circulation and variations in the Scandinavian glacier mass balance. *Quaternary Research* 47, 29–36. **Uppenbrink, J.** 1999: The North Atlantic Oscillation. *Science* 283, 948–949. **Wilby, R.L., O'Hare, G.P. and Barnsley, N.** 1997: The North Atlantic Oscillation and British Isles climate variability. *Weather* 52, 266–276.

notch An indentation at the base of a sea cliff (at about high-tide level). Notches are usually bio-erosional forms and are especially common in the tropics where the solution of LIMESTONE is important. Palaeoforms may be emergent or submergent and are indicative of SEA-LEVEL CHANGE. *HJW*

Nunn, P.D. 1995: Holocene tectonic histories for the South-central Lau group, South Pacific. *The Holocene* 5, 160–171.

nuclear accident An inadvertent release of a substantial amount of radioactive material into the environment. The most serious nuclear accident to date was the explosion at the *Chernobyl* nuclear reactor in the Ukraine in AD 1986. RADIONUCLIDES were released over a period of 10 days and spread beyond the region to include northern and western Europe, where soils, pasture and animals were contaminated. Over 130 000 people were evacuated from the local area but continue to be at risk from radiation-induced diseases. *JAM*

[See also SOIL RADIOACTIVITY]

Anspaugh, L.R., Catlin, R.J. and Goldman, M. 1988: The global impact of the Chernobyl reactor accident. *Science* 242, 1513–1519. **Boronov, A. and Bogatov, S.** 1997: *Consequences of Chernobyl*. New York: Plenum.

nuclear war The potential environmental consequences of a nuclear war are difficult to estimate. The Ambio Advisory Group to the Royal Swedish Academy of Sciences assessed the effects of a full-scale nuclear war between the USA and the former Soviet Union targeting urban, military and economic targets with particular emphasis on the effects of the enormous quantities of PARTICULATE matter that would cloak the Earth. Forests, agricultural land and oilfields would be ignited, adding PHYTOCHEMICAL SMOG to the reduction in light, short-term climatic cooling (NUCLEAR WINTER), direct blast effects and the radiation hazard, which would be particularly devastating on the terrestrial ecosystems and human populations of the Northern Hemisphere. The Group concluded that long-term and less predictable environmental effects might match or exceed these more immediate impacts. *JAM*

National Research Council 1985: *The effects on the atmosphere of a major nuclear exchange*. Washington, DC : National Academy Press. **Royal Swedish Academy of Sciences** 1983: *Nuclear war: the aftermath* [based on a Special Issue of the journal *Ambio*]. Oxford: Pergamon. **Turco, R.P., Toon, O.B., Ackerman, T.** *et al.* 1983: Global atmospheric consequences of nuclear war. *Science* 222, 1283.

nuclear winter The hypothesis that a NUCLEAR WAR would cause severe cooling of the Earth due to the injection into the atmosphere of vast quantities of DUST and soot, which would intercept SOLAR RADIATION for a prolonged time interval. Global temperatures might fall by as much as 25°C. Although precise effects are difficult to predict, analogues for nuclear winter include the contrast between day and night, the firestorms of World War II, forest fires (see BIOMASS BURNING), OIL FIRES, VOLCANIC IMPACTS ON CLIMATE and the MASS EXTINCTIONS of the geological record, possibly caused by METEORITE IMPACT.

JAM

Robock, A. 1996: Nuclear winter. In Schneider, S.H. (ed.), *Encyclopedia of climate and weather*. New York: Oxford University Press, 534–536. **Turco, R.P., Toon, O.B., Ackerman, T.P. et al.** 1990: Climate and smoke: an appraisal of nuclear winter. *Science* 247, 166–176.

nudation The removal of existing ecological communities by major disturbance prior to the initiation of ECOLOGICAL SUCCESSION. *LRW*

nuée ardente A PYROCLASTIC FLOW produced by the collapse of a LAVA DOME. The term, from the French for 'glowing cloud', was coined following the AD 1902 eruption of Mont Pelée on Martinique, in which 28 000 people died. Because of some confusion in the application of the term, it is probably best avoided. *JBH*

Cas, R.A.F. and Wright, J.V. 1987. *Volcanic successions: modern and ancient*. London: Allen and Unwin. **La Croix, A.** 1904: *La montagne Pelée et ses éruptions*. Paris: Masson. **Tanguy, J.C.** 1994: The 1902–1905 eruptions of Montagne-Pelée, Martinique – anatomy and retrospection. *Journal of Volcanology and Geothermal Research* 60, 87–107.

number of looks The number of subimages that are combined to form an output RADAR image with reduced SPECKLE. *Multilooking* (which often amounts to averaging over PIXELS in the azimuth direction), is performed as part of the data preprocessing. *PJS*

numerical-age dating A category of techniques used to obtain quantitative estimates of age and uncertainty based on an absolute timescale. It was formerly termed *absolute-age dating*. *DAR*

[See also CALIBRATED-AGE DATING, CORRELATED-AGE DATING, RELATIVE-AGE DATING]

Colman, S.M., Pierce, K.L. and Birkeland, P.W. 1987: Suggested terminology for Quaternary dating methods. *Quaternary Research* 28, 314–319.

numerical analysis Data relevant to environmental-change research are usually complex, quantitative and multivariate, consisting of many observations and many variables. Such data are often stratigraphical or geographical in character and hence the observations have a fixed order in one or two dimensions. *Biological data* (e.g. pollen stratigraphical data) usually have large numbers of variables (100–200), many zero (absence) values, are most commonly expressed as percentages and are thus 'closed' compositional data. *Geological data* (e.g. magnetic properties) usually have relatively few variables (< 50) and few zero values. *Environmental data* (e.g. lake chemistry) also have few variables (10–30), few zero values, but may have some missing unmeasured values.

Numerical analysis of such complex data-sets is now a regular part of environmental research. This is a result, in part, of the many recent developments in applied and environmental statistics (*environometrics*) and, in part, of the increasing availability of powerful computers with the advent of personal computers. Numerical analysis involves EXPLORATORY DATA ANALYSIS, *data summarisation* and CONFIRMATORY DATA ANALYSIS, the latter usually involving specific hypothesis testing.

An essential first step in any data analysis is basic exploratory data analysis. The main purpose is to provide the data analyst with 'a feel for the data'. It involves estimation of measures of central tendency (e.g. mean, median), dispersion (e.g. standard deviation, inter-quartile range), and shape (e.g. SKEWNESS, KURTOSIS) of the data, simple graphical tools such as box-and-whisker plots and scatter plots, outlier detection involving influence and leverage measures, and data transformations in an attempt to achieve a NORMAL DISTRIBUTION, which is the PROBABILITY DISTRIBUTION that is assumed by many PARAMETRIC statistical methods. Data display is an essential step, either as two- or three-dimensional scatter plots or as stratigraphical diagrams. Interactive graphical tools are indispensible in exploratory analysis. NON-PARAMETRIC regression techniques, such as locally weighted regression scatter plot smoothing (LOESS regression), are useful graphical tools for highlighting the 'signal' or major trends in data in the absence of any *a priori* model.

Patterns or 'structure' within multivariate data can be usefully detected by means of numerical CLASSIFICATION (e.g. CLUSTER ANALYSIS, TWO-WAY INDICATOR SPECIES ANALYSIS) to detect groups of observations of similar composition and/or groups of variables with similar occurrences. ORDINATION techniques (e.g. PRINCIPAL COMPONENTS ANALYSIS, CORRESPONDENCE ANALYSIS, NON-METRIC MULTIDIMENSIONAL SCALING) provide useful graphical summaries in two or three dimensions of the major patterns of variation in multivariate data, can help identify latent variables and gradients in data and can display patterns of similarity and dissimilarity between observations and variables.

When the observations come from two or more predefined groups (e.g. from two different bedrock types), DISCRIMINANT ANALYSIS provides a means of evaluating how distinct the groups are, of characterising the groups in terms of the variables and of providing a low-dimensional graphical representation of the within-group and between-group variation.

If there is some *a priori* biological or geological reason for considering some variables as responding (so-called *response variables*) to other variables, so-called *explanatory* or *predictor variables*, statistical modelling techniques such as REGRESSION ANALYSIS, *multiple linear regression* and ANALYSIS OF VARIANCE can be used to model the relationship between one response variable and one or more predictor variables. All such modelling techniques form a part of GENERALISED LINEAR MODELS. The non-parametric GENERALISED ADDITIVE MODELS can provide a useful exploratory tool. If there are two or more response variables and two or more predictor variables, constrained ordination techniques such as REDUNDANCY ANALYSIS or CANONICAL CORRESPONDENCE ANALYSIS can be used to derive a low-dimensional multivariate regression model that combines both an ordination graphical display and regression modelling. Given two sets of predictor variables

433

(e.g. climatic variables and chemical variables), it is possible by means of a series of (partial) constrained ordinations to partition the variance in the response variables into four independent components – variance due to climate independent of chemistry, variance due to chemistry independent of climate, variance due to the covariance between climate and chemistry and variance not explained by climate or chemistry. Variance decomposition can be extended for three or more groups of predictor variables.

In palaeoenvironmental research, an important role of numerical analysis involves the estimation of modern calibration or TRANSFER FUNCTIONS. These model the relationship between modern biological assemblages (e.g. pollen, diatoms) in surface sediments and contemporary environmental variables (e.g. mean July temperature, lake-water pH). Such *calibration functions* are most commonly derived by inverse regression, which assumes either a linear or a unimodal response model of organisms to the environment. A large number of techniques now exist for deriving calibration functions, including weighted-averaging regression and calibration, weighted-averaging partial least-squares regression, partial least-squares regression, GLM, modern analogue techniques and response surfaces.

Stratigraphical data are frequent in environmental research. Such data have special mathematical properties, the most important of which is that the observations are in a fixed stratigraphical and temporal order. Numerical analysis of such data requires taking account of this order. Stratigraphically constrained classification methods have been developed for partitioning or zoning such data. Individual stratigraphical variables can be partitioned by sequence splitting into segments of uniform mean and variance to provide a means of detecting and testing for consistent patterns of stratigraphical change between variables. Comparison of two or more stratigraphical records (e.g. pollen, diatoms) from the same sequence can be made numerically by one of two methods: (1) first summarising the different data-sets independently as the first few principal component-analysis or correspondence-analysis axes and then by comparing these axes by means of oscillation logs; or (2) using constrained ordination techniques such as REDUNDANCY ANALYSIS. Comparison and correlation of two or more stratigraphical sequences can be made by sequence slotting or by combined classification or ordination of the sequences. Rate-of-change analysis can be useful to estimate the amount of change per unit time in stratigraphical data. Such analyses require detailed chronologies in calendar years. MODERN ANALOGUE techniques in which modern analogues for fossil assemblages are sought numerically can provide a factual basis for interpretation of the stratigraphical data.

TIME-SERIES ANALYSIS can involve the time-domain approach based on the concept of temporal AUTOCORRELATION or the frequency-domain approach that focuses on bands of frequency or wavelength over which the variance of the time series is concentrated. Time-series analysis can be used to detect patterns of temporal autocorrelation and cross-correlation and PERIODICITY within time-ordered data. Time-series analysis makes many demanding assumptions of the data, many of which are difficult to meet in palaeoenvironmental studies.

Spatial data similarly have a fixed ordering in two dimensions. This geographical information is important in *mapping*, in trend-surface analysis, in geostatistical techniques such as KRIGING, in ordination and classification of spatial data and in the estimation of statistical parameters and the statistical testing of hypotheses in the presence of spatial autocorrelation (see SPATIAL ANALYSIS).

The testing of hypotheses about the impacts of environmental variables on biological assemblages is an important part of confirmatory data analysis in environmental research. Given the complexity of environmental data with their spatial or temporal ordering, non-normality and many zero values, hypothesis testing requires MONTE CARLO METHODS involving randomisation tests or permutation tests and constrained ordination techniques such as redundancy analysis. Such tests also provide means of testing for spatial and/or temporal trends in data.

The numerical analysis of data associated with environmental-change research is a rapidly developing field. Future developments are likely to involve classification and regression trees and neural networks, and to incorporate explicit Bayesian approaches in the statistical analysis of biological and environmental data.   *HJBB*

[See also BAYESIAN STATISTICS, CLIMATIC RECONSTRUCTION, MONTE CARLO METHODS, MULTIVARIATE ANALYSIS, STATISTICAL ANALYSIS]

Birks, H.J.B. 1987: Multivariate analysis of stratigraphical data in geology: a review. *Chemometrics and Intelligent Laboratory Systems* 2, 109–126. **Birks, H.J.B.** 1995: Quantitative palaeoenvironmental reconstructions. In Maddy, D. and Brew, J.S. (eds), *Statistical modelling of Quaternary science data*. Cambridge: Quaternary Research Association,161–254. **Birks, H.J.B.** 1998: Numerical tools in palaeolimnology – progress, potentialities, and problems. *Journal of Paleolimnology* 20, 307–332. **Birks, H.J.B and Gordon, A.D.** 1995: *Numerical methods in Quaternary pollen analysis*. London: Academic Press. **Davis, J.C.** 1986: *Statistics and data analysis in geology*. New York: John Wiley. **Jongman, R.H.G., ter Braak, C.J.F. and van Tongeren, O.F.R.** 1987: *Data analysis in community and landscape ecology*. Wageningen: Pudoc. **Legendre, P. and Fortin, M.J.** 1989: Spatial pattern and ecological analysis. *Vegetatio* 80, 107–138. **Legendre, P. and Legendre, L.** 1998: *Numerical ecology*. Amsterdam: Elsevier. **Storch, H. von and Zwiers, F.W.** 2000: *Statistical analysis in climate research*. Cambridge: Cambridge University Press.

nunatak   A mountain top protruding above an ice sheet or glacier surface. Former nunataks associated with vanished ice sheets can be identified using WEATHERING LIMITS and TRIMLINES. The term is derived from the Inuit language.   *DIB*

[See also NUNATAK HYPOTHESIS, REFUGE THEORY]

Ballantyne, C.K. 1999: An Teallach: a late Devensian nunatak in Western Ross. *Scottish Geographical Journal* 115, 249–259. **Elven, R.** 1980: The Omnsbreen glacier nunataks – a case study in plant immigration. *Norwegian Journal of Botany* 27, 1–16.

nunatak hypothesis   The idea that the patchy occurrence of mountain species is evidence of survival on the mountain tops above an ice sheet. The term is derived from the Inuit word, nunatak, for mountain tops rising above the Greenland ice sheet. The alternative is that the species migrated there after the ice sheet melted.   *BA*

[See also GLACIER VARIATIONS, GLACIAL–INTERGLACIAL CYCLE, REFUGE THEORY, REFUGIUM, TABULA RASA).

Birks, H.H. 1994: Plant macrofossils and the Nunatak theory of per-glacial survival. *Dissertationes Botanicae* 234, 129–143. **Birks,**

H.J.B. 1993: Is the hypothesis of survival on glacial nunataks necessary to explain the present-day distributions of Norwegian mountain plants? *Phytocoenologia* 23, 399–426.

nurse plant   A plant that facilitates the establishment and growth of another plant of a different species, often in ECOLOGICAL SUCCESSION, by providing direct physical shelter from, for example, excessive wind, heat or herbivory.
LRW

[See also FACILITATION]

nutrient   A raw material required by organisms for life. Nutrients include not only the essential mineral elements (MACRO- and MICRONUTRIENTS) but also water, inorganic salts and, for animals, organic compounds (carbohydrates, fats, proteins and vitamins).   *JAM*

[See also BIOGEOCHEMICAL CYCLES, MINERAL CYCLING]

Paoletti, M.G., Foissner, W. and Coleman, D.C. (eds) 1993: *Soil biota, nutrient cycling and farming systems.* Boca Raton, FL: Lewis.

nutrient pool   The store or reserve of a nutrient in an ecosystem or its component parts (in the soil, water, air or living organisms).   *RJH*

[See also BIOGEOCHEMICAL CYCLES, ECOSYSTEM CONCEPT, LIMITING FACTORS]

nutrient status   The concept of soil or other medium as a source of nutrients for plants, particularly the three major plant nutrients, nitrogen, phosphorus and potassium. The nutrient status of soils is important for natural ecosystems, agriculture, horticulture and forestry. Soils may have nutrients in sufficient quantity for plant growth, but more often than not certain elements are deficient and need to be supplied for optimum crop yields. Plant nutrients are removed from the soil naturally by LEACHING and by crops and livestock being taken and consumed elsewhere. Unless these nutrients are replaced naturally, a long-term process, or more rapidly by chemical FERTILISERS, soil fertility is reduced. The extreme situation is found in impoverished communities who have to resort to *fertility mining* for subsistence and the result is SOIL DEGRADATION. Excessive nutrient concentrations can cause imbalance of other essential nutrients and reduce yields. Certain MICRONUTRIENTS may also be in limiting supply.
EMB

[See also EUTROPHIC, EUTROPHICATION, MIRES, OLIGOTROPHIC, OLIGOTROPHICATION, SOIL DEGRADATION, SOIL QUALITY, SOIL RECLAMATION]

Tisdale, S.L., Nelson, W.L., Beaton, J.D. and Havlin, J.L. 1993: *Soil fertility and fertilizers*, 5th edn. New York: Macmillan.

O

oasis hypothesis In relation to early agriculture, this hypothesis provided an obsolete scheme to account for DOMESTICATION, whereby aggregation around reduced LATE GLACIAL water resources (oases) forged new human–animal relationships. *ARG*

[See also AGRICULTURAL HISTORY, AGRICULTURAL ORIGINS]

Childe, V.G. 1935: *New light on the most ancient East.* London: Kegan Paul.

object oriented A new approach to many fields including REMOTE SENSING and IMAGE PROCESSING based on defining classes of spatial objects in an image or in a GEOGRAPHICAL INFORMATION SYSTEM (GIS). Operators can be applied to the objects, like 'left of' or 'inside'. *HB*

objective knowledge Knowledge that has been subjected to SCIENTIFIC METHOD and hence has been researched, analysed, synthesised and debated by scientists. Although it no longer represents the subjective, untested views of individual people, it still does not imply absolute or permanent knowledge. Objective knowledge evolves as investigation continues, new information is discovered and new insights are incorporated: it is simply 'the most reliable current knowledge' (Ford, 2000). *JAM*

[See also KNOWLEDGE, SCIENCE]

Ford, E.D. 2000: *Scientific method for ecological research.* Cambridge: Cambridge University Press. **Popper, K.R.** 1982: *Objective knowledge: an evolutionary approach.* Oxford: Oxford University Press.

obligate An ENVIRONMENTAL FACTOR that is required by an organism for survival, as opposed to one that the organism may benefit from but is not necessary for survival (FACULTATIVE). Organisms may also be described as obligate in relation to particular environmental factors. Thus, an obligate HALOPHYTE is one that requires relatively high salt concentrations, whereas a facultative halophyte can tolerate a high salt concentration but does not require it. *JAM*

obliquity The angle between the plane of the Earth's orbit (the ecliptic) and the plane of the Earth's equator. Known colloquially as *axial tilt*, the full term is '*obliquity of the ecliptic*'. It affects the angle of the solar beam, which depends on time of day, time of year and latitude. For example, because of the greater obliquity in the Polar areas, INSOLATION is less concentrated and more SOLAR RADIATION is lost by atmospheric scattering, reflection and absorption. Obliquity is one of the three main orbital parameters in the MILANKOVITCH THEORY. *RDT*

[See also ECCENTRICITY, PRECESSION]

obrution Smothering by rapid burial in SEDIMENT. *LC*

[See also FOSSIL LAGERSTÄTTEN, TAPHONOMY]

observational data A record of the measurement or qualitative assessment of an environmental variable at a specified place and time. The credibility of such data depends both on the accuracy of the instruments used in measurement and the expertise of the individual in using the instrument or in making qualitative assessments. CLIMATOLOGY requires long records of daily data at the same place that have been obtained using standard instruments to produce a TIME SERIES. The ideal is a HOMOGENEOUS SERIES, but this is not always possible and a variety of statistical techniques have been developed in the last few decades to make allowances for site changes, breaks in the data series and to produce regional values. The same is true of other NATURAL ENVIRONMENTAL SCIENCES, each of which has its own criteria for the production and VALIDATION of its observational data. *BDG*

Linacre, E. 1992: *Climate data and resources.* London: Routledge. **Strangeways, I.** 1997: Ground and remotely sensed measurements. In Thompson, R. D. and Perry, A. (eds), *Applied climatology: principles and practice.* London: Routledge, 13–21.

obsidian hydration dating A technique using the rate and extent of chemical alteration of obsidian surfaces for dating purposes. Obsidian is a glassy product of VOLCANICITY, formed by the rapid cooling of silica-rich LAVAS. On exposure to water in air or surrounding SOIL, a fresh obsidian surface reacts by HYDRATION to form a rind, the thickness of which is a non-linear function of time, temperature and chemical composition. CALIBRATED-AGE DATING using this technique is possible if (1) independent evidence can be found based on NUMERICAL-AGE DATING methods such as RADIOCARBON DATING or POTASSIUM–ARGON DATING; or (2) hydration rate can be determined in the laboratory by heating experiments at different relative humidities. The technique is most widely used in archaeological studies because obsidian was widely traded in prehistoric times, although the timing of glacial events can be determined where glacial abrasion has created fresh surfaces of obsidian in GLACIAL DEPOSITS. *DAR*

[See also WEATHERING RIND]

Beck, C. and Jones, G.T. 1994: Dating surface assemblages using obsidian hydration. In Beck, C. (ed.), *Dating in exposed and surface contexts.* Albuquerque, NM: University of New Mexico Press, 47–76. **Friedman, I. and Long, W.** 1970: Hydration rate of obsidian. *Science* 191, 347–352. **Friedman, I., Trembour, F. and Hughes, R.E.** 1997: Obsidian hydration dating. In Taylor, R.E. and Aitken, M.J. (eds), *Chronometric dating in archaeology.* New York: Plenum, 297–321. **Friedman, I., Trembour, F., Smith, F.L. and Smith, G.I.** 1994: Is obsidian hydration dating affected by relative humidity? *Quaternary Research* 41, 185–190.

occult deposition Used in the context of deposition from the atmosphere of POLLUTANTS, especially ACID PRECIPITATION, it is the material deposited by INTERCEPTION. For example, hoar frost and fog droplets, which may be many times more polluted than rain, can accumulate on vegetation. In some circumstances, occult deposition combined with DRY DEPOSITION may deliver more pollutants to the Earth's surface than WET DEPOSITION. *JAM*

occupation layer A buried horizon at or near an archaeological site with evidence of human occupation. The term is widely used, not only for levels that were occupied in the strict sense (e.g. former floors or hearths and the remains of collapsed buildings), but also for various types of dumps for refuse (see MIDDEN) and the soils modified by the spreading of refuse on fields as manure (see ANTHROPEDOGENIC HORIZONS). Like *ditches* and *pits*, occupation layers tend to be rich in organic material and ARTEFACTS. *JAM*

[See also TELL]

Davidson, D.A. 1976: Processes of tell formation and erosion. In Davidson, D.A. and Shackley, M.L. (eds), *Geoarchaeology: Earth science and the past*. London: Duckworth, 255–266. **Davidson, D.A. and Simpson, I.A.** 1984: The formation of deep topsoils in Orkney. *Earth Surface Processes and Landforms* 9, 75–81. **Matthews, W.** 1995: Micromorphological characterisation and interpretation of occupation deposits and microstratigraphical sequences at Abu Salabikh, southern Iraq. In Barham, A.J. and Macphail, R.I. (eds), *Archaeological sediments and soils*. London: UCL Press, 41–74. **Needham, S. and Spence T.** 1997: Refuse and the formation of middens. *Antiquity* 71, 77–90.

ocean A continuous body of sea-water, much of it deeper than 1000 m, covering much of the Earth's surface, particularly the OCEAN BASINS, where the ocean is underlain by OCEANIC CRUST. The Earth's oceans are the *Arctic, Atlantic, Indian* and *Pacific*, with many people considering the *Southern (Antarctic) Ocean* as a fifth. Ocean water is salty because of EVAPORATION. Oceans and their fringing SEAS cover about 71% of the Earth's surface (total area about 360 million km^2), with an estimated volume of about 1.4 billion km^3 and an average depth of about 3500 m. The Pacific Ocean accounts for almost 40% of the Earth's total sea area and is the deepest ocean. It contains the deepest parts of the oceans, in the OCEANIC TRENCHES (Mindanao Trench, 11 524 m; Mariana Trench, 11 022 m). Oceans and SEAS account for about 98% by volume of the Earth's surface waters, the remainder being made up of ice (1.7%) and lakes and rivers (<1%). *GO*

[See also CONTINENTAL MARGIN, MID-OCEAN RIDGE, OCEANOGRAPHY, SEA WATER COMPOSITION]

Couper, A. (ed.) 1983: *The Times atlas of the oceans*. London: Times Books. **Duxbury, A.C. and Duxbury, A.B.** 1994: *An introduction to the world's oceans*, 4th edn. Dubuque, IA: Wm. C. Brown. **Hedgpeth, J.W.** 1957: *Classification of marine environments* [*Memoir* 671]. Boulder, CO: Geological Society of America, 17–280.

ocean basin OCEAN basins originate when extension across a RIFT VALLEY on a CONTINENT allows MAGMA from the ASTHENOSPHERE to rise and form new LITHOSPHERE bearing OCEANIC CRUST. The basin expands through the process of SEA-FLOOR SPREADING, a fundamental feature of PLATE TECTONICS. Young and mature ocean basins are exemplified by the Red Sea and the Atlantic Ocean respectively. The oldest oceanic crust in the Red Sea is about 5 million years old, while the oldest in the Atlantic Ocean is about 160 million years old. These ocean basins are characterised by a central MID-OCEAN RIDGE, where sea-floor spreading takes place, representing a CONSTRUCTIVE PLATE MARGIN (see Figure). On either side the morphology of the ocean basin is symmetrical with, in the mature ocean, an ABYSSAL plain passing into a PASSIVE CONTINENTAL MARGIN comprising CONTINENTAL RISE, CONTINENTAL SLOPE and CONTINENTAL SHELF. The true ocean–continent boundary lies close to the edge of the continental shelf (the *shelfbreak*), marking the boundary between oceanic crust and CONTINENTAL CRUST.

A DESTRUCTIVE PLATE MARGIN may develop in an ocean basin, causing oceanic crust to be destroyed at a *subduction zone*. This is marked by a deep OCEANIC TRENCH and either a volcanic ISLAND ARC or an ACTIVE CONTINENTAL MARGIN, depending on whether the subduction zone is situated within the ocean basin or at the edge of a continent. If the rate of destruction of oceanic crust exceeds its rate of formation, the ocean basin will contract (e.g. Pacific Ocean), leading ultimately to a COLLISION ZONE between continents, forming an OROGENIC BELT. The former ocean will be represented by a *suture zone*, along which fragments of oceanic crust may be preserved as *ophiolite complexes*. These simple patterns of ocean basin morphology are complicated by the presence of oceanic islands, SEAMOUNTS, GUYOTS and ATOLLS, many of which represent VOLCANOES situated at HOTSPOTS.

The different morphological elements of an ocean basin are characterised by distinct ranges of water depths, ranging from HADAL (the deepest, corresponding to trenches) through abyssal to BATHYAL (corresponding to the conti-

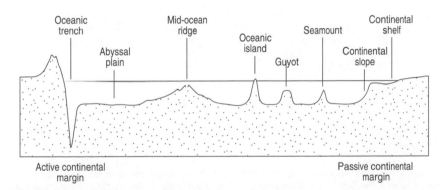

ocean basin *Cross-section of an idealised ocean basin showing the principal morphological features*

nental rise and slope). Much of the knowledge about the ocean basins has been gained since the mid-twentieth century using techniques such as DEEP-SEA DRILLING, SIDE-SCAN SONAR and SEISMIC REFLECTION SURVEYING.   *GO*

[See also OCEAN CURRENTS, THERMOHALINE CIRCULATION, WATER MASSES, WILSON CYCLE]

Menard, H.W. (ed.) 1977: *Ocean science.* San Francisco, CA: Freeman. **Open University Course Team** 1998: *The ocean basins: their structure and evolution,* 2nd edn. Oxford: Butterworth-Heinemann.

ocean currents

The surface currents of the oceans (see Figure), which move at speeds from several kilometres per day to several kilometres per hour, are driven by the surface wind systems and deflected by the distribution of continents. They form semi-closed circular patterns (GYRES), which move clockwise in the Northern Hemisphere and anticlockwise in the Southern Hemisphere. Predominantly easterly currents near the equator are driven by the TRADE WINDS, whereas the predominant westerly currents of higher latitudes are driven by the WESTERLIES. The mid-latitude gyres are centred over the subtropical high-pressure zones (see warm ANTI-CYCLONES). Depending on the relative water temperature, the surface currents may be classified as *warm* or *cold currents,* which move away from or towards the equator, respectively, and are important in the redistribution of heat in the Earth's atmosphere–ocean system. Below the surface of ocean, low 'friction coupling' and the rotation of the Earth cause a reduction in the velocity of the currents and their deflection to the right in the Northern Hemisphere and to the left in the Southern Hemisphere (see CORIOLOS FORCE), resulting in an *Ekman spiral,* which is important in explaining UPWELLING off coasts where a current is moving parallel to a coastline and towards the equator. Changes in ocean currents also reflect and cause CLIMATIC CHANGE.   *JAM*

[See also ATMOSPHERE–OCEAN INTERACTION, EKMAN MOTION, GULF STREAM, EL NIÑO, THERMOHALINE CIRCULATION]

Binkley, M.S. 1996: Oceans. In Schneider, S.H. (ed.) *Encyclopedia of climate and weather.* Oxford: Oxford University Press, 547–552. **Niiler, R.P.** 1992: The ocean circulation. In Trenberth, K.E. (ed.), *Climate system modelling.* Cambridge: Cambridge University Press, 117–148.

oceanic anoxic event

An episode of abnormally low dissolved oxygen in the OCEANS, leading to the widespread preservation of laminated, organic-rich BLACK SHALES or SAPROPELS. Factors favouring oceanic anoxic events are those associated with generally warmer ocean waters during periods of global warmth, or GREENHOUSE CONDITIONS. They include slow renewal of oxygen in bottom waters because of the absence of ICE SHEETS (see WATER MASSES); reduced levels of dissolved oxygen; increased organic PRO-DUCTIVITY; and marine TRANSGRESSION leading to the flooding of CONTINENTAL SHELF areas. Oceanic anoxic events are known from the mid-Cambrian, early mid-Ordovician, early Silurian, late Devonian, early Carboniferous, early Jurassic, late Jurassic, mid-Cretaceous and late Cretaceous.   *GO*

[See also PALAEOCEANOGRAPHY]

Schlanger, S.O. and Jenkyns, H.C. 1976: Cretaceous oceanic anoxic events: causes and consequences. *Geologie en Mijnbouw* 55, 179–184. **Stow, D.A.V., Reading, H.G. and Collinson, J.D.** 1996: Deep seas. In Reading, H.G. (ed.), *Sedimentary environments: processes, facies and stratigraphy,* 3rd edn. Oxford: Blackwell Science, 395–453.

oceanic crust

The outermost layer of the Earth that underlies the OCEANS (see Figure). Compared with CONTINENTAL CRUST, oceanic crust is relatively thin (6–8 km), dense (typically 3000 kg m^{-3}), basaltic in composition and young: because of the recycling of oceanic crust through its formation by VOLCANISM and igneous INTRUSION at CONSTRUCTIVE PLATE MARGINS and its destruction at

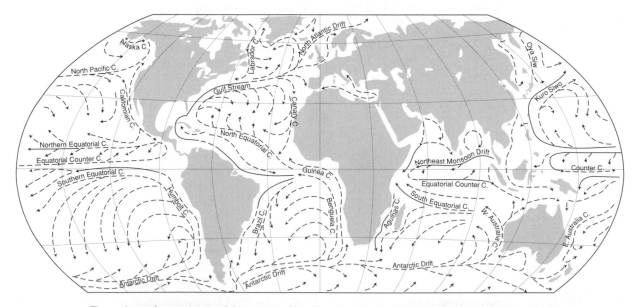

ocean currents *The major surface currents of the oceans. Note that there is seasonal variation in the positions of the currents, especially in the Indian Ocean, where the Southwest Monsoon Drift is characteristic in the Northern Hemisphere summer (after Binkley, 1996)*

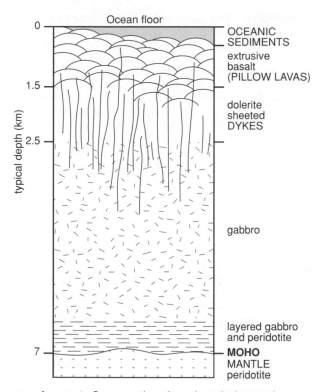

oceanic crust *Cross-section through typical oceanic crust, based on information from deep-sea drilling, seismic surveying and the study of ophiolite complexes (after Tucholke, 1998)*

DESTRUCTIVE PLATE MARGINS, there is no *in situ* oceanic crust older than 200 million years. Older oceanic crust is preserved as *ophiolite complexes* in OROGENIC BELTS. The structure and composition of oceanic crust are known from studies of ophiolite complexes, DEEP-SEA DRILLING and SEISMIC exploration. *GO*

[See also EARTH STRUCTURE]

Gass, I.G., Lippard, S.J. and Shelton, A.W. (eds) 1984: *Ophiolites and oceanic lithosphere* [*Special Publication* 13]. London: Geological Society. **Tucholke, B.E.** 1998: Discovery of 'mega-mullions' reveals gateways into the ocean crust and upper mantle. *Oceanus: reports on research from the Woods Hole Oceanographic Institution* 41(1), 15–19.

Oceanic Polar Front (OPF)

The oceanographic boundary separating warm water of high SALINITY flowing polewards from cold water of low salinity flowing towards lower latitudes. In the Southern Ocean it is defined by the northern limit of the *Antarctic Circumpolar Current* (ACC). It is approximately defined by the winter SEA ICE extent in the Arctic Ocean and is maintained by buoyant freshwater influx. It should be distinguished from the oceanic ARCTIC FRONT, which corresponds with the limit of summer sea-ice. Changes in the location of this boundary have been reconstructed on the basis of the analysis of MARINE SEDIMENTS for the period from the LAST GLACIAL MAXIMUM to the HOLOCENE (see Figure). *WENA*

[See also OCEANOGRAPHY, POLAR FRONT]

Koc, N., Jansen, E. and Haflidason, H. 1993. Paleoceanographic reconstructions of surface ocean conditions in Greenland, Iceland and Norwegian seas through the last 14 ka based on diatoms. *Quaternary Science Reviews* 12(2), 115–140. **Ruddiman, W.F. and McIntyre, A.** 1981: The North Atlantic Ocean during the last deglaciation. *Palaeogeography, Palaeoclimatology and Palaeoecology* 35, 145–214.

oceanic sediments SEDIMENTS that accumulate on the deep OCEAN floor, far from land, including PELAGIC SEDIMENT and redistributed material such as the deposits

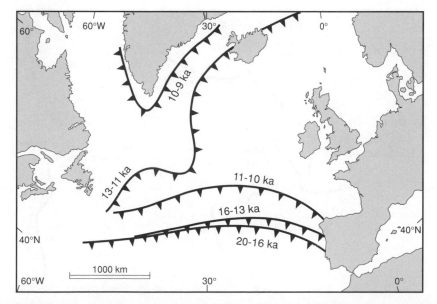

Oceanic Polar Front *Variations in the position of the Oceanic Polar Front between the Last Glacial Maximum (20–16 ka) and the early Holocene (10–9 ka), including its retreat to Iceland during the Lateglacial Interstadial (Allerød; 13–11 ka) and its re-advance to the latitude of northern Spain in the Lateglacial Stadial (Younger Dryas; 11–10 ka): dates are in radiocarbon years before present (after an early significant reconstruction by Ruddiman and McIntyre, 1981). A more recent and detailed reconstruction of palaeoceanographic conditions in the North Atlantic can be found in Koc et al. (1993)*

of TURBIDITY CURRENTS (TURBIDITES) and other forms of submarine MASS MOVEMENTS. DEEP-SEA DRILLING has demonstrated that the thickness of oceanic sediments on igneous OCEANIC CRUST increases with distance from the central axes of the MID-OCEAN RIDGES, providing evidence in support of SEA-FLOOR SPREADING and PLATE TECTONICS. Oceanic sediments hold an important record of environmental change. *GO*

[See also MARINE SEDIMENT CORES, MARINE SEDIMENTS]

Burckle, L.H., Mortlock, R. and Rudolph, S. 1996: No evidence for extreme, long term warming in early Pliocene sediments of the Southern ocean. *Marine Micropaleontology* 27, 215–226. **Jakobsson, M., Lovlie, R., Al-Hanbali, H., Arnold, E.** *et al.* 2000: Manganese and color cycles in Arctic Ocean sediments constrain Pleistocene chronology. *Geology* 28, 23–26. **Sonzogni, C., Bard, E. and Rostek, F.** 1998: Tropical sea-surface temperatures during the last glacial period: a view based on alkenones in Indian Ocean sediments. *Quaternary Science Reviews* 17, 1185–1201.

oceanic trench   A narrow, deep trough in the OCEAN floor. An oceanic trench, *deep ocean trench* or *submarine trench* is the surface expression of SUBDUCTION at a DESTRUCTIVE PLATE MARGIN and is associated with VOLCANISM and EARTHQUAKES. Oceanic trenches occur either as part of an ACTIVE CONTINENTAL MARGIN or within the ocean, with a volcanic ISLAND ARC on the over-riding plate. Trenches may be partly or completely filled by OCEANIC SEDIMENTS such as TURBIDITES (see ACCRETIONARY COMPLEX). They are particularly well developed close to the Pacific Ocean rim ('*Ring of Fire*') and include the deepest parts of the ocean such as the Mindanao Trench at 11 524 m (see Figure). *GO*

Decker, R. and Decker, B. 1997: *Volcanoes*, 3rd edn. Basingstoke: Freeman.

oceanicity   The extent to which the climate of a location is influenced by maritime influences: the equable character of climates and the converse of CONTINENTALITY. Indices of oceanicity attempt to quantify this, usually based on the annual temperature range. *Oceanic climates* (*maritime climates*) exhibit high oceanicity: these include oceanic islands and the western sides of continents in mid-latitudes, which experience the WESTERLIES throughout the year. In regions of low relief, high oceanicity may extend hundreds of kilometres inland. *JAM*

Oliver, J.E. 1996: Maritime climate. In Schneider, S.H. (ed.), *Encyclopedia of climate and weather*. New York: Oxford University Press, 491–496.

oceanography   The scientific study of the world's OCEANS. In recent years, the interdisciplinary nature of oceanography, which involves MARINE GEOLOGY, physics, chemistry, marine biology (biological oceanography) and METEOROLOGY, has alerted scientists to the complex and sensitive inter-relationships amongst the ATMOSPHERE, BIOSPHERE and oceans. Many pressing issues on the environmental agenda, including aspects of CLIMATIC CHANGE, CARBON DIOXIDE emissions, polar ICE CAP melting, POLLUTION and RADIOACTIVE WASTE disposal, involve studies of the oceans. *BTC*

[See also MARINE POLLUTION, PALAEOCEANOGRAPHY, PHYSICAL OCEANOGRAPHY]

Broecker, W. 1974: *Chemical oceanography*. New York: Harcourt Brace Jovanovich. **Garrison, T.** 1999: *Oceanography: an invitation to marine science*. Belmont, CA: Wadsworth. **Gross, M.G.** 1992: *Oceanography: a view of the Earth*, 6th edn. Englewood Cliffs, NJ: Simon and Schuster. **Moore, J.R. (ed.)** 1971: *Oceanography: readings from Scientific American*. San Francisco, CA: Freeman. **Send, U., Font, J., Krahmann, G.** *et*

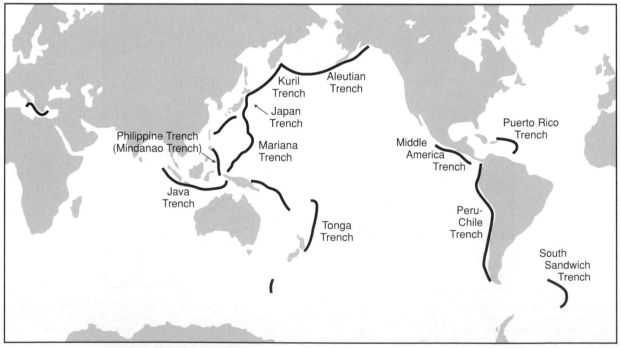

oceanic trench *The world's deep oceanic trenches (based on Decker and Decker, 1997)*

al. 1999: Recent advances in observing the physical oceanography of the western Mediterranean Sea. *Progress in Oceanography* 44, 37–64. **Summerhayes, C.P. and Thorpe, S.A.** 1996: *Oceanography: an illustrated guide.* London: Manson. **Thurman, H.V.** 1993: *Essentials of oceanography,* 4th edn. New York: Macmillan.

ochric horizon A shallow, pale coloured, surface horizon with less than 1% organic matter commonly found in REGOSOLS, ARENOSOLS and LEPTOSOLS where organic matter production is low. *EMB*

Food and Agriculture Organization of the United Nations (FAO) 1998: *World reference base for soil resources* [*Soil Resource Report* 84]. Rome: FAO, ISRIC, ISSS.

oil fire The COMBUSTION of oil is a source of ATMOSPHERIC POLLUTION and a potential ENVIRONMENTAL DISASTER. Oil fires in Kuwait during the Gulf War, 1990, which resulted from the ignition of over 700 oil wells, probably introduced more anthropogenic pollutants into the atmosphere than any other single event. *JAM*

El-Baz, F. 1992: The war for oil: effects on the land, air and sea. *Geotimes* 37, 12–15.

oil shale A BLACK SHALE or fossil SAPROPEL that is rich in carbon. Oil shales are the source rocks for conventional PETROLEUM accumulations, releasing HYDROCARBONS upon burial and heating. Oil shales at the Earth's surface that have not released their hydrocarbons can be mined as an ENERGY RESOURCE or FOSSIL FUEL (SYNFUEL), but they must be processed to yield petroleum. Their exploitation would involve large-scale opencast mining leading to ENVIRONMENTAL DEGRADATION and is not currently viable. Oil shale was worked in central Scotland from AD 1850 to 1964 and many of the spoil tips are prominent features in the landscape today. Potential reserves of oil shale exist in Colorado, Brazil and Australia. It has been estimated that the exploitation of oil shales could yield more energy than the total obtained from CRUDE OIL. *GO*

Jaber, J.O. and Probert, S.D. 1997: Exploitation of Jordanian oil-shales. *Applied Energy* 58, 161–175. **Kleinbach, M.H. and Salvagin, C.E.** 1986: *Energy technologies and conversion systems.* Englewood Cliffs, NJ: Prentice-Hall.

oil spill A leakage of oil into the marine environment: an ENVIRONMENTAL ACCIDENT. Although more oil reaches the marine environment from land-based discharges, and the largest single oil spill yet recorded (at least 400 000 tonnes) was at an offshore exploratory rig in the Gulf of Mexico, the term is normally associated with tanker accidents at sea, such as the wrecks of the *Amoco Cadiz* (220 000 tonnes) in the English Channel, the *Braer* (80 000 tonnes) off the Shetland Islands, Scotland and the *Exxon Valdez* (40 000 tonnes) in Prince William Sound, Alaska. The environmental impact depends on many factors including the properties of the oil, the characteristics of the wind, waves and currents, the sea temperature and salinity, and the air temperature. Damage tends to be intensive in the short term but not long lasting because of natural degradation ('weathering') of the oil and the capacity of wildlife to re-colonise. RESTORATION methods include chemical-assisted dispersion, containment booms, mechanical recovery, burning and BIOTECHNOLOGY (BIOREMEDIATION using oil-degrading bacteria). *JAM*

[See also MARINE POLLUTION]

Parker, J.G. 1998: Oil spill response. In Brune, D., Chapman, D.V., Gwynne, M.D. and Pacyna, J.M. (eds), *The global environment: science, technology and management.* Weinheim, Germany: VCH, 955–972. **Ritchie, W. and O'Sullivan, M.** 1994: The environmental impact of the wreck of the Braer. Edinburgh: The Scottish Office. **Teal, J.M. and Howarth, R.W.** 1984: Oil spill studies; a review of ecological effects. *Environmental Management* 8, 27–44.

Older Drift A term used in earlier literature to distinguish the relatively weathered glacial deposits produced by pre-LAST GLACIAL MAXIMUM glacial advances from the less weathered *Newer Drift* deposited during the last glacial maximum. *MHD*

[See also DRIFT]

Older Dryas Stadial A brief period of relatively cold conditions in Europe during the Late Glacial, following the BØLLING INTERSTADIAL and preceding the ALLERØD INTERSTADIAL. Evidence for this can be found from Norway to Spain. It is dated to between c. 12 and 11.8 ka BP (^{14}C years) and may be one of up to four comparable periods preceding the YOUNGER DRYAS STADIAL. *MHD*

[See also LATE GLACIAL ENVIRONMENTAL CHANGE]

Lowe, J.J., Ammann, B., Birks, H.H. *et al.* 1994: Climatic changes in areas adjacent to the North Atlantic during the last glacial–interglacial transition (14–9 ka BP). *Journal of Quaternary Science* 9, 185–198. **Mangerud, J., Andersen, S.T., Berglund, B.E. and Donner, J.J.** 1974: Quaternary stratigraphy of Norden: a proposal for terminology and classification. *Boreas* 10, 109–127.

old-growth forest An imprecise term emphasising the large, old trees in forests that have reached a mature stage of development and are particularly worthy of CONSERVATION for their economic and/or cultural value. *JAM*

[See also ANCIENT WOODLAND, PRIMARY FOREST]

Gray, A.N. 1998: Old growth/ancient forests, conservation. In Calow, P. (ed.), *The encyclopedia of ecology and environmental management.* Oxford: Blackwell Science, 485–497.

Olduvai event A short interval of NORMAL POLARITY, when there was a return to normal polarity in the Earth's magnetic field within the longer *Matuyama chron* characterised by reverse polarity. It has been dated by POTASSIUM–ARGON (K–Ar) DATING to between 1.87 and 1.67 million years ago, or 1.95 to 1.79 Ma BP based upon ORBITAL TUNING of the OXYGEN ISOTOPE record of MARINE SEDIMENTS. As it lies close to the traditional PLIOCENE–PLEISTOCENE TRANSITION, it has been an important chronological MARKER HORIZON in the study of QUATERNARY ENVIRONMENTAL CHANGE. *DH*

[See also GEOMAGNETIC POLARITY TIMESCALE]

Hilgen, F.J. 1991: Astronomical calibration of Gauss to Matuyama sapropels of the Mediterranean and implications for the geomagnetic polarity time scale. *Earth and Planetary Science Letters* 104, 226–244.

Oligocene An EPOCH of the TERTIARY period, from 35.4 to 23.3 million years ago. *GO*

[See also GEOLOGICAL TIMESCALE]

oligotrophic Nutrient-poor. The term is commonly used in relation to SOILS, surface waters (especially LAKES) and WETLANDS of low NUTRIENT STATUS. *JAM*

[See also EUTROPHIC]

oligotrophication Nutrient depletion over time. The term signifies the opposite of EUTROPHICATION and is used especially in the context of the NUTRIENT STATUS of water bodies such as lakes. *JAM*

Burkholder, J.M. 2001: Eutrophication and oligotrophication. In Levin, S.A. (ed.), *Encyclopedia of biodiversity*. Vol. 2. San Diego, CA: Academic Press, 649–670.

ombrotrophic mire A MIRE, the only water source for which is from PRECIPITATION and as a result the water is mineral poor or OLIGOTROPHIC. *MJB*

[See also RAISED MIRE]

ontogeny/ontogenesis The development of an individual organism from its earliest fertilised form (*zygote*) through to maturity. *KJW*

ontology The branch of METAPHYSICS concerned with the study of existence itself and its relationship to human consciousness. Ontology is concerned with what exists, how it does so and what can be known. It is usually contrasted with EPISTEMOLOGY. *ART*

Schatzki, T. 1991: Spatial ontology and explanation. *Annals of the Association of American Geographers* 81, 650–670.

Oort minimum The minimum in solar SUNSPOT activity between AD 1010 and 1050. *JAM*

ooze Fine-grained PELAGIC SEDIMENT (MUD) dominantly of biogenic origin. *Carbonate* or *calcareous ooze* includes *pteropod ooze*, found in water depths ranging from 1000 to 2500 m, and *globigerinid ooze* in water depths from 2000 to 4000 m. One-third of the ocean floor is covered with globigerinid ooze. *Siliceous ooze,* composed of organisms with tests of opaline silica, includes *diatom ooze*, found between 1100 and 4000 m in Antarctic regions and in low-latitude parts of the Pacific, and *radiolarian ooze* in water depths exceeding 5000 m in isolated parts of the equatorial Pacific and Indian Oceans. The distribution of calcareous and siliceous oozes is controlled by organic PRODUCTIVITY and water depth. Calcareous ooze is today found only in shallow, warm oceans, as most ocean basins have average depths of 4 to 5 km, which is below the CARBONATE COMPENSATION DEPTH. *BTC*

Boden, P. and Backman, J. 1996: A laminated sediment sequence from the northern North Atlantic Ocean and its climatic record. *Geology* 24, 507–510. Dickens, G.R. and Barron, J.A. 1997: A rapidly deposited pennate diatom ooze in Upper Miocene Lower Pliocene sediment beneath the North Pacific polar front. *Marine Micropaleontology* 31, 177–182. Kemp, A.E.S. and Baldauf, J.G. 1993: Vast Neogene laminated diatom mat deposits from the eastern equatorial Pacific Ocean. *Nature* 362, 141–144. Sigman, D.M., Altabet, M.A., Francois, R. et al.1999: The isotopic composition of diatombound nitrogen in Southern Ocean sediments. *Paleoceanography* 14, 118–134.

open system A SYSTEM that may exchange both energy and matter with its surroundings (environment);

consequently, the most realistic simplification of the real world. *CET*

Huggett, R.J. 1985: *Earth surface systems*. Berlin: Springer.

optical emission spectroscopy (OES) A technique, also known as *atomic emission spectroscopy (AES)*, used for elemental analysis in GEOCHEMISTRY and related fields. It involves analysis of the optical emission spectrum (170–780 nm wavelength) of a vapourised sample. The vapour may be held in various sources including the relatively low temperature of a flame (in *flame emission spectroscopy)* or, in more recent approaches, at much higher temperatures in a plasma (see INDUCTIVELY COUPLED PLASMA ATOMIC EMISSION SPECTROMETRY). *TY*

[See also ATOMIC ABSORPTION SPECTROPHOTOMETRY]

optical remote sensing instruments Instruments that record reflected radiation at wavelengths between 300 and 5000 nm. This region of the ELECTROMAGNETIC SPECTRUM is characterised by absorptions due to electron transitions and changes in molecular rotation and vibration within molecules. These instruments cover the wavelengths visible to the human eye and overlap at longer wavelengths with emitted radiation from high-temperature surfaces recorded by THERMAL REMOTE SENSING. The instruments that are available for optical REMOTE SENSING range from simple *cameras* to complex *hyper spectral digital scanners*.

The cameras used for AERIAL PHOTOGRAPHY were the first and simplest REMOTE SENSING instruments. They have a similar design to conventional cameras with a lens system focusing radiation from the surface on to a piece of film in the focal plane. The radiation causes a chemical response in the emulsion of the film and thus records the image. *Black-and-white films* have a single layer of emulsion, while *colour films* have three layers of emulsion separated by filters so that each layer records a different wavelength range.

The first electronic REMOTE SENSING instruments were the *return beam vidicon* (RBV) cameras, used on early METEOROLOGICAL SATELLITES (Nimbus series) and Earth resources satellites (*LANDSAT* 1–3).

Electro-optical instruments, or *multispectral scanners*, use solid-state detectors to convert radiation from the surface to electrical energy in a manner similar to *radiometers*. They build up RASTER images from lines of discrete measurements or PIXELS. Radiation from the surface is focused onto a mirror and then split up into narrow wavebands by a diffraction grating or a set of filters. The brightness of the radiation in each waveband is measured by a detector. *Whiskbroom scanners*, such as the *AVIRIS (Airborne Visible-InfraRed Imaging Spectrometer), TM (Thematic Mapper), AVHRR (Advanced Very High Resolution Radiometer) and MSS (Multispectral Scanner System)*, measure one pixel at a time with linear *photo diode arrays*. *Pushbroom scanners*, such as the *CASI (Compact Airborne Spectrographic Imager), AIS (Airborne Imaging Spectrometer) and HRV (High Resolution Visible)*, measure a line of pixels at a time with a two dimensional array of detectors or *CCD (charge coupled device)*.

Digital cameras are now in use where the film at the focal plane has been replaced by a large CCD; AERIAL PHOTOGRAPHY can therefore be acquired directly in digital form. *GMS*

442

Schowengerdt, R.A. 1997: *Remote sensing: models and methods for image processing.* New York: Academic Press. **Elachi, C.** 1987: *Introduction to the physics and techniques of remote sensing.* New York: Wiley. **Ryerson, R.A.** 1998: *Manual of remote sensing.* New York: Wiley.

optically stimulated luminescence (OSL) dating

A form of LUMINESCENCE DATING utilising the emission of luminescence from visible-light-sensitive ELECTRON TRAPS to assess time elapsed since deposition of mineral grains. It is particularly valuable because it utilises QUARTZ and allows improved measurement precision over THERMOLU-MINESCENCE DATING. *DAR*

[See also INFRARED-STIMULATED LUMINESCENCE DATING]

Haskell, E., Difley, R., Kenner, G. *et al.* 1999: A comparison of optically stimulated luminescence dating methods applied to eolian sands from the Mojave Desert in Southern Nevada. *Quaternary Science Reviews* 18, 235–242. **Huntley, D.J., Godfrey-Smith, D.I. and Thewalt, M.L.W.** 1995: Optical dating of sediments. *Nature* 313, 105–107.

optimisation

A process of searching for the 'best' solution to a problem. The search requires the *maximisation* or *minimisation* of a mathematical function (known as the *objective function*), which itself encodes what is meant by 'best' and is often subject to constraints that an acceptable solution must satisfy. *PJS*

[See also EARTH-SYSTEM ANALYSIS]

orbital forcing

The effect of the Earth's orbital parameters (see ECCENTRICITY, OBLIQUITY, PRECESSION) on the Earth's climate. In particular, orbital forcing is the main cause of GLACIAL–INTERGLACIAL CYCLES. *JAM*

[See also FORCING FACTOR, MILANKOVITCH THEORY]

orbital tuning

A method that allows global correlation by applying MILANKOVITCH THEORY to age-model development and palaeoclimatic research. It was originally developed around the observed synchroneity of changes in the OXYGEN ISOTOPE composition of FORAMINIFERA from MARINE SEDIMENT records throughout the world's oceans. Orbital tuning is based upon the assumption that variations in the isotopic composition of the oceans result from changes in global sea-ice volume that were driven by changes in solar insolation arising from the Milankovitch PERIODICITIES. These are recorded in marine sediments and can be examined using TIME-SERIES ANALYSIS, such as SPECTRAL ANALYSIS. The SPECTRAL MAPPING PROJECT TIMESCALE (SPECMAP) provides a global reference composite curve, to which other records can be compared. The method makes several assumptions, most importantly that the degree of coherency between the Milankovitch rhythms and the climate system depends on the degree of linearity of the ice volume response to external forcing. *WENA*

Ding, Z., Yu, Z., Rutter, N.W. and Liu, T. 1994: Towards an orbital timescale for Chinese loess deposits. *Quaternary Science Reviews* 13, 39–70. **Hays, J.D., Imbrie, J. and Shackleton, N.J.** 1976: Variations in earth's orbit: pacemaker of the ice ages. *Science* 194, 1121–1132. **Imbrie, J., Hays, J.D., Martinson, D.G.** *et al.* 1984: The orbital theory of Pleistocene climate: support from a revised chronology of marine δ¹⁸O record. In Berger, A. (ed.), *Milankovitch and climate.* Part I. Hingham, MA: Reidel, 269–305. **Marinson, D.G., Pisias, N.G., Hays, J.D.** *et al.*

1987: Age dating and the orbital theory of the Ice Ages: development of a high-resolution 0 to 30,000-year chronostratigraphy. *Quaternary Research* 27, 1–29. **Patience, A.J. and Kroon, D.** 1991: Oxygen isotope chronostratigraphy. In Smart, P.L. and Francis, P.D. (eds), *Quaternary dating methods* [*QRA Technical Guide* 4]. Cambridge: Quaternary Research Association, 199–228. **Shackleton, N.J.** 1967: Oxygen isotope analyses and Pleistocene temperatures re-assessed. *Nature* 215, 15–17.

ordinate

The scale value measured along the *y* axis of a Cartesian coordinate system from the origin of the system to the desired point location. *TF*

[See also CARTESIAN COORDINATES and ABSCISSA]

ordination

The arrangement of objects and/or variables along gradients on the basis of composition, occurrence or environmental preferences. The term comes from the German *Ordnung* and involves putting things into order. Ordination and CLASSIFICATION are the main approaches for the analysis of large, complex multivariate data-sets (STRUCTURING), particularly ecological and palaeoecological data and, if available, associated environmental data. Indirect ordination or GRADIENT ANALYSIS involves the biological data only and uses explicitly or implicitly a measure of the differences in composition between the objects or variables. Commonly used methods are PRINCIPAL COMPONENTS ANALYSIS and CORRESPONDENCE ANALYSIS to derive a small number of composite variables or ordination axes that represent as much of the information in the original data as possible according to stated mathematical criteria (*dimensionality* is reduced). A low (two- to three-) dimensional representation of the differences between objects and/or variables is produced as a means of summarising patterns in the data, investigating possible data structure and generating hypotheses about the environmental characteristics of the objects and possible underlying causal factors that may influence the observed patterns. Objects that are close to one another in an ordination diagram are inferred to resemble one another in terms of their composition; objects that are far apart are inferred to be dissimilar in composition; and variables that are close together are inferred to have similar environmental preferences. It is assumed that objects with similar composition reflect similar environments.

Direct ordination or GRADIENT ANALYSIS requires known environmental or other predictor variables as well as biological data. It involves modelling and summarising the responses of variables to environmental gradients by REGRESSION ANALYSIS (one response variable or species) or constrained or canonical ordination (CANONICAL CORRESPONDENCE ANALYSIS, REDUNDANCY ANALYSIS; two or more response variables). The latter simultaneously ordinates objects, variables and predictor environmental variables, thereby often simplifying the interpretation of the ordination results. Being primarily a correlative approach, ordination can aid in hypothesis generation but it can rarely demonstrate causality. *HJBB*

[See also MULTIDIMENSIONAL SCALING, NON-METRIC MULTIDIMENSIONAL SCALING, PRINCIPAL CO-ORDINATES ANALYSIS, FACTOR ANALYSIS, DISCRIMINANT ANALYSIS]

Braak, C.J.F. ter 1987: Ordination. In Jongman, R.H.G., ter Braak, C.J.F. and van Tongeren, O.F.R. (eds), *Data analysis in community and landscape ecology.* Wageningen: Pudoc, 78–173. **Braak, C.J.F. ter and Prentice, I.C.** 1988: A theory of gradient analysis. *Advances in Ecological Research* 18, 271–317. **Legendre, P. and Legendre, L.** 1998: *Numerical ecology*. Amsterdam: Elsevier.

Ordovician
A SYSTEM of rocks and a PERIOD of geological time from 495 to 442 million years ago. The late Ordovician (*Ashgill*), and possibly also the early Ordovician (*Tremadoc*), were times of GLACIATION. Land plants first evolved in the mid Ordovician. **GO**

[See also GEOLOGICAL TIMESCALE]

Eyles, N. and Young, G.M. 1994: Geodynamic controls on glaciation in Earth history. In Deynoux, M., Miller, J.M.G., Domack, E.W. *et al.* (eds), *Earth's glacial record*. Cambridge: Cambridge University Press, 1–28. **Marshall, J.D., Brenchley, P.J., Mason, P.** *et al.* 1997: Global carbon isotopic events associated with mass extinction and glaciation in the late Ordovician. *Palaeogeography, Palaeoclimatology, Palaeoecology* 132, 195–210.

ore
An aggregate of MINERALS from which one or more useful constituents, principally metals, can be extracted. The main ore of aluminium, for example, is BAUXITE. **GO**

[See also MINERAL RESOURCES, RESOURCE]

organic content
Organic matter is one of the major constituents of SOILS and is of vital importance for soil-living fauna and SOIL QUALITY. It is derived from the residues of the shoots and roots of plants that are gradually altered into HUMUS through physical fragmentation, faunal and microfaunal interactions, mineralisation and other processes of humus formation. In LAKE SEDIMENTS, the organic content reflects not only *in situ* production from plants and phytoplankton, but also inputs of eroded material (organic and mineral) from the catchment. Measurement of the organic content of soils and sediments is either by LOSS-ON-IGNITION or determination of *organic carbon* (OC) content by a wet oxidation technique. *Total organic content* (TOC) is commonly measured in the analysis of lake and MARINE SEDIMENT CORES. The terrestrial organic content of vegetation, soils and sediments is a major *carbon pool* that may be influenced by human activities to increase or decrease the atmospheric content of CARBON DIOXIDE, and hence the rate of GLOBAL WARMING. **EMB**

[See CARBON CONTENT]

Batjes, N.H. 1996: Total carbon and nitrogen in the soils of the world. *European Journal of Soil Science* 47: 151–163. **Eswaran, H., van den Bwerg, E. and Reich, P.** 1993: Organic carbon in soils of the world. *Soil Science Society of America Journal* 57, 192–194. **Post, W.M., Emmanuel, W.R., Zinke, P.J. and Stangenberger, G.** 1982: Soil carbon pools and world life zones. *Nature* 298, 1586–1589. **Tinsley, J.** 1950: The determination of organic carbon in soils by dichromate mixtures. *Transactions 4th International Congress of Soil Science* 1, 161–164.

organic farming
Farming that relies on alternatives to AGROCHEMICALS, although other non-polluting modern methods are often acceptable. The organic farming movement was founded in the AD 1930s. Lady Eve Balfour raised support for it in the UK in the 1940s and today the Soil Association (UK) supports and promotes it. A form of organic agriculture that seeks SUSTAINABILITY is *permaculture*. **CJB**

Conford, P. (ed.) 1992: *A future for the land: organic practice from a global perspective*. Bideford: Green Books.

organic sediment
A SEDIMENT or deposit containing plant and/or animal DETRITUS. **UBW**

[See also BIOGENIC SEDIMENT, LACUSTRINE SEDIMENTS, MINEROGENIC SEDIMENT, SEDIMENT TYPES]

organochlorides
Also known as *chlorinated hydrocarbons*, they are persistent, mobile, synthetic, organic PESTICIDES (e.g. DDT, dieldrin). They are highly soluble in fatty tissues and hence easily absorbed by a wide range of organisms, in addition to the main target species. **GOH**

[See also BIOLOGICAL MAGNIFICATION, HALOGENATED HYDROCARBONS, PERSISTENT ORGANIC COMPOUNDS]

organophosphates
Biodegradable, non-persistent, synthetic, organic PESTICIDES (e.g. parathion, malathion). They are not subject to BIOLOGICAL MAGNIFICATION, but are quick acting and extremely toxic (10–100 times more poisonous than ORGANOCHLORIDES) to mammals, birds and fish. **GOH**

[See also BIOLOGICAL CONTROL, PEST MANAGEMENT, PESTICIDES]

orogenesis
The process of mountain building on the Earth, associated with the DEFORMATION and UPLIFT of rocks (see TECTONICS) in an OROGENIC BELT during a period of OROGENY. Orogenesis occurs through convergent processes at DESTRUCTIVE PLATE MARGINS and continent–continent COLLISION ZONES (see PLATE TECTONICS), in contrast to EPEIROGENY. **GO**

Burg, J.-P. and Ford, M. (eds) 1997: *Orogeny through time* [*Special Publication* 121]. London: Geological Society. **Passchier, C.W.** 1995: Precambrian orogenesis: was it really different? *Geologie en Mijnbouw* 74, 141–150.

orogenic belt (orogen)
An approximately linear belt of rocks with a common history of SEDIMENTATION, DEFORMATION, VOLCANISM, igneous INTRUSION and metamorphism. Young orogenic belts comprise high mountains; older orogenic belts have undergone DENUDATION, exposing deformed and metamorphosed rocks (see GEOLOGICAL STRUCTURES, METAMORPHIC ROCKS). The classic PLATE TECTONICS model for the formation of an orogenic belt is at a continent–continent COLLISION ZONE (*collision orogen*, e.g. Himalayas, Alps). Orogenic belts can also be formed at an ocean–continent DESTRUCTIVE PLATE MARGIN, representing an ACTIVE CONTINENTAL MARGIN or *continental margin orogen* (e.g. Andes), or through TERRANE assembly (*accretionary orogen*, e.g. Rockies). These models relating OROGENY to plate tectonics have replaced GEOSYNCLINE theory. **GO**

Durand, B., Jolivet, L., Horváth, F and Séranne, M. (eds) 1999: *The Mediterranean basins: Tertiary extension within the Alpine orogen* [*Special Publication* 156]. London: Geological Society. **Harris, A.L. and Fettes, D.J. (eds)** 1987: *The Caledonian-Appalachian orogen*. [*Special Publication* 38]. London: Geological Society. **Treloar, P.J. and Searle, M.P. (eds)** 1993: *Himalayan tectonics* [*Special Publication* 74]. London: Geological Society.

orogeny
An episode of OROGENESIS (mountain building), giving rise to an OROGENIC BELT. Relatively recent episodes of orogeny are represented by high mountain

chains, such as the Himalayas (*Himalayan orogeny*, EOCENE collision), and the Alps (*Alpine orogeny*, EOCENE–OLIGOCENE deformation). The products of older episodes of orogeny have experienced DENUDATION and are represented by belts of deformed and metamorphosed rocks (see GEOLOGICAL STRUCTURES, METAMORPHIC ROCKS, THRUST FAULT). Examples include the *Hercynian* or *Variscan orogeny* (late PALAEOZOIC events in central Europe) and the *Caledonian orogeny* (early Palaeozoic events affecting northwest Europe and northeastern North America). *GO*

Prave, A.R., Kessler II, L.G., Malo, M. *et al.* 2000: Ordovician arc collision and foredeep evolution in the Gaspé Peninsula, Québec: the Taconic Orogeny in Canada and its bearing on the Grampian Orogeny in Scotland. *Journal of the Geological Society, London* 157, 393–400. **Windley, B.F.** 1995: *The evolving continents*, 3rd edn. Chichester: Wiley.

orographic Pertaining to the influence of topography, especially large-scale relief features such as mountains and hills, acting as barriers to air flow and affecting weather and climate. Orographic effects are evident from local scale to global scale in CYCLOGENESIS, LONG WAVES (Rossby waves), RAIN SHADOWS and strong, down-slope winds such as the *föhn*, *bora* and *mistral* (in Europe) and the *chinook* and *Santa Anna* (in North America). Orographic PRECIPITATION – a precipitation type generated by the forced uplift of air flow over a topographic barrier – tends to be concentrated on the windward side of mountains and uplands. *JAM*

[See also TOPOCLIMATE]

Barry, R.G. 1992: *Mountain weather and climate*, 2nd edn. London: Routledge. **Broccoli, A.J. and Manabe, S.** 1992: The effect of orography on midlatitude Northern Hemisphere dry climates. *Journal of Climate* 5, 1181–1201. **Ilide, R. and White, P.W. (eds)** 1981: *Orographic effects in planetary flows* [Global Atmospheric Research Programme Publication 23]. Geneva: World Meteorological Organization.

osmosis The DIFFUSION of a solvent through a semipermeable membrane, such as a cell wall, from a solution of relatively low concentration to one of relatively high concentration. The SOLUTES do not pass through the membrane. The process is of fundamental importance to the uptake of water by plants, where the concentration of salts (and hence '*osmotic pressure*') inside plant root cells is normally greater than in the surrounding soil water, and in *osmoregulation* of the water and salt content of organisms generally. *Reverse osmosis* is used in some DESALINATION plants and in the treatment of some wastewaters. *JAM*

osteology The study of bones. *JAM*

[See also ANIMAL REMAINS, PALAEODIET]

Gilbert, B.M. 1980: *Mammalian osteology*. Laramie, WY: B. Gilbert. **Goodman, A.H., Martin, D., Armelagos, G.J. and Clark, G.** 1984: Indications of stress from bones and teeth. In Cohen, M.N. and Armelagos, G.J. (eds), *Paleopathology at the origins of agriculture*. Orlando, FL: Academic Press. **Hesse, B. and Wapnish, P.** 1985: *Animal bone archeology: from objectives to analysis*. Washington, DC: Taraxacum. **Lambert, J.B. and Grupe, G. (eds)** 1993: *Prehistoric human bone: archaeology at the molecular level*. Berlin: Springer.

Ostracod analysis The study of a constituent of the *zooplankton* and zoobenthos (BENTHOS) in marine and terrestrial aquatic environments for BIOSTRATIGRAPHY and PALAEOENVIRONMENTAL RECONSTRUCTION. Ostracods (or ostracodes), a group of bivalved MICROFOSSILS, are crustaceans with an external calcitic or chitinous carapace (bivalved shell, typically 1 mm long, but ranging 0.3–30 mm in length), which have a well documented fossil record (early Cambrian to *Recent*). Parameters such as assemblage composition, abundance, distribution, preservation, ornamentation variation, shell shape (biometrics) and chemistry are used in their analysis. Ostracod studies are biased towards marginal to fully MARINE SEDIMENTS for reconstructing past ocean circulation and water-mass movement, SEA-SURFACE TEMPERATURES, SEA-LEVEL CHANGE, palaeobathymetry and palaeobiogeography (see BIOGEOGRAPHY). Non-marine ostracods in freshwater sediments, PEAT and soils are used for Quaternary (see QUATERNARY TIMESCALE) and HOLOCENE reconstructions of ALKALINITY and EUTROPHICATION, as well as lake-water circulation patterns and mixing (see HOLOMICTIC LAKE, MEROMICTIC LAKE). TRACE ELEMENT ratios (strontium, magnesium, calcium) and STABLE ISOTOPE ANALYSIS (oxygen, carbon) of carapaces are used as indicators of PALAEOSALINITY and lake-water temperature. *CES*

[See also MICROFOSSIL ANALYSIS].

De Deckker, P., Colin, J.P. and Peypouquet, J.P. (eds) 1988: *Ostracoda in the Earth sciences*. Amsterdam: Elsevier. **Holmes, J.A.** 1992: Nonmarine ostracods as Quaternary palaeoenvironmental indicators. *Progress in Physical Geography* 16, 405–431. **McKenzie, K.G. and Jones, P.J. (eds)** 1993: *Ostracoda in the Earth and Life sciences*. Rotterdam: Balkema. **Penney, D.N.** 1993: Northern North Sea benthic Ostracoda: modern distributions and palaeoenvironmental significance. *The Holocene* 3, 241–264. **Whatley, R.C.** 1993: Ostracoda as biostratigraphical indices in Cainozoic deep-sea sequences. In Hailwood, E.A. and Kidd, R.B. (eds), *High resolution stratigraphy* [Special Publication 70]. London: Geological Society, 155–167.

otolith A calcareous concretion from the ear of a vertebrate. Fish otoliths, also known as *ear stones*, can be indicative of the species, fish size and season caught. *JAM*

[See also SEASONALITY-OF-DEATH INDICATOR]

Mellars, P.A. and Wilkinson, M.R. 1980: Fish otoliths as indicators of seasonality in prehistoric shell middens: the evidence from Oronsay (Inner Hebrides). *Proceedings of the Prehistoric Society* 46, 19–44. **Ivany, L.C., Patterson, W.P. and Lohmann, K.C.** 2000: Cooler winters as a possible cause of mass extinction at the Eocene/Oligocene boundary. *Nature* 407, 887–890.

outbreak The occurrence of a large number of cases (in excess of BACKGROUND LEVELS) of a communicable disease in a community in a short period of time. Large-scale outbreaks, when many people in a wide geographic area are simultaneously affected, are termed EPIDEMICS. *MLW*

outgassing The release of VOLATILES (DEGASSING) from molten rocks through heating. Outgassing is believed to have been the main process in the formation of secondary planetary atmospheres. The ATMOSPHERE of the early Earth was affected by three phases of outgassing initiated by: (1) bombardment by METEORITES; (2) internal differentiation of rocks within the Earth; and (3) VOLCANISM. *JAM/GO*

[See also PRISCOAN]

Kasting, J.F. 1993: Earth's early atmosphere. *Science* 259, 920–926. **Wayne, R.P.** 1992: Atmospheric chemistry: the evolution of our atmosphere. *Journal of Photochemistry and Photobiology* 62, 379–396.

outlet glacier   A stream of fast-flowing ice discharging from an ICE SHEET or *ice cap* via a valley, or between areas of slow-moving ice as an *ice stream*.   *MJH*

[See also GLACIER]

Bickerton, R.J. and Matthews, J.A. 1993: 'Little Ice Age' variations of outlet glaciers from the Jostedalsbreen ice-cap, southern Norway: a regional lichenometric-dating study of ice-marginal moraine sequences and their climatic significance. *Journal of Quaternary Science* 8, 45–66.

outlier   (1) An area of younger rock completely surrounded by older, contrasting with the normal situation where a rock outcrop is bounded by older rocks on one side and younger on the other. Outliers can form as remnants left by EROSION or through DEFORMATION. (2) The term outlier is also used in NUMERICAL ANALYSIS for an anomalous data point, which appears not to belong to the same population as the remainder of the distribution.   *GO/JAM*

[See also INLIER]

outwash deposit   Where MELTWATER emerges at a glacier margin, the high rate of sediment input leads to the deposition of an *outwash plain* built of outwash deposits (see SANDUR). A reduction of sediment input causes downcutting and the original outwash plain forms an *outwash terrace*. Sequences of terraces and associated MORAINES in lowland valleys of rivers issuing from the Alps were the basis for the widely applied model of four Pleistocene glaciations put forward in the early twentieth century.   *JS*

[See also GLACIOFLUVIAL SEDIMENTS, GLACIOFLUVIAL LANDFORMS]

Nelson, A.R. and Shroba, R.R. 1998: Soil relative dating and outwash-terrace sequences in the northern part of the upper Arkansas Valley, central Colorado, USA. *Arctic and Alpine Research* 30, 349–361.

outwelling   The outflow of nutrients from WETLAND or ESTUARINE ENVIRONMENTS into seas of the CONTINENTAL SHELF.   *JAM*

[See also DOWNWELLING, UPWELLING]

overbank deposit   Flood sediments, usually silts and clays, but occasionally sands and gravels, deposited on river FLOODPLAINS during OVERBANK FLOW events. They tend to decrease in thickness and grain-size away from a channel.   *DML*

Walling, D.E. and He, Q. 1999: Changing rates of overbank sedimentation on the floodplains of British rivers during the past 100 years. In Brown, A.G. and Quine, T.A. (eds), *Fluvial processes and environmental change*. Chichester: Wiley, 207–222.

overbank flow   A river flow event that overtops the bank, thereby exceeding the BANKFULL stage and CHANNEL CAPACITY and spreading water across the FLOODPLAIN. Overbank flow can lead to OVERBANK DEPOSITS.   *DML*

[See also FLOOD]

Beven, K. and Carling, P.A. (eds) 1989: *Floods: hydrological, sedimentological and geomorphological implications*. Chichester: Wiley.

overcultivation   Changes in cultivation practices causing or likely to cause LAND DEGRADATION. In DRYLANDS, overcultivation is viewed as a classic cause of DESERTIFICATION. Changes may include the introduction of inappropriate cultivation techniques unsuited to the local conditions (e.g. inappropriate PLOUGHING or TILLAGE and reduction of FALLOW periods).   *RAS*

[See also OVERGRAZING]

overdeepening   The process whereby an eroding GLACIER or ICE SHEET excavates a preglacial river valley well below river BASE LEVEL. The long profiles of overdeepened valleys have steep gradients near their heads and gentler, sometimes reverse, slopes near their mouths. For example, in its overdeepened section, the Sognefjord glacial valley descends to 1308 m below sea level.   *JS*

[See also GLACIAL EROSION, U-SHAPED VALLEY]

Nesje, A., Dahl, S.O., Valen, V. and Ovstedal, J. 1992: Quaternary erosion in the Sognefjord drainage basin, western Norway. *Geomorphology* 5, 511–520.

overfishing   The reduction of fish stocks to numbers below the maximum SUSTAINABLE YIELD, that is the maximum number of fish available each year for HARVESTING without diminishing the longer-term stocks.   *JAM*

[See also FISHERIES CONSERVATION AND MANAGEMENT]

overgrazing   The rate of removal of vegetation by grazing and browsing animals that exceeds the capacity of the vegetation to recover. Consequently, there are changes in the vegetation community and typically there is soil exposure leading to SOIL EROSION. The combined impact of deteriorating pasture and TRAMPLING may reduce soil moisture and increase runoff. Overgrazing in some regions has contributed to DESERTIFICATION.   *IFS*

[See also CARRYING CAPACITY, SUSTAINABILITY, SUSTAINABLE YIELD]

Hodgson, J. and Illius, A.W. (eds) 1996: *The ecology and management of grazing systems*. Wallingford: CAB International. **Livingstone, L.** 1991; Livestock management and 'overgrazing' among pastoralists. *Ambio* 20, 80–85. **Mace, R.** 1991: Overgrazing overstated. *Nature* 349, 280–281. **Urbanska, K.M., Webb, N.R. and Edwards, P.J.** 1997: *Restoration ecology and sustainable development*. Cambridge: Cambridge University Press.

overhunting   Historically the leading cause of EXTINCTION, although now probably overtaken by HABITAT LOSS, the overexploitation of species may be for subsistence, economic or recreational purposes. Most oceanic fisheries, exploited for all three purposes, are now in steep decline and several have already collapsed, raising serious ecological, economic and food security issues. OVERFISHING rarely causes extinction of the target species, but secondary ecological consequences of massive reductions in fish abundance may be severe. While hunting by ABORIGINAL groups has popularly been viewed as far less detrimental than commercial hunting, fossil evidence suggests that traditional hunters have caused the extinction of thousands of species, including perhaps 20% of all birds, in Madagascar, North America, the Mediterranean and on many oceanic islands. Overhunting, coupled with other forms of human ENVIRONMENTAL DEGRADATION, continues to threaten many species in terrestrial, marine and aquatic ecosystems. *JTK*

[See also ECOSYSTEM COLLAPSE, FISHERIES CONSERVATION AND MANAGEMENT, SUSTAINABILITY, SUSTAINABLE YIELD, TRAGEDY OF THE COMMONS, WHALING AND SEALING]

Caughley, G. and Gunn, A. 1996: *Conservation biology in theory and practice.* Oxford: Blackwell Science. **Moulton, M.P. and Sanderson, J.** 1999: *Wildlife issues in a changing world,* 2nd edn. Boca Raton, FL: Lewis.

overkill hypothesis
As human populations expanded their geographic range across the globe, they brought with them an unparalleled ability to transform native landscapes and directly or indirectly devastate populations of NATIVE or indigenous species. According to Paul Martin and other palaeoecologists, this 'overkill' may have been responsible for the waves of EXTINCTION of native faunas (especially the MEGAFAUNA) that coincided with the arrival of environmentally significant humans in new lands. *MVL*

[See also MASS EXTINCTIONS]

Brown, J.H. and Lomolino, M.V. 1998: *Biogeography,* 2nd edn. Sunderland, MA: Sinauer. **Martin, P.S.** 1984: Prehistoric overkill. In Martin, P.S. and Klein, R.G. (eds), *Quaternary extinctions: a prehistoric revolution.* Tucson, AZ: University of Arizona Press, 345–403.

overlay analysis
The operation of stacking maps on top of each other so that information is compared at common locations. It is normally associated with GEOGRAPHICAL INFORMATION SYSTEMS (GIS), where multiple digital COVERAGES are superimposed and ENTITIES are compared at common GEOMETRIC COORDINATES (see Figure). For example, an overlay analysis examining possible environ-

mental hazards arising from the location of a new power station would involve coverages such as relief, hydrology, vegetation, soils and climatic indicators. By stacking these coverages on top of each other, and ensuring common geometric REGISTRATION, entities such as low-lying land, nearby water supplies, conservation areas of woodland, nearby agricultural land and prevailing winds could all be used within a DECISION MAKING process to determine the least cost to environmental stability.

Through time, overlays would be able to reveal changes to the environment, such as whether soils have become acidic, water supplies contaminated or trees damaged. *TVM*

[See also POLYGON ANALYSIS]

Chrisman, N.R. 1987: The accuracy of map overlays: a reassessment. *Landscape and Urban Planning* 14, 427–439. **Tomlin, C.D.** 1992: *Geographic information systems and cartographic modeling.* Englewood Cliffs, NJ: Prentice Hall. **Newcomer, J.A. and Szajgin, J.** 1984: Accumulation of thematic map error in digital overlay analysis. *American Cartographer* 11, 58–62.

overpopulation
Usually applied to animal populations, including humans, overpopulation is the condition that results where the number or density of individuals exceeds the CARRYING CAPACITY of the environment. It may be followed rapidly by a *population decline* or *population crash.*

At the end of the eighteenth century, Thomas Malthus observed the poverty and malnourishment around him and concluded that the world could only support a limited number of human beings as 'population when unchecked increases in a geometric ratio. Subsistence increases in an arithmetical ratio' (Malthus, 1798). Up to the present day, Malthus' prediction has not been proven correct as scientific development of the means of food production

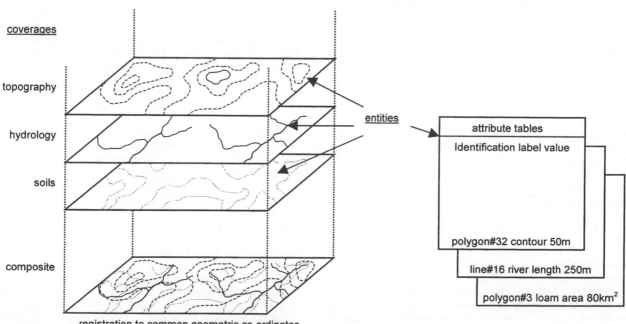

overlay analysis *The overlay principle and the fundamental units of data representation used in a geographical information system*

has kept ahead of the world's increasing population, although considerable numbers remain undernourished. Data for global population show that increases of 80 million per annum occurred in the late AD 1980s and early 1990s, mostly born in the developing countries, and it is estimated that these increases will bring the world population to over 8 billion during the first half of the twentyfirst century. At the same time, it must be appreciated that all the world's capacity is finite and that the soils most suited for agriculture are already being used.

Overpopulation implies that there are too many people present for the land to support by the provision of food and water. As a result, these people exert an undue pressure upon the land, extending cultivation into areas unsuited to cropping, removing natural vegetation, causing SOIL EROSION and initiating LAND DEGRADATION. However, the carrying capacity of the land is not a fixed value as countries have exploited their resources to enable greater or fewer people to live and at different standards of living. In Kenya, the Keita project demonstrated that increasing the rural population actually lessened the pace of land degradation. It is obvious that a balance must be made between wise use of the environment and its over-exploitation; otherwise future generations will not have the means of survival.   *EMB*

[See also DEMOGRAPHIC CHANGE, SUSTAINABLE DEVELOPMENT, LAND DEGRADATION, TECHNOLOGICAL CHANGE]

Fischer, G. and Heilig, G.K. 1997: Population momentum and the demand on land and water resources. In Greenland, D.J., Gregory, P.J. and Nye, P.H. (eds), 1997: *Land resources: on the edge of the Malthusian precipice?* Wallingford: CAB International and The Royal Society. **Food and Agriculture Organization of the United Nations (FAO)** 1995: *The Keita integrated rural development project.* Rome: FAO. **Greenland, D.J., Gregory, P.J. and Nye, P.H. (eds)** 1997: *Land resources: on the edge of the Malthusian precipice?* Wallingford: CAB International and The Royal Society. **Malthus, T.R.** 1798: *An essay on the principle of population as it affects the future improvement of society.* London: J. Johnson. **United Nations Population Fund (UNFPA)** 1991: *Population, resources and the environment: the critical challenges.* New York: UNFPA.

overtopping   The process by which coastal BARRIER BEACHES and BARRIER ISLANDS are built up by deposition from storm-generated swash flows of insufficient magnitude to extend fully across the crest. It contrasts with *overwashing*, during which higher-magnitude events may erode the barrier crest leading to *washover deposition* on the backslope and shoreward movement of the barrier. The balance between overtopping and overwashing controls the stability and migration of gravel barriers whereas, in the case of sand barriers, AEOLIAN deposition may be the more effective control.   *JAM*

Orford, J.D. and Carter, R.W.G. 1982: The structure and origins of recent sandy gravel overtopping and overwashing features at Carnsore Point, southeast Ireland. *Journal of Sedimentary Petrology* 52, 265–278.

overturning   Complete water circulation in a lake that destroys stratification and is caused by the water density differential breaking down during autumn cooling or ice melting. Overturning or *overturn* is driven by water being most dense at 4°C; it results in the sinking of oxygenated surface water and the UPWELLING of nutrient-rich bottom water, which improves lake PRODUCTIVITY.   *HHB*

[See also LIMNOLOGY, HOLOMICTIC LAKE, MEROMICTIC LAKE, LAKE STRATIFICATION]

oxidation   The chemical definition is the increase in the positive valence or the decrease in the negative valence of an element, or the loss of an *electron* from an element: ferrous iron (Fe^{2+}), for example, is oxidised to ferric (Fe^{3+}) iron. In CHEMICAL WEATHERING, this commonly means the combination of a mineral with oxygen to form *oxides* or hydroxides and usually involves atmospheric or soil oxygen dissolved in water. Well drained, AEROBIC conditions, the DECOMPOSITION of organic matter and relatively high temperatures all favour this important process of chemical weathering with iron, manganese, sulphur and titanium being among the major elements involved.   *RPDW*

[See also REDUCTION]

Brunsden, D. 1979: Weathering. In Embleton, C. and Thornes, J.B. (eds), *Process in Geomorphology.* London: Arnold, 73–129.

oxygen isotopes   Oxygen has three STABLE ISOTOPES (natural abundances, $^{16}O = 99.762\%$, $^{17}O = 0.038\%$ and $^{18}O = 0.200\%$). The ISOTOPE RATIO is expressed as $^{18}O{:}^{16}O$ relative to the Vienna Pee Dee Belemnite (VPDB) standard for CARBONATES and the Vienna Standard Mean Ocean Water (VSMOW) standard for all other samples. The $\delta^{18}O$ value of FORAMINIFERA in deep-ocean sediments is influenced primarily by global continental ice volume and thus reflects global climate (see Figure). The age of these sediments cannot be dated directly beyond the range of RADIOCARBON DATING; however, ages have been determined indirectly using palaeomagnetic stratigraphy and by matching properties of the sediments with changes in ORBITAL FORCING indicated by the MILANKOVITCH THEORY. A composite 780 ka record obtained from five MARINE SEDIMENT CORES forms the basis of the SPECTRAL MAPPING PROJECT *timescale* (SPECMAP) – the standard against which the generally fragmentary terrestrial records have been compared. Thus, ISOTOPIC STAGES have been assigned to GLACIAL and INTERGLACIAL episodes.

The terrestrial record of past climate includes the physico-chemical properties of ICE CORES taken from Antarctica, the Greenland ice sheet and high-altitude ice caps. Changes in global temperature over the last 420 ka have been reconstructed from the closely associated variations in the $\delta^{18}O$ record from the GRIP ice core from Greenland and the $\delta^{18}O$ values of O_2 occluded in the Vostok ice core from Antarctica. Although orbital forcing is evident in these cores, high-frequency chronostratigraphic markers (see CHRONOSTRATIGRAPHY) in the oxygen isotope profiles of two Greenland ICE CORES (GRIP and GISP2) illustrate the presence of DANSGAARD–OESCHGER EVENTS, which cannot be explained by the Milankovitch theory alone.

In terrestrial plants, the ISOTOPIC composition of xylem water is identical to the water absorbed by the roots. Mixing occurs as water ascends the plant to the site of PHOTOSYNTHESIS through the apoplastic or slower symplastic pathway. However, there is no ISOTOPIC FRACTIONATION until the water reaches tissues undergoing water loss, where EVAPORATION causes ISOTOPIC ENRICHMENT of ^{18}O in the remaining water. The oxygen ISOTOPE RATIO of leaf water is imparted to the intercellular CARBON DIOXIDE utilised in carbohydrate synthesis. Biochemical fractiona-

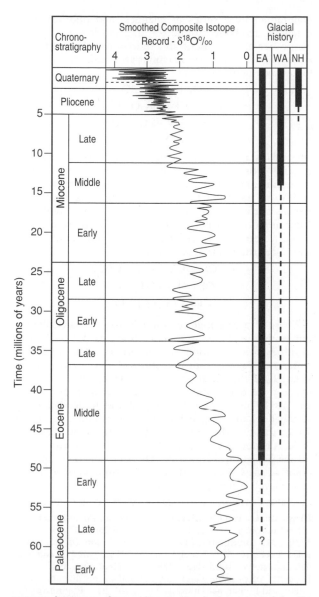

oxygen isotopes *Composite oxygen isotope record for the Cenozoic derived from marine sediment cores: the columns on the right indicate the presence of the East Antarctic Icesheet (EA) from c. 50 Ma, the West Antarctic Ice Sheet (WA) from c. 15 Ma and ice sheets in the Northern Hemisphere (NH) from around 5 Ma, based on the occurrence of ice-rafted debris (after Abreu and Anderson, 1998; Wilson et al., 2000)*

tion during photosynthesis causes an enrichment of ^{18}O of the resulting carbohydrate. Post-photosynthetic isotopic exchange occurs between the oxygen atoms of intermediate carbohydrates and metabolic water, with the extent of the exchange determined by carbohydrate recycling. Despite this complexity, the $\delta^{18}O$ values of organic matter have been used to reconstruct past climates using a TRANSFER FUNCTION or mechanistic approach. *IR*

[See also CORAL AND CORAL REEFS, LACUSTRINE SEDIMENTS, OSTRACODS, SPELEOTHEMS, PEAT]

Abreu, V.S. and Anderson, J.B. 1998: Glacial eustasy during the Cenozoic: sequence stratigraphic implications. *American Association of Petroleum Geologists Bulletin* 82, 1385–1400. **Craig, H.** 1961: Isotopic variations in meteoric waters. *Science* 133, 1702–1703. **Dansgaard, W., Johnsen, S.J., Clausen, H.B.** *et al.* 1993: Evidence for general instability of past climate from a 250-kyr ice-core record. *Nature* 364, 218–220. **Farquhar, G.D., Barbour, M.M. and Henry, B.K.** 1998: Interpretation of oxygen isotope composition of leaf material. In Griffiths, H. (ed.), *Stable isotopes: integration of biological, ecological and geochemical processes*. Oxford: Bios, 27–62. **Imbrie, J., Hays, J.D., Martinson, D.G.** *et al.* 1984: The orbital theory of Pleistocene climate: support from a revised chronology of the marine $\delta^{18}O$ record. In Berger, A., Imbrie, J., Hays, J. *et al.* (eds), *Milankovitch and climate: understanding the response to astronomical forcing*. Part I. Dordrecht: D. Reidel, 269–305. **Petit, J.R., Jouzel, J., Raynaud, D.** *et al.* 1999: Climate and atmospheric history of the past 420,000 years from the Vostok ice core, Antarctica. *Nature* 399, 429–436. **Shackleton, N.J. and Opdyke, N.D.** 1973: Oxygen isotope and palaeomagnetic stratigraphy of equatorial Pacific core V28-238: oxygen isotope temperatures and ice volumes on a 10^5 year and 10^6 year scale. *Quaternary Research* 3, 39–55. **Wilson, R.C.L., Drury, S.A. and Chapman, J.L.** 2000: *The great ice age: climate change and life*. London: Routledge.

oxygen variations The current ambient concentration of atmospheric oxygen (O_2) is 21%. The stability of this O_2 level relies on a critical balance between the consumption of O_2 by AEROBIC RESPIRATION, combustion and other processes and its production from oxygenic PHOTOSYNTHESIS and photolytic dissociation (breakdown of water and carbon dioxide molecules in the Earth's atmosphere by high-energy *ultraviolet radiation*). These essential processes have not been balanced over the whole of geological time and, as a result, O_2 concentrations have varied markedly. The earliest prebiotic atmosphere was devoid of O_2 as the small quantities produced by photolytic dissociation were rapidly consumed by ferrous iron and reduced volcanic gases. As the main source of atmospheric O_2 is from plant and algal PHOTOSYNTHESIS, the evolution of photosynthesising life on land, in conjunction with increased carbon burial and a reduction in the supply of reduced material to the atmosphere, resulted in the oxygenation of the atmosphere at the end of the Archean and into the Proterozoic. Models of oxygen variations indicate that atmospheric O_2 levels have remained between a minimum of 15% and a maximum of 35% throughout the Phanerozoic, with a large spike of 35% occurring between about 350 and 250 Ma, attributed to excessive carbon burial. *JCMc*

[See also ATMOSPHERIC COMPOSITION, OXIDATION, OXYGEN ISOTOPES, REDUCTION]

Berner, R.A. and Canfield, D.E. 1989: A new model for atmospheric oxygen over Phanerozoic time. *American Journal of Science* 289, 333–361. **Chaloner, W.G.** 1989: Fossil charcoal as an indicator of palaeoatmospheric oxygen level. *Journal of the Geological Society, London* 146, 171–174. **Lovelock, J.E. and Whitfield, M.** 1982: Life span of the biosphere. *Nature* 296, 561–563.

ozone Ozone (O_3) is a form of oxygen (O_2) that contains three atoms rather than the two atoms of ordinary oxygen. It is constantly being formed and destroyed in the STRATOSPHERE by a range of natural processes including the absorption of *ultraviolet radiation* by oxygen and ozone respectively, and by subsequent collision processes. The relatively high concentration of ozone in the stratosphere, with maximum concentrations between 20 and 25 km, is referred to as the *ozone layer*. Since the late AD 1970s significant stratospheric OZONE DEPLETION over Antarctica each Southern Hemispheric spring, has created what is commonly referred to as the *ozone hole*.

In the 1970s widespread international concern for ozone depletion centred around CHLOROFLUOROCARBONS (CFCs). These inert, non-toxic and inexpensive chemicals became widely used as a cooling agent for refrigerators and air conditioners, a propellant in aerosol sprays, a solvent in the electronics industry and a blowing agent to create packing and insulating foams. When a long-lived CFC molecule reaches the stratosphere, it is decomposed by ultraviolet radiation to produce an unattached chlorine atom. The chlorine atom initiates an ozone-destroying reaction sequence before emerging unchanged and capable of repeating the reaction many thousands of times before eventually being removed by other substances. Other ozone-depleting substances include *halons* (containing variously bromine, chlorine and fluorine), carbon tetrachloride (tetrachloromethane), methyl chloroform, methyl bromide and even HYDROCHLOROFLUOROCARBONS (HCFCs), which were introduced as substitutes for CFCs as their ozone-depletion potential was much less.

Stratospheric ozone depletion by CFCs and other compounds was first hypothesised in the early 1970s, but the proof did not come until 1985 when measurements above Halley Bay, on the Antarctic coast, revealed that large depletions had occurred there each spring since the late 1970s. By October 1987 ozone concentrations were around half the levels in the 1970s and they have continued to fall since then. Concern over ozone depletion resulted in an international CONVENTION, the 1987 Montreal Protocol on Substances which Deplete the Ozone Layer, to begin phasing out the production and consumption of ozone-depleting substances. Subsequent amendments to the Protocol have forwarded this process.

Loss of stratospheric ozone results in more *ultraviolet radiation* reaching the surface and may lead to an increase in sunburn and skin cancer, reduction in yields of some crops, damage to oceanic PLANKTON and degradation of many materials such as paints and fabrics. Given the long atmospheric residence time of CFCs, the depletion of ozone is expected to last for several decades.   *DME*

Crutzen, P.J. 1974: Estimates of possible future ozone reductions from continued use of fluorochloromethanes. *Geophysical Research Letters* 1, 205–208. **Elsom, D.M.** 1992: *Atmospheric pollution: a global problem*, 2nd edn. Oxford: Blackwell. **Stratospheric Ozone Review Group** 1996: *Stratospheric ozone 1996 [Sixth Report]*. London: Department of the Environment. **World Meteorological Organisation (WMO)** 1995: *Scientific assessment of ozone depletion: 1994 [Global Ozone Research and Monitoring Project Report* 37]. Geneva: WMO.

ozone depletion The natural and pollution-forced decrease in the ozone (O_3) layer, which is found in the STRATOSPHERE at heights between 20 and 40 km. Ozone is maintained by a series of chemical and photochemical reactions involving atomic (O) and molecular (O_2) oxygen, other gases present and short-wave *ultraviolet* (UV) SOLAR RADIATION. These reactions are usually shown in the following equations:

$$O_2 + UV \rightarrow 2O$$
$$O_2 + O + M \rightarrow O_3 + M$$
$$O_3 + UV \rightarrow O_2 + O$$
$$O + O_3 \rightarrow 2O_2$$

where M is any other gas molecule that acts as a catalyst, causing ozone to be produced and destroyed. Ozone is environmentally important because of the harmful effects of UV on most life forms. The ozone layer shields the Earth's surface from this radiation. If M is a pollutant, (such as a CHLOROFLUOROCARBON – CFC – represented by $CFCl_3$), three other reactions occur continuously and ozone destruction is accelerated:

$$CFCl_3 + UV \rightarrow Cl + CFCl_2$$
$$Cl + O_3 \rightarrow ClO + O_2$$
$$ClO + O \rightarrow Cl + O_2$$

There is a seasonal variation in ozone, particularly a diminution in late winter and early spring due to the lack of ultraviolet radiation in the winter. The loss over Antarctica is more prominent and due partly to the strength of the circumpolar WESTERLIES, which prevent ozone from being imported from lower latitudes. In the late AD 1970s and early 1980s, it was noticed that the spring depletions were increasing in amount (by up to 70%); this was the *ozone hole*. In 1993, the decrease was 99%. It is now accepted that CFCs are mainly responsible for the depletion. More important from an environmental point of view are reports that the summer recovery of ozone levels is also getting less.   *BDG*

[See also CONVENTIONS, HALOGENATED HYDROCARBONS]

Brasseur, G. and Granier, C. 1992: Mount Pinatubo aerosols, chlorofluorocarbons and ozone depletion. *Science* 257, 1239–1242. **Bridgeman, H.** 1997: Air pollution. In Thompson, R.D. and Perry, A. (eds), *Applied climatology. principles and practice.* London: Routledge: 288–303. **Crutzen, P.J.** 1974: Estimates of possible variations in total ozone due to natural causes and human activities. *Ambio* 3, 201–210. **Farman, J.C., Gardiner, B.G. and Shanklin, J.D.** 1985: Large losses of total ozone in Antarctica reveal ClO$_x$/NO$_x$ interaction. *Nature* 315, 207–210. **Gribbin, J.** 1993: *The hole in the sky*, revised edn. New York: Bantam. **Molina, M.J. and Rowland, F.S.** 1974: Stratospheric sink for chlorofluoromethanes: chlorine atom catalysed destruction of ozone. *Nature* 249, 810–812. **Roan, S.L.** 1989: *Ozone crisis. The 15-year evolution of a sudden global emergency.* Chichester: Wiley. **Urbach, F.** 1989: Potential effects of altered solar ultraviolet radiation on human skin cancer. *Photochemistry and Photobiology* 50, 507–514. **World Meteorological Organization (WMO)** 1995: *Scientific assessment of ozone depletion: 1994.* [*Global Ozone Research and Monitoring Project Report* 37]. Geneva: WMO.

P

P-forms Smooth, small-scale depressions in the form of potholes, bowls, assemblages of curved depressions (*Sichelwannen*), grooves and channels eroded into bedrock by poorly understood processes of glacial or glaciofluvial erosion. *RAS*

[See also GLACIAL LANDFORMS, GLACIOFLUVIAL LANDFORMS]

Dahl, R. 1965: Plastically sculptured detail forms on rock surfaces in northern Nordland, Norway. *Geografiska Annaler* 47, 83–140. **Kor, P.S.G., Shaw, J. and Sharpe, D.R.** 1991: Erosion of bedrock by subglacial meltwater, Georgian Bay, Ontario: a regional view. *Canadian Journal of Earth Sciences* 28, 623–642.

Pacific–North American (PNA) teleconnection

A large-scale TELECONNECTION between the climates of the North Pacific Ocean and the continent of North America, which is apparent in the 500 millibar geopotential height field. The *PNA Index* is a weighted average of 500 millibar normalised height anomaly differences between the centres of four distinct cells over the North Pacific, Hawaii, Alberta and the southeastern USA. This is correlated with the EL NIÑO–SOUTHERN OSCILLATION and accounts for some of the decadal-scale variability in the northwestern Pacific coastal salmon fisheries.

The PNA and its index cover a larger region than the *North Pacific Oscillation* and the corresponding *North Pacific Index*, which is based on areally averaged sea-level pressure over a large area of the North Pacific Ocean. The *Pacific Decadal Oscillation* and the *West Pacific Oscillation* are other relatively small North Pacific patterns. *JAM*

[See also NORTH ATLANTIC OSCILLATION]

Leathers, D.J., Yarnal, B. and Palecki, M.A. 1991: The Pacific/North American teleconnection pattern and United States climate. Part I: regional temperature and precipitation associations. *Journal of Climatology* 4, 517–528. **Mantua, N.J., Hare, S.R., Zhang, Y.** *et al.* 1997: A Pacific interdecadal climate oscillation with impacts on salmon production. *Bulletin of the American Meteorological Society* 78, 1069–1079. **National Research Council** 1999: *Global environmental change: research pathways for the next decade.* Washington, DC : National Academy Press. **Rogers, J.C.** 1981: The North Pacific Oscillation. *Journal of Climatology* 1, 39–57. **Wallace, J.M. and Gutzler, D.S.** 1981: Teleconnections in the geopotential height field during the Northern Hemisphere winter. *Monthly Weather Review* 109, 784–812.

pack ice
SEA ICE floating free under the influence of currents and wind. Pack ice comprises *ice floes* centred on relatively resistant ice, *leads* of open water, which may be subject to rapid freezing in winter, and chaotically fractured and crumpled ice-forming features (*pressure ridges* or *keels*). *RAS*

Hanna, E. 1996: The role of Antarctic sea ice in global climate change. *Progress in Physical Geography* 20, 371–401.

pahoehoe
A Hawaiian term for *ropy lava*, used to describe a LAVA FLOW with a smooth, commonly undulating surface, caused by the deformation of a surface crust formed over a fast-moving, low-viscosity BASALT flow. *GO*

[See also AA]

palaeobiology The study of FOSSILS as organisms from a biological perspective, including studies of TAXONOMY and EVOLUTION. This contrasts with the use of fossils in PALAEOECOLOGY and STRATIGRAPHY, although there is much overlap. *GO*

[See also PALAEONTOLOGY]

palaeobotany The study of FOSSIL plants. Palaeobotany is a wide field of science involving many disciplines including PALAEOCLIMATOLOGY, PALAEOECOLOGY, *palynology*, palaeofloristics, plant EVOLUTION, *phylogeny* and ONTOGENY, BIOSTRATIGRAPHY and BIOGEOGRAPHY, to name a few. Traditionally, palaeobotany research was concerned with the evolution and phylogeny of different plant groups from studies of the plant fossil record. However, over the last century the use of fossil plants as palaeoenvironmental, palaeoclimatic and palaeoatmospheric indicators has been realised and widely applied. Examples include the use of LEAF PHYSIOGONOMY, fossil wood characteristic and stable CARBON ISOTOPE composition as indicators of palaeoclimate and the use of fossil leaf STOMATA and fossil charcoal as indicators of palaeoatmospheric CARBON DIOXIDE and OXYGEN VARIATIONS, respectively. *Palaeofloristics*, the study of assemblages of fossil plants in space and time, is another important facet of palaeobotany that has provided evidence of PLATE TECTONICS, the drifting of continents and CLIMATIC CHANGE. *JCMc*

[See also CHARRED-PARTICLE ANALYSIS, STOMATAL ANALYSIS, WOOD ANALYSIS]

Knoll, A.H. and Rothwell, G.W. 1981: Palaeobotany: perspectives in 1980. *Paleobiology* 7, 7–35. **Seldon, P.A. and Edwards, D.** 1989: Colonisation of the land. In Allen, K.C. and Briggs, D.E.G. (eds), *Evolution and the fossil record.* London: Belhaven, 122–152. **Stewart, W.N. and Rothwell, G.W.** 1993: *Paleobotany and the evolution of plants.* Cambridge: Cambridge University Press.

palaeoceanography Study of the physical, chemical and biological characteristics of the OCEANS in the past (see also OCEANOGRAPHY). The aims are to determine former ocean characteristics including OCEAN BASIN size and shape, OCEAN CURRENT patterns, SEA LEVEL and events such as OCEANIC ANOXIC EVENTS, UPWELLING, desiccation and ICEBERG meltout events (HEINRICH EVENTS). Much data comes from the study of MARINE SEDIMENTS obtained by DEEP-SEA DRILLING. As well as providing direct evidence, these serve as PROXY DATA via isotopic analysis and measurements of organic CARBON CONTENT. Other data sources, including the study of magnetic stripes from SEAFLOOR SPREADING, and the known age–depth relationship for cooling OCEANIC CRUST (see MID-OCEAN RIDGE), are used to reconstruct ocean basin morphology. The reconstruction of past ocean characteristics allows the investigation of long-term relationships between ocean circulation and climate and the past record of processes in, for example, EL NIÑO–SOUTHERN OSCILLATION cyclicity and MILANKOVITCH THEORY. Most palaeoceanography is carried out on MESOZOIC and TERTIARY sediments that are still found on the ocean floor, but many of the methods can be applied to older SEDIMENTARY ROCKS preserved on land.

GO

Andel, T.H. van, Thiede, J., Sclater, J.G. and Hay, W.W. 1977: Depositional history of the S. Atlantic Ocean during the last 125 million years. *Journal of Geology* 85, 651–698. **Andrews, J.T., Austin, W.E.N., Bergsten, H. and Jennings, A.E. (eds)** 1996: *Late Quaternary palaeoceanography of the North Atlantic margin* [*Special Publication* 111]. London: Geological Society. **Berger, W.H. and Herquera, J.C.** 1992: Reading the sedimentary record of the ocean's productivity. In Falkowski, P.G. and Woodhead, A.D. (eds), *Primary productivity and biogeochemical cycles in the sea.* New York: Plenum, 455–486. **Christiansen, J.L. and Stouge, S.** 1999: Oceanic circulation as an element in palaeogeographical reconstructions: the Arenig (early Ordovician) as an example. *Terra Nova* 11, 73–78. **Hsü, K.J., Garrison, R.E., Montadert, L. *et al.*** 1977: History of the Mediterranean salinity crisis. *Nature* 267, 399–403. **Kemp, A.E.S. and Baldauf, J.G.** 1993: Vast Neogene laminated diatom mat deposits from the eastern equatorial Pacific Ocean. *Nature* 362, 141–144. **Kennett, J.P.** 1977: Cenozoic evolution of Antarctic glaciation, the circum-Antarctic ocean, and their impact on global palaeo-oceanography. *Journal of Geophysical Research* 82, 3843–3860.

Palaeocene An EPOCH of the TERTIARY period, from 65 to 56.5 million years ago. *GO*

[See also GEOLOGICAL TIMESCALE, PALAEOGENE]

palaeochannel A channel that was active in the past. Palaeochannels may represent former channel positions in systems of channels that are still active, such as channel segments on a river FLOODPLAIN that have been abandoned through channel migration or AVULSION, or whole systems of channels that no longer exist as routeways for flows, such as palaeochannels in the STRATIGRAPHICAL RECORD. Palaeochannels can be recognised by means of surface GEOMORPHOLOGY, by the geometry of BEDDING in SEDIMENTARY ROCKS or by BEDS of distinctive LITHOLOGY. The analysis of palaeochannels can yield important information about former environments and environmental change, particularly when integrated with techniques such as PALAEOHYDROLOGY, FACIES ANALYSIS and ALLUVIAL ARCHITECTURE. *GO*

[See also PALAEOFLOOD]

Brown, A.G., Keough, M.K. and Rice, R.J. 1994: Floodplain evolution in the East Midlands, United Kingdom – the Late-Glacial and Flandrian alluvial record from the Soar and Nene valleys. *Philosophical Transactions of the Royal Society of London* A348, 261–293. **Cairncross, B., Stanistreet, I.G., McCarthy, T.S. *et al.*** 1988: Paleochannels (stone-rolls) in coal seams – modern analogs from fluvial deposits of the Okavango Delta, Botswana, southern Africa. *Sedimentary Geology* 57, 107–118. **Leigh, D.S. and Feeney, T.P.** 1995: Paleochannels indicating wet climate and lack of response to lower sea-level, southeast Georgia. *Geology* 23, 687–690. **Pickup, G, Allan, G. and Baker, V.R.** 1988: History, palaeochannels, and palaeofloods of the Finke River, central Australia. In Warner, R. (ed.), *Fluvial geomorphology in Australia.* Sydney: Academic Press, 177–200. **Wu, C., Xu, Q.H., Ma, Y.H. and Zhang, X.Q.** 1996: Palaeochannels on the North China Plain: palaeoriver geomorphology. *Geomorphology* 18, 37–45.

palaeoclimatology The study of past CLIMATE. It includes CLIMATIC CHANGE throughout geological time, but most palaeoclimatic research is carried out in the context of QUATERNARY ENVIRONMENTAL CHANGE, including HOLOCENE ENVIRONMENTAL CHANGE and HISTORICAL CLIMATOLOGY. However, some confine palaeoclimatology to the study of climate prior to the period of the INSTRUMENTAL RECORD.

Model-based predictions concerning the scale and magnitude of FUTURE CLIMATIC CHANGE due to *anthropogenic impact* constitutes, as yet, an inexact science. It is clear that human-induced CLIMATIC FORCING will be superimposed on a climate system that has as one of its inherent characteristics a natural propensity for change on all timescales. Palaeoclimatology is concerned with uncovering the range of past CLIMATIC VARIATION in order that the context for present and future climate variation may be better understood and the management of climate as a resource in its own right may be optimised.

Palaeoclimatic reconstructions are most satisfactory when based on standard instrumental observations. When allowance is made for past problems such as non-standard exposure, the quality of manufacture (especially glassware), a multiplicity of measurement scales and several other difficulties, sufficient coverage can be provided from the instrumental record to permit the construction of daily weather maps, in the case of Europe, from the early eighteenth century. Elsewhere, climatic reconstructions based on direct observations are more difficult. The longest continuous record of temperature is Manley's CENTRAL ENGLAND TEMPERATURE RECORD from AD 1659, but for the Southern Hemisphere the longest record dates only from 1832 (Rio de Janeiro). INSTRUMENTAL RECORDS therefore frequently provide information on relatively poor spatial and temporal scales, and encompass a period which may be quite exceptional in the longer climatic context. Careful treatment of historical documents, such as SHIP LOG BOOK RECORDS, Mediaeval manorial records, grain prices, CHRONICLES and other early manuscripts may provide patchy information on palaeoclimates back into the first millennium AD. Before this, PROXY DATA sources from the natural world are necessary.

NATURAL ARCHIVES of the effects of CLIMATIC VARIABILITY, such as contained in TREE RINGS, LACUSTRINE SEDIMENTS and MARINE SEDIMENT CORES, CORAL analysis and ICE CORES have, with the use of advanced technologies and statistical techniques, enabled palaeoclimatic reconstruction with annual, sometimes seasonal RESOLUTION (see Figure). Changes in ocean chemistry and in ATMOSPHERIC COMPOSITION, such as that due to volcanic events, may also be recovered. Plant and animal remains, such as pollen grains (see POLLEN ANALYSIS), beetle assemblages (*Coleoptera* analysis) and planktonic foraminifera (see FORAMINIFERAL ANALYSIS), also enable TRANSFER FUNCTIONS to be derived from which palaeoclimatic information can be inferred.

On long time scales, palaeoclimatic reconstruction is complicated by the changing location of places due to PLATE TECTONICS. Before about 2.4 million years ago conditions were generally warmer than at present. Since the beginning of the Quaternary, however, the climate system has been characterised by two recurring modes: as seen in GLACIAL–INTERGLACIAL CYCLES, during which global mean temperatures may have fluctuated by about 5–7°C.

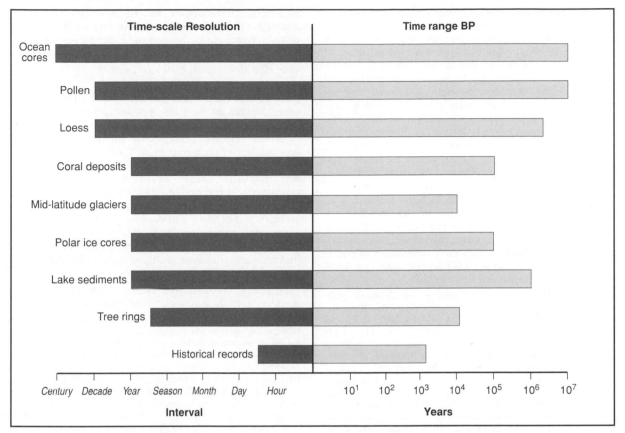

palaeoclimatology *The temporal resolution and temporal range of some palaeoclimatic data*

Considerable climatic instability has characterised glacial episodes with abrupt periods of relatively short-lived warming. From the maximum of the last glacial event, approximately 18 000 radiocarbon years BP, climate has warmed sporadically to a Holocene 'optimum' (HYPSITHERMAL) about 7000 BP when temperatures in the North Atlantic and Europe were about 2°C above those of today. During more recent times significant global warming occurred around AD 900–1200 (the MEDIAEVAL WARM PERIOD). The LITTLE ICE AGE, which reached its nadir in Europe in the late seventeenth century brought, on average, global cooling of about 1°C, which persisted into the nineteenth century. Since then, GLOBAL WARMING of about 0.6°C has occurred. Though unequivocal evidence that this is linked to anthropogenic influences has not yet been presented, the majority opinion among atmospheric scientists is that the balance of evidence suggests that a discernible human influence on global climate exists.

Modulation of past climates as a result of changes in BOUNDARY CONDITIONS is believed to result from intrinsic variations in SOLAR RADIATION and to orbitally induced changes in INSOLATION receipt. First, though the SOLAR CONSTANT has varied less than 0.2% during recent decades, greater variability in the near-ultraviolet wavelengths may produce changes in the absorption of solar radiation by stratospheric ozone. These in turn may propagate downwards to affect the LONG-WAVE pattern in the WESTERLIES and the strength of the JET STREAM. Second, orbitally induced changes in the seasonal receipt of insolation at

various latitudes are now widely accepted as the driver of the glacial–interglacial oscillations of the Quaternary. These changes occur as a result of PERIODICITIES of 21 000 years (PRECESSION of the equinoxes), 41 000 years (OBLIQUITY of the ecliptic) and 100 000 years (ECCENTRICITY), which act in conjunction to alter the receipt of insolation at key latitudes and critical seasons of the year.

On shorter time scales, SECULAR CLIMATIC CHANGES and CLIMATIC FLUCTUATIONS result from FORCING FACTORS, some internal to the present climatic system. Detecting these fluctuations and relating them to specific factors entails discriminating any 'signal' from the background 'noise' of natural CLIMATIC VARIABILITY, which can be of a similar or even greater magnitude. Among the key factors that may operate at these timescales are: changes in the composition of the atmosphere (especially trace gases such as CARBON DIOXIDE, METHANE and nitrous oxide), volcanic activity, fluctuations in the behaviour of the oceans and cryosphere, and biospheric dynamics, particularly involving biological responses to physical climatic change, which may show complex feedback relations. High-frequency variability, such as is provided by the EL NIÑO–SOUTHERN OSCILLATION (ENSO) phenomenon, is increasingly recognised as a major factor in past and present climate variability in the 2–10 year range. ENSO is influential both within the tropics and in extratropical areas where strong TELECONNECTIONS have been established in several locations. Palaeoclimatic variations may also show associations between annual, decadal and century scales as a result of ATMOSPHERE–OCEAN INTERACTION and FEEDBACK MECHANISMS.

Palaeoclimatic research, particularly that based on ICE CORES from the polar regions, has revealed the existence of several instances of ABRUPT CLIMATIC CHANGES during the last glacial–interglacial cycle. Rapid termination of glacial stages involving temperature rises of several degrees within a few decades have been identified. Equally dramatic cooling phases during early interglacial times in the northern oceans may relate to large-scale freshwater influx from greatly increased iceberg outflow (HEINRICH EVENTS), which may have induced an adjustment in the THERMOHALINE CIRCULATION. The response of the ocean to changes in the composition of the atmosphere and to insolation changes may be more rapid than hitherto believed. The lesson of palaeoclimatology is that the climatic system is probably a *quasi-transitive system* that adjusts to FORCING FACTORS by first resisting change, then moving rapidly to a new equilibrium. This renders the predictability of possible human impacts more difficult and their outcomes potentially more hazardous. *JCS*

Bradley, R.S. 1999: *Paleoclimatology: reconstructing climates of the Quaternary*, 2nd edn. San Diego, CA: Academic Press. **Cronin, T.M.** 1999: *Principles of paleoclimatology*. New York: Columbia University Press. **Crowley, T. and North, G.** 1991: *Palaeoclimatology* [Oxford Monographs on Geology and Geophysics 18]. Oxford: Clarendon Press. **Emiliani, C.** 1993: Milankovitch theory verified. *Nature* 364, 583–584. **Frakes, L.A.** 1978: *Climates throughout geological time*. Amsterdam: Elsevier. **Fritts, H.** 1976: *Tree rings and climate*. London: Academic Press. **Hecht, A.D.** 1985: Paleoclimatology: a retrospective of the past 20 years. In Hecht, A.D. (ed.), *Paleoclimate analysis and modeling*. New York: Wiley InterScience, 1–25. **Lamb, H.H.** 1982: *Climate, history and the modern world*. London: Methuen. **Lorius, C., Jouzel, J., Raynaud, D., Hansen, J. and Le Treut, H.** 1990: The ice-core record: climate sensitivity and future greenhouse warming. *Nature* 347, 139–145. **Oerlemans, J.** 1993: Evaluating the role of climate cooling in iceberg production and the Heinrich events. *Nature* 364, 783–786. **Pearce, F.** 1998: Sunny side up. *New Scientist* 159, 44–49. **Ruddiman, W.F. and Kutzbach, J.E.** 1991: Plateau uplift and climatic change. *Scientific American* 264, 66–75. **Smith, G.I. and Bischoff, J.L.** 1997: An 800,000-year paleoclimatic record from core L-92, Owens Lake, southeast California. *United States Geological Survey, Special Paper* 317, 1–172. **Stocker, T.F.** 2000: Past and future reorganizations of the climate system. *Quaternary Science Reviews* 19, 301–319.

palaeocommunity
A past COMMUNITY of organisms that may be reconstructed from FOSSIL or SUBFOSSIL evidence. *GO*

[See also DEATH ASSEMBLAGE, LIFE ASSEMBLAGE, PALAEOECOLOGY]

palaeocurrent
The direction of a past wind or water current, commonly inferred from the orientation of SEDIMENTARY STRUCTURES such as CROSS-STRATIFICATION. *MRT*

Collinson, J.D. and Thompson, D.B. 1992: *Sedimentary structures*, 2nd edn. London: Chapman and Hall. **Potter, P.E. and Pettijohn, F.J.** 1977: *Paleocurrents and basin analysis*, 2nd edn. Berlin: Springer.

palaeodata
Data relating to past environmental events and/or past environmental conditions, prior to the availability of INSTRUMENTAL RECORDS and DOCUMENTARY EVIDENCE. Palaeodata are derived from NATURAL ARCHIVES and relate primarily to the *effects* of former events or conditions, from which the events or conditions themselves are reconstructed rather than observed or measured. *JAM*

[See also ABDUCTION, PROXY DATA/PROXY EVIDENCE]

palaeodiet
The nature of the food used by humans or other animals in the past. A wide variety of sources are used to reconstruct palaeodiets. Archaeological dietary reconstruction for humans and HOMINIDS includes: the compilation of an inventory of foods known to be available, including on-site evidence such as stomach or intestine contents (rarely available), bones and teeth, COPROLITE, ARTEFACTS and RESIDUES; the application of a potentially wide range of analytical techniques (see SCIENTIFIC ARCHAEOLOGY); and the interpretation of likely environmental constraints and dietary choices. Data on palaeodiets are relevant for understanding FOOD CHAINS/WEBS, nutrition and DISEASE, CULTURAL CHANGE and palaeoenvironmental change. *JAM*

[See also ENVIRONMENTAL ARCHAEOLOGY, PALAEOETHNOBOTANY, RODENT MIDDENS, ZOOARCHAEOLOGY]

Ambrose, S.H. 1993: Isotopic analysis of paleodiets: methodological and interpretive considerations. In Sandford, M.K. (ed.), *Investigations of ancient human tissue: chemical analyses in anthropology*. Langhorne, PA: Gordon and Breach. **Gilbert Jr, R.L., Mielke, J.H. (eds)** 1985: *The analysis of prehistoric diets*. Orlando, FL: Academic Press. **Reinhardt, K.J., Greib, P.R., Callahan, M.M. and Hevly, R.H.** 1992: Discovery of colon contents in a skeletonized burial: soil sampling for dietary remains. *Journal of Archaeological Science* 19, 697–705. **Sillen, A.** 1989: Chemistry and palaeodietary research: no more easy answers. *American Antiquity* 54, 504–512. **Sobolik, K.D. (ed.)** 1994: *Paleonutrition: the diet and health of prehistoric Americans*. Carbondale, IL: Southern Illinois University Press.

palaeodose
In LUMINESCENCE DATING, the laboratory dose of *beta* (β) and/or *gamma* (γ) *radiation* needed to induce a luminescence equal to that accumulated in a SEDIMENT sample since the most recent exposure to light or significant heating ('*bleaching*' event). The unit of dose is the *gray* (1 Gy = 1 J kg^{-1}). The palaeodose is also termed the *equivalent dose* or *accumulated dose*. *DAR*

palaeodunes
Ancient, RELICT dune systems or FOSSIL dunes. Preservation of palaeodunes takes place through three processes, all of which may be associated with environmental change: (1) a reduction in wind velocity; (2) a change in wind direction; and (3) a change in rainfall regime. These process variations allow SOIL DEVELOPMENT and surface stabilisation, which can then resist surface reactivation in any later return to drier or windier conditions.

Active dunes covered large areas of present-day temperate lands during the LAST GLACIATION. Over 10^5 km^2 of Europe and America are now covered in stabilised dune areas: including much of eastern England, Sweden and Finland in Europe and the western Great Plains and eastern seaboard states of North America. The formation of SAND SEAS and *dunefields* in these now-temperate zones was stimulated by three features of the GLACIAL EPISODES in high latitudes: ARIDITY, plentiful supplies of sand from glacial outwash and strong winds blowing off the glaciers (*glacier winds*). In low latitudes, the expansion of the arid zone at the height of the glaciations brought dune-forming conditions into areas that were previously, and have since become, less arid. Around 18 000 years ago, stabilised sand seas covered as much as 50% of the land surface between 30° N and 30° S; today the proportion stands at only 10%. Extensive systems of RELICT dunes, many of which are covered by SAVANNA or grassland vegetation, have been recognised from the margins of modern, active SAND SEAS in the SAHEL, southern Africa, India and Australia and throughout the western USA. Relict dune systems have been identified from AERIAL PHOTOGRAPHS and SATELLITE REMOTE SENSING in areas where current annual precipitation is as high as 1000 mm. Fossil dunes can be recognised in the GEOLOGICAL RECORD.   *MAC*

[See also AEOLIANITE]

Clemmensen, L.B., Fornós, J.J. and Rodriquez-Perea, A. 1997: Morphology and architecture of a late Pleistocene cliff-front dune, Mallorca, Western Mediterranean. *Terra Nova* 9, 251–254. **Cooke, R. U., Warren, A. and Goudie, A.S.** 1993: Desert geomorphology. London: UCL Press. **Forman, S.L., Goetz, A.F.H. and Yuhas, R.H.** 1992: Large scale destabilized dunes on the High Plains of Colorado: understanding the landscape response to Holocene climates with the aid of images from space. *Geology* 20, 145–148. **Goring-Moris, A.N. and Goldberg, P.** 1990: Late Quaternary dune incursions in the southern Levant: archaeology, chronology and paleoenvironments. *Quaternary International* 5, 115–137. **Karpeta, W.P.** 1990: The morphology of Permian palaeodunes – a reinterpretation of the Bridgnorth sandstone around Bridgnorth, England, in the light of modern dune studies. *Sedimentary Geology* 69, 59–75.

palaeoecology   The ECOLOGY of the past. This discipline involves the study of past ecosystems and ENVIRONMENTS and, where possible, the interaction between organisms and their environment in the past. Both biological and geological evidence is used to reconstruct long-term environmental conditions and to determine the distribution and abundance of organisms in the past. Palaeoecology can be studied in any period of Earth history in which life was present. However, the main relevance of this subject to the study of contemporary ENVIRONMENTAL CHANGE lies in the most recent geological time interval, the QUATERNARY. This interval covers the past two million years or so and is characterised by a climate oscillating regularly between warmer INTERGLACIALS and colder GLACIAL EPISODES. In addition, humans evolved during this time interval.

The FLORA and/or FAUNA that lived at a particular time and place in the past can be reconstructed by analysing assemblages of fossilised organisms. Preservation of BIOTIC remains usually requires deposition in ANAEROBIC environments (such as lakes, WETLANDS, the ocean floor, etc.), desiccation or freezing. Plant MICROFOSSILS, such as pollen, and MACROFOSSILS, such as seeds, leaves and woody material, provide information on the composition of past vegetation. Similarly, analysis of FOSSIL animal remains from organisms such as OSTRACODS, CHIRONOMIDS, beetles (*Coleoptera*), CLADOCERA and MOLLUSCA, as well as larger organisms, such as mammals, allow a reconstruction of past faunal assemblages (see ANIMAL REMAINS). In addition, palaeoenvironmental conditions (e.g. climatic conditions, water chemistry and nutrient status, watershed processes, lake levels, water temperature) can be inferred from both sediments and the contained FOSSIL or SUBFOSSIL plant and animal assemblages,

assuming that the ecology and physiology of the modern species is the same as it was in the past. Additional palaeolimnological data can be inferred from the composition of diatoms, fungal remains, TESTATE AMOEBAE and chrysophytes that are found within the sedimentary matrix. In order to understand the interaction between organisms and their environment, independent environmental data are essential.

Fossil biotic data can be qualitative and/or quantitative, depending on the sampling techniques, fossil production, dispersal and preservation. Interpretation of past community composition, structure and function based on fossil material is more difficult farther back in time because ecological tolerances may change and taxa recovered may be extinct. Taphonomic processes (see TAPHONOMY) can alter the composition of fossil assemblages and must always be considered when interpreting fossil data.

Data on past environments can also be obtained from stratified sediments and landforms on the Earth's surface (LITHOLOGY and GEOMORPHOLOGY). Physical and chemical analysis of sediments can reveal the environmental history of an area and, in cases where the sediments are fossiliferous, inferences based on lithological changes can be supported by those based on fossil evidence. A wide variety of materials and methods are employed, depending on the site and nature of the stratigraphic material. These include: describing the sediment STRATIGRAPHY (for classification and correlation purposes); analysing the sediment particle-size distribution (variations in average GRAIN-SIZE or the range of sizes may reflect changes in the SEDIMENTARY ENVIRONMENT); assessing the organic content (a measure of biological productivity); techniques of MINERAL MAGNETISM (characterising sedimentary sequences and useful for core correlation); HEAVY MINERAL ANALYSIS (can reflect the provenance of inorganic deposits); elemental analyses (indication of changing erosional history of lake catchments); and STABLE ISOTOPE ANALYSIS of the sediments, or fauna contained with the sediments (indirect evidence of water temperature and hydrology). In addition, landforms that developed under a previous environmental regime, and have survived, can provide information on the nature of the environmental conditions in which they evolved, e.g. climatic, glacial and fluvial activity.

A sound CHRONOLOGY is essential in palaeoecological investigations to determine the timing of past events and rates of change, as well as for correlation among sites. Various DATING TECHNIQUES are used to provide a chronology for palaeoecological studies, including: RADIOMETRIC DATING and PALAEOMAGNETIC DATING methods; TEPHROCHRONOLOGY; DENDROCHRONOLOGY; and VARVE CHRONOLOGY.

In addition to documenting past ecosystems and environments, the palaeoecological record provides a long-term context for current phenomena, landscape patterns and processes. Many ecosystem processes (relating to both the BIOTIC and the ABIOTIC ecosystem components) occur over relatively long time scales of decades to millennia (such as ECOLOGICAL SUCCESSION, species MIGRATION, CLIMATIC CHANGE, SOIL DEVELOPMENT, geomorphological change). A long temporal perspective is therefore essential. Long-term environmental data is critical to place current GLOBAL ENVIRONMENTAL CHANGE, including CLIMATIC CHANGE, into context and to determine whether current trends are part of NATURAL ENVIRONMENTAL CHANGE. In addition, an understanding of past environmental and climatic change, and of the response of organisms to those changes, can be used in the prediction of future ecosystem dynamics. *JLF*

[See also CHRYSOPHYTES CYST ANALYSIS, DIATOM ANALYSIS, GEOMORPHOLOGY AND ENVIRONMENTAL CHANGE, PALAEO-COMMUNITY, PALAEOENVIRONMENT, PALAEOGEOGRAPHY, PALAEOHYDROLOGY, PALAEOLIMNOLOGY, POLLEN ANALYSIS]

Berglund, B.E. (ed.), 1986: *Handbook of Holocene palaeoecology and palaeohydrology*. Chichester: Wiley. **Birks, H.J.B and Birks, H.H.** 1980: *Quaternary palaeoecology*. London: Edward Arnold. **Davis, M.B.** 1994: Ecology and palaeoecology begin to merge. *Trends in Ecology and Evolution* 9, 357–358. **Delcourt, H.R. and Delcourt, P.A.** 1991: *Quaternary ecology: a paleoecological perspective*. London: Chapman and Hall. **Lowe, J.J. and Walker, M.J.C.** 1997: *Reconstructing Quaternary environments*, 2nd edn. Longman: Harlow. **Schoonmaker, P.K. and Foster, D.R.** 1991: Some implications of palaeoecology for contemporary ecology. *Botanical Review* 57, 204–245. **Walker, D.** 1990: Purpose and method in Quaternary palynology. *Review of Palaeobotany and Palynology* 64, 13–27.

palaeoentomology

The study of FOSSIL and SUBFOSSIL INSECTS, an important source of natural and anthropogenic palaeoenvironmental information. *JAM*

[See also BEETLE ANALYSIS, CHIRONOMID ANALYSIS, ELM DECLINE, INSECT ANALYSIS]

Ashworth, A.C., Buckland, P.C. and Sadler, J.P. (eds) 1997: *Studies in Quaternary entomology: an inordinate fondness for insects*. Chichester: Wiley. **Dinnin, M.H. and Sadler, J.P.** 1999: 10,000 years of change: the Holocene entomofauna of the British Isles. *Quaternary Proceedings* 7, 545–562.

palaeoenvironment

A past ENVIRONMENT. *MRT*

[See also PALAEOENVIRONMENTAL RECONSTRUCTION]

palaeoenvironmental indicator

Any biological or physical indicator of environmental conditions in the past. Interpretation of palaeoenvironmental indicators generally relies heavily on MODERN ANALOGUES, which are used to establish the sensitivity of the indicator to present environmental conditions prior to their use in reconstructing past conditions. *JAM*

[See also ANALOGUE METHOD, ENVIRONMENTAL INDICATOR, SEASONALITY INDICATOR]

Guthrie, R.D. 1982: Mammals of the mammoth steppe as palaeoenvironmental indicators. In Hopkins, D.M., Matthews Jr, J.V., Schweger, C.E. and Young, S.B. (eds), *Paleoecology of Beringia*. New York: Academic Pres, 307–326.

palaeoenvironmental reconstruction

The deduction of the characteristics of past environments from PROXY EVIDENCE such as FOSSILS, SEDIMENTOLOGY, GEOMORPHOLOGY and GEOCHEMISTRY. It is the key step in identifying past ENVIRONMENTAL CHANGE in the GEOLOGICAL RECORD. *MRT*

[See also CLIMATIC RECONSTRUCTION, FACIES ANALYSIS, HOLOCENE ENVIRONMENTAL CHANGE, MULTIPROXY APPROACH, NATURAL ARCHIVES, PALAEOCLIMATOLOGY, PALAEOECOLOGY, QUATERNARY ENVIRONMENTAL CHANGE, SEDIMENTOLOGICAL EVIDENCE OF ENVIRONMENTAL CHANGE]

Reading, H.G. (ed.) 1996: *Sedimentary environments: processes, facies and stratigraphy*, 3rd edn. Oxford: Blackwell.

palaeoethnobotany

The identification, analysis and interpretation of plant remains associated with archaeological sites. Also known as *archaeobotany* and *phytoarchaeology*, it is important in PALAEOENVIRONMENTAL RECONSTRUCTION. *JAM*

[See also ENVIRONMENTAL ARCHAEOLOGY, CEREALS, CHARRED PARTICLES, PHYTOLITHS, PLANT MACROFOSSILS]

Brooks, R.R. and Johannes, D. 1990: *Phytoarchaeology*. Leicester: Leicester University Press. **Gale, R. and Cutler, D.** 2000: *Plants in archaeology*. Otley: Westbury. **Gremillion, K.J. (ed.)** 1997: *People, plants, and landscapes: studies in paleoethnobotany*. Tuscaloosa, AL: University of Alabama Press. **Hather, J.G. (ed.)** 1994: *Tropical archaeobotany: applications and new developments*. London: Routledge. **Pearsall, D.M.** 1989: *Paleoethnobotany: a handbook of procedures*. New York: Academic Press. **Renfrew, C.** 1973: *Palaeoethnobotany: the prehistoric food plants of the Near East and Europe*. London: Methuen. **Renfrew, C. (ed.)** 1991: *New light on early farming: recent developments in palaeoethnobotany*. Edinburgh: Edinburgh University Press. **Zeist, W. van, Wasylikowa, K. and Behre, K.E. (eds)** 1991: *Progress in Old World palaeoethnobotany*. Rotterdam: Balkema.

palaeoflood

A FLOOD, usually of a river, that occurred in the past, prior to historical records. The term is usually applied to severe EVENTS, sometimes also described as *megafloods*, that may have exceeded in magnitude any that have been experienced at the same locality in historical times. Floods from the pre-Quaternary GEOLOGICAL RECORD, which are inferred from GEOLOGICAL EVIDENCE, tend not to be referred to as palaeofloods, since by definition they predate human experience. Palaeofloods can be reconstructed using evidence from SEDIMENTOLOGY and GEOMORPHOLOGY. Of particular value is the preservation of fine-grained deposits (*slackwater deposits)* in areas of reduced flow velocity. Techniques such as PALAEOHYDROLOGY can be used to reconstruct flood HYDROLOGY, DISCHARGE and magnitude, and the principles of GEOCHRONOLOGY and STRATIGRAPHY can be used to reconstruct flood chronologies.

The recognition and reconstruction of palaeofloods can provide important evidence of ENVIRONMENTAL CHANGE and CLIMATIC CHANGE. The clustering in time of large-magnitude palaeofloods, for example, may provide evidence for the HOLOCENE record of phenomena such as the EL NIÑO or MONSOON systems. Palaeoflood chronologies can be used to assess the potential hazard from severe floods in the future, and estimates of the magnitudes of palaeofloods are critical in evaluating the significance of severe floods that have occurred in recent times (see NATURAL HAZARDS).

Probably the most impressive examples of palaeofloods are those associated with the catastrophic drainage of ICE-DAMMED LAKE Missoula at the end of the last GLACIATION (see JÖKULHLAUP). Giant PALAEOCHANNELS form the CHANNELED SCABLANDS of the Columbia River Basin, Washington, USA, and are associated with giant BEDFORMS, dry waterfalls and transported blocks. It has been estimated that the peak discharge may have been up to 20 times the mean discharge of all the world's rivers today. Holocene and QUATERNARY palaeofloods are also well documented from other areas, many of which are today ARIDLANDS, including southwestern USA, Australia, South Africa, the Indian peninsula and the Middle East. Surface features on other planets have also been interpreted in terms of palaeofloods (see PLANETARY ENVIRONMENTAL CHANGE). 		GO

[See also FLOOD HISTORY]

Baker, V.R. 1987: Paleoflood hydrology and extraordinary flood events. *Journal of Hydrology* 96, 79–99. **Ely, L.L.** 1997: Response of extreme floods in the southwestern United States to climatic variations in the late Holocene. *Geomorphology* 19, 175–201. **Ely, L.L. and Baker, V.R.** 1985: Reconstructing paleoflood hydrology with slackwater deposits – Verde River, Arizona. *Physical Geography* 6, 103–126. **Kale, V.S., Singhvi, A.K., Mishra, P.K. and Benerjee, D.** 2000: Sedimentary records and luminescence chronology of Late Holocene palaeofloods in the Luni River, Thar Desert, northwest India. *Catena* 40, 337–358. **Komatsu, G. and Baker, V.R.** 1997: Paleohydrology and flood geomorphology of Ares Vallis. *Journal of Geophysical Research – Planets* 102(E2), 4151–4160. **Nott, J. and Price, D.** 1999: Waterfalls, floods and climate change: evidence from tropical Australia. *Earth and Planetary Science Letters* 171, 267–276. **Wells, L.E.** 1990: Holocene history of the El Niño phenomenon as recorded in flood sediments of northern coastal Peru. *Geology* 18, 1134–1137.

Palaeogene A subperiod of the TERTIARY, from 65 to 23.3 million years ago, comprising the PALAEOCENE, EOCENE and OLIGOCENE epochs. 		GO

[See also GEOLOGICAL TIMESCALE]

palaeogeography The geography of the geological past (see Figure). The reconstruction of palaeogeography involves many aspects of GEOLOGY and must consider the distribution of land and sea, the topography of the land surface and the sea-bed, the position of major river systems, the form of any COASTLINE and the relative configuration of the CONTINENTS and their PALAEOLATITUDE. Many SEDIMENTARY DEPOSITS and FACIES are sensitive to *palaeoclimate* and, hence, if a similar distribution of climatic belts to the present day is assumed, they can help to reconstruct several of these aspects of palaeogeography. These include COAL-bearing deposits, DESERT deposits, EVAPORITES, fossil REEFS, RED BEDS, PALAEOSOLS and TILLITES. Other aspects of palaeogeography can be reconstructed using PALAEOCURRENT indicators, FOSSIL distributions and FAUNAL PROVINCES, ISOPACHS, PALAEOCEANOGRAPHY and PALAEOMAGNETISM, as well as by matching present-day CONTINENTAL MARGINS and TECTONIC lineaments. To be strictly comparable with present-day geography, palaeogeographical maps must remove the effects of any TECTONIC shortening that has taken place through OROGENIC movement (see PALINSPASTIC MAP). The term is also applicable to the most recent geological past, on the ARCHAEOLOGICAL TIMESCALE. 		JCWC/GO

[See also CONTINENTAL DRIFT]

Christiansen, J.L. and Stouge, S. 1999: Oceanic circulation as an element in palaeogeographical reconstructions: the Arenig (early Ordovician) as an example. *Terra Nova* 11, 73–78. **Cocks, L.R.M.** 2000: The Early Palaeozoic geography of Europe. *Journal of the Geological Society, London* 157, 1–10. **Cope,**

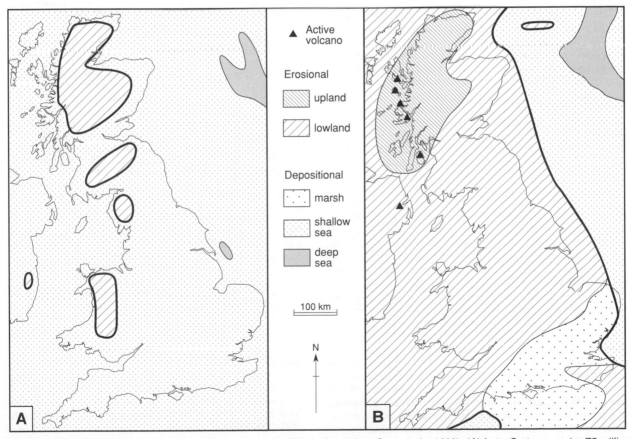

palaeogeography *Simplified palaeogeographical maps for Britain (based on Cope et al., 1992): (A) Late Cretaceous (c. 75 million years ago); (B) Late Palaeocene (c. 60 million years ago)*

J.C.W., Ingham, J.K. and Rawson, P.F. (eds)** 1992: *Atlas of palaeogeography and lithofacies* [*Memoir* 13]. London: Geological Society. **Gifford, J.A., Rapp Jr, G. and Vitali, V.** 1992: Palaeogeography of Carthage (Tunisia): coastal change during the first millennium BC. *Journal of Archaeological Science* 19, 575–596. **McKerrow, W.S. and Scotese, C.R. (eds)** 1990: *Palaeozoic palaeogeography and biogeography, Memoir* 12. London: Geological Society. **Scotese, C.R.** 1991: Jurassic and Cretaceous plate tectonic reconstructions. *Palaeogeography, Palaeoclimatology, Palaeoecology* 87, 493–501. **Smith, A.G., Smith, D.G. and Funnell, B.M.** 1994: *Atlas of Mesozoic and Cenozoic coastlines*. Cambridge: Cambridge University Press.

palaeohydrology The study of aspects of the HYDRO-LOGICAL CYCLE that operated in the past, in contrast to studies of present-day HYDROLOGY. The aims of palaeohydrology are to reconstruct hydrological parameters such as: annual PRECIPITATION, EVAPORATION, RUNOFF and DISCHARGE; variations in water storage capacity in LAKES and WETLAND areas; EXTREME EVENTS such as PALAEOFLOODS; changes in BASE LEVEL, such as LAKE-LEVEL VARIATIONS and SEA-LEVEL CHANGE; and changes in GROUNDWATER conditions. Where hydrological parameters in the past can be shown to have been different from those that exist today, palaeohydrology provides direct evidence of ENVIRONMENTAL CHANGE. Palaeohydrological relationships can also be used in GENERAL CIRCULATION MODELS to forecast the effects of future CLIMATIC CHANGE, such as changes in the frequency of FLOODS or DROUGHTS. The discipline of palaeohydrology is sometimes referred to as *palaeohydraulics*.

The concept of palaeohydrology has most significance when restricted to the reconstruction of quantitative PARAMETERS. This can be achieved through the application of empirical and theoretical relationships between hydrological parameters, such as discharge, and preserved features, such as the characteristics and thickness of river channel deposits. Variations in hydrological parameters can in turn be related to climatic and other environmental factors. The term palaeohydrology is commonly used, however, in a more general, qualitative sense, in which case there is considerable blurring of the boundaries between palaeohydrology, fluvial sedimentology, fluvial geomorphology, PALAEOLIMNOLOGY and PALAEOECOLOGY.

Methods used in the reconstruction of palaeohydrological parameters include: *geomorphometry*; SEDIMENTOLOGY, particularly FACIES ANALYSIS, GRAIN-SIZE analysis and the interpretation of SEDIMENTARY STRUCTURES; and analytical techniques such as STABLE ISOTOPE ANALYSIS. Interpretations are placed within a chronological framework using the techniques of GEOCHRONOLOGY and STRATIGRAPHY.

The important field of *continental palaeohydrology* is concerned primarily with reconstructions of river environments, such as the depth, width and morphology of PALAEOCHANNELS, flow rates and discharge regimes. The former behaviour of rivers that still exist can be reconstructed using aspects of fluvial GEOMORPHOLOGY such as RIVER TERRACES or MISFIT MEANDERS. The characteristics of ALLUVIAL DEPOSITS, particularly their GRAIN-SIZE, can be used to reconstruct hydrological parameters from deposits in present-day river basins or in the GEOLOGICAL RECORD. Analysis of the dimensions of fluvial channels has proved a fruitful line of palaeohydrological reconstruction. Empirical studies have shown consistent relationships between channel depth, channel width, MEANDER wavelength and discharge for rivers with certain CHANNEL PATTERNS. Such relationships can be applied to river channels of any age, including those preserved in SEDIMENTARY ROCKS, provided the channel pattern can be reconstructed. These approaches to reconstructions of continental palaeohydrology, using geomorphology, sediment characteristics or the geometry of sedimentary deposits, can be combined to provide detailed reconstructions of former hydrological conditions.

Palaeohydrological interpretations have been applied to landforms and deposits of all ages in the STRATIGRAPHICAL RECORD. Studies of HOLOCENE rivers have been used to determine flood chronologies and the effects of climatic change and anthropological factors on FLOOD MAGNITUDE-FREQUENCY CHANGES. In the QUATERNARY the relative effects of TECTONICS and climatic change, particularly GLACIAL–INTERGLACIAL CYCLES, on drainage patterns and drainage evolution have been investigated. Further back in the geological record, palaeohydrological methods have been applied to problems such as wetland ecology and the behaviour of rivers prior to the covering of land surfaces by terrestrial plants, with associated consequences for SOIL DEVELOPMENT. *GO*

[See also HYDROLOGICAL CYCLE]

Benito, G., Baker, V.R. and Gregory, K.J. (eds) 1998: *Palaeohydrology and environmental change*. Chichester: Wiley. **Branson, J., Brown, A.G. and Gregory, K.J. (eds)** 1996: *Global continental changes: the context of palaeohydrology* [*Special Publication* 115]. London: Geological Society. **Cross, S.L., Baker, P.A., Seltzer, G.O.** *et al.* 2000: A new estimate of the Holocene lowstand level of Lake Titicaca, central Andes, and implications for tropical palaeohydrology. *The Holocene* 10, 21–32. **Elahir, E.A.B.** 1996: El Niño and the natural variability in the flow of the Nile River. *Water Resources Research* 32, 131–137. **Gregory, K.J., Starkel, L. and Baker, V.R. (eds)** 1995: *Global continental palaeohydrology*. Chichester: Wiley. **Neut, M. van der and Eriksson, P.G.** 1999: Palaeohydrological parameters of a Proterozoic braided fluvial system (Wilgerivier Formation, Waterberg Group, South Africa) compared with a Phanerozoic example. In Smith, N.D. and Rogers, J. (eds), *Fluvial Sedimentology VI: International Association of Sedimentologists Special Publication* 28, 381–392. **Starkel, L., Gregory, K.J. and Thornes, J.B. (eds)** 1991: *Temperate palaeohydrology: fluvial processes in the temperate zone during the last 15000 years*. Chichester: Wiley. **Williams, G.P.** 1984: Paleohydrologic equations for rivers. In Costa, J.E. and Fleisher, P.J. (eds), *Developments and applications of geomorphology*. Berlin: Springer, 343–367. **Wolfe, B.B., Edwards, T.W.D., Aravena, R.** *et al.* 2000: Holocene paleohydrology and paleoclimate at treeline, north-central Russia, inferred from oxygen isotope records in lake sediment cellulose. *Quaternary Research* 53, 319–329. **Wright, V.P., Taylor, K.G. and Beck, V.H.** 2000: The paleohydrology of Lower Cretaceous seasonal wetlands, Isle of Wight, Southern England. *Journal of Sedimentary Research* 70, 619–632.

palaeolatitude The latitude of a CONTINENT at a given time in geological history. Palaeolatitude can be established using evidence from PALAEOMAGNETISM or PALAEOCLIMATOLOGY. Changes in palaeolatitude through GEOLOGICAL TIME contribute to the reconstruction of CONTINENTAL DRIFT. *DNT*

[See also APPARENT POLAR WANDER, COAL, EVAPORITES, PLATE TECTONICS]

458

palaeolimnology The study of physical and biological parameters of LAKE SEDIMENTS to reconstruct environmental and ECOSYSTEM changes in the past; usually a MULTIDISCIPLINARY study. Lake sediments generally accumulate at a rate of around 1 mm y^{-1}, forming a chronological NATURAL ARCHIVE. Sediments can be sampled by a variety of CORERS from marginal fens, open water or ice. A range of biological (FOSSIL), chemical (inorganic and organic) and physical (e.g. GRAIN-SIZE, PALAEOMAGNETISM) parameters can be analysed to reconstruct past changes in the lake ecosystem including the *catchment*, which can be interpreted in terms of climatic and ENVIRONMENTAL CHANGE, natural or human induced. CHRONOLOGY can be controlled by RADIOCARBON DATING, ^{210}Pb DATING in young sediments and VARVES if present, allowing the timing and rates of changes to be estimated. Cores within one or among several lakes can be correlated by dating (see CORRELATED-AGE DATING), PALAEOMAGNETISM, LOSS-ON-IGNITION (LOI), sedimentary features, and fossil stratigraphy.

Sediments have an AUTOCHTHONOUS component (produced within the lake) and an ALLOCHTHONOUS component (derived from the catchment or the atmosphere). Changes in the catchment are controlled largely by temperature. In combination with precipitation, temperature affects the vegetation and soil types and stability, the HYDROLOGICAL BALANCE (LAKE-LEVEL VARIATIONS), the degree of snow cover and frost activity, and sometimes glacier formation. Changes in the catchment affect the allochthonous input. The proportions of clastic (including carbonate) and organic material (measured by LOI) reflect catchment stability. The MINEROGENIC COMPONENT increases with disturbance that can result from *frost action*, glacier activity, changes in lake level and water inflow (EROSION), and human activity (DEFORESTATION, landscape modification). Base cations and the chemistry of insoluble components reflect catchment geology. N, P and *dissolved organic carbon* (DOC) (humic compounds) originate in soils. N and P increase with EUTROPHICATION, often caused by HUMAN IMPACT (FOREST CLEARANCE, FERTILISER application, SEWAGE and INDUSTRIAL WASTE). DOC increases with forest development and natural soil acidification (PODZOLISATION and PALUDIFICATION), but decreases with acidification below pH 5 resulting, for example, from ACID RAIN. Other organic material derives from vegetation or soils and comprises plant and animal MACROFOSSILS and MICROFOSSILS (POLLEN etc.) as well as decayed material. Atmospheric inputs include: POLLEN RAIN (reflecting regional vegetation); DUST (LOESS); VOLCANIC ASH (a chronological marker); *charcoal* (from vegetation burning); SPHEROIDAL CARBONACEOUS PARTICLES (yielding an industrial chronology); *radioisotopes* (^{210}Pb, ^{137}Cs etc.); strong acids in rain and HEAVY METAL residues (Pb, Zn, Cu, Cr, Mn and Hg) that are industrial POLLUTION indicators; and PESTICIDE derivatives in AEROSOLS.

Autochthonous sediments are composed of decayed remains of aquatic organisms and lake precipitates (e.g. DY, MARL, saline deposits). A lake ecosystem is affected by internal processes such as: SEDIMENT ACCUMULATION; SEDIMENT TYPE; the organisms present, their immigration, successions and extinctions and competition between them for light (plants), food (phytophagous or zoophagous), NUTRIENTS, HABITAT, dissolved O_2; seasonal OVERTURNING and degree of ANOXIA in the *hypolimnion*; and winter ice-cover (see Figure). Thus organisms are valuable INDICATOR SPECIES of past lake conditions. In a pioneer situation, aquatic organisms react very fast to climate changes, by changes in abundance if they are already present and by immigration. *Diatoms* (see DIATOM ANALYSIS) are the most abundant plant fossils, but other algae, mosses and higher plant remains can be identified (PLANT MACROFOSSILS and POLLEN). Common animal fossils include CHIRONOMIDAE, CLADOCERA, *Coleoptera*, Trichoptera, other insects and Oribatida. Fish are less common. Diatoms are sensitive pH INDICATORS and lake-water pH can be reconstructed through time by application of modern TRANSFER FUNCTIONS to the fossil assemblage. Other diatom and CHRYSOPHYTE transfer functions have been constructed for phosphorus (EUTROPHICATION) and SALINITY. Similarly, Chironomidae and Cladocera have been used for quantitative reconstruction of summer and winter temperatures. Qualitative reconstruction of lake-water quality and chemistry and lake and catchment habitats can be made from all plant and animal groups, depending upon their known ecology. *Pigments* derived from algae and cyanobacteria can be preserved in ANAEROBIC conditions and used to reconstruct EUTROPHICATION and POLLUTION. STABLE ISOTOPE fractionation, for example of C, N, O and H, in different sediment and organism components also yields environmental information.

A few lake sediment sequences span many thousand years; some include the LATE GLACIAL, and the majority cover the HOLOCENE. Long-term palaeolimnological studies can assess climate change, the reaction of lake ecosystems to natural, *climatic forcing* (e.g. Younger Dryas–Holocene transition) and natural lake ecosystem and catchment development in relation to Holocene environmental changes and internal lake processes. HUMAN IMPACT has gradually increased from mid-Holocene time, registering effects in lake sediments. Over the last century WATER POLLUTION, industrial ACIDIFICATION, EUTROPHICATION, *dams* and water removal have increasingly affected *water quality*. The continued existence of many lakes in arid climates is under severe threat. Palaeolimnology can be used to provide a baseline of natural BIODIVERSITY and ecosystem structure, against which impacts of modern technology can be registered and evaluated. ENVIRONMENTAL MONITORING programmes take several years to register trends, but high-resolution recent sediment studies can readily provide evidence of trends and indicate appropriate action for ecosystem quality improvement. Palaeolimnology thus interfaces with LIMNOLOGY and ECOSYSTEM MANAGEMENT (CONSERVATION). *HHB*

[See also LAKES AS INDICATORS OF ENVIRONMENTAL CHANGE, LAKE STRATIFICATION AND ZONATION, SEDIMENTOLOGICAL EVIDENCE OF ENVIRONMENTAL CHANGE]

Battarbee, R.W. 2000: Palaeolimnological approaches to climate change, with special regard to the biological record. *Quaternary Science Reviews* 19, 107–124. **Binford, M.W., Deevey, E.S. and Crisman, T.L.** 1983: Paleolimnology: an historical perspective on lacustrine ecosystems. *Annual Review of Ecology and Systematics* 14, 255–286. **Birks, H.H., Battarbee, R.W. and Birks, H.J.B.** 2000: The development of the aquatic ecosystem at Kråkenes Lake, western Norway, during the late glacial and early Holocene – a synthesis. *Journal of Paleolimnology* 23: 91–114. **Boomer, I., Aladin, N., Plotnikov, I. and Whatley, R.** 2000: The palaeolimnology of the Aral Sea. *Quaternary Science Reviews* 19, 1259–1278. **Charles, D.F.,**

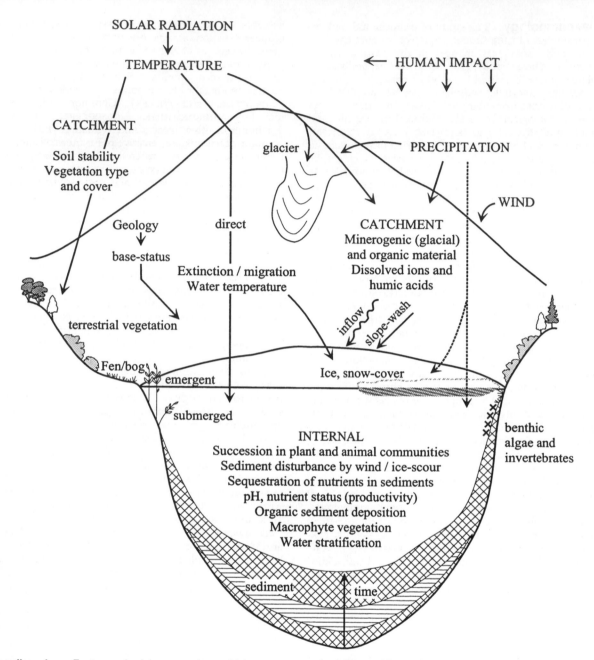

palaeolimnology *Features of a lake ecosystem, which can cause or be influenced by environmental changes that can be subsequently registered in the sediments*

Smol, J.P. and Engstrom, D.R. 1994: Paleolimnological approaches to biological monitoring. In Loeb, S.L. and Spacie, A. (editors), *Biological Monitoring of Aquatic Systems*. Boca Raton, FL: CRC Press, 233–293. **Smol, J.P., Cumming, B.F., Douglas, M.S.V. and Pienitz, R.** 1996: Inferring past climatic changes in Canada using palaeolimnological techniques. *Geoscience Canada* 21, 113–118. **Smol, J.P., Walker, I.R. and Leavitt, P.R.** 1991: IV. Paleolimnology. Paleolimnology and hindcasting climatic trends. *Verhandlungen der Internationale Vereinigung für Theroretische und Angewandte Limnologie* 24, 1240–1246.

Palaeolithic   The *Old Stone Age*, traditionally defined as the archaeological period during which stone implements were used. The period equates roughly with the end of the PLEISTOCENE, during which time HUNTING, FISHING AND GATHERING societies frequently moved location (NOMADISM) to obtain resources. Although stone was the principal material used for making tools, antler, bone and wood were commonly utilised.   *LD-P*

[See also ARCHAEOLOGICAL TIMESCALE, CAVE ART]

Cunliffe, B. 1994: *Prehistoric Europe*. Oxford: Oxford University Press. **Gamble, C.** 1986: *The Palaeolithic settlement of Europe*. Cambridge: Cambridge University Press. **Gamble, C.** 1996: *The Palaeolithic societies of Europe*. Cambridge: Cambridge University Press. **Roe, D.A.** 1986: *Studies in the upper Palaeolithic of Britain and north west Europe* [*BAR International Series* 296]. Oxford: British Archaeological Reports. **Simmons, I.G. and Tooley,**

M. 1981: *The environment in British prehistory*. London: Duckworth, 49–81.

Palaeolithic: human–environment relations

PALAEOLITHIC societies were often based on a HUNTING, FISHING AND GATHERING economy and modes of food and resource procurement which depend on the environment for their success. ENVIRONMENTAL CHANGE during the Palaeolithic had significant impacts on society. For example, CLIMATIC CHANGE and the retreat of ice sheets from many lowland areas of Europe at the end of the PLEISTOCENE influenced the MIGRATION of Palaeolithic groups. The environment would also have determined the availability of food (game, plants) and non-food (stone, wood) resources. Although the environment influenced where and how Palaeolithic groups lived, people also had the ability to modify and adapt to the environment. For example, changes in the tool types demonstrate ETHOLOGICAL (behavioural) ADAPTATION to the environment, and changing methods of food procurement and the use of fire demonstrate the ability to modify wooded landscapes. *LD-P*

Barton, N. 1991: Technological innovation and continuity at the end of the Pleistocene in Britain. In Barton, N., Roberts, A.J., Roe, D.A. *et al*. (eds), *The Late Glacial in North West Europe* [*Research Report* 77]. York: Council for British Archaeology, 234–245. Dolukhanov, P.M. 1997: The Pleistocene–Holocene transition in northern Eurasia: environmental changes and human adaptations. *Quaternary International* 41/42, 181–191. VanAndel, T.H. and Tzedakis, P.C. 1996: Palaeolithic landscapes of Europe and environs, 150,000–25,000 years ago: an overview. *Quaternary Science Reviews* 15, 481–500.

palaeomagnetic dating

A broad term that includes ARCHAEOMAGNETIC DATING and describes *age-equivalence* dating techniques based on the NATURAL REMANENT MAGNETISM (NRM) of archaeological materials, sediments or rocks. The DECLINATION, INCLINATION and MAGNETIC INTENSITY of the Earth's *geomagnetic field* (see GEOMAGNETISM) vary continually over time. Temporal variations are preserved in stratigraphic sequences. As molten volcanic rocks and baked CERAMICS cool, ferromagnetic particles within them align with the ambient geomagnetic field, acquiring THERMOREMANENT MAGNETISATION (TRM). Similarly, ferromagnetic sediment particles settling out of air, water or saturated sediments can acquire DETRITAL REMANENT MAGNETISATION (DRM), providing a proxy record of the ambient geomagnetic field at the time of formation.

On geological timescales, variations in the Earth's geomagnetic field include 180° shifts in direction of the Earth's geomagnetic pole, i.e. between NORMAL POLARITY and *reverse polarity*. GEOMAGNETIC POLARITY REVERSALS vary in duration. POLARITY CHRONS or epochs lasting 10^5–10^7 years (e.g. the Bruhnes present normal polarity) are punctuated by relatively short-lived *polarity events* (or subchrons) of 10^4–10^5 years duration (e.g. six normal polarity events occur within the Matuyama chron). The NRM signal of geomagnetic reversal boundaries in sediments, such as slowly accumulating MARINE and LAKE SEDIMENTS, LOESS and LAVA sequences, provide important age-equivalent MARKER HORIZONS. The global and broadly SYNCHRONOUS nature of geomagnetic polarity reversals means that reversal stratigraphy can be used to correlate between stratigraphic sequences on a global scale. Furthermore, POTASSIUM–ARGON DATING of field-reversal boundaries identified in lava sequences enables their use as global chronostratigraphic marker horizons. Recently, an alternative palaeomagnetic timescale has been provided by ORBITAL TUNING of the marine sediment records. Palaeomagnetic dating of field reversals has aided CORRELATION and dating of the records of QUATERNARY and pre-Quaternary ENVIRONMENTAL CHANGE preserved in different marine cores and facilitated correlation of marine records with terrestrial loess and lake sequences. For example, the GAUSS–MATUYAMA GEOMAGNETIC BOUNDARY occurs close to the PLIOCENE–PLEISTOCENE boundary.

NRM records of MAGNETIC SECULAR VARIATIONS (i.e. short term, 10^3 years or less) may be preserved in lake sediments, LOESS sequences or baked archaeological materials (e.g. pottery, kiln walls, tiles). Matching of palaeomagnetic signals from different stratigraphic sequences or samples provides a basis for establishing age equivalence. Age estimates can be derived by correlation with detailed master geomagnetic curves derived from lake sequences that have been independently dated or calibrated by, for example, RADIOCARBON DATING. Regional differences in secular geomagnetic variations mean that this palaeomagnetic dating method is limited to intraregional correlation.

A third type of palaeomagnetic dating is based on the MINERAL MAGNETISM of sediment sequences arising from variations in the magnetic mineral content. These magnetic properties are independent of the geomagnetic field and may reflect environmental changes. MAGNETIC SUSCEPTIBILITY, ISOTHERMAL REMANENT MAGNETISM (IRM) and coercivity of IRM profiles have been used to correlate marine and terrestrial stratigraphic sequences. Mineral magnetic stratigraphy has been used to correlate and establish age equivalence of Chinese loess sequences and oxygen ISOTOPIC STAGES in marine records. *MHD*

Creer, K.M. and Kopper, J.S. 1974: Paleomagnetic dating of cave paintings in Tito Bustillo Cave, Asturias, Spain. *Science* 186, 348–350. Hagstrum, J.T. and Champion, D.E. 1994: Paleomagnetic correlation of Late Quaternary lava flows in the lower east rift zone of Kilawea Volcano, Hawaii. *Journal of Geophysical Research* 99, 21679–21690. Kukla, G.J., Heller, F., Liu, X.M. *et al.* 1988: Pleistocene climates in China dated by magnetic susceptibility. *Geology* 16, 811–814. Mankinen, K.E. and Dalrymple, G.B. 1979: Revised geomagnetic polarity time scale for the interval 0–5 M.y. B.P. *Journal of Geophysical Research* 84, 615–626. Tarling, D.H. 1983: *Palaeomagnetism: principles and applications in geology, geophysics and archaeology*. London: Chapman and Hall. Tarling, D.H. 1990: Archaeomagnetism and palaeomagnetism. In Göksu, H.Y., Oberhofer, M. and Regulla, D. (eds), *Scientific dating methods*. Dordrecht: Kluwer, 217–250. Thompson, R. 1991: Palaeomagnetic dating. In: Smart, P.L. and Frances, P.D. (eds), *Quaternary dating methods – a users guide* [*QRA Technical Guide* 4]. Cambridge: Quaternary Research Association, 177–198. Thompson, R. and Oldfield, F. 1986: *Environmental magnetism*. London: Allen and Unwin. Valet, J.-P. and Meynadier, L. 1993: Geomagnetic field intensity and reversals during the past four million years. *Nature* 366, 234–238. Verosub, K.L. 1988: *Geomagnetic secular variations and the dating of Quaternary sediments* [*Special Paper* 227]. Boulder, CO: Geological Society of America, 123–138.

palaeomagnetism

The study of the characteristics of the Earth's magnetic field (see GEOMAGNETISM) in the past and its application to geological, archaeological and environmental problems. Rocks, sediment and soil retain a permanent record of the Earth's magnetic field at the time of their formation in magnetic MINERALS such as iron oxides. This MAGNETIC REMANENCE can be isolated by laboratory techniques. Of particular importance are variations in the inclination and polarity of the preserved field. The science of palaeomagnetism expanded in the AD 1950s and is now a technologically sophisticated research field of major importance to EARTH SCIENCE.

Two major applications of palaeomagnetism are in reconstructing PALAEOGEOGRAPHY and in GEOCHRONOLOGY. The INCLINATION of the preserved magnetic field is related to PALAEOLATITUDE and changes through time can be used to construct a trajectory of APPARENT POLAR WANDER. The recognition that MAGNETIC ANOMALIES in IGNEOUS ROCKS of the OCEANIC CRUST had been produced by SEA-FLOOR SPREADING was vital in the acceptance of the theory of PLATE TECTONICS. Both apparent polar wander paths and ocean-floor magnetic anomalies are invaluable in reconstructing palaeogeography and CONTINENTAL DRIFT. The dating of GEOMAGNETIC POLARITY REVERSALS has led to the development of the GEOMAGNETIC POLARITY TIMESCALE. Other applications include: the refinement of stratigraphical CORRELATION (MAGNETOSTRATIGRAPHY); PALAEOMAGNETIC DATING; determining the PROVENANCE of sediments and sedimentary rocks (see MAGNETIC SUSCEPTIBILITY); reconstructions of *palaeoclimate* and PALAEOENVIRONMENT; and the reorientation of drill CORES by comparison with present field declinations.   *DNT*

Butler, R.F. 1991: *Palaeomagnetism: magnetic domains to geologic terranes.* Oxford: Blackwell Science. **Channell, J.E.T., Stoner, J.S., Hodell, D.A. and Charles, C.D.** 2000: Geomagnetic paleointensity for the last 100 kyr from the sub-antarctic South Atlantic: a tool for inter-hemispheric correlation. *Earth and Planetary Science Letters* 175, 145–160. **Dinares-Turell, J., Orti, F., Playa, E. and Rosell, L.** 1999: Palaeomagnetic chronology of the evaporitic sedimentation in the Neogene Fortuna Basin (SE Spain): early restriction preceding the 'Messinian Salinity Crisis'. *Palaeogeography, Palaeoclimatology, Palaeoecology* 154, 161–178. **Ding, Z.L., Xiong, S.F., Sun, J.M.** *et al.* 1999: Pedostratigraphy and paleomagnetism of a similar to 7.0 Ma eolian loess-red clay sequence at Lingtai, Loess Plateau, north-central China and the implications for paleomonsoon evolution. *Palaeogeography, Palaeoclimatology, Palaeoecology* 152, 49–66. **McElhinny, M.W. and McFadden, P.L.** 1999: *Paleomagnetism: continents and oceans.* London: Academic Press. **Piper, J.D.A.** 2000: The Neoproterozoic Supercontinent: Rodinia or Palaeopangaea? *Earth and Planetary Science Letters* 176, 131–146.

palaeontology The study of FOSSILS.   *GO*

Benton, M.J. 1997: *Vertebrate palaeontology*, 2nd edn. London: Chapman and Hall. **Benton, M.J. and Harper, D.A.T.** 1997: *Basic palaeontology.* Harlow: Longman. **Clarkson, E.N.K.** 1998: *Invertebrate palaeontology and evolution*, 4th edn. Oxford: Blackwell Science.

palaeosalinity The SALINITY of any past AQUATIC

(lacustrine, lagoonal, GROUNDWATER, oceanic, etc.) or SOIL ENVIRONMENT recorded in sediments and 'measured' by PROXY EVIDENCE (biological, lithological and chemical). Past fluctuations in the salt content of water bodies reflect changing environmental conditions where net EVAPORATION exceeds precipitation. DIATOM, MOLLUSCA and OSTRACOD ANALYSIS (including TRACE ELEMENT ratio analysis of carapaces) is widely employed in palaeosalinity reconstructions. Recently, calibration datasets generated from *diatom* (and ostracod) salinity TRANSFER FUNCTIONS have been used for quantitative reconstruction of LACUSTRINE SEDIMENTS. In SALINE LAKES, stable OXYGEN ISOTOPE

RATIOS are a proxy for EVAPORATION intensity, and therefore for palaeosalinity. CHEMICAL ANALYSIS of lake sediment cores can indicate changing salinity, e.g. a sequence of freshwater carbonates through sulphates to chlorides implies increasing salinity. Oceanic palaeosalinity can be reconstructed using $\delta^{18}O$ and SEA-SURFACE TEMPERATURES (see STABLE ISOTOPE ANALYSIS). In turn, palaeosalinity can provide a history of SEA-LEVEL CHANGE, soil SALINISATION and lake evolution, e.g. inferred freshwater to saline conditions indicates a change from an open to a closed lake setting.

SALINITY is critical to the pattern of ocean THERMOHALINE CIRCULATION and past changes in salinity resulted in significant oscillations in behaviour of ocean currents. The GEOLOGICAL RECORD OF ENVIRONMENTAL CHANGE contains evidence of palaeosalinity in the form of EVAPORITE deposits that are indicative of past periods of ARIDITY. Thick Late Miocene salt deposits in the western Mediterranean identify past environments in which there was complete desiccation of the Mediterranean Sea. The effects of lower sea levels during the LAST GLACIAL MAXIMUM, coupled with environmental change to drier conditions in mid-latitudes, is believed to have caused hypersaline conditions in the Red Sea and Mediterranean. Continental evidence of palaeosalinity is also found in the geological record relating to past DESERT conditions.   *CES/MLC*

Hendry, J.P. and Kalin, R.M. 1997: Are oxygen and carbon isotopes of mollusc shells reliable palaeosalinity indicators in marginal marine environments? A case study from the Middle Jurassic of England. *Journal of the Geological Society* 154, 321–333. **Reed, J.M.** 1998: A diatom-conductivity transfer function for Spanish salt lakes. *Journal of Paleolimnology* 19, 399–416. **Rostek, F., Ruhland, G., Bassinot, F.C.** *et al.* 1993: Reconstructing sea-surface temperature and salinity using $\delta^{18}O$ and alkenone records. *Nature* 364, 319–321. **Thunnell, R.C. and Williams, D.F.** 1989: Holocene salinity changes in the Mediterranean Sea: hydrographic and depositional effects. *Nature* 338, 493–496.

palaeosciences The *'historical sciences'* or, more

appropriately, those branches of the NATURAL ENVIRONMENTAL SCIENCES (such as PALAEOCLIMATOLOGY, PALAEOECOLOGY and PALAEOCEANOGRAPHY) that focus on the reconstruction and modelling of past events rather than direct observation and experiment. However, they use MODERN ANALOGUES in the interpretation of past events and provide a long-term perspective for knowledge and understanding of present and future environmental change.   *JAM*

[See also PALAEODATA]

palaeoseismicity The study of past EARTHQUAKES

with the aim of determining their dates, EPICENTRES and magnitudes. Methods that can be used include: the study of SEDIMENTARY STRUCTURES indicative of seismically induced LIQUEFACTION or FLUIDISATION, such as *sand blows* and SOFT-SEDIMENT DEFORMATION structures; the recognition of seismically generated EVENT DEPOSITS within successions of sediments or sedimentary rocks; the analysis of GEOMORPHOLOGY; and cross-cutting features such as upward terminations of FAULTS in cross-sections. A knowledge of palaeoseismicity can contribute to the understanding of the TECTONIC evolution of fault systems and an assessment of seismic hazards, particularly in places where there is little or no historical record of large earthquakes (e.g. eastern seaboard of North America). This can allow the determination of *recurrence intervals* between large earthquakes, which is important in EARTHQUAKE PREDICTION. *GO/DNT*

Amick, D. and Gelinas, R. 1991: The search for evidence of large prehistoric earthquakes along the Atlantic seaboard. *Science* 251, 655–658. Doig, R. 1990: 2300 yr history of seismicity from silting events in Lake Tadoussac, Charlevoix, Quebec. *Geology* 18, 820–823. Jacoby, G.C., Sheppard, P.R. and Sieh, K.E. 1988: Irregular recurrence of large earthquakes along the San-Andreas Fault – evidence from trees. *Science* 241, 196–199. McCalpin, J. (ed.) 1988: *Paleoseismology*. New York: Academic Press. Obermeier, S.F. 1996: Use of liquefaction-induced features for paleoseismic analysis – an overview of how seismic liquefaction features can be distinguished from other features and how their regional distribution and properties of source sediment can be used to infer the location and strength of Holocene paleoearthquakes. *Engineering Geology* 44, 1–76. Rosetti, D.D. 1999: Soft-sediment deformation structures in late Albian to Cenomanian deposits, Sao Luis Basin, northern Brazil: evidence for palaeoseismicity. *Sedimentology* 46, 1065–1081.

palaeoslope

A slope that no longer exists. Its presence can be inferred using PALAEOCURRENT indicators and other SEDIMENTARY STRUCTURES, such as SLUMP folds. *GO*

Bradley, D. and Hanson, L. 1998: Paleoslope analysis of slump folds in the Devonian flysch of Maine. *Journal of Geology* 106, 305–318.

palaeosol

A soil formed in a past environment not representative of the present phase of soil formation. It may be a FOSSIL SOIL, completely buried by later sedimentary deposition, and possibly lithified (see LITHIFICATION), or it may be a RELICT soil at the land surface containing features indicating that it formed in a different climate or beneath other vegetation than those influencing the current conditions of soil formation. Palaeosols are clear evidence of environmental change as they usually possess characteristics that are different from present soils. The full verification of a buried soil in a single section is rarely possible and it is necessary to show that its morphology changes laterally as well as vertically, forming a *palaeocatena* (see CATENA) for positive recognition.

In southern Britain, soils in pre-Devensian materials have been described as palaeo-argillic brown earths (LUVISOLS), characterised by decalcification, clay movement and reorganisation of clayey soil matrices. Relict features of CRYOTURBATION from the Devensian are commonly seen (see INVOLUTIONS) in subsoils of CAMBISOLS in northern Europe. CHERNOZEMS have been changed to LUVISOLS and *podzoluvisols* by a change to more humid and forested conditions. Thus, the pedology of palaeosols has great potential for elucidating and reconstructing past climatic change, and possibly for predicting future pedological changes as environmental change occurs.

The most useful palaeosols are those forming time-transgressive CHRONOSEQUENCES such as occur in LOESS deposits containing several episodes of soil formation. Palaeosol sequences are also recognised in volcanic, alluvial and slope deposits. Sequences of RIVER TERRACES may have soils of successively decreasing age. Although palaeosols are less likely to be well preserved in glacial deposits, buried INTER-

GLACIAL soils of central Europe have been correlated with LOESS sequences for the middle and late Pleistocene. It is almost impossible to prove that a palaeosol has not been altered after its burial, and the RADIOCARBON DATING of palaeosols is often very difficult because the turnover of carbon in soils occurs at varying rates according to climate and other factors and the possibility of contamination by recent root material is great. *EMB*

[See also EXHUMED SOIL, GEOSOL, SOIL DATING, SOIL STRATIGRAPHY]

Catt, J.A. 1986: *Soils and Quaternary geology*. Oxford: Clarendon Press. Catt, J.A. 1989: Relict properties in soils of the central and north-west European temperate region. In Bronger, A. and Catt, J.A. (eds), *Palaeopedology: nature and application of palaeosols. Catena supplement* 16, 41–58. Catt, J.A. and Bronger, A. (eds) 1998: Reconstruction and climatic implications of palaeosols. *Catena* 34, 1–207. Ellis, S. and Matthews, J.A. 1984: Pedogenic implications of a ^{14}C-dated palaeopodzolic soil at Haugabreen, southern Norway. *Arctic and Alpine Research* 16, 77–91. Kraus, M.J. 1999: Paleosols in clastic sedimentary rocks: their geologic applications. *Earth Science Reviews* 47, 41–70. Reinhardt, J. and Sigleo, W.R. 1988: *Paleosols and weathering through geological time: principles and applications* [Special Paper 216]. Reston, VA: United States Geological Survey, 1–181. Retallack, G.J. 1990: *Soils of the past*. Boston: Unwin Hyman. Wright, V.P. (ed.) 1986: *Palaeosols: their recognition and interpretation*. Oxford: Blackwell Scientific. Yaalon, D.H. (ed.), 1971: *Palaeopedology: origin, nature and dating of palaeosols*. Jerusalem: ISSS and Jerusalem University Press.

palaeovalley

A former valley that no longer exists, or a valley that was formed in the past under different conditions than those pertaining at present (and therefore is RELICT). Examples include valleys that have been infilled by SEDIMENTARY DEPOSITS or LAVA FLOWS, or valleys whose existence in the past can be inferred from GEOLOGICAL EVIDENCE. The term is commonly used in relation to *incised palaeovalleys* in the context of SEQUENCE STRATIGRAPHY. These are river valleys that experienced major incision during a fall in BASE LEVEL and were infilled by AGGRADATION during a subsequent base-level rise such as a marine TRANSGRESSION. *GO*

Gupta, S. 1997: Tectonic control on paleovalley incision at the distal margin of the early Tertiary Alpine foreland basin, southeastern France. *Journal of Sedimentary Research* 67, 1030–1043. Leeder, M.R. and Stewart, M.D. 1996: Fluvial incision and sequence stratigraphy: alluvial responses to relative sea-level fall and their detection in the geological record. In Hesselbo, S.P. and Parkinson, D.N. (eds), *Sequence stratigraphy in British geology* [Special Publication 103]. London: Geological Society, 25–39.

palaeovelocity

The velocity of a past FLOOD flow estimated from geomorphological evidence, such as the depth of the flow, and the slope and roughness of the former channel. *JAM*

Church, M., Wolcott, J. and Maizels, J. 1990: Palaeovelocity: a parsimonious proposal. *Earth Surface Processes and Landforms* 15, 475–480.

palaeowind

A wind that operated in the past. The direction and strength of palaeowinds may be reconstructed from RELICT landforms and SEDIMENTARY STRUCTURES. There is also the possibility of reconstructing surface airflow patterns over considerable areas for specific periods of the GEOLOGICAL RECORD. Since AEOLIAN processes tend to leave their mark on the landscape under arid or semi-arid conditions, evidence of ancient airflows may provide a clear indication of CLIMATIC CHANGE. The erosional effectiveness of wind depends on weak binding agencies in the soil, a SOIL MOISTURE DEFICIT and a bare soil surface, which are all associated with ARIDITY. *RDT*

[See also PALAEODUNES, VENTIFACTS]

Carruthers, R.A. 1987: Aeolian sedimentation from the Galtymore Formation (Devonian), Ireland. In Frostick, L. and Reid, I. (eds), *Desert sediments: ancient and modern* [*Special Publication* 35]. London: Geological Society, 251–268. **Thomas, D.S.G. and Shaw, P.A.** 1991: 'Relict' desert dune systems: interpretations and problems. *Journal of Arid Environments* 20, 1–14.

Palaeozoic An ERA of geological time incorporating the CAMBRIAN, ORDOVICIAN, SILURIAN *(early/Lower Palaeozoic)*, DEVONIAN, CARBONIFEROUS and PERMIAN *(late/Upper Palaeozoic)* PERIODS. *GO*

[See also GEOLOGICAL TIMESCALE]

palimpsest A 'multilayered', RELICT feature (e.g. LANDSCAPE, CONTINENTAL MARGIN, SEDIMENT distribution, rock TEXTURE) in which the effects of successive generations of formative processes can still be deciphered. The environmental use of the term derives, by analogy, from the re-use of parchment after earlier writing has been rubbed out. *JAM/GO*

palinspastic map A reconstruction of PALAEOGEOGRAPHY that removes the effects of TECTONIC deformation and represents the original spatial disposition of features on the Earth's surface. The term is from the Greek, 'again pulling'. *JCWC*

Cope, J.C.W., Ingham, J.K. and Rawson, P.F. (eds) 1992: *Atlas of palaeogeography and lithofacies* [*Memoir* 13]. London: Geological Society. [See particularly the Lower Palaeozoic maps for northern Britain.]

palsa A dome-shaped or elongated peat-covered mound up to about 5 m high in areas of discontinuous PERMAFROST associated with the development of SEGREGATION ICE (ice lenses). It is a type of FROST MOUND. *RAS*

Seppälä, M. 1986: The origin of palsas. *Geografiska Annaler* 68A, 141–147. **Seppälä, M.** 1994: Snow depth controls palsa growth. *Permafrost and Periglacial Processes* 5, 283–288.

paludification The process whereby BLANKET BOGS or MIRES expand into formerly dry, often forested upland as a result of changes in hydrological conditions of the surrounding area. This can result from natural ENVIRONMENTAL CHANGE or ANTHROPOGENIC influence. The rising water level that causes bog expansion may result from increased precipitation or a decrease in EVAPOTRANSPIRATION as a result of a cooler and wetter or more humid ENVIRONMENT. Paludification can also be caused by a change in drainage characteristics of the area and DEFORESTATION; forest fires have also been implicated. *Sphagnum* moss species often act as pioneer communities within areas undergoing paludification, stimulating rapid colonisation by other bog or mire communities. This results in organic matter accumulation and an increase in thickness of PEAT. CLIMATIC CHANGE during the PLEISTOCENE and HOLOCENE has often been seen as responsible for paludification. *MLC*

Hulme, P.D. 1994: A paleobotanical study of a paludifying pine forest on the island of Hailuolo, Finland. *New Phytologist* 126, 153–162. **Moore, P.D. and Bellamy, D.J.** 1974 *Peatlands.* New York: Springer. **Ugolini, F.C. and Mann, D.H.** 1979: Biopedological origin of peatland in south east Alaska. *Nature* 281, 366–368.

palustrine Pertaining to the marginal areas of lakes and SWAMPS. Many lakes are sensitive to variations in HYDROLOGICAL BALANCE/BUDGET, so ancient palustrine deposits are potentially valuable indicators of environmental change. *MRT*

[See also PARALIC]

Wright, V.P. and Platt, N.H. 1995: Seasonal wetland carbonate sequences and dynamic catenas: a reappraisal of palustrine limestones. *Sedimentary Geology* 99, 65–71.

pan A closed ARIDLAND depression up to about 1000 km^2 in area, which may contain an EPHEMERAL lake or have contained a permanent lake in the past. Pans have been attributed to the combination of a variety of processes, including DEFLATION by wind, GROUNDWATER variations and SALINISATION. In some regions they are termed *nor* (Mongolia), *playa* (North America) and *sebkha* (North Africa to the Middle East) and the term pan has also been applied to SALINE LAKES and/or their basins. The term *panfan* has been used to describe an extensive plain produced by the amalgamation of ALLUVIAL FANS. *JAM*

[See also CHOTT, HARDPAN, SABKHA]

Goudie, A.S. and Wells, G.L. 1995: The nature, distribution and formation of pans in the arid zone. *Earth-Science Reviews* 38, 1–69. **Neal, J.T. (ed.)** 1975: *Playas and dried lakes.* Stroudsburg, PA: Dowden, Hutchinson and Ross. **Shaw, P.A. and Thomas, D.S.G.** 1997: Pans, playas and salt lakes. In Thomas, D.S.G. (ed.), *Arid zone geomorphology: process, form and change in drylands*, 2nd edn. Chichester: Wiley, 293–317.

Pangaea (Pangea) A single CONTINENT *(supercontinent)* assembled from all the present-day continents, that existed for up to 70 million years from the late CARBONIFEROUS or early PERMIAN to the late TRIASSIC periods. Pangaea formed when the continents of GONDWANA (Africa, Antarctica, Australia, peninsular India and South America) joined with those of *Laurasia* (Asia, Europe and North America). Its PALAEOGEOGRAPHY consisted of a continent stretching from northern to southern polar regions with a narrow equatorial part, to the east of which the TETHYS ocean formed an embayment. The concept of Pangaea ('all lands') was first proposed by Alfred Wegener as part of his theory of CONTINENTAL DRIFT. It is unclear whether such amalgamations of continents are a random or cyclical feature of Earth history, but it seems likely that at least one previous supercontinental assemblage has occurred, during the late PROTEROZOIC.

The break-up of Pangaea and its consequences dominates the MESOZOIC and CENOZOIC history of the Earth. SEA-FLOOR SPREADING first developed between North America and Africa/South America, initiating the southern part of the north Atlantic Ocean. East of Africa another rift separated Africa/South America from other Gondwanan continents and a further rift parted India from Antarctica. Progressive extension of these spreading centres through the JURASSIC and CRETACEOUS periods led to the opening of the southern Atlantic Ocean and widening of the northern Atlantic, which remained closed to the north. As Africa and India separately moved northwards, the Tethys Ocean was squeezed out of existence and during the late PALAEOGENE and early NEOGENE this compressional movement culminated in the formation of the great mountain chains that cross Europe and Asia, including the Alps and the Himalayas (see OROGENIC BELT). The Atlantic finally connected with the Arctic Ocean and continues to grow at the expense of the Pacific. In two or three hundred million years' time the Pacific may have disappeared, forming a 'neopangaean' supercontinent with Europe and Africa on its western margin. *JCWC*

[See also PLATE TECTONICS]

Embry, A.F., Beauchamp, B. and Glass, D.J. (eds) 1994: *Pangea: global environments and resources* [*Memoir 17*]. Calgary: Canadian Society of Petroleum Geologists. **Parrish, J.T.** 1993: Climate of the supercontinent Pangea. *Journal of Geology* 101, 215–233. **Piper, J.D.A. and Zhang, Q.R.** 1999: Palaeomagnetic study of Neoproterozoic glacial rocks of the Yangzi Block: palaeolatitude and configuration of South China in the late Proterozoic supercontinent. *Precambrian Research* 94, 7–10. **Scotese, C.R.** 1991: Jurassic and Cretaceous plate tectonic reconstructions. *Palaeogeography, Palaeoclimatology, Palaeoecology* 87, 493–501. **Shi, G.R. and Grunt, T.A.** 2000: Permian Gondwana–Boreal antitropicality with special reference to brachiopod faunas. *Palaeogeography, Palaeoclimatology, Palaeoecology* 155, 239–263. **Windley, B.F.** 1995: *The evolving continents*, 3rd edn. Chichester: Wiley.

parabiosphere That part of the BIOSPHERE where only dormant life exists, including areas at too high an altitude, too dry, too cold or too hot to support metabolising organisms (except technically equipped human explorers). *JAM*

Hutchinson, G.E. 1970: The biosphere. *Scientific American* 223, 45–53.

paradigm A Kuhnian term used to embrace a combination of theory and method that permits scientists to select, criticise and evaluate knowledge. For a while, scientists may operate within one paradigm, which may eventually be overthrown by another in a '*scientific revolution*'. It was so loosely used by Kuhn that he was forced into redefinition of the term and subdivision of it into exemplars and disciplinary matrixes. It is still widely used in a very loose fashion with a meaning akin to *disciplinary matrix* or *Weltanschauung* (world view or world outlook). Thus, a paradigm may be defined as a framework for the conduct of research, which pervades the problems posed and the approaches used in problem solving. *CET*

Kuhn, T. 1962: *The structure of scientific revolutions*. Chicago, IL: Chicago University Press.

paraglacial Church and Ryder (1972: 3059) applied this term to both 'nonglacial processes that are directly conditioned by GLACIATION' and the period 'during which paraglacial processes occur'. The *paraglacial concept* was developed with respect to the formation of valley-floor ALLUVIAL FANS following Late PLEISTOCENE DEGLACIATION. Although originally used to describe the gravitational (MASS MOVEMENT) and FLUVIAL reworking of GLACIAL SEDIMENTS in North America, the term has since been employed in other formerly glaciated areas and has been extended to encompass sediment reworking by AEOLIAN

and coastal-zone processes; recent and present-day analogues are also incorporated by the term.

The processes recognised as being responsible for the reworking of GLACIAL SEDIMENTS are not unique to recently deglaciated terrain, but usually occur there with greater frequency and intensity than elsewhere because of the relative abundance of readily entrainable or unstable glacial deposits. The *paraglacial readjustment* of slopes of glacigenic sediments leads to gullying, and DEBRIS CONES develop at and beyond the down-slope limit of gullies. DEBRIS FLOW activity is a particularly common paraglacial slope process and bedrock slopes are prone to FAILURE as a result of glacial steepening and pressure release. *Paraglacial activity* in FLUVIAL systems may create ALLUVIAL FANS and VALLEY FILL DEPOSITS, whilst AEOLIAN reworking of GLACIAL SEDIMENTS often leads to thick and extensive accumulations of LOESS and COVERSAND. Marine TRANSGRESSION of glaciated terrain in the coastal zone produces distinctive features reflecting the abundance and type of glacially derived SEDIMENT and the presence of glacial LANDFORMS.

The *paraglacial period*, characterised by high rates of sediment delivery, begins with the onset of DEGLACIATION and ends when SEDIMENT YIELDS fall to rates typical of unglaciated terrain. Gravitational processes may rework unstable valley-side debris within a relatively short time (decades to centuries) following ice removal, whereas FLUVIAL PROCESSES and FAILURE of rock slopes may operate over substantially longer periods (millennia). It is therefore difficult to define the end of a paraglacial period because some of these landscapes may experience delayed slope responses.

Paraglacial activity should be distinguished from PERIGLACIAL processes, which are associated with cold, non-glacial environments, although some processes and forms may be common to both zones. *PW*

[See also ALLUVIAL DEPOSITS, DENUDATION RATES, LANDSLIDE, SEA-LEVEL CHANGE, STURZSTROM, UNLOADING]

Ballantyne, C.K. and Benn, D.I. 1994: Paraglacial slope adjustment and resedimentation following recent glacier retreat, Fåbergstølsdalen, Norway. *Arctic and Alpine Research* 26, 255–269. **Benn, D.I. and Evans, D.J.A.** 1998: *Glaciers and glaciation*. London: Arnold. **Church, M. and Ryder, J.M.** 1972: Paraglacial sedimentation: a consideration of fluvial processes conditioned by glaciation. *Geological Society of America Bulletin* 83, 3059–3072. **Eyles, N., Eyles, C.H. and McCabe, A.M.** 1988: Late Pleistocene subaerial debris-flow facies of the Bow Valley, near Banff, Canadian Rocky Mountains. *Sedimentology* 35, 465–480. **Forbes, D.L. and Syvitski, J.P.M.** 1994: Paraglacial coasts. In Carter, R.W.G. and Woodroffe, C.D. (eds), *Coastal evolution: late Quaternary shoreline morphodynamics*. Cambridge: Cambridge University Press, 373–424. **Harrison, S.** 1996: Paraglacial or periglacial? The sedimentology of slope deposits in upland Northumberland. In Anderson, M.G. and Brooks, S.M. (eds), *Advances in hillslope processes*. Vol. 2. Chichester: Wiley, 1197–1218. **Matthews, J.A., Shakesby, R.A., Berrisford, M.S. and McEwen, L.J.** 1998: Periglacial patterned ground on the Styggedalsbreen glacier foreland, Jotunheimen, southern Norway: microtopographic, paraglacial and geoecological controls. *Permafrost and Periglacial Processes* 9, 147–166.

paralic Marginal marine environments such as intertidal and supratidal zones or lagoons. Important environmental controls on organisms in paralic environments include SALINITY, OXYGEN VARIATIONS, DESICCATION, temperature and unstable substrates. *LC*

[See also PALUSTRINE]

parallax The change in the apparent relative positions of two points when viewed from different positions. *TS*

parallel lamination A SEDIMENTARY STRUCTURE comprising thin sheets (usually < 1 mm) of sand or fine gravel approximately parallel to the depositional horizontal, as represented by BEDDING (cf. CROSS-STRATIFICATION). Parallel lamination implies deposition on a flat sediment surface and commonly represents upper stage PLANE BED conditions, indicative of a strong PALAEOCURRENT. *Parting lineation* is commonly preserved on bedding surfaces, giving an indication of palaeocurrent direction. This structure is sometimes called *planar lamination* or *horizontal lamination*. Thin sheets of alternating fine sand and MUD imply sediment settling from a SUSPENDED LOAD under conditions of repeatedly alternating sediment movement and still water. Such deposits accumulate over longer time intervals and are better described as *horizontal bedding*. They may represent cyclic sedimentation associated with seasons or TIDES. *GO*

[See also TIDAL RHYTHMITES]

Broadhurst, F.M. 1988: Seasons and tides in the Westphalian. In Besly, B.M. and Kelling, G. (eds), *Sedimentation in a synorogenic basin complex: the Upper Carboniferous of northwest Europe.* Glasgow: Blackie, 264–272. **Paola, C., Wiele, S.M. and Reinhart, M.A.** 1989: Upper-regime parallel lamination as the result of turbulent sediment transport and low-amplitude bedforms. *Sedimentology* 36, 47–60. **Tunbridge, I.P.** 1981: Sandy high-energy flood sedimentation – some criteria for recognition, with an example from the Devonian of S.W. England. *Sedimentary Geology* 28, 79–95.

parallel slope retreat The view that denudation of a landscape occurs by backwearing of *escarpment* slopes and hills with slope angles being maintained. Developed by L.C. King to explain the African escarpments and INSELBERGS, it is one of the three classic SLOPE EVOLUTION MODELS. *RAS*

King, L.C. 1962: *The morphology of the Earth.* Edinburgh: Oliver and Boyd.

paramagnetism This occurs when individual atoms, ions or molecules possessing a permanent magnetic dipole moment, align themselves parallel with the direction of an applied MAGNETIC FIELD. The result is a weak positive MAGNETISATION, which is lost once the field is removed. *AJP*

parameteorology The study of weather-dependent natural phenomena (such as GLACIER VARIATIONS, DROUGHTS, FLOODS, freeze-up and break-up of rivers and lakes, and the extent of SEA ICE), but usually excluding the timing of the seasonal biological phenomena, which consitutes the separate field of PHENOLOGY. *JAM*

Bradley, R.S. 1999: *Paleoclimatology*, 2nd edn. San Diego, CA: Academic Press.

parameter (1) A characteristic of a statistical POPULATION, such as the population mean (μ) or the population standard deviation (σ), as opposed to *sample estimates* of

the same, such as the SAMPLE mean ($\bar{x}$) or the sample standard deviation (s). (2) A quantity related to one or more VARIABLES in such a way that it remains constant even though the values of the variable(s) may change. *JAM*

parameterisation The technique of model CALIBRATION, *initialisation* or *tuning* whereby PARAMETERS used in the model are estimated from VARIABLES that can be measured and used to represent a complex process as a simplified function. Thus, the aim of parameterisation or *parameter estimation* is to make the model fit the data. Once the MODEL has been parameterised, CONFIRMATION or VALIDATION procedures should be applied to test it before it is used to predict the workings of the real world. Parameter estimation can be performed graphically, statistically or mathematically. *JAM*

parametric statistics Statistical procedures for testing hypotheses or estimating parameters (e.g. CONFIDENCE INTERVALS) that make assumptions about the underlying PROBABILITY DISTRIBUTION of the variables (e.g. NORMAL DISTRIBUTION) and their variability. Examples include ANALYSIS OF VARIANCE and REGRESSION ANALYSIS. *HJBB*

[See also NON-PARAMETRIC STATISTICS]

parasite An organism that obtains a material benefit from another organism, to the detriment of the second organism. *Parasitism* differs from PREDATION in that there is a long-lasting relationship with the host organism. *Endoparasites* live inside the host and include the *microparasites*, which are causative agents for many human DISEASES, such as measles, leprosy and malaria. The term *parasitoid* is used for an organism where the adult is free living but one or more immature stages of the life cycle live on or in the host. *JLI*

[See also EPIDEMIC, EPIPHYTE, PATHOGEN, PREDATOR–PREY RELATIONSHIPS, SYMBIOSIS]

Scott, M.E. and Smith, G. (eds), 1992: *Parasitic and infectious diseases.* New York: Academic Press.

parent material The mineral material from which a soil has been derived, described by Jenny as 'the initial state of the soil system' or as the C horizon. Parent material may be the bedrock below the soil profile, but often thin veneers of glacial, aeolian, volcanic or colluvial sediments are the major contributors of mineral material to soils. Young soils may have many features of the parent material, but older soils retain little influence of parent material except where it is of special composition, such as quartz sand or extremely clay-rich materials. Organic deposits are the parent materials for HISTOSOLS. *EMB*

Chesworth, W. 1973: The parent rock effect in the genesis of soil. *Geoderma* 10, 215–225. **Gellatly, A.F.** 1987: Establishment of soil covers on tills of variable texture and implications for interpreting palaeosols: a discussion. In Gardiner, V (ed.), *International geomorphology 1986.* Part II. London: Wiley, 775–784. **Paton, T.R.** 1978: *The formation of soil material.* London: George Allen and Unwin.

parietal art Paintings or engravings on rock walls as opposed to smaller, movable art objects such as sculpture (*mobiliary art*). *JAM*

[See also CAVE ART, PETROGLYPH]

parsimony The principle of parsimony, *simplicity*, or economy: an explanation that invokes the fewest assumptions or explanatory principles is to be preferred (or is more likely to be true). There are several formulations. It is often referred to as *Ockam's* (Occam's) *Razor*. *CET*

partial correlation coefficient A measure of the CORRELATION between *dependent* (response) and *independent* (explanatory) variables while the statistical effects of one or more other independent variables are 'held constant'. Partial correlation and *partial regression* are of use in differentiating the relative importance of the effects of interacting independent variables. The *order* of a partial correlation coefficient refers to the number of independent variables held constant. *JAM*

particle-induced gamma-ray emission (PIGE or PIGME) A sensitive, non-destructive technique for elemental analysis that analyses gamma rays emitted when a proton beam is applied to an object. Like PROTON INDUCED X-RAY EMISSION (PIXE), it is a method of *ion-beam analysis* (IBA). Unlike PIXE, however, it relies on excitation of the nucleus and is better for detecting light elements (lighter than sodium). It is widely used in combination with PIXE to obtain a 'total analysis' of ARTEFACTS, especially those made of flint, obsidian and pottery, with minimal sample preparation. *JAM/GO*

[See also GEOCHEMISTRY]

Ambrose, W.R., Duerden, P. and Bird, J.R. 1981: An archaeological application of PIXE–PIGME analysis to Admiralty Island obsidians. *Nuclear Instruments and Methods in Physics Research* 191, 397–402. **Reiche, I., Favre-Quattropani, L., Calligaro, T. et al.** 1999: Trace element composition of archaeological bones and postmortem alteration in the burial environment. *Nuclear Instruments and Methods in Physics Research Section B – Beam Interactions with Materials and Atoms* 150, 656–662.

particulates Small solid particles of matter; particulate matter. Particulates suspended in the ATMOSPHERE are mostly in the range 0.1–10.0 μm: the larger particles are mainly of natural origin (e.g. soil particles and sea salts), whereas the finer particles are mostly of combustion origin. *DME*

[See also AEROSOLS, DRY DEPOSITION, WET DEPOSITION]

Airborne Particles Expert Group 1999: *Source apportionment of airborne particulate matter in the United Kingdom*. London: Department of the Environment, Transport and the Regions. **Fennelly, P.F.** 1981: The origin and influence of airborne particulates. In Skinner, B.J. (ed.), *Climates, past and present*. Los Altos, CA: Kauffmann. **Groisman, P.Y.** 1992: Possible regional climate consequences of the Pinatubo eruption: an empirical approach. *Geophysical Research Letters* 19, 1603–1606.

passive continental margin A CONTINENTAL MARGIN characterised by a wide CONTINENTAL SHELF, separated by a CONTINENTAL SLOPE and CONTINENTAL RISE from the deep-ocean ABYSSAL plain and lacking active SEISMICITY or VOLCANISM that does not correspond to a PLATE MARGIN (see Figure). The continental margins of northwest Europe and eastern North America are of this type, and they are also known as *Atlantic-type* or *aseismic* continental margins. The true edge of the CONTINENT – the junction between CONTINENTAL CRUST and OCEANIC CRUST – corresponds approximately to the outer margin of the continental shelf (the *shelf break*).

A passive continental margin originates as a continental RIFT VALLEY. Spreading as a CONSTRUCTIVE PLATE MARGIN leads to the injection of oceanic crust at a MID-OCEAN RIDGE, allowing two mirror-image continental margins separated by an OCEAN to migrate to a tectonically inactive mid-plate position (see WILSON CYCLE). After the phase of rifting, the thinned continental crust of passive continental margins cools, allowing considerable SUBSIDENCE to make space for thick sequences of SEDIMENTARY DEPOSITS

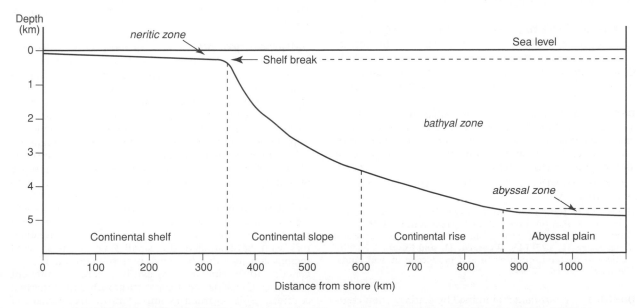

passive continental margin *Schematic section across a passive continental margin (based on Price, 1993)*

to accumulate. Modern passive continental margins have accumulated sediment piles 10 to 15 km thick over the last 200 million years. These deposits contain important SEDIMENTOLOGICAL EVIDENCE OF ENVIRONMENTAL CHANGE associated with CONTINENTAL DRIFT, as well as important PETROLEUM reserves. *CDW*

Gilchrist, A.R. and Summerfield, M.A. 1994: Tectonic models of passive margin evolution and their implications for theories of long-term landscape development. In Kirby, M. (ed.), *Process models and theoretical geomorphology*. New York: Wiley, 55–84. **Kingston, D.R., Dishroon, C.P. and Williams, P.A.** 1983: Global basin classification system. *American Association of Petroleum Geologists Bulletin* 67, 2175–2193. **Price, N.B.** 1993: Marine sediments and the ocean floor. In Duff, P.McL.D. (ed.), *Holmes' principles of physical geology*, 4th edn. London: Chapman and Hall, 543–579. **Windley, B.F.** 1996: *The evolving continents*. Chichester: Wiley.

passive microwave remote sensing   The acquisition of remotely sensed information using a microwave instrument (radiometer) that only receives energy (and does not transmit). Examples of passive microwave instruments include the SMMR (*scanning multi-channel microwave radiometer*) launched in AD 1978 and the series of SSMI (*special sensor microwave imager*) instruments launched on a range of satellites from 1987. Generally, these instruments operate over several frequencies (four in the case of the SSMI) and can also receive on different POLARISATIONS (horizontal and/or vertical). These instruments measure natural thermal emission in the microwave regime, characterised as a *brightness temperature*, and are influenced by the physical temperature of the object as well as its geometry and dielectric properties. Therefore, they have application to land (vegetation properties), cryosphere (snow cover and snow melt), oceans (surface windspeed and surface temperature) and atmosphere (water vapour, liquid water content and rain rate).   *PJS*

[See also MICROWAVE REMOTE SENSING]

Ulaby, F.T., Moore, R.K. and Fung, A.K. 1981: *Microwave remote sensing: active and passive*. Vol. 1. *Fundamentals and radiometry*. Norwood, MA: Artech House.

Pastoral Neolithic   Pre-Iron Age East African cultures based on PASTORALISM with advanced technological and economic features.   *JAM*

Collett, D.P. and Robertshaw, P.T. 1983: Dating the Pastoral Neolithic of East Africa. *African Archaeological Review* 1, 57–74.

pastoralism   An agricultural system whereby flocks and herds of animals are grazed in an open landscape or enclosed fields. The system focuses on the use of domesticated or semi-domesticated animals as a source of food and other materials such as hide.   *LD-P*

[See also DOMESTICATION, GRAZING, NOMADISM, OVERGRAZING, SEDENTISM]

Harris, D.R. (ed.) 1996: *The origins and spread of agriculture and pastoralism in Eurasia*. London: UCL Press. **Smith, A.B.** 1992: Origins and spread of pastoralism in Africa. *Annual Review of Anthropology* 21, 125–141. **Western, D. and Finch, V.** 1996: Cattle and pastoralism: survival and production in arid lands. *Human Ecology* 14, 77–94.

pasture   A *grassland* maintained by grazing, often characterised by low-growing HERBS, such as rosette GROWTH

FORMS. A pasture differs from a MEADOW, which is maintained by mowing.   *JAM*

patch dynamics   The nature and causes of change in landscape patches and patch mosaics. Some patches form mosaics of vegetation in different successional stages, which undergo cyclical changes about a *steady state*. English HEATHLANDS, for instance, contain heather (*Calluna*) patches of different age. Over time, through CYCLIC REGENERATION, the phases change places, but the peatland ecosystem as a whole remains the same. Some patches follow directional changes. Depending on the force driving the change, an abandoned field on a hillside may become a wood (through natural ECOLOGICAL SUCCESSION), a new field (through ploughing), a quarry (through digging and blasting) or a gully (through soil erosion). DISTURBANCE patches result from the action of disturbing agents. Some disturbing agents, such as fire and wind, are likely to become more frequent and more severe in certain ecosystems if GLOBAL WARMING occurs during the next century.   *RJH*

[See also FRAGMENTATION]

Forman, R.T.T. 1995: *Land mosaics: the ecology of landscapes and regions*. Cambridge: Cambridge University Press. **Pickett, S.T.A. and White, P.S. (eds)** 1985: *The ecology of disturbance and patch dynamics*. New York: Academic Press. **Shorrocks, B. and Swingland, I. R.** 1990: *Living in a patchy environment*. Oxford: Oxford University Press.

pathogens   Living organisms, usually bacteria, viruses, fungi, protozoa or parasitic worms, that cause DISEASE in humans. Exposure is usually through inhalation, ingestion, direct contact or transmission via an insect vector. The proliferation of pathogenic microbes is often linked to standards of hygiene and to changes in temperature, humidity or precipitation. The term is sometimes used also in the context of animal and plant diseases.   *MLW*

[See also HUMAN HEALTH AND ENVIRONMENTAL CHANGE, WATER-BORNE DISEASES]

Mottet, N.K. (ed.) 1985: *Environmental pathology*. New York: Oxford University Press.

patterned ground   A general term used to describe the variety of small-scale, more or less symmetrical forms, such as circles, polygons, nets and stripes, that characterise many ARCTIC and alpine regions. These forms are also found in former PERIGLACIAL environments and small forms are active down to moderate altitudes in mid-latitude temperate environments. Some distinctive forms of patterned ground also occur in hot, semi-arid regions.

Traditionally, patterned ground is classified on the basis of (a) its geometric form and (b) the presence or absence of sorting. The main geometric forms are circles, polygons, stripes, nets and steps, all of which may be sorted and unsorted with respect to particle sizes. Stripes and steps are usually confined to sloping terrain. If large-scale thermal-contraction cracking is excluded, most investigators who have studied the small-scale forms in cold regions conclude that (a) much patterned ground is polygenetic and (b) similar forms can be created by different processes. Cracking (i.e. DESICCATION rather than thermal contraction) is regarded as important in development of polygonal forms and relatively unimportant in development of circular forms. The latter are most likely the result of CRYOTURBATION, a general term referring to the complex of lateral and vertical soil displacements that accompany seasonal and/or diurnal freezing and thawing.

The wide range of cold-climate patterned-ground phenomena means that their use as a diagnostic feature of such environments is limited. Patterned ground is not limited to PERMAFROST regions or to those experiencing only seasonal frost. Moreover, lithology, grain-size, moisture availability, vegetation and a host of other site-specific factors are relevant. On the other hand, patterned-ground phenomena are best developed and most widespread in the cold regions of the world. *HMF*

[See also PERIGLACIAL LANDSCAPE EVOLUTION]

Gleason, K.J., Krantz, W.B., Caine, N. *et al.* 1986: Geometrical aspects of sorted patterned ground in recurrently frozen soil. *Science* 232, 216–220. Grab, S.W. 1997: Annually re-forming miniature sorted patterned ground in the high Drakensberg, southern Africa. *Earth Surface Processes and Landforms* 22, 733–745. Hallet, B., Anderson, S.P., Stubbs, C.W. and Gregory, E.C. 1988: Surface soil displacements in sorted circles, Western Spitzbergen. In *Proceedings, 5th International Conference on Permafrost*. Vol. 1. Trondheim: Tapir, 770–775. Matthews, J.A., Shakesby, R.A., Berrisford, M.S. and McEwen, L.J. 1998: Periglacial patterned ground on the Styggedalsbreen glacier foreland, Jotunheimen, southern Norway: micro-topographic, paraglacial and geoecological controls. *Permafrost and Periglacial Processes* 9, 147–166. Van Vliet-Lanoë, B. 1991: Differential heave, load casting and convection: converging mechanisms. A discussion of the origin of cryoturbations. *Permafrost and Periglacial Processes* 2, 123–139. Warburton, J. and Caine, N. 1999: Sorted patterned ground in the English Lake District. *Permafrost and Periglacial Processes* 10, 193–197. Washburn, A.L. 1956: Classification of patterned ground and review of suggested origins. *Geological Society of America Bulletin* 67, 823–865. Washburn, A.L. 1989: Near-surface soil displacement in sorted circles, Resolute area, Cornwallis Island, Canadian High Arctic. *Canadian Journal of Earth Sciences* 25, 941–955. Wilson, P. 1995: Forms of unusual patterned ground: examples from the Falkland Islands, South Atlantic. *Geografiska Annaler* 77A, 159–165.

payload That which a spacecraft (e.g. a REMOTE SENSING SATELLITE), an aircraft or the like carries over what is necessary for the operation of the vehicle in flight; usually the instruments that are accommodated on board. *TS*

peat The undecayed remains of vegetation that accumulates in MIRES and HEATHLAND, it can range from unhumified, yellow- or orange-brown fibrous peat of RAISED MIRES, with visible plant remains, to highly humified black amorphous heath peat. Peat is widespread in *peatlands* of north-temperate and circum-boreal regions, especially in Canada, Estonia, Finland, Ireland and Russia; in southerly latitudes in Tierra del Fuego, Falkland Islands and Tasmania; also at lower latitudes in TROPICAL PEATLANDS. In the British Isles, peat was traditionally cut (as 'turf'), dried and burnt for fuel; it is used horticulturally as a soil conditioner or growing medium. PEAT STRATIGRAPHY of OMBROTROPHIC MIRES can be interpreted as a PROXY CLIMATE RECORD. *FMC*

peat humification The analysis of peat to determine the extent to which it has degraded (or decomposed, to produce HUMIC ACID and *humin*) *in situ*. It can be determined variously, and with varying degrees of precision, by physical and chemical methods. Early field methods used a visual scale, based on the colour of water extruded from a hand sample of peat, to record PEAT STRATIGRAPHY. Aaby and Tauber (1975) later used a colorimetric laboratory technique (on an alkali extract) to determine the degree of peat humification from a Danish raised mire; they considered the technique to give a reliable indication of the (hydrological) environment at the time of peat deposition. As a consequence, peat humification is now used, particularly in northwest Europe, as a PROXY CLIMATIC INDICATOR and COLORIMETRY methods can theoretically produce continuous proxy climate records with decadal to centennial resolution, provided the peat can be dated reliably. *FMC*

Aaby, B. 1976: Cyclic climatic variations in climate over the last 5,500 years reflected in raised bogs. *Nature* 263, 281–4. Aaby, B. and Tauber, H. 1975: Rates of peat formation in relation to degree of humification and local environment, as shown by studies of a raised bog in Denmark. *Boreas* 4, 1–17. Blackford, J.J. and Chambers, F.M. 1993: Determining the degree of peat decomposition for peat-based palaeoclimatic studies. *International Peat Journal* 5, 7–24. Blackford, J.J. and Chambers, F.M. 1995: Proxy climate record for the last 1000 years from Irish blanket peat and a possible link to solar variability. *Earth and Planetary Science Letters* 133, 145–150.

peat stratigraphy The visible horizons, or layers, of a mire, such as can be seen in a PEAT core or in a vertical column of peat taken by MONOLITH SAMPLING. Peat stratigraphy is most evident in a fast-growing mire in which the peat is relatively unhumified, but in which past climatic or hydrological changes caused variations in peat growth, giving rise to changes in vegetation (which can be demonstrated through PLANT MACROFOSSIL ANALYSIS) and/or to changes in PEAT HUMIFICATION. Peat laid down in periods of dry or warmer climate will be more highly humified and so appear darker in colour than peat laid down under cooler or wetter climatic conditions.

Gross variations in the peat stratigraphy of Scandinavian bogs prompted Blytt and Sernander to postulate past climatic change and gave rise to the Blytt–Sernander scheme of postglacial climatic periods, which was then adopted over much of northwest Europe as a climatostratigraphy (or CLIMOSTRATIGRAPHY) for the HOLOCENE. This is now viewed as too simplistic, but the names used in the BLYTT–SERNANDER TIMESCALE (Pre-Boreal, Boreal, Atlantic, Sub-Boreal, Sub-Atlantic), which derived from peat stratigraphy, live on as a CHRONOSTRATIGRAPHY for Scandinavia. In northwest Europe, changes from dark peat to lighter-coloured peat in RAISED MIRES gave rise to the notion of recurrences of bog growth in times of wetter climate, producing a supposed synchronous RECURRENCE SURFACE. The major postglacial peat stratigraphic change in Europe is termed the GRENTZHORIZONT and its age in many mires may well equate with a claimed major climate shift noted at c. 2650 ^{14}C years BP. Peat stratigraphy, particularly when analysed in more detail and dated more precisely, can potentially yield valuable PROXY CLIMATIC INDICATORS. *FMC*

Barber, K.E. 1981: *Peat stratigraphy and climatic change – a palaeoecological test of the theory of cyclic bog regeneration*. Rotterdam: A.A. Balkema. Blackford, J.J. 1993: Peat bogs as sources of proxy climatic data: past approaches and future research. In Chambers, F.M. (ed.), *Climate change and human impact on the landscape*. London: Chapman and Hall, 47–56. Blackford, J.J. 2000: Palaeoclimatic records from peat bogs.

Trends in Ecology and Systematics 15, 193–198. **Blytt, A.** 1876: *Essays on the immigration of Norwegian flora during alternating rainy and dry periods.* Christiana: Cammermayer. **Dupont, L.M.** and **Brenninkmeijer, C.A.M.** 1984: Palaeobotanic and isotopic analysis of the late Sub-Boreal and early Sub-Atlantic peat from Engbertsdijksveen VII, The Netherlands. *Review of Palaeobotany and Palynology* 41, 241–271. **van Geel, B., Buurman, J. and Waterbolk, H. T.** 1996: Archaeological and palaeoecological indications of an abrupt climate change in The Netherlands, and evidence for climatological teleconnections around 2650 BP yr. *Journal of Quaternary Science* 11, 451–460. **Sernander, R.** 1908: On the evidences of Postglacial changes of climate furnished by the peat mosses of Northern Europe. *Geologiska Foreningens i Stockholm Forhandlingar* 30, 467–478.

pebble A SEDIMENT particle of GRAIN-SIZE 4–64 mm.
TY

[See also GRAVEL]

pedalfer An obsolete term for soils in which sesquioxides increase relative to silica during soil formation; a leached soil of humid climates. *EMB*

pedestal rock Relatively large caps or table-shaped rock masses supported by essentially narrow stems or shafts of rock. Many pedestal rocks are structurally homogeneous. They are thought to form in a two-stage process, the pillar or stem being formed through WEATHERING under moist conditions just beneath the surface. The caprock sometimes remains intact because of its massive structure, but its exposure, relatively dry state and protection by lichens, mosses and chemical crusts may also enhance its resistance to weathering and EROSION. Exhumation of the stem occurs through removal of the weathered bedrock, by wash and stream action and in places by wind and waves. Alternative names are *hoodoo rocks*, *mushroom rocks*, *stone* or *rock babies* and various terms in other languages, such as *Pilzfelsen*, *Tischfelsen*, *roches champignons* and *rocas fungiformas*. They differ from EARTH PILLARS, in that the latter are formed in unconsolidated sediments where coarse CLASTS provide the caprock and protect the underlying sediment from erosion. *RAS*

[See also BORNHARDTS, TOR]

Twidale, C.R. and Campbell, E.M. 1992: On the origin of pedestal rocks. *Zeitschrift für Geomorphologie NF* 36, 1–13.

pediment A gently sloping surface (0.5–7°; but usually 2–4°) connecting eroding steeper slopes or scarps with areas of sediment deposition at lower levels. A pediment is developed across bedrock that may or may not be thinly veneered with ALLUVIUM and/or an *in situ* weathering mantle. The pediment is separated from the steeper upper slope (often associated with an INSELBERG) by a relatively rapid change of slope angle in a transitional zone, sometimes termed the PIEDMONT zone, in which the change of gradient is termed the *piedmont angle*. It is an element of a piedmont belt, which may include depositional elements such as ALLUVIAL FANS and *playas*. Coalescing pediments create *pediplains*, (EROSION SURFACES produced by *pediplanation*).

Pediments are found in a wide range of climatic environments and there are numerous theories regarding the origin of pediments, only some of which invoke environmental change (normally in the form of climatic change or geological uplift or tilting) as being necessary in their for-

mation. Some consider that pediment formation and maintenance are favoured by highly seasonal, wet–dry or SEMI-ARID climates, whereas others consider that they can form in a wide range of climates. The alternate mantling and stripping of pediments are key elements of most theories invoking environmental change. In the Mojave Desert, T.M. Oberlander demonstrated the importance of more humid climates in pediment formation and extension, whilst D. Busche identified cycles of deep WEATHERING and stripping on pediment surfaces in Chad. In such theories pediments are often regarded as indicators of past humid conditions when found in arid zones and as indicators of past drier conditions when they occur in the humid tropics. *MAC*

[See also STRIPPING PHASE]

Busche, D. 1976: Pediments and climate. *Palaeoecology of Africa, the Surrounding Islands and Antarctica* 11, 20–24. **Cooke, R.V., Warren, A. and Goudie, A.** 1993: *Desert geomorphology.* London: UCL Press. **Oberlander, T.M.** 1974: Landscape inheritance and the pediment problem in the Mojave Desert of southern California. *American Journal of Science* 274, 849–875. **Thomas M.F.** 1994: *Geomorphology in the tropics: a study of weathering and denudation in low latitudes.* Chichester: Wiley. **Whitaker, C.E.** 1979: The use of the term 'pediment' and related terminology. *Zeitschrift für Geomorphologie* 23, 427–439.

pediplain/pediplanation A pediplain is an extensive, thinly alluviated EROSION SURFACE found generally in DESERT or SEMI-ARID regions. It is considered to be formed by the coalescence of two or more adjacent PEDIMENTS and occasional desert domes and to represent the end result of the mature stage of the arid erosion cycle.

Pediplanation is the action or process of formation and development of a pediplain by scarp retreat and *pedimentation*. It was proposed originally by L.C. King as an alternative to the DAVISIAN CYCLE OF EROSION and invoked on a continental scale to explain widespread multi-concave erosion surfaces with steep-sided hills in arid, SEMI-ARID and SAVANNA regions of Africa. The cycle is initiated by the uplift of a pediplain, and existing streams rapidly cut downwards to the new BASE LEVEL. Eventually, as downcutting becomes less active, small PEDIMENTS appear in valley bottoms, which become more extended as interfluve and upland areas are eroded by scarp retreat. Interfluve areas are converted to INSELBERGS and BORNHARDTS and ultimately to a landscape of low relief. *MAC*

[See also PENEPLAIN]

King L.C. 1962: *The morphology of the Earth.* Edinburgh: Oliver and Boyd. **Leroux, J.S.** 1991: Is the pediplanation cycle a useful model – evaluation in the Orange-Free-State (and elsewhere) in South Africa. *Zeitschrift für Geomorphologie* 35, 175–185. **Veldkamp, A. and Oosterom, A. P.** 1994: The role of episodic plain formation and continuous etching and stripping processes in the end-Tertiary landform development of SE Kenya. *Zeitschrift für Geomorphologie* 38, 75–90.

pedocal An obsolete term for soils in which calcium carbonate accumulates during formation; a soil that is relatively unleached. *EMB*

pedogenesis The formation and evolution of soils. Soils are formed by the operation of the processes of soil formation upon the PARENT MATERIAL over a period of time within the framework of the SOIL FORMING FACTORS.

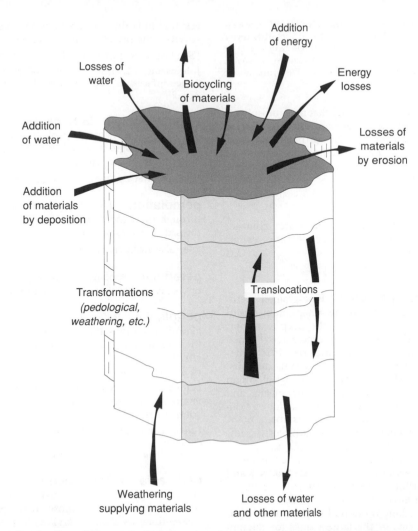

Losses of
water

Addition
of energy

Biocycling
of materials

Energy
losses

Addition
of water

Losses of
materials
by erosion

Addition
of materials
by deposition

Transformations
*(pedological,
weathering, etc.)*

Translocations

Weathering
supplying materials

Losses of water
and other materials

pedogenesis *A soil pedon showing additions, translocations, transformations and losses during pedogenesis (after Simonson, 1959)*

The processes – for example, LEACHING, PODZOLISATION, GLEYING, FERALITISATION, *calcification* and SALINISATION – are names given to 'bundles' of chemical, physical and biological actions. However, pedogenesis may be simplified as additions to, and subtractions from, a parent material within which various transformations and translocations take place (see Figure). Fresh organic matter is incorporated into the surface and raw mineral material weathered at the base of the soil. Soluble compounds are leached out of the soil in humid climates. Transformation of minerals, especially iron and manganese, occurs and translocations of organic matter, clay, aluminium and iron take place. The activities eventually result in a SOIL PROFILE with recognisable SOIL HORIZONS. Most soils, even when mature, are dynamic and continue to evolve, possibly never attaining an equilibrium state.

EMB

[See also SOIL FORMING PROCESSES]

Jenny, H. 1941: *Factors of soil formation*. New York: McGraw-Hill. **Johnson, D.L. and Watson-Stegner, D.** 1987: Evolution model of pedogenesis. *Soil Science* 143, 349–366. **Simonson, R.W.** 1959: Outline of a generalized theory of soil genesis. *Soil Science Society of America Proceedings* 23, 152–156. **Simonson, R.W.** 1978: A multi-process model of soil genesis. In Mahaney, W.C. (ed.), *Quaternary soils*. Norwich: Geoabstracts, 1–25.

pedology The scientific study of SOILS. Pedology integrates studies of soil chemistry, physics and biology to create the study of soils as a natural phenomenon in their own right. Pedology focuses upon the appearance, mode of formation, distribution, classification and use of soils. *Palaeopedology* investigates the soils of past environments (PALAEOSOLS). *EMB*

Jacobs, J.S. 1995: Archaeological pedology in the Maya Lowlands. In Collins, M.E., Carter, B.J., Gladfelter, B.G. and Southard, R.J. (eds), *Pedological perspectives in archaeological research*. Madison, WI: Soil Science Society of America, 51–80.

pedon The smallest volume that can be called a soil. It varies in area from 1 to 10 m^2 and is large enough to permit the study of the SOIL PROFILE and its SOIL HORIZONS. The depth of a pedon is governed by the depth of the genetic soil horizons and therefore is not strictly defined. Other terms that have been proposed are *soil area* and *pedo-unit*. *EMB*

Johnson, D.W. 1963: The pedon and the polypedon. *Proceedings of the Soil Science Society of America* 27, 212–215. **Soil Survey Staff** 1999: *Soil Taxonomy: a basic system of soil classification for making and interpreting soil surveys*, 2nd edn [*Agriculture Handbook* 436]. Washington, DC: United States Department of Agriculture.

pedosphere   In the same manner that the atmosphere, the biosphere and the hydrosphere are used to describe the gaseous envelope surrounding the Earth, its plants and animals, and the waters of the Earth, the term pedosphere describes the relatively thin but essential soil cover of the terrestrial parts of the planet.   *EMB*

[See also GEOSPHERE]

pelagic   Living or originating in the water column. Free-floating pelagic organisms are PLANKTONIC; swimming organisms are NEKTONIC.   *LC*

[See also BENTHIC, PELAGIC SEDIMENT]

pelagic sediment (pelagite)   Fine-grained SEDIMENT (MUD) that has its origin in the water column, usually in the OCEANS, and settles to the deep sea-floor, usually as composite particles or aggregates formed through FLOCCULATION or as faecal material. It is sometimes described as '*marine snow*' or ABYSSAL deposits. Pelagic sediment includes material of biological origin (OOZE) dominated by tests of PLANKTONIC organisms, and material of non-biological origin, which includes VOLCANIC ASH and wind-blown (AEOLIAN) ATMOSPHERIC DUST. The finest fraction of land-derived (TERRIGENOUS) sediment is termed *hemipelagic sediment (hemipelagite)*.

   The present-day distribution of pelagic sediments is controlled by organic PRODUCTIVITY, OCEAN CURRENTS and the dissolution of carbonate material below the CARBONATE COMPENSATION DEPTH. *Carbonate* or *calcareous ooze* forms the dominant sediment on shallow parts of the MID-OCEAN RIDGES and is one of the largest sinks for calcium carbonate on the Earth. *Siliceous ooze* is found in areas of deeper, colder water and high productivity. *Red clays* accumulate very slowly (0.0001–0.001 mm y^{-1}) below the carbonate compensation depth in areas of low surface productivity for siliceous plankton. Organic-rich sediment (SAPROPEL) accumulates in areas of high organic input or reduced BOTTOM WATER oxygenation. ICE-RAFTED DEBRIS is deposited in and around polar regions. Pelagic sediments are often associated with manganese *nodules*. The distribution of older pelagic sediments, derived from the analysis of CORES obtained through DEEP-SEA DRILLING, is an important tool in PALAEOCEANOGRAPHY.   *GO*

Hsu, K.J. and Jenkyns, H.C. (eds) 1974: *Pelagic sediments: on land and under the sea. International Association of Sedimentologists, Special Publication* 1. **Kumar, N., Anderson, R.F., Mortlock, R.A.** *et al.* 1995: Increased biological productivity and export production in the glacial Southern Ocean. *Nature* 378, 675–680. **Pike, J. and Kemp, A.E.S.** 1997: Early Holocene decadal-scale ocean variability recorded in Gulf of California laminated sediments. *Palaeoceanography* 12, 227–238.

peneplain   A surface of gentle relief formed through long-term DENUDATION. The term is associated in particular with the DAVISIAN CYCLE OF EROSION concept, in which it was viewed as the final stage of subaerial denudation in the absence of a change of BASE LEVEL. *Monadnocks* are

residual hills that rise above a peneplain and hence have survived the process of *peneplanation*.   *AJDF*

[See also DENUDATION CHRONOLOGY, EROSION SURFACE]

Johansson, M. 1999: Analysis of digital elevation data for palaeosurfaces in south-western Sweden. *Geomorphology* 26, 279–295.

Pennsylvanian   In North American usage, a SYSTEM of rocks and a PERIOD of geological time from 324 to 295 million years ago. It is approximately equivalent to the late CARBONIFEROUS in Europe.   *GO*

[See also GEOLOGICAL TIMESCALE, MISSISSIPPIAN]

percolation   The downward vertical flow of water through the VADOSE ZONE or *aeration zone* of soil and rock towards the WATER TABLE.   *RPDW*

[See also INFILTRATION]

perennial   A plant that lives for several to many years, or a stream that flows throughout the year.   *JLI*

periglacial   Referring to a wide range of cold non-glacial conditions that bear no necessary relationship to their temporal or spatial proximity to a GLACIER or ICE SHEET.   *RAS*

[See also FREEZE–THAW CYCLES, FROST WEATHERING, PERIGLACIAL ENVIRONMENTS, PERIGLACIAL LANDFORMS, PERIGLACIAL SEDIMENTS, PERIGLACIATION, PERMAFROST, SOLIFLUCTION]

Karte, J. and Liedtke, H. 1981: The theoretical and practical definition of the term 'periglacial' in its geographical and geological meaning. *Biuletyn Peryglacjalny* 28, 123–135.

periglacial environments   Environments with a cold, non-glacial (PERIGLACIAL) climate. There are two broad categories: (a) those that experience frequent freeze–thaw oscillations and/or deep seasonal frost (SEASONALLY FROZEN GROUND) but lack PERMAFROST; and (b) those that experience both frost action and permafrost. The *permafrost zone* of the Earth is sometimes termed the *cryolithozone*. Approximately 25% of the Earth's land surface currently experiences PERIGLACIAL conditions.   *HMF*

[See also PERIGLACIAL LANDFORMS, PERIGLACIAL LANDSCAPE EVOLUTION, PERIGLACIAL SEDIMENTS]

French, H.M. 1996: *The periglacial environment*, 2nd edn. London: Addison-Wesley Longman. **French, H.M.** 2000: Does Lozinski's periglacial realm exist today? A discussion relevant to modern usage of the term 'periglacial'. *Permafrost and Periglacial Processes* 11, 35–42.

periglacial landforms   Landforms found in present-day or former cold but essentially non-glacial climates, ranging from humid and relatively mild (e.g. Svalbard, today) to extremely cold and arid (e.g. parts of Antarctica). The distinctiveness of many of the landforms owes much to the distinctiveness of the processes. These include the growth of PERMAFROST, the development of thermal contraction cracking, the thaw of permafrost (THERMOKARST), the creep of ice-rich permafrost and the formation of wedge and injection ice. Other processes, not necessarily restricted to periglacial regions, are important because of their high magnitude or frequency. These relate to FREEZE–THAW processes affecting soils and bedrock, including the disintegration of exposed rock by frost wedging, thermal stress or a complex of poorly understood cryogenic weathering processes and, in soils, frost heaving and ice segregation.

The most distinctive periglacial landforms are associated with permafrost. Most widespread are TUNDRA polygons, 15–30 m in dimensions, formed by the thermal-contraction–cracking of the frozen ground in winter. The cracks may fill with water in early summer and after a number of years ICE WEDGES may form. Less commonly, the cracks may fill with sand. Less widespread, but equally distinctive, are various perennial ice-cored FROST MOUNDS. The largest, sometimes exceeding 40 m in height, are termed PINGOS. Other aggradational permafrost landforms associated with the preferential growth of segregated ice lenses, such as PALSAS (formed in PEAT), LITHALSAS (palsa-like features formed in mineral soil) and peat plateaus, are 1–8 m in height. Ground-ice slumps, thaw lakes and thaw depressions (e.g. ALASES) and thermokarst mounds result from PERMAFROST DEGRADATION. The *in situ* creep of permafrost, especially as it thaws in response to regional climatic change, may cause nondiastrophic structures in sedimentary strata. These include up-arching beneath valley bottoms (VALLEY BULGING) and the bending, deformation and sliding of strata on slopes, leading to CAMBERING and joint widening, or GULL formation. Within the ACTIVE LAYER, local conditions of soil moisture saturation and high pore-water pressures can induce rapid shallow movements confined to the active layer, with the top of permafrost acting as a lubricated slip plane for movement, thereby controlling the depth of the FAILURE plane. These failures, generally referred to as *active-layer detachments*, are frequent in the summer months on terrain underlain by ice-rich and unconsolidated shales and siltstones. In most cases, failure is initiated when the LIQUID LIMIT is exceeded; this happens during years of rapid spring thaw and/or following periods of unusually heavy summer precipitation.

There are also periglacial landforms related to the intense cryogenic weathering of exposed bedrock. These can occur in non-permafrost as well as permafrost environments. Coarse, angular rock-rubble, commonly termed BLOCKFIELDS in North America and Europe, and *kurums* in Siberia, occurs widely over large areas of the high-Arctic polar deserts and semi-deserts. They surround outcrops of more resistant rock that form isolated hills or TORS. In regions of extreme aridity, rectilinear, debris-veneered bedrock-controlled slopes (*Richter denudation slopes*) may evolve.

In upland areas, below bedrock cliffs, frost-shattered TALUS may accumulate as sheets and cones. Modification of talus may occur through AVALANCHE activity, DEBRIS FLOWS, ROCK GLACIER development and PRONIVAL RAMPART formation.

At a smaller scale, many periglacial landscapes possess a diversity of PATTERNED GROUND phenomena resulting from differential FROST HEAVE, *frost creep* and ice segregation in the active layer. These features include *sorted and nonsorted circles, stripes and nets*. On debris-mantled slopes, MASS WASTING processes may result in sorted and nonsorted stripes. Depending upon moisture supply, SOLIFLUCTION lobes, sheets and *terraces* may form, especially at locations below late-lying or perennial SNOWBEDS.

HMF

[See also FROST MOUND, ICE WEDGE, PERIGLACIAL LANDSCAPE EVOLUTION, PERIGLACIAL SEDIMENTS]

French, H.M. 1996: *The periglacial environment*, 2nd edn. London: Addison-Wesley Longman. **Jahn, A.** 1975: *Problems of the periglacial zone*. Warsaw: PWN Polish Scientific Publishers. **Luckman, B.H. and Fiske, C.J.** 1997: Holocene development of coarse-debris landforms in the Canadian Rocky Mountains. In Matthews, J.A., Brunsden, D., Frenzel, B. *et al.* (eds), *Rapid mass movement as a source of climatic evidence for the Holocene*. Gustav Fischer: Stuttgart, 283–297. **Washburn, A.L.** 1979: *Geocryology: a survey of periglacial processes and environments*. London: Edward Arnold.

periglacial landscape evolution The evolution of landscape under the control of cold-climate rock WEATHERING and associated MASS WASTING of the REGOLITH. Typical slope evolution is thought to involve a progressive and sequential reduction of relief with the passage of time, termed CRYOPLANATION. This is effected by slope replacement from below, with the formation of *Richter denudation slopes*, which are ultimately replaced by low-angle PEDIMENTS. Richter slopes are rectilinear, debris-veneered, bedrock surfaces on which the rate of production of weathered debris is equal to, or less than, the ability of gravity-controlled transport processes to remove it.

It is difficult to generalise about the typical sequence of PERIGLACIAL LANDFORM evolution for the following reasons. First, certain lithologies are more prone to cryogenic weathering than others; equally, some are more capable of preserving a distinctive periglacial slope morphology, once formed. Second, the variety of periglacial climates existing today means that periglacial landform assemblages may also vary. For example, depending upon the degree of aridity, running water may, or may not, be an important landscape-modifying process. Third, many areas experiencing periglacial conditions today have only recently emerged from beneath continental ice sheets. Therefore, these landscapes cannot be regarded as being in true geomorphological equilibrium. Finally, there are relatively few studies that detail, in quantitative terms, the manner and speed of slope evolution in present-day periglacial environments.

The typical slope forms found in periglacial regions today can be summarised as follows: (a) rectilinear debris-mantled (Richter) slopes; (b) free-face (i.e. exposed bedrock) and debris (i.e. TALUS) slope profiles; (c) smooth convexo-concave debris-mantled slopes; (d) stepped profiles; and (e) pediment-like forms. Whilst (b) is frequently associated with glacially over-steepened valleys, as in Svalbard and northern Scandinavia, the other slope forms fit the CRYOPLANATION concept reasonably well.

For example, Richter denudation slopes are usually described over the very arid periglacial regions, such as interior Yukon, north-central Alaska, central Siberia and Antarctica. They indicate a LANDSCAPE EVOLUTION model involving slow cryogenic bedrock WEATHERING combined with gravity-controlled free-face retreat and SLOPE REPLACEMENT from below. *Cryopediments* are the end result. In more humid environments, and especially in those recently deglaciated, not only is the rate of debris production greater as a result of the increased efficacy of frost shattering on exposed and over-steepened rock walls, but SOLIFLUCTION, SLOPEWASH, SNOW AVALANCHES and DEBRIS FLOWS are more common. As a result, slope forms are more varied.

It is sometimes assumed, though not proven, that slopes evolve more rapidly under periglacial conditions than under non-periglacial conditions. The few available data suggest that this applies only to relatively humid and recently deglaciated regions. In the equally extensive arid and unglaciated periglacial environments, there is no evidence that landscape evolution is faster than elsewhere and, instead, all the evidence points to a similarity of forms with hot, arid and semi-arid regions of the world. It is probable that many present-day slope forms in the temperate mid-latitudes of North America and Europe are essentially RELICT Pleistocene periglacial forms.   *HMF*

[See also EQUILIBRIUM CONCEPTS IN GEOMORPHOLOGICAL CONTEXTS]

Dylik, J. 1969: Slope development under periglacial conditions in the Lodz region. *Biuletyn Peryglacjalny* 18, 381–410. **Jahn, A.** 1976: Contemporaneous geomorphological processes in Longyeardalen, Vestspitsbergen (Svalbard). *Biuletyn Peryglacjalny* 26, 253–268. **French, H.M.** 1996: *The periglacial environment*, 2nd edn. London: Addison-Wesley Longman, 170–184. **French, H.M. and Harry, D.G.** 1992: Pediments and cold-climate conditions, Barn Mountain, unglaciated Yukon Territory. *Geografiska Annaler* 74, 145–157. **Pissart, A.** 1995: Deux types de versants periglaciaires de Haute Belgique. *Quaestiones Geographicae* 4, 241–245. **Priesnitz, K.** 1988: Cryoplanation. In Clark, M.J. (ed.), *Advances in periglacial geomorphology*. Chichester: Wiley, 49–67. **Rapp, A.** 1960: Recent development of mountain slopes in Karkevagge and surroundings, northern Sweden. *Geografiska Annaler* 42, 71–200. **Selby, M.J.** 1971: Slopes and their development in an ice-free, arid area of Antarctica. *Geografiska Annaler* 53, 235–245.

periglacial sediments   In the strict sense, periglacial sediments result from either the cryogenic WEATHERING of bedrock or the thawing of perennially (or seasonally) frozen ground. They include: (a) BLOCKFIELDS on expanses of level or gently-sloping terrain, blockslopes on moderate or steep slopes and blockstreams in broad valleys or gullies, all resulting from *in situ* cryogenic weathering of bedrock; (b) various unstratified and *stratified slope deposits*, that have undergone transport and subsequent deposition by frost creep, gelifluction and SLOPEWASH processes; (c) unsorted sediments that have resulted from gravity-controlled, relatively rapid, MASS-WASTING processes such as rockfalls, AVALANCHES, DEBRIS FLOWS and active-layer-detachment failures; (d) well sorted, relatively homogeneous, wind-blown silt, commonly termed LOESS; (e) locally derived AEOLIAN sand (COVERSAND); (f) NIVEO–AEOLIAN DEPOSITS; and (g) unstratified, heterogeneous assemblages of unconsolidated fine sediments (DIAMICTONS), previously ice-rich and frozen, that have experienced THERMOKARST.

Cryogenic weathering is the term used to describe the poorly understood group of physico-chemical processes thought to be responsible for the angular rock-rubble surfaces that characterise cold DESERTS. The factors influencing cold-climate rock disintegration have been examined in numerous field and experimental studies. One conclusion is that the rate of frost shattering is not only a function of the freeze–thaw frequency per year and the 9% expansion when water changes to ice, but also of the degree of water (ice) saturation and bedrock tensile strength. A different approach, more applicable to soil, is the *segregation ice model* of frost weathering. In this case, rock disintegration is the result of the progressive growth of microcracks and relatively large pores wedged open by water as it migrates towards growing ice lenses. Observations in very arid regions, such as Antarctica, suggest that fracturing can also be caused by thermal stresses generated within the rock itself, rather than by ice segregation. Finally, experimental studies have demonstrated that one of the main effects of cold-climate weathering is the production of silty particles of between 0.05 and 0.01 mm in diameter and, moreover, that QUARTZ grains are less resistant under cryogenic conditions than feldspars. Although its role is unclear, SALT WEATHERING also appears to be linked to cryogenic weathering.

Cold-climate loess is largely a homogeneous unstratified silt of wind-blown origin. It has a uniform grain-size of 0.01–0.05 mm in diameter and is particularly widespread in the marginal zones of the cold semi-arid deserts of China, Siberia and interior Yukon–Alaska. It also occurs in the temperate regions of North America and Europe. Loess deposits are often tens of metres thick and nearly always calcareous. Because (a) it shrinks when water is applied, (b) large SYNGENETIC sand-filled thermal contraction cracks are sometimes present and (c) it often contains cold-climate land snails, loess must have been deposited in arid, steppe-like environments, sufficiently cold for permafrost to occur. The majority of loess is PLEISTOCENE in age. However, because loess also forms around the margins of hot deserts, not all loess is periglacial in origin. Moreover, because it has been transported, it need not necessarily indicate a periglacial environment either in the area of deposition or, equally, in the source area.

SOLIFLUCTION and SLOPEWASH deposits are common in many present-day periglacial environments of moderate to high humidity. Pleistocene versions have been identified in many former periglacial areas. Solifluction involves frost creep and gelifluction. Solifluction deposits are heterogeneous in nature and can include a wide range of sediments, including coarse TALUS, alluvial silt and clay and DIAMICTONS of both glacial and periglacial origin. Their FABRIC is often organised such that elongate particles tend to be aligned with their long axes pointing in the direction of flow. They form lobes, tongues and sheets. Slopewash deposits are stratified sands and silts and are usually locally concentrated beneath and down-slope of SNOWBEDS. In areas of especially heavy snowfall and/or frequent FREEZE–THAW CYCLES, and where the bedrock is suited to cryogenic disintegration, the deposits can be stratified. These are sometimes referred to, especially in the PLEISTOCENE literature, as GRÈZES LITÉES.

Thermokarst sediments form when ice-rich permafrost degrades and melts. The result is a range of redeposited and heterogeneous materials, or diamictons, which often incorporate clumps of organic matter and portray deformation structures indicative of differential loading and density readjustments. Where *thaw lakes* and basins develop, LACUSTRINE SEDIMENTS form and, where the permafrost is exceptionally icy, COLLUVIAL FANS and small DELTA structures can form. It is difficult to generalise about thermokarst sediments, which are probably some of the most widespread of periglacial sediments; they are complex and have been relatively little studied.   *HMF*

[See also PERIGLACIAL LANDFORMS]

Hallet, B., Walder, J.S. and Stubbs, C.W. 1991: Weathering by segregation ice growth in micro-cracks at sustained subzero temperatures: verification from an experimental study using acoustic emissions. *Permafrost and Periglacial Processes* 2, 283–300. **Konischshev, V.N. and Rogov, V.V.** 1993: Investigations of cryogenic weathering in Europe and northern Asia. *Permafrost and Periglacial Processes* 4, 49–64. **Murton, J.M.** 1996: Thermokarst-lake-basin sediments, Tuktoyaktuk Coastlands, western Arctic Canada. *Sedimentology* 43, 737–760. **Péwé, T.L.,Tungscheng, L., Slatt, R.M. and Bingyuan, L.** 1995: Origin and character of loess-like silt in the southern Quinghai-Xizand (Tibet) Plateau, China. *United States Geological Survey Professional Paper* 1549, 1–55. **Washburn, A.L.** 1979: *Geocryology: a survey of periglacial processes and environments.* London: Edward Arnold.

periglaciation   The degree to which the processes and landforms in a landscape are affected by PERIGLACIAL (cold, non-glacial climate) conditions. Periglaciation is the equivalent of the concept of GLACIATION inasmuch as this term is used to refer to the periodic glacial modification of a landscape.   *HMF*

[See also PERIGLACIAL LANDFORMS, PERIGLACIAL LANDSCAPE EVOLUTION]

Ballantyne, C.K. and Harris, C. 1994: *The periglaciation of Great Britain*. Cambridge: Cambridge University Press.

periglaciofluvial system   The concept of a process–sediment–landform association different from both glacio-fluvial systems and the fluvial systems characteristic of temperate landscapes. The *alpine periglaciofluvial system* is distinctive in terms of channel form, bed material characteristics, sediment sources (including FROST WEATHERING in the river channel and SNOW AVALANCHE material from the valley sides) and strong linkages between the river channel and the valley slopes.   *JAM*

McEwen, L.J. and Matthews, J.A. 1998: Channel form, bed material and sediment sources of the Sprongdøla, southern Norway: evidence for a distinct periglacio-fluvial system. *Geografiska Annaler* 80(A), 17–36.

perihelion   The point on the orbit of the Earth (or that of another planet) at which it is closest to the Sun.   *JAM*

perimarine   The sedimentary environments and facies of a low-lying coastal area protected by barrier islands, comprising fluvial, lagoonal and peat marsh.   *RAS*

Hageman, B.P. 1969: Development of the western part of the Netherlands during the Holocene. *Geologie en Mijnbouw* 48, 373–388.

period   An interval of time in GEOCHRONOLOGY that is equivalent to a SYSTEM in CHRONOSTRATIGAPHY. Period and system names are the same (e.g. SILURIAN period and system).   *LC*

periodicities   Systematic recurrence of events after fixed periods of time or at a fixed FREQUENCY. Thus, the 24 h day and 365.24 days in a year repeat themselves year after year. In a similar way, TIDES can be calculated many years in advance. In the field of environmental change, the term is also used to mean recurrence at nearly but not exact intervals and with different intensities, although these are better termed quasi-periodicities (see QUASI-PERIODIC PHENOMENA). The terms '*cycle*' and '*cyclicities*' are often used in the same way.

There have been many attempts to find periodicities in

order to predict future events, but often they have been shown to be a product of the technique used (see FILTERING, harmonic analysis, *power spectrum* analysis). The main quasi-periodicities that have been found in a variety of climatic data include the MADDEN–JULIAN OSCILLATION (40–50 days), the QUASI-BIENNIAL OSCILLATION (2.5 years), SUNSPOT CYCLES (11 years), LUNAR CYCLES (18.6 years) and the *Hale cycle* (22 years). The strongest periodicities in longer-term PALAEOCLIMATOLOGY are those associated with the MILANKOVITCH THEORY. Amongst the weaker periodicities reported on intermediate or SUB-MILANKOVITCH timescales of centuries to millennia are those at c. 1600 years, 1000 years and 550 years.   *BDG*

[See also EPISODIC EVENTS, TIME SERIES]

Burroughs, W. J. 1992: *Weather cycles – real or imaginary*. Cambridge: Cambridge University Press. **Chapman, M.R. and Shackleton, N.J.** 2000: Evidence of 550 year and 1000 year cyclicities in North Atlantic circulation patterns during the Holocene. *The Holocene* 10, 287–291. **Lamb, H.H.** 1972: *Climate: present, past and future*. Vol. 1. *Fundamentals and climate now*. London: Methuen. **Stuiver, M., Grootes, P.M. and Braziunas, T.F.** 1995: The GISP δ^{18}O record of the past 16,500 years and the role of the sun, oceans and volcanoes. *Quaternary Research* 44, 341–354.

permafrost   *Perennially cryotic ground* in polar and alpine regions. It is defined on the basis of temperature rather than water/ice content; that is, ground (i.e. soil or rock) that remains at or below 0°C for at least two consecutive years.

To differentiate between the thermal (i.e. temperature) and state (i.e. frozen or unfrozen) conditions of permafrost, the terms *cryotic* and *non-cryotic* have been proposed. These terms refer solely to the temperature of the material, independent of its water or ice content. *Perennially cryotic ground* is, therefore, synonymous with permafrost, and it may be 'unfrozen', 'partially frozen' or 'frozen' depending upon the state of the ice and water content.

The PERMAFROST TABLE is the upper surface of the permafrost and the ground above the permafrost table is called the *suprapermafrost layer*. The ACTIVE LAYER is that part of the suprapermafrost layer that freezes in the winter and thaws during the summer; that is, it is seasonally frozen ground. Although seasonal frost usually penetrates to the permafrost table in most areas, in some it does not and an unfrozen zone exists between the bottom of the seasonal frost and the permafrost table. This unfrozen zone is called a TALIK. Unfrozen zones within and below the permafrost are also termed taliks.

Permafrost thickness depends on the balance between the internal heat gain with depth and the heat loss from the surface. Heat flow from the Earth's interior normally results in a temperature increase of approximately 1°C per 30–60 m increase in depth. This is known as the *geothermal gradient*. If climatic conditions at the ground surface alter, permafrost thickness will change appropriately. For example, an increase in mean surface temperature will result in a decrease in permafrost thickness, while a decrease in surface temperature will give the reverse.

Nearly 25% of the Earth's land surface is underlain by permafrost (see Figure). The majority occurs in the Northern Hemisphere. Russia possesses the largest area of permafrost (11.0 million km^2), followed by Canada (5.7 million km^2) and China (2.1 million km^2). In parts of

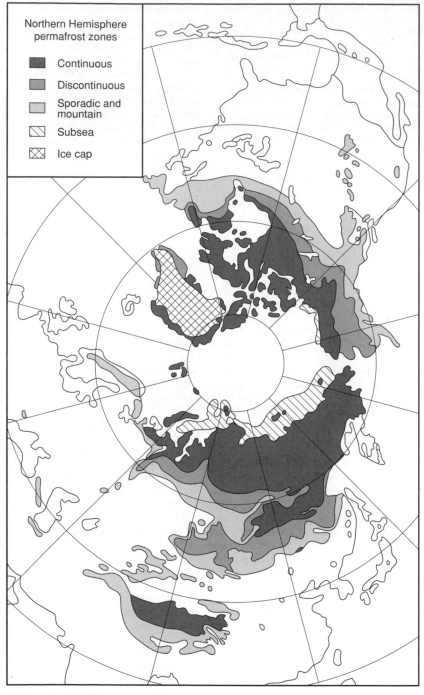

permafrost *Distribution of permafrost in the Northern Hemisphere (after French, 1996)*

Siberia and interior Alaska, permafrost has existed for several hundred thousand years; in other areas, such as the modern Mackenzie Delta, permafrost is young and currently forming under the existing cold climate.

Permafrost occurs in two contrasting and, in places, overlapping localities, namely high latitudes and high altitudes. Accordingly, permafrost can be classified into one of the following categories: (a) *polar* (or *latitudinal*) *permafrost* (i.e. permafrost in ARCTIC and Antarctic regions);

(b) *alpine permafrost* (i.e. MOUNTAIN PERMAFROST in mountainous regions); (c) *plateau permafrost* (i.e. extensive permafrost at high elevations, such as on the Tibet (Quinghai-Xizang) Plateau of China); and (d) *submarine permafrost* (i.e. on the continental shelves of the Laptev, Siberian and Beaufort Seas).

Permafrost is also classified according to whether it is continuous or discontinuous in nature. In areas of *continuous permafrost*, frozen ground is present at all localities except for localised taliks existing beneath lakes and river channels. In areas of *discontinuous permafrost*, bodies of frozen ground are separated by areas of unfrozen ground. At the southern limit of this zone permafrost becomes restricted to isolated 'islands', typically occurring beneath peaty organic sediments (sometimes termed the zone of *sporadic permafrost*). At the local level, variations in permafrost conditions are determined by a variety of terrain and other factors. Of widespread importance are the effects of relief and aspect and the nature of the physical properties of soil and rock. More complex are the controls exerted by vegetation, snow cover, water bodies, drainage and fire. GLOBAL WARMING is expected to cause widespread thaw of permafrost, especially in the discontinuous and sporadic permafrost zones.   *HMF*

[See also GEOMORPHOLOGY AND ENVIRONMENTAL CHANGE, HUMAN IMPACT ON LANDFORMS AND GEOMORPHIC PROCESSES, PERIGLACIAL SEDIMENTS, PERIGLACIAL LANDSCAPE EVOLUTION, SUBSURFACE TEMPERATURE]

Associate Committee on Geotechnical Research 1988: *Glossary of permafrost and related ground ice terms* [*Technical Memorandum* 142]. Ottawa: Permafrost Subcommittee, National Research Council of Canada. **Brown, J., Ferrians, O.J., Heginbottom, J.A. and Melnikov, E.S.** 1998: *Circum-arctic map of permafrost and ground ice*. United States Geological Survey, Map CP – 45, scale 1:10 000 000. **French, H.M.** 1996: *The periglacial environment*, 2nd edn. London: Addison-Wesley Longman. **French, H.M.** 1999: Past and present permafrost as an indicator of climate change. *Polar Research* 18, 269–274. **Haeberli, W. and Beniston, M.** 1998: Climate change and its impacts on glaciers and permafrost in the Alps. *Ambio* 27, 258–265. **Péwé, T.L.** 1991: Permafrost. In Keirsch, G.A. (ed.), *The heritage of engineering geology: the first hundred years* [*Centennial Special*. Vol. 3]. Boulder, CO: Geological Society of America, 277–298.

permafrost degradation A series of events that occur when the thermal equilibrium of PERMAFROST is disturbed either by climatic changes or by changes in ground surface condition. In the latter case, disturbance to overlying insulating vegetation and/or soil can lead to a rise in ground surface temperature and a lowering of the PERMAFROST TABLE. Subsidence occurs as the GROUND ICE thaws, the depth of subsidence depending on both the depth of thaw and the ice content of the upper layers of permafrost. Degradation can also be caused by the development of small ponds, which last until refreezing in autumn when latent heat is released locally. Human disturbance of the ground surface by fire, construction or vehicle movement can cause permafrost degradation. Ground temperature rise caused by GLOBAL WARMING is expected by many to reduce substantially areas of permafrost world-wide. For example, in Siberia, a reduction of 10% in permafrost area over a 50-year period has been predicted.   *RAS*

[See also THERMOKARST]

Koster, E.A. 1994: Global warming and periglacial landscapes. In Roberts, N. (ed.), *The changing global environment*. Oxford: Basil Blackwell, 127–149. **French, H.M.** 1996: *The periglacial environment*, 2nd edn. Harlow: Longman, 295–304. **Lunardini, V.J.** 1996: Climatic warming and the degradation of warm permafrost. *Permafrost and Periglacial Processes* 7, 311–320.

permafrost table In PERIGLACIAL environments, the boundary between the ACTIVE LAYER and the underlying PERMAFROST. Because of winter freezing penetrating

downwards at unequal rates, TALIKS may occur below the permafrost table.   *RAS*

[See also CRYOSOLS]

permanent drought A type of DROUGHT characteristic of ARIDLANDS, where there is normally insufficient moisture for agriculture without IRRIGATION, and only XEROPHYTES can grow unaided.   *JAM*

[See also CONTINGENT DROUGHT, INVISIBLE DROUGHT, SEASONAL DROUGHT]

permeable Having a structure or texture that permits the transmission of liquids or gases. In relation to rocks, sediments and soils, the distinction should be made between *primary permeability*, which allows water to pass through the pores or matrix by DIFFUSION, and *secondary permeability*, which allows the flow of water through fissures, joint or cracks. Whereas some make no distinction between permeable and PERVIOUS, others restrict usage of the term 'permeable' to what has been defined above as primary permeability and reserve pervious for secondary permeability.   *JAM/ADT*

[See also IMPERMEABLE, IMPERVIOUS, POROUS]

Stamp, L.D. (ed.) 1961: *A glossary of geographical terms*. London: Longman.

Permian A SYSTEM of rocks, and a PERIOD of geological time from 295 to 248 million years ago. The supercontinent PANGAEA had probably assembled, and a Southern Hemisphere ICE AGE that began during the CARBONIFEROUS period continued into the Permian. The end of the Permian is marked by a major MASS EXTINCTION.   *GO*

[See also GEOLOGICAL TIMESCALE, PERMO-TRIASSIC]

Eyles, N. and Young, G.M. 1994: Geodynamic controls on glaciation in Earth history. In Deynoux, M., Miller, J.M.G., Domack, E.W. *et al.* (eds), *Earth's glacial record*. Cambridge: Cambridge University Press, 1–28.

Permo-Triassic The PERMIAN and TRIASSIC geological SYSTEMS which, in northern Europe, share many characteristics resulting from a time of DESERT conditions. The term is commonly shortened to *Permo-Trias*.   *GO*

Mader, D. 1992: *Evolution of palaeoecology and palaeoenvironment of Permian and Triassic fluvial basins in Europe*. Stuttgart: Gustav Fischer.

peroxyacetylnitrate (PAN) An oxidant in urban PHOTOCHEMICAL SMOG produced from the interaction of HYDROCARBONS and NITROGEN OXIDES in vehicle exhaust. It is an eye irritant, a contributor to FOREST DECLINE and other plant damage, and a reservoir or SINK for nitrogen oxides in the upper troposphere, where it is relatively stable.   *JAM*

[See also PHOTOCHEMICAL OXIDANT, PHOTOCHEMISTRY]

McCormick, J. 1991: *Urban air pollution*. Nairobi: UNEP.

persistence (1) In the context of organisms, it refers to the developmental stage at which the organism will maintain itself indefinitely. (2) For chemicals in the environment, it is the length of time taken for the chemical to be broken down to a point at which it is no longer measurable. (3) In meteorology it describes the duration of a particular synoptic situation. (4) More generally, the term refers to the STABILITY or RESISTANCE to change of an individual, population or ecosystem in relation to a DISTURBANCE. *JLI*

[See also AUTOREGRESSIVE MODELLING, STABILITY CONCEPTS]

Blaustein A.R., Wake D.B. and Sousa W.P. 1994: Amphibian declines: judging stability, persistence, and susceptibility of populations to local and global extinctions. *Conservation Biology* 8, 60–71. **Onstad D.W. and Kornkven E.A.** 1992: Persistence and endemicity of pathogens in plant populations over time and space. *Phytopathology* 82, 561–566. **Storm G.L., Fosmire G.J. and Bellis E.D.** 1994: Persistence of metals in soil and selected vertebrates in the vicinity of the Palmerton zinc smelters. *Journal of Environmental Quality* 23, 508–514.

persistent organic compounds (POCs)

Complex organic compounds that are resistant to destruction and therefore persist in the environment without change for long periods. Most are either manufactured by humans or enter the environment inadvertently as a result of human activities. The number of synthetic organic compounds currently present in the global environment is probably 60 000–100 000 and about 1000 new ones are being produced per year. It is the chlorinated hydrocarbons (ORGANOCHLORIDES) manufactured for use as PESTICIDES or released during industrial processing that have caused most environmental concern.

The main structural classes of POCs in organochloride pesticides are *polychlorohydrocarbons* (e.g. DDT), *polychlorinated cyclodienes* (e.g. aldrin, dieldrin), *hexachlorocyclohexanes* (HCHs; e.g. lindane) and *polychlorinated monoterpenes* (e.g. toxaphene). Industrial processing produces POCs in several other structural classes including POLYCHLORINATED BIPHENYLS (PCBs), *polychlorinated dibenzo-p-dioxins* (PCDDs), *polychlorinated dibenzofurans* (PCDFs) and *polycyclic aromatic hydrocarbons* (PAHs). Most of the organochloride pesticides have been banned in industrialised countries, but they continue to be used in tropical developing countries for MALARIA control and much of past production is still involved in BIOGEOCHEMICAL CYCLES. Some PCBs are manufactured for use as liquid electrical insulators and in plastics and, although they are not highly toxic for most species, they have been implicated in the decline of seal populations through affecting reproduction. Some PCDDs and PCDFs are not manufactured for use but are produced as by-products or trace constituents and are highly toxic to fish, particularly in their early stages of development. Many PAHs are CARCINOGENS or MUTAGENS and are produced in very small quantities as a consequence of incomplete combustion of organic matter in, for example, burning FOSSIL FUELS and cooking foodstuffs.

Persistent organic compounds tend to concentrate in body-fat tissues as well as in blood and milk, exhibit BIOACCUMULATION in animals at the top of FOOD CHAINS and elicit a broad spectrum of biochemical and toxic responses. The levels of POCs found in the larger marine mammals, including whales, dolphins and porpoises, tend to be an order of magnitude greater than the levels in terrestrial birds and mammals including humans. Environmental concentrations are generally decreasing in air, water, soil, biota and food. *JAM*

[See also CHLOROFLUOROCARBONS, HALOGENATED HYDROCARBONS]

Connell, D.W. 1988: Bioaccumulation behaviour of persistent organic chemicals within aquatic organisms. *Review of Environmental Contamination and Toxicology* 102, 117–154. **Fishbein, L.** 1998: Organochlorine and polycyclic aromatic hydrocarbon contaminants. In Brune, D., Chapman, D.V., Gwynne, M.D. and Pacyna, J.M. (eds), *The global environment: science, technology and management.* Vol. 1. Weinheim: VCH, 481–497. **Kamrin, M.A. and Ringer, R.K.** 1994: PCB residues in mammals – a review. *Toxicology and Environmental Chemistry* 41, 63–84. **Korte, F.W. and Coulston, F.** 1994: Some consideration of the impact of energy and chemicals on the environment. *Ecotoxicology and Environmental Safety* 29, 243–250. **Oehme, M.** 1991: Further evidence for long-range air transport of polychlorinated aromatics and pesticides: North America and Eurasia to the Arctic. *Ambio* 20, 293–297. **Tanabe, S., Iwata, H. and Tatsukawa, R.** 1994: Global contamination by persistent organochlorides and their toxicological impact on marine mammals. *Science of the Total Environment* 15, 163–177.

perturbation

Any small-scale departure of a SYSTEM from a steady-state equilibrium. In the climatic context, the term is used for a WEATHER disturbance, such the *easterly waves* that are the precursors to TROPICAL CYCLONES. A perturbation may be described mathematically as the difference between the motion of an undisturbed wave or current in a fluid and a small superimposed motion, which is variable in space and time. It may then be viewed as the correction required to equate the actual motion in the fluid with a simple theoretical motion. *BDG*

[See also CLIMATIC FLUCTUATION]

Godske, C.L., Bergeron, T., Bjerknes, J. and Bundgaard, R.C. 1957: *Dynamic meteorology and weather forecasting.* Boston, MA: American Meteorological Society.

pervection

The mechanical movement or downwash of solid particles (especially silt) in soil. It occurs, for example, in the earliest stages of SOIL DEVELOPMENT on recently deglaciated terrain. It should be distinguished from LEACHING of chemicals in solution and ELUVIATION of clays and organic material. *JAM*

Frenot, Y., Van Vliet Lanoë, B. and Gloaguen, J.-C. 1995: Particle translocation and initial soil development on a glacier foreland, Kerguelen Islands, Subantarctic. *Arctic and Alpine Research* 27, 107–115. **Paton, T.R.** 1978: *The formation of soil material.* London: George Allen and Unwin.

pervious

(1) Synonymous with PERMEABLE. (2) The term is also used in a more restricted sense of materials, such as rocks, sediments or soils, that permit the flow of fluids through fissures, joints or cracks (*secondary permeability*), despite being non-porous. *ADT/JAM*

[See also IMPERVIOUS, POROUS]

pest

Any organism detrimental to human health, welfare and comfort. Strictly, any insect, parasite, plant or animal that reduces yield in managed ecosystems by competing with or destroying the harvestible crop. *GOH*

pest control

Reducing the abundance of a PEST and/or its EXTIRPATION. The most commonly used approach is through PESTICIDES but other methods include: (1) modification or removal of breeding sites; (2) release of irradiated males (to prevent breeding) or hormones (to prevent maturation); (3) INTRODUCTION of predators, PARASITES or DISEASE; and (4) selective breeding of pest-resistant varieties. *JAM*

[See also BIOLOGICAL CONTROL, PEST MANAGEMENT]

pest management The control and regulation of pest populations using, especially, chemical PESTICIDES and/or BIOLOGICAL CONTROL methods. *Integrated pest management* (IPM) may employ both methods, together with additional techniques, such as the use of resistant CULTIVARS, intercropping and the rationalisation of cultivation practises. *GOH*

Apple, J.L. and Smith, R.F. 1976: *Integrated pest management.* New York: Plenum.

pesticides Chemicals that kill pests. Some prefer the term *biocide*, which avoids the necessity to define PEST. Pesticides may be 'general' (*broad spectrum*) or 'specific' (*narrow spectrum*), examples of the latter including:

- *acaricide* (mites);
- *algacide* (algae);
- *antibiotic* (bacteria);
- *avicide* (birds);
- *fungicide* (fungi);
- *herbicide* (herbs, weeds);
- *insecticide* (insects);
- *molluscicide* (molluscs);
- *rodenticide* (rodents).

Use of pesticides, such as sulphur and arsenic, has a long history but modern approaches to PEST CONTROL originated in the mid-nineteenth century with the discovery of natural insecticides, such as derris dust and pyrethrum. From the AD 1930s, however, attention focused on synthetic organic pesticides, which were widely used in continuing intensification of agriculture, HORTICULTURE and SILVICULTURE. There are advantages and disadvantages of pesticide use. The main advantage is undoubtedly the increase in food production that results from the elimination of competitors for human food resources.

The non-specificity and PERSISTENCE of some pesticides remains a global problem both in the environment and in terms of HUMAN HEALTH HAZARDS, especially when accompanied by BIO-ACCUMULATION and BIOLOGICAL MAGNIFICATION. Lessons have been learned, however; use of persistent ORGANOCHLORIDES has been banned in many industrialised countries and increasing attention is being paid to alternatives, such as greater use of naturally occurring *pyrethrins*, synthetic analogues of these (*pyrethroids*), BIOLOGICAL CONTROL by natural enemies, biological control by purpose-built organisms (products of GENETIC ENGINEERING) and *integrated pest management* (IPM).

Thus, over the last 50 years, there have been changes in the types of pesticides used and the types of problems identified and the solutions adopted, at least in highly industrialised countries (see Figure). A major dilemma for third-world countries, where rapid AGRICULTURAL INTENSIFICATION is currently taking place, is whether large scale use of available, cheap but potentially environmentally damaging pesticides can be justified. *JAM*

[See also AGROCHEMICALS, DEFOLIANT, PEST MANAGEMENT]

Briggs, S.A. 1992: *Basic guide to pesticides: their characteristics and hazards.* Washington, DC: Hemisphere Publishing. **Horn, D.J.** 1988: *Ecological approach to pest management.* New York: Guilford. **Korte, F.W. and Coulson, F.** 1994: Some consideration of the impact of energy and chemicals on the environment. *Ecotoxicology and Environmental Safety* 29, 243–250. **Mannion, A.M.** 1995: *Agriculture and global change.* Chichester: Wiley. **Oehme, M.** 1991: Further evidence for long-range air transport of polychlorinated aromatics and pesticides. *Ambio* 20, 293–297. **Pettersson, O.** 1994: Swedish pesticide policy in a changing environment. In Pimentel, D. and Lehman, H. (eds), *The pesticide question: environment, economics and ethics.* New York: Chapman and Hall. **Ware, G.** 1983: *Pesticides: theory and application*, 2nd edn. New York: W.H. Freeman.

petroglyph A type of *rock art*; a carving or engraving in a cave or on an open rock surface. The carving process removes any WEATHERING RIND (*patina*) or ROCK VARNISH to reveal a fresh rock surface, on which patina forms anew. Methods such as CATION-RATIO DATING can be used to date prehistoric and historic petroglyphs. Animals or other features depicted may indicate different environmental conditions from today, such as greater moisture availability prior to 2000 years ago in the Wadi Howar region of the southeastern Sahara. *JAM*

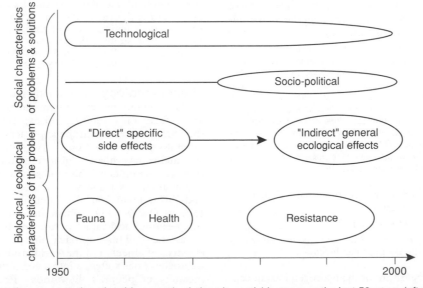

pesticides *Schematic representation of problems and solutions in pesticide use over the last 50 years (after Pettersson, 1994)*

[See also CAVE ART, GEOGLYPH, PARIETAL ART]

Chippindale, C. and Taçon, P.S.C. (eds) 1999: *The archaeology of rock-art*. Cambridge: Cambridge University Press. **Kröpelin, S.** 1993: Zur Rekonstruktion der spätquartären Umwelt amk Unteren Wadi Howar, Südöstliche Sahara (NW-Sudan). *Berliner Geographische Abhandlungen* 54. **Lanteigne, M.P.** 1991: Cation-ratio dating of rock engravings: a critical appraisal. *Antiquity* 65, 292–295.

petrography The description of rocks in hand specimens and, particularly, THIN-SECTION ANALYSIS.   *GO*

[See also PETROLOGY]

Adams, A.E., MacKenzie, W.S. and Guilford, C. 1984: *Atlas of sedimentary rocks under the microscope*. Harlow: Longman. **Harwood, G.** 1988: Microscopic techniques: II. Principles of sedimentary petrography. In Tucker, M.E. (ed.), *Techniques in sedimentology*. Oxford: Blackwell, 108–173.

petroleum Naturally-occurring HYDROCARBON compounds in liquid (CRUDE OIL), gas (NATURAL GAS) or semisolid (*asphalt, bitumen*) form that can be exploited as an ENERGY RESOURCE. Petroleum forms over geological timespans through the burial and partial decay of marine algae in organic-rich mudrocks (see BLACK SHALES, OIL SHALES): it is therefore a NON-RENEWABLE RESOURCE, or FOSSIL FUEL. Crude oil and natural gas collect in the pore spaces of SEDIMENTARY ROCKS and are extracted by drilling into rocks buried as much as several kilometres beneath the land surface or sea floor. The first oil well was drilled in Pennsylvania, USA, in AD 1859. Note that *petrol* (gasoline) is a highly processed product of petroleum: others include diesel, paraffin (kerosene) and lubricating oils.

In 1999 petroleum accounted for 65% of world energy consumption. RESERVES-to-production ratios indicate how long known supplies should last and stand at over 40 years for crude oil and 60 years for natural gas, but these estimates have changed little, or even increased, since 1989 as new discoveries and revisions to reserves estimates have matched production and consumption, and it is estimated that supplies of *conventional petroleum* should last through much of the twentyfirst century. Sources of *unconventional petroleum* include HEAVY OIL, TAR SANDS, OIL SHALES, *coalbed methane* (see COAL) and GAS HYDRATES: together these represent a far greater energy resource than the total available from conventional petroleum sources.   *GO*

Edwards, D.J. 1997: Crude oil and alternate energy production forecasts for the twenty-first century: the end of the hydrocarbon era. *American Association of Petroleum Geologists Bulletin* 81, 1292–1305. **Evans, A.M.** 1997: *An introduction to economic geology and its environmental impact*. Oxford: Blackwell Science. **Giuliano, F.A. (ed.)** 1981: *Introduction to oil and gas technology*. Boston, MA: International Human Resources Development Corporation. **Skinner, D.R.** 1981: *Introduction to petroleum processing*. Vol. 1. Houston, TX: Gulf Publishing Company. **Stoneley, R.** 1995: *An introduction to petroleum exploration for non-geologists*. Oxford: Oxford University Press.

petrology The study of rocks, principally their composition, TEXTURE and origin.   *GO*

[See also LITHOLOGY, PETROGRAPHY]

Blatt, H. and Tracy, R. 1996: *Petrology: igneous, sedimentary and metamorphic*, 2nd edn. Basingstoke: Freeman.

pH The negative logarithm of the hydrogen ion concentration in an aqueous solution:

$$pH = -\log[H^+]$$

Solutions are termed neutral when pH = 7.0; those below 7.0 are acidic, and those above alkaline.   *AWM*

[See also ACIDITY, ALKALINITY]

Phaeozems Soils with a thick, dark-coloured topsoil, rich in organic matter, but with evidence of LEACHING of carbonates from the profile and, in some cases, accumulation of clay in the lower horizons. These soils occur on the more humid plains, the forest-steppe ecotone in North America, South America and Eurasia. RELICT Phaeozems occur on the drier areas of the High Plains in Texas, formed originally under moister climatic conditions (SOIL TAXONOMY: *Udolls* or *Aquolls*).   *EMB*

[See also WORLD REFERENCE BASE FOR SOIL RESOURCES]

Phanerozoic An EON of geological time from the beginning of the CAMBRIAN period to the present day, characterised by abundant FOSSILS. The term is derived from the Greek for '*visible life*'.   *GO*

[See also GEOLOGICAL TIMESCALE]

phenology The scientific study of the life-cycle response of plants and animals to SEASONALITY and CLIMATIC CHANGE. Observations on the timing of naturally occurring phenomena, such as plant growth, flowering and fruiting, seed-times and harvesting, the migration of animals and birds and the first appearance of insects, are undertaken from year to year for seasonal comparisons. Of particular interest are the dates of first and last events such as first leaf, first flower, first and last appearance of birds and animals. For parts of Europe, long-term phenological data are available for a variety of deciduous tree species and for viticulture, in the form of VINE HARVEST records. Such observations have been used as PROXY CLIMATIC INDICATORS, thereby converting historical phenological material into CLIMATIC data.   *JBE*

Fretwell, S.D. 1972: *Populations in a seasonal environment*. Princeton, NJ: Princeton University Press. **Jefree, E.P.** 1960: Some long-term meanings from the Phenological Reports (1891–1948) of the Royal Meteorological Society. *Quarterly Journal of Royal Meteorological Society* 86, 95–103. **Pfister, C.** 1992: Monthly temperature and precipitation in central Europe 1525–1979, quantifying documentary evidence on weather and its effects. In Bradley R.S. and Jones P.D. (eds), *Climate since AD 1500*. London: Routledge, 118–142.

phenomenology Philosophies concerned with particular phenomena with an emphasis on direct intuition and a focus on meanings and experiences of the self. It therefore contrasts with the approaches developed by the natural sciences based on POSITIVISM. Phenomenological statements may thus be considered non-empirical descriptions of phenomena. This philosophical approach evolved within HUMANISM and was developed by Edward Relph and Yi-Fu Tuan from the initial formulation of Edmund Husserl (1859–1939). Phenomenology aims to uncover the meanings and values of human life by getting back to its essences. This approach aims to discover the time essences of objects and the world around us and in HUMAN GEOGRAPHY has softened into a concern with the everyday meaning of people's 'lifeworlds'. Phenomenology provides a powerful critique of POSITIVISM and throughout the

AD 1970s it was an important perspective in behavioural geography. *ART*

Relph, E. 1970: An enquiry into the relations between phenomenology and geography. *Canadian Geographer* 14, 193–201. Tuan, Y-F. 1977: *Space and place*. London: Edward Arnold.

phenotype

The characteristics of an organism, resulting from the interaction of its GENOTYPE and ENVIRONMENT. *KDB*

phi (φ)

A dimensionless expression of GRAIN-SIZE in SEDIMENTS and SEDIMENTARY ROCKS, calculated as $\varphi = -\log_2(d/d_0)$ where d is the grain diameter, and d_0 is the diameter of a 1 mm grain. High positive values on the phi scale denote small grain-sizes. *GO*

Krumbein, W.C. 1964: Some remarks on the phi notation. *Journal of Sedimentary Petrology* 34, 195–196.

-phile, -philous

Suffix for an organism that tolerates, and hence is an indicator of the *presence* of, a particular ENVIRONMENTAL FACTOR: thus, a *thermophile* is an organism tolerant of relatively high temperatures; it is *thermophilous*. *JAM*

[See also EXTREMOPHILE, -PHOBE, -PHOBOUS]

-phobe, -phobous

Suffix for an organism that is intolerant of, and hence is an indicator of the *absence* of, a particular ENVIRONMENTAL FACTOR: thus, a *chionophobe* is a plant intolerant of snow cover; it is *chionophobous*. *JAM*

[See also -PHILE, -PHILOUS]

phosphate (PO₄³⁻)

The main source of phosphorus for plants, an important FERTILISER (especially in the form of *superphosphates*) and a POLLUTANT. As a former constituent of detergents (in the form of *polyphosphates*) phosphates were a major cause of EUTROPHICATION in rivers and lakes. *JAM*

Griffiths, E.J. (ed.) 1973: *Environmental phosphorus handbook*. New York: Wiley. Jahnke, R.A. 1992: The phosphorus cycle. In Butcher, S.S., Carlson, R.J., Orians, G.H. and Wolfe, G.V. (eds), *Global biogeochemical cycles*. London: Academic Press. Toy, A.D.F. and Walsh, E.N. 1987: *Phosphorus chemistry in everyday living*. Washington, DC: American Chemical Society.

phosphate analysis

Phosphorus levels in soils can be increased markedly by human activity. This enrichment results from three sources: organic refuse (e.g. bones and plants in MIDDENS or burials); faeces and urine from people and their livestock; and deliberate manure or FERTILISER application. Samples are collected by soil augering and total soil phosphate determined, usually in the laboratory. Spatial mapping of phosphate levels may help to locate refuse dumps, cesspits, burials, areas where livestock were penned or areas where soils have been improved by manuring or fertilising. The method is, however, most effective if used in tandem with other archaeological prospection tools, such as MAGNETIC SUSCEPTIBILITY. Phosphate analysis may also be used to estimate *phosphate retention*, which has been suggested as a basis for SOIL DATING. *MJB*

[See also CHEMICAL ANALYSIS OF SOILS]

Bethell, P. and Maté, I. 1989: The use of soil phosphate analysis in archaeology: a critique. In Henderson, J. (ed.), *Scientific analysis in archaeology* [*Monograph* 19]. Oxford: Oxford University Communications for Archaeology, 1–29. Craddock, P.T., Gurney, D., Proyor, F. and Hughes, M.J. 1985: The application of phosphate analysis to the location and interpretation of archaeological sites. *Archaeological Journal* 142, 361–376. Dockrill, S.J. and Simpson, I.A. 1994: The identification of prehistoric anthropogenic soils in the Northern Isles using an integrated sampling methodology. *Archaeological Prospection* 1, 75–92. Herz, N. and Garrison, E.G., 1998: Chapter 9: Soil phosphate analysis in archaeological surveys. In *Geological methods for archaeology*. New York: Oxford University Press, 181–190. Lillios, K.T. 1993: Phosphate fractionation of soils at Agroal, Portugal. *American Antiquity* 57, 495–506. Sjöberg, A. 1976: Phosphate analysis of anthropic soils. *Journal of Field Archaeology* 3, 447–454.

phosphorite

A SEDIMENTARY DEPOSIT rich in phosphate minerals. Phosphorites are an important RESOURCE, used mainly as a raw material in fertilisers, and are significant in PALAEOENVIRONMENTAL RECONSTRUCTION, representing former CONTINENTAL SHELF sites of UPWELLING, episodes of sediment reworking, or sites of *guano* accumulation. *GO*

Abed, A.M. and Amireh, B.S. 1999: Sedimentology, geochemistry, economic potential and palaeogeography of an Upper Cretaceous phosphorite belt in the southeastern desert of Jordan. *Cretaceous Research* 20, 119–133. Gnandi, K. and Tobschall, H.J. 1999: The pollution of marine sediments by trace elements in the coastal region of Togo caused by dumping of cadmium-rich phosphorite tailing into the sea. *Environmental Geology* 38, 13–24. Notholt, A.J.G. and Jarvis, I. (eds) 1990: *Phosphorite research and development* [*Special Publication* 52]. London: Geological Society. Stoddart, D. and Scoffin, T. 1983: Phosphate rock on coral reef islands. In Goudie, A.S. and Pye, K. (eds), *Chemical sediments and geomorphology*. London: Academic Press, 369–400.

phosphorus (P)

The tenth most abundant element in the Earth's crust but only rarely found as concentrated PHOSPHATE mineral deposits, which are mined for use in fertilisers. Phosphorus is a MACRONUTRIENT but nevertheless an important source of FERTILISER POLLUTION and a major cause of EUTROPHICATION. *JAM*

[See also BIOGEOCHEMICAL CYCLES, PHOSPHATE ANALYSIS, PHOSPHORITE]

Freney, J.R. (ed.) 1982: *Cycling of carbon, nitrogen, sulfur and phosphorus in terrestrial and aquatic ecosystems*. New York: Springer.

photochemical oxidant

A SECONDARY POLLUTANT produced in the atmosphere by a complex series of chemical reactions involving volatile organic pollutants, NITROGEN OXIDES, oxygen and sunlight. Photochemical oxidants include aldehydes, nitrogen dioxide, OZONE and PEROXYACETYL NITRATE (PAN). *JAM*

[See also PHOTOCHEMISTRY, PHOTOCHEMICAL SMOG]

Seinfeld, J.H. 1986: *Atmospheric chemistry and physics of air pollution*. New York: Wiley.

photochemical smog

SMOG produced by photochemical processes. It is characterised by the presence of relatively high concentrations of various atmospheric POLLUTANTS, notably OZONE, formed by the action of sunlight on NITROGEN OXIDES and volatile hydrocarbons. The phenomenon was first recognised in Los Angeles, but is now a widespread aspect of urban pollution. The main irritants are formaldehyde, acrolein and PEROXYACETYLNITRATE (PAN). *DME*

[See also AIR POLLUTION, PHOTOCHEMISTRY, URBAN CLIMATE]

McCormick, J. 1991: *Urban air pollution*. Nairobi: UNEP. **National Academy of Sciences** 1988: *Air pollution, the automobile and human health*. Washington, DC: National Academy Press.

photochemistry   The study of chemical reactions in the ATMOSPHERE initiated, assisted or accelerated by absorption of SOLAR RADIATION, specifically photons of light in the visible or ultraviolet wavelengths. *DME*

[See also OZONE DEPLETION, PHOTOCHEMICAL SMOG]

Brasseur, G.P., Orlando, J.J. and Tyndall, G.S. (eds) 1999: *Atmospheric chemistry and global change*. New York: Oxford University Press. **Brimblecombe, P.** 1996: *Air composition and chemistry*, 2nd edn. Cambridge: Cambridge University Press.

photo-electronic erosion pin (PEEP)   An instrument, based on photo-sensitive cells, used to monitor soil or BANK EROSION and DEPOSITION automatically and quasi-continually. The PEEP system thus gives clearer pictures of the magnitude, frequency and timing of erosion and DEPOSITION events to assist process inference. *DML*

[See also EROSION PIN]

Lawler, D.M. 1991: A new technique for the automatic monitoring of erosion and deposition rates. *Water Resources Research* 27, 2125–2128. **Lawler, D.M.** 1992: Process dominance in bank erosion systems. In Carling, P.A. and Petts, G.E. (eds), *Lowland floodplain rivers: geomorphological perspectives*. Chichester: Wiley, 117–143.

photogrammetry   Techniques for making reliable and accurate environmental measurements from photographs. Although generally associated with the analysis of AERIAL PHOTOGRAPHY for mapping, the same techniques are also used to record building facades. Photogrammetry uses aerial photographs acquired as close to vertical as possible, taken with cameras calibrated for high geometric accuracy and containing *fiducial marks*, which identify the centre of the photograph. Measurements made from a single photograph require knowledge of the elevation of the points being measured to correct for terrain-induced changes in scale. If multiple photographs acquired in stereo format with a large amount of overlap are available, the PARALLAX effect caused by viewing the same point from different locations can be used to calculate elevation and therefore correct the scale. *Stereo aerial photographs* are therefore used to produce topographic maps and DIGITAL ELEVATION MODELS (DEMS). DEMS are now used to produce *orthophotographs* with terrain effects removed. *GMS*

Wolf, P.R. 1983: *Elements of photogrammetry*. McGraw-Hill: New York. **Avery, T.E.** 1992: *Fundamentals of remote sensing and airphoto interpretation*. New York: Prentice Hall.

photon   The smallest particle of electromagnetic RADIATION. *HB*

[See also ELECTROMAGNETIC SPECTRUM]

photoperiod   The duration of darkness and light, normally described in terms of day length. Many plants and animals respond and are adapted to the photoperiod. For example, a chrysanthemum blooms under short days and long nights. *RJH*

[See also LIMITING FACTORS, PHENOLOGY, PHOTOSYNTHESIS]

photosynthesis   The process by which AUTOTROPHIC ORGANISMS synthesise carbohydrates from light energy, CARBON DIOXIDE and water in the presence of CHLOROPHYLL. Oxygen is released during photosynthesis. *JAM*

Hall, D.O. and Rao, K.K. 1999: *Photosynthesis*, 6th edn. Cambridge: Cambridge University Press. **Lawlor, D.W.** 2000: *Photosynthesis*, 3rd edn. Oxford: Bios. **Raghavendra, A.S. (ed.)** 2000: *Photosynthesis : a comprehensive treatise*. Cambridge: Cambridge University Press.

photosynthetically active radiation (PAR)   The number of photons within the PAR spectrum incident per unit time on a unit surface area. The PAR spectrum is the 400–700 nm waveband. Plants use light from this waveband for PHOTOSYNTHESIS. A simple relationship exists between the number of plant molecules changed photochemically and the number of photons absorbed within the waveband; the relationship is independent of photon energy. *KJT*

Robinson, B.F. and DeWitt, D.P. 1983: Electro-optical non-imaging sensors. In Colwell, R.N., Simonett, D.S. and Ulaby, F.T. (eds), *Manual of remote sensing*, 2nd edn. Falls Church, VA: American Society of Photogrammetry.

phreatic zone   The zone below the permanent WATER TABLE, which is saturated with GROUNDWATER. *Phreatic water*, which completely fills the rock interstices and exists under HYDROSTATIC PRESSURE in the phreatic zone, is the water that feeds WELLS. *JAM*

[See also VADOSE ZONE]

Dingman, S.L. 1994: *Physical hydrology*. Englewood Cliffs, NJ: Prentice Hall.

phreatophyte   A plant that derives its water supply from GROUNDWATER: the term is from the Greek for 'well plant'. *JAM*

phyletic gradualism   EVOLUTION seen as gradual transformation of entire species populations, by means of NATURAL SELECTION. The rate of change is even and slow in explicit contrast to the concept of PUNCTUATED EQUILIBRIA. *KDB*

[See also GRADUALISM]

Eldredge, N. and Gould, S. J. 1972: Punctuated equilibria: an alternative to phyletic gradualism. In Schopf, T.J.M. (ed.), *Models in paleobiology*. San Francisco, CA: Freeman, Cooper, 82–115.

phylogeography   Study of the principles and processes governing the geographical distribution of genealogical LINEAGES, especially those of animals at the intraspecific level. *JAM*

[See also DNA, ANCIENT]

Avise, J.C. 2000: *Phylogeography: the history and formation of species*. Cambridge, MA: Harvard University Press.

physical analysis of soils   Analytical methods used to describe the properties that can be measured by physical terms or equations. The most common physical properties investigated in soils are bulk density, water-holding capacity, hydraulic conductivity, porosity, pore-size distribution and particle-size distribution.   *EMB*

[See also CHEMICAL ANALYSIS OF SOIL]

Black, C.A. (ed.) 1965: *Methods of soil analysis. Part 1. Physical and mineralogical properties, including statistics of measurement and sampling* [*Agronomy series* 9]. Madison, WI: American Society of Agronomy. **Dirksen, C.** 1999: *Soil physics measurements.* Reisenkirchen: Catena. **Menounos, B.** 1997: The water content of lake sediments and its relationship to the physical parameters: an alpine case study. *The Holocene* 7, 202–212. **Smith, K.A. and Mullins, C.E.** 1991: *Soil analysis: physical methods.* New York: Marcel Dekker.

physical geography

A subdiscipline of GEOGRAPHY and a NATURAL ENVIRONMENTAL SCIENCE that focuses on understanding the physical environments and LANDSCAPES of the Earth's surface at local to global scales. This encompasses spatial patterns in the GEOECOSPHERE and their controlling processes, the evolution of these patterns and processes, and their interactions with human activity. Physical geographers employ a wide range of methodological strategies, including FIELD RESEARCH, REMOTE SENSING, LABORATORY SCIENCE, NUMERICAL ANALYSIS, *modelling* and GEOGRAPHICAL INFORMATION SYSTEMS (GIS). Many physical geographers specialise in the pure or applied aspects of BIOGEOGRAPHY, CLIMATOLOGY or GEOMORPHOLOGY, but there are important integrative themes, one of which is ENVIRONMENTAL CHANGE. In this context, physical geography is most concerned with environmental change on timescales relevant to understanding present landscapes as the domain of human societies, past, present and future: particularly important aspects include HOLOCENE ENVIRONMENTAL CHANGE and current HUMAN IMPACTS.

JAM

[See also HUMAN GEOGRAPHY]

Gregory, K.J. 2000: *The changing nature of physical geography.* London: Arnold. **Haines-Young, R.H. and Petch, J.R.** 1986: *Physical geography: its nature and method.* London: Harper and Rowe. **Slaymaker, O. and Spencer, T.** 1998: *Physical geography and global environmental change.* London and New York: Addison-Wesley Longman.

physical oceanography

The branch of OCEANOGRAPHY dealing with physical aspects of the OCEANS including their temperature, density, OCEAN CURRENTS, UPWELLING, TIDES, waves and acoustic properties.   *BTC*

Pickard, G.L. and Emery, W.J. 1990: *Descriptive physical oceanography: an introduction,* 5th edn. Oxford: Pergamon Press.

physical weathering

The mechanical disintegration of rocks and minerals *in situ*, without chemical alteration. Also known as *mechanical weathering*, physical weathering is often closely related to CHEMICAL WEATHERING, which may reduce the mechanical resistance of rocks to physical weathering. Thus, determining the relative contribution of the two processes in overall rock breakdown is often difficult. Important physical processes include water-based mechanisms such as FREEZE–THAW CYCLES, HYDRATION and hydraulic pressure weathering. Also important are SALT WEATHERING, which involves the expansion of salt crystals, and *insolation weathering*, which reflects the thermal expansion and shrinkage of minerals, often at different, mineral-specific rates. Fire can be particularly effective in leading to rapid thermal expansion of rock. The resulting spalls have been called *ignifracts*.

UNLOADING is caused by the reduction of compressive stress owing to the removal of overlying rock. Physical weathering is particularly effective in climates of high diurnal and/or annual temperature amplitudes and areas where the temperature oscillates around 0°C. Thus, evidence of long-term physical weathering in an area may be useful in aiding PALAEOENVIRONMENTAL RECONSTRUCTION.

SHD

[See also EXFOLIATION, FREEZING INDEX, FROST WEATHERING, NIVATION]

Ballais, J.L. and Bosc, M.C. 1994: The ignifracts of the Sainte-Victoire Mountain (Lower Provence, France). In Sala, M. and Rubio, J.L. (eds), *Soil erosion and degradation as a consequence of forest fires.* Logroño: Geoforma Ediciones, 217–227. **Bland, W. and Rolls, D.** 1998: *Weathering.* London: Arnold. **Halsey, D.P., Mitchell, D.J. and Dews, S.J.** 1998: Influence of climatically induced cycles in physical weathering. *Quarterly Journal of Engineering Geology* 31, 359–367. **Matsuoka, N., Moriwaki, K, and Hirakawa, K.** 1996: Field experiments on physical weathering and wind erosion in an Antarctic cold desert. *Earth Surface Processes and Landforms* 21, 687–699.

physiography

A largely obsolete term that has been used in a variety of ways ranging from a synonym for GEOMORPHOLOGY to the holistic study of PHYSICAL GEOGRAPHY conceived in the broadest possible way.   *JAM*

Huxley, T.H. 1877: *Physiography: an introduction to the study of nature.* London: Macmillan. **Stoddart, D.R.** 1975: 'That Victorian science': Huxley's Physiography and its impact on geography. *Transactions of the Institute of British Geographers* 66, 17–40.

phytogeography

Also known as *plant geography*, phytogeography is the study of the distribution of different types of plants or TAXONOMIC groups of plants (e.g. families and genera), often with the object of analysing the geographical range of particular taxa or floras and explaining them in terms of origin, DISPERSAL and EVOLUTION of the type or group. The Figure shows the phytogeographical regions of the world.   *MJB*

[See also BIOGEOGRAPHY, FLORAL PROVINCES, ISLAND BIOGEOGRAPHY, WALLACE'S LINE, ZOOGEOGRAPHY]

Good, R. 1964: *The geography of the flowering plants.* London: Longman. **Stott, P.A.** 1981: *Historical plant geography.* London: Allen and Unwin. **Tivy, J.** 1993: *Biogeography: a study of plants in the ecosphere,* 3rd edn. London: Longman Scientific and Technical.

phytoindication

Use of plants as indicators of environmental change, for ENVIRONMENTAL MONITORING and for DATING purposes. Three main subdivisions of phytoindication methods may be recognised: (1) *phytocoenotic methods* (based on plant community properties); (2) *dendro-indication* (including DENDROCHRONOLOGY and DENDROCLIMATOLOGY); and (3) *lichenometry* (including LICHENOMETRIC DATING). Phytocoenotic methods include the use of changes in species composition and community structure through time for dating surfaces (PHYTOMETRIC DATING). This approach has been widely used in the former Soviet Union for identifying the extent and timing of debris flows, snow avalanches and landslides. Also

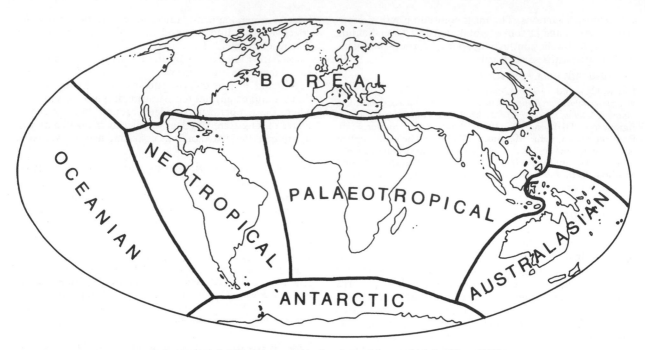

phytogeography *Phytogeographical regions of the world (after Tivy, 1993)*

included is the use of epiphytic lichens as monitors of AIR POLLUTION in cities. Such EPIPHYTES are sensitive to sulphur dioxide air pollution in urban areas, where zones of distinct lichen communities provide a semi-quantitative scale of pollution levels. *JAM*

Hawkesworth, D.L. and Rose, F. 1976: *Lichens as pollution monitors*. London: Arnold. Lekhatinov, A.M. 1988: Phytoindication methods for studying landslide and mudflow regimes. In Kozlovski, E.A. (ed.), *Landslides and mudflows*. Vol. 1. Moscow: Unesco, 94–97.

phytoliths The opalaetic bodies, also known as *silica bodies*, *biogenic opals*, *biogenic silica*, *plant opals*, *opal phytoliths*, *silica cells* or *micrometric hydrated opal-A particles* ($SiO_2.nH_2O$), which constitute the 'stony' parts of a plant and range in size from 20 to 200 μm. The term 'opal' is often used in describing phytoliths in order to distinguish them from other mineral substances secreted by plants (e.g. calcium carbonate). Soluble silica is carried passively through the plant vascular system until the evaporation of water leads to chemical PRECIPITATION of silica in or around the plant cells. Phytoliths are most abundant in the leaves where plants lose most water as a result of *transpiration*. Although the purpose of phytoliths remains unclear, it has been suggested that they function as structural skeletons, anti-wilting devices, or that they provide plant resistance to herbivory. Phytoliths are valuable BIO-INDICATORS since they are relatively inert, are extractable from SEDIMENTS and PALAEOSOLS and permit identification of the source plants to family or subfamily level. Phytolith analysis has found particular application in the reconstruction of past grasslands, where phytoliths provide complementary micromorphological evidence to GRASS CUTICLE ANALYSIS. *MJW*

[See also PALAEOECOLOGY, PALAEOBOTANY, SILICA ANALYSIS]

Pearsall, D.M. 1989: *Paleoethnobotany: a handbook of procedures.* San Diego, CA: Academic Press. Piperno, D.R. 1988: *Phytolith analysis: an archaeological and geological perspective.* San Diego, CA: Academic Press.

phytometric dating Use of plant species and community properties, especially in relation to ECOLOGICAL SUCCESSION, for dating purposes. *JAM*

[See also PHYTO-INDICATION]

Matthews, J.A. 1978: Plant colonisation patterns on a gletschervorfeld, southern Norway: a meso-scale geographical approach to vegetation change and phytometric dating. *Boreas* 7, 155–178.

phytosociology An approach to the study of vegetation, which focuses on the classification of plant communities based on their species composition, as typified by the Zürich–Montpellier school. *JLI*

[See also COMMUNITY CONCEPTS, INDIVIDUALISTIC CONCEPT]

Braun-Blanquet, J. 1964: *Pflanzensoziologie. Grundzüge der Vegetationskunde,* 3rd edn. Berlin: Springer. Keller, W., Wohlgemuth, T., Kuhn, N. *et al.* 1998: Waldgesellschaften der Schweiz auf floristischer Grundlage. *Mitteilungen der Eidgenössischen Forschungsanstalt für Wald, Schnee und Landschaft* 73, 93–357.

piedmont An area of relatively low relief at the base of an upland or mountain range. The *piedmont angle* is the sharp break of slope between the mountain and the piedmont feature (often a PEDIMENT), which is particularly prominent in semi-arid and arid INSELBERG landscapes. Piedmonts often reveal erosional and/or depositional evidence of LANDSCAPE EVOLUTION. *MAC*

[See also BAJADA]

Bourne, J.A. and Twidale, C.R. 1998: Pediments and alluvial fans: genesis and relationships in the western piedmont of the

Flinders Range, South Australia. *Australian Journal of Earth Science* 45, 123–135.

pillow lava A body of LAVA made up of bulbous, rounded masses up to a few metres across. The base of each pillow is moulded to the shapes of those beneath. Pillow lavas are characteristic of lava erupted into water, and are useful in PALAEOENVIRONMENTAL RECONSTRUCTION. One situation in which they are erupted is at MID-OCEAN RIDGES, forming the surface layer of OCEANIC CRUST.
GO

pilot survey A preliminary survey used as a basis either for an informed decision on whether a full study would be worthwhile or for deciding on a suitable SAMPLING design for a later, more definitive survey. *JAM*

[See also FIELD RESEARCH, RECONNAISSANCE SURVEY]

pine decline A mid-HOLOCENE decline in the abundance of the coniferous tree species *Pinus sylvestris* L. (Scots pine). Pollen data from sites in the north and west of the British Isles record a marked decline in the abundance of pine where it still persisted between 4800 and 4400 radiocarbon years BP. Pine stumps can be found in abundance in blanket PEAT throughout the northwestern British Isles, but pine appears not to have persisted within this habitat after about 4000 radiocarbon years BP. However, pine did not become extinct at this time, surviving at low levels in Ireland until Mediaeval times and in Scotland until the present day. The cause of this decline is still not clear. Hypotheses that have been proposed include CLIMATIC CHANGE, VOLCANIC ERUPTION, human activity and changes in FIRE FREQUENCY. *JLF*

[See also ELM DECLINE, HUMAN IMPACT ON VEGETATION HISTORY, POLLEN ANALYSIS]

Bennett, K.D. 1984: The post-glacial history of *Pinus sylvestris* in the British Isles. *Quaternary Science Reviews* 3, 133–155.
Willis, K.J., Bennett, K.D, and Birks, H.J.B. 1998: The late-Quaternary dynamics of pines in Europe. In Richardson, D.M. (ed.), *Ecology and biogeography of* Pinus. Cambridge: Cambridge University Press, 107–121.

pingo A typically dome-shaped, ice-cored hill up to 50 m high in PERMAFROST areas, formed by the freezing of water moving under a pressure gradient towards the pingo. It is the largest type of FROST MOUND. *RAS*

pinning point (1) For marine-based ICE SHEETS, a high point on the sea floor consisting of bedrock hills or upstanding masses of sediment that exert an influence on ICE SHELF dynamics by increasing locally the basal drag and reducing losses by CALVING. In applying the GLACIOMARINE HYPOTHESIS TO DEGLACIATION of the Irish Sea Basin, the Pembrokeshire peninsula, Wales and Wexford, Ireland have been viewed as pinning points preventing GLACIER SURGE and collapse of the marine-based ICE DOMES. This deglaciation model for the Irish Sea Basin is hotly disputed. (2) Evidence in marine and/or subaerial SEDIMENTS, thought to be useful in reconstructing former RELATIVE SEA LEVEL. A *pinning point curve* can provide an illustration of relative sea level history. *RAS*

Eyles, N. and McCabe, A.M. 1989: The Late Devensian (< 22,000 BP) Irish Sea Basin: the sedimentary record of a collapsed ice sheet margin. *Quaternary Science Reviews* 8, 307–351. Goldstein, R.H. and Franseen, E.K. 1995: Pinning-points – a method providing quantitative constraints on relative sea-level history. *Sedimentary Geology* 95, 1–10.

pioneer plant A plant that colonises a disturbed area, thereby initiating ECOLOGICAL SUCCESSION. *LRW*

[See also COLONISATION, RUDERAL]

piosphere An approximately circular area of land centred on a watering point (WELL or BOREHOLE) for people and/or animals in a DRYLAND area. As cattle need to drink every second day, the radius of a piosphere in theory is equivalent to one day's walk from the water source, although piospheres of up to 50 km in radius have been reported. Continual GRAZING pressure on the piospheres is seen by most researchers as leading to LAND DEGRADATION, and coalescence of piospheres as a cause of DESERTIFICATION over large areas (see Figure). It has been argued, however, that it is difficult to separate vegetation loss by grazing from that caused naturally from low rainfall, that piosphere coalescence is theoretical only and that the high nutrient input around wells and boreholes may actually lead to improved SOIL QUALITY. *RAS*

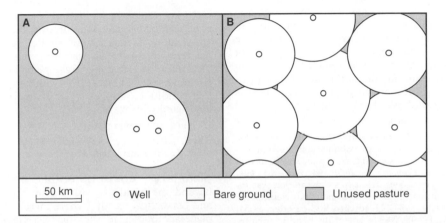

piosphere *(A) Piospheres centred around wells. (B) Theoretical coalescence of piospheres, regarded by some as a process of desertification (after Thomas and Middleton, 1994)*

Andrew, M.H. 1988: Grazing impacts in relation to livestock watering points. *Trends in Ecology and Evolution* 3, 336–339. Middleton, N. 1993: The desertification debate. *Geography Review* September 1993, 30–33. Thomas, D.S.G. and Middleton, N.J. 1994: *Desertification: exploding the myth.* Chichester: Wiley.

pipeflow The down-slope movement of water through natural tunnels or passages (*soil pipes*) in the SOIL or REGOLITH. Shallow pipes with surface connections tend to respond quickly to rainfall, but deep pipes tend to respond slowly. *Piping* may have a biotic origin, such as animal burrows or voids left after roots decay, or it may be produced by other pedological/hydrological processes, notably by cracking associated with seasonal drying or cooling cycles. Pipe development in peat in the British Isles may have been enhanced by the drying effects of AFFORESTATION. *ADT*

[See also THROUGHFLOW, RUNOFF PROCESSES]

Jones, J.A.A. 1981: *The nature of soil piping: a review of research* [*British Geomorphological Research Group Monograph* 3]. Norwich: Geobooks. Jones, J.A.A. 1997: Pipeflow contributing areas and runoff response. *Hydrological Processes* 11, 35–41.

piston sampler These samplers or CORERS are more effective than CHAMBER SAMPLERS in soft sediments or under water. They consist of a hollow tube fitted with a moveable piston. The piston creates a negative pressure within the tube, thus preventing compression of sediments and helping to keep the material within the tube when the corer is retrieved. The commonest example is probably the *Livingstone corer.* *MJB*

Nesje, A. 1992: A piston corer for lacustrine and marine sediments. *Arctic and Alpine Research* 24, 257–259. Wright, H.E. 1980: Cores of soft lake sediments. *Boreas* 9, 107–114.

pixel The smallest element of information in an image (such as that formed by REMOTE SENSING), consisting of a position and a value. Allowed values are real numbers (grey-level images), vectors of real numbers (multispectral images) and classes (CLASSIFICATION), among others. *ACF*

placer deposit A type of ORE deposit in which dense minerals of economic value have been eroded from a source material and concentrated by mechanical sorting during SEDIMENTATION. Most placer deposits occur in SEDIMENTARY DEPOSITS that accumulated in river, beach and COASTAL DUNE environments. *Palaeoplacers* are ancient placer deposits within SEDIMENTARY ROCKS and include the Witwatersrand gold deposits in South Africa of ARCHAEAN age. Much of the world's gold, tin (from tin oxide, cassiterite), titanium, diamonds and other gemstones are obtained from placer deposits. The term is derived from the Spanish *plaza*, meaning a place. *GO*

[See also MINERAL RESOURCES]

Craw, D., Youngson, J.H. and Koons, P.O. 1999: Gold dispersal and placer formation in an active oblique collisional mountain belt, Southern Alps, New Zealand. *Economic Geology and the Bulletin of the Society of Economic Geologists* 94, 605–614. Els, B.G. 1998: The auriferous late Archaean sedimentation systems of South Africa: unique palaeo-environmental conditions? *Sedimentary Geology* 120, 205–224. Jewett, S.C., Feder, H.M. and Blanchard, A. 1999: Assessment of the benthic environment following offshore placer gold mining in the northeastern Bering Sea. *Marine Environmental Research* 48, 91–122.

Thomas, M.F. and Thorpe, M.B. 1993: The geomorphology of some Quaternary placer deposits. *Zeitschrift für Geomorphologie Supplementband* 87, 183–194.

plaggic horizon A surface ANTHROPEDOGENIC HORIZON, strongly acid, brownish or blackish in colour and more than 50 cm thick, common in northwestern Europe. It was formed by long-continued *manuring* with turves cut from the forest that were used in stables or cowsheds and subsequently placed on fields, raising their level above that of surrounding areas. *EMB*

Pape, J.C. 1970: Plaggen soils in the Netherlands. *Geoderma* 4, 229–245.

plagioclimax The vegetation that wholly or partly replaces or modifies a supposed CLIMAX VEGETATION after an environmental DISTURBANCE. It may be maintained by disturbance. The alternative term *disclimax* is often used in the USA. Most of the so-called natural grasslands in the British Isles lying below 500 m are plagioclimax vegetation maintained by cattle and sheep GRAZING. If grazing were to cease, the grasslands would mostly revert to deciduous forest. In the arid and semiarid United States, ECOLOGICAL SUCCESSION around some ghost towns is permanently altered by an annual grass – cheatgrass (*Bromus tectorum*) – introduced from the Mediterranean. The cheatgrass will probably stay dominant at the expense of other, native species: SECONDARY SUCCESSION has produced a plagioclimax vegetation. Logging in the Peace River Lowlands, Wood Buffalo National Park, Canada is creating a plagioclimax vegetation in BOREAL FOREST: convergence towards the original white-spruce and mixedwood forests is not apparent and a long-term deciduous disclimax (with balsam poplar and lesser amounts of Alaska birch and aspen) is predicted. *RJH*

[See also CONVERGENCE AND DIVERGENCE, IN SUCCESSION]

Knapp, P.A. 1992: Secondary plant succession and vegetation recovery in two western Great Basin Desert ghost towns. *Biological Conservation* 6, 81–89. Timoney, K.P., Peterson, G., and Wein, R. 1997: Vegetation development of boreal riparian plant communities after flooding, fire, and logging, Peace River, Canada. *Forest Ecology and Management* 93, 101–120.

plane bed A BEDFORM comprising a flat, or nearly flat surface across which sand or fine gravel is transported. *Lower stage plane bed* forms in response to weak currents, replacing CURRENT RIPPLES in coarse sand. *Upper stage plane bed* forms in strong currents, especially where flows are shallow. Delicate flow-parallel ridges on the surface are termed *parting lineation* or *primary current lineation*. Deposition under upper stage plane bed conditions gives rise to PARALLEL LAMINATION with parting lineation. Plane bed conditions also exist for strong oscillating currents associated with WAVES. *GO*

[See also BEDFORM STABILITY DIAGRAM]

Allen, J.R.L. 1982: *Sedimentary structures: their character and physical basis.* Amsterdam: Elsevier.

planetary environmental change Discussion beyond the highly speculative on the geological history and CLIMATIC CHANGE (where there is an atmosphere) of planetary bodies in the solar system has depended on imagery and other data sent back from orbiter and lander spacecraft launched since AD 1959. The most detailed

GEOLOGICAL RECORD OF ENVIRONMENTAL CHANGE has been assembled for the *Moon*, as a result of its proximity to Earth but particularly because of repeated visits by uncrewed and crewed spacecraft and rock samples returned to Earth. According to RADIOMETRIC DATING, the heavily cratered highlands date from c. 4.4–4.0 billion years ago, but the smooth-surfaced, relatively CRATER-free *maria* (infilled by LAVA FLOWS issuing from partially molten lunar mantle) formed c. 3.9–3.2 billion years ago. The maria give an idea of rates of cratering from impacting objects (BOLIDES) over the past three million years or so, providing a means of assessing the lengths of exposure of the surfaces of other planets.

Imagery of *Mercury* resembles that obtained from the Moon with ubiquitous impact craters, though apparently no volcanism. A further difference lies in compressional cliffs or faults postdating most craters, suggesting that the planet's interior may have shrunk slightly (perhaps by only 1 km or less in diameter) many hundreds of million years after crustal solidification.

Venus and *Mars* are Earth-like in two important respects: both have dynamic atmospheres and rocky surfaces. With little EROSION and SEDIMENTATION on Venus, its cratered surface seems to have been periodically obliterated by new lava flows every few hundreds of million years. This is suggested by the comparatively small number of identified craters (c. 900). There is also clear evidence of tectonic activity in a broad belt with comparatively small-scale '*blob*' tectonics (see PLATE TECTONICS). The atmosphere may well have resembled that of Earth about four billion years ago, but it is now composed largely of CO_2 (96%) with little oxygen or water vapour. It has been argued that it is at or near a state of *unstable equilibrium* and that only moderate perturbations of radiatively active volatiles might cause climatic change. The massive CO_2 atmosphere means that Venus experiences an extreme GREENHOUSE EFFECT, leading to surface temperatures of around 450°C.

The atmospheric composition of Mars is close to that of Venus, but the low atmospheric pressure (< 1% that of Earth) combined with lower solar input mean that it is a cold (down to below −125°C at the poles) desiccated planet, probably with GROUNDWATER trapped beneath a PERMAFROST layer. Approximately half the surface is moderately cratered and, from the evidence of extensive mainly WIND EROSION, is probably at least four billion years old. Younger surfaces are volcanic in character, as in the case of the *Tharsis bulge* (a huge uplifted area the size of North America), which is thought to be between one and three billion years old. Very different climates and surface processes seem to have operated in the past. Early in its history, Mars was once apparently much warmer than now, with what appear to be relict river channels and large and small water bodies. Most valleys lie in isolated systems or clusters, suggesting supply of water from subsurface AQUIFERS, but there has been some support for the former existence of rainfall-fed streams. Recently formed small gullies on steep crater and valley slopes seen in Mars Global Surveyor satellite imagery have been interpreted in terms of flooding of aquifer water (or liquid CO_2) following the sudden release of a surface frozen barrier. Progressive cooling of Mars probably occurred through gas escape from the atmosphere, caused by the weak gravitational field and through low solar-energy input.

Different origins have been suggested for huge outflow channels resembling those of the CHANNELED SCABLANDS. They may have resulted from catastrophic floods caused by the sudden release of ARTESIAN water from beneath the frozen upper permafrost layer triggered by internal geological activity. Alternative mechanisms are DEBRIS FLOWS, liquefied crustal material and low-viscosity lava flows. Early Martian warmth and wetness have led to speculation about whether some simple forms of life may have developed. Climatic change over the last few million years may be recorded in layers apparently present throughout the polar regions, which probably represent AEOLIAN deposition of sediment transported from lower latitudes.

The so-called *giant outer planets* (Jupiter, Saturn, Uranus and Neptune) are much larger than the *inner terrestrial planets* and have thick atmospheres. *Jupiter* has no crustal surface, but has a spectacular kaleidoscopic cloud system with notable disturbances. The largest of these, the *Great Red Spot* (diameter 30 000 km), has existed at least since detailed telescopic observations began some 300 years ago, but the three *White Ovals* formed only as recently as AD 1940. No similar continuity of observations exists for the other giant outer planets.

The moons of the giant planets have proved unexpectedly varied. Images of the *Jovian moons* (the moons of Jupiter) were obtained during Voyager spacecraft flybys in 1979. *Ganymede* is composed partly of ice. Its surface has craters, but at some time in the past was warm enough for eruptions of water to flood the surface and obliterate the old craters. *Callisto* has a cratered surface modified by deformation of slightly plastic ice. The smaller inner moons, *Europa* and *Io*, are very different. Europa is geologically unique, with a smooth surface ice crust possibly floating on a global ocean. Io has a close orbit with Jupiter, resulting in a heat-generating tidal bulge causing very active volcanism, which renews the surface comparatively frequently. This explains the lack of impact craters. One of the satellites of *Saturn*, *Titan*, is unique amongst moons in the solar system in having an atmosphere. Many believe that its atmosphere will provide important insights into the early history of Earth's atmosphere and possibly into the origins of life. The largest of the moons of *Neptune*, *Triton*, is the coldest place in the solar system, such that most potential atmospheric gas is frozen. There are some impact craters but many regions have undergone resurfacing possibly through flowage of methane and nitrogen 'ice' precipitated out of Triton's thin atmosphere. A polar cap, possibly comprising frozen nitrogen, covers much of the southern hemisphere, apparently evaporating through solar warming along its northern edge via geysers or volcano-like plumes of nitrogen ejected up to 10 km from the surface. *RAS*

[See also ATMOSPHERIC COMPOSITION, GLOBAL WARMING, METEORITE IMPACT, NEAR-EARTH OBJECT, VOLCANICITY]

Bullock, M.A. and Grinspoon, D.H. 1996: The stability of climate on Venus. *Journal of Geophysical Research – Planets* 101, 7521–7529. **Carr, M.H.** 1995: The Martian drainage system and the origin of valley networks and fretted channels. *Journal of Geophysical Research – Planets* 100, 7479–7507. **Dressler, O. and Sharpton, V.L.** 1999: *Large meteorite impacts and planetary evolution* II [*Special Paper* 339]. Boulder, CO: Geological Society of America. **Greeley, R.** 1995: Geology of terrestrial planets with dynamic atmospheres. *Earth, Moon and Planets* 67, 13–29. **Gulick, V.C.** 2001: Origin of the valley networks on Mars: a hydrological perspective. *Geomorphology* 37, 241–268. **Morrison, D.** 1992: *Exploring planetary worlds.* New York: Scientific American Library. **Musselwhite, D.S., Swindle, T.D. and Lunine, J.I.** 2001: Liquid CO_2 breakout and the formation of recent small gullies on Mars. *Geophysical Research Letters* 28, 1283–1285. **Pepin, R.O.** 1994: Evolution of the Martian atmosphere. *Icarus* 111, 289–304. **Summerfield, M.A.** 1991: *Global geomorphology.* Harlow: Longman, 483–509. **Tanaka, K.L.** 1997: Sedimentary history and mass flow structures of Chryse and Acidalia Planitiae, Mars. *Journal of Geophysical Research – Planets* 102, 4131–4149. **Treiman, A.H., Fuks, K.H. and Murchie, S.** 1995: Diagenetic layers in the upper walls of Valles Marineris, Mars: evidence for drastic climate change since the mid-Hesperian. *Journal of Geophysical Research – Planets* 100, 26339–26344. **Wells, G.L. and Zimbelman, J.R.** 1997: Extraterrestrial arid surface processes. In Thomas, D.S.G. (ed.), *Arid zone geomorphology: process, form and change in drylands,* 2nd edn. Chichester: Wiley.

planetary geology

The study of the rocky celestial bodies (terrestrial planets, moons and asteroids). This enormous and recent discipline has considerable potential for studies of environmental change. First, these bodies can provide independent control on whether Earth's environmental changes are controlled by internal forces (e.g. PLATE TECTONICS) or external forces (e.g. SOLAR FORCING). Second, they provide further MODERN ANALOGUES for testing geological models for such features as the GENERAL CIRCULATION OF THE ATMOSPHERE and TECTONIC processes. *CDW*

[See also ASTROGEOLOGY, GEOLOGY, PLANETARY ENVIRONMENTAL CHANGE]

Greeley, R. 1994: *Planetary landscapes.* London: Chapman & Hall. **Shirley, J.H. and Fairbridge, R.W.** 1997: *Encyclopedia of planetary sciences.* London: Thomson.

plankton

Free-floating PELAGIC organisms, described as *planktonic* or *planktic*. *Phytoplankton* (*autotrophs*, algae) in surface waters are primary producers in the FOOD CHAIN; *zooplankton* (*heterotrophs*) are consumers. PLANKTONIC organisms such as foraminiferans, RADIOLARIA, diatoms, COCCOLITHOPHORES and the extinct graptolites have good potential as INDEX FOSSILS in BIOSTRATIGRAPHY. The term 'plankton' was first used in AD 1887 to describe organisms retained by a silk plankton net of mesh size c. 65 μm. Smaller organisms have since been described as *nannoplankton* (5–60 μm), *ultraplankton* (<5 μm) and *picoplankton* (<1 μm). *LC*

[See also DIATOM ANALYSIS, FORAMINIFERAL ANALYSIS, MARINE SEDIMENT CORES]

Planosols

Soils with a silty or loamy grey surface or shallow subsurface horizon, which are signs of periodic wetness, abruptly overlying a dense subsoil upon which downward percolating water stagnates. Breakdown of clay by FERROLYSIS leads to the abrupt change of TEXTURE in the soil profile. These soils occur on low plateau surfaces in Latin America, eastern United States, eastern Africa, southeast Asia and Australia where there is a marked alternation between wet and dry seasons. (SOIL TAXONOMY: *Albaqualfs, Albaquults* and *Argialbolls*). These soils may have been derived from VERTISOLS following a change to a wetter climate. *EMB*

[See also WORLD REFERENCE BASE FOR SOIL RESOURCES]

plant macrofossil analysis

Plant MACROFOSSILS are plant parts preserved in Quaternary sediments that are large enough to be manipulated by hand, thus contrasting with MICROFOSSILS. They usually comprise seeds and fruits, but can be vegetative parts, e.g. leaves, cuticles, bud-scales, rhizomes, twigs and wood. Macrofossils can be identified using a stereo-microscope, a high-power microscope, or SCANNING ELECTRON MICROSCOPY (SEM). Wood identification requires thin sections.

Macrofossils supplement POLLEN ANALYSIS for past vegetation and *environmental reconstruction*. They can often be identified to species level. They often occur in small quantities and hence require larger sediment samples than pollen. Percentages are not useful and stratigraphic diagrams are best presented as concentrations (numbers/unit volume sediment). Macrofossils usually have limited dispersal from their source and hence represent local vegetation development. Species that produce little or poorly preserved pollen, or cryptogams (e.g. mosses, including *Sphagnum*, Charophytes), may produce identifiable macrofossils and thus they can illuminate 'blind spots' in a pollen assemblage. Because there is no macrofossil rain equivalent to POLLEN RAIN, they have limited value for regional correlation.. However, they can indicate the local occurrence of a large pollen producer, e.g many tree species, and thus they are valuable in reconstructing TREE-LINE VARIATIONS.

Macrofossil analyses from LAKE SEDIMENTS provide information on aquatic flora and vegetation and can thus reconstruct changes in lake environment, e.g. pH, water conductivity and chemistry, water depth and EUTROPHICATION. Interpretation is aided by surface-sample representation and calibration datasets, but these are scarce for plant macrofossils. Macrofossils can demonstrate PEAT development, as plant remains form the peat. They register the local vegetation, and can reconstruct changes in surface wetness, and hence PRECIPITATION VARIATIONS. Macrofossils from archaeological contexts can reconstruct the living environment of a settlement from nearby lake or mire deposits, or directly from the settlement where they are preserved in, for example, pots, pottery, pits, latrines and fireplaces. Remains of plant foods are found in stomachs, faeces, latrines, and waste-dumps. A particularly elegant study using plant remains has been made of the lifestyle of the Italian–Austrian 5000 year-old ICEMAN. *HHB*

[See also TREE LINE/LIMIT]

Barnekow, L. 1999: Holocene treeline dynamics in the Abisko area, northern Sweden, based on pollen and macrofossil records, and the inferred climatic changes. *The Holocene* 9, 253–265. **Birks, H.H.** 1973: Modern macrofossil assemblages in lake sediments in Minnesota. In Birks, H.J.B. and West, R.G. (eds), *Quaternary Plant Ecology.* Oxford: Blackwell Scientific, 173–189. **Birks, H.H.** 1980: Plant macrofossils in Quaternary lake sediments. *Archiv für Hydrobiologie* 15, 1–60. **Birks, H.H.** 1993: The importance of plant macrofossils in late-glacial climatic reconstructions: an example from western Norway. *Quaternary Science Reviews* 12, 719–726. **Birks, H.H. and Birks, H.J.B.** 2000: Future uses of pollen analysis must include plant macrofossils. *Journal of Biogeography* 27, 31–35. **Birks, H.J.B. and Birks, H.H.** 1980: *Quaternary Palaeoecology.* London: Edward Arnold. **Bortenschlager, S. and Oeggl, K. (eds)** 2000: *The Iceman and his natural environment* [Man in the ice 4]. Vienna, NewYork: Springer. **Spindler, K.** 1994: *The Man in the Ice.* London: Weidenfeld and Nicholson. **Wasylikowa, K.** 1986: Analysis of fossil fruits and seeds. In Berglund, B.E. (ed.), *Handbook of Holocene Palaeoecology and Palaeohydrology*. Chichester: Wiley.

plant pigment analysis

Carotenoid and CHLOROPHYLL concentrations extracted from ALGAE and cyanobacteria are measured using *high-performance liquid chromatography* (HPLC). Fossil pigment compositions can discriminate between most algal groups and are, therefore, useful PALAEOECOLOGICAL indicators. Changes in *phytoplankton* composition, algal sedimentation and past glacial changes in PRIMARY PRODUCTION can be determined by pigment analysis. Chlorophyll concentrations are frequently used to estimate phytoplankton BIOMASS and PRODUCTIVITY, while DEGRADATION products are diagnostic indicators of physiological status, detrital content and GRAZING processes in natural populations of phytoplankton. When used in conjunction with *zooplankton* MICROFOSSIL ANALYSIS, plant pigments can help to reconstruct PREDATOR–PREY RELATIONSHIPS in whole lake FOOD CHAINS/WEBS. The preservation of fossil pigments varies with the water environment and is reduced under conditions of high light, oxygen, temperature and turbulence. Fossil pigments from siliceous algae are poorly preserved and therefore, fossil pigment interpretations should only be based on historical changes in pigment concentration.   *KJF*

Leavitt, P.R. and Findlay, D.L. 1994: Comparison of fossil pigments with 20 years of phytoplankton data from Eutrophic Lake 227, Experimental Lakes Area, Ontario. *Canadian Journal of Fishery and Aquatic Science* 51, 2286–2299. **Mantoura, R.F.C. and Llewellyn, C.A.** 1983: The rapid determination of algal chlorophyll and carotenoid pigments and their breakdown products in natural waters by reverse-phase high-performance liquid chromatography. *Analytica Chimica Acta* 151, 297–314.

plastic limit

The moisture content at which a sediment transforms from a brittle solid to a plastic or ductile solid. The *plasticity index* is the difference between the LIQUID LIMIT and the plastic limit and denotes the range of moisture contents over which the soil behaves as a plastic solid.   *PW*

[See also FAILURE, LIQUID LIMIT]

plastics

Substances capable of plastic flow or plastic deformation under certain conditions or at some stage of manufacture. Two main classes of plastics are manufactured: (1) *thermoplastics*, such as *polyethylene*, *polypropylene* and *polyvinyl chloride* (PVC), which retain potential plasticity after manufacture and can be remoulded by heating; (2) *thermosetting plastics*, such as *epoxy resins* and *Bakelite*, which undergo non-reversible chemical changes on heating and many of which are difficult, if not impossible, to recycle. Plastics are cheap to produce (relative to more traditional materials such as iron and steel and other metals) and they are energy efficient in terms of transport costs and durability. Lightness combined with durability has made plastics the second most important packaging material (after cardboard and paper). However, durability becomes a disadvantage when plastics are discarded. A recently invented BIODEGRADABLE plastic made from *poly-3-hydroxybutrate* (PHB), a polymer that occurs naturally in the cells of certain bacteria, may be the first of the environmentally friendly plastics of the future.   *JAM*

Porteous, A. 2000: *Dictionary of environmental science and technology*, 3rd edn. Chichester: Wiley.

plate margin

The edge of one of the Earth's plates of LITHOSPHERE. The three types of plate margin are CONSTRUCTIVE (divergent), DESTRUCTIVE (convergent) and CONSERVATIVE PLATE MARGINS (see Figure). *Plate boundary* is a broadly equivalent term, although it is sometimes restricted to the junction between the plates, whereas plate margin includes the part of the plate bordering the junction.   *GO*

[See also PLATE TECTONICS]

plate tectonics

The theory, or PARADIGM, that the Earth's LITHOSPHERE consists of discrete rigid slabs or *plates* that move laterally over the weak ASTHENOSPHERE. Three types of relative plate motion can occur at PLATE MARGINS: plates can move apart, creating new lithosphere through SEA-FLOOR SPREADING (*divergent* or CONSTRUCTIVE PLATE MARGIN); plates can move together, consuming old

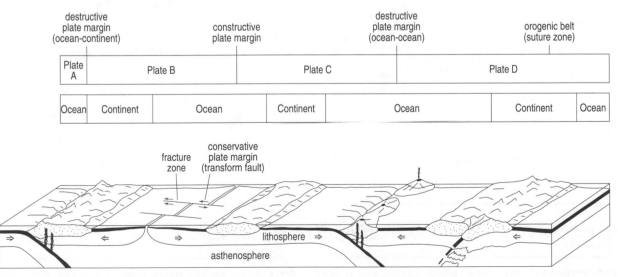

plate margin *Schematic diagram showing the different kinds of plate margins*

lithosphere through SUBDUCTION (*convergent* or DESTRUCTIVE PLATE MARGIN); or plates can slide past each other, conserving lithosphere (CONSERVATIVE PLATE MARGIN). Most of the Earth's TECTONIC activity, VOLCANOES and EARTHQUAKES are confined to plate margins, and plate interiors are essentially stable. Plate outlines can be defined using the distribution of earthquakes, volcanoes and certain major features of global GEOMORPHOLOGY: MID-OCEAN RIDGES coincide with constructive plate margins; deep OCEANIC TRENCHES, volcanic ISLAND ARCS and ACTIVE CONTINENTAL MARGINS coincide with destructive plate margins. There are seven major plates – North American, South American, Eurasian, African, Indo-Australian, Pacific and Antarctic – and numerous minor plates (see Figure). Most include areas of both CONTINENT and OCEAN, so the edges of continents (CONTINENTAL MARGINS) do not necessarily correspond to plate margins: PASSIVE CONTINENTAL MARGINS occur within plates. Plates move at rates of centimetres to tens of centimetres per year, giving rise to CONTINENTAL DRIFT.

Only lithosphere bearing relatively thin and dense OCEANIC CRUST is created and destroyed in plate tectonics. Where oceanic crust and CONTINENTAL CRUST converge, the oceanic crust is destroyed, creating an active continental margin or *continental margin orogen*. If two fragments of continental crust are brought together, a continent–continent COLLISION ZONE develops: SEDIMENTS at the continental margins are subjected to compression, DEFORMATION and UPLIFT, forming high fold mountains in an OROGENIC BELT. This plate tectonic model for OROGEN-ESIS – the WILSON CYCLE – has completely replaced GEOSYNCLINE theory.

The driving force behind plate tectonics is not fully understood. Igneous processes at plate margins contribute to a convective release of GEOTHERMAL heat, but this is unlikely to drive plate movements. More likely seems to be that the sinking of old, dense oceanic lithosphere drags the remainder of the plate behind it. It is now widely accepted that some form of plate tectonics operated in the ARCHAEAN, but because the Earth's interior was hotter then, the rates, scales and products are likely to have been significantly different (see GREENSTONE BELTS, KOMATIITES). During PERMIAN and TRIASSIC times the continental masses were assembled as a single *supercontinent* (PANGAEA) and such a configuration may have existed previously. It is not known if plate tectonics operates on any other bodies in the solar system.

Several broadly coincident discoveries culminated in the development of the plate tectonics paradigm in the early AD 1960s. After World War II, accurate echo-sounders were used to map the ocean floor, leading to the detailed charting of features such as mid-ocean ridges and oceanic trenches. RADIOMETRIC DATING of rock samples from the ocean floor gave surprisingly young ages, nowhere greater than 200 million years, and thus younger than most continental rocks. The theory of SEA-FLOOR SPREADING, developed in the late 1950s, was substantiated by the discovery of ocean-floor MAGNETIC ANOMALIES that were symmetrical either side of the mid-ocean ridges: these were interpreted in terms of ocean floor forming at

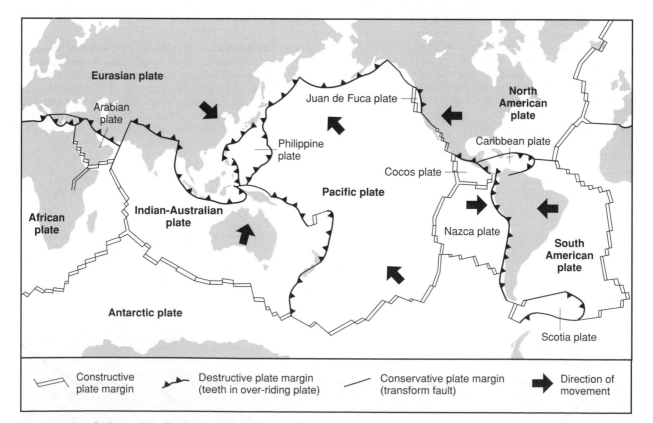

plate tectonics Outlines of the Earth's tectonic plates, showing the types of plate margins and directions of plate movements (after Decker and Decker, 1997)

mid-ocean ridges and spreading laterally while the polarity of the Earth's magnetic field repeatedly switched between normal and reverse states. DEEP-SEA DRILLING showed that the thickness of OCEANIC SEDIMENTS increased with distance from the mid-ocean ridge and hence with the age of oceanic crust. EARTHQUAKE studies showed an inclined zone of earthquake generation in the Earth's interior on the same side of oceanic trenches as active volcanoes occur: this was termed a *Benioff zone*. The locations of earthquake EPICENTRES were found to lie along narrow zones that coincided with sites of active volcanoes. Studies of SEISMIC WAVES showed that the sense of displacement on FAULTS that cut the mid-ocean ridges was consistent with crust spreading on each side of the mid-ocean ridge: these were called TRANSFORM FAULTS. The rapid development and acceptance of plate tectonics led to a genuine revolution in the EARTH SCIENCES.

Amendments to the plate tectonics theory since the 1960s have been relatively minor. Volcanic activity in plate interiors (intraplate volcanism) is attributed to HOTSPOTS that overlie MANTLE PLUMES. Some orogenic belts are now understood to have assembled over long periods of time through the accretion of small pieces of continental crust (TERRANES). The discovery of HYDROTHERMAL VENTS ('black smokers') at mid-ocean ridges has led to some revision of ideas affecting EVOLUTION and ECOLOGY.

Plate tectonics has many implications for ENVIRONMENTAL CHANGE. As a consequence of continental drift, parts of the Earth's surface have experienced changing PALAEOLATITUDE and *palaeoclimate* through GEOLOGICAL TIME, offering an explanation for climate-sensitive sediment FACIES such as COAL or EVAPORITES in temperate latitudes. Changes in PALAEOGEOGRAPHY due to the joining and splitting of land-masses have important implications for OCEAN CURRENTS and the evolution of terrestrial organisms. Long-term changes in the rate of plate activity have important consequences. Newly formed oceanic crust is elevated as a mid-ocean ridge because it has excess heat, which is lost at a steady rate, leading to a predictable *age–depth relationship for ocean floor*. Rapid spreading produces wide ridges that occupy a greater volume of the OCEAN BASINS, leading to a rise in SEA LEVEL and flooding of the continental margins. Increased volcanic activity with more rapid spreading influences the supply of VOLATILES to the ATMOSPHERE. Regional uplift during episodes of orogeny can disrupt the GENERAL CIRCULATION OF THE ATMOSPHERE and change MONSOON climates. Orogenic activity affects the volume of rocks exposed to WEATHERING: since many weathering reactions involve CARBON DIOXIDE, this can disrupt the CARBON CYCLE and influence the amount of CO_2 in the atmosphere.

In summary, the central role of plate tectonics to our understanding of the entire Earth system, including environmental change, cannot be overemphasised. *GO*

Andel, T.H. van 1994: *New views on an old planet: a history of global change*, 2nd edn. Cambridge: Cambridge University Press. **Bye, J.A.T.** 1998: Sea level change due to oscillations in seafloor spreading rate. *Physics of the Earth and Planetary Interiors* 109, 151–159. **Condie, K.C.** 1989: *Plate tectonics and crustal evolution*, 3rd edn. Oxford: Pergamon Press. **Cox, A. (ed.)** 1973: *Plate tectonics and geomagnetic reversals*. San Francisco, CA: Freeman. **Decker, R. and Decker, B.** 1997: *Volcanoes*, 3rd edn. Basingstoke: Freeman. **Gurnis, M.** 1988: Large-scale mantle convection and the aggregation and dispersal of supercontinents. *Nature* 332, 695–699. **Keary, P. and Vine, F.G.** 1990: *Global tectonics*. Oxford: Blackwell Scientific. **Kerrick, D.M. and Caldeira, K.** 1999: Was the Himalayan orogen a climatically significant coupled source and sink for atmospheric CO_2 during the Cenozoic? *Earth and Planetary Science Letters* 173, 195–203. **Komiya, T., Maruyama, S., Masuda, T. et al.** 1999: Plate tectonics at 3.8–3.7 Ga: field evidence from the Isua accretionary complex, southern West Greenland. *Journal of Geology* 107, 515–554. **Leeder, M.R.** 1999: *Sedimentology and sedimentary basins: from turbulence to tectonics*. Oxford: Blackwell Science. **McKenzie, D.** 1999: Planetary science – plate tectonics on Mars? *Nature* 399, 307–308. **Shields, O.** 1997: Rapid Earth expansion: an eclectic view. *Gondwana Research* 1, 91–94. **Summerfield, M.A.** 1991: *Global geomorphology*. Harlow: Longman. **Windley, B.F.** 1993: Uniformitarianism today: plate tectonics is the key to the past. *Journal of the Geological Society, London* 150, 7–19. **Worsley, T.R., Nance, D. and Moody, J.B.** 1984: Global tectonics and eustasy for the past two billion years. *Marine Geology* 58, 373–400.

Pleistocene   The Pleistocene is the EPOCH that follows the PLIOCENE, covers the greater part of the QUATERNARY and ended c. 10 ka BP (^{14}C years). West (1977) proposed that the term should include the whole Quaternary up to and including the HOLOCENE, since the current temperate episode may be regarded as the latest INTERGLACIAL in the Pleistocene epoch. However, this has not been generally accepted because the Holocene has other characteristics that make it distinctive. *DH/JAM*

[See also GEOLOGICAL TIMESCALE, PLIOCENE–PLEISTOCENE TRANSITION, QUATERNARY TIMESCALE]

Nilsson, T. 1993: *The Pleistocene: geology and life in the Quaternary Ice Age*. Dordrecht: Kluwer. **West, R.G.** 1977: *Pleistocene geology and biology: with especial reference to the British Isles*, 2nd edn. London: Longman.

Pleniglacial   Literally meaning 'full glacial', the term Pleniglacial or Pleni-Glacial is used by European continental stratigraphers for the part of the last cold stage (Weichselian) encompassing both the time of maximum extent of ice sheets and some INTERSTADIALS before this. For example, the Pleni-Weichselian (Pleniglacial) succeeds the Early Weichselian (up to the Odderade Interstadial) but precedes the Late Weichselian. *FMC*

Behre, K.-E. 1989: Biostratigraphy of the last glacial period in Europe. *Quaternary Science Reviews* 8, 25–44.

plinthic horizon   A clayey, iron-rich, humus-poor, mottled subsoil material that will irreversibly harden on exposure to the atmosphere, first described in AD 1807 in India by F. Buchanan. This grey clay with varying amounts of reddish iron concretions was referred to in earlier literature as LATERITE and is also known as *plinthite*. In the moist soil it can be cut with a spade, but on exposure to repeated wetting and drying it hardens as *lateritic ironstone*, in which state it is no longer considered to be plinthite. Its practical importance is that once hardened it is impossible to cultivate. Also, it is resistant to erosion and may form an abrupt edge to plateau features, particularly in Australia, Africa, India, southeast Asia and South America. It gradually weathers and releases *ironstone nodules* into contemporary soils. *EMB*

[See also DURICRUSTS]

Aleva, G.J.J. and Creutzberg D. 1994: *Laterites: concepts, geology, morphology and chemistry.*' Wageningen: International Soil Reference and Information Centre and European Commission DG XII for Science, Research and Development. Buchanan, F. 1807: *A journey from Madras through the countries of Mysore, Canara and Malabar, etc.* 3 volumes. London: East India Company. Food and Agriculture Organization of the United Nations (FAO) 1998: *World reference base for soil resources* [*Soil Resource Report* 84]. Rome: FAO, ISRIC, ISSS. MacFarlane, M. J. 1983: Laterites. In Goudie, A.S. and Pye, K. (eds), *Chemical sediments and geomorphology.* London: Academic Press. Mohr, E.C.J., Van Baren, F.A. and Schuylenborgh, J. 1972: *Tropical soils*, 3rd edn. The Hague: Mouton-Ichtiar, Baryvan Hoeve. Soil Survey Staff 1999: *Soil Taxonomy: a basic system of soil classification for making and interpreting soil surveys*, 2nd edn [*Agriculture Handbook* 436]. Washington, DC: United States Department of Agriculture.

Plinthosols   Soils affected by groundwater, containing a mixture of iron, clay and quartz, in which the iron has segregated to form a mottled horizon, the PLINTHIC HORIZON (or *plinthite*), that hardens irreversibly when dried. These soils occur mainly in the inter-tropical regions of Australia, Africa, South America, India and southeast Asia. (SOIL TAXONOMY: *Plinthaquox*). Drainage or change to a drier climate could result in the extensive hardening of these soils. Such soils were formerly referred to as LATERITES or *Latosols*.   *EMB*

[See also WORLD REFERENCE BASE FOR SOIL RESOURCES]

Pliocene   An EPOCH of the TERTIARY period, conventionally taken from 5.2 to 1.64 million years ago (Ma), although some authorities place the Pliocene–Pleistocene boundary at 2.5–2.6 Ma.   *GO*

[See also GEOLOGICAL TIMESCALE]

Pliocene–Pleistocene transition   The Pliocene–Pleistocene boundary represents the beginning of the QUATERNARY, but the exact date of this transition is the subject of debate. The boundary is usually located at the point in the STRATIGRAPHICAL RECORD where there are indications of climatic cooling, reflected by FOSSIL evidence or another climatic proxy. The reference site for the Pliocene–Pleistocene boundary accepted by the International Commission on Stratigraphy in 1985 and the International Union of Geological Sciences (IUGS), is located in Vrica in southern Italy and is based upon the first appearance of the cold-water marine OSTRACOD, *Cytheropteron testudo*. This has been dated to c. 1.64 million years ago (Ma BP) on the basis of palaeomagnetic evidence, just below the top of the OLDUVAI EVENT. The identification of *C. testudo* at this site, and hence its value as a major stratigraphic marker, has been questioned; furthermore, the dating of the top of the Olduvai event has been revised to 1.81 Ma BP. As the underlying principle for defining the boundary is to reflect the onset of a cooler climatic regime, other indicators dating as far back as 2.6 Ma BP have been suggested. Gradual cooling had already been taking place through the TERTIARY, so it is clear that any position is likely to be arbitrary. Association with a significant SYNCHRONOUS feature, such as a palaeomagnetic reversal, is therefore likely to prove the most useful form of definition.   *DH/CJC*

[See also GEOLOGICAL TIMESCALE, QUATERNARY TIMESCALE]

Aguirre, E. and Passini, G 1985: The Pliocene–Pleistocene boundary. *Episodes* 8, 116–120. Jenkins, D.G. 1987: Was the Pliocene–Pleistocene boundary placed at the wrong stratigraphic level? *Quaternary Science Reviews* 6, 41–42. Shackleton, N., Berger, A. and Peltier, W.R. 1990: An alternative astronomical calibration of the lower Pleistocene timescale based upon ODP Site 677. *Transactions of the Royal Society of Edinburgh: Earth Sciences* 81, 251–261. Suc, J.P., Bertini, A., Leroy, S.A.G. and Suballyova, D. 1997: Towards lowering of the Pliocene–Pleistocene boundary to the Gauss–Matuyama reversal. *Quaternary International* 40, 37–42. Williams, M.A.J., Dunkerley, D.L., De Deckker, P. *et al.* 1998: *Quaternary environments*, 2nd edn. London: Arnold.

plough pan   A compacted layer of soil, caused by repeated tilling of the soil to a constant depth. Plough pans are produced by TILLAGE implements such as tines, discs and moulboard ploughs, which tend to smear and compact the soil immediately below their operating depth.   *SHD*

[See also PLOUGHING]

Moore, G. (ed.) 1998: *Soil guide.* Perth: Agriculture Western Australia. Rogasik, H., Lehfeldt, J. and Morstein, K.H. 1989: Geometry of loosened zones in the plough pan – effects on recompaction. *Archiv für Acker und Pflanzenbau und Bodenkunde* 33, 661–669.

ploughing   In agriculture, the practice of cutting, turning or breaking the soil to prepare it for planting or other agricultural purposes. Ploughing can have detrimental effects on the soil, such as reducing its AGGREGATE STABILITY and increasing its ERODIBILITY, leading to accelerated losses of soil or organic matter or the development of PLOUGH PANS.   *SHD*

[See also SOIL EROSION, TILLAGE]

ploughing block   A boulder that moves down-slope, pushing soil and vegetation into a mound on its downslope side and leaving a depression or groove on its upslope side. It is a type of SOLIFLUCTION phenomenon.   *PW*

[See also GELIFLUCTION, SOIL CREEP]

Reid, J.R. and Nesje, A. 1988: A giant ploughing block, Finse, southern Norway. *Geografiska Annaler* 70A, 27–33. Wilson, P. 1993: Ploughing-boulder characteristics and associated soil properties in the Lake District and southern Scotland. *Scottish Geographical Magazine* 109, 18–26.

pluridiscipliniarity   The study of environmental problems using ideas from a number of disciplines but not retaining the methodologies of those disciplines. ENVIRONMENTAL IMPACT ASSESSMENTS, for example, are carried out in this way and generally involve the combined efforts of experts and analysts from a wide range of backgrounds working as an integral team to address a specific problem.   *JGS*

[See also INTERDISCIPLINARY, MULTIDISCIPLINARY]

O'Riordan, T. 2000: Environmental science on the move. In O'Riordan, T. (ed.), *Environmental science in environmental management*, 2nd edn. Harlow: Prentice Hall, 1–28.

pluvial   At lower latitudes, a QUATERNARY time interval characterised by relatively high moisture availability. Some arid and semi-arid regions (ARIDLANDS), such as parts of the northern Sahara and southwestern United States experienced pluvials that were synchronous with

GLACIAL EPISODES at high latitudes but, on account of regional patterns in climatic change related to the GENERAL CIRCULATION OF THE ATMOSPHERE, there was no general time equivalence between glacials and pluvials. Indeed, at the time of the LAST GLACIAL MAXIMUM, tropical Africa and Australasia experienced '*glacial aridity*'. *JAM*

[See also LAKE-LEVEL VARIATIONS, SAPROPEL]

Kallel. N., Duplessssy, J.C., Labeyrie, L. *et al.* 2000: Mediterranean pluvial periods and sapropel formation over the last 200,000 years. *Palaeogeography, Palaeoclimatology, Palaeoecology* 157, 45–58. **Street, F.A.** 1981: Tropical palaeoenvironments. *Progress in Physical Geography* 5, 157–185.

poaching   The trampling of soil by animals, leading to disturbance of the vegetation cover and modification (e.g. COMPACTION) of the soil structure.   *RAS*

podzolisation   A process of soil formation taking place in sandy soils under a moist climate associated with BOREAL FOREST and HEATHLAND. An extremely acid humus form, MOR, is developed on the soil surface. Soluble organic breakdown products from plants percolating down the profile, link with iron compounds and disrupt CLAY MINERALS to leave a sandy, bleached ALBIC HORIZON. Further down the soil profile accumulation occurs in the SPODIC HORIZON, which becomes enriched in iron, organic matter and aluminium.   *EMB*

[See also CHELATION, PODZOL]

Anderson, H.A. Berrow, M.L. Farmer, V.C. *et al.* 1982: A reassessment of podzol formation processes. *Journal of Soil Science* 33, 125–136. **Davidson, D.A.** 1987: Podzols: changing ideas on their formation. *Geography* 72, 122–128. **De Coninck, F.** 1980: Major mechanisms in the formation of spodic horizons. *Geoderma* 24,101–123. **Farmer, V.C., Russell, J.D. and Berrow, M.L.** 1980: Imogolite and proto-imogolite allophane in spodic horizons: evidence for a mobile aluminium silicate complex in podzol formation. *Journal of Soil Science* 31,673–684.

Podzols   Acid soils with distinct horizons, having a brownish or blackish subsoil, the SPODIC HORIZON, containing illuvial iron–aluminium–organic compounds, lying beneath a bleached ALBIC HORIZON. These soils occur in northern areas of North America and Eurasia under coniferous forest and heath. On deep quartz sands in the tropics, podzols with an albic horizon, many metres deep, have developed (SOIL TAXONOMY: *Spodosols*). Prehistoric FOREST CLEARANCE by humans on coarse sandy materials resulted in HEATHLAND and the development of podzols during the Holocene.   *EMB*

[See also PODZOLISATION, WORLD REFERENCE BASE FOR SOIL RESOURCES]

Dimbleby, G.W. 1962: *The development of British heathlands and their soils* [Forestry Memoir 23]. Oxford: Department of Forestry, University of Oxford. **Mackney, D.** 1961: A podzol development sequence in oakwoods and heath in central England. *Journal of Soil Science* 12, 23–40. **Muir, A.** 1961: The podzol and podzolic soils. *Advances in Agronomy* 13,1–56. **Thompson, C.H.** 1992: Genesis of podzols on coastal dunes in southern Queensland 1. Field relationships and profile morphology. *Australian Journal of Soil Research* 30, 593–613.

point   An ENTITY of zero dimension in a GEOGRAPHICAL INFORMATION SYSTEM (GIS). A single pair of COORDINATES and an ATTRIBUTE can spatially represent the location of a point (for example, a building or a spring). Together with

lines (see ARC) and areas, points make up the fundamental units of spatial representation in a GIS.   *TVM*

pointer years   In DENDROCHRONOLOGY, noticeably different TREE RINGS, identifiable in the context of those that precede and follow them. The 'difference' may be defined with respect to a variety of different parameters, e.g. total width of the ring, presence of some distinct anatomical feature, such as resin ducts, or a marked change in latewood density. This variety of parameters implies that a wide range of ecological interpretational possibilities are offered by the recording and dating of such phenomena.   *KRB*

[See also TREE RINGS, CROSS-DATING, DENDROECOLOGY]

Schweingruber, F.H. 1990: Dendroecological information in pointer years and abrupt growth changes. In Cook, E.R. and Kairiukstis, L.A. (eds), *Methods of dendrochronology. Applications in environmental sciences.* Dordrecht: Kluwer, 277–283.

polar amplification   The concept, which tends to be supported by CLIMATIC MODELS and by palaeoclimatic data (PALAEODATA), that temperature changes (warming or cooling TRENDS) tend to be exaggerated at relatively high latitudes.   *JAM*

polar desert   Areas where the cold, dry *polar climate* prohibits the growth of plants. Some very low-growing herbaceous or dwarf-shrub taxa may occur along sheltered brooksides or where water seeps from snowbeds.   *SPH*

[See also BIOMES, VEGETATION FORMATION-TYPE, TUNDRA]

Alexandrova, V.D. 1988: *Vegetation of the Soviet polar deserts.* Cambridge: Cambridge University Press.

polar front   The narrow zone that separates polar and tropical AIR MASSES. The term originated in AD 1922 with the publication of the classic Norwegian model of a mid-latitude DEPRESSION or CYCLONE. The *atmospheric polar front* is a narrow zone of potentially strong temperature, humidity and wind differences, where the warm, light tropical air is forced to rise over the dense, cold polar air. It is clearly seen on surface synoptic charts stretching for thousands of kilometres across the world's oceans between 45° and 60° latitude, but is usually disrupted over the main continental areas. With the advent of more upper-air data after World War II, it proved difficult to locate on the upper-air charts and the Norwegian model had to be modified. SATELLITE REMOTE SENSING since the 1960s has shown that it can be identified in satellite cloud photographs. It is now closely associated with the polar front JET STREAM. The OCEANIC POLAR FRONT separates cold and warm water bodies in an analogous way.
                                                   BDG

Carlson, T. N. 1991: *Mid-latitude weather systems.* London: HarperCollins. **Giles, B.D.** 1972: A three-dimensional model of a front. *Weather* 27, 352–363.

polar shore erosion   The process of FROST WEATHERING and clast removal along shores in cold-climate environments. These locations may be characterised by conspicuous ROCK PLATFORMS and backing cliffs. Elsewhere along the shores of lakes in PERIGLACIAL environments, BOULDER PAVEMENTS composed of angular boulders are indicative of bedrock EROSION due to frost-related processes. In coastal areas, the occurrence of shore platforms indicates the removal of large volumes of rock from the coastal zone and the transport of debris offshore by SEA ICE. Polar shore platforms may also be protected from erosional processes during winter by the presence of an *ice foot* that forms at the cliff-platform junction. In areas affected by GLACIO-ISOSTATIC REBOUND, shore platforms produced by polar shore erosion occur above sea level. Polar shore processes may be dominated by the effects of frost and largely unrelated to the effects of waves.   *AGD*

[See also STRANDFLAT]

Ballantyne, C.K. and Harris, C. 1994: *The periglaciation of Britain*. Cambridge: Cambridge University Press. **Dawson, A.G.** 1980: Shore erosion by frost: an example from the Scottish Lateglacial. In Lowe, J.J., Gray, J.M. and Robinson, J.E. (eds), *Studies in the Lateglacial of North-West Europe*. Oxford: Pergamon, 45–53. **Matthews, J.A., Dawson, A.G. and Shakesby, R.A.** 1986: Lake shoreline development, frost weathering and rock platform erosion in an alpine periglacial environment. *Boreas* 15, 33–50.

polarisation   The property of electromagnetic fields relating to the plane in which the electric field oscillates. A propagating electromagnetic wave has an electric field vector that oscillates in a direction at right angles to the direction of propagation. The direction of the oscillation plane is called the *polarisation plane* (or polarisation) and is determined by the orientation of the transmitting antenna. In general, polarisation can be decomposed into two orthogonal components that are often chosen to be 'horizontal' and 'vertical' with respect to the direction of propagation. The polarisation plane can also rotate continuously around the propagation direction (circular or elliptical polarisation). The propagation speed of an electromagnetic wave in a medium and the way in which it scatters from objects can depend on the polarisation of the field.   *PJS*

polarity chron   In relation to the GEOMAGNETIC POLARITY TIMESCALE, a time interval of dominantly normal or reversed polarity of the Earth's magnetic field, typically lasting approximately 10^5–10^6 years, also known as a *magnetic chron* and formerly as a *magnetic epoch*. The rock unit corresponding to a polarity chron is a MAGNETOZONE. A shorter interval of opposing polarity within a chron is a *subchron*. The four most recent polarity chrons, named after pioneering geomagnetists, are the *Brunhes* normal (to 730 000 years ago), *Matuyama* reversed (to 2.48 million years ago), *Gauss* normal (to 3.40 million years ago) and *Gilbert* reversed (to about 5 million years ago).   *DNT*

[See also GEOMAGNETISM]

polje   A large (up to several kilometres), flat-floored, partially structurally controlled, partially solutional depression in KARST terrain with subsurface drainage.   *SHD*

Bär, W.F., Fuchs, F. and Nagel, G. 1986: Lluc/Sierra Norte (Mallorca): Karst einer mediterranen Insel mit alpidischer Struktur. *Zeitschrift für Geomorphologie Supplementband* 59, 27–48.

pollen analysis   The main research technique used in investigating VEGETATION HISTORY, involving the chemical extraction of pollen (and SPORES) in stratigraphical order from a SEDIMENT sequence and their identification back to the plants that produced them. The technique is based on the fact that plants produce large quantities of pollen, which is widely dispersed, frequently identifiable to genus and sometimes to species and which is well preserved under ANAEROBIC, acid conditions (e.g. in PEAT and LACUSTRINE SEDIMENTS). The technique was first introduced by von Post in AD 1916 for investigating climatic history from Swedish MIRES. Pollen analysis is one aspect of the wider research field of *palynology* (the study of pollen grains), which also includes taxonomic and genetic studies, honey studies, forensic work and AEROBIOLOGY.

Pollen analysis, as a geologically based tool, allows the tracing through time of changes in vegetation, caused by both CLIMATE and HUMAN IMPACT. However, the exact timing of these changes requires independent dating (e.g. RADIOCARBON, TEPHROCHRONOLOGY) and the temporal RESOLUTION of the reconstruction depends on the sampling interval. The spatial resolution depends on the size of the pollen catchment area, which, in turn, is related to the size of the SEDIMENTARY BASIN. The pollen assemblage in any single sample has elements that have originated from both local and regional vegetation communities and even from quite alien communities through long-distance transportation. Pollen analysis results are illustrated in *pollen diagrams*, which show the variations of each identified pollen taxon through the sediment profile. The taxa are usually grouped ecologically and variations expressed as percentages of the POLLEN SUM or as POLLEN INFLUX. For ease of interpretation, the diagram is divided into POLLEN ASSEMBLAGE ZONES. Increasingly, stratigraphically based pollen-analysis datasets, which may even stretch back to cover several INTERGLACIALS (see Figure), are stored in a pollen DATABASE, where they are available for further analyses such as the *mapping* of vegetation communities at the regional or global scale for specific points in time. TRAINING SETS based on SURFACE POLLEN are also being developed, both to provide MODERN ANALOGUES for interpreting past vegetation more exactly and for the CALIBRATION of pollen assemblages in terms of CLIMATE parameters.   *SPH*

[See also ABSOLUTE COUNTING, CHAMBER SAMPLER, CORER, SURFACE SEDIMENT SAMPLER]

Berglund, B.E., Birks, H.J.B., Ralska-Jasiewiczowa, M. and Wright, H.E. (eds) 1996: *Palaeoecological events during the last 15000 years: regional syntheses of palaeoecological studies of lakes and mires in Europe*. Chichester: Wiley. **Berglund, B.E. and Ralska-Jasiewiczowa, M.** 1986: Pollen analysis and pollen diagrams. In Berglund, B.E. (ed.), *Handbook of Holocene palaeoecology and palaeohydrology*. Chichester: Wiley, 455–484. **Birks, H. H., Birks, H.J.B., Kaland, P.E. and Moe, D. (eds)** 1988: *The cultural landscape past, present and future*. Cambridge: Cambridge University Press. **Colinvaux, P.A., De Oliveira, P.E. and Morenõ, J.E.** 1999: *Amazon pollen manual and atlas*. New York: Harwood Academic Press. **Faegri, K. and Iversen, J.** 1989: *Textbook of pollen analysis*, 5th edn. Chichester: Wiley. **Hicks, S.** 1998: Fields, boreal forests and forest clearings as recorded by modern pollen deposition. *Paläoklimaforschung* 27, 53–66. **Hooghiemstra, H. and Ran, E.T.H.** 1994: Late Pliocene–Pleistocene high resolution pollen sequence of Colombia: an overview of climatic change. *Quaternary International* 21, 63–80. **Hooghiemstra, H. and Van't Veer, R.** 1999: A 0.6 million year pollen record from the Colombian Andes. *PAGES Newsletter* 99-3, 4–5. **Huntley, B. and Webb III, T.** 1988: *Vegetation history [Handbook of Vegetation Science*. Vol. 7]. Dordrecht: Kluwer. **Moore, P.D., Webb, J.A. and Collinson, M.E.** 1991: *Pollen analysis*, 2nd edn. Oxford: Blackwell Scientific. **Prentice, I.C.** 1985: Pollen representation, source area, and basin size; towards

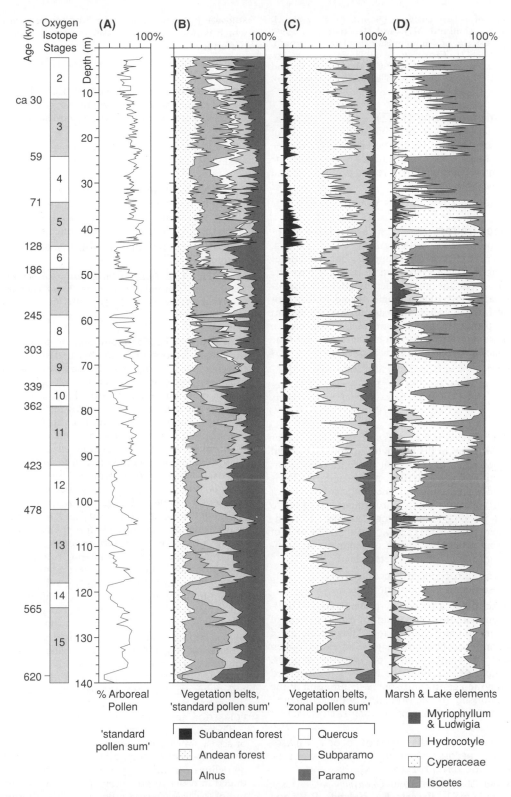

pollen analysis *Summary pollen diagram from the upper 140 m of the 357 m long core Funza 1 penetrating lacustrine sediments in the Colombian Andes. The history of the vegetation through many glacial–interglacial cycles is depicted in several different ways to the right of the timescale shown in terms of both thousands of years BP and isotopic stages: (A) tree-pollen percentage; (B) altitudinal vegetation belts based on a standard pollen sum; (C) vegetation belts based on a pollen sum excluding certain insensitive species; (D) marsh and lake species grouped according to characteristic water depth (increasing across the groups from left to right), which in general increases during glacial episodes (after Hooghiemstra and Van't Veer, 1999)*

a unified theory of pollen analysis. *Quaternary Research* 23, 76–86. **Sugita, S., Gaillard, M.J. and Broström, A.** 1999: Landscape openness and pollen records: a simulation approach. *The Holocene* 9, 409–421. **Traverse, A.** 1988: *Palaeopalynology.* Boston: Allen and Unwin. **Tzedakis, P.C., Bennett, K.D. and Magri, D.** 1994: Climate and the pollen record. *Nature* 370, 513.

pollen assemblage zone (PAZ)

pollen assemblage zone (PAZ)   A sediment package of more-or-less homogeneous composition in terms of pollen types considered to be characteristic of either the local or regional sequence of vegetation. It is used in the basic description of sections in a *pollen stratigraphy* and shown in *pollen diagrams*.   *BA*

[See also ASSEMBLAGE ZONE, POLLEN ANALYSIS]

pollen concentration   A measure of pollen abundance, usually presented as pollen grains per volume (e.g. number of grains in 1 cm^3 of sediment), but sometimes per unit of weight. *Pollen concentration diagrams* have the advantage of not creating internal dependence of percentages (closed data), in which a single strong pollen producer or a local pollen source may diminish the apparent significance of the other taxa present. *Pollen concentration values* are very sensitive to sediment changes. If good time control is available POLLEN INFLUX can be derived.   *BA*

[See also POLLEN ANALYSIS, POLLEN ASSEMBLAGE ZONE]

Prentice, I.C. and Webb III, T. 1986: Pollen percentages, tree abundances and the Fagerlind effect. *Journal of Quaternary Science* 1, 35–44.

pollen influx   Pollen influx, *pollen accumulation rate* or *absolute pollen frequency* is the estimated number of pollen grains (and spores) accumulated in a sedimentary environment per unit area per unit time. FOSSIL pollen influx is often estimated as part of palaeoecological studies of VEGETATION HISTORY and past environments. In order to estimate pollen influx, POLLEN CONCENTRATION (the number of grains per unit volume) and ACCUMULATION RATE of the fossil-bearing sediments (depth of sediment accumulated per unit time) must first be determined:

Pollen influx = pollen concentration ×
sediment accumulation rate

The main technique adopted by palynologists to estimate pollen concentrations is to add an exotic marker of known concentration and volume to a known volume of fossil-bearing material. Markers may be an exotic pollen taxon, SPORES or pollen-sized polystyrene spheres. Following sample preparation, the markers are counted with the fossil pollen grains and spores. Estimation of the age–depth relationship (based on dating of the sediments) is required to determine sediment ACCUMULATION RATES.
                                                          JLF

[See also ABSOLUTE COUNTING, PALAEOECOLOGY, PALAEO-ENVIRONMENTAL RECONSTRUCTION, POLLEN ANALYSIS, POLLEN ASSEMBLAGE ZONE, POLLEN RAIN]

Birks, H.J.B and Birks, H.H. 1980: *Quaternary palaeoecology.* London: Edward Arnold. **Faegri, K., Kaland, P.E. and Kryzwinski, K.** 1989: *Textbook of pollen analysis.* Chichester: Wiley.

pollen rain   Pollen settling from the atmosphere onto a surface, such as a SEDIMENT surface, where it becomes SURFACE POLLEN. Since 'rain' implies vertical movement, which is not always the case, the term *pollen deposition* is preferred.   *SPH*

[See also POLLEN ANALYSIS, POLLEN INFLUX].

pollen sum   The pollen count from which percentages are calculated when constructing a *pollen diagram*. Taxa included in the sum vary with the question of the investigation.   *SPH*

[See ARBOREAL POLLEN, NON-ARBOREAL POLLEN, POLLEN ANALYSIS]

pollutant   A substance introduced into a natural system by human agency *and* which impairs the system or harms organisms. A CONTAMINANT becomes a pollutant when there is damage or adverse effects.   *JAM*

[See also ECOTOXICOLOGY, POLLUTION, PRIMARY POLLUTANT, TOXIN]

Mansfield, T.A. (ed.) 1976: *Effects of air pollutants on plants.* Cambridge: Cambridge University Press. **Moriarty, F.** 1983: *Ecotoxicology: the study of pollutants in ecosystems.* London: Academic Press.

pollution   The contamination of an ENVIRONMENT to the extent that organisms are harmed or the functioning of the ecosystem is impaired. Some researchers regard natural agencies (such as volcanic eruptions) as potential causes of pollution, but normal usage is to restrict the term to the effects of substances resulting from human activities. Here, the distinction is made between *anthropogenic* pollution and *natural* DISTURBANCE of ecosystems.

Most human activities generate environmental CONTAMINANTS (potential POLLUTANTS), either deliberately in WASTE DISPOSAL or as ENVIRONMENTAL ACCIDENTS. POLLUTION HISTORY is a long one, perhaps first becoming a global environmental problem during the INDUSTRIAL REVOLUTION. Although the scale, complexity and seriousness of pollution was hardly recognised before the twentieth century, it is now a major issue of GLOBAL ENVIRONMENTAL CHANGE. Pollutants enter the environment from *point sources* (e.g. sewage outfalls) or *diffuse sources* (e.g. automobile exhaust) and as particular *pollution events* (e.g. oil spills in the oceans) or continuously (e.g. agricultural chemicals into rivers and groundwater). Pollution may involve: naturally occurring substances increased to harmful or damaging levels (e.g. HEAVY METALS from MINING activities and GREENHOUSE GASES from FOSSIL FUELS); a wide range of new, synthetic substances in gaseous, liquid or solid form (e.g. PERSISTENT ORGANIC COMPOUNDS and RADIOACTIVE FALLOUT); *biological pollutants* (e.g. the CHOLERA bacterium and other disease-causing micro-organisms); or forms of energy (as in NOISE POLLUTION and THERMAL POLLUTION).

Today, ENVIRONMENTAL PROTECTION agencies enforce or advise on various interpretations of what are acceptable levels of pollution, such as the World Health Organisation guidelines for drinking water. Measures may be introduced in order to encourage pollution reduction, such as use of the *polluter-pays principle* (PPP), although there are problems with enforcement, especially in international incidents like the Chernobyl NUCLEAR ACCIDENT. This principle, along with pollution QUOTAS – rights to release certain quantities of contaminants – acknowledges the need to minimise pollution even though it cannot be eliminated because of the ubiquity, complexity and mobility of pollutants, together with economic and political constraints on clean-up.

Integrated pollution control (IPC) recognises the need to control particular pollutants in the broader environmental context, and especially to take account of the movement of pollutants from one *environmental compartment* (e.g. atmosphere, soil and sediment, water or biota) to another. An integrated approach not only attempts *pollution prevention* at source but also optimal *pollution control* of unavoidable pollution using, for example, the '*best available technique not entailing excessive costs*' (BATNEEC). This unfortunately does not often coincide with the '*best environmental option*' (BEO). *JAM*

[See also AIR POLLUTION, DISTURBED ECOSYSTEMS, MARINE POLLUTION, PESTICIDES, SOIL POLLUTION, WATER POLLUTION]

Fellenberg, G. 1999: *The chemistry of pollution.* Chichester: Wiley. **Freedman, B.** 1995: *Environmental ecology: the ecological effects of pollution, disturbance and other stresses.* San Diego, CA: Academic Press. **Harrison, R.M. (ed.)** 1990: *Pollution: causes, effects and control.* Cambridge: Royal Society of Chemistry. **Holdgate, M.W.** 1979: *A perspective of environmental pollution.* Cambridge: Cambridge University Press. **Markham, A.** 1994: *A brief history of pollution.* London: Earthscan. **Newson, M.** 1992: The geography of pollution. In Newson, M. (ed.), *Managing the human impact on the natural environment: patterns and processes.* London: Belhaven Press, 14–35.

pollution adaptation

Pollution adaptation occurs by changes in GENE POOL composition (*genotypic adaptation*) or in outward physical appearance (*phenotypic adaptation*). Adaptation involving GENOTYPE takes two forms: (1) the development of internal physiological or structural mechanisms, which enables species to detoxify or resist pollution, e.g. pollution resistant grasslands on the hills around Manchester, UK; and (2) modifications to outward physical appearance. Genetically based phenotypic adaptation (change to PHENOTYPE can result from the effects of environment only) is displayed by many insects showing *industrial melanism*, such as the darkened melanic variety of the peppered moth (*Biston betularia*), which lives in blackened industrial areas in the UK. The camouflaged darker coloration affords protection against bird predation. Many species are unable to adapt to pollution stress except by avoiding it and growing elsewhere. Thus, highly polluted areas have low BIODIVERSITY and GENETIC DIVERSITY. Differential response/adaptation of species to pollution allows use of INDICATOR SPECIES to monitor environmental pollution. *GOH*

[See also BIO-INDICATORS, ECOLOGICAL INDICATOR, ENVIRONMENTAL INDICATORS]

Cox, C.B. and Moore, P.D. 1993: *Biogeography: an ecological and evolutionary approach*, 5th edn. Oxford: Blackwell Scientific. **Wellburn, A.** 1994: *Air pollution and climate change: the biological impact.* London: Longman.

pollution history

Trends in past POLLUTANT deposition reconstructed from NATURAL ARCHIVES or, for more recent years, measured. Industrial activities such as non-ferrous metal smelting, FOSSIL FUEL combustion and iron and steel manufacture emit a wide range of substances to the environment, and particularly the atmosphere. These include the HEAVY METALS and industrial PARTICULATES deposited and incorporated into accumulating lacustrine, peat or ice sediments. HEAVY METAL, MAGNETIC SUSCEPTIBILITY and SPHEROIDAL CARBONACEOUS PARTICLE analysis of intact sedimentary sequences may then provide evidence of past industrial activity and its EMISSIONS. DOCUMENTARY EVIDENCE may be used to ascribe causes to and dates for the changes in pollutant deposition.

Pollution histories reconstructed from Great Britain and northwest Europe usually exhibit a number of common features. Pre-INDUSTRIAL REVOLUTION episodes of elevated HEAVY METAL concentration that presumably reflect prehistoric, Roman or Mediaeval metalworking are sometimes apparent. The majority of POLLUTANTS however show marked increases in concentration from the mid to late nineteenth century, corresponding to the onset of major INDUSTRIALISATION. Concentrations rise rapidly from the turn of the twentieth century until the mid-twentieth century with widespread industrial development. Recent falls in pollutant concentration have been ascribed to declines in heavy industry and EMISSION CONTROLS. Deviations from these general trends have nonetheless been observed and are thought to reflect variations in the nature and timing of regional industrialisation. As well as elucidating trends in past industrial activity, pollution histories also provide important evidence of the causes of past ecological change. A MULTIPROXY APPROACH to reconstructing pollution histories, in conjunction with DIATOM and POLLEN ANALYSIS, has been particularly useful in assessing the causes of recent lake ACIDIFICATION.

The variable retention and unpredictable behaviour of the pollutants within the sedimentary system limits the confidence that can be placed in the reliability and accuracy of pollution histories reconstructed from natural archives. Chemical, biological and physical transformations may occur after deposition whilst the atmospheric signal may be obscured by catchment or terrestrial inputs. An appraisal of the nature of sediment accumulation processes, and of the sediments themselves, is therefore necessary for the valid interpretation of pollution histories. *DZR*

Battarbee, R.W., Anderson, N.J., Appleby, P.G. *et al.* 1988: *Lake acidification in the United Kingdom 1800–1986.* London: Ensis Publishing. **Brimblecombe, P.** 1987: *The big smoke.* London: Methuen. **Livett, E.A.** 1988: Geochemical monitoring of atmospheric heavy metal pollution: theory and applications. *Advances in Ecological Research* 18, 65–177. **Nriagu, J.O.** 1996: A history of global metal pollution. *Science* 272, 223–224. **Rippey, B.** 1990: Sediment chemistry and atmospheric contamination. *Philosophical Transactions of the Royal Society of London* B327, 311–317. **Shotyk, W., Weiss, D., Appleby, P.G.** *et al.* 1998: History of atmospheric lead deposition since 12,370 ^{14}C yr BP from a peat bog, Jura Mountains, Switzerland. *Science* 281, 1635–1640.

polychlorinated biphenyls (PCBs)

Synthetic chlorinated hydrocarbons manufactured for use in the electricity and plastics industries. The production and use of these PERSISTENT ORGANIC COMPOUNDS has been banned in most industrial countries since the AD 1970s because of adverse health effects attributed at least in part to impurities such as *polychlorinated dibenzofurans* (PCDFs). The outbreak of so-called '*Yusho disease*' in Japan in 1968, which resulted from rice oil used for cooking being contaminated by a PCB that leaked from a heat exchanger, was important in identifying such effects. *JAM*

Hutzinger, O., Safe, S. and Zitko, V. 1974: *The chemistry of PCBs*. Cleveland, OH: CRC Press.

polygenetic

polygenetic   An entity, such as a soil, sediment, sediment body, landform or landscape, is said to be polygenetic if it was formed by a sequence of different environmental conditions and processes, rather than by a single environmental condition or climate. Most entities investigated in the field of environmental change are polygenetic to some extent having, by definition, experienced varying intensities of the same processes and RATES OF ENVIRONMENTAL CHANGE, if not different processes or complete changes in environmental régime, during the timespan of their existence. It has been suggested, for example, that all soils are polygenetic and that the older they are, the more polygenetic they become. Hence, it may be more appropriate to refer to the 'evolution', rather than the 'development' of such polygenetic entities. In terms of landforms and landscapes, the larger the landform and the more stable the landscape, the more likely it is to be polygenetic in character: within-channel forms, such as CHANNEL BARS, tend to respond quickly to an environmental change, whereas the landscape of Sierra Leone (see Figure) has developed through numerous climates over geological time.   *JAM/RPDW*

[See also CLIMATOGENIC GEOMORPHOLOGY, MONOGENETIC, PALIMPSEST]

Johnson, D.L., Keller, E.A. and Rockwell, T.K. 1990: Dynamic pedogenesis: new views on some key concepts, and a model for interpreting Quaternary soils. *Quaternary Research* 33, 306–319. **Thomas, M.F.** 1994: *Geomorphology in the tropics: a study of weathering and denudation in low latitudes*. Chichester: Wiley.

polygon analysis

polygon analysis   The study and manipulation of areal ENTITIES within a GEOGRAPHICAL INFORMATION SYSTEM (GIS). A polygon is composed of at least three POINTS completely connected by at least three lines (see ARCS), along with a LABEL or IDENTIFICATION tag and some ATTRIBUTE, such as an areal measurement or landuse (agricultural field, forest, urban extent). Polygons are usually part of a GIS COVERAGE and can be interrogated by OVERLAY ANALYSIS or used to generate BUFFER ZONES. Polygon analysis examines relationships between two or more polygons and includes a number of operations, including whether one polygon is contained (completely surrounded) within another and whether one polygon is adjacent (immediately next) to another by determining shared boundaries. Consider multiple polygons representing forest stands: over time, the shape, size and location of each polygon will change; with the use of polygon analysis, changes can be summarised, evaluated and displayed using a GIS.   *TVM*

[See also ADJACENCY, SPATIAL ANALYSIS]

Baker, W.L. and Cai, Y. 1992: The role of programs for multiscale analysis of landscape structure using the GRASS geographical information system. *Landscape Ecology* 7, 291–301. **Tomlin, C.D.** 1992: *Geographic information systems and cartographic modeling*. Englewood Cliffs, NJ: Prentice-Hall.

polymorphism

polymorphism   A condition in which a population possesses more than one allele at a locus (the location in the *DNA* occupied by a particular gene). One common

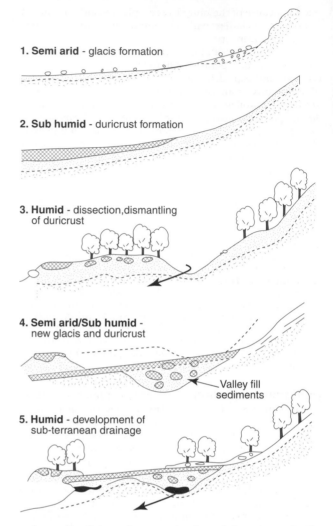

1. Semi arid - glacis formation

2. Sub humid - duricrust formation

3. Humid - dissection,dismantling of duricrust

4. Semi arid/Sub humid - new glacis and duricrust

Valley fill sediments

5. Humid - development of sub-terranean drainage

polygenetic *Schematic reconstruction of five stages in the Quaternary evolution of the present polygenetic landscape (stage 5) in Sierra Leone. Two semi-arid phases of glacis formation, two subhumid phases of* DURICRUST *(ferricrete) formation, and two humid episodes of dissection are depicted. Note that present-day* GLACIS *formation linked to the subterranean drainage channels is not shown (after Thomas, 1994)*

measure of genetic diversity is the proportion of loci in a population that are polymorphic.   *MVL*

[See also GENOTYPE, HETEROZYGOSITY]

Polynesian colonisation

Polynesian colonisation   Descendants of the early peoples who colonised Indonesia, New Guinea and Australia between 40 000 and 100 000 years ago and became highly skilled sailors, who set out on sophisticated vessels to colonise much of Micronesia and Melanesia in the western Pacific. Colonisation of the more isolated islands of Polynesia (east of Fiji) was much more recent and was probably limited to the last 2000 to 3000 years. Yet, archaeological evidence indicates that Polynesians eventually colonised the vast majority of these islands, including some of the most isolated ones on Earth. Data from ENVIRONMENTAL ARCHAEOLOGY and PALAEOECOLOGY indicate that these colonisations were far from benign, and often devastated the native flora and fauna.   *MVL*

Keast, A. and Miller, S.E. (eds) 1996: *The origin and evolution of Pacific Island biotas, New Guinea to Eastern Polynesia: patterns and processes*. Amsterdam: SPB Academic.

polyploidy   The state of having three or more sets of homologous chromosomes. The norm for eucaryotic organisms is two sets (diploid). Polyploids with odd numbers of chromosome sets (3, 5, etc.) are usually sterile, but multiplication to even numbers of sets may restore fertility and may be the basis of some rapid SPECIATION events, especially in plants.   *KDB*

population   (1) In the statistical context, the population is all the individuals of a particular sort: in practice, however, the *sampled population* (the population that is actually sampled) may differ from the *target population* (the population that is aimed at or should be sampled for valid conclusions to be reached in relation to the research problem). (2) In the ecological and human context, it is a group of individual organisms belonging to the same species: *local*, *regional* and *global populations* of species may be recognised.   *JAM*

[See also DEMOGRAPHIC CHANGE, METAPOPULATION]

population cycles   Periodic fluctuations in animal population size. All populations fluctuate as a result of relative rates of births, deaths and, in most animals, immigration and emigration. Some fluctuations show PERIODICITIES that represent a cycle, due either to *extrinsic* – largely density independent – factors (the effects of other species and the environment) or *intrinsic* – largely DENSITY DEPENDENT – factors (that relate to the life cycle and life-history strategy of the species). Without environmental constraints, a population would grow at a rate governed by its *intrinsic rate of natural increase* (*r*). However, the CARRYING CAPACITY (*K*) of the environment limits the population size, and may be exceeded only temporarily. Time lags and overcompensation in the density-dependent response mean that the size of many populations fluctuates in a specific manner (termed *stable limit cycle, damped oscillation* and *monotonic return*), with a specific periodicity (typically, cycles of 3–4 and 9–10 years for small and large New-World herbivores, respectively). Because even minor and random environmental fluctuations can induce *oscillation* of population size, the impact of widespread and rapid ENVIRONMENTAL CHANGE and/or of climatic disturbances may disrupt the established pattern of population cycles, adversely affecting crops, PESTS, PREDATOR–PREY RELATIONSHIPS and primary producers.   *SHJ*

[See also ALGAL BLOOMS, DEMOGRAPHIC CHANGE, ECOLOGICAL EXPLOSION, HOMEOSTASIS, NEGATIVE FEEDBACK, OVERHUNTING, OVERFISHING, POPULATION DYNAMICS, OVERPOPULATION]

Begon, M., Mortimer, M. and Thompson, D.J. 1996: *Population ecology: a unified study of animals and plants*, 3rd edn. Oxford: Blackwell Science. **Ricklefs, R.E.** 1990: *Ecology*, 3rd edn. New York: W.H. Freeman.

population dynamics   Variations through time in the populations of organisms, particularly in the sizes of the populations. Interplay between temporal and spatial variation is important in understanding population size, which may exhibit DENSITY DEPENDENCE and be affected by numerous ENVIRONMENTAL FACTORS.   *JLI*

Leather, S.R., Watt, A.D., Mills, N.J. and Walters, K.F.A. (eds) 1994: *Individuals, populations and patterns in ecology*. Andover: Intercept.

population ecology   The study of groups of organisms of the same species, interactions within and between POPULATIONS and relationships between populations and ENVIRONMENTAL FACTORS.   *JAM*

[See also AUTECOLOGY]

Davy, A.J., Hutchings, M.J. and Watkinnson, A.R. (eds) 1988: *Plant population ecology* [*British Ecological Society Symposium.* Vol. 28]. Oxford: Blackwell Scientific. **White, J. (ed.)** 1985: *The population structure of vegetation*. The Hague: Junk.

population viability analysis (PVA)   Given some basic information on key demographic and genetic parameters, population biologists can use a variety of PVA models to estimate the viability of a population, either in terms of expected period of PERSISTENCE or in terms of the probability that a population will persist for a given time period. PVA has become a common part of many recovery plans for ENDANGERED SPECIES.   *MVL*

[See also CONSERVATION BIOLOGY, EFFECTIVE POPULATION SIZE, MINIMUM VIABLE POPULATION, POPULATION, POPULATION CYCLES, POPULATION DYNAMICS]

Mangel, M. and Tier, C. 1994: Four facts every conservation biologist should know about persistence. *Ecology* 75, 607–14.

pore ice   Ice occurring in the pores of soil and rocks. Pore ice does not include SEGREGATED ICE and, on melting, pore ice does not yield water in excess of the pore volume of the same soil when unfrozen.   *HMF*

[See also PERMAFROST]

Associate Committee on Geotechnical Research 1988: *Glossary of permafrost and related ground-ice terms* [*Technical Memorandum* 142]. Ottawa: Permafrost Subcommittee, National Research Council of Canada.

porous   Containing interconnected voids or interstices (*pores*), which may or may not allow the passage of fluids or gases by DIFFUSION. The extent to which a rock, sediment or soil is porous (*porosity*) is measured by the ratio of the volume of the void space to the total volume of material plus voids. Porosity should be distinguished from *permeability* (the extent to which liquids or gases can pass through the material): many porous materials (such as chalk and sandstone in relation to water) are PERMEABLE; some may be porous but IMPERMEABLE (such as clay and ARGILLACEOUS rocks, in which the pore spaces are too small and water is held firmly by surface tension); others are non-porous but may be permeable or PERVIOUS (such as fissured granite). Hence useful AQUIFERS are often permeable but only moderately porous.   *JAM/RPDW*

[See also IMPERVIOUS]

Davis, S.N. 1969: Porosity and permeability of natural materials. In De Wiest, R.J.M. (ed.), *Flow through porous media*. New York: Academic Press. **Stamp, L.D.** 1961: *A glossary of geographical terms*. London: Longman.

positivism A philosophical movement, associated particularly with Auguste Compte in the nineteenth century. It maintained that observable and firmly established facts should be the only basis for belief and saw SCIENCE as the means to establishing such valid knowledge. It was part of the classical–empirical–analytical tradition of science that knowledge should be based on experience. It was explicitly applied by Compte to the study of human affairs as well as to the natural SCIENCES, and it was further developed as LOGICAL POSITIVISM in the twentieth century.

JAM/CJT

[See also EMPIRICISM, REALISM, SCIENTIFIC METHOD]

Gregory, D. 1978: *Ideology, science and human geography.* London: Hutchinson. **Hempel, C.G.** 1965: *Aspects of scientific explanation and other essays on the philosophy of science.* New York: Free Press.

possibilism Developed in the early part of the twentieth century as a counterpoint to environmental DETERMINISM, possibilism argued that environments offer a range of possibilities and that the people who occupied them had choices. Early debate centred on the limits to that choice and with identifying the constraints. At its extreme, possibilism maintained that people had the potential to control environments and override their dictats. In this latter form it became as untenable a proposition as crude determinism.

DTH

[See also ENVIRONMENTALISM]

Spate, O.H.K. 1957: How determined is possibilism? *Transactions of the Institute of British Geographers* 17, 1–12. **Tatham, G.** 1953: Environmentalism and possibilism. In Taylor, G. (ed.), *Geography in the twentieth century*, 2nd edn. London: Methuen, 128–162.

postcolonialism Not merely the investigation of the *postcolonial period*, postcolonialism is a critical perspective that highlights the misrepresentation of the nature of former colonial peoples and cultures by the colonisers both during the period of COLONIALISM and since. Postcolonialism considers: the ways in which knowledge was 'manufactured' and used by European peoples during colonisation; the continuing legacy of the widespread acceptance of 'western' views about the 'east' and 'south'; the need to investigate and celebrate alternative views; and the possibility of corrective action. It has implications for other social groups that are viewed as different and excluded and may be considered as a dimension of POSTMODERNISM.

JAM

[See also NEOCOLONIALISM]

Ashcroft, B., Griffiths, G. and Tiffin, H. (eds) 1995: *The post-colonial studies reader.* London: Routledge. **Escobar, A.** 1995: *Encountering development: the making and unmaking of the Third World.* New York: Princeton University Press. **Said, E.** 1978: *Orientalism: western conceptions of the Orient.* New York: Pantheon Books.

posthole A small pit, or mark in the ground, possibly accompanied by the remains of a wooden post (the *post pipe*) and/or packing material, which indicates the former site of an archaeological structure. Postholes are also known as *postmolds*. The term *stakehole* is reserved for smaller features produced by stakes of smaller diameter, which have simply been pushed into the ground without a prepared hole.

JAM

Shaw, I. 1999: Posthole. In Shaw, I. and Jameson, R. (eds), *A dictionary of archaeology*. Oxford: Blackwell.

postindustrialisation The general decline in MANUFACTURING INDUSTRY characteristic of developed countries in the late twentieth century, accompanied by a continuing rise in the proportion of the population employed in SERVICE INDUSTRIES.

JAM

[See also DE-INDUSTRIALISATION]

Bell, D. 1973: *The coming of post-industrial society.* New York: Basic Books. **Savitch, H.** 1988: *Post-industrial cities: politics and planning in New York, Paris and London.* Princeton, NJ: Princeton University Press.

postmodernism An ill-defined concept developed by a wide range of artists, intellectuals and scientists. It is seen as a postindustrial, postcapitalist, post-Fordian cultural logic or worldview that has replaced modernism since roughly the AD 1960s. Applied to environmental studies, it implies an adoption of a multidisciplinary or *holistic*, rather than a *reductionist* approach.

CJB

Blaikie, P. 1996: Post-modernism and global environmental change. *Global Environmental Change* 6, 81–85. **Cosgrove, D.** 1990: Environmental thought and action: pre-modern and postmodern. *Transactions of the Institute of British Geographers NS* 15, 344–358. **Harvey, D.** 1989: *The condition of postmodernity: an enquiry into the origins of cultural change.* Oxford: Basil Blackwell.

poststructuralism A philosophy that emerged during the AD 1960s as a reaction against and further development of STRUCTURALISM: it criticised the 'scientific' approach of structuralism. A leading poststructuralist was the French philosopher Jacques Derrida who stressed that the meaning of a word arises from past use in other texts and contexts and cannot be understood simply by analysing its formal position within a system of structures. French philosopher Michel Foucault strongly influenced poststructuralist thought in the 1970s, stressing that communicated knowledge is structured into 'discursive formations', and therefore helps to shape society and social institutions. Thus, poststructuralism stresses the significance of linguistic and cultural constructions, and its principal methodology is *deconstruction*.

ART

Norris, C. 1982: *Deconstruction: theory and practice.* London: Methuen. **Sarup, M.** 1989: *Post-structuralism and postmodernism.* Brighton: Harvester/Athens.

potable water Water intended for drinking, cooking or other high-quality uses. Although only two litres of water are required per day to sustain human life, it has been estimated that each individual requires a minimum of 50 L d^{-1} to remain healthy (including water for bathing and washing clothes, etc.), daily per capita usage of piped water in the UK has reached about 180 L d^{-1} and in the United States it is about 280 L d^{-1}. With > 1500 million of the world's poorest people without access to a source of safe drinking water, this is a major contributary factor to high mortality rates in developing countries, where, for example, around five million children under the age of five years of age die from WATER-BORNE DISEASES each year.

JAM

[See also DISEASE, ENVIRONMENTAL QUALITY, HUMAN HEALTH HAZARDS]

Horan, N. 1997: Collection, treatment and distribution of potable water. In Brune, D., Chapman, D.V., Gwynne, M.D. and Pacyna, J.M. (eds), *The global environment: science, technology and management*. Weinheim: VCH, 758–773. **Keller, A.Z. and Wilson, H.C.** 1992: *Hazards to drinking water supplies*. New York: Springer.

potassium–argon (K–Ar) dating

A RADIOMETRIC DATING technique based on the decay of ^{40}K to the DAUGHTER ISOTOPE ^{40}Ar, by *electron capture*. Together with ARGON–ARGON DATING, K–Ar dating is principally used to obtain ages of potassium-rich minerals in volcanic LAVAS and TUFFS. K–Ar and ^{40}Ar:^{39}Ar techniques are mainly applied to geological materials older than one million years because of the long HALF-LIFE of ^{40}K (1.31×10^9 years). However, there have been numerous applications in the field of QUATERNARY environmental change, notably, in the determination of the GEOMAGNETIC POLARITY TIMESCALE and the age of early HOMINIDS, and in dating TEPHRA layers that are found in MARINE SEDIMENT CORES, ICE CORES and LACUSTRINE SEDIMENTS. The fundamental assumptions of K–Ar and ^{40}Ar:^{39}Ar techniques are that no argon was present in the mineral immediately after crystallisation and that the system remained closed since this event. *DAR*

Leakey, L.S.B., Evernden, J.A. and Curtis, G.H. 1961: Age of bed 1, Olduvai Gorge, Tanganyika. *Nature* 191, 478–479. **McDougall, I.** 1995: Potassium–argon dating in the Pleistocene. In Rutter, N.W. and Catto. N.R. (eds), *Dating methods for Quaternary deposits*. St John's, Newfoundland: Geological Association of Canada, 1–14. **Shaeffer, O.A. and Zäringer, J.** 1966: *Potassium argon dating*. New York: Springer. **Walter, R.C.** 1997: Potassium–argon/argon–argon dating methods. In Taylor, R.E. and Aitken, M.J. (eds), *Chronometric dating in archaeology*. New York: Plenum, 97–126.

potential evapotranspiration (PET)

The maximum combined amount of EVAPORATION and *transpiration* (water loss by plants) that can take place from a surface where there is no limitation on water supply. It is a theoretical value for a particular vegetation type, which normally differs from the *actual evapotranspiration* (AET), which is the observed amount of evapotranspiration. Some definitions define PET in terms of an extensive surface of a short green crop (to facilitate comparison), in which case AET can exceed PET for vegetation types with large leaf areas, especially forests. PET may also exceed the *potential evaporation* (PE), which is the amount of evaporation from an open water surface. *JAM*

[See also EVAPOTRANSPIRATION]

Morton, F.I. 1983: Operational estimates of areal evapotranspiration and their significance to the science and practice of hydrology. *Journal of Hydrology* 66, 1–76. **Sene, K.J., Gash, J.H.C. and McBeil, D.D.** 1991: Evaporation from a tropical lake: comparison of theory with direct measurement. *Journal of Hydrology* 127, 193–217.

potential natural vegetation

The vegetation community that would develop if human activities were removed from an area and if the resulting ECOLOGICAL SUCCESSION were completed in an instant. NATURAL VEGETATION occurs where the intactness or integrity of habitats and ecosystems is free of human influence. In effect, there are few such areas remaining on Earth, but there are many areas that have been called *semi-natural*, i.e. somewhat modified by human activity. Restoring or managing vegetation to achieve 'natural vegetation' has been a common objective in CONSERVATION. Plant communities are affected by human activities but also undergo natural change. An issue therefore arises as to how natural vegetation can be defined with reference to different states throughout time. *IFS*

[See also CLIMAX VEGETATION, COMMUNITY CONCEPTS, NATURAL AREAS CONCEPT, SEMI-NATURAL VEGETATION]

Hobbs, R.J. and Norton, D.A. 1996: Towards a conceptual framework for restoration ecology. *Restoration Ecology* 4, 93–111. **Shrader-Frechette, K.S. and McCoy, E.D.** 1993: *Methods in ecology: strategies for conservation*. Cambridge: Cambridge University Press.

potentiation

The process by which a substance is made more toxic by the presence of another substance. *JAM*

[See also ANTAGONISM, SYNERGISM]

potentiometric surface

The level to which water would rise in a well tapping a confined AQUIFER: it can be viewed as a theoretical water table. The term is sometimes confused with *piezometric surface* (see WATER TABLE). *JAM*

[See also ARTESIAN]

Freeze, R.A. and Cherry, J.A. 1979: *Groundwater*. Englewood Cliffs, NJ: Prentice Hall.

Precambrian

An informal name for that part of the GEOLOGICAL RECORD older than the earliest CAMBRIAN, characterised by the absence of metazoan FOSSILS with hard parts. It is divided into two EONS: the ARCHAEAN (3900 to 2400 million years ago) and the PROTEROZOIC (2400 million years ago to the beginning of the Cambrian period). *GO*

[See also GEOLOGICAL TIMESCALE]

Coward, M.P. and Ries, A.C. (eds) 1995: *Early Precambrian processes* [*Special Publication* 95]. London: Geological Society.

precautionary principle

Where there are threats of environmental damage, the lack of scientific data should not be used to postpone measures to prevent ENVIRONMENTAL DEGRADATION. This is essentially a 'better safe than sorry' approach that has evolved with changing social, economic and political ideas. *IFS*

[See also ECOSYSTEM COLLAPSE, GLOBAL WARMING]

O'Riordan, T. and Cameron, J. (eds) 1994: *Interpreting the precautionary principle*. London: Earthscan. **O'Riordan, R., Cameron, J. and Jordan, A. (eds)** 2000: *Reinterpreting the precautionary principle*. London: Cameron May.

precession

One of the three main orbital parameters of the MILANKOVITCH THEORY of GLACIAL–INTERGLACIAL CYCLES, the 'precession of the equinoxes' relates to the combined effect of 'wobble' of the Earth about its axis and 'rotation' of the orbit around the Sun. It results in the coincidence of the PERIHELION (the position on the orbit when the Earth is closest to the Sun) with the summer EQUINOX about every 21 000 years, a phenomenon that last occurred about 11 000 years ago. *JAM*

[See also ECCENTRICITY, OBLIQUITY]

precipitation (1) A chemical reaction whereby an insoluble substance (*precipitate*) is formed from a solution (see CHEMICAL PRECIPITATION). (2) The deposition of water from the atmosphere in liquid (rain) or solid (snow, hail) form. Normally, precipitation is a term reserved for the drops or solid particles that fall to the ground; the more general term *hydrometeor* includes other forms of removal of water vapour from the atmosphere (e.g. dew, frost and rime). Three types of precipitation are recognised based on the mechanism of cooling inducing CONDENSATION: *convectional precipitation* results where moist air rises following local heating of the Earth's surface (see CONVECTION); *frontal* or *cyclonic precipitation* results from the uplift of moist air at a FRONT; OROGRAPHIC precipitation occurs where moist air is forced to rise over a topographic barrier. *JAM*

[See also ACID RAIN, CONDENSATION]

Legates, D.R. 1995: Global and terrestrial precipitation: a comparative assessment of existing climatologies. *International Journal of Climatology* 15, 237–258. **Sumner, G.** 1988: *Precipitation: process and analysis.* Chichester: Wiley.

precipitation variations Precipitation exhibits greater spatial and temporal variability than temperature. There has been a small overall increase in precipitation in the high latitudes of the Northern Hemisphere during the twentieth century and a decrease over low latitudes (notably the SAHEL since the 1960s). Whilst winter records may have been underestimates due to inefficient snow-catch, widespread evidence of wetter winters extends from the high to the mid-latitudes. Late twentieth century increases in winter precipitation have been unprecedented over recent centuries in such locations as Scotland and Norway (see Figure). Conversely, summer precipitation has declined over much of Europe and some other mid-latitude regions. These seasonal changes are consistent with climate scenarios for the twentyfirst century that incorporate a slight poleward movement of pressure zones. The larger number of high daily rainfalls evident in some recent records (notably in North America) is also replicated in scenarios of mid-twentyfirst century climate. Longer-term variations in precipitation form an important aspect of longer-term CLIMATIC CHANGE. *JCM*

[See also DROUGHT]

Dahl, S.O. and Nesje, A. 1996: A new approach to calculating Holocene winter precipitation by combining glacier equilibrium-line altitudes and pine-tree limits: a case study from Hardangerjøkulen, central southern Norway. *The Holocene* 6, 381–398. **Førland, E.J., Ashcroft, J., Dahlström, B.** *et al.* 1996: *Changes in 'normal' precipitation in the North Atlantic region* [*Report* 7/96]. Oslo: Det Norske Meteorologiske Institutt. **Hennessy, K.J., Gregory, J.M. and Mitchell, J.F.B.** 1997: Changes in daily precipitation under enhanced greenhouse conditions. *Climate Dynamics* 13, 667–680. **Jones, P., Conway, D. and Briffa, K.** 1997: Precipitation variability and drought. In Hulme, M. and Barrow, E., *Climates of the British Isles: present, past and future.* London: Routledge, 197–219. **Karl, T.R. and Knight, R.W.** 1998: Secular trends of precipitation amount, frequency and intensity in the United States. *Bulletin of the American Meteorological Society* 79, 231–241. **Nicholson, S.E.** 1994: Recent rainfall fluctuations over Africa and their relationship to past conditions over the continent. *The Holocene* 4, 121–131. **Verschuren, D., Laird, K.R. and Cumming, B.F.** 2000: Rainfall and drought in equatorial east Africa during the past 1,100 years. *Nature* 403, 410–414.

precision The degree of mutual agreement among individual measurements as a result of repeated determination under the same conditions: it is usually defined in terms of statistical uncertainty associated with the measurements and expressed in standard deviations away from the measured value. A technique may produce very precise results but they need not be an accurate representation of the real phenomenon under study, as in the case for instance of DATING TECHNIQUES. *DAR*

[See also ACCURACY]

predation The killing of one species (*prey* or victim) by another (*predator*), often but not always for food. *JAM*

Taylor, R.J. 1984: *Predation.* London: Chapman and Hall.

predator–prey relationships The interactions between prey and predators, particularly in relation to POPULATION DYNAMICS and COEVOLUTION. Such relationships illustrate mutual dependence and cyclical phenomena. *JLI*

[See also BIOLOGICAL CONTROL, POPULATION CYCLES]

Hassell, M.P. and Anderson, R.M. 1989: Predator–prey and host–pathogen interactions. In Cherrett, J.M. (ed.) 1989: *Ecological concepts.* Oxford: Blackwell Scientific, 147–96.

prediction The casting of results into the future. An example is CLIMATIC PREDICTION. The ability to predict with ACCURACY is the hallmark of a good MODEL. *JAM*

[See also ABDUCTION, FORECASTING, RETRODICTION]

prehistoric Relating to the period before written records. *SPD*

Cunliffe, B. 1998: *Prehistoric Europe.* Oxford: Oxford University Press.

prescribed fire Fire used intentionally to control or destroy vegetation, to reduce a build up of potential fuel or to control PESTS. Globally, the area being burnt in this way is estimated to be only one-fifth of what it was in the late fifteenth century. In fire-prone areas, prescribed fire is generally used under damp, cool conditions. Fire temperatures are considerably lower than those of WILDFIRES. Amounts of CO_2 released into the atmosphere by prescribed fire are thought to be negligible. *AJDF*

[See also FIRE HISTORY, FIRE FREQUENCY]

Elliott, K.J., Hendrick, R.L., Major, A.E. *et al.* 1999: Vegetation dynamics after a prescribed fire in the southern Appalachians. *Forest Ecology and Management* 114, 199–213. **Garski, C.J. and Farnsworth, A.** 2000: Fire weather and smoke management. In Whiteman, C.D. (ed.), *Mountain meteorology: fundamentals and applications.* New York: Oxford University Press, 237–272. **O'Hanlon, L.** 1995: Fighting fire with fire. *New Scientist* 1980 (15 July), 29–33.

pressure group An official or unofficial organisation that aims to stimulate public and political concern over ENVIRONMENTAL ISSUES. Pressure groups range in status from highly influential organisations, such as *Greenpeace* and *Friends of the Earth,* to less organised, radical movements that will pursue their objectives by any means necessary, including illegal activities. *JGS*

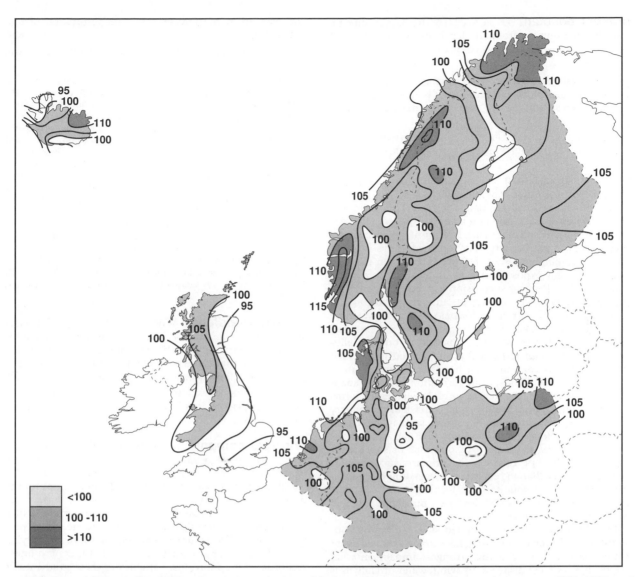

precipitation variations *Percentage change in annual precipitation over northwestern Europe, between* AD *1931–1960 and 1961–1990 (after Førland* et al.*, 1996)*

[See also ENVIRONMENTAL MOVEMENTS, GREEN POLITICS, GREENING OF SOCIETY]

pretreatment Cleaning and removal of extraneous material or CONTAMINATION, which might otherwise affect the result, from a sample to be analysed. For instance, samples for RADIOCARBON DATING should contain carbon only from the organism that is contemporary with the event being dated. Extraneous matter such as rootlets, humic substances, etc. must be excluded as this will have a younger carbon age. The ideal pretreatment chemically purifies and isolates the substance being analysed without affecting it, e.g. cellulose extraction from *wood* removes lignins and resins as well as material foreign to the wood. For radiocarbon dating of *bone*, the protein collagen is extracted, but, when the yield is low as a result of organic decay, the relative amount of contamination by soil amino acids and proteins is greater, so further extraction of the amino acid hydroxyproline, peculiar to bone, will yield a better result, albeit at the expense of a much reduced sample size. *PQD*

Hedges, R.E.M. and Law, I.A. 1989: The radiocarbon dating of bone. *Applied Geochemistry* 4, 249–253. **Hoper, S.T., McCormac, F.G., Hogg, A.G.** *et al.* 1998: Evaluation of wood pretreatments on oak and cedar. *Radiocarbon* 40, 45–50.

prevailing wind The wind direction (from where the wind is coming) with the greatest frequency, as opposed to the direction of the strongest (DOMINANT) wind. *JAM*

primary forest A forest that has never been extensively felled for timber or cleared for another LANDUSE (e.g. agriculture and estate plantations). The term is sometimes used to denote a forest perceived as undisturbed by humans. *Secondary forest* results from REGENERATION following FOREST CLEARANCE. *NDB*

[See also PRIMARY WOODLAND, TROPICAL RAIN FOREST]

primary pollutant A POLLUTANT discharged directly into the environment from an identifiable source. *JAM*

[See also SECONDARY POLLUTANT]

primary productivity The BIOMASS production rate in the producer (plant) community. It is expressed as units of energy or as units of dry organic matter per unit area per unit time. *RJH*

[See also GROSS PRIMARY PRODUCTIVITY, NET PRIMARY PRODUCTIVITY, PRODUCTIVITY, SECONDARY PRODUCTIVITY]

primary succession A sequential change in species composition (or other ecosystem characteristics) on surfaces with no residual biological legacy. This definition generally implies succession on a newly exposed, often dry and nearly sterile substrate with little or no organic matter present, low nutrient status and no seed banks or other propagules initially present. Primary succession can also occur on freshly exposed surfaces under water (e.g. overturned rocks). Primary succession can be considered the biological response to NUDATION, an extreme ALLOGENIC CHANGE, and differs from SECONDARY SUCCESSION, where the initial disturbance leaves some soil intact and can come from within (AUTOGENIC CHANGE) or outside (allogenic) the community. Typical primary SERES follow such DISTURBANCES as volcanoes, earthquakes, landslides, mining, glacial retreat, floods, dune formation and road abandonment. However, the degree of biological legacy forms a continuum between truly primary and truly secondary seres. The spatial variability of primary seres is large, from volcanoes, glaciers, floods or roads that can cover thousands of hectares to single rock outcrops or small landslides. The rate of recovery in primary succession is generally slower than in secondary succession, but predisturbance plant communities can form in < 100 y in some cases (e.g. on tropical landslides where nutrients, light and water are all abundant).

Initial COLONISATION of denuded sites is often by RUDERALS that disperse long distances. However, adjacent plant communities have the strongest impact on what establishes, as most dispersal by wind, water or animals is local. Some sites are only slowly colonised by clonal expansion from surrounding vegetation. ECECIS, or establishment on nearly sterile substrates, is often a gradual and stepwise process, although not generally predictable. The particular set of early colonists that establishes may be determined from a suite of potential colonisers by chance events. Site modification and successional trajectories are then strongly influenced by the characteristics of these first colonists. A second group of species may colonise, or grow vigorously, only after soils and vegetation develop. This process is called FACILITATION and may be most important early in ECOLOGICAL SUCCESSION in severe environments. In less harsh environments or later in succession, COMPETITION may be a more critical process than facilitation in driving successional change. A third possible scenario is minimal interaction among co-occuring plant species in succession, where growth rates and life spans of each species determine the basic pattern of species replacements. Primary succession is certainly driven by a combination of these three factors (competition, facilitation, life history attributes). Tests of the relative importance of each factor in many primary seres are under way

and will aid in the management and restoration of disturbed lands. *LRW*

[See also CLIMAX VEGETATION, LAND RESTORATION, SOIL RECLAMATION]

Burrows, C.J. 1990: *Processes of vegetation change.* London: Unwin Hyman. **Chapin, F.S. III, Walker, L.R., Fastie, C.L. and Sharman, L.C.** 1994: Mechanisms of primary succession following deglaciation at Glacier Bay, Alaska. *Ecological Monographs* 64, 149–175. **Connell, J.H. and Slatyer, R.O.** 1977: Mechanisms of succession in natural communities and their role in community stability and organization. *The American Naturalist* 111, 1119–1144. **Matthews, J.A.** 1992: *The ecology of recently-deglaciated terrain: a geoecological approach to glacier forelands and primary succession.* Cambridge: Cambridge University Press. **Miles, J. and Walton, D.H. (eds)** 1993: *Primary succession on land.* Oxford: Blackwell Scientific. **Olson, J.S.** 1958: Rates of succession and soil changes on southern Lake Michigan sand dunes. *Botanical Gazette* 119, 125–170. **Thornton, I.** 1996: *Krakatau: the destruction and reassembly of an island ecosystem.* Cambridge, MA: Harvard University Press. **Walker, L.R.** 1999. Patterns and processes in primary succession. In Walker, L.R. (ed.), *Ecosystems of disturbed ground.* Amsterdam: Elsevier, 585–610.

primary woodland Woodlands of any type that originated from *primaeval woods* (woods that established naturally prior to the onset of human interference) and which occupy sites that have been continuously wooded. True primary woodland is becoming increasingly scarce. *IFS*

[See also ANCIENT WOODLAND, CLIMAX VEGETATION, NATURAL VEGETATION, PRIMARY (RAIN) FOREST, SECONDARY WOODLANDS]

principal components analysis (PCA) An indirect ORDINATION technique that transforms the original variables in a multivariate data-set into new, composite variables or axes that are uncorrelated and account for the maximum possible proportion of the variance in the data. Axes are selected in decreasing order of importance, subject to the constraint that they are uncorrelated to all previous axes. PCA assumes that one or more underlying latent variables or gradients exist and that there is a linear response of the individual variables to these gradients. If these assumptions are valid, it can provide a useful low-dimensional representation of the original data. *HJBB*

[See also CORRESPONDENCE ANALYSIS, GRADIENT ANALYSIS, PRINCIPAL CO-ORDINATES ANALYSIS, REDUNDANCY ANALYSIS, TEMPORAL CHANGE DETECTION]

Jolliffe, I.T. 1986: *Principal components analysis.* New York: Springer.

principal components transform A technique often used in IMAGE PROCESSING and REMOTE SENSING to remove the frequently encountered problem of interband correlation of multispectral data. This problem, where data from different wavelengths convey essentially the same information, can be overcome by compressing all of the information contained in the original n channels into m components (where $m < n$). The principal components are simply linear combinations of the original data values. The first principal component (PC1) contains the greatest variation of the information in a scene. Succeeding components (PC2, PC3, . . . etc.) contain less scene variation. PRINCIPAL COMPONENTS ANALYSIS (PCA) can be used as an enhancement procedure prior to interpretation or as a preprocessing procedure prior to classification in many fields. *KJT*

Lillesand, T.M. and Kiefer, R.W. 1994: *Remote sensing and image interpretation*. New York: Wiley.

principal co-ordinates analysis An ORDINATION technique in which the distances between objects in the ordination plot are maximally correlated with the original dissimilarities or distances between objects. Almost any dissimilarity measure can be used. If Euclidean distance is used, the technique is PRINCIPAL COMPONENTS ANALYSIS. Principal co-ordinates analysis is also known as classical or *metric scaling*. *HJBB*

Gower, J.C. 1966: Some distance properties of latent root and vector methods used in multivariate analysis. *Biometrika* 53, 325–338.

Priscoan (Hadean) That part of the early history of the Earth from which no rock assemblages are definitely known. The Earth is estimated to have formed c. 4600 million years ago (Ma), and the oldest known rocks are c. 3960 Ma (see ARCHAEAN). Much evidence for conditions during the Priscoan is derived from studies of other planets and satellites (see PLANETARY GEOLOGY), the surfaces of which have not been modified by water-related and other processes, such as PLATE TECTONICS, life and WEATHERING, that occur on Earth. The Priscoan is thought to have been characterised by frequent, high-energy METEORITE IMPACT events, rapid MANTLE convection, abundant VOLCANISM and an unstable, sometimes molten, surface. An early, reducing ATMOSPHERE, possibly rich in CARBON DIOXIDE, probably developed from the release of VOLATILES during volcanism. *GO*

[See also GEOLOGICAL TIMESCALE, PLANETARY ENVIRONMENTAL CHANGE]

Bowring, S.A. and Williams, I.S. 1999: Priscoan (4.00–4.03 Ga) orthogneisses from northwestern Canada. *Contributions to Mineralogy and Petrology* 134, 3–16. **Morse, J.W. and Mackenzie, F.T.** 1998: Hadean ocean carbonate geochemistry. *Aquatic Geochemistry* 4, 301–319.

probability A quantitative description of the chance or likelihood of occurrence of a particular event expressed between 0 and 1. The higher the probability value, the more likely it is that the event will occur. If the probability is 0, the event cannot happen; if the probability is 1, the event will certainly happen. *HJBB*

probability distribution For a continuous random variable, the curve described by a mathematical formula that specifies, by the area under the curve, the PROBABILITY that the variable falls within a particular interval (e.g. NORMAL DISTRIBUTION). For a discrete random variable (e.g. counts), the probability distribution is a mathematical formula that gives the PROBABILITY of each value of the variable (e.g. *binomial distribution*). It is also known as the *probability density*. *HJBB*

prod mark A type of TOOL MARK comprising an asymmetrical gouge caused by an object transported in a strong current becoming temporarily embedded in MUD. *GO*

production The quantity of sugars manufactured by PHOTOSYNTHESIS in land plants and autotrophic algae,

some of which are used to run metabolic processes and some of which are incorporated in BIOMASS. *RJH*

[See also CARBON ASSIMILATION, NET PRIMARY PRODUCTIVITY, PRODUCTIVITY

productivity The rate at which a community of organisms manufactures new BIOMASS. It is expressed as units of energy or as units of dry organic matter per unit area per unit time (e.g. $kJ\ ha^{-1}\ y^{-1}$ or $kg\ ha^{-1}\ y^{-1}$). It consists of PRIMARY PRODUCTIVITY, the productivity of *producers*, and SECONDARY PRODUCTIVITY, the productivity of *consumers*. Primary productivity is influenced by ENVIRONMENTAL FACTORS, some of which may be LIMITING FACTORS. Light commonly limits productivity in green plants. Less than 2% of the light falling on a tropical rain-forest canopy reaches the forest floor and few plants thrive in the deep shade. Water shortage also limits primary productivity where light is not limiting. GROSS PRIMARY PRODUCTIVITY increases with increasing temperature, but so does the RESPIRATION rate. Consequently, increasing temperatures do not necessarily boost NET PRIMARY PRODUCTIVITY. In the oceans, *phytoplankton* are the powerhouse of primary PRODUCTION. Light, nutrients and the grazing rate of *zooplankton* influence their productivity. *RJH*

[See also ABOVE-GROUND PRODUCTIVITY, BELOW-GROUND PRODUCTIVITY, CARRYING CAPACITY]

Myneni, R.B., Los, S.O. and Asrar, G. 1995: Potential gross productivity of terrestrial vegetation from 1982–1990. *Geophysical Research Letters* 22, 2617–2620. **Reichle, D.E., Franklin, J.F. and Goodall, D.W.** 1975: *Productivity of world ecosystems*. Washington, DC: National Academy of Sciences.

progradation The lateral migration of environments as SEDIMENTARY DEPOSITS accumulate. Progradation is particularly characteristic of shallow subaqueous environments, where the water level sets a limit on the level to which sediment can accumulate. A classic example is the seaward migration of a DELTA as sediment is deposited, causing a local REGRESSION. In successions of sediments or SEDIMENTARY ROCKS the progradation of laterally adjacent environments gives rise to vertically stacked, gradationally bounded FACIES, a concept commonly expressed as WALTHER'S LAW. *GO*

[See also AGGRADATION, AUTOCYCLIC CHANGE, FACIES SEQUENCE]

Skene, K.I., Piper, D.J.W., Aksu, A.E. and Syvitski, P.M. 1998: Evaluation of the global oxygen isotope curve as a proxy for Quaternary sea level by modeling of delta progradation. *Journal of Sedimentary Research* 68, 1077–1092. **Walker, R.G. and James, N.P. (eds)** 1992: *Facies models: response to sea level change*. St John's, Newfoundland: Geological Association of Canada.

progressive desiccation The concept first developed in the early AD 1900s in connection with central Asia that there had been a postglacial drying-up of the environment. It was assumed that at a global scale GLACIAL EPISODES were characterised by wet conditions and aridity had increased since the waning of the last PLEISTOCENE ice sheets. The concept was also applied to southern Africa and west Africa. It is now known that glaciations, at a global scale, were characterised generally by aridity rather than increased wetness, although certain locations did indeed experience higher rainfall. *RAS*

505

[See also DESERTIFICATION, PLUVIAL]

Goudie, A.S. 1972: *The concept of post-glacial progressive desiccation* [*Research Paper* 4]. Oxford: School of Geography, University of Oxford. **Goudie, A.S.** 1990: Desert degradation. In Goudie, A.S. (ed.), *Techniques for desert reclamation.* Chichester: Wiley, 1–34.

projective approach The inferential framework of environmental change research has changed in recent decades. A sequence of three stages, from *inductive*, through *deductive* to projective modes of problem definition and research development, has been suggested. The projective approach involves responding to a research agenda set, not by what has come to light as a result of changes in the past, but by what may happen in the future. The way the theme of enhanced GLOBAL WARMING permeates research on environmental change is a prime example. The projective approach has advantages and limitations: the former including greater relevance to human society and the latter including the difficulties, if not dangers, attendant on gearing one's research to uncertain predictions. *JAM*

[See also DEDUCTION, INDUCTION, INTERNATIONALISATION OF RESEARCH]

Oldfield, F. 1993: Forward to the past: changing approaches to Quaternary palaeoecology. In Chambers, F.M. (ed.), *Climatic change and human impact on the landscape.* London: Chapman and Hall, 13–21.

pronival rampart An accumulation of debris forming a ramp or ridge(s) along the down-slope margin of a perennial or semi-permanent snowbed located typically near the base of a steep bedrock slope in PERIGLACIAL environments. The more common descriptor *protalus rampart* has been shown to be inappropriate. Other terms used for features thought to be formed in the same way as pronival ramparts include *nival ridge, nivation moraine, winter talus ridge, moraine de névé, bourrelet de névé* and *Hangblockwulst.* An assumed formation only by debris moving over a snowbed (*supranival* SNOW AVALANCHES and DEBRIS FLOWS) has been recognised as inadequate, *subnival* processes (e.g. SOLIFLUCTION, snow-push) also contributing to their formation. Relict ramparts can be mistaken for other landforms, such as glacial MORAINES, ROCK GLACIERS, LANDSLIDES and avalanche impact ridges. If correctly identified in fossil form, relict ramparts can have a limited though useful role in CLIMATIC RECONSTRUCTION, not only by indicating the former presence of long-lived snowbeds and PALAEOWIND direction, but also, where there are sufficient in an area, in reflecting any gradient in the SNOW LINE. *RAS*

[See also NIVATION, TALUS]

Shakesby, R.A. 1997: Pronival (protalus) ramparts: a review of forms, processes, diagnostic criteria and palaeoenvironmental implications. *Progress in Physical Geography* 21, 394–418. **Shakesby, R.A., Matthews, J.A., McEwen, L. and Berrisford, M.S.** 1999: Snow-push processes in pronival (protalus) rampart formation: geomorphological evidence from southern Norway. *Geografiska Annaler* 81A, 31–45.

protected area approach Conservation of BIODIVERSITY or other aspects of ecosystems and landscapes undertaken by the establishment and management of individual protected areas (such as NATURE RESERVES, BIOS-PHERE RESERVES and NATIONAL PARKS) or via networks of protected areas. *IFS*

[See also BIO-REGIONAL MANAGEMENT, CONSERVATION, CONSERVATION BIOLOGY, PROTECTED AREAS: TYPES, SINGLE LARGE OR SEVERAL SMALL RESERVES]

protected areas: types Areas of LANDSCAPES that differ greatly in the degree of protection afforded ranging from strictly protected NATIONAL PARKS (American model) and nature RESERVES, which seek to maximise BIODIVERSITY, to *multiple-use management areas* that accommodate the economic and social requirements of society in various ways while providing for SUSTAINABLE DEVELOPMENT. Protected area status implies a legislative basis and the existence of a management strategy. There are around 3000 protected areas in over 120 different countries. Ten categories of protected areas are recognised by the International Union for Conservation of Nature and Natural Resources (IUCN):

- *Biosphere Reserve* – an area containing unique communities with unusual natural features, harmonious landscapes resulting from traditional landuse, and modified or degraded ecosystems that are capable of restoration.
- *Multiple-use Management Area/Managed Resource Area* – an area designed for sustained production of or access to resources (e.g. water, timber, game, pasture, marine products or outdoor recreation).
- *National Park* – a relatively large natural area containing representative samples of landscapes and where species, habitats and scenery are of special scientific, educational or recreational interest.
- *Natural Biotic Area/Anthropological Reserve* – a natural area where modern humans have not significantly influenced the traditional ways of life of the inhabitants.
- *Natural Monument/Natural Landmark* – one or more relatively small-scale natural features rated as of outstanding national significance because of uniqueness or rarity.
- *Nature Conservation Reserve/Managed Nature Reserve/ Wildlife Sanctuary* – special sites or habitats protected for the continued well-being of resident or migratory fauna of national or global significance.
- *Protected Landscape or Seascape* – a natural or scenic area along coastlines, lakeshores or riverbanks, sometimes adjacent to visitor-use areas or population centres, with potential for development for recreational use.
- *Resource Reserve/Interim Conservation Unit* – normally a relatively isolated and inaccessible natural area under pressure for colonisation or resource development.
- *Scientific Reserve/Strict Nature Reserve* – a natural area possessing outstanding and/or fragile ecosystems, features or species of at least national scientific importance.
- *World Heritage Site* – an area of true international significance.

The UK provides an example of the many types of protected areas that may exist at the national level. Local, national and international legislation provides the basis, mostly under the broad heading of 'nature conservation', for a wide variety of statutory and non-statutory designations (32 types are listed below). Several government departments (e.g. Department of the Environment,

Ministry of Agriculture, Fisheries and Food and Department of National Heritage), conservation agencies (e.g. Countryside Council for Wales, CADW – Welsh Historic Monuments, English Heritage, English Nature, Joint Nature Conservation Committee, National Trust and Scottish Natural Heritage) and local authorities hold responsibility for the system:

- *Area of Outstanding Natural Beauty* (AONB) – in England and Wales, an area for the conservation of natural beauty, where agriculture, forestry and other rural industries are safeguarded and where the economic and social needs of local communities are also taken into account. Forty-one had been designated by 1998.
- *Area of Special Protection* (AOSP) – an area for the protection of birds (previously known as *Bird Sanctuaries*)
- *Area of Special Scientific Interest* (ASSI) – in Northern Ireland, the equivalent of a SITE OF SPECIAL SCIENTIFIC INTEREST (SSSI).
- *Biogenic Reserve* – an area for the protection of genetic diversity (part of the Council of Europe programme for a European network).
- *Biosphere Reserve* – see above (there were 13 in the UK in 1998).
- *Country Park* – an area managed by local authorities for recreation and leasure close to population centres. Country Parks do not necessarily have any nature conservation interest.
- *Environmentally Sensitive Area* (ESA) – an area of high landscape and conservation value in which farmers may enter into voluntary agreements to pursue special management procedures.
- *Geological Conservation Review Site* (GCR) – a non-statutory designated area of national or international importance for Earth science conservation (see GEOLOGICAL CONSERVATION). There were 3122 established by 1997. All are or will be designated as SSSIs.
- *Heritage Coast* – In England and Wales, a non-statutory designated section of undeveloped coast exceeding one mile in length that is of exceptionally fine scenic quality.
- *Limestone Pavement Order (LPO) land* – areas of limestone pavement considered of special interest are protected by an LPO.
- *Local Authority Nature Reserve* (LANR) – in Northern Ireland, the equivalent of a Local Nature Reserve.
- *Local Nature Reserve* (LNR) – an area of habitat of local importance because of its natural or semi-natural ecosystem. Over 620 existed in 1997.
- *Marine Nature Reserve* (MNR) – an area of special importance for marine flora and fauna and geological or geomorphological features and their study. Only three existed in 1997.
- *Marine Consultation Area* (MCA) – in Scotland, a non-statutory marine area where the quality and sensitivity of the environment is sufficient to encourage awareness of the conservation issues. There were 29 MCAs in 1998.
- *National Nature Reserve* (NNR) – a nationally or internationally important area for its natural or semi-natural ecosystems and habitats. Over 360 were in existence in 1997.
- *National Park* – in England and Wales, the 10 National Parks are each defined as an extensive area of countryside of the highest landscape quality that is preserved

and enhanced while being promoted for the enjoyment of the public.

- *National Scenic Area* (NSA) – in Scotland, an area of landscape of outstanding natural beauty that is protected in the national interest. NSAs replaced earlier designations equivalent to the AONBs and National Parks of England and Wales. There were 40 NSAs in 1997.
- *National Trust property* – land of scenic value, buildings of historical importance, etc., held for the nation by the charities, the National Trust and the National Trust for Scotland.
- *Natural Heritage Area* (NHA) – in Scotland, areas of countryside of outstanding national heritage value of wide nature conservation and landscape interest where integrated management will be encouraged (none yet designated).
- *Nature Conservation Review Site* (NCR) – a non-statutory designated area (although all are or will be designated as SSSIs) of national or international importance for biological conservation.
- *Nitrate Sensitive Area* (NSA) – in England only, an area where there are voluntary incentives to regulate landuse and/or reduce nitrate leaching.
- *Nitrate Vulnerable Zone* (NVZ) – a compulsory regulated area where agricultural production contributes to degradation of drinking water quality or eutrophication of aquifers.
- *Preferred Conservation Zone* (PCZ) – in Scotland, non-statutory coastal areas of national scenic, environmental or ecological importance where new oil- or gas-related developments would be inappropriate and where tourism and recreation take priority.
- *Ramsar Site* – an area designated in response to the Ramsar CONVENTION on Wetlands, to conserve wetlands.
- *Regional Landscape Designation* (RLD) – in Scotland, non-statutory RLDs (of which there are at least five local types) identify areas where there should be a strong presumption against development because of unexploited potential for tourism and local community benefit. As at 1997, 275 had been designated.
- *Regionally Important Geological and Geomorphological Sites* (RIGS) – a non-statutory designated site of regional or local importance to Earth science for scientific, educational, historical or aesthetic reasons.
- *Sensitive Marine Area* (SMA) – in England, a non-statutory area important nationally for its marine animal and plant communities and/or which provides ecological support to adjacent statutory sites. There were 27 SMAs in 1997.
- *Site of Importance for Nature Conservation* (SINC) – an area of local conservation value described in various ways in county and local plans but not notified as a Site of Special Scientific Interest.
- *Site of Special Scientific Interest* (SSSI) – in England, Scotland and Wales, an exceptional or representative area of nature conservation interest (defined broadly to include biological and geological/geomorphological aspects) according to a published set of quality and rarity criteria. SSSI designation (along with the ASSIs in Northern Ireland) applies to about 6500 sites in the UK, where it is the main site protection measure.
- *Special Area of Conservation* (SAC) – a site established

according to the European Directive on the Conservation of Natural Habitats and of Wild Fauna and Flora (the '*Habitats Directive*') as representative of important natural habitat types and as habitats for important species (see also Special Protection Areas, below).

- *Special Protection Area* (SPA) – a site established under the European Directive on the Conservation of Wild Birds (the '*Birds Directive*') to ensure the survival of important bird species. Together with SACs, SPAs form a network of sites termed '*Natura 2000*'.
- *World Heritage Site* – an area nominated by the government as a signatory to the Unesco Convention on the Protection of the World's Most Important Areas of Cultural and Natural Heritage. There are 14 World Heritage Cultural Sites under UK jurisdiction, including Hadrian's Wall, Ironbridge Gorge and Stonehenge, and four natural sites, St Kilda (Scotland), the Giant's Causeway (Northern Ireland), Gough Island (South Atlantic) and Henderson Island (Antarctica).   *JAM*

[See also CONSERVATION, ECOSYSTEM MANAGEMENT, ENVIRONMENTAL PROTECTION, ENVIRONMENTAL MANAGEMENT, HERITAGE, LANDSCAPE MANAGEMENT, NATIONAL PARKS, REGIONALLY IMPORTANT GEOLOGICAL AND GEOMORPHOLOGICAL SITES]

Joint Nature Conservation Committee 1998: *Nature conservation in the UK.* London: HMSO. **McNeely, J.A. and Miller, K.R. (eds)** 1984: *National parks conservation and development.* Washington, DC: Smithsonian Institution Press. **Noss, R.F.** 1987: Protecting natural areas in fragmented landscapes. *Natural Areas Journal* 7, 2–13. **Wright, R.G. and Mattson, D.J.**, 1996: The origin and purpose of National Parks and protected areas. In Wright, R.G. (ed.), *National parks and protected areas: their role in environmental protection.* Cambridge MA: Blackwell Science, 3–14.

Proterozoic

An EON of geological time from 2400 million years ago to the beginning of the CAMBRIAN period. The Proterozoic and ARCHAEAN eons make up PRECAMBRIAN time. The ATMOSPHERE became aerobic during the Proterozoic and the first evidence of metazoan invertebrate organisms is from the late Proterozoic. There may have been several ICE AGES, especially in the latest Proterozoic *(Neoproterozoic)* when all PALAEOLATITUDES appear to have been affected (see SNOWBALL EARTH).   *GO*

[See also BANDED IRON FORMATION, GEOLOGICAL TIMESCALE]

Fairchild, I.J. 1993: Balmy shores and icy wastes: the paradox of carbonates associated with glacial deposits in Neoproterozoic times. In Wright, V.P. (ed.), *Sedimentology Review* 1. Oxford: Blackwell, 1–16. **Kennedy, M.J., Runnegar, B., Prave, A.R.** *et al.* 1998: Two or four Neoproterozoic glaciations? *Geology* 26, 1059–1063. **Kershaw, S.** 1990: Evolution of the Earth's atmosphere and its geological impact. *Geology Today* 6, 55–60. **Schmidt, P.W. and Williams, G.E.** 1995: The Neoproterozoic climatic paradox: equatorial palaeolatitude for Marinoan glaciation near sea level in South Australia. *Earth and Planetary Science Letters* 134, 107–124.

Protista

Single-celled organisms with both plant and animal characteristics, which range in size from *viruses* (30–300 nm) to *protozoa* (500–50 000 nm). They may be classified as *prokaryotes* (no separation of genetic material from the rest of the cell) and *eukaryotes* (with a nucleus).   *JAM*

proton induced X-ray emission (PIXE)

A highly sensitive, non-destructive technique for elemental analysis, also known as *particle induced X-ray emission*, that analyses X-rays emitted when a proton beam is applied to an object. The method is appropriate for 'mapping' element concentrations in samples at the submicron scale and is widely used in GEOARCHAEOLOGY, often in combination with PARTICLE-INDUCED GAMMA-RAY EMISSION.   *GO*

[See also ARCHAEOLOGICAL GEOLOGY, GEOCHEMISTRY]

Montero, M.E., Aspiazu, J., Pajon, J., Miranda, S. and Moreno, E. 2000: PIXE study of Cuban Quaternary paleoclimate geological samples and speleothems. *Applied Radiation and Isotopes* 52, 289–297.

provenance

The nature of the source material or area from which a DETRITAL SEDIMENT, SEDIMENTARY ROCK or archaeological ARTEFACT has been derived. Provenance can be determined by: analysis of the rock types represented in LITHIC fragments in a SANDSTONE or RUDITE; analysing the relative proportions of QUARTZ, FELDSPAR and lithic fragments in a sandstone or rudite; or tracing the distribution of geochemical or mineral components (e.g. HEAVY MINERALS). Provenance studies can yield, for example, important information about the DENUDATION history of continental areas, the TECTONIC setting of a DEPOSITIONAL ENVIRONMENT or CULTURAL CHANGE at archaeological sites.   *GO*

[See also ARCHAEOLOGICAL GEOLOGY, PROVENIENCE]

Kilikoglou, V., Maniatus, Y. and Grimanis, A.P. 1988: The effect of purification and firing of clays on trace element provenance studies. *Archaeometry* 30, 37–46. **Ridgway, K.D., Trop, J.M. and Jones, D.E.** 1999: Petrology and provenance of the Neogene Usibelli Group and Nenana Gravel: implications for the denudation history of the central Alaska Range. *Journal of Sedimentary Research* 69, 1262–1275. **Sircombe, K.N. and Freeman, M.J.** 1999: Provenance of detrital zircons on the Western Australia coastline – implications for the geologic history of the Perth basin and denudation of the Yilgarn craton. *Geology* 27, 879–882. **Valloni, R. and Maynard, J.B.** 1981: Detrital modes of recent deep-sea sands and their relation to tectonic setting: a first approximation. *Sedimentology* 28, 75–83. **Vital, H., Stattegger, K. and Garbe-Schonberg, C.D.** 1999: Composition and trace-element geochemistry of detrital clay and heavy-mineral suites of the lowermost Amazon River: a provenance study. *Journal of Sedimentary Research* 69, 563–575. **Waelkins, M., Herz, N. and Moens, L. (eds),** 1992: *Ancient stones: quarrying, trade and provenance* [*Acta Archaeologica Lovaniensia Monographiae* 4]. Leuven: Leuven University Press.

provenience

The three-dimensional position of an archaeological find within the MATRIX at the time of discovery. *Horizontal provenience* is usually recorded in relation to a geographical grid system; *vertical provenience* is usually related to sea level or, failing this, a local datum.   *JAM*

[See also PROVENANCE]

proximate cause

Assuming a causal chain, each cause is also an effect. The proximate cause is the one immediately preceding or adjacent to the effect being studied.   *CET*

[See also ULTIMATE CAUSE]

proximity principle

In the context of WASTE MANAGEMENT, the principle that waste should be treated close to where it is generated. Waste disposal at distant LANDFILL sites should therefore be avoided. *JAM*

proxy climatic indicator A non-climatic variable or event that, because it is strongly influenced by a climatic variable, can be used to infer values of that climatic variable. In the absence of climatic records, therefore, a long series of a proxy climatic indicator can potentially be used to reconstruct a parallel series for a climatic variable. Such PROXY DATA series allow the extension backwards of INSTRUMENTAL DATA series (which generally extend back only 100–200 years) and provide, not only knowledge of climatic history, but also the long-term climatic series with which both to develop and test more rigorously theories on recent climatic change and the validity of GENERAL CIRCULATION MODELS used in the PREDICTION of FUTURE CLIMATIC CHANGE. Proxy climatic indicators may be divided into biological (grain, tree rings, phenological data) and physical (isotopes, ice formation, floods) types. Proxy data include: written records of phenomena, such as HARVEST RECORDS, phenological series (e.g. the date of appearance of cherry blossom in Japan), flood-level chronologies, LAKE-LEVEL VARIATIONS and the positions, dates of formation and disappearance of SEA ICE; and morphological or sedimentological indicators in the landscape, such as pollen records and glacial moraines. Thus, proxy indicators provide information not only for the HOLOCENE, but over the whole GEOLOGICAL TIMESCALE and are one of the most valuable investigative tools that climatologists have at their disposal. They have done much in the last two decades to further the understanding of CLIMATIC CHANGE beyond the range of INSTRUMENTAL DATA. Although some proxy records allow HIGH-RESOLUTION RECONSTRUCTIONS on an annual basis, others provide a far less detailed picture and can be used only to interpret the broader features of climatic variation. Studies in HISTORICAL CLIMATOLOGY have benefited from these sources where they have corroborated and extended DOCUMENTARY EVIDENCE of environmental change.

Proxy data can be quantitative or qualitative in nature. Many indicators (particularly those forming ANNUALLY RESOLVED RECORDS) are amenable to computerised statistical procedures, which may aid their objective use in generating reliable climatic series. An important step in all studies involving proxy data is to establish the relationship between the proxy parameter and a climatic variable or variables. This usually involves using current data and assuming that current relationships also applied in the past, i.e. the principle of ACTUALISM. The most useful indicators are those that correlate unambiguously with a single climatic parameter. Very often, however, indicators respond to more than one climatic parameter. Thus a lake-level rise may reflect either increased wetness (via increased streamflow) or decreased temperatures (and reduced evapotranspiration), a combination of the two or even a preponderance of one over a contrary change in the other. Interplay with human activities and human response may make inferences even more difficult, particularly if LANDUSE CHANGES are involved. Climatic indicators such as FAMINE and HARVEST RECORDS can be influenced more by non-climatic factors such as war, economic factors and improvements in agricultural technology than by DROUGHT and may therefore be very unreliable indicators of climate unless accompanied by more detailed historical information.

One of the most well developed proxy indicators is TREE RINGS data from the science of DENDROCHRONOLOGY. By this means some aspects of the annual climatic record, rainfall in some areas, temperatures in others, have been extended back to 3000 BC in the southwest United States, where BRISTLECONE PINES (*Pinus aristata*) have a life span measured in millennia. LAKE SEDIMENTS, including VARVE deposits, are another well established means of determining general features of rainfall and RUNOFF with annual and sometimes seasonal RESOLUTION. *Palynology* also provides information on past plant communities as environmental indicators. These techniques are used most commonly to provide information for the period since the most recent retreat of continental ice sheets. For longer time periods, those embracing, for example, the LAST GLACIATION and the preceding INTERGLACIAL, ICE CORE data based on analysis of OXYGEN ISOTOPES provide a detailed proxy temperature record over at least the past 160 000 years. These records, from Arctic and Antarctic sites, have helped to confirm the MILANKOVITCH THEORY of SOLAR FORCING. MARINE SEDIMENT CORES provide similar information, but the time-scale detail is blurred by sediment mixing and only the longer-term climatic changes tend to be discerned, rarely at less than the millennial scale. This disadvantage is, however, offset by the length of the record embraced by these slowly accumulating sea-floor deposits. *RG/DAW*

[See also HOLOCENE ENVIRONMENTAL CHANGE, NATURAL ARCHIVES, NUMERICAL ANALYSIS, PALAEOCLIMATOLOGY, QUATERNARY ENVIRONMENTAL CHANGE]

Birks, H.J.B. 1981: The use of pollen analysis in the reconstruction of past climates: a review. In Wigley, T.M.L., Ingram, M.J. and Farmer, G. (eds), *Climate and history: studies in past climates and their impact on man*. Cambridge: Cambridge University Press, 111–138. **Bradley, R.S. (ed.)** 1991: *Global changes of the past*. Boulder, CO: UCAR/Office for Interdisciplinary Earth Studies. **Duplessy, J-C.** 1978: Isotope studies. In Gribbin, J. (ed.), *Climatic change: studying the climates of the past*. Cambridge: Cambridge University Press. **Glaser, R**. 1996: Data and methods of climatological evaluation in historical climatology. *Historical Social Research* 21, 56–88. **Holzhauser, H. and Zumbühl, H.J.** 1996: The history of the Lower Grindelwald Glacier during the last 2800 years – paleosols, fossil wood and historical pictorial records – new results. *Zeitschrift für Geomorphologie NF* 104, 95–127. **Jouzel, J., Barkov, N.I., Barnola, J.M.** *et al.* 1993: Extending the Vostok ice-core record of palaeoclimate to the penultimate glacial period. *Nature* 364, 407–411. **Stuiver, M., Braziunas, T.F., Becker, B. and Kromer**, B. 1991: Climatic, solar, oceanic and geomagnetic influences on late-glacial and holocene atmospheric $^{14}C/^{12}C$-change. *Quaternary Research* 35, 1–24. **Van Geel, B. Buurman, J. and Waterbolk, H.T.** 1996: Archaeological and palaeological indications of an abrupt climate change in the Netherlands, and evidence for climatological teleconnections around 2650 BP. *Journal of Quaternary Research* 11, 451–460.

proxy data/proxy evidence/proxy records Data, evidence or records relating to an environmental variable that were derived indirectly or reconstructed from NATURAL ARCHIVES (*proxy sources*) rather than observed or measured. *JAM*

[See also PALAEOCLIMATOLOGY, CLIMATIC RECONSTRUCTION, MULTIPROXY APPROACH, PALAEOLIMNOLOGY, PROXY CLIMATIC INDICATOR]

Briffa, K. and Atkinson, T. 1997: Reconstructing Late Glacial and Holocene climates. In Hulme, M. and Jones, P.D. (eds), *Climates of the British Isles: past, present and future*. London: Routledge, 84–111.

pseudokarst Karst-like landforms produced by processes other than *solutional weathering* or solution-induced subsidence and collapse. Examples include CAVES in glaciers, which are caused by a change in phase, not dissolution, and *vulcanokarst*, which comprises caves (LAVA TUBES) within lava flows. Karst-like features in rocks of low solubility (e.g. quartzites) have in the past also been termed pseudokarst. It is, however, now commonly accepted that these features should be termed KARST, provided they are of mainly solutional origin.   *SHD*

[See also BIOKARST, CHEMICAL WEATHERING]

Doerr, S.H. 1999: Karst-like landforms and hydrology in quartzites of the Venezuelan Guyana shield: pseudokarst or 'real' karst? *Zeitschrift für Geomorphologie* 43, 1–17. **Kumpulainen, R.A.** 1997: Subsurface sediments tell about the hydrology of a volcano. *Geologiska Föreningens i Stockholm Förhandlingar* 119, 135–139. **Wray, R.A.L.** 1997: A global review of solutional weathering forms on quartz sandstones. *Earth Science Reviews* 42, 137–160.

pseudomorph The replacement of one MINERAL by another or a CAST formed by SEDIMENT replacement of a mineral, in both cases retaining the morphology of the original mineral. EVAPORITE minerals are readily dissolved and replaced during DIAGENESIS and such *halite pseudomorphs* (halite = NaCl) and *gypsum pseudomorphs* (gypsum = CaSO$_4$.2H$_2$O) are important indicators of ARIDITY in the GEOLOGICAL RECORD.   *GO*

Astin, T.R. and Rogers, D.A. 1991: 'Subaqueous shrinkage cracks' in the Devonian of Scotland reinterpreted. *Journal of Sedimentary Petrology* 61, 850–859. **Buck, B.J. and Mack, G.H.** 1995: Latest Cretaceous (Maastrichtian) aridity indicated by paleosols in the McRae Formation, south-central New Mexico. *Cretaceous Research* 16, 559–572.

pull-apart basin A SEDIMENTARY BASIN that develops along a FAULT with strike-slip (shearing) movement, commonly a TRANSFORM FAULT or CONSERVATIVE PLATE MARGIN. Bends in such faults give rise to localised areas of compression (*transpression*) and stretching (*transtension*) of the CRUST. Pull-apart basins develop as a result of SUBSIDENCE at sites of transtension, and may be supplied by rapid UPLIFT and EROSION in adjacent areas of transpression. They are characterised by rectangular shapes, small dimensions (kilometres to tens of kilometres) and rapid subsidence and infilling (several millimetres per year). Present-day examples include the Dead Sea basin and onshore and offshore basins along the San Andreas Fault in California.   *GO*

Ballance, P.F. and Reading, H.G. (eds) 1980: *Sedimentation in oblique-slip mobile zones* [*International Association of Sediment-*ologists Special Publication 4]. Oxford: Blackwell Scientific. **Crowell, J.C.** 1974: Origin of late Cenozoic basins in southern California. In Dickinson, W.R. (ed.), *Tectonics and sedimentation* [*Special Publication* 22]. Tulsa, OK: Society of Economic Paleontologists and Mineralogists, 190–204. **Holdsworth, R.E., Strachan, R.A. and Dewey, J.F. (eds)** 1998: *Continental transpressional tectonics and transtensional tectonics* [*Special Publication* 135]. London: Geological Society. **Woodcock, N.H.** 1986: The role of strike-slip fault systems at plate boundaries. *Philosophical Transactions of the Royal Society, London* A317, 13–29.

pumice VESICLE-rich VOLCANIC GLASS of high-silica composition, with up to 90% porosity and density < 1.0 g cm^{-3}. Pumice fragments float on water and hence may be transported very long distances from their source, occasionally to be utilised as tools that can later be recovered from ARCHAEOLOGICAL sites. Pumice fragments may break to produce angular glass *shards*.   *JBH*

[See also SCORIA, WELDED TUFF]

punctuated equilibria EVOLUTION seen as patterns of SPECIATION in time: periods of STASIS are 'punctuated by episodic events of allopatric speciation' (Eldredge and Gould, 1972: 96).   *KDB*

[See also ISOLATION, PHYLETIC GRADUALISM]

Eldredge, N. and Gould, S. J. 1972: Punctuated equilibria: an alternative to phyletic gradualism. In Schopf, T.J.M. (ed.), *Models in paleobiology*. San Francisco, CA: Freeman, Cooper, 82–115.

pyroclastic material Fragmentary material produced by VOLCANISM (see LAVA, VOLATILES), mostly in explosive VOLCANIC ERUPTIONS, but also by processes such as the collapse of a LAVA DOME. *Pyroclastic products* are divided by GRAIN-SIZE into ash, LAPILLI and blocks (see Table). *Pyroclastic deposits* can be emplaced by PYROCLASTIC FALL or *pyroclastic density flows*, which include PYROCLASTIC FLOW and PYROCLASTIC SURGE processes. They share many of the characteristics of SEDIMENTARY ROCKS. The term is derived from the Greek 'pur' meaning fire and 'klastos' meaning rock.   *JBH*

[See also SCORIA, TEPHRA, TUFF, VOLCANIC ASH, VOLCANIC BLOCK, VOLCANIC BOMB, VOLCANICLASTIC, WELDED TUFF]

Allen, S.R. and Cas, R.A.F. 1998: Rhyolitic fallout and pyroclastic density current deposits from a phreatoplinian eruption in the eastern Aegean Sea, Greece. *Journal of Volcanology and Geothermal Research* 86, 219–251. **Carey, S., Sigurdsson, H., Mandeville, C. and Bronto, S.** 1996: Pyroclastic flows and surges over water: An example from the 1883 Krakatau eruption. *Bulletin of Volcanology* 57, 493–511. **Cas, R.A.F. and Wright, J.V.** 1987: *Volcanic successions: modern and ancient*. London: Allen

pyroclastic material *Classification of pyroclastic fragments and deposits by grain-size (after Fisher, 1961; Orton, 1996)*

Grain-size	Pyroclastic fragments	Pyroclastic deposits	
		Unconsolidated tephra	*Consolidated pyroclastic rock*
> 64 mm	Bomb (fluidally rounded shape) or block (angular)	Block or bomb tephra; agglomerate (bombs present)	Pyroclastic breccia or agglomerate (bombs present)
2–64 mm	Lapilli	Lapilli tephra	Lapillistone (or lapilli tuff or tuff-breccia)
0.063–2 mm	Coarse ash	Coarse ash	Coarse tuff
< 0.063 mm	Fine ash	Fine ash	Fine tuff

and Unwin. **Fisher, R.V.** 1961: Proposed classification of volcaniclastic sediments and rocks. *Geological Society of America Bulletin* 72, 1395–1408. **Fisher, R.V. and Schmincke, H.U.** 1984: *Pyroclastic rocks*. Berlin: Springer. **Orton, G.J.** 1996: Volcanic environments. In Reading, H.G. (ed.), *Sedimentary environments: processes, facies and stratigraphy*, 3rd edn. Oxford: Blackwell Science, 485–567. **Walker, G.P.L.** 1971: Grain-size characteristics of pyroclastic deposits. *Journal of Geology* 79, 696–714. **Wright, J.V., Smith, A.L. and Self, S.** 1980: A working terminology of pyroclastic deposits. *Journal of Volcanology and Geothermal Research* 8, 315–336.

pyroclastic fall PYROCLASTIC MATERIAL that settles under gravity from a VOLCANIC PLUME or from fine particles separating from a PYROCLASTIC FLOW after an explosive VOLCANIC ERUPTION, forming a BED of TEPHRA or TUFF. The geometry, thickness and extent of pyroclastic fall deposits depends on GRAIN-SIZE, plume height, and wind velocity and direction. *GO*

[See also AIRFALL, ASH FALL]

pyroclastic flow A high-concentration dispersion of pyroclastic fragments and hot gases (VOLATILES) that flows as a low-turbulence, gravity-controlled DENSITY FLOW. Flows may be generated by the collapse of a VOLCANIC PLUME or a LAVA DOME (see NUÉE ARDENTE). Temperatures in pyroclastic flows are in the range 300–800°C, velocities have been estimated at up to 100 m s^{-1} and studies of the deposits of some QUATERNARY flows have shown that they travelled distances of more than 100 km. Pyroclastic flows are potentially extremely destructive and have been associated with most major explosive VOLCANIC ERUPTIONS, including Vesuvius in AD 79 and Mount St Helens in AD 1980. Pyroclastic flows are topographically controlled and their deposits tend to infill topographic depressions. They are poorly sorted, with few SEDIMENTARY STRUCTURES, and may be preserved as WELDED TUFFS. An extensive, PUMICE-rich pyroclastic flow deposit is known as an IGNIMBRITE. *JBH*

[See also PYROCLASTIC FALL, PYROCLASTIC MATERIAL, PYROCLASTIC SURGE]

Chester, D. 1993: *Volcanoes and society.* London: Edward Arnold. **Lipman, P.W. and Mullineaux, D.R. (eds)**, 1981: *The 1980 eruptions of Mount St Helens, Washington* [*Professional Paper* 1250]. Reston, VA: United States Geological Survey. **Orton, G.J.** 1996: Volcanic environments. In Reading, H.G. (ed.), *Sedimentary environments: processes, facies and stratigraphy*, 3rd edn. Oxford: Blackwell Science, 485–567.

pyroclastic surge A low-concentration dispersion of pyroclastic fragments and hot gases that flows as a turbulent DENSITY FLOW. They are hot, gas-rich, fast-moving and do not always follow topographic depressions; hence they can be very dangerous. *Base surges* are particularly associated with HYDROVOLCANIC ERUPTIONS. *Ground surges* and *ash cloud surges* are associated with VOLCANIC PLUME collapse and with PYROCLASTIC FLOWS. Pyroclastic surge deposits tend to mantle topography, although they are thickest in topographic depressions. They may contain well developed SEDIMENTARY STRUCTURES such as CROSS-STRATIFICATION. *GO*

[See also PYROCLASTIC FALL, PYROCLASTIC MATERIAL, WELDED TUFF]

Cas, R.A.F. and Wright, J.V. 1987: *Volcanic successions: modern and ancient.* London: Allen and Unwin. **Colella, A. and Hiscott, R.N.** 1997: Pyroclastic surges of the Pleistocene Monte Guardia sequence (Lipari Island, Italy): depositional processes. *Sedimentology* 44, 47–66. **Fujii, T. and Nakada, S.** 1999: The 15 September 1991 pyroclastic flows at Unzen Volcano (Japan), a flow model for associated ash-cloud surges. *Journal of Volcanology and Geothermal Research* 89, 159–172.

pyrotechnology The intentional, controlled use of fire by humans ranging from its early use in hunting and FOREST CLEARANCE to the advanced technologies of the INDUSTRIAL REVOLUTION and since. *JAM*

Head, L. 1996: Rethinking the prehistory of hunter-gatherers, fire and vegetation change in northern Australia. *The Holocene* 6, 481–487. **Perlès, C.** 1977: *Préhistoire du feu.* Paris: Masson. **Renfrew, C. and Bahn, P.** 1996: *Archaeology: theories, methods and practice*, 2nd edn. London: Thames and Hudson.

Q

qanat Especially in Iran and neighbouring regions of the Middle East, an ancient means of extracting GROUNDWATER. A qanat, *karez* or *fogarra* comprises a gently sloping tunnel (a subterranean AQUEDUCT) that conducts water from an infiltration zone beneath the WATER TABLE to the ground surface by gravity flow. Qanats are commonly located on large alluvial fans in foothill regions where annual rainfall is low (100–300 mm). Most qanats are 1–5 km long, but exceptionally are over 50 km in length.
RAS

Beaumont, P., Blake, G.H. and Wagstaff, J.M. 1988: *The Middle East. A geographical study*, 2nd edn. London: David Fulton, 93–96. **Lightfoot, D.R.** 2000: The origin and diffusion of Qanats in Arabia: new evidence from the northern and southern peninsula. *Geographical Journal* 166, 215–226.

qoz An Arabic term for an area of SAND DUNES and SAND SHEETS. It is commonly applied to the LONGITUDINAL DUNE system west of the Nile. The qoz areas are thought to have formed mainly during the earlier drier phases of the Pleistocene.
MAC

Mitchell, C.W. 1990: Physiography, geology, and soils. In Craig, G.M. (ed.), *The agriculture of the Sudan*. Oxford: Oxford University Press.

quad-tree A hierarchical data structure in which geographical space is recursively subdivided into quarters until all quadrants indicate ATTRIBUTE homogeneity, or until a predetermined cut-off depth is reached. *TVM*

[See also GEOGRAPHICAL INFORMATION SYSTEM]

quarrying (1) The open surface excavation of rocks and sediments by humans (e.g. slate, granite, sand and gravel). (2) A process of GLACIAL EROSION by which rock fragments greater than about 1 cm are removed from bedrock, involving (a) FAILURE of the rock, leading to the fragments being loosened from the bed, and (b) *evacuation*, whereby fragments are moved from their original position. *Plucking* is an alternative term for the process. (3) In the context of FLUVIAL PROCESSES, the term quarrying, or plucking, is used for erosion by the *hydraulic action* of water alone, requiring the loosening of blocks and their subsequent entrainment. This may include the process of *cavitation*, which involves the effects of shock waves generated through the collapse of the 'vapour pockets' or 'airless bubbles' that form in a fast, turbulent flow with marked pressure changes. *RAS/JAM*

[See also ABRASION, CORRASION]

Barnes, H.L. 1956: Cavitation as a geological agent. *American Journal of Science* 254, 493–505. **Benn, D.I. and Evans, D.J.A.** 1998: *Glaciers and glaciation*. London: Arnold, 188–192. **Whipple, K.X., Hanwell, G.S. and Anderson, R.S.** 2000: River incision into bedrock: mechanics and relative efficiency of plucking, abrasion and cavitation. *Geological Society of America Bulletin* 112, 490–503.

quartz A SILICATE MINERAL of composition SiO_2 *(silica)*. Because of its resistance to WEATHERING, quartz is common in SEDIMENTARY ROCKS, particularly SANDSTONE.
GO

[See also PHYTOLITH]

quartz grain surface textures High-resolution magnification made possible with SCANNING ELECTRON MICROSCOPY has shown that sand grains (especially quartz) can retain surface markings or textures diagnostic of depositional history. GLACIAL, AEOLIAN and subaqueous (FLUVIAL and littoral) PROCESSES can give rise to distinctive suites of mechanically formed textures. The evidence of more than one depositional environment may be identifiable in a sample of grains or even on a single sand grain. Analysis typically involves noting, for each of a sample of grains, the presence/absence of recognised features from a standard list. CHEMICAL WEATHERING may render any mechanically produced textures unidentifiable. *RAS*

[See also GLACIAL DEPOSITION]

Culver, S.J., Bull, P.A., Campbell, S. *et al.* 1983: A statistical investigation of operator variance in quartz grain surface texture studies. *Sedimentology* 30, 129–136. **Frihy, O.E. and Stanley, D.J.** 1987: Quartz grain surface textures and depositional interpretation, Nile Delta region, Egypt. *Marine Geology* 77, 247–255. **Tengberg, A.** 1995: Sediment sources of nebkhas in the Sahel zone, Burkino-Faso. *Physical Geography* 16, 259–275.

quartzite A SANDSTONE *(orthoquartzite)* or its derivative METAMORPHIC ROCK *(metaquartzite)* composed almost entirely of QUARTZ. A sandstone with at least 95% quartz CLASTS and little or no MATRIX, but without a quartz CEMENT, is a *quartz arenite:* therefore an orthoquartzite is a quartz arenite with a quartz cement. *TY*

Quasi-biennial Oscillation (QBO)
A well-defined oscillation of the zonal wind component of the equatorial STRATOSPHERE. The PERIODICITY of the phenomenon is about 27 months and the reversal begins at high levels, taking approximately 12 months to descend from 30 to 18 km (10–60 mbar). Recent observations of the QBO indicate that the easterlies are stronger than the westerlies, that the time between the easterly and westerly maxima is much shorter than the reverse and that there is considerable variability in the periodicity of the oscillation. For some low-latitude locations climatic variables such as rainfall and temperature demonstrate temporal variations that resemble those of the QBO. There is also evidence that the QBO may modulate other elements of tropical climate, such as EL NIÑO–SOUTHERN OSCILLATION and TROPICAL CYCLONES. For instance, the east phase inhibits hurricane formation, whereas the west phase is associated with 200% more major hurricanes. *GS*

[See also TELECONNECTIONS]

Gray, W.M. and Schaeffer, J. 1991: El Niño and QBO influences on tropical cyclone activity. In Glantz, M.H., Katz, R.W.

and Nicholls, N. (eds), *Teleconnections linking worldwide climate anomalies*. Cambridge: Cambridge University Press. **Gray, W.M., Schaeffer, J. G. and Knaff, J.A.** 1992: Influence of stratospheric QBO on ENSO variability. *Journal of the Meteorological Society of Japan* 70, 975–995. **Labitzke, K.** 1987: Sunspots, the QBO, and the stratospheric temperature of the North Polar Region. *Geophysical Research Letters* 14, 535–537.

quasi-periodic phenomena Environmental variables that display approximate cyclical patterns or PERIODICITIES in time series. For instance, an oscillation of about two years has been demonstrated in a number of climatic elements (e.g. in the CENTRAL ENGLAND TEMPERATURE RECORD) and this is believed to be related to the QUASI-BIENNIAL OSCILLATION. *GS*

[See also EL NIÑO–SOUTHERN OSCILLATION, NORTH ATLANTIC OSCILLATION]

Burroughs, W.J. 1992: *Weather cycles: real or imaginary?* Cambridge: Cambridge University Press.

Quaternary The most recent PERIOD of the GEOLOGICAL TIMESCALE, continuing to the present day and comprising the PLEISTOCENE and HOLOCENE epochs. There is currently some debate concerning the date for the start of the Quaternary period. Originally placed arbitrarily at 1 million years ago (Ma), it has been dated at a type section in southern Italy at 1.64 Ma. Some workers have suggested a revision of this date to between 1.8 and 1.9 Ma, while others prefer a different definition for the start of the Quaternary, at around 2.5 or 2.6 Ma. The latter date reflects an intensification of Northern Hemisphere GLACIATION in Eurasia and North America. *GO*

[See also GEOLOGICAL TIMESCALE, PLIOCENE–PLEISTOCENE TRANSITION, QUATERNARY TIMESCALE]

Harland, W.B., Armstrong, R.L., Cox, A.V. *et al.* 1990: *A geologic time scale 1989*. Cambridge: Cambridge University Press. **Lowe, J.J. and Walker, M.J.C.** 1997: *Reconstructing Quaternary environments*, 2nd edn. Harlow: Addison-Wesley Longman. **Williams, M., Dunkerley, D., de Deckker, P.** *et al.* 1998: *Quaternary environments*, 2nd edn. London: Arnold.

Quaternary environmental change The Quaternary Era, which covers about the last two million years, has been a time of dramatic environmental change, with the Earth's climate lurching between GLACIAL EPISODES and INTERGLACIALS. During glacials, ICE SHEETS expanded to cover large parts of the Northern Hemisphere, global sea levels fell and tropical latitudes became colder and drier. During interglacials, the ice shrank back to high latitudes and altitudes, sea levels rose and the tropics became warmer and wetter. This CYCLICITY of CLIMATIC CHANGE, recognised first in the OXYGEN ISOTOPE signal recovered from MARINE SEDIMENT CORES and later confirmed using other proxies including LOESS deposits and ICE CORES, suggests that the driving force for the GLACIAL–INTERGLACIAL CYCLES is small fluctuations in the Earth's orbit. The resulting small fluctuations in the amount of energy received by each hemisphere, with cycles of about 100 ka, 41 ka and 23 ka, have pertained throughout Earth's history, so some critical THRESHOLD must have been crossed for ORBITAL FORCING of climatic changes with the amplitude experienced during the Quaternary. A possible cause was the closure of the Isthmus of Panama once the continents had reached their present positions (see CONTINEN-

TAL DRIFT). This triggered the diversion of warm waters into the North Atlantic and so large transfers of heat into high latitudes. The warm water supplied the precipitation needed to initiate the build-up of ice sheets. A small decline in energy received could have been amplified by the high ALBEDO of the growing ice sheets, so that larger proportions of SOLAR RADIATION were reflected away, leading to further cooling. Warming and DEGLACIATION are not so easy to explain, since it is more difficult to amplify the small increase in energy received when large areas are already covered in ice. One possibility is that increased insolation delivered large amounts of MELTWATER and ICE-BERGS into the North Atlantic, severing the supply of warm water from the south and leading to reduced evaporation and starvation of the ice sheets.

The most complete records of Quaternary environmental change come from studies of deep-ocean cores, and particularly the OXYGEN ISOTOPE RATIOS of foraminifera, which are related to ocean volume and temperature. These have been used to identify more than 100 oxygen ISOTOPIC STAGES (OIS), with even numbers representing cold intervals, odd numbers warm intervals, and letters indicating smaller variations. The present interglacial is OI Stage 1, the Last Glacial Maximum Stage 2, and the last full interglacial Stage 5e. These stages have been correlated with similar fluctuations in climate reflected in loess sequences and ice cores, allowing the marine record for at least part of the Quaternary to be correlated with events on the continents.

Although these long PROXY RECORDS provide a clear framework of climatic changes during the Quaternary and give an indication of changes in the volume of ice on land, the location of the ice, the patterns of ice sheet growth and decay and the influence of climatic changes on ECOSYSTEMS are more difficult to establish. A variety of approaches have been used to reconstruct ice sheets during the LAST GLACIAL MAXIMUM, including field mapping, analysis of remotely sensed images and computer modelling. Long lake cores have provided pollen and other proxy records of changes in vegetation over long timescales, but with much higher temporal RESOLUTION than can be obtained from slowly accumulating ocean sediments. Sediments that accumulated on the continents during the last glacial–interglacial cycle are widespread and a very wide range of proxies has been applied to study climatic and environmental changes at very high resolution.

Quaternary environmental change may have been one of the driving forces in the evolution of HOMINIDS and eventually modern humans. MIGRATIONS out of Africa are thought to have occurred twice. *Homo erectus* (or possibly a predecessor) migrated early in the Pleistocene, evolving in Europe into the Neanderthals. It is generally accepted that anatomically modern humans evolved in Africa and migrated in the Late Pleistocene, reaching Europe by about 40 ka, though there are still some proponents of the multiregional hypothesis, suggesting genetic continuity with earlier hominids within regions. North America was probably first occupied by MAMMOTH hunters crossing the Bering Strait, perhaps as early as 30 ka. Australia may have been occupied earlier than Europe (50–60 ka).

The Late Pleistocene was also a time of MASS EXTINCTION, particularly of large animals such as the mammoth, the woolly rhinoceros, the sabre-tooth tiger, the giant sloth and large flightless birds. There is continuing debate over the role played by hunting, natural climatic change and human-induced vegetation changes in causing the extinctions. The fact that the MEGAFAUNA survived the many large swings in climate of the Pleistocene, only to succumb to the last when humans were actively hunting them, does suggest some degree of culpability. Human-induced extinctions and large-scale alterations of natural environments are not new, but the pace of change has accelerated enormously over the last century, coupled with changes in the chemistry of the atmosphere, which may be driving changes in global climate. Thus, the study of Quaternary environmental change may help to inform us of the likely consequences of our actions.   *DMcC*

[See also DATING TECHNIQUES, ENVIRONMENTAL CHANGE, HOLOCENE ENVIRONMENTAL CHANGE, OVERKILL HYPOTHESIS, QUATERNARY TIMESCALE]

Imbrie, J. and Imbrie, K.P. 1979: *Ice ages: solving the mystery.* London: Macmillan. **Lowe, J. and Walker, M.J.C.** 1997: *Reconstructing Quaternary environments*, 2nd edn. Harlow: Longman. **Shackleton, N.J. and Opdyke, N.D.** 1973: Oxygen isotope and palaeomagnetic stratigraphy of equatorial Pacific core V28–238: oxygen isotope temperatures and ice volumes on a 10^5 and a 10^6 scale. *Quaternary Research* 3, 39–55. **Stringer, C. and McKie, R.** 1996: *African exodus. The origins of modern humanity.* London: Pimlico. **Williams, M., Dunkerley, D., De Decker, P.** *et al.* 1998: *Quaternary environments.* London: Arnold.

Quaternary timescale

The Quaternary is the most recent major subdivision, or PERIOD, of the geological record and includes the present day. It is part of the CENOZOIC, the fourth geological era, which also includes the TERTIARY. There is some controversy about the exact timing of the beginning of the Quaternary at the PLIOCENE–PLEISTOCENE TRANSITION, with some workers favouring a 'long' Quaternary and others a much shorter period. Advocates of the shorter timescale use the Pliocene–Pleistocene boundary accepted by the International Commission on Stratigraphy in 1985 and the International Union of Geological Sciences (IUGS). It is based upon the first appearance of the cold-water marine OSTRACOD *Cytheropteron testudo* in a reference site at Vrica in Italy. This has been dated to c. 1.64 million years ago (Ma) on the basis of palaeomagnetic evidence, just below the end of the OLDUVAI EVENT (K-Ar dated to 1.87–1.67 Ma), although revisions in dating this event have led to earlier suggested dates of between 1.88 and 1.81 Ma. Workers who favour a longer timescale (see Table) argue for the boundary to be placed where there are the first global indications of cooling, such as the build-up of continental GLACIATION as defined by ICE-RAFTED DEBRIS (IRD) in MARINE SEDIMENTS and LOESS deposition. Thus, some place the boundary at around 2.4 to 2.6 Ma, close to the GAUSS–MATUYAMA BOUNDARY (c. 2.58 Ma).

The Quaternary is divided into two EPOCHS, the PLEISTOCENE, which ended at c. 10000 radiocarbon years BP (about 11500 years ago), and the HOLOCENE, which is the present warm stage. The Pleistocene may be further subdivided into: the Lower Pleistocene, 1.8–0.75 Ma; the Middle Pleistocene, 750–125 ka; and the Upper Pleistocene, 125–10 ka. These boundaries are arbitrary to a certain extent and are not sacrosanct. The Quaternary has revealed a high frequency of climatic oscillations and intensely cold phases. The exact number of GLACIAL–INTERGLACIAL CYCLES occurring during the Quaternary remains to be established, but the Earth may have experienced 30–50 GLACIAL EPISODES and the same number of INTERGLACIAL phases, depending upon the age assigned to the Pliocene–Pleistocene boundary.

Division of the Quaternary timescale is predominantly founded on CLIMATOSTRATIGRAPHIC units, distinguishing between cold glacials or STADIALS and warm interglacials or INTERSTADIALS. In areas away from glaciation, for example China, where there are extensive LOESS records, these climatostratigraphic units are defined in terms of PALAEOSOLS and LOESS STRATIGRAPHY. With increasing age, it has proved extremely difficult to differentiate individual stadials or interstadials because of their short duration, although by comparing ICE CORE records from Greenland and high-resolution records from MARINE SEDIMENTS it has been possible to subdivide the last 100 000 years into 24 different stages, mostly unnamed but given individual numbers (see ISOTOPIC STAGES).

Although the Quaternary timescale can be applied on a global basis, different countries and regions tend to use different terms or names for particular CHRONOSTRATIGRAPHIC units, thus the LAST INTERGLACIAL is known in the UK as the *Ipswichian*, in Ireland as the *Fenitan*, in Europe as the *Eemian* and in North America as the *Sangamon*. Because of the problems this causes for CORRELATION, workers now tend to relate stratigraphies to the oxygen isotope (OI) stages derived from the record found in MARINE SEDIMENT CORES, as it is globally applicable. Hence the Last Interglacial is known as OI Stage 5e.

Assigning ages to the Quaternary timescale has been achieved using PALAEOMAGNETIC DATING, whereby dates for distinctive horizons, such as GEOMAGNETIC POLARITY REVERSALS derived from K-Ar dating of associated volcanic rocks, enabled a broad temporal framework to be obtained. Ages between such MARKER HORIZONS were then estimated by extrapolation. Marine sediments, which also have a record of the palaeomagnetic timescale, then provided more detail with the development of a stratigraphy based on changes in the oxygen isotope record in FORAMINIFERA. This record has since been further refined by ORBITAL TUNING to provide a universal timescale accepted by all workers (see SPECTRAL MAPPING PROJECT TIMESCALE). Although ages still have to be extrapolated between stage boundaries or marker horizons and further refinement is still being undertaken, the level of PRECISION now available is acceptable considering the length of the Quaternary as a whole.

The Table sets out the current Quaternary timescale according to a number of sources, set against the oxygen isotope stratigraphy and the palaeomagnetic timescale. It should be emphasised that, although the general sequence and terminology shown below is widely accepted, there will still be workers who would argue for variations – the debate between the 'long' and 'short' Quaternary timescales for instance.   *CJC*

Bowen, D.Q. (ed.) 1999: *A revised correlation of Quaternary deposits in the British Isles* [*Special Report* 23]. Bath: The Geological Society. **Johnsen, J., Clausen, H.B., Dansgaard, W.** *et al.* 1992: Irregular glacial interstadials recorded in a new Greenland ice core. *Nature* 359, 311–313. **Lowe, J.J. and Walker, M.J.C.** 1997: *Reconstructing Quaternary environments*, 2nd edn. Harlow: Longman. **Prentice, A.J. and Kroon, D.**

Quaternary timescale *The Quaternary timescale (left) in millions of years before present (Ma) based on orbital tuning of the marine oxygen isotope stages (even-numbered stages are relatively cold; odd numbers denote warm stages). The palaeomagnetic timescale is represented by the three chrons that occur in the Quaternary. The remaining columns show selected regional subdivisions based on terrestrial stratigraphic records (after Šibrava, 1986; Lowe and Walker, 1997)*

Timescale Ma. BP	Marine oxygen isotope stages	Palaeo-magnetic chrons	Northern Europe	The Netherlands		British Isles	European Russia	Northern Alps		North America		Cold or temperate
0.01	1	Brunhes	Holocene	Holocene		Flandrian	Holocene	Holocene		Holocene		T
0.08	2.4d		Weichselian	Weichselian		Devensian	Devensian	Würm		Wisconsian		C
0.13	5e		Eemian	Eemian		Ipswichian	Mikulino	Riss-Würm		Sangamon		T
0.19	6		Warthe			"Wolstonian"	Moscow	Penultimate Glacial Late Riss?	Dneipr Glaciation		Late	C
0.25	7		Saale / Drenthe				Odintsovo			Illinoian		T
0.30	8		Drenthe				Dneipr	Antepenultimate Glac. Early Riss / Mindel?			Early	C
0.34	9		Domnitz (Wacken)		Holsteinian Interglacial		Romny					T
0.35	10		Fuhne (Mehleck)			Hoxnian	Pronya	Pre-Riss?		Pre-Illinoian A		C
0.43	11		Holsteinian (Muldsberg)				Lichvin					T
0.48	12		Elster 1			Anglian		Late Mindel?/ Donau		B		C
0.51	13		Elster 1/2	Elster			Oka					T
0.56	14		Elster 1			Cromerian		Early Mindel?/ Donau		C		C
0.63	15		Cromerian IV	Cromerian IV (Noordbergum)								T
0.69	16		Glacial C	Glacial C						D		C
0.72	17		Interglacial III	Interglacial III (Rosmalen)								T
0.78	18		Glacial B	Glacial B						E		C
0.79	19		Interglacial II	Interglacial II (Westerhoven)								T
	20	Matuyama	Helme (Glacial A)	Glacial A				Early Gunz?		F		C
	21		Astern Interglacial I	Interglacial I (Waardenburg)								T
0.90	22				Dorst					G		C
				Bavelian	Leerdam							T
0.97					Linge							C
					Bavel							T
				Menapian								T/C
				Waalian								T
												C
				Eburonian						H		T/C
												T
1.65	6.0					Beestonian				I		C
				Tiglian	C5-6	Pastonian						T
					C-4c	Pre-Pastonian/ Baventian						C
					Cl-4b	Bramertonian/ Antian						T
					B	Thurnian				J		C
	103				A	Ludhamian						T
	104			Praetiglian	Pre-Ludhamian							C
2.60		Gauss		Pliocene	Pliocene							

1991: Oxygen isotope chronostratigraphy. In Smart, P.L. and Frances, P.D. (eds), *Quaternary dating methods – a users guide* [*QRA Technical Guide* 4]. Cambridge: Quaternary Research Association, 199–228. **Shackleton, N., Berger, A. and Peltier, W.R.** 1990: An alternative astronomical calibration of the lower Pleistocene timescale based upon ODP Site 677. *Transactions of the Royal Society of Edinburgh: Earth Sciences* 81, 251–261. **Shackleton, N.J. and Opdyke, N.D.** 1973: Oxygen isotope and palaeomagnetic stratigraphy of equatorial Pacific core V28–238: oxygen isotope temperatures and ice volumes on a 10^5 year and 10^6 year scale. *Quaternary Research* 3, 39–55. **Šibrava, V.** 1986: Correlation of European glaciations and their relation to the deep-sea record. *Quaternary Science Reviews* 5, 433–442. **Thompson, R.** 1991: Palaeomagnetic dating. In Smart, P.L. and Frances, P.D. (eds), *Quaternary dating methods – a users guide* [*QRA Technical Guide* 4]. Cambridge: Quaternary Research Association, 177–198. **Valet, J.-P. and Meynadier, L.** 1993: Geomagnetic field intensity and reversals during the past four million years. *Nature* 366, 234–238. **Williams, M.A.J., Dunkerley, D.L., De Deckker, P. et al.** 1998: *Quaternary environments*, 2nd edn. London: Arnold. **Wilson, R.C.L., Drury, S.A. and Chapman, J.L.** 2000: *The great ice age: climate change and life*. London: Routledge.

query language A software interface for defining, accessing and interrogating data in a DATABASE. SQL (*structured query language*) is a standard interface for a RELATIONAL DATABASE. *TVM*

Lee, Y.C. and Chin, F.L. 1995: An iconic query language for topological relationships in geographical information systems. *International Journal of Geographical Information Systems* 9, 25–46. **Van der Lans, R**. 1999: *Introduction to SQL*. Harlow: Longman.

quickflow That part of PRECIPITATION that takes a rapid route to a stream channel constituting the FLOOD peak of a stream. It may be provided by a number of RUNOFF PROCESSES, notably rapid THROUGHFLOW, PIPEFLOW and HORTONIAN OVERLAND FLOW. *ADT*

[See also BASEFLOW, STORM HYDROGRAPH]

Jones, J.A.A. 1997: *Global hydrology: processes, resources and environmental management*. Harlow: Addison-Wesley Longman. **Ward, R.C. and Robinson, M.** 2000: *Principles of hydrology*, 4th edn. London: McGraw-Hill.

quota A limit on the quantity of a commodity that can be produced, exported or imported. The term is increasingly used for environmentally sensitive 'commodities', such as those associated with trade in animal species and in the 'export' of POLLUTANTS. *JAM*

[See also TRADEABLE EMISSIONS PERMITS]

R

r-selection   An evolutionary 'strategy' that maximises the intrinsic rate of increase (*r*) of an organism, allowing rapid and opportunistic colonisation of disturbed habitats.

LRW

[See also ECOLOGICAL SUCCESSION, RUDERAL, *K*-SELECTION]

Begon, M. and Mortimer, M. 1986: *Population ecology: a unified study of animals and plants*, 2nd edn. Oxford: Blackwell Scientific.

R-value, in pollen calibration   A correction factor used in PALAEOECOLOGY for FOSSIL pollen abundance data in order to reflect actual VEGETATION composition rather than pollen composition. *R*-values correct for the differences among taxa in pollen productivity, pollen dispersal, and pollen preservation.

JLF

[See also DETERIORATED POLLEN, PALAEOBOTANY, POLLEN ANALYSIS, POLLEN CONCENTRATION]

Andersen, S.T. 1973: The differential pollen productivity of trees and its significance for the interpretation of a pollen diagram from a forested region. In Birks, H.J.B and West, R.G. (eds), *Quaternary plant ecology*. Oxford: Blackwell, 109–115. **Birks, H.J.B and Birks, H.H.** 1980: *Quaternary palaeoecology*. London: Edward Arnold. **Moore, P.D., Webb, J.A. and Collinson, M.D.** 1991: *Pollen analysis*. Oxford: Blackwell.

radar   A technique for measuring the distance from a sensor to a target by using radio waves. Radars are active microwave instruments in which electromagnetic radiation (in the approximate wavelength range 1 to 100 cm) is transmitted and the waves scattered from an object are received (detected). The distance to the scattering object can be inferred from the delay time between transmission and reception. In addition to this, relative motion between the target and receiver along the line of sight can be inferred from the frequency shift of the reflected waves using the DOPPLER EFFECT. The principles of the radar (*radio detection and ranging*) technique were first developed in the early twentieth century and advanced as a result of concentrated effort during World War II. They now encompass a range of instruments including altimeters, scatterometers and synthetic aperture radars.

PJS

[See also MICROWAVE REMOTE SENSING, RADAR REMOTE SENSING INSTRUMENTS]

radar remote sensing instruments   Any of a range of REMOTE SENSING instruments based on target detection and ranging using radio waves or microwaves. There are three generic types of space-borne RADAR instruments: altimeters, scatterometers and imaging radars (e.g. *synthetic aperture radar* or SAR). *Altimeters* are nadir-(downward-) looking radars that are used to derive elevation profiles under the orbit track. These profiles are constructed by making accurate measurements of the time delay for a radar pulse to propagate to the surface and back to the sensor. This has applications in land topography and high-resolution ocean and ice mapping.

Scatterometers provide an accurate measurement of the surface reflectivity, which can be used as a measure of SURFACE ROUGHNESS. *Imaging radars* are used to acquire high-resolution (a few metres to a few tens of metres) maps over large areas using SAR from space. In this case, radar pulses are emitted as the platform moves and, by correcting the pulses for the transmission and reception times, an aperture can be synthesised that has a much greater size than the physical size of the antenna.

Radar instruments are generally composed of several parts: a transmitting source; an antenna, which shapes the transmitted energy into a beam pointing in a certain direction and collects energy from this direction; and equipment for processing and storing the data. The antenna has a BEAMWIDTH, which means that the radiation is emitted into a beam of finite angular width rather than in a single direction.

Civilian imaging radars have been carried by space-borne platforms including *Seasat* (1978), the *Shuttle Imaging Radar* experiments SIR-A (1981), SIR-B (1984) and SIR-C (1994), the *European Remote Sensing Satellites* ERS-1 (1991) and ERS-2 (1995), the Russian *Almaz Satellite* (1991), the *Japanese Earth Resources Satellite* JERS-1 (1992) and the Canadian Space Agency's *Radarsat* (1995). The active microwave instrument of the ERS satellites operates either as an imaging radar or as a scatterometer and the satellite also carries a radar altimeter.

PJS

[See also MICROWAVE REMOTE SENSING, RADAR]

Attema, E.P.W. 1991: The active microwave instrument onboard the ERS-1 Satellite. *Proceedings of the IEEE* 79, 791–799. **Elachi, C**. 1988: *Spaceborne radar remote sensing: applications and techniques*. Piscataway, NJ: IEEE Press. **Ulaby, F.T., Moore, R.K. and Fung, A.K.**, 1986: *Microwave remote sensing: active and passive*. Vol. 3. *From theory to applications*. Norwood, MA: Artech House.

radiance   The total radiant energy of a source per unit of time, unit of surface and unit of solid angle. The radiant energy per unit of time is known as *radiant flux*. The usual unit of the radiance is watt per steradian per square metre (W sr m^{-2}).

TS

radiant temperature   The BLACK-BODY equivalent temperature of a substance characterised by a given spectral RADIANCE (i.e. the only measure of an object's temperature by an instrument not in direct contact with it; for example through REMOTE SENSING). The radiant temperature is also known as the *brightness temperature*.

TS

radiation   The transmission of energy from a body to its surroundings by means of electromagnetic waves. All objects that are not at absolute zero temperature emit RADIATION, which is characterised on the basis of its *wavelength*. The Sun is the principal source of radiation and emits energy in the form of *short-wave radiation*, known as SOLAR RADIATION. Upon entering the ATMOSPHERE, most of

517

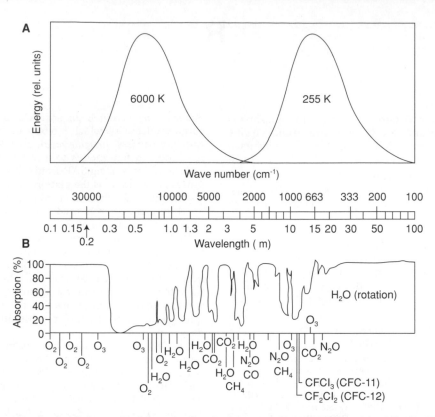

radiation *(A) The wavelength of radiation emitted from the Sun (temperature about 6000 K) and from the Earth (temperature about 255 K); (B) Percentage of incident radiation absorbed by the atmosphere and the gases responsible for that absorption at various wavelengths (after Harvey, 2000)*

this INSOLATION is transmitted without ABSORPTION. The Earth, on the other hand, emits TERRESTRIAL RADIATION of longer wavelengths (related to its temperature) and much of this re-radiated energy can be absorbed by WATER VAPOUR, CARBON DIOXIDE and OZONE in the atmosphere (see Figure). This in turn warms the atmosphere and counter-radiation is returned to the Earth's surface. This process assists in maintaining higher temperatures than would otherwise be expected. In the event of GLOBAL WARMING, the proportion of TERRESTRIAL RADIATION returned to the Earth's surface as counter radiation increases, causing an increase in overall temperature. *JBE*

[See also ABSORPTION SPECTRUM, BLACK-BODY RADIATION, ELECTROMAGNETIC SPECTRUM, RADIOACTIVITY]

Harvey, L.D.D. 2000: *Global warming: the hard science.* Harlow: Pearson Education. **Liou, K.-N.** 2002: *An introduction to atmosphere radiation, 2nd edition.* Orlando, FL: Academic Press. **Paltridge, G.W. and Platt, C.M.R.** 1976: *Radiative processes in meteorology and climatology.* Amsterdam: Elsevier.

radiation balance The balance achieved by the Earth and its atmosphere between incoming *short-wave radiation* (SOLAR RADIATION) and outgoing *long-wave radiation* (TERRESTRIAL RADIATION). The atmosphere is more transparent to the former than to the latter. Whilst most solar radiation passes through the atmosphere, only a proportion of terrestrial radiation does: the remainder is absorbed and re-emitted as long-wave radiation at a longer wavelength. If there is no ADVECTION or conduction, the temperature of

a surface will adjust itself so that as much energy is radiated as is received and, in the long term, the body will achieve an ENERGY BALANCE. The Earth–atmosphere system has a surplus of radiation between latitude 35° and the equator and a deficit between latitude 35° and the poles. The energy transfers required to achieve a global balance are made by the GENERAL CIRCULATION OF THE ATMOSPHERE and OCEAN CURRENTS. *BDG*

[See also ABSORPTION SPECTRUM, BLACK-BODY RADIATION, HEAT BALANCE]

Lockwood, J.G. 1979: *Causes of climate.* London: Arnold. **Paltridge, G.W. and Platt, C.M.R.** 1976: *Radiative processes in meteorology and climatology.* New York: Elsevier. **Seigel, R. and Howell, J.R.** 1992: *Thermal radiation heat transfer,* 3rd edn. Washington, DC: Hemisphere Publishing.

radiative transfer theory A body of theory that allows the prediction of RADIATION intensity in a medium that is able to absorb, emit and scatter radiation. Radiative transfer theory was initiated by Schuster in AD 1905 and has since then been developed mostly by astrophysicists. It is applied in REMOTE SENSING to understand better the BACKSCATTERING mechanisms on the ground when it is illuminated by radiation in the ELECTROMAGNETIC SPECTRUM. In contrast to wave theory, which uses Maxwell's equations, radiative transfer theory starts with the radiative transfer equations that govern the propagation of energy through the scattering medium. It is assumed that there is no correlation between fields and that the intensities of the fields, rather than the fields themselves, can be added. Scattering effects are considered by two MODELS: random fluctuations of permittivity and discrete particles in a homogeneous medium. Sophisticated scattering models have been developed both for optical and microwave applications. *HB*

[See also OPTICAL REMOTE SENSING INSTRUMENTS, RADAR REMOTE SENSING INSTRUMENTS]

Martonchik, J.V., Diner, D.J., Pinty, B. *et al.* 1998: Determination of land and ocean reflective, radiative, and biophysical properties using multi-angle imaging. *IEEE Transactions on Geoscience and Remote Sensing* 36, 1266–1281. **Tsang, L., Kong, J.A. and Shin, R.T.** 1985: *Theory of microwave remote sensing.* New York: Wiley.

radioactive fallout Radioactive particles that reach the Earth's surface following release into the atmosphere by nuclear explosions or NUCLEAR ACCIDENTS. *JAM*

[See also FALLOUT]

radioactive waste *Low-level radioactive waste* is waste containing small amounts of RADIONUCLIDES with a short HALF LIFE. Such material includes protective clothing, used air or water filters and nuclides from medical procedures. Treatment methods include compacting the waste (or incinerating and compacting the ashes), or immobilising liquid waste in bitumen or concrete, and storing it until radiation levels have diminished to an acceptable level for disposal. *High-level* and *intermediate-level radioactive waste* includes spent fuel from reactors, elements of the reactor structure and waste from weapon production; it contains large amounts of radioactive nuclides with long half-lives. Such waste requires careful handling and needs long-term storage in isolation from the natural environment. Spent fuel is usually stored in ponds on site before being put in interim storage or reprocessed at specialist locations. Liquid waste can be stabilised in a borosilicate glass matrix before storage. Current facilities have a finite capacity to store high-level waste and ongoing research aims to assess the feasibility of storing such material long term in deep geological repositories. However, one has to be certain that the stability and impermeability of the site is reliable enough to allow waste to decay safely over the tens or hundreds of thousands of years required. *MLW*

[See also CARCINOGEN, HAZARDOUS WASTE]

Gibb, F.G.F. 2000: A new scheme for the very deep geological disposal of high-level radioactive waste. *Journal of the Geological Society, London* 157, 27–36. **Goodess, C.M. and Palutikof, J.P.** 1989: *Future climate change and radioactive waste disposal.* Proceedings of an International Workshop held at University of East Anglia, Norwich, 1–3 November 1989. **Krausekopf, K.B.** 1988: *Radioactive waste disposal and geology.* London: Chapman and Hall. **Miller, E.W. and Miller, R.M.** 1990: *Environmental hazards: radioactive material and wastes.* Santa Barbara, CA: ABC-Clio. **OECD Nuclear Energy Agency** 1986: *Decommissioning of nuclear facilities: feasibility, needs and costs.* Paris: Organisation for Economic Co-operation and Development. **OECD Nuclear Energy Agency** 1988: *Geological disposal of radioactive waste: in situ research and investigations in OECD countries.* Paris: Organisation for Economic Co-operation and Development.

radioactivity Energetic particles or electromagnetic radiation produced on the decay of a RADIONUCLIDE. It usually comprises alpha (α), beta (β) or gamma (γ) radia-

tion: *alpha radiation* consists of helium nuclei; *beta radiation* consists of electrons; *gamma radiation* is a type of electromagnetic radiation (see ELECTROMAGNETIC SPECTRUM), with a wavelength of $< 10^{-4}$ μm, emitted by excited nuclei. *PQD*

radiocarbon dating Developed by W.F. Libby to determine the age of a variety of organic materials, radiocarbon dating depends on the interaction in the upper atmosphere of cosmic-ray-derived *neutrons* with the STABLE ISOTOPE ^{14}N to form the RADIONUCLIDE ^{14}C. This oxidises to CARBON DIOXIDE, becomes a part of atmospheric carbon dioxide and is thence incorporated into the global carbon exchange reservoir. Living organisms are presumed to have the same ^{14}C concentration as in the atmosphere and after death this is reduced by radioactive decay. By measuring the residual concentration of ^{14}C, the time elapsed since death can be calculated. The accuracy of the process depends on various factors:

1. The HALF LIFE of ^{14}C was estimated by Libby to be 5568 years, the LIBBY HALF LIFE. A more recent value was determined at 5730 ± 30 years but CALIBRATION renders half life uncertainties immaterial.

2. Atmospheric variation in ^{14}C concentration is small (up to 30 years equivalent between Northern and Southern Hemispheres), but marine samples have an APPARENT AGE of at least about 400 years because of the greater mean RESIDENCE TIME of carbon in the oceans, a RESERVOIR EFFECT.

3. ISOTOPIC FRACTIONATION causes enrichment or depletion of ^{14}C, but the effects can be corrected by MASS SPECTROMETER measurement of the stable isotope ^{13}C (c. 1% of the total carbon, the balance being ^{12}C).

4. First demonstrated by de Vries (the DE VRIES EFFECT), radiocarbon dates of timber dated by DENDROCHRONOLOGY show a non-linear deviation from CALENDAR AGES. Short-term variations superimposed on a larger long-term variation are thought to be due to variations of the MAGNETIC FIELD surrounding the Earth causing ^{14}C production rate variations.

FOSSIL FUEL combustion has diluted the concentration of atmospheric ^{14}C (the SUESS EFFECT), while *bomb carbon* was released by the atmospheric testing of nuclear weapons (the *Libby effect* or BOMB EFFECT). A comprehensive programme of HIGH-PRECISION DATING of wood of known age has produced a dataset with which radiocarbon dates can be converted to CALENDAR DATES by CALIBRATION using computer programs available on the World Wide Web from http://depts.washington.edu/qil/ and http://www.rlaha.ox.ac.uk/.

RADIOMETRIC techniques use the radioactivity of ^{14}C to estimate its concentration either by means of *gas proportional counting* or *scintillation counting* and generally require gram quantities of carbon (see CONVENTIONAL RADIOCARBON DATING). The ACCELERATOR MASS SPECTROMETRY (AMS) DATING technique uses high-voltage accelerators to measure ^{14}C ATOMS directly and requires milligram quantities of carbon. In both cases, dates are referenced to an NBS OXALIC ACID standard, calculated using the LIBBY HALF LIFE, have their isotopic fractionation normalised to −25‰ and are given in years BP, i.e. years before AD 1950. Quoted precisions of radiocarbon dates derive from counting statistics and other laboratory errors and, more recently, include a laboratory error multiplier.   *PQD*

Aitken, M.J. 1990: *Science-based dating in archaeology.* London: Longman. **Bowman, S.** 1990: *Radiocarbon dating.* Berkeley, CA: University of California Press. **Linick, T.E., Damon, P.E., Donahue, D.J. and Jull, A.J.T.** 1989: Accelerator mass spectrometry: the new revolution in radiocarbon dating. *Quaternary International* 1, 1–6. **Long, A. (ed.)** 1998: INTCAL 98: Calibration issue. *Radiocarbon* 40(3). **Lowe, J.J. (ed.)** 1991: Radiocarbon dating: recent applications and future potential. *Quaternary Proceedings* 1, 1–87. **McCormac, F.G., Hogg, A.G., Higham, T.F.G.** *et al.* 1998: Variations of radiocarbon in tree rings: Southern Hemisphere offset preliminary results. *Radiocarbon* 40, 1153–1159. **Mook, W.G. and van der Plicht, J.** 1999: Reporting ^{14}C activities and concentrations. *Radiocarbon* 41, 227–240. **Pilcher, J.R.** 1993: Radiocarbon dating and the palynologist: a realistic approach to precision and accuracy. In Chambers, F.M. (ed.), *Climatic change and human impact on the landscape.* London: Chapman and Hall, 23–32. **Renfrew, C.** 1973: *Before civilization: the radiocarbon revolution and prehistoric Europe.* New York: Alfred A. Knopf. **Stuiver, M. and Polach, H.A.** 1977: Discussion: reporting of ^{14}C data. *Radiocarbon* 19, 355–363. **Stuiver, M., Reimer, P.J., Bard, E.** *et al.* 1998: INTCAL98 Radiocarbon age calibration, 24000–0 cal BP. *Radiocarbon* 40, 1041–1083. **Taylor, R.E.** 1987: *Radiocarbon dating: an archaeological perspective.* San Diego, CA: Academic Press. **Taylor, R.E.** 1997: Radiocarbon dating. In Taylor, R.E. and Aitken, M.J. (eds), *Chronometric dating in archaeology.* New York: Plenum, 65–90. **Taylor, R.E.** 2000: Fifty years of radiocarbon dating. *American Scientist* 88, 60–67. **Taylor, R.E., Long, A. and Kra, R. (eds)** 1992: *Radiocarbon after four decades: an interdisciplinary perspective.* New York: Springer.

radiolaria   A subclass of simple amoebic protozoans (Protista) with siliceous skeletons used extensively in PALAEOCEANOGRAPHY and PALAEOCLIMATOLOGY. Many are good indicators of water depth and other factors of the marine environment.   *JAM*

[See also PLANKTON, UPWELLING RADIOLARIAN INDEX]

Casey, R.E. 1993: Radiolaria. In Lipps, J.H. (ed.), *Fossil prokaryotes and protists.* Oxford: Basil Blackwell. **Anderson, O.R.** 1983: *Radiolaria.* New York: Springer.

radioluminescence dating   A new method of LUMINESCENCE DATING in which the luminescence signal is detected during irradiation of the sample with a radioactive source. The first signal measured during the irradiation is the natural radioluminescence signal; the irradiation and the measurement of the signals continues to the point where the radioluminescence does not change; after this, the sample is bleached with artificial sunlight and the radioluminescence of the bleached sample is detected. The infrared emission peak (1.42 eV) can be used for *infrared radioluminescence dating* (IR-RL dating) of potassium feldspars, of which only 4 mg are sufficient. Radioactive irradiation is also necessary for conventional methods, where feldspars send out radioluminescence signals before thermoluminescence or optical signals are stimulated and measured. The modified physical model (*band models* are used to describe luminescence properties) that is used to explain radioluminescence stimulation and emitted spectra suggests that the infrared radioluminescence of feldspars is related to the infrared optically stimulated luminescence. It also suggests, for reasons of solid-state physics (direct measurement of the charge density of the optically sensitive traps), that radio-

luminescence can be used to scrutinise conventional optically stimulated luminescence data. The bleaching time of IR-RL is somewhat longer but close to that of OPTICALLY STIMULATED LUMINESCENCE DATING.   *AE*

[See also INFRARED-STIMULATED LUMINESCENCE DATING, THERMOLUMINESCENCE DATING]

Trautmann, T., Krbetschek, M.R., Dietrich, A. and Stolz, W. 1999: Feldspar radioluminescence: a new dating method and its physical background. *Journal of Luminescence* 85, 45–58. **Trautmann, T., Krbetschek, M.R., Dietrich, A. and Stolz, W.** 2000: The basic principle of radioluminescence dating and a first model approach. *Radiation Measurements* 32, 487–492.

radiometric dating   Techniques that use the immutable process of *radioactive decay* as the basis for the measurement of periods of time. The term radiometric dating is sometimes used in RADIOCARBON DATING to distinguish conventional methods using RADIOACTIVITY measurement (CONVENTIONAL RADIOCARBON DATING) from ACCELERATOR MASS SPECTROMETRY (AMS) DATING techniques, which directly measure ^{14}C concentrations.

Naturally occurring RADIONUCLIDES either exist because of cosmogenic origins, being formed by the interaction of cosmic rays with the Earth, or are of primaeval origins with long HALF LIVES; in the latter case they may have decay products or daughter radionuclides with shorter half lives. Generally, cosmogenic radiometric dating methods rely on knowledge of the initial level of a radionuclide and measuring the quantity remaining after a period of time, e.g. radiocarbon dating. Primaeval isotopes and their daughters exist at low concentrations in many environments and dating methods rely on some mechanism such as volcanism (e.g. POTASSIUM–ARGON DATING) or precipitation/crystallisation (e.g. URANIUM-SERIES DATING) to 'zero' the timing. A radionuclide decays to form a DAUGHTER ISOTOPE. At zero time there is none present, but in time the concentration of the daughter grows. If the daughter is itself radioactive, an equilibrium value is reached eventually where the decay of the daughter matches production. Measurement of isotope ratios before equilibrium is reached will provide the time elapsed since formation.   *PQD*

Geyh, M.A. and Schleicher, H. 1990: *Absolute age determination: physical and chemical dating methods and their application.* Berlin: Springer.

radionuclide   An ISOTOPE that is unstable and spontaneously disintegrates into a DAUGHTER ISOTOPE with the simultaneous emission of an energetic particle or electromagnetic radiation, the most common being *alpha, beta* or *gamma radiation.* The HALF LIFE is the rate of decay of the isotope. Radionuclides or *radioisotopes* either exist because of primaeval origins and have very long half lives or are formed as a result of *transmutation reactions,* during which nuclides transmute into a radionuclide of a different element. Three of the primaeval isotopes give rise to a decay series of radioactive daughters with shorter half lives, the decay series always ending with a stable isotope.   *PQD*

[See also COSMOGENIC-ISOTOPE DATING, RADIOACTIVITY]

Aitken, M.J. 1990: *Science-based dating in archaeology.* London: Longman.

radiosonde   A standard meteorological instrument carried aloft by balloon for providing upper-air data to heights of 24 km (30 hPa) or 30 km (10 hPa). Typically, elements of the instrument record temperature, pressure and relative humidity. Data are switched in turn to a small radio that transmits to a receiving station on the Earth's surface. If the balloon is tracked by radar, its displacement is used to calculate wind speed and direction: this is called a *rawinsonde*. *BDG*

Gaffen, D.J. 1993: *Historical changes in radiosonde instruments and practices: final report* [WMO Instruments and Observing Methods, Report 50; WMO/TD 541]. Geneva: World Meteorological Organization. **World Meteorological Organization** 1986: *WMO catalogue of radiosondes and upper-air wind systems in use by members, 1986* [WMO Instruments and Observing Methods, Report 27; WMO/TD 176]. Geneva: World Meteorological Organization.

radon emanation The naturally occurring radioactive ISOTOPE, radium-226 (^{226}Ra) decays to give radioactive radon-222 (HALF LIFE 3.825 days), which seeps into groundwater and soil air, sometimes leading to accumulation in mines and buildings. Such accumulation may be a health risk and can affect CONVENTIONAL RADIOCARBON DATING. A good ventilation system eliminates the problem in buildings. *JAM*

Gates, A.E. and Gundersen, L.C.S. (eds) 1992: *Geological controls on radon* [Special Paper 271]. Boulder, CO: Geological Society of America. **Nagda, N.L.** 1994: *Radon: prevalence, measurements, health risks and control*. Philadelphia, PA: ASTM.

rain flushing The enhanced removal of fine ASH grade TEPHRA from VOLCANIC PLUMES by rainwater-enhanced particle aggradation through electrostatic and surface tension effects. *JBH*

rain forest Evergreen forest formations found in both tropical and temperate areas where the climate has little or no dry season and where rainfall usually exceeds 1500 mm per annum. TROPICAL RAIN FORESTS of the equatorial regions are characterised by high precipitation from the INTERTROPICAL CONVERGENCE ZONE (ITCZ); *temperate rain forests* occur where OCEANICITY and OROGRAPHIC conditions favour high precipitation, such as northwestern North America and western South Island, New Zealand. *NDB*

[See also MONTANE FOREST, VEGETATION FORMATION-TYPE]

Sandved, K.B. and Emsley, M. 1979: *Rain forests and cloud forests*. New York: Abrams.

rain shadow The region in the lee (downwind) of a hill, mountain or mountain range where precipitation is relatively low. The cause of the rain shadow is reduced moisture content of the air (OROGRAPHIC precipitation having occurred on the windward side of the topographic obstacle) combined with descending air and adiabatic warming to the leeward of the summit. Many of the DESERTS of the world are caused or accentuated by rain-shadow effects, but similar patterns are widespread elsewhere: across the southern Norwegian mountains, for example, mean annual precipitation attains values > 4000 mm yr^{-1} close to the west coast whereas < 500 mm yr^{-1} occurs in some of the eastern valleys in the rain shadow. *JAM*

rainfall simulator A device used to investigate and simulate the effect of rainfall in the field or the laboratory.

Rainfall simulators can usually produce precipitation of a standardised intensity and kinetic energy. There are basically two main types: those that use non-pressurised droppers and those that use spraying nozzles with pressure. *SHD*

Cerdà, A., Ibanez, S. and Calvo, A. 1997: Design and operation of a small and portable rainfall simulator for rugged terrain. *Soil Technology* 11, 163–170. **Farres, P.J.** 1980: *Hardware models: simulation of rainfall* [Department of Geography Paper 5]. Portsmouth: Portsmouth Polytechnic. **Hall, M.J.** 1970: A critique of methods of simulating rainfall. *Water Resources Research* 6, 1104–1114. **Schlesinger, W.H., Abrahams, A.D., Parsons, A.J. and Wainwright, J.** 1999: Nutrient losses in runoff from grassland and shrubland habitats in southern New Mexico: 1. Rainfall simulation experiments. *Biogeochemistry* 45, 21–34. **Seuffert, O.** 1992: Vom Sinn und Unsinn der Niederschlagssimulation in der Geohydrologie und Geomorphologie. *Petermanns Geographische Mitteilungen* 136, 41–47. **Walsh, R.P.D., Coelho, C.O.A., Elmes, A. et al.** 1998: Rainfall simulation plot experiments as a tool in overland flow and soil erosion assessment, north-central Portugal. *GeoÖkoDynamik* 19, 139–152.

raised mire/bog A convex OMBROTROPHIC MIRE raised above the local landscape and groundwater, with definable boundaries including a wet marginal zone (the *lagg*), where runoff from the surrounding mineral soil mixes with water draining from the domed surface of the MIRE. *MJB*

raised shorelines LANDFORMS representing former positions of *sea level* higher than the present one, marked by erosional and depositional features (e.g. NOTCH, ROCK PLATFORM and *raised beach*). Thus, in many areas of the world, shoreline features formed during the LAST INTERGLACIAL in areas unaffected by tectonic uplift are generally found several metres above present sea level. By contrast, in areas of plate collision (e.g. New Guinea) raised shoreline features of the last interglacial age occur locally several hundred metres above present sea level. Similarly, individual raised-shoreline features can become elevated above sea level during EARTHQUAKES. For example, in western Crete, shoreline features of the fourth century AD were raised by up to 6 m during a major earthquake (COSEISMIC uplift).

In areas of the world formerly covered by ice, processes of GLACIO-ISOSTATIC REBOUND have produced raised shoreline features. Owing to the spatial variations in the thickness of former ICE SHEETS and the degree of crustal depression, raised shoreline features of a given age occur at higher altitudes closer to the former centre of glacio-isostatic uplift than near the former ice-sheet edge. The highest raised shorelines associated with the melting of the last great ice sheets occur along the shores of James Bay and Hudson Bay, Canada. Here, individual raised shorelines occur up to altitudes slightly in excess of 300 m. By contrast, in Britain, where there was a smaller ice sheet during the LAST GLACIATION, the highest raised shorelines resulting from glacio-isostatic rebound generally only reach 40 m above present sea level.

Mapping of raised shorelines in glaciated areas can also contribute greatly to an understanding of former patterns of ice-sheet recession. In many cases, individual raised shorelines can be traced as far as the position of the former ice edge. RADIOMETRIC DATING of fossil material contained within raised shoreline sediments therefore contributes to understanding the relative timing of ice retreat and former rates of glacio-isostatic rebound. *AGD*

[See also BEACH RIDGES, SEA-LEVEL CHANGE]

Johnson, L.L. and Stright, M. 1992: *Paleoshorelines and prehistory: an investigation of method.* Boca Raton, FL: CRC Press. **Lambeck, K.** 1995: Constraints on the Late Weichselian ice-sheet over the Barents Sea from observations of raised shorelines. *Quaternary Science Reviews* 14, 1–16. **Nunn, P.D.** 1995: Holocene tectonic histories of five islands in the south-central Lan group, south Pacific. *The Holocene* 5, 160–171. **Pirazzoli, P.A.** 1996: *Sea level changes: the last 20,000 years.* Chichester: Wiley. **Rose, J.** 1990: Raised shorelines. In Goudie, A.S. (ed.), *Geomorphological techniques,* 2nd edn. London: Unwin Hyman. **Sissons, J.B.** 1967: *The evolution of Scotland's scenery.* Edinburgh: Oliver and Boyd. **Sissons, J.B.** 1981: The last Scottish ice sheet: facts and speculative discussion. *Boreas* 10, 1–17.

rampa   A Brazilian term used to describe hillslope hollows fronted by ramp-like slopes of both erosion and deposition. Widespread mass movement and colluviation in Brazil have contributed to *rampa complexes,* described as amphitheatre-like hollows and associated GLACIS, which may be both colluvial (see COLLUVIUM) and alluvial (see ALLUVIUM) in origin. *MAC*

[See also ETCHPLAIN, INSELBERG, PEDIPLAIN]

De Mais, M.R.M. and De Moura, J.R. da Silva 1984: Upper Quaternary sedimentation and hillslope evolution: southeastern Brazilian plateau. *American Journal of Science* 284, 241–254.

random   Governed by chance; not completely determined by other factors; haphazard, having no recognisable patterns; non-deterministic. In statistics, having an *a priori* PROBABILITY of occurrence other than zero or one. Random SAMPLING avoids BIAS (every object in the population has an equal probability of selection). *HJBB*

[See also STOCHASTICITY]

range   (1) The geographical area occupied by a species. The boundaries of this area, also known as the *geographical range,* may be determined by ecological interactions, climatic (e.g. minimum annual temperature) or other physical factors (e.g. coastline). (2) The difference between the highest and lowest values in a dataset and therefore a measure of the full width of the variability or dispersion, but sensitive to the effects of extreme values. *JTK/JAM*

[See also ENDEMIC, HABITAT, NICHE, RANGE ADJUSTMENT, VARIOGRAM]

range adjustment   In response to environmental perturbation of natural or human origin, the geographical range occupied by a species may change. There have been many range adjustments of species following natural environmental modification, such as the climatic changes of the QUATERNARY. Human-caused CLIMATIC CHANGE is likely to cause widespread range adjustments and, if this is so, it is unlikely that all species will be able to migrate rapidly enough to remain within climatically suitable areas. Such species face greater EXTINCTION risk. Species may undergo range adjustments because of other factors, such as habitat FRAGMENTATION. Widespread fragmentation of forests in eastern North America, for example, has

allowed the white-tailed deer (*Odocoileus virginiana*) to expand its range northward. This may have helped reduce the range occupied by moose (*Alces alces*), which is severely affected by a deer parasite. In general, range adjustments occur via JUMP DISPERSAL or DIFFUSION. The latter involves a gradual expansion along the margins of a species range. *JTK*

[See also MIGRATION, NOMADISM, DISPERSAL]

Delcourt, H.R. and Delcourt, P.A. 1991: *Quaternary ecology.* London: Chapman and Hall. **Hengeveld, R.** 1990: *Dynamic biogeography.* Cambridge: Cambridge University Press.

range management   The exploitation and CONSERVATION of rangelands – extensive areas of TROPICAL or TEMPERATE GRASSLANDS. The focus may be domesticated animals (*ranching*) or wild animals (*game ranching*). Originally, the 'range' was open, but now the animals are increasingly maintained in enclosures. Range improvement techniques include the introduction of NATIVE or EXOTIC SPECIES to increase the quantity and quality of forage, reseeding, SOIL and WATER CONSERVATION and controlled use of fire. Poor range management leads to OVERGRAZING and is a major cause of DESERTIFICATION. *JAM*

[See also GAME MANAGEMENT, PRESCRIBED FIRE]

Le Houérou, H.N. 1980: The rangelands of the Sahel. *Journal of Rangeland Management* 33, 41–46.

ranker   Shallow mineral soil with an organic-rich surface horizon and low base status overlying siliceous parent material. This soil grouping has been subsumed into the LEPTOSOLS. The name ranker comes from Austria and is connotative of steep slopes. Rankers occur commonly in mountainous areas. *EMB*

Kubiena, W.L. 1953: *The soils of Europe.* London: Thomas Murby.

rapid environmental change   HIGH-RESOLUTION RECONSTRUCTION of *palaeoclimate* from, for example, ICE CORES, MARINE SEDIMENTS and LAKE SEDIMENTS provide evidence of ABRUPT CLIMATIC CHANGE occurring within decades to millennia (i.e. on SUB-MILANKOVITCH timescales). Evidence for rapid environmental change in ice cores includes 24 DANSGAARD–OESCHGER EVENTS occurring between 115 and 14 ka BP, during the LAST GLACIATION. Each cycle consists of a rapid warming of about 7°C to INTERSTADIAL conditions within a few decades (100 years or less), followed by gradual cooling to STADIAL conditions over several centuries, as for example during the Late Glacial. ELECTRICAL CONDUCTIVITY MEASUREMENTS in the GISP ice core suggest rapid oscillations between near glacial and interglacial conditions on timescales of 5–20 years within these interstadials.

Some Dansgaard–Oeschger cycles can be correlated with evidence from BIOSTRATIGRAPHY, LITHOSTRATIGRAPHY and STABLE ISOTOPES for abrupt climate changes in marine sediment cores, including ICE-RAFTED DEBRIS (IRD), low SEA-SURFACE TEMPERATURES and low SALINITY. These indicate similar patterns of climate change, namely sea surface cooling, culminating in layers of IRD (i.e. from ICEBERG armadas) during the period of maximum cooling, followed by rapid warming. Clusters of Dansgaard–Oeschger cycles have been grouped into BOND CYCLES, longer-term climate cycles of 10–15 ka year duration, characterised by temperature changes of c. 5°C over decades. The stadial conditions in each Bond cycle terminate with an abrupt shift to warmer climatic conditions, which are usually associated with evidence for a HEINRICH EVENT in the marine record.

Proxy evidence for abrupt climatic change in ocean and ice-core records have been correlated with biological, isotopic and lithological signals for interstadials (e.g. the Late Glacial Interstadial) and stadials (e.g. the YOUNGER DRYAS STADIAL) from lacustrine sequences (e.g. La Grande Pile in France).

The similarity in patterns of rapid environmental change in proxy records from ocean and ice-core records suggests atmospheric and ocean coupling. This has been explained in terms of changes in the position of the OCEANIC POLAR FRONT, and 'switching on-and-off' of the THERMOHALINE CIRCULATION as a result of periodic influxes of MELTWATER. Whether or not the pulses of icebergs are triggered by climate or ice-sheet dynamics remains unclear.

Evidence for comparable rapid events in the LAST INTERGLACIAL (*Eemian*) have been inferred from ice-core evidence, possibly confirmed by biostratigraphical evidence from long terrestrial records. There is still some debate, however, over the validity of this interpretation, for HOLOCENE ENVIRONMENTAL CHANGE has been of relatively low magnitude, including a 1.5 ka PERIODICITY in IRD in the North Atlantic and lower-frequency oscillations in PEAT HUMIFICATION and MIRE surface wetness indices. Possible rapid cooling events have, however, been identified in the Greenland ice core record at 8200 calendar years BP, in peat records at around 2650 calendar years BP, and in the DENDROCHRONOLOGICAL record at AD 560. The first of these was probably due to a pulse of meltwater from the decaying LAURENTIDE ICE SHEET. The second was coincident with a peak in radiocarbon production and the last has been attributed to either a VOLCANIC ERUPTION or METEORITE IMPACT. *MHD*

Alley, R.B., Mayewski, P.A., Sowers, T. et al. 1997: Holocene climatic instability: a prominent, widespread event 8200 yr ago. *Geology* 25, 483–486. **Andersen, E.** 1997: Younger Dryas research and its implications for understanding of abrupt climatic change. *Progress in Physical Geography* 21, 230–249. **Bond, G., Broecker, W., Johnsen, S. et al.** 1993: Correlations between climate records from North Atlantic sediments and Greenland ice. *Nature* 365, 143–147. **Blunier, T., Chappellaz, J., Schwander, J. et al.** 1998: Asynchrony of Antarctic and Greenland climate change during the last glacial period. *Nature* 394, 739–743. **Dansgaard, W., Johnsen, S.J., Clausen, H.B. et al.** 1993: Evidence for general instability of past climate from a 250-kyr ice-core record. *Nature* 364, 218–220. **Geel, B. van and Renssen, H.** 1998: Abrupt climate change around 2650 BP in North-West Europe: evidence for climatic teleconnections and a tentative explanation. In Issar, A.S. and Brown, N. (eds), *Water, environment and society in times of climatic change.* Dordrecht: Kluwer, 21–42. **Manabe, S. and Stouffer, R.J.** 2000: Study of abrupt climate change by a coupled ocean–atmosphere model. *Quaternary Science Reviews* 19, 285–299. **McManus, J.F., Bond, G.C., Broecker, W.S. et al.** 1994: High-resolution climatic records from the North Atlantic during the last interglacial. *Nature* 371, 326–329.

rapid-onset hazards Environmental HAZARDS that strike rapidly, having developed with very little advance warning relative to the RESPONSE TIME required by local populations. They include VOLCANIC ERUPTIONS, EARTHQUAKES, FLOODS, LANDSLIDES, thunderstorms, lightning and WILDFIRE. In contrast, *slow-onset hazards* may take weeks or years to develop and include heat waves, DROUGHT, insect infestations and EPIDEMICS. The high onset speed makes the human population highly vulnerable, although this also depends upon the available forecast technology and operational warning systems, as well as their own exposure and responses (such as *evacuation*). Consequently, the poorest and least educated experience the highest level of RISK. Localised hazard events such as TORNADOES, FLASH FLOODS and EARTHQUAKES usually have a shorter forecast LEAD time, if any, and carry a high mortality risk. *JGT*

[See also DISASTER, GEOINDICATOR, NATURAL HAZARD]

Burton, I., Kates, R. and White, G.F. 1993: *The environment as hazard.* New York: Guilford Press. **Smith, K**. 1996: *Environmental hazards: assessing risk and reducing disaster.* London: Routledge. **White, G.F. (ed.)** 1974: *Natural hazards: local, national, global.* Oxford: Oxford University Press.

rare species Taxa with small populations and an International Union for the Conservation of Nature RED DATA BOOK category. Rare species are usually localised and at risk but not at present an ENDANGERED SPECIES or a VULNERABLE SPECIES. Indices of rarity have been developed for assessment of the CONSERVATION importance of sites. *IFS*

[See also THREATENED SPECIES]

Gaston, K.J. 1994: *Rarity.* London: Chapman and Hall.

rarefaction analysis A numerical procedure for providing minimum-variance unbiased estimates of the expected number of variables (*t*) (e.g. pollen types) in a random sample of *n* individuals taken from a collection of *N* individuals containing *T* variables. It standardises sample counts to a common size, thereby permitting comparison of richness in different samples when the original counts are of different sizes. It can be used when individuals observed at one hierarchical level are classified into groups at a higher level, e.g. individual pollen grains classified into pollen types, species into genera, etc. No underlying hierarchical distribution is assumed. *HJBB*

[See also DIVERSITY INDICES]

Birks, H.J.B. and Line, J.M. 1992: The use of rarefaction analysis for estimating palynological richness from Quaternary pollen-analytical data. *The Holocene* 2, 1–10.

raster An image is said to be in *raster format* or conforms to a *raster data model* if it consists of a set of PIXELS arranged in a rectangular two-dimensional grid. An image is in *vector format* (VECTOR DATA MODEL) if it consists of descriptions of geometric objects, such as lines, polygons, etc. *ACF*

rates of environmental change Environmental changes occurred throughout Earth's history, but the rates of change are best detailed for the Quaternary and particularly for the last few thousand years. With evidence of climatic change available on a decadal basis from, for example ICE CORES for a period extending back into the LAST GLACIATION, it has been possible to demonstrate the existence of ABRUPT CLIMATIC CHANGES. For example, Greenland ice-core evidence has indicated that mean annual temperatures in the North Atlantic may have risen by 7°C in only 50 years at the onset of the HOLOCENE. The response of different components of the environment to abrupt Quaternary climatic changes varied. For example, evidence from Sweden suggests that the potential rate of tree colonisation across the country (about 1 km every five years) following DEGLACIATION of the last ICE SHEET was much faster than the actual rate of ice-sheet retreat. In contrast, the LAG TIME between abrupt rainfall changes and corresponding LAKE-LEVEL VARIATIONS was very brief in the tropics. Geomorphological systems can vary in their RESPONSE TIMES to change, whether caused by climatic or other (e.g. tectonic) means and whether it is slow or rapid. Small-scale systems generally equilibrate to new conditions more rapidly than large ones with, for example, soil profiles forming in as few as 10–1000 years and individual slopes on soft rocks attaining characteristic forms in 10–10 000 years. Lags can occur in geomorphological systems in response to climatic change, for example where there is an abrupt change from arid to humid conditions. In such cases, maximum erosion and sedimentation are associated with the transition period when the increase in the protective vegetative cover lags behind the increase in precipitation. Increased EROSION RATES only decline in response to increased precipitation when vegetation density increases. On a large scale, the response time for whole mid-latitude glacial and periglacial LANDSCAPES is certainly far longer than the duration of the Holocene, to judge from the extent to which many glacial and periglacial landforms and sediments are virtually unaltered by humid temperate processes. Since mid-Holocene times, magnitudes and rates of natural and human-induced environmental change have been comparable. Late Quaternary rates of environmental change can provide useful analogues for anticipated future change caused by GLOBAL WARMING. *RAS*

[See also EQUILIBRIUM CONCEPTS IN GEOMORPHOLOGICAL AND LANDSCAPE CONTEXT, LAG TIME, RAPID ENVIRONMENTAL CHANGE, RESPONSE TIME]

Allen, J.R.L. 1974: Reaction, relaxation and lag in natural systems: general principles, examples and lessons. *Earth Science Reviews* 10, 263–342. **Bluemle, J.P., Sabel, J.M. and Karlén, W.** 1999: Rate and magnitude of past global climate changes. *Environmental Geosciences* 6, 63–75. **Brunsden, D. and Thornes, J.B.** 1979: Landscape sensitivity and change. *Transactions of the Institute of British Geographers NS* 4, 463–484. **Knox, J.C.** 1972: Valley alluviation in south-western Wisconsin. *Annals of the Association of American Geographers* 62, 401–410. **Stocker, T.F.** 1999: Abrupt climate changes: from the past to the future – a review. *International Journal of Earth Science* 88, 365–374. **Street-Perrott, F.A. and Perrott, R.A.** 1990: Abrupt climatic fluctuations in the tropics: the influence of Atlantic Ocean circulation. *Nature* 343, 607–611.

rationalism A philosophy that places emphasis on prior reasoning in the acquisition of knowledge. It is a term usually contrasted with EMPIRICISM, and is most widely associated with the seventeenth- and eighteenth-century philosophers Descartes, Spinoza and Leibniz. Characteristics of rationalism include belief in the possibility of obtaining a knowledge of what exists by reason alone, that knowledge forms a single system and that everything is explicable within this system. *ART*

[See also CRITICAL RATIONALISM]

reaction time The time lapse between the occurrence of change in the environment of a SYSTEM and the *beginning* of the system response. Important factors affecting reaction times are system complexity and the mechanisms governing response. *JAM*

[See also LAG TIME, RELAXATION TIME]

reactivation surface A discontinuity within a set of CROSS-STRATIFIED sediment, commonly caused by a change in current direction or strength. *MRT*

Bristow, C. 1995: Facies analysis in the Lower Greensand using ground-penetrating radar. *Journal of the Geological Society, London* 152, 591–598. **Collinson, J.D.** 1970: Bedforms of the Tana River, Norway. *Geografiska Annaler* 52A, 31–56.

realism Realism (contrast with EMPIRICISM and *relativism*) is the philosophy that science attempts to generate true knowledge of observable and unobservable aspects of an objective world. There are multiple versions. *CET*

Rhoads, B.L. 1994: On being a 'real' geomorphologist. *Earth Surface Processes and Landforms* 19, 269–272.

reconnaissance survey The earliest phase in FIELD RESEARCH whereby sites are located and/or a preliminary investigation is carried out. It does not necessarily involve visiting the site: reconnaissance may be limited to REMOTE SENSING. If a preliminary investigation is made, this may involve a wide range of techniques, such as ARCHAEOLOGICAL PROSPECTION, GEOPHYSICAL SURVEYING, preliminary EXCAVATION or sampling of materials, or collecting specimens and making species lists, but it is normally less focused than a PILOT SURVEY. *JAM*

recreation Activities carried out for pleasure and enjoyment, beyond the necessary duties surrounding work or supporting the family. The number of people participating in recreational activities has expanded considerably in the last 30 years as a result of greater affluence, more leisure time, more disposable income, greater mobility and improved recreational technologies. Areas of natural beauty, such as NATIONAL PARKS, are increasingly popular with visitors as they are spiritually uplifting (offering fresh air, wildlife and scenery) and contain challenging terrain for energetic activities. Common outdoor recreational pastimes include hiking, horse-riding, cycling, camping, picnicking, hunting, fishing, photography, nature study, boating, skiing and off-road vehicle driving. Each has its own unique impact on land, water, air, flora and fauna, the most documented of which is TRAMPLING. ECOTOURISM is a phenomenon of the 1980s and 1990s and consists of travelling to relatively undisturbed natural areas with the object of learning while enjoying the scenery and/or wildlife; for example, a Kenyan safari holiday. A variety of ENVIRONMENTAL MANAGEMENT techniques, including controlling visitor numbers, creating temporal and spatial zones for different activities and strengthening or restoring sites, facilitate the SUSTAINABILITY of the enjoyment of natural resources. *MLW*

Cater, E. and Lowman, G. (eds) 1994: *Ecotourism: a sustainable option?* Chichester: Wiley. **France, L.** 1999: Sustainable tourism. In Pacione, M. (ed.), *Applied geography: principles and practice*. London: Routledge, 321–332. **Hammitt, W.E. and Cole, D.N.** 1987: *Wildland recreation: ecology and management*. New York: Wiley. **Kuss, F.R.** 1986: A review of major factors influencing plant responses to recreation impacts. *Environmental Management* 10, 637–650.

recurrence surface The boundary between two PEAT layers with different degrees of DECOMPOSITION (e.g. the boundary between 'black', highly humified peat and a relatively undecomposed 'white' *Sphagnum* peat layer that has formed above it). Changes in the degree of DECOMPOSITION are generally interpreted as reflecting the hydrological environment of peat deposition, with less decayed 'white' layers deposited under wetter conditions and more decayed, darker layers deposited under drier conditions. Therefore, recurrence surfaces, especially those from OMBROTROPHIC MIRES, are often interpreted as climatic signals. Granlund (1932) originally recognised five important recurrence surfaces in southern Sweden at approximately 2300 BC, 1200 BC, 600 BC, AD 400 and AD 1200, of which the middle one was the GRENTZHORIZONT. Many more have been identified in some regions. However, their extent, age and relationship to climatic variations are somewhat variable. *MJB*

[See also REGENERATION COMPLEX]

Aaby, B. 1976: Cyclic climatic variations over the past 5500 years reflected in raised bogs. *Nature* 263, 281–284. **Barber, K.E.** 1982: *Peat stratigraphy and climatic change.* Amsterdam: A.A. Balkema. **Blackford, J.** 1993: Peat bogs as sources of proxy climatic data: past approaches and future research. In Chambers, F.M. (ed.), *Climate change and human impact on the landscape.* London: Chapman & Hall, 47–56. **Granlund, E.** 1932: De svenska Hogmossarnas geologi. *Sveriges Geologiska Undersokning* 26, 1–93.

recycling The re-use of materials, usually manufactured items, once they have been discarded by a user. Recycling is undertaken to satisfy a number of objectives, which include:

- minimising overall volumes of WASTE produced (e.g. bulky paper, plastics and glass);
- reducing the exploitation of increasingly expensive non-renewable raw materials (e.g. oil, plastics and metals);
- lessening the depletion of RENEWABLE RESOURCES (e.g. paper and rubber);
- recovering valuable materials (e.g. silver from photographic film);
- making financial gains if recycling is cheaper than obtaining parts or substances from raw materials (e.g. machine parts and aluminium);
- responding to pressure from CONSERVATION groups and legislation.

Recycling can also involve energy conservation (for example, burning domestic garbage to generate heat and power) and nutrient conservation (for example, adding treated solid sewage waste or farm slurry to agricultural land). *MLW*

[See also WASTE MANAGEMENT, WASTE MINING]

Ackerman, F. 1997: *Why do we recycle? Markets, values and public policy.* Washington, DC: Island Press. **Her Majesty's Inspectorate of Pollution** 1992: *Environmental Protection Act 1990: Waste disposal and recycling.* London: HMSO. **Virtanen, Y. and Nils, S.** 1993: *Environmental impacts of waste paper recycling.* London: Earthscan. **Waite, R.** 1995: *Household waste and recycling.* London: Earthscan.

red beds Thick successions of SANDSTONES, SILTSTONES and CONGLOMERATES with a dominant red colour, that generally lack FOSSILS other than (in PHANEROZOIC successions) occasional vertebrate tracks, wood debris or rootlets. The red colour is due to the precipitation of haematite (Fe_2O_3) as a coating around CLASTS during early DIAGENESIS under conditions favoured by a hot, semi-arid climate; most red beds are attributed to DESERT or alluvial DEPOSITIONAL ENVIRONMENTS. Some of the most typical red beds formed in continental interiors after episodes of OROGENY, including European DEVONIAN and PERMO-TRIASSIC successions represented in the British Isles by the Old Red Sandstone and New Red Sandstone respectively. Red beds are unknown in the GEOLOGICAL RECORD before 2300 million years ago: prior to this, iron-rich sediments are found as BANDED IRON FORMATIONS. The changeover is attributed to a rise in atmospheric oxygen levels in the early PROTEROZOIC. *CDW*

[See also MOLASSE]

Collinson, J.D. 1996: Alluvial sediments. In Reading, H.G. (ed.), *Sedimentary environments: processes, facies and stratigraphy.* Oxford: Blackwell, 37–82. **Dubiel, R.F. and Smoot, J.P.** 1994: Criteria for interpreting paleoclimate from red beds – a tool for Pangean reconstruction. In Embry, A.F., Beauchamp, B. and Glass, D.J. (eds), *Pangean global environments and resources* [Memoir 17]. Calgary: Canadian Society of Petroleum Geologists, 295–310. **Pye, K.** 1983: Red beds. In Goudie, A.S. and Pye, K. (eds), *Chemical sediments and geomorphology.* London: Academic Press, 227–263. **Turner, P.** 1980: *Continental red beds* [Developments in Sedimentology 29]. Amsterdam: Elsevier.

Red Data books The books, established by the International Union for the Conservation of Nature (IUCN) in the late AD 1960s to draw attention to the CONSERVATION needs of species most affected by human impacts. Several countries now have Red Data books. *IFS*

[See also ENDANGERED SPECIES, GREEN LISTS, RARE SPECIES, THREATENED SPECIES]

World Conservation Monitoring Centre 1992: *Global biodiversity: status of the Earth's living resources.* London: Chapman and Hall.

redeposition The translocation or *reworking* of material and its subsequent deposition in a stratigraphically younger context, which if not recognised could result in contamination of sediment taken for DATING, or lead to incorrect inferences being made from DERIVED FOSSILS. *FMC*

Bradbury, J.P. 1996: Charcoal deposition and redeposition in Elk Lake, Minnesota, USA. *The Holocene* 6, 339–344.

reduction The addition of electrons to an *element* and, in the environment, the removal of oxygen or the addition of hydrogen to a mineral. The process should be considered together with OXIDATION as both chemical reactions are controlled by the *redox potential* (Eh). Where minerals of soils or REGOLITH are situated in an environment where oxygen is excluded, reduction will take place. Such ANAEROBIC conditions may be found, for example, below the WATER TABLE, in WETLANDS, in GLEYSOLS and in poorly oxygenated LACUSTRINE and MARINE SEDIMENTS. Environmental changes to cooler and/or wetter conditions, which tend to lead to higher water tables, more frequent waterlogging and organic accumulation, will favour reduction. *RPDW*

Brunsden, D. 1979: Weathering. In Embleton, C. and Thornes, J.B. (eds), *Process in Geomorphology.* London: Arnold, 73–129.

Gambrell, R.P. and Patrick Jr, W.H. 1978: Chemical and microbial properties of anaerobic soils and sediments. In Hook, D.D. and Crawford, R.M.M. (eds), *Plant life in anaerobic environments*. Ann Arbor, MI: Ann Arbor Science, 375–423.

reductionism The idea that all scientific theories depend on very general scientific laws, particularly those of physics and chemistry. In ecology, it frequently involves the derivation of community- or ecosystem-level concepts through the extrapolation of results from small-scale laboratory or field EXPERIMENTS. It has also been interpreted more widely as understanding complex entities from a detailed knowledge of their component parts.   *JLI*

[See also COMMUNITY ECOLOGY, HOLISTIC APPROACH]

White, J. (ed.) 1985: *The population structure of vegetation*. The Hague: Junk. **Wimsatt, W.C.** 1980: Reductionistic research strategies and their biases in the units of selection controversy. In Saarinen, E. (ed.), *Conceptual issues in ecology*. Dordrecht: D. Riedel.

redundancy analysis The canonical or constrained version of PRINCIPAL COMPONENTS ANALYSIS where the ORDINATION axes are constrained to be linear combinations of predictor or explanatory variables. The response variables (i.e. species) are assumed to have a linear response to the explanatory variables. Redundancy analysis provides a multivariate direct ordination or GRADIENT ANALYSIS of objects, variables and predictor variables. It is also known as *reduced rank regression*.   *HJBB*

[See also CANONICAL CORRELATION ANALYSIS]

Braak, C.J.F. ter 1994: Canonical community ordination. Part I: Basic theory and linear methods. *Ecoscience* 1, 127–140.

reef An elevated part of the sea-floor that rises just above sea level, including rocky outcrops, sediment shoals and mounds dominated by biological activity (see CORAL REEF). Biologically dominated reefs have a history dating back over 2000 million years in the GEOLOGICAL RECORD and can be divided into *framework-built reefs*, with an organic framework (e.g. coral reefs), and *reef mounds* in which the role of organic activity is less clear. Together these are commonly referred to as *carbonate build-ups*, being represented by carbonate SEDIMENTARY ROCKS (LIMESTONES) that lack BEDDING and are rich in FOSSILS (see also BIOHERM). Different groups of organisms have dominated reef construction through geological time, including algal STROMATOLITES in the Precambrian, *stromatoporoids* in the Devonian and *rudist bivalves* in the Cretaceous period, as well as *scleractinian corals* in the CENOZOIC up to the present day.   *GO*

Hopley, D. 1982: *The geomorphology of the Great Barrier Reef*. New York: Wiley. **Newell, N.D.** 1972: The evolution of reefs. *Scientific American* 226, 54–65. **Pomar, L. and Ward, W.C.** 1994: Response of a late Miocene Mediterranean reef platform to high-frequency eustasy. *Geology* 22, 131–134. **Stoddart, D.R.** 1969: Ecology and morphology of recent coral reefs. *Biological Reviews* 44, 433–498. **Wendt, J., Belka, Z., Kaufman, B.** *et al.* 1997: The world's most spectacular carbonate mud mounds (Middle Devonian, Algerian Sahara). *Journal of Sedimentary Research* 67, 424–436. **Wright, V.P. and Burchette, T.P.** 1996: Shallow-water carbonate environments. In Reading, H.G. (ed.), *Sedimentary environments: processes, facies and stratigraphy*, 3rd edn. Oxford: Blackwell Science, 325–394.

reference area In PALAEOECOLOGY, a region or area of relatively homogenous climate and vegetation within which one or more representative coring sites (*reference sites*) are chosen. The cores are studied using a variety of techniques (e.g. PALAEOMAGNETIC DATING) to obtain baseline PALAEOENVIRONMENTAL data.   *BA*

[See also CONTROL SITE, CORER, PALAEOBOTANY, PALAEOCLIMATOLOGY]

Berglund, B.E. 1986: Palaeoecological reference areas and reference sites. In Berglund, B.E. (ed.), *Handbook of Holocene palaeoecology and palaeohydrology*. Chichester: Wiley, 111–126. **Berglund, B.E., Birks, H.J.B., Ralska-Jasiewiczowa, M. and Wright, H.E. (eds)** 1996: *Palaeoecological events during the last 15000 years: regional syntheses of palaeoecological studies of lakes and mires in Europe*. Chichester: Wiley.

reference standard A STANDARD SUBSTANCE issued by a national or international organisation, against which measurements may be referenced by individual laboratories, ensuring that different laboratories produce results that can be directly compared. Reference standards are particularly important in the field of both stable and radioactive ISOTOPE RATIO studies.   *PQD*

[See also NBS OXALIC ACID, OXYGEN ISOTOPES]

reforestation The replanting of trees in areas previously cleared of their forest cover. Commonly, it involves planting with tree species different from the ones they replace, although forest conditions are usually retained.   *GOH*

[See also AFFORESTATION, REGENERATION]

refuge theory in temperate regions A theory suggesting that, during the GLACIAL EPISODES of the QUATERNARY, temperate FAUNA and FLORA became restricted to microenvironmentally favourable locations or 'refugia'. However, where each REFUGIUM was located, and how extensive they were, has long been a matter of debate. Past researches into full-glacial refugia has fallen into two main camps, those dealing with the REFUGE THEORY IN THE TROPICS and those dealing with refugia in higher 'temperate' latitudes, beyond the limits of GLACIERS and PERMAFROST. Most of the work on refugia in temperate latitudes has focused predominantly on where isolated pockets of plants and animals survived. In Europe, palaeoecological evidence suggests that the three southern peninsulas (Balkans, Spain and southern Italy) of Europe, plus other small microenvironmentally favourable locations closer to the ICE SHEETS, were important refugial areas. Molecular evidence is starting to suggest that full-glacial isolation in these mid- to high-latitude refuges had a significant influence on current patterns of BIODIVERSITY. Molecular examination of present-day populations of FAUNA and FLORA in Europe, for example, has revealed distinctive patterns of genetic variation within populations of grasshoppers, hedgehogs, oak trees, the common beech, black alder, the brown bear, newts, voles, silver fir and house mice. The geographical pattern of this variation can be linked to various GLACIAL refugia, suggesting that full GLACIAL isolation has left a recognisable imprint on present day BIODIVERSITY. These data have also enabled reconstruction of the major routes in the postglacial COLONISATION of northern Europe from refugia (see Figure).   *KJW*

[See also NUNATAK HYPOTHESIS]

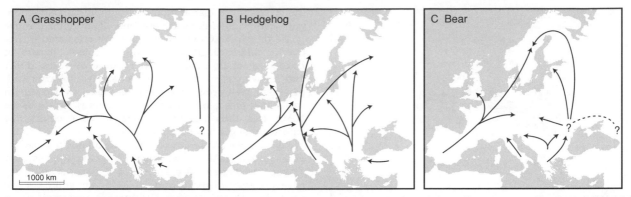

Refuge theory in temperate regions *Proposed patterns of postglacial colonisation of three European species from full-glacial refugia: (A) meadow grasshopper (*Chorthippus parallelus*); (B) hedgehog (*Erinaceus europeus/concolor*); and (C) brown bear (*Ursus arctos*). Reconstructed from molecular variations within extant populations (after Hewitt, 2000)*

Bennett, K.D., Tzedakis, P.C. and Willis, K.J. 1991: Quaternary refugia of north European trees. *Journal of Biogeography* 18, 103–115. **Hewitt, G.** 2000: The genetic legacy of the Quaternary ice ages. *Nature* 405, 907–913. **Rundgren, M. and Ingolfsson, Ó.** 1999. Plant survival in Iceland during periods of glaciation? *Journal of Biogeography* 26, 387–396. **Willis, K.J.** 1996: Where did all the flowers go? The fate of temperate European flora during glacial periods. *Endeavour* 10, 110–114. **Willis, K.J., Rudner, E. and Sümegi, P.** 2000: The full-glacial forests of central and south eastern Europe. *Quaternary Research* 53, 203–213. **Willis, K.J. and Whittaker, R.J.** 2000: The refugial debate. *Science* 287, 1406 1407.

refuge theory in the tropics

Supposedly, ARIDITY during glacial stages caused rainforests to contract into isolated blocks (see REFUGIUM) separated by SAVANNA. Closely related species of Amazonian birds have modern ranges that touch, but do not overlap. Assuming that most SPECIATION is allopatric (occurring when populations are genetically isolated), the SPECIATION of these birds from a common ancestor must have taken place when the ancestral population was divided. Glacial aridity would provide the necessary isolation. Under dry conditions the rainforest habitats of the birds would contract into isolated pockets separated by broad savannas. Isolated from other populations, speciation occurred. When moist conditions returned, species' ranges expanded until they formed the modern parapatric (touching) distributions. Other biogeographical data for lizards, frogs, plants and butterflies appear to be consistent with this hypothesis and it has been suggested that areas of high biodiversity and endemism (see ENDEMIC) indicate the locations of the former refugia (see Figure, part B). However, *ancient DNA* and protein studies indicate most of the speciation was Miocene in age (i.e. predating the last ICE AGE) and POLLEN ANALYSIS indicates that the GLACIAL EPISODES were not dry enough to cause forest isolation (see Figure, part A). *MBB*

[See also BIOGEOGRAPHY, REFUGE THEORY IN TEMPERATE REGIONS, TROPICAL RAIN FOREST]

Brown Jr, K.S. 1987: Conclusions, synthesis and alternative hypotheses. In Whitmore, T.C. and Prance, G.T. (eds), *Biogeography and Quaternary history in tropical America*. Oxford: Blackwell Scientific. **Bush, M.B.** 1994: Amazonian speciation: A necessarily complex model. *Journal of Biogeography* 21, 5–18. **Colinvaux, P.** 1979: The ice-age Amazon. *Nature* 280, 399–400. **Haffer, J.** 1969: Speciation in Amazonian forest birds. *Science* 165, 131–137. **Whitmore, T.C. and Prance, G.T. (eds)** 1987: *Biogeography and Quaternary history in tropical America*. Oxford: Clarendon Press. **Willis, K.J. and Whittaker, R.J.** 2000: The refugial debate. *Science* 287, 1406–1407.

refugium

An area of survival of species during adverse conditions, most notably during GLACIAL EPISODES when temperate trees survived in areas south of the continental ICE SHEETS and mountain species survived below the region of MOUNTAIN GLACIATION. Refugia for other types of species also existed far removed from glaciers (e.g. TROPICAL RAIN FOREST refugia). The current HOTSPOTS IN BIODIVERSITY have been termed *Holocene refugia*. *BA*

[See also BLUE-WATER REFUGIUM, NUNATAK HYPOTHESIS, REFUGE THEORY IN TEMPERATE REGIONS, REFUGE THEORY IN THE TROPICS, TABULA RASA]

Willis, K.J. and Whittaker, R.J. 2000: The refugial debate. *Science* 287, 1406–1407.

regeneration

Recovery of biotic structures or communities following damage or disturbance, as when a cut forest regenerates from seeds or previously established seedlings (*advance regeneration*). *LRW*

[See also CYCLIC REGENERATION, ECOLOGICAL SUCCESSION, SECONDARY SUCCESSION]

Cairns Jr, J. (ed.) 1980: *The recovery process in damaged ecosystems*. Ann Arbor, MI: Ann Arbor Science.

regeneration complex

The lenticular structures of variably decomposed peats seen in section in some RAISED MIRES/BOGS. They are believed to reflect a series of successive regeneration cycles of the mire ecosystem. *MJB*

[See also CYCLIC REGENERATION, RECURRENCE SURFACE]

regional climatic change

The influence of variations in the atmospheric circulation on regional weather and climate is determined largely by surface topography. A variety of modes in the atmospheric circulation (including TELECONNECTIONS) are recognised in different parts of the world. Examples include the North Pacific Oscillation and NORTH ATLANTIC OSCILLATION (NAO), in which linked oscillations of air pressure and temperature influence the frequency of surface wind directions. The latter determine

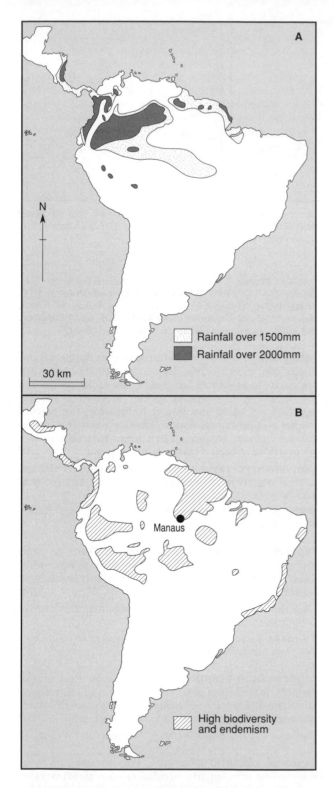

refuge theory in the tropics *Postulated refugia in Amazonia:
(A) assuming uniform reduction of present rainfall by
1000 mm y⁻1; (B) based on areas of high biodiversity and
endemism of plants and animals. The theory has largely been
refuted by palaeoecological and molecular data (after Colinvaux,
1979; Brown, 1987)*

the geographical distribution of orographic precipitation,
rain-shadow conditions and the fohn effect. In the positive
phase of the NAO, westerly winds are enhanced in frequency and strength. Northwestern coastal fringes of
Europe generally become wetter, more overcast, duller
and (in winter) milder. These ANOMALIES coincide with
persistent high pressure and dry anomalies over southern
and sometimes central Europe. Long term trends in the
NAO can thus become associated with contrasting
climatic changes in adjacent regions.

Regional climatic models (RCMs), also known as *limited
area models* (LAMs) are high-resolution models that
attempt predictions of climate at a regional scale. They
have a horizontal grid spacing of 20–50 km and a vertical
resolution of 100–1000 m. RCMs are constrained and
driven by the output from GENERAL CIRCULATION MODELS
(GCMs) and are dependent on the quality of that output.
JCM

Giorgi, F. and Mearns, L.O. 1991: Approaches to the simulation of regional climate change: a review. *Reviews of Geophysics*
29, 191–216. **Glantz, M.H. (ed.)** 1988: *Societal responses to
regional climatic change.* Boulder, CO: Westview Press. **Heino, R.**
1994: *Climate variations in Europe* [*Publication* 3/94]. Helsinki:
Academy of Finland. **Hurrell, J.W.** 1995: Decadal trends in the
North Atlantic Oscillation: regional temperatures and precipitation. *Science* 269, 676–679. **Hurrell, J.W. and van Loon, H.**
1997: Decadal variations in climate associated with the North
Atlantic Oscillation. *Climatic Change* 36, 301–326. **Wilby, R.L.
and Wigley, T.M.L.** 1997: Downscaling general circulation
model output: a review of methods and limitations. *Progress in
Physical Geography* 21, 530–548. **Wright Jr, H.E., Kutzbach,
J.E., Webb III, T. et al. (eds)** 1993: *Global climates since the Last
Glacial Maximum.* Minneapolis, MN: University of Minnesota
Press.

regional cooling The concept that some parts of the
Earth are experiencing a decrease in temperatures, despite
increasing evidence for and the widespread belief in
GLOBAL WARMING. Possible explanations include: (1) cooling in some regions being a natural consequence of global
warming, simply reflecting the spatial expression of interactions in the CLIMATIC SYSTEM; (2) additional causes of
CLIMATIC CHANGE operating at regional and local scales
and overriding any global tendency for warming; and (3)
current evidence for and future predictions of global
warming being uncertain and possibly misleading. *JAM*

[See also REGIONAL CLIMATIC CHANGE]

Allard, M., Wang, B. and Pilon, J.A. 1995: Recent cooling
along the southern shore of Hudson Strait, Quebec, Canada,
documented from permafrost temperature measurements. *Arctic
and Alpine Research* 27, 157–166. **Kullman, L.** 1996: Recent
cooling and recession of Norway spruce (*Picea abies* (L.) Karst.)
in the forest–alpine tundra ecotone of the Swedish Scandes.
Journal of Biogeography 23, 843–854. **Normile, D.** 1995: Polar
regions give cold shoulder to theories. *Science* 270, 1566.

regionalism Although it has many expressions in geographical studies, the concept of place is the richest quality of regionalism. The basic notion of subdivision of the
Earth's surface can be expressed and measured mechanically, as in Hartshorne's concept of '*areal differentiation*' or
more modern exercises in *regionalisation*, but place adds
something to regionalism. Principally, the extra dimension lies in the realms of values, meanings, traditions and
emotive ties between place, or region, and its occupants.

Vidal de la Blache's famous dictat, 'place is a medal cast in the image of its people' captures the concept in a magnificent way. Regionalism has been used as a way of organising the teaching of GEOGRAPHY, with courses on different parts of the world such as Africa, North America, Australasia and Europe on the theme of area studies. It is linked with methodological exercises in regionalisation, where the aim is to partition an area in meaningful and convenient divisions for administrative purposes such as electoral districts, school catchments and health authorities. In PHYSICAL GEOGRAPHY, similar principles can be applied to the delimitation of DRAINAGE BASINS, CLIMATIC CLASSIFICATION or the ZONATION of vegetation and soils. The concept of territory is important and underpins some of the urban regionalisations, such as the definition of standard metropolitan areas (SMAs). Regionalism has a range of applications, but its purest expression concerns people, environment and LANDSCAPES in recognisable and different territories.   *DTH*

Holt-Jensen, A. 1999: *Geography: history and concepts.* London: Sage. **Livingstone, D.** 1992: *The geographical tradition.* Oxford: Blackwell.

Regionally Important Geological and Geomorphological Sites (RIGS)

A non-statutory designation in the UK for CONSERVATION of sites that do not merit the status of a SITE OF SPECIAL SCIENTIFIC INTEREST but are worthy of protection for their educational, historical or aesthetic value with respect to MINERALS, rocks, FOSSILS, SOILS, LANDFORMS or LANDSCAPE. RIGS sites are locally managed by voluntary groups.   *GO*

[See also GEOLOGICAL CONSERVATION]

Royal Society for Nature Conservation 1999: *The RIGS handbook.* Newark: Royal Society for Nature Conservation.

regolith

A general term for all the superficial and unconsolidated material at the Earth's surface; it is not only the product of WEATHERING.   *SHD*

[See also IMPACTITE]

Gale, S.J. 1992: Regolith: the mantle of unconsolidated material on the Earth's surface. *Quaternary Research* 37, 261–262.

Regosols

Deep soils with limited profile development in unconsolidated weathered materials, excluding coarse sands and alluvial deposits. Regosols occur widely in desert areas of North America, Africa, the Near East and Australia (USDA: *Entisols*). These soils are vulnerable to GULLY EROSION.   *EMB*

[See also WORLD REFERENCE BASE FOR SOIL RESOURCES]

regression analysis A frequently used (and misused) statistical technique for describing and modelling the relationship between a *response variable* (e.g. plant species) and one or more predictor or *explanatory variables* (e.g. soil pH). The major aim is to describe the response variable as a function of one or more explanatory variables. This response function cannot be chosen so that the function will predict the responses without error. Regression tries to make these errors as small as possible and averages them to zero. The value predicted by the response function is the expected response with the error averaged out.

All regression analyses are based on a response model consisting of a systematic part, which describes the way in which the expected response depends on the explanatory variables, and an error part, which describes the way in which the observed responses deviate from the expected responses. The systematic part is specified by the regression equation, the error part by the statistical distribution of the error. In the most common type of regression, LEAST-SQUARES REGRESSION, the error distribution is assumed to follow the NORMAL DISTRIBUTION. The response variable can be quantitative or nominal (including presence/absence) and the explanatory variables can be quantitative, nominal (including presence/absence) and/or ordinal. Different types of regression analysis are used for different types of response and predictor variables (see Table). Almost all regression techniques fall in the framework of GENERALISED LINEAR MODELS. If there is more than one response variable, *multivariate regression* is required. REDUNDANCY ANALYSIS, also known as *reduced rank regression*, is a useful approach here.

Many things can go wrong in regression analysis. The type of response model or error distribution may have been incorrectly specified and there may be outliers that unduly influence the regression model. Regression diagnostics, often involving examination of RESIDUALS, are an important part of regression analysis. If a particular parametric form of the systematic part cannot be assumed, non-parametric regression techniques can be used (e.g. GENERALISED ADDITIVE MODELS). Specialised biased regression techniques exist for guarding against the effects of high correlations between explanatory variables (e.g. ridge regression, principal components regression, *partial least-squares* regression). *Inverse regression techniques*, where an environmental variable (e.g. lake pH) is modelled as a function of many biological variables (e.g. diatom species), can be used for quantitative environmental reconstruction and CLIMATIC RECONSTRUCTION.   *HJBB*

Braak, C.J.F. ter and Looman, C.W.N. 1987: Regression. In Jongman, R.H.G., ter Braak, C.J.F. and van Tongeren, O.F.R.

regression analysis *Types of regression analysis related to the characteristics of the explanatory and response variables*

Response variable	Explanatory variable			
	One		Many	
	Nominal	Quantitative	Nominal	Quantitative
Quantitative	ANOVA	Linear and non-linear regression	Multiple linear regression with nominal dummy variables	Multiple linear regression
±	χ^2 contingency table analysis	Logit regression	Multiway contingency table analysis	Multiple logit regression

(eds), *Data analysis in community and landscape ecology*. Wageningen: Podoc, 29–77. **Draper, N. and Smith, H.** 1981: *Applied regression analysis*. New York: Wiley. **Rawlings, J.O.** 1988: *Applied regression analysis: a research tool*. Pacific Grove, CA: Wadsworth.

regression, marine

A seaward shift in the position of a shoreline. Evidence of a marine regression in a succession of SEDIMENTS or SEDIMENTARY ROCKS may be preserved in the form of non-marine deposits overlying marine; an upward shallowing of FACIES; or, in TERRESTRIAL ENVIRONMENTS, a change in facies type or FACIES ARCHITECTURE that might indicate, for example, entrenchment of a river channel. A *forced regression* is caused by ALLOCYCLIC mechanisms that bring about an actual sea-level change at the coast, such as a *eustatic* sea-level fall (i.e. absolute SEA-LEVEL CHANGE) or a decrease in TECTONIC subsidence, or actual UPLIFT (a relative sea-level change). A *progradational regression* occurs through AUTOCYCLIC mechanisms under conditions of constant sea level, such as the accumulation of sediment leading to PROGRADATION of a shoreline, and can occur during a period of eustatic SEA-LEVEL RISE if the SEDIMENT ACCUMULATION RATE is sufficient. One of the challenges in modern SEDIMENTOLOGY is to separate these complex relationships amongst controls on DEPOSITIONAL ENVIRONMENTS represented in the form of regressions and TRANSGRESSIONS. *GO*

[See also COASTAL ENVIRONMENTAL CHANGE, SEQUENCE STRATIGRAPHY]

Collier, R.E.Ll., Leeder, M.R. and Maynard, J.R. 1990: Transgressions and regressions: a model for the influence of tectonic subsidence, deposition and eustasy, with application to Quaternary and Carboniferous examples. *Geological Magazine* 127, 117–128. **Hampson, G.J.** 1998: Evidence for relative sea-level falls during deposition of the Upper Carboniferous Millstone Grit, South Wales. *Geological Journal* 33, 243–266. **Havholm, K.G. and Kocurek, G.** 1994: Factors controlling aeolian sequence stratigraphy: clues from super bounding surface features in the Middle Jurassic Page Sandstone. *Sedimentology* 41, 913–934. **Leeder, M.R.** 1999: *Sedimentology and sedimentary basins: from turbulence to tectonics*. Oxford: Blackwell Science. **Leeder, M.R. and Stewart, M.D.** 1996: Fluvial incision and sequence stratigraphy: alluvial responses to relative sea-level fall and their detection in the geologic record. In Hesselbo, S.P. and Parkinson, D.N. (eds), *Sequence stratigraphy in British geology* [*Special Publication* 103]. London: Geological Society, 25–39.

rehabilitation

The partial return of a disturbed environmental SYSTEM to its predisturbance condition by human intervention. *JAM*

[See also DISTURBANCE, REMEDIATION, RESTORATION]

rejuvenation

In the geomorphological context, rejuvenation is most commonly used to describe renewed EROSION, especially with reference to FLUVIAL PROCESSES. There are many possible causes for the rejuvenation of rivers, including a fall in BASE LEVEL, tectonic UPLIFT, CLIMATIC CHANGE leading to greater rainfall, CHANNEL CHANGE or reduced load (see, for example RESERVOIRS: ENVIRONMENTAL EFFECTS), some of which are ANTHROPOGENIC. *JAM*

[See also KNICKPOINT, RIVER TERRACE]

relational database

A structured collection of data organised into two-dimensional tables known as relations, consisting of sets of records or *tuples*. Each tuple in a relation is composed of an ordered and inter-related set of FIELDS, where each set can be uniquely identified by a primary field or key. Fields are composed of ATTRIBUTES, which can be either LABELS or values. For example, a tuple on national daily precipitation would include fields of attributes measuring amount, frequency, duration, type and time. Each set of attributes would be uniquely identified by the location or code of a measuring station. Similarly, a tuple recording the state of a river would include fields of attributes measuring velocity, length, depth, water quality and load; the key field would be the name of the reach.

Relational DATABASES are usually computerised and follow a conceptual schema of a data model based on set theory. The rationale is built on the premise that data are grouped by inter-related characteristics and represented by unique fields. This allows efficient data access and data transformation through implicit relationships between tuples and between relations simply by reference to unique fields. Relationships are constructed by the user via a QUERY LANGUAGE, such as SQL (*structured query language*). The query language is a software interface that allows the user to ask various questions of the database but does not assume prior knowledge of computer architecture.

A computer database is at the heart of a GEOGRAPHICAL INFORMATION SYSTEM (GIS). Data are stored and retrieved as and when they are needed for GIS operations. Efficiency also depends on the condition of the *database management system* (such as INGRES), which oversees the link between data and GIS functions. *TVM*

Stonebraker, M. 1986: The design and implementation of INGRES. *ACM Transactions on Database Systems* 1, 189–222. **Van Roessel, J.W.** 1987: Design of a spatial data structure using the relational normal forms. *International Journal of Geographical Information Systems* 1, 33–50. **Worboys, M.F.** 1999: Relational databases and beyond. In Longley, P.A., Goodchild, M.F., Maguire, D.J. and Rhind, D.W. (eds), *Geographical information systems: principles, techniques, applications, and management*. Chichester: Wiley, 373–384.

relative age

The age of an object, a surface, a soil, a sediment or rock unit or an event, expressed as older than, younger than or equal to others, rather than in terms of an absolute standard. The order of layered (stratified) sediment or rock units is determined using the principles of STRATIGRAPHY, such as *superposition* (higher layers in a succession are younger), *original horizontality* (tilted and folded layers formed as horizontal BEDS), *cross-cutting relationships* (a feature that cuts another must be younger), *way up* (see SOLE MARKS) and UNCONFORMITIES. The relative ages of local units are compared to internationally agreed units in the STRATIGRAPHICAL COLUMN by methods of CORRELATION. The RESOLUTION of the relative, chronostratigraphical time scale varies through the PHANEROZOIC. It is generally higher in younger sediments and rocks where deformation and alteration are less, especially in the records of OCEANIC SEDIMENTS, which can be analysed using techniques such as MAGNETOSTRATIGRAPHY, STRONTIUM ISOTOPE stratigraphy and CYCLOSTRATIGRAPHY. The resolution obtained by RELATIVE-AGE DATING using BIOSTRATIGRAPHY and other techniques for correlation remains far higher than that obtained by isotopic or RADIOMETRIC DATING. *LC*

Harland, W.B., Armstrong, R.L., Cox, A.V. *et al.* 1990: *A geologic time scale 1989.* Cambridge: Cambridge University Press. McArthur, J.M. 1994: Recent trends in strontium isotope stratigraphy. *Terra Nova* 6, 331–358. Nichols, G. 1999: *Sedimentology and stratigraphy.* Oxford: Blackwell Science.

relative-age dating A range of generally inexpensive and often preliminary techniques used especially to date surfaces and sediments. The approach is distinct from NUMERICAL-AGE DATING (or *absolute-age dating*) in that the aim is to determine *age equivalence* (see CORRELATED-AGE DATING) or to place surfaces, bodies of sediment, soils or events in rank order of RELATIVE AGE. It is commonly applied to glacial deposits and to surfaces in arid environments. Techniques of relative-age dating often rely on simple measures of the degree of rock surface WEATHERING or soil development, or on *superposition*.

In the context of SURFACE DATING, common measures include: WEATHERING RIND thickness; SCHMIDT HAMMER rebound values; rock surface roughness; JOINT depth; frequency of surface boulders; surface discolouration; changes in soil depth; SOIL HORIZON development or soil chemistry; and subjective assessment of degree of weathering. Although sometimes criticised as more art than science, these techniques have been used to establish the relative ages of glacial deposits over much of North America and to distinguish glacially abraded surfaces from former NUNATAKS, allowing mapping of the upper surface of former ice sheets in a number of glaciated or formerly glaciated areas. Attempts to produce numerical ages using such measures have generally underestimated the likely degree of uncertainty. In recent years it has been possible to provide numerical exposure ages for rock surfaces using COSMOGENIC-ISOTOPE DATING. *DMcC*

[See also CATION-RATIO DATING, MARKER HORIZON, SOIL DATING, WEATHERING INDICES]

Ballantyne, C.K., McCarroll, D., Nesje, A. *et al.* 1998: The last ice sheet in north-west Scotland: reconstruction and implications. *Quaternary Science Reviews* 17, 1149–1184. McCarroll, D. 1991: Relative-age dating of inorganic deposits: the need for a more critical approach. *The Holocene* 1, 174–180.

relative density The ratio of the DENSITY of a substance at a given temperature to the density of water at −4°C. Formerly known as *specific gravity*, a substance floats on water if it has a relative density of < 1.0. *JAM*

relative sea level The local sea level in relation to the land, which may be affected by GLACIO-ISOSTASY, HYDROISOSTASY, erosional and depositional isostasy, compaction of sediments leading to subsidence, OROGENY, EPEIROGENY and continental and ice-sheet MASS ATTRACTION. Relative sea level at any location can remain zero when the rate of ABSOLUTE SEA LEVEL change is exactly balanced by a corresponding rate of local land UPLIFT or SUBSIDENCE. *MJH*

[See also SEA-LEVEL CHANGE]

Goudie, A.S. 1992: *Environmental change*, 3rd edn. Oxford: Oxford University Press, 217–254. Higgins, C.G. 1965: Causes of sea-level changes. *American Scientist* 53, 464–476.

relaxation The passage of a SYSTEM from one equilibrium state to another. The *relaxation path* is the sequence of states whereby this adjustment occurs following a PERTURBATION or DISTURBANCE. The RELAXATION TIME is the time necessary for the system reorganisation.

In the context of BIODIVERSITY, when once extensive and relatively continuous ecosystems are reduced in coverage and become fragmented, populations of NATIVE SPECIES may become stranded on shrinking patches of their natural habitats. In small and isolated habitats, EXTINCTION rates will be particularly high, while the likelihood that such patches will be recolonised will be low. As a result, the diversity of fragments of native communities will decrease, or 'relax' either until all species are lost or until extinction and immigration reach a new balance (i.e. at a relatively low number of species). *MVL*

[See also CONSERVATION BIOLOGY, FRAGMENTATION, HABITAT LOSS, ISLAND BIOGEOGRAPHY, SINGLE LARGE OR SEVERAL SMALL RESERVES]

Brown, J.H. 1971: Mammals on mountaintops: non-equilibrium insular biogeography. *American Naturalist* 105, 467–478. Chorley, R.J. and Kennedy, B.A. 1971: *Physical geography: a systems approach.* London: Prentice-Hall International. Whittaker, R.J. 1998: *Island biogeography: ecology, evolution and conservation.* Oxford: Oxford University Press.

relaxation time The time lapse between the beginning of the response of a SYSTEM to an environmental change and the attainment of a new equilibrium condition. *JAM*

[See also LAG TIME, REACTION TIME]

relict A landscape, an element of a landscape such as a soil, a soil characteristic, a particular landform, a tree stump, or any other feature that is not a product of contemporary environmental conditions is said to be a relict feature (e.g. a relict soil). *JAM*

[See also COLD-BASED GLACIERS, PALAEOSOLS]

Kleman, J., Borgström, I. and Hättestrand, C. 1994: Evidence for a relict glacial landscape in Quebec–Labrador. *Palaeogeography, Palaeoclimatology, Palaeoecology* 111, 217–228. Simpson, L.A. 1997: Relict properties of anthropogenic deep top soils as indicators of infield management in Marwick, West Mainland, Orkney. *Journal of Archaeological Science* 24, 365–380.

remediation A form of LAND RECLAMATION or regeneration, whereby soil contaminated with chemicals, radioactivity or harmful organisms is treated (chemically, physically or with biotechnology) to render it harmless. A cheaper alternative may be removal, disposal and replacement, but this simply moves the problem elsewhere and should be avoided wherever possible. There is some overlap with: REHABILITATION, applied where revegetation takes place and improves on the original cover; RESTORATION, the recreation of vegetation or landscape; and *amendment*, chemical and physical treatment to improve alkalinised or salinised soil. *CJB*

[See also CONTAMINATED LAND, LAND RESTORATION, SOIL RECLAMATION]

Fleming, G. (ed.) 1996: *Recycling derelict land.* London: Thomas Telford.

remote sensing The means by which the Earth can be studied using a device separated from it by some distance. Within the context of ENVIRONMENTAL CHANGE, the remote sensing definition can be refined to focus on the

measurement and recording of those parts of the ELEC-
TROMAGNETIC SPECTRUM emanating from the Earth's environment (surface and atmosphere) by sensors mounted
on a platform at a vantage-point above the Earth's surface.
Remote sensing provides a synoptic view that carries a
continuous and consistent record of the environment. It is
much superior to traditional representation derived by
INTERPOLATION of limited field observations.

Remote sensing through the medium of electromagnetic radiation comprises four components, all of which
are linked through the transfer of radiation. The first component is the source of electromagnetic radiation.
Radiation is either natural, originating from the Sun or the
Earth by emission, or artificial. The respective use of
either natural or artificial radiation determines whether
remote sensing is undertaken passively or actively. Most
remote sensing is passive, a familiar example being AERIAL
PHOTOGRAPHY. A common example of active remote sensing is *microwave radio direction and ranging* (RADAR) and its
use for the study of the Earth's environment is rapidly
evolving. The second component of remote sensing refers
to the interaction of radiation with the Earth's surface.
The nature of this interaction is determined by the essential physical, chemical and biological properties of the
Earth's surface, which in turn determine the amount of
radiation measured by a sensor. Electromagnetic radiation
may also interact with the atmosphere residing between
the Earth's surface and the sensor, contributing to the
radiation reaching the sensor, and this constitutes the
third component. The final component of remote sensing
for the study of environmental change is the sensor, which
measures remotely the radiation from the Earth's environment. Predominantly, the electromagnetic radiation
received by a remote sensor is measured electronically and
used to produce a digital record of radiation from the
Earth's environment. The digital record is expressed as an
image, composed of a two-dimensional array of discrete
PIXELS, each of which corresponds to the average radiation
at a given *wavelength* measured over the ground area and
intervening atmosphere corresponding to each pixel. The
radiation measured is a function of the location, time,
wavelength and the geometric relationship between the
radiation source, target of interest and sensor and can be
interpreted to provide information about the Earth's environment. The basis for this interpretation is provided
through the knowledge of how electromagnetic radiation
interacts with Earth surface and atmospheric properties,
coupled with appropriate IMAGE PROCESSING. The digital
format of the remotely sensed data is conducive for computer-based analysis and integration with other datasets
within a GEOGRAPHICAL INFORMATION SYSTEM (GIS).

Remotely sensed records for the study of the Earth's
environment are currently available from a host of sensors,
carried on *airborne* and SATELLITE platforms, which measure radiation at a range of pixel sizes (i.e. the spatial RES-
OLUTION), wavebands (i.e. the spectral resolution) and
temporal frequencies (i.e. the temporal resolution). Such
variations in resolution, along with variations in the geometric relationship of radiation measurement, may provide unique radiation measurements of the Earth's
environment, which can be exploited. This may involve
multispectral remote sensing whereby radiation measured
simultaneously in a number of wavebands is used.
Commonly, vegetation properties (e.g. leaf area) are esti-

mated using VEGETATION INDICES, which are simple mathematical expressions that combine spectral measurements
in two or more wavebands. *Multistage remote sensing*, the
use of radiance acquired at a number of spatial resolutions, offers a unique opportunity to study the Earth's
environment over a wide range of scales. The use of radiation measured at various geometric relationships (*multi-
view angle remote sensing*) provides the potential to infer
unique information on the Earth's surface through an
investigation of their angular properties. MULTITEMPORAL
ANALYSIS permits ENVIRONMENTAL MONITORING over time
periods appropriate for studying environmental change.
Moreover, the timely consistent information on the environment provided by remote sensing affords a long-term
database for the examination of the dynamics of environmental change. The most compelling argument for using
remote sensing in the study of environmental change is
that at regional to global scales remote sensing is the only
feasible means of acquiring information on significant
environmental properties at a relatively fine temporal resolution. In this instance, remote sensing may be used to
estimate directly a physical VARIABLE independently of
other data, to infer the non-measurable from the measurement of a related variable, to extrapolate local scale measurements to estimates for large areas, or to parameterise
and drive environmental MODELS.

Remotely sensed records have a huge potential to yield
useful and vital information about the Earth's environment and hence to increase our understanding of the environment in space and time. This potential will be
enhanced further by the imminent launch of more *Earth
observation satellites*, which will constitute a comprehensive
array of environmental sensors. The most noteworthy of
these are the European Space Agency's (ESA) *Envisat* and
the National Aeronautics and Space Administration's
(NASA) *Earth Observing System, Terra*. These satellites
will provide an international Earth-observing capacity for
acquiring a vast amount of data for local to global environmental-change research. These will be further comple-
mented by remotely sensed data acquired by the recent
proliferation of sensors with fine spatial resolution and, in
the far future, a new family of scientific Earth observation
missions (such as ESA's Earth Explorers). *DSB*

[See also ARCHAEOLOGICAL PROSPECTION, GEOPHYSICAL
SURVEYING]

Abbott, A. 1996: ESA wants compulsory Earth missions. *Nature*
381, 453. **Barnsley, M.J., Alison, D. and Lewis, P.** 1997: On
the information content of multiple view angle (MVA) images.
International Journal of Remote Sensing 18, 1937–1960. **Curran,
P.J.** 1985: *Principles of remote sensing*. Harlow: Longman.
Curran, P.J. and Foody, G.M. 1994: Environmental issues at
regional to global scales. In Foody, G.M. and Curran, P.J. (eds),
Environmental remote sensing from regional to global scales.
Chichester: Wiley, 1–7. **Lillesand, T.M. and Kiefer, R.W.**
1994: *Remote sensing and image interpretation*, 3rd edn. New York:
Wiley. **Stoms, D.M. and Estes, J.E.** 1993: A remote sensing
research agenda for mapping and monitoring biodiversity.
International Journal of Remote Sensing 14, 1839–1860.
Verstraete, M.M., Pinty, B. and Myeni, R.B. 1996: Potential
and limitations of information extraction on the terrestrial bio-
sphere from satellite remote sensing. *Remote Sensing of
Environment* 58, 201–214. **Wickland, D.E.** 1989: Future direc-
tions for remote sensing in terrestrial ecological research. In
Asrar, G. (ed.), *Theory and applications of optical remote sensing*.
New York: Wiley, 691–724.

Rendzina A shallow mineral soil with an organic-rich surface horizon and high base status overlying calcareous PLANT material. This soil grouping has been subsumed into the LEPTOSOLS. The name rendzina is expressive of the scratching noise of a plough on limestone (SOIL TAXONOMY: *Rendolls*).   *EMB*

Kubiena, W.L. 1953: *The soils of Europe*. London: Thomas Murby.

renewable energy Natural sources of energy, whose supply can be replaced as they are used. They include the WIND, sun, water (including TIDES and WAVES), BIOMASS and GEOTHERMAL sources. At present these are most significant in *developing countries*. Globally, their current use is limited by being technically difficult to convert into a usable form, expensive to develop, limited in their range of applications and geographically dispersed. They have considerable potential for the future and are likely increasingly to replace FOSSIL FUELS during the twentyfirst century. HYDROPOWER is the most widely used form, but often has high environmental and economic costs because of large-scale engineering. Most renewable energies have far less damaging actual or potential environmental impacts than non-renewable fossil fuels or *nuclear energy* (although the latter can be reprocessed). Many forms are unsuitable for large-scale centralised power generation but have considerable potential at present for small-scale, dispersed power needs, amenable to local control in relatively isolated places.   *JGT*

[See also ALTERNATIVE ENERGY, RESERVOIRS: ENVIRONMENTAL EFFECTS]

Boyle, G. 1996: *Renewable energy: power for a sustainable future*. Oxford: Oxford University Press. **Twidell, J. and Weir, A.** 1986: *Renewable energy resources*. London: E & F Spon.

renewable resources RESOURCES that naturally replace themselves or are replenished by human activity; they are also known as *flow resources*. They may be regarded as effectively non-renewable, however, if the rate of renewal is less than a human life span. Resources that are ubiquitous and the quantities of which are largely unaffected by humans are termed *base resources* and include water, air and solar energy. *Derivative resources* are those that rely upon and stem from base resources, and include plant and animal materials. Plants and animals have been used for food, clothing, shelter, heat, decorative and medicinal purposes for as long as humans have existed and methods of exploitation have changed from HUNTING, FISHING AND GATHERING techniques to controlled agricultural practices and uncontrolled exploitation. Ecological resources are only renewable if rates of harvesting and removal remain SUSTAINABLE and do not exceed rates of regrowth or regeneration.

The likelihood of over-exploitation grows as local and global *population growth* increases and species and habitats are increasingly being hunted or harvested to the point of EXTINCTION. Recent, widely publicised examples of non-sustainable exploitation include the depletion of fish stocks in the North Atlantic due to OVERFISHING and the DEFORESTATION of TROPICAL RAIN FOREST in South America. Even when cultivation and harvesting are undertaken in controlled environments such as agricultural land, intensive arable regimes often do not allow soil nutrient levels or

structure be maintained, and the stocking of land in excess of its CARRYING CAPACITY often leads to OVERGRAZING. Both situations can result in SOIL DEGRADATION and EROSION, along with reduced yields. Resources that are used to generate RENEWABLE ENERGY include solar, wind, geothermal, nuclear and tidal power, and these are growing in importance because of the non-renewable nature and GREENHOUSE GAS-inducing effects of FOSSIL FUELS.   *MLW*

[See also NON-RENEWABLE RESOURCES]

Clark, C.W. 1990: *Mathematical bioeconomics: the optimal management of renewable resources*. New York: Wiley. **Douglas, A.J. and Johnson, R.L.** 1993: Harvesting and replenishment policies for renewable natural resources. *Journal of Environmental Management* 38, 2742–2752. **Howe, C.W. (ed.)** 1982: *Managing renewable natural resources in developing countries*: Boulder, CO: Westview Press. **Lewis, T.R. and Schmalensee, R.** 1977: Nonconvexity and optimal exhaustion of natural resources. *International Economic Review* 18, 535–551. **Rees, J.** 1990: *Natural resources: allocation, economics and policy*, 2nd edn. London: Routledge. **Ruddle, K. and Manshard, W.** 1981: *Renewable natural resources and the environment: pressing problems in the developing world*. Dublin: Tycooly International. **World Energy Council** 1994: *New renewable energy resources: a guide to the future*. London: Kogan Page.

replication (1) The statistical procedure of repeated sampling from a class of phenomena; part of EXPERIMENTAL DESIGN. (2) The situation when two or more variables in a single correlation analysis contain the same mathematical information; consequently a notion of *redundancy*.   *CET*

reproducibility In order to be acceptable to the scientific community, an individual's experimental results or observations must be reproducible by other scientists. Implicitly, this demands that a scientist must describe the theory, methodology and techniques employed in a sufficiently comprehensive manner to permit attempts at reproduction.   *CET*

reserves The amount of a RESOURCE available for use. This depends not only on the total amount in existence, but also on factors such as new discoveries, technological advances, location of deposits, cost and demand. The supply or stock of a resource is the known reserves divided by the annual consumption. Estimates of reserves and supply are notoriously unreliable. A good illustration is given by the example of CRUDE OIL. Since little is stored, reserves-to-production ratios are an estimate of available supplies. In the AD 1960s the ratio of reserves to production indicated an available supply of no more than 30 years, partly fuelling the oil price rises and crises of the 1970s. By 1988 global consumption had doubled, but improved technology and the discovery of new reserves (including the North Sea) had increased proven reserves almost threefold, giving a supply of over 40 years. In recent years reserves-to-production ratios for crude oil have changed little: discoveries of new oilfields and improvements in technology have approximately matched global consumption each year (see PETROLEUM). Current proven reserves are estimated to be sufficient for at least 40 years and it is estimated that a similar amount of conventional petroleum remains to be found (see also SYNFUELS).   *GO*

[See also NATURE RESERVE, PROTECTED AREAS]

BP Amoco 2000: *BP Amoco statistical review of world energy, June 2000.* London: BP Amoco. **Woodcock, N.H.** 1993: Fossil fuels: a crisis of resources or effluents? *Geology Today* 9, 142–147.

reservoir effect   Global or local alteration of ^{14}C concentrations, leading to different RADIOCARBON ages for contemporary samples from different parts of the global exchange reservoir. Southern Hemisphere terrestrial sample ages are up to 30 years greater than Northern Hemisphere ones because of slow mixing between hemispheres and differences in the amount of atmospheric CARBON DIOXIDE exchange with the sea. The APPARENT AGE (c. 400 years) of marine samples compared to terrestrial samples is caused by the long mean RESIDENCE TIME of carbon in the upper mixed layer of the oceans. Another reservoir effect is the SUESS EFFECT, whereby atmospheric ^{14}C is diluted by the burning of fossil fuels, which can be locally enhanced near industrial areas. Similar effects may be found close to volcanic areas. The HARD-WATER EFFECT is a localised phenomenon in which aquatic plants derive their carbon from groundwaters containing carbon dissolved from carbonaceous rock. These effects produce greater than expected radiocarbon ages, while the opposite occurs with the BOMB EFFECT produced by the atmospheric testing of nuclear weapons, which caused a near doubling of the atmospheric ^{14}C concentration in the 1960s.   *PQD*

McCormac, F.G., Hogg, A.G., Higham, T.F.G. *et al.* 1998: Variations of radiocarbon in tree rings: Southern Hemisphere offset preliminary results. *Radiocarbon* 40, 1153–1159. **Stuiver, M., Pearson, G.W. and Braziunas, T.** 1986: Radiocarbon age calibration of marine samples back to 9000 cal year BP. *Radiocarbon* 28, 980–1021. **Stuiver, M., Reimer, P.J. and Braziunas, T.F.** 1998: High-precision radiocarbon age calibration for terrestrial and marine samples. *Radiocarbon* 40, 1127–1151.

reservoirs: environmental effects   The impact of reservoirs for flow regulation, in the management of WATER RESOURCES or during *flood control* on the local environment, river system or drainage basin. Environmental impacts can arise from initial dam construction, subsequent operation and the intrusive effects of a large water body. The magnitude and direction of impacts (hydrological, geomorphological, hydrochemical, seismic, ecological, micrometeorological and social) vary hugely with reservoir type, basin context and operational demands. Principal downstream hydrological impacts include reduction of peak-flow and flood frequency, increase in low flows and trapping of sediment (up to 95% fine sediment losses have been reported) – see Figure. This can lead to enhanced scour immediately downstream, but a reduction in width and/or depth and hence in CHANNEL CAPACITY further downstream. A downstream HYDRAULIC GEOMETRY approach is often employed to assess the magnitude and downstream effect of reservoir impacts. Reservoirs may also affect river thermal regime downstream, e.g. increase or decrease the mean river temperatures, depress the summer maximum and delay the annual thermal cycle. Seismic effects (see MICROSEISM) of large reservoirs result from loading of the Earth's crust. Freshwater ecological impacts, driven by river flow, temperature, turbidity and dissolved oxygen changes and 'dam-barrier' effects, include delay of fish spawning, fish kills associated with reservoir releases and changes in

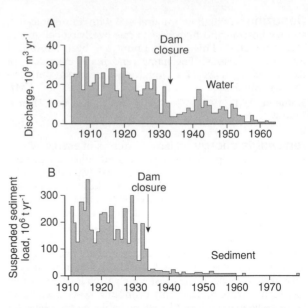

reservoirs: environmental effects *Changes in (A) discharge and (B) suspended sediment load below the Hoover Dam, Colorado River, USA, following dam closure (after Meade and Parker, 1985; Knighton, 1998)*

invertebrate behaviour and population dynamics. Local climates may become moister and thermally subdued. Social impacts, especially resettlement, are currently controversial (e.g. the Three Gorges Dam, China).   *DML*

Knighton, A.D. 1998: *Fluvial forms and processes: a new perspective.* London: Arnold. **Meade, R.H. and Parker, R.S.** 1985: Sediment in rivers of the United States. In *National Water Summary 1984. United States Geological Survey Water Supply Paper* 2275, 49–60. **Petts, G.E.** 1984: *Impounded rivers: perspectives for ecological management.* Chichester: Wiley. **Petts, G.E., Foulger, T.R., Gilvear, D.J.** *et al.* 1985: Wave movement and water quality variations during a controlled release from Kielder Reservoir, North Tyne River, UK. *Journal of Hydrology* 80, 371–389. **Surian, N.** 1999: Channel changes due to river regulation: the case of the Piave River, Italy. *Earth Surface Processes and Landforms* 24, 1135–1151.

residence time   (1) Organic matter that is incorporated into the soil in the form of HUMUS remains in the soil for a considerable period; this is its residence time. Analysis by ^{14}C reveals that, although fresh organic matter is rapidly decomposed, the average time carbon is resident in some British soil A horizons is 22 years; in subsoils the period is much longer, over 1000 years, giving rise to the possibility of CARBON SEQUESTRATION in soils. (2) Residence time is also used by hydrologists to describe the passage of water through the landscape from its input as precipitation until its re-emergence from deep percolation as springs or seepage water from soils. (3) In the context of climatology, the residence time is the length of time that a molecule spends in a reservoir or SINK before being removed. The concept is also used in other environmental contexts.   *EMB*

[See also APPARENT MEAN RESIDENCE TIME, SOIL DATING]

Jenkinson, D.S. 1990: The turnover of organic carbon and nitrogen in soil. *Philosophical Transactions of the Royal Society London* B329, 361–368. **Jenkinson, D.S. and Rayner, J.M.**

1977: The turnover of soil organic matter in some of the Rothamsted classical experiments. *Soil Science* 123, 298–305.

residual The difference between the observed value of a response variable and the value fitted or predicted by some model (e.g. a REGRESSION ANALYSIS model). Examination of residuals provides a means of checking the assumptions of the model (e.g. NORMAL DISTRIBUTION).

HJBB

Rawlings, J.O. 1988: *Applied regression analysis: a research tool.* Pacific Grove, CA: Wadsworth.

residual deposit Sediments or soils remaining as RESIDUES after removal of the majority of the material following *in situ* WEATHERING or rock disintegration. Residual deposits, which may in some cases be the products of chemical alteration, rest on the material from which they were formed. They are particularly common in the HUMID TROPICS, where they result from high rates of CHEMICAL WEATHERING, but the BLOCKFIELDS of arctic–alpine environments may also be considered as residual deposits.

JAM

[See also WEATHERING PROFILE]

Fookes, P.G. 1997: *Tropical residual soils.* London: The Geological Society.

residual landform A relatively small, upstanding LANDFORM remaining after the removal by WEATHERING and/or EROSION of the bulk of the material comprising a much larger landform.

JAM

[See also INSELBERG, KARST, TOR]

residue There are many contexts in which a relatively small quantity of material remains after the removal of the bulk. Thus, in an archaeological context, residues include small quantities of original material, such as hair, bone, blood or food, or altered chemical residues surviving on the surface of ARTEFACTS, such as weapons, tools or utensils. RESIDUAL DEPOSITS remaining after CHEMICAL WEATHERING provide another example.

JAM

[See also BOLUS, MICRORESIDUAL FRACTION]

Evershed, R.P., Heron, C., Charters, S. and Goad, L.J. 1992: The survival of food residues: new methods of analysis, interpretation and application. In Pollard, A.M. (ed.), *New developments in archaeological science.* Oxford: Oxford University Press, 187–208. **Fullagar, R., Furby, J. and Hardy, B.** 1996: Residues on stone artefacts: state of a scientific art. *Antiquity* 70, 740–744. **Gurfinkel, D.M. and Franklin, U.M.** 1988: A study in the feasibility of detecting blood residue on artefacts. *Journal of Archaeological Science* 15, 83–97. **Hillman, G., Wales, S., McLaren, F. et al.** 1993: Identifying problematic remains of ancient plant foods: a comparison of the role of chemical, histological and morphological criteria. *World Archaeology* 25, 94–124.

resilience The ability of a SYSTEM to recover to its original (equilibrium or stable) state following a PERTURBATION or DISTURBANCE. Four aspects of resilience may be recognised: (1) *elasticity* (the rate of return to the original state); (2) *amplitude* (the zone from which recovery is possible); (3) *hysteresis* (the extent to which the recovery pathway differs from the pathway of disruption); and (4) *malleability* (the degree to which a new stable or equilibrium state differs from the original state).

JAM

[See also RELAXATION, RESISTANCE, RESPONSE TIME, SENSITIVITY, SOIL RESILIENCE, STABILITY CONCEPTS]

Slaymaker, O. and Spencer, T. 1998: *Physical geography and global environmental change.* Harlow: Addison-Wesley Longman. **Westman, W.E.** 1978: *Ecology, impact assessment and environmental planning.* New York: Wiley.

resistance The ability of a system to remain unaffected by a PERTURBATION or DISTURBANCE. It may be measured by the magnitude of the disturbance necessary to invoke a response or dislodge the system from an equilibrium condition. Resistance may also be reflected in the REACTION TIME or RESPONSE TIME of the system.

JAM

[See also RESILIENCE, SENSITIVITY, STABILITY CONCEPTS]

Chorley, R.J. and Kennedy, B.A. 1971: *Physical geography: a systems approach.* London: Prentice-Hall International.

resolution, in palaeoenvironmental reconstruction Reconstruction of past environments is possible from a variety of sources of PROXY EVIDENCE (e.g. DIATOM ANALYSIS, POLLEN ANALYSIS, CHIRONOMID ANALYSIS, TREE RINGS, ICE CORES, GLACIER VARIATIONS). Different 'proxies' have different spatial and temporal resolutions. These range from the local or micro scale ($1–5 \times 10^3$ years, $1–10^6$ m^2), through the regional or macro scale ($5 \times 10^3–10^6$ years, $10^6–10^{12}$ m^2), to the global or mega scale ($> 10^6$ years, $> 10^{12}$ m^2). Palaeoenvironmental sensors include: at the local scale TREE RINGS, DIATOMS, PLANT MACROFOSSILS, CHIRONOMIDS and pollen assemblages from small (10–30 m diameter) hollows; at the regional scale pollen assemblages from lakes or bogs, STABLE ISOTOPES, glacier variations and fossil assemblages from MARINE SEDIMENT CORES; and at the global scale ICE CORES. The temporal resolution attainable from different sources depends on the sampling precision and the inherent nature of the preserved record (e.g. *annually laminated sediments*, annual TREE RINGS, SEDIMENT ACCUMULATION RATE). Different environmental processes are important at different scales, varying from extreme variations in weather and DISTURBANCE regime (local scale) to MIGRATION, SOIL DEVELOPMENT and CLIMATIC CHANGE (regional scale) and to changes in ORBITAL PARAMETERS, PLATE TECTONICS and extraterrestrial body impacts (global scale). Decisions concerning proxy source, taxonomic detail, geographical scale, sampling density in time and space, site type and NUMERICAL ANALYSIS all influence resolution in palaeoenvironmental reconstruction.

HJBB

[See also CLIMATIC RECONSTRUCTION, ENVIRONMENTAL CHANGE]

Birks, H.J.B. 1986: Late-Quaternary biotic changes in terrestrial and lacustrine environments, with particular reference to north-west Europe. In Berglund, B.E. (ed.), *Handbook of Holocene palaeoecology and palaeohydrology.* Chichester: Wiley, 3–65.

resolution, in remote sensing A measure of the size of the smallest detail identifiable on an image. There are different resolutions regarding the different aspects of an image: the *spatial resolution* is a measure of the smallest separation between two objects and is usually expressed in radians or metres, while the *spectral resolution* is the ability of a sensing system to differentiate electromagnetic radiation of different frequencies.

ACF

535

Lillesand, J. and Keiffer, R.W. 1994: *Remote sensing and image processing*, 4th edn. New York: Wiley.

resonance ionisation spectroscopy (RIS)

A spectroscopic technique for elemental analysis based on the excitation of particular atoms by a laser beam and the detection of the photo-ions produced. Small numbers of atoms can be detected, such as the rise in relatively rare rhodium (Rh) along with more common iridium (Ir) in sedimentary rocks at the K–T BOUNDARY leading to Rh:Ir ratios coincident with those in meteorites (and hence strengthening the hypothesis of MASS EXTINCTION from an extraterrestrial cause). *JAM*

Hurst, G.S. and Letokhov, V.S. 1994: Resonance ionization spectroscopy. *Physics Today* 47, 38–45.

resource

An anthropocentric concept with qualitative and quantitative meanings: in the qualitative sense, something of use to society, commonly but not necessarily a material; in the quantitative sense, the total amount that exists. Something becomes a resource once demand leads to its exploitation. The term is also used in an ecological context for material necessary for the growth and survival of any organism: *limiting resources* are those the supply of which constrains growth (see LIMITING FACTORS).

Natural resources include anything existing in nature that is capable of economic exploitation, such as plants, animals and habitats, recreational resources, MINERAL RESOURCES and ENERGY RESOURCES. The total amount of material resources in existence is considerably greater than the available RESERVES, which represent the amount available for use (that is, the location is accurately known and it is technically feasible and commercially viable to extract and supply it).

Human demands that are fulfilled by the exploitation of resources include those of food and drinking water, clothing, building materials, heat and electricity generation and manufacturing industries. Resources can be classed as either renewable or non-renewable. NON-RENEWABLE RESOURCES are those that form over geological timescales and whose reserves are finite (for example, FOSSIL FUELS, metal ores and gems) or species that, once extinct, cannot be replaced. RENEWABLE RESOURCES are those that are replaced by human activity or naturally replenish themselves (for example, water, flora and fauna). *Discrete resources* are found at isolated sites, are more likely to be non-renewable and include mineral deposits. *Continuous resources* are found ubiquitously, are more likely to be renewable and include solar energy, air and soil. Environmental change often alters the distribution and quality of resources such as vegetation and water.

What is regarded as a resource by humans changes as society changes. *Potential resources* only become resources when economic, technological or cultural conditions change. *Conditional resources* are those that are known to exist but are not suitable for exploitation at a particular time. *Undiscovered resources* have not yet been discovered but, based on experience, are expected to be discovered in the future, either in areas where the resource is known to be present (*hypothetical resources*) or in unexplored areas (*speculative resources*). *Perpetual resources*, such as solar energy, continue to be available no matter how much they are used. *GO/MLW*

[See also SUSTAINABILITY]

Cutter, S.L., Renwick, H.L. and Renwick, W.H. 1991: *Exploitation, conservation, preservation: a geographical perspective on natural resource use*. New York: Wiley. Grover, J.P. 1997: *Resource competition*. London: Chapman and Hall. Owen, O.S. and Chiras, D.D. 1990: *Natural resource conservation*, 5th edn. New York: Macmillan. Rees, J. 1990: *Natural resources: allocation, economics and policy*. London: Routledge. Simmons, I.G. 1974: *The ecology of natural resources*. London: Edward Arnold. World Resources Institute 1998: *World resources 1998–1999. A guide to the global environment: environmental change and human health*. New York: Oxford University Press. Young, M.D. 1992: *Sustainable investment and resource use: equity, environmental integrity and economic efficiency*. Carnforth: Parthenon.

respiration

A complex series of chemical reactions in organisms by which energy is made available for use. The end products are CARBON DIOXIDE, water, and energy. *RJH*

[See also CARBON ASSIMILATION, PRIMARY PRODUCTIVITY, PRODUCTIVITY]

response function

A numerical description of the way in which a dependent VARIABLE is related to a set of independent variables. In the field of DENDROCLIMATOLOGY, it is a set of correlation or regression coefficients that quantify the association between a single tree-ring chronology and multiple climate time series; usually a number of mean temperature and precipitation totals for each of a succession of months preceding and during the growing season of the tree. The sign and magnitude of the coefficients give a statistical indication of how tree growth is influenced by CLIMATIC FORCING. The statistically significant coefficients together can often indicate an optimum season (aggregation of monthly climate data) that best represents the overall influence of climate on growth and that might be used as a potential predictand in a dendroclimatic reconstruction. It is possible to summarise the results of a number of response functions analyses, e.g. in a graphical representation of the percentage of significant coefficients calculated for a given species or region, or through a PRINCIPAL COMPONENTS ANALYSIS of a number of response functions. It is also possible to explore time dependence in response functions by the simple use of moving analysis windows or the KALMAN FILTER. *KRB*

[See also TRANSFER FUNCTION]

Biondi, F. 1997: Evolutionary and moving response functions in dendroclimatology. *Dendrochronologia* 15, 139–150. Blasing, T.J., Solomon, A.M. and Duvick, D.N. 1984: Response functions revisited. *Tree-Ring Bulletin* 44, 1–15. Briffa, K.R. and Cook, E.R. 1990: Methods of response function analysis. In Cook, E.R. and Kairiukstis, L.A. (eds), *Methods of dendrochronology. Applications in environmental sciences*. Dordrecht: Kluwer/ IIASA, 240–247. Fritts, H.C., Blasing, T.J., Hayden, B.P. and Kutzbach, J.E. 1971: Multivariate techniques for specifying tree-growth and climate relationships and for reconstructing anomalies in palaeoclimate. *Journal of Applied Meteorology* 10, 845–864.

response time

The time taken for a SYSTEM to regain equilibrium after a PERTURBATION. The response of various components of the system, such as subsystems of the CLIMATIC SYSTEM (see Table) may differ, may be linear or non-linear and is often complex and poorly understood. The response time of a glacier, for example, is the time taken for a glacier to adjust to a change in mass balance. Many factors, such as the size, surface slope and thermal

response time *Response times for various components of the Earth/ocean/atmosphere system to regain an equilibrium condition after a climatic perturbation (after McGuffie and Henderson-Sellers, 1997)*

Subsystem	Response time
Atmosphere (free)	11 days
Atmosphere (boundary)	24 days
Oceans – mixed layer	7–8 years
Deep ocean	300 years
Sea ice	Days to centuries
Continents – lakes and rivers	11 days
Soil, vegetation	11 days
Snow, surface ice	24 hours
Mountain glaciers	300 years
Ice-sheet decay	1000–10 000 years
Crustal isostatic adjustment	2000–10 000 years
Mantle convection	30 000 000 years

characteristics of the glacier, are effective influences. Thus, the response time of a small glacier (thickness 150–300 m) is likely to be between 15 and 60 years: for the Greenland Ice Sheet, the response time may be about 3000 years. *JAM*

[See also LAG TIME, REACTION TIME, RELAXATION TIME]

McGuffie, K. and Henderson-Sellers, A. (eds) 1997: *A climate modelling primer*. New York: Wiley. **Paterson, W.S.B.** 1994: *The physics of glaciers*. 3rd edn. Oxford: Pergamon. **Wright Jr, H.E.** 1984: Sensitivity and response time of natural systems to climatic change in the late Quaternary. *Quaternary Science Reviews* 3, 91–131.

restoration The complete return of a disturbed environmental SYSTEM to its predisturbance condition by human intervention. *JAM*

[See also ECOLOGICAL RESTORATION, LAND RESTORATION, REHABILITATION, REMEDIATION]

Brookes, A. and Shields, F.D. (eds) 1996: *River channel restoration: guiding principles for sustainable projects*. Chichester: Wiley. **Cairns, J.** 1991: The status of the theoretical and applied science of restoration ecology. *Environmental Professional* 13, 186–194. **Urbanski, K.M., Webb, N. and Edwards, P.J. (eds)** 2000: *Restoration ecology and sustainable development*. Cambridge: Cambridge University Press.

resurgence A point where underground water emerges at the land surface. The term is often used to describe a KARST spring, the waters of which were on the surface before entering a karst AQUIFER via a *sinkhole*. *SHD*

retrodiction Retrodiction, or *postdiction*, is the process of casting results into the past rather than into the future (PREDICTION). When the additional step of embracing formal logic in the process is undertaken, retrodiction and prediction exhibit an asymmetrical relationship. DEDUCTION is the strongest form of logic: given a controlling state of affairs, plus a LAW or law-like statement, the (future) outcome is a logical necessity. However, given an outcome, plus a law or law-like statement, to infer the controlling state of affairs at the time of creation in the past is known as the logical fallacy of affirming the consequent.

ABDUCTION is a form of logic that attempts precisely this. Consequently, abduction (except in highly constrained circumstances) is rarely as powerful as deduction; nevertheless, regardless of the inherent uncertainty, retrodiction is an essential component of investigating the historical component of the environmental sciences. *CET*

Rhoads, B.L. and Thorn, C.E. 1993: Geomorphology as science: the role of theory. *Geomorphology* 6, 287–307.

retrogressive succession The controversial concept of a community succession being, in some sense, 'reversed'. There are two most common usages. First, a DISTURBANCE may drive a community 'backwards' to an earlier successional stage, as when heavily grazed grasslands show reduced cover, the spread of weeds and, eventually, soil erosion. Second, a mature community may 'deteriorate' in the sense of exhibiting a long-term decline in productivity and/or stunted growth forms. Prolonged succession without disturbance, related to such factors as low soil fertility and waterlogging, may lead to retrogression in this sense in Boreal coniferous forests. Both interpretations were present in Iversen's concept of vegetation and soil change in the later stages of an INTERGLACIAL CYCLE. *JAM/RJH*

[See also ECOLOGICAL SUCCESSION, PEDOGENESIS, PODZOLS, PODZOLISATION]

Iversen, J. 1964: Retrogressive vegetational succession in the Post-glacial. *Journal of Ecology* 52 (Supplement), 59–70. **Iversen, J.** 1969: Retrogressive development of a forest ecosystem demonstrated by pollen diagrams from fossil mor. *Oikos* 12 (Supplement), 35–49. **Matthews, J.A.** 1999: Disturbance regimes and ecosystem response on recently deglaciated terrain. In Walker, L.R. (ed.), *Disturbed ecosystems*. Amsterdam: Elsevier. **Phillips, J.D.** 1993: Progressive and regressive pedogenesis and complex soil evolution. *Quaternary Research* 40, 169–176.

return period The average time between EVENTS (such as FLOODS) of a given magnitude. Also known as the *recurrence interval*, its reciprocal is the PROBABILITY of occurrence in any one year. *JAM*

[See also EXTREME EVENTS, EXTREME CLIMATIC EVENTS, FLOOD MAGNITUDE-FREQUENCY CHANGES, MAGNITUDE-FREQUENCY CONCEPTS]

revetment A structure built along a sloping or vertical cliff or riverbank for protection against erosion. Such revetments differ from SEAWALLS by comprising an armour facing of rocks, blocks or other materials. The term is also used for free-standing, openwork structures made, for example, of wood, which may lie in front of the protected cliff. Both varieties are designed to absorb wave energy, thereby reducing erosion, but they may cause loss of beach sediment. *HJW*

[See also COASTAL ENGINEERING STRUCTURES, COASTAL (SHORE) PROTECTION]

Fletcher, C.H., Mullane, R.A. and Richmond, B.M. 1997: Beach loss along armoured shorelines on Oahu, Hawaiian Islands. *Journal of Coastal Research* 13, 209–215.

rheology The study of DEFORMATION and flow in materials, including the behaviour of ice in GLACIERS, rocks during deformation, convection in the Earth's MANTLE, and the behaviour of LAVA FLOWS and DEBRIS FLOWS. *GO*

Johnson, A.M. 1970: *Physical processes in geology*. San Francisco, CA: Freeman.

rheotrophic mire

A MIRE, the dominant water source and hence nutrition of which is derived from flowing water; for example mires formed along river floodplains. Such WETLANDS are less acidic and more nutrient-rich than OMBROTROPHIC MIRES. *MJB*

rhizocretion

Accumulation of mineral matter around roots, before or after death, occurring particularly in poorly drained and highly calcareous situations. In the immediate proximity of roots, the RHIZOSPHERE has a special environment influenced by exudates from, and respiration of, roots and associated fungi. Rhizocretions, or *rhizoliths* (including *root casts*), include sheaths of iron-rich material, ferrihydrate, which accumulate around grass roots in GLEYSOLS and of calcareous material in CALCISOLS. *EMB*

Klappa, C.F. 1980: Rhizoliths in terrestrial carbonates: classification, recognition, genesis and significance. *Sedimentology* 27, 613–629.

rhizopod analysis

The study of SUBFOSSIL tests of *amoeba* in *peatlands* and LAKE SEDIMENTS for PALAEOENVIRONMENTAL RECONSTRUCTION. Rhizopods (TESTATE AMOEBA or Testacea) are morphologically diverse freshwater unicellular (Protozoa) animals (foraminiferids) of size range 10–250 µm that are particularly well preserved in *Sphagnum* PEAT. Assemblage composition, abundance, distribution and preservation are typical parameters used in analysis. In OMBROTROPHIC MIRES (BOGS), *pseudochitinous rhizopods* are common and are resilient to HUMIFICATION and mineralisation. In *minerotrophic mires (fens)* *siliceous* or *mineralised rhizopods* are prevalent, but easily dissolved. Since moisture and nutrient content are important ecological variables in mires, rhizopods are useful for detecting HYDROLOGICAL change affecting PEAT formation and variations in mire water chemistry and pH. In lakes where dissolved oxygen content and TROPHIC LEVEL are important ecological factors, rhizopods can detect limnological change (e.g. EUTROPHICATION and ACIDIFICATION), lake *catchment* climatic change and vegetation change (e.g. presence/absence of *Sphagnum* carpets). *CES*

[See also FORAMINIFERAL ANALYSIS, MICROFOSSIL ANALYSIS, MICROSCOPY, PROXY DATA, RADIOLARIA]

Charman, D.J., Roe, H.M and Gehrels, W.R. 1998: The use of testate amoeba in studies of sea-level change: a case study from the Taf Estuary, South Wales, U.K. *The Holocene* 8, 209–218. **Hendon, D. and Charman, D.J.** 1997: The preparation of testate amoebae (Protozoa: Rhizopoda) samples from peat. *The Holocene* 7, 199–205. **Tolonen, K.** 1986: Rhizopod analysis. In Berglund, B.E. (ed.), *Handbook of Holocene palaeoecology and palaeohydrology*. Chichester: Wiley, 645–666.

rhizosphere

A zone immediately surrounding plant roots where the soil microbial and fungal population is altered by the presence of the roots. Within this zone of 1 to 2 mm thick, the bacterial population may be from two to four times more active than in the soil outside and the species present are more diverse. The specific association of plant roots and specialised fungi (MYCORRHIZAE) in the rhizosphere is normal and this symbiosis assists the plant in obtaining its nutrients from the soil. Re-establishment of the micro-organisms of the rhizosphere is a problem to be overcome in SOIL RECLAMATION. *EMB*

[See also RHIZOSPHERE]

Lynch, J.M. 1982: The rhizosphere. In Burns, R.G. and Slater, S.M. (eds), *Experimental microbial ecology*. Oxford: Blackwell Scientific, 1–23. **Tinker, P.B.** 1984: The role of microorganisms in mediating and facilitating the uptake of plant nutrients from soil. In Tinsley, J. and Darbyshire, J.F. (eds), *Biological processes and soil fertility*. The Hague: Martinus Nijhof.

rhythmite

The coupling of sedimentary laminae in sediments, indicating rhythmic sedimentation (e.g. of alternating coarse and fine layers of LACUSTRINE SEDIMENTS). Annual rhythmites are called VARVES. *FMC*

ria

A non-glaciated, V-shaped coastal valley drowned by the sea because of long-term relative SEA-LEVEL CHANGE. *HJW*

[See also ESTUARINE ENVIRONMENTS]

ridge and furrow

The ploughing of land into regular ridges and furrows was a characteristic feature of many pastures on clay-rich soils in the UK. These undulations provided better LAND DRAINAGE on the ridges in a wet season, and in a dry season the damp furrows provided better grass growth. These ridges and furrows (*selions*) were a legacy of mediaeval ploughing practices and in plan had a reversed 's' shape. With modern drainage techniques and increased mechanisation, these features are being ploughed out. However, the practice of *ridging* has been found useful in restoration of mined land, when the restored and ridged surface is planted with trees. *EMB*

[See also AGRICULTURAL IMPACT ON SOILS, HUMAN IMPACT ON SOILS, MIDDLE AGES: LANDSCAPE IMPACTS, SOIL DRAINAGE]

rift valley

A linear depression bounded by FAULTS, such as the East African Rift Valley, associated with extension of the CRUST. Rift valleys can form barriers to the MIGRATION of plants and animals and have major effects on environments. The initial stage in the break-up of a CONTINENT by PLATE TECTONICS (see WILSON CYCLE) is the development of a rift valley at the crest of an area of UPLIFT, which may lie above a MANTLE PLUME. Continued extension allows the formation of OCEANIC CRUST at a CONSTRUCTIVE PLATE MARGIN. *GO*

[See also GRABEN, HORST]

Frostick, L.E., Renaut, R., Reid, I. and Tiercelin, J.J. (eds) 1986: *Sedimentation in the African rifts* [*Special Publication* 25]. London: Geological Society. **Smith, A.** 1988: *The Great Rift: Africa's changing valley*. London: BBC Books.

rill

A small (up to several centimetres deep) ephemeral microchannel initiated on slopes by the eroding action of concentrated *overland flow*. *SHD*

[See also GULLY, SOIL EROSION]

Bryan, R.B. (ed.) 1987: *Rill erosion: processes and significance* [*Catena* 8 (Supplement)]. Cremlingen: Catena. **Moss, A.J. Green, P. and Hutka, J.** 1982: Small channels: their formation, nature and significance. *Earth Surface Processes and Landforms* 7, 401–405.

rip current A narrow seaward-flowing current in the breaker zone with a velocity of 1–5 ms⁻¹, which can scour the sea bed, especially in storms. *JAM*

Gruszczyński, M., Rudowski, S., Semil, J., Słomiński, J. and Zrobek, J. 1983: Rip currents as a geological tool. *Sedimentology* 40, 217–236.

riparian Related to river banks or the strip of land (*riparian strip*) or vegetation located beside a freshwater body. Riparian *corridors* may be important MIGRATION routes for particular species, and the management of riparian forests, and other aspects of riparian-zone conservation (such as maintenance of river banks and water quality), have become increasingly sensitive issues. *JLI*

[See also GALLERY FOREST]

Forman, R.T.T. 1995: *Land mosaics: the ecology of landscapes and regions.* Cambridge: Cambridge University Press. **Large, A.R.G and Petts, G.E.** 1994: Rehabilitation of river margins. In Calow, P. and Petts, G.E. (eds), *The rivers handbook.* Vol. 2. Oxford: Blackwell, 401–418. **Warner, R.E. and Hendrix, K.M. (eds)** 1984: *Californian riparian systems: ecology, conservation and productive management.* Berkeley, CA: University of California Press.

ripple A small flow-transverse, periodic BEDFORM, typically a few centimetres in height and tens of centimetres in spacing, produced by fluid flow over cohesionless granular material such as sand. Asymmetrical CURRENT RIPPLES are formed by a unidirectional water flow (e.g. a river); WAVE RIPPLES are symmetrical in cross-section; WIND RIPPLES are variable in form and orientation. Ripples can be preserved on the surface of a BED (*ripple mark*) and their migration produces CROSS-LAMINATION and TRANSLATENT STRATA. These SEDIMENTARY STRUCTURES are valuable in the reconstruction of PALAEOCURRENTS and PALAEOENVIRONMENTS from SEDIMENTARY DEPOSITS. *GO*

Allen, J.R.L. 1982: *Sedimentary structures: their character and physical basis.* Amsterdam: Elsevier.

riprap A collection of large rocks placed along a shore to help prevent erosion, or the rocks used. *HJW*

risk The likelihood or probability that a given activity will result in a particular set of circumstances or a series of events that have the potential to cause damage to the ENVIRONMENT or harm to its human population. *JGS*

[See also HAZARD, RISK ASSESSMENT, RISK MANAGEMENT]

Gerrard, S. 2000: Environmental risk management. In O'Riordan, T. (ed.), *Environmental science for environmental management*, 2nd edn. Harlow: Prentice Hall, 435–468.

risk assessment A procedure that identifies the likelihood that rare but possible events will occur as a result of particular human activities. Environmental risk assessment relates specifically to events that are likely to cause damage to the environment or harm to its biological populations and derives mainly from the long-established procedures adopted by engineers and health authorities. The identification of RISK is a multistaged process that may be approached from a variety of different perspectives, each with two common objectives: to estimate the probability that a particular set of circumstances will arise and to estimate the likely magnitude of the damage caused. A competent risk assessment should include as wide a variety of

potential HAZARDS as possible, an allowance for geographical variation in the extent of damage and effectiveness of limitation strategies, and cautious treatment of data sources. Lack of consideration of the above has led to inflammation of public concern and unnecessary financial hardship to industries in the past.

Increasingly, risk assessment is being used as a tool for industrial decision makers to rank potential environmental problems in order to target resource use effectively. In this sense, risk assessment is an integral part of the overall process of RISK MANAGEMENT, which is as much an economic concern as an environmental one. *JGS*

[See also ENVIRONMENTAL IMPACT ASSESSMENT, RISK]

Alexander, D.E. 1993: *Natural disasters.* London: UCL Press. **Gerrard, S.** 2000: Environmental risk management. In O'Riordan, T. (ed.), *Environmental science in environmental management*, 2nd edn. Harlow: Prentice Hall, 435–468. **Kemshall, H. and Pritchard, J. (eds)** 1996: *Good practice in risk assessment and risk management.* London: Kingsley. **Turney, R. and Pitblado, R. (eds)** 1996: *Risk assessment in the process industries.* Rugby: Institute of Chemical Engineers.

risk management The multistage process of estimating, characterising, monitoring and mitigating RISK associated with human activities, particularly those relating to industry. Risk management differs from RISK ASSESSMENT in that the latter is solely concerned with the identification of potential risks, their likelihood of occurrence and the magnitude of their effects. In contrast, management implies a broader perspective with all aspects of risk, including economic concerns in mind. The economic and socio-economic problems commonly embedded within environmental issues have prompted a diversification of risk management procedures since the mid-1990s. No assessment of risk is now completed without some input from social scientists, which means that the management process is INTERDISCIPLINARY in nature, enabling it to operate more effectively by evaluating risk from a number of viewpoints. A risk management programme progresses through a series of stages (see Figure), which does not end with the commencement of the activity in question and may include long-term monitoring and mitigation measures. *JGS*

[See also ENVIRONMENTAL IMPACT ASSESSMENT, HAZARD]

Gerrard, S. 2000: Environmental risk management. In O'Riordan, T. (ed.), *Environmental science for environmental management*, 2nd edn. Harlow: Prentice Hall, 435–468. **Kellow, A.** 1999: *International toxic risk management: ideals, interests and implementation.* Cambridge: Cambridge University Press. **Kemshall, H. and Pritchard, J. (eds)** 1996: *Good practice in risk assessment and risk management.* London: Kingsley.

river capture The acquisition by a river (or stream) of the headwaters of a second river (or stream) due to the greater erosional power of the former. The lower course of the second river thus becomes a *beheaded stream* (a type of *misfit stream* or UNDERFIT STREAM). A sharp change in the orientation of the first river at the point of 'capture' is the *elbow of capture*. The abandoned valley formerly occupied by the headwaters of the second river may become a *wind gap* (possibly also occupied by a misfit stream of insufficient size to have eroded the valley). River capture has also been termed *stream capture* or *river piracy*. The precise processes involved in river capture are not always clear,

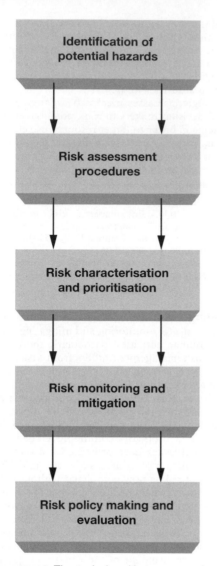

risk management *The typical multi-stage procedure for risk management*

but HEADWARD EROSION and downcutting and capture of underground and surface runoff of the second stream are usually considered the main processes involved.

JAM/RPDW

[See also DRAINAGE EVOLUTION, DENUDATION CHRONOLOGY]

Stamp, L.D. (ed.) 1961: *A glossary of geographical terms.* London: Longman.

river discharge variations Changes in river discharge on a variety of timescales in response to natural or anthropogenic controls. Variation can be divided into two types: (1) short-term variations in response to WEATHER and season; and (2) longer-term secular changes in response to changes in CLIMATE (and associated vegetation and soil changes) or LANDUSE CHANGE and other human actions (such as river diversion, water ABSTRACTION, dam construction or inter-basin transfers).

Short-term variations in river discharge occur as a result of changes in the balance between precipitation inputs, evapotranspiration losses and the dynamics of contributory RUNOFF PROCESSES. Such fluctuations occur on a variety of timescales. The most rapid are changes in discharge associated with the input of quickflow (see STORM HYDROGRAPH) during and immediately following rainstorm events. Fluctuations also occur in response to seasonal differences in rainfall and temperature. In the western UK the higher discharges (both storm peaks and BASEFLOW) of autumn and winter and low discharges of summer reflect an autumn/winter rainfall peak and reduced EVAPOTRANSPIRATION and higher WATER TABLES in the winter months. In seasonally very cold environments, the river discharges may be very low in the frozen winter months and exhibit a spring/early summer snowmelt peak (e.g. the Assiniboire River in Manitoba, Canada) or a late summer glacial meltwater peak (e.g. the Vernagt Ache meltwater stream in Austria).

Longer-term SECULAR VARIATIONS are associated both with human actions and with climatic change; average discharge, the discharge regime and FLOOD MAGNITUDE-FREQUENCY CHANGES may all be affected. Human actions that have major impacts on river discharge include: LANDUSE CHANGES (such as URBANISATION, AFFORESTATION and DEFORESTATION); the construction of dams and reservoirs (see RESERVOIRS: ENVIRONMENTAL IMPACTS); water abstraction for IRRIGATION or water supply; EFFLUENT discharges; river CHANNELISATION; and *flood control* measures.

CLIMATIC CHANGE can have major impacts on river discharge. In the Nile Basin, for example, flow along the White Nile from the Equatorial Lakes region via Lake Victoria dried up completely from 14 500 to 12 500 radiocarbon years BP in the Late Glacial period and flows from the Blue Nile were both lower and more seasonal than at present. Later, during the AFRICAN HUMID PERIOD, the Blue Nile was characterised by high floods and a less seasonal regime and the While Nile was also high.

On a shorter, more recent timescale, the last 150 years have witnessed major changes both in Blue Nile and White Nile discharge and in flood magnitude-frequency. Blue Nile flows were very high in the wet late nineteenth century before dipping markedly in the early twentieth century. Flows have dipped further (by 16.5%) in AD 1965–1987 compared with 1905–1964. In contrast, the White Nile at Mongolia increased by 84% in 1961–1982 compared with 1905–1960 in response to very high rainfall in Equatorial Lakes in the early 1960s.

In central England, the 11-years running mean of the annual discharge (as estimated using annual rainfall) has ranged between 330 mm in the AD 1740s to over 500 mm in the 1870s. Discharges in the UK are predicted to change radically with climatic change during the twenty-first century. Under the Climate Change Impacts Review Group 1996 scenario, river flows in the UK would increase in winter and decrease, especially in the south, in the summer. It is expected that there will be an increase in winter flood frequency due to increased winter storminess and wetter ANTECEDENT SOIL MOISTURE catchment conditions.

RPDW/ADT

[See also CLIMATIC CHANGE: PAST IMPACTS, CLIMATIC CHANGE: POTENTIAL FUTURE IMPACTS, HYDROLOGICAL BALANCE/BUDGET, HYDROLOGICAL CYCLE, PALAEOFLOODS, PALAEOHYDROLOGY, RIVER REGIMES]

Arnell, N.W. and Reynard, N.S. 1996: The effects of climate change due to global warming on river flows in Great Britain. *Journal of Hydrology* 183, 397–424. **Department of the Environment** 1996: *Review of the potential effects of climate change in the United Kingdom.* London: HMSO. **Williams, M.A.J.** 1985: Pleistocene aridity in tropical Africa, Australia and Asia. In Douglas, I. and Spencer, T. (eds), *Environmental change and tropical geomorphology.* London: Allen and Unwin, 219–233. **Walsh, R.P.D. and Musa, S.B.** 1990: Hydrology and Sudan's capital region. In Abu Sin, M.E. and Davies, H.R.J. (eds), *The future of Sudan's capital region: a study of development and change.* Khartoum: Khartoum University Press, 47–61.

river lake

A lake that is sufficiently small in volume relative to the river water passing through it to be substantially affected in terms of its HYDROLOGY and SEDIMENTS by the river inputs; a transitional LANDFORM between a river and a lake. *JAM*

Eyster-Smith, N.M., Wright Jr, H.E. and Cushing, E.J. 1991: Pollen studies at Lake St Croix, a river lake on the Minnesota/Wisconsin border, USA. *The Holocene* 1, 102–111.

river regime

The temporal pattern of flow or DISCHARGE of a river. Monthly averages of river flow are commonly used to describe the SEASONALITY of river flow, but the term also refers to the variability in flow in response to individual rainstorms. *RPDW*

[See also RIVER DISCHARGE VARIATIONS]

river terrace

A fluvial LANDFORM comprising a level or gently sloping surface in a river valley that is above the level of the active channel and FLOODPLAIN, from which it is separated by a steep, erosional slope. River terraces formed at river level, and represent dissected remnants of former floodplains that existed prior to one or more episodes of channel downcutting. River terraces can be confused with terrace features of different origin that can occur in river valleys, such as KAME terraces, LAKE TERRACES, PEDIMENT slopes, terraces underlain by COLLUVIUM (e.g. SOLIFLUCTION TERRACES) or terraces controlled by GEOLOGICAL STRUCTURES.

A distinction can be made between *rock-floored terraces* (STRATH TERRACES) and *alluvial terraces*: the latter are composed of ALLUVIAL DEPOSITS, whereas the former are products of EROSION. Where there are several levels of river terraces in a valley, the lowest is the youngest. They may form *paired terraces,* with matching levels on each valley side, or *unpaired terraces,* where a terrace at a particular elevation occurs on one side of the valley only. Paired terraces develop if incision was rapid relative to channel migration; unpaired terraces develop when the channel migrates rapidly relative to the rate of incision. Complex patterns of buried and partially eroded river terraces can develop after many phases of alternating ALLUVIATION and channel downcutting.

Unpaired terraces can be formed during the gradual down-cutting of a shifting river channel. In many cases, however, the preservation of river terraces may be related to ALLOCYCLIC CHANGES. These include regional UPLIFT, changes in BASE LEVEL and changes in DISCHARGE, which may in turn be due to CLIMATIC CHANGE or TECTONIC effects. River terrace formation can also occur in response to geomorphic THRESHOLD effects. The along-stream slope of river terraces gives a measure of the former CHANNEL GRADIENT. This may differ from the present-day value and

is an important quantity in the reconstruction of PALAEOHYDROLOGY. *GO*

Antoine, P. 1994: The Somme valley terrace system (northern France); a model of river response to Quaternary climatic variations since 800,000 BP. *Terra Nova* 6, 453–464. **Bridgland, D.R.** 1994: *Quaternary of the Thames* [*Geological Conservation Review Series* 7]. London: Chapman and Hall. **Green, C.P. and McGregor, D.F.M.** 1987: River terraces: a stratigraphic record of environmental change. In Gardiner, V. (ed.), *International geomorphology 1986. Proceedings of the First International Conference on Geomorphology.* Part 1. Chichester: Wiley, 977–987. **Howard, A.J., Macklin, M.G., Black, S. and Hudson-Edwards, K.A.** 2000: Holocene river development and environmental change in Upper Wharfedale, Yorkshire Dales, England. *Journal of Quaternary Science* 15, 239–252. **Molnar, P., Brown, E.T., Burchfiel, B.C.** *et al.* 1994: Quaternary climate change and the formation of river terraces across growing anticlines on the north flank of the Tien Shan, China. *Journal of Geology* 102, 583–602. **Rao, K.N. and Rao, C.U.B.** 1999: Geomorphic evidences of Pleistocene higher sea levels near Visakhapatnam, east coast of India. *Zeitschrift für Geomorphologie* 43, 19–25.

roche moutonnée

A small, streamlined, elongated rock knob or hillock shaped by GLACIAL EROSION. Its up-ice (proximal) end is usually gently sloping, smooth and polished with STRIATIONS and FRICTION CRACKS indicating glacial ABRASION. The down-ice (distal) end is steep and has a rough appearance, reflecting bedrock removal by QUARRYING. Roche moutonnée is a type of *stoss and lee feature*; large-scale versions are sometimes called *flyggbergs*. Roches moutonnées are not always very reliable palaeo-ice flow indicators because their detailed form is influenced partly by bedrock joints and foliations. *Whalebacks* are rock protuberances smoothed on both up- and down-ice ends. They differ from CRAG AND TAILS, which have steeper up-ice ends and down-ice ends that have been protected in the lee of a resistant rock mass. *JS*

[See also GLACIAL LANDFORMS, STRIATIONS]

Carol, H. 1947: The formation of roches moutonnées. *Journal of Glaciology* 1, 57–59. **Sugden, D.E., Glasser, N.F. and Clapperton, C.M.** 1992: Evolution of large roches moutonnées. *Geografiska Annaler* 74A, 253–264.

rock creep

The gradual down-slope movement of surface rock fragments under the influence of gravity. *PW*

[See also CAMBERING, MASS MOVEMENT PROCESSES, ROCK GLACIER, SOIL CREEP, SOLIFLUCTION]

rock cycle

An idealised cycle of the processes involved in the formation and destruction of rocks. MAGMA from the Earth's interior solidifies within the CRUST or at the Earth's surface to form IGNEOUS ROCKS. These are broken down by WEATHERING at the Earth's surface. The products of this may accumulate as SEDIMENTARY DEPOSITS, which may form SEDIMENTARY ROCKS upon burial and LITHIFICATION. Rocks within the crust may be subjected to elevated temperatures and stresses. They may undergo DEFORMATION to produce GEOLOGICAL STRUCTURES or complete recrystallisation to form METAMORPHIC ROCKS. UPLIFT of buried rocks, particularly in response to OROGENESIS, exposes them once more to weathering. The rock cycle provides a useful framework for considering the relationships amongst the physical and chemical processes that operate on the Earth, particularly when related to the HYDROLOGICAL CYCLE and the WILSON CYCLE. *GO*

[See also BIOGEOCHEMICAL CYCLES, CARBON CYCLE, GEOLOGY]

Skinner, B.J. and Porter, S.C. 2000: *The dynamic Earth: an introduction to physical geology*, 4th edn. New York: Wiley.

rock flour Detrital clay and especially silt produced mainly by ABRASION, attrition (particle-to-particle contact) and crushing. The term is most commonly associated with the fine powder formed by glacial action. Rock flour particularly forms much of the matrix (i.e. fine component) of subglacial TILLS. It can provide an important indicator of glacier expansion in terms of its abundance in sediment cores from lakes within catchments containing present-day or past glaciers. *JS*

[See also GLACIER MILK, GLACIER VARIATIONS]

Bischoff, J.L., Menking, K.M., Fitts, J.P. and Fitzpatrick, J.A. 1997: Climatic oscillations 10,000–155,000 yr B.P. at Owens Lake, California reflected in glacial rock flour abundance and lake salinity in core OL-92. *Quaternary Research* 48, 313–325.

rock glacier 'An accumulation of angular rock debris, usually with a distinct ridge/furrow pattern and steep front and side slopes, whose length is *generally* greater than its width; existing on a valley floor' (Hamilton and Whalley, 1995: 76). This morphological–locational definition facilitates a variety of forms, origins and processes. There is no generally accepted model of rock glacier development. Three main theories are prevalent: (1) they are PERMAFROST phenomena consisting of an ice-rock mixture; (2) they possess a core of GLACIER ice; and (3) they originate from catastrophic rock or TALUS failures (STURZSTROM). It is possible that each mechanism is capable of producing a morphologically similar form (EQUIFINALITY). A further subdivision of rock glaciers is based on level of activity: active, inactive and fossil (RELICT) forms are recognised. Fossil forms are often used in PALAEOENVIRONMENTAL RECONSTRUCTION, but the validity of interpretation depends on correct identification of the formative mechanism. *PW*

[See also PRONIVAL RAMPART]

Barsch, D. 1996: *Rock glaciers: indicators for the present and former geoecology in high mountain environments*. Berlin: Springer. Giardino, J.R., Shroder, J.F. and Vitek, J.D. 1987: *Rock glaciers*. Boston, MA: Allen and Unwin. Hamilton, S.J. and Whalley, W.B. 1995: Rock glacier nomenclature: a re-assessment. *Geomorphology* 14, 73–80. Martin, H.E. and Whalley, W.B. 1987: Rock glaciers. Part 1: rock glacier morphology: classification and distribution. *Progress in Physical Geography* 11, 260–282. Shakesby, R.A., Dawson, A.G. and Matthews, J.A. 1987: Rock glaciers, protalus ramparts and related features, Rondane, Norway: a continuum of large-scale talus-derived landforms. *Boreas* 16, 305-317. Whalley, W.B. and Martin, H.E. 1992: Rock glaciers: II. Models and mechanisms. *Progress in Physical Geography* 16, 127–186. Wilson, P. 1990: Characteristics and significance of protalus ramparts and fossil rock glaciers on Errigal Mountain, County Donegal. *Proceedings of the Royal Irish Academy* B90, 1–21.

rock platform A flat or low-angled bench or surface developed in coastal bedrock exposures, often with an associated backing cliff. Platforms are formed by hydraulic EROSION by waves, by biological processes and also in cold climate regions by FROST WEATHERING processes. Platforms cannot be produced by hydraulic processes alone and require ABRASION of platform surfaces by wave-transported debris. Their rate of formation also depends upon wave FETCH and platform LITHOLOGY. Thus, rock platforms tend to be well developed in fissile rocks (e.g. slates and mudstones), but are rare in granite areas. In cold-climate environments, rock platforms on lake margins as well as coasts can be particularly rapidly developed through POLAR SHORE EROSION. Together with other evidence of former shorelines, fossil rock platforms can be used to reconstruct former relative and absolute sea levels (see MEAN, RELATIVE and ABSOLUTE SEA LEVEL). *AGD*

[See also RAISED SHORELINE, STRANDFLAT]

Dawson, A.G., Matthews, J.A. and Shakesby, R.A. 1987: Rock platform erosion on periglacial shores: a modern analogue for Pleistocene rock platforms in Britain. In Boardman, J. (ed.), *Periglacial processes and landforms in Britain and Ireland*. Cambridge: Cambridge University Press, 173–182. Trenhaile, A.S. 1990: *The geomorphology of rock coasts*. Oxford: Clarendon Press.

rock stream A surface accumulation of coarse rock debris aligned down low- to moderate-gradient slopes or along a valley axis; a product of FROST WEATHERING and SOLIFLUCTION. *PW*

[See also BLOCKFIELDS, MASS WASTING]

rock varnish A hard rind or *patina* on rock surfaces, 5–100 µm thick, commonly found on bare rock surfaces in deserts. Rock varnish, also known as *desert varnish*, is a type of *rock coating* produced by the physico-chemical and/or biological mobilisation and subsequent deposition of manganese-rich oxides. Its presence on exposed rock is a good indicator of desert conditions since it is removed in a wet climate. *SHD*

[See also CATION-RATIO DATING, PETROGLYPH, VENTIFACTS, WEATHERING RIND]

Dorn, R.I. 1998: *Rock coatings*. Amsterdam: Elsevier. Liu, T. and Broecker, W.S. 2000: How fast does rock varnish grow? *Geology* 28, 183–186. Watson, A. and Nash, D.J. 1997: Desert crusts and varnishes. In Thomas, D.S.G. (ed.), *Arid zone geomorphology: process, form and change in drylands*. Chichester: Wiley, 69–107. Whalley, W.B. 1983: Desert varnish. In Goudie, A.S. and Pye, K. (eds), *Chemical sediments and geomorphology*. London: Academic Press, 197–226.

RockEval An analytical method for assessing the bulk composition of organic matter preserved in sediments and sedimentary rocks. Variations in composition determined by RockEval can be related to environmentally important features such as PRIMARY PRODUCTIVITY and BOTTOM WATER oxygenation. *MRT*

Talbot, M.R. and Lærdal, T. 2000: The late Pleistocene–Holocene palaeolimnology of Lake Victoria, East Africa, based upon elemental and isotopic analyses of sedimentary organic matter. *Journal of Paleolimnology* 23, 141–164.

rockshelter An overhanging cliff face, large boulder or sometimes a pile of boulders occupied in PREHISTORY by humans, normally as a temporary, seasonal shelter. Rockshelters differ from CAVES by only being enclosed from the back and ceiling: they represent semi-enclosed environments intermediate in character between open sites and cave mouths. Rockshelter floors may contain evidence of CULTURAL CHANGE and ENVIRONMENTAL CHANGE, especially in deposits at the DRIPLINE and in the space sheltered from precipitation and strong sunlight between the dripline and the backwall, where domestic activity is usually concentrated. *JAM*

[See also CAVE SEDIMENTS, ENVIRONMENTAL ARCHAEOLOGY, PETROGLYPH]

Barton, C.M. and Clark, G.A. 1993: Cultural and natural formation processes in Late Quaternary cave and rockshelter sites of western Europe and the Near East. In Goldberg, P., Nash, T.D. and Petraglia, M.D. (eds), *Formation processes in archaeological context.* Madison, WI: Prehistory Press. Laville, H., Rigaud, J. and Sackett, J. 1980: *Rock shelters of the Perigord: geological stratigraphy and archaeological succession.* New York: Academic Press. Straus, L.G. 1990: Underground archaeology: perspectives on caves and rockshelters. *Archaeological Method and Theory* 2, 255–304.

rodent midden
Rodents and several other types of small animals build accumulations of plant material cemented by urine in caves and rock crevices. Pack rat (*Neotoma* spp.), hyrax (*Procavia* spp.) and stick-nest rat (*Leoporillus* spp.) are amongst the species producing middens in North and South America, Africa, Australia and the Middle East. The organic material accumulated by midden-building animals is valuable in the reconstruction and dating of Late Quaternary VEGETATION HISTORY and other environmental changes, especially in ARIDLANDS where sedimentary sequences containing POLLEN and MACROFOSSILS are rare. Material in the middens reflects the local environment because the animals forage within a limited distance of their den. *JAM*

[See also COPROLITE, MIDDEN, FAECES ANALYSIS]

Betancourt, J.L., Van Devender, T.R. and Martin, P.S. (eds) 1990: *Packrat middens: the last 40,000 years.* Tucson, AZ: University of Arizona Press. Markgraf, V., Betancourt, J. and Rylander, K.A. 1997: Late-Holocene rodent middens from Rio Limay, Neuquen Province, Argentina. *The Holocene* 7, 325–329. McCarthy, L., Head, L. and Quade, J. 1996: Holocene palaeoecology of the northern Flinders Ranges, South Australia, based on stick-nest rat (*Lepotillus* spp.) middens: a preliminary overview. *Palaeogeography, Palaeoclimatology, Palaeoecology* 123, 205–218. Scott, L. and Bousman, C. 1990: Palynological analysis of hyrax middens from southern Africa. *Palaeogeography, Palaeoclimatology, Palaeoecology* 79, 367–379.

rogation index
The *drought rogation index* (DRI) and *excessive rain rogation index* (ERRI) are quantitative indices of climate derived from Roman Catholic ecclesiastical records. *JAM*

[See also DOCUMENTARY EVIDENCE, HISTORICAL CLIMATOLOGY]

Barriendos, M. 1997: Climatic variations in the Iberian Peninsula during the late Maunder Minimum (AD 1675–1715): an analysis of data from rogation ceremonies. *The Holocene* 7, 105–111.

Rogen moraines
The term 'Rogen moraine' was introduced to describe the distinctive MORAINE landscape around Lake Rogen in Sweden. The term, also referred to as *ribbed moraine*, is applied to fields of coalescent crescentic ridges up to about 30 m high and up to about 100 m wide lying transverse to the former ice flow direction. The composition of Rogen moraines exhibits a wide range of sediment types and structures. Several hypotheses

explaining the formation and spatial pattern of Rogen moraines have been advanced: (a) subglacial; (b) stagnant ice; (c) subglacial tectonics; (d) subglacial meltwater floods; (e) crevasse filling; and (f) marginal moraine formation. *AN*

Hättestrand, C. and Kleman, J. 1999: Ribbed moraine formation. *Quaternary Science Reviews* 18, 43–61. Lundquist, J. 1969: Problems of the so-called Rogen moraine. *Sveriges geologiske undersökelse, Series C,* 648 pp. Lundquist, J. 1989: Rogen (ribbed) moraine – identification and possible origin. *Sedimentary Geology* 62, 281–292.

Roman Period: landscape impacts
The ROMAN PERIOD brought major changes to the landscape of parts of Europe and the Mediterranean with the establishment of forts, towns and villas, interspersed with extensive field systems and connected by a network of roads. Demand for agricultural produce increased, triggering a major phase of FOREST CLEARANCE in some regions, while MINING and quarrying transformed the landscape of areas rich in minerals and building stone. The area of land suitable for agriculture was increased by DRAINAGE and IRRIGATION schemes, whilst agricultural expansion triggered SOIL EROSION and increased flooding and ALLUVIATION in many river valleys. WOODLAND MANAGEMENT was widespread, and some areas of *grassland* were managed as hay MEADOWS. Many species of plants were introduced to new areas, some to be grown in ornamental gardens. A few of these plants became naturalised, including the Sweet Chestnut (*Castanea sativa*), originally native in southern Europe but probably introduced further north, including in England. *SPD*

[See also AGRICULTURAL HISTORY, VEGETATION HISTORY]

Dark, K. and Dark, P. 1997: *The landscape of Roman Britain.* Stroud: Sutton. Dumayne, L. and Barber, K.E. 1994: The impact of the Romans on the environment of northern England: pollen data from three sites close to Hadrian's Wall. *The Holocene* 4, 165–173. Frenzel, B., Andersen, S.T., Berglund, B.E. and Gläser, B. (eds) 1994: *Evaluation of land surfaces cleared from forests in the Roman Iron Age and the time of migrating German tribes* [Paläoklimaforschung 12]. Stuttgart: Gustav Fischer. McCarthy, M.R. 1995: Archaeological and environmental evidence for the Roman impact on vegetation near Carlisle, Cumbria. *The Holocene* 5, 491–495. White, K.D. 1970: *Roman farming.* London: Thames and Hudson.

roundness
For a rock particle, an assessment of the degree of curvature of its edges and corners, increased roundness occurring through mechanical wear in transport by, for example, a river, GLACIER or the wind or through WEATHERING. Roundness may be quantified by measuring the radius of curvature of one or more corners compared with the radius of the maximum inscribed circle of the largest silhouette of the particle (maximum projection plane) or the *a*-axis length. Alternatively, estimates are commonly based on visual comparison charts of representative degrees of rounding. Roundness is a useful index to help in determining the transport history, energy of the depositional environment or distance of SEDIMENT TRANSPORT in, for example, fluvial, glacial or coastal environments. It is a more sensitive index of particle wear than SPHERICITY or SHAPE and thus usually regarded as more useful in differentiating depositional environments or in detecting greater rounding with increasing distance of transport in the same transporting environment. *RAS*

Briggs, D.J. 1977: *Sources and methods in geography. Sediments.* London: Butterworths. **Matthews, J.A.** 1987: Regional variation in the composition of Neoglacial end moraines, Jotunheimen, Norway: an altitudinal gradient in clast roundness and its possible palaeoclimatic significance. *Boreas* 16, 173–188. **Olsen, L.** 1983: A method for determining total clast roundness in sediments. *Boreas* 12, 17–21. **Powers, M.** 1953: A new roundness scale for sedimentary particles. *Journal of Sedimentary Petrology* 25, 117–119. **Shakesby, R.A.** 1980: Field measurement of roundness: a review. *Swansea Geographer* 18, 27–36. **Shakesby, R.A.** 1989: Variability in moraine morphology and composition, Storbreen, Jotunheimen, Norway: within-moraine patterns and their implications. *Geografiska Annaler* 71A, 17–29. **Trenhaile, A.S., Van Der Nol, L.V. and LaValle, P.D.** 1996: Sand grain roundness and transport in the swash zone. *Journal of Coastal Research* 12, 1017–1023.

rudaceous Adjective describing a gravel-rich SEDIMENT or SEDIMENTARY ROCK (or derivative METAMORPHIC ROCK) with a GRAIN-SIZE dominantly or entirely greater than 2 mm (i.e. GRAVEL). Rudaceous SEDIMENTARY DEPOSITS imply SEDIMENT TRANSPORT and deposition by strong currents, and commonly accumulate close to the source of the weathered material. Because LITHIC fragments are common as CLASTS, rudaceous rocks are useful in PROVENANCE studies. *TY*

Koster, E.H. and Steel, R.J. (eds) 1984: *Sedimentology of gravels and conglomerates* [Memoir 10]. Calgary: Canadian Society of Petroleum Geology.

ruderal A species with characteristics such as high seed set, widespread seed dispersal, rapid germination and fast growth that adapt it to rapid COLONISATION of recently disturbed areas, but which is intolerant of stress or competition. *LRW*

[See also *R*-SELECTION, SECONDARY SUCCESSION, WEED]

Grime, J.P. 1979: *Plant strategies and vegetation processes.* London: Wiley.

rudite A RUDACEOUS rock (i.e. rich in GRAVEL-size particles). *TY*

ruminant Of the Ruminantia, a suborder comprising those artiodactyls (even-toed mammals) possessing a complex stomach of three or four parts, in which rumination occurs. This involves a repeated digestion process and chewing of the cud. During rumination, micro-organisms produce METHANE. Some 16% of global methane production is accounted for by ruminants, especially cattle and sheep. *DCS*

runoff The part of precipitation (expressed as millimetres depth) that leaves a drainage basin as *streamflow*; a synonym for RIVER DISCHARGE. Runoff is also used as a synonym for *overland flow* in slope hydrology and EROSION PLOT studies. *ADT*

Jones, J.A.A. 1997: *Global hydrology: processes, resources and environmental management.* Harlow: Addison-Wesley Longman.

runoff coefficient The proportion of PRECIPITATION that leaves either a DRAINAGE BASIN as *streamflow* or an enclosed EROSION PLOT as *overland flow*. It is also sometimes defined as the proportion of rainstorm water that contributes to the STORM HYDROGRAPH. *ADT*

[See also RUNOFF]

Ward, R.C. and Robinson, M. 2000: *Principles of hydrology*, 4th edn. London: McGraw-Hill.

runoff processes Hydrological processes by which water travels to stream channels and generates *stream flow*. Several processes have been distinguished on the basis of their mode of origin and routeway and these have been broadly grouped into QUICKFLOW and delayed flow (BASEFLOW) processes (see Figure). In arid and semi-arid areas and catchments affected by overgrazing or devegetation,

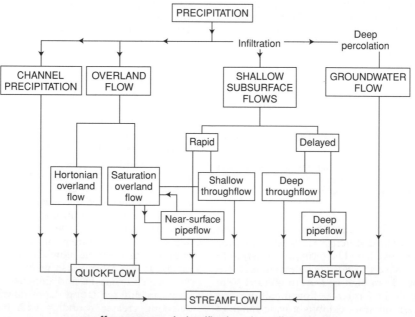

runoff processes *A classification of runoff processes*

then HORTONIAN OVERLAND FLOW is the main quickflow process, but in humid vegetated areas with more permeable topsoils, SATURATION OVERLAND FLOW and in some cases THROUGHFLOW and shallow PIPEFLOW are the chief generators of the STORM HYDROGRAPH. Baseflow is provided by GROUNDWATER flow and in some wetter environments by deeper throughflow and by more deeply seated PIPEFLOW. In vegetated areas, topography and soil properties also influence the relative importance of different quickflow processes. Steeper terrain with narrow valley bottoms favours throughflow, whereas terrain with wider valley bottoms, valley side hollows and flat interflows favours saturation overland flow. In areas with thin permeable soil or a relatively impermeable subsoil at a shallow depth, saturation overland flow can be widespread.

The different speeds of these runoff processes in delivering water to stream channels influences the hydrograph and the chemistry of the various components. The longer water has in contact with mineral and organic material in soil, bedrock and on vegetation surfaces, the greater the opportunity for desorption and therefore quickflow tends to have lower solute concentrations than subsurface flow. Thus, environmental changes affecting runoff processes will also affect the chemistry of runoff and the form of the hydrograph. The separation of precipitation into the various runoff routes is a function of the INFILTRATION capacity of the surface and modifications to groundcover will affect runoff processes. AFFORESTATION, DEFORESTATION, URBANISATION (see URBANISATION IMPACTS ON HYDROLOGY) and agricultural practices affect the infiltration characteristics of the surface and thus runoff processes. In addition to LAND COVER changes, runoff processes will be affected by CLIMATIC CHANGE. Recent predictions for the

UK indicate that winter rainfall is likely to increase with higher rainfall intensities. This is likely to increase the instances of overland flow and generate more FLOODS. Predicted drier weather in the summer months may lead to lower flows, but cracking in dried clay soils could increase the instances of by-pass flows. This rapid transmission of water through cracks in the soil structure may enhance pipeflow or throughflow, both of which contribute to quickflow, and hence may lead to the transmission of NUTRIENTS and PESTICIDES to streamwater.

ADT/RPDW

[See also HUMAN IMPACTS ON HYDROLOGY]

Department of the Environment 1996: *Review of the potential effects of climate change in the United Kingdom.* London: HMSO. **Ward, R.C. and Robinson, M.** 2000: *Principles of hydrology.* London: McGraw-Hill. **Bonell, M. with Balek, J.** 1993: Recent scientific developments and research needs in hydrological processes of the humid tropics. In Bonell, M., Hufschmidt, M.M. and Gladwell, J.S. (eds), *Hydrology and water management in the humid tropics: hydrological research issues and strategies for water management.* Cambridge: Cambridge University Press, 167–260.

runout distance The distance travelled beyond the toe of a LANDSLIDE rupture surface by the displaced material. *PW*

[See also LANDSLIDE HAZARD, STURZSTROM]

rural decline Depopulation and economic decline of rural areas in developed countries following URBANISATION. Associated in part with the *drift from the land* brought about by changes in agriculture, it has been reduced or reversed in some areas as a result of COUNTERURBANISATION. *JAM*

S

Saalian The Northern European Middle Pleistocene GLACIATION between the Holsteinian Interglacial complex and the LAST INTERGLACIAL (*Eemian*); broadly corresponding with *Wolstonian* in the British Isles, *Riss* in the Alps and *Illinoian* in North America. Now recognised as being a complex glaciation comprising at least three *substages*, the *Drente* (glacial), *Treene* (temperate) and *Warthe* (glacial), correlated with oxygen ISOTOPE STAGES 6, 7 and 8 respectively.   *MHD*

[See also QUATERNARY TIMESCALE]

Petit, J.R., Jouzel, J., Raynaud, D. *et al.* 1999: Climate and atmospheric history of the past 420,000 years from the Vostok ice core, Antarctica. *Nature* 399, 429–436.

sabkha An Arabic word for an arid SHORELINE where evaporation draws saline water through the sediment, causing the precipitation of EVAPORITE minerals within the SEDIMENTS. Sabkhas can occur on the edge of the sea, often as tidal flats (*marine sabkha*), or on the edge of a permanent or EPHEMERAL lake (*continental sabkha*). Shoreline PROGRADATION leads to extensive deposits that have distinctive features, allowing their recognition in the GEOLOGICAL RECORD.   *GO*

[See also PAN]

Amin, A. and Bankher, K. 1997: Causes of land subsidence in the Kingdom of Saudi Arabia. *Natural Hazards* 16, 57–63. **Kendall, A.C. and Harwood, G.M.** 1996: Marine evaporites: arid shorelines and basins. In Reading, H.G. (ed.), *Sedimentary environments: processes, facies and stratigraphy*, 3rd edn. Oxford: Blackwell Science, 281–324.

Sackung A type of large-scale, deep-seated MASS MOVEMENT or slope failure, also described as *rock flow*, involving gradual settlement (sag) and extension of a large rock slab or an entire hillside, producing a ridge-top trench and uphill-facing scarps known as *antislope scars*. The term, from the German for sagging, was originally proposed by Zischinsky (1966). It has been suggested that such failures in deglaciated uplands may represent PARAGLACIAL phenomena, whereby high water pressures within the rock mass during deglaciation may have led to a decrease in the interlayer SHEAR STRENGTH of the rock.   *GO/RAS*

Bisci, C., Dramis, F. and Sorriso-Valvo, M. 1996: Rock flow (sackung). In Dikau, R., Brunsden, D., Schrott, L. and Ibsen, M.-L. (eds), *Landslide recognition: identification, movement and causes*. Chichester: Wiley, 150–160. **Holmes, G. and Jarvis, J.J.** 1985: Large-scale toppling within a sackung type deformation at Ben Attow, Scotland. *Quarterly Journal of Engineering Geology* 18, 287–289. **Shroder, J.F.** 1998: Slope failure and denudation in the western Himalaya. *Geomorphology* 26, 81–105. **Zischinsky, U.** 1966: On the deformation of high slopes. *Proceedings of the First Congress of the International Society of Rock Mechanics*. Vol. 2, 179–185.

Saffir–Simpson scale A five-point scale (weak to devastating) of hurricane (TROPICAL CYCLONE) strength and damage potential based on wind speeds and the height of associated STORM SURGES.   *JAM*

Saharan dust ATMOSPHERIC DUST originating from AEOLIAN erosion in the Sahara desert. This can be transported to distant locations and has been regularly observed in northern Mediterranean locations and occasionally in northern Europe and the Caribbean. The main climatic effect is the reduction of visibility, yet in large quantities a significant influence may be exerted on the RADIATION BALANCE. SOLAR RADIATION receipts during the day are reduced by scattering and absorption, while at night the DUST VEIL will cool rapidly to enhance the subsidence of air and suppress thermal convection. As a consequence of OVERGRAZING and overcultivation on the desert margins (SAHEL), dust levels have increased as a result of the decline in SOIL STRUCTURE and moisture. The effects on the radiation balance have been modelled and the increased atmospheric stability will tend to suppress rainfall resulting in a *positive feedback*, which could further increase dust levels.   *GS*

Carlson, T.N. and Benjamin, S.G. 1980: Radiative heating rates for Saharan dust. *Journal of Atmospheric Science* 37, 193–213. **Littmann, T.** 1991: Rainfall, temperature and dust storm anomalies in the African Sahel. *Geographical Journal* 157, 136–160. **Morales, C. (ed.)** 1979: *Saharan dust: mobilization, transport and deposition* [*SCOPE Report* 14]. New York: Wiley. **Moulin, C., Lambert, C.E., Dulac, F. and Dayan, U.** 1997: Control of atmospheric export of dust from North Africa by the North Atlantic Oscillation. *Nature* 387, 691–694. **Street-Perrott, F.A., Holmes, J.A., Waller, M.P.** *et al.* 2000: Drought and dust deposition in the West African Sahel: a 5500-year record from Kajemarum Oasis, northeastern Nigeria. *The Holocene* 10, 293–302. **Swap, R., Ulanski, S., Cobbett, M. and Garstang, M.** 1996: Temporal and spatial characteristics of Saharan dust outbreaks. *Journal of Geophysical Research* 101, 4205–4220.

Sahel The drought-prone zone south of the Sahara Desert in West Africa, which was formerly characterised by *thorn woodland* but is now dominated by thorny shrubs and annual grasses. It approximately corresponds to the zone receiving a mean annual rainfall of between 150 and 750 mm but this is concentrated into a rainy season of only 2–4 months and there is high spatial and temporal variability in rainfall. The Sahel in the strict sense includes parts of six nations – Senegal, Mauritania, Mali, Burkino Faso, Niger and Chad – but the term is sometimes broadened to include the neighbouring countries also affected by DROUGHT, DESERTIFICATION and FAMINE (e.g. parts of Nigeria, Sudan, Ethiopia, Eritrea and Somalia) brought about at least in part by LAND DEGRADATION. To the south of the Sahel lies the *Soudan* or *Sudanese zone*, which receives more rainfall and is characterised by SAVANNA vegetation.   *JAM*

Adams, W.A. and Mortimore, M.J. 1997: Agricultural intensification and flexibility in the Nigerian Sahel. *Geographical Journal* 163, 150–160. **Franke, R.W. and Chasin, B.H.** 1980:

Seeds of famine: ecological destruction and the development dilemma in the West African Sahel. Montclair, NJ: Allanheld Osmun. **Gritzner, J.A.** 1988: *The West African Sahel: human agency and environmental change.* Chicago, IL: University of Chicago, Committee on Geographical Studies. **Le Houerou, H.N.** 1989: *The grazing land ecosystems of the African Sahel.* New York: Springer. **Tricart, J.** 1972: *Landforms of the humid tropics, forests and savannas.* London: Longman.

salic horizon

A surface or shallow subsurface horizon that contains a secondary enrichment of soluble salts. It must be at least 15 cm thick with at least 1% salt and have a product of the thickness (in cm) times salt percentage of 60 or more. Such horizons are found in ARIDLANDS and SEMI-ARID regions. *EMB*

Food and Agriculture Organization of the United Nations (FAO) 1998: *World reference base for soil resources* [*Soil Resource Report* 84]. Rome: FAO, ISRIC, ISSS.

saline lakes

Often called *salt lakes* or *salinas*, they occur in ENDORHEIC (internal) drainage systems, which are fed from inflowing streams, direct precipitation and groundwater. Salt concentrations exceed 5000 mg L^{-1}. Saline lakes occur in SEMI-ARID regions, ARIDLANDS and DESERTS where *potential evaporation* losses exceed precipitation input, e.g. the Dead Sea. Salts derive from both surface and GROUNDWATER sources and can be blown in from an AEOLIAN source, particularly if the lake is close to a coastal area. Where the groundwater table is close to the surface, influx of highly saline groundwater may predominate over surface waters with a low SOLUTE concentration, the inflow of which is controlled by temporally and spatially variable rainfall. EVAPORATION increases the concentration of salts in the lake, which may form vertical zonations controlled by solubility and temperature. Climate controls the relationship between surface and subsurface water inflow, and changes in the hydrological regime (as a result of increasing ARIDITY) may ultimately result in drying up of the lake, creating an EVAPORITE or salt PAN, e.g. Death Valley, California. *MLC*

[See also PALAEOSALINITY, SALINITY]

Hardie, L.A., Smoot, J.P. and Eugster, H.P. 1978: Saline lakes and their deposits: a sedimentological approach. In Matter, A. and Tucker, M. (eds), *Modern and ancient lake sediments* [*International Association of Sedimentologists Special Publication* 2]. Oxford: Blackwell Scientific, 7–41. **Bowler, J.M.** 1986: Spatial variability and hydrological evolution of Australian lake basins: analogue for Pleistocene hydrological change and evaporite formation. *Palaeogeography, Palaeoclimatology, Palaeoecology* 54, 21–41. **Herczeg, A.L. and Lyons, W.B.** 1991: A chemical model for the evolution of Australian sodium chloride lake brines. *Palaeogeography, Palaeoclimatology, Palaeoecology* 84, 43–53.

saline water

Water containing a total dissolved solids concentration > 20 000 mg L^{-1}: sea water contains around 35 000 mg L^{-1}, POTABLE WATER contains < 500 mg L^{-1} and IRRIGATION water < 1000 mg L^{-1}. *JAM*

[See also BRACKISH WATER]

salinisation

A soil process of ARIDLANDS and SEMI-ARID regions, in which soluble salts such as calcium carbonate, calcium sulphate and sodium chloride accumulate in a soil because there is insufficient rainfall to wash them out. The process occurs where CAPILLARY ACTION brings GROUNDWATER salts up into the soil. The water evaporates

leaving the salts behind to form a SOLONCHAK. Salinisation is a problem on IRRIGATION projects where drainage is inadequate, but, if good drainage is provided, salts can be leached out to give satisfactory growing conditions. Similar problems seem to have affected irrigated agriculture in early AGRARIAN CIVILISATIONS. Salinisation is also a process of DESERTIFICATION. *EMB*

Artzy, M. and Hillel, D. 1988: A defense of the theory of progressive salinization in ancient southern Mesopotamia. *Geoarchaeology* 3, 235–238. **Jacobsen, T. and Adams, R.M.** 1958: Salt and silt in ancient Mesopotamian agriculture. *Science* 128, 1251–1258. **Loyer, J.Y.** 1991: Classification des sols salés: les sols salic. *Cahiers ORSTOM, series Pédologie* XXVI(1), 51–61. **Middleton, N.J. and van Lynden, G.W.J.** 2000: Secondary salinization in South and Southeast Asia. *Progress in Environmental Science* 2, 1–19. **Szabolcs, I.** 1989: *Salt-affected soils.* Baton Rouge, FL: CRC Press. **Thomas, D.S.G. and Middleton, N.J.** 1993: Salinization: new perspectives on a major desertification issue. *Journal of Arid Environments* 24, 95–105. **Worthington, E.B. (ed.)** 1977: *Arid land irrigation in developing countries: environmental problems and effects.* Oxford: Pergamon Press.

salinity

The concentration of dissolved salts, mainly sodium chloride (NaCl), in ocean water, SALINE WATER or soil. It is measured in practical salinity units (psu) and estimated from conductivity and temperature measurements. *CES*

[See also DESALINATION, DIATOM ANALYSIS, PALAEOSALINITY, SALINISATION]

salt

(1) The mineral *halite* (sodium chloride, NaCl); common salt. (2) A COMPOUND formed when hydrogen ions in an ACID are replaced by metal ions. This occurs, for example, when an acid reacts with a BASE, which produces a salt and water. The name of the salt includes reference to a metal and an acid: sodium chloride, for example, is formed from the metal sodium (Na) and hydrochloric acid (HCl). *JAM*

[See also EVAPORITE]

salt caves

CAVES in BEDS or FORMATIONS of SALTS, which are amongst the most soluble minerals. As they are formed by *solution* of the rock, they are a type of KARST (in the broad sense) cave. Salt caves can develop rapidly by CORROSION (*chemical erosion*) and their expected lifetime is often short. In ARID regions, however, the rate of dissolution of rock-salt is slower so such caves may exist for longer. The *cave geometry* and CAVE SEDIMENTS are potential sources of information on environmental change. ANTHROPOGENIC salt caves have been formed by salt MINING and SUBSIDENCE may be a HAZARD in some areas. *JAM*

Frumkin, A., Magaritz, M., Carmi, I. and Zak, I. 1991: The Holocene climatic record of the salt caves of Mount Sedom, Israel. *The Holocene* 1, 191–200. **Wallwork, K.L.** 1960: Some problems of subsidence and land use in the mid-Cheshire industrial area. *Geographical Journal* 126, 191–199.

salt domes

Diapiric structures, composed of EVAPORITE deposits, some of which pierce the ground surface, forming mounds that may be subaerial or subaqueous. Cave passages within the salt dome of Mount Sedom, Israel, have enabled the reconstruction of Dead Sea levels and climatic changes from GROUNDWATER conditions. *HJW*

Frumkin, A., Magaritz, M., Carmi, I. and Zak, I. 1991: The Holocene climatic record of the salt caves of Mount Sedom,

Israel. *The Holocene* 1, 191–200. **Goudie, A.S.** 1989: Salt tectonics and geomorphology. *Progress in Physical Geography* 13, 597–605.

salt weathering   The breakdown of rock by the action of salts (*haloclasty*), which may be caused by the crystallisation of salt from solution, HYDRATION or the thermal expansion of crystals. It is a particularly important process in DESERTS and coastal environments, where it can be a serious hazard to buildings and other structures if salty water enters such structures by CAPILLARY ACTION.   *SHD*
Cooke, R.U. 1981: Salt weathering in deserts. *Proceedings of the Geologists' Association* 92, 1–16. **Goudie, A.S.** 1999: Experimental salt weathering of limestones in relation to rock properties. *Earth Surface Processes and Landforms* 24, 715–724.

saltation   The process in which particles are moved by either flowing water or wind in a series of bounds (as opposed to by sliding or in permanent suspension). In fluvial transport, it forms part of BEDLOAD. The particles involved in saltation are too heavy in relation to the fluid velocity to be transported in suspension in the water or air, but too light to confine movement to sliding and rolling. The term *reptation* is sometimes used to describe the associated movement of other clasts impacted by the saltating particles.   *RPDW*
[See also SUSPENDED LOAD, DEFLATION]
Abbott, J.E. and Francis, J.R.D. 1977: Saltation and suspension trajectories of solid grains in a water stream. *Philosophical Transactions of the Royal Society* A284, 225–254.

saltmarsh   A type of WETLAND characteristic of the high intertidal zone (between high NEAP and high SPRING TIDES) mainly along sheltered TEMPERATE coasts. MANGROVE SWAMPS tend to occupy the same niche in tropical environments. Environmental features include the ACCRETION of fine sediments (mud to fine sand), the presence of creeks and salt pans and a halophytic vegetation (see HALOPHYTE). Saltmarshes are sensitive environmental indicators and NATURAL ARCHIVES for palaeoenvironmental reconstruction.   *JAM*
Adam, P. 1990: *Saltmarsh ecology*. Cambridge: Cambridge University Press. **Allen, J.R.L. and Pye, K. (eds)** 1992: *Saltmarshes: morphodynamics, conservation and engineering significance*. Cambridge: Cambridge University Press. **Shaw, J. and Ceman, J.** 1999: Salt-marsh aggradation in response to late-Holocene sea-level rise at Amherst Point, Nova Scotia, Canada. *The Holocene* 9, 439–451.

saltwater intrusion   The movement of saltwater upstream beneath fresh water as a *saltwater wedge* and the flux of saltwater into coastal GROUNDWATER aquifers. It is anticipated that the projected SEA-LEVEL RISE caused by GLOBAL WARMING will lead to GROUNDWATER SALINISATION from saltwater intrusion along coasts.   *HJW*
Mulrennan, M.E. and Woodroffe, C.D. 1998: Saltwater intrusion into the coastal plains of the Lower Mary River, Northern Territory, Australia. *Journal of Environmental Management* 54, 169–188.

sample   A subset of a statistical POPULATION selected for investigation. Samples should be representative of the population from which they are drawn. *Representative* or *unbiased samples* allow accurate *sample estimates* of population PARAMETERS. RANDOM samples contain no known BIAS. *Sample size* is the number of cases or individuals

drawn from the population: it should be large enough to ensure accurate sample estimates. *Case studies* are samples with a sample size of one: even though they may be unrepresentative, the in-depth information obtained may be useful in other ways.   *JAM*
[See also SAMPLING]

sampling   The technique of characterising the properties of a population by investigating only a subset of the data. The simplest such technique is *random sampling* (where every element of the population has an equal chance of being selected). If the population is known to contain distinct classes and it is important to include the properties of these classes, *stratified sampling* can be employed (where the population is subdivided into groups by some demographic characteristic of the population and a simple random sample is taken from each group). In *systematic sampling*, every *n*th element is selected. This technique is very effective except when the dataset has periodic *fluctuations* that match the sampling frequency. More complex *sampling designs* are possible, including combinations of the above, cluster sampling and unaligned sampling. The aim of sampling is to estimate the properties of populations both efficiently and accurately.   *PJS*

sand   A SEDIMENT or particles with GRAIN-SIZE between 0.063 and 2 mm.   *TY*

sand-bed river   A river with a sediment load dominated by SAND moving as BEDLOAD. Sand-bed rivers can have a variety of CHANNEL PATTERNS, from braided to meandering. The river bed is shaped into BEDFORMS such as CHANNEL BARS, DUNES and RIPPLES. Sand-bed rivers grade into *mixed-load rivers*, which carry sufficient fine-grained sediment in suspension to develop stable banks (see COHESION) and are dominated by a meandering channel pattern.   *GO*
[See also BEDROCK CHANNEL, GRAVEL-BED RIVER, SUSPENDED-LOAD RIVER]
Bristow, C.S. 1987: Brahmaputra River: channel migration and deposition. In Ethridge, F.G., Flores, R.M. and Harvey, M.D. (eds), *Recent developments in fluvial sedimentology* [*Special Publication* 39]. Tulsa, OK: Society of Economic Paleontologists and Mineralogists, 63–74. **Collinson, J.D.** 1996: Alluvial sediments. In Reading, H.G. (ed.), *Sedimentary environments: processes, facies and stratigraphy*, 3rd edn. Oxford: Blackwell Science, 37–82.

sand bypassing   The movement of sand across an inlet or harbour, by either natural or mechanical means. Mechanical means are especially prominent where structures have been built to keep inlets from becoming shallow.   *HJW*

sand dune encroachment   Often used as an emotive descriptor, frequently criticised, of how a desert margin might advance into bordering DRYLANDS to cause DESERTIFICATION (*desert encroachment*). Alternatively, it describes the movement of coastal dunes inland, usually onto vegetated land.   *RAS*
Barker, J.R., Herlocker, D.J. and Young, S.A. 1989: Vegetal dynamics in response to sand dune encroachment within the coastal grasslands of central Somalia. *African Journal of Ecology* 27, 277–282.

sand dune reactivation   The remobilisation of sand dune sediment usually as a result of depletion of vegetation cover by people and/or animals. In temperate regions, reactivation of coastal dunes has attracted most attention, although inland PLEISTOCENE dune fields have also been reactivated following vegetation clearance, fire and grazing. The term has arguably been most associated, however, with DRYLANDS where reduction in ground cover vegetation has led to reactivation of fossil dunes (PALAEO-DUNES) formed about the LAST GLACIAL MAXIMUM. Although various techniques are used in attempts to control sand drifting and dune reactivation, in practice most solutions involve the establishment of a vegetation cover. Alternative methods in drylands include the application of high-gravity oil and salt-saturated water to promote CRUSTING and thus reduce ERODIBILITY.   *RAS*

Goudie, A.S. 2000: *The human impact on the natural environment*, 5th edn. Oxford: Blackwell, 307–311.

sand dunes   Subaerial bodies of SAND, ranging from 30 cm to 400 m high and between 1 m and 1 km wide, which have been shaped by prevailing wind conditions and the process of DEFLATION. Sand begins to accumulate where wind speed is reduced by an increase in surface roughness. The velocity of air flow within a metre or two of the Earth's surface varies with the slightest irregularity. Wind sweeps over and around an obstacle and sand grains drop out of the slowing-moving air behind the obstacle. As more sand piles up, a single, asymmetrical dune with a steep, straight lee (downwind) slope and a gentler windward slope is formed. Sand grains move up the windward slope by SALTATION to reach the crest of the dune and the straight lee slope experiences avalanching to return it to the angle of repose, which is typically 30–34°. Dunes of loose sand may be fixed to topographic obstacles or plants or they may be able to move freely.

Sand dunes are common in ARIDLANDS and in PERIGLACIAL environments and there are also extensive areas of COASTAL DUNES. Many DESERTS do not contain dunes. *Desert dunes* commonly occur in *dunefields* and as *compound dunes* rather than as isolated, simple dunes. There are many different types of sand dune. A basic classification differentiates between *transverse dunes*, LONGITUDINAL DUNES and *star dunes*, which tend to be formed by unidirectional, bidirectional and multidirectional wind regimes, respectively. *Barchans* and *parabolic dunes* are types of transverse dune: both are crescentic in plan, but, whereas the 'horns' of a barchan point in the direction of dune movement, they point upwind in a parabolic dune. Changing climate or a decline in sand supply can induce dune stabilisation by plant colonisation or soil formation on the dune surface; in this state dunes slowly lose their characteristic shapes (see Figure). Environmental changes allow ancient dune systems to be preserved as PALAEO-DUNES.   *MAC*

[See also ERG, SAND SEA, SAND SHEET]

Cooke, R., Warren, A. and Goudie, A. 1993: *Desert geomorphology*. London: UCL Press. Craig, M.S. 2000: Aeolian sand transport at the Lanphere Dunes, northern California. *Earth Surface Processes and Landforms* 25, 39–253. Hastenrath, S. 1987: The barchan sanddunes of south Peru revisited. *Zeitschrift für Geomorphologie NF* 31, 167–178. Lancaster, N. 1995: *Geomorphology of desert dunes*. London: Routledge. Muhs D.R. 1985: Age and paleoclimatic significance of Holocene sand dunes in northeastern Colorado. *Annals of the Association of American Geographers* 75, 566–582. Thomas, D.S.G. 1992: Dune activity: concepts and significance. *Journal of Arid Environments* 22, 31–38. Seppälä, M. 1971: Evolution of eolian relief of the Kaamasjoki-Kiellajoki river basin in Finnish Lapland. *Fennia* 104, 1 88.

sand ramp   A deposit predominantly of SAND either upwind or in the lee of a topographic obstacle. Sand

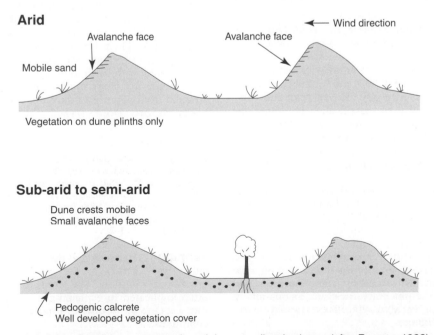

sand dunes *Possible responses of sand dunes to climatic change (after Rognon, 1982)*

ramps, which may be up to 100 m thick in DESERTS, are a potential source of palaeoenvironmental information based not only on AEOLIAN stratigraphy but also on the deposits of colluvial, fluvial or palaeosol origin that they may contain or overlie. *JAM*

Rendell, H.M. and Sheffer, N.L. 1996: Luminescence dating of sand ramps in the eastern Mojave Desert. *Geomorphology* 17, 187–197.

sand sea An extensive AEOLIAN accumulation of SAND characterised by SAND DUNES and/or SAND SHEETS; an area of sandy DESERT. Sand seas (termed ERGS in the Sahara), which should be distinguished from stony deserts (*reg*) and rocky deserts (*hamada*), require a sufficient sand supply (often provided by fluvial or coastal deposits of a preceding climatic epoch), winds strong enough to transport sand and suitable topography for sand deposition.
JAM/RPDW

McKee, E.D. (ed.) 1979: A study of global sand seas. *United States Geological Survey, Professional Paper* 1052. **Thomas, D.S.G.** 1997: Sand seas and aeolian bedforms. In Thomas, D.S.G. (ed.), *Arid zone geomorphology*. Chichester: Wiley, 371–412.

sand sheet An extensive AEOLIAN accumulation of SAND lacking DUNE development apart from *zibar dunes* (low, dome-shaped sand dunes without a slip face). Sand sheets up to about 100 000 km² are favoured by the existence of a vegetation cover, coarse sand, a high WATER TABLE, periodic flooding and the development of DURICRUSTS. *JAM*

[See also SAND SEA]

Kocurek, G. and Nielson, J. 1986: Conditions favourable for the formation of warm-climate eolian sand sheets. *Sedimentology* 33, 795–816.

sandrock A weakly or poorly cemented sand. The term is used in association with fossil PLEISTOCENE dune sands exposed at coastal sites in southwest England. *PW*

[See also AEOLIANITE, BEACHROCK, SANDSTONE]

Campbell, S., Hunt, C.O., Scourse, J.D. *et al.* (eds) 1998: *Quaternary of south-west England*. London: Chapman and Hall.

sandstone A CLASTIC SEDIMENTARY ROCK with a dominant GRAIN-SIZE between 0.063 and 2 mm (i.e. SAND). Sandstones are characterised by their TEXTURE and composition. A commonly used classification combines the absence (ARENITE) or presence (WACKE or GREYWACKE) of a MATRIX of MUD with the relative proportions of CLASTS of QUARTZ, FELDSPAR and LITHIC composition, to give a bipartite name such as *quartz arenite* or *lithic wacke* (see Figure).

Sandstones are common rocks, develop some distinctive LANDFORMS and contain much information about past environments and environmental change (see SEDIMENTOLOGICAL EVIDENCE OF ENVIRONMENTAL CHANGE). Sandstone composition provides information about PROVENANCE and TECTONIC setting; texture, FABRIC and SEDIMENTARY STRUCTURES give information about physical processes of WEATHERING, transport and deposition; and FACIES ANALYSIS of sedimentary rock successions allows the identification of PALAEOENVIRONMENTS and controls on environmental change. *TY/GO*

[See also ARKOSE, QUARTZITE]

Johnsson, M.J., Stallard, R.F. and Lundberg, N. 1991: Controls on the composition of fluvial sands from a tropical weathering environment – sands of the Orinoco River drainage basin, Venezuela and Colombia. *Geological Society of America Bulletin* 103, 1622–1647. **Mountney, N., Howell, J., Flint, S. and Jerram, D.** 1999: Climate, sediment supply and tectonics as controls on the deposition and preservation of the aeolian-fluvial Etjo Sandstone Formation, Namibia. *Journal of the Geological Society, London* 156, 771–777. **Pettijohn, F.J.** 1975: *Sedimentary rocks*, 3rd edn. New York: Harper & Row. **Pettijohn, F.J., Potter, P.E. and Siever, R.** 1987: *Sand and sandstone*. New York: Springer. **Young, R. and Young, A.** 1992: *Sandstone landforms*. Berlin: Springer.

sandur A gently sloping glacial *outwash plain* ranging in width from hundreds of metres up to several kilometres. It is typically built up of sand and gravel deposited by proglacial meltwater streams. In an enclosed valley in mountainous terrain, an outwash plain is termed a *valley train* or *valley sandur*. Where not enclosed, it is a *plain sandur*. Sandur is an Icelandic term, the plural of which is sandar. Sandar were more extensive than today beyond the limits of the ICE SHEETS during GLACIAL EPISODES. They are important sources of sand and gravel for the construction industry. *RAS*

[See also GLACIOFLUVIAL LANDFORMS, GLACIOFLUVIAL SEDIMENTS]

Church, M.A. 1972: Baffin Island sandurs: a study of Arctic fluvial processes. *Geological Survey of Canada Bulletin* 216, 1–208. **Krigström, A.** 1962: Geomorphological studies of sandur plains and their braided rivers in Iceland. *Geografiska Annaler* 44, 328–346.

sandwave A periodic, flow-transverse subaqueous BEDFORM in sand with a straight crest-line and superimposed smaller bedforms (DUNES or RIPPLES). In fluvial environments such bedforms are better referred to as compound dunes and *bars* and the term sandwave is not recommended, but it is useful in relation to *tidal sandwaves*. These are symmetrical to asymmetrical in cross-section and form in intertidal to subtidal settings, with heights of 10 m or more and spacings up to several hundred metres. Their character persists over many tides, although the superimposed bedforms may be destroyed and created with each tide. In SEDIMENTARY DEPOSITS sandwave morphology and tidal parameters can be reconstructed from patterns of CROSS-BEDDING, BOUNDING SURFACES, REACTIVATION SURFACES, MUD DRAPES and TIDAL BUNDLES. *GO*

[See also DRAA]

Allen, J.R.L. 1980: Sand waves: a model of origin and internal structure. *Sedimentary Geology* 26, 281–328. **Allen, P.A. and Homewood, P.** 1984: Evolution and mechanics of a Miocene tidal sandwave. *Sedimentology* 31, 63–81. **Berne, S., Auffret, J.-P. and Walker, P.** 1988: Internal structure of subtidal sandwaves revealed by high resolution seismic reflection. *Sedimentology* 35, 5–20.

sapping The erosion at the base of a cliff as a result of either wave action or the outflow of streams and/or groundwater outflow. The latter, termed *spring sapping*, is often implicated in explaining the development of chalk DRY VALLEYS in southern England, under PERIGLACIAL conditions. *PW*

[See also FLUVIAL PROCESSES]

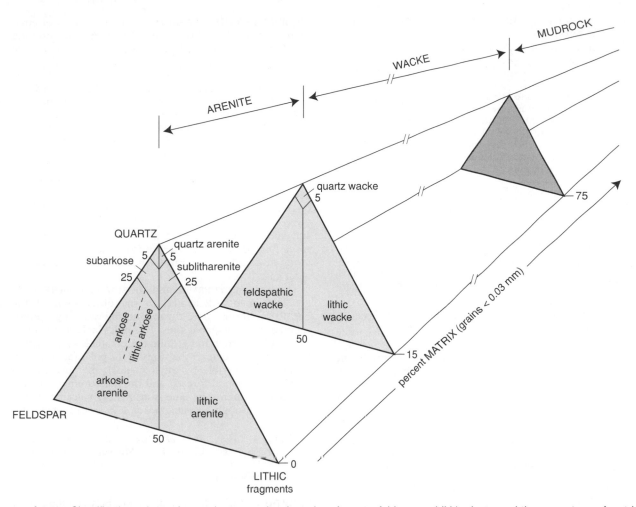

sandstone *Classification scheme for sandstones, using the ratios of quartz, feldspar and lithic clasts, and the percentage of matrix (after Pettijohn, 1975)*

saprolite A CLAY-rich, *in situ* REGOLITH that tends to preserve original bedrock structures, such as BEDDING, at least to some extent. Saprolite, from the Greek for 'rotten rock', is the product of deep CHEMICAL WEATHERING and is most commonly found in the *humid tropics*. Two broad zones may be recognised within that part of the WEATHERING PROFILE composed of saprolite. Relatively close to the bedrock is *coarse saprolite*, sometimes termed *arene*, composed largely of primary minerals in separated or partly disaggregated grains and rock fragments, many of which are of SAND size or coarser. Above this lies *fine saprolite*, in which secondary minerals of alteration predominate. The term *lithomarge* has also been used for highly altered, fine saprolite, especially where it is dominated by KAOLINITE and has a plastic consistency and light colour constituting the *pallid zone*. Saprolite has not been subject to SOIL FORMING PROCESSES (see also PEDOGENESIS). Hence, the overlying *mottled zone* of the weathering profile is part of the SOIL PROFILE B horizon, whereas saprolite constitutes the C horizon and is an *alterite*, not a SOIL. *JAM/GO*

[See also LATERITE, PLINTHIC HORIZON, PLINTHOSOLS, RESIDUAL DEPOSIT]

Scholten, T., Felix-Henningsen, P. and Schotte, M. 1997: Geology, soils and saprolites of the Swaziland Middleveld. *Soil Technology* 11, 229–246. Stolt, M.H. and Baker, J.C. 1994: *Strategies for studying saprolite and saprolite genesis* [*Special Publication* 34]. Madison, WI: Soil Science Society of America, 1–19. Tardy, Y. 1997: *Petrology of laterites and tropical soils*. Rotterdam: Balkema.

sapropel An organic-rich black MUD, composed predominantly of the remains of PLANKTON, which accumulates as PELAGIC SEDIMENTS in marine basins, lakes and estuaries under ANAEROBIC conditions. In the sediments of the eastern Mediterranean, for example, a major layer of sapropel (labelled S1) has been dated to the intervals 11 800–10 400 (subphase S1b) and 9000–8000 radiocarbon years BP (subphase S1a). Earlier sapropel layers have also been dated to about 129 000, 104 000 and 85 000 years BP. Anaerobic conditions appear here to have been caused by an increase in STRATIFICATION following increased PRECIPITATION and the delivery of an increased volume of fresh water to the Mediterranean from particularly high NILE FLOODS in PLUVIAL episodes, which would also have delivered an influx of fine sediment. *JAM/GO*

[See also AFRICAN HUMID PERIOD, BLACK SHALE, OCEANIC ANOXIC EVENTS]

Ariztegui, D., Asioli, A., Lowe, J.J. *et al.* 2000: Palaeoclimate and the formation of sapropel S1: inferences from Late Quaternary lacustrine and marine sequences in the central Mediterranean region. *Palaeogeography, Palaeoclimatology, Palaeoecology* 158, 215–240. **Cramp, A. and O'Sullivan, G.** 1999: Neogene sapropels in the Mediterranean: a review. *Marine Geology* 153, 11–28. **Rossignol-Strick, M., Nesteroff, W., Olive, P. and Vernaud-Grazzini, C.** 1982: After the deluge: Mediterranean stagnation and sapropel formation. *Nature* 295, 105–110. **Rossignol-Strick, M.** 1985: Mediterranean Quaternary sapropels, an immediate response of the African monsoon to variation of insolation. *Palaeogeography, Palaeoclimatology, Palaeoecology* 49, 237–263. **Rossignol-Strick, M.** 1999: The Holocene climatic optimum and pollen records of sapropel 1 in the eastern Mediterranean. *Quaternary Science Reviews* 18, 515–530.

saprophyte Also known as *saprotrophs* or *saprobes*, saprophytes obtain their energy requirements from dead organic matter. Fungi and BACTERIA are important groups, the most important role of which is as DECOMPOSER organisms. *JAM*

[See also BIODEGRADATION, BIOGEOCHEMICAL CYCLES, DECOMPOSITION]

sarsen A large block of silica-cemented sandstone found mainly on chalk hills (Downs) of southern England. Also known as *'grey wethers'*, *'puddingstones'* or *meulières* (France), sarsens are thought to be the remnants of a now absent DURICRUST (SILCRETE) formed under TERTIARY warm climates. *RAS*

Summerfield, M.A. and Goudie, A.S. 1980: The sarsens of southern England: their palaeoenvironmental interpretation with reference to other silcretes. In Jones, D.K.C. (ed.), *The shaping of southern England* [*Special Publication* 11]. London: Institute of British Geographers, 71–100. **Ullyot, J.S., Nash, D.J. and Shaw, P.A.** 1998: Recent advances in silcrete research and their implications for the origin and palaeoenvironmental significance of sarsens. *Proceedings of the Geologists' Association* 109, 255–270.

satellite climatology The study of CLIMATE and changing climate based primarily on data from Earth-orbiting satellites. *JBE*

[See also METEOROLOGICAL SATELLITES]

Barrett, E.C. 1974: *Climatology from satellites*. London: Methuen. **Kondratiev, Y.** 1983: *Satellite climatology*. Leningrad: Hydrometeozdat.

satellite orbit The track taken by a satellite around a planetary body. All Earth satellite orbit planes pass through the centre of the Earth. Satellite orbits may be categorised by distance from the Earth and position relative to the Earth and Sun: *low Earth orbit* relates to between 750 and 2000 km altitude; *medium Earth orbit* to about 10 000 km; and *geosynchronous Earth orbit* to 36 000 km above the Earth's surface. A satellite that always views the same part of the Earth is termed *geostationary*, whilst one that always views an area at the same Sun angle is *Sun-synchronous*. Earth observation satellites often operate using a recurrent track that repeats itself over a fixed period (for example, 16 days in the case of *LANDSAT*). *TF*

NASDA (National Space Development Agency of Japan) 1999: *Cosmic Information Center, online space notes*: http://spaceboy.nasda.go.jp/Index_e.html.

satellite remote sensing REMOTE SENSING of environmental variables carried out from instruments orbiting the Earth. The sensing from space of many geophysical quantities relevant in METEOROLOGY, CLIMATOLOGY, HYDROLOGY, OCEANOGRAPHY and GEOLOGY can be performed on a global scale and on a regular basis. In order to ensure the temporal repeatability of the measurements, a series of satellites with similar instruments have been launched. The most important limiting factor of satellite remote sensing in comparison to AIRBORNE REMOTE SENSING is the less fine RESOLUTION.

Starting from the photographs of the astronauts on board the first satellites, satellite remote sensing has become a very useful tool for surveying ENVIRONMENTAL CHANGE. Many instruments on board various satellites carry out remote monitoring of the Earth's atmosphere and surface. The observation of the atmosphere from space for METEOROLOGICAL purposes includes the monitoring of clouds, temperature, water vapour, ozone and other gases.

The OPTICAL, THERMAL and MICROWAVE REMOTE SENSING instruments dedicated to the observation of the Earth's surface are used for monitoring various geophysical variables on a global scale. Examples include: the measurements of water and ground temperature, soil moisture and snow-cover extent; the study of the polar regions (particularly inaccessible by other means); the surveying of tropical and boreal forests; and the measurement of surface displacements in land subsidence, volcanicity and seismicity through *synthetic aperture radar* INTERFEROMETRY. *TS*

Schanda, E. 1986: *Physical fundamentals of remote sensing*. Heidelberg: Springer. **Ulaby, F.T., Moore, R.K. and Fung, A.K.** 1981: *Microwave remote sensing: active and passive*. Norwood, MA: Artech House.

saturation deficit The difference between the *vapour pressure* and the *saturation vapour pressure*; i.e. the difference between the amount of moisture in the air and the amount the air could hold at the same temperature. The saturation deficit is one index of atmospheric moisture and an important concept in the understanding of HUMIDITY and PRECIPITATION processes. *BDG*

Penman, H.L. 1955: *Humidity*. London: Institute of Physics. **Sumner, G.** 1988: *Precipitation: process and analysis*. Chichester: Wiley.

saturation isothermal remanent magnetisation (SIRM) The degree of magnetisation above which any further increase in geomagnetic field will not lead to a corresponding increase in ISOTHERMAL REMANENT MAGNETISM (IRM). It is a parameter used in PALAEOMAGNETIC DATING. *MHD*

Thompson, R. 1991: Palaeomagnetic dating. In Smart, P.L. and Frances, P.D. (eds), *Quaternary dating methods – a users guide* [*QRA Technical Guide* 4]. Cambridge: Quaternary Research Association, 177–198.

saturation overland flow Water that flows over the ground surface when a WATER TABLE or temporary *perched water table* rises to the surface, producing saturated (and hence impermeable) ground. Such overland flow comprises direct *rain runoff* (rain falling onto saturated areas) and *return flow* (subsurface water from upslope returning to the surface in saturated zones). Saturation overland flow within catchments may be localised or widespread and temporary, seasonal or quasi-permanent.　　*ADT*

[See also HORTONIAN OVERLAND FLOW, RUNOFF PROCESSES]

Hewlett, J.D. and Hibbert, A.R. 1967: Factors affecting the response of small watersheds to precipitation in humid areas. In Sopper, W.E. and Lull, H.W. (eds), *Forest Hydrology*. Oxford: Pergamon, 275–290. **Ward, C. and Robinson, M.** 2000: *Principles of hydrology*, 4th edn. London: McGraw-Hill.

savanna Tropical vegetation covering a broad range of habitat types intermediate between DESERT and EVERGREEN FOREST. The commonest use of the term savanna is to describe *grasslands* with scattered trees or shrubs occurring on ancient soils that are poor in NUTRIENTS. Commonly, savanna plants are adapted to conserve water, using CAM and C-4 PHOTOSYNTHESIS (see also C-3, C-4 AND CAM PLANTS) or by possessing deep root systems. Savanna ecosystems are present in all three major tropical regions (Neotropics, Africa and Asia), but are best represented in the driest of these areas (Africa).

Four kinds of savanna may be recognised:

1. *Climatic savanna*, where rainfall is insufficient to maintain forest. Rainfall is evenly spread throughout the year with precipitation of 800 to 1800 mm being typical.
2. *Non-seasonal savanna*, where edaphic (soil) conditions prevents forest growth in a climate with no dry season. White sand soils, or soils with high concentrations of potentially toxic chemicals, e.g. aluminium, do not support tree growth. Rainfall may be high.
3. *Seasonal savanna*, where edaphic conditions combine with a strong dry season to prevent forest growth and well drained soils develop a strong moisture deficit for enough time to prevent forest development.
4. *Hyperseasonal savanna*, where strongly seasonal rainfalls cause soils to alternate between saturation and severe water deficit, preventing forest development.

Generally, the natural savannas of the Neotropics are almost all edaphic, whereas those of Australia and Africa are climatic.

In Africa, Miocene climatic change led to the replacement of woodland with savanna and may have precipitated the EVOLUTION of ground-dwelling HOMINIDS from their arboreal ancestors. All species of hominid apparently evolved within the African savanna habitats. PLEISTOCENE climate change was once thought to have led to a huge expansion of savanna (see REFUGE THEORY IN THE TROPICS).

Most natural savannas burn every three to ten years. The fires maintain the openness of the ecosystem and are important in MINERAL CYCLING. Humans have achieved long-term modification or creation of savannas, e.g. ABORIGINAL fire-stick farming in Australia. Savannas commonly support large populations of herd animals and many have been converted to grazing lands for domesticated livestock and agriculture.　　*MBB*

[See also BIOMES, FIRE HISTORY, SAHEL, STEPPE VEGETATION, TROPICAL RAIN FOREST, TROPICAL SEASONAL FOREST]

Bourlière, F. (ed.) 1983: *Tropical savannas [Ecosystems of the World*. Vol. 13]. Amsterdam: Elsevier. **Coe, M., McWilliam, N., Stone, G. and Packer, M.** 1999: *Mkomazi: the ecology, biodiversity and conservation of a Tanzanian savanna*. London: Royal Geographical Society. **Eiten, G.** 1982: Brazilian Savannas. In Huntley, B.J. and Walker, B.H. (eds), *Ecology of tropical savannas* [*Ecological Studies* 42]. Berlin: Springer, 25–47. **Huntley, B.J. and Walker, B.H. (eds)** 1982: *Ecology of tropical savannas*. Berlin: Springer. **Kellman, M. and Tackaberry, R.** 1997: *Tropical environments: the functioning and management of tropical ecosystems*. London: Routledge. **Larick, R. and Ciochon, R.L.** 1996: The African emergence and early Asian dispersals of the genus Homo. *American Scientist* 84, 538–551. **Le Houerou, H.N.** 1989: *The grazing land ecosystems of the African Sahel*. New York: Springer. **Medina, E. and Silva, J.F.** 1990: Savannas of northern South America: a steady state regulated by water–fire interactions on a background of low nutrient availability. *Journal of Biogeography* 17, 403–413. **Werner, P.A.** 1991: *Savanna ecology and management: Australian perspectives and intercontinental comparisons*. Oxford: Blackwell Scientific.

savannisation The spread of SAVANNA at the expense of TROPICAL AND SUBTROPICAL FORESTS. Analogous to DESERTIFICATION, it is currently largely caused by LAND DEGRADATION following FOREST CLEARANCE.　　*JAM*

scale concepts in environmental change The BIOSPHERE plays a central role in mediating global ENVIRONMENTAL CHANGE through its interactions with land surface ENERGY BALANCE, ATMOSPHERIC COMPOSITION and the HYDROLOGICAL and CARBON CYCLES. For these interactions to be fully understood, terrestrial ecological processes governing the exchange of energy, CARBON DIOXIDE (CO_2), water and other trace gases between the ATMOSPHERE, VEGETATION and the SOIL must be accurately quantified on regional to global scales, over periods of years to centuries. In addition, the large-scale cycling of carbon and sulphur through marine ecosystems must be determined for similar reasons. However, research in ECOLOGY has traditionally focused on individual organisms, over periods of up to years, adopting a largely reductionist approach (see REDUCTIONISM). Furthermore, experiments on the large-scale impacts of environmental change often cannot be conducted for moral and logistic reasons. The use of *upscaling*, the integration of information from small to larger scales, is therefore assuming increasing importance in research on the interactions between the biosphere and GLOBAL CHANGE and is discussed here with reference to ecological interactions with the terrestrial carbon cycle.

Variation in the processes of an ecological system and in its structural organisation both vary according to the scale at which they are observed. Accounting for control of ecological processes by the appropriate NEGATIVE and *positive* FEEDBACKS is important for upscaling, since they vary greatly in significance with scale. In the case of structural organisation of ecological systems, accounting for heterogeneity can be critical for accurate upscaling, especially if properties of the system show non-linear responses to structural changes. For example, clumps of leaves within a VEGETATION CANOPY, clustered distributions of individuals within a plant COMMUNITY and patches of different vegetation types within a LANDSCAPE all significantly modify ABSORPTION of SOLAR RADIATION by vegetation in comparison with uniform distributions of each, since absorption with respect to leaf area is non-linear. This heterogeneity has important consequences both in the upscaling of energy balance and PHOTOSYNTHESIS and in the DOWNSCALING of SATELLITE REMOTE SENSING measurements.

Mechanistic mathematical MODELS currently provide the principal method for upscaling in ecological systems. For models to be flexible and have reliable predictive value, they must contain explicit descriptions of the actions of global change, for the scale at which it acts. For example, rising atmospheric CO_2 concentration is likely to have significant impacts on the functioning of terrestrial vegetation and its interaction with the global carbon cycle during the next century, with negative feedback effects to CO_2. However, the principal direct effects of CO_2 on vegetation are on the biochemistry of photosynthesis and on the physiology of STOMATA, both of which occur at scales smaller than the leaf and vary over periods of seconds to minutes (see CARBON DIOXIDE FERTILISATION). Upscaling in this case therefore involves several orders of magnitude (see Figure), and cellular processes consequently underpin the responses of carbon storage in the terrestrial biosphere to atmospheric CO_2.

Carbon storage at progressively larger spatial scales tends to respond to atmospheric or climatic change over successively longer temporal scales, and spatial and temporal scales thereby show an approximate equivalence. A key question in this upscaling exercise is whether increasing carbon fixation in photosynthesis with atmospheric CO_2 enrichment will result in greater ecosystem carbon storage, an issue that is complicated by the concurrent rise in temperature expected in the future. Recent modelling studies suggest that carbon storage in the global terrestrial biosphere and soils will increase with CO_2 and temperature in the near future, as a consequence of higher rates of leaf photosynthesis. However, this initial response may vary between BIOMES and will saturate in the longer term as a result of compensation by processes at higher spatial and temporal scales (see Figure).

Recent technological advances have led to the development of several new approaches for investigating large-scale ecological processes that are of direct interest in the context of global change. Studies based on these approaches have also proved useful for comparisons with the results of upscaling models, which must be tested rigorously at each upscaling step, in order to avoid the propagation of errors and to produce reliable predictions. Inevitably, a model can only reproduce mechanisms that are characterised, and it therefore acts as a summary of the current state of knowledge about an ecological system, serving to highlight gaps in understanding and generating hypotheses to be tested experimentally.

The direct effects of increasing atmospheric CO_2 on intact stands of mature natural vegetation remain uncertain and experiments that address this problem have therefore been initiated at a number of sites throughout the world, following the recent development of *Free-Air Carbon Dioxide Enrichment* (FACE) *systems*. FACE systems are capable of manipulating atmospheric CO_2 concentration around stands of otherwise undisturbed vegetation growing under field conditions, up to the scale of forest trees, and with little perturbation of MICROCLIMATE. The method allows the direct study of ecosystem processes under high CO_2 concentration at larger spatial scales than have previously been possible and allows

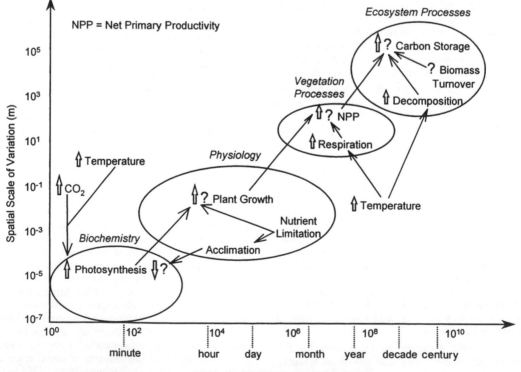

scale concepts in environmental change Upscaling the potential effects of increasing CO_2 and temperature on ecosystem carbon storage

upscaling models to be compared directly with observations at the scale of interest.

Eddy flux measurements give integrated estimates of trace gas fluxes between ecosystems and the atmosphere, and can be made from towers or aircraft, at landscape scales, over periods of several years, providing excellent opportunities for model testing. The value of such measurements is increased markedly by a complementary analysis of the fluxes from components of the ecosystem, such as canopy photosynthesis and respiration from woody tissues and the soil. This downscaling exercise reveals that relatively small net ecosystem fluxes of CO_2 are the sum of larger, but opposing photosynthetic and respiratory fluxes. A relatively small change in the balance between these can therefore change the ecosystem from a net carbon source to a net CARBON SINK, a change of key importance for the terrestrial carbon cycle.

At larger scales, measurements of the annual seasonal amplitude and CARBON ISOTOPE analysis of atmospheric carbon dioxide can provide clues about regional and global NET PRIMARY PRODUCTIVITY of the BIOSPHERE, especially at high latitudes. In addition, satellite remote sensing provides evidence about changes in the length of the growing season and the structure and activity of vegetation. Perhaps the most valuable scaling studies on terrestrial ecosystems adopt an approach where mechanistic modelling is combined and tested with experimental studies at a range of scales, and using a suite of different observation techniques. *CPO*

[See also MAGNITUDE-FREQUENCY CONCEPTS]

Amthor, J.S., Goulden, M.L., Munger, J.W. and Wofsy, S.C. 1994: Testing a mechanistic model of forest-canopy mass and energy exchange using eddy correlation: carbon dioxide and ozone uptake by a mixed oak–maple stand. *Australian Journal of Plant Physiology* 21, 623–651. **Asner, G.P., Wessman, C.A. and Archer, S.** 1998: Scale dependence of absorption of photosynthetically active radiation in terrestrial ecosystems. *Ecological Applications* 8, 1003–1021. **Baldocchi, D.B., Valentini, R., Running, S., Oechel, W. and Dahlman, R.** 1996: Strategies for measuring and modelling carbon dioxide and water vapour fluxes over terrestrial ecosystems. *Global Change Biology* 2, 159–168. **Cao, M. and Woodward, F.I.** 1998: Dynamic responses of terrestrial ecosystem carbon cycling to global climate change. *Nature* 393, 249–252. **Ehleringer, J.R. and Field, C.B. (eds)** 1993: *Scaling physiological processes. Leaf to globe.* New York: Academic Press. **Hendrey, G.R., Ellsworth, D.S., Lewin, K.F. and Nagy, J.** 1999: A free-air enrichment system for exposing tall forest vegetation to elevated atmospheric CO_2. *Global Change Biology* 5, 293–309. **Jarvis, P.G.** 1995: Scaling processes and problems. *Plant, Cell and Environment* 18, 1079–1089. **Levin, S.A.** 1992: The problem of pattern and scale in ecology. *Ecology* 73, 1943–1967. **Peterson, D.L. and Parker, V.T. (eds)** 1998: *Ecological scale: theory and applications.* New York: Columbia University Press. **Rosswell, T., Woodmonsee, R.G. and Risser, P.G.** 1988: *Scales and global change: spatial and temporal variability in biospheric and geospheric processes.* Chichester: Wiley. **Wang, Y.P. and Polglase, P.J.** 1995: Carbon balance in the tundra, boreal forest and humid tropical forest during climate change: scaling up from leaf physiology and soil carbon dynamics. *Plant, Cell and Environment* 18, 1226–1244. **Waring, R.H., Law, B.E., Goulden, M.L. et al.** 1995: Scaling gross ecosystem production at Harvard Forest with remote sensing: a comparison of estimates from a constrained quantum-use efficiency model and eddy correlation. *Plant, Cell and Environment* 18, 1201–1213. **Woodward, F.I., Lomas, M.R. and Betts, R.A.** 1998: Vegetation–climate feedbacks in a greenhouse world. *Philosophical Transactions of the Royal Society of London* B353, 29–39.

scale concepts: the palaeoecological context

PALAEOECOLOGY is concerned with changes in fauna and flora in space and time across a range of different scales, which may be viewed as hierarchical. Delcourt *et al.* (1983) devised a four-part scheme in which spatial and temporal scales are considered together and placed at different levels of complexity – the micro-, meso-, macro- and mega-scales (see Figure).

The micro-scale relates to a time scale of one to 500 years and a spatial scale of $1 m^2$ to $10^6 m^2$. This includes the study of palaeoecological change and ECOLOGICAL SUCCESSION associated with relatively recent (e.g. LITTLE ICE AGE) or short-term events (e.g. analysis of palaeovegetation change associated with short-term occupation of archaeological sites). Such events are often associated with individual localities. The meso-scale domain is defined as extending from 500 to 10 000 years and from $10^6 m^2$ to $10^{10} m^2$ and thus relates to regional palaeoecological change during the HOLOCENE. The macro-scale extends from 10 000 years to 1 000 000 years and from $10^{10} m^2$ to $10^{12} m^2$. This scale may include continent-wide changes in CLIMATE or ecosystems; examples include plant and community migrations on a subcontinental scale. In such studies, individual site investigations at the micro- and meso-scales are fitted together to give a larger-scale picture of landscape change. The pollen ISOCHRONE maps for Britain and Europe depicting palaeovegetation change during the last 13 000 years (Huntley and Birks, 1983; Birks, 1986) display both spatial and temporal patterns in pollen data, the isochrones joining sites at which pollen or vegetation events happened at the same time. The ISOCHRONE maps can be used to determine rates and directions of species and community migrations through space and time. Finally, the mega-scale domain spans 1 000 000 years to 4.6 billion years and $> 10^{12} m^2$, essentially representing changes at the global scale and long-term changes such as GLACIAL–INTERGLACIAL CYCLES or EVOLUTION. The development of GENERAL CIRCULATION MODELS and computer simulation has aided the analysis of palaeoecological change at the global scale. Different environmental and cultural processes influence the palaeoecological patterns at each scale. For example, astronomical forcing of climate may be important at a global scale, whilst EDAPHIC FACTORS are important at the micro-scale.

Spatial scale, particularly the concept of the *pollen source area* (i.e. the area from which the pollen deposited at a study was derived), is of paramount importance in *palynology*. This has been demonstrated by models that explore the relationships between pollen transport and the spatial resolution of the pollen record. Jacobson and Bradshaw (1981) suggest that the pollen deposited at a sampling site is composed of three components: *local pollen* derived from plants growing within 20 m of the sampling site; *extralocal pollen* from between 20 m and 2 km; and *regional pollen* transported to the site from distances greater than 2 km. The size of the sampling site determines the relevant contribution of each component to the pollen deposited at a site, such that the larger the site, the greater the contribution of regional pollen rain to the pollen deposited at that site. Thus, for a particular palynological investigation, the choice of site of an appropriate size is essential. The study of the ecological dynamics of a small stand of trees would require the study of a small site (e.g. forest hollow), which only reflects local

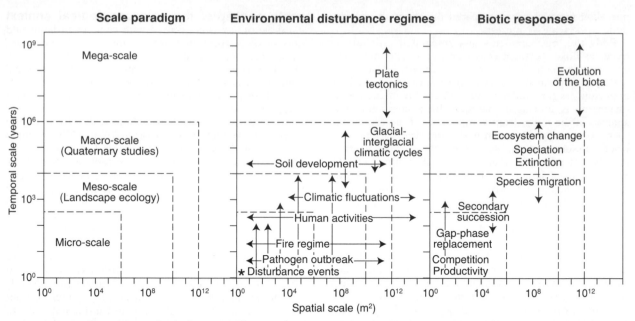

scale concepts: the palaeoecological context *The spatial and temporal scales of palaeoecological and related investigations: scale paradigms (left) within which disturbance regimes (centre – examples of disturbance events include clear cutting, earthquakes, floods, wildfire and wind throw) and (right) biotic responses are defined according to their spatial–temporal domains (after Delcourt and Delcourt, 1991)*

changes, whilst the reconstruction of landscape change associated with human settlement would require the study of a larger lake or mire.

The temporal scale provided by a palaeoecological study depends on the palaeoecological technique used, the time span of the deposit or material analysed, its accumulation rate and the interval at which the sediment is sampled for palaeoecological analysis. Palaeoecological evidence can be derived on an annual basis. For example, an analysis of CLIMATIC CHANGE with annual RESOLUTION can be obtained from the study of TREE RINGS, whilst fine resolution pollen analysis of annually laminated sediments can give yearly records of vegetation change. However, most POLLEN, *coleopteran* and MACROFOSSIL studies use coarser sediment subsampling intervals and each sample that is analysed from a sediment sequence may represent tens to hundreds of years, thus giving a broad indication of the nature of palaeoecological change. Whilst coarser subsampling leads to loss of short-term environmental information, those changes that occurred over millennia or centuries can be discerned.

In essence, spatial and temporal scales are inseparable because spatial patterns are continually changing over time.   *LD-P*

[See also LOCAL CLIMATE]

Birks, H.J.B. 1986: Holocene isochrone maps and patterns of tree-spreading in the British Isles. *Journal of Biogeography* 16, 503–540. **Birks, H.J.B.** 1986: Late-Quaternary biotic changes in terrestrial and lacustrine environments, with particular reference to north-west Europe. In Berglund, B. E. (ed.), *Handbook of Holocene palaeoecology and palaeohydrology*. Chichester: Wiley, 3–65. **Delcourt, H.R. and Delcourt, P.A.** 1991: *Quaternary ecology: a palaeoecological perspective*. London: Chapman and Hall. **Delcourt, H.R., Delcourt, P.A. and Webb III, T.** 1983: Dynamic plant ecology: the spectrum of vegetational change in space and time. *Quaternary Science Reviews* 1, 153–175.

Huntley, B. and Birks, H.J.B. 1983: *Atlas of past and present pollen maps for Europe, 0–13,000 years ago.* Cambridge: Cambridge University Press. **Jacobson, G.L. and Bradshaw, R.H.W.** 1981: The selection of sites for palaeoecological studies. *Quaternary Research* 16, 80–96. **Peglar, S.** 1993: The mid-Holocene *Ulmus* decline at Diss Mere, Norfolk, UK: a year-by-year pollen stratigraphy from annual laminations. *The Holocene* 3, 1–14. **Stein, J.K. and Linse, A.R. (eds)** 1993: *Effects of scale on archaeological and geoscientific perspectives* [*Special Paper* 283]. Boulder, CO: Geological Society of America.

scanning electron microscopy (SEM)   A form of microscopy used to obtain images of microstructures; it utilises an electron gun to provide a beam of electrons, which can be focused on a specimen: the smaller the area to be scanned, the greater the magnification. Unlike its predecessor, *transmission electron microscopy*, where the beam passed through the sample to produce an image on a flat plate, the beam scans a small area of the solid specimen and reflected and secondary electrons are detected to form an image on a screen. SEM has been used to determine surface characteristics (see QUARTZ GRAIN SURFACE TEXTURES) of individual SAND or SILT grains and thereby to identify sedimentary transport mechanisms and the PROVENANCE of deposits and ARTEFACTS. Examination of bulk materials such as TILL and LOESS can also be undertaken, allowing characterisation of structures indicative of particular depositional histories. It is also used in MICROFOSSIL ANALYSIS, such as POLLEN ANALYSIS and FORAMINIFERAL ANALYSIS, as well as in the analysis of TESTATE AMOEBAE for fine structure characteristics and to aid species identification.   *DH/CJC*

Tite, M.S. 1991: The impact of electron microscopy on ceramic studies. In Pollard, A.M. (ed.), *New developments in archaeological science*. Oxford: Oxford University Press, 111–121. **Whalley, W.B.** 1996: Scanning electron microscopy. In Menzies, J. (ed.),

Past glacial environments: sediments, forms and techniques. Oxford: Butterworth-Heinemann, 357–375.

scattering　The process by which a propagating electromagnetic field incident on an object from one direction is scattered into many other directions. The incident electromagnetic field excites an oscillating electric field inside the object (or on its surface) and this oscillating field radiates with a strength and directional pattern dictated by properties of the object (including its shape, orientation and DIELECTRIC CONSTANT).　　　　　*PJS*

Schmidt hammer　A hand-held instrument that records the distance of rebound of a spring-loaded mass. Originally designed to measure the hardness of concrete, it has been used to measure the hardness of rock surfaces and therefore the degree of WEATHERING or *case-hardening*. It thus provides a means of RELATIVE-AGE DATING applicable to rock surfaces and boulders. It has been used mainly on surfaces of HOLOCENE age in areas where datable organic material is uncommon. Surfaces that have been studied include limestone surfaces, glacial MORAINES in Arctic and alpine areas and ALLUVIAL FANS in arid and SEMI-ARID environments. It has also been used to distinguish surfaces that were glacially abraded during the LAST GLACIAL MAXIMUM from surfaces that remained exposed as NUNATAKS.　　　　　*DMcC*

Ballantyne, C.K., McCarroll, D., Nesje, A. *et al.* 1998: The last ice sheet in north-west Scotland: reconstruction and implications. *Quaternary Science Reviews* 17, 1149–1184. **Matthews, J.A. and Shakesby, R.A.** 1984: The status of the 'Little Ice Age' in southern Norway: relative-age dating of Neoglacial moraines with Schmidt hammer and lichenometry. *Boreas* 13, 333–346. **McCarroll, D.** 1994: The Schmidt hammer as a measure of degree of rock surface weathering and terrain age. In Beck, C. (ed.), *Dating in exposed and surface contexts.* Albuquerque, NM: University of New Mexico Press, 29–46.

science　'The human effort to describe and understand the natural world through passive observations, active experiments, and theoretical analysis and synthesis' (Friedlander, 1998: 38). Science is an approach to obtaining and organising knowledge distinguished by SCIENTIFIC METHOD, which currently represents the most successful way of determining what is reliable knowledge and hence what can be reliably applied to solving problems in the 'real world'. Scientific method provides an objective framework for minimising subjectivity in the highly subjective process of developing reliable knowledge (see OBJECTIVE KNOWLEDGE). Successful science is characterised by the ability to make PREDICTIONS about similar phenomena, processes and events in the world in varying places and times. To be able to do this, science involves such activities as EXPERIMENT, the construction of MODELS, testing HYPOTHESES and above all seeking to generalise and construct THEORY. The credibility of science depends on exposure of its results to REPLICATION and its hypotheses to independent testing (and *refutation* – see FALSIFICATION), and hence on the willingness of scientists to modify their conclusions in the light of new evidence. A flavour of the nature and limitations of science is provided by the following quotations from the much larger collection assembled by Wolf (1970):

- Science is knowledge gained by systematic observation, experiment, and reasoning. (Louis Pasteur)
- The great aim of all science is to cover the greatest number of empirical facts by logical deduction from the smallest number of hypotheses or axioms. (Albert Einstein)
- The scientific addition to common sense is merely a more penetrating analysis of the complex factors involved, even in seemingly simple events. (A.J. Carlson)
- The aim of science is to seek the simplest explanation of complex facts . . . seek simplicity and distrust it. (Alfred North Whitehead)
- Science is based upon the belief that the universe is reliable in its operation. (A. Standen)
- The origin of all science is in the desire to know causes. (William Hazlitt)
- The goal of research is not to have and perpetuate a contemporary theory but to approach the real truth through doubt. (C.G. Jung)
- It is not the possession of knowledge, of irrefutable truths, that constitutes the man of science, but the disinterested, incessant search for truth. (Karl Popper)
- The fundamental secret of science is that seeking truth is more important than truth itself. (Friedrich Nietzsche)
- Although this may seem a paradox, all exact science is dominated by the idea of approximation. (Bertrand Russell)
- The great tragedy of science [is] the slaying of a beautiful hypothesis by an ugly fact. (Thomas Huxley)
- The real and legitimate goal of the sciences is the endowment of human life with new inventions and riches. (Francis Bacon)
- Scientific discovery and scientific knowledge have been achieved only by those who have gone in pursuit of them without any practical purpose whatsoever in view. (Max Planck)
- People must understand that science is inherently neither a potential for good or for evil. It is a potential to be harnessed by man to do his bidding. (Glenn Seaborg)
- Research is to see what everybody else has seen, and think what nobody has thought. (Albert Szent-Gyoryi)
- Science gives us knowledge, but only philosophy can give us wisdom. (Will Durant)　　　　　*JAM*

[See also APPLIED SCIENCE]

Friedlander, M.W. 1998: *At the fringes of science.* Boulder, CO: Westview Press. **Medawar, P.B.** 1981: *Advice to a young scientist.* Oxford: Perseus Books Group (Basic Books). **Medawar, P.B.** 1984: *The limits of science.* Oxford: Oxford University Press. **Smith, J.** 2000: Nice work – but is it science? Untestable ecological theory won't help solve environmental problems [Millennium Essay]. *Nature* 408: 293. **Wolf, K.H.** 1970: A collection of scientific sayings and quotations (1–3). *Earth Science Reviews* 6, 289–296, 297–335, 353–368.

sciences　Divisions of academia with a common approach encapsulated by the term SCIENTIFIC METHOD, by which new knowledge is gained, understanding is achieved and well-tested THEORY is developed and made available for application. Variation within the sciences is recognised by such terms as *physical sciences* (such as physics and chemistry), NATURAL ENVIRONMENTAL SCIENCES, EARTH SCIENCES, *biological sciences* and SOCIAL SCIENCES. The various sciences are sometimes viewed in a hierarchical way, with the so-called 'hard sciences', especially physics, which are most highly mathematised and possess highly developed theoretical frameworks, at the top of the hierarchy. *Applied sciences*, such as engineering, forestry and agriculture are normally considered to comprise separate disciplines, although the distinction between 'pure' or 'basic' science and 'applied' science is not always clear. *JAM*

[See also LABORATORY SCIENCE, THEORY]

scientific archaeology
The application of scientific approaches, SCIENTIFIC METHOD and especially the sophisticated analytical techniques of the physical sciences in the investigation, dating and interpretation of archaeological sites and materials. Some techniques used in relation to archaeological materials are summarised in the Table. *JAM*

[See also ARCHAEOLOGICAL CHEMISTRY, ARCHAEOLOGICAL GEOLOGY, ARCHAEOMAGNETIC DATING, ARCHAEOMAGNETISM, CERAMICS, ENVIRONMENTAL ARCHAEOLOGY, GEOARCHAEOLOGY, PETROGLYPHS, ZOOARCHAEOLOGY]

Aitken, M.J. 1990: *Science-based dating in archaeology*. London: Longman. **Brothwell, D. and Higgs, E. (eds)** 1969: *Science in archaeology*. London: Thames and Hudson. **Henderson, J. (ed.)** 1989: *Scientific analysis in archaeology* [*Monograph* 19]. Oxford: Oxford University Communications for Archaeology. **Herz, N. and Garrison, E.G.** 1998: *Geological methods for archaeology*. Oxford: Oxford University Press. **Kelley, J.H. and Hanen, M.P.** 1988: *Archaeology and the methodology of science*. Albuquerque, NM: University of New Mexico Press. **Lambert, J.B.** 1998: *Traces of the past: unravelling the secrets of archaeology through chemistry*. Oxford: Perseus Books Group. **Pearsall, D.M. and Piperno, D.R. (eds)** 1993: *MASCA research papers in science and archaeology*. Philadelphia: University Museum of Archaeology and Anthropology. **Renfrew, C. and Bahn, P.** 1996: *Archaeology: theory, methods and practice*, 2nd edn. London: Thames and Hudson.

scientific method
Implicit in the term scientific method is the view that scientists follow a single, normative approach to scientific discovery and justification. Such a path is most commonly seen as an iterative sequence of tentative HYPOTHESIS postulation, testing and reformulation. Much of the underpinning of this view stems from the LOGICAL POSITIVISM versus CRITICAL RATIONALISM debate (i.e. are hypotheses verified or falsified etc.). The polar opposite view is that all SCIENCE is socially constructed (*strong relativism*) and consequently individual groups of scientists do not even speak the same scientific language. There is presently no reason to assert a single scientific methodology, nor do the majority of scientists accept strong relativism. The working reality is that individual scientists do adhere to, and are policed by, research norms shared by their particular scientific com-

scientific archaeology *Some techniques used in scientific archaeology for identifying, characterising, dating or otherwise investigating archaeological materials (based on Renfrew and Bahn, 1996)*

Archaeological material	Analytical techniques*
Bone	CARBON, NITROGEN, OXYGEN and STRONTIUM ISOTOPES, POTASSIUM–ARGON DATING, RADIOCARBON DATING
Flint	AAS, ICP-AES, NAA, PIXE/PIGME, XRF
Glasses	ICP-AES, LEAD ISOTOPES, NAA, SEM, XRF
Gold	ICP-AES, ICP-MS, PIXE, SEM, XRF
Jade	XRD
Marble	CARBON ISOTOPES, CL, NAA, OXYGEN ISOTOPES
Metal slag	ICPS-AES, LEAD ISOTOPES, NAA, SEM
Minerals and ores	ICP-AES, LEAD ISOTOPES, NAA, XRD, XRF
Obsidian	AAS, FISSION-TRACK DATING, ICP-AES, NAA, OBSIDIAN HYDRATION DATING, PIXE/PIGME, XRF
Pottery	AAS, ICP-AES, IRSL, NAA, PIXE/PIGME, TL, THIN-SECTION ANALYSIS, XRD, XRF
Pure metals and alloys of copper, silver, lead	AAS, ICP-AES, ICP-MS, LEAD ISOTOPES, NAA (not for silver), SEM, XRF
Teeth	ESR, URANIUM-SERIES DATING
Wood	DENDROCHRONOLOGY, RADIOCARBON DATING, THIN-SECTION ANALYSIS

*Dictionary entries for technique abbreviations:

AAS	ATOMIC ABSORPTION SPECTROPHOTOMETRY
CL	CATHODOLUMINESCENCE ANALYSIS
ESR	ELECTRON SPIN RESONANCE DATING
ICP-AES	INDUCTIVELY COUPLED PLASMA ATOMIC EMISSION SPECTROMETRY
ICP-MS	INDUCTIVELY COUPLED PLASMA MASS SPECTROMETRY
IRSL	INFRARED-STIMULATED LUMINESCENCE DATING
NAA	NEUTRON ACTIVATION ANALYSIS
PIGME	PARTICLE-INDUCED GAMMA-RAY EMISSION
PIXE	PROTON-INDUCED X-RAY EMISSION
SEM	Scanning ELECTRON MICROPROBE ANALYSIS
TL	THERMOLUMINESCENCE DATING
XRD	X-RAY DIFFRACTION ANALYSIS
XRF	X-RAY FLUORESCENCE ANALYSIS

munity and embracing much more than logic alone, but also enjoy some flexibility and exhibit individuality. *CET*

[See also FALSIFICATION, OBJECTIVE KNOWLEDGE, SOCIAL CONSTRUCTION, THEORY]

Chalmers, A.F. 1982: *What is this thing called science?* Milton Keynes: Open University Press. **Watson, P.J., LeBlanc, S.A. and Redman, C.L.** 1984: *Archaeological explanation: the scientific method in archaeology.* Cambridge: Cambridge University Press.

sclerochronology

The use of variations in the composition of biogenic mineralogical materials, such as shell growth banding and corals for dating purposes. It includes the linking of the banding in individual species into composite sclerochronologies. It has potential, for example, in the dating of subfossil shells from the last millennium and for the reconstruction of continuous geochemical, hydrological and climatic (*scleroclimatology*) records. With a lifespan of > 100 years, the ocean quahog (*Arctica islandica*) is a particularly suitable species for sclerochronology, which has been rather optimistically described as the marine counterpart of DENDROCHRONOLOGY. *JAM*

[See also CORAL AND CORAL REEFS, MOLLUSCA ANALYSIS]

Buddemeier, R.W. and Taylor, F.W. 2000: Sclerochronology. In Stratton Noller, J., Sowers, J.M. and Lettis, W.R. (eds), *Quaternary geochronology: methods and applications.* Washington, DC: American Geophysical Union, 25–40. **Jones, D.S.** 1983: Sclerochronology: reading the record of the molluscan shell. *American Scientist* 71, 384–391. **Marchitto Jr, T.M., Jones, G.A., Goodfriend, G.A. and Weidman, C.R.** 2000: Precise temporal correlation of Holocene mollusk shells using sclerochronology. *Quaternary Research* 53, 236–246.

sclerophyllous

Trees and shrubs which possess small, hard, thick leathery leaves with waxy cuticles as a way of adapting to climates that have a long dry season. This type of leaf reduces water loss. The plants, such as olive (*Olea europea)* and holm oak (*Quercus ilex*), tend to be evergreen, the dominant species of typical vegetation in 'Mediterranean'-type climates and a major constituent of MEDITERRANEAN-TYPE VEGETATION. *MLC*

[See also MALACOPHYLLOUS]

Reille, M. and Pons, A. 1992: The ecological significance of sclerophyllous oak forests in the western part of the Mediterranean Basin: a note on pollen analytical data. *Vegetatio* 99, 13–17.

scoria

PYROCLASTIC fragments, usually of *block* or *bomb* size, with abundant VESICLES. Scoria, sometimes referred to as *cinders*, are usually denser, more CRYSTALLINE and darker in colour than PUMICE, with a lower silica content. *GO*

[See also TEPHRA]

scour marks

A group of erosional SEDIMENTARY STRUCTURES formed when hollows are scoured on a surface of cohesive sediment (MUD) by eddies in a TURBULENT FLOW, usually of water. Like TOOL MARKS, they are commonly preserved as CASTS by later deposition of sand and seen as SOLE MARKS on the base of a bed of SANDSTONE. Examples include horseshoe-shaped *flute marks* (or *flute casts*), steep-sided linear *gutter casts*, and *obstacle scours* that form around immobile objects. Scour marks are useful indicators of PALAEOCURRENT strength and direction. *GO*

Collinson, J.D. and Thompson, D.B. 1992: *Sedimentary structures*, 2nd edn. London: Chapman and Hall. **Myrow, P.M.** 1992: Pot and gutter casts from the Chapel Island Formation, southeast Newfoundland. *Journal of Sedimentary Petrology* 62, 992–1007.

scrub

Vegetation dominated by woody plants, mostly shrubs or bushes but including small trees, which are not as tall or as closely spaced as in a woodland or forest. Scrub is commonly a SERAL COMMUNITY developing where forest is extending into grassland or woody plants are colonising abandoned agricultural land. It may also result from DEFORESTATION or other forms of ENVIRONMENTAL DEGRADATION. However, scrub may also be characteristic where climatic conditions limit the growth of trees. *JAM*

[See also HEATHLAND, SHRUBLAND]

sea

A shallow area of saltwater at the margins of an OCEAN, partly or completely surrounded by land (e.g. Caribbean Sea, Red Sea). *Epicontinental seas* are shallower than 200 m and underlain by CONTINENTAL CRUST (e.g. North Sea): they are flooded extensions of the continents (see CONTINENTAL SHELF). *Inland seas* are also recognised: they are saline because of internal drainage (i.e. no outlet to the ocean) and evaporation in high temperatures (e.g. Caspian Sea, Aral Sea) and are saline LAKES. *GO*

sea-floor spreading

The process in PLATE TECTONICS by which OCEANIC CRUST and LITHOSPHERE are formed at CONSTRUCTIVE PLATE MARGINS, which coincide with MID-OCEAN RIDGES. The separation of lithospheric plates allows decompression melting of the ASTHENOSPHERE. MAGMA rises at the ridge axis, forming new lithosphere through basaltic VOLCANISM and igneous INTRUSION. *GO*

sea ice

Ice formed in the oceans from the freezing of sea water in polar regions during cold, calm weather when sea temperatures fall below about −2°C. Floating sea ice (PACK ICE) forms when the sea freezes into *frazil ice* (fine ice particles suspended in water), *grease ice* (a soupy surface layer of ice crystals), *nilas* (a thin, elastic sheet up to 10 cm thick) and, eventually, circular pieces of ice up to 3 m in diameter (*pancake ice*). *Fast ice*, which is sea ice attached to the shore (with *tide cracks*) thickens up to 2–3 m in a single season by the addition of snowfall on top and further water freezing below. Sea ice may be crumpled, forming pressure ridges, and in the spring is broken up by strong winds and sea swell into rectangular-shaped *ice floes*, the navigation scourge of polar oceans (the largest floes off Antarctica reach 50 km in length). *Polynas* are extensive areas of open water within the area of sea ice.

The extent of sea ice is important in ATMOSPHERE–OCEAN INTERACTION, is a GEOINDICATOR of short-term CLIMATIC CHANGE and has an important role in long-term CLIMATIC CHANGE. Effects on climate include its insulating qualities, reducing the exchanges between water surface and atmosphere, increasing the surface ALBEDO of the ocean and reducing salinity as the sea ice becomes older (which affects the salinity of the seawater below). Sea ice extended towards lower latitudes during cold periods such as the LITTLE ICE AGE, YOUNGER DRYAS and GLACIAL EPISODES. Shorter-term climatic variations are also reflected in the incidence of SEA-ICE VARIATIONS (see Figure). *RDT/JAM*

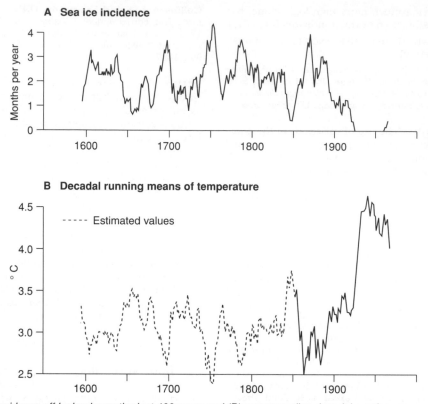

A Sea ice incidence

B Decadal running means of temperature

sea ice (A) *Sea ice incidence off Iceland over the last 400 years and (B) corresponding decadal running means of annual mean temperature. Estimated values in (B) were predicted from sea ice incidence using measured temperatures from the most recent part of the record in (A) (after Bergthórsson, 1969)*

[See also ATMOSPHERE–OCEAN INTERACTION, COUPLED OCEAN–ATMOSPHERE MODELS, DRIFTWOOD, HEINRICH EVENTS, ICE-RAFTED DEBRIS]

Barry, R.G. 1983: Arctic Ocean ice and climate: perspectives on a century of polar research. *Annals of the Association of American Geographers* 73, 485–501. **Bergthórsson, P.** 1969: An estimate of drift ice and temperature in Iceland in 1000 years. *Jökull* 19, 94–101. **Kukla, G. and Gavin, J.** 1981: Summer ice and carbon dioxide. *Science* 214, 497–503. **Mysak, L.A.** 1995: Decadal-scale variability of ice cover and climate in the Arctic Ocean and Greenland and Iceland Seas. In National Research Council, *Natural climate variability on decade-to-century time scales.* Washington, DC: National Academy Press, 253–263. **Ogilvie, A.** 1992: Documentary evidence for changes in the climate of Iceland, A.D. 1500 to 1800. In Bradley, R.S. and Jones, P.D. (eds), *Climate since A.D. 1500.* London: Routledge, 92–117. **Semtner, A.J.** 1984: The climatic response of the Arctic Ocean to Soviet river diversions. *Climatic Change* 6, 109–130. **Vernal, A. de and Hillaire-Marcel, C.** 2000: Sea-ice cover, sea-surface salinity and halo-/thermocline structure of the northwest North Atlantic: modern versus full glacial conditions. *Quaternary Science Reviews* 19, 65–85. **Walton, D.W.H. (ed.)** 1987: *Antarctic science.* Cambridge: Cambridge University Press, 140–150.

sea-ice variations Changes in the extent and spatial distribution and seasonal dates of formation and break-up of marine ice. Sea ice plays an important role in the climatic system of the Earth. It influences the production and distribution of cold water, helps control, via albedo effects, the circulation systems and is seen as an ENVIRON-MENTAL INDICATOR or GEOINDICATOR within the framework of GLOBAL WARMING. Various ice atlases provide information on the conditions in the twentieth century. During the ice season, national ice services publish weekly bulletins in which the current extent of the ice is presented – an important parameter for shipping. Longer-term changes in sea-ice parameters (such as dates of freezing and break-up) can be established for some areas through the analysis of HISTORICAL EVIDENCE. Good series have been deduced for some regions with reliable, continuous and systematic historical records, such as the lagoons of Venice, the freezing of Hudson Bay, the freezing near Iceland and the Baltic Sea. Some studies have not only established chronologies and identified spatiotemporal change, but also used the data and relationships with climatic variables to formulate prognostic models. Analyses have also demonstrated clearly that time–space changes in the appearance of sea ice correlate very well with independently developed histories of climatic change during the LITTLE ICE AGE and MEDIAEVAL WARM PERIOD.

RG

Camuffo, D. 1987: Freezing of the Venetian Lagoon since the ninth century AD in comparison to the climate of western Europe and England. *Climatic Change* 10, 43–66. **Catchpole, A.J.W. and Faurer, M.A.** 1989: Severe summer ice in Hudson Strait and Hudson Bay following major volcanic eruptions, 1751 to 1889 AD. *Climatic Change* 14, 61–79. **Koslowski, G. and Glaser, R.** 1999: Variations in reconstructed ice winter severity in the Western Baltic from 1501 to 1995 and their implications for the North Atlantic Oscillation. *Climatic Change* 41, 715–191.

sea-level change A change in absolute or relative sea level (see MEAN, RELATIVE and ABSOLUTE SEA LEVEL). It is almost impossible to describe former patterns of sea-level change for any area of the world since quantification requires measurement from a common reference level. Owing to the enormous complexity of eustatic, isostatic and tectonic processes, geological investigations for particular coastal areas only permit reconstructions of former patterns of *relative* sea-level change (see Figure). The position of relative sea level during the LAST INTERGLACIAL varies between regions. Thus, in southwest England, it is 2–5 m above present; in parts of The Netherlands –15 m; and in parts of Greece it is as high as +200 m. By contrast, during the LAST GLACIAL MAXIMUM it is estimated for many areas to have been about –120 m in the open ocean. This value may incorporate several tens of metres of hydro-isostatic unloading due to a reduced mass of ocean water caused by diminished volume.

Present estimates of SEA-LEVEL RISE as absolute values are therefore misleading if the impression is given that such estimates apply everywhere. A *global sea level* or a *global sea-level change* are not valid concepts. Thus, in Scandinavia, despite human-induced glacio-eustatic rise in sea level, relative sea level is falling in many areas as a result of continued GLACIO-ISOSTATIC REBOUND caused by the melting of the last Fennoscandian ICE SHEET. In contrast, the amount of recent relative sea-level rise in the world's major DELTAS is exceptionally high as a result of deltaic SUBSIDENCE, which is a natural phenomenon exacerbated by human impact (e.g. extraction of oil, gas and water). The *Global Sea-Level Observing System* (GLOSS) monitors long-term variations in sea level using a network of *sea-level gauges*, the positions of which are fixed by a satellite-based GLOBAL POSITIONING SYSTEM. *AGD*

[See also EUSTASY, ISOSTASY, GEOIDAL EUSTASY, GLACIO-ISOSTASY, HYDROISOSTASY]

Dawson, S. and Smith, D.E. 1997: Holocene relative sea-level changes on the margin of a glacio-isostatically uplifted area: an example from northern Scotland. *The Holocene* 7, 59–77. **Fairbanks, R.W.** 1989: A 17,000 year glacio-eustatic sea level record: influence of glacial melting rates on the Younger Dryas event and deep ocean circulation. *Nature* 342, 637–642. **Harmon, R.S., Mitterer, R.M., Kriansakal, N. et al.** 1983: U-series and amino-acid racemisation geochronology of Bermuda: implications for eustatic sea-level fluctuations over the past 250,000 years. *Palaeogeography, Palaeoclimatology, Palaeoecology* 44, 41–70. **Harvey, N., Barnett, E.J., Bourman, R.P. and Belperio, A.P.** 1999: Holocene sea-level change at Port Pirie, South Australia: a contribution to global sea-level rise estimates from tide gauges. *Journal of Coastal Research* 15, 607–615. **Lowe, J.J. and Walker, M.J.C.** 1997: *Reconstructing Quaternary environments*, 2nd edn. Harlow: Longman. **Nummedal, D., Pilkey, O.H. and Hoard, J.D. (eds)** 1987: *Sea level fluctuation and coastal evolution* [*Special Publication* 41]. Tulsa, OK: Society of Economic Paleontologists and Mineralogists. **Warwick, R.A., Barrow, E.M. and Wigley, T.M.L. (eds)** 1993: *Climate and sea level changes: observations, projections and implications.* Cambridge: Cambridge University Press.

sea-level change: past effects on coasts The rise and fall of sea level during the QUATERNARY on a global scale through glacio-eustatic causes, and the local rise and fall as a result of glacio-isostatic causes, tectonic and orogenic forces, local compaction of sediments and HYDROISOSTASY, had impacts that can still be seen along present-day coasts. RELICT shorelines now lying above sea level may be indicated on EMERGENT COASTS by *raised beach* deposits and RAISED SHORELINES backed by steep cliff-like slopes, sometimes with associated stacks and sea caves. On SUBMERGENT COASTS, RIAS, submerged dune-chains, notches and benches in the submarine topography and remnants of forests or peat layers at or below present sea level may be found. Even where local effects on RELA-

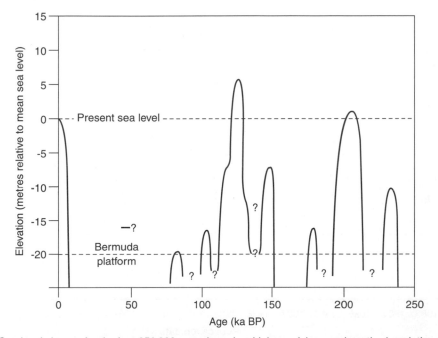

sea-level change Sea-level change for the last 250 000 years based on high-precision uranium–thorium dating of submerged corals in Bermuda (after Harmon et al., 1983; Walker and Lowe, 1997)

TIVE SEA LEVEL have been absent since mid-HOLOCENE times, sea level globally reached approximately its present position only as recently as about 6000 years ago. On emergent coasts, the present relative sea level may have been reached more recently. *RAS*

Dawson, A.G. 1992: *Ice Age Earth: late Quaternary geology and climate.* London: Routledge. **Goudie, A.** 1992: *Environmental change,* 3rd edn. Oxford: Blackwell, 217. **Lowe, J.J. and Walker, M.J.C.** 1997: *Reconstructing Quaternary environments,* 2nd edn. Harlow: Addison-Wesley Longman, 53–68. **Pirazzoli, P.A.** 1996: *Sea-level changes.* Chichester: Wiley.

sea-level change: past effects on ecosystems

Periodic SEA-LEVEL CHANGE affects coastal marine and terrestrial ecosystems in many ways. HABITAT LOSS or expansion/creation affects subsequent landward or seaward ecosystem migration. During SEA-LEVEL RISE, the CONTINENTAL SHELF (supporting a diverse marine ecosystem) is broadened, enabling diversification and landward ecosystem migration. As coastlines retreat, intrusion of saline water can lead to soil SALINISATION and flooding can threaten delicate coastal ecosystems (e.g. sea grasses, mangroves, lagoons and estuarine flats) by causing a decline in PRODUCTIVITY and BIODIVERSITY. Such coastal ecosystems may eventually drown, as seen with the periodic flooding of Carboniferous COAL swamps and the late Quaternary CORAL REEFS of Barbados. During sea-level fall, shelf ecology is restricted and coastal ecosystems expand seaward, e.g. animal-grazed vegetation on newly exposed shelf during the early Triassic, and also the late Quaternary of New England at c. 15 000 radiocarbon years BP. Lagoons may become infilled, inland seas land-locked (e.g. the temporary isolation of the Black Sea from the world ocean during the early HOLOCENE) and new alluvial habitats created by extended rivers (e.g. the Carmargue). Sea-level change also exerts some control on MIGRATION or ISOLATION of land-based floral and faunal ecosystems by the connection or dissection of dry land areas, e.g. late HOLOCENE isolation and local EXTINCTION of pine trees in Ireland. *CES*

Hallam, A. and Wignall, P.B. 2000: Mass extinctions and sea-level changes. *Earth Science Reviews* 48, 217–250. **Lowe, J. and Walker, M.J.C.** 1997: *Reconstructing Quaternary environments.* Harlow: Addison-Wesley Longman. **Pirazzoli, P.A.** 1996: *Sea level changes: the last 20,000 years.* Chichester: Wiley. **Tooley, M.J. and Jelgersma, S. (eds)** 1992: *Impacts of sea-level rise on European coastal lowlands.* Oxford: Blackwell.

sea-level rise Projections of SEA-LEVEL RISE in response to GLOBAL WARMING have been revised downwards since the early 1980s, when up to several metres of rise were predicted by some authorities (see Figure). Two major reviews by the INTERGOVERNMENTAL PANEL ON CLIMATE CHANGE (IPCC) published in 1990 and 1996 suggested progressively lower best estimate rises for AD 2100 of respectively 66 cm and 49 cm. A minority view is that sea level will fall in response to global warming. *HJW*

[See also SEA-LEVEL CHANGE, SEA-LEVEL RISE: POTENTIAL FUTURE IMPACTS ON PEOPLE, SEA-LEVEL RISE: POTENTIAL FUTURE GEOMORPHOLOGICAL IMPACTS, THERMAL EXPANSION OF THE OCEANS]

Douglas, B.C., Kearney, M.S. and Leatheman, S.P. 2000: *Sea level rise: history and consequences.* Sidcup: Academic Press –

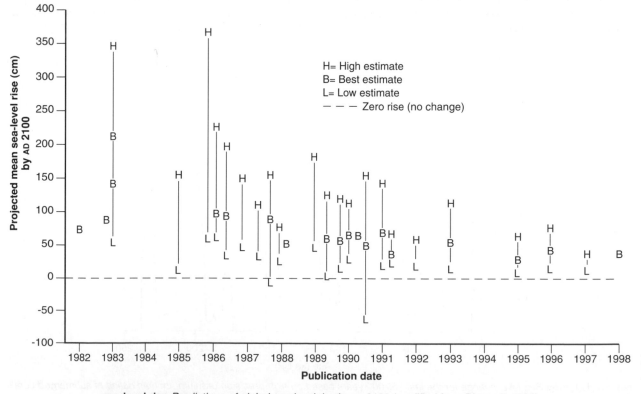

sea-level rise *Predictions of global sea-level rise by* AD *2100 (modified from Pirazzoli, 1996)*

Harcourt Publishers. **National Academy of Sciences** 1990: *Sea-level change. Studies in Geophysics.* Washington, DC: National Academy Press. **Pirazzoli, P.A.** 1989: Present and near-future sea level changes. *Palaeogeography, Palaeoclimatology, Palaeoecology* 75, 241–258. **Pirazzoli, P.A.** 1996: *Sea-level changes.* Chichester: Wiley. **Schneider, S.H.** 1992: Will sea levels rise or fall? *Nature* 356, 11–12. **Tooley, M.J. and Jelgersma, S. (eds)** 1992: *Impacts of sea level rise on European coastal lowlands* [*Institute of British Geographers Special Publication* 27]. Oxford: Blackwell. **Warrick, R. A., Barrow, E.M. and Wigley, T.M.L. (eds)** 1993: *Climate and sea level change: observations, projections and implications.* Cambridge: Cambridge University Press.

sea-level rise: potential future ecological impacts

The rise in MEAN SEA LEVEL relative to land is one aspect of SEA-LEVEL CHANGE. It excludes changes due to tidal and wave movements. Since the end of the PLEISTOCENE, sea level has risen over 100 m. GLOBAL WARMING due to the enhanced GREENHOUSE EFFECT is predicted to cause partial melting of polar ice and THERMAL EXPANSION OF THE OCEANS, raising mean sea levels some 0.5 m by about AD 2050. This may exceed the rate at which coastal ecosystems can adjust. Environmental consequences may include the inundation of important sites of agriculture and habitation on DELTAS, the erosion of coastlines and the loss of CORAL REEFS and coastal and intertidal HABITATS (e.g. MANGROVE SWAMPS and SALTMARSHES), the inland migration of which would be restricted by sea defences and the conflicting interests of industry, agriculture and human habitation. Sea-level rise will vary regionally: land masses experiencing subsidence (e.g. the Mediterranean) will suffer most, as will low-lying areas (e.g. the east coast of the USA and Bangladesh). *SHJ*

[See also ICE AGES, SALTWATER INTRUSION, SILTATION]

Raffaelli, D. and Hawkins, S. 1996: *Intertidal ecology.* London: Chapman and Hall. **Intergovernmental Panel on Climate Change (IPCC)** 1990: *Scientific assessment of climate change: report prepared for IPCC by Working Group 1.* Geneva: IPCC, WMO and UNEP. **Ince, M.** 1990: *The rising seas.* London: Earthscan.

sea-level rise: potential future geomorphological impacts

Projected SEA-LEVEL RISE caused by GLOBAL WARMING is anticipated to affect different types of coasts in different ways. The major impacts include inundation, enhanced shoreline EROSION, more frequent flooding and SALTWATER INTRUSION. For soft-rock cliffs, recession rates are expected to increase markedly. In estuaries, bays, lagoons and SABKHAS, low-lying parts can be expected to become submerged. Sea-level rise is expected to cause more rapid erosion along sandy beaches and probably add to the estimated 70% of beaches world-wide already experiencing erosion. DELTAS, where SUBSIDENCE often outpaces predicted sea-level rise, are at considerable risk. Low tropical islands may also be vulnerable, although many believe that CORAL REEF growth will keep pace with sea-level rise. Coral reefs, however, cannot tolerate water temperatures greater than about 30°C and even a rise of 1–2°C could adversely affect many shallow-water coral species. *HJW*

[See also BRUUN RULE, CORAL BLEACHING, CORAL REEFS: HUMAN IMPACT AND CONSERVATION, SALTWATER INTRUSION, SEA-LEVEL RISE: POTENTIAL FUTURE IMPACTS ON PEOPLE]

Bray, M.J. and Hooke, J.M. 1997: Prediction of soft-cliff retreat with accelerating sea-level rise. *Journal of Coastal Research* 13, 453–467. **Healy, T.** 1996: Sea level rise and impacts on nearshore sedimentation: an overview. *Geologische Rundschau* 85, 546–553. **Spencer, T.** 1995: Potentialities, uncertainties and complexities in the response of coral-reefs to future sea-level rise. *Earth Surface Processes and Landforms* 20, 49–64.

sea-level rise: potential future impacts on people

The expected major impacts on people of SEA-LEVEL RISE include permanent inundation of low-lying coastal areas, many of which (e.g. the Nile DELTA, The Netherlands and Bangladesh) are densely populated, and damage to property from increased EROSION rates. Inundation of coastal marshes and WETLANDS would reduce their value as fish nursery grounds affecting world fisheries. SALTWATER INTRUSION would affect groundwater supplies. More frequent STORM SURGES would require additional protective measures to maintain the current position of shorelines. A preferred alternative strategy for many would be MANAGED RETREAT. *HJW*

[See also SEA-LEVEL RISE: POTENTIAL FUTURE GEOMORPHOLOGICAL IMPACTS, SEA-LEVEL RISE, GLOBAL WARMING, WETLAND CLASSIFICATION]

Barth, M. C. and Titus, J. G. 1984: *Greenhouse effect and sea level rise.* New York: Van Nostrand Reinhold. **Doornkamp, J.C.** 1998: Coastal flooding, global warming and environmental management. *Journal of Environmental Management* 52, 327–333.

sea-surface temperature (SST)

The temperature of the surface water of the ocean. Although SSTs were formerly measured using a thermometer immersed in a canvas bucket or in the engine inlets of ships, estimates of SST are increasingly becoming available from SATELLITE REMOTE SENSING. Annual, seasonal and monthly mean data are readily available in atlas form. However, variation in data quantity and quality, different methods of computation employed in map preparation and comparability may vary from one atlas to another.

Charts of mean sea-surface temperature show that the expected variations with latitude are much modified by OCEAN CURRENTS. Sea temperatures in the Southern Hemisphere are generally somewhat lower than those in the Northern Hemisphere because of differences in the character of the PREVAILING WINDS and the effects of the large ice-covered Antarctic continent. The highest recorded SST was 35°C in the Persian Gulf. Minimum sea temperature may be defined as the freezing point of sea water, which is a function of SALINITY. The annual range of SST over most of the oceans of the globe is less than 10°C but ranges are higher in land-locked seas and in continental shelf waters, approaching 20°C near Korea. The waters north of the British Isles are more than 9°C warmer than is normal for the latitude and the largest positive anomaly in any ocean. Advection by currents, heat exchanges between sea and atmosphere, mixing by wave action and convective stirring can all lead to diurnal changes in SST.

Sea-surface temperature is critical in the formation of TROPICAL CYCLONES, which requires SSTs of at least 26.5°C, and is a key parameter in COUPLED OCEAN–ATMOSPHERE MODELS and in understanding the role of ATMOSPHERE–OCEAN INTERACTION in CLIMATIC CHANGE. Changes in ocean climate seem to be linked closely with large-scale atmospheric changes, and this is particularly true in the Pacific where large-scale SST increase occurs on average every 4–6 years (see EL NIÑO–SOUTHERN OSCIL-LATION). SSTs are indicators and predictors of changes within the Earth–atmosphere–ocean system, being followed by changes in pressure, wind and precipitation. Over the past century, global minimum SSTs occurred around AD 1905–1910 and maxima around 1960–1965, with a world-wide fluctuation of about 0.6°C. Renewed warming appears to have set in after the early 1970s. *AHP*

Cohen, A.L. and Tyson, P.D. 1995: Sea-surface temperature fluctuations during the Holocene off the south coast of Africa: implications for terrestrial climate and rainfall. *The Holocene* 5, 304–312. Folland, C.K., Palmer, T.N. and Parker, D.E. 1986: Sahel rainfall and worldwide sea temperatures 1901–85: observational, modelling and simulation studies. *Nature* 320, 602–607. Meteorological Office 1990: *Global ocean surface temperature atlas*. Bracknell: UK Meteorological Office and MIT. Perry, A.H. and Walker, J.M. 1978: *The ocean–atmosphere system*. Harlow: Longman.

sea water composition

Water in the world's OCEANS and SEAS contains on average 3.5% dissolved salts. This defines the SALINITY, usually expressed in parts per thousand as 35‰. BRACKISH water has a salinity less than 25‰, and *hypersaline* water has a salinity greater than about 40‰. The origin of the solutes in sea water is a combination of TERRIGENOUS products introduced by rivers and outgassing from the Earth's interior by VOLCANISM. Chloride and sodium ions make up more than 85% of the dissolved substances in sea water. Other major constituents, some essential for life in sea water, include sulphate, magnesium, calcium and potassium; there are also minor quantities of bicarbonate, bromide, borate, strontium and fluoride. Although the concentration of these ions in sea water may vary, their relative proportions remain virtually constant: they form the *conservative constituents* of sea water. The proportions of other, *non-conservative constituents* are tied to biological, seasonal or other short-term cycles and vary in space and time. NUTRIENTS include nitrogen, phosphorus and silicon: these are essential for biogenic activity and their concentrations are measured in parts per million. The quantities of gases dissolved in sea water are strongly affected by biogenic activity, particularly PHOTOSYNTHESIS. Concentrations of TRACE ELEMENTS are measured in parts per billion. They may aid or retard life in the oceans and human activity is responsible for most of their levels in shallow water depths. *BTC*

[See also EVAPORITES, HYDROTHERMAL VENTS]

Garrison, T. 1999: *Oceanography: an invitation to marine science*, 3rd edn. Belmont, CA: Wadsworth. Hill, M.N. 1963: *The sea: composition of seawater*. New York: Wiley. MacIntyre, F. 1970: Why the sea is salt. *Scientific American* November, 104–115. Open University Course Team 1995: *Seawater: its composition, properties and behaviour*, 2nd edn. Oxford: Elsevier.

seamount

An area of OCEAN floor, usually an extinct VOLCANO, rising at least 1000 m above its surroundings, usually from the ABYSSAL plain (e.g. Emperor Seamounts, northwest of Hawaii). Many seamounts are volcanic islands that subsided as their rocks cooled. *Abyssal hills* are lower features. *CDW*

[See also GUYOT]

Cousens, B., Dostal, J. and Hamilton, T.S. 1999: A near-ridge origin for seamounts at the southern terminus of the Pratt–Welker Seamount Chain, northeast Pacific Ocean. *Canadian Journal of Earth Sciences* 36, 1021–1031.

seasonal drought

A type of DROUGHT characteristic of SEMI-ARID lands, where there is normally insufficient moisture for agriculture in a distinct dry season, such as in the SAHEL. Crop growth without IRRIGATION is confined to the wet season, but is vulnerable if seasonal drought is extended by late rains or if the rains fail. It is one of four types of drought defined by C.W. Thornthwaite. *JAM*

[See also CONTINGENT DROUGHT, INVISIBLE DROUGHT, PERMANENT DROUGHT]

Thornthwaite, C.W. 1947: Climate and moisture conservation. *Annals of the Association of American Geographers* 37, 87–100.

seasonality

The main component of seasonal variation in climate is the magnitude of temperature change between winter and summer. The concept can also apply to other aspects of climate and the natural environment (such as the seasonal distribution of precipitation). *Thermal seasonality* can be quantified by a variety of indices. V. Conrad devised an index of CONTINENTALITY that expresses the annual temperature range with latitude; values range from 100 in Siberia to 0 in the Faeroes. Within the British Isles this index varies from roughly 3 around western coasts to 12 in the London area. The geographical distribution of seasonality is a function of the distribution of oceanic and continental AIR MASSES. It increases eastwards across the main land-masses of the Northern Hemisphere. Seasonality is also an important facet of global precipitation and may increase in some mid-latitude regions in the twentyfirst century as CLIMATIC SCENARIOS indicate a continuation of present trends to wetter winters and drier summers. *JCM*

[See also SEASONALITY INDICATORS]

Jones, P.D. and Briffa, K.R. 1992: Global surface air temperature variations over the twentieth century: part I. Spatial, temporal and seasonal details. *The Holocene* 2, 165–179. Troll, C. and Paffen, K.H. 1965: Seasonal climates of the Earth. In Rodenwaldt, E. and Jusatz, H.J. (eds), *World maps of climatology*. Berlin: Springer, 19–25. Walsh, R.P.D. and Lawler, D.M. 1982: Rainfall seasonality: description, spatial patterns and change through time. *Weather* 36, 201–208.

seasonality indicator

An ENVIRONMENTAL INDICATOR that is sensitive to change in seasonal environmental conditions. Seasonality indicators include many biological phenomena that are sensitive to temperature or precipitation SEASONALITY. Some are important for very HIGH-RESOLUTION RECONSTRUCTION and dating in ENVIRONMENTAL ARCHAEOLOGY and PALAEOCLIMATOLOGY (e.g. MICRODENDROCLIMATOLOGY and SCLEROCHRONOLOGY). Non-biological sources of seasonality indicators, such as VARVES and ICE CORES, should also be noted. *JAM*

[See also PHENOLOGY, SEASONALITY-OF-DEATH INDICATOR]

Bailey, G., Deith, M.R. and Shackleton, N.J. 1983: Oxygen isotope analysis and seasonality determinations: limits and potential of a new technique. *American Antiquity* 48, 390–398. Carter, R.J. 2001: Dental indicators of seasonal human presence at the Danish Boreal sites of Holmegaard I, IV and V and Mullerup and the Atlantic sites of Tybrind Vig and Ringkloster. *The Holocene* 11, 367–374. Hammer, C.U. 1989: Dating by physical and chemical seasonal variation and reference horizons. In Oeschger, H. and Langway Jr, C.C. (eds), *The environmental record in glaciers and ice sheets*. New York: Wiley, 99–121. Monks, G.G. 1981: Seasonality studies. *Advances in Archaeological Method and Theory* 4, 117–240.

seasonality-of-death indicator The presence of an organism or any characteristic of an organism that is indicative of the season in which that organism died. The presence of a migratory bird, mammal or fish, for example, may place limits on the season in which the organism died and hence the season of deposition or use. Similarly, seasonal rhythms in growth of molluscan and vertebral annuli, fish OTOLITHS or tooth cementum and dentine can provide reliable biological clocks for reconstructions in archaeological or palaeoenvironmental contexts. *JAM*

Carter, R.J. 1998: Reassessment of seasonality at the early Mesolithic site of Star Carr, Yorkshire, based on radiographs of mandibular tooth development in red deer (*Cervus elephas*). *Journal of Archaeological Science* 25, 851–856. **Gilbert, B.M. and Bass, W.M.** 1967: Seasonal dating of burials from the presence of fly pupae. *American Antiquity* 32, 534–535. **Legge, A.J. and Rowley-Conwy, P.A.** 1987: Gazelle killing in Stone Age Syria. *Scientific American* 255, 88–95. **Lieberman, D.E.** 1994: The biological basis for seasonal increments in dental cementum and their application to archaeological research. *Journal of Archaeological Science* 21, 525–539. **Monks, G.G. and Johnston, R.** 1993: Estimating season of death from growth increment data: a critical review. *ArchaeoZoologia* 2, 17–40.

seasonally frozen ground Ground that freezes annually. It is a diagnostic characteristic of those PERIGLACIAL ENVIRONMENTS without PERMAFROST. In areas with permafrost, seasonally frozen ground is termed the ACTIVE LAYER. Seasonally thawed ground thaws annually. In areas with permafrost, seasonally thawed ground can include the uppermost portion of the permafrost in places where annual thawing takes place at temperatures below 0°C, as a result of freezing-point depression due to saline pore water or a high clay content. *HMF*

[See also ACTIVE LAYER]

seawall A structure made from a variety of materials (e.g. concrete, sheet piling or wood) placed along the shore to prevent erosion. *HJW*

[See COASTAL ENGINEERING STRUCTURES]

secondary laterite Derived from primary LATERITE as a result of DENUDATION and the recementation of laterite fragments, secondary laterite is associated with benches along valley sides or basal concavities (whereas *primary laterites* are regarded as having formed *in situ* from underlying parent materials and are associated with summit features). The term is rarely used today. *MAC*

[See also FERRICRETE, DURICRUST]

Thomas, M.F. 1974: *Tropical geomorphology: a study of weathering and landform development in warm climates*. London: Macmillan.

secondary pollutant A POLLUTANT formed in the environment from a PRIMARY POLLUTANT. Secondary pollutants, such as PHOTOCHEMICAL OXIDANTS, are not discharged directly into the environment from pollution sources. *JAM*

secondary productivity The BIOMASS production rate in the consumer (animal) COMMUNITY. It is expressed as units of energy or as units of dry organic matter per unit area per unit time, but is difficult to measure. *RJH*

[See also CARBON ASSIMILATION, GROSS PRIMARY PRODUCTIVITY, PRIMARY PRODUCTIVITY, PRODUCTIVITY]

secondary succession A sequential change in species composition (or other ecosystem characteristics) on surfaces with some residual evidence of biotic activity. This definition generally implies succession where some degree of above-ground plant biomass has been removed but soils remain more or less intact. In contrast, PRIMARY SUCCESSION occurs where there is little or no residual biological legacy. Disturbances that result in secondary succession can come from both within (AUTOGENIC CHANGE) and outside (ALLOGENIC CHANGE) the biotic community. Examples include fire, ice- and wind-storms, floods, tree falls, cessation of agricultural activities (including 'old-field succession'), DEFORESTATION, insect outbreaks and various forms of aerial, aquatic or terrestrial POLLUTION.

DISPERSAL of propagules (spores, seeds or plant fragments) to a recently disturbed site is commonly augmented by propagules that survive *in situ* (e.g. seeds in the soil). Germination may be triggered by shifts in red:farred ratios of light when a gap is created in dense vegetation. The adaptation of some plants to the relatively high light and low nutrient levels common in early successional environments and of other plants to the low light and higher nutrient levels later in succession is one factor driving species change in succession. If no one species is capable of maintaining a competitive advantage in an environment that is in flux (from changes caused by growing plants), then successional change will occur. COMPETITION is now considered only one process among several (e.g. predation, mutualism, herbivory and disease) that interact to drive succession. FACILITATION can be considered the antithesis of competition. Although both processes can occur simultaneously, the net effect is generally either positive or negative. Secondary succession can be explained also by the differential arrival times, growth rates and longevities of the species without invoking positive or negative interactions. Models based on such life-history traits have successfully modelled secondary succession in forests and grasslands. Competition and facilitation may therefore be more important in altering the rate than the trajectory of succession. *LRW*

[See also CLIMAX VEGETATION]

Bazzaz, F. 1996: *Plants in a changing environment*. Cambridge: Cambridge University Press. **Connell, J.H. and Slatyer, R.O.** 1977: Mechanisms of succession in natural communities and their role in community stability and organization. *The American Naturalist* 111, 1119–1144. **Glenn-Lewin, D.C., Peet, R.K. and Veblen, T.T. (eds)** 1992: *Plant succession: theory and prediction*. London: Chapman and Hall. **Noble, I.R. and Slatyer, R.O.** 1980: The use of vital attributes to predict successional changes in plant communities subject to recurrent disturbances. *Vegetatio* 43, 5–21. **Tilman, D.** 1986: Resources, competition and the dynamics of plant communities. In Crawley, M.J. (ed.), *Plant ecology*. Oxford: Blackwell Scientific, 51–75. **Wilson, S.D.** 1999. Plant interactions during secondary succession. In Walker, L. (ed.), *Ecosystems of disturbed ground*. Amsterdam: Elsevier, 611–632.

secondary woodland Woodland re-established on land that was at some time completely cleared of trees. Although secondary woodland may occur on land that was once wooded in prehistoric times, the difference between secondary and PRIMARY WOODLAND is a consequence of the FRAGMENTATION of original primaeval forest. Thus, secondary woodlands are on land formerly used as pasture, quarries and moors, etc. and which has not been continuously wooded. Early studies of secondary woodland gave rise to theories of ECOLOGICAL SUCCESSION and CLIMAX VEGETATION. As result of ecological succession, secondary woodlands range in age and seral stage from scrub to woodland with mature trees. Most secondary woodlands have become established by themselves and have not been planted.   *IFS*

[See also ANCIENT WOODLAND, POTENTIAL NATURAL VEGETATION]

Rackham, O. 1980: *Ancient woodland: its history, vegetation and uses in England.* London: Edward Arnold. **Peterken, G.** 1993: *Woodland conservation and management,* 2nd edn. London: Chapman and Hall.

secular variation
Systematic variation through time where a persistent directional change of elements (positive or negative TREND) can be discerned over the period of record. Timescales of decades to centuries or millennia are usually implied. Secular CLIMATIC VARIATIONS may be weak tendencies only detectable in TIME-SERIES ANALYSIS by use of mean values or FILTERING and they may be obscured by shorter-term CLIMATIC FLUCTUATIONS and CLIMATIC VARIABILITY. Secular variations are recognised in other environmental contexts, such as MAGNETIC SECULAR VARIATIONS.   *RDT/BDG*

[See also PERIODICITIES, SUB-MILANKOVITCH]

Stephenson, F.R. and Wolfendale, A.W. (eds) 1988: *Secular solar and geomagnetic variations in the last 10,000 years.* Dordrecht: Kluwer. **Verosub, K.L.** 1988: Geomagnetic secular variation and the dating of Quaternary sediments. In Easterbrook, D.J. (ed.), *Dating Quaternary sediments* [*Special Paper* 227]. Boulder, CO: Geophysical Society of America, 123–139.

sedentary organism
An organism described during a non-motile stage in its life cycle.   *KDB*

sedentism
An immobile community lifestyle involving a residence in one location, in contrast to migratory or nomadic lifestyles. In early prehistory sedentism was only possible where the environment and resources were amenable throughout the year.   *LD-P*

[See also DOMESTICATION, MIGRATION, NOMADISM, PASTORALISM]

sediment
Unconsolidated material that has accumulated at the Earth's surface and may undergo LITHIFICATION to form a SEDIMENTARY ROCK. The term embraces material derived from DENUDATION of existing rocks (CLASTIC sediment), organic material, chemical precipitates and particulate material derived from volcanic activity (PYROCLASTIC sediment).   *TY*

[See also SEDIMENT TYPES]

sediment budget
The balance between the inputs, outputs and storages of sediment in a terrain unit (e.g. a river reach, a section of slope, a lake, a reservoir or a river catchment) over a period of time. The concept has become increasingly important from the 1980s with the advent of new or improved techniques (SEDIMENT FINGERPRINTING) for identifying and quantifying SEDIMENT SOURCES and the dynamics and significance of in-channel SEDIMENT STORAGES. Nested catchment SEDIMENT YIELD studies have demonstrated clearly how much SEDIMENT DELIVERY

RATIOS vary both within and between catchments, with a general tendency for ratios to fall with increasing catchment size with channel and floodplain aggradation. Studies have also demonstrated how phases of excessive slope erosion due to FOREST CLEARANCE tend to lead to low sediment delivery ratios during disturbance as rivers are unable to transport the supplied sediment effectively, but in contrast ratios increase after the slopes become revegetated and valley re-excavation by the river occurs.   *RPDW*

Phillips, J.D. 1992: Fluvial sediment budgets in the North Carolina Piedmont. *Geomorphology* 4, 231–241. **Trimble, S.W.** 1983: A sediment budget for Coon Creek basin in the Driftless Area, Wisconsin, 1853–1977. *American Journal of Science* 283, 454–474.

sediment delivery ratio
The ratio of the sediment yield at a catchment outlet to the measured or estimated off-channel (slope or landscape) erosion delivered to the channel. If the ratio is less than 1.0, it implies that net SEDIMENT STORAGE is occurring in channel margin areas, on the channel bed or on the floodplain of the river; if the ratio is higher than 1.0, it means that the river is deriving more from its bed and banks than it loses to those areas from the off-channel erosion delivered to it. In general, delivery ratios tend to fall significantly with catchment size and in a downstream direction along individual catchments.   *RPDW*

[See also SEDIMENT BUDGET]

Walling, D.E. 1983: The sediment delivery problem. *Journal of Hydrology* 65, 209–237.

sediment fingerprinting
The use of suspended sediment or deposited sediment properties to detect SEDIMENT SOURCES. This approach, which has expanded significantly since the 1980s, depends for its success on sediment sources within catchments being sufficiently distinctive in one or more aspects to enable the relative importance of contributions from the various sources or groups of sources to be deduced. A range of techniques may be involved, such as the mineralogy, the CAESIUM-137 content (derived from nuclear bomb tests and nuclear power station accidents), MINERAL MAGNETISM, colour and MICROMORPHOLOGICAL ANALYSIS of grains. The techniques can also be used to derive rates of floodplain sedimentation and aid in the construction of SEDIMENT BUDGETS for catchments.   *RPDW*

[See also HEAVY MINERAL ANALYSIS, QUARTZ GRAIN SURFACE TEXTURES, SEDIMENT YIELD, TRACERS]

Walling, D.E., Quine, T.A. and He, Q. 1992: Investigating contemporary rates of floodplain sedimentation. In Carling, P.A. and Petts, G.E. (eds), *Lowland floodplain rivers.* Chichester: Wiley, 251–327. **Walling, D.E., Woodward, J.C. and Nicholas, A.P.** 1993: A multi-parameter approach to fingerprinting suspended-sediment sources. *International Association of Hydrological Sciences Publication* 215, 329–338 [*Proceedings of the Yokohama Symposium*].

sediment flux
The total amount of material deposited on a lake or sea bottom, including MINEROGENIC, BIOGENIC and chemical components. It is expressed in g m^{-2} d^{-1}.   *UBW*

[See also ACCUMULATION RATE, HUMAN IMPACT ON SOIL, MAGNETIC SUSCEPTIBILITY, SEDIMENT FOCUSING, SEDIMENT INFLUX]

sediment focusing   A process causing LACUSTRINE SEDIMENTS to accumulate at different rates in different parts of a basin through time. The process depends on the lake's shape, water depth, total basin depth, wind exposure and so on.   *UBW*

[See also ACCUMULATION RATE, ACCUMULATION ZONE, EROSION, SEDIMENT FLUX, SEDIMENT INFLUX]

Davis, M.B., Moeller, R.E. and Ford, J. 1984: Sediment focusing and pollen influx. In Haworth, E.Y. and Lund, J.W.G. (eds), *Lake sediments and environmental history*. Leicester: Leicester University Press, 261–293. **Lehman, J.T.** 1975: Reconstructing the rate of accumulation of lake sediment: the effect of sediment focusing. *Quaternary Research* 5, 541–550.

sediment gravity flow   A category of flow in which particles are an essential part of the flow, in contrast to *fluid gravity flows (stream flow)*, in which a moving fluid may entrain particles, although *hyperconcentrated flows* represent a transitional type. Most sediment gravity flows are types of MASS MOVEMENTS. The main types of sediment gravity flow can be distinguished by the dominant particle support mechanism. In *grain flow* support is provided through particle collisions. This is the type of flow on the lee slopes of BEDFORMS that gives rise to CROSS-STRATIFICATION. Particle support in DEBRIS FLOW is also provided by strength and buoyancy attributable to a dense interstitial fluid rich in MUD. There is a continuum of processes between 'classical' muddy debris flows (*cohesive debris flow, mud flow*, LAHAR) and grain flow (*cohesionless debris flow*). Particle support in *turbidity flow* or TURBIDITY CURRENTS is provided by the upward component of eddies in a TURBULENT FLOW, giving the flowing mass a higher density than the ambient fluid and driving it as a type of DENSITY FLOW. Most sediment gravity flows move as a rapid surge and can be considered as a type of AVALANCHE. Sedimentation occurs through *solidification* ('freezing') of the entire flow, in contrast to the grain-by-grain sedimentation of particles entrained by fluid gravity flows.   *GO*

Allen, P.A. 1997: *Earth surface processes*. Oxford: Blackwell Science. **Costa, J.E.** 1988: Rheologic, geomorphic, and sedimentologic differentiation of water floods, hyperconcentrated flows, and debris flows. In Baker, V.R., Kochel, R.C. and Patton, P.C. (eds), *Flood geomorphology*. New York: Wiley, 113–122. **Leeder, M.R.** 1999: *Sedimentology and sedimentary basins: from turbulence to tectonics*. Oxford: Blackwell Science. **Lowe, D.R.** 1979: Sediment gravity flows: their classification and some problems of application to natural flows and deposits. In Doyle, L.J. and Pilkey, O.H. (eds), *Geology of continental slopes* [*Special Publication* 27]. Tulsa, OK: Society of Economic Paleontologists and Mineralogists, 75–82. **Lowe, D.R.** 1982: Sediment gravity flows: II. Depositional models with special reference to the deposits of high-density turbidity currents. *Journal of Sedimentary Petrology* 52, 279–298. **Middleton, G.V. and Hampton, M.A.** 1976: Subaqueous sediment transport and deposition by sediment gravity flows. In Stanley, D.J. and Swift, D.J.P. (eds), *Marine sediment transport and environmental management*. New York: Wiley, 197–218.

sediment influx   The amount of suspended MINEROGENIC and ORGANIC material transported into a lake by, for example, rivers or through EROSION and runoff in given periods of time.   *UBW*

[See also ACCUMULATION RATE, HUMAN IMPACT ON SOIL, MAGNETIC SUSCEPTIBILITY, SEDIMENT FLUX, SEDIMENT FOCUSING]

sediment sources   Locations (spatial and/or vertical) from which, or processes by which, transported sediment (usually either SUSPENDED SEDIMENT or floodplain or channel deposits) has been derived. The term is commonly used in studies of EROSION in DRAINAGE BASINS. Sources are usually classified into channel sources (e.g. channel banks and channel bed) and off-channel sources (i.e. material derived from slopes), but often material from surface and subsurface sources and from different parts of catchments can also be distinguished. Field monitoring of processes or SEDIMENT FINGERPRINTING techniques can be used to identify and quantify sources.   *RPDW*

[See also SEDIMENT TRANSPORT, SEDIMENT BUDGET, TRACERS]

Walling, D.E., Woodward, J.C. and Nicholas, A.P. 1993: A multi-parameter approach to fingerprinting suspended-sediment sources. *International Association of Hydrological Sciences Publication* 215, 329–338 [*Proceedings of the Yokahama Symposium*].

sediment storage   The temporary or more permanent storage of sediment eroded from the landscape at the base of slopes, in the channels of streams and rivers and in floodplains.   *RPDW*

[See also SEDIMENT BUDGET, SEDIMENT DELIVERY RATIO]

Beach, T. 1994: The fate of eroded soil: sediment sinks and sediment budgets of agrarian landscapes in southern Minnesota, 1851–1988. *Annals of the Association of American Geographers* 84, 5–28. **Warburton, J.** 1999: Environmental change and sediment yield from glacierised basins: the role of fluvial processes and sediment storage. In Brown, A.G. and Quine, T.A. (eds), *Fluvial processes and environmental change*. Chichester: Wiley, 363–384.

sediment transport   The movement of SEDIMENT from SEDIMENT SOURCES to SEDIMENTARY DEPOSITS, often interrupted by various types of temporary SEDIMENT STORAGE. Agents of sediment transport include wind, water and ice. During sediment transport by rivers, for example, material may be moved as BEDLOAD, DISSOLVED LOAD or SUSPENDED LOAD. Sediment transport distinguishes EROSION from *in situ* WEATHERING.   *JAM*

[See also MASS MOVEMENT, SALTATION]

sediment trap   (1) A site of long-term accumulation of sediment (e.g. a floodplain, reservoir, lake or basin). The determination of the amount of sediment accumulated and the period of accumulation can provide an indication of long-term EROSION RATES. (2) A settling tank to recover sediment removed from intensely cultivated slopes (e.g. vineyards). (3) A means of removing sediment from *overland flow* prior to its release into an aquifer. (4) A container or containers that collect(s) eroded sediment, which may be used in conjunction with EROSION PLOTS to allow the determination of SOIL LOSS (see GERLACH TROUGH).   *SHD*

Goudie, A. 1995: *The changing Earth: rates of geomorphological processes*. Oxford: Blackwell, 73–74. **Rice, R.J.** 1988: *Fundamentals of geomorphology*, 2nd edn. Harlow: Longman. 367–373. **Verstraeten, G. and Poesen, J.** 2000: Estimating trap efficiency of small reservoirs and ponds: methods and implications for the assessment of sediment yield. *Progress in Physical Geography* 24, 219–251.

sediment types   According to formation and DEPOSITIONAL ENVIRONMENT, composition and geotechnical properties, SEDIMENTS are divided into MINEROGENIC/CLASTIC, ORGANIC/BIOGENIC and chemical SEDIMENTS.

MINEROGENIC/CLASTIC SEDIMENTS are composed of ALLOCHTHONOUS DETRITAL material, which is derived from the physical, chemical and/or biological breakdown of rocks. Their classification is based on the size and shape of the particles and on the degree of plasticity and can be determined by among others by GRAIN-SIZE analysis. Rock fragments are transported by, for example, gravity, wind, ice or running water. Landsliding and transport by ice results in a mixture of all GRAIN-SIZES (MASS MOVEMENT PROCESSES, GLACIAL SEDIMENTS, DIAMICTON, TILL). Particles transported by wind (AEOLIAN sediments) or water (GLACIOLACUSTRINE DEPOSITS, GLACIO-FLUVIAL SEDIMENTS, FLUVIAL SEDIMENTS, LACUSTRINE SEDIMENTS, BRACKISH sediments, MARINE SEDIMENTS and DELTA sediments), in suspension (see SUSPENDED LOAD), by rolling or by SALTATION are sorted and their size reflects the speed of the transporting medium. Coarse-grained sediment indicates deposition from fast-flowing water/wind; fine-grained sediment indicates that the wind/water was slow-moving, or that only fine-grained sediment was available for transport.

Chemical sediments are formed by PRECIPITATION of AUTHIGENIC minerals from solution in water. This precipitation may be due to, for example, biochemical reactions, where activities of marine micro-organisms may lead to a decreased ACIDITY of the surrounding water, causing calcium carbonate to precipitate. INORGANIC reactions result in precipitation of, for example, opal or calcite, growth of manganese minerals on the sea floor or the evaporation of salt. Most chemical sediments and SEDIMENTARY ROCKS contain one dominant mineral and this is used as a basis for classification.

BIOGENIC and ORGANIC SEDIMENTS occur in marine, freshwater and terrestrial environments and are described and classified according to colour, amount and type of organic/minerogenic components, physical properties, humification and layer boundaries (see the Figure for an example). These AUTOCHTHONOUS and allochthonous deposits may be composed of purely organic material (PEAT, DY) or of a mixture of organic and minerogenic matter (e.g. GYTTJA, OOZE, MARL, DIATOMITE). The classification of organic sediments from lacustrine and WETLAND environments generally follows a modified version of the Troels-Smith system. In marine environments, BIOGENIC SEDIMENTS comprise mainly siliceous OOZE (large percentages of skeletons of opaline silica), calcareous/foraminifera OOZE (composed of > 30% carbonate, shells and skeletons of microorganisms) and SAPROPEL (an organic marine OOZE). *UBW*

[See also CALCRETE, CAVE SEDIMENTS, DIATOMITE, DY, GYTTJA, MARINE SEDIMENT CORES, MARL, SEDIMENT SOURCES, TUFA,]

Aaby, B. and Berglund, B.E. 1986: Characterization of peat and lake deposits. In Berglund, B.E. (ed.), *Handbook of Holocene palaeoecology and palaeohydrology.* Chichester: Wiley, 231–246. **Birks, H.J.B. and Birks, H.H.** 1980: *Quaternary palaeoecology.* London: Edward Arnold. **Reading, H.G. (ed.)** 1978: *Sedimentary environments and facies.* Oxford: Blackwell Scientific. **Troels-Smith, J.** 1955: Characterisation of unconsolidated sediments. *Danmarks Geologiske Undersøkelse* IV(3), 1–73.

sediment yield The total amount of sedimentary particles leaving a DRAINAGE BASIN over a specified time period. Strictly it comprises the sum of the SUSPENDED SEDIMENT LOAD and the BEDLOAD of a river, but in some cases it is used as a synonym for the suspended sediment yield or the bedload is estimated as a minor (10–15%) fraction of the sediment yield. When catchments are compared, sediment yields tend to be expressed in terms of per unit catchment area (*specific sediment yield*). Maximum values of $> 20\,000$ t km^{-2} y^{-1} have been recorded in severely eroded areas of the Loess Plateau, China. In general, sediment yield varies directly with relief, the ERODIBILITY of the underlying lithology and the degree of human interference, but inversely with vegetation cover and the permeability of the soil and underlying rock. Because of the extent of human disturbance over the globe, variations with climate in areas of natural vegetation are still not fully known, and although the high yields of semi-arid areas, compared with both arid areas and wetter vegetated regions with more permeable soils, is well established, how yields vary between differing forest zones as rainfall increases within wetter regions remains a subject of much debate. Comparisons between catchments are also not made easier by the fact that SEDIMENT DELIVERY RATIOS also vary greatly between catchments.

RPDW/JAM

[See also SEDIMENT STORAGE]

Laronne, J.B. and Mosley, M.P. (eds) 1982: *Erosion and sediment yield* [*Benchmark Papers in Geology* 63]. Stroudsville, PA: Hutchinson Ross. **Milliman, J.C. and Meade, R.H.** 1992: Geomorphic/tectonic control of sediment discharge to the oceans. *Journal of Geology* 100, 525–544. **Walling, D.E.** 1988: Erosion and sediment yield research – some recent perspectives. *Journal of Hydrology* 100, 113–141.

sedimentary basin A part of the Earth's surface where SEDIMENTARY DEPOSITS accumulate. Sedimentary basins can be related to three TECTONIC settings for CRUSTAL SUBSIDENCE (see PLATE TECTONICS) and divided into nine types, although there are many subdivisions. Basins in areas of crustal extension (including CONSTRUCTIVE PLATE MARGINS) include FAULT-bounded *rift basins* (see GRABEN, RIFT VALLEY), basins on CRATONS, basins at PASSIVE CONTINENTAL MARGINS and basins on OCEANIC CRUST. Basins in areas of crustal shortening, associated with SUBDUCTION and DESTRUCTIVE PLATE MARGINS, include MARGINAL BASINS in BACK-ARC and FORE-ARC settings, *intra-arc basins*, OCEANIC TRENCHES and FORELAND BASINS. PULL-APART BASINS form in areas of crustal shearing, particularly along CONSERVATIVE PLATE MARGINS. The type and distribution of DEPOSITIONAL ENVIRONMENTS represented in sediment FACIES can be used to reconstruct the type and evolution of sedimentary basins from SEDIMENTARY ROCK successions in the GEOLOGICAL RECORD. *GO*

Busby, C.J. and Ingersoll, R.V. (eds) 1995: *Tectonics of sedimentary basins.* Oxford: Blackwell Science. **Einsele, G.** 1992: *Sedimentary basins: evolution, facies and sediment budget.* Berlin: Springer. **Ingersoll, R.V.** 1988: Tectonics of sedimentary basins. *Geological Society of America Bulletin* 100, 1704–1719. **Jackson, J.A., Norris, R. and Youngson, J.** 1996: The structural evolution of active fault and fold systems in central Otago, New Zealand: evidence revealed by drainage patterns. *Journal of Structural Geology* 18, 217–234. **Leeder, M.R.** 1996: Sedimentary basins: tectonic recorders of sediment discharge from drainage catchments. *Earth Surface Processes and Landforms* 22, 229–237. **Leeder, M.R. and Gawthorpe, R.L.** 1987: Sedimentary models for extensional tiltblock/half graben basins.

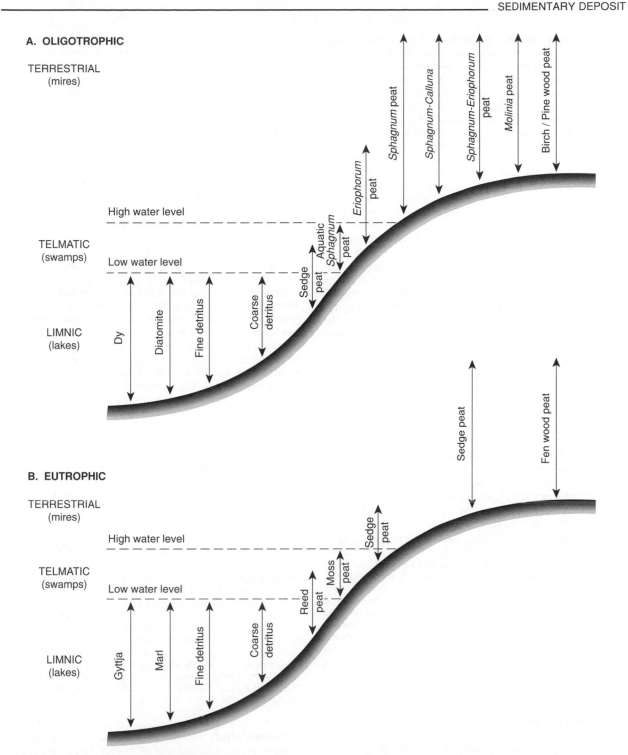

sediment types *Some sediment types formed in (A) oligotrophic and (B) eutrophic environments (after Birks and Birks, 1980)*

In Coward, M.P., Dewey, J.F. and Hancock, P.L. (eds), *Continental extensional tectonics* [*Special Publication* 28]. London: Geological Society, 139–152.

sedimentary deposit A general term for materials that have accumulated at the Earth's surface, including SOIL, SEDIMENT and PYROCLASTIC material, but with the exception of high-temperature liquids such as LAVA. Sedimentary deposits accumulate in SEDIMENTARY BASINS and can form SEDIMENTARY ROCKS after burial and DIAGENESIS. They are an important repository of information about environmental change. *GO*

[See also SEDIMENTOLOGICAL EVIDENCE OF ENVIRONMENTAL CHANGE, SEDIMENTOLOGY]

sedimentary rocks   Rocks formed at the Earth's surface, at normal temperatures and pressures. Features that are typical of sedimentary rocks include their occurrence in BEDS or *strata*, the preservation of FOSSILS and, in many cases, a fragmental, particulate (CLASTIC) texture. PYROCLASTIC deposits, formed as a result of explosive VOLCANISM, share some of these features and are commonly considered as sedimentary rocks. Stages in the formation of a sedimentary rock are: (1) the release of SEDIMENT (e.g. through WEATHERING); (2) in most cases, its transportation by gravity (see MASS MOVEMENT PROCESSES and SEDIMENT GRAVITY FLOWS) or by moving fluids (currents of water, wind or ice); (3) the accumulation of a SEDIMENTARY DEPOSIT when, for example, a current slows down, salty water evaporates or organisms die; and (4) burial and DIAGENESIS, one consequence of which may be LITHIFICATION to form a sedimentary rock.

Most sedimentary rocks fall into one of four groups, but the boundaries between them are gradational and many categories are without precise formal definition: TERRIGENOUS CLASTIC (or *siliciclastic* or DETRITAL) sediments (commonly described simply as *clastic sedimentary rocks* – e.g. MUDROCK, SANDSTONE, CONGLOMERATE and BRECCIA); biogenic and organic sediments (e.g. COAL, many types of LIMESTONE); chemical sediments (EVAPORITES); and pyroclastic and VOLCANICLASTIC deposits (e.g. TUFF). Having formed at the Earth's surface, sedimentary rocks hold a direct record of past environments and climates in their LITHOLOGY, TEXTURE, FABRIC, chemical composition and SEDIMENTARY STRUCTURES. The interpretation of successions of sedimentary rocks (see FACIES ANALYSIS) can provide a uniquely geological, long-term record of environmental change.   *TY/GO*

[See also GEOLOGICAL RECORD OF ENVIRONMENTAL CHANGE, IGNEOUS ROCKS, METAMORPHIC ROCKS, SEDIMENTOLOGICAL EVIDENCE OF ENVIRONMENTAL CHANGE, SEDIMENTOLOGY, STRATIGRAPHY]

Leeder, M.R. 1982: *Sedimentology: process and product.* London: Unwin Hyman. Nichols, G. 1999: *Sedimentology and stratigraphy.* Oxford: Blackwell. Pedley, H.M. and Frostick, L. (eds) 1999: Unravelling tectonic and climatic signals in sedimentary successions. *Journal of the Geological Society, London* 156, 747–863. Selley, R.C. 1996: *Ancient sedimentary environments and their sub-surface diagnosis,* 4th edn. London: Chapman and Hall. Tucker, M.E. 1991: *Sedimentary petrology: an introduction to the origin of sedimentary rocks,* 2nd edn. Oxford: Blackwell Science. Walker, R.G. 1992: Facies, facies models and modern stratigraphic concepts. In Walker, R.G. and James, N.P. (eds), *Facies models: response to sea level change.* St John's, Newfoundland: Geological Association of Canada, 1–14.

sedimentary structures   Distinctive morphological or geometrical features of sediments and sedimentary rocks that are crucial elements in the interpretation of DEPOSITIONAL ENVIRONMENTS using FACIES ANALYSIS. They are conventionally subdivided into *primary structures* formed during SEDIMENT TRANSPORT and DEPOSITION, *secondary structures* formed by processes affecting sediment after deposition, and *biogenic structures* resulting from the activities of organisms.

Primary structures are formed by sediment deposition (and include BEDDING and LAMINATION) or by EROSION, with subsequent deposition leading to preservation. They provide information about the type, strength and direction of PALAEOCURRENTS. Deposition from waning currents forms GRADED BEDDING. Unidirectional water flows (e.g. rivers) mould sand surfaces into BEDFORMS including asymmetrical CURRENT RIPPLES, DUNES and PLANE BEDS that can be preserved as sedimentary structures on bedding surfaces. Sediment deposition on migrating or accreting bedforms produces internal structures of CROSS-LAMINATION, CROSS-BEDDING and PARALLEL LAMINATION. Wave activity forms symmetrical WAVE RIPPLES and HUMMOCKY CROSS-STRATIFICATION, and TIDES produce FLASER BEDDING, HERRINGBONE CROSS-BEDDING and TIDAL BUNDLES. Sediment transport by wind forms WIND RIPPLES, TRANSLATENT STRATA, ADHESION structures and dunes preserved as cross-bedding. Structures produced by erosion include CHANNELS and smaller-scale SCOUR MARKS and TOOL MARKS.

Secondary structures commonly involve the disturbance of stratification. A distinctive structure is a surface pattern of sand-filled MUD CRACKS, most commonly formed by the drying and shrinkage of exposed sediments. SOFT-SEDIMENT DEFORMATION structures are formed by the near-surface deformation of unconsolidated sediment, and down-slope movements of sediment masses give rise to SLUMPS and other forms of MASS MOVEMENT. These structures can indicate the orientation of PALAEOSLOPES or point to the former activity of specific agents of disturbance, such as EARTHQUAKES or TSUNAMIS. Ploughing and dragging by GLACIERS and ICEBERGS also produce secondary deformation structures that can indicate the direction of ice movement (see GLACIOTECTONICS). A separate group of secondary structures results from the growth within sediment of ICE WEDGES or minerals formed during DIAGENESIS, in particular EVAPORITES and cemented sediment masses (CONCRETIONS).

Biogenic structures or TRACE FOSSILS include the disturbance of stratification due to burrowing and root growth (see BIOTURBATION), or tracks and trails at the sediment surface.   *MRT*

[See also SEDIMENTOLOGICAL EVIDENCE OF ENVIRONMENTAL CHANGE]

Allen, J.R.L. 1982: *Sedimentary structures: their character and physical basis.* Amsterdam: Elsevier. Allen, J.R.L. 1993: Sedimentary structures: Sorby and the last decade. *Journal of the Geological Society, London* 150, 417–425. Collinson, J.D. and Thompson, D.B. 1992: *Sedimentary structures,* 2nd edn. London: Chapman and Hall. Leeder, M.R. 1999: *Sedimentology and sedimentary basins: from turbulence to tectonics.* Oxford: Blackwell Science. Pettijohn, F.J. and Potter, P.E. 1964: *Atlas and glossary of primary sedimentary structures.* New York: Springer. Potter, P.E. and Pettijohn, F.J. 1977: *Paleocurrents and basin analysis,* 2nd edn. Berlin: Springer.

sedimentation   In the broad sense, any process that leads to the accumulation of SEDIMENT. In the strict sense, the term is restricted to the settling out of particles in the SUSPENDED LOAD from a fluid medium.   *MRT*

sedimentological evidence of environmental change   SEDIMENTS and SEDIMENTARY ROCKS provide a major source of evidence of past environmental change. All DEPOSITIONAL ENVIRONMENTS respond to externally forced change, but some are more sensitive than others and the changes are reflected in different ways. Several branches of SEDIMENTOLOGY are concerned with the identification and interpretation of evidence of environmental change preserved in sedimentary successions.

Variations in sediment ACCUMULATION RATE may reflect important changes in the *catchment* area, as SEDIMENT YIELD is a function of climatically controlled factors such as vegetation cover, rainfall intensity and amount, and SEASONALITY. It is also strongly influenced by ANTHROPOGENIC impacts, particularly LANDUSE practices. In BIOGENIC SEDIMENTS, changes in accumulation rate can result from variations in PRODUCTIVITY, which may also have natural or anthropogenic causes. Accumulation can be measured directly, typically as linear or mass accumulation rates per unit time; relative changes can also be estimated by PROXY methods such as *porosity* evaluation (e.g. GAMMA RAY ATTENUATION POROSITY EVALUATOR).

Sediment composition is affected by environmental change through its influence upon sediment supply routes, WEATHERING rates or the relative supply of BIOGENIC and MINEROGENIC components. CLAY MINERALS are a characteristic product of CHEMICAL WEATHERING and changes in the proportions of clay minerals may indicate climatic variations. Sediment GEOCHEMISTRY also carries important environmental information. Variations in the amount and type of preserved ORGANIC matter can, for example, reflect changes in vegetation, BIOMASS BURNING or BOTTOM WATER oxygen supply. The stable isotopic compositions of CARBONATE MINERALS and organic matter (see STABLE ISOTOPE ANALYSIS) are related to features such as water composition and temperature, mixing regime, nutrient availability and rates of PRIMARY PRODUCTIVITY. Some TRACE ELEMENTS are sensitive to dissolved oxygen content and can be used to track variations in the ventilation of a waterbody. In addition to mineralogical and geochemical analyses, relative changes in sediment compositions may be assessed by proxy methods, notably using colour variations (GREY-SCALE ANALYSIS), MAGNETIC SUSCEPTIBILITY and NATURAL GAMMA RADIATION.

The most widespread sedimentological evidence of environmental change is preserved as variations in FACIES, which are interpreted by FACIES ANALYSIS. Some sedimentary environments are particularly sensitive to change, especially those at the transition between zones of contrasting climate or dominant sedimentary process, or where dramatic fluctuations in water chemistry occur. On land, one such zone is the DESERT margin, where climate determines the extent of widespread AEOLIAN sediment transport. Changes in rainfall and wind intensity have led to repeated expansion and contraction of the world's desert regions. The SAHEL, for example, is characterised by extensive fields of PALAEODUNES, which testify to periods when the Sahara extended farther south. Conversely, RELICT river, lake and marsh deposits within modern deserts reveal periods of more humid climate. Where aeolian processes dominate, sediment of silt size and finer is commonly transported long distances. On land it may be deposited as LOESS, but ATMOSPHERIC DUST also makes a significant contribution to OCEANIC SEDIMENTS downwind from many major deserts. Cyclic variations in the quantity and grain-size of this dust demonstrate that QUATERNARY glaciations were accompanied by intensified winds and low-latitude ARIDITY.

Environmental change affects ALLUVIAL sedimentation principally through its influence on DISCHARGE, SEDIMENT YIELD and BASE LEVEL. The FLOODPLAINS of many modern river systems bear traces of PALAEOCHANNELS that differ in scale and form from those that are currently active, attesting to significant changes in RIVER REGIME. Buried or *incised* channels may indicate base-level fall because of SEA-LEVEL CHANGE or LAKE-LEVEL VARIATIONS. Facies changes commonly accompany changes in CHANNEL PATTERNS. Perennial streams with abundant SUSPENDED LOAD tend to MEANDER and the channel sediments are commonly associated with extensive, fine-grained floodplain deposits. A change to strongly seasonal discharge, or to conditions that favour the production of abundant BEDLOAD sediment, may cause a switch to BRAIDING in which sand and gravel dominate and little of the finer-grained sediment load is preserved. In areas of ALLUVIAL FAN sedimentation, the development of dense vegetation cover in the catchment may reduce sediment yields to the point where little or no accumulation occurs on the fans. Periods like this are commonly marked by widespread PEDOGENESIS on the fan surface.

Lakes represent another sedimentary environment that is sensitive to change, particularly those with a HYDROLOGICAL BALANCE close to unity or negative. Variations in hydrological balance typically lead to lake-level changes, which are expressed in the form of raised or drowned lacustrine and deltaic sediments, PALAEOSOLS, STROMATOLITE and *travertine* deposits. Evidence of lake-level variation is also revealed by abrupt, vertical facies changes, for example the superposition, as a result of lake-level fall, of coarse-grained, marginal deposits upon deep-water muds.

EVAPORITE-bearing sequences deposited from lakes or restricted marine waterbodies also preserve evidence of changes in hydrological balance. Variations in evaporite mineralogy, or alternations of evaporite and LIMESTONE or CLASTIC units, are commonly a result of a varying ratio of inflow to evaporation. In lakes the cause is normally climatic, notably changes in rainfall amount. This may also be the case in some marine basins, but in others the reason is variation in the rate at which normal seawater can enter the basin to replenish that lost by evaporation. Many of the world's major evaporite deposits show repeated variations in mineralogy or lithology, indicating PERIODICITIES in environmental change.

Although marine sedimentation is influenced by sediment supply from land, the most widespread changes are those associated with sea-level and OCEAN CURRENT fluctuations, and variations in the production and preservation of BIOGENIC sediments. Facies changes defining cycles of TRANSGRESSION and REGRESSION that characterise many shallow-marine sequences demonstrate the important influence exerted by sea-level change on nearshore and CONTINENTAL SHELF environments. Palaeosols and, in carbonate deposits, characteristic types of early CEMENTATION and KARST development, provide additional evidence of lowered sea-level. In the deep sea, environmental change is mainly forced by shallow-water or surface processes. Variations in grain-size may reflect fluctuations in bottom-current intensity because of changes in DEEP WATER genesis. Variable contents of biogenic components can provide evidence of changes in surface-water productivity. Increased supply of coarse clastic debris to SUBMARINE FANS may be an indication of lowered sea level, when rivers or glaciers reach the continental shelf edge.   *MRT*

Bird, M.I. and Cali, J.A. 1998: A million-year record of fire in sub-Saharan Africa. *Nature* 394, 767–769. **Blum, M.D. and Price, D.M.** 1998: Quaternary alluvial plain construction in response to glacio-eustatic and climatic controls, Texas Gulf Coastal Plain. In Shanley, K.W. and McCabe, P.J. (eds), *Relative role of eustasy, climate and tectonism in continental rocks* [*Special Publication* 59]. Tulsa, OK: Society of Economic Paleontologists and Mineralogists, 31–48. **Chamley, H.** 1989: *Clay sedimentology.* Berlin: Springer. **Demenocal, P.B., Ruddiman, W.F. and Pokras, E.M.** 1993: Influences of high-latitude and low-latitude processes on African terrestrial climate – Pleistocene eolian records from equatorial Atlantic-Ocean drilling program site 663. *Paleoceanography* 8, 209–242. **Laberg, J.S. and Vorren, T.O.** 1996: The Middle and Late Pleistocene evolution of the Bear Island Trough Mouth Fan. *Global and Planetary Change* 12, 309–330. **Lowenstein, T.K., Li, J., Brown, C., Roberts, S.M.** *et al.* 1999: 200 k.y. paleoclimate record from Death Valley salt core. *Geology* 27, 3–6. **Milligan, M.R. and Chan, M.A.** 1998: Coarse-grained Gilbert deltas: facies, sequence stratigraphy and relationships to Pleistocene climate at the eastern margin of Lake Bonneville, northern Utah. In Shanley, K.W. and McCabe, P.J. (eds), *Relative role of eustasy, climate and tectonism in continental rocks* [*Special Publication* 59]. Tulsa, OK: Society of Economic Paleontologists and Mineralogists, 177–189. **Pedley, H.M. and Frostick, L.** 1999: Unravelling tectonic and climatic signals in sedimentary successions. *Journal of the Geological Society, London* 156, 747 [Introduction to thematic set of papers, pp. 747–863]. **Reading, H.G. (ed.)** 1996: *Sedimentary environments: processes, facies and stratigraphy*, 3rd edn. Oxford: Blackwell. **Robert, C. and Kennet, J.P.** 1997: Antarctic continental weathering changes during Eocene–Oligocene cryosphere expansion: clay mineral and oxygen isotope evidence. *Geology* 25, 587–590. **Tyson, R.V.** 1995: *Sedimentary organic matter: organic facies and palynofacies.* London: Chapman and Hall.

sedimentology

The study of SEDIMENTS and SEDIMENTARY ROCKS. Important aspects include the formation of sediment particles, the description and classification of sediments and sedimentary rocks, DIAGENESIS, the processes of SEDIMENT TRANSPORT and DEPOSITION, the interpretation of SEDIMENTARY STRUCTURES, reconstructions of PALAEOGEOGRAPHY and PALAEOENVIRONMENTAL RECONSTRUCTION based on the analysis and interpretation of sedimentary deposits and FACIES. *MRT*

[See also SEDIMENTOLOGICAL EVIDENCE OF ENVIRONMENTAL CHANGE]

Leeder, M.R. 1999: *Sedimentology and sedimentary basins: from turbulence to tectonics.* Oxford: Blackwell Science. **Nichols, G.** 1999: *Sedimentology and stratigraphy.* Oxford: Blackwell Science. **Tucker, M.E.** 1991: *Sedimentary petrology: an introduction to the origin of sedimentary rocks.* Oxford: Blackwell Science.

segregated ice

Ice formed by the migration of pore water towards the freezing plane, where it forms into discrete lenses or layers. Segregated ice, sometimes termed *segregation ice*, ranges in thickness from hairline to more than 10 m. It commonly occurs in alternating layers of soil and ice. *HMF*

[See also FROST HEAVE, PATTERNED GROUND]

Associate Committee on Geotechnical Research 1988: *Glossary of permafrost and related ground-ice terms* [*Technical Memorandum* 142]. Ottawa: Permafrost Subcommittee, National Research Council of Canada.

seiche

A standing wave (resonant oscillation) in an enclosed or partly enclosed water body (such as a coastal lagoon, sound, bay, estuary or lake) generated as an abrupt response to the wind, TIDES, STORM SURGES, TSUNAMIS, sediment SLUMPS or EARTHQUAKE shaking. The waves generated often compound the NATURAL HAZARD. Seismically induced seiches can be experienced many hundreds of kilometres from the EPICENTRE of an earthquake. For example, seiches affected Loch Lomond in Scotland for over an hour after the AD 1755 Lisbon earthquake. Seiches are similar to the effect of sloshing water in a bath and can damage vessels against harbour walls. *GO*

Chapron, E., Beck, C., Pourchet, M. and Deconinck, J.-F. 1999: 1822 earthquake-triggered homogenite in Lake Le Bourget (NW Alps). *Terra Nova* 11, 86–92. **Rabinovich, A.B. and Monserrat, S.** 1998: Generation of meteorological tsunamis (large amplitude seiches) near the Balearic and Kuril Islands. *Natural Hazards* 18, 27–55. **Smith, B. and Miyaoka, E.** 1999: Frequency domain identification of harbour seiches. *Environmetrics* 10, 575–587.

seismic

Pertaining to vibrations in the Earth, including EARTHQUAKES and the artificial generation of shock waves (see GEOPHYSICAL EXPLORATION). *GO*

seismic belt

A narrow, linear zone on the Earth's surface that experiences SEISMIC activity (EARTHQUAKES). Most of the rest of the Earth's surface is free from seismic activity. The distribution of seismic belts picks out the Earth's *plate boundaries*. *DNT*

[See also PLATE TECTONICS, SEISMICITY]

seismic gap

A section of a *plate boundary* or other major FAULT that has not ruptured to produce an EARTHQUAKE for a longer time than would be expected from historical records or from evidence from PALAEOSEISMICITY, suggesting that sufficient strain energy has built up for a major earthquake to occur in the near future. EARTHQUAKE PREDICTION identifies seismic gaps as the most likely parts of SEISMIC BELTS to generate an earthquake. Recent evidence from *subduction zones*, however, does not fully support this model. *DNT*

Kagan, Y.Y. and Jackson, D.D. 1995: New seismic gap hypothesis – 5 years after. *Journal of Geophysical Research – Solid Earth* 100, 3943–3959.

seismic reflection surveying

A method of SEISMIC SURVEYING that uses the time taken for reflections of SEISMIC WAVES from prominent boundaries (*reflectors*) to travel from the Earth's surface to the reflector and back (*travel time* or *two-way travel time*) to produce images of the subsurface. Seismic reflection can produce images of the Earth's interior to many kilometres depth and is used in the exploration for geological RESOURCES, especially PETROLEUM, as well as in the scientific study of the Earth's CRUST. Seismic reflection surveys can be used to produce images of GEOLOGICAL STRUCTURES, *plate boundaries* and SEDIMENTARY BASINS, as well as in studies of the OCEAN floor, where the technique is commonly used in conjunction with SIDE-SCAN SONAR. The subdivision of cross-sections obtained by seismic reflection surveys (*seismic profiles*) into units that are considered to have time significance forms the discipline of SEISMIC STRATIGRAPHY, which strongly influenced the development of SEQUENCE STRATIGRAPHY. *Seismic tomography* and *three-dimensional seismic* (commonly referred to as '3-D seismic') are methods that use computer manipulation of large amounts of seismic reflection data for a given volume of rock to produce cross-sections or plan views of seismic properties. *GO*

[See also GEOPHYSICAL SURVEYING]

Brown, A.R. 1999: *Interpretation of three-dimensional seismic data*, 5th edn. [*Memoir 42*]. Tulsa, OK: American Association of Petroleum Geologists. **Buker, F., Green, A.G. and Horstmeyer, H.** 2000: 3-D high-resolution reflection seismic imaging of unconsolidated glacial and glaciolacustrine sediments: processing and interpretation. *Geophysics* 65, 18–34. **Cherkis, N.Z., Max, M.D., Vogt, P.R.** *et al.* 1999: Large-scale mass wasting on the north Spitsbergen continental margin, Arctic Ocean. *Geo-Marine Letters* 19, 131–142. **Morey, D. and Schuster, G.T.** 1999: Palaeoseismicity of the Oquirrh fault, Utah, from shallow seismic tomography. *Geophysical Journal International* 138, 25–35. **Torres, J., Droz, L., Savoye, B.** *et al.* 1997: Deep-sea avulsion and morphosedimentary evolution of the Rhône Fan Valley and Neofan during the Late Quaternary (north-western Mediterranean Sea). *Sedimentology* 44, 457–477.

seismic refraction surveying

A method of SEISMIC SURVEYING that uses the refraction of SEISMIC WAVES at boundaries beneath the Earth's surface to produce images of the subsurface. Seismic refraction can be used, commonly in conjunction with SEISMIC REFLECTION SURVEYING or other methods of GEOPHYSICAL SURVEYING, to distinguish boundaries in the shallow subsurface, such as the boundary between unconsolidated sediments and bedrock, and to study deep EARTH STRUCTURE. *GO*

Chroston, P.N., Jones, R. and Makin, B. 1999: Geometry of Quaternary sediments along the north Norfolk coast, UK: a shallow seismic study. *Geological Magazine* 136, 465–474. **Lanz, E., Maurer, H. and Green, A.G.** 1988: Refraction tomography over a buried waste disposal site. *Geophysics* 63, 1414–1433.

seismic stratigraphy

The subdivision of successions of sediments and/or sedimentary rocks (see STRATIGRAPHY) in the Earth's subsurface using SEISMIC REFLECTION SURVEYING techniques. Discontinuities on *seismic profiles* are interpreted as UNCONFORMITIES bounding distinct packages of *strata*. Seismic stratigraphy was developed by the PETROLEUM exploration industry in the AD 1970s and provided the opportunity to observe the geometry of strata on a very large scale. Information on LITHOLOGY from BOREHOLES and exposures on land helped to integrate seismic stratigraphy with geological field observations, including conventional FACIES ANALYSIS, leading to the development of SEQUENCE STRATIGRAPHY. *DNT*

Hart, B.S., Sibley, D.M. and Flemings, P.B. 1997: Seismic stratigraphy, facies architecture, and reservoir character of a Pleistocene shelf-margin delta complex, Eugene Island Block 330 field, offshore Louisiana. *Bulletin of the American Association of Petroleum Geologists* 81, 380–397. **Nichols, G.** 1999: *Sedimentology and stratigraphy*. Oxford: Blackwell Science. **Payton, C.E. (ed.)** 1977: *Seismic stratigraphy – applications to hydrocarbon exploration* [*Memoir 26*]. Tulsa, OK: American Association of Petroleum Geologists. **Whittaker, A.** 1998: Principles of seismic stratigraphy. In Doyle, P. and Bennett, M.R. (eds), *Unlocking the stratigraphical record: advances in modern stratigraphy*. Chichester: Wiley: 275–298.

seismic surveying

Methods of GEOPHYSICAL SURVEYING that use the passage of SEISMIC WAVES through the Earth to produce images of objects and structures in the subsurface. Seismic surveys can be carried out on land or from ships to produce an image of EARTH STRUCTURE beneath the sea floor. The source of seismic waves can be an explosion, an applied shock, or sound waves, and their arrival times at the surface, having been reflected or refracted from boundaries in the subsurface, are detected using SEISMOMETERS, *geophones* or *hydrophones*. The two methods of seismic surveying are SEISMIC REFLECTION SURVEYING and SEISMIC REFRACTION SURVEYING. *GO*

Brabham, P.J., McDonald, R.J. and McCarroll, D. 1999: The use of shallow seismic techniques to characterize sub-surface Quaternary deposits: the example of Porth Neigwl (Hells Mouth Bay), Gwynedd, N. Wales. *Quarterly Journal of Engineering Geology* 32, 119–137. **Gowda, B.M.R., Ghosh, N., Wadhwa, R.S.** *et al.* 1999: Seismic surveys for detecting scour depths downstream of the Srisailam dam, Andhra Pradesh, India. *Engineering Geology* 53, 35–46. **Kearey, P. and Brooks, M.** 1991: *An introduction to geophysical surveying*, 2nd edn. Oxford: Blackwell Science.

seismic waves

Simple periodic motions of soil, sediment and rock that transmit energy in all directions from the focus of an EARTHQUAKE or other natural or artificial SEISMIC source. There are two types of seismic wave: body waves and surface waves.

Body waves travel through the Earth and are either compressional waves (*Primary waves* or *P-waves*; displacement in the direction of wave travel) or transverse waves (*Secondary waves* or *S-waves*; displacement perpendicular to the direction of wave travel). The velocity of seismic waves (*seismic velocity*) depends on the properties of the material they travel through. P-waves travel at a greater velocity than S-waves and are the first to arrive at a SEISMOMETER after an earthquake: hence the designations Primary wave and Secondary wave. Body waves are reflected and refracted when they encounter boundaries between rocks with different properties, and S-waves are unable to travel through liquids. Thus, the arrival times of P-waves and S-waves at seismometers can be used to interpret EARTH STRUCTURE and in SEISMIC SURVEYING.

Surface waves travel on, or just below, the Earth's surface and include *Love waves*, with a shearing motion, and *Rayleigh waves* with a rolling motion. They travel more slowly than body waves and therefore arrive later at seismometers. It is surface waves that cause damaging ground shaking during a major earthquake. *DNT*

Udías, A. 2000: *Principles of seismology*. Cambridge: Cambridge University Press.

seismicity

The distribution of EARTHQUAKES in space and time. Early work suggested that earthquake frequency was inversely related to EARTHQUAKE MAGNITUDE, but recent studies have indicated that this is not strictly true for either end of the magnitude scale because of the FRACTAL nature of earthquakes. The geographical distribution of seismicity is not uniform, but shows a concentration along SEISMIC BELTS, which coincide with zones of VOLCANISM and with certain major geomorphological features such as MID-OCEAN RIDGES, OCEANIC TRENCHES and OROGENIC BELTS, picking out the PLATE MARGINS (see Figure). *DNT*

Gutenberg, B. and Richter, C.F. 1954: *Seismicity of the Earth and associated phenomena*. Princeton, NJ: Princeton University Press. **Press, F. and Siever, R.** 1986: *Earth*, 4th edn. New York: Freeman. **Tsapanos, T.M. and Christova, C.V.** 2000: Some preliminary results of a worldwide seismicity estimation: a case study of seismic hazard evaluation in South America. *Annali di Geofisica* 43, 11–22.

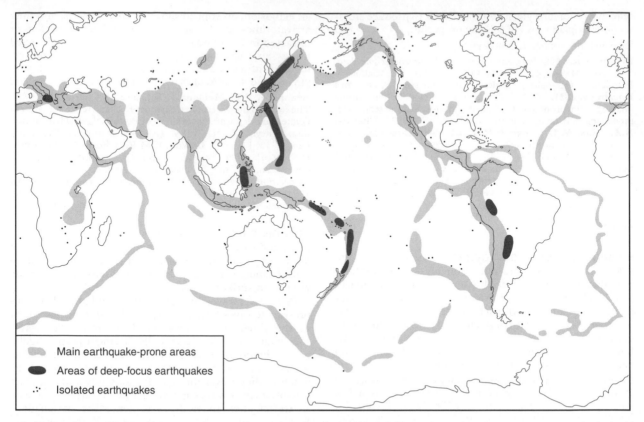

seismicity *Areas of the world that experience major earthquakes (seismic belts). The main earthquake-prone areas are shaded, with dots representing isolated events. Darker shading shows areas that experience deep-focus earthquakes (based on Press and Siever, 1986)*

seismology The study of EARTHQUAKES and other SEISMIC phenomena, including the study of EARTH STRUCTURE (*global seismology*), detailed investigation of the structure of the CRUST and outer MANTLE using artificially-generated SEISMIC WAVES (*exploration seismology*), EARTHQUAKE PREDICTION and the study of the record of earthquakes through time (*palaeoseismology*). *GO*

[See also GEOPHYSICAL EXPLORATION]

Gubbins, D. 1990: *Seismology and plate tectonics*. Cambridge: Cambridge University Press. **Lay, T. and Wallace, T.C.** 1995: *Modern global seismology*. New York: Academic Press.

seismometer An instrument that detects SEISMIC WAVES, for example generated by an EARTHQUAKE. A seismometer works on the principle of a pendulum anchored to the solid Earth: if the Earth moves, the movement of the pendulum lags behind. A recording instrument, or *seismograph*, amplifies ground movements and records them in analogue or digital form to produce a *seismogram*, or plot of vibrations against time. *DNT*

selection pressure The effectiveness of NATURAL SELECTION in modifying the genetic composition of a population. *KDB*

selective cutting A felling operation that only removes a proportion (usually selected by size and/or

species) of the trees in a block of forest. It is also termed *selective logging*. *NDB*

[See also CLEAR CUTTING, HARVESTING]

self-organisation The tendency of complex SYSTEMS to form regular patterns spontaneously by internal (*endogenous*) processes without control by external (*exogenous*) environmental factors. Examples include periglacial PATTERNED GROUND, ripple patterns on beaches and the branching patterns of erosional GULLY systems. *JAM*

[See also AUTOGENIC CHANGE, CHAOS]

Kauffman, S. 1995: *At home in the universe: the search for laws of self-organization and complexity*. Oxford: Oxford University Press. **Phillips, J.D.** 1995: Self-organization and landscape evolution. *Progress in Physical Geography* 19, 309–321. **Werner, B.T. and Hallet, B.** 1993: Numerical simulation of self-organized stone stripes. *Nature* 361, 142–145.

semi-arid A DRYLAND environment defined by the United Nations Environment Program (UNEP) (1991) on the basis of the Thornthwaite index of *available humidity*, where the ratio of annual precipitation to potential evapotranspiration averages 0.21–0.50. Using this definition, semi-arid environments cover 17.7% of the world's land surface. Semi-arid regions are usually characterised by adequate moisture for the production of livestock forage during part of the year; dryland crop production is successful in some years. *MLC*

[See also DESERTIFICATION, DROUGHT]

Beaumont, P. 1993: *Drylands: environmental management and development.* London: Routledge. **Heath, M.E., Barnes, R.F. and Metcalfe, D.S.** 1985: *Forages: the science of grassland agriculture.* Ames, IA: Iowa State University Press. **Skujins, J. (ed.)** 1991: *Semiarid lands and deserts: soil resources and reclamation.* New York: Marcel Dekker. **United Nations Environment Program** 1991: *Status of desertification and implementation of the United Nations plan of action to combat desertification* [GCSAS.III/3]. Nairobi: UNEP.

semi-natural vegetation Vegetation communities that have been affected by human activities but still retain some natural features and are mostly composed of NATIVE or indigenous species. Vegetation communities range from those that have not been changed by humans to those that are entirely the product of human activities and/or are managed intensively. This spectrum is represented by NATURAL to ARTIFICIAL. Alternatively, Peterken (1993) recognises four related concepts: (1) the *original natural state* before the impact of humans; (2) *present naturalness*, the state that would exist if humans had not become a significant factor; (3) *future naturalness*, the state that would appear if human activities were removed; and (4) *potential naturalness*, the state that would appear instantly if human activities were completely removed. *IFS*

[See also CLIMAX VEGETATION, POTENTIAL NATURAL VEGETATION, NATURAL VEGETATION, SECONDARY WOODLAND]

Bennett, K.D. 1989: A provisional map of forest types for the British Isles 5000 years ago. *Journal of Quaternary Science* 4, 141–144. **Peterken, G.** 1993: *Woodland conservation and management*, 2nd edn. London: Chapman and Hall.

sensitive clays Deposits of CLAY that undergo a loss of strength upon disturbance, causing slope failure and MASS MOVEMENTS. Those that suffer an extreme loss of strength are described as *quick clays*. The reduction in strength is caused by a breakdown of a metastable packing of the clay particles. Unlike the phenomenon of THIXOTROPY, full strength is not regained. Well known sensitive and quick clay deposits include GLACIOLACUSTRINE and GLACIOMARINE DEPOSITS in Scandinavia and Canada. *GO*

Aas, G. 1981: Stability of natural slopes in quick-clays. *Norwegian Geotechnical Institute Bulletin* 135. **Lefebvre, G., Leboeuf, D., Hornych, P. and Tanguay, L.** 1992: Slope failures associated with the 1988 Saguenay earthquake, Quebec, Canada. *Canadian Geotechnical Journal* 29, 117–130. **Locat, J., Lefebvre, G. and Ballivy, G.** 1984: Mineralogy, chemistry and physical-properties interrelationships of some sensitive clays from eastern Canada. *Canadian Geotechnical Journal* 21, 530–540.

sensitivity (1) The magnitude of the response of a SYSTEM to a PERTURBATION, DISTURBANCE or any external, environmental influence. This may depend on the nature and magnitude of the external influence and the RESISTANCE and complexity of the system. (2) The strength of the dependence of one quantity on another. High sensitivity indicates that there is a strong dependence and that small changes in the underlying quantity will produce large changes in the other. (3) *Instrument sensitivity* refers to the strength of signal that can be detected. *PJS/JAM*

[See also LANDSCAPE SENSITIVITY, RELAXATION, RESILIENCE, RESISTANCE, SENSITIVITY EXPERIMENT]

Eybergen, F.A. and Imeson, A.C. 1989: Geomorphological processes and climatic change. *Catena* 16, 307–319.

sensitivity experiments In the context of *modelling*, sensitivity concerns the scale of response of the model to a perturbing influence. It defines the degree to which the output from the model is sensitive to both the equations and PARAMETER values used in the model. Thus, sensitivity experiments, or *sensitivity analyses*, may be conducted, for example with CLIMATIC MODELS, to examine the possible effects of particular FORCING FACTORS by altering the forcing factors one at a time. *JAM/AHP*

[See also LANDSCAPE SENSITIVITY]

Broccoli, A.J. and Manabe, S. 1987: The influence of continental ice, atmospheric CO_2 and land albedo on the climate of the last glacial maximum. *Climate Dynamics* 1, 87–99. **Cess, R.D., Zhang, M.H., Potter, G.L. et al.** 1993: Uncertainties in carbon dioxide radiative forcing in atmospheric general circulation models. *Science* 262, 1252–1255. **Hansen, J., Lacis, A., Rind, D. et al.** 1984: *Climate sensitivity: analysis of feedback mechanisms* [Geophysical Monograph 29]. Washington, DC: American Geophysical Union, 130–163.

sequence stratigraphy A method of interpreting STRATIGRAPHY using packages of SEDIMENTS or SEDIMENTARY ROCKS bounded by UNCONFORMITIES, developed during the AD 1970s by geologists working for the oil company Exxon from relationships seen in subsurface SEISMIC STRATIGRAPHY. A *sequence* or *depositional sequence* is a stratigraphical unit bounded at its top and base by unconformities 'or their correlative conformities'.

Central to sequence stratigraphy is the concept of ACCOMMODATION SPACE, particularly in shallow marine environments where SEA LEVEL places a limit on the level of sediment accumulation. In such environments, accommodation space is provided by a rise in RELATIVE SEA LEVEL (although this need not be a *eustatic* change – see TRANSGRESSION). Thus a sequence represents sediment that accumulated during an episode of relative sea-level rise and the bounding unconformities are attributed to intervening periods of relative sea-level fall. A lower-order cycle of sea-level change within a sequence is a *parasequence*. These changes in relative sea level are used to divide up the stratigraphy of a SEDIMENTARY BASIN. If the sea-level changes are eustatic in character, then sequences are significant in terms of global CORRELATION, and the sequence stratigraphy approach has been integrated with BIOSTRATIGRAPHY and MAGNETOSTRATIGRAPHY to construct a curve of *global eustasy* for PHANEROZOIC time (see Figure).

A depositional *sequence* is subdivided into *systems tracts* associated with specific stages in the cycle of sea-level rise and fall, such as *highstand, transgressive* and *lowstand systems tracts*. A surface of marine transgression is called a *flooding surface* and the change from transgression to REGRESSION (PROGRADATION) or AGGRADATION represents the *maximum flooding surface* (see HIGHSTAND). Some workers prefer to use maximum flooding surfaces as sequence boundaries, but this differs from the 'classical' Exxon model for sequence stratigraphy.

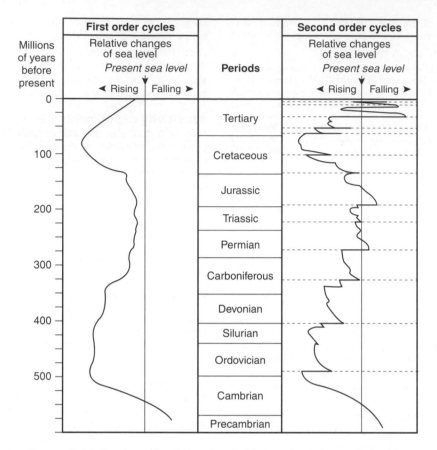

sequence stratigraphy *Curves of global eustasy, showing cycles of global sea-level change derived from sequence stratigraphical analysis (after Vail et al. in Payton, 1977, p. 84)*

Sequence stratigraphy has its proponents and opponents. A major objection is the emphasis placed by 'purists' on the eustatic control of sequences and their global significance. In some cases the sea-level curve derived from the correlation of sequences has been used to recognise sequences, which involves circular reasoning. Many cases have been documented of relative sea-level changes controlled by regional or local TECTONIC effects (see ALLOCYCLIC CHANGE). In addition, the profusion of jargon that accompanies sequence stratigraphy in some cases clouds rather than clarifies issues (see FACIES SEQUENCE).

Regardless of the global or regional significance of sequences, sequence stratigraphy has driven major advances in SEDIMENTOLOGY. It has focused research on the causes of ENVIRONMENTAL CHANGE in the GEOLOGICAL RECORD and helped to reunite sedimentology and stratigraphy in the analysis of sedimentary basins. Regional unconformities can be used as an important tool in correlation, even in areas where it is unlikely that they bound sequences of eustatic origin. Sequence stratigraphy has led to an enhanced appreciation of the effects of sea-level changes on sediment distribution patterns. With a fall in sea level, for example, part of the CONTINENTAL SHELF becomes exposed, lower reaches of rivers become entrenched and the eroded, relatively coarse sediment is supplied to offshore environments. A rise in relative sea level leads to drowning of the lower reaches of rivers as *estuaries* where sediment is trapped, starving offshore environments of sediment. Thus the effects of sea-level changes can be identified in environments remote from the coast, including DESERTS, rivers and deep-sea SUBMARINE FANS. *GO*

Catuneanu, O. and Eriksson, P.G. 1999: The sequence stratigraphic concept and the Precambrian rock record: an example from the 2.7–2.1 Ga Transvaal Supergroup, Kaapvaal craton. *Precambrian Research* 97, 215–251. **Emery, D. and Myers, K.J. (eds)** 1996: *Sequence stratigraphy.* Oxford: Blackwell Science. **Galloway, W.E.** 1989: Genetic stratigraphic sequences in basin analysis. 1: Architecture and genesis of flooding-surface bounded depositional units. *American Association of Petroleum Geologists Bulletin* 73, 125–142. **Hampson, G.J., Elliott, T. and Davies, S.J.** 1997: The application of sequence stratigraphy to Upper Carboniferous fluvio-deltaic strata of the onshore UK and Ireland: implications for the southern North Sea. *Journal of the Geological Society, London* 154, 719–733. **Havholm, K.G. and Kocurek, G.** 1994: Factors controlling aeolian sequence stratigraphy: clues from super bounding surface features in the Middle Jurassic Page Sandstone. *Sedimentology* 41, 913–934. **Hesselbo, S.P. and Parkinson, D.N. (eds)** 1996: *Sequence stratigraphy in British geology* [*Special Publication* 103]. London: Geological Society. **Miall, A.D.** 1997: *The geology of stratigraphic sequences.* Berlin: Springer. **Payton, C.E. (ed.)** 1977: *Seismic stratigraphy – applications to hydrocarbon exploration* [*Memoir* 26]. Tulsa, OK: American Association of Petroleum Geologists. **Reading, H.G. (ed.)** 1996: *Sedimentary environments: processes, facies and stratigraphy,* 3rd edn. Oxford: Blackwell Science. **Vincent, S.J., Macdonald, D.I.M. and Gutteridge, P.** 1998: Sequence stratigraphy. In Doyle, P. and Bennett, M.R. (eds), *Unlocking the stratigraphical record: advances in modern stratigraphy*. Chichester: Wiley, 299–350. **Walker, R.G. and James, N.P. (eds)** 1992: *Facies models: response to sea level change*. St John's, Newfoundland: Geological Association of Canada. **Wilgus, C.K., Hastings, B.S., Kendall, C.G.St.C. et al. (eds)** 1988: *Sea level changes: an integrated approach* [*Special Publication* 42]. Tulsa, OK: Society of Economic Paleontologists and Mineralogists.

seral community A transitional community during ECOLOGICAL SUCCESSION. *LRW*

[See also CLIMAX VEGETATION, SERE]

sere The entire sequence of developmental stages during ECOLOGICAL SUCCESSION from pioneer stage to CLIMAX VEGETATION. Modern usage usually does not imply a deterministic process to a stable endpoint through a predictable sequence of SERAL COMMUNITIES. Seral communities are simply considered transitional phases in any sequence of vegetation change. Seres can be classified by the environmental context: for example, *xeroseres* are dry, *hydroseres* are wet; *lithoseres* occur on bedrock, *psammoseres* on sand and *haloseres* in saline habitats. *LRW*

Walker, D. 1970: Direction and rates in some British postglacial hydroseres. In Walker, D. and West, R.G. (eds), *Studies in the vegetation history of the British Isles*. Cambridge: Cambridge University Press, 117–139.

seriation A RELATIVE-AGE DATING technique used by archaeologists to establish CHRONOLOGY between assemblages of ARTEFACTS. Assemblages are arranged in a series according to their similarities: serial order is used as a surrogate for order in time. *Frequency seriation* relies on the proportional abundance (frequency) of particular styles within the assemblage. This assumes that there is a systematic pattern in the adoption and decline in popularity of, for example, a pottery style through time. *JAM*

Carver, M.O.H. 1985: Theory and practice in urban pottery seriation. *Journal of Archaeological Science* 12, 353–366. **Harding, A.** 1999: Establishing archaeological chronologies. In Barker, G. (ed.), *Companion encyclopedia of archaeology*. Vol. 1. London: Routledge, 182–221. **Robinson, W.S.** 1951: A method for chronologically ordering archaeological deposits. *American Antiquity* 16, 293–301.

series A subdivision of a SYSTEM in CHRONOSTRATIGRAPHY. The base of a series is defined at a *boundary stratotype*. *LC*

service industries The progression of global economic integration has been accompanied by a gradual expansion of service employment since the late AD 1960s to a degree that tertiary activities now dominate all advanced postindustrial economies. Since that time, service employment has grown typically by rates of 40–80%, so that by 1990 service employment averaged 62% for the countries of the Organisation for Economic Co-operation and Development (OECD). Services now dominate employment in the USA and Canada to the greatest degree (70%), and the UK, Australia, Norway, Belgium and The Netherlands are close behind (69%). This has been accompanied by a higher proportion of females in the labour force.

The traditional service sector comprises the personal and public services provided by both commercial organisations and central and local government. The most numerous are the retail outlets and associated activities that are usually provided for individual customers drawn from relatively local hinterlands. Closely related to these are the distributional activities of wholesaling, warehousing and transport, although here services are provided mainly for companies and institutions. The provision of medical services and the public utilities (from schools to leisure centres, government administration and the emergency services) usually involve a much greater degree of public-sector control.

The period since the late 1960s, however, has seen the rapid growth of a wide range of specialised producer services required to support both the processes involved in economic GLOBALISATION and the development of high-technology industry. Foremost amongst these are banking, insurance, legal advice, communications, research and development, market research and advertising. The internationalisation of the financial markets and MANUFACTURING INDUSTRY and the trading of goods and services has created heavy demands for professional skills in business. Growth of such services has led to the recognition of two categories of services in additional to the 'tertiary' definition: namely the *quaternary sector* (including transactional activities) and the *quinary services* (focused on innovation).

The increase in service employment in the advanced economies has now become a more powerful agent of environmental change than the more traditional activities of agriculture, EXTRACTIVE INDUSTRY and manufacturing. This is expressed, for example, in terms of urban environmental restructuring in the central city and suburbs, traffic generation and energy consumption in order to accommodate the demands of an increasingly affluent society, and rapid technological innovation. On a global scale, both the trajectory of future environmental change in the natural environment and the human response to it are largely under the control of the service industries in terms of decision making, financing and technological innovation. *CJT*

[See also GLOBALISATION, TECHNOLOGICAL CHANGE, URBAN AND RURAL PLANNING]

Gottmann, J. 1983: *The coming of the transactional city*. Baltimore, MD: Institute for Urban Studies, University of Maryland. **Grübler, A.** 1998: *Technology and global change*. Cambridge: Cambridge University Press. **Herbert, D.T. and Thomas, C.J.** 1997: *Cities in space. City as place*. London: David Fulton. **Marshall, J.N. and Wood, P.A.** 1995: *Services and space: key aspects of urban and regional development*. London: Longman.

set-aside schemes The practice of removing a portion of agricultural land from production in return for grants or other incentives in order to protect the environment from POLLUTION or to conserve BIODIVERSITY. Set-aside schemes were initiated in the European Community in 1988, mainly to combat *nitrate pollution of groundwater* and *overproduction* of grain. Since 1993, European farmers have set aside 15% of their arable land for non-arable, non-polluting use, in return for payments based on the estimated average yield of that land. The United States' equivalent of set-aside, the *Conservation Reserve Program* (initiated 1985) seeks to combat soil erosion, but has not been very popular. A risk with set-aside is that the remaining land may be more intensively used, negating the benefit. A set-aside area may remain at one site or be moved around on an annual or other periodic cycle (*rotational set-aside*). *CJB*

Conway, G.R. and Pretty, J.N. 1991: *Unwelcome harvest: agriculture and pollution*. London: Earthscan.

sewage history Sewage inputs to water bodies often contain considerable amounts of environmental indicators, such as HEAVY METALS. These substances are incorporated into aquatic sediments and variations in their concentration may reflect the sewage POLLUTION HISTORIES of lakes and FJORDS. *DZR*

Paetzel, M., Schrader, H. and Croudace, I. 1994: Sewage history in the anoxic sediments of the fjord Nordåsvannet, western Norway: (1) Dating and trace-metal accumulation. *The Holocene* 4, 290–298.

shale A SEDIMENTARY ROCK with a dominant GRAIN-SIZE of MUD grade (i.e. a MUDROCK) that has a FISSILITY of sedimentary (i.e. non-deformational) origin. It is sometimes contrasted with MUDSTONE, where the latter is used to imply a non-fissile mudrock. *TY*

[See also BLACK SHALE, OIL SHALE]

Potter, P.E., Maynard, J.B. and Pryor, W.A. 1980: *The sedimentology of shale: study guide and reference sources*. New York: Springer.

shape In the context of a rock particle, an aspect or aspects of its external morphology. There is disagreement in the literature regarding how this term and that of *form* should be used. Although typical English-language dictionary definitions indicate that the two terms are synonymous, they have assumed specific but different meanings amongst geologists and geomorphologists when rock particles are described. Some consider that shape should be used to mean the dimensional relations of a particle – based on its mutually orthogonal long (*a*), intermediate (*b*) and short (*c*) dimensions or axes – ROUNDNESS and surface texture (e.g. STRIATIONS, roughness, QUARTZ GRAIN SURFACE TEXTURES) and that use of the term form should be restricted to the dimensional relations. An alternative view is that form is the all-embracing term that encompasses the concepts of shape (i.e. how closely the particle resembles a regular solid such as a disc, blade, rod or sphere), SPHERICITY, roundness and surface texture. Indices based on the dimensional relations of rock particles – e.g. *flatness* $((a + b)/2c)$ and *slabbiness* (c/a) – can provide useful means of reconstructing depositional 'histories' of SEDIMENTS. *RAS*

Barrett, P.J. 1980: The shape of rock particles, a critical review. *Sedimentology* 27, 291–303. Benn, D.I. and Ballantyne, C.K. 1994: Reconstructing the transport history of glacigenic sediments: a new approach based on the co-variance of clast form indices. *Sedimentary Geology* 91, 215–227. Goudie, A.S. (ed.) 1990: *Geomorphological techniques*, 2nd edn. London: Unwin Hyman, 121–127. Griffiths, J.C. 1967: *Scientific method in the analysis of sediments*. New York: McGraw-Hill. Whalley, W.B. 1972: The description and measurement of sedimentary particles and the concept of form. *Journal of Sedimentary Petrology* 42, 961–965.

shape box A device, comprising a box constructed from durable material (e.g. wood, metal) with two sides and the top removed, for determining the lengths of the three primary axes of CLASTS to the nearest millimetre. With three scales and a movable gauge, the axes can be measured without having to move the clast. The shape box allows unambiguous and reproducible measurements of the axes, which may not always be possible with alternative devices (e.g. steel tape, callipers), particularly for clasts with highly irregular outlines. *RAS*

[See also SHAPE]

Shakesby, R.A. 1979: A simple device for measuring the primary axes of clasts. *British Geomorphological Research Group, Technical Bulletin* [*Shorter Technical Methods* III], 24, 11–13.

shear strength The strength of soil or sediment is provided by the cohesive forces between particles. Shear strength is a measure of the force required to overcome the friction provided by the cohesive forces and is usually measured using the triaxial compression method, in a *shearbox* or in the field by a *penetrometer* or an instrument with vanes, which is inserted into the soil and rotated. Measurement is recorded in kPa (kilopascals) and ranges up to 30 kPa in clays, but can be nil in cohesionless sands and saturated clays. *EMB*

Atkinson, J.H. and Bramsby, P.L. 1978: *The mechanics of soils*. London: McGraw-Hill. Whalley, W.B. 1976: *Properties of materials in geomorphological explanation*. Oxford: Oxford University Press.

shell pavement A coastal LAG DEPOSIT of shells and shell fragments produced by the selective erosion of fine sedimentary MATRIX by waves, tidal currents or aeolian processes. *JAM*

Carter, R.W.G. 1976: Formation, maintenance, and geomorphological significance of an aeolian shell pavement. *Journal of Sedimentary Petrology* 46, 418–429.

shifting cultivation An agricultural system, also known as *slash-and-burn* or *swidden*, in which patches of natural forest are cleared for the temporary cultivation of crops. It commonly involves complex management of cultivated and natural plants, often characterised by intercropping and requiring detailed traditional knowledge. After a number of harvests, plots are left to develop into *secondary forest*, which provides other useful non-domesticated resources, whilst a new area is cleared for cropping elsewhere. The length of time for which a plot is cultivated will depend on rates of SOIL EXHAUSTION and colonisation by RUDERAL plants. For many communities, constraints on the area of forest available, caused by LOGGING, encroachment by outside settlers and *population growth*, mean that there is insufficient time for natural and managed *succession*. As a consequence, shifting cultivation is often accused of being a prime cause of DEFORESTATION. *NDB*

[See also AGRICULTURAL IMPACTS ON SOILS, HARVESTING]

Angelsen, A. 1995: Shifting cultivation and 'deforestation': a study from Indonesia. *World Development* 23, 1713–1729. Uhl, C. 1987: Factors controlling succession following slash-and-burn agriculture in Amazonia. *Journal of Ecology* 75, 373–383.

ship log book records Daily observations of wind direction, wind force and other weather phenomena made by ships' officers to help in navigation. Many thousands of log books exist, the earliest of which date from the mid-seventeenth century. *DAW*

[See also DOCUMENTARY EVIDENCE]

Catchpole, A.J.W. 1992: Hudson's Bay Company ships logbooks as sources of sea ice data, 1751–1870. In Bradley, R.S. and Jones, P.D. (eds), *Climate since 1500*. London: Routledge, 17–39. **Oliver, J. and Kington, J.** 1970: The usefulness of ships' log books in the synoptic analysis of past climates. *Weather*, 25, 520–528.

shore displacement   Either a shoreward or seaward change in the position of the shoreline, usually associated with human action and caused by, for example, EROSION or LAND RECLAMATION.   *RAS*

shore zone   The part of the coast that extends from the breaker zone landward across the shoreline to the upper limits of the *backshore* (see Figure). Its landward limit is often defined as the maximum height reached by storm waves and is often at the base of the stable inner dunes. Included are the *breaker, surf* and *swash zones* as part of the *nearshore* zone and the backshore zone, which is the upper part of the beach. Specific features of the shore zone include *berms* (surfaces formed by wave deposition) and *foredunes*, which are the first dunes formed, usually in rows, above the inner berm. The *coastal zone* is a broader concept and more arbitrary; it extends farther inland but is still within reach of 'coastal influences'.   *HJW*

Viles, H. and Spencer, T. 1995: *Coastal problems*. London: Edward Arnold.

shoreline   The shifting line representing the water's edge along a shore.   *HJW*

[See also COASTLINE]

shrub   A woody plant that is smaller than a tree. *JAM*

[See also SHRUBLAND]

shrubland   A vegetation and landscape type characterised by woody plants that are smaller than trees (SHRUBS). They include HEATHLANDS and Mediterranean-type shrublands, the latter characterised by small XEROPHYTIC and SCLEROPHYLLOUS bushes, characteristic of seasonally dry, Mediterranean-type climates and known as *chaparral* in California, *maquis, macchia, garrigue* or *matorral* in MEDITERRANEAN regions and *mallee* in Australia.   *MLC*

[See also MEDITERRANEAN-TYPE VEGETATION, SCRUB]

Di Castri, F., Goodall, D.W. and Specht, R.L. (eds) 1981: *Mediterranean-type shrublands* [*Ecosystems of the World*. Vol. 11]. Amsterdam: Elsevier. **Specht, R.L. (ed.)** 1979: *Heathlands and related shrublands* [*Ecosystems of the World*. Vol. 9A]. Amsterdam: Elsevier.

SI units   The international system of units (*Système International d'Unités*), derived from the mks (metre, kilogram, second) system and officially adopted in AD 1960. SI units are recommended for all scientific, industrial and commercial work. There are seven base SI quantities (see Table 1).

The two additional supplementary SI units are the radian (rad; plane angle) and the steradian (sr; solid angle). Any unit formed from two or more base units is termed 'derived' and those most relevant to the environmental change context are given in Table 2.

Quantities that are either very small or very large in relation to the base unit may be indicated by prefixes (Table 3), although those that are not multiples of three (h, da, d, c) are rarely used in scientific work.

In some cases, however, non-SI units are still used, for example in chemistry or soil science (equivalents; eq) and forestry (hectares rather than m²). Other common non-SI units are minute (min), hour (h), day (d) and degree (°).   *DZR*

SI units (1) *SI base quantities, their corresponding units and symbols*

Physical quantity	Name of unit	Symbol
Length	metre	m
Mass	kilogram	kg
Time	second	s
Electric current	ampère	A
Thermodynamic temperature	kelvin	K
Luminous intensity	candela	cd
Amount of substance	mole	mol

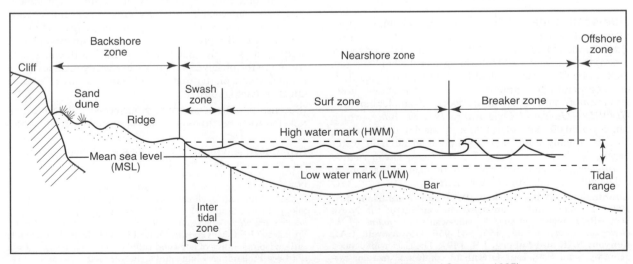

shore zone *The shore zone and its subdivisions (based on Viles and Spencer, 1995)*

SI units (2) *Some SI derived quantities and their units*

Derived quantity	Unit name	Symbol	Base units
Area	square metre		m^2
Volume	cubic metre		m^3
Density	kilogram per cubic metre		$kg\ m^{-3}$
Velocity	metre per second		$m\ s^{-1}$
Acceleration	metre per second per second		$m\ s^{-2}$
Energy	joule	J	Nm
Power	watt	W	$J\ s^{-1}$
Radioactivity	bequerel	Bq	s^{-1}

SI units (3) *Prefixes used with SI units*

Factor	Name	Symbol
10^{18}	exa-	E
10^{15}	peta-	P
10^{12}	tera-	T
10^{9}	giga-	G
10^{6}	mega-	M
10^{3}	kilo-	k
10^{2}	hecto-	h
10	deca-	da
10^{-1}	deci-	d
10^{-2}	centi-	c
10^{-3}	milli-	m
10^{-6}	micro	μ
10^{-9}	nano-	n
10^{-12}	pico-	p
10^{-15}	femto-	f
10^{-18}	atto-	a

Kaye, G.W.C. and Laby, T.H. 1995: *Tables of physical and chemical constants*, 16th edn. Harlow: Longman. **Royal Society of London** 1975: *Quantities, units and symbols*, 2nd edn. London: Royal Society.

sidereal date An age estimate based on the *sidereal year*, the exact period it takes the Earth to complete an orbit of the Sun. *DAR*

[See also CALENDAR/CALENDRICAL AGE/DATE]

side-scan sonar Ship-towed echo-sounding equipment used to survey the sea floor. The backscatter provides detailed images comparable to AERIAL PHOTOGRAPHY or SATELLITE REMOTE SENSING images of the land surface. Surveys have proved invaluable in investigating sea-floor processes and environments such as DEBRIS FLOWS, *deep-sea channels* and SUBMARINE FANS. Examples include *GLORIA (Geological Long Range Inclined Asdic)*, which can cover 20 000 km² of sea floor in one day and has a resolution roughly equivalent to a football pitch in water depths of about 5 km, and *TOBI (Towed Ocean Bottom Instrument)*, which can cover up to 600 km² of sea-floor in a day with a resolution of a few metres. *GO*

Cunningham, A.P., Barker, P.F. and Tomlinson, J.S. 1998: Tectonics and sedimentary environment of the North Scotia Ridge region revealed by side-scan sonar. *Journal of the Geological Society, London* 155, 941–956. **Dowdeswell, J.A., Kenyon, N.H. and Laberg, J.S.** 1997: The glacier-influenced Scoresby Sund Fan, East Greenland continental margin: evidence from GLORIA and 3.5 kHz records. *Marine Geology* 143,

207–221. **Masson, D.G., Canals, M., Alonso, B. *et al*.** 1998. The Canary Debris Flow: source area morphology and failure mechanisms. *Sedimentology* 45, 411–432.

sieve deposition The rapid deposition of sediment triggered by flow infiltration into the underlying substrate. It has been noted in particular on ALLUVIAL FANS and COLLUVIAL FANS in both ARIDLANDS and alpine environments, when water drains through coarse-grained material. *JAM*

Krainer, K. 1988: Sieve deposition on a small modern alluvial fan in the Lechtal Alps (Tyrol, Austria). *Zeitschrift für Geomorphologie NF 32*, 289–298.

signal strength In PALAEOCLIMATOLOGY, a measure of the influence of a predefined FORCING FACTOR as it acts upon a specified process. In most applied situations, the strength of this signal can only be measured empirically, for example by simple correlation or regression of the recorded variability in the presumed forcing against a record of process changes over time.

A good example is the use of the concepts of signal and noise in DENDROCLIMATOLOGY. Here 'signal' can be defined in many different ways, depending on the process of interest. In each of these definitions, the residual variance in tree growth (i.e. that not representing the influence of the signal) is defined as '*noise*'. Comparing one against the other provides a measure of signal strength. Hence in tree-ring studies, 'signal' has been defined as the measurable common variability contained in a number of contemporaneous tree-ring records. This may be quantified in terms of the mean correlation coefficient $\bar{r}$, obtained when the series are intercompared. The residual variance $(1 - \bar{r})$ is defined as 'noise'. By averaging multiple records, from different trees, the common variability is unaffected, while the noise cancels in proportion to the number of series being averaged. Hence, it is possible to quantify the signal strength in a chronology time series in relation to the residual noise. This is only one definition of signal (*statistical signal*). It differs from the concept of a *theoretical signal*. To continue with the dendroclimatic example, this theoretical signal might be defined as a measure of the strength of association between some specific measure of CLIMATIC VARIABILITY, such as mean summer temperature, and the variability of the mean chronology (i.e. the variance explained in a regression of CHRONOLOGY on climate series). Then the uncorrelated (non-climatic) variance would represent noise. In environmental-change studies, it is likely that many possible influences contribute to the observed variability in many processes and it is not a simple matter to distinguish or measure all of their effects unambiguously. *KRB*

[See also DENDROCLIMATOLOGY, RESPONSE FUNCTION, SIGNAL-TO-NOISE RATIO, STANDARDISATION, TRANSFER FUNCTION]

Briffa, K.R. and Jones, P.D. 1990: Basic chronology statistics and assessment. In Cook, E.R. and Kairiukstis, L.A. (eds), *Methods of dendrochronology. Applications in environmental sciences*. Dordrecht: Kluwer, 137–152. **Briffa, K.R.** 1995: Interpreting high-resolution proxy climate data – the example of dendroclimatology. In von Storch, H. and Navarra, A. (eds), *Analysis of climate variability: applications of statistical techniques*. Berlin: Springer, 77–94. **McCarroll, D. and Pawellek, F.** 1998: Stable carbon isotope ratios of latewood cellulose in *Pinus sylvestris* from northern Finland: variability and signal strength. *The Holocene* 8, 675–684. **Wigley, T.M.L., Jones, P.D. and Briffa, K.R.**

1987: Detecting the effects of acidic deposition and CO_2 fertilization on tree growth. In Kairiukstis, L., Bednarz, Z. and Filiksik, E. (eds), *Methods of dendrochronology – I. Proceedings of the task force meeting on methodology of dendrochronology: East/West approaches.* Warsaw: Polish Academy of Sciences – Systems Research Institute, 239–254.

signal-to-noise ratio (SNR)   The ratio of the power (brightness) of a signal to the noise power in the absence of the signal. It is usually measured in dB units.   *PJS*

significance level   The probability level at which it is agreed *a priori* that the null hypothesis of 'non association' or 'no difference' would be rejected in favour of the alternative hypothesis of non-zero 'association' or 'difference'. It is commonly set at $\alpha = 0.05$, 0.01 or 0.001. Hypothesis testing with small significance levels is less likely to reject wrongly the null hypothesis when it is true (*type-I error*).   *HJBB*

silcrete   A siliceous DURICRUST, commonly composed of more than 95% silica oxide. Silcrete occurs in both humid and arid TROPICAL environments, tends to develop above or at the WATER TABLE and is widespread in Australia and South Africa. It can become more than 5 m thick and is resistant to erosion.   *MAC*

[See also CALCRETE, FERRICRETE, GYPCRETE, SARSEN]

Summerfield, M.A. 1983: Silcrete. In Goudie, A.S. and Pye, K. (eds), *Chemical sediments and geomorphology: precipitates and residua in the near-surface environment.* London: Academic Press, 59–91. **Twidale, C.R.** 1983: Australian laterites and silcretes: ages and significance. *Revue de Geomorphologie Dynamique et Geographie Physique* 24, 35–45.

silica analysis   Silica (in the form of SiO_2) is the major component of Earth's crust. AMORPHOUS silica (opal-A) precipitates from natural waters inorganically as in SILCRETES, geyserites, oncoids and opal deposits. It is also present in large amounts in VOLCANIC ASH. Organic (BIOGENIC) silica is secreted by aquatic organisms (DIATOMS, RADIOLARIANS, silicoflagellates, chrysophytes, sponge spicules) and terrestrial plants (PHYTOLITHS). Siliceous MICROFOSSILS are excellent indicators of changing marine and lacustrine environmental conditions and are most abundant in nutrient-rich water. Phytoliths may record the degree of EROSION caused by HUMAN IMPACT. The amount of biogenic and INORGANIC silica in sediments can be determined by sequential leaching in 0.1 M Na_2CO_3 at 85°C for >5 h. DIATOMS and chrysophyte cysts are least resistant, phytoliths and sponge spicules occupy an intermediate position, while volcanic ash and crystalline quartz grains are most resistant. Other methods include NaOH/HF digestion or the determination of biogenic silica by infrared spectrometry.   *UBW*

[See also ACIDITY, DIATOM ANALYSIS, CHRYSOPHYTE CYST ANALYSIS, ENVIRONMENTAL INDICATOR, MARINE SEDIMENT CORES]

Ragueneau, O. and Treguer, P. 1994: Determination of biogenic silica in coastal waters: applicability and limits of the alkaline digestion method. *Marine Chemistry* 45, 43–51. **Thunell, R.C., Tappa, E. and Anderson, D.M.** 1995: Sediment fluxes and varve formation in Santa Barbara Basin, offshore California. *Geology* 23, 1083–1086.

silicate minerals   Compounds of silicon and oxygen, the dominant group of minerals in the Earth's CRUST, particularly in IGNEOUS ROCKS. Examples include QUARTZ and FELDSPAR.   *GO*

silt   A SEDIMENT or particles of GRAIN-SIZE between 0.004 and 0.063 mm.   *TY*

[See also CLAY, MUD]

siltation   The settling-out of SILT (and other fine particles) in a lake or other water body, as exemplified by the infilling with fine sediment of lakes, reservoirs, low-gradient water courses and estuaries. It is a natural process that has often been accentuated by anthropogenic effects (notably enhanced EROSION) in the landscape.   *MAB*

[See also AGGRADATION, SOIL EROSION]

Kraft, J.C., Kayan, I. and Erol, O. 1980: Geomorphic reconstructions in the environs of ancient Troy. *Science* 209, 776–782.

siltstone   A SEDIMENTARY ROCK that is not FISSILE, with a dominant GRAIN-SIZE of SILT grade (0.004–0.063 mm).   *TY*

[See also MUD, MUDROCK, MUDSTONE, SHALE]

Silurian   A SYSTEM of rocks and a PERIOD of geological time from 442 to 416 million years ago.   *GO*

[See also GEOLOGICAL TIMESCALE]

silviculture   The cultivation of trees in forests, including the management of natural forests, plantation management and forestry practices. It is essentially applied forest ecology.   *JAM*

[See also ARBORICULTURE, FOREST MANAGEMENT, HORTICULTURE, NEW FORESTRY]

Kostler, J. 1956: *Silviculture.* Edinburgh: Oliver and Boyd. **Smith, D.M., Larson, B.C., Keity, M.J. and Ashton, P.M.S.** 1997: *The practice of silviculture: applied forest ecology.* New York: Wiley.

simulation model   An approach to the mathematical representation of a SYSTEM that is used if the complex structure of the system makes analytical solutions of the underlying mathematical equations intractable. The behaviour of the model simulates the dynamics of the real system, thus helping to understand its driving mechanisms and the interactions of its VARIABLES. One of the main applications of a simulation model is to predict future states of an environmental system, given different possible *scenarios*, so as to provide decision support in ENVIRONMENTAL MANAGEMENT. A sensitivity analysis (see SENSITIVITY EXPERIMENTS) shows the variation of output parameters, depending on the variation of each input variable. The model is more sensitive to variables that cause greater changes in output parameters with smaller variations in input values. Predictions from simulation models are subject to VALIDATION. Data from REMOTE SENSING can be linked to a simulation model as input variables, or to evaluate model predictions. Simulation models are also used for understanding the interactions between electromagnetic radiation and vegetation.   *HB*

[See also MODEL, ENVIRONMENTAL MODELLING, DECISION MAKING, RADIATIVE TRANSFER THEORY]

Bossel, H. and Krieger, H. 1991: Simulation model of natural tropical forest dynamics. *Ecological Modelling* 59, 37–71. **Frohn, R.C., McGwire, K.C., Dale, V.H. and Estes, J.E.** 1996: Using satellite remote sensing analysis to evaluate a socio-economic and ecological model of deforestation in Rondonia, Brazil. *International Journal of Remote Sensing* 17, 3233–3255. **Gerard, F.F. and North, P.R.J.** 1997: Analyzing the effect of structural variability and canopy gaps on forest BRDF using a geometric-optical model. *Remote Sensing of Environment* 62, 46–62.

single large or several small reserves (SLOSS debate)

In an effort to guide the strategic and efficient design of NATURE RESERVES, conservation biologists debated whether an optimal system of reserves should be comprised of a single large reserve or several smaller reserves with the same total area as the large reserve. Drawing on different theories and empirical patterns from ISLAND BIOGEOGRAPHY, population biology and GENETICS, many scientists became embroiled in a tempestuous debate over *SLOSS* during the AD 1970s and 1980s. Many conservation biologists now view this debate as an academic exercise without a single answer. *MVL*

[See also CONSERVATION, CONSERVATION BIOLOGY, MINIMUM CRITICAL SIZE OF ECOSYSTEMS]

Lomolino, M. V. 1994: An evaluation of alternative strategies for building networks of nature reserves. *Biological Conservation* 69, 243–249. **Simberloff, D. S. and Abele, L. G.** 1982: Refuge design and island biogeography theory: effects of fragmentation. *American Naturalist* 120, 41–50. **Whittaker, R.J.** 1998: *Island biogeography: ecology, evolution and conservation.* Oxford: Oxford University Press.

singularity

An interval of distinctive WEATHER persisting for between seven and 14 days, which tends to recur in many years on or about the same date. Apparently statistically significant examples for England (based on data from AD 1873–1961) include early winter STORMS between 12 October and 22 November, renewed winter storms between 27 January and 4 February, ANTICYCLONES between 3 March and 22 March and again between 15 November and 23 November. *JAM*

Lamb, H.H. 1964: *The English climate.* London: English Universities Press.

sink

A natural *store* or *reservoir* for materials involved in BIOGEOCHEMICAL CYCLES. In the CARBON CYCLE, for example, the oceans and the BIOSPHERE are possible CARBON SINKS for some of the carbon originating from FOSSIL FUEL sources. *JAM*

[See also CARBON SEQUESTRATION]

site-catchment analysis

Assessment of the natural resources available for exploitation by the occupants of an archaeological site or settlement. The results of a site-catchment analysis, or *site exploitation territorial analysis*, represent the *potential* of the locality, not an account of the actual use of resources. *JAM*

[See also GEOARCHAEOLOGY, LANDSCAPE ARCHAEOLOGY]

Bailey, G.N. and Davidson, I. 1983: Site exploitation territories and topography: two case studies from Palaeolithic Spain. *Journal of Archaeological Science* 10, 87–115. **Vita-Finzi, C. and Higgs, E.S.** 1970: Prehistoric economy in the Mount Carmel area of Palestine: site catchment analysis. *Proceedings of the Prehistoric Society* 36, 1–37.

site formation processes

The natural processes (*non-cultural formation processes* or *n-transforms*), such as WEATHERING and EROSION, and the *cultural formation processes* (*c-transforms*) which create archaeological sites. The latter include both processes involved in establishing the site in the archaeological record (e.g. site construction and use) and those later processes involved in its modification and destruction (e.g. ploughing and looting). *JAM*

Goldberg, P., Nash, P.T. and Petraglia, M.D. (eds) 1993: *Formation processes in archaeological context* [*Monographs in World Archaeology* 17]. Madison, WI: Prehistory Press. **Schiffer, M.B.** 1987: *Formation processes of the archaeological record.* Albuquerque, NM: University of New Mexico Press. **Smith, H.** 1996: An investigation of site formation processes and geoarchaeological techniques. In Gilbertson, D., Kent, M. and Grattan, J. (eds), *The Outer Hebrides: the last 14,000 years.* Sheffield: Sheffield Academic Press, 195–206.

Sites of Special Scientific Interest (SSSIs)

In the UK, sites designated as important for CONSERVATION because of their scientific value as exceptional or representative examples of HABITATS, geological features or landforms. In 1997, there were 6382 SSSIs in England, Scotland and Wales. In Northern Ireland, *Areas of Special Scientific Interest* (ASSIs), of which there were 117 in 1998, are the equivalent. *JAM*

[See also PROTECTED AREAS, REGIONALLY IMPORTANT GEOLOGICAL AND GEOMORPHOLOGICAL SITES]

Department of the Environment and others 1982: *Code of guidance for Sites of Special Scientific Interest.* London: HMSO. **English Nature** 1999: *SSSIs: what you should know about Sites of Special Scientific Interest.* Peterborough: English Nature. **Gordon, J.E. and Campbell, S.** 1992: Conservation of glacial deposits in Great Britain: a framework for assessment and protection of sites of special scientific interest. *Geomorphology* 6, 89–97.

skewness

The lack of symmetry in a frequency or PROBABILITY DISTRIBUTION. Symmetric distributions have zero skewness. A distribution has positive skewness when it has a long thin tail to the right and has negative skewness when it has a long thin tail to the left. The coefficient of skewness is the third moment about the mean divided by the cube of the standard deviation. *HJBB*

Sokal, R.R. and Rohlf, F.J. 1995: *Biometry.* New York: W.H. Freeman.

slant range

The distance of a target from side-looking RADAR, measured in the plane joining the target and the radar. *PJS*

slope decline

One of three classic models of hillslope development, proposed by W.M. Davis, in which slope angles become progressively reduced through time. *RAS*

[See also DAVISIAN CYCLE OF EROSION, SLOPE EVOLUTION MODELS]

Davis, W.M. 1899: The geographical cycle. *Geographical Journal* 14, 481–504.

slope evolution models

DRAINAGE BASIN slopes in profile can take a variety of forms with concave, convex, rectilinear (straight) and complex (i.e. large variety of slope angles over short distances) elements. Since the development of hillslopes is in large part the result of WEATHERING and EROSION, many geomorphologists have sought to interpret their slope form in terms of change through time. Although there have been variations, three basic concepts have been proposed. In his geographical cycle, W.M. Davis suggested that in humid climates slope gradients declined through time (SLOPE DECLINE). Alternatively, W. Penck proposed that evolution occurred through replacement of steep gradients by gentler ones starting at the base of a hillslope (SLOPE REPLACEMENT). Thus, an initially simple, steep, straight slope is replaced through time by one with two elements: a steep upper slope occupying a decreasing proportion of the profile and a basal gentler slope occupying an increasing proportion of the profile. The third basic concept (PARALLEL SLOPE RETREAT), proposed by L.C. King, is that slopes can remain unchanged in form as they undergo weathering and erosion, so that slope angles are maintained. Considerable debate was generated by these models, but after the mid-1960s the focus of interest shifted to other areas of slope research (e.g. measurement of processes and rates of erosion and weathering). *RAS*

[See also SLOPE ZONATION MODELS]

Young, A. 1972: *Slopes*. Edinburgh: Oliver and Boyd. **Parsons, A.J.** 1988: *Hillslope form*. London: Routledge. **Phillips, J.D.** 1993: Instability and chaos in hillslope evolution. *American Journal of Science* 293, 25–48. **Selby, M.J.** 1993: *Hillslope materials and processes*, 2nd edn. Oxford: Oxford University Press.

slope replacement   One of three classic models of hillslope development, developed by W. Penck, in which the gentler lower slope segment extends at the expense of the upper steep segment. *RAS*

[See also SLOPE EVOLUTION MODELS]

Penck, W. 1924: *Die morphologische Analyse, ein Kapitel der physikalischen Geologie*. Stuttgart: Engelhorns.

slope zonation models   Various models of hillslope form have been devised that identify 'typical' units of a slope profile thought to behave in particular ways under the influence of WEATHERING and EROSION. The *raison d'être* for devising these models is that if the likely behaviour of each unit can be predicted, the possible development of the whole profile might be determined. Examples include the *four-unit slope model* proposed by A. Wood in AD 1942. At its simplest, this model comprises a straight TALUS slope beneath a straight free face, but Wood argued that in humid areas there would also be upper convex and lower concave units. A *five-unit alpine slope model*, proposed by N. Caine in 1974, is similar to Wood's model but adds a 'talus foot' unit between the talus and concave basal units. The addition of this unit reflects various talus-related forms (e.g. ROCK GLACIER) that occur near the bases of talus slopes in alpine PERIGLACIAL zones. Greater complexity, with a *nine-unit slope model*, was proposed by J. Dalrymple, J. Blong and A. Conacher in 1968. This provided subdivision of Wood's model and added an alluvial toe slope, river channel wall and river channel bed at the base of the slopes. *RAS*

[See also SLOPE EVOLUTION MODELS]

Caine, N. 1974: The geomorphic processes of the Alpine environment. In Ives, J.D. and Barry, R.G. (eds), *Arctic and alpine environments*. London: Methuen, 721–748. **Dalrymple, J.B., Blong, R.J. and Conacher, A.J.** 1968: A hypothetical nine-unit land-surface model. *Zeitschrift für Geomorphologie* 12, 60–76. **Small, R.J. and Clark, M.J.** 1982: *Slopes and weathering*.

Cambridge: Cambridge University Press, 69–71. **Wood, A.** 1942: The development of hillside slopes. *Proceedings of the Geologists' Association* 53, 128–140.

slopewash   (1) The process of MASS WASTING, by which rock, soil or sediment is transported down a slope as a result of *sheet erosion*. (2) The material that has been transported by the above process, often consisting of predominantly sandy particles and forming irregular, thin sediment sheets. *SHD*

[See also COLLUVIUM, SOIL EROSION]

slump   A gravity-driven MASS MOVEMENT involving significant internal DEFORMATION of the failed mass, resulting from slope failure in a subaerial or subaqueous setting. In the GEOLOGICAL RECORD slumps are preserved as a deformational SEDIMENTARY STRUCTURE, comprising a sheet of internally folded and contorted material *(slump sheet)* underlain and overlain by undisturbed BEDS. They are a common feature of certain DEPOSITIONAL ENVIRONMENTS including ALLUVIAL FANS, CONTINENTAL SLOPE deposits and SUBMARINE FANS, and *slump fold* geometry and orientation can be used to determine PALAEOSLOPE orientation. Slumps can be initiated by triggers such as rapid sediment loading, SEA-LEVEL fall or EARTHQUAKES and may develop into SEDIMENT GRAVITY FLOWS such as DEBRIS FLOWS or, in a subaqueous setting, TURBIDITY CURRENTS. *GO*

[See also LANDSLIDE]

Lomas, S.A. 1999: A Lower Cretaceous clastic slope succession, Livingston Island, Antarctica: sand-body characteristics, depositional processes and implications for slope apron depositional models. *Sedimentology* 46, 477–504. **Martinsen, O.J.** 1994: Mass movements. In Maltman, A.J. (ed.), *The geological deformation of sediments*. London: Chapman and Hall, 127–165.

smog   Originally coined in AD 1905 by Harold Des Voeux from the words 'smoke' and 'fog' to describe short-term periods of high pollution concentrations characterised by very poor visibility and epitomised by the smogs of Victorian London. The term is now commonly applied to air pollution episodes regardless of whether visibility is reduced to tens of metres or a few kilometres. *DME*

[See also INDUSTRIAL REVOLUTION, PHOTOCHEMICAL SMOG]

Brimblecombe, P. 1987: *The big smoke: a history of air pollution in London since Medieval times*. London: Methuen. **Elsom, D.M.** 1996: *Smog alert: managing urban air quality*. London: Earthscan.

snow avalanche   As snow accumulates on a mountain slope, the normal stress in the SNOWPACK increases with increasing thickness. Strong winds may increase the density in the snowpack. When the stress exceeds the strength, a gravitational FAILURE occurs. The *wind drift* of dry snow is important for snowpack thickness. Snow avalanches are most frequent on slopes of 30–40°, but wet snow avalanches can be triggered on significantly gentler SLOPES. Two types of snow avalanches are distinguished. *Dense snow avalanches* are usually subject to a line failure, leading to a translational slide. A broad slab of shearing snow then descends the slope, leaving well defined crown and flank scarps in the snowpack. Dense snow avalanches are comparable to debris flows and are considered as cohesive, cohesionless or viscoelastic. Their steep-fronted depositional lobes and levéed tracks are similar to those of DEBRIS FLOWS. Dense snow avalanches typically exhibit velocities of 10–30 m s⁻¹. *Powder snowflows* consist of fresh, powdery snow or loose older snow that has lost cohesion (dry granular snow). Powder snowflows are usually subject to a point failure, which expands quickly.

Past snow-avalanche activity can best be reconstructed from HISTORICAL and DOCUMENTARY EVIDENCE, and by the application of RADIOCARBON DATING of ORGANIC debris below and above SNOW AVALANCHE DEPOSITS either in open COLLUVIAL sections or in LAKE SEDIMENTS. *AN*

Blikra, L.H. and Fjeldstad Selvik, S. 1998: Climatic signals recorded in snow avalanche-dominated colluvium in western Norway: depositional facies successions and pollen records. *The Holocene* 8, 631–658. **Grove, J.M.** 1972: The incidence of landslides, avalanches and floods in western Norway during the Little Ice Age. *Arctic and Alpine Research* 4, 131–138. **Jonasson, C.** 1991: Holocene slope processes of periglacial mountain areas in Scandinavia and Poland. *Uppsala, UNGI Rapport* 79, 1–156. **Laternser, M. and Pfister, C.** 1997: Avalanches in Switzerland 1500–1900. In Matthews, J.A., Brunsden, D., Frenzel, B. *et al.* (eds), *Rapid mass movement as a source of climatic evidence for the Holocene* [Paläoklimaforschung. Vol. 19]. Stuttgart: Gustav Fischer, 241–266.

snow avalanche deposits AVALANCHES may be defined as rapid gravitational movements of wet or dry rock debris and/or snow occurring on steep slopes. These rapid MASS MOVEMENTS are also catastrophic or EXTREME EVENTS and large avalanches can move at speeds of 50–80 m s⁻¹. *Snow avalanches*, also termed *snow flows*, have been a subject of intense research because these avalanches may be serious NATURAL HAZARDS in mountainous terrain. Snow avalanches are able to transport large amounts of rock debris, including large boulders. However, some snow avalanches, such as powder snow avalanches during cold periods, carry little debris. Wet snow avalanches may be rich in clastic material. Dense snow avalanches have a high shear strength and their rigid plugs can transport large rock clasts. Non-turbulent powder snow avalanches are able to carry cobbles in the same way as DEBRIS FLOWS. The turbulent powder snow avalanches carry debris in turbulent suspension, in a similar way to rapid TURBIDITY CURRENTS.

Snow avalanche deposits range from blankets of scattered, unsegregated debris to irregular, thin patchy lobes of unsorted debris. The clasts are occasionally found resting upon one another as a result of settling from the downmelting snow. The debris is commonly surrounded by a blanket of waterlain sand. The clast FABRIC of snow-avalanche deposits varies on a small scale. The non-turbulent, *dense snow avalanches* may cause an internal clast fabric due to LAMINAR shear, but the fabric is destroyed when the debris melts out and settles to the ground. Characteristic features formed by snow avalanches are low-relief longitudinal grooves and ribs, where the grooves are formed by dragging of large angular clasts while the ribs are due to linear accumulation of debris, rather than erosion. Snow avalanches may also form *debris horns* on the upslope sides of large, immobile obstacles. This form is attributed to local 'freezing' of a dense snow avalanche rich in rock debris. Snow avalanches also form *debris shadows* on the down-slope sides of large boulders. These snow-avalanche impact landforms can be both erosional and depositional. At the foot of steep cliffs, where snow avalanches virtually crash-land, the snow masses splash out and eject debris from *plunge pools* (near-circular ponds

or areas of river beds) in a blast-like fashion (see SNOW-AVALANCHE IMPACT LANDFORMS). *AN*

[See also COLLUVIAL HISTORY, DEBRIS FLOW, SNOW AVALANCHE, STURZSTROM]

Blikra, L.H. and Nemec, W. 1998: Postglacial colluvium in western Norway: depositional processes, facies and palaeoclimatic record. *Sedimentology* 45, 909–959. **Jonelli, V.** 1999: Les effets de la fonte sur la sédimentation de dépôts d'avalanche de neige chargée dans le massif des Ecrins (Alpes françaises). *Géomorphologie: relief, processes, environment* 1, 39–58.

snow-avalanche impact landforms The erosional craters, pits, plunge pools or tarns and the associated depositional landforms (mounds, ramparts or tongues) produced by the impact of SNOW AVALANCHES at the foot of steep mountain slopes. *JAM*

Matthews, J.A. and McCarroll, D. 1994: Snow-avalanche impact landforms in Breheimen, southern Norway: origin, age, and paleoclimatic implications. *Arctic and Alpine Research* 26, 103–115. **Luckman, B.H., Matthews, J.A., Smith, D.J.** *et al.* 1994: Snow-avalanche impact landforms: a brief discussion of terminology. *Arctic and Alpine Research* 26, 128–129. **Smith, D.J., McCarthy, D.P. and Luckman, B.H.** 1994: Snow-avalanche impact pools in the Canadian Rocky Mountains. *Arctic and Alpine Research* 26, 116–127.

snow line (1) The temporary line (actually a zone) on GLACIERS marking the transition between the SNOW from the current year and the FIRN from older snow. (2) The altitude above which permanent snowbeds survive in the landscape (*climatic snow line* when averaged over a number of years). *AN*

[See also EQUILIBRIUM LINE, FIRN LINE, GLACIATION THRESHOLD]

snowball Earth A controversial theory that late PROTEROZOIC (Neoproterozoic) ICE AGES between about 750 and 550 million years ago affected the whole Earth, including low latitudes. Key evidence comes from SEDIMENTOLOGY and PALAEOGEOGRAPHY and includes warmwater CARBONATE rocks (*cap carbonates*) that immediately overlie GLACIAL SEDIMENTS, CARBON ISOTOPE analysis and PALAEOMAGNETISM. These peculiar palaeoclimatic conditions may be linked to the CAMBRIAN EXPLOSION of metazoan faunas. *GO*

[See also BLUE-WATER REFUGIUM]

Hoffman, P.F., Kaufman, A.J., Halverson, G.P. and Schrag, D.P. 1998: A Neoproterozoic snowball Earth. *Science* 281, 1342–1346. **Hyde, W.T., Crowley, T.J., Baum, S.K. and Peltier, W.R.** 2000: Neoproterozoic 'snowball Earth' simulations with a coupled climate/ice-sheet model. *Nature* 405, 425–429. **Jenkins, G.S. and Smith, S.R.** 1999: GCM simulations of Snowball Earth conditions during the late Proterozoic. *Geophysical Research Letters* 26, 2263–2266. **Kirschvink, J.L., Gaidos, E.J., Bertani, L.E.** *et al.* 2000: Paleoproterozoic snowball Earth: Extreme climatic and geochemical global change and its biological consequences. *Proceedings of the National Academy of Sciences of the United States of America* 97, 1400–1405. **Meert, J.G. and van der Voo, R.** 1994: The Neoproterozoic (1000–540 Ma) glacial intervals: No more snowball earth? *Earth and Planetary Science Letters* 123, 1–13.

snowbed An isolated area or patch of snow that persists throughout spring and/or summer in a site protected to some degree from INSOLATION. It may initiate NIVATION processes. *PW*

[See also PRONIVAL RAMPART, SNOWMELT, SNOWPACK]

Galen, C. and Stanton, M.L. 1995: Responses of snowbed plant-species to changes in growing-season length. *Ecology* 76, 1546–1557. **Wilson, P. and Clark, R.** 1999: Further glacier and snowbed sites of inferred Loch Lomond Stadial age in the northern Lake District, England. *Proceedings of the Geologists' Association* 110, 321–331.

snowmelt Runoff generated by melting of a SNOW-PACK; it may be concentrated in an annual event, usually restricted to a few days or weeks in spring, that may cause a snowmelt FLOOD or *freshet*. *PW*

[See also NIVAL ZONE, NIVATION, SNOWBED]

Carey, S.K. and Woo, M.K. 1999: Hydrology of two slopes in Subarctic Yukon, Canada. *Hydrological Processes* 13, 2549–2562. **Liston, G.E.** 1999: Interrelationships among snow distribution, snowmelt, and snow cover depletion: implications for atmospheric, hydrologic, and ecological modeling. *Journal of Applied Meteorology* 38, 1474–1487.

snowpack The snow that accumulates and persists in the landscape during winter. Modifications by freezing rain and diurnal melt and refreezing produce alternating layers of ice and snow. *PW*

[See also NIVAL ZONE, NIVATION, SNOWBED]

social construction The concept that people shape or 'construct' their knowledge and understanding such that it may be difficult to separate OBJECTIVE KNOWLEDGE from subjective ideas. A useful metaphor is that of a 'text'. It may be written in precise language and to a particular purpose, but its readers will put their own social construction upon it. To each individual it may convey different shades of meaning: it will be read, interpreted and experienced in different ways. Place, for example, has been described as a 'multilayered' concept and part at least of that layering of meanings comes from the ways in which place is socially constructed. With many places, such as a heritage site, social construction is involved in both the presentation (by the developer) and the interpretation (by the consumer). There have been many examples of alternative conceptualisation of ENVIRONMENT, NATURE and LANDSCAPE over the span of human history. Different human groups and cultures have had sharply contrasted perspectives on the environments they occupy. Key questions are how and why realities come to be constructed in particular ways and how shared constructions emerge and are communicated.

Reconstructions of environmental change employing SCIENTIFIC METHOD are similarly, in the first instance, representations of events and processes. Contrary to some perspectives in the humanities, however, this does not mean that the results of scientific method are representations entirely without substance: biophysical processes exhibit agency independent of humans, and humans are embedded in biophysical processes. Thus, constructed categories of human technological development (such as hunter–gatherer) and events such as the NEOLITHIC TRANSITION are generalisations with important limitations in understanding the interactions between human activity and the Earth. Deterministic trajectories of change can therefore be illusory, whereas *contingency*, defined as the

historical and local specificity that has to be accounted for in understanding and effectively manipulating environmental systems, provides a counterbalancing principle. *JAM/DTH*

Harvey, D.W. 1993: From space to place and back again: reflection on the condition of post modernity. In Bird, J., Curtis, B., Putnam, T. and Tickner, L. (eds), *Mapping the futures: local cultures, global change*. London: Routledge. **Head, L.** 2000: *Cultural landscapes and environmental change*. London: Arnold.

social sciences Academic disciplines concerned with the working of society, its structure and organisation. Economics, politics, sociology, anthropology and HUMAN GEOGRAPHY are among the main disciplines of social science. Because of their focus on the ways in which a society is organised and run, there are often strong applied derivatives of social science such as social policy and criminology. *DTH*

Cartermill 1996: *Current research in Britain: social sciences*. London: Cartermill.

sodium chloride *Halite*, or common salt (NaCl). It is a compound encountered in numerous environmental and palaeoenvironmental contexts, such as DESALINATION, DESALINISATION, SALINE LAKES, SALINISATION, SALT DOMES and SALT WEATHERING. Sodium chloride concentrations in ICE CORES provide evidence for past atmospheric circulation patterns and storminess. *DIB*

Alley, R.B., Finkel, R.C., Nishiizumi, K. *et al.* 1995: Changes in continental and sea-salt atmospheric loadings in central Greenland during the most recent deglaciation. *Journal of Glaciology* 41, 503–514.

soft engineering An engineering procedure to counter erosion without the use of permanent structures; for example by BEACH NOURISHMENT or by vegetating dunes. *HJW*

[See also HARD ENGINEERING]

Leatherman, S.P. 1996: Shoreline stabilization approaches in response to sea level rise: US experience and implications for Pacific island and Asian nations. *Water, Air and Soil Pollution* 92, 149–157.

soft-sediment deformation A category of processes and SEDIMENTARY STRUCTURES (sometimes called *sedimentary deformational structures*) involving the disturbance and deformation of mud, sand and gravel by physical forces during or shortly after DEPOSITION, during the early stages of DIAGENESIS. Such deformation of unlithified materials is commonly preserved in sediments and sedimentary rocks, but contrasts with TECTONIC deformation that formed in more deeply buried rocks, although there is some overlap.

The products of soft-sediment deformation are a variety of FOLDS and FAULTS, including *convolute lamination* and deformed CROSS-BEDDING. *Load structures* (or *load casts* and *flame structures*) are undulations of an interface (bedding surface) caused by the sinking of denser into less dense sediment (e.g. sand into mud). INVOLUTIONS are similar. A separation can be made into: (1) essentially *in situ* soft-sediment deformation structures; (2) injection or sedimentary transposition structures that involve the mobilisation of sediment (e.g. clastic *dykes*, DIAPIRS, and sand or mud VOLCANOES); and (3) MASS MOVEMENTS such as SLUMPS. However, again there is some overlap of processes and products.

For soft-sediment deformation to occur, unconsolidated sediment commonly has to undergo a sudden loss of strength: this can occur by a variety of processes such as LIQUEFACTION or FLUIDIZATION in sand, that can together be termed LIQUIDIZATION. These processes are commonly triggered by some kind of impulsive agent such as an EARTHQUAKE, TSUNAMI or STORM waves, and soft-sediment deformation structures have the potential to identify the activity of such agents in the GEOLOGICAL RECORD. *GO*

Allen, J.R.L. 1982: *Sedimentary structures: their character and physical basis.* Amsterdam: Elsevier. Allen, J.R.L. 1986: Earthquake magnitude-frequency, epicentral distance, and soft-sediment deformation in sedimentary basins. *Sedimentary Geology* 46, 67–75. Jones, M.E. and Preston, R.M.F. (eds) 1987: *Deformation of sediments and sedimentary rocks* [*Special Publication* 29]. London: Geological Society. Maltman, A.J. (ed.) 1994: *The geological deformation of sediments.* London: Chapman and Hall. Owen, G. 1996: Experimental soft-sediment deformation: structures formed by the liquefaction of unconsolidated sands and some ancient examples. *Sedimentology* 43, 279–293. Rijsdijk, K.F., Owen, G., Warren, W.P. *et al.* 1999: Clastic dykes in over-consolidated tills: evidence for subglacial hydrofracturing at Killiney Bay, eastern Ireland. *Sedimentary Geology* 129, 111–126. Ringrose, P.S. 1989: Palaeoseismic (?) liquefaction event in late Quaternary lake sediment at Glen Roy, Scotland. *Terra Nova* 1, 57–62.

soil The unconsolidated mineral and/or organic material at the surface of the Earth, capable of supporting plant growth. However, soil should be differentiated from rock, SEDIMENT and PARENT MATERIAL. Hence, it was defined by Joffe as 'a natural body of animal, mineral and organic substances differentiated into horizons of variable depth which differ from the material below in morphology, physical make-up, chemical properties and composition, and biological characteristics'. The full sequence of SOIL HORIZONS makes up the SOIL PROFILE. Soil is the basis of all terrestrial ecosystems and > 90% of human food is produced from crops grown in soil. It is the fundamental basis for the existence of life on Earth. *EMB*

[See also PALAEOSOL, PEDOGENESIS, SOIL DEVELOPMENT, SOIL FORMING PROCESSES]

Brady, N.C. 1984: *The nature and properties of soils*, 9th edn. New York: Macmillan. Ellis, S. and Mellor, A. 1995: *Soils and environment.* London: Routledge. Jacks, G.V. 1954: *Soil.* London: Thomas Nelson. Joffe, J.S. 1953: *Pedology.* New Brunswick, NJ: Pedology Publications. Nikiforoff, C.C. 1959: Reappraisal of the soil: pedogenesis consists of transactions in matter and energy between the soil and its surroundings. *Science* 129, 186–196. Simonson, R.W. 1968: Concept of soil. *Advances in Agronomy* 20, 1–47. Wild, A. 1993: *Soils and the environment: an introduction.* Cambridge: Cambridge University Press. Yaalon, D. 2000: Down to earth: why soil science matters. *Nature* 407, 301.

soil age Usually used in the sense of *absolute soil age*, or the time elapsed since the beginning of soil formation (see SOIL FORMING PROCESSSES) at the site, it may also refer to *relative soil age*, or the age of the soil constituents, particularly the organic matter component. Soils vary greatly in age and, from their morphological characteristics and carbon content, an indication of soil age may be obtained. Soil formation begins when the first fragment of organic matter is added to mineral matter and continues by development of SOIL HORIZONS of increasing clarity as the soil

progresses towards a *steady-state equilibrium* with the environment. The process may take place over a few hundred or many thousands of years, but is likely to be interrupted and a steady state may never be attained. Dates marking the beginning of soil formation can be obtained from historical events such as erosion, flooding, volcanic eruptions or aeolian deposits that provide a new PARENT MATERIAL and effectively set the clock back to zero as far as soil formation is concerned. The age of a buried PALAEOSOL commonly refers to the time elapsed since burial, which is a rather different concept. *EMB*

[See also SOIL DATING]

Gerasimov, I.P. 1980: The age of recent soils. *Geoderma* 12, 17–25. Matthews, J.A. 1980: Some problems and implications of ^{14}C dates from a podzol buried beneath an end moraine at Haugabreen, southern Norway. *Geografiska Annaler* 62A, 185–208. Proudfoot, V.B. 1958: Problems of soil history. Podzol development at Goodland and Torr Townlands, County Antrim, Northern Ireland. *Journal of Soil Science* 9, 186–198.

soil classification Classification of soils is necessary for communication and a full understanding of the range of features observed in the SOIL cover of the Earth. Soils, unlike discrete individuals in the plant and animal kingdoms, gradually change from one to another forming a continuum upon the Earth's surface. In the AD 1880s, Dokuchaev recognised that soils were independent natural bodies that could be studied from the morphology of their profiles and Sibirtzev proposed the terms ZONAL, INTRAZONAL and AZONAL as the three main classes of soils. Many systems of classification were developed in individual countries during first half of the twentieth century, based on supposed genesis, but in the 1950s emphasis turned to the morphology of the soil itself as a basis for classification (*generic classification*). During the period 1960 to 1980 the United States Department of Agriculture developed a comprehensive system of soil classification (SOIL TAXONOMY), and the legend for the Soil Map of the World was developed by the United Nations (FAO).

During the last decade, agreement has been reached amongst the soil science community regarding a WORLD REFERENCE BASE FOR SOIL RESOURCES (WRB). This system is based on a revised legend for the FAO–Unesco Soil Map of the World, but has the necessary scientific depth and background for a framework to classify world soils. Classification of soils in modern systems requires the presence of certain DIAGNOSTIC SOIL HORIZONS that are used in the allocation of a soil to the appropriate Reference Soil Group. There are 30 Reference Soil Groups in the WRB, based on the familiar soil names of the FAO system; they have been used throughout this publication and are summarised in the Table. *EMB*

Bridges, E.M., Batjes, N.H. and Nachtergaele, F.O. (eds) 1998: *World Reference Base for Soil Resources: Atlas* [ISSS Working Group RB]. Leuven: Acco. Deckers, J.A., Nachtergaele, F.O. and Spaargaren, O.C. (eds) 1998: *World Reference Base for Soil Resources: Introduction* [ISSS Working Group RB]. Leuven: Acco. Finkl Jr, C.W. (ed.) 1982: *Soil classification.* Stroudsburg, PA: Hutchinson Ross. Food and Agriculture Organization of the United Nations (FAO) 1998: *World reference base for soil resources* [*Soil Resource Report* 84]. Rome: FAO, ISRIC, ISSS. FAO–Unesco 1974: *FAO–Unesco Soil Map of the World.* Vol. 1. *Legend.* Paris: Unesco. Soil Survey Staff 1999: *Soil taxonomy: a basic system of soil classification for making and interpreting soil sur-*

soil classification *Main characteristics of the 30 soil reference groupings of the FAO/Unesco/IUSS system of soil classification and the World Reference Base for Soil Resources (see also Table under* WORLD REFERENCE BASE FOR SOIL RESOURCES*) (after Bridges* et al. *1998)*

Main characteristics	Reference soil group
Soils that are composed of organic materials	Histosols
Soils with permafrost within 1 m depth	Cryosols
Soils in which soil formation is conditioned by human influences	Anthrosols
Very shallow soils over hard rock or in unconsolidated, very gravelly material	Leptosols
Dark-coloured cracking and swelling clays	Vertisols
Young soils in alluvial deposits	Fluvisols
Strongly saline soils	Solonchaks
Soils with permanent or temporary wetness near the surface	Gleysols
Young soils in volcanic deposits	Andosols
Acid soils with a black/brown/red subsoil with illuvial iron–aluminium–organic compounds	Podzols
Wet soils with an irreversibly hardening mixture of iron, clay and quartz in the subsoil	Plinthosols
Deep, strongly weathered soils with a chemically poor, but physically stable, subsoil	Ferralsols
Soils with a bleached, temporarily water-saturated topsoil on a slowly permeable subsoil	Planosols
Soils with subsurface clay accumulation, rich in sodium	Solonetz
Soils with a thick, blackish topsoil, rich in organic matter with a calcareous subsoil	Chernozems
Soils with a thick, dark brown topsoil, rich in organic matter and a calcareous or gypsum-rich subsoil	Kastanozems
Soils with a thick, dark topsoil rich in organic matter and evidence of removal of carbonates	Phaeozems
Soils with accumulation of secondary gypsum	Gypsisols
Soils with accumulation of secondary silica	Durisols
Soils with accumulation of secondary calcium carbonates	Calcisols
Acid soils with a bleached horizon penetrating into a clay-rich subsurface horizon	Albeluvisols
Soils with subsurface accumulation of high activity clays, rich in exchangeable aluminium	Alisols
Deep, dark red, brown or yellow clayey soils having a pronounced shiny, nut-shaped structure	Nitisols
Soils with subsurface accumulation of low-activity clays and low base saturation	Acrisols
Soils with subsurface accumulation of high-activity clays	Luvisols
Soils with subsurface accumulation of low-activity clays and high base saturation	Lixisols
Acid soils with a thick, dark topsoil rich in organic matter	Umbrisols
Weakly to moderately developed soils	Cambisols
Sandy soils featuring very weak or no soil development	Arenosols
Soils with very limited soil development	Regosols

veys, 2nd edn [*Agriculture Handbook* 436]. Washington, DC: United States Department of Agriculture.

soil conservation As the soil reserves of the world are finite and within the human lifespan are not renewable except at great cost, the case for soil care is very strong. Soil is one of humanity's most precious assets; it allows plants, animals and people to live on the Earth's surface. Soil is a limited resource that can easily be destroyed by poor management or thoughtless action. Silt loams, sand, loamy sand, fine sandy loam and cultivated organic soils are vulnerable to EROSION by wind and water. Considerable effort has been expended to develop theoretical models to help counter erosion, such as the *universal soil loss equation* (USLE) in the United States and the *soil loss estimation model for southern Africa* (SLEMSA) in South Africa. If conservation techniques known since the AD 1930s were to be applied world-wide, the present impact of erosion would be greatly reduced.

Until the last decade, soil conservation was synonymous with measures to prevent SOIL EROSION, but recently the term has come to embrace protection against chemical and physical damage as well. Inappropriate use of herbicides and PESTICIDES and overuse of FERTILISERS in the developed world by factory-type agriculture is causing problems, just as poor soil management caused by ignorance and poverty affects the developing world. Society uses soils for a variety of purposes, such as industry, housing and transport. With a rising world population it is necessary to develop policies of LANDUSE and soil conservation that take into consideration the nature and properties of soils as well as the present and future demands of society.

EMB

[See also CONSERVATION, MECHANICAL SOIL CONSERVATION MEASURES, SUSTAINABILITY]

De Graaf, J. 1996: *The price of soil erosion: an economic evaluation of soil conservation and watershed development* [*Tropical Resource Management Paper* 14]. The Netherlands: Wageningen Agricultural University. **Greenland, D.J. and Szabolcs, I. (eds)** 1994: *Soil resilience and sustainable land use.* Wallingford: CAB International.

soil creep A type of MASS WASTING process involving the slow deformation (without shear failure) of the soil profile as the upper layers move imperceptibly downhill. It may involve frost creep (see SOLIFLUCTION). *SHD*

Auzet, A.V. and Ambroise, B. 1996: Soil creep dynamics, soil moisture and temperature conditions on a forested slope in the granitic Vosges Mountains, France. *Earth Surface Processes and Landforms* 21, 531–542. **Brady, N.C. and Weil, R.R.** 1999: *The nature and properties of soils.* Upper Saddle River, NJ: Prentice-Hall. **Harker, R.I.** 1996: Curved tree trunks – indicators of soil creep and other phenomena. *Journal of Geology* 104, 351–358.

soil dating The age of a soil may be estimated from the depth and degree of development of the soil horizons it contains. The content of weatherable minerals is also an indication of a soil's age. Thus, certain soils of the tropical regions with few weatherable minerals have been called old, whereas soils developed in a few hundred years on till or alluvium are said to be young. However, soil dating is complicated by different concepts of SOIL AGE and the existence of materials of different ages within most soils. An example of soil dating is provided by the RADIOCARBON DATING of included soil organic matter or carbonate carbon. Thus, CHERNOZEMS in the USA have been found to have an age of about 350 years; upper caliche layers are 2300 years old and lower caliche layers 32 000 years old. The oldest material provides a minimum estimate of time elapsed since the beginning of soil formation at the site (*absolute soil age*). *EMB*

[See also APPARENT MEAN RESIDENCE TIME, PALAEOSOL]

Birkland, P.W., Burke, R.M. and Benedict, J.B. 1989: Pedogenetic gradients for iron and aluminium accumulation and phosphorus depletion in arctic and alpine soils as a function of time and climate. *Quaternary Research* 32, 193–204. **Gellatly, A.F.** 1985: Phosphate retention: relative age dating of Holocene soil development. *Catena* 12, 227–240. **Matthews, J.A.** 1985: Radiocarbon dating of surface and buried soils: principles, problems and prospects. In Richards, K.S., Arnett, R.R. and Ellis, S. (eds), *Geomorphology and soils*. London: George Allen and Unwin, 269–288. **Scharpenseel, H.W. and Schiffmann, H.** 1977: Radiocarbon dating of soils: a review. *Zeitschrift Pflanzenernährung Düngung Bodenkunde* 140, 159–174.

soil degradation The decline in SOIL QUALITY, particularly as a result of improper use by human beings. To use the soil for almost any purpose inevitably results in some loss of its biological capability, but degradation is usually apparent in the following ways: (1) WATER EROSION (including sheet, rill, gully erosion mass movements and/or deposition); (2) WIND EROSION and/or deposition; (3) *biological degradation* (including changes in humus content of soils, vegetation disturbance or even elimination of soil biota such as worms or fungi); (4) *physical degradation* (including changes in bulk density, structure and permeability); and (5) *chemical degradation* (including acidification, alkalisation and changes in pH, salinity and chemical toxicity).

Often degradation is imperceptible, as with the annual loss of a few millimetres of topsoil by erosion, or SOIL MINING of nutrients without replenishment. The effects of both these forms of degradation can be disguised by the use of mineral FERTILISERS, but off-site sedimentation on fields and silting of water courses cause additional problems. In other cases, soils are polluted by salts from rising GROUNDWATERS or completely annihilated by industrial or constructional activity. As soil is the basis of agriculture and food production, it is in the human interest to minimise soil degradation. However, opinions differ; some socio-economists think that human ingenuity will overcome these problems, but environmental ecologists see problems of food security if soil degradation continues at present rates.

Two surveys of the impact of human-induced degradation have been undertaken recently by the International Soil Reference and Information Centre (ISRIC): GLASOD and ASSOD. GLASOD, an acronym for the '*global assessment of human-induced soil degradation*', pro-duced the *World Map of the State of Human-induced Soil Degradation*, published in AD 1990 by ISRIC and UNEP at an average scale of 1:10 000 000. It was the first attempt to present a global view of soil degradation, specifying the extent, categories, degree and causes of soil degradation. The survey revealed that 1965 million hectares of the world's soils are degraded to some degree. The major categories of degradation shown are water erosion, wind erosion and chemical and physical degradation. The degree of degradation is expressed as light, moderate, strong and extreme and the causes are attributed to DEFORESTATION, OVERGRAZING, agricultural mismanagement, overexploitation and industrial activity. The more detailed *assessment of human-induced soil degradation in south and southeast Asia* has also been completed under the acronym ASSOD; a survey of the status of land degradation at a scale of 1:5 000 000. Twenty categories of soil degradation are shown, together with estimated rates of degradation, the causative factors and the influence of management. Such surveys are important for monitoring the rate of change of soils under human influence. *EMB*

[See also LAND DEGRADATION, SOIL CONSERVATION, SOIL EROSION]

Bridges, E.M. and Oldeman, L.R. 1999: Global assessment of human-induced soil degradation. *Arid Soil Research and Rehabilitation* 13, 319–325. **Bridges, E.M. and van Baren, J.H.V.** 1997: Soil: an overlooked, undervalued and vital part of the human environment. *The Environmentalist* 17, 15–20. **Fleming, G.** 1996: *Recycling derelict land*. London: Thomas Telford. **Greenland, D.J., Gregory, P.J. and Nye, P.H. (eds)** 1997: *Land resources: on the edge of the Malthusian precipice?* Wallingford: CAB International. **Oldeman, L.R.** 1998: Soil degradation: a threat to food security? In Bindraban, P.S., van Keulen, Y., Kuyvenhoven, A. *et al.* (eds), *Food security at different scales: demographic, biophysical and socio-economic considerations*. Wageningen: AB-DLO. **Oldeman, L.R., Hakkeling, R.T.A. and Sombroek, W.G.** 1990: *Global assessment of soil degradation*. Wageningen: International Soil Reference Information Centre (ISRIC). **Oldeman, L.R., Hakkeling, R.T.A. and Sombroek, W.G.** 1991: *World map of the state of human-induced soil degradation: an explanatory note*, 2nd edn. Wageningen: UNEP and ISRIC. **Stocking, M.** 1995: Soil erosion and land degradation. In O'Riordan, T. (ed.), *Environmental science for environmental management*. Harlow: Longman, 223–237. **Van Lynden, G.W.J. and Oldeman, L.R.** 1997: *The assessment of the status of human-induced soil degradation in south and southeast Asia*. Wageningen: UNEP, FAO, ISRIC.

soil development Development of soil may be judged by the clarity and depth of the horizons that form the SOIL PROFILE. Soil development usually follows an asymptotic curve (see Figure), the shape of which depends upon the inherent characteristics of the soil and the attribute(s) being studied. There are no distinct stages in soil development, but examples may be seen of minimal soil development in solid rock, recently deposited alluvium, loess or till and complex, well developed, POLYGENETIC profiles. Soil formation begins from the first addition of organic matter to the mineral PARENT MATERIAL and continues with increasing complexity as organisms convert inert mineral material into the 'living' soil with its full complement of SOIL HORIZONS. Soil development takes place through the SOIL FORMING PROCESSES, such as LEACHING, GLEYING, PODZOLISATION, FERRALITISATION, etc., which work within a framework of SOIL FORMING FAC-

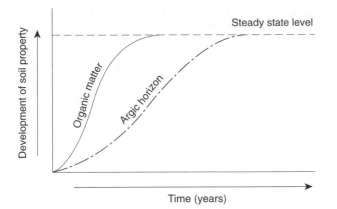

soil development *The asymptotic nature of soil development*

TORS (climate, organisms, relief, parent material and time).   *EMB*

[See also CHRONOSEQUENCE, PEDOGENESIS, SOIL AGE]

Crocker, R.L. and Major, J. 1955: Soil development in relation to vegetation and surface age at Glacier Bay, Alaska. *Journal of Ecology* 43, 427–448. **Matthews, J.A.** 1992: *The ecology of recently-deglaciated terrain.* Cambridge: Cambridge University Press, 82–144.

soil drainage Soil drainage is assessed by the morphology of the soil profile: signs of GLEYING and the relationship to the soil surface are critical. Bright soil colours normally indicate well drained conditions, whereas mottling and dull colours suggest imperfect or poorly drained conditions. Unfortunately, this simple approach is not always reliable, as soils on red Triassic mudstone PARENT MATERIALS do not readily change to grey colours and, once formed, the characteristics of gleying remain after drainage improvements have been carried out. The current approach in England and Wales allocates soils to the categories of well drained, moderately well drained, imperfectly drained, poorly drained and very poorly drained by the length of time the soil is wet within profile depth (see Figure). The scale ranges from being 'not wet within 70 cm for more than 30 days in most years' to the soil being 'wet for more than 335 days in most years'.

Conversion of forest soils to agriculture caused loss of permeability and problems of excess wetness for grain crops such as wheat. Consequently, removal of excess moisture from soils has been a necessary part of agricultural practice since at least the Middle Ages. RIDGE AND FURROW ploughing of the open fields gave better drainage on the ridges. Open ditches were dug around fields and, within fields, trenches with a layer of stones or brushwood provided lines along which water could move out. Later, deep stone-lined drains were installed in fields to remove excess water, but these were often inadequate to remove surface water. In more recent times, tile drains, mole drains and plastic pipes with a covering of porous gravel lying below the topsoil have ensured removal of moisture and allowed cultivation to proceed without hindrance.   *EMB*

[See also FIELD DRAINAGE, LAND DRAINAGE]

Corbett, W.M. and Tatler, W. 1970: *Soils in Norfolk, Sheet TM49 (Beccles North) Soil Survey Record 1.* Harpenden: Soil Survey of England and Wales. **Hodgson, J.M. (ed.)** 1974: *Soil survey field handbook.* Harpenden: Soil Survey of England and Wales. **Thomasson, A.J.** 1975: *Soils and field drainage* [Technical Monograph 7]. Harpenden: Soil Survey of England and Wales. **Food and Agriculture Organization of the United Nations (FAO)** 1990: *Guidelines for soil profile description.* Rome: FAO.

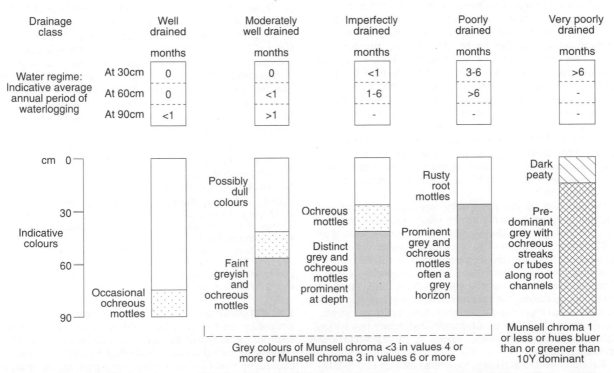

soil drainage *Soil drainage class and water regime and its relation to profile morphology (after Corbett and Tatler, 1970)*

soil emissions The role of soils in the production of the GREENHOUSE GASES, including carbon dioxide, methane and nitrous oxides, stems from a complex interaction of biological, chemical and physical processes. Changes of LANDUSE have led to increases in emissions of all three gases throughout the twentieth century. Soil management techniques are being developed to avoid these losses from the soil system.   *EMB*

[See also RADON EMANATION]

Bouwman, A.F. 1999: *Approaches to scaling of trace gas fluxes in ecosystems.* Amsterdam: Elsevier. **Bridges, E.M. and Batjes, N.H.** 1996: Soil gaseous emissions and climate change. *Geography* 81, 141–155.

soil erosion The accelerated removal of soil from the land surface, mainly by wind and water and caused directly or indirectly by human action. Although the process of EROSION of soil is an entirely natural one (sometimes called 'geological' erosion), in which weathered material is removed from parts of the landscape (and deposited elsewhere), this loss is more or less balanced in most landscapes unaffected by human action by the formation of new soil material created through WEATHERING and PEDOGENESIS. The term soil erosion has consequently become restricted in its usage to human-induced soil losses ('accelerated' erosion of soil). Soil erosion can occur on most parts of the world's land surface, but the most vulnerable areas are usually regarded as SEMI-ARID regions, where loss of a protective vegetation caused by human action (e.g. DEFORESTATION, grazing and fire) and damage to the soils caused by, for example, TILLAGE, MINING, OVERCULTIVATION and COMPACTION can leave the soils particularly prone to erosion.

Soil erosion is a two-stage process consisting of the detachment of individual particles from the soil mass and their subsequent transport. *Rainsplash* is the most important means of detachment. Raindrops hitting a bare soil surface cause particles to be thrown into the air and moved several centimetres. Wind and running water can also detach soil particles. Transport is achieved by rainsplash, *overland flow* in the form of SLOPEWASH and concentrated flow in RILLS and GULLIES, as well as by wind. Many MASS MOVEMENT processes (e.g. shallow landslides, creep) are also viewed by many researchers as processes of soil erosion. Factors controlling soil erosion are the EROSIVITY of the eroding agent and the ERODIBILITY of the soil. Slope parameters, the nature of the vegetation cover and land management and cultivation practices influence the rate of erosion.

Recognition of the potentially large quantities of soil that can be moved down-slope by tillage and land levelling have also meant that these processes are now regarded as soil erosion processes in the strict sense by many researchers. Interest has been generated recently in reconstructing past soil erosion as a means of gaining a long-term perspective on short- and medium-term EROSION RATES derived from, for example, EROSION PLOTS and EROSION PINS and CAESIUM-137 analysis. Much of the evidence comes from the build-up of datable sediments in various sinks and stores in the landscape (e.g. VALLEY FILLS and SEDIMENT TRAPS).

Strongly eroded soils are thought to lose up to 75–80% of their original organic carbon pool, so that at a global scale accelerated erosion causes large SOIL EMISSIONS of carbon into the atmosphere, contributing to atmospheric CO_2 quantities and thus to GLOBAL WARMING. It has been estimated that restoration of the estimated 250 million ha of the world affected by strong or extreme soil erosion by wind and water would be the most effective means of mitigating the GREENHOUSE EFFECT over the next 25–50 years, 'buying time' until non-carbon fuel options can take effect.   *RAS*

[See also AGRICULTURAL IMPACT ON GEOMORPHOLOGY, AGRICULTURAL IMPACT ON SOILS]

Dearing, J. 1994: Reconstructing the history of soil erosion. In Roberts, N. (ed.), *The changing global environment.* Oxford: Blackwell, 242–261. **Kirkby, M.J. and Morgan, R.P.C. (eds)** 1980: *Soil erosion.* Chichester: Wiley. **Lal, R.** 1994: *Soil erosion research methods*, 2nd edn. Delray Beach, FL: St Lucie Press. **Lal, R.** 1999: Soil management and restoration for C sequestration to mitigate the accelerated greenhouse effect. *Progress in Environmental Science* 1, 307–326. **Morgan, R.P.C.** 1995: *Soil erosion and conservation*, 2nd edn. Harlow: Longman. **O'Hara, S.L., Street-Perrott, F.A. and Burt, T.P.** 1993: Accelerated soil erosion around a Mexican highland lake caused by prehistoric agriculture. *Nature* 362, 48–51. **Pimentel, D. (ed.)** 1993: *World soil erosion and conservation.* Cambridge: Cambridge University Press.

soil erosion history Indications of past SOIL EROSION are found in archaeological and geological records. However, identifying the extent of human influence on HOLOCENE erosion in a region is not straightforward because PROXY DATA often contain ANTHROPOGENIC and climatic signals, which are difficult to separate. High natural EROSION RATES, governed for example by climate, slope angle and vegetation cover, occurred in formerly glaciated areas during the LATE GLACIAL. As the climate improved, the landscapes became vegetated and erosion decreased. Forested regions with a short AGRICULTURAL HISTORY (of a few hundred years) have been characterised by roughly constant or declining erosion rates, except for the most recent centuries when DEFORESTATION and local construction have led to increased erosion. In contrast, in a landscape with a long history of agriculture (e.g. the Mediterranean region with a history of several thousand years), erosion rates have generally increased since the time of initial DEFORESTATION: the characteristic pattern of erosion is a stepwise increase.   *UBW*

[See also ACCUMULATION RATE, ALLUVIUM, COLLUVIUM, ENVIRONMENTAL ARCHAEOLOGY, GEOLOGICAL EVIDENCE OF ENVIRONMENTAL CHANGE, HUMAN IMPACT, SEDIMENT INFLUX, SEDIMENT YIELD]

Andel, T.H. van, Zangger, E. and Demitrack, A. 1990: Land use and soil erosion in prehistoric and historical Greece. *Journal of Field Archaeology* 17, 379–396. **Dearing, J.** 1994: Reconstructing the history of soil erosion. In Roberts, N. (ed.), *The changing global environment.* Oxford: Blackwell Science, 242–263. **Berglund, B.E. (ed.)** 1991: The cultural landscape during 6000 years in southern Sweden – the Ystad Project. *Ecological Bulletins* 41, 1–495.

soil exhaustion (1) The concept that SOIL EROSION is limited by the depletion of suitable soil material (e.g. fine sediment and organic matter) available for EROSION. Thus, on very degraded terrain, a decline in SOIL LOSS may be caused by the lack of suitable material rather than by improved resistance to erosion by, for example, improved

SOIL CONSERVATION. (2) Soil that is heavily depleted of NUTRIENTS. Soils contain a limited amount of plant nutrients that, in a natural situation, are circulated between the plant and the soil. When this MINERAL CYCLING is broken, as with agriculture when crops are removed and plant nutrients are not replaced, soils steadily lose their fertility. In the Middle Ages, the three-field system of farming with fallow and the manure of grazing animals attempted to delay the exhaustion of soil for as long as possible. Soil exhaustion is a particular problem in tropical regions as most of the fertility is associated with organic matter. When forest is cleared, organic matter is quickly oxidised or eroded and the apparently high fertility under forest is replaced rapidly by an exhausted soil. The problem is exacerbated on sloping marginal lands where impoverished subsistence farmers are unable to manage erosion or to replace any nutrients removed by crops. Exhausted soils are extremely vulnerable to erosion.   *EMB/SHD*

[See also AGRICULTURAL IMPACTS ON SOILS, FIELD SYSTEMS, HUMAN IMPACT ON SOILS, SOIL DEGRADATION]

Bennett, H.H. 1939: *Soil conservation*. New York: McGraw-Hill. El-Swaify, S.A. 1999: *Sustaining the global farm: strategic issues, principles and approaches*. Honolulu, HI: International Soil Conservation Organization and University of Hawaii. Greenland, D.J., Gregory, P.J. and Nye, P.H. 1997: *Land resources: on the edge of the Malthusian precipice?* Wallingford: CABI and The Royal Society.

soil forming factors

Environmental factors that control the processes of soil formation. Five factors were suggested by Dokuchaev and these were elaborated by Jenny in the FUNCTIONAL FACTORIAL APPROACH: these are climate, organisms, parent material, relief and time. Jenny proposed the fundamental equation of soil forming factors: $s = f(cl, o, r, p, t ...)$, where a particular soil property (s) is a function of, or is dependent upon, the various factors such as climate (cl) organisms (o), relief (r), parent material (p) and time (t).

The climatic conditions of the world give an approximately latitudinal and altitudinal pattern to world soils that reflects the broad influence of temperature, but this picture is disrupted by the influence of rainfall, which supplies soil moisture. Many organisms are involved in soil formation from bacteria to mammals. Micro-organisms include bacteria, the most populous soil-living form of life. Soil *meso-fauna* include members of the arthropoda, especially springtails and mites. *Macro-fauna* include EARTHWORMS, nematodes, ants, TERMITES, millipedes, centipedes, slugs, snails and small mammals. Relief is significant at the broad scale in that high mountains and plateaux develop soils at altitude that are different from those on low-lying areas, and on plains micro-relief plays an important role. The factor of PARENT MATERIAL is usually complex as soils are rarely just formed from the rock beneath, including alluvial, colluvial and aeolian additions to the material derived from weathered rocks. Finally, the factor of time is important, because all soil processes take place over a span of time. Soils are therefore important as they usually contain a record of environmental changes that have taken place during soil formation.   *EMB*

[See also CATENA, CHRONOSEQUENCE, CHRONOFUNCTION]

Amundson, R., Harden, J. and Singer, M. (eds) 1994: *Factors of soil formation: a fiftieth anniversary retrospective* [*Special Publication* 33]. Madison, WI: Soil Science Society of America,

1–160. Crocker, R.L. 1952: Soil genesis and pedogenetic factors. *Quarterly Review of Biology* 27, 139–168. Crocker, R.L. 1959: The plant factor in soil formation. *Australian Journal of Science* 21, 180–193. Jenny, H. 1941: *Factors of soil formation*. New York: McGraw-Hill. [Republished 1994 as: *Factors of soil formation: a system of quantitative pedology*. Dover, New York.] Johnson, D.L. Keller, E.A. and Rockwell, T.K. 1990: Dynamic pedogenesis: new views on some key soil groups, and a model for interpreting Quaternary soils. *Quaternary Research* 33, 306–319.

soil forming processes

Processes that modify the REGOLITH and give it characteristics that distinguish SOIL from PARENT MATERIAL. These processes or 'bundles' of chemical, physical and biological activities react or interact with the parent material over time to produce the SOIL PROFILE. The strength of the processes varies greatly from place to place and in different parts of the world.

The most significant processes (see Figure) include accumulation of ORGANIC MATTER as fresh organic matter is actively decomposed by the soil fauna and humus is added to the upper part of the soil profile to form a MOLLIC, OCHRIC or HISTIC horizon. LEACHING occurs as water percolates through the soil, dissolving any soluble constituents and removing base cations from the cation exchange sites to form a CAMBIC horizon. Clay ELUVIATION is the mobilisation and transport of clay from the upper horizons of soils and its redeposition in the subsoil to form an ARGIC horizon. PODZOLISATION is the removal of iron and aluminium compounds from the upper part of the soil and their movement into a lower horizon with organic matter (ILLUVIATION) and deposition in a SPODIC HORIZON. FERRALITISATION is the relative accumulation of iron and aluminium oxides as silica is lost from the breakdown of clay minerals in tropical soils to form a FERRALIC horizon. GLEYING is the process of soil formation in saturated, anaerobic conditions. SALINISATION is the accumulation of soluble salts in the soil profile and SOLODISATION is the leaching of sodium from soils with *natric horizons*. *Pedoturbation* is a term for physical and biological churning of soil, resulting in homogenisation of the profile. The physical, chemical and biological changes following exposure of a fresh parent material (as in the Dutch polders) are collectively called *ripening*. WEATHERING is the alteration and breakdown of rocks to form the parent material of soils.   *EMB*

Bridges, E.M. 1998: *World soils*, 3rd edn. Cambridge: Cambridge University Press. Duchaufour, P. 1971: *Précis de pédologie*. Paris: Masson. McRae, S.G. 1988: *Practical pedology: studying soils in the field*. Chichester: Ellis Horwood. Ross, S.M. 1989: *Soil processes*. London: Routledge.

soil horizon

A layer of soil, revealed in a SOIL PROFILE, lying approximately parallel to the Earth's surface, having pedological characteristics (see Figure). Soil horizons are an expression of the action of SOIL FORMING PROCESSES upon the PARENT MATERIAL. A soil horizon may be distinguished by its morphology, chemical properties and composition and biological characteristics. Surface organic materials (MOR, MODER and MULL) are usually considered to form *soil layers* rather than horizons as they have not been formed specifically by soil forming processes. Similarly, layers in the parent material are considered to be the result of geological processes. *Topsoil* and *subsoil* are non-technical terms for soil layers that may comprise one

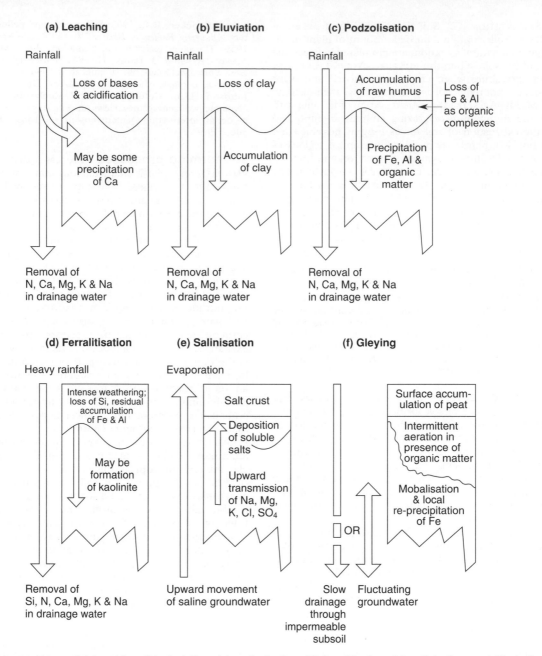

soil forming processes *(a) Leaching; (b) eluviation; (c) podzolisation; (d) ferralitisation; (e) salinisation; and (f) gleying (after McRae, 1988)*

or more horizons; the former contains the bulk of the organic matter, nutrients and living organisms.

A soil profile usually possesses one or more *master horizons*; these are designated by the letters A, B, C, and the letters E, D and R may also be used. Surface organo-mineral horizons are usually referred to as A horizons, and E represents an eluvial horizon depleted of iron and clay. The symbol B is used for illuvial horizons where accumulation of iron, clay and humus occurs (SPODIC, ARGIC, CALCIC, GYPSIC and *natric*), and B has also been used for non-illuvial (CAMBIC) forms. The letters D or R are used to signify the parent material or rock. These designations are interpretative symbols, based on horizon morphology

and implied genesis, which are used to identify and label a soil horizon. Certain horizons with quantitatively defined properties are used to identify soil units within classification systems; these are DIAGNOSTIC (or reference) SOIL HORIZONS. The WORLD REFERENCE BASE FOR SOIL RESOURCES uses 39 diagnostic horizons in combination with other diagnostic properties and materials to classify soils into 30 reference soil groups.   *EMB*

Bridges, E.M. 1993: Soil horizon designations: past use and future prospects. *Catena* 20, 363–373. **Food and Agriculture Organization of the United Nations (FAO)** 1998: *World reference base for soil resources* [*Soil Resource Report* 84]. Rome: FAO, ISRIC, ISSS.

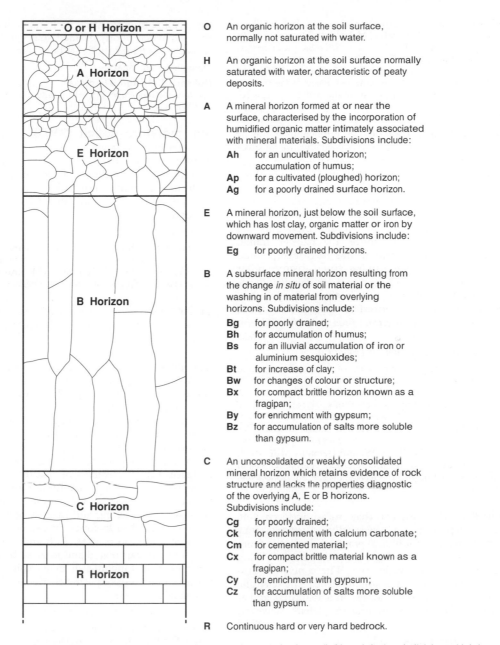

O   An organic horizon at the soil surface, normally not saturated with water.

H   An organic horizon at the soil surface normally saturated with water, characteristic of peaty deposits.

A   A mineral horizon formed at or near the surface, characterised by the incorporation of humidified organic matter intimately associated with mineral materials. Subdivisions include:

 Ah   for an uncultivated horizon; accumulation of humus;
 Ap   for a cultivated (ploughed) horizon;
 Ag   for a poorly drained surface horizon.

E   A mineral horizon, just below the soil surface, which has lost clay, organic matter or iron by downward movement. Subdivisions include:

 Eg   for poorly drained horizons.

B   A subsurface mineral horizon resulting from the change *in situ* of soil material or the washing in of material from overlying horizons. Subdivisions include:

 Bg   for poorly drained;
 Bh   for accumulation of humus;
 Bs   for an illuvial accumulation of iron or aluminium sesquioxides;
 Bt   for increase of clay;
 Bw   for changes of colour or structure;
 Bx   for compact brittle horizon known as a fragipan;
 By   for enrichment with gypsum;
 Bz   for accumulation of salts more soluble than gypsum.

C   An unconsolidated or weakly consolidated mineral horizon which retains evidence of rock structure and lacks the properties diagnostic of the overlying A, E or B horizons. Subdivisions include:

 Cg   for poorly drained;
 Ck   for enrichment with calcium carbonate;
 Cm   for cemented material;
 Cx   for compact brittle material known as a fragipan;
 Cy   for enrichment with gypsum;
 Cz   for accumulation of salts more soluble than gypsum.

R   Continuous hard or very hard bedrock.

soil horizon *Soil horizon nomenclature for a soil profile: the main horizons (left) and their subdivisions (right)*

soil loss   The amount of soil removed from part of a slope or DRAINAGE BASIN, usually expressed in terms of weight per unit area or weight per unit area per unit of time. If the BULK DENSITY of the soil is known or can be estimated, soil loss may be expressed in terms of depth of soil removed. It can be determined or estimated using a variety of techniques (e.g. EROSION PLOT, EROSION PINS, SOIL MICROPROFILING DEVICE, SEDIMENT YIELD).   *RAS*

[See also SOIL EROSION, SOIL LOSS TOLERANCE, EROSION RATE]

soil loss tolerance   The theoretical maximum acceptable rate of SOIL EROSION for a particular soil type in order to prevent deterioration in the quality of agricultural land such that it has to be abandoned. Ideally, this rate should be equivalent to the rate of soil formation, but the latter is difficult to measure and figures are crude estimates. Alternatively, they may be derived from rates of PALAEOSOL formation where data are available. In practice, quoted soil loss tolerances for cultivated lands are probably only sustainable at best over the medium term (25–50 years) before serious SOIL DEGRADATION occurs.   *RAS*

[See also DENUDATION RATES, EROSION RATES, SOIL CONSERVATION]

Alexander, E.B. 1988: Strategies for determining soil-loss tolerance. *Environmental Management* 12, 791–796. **Johnson, L.C.**

1987: Soil loss tolerance – fact or myth? *Journal of Soil and Water Conservation* 42, 155–160. **Kirkby, M.J.** 1980: The problem. In Kirkby, M.J. and Morgan, R.P.C. (eds), *Soil erosion*. Chichester: Wiley, 183–216. **Morgan, R.P.C.** 1995: *Soil erosion and conservation*, 2nd edn. Harlow: Longman, 96–97. **Ramos, M.C. and Porta, J.** 1997: Analysis of design criteria for vineyard terraces in the Mediterranean area of north-east Spain. *Soil Technology* 10, 155–166.

soil microprofiling device

A removable frame or bar mounted on stable datum points (usually a metal stake or stakes anchored in the ground), from which the micro-topography of the soil can be determined in two or three dimensions and/or soil deposition/erosion can be monitored by means of repeat-measurements. Many devices involve lowering to the ground a metal rod or rods through holes in the frame or bar and measuring the amount of rod extension above the bar or frame. Electronic distance measurement has been incorporated in some variants. For determining SOIL EROSION and/or deposition, a soil microprofiling device has a major advantage compared with EROSION PINS in that at the points of measurement it does not interfere with the processes under investigation (e.g. *overland flow* and *rainsplash erosion*). *RAS*

[See also EROSION PLOT, GERLACH TROUGH]

Hudson, N. 1964: Field measurements of accelerated erosion in localised areas. *Rhodesian Agricultural Journal* 61(3), 46–47 & 60. **McCool, D.K., Dossett, M.G. and Yecha, S.J.** 1981: A portable rill meter for field measurement of soil loss. *International Association of Hydrological Science Publication* 133, 479–484. **Shakesby, R.A.** 1993: The soil erosion bridge: a device for micro-profiling soil surfaces. *Earth Surface Processes and Landforms* 18, 823–827. **Wells, A.W. and Bennett, M.R.** 1996: A simple portable device for the measurement of ground loss and surface changes. *Zeitschrift für Geomorphologie Supplementband* 106, 255–265.

soil mining

The concept that human-induced SOIL EROSION is depleting an essentially NON-RENEWABLE RESOURCE. Thus, for agricultural practices that cause more soil to be eroded than is being renewed, the soil is being removed at an unsustainable rate. The concept is also applicable to the reduction in soil nutrients. The scientific basis for extrapolation of soil losses on agricultural land at large scales (regional, continental or global), however, may be extremely weak. *SHD*

[See also SOIL LOSS TOLERANCE]

Brown, L.R. 1981: World-population growth: soil erosion and food security. *Science* 214, 995–1002. **Mwalyosi, R.B.B.** 1992: Land-use changes and resource degradation in south-west Masailand, Tanzania. *Environmental Conservation* 19, 145–152.

soil moisture

Water contained in the soil and available for plants in the rooting zone. After excess water has drained away following rain or irrigation, the soil is said to be at FIELD CAPACITY. As a soil dries, moisture is held increasingly tightly until it becomes unavailable for plants at the permanent *wilting point*. The available *water capacity* of a soil is the moisture content between these two points. The forces necessary to draw moisture from a soil are measured in kilopascals, and are collectively referred to as the *potential of soil water* (Φ), the amount by which the free energy of water in the soil is reduced by gravitational, osmotic and matrix (capillary) potentials. The soil mois-

ture regime imposes strict limitations upon the periods in the year when poorly drained soils can be cultivated and subjected to heavy traffic without damage. *EMB*

[See also SOIL DRAINAGE]

Robson, J.D. and Thomasson, A.J. 1977: *Soil water regimes* [*Technical Monograph* 11]. Harpenden: Soil Survey of England and Wales.

soil moisture deficit (SMD)

The amount of soil moisture that is drawn upon by plants in order to make good the amount by which EVAPORATION exceeds PRECIPITATION. It can be estimated from lysimeter measurements or from evaporation estimation equations. SMD values are important in the estimation of RUNOFF and IRRIGATION requirements. *JAM*

[See also FIELD CAPACITY]

Calder, I.R., Harding, R.J. and Rosier, P.T.W. 1983: An objective assessment of soil water deficit models. *Journal of Hydrology* 60, 329–355. **Kramer, P.J. and Boyer, J.S.** 1995: *Water relations of plants and soils*. San Diego, CA: Academic Press.

soil phase

A subdivision of a SOIL SERIES indicating the influence of slope or stoniness affecting the behaviour of the soil. *EMB*

soil pollution

Concern has developed rapidly during the past decade regarding pollution of the soil by toxic substances, mainly the products of INDUSTRIALISATION. Soils have conventionally been used in WASTE MANAGEMENT. Farmyard manure and, in the past, town MIDDENS have been used to fertilise the land for crops. Unfortunately, these wastes increasingly contain HEAVY METALS that are toxic to plants and soil fauna. ACID RAIN increases the effectiveness of LEACHING, leading to ACIDIFICATION, and fallout of RADIONUCLIDES from testing and NUCLEAR ACCIDENTS has contaminated soils of most areas. Wastes from smelting and metal-working, gas works, oil refining, chemical factories and scrap yards may lead to extreme pollution of sites. Although the worst sites may have to be encapsulated and isolated by physical barriers, the biological functions of many soils polluted by toxic wastes can be restored using chemical, physical and bacteriological methods. *EMB*

[See also SOIL RADIOACTIVITY, SOIL RESTORATION]

Bridges, E.M. 1987: *Surveying derelict land*. Oxford: Clarendon Press. **Bridges, E.M.** 1991: Dealing with contaminated soils. *Soil use and management* 7, 151–158.

soil profile

The soil as revealed in a vertical section, from the organic material at the surface, through the horizons of the mineral soil, to the PARENT MATERIAL beneath (see Figure). This concept is usually extended so that the minimum amount of material that can be considered to be a soil is a three-dimensional body of soil, or PEDON. The use of a uniform system of soil profile description is desirable for international co-operation and transfer of scientific data. Descriptions are normally made under two headings: general information about the site and specific information about the soil profile. General information must include a unique profile number that refers to the location and map, such as is provided by a national grid reference system; co-ordinates will be increasingly required as global positioning equipment becomes more

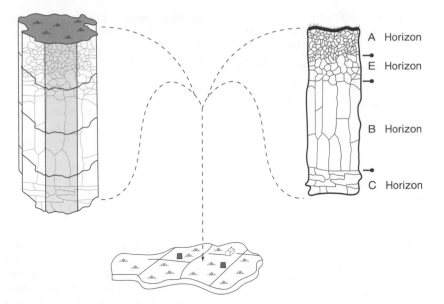

soil profile *An area of land with its soil shown as a pedon (left) and a profile (right) (after Bridges, 1997)*

widely used. SOIL HORIZON description includes horizon depth and clarity, colour, texture, structure, consistence, porosity, clay skins, cementation or compaction, nodules, roots, biological features, carbonates, gypsum and soil reaction. All features should be recorded and coded so that they may readily be transferred into a DATABASE for easy storage and retrieval. *EMB*

[See also SOIL FORMING FACTORS, SOIL FORMING PROCESSES]

Bridges, E.M. 1997: *World soils*, 3rd edn. Cambridge: Cambridge University Press. **Food and Agriculture Organization of the United Nations (FAO)** 1990: *Guidelines for soil profile description.* Rome: FAO. **Hodgson, J.M. (ed.)** 1974: *Soil survey field handbook* [*Technical Monograph* 5]. Harpenden: Soil Survey of England and Wales. **McRae, S.G.** 1988: *Practical pedology: studying soils in the field.* Chichester: Ellis Horwood. **Soil Survey Staff** 1995: *Soil survey manual.* Washington, DC: United States Department of Agriculture.

soil quality

The capacity of a soil to sustain biological productivity, for the sustainable production of healthy and nutritious crops, and its structural and biological integrity, which help resist SOIL EROSION and SOIL POLLUTION. Soil quality has been recently defined as 'the capacity of a specific type of soil to function, within natural or managed ecosystem boundaries, to sustain plant and animal productivity, maintain or enhance water and air quality and support human health and habitation' (Karlen *et al.*, 1997: 6).

The search is currently taking place for suitable criteria to use in the assessment of soil quality, but the inherent variability of soils makes this difficult. A soil of high quality for plant life should be deep, not have extreme particle-size characteristics; it should be well structured, possess adequate reserves of plant NUTRIENTS and be able to hold sufficient moisture to supply plants without stress. The characteristics that are commonly determined by soil scientists that reflect these qualities are the depth of the profile and its horizons, the texture, the structure, the chemical fertility and the water-holding capacity. *Soil fertility* may often be improved by increasing the content of organic matter, changing the pH and increasing the reserve of plant nutrients. Equally, through mismanagement these same properties can be degraded. *EMB*

[See also SOIL DEGRADATION, LAND EVALUATION]

Allan, D.L. and others, 1995: Soil Science Society of America statement on soil quality. *Agronomy News* June, 7. **Eijsackers, H.** 1998: Soil quality assessment in an international perspective: generic and land-use based quality standards. *Ambio* 27, 70–77. **Food and Agriculture Organization of the United Nations (FAO)** 1997: *Land quality indicators and their use in sustainable agriculture and rural development* [*Land and Water Bulletin* 5]. Rome: FAO. **Karlen, D.L., Mausbach, M.J., Doran, J.W. et al.** 1997: Soil quality: a concept, definition and framework for evaluation. *Soil Science Society of America Journal* 61, 4–10. **Pieri, C., Dumanski, J., Hamblin, A. and Young A.** 1995: *Land quality indicators* [*Discussion Paper* 315]. Washington, DC: World Bank.

soil radioactivity

Following the testing of nuclear weapons and NUCLEAR ACCIDENTS, contamination of the soil by RADIONUCLIDES has become a significant issue. Ten years after the *Chernobyl accident*, a large area of Ukraine remains cordoned off, and in certain areas affected by FALLOUT, grazing animals are still taking up too much radioactivity for their meat to be sold. Surveys of the amount of ^{134}Cs, ^{137}Cs and plutonium in soils have been carried out, as well as of the transfer of ^{90}Sr and ^{137}Cs from soil to crops. Small quantities of plutonium (released into the Irish Sea from Sellafield) have been involved in sea to land transfer on the north coast of Wales. The problem of soil radioactivity is greatest on shallow, organic-rich acid soils, and particularly in flushes associated with these soils in southwest Scotland, the Lake District and North Wales. On soils with calcareous clays, cation exchange locks the radionuclides into the lattice of the CLAY MINERALS and they are not circulated through the ecosystem. ^{137}Cs has been used as a marker in studies of SOIL EROSION. *EMB*

[See also CAESIUM-137, RADON EMANATION]

Cawse, P.A., Cambray, R.S., Baker, S.J. and Burton, P.J. 1988: *A survey of background levels of environmental radioactivity in Wales, 1984–1986.* Harwell: United Kingdom Atomic Energy Authority. Mayes, R.W., Atkinson, D. and Shepherd, H. 1987: *Radioactive contamination and agricultural systems.* Aberdeen: Annual Report of the Macaulay Land Use Research Institute. Walling, D.E. and Quine, T.A. 1991: Use of ^{137}Cs measurements to investigate soil erosion on arable fields in the UK. *Journal of Soil Science* 42, 147–165.

soil reclamation

In order to reclaim soils for productive use after they have been damaged by erosion, exhaustion or pollution, it is necessary to bring about the revitalisation of all soil functions. To achieve this result, it is easier to work with nature than attempt an imposed solution. The means of reclamation include raising pH where acidification has occurred (a higher pH also assists in reducing the mobility of toxic metals from soil to plants). Raising fertility through natural or synthetic manures is necessary where the soil's natural fertility has been 'mined' by overgrazing or cropping to exhaustion (*fertility mining*). Where soils have been badly eroded, reclamation may include TERRACING, grading slopes and eradicating gullies, as well as increasing soil organic matter content to reduce the risk of erosion.

Reclamation of a site requires RESTORATION of the physical form of the landscape and the CONSERVATION of any available soil material. It is necessary to ensure as far as possible that soil physical conditions (SOIL STRUCTURE) and soil chemical conditions (*soil fertility*) are satisfactory for plant growth. When soils have been in storage heaps for long periods, anaerobic conditions develop, greatly reducing the numbers of earthworms and mycorrhiza associated with plant roots are greatly diminished in number. In cases where extreme SOIL POLLUTION has occurred by organic chemicals, it may be necessary to pass the soil through a high-temperature incinerator. The resulting 'cleaned' soil is inert and completely lacking in organic matter and plant nutrients. Alternative bacterial soil-cleaning methodologies are also becoming available. It requires careful management to reintroduce organic matter and life back into the soil. *EMB*

[See also LAND RECLAMATION]

Bradshaw, A.D. 1987: The reclamation of derelict land and the ecology of ecosystems. In Jordan, W.R., Gilpin, M.E. and Aber, J.D. (eds), *Restoration ecology.* Cambridge: Cambridge University Press. Bradshaw, A.D. and Chadwick, M.J. 1980: *The restoration of land.* Oxford: Blackwell. Bridges, E.M. 1988: *Surveying derelict land [Monographs of Soil and Resources Survey* 13]. Oxford: Clarendon Press. Bridges, E.M. 1991: Dealing with contaminated soils. *Soil Use and Management* 7, 151–158.

soil regeneration

'The reformation of degraded soil through biological, chemical and/or physical agencies' (Johnson *et al.*, 1997: 586). Following SOIL DEGRADATION, soil can regenerate naturally over time during PEDOGENESIS; it may, however, be aided and accelerated artificially by humans. *JAM*

[See also ECOLOGICAL RESTORATION, LAND RECLAMATION, LAND RESTORATION, SOIL RECLAMATION]

Johnson, D.L., Ambrose, S.H., Bassett, T.J. *et al.* 1997: Meanings of environmental terms. *Journal of Environmental Quality* 26, 581–589.

soil resilience

The ability of a soil to restore its living systems after disturbance. If the DISTURBANCE is too profound, the soil may undergo irreversible change and the whole system is jeopardised. It is important to realise that the resilience of soils to different forms of change varies greatly and this should be taken into account when alternative LANDUSE is proposed. Similarly some soils will have greater resilience to the effects of CLIMATIC CHANGE than others. *EMB*

Lal, R. 1994: Sustainable land use systems and soil resilience. In Greenland, D.J. and Szabolcs, I. (eds), *Soil resilience and land use.* Wallingford: CAB International.

soil series

The main mapping unit for medium-scale soil maps, defined as soils with similar profiles derived from similar materials under similar conditions of development. The series may be divided into phases, where stoniness or slope plays a significant part in influencing soil behaviour or soil types that are differentiated by topsoil TEXTURE differences. On a generalised soil map, a number of soil series may be grouped together. *EMB*

Robinson, G.W. 1949: *Soils, their origin constitution and classification.* London: Thomas Murby.

soil stratigraphy

The use of PALAEOSOLS or GEOSOLS that evolved over a specific soil-forming interval and are traceable over a wide area as stratigraphic units or *pedostratigraphic units* (see STRATIGRAPHY). Such pedostratigraphic units represent MARKER HORIZONS, may represent a major HIATUS in the lithostratigraphic record (see LITHOSTRATIGRAPHY) and/or may provide evidence of more than one episode of SOIL DEVELOPMENT. *JAM*

[See also POLYGENETIC]

Kemp, R.A., Whiteman, C.A. and Rose, J. 1993: Palaeoenvironmental and stratigraphic significance of the Valley Farm and Barham Soils in eastern England. *Quaternary Science Reviews* 12, 833–848. Morrison, R.B. 1978: Quaternary soil stratigraphy: concepts, methods and problems. In Mahaney, W.C. (ed.), *Quaternary soils.* Norwich: GeoAbstracts, 77–108.

soil structure

Soil particles adhere to each other to form larger natural aggregates known as *peds* (see Figure). The shape of peds is described as blocky if all three axes are of equal length, subdivided into either angular or subangular blocky according to their shape. The faces of angular blocky peds intersect at sharp angles and are casts of the faces of surrounding structures; subangular blocky peds have rounded faces. Where the vertical axis is greater than the two horizontal axes, they are described as prismatic, or with rounded tops (associated with SOLONETZ) as columnar. Platy structures have a short vertical axis and the two horizontal axes are greater. Crumb or granular structures are irregular spheroids. A soil lacking structure is described as *apedal* and may be massive or single grain as in a loose sand. Soil structures are described in terms of degree of development, size and shape. Soil structure influences the INFILTRATION rate of water, rootability and the vulnerability to SOIL EROSION. It may be affected by management, especially the use of heavy farm machinery and, in the longer term, by climate and other environmental change. *EMB*

Food and Agriculture Organization of the United Nations (FAO) 1990: *Guidelines for soil description.* Rome: FAO. Soil Survey Staff 1981: *Soil Survey manual.* Washington, DC: United States Department of Agriculture.

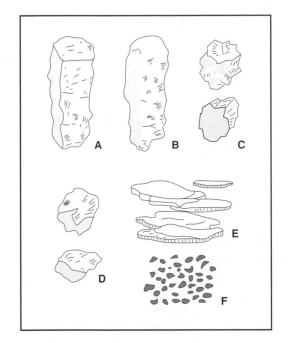

soil structure *Soil structures formed by the aggregation of sand, silt and clay particles: (a) prismatic; (b) columnar; (c) angular blocky; (d) subangular blocky; (e) platy; (f) granular or crumb (after Soil Survey Staff, 1981)*

Soil Taxonomy

A comprehensive system of classification of soils used by the Soil Survey Staff of the United States Department of Agriculture (USDA) since AD 1965. It continues to be developed up to the present day. Initially, ten Soil Orders were recognised: Entisols, Inceptisols, Mollisols, Alfisols, Ultisols, Oxisols, Vertisols, Aridisols, Spodosols and Histosols, their names being derived from Greek or Latin roots that convey the concept of the soil character or genesis. In the following example from the Mollisols, the element 'oll' is carried throughout as the order name, the prefix 'aqu' indicates the presence of poor drainage and 'argi' indicates the presence of a clay-rich subsoil horizon:

- Mollisol – *soil order*;
- Aquoll - *soil suborder*;
- Argiaquoll – *soil great group*;
- Typic Argiaquoll – *soil subgroup*.

Below the category of subgroup are further categories, the *soil family* including soils with similar texture, and SOIL SERIES indicating the geographical locality where the soil occurs and was first described. To place a soil in the system it is necessary first to identify the presence of certain DIAGNOSTIC SOIL HORIZONS, many of which are similar to those described for the WORLD REFERENCE BASE FOR SOIL RESOURCES, the system used throughout this volume. As Soil Taxonomy developed, two additional orders were introduced: *Andisols*, for soils on volcanic materials and *Gelisols* for soils with permafrost. Although developed in the USA, this comprehensive system is widely used for international comparison of soils and transfer of technical information about soils. *EMB*

Buol, S.W., Hole, F.D., McCracken, R.J. and Southward, R.J. 1997: *Soil genesis and classification*. Ames, IA: Iowa State University Press. **Soil Survey Staff** 1999: *Soil taxonomy: a basic system of soil classification for making and interpreting soil surveys*, 2nd edn. [*Agriculture Handbook* 436] Washington, DC: United States Department of Agriculture. **Soil Survey Staff** 1998: *Keys to Soil Taxonomy*, 8th edn. Washington, DC: United States Department of Agriculture.

soil texture The particle-size distribution in the soil often described in terms of the relative proportions of various size fractions. Soil texture is probably the single most important soil property influencing SOIL PROFILE morphology, water-holding capacity and water release characteristics, SOIL DRAINAGE and permeability, workability, SOIL QUALITY and fertility.

Soil particle-size distribution is determined on that proportion of a soil sample (the fine earth) that passes through a 2 mm round-hole sieve: anything larger is described as 'stones'. Early in the twentieth century, soil scientists agreed that the size limits of the three broad particle-size categories – sand, silt and clay – would be based on a logarithmic scale: sand refers to particles with an equivalent diameter of 2.0–0.2 mm; silt particles are 0.2–0.02 mm; and clay particles are < 0.002 mm. This has remained in international usage with only minor modifications, although the description of the TEXTURE of sedimentary rocks and the GRAIN-SIZE of sediments (often referred to as 'soils' by engineers) commonly placed the sand/silt boundary at 0.063 mm.

Soil texture may be determined in the field by moistening the sample, working it between finger and thumb and, with experience, recognising the three constituent grain-size categories as gritty (sand), silky (silt) and sticky (clay). In the laboratory, accurate grain-size distributions are determined, after suitable PRETREATMENT of the sample using sieving for sand fractions and pipette analysis or more sophisticated machinery (such as a Coulter counter or Sedigraph) for the finer fractions. A larger number of descriptive classes, such as silty clay, sandy clay, loam (an approximately equal mixture of sand, silt and clay), silt loam and loamy sand, are usually employed in the description of soil texture. These may be shown diagrammatically on a triangular graph (see Figure). Soil particle-size distributions may also be shown graphically as a cumulative curve or HISTOGRAM. *EMB*

Bridges, E.M. 1997: *World soils*. Cambridge: Cambridge University Press. **FAO** 1990: *Guidelines for soil description*. Rome: FAO. **McRae, S.G.** 1988: *Practical pedology: studying soils in the field*. Chichester: Ellis Horwood. **Hodgson, J.M. (ed.)** 1974: *Soil survey field handbook* [*Technical Monograph* 5]. Harpenden: Soil Survey of England and Wales. **Soil Survey Staff** 1981: *Soil survey manual*. Washington, DC: United States Department of Agriculture.

soil vulnerability Some soils can absorb, retain and circulate CONTAMINANTS better than others. Soils that are at greatest risk of being harmed by EROSION and the different forms of POLLUTION are said to be vulnerable. Soil vulnerability is described as 'the capability for the soil system to be harmed in one or more of its ecological functions' (Batjes and Bridges, 1991). Certain soils will also be at risk from the effects of climatic change. *EMB*

[See also SOIL RESILIENCE]

Batjes, N.H. and Bridges, E.M. (eds) 1991: *Mapping of soil and terrain vulnerability to specific chemical compounds in Europe at*

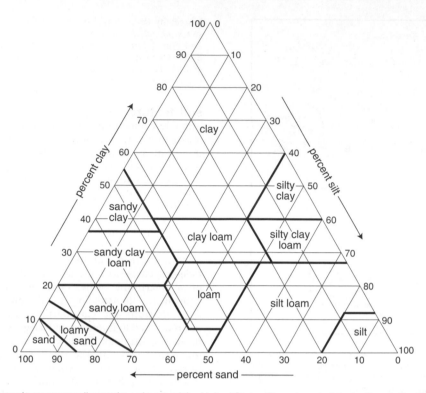

soil texture *Soil texture classes according to the scheme of the United States Department of Agriculture (after Soil Survey Staff, 1981)*

a scale of 1:5M. [*Proceedings of an international workshop organised in the framework of the chemical time bombs project of The Netherlands Ministry of Housing, Physical Planning and Environment, and the International Institute for Applied Systems Analysis*]. Wageningen: International Soil Reference and Information Centre. **Batjes, N.H. and Bridges, E.M.** 1993: Soil vulnerability to pollution in Europe. *Soil Use and Management* 9, 25–29. **Rounsevell, M.D.A. and Loveland, P.J. (eds)** 1994: *Soil response to climate change.* Berlin: Springer.

soilmarks   Patterns in bare soil that are recognised by archaeologists as indicating former modification of the LANDSCAPE by human activity. Often only detectable using AERIAL PHOTOGRAPHY or other forms of REMOTE SENSING, soilmarks include the outlines of building, field patterns and ditches obscured by later arable cultivation but nevertheless detectable because of variations in soil properties such as colour or moisture (*dampmarks*) or the differential melting of frozen ground (*frostmarks*) or snow (*snowmarks*). *JAM*

[See also CROPMARKS]

Braasch, O. 1983: *Luftbildarchäologie in Süddeutschland.* Stuttgart: Gustav Fischer.

solar constant   The rate at which energy from the Sun is received just outside the Earth's atmosphere on a surface normal to the incident radiation and at the Earth's mean distance from the Sun. It is approximately 1370 W m^{-2}. *TS*

solar cycle   A PERIODICITY in the relative activity and inactivity of the Sun leading to variations in SOLAR RADIA-TION. The best known solar cycle is associated with the 11-year SUNSPOT CYCLE. *RDT*

[See also SOLAR FORCING]

solar flare   A short-term high-energy event associated with the surface of the Sun, which is accompanied by an increase of COSMIC RADIATION as well as the emission of high-energy particles in the solar spectrum, especially in the ultraviolet wavelengths. A single sunspot may be characterised by up to 40 solar flares during its life span. *RDT*

[See also SOLAR CYCLES, SUNSPOT CYCLES]

solar forcing   CLIMATIC CHANGE caused by variation in INSOLATION received from the Sun. Although the Sun is the most important driving force of the CLIMATIC SYSTEM, the extent to which climatic change can be attributed to solar forcing remains a contentious issue. Significant variations in SOLAR RADIATION occur in the short term (diurnal/seasonal patterns), medium term (SUNSPOT CYCLES) and long term (ORBITAL FORCING associated with MILANKOVITCH THEORY). These variations relate to both the Sun's output and the variable distance of the Sun from the Earth.

There is a clear correlation between solar irradiance and the 11-year sunspot cycle, but measurements made over the last two decades suggest that the variability is too small to affect climate. There are similar difficulties with accepting solar forcing as a major cause of century to millennial-scale climatic changes. However, on these timescales intensification of solar irradiance may have been amplified to influence climatic episodes such as the MEDIAEVAL

WARM PERIOD and LITTLE ICE AGE. Possible mechanisms may involve interactions with CLOUDS, solar ultraviolet RADIATION or the SOLAR WIND (see Figure). There is growing evidence that so-called minima in solar activity (such as the MAUNDER MINIMUM) coincided with glacier advances, LAKE-LEVEL VARIATIONS and other PROXY records. The record of COSMOGENIC ISOTOPES in NATURAL ARCHIVES provides a relatively new approach to reconstructing past solar variability.

The 100 000 year Milankovitch cycle due to changes in the ellipticity (ECCENTRICITY) of the Earth's orbit around the Sun seems particularly influential in the longer term. When the Earth is closest to the Sun, greater insolation and higher surface temperatures lead to INTERGLACIAL conditions and general GLACIER retreat. Use of a non-linear regression model to separate natural and anthropogenic forcing since AD 1850 is consistent with a solar contribution of about 40% to the GLOBAL WARMING during the last 140 years. *RDT/JAM*

Beer, J., Mende, W. and Stellmacher, R. 2000: The role of the Sun in climatic forcing. *Quaternary Science Reviews* 19, 403–415. **Chambers, F.M., Ogle, M.I. and Blackford, J.J.** 1999: Palaeoenvironmental evidence for solar forcing of Holocene climate: linkages to solar science. *Progress in Physical Geography* 23, 181–204. **Friis-Christensen, E. and Lassen, K.** 1991: Length of the solar cycle: an indicator of solar activity closely associated with climate. *Science* 254, 698–700. **Geel, B. van, Raspopov, O.M., Renssen, H. *et al.*** 1999: The role of solar forcing upon climate change. *Quaternary Science Reviews* 18, 331–338. **Haigh, J.D.** 1996: The impact of solar variability on climate. *Science* 272, 981–984. **Lassen, K. and Friis-Christiansen, E.** 1995: Variability of the solar cycle during the past five centuries and the apparent association with terrestrial climate. *Journal of Atmospheric and Terrestrial Physics* 57, 835–845. **Magny, M.** 1993: Solar influences on Holocene climatic changes illustrated by correlation between past lake level fluctuations and the atmospheric ¹⁴C record. *Quaternary Research* 40, 1–9. **Nesme-Ribes, E. (ed.)** 1994: *The solar engine and its influence on terrestrial atmosphere and climate.* Berlin: Springer. **Pap, J.M., Fröhlich, C., Hudson, H.S. and Solanki, S.K. (eds)** 1994: *The Sun as a variable star.* Cambridge: Cambridge University Press. **Sonett, C.P., Giampapa, M.S. and Matthews, M.S. (eds)** 1991: *The Sun in time.* Tucson, AZ: University of Arizona Press. **Stuiver, M. and Braziunas, T.F.** 1993: Sun, ocean, climate and atmospheric ¹⁴CO₂: an evaluation of causal and spectral relationships. *The Holocene* 3, 289–305. **Svensmark, H. and Friis-Christensen, E.** 1997: Variation of cosmic ray flux and global cloud coverage – a missing link in solar–climate relationships. *Journal of Atmospheric and Terrestrial Physics* 59, 1225–1232. **Thompson, R. D.** 1998: *Atmospheric processes and systems.* London: Routledge, 19–39.

solar radiation The electromagnetic radiation emitted by the Sun. Because the temperature of the Sun is about 5800°K, some 99.9% of solar radiation is *shortwave radiation* with wave lengths in the range 0.1 to 4.0 μm and peak intensity at 0.47 μm. About one half of the energy in the *solar beam* (direct radiation) is within the visible spectrum: most of the remainder is in the near infrared with a small minority in ultraviolet wavelengths.

Globally, about 30% of the incident solar radiation reaching the Earth's atmosphere is reflected back to space, especially by CLOUDS: of the remaining 70%, some two thirds is eventually absorbed by the Earth's surface and

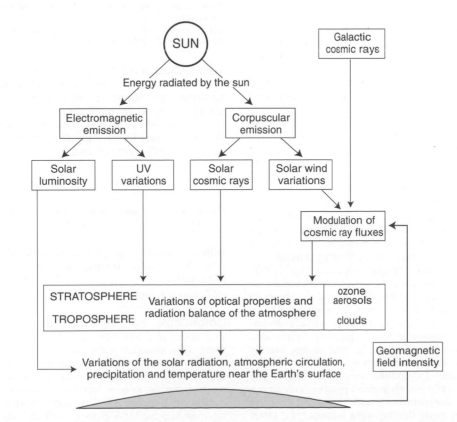

solar forcing *Possible amplifying factors in the solar forcing of the Earth's climate (after van Geel et al., 1999)*

one third is absorbed by the atmosphere (mostly by water vapour). The atmosphere reduces the *direct radiation* received at the Earth's surface to around 5% because most of the solar radiation is transmitted to the surface as *diffuse radiation* by SCATTERING. Solar radiation is measured using a *solarimeter*.   *JAM*

[See also BLACK-BODY RADIATION, ELECTROMAGNETIC SPECTRUM, ENERGY BALANCE, RADIATION BALANCE]

Coulson, K.P. 1975: *Solar and terrestrial radiation*. London: Academic Press. **Iqbal, M.** 1983: *An introduction to solar radiation*. New York: Academic Press.

solar wind The outflow of charged particles from the Sun, which interacts with the Earth's magnetic field (in the polar atmosphere) to produce magnetic storms and the spectacular, fluorescent phenomenon called an *aurora*.
                                                        RDT

sole marks SEDIMENTARY STRUCTURES that are preserved on the basal surfaces (soles) of beds of SANDSTONE or, less commonly, CONGLOMERATE or BRECCIA. Many are infills (CASTS) of marks produced by EROSION of a mud surface by a strong current and infilled by subsequent deposition (see PROD MARKS, SCOUR MARKS and TOOL MARKS). They are valuable indicators of PALAEOCURRENT directions and the original *way-up* of sequences of SEDIMENTARY ROCKS that have undergone severe TECTONIC deformation. TRACE FOSSILS are also commonly preserved as sole marks.
                                                        GO

Collinson, J.D. and Thompson, D.B. 1992: *Sedimentary structures*, 2nd edn. London: Chapman and Hall.

solifluction A widespread MASS-WASTING process occurring in periglacial regions. The term was first used to describe the slow flowage, under gravity, of near-surface water-saturated regolith observed in the Falkland Islands. Today, solifluction is generally regarded as a cold-climate process, comprising GELIFLUCTION (slow saturated flowage during thaw consolidation of ice-rich sediments) and *frost creep*. The latter is the process by which individual soil particles are moved down-slope in a ratchet-like manner, but the former accounts for most of the movement except in relatively dry environments. During a FREEZE–THAW CYCLE, expansion of the soil normal to the surface occurs, but then it settles in a more nearly vertical direction. Suitable conditions for thaw consolidation (i.e. settlement of sediment following melting) occur wherever the downward percolation of water through the soil is limited or where the melting of segregated ice lenses in the thawing soil provides excess water, which reduces internal friction and cohesion between the sediment particles.

Solifluction, in the modern sense, operates slightly differently according to whether the ground is seasonally or perennially frozen; in the former, the ground is subject to one-sided and in the latter to two-sided freezing. Under one-sided freezing (i.e. freezing from the surface downwards, typical of seasonally frozen ground) solifluction consists of only two components: (a) frost creep, usually in the autumn or in association with repeated diurnal frost throughout the year; (b) gelifluction, occurring in the spring as the seasonally frozen ground progressively thaws from the surface downwards. Under two-sided freezing (i.e. freezing from both the top and bottom, the latter because of the presence of PERMAFROST), there may be an additional mass-wasting process. This is the '*plug-like flow*' that occurs in late summer when the thawed active layer slides, *en masse*, across the lubricated slip plane provided by the ice-rich zone at the top of permafrost.

Rates of solifluction movement vary, but typically range between 0.5 and 4.0 cm y^{-1} near the surface. Total volumetric transport amounts in the western Canadian Arctic, where the active layer thickness is typically 50–70 cm, approximate 10–50 cm^3 cm^{-1} y^{-1}.

DIAMICTONS considered to be RELICT solifluction deposits have been reported from many locations throughout the world and are important indicators of ENVIRONMENTAL CHANGE. Diagnostic characteristics include: angular CLASTS (where upslope frost-weathered bedrock has been the source); low LIQUID LIMIT; a strong sediment FABRIC with clast long axes pointing in the direction of flow; and microscopic lens- or blade-shaped zones or cappings of segregated fine sediment indicative of repeated freezing and thawing.   *HMF*

[See also HEAD, PATTERNED GROUND, PERIGLACIAL LANDFORMS]

Benedict, J.B. 1970: Down-slope soil movement in a Colorado alpine region: rates, processes and climatic significance. *Arctic and Alpine Research* 2, 165–226. **Egginton, P.A. and French, H.M.** 1985: Solifluction and related processes, eastern Banks Island, N. W. T. *Canadian Journal of Earth Sciences* 22, 1671–1678. **Frenzel, B., Matthews, J.A. and Gläser, B. (eds)** 1993: *Solifluction and climatic variation in the Holocene* [*Paläoklimaforschung*. Vol. 11]. Stuttgart: Gustav Fischer. **Mackay, J.R.** 1981: Active layer slope movement in a continuous permafrost environment, Garry Island, Northwest Territories, Canada. *Canadian Journal of Earth Sciences* 18, 1666–1680. **Washburn, A.L.** 1967: Instrumental observations on mass wasting in the Mesters Vig District, Northeast Greenland. *Meddelelser öm Grønland*, 166, 1–318. **Washburn, A.L.** 1979: *Geocryology: a survey of periglacial processes and environments*. London: Edward Arnold.

solifluction terrace (1) A terrace-like accumulation of debris with steep frontal scarp and gently sloping tread resulting from SOLIFLUCTION in a PERIGLACIAL environment; essentially an amalgamation of solifluction lobes. (2) A terrace of solifluction sediments (HEAD) on the lower slopes of a river valley originating from fluvial downcutting; essentially a RELICT terrace carved by fluvial erosion.
                                                        PW

[See also GELIFLUCTION, MASS MOVEMENT PROCESSES, SOIL CREEP, SOLIFLUCTION]

Crampton, C.B. and Taylor, J.A. 1967: Solifluction terraces in south Wales. *Biuletyn Peryglacjalny* 16, 15–36.

solodisation The process of removal of sodium ions from the exchange complex of a SOLONETZ by the process of LEACHING.   *EMB*

[See also SOIL FORMING PROCESSES]

Solonchak Strongly saline soils with a SALIC HORIZON that occur in ARIDLANDS or SEMI-ARID conditions where there is a seasonal or permanent WATER TABLE near the soil surface. Salts accumulate in the upper horizon of the profile, or as an efflorescence on the soil surface, but they are washed out by rain or irrigation, only to re-form again during dry periods. An INTRAZONAL SOIL in the classification used in the USA before 1965, Solonchaks were also known as *White Alkali Soils*.   *EMB*

[See also WORLD REFERENCE BASE FOR SOIL RESOURCES]

Szabolcs, I. 1989: *Salt-affected soils*. Boca Raton, FL: CRC Press.

Solonetz

Soils with a subsurface horizon enriched with clay dominated by sodium salts (*natric horizons*). Solonetz occur in ARIDLANDS and SEMI-ARID areas where there is a predominance of sodium salts. As a result, the clay and organic matter in Solonetz is usually dispersed, making them extremely difficult to manage. Formerly, an INTRAZONAL SOIL in the classification used in the USA before AD 1965, Solonetz were known as *Black Alkali Soils*. *EMB*

[See also WORLD REFERENCE BASE FOR SOIL RESOURCES]

Szabolcs, I. 1989: *Salt-affected soils*. Boca Raton, FL: CRC Press.

solstice

The times (twice per year) when the overhead Sun is at its maximum or minimum declination and farthest from the Equator. On 21 June (*summer solstice* in the Northern Hemisphere) the Sun is directly overhead at the Tropic of Cancer (23.5°N); on 21 December (*winter solstice* in the Northern Hemisphere) the Sun is directly overhead at the Tropic of Capricorn (23.5°S). At the summer solstice, areas north of the Arctic Circle (66.5°N) receive 24 hours daylight and at the winter solstice the same area receives 24 hours darkness. *JAM*

solum

The upper part of the SOIL PROFILE including the eluvial A and illuvial B horizons. It is the genetic soil developed by the SOIL FORMING PROCESSES influenced by plant roots. *EMB*

solute

A substance in solution; the liquid in which a solute is dissolved is the *solvent*. *Solute concentration* is the amount of dissolved material (solutes), as given by the dry residue after filtering, per unit volume of water. Solute concentrations and *solute loads* (synonym: DISSOLVED LOADS) of streamwaters have been used to determine rates of CHEMICAL DENUDATION and the impact of vegetation change on nutrient losses. *ADT*

Trudgell, S.T. (ed.) 1986: *Solute processes*. Chichester: Wiley.

sorting

An attribute of the TEXTURE of a SEDIMENT or SEDIMENTARY ROCK that expresses the variation of GRAIN-SIZE, quantified as the standard deviation of the grain-size distribution. There are several established techniques for calculating a numerical value of sorting from different expressions of a grain-size distribution. In THIN-SECTION ANALYSIS and hand specimens of SANDSTONES, sorting is usually determined by visual comparison with standard charts. Standard verbal descriptions for numerical values of sorting are shown in the Table. Sorting is used in the determination of textural MATURITY, although it is commonly influenced more by the process of DEPOSITION than by the distance of SEDIMENT TRANSPORT. *TY*

[See also GREYWACKE, WACKE]

Folk, R.L. 1980: *Petrology of sedimentary rocks*. Austin, TX: Hemphill. **Folk, R.L. and Ward, W.** 1957: Brazos River bar: a study of the significance of grain size parameters. *Journal of Sedimentary Petrology*, 41, 1045–1058. **McManus, J.** 1988: Grain size determination and interpretation. In Tucker, M.E. (ed.), *Techniques in sedimentology*. Oxford: Blackwell, 63–85.

sorting *Standard verbal descriptions for numerical values of sorting, in phi (φ) units (after Folk, 1980)*

Sorting value (φ)	Descriptive term
< 0.35	Very well sorted
0.35–0.50	Well sorted
0.50–0.71	Moderately well sorted
0.71–1.0	Moderately sorted
1.0–2.0	Poorly sorted
2.0–4.0	Very poorly sorted
> 4.0	Extremely poorly sorted

Pettijohn, F.J. 1975: *Sedimentary rocks*, 3rd edn. New York: Harper & Row. **Pettijohn, F.J., Potter, P.E. and Siever, R.** 1987: *Sand and sandstone*. New York: Springer.

Southern Oscillation (SO)

An inter-annual CLIMATIC FLUCTUATION reflected in the periodic reversal (every 1–5 years) of pressure patterns above the equatorial Pacific Ocean. It is expressed by the *Southern Oscillation index*: the sea-level pressure difference between Tahiti in the eastern Pacific and Darwin in northern Australia. *JAM*

[See also EL NIÑO–SOUTHERN OSCILLATION, WALKER CIRCULATION]

Lockwood, J.G. 1984: The southern oscillation and El Niño. *Progress in Physical Geography* 8, 102–110. **Swetnam, T.W. and Betancourt, J.L.** 1990: Fire–Southern Oscillation relations in the southwestern United States. *Science* 249, 1017–1020. **Wright, P.B.** 1987: Southern oscillation. In Oliver, J.E. and Fairbridge, R.W. (eds), *The encyclopedia of climatology*. New York: Van Nostrand Reinhold, 796–800.

space agencies

National or international agencies with the mission to plan, direct and conduct aeronautical and space civil activities. *TS*

spaghetti data

Vector data composed of line segments that have no topological structure or geographical order. GIS ENTITIES need to have ATTRIBUTES and geometric structure for spatial analysis (for example OVERLAY ANALYSIS). *TVM*

spatial analysis

A collection of techniques concerned with the description and examination of data of which the positions in space, relative to each other, form an explicit part of the analysis. Such data maybe acquired from, for example, systematic or random sampling from FIELD RESEARCH, CARTOGRAPHY or REMOTE SENSING. The data may be regularly *gridded data* (as in the case of remotely sensed data), *latticed data* (such as data on postcode areas) or *distributed data* (global volcanic activity, for example). Some of the most important techniques used in spatial analysis are KRIGING, *semivariograms* and AUTOCORRELATION. Kriging (after D.G. Krige) is used to provide unbiased estimates of a variable at a given location based on the value and relative location of nearby data. It is often referred to by the acronym BLUE, for *best linear unbiased estimator*. Semivariograms and autocorrelation techniques provide measures of how data vary at different scales. *TQ*

[See also SPATIAL FILTERING]
Issaks, E.H. and Srivastava, R.M. 1989: *An introduction to applied geostatistics*. Oxford: Oxford University Press.

spatial filtering The application of a FILTERING in the spatial sense (e.g. across the two dimensions of the RASTER within IMAGE PROCESSING, often using a LOCAL OPERATOR). Such techniques can be grouped by the kind of transformation applied to the input data into *linear* (mean, for instance) and *non-linear* (median, for example). The effect of the filters also induces the classification into *low-pass* (which have a blurring effect), *high-pass* (that enhance borders and small features) and *band-pass*. The particular filters and the sequence in which they have to be applied depends on the original image and on the desired effect.

ACF

Jain, A.J. 1989: *Fundamentals of digital image processing*. Englewood Cliffs, NJ: Prentice-Hall.

spatial variation The distribution of differences between values of a random variable at spatially separated points or areas. Spatial variation can be a valuable additional source of information from REMOTE SENSING images. It can be quantified, for example, by spatial AUTO-CORRELATION, spatial autoregression (AUTOREGRESSIVE MODELLING), VARIOGRAM analysis, TEXTURE measures or the local COEFFICIENT OF VARIATION. Texture can be estimated model-free or be derived from a PROBABILITY distribution model, e.g. by fitting k-distributions with different parameters to *synthetic aperture radar* data. VEGETATION CANOPIES of different vegetation types having distinct spatial structures often result in textured remote sensing images. The use of spatial variation in the data can improve IMAGE CLASSIFICATION results.

HB

[See also SPATIAL ANALYSIS]
Luckman, A.J., Frery, A.C., Yanasse, C.C.F. and Groom, G.B. 1997: Texture in airborne SAR imagery of tropical forest and its relationship to forest regeneration stage. *International Journal of Remote Sensing* 18, 1333–1349. **Oliver, C. and Quegan, S.** 1998: *Understanding synthetic aperture radar images*. Boston, MA: Artech House. **Sheen, D. R. and Johnston, L. P.** 1992: Statistical and spatial properties of forest clutter measured with polarimetric synthetic aperture radar (SAR). *IEEE Transactions on Geoscience and Remote Sensing* 30, 578–588.

speciation The processes by which new species come into being. Scientific thinking on this topic was revolutionised by Charles Robert Darwin (1809–1882) and his theory of EVOLUTION by means of NATURAL SELECTION, the first plausible mechanism of a process by which one species might be transformed into another. Speciation might, in principle, take place gradually (PHYLETIC GRADUALISM) by the spread of new GENOTYPES through a population, or abruptly, especially following ISOLATION of populations, or by POLYPLOIDY (in plants). The contrast may also be expressed as *sympatric speciation* (development of a new species within the range of the ancestral species) or *allopatric speciation* (development of new species away from the range of the ancestral species, by transformation of an isolated population).

KDB

[See also FOUNDER EFFECT, GENETIC DRIFT, SPECIES CONCEPT]
Cronin, T.M. 1987: Speciation and cyclic climatic change. In Rampino, M.R., Sanders, J.E., Newman, W.S and Königsson,

L.K. (eds), *Climate: periodicity and predictability*. New York: Van Nostrand Reinhold, 333–342. **Darwin, C.** 1859: *On the origin of species by means of natural selection, or the preservation of favoured races in the struggle for life*. London: John Murray. **Mayr, E.** 1942: *Systematics and the origin of species from the viewpoint of a zoologist*. New York: Columbia University Press.

species–area relationship The very general tendency for diversity (as measured by the number of species) to increase with increasing area of islands or patches of terrestrial or aquatic habitats. Typically, species number increases steeply at first, but the rate of increase slows as island area increases.

MVL

[See also BIODIVERSITY, DIVERSITY CONCEPTS, ISLAND BIOGEOGRAPHY]
Rosenzweig, M.L. 1995: *Species diversity in space and time*. Cambridge: Cambridge University Press.

species concept The notion that objects, including organisms, can be collected into groups by their common and permanent characteristics, enabling distinctions to be made between such groups. When applied to organisms, a distinction can be made between the *species category* and the *species taxon*. The species category is a class whose members are species taxa. Species taxa have been delimited in many ways. Taxonomists have traditionally used an essentialist interpretation: the morphological species. However, ecologists and others have emphasised the importance of interbreeding populations, giving rise to the BIOLOGICAL SPECIES.

KDB

Mayr, E. 1982: *The growth of biological thought: diversity, evolution, and inheritance*. Cambridge, MA: Belknap Press of Harvard University Press.

species richness The number of species found within a given area or sample, also referred to as *species density*. This is the simplest and most commonly used measure of BIODIVERSITY.

JTK

[See also DIVERSITY INDICES, ENDEMIC]

species translocation The human-assisted, purposeful movement of individuals from one area followed by their free release into another. This is a general term that includes INTRODUCTION, *re-introduction* and *restocking* (movement of individuals to sites outside their native range, to sites within their historic but not current range and to sites with extant populations, respectively).

MVL

[See also EXOTIC SPECIES, NATIVE SPECIES]
Falk, D.A., Millar, C.I. and Olwell, M. (eds) 1996: *Restoring diversity: strategies for the reintroduction of endangered plants*. Covelo, CA: Island Press. **Griffith, B., Scott, J.M., Carpenter, J.W. and Reed, C.** 1989: Translocation as a species conservation tool: status and strategy. *Science* 245, 477–480.

specific heat The quantity of heat required to raise the temperature of a unit mass (1 g) of a substance by unit temperature (1°C).

JAM

speckle Statistical properties of PIXEL amplitudes in a coherent image or measurement, arising from the interactions of electromagnetic waves reflected from many objects in a resolution cell. Though speckle looks like *noise*, it is not because the same imaging configuration will produce the same speckle pattern.

PJS

spectral analysis A method for identifying characteristic frequencies in a time series (see TIME-SERIES ANALYSIS), which is also applicable to spatial series. In the frequency domain, it views a complex time series as a composite of many different PERIODICITIES. Dominant and/or other significant periodicities are identified by the *spectral density function*, which gives the proportion of the total variance in the time series accounted for at each frequency. Spectral analysis is widely used in PALAEOCLIMATOLOGY, where it has been instrumental in the testing and application of MILANKOVITCH THEORY and is currently much used to identify higher frequency (SUB-MILANKOVITCH) climatic periodicities. *JAM*

Stuiver, M. and Braziunas, T.F. 1993: Sun, ocean, climate and atmospheric $^{14}CO_2$: an evaluation of causal and spectral relationships. *The Holocene* 3, 289–305. **Yiou, P., Baert, E. and Loutre, M.F.** 1996: Spectral analysis of climate data. *Surveys in Geophysics* 17, 619–663.

Spectral Mapping Project timescale (SPECMAP) The standard, globally applicable chronology developed from OXYGEN ISOTOPE evidence in MARINE SEDIMENT records. By combining or *stacking* a series of oxygen isotope sequences to eliminate local anomalies, smoothing the record and then using ORBITAL TUNING to relate these records to known astronomical variations as predicted by MILANKOVITCH THEORY, it has proved possible to define a single chronology for the QUATERNARY utilising oxygen ISOTOPIC STAGES. It originally developed out of the successful *CLIMAP* approach. *CJC*

Imbrie, J., Hays, J.D., Martinson, D.G. *et al.* 1984: The orbital theory of Pleistocene climate: support from a revised chronology of the marine $\delta^{18}O$ record. In Berger, A., Imbrie, J., Hays, J. *et al.* (eds), *Milankovitch and climate*. Dordrecht: Reidel, 269–306.

spectrophotometer A type of instrument that measures attenuation of light passing through a solution. All spectrophotometers include an energy source, an energy spreader for choice of wavelength and an energy detector. *JAM*

[See also ATOMIC ABSORPTION SPECTROPHOTOMETRY]

specular reflection Reflection of electromagnetic or other waves in which the angle of incidence is equal to the angle of reflection. *TS*

speleology The scientific study of caves, including their formation (*speleogenesis*), CAVE deposits, cave organisms and the chemistry of cave waters. *JAM*

Chapman, P. 1993: *Caves and cave life*. London: Harper Collins. **Ford, T.D. and Cullingford, C.H.D. (eds)** 1976: *The science of speleology*. London: Academic Press.

speleothems Cave *dripstones*, including stalactites, stalagmites and other forms of cave deposit, mostly formed of calcium carbonate precipitated from cave waters that exhibit SUPERSATURATION with respect to calcite or aragonite. Supersaturation occurs by *degassing* of carbon dioxide from *dripwater* emerging in caves after previously having been in equilibrium with CARBON DIOXIDE at elevated partial pressures in the soil percolation zone. Deposition through evaporation occurs only in limited situations, usually close to the cave entrance.

Most speleothems are composed of aggregates of individual crystals, which may exhibit growth banding, and have considerable potential in HIGH-RESOLUTION RECONSTRUCTIONS of palaeoenvironments. Stalactites are generally not analysed because fluid flow may alternate between the central cavity and the outer surface such that superposition cannot be assumed. As a source of PROXY DATA in PALAEOCLIMATOLOGY, four aspects of speleothems are of particular importance: (1) OXYGEN ISOTOPE and CARBON ISOTOPE composition, which may reflect cave temperature, rainwater composition and other environmental factors; (2) recurrent luminescent banding and other LAMINATIONS, some of which are annual; (3) impurity content, including TRACE ELEMENTS and organic matter; and (4) crystallography, which is typically precipitation-evaporation related. Other advantages of speleothems in the field of PALAEOENVIRONMENTAL RECONSTRUCTION include their suitability for URANIUM-SERIES DATING and their high preservation potential. *JAM*

[See also CAVE, CAVE SEDIMENTS]

Baker, A., Smart, P.L., Edwards, R.L. and Richards, D.A. 1993: Annual growth banding in a cave stalagmite. *Nature* 364, 518–520. **Goede, A.** 1994: Continuous early last glacial palaeoenvironmental record from a Tasmanian speleothem based on stable isotope and minor element variations. *Quaternary Science Reviews* 13, 283–291. **Holmgren, K., Karlén, W., Lauritzen, S.E.** *et al.* 1999: A 3000-year high-resolution stalagmite-based record of palaeoclimate for northeastern South Africa. *The Holocene* 9, 295–309. **Lauritzen, S.E. and Lundberg, J. (eds)** 1999: Speleothems as high-resolution palaeoclimatic archives. *The Holocene* 9(6), 643–722 [Special Issue]. **Ramseyer, K., Miano, T., D'Orazio, V.** *et al.* 1997: Nature and origin of organic matter in carbonates from speleothems, marine cements and coral skeletons. *Organic Geochemistry* 26, 361–378. **Roberts, M.S., Smart, P.L. and Baker, A.** 1998: Annual trace element variations in a Holocene speleothem. *Earth and Planetary Science Letters* 154, 237–246. **Shopov, Y., Ford, D.C. and Schwarcz, H.P.** 1994: Luminescent microbanding in speleothems: high-resolution chronology and paleoclimate. *Geology* 22, 407–410.

Sphagnum Genus of bryophytes commonly known as 'bog' or 'peat' moss, adapted to retaining large quantities of water as they have no roots for absorbing soil moisture, nor tissues for internal water transport. Changes in Sphagnum content or species within PEAT, identified by BRYOPHYTE or PLANT MACROFOSSIL ANALYSIS, provide important evidence of past HYDROLOGICAL change within MIRES. The genus also has a pronounced capacity for the exchange of hydrogen and mineral ions, which is advantageous where nutrients are in short supply (e.g. on OMBROTROPHIC MIRES). It also, however, increases their susceptibility to damage from ACID RAIN and other fossil-fuel POLLUTANTS. *DZR*

Daniels, R.E. and Eddy, A. 1990: *Handbook of European Sphagna*. London: HMSO. **Ferguson, P., Robinson, R.N., Press, M.C. and Lee, J.A.** 1984: Element concentration in five *Sphagnum* species in relation to atmospheric pollution. *Journal of Bryology* 13, 107–114. **Johnson, L.C., Damman, A.H.W. and Malmer, N.** 1990: *Sphagnum* macrostructure as an indicator of decay and compaction in peat cores from an ombrotrophic south Swedish peat bog. *Journal of Ecology* 78, 633–647. **Maquoy, D. and Barber, K.** 1999: Evidence for climatic deteriorations associated with the decline of *Sphagnum imbricatum* Hornsch. ex Russ. in six ombrotrophic mires from northern England and the Scottish Borders. *The Holocene* 9, 423–437.

spheroid *World spheroids*

Name	Date	Radius (m)	Flattening
Airy	1830	6 377 563.4	1:299.32
Everest	1830	6 377 276.3	1:300.80
Clarke	1886	6 378 206.4	1:294.98
International Astronomical Union	1968	6 378 160	1:298.25
World Geodetic System – 72	1972	6 378 135.0	1:298.26
GRS (Geodetic Reference System) – 80	1980	6 378 137	1:298.26
WGS (World Geodetic System) – 84	1984	6 378 137.0	1:298.26

sphericity   For a rock particle, a measure of how closely it resembles a sphere, determined as the ratio of the true nominal diameter of a particle to the diameter of the smallest sphere that would enclose it (generally taken as the long (*a*) axis). *JS*

[See also ROUNDNESS, SHAPE]

spheroid   A shape generated by rotating an ellipse about its minor (or major) axis, forming a three-dimensional body. If the major and minor axis dimensions are equal, the result is a sphere (as used for globes); otherwise it becomes flattened at the poles (an oblate spheroid). This shape is considered acceptably close to the actual shape of the Earth, a GEOID with minor axis of 6356 km and major axis 6378 km. The spheroid is employed in the generation of MAP PROJECTIONS and translations of REMOTE SENSING imagery from the Earth's curved surface to a plane by GEOMETRIC CORRECTION. A number of variations have been calculated (see Table). *TF*

Maling, D.H. 1992: *Coordinate systems and map projections*, 2nd edn. Oxford: Pergamon Press.

spheroidal carbonaceous particles (SCP)   The spherical particles (usually 5–40 µm in diameter) produced by incomplete high-temperature coal and oil COMBUSTION, used in the reconstruction of POLLUTION HISTORY and as a DATING TECHNIQUE. Accumulating sediments incorporate SCP as they are deposited from the ATMOS-

PHERE. The SCP are extracted and counted and variations in SCP concentration with sediment depth reflect temporal trends in FOSSIL FUEL consumption. DOCUMENTARY EVIDENCE of industrial activity then provides dates for the profile features and thus a means of core correlation. The elemental carbon spheres are resistant to diagenetic change and are relatively immobile within the sediments. Thus, SCP profiles also represent surrogate pollution histories for fossil-fuel pollutants that are susceptible to postdepositional modification (e.g. sulphur). The Figure shows an idealised SCP profile and the main SCP dating horizons for the UK, although the timings of these features may vary both within and between regions. Characterising the SCP according to fuel type by SCANNING ELECTRON MICROSCOPY or ENERGY-DISPERSIVE SPECTROMETRY may provide additional dating horizons. Other applications of SCP analysis are involved in the quantification of MODERN POLLEN RAIN, the assessment of spatial variations in contemporary POLLUTANT deposition and the study of LACUSTRINE SEDIMENTS. *DZR*

Rose, N., Juggins, S., Watt, J. and Battarbee, R.W. 1994: Fuel type characterisation of spheroidal carbonaceous particles using surface chemistry. *Ambio* 23, 296–299. **Rose, N.L., Harlock, S., Appleby, P.G. and Battarbee, R.W.** 1995: Dating of recent lake sediments in the United Kingdom and Ireland using spheroidal carbonaceous particle (SCP) concentration profiles. *The Holocene* 5, 320–335. **Wik, M. and Renberg, I.** 1997: Environmental records of carbonaceous fly-ash particles from fossil fuel combustion. *Journal of Palaeolimnology* 15, 193–206.

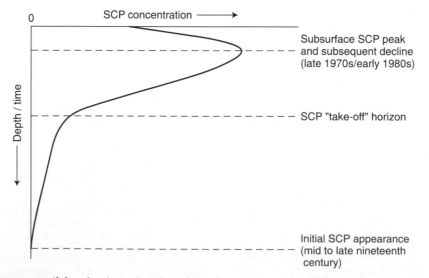

spheroidal carbonaceous particles *A schematic spheroidal carbonaceous particle (SCP) profile as used for dating in the British Isles (after Rose et al., 1995)*

spheroidal weathering The subsurface, chemical decay of concentric or spherical shells of rock (ranging in diameter from 2 cm to 2 m). It is also known as *onion-skin weathering*. The 'skins' are then loosened and separated from a block of rock by water penetrating joints or fractures and attacking the block from all sides. It is similar to the larger-scale EXFOLIATION produced usually by mechanical weathering. *MAC*

splash pedestal A small pillar (up to a few centimetres high) of soil formed where stones or roots selectively protect the underlying soil from the effect of *rainsplash erosion*, while the soil in the immediate vicinity is removed. The height of the pedestal can provide a measure of the depth of SOIL LOSS over time. *SHD*

[See also EARTH PILLAR]

Morgan, R.P.C. 1995: *Soil erosion and conservation*, 2nd edn. Harlow: Longman.

spline function A technique of curve fitting using piecewise polynomials to link some or all data points in a series. The most commonly used function is the *cubic spline*. *TQ*

[See also TIME-SERIES ANALYSIS]

Kendall, M. and Ord, K. J. 1990: *Time Series*, 3rd edn. London: Edward Arnold.

spodic horizon A black or orange brown subsurface horizon resulting from an accumulation of amorphous illuvial substances including organic matter and aluminium with or without iron. The spodic horizon is diagnostic for PODZOLS and normally lies beneath an ALBIC HORIZON. *EMB*

Food and Agriculture Organization of the United Nations (FAO) 1998: *World reference base for soil resources* [*Soil Resource Report* 84]. Rome: FAO, ISRIC, ISSS.

Spörer minimum The minimum in solar SUNSPOT activity observed between AD 1420 and 1530 and a possible influence on CLIMATIC VARIATIONS within the LITTLE ICE AGE. *JAM*

[See also MAUNDER MINIMUM]

spores Reproductive bodies of the lower plants (cryptogams) such as ferns and mosses. As with pollen of flowering plants (Spermatophytes), the walls of spores of mosses, hepatics and ferns consist of highly polymerised terpens (*sporopollenin*). These are therefore very resistant to corrosion. In common with pollen, spores have a rich wall morphology and commonly occur as FOSSILS and SUBFOSSILS, which can be used in *palaeoecological reconstruction*. *BA*

[See also MICROFOSSIL, POLLEN ANALYSIS]

Dickson, J.H. 1986: Bryophyte analysis. In Berglund, B.E. (ed.), *Handbook of Holocene palaeoecology and palaeohydrology*. Chichester: Wiley, 627–643.

spread, vectors of Those agents (e.g. wind, water, birds) that *effect* (i.e. bring about) the long-distance transfer of plant progagules from one place to another. *FMC*

[See also ESTABLISHMENT, AGENCIES OF, TREE MIGRATION/ SPREADING]

Chambers, F.M. and Elliott, L. 1989: Spread and expansion of *Alnus* Mill. in the British Isles: timing, agencies and possible vectors. *Journal of Biogeography* 16, 541–550.

spring tide The highest TIDE occurring at new and full moon. *HJW*

[See also NEAP TIDE]

stability concepts: ecological contexts There are various perspectives on what is meant by stability in the context of ECOSYSTEMS. RESISTANCE to and rate of recovery from disturbance reflect stability of the biotic communities of the ecosystem and are commonly referred to as RESILIENCE. Stability may also refer to features of the physical environment of ecosystems. An ecosystem that is disturbed to the degree that it is unable to return to its original state may undergo structural alteration, perhaps changing from grassland to desert, for example. There is evidence that the BIODIVERSITY of an ecosystem is related to its resilience. This phenomenon may be studied in the context of the effects of biodiversity on ecosystem function. Ecologists have gained particularly valuable insights into aspects of stability by investigating the effects of biodiversity on specific aspects of ecosystem function, such as nutrient cycling or productivity. Studies suggest that many aspects of ecosystem function may be impaired by species losses. As losses continue, the stability of the ecosystem will be surpassed and its function or structure will change. *JTK*

[See also DIVERSITY–STABILITY HYPOTHESIS, EQUILIBRIUM CONCEPTS, INTERMEDIATE DISTURBANCE HYPOTHESIS, PERSISTENCE]

Goodman, D. 1975: The theory of diversity stability relationships in ecology. *Quarterly Review of Biology* 50, 237–266. Huston, M.A. 1994: *Biological diversity: the coexistence of species on changing landscapes*. Cambridge: Cambridge University Press. Pimm, S.L. 1991: *The balance of nature? Ecological issues in the conservation of species and communities*. Chicago, IL: University of Chicago Press. Westman, W.E. 1978: Measuring the inertia and resilience of ecosystems. *Bioscience* 28, 705–710.

stable isotope An isotope that does not undergo spontaneous radioactive decay. For an ISOTOPE to be stable, the *neutron-to-proton ratio* must be approximately equal to or greater than unity. ATOMS with an even number of protons are more abundant than atoms with an odd number. *IR*

[See also RADIONUCLIDE]

stable isotope analysis The determination of the ISOTOPE RATIO of a sample using a MASS SPECTROMETER. The stable isotope composition is determined for the two main STABLE ISOTOPES of an element. The difference between a sample and a REFERENCE STANDARD can be measured far more precisely than absolute ratios and this approach allows very small differences in the isotopic composition of two samples to be determined. The differences between samples and reference standards are usually small and expressed as parts per thousand or *per mille* (‰). *IR*

[See also CARBON ISOTOPES, HYDROGEN ISOTOPES, ISOTOPES AS INDICATORS OF ENVIRONMENTAL CHANGE, LEAD ISOTOPES, NITROGEN ISOTOPES, OXYGEN ISOTOPES, STRONTIUM ISOTOPES, SULPHUR ISOTOPES]

Ehleringer, J.R. and Rundel, P.W. 1989: Stable isotopes: history, units, and instrumentation. In Rundel, P.W., Ehleringer, J.R. and Nagy, K.A. (eds), *Stable isotopes in ecological research* [*Ecological Studies*. Vol. 68]. Berlin: Springer, 1–15. **Friedman, I. and O'Neil, J.R.** 1977: Compilation of stable isotope fractionation factors of geochemical interest. In Fleischer, M. (ed.), *Data of geochemisty*, 6th edn [*Geological Survey Professional Paper* 440-KK]. Washington, DC: United States Government Printing Office, 1–12.

stadial A short cold episode during which LOCAL GLACIATION may occur or a subdivision of a GLACIAL EPISODE characterised by a relative deterioration of climate. The last stadial event occurred at the end of the LAST GLACIATION and was known as the YOUNGER DRYAS STADIAL, lasting just over 1000 years. *DH*

[See also LOCH LOMOND STADIAL]

Peteet, D.M. 1995: Global Younger Dryas? *Quaternary International* 28, 93–104.

stage A subdivision of a SERIES in CHRONOSTRATIGRAPHY and the basic unit for stratigraphical CORRELATION on regional and global scales. Stage names end with -ian (e.g. MESSINIAN). *LC*

stand 'A spatially continuous group of trees and associated vegetation having similar structures and growing under similar soil and climatic conditions' (Oliver and Larson, 1996: 1). So-called homogeneous stands are the fundamental field units of vegetation from which samples are drawn in PHYTOSOCIOLOGY. *JLI*

Oliver, C.D. and Larson, B.C. 1996: *Forest stand dynamics*, update edn. New York: John Wiley.

standard method A generally accepted method or set of analytical procedures that is recognised as producing results within the limits of ACCURACY and PRECISION acceptable for particular purposes. *JAM*

standard substance A substance, such as a powder or solution, of known composition or strength, which is used as a basis for comparison with samples and/or for INSTRUMENT CALIBRATION. *JAM*

[See also REFERENCE STANDARD]

standardisation A statistical procedure for improving data comparability and amenability to analysis. In DENDROCHRONOLOGY and DENDROCLIMATOLOGY, it is the removal of part of the variance in a TREE-RING time series with the intention of highlighting some statistical 'signal' of interest in the data by removing the variance believed to be due to some other process (and hence considered in this application to be 'noise'). An example would be the removal of the long-term trend in a series of measured ring widths, which represents the progressive thinning of ring widths that occurs as the tree increases in girth (i.e. roughly constant growth being distributed around bands of ever increasing diameter). In this example, the evidence of climatic change is deemed to be superimposed on this non-climate trend. Removing the trend will thus prevent it being inappropriately interpreted as evidence of a trend in climate. Standardisation is most frequently achieved by fitting some function or by applying a low-pass filter to the data and subtracting the function or filtered values from

the measurement series – a form of high-pass FILTERING. In fact, there are many approaches and techniques used to standardise palaeoclimatic data series, but, in practice, their use is highly problematic because they can lead to a loss of recoverable climatic variance at longer timescales. *KRB*

[See also CHRONOLOGY, NORMALISATION, TREE-RING INDEX]

Briffa, K.R., Jones, P.D., Schweingruber, F.H. *et al.* 1996: Tree-ring variables as proxy indicators: Problems with low-frequency signals. In Jones, P.D., Bradley, R.S. and Jouzel, J. (eds), *Climatic variations and forcing mechanisms of the last 2000 years.* Berlin: Springer, 9–141. **Cook, E.R., Briffa, K.R., Meko, D.M.** *et al.* 1995: The 'segment length curse' in long tree-ring chronology development for paleoclimatic studies. *The Holocene* 5, 229–237. **Cook, E.R., Briffa, K.R., Shiyatov, S. and Mazepa, V.** 1990: Tree-ring standardization and growth-trend estimation. In Cook, E.R. and Kairiukstis, L.A. (eds), *Methods of dendrochronology: applications in environmental sciences.* Dordrecht: Kluwer, 104–123.

standing crop The BIOMASS present in a defined area at a particular time. The term may be used with reference to the VEGETATION component only and, sometimes, only to the *above-ground biomass*. *JAM*

stasis Used in descriptions of the process of EVOLUTION to refer to periods of little or no evolutionary change in species. *KDB*

[See also PHYLETIC GRADUALISM, PUNCTUATED EQUILIBRIA]

statistical analysis A general term for that branch of mathematics dealing with data summarisation, estimation, generalisation, prediction and/or hypothesis testing in situations where some degree of uncertainty usually exists. *HJBB*

[See also DESCRIPTIVE STATISTICS, INFERENTIAL STATISTICS, NUMERICAL ANALYSIS, MULTIVARIATE ANALYSIS]

statistical uncertainty The inability to determine results exactly because of the combined effects of natural variability in populations and sampling errors. *JAM*

[See also SIGNAL-TO-NOISE RATIO]

stemflow The drainage of intercepted precipitation that is directed down the shoot systems of plants. When PRECIPITATION is subject to INTERCEPTION by a vegetation *canopy*, the water may drip towards the ground (*throughfall*), may not reach the ground because of EVAPORATION or may reach the ground as stemflow. *JAM/RAS*

Tagajki, M., Sasaki, S., Gyokusen, K. and Saito, A. 1997: Stemflow chemistry of urban street trees. *Environmental Pollution* 96, 107–109.

steno- A prefix given to organisms that have a narrow TOLERANCE of specific environmental factors (such as *stenothermic* for narrow tolerance of temperature conditions). *JLI*

[See also EURY-]

steppe vegetation A vegetation type dominated by grasses existing in temperate regions where the PRECIPITATION:EVAPORATION ratio is too low to enable trees to grow. So-called '*cold steppe*' vegetation covered wide areas at the margins of the ICE SHEETS during the LAST GLACIATION. *Artemisia* is a characteristic species of the steppe and its pollen is frequently found in Late Glacial POLLEN ASSEMBLAGES, along with *Rumex* and Chenopodiaceae pollen. Today, steppe vegetation is found on the plains of central Europe and Eurasia. *SPH*

[See also BIOMES, TEMPERATE GRASSLANDS, VEGETATION FORMATION-TYPE]

Archibold, O.W. 1995: *Ecology of world vegetation*. London: Chapman and Hall. Laurenko, E.M. and Karamysheva, Z.V. 1993: Steppes of the former Soviet Union and Mongolia. In Coupland, R.J. (ed.), *Natural grasslands: Eastern Hemisphere and résumé* [*Ecosystems of the World*. Vol. 8B]. Amsterdam: Elsevier, 3–60.

stereoscopic vision

The process of observing a scene from each of two separate simultaneous viewpoints and thereby allowing the perception of depth. In humans, the brain interprets the variations between the images seen by the eyes and introduces an awareness of three dimensionality to the scene. Normal human capabilities can be extended using instruments such as: a *stereoscope* to view overlapping AERIAL PHOTOGRAPHS and thereby derive CONTOUR information; optical filter systems to view images on a screen; and special headsets, as used in *virtual reality systems*. *TF*

sterilisation

The process of destroying all microbial life. It differs from *disinfection* in that BACTERIA are destroyed as well as the other types of micro-organisms. Thus, *pasteurisation* of milk disinfects but does not sterilise. *JAM*

stewardship concept

In *natural resources management*, voluntary actions taken by individuals or organisations to promote ecological goals and CONSERVATION. The concept covers the management of both tangibles and intangibles to ensure their availability for others, especially for future generations. *JAM*

[See also ENVIRONMENTAL MANAGEMENT, SUSTAINABILITY]

Alpert, P. 2001: Stewardship, concept of. In Levin, S.A. (ed.), *Encyclopedia of biodiversity*. Vol. 5. San Diego, CA: Academic Press, 481–484. Johnson, N.C., Malk, A.J., Sexton, W.T. and Szave, R. (eds) 1999: *Ecological stewardship: a common reference for ecosystem management*. Oxford: Oxford University Press.

stillstand

(1) A period of stability between two phases of TECTONIC activity. (2) A period in which GLACIER dimensions remain constant and hence the snout is stationary. *PW*

[See also HIGHSTAND, EQUILIBRIUM CONCEPTS IN GEOMORPHOLOGY]

stochasticity

An attribute of concepts, models and phenomena that denotes an element of randomness and limited predictability, and hence the appropriateness of the ideas and techniques of probability and statistical inference. It is commonly encountered in the context of environmental change, especially in terms of stochastic processes and stochastic models. A *stochastic process* is one in which changes in time (or space) following statistical or probabilistic laws, in contrast to a *deterministic process* in which the outcome is predictable with certainty. The essential characteristic of a stochastic process is that uncertainty is propagated over time or space. The degree of stochasticity in nature may, however, be less than appears at first sight because of inaccurate measurement and imperfect knowledge. *Stochastic models* attempt to represent stochastic processes numerically by characterising process uncertainty as PROBABILITY DISTRIBUTIONS. *JAM*

Mann, C.J. 1970: Randomness in nature. *Bulletin of the Geological Society of America* 81, 95–104. Richards, K.S. 1979: *Stochastic processes in one-dimensional series* [CATMOG 23]. Norwich: Geo Abstracts.

stomata

Pores in leaves, which permit gaseous exchange between the leaf interior and the atmosphere. *JAM*

stomatal analysis

The study of stomatal pore numbers on the leaf surfaces of SUBFOSSIL and FOSSIL leaves in order to reconstruct atmospheric CARBON DIOXIDE (CO_2) concentrations in the recent and geological past. This BIOINDICATOR or PROXY CLIMATIC INDICATOR method is based on a well established inverse relationship that exists between atmospheric CO_2 concentration and the *stomatal density* (number of stomata per mm^2 area of leaf surface) and *stomatal index* (a ratio of the number of stomatal cells to the total number of epidermal plus stomatal cells) of leaves. Quantitative atmospheric CO_2 estimates for the Quaternary and Cenozoic are obtained using inverse regression analysis of stomatal density/index responses of leaves to known atmospheric CARBON DIOXIDE VARIATIONS. For times earlier than the Tertiary, for which there are no extant species, CO_2 reconstructions are based on *stomatal ratio analysis* (a ratio of the stomatal index of the nearest living ecological/morphological equivalent of the fossil divided by the stomatal index of the fossil taxa), which are standardised against long-term CARBON CYCLE models. *JCMc*

[See also EVAPOTRANSPIRATION]

McElwain, J.C. 1998: Do fossil plants signal palaeoatmospheric CO_2 concentration in the geological past? *Philosophical Transactions of the Royal Society of London* B353, 83–96. Rundgren, M. and Beerling, D. 1999: A Holocene CO_2 record from the stomatal index of subfossil *Salix herbacea* L. leaves from northern Sweden. *The Holocene* 9, 509–513. Van der Burgh, J., Visscher, J., Dilcher, H. and Kurschner, W.M. 1993: Palaeoatmospheric signatures in Neogene fossil leaves. *Science* 260, 1788–1790. Woodward, F.I. 1987: Stomatal numbers are sensitive to CO_2 increases from pre-industrial levels. *Nature* 327, 617–618.

stomatal conductance

The ease with which gases diffuse through stomata. *Stomatal resistance* is the reciprocal of stomatal conductance. *JAM*

stone line

A subsurface layer of CLASTS encountered in natural exposures, soil pits and road cuttings in SAPROLITE landscapes of the tropics and subtropics. Ideas on the formation of stone lines include: (1) relict *stone pavements* (or LAG DEPOSITS) formed by removal of fine sedimentary MATRIX; (2) a natural consequence of BIOTURBATION by, for example, TERMITES and mole rats, involving the downward movement in a soil of clasts to the lower limit of faunal activity; (3) derivation from WEATHERING PROFILES; and (4) residual deposits associated with a WEATHERING FRONT. Fundamentally, there is controversy between those who favour an ALLOCHTHONOUS origin (i.e. the stone lines are formed by an accumulation of transported material) and those who adopt an AUTOCHTHONOUS hypothesis (i.e. they are formed by *in situ* processes). The term has also been used in connection with layers of rock fragments on interfluves in mid-latitudes. *Stone layer* is an alternative term, which is viewed by some authorities as the correct one. *RAS*

[See also PEDIMENT, PEDOGENESIS, WEATHERING]

Alexandre, J. and Symoens, J.J. (eds) 1989: Stone lines. *Geo Eco Trop* 11, 1–237. Johnson, D.L. and Balek, C.L. 1991: The genesis of Quaternary landscapes with stone-lines. *Physical Geography* 12, 385–395. Paton, T.R., Humphreys, G.S. and Mitchell, P.B. 1995: *Soils: a new global view*. London: UCL Press.

storm beach Along some shorelines, storms deposit coarse-grained sediment (sand, shingle and gravel) on the upper beach. These high *berms*, or storm beaches, are usually durable, unlike those formed lower on the beach profile. Storm waves also carry sand from the beach to the bottom near-shore. *HJW*

storm hydrograph A graphical representation of channel DISCHARGE from a catchment in response to a rainfall event. Rainfall and RUNOFF are usually plotted together and thus the storm hydrograph is a useful tool in examining rainfall–runoff relationships and in separating QUICKFLOW and BASEFLOW components. Storm hydrograph characteristics, such as time-to-peak, storm runoff as a percentage of rainfall and the form of the recession limb, vary considerably both between catchments and temporally in the same catchment. Runoff responses vary with fixed factors (such as topography and geology), transient factors (such as rainfall characteristics and antecedent soil moisture and weather) and factors such as LAND COVER and soil properties, which are also highly susceptible to radical alteration as a result of human activities or climatic change. *ADT*

[See also HYDROGRAPH]

Hewlett, J.D. and Hibbert, A.R. 1967: Factors affecting the response of small watersheds to precipitation in humid areas. In Supper, W.E. and Lull, H.W. (eds), *Forest hydrology*. Oxford: Pergamon, 275–290.

storm surge A type of short-term SEA-LEVEL CHANGE caused by EXTREME CLIMATIC EVENTS. In tropical regions, where the main cause is TROPICAL CYCLONES, storm surges are confined to within a few tens of kilometres of where the storms impinge on the coast: in extratropical regions they may be an order of magnitude more extensive. They are most severe and damaging on low-lying coasts with shallow water and when they coincide with the high sea levels of SPRING TIDES. *JAM*

[See also EXPLORATORY DATA ANALYSIS, SEICHE]

Murty, T.S., Flather, R.A. and Henry, R.F. 1986: The storm surge problem in the Bay of Bengal. *Progress in Oceanography* 16, 195–233. Pugh, D.T. 1987: *Tides, surges and mean sea level*. Chichester: Wiley.

storms A wide variety of severe weather events involving strong winds are colloquially termed storms in different parts of the world. These are often associated with precipitation because the strong air-pressure gradients that usually cause the strong winds are often associated with areas of low pressure (CYCLONES, DEPRESSIONS). A *storm track* is the trajectory of a storm across the Earth's surface or the average trajectory of similar types of storms.

Gales are defined as winds having an average velocity measured 10 m above the surface of 34 knots (17.2 ms^{-1}) or more sustained over at least 10 minutes. This corresponds to wind force 8 or more on the BEAUFORT SCALE. Mid-latitude depressions are most likely to lead to such wind speeds in the winter when thermal gradients are largest. Rapid convergence uplift occurs at such temperature gradients (baroclinic conditions). Air pressure falls as a result – if it drops at a rate exceeding 12 millibars within a 12 hour period, *explosive cyclogenesis* is said to occur.

Fluctuations of a *gale index* for the British Isles correspond to changes between progressive and blocked phases. The early 1990s were notable for an abrupt increase in frequency of severe gales, attributable to the positive phase of the NORTH ATLANTIC OSCILLATION. It does not necessarily follow that a general increase in wind speeds will have occurred. This is reflected in the frequency distribution of deep low-pressure systems over the North Atlantic, whereby the largest proportionate increase has been in the case of the deepest depressions.

The impact of strong winds depends not just on absolute wind speed but also on the relative severity for a location. It follows that storm tracks are thus important variables. The storm of October 1987 caused £1.2 billion of damage in the UK, largely because it passed over southern England, an area that has a greater sensitivity to strong winds than Scotland, because of their relative infrequency in the former region. World-wide, roughly 70% of weather-related insurance losses are attributable to storms. The single weather events to have caused the largest losses in financial terms are TROPICAL CYCLONES that have made landfall over densely populated areas of the southern USA. Hurricane Andrew (1992) caused the largest damage to date (US $17 billion), the losses being generated from a very small area in southern Florida. *JCM*

[See also CLIMATIC VARIABILITY, EXTREME CLIMATIC EVENTS]

Agustsdottir, A.M., Barron, E.J., Bice, K.L. *et al.* 1999: Storm activity in ancient climates 1 and 2. *Journal of Geophysical Research D: Atmospheres* 104, 27277–27320. Dawson, A.G., Hickey, K., McKenna, J. and Foster, I.D.L. 1997: A 200-year record of gale frequency, Edinburgh, Scotland: possible link with high-magnitude volcanic eruptions. *The Holocene* 7, 337–341. Franzen, L.G. 1991: The changing frequency of gales on the Swedish west-coast and its possible relation to the increased damage to coniferous forests of southern Sweden. *International Journal of Climatology* 11, 769–793. Hoskins, B.J. and Valdes, P.J. 1990: On the existence of storm tracks. *Journal of Atmospheric Sciences* 47, 1854–1864. Lamb, H.H. 1991: *Historic storms of the North Sea, British Isles and northwest Europe*. Cambridge: Cambridge University Press. Palutikof, J., Holt, T. and Skellern, A. 1997: Wind: resource and hazard. In Hulme, M. and Barrow, E. (eds), *Climates of the British Isles, past, present and future*. London: Routledge, 220–242. Peilke Sr, R.A. and Peilke Jr, R.A. (eds) 1999: *Storms*. 2 vols. London: Routledge. Royal Academy of Engineering 1995: *Windstorm; coming to terms with mankind's worst natural hazard*. London: Royal Academy of Engineering. Schinke, H. 1993: On the occurrence of deep cyclones over Europe and the North Atlantic in the period 1930–1991. *Beiträge zur Physik der Atmosphäre* 66, 223–238. Waves and Storms in the North Atlantic (WASA) Group 1998: Changing waves and storms in the Northeast

Atlantic? *Bulletin of the American Meteorological Society* 79, 741–760.

strandflat First used by the Norwegian explorer and scientist Fridtjof Nansen to describe extensive, relatively flat areas (ROCK PLATFORMS and conspicuous ice-moulded bedrock) and adjacent cliffs of the coastlines of western Norway, southern Alaska and the Antarctic Peninsula. In western Norway, individual areas of strandflat are up to 80 km wide. Several theories have been advocated to account for the origin of strandflat areas. One view is that the rock surfaces are the product of marine planation during the CENOZOIC era. Another view is that they are produced as a result of GLACIAL EROSION in coastal areas. Yet others consider that cold-climate coastal FROST WEATHERING and POLAR SHORE EROSION of bedrock are important in strandflat formation. A curious aspect of many strandflat areas is that, despite their presence in areas formerly affected by QUATERNARY GLACIATION, many of these areas do not exhibit evidence of glacio-isostatic deformation.

AGD

[See also GLACIO-ISOSTATIC REBOUND]

Dawson, A.G. 1994: Strandflat development and Quaternary shorelines on Tiree and Coll, Scottish Hebrides. *Journal of Quaternary Science* 9, 349–356. **Holtedahl, H.** 1998: The Norwegian strandflat – a geomorphological puzzle. *Norsk Geologisk Tidsskrift* 78, 47–66. **Nansen, F.** 1922: The strandflat and isostasy. *Videnskapselkapets Skrifter* 1921 I *Math.-Naturw. Kl* 2, 1-313 [*Norges Videnskaps Akademie i Kristiana*].

strandline (1) The upper limit of the LITTORAL ZONE, where flotsam and jetsam accumulates either due to ebbing of the high tide or to storm wave action. It is also known as the *driftline*, where stranded organic matter may support diverse animal communities (e.g. invertebrates and birds). The upper strandline may be colonised by sea-dispersed plants. Evidence of prehistoric strandlines (both marine and fresh water) assist in the reconstruction of PALAEOENVIRONMENTS. The term is also applied in this sense to lakes and rivers.

(2) A strandline of snow and ice may occur in the ACCUMULATION AREA of a glacier where, after the occurrence of a GLACIER SURGE, it marks the former position of the glacier surface. After the rapid drainage of an ICE-DAMMED LAKE, stranded ICEBERGS may also produce a strandline for a short time.

SHJ/JAM

[See also DISPERSAL, ECOTONE, ISLAND BIOGEOGRAPHY, LIMNOLOGY, MANGROVE, PALAEOCEANOGRAPHY, SALT-MARSH]

strange attractor One form of ATTRACTOR in DYNAMICAL SYSTEMS that exhibits particularly complex behaviour; also called a *chaotic attractor*.

CET

Phillips, J.D. 1999: *Earth surface systems: complexity, order and scale*. Oxford: Blackwell.

strath terrace *Strath* is a Scottish term meaning wide, flat-floored valley. In 1932, W. Bucher introduced the term to distinguish rock-floored or *degradational fluvial terraces* from aggradational ones. Such terraces are common in tectonically active mountains.

RAS

Bucher, W.H. 1932: Strath as a geomorphic term. *Science* 75, 130–131. **Bull, W.B.** 1991: *Geomorphic responses to climatic change*. Oxford: Oxford University Press.

stratification The layered character of many soils, sediments and rocks, principally sedimentary rocks but also others such as LAVA FLOWS and some igneous INTRUSIONS. The layers are *strata* (singular: *stratum*) and their orientation is described by STRIKE and DIP. Stratification in sediments and sedimentary rocks is separated by scale into BEDDING (units thicker than 1 cm) and LAMINATION (units thinner than 1 cm), although 'bedding' is sometimes used, regardless of thickness, to imply strata deposited by distinct episodes of sedimentation as opposed to unsteady sedimentation within one episode (see PARALLEL LAMINATION).

GO

[See also CROSS-STRATIFICATION]

Collinson, J.D. and Thompson, D.B. 1992: *Sedimentary structures*, 2nd edn. London: Chapman and Hall. **McKee, E.D. and Weir, G.W.** 1953: Terminology for stratification and cross-stratification in sedimentary rocks. *Geological Society of America Bulletin* 64, 381–390.

stratigraphical column The subdivision of the STRATIGRAPHICAL RECORD into units of global significance (primarily SYSTEMS) that correspond to units of time (PERIODS), establishing the RELATIVE AGES of SEDIMENTARY DEPOSITS or events. *Absolute dates* may be added to the stratigraphical column (*chronostratic scale*) through GEOCHRONOLOGY to produce a GEOLOGICAL TIMESCALE (*geochronologic scale*). Units of sediment or rock with time significance (*time-rock units*) can be thought of as being *imposed* on the stratigraphical record and each basal boundary is *agreed* by international discussion with respect to a *Global Stratotype Section and Point* (GSSP). The correct calibration of a locally determined geological succession with the global standards depends on accurate CORRELATION. BIOSTRATIGRAPHY is integral to the definition of units in the global standard STRATIGRAPHY for the PHANEROZOIC and can be used also for the uppermost PRECAMBRIAN (*Vendian*), where there are soft-bodied FOSSILS. For older Precambrian rock successions, the boundaries of units are based on tectonic, igneous and metamorphic histories and defined by geochronology. The Phanerozoic is divided into three *erathems*, which correspond to gross evolutionary changes in fossil faunas – in broad terms, the PALAEOZOIC radiation of shelly marine invertebrates, MESOZOIC age of dinosaurs and CENOZOIC age of mammals.

LC

Conway-Morris, S. 1990: Late Precambrian – early Cambrian metazoan diversification. In Briggs, D.E.G. and Crowther, P.R. (eds), *Palaeobiology: a synthesis*. Oxford: Blackwell Scientific, 30–36. **Holland, C.H.** 1998: Chronostratigraphy (global standard stratigraphy): a personal perspective. In Doyle, P. and Bennett, M.R. (eds), *Unlocking the stratigraphical record: advances in modern stratigraphy*. Chichester: Wiley, 383–392. **Haq, B.U. and van Eysinga, W.B.** 1998: *Geological time table*, 5th edn. Amsterdam: Elsevier. **Harland, W.B., Armstrong, R.L., Cox, A.V.** *et al.* 1990: *A geologic time scale 1989*. Cambridge: Cambridge University Press. **Whittaker, A., Cope, J.C.W., Cowie, J.W.** *et al.* 1991: A guide to stratigraphical procedure. *Journal of the Geological Society, London* 148, 813–824.

stratigraphical record That part of the GEOLOGICAL RECORD comprising stratified materials (*strata*), principally SEDIMENTS, SEDIMENTARY ROCKS, PYROCLASTIC rocks and LAVAS. These materials can be divided into units that have temporal significance, using the principles of STRATIGRA-

609

PHY, and provide a historical record of processes and events affecting the Earth's surface. *GO*

[See also FOSSIL RECORD, GEOLOGICAL RECORD]

Ager, D.V. 1993: *The nature of the stratigraphical record*, 3rd edn. Chichester: Wiley.

stratigraphy The placing of events and processes into a time sequence through the study of sediment and rock successions. By ordering and dating the STRATIGRAPHICAL RECORD into an internationally agreed GEOLOGICAL TIMESCALE, stratigraphy provides the means for comparing events in Earth history on a global scale: it is the key to the interpretation of Earth history.

Modern stratigraphical analysis relies on a synthesis of data from diverse fields, including SEDIMENTOLOGY, PALAEONTOLOGY, igneous and metamorphic PETROLOGY, GEOCHEMISTRY, GEOPHYSICS and structural geology. Traditionally, the three main approaches to stratigraphy are LITHOSTRATIGRAPHY (sediments and rocks), BIO-STRATIGRAPHY (FOSSILS and SUBFOSSILS) and CHRONO-STRATIGRAPHY (RELATIVE AGES). Nowadays, these are supplemented by high-resolution techniques such as MAG-NETOSTRATIGRAPHY, SEISMIC STRATIGRAPHY, SEQUENCE STRATIGRAPHY, CYCLOSTRATIGRAPHY, EVENT STRATIGRAPHY, CHEMOSTRATIGRAPHY and STRONTIUM ISOTOPE stratigraphy. By using the radioactive decay of isotopes with various half-lives, an absolute *chronometric timescale* has been calibrated with the *chronostratic scale* of the STRATIGRAPHI-CAL COLUMN to give a *geochronologic scale* or geological timescale (see GEOCHRONOLOGY).

Stratigraphy relies on first establishing the succession of sediments or rocks, and then the history of events represented in them. Eighteenth and nineteenth century geologists including William Smith (1769–1839), Adam Sedgwick (1785–1873) and Roderick Impey Murchison (1792–1871) mapped the space and time relationships of rock units to build up a relative geological timescale (the stratigraphical column), establishing the major ERAS and SYSTEMS of chronostratigraphy. Lithostratigraphical principles such as the *original horizontality* of BEDS, *superposition* (higher beds are younger), *cross-cutting relationships*, *way up*, DEFORMATION and angular UNCONFORMITIES were important in ordering sequences, the value of fossils in characterising sedimentary strata was recognised, and stratigraphy was used to reconstruct past environments and document changing life forms. UNIFORMITARIANISM was important in the recognition that geological processes operating today were responsible for the GEOLOGICAL RECORD. However, although angular unconformities and OROGENIC BELTS were recognised as indicating episodes of EROSION, gaps in the geological record, periods of mountain building and SEA-LEVEL CHANGES, it was only with the acceptance of SEA-FLOOR SPREADING in the AD 1960s that PLATE TECTONICS provided a mechanism for explaining the larger-scale changes in PALAEOGEOGRAPHY and PALAEO-CLIMATOLOGY that had affected the continents. Stratigraphical studies provided the means for interpreting and constraining the dynamic history of ENVIRONMENTAL CHANGE associated with CONTINENTAL DRIFT.

Problems of CORRELATION arose through the historical establishment of geological systems in different areas of Europe, some where the lower, upper or both boundaries were missing, and with the potential for overlap. The International Commission on Stratigraphy of the International Union of Geological Sciences (IUGS), which has a coordinating role, has concentrated on the formal selection and definition of *boundary stratotypes*. These are outcrop sections where a boundary is designated and marked as the *Global Stratotype Section and Point*, or '*golden spike*', a reference point for global correlation, and where by definition time is tied into a rock succession. Standard procedures for stratigraphical nomenclature include systems of terminology and a hierarchy for stratigraphical units in the principal types of stratigraphy.

FACIES ANALYSIS is a key interpretive method in stratigraphy to investigate environmental change through time. Descriptive FACIES and FACIES MODELS are interpreted in terms of depositional processes and DEPOSITIONAL ENVI-RONMENTS. Linked within a stratigraphical framework, FACIES ASSOCIATIONS and their geometry allow interpretations to be made of the controls on environmental change in the geological record.

Unconformities indicate gaps in the continuity of the geological record. Estimates of its completeness have been based on SEDIMENT ACCUMULATION RATES and volumes in modern settings but there is little agreement that such values can be measured accurately. Even in the deep oceans, where sediment accumulation is relatively continuous and slow, sedimentation rates are variable and there are gaps (see HIATUS). Fossils provide the most important stratigraphical tool for detecting missing time intervals and correlating rocks between locations. The completeness of the FOSSIL RECORD has been estimated based on preservation probabilities and gaps in the stratigraphical ranges of taxa and evolutionary lineages. *LC*

[See also GEOLOGICAL CONTROLS ON ENVIRONMENTAL CHANGE, GEOLOGICAL RECORD OF ENVIRONMENTAL CHANGE, SEDIMENTOLOGICAL EVIDENCE OF ENVIRONMENTAL CHANGE]

Doyle, P. and Bennett, M.R. (eds) 1998: *Unlocking the stratigraphical record: advances in modern stratigraphy*. Chichester: Wiley. **Doyle, P., Bennett, M.R. and Baxter, A.N.** 1994: *The key to Earth history: an introduction to stratigraphy*. Chichester: Wiley. **Foote, M. and Sepkoski Jr, J.J.** 1999: Absolute measures of the completeness of the fossil record. *Nature* 398, 415–417. **Hailwood, E.A. and Kidd, R.B. (eds)** 1993: *High resolution stratigraphy* [*Special Publication* 70]. London: Geological Society. **Hedberg, H.D. (ed.)** 1976: Introduction to an international guide to stratigraphic classification, terminology, and usage. Report No. 7a of International Subcommission on Stratigraphic Classification. *Lethaia* 5, 283–295. **Holland, C.H.** 1986: Does the golden spike still glitter? *Journal of the Geological Society, London* 143, 3–21. **McShea, D. and Raup, D.M.** 1986: Completeness of the geological record. *Journal of Geology* 94, 569–574. **Nichols, G.** 1999: *Sedimentology and stratigraphy*. Oxford: Blackwell Science. **Posamentier, H.W., Summerhayes, C.P., Haq, B.U. and Allen, G.P. (eds)** 1993: *Sequence stratigraphy and facies associations* [*International Association of Sedimentologists Special Publication* 18]. Oxford: Blackwell Science. **Ramsay, A.T.S. and Baldauf, J.G.** 1999: *A reassessment of the Southern Ocean biochronology, Memoir* 18. London: Geological Society. **Whittaker, A., Cope, J.C.W., Cowie, J.W. et al.** 1991: A guide to stratigraphical procedure. *Journal of the Geological Society, London* 148, 813–824.

stratosphere The layer of the ATMOSPHERE above the TROPOSPHERE, between the tropopause (8–15 km above the surface of the Earth) and the *stratopause* (at about 50 km). It consists of two parts: a lower isothermal layer up to about 20 km, beyond which temperatures rise from around −50°C to about 0°C due to the presence of OZONE, which absorbs ultraviolet RADIATION. With little vertical mixing, the residence time of any particles that reach the stratosphere is much longer than in the troposphere. While the troposphere has been warming in recent decades, the stratosphere has been cooling. *JAM*

[See also OZONE DEPLETION]

Kodera, K. and Koide, H. 1997: Spatial and seasonal characteristics of recent decadal trends in the Northern Hemisphere troposphere and stratosphere. *Journal of Geophysical Research* 102, 19433–19447.

striation A scratch or minute line on bedrock, often just fractions of a millimetre deep, but sometimes up to a few millimetres deep and wide. In fluvial environments, striations are caused by CLASTS impacting the bedrock surface, whilst in glacial environments they are caused by clasts being dragged across it. They are also found on the faces of geological faults, along the paths of PYROCLASTIC flows and associated with sliding snow. Striations are, however, most commonly associated with basal sliding of GLACIERS and ICE SHEETS; glacial striations may be distinctive. Sets of parallel and subparallel striations or *striae* can be used to reconstruct ice-flow direction and, where more than one set of directions is found (so-called *cross-cutting striations*), they may be used to infer more than one glacial episode involving different ice movement directions. *JS*

[See also ABRASION, GLACIAL EROSION]

Demorest, M. 1938: Ice flowage as revealed by glacial striae. *Journal of Geology* 46, 700–725. **Dyson, J.L.** 1937: Snowslide striations. *Journal of Geology* 45, 549–557. **Iverson, N.R.** 1991: Morphology of glacial striae: implications for abrasion of glacier beds and fault surfaces. *Geological Society of America Bulletin* 103, 1308–1316. **McCarroll, D., Matthews, J.A. and Shakesby, R.A.** 1989: 'Striations' produced by catastrophic subglacial drainage of a glacier-dammed lake, Mjølkedalsbreen, southern Norway. *Journal of Glaciology* 35, 193–196.

strike The 'trend' of a sloping or tilted surface or planar feature, defined as the orientation of a horizontal line on the surface, and used (along with DIP) to describe the orientation of sediment or rock *strata* and other geological features, including GEOLOGICAL STRUCTURES. *GO*

strip cropping An agricultural practice in which elongated bands of crops are aligned in rows and strips across a slope, making it effective as a SOIL CONSERVATION measure. *SHD*

Food and Agriculture Organisation (FAO) 1965: *Soil erosion by water.* Rome: United Nations. **Hudson, N.W.** 1971: *Soil conservation.* London: Batsford. **Sharma, N.N., Paul, S.R., Dey, J.K.** *et al.* 1997: Effect of contour strip cropping of pineapple (*Ananas comosus*) on rice (*Oryza sativa*), sesame (*Sesamum indicum*) and maize (*Zea mays*) cropping sequences and their effect on soil properties. *Indian Journal of Agricultural Science* 67, 20–22.

strip lynchet A bench-like modification on a hillslope in Britain to enable the cultivation of hillsides and constructed in Mediaeval times. Strip lynchets tend to be longer than their prehistoric forebears (LYNCHETS), sometimes being as much as 200 m in length. Where whole flights occur together on a hillslope, they are often linked by ramps giving access to the treads. *SHD*

Bell, M. and Walker, M.J.C. 1992: *Late Quaternary environmental change: physical and human perspectives.* Harlow: Longman. **Taylor, C.** 1975: *Fields in the English landscape.* London: Dent.

stripping phase A stage in the erosion of a deep-weathered profile. CORESTONES are initially exposed, and eventually the *basal surface of weathering* (WEATHERING FRONT) may be uncovered. BORNHARDTS, ETCHPLAINS, INSELBERGS and TORS are all formed by the process of stripping. Stripping phases may result from uplift or tilting of landsurfaces or from a climatic change that enhances erosion (such as from humid and vegetated to semi-arid conditions or from humid to more humid conditions). *MAC*

Mabbutt, J.A. 1961: A stripped land surface in Western Australia. *Transactions of the Institute of British Geographers* 29, 101–114.

stromatolite A laminated SEDIMENTARY STRUCTURE of organic origin, produced by the trapping, binding and (sometimes) precipitation of fine particles by microbial mats of CYANOBACTERIA (blue–green algae). FOSSIL stromatolites (they are actually TRACE FOSSILS) occur mainly in LIMESTONES and dolomites as centimetre- to metre-high domes and columns, although wavy and buckled sheets and spheroidal forms (*oncolites*) also occur. They are known from as early as c. 3500 million years ago, representing some of the earliest evidence of life on Earth. They reached their greatest diversity and abundance during the PROTEROZOIC, but the rise of multicellular organisms from the CAMBRIAN period onwards has resulted in their restriction today to environments that are inhospitable to more complex life-forms (e.g. hypersaline lagoons). *CDW*

Awramik, S.M. 1990: Stromatolites. In Briggs, D.E.G. and Crowther, P.R. (eds), *Palaeobiology: a synthesis.* Oxford: Blackwell Science, 336–341. **Buick, R.** 1992: The antiquity of oxygenic photosynthesis: evidence from stromatolites in sulphate-deficient Archaean lakes. *Science* 255, 74–77. **Nisbet, E.** 1987: *The young Earth: an introduction to Archaean geology.* London: Allen and Unwin. **Vanyo, J.P. and Awramik, S.M.** 1985: Stromatolites and Earth–Sun–Moon dynamics. *Precambrian Research* 29, 121–142. **Walter, M.R. (ed.)** 1976: *Stromatolites.* Amsterdam: Elsevier. **Walter, M.R. and Heys, G.R.** 1985: Links between the rise of the Metazoa and the decline of the stromatolites. *Precambrian Research* 29, 149–174.

strontium (Sr) isotope ratio The ratio between radiogenic ^{87}Sr and non-radiogenic ^{86}Sr (^{87}Sr:^{86}Sr). In sediments poor in ^{87}Rb, the parent isotope of ^{87}Sr, strontium isotopes are highly conservative and are preserved unchanged in synsedimentary minerals such as CARBONATES and EVAPORITES. They provide a powerful tool for tracing water sources and distinguishing ancient marine from non-marine deposits. The ^{87}Sr:^{86}Sr of seawater is spatially homogeneous and has varied systematically through time. Variations over the last c. 500 million years are known with a reasonable degree of certainty and can be used for the CORRELATION of marine deposits. The ^{87}Sr:^{86}Sr of continental waters, on the other hand, varies locally, depending upon catchment geology. Rocks and minerals with an original ^{87}Sr:^{86}Sr value different from that of contemporaneous sea-water cannot have formed in a marine environment. *MRT*

McArthur, J.M. 1994: Recent trends in strontium isotope stratigraphy. *Terra Nova* 6, 331–358. **McArthur, J.M.,**

Howarth, R.J. and Bailey, T.R. 2001: Strontium isotope stratigraphy: LOWESS Version 3: Best Fit to the marine Sr-isotope curve for 0–509 Ma and accompanying look-up table for deriving numerical age. *Journal of Geology* 109, 155–170. **Palmer, M.R. and Edmond, J.M.** 1992: Controls over the strontium isotope composition of river water. *Geochimica et Cosmochimica Acta* 56, 2099–2111. **Poyato-Ariza, F.J., Talbot, M.R., Fregenal-Martínez, M.A.** *et al.* 1998: First isotopic and multidisciplinary evidence for nonmarine coelacanths and pycnodont fishes: palaeoenvironmental implications. *Palaeogeography, Palaeoclimatology, Palaeoecology* 144, 65–84.

strontium isotopes

Strontium has four naturally occurring STABLE ISOTOPES (natural abundances, $^{84}Sr = 0.56\%$, $^{86}Sr = 9.86\%$, $^{87}Sr = 7.00\%$ and $^{88}Sr = 82.58\%$). The ISOTOPE RATIO is expressed as the absolute $^{87}Sr:^{86}Sr$ value. As ^{87}Rb decays to form ^{87}Sr, the natural abundance of strontium isotopes is not constant. The formation of rubidium-bearing minerals may be dated by determining ^{87}Sr values. The major inputs of strontium into the oceans are from HYDROTHERMAL activity at the mid-ocean ridges and from continental WEATHERING. Over the past 40 million years, the oceanic $^{87}Sr:^{86}Sr$ value has increased owing to weathering of PRECAMBRIAN rocks. Strontium readily replaces calcium in the crystal lattice of many rocks and minerals, especially sulphates and CARBONATES. $^{87}Sr:^{86}Sr$ are not fractionated during biological processes and, if diagenetic strontium is removed, $^{87}Sr:^{86}Sr$ values may be used to determine the contribution to the diet from marine sources. *IR*

[See also CARBON ISOTOPES, NITROGEN ISOTOPES, ISOTOPES AS INDICATORS OF ENVIRONMENTAL CHANGE]

Harris, N., Bickle, M., Chapman, H. *et al.* 1998: The significance of Himalayan rivers for silicate weathering rates: evidence from the Bhote Kosi tributary. *Chemical Geology* 144, 205–220. **Krishnaswami, S., Trivedi, J.R., Sarin, M.M.** *et al.* 1992: Strontium isotopes and rubidium in the Ganga–Brahmaputra river system: weathering in the Himalaya, fluxes to the Bay of Bengal and contributions to the evolution of oceanic $^{87}Sr/^{86}Sr$. *Earth and Planetary Science Letters* 109, 243–253. **Sealy, J.C., van der Merwe, N.J., Sillen, A.** *et al.* 1991: $^{87}Sr/^{86}Sr$ as a dietary indicator in modern and archaeological bone. *Journal of Archaeological Science* 18, 399–416.

structuralism

A range of principles and practices, which developed during the AD 1950s and 1960s, later exemplified by the approaches of Claude Lévi-Strauss and Louis Althusser during the late 1960s and early 1970s. Within the Marxist philosophies from the early 1970s there has been a strong emphasis upon the significance of the structural dimensions of society. This approach asserts that wider underlying societal structures and structural relationships determine the empirical world of observable phenomena. The explanation of environmental phenomena is therefore achieved through a description and understanding of these underlying structures, and individual social behaviour can be interpreted in terms of these structures. Structuralism offers another approach to at least the societal aspects of environmental change in contrast to the traditions of LOGICAL POSITIVISM and CRITICAL RATIONALISM. *ART*

Gregory, D. 1978: *Ideology, science and human geography*. London: Hutchinson. **Harland, R.** 1987: *Superstructuralism: the philosophy of structuralism and post-structuralism*. New York: Methuen.

structuration

An approach to social theory associated primarily with the British sociologist Anthony Giddens, who developed structuration theory during the late 1970s and early 1980s. It seeks to transcend *hermeneutics* (the study of interpretation and meaning), *functionalism* (form and function of the world) and STRUCTURALISM. Structuration stresses the interconnection between knowledgeable and capable human agents and the wider social systems and structures within which they operate, thus addressing the dualism between structure and agency. An important theme of such theory is that the wider structural characteristics of social systems are both the medium and the outcome of the social practices that constitute the systems. In the context of environmental change, humans may be seen as agents in modifying the natural environment. *ART*

Cohen, I. 1989: *Structuration theory: Anthony Giddens and the constitution of social life*. London: Macmillan. **Gregory, D.** 1981: Human agency and human geography. *Transactions of the Institute of British Geographers* 6, 1–18.

structuring

Simplification of a complex dataset to reveal an underlying simpler structure. CLASSIFICATION and ORDINATION are alternative approaches to structuring. *JAM*

[See also NUMERICAL ANALYSIS]

Lambert, J.M. and Dale, M.B. 1964: The use of statistics in phytosociology. *Advances in Ecological Research* 2, 59–99.

sturzstrom

A catastrophic FAILURE of bedrock slopes leading to the very rapid down-slope transfer of debris; also known as *bergsturz* or *rock avalanche*. Sturzstroms commonly result from UNLOADING, VOLCANICITY and/or SEISMICITY. *PW*

[See also EARTHQUAKE, LANDSLIDE, MASS MOVEMENT PROCESSES]

Dawson, A.G., Matthews, J.A. and Shakesby, R.A. 1986: A catastrophic landslide (sturzstrom) in Verkilsdalen, Rondane National Park, southern Norway. *Geografiska Annaler* 68A 77–87. **Hsü, K.J.** 1975: Catastrophic debris streams (sturzstroms) generated by rockfalls. *Geological Society of America Bulletin* 86, 129–140.

Subarctic

Those areas where the mean monthly air temperature does not exceed +10°C for more than four months and where the coldest month is below 0°C. The northern boundary of the Subarctic approximates the TREE LINE. The Subarctic is characterised by the presence of the BOREAL FOREST, a zone of continuous forest extending across both North America and Eurasia. In North America, the dominant species are spruce (*Picea glauca, Picea maraiana*) whereas in Eurasia the dominant species are Scots pine (*Pinus silvestris*) and larch (*Larix dahurica*). At its southern margins, the subarctic forests merge into the northern TEMPERATE FOREST or into TEMPERATE GRASSLANDS and SEMI-ARID woodlands in more continental areas. *HMF*

[See also ARCTIC, LOW ARCTIC, HIGH ARCTIC]

French, H.M. 1999: Arctic environments. In Alexander, D.E. and Fairbridge, R.W. (eds), *Encyclopedia of environmental science*. Dordrecht: Kluwer, 29–33. **Unesco** 1970: *Ecology of the Subarctic regions* [*Proceedings of the Helsinki Symposium*]. Paris: Unesco.

subclimax The penultimate stage of ECOLOGICAL SUC-CESSION in all complete primary and secondary SERES. It persists for a long time but is expected eventually to be replaced by the CLIMAX VEGETATION. In eastern North America, the early Holocene coniferous forest that eventually gave way to a deciduous forest might be thought of as a subclimax community.   *RJH*

[See also EQUILIBRIUM CONCEPTS]

subduction The process by which LITHOSPHERE bearing OCEANIC CRUST is destroyed in PLATE TECTONICS. A *subduction zone* is equivalent to a DESTRUCTIVE PLATE MARGIN.
GO

subfossil A dead organism that is not truly fossilised (*fossilisation* being the replacement of the carbon content by silica or other mineral compounds). In QUATERNARY studies subfossils are often referred to as fossils.   *BA*

[See also FOSSIL, MACROFOSSIL, MICROFOSSIL, PALAEO-BOTANY]

subglacial Pertaining to the environment beneath a GLACIER or ice sheet and the sediments, processes or landforms originating at the base of glaciers.   *JS*

[See also GLACIAL SEDIMENTS, LANDSYSTEM, LODGEMENT TILL]

Eyles, N. and Menzies, J. 1983: The subglacial landsystem. In Eyles, N. (ed.), *Glacial geology*. Oxford: Pergamon Press, 19–70. **Menzies, J. and Rose, J. (eds)** 1989: Subglacial bedforms – drumlins, rogen moraine and associated subglacial bedforms. *Sedimentary Geology* 62, 117–430.

sublimation The change of state of a solid directly into a gas. In meteorology this transformation refers to the change of ice directly into water vapour without becoming liquid water first. The process is common in subsaturated air. The reverse process (water vapour to ice) is sometimes incorrectly referred to as sublimation; the correct term for this is *deposition*, which dominates under saturated or supersaturated conditions. The other transformations of water are CONDENSATION, EVAPORATION, freezing and melting. *Sublimation till* is the product of GLACIAL DEPOSITION where ice loss is dominated by sublimation rather than melting.   *BDG*

Linacre, E. and Geert, B. 1997: *Climates and weather explained*. London: Routledge.

submarine canyon A steep-sided sea-floor valley that cuts across the CONTINENTAL SHELF and CONTINENTAL SLOPE, losing topographic expression at the base of the slope or on the CONTINENTAL RISE, where it may evolve into a levéed *deep-sea channel*. Submarine canyons resemble river-cut subaerial canyons, with sinuous or straight planform expressions and tributary systems, whereas deep-sea channels usually have distributary patterns. Many submarine canyons feed large SUBMARINE FANS. Submarine canyons may be cut into basement rocks or continental shelf sediments and their spacing is related to slope gradient. Their length and size is related to their age: canyons incise headwards and thus older canyons cross the continental shelves, whereas younger canyons are little more than excavated gullies at or beyond the *shelfbreak*. In

some countries, submarine canyons have been chosen as sites for *waste disposal*.   *BTC*

Davies, J.R., Fletcher, C.J.N., Waters, R.A. *et al.* 1997: *Geology of the country around Llanilar and Rhayader: memoir for 1:50,000 geological sheets 178 and 179 (England and Wales)*. London: The Stationery Office. **Harris, P.T., O'Brien, P.E., Quilty, P. *et al.*** 1999: Sedimentation and continental slope processes in the vicinity of an ocean waste-disposal site, southeastern Tasmania. *Australian Journal of Earth Sciences* 46, 577–591. **Kudrass, H.R., Michels, K.H., Wiedicke, M. and Suckwo, A.** 1998: Cyclones and tides as feeders of a submarine canyon off Bangladesh. *Geology* 26, 715–718. **Pickering, K.T., Hiscott, R.N. and Hein, F.J.** 1989: *Deep-marine environments: clastic sedimentation and tectonics*. London: Unwin Hyman. **Shepard, F.P.** 1977: *Geological oceanography: evolution of coasts, continental margins, and the deep-sea floor*. New York: Crane, Russak and Co.

submarine fan An accumulation of land-derived (TERRIGENOUS) sediment on the OCEAN floor beyond the edge of the CONTINENTAL SHELF and extending to ABYSSAL depths, usually offshore from the mouth of a major river system. Current submarine fan FACIES MODELS also encompass deep-water CLASTIC systems that do not have the conical shape implied by the term. Submarine fans range in size from the Bengal Fan, which covers 3 million km² of the Indian Ocean floor, to smaller systems like the Almeria Fan off southern Spain, or the Crati Fan off southern Italy, which are less than 50 km in diameter. The surfaces of submarine fans are characterised by *deep-water channels* and lobes.

Sediment DEPOSITION is dominated by EVENT processes such as TURBIDITY CURRENTS, DEBRIS FLOWS and SLUMPS and slides (MASS MOVEMENTS). There is usually a progressive decrease in GRAIN-SIZE from proximal to distal parts of the fan. The level of activity on submarine fans may be dependent on SEA-LEVEL CHANGE: when sea-level is low, land-derived sediment is fed through SUBMARINE CANYONS into submarine fan systems, bypassing the continental shelf; when sea-level is high (as today), sediment is trapped in DELTAS and other continental shelf systems, allowing submarine fan surfaces to be draped with MUD. Many ancient TURBIDITE successions in the GEOLOGICAL RECORD have been interpreted as submarine fan systems, and similar relationships with *eustatic* sea-level change have been inferred.   *BTC*

Bouma, A.H., Normark, W.R. and Barnes, N.E. (eds) 1985: *Submarine fans and related submarine systems*. New York: Springer. **Carlson, P.R., Cowan, E.A., Powell, R.D. and Cai, J.** 1999: Growth of a post-Little Ice Age submarine fan, Glacier Bay, Alaska. *Geo-Marine Letters* 19, 227–236. **Normark, W.R., Piper, D.J.W. and Hiscott, R.N.** 1998: Sea level controls on the textural characteristics and depositional architecture of the Hueneme and associated submarine fan systems, Santa Monica Basin, California. *Sedimentology* 45, 53–70. **Pickering, K.T., Hiscott, R.N. and Hein, F.J.** 1989: *Deep-marine environments: clastic sedimentation and tectonics*. London: Unwin Hyman. **Shanmugam, G. and Moiola, R.J.** 1988: Submarine fans: characteristics, models, classification and reservoir potential. *Earth-Science Reviews* 24, 383–428. **Stow, D.A.V., Reading, H.G. and Collinson, J.D.** 1996: Deep seas. In Reading, H.G. (ed.), *Sedimentary environments: processes, facies and stratigraphy*, 3rd edn. Oxford: Blackwell Science, 395–453. **Weimer, P. and Link, M.H. (eds)** 1991: *Seismic facies and sedimentary processes of submarine fans and turbidite systems*. Berlin: Springer.

submergent coast A coast that has become covered by water because of a rise in sea level, a lowering of land or a combination of the two. Submergent coasts may be recognised by submerged forests or other drowned terrestrial features. *HJW*

[See also COASTAL CHANGE, EMERGENT COAST]

sub-Milankovitch CLIMATIC VARIATIONS on millennial to decadal timescales *or less*, which are too short-term to be explained by ORBITAL FORCING as described by MILANKOVITCH THEORY, are said to be sub-Milankovitch in scale. By means of a non-linear response, *climatic forcing* at sub-Milankovitch frequencies may nevertheless contribute to larger-scale events, such as the initiation of Northern Hemisphere GLACIATION about 2.75 million years ago. *JAM*

Murray, D. and Overpeck, J. (eds) 1993: Decadal to millennial-scale variability in the climatic system. *Quaternary Science Reviews* 12(6) [Special Issue]. Roberts, N. 1993: Sub-Milankovitch palaeoclimatic events: their recognition and correlation. *Climatic Change* 24, 175–178. Willis, K.J., Kleczkowski, A., Briggs, K.M. and Gilligan, C.A. 1999: The role of sub-Milankovitch climatic forcing in the initiation of the Northern Hemisphere glaciation. *Science* 285, 568–571.

subsidence The downward movement of material *en masse* relative to its surroundings, including LAND SUBSIDENCE, AIR SUBSIDENCE and CRUSTAL SUBSIDENCE. *GO*

subsistence agriculture Growing crops for food and other necessities, without a surplus for sale. Such systems may be vulnerable to crop failure because there are inadequate reserves. *JAM*

[See also SHIFTING CULTIVATION]

subsurface temperature Temperature changes at the Earth's surface propagate slowly into the subsurface and hence provide evidence of past surface temperatures. Climatic changes can therefore be estimated from subsurface temperature profiles measured in BOREHOLES. For example, underground temperature measurements from boreholes in bedrock in North America, Europe, Africa and Australia indicate mean surface GROUND TEMPERATURE (GST) has increased about 1.0°C over the past five centuries and 0.5°C during the twentieth century (see Figure). *Borehole temperatures* from the Greenland Ice Sheet have similarly enabled direct estimates of air temperatures during the HYPSITHERMAL, MEDIAEVAL WARM PERIOD and LITTLE ICE AGE. *JAM*

Dahl-Jensen, D., Mosegaard, K., Gundestrup, N. *et al.* 1998: Past temperatures directly from the Greenland Ice Sheet. *Science* 282, 268–271. Huang, S., Pollock, H.N. and Shen, P.-Y. 2000: Temperature trends over the past five centuries reconstructed from borehole temperatures. *Nature* 403, 756–758. Pollack, H.N., Huang, S. and Shen, P.-Y. 1997: Climate change record in subsurface temperatures: a global perspective. *Science* 281, 1635–1639.

subtropical The environments, especially the warm climates, immediately poleward of the tropics to about 40° latitude; to some 'subtropical' is synonymous with *warm temperate*. The *dry-summer subtropical* climates of west coasts (Mediterranean-type regions) contrast with the *humid subtropical* climates of east coasts (which lack a distinct dry season) and the DESERTS and ARIDLANDS of the continental interiors. *Mediterranean-type climates* are uniquely characterised by winter rains (brought by mid-latitude DEPRESSIONS) and summer drought (caused by the dominance of the subtropical ANTICYCLONES). Semi-permanent high-pressure cells occur over the subtropical oceans (centred on the so-called *horse latitudes*), which are in turn characterised by the subtropical GYRES. *JAM*

[See also CLIMATIC CLASSIFICATION, TEMPERATE, TROPICAL]

Nicholls, N. 1998: Subtropical climate. In Schneider, S.H. (ed.), *Encyclopedia of climate and weather*. New York: Oxford University Press, 733–734. Perry, A.H. 1981: Mediterranean climate: a synoptic reappraisal. *Progress in Physical Geography* 5, 105–113.

Suess effect The dilution, since the start of the INDUSTRIAL REVOLUTION, of atmospheric ^{14}C by CARBON DIOXIDE from combustion of FOSSIL FUELS containing no ^{14}C. The atmospheric ^{14}C concentration has decreased by about an average 0.03% per annum. Because of this, 'modern' references originally used in RADIOCARBON DATING were materials such as tree rings grown in AD 1850, but the modern REFERENCE STANDARD, NBS OXALIC ACID,

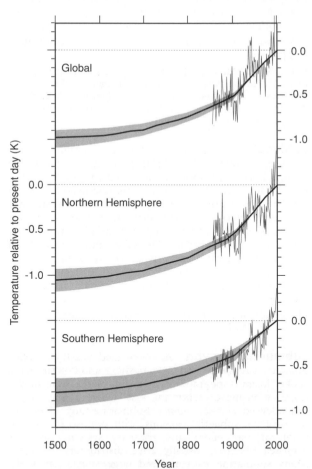

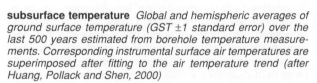

subsurface temperature *Global and hemispheric averages of ground surface temperature (GST ±1 standard error) over the last 500 years estimated from borehole temperature measurements. Corresponding instrumental surface air temperatures are superimposed after fitting to the air temperature trend (after Huang, Pollack and Shen, 2000)*

was rapidly introduced. The Suess effect is enhanced downwind of industrial areas and a natural version may exist in close proximity to volcanic vents where ^{14}C-free CO_2 is produced. *PQD*

Suess, H.E. 1955: Radiocarbon concentration in modern wood. *Science* 122, 415–417.

suffosion The subsurface evacuation of fine material by a combination of downwashing and solution. Essentially a category of erosion by *piping*, suffosion can lead to closed depressions in unconsolidated material overlying limestone bedrock. *SHD*

Motyka, J. 1998: A conceptual model of hydraulic networks in carbonate rocks, illustrated by examples from Poland. *Hydrogeology Journal* 6, 469–482.

sulphation In an environmental context, the chemical reaction between calcium carbonate and SULPHUR DIOXIDE exemplified by the weathering of LIMESTONE building materials in polluted urban environments. *JAM/RAS*

Nord, A.G. and Holenyi, K. 1999: Sulphur deposition and damage on limestone and sandstone in Stockholm city buildings. *Water Air and Soil Pollution* 109, 147–162.

sulphur cycle The cyclical progress of sulphur through living things, air, rocks, soil and water. Atmospheric sulphur, occurring as various oxides and sulphides, dissolves in rainwater. Once on land and in the oceans, it is assimilated and metabolised by animals, plants, and micro-organisms. It returns to soil and water bodies in dead organisms. Soil sulphates are subject to REDUCTION and soil sulphides to OXIDATION. Under reducing conditions, hydrogen sulphide may be liberated, which returns to the atmosphere, and elemental sulphur may be produced. Under oxidising conditions, sulphates may form, including sulphuric acid. ACIDIFICATION in sulphide-rich soils, as found on mine spoil heaps, is a serious problem. The burning of FOSSIL FUELS releases SULPHUR DIOXIDE into the air (as do volcanoes). This quickly oxidises and falls as ACID RAIN. Steps to remedy the acid-rain problem may reduce atmospheric sulphur concentrations. In consequence, it may become necessary to supplement soil sulphur through FERTILISER applications. *RJH*

[See also BIOGEOCHEMICAL CYCLES, CARBON CYCLE, NITROGEN CYCLE, SULPHUR ISOTOPES]

Charlson, R. J. Anderson, T. L. and McDuff, R. E. 2000: The sulphur cycle. In Jacobson, M.C., Charleson, R.J., Rodhe, H. and Orians, G.H. (eds), *Earth system science: from global biogeochemical cycles to global changes*. London: Academic Press, 343–359. Lovelock, J. 1997: A geophysiologist's thought on the natural sulphur cycle. *Philosophical Transactions of the Royal Society of London* B352, 143–147.

sulphur dioxide The most common source of this gas in the environment arises from the burning of FOSSIL FUELS, followed by ore smelting. In strong concentrations it is an irritant, especially on eyes, and has phytotoxic effects on plants. The gas (SO_2) combines with water in the atmosphere to form weak sulphuric acid, which has a corrosive effect on stonework, especially limestone (see SULPHATION), and is a factor in FOREST DECLINE. Concentrations of up to 3 ppmv (parts per million by volume) can occur in urban areas. Volcanic EMISSIONS are the major natural source of sulphur dioxide, others being

FOREST FIRES and planktonic ALGAL BLOOMS. Sulphur dioxide has a short lifetime in the atmosphere of 2–4 days, being removed by both WET and DRY DEPOSITION. Sulphate PARTICULATES form the most abundant group of secondary AEROSOLS in the atmosphere, where they have increased markedly over the past 50–100 years and act to cool the Earth by scattering incoming SOLAR RADIATION. *AHP*

[See also ACID RAIN, AIR POLLUTION, SULPHUR CYCLE]

Bridgman, H. 1990: *Global air pollution*. London: Belhaven. Charlson, R.J. and Wigley, T.M.L. 1994: Sulfate aerosol and climatic change. *Scientific American* 270, 48–57.

sulphur isotopes Sulphur has four STABLE ISOTOPES (natural abundance: ^{32}S = 95.02%, ^{33}S=0.75%, ^{34}S = 4.21% and ^{36}S = 0.02%). The ISOTOPE RATIO is expressed as ^{34}S:^{32}S relative to sulphur in *troilite* (FeS) from the meteorite Canyon Diablo (CD). Sulphur occurs in most natural systems; however, biological isotope effects have only been studied in a few species of microorganism. In the ANOXIC environment of the LITHOSPHERE, sulphur occurs as metal sulphide minerals, the most common of which, pyrite (FeS_2), is a major sink of sulphur in the SULPHUR CYCLE. The largest ISOTOPIC FRACTIONATION occurs during the reduction of sulphate (SO_4^{2-}) to hydrogen sulphide (H_2S) by ANAEROBIC bacteria, causing an ISOTOPIC DEPLETION in ^{34}S. Most vegetation obtains sulphur from either atmospheric SULPHUR DIOXIDE (SO_2) or from aqueous solutions in the SOIL; hence the wide range and complexity of plant δ^{34}S values. The ISOTOPE RATIO of atmospheric SO_2 originating from ANTHROPOGENIC activities is related to the source from which it was originally derived and can be used to monitor ATMOSPHERIC POLLUTION. At each TROPHIC LEVEL in a FOOD CHAIN there is little enrichment in ^{34}S. Distinct differences between marine and terrestrial δ^{34}S values enable the use of δ^{34}S values in dietary reconstruction. *IR*

Chambers, L. A. and Trudinger, P. A. 1979: Microbiological fractionation of stable sulfur isotopes: a review and critique. *Geomicrobiology Journal* 1, 249–293. Krouse, H. R. 1989: Sulfur isotope studies of the pedosphere and biosphere. In Rundel, P. W., Ehleringer, J. R., Nagy, K.A. (eds), *Stable isotopes in ecological research* [*Ecological Studies*. Vol. 68]. Berlin: Springer, 424–444. Ohmoto, H. and Goldhaber, M. B. 1997: Sulfur and carbon isotopes. In Barnes, H. L. (ed.), *Geochemistry of hydrothermal ore deposits*, 3rd edn. New York: Wiley, 517–611. Zhao, F.J., Spiro, B., Poulton, P.R. and McGrath, S.P. 1998: Use of sulfur isotope ratios to determine anthropogenic sulfur signals in a grassland ecosystem. *Environmental Science and Technology* 32, 2288–2291.

sunspot cycles *Sunspots* are relatively dark areas on the surface of the Sun, which are the visible manifestation of convection cells near the Sun's surface. Sunspot cycles are periodic alternations between *sunspot maxima* (when the number of sunspots is high and, paradoxically, SOLAR RADIATION is relatively high) and contrasting phases of *sunspot minima*. The 11-year cycle is well known (see Figure) and others have been recognised, such as about 22 years (the *Hale cycle*), around 88 years (the *Gleissberg cycle*) and around 200 years.

Sunspot cycles are part of the CLIMATIC CHANGE debate, but their control on global climates is both confusing and contradictory. However, when it is realised that sunspot cycles represent a periodic change of Sun-surface temperatures, of the order of 1100°C, then it does appear perti-

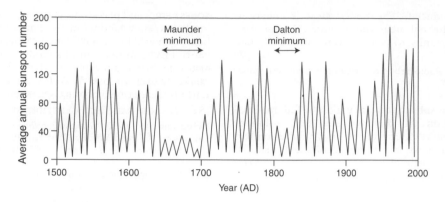

sunspot cycles *The prominent 11-year sunspot cycle and evidence of lower-frequency variations in average annual sunspot numbers between* AD *1500 and 1993 (after Bryant, 1997)*

nent to consider them in this context, even though associated SOLAR CONSTANT changes are only about 0.1%. The peak of the MEDIAEVAL WARM PERIOD (twelfth century), and early and late twentieth century global warming, have been correlated with sunspot maxima and, conversely, the most severe winters of the LITTLE ICE AGE (AD 1645–1705) are said to correspond with a sunspot minimum (see MAUNDER MINIMUM). However, sunspot–temperature correlations are not precise and research in the tropics has revealed positive correlations over the period 1930–1950 and negative correlations for the period 1875–1920. *RDT*

[See also COSMOGENIC-ISOTOPE DATING, DALTON MINIMUM, OORT MINIMUM, WOLF MINIMUM]

Bryant, E. 1997: *Climate process and change.* New York: Cambridge University Press. **Grove, J. M.** 1988: *The Little Ice Age.* London: Methuen. **Schove, D.J.** 1983: Sunspot, auroral, radiocarbon and climatic fluctuations since 7000 BC. *Annales Geophysicae* 4. **Schove, D.J. (ed.)** 1983: Sunspot cycles. Stroudsburg, PA: Hutchinson Ross.

super bounding surface A major BOUNDING SURFACE in a succession of AEOLIAN deposits, taken to represent a HIATUS in ERG development, and sometimes referred to as a *super surface.* *GO*

Kocurek, G. 1988: First-order and super bounding surfaces in eolian sequences – bounding surfaces revisited. *Sedimentary Geology* 56, 193–206. **Mountney, N., Howell, J., Flint, S. and Jerram, D.** 1999: Climate, sediment supply and tectonics as controls on the deposition and preservation of the aeolian–fluvial Etjo Sandstone Formation, Namibia. *Journal of the Geological Society, London* 156, 771–777.

supercontinent A former CONTINENT subsequently split into two or more continents. Parts of the continent may have been submerged beneath shallow seas. For example, Africa, Antarctica, Australia and South America, together with India and Arabia, were formerly united as GONDWANA, and North America and Eurasia were joined forming *Laurasia.* Between about 260 and 180 million years ago, these two supercontinents formed one supercontinent (PANGAEA). An earlier supercontinent that may have existed during the PROTEROZOIC has been given the name *Rodinia.* During Earth's history there may have been as many as seven *supercontinent cycles*, each lasting around 500 million years, during which supercontinents were formed and split up. *JAM/GO*

[See also CONTINENTAL DRIFT, PALAEOGEOGRAPHY, PLATE TECTONICS]

Dalziel, I.W.D. 1995: Earth before Pangaea. *Scientific American* 272, 58–73. **Murphy, J.B. and Nance, R.D.** 1992: Mountain belts and the supercontinent cycle. *Scientific American* 266, 84–91. **Piper, J.D.A.** 2000: The Neoproterozoic supercontinent: Rodinia or Palaeopangaea? *Earth and Planetary Science Letters* 176, 131–146.

supercooling The cooling of a liquid below its freezing point without transformation into a solid. Water droplets in clouds are typically supercooled to about –20°C before freezing, but exceptionally reach –40°C in the absence of CONDENSATION NUCLEI. *JAM*

supergroup A unit in LITHOSTRATIGRAPHY comprising two or more GROUPS with related characteristics of LITHOLOGY or within major bounding UNCONFORMITIES (e.g. Dalradian Supergroup). *LC*

superimposed drainage A river drainage system that developed on one set of rock structures or sedimentary deposits and was subsequently 'let down' through erosion onto another, underlying set of structures, with which it is has no apparent relationship. Superimposed or *epigenetic drainage* is a type of DISCORDANT DRAINAGE that was inherited and hence cannot be explained in terms of contemporary adjustments between rocks and relief. The courses of many large, eastward-flowing rivers in England and Wales, for example, are believed to have been superimposed onto underlying older rocks from chalk strata uplifted and tilted towards the east at the end of the CRETACEOUS period. *JAM*

[See also ANTECEDENT DRAINAGE]

Sparks, B.W. 1986: *Geomorphology*, 3rd edn. London: Longman.

superimposed ice Ice formed from water that percolates down through snow near the EQUILIBRIUM LINE of a GLACIER and freezes in contact with cold snow or ice. *RAS*

[See also MASS BALANCE, MELTWATER, SUPRAGLACIAL]

supersaturation The state existing where (1) a solution contains more of a solute than is necessary to saturate it, or (2) the atmosphere contains more water vapour than is necessary to produce saturation. In supersaturated air, the vapour pressure is greater than the saturation vapour pressure (see SATURATION DEFICIT). Because of the differences between saturation vapour pressures over ice and water, it is possible for air to be supersaturated with respect to ice while being only saturated with respect to water. Thus, in CLOUD formation processes, ice crystals may grow at the expense of water droplets.   *BDG*

Ahrens, C.D. 1994: *Meteorology today: an introduction to weather, climate and the environment.* Minneapolis/St Paul, MN: West Publishing Company. **Sumner, G.** 1988: *Precipitation: process and analysis.* Chichester: Wiley.

supraglacial   Relating to the upper surface of a glacier or ice sheet. The term *superglacial* is a little used synonym.   *JS*

[See also FLOW TILL, GLACIAL SEDIMENTS, LANDSYSTEM]

Paul, M.A. 1983: The supraglacial landsystem. In Eyles, N. (ed.), *Glacial geology.* Oxford: Pergamon Press, 71–90.

surface dating   The dating of exposed surfaces in the landscape as opposed to subsurface material; also known as EXPOSURE-AGE DATING, this type of DATING TECHNIQUE is of particular interest in the investigation of archaeological and geomorphological change. Included are some established methods that have been recently improved (such as those based on rock WEATHERING RINDS), some methods that are adaptations of those traditionally applied in the subsurface context (such as THERMOLUMINESCENCE DATING) and some new methods specifically designed for exposed features (such as COSMOGENIC-ISOTOPE DATING).   *JAM*

[See also LICHENOMETRIC DATING, OBSIDIAN HYDRATION DATING, SCHMIDT HAMMER]

Beck, C. (ed.) 1994: *Dating in exposed and surface contexts.* Albuquerque, NM: University of New Mexico Press. **McCarroll, D.** 1994: A new approach to lichenometry: dating single-age and diachronous surfaces. *The Holocene* 4, 383–396.

surface displacement   Geomorphological changes at the surface of the Earth whereby a mass of sediment or rock is moved down-slope, generally in the absence of running water. Small-scale surface displacement processes include *creep* and SOLIFLUCTION, whereas larger-scale MASS MOVEMENT PROCESSES include flow (e.g. DEBRIS FLOW) and slide (e.g. rockslide) phenomena. Rarer examples of surface displacement include fault scarps caused by EARTHQUAKES and SUBSIDENCE.   *AGD*

Skinner, B.S. and Porter, S.C. 1987: *Physical geology.* Chichester: Wiley.

surface pollen   Pollen contained in the surface layers of a SEDIMENT (*moss* polsters and PEATS, leaf litter, soil or lake sediment). The POLLEN ASSEMBLAGE in a surface sample gives a MODERN ANALOGUE for the VEGETATION (local and regional) from within which the sample was taken. Modern analogues are used to reconstruct vegetation from FOSSIL pollen assemblages, while training sets of surface pollen samples numerically calibrated with environmental parameters allow a reconstruction of those environmental factors that affect vegetation. Surface pollen is also sampled as *pollen deposition* (POLLEN RAIN) by

means of *pollen traps*. In this instance, the time period represented by the sample is controlled and POLLEN INFLUX (grains cm^{-2} y^{-1}) can be calculated. Annual variations in pollen deposition reflect climate while the long-term average pollen influx of a taxon reflects the density and abundance of that taxon in the catchment area.   *SPH*

[See also PALAEOENVIRONMENTAL RECONSTRUCTION, POLLEN ANALYSIS, *R*-VALUES, VEGETATION HISTORY]

Gaillard, M.-J., Berglund, B.E., Frenzel, B. and Huckreide, U. (eds) 1998: *Quantification of land surfaces cleared of forests during the Holocene – modern pollen/vegetation/landscape relationships as an aid to the interpretation of fossil pollen data* [Paläoklimaforschung Vol. 27]. Stuttgart: Gustav Fischer. **Hicks, S.** 1994: Present and past records of Lapland forests. *Review of Palaeobotany and Palynology* 82, 17–35. **Tauber, H.** 1977: Investigations of aerial pollen transport in a forested area. *Dansk Botanisk Arkiv* 32, 1–121.

surface roughness   The variation in height of a surface. Qualitatively, surfaces can be considered rough (or smooth) where the fluctuations in surface height are larger (or smaller) than a fraction of the wavelength of scattered radiation.   *PJS*

surface sediment sampler   An implement for taking undisturbed samples from the uppermost part of waterlaid (lake and marine) sediments. Surface samplers are either gravity samplers or freeze core samplers. Both take short cores across the sediment/water interface, from which, for example, the surface 0–1 cm can be subsampled. Such surface samples are used for calibrating data for *environmental reconstructions* and/or for producing MODERN ANALOGUES, while the analysis of consecutive samples through surface short cores allows recent ENVIRONMENTAL CHANGES (e.g. POLLUTION and ACIDIFICATION) to be followed. *Gravity samplers* include Kajak, Limnos and Glew samplers. All take a circular core measuring 5–10 cm in diameter and the Limnos sampler automatically subsamples in 1 cm thick slices in the field. *Freeze core samplers* (see FREEZE CORING) take a frozen sample either around the outside of a cylindrical tube or on the surface of a rectangular plate. The latter provides a bigger and more easily handled sample.   *SPH*

[See also CHAMBER SAMPLER, PISTON SAMPLER, CORERS]

Glew, J. 1995: Conversion of shallow water gravity coring equipment for deep water operation. *Journal of Palaeolimnology* 14, 83–88. **Håkanson, L. and Jansson, M.** 1983: *Principles of lake sedimentology.* Berlin: Springer. **Renberg, I.** 1991: The Hon-Kajak sediment corer. *Journal of Paleolimnology* 6, 167–170.

surface tension   The tension due to the forces of attraction between the molecules of a liquid, which act in a way that minimises the surface area. With pure water, surface tension is inversely related to temperature. It is important in many environmental processes, such as EVAPORATION from water surfaces, *nucleation* (growth of droplets) in clouds and CAPILLARY ACTION in soils.   *BDG*

surveying   The procedure of measuring and locational recording of data pertaining to features of the Earth. The result of the process is often a map-type image on paper or in digital form. Surveys are normally based on a well defined network of control points fixed in two- or three-dimensional space with respect to the Earth.

Measurements are related to a datum. This may be located on the Earth's surface, for example, *latitude* and *longitude* (origin: Equator and Greenwich Meridian) or within the Earth, for example, the *World Geodetic System* (WGS), the origin of which is the centre of a reference ellipsoid. The observed, measured or plotted information is thus located in a geodetic reference system, a MAP PROJECTION or a local COORDINATE SYSTEM, such as WGS84 (see Table in SPHEROID), Transverse Mercator projection or Ordnance Survey National Grid, respectively.

The information gathering methods used in surveying for environmental purposes range widely. They may involve on-site survey techniques and instrumentation such as GPS receivers, theodolites, electronic distance measuring (EDM) equipment, nowadays often combined in one system known as a total station. Alternatively, remotely sensed images (see REMOTE SENSING) may only require ground truth checks to compare and confirm features on the ground with their recorded characteristics on the image, thus allowing extended interpretation and data extraction using computer software. Remotely sensed imagery, from AERIAL PHOTOGRAPHY to satellite images, can be GEOMETRICALLY CORRECTED to fit with, say, topographic base maps on specific projection or coordinate systems. The integrated study of all elements of spatial information handling including surveying, data acquisition, processing, manipulation and presentation is now termed *geomatics*. *TF*

Bannister, A., Raymond, S. and Baker, R. 1998: *Surveying*, 7th edn. Harlow: Longman. **Kavanagh, B.F. and Bird, S.J.G.** 1996: *Surveying: principles and applications*. Englewood Cliffs, NJ: Prentice-Hall.

suspended load That proportion of the SEDIMENT transported in suspension by a flow. In rivers and streams it usually consists of particles < 0.2 mm in diameter. The particles are prevented from falling out of suspension by the upward component of movement in the turbulence created by eddies in the main flow. The suspended sediment may be derived by the river through bed and/or BANK EROSION, or it may be applied to the river by RUNOFF processes. *MAB*

suspended-load river A river with a sediment load that is dominated by fine-grained sediment carried as SUSPENDED LOAD. Such rivers are typical of low CHANNEL GRADIENTS. Because of the cohesive nature of the sediment (see COHESION), rivers develop stable channel banks and typically have a high-sinuosity, anastomosing CHANNEL PATTERN. *GO*

[See also GRAVEL-BED RIVER, SAND-BED RIVER]

Rust, B.R. and Legun, A.S. 1983: Modern anastomosing–fluvial deposits in arid Central Australia and a Carboniferous analogue in New Brunswick, Canada. In Collinson, J.D. and Lewin, J. (eds), *Modern and ancient fluvial systems* [*International Association of Sedimentologists Special Publication* 6]. Oxford: Blackwell Scientific, 385–392.

sustainability The use of the biophysical environment by humans in such a way that its productive functions remain indefinitely available. As expressed by FAO (1991), *sustainable land management* is the management and conservation of the natural resource base and the orientation of technological and institutional change in such

a manner as to ensure the attainment and continued satisfaction of human needs for present and future generations. Such sustainability in the agriculture, forestry and fisheries sectors conserves land, water and plant and animal genetic resources; it is both economically viable and socially acceptable. *Destructive exploitation* (the antithesis of sustainability) of any natural resource, particularly water and soil, will have implications for future generations as without water to drink and soil in which to grow crops the outlook for humanity is bleak. This concept of sustainability was further developed following the United Nations Conference on Environment and Development held in Rio de Janeiro in AD 1992, and its action plan (*Agenda 21*), which should be implemented at global to local scales.

Douglas (1984) proposed three basic components of sustainability, although the concept is applicable beyond agricultural systems.

- Sustainability as long-term *food sufficiency*, which requires agricultural systems that are more ecologically based and that do not destroy their natural resources.
- Sustainability as *stewardship*, requiring agricultural systems that are based on a conscious ethic regarding humankind's relationship to future generations and to other species (see STEWARDSHIP CONCEPT).
- Sustainability as *community*, requiring agricultural systems that are equitable.

To attain SUSTAINABLE DEVELOPMENT, it is necessary for national governments to rethink policies so that they may achieve the most fundamental goals of society, namely people-centred, equitable and sustainable development, and ultimately democracy and peace. Education for sustainable development needs to involve all sectors of society and to include basic messages regarding the environment and development, giving attention to scientific accuracy, interdisciplinarity and a global perspective. It is highly desirable for the whole of Agenda 21 to be implemented to ensure that natural resources remain for our children. *MGC*

[See also CONSERVATION]

Commission on Sustainable Development 1996: *Promoting education, public awareness and training*. New York: United Nations Department for Policy Coordination and Sustainable Development. **Douglas, G.K.** 1984: *Agricultural sustainability in a changing world order*. Boulder, CO: Westview Press. **Food and Agricultural Organization of the United Nations (FAO)** 1991: *The den Bosch declaration and agenda for action on sustainable agriculture and rural development*. Rome: FAO. **Goodland, R.** 1995: The concept of environmental sustainability. *Annual Review Ecological Systems* 26, 1–24. **Hueting, R. and Reijnders, L.** 1998: Sustainability is an objective concept. *Ecological Economics* 27, 139–147. **Lafferty, W.** 1998: *From Earth summit to local Agenda 21*. Carbondale, IL: Earthscan. **Newman, P and Renworthy, J.** 1999: *Sustainability and cities*. Washington, DC: Island Press. **Redclift, M. (ed.)** 1999: *Sustainability: life chances and livelihoods*. London: Routledge. **United Nations Conference on Environment and Development** 1992: *Agenda 21: programme of action for sustainable development*. New York: United Nations.

sustainable development Not a new concept, nor one that has a precise definition. Most accept that it means 'marrying' environmental care to development and bequeathing the same resource endowment enjoyed by the present generation to the future ('inter-generational equity'). The concept has become of increasing importance with the realisation that standards of living in developed countries, which many other nations would like to emulate, cannot be maintained with likely populations and technology. In this form, it appeared in the 1970s, and was promoted by *The World Conservation Strategy* (IUCN, UNEP and WWF, 1980) and the *Brundtland Report* (World Commission on Environment and Development, 1987). Problems that beset it include a failure to clarify what 'development' means and the conflict between opposing goals of LIMITS TO GROWTH and hope of ever-increasing consumption. There is debate as to whether it is a valuable guide-rail, a distracting shibboleth or something that can spawn practical strategies and techniques for sustained improved living standards without ENVIRONMENTAL DEGRADATION. *CJB*

[See also SUSTAINABILITY]

Beckerman, W. 1994: Sustainable development: is it a useful concept? *Environmental Values* 3, 191–209. Elliott, J.A. 1994: *An introduction to sustainable development.* London: Routledge. Hanley, N. 2000: The concept of sustainable development: an economic perspective. *Progress in Environmental Science* 2, 181–203. IUCN, UNEP and WWF 1980: *The World Conservation Strategy.* Gland, Switzerland: International Union for the Conservation of Nature and Natural Resources. Redclift, M. 1987: *Sustainable development: exploring the contradictions.* London: Methuen. Reid, D. 1995: *Sustainable development: an introductory guide.* London: Earthscan. World Commission on Environment and Development 1987: *Our common future* [*The Bruntland Report*]. Oxford: Oxford University Press.

sustainable yield Removing resources from an environment (HARVESTING) at a rate that can be maintained without the onset of ENVIRONMENTAL DEGRADATION. The *maximum sustainable yield* is the maximum rate of removal that can be permitted while ensuring that the harvest will be maintained in the long run. *JAM*

[See also SUSTAINABILITY]

suture The boundary in an OROGENIC BELT between two former CONTINENTS that joined at a continent–continent COLLISION ZONE, representing the site of a former OCEAN. *GO*

[See also WILSON CYCLE]

Burke, K., Dewey, J.F. and Kidd, W.S.F. 1977: World distribution of sutures: the sites of former oceans. *Tectonophysics* 40, 69–99. Woodcock, N.H., Quirk, D.G., Fitches, W.R. and Barnes, R.P. (eds) 1999: *In sight of the suture: the Palaeozoic geology of the Isle of Man in its Iapetus Ocean context* [*Special Publication* 160]. London: Geological Society.

swamp A general term for a WETLAND, including PEATlands and also seasonally inundated wetlands that do not accumulate PEAT. Some use the term to differentiate wetlands with trees from those without (*marsh*), especially in tropical environments. *MBB*

[See also MANGROVE SWAMP, MIRES, SEDIMENT TYPES, WETLAND CLASSIFICATION]

swath The width of an imaged scene swept out by a moving beam. *PJS*

symbiosis A relationship between two or more living organisms (*symbionts*). The term is often restricted to

cases where all organisms benefit (more correctly termed MUTUALISM), but it may also involve cases where one organism benefits either to the cost of another (*parasitism*) or without any significant effect (COMMENSALISM). *JLI*

[See also EPIPHYTE, MYCORRHIZA, PARASITE]

synchronous Formed or occurring at the same time; contemporary, such as a rock surface on which every point is the same age. The extent to which events can be described as synchronous is often restricted by the PRECISION of age estimates. *DAR*

[See also DIACHRONOUS, TIME TRANSGRESSIVE]

synergism The combined environmental effect of two or more factors, such as POLLUTANTS, (inter)acting together, when the combined effect is greater than the sum of the individual effects: *anergism* occurs when the combined effects are less. *GOH*

Wellburn, A. 1988: *Air pollution and acid rain.* London: Longman.

synfuel A FOSSIL FUEL that must be processed (i.e. energy must be put in) to yield PETROLEUM. Examples include OIL SHALES and TAR SANDS. *GO*

Barbiroli, G. and Mazzaracchio, P. 1995: Synthetic fuel technologies as strategic pathways. *Energy Sources* 17, 595–604.

syngenetic Refers to an ICE WEDGE that grows upwards as the surface of the PERMAFROST rises as a result of the addition of SEDIMENT at the growth site. Such a feature often develops as a vertically nested chevron form. *PW*

[See also EPIGENETIC, PERIGLACIAL]

synoptic catalogues The aim of a synoptic catalogue is to identify a relatively small number of modes or patterns in the atmospheric circulation of a particular region for a stated time interval (most typically 24 hours). Catalogues thus consist of a series of codes that indicate the incidence of defined patterns of air pressure, AIR MASS or AIRFLOW TYPE. Synoptic catalogues can be compiled by subjective or objective (automated) methods. The most well known exponent of the subjective catalogue was H.H. Lamb. Other subjective classifications have adopted a larger study region (e.g. Grosswetterlagen – see WEATHER TYPES). Objective (automated or computer-assisted) schemes have been widely used over recent years to achieve a greater consistency of weather type generation. These include the pressure correlation based method of Lund and the sum of squares analysis of surface or upper air pressure patterns associated with Kirchofer. *JCM*

El-Kadi, A.K.A. and Smithson, P.A. 1992: Atmospheric classifications and synoptic climatology. *Progress in Physical Geography* 16, 432–455. Lamb, H.H. 1972: *British Isles weather types and a register of the daily sequence of circulation patterns, 1861–1971* [*Meteorological Office Geophysical Memoir* 116]. London: HMSO. Yarnal, B. 1993: *Synoptic climatology in environmental analysis: a primer.* London: Belhaven.

synoptic climatology The study of the relationship between the atmospheric circulation and local or regional climates. *Synoptic scale* analysis typically ranges from 500 to 10 000 km and involves two stages: first, the classification of the atmospheric circulation into a series of distinctive WEATHER TYPES; and, second, the assessment of these categories in relation to the weather elements of the region. Until the AD 1960s, synoptic classifications were developed through subjective procedures, but with an increase in the availability of suitable data banks, data from METEOROLOGICAL SATELLITES and automated classification procedures, more objective numerical techniques are now commonly employed. A *synoptic map* summarises the state of the atmosphere over a large area at a given time. Synoptic climatology is a dominant approach within the field of *geographical climatology*. *AHP*

Barry, R.G. and Perry, A.H. 1973: *Synoptic climatology: methods and applications.* London: Methuen. **Smithson, P.A.** 1986: Dynamic and synoptic climatology. *Progress in Physical Geography* 12, 90–96. **Yarnal, B.** 1993: *Synoptic climatology in environmental analysis.* London: Belhaven.

synthesis indicator
An ENVIRONMENTAL INDICATOR that provides information on the *links between* three types of indicator: ECOSYSTEM INDICATORS, INTERACTION INDICATORS and indicators of the human condition. Synthesis indicators provide a basis for anticipatory policy development and decision making. *JAM*

[See also GEOINDICATOR]

Hodge, R.A. 1996: Indicators and their role in assessing progress towards sustainability. In Berger, A.R. and Iams, W.J. (eds), *Geoindicators: assessing rapid environmental changes in earth systems.* Rotterdam: Balkema, 19–24.

system
A system is a set of objects, together with the relationships between the objects and between their attributes. The definition implies that a system must have a clearly demarcated boundary and that what is inside the system may be sharply distinguished from what is outside the system (in systems jargon what is outside the system is called the ENVIRONMENT). Virtually all sciences employ a systems approach (explicitly or implicitly) and quantitative analysis must embrace systems. The systems approach predates GENERAL SYSTEM(S) THEORY, which is primarily an attempt to generalise and systematise the components of systems thinking that have interdisciplinary utility. Use of systems thinking emphasises the interaction between things, as is illustrated at a simple level by a static correlation model and at a sophisticated level by a DYNAMICAL SYSTEMS model. Some types of systems are defined in the Table; there is no single, or correct, way to classify systems: rather there are innumerable ways, any one of which may be appropriate for a specific purpose. The classifications listed are extremely simple and represent merely an easy starting point. *CET*

Bennett, R.J. and Chorley, R.J. 1978: *Environmental systems: philosophy, analysis and control.* Princeton, NJ: Princeton University Press. **Chorley, R.J. and Kennedy, B.** 1971: *Physical geography: a systems approach.* London: Prentice-Hall International. **Dearing, J.A. and Zolitscha, B.** 1999: System dynamics and environmental change: an exploratory study of Holocene lake sediments at Holzmaar, Germany. *The Holocene* 9, 531–540. **Huggett, R.J.** 1985: *Earth surface systems.* Berlin: Springer. **Phillips, J.D.** 1999: *Earth surface systems: complexity, order and scale.* Oxford: Blackwell. **Smalley, I.J. and Vita-Finzi, C.** 1969: The concept of 'system' in the Earth sciences. *Bulletin of the Geological Society of America* 80, 1591. **Weaver, W.** 1958: *A quarter century in the natural sciences. Annual Report of the Rockefeller Foundation.* New York: Rockefeller Foundation, 7–122.

system (stratigraphical)
A major unit of global application in the hierarchy of CHRONOSTRATIGRAPHY (e.g. Silurian System). *LC*

[See also PERIOD]

system *A selected listing of simple classification schemes for systems*

Name	Primary attributes	Source*
Functional classification		
Isolated system	Boundary closed to both energy and mass transfer	C & K
Closed system	Boundary closed to transfer of mass, but open to energy transfer	C & K
Open system	Boundary open to transfer of both energy and mass	C & K
Structural classification		
Morphological system	Limited to a network of structural relationships between parts	C & K
Cascading system	The system is defined by the path of energy or mass throughputs	C & K
Process–response system	At least one morphological and one cascading system linked so that the manner of the form–process link is demonstrated	C & K
Control system (transducer)	A process–response system controlled by an intelligence	C & K
Self-maintaining system	The lowest form of life (e.g. a cell)	C & K
Plant	A living structure	C & K
Animal	A living structure	C & K
Ecosystem	Plants, animals, and their habitat	C & K
Man		C & K
Social system		C & K
Human ecosystem	Interlocking social system and ecosystem	C & K
Complexity and organisational classification		
Simple system	A small number of components described by a small number of variables	W
Complex, disorganised system	A large number of components weakly or randomly coupled	W
Complex, organised system	A large system with strong couplings between elements	W

* C & K: Chorley and Kennedy (1971); W: Weaver (1958)

T

tabula rasa Latin for a flat, clean table; the term refers to a hypothesis in BIOGEOGRAPHY that explains the present disjunct ranges of species distributions in terms, first, of local extinction (EXTIRPATION), caused by the expansion of PLEISTOCENE ice and, second, by subsequent re-immigration. It contrasts with the NUNATAK HYPOTHESIS. *BA*

taele gravels Dissected and cryoturbated SOLIFLUCTION gravels on the chalklands of southern and eastern England, the ERRATIC content of which suggests derivation from the penultimate GLACIATION. *PW*

[See also HEAD, PERIGLACIAL, PERIGLACIATION]

tafoni Holes or depressions (singular: tafone), usually less than a few metres in width, with arch-shaped entrances, concave inner walls and overhanging margins (*visors*). These 'cavernous' or 'honeycomb' forms of WEATHERING, commonly occurring on steep rock faces of chiefly coarse- to medium-grained lithologies, may be accentuated by *case hardening* of the surface of the rock outcrop. The term is of Corsican origin. *SHD*

[See also CHEMICAL WEATHERING, PHYSICAL WEATHERING]

Matsukura, Y. and Tanaka, Y. 2000: Effect of rock hardness and water content on tafoni weathering in the granite of Mount Doeg-Sung, Korea. *Geografiska Annaler* 82A, 59–67. **Mellor, A., Short, J. and Kirkby, S.J.** 1997: Tafoni in the El Charro area, Andalucia, southern Spain. *Earth Surface Processes and Landforms* 22, 817–833. **Sunamura, T.** 1996: A physical model for the rate of coastal tafoni development. *Journal of Geology* 104, 741–748.

talik A layer or body of unfrozen ground in a PERMAFROST region. *HMF*

talus The accumulation of coarse debris at the foot of rockwalls, and the steep valley-side slopes formed by such material (often at or close to the angle of repose). Talus or *scree* is prominent in mountain areas that experience, or previously experienced, a PERIGLACIAL climate. *PW*

[See also FROST WEATHERING, MASS WASTING]

McCarroll, D., Shakesby, R.A. and Matthews, J.A. 1998: Spatial and temporal patterns of Late Holocene rockfall activity on a Norwegian talus slope: a lichenometric and simulation modeling approach. *Arctic and Alpine Research* 30, 51–60.

taphonomy Biological, physical and chemical processes of FOSSIL preservation, including decay, disarticulation, transport, burial, flattening, compaction and DIAGENESIS. Taphonomic studies can yield valuable palaeoenvironmental information. *LC*

[See also FOSSIL LAGERSTÄTTEN]

Allison, P.A. and Briggs, D.E.G. (eds) 1991: *Taphonomy: releasing the data locked in the stratigraphic record*. New York: Plenum. **Brenchley, P.J. and Harper, D.A.T.** 1998: *Palaeoecology: ecosystems, environments and evolution*. London: Chapman and Hall. **Briggs, D.E.G.** 1995: Experimental taphonomy. *Palaios* 10, 539–550. **Marean, C.W., Abe, Y., Frey, C.J. and Randall, R.C.** 2000: Zooarchaeological and taphonomic analysis of the Die Kelders Cave 1 Layers 10 and 11 Middle Stone Age larger mammal fauna. *Journal of Human Evolution* 38, 197–233. **Taylor, W.L. and Brett, C.E.** 1996: Taphonomy and paleoecology of echinoderm *Lagerstätten* from the Silurian (Wenlockian) Rochester Shale. *Palaios* 11, 118–140.

tar sands Deposits of sand or sandstone in which a near-solid form of PETROLEUM acts as a CEMENT. Tar sands can be worked by opencast mining and heated to release CRUDE OIL (see SYNFUEL), but large areas need to be exploited to be commercially viable. The potential yield of oil from tar sands exceeds the known reserves of *conventional petroleum*. The Athabasca Tar Sands in Canada contribute a significant proportion of Canada's oil consumption. *GO*

Hubbard, S.M., Pemberton, S.G. and Howard, E.A. 1999: Regional geology and sedimentology of the basal Cretaceous Peace River Oil Sands deposit, north-central Alberta. *Bulletin of Canadian Petroleum Geology* 47, 270–297.

taxon cycle A theory developed by E.O. Wilson in the early 1960s that predicts a regular, progressive change in the ecological and evolutionary characteristics of a species once it has colonised an isolated *oceanic island*. Over many generations the species is transformed, primarily through NATURAL SELECTION, from one comprising generalist (broad-niched) individuals occupying the beachfronts to ever more specialised populations, which penetrate deeper and deeper into the island's interior. Eventually, the descendants of the original colonists form new species that become so specialised and so limited in their distribution that they become extinct and are eventually replaced by a new wave of colonists and their descendants. *MVL*

[See also EVOLUTION, ISLAND BIOGEOGRAPHY, NICHE]

Wilson, E. O. 1961: The nature of the taxon cycle in the Melanesian ant fauna. *American Naturalist* 95, 169–193. **Whittaker, R.J.** 1998: *Island biogeography: ecology, evolution and conservation*. Oxford: Oxford University Press.

taxonomy The theory and practice of describing, naming and classifying features or objects according to general principles and laws; for example, the Linnaean classification of plants. *MJB*

technocentrism The environmental philosophy that considers technology as able to provide the solution to all human-environmental problems; a form of *anthropocentrism*. *JAM*

[See also ECOCENTRISM, ENVIRONMENTALISM, TECHNOLOGICAL CHANGE, TECHNOLOGICAL PARADOX]

technological change Change in 'technological hardware', the knowledge required to produce and use such hardware and the organisation of human activities that such hardware makes possible. Invention (discovery) is followed by development (use or application) and spread (adoption or diffusion); newer technologies typically supersede older ones. The spread of technologies is at the core of the historical technological changes relevant to GLOBAL CHANGE. There are complex interdependencies between technologies. Grübler (1998), for example, recognises five overlapping 'technological clusters' since the onset of the INDUSTRIAL REVOLUTION: (1) AD 1750–1820, (2) 1800–1870, (3) 1850–1940, (4) 1920–2000, (5) 1980 and currently emerging. Each was characterised by particular 'dominant' and 'emerging' technologies involving energy production, transport and communications, materials, industrial processes and consumer products.

JAM

[See also GLOBAL ENVIRONMENTAL CHANGE, TECHNOCENTRISM]

Ausubel, J.H. and Sladovich, H.E. (eds) 1989: *Technology and environment*. Washington, DC: National Academy Press. **Grübler, A.** 1998: *Technology and global change*. Cambridge: Cambridge University Press. **Marcus, A.I. and Segal, H.P.** 1999: *Technology in America: a brief history*. Fort Worth, TX: Harcourt Brace. **Mokyr, J.** 1990: *The lever of riches: technological creativity and economic progress*. Oxford: Oxford University Press. **Technological Forecasting and Social Change** 1996: Special Issue: 'Technology and the environment'. *Technological Forecasting and Social Change* 53(1).

technological paradox The 'paradox' of technological development: that technology is a cause of environmental change, a means of assessing its impacts and a source of remedies for adverse impacts. *JAM*

Gray, P.E. 1989: The paradox of technological development. In Ausubel, J.H. and Sladovich, H.E. (eds), *Technology and environment*. Washington, DC: National Academy Press, 192–204.

tectonics Processes that relate to the DEFORMATION, UPLIFT and SUBSIDENCE of the Earth's CRUST, exerting a major influence on GEOMORPHOLOGY, LANDSCAPE EVOLUTION, ENVIRONMENTAL CHANGE and patterns of SEDIMENTATION. Tectonic history is recorded in GEOLOGICAL STRUCTURES. NEOTECTONICS refers to tectonic processes operating during the CENOZOIC. *GO*

[See also SEDIMENTARY BASIN]

Crowley, T.J. and Burke, K.C. 1999: *Tectonic boundary conditions for climatic reconstructions* [*Oxford Monographs on Geology and Geophysics* 39]. Oxford: Oxford University Press. **Frostick, L.E. and Steel, R.J. (eds)** 1994: *Tectonic controls and signatures in sedimentary successions* [*International Association of Sedimentologists Special Publication* 20]. Oxford: Blackwell Scientific. **Mack, G.H. and Leeder, M.R.** 1999: Climatic and tectonic controls on alluvial-fan and axial–fluvial sedimentation in the Plio–Pleistocene Palomas half graben, southern Rio Grande rift. *Journal of Sedimentary Research* 69, 635–652. **MacNiocaill, C. and Ryan, P.D. (eds)** 1999: *Continental tectonics* [*Special Publication* 164]. London: Geological Society. **Stokes, M. and Mather, A.E.** 2000: Response of Plio–Pleistocene alluvial systems to tectonically induced base-level changes, Vera Basin, SE Spain. *Journal of the Geological Society, London* 157, 303–316.

tektite A small spherule of glass formed by melting and rapid cooling during flight through the air. Tektites are generally taken to represent EJECTA from METEORITE IMPACTS. Tektites with a common mineralogy and age are generally found within a limited area known as a *strewnfield*: examples include parts of Libya and the Ivory Coast. *GO*

[See also K–T BOUNDARY]

King, E.A. 1977: The origin of tektites: a brief review. *American Scientist* 65, 212–218. **Smit, J.** 1999: The global stratigraphy of the Cretaceous–Tertiary boundary impact ejecta. *Annual Review of Earth and Planetary Sciences* 27, 75–113.

teleconnection The linking of environmental events in time and place, especially the connection of CLIMATIC VARIATIONS at locations that are geographically remote. Teleconnection is possible because of the dynamic nature of the Earth–atmosphere system, changes in one part of the system affecting other locations after a time lag. The variations at one location are usually considered as entirely atmospheric, while those at another may be atmospheric or related to BOUNDARY CONDITIONS, for instance fluctuations in SEA-SURFACE TEMPERATURE. Teleconnections can be demonstrated using statistical analysis, but explanation requires a physical understanding that may be elucidated by GENERAL CIRCULATION MODELS. The most pronounced teleconnections globally are those associated with EL NIÑO–SOUTHERN OSCILLATION (ENSO) phenomena. ENSO dominates the interannual CLIMATIC VARIABILITY in tropical latitudes and there is a pronounced teleconnection between it and surface pressure over the North Pacific Ocean and between it and monsoon rainfall in southeast Asia. There are even suggestions that ENSO can be detected in European climate. *GS*

[See also NORTH ATLANTIC OSCILLATION, PACIFIC–NORTH AMERICAN TELECONNECTION, PERIODICITY]

Fraedrich, K. and Muller, K. 1992: Climate anomalies in Europe associated with ENSO extremes. *International Journal of Climatology* 12, 25–31. **Glantz, M.H., Katz, R.W. and Nicholl, N. (eds)** 1991: *Teleconnections linking worldwide climate anomalies: scientific basis and societal impact*. Cambridge: Cambridge University Press. **Leathers, D.J. and Palecki, M.A.** 1991: The Pacific/North American teleconnection pattern and United States climate, part II: temporal characteristics and index specifications. *Journal of Climate* 4, 707–716. **Singh, N. and Sontakke, N.A.** 1999: On the variability and prediction of the rainfall of the post-monsoon season over India. *International Journal of Climatology* 19, 309–339. **Wilby, R.** 1993: Evidence of ENSO in the synoptic climate of the British Isles. *Weather* 48, 234–239.

teleology A form of explanation that asserts that phenomena and observations are best explained and understood not by means or causes, but by outcomes or aims, intentions or purposes. Hence it focuses on the ends as a means to understanding the cause. *ART*

tell A large, low mound or hill-like LANDFORM built up over thousands of years from the massive accumulation of debris from ancient settlements. Tells are characteristic of Eurasia, especially the Near and Middle East, and often made largely of successive generations of mud and/or wattle houses. The term *kom* has also been used for these features. An outstanding example is Tell Abu Hureya, which occupies an area of 11.5 ha on the bank of the Euphrates in northern Syria and has been dated from at least 8000 to 5000 BC. The ENVIRONMENTAL ARCHAEOLOGY of the site has yielded important information relating to the early and gradual transition from hunting to agriculture. *JAM/RAS*

[See also AGRICULTURAL ORIGINS]

Bruins, H.J. and VanDerPflicht, J. 1995: Tell es-Sultan (Jericho): radiocarbon results of short-lived cereal and multiyear charcoal samples from the end of the Middle Bronze Age. *Radiocarbon* 37, 213–220. **Holz, R.K.** 1969: Man-made landforms in the Nile Delta. *Geographical Review* 59, 253–269. **Moore, A.M.T.** 1979: A pre-Neolithic farmers' village on the Euphrates. *Scientific American* 241, 50–58. **Moore, A.M.T., Hillman, G.C. and Legge, A.J.** 1998: *Abu Hureya and the advent of agriculture.* New York: Oxford University Press. **Rosen, A.M.** 1986: *Cities of clay: the geoarchaeology of tells.* Chicago: University of Chicago Press.

temperate Normally applied to the 'mild' or 'moderate' mid-latitude climates that are intermediate, especially in terms of their thermal character, between hot (TROPICAL) and cold climates. *Warm temperate* and *cool temperate* climates are sometimes subdivided into oceanic and continental types, leading to four main types of temperate climate. However, the term 'temperate' is less appropriate for the extremely seasonal, mid-latitude climates of continental interiors, where neither the summer nor the winter season can be described as mild or moderate. Thus, the most commonly recognised temperate climates range from the warm temperate oceanic climates of mediterranean-type regions to the cool temperate climates of higher-latitude west coasts (such as that characteristic of the British Isles). *JAM*

[See also CLIMATIC CLASSIFICATION, CONTINENTALITY, GLACIER THERMAL REGIME, SUBTROPICAL]

Bailey, H.P. 1964: Towards a unified concept of the temperate climate. *Geographical Review* 54, 516–545. **Diaz, H.F.** 1996: Temperate climate. In Schneider, S.H. (ed.), *Encyclopedia of climate and weather.* New York: Oxford University Press, 744–747.

temperate forests Communities in TEMPERATE zones dominated by DECIDUOUS trees and shrubs, but with some EVERGREEN species. The latter are conifers in the mountains, sclerophyll taxa (see SCLEROPHYLLOUS) in Mediterranean areas (i.e. near west coasts), and broadleaved, laurophyllous evergreens (*laurisilva*) where the climate is both warm and humid enough (i.e. near east coasts or in low-latitude mountain areas).

Temperate forests originated from Tertiary tropical to subtropical floras. During the PLEISTOCENE, multiple climatic changes linked to the GLACIAL–INTERGLACIAL CYCLES resulted in the stepwise loss of THERMOPHILOUS and hygrophilous (see HYGROPHYTE) taxa during each GLACIAL EPISODE (cold phase). This process was: (1) strongest in Europe, where mountain systems trending west–east inhibited re-immigration from ice-age REFUGIA and where the climate was dominated by cold, dry Atlantic air masses during glacial phases; (2) intermediate in eastern North America, where mountain chains trend north–south and where the climate was affected by subtropical air masses, even during glacial intervals; and (3) weakest in southeast Asia, where subtropical air masses prevailed in the absence of glacial effects on the climate. These differences resulted in species diversities among woody taxa increasing from Europe to southeast North America to southeast Asia.

During each GLACIATION deciduous trees in Europe became regionally extinct at higher latitudes and survived in refugia in lower-latitude mountains, where orographic precipitation permitted tree growth. In these lower latitudes, higher intra-specific genetic diversity is found, a fact important for conservation of BIODIVERSITY. Despite this impoverishment of the flora, SPECIATION occurred in certain families during the PLEISTOCENE. During the HOLOCENE, characteristic vegetational sequences, usually inferred from POLLEN ASSEMBLAGE ZONES for REFERENCE AREAS, are controlled by distances to the refuge areas, migrational speeds of taxa (with or without geographical barriers), CLIMATIC CHANGE, PEDOGENESIS and HUMAN IMPACT. *BA*

Bennett, K.D., Tzedakis, P.C. and Willis, K.J. 1991: Quaternary refugia of north European trees. *Journal of Biogeography* 18, 103–115. **Birks, H.J.B.** 1986: Late-Quaternary biotic changes in terrestrial and lacustrine environments, with particular reference to north-west Europe. In Berglund, B.E. (ed.), *Handbook of Holocene palaeoecology and palaeohydrology.* Chichester: Wiley, 3–65. **Ellenberg, H.** 1988: *Vegetation ecology of Central Europe.* Cambridge: Cambridge University Press. **Ovington, J.D. (ed.)** 1981: *Temperate broad-leaved evergreen forests* [Ecosytems of the World. Vol. 10]. Amsterdam: Elsevier. **Röhrig, E. and Ulrich, B. (eds)** 1991: *Temperate deciduous forests* [Ecosytems of the World. Vol. 7]. Amsterdam: Elsevier. **Latham, R.E. and Ricklefs, R.E.** 1993: Continental comparisons of temperate-zone tree species diversity. In Ricklefs, R.E. and Schluter, D. (eds), *Species diversity in ecological communities: historical and geographical perspectives.* Chicago, IL: University of Chicago Press, 294–314.

temperate grasslands A vegetation FORMATION TYPE, most extensive in the Northern Hemisphere, and including the *prairies* of the Great Plains of North America, the STEPPES of Eurasia, the *veld* of South Africa and the *pampas* of South America. The natural climatic limits of these grasslands are determined by moisture and ARIDITY, but are difficult to define precisely because they are also affected by soil type, natural herbivore populations, and natural and anthropogenic fires. They differ from tropical grassland (SAVANNA) in the abundance of C-4 grasses (see C-3, C-4 AND CAM PLANTS).

HUMAN IMPACT ON VEGETATION has been extensive in temperate grassland regions to the extent that little if any remains in a NATURAL state, especially through GRAZING by domestic livestock and the expansion of ARTIFICIAL grasslands. Natural or semi-natural temperate grasslands are subject to seasonal and periodic DROUGHTS, which also affect these 'substitute' communities. *JAM*

Bamforth, D.B. 1988: *Ecology and human organization on the Great Plains.* New York: Plenum. **Coupland, R.T. (ed.)** 1992: *Natural grasslands: introduction and Western Hemisphere* [Ecosystems of the World. Vol. 8A]. Amsterdam: Elsevier. **Coupland, R.T. (ed.)** 1993: *Natural grasslands: Eastern Hemisphere and résumé* [Ecosystems of the World. Vol. 8B]. Amsterdam: Elsevier.

temporal change detection ENVIRONMENTAL MONITORING attempts to identify temporal change of environmental variables. Fundamental understanding of such short-term environmental change is needed for SUSTAINABLE management of natural RESOURCES. SIMULATION MODELS can be used to predict expected changes for different *scenarios.* A VALIDATION of the model should be carried out to assess its reliability.

REMOTE SENSING provides techniques for change detection from MULTITEMPORAL ANALYSIS. Changing environmental variables result in different reflectance or scattering properties, which can be monitored by optical and radar sensors. For instance, burnt areas in boreal forests show clearly on optical imagery, while flooded areas in forests are easier to detect from radar, which penetrates the canopy.

Various methods for detecting temporal changes from repeated imaging are available. *Image ratioing* calculates the pixel-wise ratios of values of the same band at two different acquisition times. *Image differencing* calculates differences between values acquired at different times to detect changes. Image differencing using two images of *Normalised Difference Vegetation Index* has been used to detect vegetation change. PRINCIPAL COMPONENTS ANALYSIS of multitemporal imagery extracts a number of principal components that are linear combinations of the data vectors. The weight factor of each variable contributing to a principal component is called a *loading*. The signs of the loadings of variables representing the same image channel at different times indicate whether a change has occurred. IMAGE CLASSIFICATION techniques like the *spectral temporal change classifier* (Michener and Houhoulis, 1997) have been applied to derive a thematic map of changes in LAND COVER. In this approach a class is not defined as a land-cover type but as a transition between two land-cover types.

Change-detection analyses have been applied in many fields, such as soil SALINITY dynamics, tropical FOREST REGENERATION and tropical DEFORESTATION, as well as vegetation responses to DROUGHT, FLOODS, insect outbreaks and DUST STORMS. *HB*

[See also CLIMATIC CHANGE, TIME SERIES ANALYSIS, TROPICAL RAIN FOREST]

Dwivedi, R.S. and Sreenivas, K. 1998: Image transforms as a tool for the study of soil salinity and alkalinity dynamics. *International Journal of Remote Sensing* 19, 605–619. Guirguis, S.K., Hassan, H.M., El Raey, M.E. and Hussain, M.M.A. 1996: Multi-temporal change of Lake Brullus, Egypt, from 1983 to 1991. *International Journal of Remote Sensing* 17, 2915–2921. Michener, W.K. and Houhoulis, P.F. 1997: Detection of vegetation changes associated with extensive flooding in a forested ecosystem. *Photogrammetric Engineering and Remote Sensing* 63, 1363–1374. Yanasse, C.D.F., SantAnna, S.J.S., Frery, A.C. et al. 1997: Exploratory study of the relationship between tropical forest regeneration stages and SIR-C L and C data. *Remote Sensing of Environment* 59, 180–190.

ten-to-twelve-year oscillation (TTO) A QUASI-PERIODIC PHENOMENON reported, for example, from the Arctic stratosphere during the winter and more widely in the middle and upper troposphere. The TTO may provide a link between the QUASI-BIENNIAL OSCILLATION (QBO) of the tropics and higher latitudes. It has been in phase with the 11-year SOLAR CYCLE for the last four recorded cycles, but there is, as yet, no known mechanism whereby the 11-year solar cycle influences the Earth's stratosphere and troposphere. *JAM*

Loon, H. van, and Labitzke, K. 1994: The 10–12-year atmospheric oscillation. *Meteorologische Zeitschrift NS* 3, 259–266.

tephra A collective term for all the unconsolidated, primary pyroclastic products of a volcanic eruption, independent of grain size. *JBH*

[See also TEPHRA ANALYSIS, TEPHROSTRATIGRAPHY and TEPHROCHRONOLOGY]

Thorarinsson, S. 1981: Tephra studies and tephrochronology: A historical review with special reference to Iceland. In Self, S. and Sparks, R.S.J. (eds), *Tephra studies*. Dordrecht: Reidel, 1–12.

tephra analysis Instrumental methodology to determine the (geochemical) composition of TEPHRA. Analysis of both VOLCANIC GLASS and mineral phases (e.g. Fe–Ti oxides, mafic minerals) can reveal potentially unique fingerprints for particular tephras, providing the foundation for TEPHROCHRONOLOGY. Grain-discrete beam methods are favoured over bulk methods as the former enable detection of trends within a single eruptive phase and postdepositional mixing, and allow for avoidance of contaminants or inclusions. The most widely used approach is that of wavelength-dispersive ELECTRON MICROPROBE MICROANALYSIS. However, when applied to glass shards, it is prone to analytical errors necessitating high precision and interlaboratory standardisation. Supplementing the electron microprobe, laser-ablation INDUCTIVELY-COUPLED PLASMA MASS SPECTROMETRY allows assay of major, trace and rare-earth elements in individual glass shards. *JBH*

Hunt, J.B. and Hill, P.G. 1993: Tephra geochemistry: a discussion of some persistent analytical problems. *The Holocene* 3, 271–278. Larsen, G. 1981: Tephrochronology by microprobe glass analysis. In Self, S. and Sparks, R.S.J. (eds), *Tephra studies*. Dordrecht: Reidel, 95–102. Pearce, N.J.G., Westgate, J.A. and Perkins, W.T. 1996: Developments in the analysis of volcanic glass shards by laser ablation ICP-MS: quantitative and single internal standard–multi-element methods. *Quaternary International* 34–36, 213–227.

tephrochronology Broadly, a means of dating or correlating signals of (palaeo)environmental change based upon the near-instantaneous deposition of TEPHRA. Tephra is dispersed from its volcanic source by atmospheric or oceanic circulation and, within hours to weeks, may be deposited on a wide variety of environmental surfaces, including lakes, mires, soils, loess, deltas, moraines, fluvioglacial deposits, ice sheets, glaciers, solifluction and gelifluction lobes, estuaries and salt marshes, and deep and shallow oceans. Tephras may be transported many thousands of kilometres, so tephrochronology is not restricted to volcanic areas. In the palaeoenvironmental record, a tephra layer defines an isochronous surface or time-plane within sediments forming potential sources of PROXY DATA for PALAEOENVIRONMENTAL RECONSTRUCTION.

Geochemical TEPHRA ANALYSIS frequently links tephra, uniquely, to a particular eruption of an identifiable volcano. In this manner, tephras possess a geochemical fingerprint. Additional properties of the tephra such as mineralogy, grain-size and stratigraphical associations are also important in characterising a particular layer.

Thorarinsson (1981) defined tephrochronology as a dating method based on the 'identification, correlation and dating of tephra layers'. In initial studies, age controls were obtained for the youngest tephras primarily though reference to the historical record (e.g. religious documents and sagas in Iceland). Following the development of radiocarbon dating, numerical ages for tephras were obtained by dating-associated organic sediments. In both instances, chronology is transferred between sites using TEPHROSTRATIGRAPHY, i.e. depositional sequences, landscapes or geomorphic/archaeological events are dated using their stratigraphic relationship to tephra layers of known age. This spatial transference of *chronology* through TEPHROSTRATIGRAPHY defines tephrochronology in the strict sense. *JBH*

Dugmore, A.J., Larsen, G. and Newton, A.J. 1995: Seven tephra isochrones in Scotland. *The Holocene* 5, 257–266. **Einarsson, T**. 1986: Tephrochronology. In Berglund, B.E. (ed.), *Handbook of Holocene palaeoecology and palaeohydrology*. Chichester: Wiley, 329–342. **Schmincke, H.-U. and Bogaard, P.v.d.** 1991: Tephra layers and tephra events. In Einsele, G. Ricken, W. and Seilacher, A. (eds), *Cycles and events in stratigraphy*. Berlin: Springer, 392–429. **Self, S. and Sparks, R.S.J. (eds)** 1981: *Tephra studies*. Dordrecht: Reidel. **Sheets, P.D. and Grayson, D.K.** 1979: *Volcanic activity and human ecology*. New York: Academic Press. **Thorarinsson, S**. 1981: Tephra studies and tephrochronology: A historical review with special reference to Iceland. In Self, S. and Sparks, R.S.J. (eds), *Tephra studies*. Dordrecht: Reidel, 1–12.

tephrochronometry The direct dating of TEPHRA layers to provide an absolute chronology, e.g. by LUMINESCENCE DATING, FISSION TRACK DATING, or RADIOMETRIC DATING (K–Ar, Ar–Ar or U) methods.   *JBH*

tephrology An all-embracing term for studies of TEPHRA, including TEPHRA ANALYSIS, TEPHROCHRONOLOGY, TEPHROSTRATIGRAPHY and TEPHROCHRONOMETRY.   *JBH*

Lowe, D.J. (ed.) 1996: Tephra, loess, and paleosols – an integration. *Quaternary International* 34–36, 1–261.

tephrostratigraphy The use of TEPHRA layers to help correlate sediment sequences in the absence of a tephra-based absolute chronology derived either by TEPHROCHRONOMETRY or by dating of tephra-associated sediments (see TEPHROCHRONOLOGY).   *JBH*

Hunt, J.B. and Lowe, D.L. 2001: Tephra Nomenclatura. *Journal of Archaeological Sciences*. In press.

teratology The study of congenital malformation. Viral infections, *ionising radiation*, and chemicals are amongst the areas of environmental interest.   *JAM*

termination The abrupt ending of a GLACIAL EPISODE, especially as reflected in OXYGEN ISOTOPE records from MARINE SEDIMENT CORES. The seven terminations commonly recognised during the last 600 000 years, which are useful for CORE CORRELATION, were each shortly followed by the relatively slow build-up of ice volumes towards the next glaciation. Terminations are numbered consecutively with the youngest, Termination I, referring to the end of the LAST GLACIATION.   *JAM*

Broecker, W.S. 1984: Terminations. In Berger, A., Imbrie, J., Hays, J., Kukla, G. and Saltzman, B. (eds), *Milankovitch and climate*. Dordrecht: Reidel, 687–698. **Mix, A.C. and Ruddiman, W.F.** 1985: Structure and timing of the last deglaciation: oxygen isotope evidence. *Quaternary Science Reviews* 4, 59–108. **Raymo, M.E.** 1997: The timing of major climatic terminations. *Paleoceanography* 12, 577–585.

termites Insects of the order Isoptera. Some produce conspicuous mounds (*termitaria*). All are important for BIOTURBATION, DECOMPOSITION and MINERAL CYCLING in the soils of the tropics and subtropics, although some extend to temperate latitudes. Termites are a very diverse group with > 2600 species and comprise an estimated 10% of tropical animal BIOMASS. They are also important PESTS, especially those that consume wood, and are estimated to cause damage exceeding two million US dollars per year.   *JAM*

Collins, N.M. 1989: Termites. In Lieth, H. and Werger, M.J.A. (eds), *Tropical rain forest ecosystems*. Amsterdam: Elsevier, 455–471. **Edwards, R. and Mill, A.E.** 1986: *Termites in buildings: their biology and control*. East Grinstead: Rentokil. **Goudie, A.S.** 1988: The geomorphological role of termites and earthworms in the tropics. In Viles, H.A. (ed.), *Biogeomorphology*. Oxford: Blackwell, 166–192. **Harris, W.V.** 1961: *Termites: their recognition and control*. London: Longman. **Lee, K.E. and Wood, T.G.** 1971: *Termites and soils*. London: Academic Press.

terra fusca A term used by W.L. Kubiena for shallow, decalcified, humus-deficient loamy soils with a brown to reddish brown colour developed over limestone rock.

EMB

Kubiena, W.L. 1953: *The soils of Europe*. London: Thomas Murby.

terra preta do indio An ANTHROPOGENIC, organic rich '*Black Earth*' soil found on terraces along the Amazon and its tributaries under degraded rainforest. Its formation is attributed to pre-Colombian times. American Indian tribes modified these soils sufficiently for them to be considered as hortic ANTHROSOLS and their activities are still reflected in the lack of regrowth to fully developed tropical rainforest.   *JAM/EMB*

Eden, M.J., Bray, W., Herrera, L. and McEwan, C. 1984: Terra preta soils and their archaeological context in the Caquetá Basin of southeast Colombia. *American Antiquity* 49,125–140. **Sombroek, W.G.** 1992: Biomass and carbon storage in the Amazon ecosystem. *Interciencia* 17, 269–272.

terra rossa A term used by W.L. Kubiena for shallow, decalcified, humus-deficient clayey soils with a bright red colour developed over pure limestone, especially in KARST landscapes in the Mediterranean region. There is usually a high content of wind-blown mineral grains and the iron compounds that coat the clays give the bright red coloration. Some may be RELICT soils.   *EMB/JAM*

Kubiena, W.L. 1953: *The soils of Europe*. London: Thomas Murby. **Macleod, D.A.** 1980: The origin of red Mediterranean soils in Epirus, Greece. *Journal of Soil Science* 31, 125–136. **Nihlén, T. and Olsson, S.** 1995: Influence of aeolian dust on soil formation in the Aegian area. *Zeitschrift für Geomorphologie NF* 39, 341–361.

terracette A small step, generally less than about 20 cm wide, along the contour usually of steep grassy slopes. Terracettes can form through SOIL CREEP, animal activity and/or build-up of sediment around vegetation.   *AJDF*

Bergkamp, G. 1998: A hierarchical view of the interactions of runoff and infiltration with vegetation and microtopography in semiarid shrublands. *Catena* 33, 201–220.

terracing In agriculture, a type of MECHANICAL SOIL CONSERVATION MEASURE. There are three different types of terracing depending on their main function. *Diversion or channel terraces* comprise earth embankments built along the contour on a slope to intercept surface runoff and convey it to an established outlet at a sufficiently low velocity so that it does not promote SOIL EROSION. They are usually built on slopes of no more than 7°. *Reverse-slope terraces* on slopes of less than 4–5° help to promote infiltration of rainfall and *overland flow*. These are also known as *retention terraces*. BENCH TERRACES convert a steep slope (up to

625

30°) into a series of steps with horizontal or near-horizontal ledges and steep or vertical walls between the ledges. Near-horizontal terraces with raised lips at the outer edge are used to retain IRRIGATION water for growing rice, tea, fruit trees and other high-value crops. *Fanya juu terraces* comprise narrow shelves formed by digging a ditch along the contour and throwing the soil upslope to form an embankment stabilised by planting grass. *RAS*

[See also RIVER TERRACE, SOIL CONSERVATION]

Beach, T. and Dunning, N. 1995: Ancient Maya terracing and modern conservation. *Journal of Soil and Water Conservation* 50, 138–145. **Denevan, W.M., Mathewson, K. and Knapp, G. (eds)** 1987: *Pre-Hispanic agricultural terraces in the Andean region* [*BAR International Series* 359]. Oxford: British Archaeological Reports. **Morgan, R.P.C.** 1995: *Soil erosion and conservation*, 2nd edn. Harlow: Longman, 138–139. **Thomas, D.B. and Biannah, E.K.** 1989: Origin, application and design of the fanya juu terrace. In Moldenhauer, W.C., Hudson, N.W., Sheng, T.C. and Lee, S.W. (eds), *Development of conservation farming on hillslopes*. Ankeny, IA: Soil and Water Conservation Society, 185–194. **Wagstaff, M.** 1992: Agricultural terraces: the Vasilikos Valley, Cyprus. In Bell, M. and Boardman, J. (eds), *Past and present soil erosion.* Oxford: Oxbow Books, 155–161.

terrain A general geomorphological term for the surface characteristics of a tract of land or LANDSCAPE. *Terrain analysis,* an outdated term for GEOMORPHOLOGY with military connotations and still used in the context of applied geomorphology, has been recently adopted in the context of GEOGRAPHIC INFORMATION SYSTEMS, especially in relation to DIGITAL TERRAIN MODELS. *JAM*

[See also DIGITAL ELEVATION MODEL, TERRANE]

Mitchell, C.W. 1991: *Terrain evaluation: an introductory handbook to the history, principles and methods of practical terrain assessment.* Harlow: Longman Scientific and Technical. **Ollier, C.D.** 1977: Terrain classification: methods, applications and principles. In Hails, J.R. (ed.), *Applied geomorphology.* Amsterdam: Elsevier, 277–316. **Townshend, J.R.G. (ed.)** 1981: *Terrain analysis and remote sensing.* London: Allen and Unwin. **Wilson, J.P. and Gallant, J. (eds)** 2000: *Terrain analysis: principles and applications.* Chichester: Wiley.

terrane A FAULT-bounded area of rocks in an OROGENIC BELT that has a history of SEDIMENTATION or DEFORMATION distinct from adjacent areas. *Terrane tectonics* is a modification to PLATE TECTONICS that allows for some orogenic belts (e.g. North American Cordillera) to have been assembled over a long period of time through small fragments of CONTINENTAL CRUST (*terranes*) individually meeting the CONTINENTAL MARGIN at an ocean–continent DESTRUCTIVE PLATE MARGIN, forming an *accretionary orogen.* Terranes may migrate for hundreds of kilometres along the continental margin before coming to rest (*docking*) at a great distance from where their SEDIMENTARY ROCKS originally formed: these may be known as *suspect terranes* or *allochthonous terranes.* *GO*

[See also TERRAIN]

Aberhan, M. 1999: Terrane history of the Canadian Cordillera: estimating amounts of latitudinal displacement and rotation of Wrangellia and Stikinia. *Geological Magazine* 136, 481–492. **Coney, P.J., Jones, D.L. and Monger, J.W.H.** 1980: Cordilleran suspect terranes. *Nature* 288, 329–333. **Howell, D.** 1994: *Principles of terrane analysis: new applications for global tectonics.* New York: Chapman and Hall. **Sengör, A.M.C. and Dewey, J.F.** 1990: Terranology: vice or virtue? *Philosophical Transactions of the Royal Society of London* A331, 457–478.

terrestrial environment Abiotic and biotic conditions of ecosystems on land. *BA*

[See also AQUATIC ENVIRONMENT]

Aber, J.D. and Meliko, J.M. 1991: *Terrestrial ecosystems.* Philadelphia, PA: Saunders.

terrestrial radiation RADIATION emitted by the Earth, often referred to as *long-wave radiation.* This radiation lies in the INFRARED range of the ELECTROMAGNETIC SPECTRUM. *JBE*

[See also BLACK-BODY RADIATION]

Paltridge, G.W. and Platt, C.M.R. 1976: *Radiative processes in meteorology and climatology.* Amsterdam: Elsevier.

terric horizon An ANTHROPEDOGENIC surface SOIL HORIZON, the surface of which is raised by additions of earthy material, compost or farmyard manure, having signs of considerable biological activity. *EMB*

terricolous Ground dwelling plants or animals that live wholly or predominantly on the ground surface, or organisms that inhabit the soil. Terricolous lichens, for example, may be distinguished from those that grow on rock surfaces (*epilithic*) or on other plants, especially trees (*epiphytic*). *JAM*

terrigenous SEDIMENT derived from land areas through WEATHERING and EROSION, although terrigenous deposits may accumulate in terrestrial or subaqueous settings. Applications of the term include the recognition of *terrigenous clastic* SEDIMENTARY ROCKS and, in the context of OCEANIC SEDIMENTS, the distinction between terrigenous sediments and PELAGIC SEDIMENTS. *GO*

Tertiary A SYSTEM of rocks and a PERIOD of geological time from 65 million years ago (Ma) to the beginning of the QUATERNARY period. It is subdivided into two subperiods, the PALAEOGENE (65–24 Ma; *Palaeocene, Eocene* and *Oligocene* EPOCHS) and the NEOGENE (24–1.8 Ma; *Miocene* and *Pliocene* epochs). Primates, grasses and horses first appeared in the Palaeogene. During the Eocene, India collided with Asia to begin the uplift of the Himalayas (see PLATE TECTONICS). HOMINIDS first appeared in the Neogene, and major environmental events included the enlargement of the Antarctic ICE SHEET and the first evidence of Arctic GLACIATION. *GO*

[See also GEOLOGICAL TIMESCALE]

testate amoebae Unicellular protozoa (Rhizopoda), the tests of which are preserved well in PEAT, LAKES and SOILS and can be used to reconstruct palaeohydrological conditions (moisture and depth to WATER TABLE) using MODERN ANALOGUES and TRANSFER FUNCTIONS. *DH*

[See also RHIZOPOD ANALYSIS]

Charman, D.J., Hendon, D. and Woodland, W.A. 2000: *The identification of testate amoebae (Protozoa: Rhizopoda) from British oligotrophic peatlands.* Cambridge: Quaternary Research Association.

Tethys An equatorial OCEAN that lay between the northern part of PANGAEA (the continents of *Laurasia*) and the southern continents that comprised GONDWANA. It was closed at its western end between Spain and North Africa and widened progressively eastwards. Tethys is first recognisable as a palaeogeographical entity (see PALAEO-GEOGRAPHY) after the late CARBONIFEROUS episode of OROGENY that assembled Pangaea and it persisted through the MESOZOIC era. Its closure during the CENOZOIC era produced a series of OROGENIC BELTS extending from the Pyrenees and the Alps to the Himalayas. Some parts of the Mediterranean Sea, such as the Levantine and Ionian basins, may represent relics of Tethys that may disappear as the African plate continues to move northwards against the European plate (see PLATE TECTONICS). It is named after an ancient Greek sea goddess.   *JCWC*

[See also CONTINENTAL DRIFT]

Audley-Charles, M.G. and Hallam, A. (eds) 1988: *Gondwana and Tethys* [*Special Publication* 37]. London: Geological Society. **Sakinc, M., Yaltirak, C. and Oktay, F.Y.** 1999: Palaeogeographical evolution of the Thrace Neogene Basin and the Tethys–Paratethys relations at northwestern Turkey (Thrace). *Palaeogeography, Palaeoclimatology, Palaeoecology* 153, 17–40. **Wortmann, U.G., Hesse, R. and Zacher, W.** 1999: Major-element analysis of cyclic black shales: paleoceanographic implications for the Early Cretaceous deep western Tethys. *Paleoceanography* 14, 525–541.

tetrapod   A concrete armour unit with four legs, designed to be used in numbers for protecting the shore-line against erosion.   *HJW*

[See also HEXAPOD, COASTAL ENGINEERING STRUCTURES]

texture   The size, shape and arrangement of the constituents of a soil, sediment or rock (e.g. CLASTS in a SEDIMENTARY ROCK or crystals in an IGNEOUS ROCK). The main attributes of texture in sediments and sedimentary rocks are GRAIN-SIZE, SORTING, grain SHAPE (described by particle form and indices of ROUNDNESS and SPHERICITY), FABRIC and surface textures of particles (see QUARTZ GRAIN SURFACE TEXTURES, STRIATIONS). Many textural attributes are progressively modified as sediment released by WEATHERING undergoes SEDIMENT TRANSPORT, and several are combined in the concept of textural MATURITY. Relationships are recognised between textural attributes and transport processes or DEPOSITIONAL ENVIRONMENTS.   *GO/TY*

[See also IMAGE TEXTURE, SOIL TEXTURE]

Friedman, G.M. 1961: Distinction between dune, beach and river sands from their textural characteristics. *Journal of Sedimentary Petrology* 31, 514–529. **Pettijohn, F.J.** 1975: *Sedimentary rocks*, 3rd edn. New York: Harper and Row. **Tucker, M.E. (ed.)** 1988: *Techniques in sedimentology*. Oxford: Blackwell.

theory   A single definition of theory is not possible. Simplistically, this is true because the word is commonly used in two different ways; at a more sophisticated level in one of the uses the term has varying portent.

The simpler and more restricted use of theory may be seen in the sequence: conjecture, HYPOTHESIS, theory. Here the word is used to identify a construct that is assumed to be true and forms the basis for PREDICTION and/or explanation – thereby forming the basis for accumulation of further empirical knowledge. In the conjecture, hypothesis, theory sequence there is no formal distinction between the terms, but there is widespread, informal recognition that it reflects increasing certainty or confidence in the scientific underpinning of the idea. It is

embraced colloquially in the expression 'I have a theory about that'.

The infinitely more complex use of the word emerges in expressions such as 'a body of theory' or 'scientific theory'. Here theory is being used to designate sets of linked ideas (some of which, at least, are often called LAWS) that underpin and pervade the entire scientific endeavour or some distinctive and substantive portion of it, such as thermodynamics or an entire discipline. This use of the term has not only been the focus of multiple attempts to create a satisfactory definition in the immediate sense, but the very content, nature and role of the concept has been, and continues to be, open to debate.

One useful way of grasping the general purpose of theory is to view it as any generalisation that goes beyond available data. Science is not just about accumulating information, nor indeed just about explaining (predicting and/or retrodicting) experiences open to the human senses; it seeks to generalise, simplify, unify the known and predict and prepare for the, as yet, unknown. At its deepest level, the debate about the nature and role of theory has been conducted within the philosophy of science.

Most natural or Earth scientists have not received any training in philosophy, but have grown up in scientific cultures in which there has been tremendous emphasis on FIELD RESEARCH. Consequently, there is a widespread tendency to see theory in terms of the old debate between logical empiricists and critical rationalists. The simplified version of this debate common in the NATURAL ENVIRONMENTAL SCIENCES tends to emphasise its normative nature and its distinctions between several couplets: (1) observation and theory; (2) VERIFICATION and FALSIFICATION; and (3) the role of theory in scientific discovery and scientific justification. Coupled with a preference for field research, these ideas have tended to foster a dislike for theory among many environmental scientists and a belief that it is an arcane and mischievous activity that may be safely left to a few armchair dilettantes in their discipline. For most philosophers of science the LOGICAL POSITIVISM/CRITICAL RATIONALISM debate is long over. In any event, such a debate would be less significant than one focused upon the role of theory in something such as *relativism*, where it is seen as socially constructed and changes in it do not imply convergence on some external truth, versus its role in REALISM, where truth is seen as an external reality upon which science is converging.

Acknowledging that a definition of theory is dependent upon one's philosophy of science, it is possible to offer some useful ideas stemming from modern naturalised philosophies of science. Foremost, theory is pervasive and hierarchical. At a foundational level, theory rests upon embedded ONTOLOGY, '*Weltenschauung*' (world view or world outlook), regulative principles or myths – metaphysical elements that are unfalsifiable and empirically untestable. At a more superficial level all observations are made in the light of theory, and consequently are theory-laden. However, only in its most extreme form does the theory-laden nature of observation preclude adequate, independent testing of theory. Contextual distinction between *exploratory theories* (e.g. those that underpin belief in the utility of a radio telescope) and *explanatory theories* (e.g. the astronomical beliefs being investigated using the radio telescope) not only highlights the importance of theory; it often also assuages the concern about theory-laden observation.

At the level of metatheory, as well as at less fundamental levels, it is important to appreciate that most environmental sciences do not seek to challenge the fundamental sciences (such as physics and chemistry), but rather take their findings as given. This raises two important issues. First, as scientific theory requires, at a minimum, internal consistency, theory in the environmental sciences needs to be consistent with the fundamental sciences. Second, perhaps the environmental sciences can be reduced to the fundamental sciences: this view is generally seen as fallacious, as the environmental sciences contain many elaborate and integrative structures that represent distinctive content, and thereby necessitate additional theory.

Some philosophers of science pursuing naturalised approaches see an important distinction between 'data' and 'phenomena' and not between observation and theory. *Data* are the product of procedures designed to gather information about the world, are accessible to the human perceptual system and are thus available for public inspection. *Phenomena* are objects, entities, events and processes that exist in the world and, while they cannot be observed, they are the foci of explanatory theories.

Scientific theory is dependent upon the regulative principles and types of scientific argument employed. Scientific arguments are not exclusively or purely logical; other kinds of judgement are also relevant. In the environmental sciences, ABDUCTION (arguing from a resulting state of affairs to a possible cause) plays an important role because of the desire to explain the present in terms of the past. However, in terms of pure logic, abduction embraces the logical fallacy of affirming the consequent and is inherently weaker than DEDUCTION (given a controlling state of affairs, a law or law-like statement, the resulting state of affairs is inevitable). As the time scale of interest increases in the environmental sciences, what constitutes acceptable theory becomes more conjectural, and historical contingency must also be embraced. *CET*

[See also SCIENTIFIC METHOD, SCIENCE]

Rhoads, B.L. and Thorn, C.E. 1993: Geomorphology as science: the role of theory. *Geomorphology* 6, 287–307. **Rhoads, B.L. and Thorn, C.E.** 1994: Contemporary philosophical perspectives on physical geography with emphasis on geomorphology. *Geographical Review* 84, 90–101. **Rhoads, B.L. and Thorn, C.E.** 1996: *The scientific nature of geomorphology*. Chichester: Wiley. **Suppe, F.** 1977: *The structure of scientific theories*, 2nd edn. Urbana, IL: University of Illinois Press. **Von Englehardt, W. and Zimmermann, J.** 1988: *Theory of Earth science*, translated by L. Fischer (first published in German in 1982). Cambridge: Cambridge University Press.

thermal analysis Determination of the firing temperature of CERAMICS by measuring the effects of refiring (reheating), such as thermal expansion, shrinkage or weight loss (*thermogravimetric analysis*). *JAM*

Rice, P.M. 1987: *Pottery analysis: a sourcebook*. Chicago, IL: University of Chicago Press. **Tite, M.S.** 1969: Determination of the firing temperature of ancient ceramics by measurement of thermal expansion: a reassessment. *Archaeometry* 11, 131–143.

thermal capacity The quantity of heat required to raise the temperature of a system by one degree either at constant volume or at constant pressure. The SPECIFIC HEAT is the thermal capacity referred to a unit mass. *TS*

thermal expansion of the oceans The expansion of ocean volume due to heating, probably only of the upper part of the oceans rather than the entire water depth. For example, it has been estimated that heating of the upper 100 m by 10°C would cause a world *eustatic* SEA-LEVEL RISE of 10 cm. *RAS*

[See also SEA-LEVEL CHANGE, SEA-LEVEL RISE: POTENTIAL FUTURE HUMAN IMPACTS]

thermal-ionisation mass spectrometry (TIMS) An analytical technique used to obtain measurements of ISOTOPE RATIOS and abundance. Thermally ionised particles are accelerated towards collectors and separated in a magnetic field by deflection proportional to their mass. Isotope ratios measured by this method are used in NUMERICAL-AGE DATING, particularly in URANIUM–THORIUM DATING. *DAR*

Bard, E., Hamelin, B., Fairbanks, R.G. and Zindler, A. 1990: Calibration of the ^{14}C timescale over the past 30,000 years using mass spectrometric U–Th ages from Barbados corals. *Nature* 345, 405–410. **Beck, J.W., Richards, D.A., Edwards, R.L. et al.** 2001: Extremely large variations of atmospheric ^{14}C concentrations during the last glacial period. *Science* 292, 2453–2458. **Edwards, L., Chen, J.H. and Wasserburg, G.J.** 1986: ^{238}U–^{234}U–^{230}Th–^{232}Th systematics and the precise measurement of time over the past 500,000 years. *Earth and Planetary Science Letters* 81, 175–192.

thermal pollution Artificial release of heat into the environment at a level sufficient to harm organisms or damage ecosystems. Although it occurs more widely, thermal pollution is most frequently discussed in relation to aquatic environments and particularly in association with the discharge of cooling water from power stations. Thermal pollution may favour the spread of EXOTIC SPECIES of fish, encourage the growth of micro-organisms with potential human-health implications or create problems for fish and other organisms by reducing the availability of dissolved oxygen. *JAM*

[See also URBAN CLIMATE]

Langford, T.E.L. 1990: *Ecological effects of thermal discharges*. London: Elsevier.

thermal remote sensing REMOTE SENSING utilising *thermal radiation* at wavelengths between 3000 and 14 000 nm emitted from a surface. Objects at temperatures above absolute zero will emit RADIATION with a peak emission wavelength inversely proportional to temperature (the Earth's surface emits at around 10 000 nm; the Sun emits at around 500 nm). Instruments to record emitted thermal radiation have similar designs to OPTICAL REMOTE SENSING INSTRUMENTS such as *whiskbroom scanners*. The detectors are selected to record at longer wavelengths and must be cooled to increase their sensitivity. The detectors also require continual CALIBRATION, by allowing the detectors to view objects of known temperature. Instruments actually record *brightness temperature*, which is a function of the actual temperature of the surface and its *emissivity*. Applications include: SEA-SURFACE TEMPERATURE measurement from METEOROLOGICAL SATELLITES, THERMAL POLLUTION mapping of buildings and outfalls, and differential heating and cooling measurements from day and night images, all of which have significance for environmental change. *GMS*

Elachi, C. 1987: *Introduction to the physics and techniques of remote sensing*. New York: Wiley.

thermocline   The layer within an ocean or lake that separates the warmer surface water from deeper colder water. The temperature gradient in the thermocline exceeds that in the layers above and below. In temperate lakes it may be only a few metres thick and seasonal, being destroyed by *overturning* driven by seasonal changes in water temperature and density. In the tropical ocean it is permanent and up to several hundred metres thick.   *JAM*

[See LAKE STRATIFICATION AND ZONATION]

thermohaline circulation   Sometimes termed the global *ocean conveyor belt*, the thermohaline circulation arises from density differences produced by temperature and SALINITY changes in seawater. Essentially, salty DEEP WATER is carried towards the Pacific Ocean where it upwells (see UPWELLING) towards the surface and is then transported back to regions of DOWNWELLING in the North Atlantic (primarily the Norwegian–Greenland Sea and the Labrador Sea) and in the Weddell Sea (see Figure). The system operates in today's ocean to compensate for the transport of water vapour through the atmosphere from the Atlantic to the Pacific Ocean. It plays a very important role in the GLOBAL CLIMATIC SYSTEM, annually releasing some 5×10^{21} calories of heat into the atmosphere over the North Atlantic.

Thermohaline circulation provides an important stabilising effect on climate, but also has the potential to cause ABRUPT CLIMATIC CHANGE in the space of a few decades if perturbed. Modest changes to the oceanic water balance can have profound impact on thermohaline circulation. *Modelling* experiments have shown that the thermohaline circulation of the Atlantic is sensitive to salinity forcing resulting from MELTWATER release. The resultant changes in deep-water formation can be monitored through nutrient proxy data, such as CADMIUM:CALCIUM RATIOS and $\delta^{13}C$ in benthic FORAMINIFERA from MARINE SEDIMENT CORES. It appears that much of the climate variability experienced in the North Atlantic region over the last GLACIAL–INTERGLACIAL CYCLE can be explained by changes in *North Atlantic Deep Water* formation.   *WENA*

[See also CONTOUR CURRENT, WATER MASS, YOUNGER DRYAS]

Boyle, E.A. and Keigwin, L. 1987: North Atlantic thermohaline circulation during the past 20,000 years linked to high-latitude surface temperature. *Nature* 330, 35–40. **Broecker, W.S., Peteet, D.M. and Rind, D.** 1985: Does the ocean–atmosphere system have more than one stable mode of operation? *Nature* 315, 21–25. **Street-Perrott, F.A. and Perrott, R.A.** 1990: Abrupt climatic fluctuations in the tropics: the influence of the Atlantic Ocean circulation. *Nature* 343, 607–612. **Wood, R.A., Keen, A.B., Mitchell, J.F.B. and Gregory, J.M.** 1999: Changing spatial structure of the thermohaline circulation in response to atmospheric CO_2 forcing in a climatic model. *Nature* 399, 572–575. **Yu, E.F., François, R. and Bacon, M.P.** 1996: Similar rates of modern and last-glacial ocean thermohaline circulation inferred from radiochemical data. *Nature* 379, 689–694.

thermokarst   The unique complex of permafrost-related processes associated with the thaw of frozen, ice-rich ground. Subsidence, slumping and resedimentation occur, resulting in irregular, hummocky topography. The term usually includes *thermal erosion* (i.e. thaw and erosion of frozen ground by running water).

Thermokarst results when disturbance to the ground thermal regime initiates thaw of permafrost. Natural disturbances include: destruction of the shrub TUNDRA or BOREAL FOREST vegetation, often by lightning-induced fire;

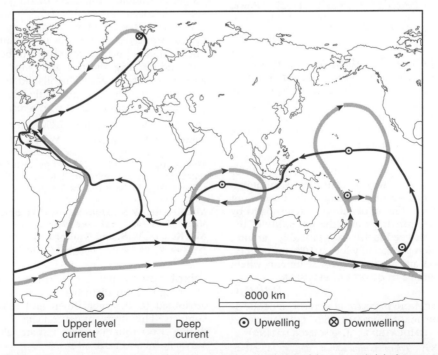

thermohaline circulation *Schematic representation of the thermohaline circulation of the ocean ('global conveyor belt'): note the difference between surface and deep-water currents and the formation of the deep current by downwelling in the North Atlantic (after Schmitz, 1995)*

the occurrence of *active-layer detachment* failures on slopes; undercutting and slumping along river banks by thermal erosion; and regional climatic warming. Much thermokarst activity is human-induced. For example, extensive human-induced thermokarst occurred in central Alaska in the AD 1930s and 1940s following clearance of the land for agricultural purposes. Not only was much of the permafrost composed of ice-rich unconsolidated silty materials but it also contained large ice-wedge bodies. Their preferential melt and general subsidence of the ground quickly resulted in hummocky topography, typically 3–5 m in relative relief. These *permafrost mounds* are more correctly termed *thermokarst mounds*.

Thermokarst is an important sedimentary process, initiating widespread resedimentation and SOFT-SEDIMENT DEFORMATION. Thermokarst sediments (occurring in, for example, thermokarst basins, lakes, slumps and deepening thaw layers and sedimentary structures, e.g. ice wedge pseudomorphs and thermokarst involutions) may be preserved in PLEISTOCENE sediments, providing evidence for previous episodes of thermokarst. The prevention of human-induced thermokarst and the reparation of thermokarst-induced terrain damage can be of considerable local economic and environmental significance. Thermokarst-related problems are likely to intensify and spread under conditions of GLOBAL WARMING. *HMF*

[See also ALAS, PERMAFROST]

Ferrians, O., Kachadoorian, R. and Green, G.W. 1969: Permafrost and related engineering problems in Alaska. *United States Geological Survey* [*Professional Paper* 678]. Reston, VA: United States Geological Survey, 1–37. **Murton, J.B.** 1996: Thermokarst-lake–basin sediments, Tuktoyaktuk Coastlands, western Arctic Canada. *Sedimentology* 43, 737–760. **Murton, J.B. and French, H.M.** 1993: Thermokarst involutions, Summer Island, Pleistocene Mackenzie Delta, western Canadian Arctic. *Permafrost and Periglacial Processes* 4, 217–227. **Péwé, T.L.** 1954: Effect of permafrost upon cultivated fields. *United States Geological Survey Bulletin* 989, 315–351.

thermoluminescence (TL) dating

A dating technique utilising the emission of photons from luminescence centres in a crystal lattice when it is heated. The intensity of emission spectra is related to accumulated radiation damage, which increases with mineral age such that the greater the luminescence, the older the sample. TL dating is particularly useful for dating material such as pottery, hearths or tephra that were heated strongly immediately prior to deposition and was originally developed for archaeological dating of pottery and burnt flints. Although pottery only 100 years old may be dated, the lower limit for sediments is older. The upper limit is determined by saturation of the electron traps and the instability of the TL signal, and for sediments is probably around 100–150 ka. Precision is relatively low at 10% for pottery, hearths and TEPHRAS and up to 15% for sediments where dating is based on grains of QUARTZ or FELDSPAR.

DAR/CJC

[See also LUMINESCENCE DATING, ELECTRON SPIN RESONANCE DATING, OPTICALLY STIMULATED LUMINESCENCE DATING, INFRARED-STIMULATED LUMINESCENCE DATING]

Aitken, M.J. 1985: *Thermoluminescence dating*. London: Academic Press. **Berger, G.W.** 1994: Thermoluminescence dating of sediments older than 100 ka. *Quaternary Science Reviews* 13, 445–455. **Fleming, S.J.** 1979: *Thermoluminescence techniques*

in archaeology. Oxford: Clarendon Press. **Van den Haute, P., Vancraeynest, L. and de Corte, F.** 1998: The Late Pleistocene loess deposits of eastern Belgium: new TL age determinations. *Journal of Quaternary Science* 13, 487–497. **Wintle, A.G. and Huntley, D.J.** 1982: Thermoluminescence dating of sediments. *Quaternary Science Reviews* 1, 31–53. **Zöller, L. and Wagner, G.A.** 1990: Thermoluminescence dating of loess: recent developments. *Quaternary International* 7/8, 119–128.

thermophilous

From the Greek for warmth-loving. Thermophilous taxa or communities require relatively high air or water temperatures (such as mean annual or growing-season temperatures). *BA*

Kullman, L. 1998: The occurrence of thermophilous trees in the Scandes Mountains during the early Holocene: evidence for a diverse tree flora from macroscopic remains. *Journal of Ecology* 86, 421–428. **Salvigsen, O., Forman, S.L. and Miller, G.H.** 1992: Thermophilous molluscs on Svalbard during the Holocene and their paleoclimatic implications. *Polar Research* 11, 1–10.

thermoremanent magnetisation (TRM)

Magnetic properties acquired by certain materials following heating above the Curie Point. On cooling, ferromagnetic particles align with the Earth's ambient geomagnetic field, providing a basis for PALAEOMAGNETIC and ARCHAEOMAGNETIC DATING. *MHD*

[See also FERROMAGNETISM]

Thompson, R. 1991: Palaeomagnetic dating. In Smart, P.L. and Frances, P.D. (eds), *Quaternary dating methods – a users guide* [*QRA Technical Guide* 4]. Cambridge: Quaternary Research Association, 177–198.

thin-section analysis

The study of a sample in the form of a thin glass-mounted slice using an optical microscope. A standard thickness of 0.03 mm is used in the study of rocks (PETROGRAPHY). At this thickness most common MINERALS allow the transmission of light. The optical properties of minerals in plane-polarised light and under crossed polars enable them to be accurately identified, and the magnification allows for the detailed examination of TEXTURE. Unconsolidated material (e.g. SEDIMENT or SOIL) can be studied in thin section if the sample is first impregnated with a resin. *TY/GO*

[See also MICROMORPHOLOGICAL ANALYSIS]

Adams, A.E., MacKenzie, W.S. and Guilford, C. 1984: *Atlas of sedimentary rocks under the microscope*. Harlow: Longman. **Harwood, G.M.** 1988: Microscopic techniques: II. Principles of sedimentary petrography. In Tucker, M.E. (ed.), *Techniques in sedimentology*. Oxford: Blackwell Scientific, 108–173. **Kerr, P.F.** 1977: *Optical mineralogy*, 4th edn. New York: McGraw Hill. **MacKenzie, W.S., Donaldson, C.H. and Guilford, C.** 1982: *Atlas of igneous rocks and their textures*. Harlow: Longman. **MacKenzie, W.S. and Guilford, C.** 1980: *Atlas of rock-forming minerals in thin section*. Harlow: Longman.

thixotropy

A property of some CLAYS involving a loss of strength upon disturbance that is gradually restored, attributed to the breakdown and reforming of particle contacts. Thixotropy of natural clay deposits, for example during EARTHQUAKES, can initiate slope failure, MASS MOVEMENTS and LANDSLIDES. *GO*

Perret, D., Locat, J. and Martignoni, P. 1996: Thixotropic behavior during shear of a fine-grained mud from Eastern Canada. *Engineering Geology* 43, 31–44.

threatened species A species the numbers, geographical distribution or survival of which is under threat from human activities. *Threat numbers* have been used to quantify the status of threatened species. *IFS*

[See also ENDANGERED SPECIES, RARE SPECIES, RED DATA BOOKS, VULNERABLE SPECIES]

threshold A critical level of disturbance or size and/or frequency of event that produces a sudden, and in some cases irreversible, change to a natural system, landform or landscape. In a systems context, threshold has also been defined as a condition marking the transition from one state to another. It is also used as an adjective to describe the magnitude of an event or the value of a process necessary for significant erosion or geomorphic work to result, e.g. the threshold (or critical) river discharge for bedload entrainment to occur.

Thresholds of many different types are considered to be of great relevance in bringing about radical change in geomorphic systems. Change can occur because of the crossing of either *extrinsic* or *intrinsic thresholds*, where the former is initiated by a change in an external factor such as climate or tectonic activity and the latter represents an inherent property (EXTREME EVENT) of the geomorphic system itself. When thresholds are crossed, a system in a *stable equilibrium* may be able to accommodate the disturbance and return to its predisturbance state, whereas one in *unstable equilibrium* will not return to its predisturbance state, but will move out of equilibrium or approach a new equilibrium. The times taken for an unstable system to react to a change in conditions and to attain a new characteristic equilibrium time are termed the RESPONSE TIME (or *reaction time*) and the RELAXATION TIME, respectively.

In fluvial geomorphology, *threshold theory* is based on the assumption that all particles on the channel boundary are on the verge of movement at bankfull discharge, though it is known that in many rivers significant bedload movement and bank erosion occur at discharges significantly below bankfull. *RPDW*

[See also EQUILIBRIUM CONCEPTS IN GEOMORPHOLOGY, LANDSCAPE EVOLUTION]

Brunsden, D. 1980: Applicable models of long term landform evolution. *Zeitschrift für Geomorphologie Supplementband* 36, 16–26. **Coates, D.R. and Vitek, J.D. (eds)** 1980: *Thresholds in geomorphology.* Boston, MA: George, Allen and Unwin. **Schumm, S.A.** 1973: Geomorphic thresholds and complex response of drainage systems. In Morisawa, M. (ed.), *Fluvial geomorphology*, Binghampton, NY: New York State University Publications in Geomorphology, 299–309.

throughflow The down-slope movement of water laterally through the SOIL or REGOLITH either as unsaturated flow or, more importantly, saturated flow above the main groundwater table (see WATER TABLE). It is usually produced by a sharp or progressive decline in permeability (and resulting ponding of water) down the soil profile. *RPDW*

[See also PIPEFLOW]

thrust fault A low-angle reverse FAULT formed by compression and shortening of the CRUST. Thrust faults are an important feature of the DEFORMATION in many OROGENIC BELTS, where they are associated with FOLDS in *fold–thrust belts*. The relative displacement across the *thrust*

plane may be many tens of kilometres and the overlying, tectonically transported sheet of rocks is known as a *nappe*. *GO*

[See also FORELAND BASIN, IMBRICATION]

Alvarez, W. 1999: Drainage on evolving fold–thrust belts: a study of transverse canyons in the Apennines. *Basin Research* 11, 267–284. **Butler, R.W.H.** 1982: The terminology of structures in thrust belts. *Journal of Structural Geology* 4, 239–245. **Shaw, J.H., Bilotti, F. and Brennan, P.A.** 1999: Patterns of imbricate thrusting. *Geological Society of America Bulletin* 111, 1140–1154.

thufur A perennial, well vegetated, earth-cored hummock up to c. 0.5 m high and c. 1.0 m in diameter, formed in either the ACTIVE LAYER in PERMAFROST areas or in SEASONALLY FROZEN GROUND in non-permafrost areas. The term is of Icelandic origin. *HMF*

[See also PATTERNED GROUND]

Grab, S.W. 1994: Thufur in the Mohlesi Valley, Lesotho, southern Africa. *Permafrost and Periglacial Processes* 5, 111–118. **Thorarinsson, A.** 1951: Notes on patterned ground in Iceland, with particular reference to the Icelandic 'flas'. *Geografiska Annaler* 33, 144–156. **Schunke, E. and Zoltai, S.C.** 1988: Earth hummocks (thufur). In Clark, M.J. (ed.), *Advances in periglacial geomorphology*. Chichester: Wiley, 231–245. **Scotter, G.W. and Zoltai, S.C.** 1982: Earth hummocks in the Sunshine area of the Rocky Mountains, Alberta and British Columbia. *Arctic* 35, 411–416.

tidal bore In ESTUARINE ENVIRONMENTS with exceptional tidal ranges (e.g. Bay of Fundy, located between New Brunswick and Nova Scotia, Canada; up to 15 m), the wave of advancing water (up to 5 m in the Amazon River) formed on the rising tide. *RAS*

tidal bundles A style of CROSS-BEDDING in which sandy *foresets* are interrupted by MUD DRAPES or REACTIVATION SURFACES, representing increments of sediment deposited by one ebb–flood cycle of TIDES. Subtidal SANDWAVES commonly develop in areas dominated by either the *flood tide* (incoming) or the *ebb tide* (outgoing). At *slack water* (a period of still water at high tide or low tide) the BEDFORM surface is covered with mud. During the subordinate tide a small amount of sand is moved before another mud drape is deposited at the next slack-water period. Significant sand movement resumes with the next dominant tide, giving rise to cross-bedding. In intertidal settings tidal bundles are each characterised by a single mud drape. Tidal bundles preserved in the geological record are a valuable indicator of PALAEOENVIRONMENT. Their spacing commonly shows a regular PERIODICITY, which has been interpreted in terms of fortnightly cycles of SPRING TIDES and NEAP TIDES. Such datasets provide information about rates of bedform migration and the number of days in the lunar month. *GO*

[See also TIDAL RHYTHMITE]

Allen, J.R.L. 1982: Mud-drapes in sandwave deposits: a physical model with application to the Folkestone Beds (early Cretaceous, southeast England). *Philosophical Transactions of the Royal Society, London* A306, 291–345. **Allen, P.A. and Homewood, P.** 1984: Evolution and mechanics of a Miocene tidal sandwave. *Sedimentology* 31, 63–81. **Yang, C.-S. and Nio, S.-D.** 1985: The estimation of palaeohydrodynamic processes from subtidal deposits using time series analysis methods. *Sedimentology* 32, 41–57.

tidal rhythmite A style of BEDDING comprising couplets of sand and mud, each deposited during one ebb–flood tidal cycle, which commonly show cyclical variations in thickness related to PERIODICITY in SPRING TIDES and NEAP TIDES. If the sand beds are rippled, FLASER BEDDING is produced. *GO*

[See also RHYTHMITE]

Kvale, E.P., Johnson, H.W., Sonett, C.P. *et al.* 1999: Calculating lunar retreat rates using tidal rhythmites. *Journal of Sedimentary Research* 69, 1154–1168. **Williams, G.E.** 1989: Late Precambrian tidal rhythmites in South Australia and the history of the Earth's rotation. *Journal of the Geological Society, London* 146, 97–111.

tidalite A sediment or sedimentary rock that accumulated under the influence of TIDES. Diagnostic features include HERRINGBONE CROSS-BEDDING, FLASER BEDDING, TIDAL BUNDLES and TIDAL RHYTHMITES. The characteristics of tidalites can be used to reconstruct the cyclicity of tidal processes in the Earth's past. *GO*

Alexander, C., Davis, R.A., Jr. and Henry, V.J. (eds) 1988: *Tidalites: processes and products* [*Special Publication* 61]. Tulsa, OK: Society of Economic Paleontologists and Mineralogists.

tide gauge records Instrumental records of relative sea level (see MEAN, RELATIVE and ABSOLUTE SEA LEVEL). The longest tide gauge records (e.g. Amsterdam) provide a record of relative sea-level change covering the last c. 200 years. Data from areas experiencing GLACIO-ISOSTATIC UPLIFT indicate a progressive fall in relative sea level during recent decades. By contrast, tide gauge records from areas experiencing crustal SUBSIDENCE show a rise over the same period. Analysis of tide gauge records worldwide is particularly important for detecting whether SEA-LEVEL RISE due to GLOBAL WARMING effects (especially THERMAL EXPANSION OF THE OCEANS and melting of ICE SHEETS and GLACIERS) is occurring. The most recent analysis of the records suggests an average rise in sea level worldwide of about 1 mm y^{-1}. *AGD*

[See also GLACIO-EUSTASY, INSTRUMENTAL RECORDS, SEA-LEVEL CHANGE]

Pirazzoli, P.A. 1991: *World atlas of sea level changes*. Amsterdam: Elsevier. **Shennan, I. and Woodworth, P.L.** 1992: A comparison of late Holocene and twentieth-century sea-level trends from the UK and North Sea region. *Geophysical Journal International* 109, 96–105.

tides Periodic rises and falls of Earth materials (water, air, land) resulting from the gravitational attraction of the Moon, Sun and other astronomical bodies as the Earth rotates. Tides are especially conspicuous in the oceans, where their periodicity and range vary with coastal configuration. *Tidal range* (the vertical distance tides rise and fall) varies up to as much as 15 m. Tides may be diurnal (one high and one low tide each lunar day), semidiurnal (highs and lows twice a day), mixed or occur as a double tide (a tidal day with two distinct maximum or minimum tides separated by a single small minimum or maximum). The highest and lowest ranges are known as SPRING TIDES (when the Sun, Moon and Earth are in alignment) and NEAP TIDES (when the directions to the Sun and Moon from Earth are nearly at right angles). Rising tides in some river mouths and estuaries cause a wave of water translation upstream known as a TIDAL BORE. *HJW*

[See also TIDAL BUNDLES]

Cartwright, D.E. 1999: *Tides: a scientific history*. Cambridge: Cambridge University Press. **Pugh, D.T.** 1987: *Tides, surges and mean sea level*. Chichester: Wiley.

tiering The vertical segregation of organisms in a COMMUNITY according to their height above the substrate, or depth within it, commonly as a response to competition for RESOURCES such as food or light. *LC*

[See also EPIFAUNA, INFAUNA]

till A poorly sorted, typically unstratified glacigenic deposit containing particles potentially ranging from clay to boulders in size, though it may contain only a part or parts of this size range. Frequently in older literature and less frequently in more recent publications, PLEISTOCENE till is referred to by the outdated terms '*boulder clay*' and DRIFT. Today, DIAMICTON is the preferred non-genetic, descriptive term for undifferentiated, poorly sorted *glacigenic sediments*. Nevertheless, the term till remains ingrained in the literature. Till is regarded by most workers as being deposited directly by or from GLACIER ice. Its deposition is either observed in operation around modern glaciers or reconstructed from observation and measurements (e.g. till FABRIC ANALYSIS) of fossil deposits combined with, where possible, the landscape and LANDFORM context. Genetic till subtypes (e.g. LODGEMENT, MELTOUT, SUBLIMATION, DEFORMATION and FLOW) in former glacial landscapes can be difficult to identify. Nevertheless, careful study of till yields important clues on thermal regime, modes of debris entrainment, glacier dynamics and flow patterns. In the context of FACIES ARCHITECTURE and in association with sorted sediment, this gives insight into processes and DEPOSITIONAL ENVIRONMENTS. *JS*

[See also GLACIAL SEDIMENTS, GLACIOFLUVIAL SEDIMENTS]

Boulton, G.S. 1978: Boulder shapes and grain-size distributions of debris as indicators of transport paths through a glacier and till genesis. *Sedimentology* 25, 773–799. **Geikie, A.** 1863: On the phenomena of the glacial drift of Scotland. *Transactions of the Geological Society Glasgow* 1, 190 pp. **Goldthwait, R.P and Matsch, C.L. (eds)** 1988: *Genetic classification of glacigenic deposits*. Rotterdam: Balkema. **Kjaer, K.H.** 1999: Mode of subglacial transport deduced from till properties, Myrdalsjökull, Iceland. *Sedimentary Geology* 128, 271–292.

tillage The mechanical manipulation of soil in agriculture, usually involving the modification of soil conditions for crop production and weed control. *Conventional tillage* practices often involve PLOUGHING of the soil, whereas *conservation*, *minimum* or *zero tillage* practices, which are used as SOIL CONSERVATION measures, involve lower levels of disturbance of the soil. *SHD*

Hudson, N.W. 1971: *Soil conservation*. London: Batsford. **Moran, C.J., Koppi, A.J., Murphy, B.W. and McBratney, A. B.** 1988: Comparison of the macropore structure of a sandy loam surface soil horizon subjected to two tillage treatments. *Soil Use and Management* 4, 96–102. **Morgan, R.P.C.** 1995: *Soil erosion and conservation*, 2nd edn. Harlow: Longman.

tillite Lithified rather than unconsolidated TILL, usually of pre-PLEISTOCENE, rather than Pleistocene, age. *JS*

Evans, J.A., Fitches, W.R. and Muir, R.J. 1998: Laurentian clasts in a Neoproterozoic tillite from Scotland. *Journal of Geology* 106, 361–366.

time-series analysis The detection of a signal against the background noise in a time-dependent dataset and the estimation of the parameters and properties characterising the signal. Signals can be classified broadly into DETERMINISTIC and STOCHASTIC signals. A deterministic signal (e.g. a periodic signal) can be predicted for arbitrary spaces of time. For a stochastic (occurring by chance) signal, no such prediction can be made beyond a certain time interval. For any finite time series the classification into these two categories is ambiguous. Usually, processes in the source of the signal (e.g. the nucleus of an active galaxy) and observational errors introduce a random component into the series, called *noise*. The analysis of such series usually aims at removing the noise and fitting a model to the remaining component of the series. Suitable models can be obtained by shifting a known series by some time lag or by repeating fragments of it with some frequency. Stochastic signals are analysed more appropriately in the time domain and periodic signals in the frequency domain (e.g. by using the *Fourier transform*). Models usually depend on several parameters. Fitting of the model to the signal means choosing the best set of these parameters. In the field of environmental change, time-series analysis is widely applied in, for example, DENDROCHRONOLOGY, HYDROLOGICAL change and CLIMATIC CHANGE.   *KJT*

[See also AUTOREGRESSIVE MODELLING, FILTERING, SPECTRAL ANALYSIS]

Chatfield, C. 1980: *The analysis of time series: an introduction.* London: Chapman and Hall. **Kendall, M. and Ord, K.J.** 1990: *Time series*, 3rd edn. London: Arnold. **Tsonis, A.A., Roebber, P.J. and Elsner, J.B.** 1998: Characteristic time scale in the global temperature record. *Geophysical Research Letters* 25(15), 2821–2823.

time transgressive A stratigraphic boundary or unit that crosses time planes; not having the same age at all points on the surface.   *DAR*

[See also DIACHRONOUS]

tin (Sn) A non-ferrous HEAVY METAL and TRACE ELEMENT that is largely non-toxic except in some organic compounds used as PESTICIDES and anti-fouling agents on the keel of ships. It is widely used in inorganic form as coatings, solders and alloys. MINING of tin has a long history and, in the form of bronze (an alloy with COPPER), it was indirectly responsible for some major PREHISTORIC human impacts on the environment, such as FOREST CLEARANCE and SOIL DEGRADATION.   *JAM*

[See also BRONZE AGE LANDSCAPE IMPACTS]

Franklin, A., Olin, J. and Wertime, T. 1978: *The search for ancient tin.* Washington, DC: Smithsonian Institution. Penhallurick, R.D. 1986: *Tin in antiquity.* London: Institute of Metals.

tipline A stratigraphic layer at an archaeological site, made up of material, such as dumped *garbage*, which accumulated after moving down a slope.   *JAM*

tolerance The ability of organisms to withstand inhospitable conditions through genetic change. For example, some plant species (or populations within species) can tolerate higher concentrations of heavy metals in the soil than other species (populations) and this tolerance has been shown to have a genetic basis.   *KDB*

[See also SOIL LOSS TOLERANCE]

Grime, J.P. 1979: *Plant strategies and vegetation processes.* Chichester: Wiley.

tolerance model A model of ECOLOGICAL SUCCESSION suggesting that early colonising species neither facilitate nor inhibit the establishment or growth of subsequent arrivals; subsequent species replacement is dependent on tolerance of increasingly lower levels of resources.   *LRW*

[See also COLONISATION, FACILITATION, INHIBITION MODEL]

tool marks A group of erosional SEDIMENTARY STRUCTURES formed when debris transported by a current gouges hollows on a surface of cohesive sediment (MUD). Like SCOUR MARKS, they are commonly preserved as CASTS by deposition of sand and seen as SOLE MARKS on the base of a bed of SANDSTONE. Examples include long, straight *groove marks* and a variety of intricate *bounce, skip* and *prod marks*. Tool marks are useful indicators of PALAEOCURRENT strength and direction.   *GO*

[See also ICEBERG PLOUGH MARKS]

Beukes, N.J. 1996: Sole marks and combined-flow storm event beds in the Brixton Formation of the siliciclastic Archean Witwatersrand Supergroup, South Africa. *Journal of Sedimentary Research* 66, 567–576. **Potter, P.E. and Pettijohn, F.J.** 1977: *Paleocurrents and basin analysis*, 2nd edn. Berlin: Springer.

topoclimate Differences in the climate from place to place caused by topography. Topoclimate, *topographic climate or orographic microclimate* is an aspect of LOCAL CLIMATE or MESOCLIMATE including, for example, the diurnal wind systems that may develop in mountain valleys: upslope (*anabatic*) and up-valley (*valley*) *winds* develop during the day in response to local differential heating, whereas stronger down-slope (*katabatic*) and down-valley (*mountain*) *winds* occur at night because of cold-air drainage downhill.   *JAM*

toponymy The study of *place-names*. In the reconstruction of environmental change, place-names are useful in particular for the clues they provide to the origin of settlements and the history of LANDUSE. There is a relationship, for example, between soils and the CHRONOLOGY of settlements in the distribution of Scandinavian place-names in eastern England: earlier names and larger settlements tend to occur on the more fertile soils.   *JAM*

Darby, H.C. 1956: The clearing of the woodlands in Europe. In Thomas Jr, W.I. (ed.) *Man's role in changing the face of the Earth.* Chicago: Chicago University Press, 183–216. **Evans, J. and O'Connor, T.** 1999: *Environmental archaeology: principles and methods.* London: **Gelling, M.** 1984: *Place-names in the landscape.* London: Dent. **Gover, J.E.B., Mawer, A. and Stenton, F.M.** 1942: *The place-names of Middlesex.* Cambridge: Cambridge University Press.

tor An exposure of bedrock rising above its surroundings and formed by: (1) subaerial weathering causing spheroidal modification to the form of outcrops; (2) differential WEATHERING and removal of weathered debris by MASS WASTING and stripping; (3) scarp retreat of larger INSELBERGS, or (4) freeze-thaw weathering in PERIGLACIAL conditions followed by SOLIFLUCTION. Marked jointing usually controls the detailed form of a tor. Tors are associated with TROPICAL and periglacial environments. One view of British examples is that they formed in two stages by subsurface weathering during TERTIARY tropical conditions followed by removal of weathering products during PLEISTOCENE periglacial processes. Alternatively, they are regarded as essentially RELICT periglacial forms. The *koppies* or *kopjes* of Africa are similar in form and usually interpreted in terms of theory 3.   *PW*

[See also CLITTER, CORESTONE, GROWAN, GRUSSIFICATION, INSELBERG]

Gerrard, J. 1994: Classics in physical geography revisited – the problem of tors. *Progress in Physical Geography* 18, 559–563. **Linton, D.L.** 1955: The problem of tors. *Geographical Journal* 121, 470–487. **Palmer, J. and Neilson, R.A.** 1962: The origin of granite tors, Dartmoor, Devonshire. *Proceedings of the Yorkshire Geological Society* 33, 315–340.

tornado An intense rotating STORM, usually 100–200 m in diameter with a characteristic funnel cloud and winds spiralling upwards at speeds of the order of 100 m s^{-1}. Tornadoes are associated with violent thunderstorms and steep temperature gradients between warm and cold moist AIRMASSES and are a major NATURAL HAZARD in the mid-west and Great Plains of the United States, especially in spring. There are usually between 500 and 1000 tornadoes per year in the USA, causing 25 to 500 deaths per year. They are less common elsewhere (about 15 per year are reported in Australia). *Tornado outbreaks* are defined as five or more tornadoes associated with a single weather system: the largest recorded outbreak in the mid-west involved over 100 individual tornadoes on 11 April 1965 (the 'Palm Sunday' outbreak). Individual tornadoes usually consist of a solitary funnel but two or more smaller funnels (*suction vortices*) may rotate around a common centre. Intensity is measured on the six-point *Fujita–Pearson scale* based on three easily measured characteristics: maximum wind speed, damage path length (across the ground surface) and average path width. Related, more frequent but weaker *waterspouts* occur over water.   *JAM*

Bluestein, H.B. 1999: *Tornado alley: monster storms of the Great Plains.* New York: Oxford University Press. **Church, C., Burgess, D., Doswell, C. and Davies-Jones, R. (eds)** 1993: *The tornado: its structure, dynamics, prediction and hazards* [*Geophysical Monograph* 79]. Washington, DC: American Geophysical Union. **Fujita, T.T. and Pearson, A.D.** 1972: *FPP tornado scale and its application* [*Satellite and Mesometeorology Research Project, Research Paper* 98]. Chicago, IL: University of Chicago. **Sadowski, A.F.** 1966: Tornadoes with hurricanes. *Weatherwise* 19, 71–75. **Snow, J.T.** 1987: Tornadoes. In Oliver, J.E. and Fairbridge, R.W. (eds), *The encyclopedia of climatology.* New York: Van Nostrand Reinhold, 845–856. **Stowe, J.T.** 1984: The tornado. *Scientific American* 250, 86–96.

tower karst A KARST landscape of residual hills scattered across a plain. A well known example is the karst towers of Guizhou, China. Tower karst, known as *turmkarst* in Germany, was considered at one time to be characteristic of humid tropical environments, but it is now considered that it can be produced in a variety of climatic environments with examples in northern Canada and seasonal tropical northern Australia. Classic tower karst has sheer cliffs (a product of uplift and fluvial COR-RASION and undercutting), but many tower slopes are much less steep.   *SHD*

[See also TROPICAL KARST]

McDonald, R.C. 1979: Tower karst geomorphology in Belize. *Zeitschrift für Geomorphologie* 32, 35–45. **Williams P.W.** 1987: Geomorphic inheritance and the development of tower karst. *Earth Surface Processes and Landforms* 12, 453–465. **Yuan, D.** 1987: New observations on tower karst. In Gardiner, V. (ed.), *International geomorphology 1986.* Vol. 2. Chichester: Wiley, 1109–1123.

toxicant An artificial (synthetic) substance that is toxic or poisonous.   *JAM*

[See also ECOTOXICOLOGY, TOXIN]

toxin A natural substance that is toxic or poisonous. A substance can only be considered toxic in relation to a specific biological system and with reference to a response that is considered deleterious.   *JAM*

[See also TOXICANT]

trace element An element that is a minor constituent of the Earth's materials. Eight elements – aluminium, calcium, iron, magnesium, oxygen, potassium, silicon and sodium – comprise 99% of the Earth's crust; the remainder constitute trace elements, which may nevertheless attain high concentrations locally where released into the environment by WEATHERING or human activity, such as the MINING of ores, metal refining and the combustion of FOSSIL FUELS. The trace metals include antimony, arsenic, CADMIUM, chromium, cobalt, COPPER, LEAD, MERCURY, molybdenum, nickel, selenium, TIN, titanium, vanadium and zinc. Where trace elements are released, they may be useful ENVIRONMENTAL INDICATORS both in modern and in palaeoenvironmental contexts. Some trace elements are required in small quantities by organisms (MICRONUTRIENTS), but others perform no known biological function; many are toxic even at low concentrations.   *JAM*

[See also HEAVY METALS, TRACE GASES]

Benjamin, M.M. and Honeyman, B.D. 1992: Trace metals. In Butcher, S.S., Charleson, R.J., Orians, G.H. and Wolfe, G.V. (eds), *Global biochemical cycles.* London: Academic Press, 317–352. **Hölzer, A. and Hölzer, A.** 1998: Silicon and titanium in peat profiles as indicators of human impact. *The Holocene* 8, 685–696. **Nriagu, J.O.** 1989: A global assessment of the natural sources of atmospheric trace metals. *Nature* 338, 47–49.

trace fossil Evidence of biological activity preserved in SEDIMENTS and SEDIMENTARY ROCKS as SEDIMENTARY STRUCTURES, including tracks and trails, burrows, borings and COPROLITES. The study of trace fossils (also known as *ichnofossils* or *Lebensspuren*) is *ichnology*. Trace fossils are formed at BED contacts (e.g. footprints) or within beds (e.g. burrows of INFAUNA) and may become enhanced during DIAGENESIS, for example by mineralisation of burrow linings or differential COMPACTION. Trace fossils are named (ichnogenus, ichnospecies) according to morphology, but without interpretation of behaviour or the tracemaker. A useful classification of trace fossils is according to behaviour, including feeding, locomotion and dwelling traces. Preservation *in situ* (see *body fossils*; DEATH ASSEMBLAGE) means that trace fossils are particularly useful for PALAEOECOLOGY and PALAEOENVIRONMENTAL RECONSTRUCTION. ICHNOFACIES are characterised by an assemblage of trace fossils and types of behaviour representative of a bathymetric or sedimentary zone, ranging from non-marine to ABYSSAL depths. The term is also used in the archaeological context. *LC*

Brenchley, P.J. and Harper, D.A.T. 1998: *Palaeoecology: ecosystems, environments and evolution.* London: Chapman and Hall. **Bromley, R.G.** 1996: *Trace fossils: biology, taphonomy and applications,* 2nd edn. London: Chapman and Hall. **Ekdale, A.A., Bromley, R.G. and Pemberton, S.G.** 1984: *Ichnology: trace fossils in sedimentology and stratigraphy* [Short Course Notes 15]. Tulsa, OK: Society of Economic Paleontologists and Mineralogists. **Frey, R.W., Pemberton, S.G. and Saunders, T.D.A.** 1990: Ichnofacies and bathymetry: a passive relationship. *Journal of Paleontology* 64, 155–158. **Gautier, A.** 1993: Trace fossils in archaeozoology. *Journal of Archaeological Science* 20, 511–523. **Savrda, C.E.** 1995: Ichnologic applications in paleoceanographic, paleoclimatic, and sea-level studies. *Palaios* 10, 565–577.

trace gas A gas found in the atmosphere in very low quantities. Trace gases may be NATURAL and/or ANTHROPOGENIC in origin and include (1) *inert gases* (e.g. helium, krypton, neon, xenon), and (2) gases that take part in natural environmental processes (e.g. ammonia, CARBON DIOXIDE, CHLOROFLUOROCARBONS and METHANE). *JAM*

[See also TRACE ELEMENT]

tracer A substance that is introduced deliberately or accidentally into the environment, or occurs naturally in the environment, and the progress of which can be monitored to determine some aspect of the behaviour of the system. In AQUIFERS or SOILS, for example, tracers such as fluorescent dyes or radioactive isotopes can be used to follow the movement of water, solutes and sediment. *SHD*

[See also CAESIUM-137]

Foster, I.D. (ed.) 2000: *Tracers in geomorphology.* Chichester: Wiley. **Foth, H.D.** 1990: *Fundamentals of soil science.* New York: Wiley. **Nativ, R., Gunay, G., Hotzl, H.** *et al.* 1999: Separation of groundwater-flow components in a karstified aquifer using environmental tracers. *Applied Geochemistry* 14, 1001–1014.

tracheidogram A series of measurements of the sizes of xylem cells (*tracheids*) in a radial file of cells across the annual growth ring of a conifer tree. The variation of cell-size dimension, in what is virtually a cross-section of an *annual ring*, is influenced by changing growth conditions (such as climate variability), so tracheidograms represent a source of information on intra-annual tree-growth forcing. *KRB*

[See also TREE RINGS, CAMBIUM, DENDROCHRONOLOGY]

Vaganov, E.A. 1990: The tracheidogram method in tree-ring analysis and its application. In Cook, E.R. and Kairiukstis, L.A. (eds), *Methods of dendrochronology: applications in environmental sciences.* Dordrecht: Kluwer, 63–76.

trackway Typically a wooden structure constructed across a low-lying, WETLAND area that would be otherwise difficult to access. Trackways were used to provide ease of passage (e.g. to a settlement). Trackways can vary in design and composition from sturdy cut timbers laid longitudinally or transversely and fastened by stakes, to less substantial bundles of heather or brushwood laid on the ground surface. Trackways were often constructed from alder, hazel, oak, ash or lime in areas susceptible to waterlogging, such as ESTUARINE ENVIRONMENTS and *peatlands*,

and may date from the NEOLITHIC to the Mediaeval period. Many date to the BRONZE AGE and it is thought that they provide evidence for a human response to climatic deterioration or marine TRANSGRESSION. *LD-P*

[See also CLIMATIC CHANGE: PAST IMPACTS ON HUMANS, MIRES, SALTMARSH]

Bell, M. and Neumann, H. 1997: Prehistoric intertidal archaeology and environments in the Severn Estuary, Wales. *World Archaeology* 29, 95–113. **Caseldine, C. and Hatton, J.** 1996: Early land clearance and wooden trackway construction in the third and fourth millennia BC at Corlea, Co Longford. *Biology and Environment – Proceedings of the Royal Irish Academy* 96B, 11–19. **Coles, J.M. and Lawson, A.** 1987: *European wetlands in prehistory.* Oxford: Clarendon Press.

Trade Winds Surface planetary winds that blow with an easterly component of flow from the subtropical high-pressure areas towards the Equator (*Northeast Trades* in the Northern Hemisphere; *Southeast Trades* in the Southern Hemisphere). They are the most persistent of the surface winds of the GENERAL CIRCULATION OF THE ATMOSPHERE, exhibiting a regularity of 60–70% over the oceans (90–95% towards the middle of their course). *JAM*

[See also ANTITRADES, HADLEY CELL]

tradeable emissions permits In an air quality management programme, a market-based system for controlling total aggregate EMISSIONS of a POLLUTANT from a collective community (a specific economic sector, a region or a country) based on the allocation of permits to community members for the right to release specified quantities of a pollutant. These permits can be traded within the community as a marketable commodity. For instance, Region A could reduce SULPHUR DIOXIDE emissions to a level below that required by air quality standards while transferring the 'right to pollute' to another region. Region B would then have the right not to meet the standards imposed on it, provided that the net rate of emissions from the two regions was within the protocol requirements. *GS*

Heggelund, M. 1991: *Emissions permit trading: a policy tool to reduce the atmospheric concentration of greenhouse gases.* Calgary: Canadian Energy Research Institute. **Rapaport, R.** 1986: Trading dollars for dirty air. *Science* 86, 75.

tradition In the context of archaeology: (1) the evolutionary history of an INDUSTRY or of related industries; or (2) a set of historically connected industries; or simply (3) a set of culturally connected industries. *JAM*

trafficability The ease with which terrain can support or permit the passage of vehicles, humans or animals. Environmental factors that influence or hinder a vehicle's progress include surface topography and obstacles, soil strength and moisture content, vegetation type and cover, snow thickness and the presence of surface water. *MLW*

Laut, P. and Davis, R. 1989: Cross-country trafficability ratings for Cape York Peninsula: a method based on simple landscape data and expert systems. In Ball, D. and Babbage, R. (eds), *Geographic information systems: defence applications.* Sydney, Australia: Pergamon-Brasseys.

tragedy of the commons Overexploitation of 'international' or 'commonly' owned resources (such as open oceans, global atmosphere, TROPICAL RAIN FOREST) by unregulated use. The phrase derives from the overuse of COMMON LAND historically: tragic consequences are attributed to the gains of individuals exceeding communal losses when CARRYING CAPACITY is exceeded. Disruption of *global commons* has repercussions for all the world's ecosystems.   *GOH*

[See also OVERFISHING, OVERGRAZING, OVERHUNTING, SUS-TAINABILITY, SUSTAINABLE DEVELOPMENT]

Hardin, G. 1968: The tragedy of the commons. *Science* 162, 1243–1248. **Hardin, G. and Baden, J. (eds)** 1977: *Managing the commons.* San Francisco, CA: Freeman. **World Commission on Environment and Development** 1987: *Our common future.* Oxford: Oxford University Press.

training area   A user-defined area in supervised IMAGE CLASSIFICATIONS in REMOTE SENSING for deriving spectral signatures of individual classes from a small sample of PIX-ELS with known class membership.   *HB*

[See also TRAINING SET]

training set   In the context of PALAEOENVIRONMENTAL RECONSTRUCTION, a dataset used to derive the environmental relationships from modern organisms or features for later use as ENVIRONMENTAL INDICATORS. The relationships are modelled statistically and the resulting TRANSFER FUNCTION is used to transform similar data from FOSSILS, SUBFOSSILS or other PROXY EVIDENCE into quantitative estimates of the past environmental variables. The training set, or *calibration set*, which is used in model derivation, should be independent of data used for model CONFIRMA-TION or VALIDATION and for the reconstruction in the strict sense.   *DH/JAM*

Fritz, S.C., Juggins, S., Battarbee, R.W. and Engstrom, D.R. 1991: Reconstruction of past changes in salinity and climate using a diatom-based transfer function. *Nature* 352, 702–704. **Gordon, G.A.** 1982: Verification of dendroclimatic reconstructions. In Hughes, M.K., Kelly, P.M., Pilcher, J.R. and LaMarche, V.C. (eds), *Climate from tree rings.* Cambridge: Cambridge University Press, 58–62. **Guiot, J.** 1990: Methods of calibration. In Cool, E.R. and Kairiukis, L.A. (eds), *Methods of dendrochronology: applications in the environmental sciences.* Dordrecht: Kluwer, 165–178.

trampling   Impacts on soils and vegetation produced by the repeated passage of humans, animals or off-road vehicles. Impacts on soil include COMPACTION (characterised by an increase in bulk density and penetration resistance, reduction in porosity and collapse of soil structure) and a reduced ability for water to infiltrate and percolate. This combination often leads to an increase in the velocity and quantity of surface runoff and can accelerate soil loss by EROSION. Erosion is facilitated by the impacts that trampling imposes on plants, including a reduction in the height, percentage cover and overall biomass of vegetation. Trampling can also alter the composition of species in a plant community as opportunist species invade newly created bare ground and plants that have ADAPTATIONS to trampling (including roseate and creeping morphologies, low-growing woody stems and basal growth points) become dominant. Severe compaction reduces soil aeration and can induce changes in microorganism populations, favouring anaerobic species. A reduction in soil arthropod and earthworm numbers is

another consequence of compaction. Trampling has become a concern in many National Parks and fragile ecosystems such as sand dunes, where increasing numbers of visitors have contributed to vegetation loss and footpath erosion. Good visitor and ENVIRONMENTAL MANAGEMENT practices help to alleviate trampling impacts.   *MLW*

Cole, D.N. 1995: Experimental trampling of vegetation I: relationship between trampling intensity and vegetation response. *Journal of Applied Ecology* 32, 203–214. **Liddle, M.J.** 1991: Recreation ecology: effects of trampling on plants and corals. *Trends in Ecology and Evolution* 6, 13–17. **Weaver, T. and Dale, D.** 1978: Trampling effects of hikers, motorcycles and horses in meadows and forests. *Journal of Applied Ecology* 15, 451–457.

transfer function   A set of numerical weights (equivalent to a set of regression coefficients) that, when multiplied by the appropriate predictor time series and summed, provide equivalent time-series estimates of the predictand (in simple regression) or predictands (in multiple regression). In other words, the transfer function represents a weighting function by which measurements or observations of the behaviour of one phenomenon may be 'transferred' into estimates of another. The term is commonly used in various branches of PALAEOCLIMATOLOGY (e.g. to reconstruct climate variability from various biological archives such as are found in TREE RINGS or LAKE and MARINE SEDIMENTS).   *KRB*

[See also NUMERICAL ANALYSIS, TRAINING SET]

Birks, H.J.B. and Gordon, A.D. 1985: *Numerical methods in Quaternary pollen analysis.* London: Academic Press. **Bradley, R.S.** 1999: *Paleoclimatology: reconstructing climates of the Quaternary,* 2nd edn. London: Academic Press. **Cook, E.R., Briffa, K.R. and Jones, P.D.** 1994: Spatial regression methods in dendroclimatology: a review and comparison of two techniques. *International Journal of Climatology* 14, 379–402. **Webb III, T., Howe, S.E., Bradshaw, R.H.W. and Heide, K.M.** 1981: Estimating plant abundances from pollen percentages: the use of regression analysis. *Review of Palaeobotany and Palynology* 34, 269–300.

transform fault   A strike-strip FAULT linking CON-STRUCTIVE or DESTRUCTIVE PLATE MARGINS as a CONSERVA-TIVE PLATE MARGIN. Examples include the San Andreas Fault in California and the offsets of MID-OCEAN RIDGES.
                              GO

[See also PLATE TECTONICS]

Wilson, J.T. 1965: A new class of faults and their bearing on continental drift. *Nature* 207, 343–347.

transgression, marine   Flooding of the land by the sea; a relative rise in SEA LEVEL. Evidence of a marine transgression in SEDIMENTARY DEPOSITS may be preserved in the form of marine deposits (e.g. MARINE BAND) overlying non-marine, in an upward deepening of FACIES or, in the deposits of TERRESTRIAL ENVIRONMENTS, as a change in facies type or FACIES ARCHITECTURE. A surface of marine transgression is termed a *flooding surface* in SEQUENCE STRATIGRAPHY.

A marine transgression may be caused by one or several of the following factors: (1) *eustatic* SEA-LEVEL RISE (i.e. absolute sea-level change; a mechanism of ALLOCYCLIC CHANGE); (2) TECTONIC subsidence (a relative sea-level change; an allocyclic change); or (3) a decrease in sediment supply (relative sea-level change), which could be caused by an allocyclic change (e.g. CLIMATIC CHANGE in the source area) or an AUTOCYCLIC CHANGE (e.g. abandonment of a DELTA lobe due to AVULSION). One of the challenges for modern SEDIMENTOLOGY is to clarify the complex relationships between these controlling factors in the GEOLOGICAL RECORD. *GO*

[See also COASTAL ENVIRONMENTAL CHANGE, HIGHSTAND, REGRESSION, SEA-LEVEL CHANGE]

Collier, R.E.Ll., Leeder, M.R. and Maynard, J.R. 1990: Transgressions and regressions: a model for the influence of tectonic subsidence, deposition and eustasy, with application to Quaternary and Carboniferous examples. *Geological Magazine* 127, 117–128. **Davey, S.D. and Jenkyns, H.C.** 1999: Carbon-isotope stratigraphy of shallow-water limestones and implications for the timing of Late Cretaceous sea-level rise and anoxic events (Cenomanian–Turonian of the peri-Adriatic carbonate platform, Croatia). *Eclogae Geologicae Helvetiae* 92, 163–170. **Muto, T. and Steel, R.J.** 1997: Principles of regression and transgression: the nature of the interplay between accommodation and sediment supply. *Journal of Sedimentary Research* 67, 994–1000. **Smith, D.E., Firth, C.R., Brooks, C.L.** *et al.* 1999: Relative sea-level rise during the Main Postglacial Transgression in NE Scotland, UK. *Transactions of the Royal Society of Edinburgh – Earth Sciences* 90, 1–27.

transhumance The biannual movement of domesticated animals between lowland areas and a mountain area of contrasting climate and resources. *Ascending transhumance* is the most common form, where most of the time is spent in the lowlands with uphill movement to greener summer pastures and available water; in *descending transhumance*, the coldest season of the year is spent in the lowlands. It reflects a semi-nomadic lifestyle that seems to date back at least to the Bronze Age. Networks of trails used for transhumance in the Alps and Mediterranean regions reflect the importance of seasonal *drives* of typically no more than 100–300 km in southern France and northern Italy but as much as 700–1000 km in semi-arid Spain. This traditional practice is in decline: at its peak in the sixteenth century, transhumance in central Spain involved up to about five million sheep. *JAM*

Brisebarre, A.M. 1978: *Bergers des Cévennes.* Paris: Berger-Levrault. **Gintzburger, G., Rochon, J.J. and Conesa, A.P.** 1990: The French Mediterranean zones: sheep rearing systems and the present and potential role of pasture legumes. In Osman, A.E., Ibrahim, M.H. and Jones, M.A. (eds), *The role of legumes in the farming systems of the Mediterranean area.* Aleppo, Syria: Icardia, 179–194. **Ruíz, M. and Ruíz, J.P.** 1986: Ecological history of transhumance in Spain. *Biological Conservation* 37, 73–86.

transient response The time-dependent change in a SYSTEM during the transition from one equilibrium or steady-state condition to another. *JAM*

translatent strata A form of STRATIFICATION representing the boundaries between units of sediment (*sets*), each formed by the migration of a single RIPPLE (see CLIMBING RIPPLE CROSS-LAMINATION). The migration of some kinds of ripple, such as WIND RIPPLES, does not form CROSS-LAMINATION and gently inclined translatent strata may be their main preserved SEDIMENTARY STRUCTURE. *GO*

Hunter, R.E. 1977: Basic types of stratification in small eolian dunes. *Sedimentology* 24, 361–387.

transport impacts Each stage of cultural development has been accompanied by TECHNOLOGICAL CHANGE, which has demanded numerous different and more sophisticated ways to transport goods and passengers. The period prior to the INDUSTRIAL REVOLUTION was characterised by basic transport means, such as foot, horseback and horse-drawn canal barges with limited environmental damage. During the industrial revolution, however, rail transport became characteristic in industrialised countries. Steam trains had indirect impacts on the environment through a growing demand for metal and wood to construct tracks, and direct impacts through pollution in the form of particulate emissions and gases.

Even more significant has been the emergence of terrestrial motor vehicles as the dominant form of transportation of both goods and passengers in the past few decades. This has led to road transport becoming a major source of AIR POLLUTION. Indeed, in most industrialised countries, where the impacts are most intense, motor vehicles are responsible for around 5% of total emissions of SULPHUR DIOXIDE, 10% of PARTICULATE emissions, 45% of HYDROCARBONS, over 50% of NITROGEN OXIDES and about 80% of CARBON MONOXIDE emissions. In addition, motor vehicles account for approximately one quarter of the anthropogenic CARBON DIOXIDE emissions for the world, which is a significant contribution to the GREENHOUSE EFFECT and GLOBAL WARMING. Particulate matter emitted from vehicles is also of great concern on account of the long residence time of dust particles in the atmosphere and their contribution to the formation of PHOTOCHEMICAL SMOG and DUST VEILS. Moreover, road transport is the fastest growing source of these pollutants, in line with the increase in car ownership world wide and its associated impacts (e.g. traffic congestion, noise pollution and detriment to human health). Road construction schemes have been unsuccessful in keeping pace with the increase in private car ownership and road freight, especially in the developing world, where many urban areas are now hyper-congested and very polluted.

Transportation by sea-going vessels is a source of MARINE POLLUTION but is relatively less of a problem than road transport. The transportation of oil in tankers has resulted in numerous OIL SPILLS, several of which have had disastrous impacts on the regional environment. Notable examples are the *Exxon Valdez* spill off Alaska, 1989, and the *Sea Empress* spill off Milford Haven, South Wales in 1996, in which 72 000 tonnes of crude oil was released into the sea. The ecological impacts of such events are devastating in the short and medium term as oil remains in the water column and bottom sediments for some time after the initial event.

Air transport has received much less attention as a potential source of pollution, but the release of gases and particles from aircraft engines does cause pollution in the higher levels of the atmosphere. Aircraft are also a major source of NOISE POLLUTION. *JGS*

[See also ESTUARINE ENVIRONMENTS]

Chin, A.T.H. 1995: Containing air pollution and traffic congestion: transport policy and the environment in Singapore. *Atmospheric environment* 30, 787–801. **Dyrynda, P.** 1996: *An appraisal of the early impacts of the 'Sea Empress' oil spill on shore ecology within south-west Wales.* Swansea: University of Wales Swansea. **Goodwin, P.** 1999: Transformation of transport policy in Great Britain. *Transportation Research Part A: Policy and Practice* 33, 7–8, 655–669. **Johnson, D.L. and Lewis, L.A.** 1995: *Land degradation: creation and destruction.* Oxford:

Blackwell. **Penner, J.E., Lister, D., Griggs, D.J.** *et al.* 1999: *Aviation and the global atmosphere*. Cambridge: Cambridge University Press.

transport-limited erosion A state in which more material is supplied as a result of EROSION than can be transported away from the site by agents (e.g. wind, water) is described as being transport limited. *SHD*

[See also DETACHMENT-LIMITED EROSION, EROSION, HORTONIAN OVERLAND FLOW, SATURATION OVERLAND FLOW, RUNOFF PROCESSES].

Morgan, R.P.C. 1995. *Soil erosion and conservation*, 2nd edn. Harlow: Longman.

trap (1) A subsurface geological feature (e.g. an antiform) that allows PETROLEUM in the pore spaces of sedimentary rocks to accumulate in one area. (2) A term formerly used to describe the sheet-like character of many IGNEOUS ROCKS, particularly LAVAS, derived from the Swedish 'trappa', meaning a staircase. The term is now obsolete in a descriptive sense, but persists in some names such as the Cretaceous–Tertiary *Deccan Traps* of west-central India. *GO*

[See also K–T BOUNDARY, LAVA PLATEAU]

tree line/limit Tree lines in the broad sense are generalisations of the limits of tree distribution in the landscape and are altitudinal (upper, lower), latitudinal, or artificial. Various definitions (see Figure) describe different 'lines' or 'limits' associated with the ECOTONE at the tree line.

- *Forest*. A continuous stand with trees closer than 30 m.
- *Climatic forest-limit*. The potential occurrence of forest limited by regional climate and optimal local climatic and soil conditions.
- *Empirical forest-limit*, or limit of continuous forest and isolated forest stands (minimum 15 trees). The observed limit. It is frequently reduced from the climatic limit by local environmental factors or human disturbance.
- *Tree*. A tree is a growth form that attains a minimum height (conventionally 5 m for conifers, 2.5 m for birch; some use 2 m).
- *Tree line*. This includes the individual trees and isolated tree clusters that extend beyond the forest limit. Tree line in the strict sense refers to the limit of the tree growth forms. Beyond is shrub tundra and/or alpine vegetation.
- *Tree-species line/limit*. The extent of the tree species, often stunted (KRUMMHOLZ). The zone between the

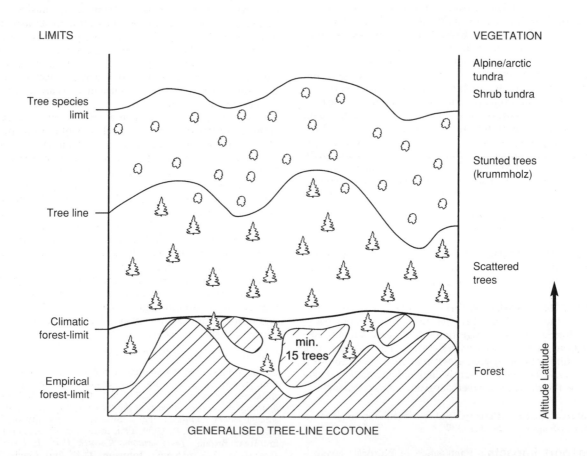

tree line/limit *Illustration of the definitions of tree line and related concepts across the tree-line ecotone*

tree-species limit and the tree line in the strict sense is the KAMPFZONE.

Lower altitudinal/latitudinal tree lines border on *grassland* or STEPPE. Tree growth is usually limited by drought associated with fire and human disturbance.

Artificial tree lines result from logging, grazing, burning, etc.   *HHB*

[See also TREE-LINE VARIATIONS]

Aas, B. and Faarlund, T. 1996: The present and the Holocene subalpine birch belt in *Norway. Paläoklimaforschung* 20, 19–42.
Payette, S. 1983: The forest tundra and present tree-lines of the northern Québec-Labrador Peninsula. In Morisset, P. and Payette, S. (eds), *Tree-line Ecology. Proceedings of the Northern Québec Tree-line Conference. Collection Nordicana* 47, 3–23.

tree-line variations

TREE LINES vary spatially, temporally and in species composition. Tree lines occur world-wide where the growing-season mean temperature is 5–7.5°C. Within this range, tree lines vary with local climatic conditions and can be 100 m higher on sheltered south-facing slopes. Tree lines are higher in continental than in oceanic areas and in high massifs (MASSENERHEBUNG EFFECT). Following climatic gradients, tree lines descend with increasing latitude towards the *polar tree line*.

The limitation of tree growth at tree lines is complex. At low temperatures, photosynthesis is adequate, but carbon allocation (biosynthesis) requiring energy production is limited. Branches and apices exposed above the microclimatic boundary layer cool rapidly and cannot grow. Canopy shade delays snow melting and restricts radiative warming of the roots, and this is an important factor acting at a tree line. Various deciduous and coniferous species form tree lines world-wide. *Betula* spp. are particularly adapted to withstand deep winter snow that would snap less flexible trunks. *Picea* spp. regenerate and spread from low branches if the trunks are damaged.

Tree lines may react slowly to climatic change. Amelioration allows trunk development from pre-existing KRUMMHOLZ and seedling establishment (Payette, 1985). Seedling establishment depends on summer soil moisture and winter survival of saplings once the height exceeds the BOUNDARY LAYER. Deterioration may not kill trees directly, but will prevent regeneration and may cause krummholz formation, in which form tree species can survive for centuries.

Logging, grazing and burning destroy climatic tree lines. The resulting soil degeneration and nutrient depletion may inhibit regeneration (e.g. by MOR HUMUS formation, PODZOLISATION, and BLANKET MIRE development in western Britain). Elsewhere, reduction in human pressure may result in forest regeneration (e.g. abandoned coastal heaths and mountain summer farms in Norway) and tree-line advance.

Temporal tree-line fluctuations can have climatic significance, but with up to several centuries lag-time, although mountain birch in northern Norway has shown remarkable expansion during the last 50 years. Pollen percentage analysis is insensitive to tree-line fluctuations involving anemophilous trees. POLLEN INFLUX is more sensitive. Improved reconstructions can be made by the additional study of macrofossils (including stomata from conifer needles), as these are more locally deposited. Ideally a transect across the ECOTONE should be studied. MEGAFOSSILS (roots, trunks) indicate the point of tree growth and they can be used to reconstruct former tree lines.   *HHB*

[See also TREE LINE/LIMIT, MACROFOSSIL, POLLEN ANALYSIS]

Aas, B. and Faarlund, T. 1996. The present and the Holocene subalpine birch belt in *Norway. Paläoklimaforschung* 20, 19–42.
Birks, H.H., Vorren, K.-D. and Birks, H.J.B. 1996: Holocene treelines, dendrochronology and palaeoclimate. *Paläoklimaforschung* 20, 1–18. **Dahl, E.** 1986: Zonation in arctic and alpine tundra and fellfield ecobiomes. In Polunin, N. (ed.), *Ecosystem Theory and Application; Environmental Monographs and Symposia.* Chichester: Wiley, 35–62. **Körner, C.** 1998: A reassessment of high elevation treeline positions and their explanation. *Oecologia* 115, 445–459. **Odland, A.** 1996: Differences in the vertical distribution pattern of *Betula pubescens* in Norway and its ecological significance. *Paläoklimaforschung* 20, 43–59. **Payette, S.** 1985: Secular climate change in old-growth tree-line vegetation of northern Quebec. *Nature* 315, 135–138. **Sveinbjörnson, B.** 1992: Arctic tree line in a changing climate. In Chapin, F.S., Jefferies, R.L., Reynolds, J.F. *et al.* (eds), *Arctic ecosystems in a changing climate – an ecophysiological perspective.* San Diego, CA: Academic Press, 239–256. **Wick, L. and Tinner, W.** 1997: Vegetation changes and timberline fluctuation in the central Alps as indicators of Holocene climatic oscillations. *Arctic and Alpine Research* 29, 445–458.

tree migration/spreading

Tree MIGRATION describes the spread of tree populations across a landscape. Land plants, such as trees, are not able to move in response to a change in environmental conditions and their survival options are either to adapt or to migrate; if they fail to do either, they will undergo EXTIRPATION or EXTINCTION. Migration of plant populations refers to their spreading outside of their former geographical RANGE, via propagules (sexual or vegetative) that establish and reach maturity under favourable environmental conditions. For any given species, migration tends to occur in response to an ENVIRONMENTAL CHANGE, which results in either unfavourable climatic conditions locally and/or suitable climatic conditions in an area previously unfavourable. As conditions become favourable for establishment outside of the range of a population, new individuals will establish themselves by DISPERSAL into the area or by the spread of VEGETATIVE GROWTH. Under these conditions an advancing range margin will develop. In contrast, if unfavourable conditions develop and established plants die, they will not be replaced by further seedlings or vegetative spread. Under these conditions a retreating range margin develops. The relative speeds of the retreating and advancing range margins of a plant population in a changing environment determine whether survival or extinction of that population occurs.

Because of the long generation time of most tree species (decades to centuries), tree migration cannot be observed directly and retrospective studies are required. Various approaches can be used to examine the spread of tree populations over time, including the approaches used in PALAEOECOLOGY (such as the pollen records from a number of sites distributed across a region), DENDROECOLOGY (e.g. examination of tree remains that can indicate either when a species first became established within a region or the distribution of that species relative to its current range), historical (e.g. AERIAL PHOTOGRAPHY showing range expansion or contraction) and ecological studies (e.g. examination of seedling dynamics at the *migration front* and of established trees within the range boundaries). Only the stages of migration (establishment and growth of seedlings and saplings away from the parent population) can be studied by ecologists and foresters. The study of plant migration is essential to aid in the prediction of the abundance and distribution of tree populations in response to future GLOBAL CLIMATIC CHANGE. *JLF*

Bennett, K.D. 1998: The power of movement in plants. *Trends in Ecology and Evolution* 13, 339–340. Bennett, K.D. 1986: The rate of spread and population increase of forest trees during the postglacial. *Philosophical Transactions of the Royal Society of London* B314, 523–531. Delcourt, H.R. and Delcourt, P.A. 1991: *Quaternary ecology – a palaeoecological perspective*. London: Chapman and Hall. MacDonald, G.M. 1993: Fossil pollen analysis and the reconstruction of plant invasions. *Advances in Ecological Research* 24, 67–109. Tallis, J.H. 1991: *Plant community history*. London: Chapman and Hall.

tree ring The product of incremental radial growth in the stems, branches and roots of trees. The familiar concentric rings revealed in cross-sections of many extratropical trees are laid down year on year as the meristem cells in the vascular CAMBIUM near the outer circumference of the tree divide to form the tree's conducting tissue (*phloem*, within which sugars are transported, and *xylem*, by which water is transported). The xylem cells build continuously upon each other from the pith outwards, forming an ever-increasing amount of wood. In regions where the thermal climate oscillates with a strong annual rhythm, trees are generally dormant in the cold season and begin to grow as it warms in the spring. During the early growing season, cell division is relatively rapid and large thin-walled cells are often produced. As the year progresses into summer, cell division continues, but smaller, thicker-walled cells may be produced until, eventually, cell division ceases for the winter. The boundary between the, often darker and denser, late-summer wood and the lighter spring wood clearly delineates one year's growth from the next. The varying width of the rings provides an indication of the changing conditions averaged over the year. In generally cool environments, the rings may represent a record of average temperatures. In drier environments it may be the annual cycle of moisture availability that controls tree growth, with wider rings indicating wetter years and thin ones showing drought. Hence, provided the seasonal limitation is not too great or too irregular, it is possible to date and measure the widths of tree rings precisely and use this information to provide a continuous record of the INTERANNUAL VARIABILITY, not only of the growth of the tree over its lifetime, but also of specific limiting factors that influenced it. By careful selection of particular groups of trees, it is possible to study the changing behaviour of numerous factors that influence tree growth. *KRB*

[See also DENDROCHRONOLOGY, DENDROCLIMATOLOGY, DENDROECOLOGY, CHRONOLOGY, CROSS-DATING, FALSE RING, LIGHT RING, RESPONSE FUNCTION]

Briffa, K.R. 2000: Annual climate variability in the Holocene: interpreting the message of ancient trees. *Quaternary Science Reviews* 19, 87–105. Dean, J.S., Meko, D.M. and Swetnam, T.W. (eds) 1996: *Tree rings, environment and humanity*. Tucson, AZ: University of Arizona Press. Esper, J. 2000: Long-term tree-ring variations in *Juniperus* at the upper timber line in the Karakorum (Pakistan). *The Holocene* 10, 253–260. Fritts, H.C.

1976: *Tree rings and climate*. London: Academic Press. Schweingruber, H.C. 1988: *Tree rings. Basics and applications of dendrochronology*. Dordrecht: Kluwer. Wimmer, R. and Vetter, R.E. (eds) 1999: *Tree-ring analysis: biological, methodological and environmental aspects*. Wallingford: CAB International.

tree-ring index The annual value in a tree-ring CHRONOLOGY that has been dendrochronologically 'standardised'. This transforms the series of annual measurements of whatever tree-ring parameter is represented into a series of relative indices, with a prescribed mean (and perhaps standard deviation). *KRB*

[See also DENDROCHRONOLOGY, STANDARDISATION]

Cook, E.R., Briffa, K.R., Shiyatov, S., Mazepa, V. 1990: Tree-ring standardization and growth-trend estimation. In Cook, E.R. and Kairiukstis, L.A. (eds), *Methods of dendrochronology: applications in environmental sciences*. Dordrecht: Kluwer, 104–123. Fritts, H.C. 1976: *Tree rings and climate*. London: Academic Press.

trend A tendency for the average value of a variable to exhibit a monotonic rise or fall. *JAM*

[See also CLIMATIC FLUCTUATION]

triangular irregular network (TIN) A technique used in the INTERPOLATION process to create a set of regularly spaced values from irregularly spaced data points. Original data points are linked to form a pattern of triangles, thus ensuring that all values are included. INTERPOLATION can then be carried out by mathematically fitting a grid to the TIN surface. *TF*

[See also DELAUNAY TRIANGULATION, TRIANGULATION SCHEME]

triangulation scheme A system in which survey points or data points on a surface are linked by straight lines to form triangles. This structure allows simple trigonometric calculation of position and INTERPOLATION of intermediate values on the surface. It is used, for example, in contouring. *TF*

[See also CONTOUR, TRIANGULAR IRREGULAR NETWORK]

Triassic A SYSTEM of rocks and a PERIOD of geological time from 248 to 205 million years ago. The first dinosaurs and mammals appeared and, in the late Triassic, PANGAEA began to break apart. *GO*

[See also CONTINENTAL DRIFT, GEOLOGICAL TIMESCALE, PERMO-TRIASSIC]

Parrish, J.T. 1993: Climate of the supercontinent Pangea. *Journal of Geology* 101, 215–233.

tributary A channel that joins with others in the downstream direction. A tributary pattern characterises most river systems. *GO*

[See also DISTRIBUTARY, DRAINAGE BASIN]

trimline A boundary line or limit in the LANDSCAPE produced by EROSION and separating landscapes of different age and maturity. The advance and subsequent retreat of a glacier, for example, may produce a trimline, above which VEGETATION is relatively mature, rock surfaces are relatively well weathered and PERIGLACIAL LANDFORMS, such as BLOCKFIELDS, are relatively well developed. The sharpness of the trimline depends on the age and effectiveness of the erosional event, the nature and extent of landscape modification since formation and the age difference between the land surfaces above and below the trimline.   *JAM*

[See also WEATHERING LIMITS]

Ballantyne, C.K. 1997: Periglacial trimlines – the Scottish Highlands. *Quaternary International* 38/39, 119–136.

trophic–dynamic approach

An approach to understanding how ECOSYSTEMS function based on ENERGY FLOW and MINERAL CYCLING between TROPHIC LEVELS.   *JAM*

[See also ECOLOGICAL ENERGETICS]

Lindeman, R.L. 1942: The trophic–dynamic aspect of ecology. *Ecology* 23, 399–419.

trophic level

A feeding level in an ecosystem. The four trophic levels are *autotrophs* (primary producers) plus the three *heterotroph* groups – HERBIVORES (*primary consumers*), CARNIVORES (*secondary consumers*) and *top carnivores* (tertiary consumers).   *RJH*

[See also ECOSYSTEM CONCEPT, FOOD CHAIN/WEB, TROPHIC–DYNAMIC APPROACH]

tropical

In the strict sense, 'tropical' refers to those parts of the world that lie between the Tropics of Cancer (23.5°N) and Capricorn (23.5°S), although this astronomical definition is sometimes relaxed for environmental, ecological and geographical purposes. *Tropical climates* include the zone of deep easterly flow bounded at the Earth's surface by the large, warm, semi-permanent subtropical ANTICYCLONES. The TRADE WINDS converge near the Equator in the INTERTROPICAL CONVERGENCE ZONE (ITCZ). The primary subdivision of the tropical climates is into the wet or *humid tropics* and the *dry tropics*, the former including most of the equatorial zone and coastal regions with on-shore winds all year around, the latter including most continental interiors and west coasts outside the equatorial zone. This simplified model is modified by the distribution of the oceans and continents, which introduces the monsoonal circulation (see MONSOON).   *RDT/JAM*

Garstang, M. 1998: Tropical climates. In Schneider, S.H. (ed.), *Encyclopedia of climate and weather*. New York: Oxford University Press, 778–780. **McGregor, G.R. and Nieuwolt, S. (eds)** 1998: *Tropical climatology*. Chichester: Wiley.

tropical and subtropical forests

These ecosystems are sensitive to both the amount and seasonal distribution of rainfall. Seasonality increases and PRECIPITATION decreases away from the Equator, leading to a progressive change from EVERGREEN to xeromorphic (XEROPHYTE), SCLEROPHYLLOUS woodland. This general pattern is modified by relief, soils and biological history. TROPICAL RAIN FOREST is replaced by TROPICAL SEASONAL FOREST, where annual rainfall drops below 1500 mm or there are more than two dry months per year. *Semi-evergreen seasonal forest* gives way to *deciduous tropical forest* and arid *thorn forest*, where the annual dry season exceeds six months. Above 900 m, MONTANE FOREST formations dominate. Fresh-water *swamp forests* (including the *igapo* and *varzea* forest of Amazonia) and MANGROVES occur where soils are flooded by fresh and saline water, respectively. *Peat swamp*

forest forms on PEAT deposits and *heath forest* (known also as *caatinga*, *campina* and *kerangas*) occurs on podsolised white sands.   *NDB*

Brown, S. and Lugo, A.E. 1990: Tropical secondary forests. *Journal of Tropical Ecology* 6, 1–32. **Whitmore, T.C.** 1998: *An introduction to tropical rain forests*, 2nd edn. Oxford: Oxford University Press.

tropical cyclone

An intense, circular, low-pressure vortex that develops over warm (> 26.5°C) tropical oceans where the latent heat flux, the CORIOLIS FORCE and upper air divergence allow *easterly waves* to intensify. Thus, tropical cyclones do not form outside the tropics, over land or within 5° latitude of the Equator. They can develop rapidly through several stages from the *tropical wave* via a *tropical depression* (wind speed < 56 km h^{-1}) and a *tropical storm* (> 56 km h^{-1}) to a tropical cyclone (> 103.6 km h^{-1}), the whole developmental sequence and lifespan averaging about 10 days. *Hurricanes* are sometimes defined as including the tropical storm and tropical cyclone classes.

Tropical cyclones differ from extratropical CYCLONES in having a warm core, an intense ring of CONVECTION (the eye wall) where wind speeds are greatest (up to about 100 m s^{-1}) and precipitation heaviest (up to 50 cm per day), and a clear, calm, central eye (25–50 km in diameter). Typically around 650 km in diameter, tropical cyclones are smaller than *extratropical cyclones*.

The 'hurricane season' is mainly summer and especially autumn, when SEA-SURFACE TEMPERATURES are highest. The global annual average number of tropical cyclones is about 80 but there is considerable variation in interannual frequency, possibly linked to the EL NIÑO–SOUTHERN OSCILLATION phenomenon. They are most common in the northwestern Pacific (about one third of the annual total), which is attributed to the greater area of warm surface water. Tropical cyclones are known as *typhoons* in the northwestern Pacific, *baquios* in the Philippines, *willy-willies* in northern Australia and *hurricanes* in the northwestern Atlantic. There is considerable evidence of changes in frequency and spatial distribution over the past few centuries and between the glacial and interglacial episodes of the Quaternary. Some are predicting more frequent and severe tropical cyclones with GLOBAL WARMING.   *RDT/JAM*

[See also CORAL REEFS, MANGROVE SWAMPS: IMPACTS OF TROPICAL CYCLONES, SAFFIR–SIMPSON SCALE]

Anthes, R.A. 1982: *Tropical cyclones: their evolution, structure and effects* [*Meteorological Monographs* 19(41)]. Boston: American Meteorological Society. **Eichler, T.P.** 1996: Hurricanes. In Schneider, S.H. (ed.), *Encyclopedia of climate and weather*. New York: Oxford University Press, 407–411. **Fendell, F.E.** 1974: Tropical cyclones. *Advances in Geophysics* 17, 2–100. **Pieke, R.A.** 1990: *The hurricane*. London: Routledge. **Riehl, H.** 1987: Hurricanes. In Oliver, J.E. and Fairbridge, R.W. (eds), *The encyclopedia of climatology*. New York: Van Nostrand Reinhold, 483–496. **Walsh, R.P.D. and Reading, A.J.** 1991: Historical changes in tropical cyclone frequency within the Caribbean since 1500. *Würzburger Geographische Arbeiten* 80, 199–240.

tropical cyclones: impact on people

The main threats to population are from extreme winds, coastal FLOODS and torrential rainfall. People living along low-lying coastal areas, and in isolated island groups, are particularly susceptible to RISK. Violent winds cause death and injury from flying objects, structural damage and crop defoliation, as well as being a serious shipping hazard, while torrential rains result in problems of river flooding and reservoir management. However, the greatest loss of life (90% of some 20 000 fatalities per year) and the most extensive property damage result from STORM SURGES, with the only remedy being for people to seek higher ground immediately. Financially, the highest losses are experienced by developed countries, although *developing countries* are hardest hit in terms of loss of life, and intermediate societies in terms of damage to economic infrastructure. A tropical cyclone that struck coastal Bangladesh in 1991 caused over 130 000 deaths from drowning and left over 10 million people homeless. METEOROLOGICAL SATELLITES give advanced warnings of potential threats posed and have led to advances in FORECASTING. Tropical cyclones sometimes bring benefits by acting as economic catalysts and by playing a supporting role in regional ecological REGENERATION. GLOBAL WARMING may trigger additional risks posed to people by increasing the frequency and intensity of such STORMS, which are likely to become more hazardous as population growth leads to more settlement in areas of greater risk.   *JBE/GS*

Diaz, H.F. and Pulwarty, R.S. (eds) 1997: *Hurricanes: climate and socioeconomic impacts.* Berlin: Springer. Elsner, J.B. and Kara, A.B. 1999: *Hurricanes of the North Atlantic: climate and society.* New York: Oxford University Press. Hammerton, J.L., George, C. and Pilgrim, R. 1984: Hurricanes and agriculture: losses and remedial actions. *Disasters* 8, 279–286. Simpson, R.H. and Riehl, H. 1981: *The hurricane and its impact.* Oxford: Basil Blackwell. Southern, R.L. 1979: The global socio-economic impact of tropical cyclones. *Australian Meteorological Magazine* 27, 175–195. Sugg, A.L. 1968: Beneficial aspects of the tropical cyclone. *Journal of Applied Meteorology* 7, 39–45. Walsh, R.P.D. 1999: Extreme weather events. In Pacione, M. (ed.), *Applied geography: principles and practice.* London: Routledge, 51–65.

tropical forest fires The majority of tropical forests, with the exception of ever-wet RAIN FORESTS, have burned frequently over many thousands of years. The structure and composition of many TROPICAL SEASONAL FORESTS reflects adaptation to the prevailing FIRE FREQUENCY. FOSSIL and SUBFOSSIL evidence suggests that extensive forest fires occurred in TROPICAL RAIN FORESTS during the PLEISTOCENE and there is evidence of widespread natural WILDFIRES in historical times. Fire is used extensively in the tropics for DEFORESTATION. During recent severe EL NIÑO drought events, fire was used to accelerate the rate of forest clearance for agriculture and often spread out of control. During the El Niño event of 1982–1983, an estimated 4 million ha of rain forest burned on the island of Borneo. Nearly six times more logged forest than PRIMARY FOREST burned. At least 2 million ha burned again in 1997–1998, whilst a further 2 million ha burned in the Amazon. Such fires may contribute substantially to rising levels of atmospheric CARBON DIOXIDE.   *NDB*

[See also BIOMASS BURNING, DISTURBANCE, FIRE IMPACTS, GLOBAL WARMING, GREENHOUSE GASES]

Goldammer, J.G. and Price, C. 1998: Potential impacts of climate change on fire regimes in the tropics based on MAGICC and a GIS GMC-derived lightning model. *Climate Change* 39, 273–296. Stott, P.A., Goldammer, J.G., and Werner, W.L. 1990:The role of fire in tropical lowland deciduous forest of Asia. In Goldammer, J.G. (ed.), *Fire in the tropical biota: ecosystem processes and global challenges.* Berlin: Springer. Kauffman, J.B., Uhl, C. and Cummings, D.L. 1988: Fire in the Venezuelan Amazon 1: fuel biomass and fire chemistry in the evergreen rainforest of Venezuela. *Oikos* 53, 167–175. Stott, P.A. 2000: Combustion in tropical biomass fires: a critical review. *Progress in Physical Geography* 24, 355–377.

tropical grassland Typically, a tall grassland sustained by fire and grazing. This major VEGETATION FORMATION-TYPE has in many areas replaced former TROPICAL RAIN FOREST, TROPICAL SEASONAL FOREST and tropical SCRUB.   *GOH*

[See also PLAGIOCLIMAX, SAVANNA]

tropical karst The LANDFORMS of areas of LIMESTONE rock in tropical environments. The concept of a distinctive tropical karst landscape has been much discussed. COCKPIT KARST and TOWER KARST were regarded by some workers as distinctively tropical and also as comprising successive evolutionary stages in karst LANDSCAPE EVOLUTION, although both views have now been questioned. Both the higher density and star-shaped (vallied) nature of DOLINES in some tropical landscapes compared with in temperate areas have been linked to higher rainfall intensities and shallow subsurface runoff rates (subcutaneous flows) in the humid tropics. Cockpit and tower karst, however, have been found in extratropical as well as tropical regions and in the Giulin area of southern China have been found to represent alternative rather than successive landscape forms. Also, the great height attained by limestone pinnacles in tropical areas such as Mulu, Sarawak, linked by some to high rainfall, probably results instead from a combination of a history of uplift and the strength and joint spacing of the limestone involved. Rates of solution and DENUDATION in tropical karst landscapes are somewhat higher than in colder climates, despite the lower solubility of CARBON DIOXIDE in warmer water. This has been linked to two groups of factors: (1) higher availability of soil carbon dioxide and organic acids and heightened microbial activity linked to the greater vegetation and soil cover and continuously high temperatures of the humid tropics; and (2) higher rainfall of the humid tropics compared with most cooler humid environments.

                                    JAM/RPDW

Ford, D.C. and Williams, P.W. 1989: *Karst geomorphology and hydrology.* London: Chapman and Hall. Jennings, J.N. 1972: The character of tropical humid karst. *Zeitschrift für Geomorphologie NF* 16, 336–341. Sweeting, M.M. 1972: *Karst landforms.* London: Macmillan. Tang, T. and Day M.J. 2000: Field survey and analysis of hillslopes on tower karst in Guilin, southern China. *Earth Surface Processes and Landforms* 25, 1221–1235. Thomas, M.F. 1994: *Geomorphology in the tropics: a study of weathering and denudation in low latitudes.* Chichester: Wiley. Williams, P.J. 1993: Climatological and geological factors controlling the development of polygonal karst. *Zeitschrift für Geomorphologie Supplementband* 93, 159–193.

tropical peatlands PEAT may be defined as soils with an organic content >65% by volume and where the organic-rich horizon is >50 cm depth. Such soils form when organic DETRITUS fails to decompose and accumulates at the soil surface. Processes that slow DECOMPOSITION are severe cold and waterlogging. Two basic types of peatland exist in the tropics: OMBROTROPHIC MIRES and RHEOTROPHIC (*topogenous*) MIRES.

OMBROTROPHIC MIRES have a surface higher than the general land surface and nutrient cycling is AUTOCHTHONOUS. These peats are highly acidic, with very low NUTRIENT availability. High sulphide content and low oxygen availability inhibit decomposition. Such peats commonly form deep (> 20 m) deposits on the landward fringe of MANGROVES. Peat accumulation rates can be as high as 0.5 m per century in these environments.

RHEOTROPHIC mires are ALLOCHTHONOUS, deriving their nutrients from river water or from proximate mineral soils. Abandoned ox-bows, deltaic peatlands, backswamps and depressional wetlands contain topogenous peat. Frequently, the surface of the peat is a hard crust overlying a semi-liquid interior. The crust may be strong enough to support a wetland forest with trees > 20 m in height.

Ice scouring and corrie formation of past glacial advances has left depressions suitable for peat development at elevations above 3000 m on many tropical mountains. As in TEMPERATE regions, the tropical peats of all elevations are a rich source of MICROFOSSILS such as pollen, phytoliths, diatoms and charcoal, as well as MACROFOSSILS such as wood and beetles. These SUBFOSSILS are used in *environmental reconstruction* and to evaluate past CLIMATIC CHANGE. In strongly seasonal environments there may be no peat accumulation, because during the dry season the wetland dries out and all surface organic material undergoes OXIDATION. *MBB*

[See also ACIDITY, ANAEROBIC, BIOMES, LAND DRAINAGE, MIRES, SWAMP, WETLAND, WETLAND CLASSIFICATION]

Clapperton, C.M. 1993: *Quaternary geology and geomorphology of South America*. Amsterdam: Elsevier. **Whitten, A.J., Damanik, S.J., Anwar, J. and Hisyam, N.** 1987: *The ecology of Sumatra*. Yogyakarta, Indonesia: Gadjah Mada University Press.

tropical rain forest

An EVERGREEN forest VEGETATION FORMATION-TYPE found in those tropical areas that have a climate with little or no dry season and where rainfall usually exceeds 1500 mm per annum. Tropical rain forest is found in three main VEGETATION FORMATIONS. The largest of these, 4×10^6 km^2 in extent, is in Central and South America and includes the vast basin of the Amazon. Central Africa has some 1.8×10^6 km^2 of rain forest with fragments surviving on the east coast of Madagascar. Southeast Asia has a total of about 2.5×10^6 km^2 of tropical rain forest, distributed across the Indo–Malay archipelago. Although these forests have very few plant species in common, they show remarkable physiognomic similarity, with a multilayered forest *canopy*, trees of at least 30 m height and being rich in EPIPHYTES and lianes. The understorey is often sparse because of very low light levels. This BIOME is believed to support several million species of organism, although a minority of them have been scientifically described. Peninsular Malaysia alone has in excess of 8000 species of vascular plant. Many plant species belong to large congeneric series that are genetically very similar. Most plant species are of restricted distribution and occur at low population densities, making them vulnerable to EXTINCTION.

CONTINENTAL DRIFT and CLIMATIC CHANGE have determined the biogeography of tropical rain forests. The floras of all three tropical regions have a common evolutionary origin in GONDWANA. They have subsequently diverged and mixed with species from Laurasia. During the

QUATERNARY, tropical climates have been extremely variable. Rainfall may have been 30% less during much of the PLEISTOCENE and sea level fell by over 100 m during GLACIAL EPISODES. Rain forest may have been periodically restricted to REFUGIA of higher rainfall.

Tropical rain forests have rarely supported large human populations, though at the time of first European contact, in the sixteenth century, it is estimated that Amazonia had a population of several million people. However, millennia of use by humans have had a substantial impact on forest composition. Present-day populations mostly practise SHIFTING CULTIVATION. The widespread DEGRADATION of tropical rain forests by commercial logging and rapid DEFORESTATION are causing global concern. *NDB*

[See also BIODIVERSITY, REFUGE THEORY IN THE TROPICS, TROPICAL FOREST]

Brown, N. 1998: Degeneration versus regeneration – logging in tropical rain forests. In Goldsmith, F.B. (ed.), *Tropical rain forest: a wider perspective*. London: Chapman & Hall. **Flenley, J.** 1979: *The equatorial rain forest: a geological history*. London: Butterworth. **Golley, F.B., Lieth, H. and Werger, M.J.A.** (eds) *Tropical rain forest ecosystems* [*Ecosystems of the World*. Vol. 14]. Amsterdam: Elsevier. **Mabberley, D.J.** 1992: *Tropical rain forest ecology*, 2nd edn. Glasgow: Blackie. **Parks, C.C.** 1992: *Tropical rain forests*. London: Routledge. **Primack, R.B. and Lovejoy, T.E.** (eds) 1995: *Ecology, conservation and management of Southeast Asian rainforests*. New Haven, CT: Yale University Press. **Richards, P.W.** 1999: *The tropical rain forest*, 3rd edn. Cambridge: Cambridge University Press. **Whitmore, T.C.** 1998: *An introduction to tropical rain forests*, 2nd edn. Oxford: Oxford University Press.

tropical seasonal forest

A forest VEGETATION FORMATION-TYPE, sometimes referred to as *monsoon forest*, that occurs in those part of the tropics with a strong dry season, including most of the African tropical forests. Tropical seasonal forests comprise approximately 28% of the Earth's forested area. As the severity of seasonal drought increases, so does the frequency of DECIDUOUS trees, especially among those that form the upper *canopy* of the forest. The forests are of lower structure and typically less diverse than those found in ever-wet areas and may be dominated by a small number of tree species. Many species are fire tolerant. The FLORA of seasonal forests contains many families and genera that are common in the ever-wet tropics, implying a comparatively recent evolutionary adaptation to increasing SEASONALITY by many tropical plants. The boundary between tropical seasonal forest and SAVANNA is highly dynamic and is dependent on climate, soils, fire regime, grazing and human management. *NDB*

[See also EVOLUTION, FIRE FREQUENCY, FIRE IMPACTS (ECOLOGICAL), TROPICAL AND SUBTROPICAL FORESTS, TROPICAL RAIN FOREST]

Longman, K.A. and Jenik, N. 1987: *Tropical forest and its environment*. London: Longman. **Redford, K.H. and Padoch, C.** 1992: *Conservation of Neotropical forests: working from traditional resource use*. New York: Columbia University Press.

troposphere

The lowest layer of the ATMOSPHERE, in which temperature decreases in altitude with a LAPSE RATE of about 6.5°C km^{-1} up to the *tropopause*, which occurs at about 8 km above the poles and about 16 km above the Equator and where the temperature is about −50 to −60°C. The troposphere contains about 75% of the mass of the atmosphere, almost all the weather systems affecting the Earth's CLIMATE and WEATHER, and is the part of the atmosphere most affected by human impact.   *JAM*

tsunami   An ocean wave (in Japanese, literally 'harbour wave') that travels at exceptionally high velocities and is characterised by long wavelengths and periods. When the *wave train* reaches the coast, individual tsunami waves frequently reach heights in excess of 5 m and result in extensive loss of life and damage to property. In 1998, an earthquake off the coast of Papua New Guinea produced a destructive tsunami that flooded coastal villages and killed several thousand people. Tsunamis normally originate as a result of offshore EARTHQUAKE activity, where a seabed fault causes disturbance of the ocean water column. They can also be caused by underwater slides and by terrestrial LANDSLIDES that enter the sea and generate waves. It is also thought that large tsunamis can be generated by BOLIDE impact. For example, the Cretaceous–Tertiary bolide impact in Yucatan, Mexico c. 65 million years ago is thought to have produced an enormous tsunami.   *AGD*

Bernard, E.N. 1991: *Tsunami hazard: a practical guide to tsunami hazard reduction.* Dordrecht: Kluwer. **Dawson, A.G.** 1994: Geomorphological effects of tsunami runup and backwash. *Geomorphology* 10, 83–94. **Dawson, A.G., Long, D. and Smith, D.E.** 1987: The Storegga Slides: evidence from eastern Scotland of a possible tsunami. *Marine Geology* 82, 271–276. **Hemphill-Haley, E.** 1996: Diatoms as an aid in identifying late-Holocene tsunami deposits. *The Holocene* 6, 439–448. **Lowe, D.J. and de Lange, W.P.** 2000: Volcano-meteorological tsunamis, the c. AD 200 Taupo eruption (New Zealand) and the possibility of a global tsunami. *The Holocene* 10, 401–407.

tufa   Calcareous tufa, *travertine* or *flowstone* is a secondary, freshwater, carbonate deposit formed primarily from the above-ground DEGASSING of calcium-carbonate-rich waters. Tufa deposits can be divided into *valley-side tufa* associated with springs and *valley-bottom tufa* associated with rivers, waterfalls and barrage systems. A late-Holocene decline in tufa deposition over much of Europe has been suggested, possibly related to natural or anthropogenic environmental changes, but there may have been significant under-reporting of the extent and rate of contemporary deposition.   *JAM*

[See also SPELEOTHEMS]

Baker, A. and Simms, M.J. 1998: Active deposition of calcareous tufa in Wessex, UK, and its implications for the 'late-Holocene tufa decline'. *The Holocene* 8, 359–365. **Goudie, A.S., Viles, H.A. and Pentecost, A.** 1993: The late-Holocene tufa decline in Europe. *The Holocene* 3, 181–186. **Pentecost, A.** 1995: The Quaternary travertine deposits of Europe and Asia Minor. *Quaternary Science Reviews* 14, 1005–1028.

tuff   A PYROCLASTIC rock; specifically a lithified ASH.   *JBH*

[See also WELDED TUFF]

tumulus   An earth or rubble monument in prehistoric Europe (Early NEOLITHIC to Mediaeval), usually covering a burial. Tumuli range from heaps of stones (*cairns*) to various types of circular or elongate earth mounds (*barrows*). In addition to their archaeological significance, the buried land surfaces beneath barrows are of importance for reconstructing environmental changes, such as the evolution of soils and vegetation.   *JAM*

[See also TELL]

Dimbleby, G.W. 1962: *The development of British heathlands and their soils.* Oxford: Clarendon Press. **Dimbleby, G.W.** 1985: *The palynology of archaeological sites.* London: Academic Press.

tundra   The generally flat treeless area of the Arctic, north of the BOREAL FOREST and to the south of the POLAR DESERT. Tundra tends to occur where the ambient air temperature is less than 0°C for at least seven months of the year and where there is continuous PERMAFROST. This circum-Arctic zone, therefore, extends much further south in North America than it does in Fennoscandia. The tundra–boreal forest boundary coincides with the summer position of the Arctic front. The very short growing season, coupled with shallow soils and the prevalence of CRYOTURBATION and SOLIFLUCTION, inhibits tree growth and the vegetation consists of low-growing woody herbaceous PERENNIALS, grasses, sedges, mosses and lichens.

Tundra areas developed in the early PLEISTOCENE with the vegetation probably evolving from pre-adapted plants of boreal forest and/or alpine areas as land masses moved into higher latitudes. During the GLACIAL EPISODES, tundra existed around the southern margins of the ice sheets, where it graded into STEPPE VEGETATION. Its distribution at those times was, therefore, much further south than it is today and it covered greater areas.

Tundra vegetation is characterised by a diversity of communities, which reflect the edaphic, topographic and microclimate conditions, with local variation often being related to the length of the snow cover. SPECIES DIVERSITY, the continuity of the vegetation cover and the height of the plants all decrease towards the pole. In the LOW ARCTIC, shrub tundra is characteristic and vegetation covers 80–100% of the land surface. Here, mosaics of *mosses* and sedges in low-lying wet areas, and HEATHLAND and dwarf shrubs in more elevated drier areas, are interspersed with lichen covered fellfields and exposed rocky ridges. In the HIGH ARCTIC, vegetation covers less than 50% of the land, bryophtyes and lichens are important and snow-patch catenas are characteristic. As a result of the climatic conditions, the rate of decomposition is very slow, so carbon reserves can be quite high (see CARBON CYCLE). MIRES are common and, in the low Arctic, PEAT accumulation can reach several metres. The wildlife is dominated by mammalian grazers (caribou, reindeer, musk-ox, voles and lemmings) and herbivorous birds (geese and ptarmigan).   *SPH*

[See also BIOMES, TUNDRA VEGETATION: HUMAN IMPACT, VEGETATION FORMATION-TYPE]

Bliss, L.C., Heal, O.W. and Moore, J.J. 1981: *Tundra ecosystems: a comparative analysis.* Cambridge: Cambridge University Press. **Chernov, Y.I.** 1985: *The living tundra.* Cambridge: Cambridge University Press. **Griggs, R.F.** 1934: The problem of arctic vegetation. *Journal of the Washington Academy of Sciences* 25, 153–175. **Heal, O.W., Callaghan, T.V., Cornelissen, J.H.C. et al.** 1998: *Global change in Europe's cold regions. Ecosystems Research Report* 27, EUR 18178 EN. **Journal of Vegetation Science** 1994: Special Issue on Circumpolar Arctic Vegetation. *Journal of Vegetation Science* 5(6), 758–920. **Wielgolaski, F.E. (ed.)** 1997: *Polar and alpine tundra* [*Ecosystems of the World.* Vol. 3]. Amsterdam: Elsevier.

tundra vegetation: human impact

TUNDRA vegetation, existing under the harsh climate conditions of a MARGINAL AREA, is highly sensitive to HUMAN IMPACT. Not only can a small amount of impact cause extensive changes to the vegetation, but recovery from the impact is slow. Impact was relatively limited up until World War II since, until then, the main occupants of the tundra had followed a nomadic way of life with a HUNTING, FISHING AND GATHERING or a PASTORAL economy. Since then, exploration for and exploitation of oil, gas and mineral ores and the establishment of industrial and domestic buildings, together with the increase in off-road vehicles, have, locally, caused dramatic changes. Disruption of the insulating soil and vegetation cover has led to a lowering of the summer PERMAFROST table and the development of a THERMOKARST surface. Shallow LANDSLIDES and gullying have increased, whilst mineral extraction and processing have resulted in severely polluted areas (see POLLUTION).

SPH

[See also ARCTIC ENVIRONMENTAL CHANGE]

Arctic and Alpine Research 1987: Special Issue on Restoration and Vegetation Succession in Circumpolar Lands. *Arctic and Alpine Research* 19(4), 342–583. **Arnalds, A.** 1987: Ecosystem and recovery in Iceland. *Arctic and Alpine Research* 19, 508–513. **French, H.M.** 1996: *The periglacial environment*, 2nd edn. Cambridge: Cambridge University Press. **Komárková, V. and Wielgolaski, F.E.** 1999: Stress and disturbance in cold region ecosystems. In Walker, L.R. (ed.), *Ecosystems of disturbed ground* [*Ecosystems of the World*. Vol. 16]. Amsterdam: Elsevier, 39–122. **Williams, P.J. and Smith, M.W.** 1989: *The frozen Earth*. Cambridge: Cambridge University Press.

tunnel valley

A large channel up to 100 km long and 4 km wide, often exhibiting an irregular long profile and OVERDEEPENING and infilled with later sediments. The origin of tunnel valleys is controversial, but they are thought to be excavated by SUBGLACIAL meltwaters, possibly in catastrophic outburst floods or JÖKULHLAUPS. Also known as *tunnel channels*, tunnel valleys occur singly or in groups and are normally oriented at right angles to former ice margins.

DIB

[See also URSTROMTÄLER]

Ehlers, J. 1996: *Quaternary and glacial geology*. Chichester: Wiley. **O'Cofaigh, C.** 1996: Tunnel valley genesis. *Progress in Physical Geography* 20, 1–19.

turbidite

A BED of sediment deposited by a TURBIDITY CURRENT in a deep-water SEDIMENTARY BASIN, including ocean basins, deep lakes and deep marine basins on continental crust. Turbidites range from millimetres to metres in thickness and are characterised by moderate to poor sorting (turbidite sandstones in the geological record are commonly GREYWACKES), GRADED BEDDING, tabular geometry, extreme lateral continuity, an erosional base commonly preserving SCOUR MARKS, TOOL MARKS or TRACE FOSSILS and a regular sequence of internal SEDIMENTARY STRUCTURES (the '*Bouma sequence*') from structureless sediment at the base, through parallel lamination, cross-lamination and laminated sand and mud, to mud. Most turbidites preserve an incomplete Bouma sequence, but those divisions that are present are in a consistent order.

A distinction can be made between: *fine-grained turbidites* (*distal turbidites*) with thin, fine-grained turbidites and mud-dominated successions; *classical turbidites*, which conform to the Bouma sequence, with approximately equal proportions of sand and mud; and *coarse-grained turbidites* (*proximal turbidites*) with thick, coarse-grained turbidites and sand- or gravel-dominated successions. There is some overlap between the processes responsible for coarse-grained turbidites and DEBRIS FLOWS. The type of turbidite depends not only on proximity to the source, but also on the setting of the DEPOSITIONAL ENVIRONMENT: a turbidite deposited in a deep-sea channel may be thick and coarse-grained, whereas the overbank portion of the same event may deposit a fine-grained turbidite.

Turbidite successions are characterised by monotonously interbedded mud and turbidite sandstones. The mud is interpreted as *background sediment* that accumulated between turbidity current EVENTS. Turbidites are deposited in a range of deep-water environments, including SUBMARINE FANS and ABYSSAL plains. Most are inferred to have been deposited by turbidity currents triggered by EARTHQUAKES, slope-wasting or rivers in FLOOD. Turbidite sandstones in deep-water basins, such as offshore Brazil, Angola, Nigeria, the USA and the UK, make up a large proportion of recently discovered giant oilfields and hence interest in their study has increased in recent years. *BTC*

[See also FLYSCH]

Bouma, A.H. 1962: *Sedimentology of some flysch deposits: a graphic approach to facies interpretation*. Amsterdam: Elsevier. **Bouma, A.H.** 2000: Coarse-grained and fine-grained turbidite systems as end member models: applicability and dangers. *Marine and Petroleum Geology* 17, 137–143. **Kuenen, Ph.H. and Migliorini, C.I.** 1950: Turbidity currents as a cause of graded bedding. *Journal of Geology* 58, 91–127. **Pickering, K.T., Hiscott, R.N. and Hein, F.J.** 1989: *Deep-marine environments: clastic sedimentation and tectonics*. London: Unwin Hyman. **Shanmugam, G.** 2000: 50 years of the turbidite paradigm (1950s–1990s): deep-water processes and facies models – a critical perspective. *Marine and Petroleum Geology* 17, 285–342. **Stow, D.A.V., Reading, H.G. and Collinson, J.D.** 1996: Deep seas. In Reading, H.G. (ed.), *Sedimentary environments: processes, facies and stratigraphy*, 3rd edn. Oxford: Blackwell Science, 395–453. **Weltje, G.J. and deBoer, P.L.** 1993: Astronomically induced paleoclimatic oscillations reflected in Pliocene turbidite deposits on Corfu (Greece): implications for the interpretation of higher order cyclicity in ancient turbidite systems. *Geology* 21, 307–310.

turbidity

The extent to which there is interference with the passage of light through air or water. *Atmospheric turbidity* may be caused by water droplets, PARTICULATES or gases, which may be of natural origin or POLLUTANTS. Suspended particles from a similarly wide range of origins affect the turbidity of water in rivers and lakes. *JAM*

[See also AEROSOLS, SUSPENDED LOAD]

turbidity current

A turbulent mixture of water and suspended sediment that flows as a DENSITY FLOW. Turbidity currents, or *turbidity flows*, are a type of SEDIMENT GRAVITY FLOW. In OCEAN BASINS they develop at the upper part of the CONTINENTAL SLOPE and flow as a bottom-current down the slope and across the ABYSSAL plain. They can develop from SLUMPS or DEBRIS FLOWS or can be generated directly by EARTHQUAKES, storm waves, input from river floods or sediment instability in SUBMARINE CANYONS. They are very effective in transporting large volumes of terrigenous or carbonate sediment into deep-water environments, where they contribute to the construction of SUBMARINE FANS. They also occur in deep lakes. Sediment is deposited in the form of a TURBIDITE as the turbidity current slows as a result of a combination of momentum loss and mixing (dilution) with water. Deposition also occurs at changes in gradient, at opposing topography or beyond channel mouths where flows can expand and collapse. Large, vigorous turbidity currents have been known to break submarine cables.   *BTC*

[See also MEGATURBIDITE]

Beattie, P.D. and Dade, W.B. 1996: Is scaling in turbidite deposition consistent with forcing by earthquakes? *Journal of Sedimentary Research* 66, 909–915. **Heezen, B.C. and Ewing, M.** 1952: Turbidity currents and submarine slumps, and the 1929 Grand Banks earthquake. *American Journal of Science* 250, 849–873. **Johnson, D.W.** 1939: *The origin of submarine canyons: a critical review of hypotheses.* New York: Columbia University Press. **Leeder, M.R.** 1999: *Sedimentology and sedimentary basins: from turbulence to tectonics.* Oxford: Blackwell Science. **Walker, R.G.** 1992: Turbidites and submarine fans. In Walker, R.G. and James, N.P. (eds), *Facies models: response to sea level change.* St John's, Newfoundland: Geological Association of Canada.

turbulent flow   Flow dominated by inertial forces and characterised by *turbulent eddies*: pockets of fluid moving in directions and at speeds that differ from the mean flow velocity. Turbulent flows occur at high values of the *Reynolds number* (ratio of inertial to viscous forces), the change from LAMINAR FLOW occurring at values between 500 and 2000. Most water flows and all natural air flows are turbulent.   *GO*

Clifford, N.J., French, J.R. and Hardisty, J. (eds) 1993: *Turbulence: perspectives on flow in sedimentary transport.* Chichester: Wiley. **Leeder, M.R.** 1999: *Sedimentology and sedimentary basins: from turbulence to tectonics.* Oxford: Blackwell Science.

turnover rate   The rate of change in the species composition (types of species) in an isolate or community over time. *Turnover* may occur despite relative constancy in the number of species detected. While turnover may be an important and relatively common phenomenon, it can be difficult to quantify accurately. *Cryptoturnover* describes the situation where, with irregular surveys, turnover occurring in the interval between two samples is not recorded; *pseudoturnover* occurs with incomplete censuses, where species appear to have come and gone from an isolate when they have not.   *MVL*

[See also ISLAND BIOGEOGRAPHY]

Brown, J.H. and Lomolino, M.V. 1998: *Biogeography,* 2nd edn. Sunderland, MA: Sinauer. **MacArthur, R.H. and Wilson, E.O.** 1967: *The theory of island biogeography.* Princeton, NJ: Princeton University Press. **Whittaker, R.J.** 1998: *Island biogeography: ecology, evolution and conservation.* Oxford: Oxford University Press.

two-way indicator species analysis (TWINSPAN)   A fast, robust, and effective numerical method for the simultaneous CLASSIFICATION of objects and variables in large, heterogeneous datasets. It uses CORRESPONDENCE ANALYSIS as the initial basis for the object classification. It also derives indicators for the different clusters. It is implemented by the computer program TWINSPAN.   *HJBB*

Hill, M.O. 1979: TWINSPAN – A FORTRAN program for arranging multivariate data in an ordered two-way table by classification of individuals and attributes. Ithaca, NY: Cornell University (available from http://cc.oulu.fi/~jarioska).

typological sequencing   RELATIVE-AGE DATING of ARTEFACTS based on their similarities. It is based on two principles: first, that artefacts of a particular period are distinctive; and second, that changes in style are gradual and directional. Particular artefacts may therefore be assigned to a unique position in a series representative of a time sequence.   *JAM*

[See also SERIATION]

Renfrew, C. and Bahn, P. 1996: *Archaeology: theories, methods and practice,* 2nd edn. London: Thames and Hudson.

U

U-shaped valley Cross-sections of glacial valleys or troughs commonly appear U-shaped, with steep bedrock sides and comparatively flat floors, compared with the typically V-SHAPED VALLEYS of rivers. If, however, surficial deposits on the floor of the glacial valley are excluded, the eroded bedrock cross-sectional form approximates more closely to that of a parabola or similar mathematically defined curve, suggesting that GLACIER EROSION modifies the valley form to provide the most efficient shape for evacuating GLACIER ice. Research in the Ben Ohau Range, New Zealand, has suggested that 70 000 years may be required for the development of a recognisable parabolic form from a V-shaped valley, while it may take some 300 000 years to produce a mature glacial trough.   *JS*

[See also CIRQUE, GLACIER, OVERDEEPENING]

Harbor, J.M. 1992: Numerical modelling of the development of U-shaped valleys by glacial erosion. *Geological Society of America Bulletin* 104, 1364–1375. **Harbor, J.M., Hallet, B. and Raymond, C.F.** 1988: A numerical model of landform development by glacial erosion. *Nature* 333, 347–349. **Kirkbride, M.P. and Mathews, D.** 1997: The role of fluvial and glacial erosion in landscape evolution: the Ben Ohau Range, New Zealand. *Earth Surface Processes and Landforms* 22, 317–327.

ultimate causes Explanations that require no further explanation of the premises used in the explanation. A reductionist, and perhaps illusory, concept.   *CET*

[See also PROXIMATE CAUSE]

Diamond, J. 1998: *Guns, germs and steel.* London: Vintage.

umbric horizon A deep, dark-coloured humus-rich horizon with a base saturation of <50%, poor in plant nutrient reserves and characteristic of UMBRISOLS.   *EMB*

umbrisols Acid soils with a thick dark surface horizon, an UMBRIC horizon, rich in organic matter, that occur in cool humid, often mountainous, regions. These soils are found, often in mid-slope positions, in the northwestern part of the Iberian peninsula, the northwest coast of USA and Canada and the mountain ranges of the Himalayas and Andes. This reference soil grouping brings together all deep, free-draining immature soils in which between 2% and 5% desaturated organic matter has accumulated. If limed and cultivated, the umbric horizon begins to resemble a MOLLIC HORIZON of the PHAEOZEMS.   *EMB*

[See also WORLD REFERENCE BASE FOR SOIL RESOURCES]

uncertainty principle The concept, based on Heisenberg's uncertainty principle from atomic physics, that it is impossible to measure exactly two *state variables* of a SYSTEM (e.g. position, momentum) at the same time. Application in its original form to environmental and other *macroscopic systems* is questionable but there are implications.   *JAM*

[See also INDETERMINACY]

Harrison, S. and Dunham, P. 1998: Decoherence, quantum theory and their implications for the philosophy of geomorphology. *Transactions of the Institute of British Geographers* 23, 501–514.

unconformity A GEOLOGICAL STRUCTURE comprising a surface between units in a rock succession separated in time by a significant interval that is not represented in that succession, i.e. a gap in the GEOLOGICAL RECORD. In the missing time interval (HIATUS), which commonly represents intervals of tens to hundreds of millions of years, the older rocks may have been subjected to burial, TECTONIC deformation, UPLIFT and EROSION, before accumulation of the upper succession. Rocks or sediments formed during this interval may be preserved elsewhere. Different styles of unconformity include: an *angular unconformity*, where there is a difference in the orientation of BEDDING at the surface of the unconformity; a *disconformity* or *parallel unconformity*, where there is no angular discordance; a *nonconformity*, where the older rocks are IGNEOUS or META-MORPHIC ROCKS; and a *landscape unconformity*, where LANDSCAPE features (e.g. hills or PALAEOSOLS) are preserved. Unconformities may indicate periods of OROGENY or reduced SEA LEVEL. If traced towards the centre of a SEDIMENTARY BASIN, an unconformity may be replaced by a conformable junction: this forms the basis for the separation of sequences in SEQUENCE STRATIGRAPHY.   *JCWC/GO*

[See also DIASTEM]

Lisle, R.J. 1995: *Geological structures and maps.* Oxford: Butterworth-Heinemann. **Whitten, D.G.A. and Brooks, J.R.V.** 1972: *The Penguin dictionary of geology.* London: Penguin. **Williams, G.E. and Schmidt, P.W.** 1997: Palaeomagnetic dating of sub-Torridon Group weathering profiles, NW Scotland: verification of Neoproterozoic palaeosols. *Journal of the Geological Society, London* 154, 987–997. **Wilson, R.C.L.** 1992: Sequence stratigraphy: an introduction. In Brown, G.C., Hawkesworth, C.J. and Wilson, R.C.L. (eds), *Understanding the Earth: a new synthesis.* Cambridge: Cambridge University Press, 388–414.

underfit stream A stream apparently too small for its valley. Underfit streams have been used to infer CLIMATIC CHANGE, in which much higher discharges and rainfalls in a postglacial, valley-cutting period, gave way to channel 'shrinkage' and 'underfitness' as drier conditions returned. Such simple interpretations have since been criticised strongly.   *DML*

[See also MISFIT MEANDER, RIVER CAPTURE]

Dury, G.H. 1977: Underfit streams: retrospect, perspect and prospect. In Gregory, K.J. (ed.), *River channel changes.* Chichester: Wiley, 281–293.

uniclinal shifting The process of ASYMMETRICAL VALLEY formation controlled by the geological structure whereby a stream following the line of the geological STRIKE tends to erode sideways in the direction of the DIP.   *JAM*

Stamp, L.D. (ed.) 1961: *A glossary of geographical terms.* London: Longman.

uniformitarianism The principle that physical, chemical and biological laws operate in the same way today as they have throughout GEOLOGICAL TIME. Because the Earth has changed through time (e.g. mean surface temperature, atmospheric composition, rates and composition of volcanic activity), this does not imply that the precise conditions that can be observed today are representative of all geological time, although this is sometimes the interpretation given to uniformitarianism, especially when expressed in its popular form that 'the present is the key to the past'. Perhaps more than any other development, the acceptance of uniformitarianism has allowed geological problems to be solved using SCIENTIFIC METHODS.

The history of this critical but disputed term is important in understanding its meaning. James Hutton (1726–1797) first suggested in his 'Theory of the Earth' that the geological past was similar to the present and argued for the interpretation of the GEOLOGICAL RECORD in terms of the products of slow, persistent processes operating over unimaginably long spans of time, as opposed to the cataclysmic EVENTS favoured by proponents of CATASTROPHISM. Hutton's idea that the past had a 'uniformity' with the present was taken up forcefully by Charles Lyell (1797–1875) in his *Principles of Geology* with the term 'uniformitarianism' being used in a review of the work by William Whewell in AD 1832. Lyell's 'uniformity' is complex, with four distinct meanings. *Uniformity of law* considered that natural laws are constant in time and space. *Uniformity of process* held that, if a past event can be explained by processes now known to be operating, then additional, unknown causes should not be invoked: this is merely an extension of the philosophical principle of economy. These two meanings, constituting '*methodological uniformitarianism*' are untestable methodological assumptions common to all sciences. Lyell also affirmed *Uniformity of rate,* asserting that natural processes occur by slow, steady and gradual means, and *Uniformity of state,* which held that although change occurs on the Earth it is fundamentally directionless, with the Earth being in a more-or-less steady state. These last two definitions are testable theories about how the Earth works ('*substantive uniformitarianism*'); they have respectively been partially and completely falsified. The widespread acceptance of 'the present is the key to the past' as a definition of uniformitarianism, when it is little more than a scientific platitude, epitomises the elusive and ambiguous nature of Lyell's definition. Modern uniformitarianism's assumption of the operation in the past of identical natural laws to those observed at present allows for the occurrence of occasional cataclysmic events.   *CDW*

[See also ACTUALISM, GRADUALISM, NEOCATASTROPHISM]

Albritton, C.C. (ed.) 1967: *Uniformity and simplicity* [Special Paper 89]. Boulder, CO: Geological Society of America. **Blundell, D. and Scott, A.C. (eds)** 1998: *Lyell: the past is the key to the present* [Special Publication 143]. London: Geological Society. **Craig, G.Y. and Hull, J.H. (eds)** 1999: *James Hutton – present and future* [Special Publication 150]. London: Geological Society. **Gould, S.J.** 1965: Is uniformitarianism necessary? *American Journal of Science* 263, 223–228. **Gould, S.J.** 1987: *Time's arrow, time's cycle: myth and metaphor in the discovery of geological time.* Cambridge, MA: Harvard University Press. **Hallam, A.** 1989: *Great geological controversies*, 2nd edn. Oxford: Oxford

University Press. **Hooykaas, R.** 1963: *Natural law and divine miracle: the principle of uniformity in geology, biology and theology.* Leiden: E.J. Brill. **Hutton, J.** 1788: Theory of the earth: or an investigation of the laws observable in the composition, dissolution and restoration of land upon the globe. *Transactions of the Royal Society of Edinburgh* 1, 209–304. **Lyell, C.** 1830–1833: *Principles of geology, being an attempt to explain the former changes of the Earth's surface by reference to causes now in operation.* London: John Murray. **Rudwick, M.J.S.** 1972: *The meaning of fossils: episodes from the history of palaeontology.* London: Macdonald. **Shea, J.** 1982: Twelve fallacies of uniformitarianism. *Geology* 10, 455–460.

unloading The removal of mass from an area of the Earth's surface over any timescale. At the shortest timescale, a landslip may induce unloading of the ground surface where the sediment or rock has been removed. Unloading can lead to expansion of a rock, to rock fracture and to the weakening of rock joints (DILATION), contributing sometimes to the process of EXFOLIATION. During the QUATERNARY, the melting and consequent unloading of ICE SHEETS resulted in GLACIO-ISOSTATIC REBOUND. Over millions of years, long-term tectonic and EPEIROGENIC uplift may also cause unloading.   *AGD*

[See also HYDROISOSTASY, ISOSTASY]

Fairbridge, R.W. 1961: Eustatic changes in sea level. *Physics and Chemistry of the Earth* 4, 99–185. **Lambeck, K.** 1995: Late Devensian and Holocene shorelines of the British Isles and North Sea from models of glacio-hydro-isostatic rebound. *Journal of the Geological Society* 152, 437–448.

uplift Upward movement of the Earth's surface, driven by compressional stresses during OROGENY (see, for example, COLLISION ZONE); by EPEIROGENY, for example above an INTRUSION; or by ISOSTASY.   *GO*

[See also HIMALAYAN UPLIFT, PLATE TECTONICS, SUBSIDENCE]

Burbank, D.W. 1992: Causes of recent Himalayan uplift deduced from deposited patterns in the Ganges basin. *Nature* 357, 680–683. **Smith, B.J., Whalley, W.B. and Warke, P.A. (eds)** 1999: *Uplift, erosion and stability: perspectives on long-term landscape development* [Special Publication 162]. London: Geological Society.

upwelling The vertical movement of water from the deep ocean to the sea surface. There are three types: (1) *coastal upwelling* occurs as surface waters move away from coasts in response to equatorward winds along coasts oriented north–south (e.g. on the eastern side of the Pacific and Atlantic Oceans off the coasts of California, Peru, Morocco and South Africa); (2) *equatorial upwelling* results from easterly winds blowing along the Equator; (3) *open-ocean upwelling* occurs under atmospheric cyclones. In all cases upwelling produces relatively cold and nutrient-rich surface waters, many of which support important fisheries, such as the Peruvian anchovy fishery, its associated sea-bird populations and guano deposits. Breakdown of the Peruvian upwelling occurs in EL NIÑO years.

JAM/AHP

[See also OCEAN CURRENTS, EL NIÑO–SOUTHERN OSCILLATION]

Bakun, A. 1990: Global climate change and the intensification of coastal ocean upwelling. *Science* 247, 198–201. **Berger, W.H., Diester-Haas, L. and Killingley, J.S.** 1978: Upwelling off north-west Africa: the Holocene decrease as seen in carbon isotopes and sedimentological indicators. *Oceanologia Acta* 1, 3–7. **Hsieh, W.W. and Boer, G.J.** 1992: Global climate change and ocean upwelling. *Fisheries Oceanography* 1, 333–338. **Murphy, R.C.** 1981: The guano and anchoveta fishery. In Glantz, M.H. and Thompson, J.D. (eds), *Resource management and environmental uncertainty*. New York: Wiley, 81–106. **Richards, F.A. (ed.)** 1981: *Coastal upwelling*. Washington, DC: American Geophysical Union. **Skaggs, J.M.** 1994: *The great guano rush*. New York: St Martin's Griffin.

upwelling radiolarian index (URI)　The number of species of Radiolaria in a sample of marine sediment comprising the UPWELLING assemblage expressed as a percentage of all the Radiolaria species present. In core-top sediments, URI values are positively correlated with THERMOCLINE depth and can be used as a proxy indicator in PALAEOCEANOGRAPHY and PALAEOCLIMATOLOGY.　*JAM*

Haslett, S.K. 1995: Mapping Holocene upwelling in the eastern equatorial Pacific using Radiolaria. *The Holocene* 5, 470–478.

uranium-series dating　A range of DATING techniques based on the radioactive-decay series of uranium-238 (^{238}U)and uranium-235 (^{235}U). Ages are based on the extent to which ISOTOPES in the decay chains have reached isotopic equilibrium after initial disturbance, such as chemical fractionation. With increasing time since deposition, the activity ratio between daughter and parent nuclides tends to unity. The technique has been most widely applied to the dating of CARBONATES precipitated from waters that have negligible protactinium-231 (^{231}Pa) and thorium-230 (^{230}Th) (insoluble daughter nuclides of ^{235}U and ^{238}U respectively) and significant abundance of uranium isotopes (soluble parent nuclides) (see SPELEOTHEMS, CORAL TERRACES, MARINE SEDIMENT CORES) Approximate dating limits vary according to the parent/DAUGHTER ISOTOPE pairs and their respective HALF LIVES: for ^{230}Th:^{234}U the half life is 500 ka; for ^{231}Pa:^{235}U the half life is 250 ka). An age estimate with PRECISION less than 1 ka (two standard deviations) can be obtained for corals formed during the last INTERGLACIAL using URANIUM–THORIUM DATING and THERMAL-IONISATION MASS SPECTROMETRY. Uranium-series dating has also been applied to bones, molluscs and peat, but attempts are complicated by open-system behaviour or detrital contamination.　*DAR*

Bard, E., Arnold, M., Hamelin, B. *et al.* 1998: Radiocarbon calibration by means of mass spectrometric ^{230}Th/^{234}U and ^{14}C ages of corals: an updated database including samples from Barbados, Mururoa and Tahiti. *Radiocarbon* 40, 215–220. **Edwards, R.L., Chen, J.H., Ku, T-L. and Wasserburg, G.J.** 1987: Precise timing of the last interglacial period from mass-spectrometric determination of ^{230}Th in corals. *Science* 236, 1547–1553. **Ivanovich, M. and Harmon, R.S.** 1992: *Uranium-series disequilibrium: applications to earth, marine and environmental sciences*, 2nd edn. Oxford: Oxford University Press. **Schwarcz, H.P.** 1997: Uranium series dating. In Taylor, R.E. and Aitken, M.J. (eds), *Chronometric dating in archaeology*. New York: Plenum, 159–182. **Smart, P.L.** 1991: Uranium series dating. In Smart, P.L. and Frances, P.D. (eds), *Quaternary dating methods: a user's guide* [*QRA Technical Guide* 4]. London: Quaternary Research Association, 45–83.

uranium–thorium dating　Used, principally, to date CORALS, SPELEOTHEMS, TUFA and MARLS, this technique is based on the decay of uranium-234 (^{234}U) to thorium-230 (^{230}Th). The approximate age limit is 350 ka using alpha-

spectrometric techniques but can be improved to 500 ka using THERMAL-IONISATION MASS SPECTROMETRY (TIMS). It is the most commonly applied type of URANIUM-SERIES DATING.　*DAR*

Bard, E., Hamelin, B., Fairbanks, R.G. and Zindler, A. 1990: Calibration of the ^{14}C timescale over the past 30,000 years using mass spectrometric U–Th ages from Barbados corals. *Nature* 245, 405–410.

urban and rural planning　The planning of human settlement has distant origins, but the long historic record is punctuated with more spectacular expressions such as the Roman grid-iron towns or the aesthetic forms of Baroque European cities. *City planning* was the clearest form of planning because cities were designed to fulfil purposes such as means of control, expressions of power, forms of defence or reflections of some ideology. *Rural planning* was less obvious historically and the formal arrival of *town and country planning* and, more recently, *city and regional planning* serve to blur a division that in many societies has become less meaningful. It has been argued that town and country planning as a task of government developed from public health and housing policies. The anarchic movement has been seen as the source of many visions of planning and twentieth-century urban planning as a reaction against the evils of the nineteenth-century city.

Some forms of planning are visionary. *New towns*, *garden cities*, the city beautiful, *green belts* and NATIONAL PARKS all fall within that definition. Others are essential but more mundane, such as development control, traffic management and aspects of CONSERVATION and listed buildings. Planning affects both LANDUSE and land values and these values add a political dimension. Planning must cater for the needs of people and the social aspects have become significant. Again, planning involves how to conserve and preserve NATURAL and SEMI-NATURAL environments, so issues such as '*greenfield*' and '*brownfield*' *sites* are now at the forefront of public debate. Many global problems exemplify the need to plan, but different societies have varying abilities to meet that need.　*DTH*

[See also URBAN ENVIRONMENTAL CHANGE]

Cullingworth, J.B. and Nadim, V. 1994: *Town and country planning in Britain*. London: Routledge. **Hall, P.** 1996: *Cities of tomorrow: an intellectual history of urban planning and design in the twentieth century*. Oxford: Blackwell. **Morgan, P. and Nott, S.** 1995: *Development control, law, policy and practice*. London: Butterworth.

urban climate　The distinctive LOCAL CLIMATE of large urban areas, which may be viewed as the cumulative effects of numerous anthropogenic modifications of MICROCLIMATE. Urban climates exhibit distinct contrasts with the surrounding rural areas. Characteristics of URBANISATION leading to urban climates include the properties of urban surfaces and materials (such as albedo, heat capacity, permeability and geometry), anthropogenic heat production and *air quality*. These affect energy, moisture and wind movement patterns within the urban BOUNDARY LAYER, especially in the *urban canopy layer* (below roof level). Although there are differences depending on such factors as the extent of the built-up area, relief, LANDUSE, season and location, urban climates tend to exhibit higher air temperatures, greater cloudiness and precipitation, more thunderstorms, less snowfall, lower relative humidity, lower wind speeds, greater frequencies of fog and higher levels of gaseous and particulate pollution, in comparison with adjacent rural environments.

JAM

[See also HEAT ISLAND, PHOTOCHEMICAL SMOG, SMOG]

Chandler, T.J. 1965: *The climate of London.* London: Hutchinson. **Goldreich, Y.** 1984: Urban topoclimatology. *Progress in Physical Geography* 8, 336–364. **Landsberg, H.** 1981: *The urban climate.* New York: Academic Press. **Lowry, W.P.** 1998: Urban effects on precipitation amount. *Progress in Physical Geography* 22, 477–520. **Oke, T.R. (ed.)** 1986: *Urban climatology and its applications with special regard to tropical areas* [*Publication* 652]. Geneva: World Meteorological Organisation.

urban environmental change There are many forms of environmental change, but the clearest common strand relates to the modification of the built form of the city environment. Many changes arise from the ageing process as urban fabric becomes outdated; others arise from functional adaptations as the activity for which a structure was created becomes redundant; and yet others stem from the pressures of growth and congestion in expanding cities. The abandoned warehouses of old docklands and the decline of inner-city MANUFACTURING INDUSTRY are examples of functional changes, whereas inadequate transport systems exemplify the impacts of growth and congestion. The expansion of urban areas has also led to distinctive natural environmental problems associated, for example, with URBAN CLIMATE, AIR POLLUTION, FLOODPLAIN MANAGEMENT and SUBSIDENCE.

Planners respond to urban environmental deterioration with a variety of measures. *Urban renewal* involves the clearance of old, substandard buildings and their replacement. In the United States, the term 'federal bulldozer' was used to characterise the main period of slum clearance. Urban improvement focuses less on clearance and more on refurbishment and the upgrading of older structures and has been strongly favoured wherever possible. Most urban environmental change, however, has occurred at the edges of cities with *suburban growth*. This in turn has created urban/rural LANDUSE conflict, greater planning controls and the preferences for re-using older urban places, the '*brownfield sites*', rather than further impingement upon rural areas (the '*greenfield sites*'). *Development controls*, together with specific instruments such as '*green belts*', have been the main arms of planners. *DTH*

[See also URBAN AND RURAL PLANNING]

Douglas, I. 1983: *The urban environment.* London: Arnold. **Douglas, I.** 1999: Physical problems of the urban environment. In Pacione, M. (ed.), *Applied geography: principles and practice.* London: Routledge, 124–134. **Herbert, D.T. and Thomas, C.J.** 1997: *Cities in space: city as place.* London: Fulton. **Hall, P.** 1996: *Cities of tomorrow: an intellectual history of planning and design.* Oxford: Blackwell.

urban revolution A theory of the origin of cities: namely that they emerged from a process of agricultural change. In a more generic form it suggests a more general process of relatively rapid urban development. *DTH*

[See also URBANISATION]

Carter, H. 1977: Urban origins: a review. *Progress in Human Geography* 1, 12–32.

urbanisation The shift towards a higher proportion of a society's population being classed as urban. England and Wales were the early examples of 'modern' urbanisation, linked with INDUSTRIALISATION and DEMOGRAPHIC CHANGE. Whereas 26% of the population could be classed as urban in AD 1801, this rose to 45% by 1851 and 75% by 1911. This process of urbanisation is sometimes modelled as a logistic curve with a slow start, sharp rise and levelling off. The curve can be fitted to other western societies to varying degrees. All started on the curve later than England and Wales, some markedly so, and some have not reached the levelling-out stage. Once a levelling-out stage is reached, the curve can decline and a society may appear to become less urbanised. This is more apparent than real as COUNTERURBANISATION moves clusters of urban dwellers away from cities into smaller towns and rural areas. As information technology networks become more efficient, this dispersal may accentuate. The term *urban growth* is more commonly applied to the physical spread of cities.

The experience of urbanisation in less developed countries has been different. Rural populations have continued to grow, so the rate of urbanisation is lower but urban growth may be greater. As *economic development* has not accompanied urbanisation, the term '*pseudo-urbanisation*' has been applied, though this ignores the societal and temporal differences that tend to undermine the value of such comparisons with Western experience. Less developed countries remain the least urbanised parts of the world, but they are also those where the largest cities are emerging. *DTH*

Drakakis-Smith, D. (ed.) 1986: *Urbanisation in the developing world.* London: Croom Helm. **Gilbert, A.** 1996: Third world urbanization. In Douglas, I., Huggett, R. and Robinson, M. (eds), *Companion encyclopedia of geography.* London: Routledge, 391–407. **Herbert, D.T. and Thomas, C.J.** 1997: *Cities in space: city as place.* London: Fulton.

urbanisation impacts on hydrology Effects on water movement through DRAINAGE BASINS produced by the growth of urban areas within catchments. URBANISATION of drainage basins results in: replacement of rural surfaces with less pervious and more hydraulically smooth ones; the introduction of piped water supply systems, storm and wastewater drainage networks and other hydraulic conduits; an increase in waste and pollution generation; a possible expansion of industrial operations; and in some cases a modest increase in precipitation (the so-called *urban rainfall island*). Hydrological effects depend on the character of the urban area LANDUSE patchwork and its size and location with respect to the drainage network and hydrologically sensitive areas. Impacts include: increases in runoff coefficient and river flood magnitude (up to eightfold increases have been reported), frequency (threefold increases have been reported) and flashiness, which can increase channel erosion and CHANNEL CAPACITY downstream of urban areas; decreases in INFILTRATION, GROUNDWATER recharge and river low flows (except in arid areas, where the reverse can occur); deterioration in water and sediment quality, with rises in concentrations of CONTAMINANTS associated with industrial effluents, sewerage additions and transport networks (e.g. ammonia, LEAD and CADMIUM); and a deterioration in freshwater habitat quality and invertebrate and fish populations. Groundwater tables are now rising under many Western cities, associated with a recent decline in water-abstracting industries.   *DML*

[See also LANDUSE IMPACTS ON HYDROLOGY]

Hollis, G.E. 1975: The effect of urbanisation on floods of different recurrence intervals. *Water Resources Research* 11, 431–434.
Niemczynowicz, J. 1999: Urban hydrology and water management – present and future challenges. *Urban Water* 1, 1–14.
Roberts, C.R. 1989: Flood frequency and urban-induced channel change: some British examples. In Beven, K. and Carling, P.A. (eds), *Floods: hydrological, sedimentological and geomorphological implications*. Chichester: Wiley, 57–82.

Urstromtäler Wide, shallow valleys (up to several hundreds of kilometres in length and 100 m wide) in the North German Lowland. Also termed ice-marginal valleys, they were excavated largely by MELTWATER flowing along the former margins of the FENNOSCANDIAN ICE SHEET and draining towards the west. The present-day courses of the rivers Elbe, Havel, Oder, Spree, Vistula and Weser follow, in part, the five principal Urstromtäler, which are termed *Pradoliny* in Poland.   *JAM*

[See also TUNNEL VALLEY]

Ehlers, J. 1996: *Quaternary and glacial geology*. Chichester: Wiley. **Embleton, C. and King, C.A.M.** 1975: *Glacial geomorphology*, 2nd edn. London: Arnold.

V

V-shaped valley A valley, the cross-profile of which has the shape of a 'V'. It is generally considered to be typical of a landscape characterised by active downcutting by a river combined with the effects of active valley-side slope processes.   *JAM/RPDW*

[See also GORGE, U-SHAPED VALLEY]

vadose zone The unsaturated GROUNDWATER zone above the permanent WATER TABLE; also known as the *aeration zone*. *Vadose water* occurs in the vadose zone, where it circulates more-or-less freely under the influence of gravity, depending on the *porosity* (the ratio of the voids to the total volume of material) and *permeability* (the degree of continuity between voids) of the rocks, and on the extent to which the rocks are PERVIOUS (with drainage along fissures in otherwise IMPERMEABLE rocks).   *JAM*

[See also CAVES, KARST, PERMEABLE, PHREATIC ZONE, POROUS]

Parlange, M.B. and Hopmans, J.W. 1999: *Vadose zone hydrology: cutting across disciplines.* New York: Oxford University Press.

validation An assessment of the reliability of a MODEL, usually carried out by comparing the model predictions with an independent dataset representing the real world. This may involve various methods, such as PRINCIPAL COMPONENTS ANALYSIS, Andrews' curves, distance measures (e.g. root mean square error, squared Euclidian distance, Mahalanobis distance), CORRELATION ANALYSIS, coefficients of correspondence (κ, τ), or statistical tests (Chi-squared test, Wilcoxon's signed ranks test, t-test, ANALYSIS OF VARIANCE). In REMOTE SENSING, validation is mostly important for: (1) the comparison of BACKSCATTERING and RADIATIVE TRANSFER MODELS with observed values; (2) the assessment of PRECISION and ACCURACY of estimated biophysical parameters (e.g. LEAF-AREA INDEX, BIOMASS); and (3) the determination of the accuracy of IMAGE CLASSIFICATION. To perform such a validation, GROUND MEASUREMENTS are needed. These are independent data, which have not been used for parameter estimation, spectral signature identification or any other previous processing steps. Without a validation, the degree of correspondence of a model to reality, and hence its usefulness, remain unknown. It has been suggested that *evaluation* would be a more appropriate term.   *HB*

[See also CALIBRATION, CONFIRMATION, VERIFICATION]

Brown, T.N. and Kulasiri, D. 1996: Validating models of complex, stochastic, biological systems. *Ecological Modelling* 86, 129–134. **Carlstrom, A. and Ulander, L.M.H.** 1995: Validation of backscatter models for level and deformed sea-ice in ERS-1 SAR images. *International Journal of Remote Sensing* 16, 3245–3266.

valley bulging Anticlinal deformation of soft strata underlying a valley floor. Valley bulging is attributed to pressures generated by the weight of competent strata on the valley sides and is often associated with CAMBERING and GULL development.   *PW*

[See also MASS MOVEMENT PROCESSES]

Hutchinson, J.N. 1992: Engineering in relict periglacial and extraglacial areas in Britain. *Quaternary Proceedings* 2, 49–65.

valley fill deposits Accumulations of SEDIMENT (alluvial and colluvial) in valley floor positions (see Figure), representing a phase or phases of accelerated MASS WASTING on valley slopes and/or stream AGGRADATION commonly in response to CLIMATIC CHANGE, accelerated SOIL EROSION or SEA-LEVEL RISE in coastal locations.

RAS

[See also ALLUVIAL FILL, ALLUVIUM, ENVIRONMENTAL ARCHAEOLOGY, COLLUVIUM, HUMAN IMPACT ON LANDFORMS AND GEOMORPHIC PROCESSES, MEDITERRANEAN ENVIRONMENTAL CHANGE AND HUMAN IMPACT, SOIL EROSION HISTORY]

Knox, J.C. 1987: Historical valley floor sedimentation in the upper Mississippi valley. *Annals of the Association of American Geographers* 77, 224–244. **Meulen, S. van der** 1995: Younger Dryas deposits of the Tjonger valley fill in the NE Netherlands. *Geologie en Mijnbouw* 74, 257–260. **Vita-Finzi, C.** 1969: *The Mediterranean valleys.* Cambridge: Cambridge University Press. **Willis, B.J.** 1997: Architecture of fluvial-dominated valley-fill deposits in the Cretaceous Fall River Formation. *Sedimentology* 44, 735–757.

Van't Hoff's rule The rate of a chemical reaction will approximately double for a 10°C rise in temperature.

JAM

variable A characteristic or quantity that may take on a range of values. Common *environmental variables* include, for example, AIR TEMPERATURE, soil moisture and ACIDITY, and leaf area. These examples are *continuous variables*, which exhibit no sharp breaks between values. *Discontinuous* or *discrete variables*, such as the number of cases in a disease outbreak, or landuse categories, can only be measured in integers (whole numbers). In REGRESSION ANALYSIS, the *dependent* or *response variable* (y) is predicted from the *independent* or *explanatory variable* (x).   *JAM*

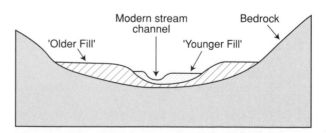

valley fill deposits *Schematic representation of the relationship between the Older and Younger Fills in Mediterranean valleys (based on Vita-Finzi, 1969)*

variance A measure of the degree of variation or dispersion around the mean or average: it is the square of the standard deviation. *JAM*

[See also ANALYSIS OF VARIANCE]

variogram The variogram $2\gamma(h)$ is the VARIANCE of the increments of spatially separated measurement points given a random variable $Z(X)$ measured at locations x_1, $x_2, \ldots, x_n$. The empirical variogram can be estimated as the mean squared difference of the measurements at all sampling locations x_i and $x_i + h$ with distance h. However, this estimator is sensitive against outliers and Cressie (1993) and Haslett (1997) suggest more robust estimators. The variogram is not dependent on the location of the points but on the distance between the points. Plotting the empirical variogram against distance h shows the spatial dependence of the measurements. A *theoretical variogram model* can be fitted to the empirical variogram. Theoretical variogram models need to fulfil certain mathematical restrictions. Suitable variogram models include the linear, spherical, exponential, rational quadratic, wave and power variogram. A variogram model is characterised by its *range, nugget* and *sill*. The range is the distance h for which $2\gamma(h)$ reaches an upper limit (the sill), which is the overall variance. A nugget effect is observed if $2\gamma(0) \neq 0$. It is caused by microscale variation and measurement error. Variograms are used in the spatial INTERPOLATION methods known as KRIGING. The *semivariogram* $\gamma(h)$ is defined as the variogram divided by two. *HB*

Cressie, N.A.C. 1993: *Statistics for spatial data.* New York: Wiley. **Haslett, J.** 1997: On the sample variogram and the sample autocovariance for non stationary time series. *The Statistician* 46, 475–486.

varve Derived from the Swedish for a lap (*varv*), varves are sediment laminations or beds deposited on an annual cycle, forming couplets. Varves arise from seasonal variations in sediment supply producing alternating coarse (summer) and fine (winter) layers, and are of value in dating through the construction of VARVE CHRONOLOGIES. *MHD*

[See also RHYTHMITE]

Lotter, A.F. and Lemcke, G. 1999: Methods for preparing and counting biochemical varves. *Boreas* 28, 243–252. **Saarnisto, M.** 1986: Annually laminated sediments. In Berglund, B.E. (ed.), *Handbook of Holocene palaeoecology and palaeohydrology.* Chichester: Wiley, 343–370. **Schlüchter, Ch. (ed.)** 1979: *Moraines and varves.* Rotterdam: Balkema.

varve chronology VARVES are annually deposited couplets of sedimentary laminations or beds: hence the duration of the period of sediment deposition can be determined by counting these annual couplets. Correlation between varve sequences, based on cross-matching patterns in the relative thicknesses of individual varves within sequences, enables the construction of a varve MASTER CHRONOLOGY for a region (analogous to the construction of master chronologies in DENDROCHRONOLOGY).

De Geer's pioneering work on glaciolacustrine varves arising from seasonal variations in glacial MELTWATER and sediment supply into *proglacial* LAKES enabled the construction of a varve chronology for Sweden. De Geer reconstructed the regional pattern of DEGLACIATION by correlating varve sequences from former and existing proglacial lakes across Sweden, and by assuming that the basal varve represented the onset of deglaciation at a site. Subsequently, De Geer's Swedish varve chronology has been revised and the uppermost varve in the chronology linked to the present day, providing CALENDAR AGES for individual varve couplets extending back to about 11.5 ka BP (^{14}C years). Varve chronologies have formed the basis for reconstructing the spatial and temporal pattern of deglaciation in the Baltic region and this application has been used with varying success in other parts of the world, most notably in reconstructing the decay of the LAURENTIAN ICE SHEET in North America.

Varve counting estimates of the duration of the BØLLING and ALLERØD INTERSTADIALS and the YOUNGER DRYAS STADIAL have helped to identify patterns in the timing of LATE GLACIAL ENVIRONMENTAL CHANGE in Europe. RADIOCARBON DATING of organic material from varve chronologies with an estimated CALENDAR AGE enables the CALIBRATION OF RADIOCARBON DATES. This application, analogous to calibration in dendrochronology, identifies discrepancies in radiocarbon dates, which may be attributed to past variations in ^{14}C activity.

In addition to glaciolacustrine varves, annual couplets resulting from seasonal variations in organic or chemical precipitation and from biological productivity can be used to estimate the duration of other QUATERNARY time periods or events. Such couplets are more normally referred to as *annually laminated sediments.* *MHD*

[See also RHYTHMITES]

Anderson, R.Y. 1996: Seasonal sedimentation: framework for reconstructing climate and environmental change. In Kemp, A.E.S. (ed.), *Palaeoecology and palaeoceanography from laminated sediments* [Geological Society Special Publication 116]. Bath: The Geological Society, 1–15. **Cato, I.** 1985: The definitive connection of the Swedish geochronological time scale with the present, and the new date of the zero year in Döviken, northern Sweden. *Boreas* 14, 117–122. **Geer, G. de** 1912: A geochronology of the last 12,000 years. *XIth International Geological Congress, Stockholm.* 1, 241–253. **Gulliksen, S., Birks, H.H., Possnert, G. and Mangerud, J.** 1998: The calendar age of the Younger-Dryas–Holocene transition at Krakenes, western Norway. *The Holocene* 8, 249–260. **Leeman, A. and Niessen, F.** 1994: Varve formation and the climatic record in an Alpine proglacial lake: calibrating annually-laminated sediments against hydrological and meteorological data. *The Holocene* 4, 1–8. **Lundqvist, J.** 1986: Late Weischselian glaciation and deglaciation in Scandinavia. *Quaternary Science Reviews* 5, 269–292. **O'Sullivan, P.E.** 1983: Annually laminated sediments and the study of Quaternary environmental changes: a review. *Quaternary Science Reviews* 1, 245–313. **Stromberg, B.** 1994: Younger Dryas deglaciation at Mt Billinen and clay varve dating at the Younger Dryas/Preboreal transition. *Boreas* 23, 177–193. **Walker, M.J.C., Coope, G.R. and Lowe, J.J.** 1993: The Devensian (Weichselian) Lateglacial palaeoenvironmental record from Gransmoor, East Yorkshire, England. *Quaternary Science Reviews* 12, 659–680. **Wohlfarth, B., Björck, S., Possnert, G. et al.** 1993: AMS dating Swedish varved clays of the last glacial/interglacial transition and the potential/difficulties of calibrating Late Weichselian 'absolute' chronologies. *Boreas* 22, 113–128.

vector An organism that carries disease from host to host, such as the *Anopheles* mosquito (see MALARIA) and rats. *JAM*

[See also WATER-RELATED VECTOR]

vector data model   The spatial representation of data structures in a GEOGRAPHICAL INFORMATION SYSTEM by assigning continuous coordinates, lengths and dimensions to objects, as opposed to the RASTER data model, which uses discrete PIXEL values. A *point entity* is any object positioned at a coordinate pair *x* and *y* (e.g. a symbol or a string of characters to be printed on a map). A *line entity* is defined as all linear features made up of straight-line segments between two or more points. A line has a certain length and may have additional properties. A curve can be approximated by using many points with short line segments between them. An *area entity* (or polygon or region) is built of several lines and is characterised by a unique perimeter, area and shape. Each polygon has a unique identifier that points to an ATTRIBUTE table containing information about the area entity. The entities are interrelated with neighbouring entities and build a complex topography.   *HB*

[See also OBJECT ORIENTED, POLYGON ANALYSIS]

Burrough, P.A. 1986: *Principles of geographical information systems for land resources assessment* [*Monographs on Soil and Resources Survey* 12]. Oxford: Oxford University Press.

vegetation   The plant individuals considered *en masse* at a specific locality. Many studies of the impacts of climatic change are based on the analysis of *vegetation change*.   *JLI*

[See also FAUNATION]

Eyre, S.R. 1963. *Vegetation and soils: a world picture*. London: Arnold. **Woodward, F.I.** 1987: *Climate and plant distribution*. Cambridge: Cambridge University Press.

vegetation canopy   The layers of a continuous area of vegetation, for instance in a forest consisting of trunks, branches and foliage. The three-dimensional structure of a vegetation canopy affects the BACKSCATTERING of RADIATION in the ELECTROMAGNETIC SPECTRUM, enabling REMOTE SENSING to retrieve biophysical parameters from the canopy. Herbaceous (grasses and herbs) and woody vegetation (shrubs, coniferous and deciduous trees) may be distinguished. Whereas in dense vegetation short optical waves interact mainly with the foliage, microwaves are able to penetrate the vegetation canopy to an extent related to the *wavelength* and interact directly with branches and trunks. Different SCATTERING mechanisms can be separated and used for IMAGE CLASSIFICATION of structural vegetation types.   *HB*

[See also BIOMASS, OPTICAL REMOTE SENSING INSTRUMENTS, RADAR REMOTE SENSING INSTRUMENTS]

Dobson, M.C., Ulaby, F.T. and Pierce, L.E. 1995: Land-cover classification and estimation of terrain attributes using synthetic-aperture radar. *Remote Sensing of Environment* 51, 199–214. **Freeman, A. and Durden, S.L.** 1998: A three-component scattering model for polarimetric SAR data. *IEEE Transactions on Geoscience and Remote Sensing* 36, 963–973. **Imhoff, M.L.** 1995: A theoretical analysis of the effect of forest structure on synthetic aperture radar backscatter and the remote sensing of biomass. *IEEE Transactions on Geoscience and Remote Sensing* 33, 341–352.

vegetation formation   A subdivision of a VEGETATION FORMATION-TYPE distinguished by its geographical distribution. Thus, the Amazonian Tropical Rain Forest is a formation within the Tropical Rain Forest formation-type.

Formations are continental-scale units with similar physiognomy or dominant LIFE FORM.   *JLI*

[See also BIOME]

Collinson, A.S. 1968: *Introduction to world vegetation*. London: George Allen and Unwin. **Holdridge, L.R.** 1947: Determination of world plant formations from simple climatic data. *Science* 105, 367–368.

vegetation formation-type   A classification unit used to describe physiognomically distinct forms of vegetation at the global scale (e.g. the Tropical Rain Forest formation-type). Each formation-type contains several VEGETATION FORMATIONS.   *JLI*

[See also BIOME, LIFE ZONE]

Holdridge, L.R. 1967 *Life zone ecology*. San José, Costa Rica: Tropical Science Center.

vegetation history   The reconstruction of changes in VEGETATION through time at different temporal and spatial scales. The most commonly used research technique is POLLEN ANALYSIS supplemented by PLANT MACROFOSSIL ANALYSIS and, for the recent past, HISTORICAL EVIDENCE. MEGAFOSSILS also provide evidence that individual tree species have, in the past, thrived beyond their present range. Vegetation reflects a range of factors including topography, bedrock and soil, but is basically influenced by CLIMATE. Past vegetation, therefore, is often used as a PROXY for past climate. Much knowledge of QUATERNARY GLACIAL, INTERGLACIAL and INTERSTADIAL conditions comes from vegetation history studies, some utilising very long sedimentary sequences such as La Grande Pile in northeastern France and Funza in the Colombian Andes. The vegetation of the different interglacials is determined by the rate of MIGRATION of plant species from their GLACIAL REFUGIA and the COMPETITION between species, as much as by the rapidity of the warming and the changing land/sea configuration. The *succession* of vegetation units through an INTERGLACIAL CYCLE follows a general pattern and one aspect of vegetation history involves comparing and contrasting the species composition, order of species arrival and possible extinction of species from one interglacial to the next.

For the past 5000 years, however, HUMAN IMPACT ON TERRESTRIAL VEGETATION has led to an increasing obscuration of the underlying climatic signal. Vegetation history for this period thus includes tracing the spatial cover and duration of different ANTHROPOGENIC communities. The temporal detail with which vegetation history can be followed depends on the sampling interval of the SEDIMENT containing the record; the spatial detail depends on the density of sampling points; and the precision of vegetation community reconstruction depends on the identification level of the FOSSIL or SUBFOSSIL. A lower taxonomic level can be achieved with plant macro- and megafossils than with pollen. MODERN ANALOGUES are useful in reconstructing past vegetation, although not all past communities have modern counterparts. Numerical methods are being increasingly used to quantify vegetation changes through time, and mapped reconstructions of the spatial variations in vegetation at the regional, continental and global scales at specific points in time during the Holocene are being made, utilising fossil records archived in DATABASES.   *SPH*

[See also CLIMATIC RECONSTRUCTION, FOREST CLEARANCE, HUMAN IMPACT ON VEGETATION HISTORY, PALAEOCLIMATOLOGY]

Agnoletti, M. and Anderson, S. 2000: *Methods and approaches in forest history.* Wallingford: CABI Publishing. Berglund, B.E., Birks, H.J.B., Ralska-Jasiewiczova, M. and Wright, H.E. (eds) 1996: *Palaeoecological events during the last 15000 years: regional syntheses of palaeoecological studies of lakes and mires in Europe.* Chichester: Wiley. De Beaulie, J. and Reille, M. 1992: The last climatic cycle at La Grande Pile (Vosges, France): a new pollen profile. *Quaternary Science Reviews* 11, 431–438. Birks, H.H., Birks, H.J.B., Kaland, P.E. and Moe, D. (eds) 1988: *The cultural landscape past, present and future.* Cambridge: Cambridge University Press. Godwin, H. 1975: *History of the British flora*, 2nd edn. Cambridge: Cambridge University Press. Hooghiemstra, H. 1984: *Vegetational and climatic history of the high plain of Bogotá, Colombia: a continuous record of the last 3.5 million years* [Dissertationes Botanical 79]. Valdus: J. Cramer. Huntley, B. and Webb III, T. 1988: *Vegetation history* [Handbook of Vegetation Science. Vol. 7]. Dordrecht: Kluwer. Prentice, I.C., Jolly, D. and BIOME 6000 participants 2000: Mid-Holocene and glacial maximum vegetation geography of the northern continents and Africa [Introduction to the BIOME 6000 Special Issue]. *Journal of Biogeography* 27, 506–520. Smith, A.G. and Cloutman, E.W. 1988: Reconstruction of Holocene vegetation history in three dimensions at Waun Fignen Felen, an upland site in South Wales. *Philisophical Transactions of the Royal Society of London* B322, 159–219. Tallis, J.H. 1991: *Plant community history.* London: Chapman and Hall. Walker, D. and West, R.G. 1970: *Studies in the vegetation history of the British Isles.* Cambridge: Cambridge University Press.

vegetation indices

Simple mathematical expressions (summing, differencing, ratioing) that combine two or more remotely sensed spectral measurements (see REMOTE SENSING) to provide non-destructive estimates of a number of properties of green vegetation. Vegetation indices may be computed from DIGITAL NUMBERS (DN), at sensor radiances, apparent *reflectances*, land-leaving RADIANCES or surface *reflectances* to provide dimensionless diagnostic information about the LEAF AREA INDEX (LAI), percentage cover, green BIOMASS and absorbed PHOTOSYNTHETICALLY ACTIVE RADIATION (APAR) of VEGETATION CANOPIES over space and time.

The most widely used vegetation indices are those that use remotely sensed data recorded in discrete red and near-infrared spectral wavebands. These attempt to exploit the contrast between the high photosynthetic pigment absorption at red *wavelengths* and the high reflectivity of plant material at near-infrared wavelengths. These vegetation indices fall principally into two categories, namely ratio-based and linear-based. *Ratio-based indices* are constructed by ratioing a combination of the red and near-infrared wavebands by another combination of the same wavebands. The most common index of this form is the *normalised difference vegetation index* (NDVI). *Linear-based indices* are orthogonal sets of *n* linear equations calculated using data from *n* spectral bands. Examples of linear-based indices include the *perpendicular vegetation index* (PVI) and *tasseled cap transformations*. Vegetation indices are popular as they are computationally simple. However, contributions to the remotely sensed signal from soil background, atmospheric attenuation and solar irradiance variability have proved problematic.

Modified vegetation indices have been developed to overcome these inherent problems. *Soil adjusted vegetation indices* (SAVI) are designed to diminish soil background effects where soil brightness contributes to the remotely-sensed signal, particularly from VEGETATION CANOPIES with low cover. Other indices are designed to reduce the relative effects of atmospheric perturbations. The *atmospherically resistant vegetation index* (ARVI), for example, incorporates a self-correction process for atmospheric influences at red *wavelengths* by using the difference in RADIANCE between the blue and red wavebands. Other indices combine the properties of the SAVI and ARVI indices (e.g. SARVI). The acquisition of information on vegetation properties at global scales has encouraged the development of vegetation indices such as the *global environmental monitoring index* (GEMI). Furthermore, a new breed of vegetation indices is available to utilise the information provided by sensors with narrow wavebands and wider spectral resolution. Since their advent, vegetation indices have evolved to become an important step in the retrieval of vegetation properties from remotely sensed data for use in the study of environmental change, particularly at regional to global scales.   *DSB*

Asrar, G., Fuchs, M., Kanesmau, E.T. and Hatfield, J.L. 1984: Estimating absorbed photosynthetic radiation and leaf area index from spectral reflectance in wheat. *Agronomy Journal* 76, 300–306. Boyd, D.S. and Ripple, W.J. 1997: Potential vegetation indices for determining global forest cover. *International Journal of Remote Sensing* 18, 1359–1410. Crist, E.P. and Cicone, R.C. 1984: A physically-based transformation of Thematic Mapper – the TM tasseled cap. *IEEE Transactions on Geoscience and Remote Sensing* 22, 256–263. Jackson, R.D. and Huete, A.R. 1991: Interpreting vegetation indices. *Preventative Veterinary Medicine* 11, 185–200. Malthus, T.M., Andrieu, B., Danson, F.M. et al. 1993: Candidate high spectral resolution infrared indices for crop cover. *Remote Sensing of Environment* 46, 204–212. Pinty, B. and Verstraete, M.M. 1992: GEMI: a non-linear index to monitor global vegetation from satellites. *Vegetatio* 101, 15–20.

ventifact

A wind-abraded CLAST with a fluted, faceted or grooved appearance. Ventifacts are common in arid and polar areas where the absence of vegetation exposes the clast to abrasion by dust, sand or snow containing these particles. The clast may have three abraded faces (the so-called *dreikanter*). Under certain circumstances, ventifacts can be used to infer local and regional present day wind or PALAEOWIND patterns.   *HMF*

Christiansen, H.H. and Svensson, H. 1998: Windpolished boulders as indicators of Late Weichselian wind regime in Denmark in relation to neighbouring areas. *Permafrost and Periglacial Processes* 9, 1–21. Fristrup, B. 1952/3: Wind erosion within the Arctic deserts. *Geografisk Tidsskrift* 52, 51–56.

verification

Establishment of the truth or verity of a proposition. This is a very difficult task as it is logically possible only in a CLOSED SYSTEM. *Proof* of the truth of a HYPOTHESIS or the reliability of the predictions of a MODEL is therefore rarely possible, especially in the investigation of natural systems, which are never closed. Hence, it is more appropriate to describe the apparent success of attempts to establish the truth of hypotheses and successful demonstrations of the reliability of models, as CONFIRMATION rather than verification. However, complete confirmation is logically precluded by the fallacy of affirming the consequent and practically precluded by incomplete access to natural phenomena. *JAM*

[See also FALSIFICATION, VALIDATION]

Oreskes, N., Shrader-Frechette, K. and Belitz, K. 1994: Verification, validation and confirmation of numerical models in the Earth sciences. *Science* 263, 641–646.

vermiculite

A 2:1 CLAY MINERAL with expanding characteristics when wet, having two sheets of silicon tetrahedra with between them a sheet of aluminium octahedra. Substitution of Mg for Al and Al for Si in the clay lattice gives the high negative charge associated with these minerals. Vermiculites are weathered from micas or illites. *EMB*

Mellor, A. 1986: Hydrobiotite formation in some Norwegian arctic-alpine soils developing in Neoglacial till. *Norsk Geologisk Tidsskrift* 66, 183–185.

Vertisols

Dark coloured, cracking and swelling soils with more than 30% clay usually, but not always, of the *smectite* group, weathered from base-rich PARENT MATERIALS. Vertisols commonly occur in subtropical regions where there is a marked alternation of wet and dry seasons as in parts of East Africa, India and Australia. *EMB*

[See also GILGAI, WORLD REFERENCE BASE FOR SOIL RESOURCES]

Ahmad, N. and Mermut, A. 1996: *Vertisols and technologies for their management* [*Developments in Soil Science* 24]. Amsterdam: Elsevier. **Blokhuis, W.A.** 1993: *Vertisols in the central clay plain of the Sudan.* Wageningen: Wageningen Agricultural University.

vesicle

A cavity caused by trapped gas. Spherical to elongate vesicles are common in LAVA, formed by the trapping of VOLATILES exsolved during the decompression of MAGMA. Once filled by mineral material they are known as *amygdales*. Vesicles also occur in other types of materials, especially as trapped air bubbles. *JBH*

vicariance

The separation of a population or species into disjunct groups resulting from the formation of a physical barrier. A vicariance event can lead to the evolution of new LINEAGES if the separation is permanent or occurs over a long period of time. *KJW*

vine harvests

Dates of European vine harvests were recorded from the late fifteenth century in regions of France, Germany, Switzerland and Austria. These data have been used to provide a detailed, annual record of the climate of western Europe during the LITTLE ICE AGE and later. They are especially valuable for the pre-instrumental period. Until the present century and the introduction of new methods and farming practices, the dates at which the harvests took place were largely dependent upon climatic conditions during the growing season (April to September) and were particularly sensitive to temperatures in the early part of the year. Higher temperatures led to earlier harvests, but dry conditions also tended to bring the dates forward. The series of harvest dates confirm the evidence of contemporary GLACIER VARIATIONS and of DENDROCHRONOLOGY, although the tree ring variations appear to be more dependent on late summer conditions. They also reveal the more detailed patterns of CLIMATIC VARIATIONS at the short temporal and the small spatial scales.

The AD 1690s stand out as particularly cold years, with late and poor harvests of both vines and wheat. *DAW*

[See also PHENOLOGY, WINE QUALITY]

Bray, J.R. 1982: Alpine glacier advance in relation to a proxy summer temperature index based mainly on wine harvest dates. *Boreas* 11, 1–10. **Le Roy Ladurie, E.** 1971: *Times of feast, times of famine: a history of climate since the year 1000.* London: George Allen and Unwin. **Pfister, C.** 1981: An analysis of the Little Ice Age climate in Switzerland and its consequences for agricultural production. In Wigley, T.M.L., Ingram, M.J. and Farmer, G. (eds), *Climate and history: studies in past climates and their impact on man.* Cambridge: Cambridge University Press, 214–248. **Lauscher, F.** 1983: Weinlese in Frankreich und Jahrestemperatur in Paris seit 1453. *Wetter und Leben* 35, 39–42.

vitrinite reflectance

A technique that uses the optical properties of organic debris in SEDIMENTARY DEPOSITS to infer their maximum burial temperature and hence maximum burial depth. Together with other techniques of *palaeothermometry* such as FISSION TRACK ANALYSIS, vitrinite reflectance can be used to infer the DENUDATION CHRONOLOGY and LANDSCAPE EVOLUTION of a region. The technique is also used to determine the rank of COAL and to assess the MATURITY of potential petroleum source rocks. *GO*

O'Sullivan, P.B. 1999: Thermochronology, denudation and variations in palaeosurface temperature: a case study from the North Slope foreland basin, Alaska. *Basin Research* 11, 191–204. **Pearson, M.J. and Russell, M.A.** 2000: Subsidence and erosion in the Pennine Carboniferous Basin, England: lithological and thermal constraints on maturity modelling. *Journal of the Geological Society, London* 157, 471–482.

volatiles

Substances that change readily into a vapour. In the context of VOLCANISM, volatiles are elements and compounds dissolved in MAGMA that are released as gases as magma decompresses on approaching the Earth's surface (see OUTGASSING). Their expansion contributes to the explosivity of some VOLCANIC ERUPTIONS. Common volatiles are water, CARBON DIOXIDE, hydrochloric acid, chlorine, fluorine and SULPHUR DIOXIDE. Their release to the ATMOSPHERE contributes to VOLCANIC IMPACTS ON CLIMATE. They may be trapped in IGNEOUS ROCKS, forming VESICLES. *Volatile organic compounds* (VOCs) are widely used as fuels and solvents and can be important POLLUTANTS. *JBH*

[See also COMET, DEGASSING]

Sparks, R.S.J., Bursik, M.I., Carey, S.N. *et al.* 1997: *Volcanic plumes.* Chichester: John Wiley.

volcanic aerosols

Microscopic solid and liquid particles that are ejected by volcanic eruptions into the TROPOSPHERE and STRATOSPHERE, where they are suspended for periods lasting from days to months, depending on their altitude and weather patterns. They reduce INSOLATION, add to the atmosphere's TURBIDITY, act as CONDENSATION NUCLEI, contribute to OZONE DEPLETION and modify SOLAR RADIATION fluxes through ABSORPTION and SCATTERING. In the troposphere aerosols are usually removed by rainfall within a few days or weeks. However, those entering the stratosphere (usually *sulphates*) may linger for months at heights of 20–30 km or more. They spread out as a DUST VEIL, travelling round the Earth in 10 days to a few weeks. This becomes increasingly uniform and may cover an entire hemisphere in about six months. Thereafter, the AEROSOLS may remain in the stratosphere for 1–7 years. Arrhenius was an early exponent of volcanic activity as a principal factor in climatic change. *JGT*

[See also JUNGE LAYER, VOLCANIC IMPACTS]

Arrhenius, G.O.S. 1896: On the influence of carbonic acid in the air upon the temperature of the ground. *Philosophical Magazine ser. 5* 4, 237–276. **Bradley, R.S. and Jones, P.D.** 1992: Records of explosive volcanic eruptions over the last 500 years. In Bradley, R.S. and Jones, P.D. (eds), *Climate since AD 1500*. London: Routledge, 606–622. **Rampino, M. and Self, S.** 1984: Sulfur-rich volcanic eruptions and stratospheric aerosols. *Nature* 310, 677–679.

volcanic ash   Unconsolidated PYROCLASTIC material finer in GRAIN-SIZE than 2 mm. The lithified equivalent is a TUFF. The term is commonly, but less correctly, applied to all pyroclastic deposits (see TEPHRA). *JBH*

[See also VOLCANIC BLOCK, VOLCANIC BOMB, LAPILLI]

volcanic block   A PYROCLASTIC fragment greater than 64 mm in GRAIN-SIZE, with an angular shape. *GO*

[See also VOLCANIC BOMB]

volcanic bomb   A PYROCLASTIC fragment greater than 64 mm in GRAIN-SIZE, with a rounded shape, having been molten and plastic when thrown from a VOLCANO. *GO*

[See also VOLCANIC BLOCK]

volcanic eruption   The effusive or explosive expulsion of MAGMA at a VOLCANO on the surface of the Earth (or another planet) as liquid (LAVA), solid (PYROCLASTIC) and gaseous (VOLATILES) materials. There is a range of styles of volcanic eruptions (see Table), which can be separated into eruptions from an approximately circular opening or *vent* (*central eruption*) and those from a linear opening (FISSURE ERUPTION). The explosiveness of an eruption is influenced by the gas content of the magma, the ease with which gas can be exsolved from the magma and contact between

volcanic eruption *Names and characteristics of the principal types of volcanic eruptions*

Named eruption type	Eruption characteristics and products
Bandaian	Collapse of volcanic flanks/edifice, leading to massive landslides or rock avalanches
Hawaiian	Quiet to moderate activity; caldera, fissure and crater eruptions; thin, mobile lava flows building lava domes; minor ash; slightly higher gas contents can lead to occasional lava fountains
Hydrovolcanic: Maar	Low cones with bowl-shaped craters and high width:rim-height ratio; explosive ejection of tephra and surrounding bedrock
Hydrovolcanic: Surtseyan	Violently explosive eruption caused by rapid cooling effect of water on near-surface magma (phreatomagmatic); ejection of glass tephra (often as large bombs)
Icelandic	Quiet fissure eruptions releasing large-volume, free-flowing and gas-poor basalt sheets with minor tephra
Katmaian	Large-volume ignimbrite production with welded tuffs
Lateral blast	Although not taking its name from Mount St Helens in 1980, this volcano provided the first evidence of this type of eruption; seismicity created flank instability and collapse with rapid depressurization; Plinian-type products ejected laterally at great speed (> 600 km h⁻¹); blast not counter to gravitational force
Maar	(See Hydrovolcanic: Maar)
Peléean	High-viscosity magma results in blocking of stratovolcano conduit by dome or plug, delaying explosivity; gas/lava escape on flanks or by uplift/destruction of plug; gas blasts drive blocks and ash down-slope as nuées ardentes and pyroclastic flows
Plinian	A more violent form of Vesuvian type; can be associated with caldera collapse; final major phase is uprush of gas and pyroclasts in vertical column, narrow at base but laterally expansive at altitude; wide dispersal of tephra and pumice; named after Pliny the Younger and observations of AD 79 eruption at Vesuvius
Strombolian	Summit crater edifices generating rhythmic to near-continuous explosions driven by gas-release; lava clasts ejected as bombs and scoria; thicker outpourings of relatively fluid lava; limited altitude steam clouds
Surtseyan	(See Hydrovolcanic: Surtseyan)
Ultra-plinian	More violent form of Plinian eruption (sometimes known as Krakatoan)
Vesuvian	After long quiet or dormant interval, extremely violent explosion of gas-charged magma from stratocone vent; emptying of conduit to considerable depth; glowing lava spray above vent, with ash-laden clouds repeatedly extending to altitude
Vulcanian	Central vent stratocones ejecting short, thick and viscous lava flows; lava lake in vent develops gas-trapping crusts; eruptions increase in violence ejecting large lava-crust bombs, pumice and tephra; ash-laden 'cauliflower' clouds deposit tephra over the volcano's flanks, interspersed with longer periods of quiescence until lava crust is broken, clearing vent; lava flows from summit after main explosive phase

magma and water (see HYDROVOLCANIC ERUPTION), and may be measured using the VOLCANIC EXPLOSIVITY INDEX. The sizes of eruptions can be compared by estimating the volume of eruption products. Mount St Helens in AD 1980 erupted less than 1 km³ and Pinatubo in the Philippines erupted about 10 km³ in AD 1991. Records of past eruptions are preserved in IGNEOUS ROCKS, especially spreads of TEPHRA. The largest known eruption of the last 2000 years was Taupo, New Zealand, in AD 186 (50 km³) and the largest eruption in the Quaternary was possibly that of Toba, Sumatra, approximately 74 000 years ago, which deposited an IGNIMBRITE of volume greater than 2000 km³, forming a CALDERA. Volcanic eruptions can have major environmental effects on climate, landscape, people and vegetation, including LAHARS, volcanic EARTHQUAKES and TSUNAMIS, and they may contribute to MASS EXTINCTIONS (see K–T BOUNDARY). *JBH*

[See also VOLCANIC IMPACTS]

Francis, P. 1993: *Volcanoes: a planetary perspective.* Oxford: Oxford University Press. **Gilbert, J.S. and Sparks, R.S.J. (eds)** 1998:*The physics of explosive volcanic eruptions* [*Special Publication* 145]. London: Geological Society. **Huppert, H.E. and Dade, W.B.** 1998: Natural disasters: explosive volcanic eruptions and gigantic landslides. *Theoretical and Computational Fluid Dynamics* 10, 201–212. **Kristmannsdottir, H., Bjornsson, A., Palsson, S. and Sveinbjornsdottir, A.E.** 1999: The impact of the 1996 subglacial volcanic eruption in Vatnajokull on the river Jokulsa a Fjollum, north Iceland. *Journal of Volcanology and Geothermal Research* 92, 359–372. **Kusanagi, T. and Matsui, T.** 2000: The change of eruption styles of Martian volcanoes and estimates of the water content of the Martian mantle. *Physics of the Earth and Planetary Interiors* 117, 437–447. **MacDonald, G.A.** 1972: *Volcanoes.* Engelwood Cliffs, NJ: Prentice-Hall. **Pyle, D.M.** 1998: Forecasting sizes and response times of future extreme volcanic events. *Geology* 26, 367–370.

volcanic explosivity index (VEI)

An index (on a scale of 1 to 8) based entirely on geological (volcanological) criteria for assessing the magnitude, intensity, dispersive power and destructiveness of volcanic eruptions. More than 110 eruptions with an index of 4 or more (consistent with the injection of material into the STRATOSPHERE) have occurred over the last 500 years. *AHP/JAM*

[See also DUST VEIL INDEX, GLACIOLOGICAL VOLCANIC INDEX]

Newall, G.C. and Self, S. 1982: The volcanic explosivity index (VEI): an estimate of the explosive magnitude for historical vulcanism. *Journal of Geophysical Research* 87, 1231–1238. **Simkin, T., Siebert, L., McClelland, L.** *et al.* 1981: *Volcanoes of the world.* Stroudsberg, PA: Hutchinson Ross.

volcanic gases

One of the three types of substances ejected from a volcano – the other two being solids (TEPHRA) and liquids (LAVA). The commonest gas released is water vapour in the form of steam (60–90%); the other gases in order of abundance are: CARBON DIOXIDE, nitrogen, SULPHUR DIOXIDE, hydrogen, CARBON MONOXIDE, sulphur and chlorine, some of which are GREENHOUSE GASES and of importance in CLIMATIC CHANGE. *BDG*

[See also ACIDITY RECORD OF VOLCANIC ERUPTIONS, GLACIOLOGICAL VOLCANIC INDEX, VOLCANIC IMPACTS]

Holmes, A. 1965: *Principles of physical geology.* London: Nelson. **Kelly, P.M. and Jones, P.D.** 1996: The spatial response of the climatic system to explosive eruptions. *International Journal of Climatology* 16, 537–550.

volcanic glass

An IGNEOUS ROCK that lacks MINERALS, retaining the atomic structure of a liquid, as a result of rapid cooling of MAGMA. An example is *obsidian*. Many PYROCLASTIC fragments have the structure of volcanic glass and these may contribute up to 100% of a TEPHRA layer. *GO/JBH*

[See also WELDED TUFF]

Fisher, R.V. and Schminke, H.-U. 1984: *Pyroclastic rocks.* Berlin: Springer.

volcanic impacts on climate

Comments in ancient manuscripts during times of increased volcanic activity that 'the fruits did not ripen' or 'the Sun's light was dimmed and its ray enfeebled for periods up to one year' indicate a long-standing awareness of links between VOLCANIC ERUPTIONS and climate. Some of the climatic effects are now known to be global in nature with major eruptions such as Laki (AD 1783) and Krakatoa (1883) producing a planetary-scale DUST VEIL and reductions in INSOLATION at the surface of over 20%.

To generate significant climatic impacts, an eruption must first be explosive enough to eject material into the lower STRATOSPHERE (above c. 20 km). Second, the composition of the materials ejected is significant. Sulphur compounds (later oxidising to sulphate AEROSOLS) and fine dusts are the climatically active components and may have a RESIDENCE TIME in the atmosphere of months to several years. Particles of diameter 0.5–2 μm are particularly effective reflectors, being comparable in size to typical wavelengths of incoming SOLAR RADIATION. Third, the location of the volcano is crucial. An eruption near the Equator maximises the poleward spread of the dust veil by the planetary wind circulation. When these requirements are met, as in the cases of El Chichón (1982) and Mount Pinatubo (1991), global surface cooling of up to 0.5°C occurs approximately 1.0–1.5 years after the event and cooling tendencies predominate for about three years.

Major eruptions seem to be associated with a warming of the landmasses of the high northern latitudes during the following winter. A poleward displacement of the North Atlantic WESTERLIES appears to be responsible, driving more vigorous depressions deeper into Eurasia. This strengthening of the polar vortex may result from the heightened Equator–pole temperature contrast created as the volcanic aerosol in the high tropical atmosphere is warmed by the Sun. During subsequent years, general cooling predominates at all latitudes with volcanically related temperature ANOMALIES gradually disappearing over approximately three years. The volcanic signal is strongest in the summer and autumn and less discernible in winter. The signatures of past eruptions are often well displayed in ICE CORE and TREE RING data and show strong correlations with trends in global temperature.

Though individual volcanic eruptions may mask, for a time, the effects of anthropogenic GLOBAL WARMING, their cooling effect is not sufficient to reverse a greenhouse signal expected to be +0.1–0.2°C per decade early in the twentyfirst century, unless a major increase in volcanic activity occurs. *JCS*

[See also DUST VEIL INDEX, GLACIOLOGICAL VOLCANIC INDEX, VOLCANIC AEROSOLS, VOLCANIC EXPLOSIVITY INDEX]

Baillie, M.G.L. and Nunro, M.A.R. 1988: Irish tree rings, Santorini and volcanic dust veils. *Nature* 322, 344–346.

Caseldine, C., Hatton, J., Huber, U. *et al.* 1998: Assessing the impact of volcanic activity on mid-Holocene climate in Ireland: the need for replicate data. *The Holocene* 8, 105–111. **Kelly, P.M., Jones, P.D. and Jia, P.Q.** 1996: The spatial response of the climate system to explosive volcanic eruptions. *International Journal of Climatology* 16, 537–550. **Kondratyev, K. and Galindo, I.** 1997: *Volcanic activity and climate.* Williamsburg, VA: Deepak. **Porter, S.C.** 1981: Recent glacier variations and volcanic eruptions. *Nature* 291, 139–142. **Scuderi, L.A.** 1990: Tree ring evidence for climatically effective volcanic eruptions. *Quaternary Research* 34, 67–86.

volcanic impacts on people

In the short term, *volcanic hazards* pose threats to life, property, communications, utilities and agriculture: in the long term, however, subsequent economic benefits from volcanoes may outweigh these losses. PYROCLASTIC flows account for the greatest loss of life, owing to their high velocities and temperature. Risks also arise from flooding associated with LAHARS and TSUNAMIS. Crops and animals are most affected by TEPHRA falls, which can cover large areas. LAVA FLOWS are negligible causes of death, although they may cause much damage and destruction to buildings. VOLCANIC GASES and ACID RAINS generally have limited direct impact on people.

The aftermath of a VOLCANIC ERUPTION can, however, bring considerable benefits to an area. Volcanic soil is extremely fertile, GEOTHERMAL activity can be harnessed for electricity generation and most of the metallic minerals mined in the world are associated with past volcanic activity. Many volcanoes also prove to be tourist attractions world-wide. *JBE*

Blong, R.J. 1984: *Volcanic hazards: a sourcebook on the effects of eruptions.* London and New York: Academic Press. **Chester, D.** 1993: *Volcanoes and society.* London: Edward Arnold. **Payson, D.S. and Grayson, D.K. (eds)** 1979: *Volcanic activity and human ecology.* London: Academic Press.

volcanic impacts on vegetation

VEGETATION changes temporally associated with VOLCANIC ERUPTIONS. Changes may result directly from the deposition of volcanic ash (TEPHRA), lapilli and bombs, causing plants close to the source to become smothered, or as a result of poisonous substances contained in the ash. Indirect effects may be short term, due to the presence of the ash cloud, or longer term, due to sulphate aerosol clouds that can cover wide areas and persist in the atmosphere for a number of years. Both of these intercept SOLAR RADIATION, resulting in a shift of climate that may be reflected in changes in species composition. Past changes of this type may be revealed by combining POLLEN ANALYSIS and TEPHROCHRONOLOGY from PEAT and lake SEDIMENTS with DENDROCHRONOLOGICAL studies. Indirect effects may also be due to short-term OZONE DEPLETION caused by the injection of millions of tons of SO_2 into the atmosphere, where it may activate chlorine into forms that catalyse ozone destruction. Long-distance impacts on vegetation are more controversial than local impacts. *SPH*

[See also CLIMATIC CHANGE, ECOLOGICAL SUCCESSION]

Birks, H.J.B. 1994: Did Icelandic volcanic eruptions influence the postglacial vegetational history of the British Isles? *Trends in Ecology and Evolution* 9, 312–314. **Blackford, J.J., Edwards, K.J., Dugmore, A.J.** *et al.* 1992: Icelandic volcanic ash and the mid-Holocene Scots pine (*Pinus sylvestris*) pollen decline in northern Scotland. *The Holocene* 2, 260–265. **Hall, V.A.,**

Pilcher, J.R. and McCormac, F.G. 1994: Icelandic volcanic ash and the mid-Holocene Scots pine (*Pinus sylvestris*) decline in the north of Ireland: no correlation. *The Holocene* 4, 79–83. **Hansen, J., Sato, M., Lacis, A. and Ruedy.** 1997: The missing climate forcing. *Philosophical Transactions of the Royal Society of London* B352, 231–240. **Scuderi, L.A.** 1990: Tree-ring evidence for climatically effective volcanic eruptions. *Quaternary Research* 34, 67–85. **Vogelmann, A.M., Ackerman, T.P. and Turco, R.P.** 1992: Enhancements in biologically effective ultraviolet radiation following volcanic eruptions. *Nature* 359, 47–49.

volcanic plume

A buoyant column of PYROCLASTIC debris (TEPHRA) and volcanic gas (VOLATILES) that rises from an explosive VOLCANIC ERUPTION. Volcanic plumes, which are also known as *eruption columns*, can carry dust and gases to heights of > 50 km in the ATMOSPHERE. *GO*

Sparks, R.S.J., Bursik, M.I., Carey, S.N. *et al.* 1997: *Volcanic plumes.* Chichester: Wiley.

volcanicity

The distribution of volcanic activity (VOLCANISM) in space and time. ACTIVE VOLCANOES are confined to narrow zones that coincide with SEISMIC BELTS and define the PLATE MARGINS (see Figure). *Vulcanicity* is a synonym, and both terms are commonly used as synonyms for volcanism. However, it is recommended that the distinction between volcanicity and volcanism made in this dictionary is maintained. *GO*

[See also SEISMICITY]

Brothers, R.N. and Delaloye, M. 1982: Obducted ophiolites of North Island, New Zealand: origin, age, emplacement and tectonic implications for Tertiary and Quaternary volcanicity. *New Zealand Journal of Geology and Geophysics* 25, 257–274. **Orton, G.J.** 1996: Volcanic environments. In Reading, H.G. (ed.), *Sedimentary environments: processes, facies and stratigraphy*, 3rd edn. Oxford: Blackwell Science, 485–567. **Straub, S.M. and Schmincke, H.U.** 1998: Evaluating the tephra input into Pacific Ocean sediments: distribution in space and time. *Geologische Rundschau* 87, 461–476.

volcaniclastic

Fragmental (CLASTIC) sediments and rocks composed of volcanic materials. As well as material of PYROCLASTIC origin (formed by explosive VOLCANIC ERUPTIONS), the term covers fragments of *epiclastic* origin (formed by the WEATHERING of volcanic rocks) and *autoclastic* origin (formed by the fragmentation of LAVA during flow). *GO*

Lajoie, J. and Stix, J. 1992: Volcaniclastic rocks. In Walker, R.G. and James, N.P. (eds), *Facies models: response to sea level change.* St John's, Newfoundland: Geological Association of Canada, 101–118.

volcanism

Activity associated with the expulsion at the surface of the Earth (or another planet) of MAGMA and associated VOLATILES. Volcanism covers all aspects of *volcanic activity* including VOLCANOES, VOLCANIC ERUPTIONS, and their distribution in space and time (VOLCANICITY). *Vulcanism* is a synonym. Long-term records of volcanism may be recorded in ICE CORES, and the magnitude of former explosive volcanism calculated using the VOLCANIC EXPLOSIVITY INDEX or the DUST VEIL INDEX. Volcanism can contribute to CLIMATIC CHANGE through the release of volatiles or EJECTA to the ATMOSPHERE. Conversely, climatic change can trigger volcanism through changes in HYDROSTATIC PRESSURE or CRYOSTATIC PRESSURE regimes. *JBH*

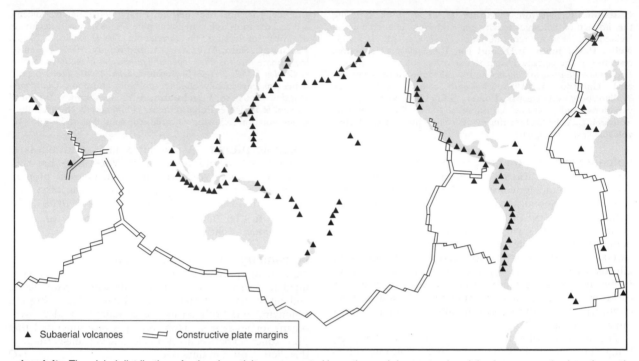

volcanicity *The global distribution of volcanic activity, represented by active and dormant subaerial volcanoes, and submarine activity at constructive plate margins (after Orton, 1996)*

[See also GLACIOLOGICAL VOLCANIC INDEX, IGNEOUS ROCKS, VOLCANIC IMPACTS]

Cas, R.A.F. and Wright, J.V. 1987: *Volcanic successions: modern and ancient.* London: Allen and Unwin. **Orton, G.J.** 1996: Volcanic environments. In Reading, H.G. (ed.), *Sedimentary environments: processes, facies and stratigraphy*, 3rd edn. Oxford: Blackwell, 485–567. **Summerfield, M.A.** 1991: *Global geomorphology.* Harlow: Longman. **Upton, B.G.J.** 1993: Volcanoes and their products. In Duff, P.McL.D. (ed.), *Holmes' principles of physical geology*, 4th edn. London: Chapman and Hall, 207–258. **Wilson, M.** 1989: *Igneous petrogenesis.* London: Unwin Hyman.

volcano An opening in the surface of the Earth (or another planet) through which MAGMA can be expelled in a VOLCANIC ERUPTION as LAVA, PYROCLASTIC material or VOLATILES, and the LANDFORM produced by such activity. Volcanoes can be divided into two categories: those associated with eruptions from a more-or-less circular *vent* above a cylindrical conduit (*central volcanoes*) and those associated with a linear opening above a sheet-like conduit (*fissure volcanoes*).

The form of a volcano depends on magma properties and the eruption pattern and size (see Figure). FISSURE ERUPTIONS give rise to sheet-like LAVA FLOWS and FLOOD BASALTS and may construct LAVA PLATEAUX. Although not normally considered as volcanoes, fissure eruption activity along MID-OCEAN RIDGES creates OCEANIC CRUST. The accumulation of volcanic products around a CRATER forms a variety of types of central volcanoes, including low rings (see MAAR) and large, gently sloping *shield volcanoes* (e.g. Hawaii). *Cinder cones* are conical hills of pyroclastic EJECTA. The classic conical form of large *stratovolcanoes*, also known as *composite volcanoes*, is exemplified by Mounts Fuji (Japan), Egmont (New Zealand), Mayon

(Philippines) and Adams (USA). They are steep-sided cones of alternating LAVA FLOWS and pyroclastic material lying close to its angle of repose. Large explosive eruptions may destroy the volcanic edifice through explosion and subsidence, forming a CALDERA. The largest volcano on Earth is Mauna Loa (Hawaii), a shield volcano rising 10 000 m from the ocean floor. Olympus Mons on Mars rises 26 km above its surrounding plain and is probably the largest volcano in the solar system.

Volcanoes on the Earth are described as ACTIVE, DORMANT or EXTINCT VOLCANOES, depending on the likelihood of further volcanic eruptions. Their environmental impact includes volcanic eruptions and associated effects such as LAHARS, LANDSLIDES and HYDROTHERMAL activity.

Similar morphological features associated with the expulsion of SEDIMENT that has been subjected to FLUIDISATION or some other form of LIQUIDISATION are described as *sand volcanoes, mud volcanoes, sand boils* and *sand blows* (see SOFT-SEDIMENT DEFORMATION). *JBH*

[See also METEORITE IMPACT]

Chester, D. 1993: *Volcanoes and society.* London: Edward Arnold. **Decker, R. and Decker, B.** 1997: *Volcanoes*, 3rd edn. Basingstoke: Freeman. **Firth, C.R. and McGuire, W.J. (eds)** 1999: *Volcanoes in the Quaternary* [*Special Publication* 161]. London: Geological Society. **Francis, P.** 1993: *Volcanoes: a planetary perspective.* Oxford: Oxford University Press. **Ollier, C.D.** 1988: *Volcanoes.* Oxford: Blackwell. **Sigurdsson, H., Houghton, B., Rymer, H.** *et al.* **(eds)** 1999: *Encyclopedia of volcanoes.* New York: Academic Press.

volume scattering The integrated effect of multiple scattering from a medium (as opposed to *surface scattering*). Volume scattering media include vegetation canopies

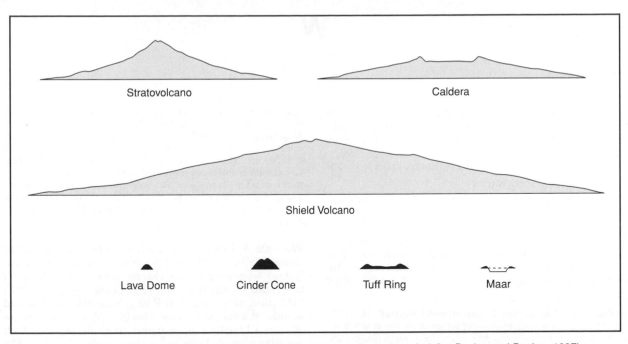

volcano *Simplified morphology of the main types of volcanoes, drawn to scale (after Decker and Decker, 1997)*

and layers of snow or soil that are penetrated by incident radiation. *PJS*

vulnerability The degree to which an ecological, human or economic SYSTEM may react adversely to the occurrence of a PERTURBATION, DISTURBANCE or hazardous environmental change: the potential of a system to experience adverse environmental impacts; an antonym of RESILIENCE and a corollary of RISK. Vulnerability is an important aspect of SUSTAINABILITY, which may be enhanced by reducing vulnerability through *mitigation* (reducing the external HAZARD) or ADAPTATION (system change to improve viability). *JAM*

Blaikie, P., Cannon, T., Davis, I. and Wisner, B. 1994: *At risk: natural hazards, people's vulnerability and disasters*. London: Routledge. **Smit, B. (ed.)** 1993: *Adaptation to climatic variability and change* [*Occasional Paper* 19]. Guelph, Ontario: University of Guelph, Department of Geography. **Timmerman, P.** 1981: *Vulnerability, resilience and the collapse of society: a review of models and possible applications* [*Monograph* 1]. Toronto: Institute of Environmental Studies. **United Nations Disaster Relief Organization** 1982: *Natural disasters and vulnerability analysis*. Geneva: United Nations Disaster Relief Co-ordinator.

vulnerable species Species believed likely to become ENDANGERED SPECIES if present changes continue, and an International Union for the Conservation of Nature RED DATA BOOK category; includes species with at least 10% probability of extinction within 100 years. *IFS*

[See also RARE SPECIES, THREATENED SPECIES]

W

wacke A poorly sorted SANDSTONE, with between 15 and 75% MATRIX, typical either of deposition in a low-energy DEPOSITIONAL ENVIRONMENT or of a process of deposition that results in rapid SEDIMENTATION (e.g. deposition from a TURBIDITY CURRENT or DEBRIS FLOW).   *TY*

[See also FLYSCH, GREYWACKE]

Pettijohn, F.J., Potter, P.E. and Siever, R. 1987: *Sand and sandstone*. New York: Springer.

wadi An EPHEMERAL stream course in a DESERT or semi-desert area. Streamflow is occasional, sometimes discontinuous and may be torrential when it does occur. The term is of Arabic origin.   *RAS*

[See also ARROYO, GULLY]

Smith, S.E., El-Shamy, I. and Abd-El Monsef, H. 1997: Locating regions of high probability for groundwater in the Wadi El-Arish Basin, Sinai, Egypt. *Journal of African Earth Sciences* 25, 253–262.

Walker circulation A series of large-scale zonal cells in the equatorial atmosphere that, in contrast to the meridional HADLEY CELLS, move air latitudinally. Each *Walker cell* consists of rising warm air and subsiding cold air driven thermally by LATENT HEAT and is confined to the tropical zone by rotational forces. The cells exist permanently and move from east to west. The average position of the strongest cell of the Walker circulation consists of a broad region of rising air over the warm surface of the equatorial western Pacific and eastern Indian Oceans and widespread sinking air over the cold waters of the eastern Pacific. This cell was first recognised by Sir Gilbert Walker in the AD 1920s during his attempts to forecast the rainfall of the Indian MONSOON. Other cells are associated with ascending air over northeastern Brazil and eastern Africa.

Seasonal variations in the Walker circulation are linked to the larger-scale MONSOON circulation and inter-annual variations to EL NIÑO–SOUTHERN OSCILLATION (ENSO) phenomena. Indeed, the SOUTHERN OSCILLATION, defined as the oscillating sea-level pressure difference between Tahiti and Darwin, provides a record of variations in the Walker circulation. During an ENSO event, the position of the rising air of the main Walker cell shifts to the central and eastern Pacific, pressure is low in Tahiti and high in Darwin and the circulation is reversed. This brings enhanced rainfall over the eastern Pacific and western coastal South America, while drought conditions affect northern Australia and Indonesia. At the same time, the Indian monsoon weakens, there is a decrease in rainfall over Brazil and two of the more far-reaching ANOMALIES include floods in California and cold weather in northeastern North America.   *JAM*

[See also MADDEN–JULIAN OSCILLATION]

Julian, P.R. and Chervin, R.M. 1978: A study of the southern oscillation and Walker circulation phenomenon. *Monthly Weather Review* 106, 1433–1451. **Lau, W.K.-M.** 1996: Tropical circula-

tions. In Schneider, S.H. (ed.), *Encyclopedia of climate and weather*. New York: Oxford University Press, 775–778. **Shulmeister, J. and Lees, B.G.** 1995: Pollen evidence from tropical Australia for the onset of an ENSO-dominated climate at c. 4000 BP. *The Holocene* 5, 10–18. **Walker, G.T.** 1924: Correlation in seasonal variation of weather IX: a further study of world weather. *Memoirs of the Indian Meteorological Department* 24, 275–332. **Wyrthi, K.** 1982: The southern oscillation, ocean–atmosphere interaction and El Niño. *Marine Technology Society Journal* 16, 2–10.

Wallace's Line A boundary, described by Alfred Russel Wallace (1823–1913), across the Indonesian archipelago separating faunas characteristic of the islands of the Sunda continental shelf (e.g. Borneo, Java and Bali) from those of the Sahul shelf (e.g. Sulawesi, Lombok and islands further east and south). Many variants of Wallace's Line have been suggested in this region, which has been termed *Wallacea*.   *KDB*

[See also BIOGEOGRAPHY]

Wallace, A. R. 1869: *The Malay archipelago: the land of the orang-utan and the bird of paradise. A narrative of travel with studies of man and nature*. London: Macmillan.

Walther's law 'Only those facies and facies-areas can be superimposed primarily which can be observed beside each other at the present time' (Walther, 1894; translation in Blatt, Middleton and Murray, 1980). This 'law' embodies the principle that in many environmental systems characterised by subenvironments, each of which is preserved as a distinct FACIES, the accumulation of sediment over time causes the subenvironments to migrate laterally in space such that the facies become stacked vertically in the GEOLOGICAL RECORD, separated by gradational boundaries. The lateral migration of environment belts with sediment accumulation is PROGRADATION and the classic illustration of Walther's Law is the behaviour of a rivermouth DELTA where, over time, finer-grained off-shore deposits come to be overlain by coarser-grained nearshore deposits. The application of Walther's Law was vital to the development of FACIES MODELS from the AD 1960s to the 1980s, but its relevance to interpreting sedimentary rock successions is dependent on the validity of UNIFORMITARIANISM, which may not be so rigidly applicable where environmental change was rapid.   *GO*

[See also SEQUENCE STRATIGRAPHY]

Blatt, H., Middleton, G.V. and Murray, R.C. 1980: *Origin of sedimentary rocks*, 2nd edn. Englewood Cliffs, NJ: Prentice-Hall. **Reading, H.G. and Levell, B.K.** 1996: Controls on the sedimentary rock record. In Reading, H.G. (ed.), *Sedimentary environments: processes, facies and stratigraphy*, 3rd edn. Oxford: Blackwell Science, 5–36. **Soreghan, G.S.** 1997: Walther's Law, climate change, and Upper Paleozoic cyclostratigraphy in the ancestral Rocky Mountains. *Journal of Sedimentary Research* 67, 1001–1004.

washload A term sometimes used to describe that finest-grained part of the SUSPENDED LOAD in water, which remains almost permanently suspended, and would settle very slowly if the flow stopped. The equivalent sediment load in air can be termed *dustload*. *GO*

waste An object or substance that is no longer required. Some examples include: (1) RESIDUES from raw-materials extraction, manufacturing and food consumption; (2) products that are surplus to requirement, off-specification, worn out, or out-of-date; (3) contaminated materials; (4) hazardous by-products of processes (see HAZARDOUS WASTE); and (5) substances, the use of which has been made illegal. Waste is an outcome of every production process and in that context it is commonly classified according to origin, such as AGRICULTURAL, DOMESTIC and INDUSTRIAL WASTE. *JAM*

[See also COMMON LAND, FIELD SYSTEMS]

waste management Towards the end of the twentieth century, it has become clear to all governments that the quantity of waste generated by the human population, about 10 kg per week per household, requires careful management. Disasters related to waste disposal, such as CONTAMINATED LAND, methane explosions, GROUNDWATER POLLUTION, NUCLEAR ACCIDENTS and transport of hazardous materials through areas of dense population, have drawn attention to the necessity for clear policy guidelines on the subject. Strategies for waste minimisation include the reduction of waste at source, treatment, *product substitution*, WASTE MINING and WASTE RECYCLING. Systems of *integrated waste management* have, however, rarely been fully implemented. *EMB*

[See also AGRICULTURAL WASTE, DOMESTIC WASTE, HAZARDOUS WASTE, INDUSTRIAL WASTE, LANDFILL, RADIOACTIVE WASTE, SUSTAINABILITY]

Harrison, R.M. and Hester, R.E. 1995: *Waste treatment and disposal.* Cambridge: Royal Society of Chemistry. **Kharbanda, O.P. and Stallworthy, E.A.** 1990: *Waste management: towards a sustainable society.* New York: Auburn House. **National Research Council** 1978: *Multimedium management of municipal sludge.* Washington, DC: National Academy Press. **Oweis, I.S. and Khera, R.P.** 1990: *Geotechnology of waste management.* London: Butterworths. **Thomas-Hope, E. (ed.)** 1998: *Solid waste management: critical issues for developing countries.* Kingston, Jamaica: Canoe Press, University of the West Indies.

waste mining The re-use of waste produced by the processes of MANUFACTURING INDUSTRY. It includes *coprocessing* or recovering secondary materials, *material recovery* from liquid and airborne waste streams and material recovery from the wastes of historical manufacturing. *JAM*

[See also WASTE RECYCLING]

waste recycling Strictly the reprocessing of the disused products of MANUFACTURING INDUSTRY, as opposed to the processing of manufacturing waste (WASTE MINING). *JAM*

[See also WASTE MANAGEMENT]

water-based disease A DISEASE that is spread by an infecting agent, at least part of the life cycle of which takes place in an aquatic animal. Unlike WATER-BORNE DISEASES,

water-based diseases do not necessarily involve the ingestion of water. *JAM*

[See also WATER-RELATED VECTOR]

water-borne diseases Diseases transmitted by the ingestion of contaminated water (POTABLE WATER, bathing water, etc.) including CHOLERA, *dysentery*, *schistomiasis* and *typhoid*. Most water-borne diseases occur when pathogenic bacterial, viral, protozoal or helminthic microorganisms are transmitted by water contaminated by faeces from humans or domesticated animals. Maintenance of sanitation and chemical water treatments (e.g. *chlorination*) are the most effective prevention and control measures. *JAM*

[See also DISEASE, EPIDEMIC, HUMAN HEALTH HAZARDS, WATER-BASED DISEASE]

National Academy of Sciences 1977: *Drinking water and health.* Washington, DC: National Academy Press. **Salvato, J.A.** 1992: *Environmental engineering and sanitation.* New York: Wiley.

water erosion The removal of soil by *rainsplash*, SLOPEWASH and movement of water flowing in discrete hillslope channels (RILLS and GULLIES). The term is often used to distinguish this form of SOIL EROSION by water from WIND EROSION. *RAS*

[See also ERODIBILITY, EROSIVITY, FLUVIAL PROCESSES]

Bryan, R.B. 2000: Soil erodibility and processes of water erosion on slopes. *Geomorphology* 32, 385–415.

water mass A body of water in the OCEAN that can be characterised and distinguished by its range of temperature and SALINITY and, therefore, by its density, defined by a temperature–salinity curve. These properties are acquired by near-surface water at specific latitudes and are retained as water masses move through the oceans as OCEAN CURRENTS. The temperature of 75% of all ocean water is in the range 0–5°C, whereas the average temperature of surface waters is 17.5°C: this indicates that most water masses originate at high latitudes. The uniformity of temperature and salinity implies that water masses move between oceans (see THERMOHALINE CIRCULATION). Three groups of global ocean water masses are found in the Atlantic, Indian and Pacific Oceans: *surface water* (0–1 km depth), *intermediate water* (1–2 km depth) and DEEP WATER (deeper than 2 km). The deepest water, in contact with the sea bed, is BOTTOM WATER. *BTC*

[See also CONTOUR CURRENT]

Pinet, R. 1998: *Invitation to oceanography.* Sudbury, MA: Jones and Bartlett. **Open University Course Team** 1989: *Ocean circulation.* Oxford: Pergamon. **Stephens, J.C. and Marshall, D.P.** 2000: Dynamical pathways of Antarctic Bottom Water in the Atlantic. *Journal of Physical Oceanography* 30, 622–640. **Summerhayes, C.P. and Thorpe, S.A.** 1996: *Oceanography: an illustrated guide.* London: Manson.

water pollution The artificial introduction of substances to the AQUATIC ENVIRONMENT that are likely to be a hazard to human health or the ecological system. Water pollution can lead to DISEASE transmission and reductions in water potability. It may also be poisonous to humans, animals and plants and it can be unsightly and smelly. Categories of pollutants are numerous and include NUTRIENTS, PESTICIDES, RADIONUCLIDES, PATHOGENS, heat, oil and metals. Pollution sources can be classified into point and diffuse: pollution from *point sources*, such as sewage or industrial effluent outfalls, is the easiest to monitor, predict and thus control. Many of the recent improvements in *water quality* in Europe and North America have been accomplished through reduced or improved quality of such discharges. However, outfalls discharging untreated or partly treated sewage into coastal or riverine waters remain a major source of water pollution.

The application of FERTILISERS and pesticides to agricultural land has increased exponentially since the 1950s. This has led to increasing concern over *diffuse sources* from agricultural land leading to EUTROPHICATION or BIOLOGICAL MAGNIFICATION in the FOOD CHAIN and to pollution of GROUNDWATER. Diffuse sources enter watercourses via *overland flow*, THROUGHFLOW or LAND DRAINAGE and are harder to prevent because occurrences tend to be episodic (see EPISODIC EVENTS) and dependent on the hydrological pathway taken during STORMS. Fertilisers, manure and slurry are potential sources of NITRATES, which are readily desorbed and transported into streams in solution. PHOSPHORUS, which is much less soluble than nitrogen, can be transported to streams from the same sources, but is predominantly adsorbed to eroded sediment.

Urban runoff is a major contributor to water pollution in estuaries, along coasts and in the lower reaches of rivers. It can contain high levels of metals, oil and chemicals from roads, as well as industrial effluents.   *ADT*

[See also MARINE POLLUTION, METAL POLLUTION, OIL SPILL, THERMAL POLLUTION, WASTE MANAGEMENT]

Edwards, A.C. and Withers, P.J.A. 1998: Soil phosphorus management and water quality: a UK perspective. *Soil Use and Land Management* 14, 124–130. **Goulding, K.W.T., Matchett, L.S., Heckrath, G. et al.** 1996: Nitrogen and phosphorus flows from agricultural hillslopes. In Anderson, M.G. and Brooks, S.M. (eds), *Advances in hillslope processes*. Vol. I. Chichester: Wiley 231–227. **Heathwaite, A.L. and Johnes, P.J.** 1996: The contribution of nitrogen species and phosphorus fractions to stream water quality in agricultural catchments. *Hydrological Processes* 10, 971–983. **Mason, C.F. and Macdonald, S.M.** 1993: Impact of organochlorine pesticide residues and PCBs on otters (*Lutra lutra*) in eastern England. *Science of the Total Environment* 138, 147–160. **Organisation for Economic Co-operation and Development (OECD)** 1986: *Water pollution by fertilizers and pesticides*. Paris: OECD. **Sharp, J.J.** 1990: The use of ocean outfalls for efficient disposal in small communities and developing countries. *Water International* 15, 35–43. **Smol, J.P.** 2002: *Pollution of lakes and rivers: a palaeoenvironmental perspective*. London: Arnold.

water-related vector

An insect that lives and breeds close to water and spreads a disease, such as *dengue fever*, *haemorrhagic fever* and malaria, by biting. Thus, the female *Anopheles* mosquito spreads MALARIA by infecting the host with the protozoan *Plasmodium vivax*.   *JAM*

Harrison, G.A. 1978: *Mosquitoes, malaria and man*. New York: Dutton.

water repellency

The reduced affinity of a material for water. In soils, water repellency or *hydrophobicity* can delay the penetration of water drops sometimes in excess of several hours. Soil water repellency is usually caused by a coating of long-chained organic molecules on individual soil particles. These substances are released from a range of plants or micro-organisms either naturally or during burning. Soil water repellency is temporally variable and often most prominent after prolonged dry spells. Owing to the cultivation of certain frequently introduced plant species and the increase in WILDFIRES in some regions, water repellency has developed in previously unaffected areas. Amongst the effects of water repellency are inhibited plant growth, increased *overland flow* and SOIL EROSION, uneven spatial and vertical wetting patterns and enhanced risk of GROUNDWATER POLLUTION due to the generation of preferential flow pathways.   *SHD*

Dekker, L.W. and Ritsema, C.J. 1994: How water moves in a water repellent sandy soil. 1. Potential and actual water-repellency. *Water Resources Research* 30, 2507–2517. **Doerr, S.H., Shakesby, R.A. and Walsh, R.P.D.** 2000: Soil water repellency: its origin, characteristics and hydro-geomorphological implications. *Earth-Science Reviews* 51, 33–65. **Shakesby, R.A., Doerr, S.H. and Walsh, R.P.D.** 2000: The erosional impact of soil hydrophobicity: current problems and future research directions. *Journal of Hydrology* 231/232, 178–191.

water resources

Fresh water is vital to human life and access to water and protection of resources are issues on many national security agendas. Although 71% of the Earth's surface is covered by water, only 0.008% of this is available for human consumption. Water ABSTRACTION has risen by 3500% in the last 300 years (four times as fast as population growth) and demand continues to rise steeply. Water is required in large volumes for drinking, washing, irrigation, industrial processes and cooling. Most industrialised nations have legislation to protect *water quality*, but hydrological processes do not obey political boundaries, and nations or authorities that share large river catchments may have conflicting laws governing quality and abstraction. GROUNDWATER resources are even more difficult to monitor and allocate than surface water resources.   *MLW*

[See also ENVIRONMENTAL SECURITY, RESOURCE]

Burmaster, D.E. 1986: Groundwater – saving the unseen resource. *Environment* 28, 25–28. **Hillel, D.** 1994: *Rivers of Eden: the struggle for water and the quest for peace in the Middle East*. New York: Oxford University Press. **Shiklomanov, I.A.** 1993: World fresh water resources. In Gleick, P.H. (ed.), *Water in crisis*. New York: Oxford University Press. **Van der Leeden, F.** 1975: *Water resources of the world*. Port Washington, NY: Water Information Centre.

water table

The upper limit of the water-saturated or PHREATIC ZONE of GROUNDWATER in an unconfined AQUIFER. It normally varies seasonally and may be lowered by water ABSTRACTION. Also known as the *phreatic surface* (or sometimes as the *piezometric surface*), it can be located by observing the level of the water surface in WELLS and natural fissures. In areas where the rocks are PERMEABLE, the topography of the water table tends to follow the topography of the land surface in a general way but with less relief. *Springs*, sometimes forming a *spring-line*, occur where the water table intersects the land surface, the DISCHARGE of the springs usually reflecting seasonal changes in the height of the water table. A *perched watertable* is a suspended watertable that lies above a local, impermeable stratum (AQUICLUDE) and hence is isolated from the main body of groundwater.   *JAM*

[See also POTENTIOMETRIC SURFACE]

water use efficiency (WUE) The ratio of carbon assimilation by PHOTOSYNTHESIS to water loss by *transpiration* (see EVAPOTRANSPIRATION). WUE tends to increase in response to an increase in atmospheric CARBON DIOXIDE concentration. *JAM*

water vapour The non-visible, gaseous form of water and an important constituent of the ATMOSPHERIC COMPOSITION. It plays a vital role in the formation of CLOUDS and PRECIPITATION and in controlling the long-wave RADIATION BALANCE of the atmosphere. Water vapour is a principal GREENHOUSE GAS and, although not a main factor in enhancing the GREENHOUSE EFFECT, it tends to reinforce the process by its ability to absorb TERRESTRIAL RADIATION. With higher temperatures, the result of enhanced GLOBAL WARMING, greater EVAPORATION from the oceans, will occur and amounts of WATER VAPOUR in the atmosphere will increase. Warmer air is able to hold more moisture. This will add to the greenhouse effect and is an example of a positive FEEDBACK MECHANISM. Conversely, the additional moisture in the atmosphere will also lead to increased cloud formation. This, in turn, will reduce the amount of SOLAR RADIATION reaching the Earth's surface, thereby producing a negative feedback. *JBE*

Ludlam, F.H. 1980: *Clouds and storms: the behavior and effect of water in the atmosphere.* University Park, PA: Pennsylvania State University Press. **Rind, D., Chiou, E.W. and Chu, W.** *et al.* 1991: Positive water vapour feedback in climate models confirmed by satellite data. *Nature* 349, 500–503.

waterfall Strictly a vertical descent of water along a watercourse, but the term is often applied to a variety of other less vertical forms that merge with rapids. *JAM*

[See also KNICKPOINT]

Young, R.W. 1985: Waterfalls: form and process. *Zeitschrift für Geomorphologie Supplementband* 55, 81–95.

watershed (1) A line delimiting the boundary of a DRAINAGE BASIN or *catchment*. Also known as a *water divide*, a watershed separates the area draining into one river system from that draining into neighbouring systems. For GROUNDWATER, the equivalent term is *phreatic divide*, which may or may not coincide with the surface watershed. (2) In America, watershed is a synonym for the catchment itself. *JAM*

wave base The water depth below which the water column in lakes and shallow seas is not disturbed by waves on the water surface. If the water depth is less than the wave base, water movement is transformed into an oscillating current at the BED, which may be sufficient to transport SEDIMENT. The depth of wave base depends on wave characteristics. Under normal conditions the *fairweather wave base* is typically between 5 and 15 m; during STORMS the *storm wave base* may be 200 m or more. Thus BEDFORMS and SEDIMENTARY STRUCTURES of waves are normally a shallow-water phenomenon, but storms may affect the entire CONTINENTAL SHELF. *GO*

[See also BEDFORM STABILITY DIAGRAM, HUMMOCKY CROSS-STRATIFICATION]

Johnson, H.D. and Baldwin, C.T. 1996: Shallow clastic seas. In Reading, H.G. (ed.), *Sedimentary environments: processes, facies and stratigraphy*, 3rd edn. Oxford: Blackwell Science, 232–280.

wave energy The potential and kinetic energy present in waves. The bulk of this energy is that received as wind blows over a water surface. *High-energy coasts* are those that have a large FETCH. On particular coasts wave energy may vary through time in response to SEA-LEVEL CHANGE or other factors, such as the growth of CORAL REEFS. Thus, there was a *Holocene high-energy window* along parts of the Great Barrier Reef following the mid-Holocene TRANSGRESSION, but prior to the development of a protective reef. *HJW*

Hopley, D. 1984: The Holocene 'high energy window' on the central Great Barrier Reef. In Thom, B.G. (ed.), *Coastal geomorphology in Australia.* Sydney: Academic Press, 135–150.

wave refraction The change in direction of a wave due to bottom interaction as it moves at an angle progressively into shallower water. *HJW*

wave ripple A periodic, flow-transverse BEDFORM in sand or fine gravel, typically of height 1–5 cm and spacing 5–50 cm, characterised by symmetrical cross-sections and straight crest lines that occasionally bifurcate ('tuning-fork junctions'). Such *symmetrical ripples* are produced by the oscillating currents that develop above WAVE BASE in lakes and the sea and can be preserved as a SEDIMENTARY STRUCTURE. Wave ripples form by a variety of processes and, unlike current ripples, can form in coarse sand and fine gravel. Sediment accumulation produces complex patterns of CROSS-LAMINATION. Wave-ripple geometry and size is related to sediment GRAIN-SIZE and wave characteristics and can be used to reconstruct palaeowave conditions in the geological record. Wave ripples are also useful in reconstructing PALAEOCURRENTS and PALAEOGEOGRAPHY. Some ripples combine the characteristics of wave ripples and CURRENT RIPPLES (e.g. asymmetrical cross-section, close spacing, straight bifurcating crests). These *combined-flow ripples* or *wave–current ripples* are indicative of complex flows that combine an oscillatory and a unidirectional component. *GO*

[See also BEDFORM STABILITY DIAGRAM]

Allen, J.R.L. 1982: *Sedimentary structures: their character and physical basis.* Amsterdam: Elsevier. **Allen, P.A.** 1984: Reconstruction of ancient sea conditions with an example from the Swiss Molasse. *Marine Geology* 60, 455–473. **Allen, P.A.** 1997: *Earth surface processes.* Oxford: Blackwell Science. **Diem, B.** 1985: Analytical method for estimating palaeowave climate and water depth from wave ripple marks. *Sedimentology* 32, 685–704. **Leeder, M.R.** 1999: *Sedimentology and sedimentary basins: from turbulence to tectonics.* Oxford: Blackwell Science.

weather The instantaneous state of the atmosphere. It contrasts with CLIMATE, where the focus is on a timescale of years rather than days. Weather is a multivariate phenomenon involving a range of elements, including CLOUDS, HUMIDITY, PRECIPITATION, pressure, AIR TEMPERATURE, visibility and wind. The science of METEOROLOGY studies the weather. The prediction of future weather is the task of weather FORECASTING. *AHP*

[See also EXTREME WEATHER EVENTS]

Barry, R.G. and Chorley, R.J. 1992: *Atmosphere, weather and climate*, 6th edn. London: Routledge. **Battan, L.J.** 1983: *Weather in your life.* San Francisco, CA: Freeman.

weather diaries Systematic daily observations of the WEATHER made without meteorological instruments and recorded in diary format, often with other information. One was kept briefly by Ptolemy in AD 120, although William Merle of Merton College, Oxford, was the first to keep an extended daily weather journal, from AD 1337 to 1344. Over the next 200 years this practice spread slowly across Europe and later beyond, continuing well into the period of INSTRUMENTAL RECORDS. Writing calendars (*Schreibkalender*) were sometimes used to record weather next to planetary, lunar and zodiacal constellations, thereby relating weather to zodiacal configurations. Wind direction (and sometimes measures of force), forms of PRECIPITATION and general weather comments were usually included and, as instruments became available, pressure and temperature were sometimes added (see EARLY INSTRUMENTAL METEOROLOGICAL RECORDS). It is also sometimes possible to relate systematic observations of other weather elements, such as thunderstorm occurrence, TROPICAL CYCLONES or the number of days with precipitation, to their modern-day comparative values.

The resulting data series is often chronologically exact by day, month and year, although comparability between records is sometimes hard to establish. STATISTICAL ANALYSIS of frequencies may allow numerical comparisons with the present climate. Derivative parameters can be used to give other indications of temperature change, for example by using the proportion of snow days among the days with precipitation. Particularly where diaries from a spatial cluster of locations are available, synoptic reconstructions (see SYNOPTIC CLIMATOLOGY) may be possible from them. Diaries tend to be restricted to Europe, North America and regions of European colonisation, the reasons being the special technical prerequisites, such as the spread of book printing, the taking of astrometeorological observations and the resulting attempts at WEATHER FORECASTING.

JGT/RG

[See also HISTORICAL CLIMATOLOGY]

Frisinger, H.H. 1977: *The history of meteorology to 1800.* New York: American Meteorological Society. **Le Roy Ladurie, E.** 1971: *Times of feast, times of famine: a history of climate since the year 1000*, translated by B. Bray. New York: Doubleday. **Merle, W.** 1891: *Consideraciones temperici pro 7 annis (1337–1344).* Reproduced and translated under the supervision of G.J. Symons and printed as *The earliest known Journal of Weather.* London: E. Stanford. **Pfister, C., Brazdil, R. and Glaser, R. (eds)** 1999: Climatic variability in sixteenth-century Europe and its social dimensions. *Climatic Change* 43, 1–351 [Special Issue]. **Walsh, R.P.D., Glaser, R. and Militzer, S.** 1999: The climate of Madras in the eighteenth century. *International Journal of Climatology* 19, 1025–1047.

weather satellites Earth-orbiting space vehicles, travelling outside the Earth's ATMOSPHERE, designed for the study and prediction of WEATHER. Now an integral part of the meteorological observation system, the first purpose-built meteorological satellite (TIROS-1) was launched on 1 April 1960. Today, using the technique of SATELLITE REMOTE SENSING, satellites provide 'real-time' monitoring of the operations of the atmosphere. All meteorological satellites carry *radiometers*, which transmit cloud images and also measure TERRESTRIAL and SOLAR RADIATION. There are two types of meteorological satellites, classified according to their SATELLITE ORBITS. *Polar-orbit-*

ing satellites monitor cloud systems directly beneath them and make measurements of atmospheric temperature and humidity and ground and sea-surface temperatures. *Geostationary satellites* provide a broader perspective over a given fixed point and can monitor the progress of large-scale weather systems. Meteorological satellites have proved invaluable in providing advanced warnings of impending TROPICAL CYCLONES and other severe storms. Other useful meteorological applications include long-range weather forecasts. *JBE*

[See also METEOROLOGICAL SATELLITES]

Carleton, A.M. 1991: *Satellite remote sensing in climatology.* London: Belhaven. **Mason, B. and Schmetz, J.** 1992: Meteorological satellites. *International Journal of Remote Sensing* 13, 1153–1172. **NASA** 1982: *Meteorological satellites: past, present and future* [*NASA Conference Publication* 2227]. Washington, DC: National Aeronautics and Space Administration.

weather types Recurrent patterns in regional atmospheric circulation normally based on sea-level isobaric charts. Weather-type classification is central to studies in SYNOPTIC CLIMATOLOGY. Pioneered by Baur in the AD 1930s (the *Grosswetterlagen*), many manual and automated weather-type classifications (SYNOPTIC CATALOGUES) have since been developed for a number of regions. The most well known and frequently applied manual method is Lamb's Weather Type (LWT) Catalogue for the British Isles. This classifies daily surface isobaric charts into either eight directional types (N – north, NE – northeast, E – east, and so on), three non-directional types (A – anicyclonic, C – cyclonic and U – unknown) or 16 hybrids (e.g. ANE). Each type has its own characteristic weather depending on season. Weather type catalogues have been used for the analysis of a large number of environmental variables, for instance PRECIPITATION VARIATIONS, tropospheric OZONE amounts and long-term variations in AIR TEMPERATURE. Automated classifications are derived using statistical techniques such as PRINCIPAL COMPONENTS ANALYSIS. *GS*

[See also AIRFLOW TYPES]

Lamb, H.H. 1972: *British Isles weather types and a register of the daily sequences of circulation patterns 1861–1971* [*Meteorological Office Geophysical Memoir* 118]. London: HMSO. **O'Hare, G.P. and Sweeney, J.** 1993: Lamb's circulation types and British weather: an evaluation. *Geography* 78, 43–60. **O'Hare, G.P. and Wilby, R.L.** 1995: A review of ozone pollution in the United Kingdom and Ireland with an analysis using Lamb Weather Types. *Geographical Journal* 161, 1–20. **Sowden, I.P. and Parker, D.E.** 1981: A study of climatic variability of Central England Temperature in relation to Lamb synoptic types. *Journal of Climatology* 1, 3–10. **Sweeney, J. and O'Hare, G.P.** 1992: Geographical variations in precipitation yields and circulation types in Britain and Ireland. *Transactions of the Institute of British Geographers* 17, 448–463.

weathering The *in situ* alteration and fragmentation of material exposed to the ATMOSPHERE or HYDROSPHERE. The general term weathering can be subdivided into PHYSICAL, CHEMICAL and BIOLOGICAL WEATHERING. Weathering is a fundamental process in the formation of, for example, clay minerals and soils in general. Type and rate of weathering are often closely related to the prevailing environmental conditions (e.g. climate and host material). Thus, weathering products can sometimes be used for either CLIMATIC RECONSTRUCTION, provided their age is known, or RELATIVE-AGE DATING, if their environmental history is known. *SHD*

[See also WEATHERING RATES]

Bland, W. and Rolls, D. 1998: *Weathering*. London: Arnold. **Kittrick, J.A.** 1986: *Soil mineral weathering*. New York: Van Nostrand Reinhold. **Nord, A.G. and Holenyi, K.** 1999: Sulphur deposition and damage on limestone and sandstone in Stockholm city buildings. *Water, Air and Soil Pollution* 109, 147–162. **Yatsu, E.** 1988: *The nature of weathering*. Tokyo: Sozosha.

weathering front The interface between fresh and weathered rock beneath the ground surface; a term proposed by J.A. Mabbutt to replace the term *basal surface of weathering*. *MAC*

[See also WEATHERING PROFILE]

Mabbutt, J.A. 1961: A stripped land surface in Western Australia. *Transactions of the Institute of British Geographers* 29, 101–114.

weathering indices Measures used in RELATIVE-AGE DATING that indicate the degree of WEATHERING of rock surfaces or the degree of SOIL DEVELOPMENT on a surface. Indices, often semi-quantitative, include a range of usually simple and inexpensive measurements, such as: WEATHERING RIND thickness; SCHMIDT HAMMER rebound values; rock surface roughness; JOINT depth; frequency of surface boulders; surface discolouration; changes in soil depth; SOIL HORIZON development or soil chemistry; and subjective assessment of degree of weathering. *DMcC*

Birkeland, P.W., Berry, M.E. and Swanson, D.K. 1991: Use of soil catena field data for estimating relative ages of moraines. *Geology* 19, 281–283.

weathering limits Abrupt spatial boundaries between land surfaces exhibiting different degrees of WEATHERING. They occur within glaciated mountain regions, where part of the landscape has been occupied by glaciers, leaving other parts exposed to PERIGLACIAL processes. Glacier occupancy may result in the removal of weathered material by EROSION, in which case the weathering limit is known as a TRIMLINE. Alternatively, in the case of COLD-BASED GLACIERS, ice cover may serve to protect a land surface from further weathering. Weathering limits may also reflect altitudinal variations in the degree of weathering or spatial variations in GLACIER THERMAL REGIME; care must be taken to distinguish these cases from weathering limits that mark glacier limits. *DIB*

[See also BLOCKFIELD]

Ballantyne, C.K., McCarroll, D., Nesje, A. *et al.* 1998: The last ice sheet in north-west Scotland: reconstruction and implications. *Quaternary Science Reviews* 17, 1149–1184. **Ives, J.** 1978: The maximum extent of the Laurentide Ice Sheet along the eastern coast of North America during the last glaciation. *Arctic* 31, 24–53. **McCarroll, D., Antio, J., Heikkinen, O. and Kontanieni, L.** 1996: Degree of rock surface weathering on fjell summits in northern Finland: implications for the thermal regime of the last ice sheet. *Boreas* 25, 1–7.

weathering profile A vertical cross-section through the zone immediately beneath the Earth's surface affected by *in situ* CHEMICAL WEATHERING. Its lower limit is defined by the WEATHERING FRONT, below which there is no visible sign of weathering in bedrock or rock particles, and its

subdivisions normally reflect a progressive increase in the degree of weathering towards the surface. It may contain CORESTONES of relatively unweathered rock and it may be capped by a SOIL, RESIDUAL DEPOSIT or DURICRUST. *JAM*

[See also REGOLITH, SAPROLITE, SOIL PROFILE]

Fookes, P.G. 1997: *Tropical residual soils*. London: The Geological Society. **Thomas, M.F.** 1996: *Geomorphology in the tropics: a study of weathering and denudation in low latitudes*. Chichester: Wiley.

weathering rates Weathering proceeds at different rates depending on the type of WEATHERING, host material and environmental conditions. While weathering may be sufficiently rapid for direct measurement in short-term laboratory experiments, in most cases rates have been determined using indirect approaches. For example, average surface weathering rates of rocks can be determined from gravestones since their exact period of exposure to the atmosphere is known. Owing to differences in technique as well as to local conditions, average weathering rates reported for particular rock types and environments vary considerably between studies. Nevertheless, 'typical' weathering rates are often used as GEOINDICATORS in landscape evolution studies to provide a crude estimate of the time of exposure of certain features and are thus useful for PALAEOENVIRONMENTAL RECONSTRUCTION. *SHD*

Bland, W. and Rolls, D. 1998: *Weathering*. London: Arnold. **Colman, S.M. and Dethier, D.P. (eds)** 1986: *Rates of chemical weathering of rocks and minerals*. Orlando, FL: Academic Press. **White, A.F. and Brantley, S.L.** 1995: Chemical weathering of silicate minerals: an overview. *Reviews in Mineralogy* 31, 1–22.

weathering rind A zone of discolouration penetrating a rock surface (bedrock or boulders), which is used to indicate the degree of WEATHERING. Weathering rind thickness has been used in RELATIVE-AGE DATING of surfaces and, by constructing a CALIBRATION curve using surfaces of known age, to suggest absolute dates. *DMcC*

[See also PETROGLYPH, ROCK VARNISH]

Knuepfer, P.L.K. 1994: Use of rock weathering rinds in dating geomorphic surfaces. In Beck, C. (ed.), *Dating in exposed and surface contexts*. Albuquerque, NM: University of New Mexico Press, 15–28.

weed Any plant unwanted by humans in an ecosystem; especially a plant that colonises an agricultural ecosystem in an aggressive way and is difficult to control. *JAM*

[See also ALIEN SPECIES, RUDERAL]

Holm, L., Doll, J., Holm, E. *et al.* 1997: *World weeds: natural histories and distributions*. New York: Wiley. **Salisbury, E.** 1964: *Weeds and aliens*, 2nd edn. London: Collins.

welded tuff A PYROCLASTIC deposit that was hot enough for fragments of PUMICE and shards of VOLCANIC GLASS to deform and fuse together under the compressional load of the deposit, creating an indurated deposit (see LITHIFICATION). Flattened fragments of pumice known as *fiamme* define a streaky *eutaxitic texture*. Welded tuffs are evidence of hot pyroclastic deposits (600°C), principally from PYROCLASTIC FLOWS and PYROCLASTIC SURGES, but also from some PYROCLASTIC FALLS. *GO*

[See also IGNIMBRITE]

Fritz, W.J. and Stillman, C.J. 1996: A subaqueous welded tuff from the Ordovician of County Waterford, Ireland. *Journal of Volcanology and Geothermal Research* 70, 91–106. **Calderone, G.M., Grönvold, K. and Oskarsson, N.** 1990: The welded air-fall tuff layer at Krafla, northern Iceland – a composite eruption triggered by injection of basaltic magma. *Journal of Volcanology and Geothermal Research* 44, 303–314.

well A pit dug or a shaft bored beneath the ground surface to exploit groundwater AQUIFERS. The principle behind a permanent well is to tap the saturated or PHREATIC ZONE below the WATER TABLE. If there is a strong seasonal lowering of the water table, the well may run dry. In semi-arid and subhumid parts of the SAHEL, shallow wells are dug in EPHEMERAL river beds each year following the wet season to tap shallow GROUNDWATER aquifers; these tend to run dry in drought years following poor wet seasons. There are numerous methods for raising the water to the surface. In the case of ARTESIAN wells, the water flows naturally to the surface under considerable HYDROSTATIC PRESSURE.   *JAM/RPDW*

Walsh, R.P.D. 1991: Climate, hydrology and water resources. In Craig, G.M. (ed.), *Agriculture in the Sudan*. Oxford: Oxford University Press, 19–53.

Westerlies The zone of mid-latitude winds, both at the surface and in the upper air, polewards from the subtropical high-pressure belt. Sometimes known as the *Ferrel Westerlies*, they are a dynamic response to the thermal gradient between low and high latitudes on a rotating Earth. Westerlies are characterised at the surface by migratory DEPRESSIONS and ANTICYCLONES, travelling generally from west to east, and aloft (the *upper westerlies*) by the circumpolar vortex of LONG WAVES centred on a core of high-velocity JET STREAMS.

The average position of the Westerlies is approximately between latitudes 35° and 65°. However, in the Northern Hemisphere the Westerlies exhibit strong seasonal variations in position and strength: they are closer to the poles in winter and closer to the Equator in summer and they strengthen in winter because of the enhanced thermal contrast between the Arctic and the tropics (in turn caused by the general increase in the area of land with latitude in the Northern Hemisphere). The Westerlies of the Southern Hemisphere are up to 60% stronger than those of the Northern Hemisphere because the broader oceanic expanses do not favour the BLOCKING activity that can disrupt flow. At the Last Glacial Maximum, the Westerlies appear to have strengthened, those of the Northern Hemisphere migrating towards the tropics but those of the Southern Hemisphere being displaced polewards.

AHP/JAM

[See also GENERAL CIRCULATION OF THE ATMOSPHERE, ZONAL INDEX]

Hare, F.K. 1960: The Westerlies. *Geographical Review* 50, 345–367. **Sellers, W.D.** 1965: *Physical climatology*. Chicago, IL: University of Chicago Press. **Wyrwoll, K.-H., Dong, B. and Valdes, P.** 2000: On the position of the Southern Hemisphere Westerlies at the Last Glacial Maximum: an outline of AGCM simulation results and evaluation of their implications. *Quaternary Science Reviews* 19, 881–898.

wet deposition Used in the context of deposition from the atmosphere of POLLUTANTS, especially ACID PRECIPITATION, wet deposition includes material dissolved in water droplets, serving as CONDENSATION NUCLEI for the droplets, or PARTICULATES captured by descending droplets (including rain, hail and snow). Rain is the most effective of the wet deposition types (*rainout*) at removing pollutants from the atmosphere; snow is the least effective.

JAM

[See also DRY DEPOSITION, OCCULT DEPOSITION]

Norton, S.A. 1999: Acid precipitation: sources to effects. In Alexander, D.E. and Fairbridge, R.H. (eds), *Encyclopedia of environmental science*. Dordrecht: Kluwer, 1–6.

wetland An ecosystem and/or landscape where water is the primary control on environment and associated biota, characterised by a WATER TABLE at or near the land surface, or where the land surface is covered by a shallow water body (less than 6 m deep).   *MJB*

[See also LAND DRAINAGE, MANGROVE, MIRE, TROPICAL PEATLANDS, WETLAND CLASSIFICATION, WETLAND CONSERVATION]

Hughes, J.M.R. and Heathwaite, A.L. (eds) 1995: *Hydrology and hydrochemistry of British wetlands*. Chichester: Wiley. **Lewis, W.M. (ed.)** 1995: *Wetlands: characteristics and boundaries*. Washington, DC: National Research Council.

wetland classification Many different wetland classification schemes have been devised, but none has been universally adopted. There are strong regional differences in both the terminology used and the basis of the classification. Wetlands are often classified on the basis of their surface vegetation communities, which is relatively straightforward: the vegetation types present reflect the underlying wetland properties; they are immediately visible and therefore simple to measure. However, other classification bases are also used, for example: division by water source (e.g. RHEOTROPHIC or OMBROTROPHIC); division by trophic status (EUTROPHIC or OLIGOTROPHIC); or division on physiographic grounds (e.g. basin or TOPOGENOUS mire; valley mire, RAISED MIRE/BOG). A detailed scheme that attempts to integrate these different approaches has been developed for Canadian wetlands.

MJB

[See also MANGROVE, MIRES, PEAT, SALTMARSH, SWAMP, TROPICAL PEATLANDS]

Heathwaite, A.L. and Göttlich, Kh. 1993: *Mires: process, exploitation and conservation*. Chichester: Wiley. **National Wetlands Working Group** 1988: *Wetlands of Canada*. Montreal: Polyscience Publications. **Rodwell, J.S. (ed.)** 1995: *British plant communities*. Vol. 4. *Aquatic communities, swamps and tall-herb fens*. Cambridge: Cambridge University Press. **Wheeler, B.D. and Proctor, M.C.F.** 2000: Ecological gradients, subdivisions and terminology of north-west European mires. *Journal of Ecology* 88, 187–203.

wetland conservation Until recently, WETLANDS were viewed as ecosystems to be avoided and were prime targets for LAND DRAINAGE. However, conservation and restoration of wetlands has followed increasing recognition of their role in maintaining WATER TABLES, improving *water quality*, moderation of downstream FLOODS, sustaining fish and wildlife, stabilising shorelines and supporting recreational activities. These functions often make wetlands more valuable over the long term for the ECOLOGICAL GOODS AND SERVICES that they provide than if they were drained. Conservation is important in slowing the loss and degradation of wetlands. *Wetland restoration* is complicated by the interactions between hydrology, soils and ecology that commonly developed over thousands of years in natural systems prior to human impacts.

Many wetlands are exceptionally important in terms of the BIODIVERSITY that they support. This function is clear in relation to vertebrates in arid regions, where wetlands provide water and shady habitats; in urban areas, where other refuges are in short supply; and in coastal wetlands, which shelter large populations of fish, invertebrates and birds. The importance of wetland conservation has been recognised internationally in the Ramsar Convention on Wetlands of International Importance Especially as Wildfowl Habitat, 1971. By 1989, over 400 wetlands covering > 93 million hectares had been designated as worthy of protection under the terms of this convention, including major wetlands under threat such as the Florida Everglades (USA), the Camargue (France) and the Okavango Delta (Botswana). *JAM*

[See also CONSERVATION]

Dugan, P.J. (ed.) 1990: *Wetland conservation: a review of current issues and required action*. Gland, Switzerland: IUCN. **Greeson, P.E., Clark, J.R. and Clark, J.E.** 1979: *Wetland functions and values: the state of our understanding*. Minneapolis, MN: American Water Resources Association. **Maltby, E.** 1986: *Waterlogged wealth*. London: Earthscan. **Maltby, E. and Lucas, E.** 1997: Wetland restoration. In Brune, D., Chapman, D.V., Gwynne, M.D. and Pacyna, J.M. (eds), *The global environment: science, technology and management*. Vol. 2. Weinheim, Germany: VCH, 946–954. **Mitsch, W.J. and Gosselink, J.G.** 1986: *Wetlands*. New York: Van Nostrand Reinhold. **Wheeler, B.D. and Shaw, S. (eds)** 1995: *Restoration of temperate wetlands*. Chichester: Wiley. **Williams, M. (ed.)** 1990: *Wetlands: a threatened landscape*. Oxford: Blackwell.

wetting front The leading edge of infiltrating rainwater within a previously dry soil. *RPDW*

[See also INFILTRATION]

whaling and sealing Hunting of whales began along the Atlantic coast of Europe. The Basques hunted whales in the Bay of Biscay in the eleventh century and the Norwegians hunted whales in fjords. The development of seaworthy vessels and reports from early EXPLORATION led to widespread whaling in the Arctic, north Atlantic and north Pacific. In the seventeenth century, the Dutch were the leading whaling nation and by the nineteenth century the Americans led a global-scale industry based on whale oil. Important technological advances included the adoption of steam-driven vessels and explosive harpoons in the AD 1880s. The first whaling in Antarctic waters occurred in 1904. These developments led to reductions in the stocks, especially of the largest baleen whales (blue, humpback and right whales) and eventually to their ENDANGERED SPECIES status in the late twentieth century. In 1983, all commercial whaling was banned by the International Whaling Commission (IWC) and, in 1994, most of the Southern Ocean south of 40°S was declared a sanctuary for whales, from which all whaling is banned. Although populations of most species of large whales have recovered, continued international vigilance by the IWC and others is necessary to ensure that the Earth's largest animals are not, once again, brought to the verge of extinction.

Sealing has followed a similar pattern. Hunting for subsistence was followed by commercial exploitation of harp and fur seals for oil and fur. The first International Convention for managing stocks of a marine organism (the North Pacific Seal Fur Convention, 1911) was established for fur seals of the Pribilou Islands in the Bering Sea. Many species of seals are, however, in a similar depleted state to the whales. *JAM*

[See also FISHERIES CONSERVATION AND MANAGEMENT]

Cooke, J.G. 1994: The management of whaling. *Aquatic Mammals* 20, 129–135. **Cushing, D.H.** 1988: *The provident sea*. Cambridge: Cambridge University Press. **Francis, D.** 1991: *The great chase: a history of world whaling*. London: Penguin Books. **McIntyre, A.D.** 1997: Marine and coastal systems. In Brune, D., Chapman, D.V., Gwynne, M.D. and Pacyna, J.M. (eds), *The global environment: science, technology and management*. Vol. 1.Weinheim, Germany: VCH, 253–263.

width–depth ratio An index of river channel cross-section shape (occasionally termed the *form ratio F*). The ratio is usually derived for bankfull stage, thus: $F = w/d$, where w is bankfull channel width and d is mean bankfull depth. Values vary with the silt-clay percentage of the bank material and with the presence/absence and nature of bank vegetation, with low ratios in humid vegetated environments with silt-clay soils and high ratios in SEMI-ARID and ARID regions, especially with coarser bank material (see CHANNEL SHAPE). The width–depth ratio is important for flow patterns in river bends and a high value may indicate potential lateral instability. *DML*

[See also BANKFULL DISCHARGE]

Knighton, A.D. 1998: *Fluvial forms and processes; a new perspective*. London: Arnold, Chapter 5.

wiggle matching The application of RADIOCARBON DATING to a floating DENDROCHRONOLOGY sequence (or any other sequence of samples separated by known time intervals) with the object of matching the sequence of dates obtained to the radiocarbon CALIBRATION curve, thereby linking the FLOATING CHRONOLOGY to a MASTER CHRONOLOGY and providing an accurate date for the floating sequence. When sites under investigation provide quantities of timber, floating dendrochronological sequences may be established that, as a result of local environmental effects or for other reasons, cannot be matched to a master chronology. Radiocarbon dates do not have the PRECISION of dendrochronological dates and calibration sometimes yields a wide range of possible CALENDAR DATES. By obtaining several radiocarbon dates for wood from a floating chronology at precisely known intervals, a pattern of 'carbon' age versus calendar time interval should emerge, which can be matched with the '*wiggles*' of the radiocarbon calibration curve. *PQD*

Bayliss, A., Groves, C., McCormac, G. *et al.* 1999: Precise dating of the Norfolk timber circle. *Nature* 402, 479. **Pearson, G.W.** 1986: Precise calendrical dating of known growth-period samples using a 'curve fitting' technique. *Radiocarbon* 28, 292–299.

wilderness The concept of land that retains its primitive features. According to the 1994 Wilderness Act (USA), it is land that does not have permanent improvements or human habitation and which is protected and managed to preserve its NATURAL conditions. It is land that has no mining, agriculture, settlement or other human impact. Aldo Leopold (1921) referred to *wilderness areas* as 'big enough to absorb a two weeks' pack trip' and McCloskey and Spalding (1989) referred to 'an undeveloped land still primarily shaped by the forces of nature'. A gradient in conditions from wilderness to totally human-made surfaces and structures could be used to assess environmental integrity or the extent to which change has occurred. *IFS*

[See also BIOREGIONAL MANAGEMENT, CONSERVATION BIOLOGY, NATURAL AREAS CONCEPT, NATURAL VEGETATION, PRIMARY WOODLAND]

Leopold, A. 1921: The wilderness and its place in forest recreation policy. *Journal of Forestry* 19, 718–721. McCloskey, J.M. and Spalding, H. 1989: A reconnaissance-level inventory of the amount of wilderness remaining in the world. *Ambio* 18, 221–227. Oelschlager, M. 1991: *The idea of wilderness: from prehistory to the age of ecology.* New Haven, CN: Yale University Press.

wildfire
An uncontrolled fire, usually on uncultivated land (e.g. scrub or forest). Reduced land management leading to a build-up of flammable vegetative matter and landuse change to flammable plant and tree species have increased the incidence of wildfires. Wildfires can cause important environmental changes, including changes in plant and animal communities, the depletion of soil nutrients and the enhancement or development of WATER REPELLENCY in the soil, which can exacerbate the enhanced *overland flow* and SOIL EROSION, which often follow wildfire. *AJDF*

[See also LAND DEGRADATION, PRESCRIBED FIRE]

Baird, M., Zabowski, D. and Everett, R.L. 1999: Wildfire effects on carbon and nitrogen in inland coniferous forests. *Plant and Soil* 209, 233–243. Shakesby, R.A., Coelho, C.de O.A., Ferreira, A.D. et al. 1993: Wildfire impacts on soil erosion and hydrology in wet Mediterranean forest, Portugal. *International Journal of Wildland Fire* 3, 95–110.

wildlife conservation and management
The controlled use and systematic protection of indigenous fauna. It is broader than GAME MANAGEMENT, which focuses on HARVESTING the animal populations. Wildlife conservation and management may have other purposes than harvesting for food or sport, including preservation or protection for moral, aesthetic or scientific reasons. The conservation of ENDANGERED SPECIES is a particularly important purpose. In principle rather than practice, flora as well as fauna may be classified as 'wildlife': however, plant species conservation would normally be considered as part of *biological conservation*.

Wildlife conservation has evolved from a local to a national government-controlled enterprise and is now international in scope. The international movement to protect wildlife began with the bilateral Migratory Bird Treaty Act of 1916 signed by the USA and Great Britain (acting for Canada). Since then numerous other bilateral and multilateral treaties and CONVENTIONS have involved single species, groups of species, habitat protection and/or trade in endangered species. There is, however, increasing recognition of a role for the private sector. In developing countries in particular, conservation efforts tend to succeed where local people retain ownership rights and can profit from the preservation effort. *JAM*

[See also BUSH MEAT, CONSERVATION]

Anderson, T.L. and Hill, P.J. (eds) 1995: *Wildlife in the market place.* Lanham, MD: Rowman and Littlefield. Bonner, R. 1993: *At the hand of man: peril and hope for Africa's wildlife.* New York: Knopf. Gilbert, F.F. and Dodds, D.G. 1992: *The philosophy and practise of wildlife management,* 2nd edn. Melbourne, FL: Krieger. Moulton, M.P. and Sanderson, J. 1999: *Wildlife issues in a changing world,* 2nd edn. Boca Raton, FL: Lewis. Robinson, J.G. and Redford, K.H. (eds) 1991: *Neotropical wildlife use and conservation.* Chicago, IL: University of Chicago Press.

wildlife corridor
A linear landscape feature that acts as a conduit for the MIGRATION or DISPERSAL of organisms between fragmented habitat; hence the alternative term *habitat corridor*. The concept has been prompted as a way of counteracting the process of habitat FRAGMENTATION and isolation. It has often been used as a popular concept in environmental planning when there is no evidence that such features do in fact act as conduits for dispersion. There are some well documented examples of linear features acting as conduits, but generally little is known about the structure and characteristics of linear features that promote dispersion of individual plants and animals. *IFS*

[See also CONSERVATION BIOLOGY, HABITAT ISLAND, HEDGEROW REMOVAL, METAPOPULATION MODEL, SINGLE LARGE OR SEVERAL SMALL RESERVES]

Spellerberg, I.F. and Gaywood, M.J. 1993: *Linear features: linear habitat or wildlife corridors* [English Nature Research Reports 60]. Peterborough: English Nature.

Wildwood
As defined by Rackham (1976: 233), the Wildwood is the wholly natural woodland, unaffected by NEOLITHIC or later civilisation; it no longer exists in the British Isles. Indeed, increasing evidence for MESOLITHIC impacts on early- and mid-HOLOCENE British woodlands suggest that 'wholly natural' woodland may never have developed in Britain during this INTERGLACIAL. Maps summarising woodland types in Britain at 5000 radiocarbon years BP, just before the ELM DECLINE and the onset of widespread evidence for NEOLITHIC LANDSCAPE IMPACTS, have been based on POLLEN ANALYSIS. These demonstrate the dominance of woodland and the wide variety of woodland types that were characteristic of the British landscape with minimal HUMAN IMPACT. *MJB*

[See also CLIMAX VEGETATION, FOREST CLEARANCE, NATURAL VEGETATION, PRIMARY WOODLAND]

Bennett, K.D. 1989: A provisional map of forest types for the British Isles 5000 years ago. *Journal of Quaternary Science* 4, 141–144. Rackham, O. 1976: *Trees and woodlands in the British landscape.* London: Dent.

Wilson cycle
An idealised cycle of TECTONIC processes, also known as the *tectonic cycle*, in which an OCEAN forms through the stretching of a RIFT VALLEY on a CONTINENT, allowing SEA-FLOOR SPREADING and the emplacement of OCEANIC CRUST to develop, then closes through SUBDUCTION at one or more DESTRUCTIVE PLATE MARGINS, leading to OROGENY at a continent–continent COLLISION ZONE (see Figure). The Wilson cycle is a model that relates the interpretation of OROGENIC BELTS to PLATE TECTONICS concepts, replacing GEOSYNCLINE theory. *GO*

[See also ROCK CYCLE]

Mitchell, A.H.G. and Reading, H.G. 1986: Sedimentation and tectonics. In Reading, H.G. (ed.), *Sedimentary environments*

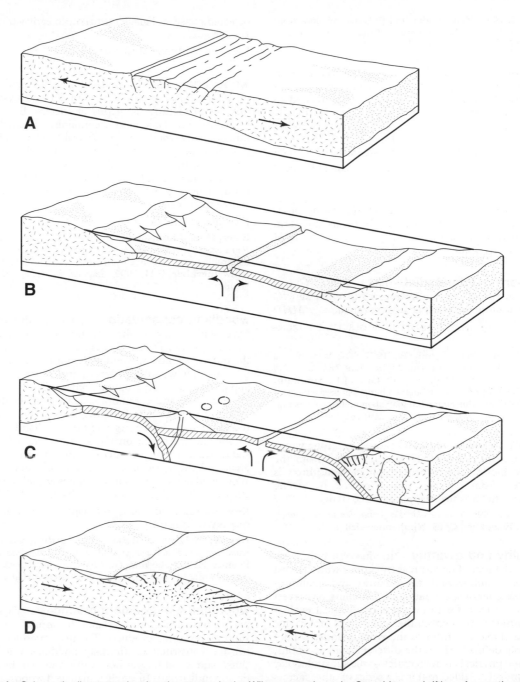

Wilson cycle *Schematic diagrams showing the stages in the Wilson tectonic cycle. Stretching and rifting of a continent (A) allows emplacement of oceanic crust and formation of an ocean with a constructive plate margin (B). Development of destructive plate margins (C) causes contraction of the ocean, leading to continent-continent collision and formation of an orogenic belt (D) (after Nichols, 1993)*

and facies, 2nd edn. Oxford: Blackwell, 471–519. **Nichols, G.** 1993: Continental margins and basin evolution. In Duff, P.McL.D. (ed.), *Holmes' principles of physical geology*, 4th edn. London: Chapman and Hall, 698–723. **Wilson, J.T.** 1966: Did the Atlantic close and then re-open? *Nature* 211, 676–681.

wind erosion   Erosion caused by wind or AEOLIAN erosion; it can be divided into DEFLATION (ENTRAINMENT of non-cohesive sediment) and ABRASION (wearing down of consolidated, cohesive material), in which sand particles carried in the air stream act as abrasive agents.   Abrasion may form streamlined bedrock outcrops (*yardangs*) and a range of other landforms. Deflation is important in SOIL EROSION and DESERTIFICATION.   *SHD*

[See also AEOLIAN, PALAEOWIND, VENTIFACT]

Greeley, R. and Iversen, J.D. 1985: *Wind as a geological process*. Cambridge: Cambridge University Press. **Stetler, L.D. and**

Saxton, K.E. 1996: Wind erosion and PM(10) emissions from agricultural fields on the Columbia Plateau. *Earth Surface Processes and Landforms* 21, 673–685.

wind ripple A variety of types of RIPPLES can form in sand under transport by wind. They include *impact ripples* related to sand SALTATION, *aerodynamic ripples* related to secondary flows, and ADHESION ripples. They are mostly low-relief BEDFORMS with little internal structure other than TRANSLATENT STRATA and are thus poorly preserved as SEDIMENTARY STRUCTURES in AEOLIAN SEDIMENTS. *GO*

Allen, P.A. 1997: *Earth surface processes.* Oxford: Blackwell Science.

wind shadow The sheltered region downwind of an obstacle that is acting as a *windbreak*. In theory, the term can be applied at any scale. *JAM*

[See also RAIN SHADOW]

Windermere Interstadial The *Late Glacial Interstadial* in Britain, lasting from c. 13 to 11 ka BP (^{14}C years), followed by the LOCH LOMOND STADIAL. *MHD*

[See also ALLERØD INTERSTADIAL, LATE GLACIAL ENVIRONMENTAL CHANGE]

Coope, C.R. and Pennington, W. 1977: The Windermere Interstadial of the Late Devensian. *Philosophical Transactions of the Royal Society, London* B280, 337–339. Lowe, J.J., Ammann, B., Birks, H.H. *et al.* 1994: Climatic changes in areas adjacent to the North Atlantic during the last glacial–interglacial transition (14–9 ka BP). *Journal of Quaternary Science* 9, 185–198.

windthrow Damage to forests caused by strong winds and involving the roots of trees being pulled from the ground and the intact tree being toppled. Windthrow is distinguished from *windbreak*, also caused by strong winds, which involves the snapping of tree stems. *JLI*

Kimmins, J.P. 1996: *Forest ecology: a foundation for sustainable management.* Englewood Cliffs, NJ: Prentice-Hall.

wine quality and quantity 'In vino' not only 'veritas' but also climate. For this reason various wine-harvest parameters, including both the quantity and quality of wine and the phenological data on blossoming or harvest season, can be used for historical climatological evaluation. In contrast to the explicit thermal control of the wine quality, the climatic control of the wine quantity cannot be so sharply defined. Here, the different site conditions, especially the physical characteristics of the substrata, play an important role. Added to this is a series of other factors – which in retrospect can only be partially solved – such as weather conditions at the start of flowering and during flowering. In summary, in spite of the other factors, hygrometric control of wine quantity can be detected. Especially close connections result between wine quality, the amount of warmth in late summer and early autumn and the VINE HARVEST parameters. *RG*

Lauer, W. and Frankenberg, P. 1986: *Zur Rekonstruktion des Klimas im Bereich der Rheinpfalz seit Mitte des 16. Jahrhunderts mit hilfe von Zeitreihen der Weinquantität und Weinqualität.* Stuttgart: Fischer.

Wolf minimum The minimum in solar SUNSPOT activity observed between AD 1280 and 1340; a possible cause of a cold phase and glacier HIGHSTAND early within the LITTLE ICE AGE. *JAM*

[See also MAUNDER MINIMUM]

wood analysis In archaeological or palaeoenvironmental studies, this involves microscopic analysis of wood samples to identify the genus or species. Extensive data banks and reference works, containing detailed information on many wood types are available. Correct identification of wood remnants, in conjunction with information on their particular geographical ranges or ecological preferences, may enable researchers to make deductions about environmental conditions prevailing at the time the source trees were growing. *KRB*

[See also PLANT MACROFOSSILS]

Biger, G. and Liphschitz, N. 1991: The recent distribution of *Pinus brutia*: a reassessment based on dendroarchaeological and dendrohistorical evidence from Israel. *The Holocene* 1, 157–161. Schweingruber, F.H. 1992: *Anatomie europäscher Hölzer.* Bern: Haupt.

woodland conservation There are different spatial scales with respect to the objectives of woodland conservation. One objective is to ensure suitable examples of all the different kinds of woodland communities. Another is to ensure that woodland communities retain their integrity and include self-perpetuating populations. The CONSERVATION of woodland communities also includes the conservation of NATIVE or indigenous plants and animals within the woodlands. The assessment of woodlands for conservation may be based on the species, type of woodland, stand structure, management, HABITATS within the wood, area, and the wood in relation to the wider LANDSCAPE. Woodland conservation in Europe is an old idea and specific records of conservation date back to AD 1100. *IFS*

[See CONSERVATION BIOLOGY, PRIMARY WOODLAND, SECONDARY WOODLAND]

Hunter, M.L. 1990: *Wildlife, forests and forestry. Principles of managing forests for biological diversity.* Englewood Cliffs, NJ: Prentice Hall. Peterken, G. 1993: *Woodland conservation and management*, 2nd edn. London: Chapman and Hall.

woodland management practices Ways in which people have utilised woodlands to ensure a continuous supply of essential products. The main products were timber for constructions (initially buildings but later also ships) and wood for hurdles, wattle and fuel (both domestic and industrial). In English managed woods, the timber trees, 90% of which were oak (*Quercus* spp.), were left standing (standards in *coppice-with-standards*) and cut only when fully grown, while the *underwood* was subjected to COPPICING or pollarding and the whole area used for *wood pasture*. Since animals eat the young shoots of coppiced trees, *pollarding* (cutting off at some height above the ground) was usually practised in wood pasture. A managed woodland provided timber and wood for centuries and was a permanent feature in the landscape. Woodland management is known from the NEOLITHIC, was particularly important during Anglo-Saxon and MEDIAEVAL times and continued until the early nineteenth century. In the eighteenth century, oak bark, from the timber trees, was in great demand for the tanning industry. *SPH*

[See also MIDDLE AGES: LANDSCAPE IMPACTS, NEOLITHIC LANDSCAPE IMPACTS, ROMAN PERIOD: LANDSCAPE IMPACTS]

Edwards, K.J. 1993: Models of mid-Holocene forest farming for north-west Europe. In Chambers, F.M. (ed.), *Climatic change and human impact on the landscape*. London: Chapman and Hall, 133–145. **Haas, J.N., Karg, S. and Rasmussen, P.** 1998: Beech leaves and twigs used as winter fodder: examples from historic and prehistoric times. *Environmental Archaeology* 1, 81–86. **Rackham, O.** 1986: *The history of the countryside*. London: Dent. **Rasmussen, P.** 1990: Pollarding of trees in the Neolithic: often presumed – difficult to prove. In Robinson, D.E. (ed.), *Experimentation and reconstruction in environmental archaeology*. Oxford: Oxbow, 77–99.

World Reference Base for Soil Resources (WRB)

After a century of discussion, an agreed basis for classification of the soils of the world has been reached by the International Union of Soil Sciences (IUSS). The system is based upon the legend developed for the FAO–Unesco *Soil Map of the World*. DIAGNOSTIC SOIL HORIZONS are used to place soils in their correct category in the system that has 30 *Soil Reference Groupings*. A simplified key, in the form of a flow chart, to the reference soil groups (named in this volume) is shown in the Table.

EMB

[See also SOIL CLASSIFICATION]

Bridges, E.M., Batjes, N.H. and Nachtergaele, F.O. (eds) 1998: *World reference base for soil resources: Atlas* [ISSS Working Group RB]. Leuven: Acco. **Deckers, J.A., Nachtergaele, F.O. and Spaargaren, O.C. (eds)** 1998: *World reference base for soil resources: Introduction* [ISSS Working Group RB]. Leuven: Acco. **ISSS Working Group RB** 1998: *World reference base for soil resources* [*World Soil Resources Report* 84]. Rome: FAO, ISRIC, ISSS.

World Reference Base for Soil Resources *Simplified key to the 30 Soil Reference Groupings of the FAO/Unesco/IUSS system of soil classification and the World Reference Base for Soil Resources*

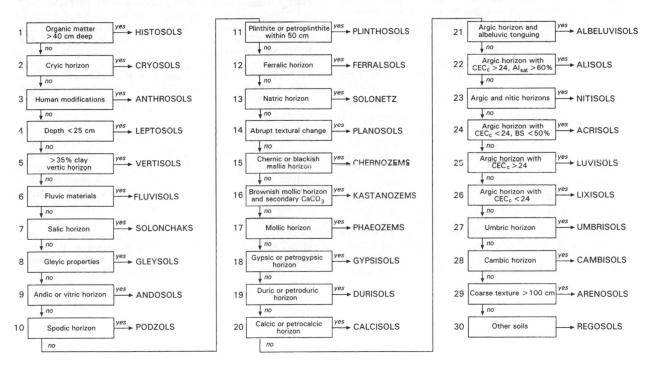

X

X-radiography   Based on the extent to which X-rays penetrate materials of different composition and density, the technique is widely used in tree-ring DENSITOMETRY and to detect stratigraphic variations in sediment cores. It is particularly useful for investigating small-scale and faint structures that are not clearly visible to the naked eye.

JAM

Axelsson, V. and Händel, K. 1972: X-radiography of unextruded cores. *Geografiska Annaler* 54(A), 34–37. **Butler, S.** 1992: X-radiography of archaeological soil and sediment profiles. *Journal of Archeological Science* 19, 151–161. **Dugmore, A.J. and Newton, A.J.** 1992: Thin tephra layers in peat revealed by X-radiography. *Journal of Archaeological Science* 19, 163–170. **Koivisto, S. and Saarnisto, M.** 1978: Conventional radiography, xeroradiography, tomography and contrast enhancement in the study of laminated sediments. *Geografiska Annaler* 60(A), 55–61.

X-ray diffraction analysis (XRD)

A technique for the identification of crystalline materials used in GEOCHEMISTRY and mineralogical investigations generally, particularly of fine-grained SEDIMENTS and SEDIMENTARY ROCKS. The spacing of lattice planes in the sample (d) is determined from the diffraction of X-rays using *Bragg's Law* ($n\lambda = 2d \sin\theta$), where n is an integer, λ is the wavelength of the radiation and θ is the angle of incidence). Sample preparations commonly include random powder mounts and oriented smears of the CLAY fraction.   *TY*

Ferrell, R.E., Hart, G.F., Swamy, S. and Murthy, B. 1998: X-ray mineralogical discrimination of depositional environments of the Krishna Delta, peninsular India. *Journal of Sedimentary Research* 68, 148–154. **Hardy, R.G. and Tucker, M.E.** 1988: X-ray powder diffraction of sediments. In Tucker, M.E. (ed.), *Techniques in sedimentology.* Oxford: Blackwell Scientific, 191–228. **Minzoni-Déroche, A.** 1981: X-ray diffraction analysis and petrography as useful methods for ceramic typology. *Journal of Field Archaeology* 8, 511–513.

X-ray fluorescence analysis (XRF)

A technique for elemental analysis used in GEOCHEMISTRY and related fields such as GEOARCHAEOLOGY, involving the measurement of secondary X-rays emitted by a target bombarded with high-energy X-rays. It is generally used on samples prepared either as beads (typically around 0.75 g) fused with a flux (for major elements), or pressed pellets of powder (typically around 7 g) with a binder (for minor and trace elements).   *TY*

Fairchild, I.J., Hendry, G., Quest, M. and Tucker, M.E. 1988: Chemical analysis of sedimentary rocks. In Tucker, M.E. (ed.), *Techniques in sedimentology.* Oxford: Blackwell Scientific, 274–354. **Janssens, K., Adams, T. and Rindby, A.** 2000: *Microscopic X-ray fluorescence.* Chichester: Wiley. **Potts, P.J., Ellis, A.T., Kregsamer, P. et al.** 1999: X-ray fluorescence spectrometry. *Journal of Analytical Atomic Spectrometry* 14, 1773–1799. **Vittiglio, G., Janssens, K., Vekemans, B. et al.** 1999: A compact small-beam XRF instrument for in-situ analysis of objects of historical and/or artistic value. *Spectrochimica Acta. Part B – Atomic Spectroscopy* 54, 1697–1710.

xerophyte

A plant that is tolerant of drought conditions. Various climatic ADAPTATIONS include: leaf adaptations; long tap roots; the development of water-storing tissues (*succulence*); and life cycles that allow survival during a hot, dry season as bulbs, corms or seeds. Xerophytes are commonly found in DESERTS and DRYLANDS and in coastal and other windy environments.   *MLC*

Crawford, R.M.M. 1989: *Studies in plant survival.* Oxford: Blackwell Scientific. **Walter, H. and Stadelman, E.** 1974: A new approach to the water relations of desert plants. In Brown, G.W. (ed.), *Desert biology.* New York: Academic Press.

Y

Younger Dryas Stadial The *Late Glacial Stadial* lasting c. 11–10 ka BP (^{14}C years) that followed the ALLERØD INTERSTADIAL, culminating in the termination of the *Weischselian* GLACIATION in Northern Europe around 11 500 years ago (calibrated ^{14}C years). It is broadly equivalent to the LOCH LOMOND STADIAL in Britain and is thought to have lasted at least 1150 years on VARVE counts, but between 1200 and 1300 years according to evidence from the GISP2 Greenland ICE CORE. Summer temperatures were up to about 10°C lower than today, but there were large regional variations and the extent to which the Younger Dryas was a truly global phenomenon is still controversial.   *MHD*

[See also LATE GLACIAL ENVIRONMENTAL CHANGE]

Alley, R.B. 2000: The Younger Dryas cold interval as viewed from central Greenland. *Quaternary Science Reviews* 19, 213–226.

Amman, B. (ed.) 2000: Biotic response to rapid climatic changes around the Younger Dryas. *Palaeogeography, Palaeoclimatology, Palaeoecology* 159(3–4), 191–361 [Special Issue]. **Bennett, K.D., Haberle, S.G. and Lumley, S.H.** 2000: The Last Glacial–Holocene transition in southern Chile. *Science* 290, 325–328. **Lowe, J.J., Ammann, B., Birks, H.H.** *et al.* 1994: Climatic changes in areas adjacent to the North Atlantic during the last glacial–interglacial transition (14–9 ka BP) *Journal of Quaternary Science* 9, 185–198. **Peteet, D.M.** 1995: Global Younger Dryas. *Quaternary International* 28, 93–104. **Roberts, N., Taieb, M., Barker, P.** *et al.* 1993: Timing of the Younger Dryas event in East Africa from lake-level changes. *Nature* 366, 146–148. **Wright Jr, H.E.** 1989: The amphi-Atlantic distribution of the Younger Dryas palaeoclimatic oscillation. *Quaternary Science Reviews* 8, 295–306.

Z

zonal index   A measure of the zonal (longitudinal or west–east) component of the atmospheric circulation of mid-latitudes expressed as the horizontal pressure difference between latitudes 35° and 55°. Over an interval of three to eight weeks, the zonal index varies from about 3 (low index) to 8 millibars (high index). The *index cycle* describes the variation from low to high and back to low zonal index, the latter being characterised by well developed LONG WAVES, significant meridional (latitudinal) energy transport towards the poles and BLOCKING situations.   *JAM*

Namias, J. 1950: The zonal index and its role in the general circulation. *Journal of Meteorology* 7, 130–139. **Oliver, J.E.** 1987: Zonal circulation and index. In Oliver, J.E. and Fairbridge, R.W. (eds), *The encyclopedia of climatology*. New York: Van Nostrand Reinholt, 942–946.

zonal soil   Used originally by Dokuchaev for a soil that is characteristic of a latitudinal area of the Earth's surface, zonal soils have well developed profiles and are more-or-less in equilibrium with the environment, particularly climate, of the zone in which they occur. It was a *soil order* in the system of classification used in the USA before AD 1965; but is not used in SOIL TAXONOMY.   *EMB*

Baldwin, M., Kellogg, C.E. and Thorp, J. 1938: Soil classification. In *Soils and men, yearbook of agriculture*. Washington, DC: United States Department of Agriculture.

zonation, spatial   The recognition of spatial zones in the LANDSCAPE at local to global scales may be considered as either (1) a methodological device or (2) a representation of reality. It may be methodologically appropriate to summarise spatial variation of, for example, vegetation and soils as relatively homogeneous areas separated by sharp boundaries (implied by the term zone) even though the landscape exhibits an intergrading continuum of variation. Where zones exist in the landscape, their boundaries often coincide with discontinuities in environmental gradients. Similar concepts apply to the nature and recognition of temporal STAGES.   *JAM*

[See also ALTITUDINAL ZONATION, ECOCLINE, ECOTONE, LANDSCAPE MOSAIC, ZONAL SOIL]

zooarchaeology   The identification, analysis and interpretation of animal bones, teeth and other resistant animal tissues associated with archaeological sites.   *JAM*

[See also ENVIRONMENTAL ARCHAEOLOGY, SCIENTIFIC ARCHAEOLOGY]

Davis, S.J.M. 1987: *The archaeology of animals*. Batsford: London. **Klein, R.G. and Cruz-Uribe, K.** 1984: *The analysis of animal bones from archaeological sites*. Chicago, IL: University of Chicago Press. **Reitz, J. and Wing, E.S.** 1999: *Zooarchaeology*. Cambridge: Cambridge University Press.

zoogeography   The study of the geographical distributions of animals and their explanation in terms of combinations of geographical, ecological and historical factors, which affect their physical, climatic or time range. Diverse factors regulate species distribution, including competition, predation, a lack of suitable food, adverse climate and physical conditions. The spread of a species is facilitated by LAND BRIDGES and prevented by physical barriers (such as oceans, deserts or mountain ranges), climatic barriers (such as temperature, daylight length or moisture) and biological factors (such as disease).   *DCS*

Darlington, P.J. 1957: *Zoogeography: the geographical distribution of animals*. New York: Wiley. **Illies, J.** 1974: *Introduction to zoogeography*. London: Macmillan. **Udvardy, M.D.F.** 1969: *Dynamic zoogeography*. New York: Van Nostrand Reinhold. **Whittaker, R.J.** 1998: *Island biogeography : ecology, evolution, and conservation*. New York: Oxford University Press.

zoogeomorphology   The study of animals as agents in creating and modifying LANDFORMS and geomorphic processes. Animals excavate, burrow, wallow, trample, build mounds and dams, graze vegetation, eat earth (*geophagy*) and excrete. The direct and indirect effects of wild and domesticated animals are often underestimated.   *JAM*

[See also GEOMORPHOLOGY, COPROLITE]

Butler, D.R. 1995: *Zoogeomorphology: animals as geomorphic agents*. Cambridge: Cambridge University Press. **Trimble, S.W. and Mendel, A.C.** 1995: The cow as a geomorphic agent: a critical review. *Geomorphology* 13, 233–253.

zoom   In IMAGE PROCESSING, an operation that enlarges (*zoom in*) or reduces (*zoom out*) an image or a portion of an image in an output device. It may consist of the mere replication or discarding of values or it may rely on INTERPOLATION or re-sampling techniques.   *ACF*

Index

The index should be viewed as an indispensable aid to finding a topic or term if the headword is not the term with which the reader is most familiar. It includes terms, often given in italic in the main text, that are not included as headwords; the headwords themselves are not repeated in the index. The index includes expansions of many of the acronyms and abbreviations used in the text, as well as common North American spellings.

677